Solutions Manual

to accompany

BEDFORD•FOWLER ENGINEERING MECHANICS

Dynamics

Solutions Manual

to accompany

BEDFORD•FOWLER ENGINEERING MECHANICS

Dynamics

Second Edition

by

Wallace Fowler
University of Texas, Austin

Anthony Bedford
University of Texas, Austin

Eugene L. Davis
University of Texas, Austin

Prentice Hall, Upper Saddle River, NJ 07458

Solutions Manual to accompany Bedford/Fowlers's **ENGINEERING MECHANICS**: Dynamics, Second Edition

Printed in the United States of America

ISBN 0-201-30869-X

10 9 8 7 6 5 4 3 2 1

Prentice-Hall International (UK) Limited, *London*
Prentice-Hall of Australia Pty. Limited, *Sydney*
Prentice-Hall Canada, Inc., *London*
Prentice-Hall Hispanoamericana, S.A., *Mexico*
Prentice-Hall of India Private Limited, *New Delhi*
Prentice-Hall of Japan, Inc., *Tokyo*
Prentice-Hall (Singapore) Pte Ltd
Editora Prentice-Hall do Brazil, Ltda., *Rio de Janeiro*

CHAPTER 1 - Solutions

=====================================<>=====================================

Problem 1.1 The value of is 3.141592654... What is the value of π to seven significant digits?

Solution: Start with the first non-zero digit on the left, record the next four digits. Round the last digit to the nearest integer. The resultis $\pi = 3.141593$ to seven significant digits.

=====================================<>=====================================

Problem 1.2 What is the value of e (the base of the natural logarithms) to five significant digits?

Solution: The value of e is 2.718281828.... Record the first five digits, and round the last digit to the nearest integer. The result is $e = 2.7183$ to five significant digits.

=====================================<>=====================================

Problem 1.3 Determine the value of the expression $\frac{1}{(2-\pi)}$ to three significant digits.

Solution: The value of the expression is $\frac{1}{(2-\pi)} = -0.87596919....$ Start with the first non-zero digit on the left, and record the next three digits left to right. Round the last digit to the nearest integer. The result is $\frac{1}{(2-\pi)} = -0.876$ to three significant digits.

=====================================<>=====================================

Problem 1.4 The opening in a soccer goal is 24 ft wide and 8 ft high. What are its dimensions in meters, to three significant digits.

Solution: The conversion between feet and meters, found inside the front cover of the textbook, is $1\ m = 3.281\ ft$. The goal width, $w = 24\,ft\left(\frac{1\,m}{3.281\,ft}\right) = 7.3148\ m = 7.31\ m$. The goal height is given by $h = 8\,ft\left(\frac{1\ m}{3.281\ ft}\right) = 2.438\ m = 2.44\ m$.

=====================================<>=====================================

Problem 1.5: The dimensions of the Boeing 777-200 aircraft are length = 209 ft 1 in., wingspan = 199 ft 11 in., and height = 60 ft 6 in. Express these dimensions in meters to three significant digits.

Solution: The conversion between feet and meters, found inside the front cover of the textbook, is $1\ m = 3.281\ ft$. $Length = 209\ ft\ 1\ in = 209.0833\ ft\left(\frac{1\ m}{3.281\ ft}\right) = 63.7\ m$,

$$Wingspan = 199\ ft\ 11\ in = 199.917\ ft\left(\frac{1\ m}{3.281\ ft}\right) = 60.9\ m,$$

$$Height = 60\ ft\ 6\ in = 60.5\ ft\left(\frac{1\ m}{3.281\ ft}\right) = 18.4\ m.$$

=====================================<>=====================================

==================================<>==================================

Problem 1.6: Suppose that you have just purchased a Ferrari Dino 246GT coupe and you want to know whether you can use your set of SAE (U.S. Customary Units) wrenches to work on it. You have wrenches with widths w = 1/4 in., 1/2 in., 3/4 in. and 1 in., and the car has nuts with dimensions n = 5 mm, 10 mm, 15 mm, 20 mm, and 25 mm. Defining a wrench to fit if w is no more than 2 percent larger than n, which of your wrenches can you use?

Solution: Convert the metric size n to inches, and compute the percentage difference between the metric sized nut and the SAE wrench. The results are:

$$5\ mm\left(\frac{1\ inch}{25.4\ mm}\right)=0.19685..in,\quad \left(\frac{0.19685-0.25}{0.19685}\right)100=-27.0\%$$

$$10\ mm\left(\frac{1\ inch}{25.4\ mm}\right)=0.3937..in\ ,\quad \left(\frac{0.3937-0.5}{0.3937}\right)100=-27.0\%$$

$$15\ mm\left(\frac{1\ inch}{25.4\ mm}\right)=0.5905..in,\quad \left(\frac{0.5905-0.5}{0.5905}\right)100=+15.3\%$$

$$20\ mm\left(\frac{1\ inch}{25.4\ mm}\right)=0.7874..in,\quad \left(\frac{0.7874-0.75}{0.7874}\right)100=+\ 4.7\%$$

$$25\ mm\left(\frac{1\ inch}{25.4\ mm}\right)=0.9843..in\ ,\quad \left(\frac{0.9843-1.0}{0.9843}\right)100=-\ 1.6\%$$

A negative percentage implies that the metric nut is smaller than the SAE wrench; a positive percentage means that the nut is larger then the wrench. Thus within the definition of the 2 % fit, the 1 in. wrench will fit the 25 mm nut. **The other wrenches cannot be used.**

==================================<>==================================

Problem 1.7 The 1829 *Rocket,* shown in example 1.3, could draw a carriage with 30 passengers at 25 miles/hr. Determine its velocity to three significant digits: (a) in ft/s, (b) in km/hr.

Solution: Convert the units using the Unit Conversions given in Table 1.2. The results, to three significant digits:

$$25\left(\frac{\text{miles}}{\text{hr}}\right)\left(\frac{5280\text{ ft}}{1\text{ mile}}\right)\left(\frac{1\text{ hr}}{3600\text{ s}}\right)=36.666...\left(\frac{\text{ft}}{\text{s}}\right)=36.7\left(\frac{\text{ft}}{\text{s}}\right)$$

$$25\left(\frac{\text{miles}}{\text{hr}}\right)\left(\frac{5280\text{ ft}}{1\text{ mile}}\right)\left(\frac{0.3048\text{ m}}{1\text{ ft}}\right)\left(\frac{1\text{ km}}{1000\text{ m}}\right)=40.2336...\left(\frac{\text{km}}{\text{hr}}\right)=40.2\left(\frac{\text{km}}{\text{hr}}\right)$$

==================================<>==================================

Problem 1.8 High speed "bullet trains" began running between Tokyo and Osaka, Japan in 1964. If a bullet train travels at 240 km/hr, what is its speed in miles/hr to three significant digits?

Solution: Convert the units using Table 1.2. The results are:

$$240\left(\frac{\text{km}}{\text{hr}}\right)\left(\frac{1\text{ mile}}{5280\text{ ft}}\right)\left(\frac{1\text{ ft}}{0.3048\text{ m}}\right)\left(\frac{1000\text{ m}}{1\text{ km}}\right)=149.12908...\left(\frac{\text{mile}}{\text{hr}}\right)=149\left(\frac{\text{mile}}{\text{hr}}\right)$$

==================================<>==================================

=================================<>=================================

Problem 1.9 In December, 1986, Dick Rutan and Jeana Yeager flew the Voyager aircraft around the world nonstop. They flew a distance of 40,212 km in 9 days, 3 minutes, and 44 seconds. (a) Determine the distance they flew in miles to three significant digits, (b) Determine their average speed in kilometers per hour, miles per hour, and knots, to three significant digits.

Solution: Convert the units using Table 1.2.

(a) $40212\ km\left(\frac{1000\ m}{1\ km}\right)\left(\frac{1\ ft}{0.3048\ m}\right)\left(\frac{1\ mile}{5280\ ft}\right) = 24987\ mi = 25000\ mi = 2.50\times10^4\ mi$

(b) The time of flight is 9 days 3 min 44 sec $= \left((9)(24)+\left(\frac{3}{60}\right)+\left(\frac{44}{3600}\right)\right)$hours = 216.062 hours.

The average speed is $\left(\frac{40212\ km}{216.062\ hours}\right) = 186.11\ \frac{km}{hr} = 186\frac{km}{hr}$. Converting,

$\left(186.11\ \frac{km}{hr}\right)\left(\frac{1\ mile}{1.609\ km}\right) = 115.7\frac{mi}{hr} = 116\frac{mi}{hr}$, or

$\left(186.11\ \frac{km}{hr}\right)\left(\frac{1\ nautical\ mile}{1.852\ km}\right) = 100.49\ knots = 100\ knots = 1.00\times10^2\ knots$, to three significant digits.

=================================<>=================================

Problem 1.10 Engineers who study shock waves sometimes express velocity in millimeters per microsecond (mm/s). Suppose that the velocity of a wavefront is measured and determined to be 5 mm/s. Determine its velocity (a) in m/s; (b) in mi/s.

Solution: Convert units using Tables 1.1 and 1.2. The results:

(a) $5\left(\frac{mm}{\mu s}\right)\left(\frac{1\ m}{1000\ mm}\right)\left(\frac{10^6\ \mu s}{1\ s}\right) = 5000\left(\frac{m}{s}\right)$. Next, use this result to get (b):

(b) $5000\left(\frac{m}{s}\right)\left(\frac{1\ ft}{0.3048\ m}\right)\left(\frac{1\ mi}{5280\ ft}\right) = 3.10685...\left(\frac{mi}{s}\right) = 3.11\left(\frac{mi}{s}\right)$

=================================<>=================================

Problem 1.11 Geophysicists measure the motion of a glacier and discover that it is moving at 80 mm/year. What is its velocity in m/s?

Solution: Convert units using Tables 1.1 and 1.2. The result is:

$$80\left(\frac{mm}{year}\right)\left(\frac{1\ m}{1000\ mm}\right)\left(\frac{1\ hr}{3600\ s}\right)\left(\frac{1\ day}{24\ hr}\right)\left(\frac{1\ year}{365.25\ days}\right) = 2.535...(10^{-9})\left(\frac{m}{s}\right) = 2.54\ (10^{-9})\left(\frac{m}{s}\right)$$

=================================<>=================================

Problem 1.12 The acceleration due to gravity at sea level in SI units is $g = 9.81$ m/s. By converting units, use this value to determine the acceleration due to gravity at sea level in U. S. Customary units.

Solution: Use Table 1.2. The result is:

$$g = 9.81\left(\frac{m}{s^2}\right)\left(\frac{1\ ft}{0.3048\ m}\right) = 32.185...\left(\frac{ft}{s^2}\right) = 32.2\left(\frac{ft}{s^2}\right)$$

=================================<>=================================

================================<>================================

Problem 1.13 A *furlong per fortnight* is a facetious unit of velocity, perhaps made up by a student as a satirical comment on the bewildering variety of units engineers must deal with. A furlong is 660 ft (1/8 mile). A fortnight is two weeks (fourteen days). If you walk to class at 5 ft/s, what is your velocity in furlongs per fortnight to three significant digits?

Solution: Convert the units using the given conversions. Record the first three digits on the left, and add zeros as required by the number of tens in the exponent. The result is

$$:\left(5\frac{ft}{s}\right)\left(\frac{1furlong}{660ft}\right)\left(\frac{3600s}{1hr}\right)\left(\frac{24hr}{1day}\right)\left(\frac{14day}{1fortnight}\right)=\left(9160\frac{furlongs}{fortnight}\right)=9.16\times10^3\frac{furlongs}{fortnight}$$

================================<>================================

Problem 1.14 The cross sectional area of a beam is 480 in^2. What is its cross section in m^2 ?

Solution: Convert units using Table 1.2. The result:

$$480\ in^2\left(\frac{1\ ft}{12\ in}\right)^2\left(\frac{0.3048\ m}{1\ ft}\right)^2 = 0.30967...m^2 = 0.310\ m^2$$

================================<>================================

Problem 1.15 At sea level, the weight density (weight per unit volume) of water is approximately 62.4 lb/ft^3. Use this value to determine the mass density of water in kg/m^3.

Solution: Convert the units using Table 1.2 and the fact that weight equals mass times the acceleration of gravity. First, convert to N/m^3. $62.4\left(\frac{lb}{ft^3}\right)\left(\frac{1\ N}{0.2248\ lb}\right)\left(\frac{3.281\ ft}{1\ m}\right)^3=9804.1\left(\frac{N}{m^3}\right)=9.80\times10^3\left(\frac{N}{m^3}\right).$

Divide by $g=9.81\left(\frac{m}{s^2}\right)$ and recall that $1\ kg = 1\left(\frac{N\ s^2}{m}\right)$. Hence,

$$9804.1\left(\frac{N}{m^3}\right)\left(\frac{1}{9.81}\right)\left(\frac{s^2}{m}\right)=999\left(\frac{N\ s^2}{m}\right)\left(\frac{1}{m^3}\right)=999\left(\frac{kg}{m^3}\right).$$

================================<>================================

Problem 1.16 A pressure transducer measures a value of 300 lb/in^2 Determine the value of the pressure in Pascals. A Pascal (Pa) is 1 N/m^2.

Solution: Convert the units using Table 1.2 and the definition of the Pascal unit. The result:

$$300\left(\frac{\text{lb}}{\text{in}^2}\right)\left(\frac{4.448\ \text{N}}{1\ \text{lb}}\right)\left(\frac{12\ \text{in}}{1\ \text{ft}}\right)^2\left(\frac{1\ \text{ft}}{0.3048\ \text{m}}\right)^2=2.0683...(10^6)\left(\frac{\text{N}}{\text{m}^2}\right) = 2.07(10^6)\ \text{Pa}$$

================================<>================================

Problem 1.17 A horsepower is 550 ft-lb/s. A watt is 1 N-m/s. Determine the number of watts generated by (a) the Wright brothers' 1903 airplane, which had a 12-horsepower engine; (b) a modern passenger jet with a power of 100000 horsepower at cruising speed.

Solution: Convert units using Table 1.2 and the definition of a horsepower to derive the conversion between horsepower and watts. The result

(a) $12\ \text{hp}\left(\frac{746\ \text{watt}}{1\ \text{hp}}\right) = 8950$ wat .(b) $10^5\ \text{hp}\left(\frac{746\ \text{watt}}{1\ \text{hp}}\right)=7.46\left(10^7\right)$ watt

================================<>================================

==================================<>==================================

Problem 1.18 In SI units, the universal gravitational constant G has the value, $G = 6.67(10^{-11})\left(\frac{\text{N-m}^2}{\text{kg}^2}\right)$. Determine the value of G in U.S. Customary units.

Solution: Convert units using Table 1.2. The result:

$$6.67(10^{-11})\left(\frac{\text{N-m}^2}{\text{kg}^2}\right)\left(\frac{1\text{ lb}}{4.448\text{ N}}\right)\left(\frac{1\text{ ft}}{0.3048\text{ m}}\right)^2\left(\frac{14.59\text{ kg}}{1\text{ slug}}\right)^2 = 3.43590\ldots(10^{-8})\left(\frac{\text{lb-ft}^2}{\text{slug}^2}\right)$$

$$= 3.44(10^{-8})\left(\frac{\text{lb-ft}^2}{\text{slug}^2}\right)$$

==================================<>==================================

Problem 1.19 If the Earth is modeled as a homogenous sphere, the velocity of a satellite in a circular orbit is $v = \sqrt{\frac{gR_E^2}{r}}$ where R_E is the radius of the Earth and r is the radius of the orbit. (a) If g is in m/s^2 and R_E and r are in meters, what are the units of v? (b) If R_E= 6370 km and r = 6670 km what is the value of v to three significant digits? (c) For the orbit described in Part (b), what is the value of v in miles/s to three significant digits?

Solution: For (a), substitute the units into the expression and reduce:

(a) $$\sqrt{\frac{g\left(\frac{m}{s^2}\right)(R_E\ m)^2}{(r\ m)}} = \sqrt{\frac{gR_E^2}{r}\left(\frac{m^3}{m\ s^2}\right)} = v\left(\frac{m}{s}\right)$$ Hence, the units are m/s

For (b), substitute the numerical values into the expression, using $g = 9.81\left(\frac{m}{s^2}\right)$.

$$v = \sqrt{\frac{(9.81\ \frac{m}{s^2})((6370\ km)(10^3\ \frac{m}{km}))^2}{(6670\ km)(10^3\ \frac{m}{km})}} = \sqrt{59.679\ldots(10^6)}\left(\frac{m}{s}\right) = 7.7252\ldots(10^3)\left(\frac{m}{s}\right)$$

(b) $v = 7730\left(\frac{m}{s}\right)$

For (c), convert units using Table 1.2. The result:

(c) $$v = 7730\left(\frac{m}{s}\right)\left(\frac{1\ ft}{0.3048\ m}\right)\left(\frac{1\ mile}{5280\ ft}\right) = 4.803\ldots\left(\frac{mile}{s}\right) = 4.80\left(\frac{mile}{s}\right)$$

==================================<>==================================

Problem 1.20 In the equation $T = \frac{1}{2}\ I\ \omega^2$ the term I is in kg-m^2 and is in s^{-1} .(a) What are the SI units of T? (b) The value of T is 100 when I is in kg-m^2 and is in s^{-1}. What is the value of T when it is expressed in U. S. Customary base units?

Solution: For (a), substitute the units into the expression for T:

(a) $T = \left(\frac{1}{2}\right)(I\ kg\text{-}m^2)(w\ s^{-1})^2 \quad T = \left(\frac{1}{2}\right)(I\ kg\text{-}m^2)(w\ s^{-1})^2$

For (b), convert units using Table 1.2. The result:

(b) $$100\left(\frac{kg\text{-}m^2}{s^2}\right)\left(\frac{1\ slug}{14.59\ kg}\right)\left(\frac{1\ ft}{0.3048\ m}\right)^2 = 73.7759\ldots\left(\frac{slug\text{-}ft^2}{s^2}\right) = 73.8\left(\frac{slug\text{-}ft^2}{s^2}\right)$$

==================================<>==================================

Problem 1.21 The aerodynamic drag force D exerted on a moving object by a gas is given by the expression $D = C_D S \frac{1}{2} \rho v^2$, where the drag coefficient, C_D, is dimensionless, S is reference area, ρ is the mass per unit volume of the gas, and v is the velocity of the object relative to the gas.

(a) Suppose that D is 800 when S, ρ, and v are expressed in SI base units. By converting units, find the value of D when S, ρ, and v are expressed in U.S. Customary base units.

(b) The drag force D is in newtons when the expression is evaluated in SI base units and is in pounds when the expression is evaluated in U.S. Customary base units. By using the result from part (a), determine the conversion factor from newtons to pounds.

Solution: For (a), we just carry out the conversion unit by unit. We get:

(a) $$800(m^2)\left(\frac{kg}{m^3}\right)\left(\frac{m}{s}\right)^2 = 800\left(\frac{kg\ m}{s^2}\right) = 800\left(\frac{0.0685\ slug}{1\ kg}\right)\left(\frac{3.281\ ft}{1\ m}\right)\left(\frac{1}{s^2}\right) = 1.80\times10^2\left(\frac{slug\ ft}{s^2}\right)$$

(b) From (a), 800 N = 180 lb. Hence, 1 N = 0.225 lb.

Problem 1.22 The Lockheed-Martin X-33 reusable launch test vehicle, when fully fueled, will weigh 273,300 lb at sea level. Its weight at sea level with its fule expended will be 62,700 lb.

(a) Determine its mass in slugs when fully fueled and with its fuel expended.

(b) Determine its weight in meganewtons at sea level when it is fully fueled and with its fuel expended.

Solution: For (a), we use $m = W/g$, where $g = 32.2\left(\frac{ft}{s^2}\right)$. Our results are:

$$m_{loaded} = \left(\frac{273{,}300\ lb}{32.2\ ft/s^2}\right) = 8490\left(\frac{lb\ s^2}{ft}\right) = 8490\ slugs, \text{ and } m_{empty} = \left(\frac{62{,}700\ lb}{32.2\ ft/s^2}\right) = 1950\left(\frac{lb\ s^2}{ft}\right) = 1950\ slugs$$

For (b), we need to convert the given weights to newtons and then to meganewtons.

$$W_{loaded} = 273{,}300\ lb\left(\frac{1\ N}{0.2248\ lb}\right) = 1{,}216{,}000\ N = 1.22\ MN$$

$$W_{empty} = 62{,}700\ lb\left(\frac{1\ N}{0.2248\ lb}\right) = 279{,}000\ N = 0.279\ MN$$

Problem 1.23 The acceleration due to gravity is 13.2 ft/s^2 on the surface of Mars and 32.2 ft/s^2 on the surface of the earth. If a woman weighs 125 lb on Earth, To survive and work on the surface of Mars, she must wear life-support and carry tools. What is the maximum allowable weight on Earth of the woman's clothing, equipment, and tools if the engineers don't want the total weight on Mars of the woman, her clothing, equipment, and tools to exceed 125 lb?

Solution: Note that the mass is the same on earth and Mars. Thus $W_m = g_m m = 125\ lb$, and $W_e = g_e m$ Take the ratio, $\left(\frac{W_e}{W_m}\right) = \left(\frac{m\ g_e}{m\ g_m}\right) = \left(\frac{g_e}{g_m}\right)$ Solving for the weight on Earth corresponding to 125 lbs on Mars, we get:

$$W_e = W_m\left(\frac{g_e}{g_m}\right) = 125(lb)\left(\frac{32.2\ ft/s^2}{13.2\ ft/s^2}\right) = 305\ lb.$$ Next, we subtract the woman's weight to get the Earth weight of the equipment, tools, and clothing. We get $W_e = 305\ lb - 125\ lb = 180\ lb$

=====================================<>=====================================

Problem 1.24 A person has a mass of 50 kg. (a) The acceleration due to gravity at sea level is $g = 9.81\ m/s^2$. What is the person's weight at sea level? (b) The acceleration due to gravity on the Moon is $g = 1.62\ m/s^2$. What would the person weigh on the Moon?

Solution: Use Eq (1.6). (a) $W_e = 50\ kg\left(9.81\dfrac{m}{s^2}\right) = 490.5\ N = 491\ N$, and

$$\text{(b)}\quad W_{moon} = 50\ kg\left(1.62\frac{m}{s^2}\right) = 81\ N.$$

=====================================<>=====================================

Problem 1.25 The acceleration due to gravity at sea level is $g = 9.81\ m/s^2$. The radius of the earth is 6370 km. The Universal Gravitation Constant is $G = 6.67(10^{-11})\ N\ m^2/kg^2$. Use this information to obtain the mass of the earth.

Solution: Use Eq (1.3) $a = \dfrac{Gm_E}{R^2}$. Solve for the mass,

$$m_E = \frac{gR^2}{G} = \frac{(9.81\ m/s^2)(6370\ km)^2\left(10^3\ \dfrac{m}{km}\right)^2}{6.67(10^{-11})\left(\dfrac{N\text{-}m^2}{kg^2}\right)} = 5.9679...(10^{24})kg\ =\ 5.97(10^{24})\ kg$$

=====================================<>=====================================

Problem 1.26 A person weighs 180 lb at sea level. The radius of the Earth is 3960 miles. What force is exerted on the person by the gravitational attraction of the Earth if he is in a space station in near-Earth orbit 200 miles above the surface of the Earth?

Solution: Use Eq (1.5).

$$W = mg\left(\frac{R_E}{r}\right)^2 = \left(\frac{W_E}{g}\right)g\left(\frac{R_E}{R_E+H}\right)^2 = W_E\left(\frac{3960}{3960+200}\right)^2 = (180)(0.90616) = 163\ lb$$

=====================================<>=====================================

Problem 1.27 The acceleration due to gravity on the surface of the Moon is 1.62 m/s^2. The radius of the Moon is $R_M = 1738$ km. Determine the acceleration due to gravity at a point 1738 km above its surface.

Solution: Use Eq (1.4), rewritten to apply to the Moon...$a = g_M\left(\dfrac{R_M}{r}\right)^2$

$$a = (1.62\ m/s^2)\left(\frac{R_M}{R_M+R_M}\right)^2 = \left(1.62\ m/s^2\right)\left(\frac{1}{2}\right)^2 = 0.405\ m/s^2$$

=====================================<>=====================================

Problem 1.28 If an object is near the surface of the Earth, the variation of its weight with distance from the center of the Earth can often be neglected. The acceleration due to gravity at the surface of the Earth is 9.81 m/s^2 .The radius of the Earth is 6370 km. The weight of an object at sea level is ***mg***, where ***m*** is its mass. At what height above the Earth does the weight decrease to 0.99 ***mg***?

Solution: Use a variation of Eq (1.5). $W = mg\left(\dfrac{R_E}{R_E+h}\right)^2 = 0.99mg$ Solve for the radial height,

$$h = R_E\left(\frac{1}{\sqrt{0.99}} - 1\right) = (6370)(1.0050378 - 1.0) = 32.09...\ km\ =\ 32{,}100\ m = 32.1\ km$$

=====================================<>=====================================

Problem 1.29 The centers of two oranges are one meter apart. The mass of each orange is 0.2 kg. What gravitational force do they exert on each other? (The Universal Gravitational Constant is $G = 6.67 \times 10^{-11} \dfrac{N\text{-}m^2}{kg^2}$.)

Solution: Use Eq (1.1) $F = \dfrac{G\, m_1 m_2}{r^2}$. Substitute:

$$F = \frac{(6.67)(10^{-11})(0.2)(0.2)}{1^2} = 2.668(10^{-12})\ N$$

Problem 1.30 At a point between the Earth and the Moon, the magnitude of the Earth's gravitational acceleration equals the magnitude of the Moon's gravitational acceleration. What is the distance from the center of the Earth to that point to three significant digits. The distance from the center of the Earth to the center of the Moon is 383,000 km and the radius of the Earth is 6370 km. The radius of the Moon is 1738 km and the acceleration of gravity at its surface is 1.62 m/s^2.

Solution: Let r_{Ep} be the distance from the Earth to the point where the gravitational accelerations are the same and let r_{Mp} be the distance from the Moon to that point. Then, $r_{Ep} + r_{Mp} = r_{EM} = 383000\ km$. The fact that the gravitational attractions by the Earth and the Moon at this point are equal leads to the equation

$g_E \left(\dfrac{R_E}{r_{Ep}} \right)^2 = g_M \left(\dfrac{R_M}{r_{Ep}} \right)^2$, where $r_{EM} = 383{,}000\ km$. Substituting the correct numerical values leads to the

equation $9.81 \left(\dfrac{m}{s^2} \right) \left(\dfrac{6370\ km}{r_{Ep}} \right)^2 = 1.62 \left(\dfrac{m}{s^2} \right) \left(\dfrac{1738\ km}{r_{EM} - r_{Ep}} \right)^2$, where r_{Ep} is the only unknown. Solving, we get $r_{Ep} = 344{,}770\ km. = 345{,}000\ km.$

================================<>================================

The following problems (Problem 2.1 though Problem 2.35) involve straight line motion. The time t is in seconds unless otherwise stated.

================================<>================================

Problem 2.1 The graph of the position s of a point as a function of time is a straight line. When $t = 4$ seconds, $s = 34\ m$, and when $t = 20$ seconds, $s = 72\ m$. (a) Determine the velocity of the point by calculating the slope of the straight line. (b) Obtain the equation for s as a function of time and use it to determine the velocity of the point.

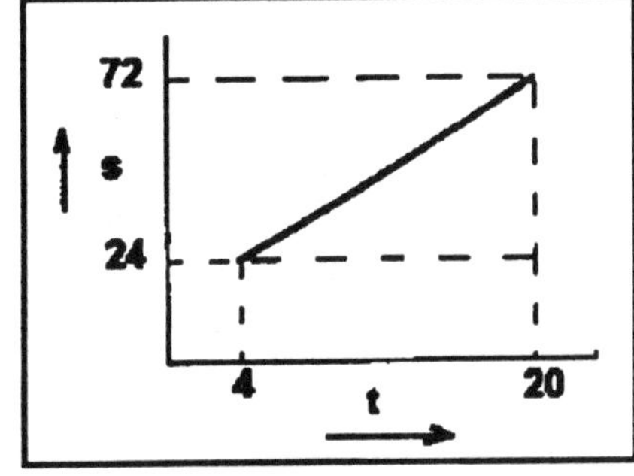

Solution (a) By definition, the slope of the line is $v = \frac{\Delta s}{\Delta t}$, or

$$\boxed{v = \frac{72-24}{20-4} = \frac{48}{16} = 3\ m/s}$$

(b) The equation of a straight line in canonical form is $s = v(t - t_o) + s_o$, where v is the slope, t is the time, and s_o is the value at $t = t_o$. Substituting $t_o = 4$ seconds, $s_o = 24\ m$, and the equation is $\boxed{s = 3(t-4) + 24\ m}$. Transform this equation to the origin, $\boxed{s = 3t - (3)(4) + 24 = 3t + 12\ m}$. The velocity of the point by definition is $\boxed{v = \frac{ds}{dt} = 3\ m/s}$

================================<>================================

Problem 2.2 The graph of the position s of a milling machine as a function of time is a straight line. When $t = 0.2$ seconds, $s = 90\ mm$. During the interval of time from $t = 0.6$ seconds to $t = 1.2$ seconds, the displacement of the point is $\Delta s = -180\ mm$. (a) Determine the equation for s as a function of time. (b) What is the velocity of the point?

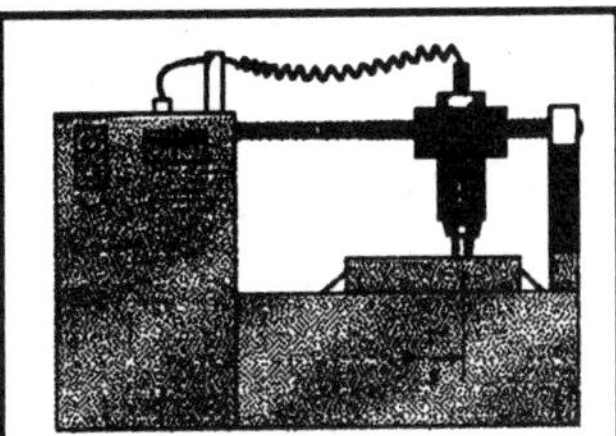

Solution (a) The equation of the point displacement is $s = \left(\frac{\Delta s}{\Delta t}\right)(t - t_o) + s_o$, from which $s = \left(\frac{-180}{1.2-0.6}\right)(t - 0.2) + 90\ mm$, or $\boxed{s = -300(t-0.2) + 90 = -300t + 150\ mm}$ Convert to meters, $s = -0.3t + 0.15\ m$. (b) The velocity of the point: $\boxed{v = \frac{ds}{dt} = -0.3\ m/second}$

================================<>================================

Problem 2.3 The graph of the velocity v of a point as a function of time is a straight line. When $t = 2\ s$, $v = 4\ ft/s$, and when $t = 4\ s$, $v = -10\ ft/s$. (a) Determine the acceleration of the point by calculating the slope of the straight line. (b) Obtain the equation for v as a function of time and use it to determine the acceleration of the point.

Solution (a) The acceleration is the slope of the straight line: $\boxed{a = \frac{\Delta v}{\Delta t} = \frac{-10-4}{4-2} = \frac{-14}{2} = -7\ ft/s^2}$ (b) The equation of the straight line velocity is $\boxed{v = \frac{\Delta v}{\Delta t}(t - t_o) + v_o = -7(t-2) + 4 = -7t + 18\ ft/s}$. The acceleration is $a = \frac{dv}{dt} = -7\ ft/s^2$

Problem 2.4 The position of a point is $s = 2t^2 - 10\ ft$. (a) What is the displacement of the point from $t = 0$ to $t = 4$ seconds? (b) What are the velocity and acceleration at $t = 0$? (c) What are the velocity and acceleration at $t = 4$ seconds?

Solution The displacement is $\boxed{\Delta s = s(4) - s(0) = 2(16) - 10 - (2(0) - 10) = 32\ ft}$.

(b) Velocity at $t = 0$ is $\boxed{\left.\frac{ds}{dt}\right|_{t=0} = [4t]_{t=o} = 0}$. Acceleration at $t = 0$ is $\boxed{\left.\frac{d^2s}{dt^2}\right|_{t=0} = [4]_{t=0} = 4\ ft/s^2}$

c) Velocity at $t = 4$ seconds is $\boxed{\left.\frac{ds}{dt}\right|_{t=4} = [4t]_{t=4} = 16\ ft/s}$. Acceleration at $t = 4$ seconds is $\boxed{\frac{d^2s}{dt^2} = [4]_{t=4} = 4\ ft/s^2}$

Problem 2.5 A rocket starts from rest and travels straight up. Its height above the ground is measured by radar from $t = 0$ to $t = 4$ seconds and is found to be approximated by the function $s = 10t^2\ m$. (a) What is the displacement during this period of time? (b) What is the velocity at $t = 4$ seconds? (c) What is the acceleration during the first 4 seconds?

Solution

a) The displacement is $\boxed{\Delta s = s(4) - s(0) = 10(4^2) - 10(0) = 160\ m}$.

(b) The velocity at $t = 4$ seconds is $\boxed{\left.\frac{ds}{dt}\right|_{t=4} = [20t]_{t=4} = 80\ m/s}$.

(c) The acceleration over the interval is $\boxed{\frac{d^2s}{dt^2} = 20\ m/s^2}$

==================================<>==================================

Problem 2.6 The position of a point during the interval of time from $t = 0$ to $t = 6$ seconds is $s = -\frac{1}{2}t^3 + 6t^2 + 4t\ m$. (a) What is the displacement of the point during this interval of time? (b) What is the maximum velocity during this interval of time, and at what time does it occur? (c) What is the acceleration when the velocity is a maximum?

Solution (a) The displacement is $\boxed{\Delta s = s(6) - s(0) = -\frac{1}{2}6^3 + 6(6^2) + 4(6) - 0 = 132\ m}$. (b) The velocity is $v = \frac{ds}{dt} = -\frac{3}{2}t^2 + 12t + 4\ m/s$. The maximum (or minimum) occurs when $\frac{dv}{dt} = 0$, or when $-3t + 12 = 0$, from which $\boxed{t = 4}$ seconds. This point is a maximum if the second derivative is negative, $\left.\frac{d^2v}{dt^2}\right|_{t=4} = -3 < 0$, which is indeed the case. The value of the maximum velocity is $\boxed{\left.\frac{ds}{dt}\right|_{t=4} = -\frac{3}{2}(4^2) + 12(4) + 4 = 28\ m/s}$ (c) The value of the acceleration is $\boxed{\left.\frac{d^2s}{dt^2}\right|_{t=4} = [-3t + 12]_{t=4} = 0}$

[*Note:* If the value extremes of a function occur at the beginning or end of an interval, in the strict sense they are called the "greatest value/least value", and not "maximum value/minimum value". However, since the engineer's interest in extreme values is usually independent of where they occur over the interval, *the text does not make the distinction between "maximum/minimum" values and "greatest/least" values.* Since the vanishing first derivative method cannot in general be used to find greatest/least values, a search of function values may be required to find the value extremes. *End of Note.*]

==================================<>==================================

Problem 2.7 The position of a point during the interval of time from $t = 0$ to $t = 3$ seconds is $s = 12 + 5t^2 - t^3\ ft$. (a) What is the maximum velocity during this interval of time, and at what time does it occur? (b) What is the acceleration when the velocity is a maximum?

Solution (a) The velocity is $\frac{ds}{dt} = 10t - 3t^2$. The maximum occurs when $\frac{dv}{dt} = 10 - 6t = 0$, from which $\boxed{t = \frac{10}{6} = 1.667}$ seconds. This is indeed a maximum, since $\frac{d^2v}{dt^2} = -6 < 0$. The maximum velocity is $\boxed{v = \left[10t - 3t^2\right]_{t=1.667} = 8.33\ ft/s}$. (b) The acceleration is $\boxed{\frac{dv}{dt} = 0}$ when the velocity is a maximum.

[See Note under solution to Problem 2.6.]

==================================<>==================================

Problem 2.8 The mechanism causes the displacement of point P to be $s = 0.2\sin(\pi t)$ meters, where t is in seconds and the argument πt of the sine function is in radians. Determine the velocity and acceleration of a point P at $t = 3.8\ s$.

Solution The velocity is $v = \frac{ds}{dt} = 0.2\ \pi\ \cos(\pi t)$, and acceleration is $a = \frac{dv}{dt} = \frac{d^2s}{dt^2} = -0.2\ \pi^2 \sin(\pi t)$.

Evaluating at $t = 3.8\ s$, [$(\pi t)_{t=3.8s} = 11.94\ rad$], we get $v_{t=3.8s} = 0.508\ m/s$ and *a =1.16 m/s/s.*

==================================<>==================================

==================================<>==================================

Problem 2.9 For the mechanism in Problem 2.8, draw graphs of the position s, velocity v, and accleration a of point P as a function of time for $0 < t < 4\ s$. Using your graphs, confirm that the slope of the graph of the position s is zero at times for which the velocity $v = {ds}/{dt}$ is zero, and the slope of the graph of the velocity v, is zero at times for which the acceleration, $a = \frac{dv}{dt}$ is zero.

Solution

We can use a plotting calculator or very simple computer program to plot the curves. We get the curves below.

From the graph, we see that the slope of the s curve is zero where the velocity is zero (t = 0.5 s, 1.5 s, 2.5s, and 3.5s). Also, the slope of the velocity curve is zero where the acceleration is zero (At t = 0 s, 1 s, 2 s, 3 s, and 4 s.

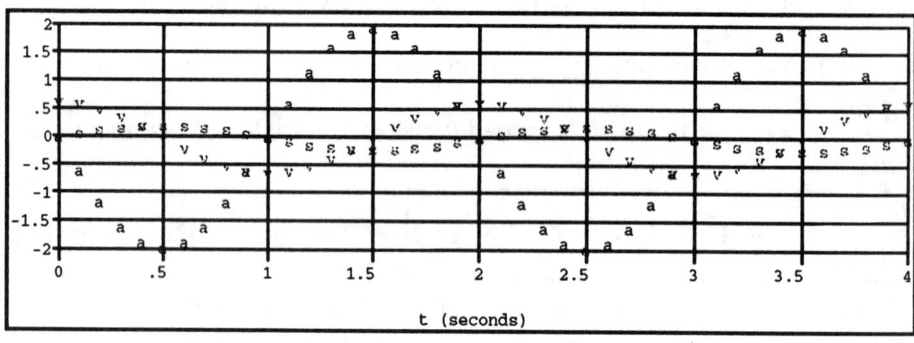

==================================<>==================================

Problem 2.10 A seismograph measures the horizontal motion of the ground during an earthquake. An engineer analyzing the data determines that for a ten-second interval of time beginning at $t = 0$, the position is approximated by $s = 100\cos(2\pi t)\ mm$. What are (a) the maximum velocity and (b) maximum acceleration of the ground during the ten second interval?

Solution (a) The velocity is $\frac{ds}{dt} = -(2\pi)100\sin(2\pi t)\ mm/s = -0.2\pi\sin(2\pi t)\ \ m/s$. The velocity maxima occur at $\frac{dv}{dt} = -0.4\pi^2\cos(2\pi t) = 0$, from which $2\pi t = \frac{(2n-1)\pi}{2}$, or

$t = \frac{(2n-1)}{4}$, $n = 1, 2, 3, \ldots. M$, where $\frac{(2M-1)}{4} \le 10$ seconds. These velocity maxima have the absolute value $\left|\frac{ds}{dt}\right|_{t=\frac{(2n-1)}{4}} = [0.2\pi] = 0.628\ \ m/s$.

(b) The acceleration is $\frac{d^2s}{dt^2} = -0.4\pi^2\cos(2\pi t)$. The acceleration maxima occur at

$\frac{d^3s}{dt^3} = \frac{d^2v}{dt^2} = 0.8\pi^3\sin(2\pi t) = 0$, from which $2\pi t = n\pi$, or

$t = \frac{n}{2}$, $n = 0, 1, 2, \ldots K$, where $\frac{K}{2} \le 10$ seconds. These acceleration maxima have the absolute value $\left|\frac{dv}{dt}\right|_{t=\frac{n\pi}{2}} = 0.4\pi^2 = 3.95\ \ m/s^2$. [See Note under solution to Problem 2.6.]

==================================<>==================================

=====================================<>=====================================

Problem 2.11 During an assembly operation, a robot's arm moves along a straight line. During an interval of time from $t = 0$ to $t = 1$ second, its position is given by $s = 3t^2 - 2t^3$ *in*. During this 1 second interval, determine: (a) the displacement of the arm; (b) the maximum and minimum values of the velocity; (c) the maximum and minimum values of the acceleration.

Solution (a) The displacement is $\Delta s = s(1) - s(0) = 3(1)^2 - 2(1)^3 = 1\ in.$, or $\Delta s = \frac{1}{12}\ ft = 0.0833\ ft$.

(b) The velocity is $\frac{ds}{dt} = v = 6t - 6t^2$. The extremum is at $\frac{dv}{dt} = 6 - 12t = 0$, or $t = 0.5$ s This is a maximum, since $\left.\frac{d^2v}{dt^2}\right|_{t=0.5} = -12 < 0$. Maximum velocity is $v_{t=0.5} = \left[6t - 6t^2\right]_{t=0.5} = 1.5$ in / s . There is no true minimum, since there is only one value of *t* for which $\frac{dv}{dt} = 0$, but the *least value* of the velocity occurs at $t = 0$ and $t = 1\ s$, where $v = 0$. (c) The acceleration is $\frac{dv}{dt} = a = 6 - 12t$. The maximum and minimum occur when $\frac{d^2v}{dt^2} = \frac{da}{dt} = 0$. This does not occur, since $\frac{da}{dt} = -12 \neq 0$. The *greatest value* of acceleration is at $t = 0$, where $a = \frac{dv}{dt} = 6\ in/s^2 = 0.5\ ft/s^2$. The *least value* is at $t = 1$ second, where $a = \frac{dv}{dt} = -6\ in/s^2 = -0.5\ ft/s^2$. [See Note under solution to Problem 2.6.]

=====================================<>=====================================

Problem 2.12 In the test of a prototype car, the driver starts the car from rest at $t = 0$, accelerates, and then applies the brakes. Engineers measuring the position of the car find that from $t = 0$ to $t = 18$ seconds it is approximated by $s = 5t^2 + \frac{1}{3}t^3 - \frac{1}{50}t^4$ *ft*. (a) What is the maximum velocity, and at what time does it occur? (b) What is the maximum acceleration, and at what time does it occur?

Solution Assume that *s* is measured in feet, and that *t* is measured in seconds. (a) The velocity is $\frac{ds}{dt} = v = 10t + t^2 - \frac{2}{25}t^3\ ft/s$. The maximum occurs at $\frac{dv}{dt} = 10 + 2t - \frac{6}{25}t^2 = 0$. In canonical form this quadratic equation is $t^2 + 2bt + c = 0$, where $b = -\frac{25}{6}$, $c = -\frac{125}{3}$, with the solution $t = -b \pm \sqrt{b^2 - c} = 11.85,\ = -3.52$ seconds. The negative value is outside the interval of interest. At $t = 11.85$ s there is a maximum since $\left.\frac{d^2v}{dt^2}\right|_{t=11.85} = \left[2 - \frac{12}{25}t\right]_{t=11.85} = -3.69 < 0$. The maximum velocity is $v(11.85) = \left[10t + t^2 - \frac{2}{25}t^3\right]_{t=11.85} = 125.8\ ft/s$. (b) Acceleration is $a = \frac{dv}{dt} = 10 + 2t - \left(\frac{6}{25}\right)t^2$. The maximum occurs at $\frac{da}{dt} = 2 - \frac{12}{25}t = 0$, from which $t = \frac{50}{12} = 4.17$ seconds. This is a maximum since $\left.\frac{d^2a}{dt^2}\right|_{t=4.17} = -\frac{12}{25} < 0$. The maximum acceleration is $a(4.17) = \left[10 + 2t - \frac{6}{25}t^2\right]_{t=4.17} = 14.17\ ft/s^2$

=====================================<>=====================================

================================<>================================

Problem 2.13 Suppose you want to approximate the position of a vehicle you are testing by the power series $s = A + Bt + Ct^2 + Dt^3$ where *A, B, C,* and *D* are constants. The vehicle starts from rest at $t = 0$ and $s = 0$. At $t = 4$ seconds, $s = 176\,ft$, and at $t = 8$ seconds, $s = 448\,ft$. (a) Determine *A,B,C,* and *D.* (b) What are the approximate velocity and acceleration of the vehicle at $t = 8$ seconds?

Solution At $t = 0$, $s = 0$, hence $\boxed{0 = A}$, and $s = Bt + Ct^2 + Dt^3$. Since the vehicle starts from rest at $t = 0$, the velocity is zero, and $\boxed{0 = B}$ At $t = 4$ seconds, $s = 176\,ft$, from which $176 = 16C + 64D$. At $t = 8$ seconds $s = 448\,ft$, from which $448 = 64C + 512D$. Solve these two simultaneous equations to obtain $\boxed{C = 15}$ and $\boxed{D = -1}$. (b) The position is given by $s = 15t^2 - t^3$, from which the velocity is $v = \frac{ds}{dt} = 30t - 3t^2$. The velocity at $t = 8$ seconds is $\boxed{v(8) = \left[30t - 3t^2\right]_{t=8} = 48\,ft/s}$. The acceleration is $a = \frac{dv}{dt} = 30 - 6t$. The acceleration at $t = 8$ seconds is $\boxed{a(8) = [30 - 6t]_{t=8} = -18\,ft/s^2}$.

================================<>================================

Problem 2.14 The acceleration of a point is $a = 20t\,m/s^2$. When $t = 0$, $s = 40\,m$, and $v = -10\,m/s$. What are the position and velocity at $t = 3\,s$?

Solution The velocity is $v = \int a\,dt + C_1$, where C_1 is the constant of integration. Thus

$v = \int 20\,t\,dt + C_1 = 10t^2 + C_1$. At $t = 0$, $v = -10\,m/s$, hence $C_1 = -10$ and the velocity is

$v = 10t^2 - 10\;m/s$. The position is $s = \int v\,dt + C_2$, where C_2 is the constant of integration.

$s = \int (10t^2 - 10)\,dt + C_2 = \left(\frac{10}{3}\right)t^3 - 10t + C_2$. At $t = 0$, $s = 40\,m$, thus $C_2 = 40$. The position is

$s = \left(\frac{10}{3}\right)t^3 - 10t + 40\;m$. At $t = 3$ seconds, $\boxed{s(3) = \left[\frac{10}{3}t^3 - 10t + 40\right]_{t=3} = 100\;m}$. The velocity at

$t = 3$ seconds is $\boxed{v(3) = \left[10t^2 - 10\right]_{t=3} = 80\,m/s}$

================================<>================================

Problem 2.15 The acceleration of a point is $a = 60t - 36t^2\;ft/s^2$. When $t = 0$, $s = 0$ and $v = 20\;ft/s$. What are position and velocity as a function of time?

Solution The velocity is $v = \int a\,dt + C_1 = \int (60t - 36t^2) + C_1 = 30t^2 - 12t^3 + C_1$. At $t = 0$,

$v = 20\;ft/s$, hence $C_1 = 20$, and the velocity as a function of time is $\boxed{v = 30t^2 - 12t^3 + 20\;ft/s}$. The position is $s = \int v\,dt + C_2 = \int \left(30t^2 - 12t^3 + 20\right) + C_2 = 10t^3 - 3t^4 + 20t + C_2$. At $t = 0$, $s = 0$, hence

$C_2 = 0$, and the position is $\boxed{s = 10t^3 - 3t^4 + 20t\;ft}$

================================<>================================

==========================<>==========================

Problem 2.16 Suppose that during the preliminary design of a car, you assume that its maximum acceleration is approximately constant. What constant acceleration is necessary if you want the car to be able to accelerate from rest to a velocity of 55 mi/hr in 10 seconds? What distance would the car travel during that time?

Solution The velocity is $v = \int a\, dt + C_1 = at + C_1$. Since the car starts from rest at $t = 0$, $C_1 = 0$, and the velocity is $v = at$. At the end of 10 seconds, $v = 55\ mi/hr\left(\frac{1\ hr}{3600\ s}\right)\left(\frac{5280\ ft}{1\ mi}\right) = 80.67\ \ ft/s$, from which $10a = 80.67$, or the required acceleration is $\boxed{a = 8.07\ ft/s^2}$. The distance traveled is

$$\boxed{s = \int_0^{10} 8.07t\, dt = \left[4.033t^2\right]_0^{10} = 403\ ft}$$

==========================<>==========================

Problem 2.17 An entomologist estimates that a flea one millimeter in length attains a velocity of 1.3 *m/s* in a distance of one body length when jumping. What constant acceleration is necessary to achieve that velocity?

Solution The velocity is $v = \int a\, dt + C_1 = at + C_1$. The flea jumps from rest at $t = 0$, hence $C_1 = 0$, and $v = at$. The position is given by $s = \int v\, dt + C_2 = \frac{a}{2}t^2 + C_2$. Assume that $s = 0$ at $t = 0$, thus $C_2 = 0$, and the position is $s = \frac{a}{2}t^2$. Let t_1 be the time after the jump that the flea has moved one body length. Then $t_1 = \sqrt{\frac{2(0.001)}{a}} = \sqrt{\frac{2 \times 10^{-3}}{a}}$. Substitute into the velocity to obtain

$v = a\sqrt{\frac{2 \times 10^{-3}}{a}} = \sqrt{2a \times 10^{-3}}\ \ m/s$. The velocity at $t = t_1$ is $v = 1.3\ m/s$, from which

$a = \frac{(1.3)^2 \times 10^3}{2} = 845\ \ m/s^2$. The acceleration in g's is $\boxed{a_g = \frac{845}{9.81} = 86.1\ \ g}$.

==========================<>==========================

Problem 2.18 Missiles designed for defense against ballistic missiles achieve accelerations in excess of 100 *g's*, or one hundred times the acceleration of gravity. If a missile has a constant acceleration of 100 *g's*, how long does it take to go from rest to 60 mi/hr? What is its displacement during that time?

Solution Use $g = 32.2\ \ ft/s^2$ for the acceleration of gravity. Since the missile starts from

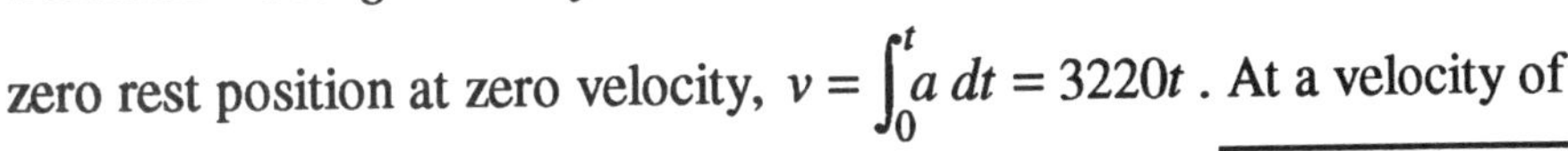

zero rest position at zero velocity, $v = \int_0^t a\, dt = 3220t$. At a velocity of

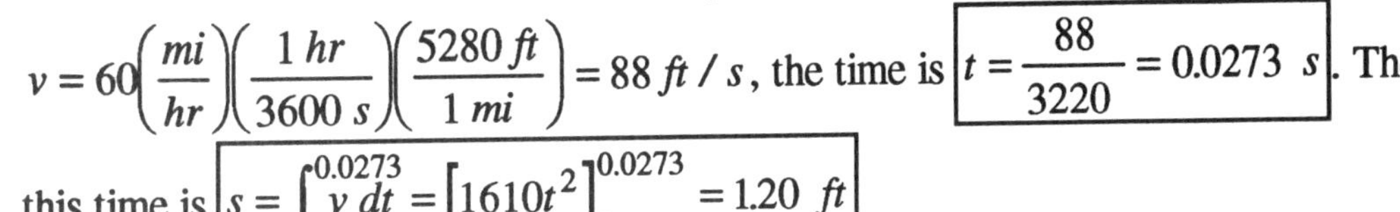

$v = 60\left(\frac{mi}{hr}\right)\left(\frac{1\ hr}{3600\ s}\right)\left(\frac{5280\ ft}{1\ mi}\right) = 88\ ft/s$, the time is $\boxed{t = \frac{88}{3220} = 0.0273\ \ s}$. The displacement during this time is $\boxed{s = \int_0^{0.0273} v\, dt = \left[1610t^2\right]_0^{0.0273} = 1.20\ \ ft}$

==========================<>==========================

==================================<>==================================

Problem 2.19 Suppose you want to throw some keys to a friend standing on a second floor balcony. If you release the keys at 1.5 *m* above ground, what vertical velocity is required for them to reach your friend's hand 6 meters above ground?

Solution The acceleration of gravity opposes the upward travel. The velocity is $v = \int -g\,dt + v(0) = -gt + v(0)$, where $v(0)$ is the vertical velocity required. The vertical position of the keys is $s = \int (-gt + v(0))dt + s(0) = -\frac{g}{2}t^2 + v(0)t + s(0)$. Let $t = t_1$ be time required to reach the required height, and assume that the velocity at that height is zero. Then $0 = -gt_1 + v(0)$, from which $v(0) = gt_1$. Substituting into the position: $6 = -\frac{g}{2}t_1^2 + gt_1^2 + 1.5$, from which $t_1 = \sqrt{\frac{2(4.5)}{g}} = \sqrt{\frac{9}{9.81}} = 0.958\ s$. Substituting, $\boxed{v(0) = gt_1 = (9.81)(0.958) = 9.4\ m/s}$

==================================<>==================================

Problem 2.20 The lunar module descends toward the surface of the Moon at 1 *m/s* when its landing probes, which extend 2 *m* below the landing gear, touch the surface, automatically shutting off the engines. Determine the velocity at which the landing gear contact the surface. (The acceleration of gravity at the surface of the Moon is $1.62\ m/s^2$.

Solution The acceleration acts to increase the velocity. The velocity is $v = \int g_m dt + v(0) = g_m t + v(0)$ where $v(0) = 1\ m/s$ is the initial velocity. The distance traveled after probe contact is $s = \int v dt + s(0) = \int (g_m t + v(0))dt + s(0) = \frac{g_m}{2}t^2 + v(0)t + s(0)$, where $s(0) = 0$ is the initial position at probe contact, from which $s = \frac{g_m}{2}t^2 + 1t$. At landing gear contact, $s = 2\ m$, from which $\frac{g_m}{2}t^2 + t - 2 = 0$. In canonical form this quadratic equation is $t^2 + 2bt + c = 0$, where $b = \frac{1}{g_m}$ and $c = -\frac{4}{g_m}$, with the solution $t = -b \pm \sqrt{b^2 - c} = 1.07, \quad = -2.3$. The negative value has no meaning here, $t = 1.07$ seconds, and the velocity at contact is $\boxed{v(1.07) = [g_m t + 1]_{t=1.07} = (1.62)(1.07) + 1 = 2.73\ m/s}$

==================================<>==================================

================================<>================================

Problem 2.21 In 1960, R. C. Owens of the Baltimore Colts blocked a Washington Redskins field goal attempt by jumping up and knocking the ball away in front of the cross bar at a point 11 *ft* above the field. If he was 6 *ft* 3 *in.* tall and could reach 1 *ft* 11 *in.* above his head, what was his initial velocity as he left the ground?

Solution Assume that at the maximum height reached (11 *ft*) the velocity is zero. The distance traveled is $s = 11 - 8.1667 = 2.8333\,ft$. The initial velocity is $v(t_o) = \sqrt{2gs}$, where $g = 32.17\,ft/s^2$ is the acceleration of gravity, from which $\boxed{v(t_o) = 13.5\,ft/s}$. *.Check.* The equation for vertical straight line motion is $s(t) = -\frac{g}{2}t^2 + v(t_o)t + s(t_o)$, where $v(t_o)$, $s(t_o)$ are the initial velocity and displacement. The time when the velocity is zero is given by $\frac{ds(t)}{dt} = -gt + v(t_o) = 0$, from which $t = \frac{v(t_o)}{g}$. Substitute into the equation for straight line vertical motion to obtain $v^2(t_o) = 2gs(t_o)$, from which $v(t_o) = 13.5\,ft/s$. *check.*

================================<>================================

Problem 2.22 The velocity of a bobsled is $v = 10t\,ft/s$. When $t = 2$ seconds, its position is $s = 25\,ft$. What is its position at $t = 10\,s$?

Solution: . The equation for straight line displacement under constant acceleration is

$$s(t - t_o) = \frac{a(t - t_o)^2}{2} + v(t_o)(t - t_o) + s(t_o).$$

Choose $t_o = 0$. At $t = 2$, the acceleration is

$$a = \left[\frac{dv(t)}{dt}\right]_{t=2} = 10\,ft/s^2,$$

the velocity is $v(t_o) = 10(2) = 20\,ft/s$, and the initial displacement is $s(t_o) = 25\,ft$. At $t = 10$ seconds, the displacement is $\boxed{s(10-2) = \frac{10}{2}(10-2)^2 + 20(10-2) + 25 = 505\,ft}$

================================<>================================

Problem 2.23 The acceleration of an object is $a = 30 - 6t\,ft/s^2$. When $t = 0$, $s = 0$ and $v = 0$. What is the maximum velocity during the interval of time $t = 0$ to 10 seconds?

Solution: The velocity is $v(t) = \int a\,dt + v(0) = \int (30 - 6t)dt = 30t - 3t^2$, since $v(0) = 0$. The velocity occurs at $\frac{dv(t)}{dt} = 30 - 6t = 0$, from which $t = 5$ seconds. This is a maximum, since $\frac{d^2v(t)}{dt^2} = -6 < 0$. Substitute into the velocity equation: $\boxed{v(5) = \left[30t - 3t^2\right]_{t=5} = 75\,ft/s}$ is the maximum velocity in the interval. [See Note under the solution to Problem 2.6.]

================================<>================================

==================================<>==================================

Problem 2.24 The velocity of an object is $v = 200 - 2t^2\ m/s$. When $t = 3$ seconds, its position is $s = 600\ m$. What are the position and acceleration of the object at $t = 6$ seconds?

Solution: The acceleration is $\frac{dv(t)}{dt} = -4t\ m/s^2$. At $t = 6$ seconds, the acceleration is $a = -24\ m/s^2$. Choose the initial conditions at $t_o = 3$ seconds. The position is obtained from the velocity: $\boxed{s(t - t_o) = \int_3^6 v(t)\,dt + s(t_o) = \left[200t - \frac{2}{3}t^3\right]_3^6 + 600 = 1074\ m}$.

==================================<>==================================

Problem 2.25 The acceleration of a part undergoing a machining operation is measured and determined to be $a = 12 - 6t\ mm/s^2$. When $t = 0, v = 0$. For the interval of time from $t = 0$ to $t = 4$ seconds, determine (a) the maximum velocity, (b) the displacement.

Solution: (a) The velocity is $v(t) = \int a(t)dt + v(0) = 12t - 3t^2$, since $v(0) = 0$. The maximum occurs at time obtained from $\frac{dv(t)}{dt} = 0 = 12 - 6t$, from which $t = 2$ seconds. This is a maximum since $\frac{d^2v(t)}{dt^2} = -6 < 0$. [See Note under the solution to Problem 2.6.] Substitute into the velocity equation: $\boxed{v(2) = \left[12t - 3t^2\right]_{t=2} = 12\ mm/s}$. (b) The displacement is

$$\boxed{s(4) = \int_0^4 v(t)dt = \int_0^4 \left(12t - 3t^2\right)dt = \left[6t^2 - t^3\right]_0^4 = 32\ mm}$$

==================================<>==================================

Problem 2.26 The missile shown in Problem 2.18 starts from rest and accelerates straight up for 3 seconds at 100*g*'s. After 3 seconds, its weight and aerodynamic drag cause it to have a nearly constant deceleration of 4*g*'s. How long does it take the missile to go from the ground to an altitude of 50,000 *ft*?

Solution: Divide the time of flight into two intervals: the period of 100*g* acceleration (3 seconds) and the period of 4*g* deceleration. The velocity after 3 seconds is $v(3) = \int_0^3 100(32.17)dt = [3217t]_0^3 = 9651\,ft/s$, where the missile flight starts from rest. The distance traveled is $s(3) = \int_0^3 v(t)dt = \int_0^3 3217t\,dt = \left[1608.5t^2\right]_0^3 = 14{,}476.5\,ft$. The equation for the displacement during the second period is $s(t-3) = -\frac{4(32.17)}{2}(t-3)^2 + v(3)(t-3) + s(3)$, where $t \ge 3$ seconds. For $s(t-3) = 50000\,ft$, this becomes a quadratic equation:, $-64.34(t-3)^2 + 9651(t-3) - 35523.5 = 0$. In canonical form $(t-3)^2 + 2b(t-3) + c = 0$, where $b = -75$ and $c = 552.12$, with the solution $(t-3) = -b \pm \sqrt{b^2 - c} = 3.776\ ,\ = 146.2$. The total time of flight is $\boxed{t = 6.776}$ seconds.

==================================<>==================================

================================<>================================

Problem 2.27 The graph describes an airplane's acceleration during its take-off run. Determine the airplane's velocity when it rotates (lifts off) at t = 20 s.

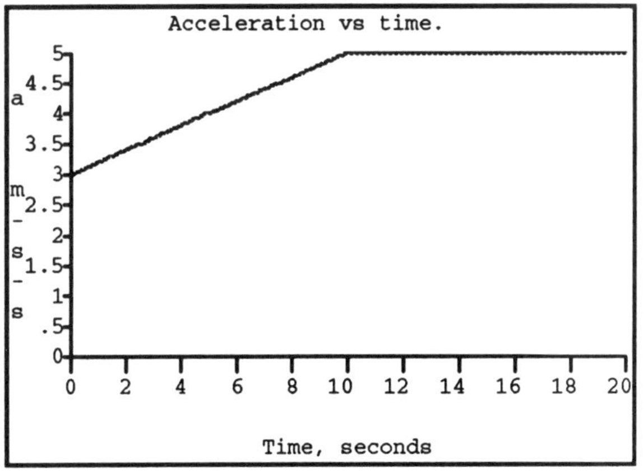

Solution: The acceleration of the airplane in the interval (0,10 s) is given by $a_1(t) = 3 + 0.2\ t\ \text{m/s}^2$ and the acceleration over the interval (10 s,20 s) is given by $a_2 = 5\ \text{m/s}^2$. The velocity of the aircraft at lift-off is given by $v(20) = \int_0^{10} a_1\, dt + \int_{10}^{20} a_2\, dt$. Carrying out the indicated integrations, the result is $v(20) = \left[3t + 0.2\ t^2/2\right]_0^{10} + [5\ t]_{10}^{20}\ \text{m/s}$. Evaluating the result, we get $v(20) = 30 + 0.2(50) + 50 = 90\ \text{m/s}$.

================================<>================================

Problem 2.28 In problem 2.27, determine the distance that the airplane has traveled when it rotates at t = 20 s.

Solution: From 2.27, the velocity is given by $v(20) = \int_0^{10} a_1\, dt + \int_{10}^{20} a_2\, dt$, where $a_1(t) = 3 + 0.2\ t\ \text{m/s}^2$ and the acceleration over the interval (10 s,20 s) is given by $a_2 = 5\ \text{m/s}^2$. From Problem 2.27, the velocity at 10 s is given by $v(10) = \left[3t + 0.2\ t^2/2\right]_0^{10}\ \text{m/s} = \left[30 + 0.2(10)^2/2\right]\text{m/s} = 40\ \text{m/s}$. The distance traveled during the first 10 seconds of the motion is given by $s(10) = \left[3\ t^2/2 + 0.2\ t^3/6\right]_0^{10}\ \text{m} = \left[3\ (10)^2/2 + 0.2\ (10)^3/6\right]\text{m} = 183.3\ \text{m}$. The second phase of the motion takes place in the interval from 10 seconds to 20 seconds, during which the acceleration is constant. The velocity and position at 20 seconds are given by $v(20) = v(10) + 5(t-10)_{t=20\ s}\ \text{m/s} = 90\ \text{m/s}$ and $s(20) = [s(10) + v(10)(t-10) + 5(t-10)^2/2]_{t=20s}\ \text{m}$. Evaluating, we get $s(20) = [183.3 + 40(10) + 5(20-10)^2/2]\ \text{m} = 833.3\ \text{m}$.

Discussion: Note that the airplane gains more speed and goes much farther during the last 10 s of the 20 s takeoff run than it does during the first 10 s. The acceleration is lower during the first 10 s and the airplane is moving much slower during the initial 10 s.

================================<>================================

Problem 2.29 A car is traveling at 30 *mi/hr* when a traffic light 295 feet ahead turns yellow. The light will remain yellow for five seconds. (a) What constant acceleration will cause the car to reach the light at the instant it turns red, and what will the velocity of the car be when it reaches the light? (b) If the driver decides not to try to make the light, what constant rate of acceleration will cause the car to come to a stop just as it reaches the light?

Solution: The displacement for straight line motion under constant acceleration is $s(t) = \frac{a}{2}t^2 + v(0)t + s(0)$. Assume that $s(0) = 0$. (a) The velocity at $t = 0$ is given:

$v(0) = 30\ mi/hr = 30\left(\frac{5280\ ft}{1\ mi}\right)\left(\frac{1\ hr}{3600\ s}\right) = 44\ ft/s$. The distance traveled in 5 seconds is $s(5) = 295\ ft$,

from which $295 = \frac{a}{2}(5)^2 + 44(5)$. Solve: $\boxed{a = 6\ ft/s^2}$ The velocity for straight line motion under

Solution continued on next page

Continuation of Solution to Problem 2.29

constant acceleration is $\frac{ds(t)}{dt} = v(t) = at + v(0)$. At $t = 5$ seconds $v(5) = 6(5) + 44 = 74\ ft/s = 50.45\ mi/hr$.

(b) Two conditions are to be met: the car comes to a stop as it reaches the light, therefore the distance traveled is $295\ ft$ and the velocity is zero at the end of that travel. The distance traveled is $s(t) = 295 = \frac{a}{2}t^2 + 44t$, and the velocity is $v(t) = 0 = at + 44$. The time required is $t = -\frac{44}{a}$. Substitute into the displacement equation to obtain $a = -\frac{(44)^2}{2(295)} = -3.28\ ft/s^2$. The time required is $t = 13.4\ s$.

==================================<>==================================

Problem 2.30 At $t = 0$ a motorist traveling at $100\ km/hr$ sees a deer standing in the road $100\ m$ ahead. After a reaction time of 0.3 seconds, he applies the brakes and decelerates at a constant rate of $4\ m/s^2$. If the deer takes 5 seconds from $t = 0$ to react and leave the road, does the motorist miss him?

Solution: Divide the time into two intervals, the reaction time of the motorist (0.3 s) and the time before the deer leaves the road ($5 - 0.3 = 4.7\ s$). The initial velocity is $v(0) = 100\frac{\text{km}}{\text{hr}} = 27.8\ \text{m/s}$. The distance traveled in the first interval is $s = 27.8(0.3) = 8.33\ m$. The distance traveled in the second interval is $s(t) = -\frac{a}{2}t^2 + v(0)t = -2(4.7)^2 + 27.8(4.7) = 86.4\ m$. The total distance traveled is $s(5) = 8.33 + 86.4 = 94.7\ m$, which is less than the $100\ m$ from motorist to deer at $t = 0$. Yes, the motorist misses the deer.

==================================<>==================================

Problem 2.31 A high speed rail transportation system has a top speed of $100\ m/s$. For the comfort of the passengers, the magnitude of the acceleration and deceleration is limited to $2\ m/s^2$. Determine the time required for a trip of $100\ km$.

Solution: Divide the time of travel into three intervals: The time required to reach a top speed of $100\ m/s$, the time traveling at top speed, and the time required to decelerate from top speed to zero. From symmetry, the first and last time intervals are equal, and the distances traveled during these intervals are equal. The initial time is obtained from $v(t_1) = at_1$, from which $t_1 = 100/2 = 50\ s$. The distance traveled during this time is $s(t_1) = at_1^2/2$ from which $s(t_1) = 2(50)^2/2 = 2500\ \text{m}$. The third time interval is given by $v(t_3) = -at_3 + 100 = 0$, from which $t_3 = 100/2 = 50\ s$. *Check.* The distance traveled is $s(t_3) = -\frac{a}{2}t_3^2 + 100t_3$, from which $s(t_3) = 2500\ m$. *Check.* The distance traveled at top speed is $s(t_2) = 100000 - 2500 - 2500 = 95000\ m = 95\ km$. The time of travel is obtained from the distance traveled at zero acceleration: : $s(t_2) = 95000 = 100t_2$, from which $t_2 = 950$. The total time of travel is $t_{total} = t_1 + t_2 + t_3 = 50 + 950 + 50 = 1050\ s$ $\boxed{= 17.5\ minutes}$.

A plot of velocity versus time can be made and the area under the curve will be the distance traveled. The length of the constant speed section of the trip can be adjusted to force the length of the trip to be the required $100\ km$.

==================================<>==================================

=================================<>=================================

Problem 2.32 The nearest star, Proxima Centauri, is 4.22 light years from the Earth. Ignoring relative motion between the solar system and Proxima Centauri, suppose that a spacecraft accelerates from the vicinity of the Earth at $0.01g$ (0.01 times the acceleration due to gravity at sea level) until it reaches one-tenth the speed of light, coasts until time to decelerate, then decelerates at $0.01g$ until it comes to rest in the vicinity of Proxima Centauri. How long does the trip take? (Light travels at $3\times10^8\ m/s$. A solar year is 365.2422 solar days.)

Solution: The distance to Proxima Centauri is

$d = (4.22\ light\text{-}year)(3\times10^8\ m/s)(365.2422\ day)\left(\frac{86400\ s}{1\ day}\right) = 3.995\times10^{16}\ m$. Divide the time of flight into the three intervals. The time required to reach 0.1 times the speed of light is

$t_1 = \frac{v}{a} = \frac{3\times10^7\ m/s}{0.0981\ m/s^2} = 3.0581\times10^8$ seconds. The distance traveled is $s(t_1) = \frac{a}{2}t_1^2 + v(0)t + s(0)$, where $v(0) = 0$ and $s(0) = 0$ (from the conditions in the problem), from which $s(t_1) = 4.587\times10^{15}\ m$. From symmetry, $t_3 = t_1$, and $s(t_1) = s(t_3)$. The length of the middle interval is

$s(t_2) = d - t_1 - t_3 = 3.0777\times10^{16}\ m$. The time of flight at constant velocity is

$t_2 = \frac{3.0777\ \times10^{16} m}{3\times10^7} = 1.026\times10^9$ seconds. The total time of flight is

$t_{total} = t_1 + t_2 + t_3 = 1.63751\times10^9$ seconds. In solar years:

$$t_{total} = (1.63751\times10^9\ sec)\left(\frac{1\ solar\ years}{365.2422\ days}\right)\left(\frac{1\ days}{86400\ sec}\right) = 51.89\ solar\ years$$

=================================<>=================================

Problem 2.33 A race car starts from rest and accelerates at $a = 5 + 2t\ ft/s^2$ for 10 seconds. The brakes are then applied and the car has a constant acceleration $a = -30\ ft/s^2$ until it comes to rest. Determine (a) the maximum velocity, (b) the total distance traveled; (c) the total time of travel.

Solution: (a) For the first interval, the velocity is $v(t) = \int(5+2t)dt + v(0) = 5t + t^2$ since $v(0) = 0$. The velocity is an increasing monotone function; hence the maximum occurs at the end of the interval, $t = 10\ s$, from which $\boxed{v_{\max} = 150\ ft/s}$. (b) The distance traveled in the first interval is

$s(10) = \int_0^{10}(5t + t^2)dt = \left[\frac{5}{2}t^2 + \frac{1}{3}t^3\right]_0^{10} = 583.33\ ft$. The time of travel in the second interval is

$v(t_2 - 10) = 0 = a(t_2 - 10) + v(10)$, $t_2 \geq 10\ s$, from which $(t_2 - 10) = -\frac{150}{-30} = 5$, and (c) the total time of travel is $t_2 = 15$ The total distance traveled is $s(t_2 - 10) = \frac{a}{2}(t_2 - 10)^2 + v(10)(t_2 - 10) + s(10)$,

from which (b) $\boxed{s(5) = \frac{-30}{2}5^2 + 150(5) + 583.33 = 958.33\ ft}$

=================================<>=================================

==================================<>==================================

Problem 2.34 When $t = 0$ the position of a point is $s = 6\ m$ and its velocity is $v = 2\ m/s$. From $t = 0$ to $t = 6$, its acceleration is $a = 2 + 2t^2\ m/s^2$. From $t = 6$ until it comes to rest, its acceleration is $a = -4\ m/s^2$. (a) What is the total time of travel? (b) What total distance does it move?

Solution: For the first interval the velocity is $v(t) = \int (2 + 2t^2) dt + v(0) = \left[2t + \frac{2}{3}t^3\right] + 2\ m/s$. The velocity at the end of the interval is $v(6) = 158\ m/s$. The displacement in the first interval is $s(t) = \int \left(2t + \frac{2}{3}t^3 + 2\right) dt + 6 = \left[t^2 + \frac{1}{6}t^4 + 2t\right] + 6$. The displacement at the end of the interval is $s(6) = 270\ m$. For the second interval, the velocity is $v(t-6) = a(t-6) + v(6) = 0$, $t \ge 6$, from which $(t-6) = -\frac{v(6)}{a} = -\frac{158}{-4} = 39.5$. The total time of travel is (a) $\boxed{t_{total} = 39.5 + 6 = 45.5}$ seconds.(b) The distance traveled is $s(t-6) = \frac{-4}{2}(t-6)^2 + v(6)(t-6) + s(6) = -2(39.5)^2 + 158(39.5) + 270$, from the total distance is $\boxed{s_{total} = 3390.5\ m}$

==================================<>==================================

Problem 2.35 Zoologists studying the ecology of the Serengeti Plain estimate that the average adult cheetah can run $100\ km/hr$, and that the average springbuck can run $65\ km/hr$. If the animals run along the same straight line, start at the same time, and are each assumed to have constant acceleration and reach top speed in 4 seconds, how close must the a cheetah be when the chase begins to catch a springbuck in 15 seconds.

Solution: The top speeds are $V_c = 100\ km/hr = 27.78\ m/s$ for the cheetah, and $V_s = 65\ km/hr = 18.06\ m/s$. The acceleration is $a_c = \frac{V_c}{4} = 6.94\ m/s^2$ for the cheetah, and $a_s = \frac{V_s}{4} = 4.513\ m/s^2$ for the springbuck.Divide the intervals into the acceleration phase and the chase phase. For the cheetah, the distance traveled in the first is $s_c(t) = \frac{6.94}{2}(4)^2 = 55.56\ m$. The total distance traveled at the end of the second phase is $s_{total} = V_c(11) + 55.56 = 361.1\ m$. For the springbuck, the distance traveled during the acceleration phase is $s_s(t) = \frac{4.513}{2}(4)^2 = 36.11\ m$. The distance traveled at the end of the second phase is $s_s(t) = 18.06(11) + 36.1 = 234.7\ m$. The permissible separation between the two at the beginning for a successful chase is $\boxed{d = s_c(15) - s_s(15) = 361.2 - 234.7 = 126.4\ m}$

==================================<>==================================

===================================<>===================================

Problem 2.36 Suppose that a person unwisely drives 75 mi/hr in a 55 mi/hr zone and passes a police car going 55 mi/hr in the same direction. If the police begin constant acceleration at the instant they are passed and increase their speed to 80 mi/hr in 4 seconds, how long does it take them to be even with the pursued car?

Solution: The conversion from *mi/hr* to *ft/s* is $1.467\left(\frac{ft\text{ - }hr}{mi\text{ - }second}\right)$. The acceleration of the police car is $a = \frac{(80-55)(1.467)\text{ft/s}}{4\text{ s}} = 9.169\text{ ft/s}^2$. The distance traveled during acceleration is $s(t_1) = \frac{9.169}{2}(4)^2 + 55(1.467)(4) = 396\text{ ft}$. The distance traveled by the pursued car during this acceleration is $s_c(t_1) = 75(1.467)t_1 = 110(4) = 440\,ft$. The separation between the two cars at 4 seconds is $d = 440 - 396 = 44\,ft$. This distance is traversed in the time $t_2 = \frac{44}{(80-75)(1.467)} = 6$. The total time is $\boxed{t_{total} = 6 + 4 = 10}$ seconds.

===================================<>===================================

Problem 2.37 If $\theta = 1$ radian and $\frac{d\theta}{dt} = 1$ radian/second, what is the velocity of P relative to O?

Strategy: The position of P relative to O is $s = (2\,ft)\cos\theta + (2\,ft)\cos\theta$. Take the derivative of this expression with respect to time to determine the velocity.

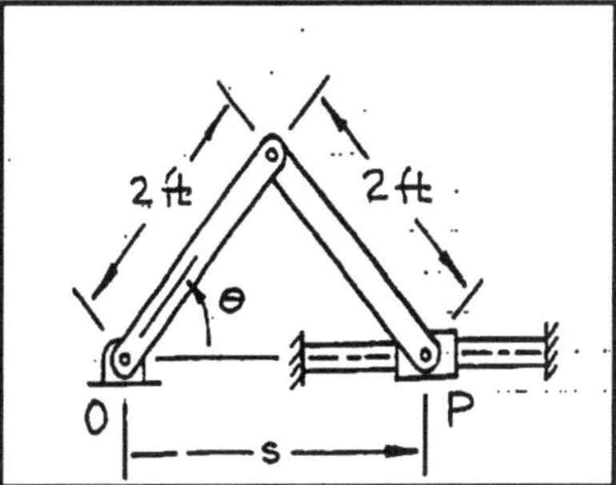

Solution: The distance s from point O is

$s = (2\,ft)\cos\theta + (2\,ft)\cos\theta$. The derivative is $\frac{ds}{dt} = -4\sin\theta\frac{d\theta}{dt}$.

For $\theta = 1$ radian and $\frac{d\theta}{dt} = 1$ radian/second,

$$\boxed{\frac{ds}{dt} = v(t) = -4(\sin(1\,rad)) = -4(0.841) = -3.37\,ft/s}$$

===================================<>===================================

Problem 2.38 In Problem 2.37, if $\theta = 1$ radian, $\frac{d\theta}{dt} = -2$ radian per second and $\frac{d^2\theta}{dt^2} = 0$, what are the velocity and acceleration of P relative to O?

Solution: The velocity is $\boxed{\frac{ds}{dt} = -4\sin\theta\frac{d\theta}{dt} = -4(\sin(1\,rad))(-2) = 6.73\,ft/s}$. The acceleration is

$\frac{d^2s}{dt^2} = -4\cos\theta\left(\frac{d\theta}{dt}\right)^2 - 4\sin\theta\left(\frac{d^2\theta}{dt^2}\right)$, from which $\boxed{\frac{d^2s}{dt^2} = a = -4\cos(1\,rad)(4) = -8.64\,ft/s^2}$

===================================<>===================================

===================================<>===================================

Problem 2.39 If $\theta = 1$ radian and $\frac{d\theta}{dt} = 1$ radian per second, what is the velocity of P relative to O?

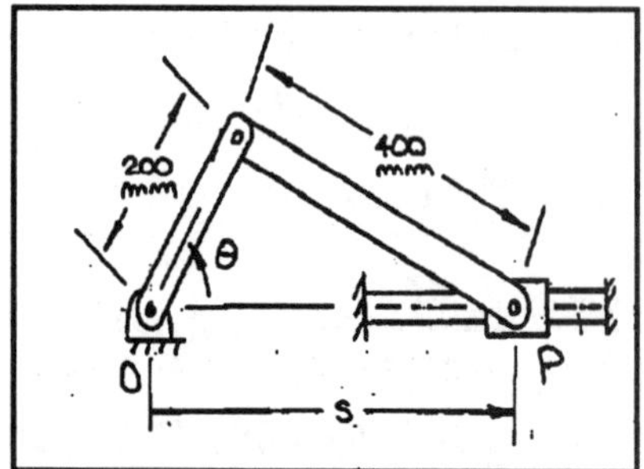

Solution: The acute angle formed by the 400 mm arm with the horizontal is given by the sine law: $\frac{200}{\sin\alpha} = \frac{400}{\sin\theta}$, from which $\sin\alpha = \left(\frac{200}{400}\right)\sin\theta$. For $\theta = 1$ radian, $\alpha = 0.4343$ radians. The position relative to O is . $s = 200\cos\theta + 400\cos\alpha$. The velocity is $\frac{ds}{dt} = v(t) = -200\sin\theta\left(\frac{d\theta}{dt}\right) - 400\sin\alpha\left(\frac{d\alpha}{dt}\right)$. From the expression for the angle α, $\cos\alpha\left(\frac{d\alpha}{dt}\right) = 0.5\cos\theta\left(\frac{d\theta}{dt}\right)$, from which the velocity is $v(t) = (-200\sin\theta - 200\tan\alpha\cos\theta)\left(\frac{d\theta}{dt}\right)$. Substitute: $\boxed{v(t) = -218.4\ mm/s}$.

===================================<>===================================

Problem 2.40 An engineer designing a system to control a router for a machining process models it so that the router's acceleration during an interval of time is $a = -2v m/s^2$. When $t = 0$, its position is $s = 0$ and its velocity is $v = 2m/s$. Determine the router's velocity as a function of time.

Solution: we are given $a = \frac{dv}{dt} = -2v$ Integrating, we get $\int_2^v \frac{dv}{v} = -2\int_0^t dt$ or $\ln(v)]_2^v = -2t$, or $\ln\left(\frac{v}{2}\right) = -2t$. Converting to exponential form, we get $\frac{v}{2} = e^{-2t}$, or $v = 2e^{-2t} m/s$.

===================================<>===================================

Problem 2.41 In Problem 2.40, determine the router's position as a function of time.

Solution: From Problem 2.40, we have $a = \frac{dv}{dt} = -2v$. Changing variables, we can write this as $a = \frac{dv}{ds}\frac{ds}{dt} = v\frac{dv}{ds} = -2v$ or $\frac{dv}{ds} = -2$. Integrating, we get $\int_2^v dv = -2\int_0^s ds$, or $v - 2 = -2s$. Solving for s, we get $s = 1 - v/2\ m$. Recall from Problem 2.40, we had $v = 2e^{-2t} m/s$. substituting into the expression for s we get $s = 1 - e^{-2t} m$.

===================================<>===================================

Problem 2.42 A boat is moving at 20 *ft/s* when its engine is shut down. Due to hydrodynamic drag, its acceleration is $a = -0.1v^2\ ft/s^2$. What is the boat's velocity 2 seconds later?

Solution The strategy is to solve the differential equation for the velocity. The equation is $\frac{dv}{dt} = -0.1v^2$. Use method of separation of variables: $\frac{dv}{v^2} = -0.1dt$. Integrate both sides: $-\frac{1}{v} = -0.1t + C$. At $t = 0$, $v(0) = 20\,ft/s$, from which $C = -\frac{1}{20}$. The solution: $v(t) = \frac{20}{1+2t}$. At $t = 2\ s$, $\boxed{v(2) = 4\,ft/s}$

===================================<>===================================

==================================<>==================================

Problem 2.43 In Problem 2.42, what distance does the boat move in the 2 seconds following the shutdown of its engine?

Solution Use the solution to Problem 2.42. The strategy is to solve the differential equation for the position. The equation is $\frac{ds}{dt} = \frac{20}{1+2t}$. The solution is $s(t) = (10)\ln(1+2t)$. At $t = 2$, $\boxed{s(2) = 16.1\ ft}$.

==================================<>==================================

Problem 2.44 A steel ball is released from rest in a container of oil. Its downward acceleration is $a = 0.9g - cv$, where g is the acceleration due to gravity at sea level, and c is a constant. What is the velocity of the ball as a function of time?

Solution The acceleration is $\frac{dv}{dt} = 0.9g - cv$. Separate variables, $\frac{dv}{0.9g - cv} = dt$. Integrate: $v = \frac{0.9g}{c} - Ce^{-ct}$, where $cv < 0.9g$. Note $v(0) = 0$, $C = \frac{0.9g}{c}$, from which $\boxed{v = \frac{0.9g}{c}\left(1 - e^{-ct}\right)}$.

==================================<>==================================

Problem 2.45 In Problem 2.44, determine the position of the ball relative to its intitial position as a function of time.

Solution From Problem 2.140 $\frac{ds}{dt} = \frac{0.9g}{c}\left(1 - e^{-ct}\right)$. Integrate: $s(t) = \frac{0.9g}{c}(t + \frac{1}{c}e^{-ct}) + C$. Note $s(0) = 0$ from which $C = -\frac{0.9g}{c^2}$, and $\boxed{s(t) = \frac{0.9g}{c^2}(ct + e^{-ct} - 1)}$

==================================<>==================================

Problem 2.46 The greatest depth yet discovered in the oceans is the Marianas Trench in the western Pacific Ocean. A steel ball released at the surface requires 64 minutes to reach the bottom. The ball's downward acceleration is $a = 0.9g - cv$, where g is the acceleration due to gravity at sea level and the constant $c = 3.02s^{-1}$. What is the depth of the Marianas Trench in miles?

Solution Use the solution to Problem 2.40, with $g = 32.17\ ft/s^2$. The velocity is $\frac{ds}{dt} = \frac{0.9g}{c}\left(1 - e^{-ct}\right)$. Integrate: $s(t) = \frac{0.9g}{c}\left(t + \frac{1}{c}e^{-ct}\right) + C$. At $t = 0$, $s(0) = 0$, hence $C = -\frac{0.9g}{c^2}$, from which $s(t) = \frac{0.9g}{c}\left(t + \frac{1}{c}e^{-ct} - \frac{1}{c}\right)ft$. For $t = 64\ min = 3480\ s$, the depth is

$$\boxed{d = \left(\frac{36811.2\ ft}{5280\ ft/mi}\right) = 6.972\ miles}.$$

==================================<>==================================

=================================<>=================================

Problem 2.47 To study the effects of meteor impacts on satellites, engineers use a *rail gun* to accelerate a plastic pellet to a high velocity. They determine that when the pellet has traveled 1 *m* from the gun, its velocity is 2.25 km/s, and when it has traveled 2 *m* from the gun, its velocity is 1 km/s. Assume that the acceleration of the pellet after it leaves the gun is given by $a = -cv^2$, where *c* is a constant. (a) What is the value of *c*, and what are its SI units? (b) What was the velocity of the pellet as it left the rail gun?

Solution The acceleration is $\frac{dv}{dt} = -cv^2$, from which $\frac{dv}{-cv^2} = dt$. The velocity is $\frac{ds}{dt} = v$, from which $\frac{ds}{v} = dt$. Equate: $\frac{dv}{-cv} = ds$. Integrate: $-\frac{1}{c}\ln v = s + C$. (a) When $s = 1\ m$, $v(1) = 2.25\times 10^3\ m/s$, from which (1) $-\left(\frac{1}{c}\right)\ln(2.25\times 10^3) = 1 + C$. When $s = 2\ m$, $v(2) = 1000\ m/s$, from which (2) $-\left(\frac{1}{c}\right)\ln(1000) = 2 + C$. Solve: $\boxed{c = 0.8109\ m^{-1}}$, $C = -10.518\ m$. (b) Invert the velocity: $v(s) = 5062.5e^{-cs}$ .The velocity at the exit is $\boxed{v(0) = 5062.5\ m/s}$

=================================<>=================================

Problem 2.48 If aerodynamic drag is taken into account, the acceleration of a falling object can be approximated by $a = g - cv^2$, where *g* is the acceleration due to gravity at sea level and *c* is a constant. (a) If an object is released from rest, what is its velocity as a function of the distance s from the point of release? (b) Determine the limit of your answer to Part (a) as $c \to 0$ and show that it agrees with the solution you obtain by assuming that the acceleration $a = g$.

Solution The differential equation for the velocity is $\frac{dv}{dt} = g - cv^2$, where the distance traveled is positive downward, and $c \ge 0$, $g > 0$, and $g > cv^2$. Use method of separation of variables, $\frac{dv}{g - cv^2} = dt$. The equation for the distance is $\frac{ds}{dt} = v$, and $\frac{ds}{v} = dt$. Equate, to obtain $\frac{vdv}{g - cv^2} = ds$. Integrate both sides: $-\frac{1}{2c}\ln(g - cv^2) = s + C$. For $s = 0$, $v(0) = 0$, from which $C = -\frac{1}{2c}\ln(g)$. Reduce algebraically: $\ln\left(\frac{g - cv^2}{g}\right) = -2cs$ (a) Invert the solution to obtain $v^2 = \left(\frac{g}{c}\right)(1 - e^{-2cs})$ where $(g > cv^2)$. (b) The limit as $c \to 0$ is obtained by expanding the exponential in a series: $e^{-2cs} = 1 - 2cs + \frac{(2cs)^2}{2!} - \ldots$ from which $\boxed{v^2|_{c\to 0} = 2gs}$. For free fall downward toward a flat earth, $\frac{dv}{dt} = g$ and $\frac{ds}{dt} = v$. The solution: $v(t) = gt + C_1$, and $s(t) = \frac{gt^2}{2} + C_1 t + C_2$. At $t = 0$, $v = 0$ and $s = 0$, hence $C_1 = 0$ and $v(t) = gt$, and $v = gt$ and $s(t) = \frac{gt^2}{2}$. Solve for *t*, $t = \sqrt{\frac{2s}{g}}$ and substitute: $v = \sqrt{2gs}$, from which $\boxed{v^2 = 2gs}$

=================================<>=================================

==================================<>==================================

Problem 2.49 A skydiver jumps from a helicopter and is falling straight down at 30 *m/s* when her parachute opens. From then on her downward acceleration is approximately $a = g - cv^2$, where $g = 9.81\ m/s^2$ and c is a constant. After an initial "transient" period she descends at a nearly constant velocity of $5\ m/s$. (a) What is the value of c and what are its SI units? (b) What maximum deceleration is she subjected to? (c) What is her downward velocity when she has fallen 2 meters from the point where her parachute opens?

Solution Assume $c > 0$. (a) After the initial transient, she falls at a constant velocity, so that the acceleration is zero and $cv^2 = g$, from which $\boxed{c = \frac{g}{v^2} = \frac{9.81\ m/s^2}{(5)^2\ m^2/s^2} = 0.3924\ m^{-1}}$ (b) The maximum acceleration (in absolute value) occurs when the parachute first opens, when the velocity is highest: $\boxed{a_{\max} = |g - cv^2| = |g - c(30)^2| = 343.4\ m/s^2}$ (c) Choose coordinates such that distance is measured positive downward. The velocity is related to position by the chain rule: $\frac{dv}{dt} = \frac{dv}{ds}\frac{ds}{dt} = v\frac{dv}{ds} = a$, from which $\frac{vdv}{g - cv^2} = ds$. Integrate: $\left(-\frac{1}{2c}\right)\ln|g - cv^2| = s + C$. When the parachute opens $s = 0$ and $v = 30\ m/s$, from which $C = -\left(\frac{1}{2c}\right)\ln|g - 900c| = -7.4398$. The velocity as a function of distance is $\ln|g - cv^2| = -2c(s + C)$. For $s = 2\ m$, $\boxed{v = 14.4\ m/s}$ [*Note:* The absolute magnitude of the argument of the natural logarithm is required by the formal integral (See No. 141.1, page 35, Dwight's "Tables of Integrals and Other Mathematical Data" 4th Ed. 1961.). From an engineering viewpoint it is required because of the possibility that the $cv^2 > g$ during the "transient" period. (Distance is positive downward, and $c \geq 0,\ g > 0$.)]

==================================<>==================================

Problem 2.50 A rocket sled starts from rest and accelerates at $a = 3t^2\ m/s^2$ until its velocity is 1000 *m/s*. It then hits a water brake and its acceleration is $a = -0.001v^2\ m/s^2$ until its velocity decreases to $500\ m/s$. What total distance does the sled travel?

Solution Divide the interval into two phases. *In the first phase:* $\frac{dv}{dt} = 3t^2$,from which: $v(t) = t^3 + C_1$. At $t = 0$, $v(0) = 0$, from which $C_1 = 0$. The distance is $\frac{ds}{dt} = v(t) = t^3$, from which $s(t) = \frac{t^4}{4} + C_2$. At $t = 0$, $s(0) = 0$, from which $C_2 = 0$. At $v = 1000\ m/s$, the time elapsed is $t_1 = v^{\frac{1}{3}} = (1000)^{\frac{1}{3}} = 10$ seconds. The distance traveled is $s(t_1) = \frac{10^4}{4} = 2500\ m$. *In the second phase:* Use the chain rule: $v\frac{dv}{ds} = -0.001v^2$, from which $\frac{dv}{0.001v} = -ds$. The solution: $\ln(v) = -(10^{-3})s + C$. At $v = 1000\ m/s$, $s = 0$, from which $C = \ln(10^3) = 6.9078$. From which $\ln(v) = -(10^{-3})s + 6.9078$. At $v = 500\ m/s$, $s = 693.1\ m$. The total distance traveled is $\boxed{d = 2500 + 693.1 = 3193\ m}$

==================================<>==================================

==================================<>==================================

Problem 2.51 The velocity of a point is given by the equation $v = \left(24 - 2s^2\right)^{\frac{1}{2}}\ m/s$. What is its acceleration when $s = 2\ m$?

Solution The velocity as a function of distance is given by the chain rule: $v\frac{dv}{ds} = \frac{dv}{dt} = a$. Solve for the acceleration and carry out the differentiation: $a = v\frac{dv}{ds} = \left(24 - 2s^2\right)^{\frac{1}{2}}\left(\frac{1}{2}\right)\left(24 - 2s^2\right)^{-\frac{1}{2}}(-4s) = -2s$. At $s = 2\ m$, the acceleration is $\boxed{a = -4\ m/s^2}$

==================================<>==================================

Problem 2.52 The velocity of an object subjected to the earth's gravitational field is $v = \left[v_o^2 + 2gR_E^2\left(\frac{1}{s} - \frac{1}{s_o}\right)\right]^{\frac{1}{2}}$, where v_o is the velocity at position s_o and R_E is the radius of the earth. Using this equation, show that the object's acceleration is $a = \frac{gR_E^2}{s^2}$.

Solution Using the chain rule, the velocity as a function of distance is given by $v\frac{dv}{ds} = a$. Solve for a and carry out the differentiation.

$$a = \left[v_o^2 + 2gR_E^2\left(\frac{1}{s} - \frac{1}{s_o}\right)\right]^{\frac{1}{2}}\left(\frac{1}{2}\right)\left[v_o^2 + 2gR_E^2\left(\frac{1}{s} - \frac{1}{s_o}\right)\right]^{-\frac{1}{2}}(-2gR_E^2 s^{-2}) = \boxed{-\frac{gR_E^2}{s^2}}$$

==================================<>==================================

Problem 2.53 Engineers analyzing the motion of a linkage determine that the velocity of an attachment point is given by $v = A + 4s^2\ ft/s$, where A is a constant. When $s = 2\ ft$, its acceleration is measured and determined to be $a = 320\ ft/s^2$. What is its velocity when $s = 2\ ft$?

Solution The velocity as a function of the distance is $v\frac{dv}{ds} = a$. Solve for a and carry out the differentiation. $a = v\frac{dv}{ds} = \left(A + 4s^2\right)(8s)$. When $s = 2\ ft$, $a = 320\ ft/s^2$, from which $A = 4$. The velocity at $s = 2\ ft$ is $\boxed{v = 4 + 4(2^2) = 20\ ft/s}$

==================================<>==================================

Problem 2.54 The acceleration of an object is given by the function $a = 2s\ ft/s^2$. When $t = 0$, $v = 1\ ft/s$. What is the velocity when the object has moved $2\ ft$ from its initial position?

Solution The differential equations for the velocity and distance are $\frac{dv}{dt} = 2s$, and $\frac{ds}{dt} = v$. Use the chain rule, separate variables and integrate: $v\frac{dv}{ds} = 2s$, from which $v^2 = 2s^2 + C_1$. At $t = 0$, $s = 0$, and $v = 1\ ft/s$, from which $C_1 = 1$ and $v^2 = 2s^2 + 1$. At $s = 2\ ft$, $v^2 = 9$, $\boxed{v = 3\ ft/s}$ where the positive sign on the square root is chosen because the velocity increases with distance.

==================================<>==================================

====================================<>====================================

Problem 2.55 The acceleration of an object is given by $a = 3s^2\ ft/s^2$. At $s = 0$, its velocity is $v = 10\ ft/s$. What is its velocity when $s = 44\ ft$?

Solution The velocity as a function of the position is given by $v\frac{dv}{ds} = \frac{dv}{dt} = a$. Substitute the given acceleration and separate variables: $vdv = 3s^2 ds$, from which $\frac{v^2}{2} = s^3 + C$. At $s = 0$, $v(0) = 10\ ft/s$, from which $C = \frac{10^2}{2} = 50$. The solution: $v^2 = 2s^2 + 100$. At $s = 4\ ft$,

$v(4) = \pm\sqrt{2(4^2) + 100} = \pm 15.1\ ft/s$. Take the positive root: $\boxed{v(4) = 15.1\ ft/s}$

====================================<>====================================

Problem 2.56 A gas gun used by engineers to investigate the properties of materials subjected to high pressure subjects a projectile to an acceleration $a = 3500/s^{1.4}\ m/s^2$, where s is the projectile's position in meters. If the projectile starts from rest at $s = 1.5\ m$, what is its velocity when $s = 3\ m$?

Solution: The acceleration can be written as $a = \frac{dv}{ds}\frac{ds}{dt} = v\frac{dv}{ds} = 3500/s^{1.4}$ or $\int_0^v vdv = 3500\int_{1.5}^3 s^{-1.4} ds$.

Integrating, we get $\frac{v^2}{2} = 3500\left(\frac{s^{-0.4}}{-0.4}\right)$ or $v^2 = \left(\frac{-7000}{0.4}\right)\left[3^{-0.4} - 1.5^{-0.4}\right] = 3603\ m^2/s^2$.; $v = 60.0 m/s$.

====================================<>====================================

Problem 2.57 A *spring-mass* oscillator consists of a mass and a spring connected as shown. The coordinate s measures the displacement of the mass relative to its position when the spring is unstretched, If the spring is linear, the mass is subjected to a deceleration proportional to s. Suppose that $a = -4s\ m/s^2$, and you give the mass a velocity $v = 1\ m/s$ in the position $s = 0$. (a) How far will the mass move to the right before the spring brings it to a stop? (b) What will be the velocity of the mass when it has returned to the position $s = 0$?

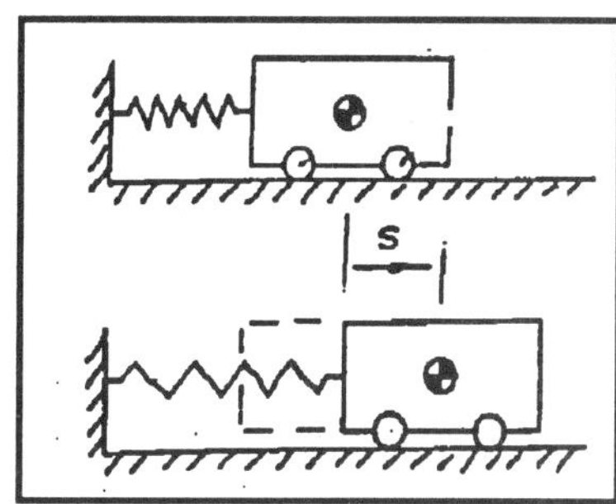

Solution

The velocity of the mass as a function of its position is given by $v\,dv/ds = a$. Substitute the given acceleration, separate variables, and integrate: $vdv = -4s\ ds$, from which $v^2/2 = -2s^2 + C$. The initial velocity $v(0) = 1\ m/s$ at $s = 0$, from which $C = 1/2$. The velocity is $v^2/2 = -2s^2 + 1/2$.

(a) The velocity is zero at the position given by $0 = -2(s_1)^2 + \frac{1}{2}$, from which $s_1 = \pm\sqrt{\frac{1}{4}} = \pm\frac{1}{2}\ m$. Since the displacement has the same sign as the velocity, $\boxed{s_1 = +1/2\ \text{m}}$ is the distance traveled before the spring brings it to a stop.

(b) At the return to $v = \pm\sqrt{\frac{2}{2}} = \pm 1\ m/s$, the velocity is $v = \pm\sqrt{\frac{2}{2}} = \pm 1\ m/s$. From the physical situation, the velocity on the first return is negative (opposite the sign of the initial displacement), $\boxed{v = -1\ m/s}$.

====================================<>====================================

===================================<>===================================

Problem 2.58 In Problem 2.57, suppose that at $t = 0$ you release the mass from rest in the position $s = 1\ m$. Determine the velocity of the mass as a function of s as it moves from the initial position to $s = 0$.

Solution From the solution to Problem 2.57, the velocity as a function of position is given by $\frac{v^2}{2} = -2s^2 + C$. At $t = 0$, $v = 0$ and $s = 1\ m$, from which $C = 2(1)^2 = 2$. The velocity is given by $v(s) = \pm\left(-4s^2 + 4\right)^{\frac{1}{2}} = \pm 2\sqrt{1-s^2}\ m/s$. From the physical situation, the velocity is negative (opposite the sign of the initial displacement) : $\boxed{v = -2\sqrt{1-s^2}\ m/s}$. [*Note*: From the initial conditions, $s^2 \le 1$ always.]

===================================<>===================================

Problem 2.59 In Problem 2.57, suppose that at $t = 0$ you release the mass from rest in the position $s = 1\ m$. Determine the position of the mass as a function of time as it moves from its initial position to $s = 0$.

Solution The differential equations for the velocity and position are $\frac{dv}{dt} = -4s$, and $\frac{ds}{dt} = v$. Use the chain rule: $v\frac{dv}{ds} = -4s$. Separate variables and integrate: $v^2 = -4s^2 + C$.At $t = 0$, $s = 1\ m$, and $v = 0$, from which $C = 4$, and $v^2 = 4\left(1-s^2\right)$. Substitute: $\frac{ds}{dt} = \pm 2\sqrt{1-s^2}$. Separate variables and integrate: $\frac{ds}{\sqrt{1-s^2}} = \pm 2dt$, $-\cos^{-1}(s) = \pm 2t + C$. At $t = 0$, $s = 1\ m$, from which $C = \cos^{-1}(1) = 0$, from which $\boxed{s(t) = \cos 2t\ m}$ where the negative sign for the square root is chosen because s decreases with increasing t at $t = 0+ \ldots$

===================================<>===================================

==================================<>==================================

Problem 2.60 The mass is released from rest with the springs unstretched. Its downward acceleration is $a = 32.2 - 50s\ ft/s^2$, where s is the position of the mass measured from the position in which it is released. (a) How far does the mass fall? (b) What is the maximum velocity of the mass as it falls?

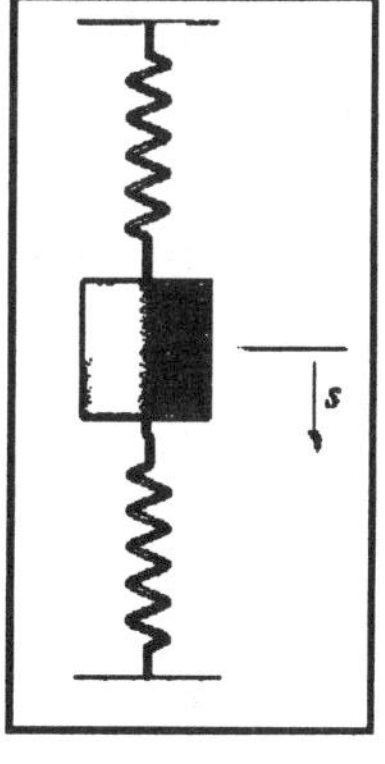

Solution: The acceleration is given by $a = \dfrac{dv}{dt} = \dfrac{dv}{ds}\dfrac{ds}{dt} = v\dfrac{dv}{ds} = 32.2 - 50s\ ft/s^2$.

Integrating, we get $\int_0^v v\,dv = \int_0^s (32.2 - 50s)ds$ or $\dfrac{v^2}{2} = 32.2s - 25s^2$.

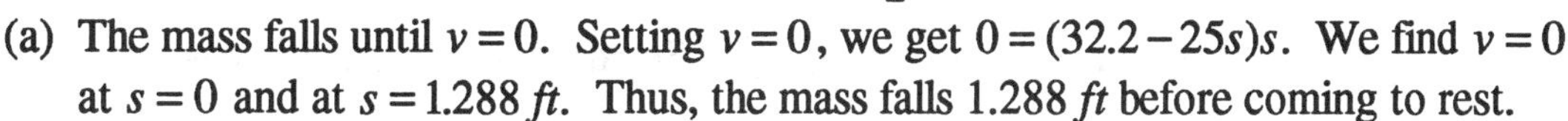

(a) The mass falls until $v = 0$. Setting $v = 0$, we get $0 = (32.2 - 25s)s$. We find $v = 0$ at $s = 0$ and at $s = 1.288\ ft$. Thus, the mass falls $1.288\ ft$ before coming to rest.

(b) From the integration of the equation of motion, we have $v^2 = 2(32.2s - 25s^2)$. The maximum velocity occurs where $\dfrac{dv}{ds} = 0$. From the original equation for acceleration, we have $a = v\dfrac{dv}{ds} = (32.2 - 50s)\ ft/s^2$. Since we want maximum velocity, we can assume that $v \neq 0$ at this point. Thus, $0 = (32.2 - 50s)$, or $s = (32.2/50)\ ft$ when $v = v_{MAX}$. Substituting this value for s into the equation for v, we get $v_{MAX}^2 = 2\left(\dfrac{(32.2)^2}{50} - \dfrac{(25)(32.2)^2}{50^2}\right)$, or $v_{MAX} = 4.55\ ft/s$

==================================<>==================================

Problem 2.61 The position of the mass in Problem 2.60 as a function of time is $s = \left(\dfrac{32.2}{50}\right)\left[1 - \cos\left(\sqrt{50}t\right)\right]$. Prove that this is the correct solution for s; that is, prove that it satisfies the initial conditions and give the correct acceleration.

Strategy: We need to take two derivatives and then reintroduce the definition of s into the equation for acceleration. We then need to evaluate s and v at $t = 0$.

Solution: From the given expression for position, the velocity can be written as

$v = \dfrac{ds}{dt} = \left(\dfrac{32.2}{50}\right)\left[\sqrt{50}\sin\left(\sqrt{50}t\right)\right]$ and the acceleration can be written as

$a = \dfrac{dv}{dt} = \left(\dfrac{32.2}{50}\right)\left[\left[\sqrt{50}\right]^2\cos\left(\sqrt{50}t\right)\right]$ or $a = (32.2)\left[\cos\left(\sqrt{50}t\right)\right]$. Rewriting the acceleration, we form

$a = (32.2)\left[1 - 1 + \cos\left(\sqrt{50}t\right)\right]$ or $a = 32.2 - 32.2\left[1 - \cos\left(\sqrt{50}t\right)\right]$ again,

$a = 32.2 - 32.2\left(\dfrac{50}{50}\right)\left[1 - \cos\left(\sqrt{50}t\right)\right]$ Rearranging terms, we have

$a = 32.2 - 50\left(\dfrac{32.2}{50}\right)\left[1 - \cos\left(\sqrt{50}t\right)\right]$ or $a = 32.2 - 50s$. Thus, the original equation for acceleration is verified. Evaluating s and v at $t = 0$, we get $s = 0$ and $v = 0$. Hence, the solution is verified

==================================<>==================================

==================================<>==================================

Problem 2.62 If a spacecraft is 100 miles above the surface of the Earth, what initial velocity v_o straight away from the Earth would be required for it to reach the Moon's orbit 238,000 miles from the center of the Earth? The radius of the earth is 3960 miles. Neglect the effect of the Moon's gravity. (See Example 2.5.)

Solution For computational convenience, convert the acceleration due to Earth's gravity into the units given in the problem, namely miles and hours: $g = \left(\frac{32.17\,ft}{1\,s^2}\right)\left(\frac{1\,mile}{5280\,ft}\right)\left(\frac{3600^2\,s^2}{1\,hr^2}\right) = 78962.7\,mi/hr^2$.

The velocity as a function of position is given by $v\frac{dv}{ds} = a = -\frac{gR_E^2}{s^2}$. Separate variables, $vdv = -gR_E^2\frac{ds}{s^2}$.

Integrate: $v^2 = -2gR_E^2\left(-\frac{1}{s}\right) + C$.. Suppose that the velocity at the distance of the Moon's orbit is zero.

Then $0 = 2(78962.7)\left(\frac{3960^2}{238{,}000}\right) + C$, from which $C = -10405562\,mi^2/hr^2$. At the 100 mile altitude,

the equation for the velocity is $v_o^2 = 2g\left(\frac{R_E^2}{R_E + 100}\right) + C$. From which

$\boxed{v_o = \sqrt{599575671} = 24486.2\,mi/hr}$ Converting:

$\boxed{v_o = \left(\frac{24486.2\,mi}{1\,hr}\right)\left(\frac{5280\,ft}{1\,mi}\right)\left(\frac{1\,hr}{3600\,s}\right) = 35913.1\,ft/s}$. *Check:* Use the result of Example 2.5,

$v_o = \sqrt{2gR_E^2\left(\frac{1}{s_o} - \frac{1}{h}\right)}$, (where $h > s_o$ always), and $h = 238{,}000$, from which

$v_o = 24486.2\,mi/hr$.*check.*

==================================<>==================================

Problem 2.63 The radius of the Moon is $R_M = 1738\,km$. The acceleration of gravity at its surface is $1.62\,m/s^2$. If an object is released from rest 1738 *km* above the surface of the Moon, what is the magnitude of its velocity just before it impacts the surface?.

Solution The velocity as a function of position is $v\frac{dv}{ds} = -\frac{g_m R_M^2}{s^2}$. Separate variables and integrate to obtain (see Problem 2.56) $v^2 = 2g_m\left(\frac{R_M^2}{s}\right) + C$. At a distance $s_o = R_M + 1738 = 3476\,km$ above the surface, $v = 0$, from which $C = -2815560\,m^2/s^2$. The velocity at the surface $s = R_M$ is

$\boxed{v = \sqrt{2g_m R_M + C} = 1678\,m/s}$

==================================<>==================================

================================<>================================

Problem 2.64 Using the data in Problem 2.63, determine the escape velocity from the surface of the Moon. (See Example 2.5.)

Solution From the solution to Problem 2.63, the velocity is given by $v^2 = \frac{2g_m R_m^2}{s} + C$. Assume that as $s \to \infty$, the velocity goes to zero, from which $C = 0$. At $s_o = R_M$, $v^2 = 2g_m R_M$, from which

$\boxed{v_{escape} = \sqrt{2g_m R_M} = 2373 \; m/s}$ *Check:* From Example 2.5, $v_o = \sqrt{2g_m R_M^2\left(\frac{1}{s_o} - \frac{1}{h}\right)}$, (where $h > s_o$ always), choose $s_o = R_M$ and $h \to \infty$, $v_o = \sqrt{2g_m R_M}$, *check.*

================================<>================================

Problem 2.65 Suppose that a tunnel could be drilled straight through the earth from the north pole to the south pole and the air evacuated from it. An object dropped from the surface would fall with the acceleration $a = -\frac{gs}{R_E}$, where g is the acceleration of gravity at sea level, R_E is radius of the Earth, and s is the distance of the object from the center of the Earth. (Gravitational acceleration is equal to zero at the center of the Earth and increases linearly with the distance from the center.) What is the magnitude of the velocity of the dropped object when it reaches the center of the Earth?

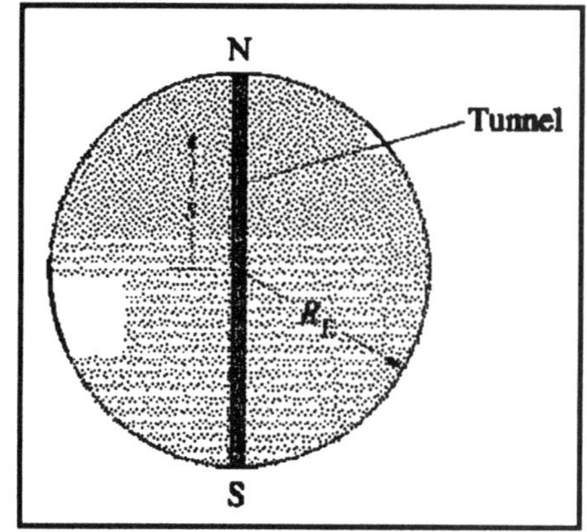

Solution The velocity as a function of position is given by $v\frac{dv}{ds} = -\frac{gs}{R_E}$. Separate variables and integrate: $v^2 = -\left(\frac{g}{R_E}\right)s^2 + C$. At $s = R_E$, $v = 0$, from which $C = gR_E$. Combine and reduce:

$v^2 = gR_E\left(1 - \frac{s^2}{R_E^2}\right)$ At the center of the earth $s = 0$, and the velocity is $\boxed{v = \sqrt{gR_E}}$

================================<>================================

===================================<>===================================

Problem 2.66 The acceleration of gravity of a hypothetical two-dimensional planet would depend upon the distance s from the center of the planet according to the relation $g = -\frac{k}{s}$, where k is a constant. Let the radius of the planet be R_T and let the magnitude of the acceleration due to gravity at its surface by g_T. (a) If an object is given an outward velocity v_o at a distance s_o from the center of the planet, determine its velocity as a function of s. (b) Show that there is no escape velocity from a two dimensional planet, thereby explaining why we have never been visited by any two-dimensional beings.

Solution At the surface of the planet the acceleration is $-\frac{k}{R_T} = -g_T$, from which $k = g_T R_T$. The velocity as a function of position is $v\frac{dv}{ds} = -\frac{g_T R_T}{s}$. Separate variables and integrate: $v^2(s) = -2g_T R_T \ln(s) + C$. At a distance s_o the velocity is v_o, from which $C = v_o^2 + 2g_T R_T \ln(s_o)$. Substitute and rearrange algebraically to obtain (1) $v^2(s) = -2g_T R_T \ln\left(\frac{s}{s_o}\right) + v_o^2$, from which

$\boxed{v(s) = \pm\sqrt{v_o^2 - 2g_T R_T \ln\left(\frac{s}{s_o}\right)}}$. (b) Suppose that $s_o = R_T$ and $s \to \infty$, so that v_o is the escape velocity. Since $s > s_o$, then, since $\ln(x) > 0$ if $x > 1$, and $2g_T R_T > 0$, the term on the right is negative. Since $\lim_{x\to\infty} [\ln(x)] \to \infty$ then the term under the radical will become negative unless v_o^2 also grows at least as fast as $\ln\left(\frac{s}{s_o}\right)$ as $s \to \infty$.Thus the escape velocity is infinite. *Note:* We could use equation (1) above, and take s to be a great distance, from which $v_o^2 = \lim_{s\to\infty} 2g_T R_T \ln\left(\frac{s}{s_o}\right) + \lim_{s\to\infty} v^2(s)$. which demonstrates that the escape velocity is infinite, regardless of the value of $\lim_{s\to\infty} v^2(s)$ (which cannot be negative). *Note:* We cannot choose $\lim_{s\to\infty} v^2(s) \to 0$, as is customary for a 3-dimensional spherical planet, because there is no reason to believe that the velocity at a great distance from a two dimensional planet can be made to approach zero in the limit.

===================================<>===================================

Problem 2.67 The Cartesian coordinates of a point (in meters) are $x = 2t + 4$, $y = t^3 - 2t$, and $z = 4t^2 - 4$, where t is in seconds. What are its velocity and acceleration at $t = 4$ seconds?

Solution The components of the velocity are $v_x = \frac{d}{dt}(2t+4) = 2\ m/s$.

$v_y = \frac{d}{dt}\left(t^3 - 2t\right) = 3t^2 - 2\ m/s$, $v_z = \frac{d}{dt}\left(4t^2 - 4\right) = 8t\ m/s$. At $t = 4$, $\boxed{\vec{v} = 2\vec{i} + 46\vec{j} + 32\vec{k}\ (m/s)}$.

The acceleration components are $a_x = \frac{dv_x}{dt} = \frac{d}{dt}(2) = 0$, $a_y = \frac{dv_y}{dt} = \frac{d}{dt}\left(3t^2 - 2\right) = 6t$,

$a_z = \frac{dv_z}{dt} = \frac{d}{dy}(8t) = 8$. At $t = 4$, $\boxed{\vec{a} = 0\vec{i} + 24\vec{j} + 8\vec{k}\ \left(m/s^2\right)}$

===================================<>===================================

====================================<>====================================

Problem 2.68 The velocity of a point is $\vec{v} = 2\vec{i} + 3t^2\vec{j}\ (ft/s)$. At $t = 0$, its position is $\vec{r} = -\vec{i} + 2\vec{j}\ (ft)$. What is its position at $t = 2$ seconds.

Solution The strategy is to solve the differential equations for the position as a function of time. The differential equations are $\frac{dx}{dt} = v_x = 2$, $\frac{dy}{dt} = v_y = 3t^2$, and $\frac{dz}{dt} = v_z = 0$. Integrate to obtain $x = 2t + C_x$, $y = t^3 + C_y$, and $z = C_z$, where C_x, C_y and C_z are constants of integration. At $t = 0$, $\vec{r} = -\vec{i} + 2\vec{j}\ (ft)$ from which $C_x = -1$, $C_y = 2$, and $C_z = 0$. The position of the point as a function of time is $\vec{r}(t) = (2t-1)\vec{i} + (t^3+2)\vec{j} + 0\vec{k}$. At $t = 2$ seconds, the position is $\boxed{\vec{r}(2) = 3\vec{i} + 10\vec{j}\ (ft)}$

====================================<>====================================

Problem 2.69 The acceleration components of a point (in ft/s^2) are $a_x = 3t^2$, $a_y = 6t$, and $a_z = 0$. At $t = 0$, $x = 5\,ft$, $v_x = 3\,ft/s$, $y = 1\,ft$, $v_y = -2\,ft/s$, $z = 0$, and $v_z = 0$. What are its position and velocity vector at $t = 3$ seconds?

Solution The differential equations for the velocity are: $\frac{dv_x}{dt} = a_x = 3t^2$, from which $v_x = t^3 + V_x$, $\frac{dv_y}{dt} = a_y = 6t$, from which $v_y = 3t^2 + V_y$, and $\frac{dv_z}{dt} = a_z = 0$, from which $v_z = V_z$, , where V_x, V_y, V_z are constants of integration. At $t = 0$, $v_x = 3\,ft/s$, $v_y = -2\,ft/s$, $v_z = 0$, from which $V_x = 3$, $V_y = -2$, and $V_z = 0$. The velocity vector is $\vec{v}(t) = (t^3+3)\vec{i} + (3t^2-2)\vec{j} + 0\vec{k}$. At $t = 3$ seconds, $\boxed{\vec{v}(3) = 30\vec{i} + 25\vec{j}\ (ft/s)}$. The differential equations for the position are: $\frac{dx}{dt} = v_x = t^3 + 3$, from which $x(t) = \frac{t^4}{4} + 3t + C_x$, $\frac{dy}{dt} = v_y = 3t^2 - 2$, from which $y(t) = t^3 - 2t + C_y$. $\frac{dz}{dt} = v_z = 0$, from which $z = C_z$. At $t = 0$, $x = 5\,ft$, $y = 1\,ft$, and $z = 0$, from which $C_x = 5$, $C_y = 1$, and $C_z = 0$. The position vector is $\vec{r}(t) = \left(\frac{t^4}{4} + 3t + 5\right)\vec{i} + (t^3 - 2t + 1)\vec{j}$. At $t = 3$, $\boxed{\vec{r}(3) = 34.25\vec{i} + 22\vec{j}\ (ft)}$

====================================<>====================================

Problem 2.70 The acceleration components of an object (in m/s^2) $a_x = 2t$, $a_y = 4t^2 - 2$, and $a_z = -6$. At $t = 0$, the position of the object is $\vec{r} = 10\vec{j} - 10\vec{k}\ (m)$ and its velocity is $\vec{v} = 2\vec{i} - 4\vec{j}\ (m/s)$. Determine its position when $t = 4$ seconds.

Solution The equations: $\frac{dv_x}{dt} = 2t$, from which $v_x(t) = t^2 + V_x$. From the initial conditions, $V_x = 2$, and $v_x(t) = t^2 + 2$. $\frac{dv_y}{dt} = 4t^2 - 2$, from which $v_y(t) = \frac{4}{3}t^3 - 2t + V_y$. From the initial conditions $V_y = -4$, and $v_y(t) = \frac{4}{3}t^3 - 2t - 4$.. $\frac{dv_z}{dt} = -6$, from which $v_z(t) = -6t + V_z$. From the initial conditions, $V_z = 0$, and $v_z(t) = -6t$.

Solution continued on next page

The position equations are:

$\frac{dx}{dt} = t^2 + 2$, from which $x(t) = \frac{1}{3}t^3 + 2t + C_x$.. From the initial conditions, $C_x = 0$, and

$x(t) = \frac{1}{3}t^3 + 2t$.

$\frac{dy}{dt} = \frac{4}{3}t^3 - 2t - 4$, from which $y(t) = \frac{1}{3}t^4 - t^2 - 4t + C_y$. From the initial conditions, $C_y = 10$, and

$y(t) = \frac{1}{3}t^4 - t^2 - 4t + 10$.

$\frac{dz}{dt} = -6t$., from which $z(t) = -3t^2 + C_z$. From the initial conditions $C_z = -10$, and $z(t) = -3t^2 - 10$.

At $t = 4$ seconds: $\boxed{x = 29.33\ m}$. $\boxed{y = 63.33\ m}$. $\boxed{z = -58\ m}$

==================================<>==================================

Problem 2.71 Suppose that you are designing a mortar to send a rescue line from a coast guard boat to ships in distress. The line is attached to a weight that is fired by the mortar. The mortar is to be mounted so that it fires at 45° above the horizontal. If you neglect aerodynamic drag and the weight of the line for your preliminary design and assume a muzzle velocity of $100\ ft/s$ at $t = 0$, what are the x and y coordinates of the weight as a function of time?

Solution Neglect the height of the mortar above sea level.. At $t = 0$ the velocities are: $v_x = 100\cos(45^o)$, $v_y = 100\sin(45^o)$. The accelerations are: $a_x = 0$, $a_y = -g$. Denote the constants of integration by V_x, V_y, C_x, C_y. The equations are $\frac{dv_x}{dt} = a_x = 0$, from which $v_x = V_x = 100\cos(45^o)$.
$\frac{dv_y}{dt} = a_y = -g$, from which $v_y = -gt + V_y$. From the initial conditions, $V_y = 100\sin(45^o)$. The positions are given by: $\frac{dx}{dt} = V_x$, from which $x(t) = V_x t + C_x$.. From the initial conditions $C_x = 0$. $\frac{dy}{dt} = -gt + V_y$, from which $y(t) = -\frac{g}{2}t^2 + V_y t + C_y$. From the initial conditions $C_y = 0$. Substitute and collect results:

$\boxed{x(t) = 100\cos(45^o)t = 70.7t}$. $\boxed{y(t) = -\frac{g}{2}t^2 + 100\sin(45^o)t = -16.1t^2 + 70.7t}$

==================================<>==================================

Problem 2.72 In Problem 2.71, what must the mortar's muzzle velocity be to reach ships 1000 feet away.?

Solution The condition is: $y(t) \geq 0$ for $x \leq 1000\ ft$. From Problem 2.65, $y(t) = -\frac{g}{2}t^2 + 0.707V_M t$, where V_M is the muzzle velocity, from which $0 = \left(-\frac{g}{2}t + 0.707V_M\right)t$, from which $t_{flight} = \frac{\sqrt{2}\,V_M}{g}$. For the x-coordinate: $1000 = \frac{V_M}{\sqrt{2}}t_{flight} = \frac{V_M^2}{g}$, from which $\boxed{V_M = \sqrt{1000g} = 179.4\ ft/s}$

==================================<>==================================

==================================<>==================================

Problem 2.73 If a stone is thrown horizontally from the top of a 100 *ft* tall building at 50 *ft/s*, what horizontal distance from the point at which it is thrown does it hit the ground? (Assume level ground.) What is the magnitude of its velocity just before it hits?

Solution The equations for the velocity: $\frac{dv_x}{dt} = 0$, from which $v_x = V_x = 50\,ft/s$. $\frac{dv_y}{dt} = -g$, from which $v_y = -gt + V_y$. At $t = 0$, $v_y = 0$, from which $V_y = 0$. The equations for the position: $\frac{dx}{dt} = 50$, from which $x(t) = 50t + C_x$. From the initial conditions, $C_x = 0$. $\frac{dy}{dt} = v_y = -gt$, from which $y(t) = -\frac{g}{2}t^2 + C_y$. At $t = 0$, $y(0) = 100\,ft$, from which $C_y = 100\,ft$. The x, y coordinates of the path are $x(t) = 50t$, $y(t) = -\frac{g}{2}t^2 + 100$. The time of impact is given by: $0 = -\frac{g}{2}t^2_{impact} + 100$, from which $t_{impact} = \sqrt{\frac{200}{g}} = 2.493$ seconds. Substitute into the equation for the x-coordinate: $\boxed{x_{impact} = 50(2.493) = 124.67\,ft}$. The magnitude of the velocity at time of impact is

$$\boxed{V_{impact} = \sqrt{50^2 + (gt_{impact})^2} = 94.52\,ft/s}$$

==================================<>==================================

Problem 2.74 A projectile is launched from ground level with an initial velocity v_o What initial angle θ_o above the horizontal causes the range R to be a maximum, and what is the maximum range?

Solution The equations for the velocities: $\frac{dv_x}{dt} = 0$, from which $v_x(t) = V_x$. At $t = 0$, $v_x(0) = V_M \cos\theta_o$, where V_M is the initial muzzle velocity, from which $v_x(t) = V_M \cos\theta_o$. $\frac{dv_y}{dt} = -g$, from which $v_y(t) = -gt + V_y$. At $t = 0$, $v_y(0) = V_M \sin\theta_o$, from which $v_y(t) = -gt + V_M \sin\theta_o$. The equations for the x-y coordinates of the path: $\frac{dx}{dt} = v_x(t) = V_M \cos\theta_o$, from which $x(t) = V_M t \cos\theta_o + C_x$. At $t = 0$, $x(0) = 0$, from which $C_x = 0$ and $x(t) = V_M t \cos\theta_o$. $\frac{dy}{dt} = v_y(t) = -gt + V_M \sin\theta_o$, from which $y(t) = -\frac{g}{2}t^2 + V_M t \sin\theta_o + C_y$. At $t = 0$, $y(0) = 0$, from which $y(t) = -\frac{g}{2}t^2 + V_M t \sin\theta_o$. The range is determined by the maximum value of *x* for $y \ge 0$. The time of impact is obtained from $y(t) = 0 = \left(-\frac{g}{2}t_{imp} + V_M \sin\theta_o\right)t_{imp}$, Discarding $t_{imp} = 0$, $t_{imp} = \frac{2V_M \sin\theta_o}{g}$. Substitute into the equation for the x-coordinate:

$$x(t_{imp}) = V_M \cos\theta_o\left(\frac{2V_M \sin\theta_o}{g}\right) = \left(\frac{2V_M^2}{g}\right)\sin\theta_o \cos\theta_o.$$

For a given muzzle velocity, this is a maximum when $f(\theta_o) = \sin\theta_o \cos\theta_o$ is a maximum.

Solution continued on next page

The minimum (maximum) occurs when the first derivative vanishes:
$\frac{\partial f(\theta_o)}{\partial\theta_o} = \cos^2\theta_o - \sin^2\theta_o = 0 = \cos^2\theta_o\left(1-\tan^2\theta_o\right) = 0$. This has two possible solutions: $\cos^2\theta_o = 0$, from which $\theta_o = \frac{\pi(2n+1)}{2}$, $n = 0, 1, 2\ldots$ (the muzzle is pointed straight up or straight down!) or $\tan^2\theta_o = 1$, from which $\theta_o = \tan^{-1}(1)$. The first solution can be disregarded for physical reasons, and the principal value of the second is $\boxed{\theta_o = 45^o}$ The maximum range is $\boxed{x_{\max} = \left(\frac{2V_m^2}{g}\right)\left(\frac{1}{2}\right) = \left(\frac{V_m^2}{g}\right)}$ where V_M is the muzzle velocity. *Note:* This result may be used to solve Problem 2.66 by direct substitution of the data given in that problem.

==================================<>==================================

Problem 2.75 A pilot wants to drop supplies to remote locations in the Australian outback. He intends to fly horizontally and release the packages with no vertical velocity. Derive an equation for the horizontal distance *d* at which he should release the package in terms of the airplane's velocity v_o and altitude h.

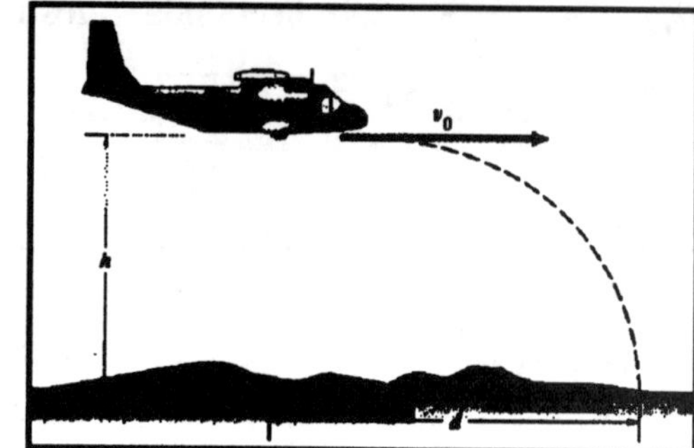

Solution The equations for the velocity are: (1) $\frac{dv_x}{dt} = 0$, from which $v_x = V_x$, where V_x is a constant of integration. At $t = 0$, $v_x(0) = v_o$, from which $v_x(t) = v_o$. (2) $\frac{dv_y}{dt} = -g$, from which $v_y(t) = -gt + V_y$. At $t = 0$, $v_y(0) = 0$, from which $V_y = 0$.

The equations for the x-y coordinates of the path: (3) $\frac{dx}{dy} = v_x = v_o$, from which $x(t) = v_o t + C_x$. At $t = 0$, $x(0) = 0$, from which $C_x = 0$, and $x(t) = v_o t$. (4) $\frac{dy}{dt} = v_y = -gt$, from which $y(t) = -\frac{g}{2}t^2 + C_y$. At $t = 0$, $y(0) = h$, from which $C_y = h$, and $y(t) = -\frac{g}{2}t^2 + h$. At time of impact, $y(t_{impact}) = 0 = -\frac{g}{2}t_{impact}^2 + h$, from which $t_{impact} = \sqrt{\frac{2h}{g}}$. Substitute in the x-coordinate equation,

$$\boxed{x(t_{impact}) = d = v_o\sqrt{\frac{2h}{g}}}$$

==================================<>==================================

==================================<>==================================

Problem 2.76 If the pitching wedge the golfer is using gives the ball an initial angle $\theta_0 = 50^\circ$, what range of velocities v_0 will cause the ball to land within $3 ft$ of the hole? (Assume the hole lies in the plane of the ball's trajedory). Strategy: We need to find the velocities which cause the ball to pass through the points (27,3) ft (3 feet short of the hole) and (33,3) feet (3 feet beyond the hole). Solution: set the coordinate origin at the point where the golfer strikes the ball. The motion in the horizontal (x) direction is given by $a_x = 0$, $V_x = V_0 \cos\theta_0$, $x = (V_0 \cos\theta_0)t$. The motion in the vertical (y) direction is given by $a_y = -g$, $V_y = V_0 \sin\theta_0 - gt, y = (V_0 \sin\theta_0)t - \frac{gt^2}{2}$. From the x equation, we can find the time at which the ball reaches the required value of x (27 or 33 feet). This time is $t_f = x_f / (V_0 \cos\theta_0)$. We can substitute this information the equation for Y with $Y_f = 3 ft$ and solve for V_0. The results are: For hitting (27,3) feet, $V_0 = 31.19 ft/s$. For hitting (33,3) feet, $V_0 = 34.16 ft/s$.

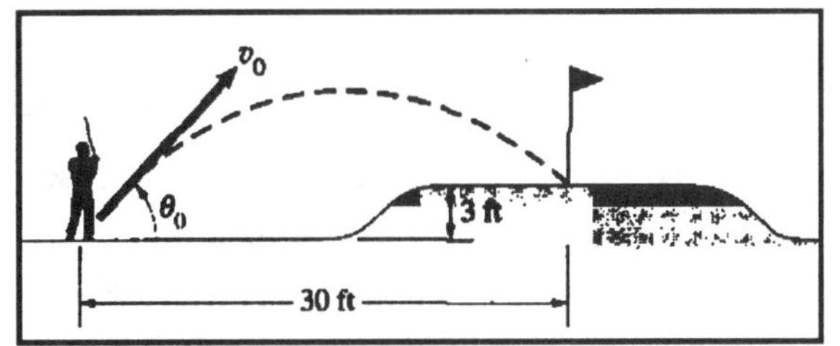

==================================<>==================================

Problem 2.77 A baseball batter strikes the ball at 3 feet above home plate and pops it up at an angle of 60° above the horizontal. The second baseman catches it at 6 feet above second base. What was the ball's initial velocity.

Solution A professional baseball "diamond" is a square $90 ft$ on a side. The distance from home plate to second base is $d = 2(90)\cos 45^\circ = 127.3 ft$. The equations for the velocity are: (1) $\frac{dv_x}{dt} = 0$, from which $v_x = V_B \cos 60^\circ$. (2) $\frac{dv_y}{dy} = -g$, from which $v_y = -gt + V_B \sin 60^\circ$. (3) $\frac{dx}{dt} = V_B \cos 60^\circ$, from which $x(t) = V_B \cos 60^\circ t$, since the initial value of x(t) is zero. (4) $\frac{dy}{dx} = -gt + V_B \sin 60^o$, from which $y(t) = -\frac{g}{2}t^2 + V_B \sin 60^o t + 3$, since $y(0) = 3 ft$. The time of the catch is given by $x(t_c) = d = V_B \cos 60^o t_c$, from which $t_c = \frac{d}{V_B \cos 60^o}$. Substitute into the y-coordinate:

$$y(t_c) = 6 = -\frac{g}{2}\left(\frac{d}{V_B \cos 60^o}\right)^2 + V_B \sin 60^o \left(\frac{d}{V_B \cos 60^o}\right) + 3.$$

Solve:

$$\boxed{V_B = \sqrt{\frac{d^2 g}{2\cos^2 60^o (d \tan 60^o - 3)}} = 69.2 ft/s}.$$

==================================<>==================================

================================<>================================

Problem 2.78 A baseball pitcher releases a fastball with an initial velocity $v_o = 90\ mi/hr$. Let θ be the initial angle of the ball's velocity vector above the horizontal. When it is released, the ball is 6 *ft* above the ground and 58 *ft* from the batter's plate. The batter's strike zone (between his knees and shoulders) extends from 1 *ft* 10 *in.* above the ground to 4 *ft* 6 *in.* above the ground. Neglecting aerodynamic effects, determine whether the ball will hit the strike zone (a) if $\theta = 1^o$; (b) if $\theta = 2^o$.

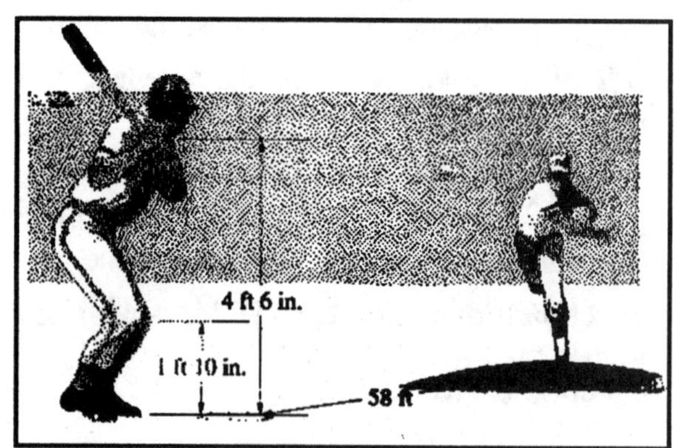

Solution The initial velocity is $v_o = 90\ mi/hr = 132\ ft/s$. The velocity equations are (1) $\frac{dv_x}{dt} = 0$, from which $v_x = v_o \cos\theta$. (2) $\frac{dv_y}{dt} = -g$, from which $v_y = -gt + v_o \sin\theta$. (3) $\frac{dx}{dt} = v_o \cos\theta$, from which $x(t) = v_o \cos\theta\ t$, since the initial position is zero. (4) $\frac{dy}{dt} = -gt + v_o \sin\theta$, from which $y(t) = -\frac{g}{2}t^2 + v_o \sin\theta\ t + 6$, since the initial position is $y(0) = 6\ ft$. At a distance $d = 58\ ft$, the height is *h*. The time of passage across the home plate is $x(t_p) = d = v_o \cos\theta\ t_p$, from which $t_p = \frac{d}{v_o \cos\theta}$.

Substitute: $y(t_p) = h = -\frac{g}{2}\left(\frac{d}{v_o \cos\theta}\right)^2 + d\tan\theta + 6$. For $\theta = 1^o$, $\boxed{h = 3.91\ ft}$, <u>Yes</u>, the pitcher hits the strike zone. For $\theta = 2^o$, $\boxed{h = 4.92\ ft}$ <u>No</u>, the pitcher misses the strike zone.

================================<>================================

=================================<>=================================

Problem 2.79 In Problem 2.78, assume that the pitcher releases the ball at an angle $\theta = 1^o$ above the horizontal and determine the range of velocities v_o (in ft/s) within which he must release the ball to hit the strike zone.

Solution From the solution to Problem 2.78, $h = -\frac{g}{2}\left(\frac{d}{v_o \cos\theta}\right)^2 + d\tan\theta + 6$, where $d = 58\,ft$, and $4.5 \ge h \ge 1.833\,ft$. Solve for the initial velocity: $v_o = \sqrt{\frac{gd^2}{2\cos^2\theta(d\tan\theta + 6 - h)}}$. For $h = 4.5$, $v_o = 146.8\,ft/s$. For $h = 1.833$, $v_o = 102.2\,ft/s$. The pitcher will hit the strike zone for velocities of release of $\boxed{102.2 \le v_o \le 146.8\,ft/s}$, and a release angle of $\theta = 1^o$. *Check:* The range of velocities in miles per hour is $69.7\,mph \le v_o \le 100.1\,mph$, which is within the range of major league pitchers, although the 100 mph upper value is achievable only by a talented few (Nolan Ryan, while with the Houston Astros, would occasionally in a game throw a 105 mph fast ball, as measured by hand held radar from behind the plate).

=================================<>=================================

Problem 2.80 A zoology student is provided with a bow and arrow tipped with a syringe of sedative and is assigned to measure the temperature of a black rhino (*Diceros bicornis*). The range of his bow, when it is fully drawn and aimed at an angle of 45 degrees above the horizontal is 100 meters. If a truculent rhino charges straight toward him at 30 *km/hr*, and he aims his bow 20° above the horizontal, how far away should the rhino be when he releases the arrow?

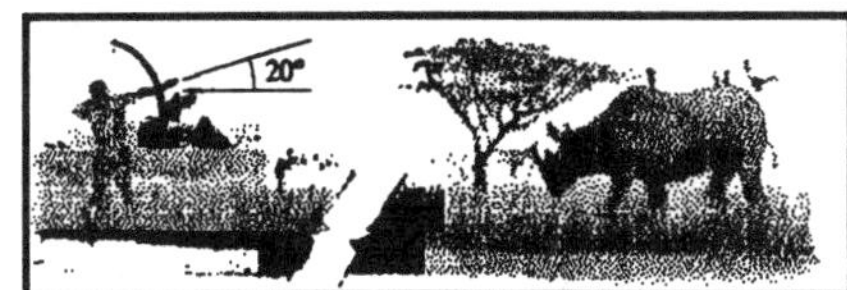

Solution The strategy is (a) to determine the range and flight time of the arrow when aimed 20° above the horizontal, (b) to determine the distance traveled by the rhino during this flight time, and then (c) to add this distance to the range of the arrow. Neglect aerodynamic drag on the arrow. The equations for the trajectory are: Denote the constants of integration by V_x, V_y, C_x, C_y, and the velocity of the arrow by V_A.. (1) $\frac{dv_x}{dt} = 0$,. from which $v_x = V_x$. At $t = 0$, $V_x = V_A\cos\theta$. (2) $\frac{dv_y}{dt} = -g$, from which $v_y = -gt + V_y$. At $t = 0$, $V_y = V_A \sin\theta$. (3) $\frac{dx}{dt} = v_x = V_A\cos\theta$, from which $x(t) = V_A\cos\theta\, t + C_x$. At $t = 0$, $x(0) = 0$, from which $C_x = 0$. (4) $\frac{dy}{dt} = v_y = -gt + V_A\sin\theta$, from which $y = -\frac{g}{2}t^2 + V_A\sin\theta\, t + C_y$. At $t = 0$, $y = 0$, from which $C_y = 0$. The time of flight is given by $y(t_{flight}) = 0 = (-\frac{g}{2}t_{flight} + V_A\sin\theta)\, t_{flight}$, from which $t_{flight} = \frac{2V_A\sin\theta}{g}$. The range is given by $x(t_{flight}) = R = V_A\cos\theta\, t_{flight} = \frac{2V_A^2\cos\theta\sin\theta}{g}$. The maximum range (100 meters) occurs when the arrow is aimed 45° above the horizon. Solve for the arrow velocity: $V_A = \sqrt{gR_{\max}} = 31.3\,m/s$. The time of flight when the angle is 20° is $t_{flight} = \frac{2V_A\sin\theta}{g} = 2.18$ s, and the range is $R = V_A\cos\theta\, t_{flight} = 64.3\,m$. The speed of the rhino is $30\,km/hr = 8.33\,m/s$. The rhino travels a distance $d = 8.33(2.18) = 18.2\,m$. The required range when the arrow is released is $\boxed{d + R = 82.5\,m}$.

=================================<>=================================

==================================<>==================================

Problem 2.81 The crossbar of the goal posts in American football is $y_c = 10\,ft$ above the ground. To kick a field goal, the ball must go between the two uprights supporting the crossbar and be above the crossbar when it does so. Suppose the kicker attempts a 40-yard field goal $(x_c = 120\,ft)$, and kicks the ball with an initial velocity $v_0 = 70\,ft/s$ and $\theta_0 = 40°$. By what vertical distance does the ball clear the crossbar? Solution: Set the coordinate origin at the point where the ball is kicked. The x (horizontal) motion of the ball is given by $a_x = 0$, $V_x = V_0 \cos\theta_0, x = (V_0 \cos\theta_0)t$. The y motion is given by $a_y = -g$ $V_y = V_0 \sin\theta_0 - gt$, $y = (V_o \sin\theta_0)t - \frac{gt^2}{2}$. Set $x = x_c = 120\,ft$ and find the time t_c at which the ball crossed the plane of the goal posts. Substitute this time into the y equation to find the y coordinate Y_B of the ball as it passes over the crossbar. Substituting in the numbers $(g = 32.2\,ft/s^2)$, we get $t_c = 2.24s$ and $y_B = 20.06\,ft$. Thus, the ball clears the crossbar by 10.06 feet.

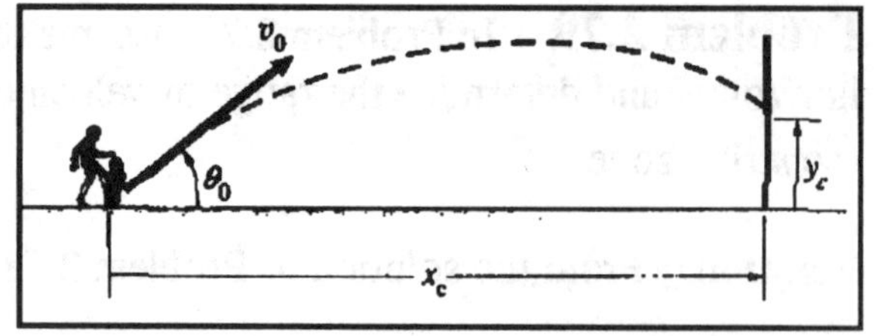

==================================<>==================================

Problem 2.82 In Problem 2.81, suppose you want to determine the minimum initial velocity of the football and the corresponding initial angle needed to make a field goal.

(a) Show that the initial angle satisfies the equation

$$\tan\theta_0 = \frac{y_C}{x_C} + \sqrt{\left(\frac{y_C}{x_C}\right)^2 + 1}$$

(b) Determine the minimum velocity and initial angle needed to make a 40 yard field goal.

Strategy: There are numerous combinations of initial velocity and angle which will *just make* the field goal. Let us assume the limiting case – where the ball hits the crossbar [goes through the point (x_C, y_C)]. We will write the equations of motion for this case and will eliminate the time from the y equation using the x equation. We will then have an equation in the initial velocity, the initial angle, and the coordinates of the crossbar. We will then use implicit differentiation to find the derivative $(dV_0/d\theta_0)$. We will set this derivative to zero.

Solution: The equations of motion for a football which hits the bar at time t_C are $x_C = V_0 \cos\theta_0 t_C$ and $y_C = V_0 \sin\theta_0 t_C - gt_C^2/2$. Eliminating t_C from the y-equation yields $y_C = \frac{\sin\theta_0 x_C}{\cos\theta_0} - \frac{gx_C^2}{2V_0^2\cos^2\theta_0}$. Multiplying the whole equation by $V_0^2\cos^2\theta_0$, we get $y_C V_0^2 \cos^2\theta_0 = x_C V_0^2 \sin\theta_0 \cos\theta_0 - gx_C^2/2$. We now differentiate this equation with respect to θ_0 and set $(dV_0/d\theta_0) = 0$. Taking the derivative, we get $\left(\frac{dV_0}{d\theta_0}\right)(2V_0 y_C \cos^2\theta_0 - 2V_0 x_C \sin\theta_0 \cos\theta_0) = V_0^2 x_C(\cos^2\theta_0 - \sin^2\theta_0) + 2V_0^2 y_C \sin\theta_0 \cos\theta_0$. Setting $(dV_0/d\theta_0) = 0$, we get $V_0^2 x_C(\cos^2\theta_0 - \sin^2\theta_0) + 2V_0^2 y_C \sin\theta_0 \cos\theta_0 = 0$. After algebraic manipulation, this becomes $\tan^2\theta_0 - 2\frac{y_C}{x_C}\tan\theta_0 - 1 = 0$. Solving, we get $\tan\theta_0 = \frac{y_C}{x_C} + \sqrt{\left(\frac{y_C}{x_C}\right)^2 + 1}$. We actually get the relation with a ± in front of the radical, but the negative sign is meaningless.

(b) Solving for the 40 yard field goal, we merely substitute numbers into our relations. We get $V_0 = 64.8\,ft/s$ and $\theta_0 = 47.38°$.

==================================<>==================================

==================================<>==================================

Problem 2.83 The cliff divers of Acapulco, Mexico must time their dives so that they enter the water at the crest (high point) of a wave. The crests of the waves are 2 ft above the mean water depth $h = 12\ ft$, and the horizontal velocity of the waves is $\sqrt{gh}$. The diver's aiming point is 6 ft out from the base of the cliff. Assume that his velocity is horizontal when he begins the dive. (a) What is the magnitude of his velocity in miles per hour when he enters the water? (b) How far from his aiming point should a wave crest be when he dives in order for him to enter the water at the crest?

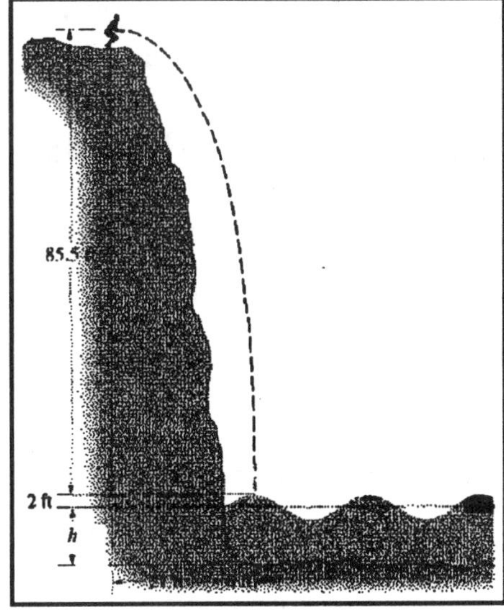

Solution The strategy is to (a) determine the time from the dive to water entry, (b) the horizontal velocity required to travel the required horizontal distance, (c) the vertical velocity at time of entry, (d) the distance traveled by the wave crest during the dive. The horizontal distance to be traveled is $d_x = 21 + 6 = 27\ ft$. The vertical distance is $d_y = 85.5\ ft.$. The trajectory equations are given by (1) $\frac{dv_x}{dt} = 0$, from which $v_x = V_x$, where V_x is the unknown horizontal velocity at $t = 0$. (2) $\frac{dv_y}{dt} = -g$, from which $v_y = -gt$, since the vertical velocity is zero at $t = 0$. (3) $\frac{dx}{dt} = v_x = V_x$, from which $v_x = V_x t$, since $x(0) = 0$. (4) $\frac{dy}{dt} = v_y = -gt$, from which $y = -\frac{g}{2}t^2$, since $y(0) = 0$. The time of the dive is $y(t_{dive}) = -85.5 = -\frac{g}{2}t_{dive}^2$, from which $t_{dive} = \sqrt{\frac{2(85.5)}{g}} = 2.3$ seconds. The horizontal velocity is given by $V_x = \frac{27}{2.3} = 11.7\ ft/s$. The vertical velocity at entry is $v_y(t_{dive}) = -gt_{dive} = -74.2\ ft/s$. (a) The magnitude of the velocity is $V = \sqrt{V_x^2 + v_y(t_{dive})^2} = 75.1\ ft/s \boxed{= 51.2\ mi/hr}$. (b) The velocity of the wave crest is $v_{crest} = \sqrt{gh} = 19.65\ ft/s$. The distance traveled by the wave crest is $\boxed{d_{crest} = v_{crest}t_{dive} = 2.3(19.65) = 45.3\ ft}$, which is the distance the wave crest must be from his aiming point at the beginning of the dive.

==================================<>==================================

=================================<>=================================

Problem 2.84 A projectile is launched at 10 m/s from a sloping surface. Determine the range R.

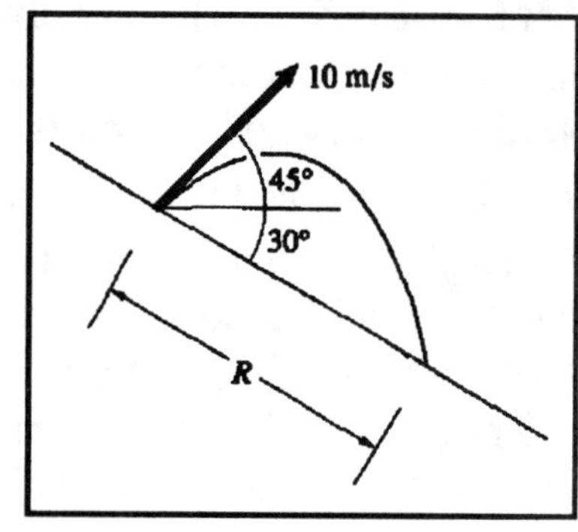

Solution Choose a coordinate system with the x axis parallel to the surface. Denote $\alpha = 30^o$, $\theta = 45^o$, $\beta = \theta + \alpha = 75^o$, and $V = 10\ m/s$.The trajectory equations are: (1) $\frac{dv_x}{dt} = g\sin\alpha$, from which $v_x = (g\sin\alpha)t + V\cos\beta$

(2) $\frac{dv_y}{dt} = -g\cos\alpha$, from which $v_y = -(g\cos\alpha)t + V\sin\beta$.

(3) $\frac{dx}{dt} = v_x = (g\sin\alpha)t + V\cos\beta$, from which $x(t) = \frac{(g\sin\alpha)}{2}t^2 + (V\cos\beta)\,t$, since $x(0) = 0$

(4) $\frac{dy}{dt} = v_y = -(g\cos\alpha)t + V\sin\beta$, from which $y(t) = -\frac{g\cos\alpha}{2}t^2 + (V\sin\beta)\,t$ since $y(0) = 0$:The time of flight is determined from $y(t_{flight}) = 0 = \left(-\frac{g\cos\alpha}{2}t_{flight} + V\sin\beta\right)t_{flught}$, from which

$t_{flight} = \frac{2V\sin\beta}{g\cos\alpha} = 2.27\ s$. The range is given by $x(t_{flight}) = R = \frac{g\sin\alpha}{2}t_{flight}^2 + (V\cos\beta)t_{flight}$, from which $\boxed{R = 18.57\ m}$ *Check:* An alternate approach: Choose a coordinate system with the origin at O and the x axis horizontal. The equation for the sloping surface (in two dimensions) is $y = -x\tan\alpha$. The trajectory equations in the new system are taken from above with $\alpha = 0$: $x(t) = (V\cos 45^o)t$ and $y(t) = -\frac{g}{2}t^2 + (V\sin\theta)t$. For impact with the sloping surface, $y(t) = -x(t)\tan\alpha$, from which $-\frac{g}{2}t^2 + V(\sin\theta + \cos\theta\tan\alpha)t = 0$, and the time of flight is $t_{flight} = \frac{2V(\sin\theta + \cos\theta\tan\alpha)}{g} = 2.27\ s$ seconds. *check.* The coordinates at impact: $x(t_{flight}) = (V\cos\theta)t_{flight} = 16.1\ m$, and $y(t_{flight}) = -x(t_{flight})\tan\alpha = -9.28\ m$. The range along the slope is $\boxed{R = \sqrt{x^2 + y^2} = 18.57\ m}$. *check.*

=================================<>=================================

Problem 2.85 A projectile is launched at 100 ft/s at 60° above the horizontal. The surface on which it lands is described by the equation shown. Determine the point of impact.

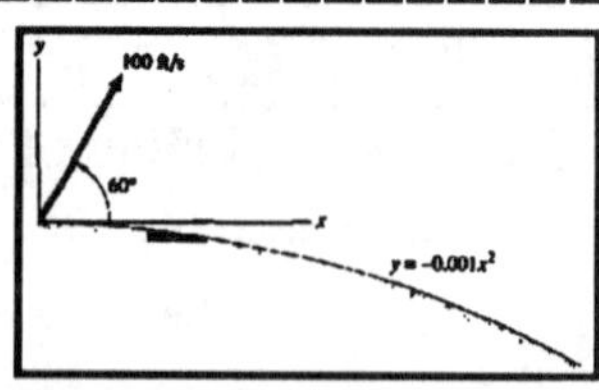

Strategy: Find the equations for the x and y coordinates of the projectile and substitute them into the equation for the surface. Solve for the time of impact then substitute this time back into the equations for the x and y coordinates of the projectile.

Solution: The motion in the x direction is $a_x = 0, v_x = V_0\cos\theta_0$, $x = (V_0\cos\theta_0)t$, and the motion in the y direction is given by $a_y = -g$, $v_y = (V_0\sin\theta_o) - gt$, $y = (V_0\sin\theta_0)t - gt^2/2$. We know that $V_0 = 100\ ft/s$ and $\theta_o = 60°$. The equation of the surface upon which the projectile impacts is $y = -0.001x^2$. Thus, the time of impact, t_I, can be determined by substituting the values of x and y from the motion equations into the equation for the surface. Hence, we get $(V_o\sin\theta_o)t_I - g\frac{t_I^2}{2} = -0.001(V_o\cos\theta_o)^2 t_I^2$. Evaluating with the known values, we get $t_I = 6.37s$ Substituting this value into the motion equations reveals that impact occurs at $(x, y) = (318.4,\ -101.4)ft$.

=================================<>=================================

====================================<>====================================

Problem 2.86 At $t = 0$, a steel ball in a tank of oil is given a horizontal velocity $\vec{v} = 2\vec{i}\ m/s$. The components of the acceleration in m/s^2 are $a_x = -1.2v_x$, $a_y = -8 - 1.2v_y$, $a_z = -1.2v_z$. What is the velocity of the ball at $t = 1$ second?

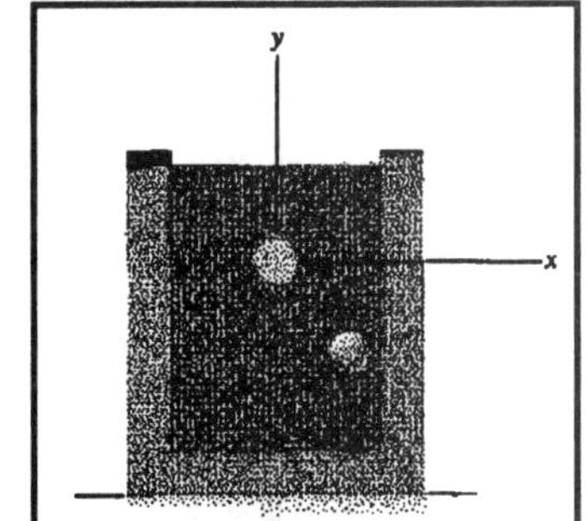

Solution Assume that the effect of gravity is included in the given accelerations. The equations for the path are obtained from:

(1) $\frac{dv_x}{dt} = a_x = -1.2v_x$. Separate variables and integrate: $\frac{dv_x}{v_x} = -1.2\,dt$,

from which $\ln(v_x) = -1.2t + V_x$. At $t = 0$, $v_x(0) = 2$, from which $\ln\left(\frac{v_x}{2}\right) = -1.2t$. Inverting:

$v_x(t) = 2e^{-1.2t}$. (2) $\frac{dv_y}{dt} = a_y = -8 - 1.2v_y$. Separate variables and integrate: $\frac{dv_y}{\frac{8}{1.2} + v_y} = -1.2dt$, from

which $\ln\left(\frac{8}{1.2} + v_y\right) = -1.2t + V_y$. At $t = 0$, $v_y(0) = 0$, from $\ln\left(1 + \frac{1.2}{8}v_y\right) = -1.2t$. Inverting:

$v_y(t) = \frac{8}{1.2}\left(e^{-1.2t} - 1\right)$. (3) $\frac{dv_z}{dt} = a_z = -1.2v_z$, from which $\ln(v_z) = -1.2t + V_z$. Invert to obtain

$v_z(t) = V_z e^{-1.2t}$. At $t = 0$, $v_z(0) = 0$, hence $V_z = 0$ and $v_z(t) = 0$. At $t = 1$ second,

$\boxed{v_x(1) = 2e^{-1.2} = 0.6024\ m/s}$, and $\boxed{v_y(1) = -\left(\frac{8}{1.2}\right)\left(1 - e^{-1.2}\right) = -4.66\ m/s}$, or $\boxed{\vec{v} = 0.602\vec{i} - 4.66\vec{j}}$

====================================<>====================================

Problem 2.87 In Problem 2.86, what is the position of the ball at $t = 1$ seconds relative to its position at $t = 0$?

Solution Use the solution for the velocity components from Problem 2.78. The equations for the coordinates: (1) $\frac{dx}{dt} = v_x = 2e^{-1.2t}$, from which $x(t) = -\left(\frac{2}{1.2}\right)e^{-1.2t} + C_x$. At $t = 0$, $x(0) = 0$, from which $x(t) = \left(\frac{2}{1.2}\right)\left(1 - e^{-1.2t}\right)$. (2) $\frac{dy}{dt} = \left(\frac{8}{1.2}\right)\left(e^{-1.2t} - 1\right)$, from which $y(t) = -\left(\frac{8}{1.2}\right)\left(\frac{e^{-1.2t}}{1.2} + t\right) + C_y$. At $t = 0$, $y(0) = 0$, from which $y(t) = -\left(\frac{8}{1.2}\right)\left(\frac{e^{-1.2t}}{1.2} + t - \frac{1}{1.2}\right)$. (3) Since $v_z(0) = 0$ and $z(0) = 0$, then $z(t) = 0$. At $t = 1$, $\boxed{x(1) = \left(\frac{2}{1.2}\right)\left(1 - e^{-1.2}\right) = 1.165\ m}$. $\boxed{y(1) = -\left(\frac{8}{1.2}\right)\left(\frac{e^{-1.2}}{1.2} + 1 - \frac{1}{1.2}\right) = -2.784\ m}$, or

$\boxed{\vec{r} = 1.165\vec{i} - 2.784\vec{j}\ (m)}$

====================================<>====================================

Problem 2.88 You must design a device for an assembly line that launches small parts through the air into a bin. The launch point is $x = 200\ mm,\ y = -50\ mm,\ z = -100\ mm$. (The y-axis is vertical and positive upward.) To land in the bin, the parts must pass through the point $x = 600\ mm, y = 200\ mm$, and $z = 100\ mm$ *moving horizontally.* Determine the components of velocity the launcher must give the parts.

Solution The strategy is to treat the point (600, 200, 100) as an impact point with $v_y(t_{impact}) = 0$. Denote the initial velocities by V_x, V_y, V_z. The path is abstained from: $\frac{dv_x}{dt} = a_x = 0$, from which $v_x(t) = V_x$. $\frac{dx}{dt} = v_x = V_x$, from which $x(t) = V_x t + C_x$. At $t = 0$, $x(0) = 200\ mm$, from which $C_x = 200$. $\frac{dv_y}{dt} = -g$, from which $v_y(t) = -gt + V_y$. $\frac{dy}{dt} = v_y = -gt + V_y$, from which $y(t) = -\frac{g}{2}t^2 + V_y t + C_y$. At $t = 0$, $y(0) = -50\ mm$, from which $C_y = -50$. $\frac{dv_z}{dt} = a_z = 0$, from which $v_z(t) = V_z$. $\frac{dz}{dt} = v_z = V_z$, from which $z(t) = V_z t + C_z$. At $t = 0$, $z(0) = -100\ mm$, from which $C_z = -100$. At the point (600, 200, 100) (1) $v_y(t_{impact}) = 0 = -g t_{impact} + V_y$ and (2) $y(t_{impact}) = 200 = -\frac{g}{2}t_{impact}^2 + V_y t_{impact} - 50$. Substitute $t_{impact} = \frac{V_y}{g}$ from the first condition into the second and solve for V_y: $\boxed{V_y = \sqrt{\frac{g}{2}} = 2.21\ m/s}$, from which $t_{impact} = \frac{V_y}{g} = 0.2258$ seconds.

Substitute: $x(t_{impact}) = 600 = V_x t_{impact} + 200$, from which $\boxed{V_x = \frac{400\ mm}{t_{impact}} = 1.77\ m/s}$.

$z(t_{impact}) = 100 = V_z t_{impact} - 100$, from which $\boxed{V_z = \frac{200\ mm}{t_{impact}} = 0.89\ m/s}$.

Problem 2.89 If $y = 150\ mm$, $\frac{dy}{dt} = 300\ mm/s$, and $\frac{d^2y}{dt^2} = 0$, what are the magnitudes of the velocity and acceleration of point P?

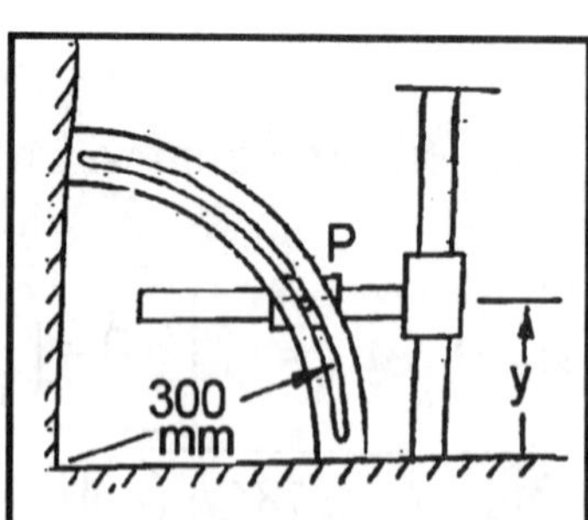

Solution The equation for the location of the point P is $R^2 = x^2 + y^2$, from which (1) $x = (R^2 - y^2)^{\frac{1}{2}} = 0.2598\ m$, and

(2) $\frac{dx}{dt} = -\left(\frac{y}{x}\right)\left(\frac{dy}{dt}\right) = -0.1732\ m/s$,

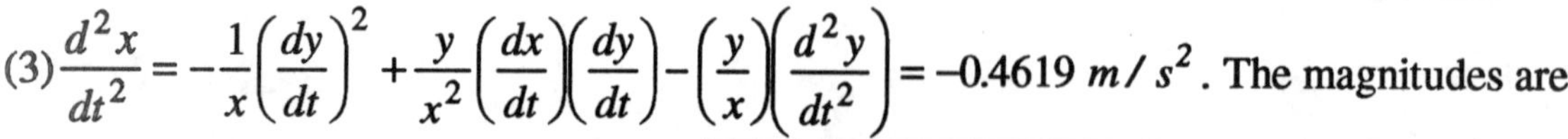

(3) $\frac{d^2x}{dt^2} = -\frac{1}{x}\left(\frac{dy}{dt}\right)^2 + \frac{y}{x^2}\left(\frac{dx}{dt}\right)\left(\frac{dy}{dt}\right) - \left(\frac{y}{x}\right)\left(\frac{d^2y}{dt^2}\right) = -0.4619\ m/s^2$. The magnitudes are:

$\boxed{|v_P| = \sqrt{\left(\frac{dx}{dt}\right)^2 + \left(\frac{dy}{dt}\right)^2} = 0.3464\ m/s}$ $\boxed{|a_P| = \sqrt{\left(\frac{d^2x}{dt^2}\right)^2 + \left(\frac{d^2y}{dt^2}\right)^2} = 0.4619\ m/s^2}$

====================================<>==================================

Problem 2.90 A car travels at a constant speed of 100 *km/hr* on a straight road of increasing grade whose vertical profile can be approximated by the equation shown. When $x = 400\ m$, what is the car's acceleration?

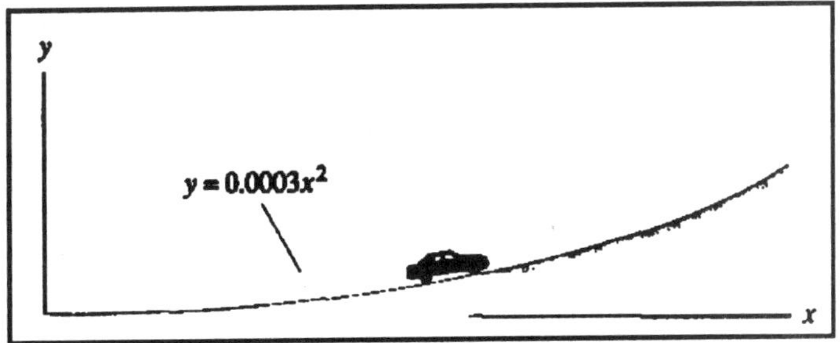

Solution Denote $C = 0.0003$ and $V = 100\ km/hr = 27.78\ m/s$. The magnitude of the constant velocity is

$V = \sqrt{\left(\frac{dy}{dt}\right)^2 + \left(\frac{dx}{dt}\right)^2}$ The equation for the road is $y = Cx^2$ from which $\frac{dy}{dt} = 2Cx\left(\frac{dx}{dt}\right)$. Substitute and solve: $\left|\frac{dx}{dt}\right| = \frac{V}{\sqrt{(2Cx)^2 + 1}} = 27.01\ m/s$. $\frac{dx}{dt}$ is positive (car is moving to right in sketch). The acceleration is

$$\boxed{\frac{d^2x}{dt^2} = \frac{d}{dt}\left(\frac{V}{\sqrt{(2Cx)^2+1}}\right) = \frac{-4C^2Vx}{\left((2Cx)^2+1\right)^{\frac{3}{2}}}\left(\frac{dx}{dt}\right) = -0.0993\ m/s^2}.$$

$$\boxed{\frac{d^2y}{dt^2} = 2C\left(\frac{dx}{dt}\right)^2 + 2Cx\left(\frac{d^2x}{dt^2}\right) = 0.4139\ m/s^2}, \text{ or } \boxed{\vec{a} = -0.099\vec{i} + 0.414\vec{j}\ \left(m/s^2\right)}$$

====================================<>==================================

Problem 2.91 Suppose that a projectile has the initial conditions shown in Figure 2.18. Show that in terms of the x', y' coordinate system with its origin at the highest point of the trajectory, the equation describing the trajectory is

$$y' = -\frac{g}{2v_o^2\cos^2\theta_o}(x')^2.$$

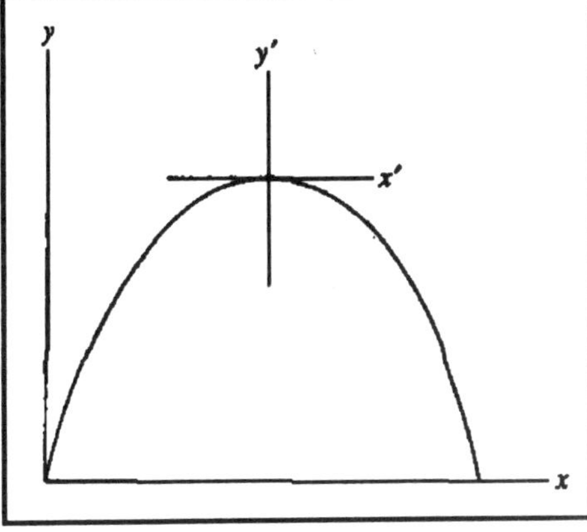

Solution The initial conditions are $t = 0$, $x(0) = 0$, $y(0) = 0$, $v_x(0) = v_o\cos\theta_o$, and $v_y(0) = v_o\sin\theta_o$. The accelerations are $a_x(t) = 0$, $a_y(t) = -g$. The path of the projectile in the *x, y* system is obtained by solving the differential equations subject to the initial conditions: $x(t) = (v_o\cos\theta_o)t$, $y(t) = -\frac{g}{2}t^2 + (v_o\sin\theta_o)t$. Eliminate t from the equations by substituting $t = \frac{x}{v_o\cos\theta_o}$ to obtain $y(x) = -\frac{gx^2}{2v_o^2\cos^2\theta_o} + x\tan\theta_o$. At the peak, $\left.\frac{dy}{dx}\right|_{peak} = 0$, from which $x_p = \frac{v_o^2\cos\theta_o\sin\theta_o}{g}$, and $y_p = \frac{v_o^2\sin^2\theta_o}{2g}$. The primed coordinates: $y' = y - y_p$ $x' = x - x_p$.

Substitute and reduce: $y' = -\frac{g(x'+x_p)^2}{2v_o^2\cos^2\theta_o} + (x'+x_p)\tan\theta_o - y_p$.

Solution continued on next page

Continuation Solution to Problem 2.91.

$$y' = -\frac{g}{2v_o^2\cos^2\theta_o}((x')^2 + x_p^2 + 2x'x_p) + (x' + x_p)\tan\theta_o - \frac{v_o^2\sin^2\theta_o}{2g}.$$

Substitute $x_p = \dfrac{v_o^2\cos\theta_o\sin\theta_o}{g}$,

$$y' = \frac{-g(x')^2}{2v_o^2\cos^2\theta_o} - \frac{v_o^2\sin^2\theta_o}{2g} - x'\tan\theta_o + x'\tan\theta_o + \frac{v_o^2\sin^2\theta_o}{g} - \frac{v_o^2\sin^2\theta_o}{2g}.$$

$$\boxed{y' = -\frac{g}{2v_o^2\cos^2\theta_o}(x')^2}$$

==================================<>==================================

Problem 2.92 The acceleration components of a point are $a_x = -4\cos(2t)$, $a_y = -4\sin(2t)$, $a_z = 0$. At $t = 0$, its position and velocity are $\vec{r} = \vec{i}$, $\vec{v} = 2\vec{j}$. Show that (a) the magnitude of the velocity is constant; (b) the velocity and acceleration vectors are perpendicular; (c) the magnitude of the acceleration is constant and points toward the origin; (d) the trajectory of a point is a circle with its center at the origin.

Solution The equations for the path are (1) $\dfrac{dv_x}{dt} = a_x = -4\cos(2t)$, from which $v_x(t) = -2\sin(2t) + V_x$. At $t = 0$, $v_x(0) = 0$, from which $V_x = 0$. $\dfrac{dx}{dt} = v_x = -2\sin(2t)$, from which $x(t) = \cos(2t) + C_x$. At $t = 0$, $x(0) = 1$, from which $C_x = 0$. (2) $\dfrac{dv_y}{dt} = a_y = -4\sin(2t)$, from which $v_y(t) = 2\cos(2t) + V_y$. At $t = 0$, $v_y(0) = 2$, from which $V_y = 0$. $\dfrac{dy}{dt} = v_y = 2\cos(2t)$, from which $y(t) = \sin(2t) + C_y$. At $t = 0$, $y(0) = 0$, from which $C_y = 0$. (3) For $a_z = 0$ and zero initial conditions, it follows that $v_z(t) = 0$ and $z(t) = 0$. (a) The magnitude of the velocity is $|\vec{v}| = \sqrt{(-2\sin(2t))^2 + (2\cos(2t))^2} = 2 = const$. (b) The velocity is $\vec{v}(t) = -\vec{i}\,2\sin(2t) + \vec{j}\,2\cos(2t)$. The acceleration is $\vec{a}(t) = -\vec{i}\,4\cos(2t) - \vec{j}\,4\sin(2t)$. If the two are perpendicular, the dot product should vanish: $\vec{a}(t)\cdot\vec{v}(t) = (-2\sin(2t))(-4\cos(2t)) + (2\cos(2t))(-4\sin(2t)) = 0$, and it does (c) The magnitude of the acceleration: $\boxed{|\vec{a}| = \sqrt{(-4\cos(2t))^2 + (-4\sin(2t))^2} = 4 = const}$. The unit vector parallel to the acceleration is $\vec{e} = \dfrac{\vec{a}}{|\vec{a}|} = -\vec{i}\cos(2t) - \vec{j}\sin(2t)$, which always points to the origin. (d) The trajectory path is $x(t) = \cos(2t)$ and $y(t) = \sin(2t)$. These satisfy the condition for a circle of radius 1: $\boxed{1 = x^2 + y^2}$

==================================<>==================================

==================================<>==================================

Problem 2.93 Suppose that the jet engine in Example 2.8 starts from rest and has a constant angular acceleration $\alpha = 5\ rad/s^2$. (a) How long does it take to reach an angular velocity of 10,000 rpm? (b) How many revolutions does it turn in that time?

Solution: Convert 10000 rpm to *rad/s*. $10000rpm = 10000\left(\frac{rev}{\min}\right)\left(\frac{1\min}{60s}\right)\left(\frac{2\pi\, rad}{1\, rev}\right) = 1047.2 rad/s$.

The motion of the rotor is characterized by $\alpha = 5rad/s^2$, $\omega = 5t\ rad/s$, $\theta = 2.5t^2\ rad$. The time at which $\omega = 1047.2 rad/s$ is $t_{10k} = 1047.2/5\ s = 209.4s$. The number of revolutions turned in this time is given by $\theta = 2.5t_{10k}^2 = 109662 rad = 17453 rev$.

==================================<>==================================

Problem 2.94 Let L be a line from the center of the earth to a fixed point on the equator and let L_O denote a fixed reference direction. The figure shows the earth seen from the north pole. (a) Is $\frac{d\theta}{dt}$ positive or negative? (b) What is the magnitude of $\frac{d\theta}{dt}$ in rad/s?.

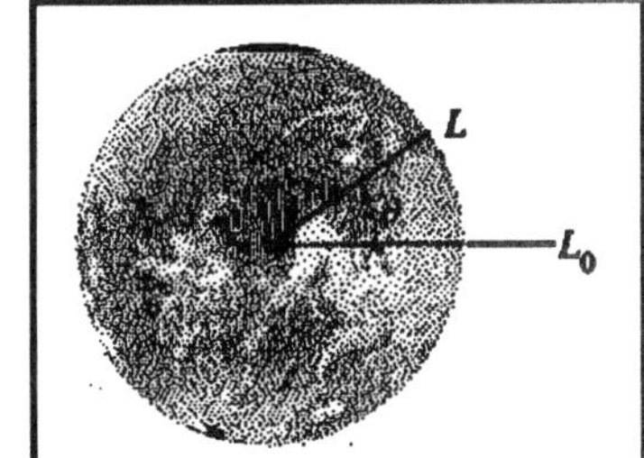

Solution (a) The earth rotates to the east, which viewed from the North pole is a counterclockwise rotation. Thus $\frac{d\theta}{dt}$ is positive. (b) The earth rotates about its on axis 2π radians per day relative to the sun, plus one revolution per year relative to a fixed direction, since the annual rotation about the sun is in a counterclockwise direction viewed from the north pole of the ecliptic. Take the length of the year to be $year = 365.242198781... = 365.2422\ days$. The revolutions per day is $rpd = 1 + \frac{1}{year} = 1.002737909... = 1.0027379\ rev/day$. Thus

$$\boxed{\frac{d\theta}{dt} = 1.0027379\left(\frac{rev}{day}\right)\left(\frac{2\pi\, rad}{1\, rev}\right) = 6.300\ rad/day}$$

or one revolution in $23^h 56^m 4^s$. *This is approximately one revolution in 24 hours..Check:*

$\frac{d\theta}{dt} = rpd\left(\frac{2\pi\, rad}{1\, day}\right)\left(\frac{1\, day}{86400\, s}\right) = 7.292116 \times 10^{-5}\ rad/s$. For the year 1979 NASA adopted the value $\omega_P = 7.2921158491837 \times 10^{-5}\ rad/s$ with respect to a precessing equinox. *check.*

==================================<>==================================

Problem 2.95 The angle between L and a reference line L_O is $\theta = 2t^2$ rad. (a) What are the angular velocity and angular acceleration of L relative to L_O at $t = 6$ seconds? (b) How many revolutions does L rotate relative to L_O during the interval of time from $t = 0$ to $t = 6$ seconds?.

Solution (a)The angular velocity is $\boxed{\frac{d\theta}{dt} = 4t|_{t=6} = 24\ rad/s}$. The angular acceleration is

$\boxed{\frac{d^2\theta}{dt^2} = 4\ rad/s^2}$ (b) The number of rotations is $\boxed{R = \left[\frac{2t^2}{2\pi}\right]_{t=6} = \frac{72}{2\pi} = 11.459\ revolutions}$

==================================<>==================================

=================================<>=================================

Problem 2.96 The angle θ between the bar and the horizontal line is $\theta = t^3 - 2t^2 + 4$ degrees. Determine the angular velocity and angular acceleration of the bar at $t = 10$ seconds.

Solution The angular velocity is

$$\frac{d\theta}{dt} = \left[3t^2 - 4t\right]_{t=10} = 260\left(\frac{deg}{s}\right)\left(\frac{\pi\, rad}{180\, deg}\right) = 4.538\ rad/s$$

The acceleration is $$\frac{d^2\theta}{dt^2} = \left[6t - 4\right]_{t=10} = 56\left(\frac{deg}{s}\right)\left(\frac{\pi\, rad}{180\, deg}\right) = 0.9774\ rad/s^2$$

=================================<>=================================

Problem 2.97 The angular acceleration of a line L relative to a line L_0 is $\alpha = 30 - 6t\ rad/s^2$. When $t = 0$, $\theta = 0$ and $\omega = 0$. What is the maximum angular velocity of L relative to L_0 during the interval of time from $t = 0$ to $t = 10$ seconds?

Solution The angular velocity is $\omega(t) = \int \alpha\, dt + C = 30t - 3t^2 + C$. At $t = 0$, $\omega(0) = 0$ from which $C = 0$. The maximum(minimum) occurs when the slope is zero, $\frac{d\omega}{dt} = 0 = 30 - 6t = 0$, from which $t = \frac{30}{6} = 5$ seconds. The value of the angular velocity is $\omega_{max} = \left[30t - 3t^2\right]_{t=5} = 75\ rad/s$. This is indeed a maximum since $\left[\frac{d^2\omega}{dt^2}\right]_{t=5} = -6 < 0$. [See Note under solution to Problem 2.6.]

=================================<>=================================

Problem 2.98 A gas turbine starts rotating from rest and has angular acceleration $\alpha = 6t\ rad/s^2$ for 3 seconds. It then slows down with constant angular deceleration $\alpha = -3\ rad/s^2$ until it stops. (a) What maximum angular velocity does it attains? (b) Through what total angle does it turn?

Solution (a) For $0 \le t \le 3$ the angular velocity is $\omega(t \le 3) = \int \alpha\, dt + C = 3t^2 + C$. At $t = 0$, $\omega(0) = 0$, from which $C = 0$. The maximum is $\left[\omega(t)\right]_{t=3} = 27\ rad/s$. The angular travel is $\theta(t \le 3) = \int \omega\, dt + C = t^3 + C$. At $t = 0$, $\theta(0) = 0$, from which $C = 0$. The total travel during the first interval is $\left[\theta(t)\right]_{t=3} = 3^3 = 27\ rad$. (b) For $t > 3$, the angular velocity is $\omega(t > 3) = \int \alpha\, dt + C = -3t + C$. At $t = 3$, $\omega(3) = 27\ rad/s$, from which $C = 36$. The angular travel is $\theta(t > 3) = \int \omega dt + C = -\frac{3}{2}t^2 + 36t + C$. At $t = 3$, $\theta(3) = 27\ rad$, from which $C = -67.5$. The time of travel is obtained from $\omega(t_s) = 0 = -3t_s + 36$, from which the stopping time is $t_s = \frac{36}{3} = 12$ seconds.

The total travel is $$\theta(t = 12) = \left[-\frac{3}{2}t^2 + 36t - 67.5\right]_{t=12} = 148.5\ rad$$

=================================<>=================================

==================================<>==================================

Problem 2.99 The rotor of an electric generator is rotating at 200 rpm (revolutions per minute) when the motor is turned off. Due to frictional effects, the angular deceleration of the rotor after it is turned off is $\alpha = -0.01\omega \ rad/s^2$, where ω is the angular velocity in rad/s. How many revolutions does the rotor turn after the motor is turned off?

Solution The angular velocity at $t = 0$ is $\omega(0) = 200\left(\frac{rev}{min}\right)\left(\frac{1\ min}{60\ s}\right)\left(\frac{2\pi\ rad}{rev}\right) = 20.944\ rad/s$. The angular velocity at time t is given by $\frac{d\omega}{dt} = a = -0.01\omega$. Separate variables and integrate: $\frac{d\omega}{\omega} = -0.01dt$, from which $\ln(\omega) = -0.01t + C$. Invert to obtain $\omega(t) = Ce^{-0.01t}$. At $t = 0$, $\omega(0) = 20.944\ rad/s$, from which $C = 20.944$. The angular travel is $\frac{d\theta}{dt} = \omega = 20.944e^{-0.01t}$, from which $\theta(t) = -2094.4e^{-0.01t} + C$. Count the angular travel from the time the motor is turned off, from which $\theta(0) = 0$, and $C = 2094.4$, and $\theta(t) = 2094.4\left(1 - e^{-0.01t}\right)$. The rotor comes to rest at a time so great that

$\omega \to 0 = \lim_{t\to\infty}\left(20.944e^{-0.01t}\right) \to 0$. Substitute $e^{-0.01t} \to 0$ into the angular travel to obtain

$$\boxed{\theta_{total} = 2094.4\ rad\left(\frac{revs}{2\pi\ rad}\right) = 333.33\ revs}$$

==================================<>==================================

Problem 2.100 The needle of a measuring instrument is connected to a torsional spring that subjects it to an angular acceleration $\alpha = -4\theta\ rad/s^2$, where θ is the needle's angular position in radians relative to a reference position. If the needle is released from rest at $\theta = 1$ rad, what is its angular velocity at $\theta = 0$?

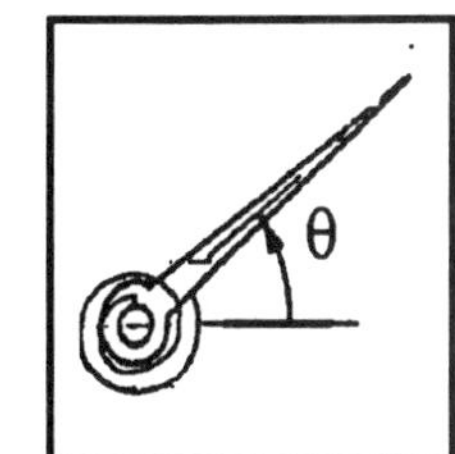

Solution From the chain rule, $\omega\frac{d\omega}{d\theta} = \alpha = -4\theta$. Separate variables and integrate: $\omega\ d\omega = -4\theta\ d\theta$, from which $\omega^2 = -4\theta^2 + C$. At $\theta = 1$, $\omega = 0$, from which $C = 4$, and $\omega^2 = 4\left(1 - \theta^2\right)$. At $\theta = 0$, $\omega^2(0) = 4$, and $\boxed{\omega(0) = -2\ rad/s}$ where the negative sign is chosen because the system is released from rest and $\alpha < 0$.

==================================<>==================================

Problem 2.101 The angle θ measures the direction of the unit vector $\vec{e}$ relative to the x-axis. Given that $\omega = \frac{d\theta}{dt} = 2\ rad/s$, determine the vector $\frac{d\vec{e}}{dt}$ (a) when $\theta = 0$; (b) when $\theta = 90^o$; (c) when $\theta = 180^o$.

Solution The solution: $\frac{d\vec{e}}{dt} = \frac{d\theta}{dt}\vec{n}$, where $\vec{n} = \left(\vec{i}\cos\left(\theta + \frac{\pi}{2}\right) + \vec{j}\sin\left(\theta + \frac{\pi}{2}\right)\right)$ Thus (a) $\boxed{\left[\frac{d\vec{e}}{dt}\right]_{\theta=0} = 2\vec{j}}$. (b) $\boxed{\left[\frac{d\vec{e}}{dt}\right]_{\theta=90^\circ} = -2\vec{i}}$. (c) $\boxed{\left[\frac{d\vec{e}}{dt}\right]_{\theta=180^\circ} = -2\vec{j}}$

==================================<>==================================

=================================<>=================================

Problem 2.102 In Problem 2.101, suppose that the angle $\theta = 2t^2$ rad. What is the vector $\frac{d\vec{e}}{dt}$ at $t = 4$ seconds?

Solution By definition: $\frac{d\vec{e}}{dt} = \left(\frac{d\theta}{dt}\right)\vec{n}$, where $\vec{n} = \vec{i}\cos\left(\theta + \frac{\pi}{2}\right) + \vec{j}\sin\left(\theta + \frac{\pi}{2}\right)$ is a unit vector in the direction of positive θ. The angular rate of change is $\left[\frac{d\theta}{dt}\right]_{t=4} = [4t]_{t=4} = 16\ rad/s$. The angle is $\theta = \left[\text{mod}(2t^2, 2\pi)\right]_{t=4} = \text{mod}(32, 2\pi) = 0.5841\ rad$, where mod(x,y) ("modulus") is a standard function that returns the remainder of division of the first argument by the second. From which,

$$\left[\frac{d\vec{e}}{dt}\right]_{t=4} = 16\left(\vec{i}\cos\left(0.5841 + \frac{\pi}{2}\right) + \vec{j}\sin\left(0.5841 + \frac{\pi}{2}\right)\right) = -8.823\vec{i} + 13.35\vec{j}$$

=================================<>=================================

Problem 2.103 The line OP is of constant length R. The angle $\theta = \omega_o t$, where ω_o is a constant.

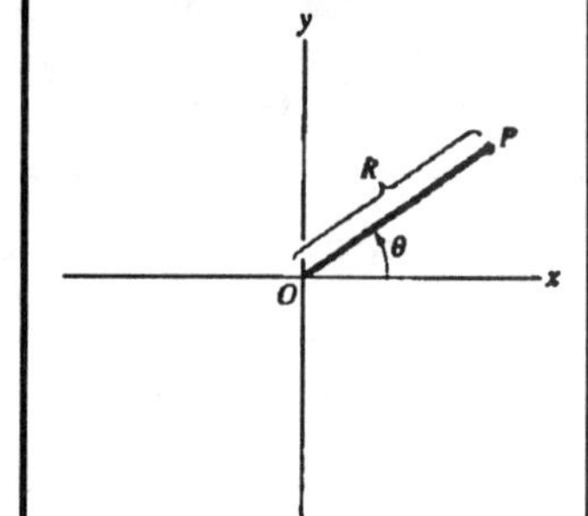

(a) Use the relations $v_x = \frac{dx}{dt}$, $v_y = \frac{dy}{dt}$ to determine the velocity of P relative to O. (b) Use Eq (2.33) to determine the velocity of P relative to O and confirm that your result agrees with the result of Part (a).

Solution (a) The point P is described by $\vec{P} = \vec{i}x + \vec{j}y$. Take the derivative: $\frac{d\vec{P}}{dt} = \vec{i}\left(\frac{dx}{dt}\right) + \vec{j}\left(\frac{dy}{dt}\right)$. The coordinates are related to the angle θ by $x = R\sin\theta$, $y = R\cos\theta$. Take the derivative and note that R is a constant and $\theta = \omega_o t$, so that $\frac{d\theta}{dt} = \omega_o$: $\frac{dx}{dt} = -R\cos\theta\left(\frac{d\theta}{dt}\right)$, $\frac{dy}{dt} = R\sin\theta\left(\frac{d\theta}{dt}\right)$. Substitute into the derivative of the vector $\vec{P}$,

$$\frac{d\vec{P}}{dt} = R\left(\frac{d\theta}{dt}\right)\left(-\vec{i}\sin\theta + \vec{j}\cos\theta\right) = R\omega_o\left(-\vec{i}\sin(\omega_o t) + \vec{j}\cos(\omega_o t)\right)$$

which is the velocity of the point P relative to the origin O. (b) Note that $\vec{P} = R\vec{e}$, and $\frac{d\vec{P}}{dt} = R\frac{d\vec{e}}{dt}$ when R is constant. Use the definition (Eq (2.33)), $\frac{d\vec{e}}{dt} = \left(\frac{d\theta}{dt}\right)\vec{n}$, where $\vec{n}$ is a unit vector in the direction of positive θ, (i.e., perpendicular to $\vec{e}$). Thus $\vec{n} = \vec{i}\cos\left(\theta + \frac{\pi}{2}\right) + \vec{j}\sin\left(\theta + \frac{\pi}{2}\right)$. Use the trigonometric sum-of-angles identities to obtain: $\vec{n} = -\vec{i}\sin\theta + \vec{j}\cos\theta$. Substitute,

$$\frac{d\vec{P}}{dt} = R\omega_o\left(-\vec{i}\sin(\omega_o t) + \vec{j}\cos(\omega_o t)\right)$$

The results are the same.

=================================<>=================================

=====================================<>=====================================

Problem 2.104 The armature of an electric motor rotates at a constant rate. The magnitude of the velocity of point P relative to O is 4 *m/s*. (a) What are the normal and tangential components of the acceleration of P relative to O? (b) What is the angular velocity of the armature?

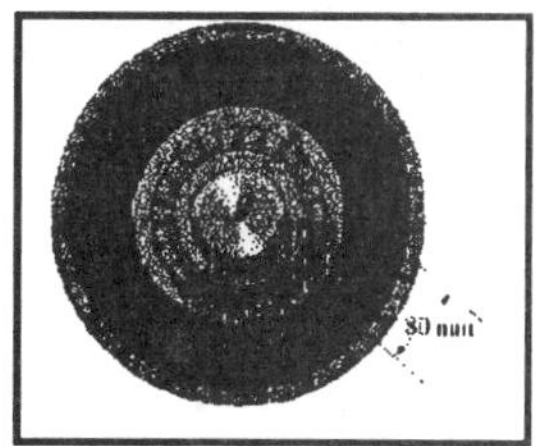

Solution The distance traveled by point P is $s = R\theta$. The velocity is given to be $v = \frac{ds}{dt} = R\frac{d\theta}{dt} = 4\ m/s$, where R is constant, from which (b) the angular velocity of the armature is $\boxed{\frac{d\theta}{dt} = \omega = \frac{v}{R} = \frac{4}{0.08} = 50\ rad/s}$. (a) The tangential acceleration is zero, i.e., $\boxed{a_t = R\frac{d\omega}{dt} = 0}$, since rotation rate is constant. The normal acceleration is $\boxed{a_n = \frac{v^2}{R} = R\omega^2 = 200\ m/s^2}$

=====================================<>=====================================

Problem 2.105 The armature in Problem 2.104 starts from rest and has constant angular acceleration $\alpha = 10\ rad/s^2$. What are the velocity and acceleration of P relative to O in terms of the normal and tangential components after 10 seconds?

Solution The angular velocity is $\omega(t) = \int \alpha\, dt + C = 10t + C$. Since the armature starts from rest, $\omega(0) = 0$, and $C = 0$. The velocity of the point P is $\boxed{\vec{v} = R\omega\,\vec{e}_t = 0.08(100)\vec{e}_t = 8\vec{e}_t\ m/s}$. The acceleration is $\boxed{\vec{a} = (R\alpha)\vec{e}_t + (R\omega^2)\vec{e}_n = 0.08\times10^1\vec{e}_t + 0.08\times10^4\vec{e}_n = 0.8\vec{e}_t + 800\vec{e}_n\ m/s^2}$

=====================================<>=====================================

Problem 2.106 Suppose you want to design a medical centrifuge to subject samples to a normal acceleration of 1000 *g*'s. (a) If the distance from the center of the centrifuge to the sample is 300 *mm*, what speed of rotation in rpm (revolutions per minute) is necessary? (b) If you want the centrifuge to reach its design rpm in one minute, what constant angular acceleration is necessary?

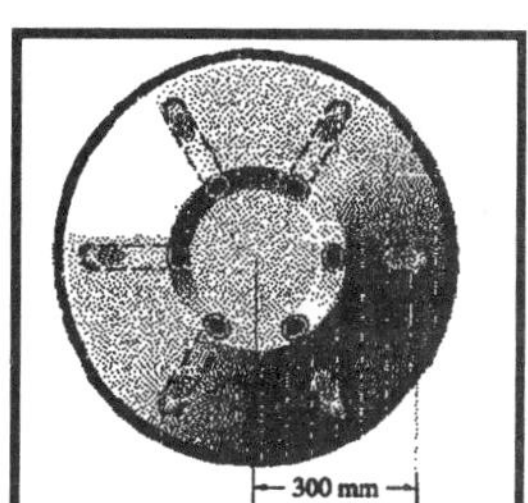

Solution (a) The normal acceleration at a constant rotation rate is $a_n = R\omega^2$, giving $\omega = \sqrt{\frac{a_n}{R}} = \sqrt{\frac{(1000)9.81}{0.3}} = 180.83\ rad/s$. The speed in rpm is $\boxed{N = \omega\left(\frac{rad}{s}\right)\left(\frac{1\ rev}{2\pi\ rad}\right)\left(\frac{60\ s}{1\ min}\right) = 1726.8\ rpm}$. (b) The angular acceleration is $\boxed{\alpha = \frac{\omega}{t} = \frac{180.83}{60} = 3.01\ rad/s^2}$

=====================================<>=====================================

Problem 2.107 In Example 2.12, what are the magnitudes of the normal and tangential components of acceleration to which the samples are subjected at the instant the centrifuge is turned on.

Solution: From Example 2.12 we have $\alpha = A - B\omega^2$ and $\omega = \omega_{MAX}\left(\frac{e^{2\sqrt{(AB)}\,t} - 1}{e^{2\sqrt{(AB)}\,t} + 1}\right)$ where $A = 7.69 rad/s^2$, $B = 1.96\times10^{-5} rad^{-1}$, and $\omega_{MAX} = 626 rad/s$. The radius is $r = 300mm = 0.3m$. Setting $t = 0s$, we find $\omega_o = 0$. Thus, the normal acceleration is $a_N = r\omega_o^2 = 0$, and the tangential acceleration is $a_T = r\alpha = (0.3)A = 2.31 m/s^2$.

=====================================<>=====================================

===================================<>===================================

Problem 2.108 In Example 2.12, what are the magnitudes of the normal and tangential components of acceleration to which the samples are subjected 10 s after the centrifuge is turned on.

Solution: From Example 2.12 we have $\alpha = A - B\omega^2$ and $\omega = \omega_{MAX}\left(\dfrac{e^{2\sqrt{(AB)}\,t}-1}{e^{2\sqrt{(AB)}\,t}+1}\right)$ where

$A = 7.69 rad/s^2$, $B = 1.96\times10^{-5} rad^{-1}$, and $\omega_{MAX} = 626 rad/s$. The radius is $r = 300mm = 0.3m$. Setting $t = 10s$, we find $\omega = 76.47 rad/s$. The normal acceleration is $a_N = r\omega^2 = 1754.3 m/s^2$. The value for α at 10 seconds is $a_{10_s} = 7.575 rad/s^2$ and the tangential acceleration is $a_T = r\alpha = 2.27 m/s^2$.

===================================<>===================================

Problem 2.109 A powerboat being tested for maneuverability is started from rest and driven in a circular path of 40 ft radius. The magnitude of its velocity is increased at a constant rate of 2 ft/s^2. In terms of normal and tangential components, (a) determine the velocity as a function of time; (b) the acceleration as a function of time.

Solution The tangential acceleration is given to be $\boxed{a_t = 2 ft/s^2}$. (a) The tangential velocity is $v = \int a_t dt + C = 2t\ ft/s$ since the motion starts from rest. (The normal velocity is zero since the path is a circle.) The angular velocity is $\omega = \dfrac{v_t}{R} = \dfrac{t}{20}\ rad/s$. *Check:* The tangential acceleration is $a_t = R\dfrac{d\omega}{dt}$, from which $\dfrac{d\omega}{dt} = \dfrac{a_t}{R}$. Integrating, $\omega(t) = \left(\dfrac{1}{R}\right)\int a_t dt = \dfrac{t}{20}\ rad/s$, since motion starts from rest. *check.*

The normal acceleration is $\boxed{a_n = R\omega^2 = \left(\dfrac{40}{400}\right)t^2 = 0.1t^2\ ft/s^2}$

===================================<>===================================

Problem 2.110 The angle $\theta = 2t^2\ rad$. (a) What are the magnitudes of the velocity and acceleration of P relative to O at $t = 1$ second. (b) What distance along the circular path does P move from $t = 0$ to $t = 1$ second?

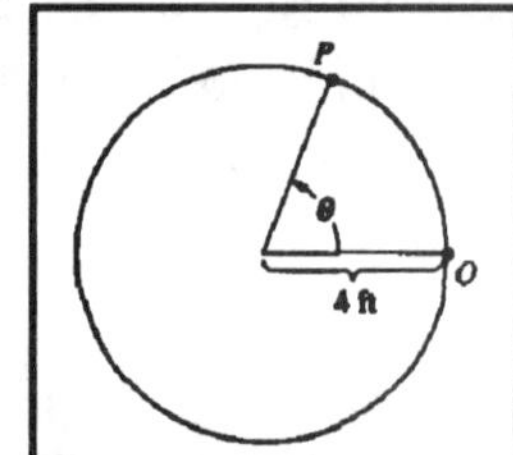

Solution Choose a coordinate system with the x-axis coinciding with $\theta = 0$, positive toward O. The position vector of P is $\vec{P} = \vec{i}x + \vec{j}y$. On the circle, $x = R\cos\theta$, and $y = R\sin\theta$. The velocity of the point P is

$\dfrac{d\vec{P}}{dt} = R\left(\dfrac{d\theta}{dt}\right)\left(-\vec{i}\sin\theta + \vec{j}\cos\theta\right)$. The acceleration is

$\dfrac{d^2\vec{P}}{dt^2} = R\left(\dfrac{d^2\theta}{dt^2}\right)\left(-\vec{i}\sin\theta + \vec{j}\cos\theta\right) + R\left(\dfrac{d\theta}{dt}\right)^2(-\vec{i}\cos\theta - \vec{j}\sin\theta)$. The derivatives are:

$\dfrac{d\theta}{dt} = \dfrac{d}{dt}\left(2t^2\right) = 4t$, $\dfrac{d^2\theta}{dt^2} = 4$, from which $\left[\dfrac{d\vec{P}}{dt}\right]_{t=1} = 16\left(-\vec{i}\sin(2) + \vec{j}\cos(2)\right) = -14.55\vec{i} - 6.66\vec{j}$. (a)

The magnitude of the velocity: : $\boxed{\left|\dfrac{d\vec{P}}{dt}\right|_{t=1} = \sqrt{14.55^2 + 6.66^2} = 16\ ft/s}$. The acceleration is:

$\left[\dfrac{d^2\vec{P}}{dt^2}\right]_{t=1} = (-16\sin(2) - 64\cos(2))\vec{i} + (16\cos(2) - 64\sin(2))\vec{j} = 12.08\vec{i} - 64.85\vec{j}$. The magnitude of the

Solution continued on next page

acceleration: $\left|\dfrac{d^2\vec{P}}{dt^2}\right| = \sqrt{12.08^2 + 64.85^2} = 65.97 = 66\ ft/s^2$ (b) The distance moved along the path is $s = R\theta = 4\left[2t^2\right]_{t=1} = 8\ ft$

==================================<>==================================

Problem 2.111 In Problem 2.110, what are the magnitudes of the velocity and acceleration of P relative to O when P has gone one revolution around the circular path starting from $t = 0$?

Solution The time lapsed is $\theta = 2t^2 = 2\pi$ from which $t = \sqrt{\pi} = 1.77$ seconds. The angular velocity is $\dfrac{d\theta}{dt} = \left[\dfrac{d}{dt}2t^2\right]_{t=1.77} = \omega = 7.09 = 7.1\ rad/s$. The angular acceleration is $\dfrac{d^2\theta}{dt^2} = 4\ rad/s^2$. Use the solution from Problem 2.100: The velocity is

$\dfrac{d\vec{P}}{dt} = R\left(\dfrac{d\theta}{dt}\right)\left(-\vec{i}\sin\theta + \vec{j}\cos\theta\right) = 4(7.1)\left(-\vec{i}\sin(2\pi) + \vec{j}\cos(2\pi)\right) = 28.36\vec{j}\ ft/s$, from which the magnitude is $\left|\dfrac{d\vec{P}}{dt}\right|_{t=1.77} = 28.36\ ft/s$. *Check:* The velocity is $v = R\omega = 4(7.09) = 28.36\ ft/s$ along the path, which coincides with the positive *y*-axis at $\theta = 2\pi$. *check.* The acceleration (from Problem 2.100)

$\dfrac{d^2\vec{P}}{dt^2} = \left(-R\left(\dfrac{d^2\theta}{dt^2}\right)\sin\theta - R\left(\dfrac{d\theta}{dt}\right)^2\cos\theta\right)\vec{i} + \left(R\left(\dfrac{d^2\theta}{dt^2}\right)\cos\theta - R\left(\dfrac{d\theta}{dt}\right)^2\sin\theta\right)\vec{j}$, from which

$\left[\dfrac{d^2\vec{P}}{dt^2}\right]_{t=1.77} = -201.06\vec{i} + 16\vec{j}$. *Check:* The tangential acceleration is $a_t = R\left(\dfrac{d\omega}{dt}\right) = 4(4) = 16\ ft/s^2$, which coincides with the positive *y*-axis at $\theta = 2\pi$. The normal acceleration is $a_n = R\omega^2 = 4(7.09)^2 = 201.06$, which coincides with the negative *x*-axis at $\theta = 2\pi$. *check.*

The magnitude is $\left|\dfrac{d^2\vec{P}}{dt^2}\right|_{t=1.77} = \sqrt{201.06^2 + 16^2} = 201.7\ ft/s^2$.

==================================<>==================================

Problem 2.112 The radius of the earth is 3960 miles. If you are standing at the equator, what is the magnitude of your velocity relative to a nonrotating reference frame with its origin at the center of the earth?

Solution The angular velocity of the earth in inertial space is 2π radians per sidereal day. A sidereal day is $t = 23\ hr\ 56\ min\ 4\ s = 86164\ s$ long, from which $\omega = 7.29212 \times 10^{-5}\ rad/s$. The velocity is

$$v = R\omega = 3960\ \text{mile}\left(\frac{5280\ \text{ft}}{1\ \text{mile}}\right)(7.29212\times10^{-5}) = 1525\ \text{ft/s}$$

$$= 1525\left(\frac{1\ \text{mile}}{5280\ \text{ft}}\right)\left(\frac{3600\ \text{s}}{1\ \text{hr}}\right) = 1039.6\ \text{mile/hr}$$

==================================<>==================================

================================<>================================

Problem 2.113 The radius of the earth is 6370 *km*. If you are standing at the equator, what is the magnitude of your acceleration relative a nonrotating reference frame with its origin at the center of the earth?

Solution From Problem 2.112, the angular velocity of the earth in inertial space is $\omega = 7.29212\times 10^{-5}\ rad/s$. The earth rotates at a constant rate, so that the tangential acceleration is zero. The normal acceleration is $a_n = R\omega^2 = 6370(7.29212\times 10^{-5})^2 = 3.3873\times 10^{-5}\ km/s^2$

================================<>================================

Problem 2.114 Suppose that you are standing at point P at 30° north latitude. (That is a point that is 30° north of the equator.) The radius of the earth is $R_E = 3960$ miles. What are the magnitudes of your velocity and acceleration relative to a nonrotating reference frame with its origin at the center of the earth?

Solution Assume a spherical earth. From Problem 2.113, the angular velocity of the earth in inertial space is $\omega = 7.29212\times 10^{-5}\ rad/s$. The linear velocity is

$$v = R\omega\cos 30^o = 3960\ mile\left(\frac{5280\ ft}{1\ mile}\right)(7.29212\times 10^{-5})(0.866) = 1320\ ft/s$$

The tangential acceleration is zero, since the earth rotates at a constant rate. The normal acceleration is

$$a_n = R\omega^2\cos(30^o) = 3960\ mile\left(\frac{5280\ ft}{1\ mile}\right)(7.29212\times 10^{-5})^2(0.866) = 0.0963\ ft/s^2$$

================================<>================================

Problem 2.115 The magnitude of the velocity of the airplane is constant and equal to 400 *m/s*. The rate of change of the path angle θ is constant and equal to 5° per second. (a) What are the velocity and acceleration of the airplane in terms of the tangential and normal coordinates? (b) What is the instantaneous radius of curvature of the airplane's path?

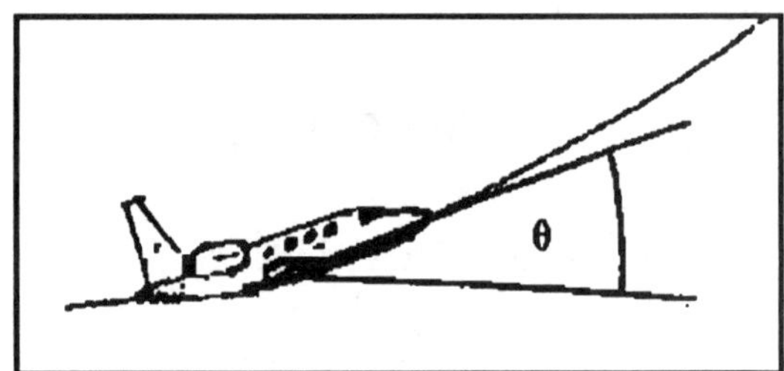

Solution The tangential component of the velocity is $\vec{v} = 400\vec{e}_t\ m/s$ The angular velocity is $\frac{d\theta}{dt} = 5\left(\frac{\deg}{s}\right)\left(\frac{\pi\ rad}{180\ deg}\right) = 0.08727\ rad/s$. The instantaneous radius of curvature is defined by $\frac{ds}{dt} = \rho\frac{d\theta}{dt}$, from which $\rho = \frac{\left(\frac{ds}{dt}\right)}{\left(\frac{d\theta}{dt}\right)} = \frac{400}{0.08727} = 4583.7\ m$ The path velocity is constant, so the tangential component of the acceleration is zero. The normal component of the acceleration is $\vec{a} = \frac{v^2}{\rho}\vec{e}_n = 34.9\vec{e}_n\ m/s$

================================<>================================

================================<>================================

Problem 2.116 At $t = 0$ a car starts from rest at point A. It moves toward the right and the tangential component of its acceleration is $a_t = 0.4t\ m/s^2$. What is the magnitude of the car's acceleration when it reaches point B?

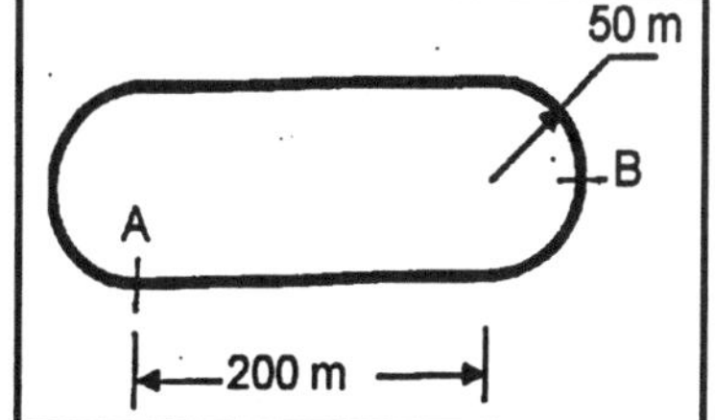

Solution The velocity of the car along the path as a function of time is $v(t) = \int_0^t a_t dt = 0.2t^2$, since the car starts from rest. The distance traveled from point A is $s = \int_0^t v(t)dt = \frac{0.2}{3}t^3$. The distance is known from the sketch, $s = 200 + R\theta = 200 + 50\pi/2 = 278.54$ m.

The time of travel is $t = \left(\frac{3s}{0.2}\right)^{\frac{1}{3}} = 16.11\ s$. The tangential acceleration at point B is

$\vec{a}_t = \left[0.4t\vec{j}\right]_{t=16.11} = 6.44\vec{j}\ m/s^2$ The velocity at point B is $\vec{v} = \left[0.2t^2\right]_{t=16.11}\vec{j} = 51.88\vec{j}\ \ m/s$. The normal acceleration at point B is $\vec{a}_n = -\frac{|\vec{v}|^2}{R}\vec{i} = -\frac{(51.88)^2}{50}\vec{i} = -53.83\vec{i}\ m/s^2$. The magnitude of the acceleration at point B is $\boxed{|\vec{a}| = \sqrt{6.44^2 + 53.83^2} = 54.22\ m/s^2}$

================================<>================================

Problem 2.117 A group of engineering students constructs a sun-powered car and tests it on a circular track of 1000 ft radius. If the car starts from rest and the tangential acceleration component of its acceleration is given in terms of the car's velocity as $a_t = 2 - 0.1v\ \ ft/s^2$. Determine v and the magnitude of the car's acceleration 15 s after it starts.

Solution: The acceleration is given by $a_t = \frac{dv}{dt} = 2 - 0.1v\,ft/s^2$. Setting up the integral, we get $\int_0^v \frac{dv}{2-0.1v} = \int_0^t dt$. Integrating, we have $-\left(\frac{1}{0.1}\right)\ln(2-0.1\ v) + \left(\frac{1}{0.1}\right)\ln(2) = 15$. Solving for v, we get $v = 15.54\,ft/s$. Substititing this back into the equation for the acceleration, we get $a_t = 0.446\,ft/s^2$. In the radial direction, $a_n = v^2/r = (15.54)^2/1000 = 0.241\,ft/s^2$. The total acceleration is then given by $|a| = \sqrt{a_t^2 + a_n^2} = \sqrt{(0.241)^2 + (0.446)^2} = 0.507\,ft/s^2$

================================<>================================

Problem 2.118 Suppose that the tangential component of acceleration in Problem 2.117 is $a_t = 2 - 0.008s\,ft/s^2$, where s is the distance the car travels along the track from the point where it starts from rest. Determine the velocity v and the magnitude of the car's acceleration when it has traveled a distance $s = 100\,ft$.

Solution: The acceleration is given by $a_t = \frac{dv}{dt} = v\frac{dv}{ds} = 2 - 0.008s\,ft/s^2$. Setting up the integral, we get $\int_0^v vdv = \int_0^s (2 - 0.008s)ds$. Integrating, we get $\frac{v^2}{2} = \left(2s - 0.008\frac{s^2}{2}\right)$. Evaluating at $s = 100\,ft$, we get $v = 17.89\,ft/s$. Substituting $s = 100\,ft$ into the acceleration relationship, we get $a_t = 1.2\,ft/s^2$. In the radial direction, $a_n = v^2/r = (17.89)^2/1000 = 0.320\,ft/s^2$. The total acceleration is then given by $|a| = \sqrt{a_t^2 + a_n^2} = \sqrt{(0.320)^2 + (1.2)^2} = 1.24\,ft/s^2$

================================<>================================

================================<>================================

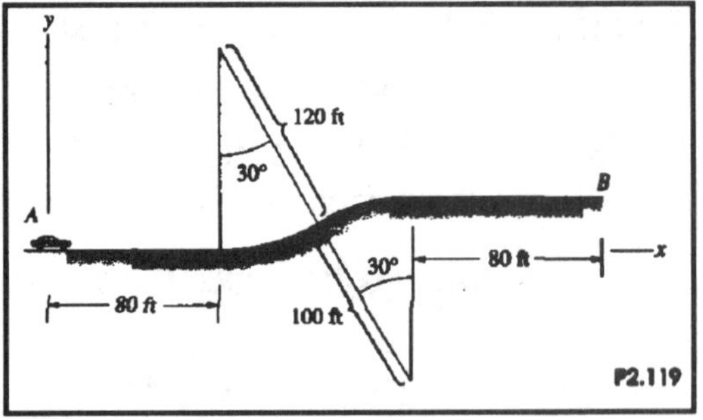

Problem 2.119 A car increases its speed at a constant rate from 40 *mph* at A to 60 *mph* at B. What is the magnitude of its acceleration 2 seconds after it passes point A?

Solution Use the chain rule to obtain $v\frac{dv}{ds} = a$, where a is constant. Separate variables and integrate: $v^2 = 2as + C$. At $s = 0$, $v(0) = 40\left(\frac{5280}{3600}\right) = 58.67\ ft/s$, from which $C = 3441.78$. The acceleration is $a = \frac{v^2 - C}{2s}$ The distance traveled from A to B is $s = 2(80) + (30)\left(\frac{\pi}{180}\right)(120 + 100) = 275\ ft$, and the speed in $[v(s)]_{s=275} = 60\left(\frac{5280}{3600}\right) = 88\ ft/s$, from the constant acceleration is $a = \frac{(88)^2 - 3441.78}{2(275)} = 7.817\ ft/s^2$. The velocity is as a function of time is $v(t) = v(0) + at = 58.67 + 7.817t\ \ ft/s$. The distance from A is $s(t) = 58.67t + \frac{7.817}{2}t^2$. At a point 2 seconds past A, the distance is $s(2) = 132.97\ ft$, and the velocity is $v(2) = 74.3\ ft/s$. The first part of the hill ends at 142.83, so that at this point the car is still in the first part of the hill. The tangential acceleration is $a_t = 7.817\ ft/s^2$. The normal acceleration is $a_n = \frac{v^2}{R} = \frac{(74.3)^2}{120} = 46.0\ ft/s^2$. The magnitude of the acceleration is $\boxed{|\vec{a}| = \sqrt{7.817^2 + 46.0^2} = 46.66\ ft/s^2}$ *Note:* This is a rather high value of acceleration-- the driver (and passengers) would no doubt be uncomfortable. *End of note.*

================================<>================================

Problem 2.120 Determine the acceleration of the car in Problem 2.119 when it has traveled along the road a distance (a) 120 *ft* from A; (b) 160 *ft* from A.

Solution Use the solution in Problem 2.119. (a) The velocity at a distance 120 *ft* from A is $v(120) = \sqrt{2as + C} = \sqrt{(2)(7.817)(120) + 3441.78} = 72.92\ ft/s$. At 120 *ft* the car is in the first part of the hill. The tangential acceleration is $a_t = 7.817\ ft/s^2$ from Problem 2.107. The normal acceleration is $a_n = \frac{(v(120))^2}{R} = \frac{(72.92)^2}{120} = 44.3\ ft/s^2$. The magnitude of the acceleration is $\boxed{|\vec{a}| = \sqrt{7.817^2 + 44.3^2} = 45\ ft/s^2}$ (b) The velocity at distance 160 *ft* from A is $v(160) = \sqrt{2(7.817)(160) + C} = 77.1\ \ ft/s$. At 160 *ft* the car is on the second part of the hill. The tangential acceleration is unchanged: $a_t = 7.817\ ft/s^2$. The normal acceleration is $a_n = \frac{(v(160))^2}{R} = \frac{5943.1}{100} = 59.43\ ft/s^2$. [*Note:* The car will "lift off" from the road *End of note.*] The magnitude of the acceleration is $\boxed{|\vec{a}| = \sqrt{7.817^2 + 59.43^2} = 59.94\ ft/s^2}$

================================<>================================

================================<>================================

Problem 2.121 An astronaut candidate is to be tested in a centrifuge with a radius of 10 *m*. He will lose consciousness if his total horizontal acceleration reaches 14 *g*'s. What is the maximum constant angular acceleration of the centrifuge, starting from rest, if he is not to lose consciousness within one minute?

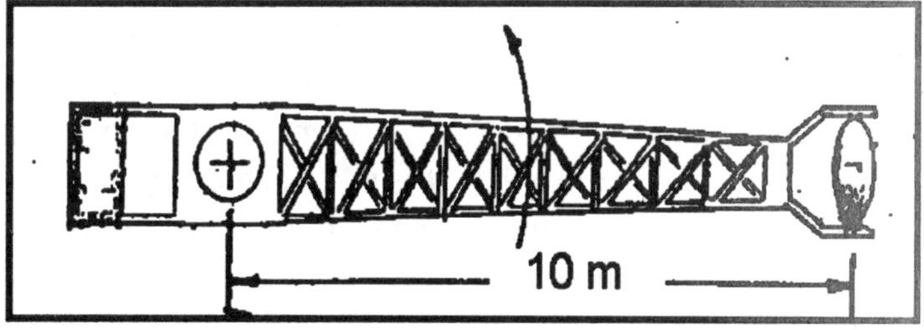

Solution The angular velocity is $\omega(t) = \alpha t$ for constant acceleration from rest. The tangential acceleration at the astronaut is $a_t = R\alpha$. The normal acceleration is $a_n = R\omega^2 = R\alpha^2 t^2$. The magnitude of the acceleration is $|\vec{a}| = \sqrt{R^2\alpha^2 + R^2\alpha^4 t^4} = 14(9.81)\ m/s^2$. Square both sides and reduce: $R^2 t^4 \alpha^4 + R^2 \alpha^2 - 18862.3 = 0$. In canonical form: $\alpha^4 + 2b\alpha^2 + c = 0$, where $b = 3.858 \times 10^{-8}$ and $c = -1.455 \times 10^{-5}$. The solution is $\alpha^2 = \pm 0.003815$, from which $\boxed{\alpha = \sqrt{0.003815} = 0.06177\ rad/s^2}$ $\boxed{= 3.54\ deg/s^2}$ is the maximum constant acceleration from rest allowed.

================================<>================================

Problem 2.122 After first-stage separation and before the second-stage engines have fired, a rocket is moving at $v = 3000 m/s$ and the angle between its velocity vector and the vertical is 60°. Because aerodynamic forces are negligible, the rocket's acceleration is that due to gravity, which is $9.50 m/s^2$ at the rocket's altitude. Determine (a) the normal and tangential components of the rocket's acceleration, and (b) the instantaneous radius of curvature of the rocket's path.

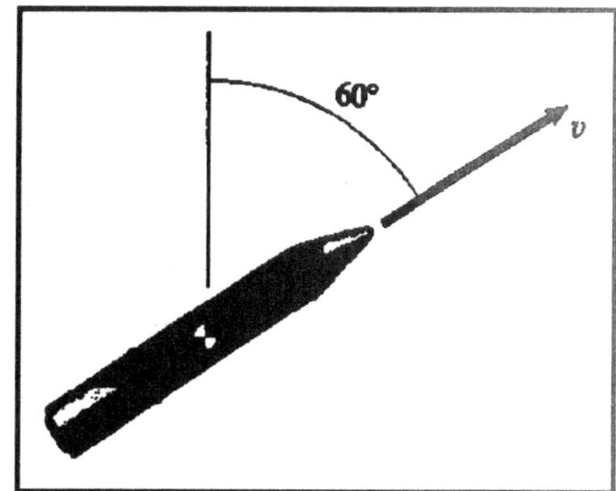

Solution:

The components of the acceleration are $a_T = g\cos(60°)$ toward the rear of the rocket, and $a_N = g\sin(60°)$ normal to the axis of the rocket directed 30 degrees away from straight down.

The normal acceleration is also given by $a_N = v^2/r$, where *r* is the radius of curvature of the path. Substituting $g = 9.50 m/s^2$ and $v = 3000 m/s$ into these relations, we get $a_T = 4.75 m/s^2$, $a_N = 8.23 m/s^2$, and $r = 1094\ km$.

================================<>================================

===================================<>===================================

Problem 2.123 A projectile has an initial velocity of 20 ft/s at 30° above the horizontal. (a) What are the velocity and acceleration of the projectile in terms of the normal and tangential components when it is at its highest point of its trajectory? b) What is the instantaneous radius of curvature of the projectile's path when it is at its highest point of its trajectory?

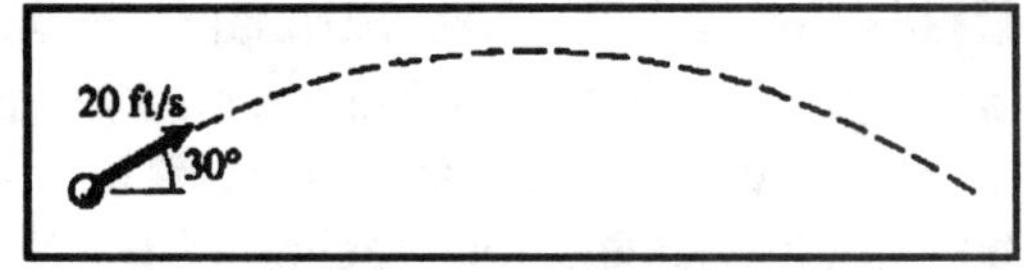

Strategy: In Part (b) you can determine the radius of curvature from the relation $a_n = \frac{v^2}{\rho}$.

Solution Denote $V_m = 20\, ft/s$ and $\theta = 30^o$. The initial components of velocity are $v_x(0) = V_m \cos\theta$ and $v_y(0) = V_m \sin\theta$. The equations for the path are obtained from: (1) $\frac{dv_x}{dt} = 0$, from which. $v_x(t) = V_m \cos\theta$. (2) $\frac{dv_y}{dt} = -g$, from which $v_y(t) = -gt + V_m \sin\theta$. The maximum point in the trajectory occurs at $v_y(t_{max}) = 0$, from which the tangential velocity at this point is $\boxed{v = v_x(t_{max}) = V_m \cos\theta = 17.32 \text{ ft/s}}$. (a) The tangential acceleration is zero at this point, since the acceleration parallel to the x-axis is zero. The normal acceleration is $\boxed{a_n = g = 32.17\, ft/s^2}$ (b) The radius of curvature is $\boxed{\rho = \frac{v^2}{a_n} = \frac{(20\cos(30^o))^2}{32.17} = 9.325\, ft}$

===================================<>===================================

Problem 2.124 In Problem 2.123, let $t = 0$ be the instant at which the projectile is launched. (a) What are the velocity and acceleration in terms of normal and tangential coordinates at $t = 0.2$ seconds? (b) Use the relation $a_n = \frac{v^2}{\rho}$ to determine the instantaneous radius of curvature of the path at $t = 0.2$ seconds?

Solution From the solution to Problem 2.123 the components of velocity are $\vec{v} = \vec{i} V_m \cos\theta + \vec{j}(-gt + V_m \sin\theta)$, where θ is the launch angle. At $t = 0.2\, s$, the tangential velocity is $\boxed{[v_t(t)]_{t=0.2} = \left[\sqrt{(V_m \cos\theta)^2 + (V_m \sin\theta - gt)^2}\right]_{t=0.2} = 17.684\, ft/s}$. The path angle is given by $\tan\beta = \frac{v_y(t)}{v_x(t)} = \frac{-gt + V_m \sin\theta}{V_m \cos\theta}$. At $t = 0.2$ seconds $\tan\beta = \frac{3.566}{17.32} = 0.2059$, from which $\beta = 0.2030\ rad$. The velocity is $\vec{v} = v_t \vec{e}_t$, where $\vec{e}_t = (\vec{i}\cos\beta + \vec{j}\sin\beta)$. The unit vector normal to $\vec{e}_t$ pointing toward the instantaneous radial center is $\vec{e}_n = \vec{i}\cos\left(\beta - \frac{\pi}{2}\right) + \vec{j}\sin\left(\beta - \frac{\pi}{2}\right) = \vec{i}\sin\beta - \vec{j}\cos\beta$ The acceleration is $\vec{a} = -g\vec{j}$. From geometry, the component parallel to the unit vector $\vec{e}_t$ is $\boxed{a_t = -g\sin\beta = -6.487\, ft/s^2}$, and the component normal to the unit vector $\vec{e}_t$ is $\boxed{a_n = g\cos\beta = 31.51\, ft/s^2}$

Solution continued on the next page

Continuation of Solution to Problem 2.124

Check: By definition the parallel component of two vectors is the dot product: $a_t = \vec{a}\cdot\vec{e}_t = -g\sin\beta = -6.487\ ft/s^2$.*check.* The normal component is : $a_n = \vec{a}\cdot\vec{e}_n = g\cos\beta = 31.51\ ft/s^2$. *check..* (b) The instantaneous radius of curvature at $t = 0.2\ s$ is

$$\boxed{\rho = \frac{(v_t(0.2))^2}{a_n} = \frac{(17.68)^2}{31.51} = 9.92\ ft}$$

Note: The path of the projectile is a parabola. An insight into the definition of the radius of curvature is provided by the observation that the parabola has the highest curvature (smallest ρ) at its vertex, and as you move away from its vertex the curvature will steadily lessen (ρ will become larger). *End of note.*

===============================<>===============================

Problem 2.125 In Problem 2.123, let $t = 0$ be the instant at which the projectile is launched. Use Equation (2.42) to determine the instantaneous radius of curvature of the path at $t = 0.2s$.

Solution: Equation (2.42) is $\rho = \dfrac{\left[1+\left(\dfrac{dy}{dx}\right)^2\right]^{3/2}}{\left|\dfrac{d^2y}{dx^2}\right|}$. We need *y(x)* for the path of the projectile of Problem 2.123. We have $x = (V_0\cos\theta_0)\,t$ and $y = (V_0\sin\theta_0)t - gt^2/2$. Solving the x relation for t and substituting this into the y relation, we get $y = \dfrac{V_0\sin\theta_0}{V_0\cos\theta_0}x - \dfrac{gx^2}{2(V_0\cos\theta_0)^2}$. The derivatives required to evaluate the radius of curvature are $\dfrac{dy}{dx} = \tan\theta_0 - \dfrac{gx}{(V_0\cos\theta_0)^2}$, and $\dfrac{d^2y}{dx^2} = -\dfrac{g}{(V_0\cos\theta_0)^2}$. Substituting these into Equation (2.42) and using $g = 32.17\,ft/s^2$, we get $x = 3.464\,ft$, $\dfrac{dy}{dx} = 0.2059$, $\dfrac{d^2y}{dx^2} = -0.1072$, and $\rho = 9.92\,ft$. This agrees with the value found in Problem 2.124 above.

===============================<>===============================

Problem 2.126 The cartesian coordinates of a point are $x = 20 + 4t^2\ m$, $y = 10 - t^3\ m$. What is the instantaneous radius of curvature of the path at $t = 3$ seconds?

Solution The components of the velocity: $\vec{v} = 8t\,\vec{i} - (3t^2)\vec{j}$. At $t = 3$ seconds, the magnitude of the velocity is $|\vec{v}|_{t=3} = \sqrt{(8t)^2 + (-3t^2)^2} = 36.12\ m/s$. The components of the acceleration are $\vec{a} = 8\vec{i} - (6t)\vec{j}$. The instantaneous path angle is $\tan\beta = \dfrac{v_y}{v_x} = \dfrac{-3t^2}{8t}$. At $t = 3$ seconds, $\beta = -0.8442\ rad$. The unit vector parallel to the path is $\vec{e}_t = \vec{i}\cos\beta + \vec{j}\sin\beta$. The unit vector normal to the path pointing toward the instantaneous radial center is

$\vec{e}_n = \vec{i}\cos\left(\beta - \dfrac{\pi}{2}\right) + \vec{j}\sin\left(\beta - \dfrac{\pi}{2}\right) = \vec{i}\sin\beta - \vec{j}\cos\beta$. The normal acceleration is the component of acceleration in the direction of $\vec{e}_n$. Thus, $a_n = \vec{e}_n \bullet \vec{a}$ or $a_n = 8\sin\beta + (6t)\cos\beta$. At $t = 3$ seconds, $a_n = 5.98\ m/s^2$. The radius of curvature at $t = 3$ seconds is $\boxed{\rho = \dfrac{|\vec{v}|^2}{a_n} = 218.3\ m}$

===============================<>===============================

====================================<>====================================

Problem 2.127 In Example 2.11, determine the tangential and normal components of the helicopter's acceleration at $t = 6s$.

Solution: The solution will follow that of Example 2.11, with the time changed to $t = 6s$. The helicopter starts from rest $(v_x, v_y) = (0,0)$ at $t = 0$. Assume that motion starts at the origin (0,0). The equations for the motion in the x direction are $a_x = 0.6\ t\ m/s^2, v_x = 0.3\ t^2 m/s,\ x = 0.1\ t^3 m$, and the equations for motion in the y direction are $a_y = 1.8 - 0.36\ t\ m/s^2,\ v_y = 1.8t - 0.18\ t^2\ m/s$, and $y = 0.9\ t^2 - 0.06\ t^3\ m$. At $t = 6\ s$, the variables have the values $a_x = 3.6 m/s^2,\ a_y = -0.36 m/s^2$, $v_x = 10.8 m/s, v_y = 4.32 m/s, x = 21.6 m$, and $y = 19.44 m$. The magnitude of the velocity is given by $|\mathbf{v}| = \sqrt{v_x^2 + v_y^2} = 11.63 m/s$. The unit vector in the tangential direction is given by

$\mathbf{e_T} = \frac{\mathbf{v}}{|\mathbf{v}|} = \frac{v_x\mathbf{i} + v_y\mathbf{j}}{|\mathbf{v}|} = 0.928\ \mathbf{i} + 0.371\ \mathbf{j}$. The tangential acceleration component is given by

$a_T = \mathbf{a} \bullet \mathbf{e_T} = 0.928 a_x + 0.371 a_y = 3.21 m/s^2$.

The magnitude of the acceleration is given by $|\mathbf{a}| = \sqrt{a_x^2 + a_y^2} = 3.62 m/s^2$.

The normal acceleration component is given by $a_N = \sqrt{|\mathbf{a}|^2 - a_T^2} = 1.67 m/s^2$

====================================<>====================================

Problem 2.128 In Example 2.11, use Equation (2.42) to determine the instantaneous radius of curvature of the helicopter's path at t = 6 s.

Solution: Equation (2.42) is $\rho = \frac{\left[1 + \left(\frac{dy}{dx}\right)^2\right]^{3/2}}{\left|\frac{d^2y}{dx^2}\right|}$.

We need *y(x)* for the path of the helicopter of Example 2.11.

We have $x = 0.1\ t^3$ and $y = 0.9t^2 - 0.06t^3$. Solving the x relation for t and substituting this into the y relation, we get $y = (0.9)(10x)^{2/3} - 0.06(10x)$.

The derivatives required to evaluate the radius of curvature are $\frac{dy}{dx} = \left(\frac{2}{3}\right)(0.9)(10)^{2/3} x^{-1/3} - 0.6$, and $\frac{d^2y}{dx^2} = \left(-\frac{1}{3}\right)\left(\frac{2}{3}\right)(0.9)(10)^{2/3} x^{-4/3}$. Substituting these into Equation (2.42), we get $x = 21.6m$, $\frac{dy}{dx} = 0.4$, $\frac{d^2y}{dx^2} = -0.0154$, and $\rho = 80.96m$. This agrees with the value found from $\rho = \frac{|\mathbf{v}|^2}{a_N} = 80.96m$ using the values in Problem 2.127

====================================<>====================================

================================<>================================

Problem 2.129 For astronaut training, the airplane shown is to achieve "weightlessness" for a short period of time by flying along a path such that its acceleration is $a_x = 0$, $a_y = -g$. If its velocity at O at $t = 0$ is $\vec{v} = v_o\vec{i}$, show that the autopilot must fly the airplane so that its tangential component of the acceleration as a function of time is

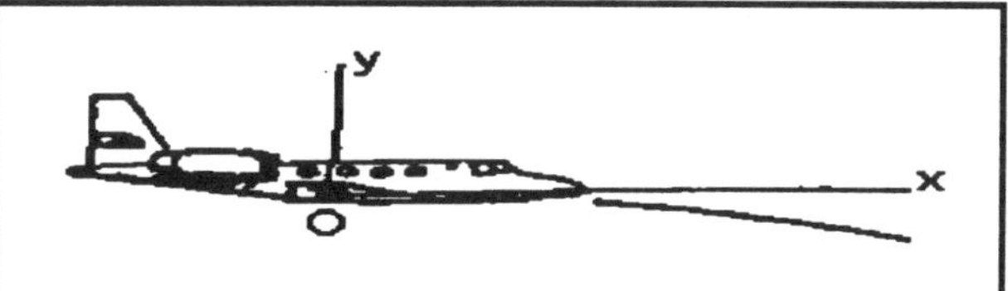

$$a_t = g\frac{\left(\frac{gt}{v_o}\right)}{\sqrt{1+\left(\frac{gt}{v_o}\right)^2}}.$$

Solution The velocity of the path is $\vec{v}(t) = v_o\vec{i} - gt\vec{j}$. The path angle is β: $\tan\beta = \frac{v_y}{v_x} = \frac{-gt}{v_o}$,

$\sin\beta = \frac{-gt}{\sqrt{v_o^2 + (gt)^2}}$. The unit vector parallel to the velocity vector is $\vec{e} = \vec{i}\cos\beta + \vec{j}\sin\beta$. The acceleration vector is $\vec{a} = -\vec{j}g$. The component of the acceleration tangent to the flight path is

$a_t = -g\sin\beta.$, from which $a_t = g\frac{gt}{\sqrt{v_o^2 + (gt)^2}}$. Divide by v_o, $\boxed{a_t = g\left(1+\left(\frac{gt}{v_o}\right)^2\right)^{-\frac{1}{2}}\left(\frac{gt}{v_o}\right)}$

================================<>================================

Problem 2.130 In Problem 2.129, what is the airplane's normal component of acceleration as a function of time?

Solution From Problem 2.129, the velocity is $\vec{v}(t) = v_o\vec{i} - gt\vec{j}$. The flight path angle is β, from which $\cos\beta = \frac{v_o}{\sqrt{v_o^2 + (gt)^2}}$. The unit vector parallel to the flight path is $\vec{e} = \vec{i}\cos\beta + \vec{j}\sin\beta$. The unit vector normal to $\vec{e}$ is $\vec{e}_n = \vec{i}\cos\left(\beta - \frac{\pi}{2}\right) + \vec{j}\sin\left(\beta - \frac{\pi}{2}\right) = \vec{i}\sin\beta - \vec{j}\cos\beta$, pointing toward the instantaneous radial center of the path. The acceleration is $\vec{a} = -\vec{j}g$. The component parallel to the normal component is $a_n = g\cos\beta$, from which $\boxed{a_n = g\frac{v_o}{\sqrt{v_o^2 + (gt)^2}} = g\left(1+\left(\frac{gt}{v_o}\right)^2\right)^{-\frac{1}{2}}}$

================================<>================================

==================================<>==================================

Problem 2.131 If $y = 100\ mm$, $\dfrac{dy}{dt} = 200\ mm/s$, and $\dfrac{d^2y}{dt^2} = 0$, what are the velocity and acceleration of P in terms of normal and tangential components?

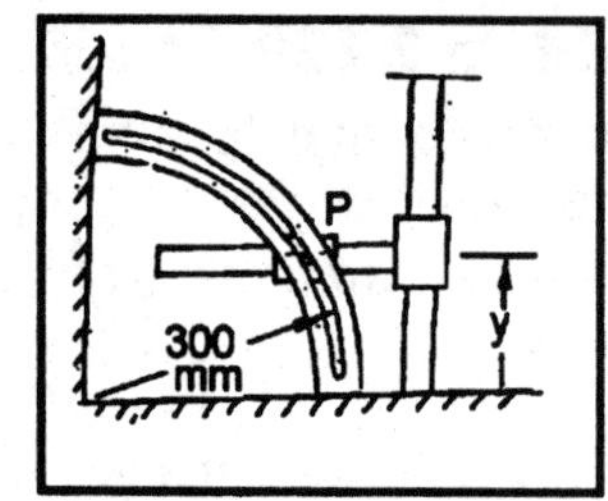

Solution The equation for the circular guide is $R^2 = x^2 + y^2$, from which $x = \sqrt{R^2 - y^2} = 0.283\ m$, and $\dfrac{dx}{dt} = -\left(\dfrac{y}{x}\right)\dfrac{dy}{dt} = v_x = -0.707\ m/s$.

The velocity of point P is $\vec{v}_p = \vec{i}v_x + \vec{j}v_y$, from which the velocity is $|\vec{v}| = \sqrt{v_x^2 + v_y^2} = 0.212\ \ m/s$. The angular velocity $\omega = \dfrac{|\vec{v}|}{R} = 0.7071\ rad/s$. The angle is $\beta = \tan^{-1}\left(\dfrac{y}{x}\right) = 19.5^o$

$a_x = \dfrac{dv_x}{dt} = \dfrac{d}{dt}\left(-\dfrac{y}{x}\dfrac{dy}{dt}\right) = -\dfrac{1}{x}\left(\dfrac{dy}{dt}\right)^2 + \dfrac{y}{x^2}\left(\dfrac{dx}{dt}\right)\left(\dfrac{dy}{dt}\right) - \left(\dfrac{y}{x}\right)\left(\dfrac{d^2y}{dt^2}\right) = -0.0707\ \ m/s^2$ The unit vector tangent to the path (normal to the radius vector *for a circle*) is $\vec{e}_p = -\vec{i}\sin\beta + \vec{j}\cos\beta$, from which $\boxed{a_t = -a_x\sin\beta = 53.0\ mm/s^2}$ since $a_y = 0$ $\boxed{a_n = -R\omega^2 = -0.150\ m/s^2}$. *Check:* $a_n = a_x\cos\beta = -0.15\ \ m/s^2$ *check.*

==================================<>==================================

Problem 2.132 In Problem 2.131, the point P moves upward in the slot with velocity $\vec{v} = 300\vec{e}_t\ \ mm/s$. When $y = 150\ mm$, what are $\dfrac{dy}{dt}$ and $\dfrac{d^2y}{dt^2}$?

Solution The position in the guide slot is $y = R\sin\theta$, from which $\theta = \sin^{-1}\left(\dfrac{y}{R}\right) = \sin^{-1}(0.5) = 30^o$. $x = R\cos\theta = 259.8\ mm$. From the solution to Problem 2.131, $v_x = -\left(\dfrac{y}{x}\right)\dfrac{dy}{dx} = -\left(\dfrac{y}{x}\right)v_y$. The velocity is

$|\vec{v}| = 300 = \sqrt{{v_x}^2 + {v_y}^2} = v_y\sqrt{\left(\dfrac{y}{x}\right)^2 + 1}$, from which $\boxed{v_y = 300\left(\left(\dfrac{y}{x}\right)^2 + 1\right)^{-\frac{1}{2}} = 259.8\ mm/s}$. and $v_x = -150\ mm/s$ (Since the point is moving upward in the slot, v_y is positive.). The velocity along the path in the guide slot is constant, hence $a_t = 0$. The normal acceleration is

$a_n = \dfrac{|\vec{v}|^2}{R} = 300\ mm/s^2$ directed toward the radius center, from which $\boxed{\dfrac{d^2y}{dt^2} = -a_n\sin\theta = -150\ mm/s^2}$

==================================<>==================================

==================================<>==================================

Problem 2.133 A car travels at 100 *km/hr* on a straight road of increasing grade whose vertical profile can be approximated by the equation shown. When $x = 400\ m$, what are the tangential and normal components of the car's acceleration?

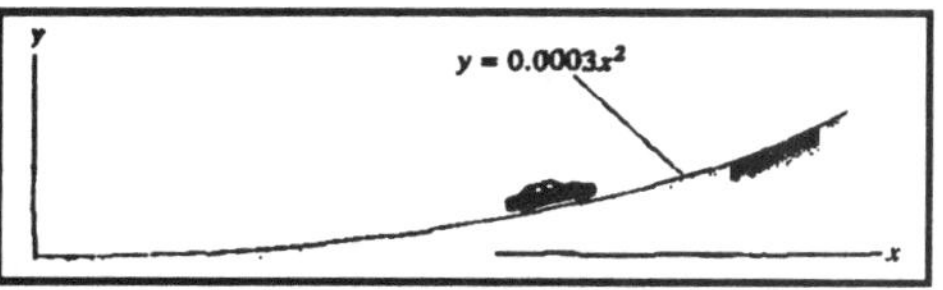

Solution The strategy is to use the acceleration in cartesian coordinates found in the solution to Problem 2.90, find the angle with respect to the *x*-axis, $\theta = \tan^{-1}\left(\frac{dy}{dx}\right)$, and use this angle to transform the accelerations to the tangential and normal components. From the solution to Problem 2.90 the accelerations are $\vec{a} = -0.0993\vec{i} + 0.4139\vec{j}\ \left(m/s^2\right)$. The angle at

$\theta = \tan^{-1}\left(\frac{d}{dx}Cx^2\right)_{x=400} = \tan^{-1}\left(6x \times 10^{-4}\right)_{x=400} = 13.5^o$ From trigonometry (see figure) the transformation is $a_t = a_x \cos\theta + a_y \sin\theta$, $a_n = -a_x \sin\theta + a_y \cos\theta$, from which $\boxed{a_t = 0.000035... = 0}$. $\boxed{a_n = 0.4256\ m/s^2}$ *Check:* The velocity is constant along the path, so the tangential component of the acceleration is zero, $a_t = \frac{dv}{dt} = 0$, *check.*

==================================<>==================================

Problem 2.134 A boy rides a skateboard on the concrete surface of an empty drainage canal described by the equation shown. He starts at $y = 20\ ft$, and the magnitude of his velocity is approximated by $v = \sqrt{2(32.2)(20-y)}\ ft/s$. (a) Use Equation 2.42 to determine the instantaneous radius of curvature of his path when he reaches the bottom. (b) What is the normal component of his acceleration when he reaches the bottom?

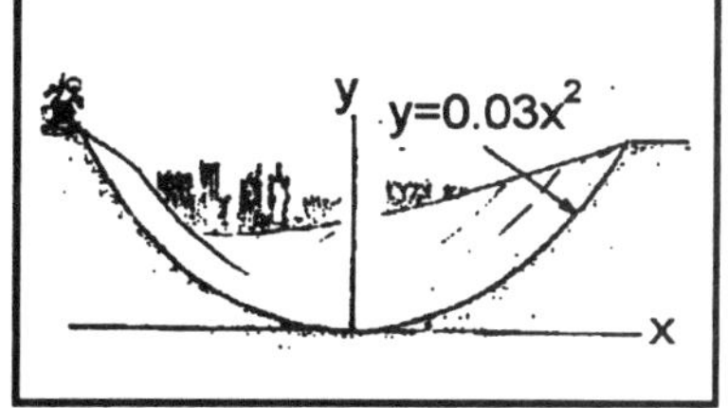

Solution The magnitude of the velocity is $\sqrt{\left(\frac{dy}{dt}\right)^2 + \left(\frac{dx}{dt}\right)^2} = v = K(20-y)^{\frac{1}{2}} = K\left(20 - Cx^2\right)^{\frac{1}{2}}$,

where $K = 8.025$, $C = 0.03$. From $y = Cx^2$, $\frac{dy}{dt} = 2Cx\left(\frac{dx}{dt}\right)$, $\frac{d^2y}{dt^2} = 2C\left(\frac{dx}{dt}\right)^2 + 2Cx\left(\frac{d^2x}{dt^2}\right)$.

Substitute: $\left|\frac{dx}{dt}\right| = \frac{K\left(20-Cx^2\right)^{\frac{1}{2}}}{\left(4C^2x^2+1\right)^{\frac{1}{2}}}$. Since the boy is moving the the right, $\frac{dx}{dt} > 0$, and $\left|\frac{dx}{dt}\right| = \frac{dx}{dt}$. The acceleration is $\frac{d^2x}{dt^2} = \frac{-KCx}{\left(20-Cx^2\right)^{\frac{1}{2}}\left(4C^2x^2+1\right)^{\frac{1}{2}}}\left(\frac{dx}{dt}\right) - \frac{K\left(4C^2x\right)\left(20-Cx^2\right)^{\frac{1}{2}}}{\left(4C^2x^2+1\right)^{\frac{3}{2}}}\left(\frac{dx}{dt}\right)$. At the bottom of the canal the values are $\left(\frac{dx}{dt}\right)_{x=0} = K\sqrt{20} = 35.89$ ft/s. $\left(\frac{dy}{dt}\right)_{x=0} = 0$, $\left(\frac{d^2x}{dt^2}\right)_{x=0} = 0$,

Solution continued on next page

$\left(\frac{d^2y}{dt^2}\right)_{x=0} = 2C\left(\frac{dx}{dt}\right)^2\Big|_{x=0} = 77.28$ ft/s^2. The angle with respect to the x axis at the bottom of the canal is

$\theta = \tan^{-1}\left(\frac{dy}{dx}\right)_{x=0} = 0$. From the solution to Problem 2.133, the tangential and normal accelerations are

$a_t = a_x\cos\theta + a_y\sin\theta$, $a_n = -a_x\sin\theta + a_y\cos\theta$, from which $\boxed{a_t = 0}$, and $\boxed{a_n = 77.28\ ft/s^2}$. *Check:* The velocity is constant along the path, so the tangential component of the acceleration is zero, $a_t = \frac{dv}{dt} = 0$. *check.* By inspection, the normal acceleration at the bottom of the canal is identical to the y component of the acceleration. *check.*

==================================<>==================================

Problem 2.135 In Problem 2.134, what is the normal component of the boy's acceleration when he has passed the bottom and reached $y = 10\,ft$?

Solution Use the results of the solutions to Problem 2.133 and 2.134. From the solution to Problem 2.134, at $y = 10\,ft$, $x_{y=10} = \left(\sqrt{\frac{y}{C}}\right)_{y=10} = 18.257\,ft$, from which

$$\left(\frac{dx}{dt}\right)_{y=10} = \left(K(20 - Cx^2)^{\frac{1}{2}}(4C^2x^2+1)^{-\frac{1}{2}}\right)_{y=10} = 17.11 \text{ ft/s.}$$

$$\left(\frac{d^2x}{dt^2}\right)_{y=10} = -K\left(\frac{dx}{dt}\right)_{y=10}\left(\frac{Cx}{(20-Cx^2)^{\frac{1}{2}}(4C^2x^2+1)^{\frac{1}{2}}} + \frac{(4C^2x)(20-Cx^2)^{\frac{1}{2}}}{(4C^2x^2+1)^{\frac{3}{2}}}\right)_{y=10} = -24.78 \text{ ft/s}\cdot$$

$$\left(\frac{d^2y}{dt^2}\right)_{y=10} = \left(2C\left(\frac{dx}{dt}\right)^2 + 2Cx\left(\frac{d^2x}{dt^2}\right)\right)_{y=10} = -9.58\ ft/s^2.$$ The angle is $\theta = \tan^{-1}\left(\frac{dy}{dx}\right)_{y=10} = 47.61^o$.

From the solution to Problem 2.133, $a_t = a_x\cos\theta + a_y\sin\theta$, $a_n = -a_x\sin\theta + a_y\cos\theta$, from which

$\boxed{a_t = -23.78\ ft/s^2}$, $\boxed{a_n = 11.84\ ft/s^2}$

==================================<>==================================

Problem 2.136 By using Eqs (2.41): (a) Show that the relations between the cartesian unit vectors and the unit vectors $\mathbf{e_T}$ and $\mathbf{e_N}$ are $\mathbf{i} = \cos\theta\,\mathbf{e_T} - \sin\theta\,\mathbf{e_N}$ and $\mathbf{j} = \sin\theta\,\mathbf{e_T} + \cos\theta\,\mathbf{e_N}$ (b) Show that $d\mathbf{e_T}/dt = (d\theta/dt)\mathbf{e_N}$ and $d\mathbf{e_N}/dt = -(d\theta/dt)\mathbf{e_T}$.

Solution: Equations (2.41) are $\mathbf{e_T} = \cos\theta\,\mathbf{i} + \sin\theta\,\mathbf{j}$ and $\mathbf{e_N} = -\sin\theta\,\mathbf{i} + \cos\theta\,\mathbf{j}$.

(a) Multiplying the equation for $\mathbf{e_T}$ by $\cos\theta$ and the equation for $\mathbf{e_N}$ by $(-\sin\theta)$ and adding the two equations, we get $\mathbf{i} = \cos\theta\,\mathbf{e_T} - \sin\theta\,\mathbf{e_N}$. Similarly, by multiplying the equation for $\mathbf{e_T}$ by $\sin\theta$ and the equation for $\mathbf{e_N}$ by $\cos\theta$ and adding, we get $\mathbf{j} = \sin\theta\,\mathbf{e_T} + \cos\theta\,\mathbf{e_N}$.

(b) Taking the derivative of $\mathbf{e_T} = \cos\theta\,\mathbf{i} + \sin\theta\,\mathbf{j}$, we get $\frac{d\mathbf{e_T}}{dt} = (-\sin\theta\,\mathbf{i} + \cos\theta\,\mathbf{j})\frac{d\theta}{dt} = \mathbf{e_N}\frac{d\theta}{dt}$.
Similarly, taking the derivative of $\mathbf{e_N} = -\sin\theta\,\mathbf{i} + \cos\theta\,\mathbf{j}$, we get $d\mathbf{e_N}/dt = -(d\theta/dt)\mathbf{e_T}$

==================================<>==================================

=================================<>=================================

Problem 2.137 At a particular time, the polar coordinates of a point P moving in the x-y plane are $r = 4\ ft$, $\theta = 0.5$ radians, and their time derivatives are $\frac{dr}{dt} = 8\ ft/s$, and $\frac{d\theta}{dt} = -2\ rad/s$. (a) What is the magnitude of the velocity of P? (b) What are the cartesian components of the velocity of P?

Solution (a) The polar components of the velocity are $\vec{v} = \left(\frac{dr}{dt}\right)\vec{e}_r + r\left(\frac{d\theta}{dt}\right)\vec{e}_\theta = 8\vec{e}_r - 8\vec{e}_\theta$. The magnitude of the velocity is $\boxed{|\vec{v}| = \sqrt{8^2 + 8^2} = 11.31\ ft/s}$ (b) The cartesian components of the velocity are $v_x = \left(\frac{dr}{dt}\right)\cos\theta - r\sin\theta\left(\frac{d\theta}{dt}\right) = 10.86\ ft/s$. $v_y = \left(\frac{dr}{dt}\right)\sin\theta + r\cos\theta\left(\frac{d\theta}{dt}\right) = -3.19\ ft/s$, from which $\boxed{\vec{v} = 10.86\vec{i} - 3.19\vec{j}}$

=================================<>=================================

Problem 2.138 In Problem 2.137, suppose that $\frac{d^2r}{dt^2} = 6\ ft/s^2$, and $\frac{d^2\theta}{dt^2} = 3\ rad/s^2$. At the instant described, determine (a) the magnitude of the acceleration of P; (b) the instantaneous radius of curvature of the path.

Solution From Problem 2.137, $r = 4\ ft$, $\theta = 0.5$ radians, $\frac{dr}{dt} = 8\ ft/s$. The accelerations are:

$a_r = \frac{d^2r}{dt^2} - r\left(\frac{d\theta}{dt}\right)^2 = -10\ ft/s^2$, $a_\theta = r\frac{d^2\theta}{dt^2} + 2\left(\frac{dr}{dt}\right)\left(\frac{d\theta}{dt}\right) = -20\ ft/s^2$ (a) The magnitude: $\boxed{|\vec{a}| = \sqrt{a_r^2 + a_\theta^2} = 22.36\ ft/s^2}$. (b) From geometry: $a_n = -a_r\cos 45^o + a_\theta \sin 45^o = \frac{v^2}{\rho}$, from which

$$\boxed{\rho = \frac{\sqrt{2}(8^2 + 8^2)}{(10 + 20)} = 6.03\ ft}$$

=================================<>=================================

Problem 2.139 The polar coordinates of a point P moving in the x-y plane are $r = t^3 - 4t\ m$, and $\theta = t^2 - t\ rad$. Determine the velocity of P in terms of radial and transverse components at $t = 1$ second.

Solution The velocity: $\vec{v} = (3t^2 - 4)\vec{e}_r - (t^3 - 4t)(2t - 1)\vec{e}_\theta$. At $t = 1$ second, $\boxed{\vec{v} = -1\vec{e}_r - 3\vec{e}_\theta}$

=================================<>=================================

Problem 2.140 In Problem 2.139, what is the acceleration of P in terms of its radial and transverse components at $t = 1$ second?

Solution The acceleration vector is $\vec{a} = \left(\frac{d^2r}{dt^2} - r\left(\frac{d\theta}{dt}\right)^2\right)\vec{e}_r + \left(r\left(\frac{d^2\theta}{dt^2}\right) + 2\left(\frac{dr}{dt}\right)\left(\frac{d\theta}{dt}\right)\right)\vec{e}_\theta$. Substitute:

$r = \left[t^3 - 4t\right]_{t=1} = -3\ m$, $\frac{dr}{dt}\Big|_{t=1} = -1\ m/s$. $\frac{d^2r}{dt^2} = [6t]_{t=1} = 6\ m/s^2$. $\frac{d\theta}{dt}\Big|_{t=1} = [2t - 1]_{t=1} = 1\ rad/s$.

$\frac{d^2\theta}{dt^2} = 2$. From which $\boxed{\vec{a} = [6 - (-3)(1)]\vec{e}_r + [-3(2) + 2(-1)(1)]\vec{e}_\theta = 9\vec{e}_r - 8\vec{e}_\theta}$

=================================<>=================================

==================================<>==================================

Problem 2.141 The radial line rotates with a constant angular velocity of 2 *rad/s*. Point P moves along the line at a constant speed of 4 *m/s*. Determine the magnitude of the velocity and acceleration of P when $r = 2$.

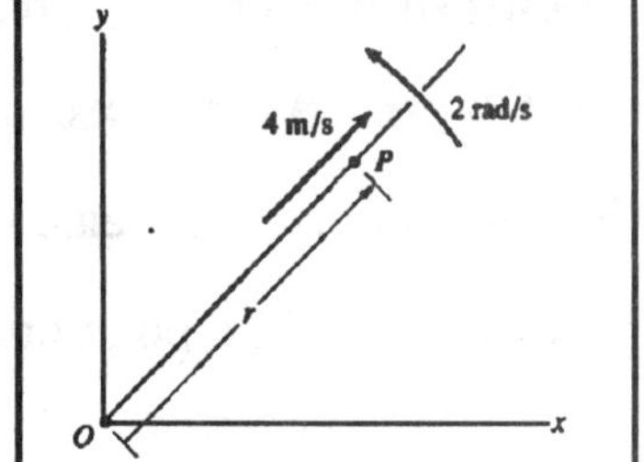

Solution The angular velocity of the line is $\frac{d\theta}{dt} = \omega = 2\ rad/s$, from which $\frac{d^2\theta}{dt^2} = 0$. The radial velocity of the point is $\frac{dr}{dt} = 4\ m/s$, from which $\frac{d^2r}{dt^2} = 0$. The vector velocity is $\vec{v} = \left(\frac{dr}{dt}\right)\vec{e}_r + r\left(\frac{d\theta}{dt}\right)\vec{e}_\theta = 4\vec{e}_r + 4\vec{e}_\theta$. The magnitude is $\boxed{|\vec{v}| = \sqrt{4^2 + 4^2} = 5.66\ m/s}$ The acceleration is $\vec{a} = [-2(4)]\vec{e}_r + [2(4)(2)]\vec{e}_\theta = -8\vec{e}_r + 16\vec{e}_\theta$. The magnitude is $\boxed{|\vec{a}| = \sqrt{8^2 + 16^2} = 17.89\ m/s^2}$

==================================<>==================================

Problem 2.142 At the instant shown, the coordinates of the slider A are $x = 1.6\,ft, y = 1.0\,ft$, and its velocity and acceleration are $\mathbf{v} = 10\mathbf{j}\ ft/s$ and $\mathbf{a} = -32.2\mathbf{j}\ ft/s^2$. Determine the slider's velocity and acceleration in terms of polar coordinates.

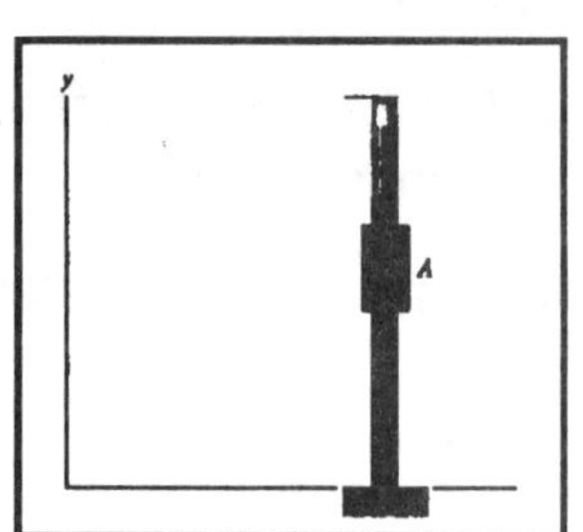

Solution: The equation relating the unit vectors in polar coordinates to those in cartesian coordinates are $\mathbf{e}_r = \cos\theta\,\mathbf{i} + \sin\theta\,\mathbf{j}$ and $\mathbf{e}_\theta = -\sin\theta\,\mathbf{i} + \cos\theta\,\mathbf{j}$. The inverse relationships are $\mathbf{i} = \cos\theta\,\mathbf{e}_r - \sin\theta\,\mathbf{e}_\theta$ and $\mathbf{j} = \sin\theta\,\mathbf{e}_r + \cos\theta\,\mathbf{e}_\theta$. The angle θ is determined from $\tan\theta = \frac{y}{x} = \frac{1.0}{1.6}$. Hence, $\theta = 32.01°$. Then, in the general case, $v_r = v_x\cos\theta + v_y\sin\theta$ and $v_\theta = -v_x\sin\theta + v_y\cos\theta$. Similar equations hold for the accelerations. Substituting in numbers, and noting that $v_x = a_x = 0$, we get $v_r = 5.30\,ft/s$, $v_\theta = 8.48\,ft/s$, $a_r = -17.07\,ft/s^2$, and $a_\theta = -27.31\,ft/s^2$

==================================<>==================================

Problem 2.143 In Problem 2.142, determine $\frac{d^2r}{dt^2}$ and $\frac{d^2\theta}{dt^2}$.

Solution: The total radial acceleration is $a_r = \frac{d^2r}{dt^2} - r\left(\frac{d\theta}{dt}\right)^2$ and the total circumferential acceleration is $a_\theta = r\frac{d^2\theta}{dt^2} + 2\frac{dr}{dt}\frac{d\theta}{dt}$. The velocity relationships are $v_r = \frac{dr}{dt}$ and $v_\theta = r\frac{d\theta}{dt}$. From Problem 2.142, we have that $v_r = 5.30\,ft/s$, $v_\theta = 8.48\,ft/s$, $a_r = -17.07\,ft/s^2$, and $a_\theta = -27.31\,ft/s^2$. Also, note that $r = \sqrt{x^2 + y^2} = 1.89\,ft$. From the equations for the velocity components, we get $\frac{dr}{dt} = v_r = 5.30\,ft/s$ and $r\frac{d\theta}{dt} = v_\theta = 8.48\,ft/s$. Since we know *r*, we can determine $\frac{d\theta}{dt} = 4.49\ rad/s$. We now go to the acceleration equations. We know everything in the a_r equation except $\frac{d^2r}{dt^2}$ and we know everything in the a_θ equation except $\frac{d^2\theta}{dt^2}$. Evaluating these, we get $\frac{d^2r}{dt^2} = 21.0\,ft/s^2$ and $\frac{d^2\theta}{dt^2} = -39.6\ rad/s^2$

==================================<>==================================

Problem 2.144 A boat searching for underwater archaeology sites in the Aegean Sea moves at 4 knots and follows the path $r = 10\theta$ meters, where θ is in radians. (A knot is one nautical mile, or 1852 meters, per hour.) When $\theta = 2\pi\ rad$, determine the boat's velocity: (a) in terms of polar coordinates; (b) in terms of cartesian coordinates.

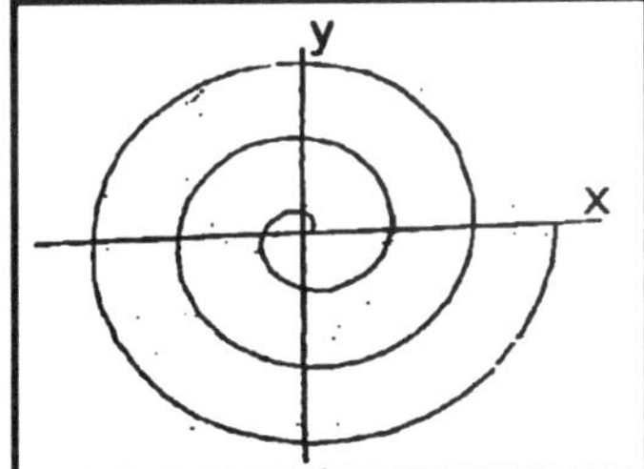

Solution The velocity along the path is

$v = 4\left(\frac{1852\ m}{1\ knot}\right)\left(\frac{1\ hr}{3600\ s}\right) = 2.06\ m/s$. (a) The path is $r = 10\theta$. The velocity $v_r = \frac{dr}{dt} = \frac{d}{dt}(10\theta) = 10\frac{d\theta}{dt}\ m/s$. The velocity along the path is related to the components by

$$v^2 = v_r^2 + v_\theta^2 = \left(\frac{dr}{dt}\right)^2 + r^2\left(\frac{d\theta}{dt}\right)^2 = 2.06^2.$$

At $\theta = 2\pi$, $r = 10(2\pi) = 62.8\ m$. Substitute: $2.06^2 = \left(10\frac{d\theta}{dt}\right)^2 + r^2\left(\frac{d\theta}{dt}\right)^2 = \left(100 + 62.8^2\right)\left(\frac{d\theta}{dt}\right)^2$,

from which $\frac{d\theta}{dt} = 0.0323\ rad/s$, $\boxed{v_r = 10\frac{d\theta}{dt} = 0.323\ m/s}$, $\boxed{v_\theta = r\frac{d\theta}{dt} = 2.032\ m/s}$

(b) From geometry, the cartesian components are $v_x = v_r\cos\theta + v_\theta\sin\theta$, and $v_y = v_r\sin\theta + v_\theta\cos\theta$.

At $\theta = 2\pi$, $\boxed{v_x = v_r}$, and $\boxed{v_y = v_\theta}$

Problem 2.145 In Problem 2.144, what is the boat's acceleration in terms of polar coordinates?

Strategy: The magnitude of the boat's velocity is constant, so you know that the tangential component of the acceleration equals zero.

Solution . From the solution of Problem 2.126, $\beta = \tan^{-1}\left(\frac{v_\theta}{v_r}\right) = \tan^{-1}\left(\frac{2.032}{0.323}\right) = 80.96^o$,

$(1)\ \frac{d^2r}{dt^2} = 10\frac{d^2\theta}{dt^2}$. The acceleration components are: $(2)\ a_r = \frac{d^2r}{dt^2} - r\left(\frac{d\theta}{dt}\right)^2$, and

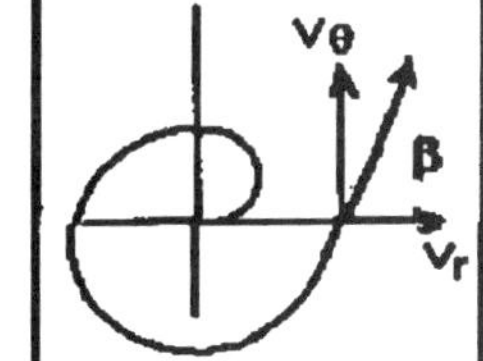

$(3)\ a_\theta = r\frac{d^2\theta}{dr^2} + 2\left(\frac{dr}{dt}\right)\left(\frac{d\theta}{dt}\right)$. The tangential acceleration is known to be zero:

$(4)\ a_t = a_r\cos\beta + a_\theta\sin\beta = 0$. From the solution to Problem 2.126, $\frac{dr}{dt} = 0.323\ m/s$,

$\frac{d\theta}{dt} = 0.0323\ m/s$, $r = 62.8\ m$. Substitute into the four equations in four unknowns and solve:

$\boxed{a_r = -0.674\ m/s^2}$, $\frac{d^2r}{dt^2} = -0.00162\ m/s^2$, $\boxed{a_\theta = 0.01072\ m/s^2}$, $\frac{d^2\theta}{dt^2} = -0.0001624\ rad/s^2$:

Problem 2.146 A point P moves in the x-y plane along the path described by the equation $r = e^{\theta}$ where θ is in radians. The angular velocity $\frac{d\theta}{dt} = \omega = const$, and $\theta = 0$ at $t = 0$. (a) Draw a polar graph of the path for values of θ from zero to 2π. (b) Show that the velocity and acceleration as functions of time are $\vec{v} = \omega_o e^{\omega_o t}(\vec{e}_r + \vec{e}_\theta)$, and $\vec{a} = 2\omega_o^2 e^{\omega_o t}\vec{e}_\theta$.

Solution (a) The graph is shown (the $r = 300$ ticks indicate the scale).

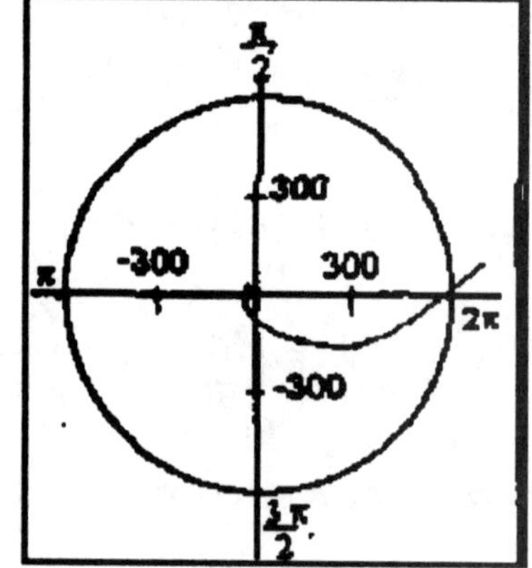

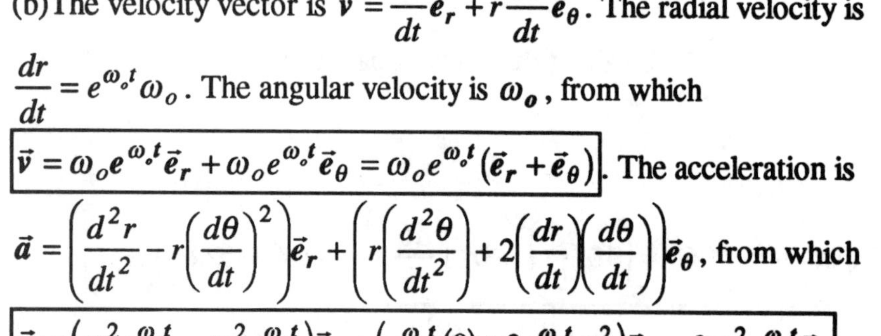

(b)The velocity vector is $\vec{v} = \frac{dr}{dt}\vec{e}_r + r\frac{d\theta}{dt}\vec{e}_\theta$. The radial velocity is $\frac{dr}{dt} = e^{\omega_o t}\omega_o$. The angular velocity is ω_o, from which $\boxed{\vec{v} = \omega_o e^{\omega_o t}\vec{e}_r + \omega_o e^{\omega_o t}\vec{e}_\theta = \omega_o e^{\omega_o t}(\vec{e}_r + \vec{e}_\theta)}$. The acceleration is

$$\vec{a} = \left(\frac{d^2 r}{dt^2} - r\left(\frac{d\theta}{dt}\right)^2\right)\vec{e}_r + \left(r\left(\frac{d^2\theta}{dt^2}\right) + 2\left(\frac{dr}{dt}\right)\left(\frac{d\theta}{dt}\right)\right)\vec{e}_\theta,$$ from which

$$\boxed{\vec{a} = \left(\omega_o^2 e^{\omega_o t} - \omega_o^2 e^{\omega_o t}\right)\vec{e}_r + \left(e^{\omega_o t}(0) + 2e^{\omega_o t}\omega_o^2\right)\vec{e}_\theta = 2\omega_o^2 e^{\omega_o t}\vec{e}_\theta}$$

Problem 2.147 In Problem 2.146, show that the instantaneous radius of curvature of the path as a function of time is $\rho = \sqrt{2}e^{\omega_o t}$.

Solution Use the results of the solution to Problem 2.128, $\vec{v} = \omega_o e^{\omega_o t}(\vec{e}_r + \vec{e}_\theta)$, $\vec{a} = 2\omega_o^2 e^{\omega_o t}\vec{e}_\theta$, from which $|\vec{v}| = \sqrt{\vec{v}\cdot\vec{v}} = \sqrt{2}\omega_o e^{\omega_o t}$. From the dot product $|\vec{a}_{tangent}| = \frac{\vec{a}\cdot\vec{v}}{|\vec{v}|} = \sqrt{2}\omega_o^2 e^{\omega_o t}$. Noting that

$$a_{normal} = \sqrt{|\vec{a}|^2 - |\vec{a}_{tangent}|^2} = \sqrt{2}\omega_o^2 e^{\omega_o t} = \frac{|\vec{v}|^2}{\rho},$$ from which $\boxed{\rho = \sqrt{2}e^{\omega_o t}}$

================================<>================================

Problem 2.148 In Example 2.14, determine the acceleration of point P at $t = 0.8$; (a) in terms of radial and transverse components; (b) in terms of cartesian components.

Solution The path of P is given as $r = 1 - 0.5\cos 2\pi t\ m$, $\theta = 0.5 - 0.2\sin 2\pi t\ rad$, from which at $t = 0.8$ seconds: $r = 0.8455\ m$, $\frac{dr}{dt} = 0.5(2\pi)\sin 2\pi t = -2.988\ m/s$, $\frac{d^2r}{dt^2} = 2\pi^2\cos 2\pi t = 6.1\ m/s^2$, $\theta = 0.6902\ rad$, $\frac{d\theta}{dt} = -\pi\cos 2\pi t = -0.3883\ rad/s$, $\frac{d^2\theta}{dt^2} = 2\pi^2\sin 2\pi t = -7.51\ rad/s^2$. (a) The radial acceleration is $[a_r]_{t=0.8} = \left(\frac{d^2r}{dt^2} - r\left(\frac{d\theta}{dt}\right)^2\right)_{t=0.8} = 5.97\ m/s^2$ The transverse acceleration is

$[a_\theta]_{t=0.8} = \left(r\left(\frac{d^2\theta}{dt^2}\right) + 2\left(\frac{dr}{dt}\right)\left(\frac{d\theta}{dt}\right)\right)_{t=0.8} = -4.03\ m/s^2$, from which $\boxed{\vec{a} = 5.972\vec{e}_r - 4.03\vec{e}_\theta\ (m/s^2)}$

(b) The cartesian coordinates are $x = r\cos\theta$, $y = r\sin\theta$. Differentiate twice and collect terms:

$$\frac{d^2x}{dt^2} = \left(\left(\frac{d^2r}{dt^2}\right) - r\left(\frac{d\theta}{dt}\right)^2\right)\cos\theta - \left(r\left(\frac{d^2\theta}{dt^2}\right) + 2\left(\frac{dr}{dt}\right)\left(\frac{d\theta}{dt}\right)\right)\sin\theta = a_r\cos\theta - a_\theta\sin\theta = 7.170\ m/s^2.$$

$$\frac{d^2y}{dt^2} = \left(\left(\frac{d^2r}{dt^2}\right) - r\left(\frac{d\theta}{dt}\right)^2\right)\sin\theta + \left(r\left(\frac{d^2\theta}{dt^2}\right) + 2\left(\frac{dr}{dt}\right)\left(\frac{d\theta}{dt}\right)\right)\cos\theta = a_r\sin\theta + a_\theta\cos\theta = 0.6961\ m/s^2,$$

from which $\boxed{\vec{a} = 7.170\vec{i} + 0.696\vec{j}\ (m/s^2)}$

================================<>================================

Problem 2.149 A bead slides along a wire that rotates in the *x-y* plane with constant angular velocity ω_o. The radial component of the bead's acceleration is zero. The radial component of its velocity is v_o when $r = r_o$. Determine the radial and transverse components of the bead's velocity as a function of *r*. *Strategy* The radial component of the velocity is $v_r = \frac{dr}{dt}$ and the radial component of the acceleration is

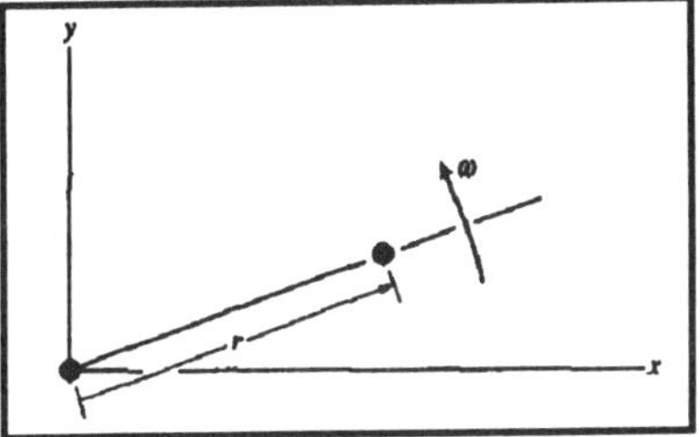

$a_r = \frac{d^2r}{dt^2} - r\left(\frac{d\theta}{dt}\right)^2 = \left(\frac{dv_r}{dt}\right) - \omega_o^2 r$. By the chain rule: $\frac{dv_r}{dt} = \frac{dv_r}{dr}\frac{dr}{dt} = v_r\frac{dv_r}{dr}$. You can express the radial component of the acceleration in the form $a_r = v_r\frac{dv_r}{dr} - r\omega_o^2$.

Solution From the strategy: $a_r = 0 = v_r\frac{dv_r}{dr} - \omega_o^2 r$. Separate variables and integrate: $v_r dv_r = \omega_o^2 r dr$, from which $\frac{v_r^2}{2} = \omega_o^2\frac{r^2}{2} + C$. At $r = r_o$, $v_r = v_o$, from which $C = \frac{v_o^2 - \omega_o^2 r_o^2}{2}$, and $v_r = \sqrt{v_o^2 + \omega_o^2(r^2 - r_o^2)}$. The transverse component is $v_\theta = r\left(\frac{d\theta}{dt}\right) = r\omega_o$, from which

$\boxed{\vec{v} = \sqrt{v_o^2 + \omega_o^2(r^2 - r_o^2)}\,\vec{e}_r + r\omega_o\vec{e}_\theta}$

================================<>================================

==================================<>==================================

Problem 2.150 The motion of a point in the *x-y* plane is such that its transverse component of acceleration, a_θ, is zero, show that the product of its radial position and its transverse velocity is constant; $rv_\theta = constant$.

Solution: We are given that $a_\theta = r\alpha + 2v_r\omega = 0$. Multiply the entire relationship by *r*. We get$0 = (r^2\alpha + 2rv_r\omega) = \left(r^2\left(\frac{d\omega}{dt}\right) + 2r\left(\frac{dr}{dt}\right)_r \omega \right) = \frac{d}{dt}(r^2\omega)$. Note that if $\frac{d}{dt}(r^2\omega) = 0$, then $r^2\omega = constant$. Now note that $v_\theta = r\omega$. We have $r^2\omega = r(r\omega) = rv_\theta = constant$. This was what we needed to prove.

==================================<>==================================

Problem 2.151 From astronomical data, Kepler deduced that the line from the sun to a planet traces out equal areas in equal times. Show that this result follows from the fact that the transverse component of the planet's acceleration is zero. [When *r* changes by an amount *dr* and θ changes by an amount $d\theta$, the resulting differential area is given by $dA = \frac{1}{2}r(rd\theta)$].

Solution: From the solution to Problem 2.150, $a_\theta = 0$ implies that $r^2\omega = r^2\frac{d\theta}{dt} = constant$. The element of area is $dA = \frac{1}{2}r(rd\theta)$, or $\frac{dA}{dt} = \frac{1}{2}r\left(r\frac{d\theta}{dt}\right) = \frac{1}{2}r^2\omega = constant$. Thus, if $\frac{dA}{dt} = constant$, then equal areas are swept out in equal times.

==================================<>==================================

Problem 2.152 The bar rotates in the *x-y* plane with constant angular velocity ω_o. The radial component of acceleration of the collar C is $a_r = -Kr$, where *K* is a constant. When $r = r_o$, the radial component of velocity is v_o.Determine the radial and transverse components of the velocity as a function of r.

Solution Use the same strategy used in Problem 2.131. The radial acceleration is given by $a_r = v_r\frac{dv_r}{dr} - r\omega_o^2$, where the chain rule has been used to obtain: $\frac{dv_r}{dt} = v_r\frac{dv_r}{dr}$. From which $a_r = -Kr = v_r\frac{dv_r}{dr} - \omega_o^2 r$. Separate variables and integrate: $v_r dv_r = (\omega_o^2 - K)rdr$, from which $\frac{v_r^2}{2} = (\omega_o^2 - K)\frac{r^2}{2} + C$. At $r = r_o$, $v_r = v_o$, from which $C = \frac{v_o^2 - (\omega_o^2 - K)r_o^2}{2}$, and $v_r = \sqrt{v_o^2 + (\omega_o^2 - K)(r^2 - r_o^2)}$. The transverse velocity is $v_\theta = \omega_o r$, from which

$$\boxed{\vec{v} = \sqrt{v_o^2 + (\omega_o^2 - K)(r^2 - r_o^2)}\vec{e}_r + r\omega_o\vec{e}_\theta}$$

==================================<>==================================

=================================<>=================================

Problem 2.153 The hydraulic actuator moves the pin P upward with constant velocity $\mathbf{v} = 0.2\mathbf{j}\, ft/s$. Determine the velocity components v_r and v_θ of the pin and the angular velocity of the slotted bar when $\theta = 30°$

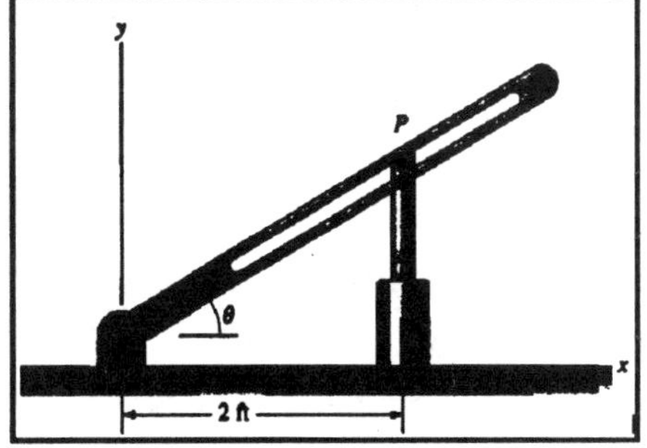

Solution: From the geometry, $\sin\theta = y/r$ and $\cos\theta = 2/r$, where $r = \sqrt{2^2 + y^2}$. The velocity in polar coordinates is given by $\mathbf{v} = v_r\,\mathbf{e_r} + r\omega\,\mathbf{e_\theta}$, where $v_r = dr/dt$ and $\omega = d\theta/dt$. We need to get the components of velocity in terms of the known value of the velocity in rectangular cartesian coordinates. We are given that $\mathbf{v} = 0\,\mathbf{i} + 0.2\,\mathbf{j}\, ft/s$ and that the pin P moves upward at constant velocity (its acceleration is zero). Thus, $x = 2$, $v_x = 0$, $a_x = 0$, and y varies with $v_y = 0.2 ft/s$, $a_y = 0$. We want to change these values into polar coordinates. From $r^2 = 2^2 + y^2$, we get $r\dot{r} = y\dot{y}$, where the dot over the symbol stands for the derivative with respect to time. From $r\cos\theta = 2$, we get $\dot{r}\cos\theta - r\dot{\theta}\sin\theta = 0$. From the first expression, we get $\dot{r} = y\dot{y}/r$ and these two expressions, we get $\dot{\theta} = 2\dot{y}/r^2$. Writing the expression for velocity in polar coordinates, i.e., $\mathbf{v} = v_r\,\mathbf{e_r} + r\omega\,\mathbf{e_\theta}$, and substituting in the values for $v_r = \dot{r}$, and for r and $\dot{\theta}$, we get$\mathbf{v} = (y\dot{y}/r)\mathbf{e_r} + \left(2\dot{y}/r\right)\mathbf{e_\theta}$. Evaluating this at $\theta = 30°$, we get $r = 2.31 ft, v_r = 0.1 ft/s$ $v_\theta = 0.1732 ft/s$.

Note: If we add the two velocity components vectorially, we get $|\mathbf{v}| = 0.2 ft/s$ *as expected.*

=================================<>=================================

Problem 2.154 In Problem 2.153, determine the acceleration components a_r and a_θ of the pin and the angular velocity of the slotted bar when $\theta = 30°$

Solution: We will use all of the solution to problem 2.153 in this solution. What we need to do is to form all of the elements necessary to evaluate the acceleration components a_r and a_θ. We know that $a_r = \ddot{r} - r\dot{\theta}^2$ and that $a_n = r\ddot{\theta} + 2\dot{r}\dot{\theta}$. Given the information found in solving Problem 2.153, we need only be able to evaluate $\ddot{r}$ and $\ddot{\theta}$ to be able to form the expressions for a_r and a_θ. From $r\dot{r} = y\dot{y}$, we get that $r\ddot{r} + \dot{r}^2 = y\ddot{y} + \dot{y}^2$. However, we know that $\ddot{y} = 0$. Hence, $\ddot{r} = \left(\dot{y}^2 - \dot{r}^2\right)/r$, and from $\dot{\theta} = 2\dot{y}/r^2$, we get $\ddot{\theta} = -\dot{\theta}/\left(r\dot{r}\right)$. Substituting these values into the expressions for the radial and normal acceleration components, we get $a_n = a_r = 0$. Also, $\ddot{\theta} = -2\dot{\theta}\,\dot{r}/r = -4\dot{y}\dot{r}/r^3 = \dfrac{-4 ft(0.2 ft/s)(-1 ft/s)}{(2.31 ft)^3}$, or $\ddot{\theta} = -0.0065 rad/s^2$

=================================<>=================================

====================================<>====================================

Problem 2.155 In Example 2.15, determine the velocity of the cam follower when $\theta = 135^\circ$: (a) in terms of polar coordinates; (b) in terms of cartesian coordinates.

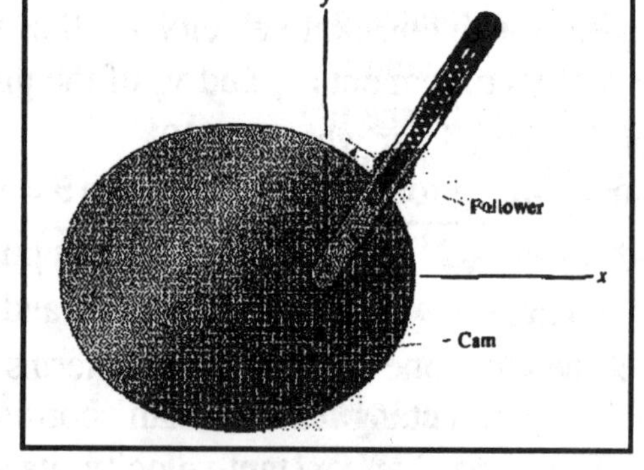

Solution:

(a) From the solution to Example 2.15, we have

$$\mathbf{v} = \left[\frac{0.075\sin\theta}{(1+0.5\cos\theta)^2}\right]\frac{d\theta}{dt}\mathbf{e}_r + \left[\frac{0.15}{(1+0.5\cos\theta)}\right]\frac{d\theta}{dt}\mathbf{e}_\theta.$$ Evaluating at $\theta = 135^\circ$, we get $\mathbf{v} = 0.508\ \mathbf{e}_r + 0.928\ \mathbf{e}_\theta ft/s$

(b) The transformation to cartesian coordinates can be derived from $\mathbf{e}_r = \cos\theta\,\mathbf{i} + \sin\theta\,\mathbf{j}$, and $\mathbf{e}_\theta = -\sin\theta\,\mathbf{i} + \cos\theta\,\mathbf{j}$.

Substituting these into $\mathbf{v} = v_r\,\mathbf{e}_r + v_\theta \mathbf{e}_\theta$, we get $\mathbf{v} = (v_r\cos\theta - v_\theta\sin\theta)\mathbf{i} + (v_r\sin\theta + v_\theta\cos\theta)\mathbf{j}$.

Substituting in the numbers, we get $\mathbf{v} = \ -1.02\,\mathbf{i} - 0.297\,\mathbf{j}\,ft/s$

====================================<>====================================

Problem 2.156 In Example 2.15, determine the acceleration of the cam follower when $\theta = 135^\circ$: (a) in terms of polar coordinates; (b) in terms of cartesian coordinates.

Solution: The solution to Problem 2.155 will be used in this solution. In order to determine the components of the acceleration in polar coordinates, we need to be able to determine all of the variables in the right hand sides of $a_r = \ddot{r} - r\dot{\theta}^2$ and that $a_\theta = r\ddot{\theta} + 2\dot{r}\dot{\theta}$. We already know everything except $\ddot{r}$ and $\ddot{\theta}$. Since ω is constant, $\ddot{\theta} = \dot{\omega} = 0$. We need only to find the value for r and the value for $\ddot{r}$ at $\theta = 135^\circ$. Substituting into the original equation for r, we find that $r = 0.232\ m$ at this position on the cam. To find $\ddot{r}$, we start with $\dot{r} = v_r$. Taking a derivative, we get $\ddot{r} = \frac{d}{dt}(v_r) = \frac{d}{dt}\left[\left(\frac{0.075\sin\theta}{(1+0.5\cos\theta)^2}\right)\dot{\theta}\right]$,

or

$$\text{or } \ddot{r} = \frac{d}{d\theta}[v_r]\frac{d\theta}{dt} = \frac{d}{d\theta}\left[\left(\frac{0.075\sin\theta}{(1+0.5\cos\theta)^2}\right)\dot{\theta}\right]\frac{d\theta}{dt}$$

$$\ddot{r} = \left[\left(\frac{(1+0.5\cos\theta)^2(0.075\cos\theta) - (0.075\cos\theta)2(1+0.5\cos\theta)(-0.5\sin\theta)}{(1+0.5\cos\theta)^4}\right)\dot{\theta}^2\right].$$

Evaluating, we get $\ddot{r} = -4.25 ft/s^2$. Substituting this into the equation for a_r and evaluating a_n, we get $a_r = -7.96 ft/s^2$ and $a_\theta = 4.06 ft/s^2$

(b) The transformation to cartesian coordinates can be derived from $\mathbf{e}_r = \cos\theta\,\mathbf{i} + \sin\theta\,\mathbf{j}$, and $\mathbf{e}_\theta = -\sin\theta\,\mathbf{i} + \cos\theta\,\mathbf{j}$. Substituting these into $\mathbf{a} = a_r\,\mathbf{e}_r + a_\theta \mathbf{e}_\theta$, we get $\mathbf{a} = (a_r\cos\theta - a_\theta\sin\theta)\mathbf{i} + (a_r\sin\theta + a_\theta\cos\theta)\mathbf{j}$. Substituting in the numbers, we get $\mathbf{a} = \ 2.76\,\mathbf{i} - 8.50\,\mathbf{j}\,ft/s^2$

====================================<>====================================

==================================<>================================

Problem 2.157 In the cam follower mechanism, the slotted bar rotates with a constant angular velocity $\omega = 10 rad/s$ and the radial position of the follower A is determined by the profile of the stationary cam. The path of the follower is described by the polar equation $r = 1 + 0.5\cos(2\theta)\ ft$.

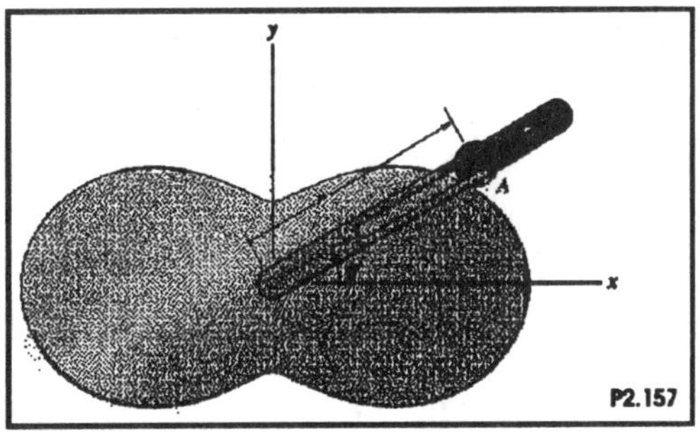

Determine the velocity of the cam follower when $\theta = 30°$ (a) in terms of polar coordinates; (b) in terms of cartesian coordinates.

P2.157

Solution: The solution for this problem follows the solution for Problem 2.155 and Example 2.15 exactly. The only difference is the form of the function $r(\theta)$ which describes the shape of the stationary cam. Here, $r = 1 + 0.5\cos(2\theta)\ ft$. We proceed as in Example 2.15. (a) We find the value for $v_r = \dot{r} = \frac{dr}{dt}$ by $\frac{dr}{dt} = \frac{dr}{d\theta}\frac{d\theta}{dt} = \frac{dr}{d\theta}\omega = -0.5(2)\sin(2\theta)\omega = -\omega\sin(2\theta)$. Note that $\omega = \dot{\theta} = constant$ and $\ddot{\theta} = 0$ as before.

Hence, $\mathbf{v} = v_r\mathbf{e}_r + r\omega\,\mathbf{e}_\theta = -\omega\sin(2\theta)\mathbf{e}_r + r\omega\,\mathbf{e}_\theta$. At $\theta = 30°$, $r = 1.25 ft$, and $\omega = 10 rad/s$. Substituting in values, we get $\mathbf{v} = -(10)\sin(60)\mathbf{e}_r + (1.25)(10)\,\mathbf{e}_\theta = -8.66\mathbf{e}_r + 12.5\,\mathbf{e}_\theta ft/s \pm$.

(b) The transformation to cartesian coordinates can be derived from $\mathbf{e}_r = \cos\theta\,\mathbf{i} + \sin\theta\,\mathbf{j}$, and $\mathbf{e}_\theta = -\sin\theta\,\mathbf{i} + \cos\theta\,\mathbf{j}$. Substituting these into $\mathbf{v} = v_r\,\mathbf{e}_r + v_\theta\mathbf{e}_\theta$, we get $\mathbf{v} = (v_r\cos\theta - v_\theta\sin\theta)\mathbf{i} + (v_r\sin\theta + v_\theta\cos\theta)\mathbf{j}$. Substituting in the numbers, we get

$\mathbf{v} = -13.75\,\mathbf{i} + 6.50\,\mathbf{j}\ ft/s$

==================================<>================================

Problem 2.158 In Problem 2.157, determine the acceleration of the cam follower when $\theta = 30°$: (a) in terms of polar coordinates; (b) in terms of cartesian coordinates.

Solution: The solution to this problem follows that of Problem 2.157 in the same way that the solution to Problem 2.156 followed that of Problem 2.155. Information from the solution to Problem 2.157 will be used in this solution. In order to determine the components of the acceleration in polar coordinates, we need to be able to determine all of the variables in the right hand sides of $a_r = \ddot{r} - r\dot{\theta}^2$ and that $a_\theta = r\ddot{\theta} - 2\dot{r}\dot{\theta}$. We already know everything except $\ddot{r}$ and $\ddot{\theta}$. Since ω is constant, $\ddot{\theta} = \dot{\omega} = 0$. We need only to find the value for r and the value for $\ddot{r}$ at $\theta = 30°$. Substituting into the original equation for r, we find that $r = 0.232\ m$ at this position on the cam. To find $\ddot{r}$, we start with $\dot{r} = v_r$. Taking a derivative, we get $\ddot{r} = \frac{d}{dt}(v_r) = \frac{d}{dt}\left[-\sin(2\theta)\dot{\theta}\right]$, or

or $\ddot{r} = \frac{d}{d\theta}[v_r]\frac{d\theta}{dt} = \frac{d}{d\theta}[-\sin(2\theta)\omega]\omega = -2\cos(2\theta)\omega^2$

Evaluating, we get $\ddot{r} = -100 ft/s^2$. Substituting this into the equation for a_r and evaluating a_n, we get $a_r = 123.2 ft/s^2$ and $a_\theta = -173.2 ft/s^2$

(b) The transformation to cartesian coordinates can be derived from $\mathbf{e}_r = \cos\theta\,\mathbf{i} + \sin\theta\,\mathbf{j}$, and $\mathbf{e}_\theta = -\sin\theta\,\mathbf{i} + \cos\theta\,\mathbf{j}$. Substituting these into $\mathbf{a} = a_r\,\mathbf{e}_r + a_\theta\mathbf{e}_\theta$, we get $\mathbf{a} = (a_r\cos\theta - a_\theta\sin\theta)\mathbf{i} + (a_r\sin\theta + a_\theta\cos\theta)\mathbf{j}$. Substituting in the numbers, we get

$\mathbf{a} = -20.1\,\mathbf{i} - 211.6\,\mathbf{j}\ ft/s^2$

==================================<>================================

=================================<>=================================

Problem 2.159 The cartesian coordinates of a point P in the x-y plane are related to its polar coordinates by $x = r\cos\theta$, $y = r\sin\theta$. (a) Show that the unit vectors $\vec{i}$, $\vec{j}$ are related to the unit vectors $\vec{e}_r$, $\vec{e}_\theta$ by $\vec{i} = \vec{e}_r\cos\theta - \vec{e}_\theta\sin\theta$, $\vec{j} = \vec{e}_r\sin\theta + \vec{e}_\theta\cos\theta$. (b) Beginning with the expression for the position vector $\vec{r} = \vec{i}x + \vec{j}y$, derive Eq (2.44) for the position vectors in terms of polar coordinates. (c) By taking the time derivative of the position vector of point P expressed in terms of cartesian coordinates, derive Eq(2.47) for the velocity in terms of polar coordinates.

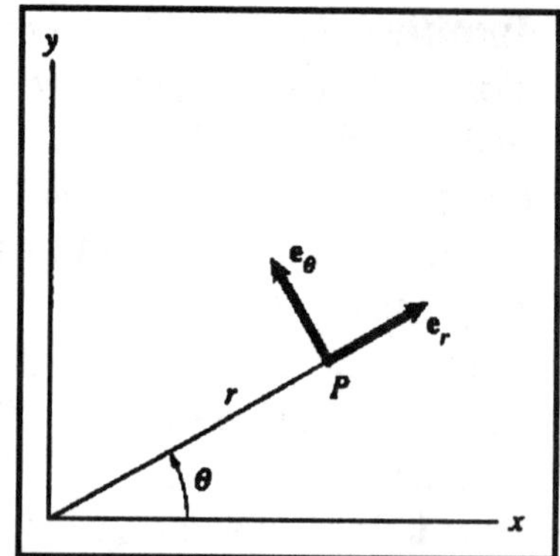

Solution (a) From geometry (see Figure), the radial unit vector is $\vec{e}_r = \vec{i}\cos\theta + \vec{j}\sin\theta$, and since the transverse unit vector is at right angles: $\vec{e}_\theta = \vec{i}\cos\left(\theta + \frac{\pi}{2}\right) + \vec{j}\sin\left(\theta + \frac{\pi}{2}\right) = -\vec{i}\sin\theta + \vec{j}\cos\theta$. Solve for $\vec{i}$ by multiplying $\vec{e}_r$ by $\cos\theta$, $\vec{e}_\theta$ by $\sin\theta$, and subtracting the resulting equations: $\boxed{\vec{i} = \vec{e}_r\cos\theta - \vec{e}_\theta\sin\theta}$. Solve for $\vec{j}$ by multiplying $\vec{e}_r$ by $\sin\theta$, and $\vec{e}_\theta$ by $\cos\theta$, and the results: $\boxed{\vec{j} = \vec{e}_r\sin\theta + \vec{e}_\theta\cos\theta}$ (b) The position vector is $\vec{r} = x\vec{i} + y\vec{j} = (r\cos\theta)\vec{i} + (r\sin\theta)\vec{j} = r(\vec{i}\cos\theta + \vec{j}\sin\theta)$. Use the results of Part (a) expressing $\vec{i}$, $\vec{j}$ in terms of $\vec{e}_r$, $\vec{e}_\theta$: $\boxed{\vec{r} = r(\vec{e}_r\cos^2\theta - \vec{e}_\theta\cos\theta\sin\theta + \vec{e}_r\sin^2\theta + \vec{e}_\theta\sin\theta\cos\theta) = r\vec{e}_r}$ (c) The time derivatives are: $\frac{d\vec{r}}{dt} = \vec{v} = \vec{i}\left(\frac{dr}{dt}\cos\theta - r\sin\theta\frac{d\theta}{dt}\right) + \vec{j}\left(\frac{dr}{dt}\sin\theta + r\cos\theta\frac{d\theta}{dt}\right)$, from which $\vec{v} = \frac{dr}{dt}(\vec{i}\cos\theta + \vec{j}\sin\theta) + r\frac{d\theta}{dt}(-\vec{i}\sin\theta + \vec{j}\cos\theta)$. Substitute the results of Part(a)

$$\boxed{\vec{v} = \frac{dr}{dt}\vec{e}_r + r\frac{d\theta}{dt}\vec{e}_\theta = \frac{dr}{dt}\vec{e}_r + r\omega\vec{e}_\theta}$$

=================================<>=================================

Problem 2.160 The airplane flies in a straight line at 400 mi/hr. The radius of its propellor is 5 ft and it turns at 2000 rpm (revolutions per minute) in the counterclockwise direction when seen from the front of the airplane. Determine the velocity and acceleration of a point on the tip of the propeller in terms of cylindrical coordinates. (Let the z axis be oriented as shown in the figure.)

Solution The speed is $v = 400\left(\frac{mile}{hr}\right)\left(\frac{1\ hr}{3600\ s}\right)\left(\frac{5280\ ft}{1\ mile}\right) = 586.7\ ft/s$

The angular velocity is $\omega = 2000\left(\frac{2\pi\ rad}{1\ rev}\right)\left(\frac{1\ min}{60\ s}\right) = 209.4\ rad/s$. The radial velocity at the propeller tip is zero. The transverse velocity is $v_\theta = \omega r = 1047.2\ ft/s$. The velocity vector in cylindrical coordinates is $\boxed{\vec{v} = 1047.2\vec{e}_\theta + 586.7\vec{e}_z\ ft/s}$. The radial acceleration is

$a_r = -r\omega^2 = -5(209.4)^2 = -219324.5\ ft/s^2$. The transverse acceleration is

$a_\theta = r\frac{d^2\theta}{dt^2} + 2\left(\frac{dr}{dt}\right)\left(\frac{d\theta}{dt}\right) = 0$, since the propeller rotates at a constant angular velocity. The acceleration

$a_z = 0$, since the airplane travels at constant speed. Thus $\vec{a} = -219324.5\vec{e}_r\ ft/s^2$

=================================<>=================================

==================================<>==================================

Problem 2.161 A charged particle P in a magnetic field moves along the spiral path described by $r = 1\ m$, $\theta = 2z\ rad$, where z is in meters. The particle moves along the path in the direction shown with constant speed $|\vec{v}| = 1\ km/s$. What is the velocity of the particle in terms of cylindrical coordinates?

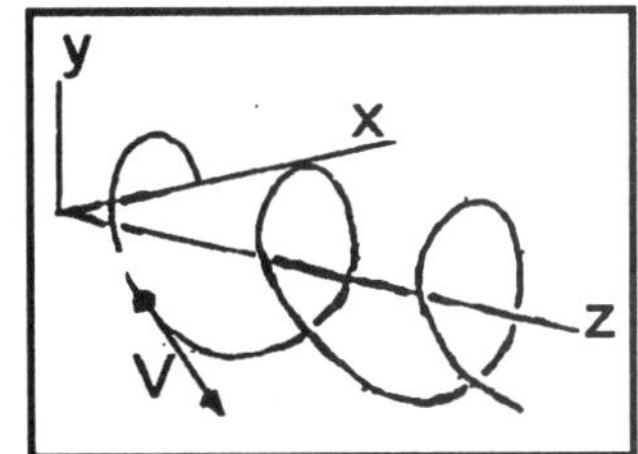

Solution The radial velocity is zero, since the path has a constant radius.

The magnitude of the velocity is $v = \sqrt{r^2\left(\frac{d\theta}{dt}\right)^2 + \left(\frac{dz}{dt}\right)^2} = 1000\ m/s$. The angular velocity is

$\frac{d\theta}{dt} = 2\frac{dz}{dt}$. Substitute: $v = \sqrt{r^2\left(\frac{d\theta}{dt}\right)^2 + \frac{1}{4}\left(\frac{d\theta}{dt}\right)^2} = \left(\frac{d\theta}{dt}\right)\sqrt{r^2 + \frac{1}{4}} = \sqrt{1.25}$, from which

$\frac{d\theta}{dt} = \frac{1000}{\sqrt{1.25}} = 894.4\ rad/s$, from which the transverse velocity is $v_\theta = r\left(\frac{d\theta}{dt}\right) = 894.4\ \ m/s$. The velocity along the cylindrical axis is $\frac{dz}{dt} = \frac{1}{2}\left(\frac{d\theta}{dt}\right) = 447.2\ m/s$. The velocity vector:

$\boxed{\vec{v} = 894.4\vec{e}_\theta + 447.2\vec{e}_z}$

==================================<>==================================

Problem 2.162 Two cars approach an intersection. Car A is going 20 *m/s* and is accelerating at 2 m/s^2, and car B is going 10 *m/s* and is decelerating at 3 m/s^2. In terms of the earth fixed coordinate system shown, determine the velocity of car A relative to car B and the velocity of car B relative to car A.

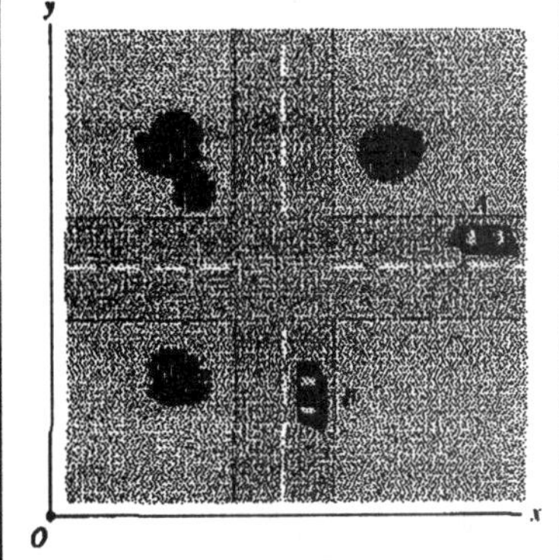

Solution The velocity of car A is . The velocity of car B is $\vec{v}_B = \vec{j}10\ m/s$. The relative velocity is

$\boxed{\vec{v}_{A/B} = \vec{v}_A - v_B = -\vec{i}\,20 - \vec{j}\,10\ m/s}$. $\boxed{\vec{v}_{B/A} = \vec{v}_B - \vec{v}_A = 20\vec{i} + 10\vec{j}}$

==================================<>==================================

Problem 2.163 In Problem 2.162, determine the accelerations of car A relative to car B, and the accelerations of car B relative car A.

Solution Car A is accelerating in the positive x direction; car B is accelerating in the negative y direction: $\boxed{\vec{a}_{A/B} = \vec{a}_A - \vec{a}_B = 2\vec{i} - (-3\vec{j}) = 2\vec{i} + 3\vec{j}\ \ ms^2/}$. $\boxed{\vec{a}_{B/A} = \vec{a}_B - \vec{a}_A = -2\vec{i} - 3\vec{j}\ \ m/s^2}$

==================================<>==================================

====================================<>====================================

Problem 2.164 Suppose that the two cars in Problem 2.162 approach the intersection with constant velocities. Prove that the cars will reach the intersection at the same time if the velocity of car A relative to car B points from car A toward car B

Solution A vector pointing from A to B is $\vec{r}_{A/B} = \vec{r}_A - \vec{r}_B$. The strategy is to prove that $\vec{v}_{A/B} = \vec{v}_A - \vec{v}_B$ is collinear with $\vec{r}_{A/B}$. The location of car A is $\vec{r}_A(t) = \int \vec{v}_A dt + \vec{R}_A = \vec{v}_A t + \vec{R}_A$, where $\vec{R}_A$ is a vector constant of integration, and the velocity is constant. The location of car B is $\vec{r}_B(t) = \int \vec{v}_B\, dt + \vec{R}_B = \vec{v}_B t + \vec{R}_B$.The time required for car A to reach the intersection is t_A. The time required for B to arrive at the intersection is t_B. If they both arrive at the same time, then $t_A = t_B = T$, and the relative location $\vec{r}_{A/B} = 0$, or $0 = \vec{v}_A T - \vec{v}_B T + (\vec{R}_A - \vec{R}_B)$, from which $(\vec{R}_A - \vec{R}_B) = -\vec{v}_A T + \vec{v}_B T$. Thus $\boxed{\vec{r}_{A/B} = \vec{r}_A - \vec{r}_B = \vec{v}_A(t-T) - \vec{v}_B(t-T) = \vec{v}_{A/B}(t-T)}$, which shows that the two vectors are collinear if they arrive at the intersection at the same time. The converse, namely, that if they are not collinear they will not arrive at the same time follows immediately from the supposition that $t_A \neq t_B$ and the fact that the velocities are constant.

====================================<>====================================

Problem 2.165 Two sailboats have constant velocities $\vec{v}_A$ and $\vec{v}_B$ relative to the earth. The skipper of boat A sights a compass point on the horizon behind boat B. Seeing that boat B remains stationary relative to that point, he knows that he must change course to avoid a collision. Use Eq (2.72) to explain why.

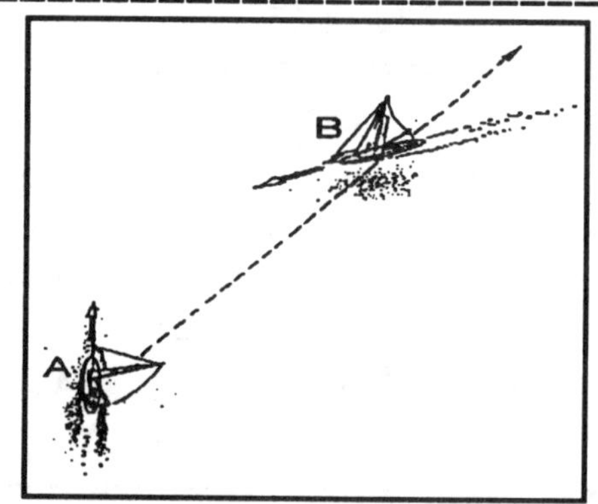

Solution Eq (2.72) is $\vec{v}_A = \vec{v}_B + \vec{v}_{A/B}$. (See Problem 2.152). If the location of B remains in a fixed direction relative to A, then the relative velocity vector must be collinear with the straight line connecting A and B, and after a time *T* the gap will close to zero.

====================================<>====================================

Problem 2.166 Two projectiles A and B are launched at the same time with the initial velocities and elevation angles shown relative to the earth-fixed coordinates. At the instant B reaches is highest point, determine (a) the acceleration of A relative to B; (b) the velocity of A relative to B; (c) the position vector of A relative to B.

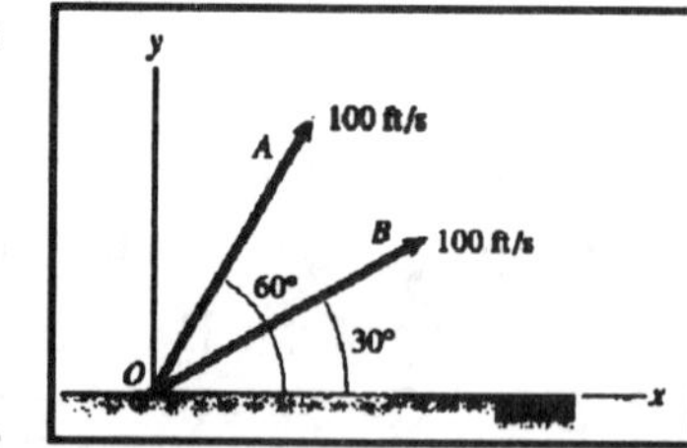

Solution For brevity, denote $V_A = 100\, ft/s$, $V_B = 100\, ft/s$, $\theta_A = 60^o$, $\theta_B = 30^o$. The differential equations are: $\frac{d^2y}{dt^2} = -g$, $\frac{d^2x}{dt^2} = 0$. Integrating twice, with the initial conditions applied: For A: $\frac{dy_A}{dt} = -gt + V_A \sin\theta_A$, $y_A(t) = -gt^2/2 + (V_A \sin\theta_A)t$, $dx_A/dt = V_A\cos\theta_A$, $x_A(t) = (V_A \cos\theta_A)t$. For B: $y_B(t) = -gt^2/2 + (V_B \sin\theta_B)t$, $dy_B/dt = -gt + V_B\sin\theta_B$, $dx_B/dt = V_B\cos\theta_B$, $x_B(t) = (V_B\cos\theta_B)t$. B reaches its highest point when $dy_B/dt = 0 = -gt_B + V_B\sin\theta_B$, from which $t_B = V_B \sin\theta_B / g = 1.554$ s. (a) The acceleration of A relative to B is $\boxed{\vec{a}_{A/B} = -\vec{j}g + \vec{j}g = 0}$ (b) The velocity of A relative to B is $\boxed{\vec{v}_{A/B} = (V_A\cos\theta_A - V_B\cos\theta_B)\vec{i} + (V_A\sin\theta_A - V_B\sin\theta_B)\vec{j} = -36.6\vec{i} + 36.6\vec{j}}$. (c) The position vector of A relative to B $\boxed{\vec{r}_{A/B} = (V_A\cos\theta_A - V_B\cos\theta_B)t_B\vec{i} + (V_A\sin\theta_A - V_B\sin\theta_B)t_B\vec{j} = -56.9\vec{i} + 56.9\vec{j}}$

====================================<>====================================

==================================<>==================================

Problem 2.167 In a machining process, the disk rotates about the fixed point O with a constant angular velocity of 10 *rad/s*. In terms of the non-rotating coordinate system shown, what is the magnitude of the velocity of A relative to B?

Solution The position of point A is

$\vec{r}_A = \vec{i}\,2\cos(\omega t + \frac{\pi}{2}) + \vec{j}\,2\sin(\omega t + \frac{\pi}{2}) = -\vec{i}\,2\sin\omega t + \vec{j}\,2\cos\omega t \ (ft)$. The position of point B is $\vec{r}_B = \vec{i}\,2\cos\omega t + \vec{j}\,2\sin\omega t \ (ft)$. The velocities are

$\vec{v}_A = -\vec{i}\,2\omega\cos\omega t - \vec{j}\,2\omega\sin\omega t \ (ft/s)$,

$\vec{v}_B = -\vec{i}\,2\omega\sin\omega + \vec{j}\,2\omega\cos\omega t \ (ft/s)$ The relative velocity is

$\vec{v}_{A/B} = \vec{v}_A - \vec{v}_B = 2\omega(-\cos\omega t + sin\omega t)\vec{i} + 2\omega(-\sin\omega t - \cos\omega t)\vec{j} \ (ft/s)$. The magnitude is

$$\boxed{|\vec{v}_{A/B}| = 2\omega\sqrt{(-\cos\omega t + \sin\omega t)^2 + (-\sin\omega t - \cos\omega t)^2} = 2\sqrt{2}\omega = 28.28\ ft/s}$$

==================================<>==================================

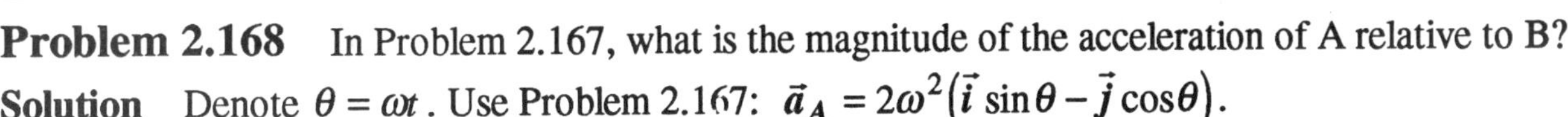

Problem 2.168 In Problem 2.167, what is the magnitude of the acceleration of A relative to B?

Solution Denote $\theta = \omega t$. Use Problem 2.167: $\vec{a}_A = 2\omega^2(\vec{i}\sin\theta - \vec{j}\cos\theta)$.

$\vec{a}_B = 2\omega^2(-\vec{i}\cos\theta - \vec{j}\sin\theta)$. The relative acceleration:

$\vec{a}_{A/B} = 2\omega^2[(\sin\theta + \cos\theta)\vec{i} + (-\cos\theta + \sin\theta)\vec{j}]$. $\boxed{|\vec{a}_{A/B}| = 2\sqrt{2}\omega^2 = 282.8\ ft/s^2}$

==================================<>==================================

====================================<>====================================

Problem 2.169 The bar rotates about the fixed point O with a constant angular velocity of 2 *rad/s*. Point A moves outward along the bar at a constant rate of 100 *mm/s*. Point B is a fixed point on the bar. In terms of the nonrotating reference frame with origin *O*, what is the magnitude of the velocity of point A relative to point B?

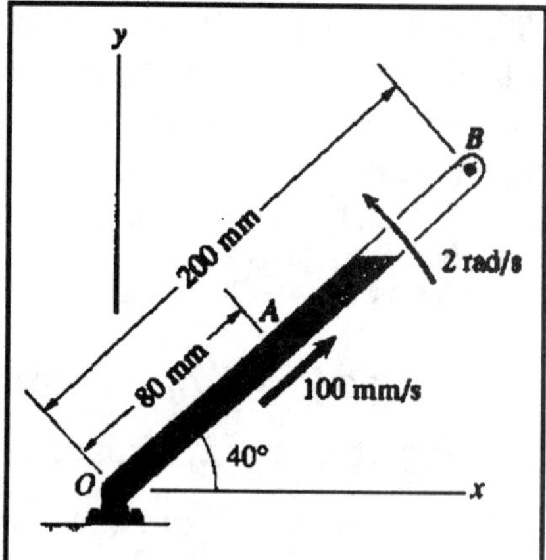

Solution Use polar coordinates: The radial velocities:

$$\vec{v}_A = \frac{dr}{dt}\vec{e}_r + r\frac{d\theta}{dt}\vec{e}_\theta = 100\vec{e}_r + 80(2)\vec{e}_\theta = 100\vec{e}_r + 160\vec{e}_\theta \ (mm/s).$$

$\vec{v}_B = r\dfrac{d\theta}{dt} = 200(2)\vec{e}_\theta = 400\vec{e}_\theta \ mm/s$. The relative velocity in polar coordinates: $\vec{v}_{A/B} = \vec{v}_A - \vec{v}_B = 100\vec{e}_r - 240\vec{e}_\theta \ (mm/s)$. The magnitude of the relative velocity: $\boxed{|\vec{v}_{A/B}| = \sqrt{100^2 + (240)^2} = 260 \ mm/s}$

====================================<>====================================

Problem 2.170 In Problem 2.169, what is the magnitude of the acceleration of point B relative to point A at the instant shown?

Solution Use polar coordinates. the radial accelerations:

$$\vec{a}_A = \left(\frac{d^2r}{dt^2} - r\left(\frac{d\theta}{dt}\right)^2\right)\vec{e}_r + \left(r\frac{d^2\theta}{dt^2} + 2\left(\frac{dr}{dt}\right)\left(\frac{d\theta}{dt}\right)\right)\vec{e}_\theta = -80(4)\vec{e}_r + 2(100)2\vec{e}_\theta.$$

$$\vec{a}_B = \left(\frac{d^2r}{dt^2} - r\left(\frac{d\theta}{dt}\right)^2\right)\vec{e}_r + \left(r\frac{d^2\theta}{dt^2} + 2\left(\frac{dr}{dt}\right)\left(\frac{d\theta}{dt}\right)\right)\vec{e}_\theta = -200(4)\vec{e}_r.$$

The relative acceleration is $\vec{a}_{B/A} = (-800 + 320)\vec{e}_r - (400)\vec{e}_\theta$. The magnitude is $\boxed{|\vec{a}_{B/A}| = \sqrt{480^2 + 400^2} = 625 \ mm/s^2}$

====================================<>====================================

Problem 2.171 The bars OA and OB are each 400 *mm* long and rotate in the *x-y* plane. OA has a counterclockwise angular velocity of 10 *rad/s* and a counterclockwise angular acceleration of 2 *rad/s*^{2}. AB has a counterclockwise angular velocity of 5 *rad/s* relative to the nonrotating coordinate system. What is the velocity of point B relative to point A?

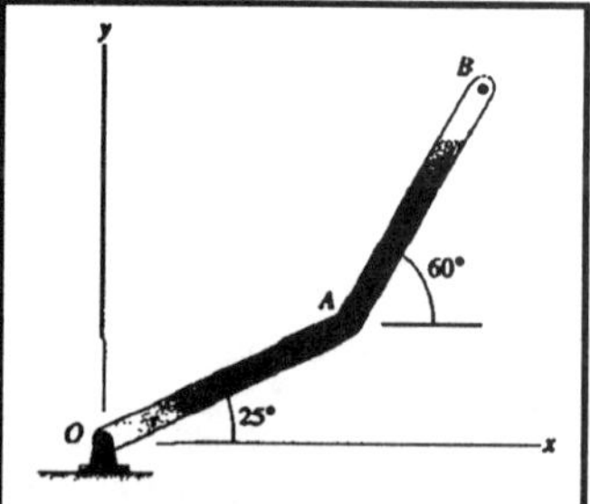

Solution Use polar coordinates. The velocity of B relative to A is

$$\vec{v}_{B/A} = \left(\frac{dr}{dt}\right)\vec{e}_r + \left(r\frac{d\theta}{dt}\right)\vec{e}_\theta = 400(5)\vec{e}_\theta \ (mm/s).$$

In cartesian coordinates: (See Problem 2.133) $\vec{e}_\theta = -\vec{i}\sin\theta + \vec{j}\cos\theta$, from which the relative velocity in cartesian coordinates is

$$\boxed{\vec{v}_{B/A} = (400)(5)\left(-\vec{i}\sin 60^o + \vec{j}\cos 60^o\right) = -1732.1\vec{i} + 1000\vec{j} \ (mm/s)}$$

====================================<>====================================

Problem 2.172 In Problem 2.171, what is the acceleration of point B relative to point A?

Solution Use polar coordinates: The acceleration of B relative to A is

$$\vec{a}_{B/A} = \left(\frac{d^2r}{dt^2} - r\left(\frac{d\theta}{dt}\right)^2\right)\vec{e}_r + \left(r\frac{d^2\theta}{dt^2} + 2\left(\frac{dr}{dt}\right)\left(\frac{d\theta}{dt}\right)\right)\vec{e}_\theta = -400(25)\vec{e}_r.$$

In cartesian coordinates: $\vec{e}_r = \vec{i}\cos\theta + \vec{j}\sin\theta$, from which $\boxed{\vec{a}_{B/A} = -5000\vec{i} - 8660\vec{j} \ (mm/s^2)}$

====================================<>====================================

==================================<>==================================

Problem 2.173 In Problem 2.171, what is the velocity of point B relative to the fixed point O?

Solution The strategy is to find the velocity of point A relative to O, and the add the velocity of B relative to A. Use polar coordinates. The velocity of A: $\vec{v}_{A/O} = \left(\frac{dr}{dt}\right)\vec{e}_r + \left(r\frac{d\theta}{dt}\right)\vec{e}_\theta = 400(10)\vec{e}_\theta$. In cartesian coordinates, $\vec{e}_\theta = -\vec{i}\sin\theta + \vec{j}\cos\theta$, from which

$$\vec{v}_{A/O} = -(4000\sin 25^o)\vec{i} + \left(4000\cos 25^o\right)\vec{j} = -1690.5\vec{i} + 3625.2\vec{j} \ (mm/s)$$

From the solution to Problem 2.159, $\vec{v}_{B/A} = -1732.1\vec{i} + 1000\vec{j} \ (mm/s)$. The relative velocity of B is

$$\boxed{\vec{v}_{B/O} = \vec{v}_{A/O} + \vec{v}_{B/A} = -3422.6\vec{i} + 4625.2\vec{j} \ (mm/s)}.$$

==================================<>==================================

Problem 2.174 In Problem 2.171, what is the acceleration of point B relative to the fixed point O?

Solution The strategy is to find the acceleration of A relative to O, and then add the acceleration of B relative to A. The acceleration of A:

$$\vec{a}_{A/O} = \left(\frac{d^2r}{dt^2} - r\left(\frac{d\theta}{dt}\right)^2\right)\vec{e}_r + \left(r\frac{d^2\theta}{dt^2} + 2\left(\frac{dr}{dt}\right)\left(\frac{d\theta}{dt}\right)\right)\vec{e}_\theta = -0.4\left(10^2\right)\vec{e}_r + 2(0.4)\vec{e}_\theta,$$

$\vec{a}_{A/O} = -40\vec{e}_r + 0.8\vec{e}_\theta \ \left(m/s^2\right)$ In cartesian coordinates, $\vec{e}_r = \vec{i}\cos\theta + \vec{j}\sin\theta$,

$\vec{e}_\theta = -\vec{i}\sin\theta + \vec{j}\cos\theta$,from which $\vec{a}_{A/O} = (-40\cos 25^o - 0.8\sin 25^o)\vec{i} + \left(0..8\cos 25^o - 40\sin 25^o\right)\vec{j}$

$\vec{a}_{A/O} = -36.59\vec{i} - 16.18\vec{j} \ \left(m/s^2\right)$ From Problem 2.160, the acceleration of B relative to A is

$\vec{a}_{B/A} = -5\vec{i} - 8.66\vec{j} \ \left(m/s^2\right)$, from which $\boxed{\vec{a}_{B/O} = \vec{a}_{A/O} + \vec{a}_{B/A} = -41.59\vec{i} - 24.84\vec{j} \ \left(m/s^2\right)}$

==================================<>==================================

Problem 2.175 Points O and O' are fixed relative to the x-y reference frame. At the instant shown,
$r_A = 1.8m$, $\theta_A = 50°$, $dr_A/dt = 12 m/s$, $d\theta_A/dt = 4\ rad/s$,
$r_B = 1.8m$, $\theta_B = 40°$, $dr_B/dt = -8 m/s$, $d\theta_B/dt = 6\ rad/s$

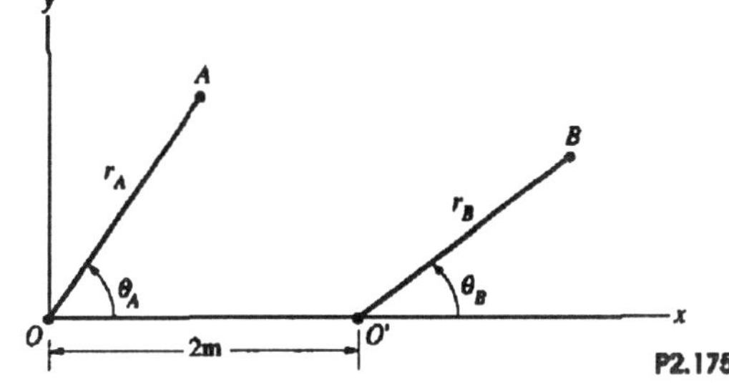

Solution: It is convenient to evaluate the velocities in polar coordinates for points A and B separately and then to transform to rectangular cartesian coordinates before finding the velocity of A relative to B. If we did it in polar coordinates, the unit vectors for the OA and O'B would not match. For either point A or point B, the velocity is given by $\mathbf{v}_{A,B} = \frac{dr_{A,B}}{dt}\mathbf{e}_{r_{A,B}} + r_{A,B}\frac{d\theta_{A,B}}{dt}\mathbf{e}_{\theta_{A,B}}$. Substituting in the known values, we get $\mathbf{v}_A = 12.0\mathbf{e}_{r_A} + 7.2\mathbf{e}_{\theta_A}\ m/s$ and $\mathbf{v}_B = -8.0\mathbf{e}_{r_B} + 10.8\mathbf{e}_{\theta_B}\ m/s$. Recall that $\theta_A = 50°$ and $\theta_B = 40°$ The transformation to cartesian coordinates can be derived from $\mathbf{e}_r = \cos\theta\,\mathbf{i} + \sin\theta\,\mathbf{j}$, and $\mathbf{e}_\theta = -\sin\theta\,\mathbf{i} + \cos\theta\,\mathbf{j}$. Substituting these into $\mathbf{v}_{A,B} = v_{r_{A,B}}\mathbf{e}_{r_{A,B}} + v_{\theta_{A,B}}\mathbf{e}_{\theta_{A,B}}$, we get $\mathbf{v} = (v_r\cos\theta - v_\theta\sin\theta)\mathbf{i} + (v_r\sin\theta + v_\theta\cos\theta)\mathbf{j}$. Substituting in the numbers for points A and B, we get $\mathbf{v}_A = 2.20\,\mathbf{i} + 13.82\,\mathbf{j}\ m/s$ and $\mathbf{v}_B = -13.07\,\mathbf{i} + 3.13\,\mathbf{j}\ m/s$. The velocity of point A relative to point B is given by $\mathbf{v}_{A/B} = \mathbf{v}_A - \mathbf{v}_B$. Thus, $\mathbf{v}_{A/B} = 15.17\,\mathbf{i} + 10.69\,\mathbf{j}\ m/s$, and $|\mathbf{v}_{A/B}| = 18.64\ m/s$

==================================<>==================================

==================================<>==================================

Problem 2.176 At the instant shown in Problem 2.175, $d^2r_A/dt^2 = 20 m/s^2$, $d^2\theta_A/dt^2 = 0$, $d^2r_B/dt^2 = 10 m/s^2$, $d^2\theta_B/dt^2 = 0$. What is the magnitude of the acceleration of point A relative to point B?

Solution: We will use the same procedure as in Problem 2.175. We will evaluate the accelerations in polar coordinates for points A and B separately and then transform to rectangular cartesian coordinates. Then we will find the acceleration of point A relative to point B in rectangular cartesian coordinates. Finally we will find the magnitude of this relative acceleration. The acceleration components for point A or for point B are given as $a_{r_{A,B}} = (d^2r/dt^2)_{A,B} - r_{A,B}(d\theta/dt)^2_{A,B}$, and $a_{\theta_{A,B}} = r_{A,B}(d^2\theta/dt^2)_{A,B} + 2(dr/dt)_{A,B}(d\theta/dt)_{A,B}$. Substituting in the known values, we get $\mathbf{a}_A = -8.8\mathbf{e}_{r_A} + 96\mathbf{e}_{\theta_A}\ (m/s^2)$, and $\mathbf{a}_B = -54.8\mathbf{e}_{r_B} - 96.0\mathbf{e}_{\theta_B}\ (m/s^2)$. Using the same transformations as in Problem 2.175, we get $\mathbf{a}_A = -79.2\,\mathbf{i} + 55.0\,\mathbf{j}\ (\mathrm{m/s^2})$ and $\mathbf{a}_B = 19.73\,\mathbf{i} - 108.77\,\mathbf{j}\ (\mathrm{m/s^2})$. The relative acceleration is given by $\mathbf{a}_{A/B} = \mathbf{a}_A - \mathbf{a}_B = -98.9\,\mathbf{i} + 163.8\,\mathbf{j}\ (m/s^2)$. The magnitude of the relative acceleration is given by $|\mathbf{a}_{A/B}| = 193.6\ (m/s^2)$.

==================================<>==================================

Problem 2.177 The train on the circular track is traveling at a constant speed of 50 *ft/s*. The train on the straight track is traveling at 20 *ft/s* and is increasing its speed at 2 *ft/s*^{2}. In terms of the coordinate system shown, what is the velocity of passenger A relative to Passenger B?

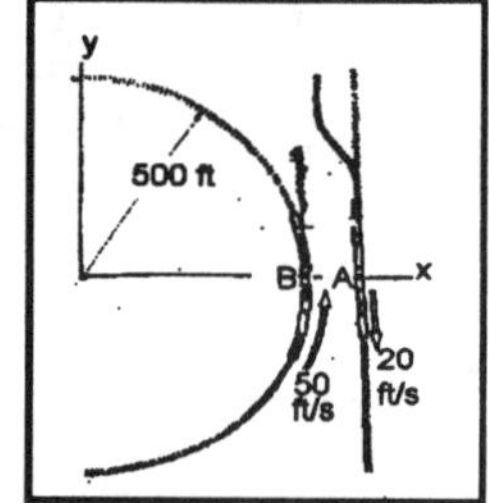

Solution The velocity of the train on the circular track is

$$\vec{v}_B = r\left(\frac{d\theta}{dt}\right)\vec{e}_\theta = V\vec{e}_\theta = 50\vec{e}_\theta = (-50\sin\theta)\vec{i} + (50\cos\theta)\vec{j} = 50\vec{j}\ \ (ft/s)$$

since the angle θ is zero. The velocity of the train on the straight track is $\vec{v}_A = -20\vec{j}\ \ (ft/s)$. The relative velocity is $\boxed{\vec{v}_{A/B} = \vec{v}_A - \vec{v}_B = (-20-50)\vec{j} = -70\vec{j}\ (ft/s)}$

==================================<>==================================

Problem 2.178 In Problem 2.177, what is the acceleration of passenger A relative to passenger B?

Solution The acceleration of B is

$$\vec{a}_B = \left(\frac{d^2r}{dt^2} - r\left(\frac{d\theta}{dt}\right)^2\right)\vec{e}_r + \left(r\frac{d^2q}{dt^2} + 2\left(\frac{dr}{dt}\right)\left(\frac{d\theta}{dt}\right)\right)\vec{e}_\theta = -\frac{V^2}{r}\vec{e}_r = -\frac{(50)^2}{500}\vec{e}_r = -5\vec{e}_r$$

$\vec{a}_B = -5(\vec{i}\cos\theta + \vec{j}\sin\theta) = -5\vec{i}\ \ (ft/s^2)$. The acceleration of A is $\vec{a}_A = -2\vec{j}\ (ft/s^2)$. The relative acceleration is $\boxed{\vec{a}_{A/B} = \vec{a}_A - \vec{a}_B = 5\vec{i} - 2\vec{j}\ \ (ft/s^2)}$.

==================================<>==================================

====================================<>====================================

Problem 2.179 The velocity of the boat relative to the earth-fixed coordinate system is $40\vec{i}$ (ft/s) and is constant. The length of the tow rope is 50 *ft*. The angle θ is 30° and is increasing at a constant rate of 10° /second. What are the velocity and acceleration of the skier relative to the boat?

Solution The velocity of the skier relative to the boat is

$$\vec{v}_{S/B} = r\left(\frac{d\theta}{dt}\right)\vec{e}_\theta = 50(10)\left(\frac{\pi}{180}\right)\vec{e}_\theta = 8.727\vec{e}_\theta \ ft/s$$

$\vec{v}_{S/B} = 8.727\left(-\vec{i}\sin\theta + \vec{j}\cos\theta\right)$, where $\theta = 210^o$, from which $\boxed{\vec{v}_{S/B} = 4.363\vec{i} - 7.557\vec{j}}$. The acceleration of the skier relative to the boat is

$$\vec{a}_{S/B} = -r\left(\frac{d\theta}{dt}\right)^2\vec{e}_r + r\left(\frac{d^2\theta}{dt^2}\right)\vec{e}_\theta = -50(10)^2\left(\frac{\pi}{180}\right)^2\vec{e}_r = -1.523\vec{e}_\theta \ (ft/s^2),$$

$$\boxed{\vec{a}_{S/B} = -1.523(\vec{i}\cos\theta + \vec{j}\sin\theta) = 1.32\vec{i} + 0.762\vec{j} \ \left(ft/s^2\right)}$$

====================================<>====================================

Problem 2.180 In Problem 2.179, what are the velocity and acceleration of the skier relative to the earth?

Solution Use the solution to Problem 2.179. The velocity of the skier relative to the earth is $\boxed{\vec{v}_{S/E} = \vec{v}_{S/B} + \vec{v}_{B/E} = 40\vec{i} + 4.36\vec{i} - 7.56\vec{j} = 44.36\vec{i} - 7.56\vec{j}}$. The acceleration is $\boxed{\vec{a}_{S/E} = \vec{a}_{S/B} + \vec{a}_{B/E} = \vec{a}_{S/B} + 0 = 1.32\vec{i} + 0.76\vec{j}}$

====================================<>====================================

Problem 2.181 A hockey player is skating with velocity components $v_x = 4\,ft/s$, $v_z = -20\,ft/s$ when he hits a slap shot with a velocity of magnitude $100\,ft/s$ relative to him. The position of the puck when he hits it is $x = 12\,ft$, $z = 12\,ft$. If he hits the puck so that its velocity relative to him is directed toward the center of the goal, where will the puck intersect the *z* axis? Will it enter the $6\,ft$ wide goal?

Solution The strategy is to integrate the equations for the velocity of the puck in rink-fixed coordinates to determine the path of the puck as a function of time. The instantaneous angle of the puck path relative to the player is $\theta = \tan^{-1}\left(\frac{12}{12}\right) = 45^o$. The velocity of the puck relative to the player is $\vec{v}_{P/H} = -V\vec{i}\cos\theta - V\vec{k}\sin\theta = 100\left(-\vec{i}\cos 45^o - \vec{k}\sin 45^o\right) = -70.71\left(\vec{i} + \vec{k}\right)$.

The velocity of the player is $\vec{v}_{H/R} = 4\vec{i} - 20\vec{k}$. The velocity of the puck relative to the rink is $\vec{v}_{P/R} = \vec{v}_{P/H} + \vec{v}_{H/R} = -66.71\vec{i} - 90.71\,ft/s$. The path is $x(t) = -66.71t + 12$. $z(t) = -90.71t + 12$. When the puck crosses the *x* axis, $z(T) = 0$, from which $T = 0.132\ s$, from which $\boxed{x = 3.17\,ft}$. No, the *x*-axis crossing exceeds the half-width of the goal.

====================================<>====================================

=====================================<>=====================================

Problem 2.182 In Problem 2.181, at what point on the x axis should the player aim the puck's velocity vector relative to him so that it enters the goal?

Solution Use the solution to Problem 2.181. The velocity of the puck relative to the player is $\vec{v}_{P/H} = -V(\vec{i}\cos\theta + \vec{k}\sin\theta)$ (ft/s). The velocity of the player relative to the rink is $\vec{v}_{H/R} = 4\vec{i} - 20\vec{k}$ (ft/s). The velocity of the puck relative to the rink is $\vec{v}_{P/R} = \vec{v}_{P/H} + \vec{v}_{H/R} = (4 - V\cos\theta)\vec{i} - (20 + V\sin\theta)\vec{k}$. The path of the puck is $x(t) = (4 - V\cos\theta)t + 12$ ft, $z(t) = -(20 + V\sin\theta)t + 12$. The x axis crossing occurs when $z(T) = 0$, from which $T = \dfrac{12}{20 + 100\sin\theta}$ s, from which $x = \left(\dfrac{(4 - 100\cos\theta)}{20 + 100\sin\theta} + 1\right)12$. *Check:* This should reduce to $x = 3.17$ ft (the solution to Problem 2.181) when $\theta = 45^o$, and it does. *check.* To hit the center of the goal, $x = 0$, from which $(\cos\theta - \sin\theta) = 0.24$. This can be solved by iteration to obtain $\theta = 35.23^o$. The aiming point is the point that the puck would strike if the player were stationary, hence the velocity relative to the player is used. The path relative to the player is $x_{P/H}(t) = -100\cos 35.23^o t + 12$, and $z_{P/H}(t) = -100\sin 35.23^o t + 12$. At $z_{P/H}(T) = 0$, $T = \dfrac{12}{100\sin 35.23^o} = 0.208$ s, from which $\boxed{x_{P/H}(T) = -4.993\ ft}$ is the aiming point.

=====================================<>=====================================

Problem 2.183 An airplane flies in a jet stream flowing east at 100 miles per hour. The airplane's airspeed (its velocity relative to the air) is 500 miles per hour toward the northwest. What are magnitude and direction of airplane's velocity relative to the earth?

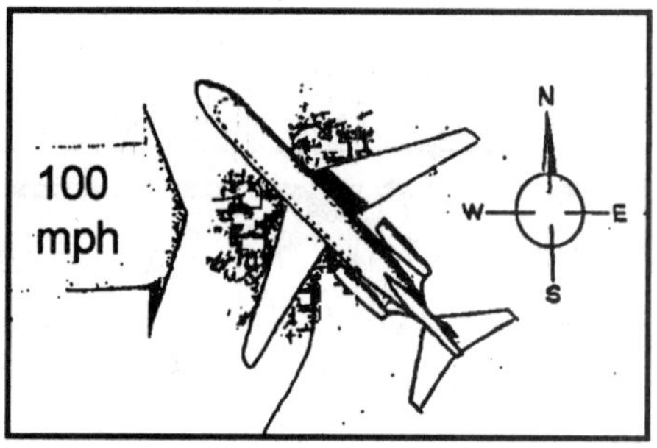

Solution The airplane's relative velocity is $\vec{v}_{A/R} = V(\vec{i}\cos\theta + \vec{j}\sin\theta)$, where $V = 500$ mi/hr, and $\theta = 135^o$. The velocity of the air relative to the ground is $\vec{v}_{W/G} = 100\vec{i}$ (mi/hr). The velocity of the airplane relative to the ground is $\vec{v}_{A/G} = \vec{v}_{A/R} + \vec{v}_{W/G} = (100 + 500\cos\theta)\vec{i} + 500\sin\theta\vec{j} = -253.6\vec{i} + 353.6\vec{j}$. The magnitude is $\boxed{|\vec{v}_{W/G}| = \sqrt{253.6^2 + 353.6^2} = 435.1\ mi/hr}$ The direction is $\beta = \tan^{-1}\left(\dfrac{353.6}{-253.6}\right) + 180^o = 125.65^o$, where 180° must be added because the angle is in the second quadrant. (The compass direction on a four point compass is $125.65^o - 90^o = 35.65^o$ West of North.)

=====================================<>=====================================

Problem 2.184 In Problem 2.183, if the pilot wants to fly toward a city that is northwest of his current position, what direction must he point the airplane, and what will be the magnitude of his velocity relative to the earth?

Solution The velocity relative to the earth: $\vec{v}_{A/G} = \vec{v}_{A/W} + \vec{v}_{W/G}$

$\vec{v}_{A/G} = 500(\vec{i}\cos\theta + \vec{j}\sin\theta) + 100\vec{i} = V(\vec{i}\cos 135^o + \vec{j}\sin 135^o)$, from which $\sin\theta + \cos\theta = -0.2$ Solve: $\boxed{\theta = 143.1^o}$ (53.1° West of North), and $\boxed{V = 424.26\ mi/hr}$

=====================================<>=====================================

====================================<>====================================

Problem 2.185 A river flows north at 3 *m/s*. (Assume that the current is uniform.) If you want to travel in a straight line from point C to point D in a boat that moves a constant speed of 10 *m/s* relative to the water, in what direction do you point the boat? How long does it take to make the crossing?

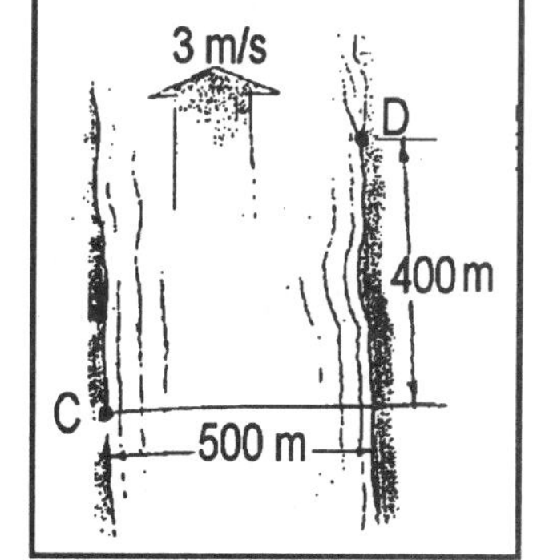

Solution The direction of travel from C to D is $\beta = \tan^{-1}\left(\frac{400}{500}\right) = 38.66^o$ north of east. The velocity relative to the earth is $\vec{v}_{B/E} = \vec{v}_{B/W} + \vec{v}_{W/E}$,

$\vec{v}_{B/E} = V_{B/E}(\vec{i}\cos\beta + \vec{j}\sin\beta) = V_{B/W}(\vec{i}\cos\theta + \vec{j}\sin\theta) + 3\vec{j}\ (m/s)$

Reduce: $V_{B/E}\cos\beta = V_{B/W}\cos\theta$, $V_{B/E}\sin\beta = V_{B/W}\sin\theta + 3$, from which

$\tan\beta = \frac{V_{B/W}\sin\theta + 3}{V_{B/W}\cos\theta}$. Solve: $\boxed{\theta = \beta - \sin^{-1}\left(\frac{3\cos\beta}{10}\right) = 38.66 - \sin^{-1}(0.234) = 25.1^o}$ north of east (64.9° east of north) and $V_{B/E} = V_{B/W}\left(\frac{\cos\theta}{\cos\beta}\right) = 10\left(\frac{0.9054}{0.7809}\right) = 11.6\ m/s$. The distance is

$L = \sqrt{400^2 + 500^2} = 640.3\ m$, and the time of travel is $\boxed{t = \frac{L}{V} = 55.2\ s}$

====================================<>====================================

Problem 2.186 In Problem 2.185, what is the minimum boat speed relative to the water necessary to make the trip from point C to D?

Solution . The strategy is (a) to show that when the boat must follow a given straight line path, the magnitude of the boat velocity relative to the water is a minimum when the boat heading is normal to the desired path, and then (b) to determine the value of this minimum.

From the solution to Problem 2.185, the velocity along the desired path relative to the earth is $\vec{v}_{B/E} = V_{B/E}(\vec{i}\cos\beta + \vec{j}\sin\beta) = V_{B/W}(\vec{i}\cos\theta + j\sin\theta) + 3\vec{j}$. Equate the components, and take the ratio to eliminate $V_{B/E}$ to obtain $\tan\beta = \frac{V_{B/W}\sin\theta + 3}{V_{B/W}\cos\theta}$. Solve: $V_{B/W} = \frac{3\cos\beta}{\sin(\beta - \theta)}$. The minimum occurs when $\frac{dV_{B/W}}{d\theta} = \frac{\cos(\beta-\theta)}{\sin^2(\beta-\theta)}(3\cos\beta) = 0$. Since β is a constant angle along the desired path, this minimum is satisfied when $\cos(\beta - \theta) = 0$, or $\beta - \theta = \pm\frac{\pi}{2}$. From physical considerations, the boat will cross only for an easterly heading, hence the positive sign is applicable, and

$\theta = \beta - \frac{\pi}{2} = 38.66^o - 90^o = -51.34^o$, *which is perpendicular to the path.* That this is indeed a minimum:

$$\left[\frac{d^2V_{B/W}}{d\theta^2}\right]_{\beta-\theta=\frac{\pi}{2}} = \left[-\frac{6\cos\beta\cos^2(\beta-\theta)}{\sin^3(\beta-\theta)} + \frac{3\cos\beta}{\sin(\beta-\theta)}\right]_{\beta-\theta=\frac{\pi}{2}} = 3\cos\beta > 0$$

The value of the minimum for this heading is $\boxed{V_{B/W} = \left[\frac{3\cos\beta}{\sin(\beta-\theta)}\right]_{\beta-\theta=\frac{\pi}{2}} = 3\cos\beta = 2.34\ m/s}$

====================================<>====================================

Problem 2.187 Relative to the earth, a sailboat sails north at velocity v_o and then sails east at the same velocity. The velocity of the wind is uniform and constant. A "telltale" on the boat points in the direction of the velocity of the wind relative to the boat. What are the direction and magnitude of the wind's velocity relative to the earth? Your answer for the magnitude velocity of the wind will be in terms of v_o.

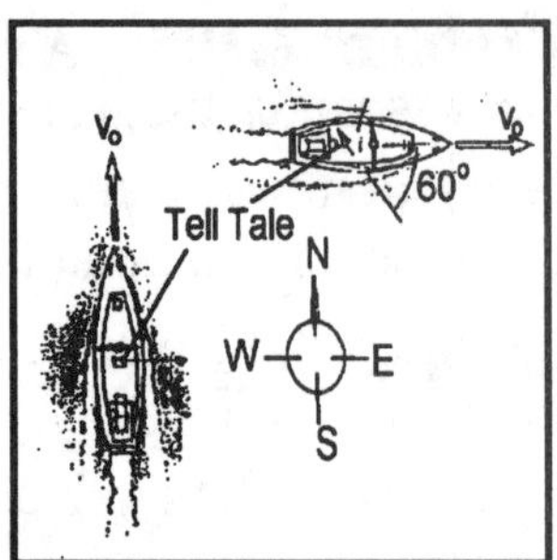

Solution The relative wind diagrams are shown for the two legs of the path.

On each leg, $|\vec{v}_B| = v_o$. From geometry and the two legs, the horizontal component of the wind's velocity is:

$v_{W_x} = v_o - v_o \cot 60^o = 0.423 v_o$, Solve the right triangle with base v_{Wx}, altitude v_o, and hypotenuse v_W,

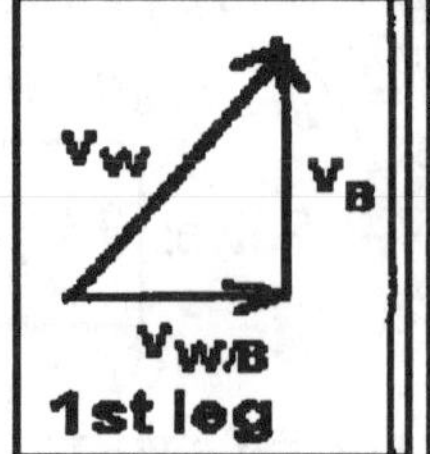

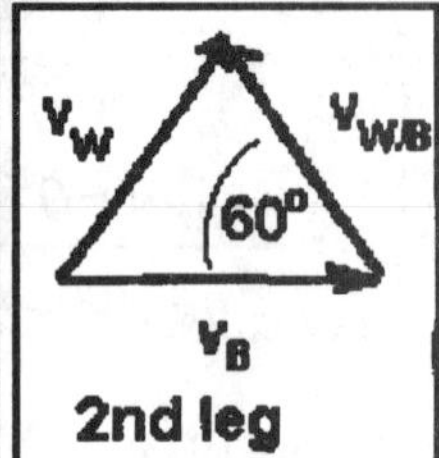

$|\vec{v}_W| = \sqrt{(0.423 v_o)^2 + v_o^2} = 1.086 v_o$. The angle of the wind is found from the first leg: $\beta = \sin^{-1}\left(\frac{0.423 v_o}{1.086 v_o}\right) = 22.9^o$ east of north.

Problem 2.188 The origin O of the non-rotating coordinate system is at the center of the earth and the y axis points north. The satellite A on the x axis is in a circular polar orbit of radius R and its velocity is $v_A \vec{j}$. Let ω be the angular velocity of the earth. What is the satellite's velocity relative to the point B on the earth directly under the satellite?

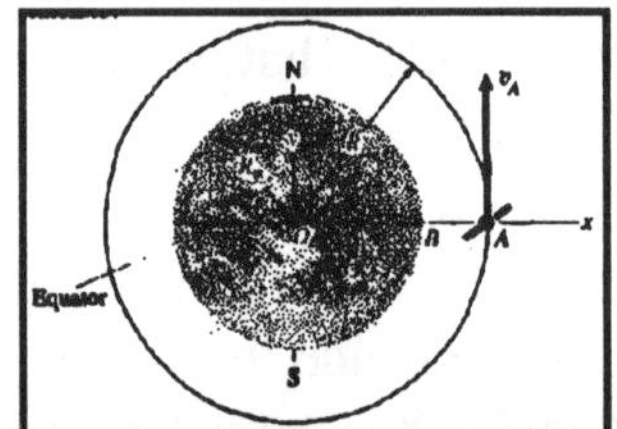

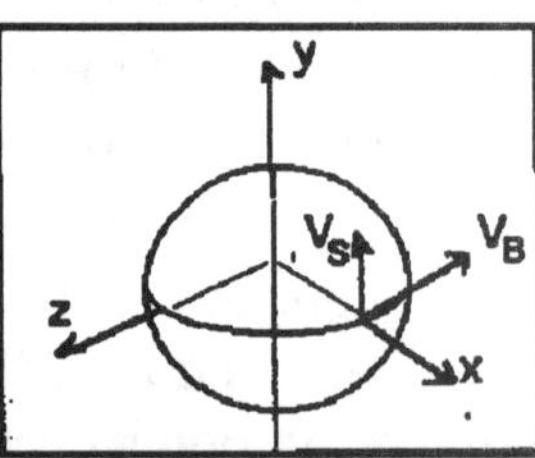

Solution The velocity of the satellite given in the problem is $\vec{v}_S = v_A \vec{j}$. The velocity of the point B is $\vec{v}_B = -R_E \omega \vec{k}$, since it is along the negative z axis. The relative velocity of the satellite is $\vec{v}_{S/B} = \vec{v}_S - \vec{v}_B$, from which $\vec{v}_{S/B} = v_A \vec{j} + R_E \omega \vec{k}$

Problem 2.189 In Problem 2.188, what is the satellite's acceleration relative to the point B on the earth directly below the satellite?

Solution The relative acceleration is $\vec{a}_{S/B} = \vec{a}_S - \vec{a}_B$. The acceleration of the satellite is

$\vec{a}_S = -\omega^2 R \vec{i} = -\frac{v_A^2}{R}\vec{i}$. The acceleration of B is $\vec{a}_B = -R_E \omega^2 \vec{i}$, from which $\vec{a}_{S/B} = \vec{i}\left(\omega^2 R_E - \frac{v_A^2}{R}\right)$

==================================<>==================================

Problem 2.190 An engineer analyzing a machining process determines that from $t = 0$ to $t = 4$ seconds, the workpiece starts from rest and moves in a straight line with acceleration $a = 2 + t^{0.5} - t^{1.5}\ ft/s^2$. (a) Draw a graph of the position of the workpiece relative to its position at $t = 0$ for values of time from $t = 0$ to $t = 4$ seconds. (b) Estimate the maximum velocity during this time interval and the time at which it occurs.

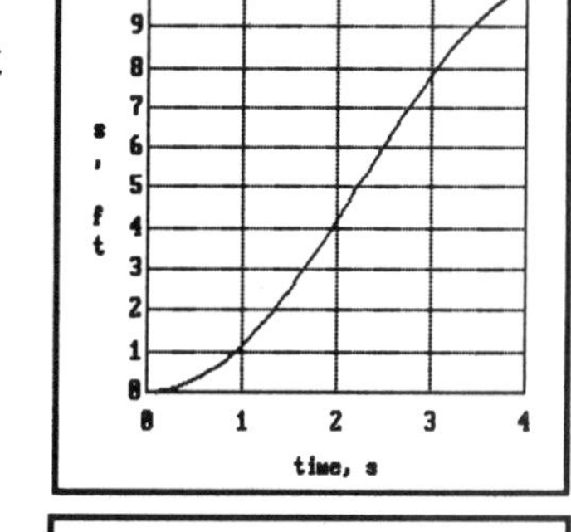

Solution The velocity is:

$v(t) = \int (2 + t^{0.5} - t^{1.5})dt = 2t + \left(\frac{2}{3}\right)t^{1.5} - \left(\frac{2}{5}\right)t^{2.5}\ ft/s$, where the constant of integration is zero because the system starts from rest. (a) The position:

$s(t) = \int v(t)dt = t^2 + \left(\frac{4}{15}\right)t^{2.5} - \left(\frac{4}{35}\right)t^{3.5}\ ft$. A graph of the position over the interval $0 \le t \le 4$ seconds is shown.

(b) The maximum velocity occurs at $\frac{dv(t)}{dt} = 0 = 2 + t^{0.5} - t^{1.5} = 0$. A graph of the zero crossing of $f(a) = 2 + t^{0.5} - t^{1.5}$ is shown. The graph crosses the zero axis at approximately $t = 2.31\ s$. (This was refined by iteration to $t = 2.3146\ s$). The value of the velocity at this maximum is

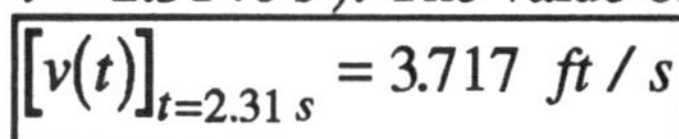

$$[v(t)]_{t=2.31\ s} = 3.717\ ft/s$$

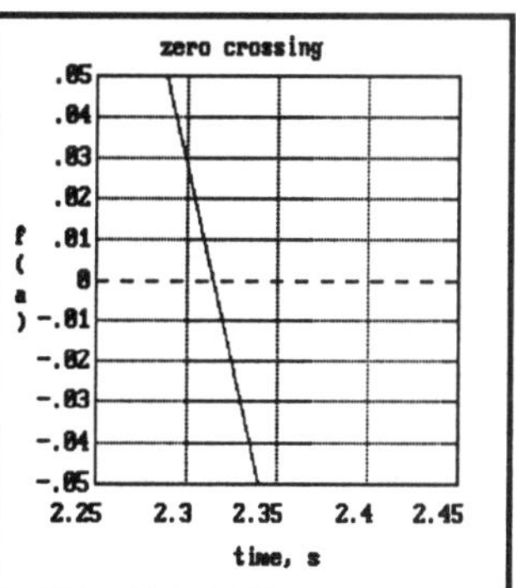

==================================<>==================================

================================<>================================

Problem 2.191 In Problem 2.78, determine the range of angles determine the range of angles θ for which the pitched must release the ball the ball to hit the strike zone.

Solution Problem 2.78 statement is: A baseball pitcher releases a fastball with an initial velocity $v_o = 90\ mi/hr$. Let θ be the initial angle of the ball's velocity vector above the horizontal. When it is released, the ball is 6 *ft* above the ground and 58 *ft* from the batter's plate. The batter's strike zone (between his knees and shoulders) extends from 1 *ft* 10 *in.* above the ground to 4 *ft* 6 *in.* above the ground. Neglect aerodynamic effects.

Use the solution to Problem 2.78. The initial velocity is $v_o = 90\ mi/hr = 132\ ft/s$. The velocity equations are (1) $\frac{dv_x}{dt} = 0$, from which $v_x = v_o \cos\theta$. (2) $\frac{dv_y}{dt} = -g$, from which $v_y = -gt + v_o \sin\theta$. (3) $\frac{dx}{dt} = v_o \cos\theta$, from which $x(t) = v_o \cos\theta\, t$, since the initial position is zero. (4) $\frac{dy}{dt} = -gt + v_o \sin\theta$, from which $y(t) = -\frac{g}{2}t^2 + v_o \sin\theta\, t + 6$, since the initial position is $y(0) = 6\ ft$. At a distance $d = 58\ ft$, the height is *h*. The time of passage across the home plate is $x(t_p) = d = v_o \cos\theta\, t_p$, from which $t_p = \frac{d}{v_o \cos\theta}$. Substitute: $y(t_p) = h = -\frac{g}{2}\left(\frac{d}{v_o \cos\theta}\right)^2 + d\tan\theta + 6$.. A graph of *h* against θ is shown; a rough estimate of the angle range was obtained from this graph, and then improved by iteration. The result

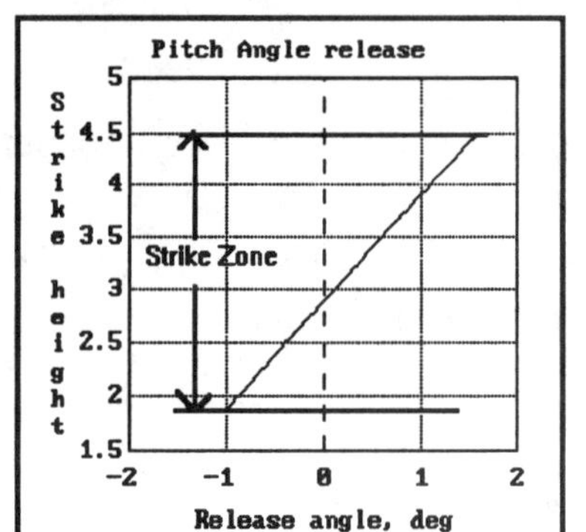

$$\boxed{-1.047^o \le \theta \le 1.588^o}$$

================================<>================================

Problem 2.192 A catapult designed to throw a line to ships in distress throws a projectile with initial velocity $v_o(1 - 0.4\sin\theta_o)$, where θ_o is the angle above the horizontal. Determine the value of θ_o. for which the distance the projectile is thrown is a maximum, and show that the maximum distance is $\frac{0.559 v_o^2}{g}$.

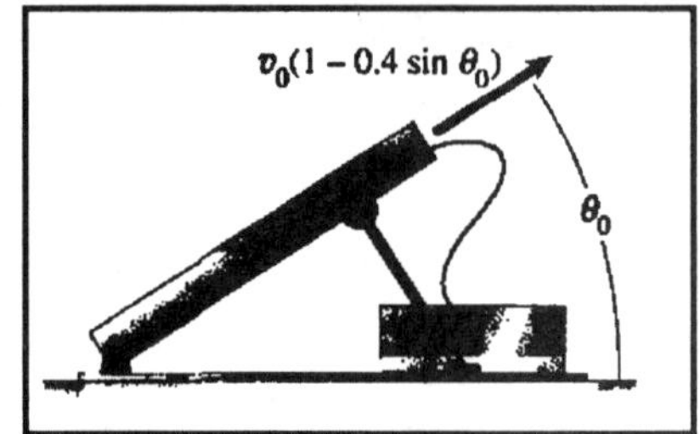

Solution The path is obtained by integrating the equations of motion: $\frac{dv_y}{dt} = -g$, and $\frac{dv_x}{dt} = 0$, from which $v_y(t) = -gt + V_o \sin\theta_o$, and $v_x(t) = V_o \cos\theta_o$. $y(t) = -\frac{g}{2}t^2 + (V_o \sin\theta_o)t$, and $x(t) = (V_o \cos\theta_o)t$. At the end of flight, $y(t_{flight}) = 0$, from which $t_{flight} = \frac{2V_o \sin\theta_o}{g}$. The range is $R = x(t_{flight}) = \frac{2V_o^2 \sin\theta_o \cos\theta_o}{g}$. Noting $V_o = v_o(1 - 0.4\sin\theta_o)$, the maximum range occurs when

$$\frac{dR}{d\theta_o} = 0 = \frac{d}{d\theta_o}\left(\frac{2V_o^2 \sin\theta_o \cos\theta_o}{g}\right) = 0, \text{ from which:}$$

Solution continued on next page

==============================<>==============================

Continuation Solution to Problem 2.192

after some algebraic reduction, $\frac{dR}{d\theta_o} = \left(\frac{-0.8\cos^2\theta\sin\theta}{(1-0.4\sin\theta_o)}\right) + \cos(2\theta) = 0$,

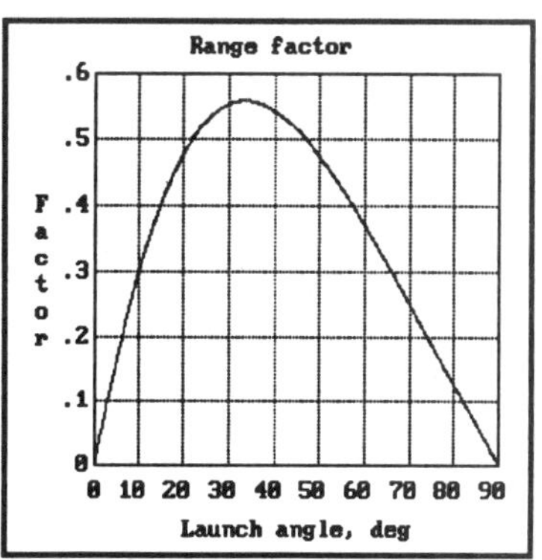

from which $\boxed{\theta_o = 33.41^o}$. This is confirmed as a maximum by graphing $R\left(\frac{g}{v_o^2}\right)$ as shown. The maximum range is

$$\boxed{R_{max} = \left[\frac{v_o^2(1-0.4\sin\theta_o)^2\sin\theta_o\cos\theta_o}{g}\right]_{\theta_o=33.41^o} = 0.5589\frac{v_o^2}{g}}$$

==============================<>==============================

Problem 2.193 At $t = 0$, a projectile is located at the origin and has a velocity of $20\ m/s$ at 40° above the horizontal. The profile of the ground surface it strikes can be approximated by the equation $y = 0.4x - 0.006x^2\ m$, where x and y are in meters. Determine the approximate coordinates of the point where it hits the ground.

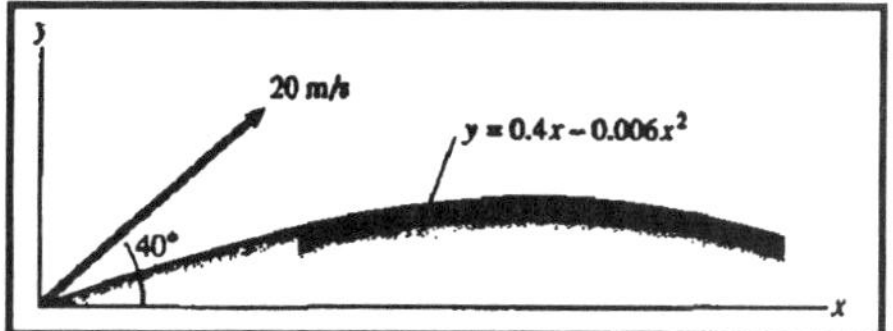

Solution The path of the projectile is obtained by integrating the equations of motion: $\frac{d^2y}{dt^2} = -g$ and $\frac{d^2x}{dt^2} = 0$ using the initial conditions. The result: $y(t) = -\frac{g}{2}t^2 + (V_o\sin\theta)t\ m$, and $x(t) = (V_o\cos\theta)t\ m$.

At impact, $y(t_{flight}) = y_{impact} = 0.4x_{impact} - 0.006x_{impact}^2$. Solve for t_{flight}:

$t_{flight} = \frac{V_o\sin\theta}{g}\left(1 \pm \sqrt{1-\frac{2gy_{impact}}{V_o^2\sin^2\theta}}\right)$. Substitute: $x_{impact} = \frac{V_o^2\cos\theta\sin\theta}{g}\left(1 \pm \sqrt{1-\frac{2gy_{impact}}{V_o^2\sin^2\theta}}\right)$. The two functions $f(x_{impact}) = \frac{V_o^2\cos\theta\sin\theta}{g}\left(1 \pm \sqrt{1-\frac{2g(0.4x_{impact}-0.006x_{impact}^2)}{V_o^2\sin^2\theta}}\right) - x_{impact}$ were graphed against values of x_{impact} to determine the zero crossings. Only one crossing is of interest (the other was $x_{impact} = 0$), and this value was refined by iteration (using **TK Solver Plus**) to yield $\boxed{x_{impact} = 29.477\ m}$ from which $\boxed{y_{impact} = 6.577\ m}$

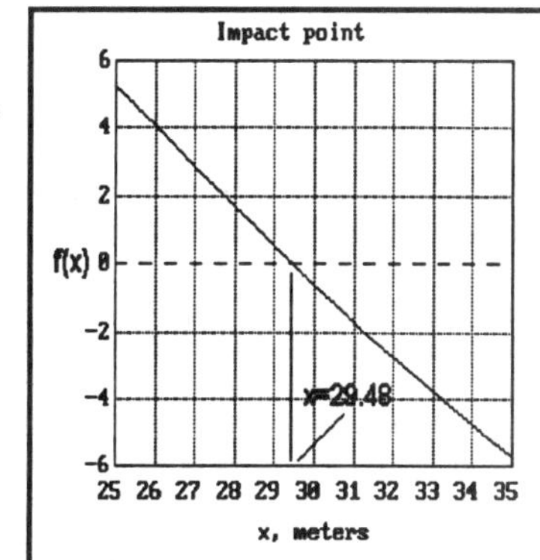

==============================<>==============================

Problem 2.194 A carpenter working on a house asks his apprentice to throw him an apple. The apple is thrown at 32 *ft/s*. What two values of θ_o will cause the apple to land in the carpenter's hand, 12 *ft* horizontal and 12 *ft* vertically from the point where it is thrown?

Solution The path obtained from the equations of motion is given by $y(t) = -\frac{g}{2}t^2 + (V_o \sin\theta_o)t$, and $x(t) = (V_o \cos\theta_o)t$. When the apple reaches the hand $y(t_{flight}) = -\frac{g}{2}t_{flight}^2 + (V_o \sin\theta_o)t_{flight} = 12 \text{ ft}$. Solve for the time of flight: $t_{flight} = \frac{V_o \sin\theta_o}{g}\left(1 \pm \sqrt{1 + \frac{24g}{(V_o \sin\theta_o)^2}}\right)$, from which

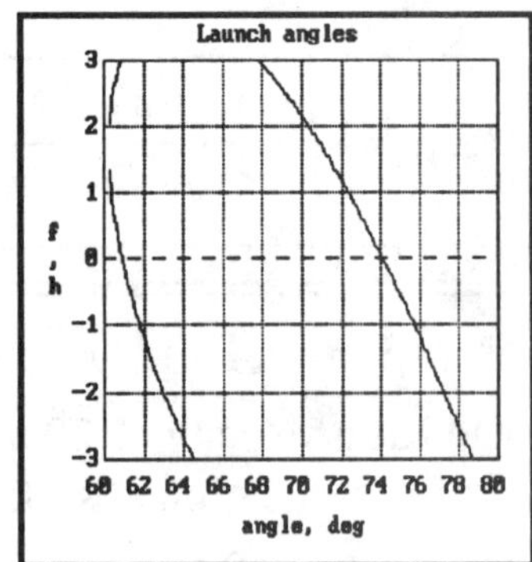

$$x(t_{flight}) = 12 = \frac{V_o^2 \sin\theta_o \cos\theta_o}{g}\left(1 \pm \sqrt{1 - \frac{24g}{(V_o \sin\theta_o)^2}}\right)$$ The two functions

$$f(\theta_o) = \frac{V_o^2 \sin\theta_o \cos\theta_o}{g}\left(1 \pm \sqrt{1 - \frac{24g}{(V_o \sin\theta_o)^2}}\right) - 12$$ were graphed to determine the zero crossings and the results were refined by iteration (using **TK Solver Plus**) to obtain $\theta_o = 60.9^o$ and $\theta_o = 74.08^o$

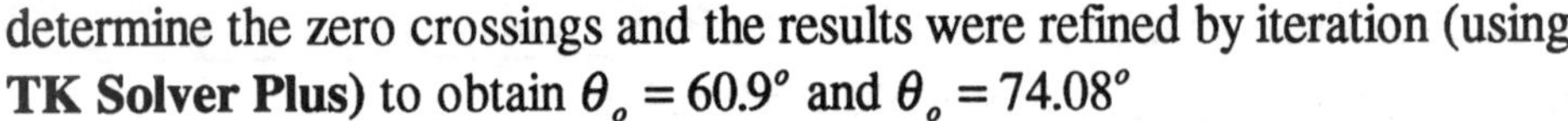

Problem 2.195 A motorcycle starts from rest at $t = 0$ and moves along a circular track with a 400 *m* radius. The tangential component of its acceleration is $a_t = 2 + 0.2t \text{ m/s}^2$. When the magnitude of its total acceleration reaches 6 *m/s²*, friction can no longer keep it on the circular track, and it spins out. How long after it starts does it spin out, and how fast is it going?

Solution The tangential component of the acceleration can be integrated independently of the radial acceleration to determine the angular velocity:

$$a_\theta = r\frac{d^2\theta}{dt^2} + 2\left(\frac{dr}{dt}\right)\left(\frac{d\theta}{dt}\right) = r\frac{d^2\theta}{dt^2} = 2 + 0.2t \text{ m/s}^2,$$ since the radius is a constant.

From which $r\omega = 2t + 0.1t^2 \text{ m/s}$. The radial acceleration is

$$a_r = \frac{d^2r}{dt^2} - r\omega^2 = -\left(\frac{1}{r}\right)(2t + 0.1t^2)^2 \text{ m/s}^2$$. The magnitude of the acceleration is

$$|\vec{a}| = \sqrt{(2 + 0.2t)^2 + \left(\frac{(2t + 0.1t^2)^2}{r}\right)^2}$$, from which, at time of spin out: $36 = (2 + 0.2t)^2 + \left(\frac{(2t + 0.1t^2)^2}{r}\right)^2$.

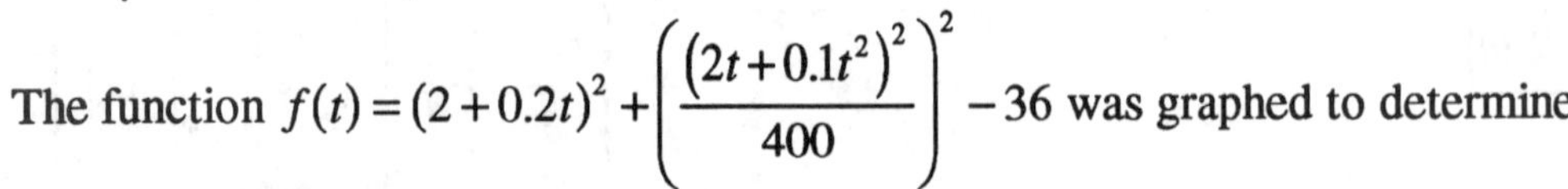

The function $f(t) = (2 + 0.2t)^2 + \left(\frac{(2t + 0.1t^2)^2}{400}\right)^2 - 36$ was graphed to determine the zero crossing and the result refined by iteration.

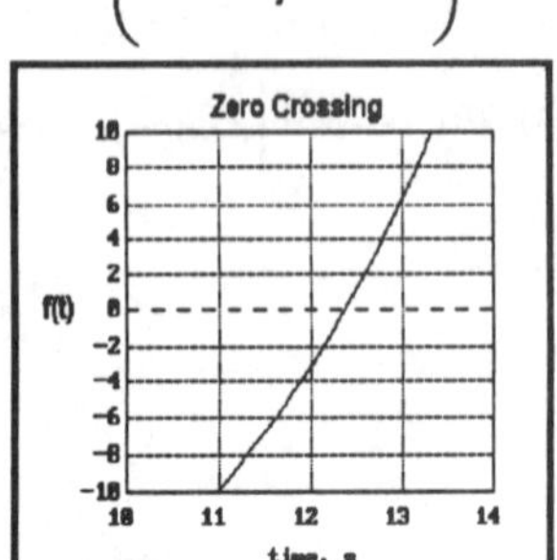

The result: $t = 12.36$ s (A zero crossing also occurs at a negative time; it is ignored.)

The velocity at spin out is $\boxed{V = [r\omega]_{t=12.36\,s} = \left[2t + 0.1t^2\right]_{t=12.36\,s} = 40 \, m/s}$

$$V = 40\left(\frac{m}{s}\right)\left(\frac{1 \, km}{10^3 \, m}\right)\left(\frac{3600 \, s}{1 \, hr}\right) = 144 \, km/hr$$

==================================<>==================================

Problem 2.196 At $t = 0$ a steel ball in a tank of oil is given a horizontal velocity $\vec{v} = 2\vec{i}$ m/s. The components of its acceleration are $a_x = -cv_x$, $a_y = -0.8g - cv_y$, $a_z = -cv_z$, where c is a constant. When the ball hits the bottom of the tank, its position relative to its position at $t = 0$ is $\vec{r} = 0.8\vec{i} - \vec{j}$ m. What is the value of c?

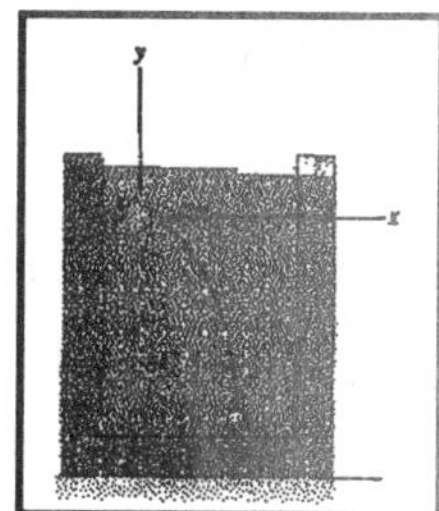

Solution Choose a coordinate system with the origin at O and the x axis parallel to the plane surface.The path is obtained by integrating the equations of motion: $\frac{dv_x}{dt} = -cv_x$, $\frac{dv_y}{dt} = -0.8g - cv_y$, and $\frac{dv_z}{dt} = -cv_z$. Separating variables and integrating: $v_x(t) = C_{vx}e^{-ct}$, $v_y(t) = -\frac{0.8g}{c} + C_{vy}e^{-ct}$, and $v_z(t) = C_{vz}e^{-ct}$. From the initial conditions, $C_{vx} = 2$, $C_{vy} = \frac{0.8g}{c}$, $C_{vz} = 0$, from which $v_x = 2e^{-ct}$, $v_y(t) = \frac{0.8g}{c}(e^{-ct} - 1)$, and $v_z(t) = 0$. Integrating: $x(t) = -\frac{2}{c}e^{-ct} + C_x$, $y(t) = -\frac{0.8g}{c}\left(\frac{1}{c}e^{-ct} + t\right) + C_y$, $z(t) = 0$. From the initial conditions, $x(0) = 0$, $y(0) = 0$, from which $C_x = \frac{2}{c}$, $C_y = \frac{0.8g}{c^2}$, from which $x(t) = \frac{2}{c}(1 - e^{-ct})$ m, $y(t) = \frac{0.8g}{c^2}(1 - e^{-ct}) - \frac{0.8g}{c}t$ m, and $z(t) = 0$. When the ball strikes the bottom $y(t) = -1$, $x(t) = 0.8$. From the equation for x, at time of impact, $e^{-ct} = 1 - 0.4c$, or $ct = -\ln(1 - 0.4c)$. Substitute these into the equation for y to obtain $0 = (0.8g)(0.4c + \ln(1 - 0.4c)) + c^2$. A graph of $f(c) = 0.8g(0.4c + \ln(1 - 0.4c)) + c^2$ was used to find the zero crossing, and the result refined by iteration to obtain $c = 1.313\ s^{-1}$

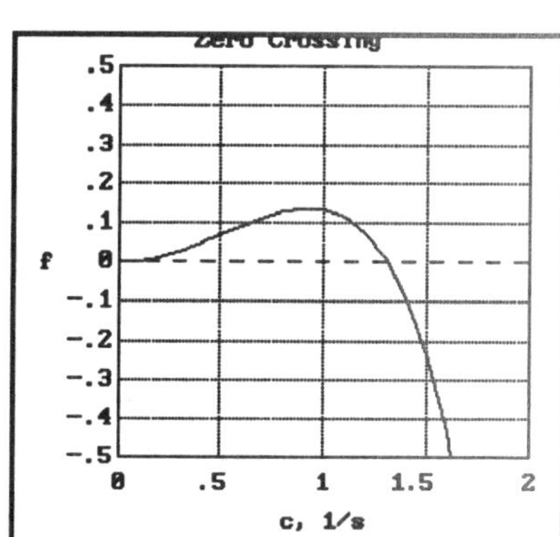

==================================<>==================================

Problem 2.197 The polar coordinates of a point P moving in the x-y plane are $r = t^3 - 4t$ meters, $\theta = t^2 - t$ radians. (a) Draw a graph of the magnitude of the velocity of P from $t = 0$ to $t = 2$ seconds. (b) Estimate the minimum magnitude of the velocity and the time at which it occurs.

Solution (a) The velocity is $\vec{v} = \frac{dr}{dt}\vec{e}_r + r\left(\frac{d\theta}{dt}\right)\vec{e}_\theta = (3t^2 - 4)\vec{e}_r + (t^3 - 4t)(2t - 1)\vec{e}_\theta$. The magnitude is $|\vec{v}_P| = \sqrt{(3t^2 - 4)^2 + (t^3 - 4t)^2(2t - 1)^2}$ m/s. The graph of the magnitude is shown. (b) The minimum magnitude is found from a search of the tabulated values of the magnitude: $|\vec{v}_P| = 2.63$ m/s at $t = 0.79$ s

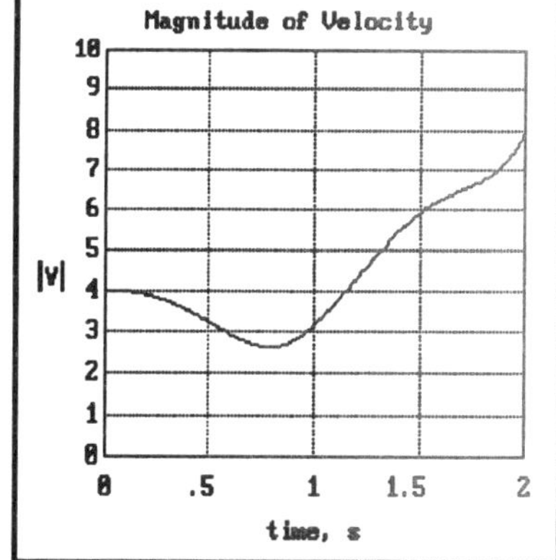

Note: **Mathcad, TK Solver** and similar programs have built-in functions for automatically determining the max or min of an array. For the users of **TK Solver Plus:** After the construction of the table (lists) of v and t, add the following rules to the Rule Sheet: (a) **vmin=MIN('v)** (find the minimum value in the list v); (b) **count=MEMBER('v, vmin)** (find the location or index number of vmin); (c) **time=ELT('t, count)** (find the time associated with that index number); (d) **Solve** (F9 key). This can be extended to any number of table lists. *End of Note.*

==================================<>==================================

Problem 2.198 (a) Draw a graph of the magnitude of the acceleration of the point P in Problem 2.197 from $t = 0$ to $t = 2$ seconds.

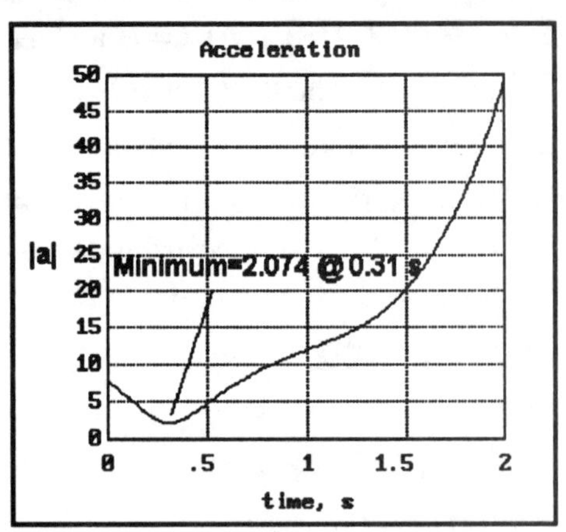

Solution The acceleration is

$$\vec{a} = \left(\frac{d^2r}{dt^2} - r\left(\frac{d\theta}{dt}\right)^2\right)\vec{e}_r + \left(r\left(\frac{d^2\theta}{dt^2}\right) + 2\left(\frac{dr}{dt}\right)\left(\frac{d\theta}{dt}\right)\right)\vec{e}_\theta.$$ From Problem 2.183, $r = (t^3 - 4t)$, $\frac{dr}{dt} = 3t^2 - 4$, $\frac{d^2r}{dt^2} = 6t$; $\theta = t^2 - t$, $\frac{d\theta}{dt} = 2t - 1$, $\frac{d^2\theta}{dt^2} = 2$. Substitute: $a_r = 6t - (2t-1)^2$, $a_\theta = (t^3 - 4t)(2) + 2(3t^2 - 4)(2t - 1)$. The magnitude is

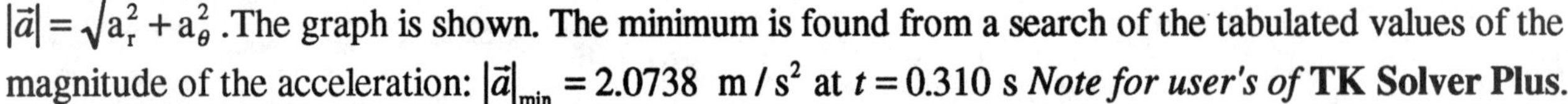

$|\vec{a}| = \sqrt{a_r^2 + a_\theta^2}$. The graph is shown. The minimum is found from a search of the tabulated values of the magnitude of the acceleration: $|\vec{a}|_{min} = 2.0738 \ \text{m/s}^2$ at $t = 0.310$ s *Note for user's of* **TK Solver Plus.**

Problem 2.199 The robot is programmed so that the point P describes the path: $r = 1 - 0.5\cos(2\pi t)$ meters, $\theta = 0.5 - 0.2\sin[2\pi(t - 0.1)]$ radians. Determine the values of r and θ at which the magnitude of the velocity of P attains its maximum value.

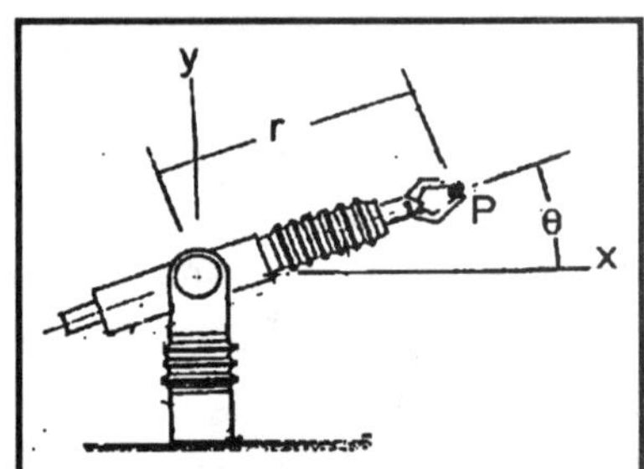

Solution The velocity is $\vec{v}_P = \frac{dr}{dt}\vec{e}_r + r\left(\frac{d\theta}{dt}\right)\vec{e}_\theta$

$\frac{dr}{dt} = \pi\sin(2\pi t)$ m, $\frac{d\theta}{dt} = -0.4\pi\cos[2\pi(t - 0.1)]$

The magnitude of the velocity is

$|\vec{v}_P| = \sqrt{(\pi\sin(2\pi t))^2 + (1 - 0.5\cos(2\pi t))^2(-0.4\pi\cos[2\pi(t - 0.1)])^2}$ The maximum value was found from a search of the tabulated values: $|\vec{v}_P|_{max} = 3.248 \ \text{m/s}$. The corresponding values of r and θ are $r = 1.066$ m and $\theta = 37^o$

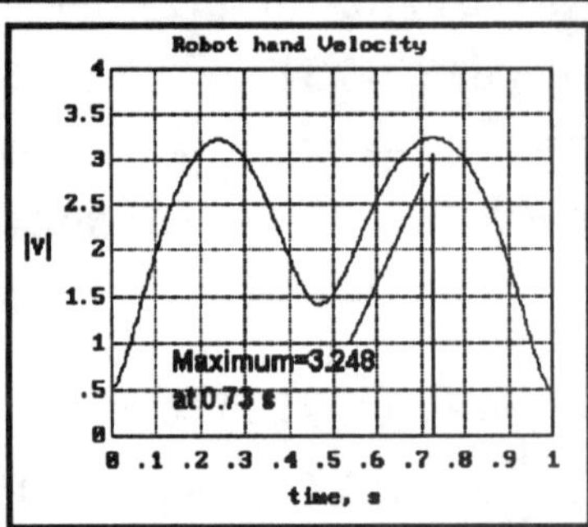

==================================<>==================================

Problem 2.200 In Problem 2.199, determine the values of r and θ at which the magnitude of the acceleration of P attains its maximum value.

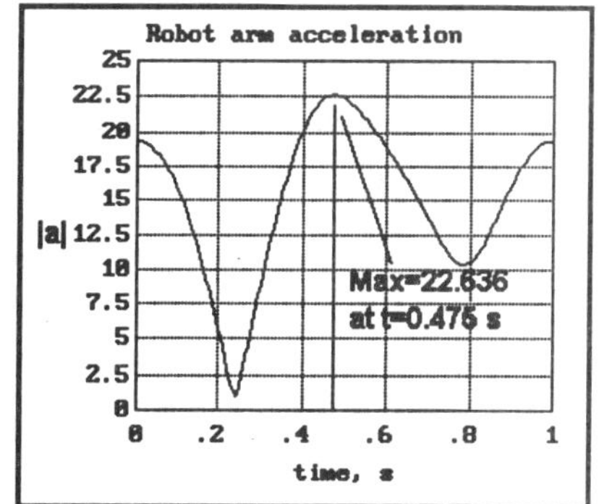

Solution The acceleration components are: $a_r = \frac{d^2r}{dt^2} - r\left(\frac{d\theta}{dt}\right)^2$, from which:

$a_r = 2\pi^2\cos(2\pi t) - (1 - 0.5\cos(2\pi t))(-0.4\pi\cos(2\pi(t - 0.1)))^2$

$a_\theta = r\left(\frac{d^2\theta}{dt^2}\right) + 2\left(\frac{dr}{dt}\right)\left(\frac{d\theta}{dt}\right)$, from which:

$a_\theta = (1 - 0.5\cos(2\pi t))(0.8\pi^2\sin(2\pi(t - 0.1)) + 2(\pi\sin(2\pi t))(-0.4\pi\cos(2\pi(t - 0.1)))$. The magnitude is $|\vec{a}_p| = \sqrt{a_r^2 + a_\theta^2}$. Although not required by the problem, a graph of the acceleration as a function of time is shown. The magnitude is found by a search of the tabulated values: $|\vec{a}_p|_{max} = 22.64 \text{ m/s}^2$. The values of the radius and angle are $r = 1.494$ m and $\theta = 20.5^o$.

==================================<>==================================

Problem 2.201 In the cam follower mechanism, the slotted bar rotates with a constant angular velocity $\omega = 10 rad/s$, and the radial position of the follower A is determined by the profile of the stationary cam. The path of the follower is described by the ploar equation $r = 1 + 0.5\cos(2\theta)$.

(a) Draw a graph of the magnitude of the follower's acceleration as a function of θ for $0 \le \theta \le 360°$.

(b) Use your graph to estimate the maximum magnitude of the follower's acceleration and the angle(s) at which it occurs.

Solution: Refer to the solutions of Problems 2.157 and 2.158. In these problems, the acceleration components for the follower were derived as functions of the angle θ. We proceed as in Example 2.15.

(a) We find the value for $v_r = \dot{r} = \frac{dr}{dt}$ by $\frac{dr}{dt} = \frac{dr}{d\theta}\frac{d\theta}{dt} = \frac{dr}{d\theta}\omega = -0.5(2)\sin(2\theta)\omega = -\omega\sin(2\theta)\int$. Note that $\omega = \dot{\theta} = constant$ and $\ddot{\theta} = 0$. Hence, $\mathbf{v} = v_r\mathbf{e}_r + r\omega\,\mathbf{e}_\theta = -\omega\sin(2\theta)\mathbf{e}_r + r\omega\,\mathbf{e}_\theta$ In order to determine the components of the acceleration in polar coordinates, we need to be able to determine all of the variables in the right hand sides of $a_r = \ddot{r} - r\dot{\theta}^2$ and that $a_\theta = r\ddot{\theta} + 2\dot{r}\dot{\theta}$. We already know everything except $\ddot{r}$ and $\ddot{\theta}$. Since ω is constant, $\ddot{\theta} = \dot{\omega} = 0$. We need only to find $\ddot{r}$. To find $\ddot{r}$, we start with $\dot{r} = v_r$. Taking a derivative, we get $\ddot{r} = \frac{d}{dt}(v_r) = \frac{d}{dt}\left[-\sin(2\theta)\dot{\theta}\right]$, or

or $\ddot{r} = \frac{d}{d\theta}\left[v_r\right]\frac{d\theta}{dt} = \frac{d}{d\theta}\left[-\sin(2\theta)\omega\right]\omega = -2\cos(2\theta)\omega^2$.

We can now solve for the accelerations and the magnitude is given as $|\mathbf{a}| = \sqrt{a_r^2 + a_\theta^2}$. Using an automatic numerical solver to plot the results, we get the graph at the right.

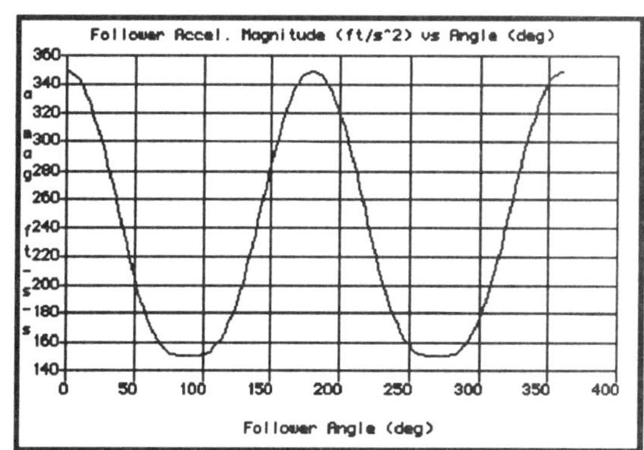

(b) From the graph, the maximum magnitudes of acceleration occur at $\theta = 0°$ and $\theta = 180°$, and the value for the maximum is $|a|_{MAX} \cong 350 ft/s^2$

==================================<>==================================

==================================<>==================================

==================================<>==================================

Problem 2.202 Suppose you throw a ball straight up at *10 m/s* and release it *2 m* above the ground. (a) What maximum height above the ground does it reach? (b) How long after release does it hit the ground? (c) What is the magnitude of its velocity just before it hits the ground?

Solution: The equations of motion for the ball are $a_y = -g = -9.81 m/s^2$, $v_y = v_{y0} - gt = 10 - 9.81t\ (m/s)$, and $y = y_0 + v_{y0}t - gt^2/2 = 2 + 10t - 9.81t^2/2\ (m)$.

(a) The maximum height occurs when the velocity is zero. Call this time $t = t_1$. It can be obtained by setting velocity to zero, i.e., $v_y = 0 = 10 - 9.81t_1\ (m/s)$. Solving, we get $t_1 = 1.02s$. Substititing this time into the y equation, we get a maximum height of $y_{MAX} = 7.10m$.

(b) The ball hits the ground when $y = 0m$. To find out when this occurs, we set $y = 0m$ into the y equation and solve for the time(s) when this occurs. There will be two times, one positive and one negative. Only the positive time has meaning for us. Let this time be $t = t_2$. The equation for t_2 is $y = 0 = 2 + 10t_2 - 9.81t_2{}^2/2\ (m)$. Solving , we get $t_2 = 2.22s$.

(c) The velocity at impact is determined by substituting $t_2 = 2.22s$ into the equation for v_y. Doing this, we find that at impact, $v_y = -11.8m/s$

==================================<>==================================

Problem 2.203 Suppose that you must determine the duration of the yellow light at a highway intersection. Assume that cars will be approaching the intersection traveling as fast as 65 *mi/hr*, that the driver's reaction times are as long as 0.5 seconds, and that cars can safely achieve a deceleration of at least 0.4*g*. (a) How long must the yellow light be on to allow drivers to safely come to a stop before the light turns red? (b) What is the minimum distance cars must be from the intersection when the light turns yellow to safely come to a stop?

Solution The speed-time equation from initial speed to stop is given by integrating the equation $\frac{d^2s}{dt^2} = -0.4g$. From which $\frac{ds}{dt} = -0.4gt + V_o$, and $s(t) = -0.2gt^2 + V_o t$, where V_o is the initial speed and the distance is referenced from the point where the brakes are applied. The initial speed is: $V_o\left(\frac{65 \text{ mi}}{1 \text{ hr}}\right)\left(\frac{5280 \text{ ft}}{1 \text{ mi}}\right)\left(\frac{1 \text{ hr}}{3600 \text{ s}}\right) = 95.33 \text{ ft/s}$. (a)The time required to come to a full stop $\frac{ds(t_o)}{dt} = 0$ is $t_o = \frac{V_o}{0.4g} = \frac{95.33}{(0.4)(32.17)} = 7.40 \text{ s}$. The driver's reaction time increases this by 0.5 second, hence the total time to stop after observing the yellow light is $T = t_o + 0.5 = 7.90$ s (b) The distance traveled after brake application is traveled from brake application to full stop is given by $s(t_o) = -0.2gt_o^2 + V_o t_o$, from which $s(t_o) = 353.14$ ft. The distance traveled during the reaction time is $d = V_o(0.5) = 95.33(0.5) = 47.66$ ft, from which the total distance is $d_t = 353.14 + 47.66 = 400.8$ ft

==================================<>==================================

==================================<>==================================

Problem 2.204 The acceleration of a point moving along a straight line is $a = 4t + 2 \ \text{m/s}^2$. When $t = 2$ seconds, its position is $s = 36$ meters, and when $t = 4$ seconds, its position is $s = 90$ meters. What is its velocity when $t = 4$ seconds?

Solution The position-time equation is given by integrating $\frac{d^2s}{dt^2} = 4t + 2$, from which $\frac{ds}{dt} = 2t^2 + 2t + V_o$, and $s(t) = \left(\frac{2}{3}\right)t^3 + t^2 + V_o t + d_o$, where V_o, d_o are the initial velocity and position. From the problem conditions: $s(2) = \left(\frac{2}{3}\right)2^3 + (2^2) + V_o(2) + d_o = 36$, from which (1) $2V_o + d_o = \left(\frac{80}{3}\right)$. $s(4) = \left(\frac{2}{3}\right)4^3 + (4^2) + V_o(4) + d_o = 90$, from which (2) $4V_o + d_o = \left(\frac{94}{3}\right)$. Subtract (1) from (2) to obtain $V_o = \left(\frac{94 - 80}{6}\right) = 2.33 \ \text{m/s}$. The velocity at $t = 4$ seconds is

$$\left[\frac{ds(t)}{dt}\right]_{t=4} = \left[2t^2 + 2t + V_o\right]_{t=4} = 32 + 8 + 2.33 = 42.33 \ \text{m/s}$$

==================================<>==================================

Problem 2.205 A model rocket takes off straight up. Its acceleration during the two seconds its motor burns is $25 \ \text{m/s}^2$. Neglect aerodynamic drag. Determine (a) The maximum velocity during the flight, and (b) the maximum altitude reached.

Solution The strategy is to solve the equations of motion for the two phases of the flight: during burn $0 \le t \le 2$ s seconds, and after burnout: $t > 2$ s. Phase 1: The acceleration is: $\frac{d^2s}{dt^2} = 25$, from which $\frac{ds}{dt} = 25t$, and $s(t) = 12.5t^2$, since the initial velocity and position are zero. The velocity at burnout is $V_{burnout} = (25)(2) = 50 \ \text{m/s}$. The altitude at burnout is $h_{burnout} = (12.5)(4) = 50 \ \text{m}$. Phase 2. The acceleration is: $\frac{d^2s}{dt^2} = -g$, from which $\frac{ds}{dt} = -g(t-2) + V_{burnout}$ $(t \ge 2)$, and $s(t) = -g(t-2)^2/2 + V_{burnout}(t-2) + h_{burnout}$, $(t \ge 2)$. The velocity during phase 1 is constantly increasing because of the rocket's positive acceleration. Maximum occurs at burnout because after burnout, the rocket has negative acceleration and velocity constantly decreses until it reaches zero at maximum altitude. The velocity from maximum altitude to impact must be constantly increasing since the rocket is falling straight down under the action of gravity. Thus the maximum velocity during phase 2 occurs when the rocket impacts the ground. The issue of maximum velocity becomes this: is the velocity at burnout greater or less than the velocity at ground impact? The time of flight is given by $0 = -g\left(t_{flight} - 2\right)^2/2 + V_{burnout}\left(t_{flight} - 2\right) + h_{burnout}$, from which, in canonical form: $\left(t_{flight} - 2\right)^2 + 2b(t_{flight} - 2) + c = 0$, where $b = -(V_{burnout}/g)$ and $c = -(2h_{burnout}/g)$ The solution $\left(t_{flight} - 2\right) = -b \pm \sqrt{b^2 - c} = 11.11, \ = -0.92$ s. Since the negative time is not allowed, the time of flight is $t_{flight} = 13.11$ s. The velocity at impact is $V_{impact} = -g\left(t_{flight} - 2\right) + V_{burnout} = -59 \ \text{m/s}$ which is <u>higher in magnitude</u> than the velocity at burnout. The time of maximum altitude is given by $\frac{ds}{dt} = 0 = -g\left(t_{\max alt} - 2\right) + V_{burnout}$, from which $t_{\max alt} - 2 = \frac{V_{burnout}}{g} = 5.1$ s, from which $t_{\max alt} = 7.1$ s. The maximum altitude is $h_{\max} = -\frac{g}{2}\left(t_{\max alt} - 2\right)^2 + V_{burnout}\left(t_{\max alt} - 2\right) + h_{burnout} = 177.42 \ \text{m}$

==================================<>==================================

===================================<>===================================

Problem 2.206 In Problem 2.205, if the rocket's parachute fails to open, what is the total time of flight from takeoff until the rocket hits the ground?

Solution The solution to Problem 2.205 was (serendipitously) posed in a manner to yield the time of flight as a peripheral answer. The time of flight is given there as $t_{flight} = 13.11$ s

===================================<>===================================

Problem 2.207 The acceleration of a point moving along a straight line is $a = -cv^3$, where c is a constant. If the velocity of the point is v_o, what distance does it move before its velocity decreases to $\frac{v_o}{2}$?

Solution The acceleration is $\frac{dv}{dt} = -cv^3$. Using the chain rule, $\frac{dv}{dt} = \frac{dv}{ds}\frac{ds}{dt} = v\frac{dv}{ds} = -cv^3$. Separating variables and integrating: $\frac{dv}{v^2} = -cds$, from which $-\frac{1}{v} = -cs + C$. At $s = 0$, $v = v_o$, from which $-\frac{1}{v} = -cs - \frac{1}{v_o}$, and $v = \frac{v_o}{1 + v_o cs}$. Invert: $v_o cs = \frac{v_o}{v} - 1$. When $v = \frac{v_o}{2}$, $\boxed{s = \left(\frac{1}{cv_o}\right)}$

===================================<>===================================

Problem 2.208 Water leaves the nozzle at 20° above the horizontal and strikes the wall at the point indicated. What was the velocity of the water as it left the nozzle? *Strategy:* Determine the motion of the water by treating each particle of water as a projectile.

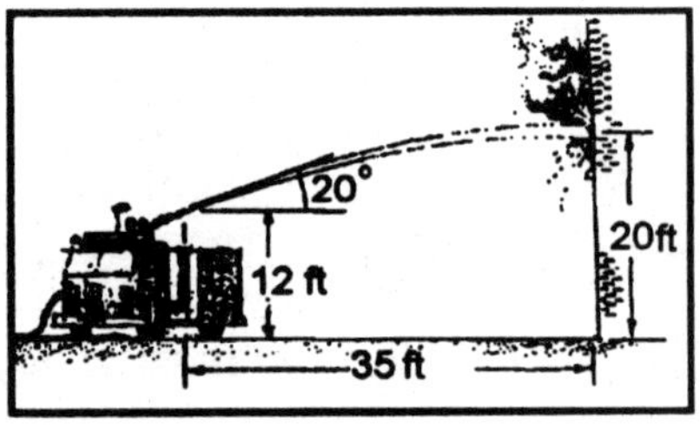

Solution Denote $\theta = 20^o$. The path is obtained by integrating the equations: $\frac{dv_y}{dt} = -g$ and $\frac{dv_x}{dt} = 0$, from which $\frac{dy}{dt} = -gt + V_n \sin\theta$, $\frac{dx}{dt} = V_n \cos\theta$. $y = -\frac{g}{2}t^2 + (V_n \sin\theta)t + y_o$. $x = (V_n \cos\theta)t + x_o$.

Choose the origin at the nozzle so that $y_o = 0$, and $x_o = 0$. When the stream is $y(t_{impact}) = (20 - 12) = 8\,ft$, the time is $0 = -\frac{g}{2}(t_{impact})^2 + (V_n \sin\theta)t_{impact} - 8$. At this same time the horizontal distance is $x(t_{impact}) = 35 = (V_n \cos\theta)t_{impact}$, from which $t_{impact} = \frac{35}{V_n \cos\theta}$. Substitute:

$$0 = -\frac{g}{2}\left(\frac{35}{V_n \cos\theta}\right)^2 + 35\tan\theta - 8, \text{ from which } \boxed{V_n = \left(\frac{35}{\cos\theta}\right)\sqrt{\frac{g}{2(35\tan\theta - 8)}} = 68.62\ ft/s}$$

===================================<>===================================

===================================<>===================================

Problem 2.209 In practice, a quarterback throws the football with a velocity v_o at 45° above the horizontal. At the same instant, a receiver standing $20\,ft$ in front of him starts running straight down field at $10\,ft/s$ and catches the ball. Assume that the ball is thrown and caught at the same height above the ground. What is the velocity v_o?

Solution Denote $\theta = 45^o$. The path is determined by integrating the equations; $\frac{d^2y}{dt^2} = -g$, $\frac{d^2x}{dt^2} = 0$, from which $\frac{dy}{dt} = -gt + v_o \sin\theta$, $\frac{dx}{dt} = v_o \cos\theta$. $y = -\frac{g}{2}t^2 + (v_o \sin\theta)t$, $x = (v_o \cos\theta)t$ where the origin is taken at the point where the ball leaves the quarterback's hand.

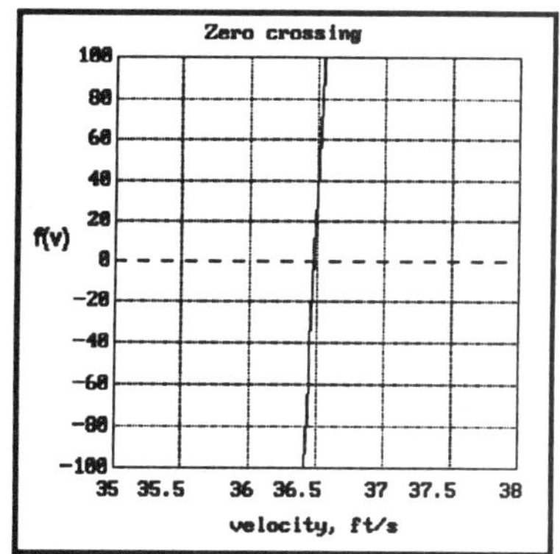

When the ball reaches the receiver's hands, $y = 0$, from which $t_{flight} = \sqrt{\frac{2v_o \sin\theta}{g}}$. At this time the distance down field is the distance to the receiver: $x = 10t_{flight} + 20$. But also $x = (v_o \cos\theta)t_{flight}$, from which $t_{flight} = \frac{20}{(v_o \cos\theta - 10)}$. Substitute: $\frac{20}{(v_o \cos\theta - 10)} = \sqrt{\frac{2v_o \sin\theta}{g}}$, from which $400g = 2v_o \sin\theta(v_o \cos\theta - 10)^2$. The function $f(v_o) = 2v_o \sin\theta(v_o \cos\theta - 10)^2 - 400g$ was graphed to find the zero crossing, and the result refined by iteration: $\boxed{v_o = 36.48\,ft/s}$. *Check:* The time of flight is $t = 1.27\ s$ and the distance down field that the quarterback throws the ball is $d = 12.7 + 20 = 32.7\,ft = 10.6\,yds$, which seem reasonable for a short, "lob" pass. *check.*

===================================<>===================================

Problem 2.210 The constant velocity $v = 2\ m/s$. What are the magnitudes of the velocity and acceleration of point P when $x = 0.35\ m$?

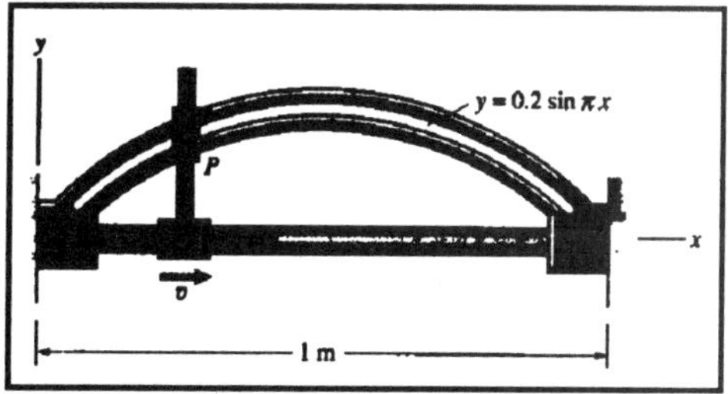

Solution The path is obtained by integrating $\frac{dx}{dt} = 2$, from which $x = 2t + x_o$. Choose the origin at $x(0) = 0$, from which $x(t) = 2t\ m$, from which $y(t) = 0.2\sin(2\pi t)$. The velocity of P: $\vec{v}_P = 2\vec{i} + (0.4\pi\cos(2\pi t))\vec{j}$. At $x = 0.25\ m$, $t = 0.125\ s$, from which the magnitude of the velocity of P is $\boxed{|\vec{v}_P| = \sqrt{2^2 + \left(0.2\pi\cos\left(\frac{\pi}{4}\right)\right)^2} = 2.19\ m/s}$ The acceleration components are $\frac{d^2x}{dt^2} = 0$, $\frac{d^2y}{dt^2} = -(0.8\pi^2\sin(2\pi t))$. At $x = 0.25\ m$, $\frac{d^2y}{dt^2} = -\left(0.8\pi^2\sin\left(\frac{\pi}{4}\right)\right) = -5.58\ m/s^2$. The magnitude is $\boxed{|\vec{a}| = 5.58\ m/s^2}$

===================================<>===================================

================================<>================================

Problem 2.211 In Problem 2.210, what is the acceleration of point P in terms of normal and tangential components when $x = 0.25\ m$? What is the instantaneous radius of curvature of the path?

Solution From Problem 2.210, $y = 0.2\sin 2\pi t\ m, x = 2t\ m$; $v_y = 0.4\pi\cos(2\pi t)\ m/s$, $v_x = 2\ m/s$; $a_x = 0$, $a_y = -\left(0.8\pi^2\sin(2\pi t)\right)$. The instantaneous path angle at $x = 0.25\ m$ is

$$\theta = \tan^{-1}\left(\frac{v_y}{v_x}\right) = \tan^{-1}\left(\frac{0.4\pi\cos\left(\frac{\pi}{4}\right)}{2}\right) = 23.96^o$$. The components of the acceleration are:

$$\boxed{a_t = a_x\cos\theta + a_y\sin\theta = -5.58(\sin\theta) = -2.267\ m/s^2}$$,

$$\boxed{a_n = a_x\sin\theta - a_y\cos\theta = -(-5.58)\cos\theta = 5.10\ m/s}$$. The instantaneous radius of curvature is

$$\boxed{\rho = \frac{v^2}{a_n} = \frac{v_x^2 + v_y^2}{5.10} = \frac{2.1886^2}{5.10} = 0.939\ m}$$.

================================<>================================

Problem 2.212 In Problem 2.210, what is the acceleration of point P in terms of the radial and transverse components (polar coordinates) when $x = 0.25\ m$?

Solution From Problem 2.210, $\vec{a} = -\vec{j}\left(0.8\pi^2\sin\left(\frac{\pi}{4}\right)\right) = -5.58\vec{j}$. Noting $\vec{j} = \vec{e}_r\sin\theta + \vec{e}_\theta\cos\theta$,

where $\theta = \tan^{-1}\left(\frac{y}{x}\right) = \tan^{-1}\left(\frac{0.2\sin\left(\frac{\pi}{4}\right)}{0.25}\right) = 29.5^o$. From which

$$\boxed{\vec{a} = -\left(0.8\pi^2\cos\left(\frac{\pi}{4}\right)\right)(\vec{e}_r\sin\theta + \vec{e}_\theta\cos\theta) = -2.75\vec{e}_r - 4.86\vec{e}_\theta}$$

================================<>================================

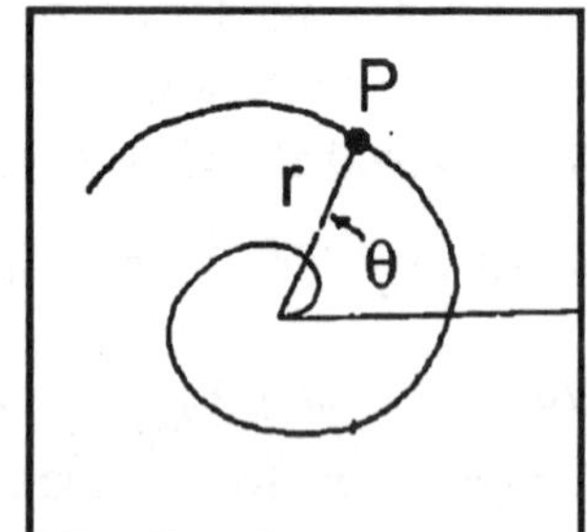

Problem 2.213 A point P moves in the *x-y* plane along the spiral path $r = (0.1)\theta\ ft$, where θ is in radians. The angular position $\theta = 2t\ rad$, where t is in seconds. Determine the magnitudes of the velocity and acceleration of P at $t = 1$ seconds.

Solution The path: $r = 0.2t\ ft$, $\theta = 2t\ rad$. The velocity components are

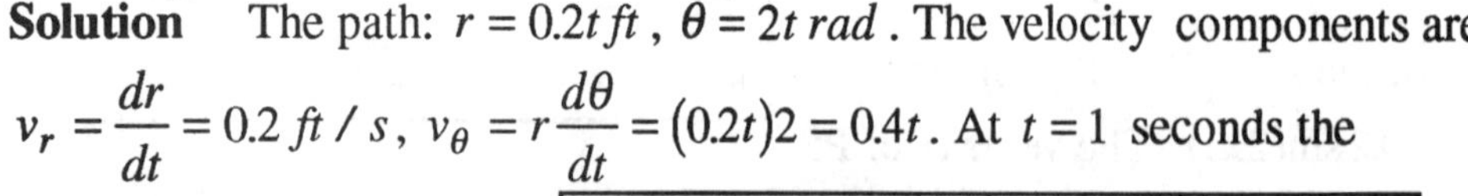

$v_r = \frac{dr}{dt} = 0.2\ ft/s$, $v_\theta = r\frac{d\theta}{dt} = (0.2t)2 = 0.4t$. At $t = 1$ seconds the magnitude of the velocity is $\boxed{|\vec{v}| = \sqrt{v_r^2 + v_\theta^2} = \sqrt{0.2^2 + 0.4^2} = 0.447\ ft/s}$ The acceleration components are: $a_r = \frac{d^2r}{dt^2} - r\left(\frac{d\theta}{dt}\right)^2 = -(0.2t)(2^2)\ ft/s^2$,

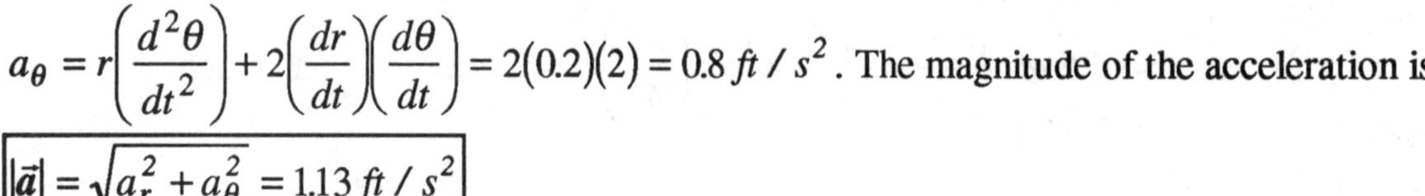

$a_\theta = r\left(\frac{d^2\theta}{dt^2}\right) + 2\left(\frac{dr}{dt}\right)\left(\frac{d\theta}{dt}\right) = 2(0.2)(2) = 0.8\ ft/s^2$. The magnitude of the acceleration is

$$\boxed{|\vec{a}| = \sqrt{a_r^2 + a_\theta^2} = 1.13\ ft/s^2}$$

================================<>================================

==================================<>==================================

Problem 2.214 In the cam-follower mechanism, the slotted bar rotates with a constant angular velocity $\omega = 12 rad/s$, and the radial position of the follower A is determined by the profile of the stationary cam. The slotted bar is pinned a distance $h = 0.2m$ to the left of the center of the circular cam. The follower moves in a circular path of *0.42 m* radius. Determine the velocity of the cam when $\theta = 40°$ (a) in terms of polar coordinates, and (b) in terms of cartesian coordinates.

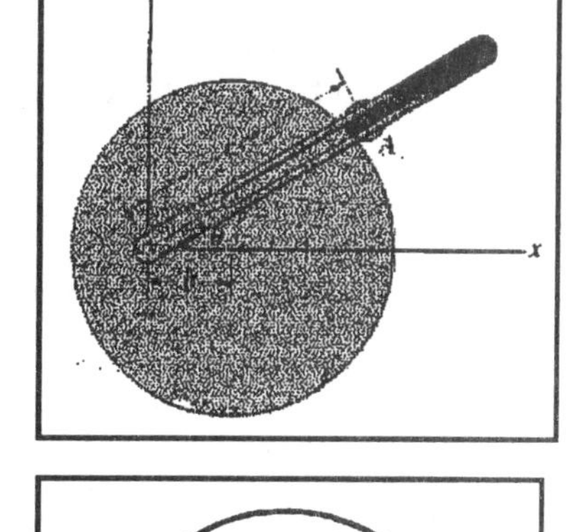

Solution:

(a) The first step is to get an equation for the path of the follower in terms of the angle θ. This can be most easily done by referring to the diagram at the right. Using the law of cosines, we can write $R^2 = h^2 + r^2 - 2hr\cos\theta$. This can be rewritten as $r^2 - 2hr\cos\theta + (h^2 - R^2) = 0$. We need to find the components of the velocity. These are $v_r = \dot{r}$ and $v_\theta = r\dot{\theta}$. We can differentiate the relation derived from the law of cosines to get $\dot{r}$. Carrying out this differentiation, we get $2r\dot{r} - 2h\dot{r}\cos\theta + 2hr\dot{\theta}\sin\theta = 0$. Solving for $\dot{r}$, we get $\dot{r} = \dfrac{h\,r\dot{\theta}\sin\theta}{(h\cos\theta - r)}$. Recalling that $\omega = \dot{\theta}$ and substituting in the numerical values,i.e., $R = 0.42m$, $h = 0.2m$, $\omega = 12 rad/s$, and $\theta = 40°$, we get $r = 0.553m$, $v_r = -2.13 m/s$, and $v_\theta = 6.64 m/s$

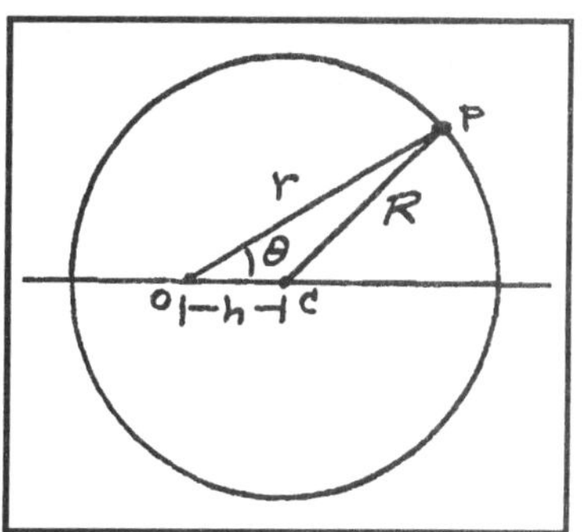

(b) The transformation to cartesian coordinates can be derived from $\mathbf{e_r} = \cos\theta\,\mathbf{i} + \sin\theta\,\mathbf{j}$, and $\mathbf{e}_\theta = -\sin\theta\,\mathbf{i} + \cos\theta\,\mathbf{j}$. Substituting these into $\mathbf{v} = v_r\,\mathbf{e_r} + v_\theta \mathbf{e}_\theta$, we get $\mathbf{v} = (v_r\cos\theta - v_\theta\sin\theta)\mathbf{i} + (v_r\sin\theta + v_\theta\cos\theta)\mathbf{j}$. Substituting in the numbers, $\mathbf{v} = -5.90\,\mathbf{i} + 3.71\,\mathbf{j}\ m/s$

==================================<>==================================

Problem 2.215 In Problem 2.214, determine the acceleration of the cam follower when $\theta = 40°$: (a) in terms of polar coordinates; (b) in terms of cartesian coordinates.

Solution: The solution to this problem follows that of Problem 2.214 in the same way that the solution to Problem 2.156 follows that of Problem 2.155. Information from the solution to Problem 2.214 will be used in this solution. In order to determine the components of the acceleration in polar coordinates, we need to be able to determine all of the variables in the right hand sides of $a_r = \ddot{r} - r\dot{\theta}^2$ and that $a_\theta = r\ddot{\theta} + 2\dot{r}\dot{\theta}$. We already know everything except $\ddot{r}$ and $\ddot{\theta}$. Since ω is constant, $\ddot{\theta} = \dot{\omega} = 0$. We need only to find the value for r and the value for $\ddot{r}$ at $\theta = 40°$. Substituting into the original equation for *r*, we find that $r = 0.553\ m$ at this position on the cam. To find $\ddot{r}$, we start with $\dot{r} = v_r$. Taking a derivative, we start with $r\dot{r} - h\dot{r}\cos\theta + hr\dot{\theta}\sin\theta = 0$ from Problem 2.214 (we divided through by 2).

Taking a derivative with respect to time, we get $\ddot{r} = \dfrac{\dot{r}^2 + 2hr\dot{\theta}\sin\theta + hr\dot{\theta}^2\cos\theta + hr\ddot{\theta}\sin\theta}{(h\cos\theta - r)}$.

Evaluating, we get $\ddot{r} = -46.17 m/s^2$. Substituting this into the equation for a_r and evaluating a_n, we get $a_r = -125.81 m/s^2$ and $a_\theta = -51.2 m/s^2$

Soluton continues next page

(b) The transformation to cartesian coordinates can be derived from $\mathbf{e}_r = \cos\theta\,\mathbf{i} + \sin\theta\,\mathbf{j}$, and $\mathbf{e}_\theta = -\sin\theta\,\mathbf{i} + \cos\theta\,\mathbf{j}$. Substituting these into $\mathbf{a} = a_r\,\mathbf{e}_r + a_\theta\mathbf{e}_\theta$, we get $\mathbf{a} = (a_r\cos\theta - a_\theta\sin\theta)\mathbf{i} + (a_r\sin\theta + a_\theta\cos\theta)\mathbf{j}$. Substituting in the numbers, we get $\mathbf{a} = -63.46\,\mathbf{i} - 120.1\,\mathbf{j}\ m/s^2$

==================================<>==================================

Problem 2.216 A manned vehicle (M) attempts to rendezvous with a satellite (S) to repair. (They are not shown to scale.) The magnitude of the satellite's velocity is $|\vec{v}_S| = 6\ km/s$, and a sighting determines that the angle $\beta = 40^o$. If you assume that their velocities remain constant and the vehicles move along the straight lines shown, what should be the magnitude of v_M to achieve rendezvous?

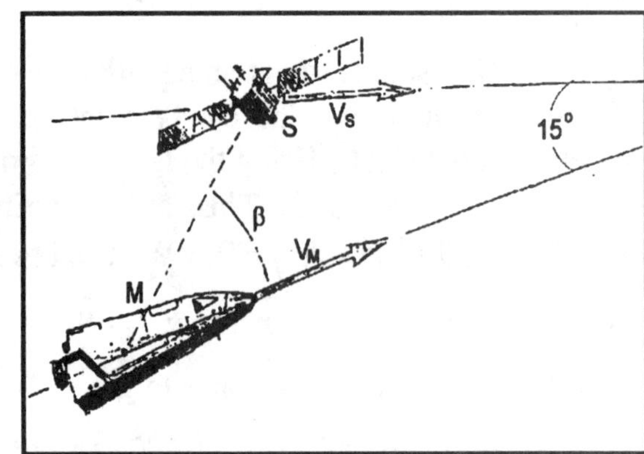

Solution Assume motion in the *x-y* plane. The velocities and directions of travel are constant; the magnitudes of the velocities are constrained by the condition that the instantaneous triangles formed by the vehicles as they move must be similar if the two vehicles are to meet at the same point. The interior angles are constant for similar triangles, and the two unknown velocities may be determined from the law of sines everywhere along the path. The velocity of the satellite is related to the velocity of the manned vehicle, by :

$$\frac{|\vec{v}_M|}{\sin(180-\beta-15)} = \frac{|\vec{v}_S|}{\sin\beta},$$ from which $$\boxed{|\vec{v}_M| = \frac{|\vec{v}_S|\sin(125^o)}{\sin(40^o)} = 6\left(\frac{0.8192}{0.6428}\right) = 7.65\ km/s}.$$

==================================<>==================================

Problem 2.217 In Problem 2.216, what is the magnitude of the velocity of the manned vehicle relative to the spacecraft once the magnitude of $\vec{v}_M$ has been adjusted to achieve rendezvous?

Solution From the law of sines (see the solution to Problem 2.216) $\frac{|\vec{v}_{SM}|}{\sin(15^o)} = \frac{|\vec{v}_S|}{\sin(40^o)}$, from which

$$\boxed{|\vec{v}_{SM}| = |\vec{v}_S|\left(\frac{\sin(15^o)}{\sin(40^o)}\right) = 2.416\ km/s}$$

==================================<>==================================

================================<>================================

Problem 2.218 The three 1-*ft* bars rotate in the x-y plane with constant angular velocity ω. If $\omega = 20\ rad/s$, what is the magnitude of the velocity of point C relative to point A in terms of the non-rotating coordinates system?

Solution Denote $\theta = 45^o$, $\beta = 90^o$, $\gamma = 0$. The velocity of point C relative to B is

$\vec{v}_{CB} = r\omega\vec{e}_\theta = 20\vec{e}_\theta = 20(-\vec{i}\sin\theta + \vec{j}\cos\theta) = 14.14(-\vec{i} + \vec{j})$. The velocity of point B relative to A is $\vec{v}_{BA} = r\omega\vec{e}_\theta = 20\vec{e}_\theta = 20(-\vec{i}\sin\beta + \vec{j}\cos\beta) = -20\vec{i}\ ft/s$ The velocity of point C relative to A is $\vec{v}_{CA} = \vec{v}_{CB} + \vec{v}_{BA} = -34.14\vec{i} + 14.14\vec{j}$. The magnitude is

$|\vec{v}_{CA}| = \sqrt{34.14^2 + 14.14^2} = 36.95... = 37\ ft/s$.

================================<>================================

Problem 2.219 In Problem 2.218, What is the velocity of point C relative to the fixed point O?

Solution From the solution to Problem 2.218, the velocity of point C relative to A is $\vec{v}_{CA} = -28.28\vec{i} + 14.14\vec{j}\ ft/s$. The velocity of point A relative to O is $\vec{v}_{AO} = r\omega\vec{e}_\theta = +20\vec{j}\ ft/s$, from which the velocity of C relative to O is $\vec{v}_{CO} = \vec{v}_{CA} + \vec{v}_{AO} = -34.14\vec{i} + 34.14\vec{j}\ \ ft/s$

================================<>================================

Problem 2.220 In Problem 2.218, accelerometers mounted at C indicate that the acceleration of point C relative to fixed point O is $\vec{a}_C = -1500\vec{i} - 1500\vec{j}\ ft/s^2$. What is the angular velocity ω? Can you tell from this information whether ω is counterclockwise or clockwise?

Solution The acceleration of C relative to B is

$\vec{a}_{CB} = \left(\frac{d^2r}{dt^2} - r\left(\frac{d\theta}{dt}\right)^2\right)\vec{e}_r + \left(r\left(\frac{d^2\theta}{dt^2}\right) + 2\left(\frac{dr}{dt}\right)\left(\frac{d\theta}{dt}\right)\right)\vec{e}_\theta = -\omega^2\vec{e}_r\ ft/s^2$, since $r = 1\,ft$. In cartesian coordinates: $\vec{a}_{CB} = -\omega^2(\vec{i}\cos 45^o + \vec{j}\sin 45^o) = -\frac{\omega^2}{\sqrt{2}}(\vec{i} + \vec{j})\ \ ft/s^2$ The acceleration of B relative to A is $\vec{a}_{BA} = -\omega^2\vec{e}_r$, from which $\vec{a}_{BA} = -\vec{j}\omega^2\ ft/s^2$. The acceleration of A relative to O is $\vec{a}_{AO} = -\omega^2\vec{e}_r = -\omega^2\vec{i}\ ft/s^2$. The acceleration of C relative to O is

$\vec{a}_{CO} = \vec{a}_{CB} + \vec{a}_{BA} + \vec{a}_{AO} = -\omega^2\left(\frac{1}{\sqrt{2}} + 1\right)\vec{i} - \omega^2\left(\frac{1}{\sqrt{2}} + 1\right)\vec{j} = -1500\vec{i} - 1500\vec{j}\ ft/s^2$.

Solve: $\omega = \pm\sqrt{\frac{1500}{\left(\frac{1}{\sqrt{2}} + 1\right)}} = \pm 29.64\ \ rad/s$. No, because information indicating direction of rotation is lost when the angular velocity is squared.

================================<>================================

==============================<>==============================

Problem 3.1 The *2-kg* collar A is initially at rest on the smooth horizontal bar. At $t = 0$ it is subjected to a constant force $F = 4N$. (a) How fast is the collar moving at $t = 1s$? (b) What distance has the collar moved at $t = 1s$?

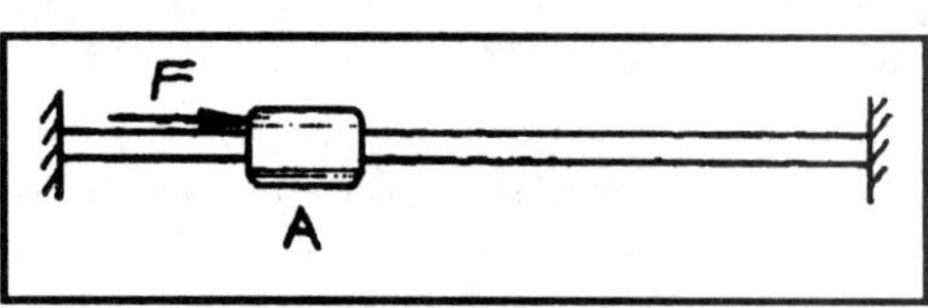

Solution: Draw the free body diagram of the collar and write the equations of motion using the information on the free body diagram. Then integrate the equations of motion using the known initial conditions. Assume x is horizontal and y is vertical. The equations of motion are $\sum F_X = ma_X = 4Newtons = (2kg)(a_X)$ and $\sum F_Y = N - mg = 0$. Thus $a_X = 2m/s^2$, and $N = 19.62Newtons$. Integrating the acceleration in the x direction,we get $v_X = a_X t$, and $x = a_X t^2/2$. Substititing in $t = 1s$, we get $v_X = 2m/s$ and $x = 1m$.

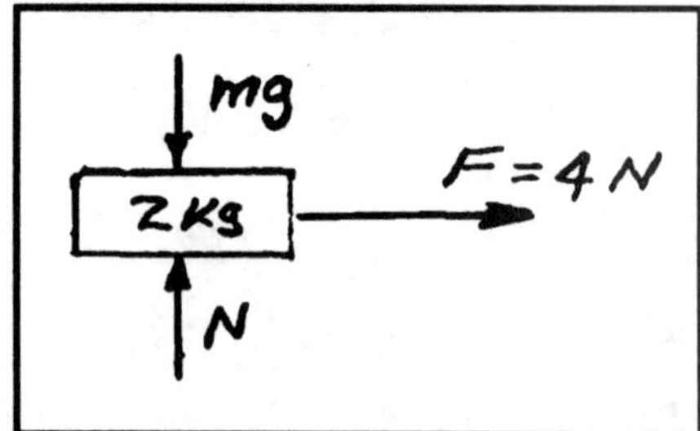

==============================<>==============================

Problem 3.2 Solve Problem 3.1 if the coefficient of kinetic friction between the collar and bar is $\mu_k = 0.1$.

Solution: Draw the free body diagram of the collar and write the equations of motion using the information on the free body diagram. Then integrate the equations of motion using the known initial conditions. Assume x is horizontal and y is vertical. The equations of motion are $\sum F_X = ma_X = (4 - \mu_k N)Newtons = (2kg)(a_X)$ and $\sum F_Y = N - mg = 0$. Thus $a_X = 1.02m/s^2$, and $N = 19.62Newtons$. Integrating the acceleration in the x direction,we get $v_X = a_X t$, and $x = a_X t^2/2$. Substititing in $t = 1s$, we get $v_X = 1.02m/s$ and $x = 0.51m$.

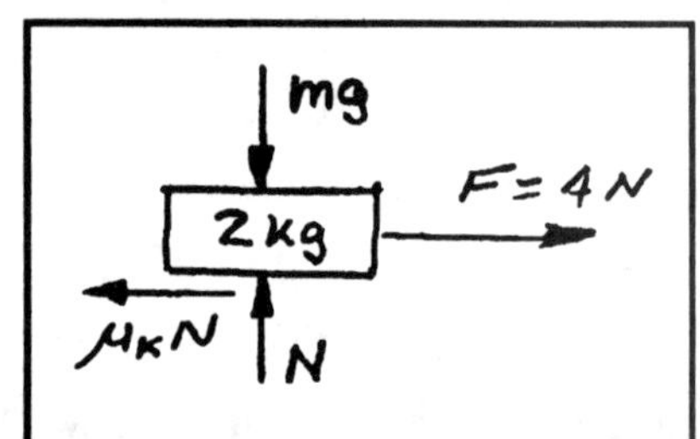

==============================<>==============================

Problem 3.3 The *20 lb* collar A starts at rest on the smooth bar at $t = 0$ and is subjected to a constant force $F = 10\ lb$. (a) How fast is the collar moving at $t = 1s$? (b) What distance has the collar moved along the bar at ?

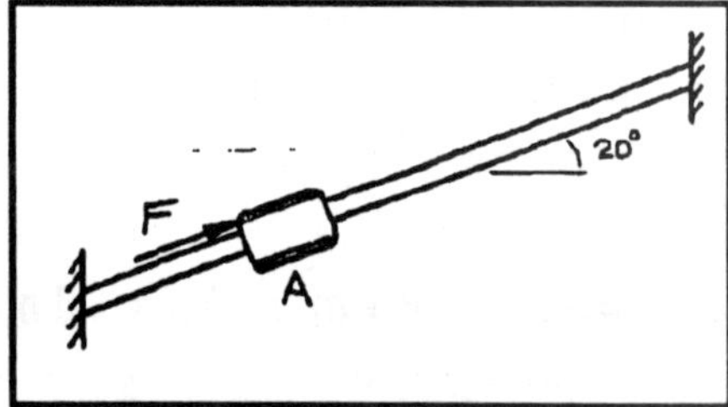

Solution: Draw the free body diagram of the collar and write the equations of motion using the information on the free body diagram. Then integrate the equations of motion using the known initial conditions. Assume x is parallel to the bar and y is normal to the bar. Note that $m = W/g = 20/32.2 = 0.621slug$. The equations of motion are $\sum F_X = ma_X = F - W\sin\theta$ and $\sum F_Y = N - W\cos\theta = 0$. Thus $a_X = 5.09ft/s^2$, and $N = 18.79lb$. Integrating the acceleration in the x direction,we get $v_X = a_X t$, and $x = a_X t^2/2$. Substititing in $t = 1s$, we get $v_X = 5.09ft/s$ and $x = 2.54ft$.

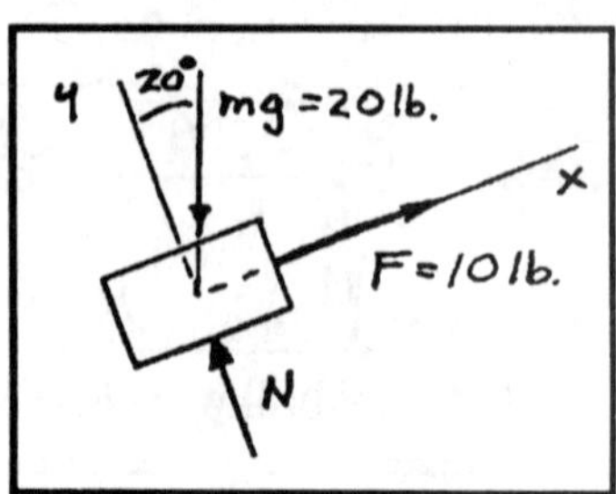

==============================<>==============================

==================================<>==================================

Problem 3.4 Solve Problem 3.3 if the coefficients of static and kinetic friction between the collar and bar are $\mu_s = \mu_k = 0.1$.

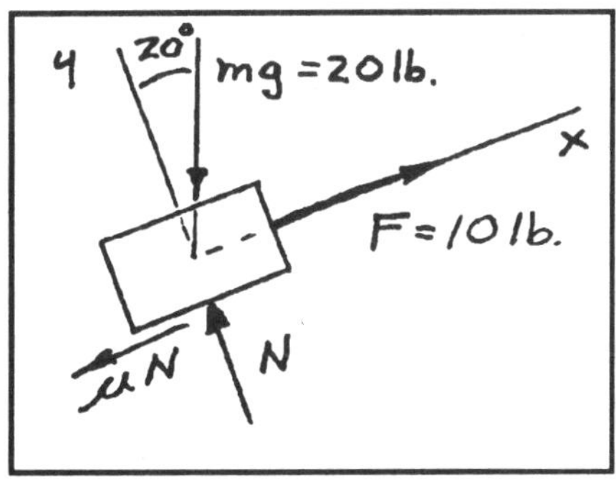
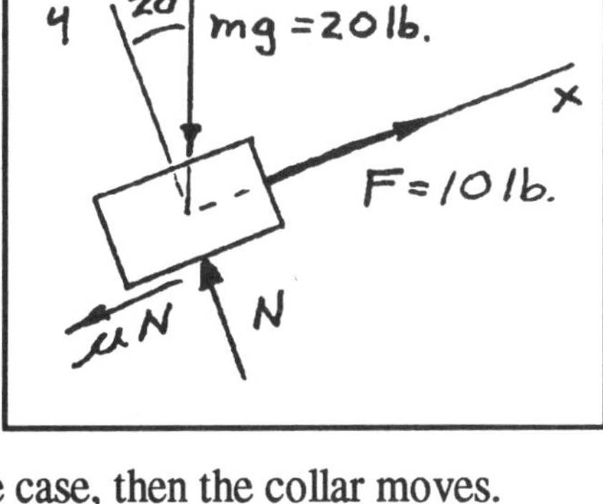

Solution: Draw the free body diagram of the collar and write the equations of motion using the information on the free body diagram. Assume x is parallel to the bar and y is normal to the bar. Note that $m = W/g = 20/32.2 = 0.621 slug$. In this case, since the collar starts from rest, we must determine whether the collar moves before writing equations of motion. To do this, we assume a static collar and determine whether the friction force necessary to keep the collar stationary is greater than the static friction force available. If this is the case, then the collar moves. Otherwise, the collar remains stationary.

The equations of equilibrium for the collar are $\sum F_X = 0 = F - W\sin\theta - f_{REQ}$ and $\sum F_Y = N - W\cos\theta = 0$. Substituting in the values for the variables, we find that $N = 18.79 lb$, $f_{REQ} = 3.16\ lb$, and $f_{AVAIL} = 1.88\ lb$. Hence, the collar moves and we must analyze the motion.

The equations of motion are $\sum F_X = ma_X = F - W\sin\theta - \mu_k N$ and $\sum F_Y = N - W\cos\theta = 0$. Thus $a_X = 2.06 ft/s^2$. Integrating the acceleration in the x direction,we get $v_X = a_X t$, and $x = a_X t^2/2$. Substititing in $t = 1s$, we get $v_X = 2.06 ft/s$ and $x = 1.03 ft$

==================================<>==================================

Problem 3.5 Suppose a person conducts an experiment in which he rides on an elevator while standing on a set of scales (Fig (a)). Let a be the elevator's upward acceleration relative to the earth. The forces acting on the person are his weight $W = 150\ lb$ and the force N exerted on him by the scales (Fig(b)). (a) If the scales read $155\ lb$, what is the elevator's acceleration? (b) If the elevator's acceleration is $a = -2 ft/s^2$, what do the scales read?.

a
W
N
(a)
(b)

Solution: The motion is in the y direction. Figure (b) is the free body diagram of the man. The equation of motion is

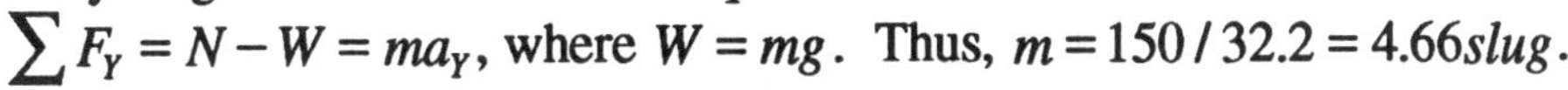

$\sum F_Y = N - W = ma_Y$, where $W = mg$. Thus, $m = 150/32.2 = 4.66 slug$.

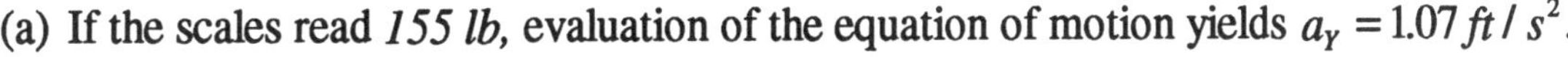

(a) If the scales read *155 lb,* evaluation of the equation of motion yields $a_Y = 1.07 ft/s^2$.

(b) If acceleration $a_Y = -2 ft/s^2$, then the equation of motion yields $N = 140.68\ lb$ for the scale reading.

==================================<>==================================

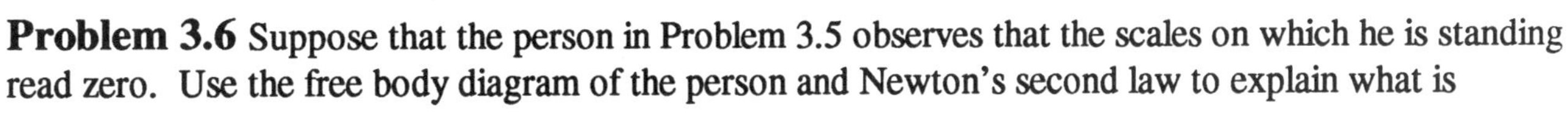

Problem 3.6 Suppose that the person in Problem 3.5 observes that the scales on which he is standing read zero. Use the free body diagram of the person and Newton's second law to explain what is happening.

Solution: The equation of motion is $\sum F_Y = N - W = ma_Y$, where $W = mg$. If the scale reading, *N,* is zero, the equation gives the acceleration as $a_Y = -32.2 ft/s^2$. The cables have broken on the elevator and the emergency brakes have failed. The man and elevator are freely falling toward the bottom of the elevator shaft.

==================================<>==================================

================================<>================================

Problem 3.7 Suppose the person in Problem 3.5 is standing on a European scale that displays his mass in kilograms, and that his actual mass is 68 kg. (Notice that to determine the force exerted on the scale, you must convert its reading in kilograms into newtons by multiplying by $9.81 m/s^2$)

(a) If the scales read *70 kg*, what is the elevator's acceleration in m/s^2?

(b) If the elevator's acceleration is $a = -0.5 m/s^2$, what do the scales read?

Solution: The motion is in the y direction. The equation of motion is $\sum F_Y = N - W = ma_Y$, where $W = mg$. Thus, the weight of the man in force units (newtons) is given by $W = mg = (68kg)(9.81 m/s^2) = 667.1 Newtons$. The scale reading, *S*, in *kg* will be $S = N/g$.

(c) If the scales read 70 *kg*, evaluation of the equation of motion yields $N = 686.7 newtons$ and $a_Y = 0.289 m/s^2$.

If acceleration $a_Y = -0.5 m/s^2$, then the equation of motion yields $N = 633.1 newtons$, and the corresponding scale reading is $S = 64.5$ *kg*.

================================<>================================

Problem 3.8 The total external force on a 10 *kg* object is $90\vec{i} - 60\vec{j} + 20\vec{k}$ (N). What is the magnitude of the acceleration relative to an inertial frame?

Solution The acceleration is $\vec{a} = \frac{\vec{F}}{m} = 9\vec{i} - 6\vec{j} + 2\vec{k}$ m/s^2. The magnitude is

$$\boxed{|\vec{a}| = \sqrt{9^2 + 6^2 + 2^2} = 11\ m/s^2}$$

================================<>================================

Problem 3.9 The total external force on a 20 *lb.* object is $\sum \vec{F} = 10\vec{i} + 20\vec{j}$ *lb*. When $t = 0$ its position vector is $\vec{r} = 0$ and its velocity vector is $\vec{v} = 20\vec{i} - 10\vec{j}$ ft/s. Determine the position and velocity of the object when $t = 2$ *s*.

Solution The mass is $m = \frac{W}{g} = 0.6217$ *slug*. The equation for the path is obtained by integrating $m\frac{d\vec{v}}{dt} = 10\vec{i} + 20\vec{j}$, from which $\vec{v}(t) = \left(\frac{1}{m}\right)(\vec{i}10t + \vec{j}20t) + \vec{i}C_x + \vec{j}C_y$. From the initial conditions for the velocity: $C_x = 20$ and $C_y = -10$, from which the velocity is

$\vec{v}(t) = \left(\left(\frac{10}{m}t + 20\right)\vec{i} + \left(\frac{20}{m}t - 10\right)\vec{j}\right)$ (ft/s). The position is

$\vec{r}(t) = \left(\left(\frac{5}{m}t^2 + 20t\right)\vec{i} + \left(\frac{10}{m}t^2 - 10t\right)\vec{j}\right)$ (ft), since $\vec{r}(0) = 0$. At $t = 2$ *s*,

$\boxed{\vec{v}(2) = 52.17\vec{i} + 54.34\vec{j}\ \ (ft/s)}$, $\boxed{\vec{r}(2) = 72.17\vec{i} + 44.34\vec{j}}$

================================<>================================

==================================<>==================================

Problem 3.10 The total external force on an object is $\vec{F} = 10t\vec{i} + 60\vec{j}$ (lb). When $t = 0$, its position vector relative to an inertial frame is $\vec{r} = 0$ and its velocity is $\vec{v} = 0.2\vec{j}$ (ft/s). When $t = 5\ s$, the magnitude of its position is measured and found to be 8.75 *ft*. What is the mass of the object?

Solution The differential equation is $m\frac{d\vec{v}}{dt} = 10t\vec{i} + 60\vec{j}$. Integrating:

$\vec{v} = (\frac{5}{m}t^2 + C_{vx})\vec{i} + \left(\frac{60}{m}t + C_{vy}\right)\vec{j}$. From the initial conditions: $C_{vx} = 0$ and $C_{vy} = 0.2$, from which

$\vec{v}(t) = \left(\frac{5}{m}t^2\right)\vec{i} + \left(\frac{60}{m}t + 0.2\right)\vec{j}$ (ft/s). The position is

$\vec{r}(t) = \left(\frac{5}{3m}t^3 + C_x\right)\vec{i} + \left(\frac{30}{m}t^2 + 0.2t + C_y\right)\vec{j}$ (ft). From the initial conditions: $C_x = 0$, $C_y = 0$. At

$t = 5\ s$, the position is $\vec{r}(5) = \frac{1}{m}\left(\frac{5^4}{3}\right)\vec{i} + \left(\frac{750}{m} + 1.0\right)\vec{j}$ (ft). The magnitude is

$8.75 = \sqrt{\left(\frac{208.33}{m}\right)^2 + \left(\frac{750}{m} + 1.0\right)^2}$. Carry out the operations and reduce to obtain $m^2 + 2bm + c = 0$,

where $b = -9.926$, $c = -8018.6$. The solution $m = -b \pm \sqrt{b^2 - c} = 100.0, \quad = -80.2$ *slugs*. The negative root has no meaning, hence $\boxed{m = 100.0\ slugs}$

==================================<>==================================

Problem 3.11 The position of a 10 *kg* object relative to an inertial reference frame is $\vec{r} = \frac{1}{3}t^3\vec{i} + 4t\vec{j} - 30t^2\vec{k}$ (m). What are the components of the total external force acting on the object at $t = 10$ seconds?

Solution The acceleration is found by differentiating twice: $v(t) = t^2\vec{i} - 4\vec{j} - 60t\vec{k}$ (m/s), $\vec{a}(t) = 2t\vec{i} - 60\vec{k}$ (m/s^2), from which $\vec{a}(10) = 20\vec{i} - 60\vec{k}$ (m/s^2). The applied force is

$\boxed{\vec{F}(10) = m\vec{a}(10) = 200\vec{i} - 600\vec{k}\ (N)}$

==================================<>==================================

Problem 3.12 If the 15,000 *lb.* helicopter starts from rest and its rotor exerts a constant 20,000 *lb.* vertical force, how high does it rise in two seconds?

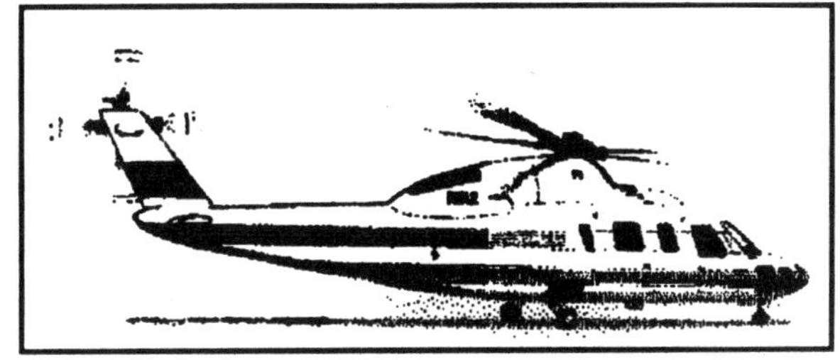

Solution The mass of the helicopter is $m = \frac{W}{g} = 466.27$ *slugs*. The acceleration is $\vec{a} = \frac{\vec{F}}{m} - g\vec{j} = (42.89 - g)\vec{j} = 10.72\vec{j}$ (ft/s^2). The velocity is $\vec{v}(t) = 10.72t\vec{j}$ (ft/s), since $\vec{v}(0) = 0$. The distance is $\vec{r}(t) = \frac{10.72}{2}t^2\vec{j}$ (ft), since $\vec{r}(0) = 0$. At $t = 2\ s$, $\boxed{\vec{r}(2) = 21.45\vec{j}\ (ft)}$

==================================<>==================================

Problem 3.13 A 1 *lb.* collar A is initially at rest in the position shown on the smooth horizontal bar. At $t = 0$, a force $\vec{F} = \left(\frac{1}{20}\right)t^2\vec{i} + \left(\frac{1}{10}\right)t\vec{j} - \left(\frac{1}{30}\right)t^3\vec{k}$ (lb) is applied to the collar, causing it to slide along the bar. What is the velocity of the collar when it reaches the right end of the bar?

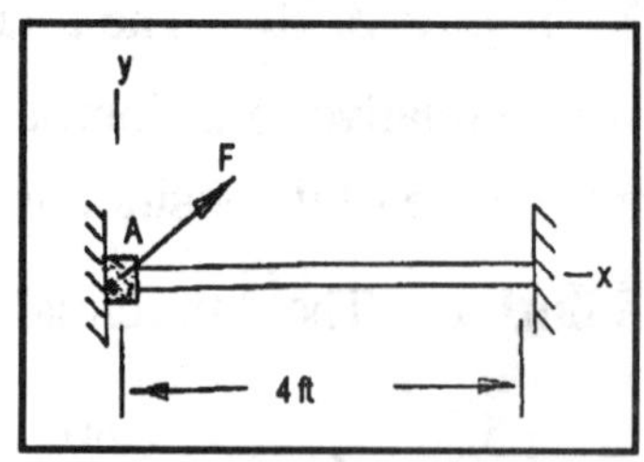

Solution The mass is $m = \frac{W}{g} = 0.03108$ *slugs*. From Newton's second law, the component of the acceleration of the collar parallel to the bar is $\vec{a} = \left(\frac{1}{m}\right)\left(\left(\frac{1}{20}\right)t^2\vec{i}\right)$ (ft/s^2).

The velocity is $\vec{v}(t) = \left(\frac{1}{m}\right)\left(\frac{1}{60}t^3\vec{i}\right)$ (ft/s), since the collar starts from rest. The displacement:

$\vec{r}(t) = \left(\frac{1}{m}\right)\left(\frac{1}{240}t^4\vec{i}\right)$ (ft). At $\vec{r} = 4\vec{i}$, the time is $t = (240m(4))^{0.25} = 2.337$ s. The velocity at the end of the bar is $\boxed{\vec{v} = 6.846\vec{i} \ ft/s}$

Problem 3.14 The airplane weighs 20, 000 *lb*. At the instant shown, the pilot increases the thrust *T* of the engine by 5000 *lb*. The horizontal component of the airplane's acceleration the instant before the thrust is increased is 20 *ft/s*^{2} What is the horizontal component of the airplane's acceleration the instant after the thrust is increased?

Solution The horizontal component of thrust is $T_{hor} = T\cos(15^o)$. Before the increase, the thrust is

$T_{hor} = \left(\frac{20000}{32.17}\right)20 = 12{,}434$ lb. The horizontal component after the increase is

$T_{after} = 12{,}434 + (5000)\cos(15^o) = 17263.6$ lb. The new horizontal component of the acceleration is

$$\boxed{a_{after} = \left(\frac{32.17}{20{,}000}\right)T_{after} = 27.77 \ ft/s^2}$$

Problem 3.15 The rocket travels straight up at low altitude. Its weight at the present time is 200 *kip*, and the thrust of its engine is 270 *kip*. An onboard accelerometer indicates that its acceleration is 10 *ft/s*^{2} upward. What is the magnitude of the aerodynamic drag on the rocket?

Solution The mass of the rocket is $m = \frac{200000}{32.17} = 6217$ *slugs*. The force on the rocket is $\vec{F} = \vec{F}_{thrust} + \vec{F}_{drag} - \vec{W} = m\vec{a}$, where $\vec{a}$ is positive upward. From which $\vec{F}_{drag} = -\vec{F}_{thrust} + m\vec{a} + \vec{W} = (-270{,}000 + 6217(10) + 200{,}000)\vec{j} = -7830.3\vec{j}$ (lb) from which the magnitude is $\boxed{|\vec{F}_{drag}| = 7830 \ lb}$ [*Note:* If a value $g = 32.2$ ft/s^2 is adopted, $|\vec{F}_{drag}| = 7888$ (lb)].

================================<>================================

Problem 3.16 A cart partially filled with water is initially stationary (Fig. a). The total mass of the cart and water is m. The cart is subjected to a time dependent force (Fig. b). If the horizontal forces exerted on the wheels by the floor and no water sloshes out, what is the x coordinate of the center of the cart after the motion of the water has subsided?

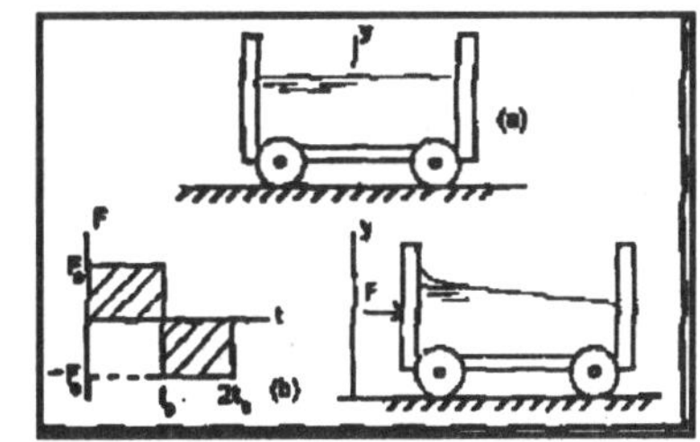

Solution The acceleration during the first time interval is $a = \frac{F_o}{m}$. The velocity is $v(0 \le t \le t_o) = \frac{F_o t}{m}$, since the cart starts from rest. The position: $s(0 \le t \le t_o) = \frac{F_o t^2}{2m}$. In the second interval, $a = -\frac{F_o}{m}$. The velocity: $v(t_o < t \le 2t_o) = \frac{-F_o t}{m} + \frac{2F_o t_o}{m}$. The position: $s(t_o < t \le 2t_o) = -\frac{F_o t^2}{2m} + \frac{2F_o t_o}{m}t - \frac{F_o t_o^2}{m}$. At the end of the second interval the position is $s(t = 2t_o) = -\frac{2F_o t_o^2}{m} + \frac{4F_o t_o^2}{m} - \frac{F_o t_o^2}{m} = \frac{F_o t_o^2}{m}$, and the velocity is $v(t = 2t_o) = -\frac{F_o(2t_o)}{m} + \frac{2F_o t_o}{m} = 0$. Since the velocity is zero when $t > 2t_o$, and the forces are zero when $t > 2t_o$, the center of mass of the cart remains stationary, and eventually the water will settle due to internal losses. The position of the cart is $\boxed{s(t > 2t_o) = s(t = 2t_o) = \frac{F_o t_o^2}{m}}$

================================<>================================

Problem 3.17 The combined weight of the motorcycle and rider is 360 *lb*. The coefficient of kinetic friction between the motorcycle's tires and the road is $\mu_k = 0.8$. If he spins the rear (drive) wheel, the normal force between the rear wheel and the road is 250 *lb*., and the horizontal force exerted on the front wheel by the road is negligible, what is the resulting horizontal acceleration?

Solution The horizontal component of the force is $F = \mu_k N = (0.8)(250) = 200\ lb$. The acceleration is

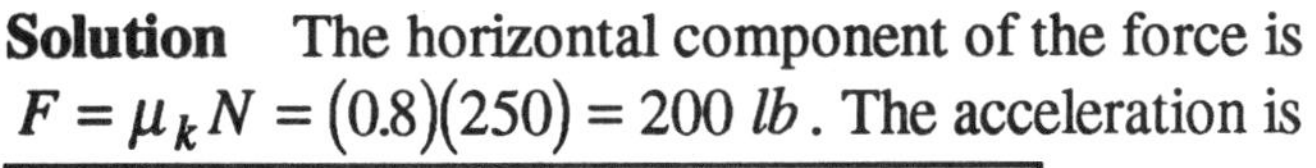

$$\boxed{a = \left(\frac{g}{W}\right)F = \left(\frac{32.17}{360}\right)200 = 17.87\ ft/s^2}$$

================================<>================================

Problem 3.18 The bucket B weighs 400 *lb*. and the acceleration of its center of mass is $\vec{a} = -30\vec{i} - 10\vec{j}\ \ (ft/s^2)$ Determine the x and y components of the total force exerted on the bucket by its supports.

Solution The force on the bucket is

$\vec{F} - \vec{W} = \left(\frac{400}{32.17}\right)(-30\vec{i} - 10\vec{j}) = -373.0\vec{i} - 124.34\vec{j}$, from which

$\boxed{\vec{F} = -373.0\vec{i} - (124.34 - 400)\vec{j} = -373.0\vec{i} + 275.7\vec{j}\ \ (lb)}$ [*Note:* If the value $g = 32.2\ ft/s^2$ is adopted, $\vec{F} = -372.7\vec{i} + 275.8\vec{j}\ \ (lb)$.]

================================<>================================

==================================<>==================================

Problem 3.19 During a test flight in which a 9000 *kg* helicopter starts from rest at $t = 0$, the acceleration of its center of mass from $t = 0$ to $t = 10$ seconds is $\vec{a} = (0.6t)\vec{i} + (1.8 - 0.36t)\vec{j}\ m/s^2$. What is the magnitude of the total external force on the helicopter (including its weight) at $t = 6\ s$?

Solution From Newton's second law: $\sum \vec{F} = ma$. The sum of the external forces is

$\sum \vec{F} = \vec{F} - \vec{W} = 9000\left[(0.6t)\vec{i} + (1.8 - 0.36t)\vec{j}\right]_{t=6} = 32400\vec{i} - 3240\vec{j}$, from which the magnitude is

$\left|\sum \vec{F}\right| = \sqrt{32400^2 + 3240^2} = 32.562\ (N)$.

==================================<>==================================

Problem 3.20 The engineers conducting the test described in Problem 3.19 want to express the total force on the helicopter at $t = 6$ seconds in terms of three forces: the weight *W*, a component *T* tangent to the path, and a component *L* normal to the path. What are the values of *W*, *T*, and *L*?

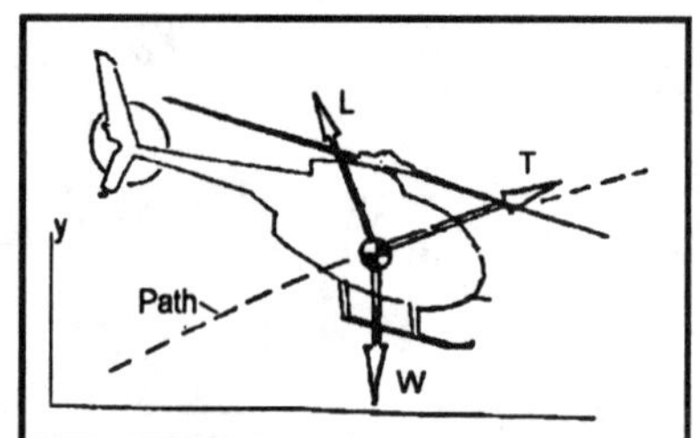

Solution Integrate the acceleration: $\vec{v} = (0.3t^2)\vec{i} + (1.8t - 0.18t^2)\vec{j}$, since the helicopter starts from rest. The instantaneous flight path angle is

$$\tan\beta = \frac{dy}{dx} = \left(\frac{dy}{dt}\right)\left(\frac{dx}{dt}\right)^{-1} = \frac{(1.8t - 0.18t^2)}{(0.3t^2)}.$$ At $t = 6\ s$, $\beta_{t=6} = \tan^{-1}\left(\frac{(1.8(6) - 0.18(6)^2)}{0.3(6)^2}\right) = 21.8^o$. A unit vector tangent to this path is $\vec{e}_t = \vec{i}\cos\beta + \vec{j}\sin\beta$. A unit vector normal to this path $\vec{e}_n = -\vec{i}\sin\beta + \vec{j}\cos\beta$. The weight acts downward: $\vec{W} = -\vec{j}(9000)(9.81) = -88.29\vec{j}\ (kN)$. From Newton's second law, $\vec{F} - \vec{W} = m\vec{a}$, from which $\vec{F} = \vec{W} + m\vec{a} = 32400\vec{i} + 85050\vec{j}\ (N)$. The component tangent to the path is $T = \vec{F}\cdot\vec{e}_t = 32400\cos\beta + 85050\sin\beta = 61669.4\ (N)$ The component normal to the path is $L = \vec{F}\cdot\vec{e}_n = -32400\sin\beta + 85050\cos\beta = 66934\ (N)$

==================================<>==================================

Problem 3.21 At the instant shown, the *11000 kg* airplane's velocity is $\mathbf{v} = 270\ \mathbf{i}\ m/s$. The forces acting on the aircraft are its weight, the thrust, $T = 110\ kN$, the lift $L = 260\ kN$, and the drag, $D = 34\ kN$. (The x-axis is paralel to the airplane's path.) Determine the magnitude of the airplane's acceleration.

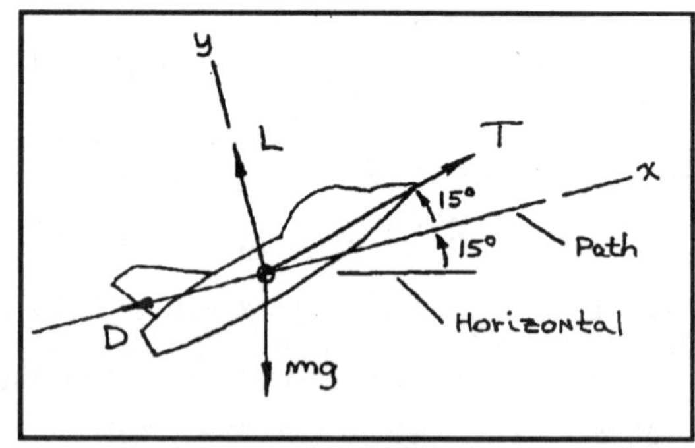

Solution: Let us sum forces and write the acceleration components along the x and y axes as shown. After the acceleration components are known, we can determine its magnitude. The equations of motion, in the coordinate directions, are $\sum F_X = T\cos 15^\circ - D - W\sin 15^\circ = ma_X$, and $\sum F_Y = L + T\sin 15^\circ - W\cos 15^\circ = ma_Y$. Substituting in the given values for the force magnitudes, we get $a_X = 4.03 m/s^2$ and $a_Y = 16.75 m/s^2$. The magnitude of the acceleration is $|\mathbf{a}| = \sqrt{a_X^2 + a_Y^2} = 17.23 m/s^2$

==================================<>==================================

===================================<>==================================

Problem 3.22 At the instant shown in Problem 3.21, determine: (a) the rate of change of the magnitude of the airplane's velocity dv_r/dt; and (b) the instantaneous radius of curvature of the airplane's path.

Soluton: To find the needed quantities, we go to the equations of motion in normal and tangential coordinates. From Eq. (2.40), we have $a_t = dv/dt$ and $a_n = v^2 / \rho$. We know that $a_X = a_t$ and $a_Y = a_n$ because of the directions of our coordinates (x is along the velocity vector). From the solution to Problem 3.21, we have $a_t = a_X = 4.03 m/s^2$ and $a_n = a_Y = 16.75 m/s^2$. The velocity is $\mathbf{v} = 270\ \mathbf{i}\ m/s$ and from this, we can determine the radius of curvature of the path $\rho = v^2 / a_n = 4353m$

===================================<>==================================

Problem 3.23 The coordinates of the *360 kg* sport plane's center of mass relative to an earth-fixed reference frame during an interval of time are $x = 20t - 1.63t^2$, $y = 35t - 0.15t^3$, and $z = -20t + 1.38t^2$, where t is in seconds (and distances are in meters). The y axis points upward. The forces exerted on the plane are its weight, the thrust vector **T** exerted by its engine, the lift force vector **L**, and the drag force vector **D.** At t = 4 s, determine **T + L + D.**

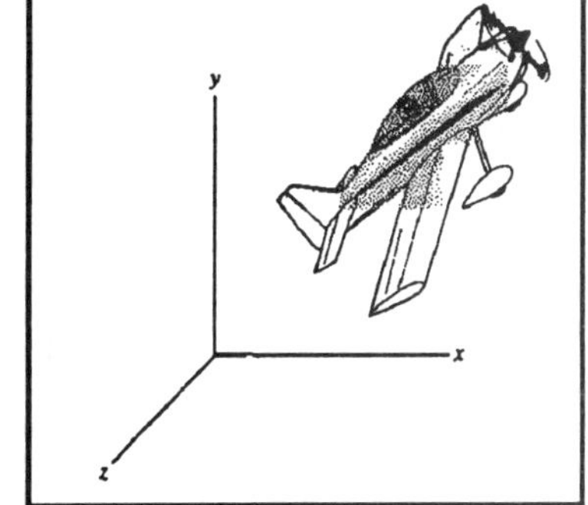

Soluton: There are four forces acting on the airplane. Newton's second law, in vector form, given $\mathbf{T}+\mathbf{L}+\mathbf{D}+\mathbf{W} = (\mathbf{T}+\mathbf{L}+\mathbf{D}) - mg\mathbf{j} = m\mathbf{a}$. Since we know the weight of the airplane and can evaluate the total acceleration of the airplane from the information given, we can evaluate the **(T + L + D)** (but we cannot evaluate these forces separately without more information. Differentiating the position equations twice and evaluating at *t = 4.0 s* , we get $a_X = -3.26 m/s^2$, $a_Y = -3.60 m/s^2$, and $a_Z = 2.76 m/s^2$. (Note that the acceleration components are constant over this time interval. Substituting into the equation for acceleration,we get $(\mathbf{T}+\mathbf{D}+\mathbf{L}) = m\mathbf{a} + mg\mathbf{j}$. The mass of the airplane is 360 kg. Thus,

$(\mathbf{T}+\mathbf{D}+\mathbf{L}) = \ -1174\,\mathbf{i} + 2236\,\mathbf{j} + 994\mathbf{k}\ (newtons)$

===================================<>==================================

Problem 3.24 the force in newtons exerted on the *360 kg* sport plane in Problem 3.23 by its engine , the lift force, and the drag force during an interval of time is

$(\mathbf{T}+\mathbf{D}+\mathbf{L}) = \ (-1000 + 280t)\ \mathbf{i} + (4000 - 430t)\ \mathbf{j} + (720 + 200t)\mathbf{k}\ (newtons)$, where t is the time in seconds. If the coordinates of the plane's center of mass are *(0,0,0)* and its velocity is $\mathbf{v} = 20\ \mathbf{i} + 35\ \mathbf{j} - 20\mathbf{k}\ m/s$ at *t=0s,*what are the coordinates of the center of mass at $t = 4s$?

Soluton: Since we are working in nonrotating rectangular cartesian coordinates, we can consider the motion in each axis separately. From Problem 3.23, we have $(\mathbf{T}+\mathbf{D}+\mathbf{L}) = m\mathbf{a} + mg\mathbf{j}$. Separating the information for each axis, we have $ma_X = -1000 + 280t$, $ma_Y = 4000 - 430t - mg$, and $ma_Z = 720 + 200t$

Integrating the x equation, we get $v_x = v_{x0} + (1/m)(-1000t + 280t^2/2)$ and

$x = v_{X0}t + (1/m)\left(-1000t^2/2 + 280t^3/6\right)$.

Integrating the y equation, we get $v_Y = v_{Y0} + (1/m)((4000 - mg)t - 430t^2/2)$ and

$y = v_{Y0}t + (1/m)\left((4000 - mg)t^2/2 - 430t^3/6\right)$

Integrating the z equation, we get $v_Z = v_{Z0} + (1/m)(720t + 200t^2/2)$ and

$z = v_{Z0}t + (1/m)\left(720t^2/2 + 200t^3/6\right)$. Evaluating at *t = 4s* we find the aircraft at (66.1, 137.7, $-58.1)m$ relative to its initial position at *t=0.*

===================================<>==================================

==================================<>==================================

Problem 3.25 The robot manipulator is programmed so that $x = 4 + t^2$ *in.*, $y = \frac{1}{4}x^2$ *in.*, $z = 0$ during the interval of time from $t = 0$ to $t = 4$ seconds. What are the *x* and *y* components of the total force exerted by the jaws of the manipulator on the 10 *lb* widget A at $t = 2$ seconds?

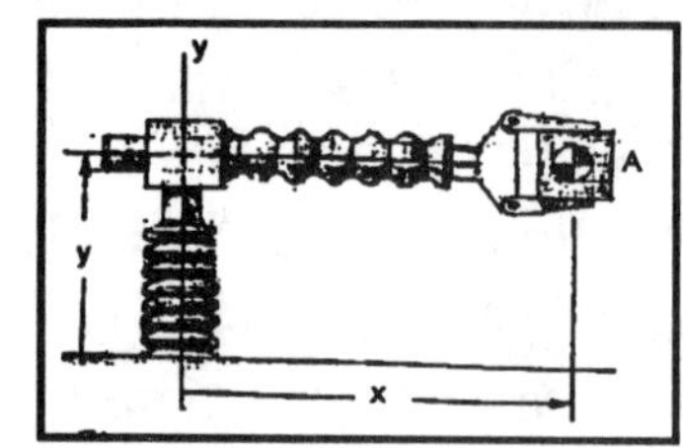

Solution The velocity is $\frac{dx}{dt} = 2t \ in/s$, from which

$\frac{d^2x}{dt^2} = 2 \ in/s^2 = \frac{1}{6} \ ft/s^2$. $\frac{dy}{dt} = \frac{dy}{dx}\frac{dx}{dt} = xt \ in/s$.

$\frac{d^2y}{dt^2} = \left(\frac{d^2y}{dx^2}\right)\left(\frac{dx}{dt}\right)^2 + \left(\frac{dy}{dx}\right)\left(\frac{d^2x}{dt^2}\right) = 3t^2 + 4 \ in/s^2 = \frac{t^2}{4} + \frac{1}{3} \ ft/s^2$. The force exerted by the jaws is

$$\vec{F} = \left(\frac{W}{g}\right)\left(\frac{1}{6}\vec{i} + \frac{4}{3}\vec{j}\right) + W\vec{j} = 0.0518\vec{i} + 10.414\vec{j} \ (lb)$$

==================================<>==================================

Problem 3.26 The robot manipulator in Problem 3.25 is stationary at $t = 0$ and is programmed so that $a_x = 2 - 0.4v_x \ in/s^2$, $a_y = 1 - 0.2v_y \ in/s^2$, $a_z = 0$ during the time from $t = 0$ to $t = 4$ seconds. What are the *x* and *y* components of the total force exerted by the jaws of the manipulator on the 10 *lb.* widget A at $t = 2$ seconds?

Solution The force exerted by the jaws is obtained by integrating the acceleration to obtain an expression for the velocity as a function of time, and then differentiating the velocity to obtain the accelerations as a function of time. From $\frac{dv}{dt} = a$, each of the expressions for the acceleration can be integrated by separation of variables.(1) $\frac{dv_x}{2 - 0.4v_x} = dt$, from which $v_x = \frac{1}{0.4}\left(2 - Ce^{-0.4t}\right)$. From $v_x(0) = 0$, $C = 2$, from which $v_x(t) = 5\left(1 - e^{-0.4t}\right) \ in/s$, and the acceleration is $a_x(t) = 2e^{-0.4t} \ in/s^2$.

(2) $\frac{dv_y}{1 - 0.2v_y} = dt$, from which $v_y(t) = \left(\frac{1}{0.2}\right)\left(1 - Ce^{-0.2t}\right)$. From $v_y(0) = 0$, $C = 1$, from which $v_y(t) = 5\left(1 - e^{-0.2t}\right) \ in/s$, and the acceleration is $a_y(t) = e^{-0.2t} \ in/s^2$. The force exerted by the jaws is

$$\vec{F} = \left(\frac{W}{12g}\right)\left(a_x(2)\vec{i} + a_y(2)\vec{j}\right) + W\vec{j} = 0.0233\vec{i} + 10.0174\vec{j} \ (lb)$$

==================================<>==================================

===================================<>===================================

Problem 3.27 In the sport of curling, the object is to slide a "stone" weighing 44 *lb.* onto the center of a target located 31 yards from the point of release. If $\mu_k = 0.01$ and the stone is thrown directly toward the target, what initial velocity would result in a perfect shot?

Solution From Newton's second law, $-\mu_k N = \left(\frac{W}{g}\right)a$. The normal force exerted on the stone is $N = W$, from which $\frac{d^2s}{dt^2} = -g\mu_k$,

$\frac{ds}{dt} = v(t) = -g\mu_k t + V_o$, since $v(0) = V_o$, and $s(t) = -\frac{g\mu_k}{2}t^2 + V_o t\ ft$,

since $s(0) = 0$. At the target, $v(t_{target}) = 0 = -g\mu_k t_{target} + V_o$, from which

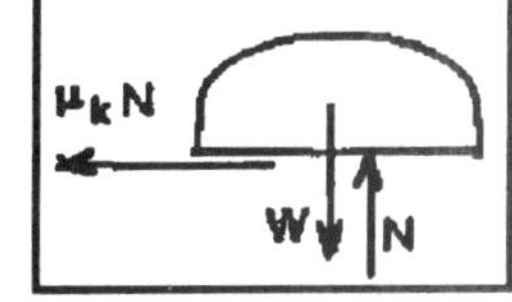

$t_{target} = \frac{V_o}{g\mu_k}$. The distance to the target is $s(t_{target}) = d = 3(31) = 93\,ft$, from

which $s(t_{target}) = -\frac{g\mu_k}{2}t_{target}^2 + V_o t_{target} = d$, from which

$-\frac{1}{2g\mu_k}V_o^2 + \frac{1}{g\mu_k}V_o^2 = d$. The initial velocity is $\boxed{V_o = \sqrt{2g\mu_k d} = \sqrt{2(32.17)(0.01)(93)} = 7.735\ ft/s}$

Note: If the value $g = 32.2\ ft/s^2$ is adopted, $V_o = 7.739\ ft/s$.

===================================<>===================================

Problem 3.28 The two weights are released from rest. How far does the 50 *lb.* weight fall in one-half second?

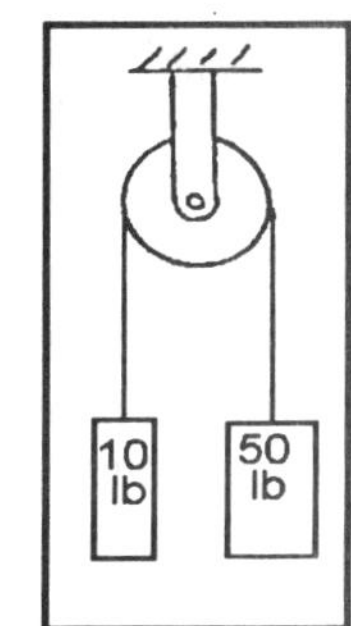

Solution Write the equations for the position of the 50 *lb.* weight, with y positive downward. From Newton's second law, $T - 10 = \left(\frac{10}{g}\right)a_{10}$, and $T - 50 = \left(\frac{50}{g}\right)a_{50}$, and since the pulley is one-to-one: $a_{10} = -a_{50}$, from which $a_{50} = \left(\frac{g}{60}\right)F = \left(\frac{2}{3}\right)g\ ft/s^2$.

The velocity is $v(t) = \left(\frac{2}{3}\right)gt\ ft/s$ since $v(0) = 0$. The position is $y(t) = \frac{gt^2}{3}\ ft$, since $y(0) = 0$. The distance fallen in one-half second is $\boxed{y(t = 0.5) = \frac{g}{12} = 2.68\ ft}$

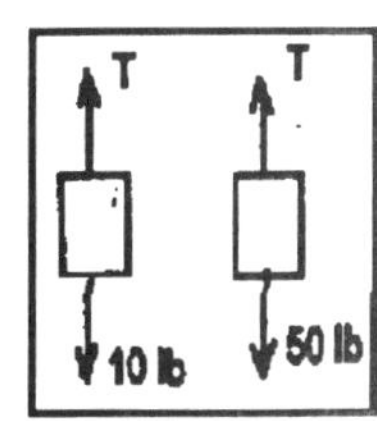

===================================<>===================================

================================<>================================

Problem 3.29 In Example 3.2, what is the ratio of the tension in the cable to the weight of crate B after the crates are release from rest?

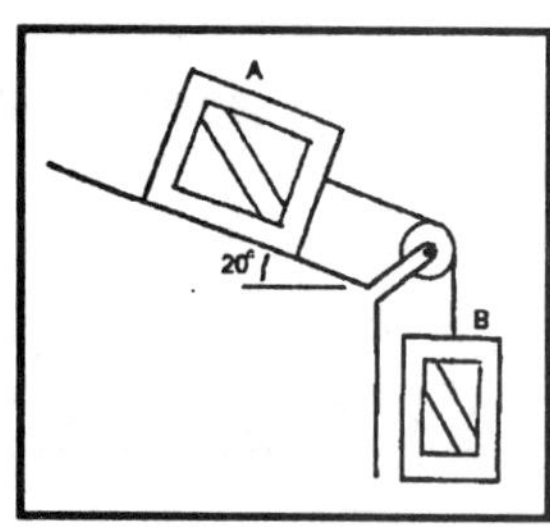

Solution Assume that crate A will slip, as in the example. From Newton's second law the sum of the forces on crate A is

$\sum F_x = T + m_a g \sin\theta - \mu_k N = m_a a_x$, $\sum F_y = N - m_a g \cos\theta = 0$. From Example 3.2 $m_a = 40\ kg$, $m_b = 30\ kg$, $\mu_k = 0.15$, $\theta = 20^o$. From the second equation, $N = 368.7.7\ N$. The equation of motion for crate B is $\sum F_x = m_b g - T = m_b a_x$. Substitute to obtain the two simultaneous equations: $T - 40a_x = -78.9$, $T + 30a_x = 294.3$. Solve: $T = 134.36\ N$,

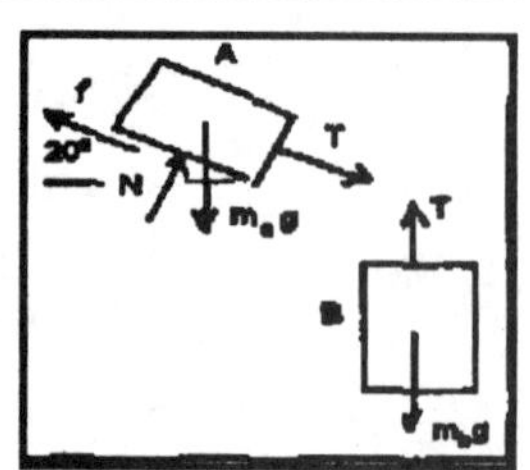

$a_x = \dfrac{294.3 - T}{30} = 5.33\ m/s^2$. The weight of crate B is $W_b = m_b g = 294.3\ N$, from which $\boxed{\dfrac{T}{W_b} = \dfrac{134.36}{294.3} = 0.4565}$

================================<>================================

Problem 3.30 Each box weighs 50 pounds and friction can be neglected. If the boxes start from rest at $t = 0$, determine the magnitude of their velocity and the distance they have moved from their initial position at $t = 1$ second.

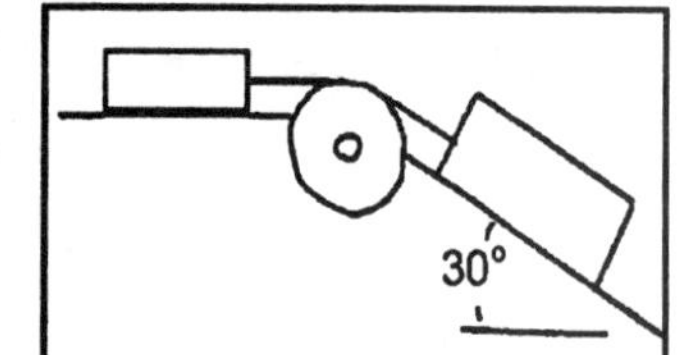
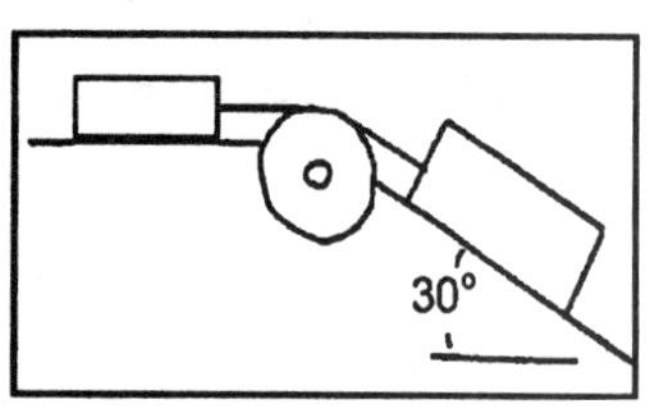

Solution From Newton's second law for the two boxes: $T = \dfrac{W}{g}a$, and $-T + W\sin\theta = \dfrac{W}{g}a$. Solve: $a = \dfrac{g\sin\theta}{2} = \dfrac{g}{4}$, from which the velocity is $v(t) = \dfrac{gt}{4}\ ft/s$, since $v(0) = 0$. At $t = 1\ s$, $\boxed{v(t=1) = \dfrac{g}{4} = 8.043\ ft/s}$. The

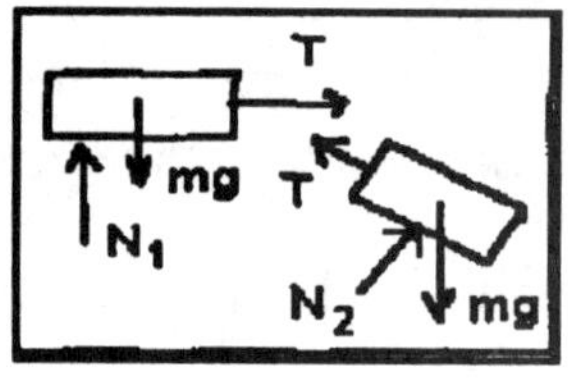

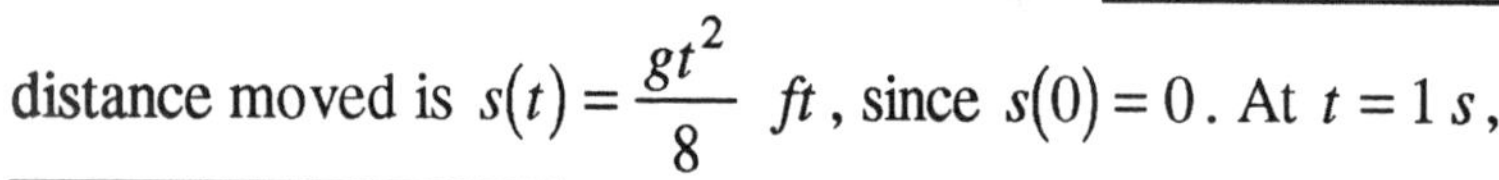

distance moved is $s(t) = \dfrac{gt^2}{8}\ ft$, since $s(0) = 0$. At $t = 1\ s$,

$\boxed{s(t=1) = \dfrac{g}{8} = 4.02\ ft}$

================================<>================================

Problem 3.31 In Problem 3.30, determine the magnitude of the velocity of the boxes and the distance they have moved from their initial position at $t = 1\ s$ if the coefficient of kinetic friction between the boxes and the surface is $\mu_k = 0.15$.

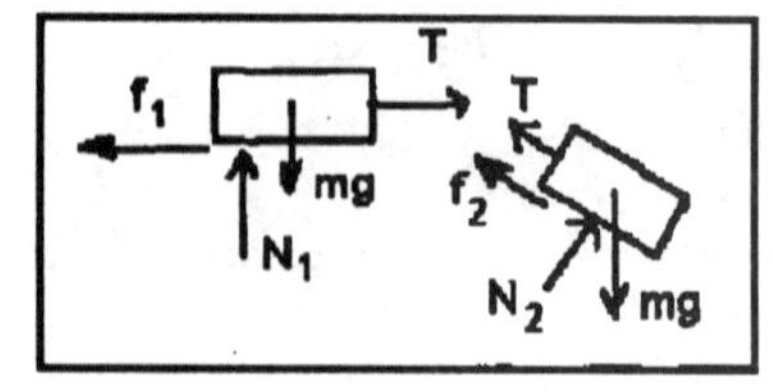

Solution Newton's second law for block 1 is $T - \mu_k N_1 = \dfrac{W}{g}a_1$, and for block 2 $-T - \mu_k N_2 + W\sin\theta = \left(\dfrac{W}{g}\right)a_2$. From the summation of forces: $N_1 = W$, $N_2 = W\cos\theta$.

Since the pulley is one-to-one $a_1 = a_2 = a$, from which $T - \mu_k W = \left(\dfrac{W}{g}\right)a$, and

$-T + W(\sin\theta - \mu_k \cos\theta) = \left(\dfrac{W}{g}\right)a$. Solve: $a = \dfrac{g}{2}(\sin\theta - \mu_k(1 - \cos\theta)) = \dfrac{g}{4}(1 - \mu_k(2 + \sqrt{3}))$

Solution continued on next page

Continuation of Solution to Problem 3.31

Check: This should reduce to the expression for the acceleration in Problem 3.30 when $\mu_k = 0$. *check.*

The velocity is $v(t) = \left(\frac{g}{4}\right)\left(1 - \mu_k\left(2+\sqrt{3}\right)\right)t$, since $v(0) = 0$. At $t = 1\ s$, $\boxed{v(t=1) = 3.54\ ft/s}$. The distance is $s(t) = \left(\frac{g}{8}\right)\left(1 - \mu_k\left(2+\sqrt{3}\right)\right)t^2$, since $s(0) = 0$. At $t = 1\ s$, $\boxed{s(t=1) = 1.770\ ft}$

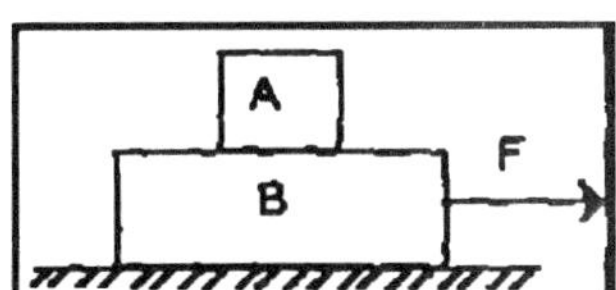

Problem 3.32 The masses $m_A = 15\ kg$, $m_B = 30\ kg$, and the coefficients of friction between all of the surfaces are $\mu_s = 0.4$, $\mu_k = 0.35$. What is the largest force *F* that can be applied without causing A to slip relative to B? What is the resulting acceleration?

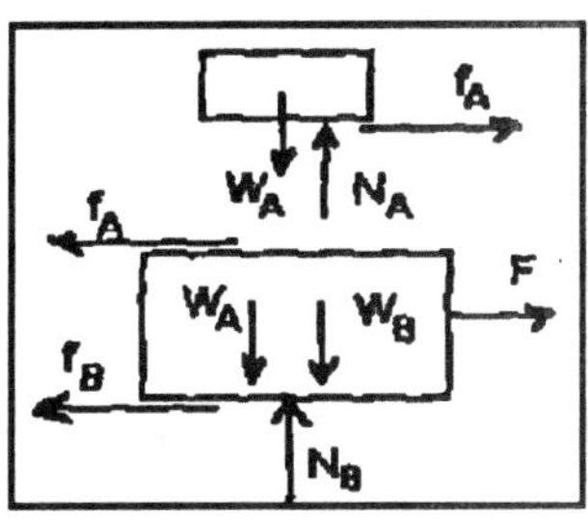

Solution From the summation of forces $N_A = W_A$, $N_B = W_A + W_B$ From Newton's second law, the equation of motion for the upper block at impending slip is $\mu_s N_A = m_A a$, from which the condition for slip is $\boxed{a = \mu_s \frac{W_A}{m_A} = \mu_s g = (0.4)(9.81) = 3.924\ \ m/s^2}$. When the system is accelerating the bottom block has slipped relative to the lower surface. From Newton's second law, $F - \mu_k N_B = (m_A + m_B)a$, from which

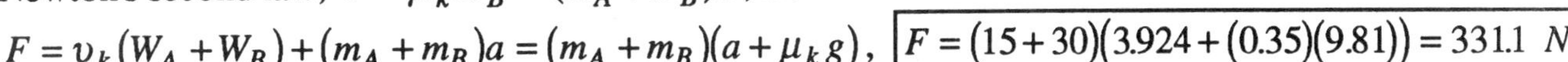

$F = \upsilon_k(W_A + W_B) + (m_A + m_B)a = (m_A + m_B)(a + \mu_k g)$, $\boxed{F = (15+30)(3.924 + (0.35)(9.81)) = 331.1\ \ N}$

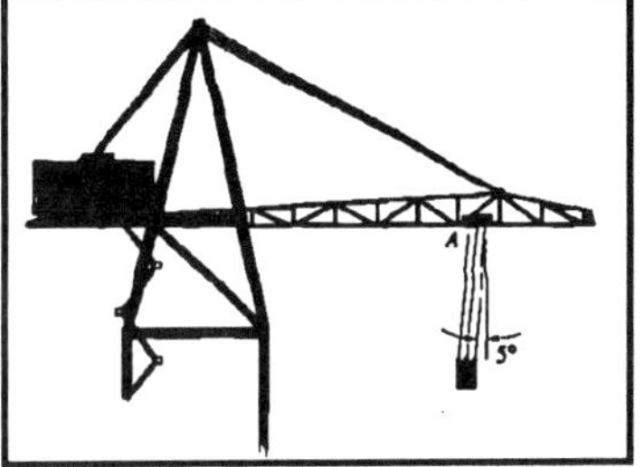

Problem 3.33 The crane's trolley at A moves to the right with constant acceleration and the 800 *kg* load moves without swinging. (a) What is the acceleration of the trolley and load? (b) What is the sum of the tensions in the parallel cables supporting the load?

Solution (a) From Newton's second law, $T\sin\theta = ma_x$, and $T\cos\theta - mg = 0$. Solve

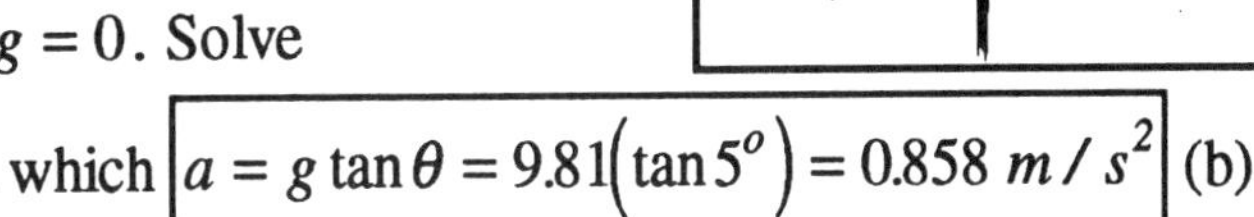

$a_x = \frac{T\sin\theta}{m}$, $T = \frac{mg}{\cos\theta}$, from which $\boxed{a = g\tan\theta = 9.81\left(\tan 5^o\right) = 0.858\ m/s^2}$ (b) $\boxed{T = \frac{800(9.81)}{\cos 5^o} = 7878\ \ N}$

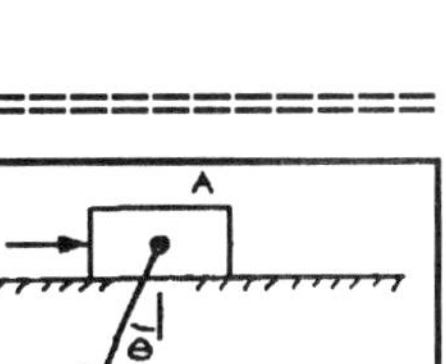

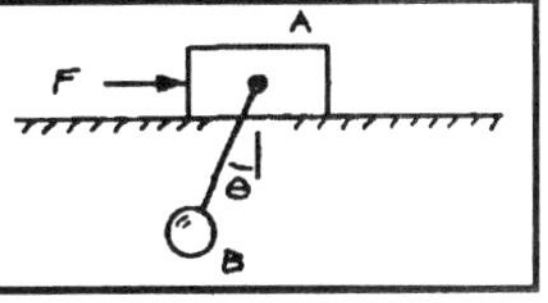

Problem 3.34 The mass of A is *30 kg* and the mass of B is *5 kg*. A slides on the smooth surface and the angle $\theta = 20^\circ$ is constant. What is the force *F?*

Solution: Draw separate free body diagrams of objects A and B. Write the equations of motion using the diagrams. Note that the acceleration for A and B are the same. The equations for A are $\sum F_X = F - T\sin 20^\circ = m_A a$ and $\sum F_Y = N - m_A g - T\cos 20^\circ = 0$. The equations for B are $\sum F_X = T\sin 20^\circ = m_B a$ and $\sum F_Y = T\cos 20^\circ - m_B g = 0$. Inserting the numbers, we get $a = 3.57 m/s^2$, $T = 52.2$ *newtons*, and $F = 125.0$ *newtons*.

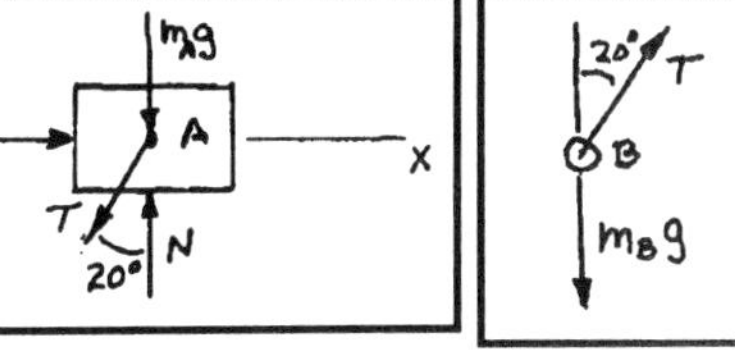

==================================<>==================================

Problem 3.35 In Problem 3.34, what is the force F if the coefficient of kinetic friction between A and the surface is $\mu_k = 0.24$?

Solution: Draw separate free body diagrams of objects A and B. Write the equations of motion using the diagrams. Note that the acceleration for A and B are the same. The equations for A are $\sum F_X = F - T\sin 20^\circ - \mu_k N = m_A a$ and $\sum F_Y = N - m_A g - T\cos 20^\circ = 0$. The equations for B are $\sum F_X = T\sin 20^\circ = m_B a$ and $\sum F_Y = T\cos 20^\circ - m_B g = 0$. Inserting the numbers, we get $a = 3.57 m/s^2$ and $F = 207.4$ *newtons*. Note that the acceleration stayed the same. This is because the acceleration of B had to remain the same. The force F just increased enough to overcome the friction force.

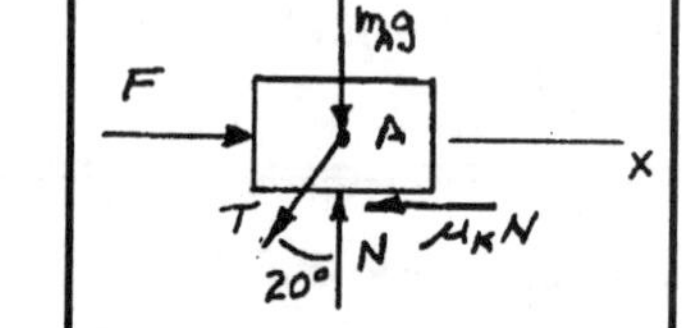

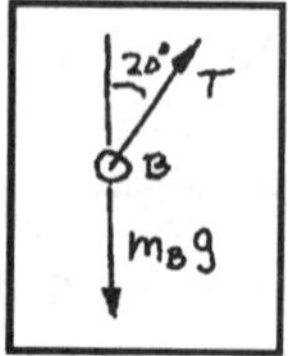

==================================<>==================================

Problem 3.36 The 100 lb. crate is initially stationary. The coefficients of friction between the crate and the inclined surface are $\mu_s = 0.2$, $\mu_k = 0.16$. Determine how far the crate moves from its initial position in 2 seconds if the horizontal force $F = 90\ lb$.

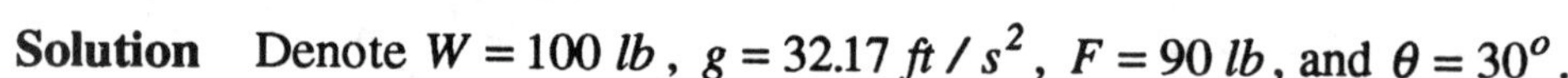

Solution Denote $W = 100\ lb$, $g = 32.17\ ft/s^2$, $F = 90\ lb$, and $\theta = 30^o$. Choose a coordinate system with the positive x axis parallel to the inclined surface. (See free body diagram next page.) The normal force exerted by the surface on the box is $N = F\sin\theta + W\cos\theta = 131.6\ \ lb$. The sum of the non-friction forces acting to move the box is $F_c = F\cos\theta - W\sin\theta = 27.9\ \ lb$. Slip only occurs if

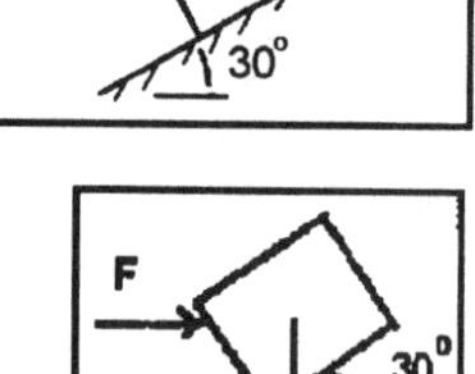

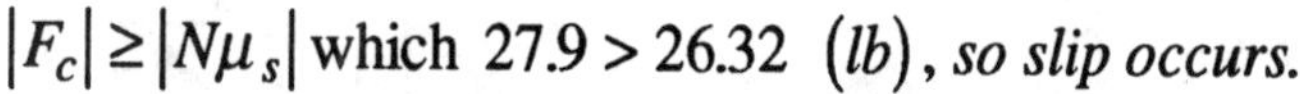

$|F_c| \geq |N\mu_s|$ which $27.9 > 26.32\ \ (lb)$, *so slip occurs.*

The direction of slip is determined from the sign of the sum of the non friction forces: $F_c > 0$, implies that the box slips up the surface, and $F_c < 0$ implies that the box slips down the surface (if the condition for slip is met). Since $F_c > 0$ *the box slips up the surface.* After the box slips, the sum of the forces on the box parallel to the surface is $\sum F_x = F_c - \text{sgn}(F_c)\mu_k N$, where $\text{sgn}(F_c) = \frac{F_c}{|F_c|}$. From Newton's second law, $\sum F_x = \left(\frac{W}{g}\right)a$, from which $a = \frac{g}{W}(F_c - \text{sgn}(F_c)\mu_k N) = 2.125\ \ ft/s^2$. The velocity is $v(t) = at\ \ ft/s$, since $v(0) = 0$. The displacement is $s = \frac{a}{2}t^2\ \ ft$, since $s(0) = 0$. The position after 2 seconds is $\boxed{s(2) = 4.43\ \ ft}$ up the inclined surface.

==================================<>==================================

Problem 3.37 In Problem 3.36, determine how far the crate moves from its initial position in 2 seconds if the horizontal force $F = 30\ lb$.

Solution Use the definitions of terms given in the solution to Problem 3.36. For $F = 30\ lb$, $N = F\sin\theta + W\cos\theta = 101.6\ lb$, and $F_c = F\cos\theta - W\sin\theta = -24.0\ lb$ from which, $|F_c| = 24.0 > |\mu_s N| = 20.3\ lb$, so slip occurs. Since $F_c < 0$, *the box will slip down the surface.* From the solution to Problem 3.24, after slip occurs, $a = \left(\frac{g}{W}\right)(F_c - \text{sgn}(F_c)\mu_k N) = -2.497\ \ ft/s^2$. The position is $s(t) = \frac{a}{2}t^2$. At 2 seconds, $\boxed{s(2) = -4.995... = -5\ \ ft}$ down the surface.

==================================<>==================================

================================<>================================

Problem 3.38 The crate has a mass of 120 *kg* and the coefficients of friction between it and the sloping dock are $\mu_s = 0.6$, $\mu_k = 0.5$. (a) What tension must the winch exert on the cable to start the stationary crate sliding up the dock? (b) If the tension is maintained at the value determined in Part (a), what is the magnitude of the crate's velocity when it has moved 10 ft up the dock?

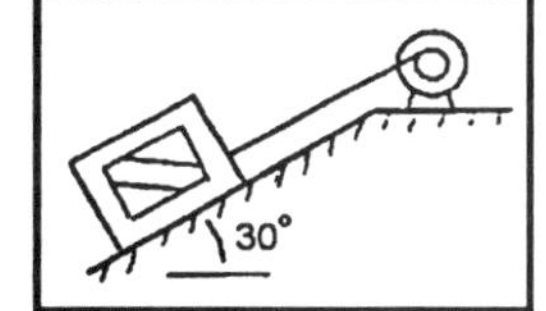

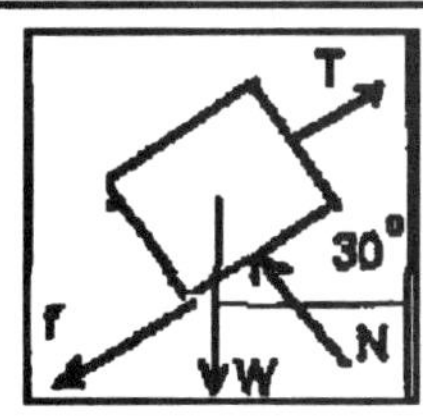

Solution Choose a coordinate system with the *x* axis parallel to the surface. Denote $\theta = 30^o$. (a) The normal force exerted by the surface on the crate is $N = W\cos\theta = 120(9.81)(0.866) = 1019.5\ N$. The force tending to move the crate is $F_c = T - W\sin\theta$, from which the tension required to start slip is $\boxed{T = W(\sin\theta) + \mu_s N\ = 1200.3\ N}$. (b) After slip begins, the force acting to move the crate is $F = T - W\sin\theta - \mu_k N = 101.95\ N$. From Newton's second law, $F = ma$, from which $a = \left(\frac{F}{m}\right) = \frac{101.95}{120} = 0.8496\ m/s^2$. The velocity is $v(t) = at = 0.8496t\ m/s$, since $v(0) = 0$. The position is $s(t) = \frac{a}{2}t^2$, since $s(0) = 0$. When the crate has moved 10 *ft* up the slope, $t_{10} = \sqrt{\frac{2(10)}{a}} = 4.85\ s$. and the velocity is $\boxed{v(t = 4.85) = a(4.85) = 4.122\ m/s}$

================================<>================================

Problem 3.39 The utility vehicle is moving forward at 10 *ft/s*. The coefficients of friction between its load A and the bed of the vehicle are $\mu_s = 0.5$, $\mu_k = 0.45$. If $\alpha = 0$, determine the shortest distance in which the vehicle can be brought to a stop without causing the load to slide on the bed.

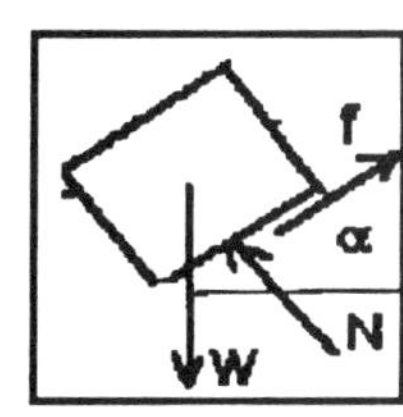

Solution The force on the load required to induce slip forward on the bed is $F = \mu_s W\cos\alpha + W\sin\alpha$. From Newton's second law, $F = ma_L$, from which the acceleration of the load relative to the vehicle is $a_L = \frac{F}{m} = g(\mu_s\cos\alpha + \sin\alpha)$. The acceleration required of the vehicle is $a_v = -a_L = -g(\mu_s\cos\alpha + \sin\alpha) = -16.1\ ft/s^2$. The velocity is $v(t) = a_v t + 10\ ft/s$, since $v(0) = 10\ ft/s$. Substitute and solve: the vehicle will come to a stop at $t = \frac{10}{a_v} = 0.6217\ s$. The distance traveled in this time is $s(t) = -a_v t^2 + 10t\ ft$, since $s(0) = 0$, from which $\boxed{s(t = 0.6217) = 3.11\, ft}$ This is the shortest possible distance that the vehicle can be stopped without the load slipping.

================================<>================================

Problem 3.40 In Problem 3.39, determine the shortest distance if the angle is (a) 15°; (b) −15°.

Solution Use the solution to Problem 3.39: $a_v = -a_L = -g(\mu_s\cos\alpha + \sin\alpha), t = \frac{10}{a_v}$, and $s(t) = \frac{a_v}{2}t^2 + 10t$. (a) For $\alpha = 15^o$, $a_v = -23.9\ ft/s^2$, $t = 0.4191\ s$, $\boxed{s(t = 0.4191) = 2.1\, ft}$. (b) For $\alpha = -15^o$, $a_v = -7.21\, ft/s^2$, $t = 1.387\ s$, and $\boxed{s(t = 1.387) = 6.93\, ft}$

================================<>================================

==================================<>==================================

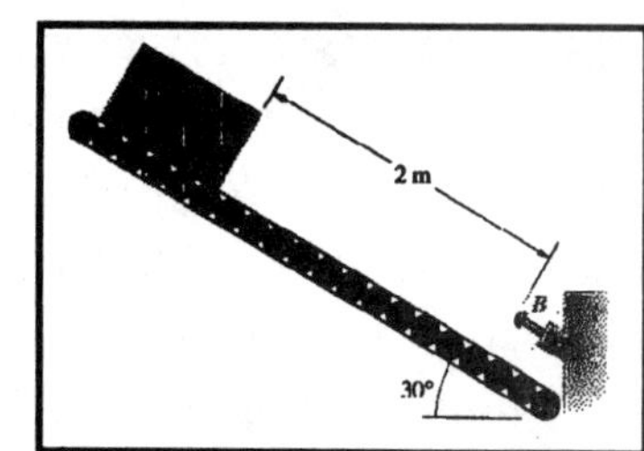

Problem 3.41 In an assembly line process, the 20 *kg* package A starts from rest and slides down the smooth ramp. Suppose that you want to design the hydraulic device B to exerts a constant force of magnitude F on the package and bring it to rest in a distance of 100 *mm*. What is the required force F?

Solution The force accelerating the package down the ramp is $F = mg\sin\theta\ N$. From Newton's second law, $F = ma$, from which the

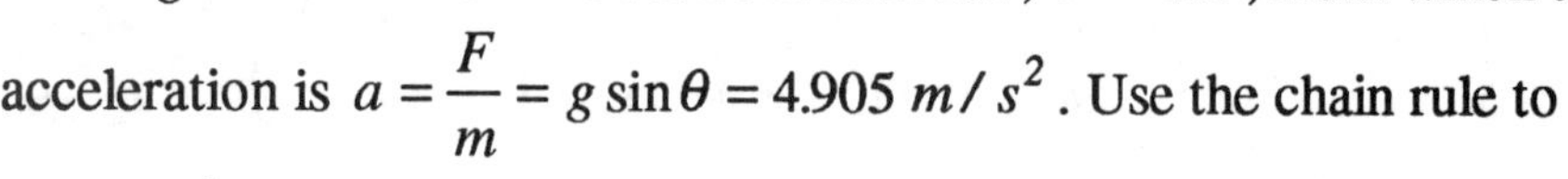

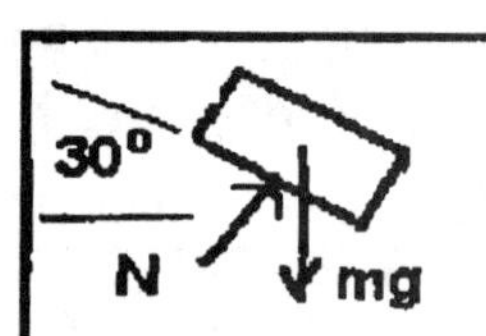

acceleration is $a = \frac{F}{m} = g\sin\theta = 4.905\ m/s^2$. Use the chain rule to write $v\frac{dv}{ds} = a$, from which $v^2(s) = 2as$, since $v(0) = 0$. At $s = 2\ m$, $v^2(s = 2) = 19.62\ \ m^2/s^2$ The deceleration of the package at the bottom must stop the package within 100 *mm*. Use the chain rule: $v\frac{dv}{ds} = -\frac{F}{m} + g\sin 30^o$, from which $v(s)^2 = 2\left(-\frac{F}{m} + g\sin 30^o\right)s + C$. Since $v^2(0) = 19.62\ m^2/s^2$, then $C = 19.62$, from which $v^2(s) = \left(-\frac{2F}{m} + g\right)s + (19.62)$. When $s = 0.1\ m$, $v(s = 0.1) = 0$. Solve: $\boxed{F = \left(\frac{g + 10(19.62)}{2}\right)m = 2060.1\ \ N}$

==================================<>==================================

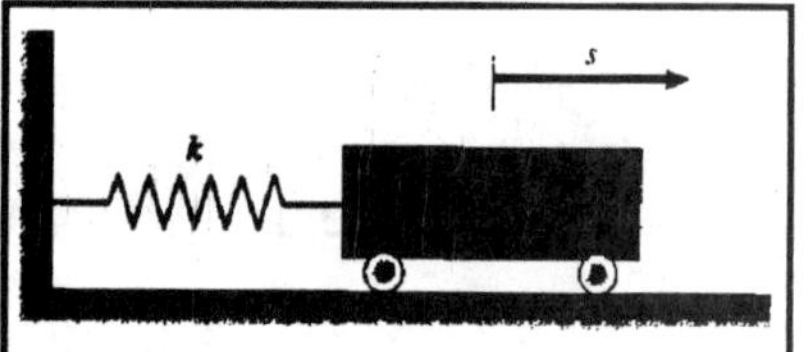

Problem 3.42 The force exerted on the 10 *kg* mass by the linear spring is $F = -ks$, where k is the "spring constant" and s is the displacement of mass from its position when the spring is unstretched. The value of k is 50 *N/m*. The mass is released from rest in the position $s = 1\ m$. (a) What is the acceleration of the mass at the instant it is released? (b) What is the velocity of the mass when it reaches the position $s = 0$?

Solution At the instant of release, the force on the mass is $F = -ks = -50(1)\ N$. The acceleration is $\boxed{a = \frac{F}{m} = -\frac{50}{10} = -5\ \ m/s^2}$. Use the chain rule to write $v\frac{dv}{ds} = -\frac{ks}{m}$, from which $v^2(s) = -\frac{k}{m}s^2 + C$. Note that $v(s = 1) = 0$, from which $C = \frac{k}{m}$, and $v^2(s) = \left(\frac{k}{m}\right)(1 - s^2)\ \ m^2/s^2$. When $s = 0$,

$\boxed{v(0) = \pm\sqrt{\frac{k}{m}} = \pm 2.236\ \ m/s}$.

==================================<>==================================

================================<>================================

Problem 3.43 A skydiver and his parachute weigh 200 pounds. He is falling vertically at 100 *ft/s* when his parachute opens. With the parachute open, the magnitude of the drag force is $0.5v^2$. (a) What is the magnitude of his acceleration at the instant the parachute opens? (b) What is the magnitude of his velocity when he has descended 20 feet from the point where the parachute opens?

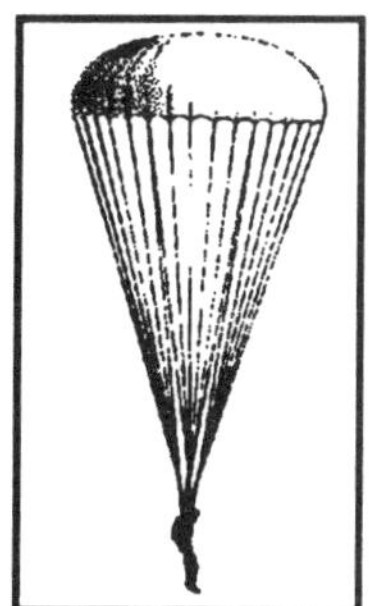

Solution: Choose a coordinate system with s positive downward. For brevity denote $C_d = 0.5$, $W = 200\ lb$, $g = 32.17\ ft/s^2$. From Newton's second law $W - D = \left(\frac{W}{g}\right)\left(\frac{dv}{dt}\right)$, where $D = 0.5v^2$. Use the chain rule to write $v\frac{dv}{ds} = -\frac{0.5v^2 g}{W} + g = g\left(1 - \frac{C_d v^2}{W}\right)$.

(a) At the instant of full opening, the initial velocity has not decreased, and the magnitude of the acceleration is $\boxed{|a_{init}| = \left|g\left(1 - \frac{C_d}{W}v^2\right)\right| = 24g = 772.08\ ft/s}$.

(b) Separate variables and integrate: $\dfrac{vdv}{1 - \dfrac{C_d v^2}{W}} = gds$, from which

$\ln\left(1 - \frac{C_d v^2}{W}\right) = -\left(\frac{2C_d g}{W}\right)s + C$. Invert and solve: $v^2(s) = \left(\frac{W}{C_d}\right)\left(1 - Ce^{-\frac{2C_d g}{W}s}\right)$. At $s = 0$,

$v(0) = 100\ ft/s$, from which $C = 1 - \frac{C_d 10^4}{W} = -24$, and $v^2(s) = \left(\frac{W}{C_d}\right)\left(1 + 24e^{-\frac{2C_d g}{W}s}\right)$. At $s = 20\ ft$

the velocity is $\boxed{v(s = 20) = \sqrt{2W\left(1 + 24e^{-\frac{g}{W}(20)}\right)} = 28.0\ ft/s}$

================================<>================================

Problem 3.44 The horizontal force exerted on the C-17 cargo plane during its takeoff run is $T - 127v^2$ *N*, where $T = 740{,}000\ N$ is the thrust of its engines and v is the plane's velocity in meters per second. The plane's mass is 265,000 *kg*. How long does it take the plane to reach its takeoff velocity of 72 *m/s*?

Solution: The equation of motion parallel to the runway is given by $m\frac{dv}{dt} = T - 127v^2$. From the integral tables, we find that $m\int_0^{V_f} \frac{dv}{a^2 - b^2v^2} = \left(\frac{m}{2ab}\right)\left(\ln\left(\frac{a + bv}{a - bv}\right)\right)\Bigg|_0^{Vf}$. Applying this to our problem, we see that $a^2 = T = 740{,}000\ N$, $b^2 = 127$, and $m = 265{,}000\ kg$. Integrating, we get

$\int_0^{t_{TO}} dt = m\int_0^{V_{TO}} \frac{dv}{T - 127v^2} = \left(\frac{m}{2\sqrt{127T}}\right)\left(\ln\left(\frac{\sqrt{T} + \sqrt{127}v}{\sqrt{T} - \sqrt{127}v}\right)\right)\Bigg|_0^{V_{TO}}$. Evaluating, we get $t_{TO} = 48.3s$

================================<>================================

==================================<>==================================

Problem 3.45 In Problem 3.44, what runway length is required for the C-17 to take off?

Solution:

Beginning with the equation of motion along the runway, we have $m\frac{dv}{dt} = mv\frac{dv}{dx} = T - 127v^2$. Setting up the integral, we get $\int_0^{x_{TO}} dx = m\int_0^{V_{TO}} \frac{vdv}{T-127v^2}$. From the integral tables, we have $m\int_0^{V_{TO}} \frac{vdv}{a^2 - b^2v^2} = -\left(\frac{m}{2b^2}\right)\ln\left(a^2 - b^2v^2\right)\Big|_0^{V_{TO}}$. As in Probem 3.45, $a^2 = T = 740{,}000\ N$, $b^2 = 127$, and $m = 265{,}000\ kg$. Evaluating, we get $x_{TO} = 2300\ m$

==================================<>==================================

Problem 3.46 A 200 *lb.* "bungee jumper" jumps from a bridge 130 *ft* above a river. The bungee cord has an unstretched length of 60 *ft* and has a spring constant $k = 14\ lb/ft$. (a) How far above the river is he when the cord brings him to a stop? (b) What maximum force does the cord exert on him?

Solution:

Choose a coordinate system with s positive downward. Divide the fall into two parts: (1) the free fall until the bungee unstretched length is reached, (2) the fall to the full extension of the bungee cord. For Part (1): From Newton's second law $\frac{ds}{dt} = g$. Use the chain rule to write: $v\frac{dv}{ds} = g$. Separate variables and integrate: $v^2(s) = 2gs$, since $v(0) = 0$. At $s = 60\,ft$, $v(s=60) = \sqrt{2gs} = 62.1\ ft/s$. For Part (2):

T

W

From Newton's second law $W - T = \frac{W}{g}\left(\frac{dv}{dt}\right)$, where $T = k(s-60)$.

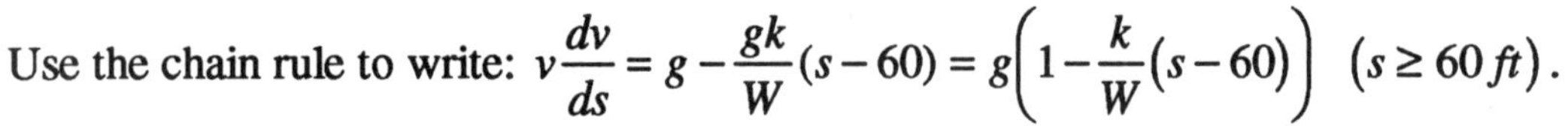

Use the chain rule to write: $v\frac{dv}{ds} = g - \frac{gk}{W}(s-60) = g\left(1 - \frac{k}{W}(s-60)\right)$ $(s \ge 60\,ft)$.

Separate variables and integrate: $v^2(s) = 2gs\left(1 - \frac{k}{2W}(s-120)\right) + C$. At $s = 60$, $v^2(s=60) = [2gs]_{s=60} = 120g$, from which $C = -\frac{gk}{W}(60^2) = -8106.8$. The velocity is $v^2(s) = -\frac{gk}{W}s^2 + 2g\left(1 + \frac{60k}{W}\right)s - \frac{gk}{W}(60^2)$. When the jumper is brought to a stop, $v(s_{stop}) = 0$, from which $s^2 + 2bs + c = 0$, where $b = -\left(\frac{W}{k} + 60\right)$, and $c = 60^2$. The solution:

$s_{stop} = -b \pm \sqrt{b^2 - c} = 118.08, \quad = 30.49\ ft$. (a) The first value represents the maximum distance on the first excursion, from which $\boxed{h = 130 - 118.1 = 11.92\ ft}$ is the height above the river at which he comes to a stop. (b) The maximum force exerted by the bungee cord is $\boxed{F = k(s-60) = 14(118.1-60) = 813\ lb}$

==================================<>==================================

==================================<>==================================

Problem 3.47 In Problem 3.46, what maximum velocity does the jumper reach, and at what height above the river does it occur?

Solution:

Use the solution to Problem 3.46 for the velocity: $v^2(s) = -\frac{gk}{W}s^2 + 2g\left(1 + \frac{60k}{W}\right)s - \frac{gk}{W}(60^2)$. The maximum occurs when $\frac{d(v^2)}{ds} = 0$, since $v(s) \geq 0$, $\frac{d(v^2)}{ds} = -2\frac{gk}{W}s_{max} + 2g\left(1 + 60\frac{k}{W}\right) = 0$, from which $s_{max} = \left(\frac{W}{k} + 60\right) = 74.29\ ft$. The height above the river is $\boxed{h_{max} = 130 - s_{max} = 55.71\ ft}$ with a maximum velocity of $\boxed{v_{max} = \sqrt{-\frac{gk}{W}s_{max}^2 + 2g\left(1 + \frac{60k}{W}\right)s_{max} - \left(\frac{gk}{W}\right)(60^2)} = 65.73\ ft/s}$

==================================<>==================================

Problem 3.48 In a cathode ray tube, an electron (mass $m_e = 9.11\times10^{-31}\ kg$) is projected at O with velocity $\vec{v} = \left(2.2\times10^{7}\right)\vec{i}\ \ m/s$. While it is between the charged plates, the electric field generated by the plates subjects it to a force $\vec{F} = -eE\vec{j}$, where the charge of the electron and the electric field strength $E = 15\ kN/C$. External forces on the electron are negligible when it is not between the plates. Where does it strike the screen?

Solution For brevity denote $L = 0.03\ m$, $D = 0.1\ m$. The time spent between the charged plates is $t_p = \dfrac{L}{V} = \dfrac{3\times10^{-2}m}{2.2\times10^{7}m/s} = 1.3636\times10^{-9}s$.

From Newton's second law, $\vec{F} = m_e\vec{a}_p$. The acceleration due to the charged plates is

$\vec{a}_p = \dfrac{-eE}{m_e}\vec{j} = -\dfrac{\left(1.6\times10^{-19}\right)\left(15\times10^{3}\right)}{9.11\times10^{-31}}\vec{j} = \vec{j}2.6345\times10^{15}\ m/s^2$. The velocity is $\vec{v}_y = -\vec{a}_p t$ and the displacement is $\vec{y} = \dfrac{\vec{a}_p}{2}t^2$. At the exit from the plates the displacement is

$\vec{y}_p = -\dfrac{\vec{a}_p t_p^2}{2} = -\vec{j}2.4494\times10^{-3}\ m$. The velocity is $\vec{v}_{yp} = -\vec{a}_p t = -\vec{j}3.59246\times10^{6}\ m/s$. The time spent in traversing the distance between the plates and the screen is

$t_{ps} = \dfrac{D}{V} = \dfrac{10^{-1}m}{2.2\times10^{7}m/s} = 4.5455\times10^{-9}\ s$. The vertical displacement at the screen is

$$\boxed{\vec{y}_s = \vec{v}_{yp}t_{ps} + \vec{y}_p = -\vec{j}\left(3.592456\times10^{6}\right)\left(4.5455\times10^{-9}\right) - \vec{j}2.4494\times10^{-3} = -\vec{j}18.78\ mm}$$

Problem 3.49 In Problem 3.48, determine where the electron strikes the screen if the electric field strength is $E = 15\sin\omega t\ \ kN/C$, where $\omega = 2\times10^{9}\ \ rad/s$.

Solution Use the solution to Problem 3.48. Assume that the electron enters the space between the charged plates at $t = 0$, so that at that instant the electric firld strength is zero. The acceleration due to the charged plates is

$\vec{a} = -\dfrac{eE}{m_e}\vec{j} = -\dfrac{\left(1.6\times10^{-19}\ C\right)\left(15000\sin\omega t\ N/C\right)}{9.11\times10^{-31}\ kg}\vec{j} = -\vec{j}\left(2.6345\times10^{15}\right)\sin\omega t\ \ m/s^2$. The velocity is

$\vec{v}_y = \vec{j}\dfrac{\left(2.6345\times10^{15}\right)}{\omega}\cos\omega t + \vec{C}$. Since $\vec{v}_y(0) = 0$ $\vec{C} = -\dfrac{2.6345\times10^{15}}{2\times10^{9}}\vec{j} = -\vec{j}1.3172\times10^{6}$. The

displacement is $\vec{y} = \vec{j}\dfrac{\left(2.6345\times10^{15}\right)}{\omega^2}\sin\omega t + \vec{C}t$, since $\vec{y}(0) = 0$. The time spent between the charged plates is (see Problem 3.34) $t_p = 1.3636\times10^{-9}\ \ s$, from which $\omega t_p = 2.7273\ \ rad$. At exit from the plates, the vertical velocity is $\vec{v}_{yp} = \vec{j}\dfrac{2.6345\times10^{15}}{2\times10^{9}}\cos\left(\omega t_p\right) + \vec{C} = -\vec{j}2.523\times10^{6}\ \ m/s$.

Solution continued on next page

The displacement is $\vec{y}_p = \vec{j}\dfrac{2.6345\times 10^{15}}{4\times 10^{18}}\sin(\omega t_p)+\vec{C}t_p = -\vec{j}1.531\times 10^{-3}\ m$. The time spent between the plates and the screen is $t_{ps} = 4.5455\times 10^{-9}\ s$. The vertical deflection at the screen is

$$\boxed{\vec{y}_s = \vec{y}_p + \vec{v}_{yp}t_{ps} = -\vec{j}13\ mm}$$

=================================<>=================================

Problem 3.50 An astronaut needs to travel from a space station to a satellite that needs repair. He departs the space station at O. A spring-loaded device gives his maneuvering unit an initial velocity of 1 *m/s* (relative to the space station) in the *y* direction. At that instant, the position of the satellite is $x = 70\ m$, $y = 50\ m$, $z = 0$, and it is drifting at 2 *m/s* (relative to the station) in the *x* direction. The astronaut intercepts the satellite by applying a constant thrust parallel to the *x* axis. The total mass of the of the astronaut and his maneuvering unit is 300 *kg*. (a) How long does it take him to make the intercept? (b) What is the magnitude of the thrust he must apply to make his intercept? (c) What is his velocity relative to the satellite when he reaches it?

Solution The path of the satellite relative to the space station is $x_s(t) = 2t + 70\ m$, $y_s(t) = 50\ m$. From Newton's second law, $T = ma_x$, $0 = ma_y$. Integrate to obtain the path of the astronaut, using the initial conditions $v_x = 0, v_y = 1\ m/s$, $x = 0, y = 0$. $y_a(t) = t$, $x_a(t) = \dfrac{T}{2m}t^2$. (a) When the astronaut intercepts the *x* path of the satellite, $y_a(t_{\text{int}}) = y_s(t_{\text{int}})$, from which $\boxed{t_{\text{int}} = 50\ s}$. (b) The intercept of the *y*-axis path occurs when $x_a(t_{\text{int}}) = x_s(t_{\text{int}})$, from which $\dfrac{T}{2m}t_{\text{int}}^2 = 2t_{\text{int}} + 70$, from which

$$\boxed{T = (2m)\left(\frac{2t_{\text{int}}+70}{t_{\text{int}}^2}\right) = 2(300)\left(\frac{170}{2500}\right) = 40.8\ N}$$

(c) The velocity of the astronaut relative to the space station is $\vec{v} = \vec{i}\left(\dfrac{T}{m}\right)t_{int} + \vec{j} = 6.8\vec{i} + \vec{j}$. The velocity of the satellite relative to the space station is $\vec{v}_s = 2\vec{i}$. The velocity of the astronaut relative to the satellite is $\boxed{\vec{v}_{a/s} = \vec{i}(6.8-2)+\vec{j} = 4.8\vec{i} + \vec{j}\ m/s}$

=================================<>=================================

===================================<>===================================

Problem 3.51 What is the acceleration of the 8 kg collar A relative to the smooth bar?

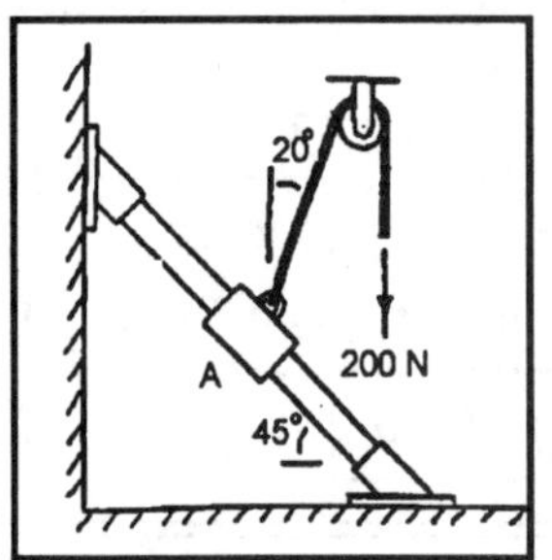

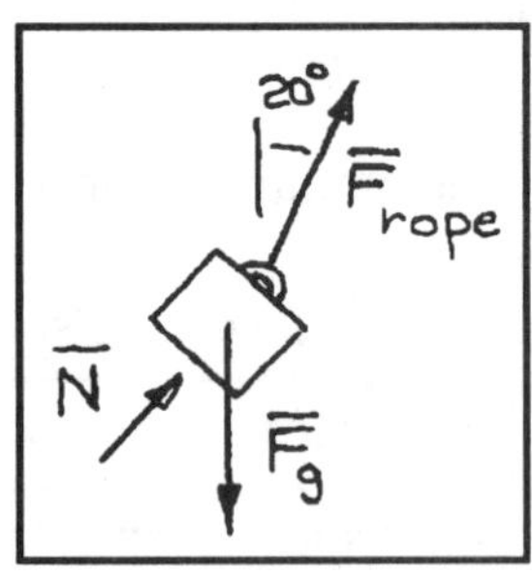

Solution For brevity, denote $\theta = 20^o$, $\alpha = 45^o$, $F = 200\ N$, $m = 8\ kg$. The force exerted by the rope on the collar is $\vec{F}_{rope} = 200(\vec{i}\sin\theta + \vec{j}\cos\theta) = 68.4\vec{i} + 187.9\vec{j}\ N$. The force due to gravity is $\vec{F}_g = -gm\vec{j} = -78.5\vec{j}\ N$. The unit vector parallel to the bar, positive upward, is $\vec{e}_B = -\vec{i}\cos\alpha + \vec{j}\sin\alpha$. The sum of the forces acting to move the collar is $\sum F = F_c = \vec{e}_B \cdot \vec{F}_{rope} + \vec{e}_B \cdot \vec{F}_g = |\vec{F}_{rope}|\sin(\alpha - \theta) - gm\sin\alpha = 29.03\ N$. The collar tends to slide up the bar since $F_c > 0$. From Newton's second law, the acceleration is

$$\boxed{a = \frac{F_c}{m} = 3.63\ m/s^2}.$$

===================================<>===================================

Problem 3.52 In Problem 3.51 determine the acceleration of the collar A relative to the bar if the coefficient of kinetic friction between the collar and the bar is $\mu_k = 0.1$.

Solution Use the solution to Problem 3.51. $F_c = |\vec{F}_{rope}|\sin(\alpha - \theta) - gm\sin\alpha = 29.03\ N$. The normal force is perpendicular to the bar, with the unit vector $\vec{e}_N = \vec{i}\sin\alpha + \vec{j}\cos\alpha$. The normal force is $N = \vec{e}_N \cdot \vec{F}_{rope} + \vec{e}_N \cdot \vec{F}_g = |\vec{F}_{rope}|\cos(\alpha - \theta) - gm\cos\alpha = 125.77\ N$. The collar tends to slide up the bar since $F_c > 0$. The friction force opposes the motion, so that the sum of the forces on the collar is $\sum F = F_c - \mu_k N = 16.45\ N$. From Newton's second law, the acceleration of the collar is

$$\boxed{a = \frac{16.45}{8} = 2.06\ m/s^2}$$ up the bar.

===================================<>===================================

Problem 3.53 The acceleration of the 20 $lb.$ collar A is $2\vec{i} + 3\vec{j} - 3\vec{k}\ ft/s^2$. What is the force F?

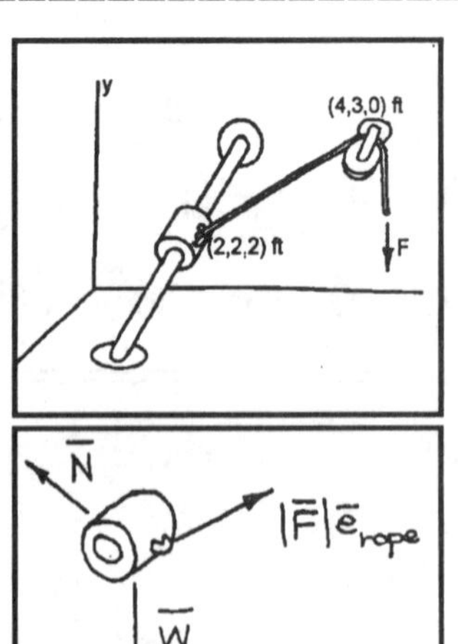

Solution The magnitude of the force acting on the collar parallel to the bar is $|\vec{F}_B| = |\vec{F}|\vec{e}_{rope} \cdot \vec{e}_B + \vec{W} \cdot \vec{e}_B$. From Newton's second law, the components of the force acting parallel to the bar are

$\vec{F}_B = \left(\frac{W}{g}\right)\vec{a} = (0.6217)(2\vec{i} + 3\vec{j} - 3\vec{k}) = 1.2434\vec{i} + 1.8651\vec{j} - 1.8651\vec{k}$, with magnitude $|\vec{F}_B| = \sqrt{F_x^2 + F_y^2 + F_z^2} = 2.9160\ lb$. The weight is $\vec{W} = -|\vec{W}|\vec{j}$. The unit vector parallel to the rope attached to the collar is

$$\vec{e}_{rope} = \frac{\vec{r}_w - \vec{r}_c}{|\vec{r}_w - \vec{r}_c|} = \frac{(4-2)\vec{i} + (3-2)\vec{j} + (0-2)\vec{k}}{3}$$

$= 0.6667\vec{i} + 0.3333\vec{j} - 0.6667\vec{k}$. The unit vector parallel to the bar is

$$\vec{e}_B = \frac{\vec{a}}{|\vec{a}|} = 0.4264\vec{i} + 0.6396\vec{j} - 0.6396\vec{k}.$$

Solution continued on next page

The component of gravity parallel to the bar is $\left|\vec{W}_B\right| = \vec{W}\cdot\vec{e}_B = -12.7920\ lb$. Substitute and solve:

$\left|\vec{F}_B\right| = \left|\vec{F}\right|\vec{e}_{rope}\cdot\vec{e}_B + \vec{W}\cdot\vec{e}_B = \left|\vec{F}\right|(0.9239) - 12.7920 = 2.9160$, from which

$$\left|\vec{F}\right| = \frac{2.9160 + 12.7920}{0.9239} = 17.0\ lb$$

==================================<>==================================

Problem 3.54 In Problem 3.53, determine the force *F* if the coefficient of kinetic friction between the collar and the bar is $\mu_k = 0.1$

Solution Use the results of the solution to Problem 3.53. The force acting on the collar parallel to the bar is $\left|\vec{F}_B\right| = \left|\vec{F}\right|\vec{e}_{rope}\cdot\vec{e}_B + \vec{W}\cdot\vec{e}_B - \mu_k\left|\vec{N}\right|$. The unit vectors $\vec{e}_{rope}$ and $\vec{e}_B$ are determined in the solution to Problem 3.53. The unit vector parallel to the normal force is perpendicular to $\vec{e}_B$. The components of the rope tension normal to the bar is $\left|\vec{F}_N\right| = \left|\vec{F}\right|\left|\vec{e}_{rope}\times\vec{e}_B\right| = \left|\vec{F}\right|\sin\theta$, where $\theta = \cos^{-1}\left(\vec{e}_{rope}\cdot\vec{e}_B\right) = 22.5^o$ The components of gravity normal to the bar are: $\left|\vec{W}_N\right| = \left|\vec{W}\times\vec{e}_B\right| = W\sin\alpha$, where $\alpha = \cos^{-1}\left(-\vec{j}\cdot\vec{e}_B\right) = 129.8^o$. The magnitude of the normal force is $\left|\vec{N}\right| = \left|\vec{F}\right|\cos\theta + W\sin\alpha$. The force parallel to the bar is

$\left|\vec{F}_B\right| = \left|\vec{F}\right|(\cos\theta - \mu_k\sin\theta) + W(\cos\alpha - \mu_k\sin\alpha) = \left|\vec{F}\right|(0.8856) - 14.3295$. From Problem 3.53

$\left|\vec{F}_B\right| = 2.9160\ \ lb$. Solve: $\left|\vec{F}\right| = \dfrac{2.9160 + 14.3295}{0.8856} = 19.4\ lb$

==================================<>==================================

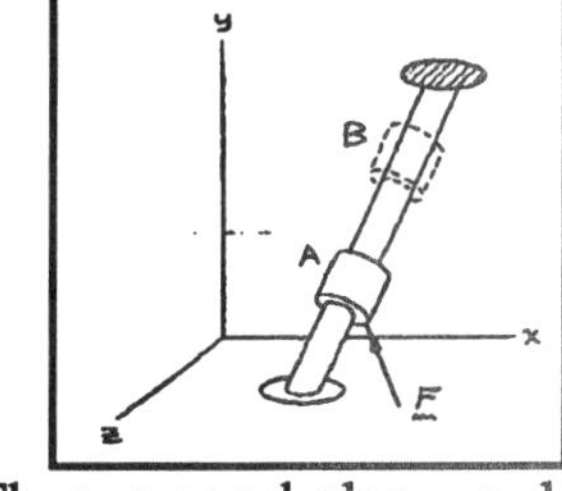

Problem 3.55 The *6-kg* collar starts from rest at position A, where the coordinates of its center of mass are *(400,200,200) mm,* and slides up the smooth bar to positon B, where the coordinates of its center of mass are *(500,400,0) mm* under the action of a constant force **F** = $-40\,\mathbf{i} + 70\,\mathbf{j} - 40\mathbf{k}$ (*newtons*). How long does it take to go from A to B.

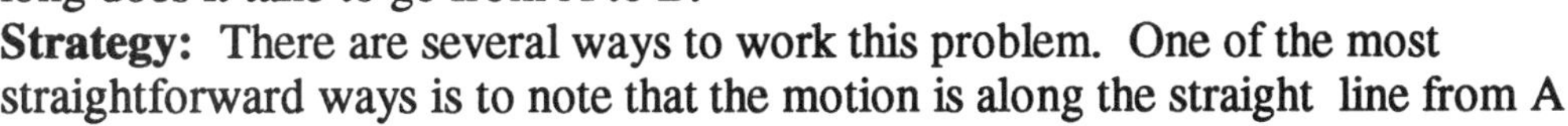

Strategy: There are several ways to work this problem. One of the most straightforward ways is to note that the motion is along the straight line from A

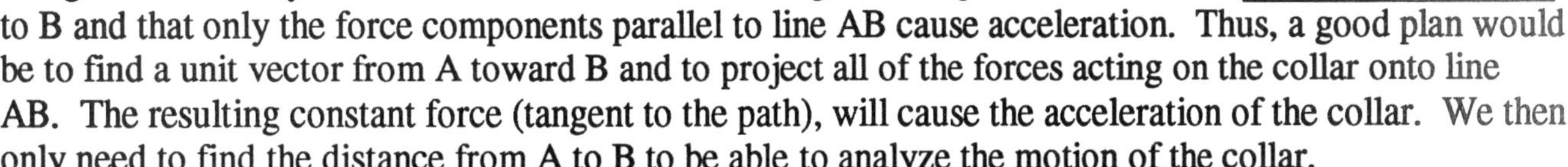

to B and that only the force components parallel to line AB cause acceleration. Thus, a good plan would be to find a unit vector from A toward B and to project all of the forces acting on the collar onto line AB. The resulting constant force (tangent to the path), will cause the acceleration of the collar. We then only need to find the distance from A to B to be able to analyze the motion of the collar.

Solution: The unit vector from A toward B is $\mathbf{e}_{AB} = \mathbf{e}_t = 0.333\,\mathbf{i} + 0.667\,\mathbf{j} - 0.667\mathbf{k}$ and the distance from A to B is *0.3 m.* The free body diagram of the collar is shown at the right. There are three forces acting on the collar. These are the applied force **F**, the weight force $\mathbf{W} = -m g\mathbf{j} = -58.86\mathbf{j}$ (*newtons*), and the force **N** which acts normal to the smooth bar. Note that **N**, by its definition, will have no component tangent to the bar. Thus, we need only consider **F** and **W** when finding force components tangent to the bar. Also note that **N** is the force that the bar exerts on the collar to keep it in line with the bar. This will be important in the next problem.

The equation of motion for the collar is $\sum\mathbf{F}_{collar} = \mathbf{F} + \mathbf{W} + \mathbf{N} = m\mathbf{a}$. In the direction tangent to the bar, the equation is $(\mathbf{F} + \mathbf{W})\bullet\mathbf{e}_{AB} = ma_t$.

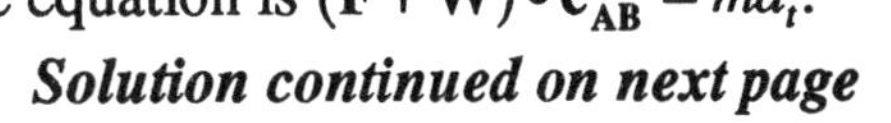

Solution continued on next page

The projection of **(F+W)** onto line AB yields a force along AB which is $|\mathbf{F}_{AB}| = 20.76$ (*newtons*). The acceleration of the *6-kg* collar caused by this force is $a_t = 3.46 m/s^2$. We now only need to know how long it takes the collar to move a distance of *0.3m,* starting from rest, with this acceleration. The kinematic equations are $v_t = a_t t$, and $s_t = a_t t^2/2$. We set $s_t = 0.3m$ and solve for the time. The time required is $t = 0.416s$

==================================<>==================================

Problem 3.56 In Problem 3.55, how long does it take for the collar to go from A to B if the coefficient of kinetic friction between the bar and the collar is $\mu_k = 0.2$?

Strategy: This problem is almost the same as problem 3.55. The major difference is that now we must calculate the magnitude of the normal force, **N,** and then must add a term $\mu_k|\mathbf{N}|$ to the forces tangent to the bar (in the direction from B toward A – opposing the motion). This will give us a new acceleration, which will result in a longer time for the collar to go from A to B.

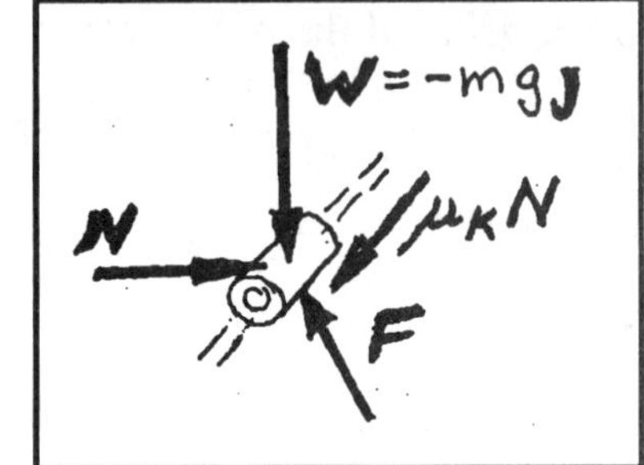

Solution: We use the unit vector $\mathbf{e}_{AB}$ from Problem 3.55. The free body diagram for the collar is shown at the right. There are four forces acting on the collar. These are the applied force **F,** the weight force $\mathbf{W} = -mg\mathbf{j} = -58.86\mathbf{j}$ (*newtons*)**,** the force **N** which acts normal to the smooth bar, and the friction force $\mathbf{f} = -\mu_k|\mathbf{N}|\mathbf{e}_{AB}$. The normal force must be equal and opposite to the components of the forces **F** and **W** which are perpendicular (not parallel) to AB. The friction force is parallel to AB. The magnitude of $|\mathbf{F}+\mathbf{W}|$ is calculate by adding these two known forces and then finding the magnitude of the sum. The result is that $|\mathbf{F}+\mathbf{W}| = 57.66$ (*newtons*). From Problem 3.55, we know that the component of $|\mathbf{F}+\mathbf{W}|$ tangent to the bar is $|\mathbf{F}_{AB}| = 20.76$ (*newtons*). Hence, knowing the total force and its component tangent to the bar, we can find the magnitude of its component normal to the bar. Thus, the magnitude of the component of $|\mathbf{F}+\mathbf{W}|$ normal to the bar is *53.79 newtons.* This is also the magnitude of the normal force **N.** The equation of motion for the collar is $\sum \mathbf{F}_{collar} = \mathbf{F}+\mathbf{W}+\mathbf{N}-\mu_k|\mathbf{N}|\mathbf{e}_{AB} = m\mathbf{a}$. In the direction tangent to the bar, the equation is $(\mathbf{F}+\mathbf{W})\bullet\mathbf{e}_{AB} - \mu_k|\mathbf{N}| = ma_t$.

The force tangent to the bar is $F_{AB} = (\mathbf{F}+\mathbf{W})\bullet\mathbf{e}_{AB} - \mu_k|\mathbf{N}| = 10.00$ (*newtons*) . The acceleration of the *6-kg* collar caused by this force is $a_t = 1.667 m/s^2$. We now only need to know how long it takes the collar to move a distance of *0.3m,* starting from rest, with this acceleration. The kinematic equations are $v_t = a_t t$, and $s_t = a_t t^2/2$. We set $s_t = 0.3m$ and solve for the time. The time required is $t = 0.600s$

==================================<>==================================

Problem 3.57 The crate is drawn across the floor by a winch that retracts the cable at a constant rate of 0.2 *m/s*. The crate's mass is 120 *kg* and the coefficient of kinetic friction between the crate and the floor is $\mu_k = 0.24$. (a) At the instant shown, what is the tension in the cable? (b) Obtain a "quasi-static" solution for the tension in the cable by ignoring the crate's acceleration and compare it to your result in Part (a).

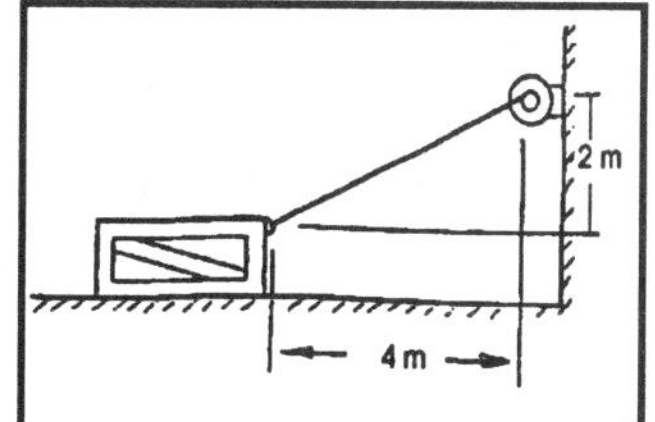

Solution The length of the cable is $L = \sqrt{2^2 + x^2}$. The rate of change of length is $\frac{dL}{dt} = \frac{x}{L}\frac{dx}{dt} = 0.2\ m/s$, from which the velocity of the crate is $\frac{dx}{dt} = \left(\frac{0.2L}{x}\right)$.

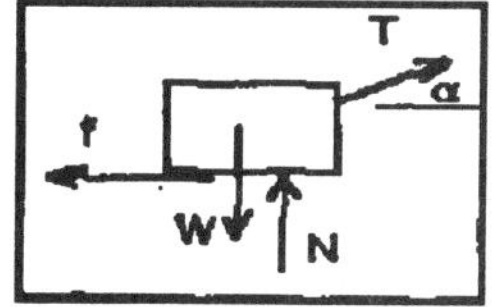

The acceleration of the crate is

$\frac{d^2x}{dt^2} = \frac{d}{dt}\frac{dx}{dt} = \frac{d}{dt}\left(\frac{0.2L}{x}\right) = \frac{0.2}{x}\frac{dL}{dt} - \frac{0.2L}{x^2}\frac{dx}{dt} = \frac{0.2^2}{x} - \frac{0.2^2 L}{x^2} = -\frac{0.64}{x^3}$. The normal force exerted by the floor on the crate is reduced by the lifting action of the cable: $N = mg - T\sin\alpha = mg - T\frac{2}{L}$. From Newton's second law, $\sum F = T\cos\alpha - \mu_k N = m\frac{d^2x}{dt^2}$.

(a) Substitute and solve: $T\left(\frac{x}{L}\right) - \mu_k\left(mg - \frac{2T}{L}\right) = m\left(-\frac{0.64}{x^3}\right)$, from which

$$T = \frac{-m\left(\frac{0.64}{x^3}\right) + \mu_k(mg)}{\frac{x}{L} + \mu_k\left(\frac{2}{L}\right)} = 280.8\ N$$

(b) Neglect the acceleration, so that

$$T = \frac{\mu_k gm}{\left(\frac{x}{L} + \mu_k\left(\frac{2}{L}\right)\right)} = \frac{\mu_k gmL}{x + 2\mu_k} = 282.0\ N$$

Check: A reasonableness test is that since the crate is *decelerating,* $(a < 0)$ the quasi-static tension should be greater than the correct tension.

Problem 3.58 If $y = 100\ mm$, $\frac{dy}{dt} = 600\ mm/s$, and $\frac{d^2y}{dt^2} = -200\ mm/s^2$, what horizontal force is exerted on the 0.4 *kg* slider A by the smooth circular slot?

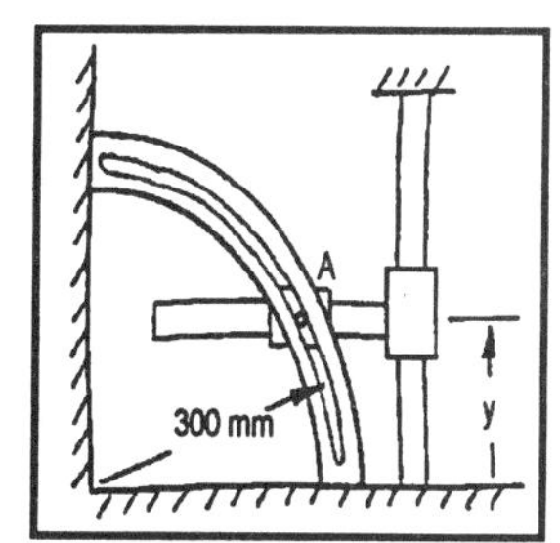

Solution The horizontal displacement is $x^2 = R^2 - y^2$. Differentiate twice

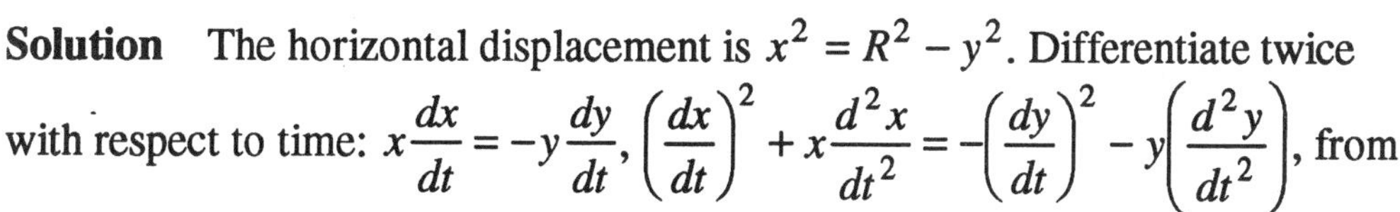

with respect to time: $x\frac{dx}{dt} = -y\frac{dy}{dt}$, $\left(\frac{dx}{dt}\right)^2 + x\frac{d^2x}{dt^2} = -\left(\frac{dy}{dt}\right)^2 - y\left(\frac{d^2y}{dt^2}\right)$, from which. $\frac{d^2x}{dt^2} = -\left(\frac{1}{x}\right)\left(\left(\frac{y}{x}\right)^2 + 1\right)\left(\frac{dy}{dt}\right)^2 - \left(\frac{y}{x}\right)\frac{d^2y}{dt^2}$. Substitute: $\frac{d^2x}{dt^2} = -1.3612\ m/s^2$. From Newton's second law, $F_h = ma_x = -1.361(0.4) = -0.544\ N$

================================<>================================

Problem 3.59 A *3000-lb* car travels at a constant speed of *60 mi/hr*on a straight road of increasing grade whose profile can be approximated by the equation shown $(y = 0.0003x^2)$. When x = *400 ft,* what are the x and y components of the total force acting on the car (including its weight).

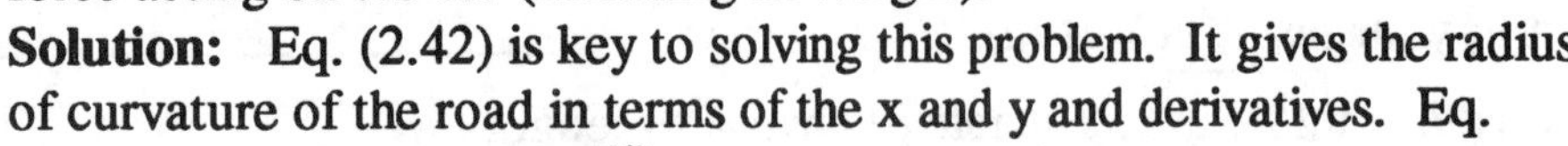

Solution: Eq. (2.42) is key to solving this problem. It gives the radius of curvature of the road in terms of the x and y and derivatives. Eq. (2.42) is $\rho = \frac{\left[1+(dy/dx)^2\right]^{3/2}}{\left|d^2y/dx^2\right|}$. For our problem, $(dy/dx) = 0.0006x$ and $(d^2y/dx^2) = 0.0006$. When x = *400 ft,* y = *48ft*, and $\rho = 1813 ft$.

We can now use the equations for accelerations in normal and tangential coordinates to find the total acceleration of the car. We have $a_t = dv/dt = 0$ since $v = 60\ mi/hr = 88 ft/s = constant$. The normal acceleration is given by $a_n = v^2/\rho = 4.27 ft/s^2$. We now need to transform coordinates back to the original *x-y* coordinates. To do this, we need the angle θ between the instantaneous velocity vector and the horizontal x axis. From calculus, recall that $\tan\theta = dy/dx = 0.0006x$. For x = *400 ft,* we have $\theta = 13.5°$. The total acceleration can now be written as $\mathbf{a} = a_t\,\mathbf{e}_t + a_n\mathbf{e}_n = 0\ \mathbf{e}_t + 4.27\mathbf{e}_n\ ft/s^2$ where $\mathbf{e}_n$ points upward with respect to the road. The transformation equations which take us back to *x-y* coordinates are $\mathbf{e}_t = \cos\theta\,\mathbf{i} + \sin\theta\,\mathbf{j}$ and $\mathbf{e}_n = -\sin\theta\,\mathbf{i} + \cos\theta\,\mathbf{j}$. Substituting for the unit vectors in the acceleration equation, we get $\mathbf{a} = (a_t\cos\theta - a_n\sin\theta)\,\mathbf{i} + (a_t\sin\theta + a_n\cos\theta)\,\mathbf{j} = a_X\,\mathbf{i} + a_Y\,\mathbf{j}$. Separating the components and evaluating the numbers, we get $a_X = -0.997 ft/s^2$ and $a_Y = 4.15 ft/s^2$. Finally, there are three forces acting on the car (the weight, the normal force, and friction). The sum of these forces is the total force acting on the car and this sum appears in Newton's second law

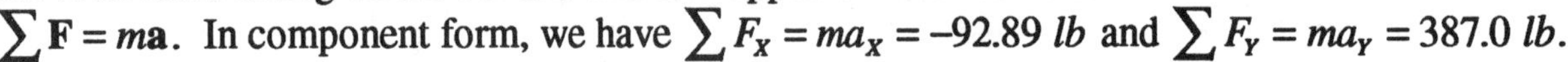

$\sum\mathbf{F} = m\mathbf{a}$. In component form, we have $\sum F_X = ma_X = -92.89\ lb$ and $\sum F_Y = ma_Y = 387.0\ lb$.

================================<>================================

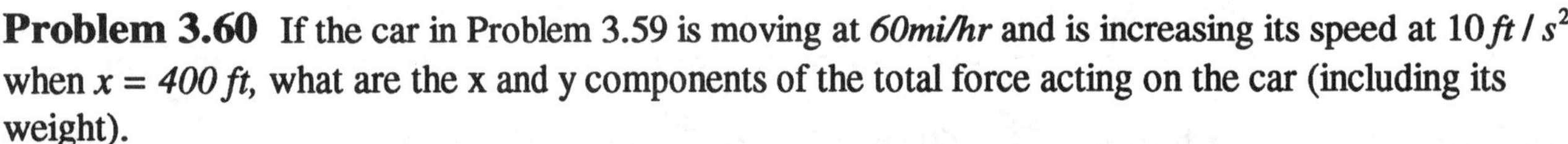

Problem 3.60 If the car in Problem 3.59 is moving at *60mi/hr* and is increasing its speed at $10 ft/s^2$ when x = *400 ft,* what are the x and y components of the total force acting on the car (including its weight).

Solution: The solution to this problem is identical to that of Problem 3.59 except for the number given to the tangential acceleration. In Problem 3.59, $a_t = 0$, whereas here, $a_t = 10 ft/s^2$ when x = *400 ft.* Substituting in values, we get $a_X = 8.73 ft/s^2$, $a_Y = 6.49 ft/s^2$, $\sum F_X = 813.1\ lb$, and $\sum F_Y = 604.5\ lb$.

================================<>================================

==================================<>==================================

Problem 3.61 The two 100 *lb.* blocks are released from rest. Determine the magnitudes of their accelerations if friction at all contacting surfaces is negligible. *Strategy*: Use the fact that the components of the accelerations of the blocks perpendicular to their mutual interface must be equal.

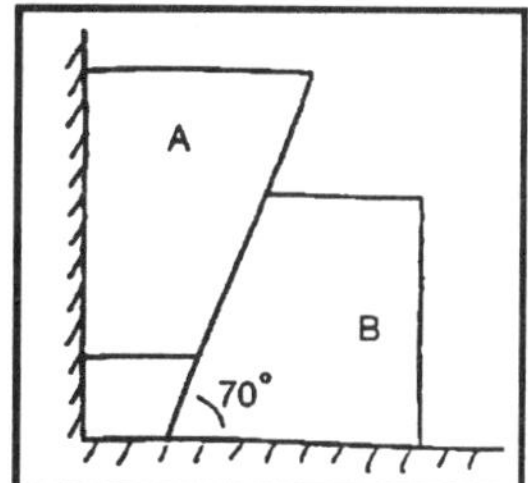

Solution The relative motion of the blocks is constrained by the surface separating the blocks. The equation of the line separating the blocks is $y = x\tan 70^o$, where y is positive upward and x is positive to the right. A positive displacement of block A results in a negative displacement of B (as contact is maintained) from which $s_A = -s_B \tan 70^o$, and from which

$$\frac{d^2 s_A}{dt^2} = -\frac{d^2 s_B}{dt^2}\tan 70^o.$$ Thus (1) $a_A = -a_B \tan 70^o$.

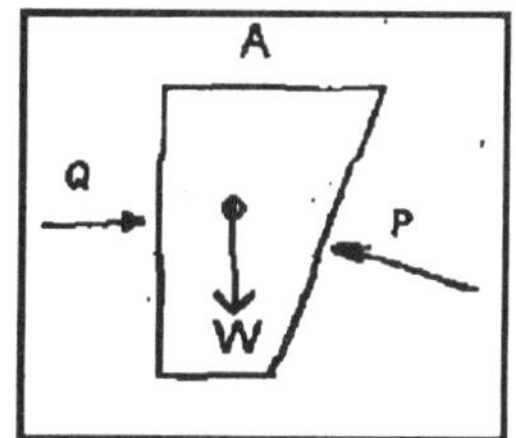

From Newton's second law: for block A, (2) $\sum F_y = -W + P\cos 70^o = ma_A$, for block B, (3) $\sum F_x = P\sin 70^o = ma_B$, from which $a_A = -\frac{W}{m} + \frac{a_B}{\tan 70^o}$ ft / s². Use (1) to obtain $a_A = -\frac{g}{1+\cot^2 70^o} = -28.4$ ft / s² and $a_B = -\frac{a_A}{\tan 70^o} = 10.34$ ft / s², where a_A is positive upward and a_B is positive to the right.

==================================<>==================================

Problem 3.62 In Problem 3.61, determine how long it takes for block A to fall 1 *ft* if $\mu_k = 0.1$ at all contacting surfaces.

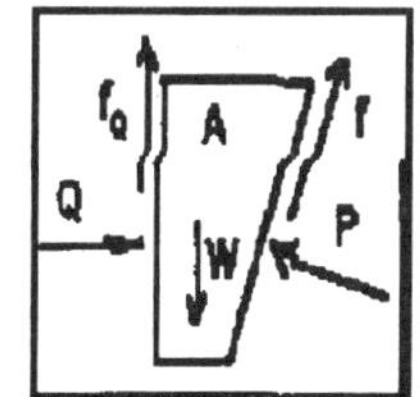

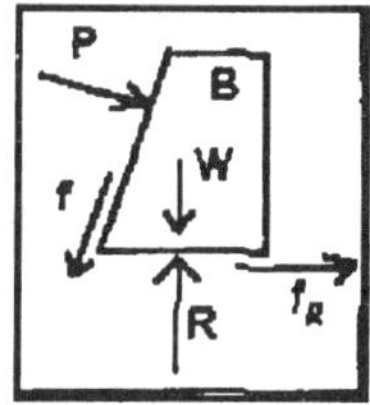

Solution Use the results of the solution to Problem 3.43. Denote by Q the normal force at the wall, and by P the normal force at the contacting surface, and R the normal force exerted by the floor on block B. For a_A positive upward and a_B positive to the right, (1) $a_A = -a_B \tan 70^o$ so long as contact is maintained. From Newton's second law for block A, (2) $\sum F_x = Q - P\sin 70^o + f\cos 70^o = 0$, (3) $\sum F_y = -W + f_Q + f\cos 70^o + P\cos 70^o = ma_A$. For block B: (4) $\sum F_x = P\sin 70^o - f\cos 70^o - f_R = ma_B$, (5) $\sum F_y = -W + R - P\cos 70^o - f\sin 70^o = 0$. In addition: (6) $f = \mu_k P$, (7) $f_R = \mu_k R$, (8) $f_Q = \mu_k Q$. Solve these eight equations by iteration: $a_A = -24.7$ ft / s², $a_B = 9$ ft / s². *Check:* (1) The effect of friction should reduce the downward acceleration of A in Problem 3.61, and (2) for $\mu_k = 0$, this should reduce to the solution to Problem 3.61. *check.* The displacement is $y = \frac{a_A}{2}t^2$ *ft*, from which, for $y = -1$ ft, $t = \sqrt{-\frac{2}{a_A}} = 0.284$ s

==================================<>==================================

Problem 3.63 In Example 3.3, a sport utility vehicle moves through the air and impacts the ground. After the impact, the vehicle would rebound and leave the ground again. Determine the vertical component of velocity of the vehicle's center of mass at the instant its wheels leave the ground. (the wheels leave the ground when the center of mass is at *y=0*).

Solution: This analysis follows that of Example 3.3. The equation for velocity used to determine how far down the vehicle compresses its springs also applies as the vehicle rebounds. From Example 3.3, we know that the vehicle comes to rest with $v_Y = 0$ and $y = 1ft$. Following the Example, the velocity on the rebound is given by $\int_0^{v_Y} v_Y dv_Y = \int_1^0 32.2(1-6y)dy$. Evaluation, we get $v_Y = -11.3 ft/s$. (+y is down). Note that the vertical velocity component on rebound is the negative of the vertical velocity of impact.

==================================<>==================================

===================================<>===================================

Problem 3.64 In Example 3.3, the duration of the vehicle's impact with the ground is 0.255 s and its wheels leave the ground when the center of mass is at *y=0*. Determine the horizontal distance the vehicle travels in the air when it rebounds and leaves the ground again. (Assume that the vehicle remains horizontal.)

Solution: From Problem 3.64, we know the vertical velocity component as the vehicle leaves the ground is $v_Y = -11.3 ft/s$. To find the details of the post impact trajectory, we must know v_X after the impact. From Example 3.3, the horizontal force during the impact is *–2400 lb*. The mass of the vehicle is $m = W/g = 3000/32.2 = 93.17 slug$. During the impact, the acceleration in the x direction is constant and is $a_x = F_x/m = -2400/m = -25.76 ft/s^2$. The impact lasts for *0.255 s* and the velocity in the x direction after impact is given by $v_{X1} = v_{X0} + a_X t_{IMPACT}$. From Ex. 3.3, $v_{X0} = 44 ft/s$. Evaluating numerically, we get $v_{X1} = 37.4 ft/s$.

We now have a situation in which the vehicle leaves the ground with $v_{Y1} = -11.3 ft/s$ and $v_{X1} = 37.4 ft/s$. Recall that +y is downward in this problem. We now have a projectile problem, leaving *(0,0) ft* with the given initial velocity components. The equations are: $a_X = 0$, $a_Y = 32.2 ft/s^2$, $v_X = v_{X1}$, $v_Y = v_{Y1} + gt$, $x = v_{X1}t \ (ft)$, and $y = v_{Y1}t + 32.2t^2/2 \ (ft)$. Setting *y=0* and solving for the time of impact (not the zero root), we get $t_{IMPACT} = 0.701 s$. The distance that the vehicle travels in the air is the x coordinate at impact. Evaluating, we get $x_{IMPACT} = 26.3 ft$

===================================<>===================================

Problem 3.65 The sum of the forces acting on an object of mass *m* is $\sum \mathbf{F}$ and the *xyz* reference frame is inertial or Newtonian; that is $\sum \mathbf{F} = m\dfrac{d^2\mathbf{r}}{dt^2} = m\left(\dfrac{d^2x}{dt^2}\mathbf{i} + \dfrac{d^2y}{dt^2}\mathbf{j} + \dfrac{d^2z}{dt^2}\mathbf{k}\right)$, where *x, y,* and *z* are the coordinates of the object's center of mass. The parallel *x'y'z'* reference frame is moving at a constant velocity relative to the *xyz* reference frame; that is, $d\mathbf{R}/dt$ is constant. Prove that the *x'y'z'* reference frame is inertial. A reference frame that translates at a constant velocity relative to an inertial reference frame is also inertial.

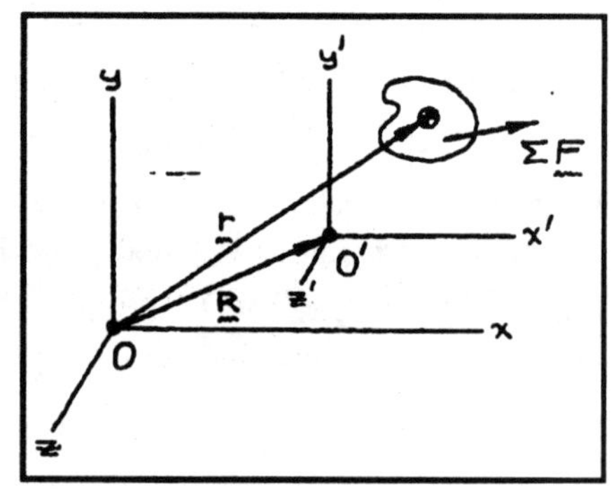

Solution: Let the vector from *O'* to the object be $\mathbf{r}'$. We are given that $\sum \mathbf{F} = m\dfrac{d^2\mathbf{r}}{dt^2}$ for the *xyz* reference frame and we want to show that $\sum \mathbf{F} = m\dfrac{d^2\mathbf{r}'}{dt^2}$ for the *x'y'z'* reference system. Note that $\mathbf{r} = \mathbf{R} + \mathbf{r}'$. Taking a derivative, we have $\dfrac{d\mathbf{r}}{dt} = \dfrac{d\mathbf{R}}{dt} + \dfrac{d\mathbf{r}'}{dt}$, (where $\dfrac{d\mathbf{R}}{dt}$ is constant). Taking a second derivative, we have $\dfrac{d^2\mathbf{r}}{dt^2} = \dfrac{d^2\mathbf{r}'}{dt^2}$, because the derivative of the constant term dropped out. Going back to the given relationship, we can now writen that $\sum \mathbf{F} = m\dfrac{d^2\mathbf{r}}{dt^2} = m\dfrac{d^2\mathbf{r}'}{dt^2}$. Hence, the *x'y'z'* reference frame is inertial.

===================================<>===================================

==================================<>==================================

Problem 3.66 The boat weighs *2600-lb* with its passengers. It is moving in a circular path of radius *R=80 ft* at a constant speed of *15 mi/hr*. Determine the magnitudes of: (a) the total horizontal force on the boat in the direction tangent to its path; (b) the total horizontal force on the boat in the direction perpendicular to its path; (c) the total vertical force acting on the boat (inculding its weight).

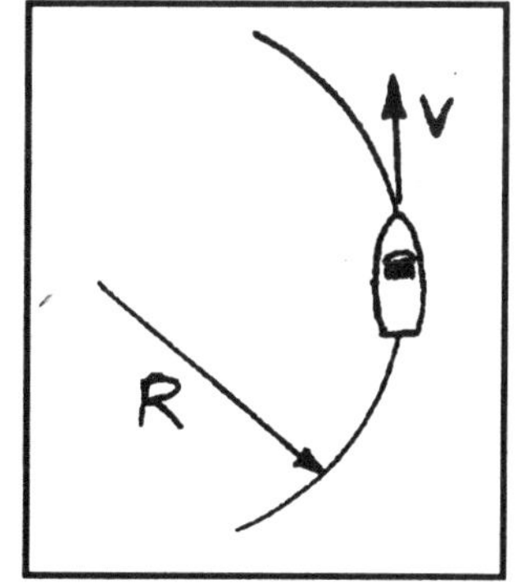

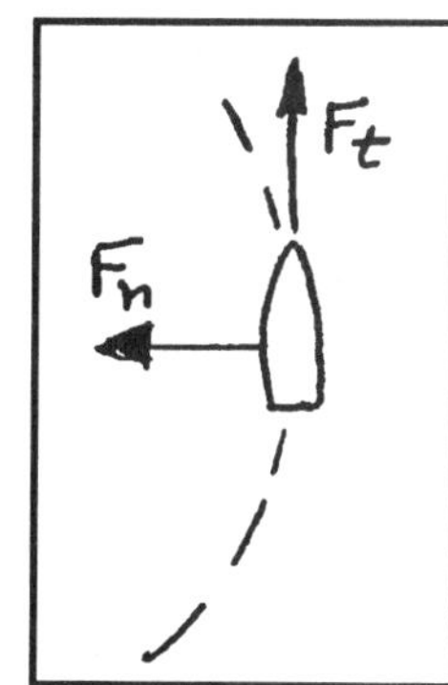

Solution: First, we need to draw free body diagrams of the boat, showing forces in all three coordinate directions. A top view and a rear view are shown. The mass of the boat is determined from *W=mg*. The equations of motion in the three coordinate directions are $\sum F_t = ma_t = m\,dv/dt$, $\sum F_n = ma_n = mv^2/R$, and $\sum F_{vert} = B - mg = 0$. For this problem, we have $dv/dt = 0$, $R = 80ft$, and $v = 15mi/hr = 22ft/s$. Substituting these values into the equations of motion, we get $a_t = 0$, $a_n = 6.05ft/s^2$, and $a_{vert} = 0$. Hence, the associated forces are $\sum F_t = 0$, $\sum F_{vert} = 0$, and $\sum F_n = 488.5\ lb$ (inward toward the center of curvature of the path.)

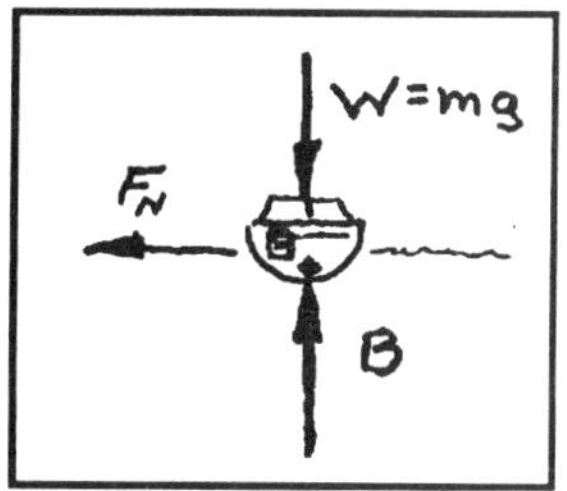

==================================<>==================================

Problem 3.67 Suppose that the mass of the boat in Problem 3.66 is 1200 kg with its passengers. At the present instant, it is moving in a circular path of radius *R = 30 m* at *4 m/s* and its speed is increasing at $2m/s^2$. Determine the magnitudes of: (a) the total horizontal force on the boat in the direction tangent to its path; (b) the total horizontal force on the boat in the direction perpendicular to its path.

Solution: The solution to this problem is the same as the solution of Problem 3.66 except that the units have changed and in this problem, *dv/dt* is NOT zero (hence we have an acceleration and force component in the tangential direction). We have $m = 1200kg$, $dv/dt = 2m/s^2$, $R = 30m$, and $v = 4m/s$. Substituting these numbers into the solution of Problem 3.66, we get $a_t = 2m/s^2$, $a_n = 0.533m/s^2$, and $a_{vert} = 0$. Hence, the forces are $\sum F_t = 2400\ (newtons)$, $\sum F_{vert} = 0$, and $\sum F_n = 640\ (newtons)$.

==================================<>==================================

Problem 3.68 If you choose the velocity of the train in Example 3.5 properly, the lateral force *S* exerted on it as it travels along the circular track is zero. What is the necessary velocity?

Solution: Refer to Example 3.5. The equations of motion in the normal and vertical directions are $M\sin 40^\circ - S\cos 40^\circ = mv^2/R$ and $M\cos 40^\circ + S\sin 40^\circ = mg$. Here the bank angle of the track is 40°. To make the analysis more general, we could replace the bank angle with a literal value, say ϕ. We will do this when we complete the analysis. We want the side force to be zero so we set $S = 0$ in both equations. We get $M\cos\phi = M\cos 40^\circ = mg$ and $M\sin\phi = M\sin 40^\circ = mv^2/R$. Divide the second equation by the first and we get $\tan\phi = \tan 40^\circ = v^2/(gR)$. Hence, the necessary velocity for Example 3.5 is given by $v^2 = gR\tan 40^\circ$ and in the more general case, with a track bank angle of ϕ, the velocity for no side force is given by $v^2 = gR\tan\phi$. For the data given, $v = 35.1m/s$.

==================================<>==================================

=================================<>=================================

Problem 3.69 An astronaut candidate is tested in a centrifuge with a raduis of *10 m*. If the centrifuge rotates in the horizontal plane with a constant angular velocity of one revolution every two seconds, what horizontal force is exerted on the astronaut. His mass is *72 kg*.

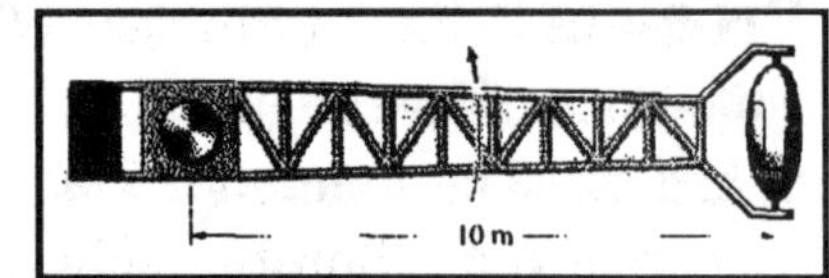

Strategy: If a point moves in a circular path of radius *R*, the magnitude of its velocity *v* is related to the angular velocity ω by $v = R\omega$, so the normal component of its acceleration can be written $a_n = v^2 / R = \omega^2 R$ (See Eq. 2.45). You can use this expression to determine the normal component of the force acting on the astronaut.

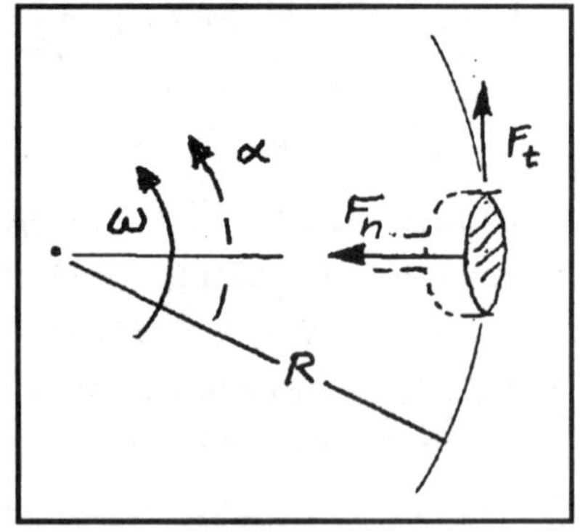

Solution: The equations of motion in the two horizontal coordinate directions are $\sum F_t = ma_t = m\,dv/dt = mR(d\omega / dt)$ and $\sum F_n = ma_n = mv^2 / R = m\omega^2 R$. For this problem, we have $dv/dt = 0$, $\omega = 0.5 rev / s = \pi\ (rad / s)$. Substituting values into the equations, we get $\sum F_t = 0$ and $\sum F_n = 7106\ (newtons)$ radially inward.

=================================<>=================================

Problem 3.70 The centrifuge in Problem 3.69 starts from rest at *t=0* and is subjected to a constant angular acceleration $\alpha = 0.1 rad / s^2$. What is the magnitude of the total horizontal force exerted on the astronaut at *t = 30s?*

Solution: The solution to this problem is much like that of Problem 3.69. In this case, we do have an angular acceleration and thus we do have a tangential force acting on the astronaut. The equation for ω is $\omega = \alpha t\ (rad / s)$. At *t = 30s*, $\omega = 3\ rad / s$. The equations of motion in the two horizontal coordinate directions are $\sum F_t = ma_t = m\,dv/dt = mR(d\omega / dt) = mR\alpha$ and $\sum F_n = ma_n = mv^2 / R = m\omega^2 R$. Substituting values into the equations, we get $\sum F_t = 72\ (newtons)$ in the direction of the velocity and $\sum F_n = 6480\ (newtons)$ radially inward. Adding the two force components vectorially, we get $\sum F_{total} = 6480.4\ (newtons)$. (Almost all of the force is radial).

=================================<>=================================

Problem 3.71 The circular disk lies in the horizontal plane and rotates with a constant counterclockwise angular velocity of *6 rad/s*. The *4-lb* slider is supported horizontally by the smooth slot and the string attached at B. What is the tension in the string.

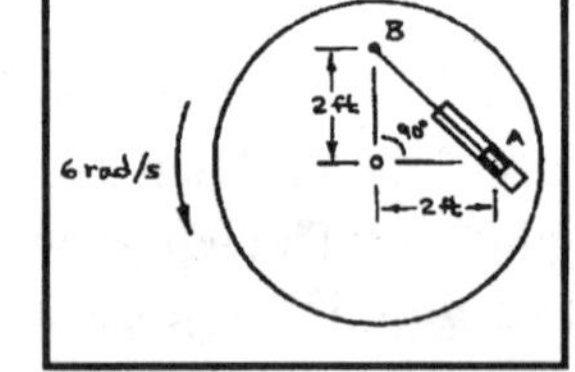

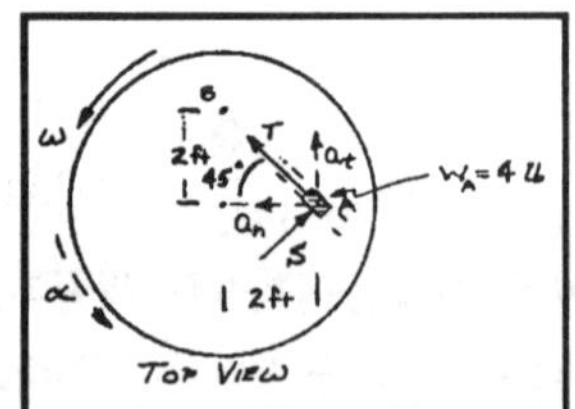

Solution: We will first find the normal and tangential forces necessary to cause the motion described and then will resolve these forces into components parallel and perpendicular to the slot. The equations of motion in the radial and normal coordinate directions are

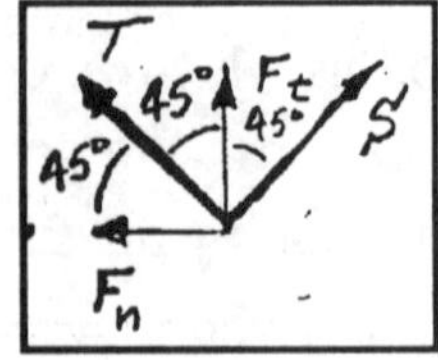

$\sum F_t = ma_t = m\,dv/dt = mR(d\omega / dt) = mR\alpha$ and $\sum F_n = ma_n = mv^2 / R = m\omega^2 R$.
In this situation, $R = 2 ft$, $\omega = 6 rad / s$, $\alpha = 0$, and $m = W / g = 4 / 32.2 = 0.124 slug$. Substituting values into the equations, we get $\sum F_t = 0$ and $\sum F_n = 8.94\ lb$ radially inward. We now need to resolve this force into components parallel and perpendicular to the slot. The equations for the transformation are $T = F_t \cos 45^\circ + F_n \sin 45^\circ = 6.32\ lb$ and $S = F_t \sin 45^\circ - F_n \cos 45^\circ = -6.32\ lb$

=================================<>=================================

====================================<>====================================

Problem 3.72 In Problem 3.71, determine the tension in the string if the circular disk has a counterclockwise angular acceleration of 10 rad/s^2 at the instant shown.

Solution: The solution is identical to that of Problem 3.71 except for the fact that here $\alpha = 10\ rad/s^2$ instead of zero. Substituting this new value into the solution of 3.71, we get $\sum F_t = 2.48\ lb$ in the direction of the velocity and $\sum F_n = 8.94\ lb$ radially inward. Transforming coordinates, we get $T = 8.08\ lb$ and $S = -4.57\ lb$.

====================================<>====================================

Problem 3.73 The 2 *kg* slider A starts from rest and slides *in the horizontal plane* along the smooth circular bar under the action of a tangential force $F_t = 4t$ newtons. At $t = 4$ seconds, determine (a) The magnitude of the velocity of A; (b) the magnitude of the horizontal force exerted on A by the bar.

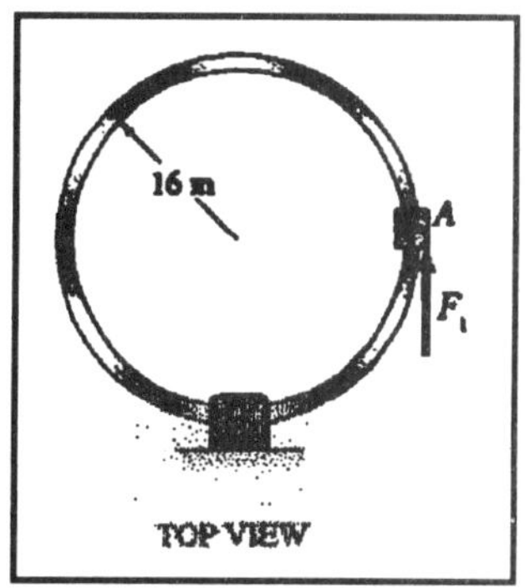

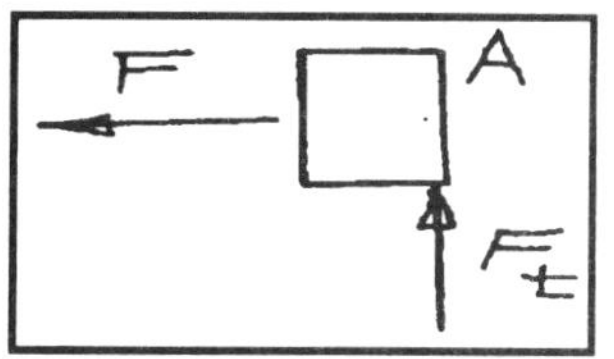

Solution The tangential acceleration is $a_t = \frac{dv}{dt}$. The equation of motion s $F_t = m\frac{dv}{dt}$, from which $v = \frac{1}{m}\int F_t dt + C = \left(\frac{1}{2}\right)(2)t^2$, since $v(0) = 0$. (a) At $t = 4$ s, $v = 16$ m/s. The horizontal force exerted on A by the bar is $F = m\frac{v^2}{R} = 2\left(\frac{16^2}{16}\right) = 32$ N

====================================<>====================================

Problem 3.74 Small parts on a conveyor belt moving with constant velocity v are allowed to drop into a bin. Show that the angle α at which the parts start sliding on the belt satisfies the equation $\cos\alpha - \frac{1}{\mu_s}\sin\alpha = \frac{v^2}{gR}$, where μ_s is the coefficient of static friction between the parts and the belt.

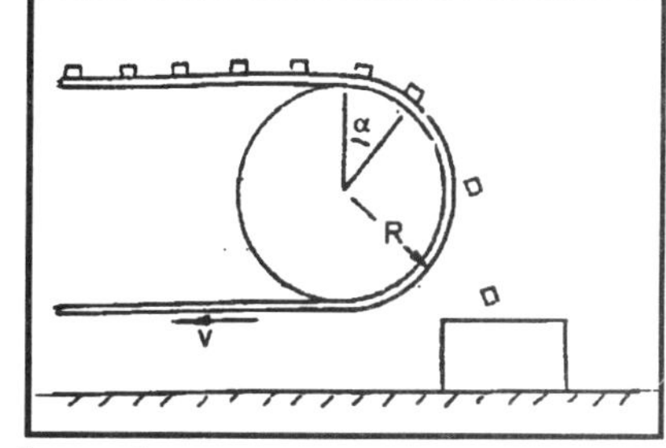

Solution The condition for sliding is $\sum F_t = -mg\sin\alpha + f = 0$, where $-mg\sin\alpha$ is the component of weight acting tangentially to the belt, and $f = \mu_s N$ is the friction force tangential to the belt. From Newton's second law the force perpendicular to the belt is $N - mg\cos\alpha = -m\frac{v^2}{R}$, from which the condition for slip is $-mg\sin\alpha + \mu_s mg\cos\alpha - \mu_s m\frac{v^2}{R} = 0$. Solve: $\cos\alpha - \frac{1}{\mu_s}\sin\alpha = \frac{v^2}{gR}$

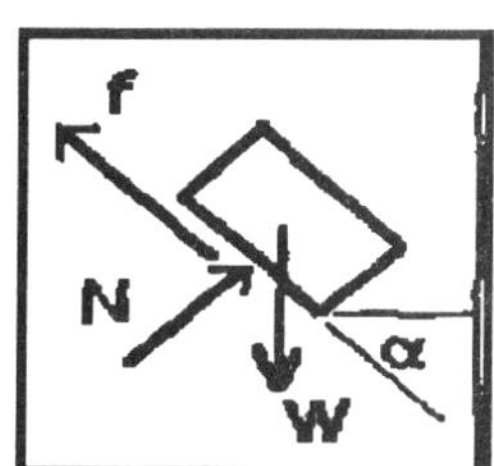

====================================<>====================================

==================================<>==================================

Problem 3.75 The mass m rotates around the vertical pole in a horizontal circular path. Determine the magnitude of its velocity in terms of θ.

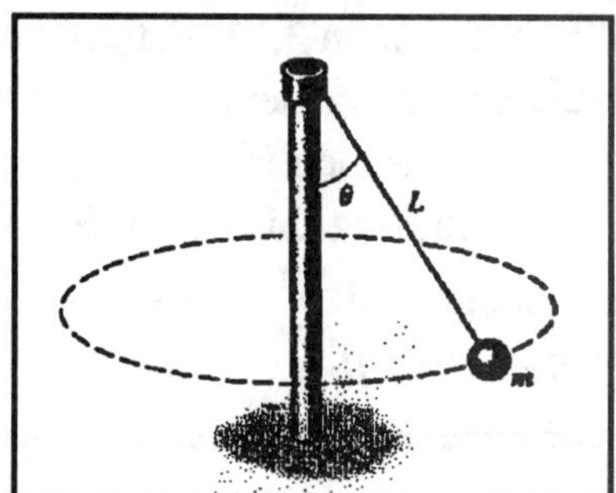

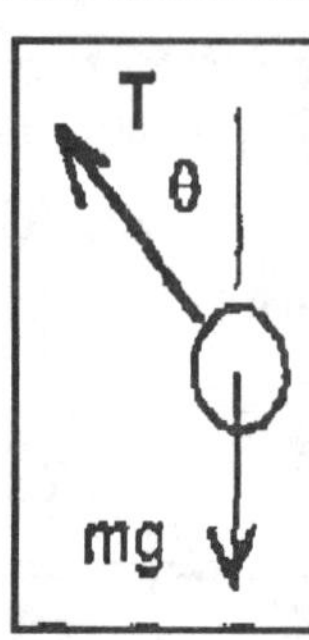

Solution From Newton's second law, $T\sin\theta = m\dfrac{v^2}{R}$, and $T\cos\theta = mg$. The radius is $R = L\sin\theta$, from which $\dfrac{mg\sin\theta}{\cos\theta} = m\dfrac{v^2}{L\sin\theta}$. Solve: $v = \sqrt{\dfrac{gL\sin^2\theta}{\cos\theta}}$

==================================<>==================================

Problem 3.76 In Problem 3.75, if $m = 1$ slug, $L = 4$ ft, and the velocity $v = 15$ ft/s, what is the tension in the string?

Solution From the solution to Problem 3.75, the velocity and the angle are related by the equation: $\tan\theta\sin\theta = \dfrac{v^2}{gL}$. The angle was determined from the zero crossing for $f(\theta) = -\tan\theta\sin\theta + \dfrac{v^2}{gL}$ and the result refined by iteration. The result: $\theta = 62.998\ldots = 63^\circ$. The radius of the horizontal path is $R = L\sin\theta = 3.56$ ft The tension in the string is the resultant of the two forces: The weight of the ball and the force due to centripetal acceleration: $T = \sqrt{W^2 + \left(m\dfrac{v^2}{R}\right)^2} = 70.86$ lb

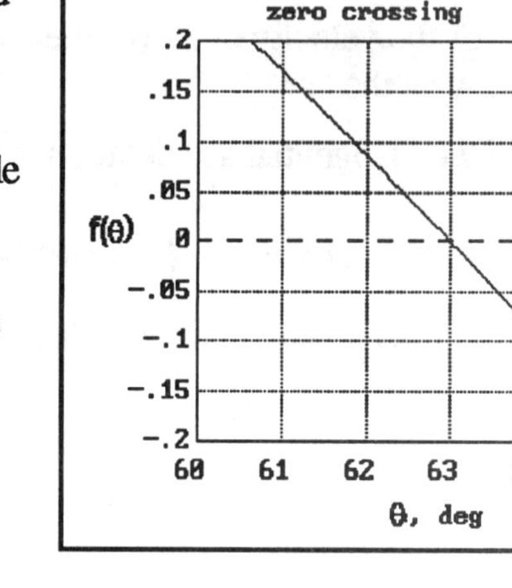

==================================<>==================================

==================================<>==================================

Problem 3.77 The 10 *kg* mass *m* rotates around the vertical pole in a horizontal circular path of radius $R = 1\ m$. If the magnitude of the velocity is $v = 3\ m/s$, what are the tensions in the strings A and B?

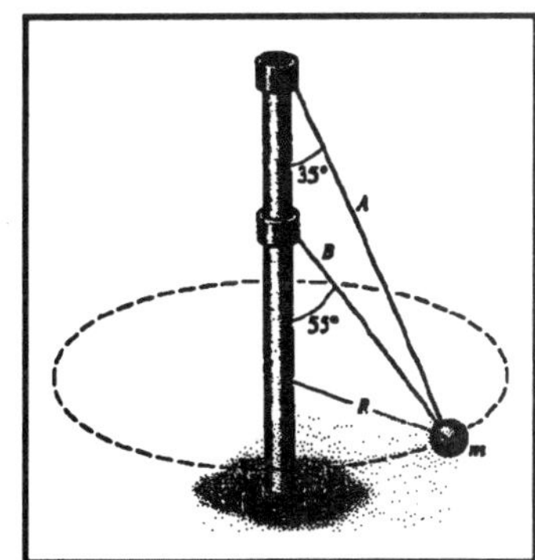

Solution Choose a cartesian coordinate system in the vertical plane that rotates with the mass. The weight of the mass is $\vec{W} = -\vec{j}mg = -\vec{j}98.1\ N$. The radial acceleration is by definition directed inward:

$\vec{a}_n = -\vec{i}\left(\frac{v^2}{R}\right) = -9\vec{i}\ \ m/s^2$. The angles from the horizontal are

$\theta_A = 90^o + 35^o = 125^o$, $\theta_B = 90^o + 55^o = 145^o$ The unit vectors parallel to the strings, from the pole to the mass ,are: $\vec{e}_A = +\vec{i}\cos\theta_A + \vec{j}\sin\theta_A$. $\vec{e}_B = +\vec{i}\cos\theta_B + \vec{j}\sin\theta_B$. From Newton's second law for the mass, $\vec{T} - \vec{W} = m\vec{a}_n$, from which $|\vec{T}_A|\vec{e}_A + |\vec{T}_B|\vec{e}_B - \vec{j}mg = -\vec{i}\left(m\frac{v^2}{R}\right)$. Separate components to obtain the two simultaneous equations:

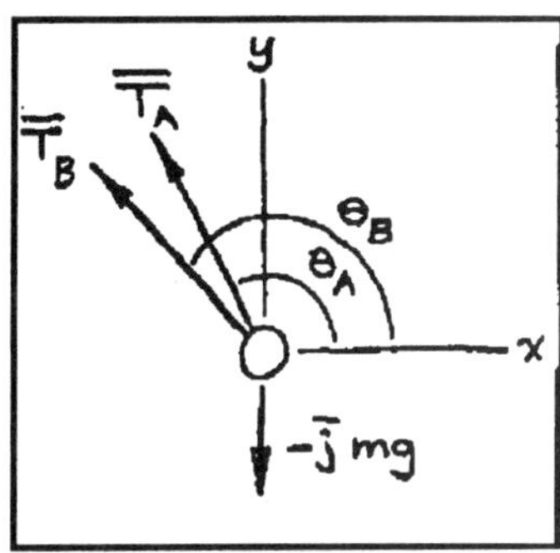

$|\vec{T}_A|\cos 125^o + |\vec{T}_B|\cos 145^o = -90\ \ N$ $|\vec{T}_A|\sin 55^o + |\vec{T}_B|\sin 35^o = 98.1\ \ N$.

Solve: $\boxed{|\vec{T}_A| = 84\ N}$. $\boxed{|\vec{T}_B| = 51\ \ N}$

==================================<>==================================

Problem 3.78 In Problem 3.77, what is the range of values of *v* for which the mass will remain in the circular path described?

Solution The minimum value of *v* will occur when the string B is impending zero, and the maximum will occur when string A is impending zero. From the solution to Problem 3.77,

$|\vec{T}_A|\cos 125^o + |\vec{T}_B|\cos 145^o = -m\left(\frac{v^2}{R}\right)$, $|\vec{T}_A|\sin 125^o + |\vec{T}_B|\sin 145^o = mg$. These equations are to be solved for the velocity when one of the string tensions is set equal to zero. For $|\vec{T}_A| = 0$, $v = 3.743\ m/s$.For $|\vec{T}_B| = 0$, $v = 2.621\ \ m/s$. The range: $\boxed{2.62 \le v \le 3.74\ \ m/s}$

==================================<>==================================

Problem 3.79 Suppose you are designing a monorail transportation system that will travel at 50 *m/s*, and decide that the angle α that the cars swing out from the vertical when they go through a turn must not be larger than 20°. If the turns in the track consist of circular arcs of constant radius R, what is the minimum allowable value of R?

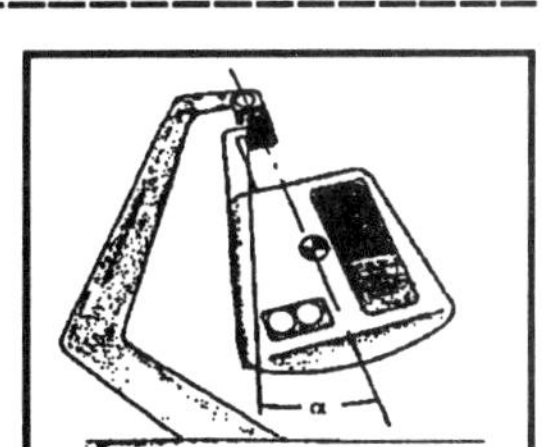

Solution Choose a coordinate system with the origin at the center of the circular arc with the *x* axis horizontal. The weight is $\vec{W} = -\vec{j}mg$. The support reaction is $\vec{T} = T(-\vec{i}\sin a + \vec{j}\cos\alpha)$.

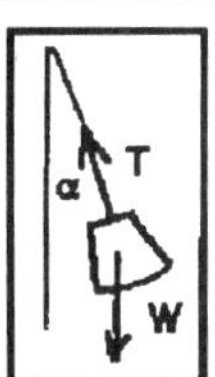

Solution continued on next page

Continuation of Solution to Problem 3.79

The acceleration is $\vec{a}_n = -\vec{i}\left(\frac{v^2}{R}\right)$. From Newton's second law, $\vec{T} + \vec{W} = m\vec{a}_n$. Separate components to obtain: $T\cos\alpha = mg$, $-T\sin\alpha = -m\left(\frac{v^2}{R}\right)$. Solve: $\boxed{R = \frac{v^2}{g\tan\alpha} = 700.2\ m}$

=====================================<>=====================================

Problem 3.80 An airplane of weight $W = 200{,}000\ lb$ makes a turn at constant altitude and at constant velocity $v = 600\ ft/s$. The bank angle is 15 . (a) Determine the lift force L. (b) What is the radius of curvature of the plane's path?

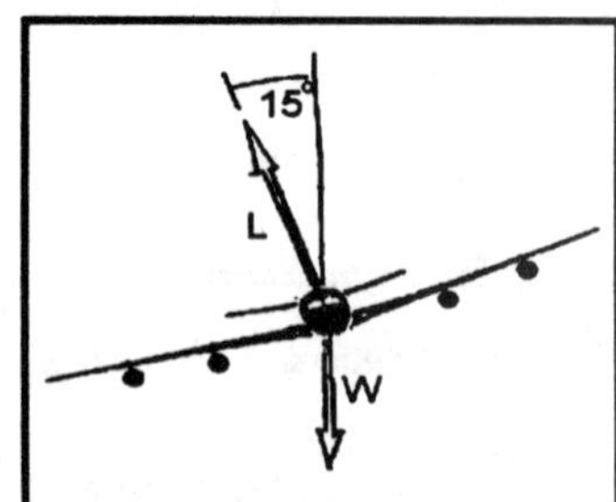

Solution The weight is $\vec{W} = -\vec{j}W = -\vec{j}(2\times10^5)\ lb$. The normal acceleration is $\vec{a}_n = \vec{i}\left(\frac{v^2}{\rho}\right)$. The lift is

$\vec{L} = |\vec{L}|(\vec{i}\cos 105^o + \vec{j}\sin 105^o) = |\vec{L}|(-0.2588\vec{i} + 0.9659\vec{j})$.

(a) From Newton's second law, $\sum\vec{F} = \vec{L} + \vec{W}_n = m\vec{a}_n$, from which, substituting values and separating the $\vec{j}$ components: $|\vec{L}|(0.9659) = 2\times10^5$, $\boxed{|\vec{L}| = \frac{2\times10^5}{0.9659} = 207055\ lb}$. (b) The radius of curvature is obtained from Newton's law: $|\vec{L}|(-0.2588) = -m\left(\frac{v^2}{\rho}\right)$, from which $\boxed{\rho = \left(\frac{W}{g}\right)\left(\frac{v^2}{|\vec{L}|(0.2588)}\right) = 41763.7\ ft}$.

Note: If the value $g = 32.2\ ft/s^2$ is adopted, $\rho = 41724.8\ ft$

=====================================<>=====================================

Problem 3.81 The suspended mass m is stationary. (a) What are the tensions in the strings? (b) If string A is cut, what is the tension in string B immediately afterwards?

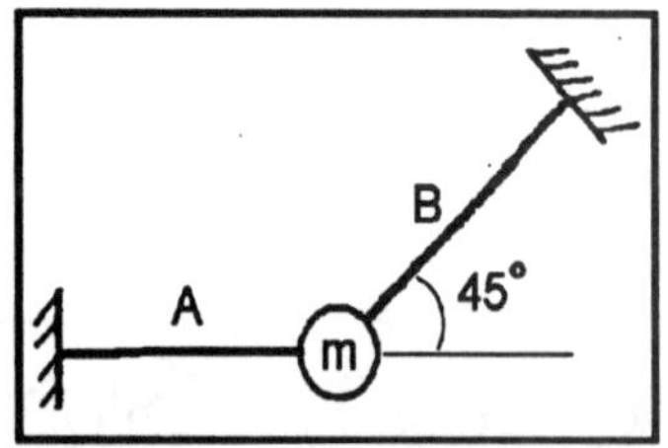

Solution (a) With the mass stationary, the external forces are the tensions in the strings and the weight. The tension in A is $\vec{T}_A = |\vec{T}_A|\vec{e}_A = |\vec{T}_A|(-\vec{i})$.

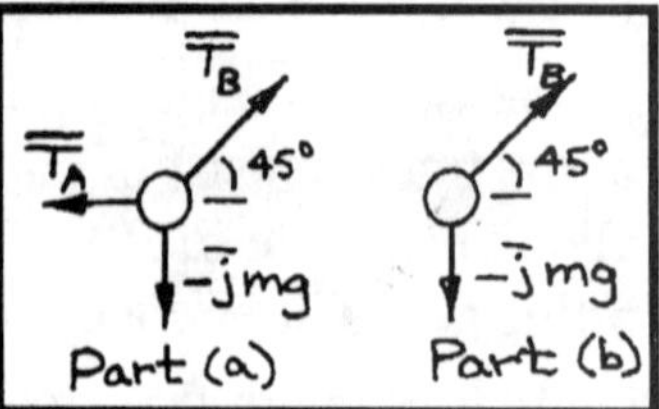

The tension in B is

$\vec{T} = |\vec{T}_B|\vec{e}_B| = \vec{T}_B|(\vec{i}\cos(45^o) + \vec{j}\sin(45^o)) = \frac{1}{\sqrt{2}}|\vec{T}_B|(\vec{i} + \vec{j})$. The weight is $\vec{W} = -mg\vec{j}$. The equilibrium condition is $\sum\vec{F} = \vec{T}_A + \vec{T}_B + \vec{W} = 0$, from which the two equations: $|\vec{T}_B|\left(\frac{1}{\sqrt{2}}\right) = mg$, and $|\vec{T}_B|\left(\frac{1}{\sqrt{2}}\right) = |\vec{T}_A|$. Solve:

$\boxed{|\vec{T}_B| = \sqrt{2}mg}$ and $\boxed{|\vec{T}_A| = mg}$.

(b) Immediately after string A is cut, the velocity is zero.

Solution continued on next page

Newton's second law for the mass is $\vec{T}_B - \vec{j}mg = m\vec{a}$. Since the motion is constrained to move in a circle with string B the radius, the acceleration has components normal to and tangential to the circular path. The tangential component contributes nothing to the tension in B. The normal component *immediately after the string is cut* is $\frac{v^2}{R} = 0$ since the velocity is zero. The normal component is also equal to the dot product: $m\vec{a}\cdot\vec{e}_B = 0 = \vec{T}_B\cdot\vec{e}_B + \left(-\vec{j}mg\right)\cdot\left(\frac{1}{\sqrt{2}}\left(\vec{i}+\vec{j}\right)\right) = \left|\vec{T}_B\right| - \frac{mg}{\sqrt{2}}$. From which

$$\boxed{\left|\vec{T}_B\right| = \frac{mg}{\sqrt{2}}}$$

==================================<>==================================

Problem 3.82 An airplane flies with constant velocity along a circular path in the vertical plane. The radius of its circular path is 5000 *ft*. The pilot weighs 150 *lb*. (a) The pilot will experience "weightlessness" at the top of the circular path if the airplane exerts no net force on him at that point. Draw a free body diagram of the pilot and use it to determine the velocity v necessary to achieve this condition. (b) Determine the force exerted on the pilot by the airplane at the top of the circular path if the airplane is traveling at twice the velocity determined in Part (a).

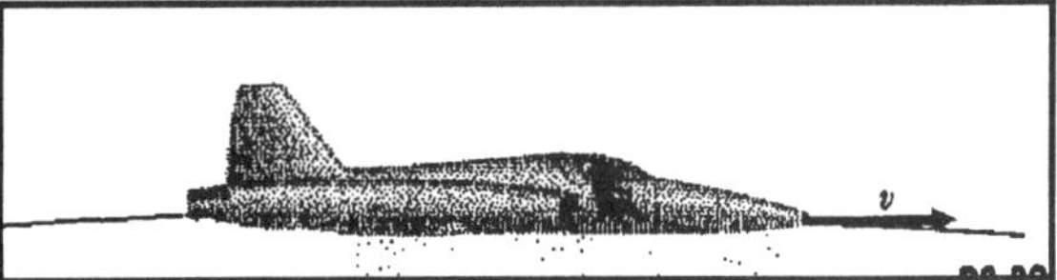

Solution Choose a coordinate system with the origin at the center of the circular path, with the y axis upward.. (a) The normal acceleration is $\vec{a}_n = -\vec{j}\frac{v^2}{R}$, and the pilot's weight is $\vec{W} = -\vec{j}mg$. The normal force exerted on the pilot by the airplane is $\vec{N} = \vec{j}N$. From Newton's second law, $\sum\vec{F} = \vec{N} + \vec{W} = m\vec{a}_n$, from which, for $\vec{N} = 0$,

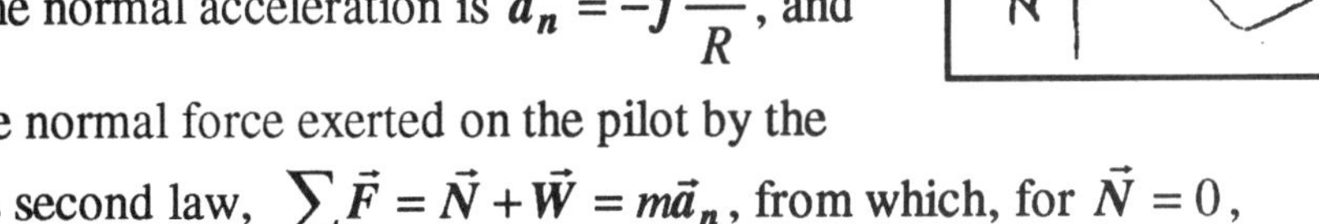

$-mg = -m\left(\frac{v^2}{R}\right)$, from which $\boxed{v = \sqrt{Rg} = \sqrt{5000(32.17)} = 401.1\ ft/s}$ (b) If the velocity is doubled, the force exerted on the pilot is $\vec{N} = -m\frac{v^2}{R}\vec{j} + mg\vec{j}$, from which, substituting $v = 2\sqrt{Rg}$,

$N = -4mg + mg = -3mg = -3\left(\frac{W}{g}\right)g = -3W$, where W is the weight of the pilot, from which

$\boxed{N = -3(150) = -450\ lb}$. The force is downward (exerted by the seatbelt constraints, or possibly by the overhead canopy), tending to keep the pilot in the airplane.

==================================<>==================================

==================================<>==================================

Problem 3.83 The smooth circular bar rotates with constant angular velocity ω_o about the vertical axis AB. Determine the angle β at which the slider of mass *m* will remain stationary relative to the circular bar.

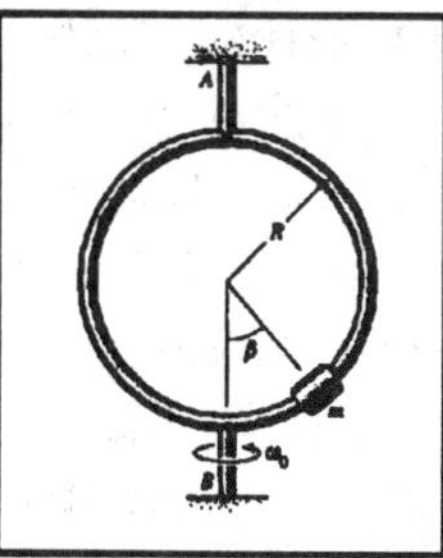

Solution Choose a coordinate system in the plane of the ring, with the *y* axis coincident with the axis AB and the origin at the center of the ring. The horizontal distance of the slider from the axis AB is *x*. The acceleration normal to the AB axis is $a_n = -\omega_o^2 x$. The distance $x = R\sin\beta$, from which $a_n = -\omega_o^2 R\sin\beta$. The slider is constrained to move on the circular ring, and the component of a_n tangential to the ring is $a_n\cos\beta = -\omega_o^2 R\sin\beta\cos\beta$. The component of the weight tangential to the ring is $-mg\sin\beta$. From Newton's second law, $-mg\sin\beta = -m\omega_o^2 R\cos\beta\sin\beta$. This is satisfied by $\sin\beta = 0$, $\boxed{\beta = 0,\ \beta = \pi}$ when the acceleration normal to the AB axis is zero, and by $\cos\beta = \dfrac{g}{R\omega_o^2}$, $\boxed{\beta = \cos^{-1}\left(\dfrac{g}{R\omega_o^2}\right)}$ for all other angles.

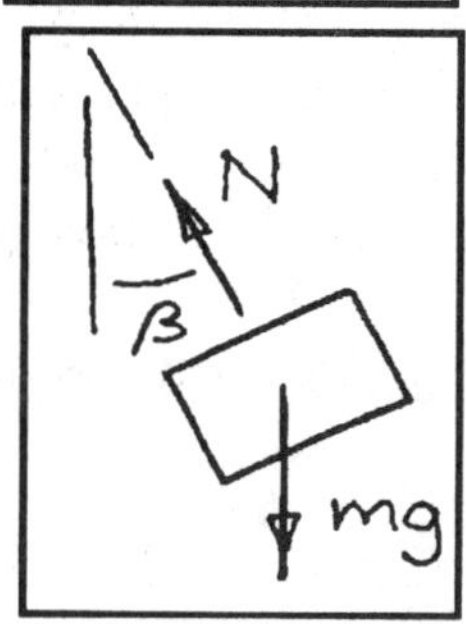

==================================<>==================================

Problem 3.84 The force exerted on a charged particle by a magnetic field is $\vec{F} = q\vec{v}\times\vec{B}$, where *q* and $\vec{v}$ are the charge and velocity of the particle, and $\vec{B}$ is the magnetic field vector. A particle of mass *m* and positive charge *q* is projected at O with velocity $\vec{v} = v_o\vec{i}$ into a uniform magnetic field $\vec{B} = B_o\vec{k}$. Using normal and tangential components, show that: (a) the magnitude of the particle's velocity is constant,; (b) the particle's path is a circle of radius $m\dfrac{v_o}{qB_o}$.

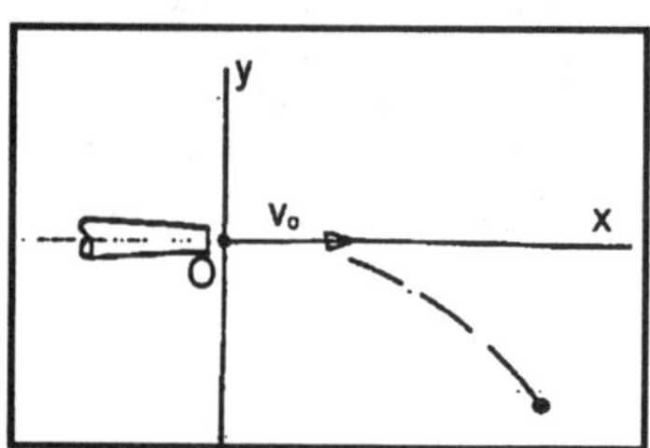

Solution (a) The force $\vec{F} = q\left(\vec{v}\times\vec{B}\right)$. is everywhere normal to the velocity and the magnetic field vector on the particle path. Therefore the tangential component of the force is zero, hence from Newton's second law the tangential component of the acceleration $\dfrac{dv}{dt} = 0$, from which $\boxed{v(t) = C = v_o}$, and the velocity is a constant. Since there is no component of force in the z-direction, and no initial z-component of the velocity, the motion remains in the *x-y* plane. The unit vector $\vec{k}$ is positive out of the paper in the figure; by application of the right hand rule the cross product $\vec{v}\times\vec{B}$ is directed along a unit vector toward the instantaneous center of the path at every instant, from which $\vec{F} = -\left|\vec{F}\right|\vec{e}_n$, where $\vec{e}_n$ is a unit vector normal to the path. The normal component of the acceleration is $\vec{a}_n = -\left(v_o^2/\rho\right)\vec{e}_n$, where ρ is the radius of curvature of the path. From Newton's second law, $\vec{F} = m\vec{a}_n$, from which $-\left|\vec{F}\right| = -m\left(v_o^2/\rho\right)$. The magnitude of the cross product can be written as $\left|\vec{v}\times\vec{B}\right| = v_o B_o\sin\theta = v_o B_o$, since $\theta = 90^o$ is the angle between $\vec{v}$ and $\vec{B}$. From which: $qv_oB_o = m\dfrac{v_o^2}{\rho}$, from which the radius of curvature is $\boxed{\rho = \dfrac{mv_o}{qB_o}}$. Since the term on the right is a constant, the radius of curvature is a constant, and the path is a circle with radius $\dfrac{mv_o}{qB_o}$.

==================================<>==================================

==================================<>==================================

Problem 3.85 A mass m is attached to a string that is wrapped around a fixed post of radius R. At $t = 0$, the object is given a velocity v_o as shown. Neglect external forces on m other that the force exerted by the string. Determine the tension in the string as a function of the angle θ. *Strategy*: The velocity vector of the mass is perpendicular to the string. Express Newton's second law in terms of normal and tangential components.

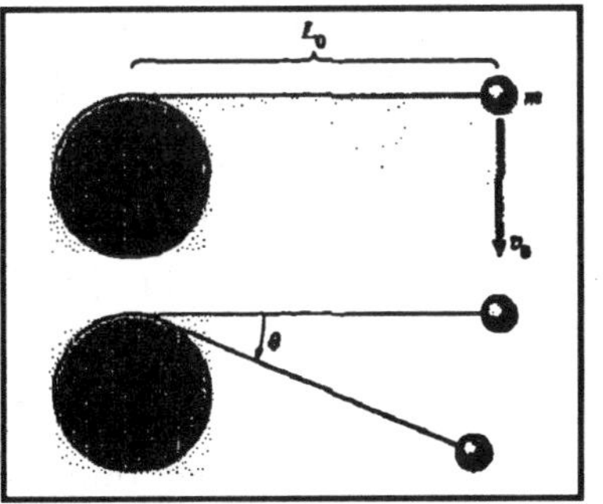

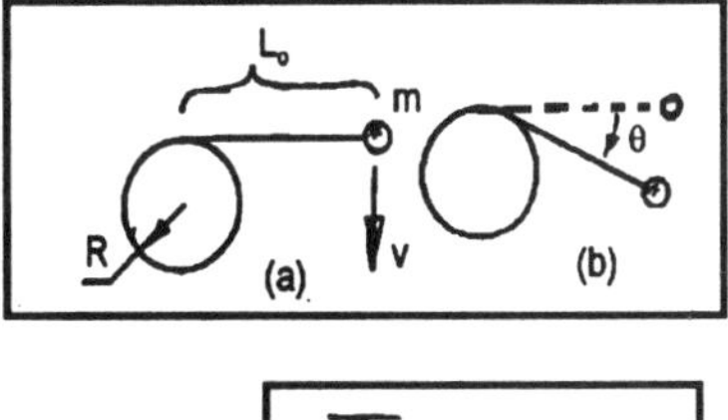

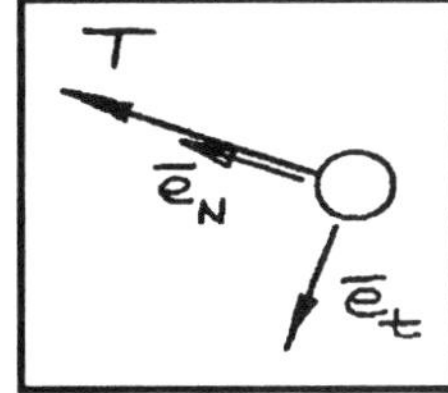

Solution Make a hypotheitical cut in the string and denote the tension in the part connected to the mass by T. The acceleration normal to the path is $\frac{v^2}{\rho}$. The instantaneous radius of the path is $\rho = L_o - R\theta$, from which by Newton's second law, $\sum F_n = T = m\frac{v_o^2}{L_o - R\theta}$, from which

$$\boxed{T = m\frac{v_o^2}{L_o - R\theta}}$$

==================================<>==================================

Problem 3.86 In Problem 3.85 determine θ as a function of time.

Solution Use the solution to Problem 3.85. The angular velocity is $\frac{d\theta}{dt} = \frac{v_o}{\rho}$. From Problem 3.85, $\rho = L_o - R\theta$, from which $\frac{d\theta}{dt} = \frac{v_o}{(L_o - R\theta)}$. Separate variables: $(L_o - R\theta)d\theta = v_o dt$. Integrate: $L_o\theta - \frac{R}{2}\theta^2 = v_o t$, since $\theta(0) = 0$. In canonical form $\theta^2 + 2b\theta + c = 0$, where $b = -\frac{L}{R}$, and $c = \frac{2v_o t}{R}$. The solution: $\theta = -b \pm \sqrt{b^2 - c} = \frac{L_o}{R} \pm \sqrt{\left(\frac{L_o}{R}\right)^2 - \frac{2v_o t}{R}}$. The angle increases with time, so the negative sign applies. Reduce: $\boxed{\theta = \frac{L_o}{R}\left(1 - \sqrt{1 - \frac{2Rv_o t}{L_o^2}}\right)}$ *Check:* When $R\theta = L_o$, the string has been fully wrapped around the post. Substitute to obtain: $\sqrt{1 - \frac{2Rv_o t}{L_o^2}} = 0$, from which $\frac{2Rv_o t}{L_o^2} = 1$, which is the value for impending failure as t increases because of the imaginary square root. Thus the solution behaves as expected. *check.*

==================================<>==================================

==================================<>==================================

Problem 3.87 The sum of the forces in newtons exerted on the *360 kg* sport plane (including its weight) during an interval of time is $(-1000+280t)\,\mathbf{i}+(480-430t)\,\mathbf{j}+(720+200t)\mathbf{k}$, where t is the time in seconds. At $t=0$ the velocity of the plane's center of gravity is $20\,\mathbf{i}+35\,\mathbf{j}-20\mathbf{k}$ (m/s). If you resolve the sum of the forces on the plane into components tangent to the plane's flight path at $t=2s$, what are their values $\sum F_t$ and $\sum F_n$?

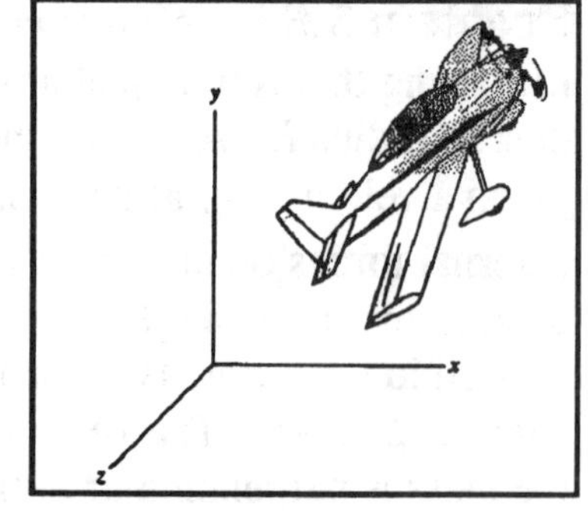

Solution: This problem has several steps. First, we must write Newton's second law and find the acceleration of the aircraft. We then integrate the components of the acceleration (separately) to find the velocity components as functions of time. Then we evaluate the velocity of the aircraft and the force acting on the aircraft at *t = 2s.* Next, we find a unit vector along the velocity vector of the aircraft and project the total force acting on the aircraft onto this direction. Finally, we find the magnitude of the total force acting on the aircraft and the force component normal to the direction of flight. We have $a_X=(1/m)(-1000+280t)$, $a_Y=(1/m)(480-430t)$, and $a_Z=(1/m)(720+200t)$. Integrating and inserting the known initial velocities, we obtain the relations $v_X=v_{X0}+(1/m)(-1000t+280t^2/2)\ (m/s)=20+(1/m)(-1000t+140t^2)\ (m/s)$, $v_Y=35+(1/m)(480t-215t^2)\ (m/s)$, and $v_Z=-20+(1/m)(720t+100t^2)(m/s)$. The velocity at *t = 2s* is $\mathbf{v}=16\,\mathbf{i}+35.3\,\mathbf{j}-14.9\mathbf{k}\ (m/s)$ and the unit vector parallel to **v** is $\mathbf{e}_v=0.386\,\mathbf{i}+0.850\,\mathbf{j}-0.359\mathbf{k}$. The total force acting on the aircraft at *t=2s* is $\mathbf{F}=-440\,\mathbf{i}-380\,\mathbf{j}+1120\mathbf{k}$ *newtons*. The component of **F** tangent to the direction of flight is $\sum F_t=\mathbf{F}\bullet\mathbf{e}_v=-894.5$ *newtons*. The magnitude of the total force acting on the aircraft is $|\mathbf{F}|=1261.9$ *newtons.* The component of **F** normal to the direction of flight is given by $\sum F_n=\sqrt{|\mathbf{F}|^2-\left(\sum F_t\right)^2}=890.1$ *newtons*

==================================<>==================================

Problem 3.88 In problem 3.87, what is the instantaneous radius of curvature of the plane's path at *t = 20 s*? The vector components of the sum of the forces in the directions tangent and normal to the path lie in the osculating plane. At t=2s, determine the components of a unit vector perpendicular to the osculating plane.

Strategy: From the solution to problem 3.87, we know the total force vector and acceleration vector acting on the plane. We also know the direction of the velocity vector. From the velocity and the magnitude of the normal acceleration, we can determine the radius of curvature of the path.

The cross product of the velocity vector and the total force vector will give a vector perpendicular to the plane containing the velocity vector and the total force vector. This vector is perpendicular to the plane of the osculating path. We need then only find a unit vector in the direction of this vector.

Solution: From Problem 3.87, we know at *t=2s*, that $a_n=\sum F_n/m=2.47m/s^2$. We can find the magnitude of the velocity $|\mathbf{v}|=41.5m/s$ at this time. The radius of curvature of the path can then be found from $\rho=|\mathbf{v}|^2/a_n=696.5m$.

The cross product yields the desired unit vector, i.e., $\mathbf{e}=(\mathbf{F}\times\mathbf{v})/|\mathbf{F}\times\mathbf{v}|=-0.916\,\mathbf{i}+0.308\,\mathbf{j}-0.256\mathbf{k}$

==================================<>==================================

=================================<>=================================

Problem 3.89 A car is traveling on a straight level road when the driver perceives a hazard ahead. After a reaction time of 0.5 seconds, he applies the brakes, locking the wheels. The coefficient of kinetic friction between the tires and the road is $\mu_k = 0.6$. Determine the total distance the car travels before coming to rest, including the distance traveled before the brakes are applied, if it is traveling at (a) 55 *mi/hr*; (b) 65 *mi/hr*.

Solution Use the chain rule: $v\dfrac{dv}{dx} = -\dfrac{\mu_k W}{m} = -g\mu_k$, from which $v^2(x) = -2g\mu_k x + V_o^2$. The car comes to a rest when $v_o = 0$, from which $x = \dfrac{V_o^2}{2\mu_k g}$. The initial distance traveled before the brakes are applies is $x_o = V_o t_d = \dfrac{V_o}{2}$. The total distance: $x = \dfrac{V_o}{2}\left(1 + \dfrac{V_o}{\mu_k g}\right)$. (a) For $V_o = 55\ mi/hr = 55\left(\dfrac{5280}{3600}\right) = 80.67\ ft/s$, $\boxed{x = 208.9\ ft}$ (b) For $V_o = 65\ mi/hr = 65\left(\dfrac{5280}{3600}\right) = 95.33\ ft/s$, $\boxed{x = 283.1\ ft}$ *Note*: If the value $g = 32.2\ ft/s^2$ is adopted, the results are (a) $x = 208.7\ ft$, (b) $282.9\ ft$

=================================<>=================================

Problem 3.90 If the car in Problem 3.89 is traveling at 65 *mi/hr* and rain decreases the value of μ_k to 0.4, what total distance does the car travel before coming to rest?

Solution Use the solution to Problem 389: $x = \dfrac{V_o}{2}(1 + \dfrac{V_o}{\mu_k g})$. For $65\ mi/hr = 95.3\ ft/s$,

$\boxed{x = 400.8\ ft}$

=================================<>=================================

Problem 3.91 A car traveling at 30 *m/s* is at the top of a hill. The coefficient of kinetic friction between the tires and the road is $\mu_k = 0.8$ and the instantaneous radius of curvature of the car's path is 200 meters. If the driver applies the brakes and the car's wheels lock, what is the resulting deceleration of the car tangent to its path?

Solution From Newton's second law; $N - W = -m\dfrac{v^2}{R}$ from which $N = m\left(g - \dfrac{v^2}{R}\right)$. The acceleration tangent to the path is $\dfrac{dv}{dt}$, from which $\dfrac{dv}{dt} = -\dfrac{\mu_k N}{m}$., and $\boxed{\dfrac{dv}{dt} = -\mu_k\left(g - \dfrac{v^2}{R}\right) = 4.25\ m/s^2}$

=================================<>=================================

==================================<>==================================

Problem 3.92 Suppose that the car in Problem 3.91 is at the bottom of a depression whose radius of curvature is 200 meters when the driver applies the brakes. What is resulting deceleration of the car tangent to its path?

Solution Use the solution to Problem 3.91: $\frac{dv}{dt} = -\frac{\mu_k N}{m}$. From Newton's second law, $N - W = m\left(\frac{v^2}{R}\right)$, from which $N = m\left(g + \left(\frac{v^2}{R}\right)\right)$, and

$$\boxed{\frac{dv}{dt} = -\mu_k\left(g + \frac{v^2}{R}\right) = -11.45\ m/s^2}$$

==================================<>==================================

Problem 3.93 A freeway off-ramp is circular with radius R, and the roadway is banked at and angle β. Show that the maximum constant velocity at which a car can travel the off ramp without losing traction is $v = \sqrt{gR\left(\frac{\sin\beta + \mu_s\cos\beta}{\cos\beta - \mu_s\sin\beta}\right)}$.

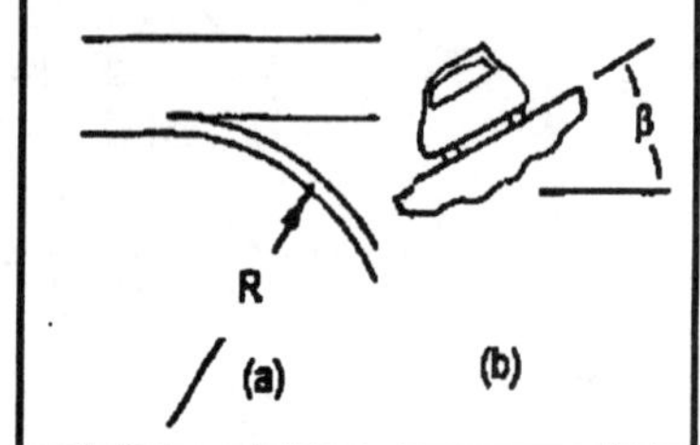

Solution Choose a coordinate system with the origin at the vehicle and *the x axis parallel to the surface of the road.* In this system the normal force is $\vec{N} = |\vec{N}|\vec{j}$. The weight is $\vec{W} = mg(-\vec{i}\sin\beta - \vec{j}\cos\beta)$. The friction force exerted on the vehicle is $\vec{f} = -\mu_s|\vec{N}|\vec{i}$. The acceleration tangent to the surface of the road is $\vec{a} = -\left(\frac{v^2}{R}\right)(\vec{i}\cos\beta - \vec{j}\sin\beta)$. From Newton's second law, $\vec{N} + \vec{f} + \vec{W} = m\vec{a}$. Equate components:

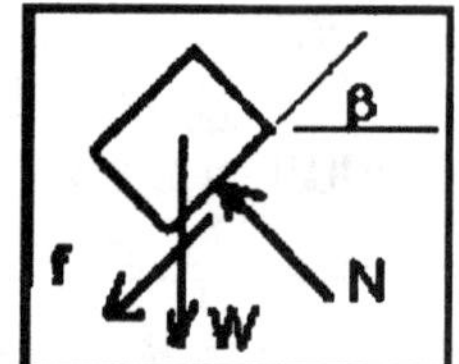

$\sum F_x = -\mu_s N - mg\sin\beta = -m\left(\frac{v^2}{R}\right)\cos\beta$. $\sum F_y = N - mg\cos\beta = m\left(\frac{v^2}{R}\right)\sin\beta$.

Solve: $\frac{v^2}{R}(\cos\beta - \mu_s\sin\beta) = g(\sin\beta + \mu_s\cos\beta)$, from which $\boxed{v = \sqrt{gR\left(\frac{\sin\beta + \mu_s\cos\beta}{\cos\beta - \mu_s\sin\beta}\right)}}$:

==================================<>==================================

=================================<>=================================

Problem 3.94 The polar coordinates of the center of mass of an object are $r = t^2 + 2\ ft$, $\theta = 2t^3 - t^2\ \ rad$, and its mass is 3 slugs. What are the radial and transverse components of the total external force on the object at $t = 1\ s$?

Solution The radial component of the acceleration is $a_r = \dfrac{d^2r}{dt^2} - r\left(\dfrac{d\theta}{dt}\right)^2$. The derivatives:

$\dfrac{dr}{dt} = \dfrac{d}{dt}\left(t^2 + 2\right) = 2t$, $\dfrac{d^2r}{dt^2} = 2$, $\dfrac{d\theta}{dt} = \dfrac{d}{dt}\left(2t^3 - t^2\right) = 6t^2 - 2t$, $\dfrac{d^2\theta}{dt^2} = \dfrac{d}{dt}\left(6t^2 - 2t\right) = 12t - 2$, from which $[a_r]_{t=1} = \left[2 - \left(t^2 + 2\right)\left(6t^2 - 2t\right)^2\right]_{t=1} = -46\ ft/s^2$. $\boxed{F_r = 3a_r = -138\ lb}$. The transverse component of the acceleration is $a_\theta = r\left(\dfrac{d^2\theta}{dt^2}\right) + 2\left(\dfrac{dr}{dt}\right)\left(\dfrac{d\theta}{dt}\right)$, from which:

$[a_\theta]_{t=1} = \left[\left(t^2 + 2\right)(12t - 2) + 2(2t)\left(6t^2 - 2t\right)\right]_{t=1} = 46\ ft/s^2$. From which $\boxed{F_\theta = 3(46) = 138\ \ lb}$

=================================<>=================================

Problem 3.95 The polar coordinates of the center of mass of an object are $r = 2t^3 + 4t$ m, $\theta = t^2 - t\ \ rad$, and its mass is $20\ \ kg$. What are the radial and transverse components of the total external force at $t = 1\ \ s$?

Solution The radial component of the acceleration is $a_r = \dfrac{d^2r}{dt^2} - r\left(\dfrac{d\theta}{dt}\right)^2$. The derivatives:

$\dfrac{dr}{dt} = \dfrac{d}{dt}\left(2t^3 + 4t\right) = 6t^2 + 4$, $\dfrac{d^2r}{dt^2} = \dfrac{d}{dt}\left(6t^2 + 4\right) = 12t$; $\dfrac{d\theta}{dt} = \dfrac{d}{dt}\left(t^2 - t\right) = 2t - 1$,

$\dfrac{d^2\theta}{dt^2} = \dfrac{d}{dt}(2t - 1) = 2$, from which $[a_r]_{t=1} = \left[12t - \left(2t^3 + 4t\right)(2t - 1)^2\right]_{t=1} = 6\ \ m/s^2$, from which

$\boxed{F_r = 20(6) = 120\ \ N}$. The transverse component of the acceleration is $a_\theta = r\dfrac{d^2\theta}{dt^2} + 2\left(\dfrac{dr}{dt}\right)\left(\dfrac{d\theta}{dt}\right)$, from which $[a_\theta]_{t=1} = \left[\left(2t^3 + 4t\right)(2) + 2\left(6t^2 + 4\right)(2t - 1)\right]_{t=1} = 32\ m/s^2$, and $\boxed{F_\theta = 20(32) = 640\ N}$

=================================<>=================================

Problem 3.96 The robot is programmed so that the 0.4 *kg* part A describes the path $r = 1 - 0.5\cos(2\pi t)$ meters, $\theta = 0.5 - 0.2\sin(2\pi t)$ radians. At $t = 2$ seconds, determine the radial and transverse components of the force exerted on A by the robot's jaws.

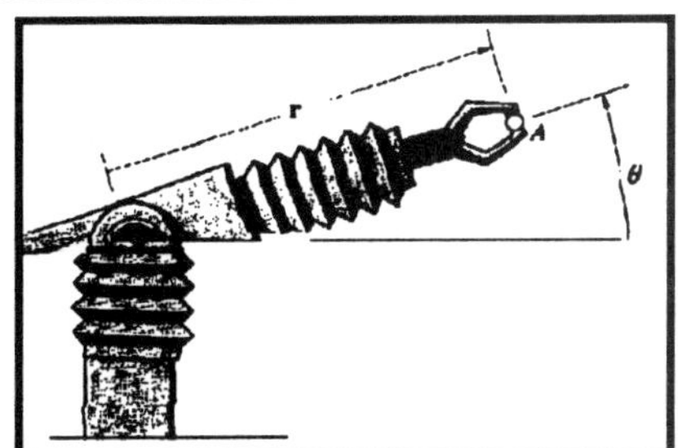

Solution The radial component of the acceleration is

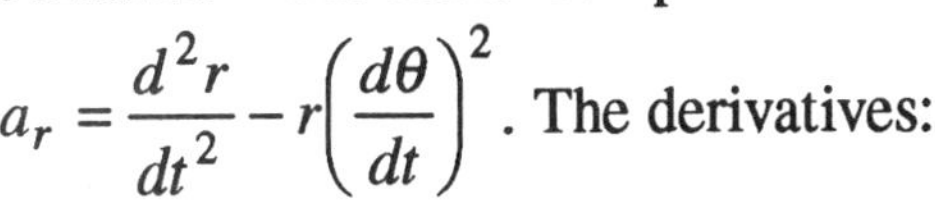

$a_r = \dfrac{d^2r}{dt^2} - r\left(\dfrac{d\theta}{dt}\right)^2$. The derivatives:

$\dfrac{dr}{dt} = \dfrac{d}{dt}(1 - 0.5\cos 2\pi t) = \pi\sin 2\pi t$, $\dfrac{d^2r}{dt^2} = \dfrac{d}{dt}(\pi\sin 2\pi t) = 2\pi^2\cos 2\pi t$;

$\dfrac{d\theta}{dt} = \dfrac{d}{dt}(0.5 - 0.2\sin 2\pi t) = -0.4\pi\cos 2\pi t$.

Solution continued on next page

$\dfrac{d^2\theta}{dt^2} = \dfrac{d}{dt}(-0.4\pi\cos 2\pi t) = 0.8\pi^2 \sin 2\pi t$. From which

$[a_r]_{t=2} = 2\pi^2 \cos 4\pi - (1 - 0.5\cos 4\pi)(-0.4\pi\cos 4\pi)^2$,

$= 2\pi^2 - 0.08\pi^2 = 18.95\ m/s^2$; $\theta(t=2) = 0.5\ rad$. From Newton's second law,

$F_r - mg\sin\theta = ma_r$, and $F_\theta - mg\cos\theta = ma_\theta$, from which $\boxed{F_r = 0.4a_r + 0.4g\sin\theta = 9.46\ N}$. The transverse component of the acceleration is $a_\theta = r\left(\dfrac{d^2\theta}{dt^2}\right) + 2\left(\dfrac{dr}{dt}\right)\left(\dfrac{d\theta}{dt}\right)$, from which

$[a_\theta]_{t=2} = (1 - 0.5\cos 4\pi)(0.8\pi^2 \sin 4\pi) + 2(\pi\sin 4\pi)(-0.4\pi\sin 4\pi) = 0$, and $\boxed{F_\theta = 3.44\ N}$

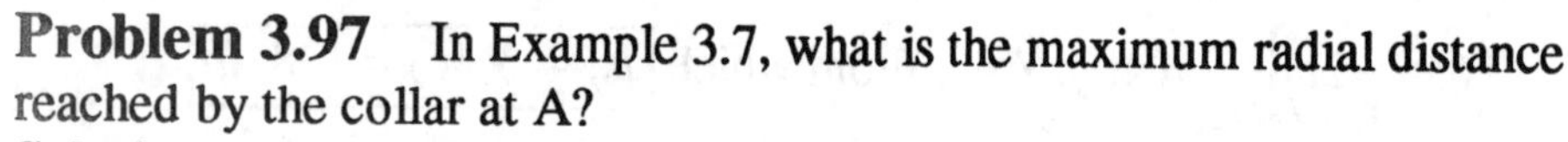

Problem 3.97 In Example 3.7, what is the maximum radial distance reached by the collar at A?

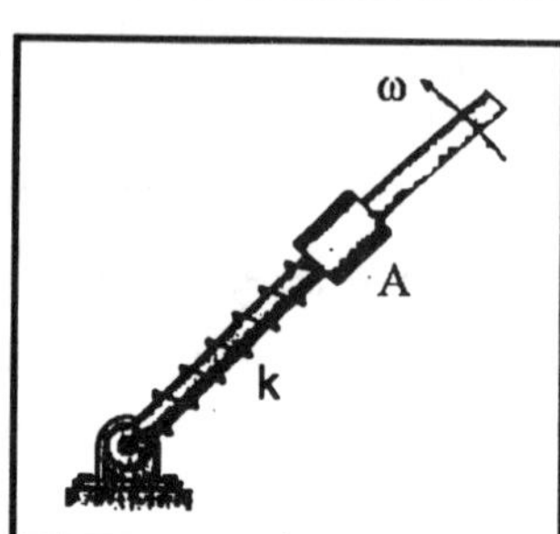

Solution Follow the strategy outlined in Example 3.7. The smooth bar rotates in the horizontal plane with constant angular velocity ω_0. The unstretched length of the spring is r_o. The collar A has mass m and is released at $r = r_o$ with no radial velocity. The spring exerts a radial force on the mass $-k(r - r_o)$, where the negative sign implies that the force is directed toward the origin. The radial component of the acceleration is $a_r = \dfrac{d^2 r}{dt^2} - r\omega^2$, from which

$-k(r - r_o) = m\left(\dfrac{dv_r}{dt} - r\omega^2\right)$. Rearrange: $\dfrac{dv_r}{dt} = r\omega^2 - \dfrac{k}{m}(r - r_o)$. Use the chain rule:

$\dfrac{dv_r}{dt} = \dfrac{dv_r}{dr}\dfrac{dr}{dt} = v_r\dfrac{dv_r}{dr}$, from which $v_r\dfrac{dv_r}{dr} = r\left(\omega^2 - \dfrac{k}{m}\right) + \dfrac{k}{m}r_o$. Separate variables and integrate:

$v_r dv_r = \left(r\left(\omega^2 - \dfrac{k}{m}\right) + \dfrac{k}{m}r_o\right)dr$, from which $\dfrac{v_r^2}{2} = \left(\omega^2 - \dfrac{k}{m}\right)\dfrac{r^2}{2} + \dfrac{k}{m}r_o r + C$. At $r = r_o$, $v_r(r_o) = 0$,

from which $C = \left(\omega^2 + \dfrac{k}{m}\right)\dfrac{r_o^2}{2}$, and (1) $v^2 = \left(\omega^2 - \dfrac{k}{m}\right)(r^2 - r_o^2) + \dfrac{2k}{m}r_o(r - r_o)$. At the maximum extension of the spring, $r = r_{\max}$, and the velocity is zero. Two conditions apply: (a) If $\omega^2 - \dfrac{k}{m} = 0$, the *maximum extension* $r_{\max}$ *is undefined.* (b) We suppose that $\omega^2 \neq \dfrac{k}{m}$. In canonical form:

$r^2 + 2br + c = 0$, where $b = \dfrac{kr_o}{m\left(\omega^2 - \dfrac{k}{m}\right)}$ and $c = \left(\omega^2 - \dfrac{k}{m}\right)^{-1}\left(\omega^2 + \dfrac{k}{m}\right)r_o^2$. The solution:

$r = -b \pm \sqrt{b^2 - c}$, which, after rearrangement, becomes:

$r = r_o\left(-\dfrac{k}{m} \pm \omega^2\right)\left(\omega^2 - \dfrac{k}{m}\right)^{-1} = r_o,\ = -r_o\left(\omega^2 - \dfrac{k}{m}\right)^{-1}\left(\omega^2 + \dfrac{k}{m}\right)$.

Solution continued on next page

The root $r = r_o$ is a trivial solution, from which $\boxed{r = r_o\left(\dfrac{1+\dfrac{\omega^2 m}{k}}{1-\dfrac{\omega^2 m}{k}}\right)}$ [*Note:* An alternate approach is to divide (1) by $r - r_o$, use $\dfrac{r^2 - r_o^2}{r - r_o} = r + r_o$ and solve. *End Note.*]

==================================<>==================================

Problem 3.98 The smooth bar rotates in the horizontal plane with constant angular velocity $\omega_o = 60\ rpm$ (revolutions per minute). If the 2 *lb.* collar A is released at $r = 1\,ft$ with no radial velocity, what is the magnitude of its velocity when it reaches the end of the bar?

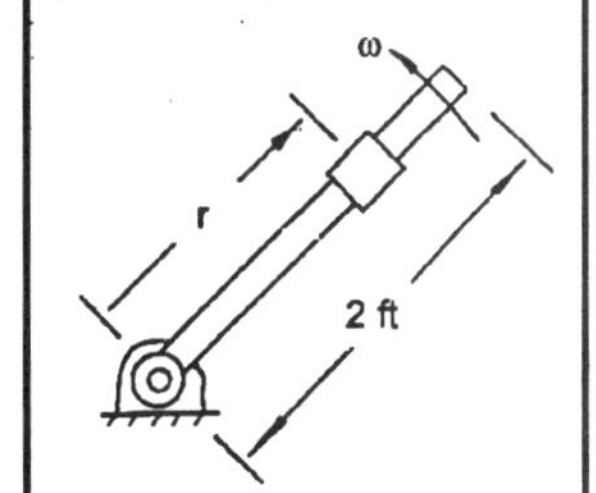

Solution The radial component of the acceleration of the mass is $a_r = \left(\dfrac{dv_r}{dt}\right) - r\omega_o^2$. The external radial forces (parallel to the smooth bar) are zero. From Newton's second law, $ma_r = 0 = \dfrac{dv_r}{dt} - \omega_o^2 r$. Use the chain rule $\dfrac{dv_r}{dt} = \dfrac{dv_r}{dr}\dfrac{dr}{dt} = v_r\dfrac{dv_r}{dr}$.

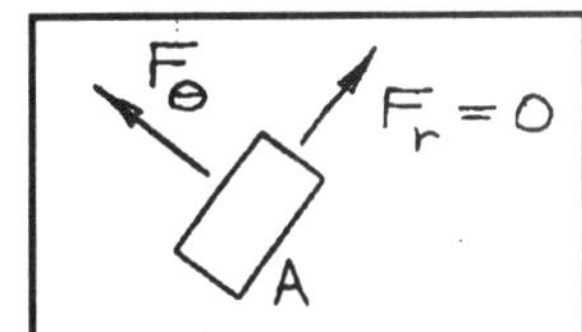

Substitute and rearrange: $v_r\dfrac{dv_r}{dr} = r\omega_o^2$. Separate variables and integrate: $v_r dv_r = r\omega_o^2 dr$, from which $v_r{}^2 = \omega_o^2 r^2 + C$. At $r = 1$, $v(1) = 0$, from which $C = -\omega_o^2$, and $v_r{}^2 = \omega_o^2(r^2 - 1)$. *Check*: If $k = 0$, the expression for the velocity in Example 3.6 reduces to $v_r = \omega_o\sqrt{r^2 - r_o^2}$, which is the same result when $r_o = 1$, as expected. *check.* When the mass reaches the end of the bar, $v_r = \omega_o\sqrt{2^2 - 1} = \sqrt{3}\omega_o$. The angular velocity is $\omega_o = 60\ \dfrac{rev}{min}\left(\dfrac{2\pi\ rad}{1\ rev}\right)\left(\dfrac{1\ min}{60\ s}\right) = 2\pi\ rad/s$, and the radial component of the velocity is $v_r = \sqrt{3}(2\pi) = 10.88\ ft/s$. The transverse component of the velocity of the mass at the end of the bar is $v_\theta = r\omega_o = 2(2\pi) = 4\pi\ ft/s$. The magnitude of the velocity: $\boxed{v = \sqrt{v_r^2 + v_\theta^2} = 16.62\ ft/s}$

==================================<>==================================

Problem 3.99 In Problem 3.98, what is the maximum horizontal force exerted on the collar by the bar?

Solution The radial force is zero, since the bar is smooth. The transverse acceleration is $a_\theta = r\dfrac{d^2\theta}{dt^2} + 2\left(\dfrac{dr}{dt}\right)\left(\dfrac{d\theta}{dt}\right) = 2v_r\omega_o$. The force is $F_\theta = 2mv_r\omega_o$. From the solution to Problem 3.70, $v_r = \omega_o\sqrt{2^2 - 1} = \sqrt{3}\omega_o$ at the end of the bar, from which $\boxed{F_\theta = 2\sqrt{3}\left(\dfrac{2}{32.17}\right)(2\pi)^2 = 8.50\ lb}$

==================================<>==================================

==================================<>==================================

Problem 3.100 The mass m is released from rest with the string horizontal. By using Newton's second law in terms of polar coordinates, determine the magnitude of the velocity of the mass and the tension in the string as a function of θ.

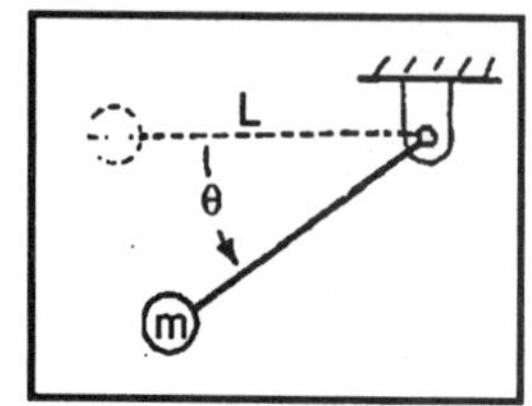

Solution The components of the weight in the radial and transverse direction are $\vec{W} = W(\vec{e}_r \sin\theta + \vec{e}_\theta \cos\theta)$. The acceleration is

$$\vec{a} = \left(\left(\frac{d^2r}{dt^2}\right) - r\left(\frac{d\theta}{dt}\right)^2\right)\vec{e}_r + \left(r\frac{d^2\theta}{dt^2} + 2\left(\frac{dr}{dt}\right)\left(\frac{d\theta}{dt}\right)\right)\vec{e}_\theta = -L\left(\frac{d\theta}{dt}\right)^2 \vec{e}_r + L\left(\frac{d^2\theta}{dt^2}\right)\vec{e}_\theta .$$

From Newton's second law $-\vec{T} + \vec{W} = m\vec{a}$. (a) For the transverse component: $-\vec{e}_\theta \cdot \vec{T} + \vec{e}_\theta \cdot \vec{W} = m\vec{e}_\theta \cdot \vec{a}$, from which $W\cos\theta = mL\frac{d\omega}{dt}$. Use the chain rule: $\frac{d\omega}{dt} = \frac{d\omega}{d\theta}\frac{d\theta}{dt} = \omega\frac{d\omega}{d\theta}$. Substitute:

$\omega\frac{d\omega}{d\theta} = \frac{g}{L}\cos\theta$. Separate variables and integrate: $\omega d\omega = \left(\frac{g}{L}\cos\theta\right)d\theta$, from which

$\omega^2 = \frac{2g}{L}\sin\theta$. The velocity is $v_\theta = L\omega$, from which $v_\theta{}^2 = 2gL\sin\theta$,

$\boxed{v_\theta = \sqrt{2gL\sin\theta}}$ (b) For the radial component, $-\vec{e}_r \cdot \vec{T} + \vec{e}_r \cdot \vec{W} = m\vec{e}_r \cdot \vec{a}$, from which

$-T + mg\sin\theta = -mL\omega^2$. Substitute the expression $\omega^2 = \frac{2g}{L}\sin\theta$, to obtain

$\boxed{T = 2mg\sin\theta + mg\sin\theta = 3mg\sin\theta}$

==================================<>==================================

Problem 3.101 The *1-lb* block A is given an initial velocity $v_0 = 14\,ft/s$ to the right when it is the position $\theta = 0$, causing it to slide up the smooth circular surface. By using Newton's second law in terms of polar coordinates, determine the magnitude of the velocity of the block when $\theta = 60°$

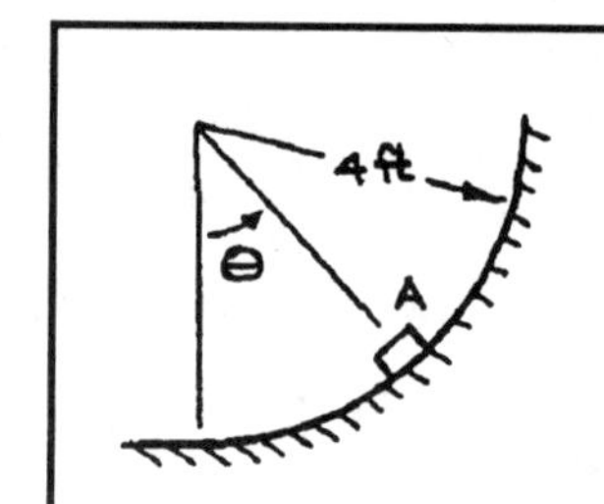

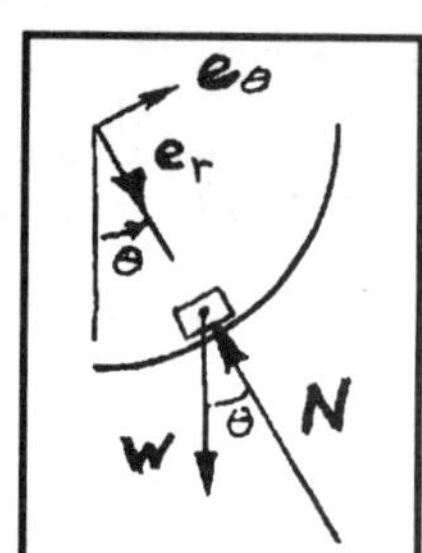

Solution: For this problem, Newton's second law in polar coordinates states

$\sum F_r = mg\cos\theta - N = m(d^2r/dt^2 - r\omega^2)$ and $\sum F_\theta = -mg\sin\theta = m(r\alpha + 2\omega(dr/dt))$. In this problem, r is constant. Thus $(dr/dt) = (d^2r/dt^2) = 0$, and the equations reduce to $N = mr\omega^2 + mg\cos\theta$ and $r\alpha = -g\sin\theta$. The first equation gives us a way to evaluate the normal force while the second can be integrated to give $\omega(\theta)$. We rewrite the second equation as $\alpha = \frac{d\omega}{dt} = \frac{d\omega}{d\theta}\frac{d\theta}{dt} = \omega\frac{d\omega}{d\theta} = -\left(\frac{g}{r}\right)\sin\theta$ and then integrate $\int_{\omega_0}^{\omega_{60}} \omega d\omega = -\left(\frac{g}{r}\right)\int_0^{60°}\sin\theta d\theta$. Carrying out the integration, we get

$\frac{\omega_{60}^2}{2} - \frac{\omega_0^2}{2} = -\left(\frac{g}{r}\right)(-\cos\theta)\Big|_0^{60°} = -\left(\frac{g}{r}\right)(1 - \cos 60°)$. Noting that $\omega_0 = v_0/R = 3.5\,rad/s$, we can solve for $\omega_{60} = 2.05\,rad/s$ and $v_{60} = R\omega_{60} = 8.20\,ft/s$.

==================================<>==================================

Problem 3.102 In Problem 3.101, determine the normal force exerted on the block by the smooth surface when $\theta = 60°$.

Solution: From the solution to Problem 3.101, we have $N_{60} = mr\omega_{60}{}^2 + mg\cos 60°$ or $N = 1.02\ lb$.

==================================<>==================================

==================================<>==================================

Problem 3.103 The skier passes point A going 17 *m/s*. From A to B, the radius of his circular path is 6 meters. By using Newton's second law in terms of polar coordinates, determine the magnitude of his velocity as he leaves the jump at B. Neglect transverse forces other than the transverse component of his weight.

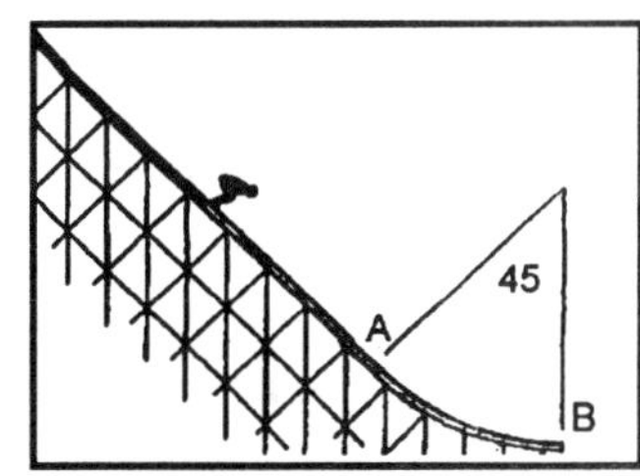

Solution The transverse component of the skier's weight (between points A and B) is $W_\theta = W\sin(45^o - \theta)$ $(0 \le \theta \le 45^o)$. The Newton's second law

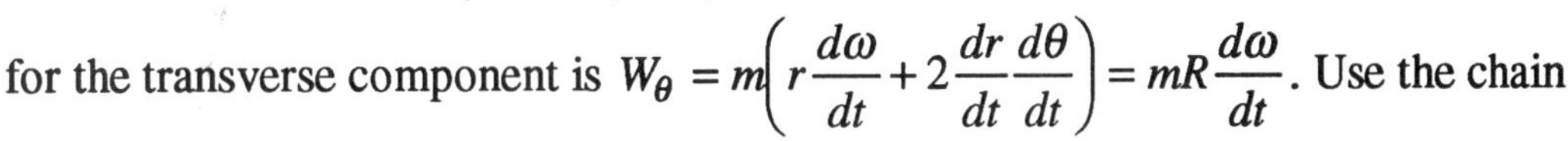

for the transverse component is $W_\theta = m\left(r\frac{d\omega}{dt} + 2\frac{dr}{dt}\frac{d\theta}{dt}\right) = mR\frac{d\omega}{dt}$. Use the chain

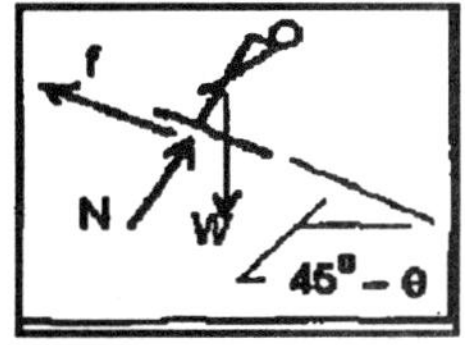

rule: $\frac{d\omega}{dt} = \frac{d\omega}{d\theta}\frac{d\theta}{dt} = \omega\frac{d\omega}{d\theta}$. Substitute and rearrange: $\omega\frac{d\omega}{d\theta} = \frac{g}{R}\sin(45^o - \theta)$.

Separate variables and integrate: $\omega^2 = \frac{2g}{R}\cos(45^o - \theta) + C$. When $\theta = 0$, $R\omega = 17\ m/s = V_A$, from which $C = \frac{V_A^2}{R^2} + \frac{\sqrt{2}g}{R}$, and $\omega^2 = \frac{2g}{R}\cos(45^o - \theta) + \frac{V_A^2}{R^2} - \frac{\sqrt{2}g}{R}$. When $\theta = 45^o$,

$\omega_B^2 = \frac{2g}{R} + \frac{V_A^2}{R^2} - \frac{\sqrt{2}g}{R}$. The transverse velocity squared is $v_A^2 = R^2\omega_B^2 = 2gR(1 - \frac{1}{\sqrt{2}}) + V_A^2$,

$$\boxed{v_A = \sqrt{2gR\left(1 - \frac{1}{\sqrt{2}}\right) + V_A^2} = \sqrt{2(9.81)(6)(1 - 0.707) + 17^2} = 17.986.. = 18\ m/s}.$$

==================================<>==================================

Problem 3.104 A 2 *kg* mass rests on a flat horizontal bar. The bar begins rotating in the vertical plane about O with a constant angular acceleration of 1 *rad/s*^{2}. The mass is observed to slip relative to the bar when the bar is 30 above the horizontal. What is the static coefficient of friction between the mass and the bar? Does the mass slip toward or away from O?

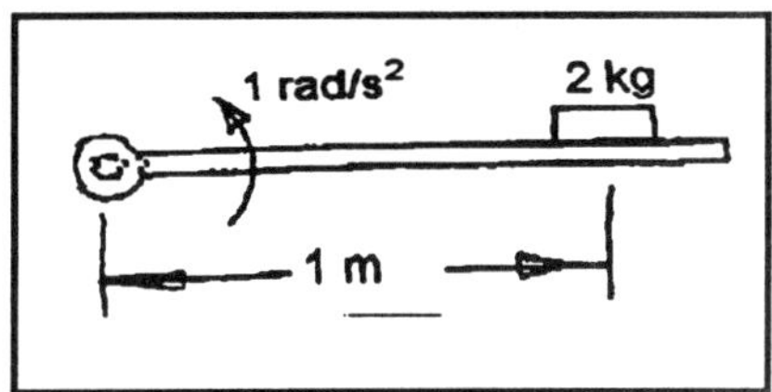

Solution From Newton's second law for the radial component $-mg\sin\theta \pm \mu_s N = -mR\omega^2$, and for the normal component: $N - mg\cos\theta = mR\alpha$. Solve, and note that $\alpha = \frac{d\omega}{dt} = \omega\frac{d\omega}{d\theta} = 1 = const$,

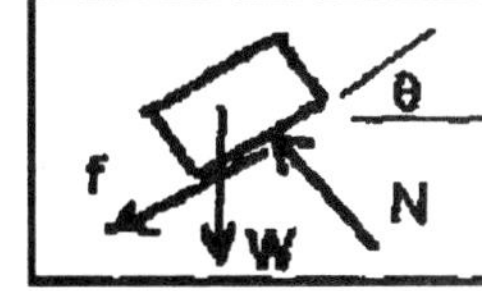

$\omega^2 = 2\theta$, since $\omega(0) = 0$, to obtain $-g\sin\theta \pm \mu_s(g\cos\theta + R\alpha) = -2R\theta$. For $\alpha = 1$, $R = 1$, this reduces to $\pm\mu_s(1 + g\cos\theta) = -2\theta + g\sin\theta$. Define the quantity $F_R = 2\theta - g\sin\theta$. If $F_R > 0$, the block will tend to slide away from O, the friction force will oppose the motion, and the negative sign is to be chosen. If $F_R < 0$, the block will tend to slide toward O, the friction force will oppose the motion, and the positive sign is to be chosen. The equilibrium condition is derived from the equations of motion: $\text{sgn}(F_R)\mu_s(1 + g\cos\theta) = (2\theta - g\sin\theta)$. from which

$\boxed{\mu_s = \text{sgn}(F_R)\frac{2\theta - g\sin\theta}{1 + g\cos\theta} = 0.406}$. Since $F_r = -3.86 < 0$, the block will slide toward O.

==================================<>==================================

=================================<>=================================

Problem 3.105 The 1/4 *lb.* slider A is pushed along the circular bar by the slotted bar. The circular bar rotates in the horizontal plane. The angular position of the slotted bar is $\theta = 10t^2\ rad$. Determine the radial and transverse components of the total external force on A at $t = 0.2$ seconds.

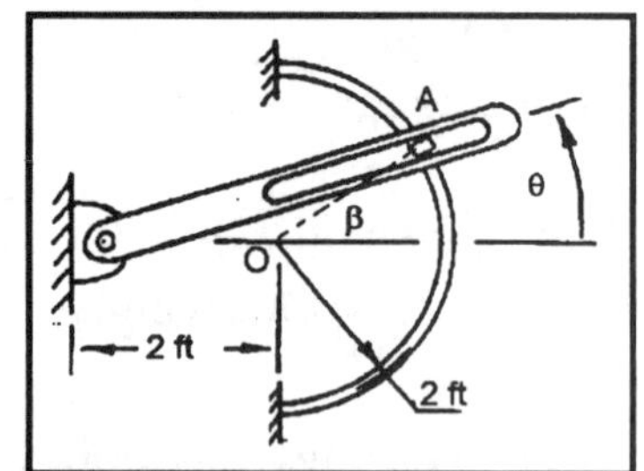

Solution The interior angle β is between the radius from O to the slider A and the horizontal, as shown in the figure. The interior angle formed by the radius from C to the slider A and the line from O to the slider is $\beta - \theta$.

The angle β is found by applying the law of sines: $\dfrac{2}{\sin\theta} = \dfrac{2}{\sin(\beta-\theta)}$ from which $\sin\theta = \sin(\beta - \theta)$ which is satisfied by $\beta = 2\theta$. The radial distance of the slider from the hinge point is also found from the sine law: $\dfrac{r}{\sin(180-\beta)} = \dfrac{2}{\sin\theta}$, $r = \dfrac{2\sin 2\theta}{\sin\theta}$, from which $r = 4\cos\theta$. The radial component of the acceleration is $a_r = \dfrac{d^2r}{dt^2} - r\left(\dfrac{d\theta}{dt}\right)^2$. The derivatives: $\dfrac{d\theta}{dt} = \dfrac{d}{dt}\left(10t^2\right) = 20t$. $\dfrac{d^2\theta}{dt^2} = 20$.

$\dfrac{dr}{dt} = -4\sin\theta\left(\dfrac{d\theta}{dt}\right) = -(80\sin\theta)t$. $\dfrac{d^2r}{dt^2} = -80\sin\theta - (1600\cos\theta)t^2$. Substitute:

$$[a_r]_{t=0.2} = \left[a_r = -80\sin\left(10t^2\right) - \left(1600\cos\left(10t^2\right)\right)t^2 - \left(4\cos\left(10t^2\right)\right)(20t)^2\right]_{t=0.2} = -149.0\ ft/s^2.$$

From Newton's second law, the radial component of the external force is $\boxed{F_r = \left(\dfrac{W}{g}\right)a_r = -1.158\ lb}$. The transverse acceleration is $a_\theta = r\left(\dfrac{d^2\theta}{dt^2}\right) + 2\left(\dfrac{dr}{dt}\right)\left(\dfrac{d\theta}{dt}\right)$. Substitute:

$$[a_\theta]_{t=0.2} = \left[a_\theta = \left(4\cos\left(10t^2\right)\right)(20) + 2\left(-80\sin\left(10t^2\right)\right)(t)(20)(t)\right]_{t=0.2} = 23.84\ ft/s^2.$$ The transverse component of the external force is $\boxed{F_\theta = \left(\dfrac{W}{g}\right)a_\theta = 0.185\ lb}$

=================================<>=================================

Problem 3.106 In Problem 3.105, suppose that the circular bar lies in the vertical plane. Determine the radial and transverse components of the total force exerted on A by the circular and slotted bars at $t = 0.25$ seconds.

Solution Assume that the orientation in the vertical plane is such that the $\theta = 0$ line is horizontal. Use the solution to Problem 3.105. For positive values of θ the radial component of acceleration due to gravity acts toward the origin, which by definition is the same direction as the radial acceleration. The transverse component of the acceleration due to gravity acts in the same direction as the transverse acceleration. From which the components of the acceleration due to gravity in the radial and transverse directions are $g_r = g\sin\theta$ and $g_\theta = g\cos\theta$. These are to be added to the radial and transverse components of acceleration due to the motion. From Problem 3.105, $\theta = 10t^2\ rad$

$$[a_r]_{t=0.25} = \left[-80\sin\theta - (1600\cos\theta)t^2 - (4\cos\theta)(20t)^2\right]_{t=0.25} = -209\ ft/s^2.$$

Solution continued on next page

From Newton's second law for the radial component $F_r - mg\sin\theta = \left(\frac{W}{g}\right)a_r$, from which $F_r = -1.478\ lb$ The transverse component of the acceleration is (from Problem 3.105) $[a_\theta]_{t=0.25} = [(4\cos\theta)(20) + 2(-80\sin\theta)(t)(20)(t)]_{t=0.25} = -52.14\ ft/s^2$. From Newton's second law for transverse component $F_\theta - mg\cos\theta = \left(\frac{W}{g}\right)a_\theta$, from which $\boxed{F_\theta = -0.2025\ lb}$

====================================<>====================================

Problem 3.107 The slotted bar rotates in the horizontal plane with constant angular velocity ω_o. The mass m has a pin that fits into the slot of the bar. A spring holds the pin against the surface of the fixed cam. The surface of the cam is described by $r = r_o(2 - \cos\theta)$. Determine the radial and transverse components of the total external force exerted on the pin as functions of θ.

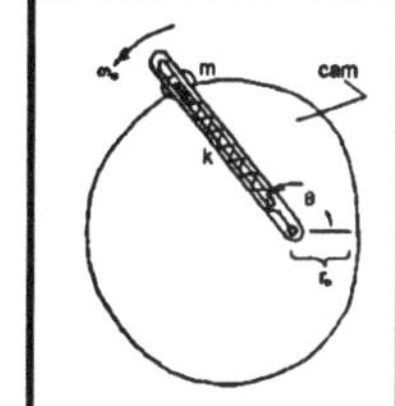

Solution The angular velocity is constant, from which $\theta = \int \omega_o dt + C = \omega_o t + C$.

Assume that $\theta(t=0) = 0$, from which $C = 0$. The radial acceleration is $a_r = \frac{d^2r}{dt^2} - r\left(\frac{d\theta}{dt}\right)^2$. The derivatives: $\frac{d\theta}{dt} = \frac{d}{dt}(\omega_o t) = \omega_o$, $\frac{d^2\theta}{dt^2} = 0$. $\frac{dr}{dt} = \frac{d}{dt}(r_o(2-\cos\theta)) = r_o \sin\theta\left(\frac{d\theta}{dt}\right) = \omega_o r_o \sin\theta$, $\frac{d^2r}{dt^2} = \frac{d}{dt}(\omega_o r_o \sin\theta) = \omega_o^2 r_o \cos\theta$. Substitute: $a_r = \omega_o^2 r_o \cos\theta - r_o(2-\cos\theta)(\omega_o^2) = 2r_o\omega_o^2(\cos\theta - 1)$.

From Newton's second law the radial component of the external force is $\boxed{F_r = ma_r = 2mr_o\omega_o^2(\cos\theta - 1)}$.

The transverse component of the acceleration is $a_\theta = r\frac{d^2\theta}{dt^2} + 2\left(\frac{dr}{dt}\right)\left(\frac{d\theta}{dt}\right)$. Substitute:

$a_\theta = 2r_o\omega_o^2 \sin\theta$. From Newton's second law, the transverse component of the external force is $\boxed{F_\theta = 2mr_o\omega_o^2 \sin\theta}$

====================================<>====================================

Problem 3.108 In Problem 3.107 suppose that the unstretched length of the spring is r_o. Determine the smallest value of the spring constant *k* for which the pin will remain on the surface of the cam.

Solution The spring force holding the pin on the surface of the cam is $F_r = k(r - r_o) = k(r_o(2-\cos\theta) - r_o) = kr_o(1-\cos\theta)$. This force acts toward the origin, which by definition is the same direction as the radial acceleration, from which Newton's second law for th pin is $\sum F = kr_o(1-\cos\theta) = -ma_r$. From the solution to Problem 3.107, $kr_o(1-\cos\theta) = -2mr\omega_o^2(\cos\theta - 1)$.

Reduce and solve: $k = 2m\omega_o^2$. Since $\cos\theta \le 1$, $kr_o(1-\cos\theta) \ge 0$, and $2mr_o\omega_o^2(\cos\theta - 1) \le 0$. If $k > 2m\omega_o^2$, Define $F_{eq} = kr_o(1-\cos\theta) + 2mr\omega_o^2(\cos\theta - 1)$. If $F_{eq} > 0$ the spring force dominates over the range of θ, so that the pin remains on the cam surface. If $k < 2m\omega_o^2$, $F_{eq} < 0$ and the radial acceleration dominates over the range of θ, so that the pin will leave the cam surface at some value of θ. Thus $\boxed{k = 2m\omega_o^2}$ is the minimum value of the spring constant required to keep the pin in contact with the cam surface.

====================================<>====================================

====================================<>====================================

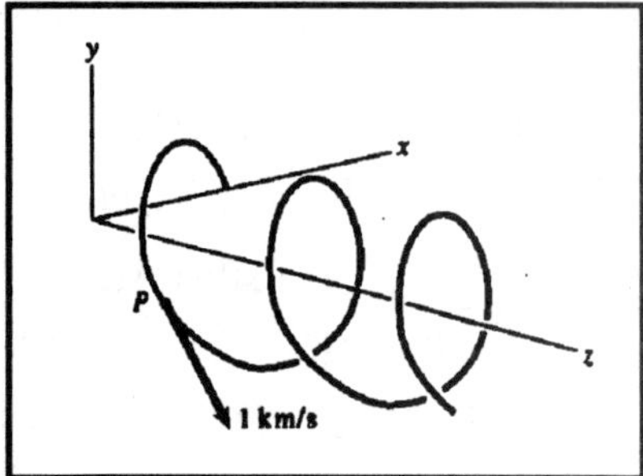

Problem 3.109 A charged particle P in a magnetic field moves along the spiral path described by $r = 1m,\ \theta = 2z\ rad$, where z is in meters. The particle moves along the path in the direction shown with a constant speed $|\mathbf{v}| = 1km/s$. The mass of the particle is $1.67\times10^{-27}kg$. Determine the sum of the forces on the particle in terms of cylindrical coordinates.

Solution: The force components in cylindrical coordinates are given by $\sum F_r = ma_r = m\left(\frac{d^2r}{dt^w} - r\omega^2\right)$, $\sum F_\theta = ma_\theta = m\left(r\alpha + 2\left(\frac{dr}{dt}\right)\omega\right)$, and $\sum F_Z = ma_Z = m\frac{d^2z}{dt^2}$. From the given information, $\frac{dr}{dt} = \frac{d^2r}{dt^2} = 0$. We also have that $\theta = 2z$. Taking derivatives of this, we see that $\frac{d\theta}{dt} = \omega = 2\frac{dz}{dt} = 2v_Z$. Taking another derivative, we get $\alpha = 2a_Z$There is no radial velocity component so the constant magnitude of the velocity $|\mathbf{v}|^2 = v_\theta^2 + v_Z^2 = r^2\omega^2 + v_Z^2 = (1000m/s)^2$. Taking the derivative of this expression with respect to time, we get $r^2\left(2\omega\frac{d\omega}{dt}\right) + 2v_Z\frac{dv_Z}{dt} = 0$. Noting that $\frac{dv_Z}{dt} = \frac{d^2z}{dt^2}$ and that $\alpha = \frac{d\omega}{dt}$, we can eliminate $\frac{dv_z}{dt}$ from the equation. We get $2r^2\omega\alpha + 2\left(\frac{\omega}{2}\right)\left(\frac{\alpha}{2}\right)$, giving $(2r^2 + 1/2)\omega\alpha = 0$. Since $\omega \neq 0,\ \alpha = 0$, and $a_Z = 0$. Substituting these into the equations of motion, we get $\omega^2 = 4(1000)^2 5(rad/s)^2$, and $\sum F_r = -mr\omega^2 = -1.34\times10^{-21}m/s^2$, $\sum F_\theta = 0$ and $\sum F_Z = 0$

====================================<>====================================

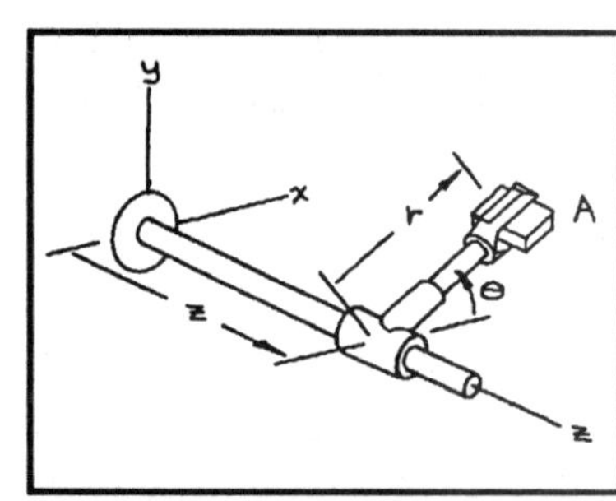

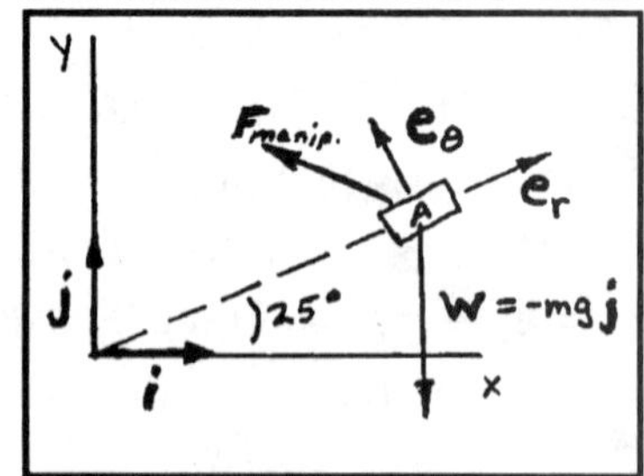

Problem 3.110 At the instant shown, the cylindrical coordinates of the *4-kg* part A held by the robotic manipulator are $r = 0.6m,\ \theta = 25°$, and $z = 0.8m$. (The coordinate system is earth-fixed and the y axis points upward). A's radial position is increasing at $\frac{dr}{dt} = 0.2m/s$ and $\frac{d^2r}{dt^2} = -0.4m/s^2$. The angle θ is increasing at $\frac{d\theta}{dt} = 1.2rad/s$ and $\frac{d^2\theta}{dt^2} = 2.8rad/s^2$. The base of the manipulator is accelerating in the z direction at $\frac{d^2z}{dt^2} = 2.5m/s^2$. Determine the force vector exerted on A by the manipulator in cylindrical coordinates.

Solution: The total force acting on part A in cylindrical coordinates is

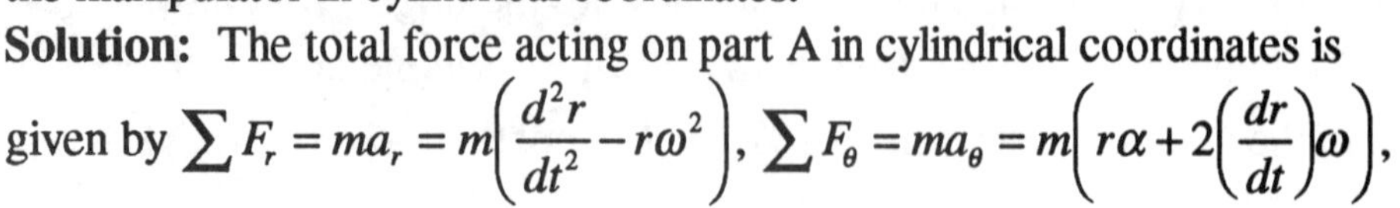

given by $\sum F_r = ma_r = m\left(\frac{d^2r}{dt^2} - r\omega^2\right)$, $\sum F_\theta = ma_\theta = m\left(r\alpha + 2\left(\frac{dr}{dt}\right)\omega\right)$,

and $\sum F_Z = ma_Z = m\frac{d^2z}{dt^2}$. We are given the values of every term in the right hand side of these equations. (Recall the definitions of ω and α. Substituting in the known values, we get $\sum F_r = -5.06$ (*newtons*), $\sum F_\theta = 8.64$(*newtons*), and $\sum F_Z = 10.0$ (*newtons*). These are the total forces acting on Part A, including the weight.

Solution continued on next page

To find the forces exerted on the part by the manipulator, we need to draw a free body diagram of the part and resolve the weight into components along the various axes. We get $\sum \mathbf{F} = \sum \mathbf{F}_{manip} + \mathbf{W} = m\mathbf{a}$ where the components of $\sum \mathbf{F}$ have already been determined above. In cylindrical coordinates, the weight is given as $\mathbf{W} = -mg\sin\theta\,\mathbf{e}_r - mg\cos\theta\,\mathbf{e}_\theta$. From the previous equation, $\sum \mathbf{F} = \sum \mathbf{F}_{manip} - mg\sin\theta\,\mathbf{e}_r - mg\cos\theta\,\mathbf{e}_\theta$ Substituting in terms of the components, we get $\left(\sum F_{manip}\right)_r = 11.5\ (newtons)$, $\left(\sum F_{manip}\right)_\theta = 44.2\ (newtons)$ and $\left(\sum F_{manip}\right)_Z = 10.0\ (newtons)$.

==================================<>==================================

Problem 3.111 Suppose that the robotic manipulator in Problem 3.110 is used in a space station to investigate zero-*g* manufacturing techniques. During an interval of time, it is programmed so that the cylindrical coordinates of the *4-kg* part A are $\theta = 0.15t^2 rad$, $r = 0.5(1+\sin\theta)m$, and $z = 0.8(1+\theta)m$ Determine the force vector exerted on A by the manipulator at $t = 2s$ in terms of cylindrical coordinates.

Solution: The contact forces exerted by the manipulator on part A in cylindrical coordinates are given by $\sum F_r = ma_r = m\left(\frac{d^2r}{dt^w} - r\omega^2\right)$, $\sum F_\theta = ma_\theta = m\left(r\alpha + 2\left(\frac{dr}{dt}\right)\omega\right)$, and $\sum F_Z = ma_Z = m\frac{d^2z}{dt^2}$. The manipulator and part A are both in free fall ("zero-g") around the earth with the rest of the space station. The weight force is not part of the contact forces experienced by Part A. From the given information, we determine the following: $\theta = 0.15t^2 rad$, $\omega = 0.30t\ rad/s$, $\alpha = 0.30 rad/s^2$, $r = 0.5(1+\sin\theta)m$, $\frac{dr}{dt} = 0.5\omega\cos\theta\ m/s$, $\frac{d^2r}{dt^2} = 0.5\alpha\cos\theta - 0.5\omega^2\sin\theta\ m/s^2$, $z = 0.8(1+\theta)m$, $\frac{dz}{dt} = 0.8\omega\ m/s$, and $\frac{d^2z}{dt^2} = 0.8\alpha\ m/s^2$ Substituting these expressions into the equations for the force components, we get $\sum F_r = -2868\ (newtons)$, $\sum F_\theta = -13{,}673\ (newtons)$ and $\sum F_Z = 96\ (newtons)$.

==================================<>==================================

Problem 3.112 In problem 3.111, draw a graph of the magnitude of the force exerted on part A by the manipulator as a function of time from $t = 0$ to $t = 5s$ and use it to estimate the maximum force during that interval of time.

Solution: Use a numerical solver to work problem 3.111 for a series of values of time during the required interval and plot the magnitude of the resulting force as a function of time. From the graph, the maximum force magnitude is approximately *8.4 newton* and it occurs at a time of about 4.4 seconds.

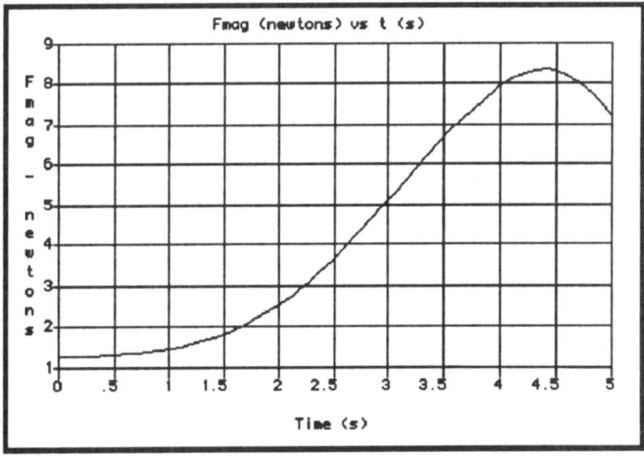

==================================<>==================================

==================================<>==================================

For the following problems use the values $R_E = 6370\ km = 3960\ mi$ for the radius of the earth.

==================================<>==================================

Problem 3.113 A satellite is in a circular orbit 200 miles above the earth's surface. (a) What is the magnitude of its velocity? (b) How long does it take to complete one orbit?

Solution (a) For $g_o = 32.17\ ft/s^2$ on the surface of the earth, the magnitude of the acceleration due to the earth's gravity at a height $h = r - R_E = 200\ mi$ above the surface is $g_h = g\left(\frac{R_E}{r}\right)^2 = 29.15\ ft/s^2$. For a satellite in a circular orbit of radius r, this must be balanced by the radial acceleration of the path, $a_r = \omega^2 r = \frac{V^2}{r}$, where $r = (3960 + 200)\left(\frac{5280\ ft}{mi}\right) = 21{,}964{,}800\ ft$, from which

$$\boxed{V = \sqrt{r g_h} = \sqrt{\frac{g_o R_E^2}{r}} = 25304.1\ ft/s}$$ (see Eq(2.68).

(b) The length of a circuit is $L_{orbit} = 2\pi r = 138{,}008{,}909\ ft$. The time required to complete a circuit is $T = \frac{L_{orbit}}{V} = 5454.0$ seconds, $\boxed{T = 90\ min\ 54\ s}$. *Note:* If the value $g_o = 32.2\ ft/s^2$ is adopted, then $V = 25316\ ft/s$, and $T = 5451.5\ s$, $T = 90\ min\ 51\ s$.

==================================<>==================================

Problem 3.114 The moon is approximately 238,000 miles from earth. Assuming that the moon's orbit around the earth is circular with velocity given by Eq(3.24), determine the time for the moon to make one revolution around the earth.

Solution Assume that the distance given is from the center of the earth. From Eq (3.24), $v_o = \sqrt{\frac{g R_E^2}{r_o}(5280)} = 2500.6\ ft/s = 0.4736\ mi/sec$. The length of the circuit is $L_{orbit} = 2\pi r = 1{,}495{,}398\ mile$. The time required for a circuit is $T = \frac{L_{orbit}}{V} = 2{,}360{,}161\ s$. The length of a solar day is 86,400 seconds, from which $\boxed{T = 27.3\ solar\ days}$

==================================<>==================================

Problem 3.115 A satellite is given an initial velocity $v_o = 22{,}000\ ft/s$ at a distance $r_o = 2R_E\ ft$ from the center of the earth as shown in Figure3.17a. (a) What is the maximum radius of the resulting elliptical orbit? (b) What is the magnitude of the velocity of the satellite when it is at its maximum radius?

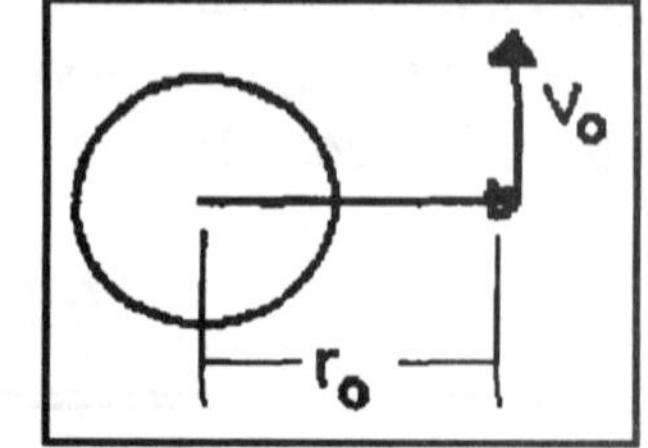

Solution From the Figure 3.17a, the velocity vector is perpendicular to the radius vector, from which the initial point is either the apogee of the perigee of the orbit. The eccentricity of the resulting elliptic obit is $\varepsilon = \frac{r_o v_o^2}{g R_E^2} - 1$. The radius of the earth is $R_E = (3960\ mi)\left(\frac{5280\ ft}{1\ mi}\right) = 20{,}908{,}800\ ft$. Substitute, and use $g = 32.17\ ft/s^2$. The result: $\varepsilon = 0.4391$, from which the initial condition is perigee.

Solution continued on next page

The maximum radius is $\boxed{r_{max} = r_o\left(\frac{1+\varepsilon}{1-\varepsilon}\right) = 5.132R_E}$ *Note:* If $g = 32.2\ ft/s^2$ is adopted, $r_{max} = 5.11R_E$. *End Note.* (b) The velocity at apogee is $v_{ap} = r_{max}\left(\frac{d\theta}{dt}\right)$. From Eq (2.58), $r_{max}v_{ap} = r_o v_o$, from which $v_{ap} = \left(\frac{r_o}{r_{max}}\right)v_o = \frac{22000}{\left(\frac{5.13}{2}\right)} = 8574.4\ ft/s$. *Note:* If the value $g = 32.2\ ft/s^2$ is adopted, $v_{ap} = 8603\ ft/s$. *End Note.*

=================================<>=================================

Problem 3.116 Draw a graph of the elliptic orbit described in Problem 3.115.
Solution The graph is shown.

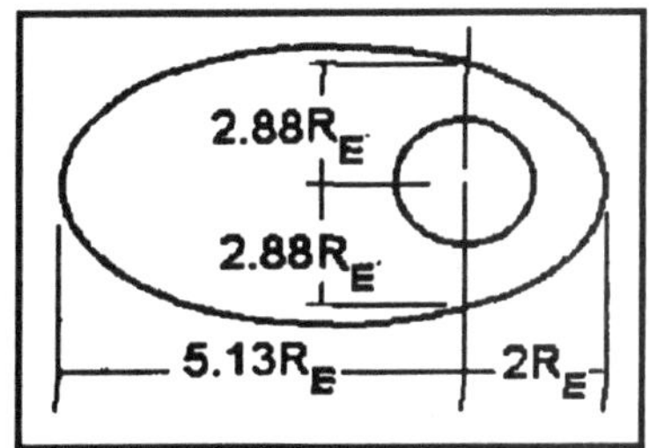

=================================<>=================================

Problem 3.117 A satellite is given an initial velocity v_o at a distance $r_o = 6800$ km as shown in Figure 3.17a. The resulting orbit has a maximum radius of 20,000 *km*. What is v_o?
Solution The initial point is perigee, and the maximum distance is at apogee. The two distances are related by $r_{max} = r_o\left(\frac{1+\varepsilon}{1-\varepsilon}\right)$. Invert, to obtain $\varepsilon = \left(\frac{r_{max} - r_o}{r_{max} + r_o}\right) = 0.4925$. The velocity at perigee is given by

$$v_o = \sqrt{\frac{gR_E^2(1+\varepsilon)}{r_o}} = 9.347\ \text{km/s}$$

=================================<>=================================

Problem 3.118 In Problem 3.117, what velocity v_o would be necessary to put the satellite into a parabolic escape orbit?
Solution The escape velocity (see Eq (2.67)) is given by

$$v_o|_{\varepsilon=1} = \left[\sqrt{\frac{gR_E^2(1+\varepsilon)}{r_o}}\right]_{\varepsilon=1} = \sqrt{\frac{2gR_E^2}{r_o}} = 10.82\ \text{km/s}$$

=================================<>=================================

==================================<>==================================

Problem 3.119 At $t = 0$, an earth satellite is a distance r_o from the center of the earth and has an initial velocity v_o in the direction shown. Show that the polar equation for the resulting orbit is

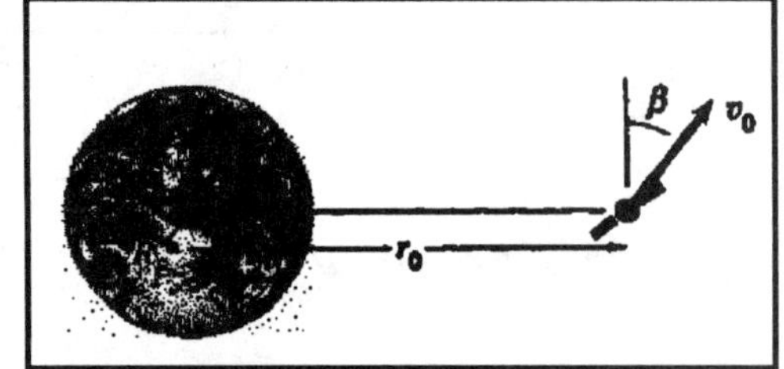

$$\frac{r}{r_o} = \frac{(\varepsilon+1)\cos^2\beta}{\left[(\varepsilon+1)\cos^2\beta - 1\right]\cos\theta - (\varepsilon+1)\sin\beta\cos\beta\sin\theta + 1},$$

where $\varepsilon = \left(\dfrac{r_o v_o^2}{g R_E^2}\right) - 1$.

Solution The strategy is to start with Eq(3.12) and Eq (3.13) and solve, using the initial conditions for the transverse velocity and radial velocity. For brevity, denote (1) $\mu = gR_E^2$.The radial acceleration is (Eq (3.12)) $a_r = \left(\dfrac{d^2r}{dt^2} - r\left(\dfrac{d\theta}{dt}\right)^2\right) = -\dfrac{\mu}{r^2}$. The transverse acceleration is (Eq (3.13)).

$a_\theta = \dfrac{1}{r}\dfrac{d}{dt}\left(r^2\dfrac{d\theta}{dt}\right) = 0$. Integrate the second equation, to obtain:(2) $r^2\left(\dfrac{d\theta}{dt}\right) = const$. The transverse velocity is $v_\theta = r\left(\dfrac{d\theta}{dt}\right)$. At $t = 0$, $v_\theta = v_o\cos\beta$, $r = r_o$, from which $r^2\left(\dfrac{d\theta}{dt}\right) = const = r_o v_o \cos\beta$.

For brevity denote (3) $p = r^2\left(\dfrac{d\theta}{dt}\right) = r_o v_o\cos\beta$. Substitute into the radial acceleration to obtain: (4) $\dfrac{d^2r}{dt^2} - \dfrac{p^2}{r^3} = -\dfrac{\mu}{r^2}$. Make the usual transformation, $r = \dfrac{1}{u}$, from which

$\dfrac{dr}{dt} = v_r = -\dfrac{1}{u^2}\dfrac{du}{dt} = -\dfrac{1}{u^2}\dfrac{du}{d\theta}\dfrac{d\theta}{dt} = -\left(\dfrac{du}{d\theta}\right)\left(r^2\dfrac{d\theta}{dt}\right) = -p\left(\dfrac{du}{d\theta}\right)$, and

$\dfrac{d^2r}{dt^2} = -p\left(\dfrac{d^2u}{d\theta^2}\dfrac{d\theta}{dt}\right) = -p^2u^2\left(\dfrac{d^2u}{d\theta^2}\right)$. Substitute into (4) to obtain: (5) $\dfrac{d^2u}{d\theta^2} + u = \left(\dfrac{\mu}{p^2}\right)$. This is a inhomogeneous second order differential equation with the solution: $u(\theta) = A\sin\theta + B\cos\theta + \left(\dfrac{\mu}{p^2}\right)$.

At $\theta = 0$, $u(0) = \dfrac{1}{r_o}$, and $\dfrac{du(0)}{d\theta} = -\dfrac{v_r}{p} = -\dfrac{v_o\sin\beta}{p}$. Solve for the initial conditions: $B = \dfrac{1}{r_o} - \left(\dfrac{\mu}{p^2}\right)$, and

$A = -\dfrac{v_o\sin\beta}{p}$. The solution: $u(\theta) = -\dfrac{v_o\sin\beta}{p}\sin\theta + \left(\dfrac{1}{r_o} - \dfrac{\mu}{p^2}\right)\cos\theta + \left(\dfrac{\mu}{p^2}\right)$. Substitute $r = \dfrac{1}{u}$:

$$r(\theta) = \frac{p^2}{-v_o p\sin\beta\sin\theta + \left(\dfrac{p^2}{r_o} - \mu\right)\cos\theta + \mu}.$$ Substitute the definition of p and μ:

$$r(\theta) = \frac{r_o^2 v_o^2\cos^2\beta}{-v_o^2 r_o\cos\beta\sin\beta\sin\theta + \left(r_o v_o^2\cos^2\beta - gR_E^2\right)\cos\theta + gR_E^2}.$$

Solution continued on next page

Divide both sides by r_o, and the top and bottom of the right side by gR_E^2 to obtain

$$\frac{r(\theta)}{r_o}=\frac{\left(\dfrac{r_o v_o^2}{gR_E^2}\right)\cos^2\beta}{-\left(\dfrac{r_o v_o^2}{gR_E^2}\right)\cos\beta\sin\beta\sin\theta+\left(\dfrac{r_o v_o^2}{gR_E^2}\cos^2\beta-1\right)\cos\theta+1}$$

. Define $\varepsilon=\left(\dfrac{r_o v_o^2}{gR_E^2}-1\right)$, and substitute:

$$\boxed{\frac{r(\theta)}{r_o}=\frac{(1+\varepsilon)\cos^2\beta}{\left[(1+\varepsilon)\cos^2\beta-1\right]\cos\theta-(1+\varepsilon)\sin\beta\cos\beta\sin\theta+1}}$$

. *Check:* For $\beta=0$, this should reduce to Eq(2.66), and it does. *check. Note:* The angle θ above is not the standard *true anomaly*, (which is zero at perigee (periapsis)) but is instead an angle artificially set to zero at $t=0$ for convenience in solving the problem. However, the angle θ that appears in Eq (3.22) is *the true anomaly.*

==================================<>==================================

Problem 3.120 Draw the graphs of the orbits given by the polar equation obtained in Problem 3.117 for $\varepsilon=0$ and $\beta=0,\ 30^o$, and 60°.

Solution The table of values for $\varepsilon=0$ is shown.

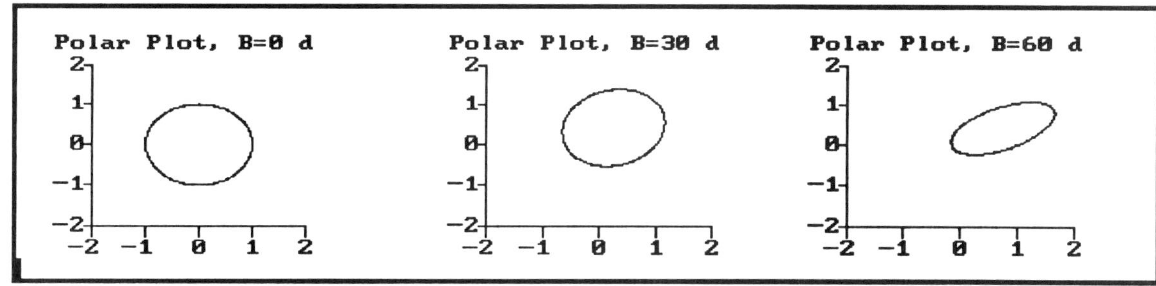

Table of $\dfrac{r(\theta)}{r_o}$

θ,deg	β=0	β=30°	β=60°		θ,deg	β=0	β=30°	β=60°
0	1	1	1		190	1	.5676	.1378
10	1	1.105	1.343		200	1	.5423	.1349
20	1	1.216	1.699		210	1	.5234	.134
30	1	1.323	1.866		220	1	.5103	.1349
40	1	1.415	1.699		230	1	.5025	.1378
50	1	1.478	1.343		240	1	.5	.1429
60	1	1.5	1		250	1	.5025	.1503
70	1	1.478	.7428		260	1	.5103	.1606
80	1	1.415	.5639		270	1	.5234	.1745
90	1	1.323	.4409		280	1	.5423	.1929
100	1	1.216	.3552		290	1	.5676	.2173
110	1	1.105	.2943		300	1	.6	.25
120	1	1	.25		310	1	.6405	.2943
130	1	.9047	.2173		320	1	.6901	.3552
140	1	.8213	.1929		330	1	.75	.4409
150	1	.75	.1745		340	1	.8213	.5639
160	1	.6901	.1606		350	1	.9047	.7428
170	1	.6405	.1503		360	1	1	1
180	1	.6	.1429					

==================================<>==================================

==================================<>==================================

Problem 3.121 A 1 kg object moves along the x axis under the action of the force $F_x = 6t\ N$. At $t = 0$, its position and velocity are $x = 0$ and $v_x = 10\ m/s$. Using numerical integration with $\Delta t = 0.1\ s$, determine the position and velocity for the first five time steps.

Strategy: At the initial time $t_o = 0$, $x(t_o) = 0$ and $v_x(t_o) = 10\ m/s$. You can use Eqs (3.13) and (3.14) to determine the velocity and position at time $t_o + \Delta t = 0.1\ s$. The position is $x(t_o + \Delta t) = x_o(t_o) + v_x(t_o)\Delta t$: $x(0.1) = x(0) + v_x(0)\Delta t = 0 + 10(0.1) = 1\ m$. and the velocity is $v_x(t_o + \Delta t) = v_x(t_o) + \frac{1}{m}F_x(t_o)\Delta t$: $v_x(0.1) = 10 + \frac{1}{1}6(0)(0.1) = 10\ m/s$. Use these values of position and velocity as the initial conditions for the next time step.

Solution The integration was carried out using **TK Solver 2.** The time was expressed as an array (list) such that $t[i] = t_o + i\Delta t$ $(0 \le i \le 5)$. A corresponding array for the force: $F_x[i] = (6)t[i]$ $(0 \le i \le 5)$.

The integration was done in a called procedure: The first values in the arrays for x and v were set equal to the starting conditions: $x[1] = 0$, $v[1] = 10$. The following values were computed in the loop:

For $i = 1$ to length('t)

$x[i+1] := x[i] + v[i]\Delta t$

$v[i+1] := v[i] + F_x[i]\Delta t$

next i.

The tabulated results are shown.

Table p3d79

Euler Integration

Time, s	Position, m	Velocity, m/s
0.000	0.000	10.000
0.100	1.000	10.000
0.200	2.000	10.060
0.300	3.006	10.180
0.400	4.024	10.360
0.500	5.060	10.600

==================================<>==================================

Problem 3.122 For the 1 kg object in Problem 3.121, draw a graph comparing the exact solution from $t = 0$ to $t = 10\ s$ with the solutions obtained using numerical integration with $\Delta t = 2\ s$, $\Delta t = 0.5\ s$ and $\Delta t = 0.1\ s$.

Solution The integration was done using an algorithm as described earlier. The results are shown. The exact values were obtained by an analytical integration of the equations $\frac{dv}{dt} = 6t$, $\frac{dx}{dt} = v$, subject to the initial conditions, from which $v(t) = 3t^2 + 10\ m/s$, $x(t) = t^3 + 10t\ m$ The exact values are always higher than the values obtained using numerical integration. For $\Delta t = 0.1\ s$, the difference between the exact values and the values obtained by numerical integration are difficult to see on the graph; but for the other time steps the differences can be seen.

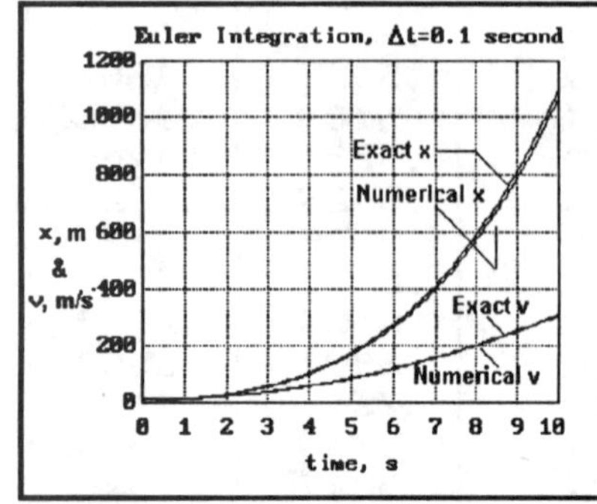

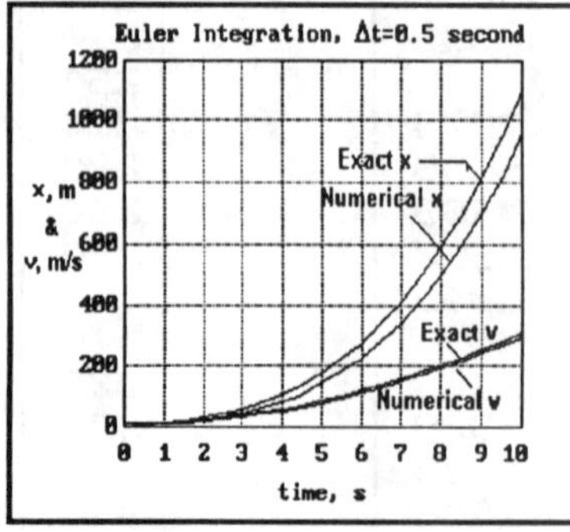

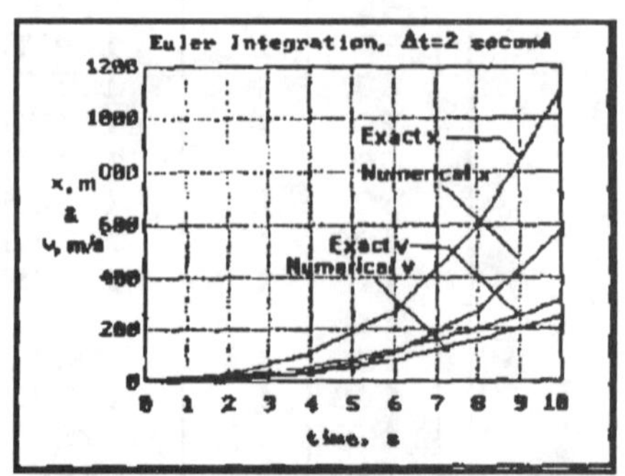

==================================<>==================================

==================================<>==================================

Problem 3.123 At $t = 0$, an object released from rest falls with constant acceleration $g = 9.81\ m/s^2$. (a) Using the closed form solution, determine the velocity of the object and the distance is has fallen at $t = 2$ seconds. (b) Approximate the answers to Part (a) using numerical integration with $\Delta t = 0.2$ seconds.

Solution The closed form solution is $\boxed{[v(t)]_{t=2} = [gt]_{t=2} = 19.62\ m/s}$, and $\boxed{[x(t)]_{t=2} = \left[\frac{g}{2}t^2\right]_{t=2} = 19.62\ m}$. (b) The numerical integration was done with an algorithm described in Problem 3.79. The results are shown, with $\boxed{[x]_{t=2} = 17.658\ m}$ and $\boxed{[v]_{t=2} = 19.620\ m/s}$

Table p3d81

Euler Integration, Δt=0.2 s

Time, s	Position, m	Velocity, m/s
0.000	0.000	0.000
0.200	0.000	1.962
0.400	0.392	3.924
0.600	1.177	5.886
0.800	2.354	7.848
1.000	3.924	9.810
1.200	5.886	11.772
1.400	8.240	13.734
1.600	10.987	15.696
1.800	14.126	17.658
2.000	17.658	19.620

==================================<>==================================

Problem 3.124 In problem 3.123, draw a graph of the distance the object falls as a function of time from $t = 0$ to $t = 4\ s$, comparing the closed form solution, the numerical solution using $\Delta t = 0.5\ s$, and the numerical solution using $\Delta t = 0.05\ s$.

Solution The results are shown. For $\Delta t = 0.05\ s$, it is impossible to distinguish the closed form solution and the results of the numerical integration. For $\Delta t = 0.5\ s$, it is easy to distinguish the closed form solution for the distance x and the numerical solution for the distance x. However, for both cases the differences in the two solutions are impossible to distinguish.

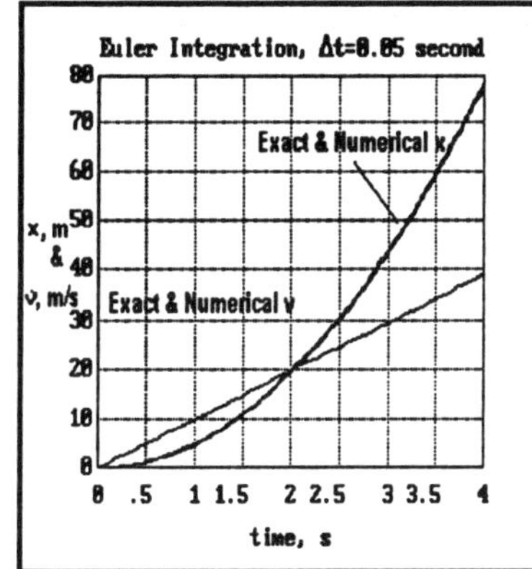

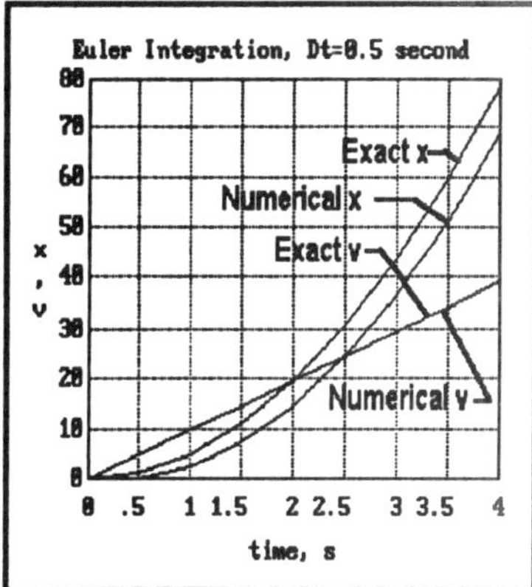

An examination of the tabulated results shows that the two solutions for the velocity agree exactly to 5 significant figures. (The velocity is a straight line function of the acceleration here, so this latter result is to be expected: the Euler method is exact for a straight line function.)

==================================<>==================================

====================================<>====================================

Problem 3.125 A 1000 *slug* rocket starts from rest and travels straight up. The total force exerted on it is $F = 100{,}000 + 10{,}000t - v^2\ lb$. Using numerical integration with $\Delta t = 0.1\ s$, determine the rocket's height and velocity for the first five time steps. (Assume that the change in the weight is negligible over this time period.)

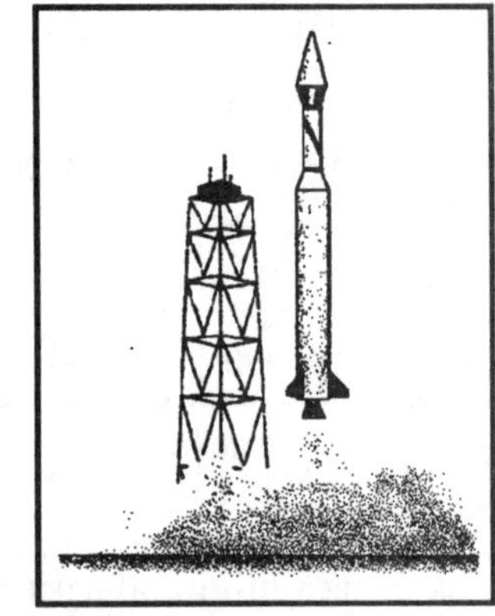

Solution Assume that the "total force" given in the problem includes the weight due to gravity. The acceleration of the rocket is $a = \frac{F}{m} = \frac{(1\times 10^5 + 1\times 10^4 t - v^2)}{10^3} = 100 + 10t - (10^{-3})v^2$. The equations to be integrated are $\frac{dv}{dt} = 100 + 10t - (10^{-3})v^2$, $\frac{dx}{dt} = v$, $x(0) = 0$, $v(0) = 0$. The algorithm used to implement the Euler method is (1) $t[i] = t_o + i\Delta t = i\Delta t \quad (0 \le i \le 5)$, (2) $x[1] = 0$, $v[1] = 0$, (3) For $i = 1$ to length('t), $a = 100 + 10t[i] - (10^{-3})(v[i])^2$, $x[i+1] = x[i] + v[i]\Delta t$, $v[i+1] = v[i] + a\Delta t$, next *i*. This algorithm was implemented in **TK Solver Plus.** The results are tabulated:

Table p3d83

Euler Integration, t = 0.1 s

time,s	height,	velocity,ft/s
0.000	0.000	0.000
.100	0.000	10.000
.200	1.000	20.090
.300	3.009	30.250
.400	6.034	40.458
.500	10.080	50.694

====================================<>====================================

Problem 3.126 The force exerted on the 50 *kg* mass by the linear spring is $F = -kx$, where x is the displacement of the mass from its position when the spring is unstretched. The spring constant is $50\ N/m$. The mass is released from rest in the position $x = 1\ m$. Use numerical integration with $\Delta t = 0.01\ s$ to determine the position and velocity of the mass for the first five time steps.

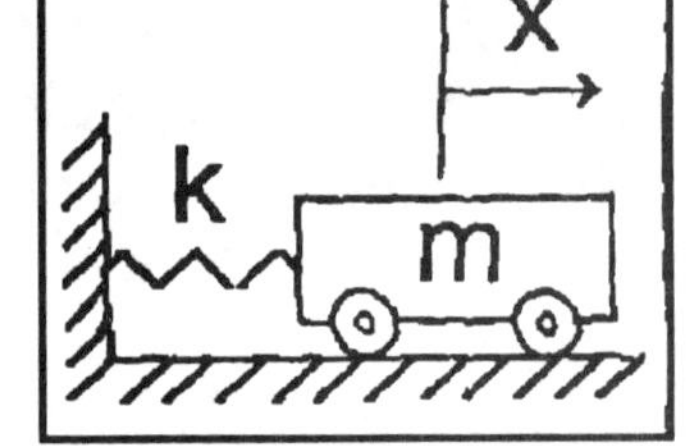

Solution The acceleration of the mass is $a = -\frac{k}{m}x = -x$. The equations are: $\frac{dv}{dt} = -x$, $\frac{dx}{dt} = v$, $x(0) = 1$, $v(0) = 0$. The algorithm used to implement the Euler method is $t[i] = t_o + i\Delta t = i\Delta t \quad (0 \le i \le 5)$, (2) $x[1] = 1$, $v[1] = 0$, For $i = 1$ to length('t), $a = -x[i]$, $x[i+1] = x[i] + v[i]\Delta t$, $v[i+1] = v[i] + a\Delta t$, next *i*. The results are tabulated.

Table p3d84

Euler Integration, Δt = 0.01 s

time,s	position, m	velocity,m/s
0.0000	1.0000	0.0000
.0100	1.0000	-.0100
.0200	.9999	-.0200
.0300	.9997	-.0300
.0400	.9994	-.0400
.0500	.9990	-.0500

====================================<>====================================

=================================<>================================

Problem 3.127 In Problem 3.126 use numerical integration with $\Delta t = 0.01\ s$ to determine the position and velocity of the mass in terms of time from $t = 0$ to $t = 10\ s$ and draw graphs of your results.

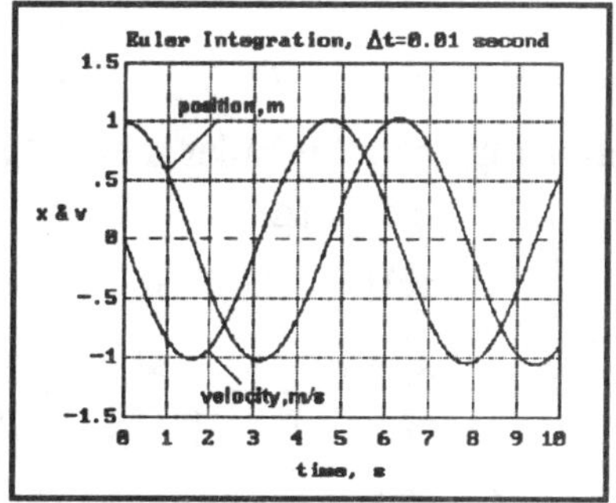

Solution The algorithm is described in the solution to Problem 3.126. The graph is shown.

=================================<>================================

Problem 3.128 At $t = 0$, the velocity of a 50 *slug* machine element that moves along the *x* axis is $v_x = 22\ ft/s$. Measurements of the total force $\sum F_x$ acting on the element at 0.1 *s* intervals from $t = 0$ to $t = 0.9\ s$ give the following values:

Time, s	force, lb.	Time, s	force, lb.
0.0	50.0	0.5	58.8
0.1	51.1	0.6	57.6
0.2	56.0	0.7	55.4
0.3	57.2	0.8	52.1
0.4	58.5	0.9	49.9

Determine approximately how far the element moves from $t = 0$ to $t = 1\ s$ and the approximate velocity at $t = 1\ s$.

Solution The algorithm is (1) $t[i] = t_o + i\Delta t = i\Delta t \quad (0 \le i \le 5)$, $\Delta t = 0.1\ s$, (2) $x[1] = 0$, $v[1] = 22$, $a = \dfrac{F[i]}{50}$, $x[i+1] = x[i] + v[i]\Delta t$, $v[i+1] = v[i] + a\Delta t$, next *i*, where $F[i]$ are the table entries above. The results are tabulated. At $t = 1\ s$, $\boxed{x \cong 22.5\ ft}$, $\boxed{v \cong 23.1\ ft/s}$. The last entry illustrates that the Euler method is a "predictor". (A neat example.)

Table p3d86

Euler Integration, Using Table entries

time,s	force,lb	position, ft	velocity,ft/s
0.000	50.0	0.000	22.000
.100	51.1	2.200	22.100
.200	56.0	4.410	22.202
.300	57.2	6.630	22.314
.400	58.5	8.862	22.429
.500	58.8	11.105	22.546
.600	57.6	13.359	22.663
.700	55.4	15.625	22.778
.800	52.1	17.903	22.889
.900	49.9	20.192	22.993
1.000		22.491	23.093

=================================<>================================

Problem 3.129 The lateral supports of a 100 *kg* structural element exert the horizontal force components $F_x = -2000x$, $F_y = -2000y$, where *x* and *y* are the coordinates of the center of mass in meters. At $t = 0$, the coordinates and component of velocity at the center of mass are $x = 0.1\ m$, $y = 0$, $v_x = 0$, and $v_y = 1\ m/s$. Using $\Delta t = 0.1\ s$, determine the approximate position and velocity of the center of mass for the first five time steps.

Solution The equations are $\dfrac{dv_y}{dt} = -\dfrac{2000}{100} = -20y$, $\dfrac{dy}{dt} = v_y$,

$v_y(0) = 1\ m$, $y(0) = 0$; $\dfrac{dv_x}{dt} = -\dfrac{2000x}{100} = -20x$, $\dfrac{dx}{dt} = v_x$, $v_x(0) = 0$, $x(0) = 0.1\ m$.

Solution continued on next page

Denoting by u, v the v_x, v_y components, the algorithm for implementing the Euler method is (1) $t[i]=t_o+i\Delta t=i\Delta t\ \ (0\le i\le 5)$, (2) $x[1]=0.1$, $u[1]=0$, $y[1]=0$, $v[1]=1$.(3) For $i=1$ to length('t), $a_x=-20x[i]$, $x[i+1]=x[i]+u[i]\Delta t$, $u[i+1]=u[i]+a_x\Delta t$; $a_y=-20y[i]$, $y[i+1]=y[i]+v[i]\Delta t$, $v[i+1]=v[i]+a_y\Delta t$, next i. The results are tabulated.

Table p3d87

Euler Integration, four variables

time,s	x, m	u, m s	y,m	v, m s
0.0000	.1000	0.0000	0.0000	1.0000
.1000	.1000	-.2000	.1000	1.0000
.2000	.0800	-.4000	.2000	.8000
.3000	.0400	-.5600	.2800	.4000
.4000	-.0160	-.6400	.3200	-.1600
.5000	-.0800	-.6080	.3040	-.8000

==================================<>==================================

Problem 3.130 In Problem 3.129, use numerical integration with $\Delta t=0.001\ s$ to determine the elliptical path described by the center of mass and draw a graph of the path.

Solution From the mass and the restoring forces, an estimate of the period is $T=\dfrac{2\pi}{\omega}=2\pi\sqrt{\dfrac{200}{2000}}=1.40\ s$ The algorithm for implementing the Euler method is given in the solution to Problem 3.87. A graph of the y value as a function of x illustrates the elliptical path of the center of mass.

Although not required by the problem, the graph of the x,y, wave forms over a full period is also shown as a reasonableness check.

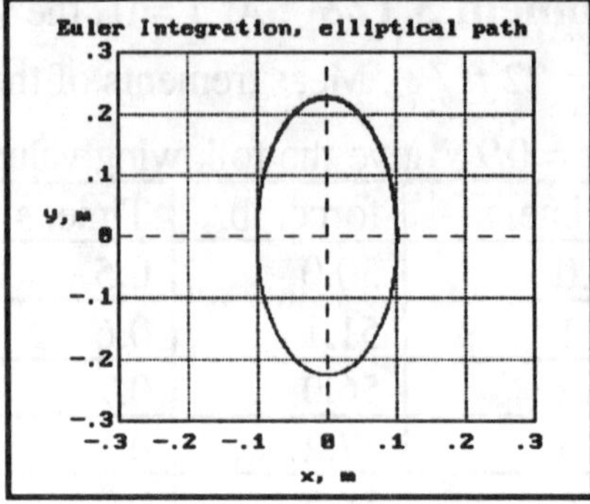

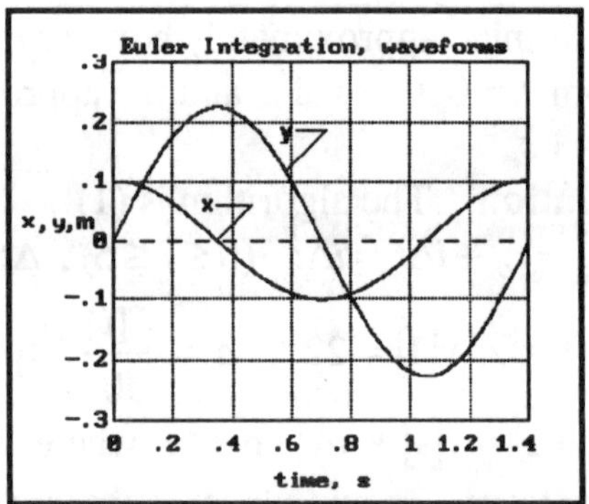

==================================<>==================================

Problem 3.131 A car starts from rest at $t=0$. Its acceleration is $a=10+2t-0.0185t^3\ ft/s^2$. (a) Using the closed form solution, determine the distance the car has traveled and its velocity at $t=6\ s$. (b) Use numerical integration with $\Delta t=0.1\ s$ to approximate the answers in Part (a). (c) Use numerical integration with $\Delta t=0.01\ s$ to approximate the answers in Part (a).

Solution (a) The closed form solution: $v(t)=\int a\,dt+C=10t+t^2-0.004625t^4+C$. Since $v(0)=0$, $C=0$. $x(t)=\int v\,dt+C_x=5t^2+\left(\frac{1}{3}\right)t^3-\left(9.25\times10^{-4}\right)t^5+C_x$. Since $x(0)=0$, $C_x=0$. At $t=6\ s$, $\boxed{[v(t)]_{t=6}=90.006\ ft/s}$. $\boxed{[x(t)]_{t=6}=244.807\ ft}$ (b) The standard algorithm (see, for example, Problem 3.83, with, of course, a different acceleration) is used to implement the Euler method of integration. For $\Delta t=0.1\ s$, $[v]_{t=6}=89.6041\,ft/s$. $\boxed{[x]_{t=6}=238.8301\ ft}$ (c) For $\boxed{[v]_{t=6}=89.9660\,ft/s}$, $\boxed{[x]_{t=6}=244.2074\ ft}$

==================================<>==================================

Problem 3.132 A 20 *kg* projectile is launched from the ground with velocity components $v_x = 100 \ m/s$, $v_y = 49 \ m/s$. The magnitude of the aerodynamic drag is $C|\vec{v}|^2$., where C is a constant. If the range of the projectile is 600 *m*, what is the constant C? (Use numerical integration with $\Delta t = 0.01 \ s$ to compute the trajectory.)

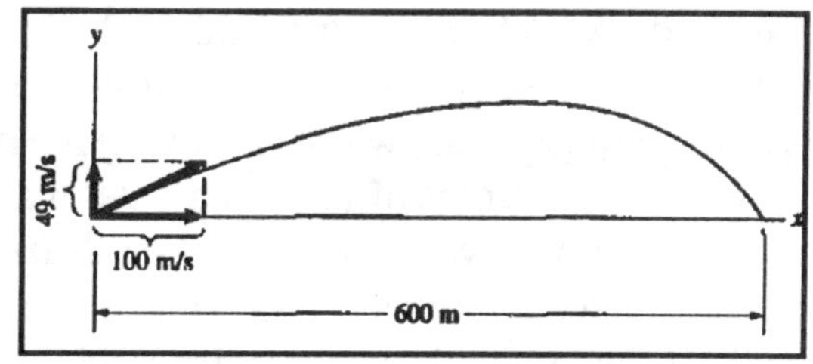

Solution Denote v_x, v_y by u, v. The strategy is a modification of that used in Example 3.7. The range R is determined as a function of C; from which the value of C is determined from R(C)=600 *m*. This requires a numerical integration for every value of C in an array. The equations are (see Eq (3.16)): $\sum F_x = -C\sqrt{u^2+v^2}u$, $\sum F_y = -mg - C\sqrt{u^2+v^2}v$, from which the accelerations to be integrated are $a_x = -\left(\frac{C}{m}\right)\sqrt{u^2+v^2}u$, $a_y = -g - \left(\frac{C}{m}\right)\sqrt{u^2+v^2}v$. For **TK Solver Plus** the following algorithm was implemented as a called procedure. The procedure call is the only rule on the Rule Sheet. (An equivalent algorithm can be developed for **Mathcad** and other similar packages). The initial values are $x_i, u_i \ y_i, v_i$, which are "placed" in the first element of the arrays x, u, y, v. The input variable C is an array of values (called a "list" in **TK** jargon), and the output R is stored in an array. The integration is done in an interior loop, with the values of C controlled by an outer loop. The range x is determined when the projectile strikes the ground, thus when the integration yields $y \le 0$, $t > 0$, the integration halts, and the corresponding value of x is stored in the range R. The graph of R as a function of C is shown, from which a coarse estimate of C was taken. The tabulated values near $R = 600 \ m$ yield the refined value $\boxed{C = 0.021655}$. As a final check, the trajectory height was graphed as a function of the range for the value of C. The graph is shown.

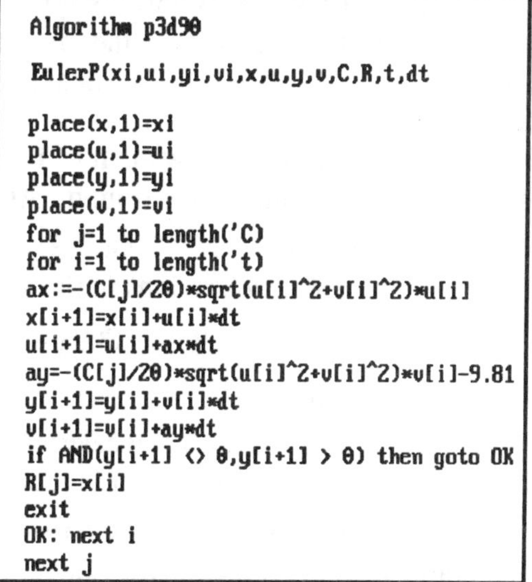

```
Algorithm p3d90
EulerP(xi,ui,yi,vi,x,u,y,v,C,R,t,dt
place(x,1)=xi
place(u,1)=ui
place(y,1)=yi
place(v,1)=vi
for j=1 to length('C)
for i=1 to length('t)
ax:=-(C[j]/20)*sqrt(u[i]^2+v[i]^2)*u[i]
x[i+1]=x[i]+u[i]*dt
u[i+1]=u[i]+ax*dt
ay=-(C[j]/20)*sqrt(u[i]^2+v[i]^2)*v[i]-9.81
y[i+1]=y[i]+v[i]*dt
v[i+1]=v[i]+ay*dt
if AND(y[i+1] <> 0,y[i+1] > 0) then goto OK
R[j]=x[i]
exit
OK: next i
next j
```

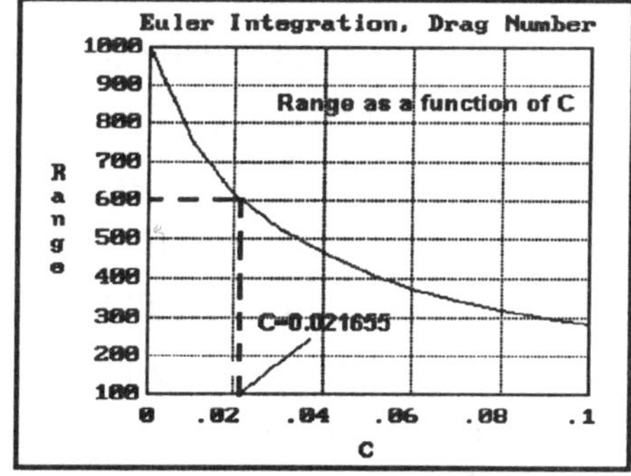

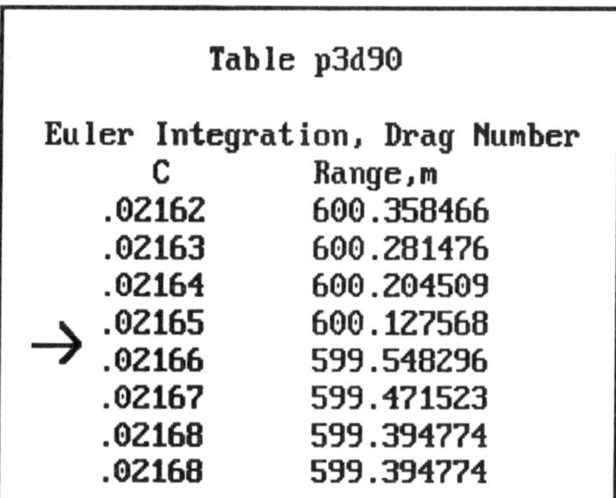

Table p3d90

Euler Integration, Drag Number

C	Range,m
.02162	600.358466
.02163	600.281476
.02164	600.204509
.02165	600.127568
→ .02166	599.548296
.02167	599.471523
.02168	599.394774
.02168	599.394774

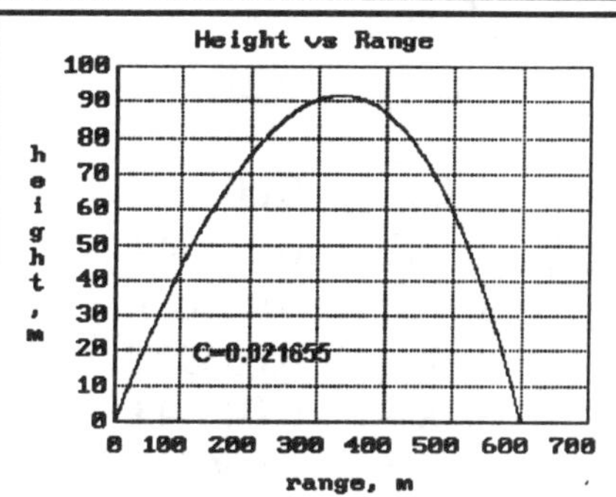

====================================<>====================================

Problem 3.133 The sport utility vehicle, which weighs 3000 lb with its driver, has left the ground after driving over a rise. At the instant shown, it is moving horizontally at 30 mi/hr and the bottoms of its tires are 24 in above the (approximately) level ground. The earth-fixed coordinate system is placed with its origin 30 in above the ground, at the height of the vehicle's center of mass when the tires first contact the ground. (Assume that the vehicle remains horizontal). When that occurs, the vehicle's center of mass initially continues moving downward and then rebounds upward due to the flexure of the suspension system. While the tires are in contact with the ground, the force exerted on the by the road is $-R\bar{i}-N\bar{j}=-2400\bar{i}-(18{,}000y+6000y^3+600v_y)\bar{j}\ (lb)$, where y is the vertical position of the center of mass in feet. Letting $t=0$ be the instant that the tires contact the ground, use numerical integration with $\Delta t=0.001s$ to determine the vertical position y and velocity for the first five time steps.

Solution:

Most of this solution follows that of Example 3.2. The problem set-up and motion are exactly the same until the wheels contact the ground. At the instant of contact, $t=0$, the following conditions are present. $x_0=0,\ y_0=0,\ v_{x0}=44\,ft/s,\ v_{y0}=11.3\,ft/s$. These conditions are derived in Ex. 3.2. Following Example 3.2, the accelerations in the x and y directions after the impact are given by $a_x=-2400(g/W)$ and $a_y=g\left(1-\frac{N}{W}\right)=g\left(\frac{W-18000y-6000y^3-600v_y}{W}\right)$, or $a_y=\left(\frac{g}{W}\right)(W-18000y-6000y^3-600v_y)$.

Obviously, this form for the acceleration in the y direction is more complicated than that of Ex. 3.2. Note that the y acceleration is a function of time since y and v_y are functions of time.

We will use numerical integration to solve it. Setting this up on a computer is a very good idea. We begin with the at $t=0$ given above. The equations for the Euler numerical integration shown in the examples just preceding this problem will be used. Starting with x_0, v_{x0}, y_0, v_{y0} at t_0 (all these values are known from the initial conditions, we find the subsequent values of these variables from the relations

$$t_{i+1}=t_i+\Delta t,\quad x_{i+1}=x_i+v_{xi}\Delta t$$
$$x(t_{i+1})=x(t_i)+v_x(t_i)\Delta t,$$
$$y(t_{i+1})=y(t_i)+v_y(t_i)\Delta t,$$
$$v_x(t_{i+1})=v_x(t_i)+a_x(t_i)\Delta t,$$
$$v_y(t_{i+1})=v_y(t_i)+a_y(t_i)\Delta t$$

Starting with $I=0$ and going out five steps, evaluating the accelerations at every step, we get

Time (sec)	y (ft)	v_y (ft/s)
0.00	0.00	11.34901
0.001	0.011349	11.30812
0.002	0.022657	11.26530
0.003	0.033922	11.22058
0.004	0.045143	11.17396
0.005	0.056317	11.12547

====================================<>====================================

Problem 3.134 In Problem 3.133, use numerical integration with $\Delta t = 0.001s$ to estimate the magnitude of the maximum acceleration to which the vehicle is subjected during its impact with the ground.

Solution:

Use the computer program set up to solve Problem 3.133. At each step, calculate the value of the total acceleration using $a_{TOTAL} = \sqrt{a_x^2 + a_y^2}$ and plot a_{TOTAL} versus the time, t, for the duration of the impact. To find the duration of the impact, plot y versus time and find the time at which $y = 0$ occurs again. The plots for this variables are shown below.

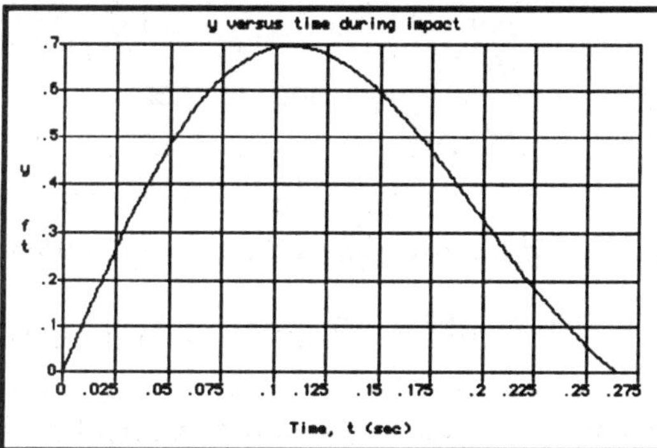

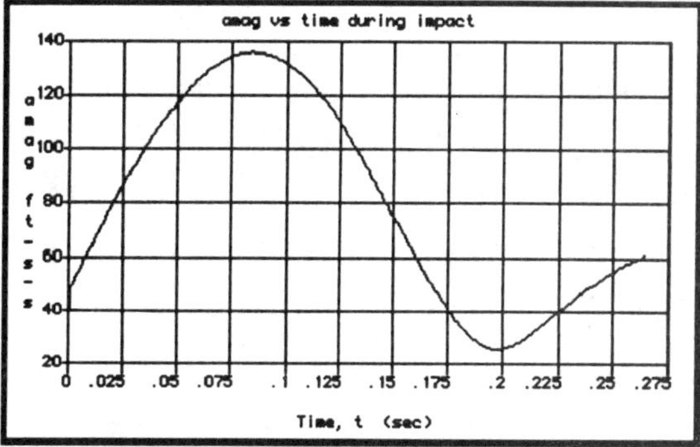

From the first plot, we see that y goes back to zero at $t = 0.26s$. (From the computer output, y reaches zero again between 0.264 *s* and 0.265 *s.).* From the second curve, the maximum acceleration occurs at about t = 0.85 *s*. The magnitude of this acceleration is about *135 ft/s*^{2}. (From the computer output, the maximum acceleration is 137.0 *ft/s*^{2} at t = 0.85 *s*).

Problem 3.135 In Problem 3.133, use numerical integration with $\Delta t = 0.001s$ to estimate the components of velocity of the vehicle's center of mass at the instant its wheels leave the ground. (The wheels leave the ground at the instant that y = 0.)

Solution:

We use the equations of Problem 3.133 and integrate out to the value t = *0.264 s.*, where y goes to zero again. At that point, we note the values of the two velocity components, which we are integrating anyway. Since we have integrated these values, they are easy to plot. The plots of the velocity components during the impact are shown below.

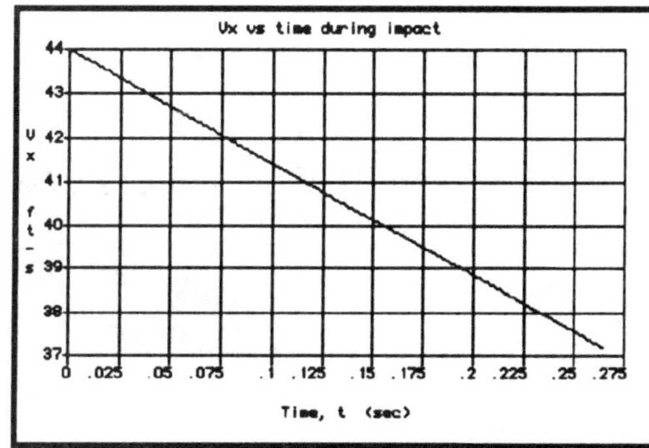

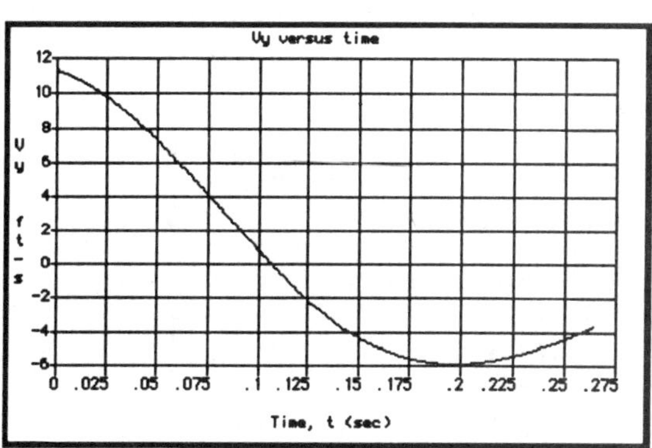

From the plots, v_x is about 37 *ft/s* at the instant $y = 0$ and v_y is about –3.8 *ft/s* (recall that positive is down in this problem). {From the computer output, $v_x = 37.2 ft/s$ and $v_y = -3.68 ft/s$. }

====================================<>====================================

Problem 3.136 The Acura NSX, which weighs *3250 lb* with its driver, can brake from *60 mi/hr* to a stop in a distance of *112 ft.* (a) If you assume its deceleration is constant, what is its deceleration and the magnitude of the horizontal force its tires exert on the road? (b) If the tires are at the limit of adhesion (slip is impending) and the normal force exerted on the car by the road equals the car's weight, what is the coefficient of static friction μ_s? (This analysis neglects the effects of horizontal and vertical aerodynamic forces).

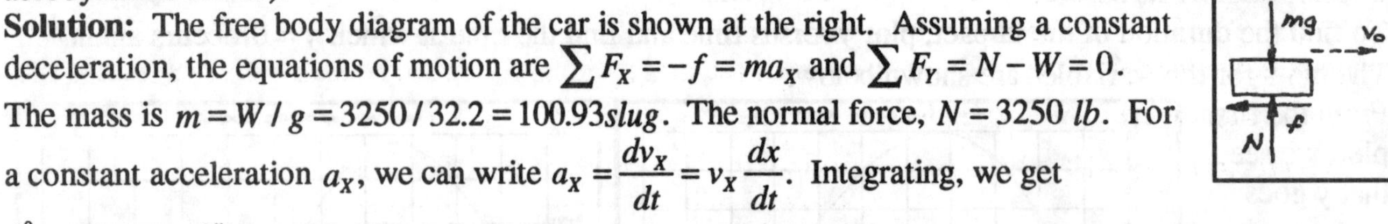

Solution: The free body diagram of the car is shown at the right. Assuming a constant deceleration, the equations of motion are $\sum F_X = -f = ma_X$ and $\sum F_Y = N - W = 0$. The mass is $m = W / g = 3250 / 32.2 = 100.93 slug$. The normal force, $N = 3250\ lb$. For a constant acceleration a_X, we can write $a_X = \dfrac{dv_X}{dt} = v_X \dfrac{dx}{dt}$. Integrating, we get $\int_{v_0}^{0} v_X dv_X = a_X \int_0^x dx$ or $v_0^2 = -2a_x x_{stop}$. Note that a_X is negative. Setting $v_0 = 60 mi / hr = 88 ft / s$ and $x_{stop} = 112 ft$, we get $a_X = -17.29 ft / s^2$. The horizontal force acting on the car is the friction force, which is $f = 1745\ lb$. If we assume that slip is impending, we know that $f = \mu_S N$. Evaluating, we get $\mu_S = 0.537$.

====================================<>====================================

Problem 3.137 Using the coefficient of friction obtained in Problem 3.136, determine the highest speed at which the NSX could drive on a flat circular track of 600 ft radius without skidding.

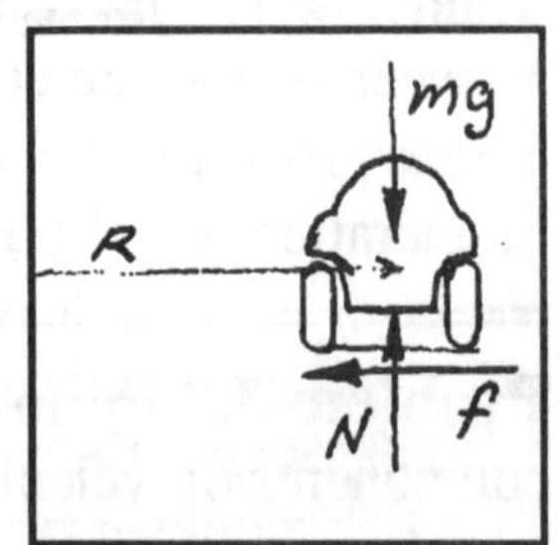

Solution: The free body diagram is at the right. The normal force is equal to the weight and the friction force has the same magnitude as in Problem 3.136 since $f = \mu_S N$. The equation of motion in the radial direction (from the center of curvature of the track to the car) is $\sum F_r = mv^2 / R = f = \mu_S N = \mu_S mg$. Thus, we have that $mv^2 / R = \mu_S mg$ or $v^2 = \mu_S Rg$. Inserting the numbers, we obtain $v = 101.8 ft / s = 69.4 mi / hr$

====================================<>====================================

Problem 3.138 A cog engine hauls three cars of sightseers to a mountain top in Bavaria. The mass of each car including its passengers is 10 *Mg* and the friction forces exerted by the wheels of the cars are negligible. Determine the forces in couplings 1, 2, and 3 if: (a) the engine is moving at constant velocity; (b) the engine is accelerating up the mountain at 1.2 *m/s²*.

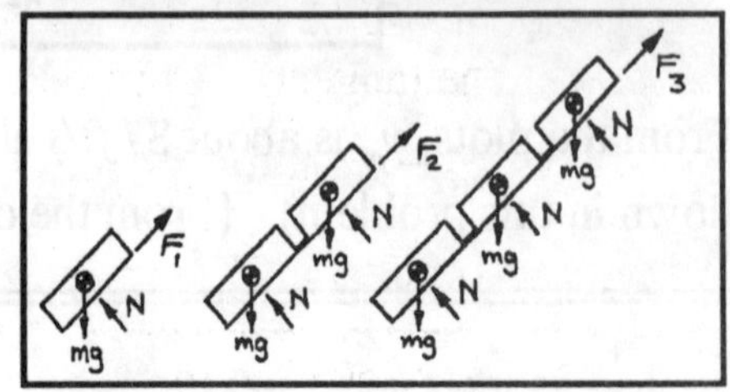

Solution (a) The force in coupling 1 is $\boxed{F_1 = 10{,}000 g \sin 40^o = 63.1\ kN}$. The force on coupling 2 is $\boxed{F_2 = 20{,}000 g \sin(40) = 126.1\ kN}$. The force on coupling 3 is $\boxed{F_3 = 30{,}000 g \sin 40^o = 189.2\ kN}$.(b) From Newton's second law, $F_{1a} - mg \sin 40^o = ma$. Under constant acceleration up the mountain, the force on coupling 1 is $\boxed{F_{1a} = 10{,}000 a + 10{,}000 g \sin 40^o = 75.1\ kN}$. The force on coupling 2 is $\boxed{F_{2a} = 2F_{1a} = 150.1\ kN}$ The force on coupling 3 is $\boxed{F_{2a} = 3F_{1a} = 225.2\ kN}$

====================================<>====================================

================================<>================================

Problem 3.139 In a future mission, a spacecraft approaches the surface of an asteroid passing near the earth. Just before it touches down, the spacecraft is moving downward at constant velocity relative to the surface of the asteroid and its downward thrust is 0.01 *N*. The computer decreases the downward thrust to 0.005 *N*, and an onboard laser interferometer determines that the acceleration of the spacecraft relative to the surface becomes $5\times10^{-6}\ N/kg^2$ downward. What is the gravitational acceleration of the asteroid near its surface?

Solution Assume that the mass of the spacecraft is negligible compared to mass of the asteroid. With constant downward velocity, the thrust balances the gravitational force: $0.01-mg_s=0$, where *m* is the mass of the space craft. With the change in thrust, this becomes $0.005-mg_s=m\left(-5\times10^{-6}\right)N/kg^2$. Multiply the first equation by 0.005, the second by 0.01, and subtract: The result: $\boxed{g_s=\left(\frac{0.01\left(5\times10^{-6}\right)}{(0.01-0.005)}\right)=1\times10^{-5}\ N/kg^2}$

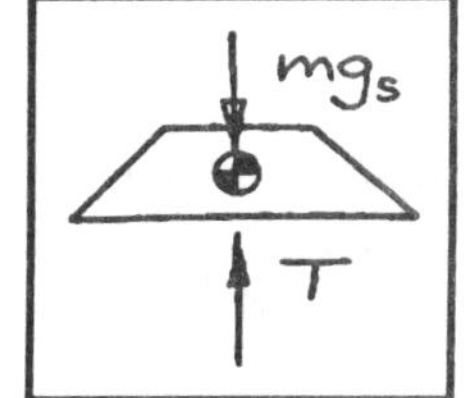

================================<>================================

Problem 3.140 A car with a mass of *1470-kg* with its driver, is driven at *130 km/hr* over a slight rise in the road. At the top of the rise, the driver applies the brakes. The coefficient of static friction between the tires and the road is $\mu_S=0.9$ and the radius of curvature of the rise is *160 m.* Determine the car's deceleration at the instant the brakes are applied and compare this to the deceleration on a flat road.

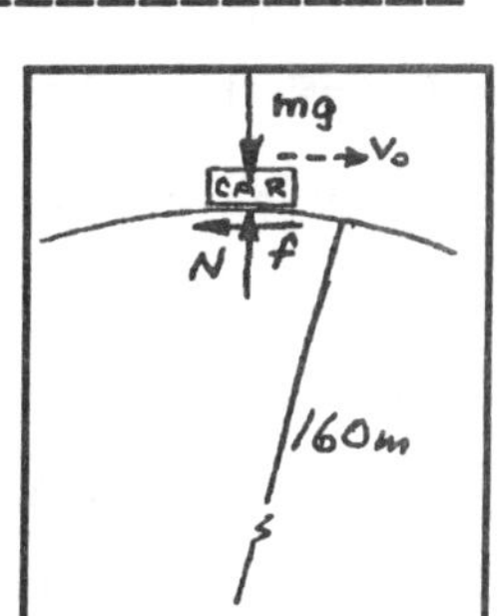

Solution: First, note that $130km/hr=36.11m/s$. We have a situation in which the car going over the rise reduces the normal force exerted on the car by the road and also reduces the braking force. The free body diagram of the car braking over the rise is shown at the right along with the free body diagram of the car braking on a level surface. For the car going over the rise, the equations of motion are $\sum F_t=-f=ma_t$, where *f* is the friction force. The normal equation is $\sum F_n=N-mg=mv^2/R$. The relation between friction and normal force is given as $f=\mu_S N$. Solving, we get $a_t=-1.49m/s^2$.

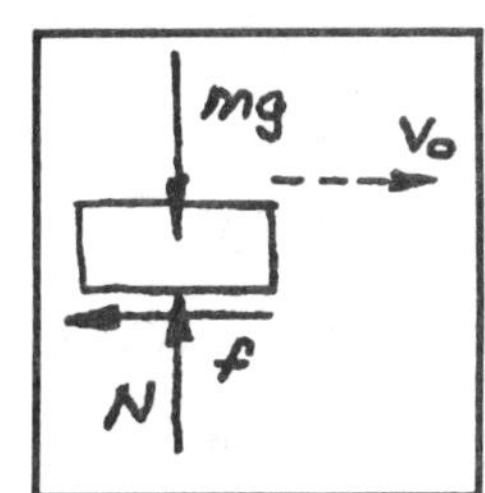

For the car braking on a level surface, the equations are $N-mg=0,\ f=\mu_S N,$ and $-f=ma_X$. Evaluating, we get $a_X=8.83m/s^2$. Note that the accelerations are VERY different. We conclude that at 130 km/hr, a rise in the road with a radius of *160 m* is not "slight". The car does not become airborne, but if the radius of curvature were smaller, the car would leave the road.

================================<>================================

Problem 3.141 The car drives at constant velocity up the straight segment of road on the left. If the car's tires continue to exert the same tangential force on the road after the car has gone over the crest of the hill and is on the straight segment of road on the right, what will be the car's acceleration?

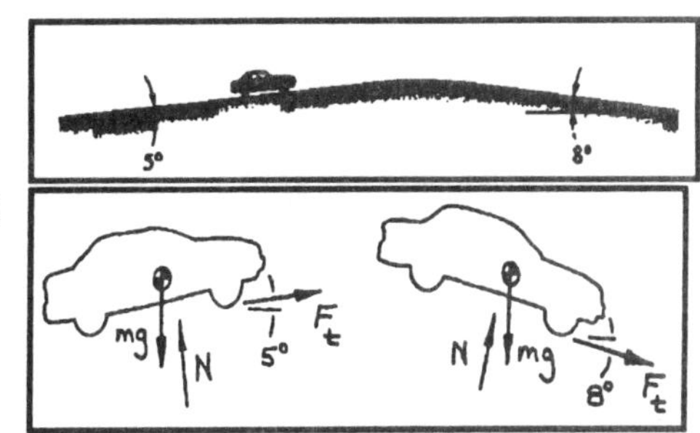

Solution The tangential force on the left is, from Newton's second law, $F_t-mg\sin\left(5^o\right)=ma=0$. On the right, from Newton's second law,: $F_t+mg\sin\left(8^o\right)=ma$ from which the acceleration is

$\boxed{a=g\left(\sin5^o+\sin8^o\right)=0.2264g}$

================================<>================================

================================<>================================

Problem 3.142 The aircraft carrier Nimitz weighs 91,000 tons (a ton is 2000 lb.) Suppose that it is traveling at its top speed of approximately 30 knots (a knot is approximately 6076 feet per hour) when its engines are shut down. If the water exerts a drag force of magnitude 20,000v pounds, where v is the carrier's velocity in *ft/s*, what distance does the carrier move before coming to rest?

Solution The force on the carrier is $F = -Kv$, where $K = 20{,}000$. The acceleration is $a = \frac{F}{m} = -\frac{gK}{W}v$. Use the chain rule to write $v\frac{dv}{dx} = -\frac{gK}{W}v$,. Separate variables and integrate: $dv = -\frac{gK}{W}dx$, $v(x) = -\frac{gK}{W}x + C$. The initial velocity: $v(0) = 30\left(\frac{6076\ ft}{1\ hr}\right)\left(\frac{1\ hr}{3600\ s}\right) = 50.63\ ft/s$, from which $C = v(0) = 50.63$, and $v(x) = v(0) - \frac{gK}{W}x$, from which, at rest,

$$\boxed{x = \frac{Wv(0)}{gK} = 14{,}321\ ft = 2.71\ mi}$$

================================<>================================

Problem 3.143 If $m_A = 10\ kg$, $m_B = 40\ kg$, and the coefficient of kinetic friction between all surfaces is $\mu_k = 0.11$, what is the acceleration of B down the inclined surface?

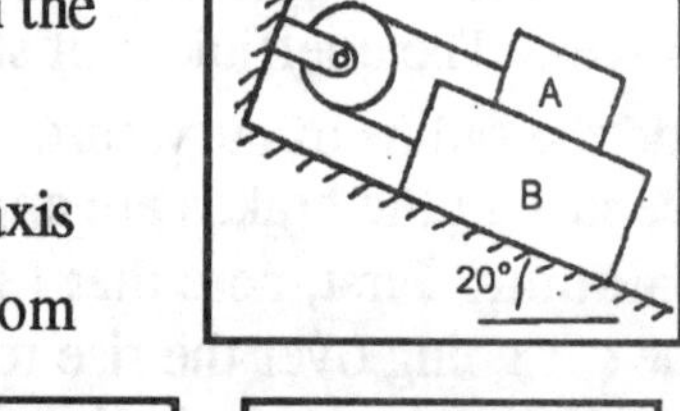

Solution Choose a coordinate system with the origin at the wall and the x axis parallel to the plane surface. Denote $\theta = 20^o$. Assume that slip has begun. From Newton's second law for block A:

(1) $\sum F_x = -T + m_A g \sin\theta + \mu_k N_A = m_A a_A$,

(2) $\sum F_y = N_A - m_A g \cos\theta = 0$. From Newton's second law for block B: (3) $\sum F_x = -T - \mu_k N_B - \mu_k N_A + m_B g \sin\theta = m_B a_B$,

(4) $\sum F_y = N_B - N_A - m_B g \cos\theta = 0$. Since the pulley is one-to-one, the sum of the displacements is $x_B + x_A = 0$. Differentiate twice:

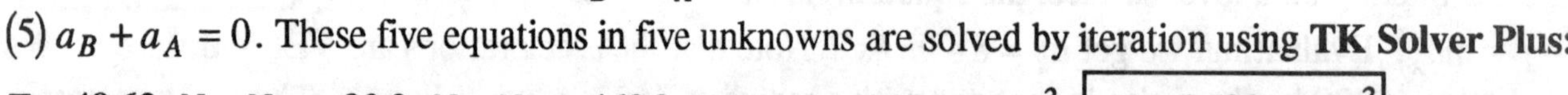

(5) $a_B + a_A = 0$. These five equations in five unknowns are solved by iteration using **TK Solver Plus:**

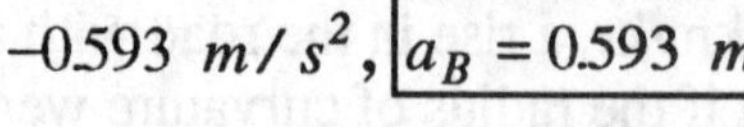

$T = 49.63\ N$, $N_A = 92.2\ N$, $N_B = 460.9\ N$, $a_A = -0.593\ m/s^2$, $\boxed{a_B = 0.593\ m/s^2}$

================================<>================================

==================================<>==================================

Problem 3.144 In Problem 3.143, if A weighs 20 *lb.*, B weighs 100 *lb.*, and the coefficient of kinetic friction between all surfaces is $\mu_k = 0.15$, what is the tension in the cord as B slides down the inclined surface?

Solution From the solution to Problem 3.143,

(1) $\sum F_x = -T + m_A g \sin\theta + \mu_k N_A = m_A a_A$,

(2) $\sum F_y = N_A - m_A g \cos\theta = 0$. For block B:

(3) $\sum F_x = -T - \mu_k N_B - \mu_k N_A + m_B g \sin\theta = m_B a_B$,

(4) $\sum F_y = N_B - N_A - m_B g \cos\theta = 0$.

(5) $a_B + a_A = 0$ Solve by iteration: $\boxed{T = 10.46\ lb}$, $N_A = 18.8\ lb$, $N_B - 112.8\ lb$, $a_A = -1.29\ ft/s^2$, $a_B = 1.29\ ft/s^2$

==================================<>==================================

Problem 3.145 A gas gun is used to accelerate projectiles to high velocities for research on material properties. The projectile is held in place while gas is pumped into the tube to a high pressure p_o on the left and the tube is evacuated on the right. Assume that the pressure *p* of the gas is related to the volume V it occupies by $pV^\gamma = const$, where γ is a constant. If friction can be neglected, show that the velocity of the projectile after it has traveled a distance *x* is $\sqrt{\dfrac{2p_o A x_o^\gamma}{m(\gamma-1)}\left(\dfrac{1}{x_o^{\gamma-1}} - \dfrac{1}{x^{\gamma-1}}\right)}$, where *m* is the mass of the projectile and *A* is the cross-sectional area of the tube.

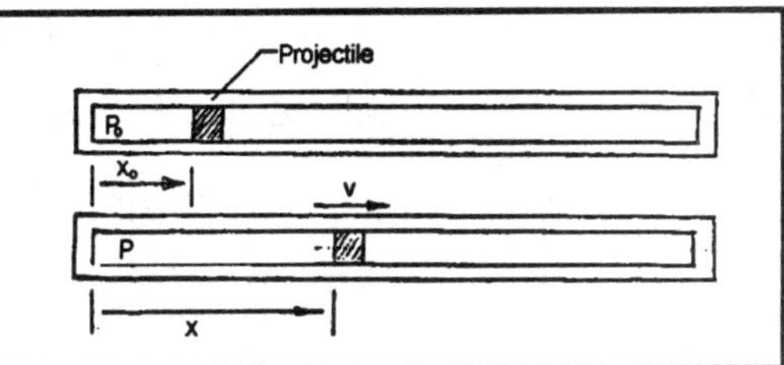

Solution The force acting on the projectile is $F = pA$ where *p* is the instantaneous pressure and A is the area. From $pV^\gamma = K$, where $K = p_o V_o^\gamma$ is a constant, and the volume $V = Ax$, it follows that $p = \dfrac{K}{(Ax)^\gamma}$, and the force is

$F = KA^{1-\gamma}x^{-\gamma}$. The acceleration is $a = \dfrac{F}{m} = \dfrac{K}{m}A^{1-\gamma}x^{-\gamma}$. The equation to be integrated:

$v\dfrac{dv}{dx} = \dfrac{K}{m}A^{1-\gamma}x^{-\gamma}$, where the chain rule $\dfrac{dv}{dt} = \dfrac{dv}{dx}\dfrac{dx}{dt} = v\dfrac{dv}{dx}$ has been used. Separate variables and integrate: $v^2 = 2\left(\dfrac{K}{m}\right)A^{1-\gamma}\int x^{-\gamma}dx + C = 2\left(\dfrac{K}{m}\right)A^{1-\gamma}\left(\dfrac{x^{1-\gamma}}{1-\gamma}\right) + C$. When $x = x_o$, $v_o = 0$, therefore

$v^2 = 2\left(\dfrac{K}{m}\right)\left(\dfrac{A^{1-\gamma}}{1-\gamma}\right)\left(x^{1-\gamma} - x_o^{1-\gamma}\right)$. Substitute $K = p_o V_o^\gamma = p_o A^\gamma x_o^\gamma$ and reduce:

$v^2 = \dfrac{2p_o A x_o^\gamma}{m(1-\gamma)}\left(x^{1-\gamma} - x_o^{1-\gamma}\right)$. Rearranging: $\boxed{v = \sqrt{\dfrac{2p_o A x_o^\gamma}{m(\gamma-1)}\left(\dfrac{1}{x_o^{\gamma-1}} - \dfrac{1}{x^{\gamma-1}}\right)}}$

==================================<>==================================

==================================<>==================================

Problem 3.146 The weights of the blocks are $W_A = 120\ lb$ and $W_B = 20\ lb$ and the surfaces are smooth. Determine the acceleration of block A and the tension in the cord.

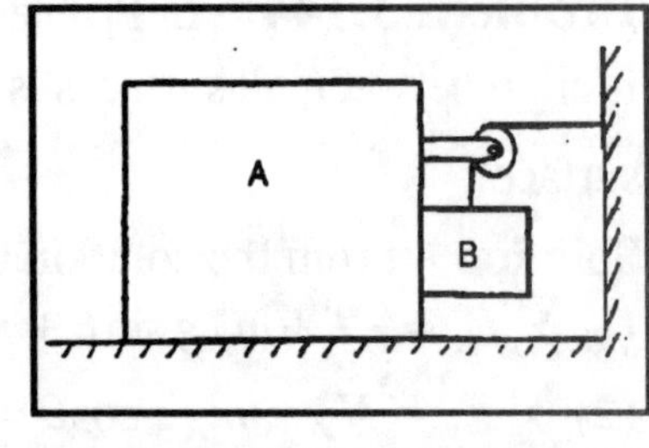

Solution Denote the tension in the cord near the wall by T_A. From Newton's second law for the two blocks: $\sum F_x = T_A = \left(\frac{W_A}{g} + \frac{W_B}{g}\right) a_A$.

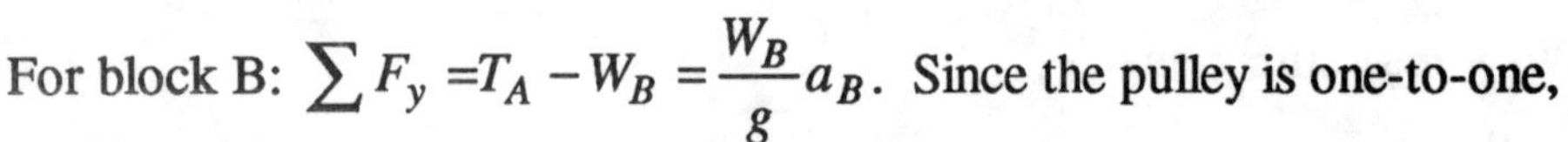

For block B: $\sum F_y = T_A - W_B = \frac{W_B}{g} a_B$. Since the pulley is one-to-one, as the displacement of B increases *downward* (negatively) the displacement of A increases *to the right* (positively), from which $x_A = -x_B$. Differentiate twice to obtain $a_A = -a_B$. Equate the expressions to obtain:

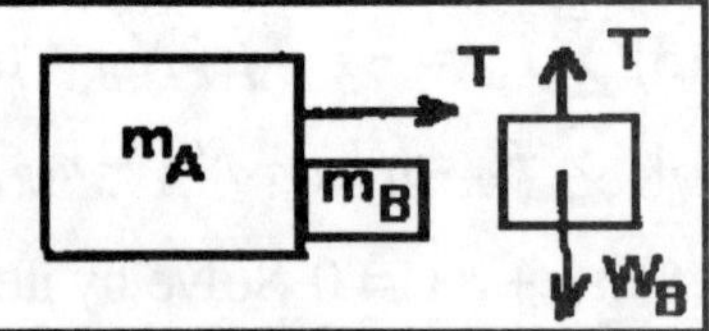

$$a\left(\frac{W_A}{g} + \frac{W_B}{g}\right) = W_B + \frac{W_B}{g} a$$, from which $$\boxed{a = g\left(\frac{W_B}{W_A + 2W_B}\right) = g\left(\frac{20}{160}\right) = \frac{32.17}{8} = 4.02\ ft/s^2}$$

==================================<>==================================

Problem 3.147 The 100 Mg space shuttle is in orbit when its engines are turned on, exerting a thrust force $\vec{T} = 10\vec{i} - 20\vec{j} + 10\vec{k}\ \ (kN)$ for 2 seconds. Neglect the resulting change in its mass. At the end of the two-second burn, fuel is still sloshing back and forth in the shuttle's tanks. What is the change in the velocity of the center of mass of the shuttle (including the fuel it contains) due to the two-second burn.

Solution At the completion of the burn, there are no external forces on the shuttle (it is in free fall) and the fuel sloshing is caused by internal forces that cancel, and the center of mass is unaffected. The change in velocity is $$\boxed{\Delta\vec{v} = \int_0^2 \frac{\vec{T}}{m} dt = \frac{2(10^4)}{10^5}\vec{i} - \frac{2(2\times 10^4)}{10^5}\vec{j} + \frac{2(10^4)}{10^5}\vec{k} = 0.2\vec{i} - 0.4\vec{j} + 0.2\vec{k}\ (m/s)}$$

==================================<>==================================

Problem 3.148 The water skier contacts the ramp with a velocity of 25 *mi/hr* parallel to the surface of the ramp. Neglecting friction and assuming that the tow rope exerts no force on him once he touches the ramp, estimate the horizontal length of his jump from the end of the ramp.

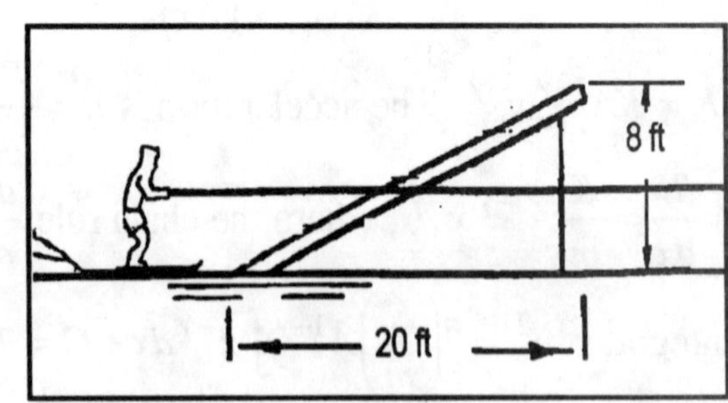

Solution Break the path into two parts: (1) The path from the base to the top of the ramp, and (2) from the top of the ramp until impact with the water. Let u be the velocity parallel to the surface of the ramp, and let z be the distance along the surface of the ramp.

From the chain rule, $u\frac{du}{dz} = -g\sin\theta$, where $\theta = \tan^{-1}\left(\frac{8}{20}\right) = 21.8^o$. Separate variables and integrate:

$u^2 = -(2g\sin\theta)z + C$. At the base of the ramp $u(0) = 25\frac{mi}{hr}\left(\frac{5280\ ft}{1\ mi}\right)\left(\frac{1\ hr}{3600\ s}\right) = 36.67\ ft/s$, frOm

Problem continued on next page

which $C = \left(36.67^2\right) = 1344.4$ and $u = \sqrt{C - (2g\sin\theta)z}$. At $z = \sqrt{8^2 + 20^2} = 21.54\ ft$ $u = 28.8\ ft/s$.(2) In the second part of the path the skier is in free fall. The equations to be integrated are $\frac{dv_y}{dt} = -g$, $\frac{dy}{dt} = v_y$, with $v(0) = u\sin\theta = 28.8(0.3714) = 10.7\ ft/s$, $y(0) = 8\ ft$. $\frac{dv_x}{dt} = 0$, $\frac{dx}{dt} = v_x$, with $v_x(0) = u\cos\theta = 26.7\ ft/s$, $x(0) = 0$. Integrating: $v_y(t) = -gt + 10.7\ ft/s$.

$y(t) = -\frac{g}{2}t^2 + 10.7t + 8\ ft$. $v_x(t) = 26.7\ ft/s$, $x(t) = 26.7t$. When $y(t_{impact}) = 0$, the skier has hit the water. The impact time is $t_{impact}^2 + 2bt_{impact} + c = 0$ where $b = -\frac{10.7}{g}$, $c = -\frac{16}{g}$. The solution $t_{impact} = -b \pm \sqrt{b^2 - c} = 1.11\ s,\ = -0.45\ s$. The negative values has no meaning here. The horizontal distance is $\boxed{x(t_{impact}) = 26.7t_{impact} = 29.7\ ft}$

==================================<>==================================

Problem 3.149 Suppose that you are designing a roller coaster track that will take the cars through a vertical loop of 40 *ft* radius. If you decide that for safety the downward force exerted on a passenger by his seat at the top of the loop should be at least one-half his weight, what is the minimum safe velocity of the cars at the top of the track?

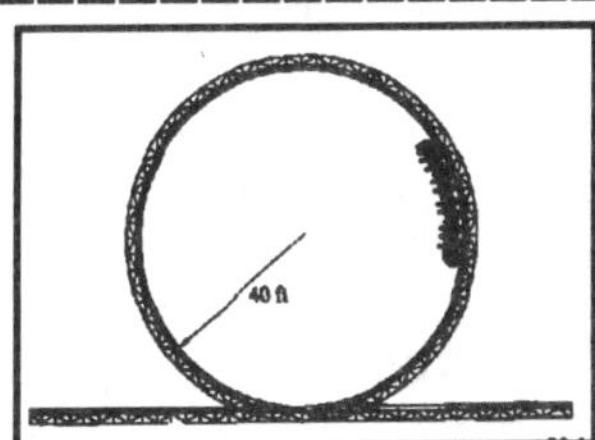

Solution Denote the normal force exerted on the passenger by the seat by *N*. From Newton's second law, at the top of the loop $-N - mg = -m\left(\frac{v^2}{R}\right)$, from which $-\frac{N}{m} = g - \frac{v^2}{R} = -\frac{g}{2}$. From which :

$$\boxed{v = \sqrt{\frac{3Rg}{2}} = 43.93.. = 44\ ft/s}$$

==================================<>==================================

Problem 3.150 As the smooth bar rotates in the horizontal plane, the string winds up on the fixed cylinder and draws the 1 *kg* collar A inward. The bar starts from rest at $t = 0$ in the position shown and rotates with constant acceleration. What is the tension at $t = 1\ s$?

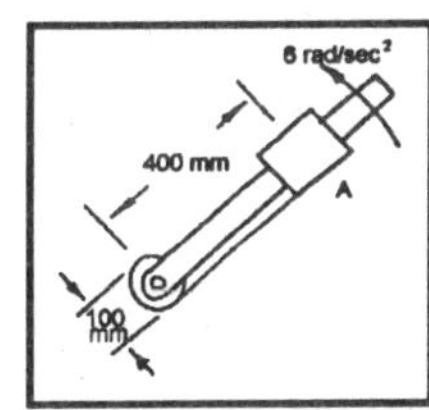

Solution The angular velocity of the spool relative to the bar is $\alpha = 6\ rad/s^2$. The acceleration of the collar relative to the bar is

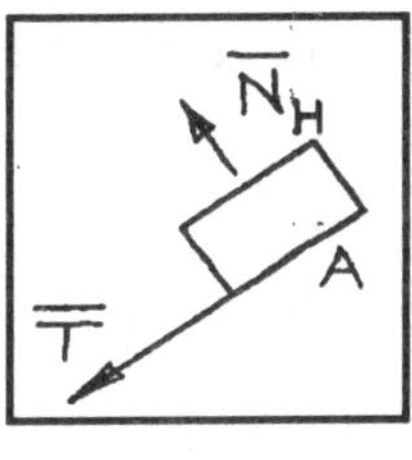

$\frac{d^2r}{dt^2} = -R\alpha = -0.05(6) = -0.3\ m/s^2$ The take up velocity of the spool is $v_s = \int R\alpha dt = -0.05(6)t = -0.3t\ m/s$. The velocity of the collar relative to the bar is $\frac{dr}{dt} = -0.3t\ m/s$.

Solution continued on next page

Continuation of solution to Problem 3.150

The velocity of the collar relative to the bar is $dr/dt = -0.3t$ m/s.

The position of the collar relative to the bar is $r = -0.15t^2 + 0.4$ m. The angular acceleration of the collar is $\frac{d^2\theta}{dt^2} = 6\ rad/s^2$. The angular velocity of the collar is $\frac{d\theta}{dt} = 6t\ \ rad/s$. The radial acceleration is

$a_r = \frac{d^2r}{dt^2} - r\left(\frac{d\theta}{dt}\right)^2 = -0.3 - \left(-0.15t^2 + 0.4\right)(6t)^2$. At $t = 1\ s$ the radial acceleration is $a_r = -9.3\ \ m/s^2$, and the tension in the string is $\boxed{|T| = |ma_r| = 9.3\ N}$

==================================<>==================================

Problem 3.151 In Problem 3.150, suppose that the coefficient of kinetic friction between the collar and the bar is $\mu_k = 0.2$. What is the tension in the string at $t = 1\ s$?

Solution Use the results of the solution to Problem 3.150. At $t = 1\ s$, the horizontal normal force is

$N_H = |ma_\theta| = m\left|\left(r\frac{d^2\theta}{dt^2} + 2\left(\frac{dr}{dt}\right)\left(\frac{d\theta}{dt}\right)\right)\right| = 2.1\ \ N$, from which the total normal force is

$N = \sqrt{N_H^2 + (mg)^2}$ From Newton's second law: $\left(-T + \mu_k\sqrt{N_H^2 + \left(mg^2\right)}\right)\vec{e}_r + N_H\vec{e}_\theta = ma_r\vec{e}_r + ma_\theta\vec{e}_\theta$,

from which $-T + \mu_k\sqrt{N_H^2 + (mg)^2} = ma_r$. From the solution to Problem 3.152, $a_r = -9.3\ \ m/s^2$.

Solve: The tension is $\boxed{T = 11.306\ \ N}$

==================================<>==================================

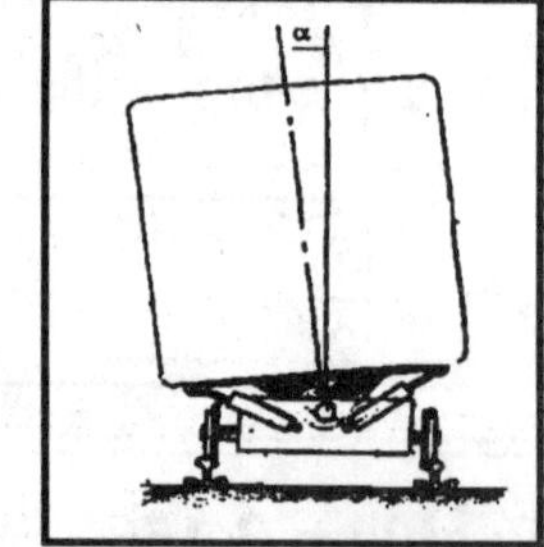

Problem 3.152 If you want to design the cars of a train to tilt as the train goes around curves to achieve maximum passenger comfort, what is the relationship between the desired tilt and α, the velocity v of the train, and the instantaneous radius of curvature of the track?

Solution For comfort, the passenger should feel the total effects of acceleration down toward his feet, that is, apparent side (radial) accelerations should not be felt. This condition is achieved when the tilt α is such that $mg\sin\alpha - m\left(v^2/\rho\right)\cos\alpha = 0$, from which

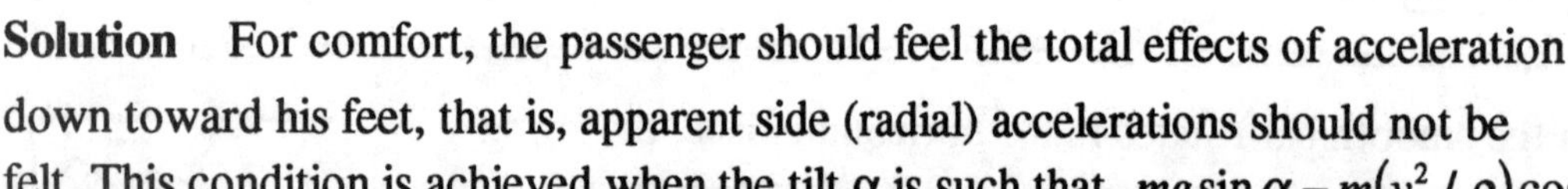

$\boxed{\tan\alpha = \frac{v^2}{\rho g}}$.

==================================<>==================================

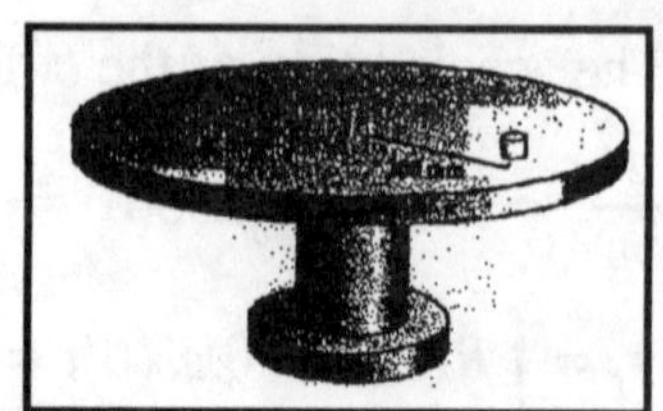

Problem 3.153 To determine the coefficient of static friction between two materials, an engineer at the U. S. Bureau of Standardsplaces a small sample of one material on a horizontal disk surfaced with the other one, then rotates the disk from rest with a constant angular acceleration of 0.4 rad/s^2. If he determines that the small sample slips on the disk after 9.903 seconds, what is the coefficient of friction?

Solution continued on next page

Solution The angular velocity after $t = 9.903\ s$ is $\omega = 0.4t = 3.9612\ rad/s$. The radial acceleration is $a_n = 0.2\omega^2 = 3.138\ m/s^2$. The tangential acceleration is $a_t = (0.2)0.4 = 0.08\ m/s^2$. At the instant before slip occurs, Newton's second law for the small sample is $\sum F = \mu_s N = \mu_s m g = m\sqrt{a_n^2 + a_t^2}$, from which $\boxed{\mu_s = \frac{\sqrt{a_n^2 + a_t^2}}{g} = 0.320}$

=================================<>=================================

Problem 3.154 The 1 *kg* slider A is pushed along the curved bar by the slotted bar. The curved bar lies in the horizontal plane and its profile is described by $r = 2\left(\frac{\theta}{2\pi} + 1\right)$ meters, where θ is in radians. The angular position of the slotted bar is $\theta = 2t\ rad$. Determine the radial and transverse components of the total external force exerted on A when $\theta = 120^o$.

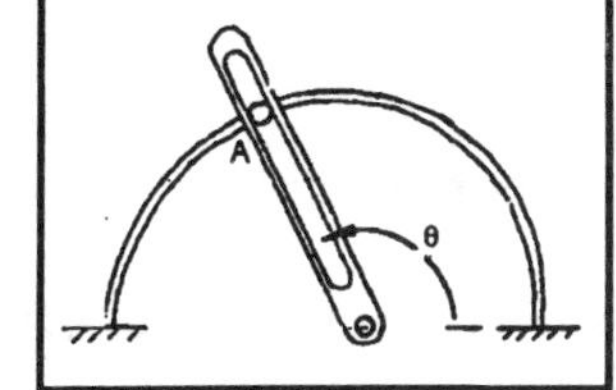

Solution The radial position is $r = 2\left(\frac{t}{\pi} + 1\right)$. The radial velocity: $\frac{dr}{dt} = \frac{2}{\pi}$. The radial acceleration is zero. The angular velocity: $\frac{d\theta}{dt} = 2$. The angular acceleration is zero. At $\theta = 120^o = 2.09\ rad$. From Newton's second law, the radial force is $F_r = ma_r$, from which $\boxed{\vec{F}_r = -\left[r\left(\frac{d\theta}{dt}\right)^2\right]\vec{e}_r = -10.67\vec{e}_r\ N}$ The transverse force is $F_\theta = ma_\theta$, from which $\boxed{\vec{F}_\theta = 2\left[\left(\frac{dr}{dt}\right)\left(\frac{d\theta}{dt}\right)\right]\vec{e}_\theta = 2.55\vec{e}_\theta\ N}$

=================================<>=================================

Problem 3.155 In Problem 3.154, suppose that the curved bar lies in the vertical plane. Determine the radial and transverse components of the total force exerted on A by the curved and slotted bars at $t = 0.5\ s$.

Solution Assume that the curved bar is vertical such that the line $\theta = 0$ is horizontal. The weight has the components: $\vec{W} = (W\sin\theta)\vec{e}_r + (W\cos\theta)\vec{e}_\theta$. From Newton's second law: $F_r - W\sin\theta = ma_r$, and $F_\theta - W\cos\theta = ma_\theta$., from which $\vec{F}_r - g\sin 2t\,\vec{e}_r = -r(d\theta/dt)^2\vec{e}_r$, from which $F_r = (-2\left(\frac{t}{\pi} + 1\right)(2^2) + g\sin 2t)$, at $t = 0.5\ s$, $\boxed{F_r = -1.02\ N}$. The transverse component $F_\theta = 2\left(\frac{2}{\pi}\right)(2) + g\cos 2t = \left(\frac{8}{\pi} + g\cos 2t\right)$. At $t = 0.5\ s$, $\boxed{F_\theta = 7.85\ N}$

=================================<>=================================

================================<>================================

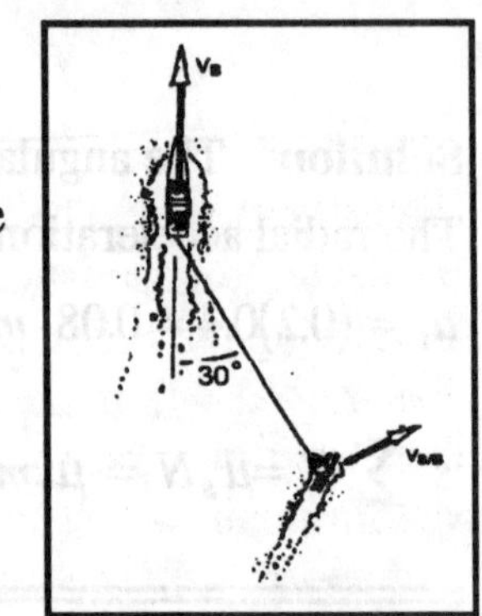

Problem 3.156 The ski boat moves relative to the water with a constant velocity of magnitude $|\vec{v}_B| = 30\ ft/s$. The magnitude of the 170 *lb.* skier's velocity relative to the boat is $|\vec{v}_{S/B}| = 10\ ft/s$. The tension in the 36 *ft* tow rope is 40 *lb.*, and the horizontal force exerted on the skier by the water is perpendicular to the direction of his motion relative to the water. If you can neglect other horizontal forces, what is the skier's acceleration in the direction of his motion relative to the water?

Solution . Denote $\theta = 30^o$ The boat's acceleration is zero, so the skier's acceleration is $\vec{a}_S = \vec{a}_B + \vec{a}_{S/B} = \vec{a}_{S/B}$. Relative to the boat, the skier moves in a circular path and the boat's velocity and the skier's velocity relative to the boat are known. The skier's velocity is $\vec{v}_S = |\vec{v}_S|(\vec{i}\cos 30^o + \vec{j}\sin 30^o) = 0.866\vec{i} + 0.5\vec{j}\ ft/s$. Since the force exerted on the water is perpendicular to the skier's direction of motion, the only force on the skier in the direction of motion is the component of rope tension in that direction. This component is $40\cos 43.9^o = 28.8\ lb$, so the acceleration in the direction of motion is $\boxed{a_{path} = 28.8\left(\frac{32.17}{170}\right) = 5.46\ ft/s^2}$.

================================<>================================

Problem 3.157 In Problem 3.156, what is the magnitude of the horizontal force exerted on the skier by the water?

Solution Use the solution to Problem 3.156. In terms of the tangential and normal accelerations, the total acceleration along the path (from the solution to Problem 3.156) is $a_n\cos 43.9^o + a_t\sin 43.9^o = 5.46\ ft/s^2$. The normal component of the acceleration is $a_n = \frac{v_o^2}{R} = \frac{10^2}{36} = 2.78\ ft/s^2$, from which $a_t = 4.99\ ft/s^2$. The force components are: $\sum\vec{F} = m\vec{a}$, $(F_N + 40)\vec{e}_n + F_t\vec{e}_t = \left(\frac{170}{32.17}\right)(a_n\vec{e}_n + a_t\vec{e}_t)$, from which $F_n = -25.3\ lb$, $F_t = 26.3\ lb$, $\boxed{|\vec{F}| = 36.5\ lb}$

================================<>================================

CHAPTER 4 - Solutions

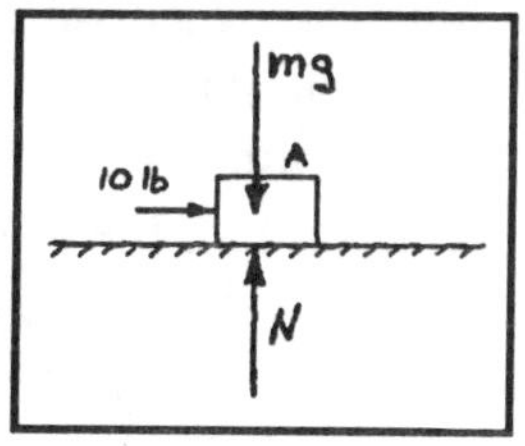

Problem 4.1 The *20-lb* box A is at rest on the smooth surface when it is subjected to a constant *10-lb* force. (a) By using Newton's second law to determine the acceleration of the box, determine how fast it is traveling when it has moved *4 ft* to the right. (b) Use Eq(4.7) to determine the work done on the box when it moves *4 ft* to the right, (c) Use the principle of work and energy to determine how fast the box is traveling when it has moved *4 ft* to the right.

Solution: (a) The mass is $m = W / g = 0.621 slug$ .From Newton's second law, we have $a_x = F_x / m = 10 / m = 16.1 ft / s^2$, Integrate twice and find the time when $x = 4 ft$. Doing this, we get $v_x = 16.1t$, $x = 8.05t^2$, and when $x = 4 ft$, $t = 0.705 s$. Substituting this into the velocity equation, we get $v_x = 11.3 ft / s$. (b) The work done by the *10 lb* force is $W = \int_0^4 F_x dx = 4F_x = 40 ft - lb$. (c) $W = mv_x^2 / 2 = 40 ft - lb$. Substituting in the values value for mass and solving, we get $v_x = 11.3 ft / s$.

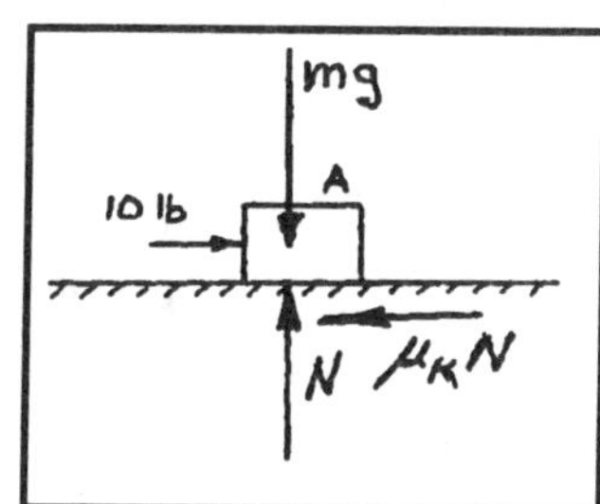

Problem 4.2 Solve Problem 4.1 if the coefficient of kinetic friction between the box and the surface is $\mu_k = 0.2$.

Solution: The work done by the horizontal forces acting on the box is $Work = \int_0^4 (F_x - \mu_k N) dx$ where $N = mg$. This is equal to the change in kinetic energy of the box. Hence $Work = \int_0^4 (10 - 4) dx = 24 ft - lb = mv_x^2 / 2$. Solving, we get $v_x = 8.79 ft / s$

Problem 4.3 The *4-kg* box A is released from rest on the smooth inclined surface. Determine how fast the box is sliding when it has moved *2 m* in the direction parallel to the surface. (a) by using Newton's second law to determine the acceleration of the box; (b) by using Eq. (4.7) and the principle of work and energy.

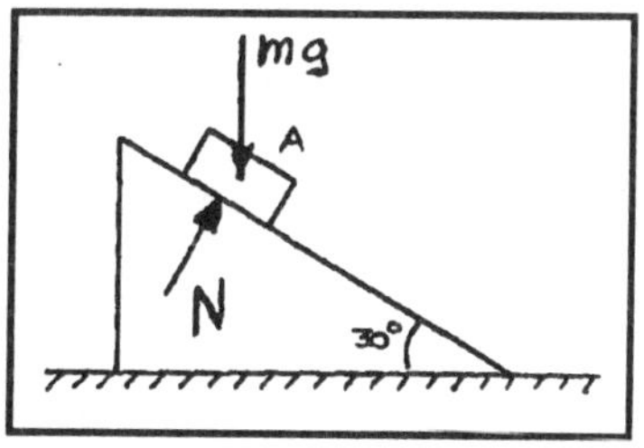

Solution: Choose the x axis parallel to the surface and the y axis perpendicular to it. The component of the weight parallel to the surface is the constant force which accelerates the box down the slope. See the free body diagram. The acceleration of the box is $a_x = F_x / m = mg \sin 30^\circ / m$. We next integrate twice and find the time at which $x = 2$ *m*. The time is $t = 0.903$ *s*, and the corresponding velocity is *4.43 m/s*. (b) The work done by the force $F_x = mg \sin 30^\circ$ is $Work = (F_x)(\Delta x) = (19.62)(2) = 39.24 newton - m$. This is equal to the change in kinetic energy. Hence, $mv_x^2 / 2 = Work$. Solving, we get $v_x = 4.43 m / s$ as before.

==================================<>==================================

Problem 4.4 Solve Problem 4.3 if the coefficient of friction between the box and the inclined surface is $\mu_k = 0.2$.

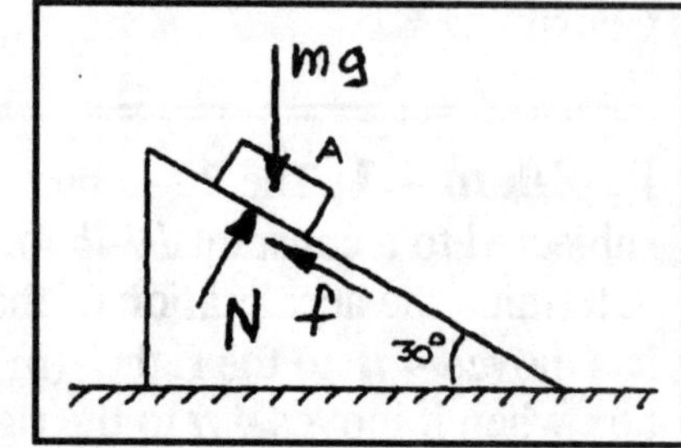

Solution: Choose the x axis parallel to the surface and the y axis perpendicular to it. The component of the weight parallel to the surface minus the friction force is the constant force which accelerates the box down the slope. See the free body diagram. The acceleration of the box down the plane is

$a_x = (F_x - friction)/m = (mg\sin 30^\circ - \mu_k N)/m$, where $N = mg\cos 30^\circ$.

We next integrate twice and find the time at which $x = 2\ m$. The time is $t = 1.117\ s$, and the corresponding velocity is *3.58 m/s.* (b) The work done by the force $(F_x - friction) = (mg\sin 30^\circ - \mu_k N) = 12.82 newtons$ is $Work = (F_x - \mu_k N)(\Delta x) = (12.82)(2) = 25.65 newton - m$. This is equal to the change in kinetic energy. Hence, $mv_x^2/2 = Work$. Solving, we get $v_x = 3.58 m/s$ as before.

==================================<>==================================

Problem 4.5 The fictional starship *Enterprise* obtains its power by combining matter and antimatter, achieving complete conversion of mass into energy. The amount of energy in an amount of matter of mass *m* is given by Einstein's equation $E = mc^2$, where *c* is the speed of light $(3\times 10^8\ m/s)$. (a) The mass of the *Enterprise* is approximately $5\times 10^9\ kg$. How much mass must be converted into kinetic energy to accelerate it to one tenth the speed of light? (b) How much mass must be converted into kinetic energy to accelerate a 200,000 *lb.* airliner from rest to 600 *mi/hr*?

Solution (a) The work done by converting mass is equal to the gain in kinetic energy. The kinetic energy of the *Enterprise* is $\frac{1}{2}mv^2 = \int_0^{0.1c} mv\,dv = \frac{m(0.1c)^2}{2} = \frac{(5\times 10^9)(9\times 10^{14})}{2} = 2.25\times 10^{24}\ N\text{-}m$.

The mass to be converted is $\boxed{m = \frac{2.25\times 10^{24}}{c^2} = 2.5\times 10^7\ kg}$. The percentage of the mass of the *Enterprise* is $R\% = \frac{2.5\times 10^7}{5\times 10^9}(100) = 0.5\%$. (b) The mass of the airliner is

$M_a = \frac{W}{g} = \frac{200{,}000}{32.17} = 6216.97\ slugs$. The velocity of the airliner is

$v_a = 600\frac{mi}{hr}\left(\frac{1\ hr}{3600\ s}\right)\left(\frac{5280\ ft}{1\ mi}\right) = 880\ ft/s$. The kinetic energy of the airliner is

$m\int_0^{v_a} v\,dv = \frac{mv_a^2}{2} = \frac{6217(880)^2}{2} = 2.4\times 10^9\ ft-lb$. The speed of light in common units is

$c = 3\times 10^8\ \frac{m}{s}\left(\frac{1\ ft}{0.3048\ m}\right) = 9.84\times 10^8\ ft/s$. The mass to be converted is

$m_a = \frac{2.4\times 10^9}{c^2} = 2.48\times 10^{-9}\ slugs$, from which $R_a\% = \frac{m_a}{M_a}(100) = 4\times 10^{-11}\%$ of the airliner mass.

==================================<>==================================

================================<>================================

Problem 4.6 A 2000 *lb.* drag racer can accelerate from rest to 300 *mi/hr* in a quarter of a mile. (a) How much work is done on the car? (b) If you assume as a first approximation that the tangential force exerted on the car is constant, what is the magnitude of the force?

Solution

(a) The kinetic energy of the car at the end of the quarter mile is

$$\boxed{\frac{1}{2}\left(\frac{W}{g}\right)\left((300)\left(\frac{5280}{3600}\right)\right)^2 = 6.01\times 10^6 \ ft\text{-}lb.}$$

(b) From the principle of work and energy, the work done on the car is

$\int_0^{1320} F dr = F(1320) = 6.01\times 10^6 \ ft\text{-}lb$, from which $\boxed{F = \frac{6.01\times 10^6}{1320} = 4559.1 \ lb}$

================================<>================================

Problem 4.7 Assume that all of the weight of the drag racer in Problem 4.6 acts on its rear (drive) wheels and that the coefficients of friction between the wheels and the road are $\mu_s = \mu_k = 0.9$. Use the principle of work and energy to determine the maximum velocity in miles per hour that the car can theoretically reach in a quarter mile. What do you think might account for the discrepancy between your answer and the car's actual velocity of 300 *mi/hr*?

Solution The maximum force that the wheels can exert is $F_w = 0.9(2000) = 1800 \ lb$.

The work done to increase the energy if the force is constant is

$$\frac{1}{2}mv^2 = \int F dr = Fr = 1800\left(\frac{5280}{4}\right) = 2.376\times 10^6 \ ft\text{-}lb.$$

The maximum velocity if the force is constant: $\boxed{v = \sqrt{\frac{2g(2.376\times 10^6)}{W}}\left(\frac{3600}{5280}\right) = 188.5 \ mi/hr}$.

Check: From the chain rule: $v\frac{dv}{dr} = \frac{F}{m} = \frac{g\mu_k W}{W} = g\mu_k$, from which

$$v = \sqrt{2g\mu_k r}\left(\frac{3600}{5280}\right) = 188.5 \ mi/hr. \ check.$$

NOTE: Obviously the tires are capable of exerting a much higher thrust than that obtained using the friction model. A possible explanation is that the car was designed such that aerodynamic force greatly increases the normal force between the car and the road, thereby obviating a friction model based on weight alone. This design feature is found in modern race cars. Also, drag racers use special tires which are heated up just before the race by "smoking the tires" on a track wet down with a liquid bleach. This causes the hot tires to stick to the track and tear away as they turn, effectively gearing the car to the track and increasing the effective coefficient of friction.

================================<>================================

Problem 4.8 Assuming as a first approximation that the tangential force exerted on the drag racer in Problem 4.6 is constant. Determine (a) the maximum power and (b) the average power transferred to the car as it accelerates from rest to 300 *mi/hr* in *4.91 s*?

Solution (a) The instantaneous power is $P = Fv$. From Newton's second law, $F = \left(\frac{W}{g}\right)\left(\frac{dv}{dt}\right)$.

Rearrange and use the chain rule: $v\frac{dv}{dx} = \frac{gF}{W}$, $v = \sqrt{\frac{2gFx}{W}}$ from which $P = Fv = \sqrt{\frac{2gF^3x}{W}}$. At $x = 1320\ ft$, $P = 2.006\times10^6 \text{ ft-lb/s} = \frac{2.004\times10^6}{550} = 3647.3 \text{ hp}$.

(b) . The average power is $P_{av} = \frac{U_{12}}{t_2 - t_1}$ The work is(from Problem 4.6) $U_{12} = 6.01\times10^6\ ft-lb$.

The car's acceleration is $a = \frac{F}{m} = \frac{4559}{(2000/32.2)} = 73.3\ ft/s^2$ so the time to reach its maximum velocity is $t = \frac{v}{a} = \frac{440}{73.3} = 6.00s$ therefore the average power is $P_{av} = \frac{6.01\times10^6}{6.00} = 1.00\times10^6\ ft-lb/s$.

Problem 4.9 A 10,000 *kg* airplane must reach a velocity of 60 *m/s* to take off. If the horizontal force exerted by its engine is 60 *kN* and you neglect other horizontal forces, what length runway is needed?

Solution From the principle of work and energy $\int_0^r Fds = Fr = \frac{1}{2}mv^2$, from which $\boxed{r = \frac{mv^2}{2F} = 300\ m}$.

Problem 4.10 Suppose that you want to design an auxiliary rocket unit that will allow the airplane in Problem 4.9 to reach its takeoff speed using only 100 *m* of runway. For your preliminary design calculation, you can assume that the combined mass of the rocket and the airplane is constant and equal to 10,500 *kg*. What horizontal component of thrust must the rocket unit provide?

Solution: The kinetic energy at takeoff is $\frac{1}{2}mv^2 = \frac{10{,}500}{2}(60^2) = 1.89\times10^7\ N-m$. The work done by the rocket is $\int_0^r F_R ds = F_R r$,. and the work done by the engines is $\int_0^r Fds = 60{,}000r$. From the principle of work and energy: $F_R r + 60{,}000r = \frac{1}{2}mv^2$, from which, for $r = 100\ m$,

$$\boxed{F_R = \frac{1.89\times10^7}{10^2} - 60{,}000 = 1.29\times10^5\ N}$$

=====================================<>=====================================

Problem 4.11 Determine (a) the maximum power and (b) the average power transferred to the airplane in Problem 4.9 while it takes off.

Solution: (a) The maximum power is $P = (60,000N)(60m/s) = 3.6 \times 10^6 N-m/s(w)$ (b) The airplane's acceleration is $a = \frac{F}{m} = \frac{60,000}{10,000} = 6m/s^2$ so the time required to take off is $t = \frac{v}{a} = \frac{60}{6} = 10s$ The average power is $P_{av} = \frac{\frac{1}{2}mv_2^2 - \frac{1}{2}mv_1^2}{t_2 - t_1} = \frac{\frac{1}{2}(10,000)(60)^2}{10} = 1.8 \times 10^6 W.$

=====================================<>=====================================

Problem 4.12 (a) Suppose that you hold a *0.15-kg* ball *2 m* above the ground and release it from rest. Use Eq. (4.8) to determine the work done on the ball by gravity as it falls to the ground. (b) Suppose that you then throw the ball upward at *3 m/s*, releasing it *2 m* above the ground. Use Eq. (4.8) to determine the work done on the ball by gravity from the time you release it until it hits the ground. (c) Use the principle of work and energy to determine the magnitude of the ball's velocity just before it hits the ground in cases (a) and (b).

Solution: (a) Let s be measured positive downward from the point of the ball's release

$U_{12} = \sum F_t = (s_2 - s_1) = (0.15)(9.81)(2-0) = 2.94N - m.$

(b) $U_{12} = \sum F_t = (s_2 - s_1) = (0.15)(9.81)(2-0) = 2.94N - m.$

(c) (c) $U_{12} = \frac{1}{2}mv_2^2 - \frac{1}{2}mv_1^2$. $2.94 = \frac{1}{2}(0.15)v^2 - 0$, $v = 6.26m/s$ In case (a),

$2.94 = \frac{1}{2}(0.15)v^2 - 0$, $v = 6.26m/s$. In case (b) $2.94 = \frac{1}{2}(0.15)v^2 - \frac{1}{2}(0.15)(3)^2$,

$v = 6.95m/s$.

=====================================<>=====================================

Problem 4.13 The force exerted on a car by a prototype crash barrier as the barrier crushes is $F = -(1000 + 10,000s)$ *lb.*, where *s* is the distance from the initial contact. Suppose that you want to design the barrier so that it can stop a 5000 *lb.* car traveling at 80 *mi/hr*. What is the necessary effective length of the barrier? That is, what is the distance required for the barrier to bring the car to a stop?

Solution The kinetic energy of a 5000 *lb.* car at 80 *mi/hr* is:

$\frac{1}{2}\left(\frac{W}{g}\right)v^2 = \frac{5000}{2(32.17)}\left(80\left(\frac{5280}{3600}\right)\right)^2 = 1.07 \times 10^6$ *ft - lb*. The work done by the barrier is

$\int Fds = (1000s + 5000s^2) = \frac{1}{2}\left(\frac{W}{g}\right)v^2 = 1.07 \times 10^6$ *ft - lb*. In canonical form $s^2 + 2bs + c = 0$, where $b = 0.1$ and $c = -214$. The solution: $s = -b \pm \sqrt{b^2 - c} = 14.53, -14.7$ *ft*. $\boxed{s = 14.3 \text{ ft}}$ since the negative root has no meaning here.

=====================================<>=====================================

Problem 4.14 The component of the total external force tangent to a 2 *lb.* object's path is $\sum F_t = 4s - s^2$ *lb.*, where *s* is its position measured along the path in feet. At $s = 0$, the object's velocity is $v = 10\,ft/s$. (a) How much work is done on the object as it moves from $s = 0$ to $s = 4\,ft$? (b) What is its velocity when it reaches $s = 4\ ft$?

Solution (a) The work done: $U = \int_0^4 F ds = \int_0^4 \left(4s - s^2\right) ds = \left[2s^2 - \frac{s^3}{3}\right]_0^4$ $\boxed{= 10.67\ ft\text{-}lb.}$. (b) From the principle of work and energy (see definition near top of page 141 in text)

$\int_0^s F ds = \frac{1}{2}\left(\frac{W}{g}\right) v(s)^2 - \frac{1}{2}\left(\frac{W}{g}\right) v_o^2$, from which $v^2(s) = \frac{2g}{W}\left(2s^2 - \frac{s^3}{3}\right) + v_o^2$ where $v_o = 10\,ft/s$. At $s = 4$, $\boxed{v(4) = 21.05\ ft/s}$

Problem 4.15 The component of the total external force tangent to a 10 *kg* object's path is $\sum F_t = 100 - 20t\ N$, where *t* is in seconds. When $t = 0$, its velocity is $v = 4\ m/s$. How much work is done on the object from $t = 2\ s$ to $t = 4\ s$?

Solution From Newton's second law, the acceleration is $\frac{dv}{dt} = \frac{F}{m} = \frac{100 - 20t}{10} = 10 - 2t\ m/s^2$. The velocity is $v(t) = 10t - t^2 + 4\ m/s$, since $v(0) = 4\ m/s$. The work done is

$U = \int_{s_1}^{s_2} F ds = \int_{t_1}^{t_2} F\left(\frac{ds}{dt}\right) dt = \int_2^4 Fv\, dt = \int_2^4 (100 - 20t)(10t - t^2 + 4) dt$, from which

$$\boxed{U = \int_2^4 \left(20t^3 - 300t^2 + 920t + 400\right) dt = \left[5t^4 - 100t^3 + 460t^2 + 400t\right]_2^4 = 1920\ N\text{-}m}$$

Problem 4.16 A group of engineering students constructs a sun-powered car and tests it on a circular track of *1000-ft* radius. The car, which weighs *450 lb* with its occupant, starts from rest. The total tangential component of force on the car is $\sum F_t = 28 - 0.1s$ (*lb*), where *s* is the distance the car travels along the track from the point where it starts. (a) Determine the work done on the car when it has gone a distance of $s = 100\,ft$. (b) Use the principle of work and energy to determine the magnitude of the car's velocity when $s = 100\,ft$.

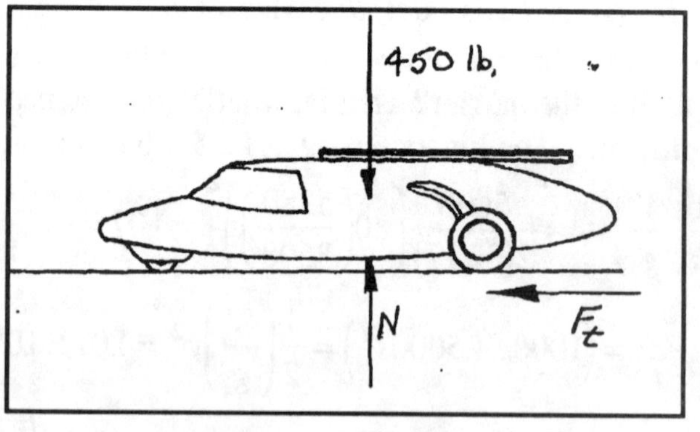

Solution: (a) $U_{12} = \int_{s_1}^{s_2} \sum F_t ds = \int_0^{100} (28 - 0.1s) ds = 28(100) - 0.1\frac{(100)^2}{2} = 2300\,ft - lb.$

(b) $U_{12} = \frac{1}{2} m v_2^2 - \frac{1}{2} m v_1^2$: $2300 = (450/32.2) v^2/2 - 0$, and $v = 18.1\,ft/s$.

=====================================<>=====================================

Problem 4.17 At the instant shown, the *160-lb* vaulter's center of mass is *8.5 ft* above the ground and the vertical component of his velocity is *4 ft/s*. As his pole straightens, it exerts a vertical force on him of magnitude $180+2.8y^2 lb$, where y is the vertical position of his center of mass *relative to its position at the instant shown.* This force is exerted on him from $y = 0$ to $y = 4 ft$, when he releases the pole. What is the maximum height above the ground reached by his center of mass?

Solution: The work done on him by the pole is

$U_{pole} = \int_0^4 (180+2.8y^2)dy = 180(4) + 2.8\frac{(4)^3}{3} = 780 ft-lb$. Let y_{max} be his maximum height above the ground. The work done by his weight from the instant shown to the maximum height is $-160(y_{max}-8.5) = U_{weight}$, or $U_{weight} + U_{pole} = mv_2^2/2 - mv_1^2/2$

$780 - 160(y_{max} - 8.5) = 0 - \frac{1}{2}(160/32.2)(4)^2$. Solving, $y_{max} = 13.6 ft$

=====================================<>=====================================

Problem 4.18 The component of the total external force tangent to the path of an object of mass m is $\sum F_t = -cv$, where v is the magnitude of the object's velocity and c is a constant. When the position $s = 0$, its velocity is $v = v_o$. How much work is done on the object as it moves from $s = 0$ to position $s = s_f$?

Solution From Newton's second law $F = m\frac{dv}{dt}$. Rearrange and use the chain rule: $v\frac{dv}{ds} = \frac{F}{m} = -\frac{c}{m}v$,

from which $\frac{dv}{ds} = -\frac{c}{m}$, $v = -\frac{c}{m}s + v_o$. The work done is

$U = \int_0^{s_f} Fds = -c\int_0^{s_f}\left(-\frac{c}{m}s + v_o\right)ds = -c\left[-\frac{c}{2m}s^2 + v_o s\right]_0^{s_f}$, $\boxed{U = cs_f\left(\frac{c}{2m}s_f - v_o\right)}$.

=====================================<>=====================================

Problem 4.19 The coefficients of friction between the 160 *kg* crate and the ramp are $\mu_s = 0.3$, and $\mu_k = 0.28$. (a) What tension T_o must the winch exert to start the crate moving up the ramp? (b) If the tension remains at the value T_o after the crate starts sliding what total work must be done on the crate as it slides a distance $s = 3\ m$ up the ramp, and what is the resulting velocity of the crate?

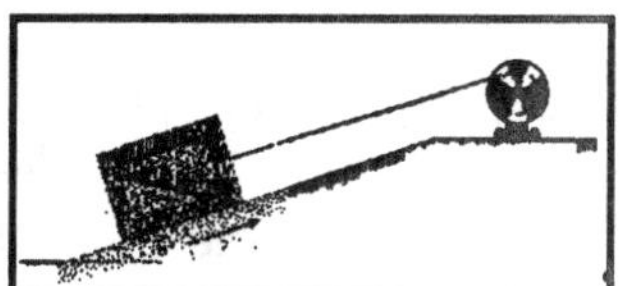

Solution (a) The tension is $T_o = W\sin\theta + \mu_s N$, from which $\boxed{T_o = mg(\sin\theta + \mu_s\cos\theta) = 932.9\ N}$. (b) The work done on the crate by (non-friction) external forces is

$U_{weight} = \int_0^3 T_o ds - \int_0^3 (mg\sin\theta)ds = 939.1(3) - 1455.1 = 1343.5\ N\text{-}m$. The work done on the crate by friction is $U_f = \int_0^3 (-\mu_k N)ds = -3\mu_k mg\cos\theta = -1253.9\ N\text{-}m$.

From the principle of work and energy is $U_{weight} + U_f = \frac{1}{2}mv^2$,

from which $v = \sqrt{\frac{6(T_o - mg(\sin\theta + \mu_k\cos\theta))}{m}}$ $\boxed{v = 1.06\ m/s}$

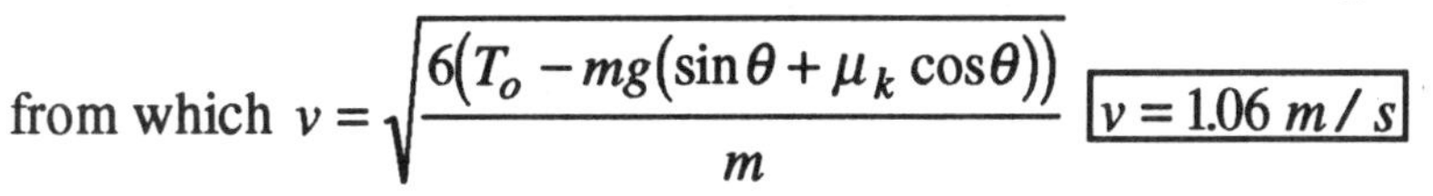

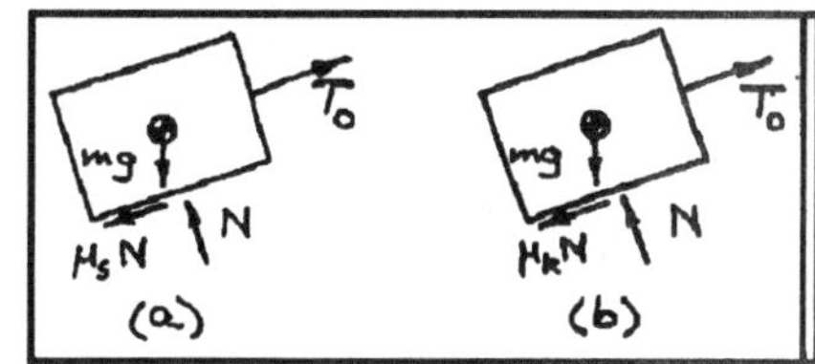

=====================================<>=====================================

==================================<>==================================

Problem 4.20 In Problem 4.19, if the winch exerts a tension $T = T_o(1+0.1s)$ after the crate starts sliding, what total work is done on the crate as it slides a distance $s = 3\ m$ up the ramp, and what is the resulting velocity of the crate?

Solution:

The work done on the crate is $U = \int_0^3 T ds - \int_0^3 (mg\sin\theta)ds - \mu_k \int_0^3 (mg\cos\theta)ds$, from which

$$U = T_o\left[\left(s+0.05s^2\right)\right]_0^3 - (mg\sin\theta)(3) - \mu_k(mg\cos\theta)(3).$$

From the solution to Problem 4.19, $T_o = 932.9\ N\text{-}m$, from which the total work done is $\boxed{U = 3218.4 - 1455.1 - 1253.9 = 509.36\ N\text{-}m}$.

From the principle of work and energy, $U = \frac{1}{2}mv^2$, from which $\boxed{v = \sqrt{\frac{2U}{m}} = 2.52\ m/s}$

==================================<>==================================

Problem 4.21 The 200 mm gas gun is evacuated on the right of the 8 kg projectile. On the left of the piston the tube contains gas with pressure $p_o = 1\times10^5\ Pa\ (N/m^2)$. The force F is slowly increased, moving the piston 0.5 m to the left from the position shown. The force is then removed and the piston accelerates to the right. If you neglect friction and assume that the pressure of the gas is related to its volume by $pV = const.$, what is the velocity of the piston when it has returned to its original position?

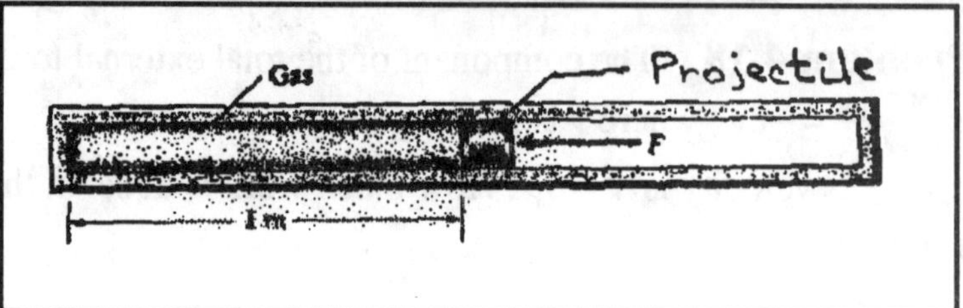

Solution:

The constant is $K = pV = 1\times10^5(1)(0.1)^2\pi = 3141.6\ N\text{-}m$. The force is $F = pA$. The volume is $V = As$, from which the pressure varies as the inverse distance: $p = \frac{K}{As}$, from which $F = \frac{K}{s}$. The work done by the gas is $U = \int_{0.5}^{1} F ds = \int_{0.5}^{1} \frac{K}{x} ds = \left[K\ln(s)\right]_{0.5}^{1.0} = K\ln(2)$.

From the principle of work and energy, the work done by the gas is equal to the gain in kinetic energy: $K\ln(2) = \frac{1}{2}mv^2$, and $v^2 = \frac{2K}{m}\ln(2)$, $\boxed{v = \sqrt{\frac{2K}{m}\ln(2)} = 23.33\ m/s}$

Note: The argument of $\ln(2)$ is dimensionless, since it is ratio of two distances.

==================================<>==================================

=================================<>=================================

Problem 4.22 In Problem 4.21, if you assume that the pressure of the gas is related to its volume by $pV = const.$ while it is compressed (an isothermal process) and by $pV^{1.4} = const.$ while it is expanding (an isentropic process), what is the velocity of the piston when it has returned to its previous position?

Solution

The isothermal constant is $K = 3141.6\ N\text{-}m$ from the solution to Problem 4.21.

The pressure at the leftmost position is $p = \dfrac{K}{A(0.5)} = 2\times 10^5\ N/m^2$.

The isentropic expansion constant is $K_e = pV^{1.4} = (2\times 10^5)(A^{1.4})(0.5^{1.4}) = 596.5\ N\text{-}m$

The pressure during expansion is $p = \dfrac{K_e}{(As)^{1.4}} = \dfrac{K_e}{A^{1.4}} s^{-1.4}$.

The force is $F = pA = K_e A^{-0.4} s^{-1.4}$. The work done by the gas during expansion is

$$U = \int_{0.5}^{1.0} F ds = \int_{0.5}^{1.0} K_e A^{-0.4} s^{-1.4} ds = K_e A^{-0.4}\left[\frac{s^{-0.4}}{-0.4}\right]_{0.5}^{1.0} = 1901.8\ N\text{-}m.$$

From the principle of work and energy, the work done is equal to the gain in kinetic energy,

$\int_{0.5}^{1} F ds = \dfrac{1}{2} mv^2$, from which the velocity is $\boxed{v = \sqrt{\dfrac{2(1901.8)}{m}} = 21.8\ m/s}$.

=================================<>=================================

Problem 4.23 The system is released from rest. By applying the principle of work and energy to each weight, determine the magnitude of the velocity of the weights when they have moved 1 ft.

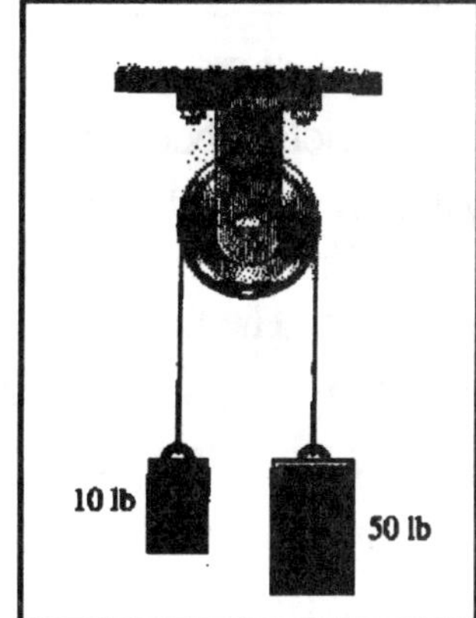

Solution

Since the pulley is one-to-one, the magnitudes of the velocities of the weights are equal.. Let s denote the upward displacement of weight A and the downward displacement of weight B. The principle of work and energy for weight A is $U_A = \int_0^s (T - W_A) ds = \dfrac{1}{2}\left(\dfrac{W_A}{g}\right) v^2$, and for the weight B is

$$U_B = \int_0^s (W_B - T) ds = \frac{1}{2}\left(\frac{W_B}{g}\right) v^2.$$

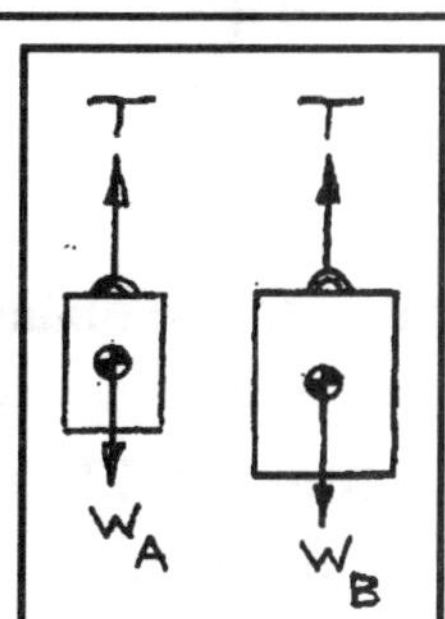

Weight A moves upward: the work done by A is $U_A = \int_0^s (-W_A) ds = -W_A s$.

Add: $U_A + U_B = W_B s - W_A s = \dfrac{1}{2}\left(\dfrac{W_A}{g}\right) v^2 + \dfrac{1}{2}\left(\dfrac{W_B}{g}\right) v^2$,

$$\boxed{|v| = \sqrt{\frac{2g(W_B - W_A)s}{(W_A + W_B)}} = 6.55\ ft/s}.$$

=================================<>=================================

=================================<>=================================

Problem 4.24 In Problem 4.23, what is the average power transmitted to the system during its motion?

Solution: The free body diagrams of the weights are as showin in Problem 3.23. Writing Newton's second law for each weight, $50 - T = (50/32.2)a$, $T - 10 = (10/32.2)a$, and solving, we obtain the acceleration: $a = 21.5\,ft/s^2$ We determine the time required to move 1ft:

$$1 = \frac{1}{2}at^2 = \frac{1}{2}(21.5)t^2,\ t = 0.305s.$$

The average power is $P_{av} = \dfrac{U_{12}}{t_2 - t_1} = \dfrac{(50-T)(1) + (T-10)(1)}{0.305} = 131\,ft - lb/s$

=================================<>=================================

Problem 4.25 Solve Problem 4.23 by applying the principle of work and energy to the system consisting of the two weights, the cable, and the pulley.

Solution Since the pulley is one-to-one, the magnitudes of the velocities of the weights are equal. The work done on the system is $U = \int_0^s F ds = \int_0^s (W_B - W_A) ds = (W_B - W_A)s$, where s is the displacement of block B. From the principle of work and energy, the work done by the system is equal to the gain in kinetic energy for the system: $U = \frac{1}{2}\left(\frac{W_A}{g} + \frac{W_B}{g}\right)v^2$. From which $v = \pm\sqrt{\frac{2g(W_B - W_A)s}{(W_B + W_A)}}$. For $s_A = 1\,ft$,

$\boxed{|v| = 6.55\ ft/s}$ From physical reasons, the velocity of B is downward, and the velocity of A is upward.

=================================<>=================================

Problem 4.26 The mass of each box is *12 kg* and the coefficient of kinetic friction between the boxes and the surface is $\mu_k = 0.05$. The system is released from rest. Determine the magnitude of the velocity of the weights when they have moved *1 m*.

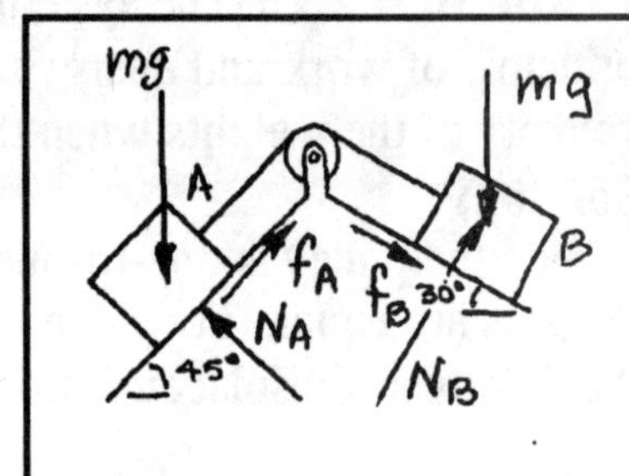

Solution: The free body diagrams are as shown. The normal forces are

$$N_A = mg\cos 45^\circ = (12)(9.81)\cos 45^\circ = 83.2N,$$

$$N_B = mg\cos 30^\circ = (12)(9.81)\cos 30^\circ = 101.9N,$$

The principle of work and energy for each box is

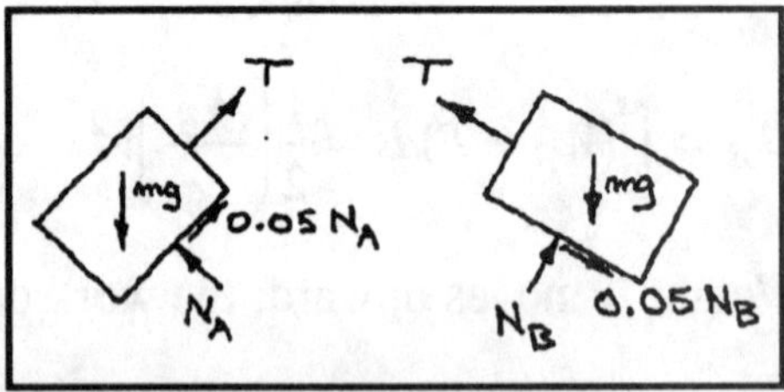

$$(mg\sin 45^\circ - 0.05N_A - T)(1) = \frac{1}{2}mv^2$$

$$(T - mg\sin 30^\circ - 0.05N_B)(1) = \frac{1}{2}mv^2$$ Solving these equations, we obtain $T = 71.5N$ $v = 1.12 m/s$.

=================================<>=================================

Problem 4.27 In Problem 4.26, what average power is transmitted to the system during the motion?

Solution: From the solution of Problem 4.26, the boxes are moving at *1.12 m/s* when they have moved *1 m*. Let a be the magnitude of the acceleration and t the time required to move *1 m*. Then, $v = at$; $1.12 = a\,t$, and $s = at^2/2$; $1 = at^2/2$. Solving, we obtain $t = 1.78s$. Then, the average power is

$$P_{av} = \frac{mv_2^2/2 - mv_1^2/2}{t_2 - t_1} = \frac{2\left[(12)(1.12)^2/2\right]}{1.78} = 8.49\ W.$$

=================================<>=================================

====================================<>====================================

Problem 4.28 The masses of the three blocks are $m_A = 40\ kg$, $m_B = 16\ kg$, and $m_C = 12\ kg$. Neglect the mass of the bar holding C. Friction is negligible. By applying the principle of work and energy to A and B individually, determine the magnitude of their velocity when they have moved 500 *mm*.

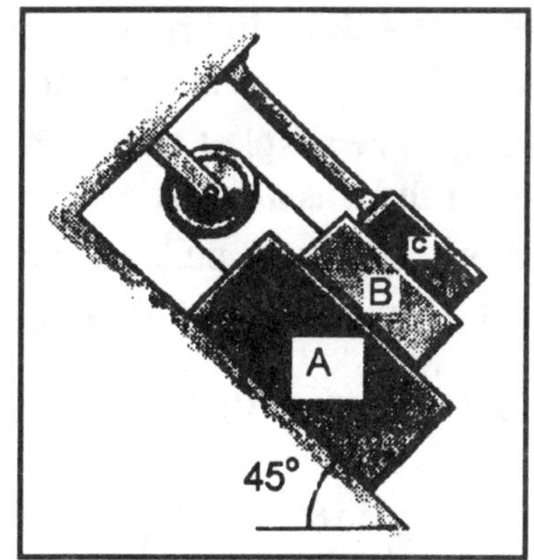

Solution

Denote $b = 0.5\ m$. Since the pulley is one-to-one, denote $|v_A| = |v_B| = v$. The principle of work and energy for weight A is

$\int_0^b (m_A g \sin\theta - T)ds = \frac{1}{2}m_A v^2$, and for weight B

$\int_0^b (T - m_B g \sin\theta)ds = \frac{1}{2}m_B v^2$. Add the two equations:

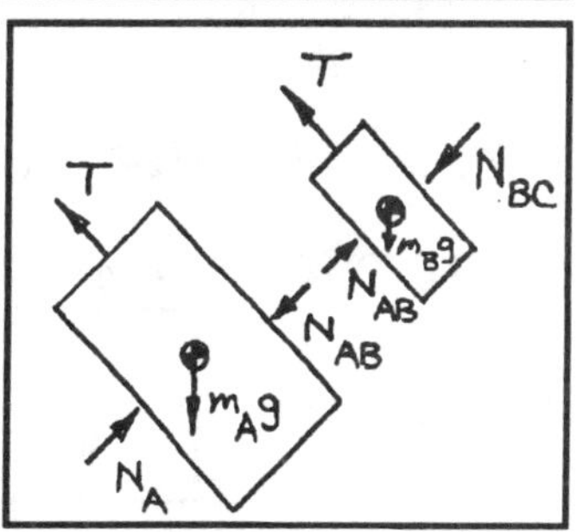

$(m_A - m_B)gb\sin\theta = \frac{1}{2}(m_A + m_B)v^2$. Solve:

$$|v_A| = |v_B| = \sqrt{\frac{2(m_A - m_B)gb\sin\theta}{(m_A + m_B)}} = 1.72\ m/s$$

====================================<>====================================

Problem 4.29 Solve Problem 4.28 by applying the principle of work and energy to the system consisting of A, B, the cable connecting them, and the pulley.

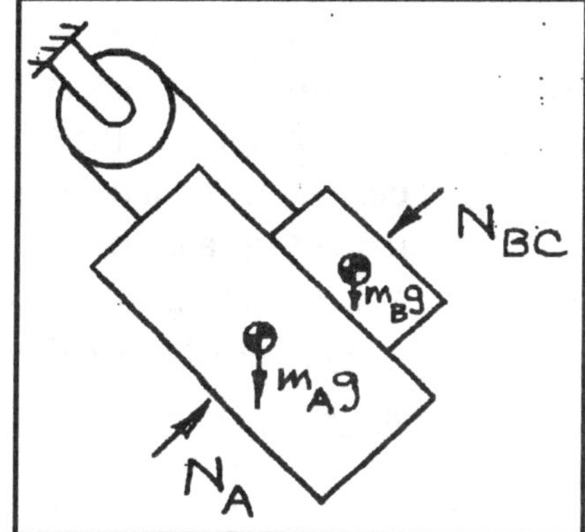

Solution

The strategy is a modification of that followed in Example 4.5. Choose a coordinate system with the origin at the pulley axis and the positive *x* axis parallel to the inclined surface. Since the pulley is one-to-one, $x_A = -x_B$. Differentiate to obtain $v_A = -v_B$. Denote $b = 0.5\ m$. From the principle of work and energy the work done by the external forces on the complete system is equal to the gain in kinetic energy,

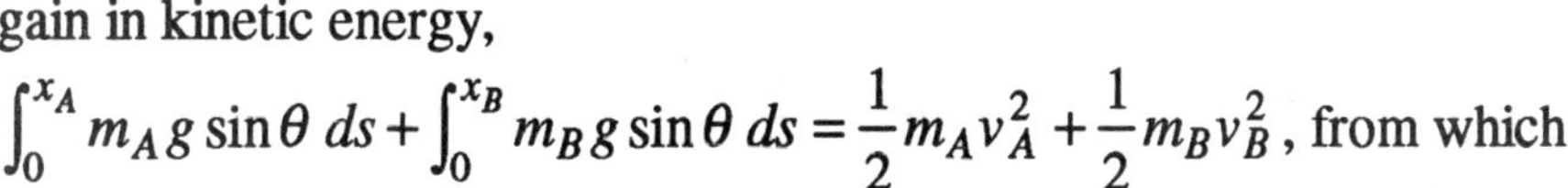

$\int_0^{x_A} m_A g \sin\theta\, ds + \int_0^{x_B} m_B g \sin\theta\, ds = \frac{1}{2}m_A v_A^2 + \frac{1}{2}m_B v_B^2$, from which

$(m_B - m_A)gb\sin\theta = \frac{1}{2}(m_A + m_B)v_A^2$ and $|v_A| = |v_B| = \sqrt{\frac{(m_A - m_B)}{(m_A + m_B)}2gb\sin\theta} = 1.72\ m/s$.

====================================<>====================================

==================================<>==================================

Problem 4.30 In Problem 4.28, determine the magnitude of the velocity of A and B when they have moved 500 *mm* if the coefficient of kinetic friction between all surfaces is $\mu_k = 0.1$.

Strategy: The simplest approach is to apply the principle of work and energy to A and B individually. If you treat them as a single system, you must account for the work done by internal friction forces.

Solution Since the pulley is one-to-one, $|v_A| = |v_B| = v$. Denote $b = 0.5\ m$. The principle of work and energy applied to weight A is $\int_0^b (m_A g \sin\theta - T - \mu_k N_A - \mu_k N_{AB}) ds = \frac{1}{2} m_A v^2$, and for weight B $\int_0^b (T - m_B g \sin\theta - \mu_k N_{BC} - \mu_k N_{AB}) ds = \frac{1}{2} m_B v^2$. Add the two equations,

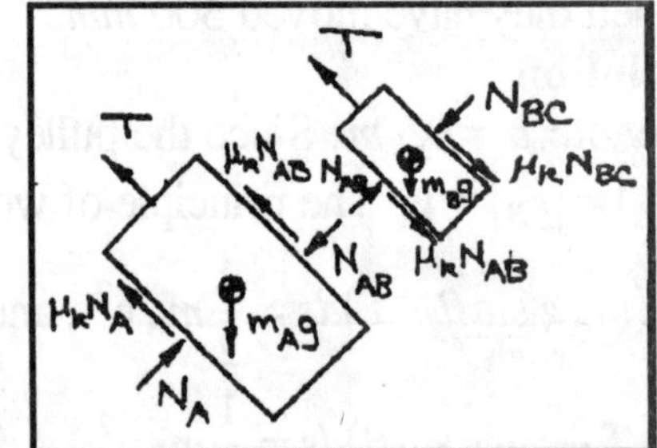

$$(m_A - m_B) g b \sin\theta - \mu_k b (N_A + 2N_{AB} + N_{BC}) = \frac{1}{2}(m_A + m_B) v^2.$$

The normal forces are $N_{BC} = m_C g \cos\theta$, $N_{AB} = (m_B + m_C) g \cos\theta$, and $N_A = (m_A + m_B + m_C) g \cos\theta$.

Substitute and solve, $v = \sqrt{\dfrac{2gb(m_A - m_B)\sin\theta - \mu_k (m_A + 3m_B + 4m_C)\cos\theta}{(m_A + m_B)}} = 1.135\ m/s$

==================================<>==================================

Problem 4.31 (a) Suppose that you hold a *0.15 kg* ball *2 m* above the ground and release it from rest. Use Eq (4.15) to determine the work done on the ball by gravity as it falls to the ground. (b) Suppose that you then throw the ball upward at *3 m/s*, releasing it at *2 m* above the ground. Use Eq (4.15) to determine the work done on the ball by gravity from the time you release it until it hits the ground. (c) Use the principle of work and energy to determine the magnitude of the ball's velocity just before it hits the ground in cases (a) and (b).

Solution: (a) Let the origin of the coordinate system (datum) be at ground level with the y axis upward.

$U_{12} = -mg(y_2 - y_1) = -(0.15)(9.81)(0-2) = 2.94 N-m$

(b) $U_{12} = -mg(y_2 - y_1) = -(0.15)(9.81)(0-2) = 2.94 N-m$

(c) In case (a) $U_{12} = \frac{1}{2} m v_2^2 - \frac{1}{2} m v_1^2$: $2.94 = \frac{1}{2}(0.15) v^2 - 0$, $v = 6.26 m/s$

In case (b), $U_{12} = \frac{1}{2} m v_2^2 - \frac{1}{2} m v_1^2$: $2.94 = \frac{1}{2}(0.15) v^2 - \frac{1}{2}(0.15)(3)^2$ $v = 6.95 m/s$.

==================================<>==================================

Problem 4.32 Suppose that you throw rocks from the top of a 200 *m* cliff with a velocity of 10 *m/s* in the direction shown. Neglecting aerodynamic drag, use the principle of work and energy to determine the magnitude of the velocity of the rock just before it hits the ground in each case.

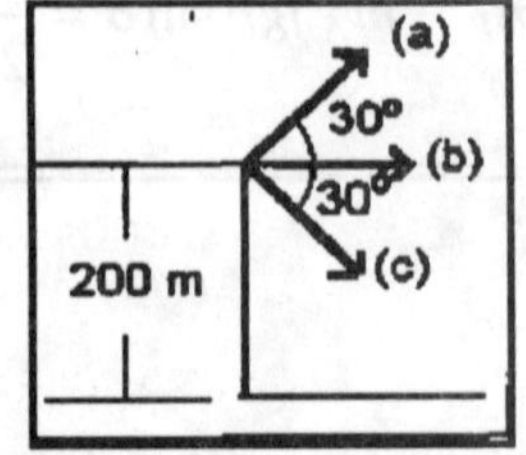

Solution Denote the initial velocity by v_o, The work done by gravity in each case will be the same, $U_{gravity} = \int_0^{-h} (-mg) ds = mgh$. From the principle of work and energy, at impact: $+mgh = \frac{1}{2} m v_{imp}^2 - \frac{1}{2} m v_o^2$, from which $v_{imp}^2 = 2gh + v_o^2$. The magnitude of the velocity at impact is $v_{imp} = \sqrt{10^2 + 2(9.81)(200)} = 63.4\ m/s$, in all three cases.

==================================<>==================================

==============================<>==============================

Problem 4.33 The 30 *kg* box starts from rest at position 1. Neglect friction. For cases (a) and (b) determine the work done on the box from position 1 to position 2 and the magnitude of the velocity of the box at position 2.

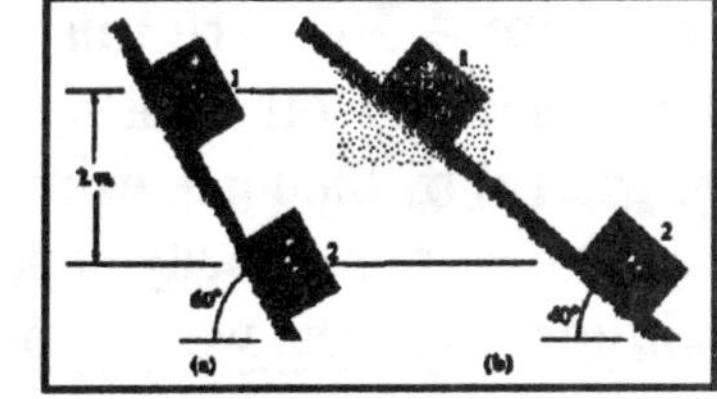

Solution Choose a coordinate system with the y axis positive downward. The work done by the weight is the same in both cases:

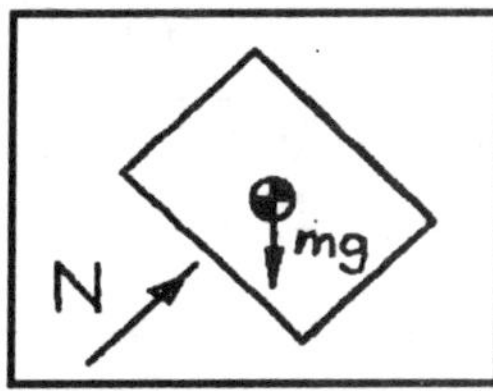

$U_{1-2} = \int_0^h (mg)dy = mgh = 30(9.81)(2) = 588.6\ N\text{-}m$. From the principle of work and energy the gain in kinetic energy is $U_{1-2} = \frac{1}{2}mv^2$, from which

$$\boxed{v = \sqrt{\frac{2U_{1-2}}{m}} = \sqrt{2gh} = 6.26\ m/s}$$

==============================<>==============================

Problem 4.34 Solve Problem 4.33 if the coefficient of kinetic friction between the box and the inclined surface is $\mu_k = 0.2$.

Solution Choose a coordinate system with the y axis positive downward. The work done by friction in the two cases will be different, since the normal forces will be different, and the total travel will be different. The work done by the weight will be the same: $U_{1-2} = \int_0^h (mg)dy = mgh = 588.6\ N\text{-}m$. The work done

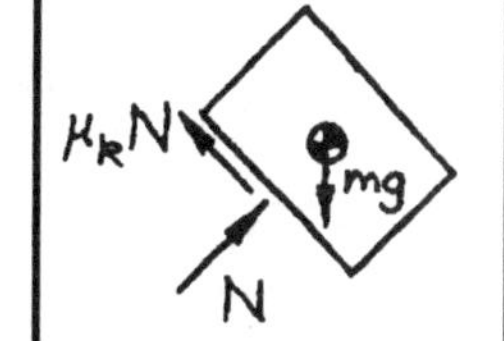

by friction is $U_{friction} = \int_0^{\frac{h}{\sin\theta}} F\,ds = \int_0^{\frac{h}{\sin\theta}} (-\mu_k N)ds$. From the principle of work and energy, the work done is equal to the kinetic energy: $U_{1-2} + U_{friction} = \frac{1}{2}mv^2$. For an inclined surface at an angle θ, and for a vertical displacement *h*, the kinetic energy will be

$$\frac{1}{2}mv^2 = mgh - \int_0^{\frac{h}{\sin\theta}} \mu_k N ds = mgh - \mu_k mg\cos\theta\left(\frac{h}{\sin\theta}\right) = mgh\left(1 - \frac{\mu_k}{\tan\theta}\right).$$

The velocity is obtained from the kinetic energy: $v = \sqrt{2gh\left(1 - \frac{\mu_k}{\tan\theta}\right)}$. For case (a) $\theta = 60^o$, the work done is $\boxed{\frac{1}{2}mv^2 = 521\ N\text{-}m}$ the velocity is $\boxed{v = 5.89\ m/s}$. For case (b), $\theta = 40^o$, $\boxed{\frac{1}{2}mv^2 = 448\ N\text{-}m}$ and $\boxed{v = 5.47\ m/s}$

==============================<>==============================

Problem 4.35 In case (a), the *5-oz.* ball [*16 ounces (oz)* equal *1 pound*] is released from rest at position 1 and falls to position 2. In case (b), the ball is released from rest at position 1 and swings to positon 2. For cases (a) and (b), determine the work done on the ball from position 1 to position 2 and the velocity of the ball at position 2.

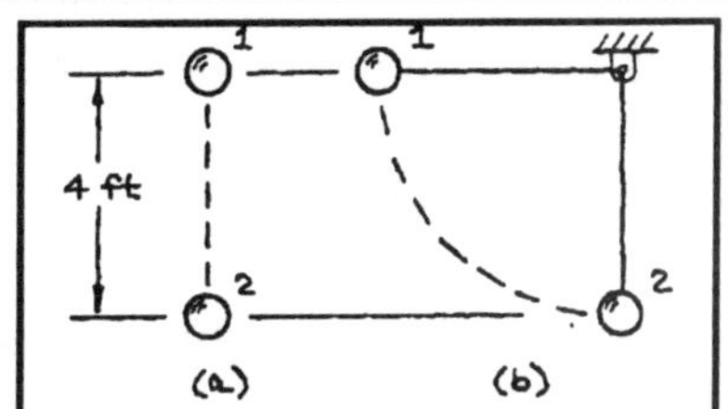

Solution:

In case (b) the work done by the string tension is zero since it acts perpendicular to the ball's motion. So in both cases, $U_{12} = -mg(y_2 - y_1) = -(5/16)(0-4) = 1.25 ft-lb$. Also, $U_{12} = mv_2^2/2 - mv_1^2/2$. Equating the two values for U_{12} and noting that $v_1 = 0$, we get $v_2 = 16.0 ft/s$

==============================<>==============================

Problem 4.36 The ball of mass m is released from rest in position 1. Determine the work done on the ball as it swings to position 2 (a) by its weight, (b) by the force exerted on it by the string. (c) What is the magnitude of its velocity at position 2?

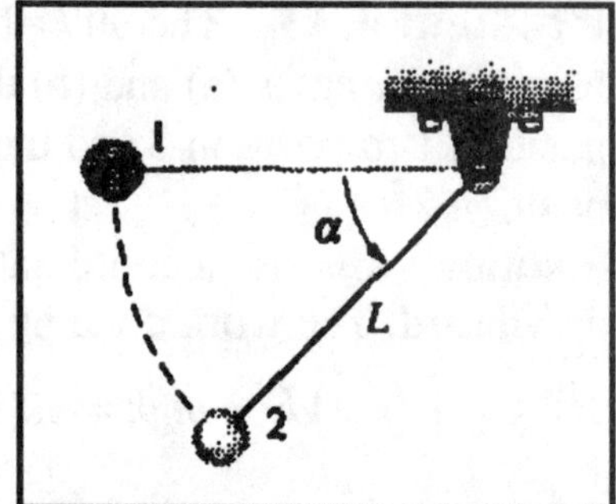

Solution (a) The work done by gravity is

$$\boxed{\int_0^s F ds = mg\int_0^{L\sin\alpha} ds = mgL\sin\alpha}$$

(b) The force exerted by the string is perpendicular to the motion, so no work is done by the string tension. (c) From the principle of work and energy: $mgL\sin\alpha = \frac{1}{2}mv^2$, from which $\boxed{v = \sqrt{2gL\sin\alpha}}$

Problem 4.37 In Problem 4.36, what is the tension in the string at position 2?

Solution From Newton's second law, $mL\left(\frac{d\alpha}{dt}\right)^2 = -mg\sin\alpha + T$, from which

$T = mg\sin\alpha + mL\left(\frac{d\alpha}{dt}\right)^2 = mg\sin\alpha + m\frac{v^2}{L}$. From the solution to Problem 4.37, $v^2 = 2gL\sin\alpha$, from which $\boxed{T = 3mg\sin\alpha}$

Problem 4.38 The 200 kg wrecker's ball hangs from a 6 m cable. If it is stationary in position 1, what is the magnitude of its velocity before it hits the wall in position 2?

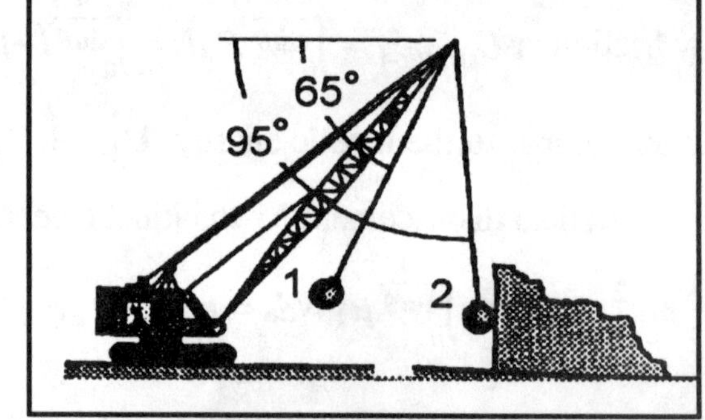

Solution The vertical distance traveled is $s = L\sin\alpha$, from which $ds = L\cos\alpha\, d\alpha$. The work done is

$$U_{gravity} = \int mgds = \int_{\alpha_1}^{\alpha_2} mgL\cos\alpha\, d\alpha$$

$U_{gravity} = mgL[\sin\alpha]_{65^\circ}^{95^\circ} = 1058.1\ N\text{-}m$. From the principle of work and energy: $U_{gravity} = \frac{1}{2}mv_1^2$. The velocity is $\boxed{v = \sqrt{\frac{2U_{gravity}}{m}} = 3.25\ m/s}$

Problem 4.39 In Problem 4.38, what is the maximum tension in the cable during the motion of the ball from position 1 to position 2?

Solution From the solution to Problem 4.38, the tension in the cable is $T = mg\sin\alpha + m\frac{v^2}{L}$. From the solution to Problem 4.38, $v^2 = 2gL\left[\sin\alpha - \sin 65^o\right]\ \left(65^o \le \alpha \le 95^o\right)$, from which

$T = 3mg\sin\alpha - 2mg\sin 65^o$. The maximum tension occurs when $\sin\alpha$ is a maximum in the interval $\left(65^o \le \alpha \le 95^o\right)$, from which $\boxed{T = 3mg\sin 90^o - 2mg\sin 65^o = 2329.6\ N}$.

==================================<>==================================

Problem 4.40 A stunt driver wants to drive a car through a circular loop of radius R and hires you as a consultant to tell him the necessary velocity v_o at which the car must enter the loop so that it can coast through without losing contact with the track. (a) What is v_o if you neglect friction and the aerodynamic drag for your first rough estimate? (b) What is the resulting velocity of the car at the top of the loop?

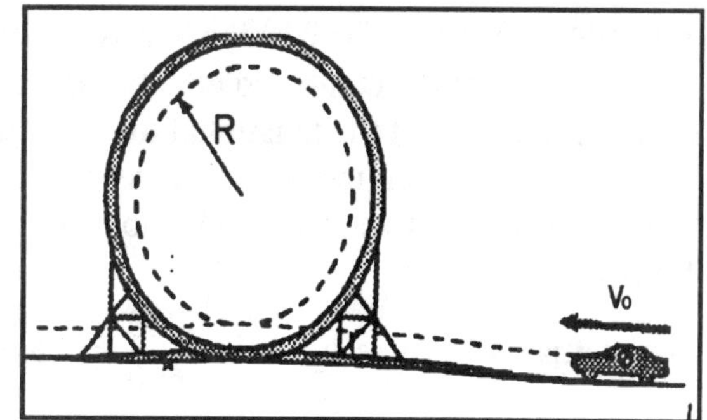

Solution The mass times the radial acceleration must be equal to or greater than the weight of the car at the top of the loop. $m\frac{v_{top}^2}{R} \geq mg$, from which (b) $\boxed{v_{top} \geq \sqrt{gR}}$. The work done in moving the car from ground level to the top of the loop is $U_{gravity} = \int_0^{2R}(-mg)ds = -2mgR$. From the principle of work and energy, at the top of the loop $-2mgR = \frac{1}{2}mv_{top}^2 - \frac{1}{2}mv_o^2$. Reduce: $\frac{1}{2}mv_o^2 = 2mgR + \frac{1}{2}mgR = \frac{5}{2}mgR$, from which (a) $\boxed{v_o = \sqrt{5gR}}$

==================================<>==================================

Problem 4.41 The 2 *kg* collar starts from rest at position 1 and slides down the smooth rigid wire. The *y* axis points upward. What is the collar's velocity when it reaches position 2?

Solution The work done by the weight is $U_{weight} = mgh$, where $h = y_1 - y_2 = 5 - (-1) = 6\ m$. From the principle of work and energy, $mgh = \frac{1}{2}mv^2$, from which $\boxed{v = \sqrt{2gh} = 10.85\ m/s}$

==================================<>==================================

Problem 4.42 The forces acting on the *24,000 lb* airplane are the thrust *T* and drag *D*, which are parallel to the plane's path, the lift *L*, which is perpendicular to the path, and the weight *W*. The airplane climbs from a *5000 ft* altitude to *10,000 ft*. During the climb the magnitude of the velocity decreases from *800 ft/s* to *600 ft/s*. (a) What work is done on the airplane by its lift during the climb? (b) What work is done by the thrust and drag combined?

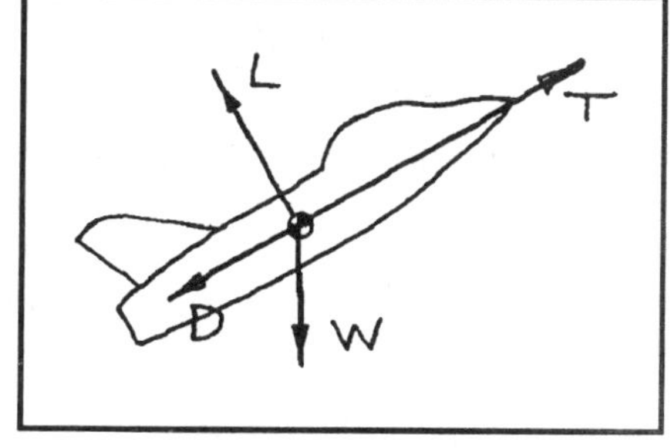

Solution: (a) The lift, *L*, being perpendicular to the path, does no work. (b) Let U_C be the work done by the thrust and drag (working together). The sum of U_C and the work done by the weight equals the change in the kinetic energy: $U_{12} = mv_2^2/2 - mv_1^2/2$:
$U_C - 24{,}000(10{,}000 - 5{,}000) = (24{,}000/32.2)\left[(600)^2 - (800)^2\right]/2$. Solving, $U_C = 1.57\times10^7\ ft-lb$.

==================================<>==================================

Problem 4.43 If the airplane in Problem 4.42 is moving at *600 ft/s* when it starts its climb at *5000 ft*, and the work done by the thrust and drag forces combined as it climbs to *10,000 ft* is $1.8\times10^8\ ft-lb$, what is the magnitude of the airplane's velocity when it reaches *10,000 ft* ?

Solution: $U_{12} = mv_2^2/2 - mv_1^2/2 = 1.8\times10^8 - 24{,}000(10{,}000-5{,}000) = \frac{(24{,}000/32.2)}{2}\left[v^2 - (600)^2\right]$.

Solving, we get $v = 722\ ft/s$

==================================<>==================================

==<>==

Problem 4.44 The 2400 *lb.* car is traveling 40 *mi/hr* at position 1. If the combined effect of the aerodynamic drag on the car and the tangential force exerted on the road by the wheels is that they exert no net tangential force on the car, what is its velocity at position 2?

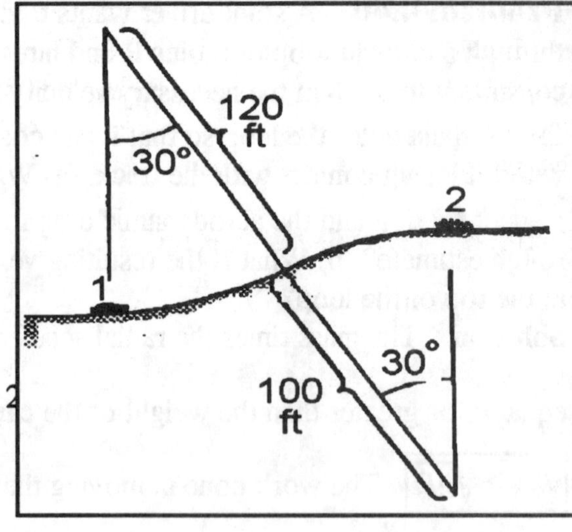

Solution

The initial velocity is $v_1 = 40\left(\dfrac{5280}{3600}\right) = 58.67\ ft/s$

The change in elevation of the car is

$h = 120(1-\cos 30^o) + 100\left(1-\cos 30^o\right) = 220\left(1-\cos 30^o\right) = 2$

The initial kinetic energy is $\dfrac{1}{2}\left(\dfrac{W}{g}\right)v_1^2 = 128{,}384\ ft\text{-}lb$

The work done by gravity is $U_{gravity} = \int_0^h (-W)ds = -Wh = -2400(h) = -70738.6\ ft\text{-}lb$.

From the principle of work and energy the work done is equal to the gain in kinetic energy:

$U_{gravity} = \dfrac{1}{2}\left(\dfrac{W}{g}\right)v_2^2 - \dfrac{1}{2}\left(\dfrac{W}{g}\right)v_1^2$, from which $v_2 = \sqrt{\dfrac{2g(-70{,}738.6 + 128{,}384.6)}{W}} = 39.3\ ft/s = 26.8\ mph$

==<>==

Problem 4.45 In Problem 4.44, if the combined effect of aerodynamic drag on the car and the tangential force exerted on the road by its wheels is that they exert a constant 400 *lb.* tangential force on the car in the direction of its motion, what is its velocity at position 2?

Solution

From the solution to Problem 4.44, the work done by gravity is $U_{gravity} = -70{,}738.6\ N\text{-}m$ due to the change in elevation of the car of $h = 29.47\ ft$, and $\dfrac{1}{2}\left(\dfrac{W}{g}\right)v_1^2 = 128{,}384\ ft\text{-}lb$.

The length of road between positions 1 and 2 is $s = 120\left(30^o\right)\left(\dfrac{\pi}{180^o}\right) + 100\left(30^o\right)\left(\dfrac{\pi}{180^o}\right) = 115.2\ ft$. The work done by the tangential force is $U_{tgt} = \int_0^s 400 ds = 400(115.2) = 46076.7\ ft\ lb$.

From the principle of work and energy $U_{gravity} + U_{tgt} = \dfrac{1}{2}\left(\dfrac{W}{g}\right)v_2^2 - \dfrac{1}{2}\left(\dfrac{W}{g}\right)v_1^2$, from which

$v = \sqrt{\dfrac{2(32.17)(-70{,}738.6 + 46{,}076.7 + 128{,}384.6)}{2400}} = 52.73\ ft/s = 36\ mph$

==<>==

==================================<>==================================

Problem 4.46 The mass of the rocket is 250 *kg*, and it has a constant thrust of 6000 *N*. The total length of the launching ramp is 10 *m*. Neglecting friction, drag, and the change in mass of the rocket, determine the magnitude of its velocity when it reaches the end of the ramp.

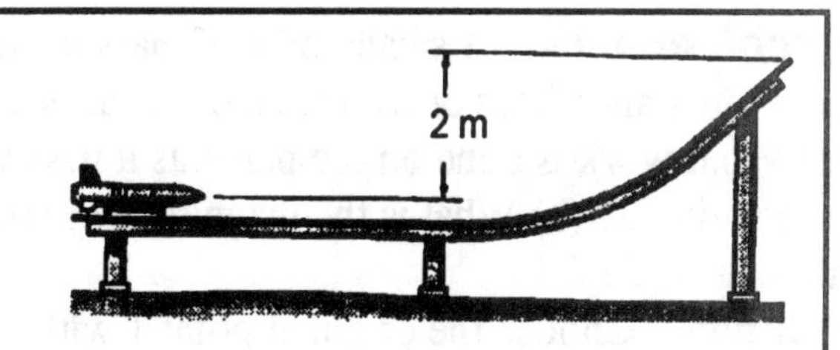

Solution The thrust is tangential to the path of the rocket so that the work done is $U_{thrust} = \int_0^{10} 6000 ds = 60{,}000 \ N\text{-}m$. The work done by gravity on the weight is $U_{weight} = \int_0^h (-mg)ds = -mgh = -(250)(9.18)(2) = -4590 \ N\text{-}m$. From the principle of work and energy $U_{thrust} + U_{weight} = \frac{1}{2}mv^2$, from which $\boxed{v = \sqrt{\frac{2(6000-4590)}{250}} = 21.05 \ m/s}$

==================================<>==================================

Problem 4.47 A bioengineer interested in energy requirements of sports determines from videotape that when the athlete begins his motion to throw the *7.25-kg* shot (Fig. a), the shot is stationary and *1.50 m* above the ground. At the instant he releases it, (Fig. B), the shot is *2.10-m* above the ground. The shot reaches a maximum height of *4.60-m* above the ground and travels a horizontal distance of *18.66-m* from the point where it was released. How much work does the athlete do on the shot from the beginning of his motion to the instant he releases it?

Solution: Let v_{xo} and v_{yo} be the velocity components at the instant of release. Using the chain rule,

$$a_y = \frac{dv_y}{dt} = \frac{dv_y}{dy}\frac{dy}{dt} = \frac{dv_y}{dy}v_y = -g, \text{ and}$$

integrating, $\int_{v_{yo}}^{o} v_y dv_y = -g\int_{2.1}^{4.6} dy$.

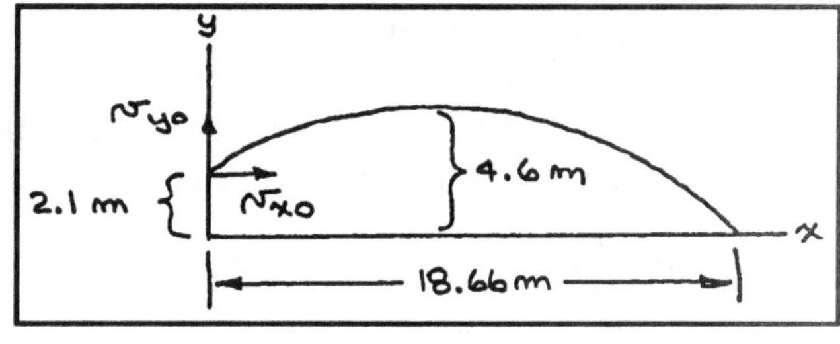

$-\frac{1}{2}v_{yo}^2 = -g(4.6-2.1)$, we find that

$v_{yo} = 7.00 m/s$. The shot's x and y coordinates are given by $x = v_{xo}t \quad y = 2.1 + v_{yo}t - \frac{1}{2}gt^2$

Solving the first equation for t and substituting it into the second, $y = 2.1 + v_{yo}\left(\frac{x}{v_{xo}}\right) - \frac{1}{2}g\left(\frac{x}{v_{xo}}\right)^2$

Setting $x = 18.66m, y = 0$ in this equation and solving for v_{xo} gives $v_{xo} = 11.1 m/s$. The magnitude of the shot's velocity at release is $U_2 = \sqrt{v_{xo}^2 + v_{yo}^2} = 13.1 m/s$

Let U_A be the work he does $U_A - mg(y_2 - y_1) = \frac{1}{2}mv_2^2 - \frac{1}{2}mv_1^2$

$U_A - mg(7.25)(9.81)(2.1-1.5) = \frac{1}{2}(7.25)(13.1)^2 - 0$, or $U_A = 666 N-m$

==================================<>==================================

==================================<>==================================

Problem 4.48 A small pellet of mass m starts from rest at position 1 and slides down the smooth surface of the cylinder. (a) What work is done on the pellet as it moves from position 1 to position 2? (b) What is the magnitude of the pellet's velocity at position 2?

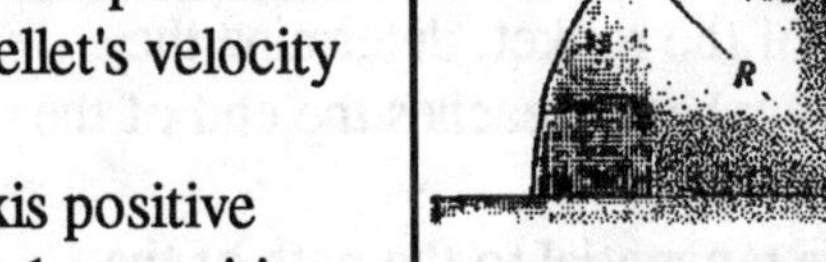

Solution Choose the origin at point 1 with y axis positive upward. (a) The change in height h from position 1 to position 2 is $h = -(R - R\cos\alpha) = -R(1-\cos\alpha)$, from which $U_{gravity} = \int_0^h (-mgh)ds = mgR(1-\cos\alpha)$. (b) From the principle of work and energy as the pellet slides down $U_{gravity} = \frac{1}{2}mv^2$, from which

$$\boxed{v = \sqrt{2gR(1-\cos\alpha)}}$$

==================================<>==================================

Problem 4.49 In Problem 4.48, what is the angle at which the pellet leaves the surface of the cylinder?

Solution The radial component of the weight of the pellet is $w_r = mg\cos\alpha$. From Newton's second law for the pellet at impending loss of contact with the cylinder, $m\frac{v^2}{R} = mg\cos\alpha$. From the solution to Problem 4.43, $v^2 = 2gR(1-\cos\alpha)$, from which $+2g(1-\cos\alpha) = g\cos\alpha$. Solve: $\cos\alpha = \frac{2}{3}$, from which

$\boxed{\alpha = 48.2^o}$.

==================================<>==================================

Problem 4.50 Suppose that you want to design a "bumper" that will bring a 50 *lb.* package moving at 10 *ft/s* to rest in 6 *in.* from the point of contact. If friction is negligible, what is the necessary spring constant k?

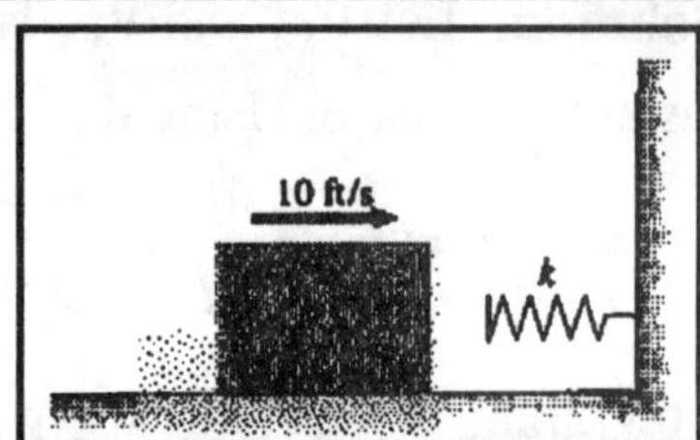

Solution From the principle of work and energy, the work done on the spring must equal the change in kinetic energy of the package within the distance 6 *in.* $\frac{1}{2}kS^2 = \frac{1}{2}\left(\frac{W}{g}\right)v^2$ from which

$$\boxed{k = \left(\frac{W}{g}\right)\left(\frac{v}{S}\right)^2 = \left(\frac{50}{32.17}\right)\left(\frac{10}{0.5}\right)^2 = 621.7 \ lb/ft}$$

==================================<>==================================

Problem 4.51 In Problem 4.50, what spring constant is necessary if the coefficient of friction between the package and the floor is $\mu_k = 0.3$ and the package contacts the spring moving at 10 *ft/s*?

Solution The work done on the spring over the stopping distance is $U_S = \int_0^S F ds = \int_0^S ks\, ds = \frac{1}{2}kS^2$.

The work done by friction over the stopping distance is $U_f = \int_0^S F\, ds = \int_0^S \mu_k W ds = \mu_k WS$. From the principle of work and energy the work done must equal the kinetic energy of the package: .

$\frac{1}{2}kS^2 + \mu_k WS = \frac{1}{2}\left(\frac{W}{g}\right)v^2$, from which, for $S = \frac{1}{2}\ ft$, $\boxed{k = \left(\frac{W}{g}\right)\frac{(v^2 - 2g\mu_k S)}{S^2} = 561.7 \ lb/ft}$ *Note:*

1==================================<>==================================

1==================================<>==================================

Problem 4.52 An astronaut in an excursion module approaches a space station docking collar. The designer of the collar incorporated a spring to attenuate the shock due to docking. The spring constant is $k = 4800 N/m$. The mass of the astronaut and module is *780-kg*. If the module contacts the docking collar moving at *0.1 m/s* relative to the collar, what distance is required for the spring to decrease its relative velocity to zero? What is the module's maximum relative deceleration?

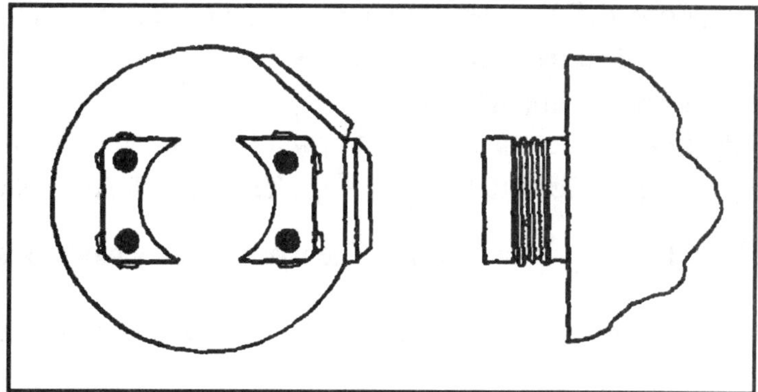

Solution: Assume that the spring is initially unstretched. $U_{12} = -\frac{1}{2}k\left(S_2^2 - S_1^2\right) = \frac{1}{2}mv_2^{\ 2} - \frac{1}{2}mv_1^{\ 2}$: or

$$-\frac{1}{2}(4800)\left(S_2^2 - 0\right) = 0 - \frac{1}{2}(780)(0.1)^2$$

Solving, $S_2 = 0.0403 m$.

The maximum deceleration is $a = \dfrac{-kS_2}{m} = \dfrac{(4800)(0.0403)}{780} = -0.248 m/s^2$.

1==================================<>==================================

Problem 4.53 In Problem 4.52, suppose that you design the docking collar so that a *10,000-kg* vehicle moving at *0.2 m/s* relative to the collar will be brought to rest in a distance of *0.15 m*. If the module described in Problem 4.52 contacts the docking collar moving at *0.1 m/s*, what is its maximum relative acceleration?

Solution: First determine the spring constant. $U_{12} = -\frac{1}{2}k\left(S_2^2 - S_1^2\right) = \frac{1}{2}mv_2^{\ 2} - \frac{1}{2}mv_1^{\ 2}$:

$$-\frac{1}{2}k\left[(0.15)^2 - 0\right] = 0 - \frac{1}{2}(10,000)(0.2)^2 \quad k = 17,800 N/m$$

Now determine the distance required to stop the *789-kg* vehicle.

$$-\frac{1}{2}(17,800)\left(S_2^2 - 0\right) = 0 - \frac{1}{2}(780)(0.1)^2 \, S_2 = 0.0209 m$$

The maximum deceleration is $a = \dfrac{-kS_2}{m} = \dfrac{(17,800)(0.0209)}{780} = -0.477 m/s^2$

1==================================<>==================================

Problem 4.54 The system is released from rest with the spring unstretched. If the spring constant is $k = 30\ lb/ft$, what maximum velocity do the weights attain?

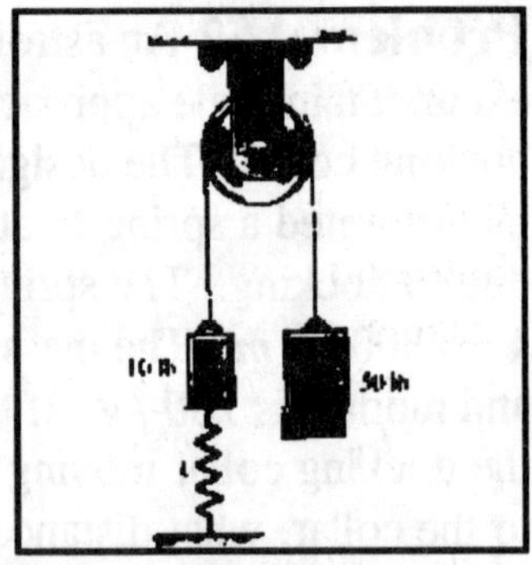

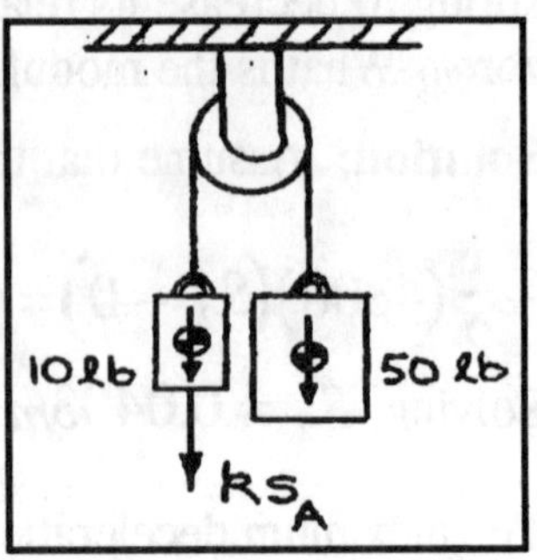

Solution Choose s_A positive upward. Since the pulley is one-to-one, $s_A = -s_B$. Differentiate: $v_A = -v_B$. The work done by block A is

$U_A = \int_0^{s_A}(-W_A)ds = -W_A s_A$. The work done by block B is

$U_B = \int_0^{s_B}(-W_B)ds = -W_B s_B = W_B s_A$. The work done *by the spring* is

$U_S = \int_0^S F\,ds = \int_0^S -ks\,ds = -\frac{1}{2}kS^2 = -\frac{1}{2}ks_A^2$, where the stretch of the spring is $S = s_A$. From the principle of work and energy

$(W_B - W_A)s_A - \frac{1}{2}ks_A^2 = \frac{1}{2}\left(\frac{W_A}{g}\right)v_A^2 + \frac{1}{2}\left(\frac{W_B}{g}\right)v_B^2$, from which

$v_A^2 = \frac{2g(W_B - W_A)s_A}{(W_B + W_A)} - \frac{gks_A^2}{(W_A + W_B)}$. The maximum occurs at

$\frac{d(v_A^2)}{ds_A} = \frac{2g(W_B - W_A)}{(W_B + W_A)} - \frac{2gks_A}{(W_B + W_A)} = 0$. Solve: $s_A = \frac{(W_B - W_A)}{k} = \frac{40}{30} = 1.3333\ ft$, from which

$\boxed{|v_{max}| = 5.35\ ft/s}$. This is indeed a maximum, since $\frac{d^2(v_A^2)}{ds_A^2} = -\frac{2gk}{(W_B + W_A)} < 0$

Problem 4.55 Suppose that you don't know the spring constant *k* of the system in Problem 4.54. If you release the system from rest with the spring unstretched until you observe that the 50 *lb.* falls 2 *ft* before rebounding, what is k?

Solution The stretch of the spring is $S = s_A$. From the solution to Problem 4.20,

$v_A^2 = \frac{2g(W_B - W_A)s_A}{(W_B + W_A)} - \frac{gks_A^2}{(W_A + W_B)}$. At the point of rebound of block B, the velocity is zero, from which $0 = s_A(2(W_B - W_A) - ks_A)$, and for $s_A = 2\ ft$ the spring constant is

$$\boxed{k = \frac{2(W_B - W_A)}{s_A} = \frac{2(40)}{2} = 40\ lb/ft}$$

=================================<>=================================

Problem 4.56 In Example 4.5, suppose that the unstretched length of each spring is 200 *mm* and you want to design the device so that the hammer strikes the workpiece at 5 *m/s*. Determine the necessary spring constant k.

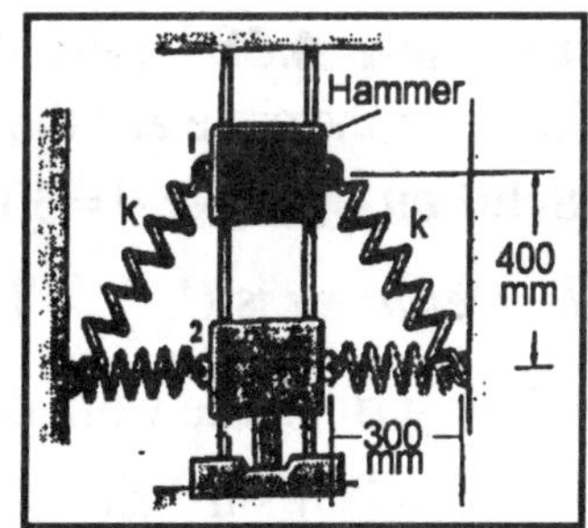

Solution From Example 4.5, the mass of the hammer is 40 *kg*. The distance in position 2 is 0.3 *m*. The stretched length of the spring is $S_2 = 0.3 - 0.2 = 0.1\ \text{m}$. In position 1 the stretched length of the spring is $S_1 = \sqrt{0.4^2 + 0.3^2} - 0.2 = 0.3\ \text{m}$. The work done by the springs from position 1 to position 2 is $U_{springs} = \int_1^2 F ds = 2\left(\frac{1}{2}kS_1^2 - \frac{1}{2}kS_2^2\right) = k(0.3^2 - 0.1^2) = 0.08k$. The work done by the weight is $U_{weight} = mg(0.4) = 157\ \text{N-m}$. The work done is equal to the gain in kinetic energy: $U_{springs} + U_{weight} = 0.08k + 157 = \frac{1}{2}mv^2$. Solve for the spring constant when :$v = 5\ \text{m/s}$,

$$k = \frac{\frac{1}{2}(40)(5^2) - 157}{0.08} = 4288\ \text{N/m}$$

=================================<>=================================

Problem 4.57 The 20 *kg* crate is released from rest with the spring unstretched. The spring constant is $k = 100\ \text{N/m}$. Neglect friction. (a) How far down the inclined surface does the crate slide before it stops? (b) What maximum velocity does it attain on the way down?

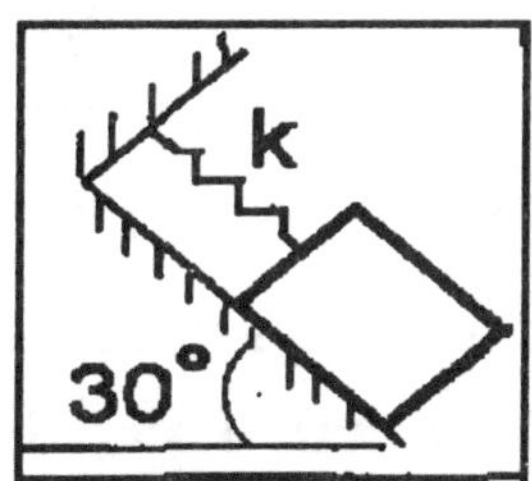

Solution Denote $\theta = 30^\circ$. Choose the *x* axis parallel to the incline, with the starting point $x = 0$. The work done by the spring as the crate slides down the incline is $U_{spring} = \int_0^S F\,ds = \int_0^S -ks\,ds = -\frac{1}{2}kS^2$. The stretch of the spring is $S = x$.

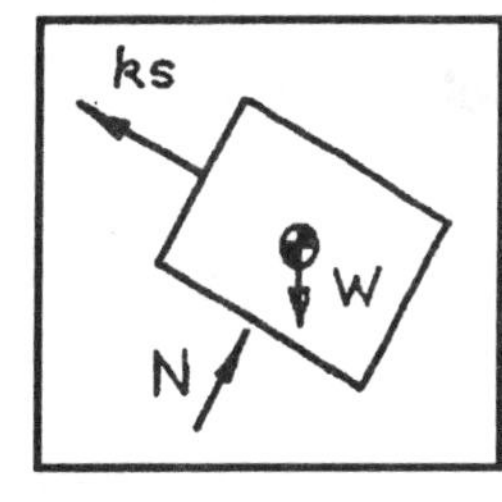

The work done by the weight is $U_{weight} = \int_0^x F\,ds = \int_0^x (W\sin\theta)ds = mgx\sin\theta$. (a) From the principle of work and energy, $U_{spring} + U_{weight} = \frac{1}{2}mv^2$, from which $mv^2 + kx^2 = 2mgx\sin\theta$. When the crate comes to rest $v = 0$, from which

$$x_{\max} = \frac{2mg}{k}\sin\theta = 1.96\ \text{m}.$$

(b) The max velocity occurs when $\frac{d(v^2)}{dx} = 2mg\sin\theta - 2kx = 0$, from which $x = \frac{mg\sin\theta}{k} = 0.981\ \text{m}$. The maximum velocity: $v(x = 0.981) = \sqrt{\frac{2mgx\sin\theta - kx^2}{m}}\Big|_{x=0.981} = 2.19\ \text{m/s}$

=================================<>=================================

===================================<>===================================

Problem 4.58 Solve Problem 4.57 if the coefficient of kinetic friction between the crate and the surface is $\mu_k = 0.12$.

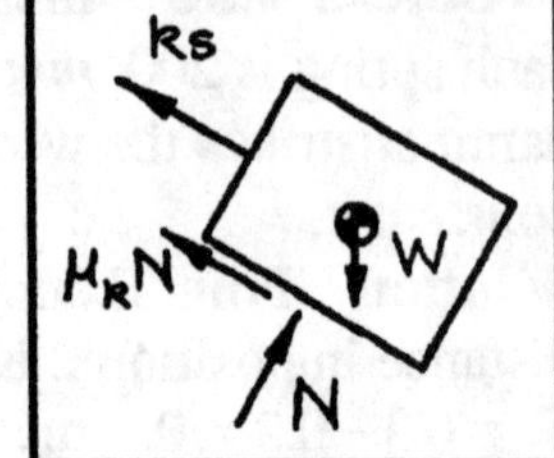

Solution Denote $\theta = 30^o$. From the solution to Problem 4.57, the work done *by the spring* is $U_{spring} = \int_0^S F\,ds = \int_0^S -ks\,ds = -\frac{1}{2}kS^2$, where $S = x$ is the stretch of the spring. The work done by the weight is $U_{weight} = \int_0^x (W\sin\theta)dx = mgx\sin\theta$. The work done by friction is $U_{friction} = \int_0^x f\,ds = \int_0^x -\mu_k N dx = -\mu_k mgx\cos\theta$.

(a) The total work done is $-\frac{1}{2}kx^2 - \mu_k mgx\cos\theta + mgx\sin\theta = \frac{1}{2}mv^2$, from which $mv^2 + kx^2 = 2mgx(\sin\theta - \mu_k\cos\theta)$. When the box comes to rest, $v = 0$,

$$x_{max} = \frac{2mg(\sin\theta - \mu_k\cos\theta)}{k} = 1.55 \text{ m}$$

(b) The maximum velocity occurs when $\frac{dv^2}{dx} = 0 = 2g(\sin\theta - \mu_k\cos\theta) - 2\left(\frac{k}{m}\right)x = 0$, from which $x = \left(\frac{m}{k}\right)(g(\sin\theta - \mu_k\cos\theta)) = 0.777$ m. The maximum velocity is

$$v(x = 0.777) = \left[\sqrt{2gx(\sin\theta - \mu_k\cos\theta) - \left(\frac{k}{m}\right)x^2}\right]_{x=0.777} = 1.738 \text{ m/s}$$

===================================<>===================================

Problem 4.59 Solve Problem 4.57 if the coefficient of kinetic friction between the crate and the surface is $\mu_k = 0.16$ and the tension in the spring when the crate is released is 20 *N*.

Solution Denote $\theta = 30^o$. The initial stretch of the spring is $S_o = \frac{F_o}{k} = 0.2$ m. The work done by the spring is $U_{spring} = U(S) + U(S_o) = -\frac{1}{2}k(S + S_o)^2 + \frac{1}{2}kS_o^2$. From the solution to Problem 4.58 the work done by the weight is $U_{weight} = mgx\sin\theta$, and the work done by friction is $U_{friction} = -\mu_k mg\cos\theta$. From the principle of work and energy, $U_{spring} + U_{weight} + U_{friction} = \frac{1}{2}mv^2$. Note from Problem 4.57 $x = S$, $x_o = S_o$, from which $-\frac{1}{2}k(x + x_o)^2 + \frac{1}{2}kx_o^2 + mgx(\sin\theta - \mu_k\cos\theta) = \frac{1}{2}mv^2$.

(a) When the crate comes to rest, $v = 0$, from which $k(x + x_o)^2 = 2mgx(\sin\theta - \mu_k\cos\theta) + kx_o^2$, from which $x_{max} = 2\left(\frac{m}{k}\right)g(\sin\theta - \mu_k\cos\theta) - 2x_o = 1.018$ m

(b) The maximum velocity occurs when $\frac{dv^2}{dx} = 0 = 2g(\sin\theta - \mu_k\cos\theta) - 2\left(\frac{k}{m}\right)(x + x_o)$, from which $x = \left(\frac{m}{k}\right)g(\sin\theta - \mu_k\cos\theta) - x_o = 0.509$ m. The maximum velocity:

$$v(x = 0.509) = \left[\sqrt{2gx(\sin\theta - \mu_k\cos\theta) - \left(\frac{k}{m}\right)(x^2 + 2xx_o)}\right]_{x=0.509} = 1.138 \text{ m/s}$$

===================================<>===================================

================================<>================================

Problem 4.60 The 20 *lb.* collar starts from rest at position 1 with the spring unstretched. The spring constant is $k = 40$ lb / ft. Neglect friction. How far does the collar fall relative to position 1?

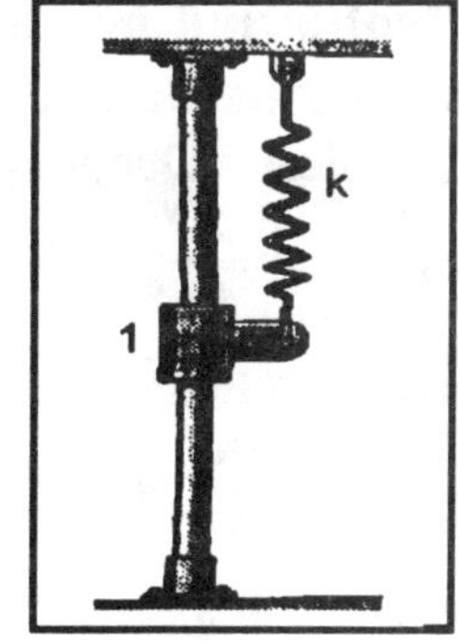

Solution The work done by gravity is $U_{gravity} = \int_0^{-x}(-W)ds = Wx$. The work done by the spring is $U_{spring} = \int_0^{-S}(ks)ds = -\frac{1}{2}kS^2$. From the principle of work and energy, $Wx - \frac{1}{2}kS^2 = \frac{1}{2}\left(\frac{W}{g}\right)v^2$, where the spring stretch is $S = x$. At the maximum excursion, $v = 0$, and $\frac{1}{2}kx^2 = Wx$, from which

$x = 2\left(\frac{W}{k}\right) = 2\left(\frac{20}{40}\right) = 1$ ft.

================================<>================================

Problem 4.61 In Problem 4.60, what maximum velocity does the collar attain?

Solution From the solution to Problem 4.46 $Wx - \frac{1}{2}kS^2 = \frac{1}{2}\left(\frac{W}{g}\right)v^2$, where the spring stretch $S = x$.

From which $v^2 = 2gx - \frac{gk}{W}x^2$. The maximum occurs at $\frac{dv^2}{dx} = 0 = 2g - 2\frac{gk}{W}x$, from which $x = \frac{W}{k}$. The maximum velocity is $v = \sqrt{\frac{gW}{k}} = 4.01$ ft / s

================================<>================================

Problem 4.62 What is the solution of Problem 4.60 if the tension in the spring in position 1 is 4 *lb.*?

Solution The work done by the spring from position 1 to position 2 is $U_{spring} = \int_0^{S}(-T - ks)ds = -TS - \frac{1}{2}kS^2$. The work done by gravity is $U_{gravity} = \int_0^{-x}(-W)ds = Wx$. From the principle of work and energy $U_{spring} + U_{gravity} = \frac{1}{2}\left(\frac{W}{g}\right)v^2$. The stretch of the spring is $S = x$, and the velocity is zero at the maximum excursion, from which $Wx - Tx - \frac{1}{2}kx^2 = 0$. Solve: $x = 0.8$ ft

================================<>================================

Problem 4.63 The 4 *kg* collar is released from rest at position 1. Neglect friction. If the spring constant is $k = 6$ kN / m and the spring is unstretched in position 2, what is the velocity of the collar when it has fallen to position 2?

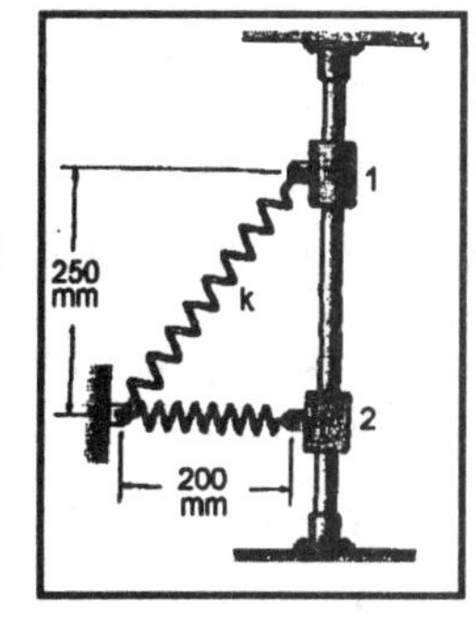

Solution Denote $d = 200$ mm, $h = 250$ mm. The stretch of the spring in position 1 is $S_1 = \sqrt{h^2 + d^2} - d = 0.120$ m and at 2 $S_2 = 0$. The work done by the spring on the collar is $U_{spring} = \int_{0.12}^{0}(-ks)ds = \left[-\frac{1}{2}ks^2\right]_{0.120}^{0} = 43.31$ N - m. The work done by gravity is $U_{gravity} = \int_0^{-h}(-mg)ds = mgh = 9.81$ N - m.

From the principle of work and energy $U_{spring} + U_{gravity} = \frac{1}{2}mv^2$, from which

$v = \sqrt{\left(\frac{2}{m}\right)(V_{spring} + U_{gravity})} = 5.15$ m / s

================================<>================================

===================================<>===================================

Problem 4.64 In Problem 4.63, if the spring constant is $k = 4$ kN/m and the tension in the spring in position 2 is 500 *N*, what is the velocity of the collar when it has fallen to position 2?

Solution Denote $d = 200$ mm, $h = 250$ mm. The stretch of the spring at position 2 is $S_2 = \frac{T}{k} = \frac{500}{4000} = 0.125$ m. The unstretched length of the spring is $L = d - S_2 = 200 - 0.125 = 0.075$ m. The stretch of the spring at position 1 is $S_1 = \sqrt{h^2 + d^2} - L = 0.245$ m. The work done by the spring is $U_{spring} = \int_{S_1}^{S_2} (-ks)ds = \frac{1}{2}k(S_1^2 - S_2^2) = 88.95$ N-m. The work done by gravity is $U_{gravity} = mgh = 9.81$ N-m. From the principle of work and energy is $U_{spring} + U_{gravity} = \frac{1}{2}mv^2$, from which $v = \sqrt{\frac{2(U_{spring} + U_{gravity})}{m}} = 7.027$ m/s

===================================<>===================================

Problem 4.65 In Problem 4.63, suppose that you don't know the spring constant *k*. If the spring is unstretched in position 2 and the velocity of the collar when it has fallen to position 2 is 4 *m/s*, what is k?

Solution The kinetic energy at position 2 is $\frac{1}{2}mv^2 = 32$ N-m. From the solution to Problem 4.63, the stretch of the spring in position 1 is $S_1 = \sqrt{h^2 + d^2} - d = 0.120$ m. The potential of the spring is $U_{spring} = \int_{S_1}^{0} (-ks)ds = \frac{1}{2}kS_1^2$. The work done by gravity is $U_{gravity} = mgh = 9.81$ N-m. From the principle of work of work and energy, $U_{spring} + U_{gravity} = \frac{1}{2}mv^2$. Substitute and solve:

$$k = \frac{2(\frac{1}{2}mv^2 - U_{gravity})}{S_1^2} = 3074 \text{ N/m}$$

===================================<>===================================

Problem 4.66 The 10 *kg* collar starts from rest at position 1 and slides along the smooth bar. The y-axis points upward. The spring constant is $k = 100$ N/m and the unstretched length of the spring is 2 *m*. What is the velocity of the collar when it reaches position 2?

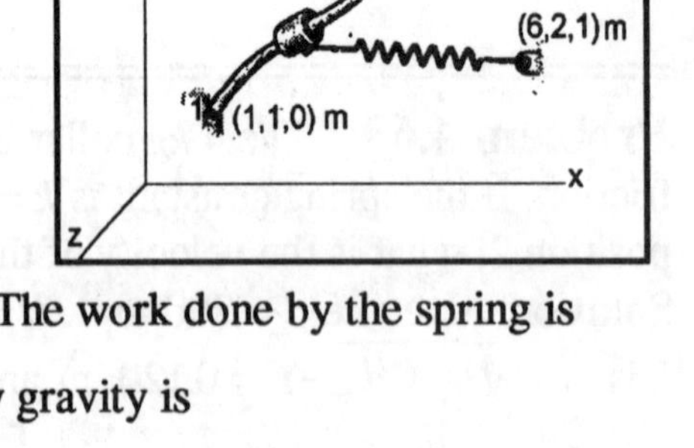

Solution The stretch of the spring at position 1 is $S_1 = \sqrt{(6-1)^2 + (2-1)^2 + (1-0)^2} - 2 = 3.2$ m. The stretch of the spring at position 2 is $S_2 = \sqrt{(6-4)^2 + (2-4)^2 + (1-2)^2} - 2 = 1$ m. The work done by the spring is $U_{spring} = \int_{S_1}^{S_2} (-ks)ds = \frac{1}{2}k(S_1^2 - S_2^2) = 460.8$ N-m. The work done by gravity is $U_{gravity} = \int_0^h (-mg)ds = -mgh = -(10)(9.81)(4-1) = -294.3$ N-m. From the principle of work and energy: $U_{spring} + U_{gravity} = \frac{1}{2}mv^2$, from which $v = \sqrt{\frac{2(U_{spring} + U_{gravity})}{m}} = 5.77$ m/s

===================================<>===================================

==================================<>==================================

Problem 4.67 A spring powered mortar is used to launch 10 *lb.* packages of fireworks into the a The package starts from rest with the spring compressed to a length of 6 *in.*; The unstretched length of the spring is 30 *in.* If the spring constant is $k = 1300\ lb/ft$, what is the magnitude of the velocity of the package as it leaves the mortar?

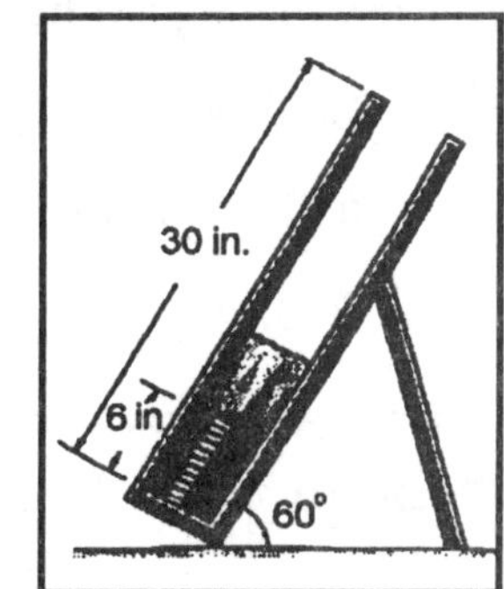

Solution The work done by the spring is

$U_{spring} = \int_0^{2.0}(ks)ds = \frac{1}{2}k2^2 = 2600\ ft\text{-}lb$. The work done by gravity is

$U_{gravity} = \int_0^{2}(-W\sin\theta) = -Wh\sin\theta = -10(2)(\sin 60^o) = -17.3\ ft\text{-}lb$. From the

principle of work and energy: $U_{spring} + U_{gravity} = \frac{1}{2}\frac{W}{g}v^2$, from which

$$v = \sqrt{\frac{2g(V_{spring} + U_{gravity})}{W}} = 128.9\ ft/s$$

==================================<>==================================

Problem 4.68 Suppose that you want to design the mortar in Problem 4.67 to throw the package to a height of 150 *ft* above its initial position. Neglecting friction and drag, determine the necessary spring constant.

Solution Divide the path of the package into (a) the path exterior to the barrel, (b) the path interior to the barrel. (a) *Exterior to the barrel.* The work done by gravity at $h = 150\ ft$ is

$U_{gravity} = \int_0^{150}(-W)ds = -Wh = -10(150) = -1500\ ft\ lb$. Denote the kinetic energy *due to the vertical component of the velocity* by $\frac{1}{2}\left(\frac{W}{g}\right)(v\sin\theta)^2$. From the principle of work and energy

$U_{gravity} = \left[\frac{1}{2}\left(\frac{W}{g}\right)(v\sin\theta)^2\right]_{h=150\ ft} - \left[\frac{1}{2}\left(\frac{W}{g}\right)(v\sin\theta)^2\right]_{muzzle}$. At 150 *ft* the vertical component of the velocity is zero, from which the muzzle velocity is $v = \sqrt{\frac{2gU_{gravity}}{W\sin^2\theta}} = \sqrt{\frac{2(32.17)(1500)}{10(\sin^2 60^o)}} = 113.4\ ft/s$

(b) *Interior to the barrel:* The spring stretch is $S = 2\ ft$. The work done by the spring is

$U_{spring} = \int_0^{2}(F)ds = \int_0^{2}(ks)\,ds = \frac{1}{2}k(2^2) = 2k$. The work done by gravity over the length of the barrel is

$U_{barrel} = \int_0^{2}(-W\sin\theta)ds = -2W\sin\theta$. From the principle of work and energy:

$U_{barrel} + U_{spring} = \left[\left(\frac{1}{2}\right)\left(\frac{W}{g}\right)(v)^2\right]_{muzzle}\ ft\text{-}lb$, from which

$U_{spring} = 2W\sin\theta + \left(\frac{1}{2}\frac{W}{g}v^2\right) = 2017.3\ ft\text{-}lb$. The spring constant is $k = \frac{2(2017.3)}{(2^2)} = 1008.7\ lb/ft$

==================================<>==================================

Problem 4.69 Suppose an object has a string or cable with constant tension T attached as shown. The force exerted on the object can be expressed in terms of polar coordinates as $\vec{F} = -T\vec{e}_r$. Show that the work done on the object as it moves along an arbitrary plane path from radial position r_1 to radial position r_2 is $U = -T(r_1 - r_2)$.

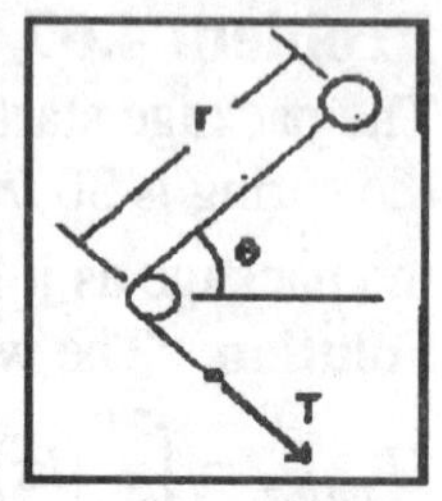

Solution The work done on the object is $U = \int_{r_1}^{r_2} \vec{F} \cdot d\vec{s}$. Suppose that the arbitrary path is defined by $d\vec{r} = (dr\vec{e}_r + rd\theta\vec{e}_\theta)$, and the work done is

$$U = \int_{r_1}^{r_2} \vec{F} \cdot d\vec{r} = \int_{r_1}^{r_2} -T(\vec{e}_r \cdot \vec{e}_r)dr + \int_{r_1}^{r_2} T(r\vec{e}_r \cdot \vec{e}_\theta)rd\theta = -\int_{r_1}^{r_2} Tdr = -T(r_2 - r_1) \text{ since } \vec{e}_r \cdot \vec{e}_\theta = 0 \text{ by definition.}$$

Problem 4.70 The 2 *kg* collar is initially at rest at position 1. A constant 100 *N* force is applied to the string, causing the collar to slide up the smooth vertical bar. What is the velocity of the collar when it reaches position 2?

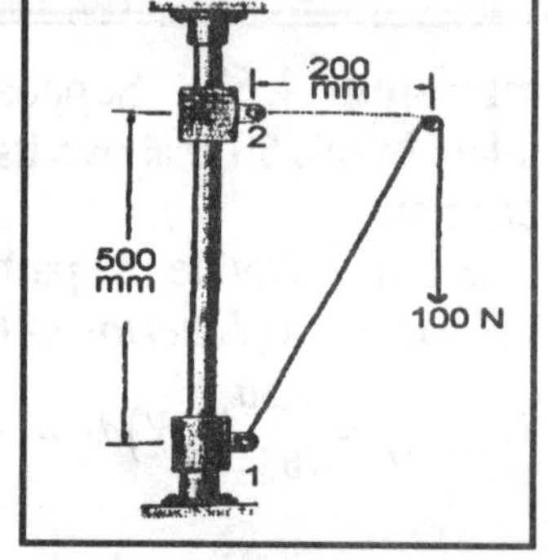

Solution The constant force on the end of the string acts through a distance $s = \sqrt{0.5^2 + 0.2^2} - 0.2 = 0.3385 \text{ m}$. The work done by the constant force is $U_F = Fs = 33.85 \text{ N-m}$. The work done by gravity on the collar is $U_{gravity} = \int_0^h (-mg)ds = -mgh = -(2)(9.81)(0.5) = -9.81 \text{ N-m}$. From the principle of work and energy:, $U_F + U_{gravity} = \frac{1}{2}mv^2$, from which

$$v = \sqrt{\frac{2(U_F + U_{gravity})}{m}} = 4.90 \text{ m/s}$$

Problem 4.71 The 10 *kg* collar starts from rest at position 1. The tension in the string is 200 *N*, and the *y* axis points upward. If friction is negligible, what is the magnitude of the collar's velocity when it reaches position 2?

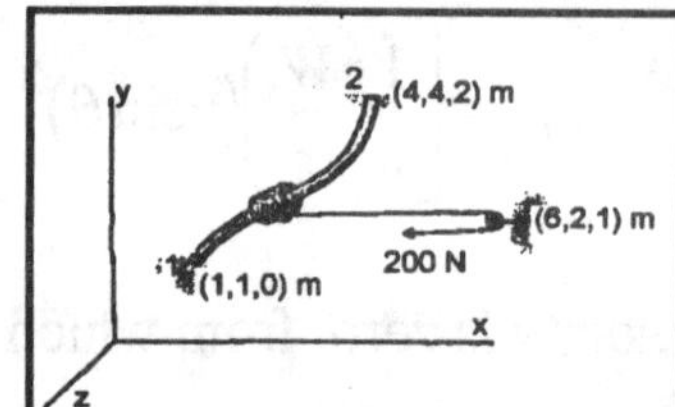

Solution

The constant force moves a distance

$$s = \sqrt{(6-1)^2 + (2-1)^2 + (1-0)^2} - \sqrt{(6-4)^2 + (2-4)^2 + (1-2)^2} = 2.2 \; m$$

The work done by the constant force is $U_F = \int_0^s Fds = Fs = 439.2 \; N\text{-}m$.

The work done by gravity is $U_{gravity} = \int_0^h (-mg)ds = -mgh = -(10)(9.81)(3) = -294.3 \; N\text{-}m$.

From the principle of work and energy $U_F + U_{gravity} = \frac{1}{2}mv^2$, from which

$$\boxed{v = \sqrt{\frac{2(U_F + U_{gravity})}{10}} = 5.38 \; m/s}$$

====================================<>====================================

Problem 4.72 The cable extending from A to B engages the arresting hook of the F/A-18 at C. The arresting mechanism maintains the tension in the cable at a constant value of *880-kN*, bringing the *11,800-kg* airplane to rest in a distance of *22-m*. What was the airplane's initial velocity?

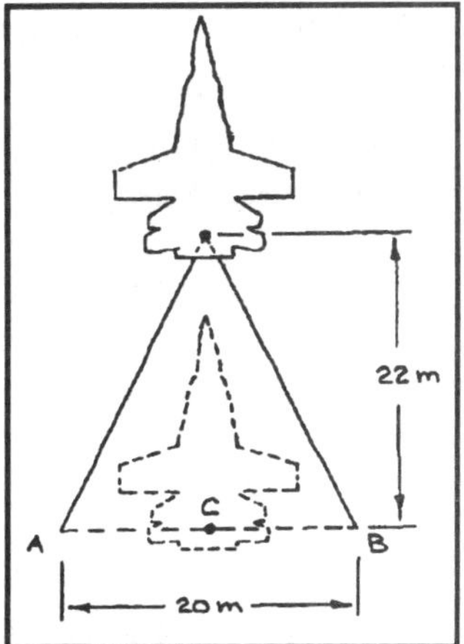

Solution:

The final length of each cable is $L_2 = \sqrt{(22)^2 + (10)^2} = 24.2m$ so the work done by each cable (see problem 4.69) is

$$-T(L_2 - L_1) = -(880{,}000)(24.2 - 10); \quad U_{12} = \frac{1}{2}mv_2^{\,2} - \frac{1}{2}mv_1^{\,2}:$$

$$-2(880{,}000)(24.2 - 10) = 0 - \frac{1}{2}(11{,}800)v_1^{\,2}; \quad v_1 = 65.0m/s$$

====================================<>====================================

Problem 4.73 In Problem 4.72, what is the airplane's maximum deceleration? If the *82-kg* pilot is subjected to the same deceleration, what is the maximum force exerted on him by his restraints?

Solution: The maximum deceleration occurs at the end of the motion, when the components of force exerted by the cables parallel to the plane's motion are greatest. Newton's second law for the plane at the end of the motion is $-2T\cos\beta = ma,$

where $\beta = \arctan(10/22) = 24.4^\circ$. Therefore $a = \dfrac{-2(880{,}000)\cos 24.4^\circ}{11{,}800} = -136m/s^2$

Newton's second law for the pilot is $F = ma = (82)(-136) = -11{,}100N$

====================================<>====================================

Problem 4.74 A spacecraft 200 *mi* above the earth has escape velocity $v_{esc} = \sqrt{\dfrac{2gR_E^2}{r}}$, where r is its distance from the center of the earth and $R_E = 3960\ mi$ is the radius of the earth. What is the magnitude of the spacecraft's velocity when it reaches the Moon's orbit 238,000 *mi* from the center of the earth?

Solution The escape velocity is $v_{esc} = \sqrt{2gR_E^2/r} = 35785.4\ \text{ft/s}$. The force on the satellite is $\vec{F} = -\dfrac{mgR_E^2}{r^2}\vec{e}_r$. The work done on the satellite as it moves outward is

$$U = \int_{r_1}^{r_2} F dr = -mg\int_{r_1}^{r_2}\frac{R_E^2}{r^2}dr = -\left[mgR_E^2\left(\frac{1}{r_2} - \frac{1}{r_1}\right)\right]$$

$$= -mgR_E^2\left(\frac{1}{238000} - \frac{1}{4160}\right)(5280) = 6.29m\times 10^8\ ft\text{-}lb.$$ From the principle of work and energy:

$U = \dfrac{1}{2}mv^2 - \dfrac{1}{2}mv_{esc}^2$, from which $v = \sqrt{v_{esc}^2 - \dfrac{2U}{m}}$, $\boxed{v = \sqrt{v_{esc}^2 - 2(6.29\times 10^8)} = 4731\ ft/s}$. *Check:* A satellite moving at escape velocity near the earth will be moving at escape velocity at all subsequent points in its path, in the absence of drag or propulsion. From the problem statement, $v_{esc} = \sqrt{\dfrac{2gR_E^2}{r}}$, from which the velocity of a satellite varies inversely as the radial distance from the center of the earth:

$$v = v_{esc}\sqrt{\frac{R_e + h}{R_M}} = v_{esc}\sqrt{\frac{4160}{238{,}000}} = 35{,}785(\sqrt{0.0175}) = 4731.1\ ft/s.\ check.$$

====================================<>====================================

================================<>================================

Problem 4.75 A piece of ejecta thrown up by the impact of a meteorite on the Moon has a velocity of 200 *m/s* magnitude relative to the center of the Moon when it is 1000 *km* above the Moon's surface. What is the magnitude of its velocity just before it strikes the Moon's surface? (The acceleration due to gravity at the Moon's surface is 1.62 m/s^2 and the Moon's radius is 1738 *km*.)

Solution The kinetic energy at $h = 1000\ km$ is $\left[\frac{m}{2}v^2\right]_{R_M+h} = 2m\times10^4\ N\text{-}m.$

The work done on the ejecta as it falls from 1000 *km* is

$$U_{ejecta} = \int_{R_M+h}^{R_M}(-W_{ejecta})ds = \int_{R_M+h}^{R_M}\left(-mg_M\frac{R_M^2}{s^2}ds\right) = \left[mg_M\frac{R_M^2}{s}\right]_{R_M+h}^{R_M} = mg_M R_M\frac{h}{R_M+h},$$

$U_{ejecta} = 1.028m\times10^6\ N\text{-}m$. From the principle of work and energy, at the Moon's surface:

$U_{ejecta} = \left[\frac{m}{2}v^2\right]_{surface} - \left[\frac{m}{2}v^2\right]_{R_M+h}$ from which $\boxed{v_{surface} = \sqrt{2\left(1.028\times10^6 + 2\times10^4\right)} = 1448\ m/s}$

================================<>================================

Problem 4.76 A satellite in a circular orbit of radius *r* around the earth has velocity $v = \sqrt{gR_E^2/r}$, where $R_E = 6370\ km$ is the radius of the earth. Suppose that you are designing a rocket to transfer a 900 *kg* communication satellite from a parking orbit with 6700 *km* radius to a geosynchronous orbit with 42,222 *km* radius. How much work must the rocket do on the satellite?

Solution Denote the work to be done by the rocket by U_{rocket}. Denote $R_{park} = 6700\ km$, $R_{geo} = 42222\ km$. The work done by the satellite's weight as it moves from the parking orbit to the geosynchronous orbit is

$$U_{transfer} = \int_{R_{park}}^{R_{geo}} Fds = \int_{R_{park}}^{R_{geo}}\left(-mg\frac{R_E^2}{s^2}ds\right) = \left[mg\frac{R_E^2}{s}\right]_{R_{park}}^{R_{geo}} = mgR_E^2\left(\frac{1}{R_{geo}} - \frac{1}{R_{park}}\right),$$

$U_{transfer} = -4.5\times10^9\ N\text{-}m$. From the principle of work and energy:

$U_{transfer} + U_{rocket} = \left[\frac{1}{2}mv^2\right]_{geo} - \left[\frac{1}{2}mv^2\right]_{park}$. from which

$U_{rocket} = \left[\frac{1}{2}mv^2\right]_{geo} - \left[\frac{1}{2}mv^2\right]_{park} - U_{transfer}$. Noting $\left[\frac{1}{2}mv^2\right]_{geo} = \frac{m}{2}\left(\frac{gR_E^2}{R_{geo}}\right) = 4.24\times10^9\ N\text{-}m$,

$\left[\frac{1}{2}mv^2\right]_{park} = \frac{m}{2}\left(g\frac{R_E^2}{R_{park}}\right) = 2.67\times10^{10}\ N\text{-}m$, from which $\boxed{U_{rocket} = 2.25\times10^{10}\ N\text{-}m}$

================================<>================================

Problem 4.77 The force exerted on a charged particle by a magnetic field is $\mathbf{F} = q\mathbf{v}\times\mathbf{B}$ where q and **v** are the charge and velocity of the particle and **B** is the magnetic field vector. If other forces on the particle are negligible, use the principle of work and energy to show that the magnitude of the particle's velocity is constant.

Solution: The force vector **F** is given by a cross product involving **v**. This means that the force vector is ALWAYS perpendicular to the velocity vector. Hence, the force field does no work on the charged particle – it only changes the direction of its motion. Hence, if work is zero, the change in kinetic energy is also zero and the velocity of the charged particle is constant.

================================<>================================

==================================<>==================================

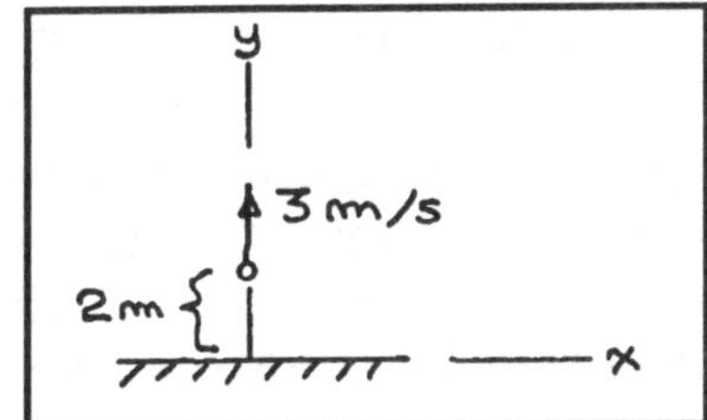

Problem 4.78 Suppose that you throw a ball straight up at *3 m/s*, releasing it *2-m* above the ground. (a) Use the principle of work and energy to determine the ball's velocity just before it hits the ground. (b) After confirming that the system is conservative, use conservation of energy to determine the ball's velocity just before it hits the ground.

Solution: (a) $U_{12} = \frac{1}{2}mv_2^{\;2} - \frac{1}{2}mv_1^{\;2}$:

$$-mg(y_2 - y_1) = \frac{1}{2}mv_2^{\;2} - \frac{1}{2}mv_1^{\;2};\quad -(9.81)(0-2) = \frac{1}{2}v_2^{\;2} - \frac{1}{2}(3)^2,\ v_2 = 6.95 m/s$$

(b) $V_1 + \frac{1}{2}mv_1^{\;2} = V_2 + \frac{1}{2}mv_2^{\;2}$; $mgy_1 + \frac{1}{2}mv_1^{\;2} = mgy_2 + \frac{1}{2}mv_2^{\;2}$,

$$(9.81)(2) + \frac{1}{2}(3)^2 = 0 + \frac{1}{2}v_2^{\;2},\ v_2 = 6.95 m/s$$

==================================<>==================================

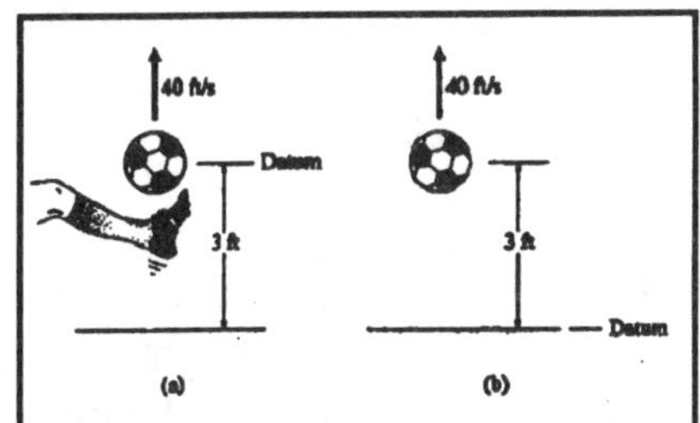

Problem 4.79 Suppose that you kick a soccer ball straight up. When it leaves your foot, it is 3 *ft* above the ground and moving at 40 *ft/s*. Neglecting drag, use conservation of energy to determine how high the ball goes and how fast it will be going just before it hits the ground. Obtain the answers by expressing the potential energy in terms of a datum (a) at the level of the ball's initial position and (b) at ground level.

Solution (a) Choose the datum at the height of the foot. The total energy at that point is $\frac{1}{2}mv_1^2$. At the maximum height, the velocity is zero, and the total energy is mgh_{max}. From conservation of energy $\frac{1}{2}mv_1^2 = mgh_{max}$, from which $h_{max} = \frac{v_1^2}{2g} = \frac{40^2}{2(32.17)} = 24.9\ ft$ is the height above the initial point, and $\boxed{h_{max} + 3 = 27.9\ ft}$ is the height above the ground. The conservation of energy condition when the ball hits the ground is $\frac{1}{2}mv_1^2 = \frac{1}{2}v_g^2 - mgh_1$, from which $\boxed{v_g = \sqrt{v_1^2 + 2gh_1} = 42.34\ ft/s}$. (b) Use the ground as a datum. The initial total energy is $\frac{1}{2}mv_1^2 + mgh_1$. At the maximum height the velocity is zero, and from conservation of energy, $\frac{1}{2}mv_1^2 + mgh_1 = mgh_{max}$, from which $\boxed{h_{max} = \frac{v_1^2}{2g} + h_1 = 27.9\ ft}$ above the ground. The total energy as it strikes the ground is $\frac{1}{2}mv_g^2$, from which $\frac{1}{2}mv_1^2 + mgh_1 = \frac{1}{2}mv_g^2$, and $\boxed{v_g = \sqrt{v_1^2 + 2gh_1} = 42.34\ ft/s}$

==================================<>==================================

====================================<>====================================

Problem 4.80 The lunar module could make a safe landing if its vertical velocity at impact was 5 *m/s* or less. Suppose that you want to determine the greatest height *h* at which the pilot could shut off the engine if the velocity of the lander relative to the surface was (a) zero; (b) 2 *m/s* downward, (c) 2 *m/s* upward. Use conservation of energy to determine *h* in each case. The acceleration due to gravity at the surface of the Moon is 1.62 m/s^2.

Solution (a) At the height *h* the total energy is $V_h = mgh$. At the surface the conservation of energy condition is $\frac{1}{2}mv_s^2 = V_h$, from which $\boxed{h = \frac{v_s^2}{2g} = \frac{(5^2)}{2(1.62)} = 7.72\ m}$. (b) At height *h* the total energy is $mgh + \frac{1}{2}mv_h^2$. At the surface the conservation of energy condition is $\frac{1}{2}mv_h^2 + mgh = \frac{1}{2}mv_s^2$. From which $\boxed{h = \frac{v_s^2}{2g} - \frac{v_h^2}{2g} = 6.48\ m}$ (c) The answer is the same as in Part (b); the direction of the velocity is not significant: it appears squared in the energy equations: a downward velocity increases the velocity of impact from a given height; and upward velocity temporarily increases the height, so that at impact the velocity is equivalent to the downward velocity case. The effect is the same.

====================================<>====================================

Problem 4.81 The *2-kg* collar starts from rest at position 1 and slides down the smooth rigid wire. The *y* axis points upward. Use conservation of energy to determine the magnitude of the collar's velocity as a function of its *y* coordinate.

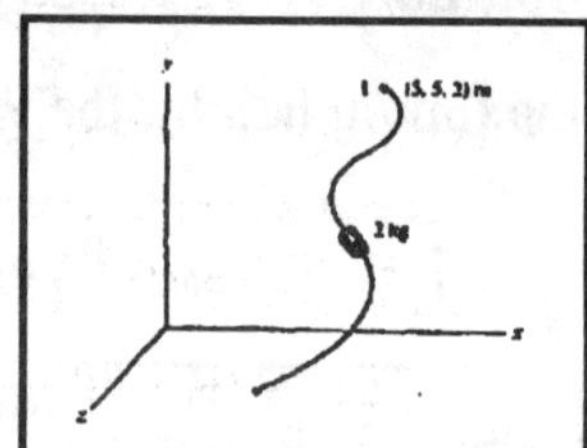

Solution: Conservation of energy gives: $V_1 + \frac{1}{2}mv_1{}^2 = V_2 + \frac{1}{2}mv_2{}^2$: or $mgy_1 + mv_1{}^2 = mgy + \frac{1}{2}mv^2$. Solving, $(9.81)(5) + 0 = 9.81y + \frac{1}{2}v^2$, This results in $v = \sqrt{2(9.81)(5-y)}$

====================================<>====================================

Problem 4.82 The spring constant $k = 40\ N/m$ and the mass $m = 12\ kg$. The surface is smooth. With the spring unstretched, the mass is given an initial velocity $v_1 = 2m/s$ to the right. The mass moves to the right, stretching the spring, until its velocity decreases to zero and it begins moving back toward its initial position. Determine the sum of the kinetic energy of the mass and the potential energy of the spring; (a) immediately after the mass is given its initial velocity; (b) at the instant the velocity of the mass has decreased to zero; (c) at the instant the mass returns to its original position.

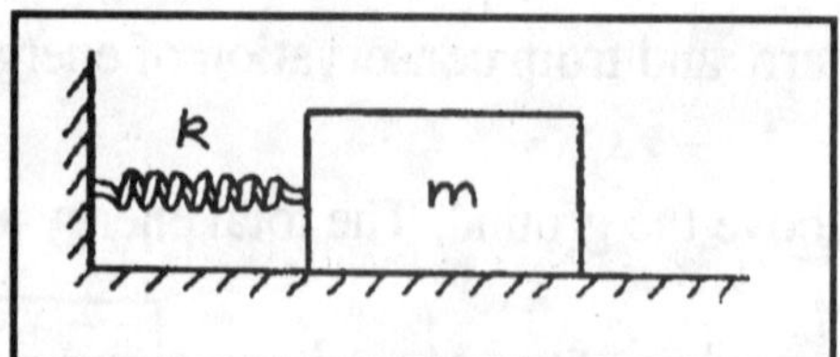

Solution: For part (a) $\frac{1}{2}mv_1{}^2 + \frac{1}{2}ks_1^2 = \frac{1}{2}(12)(2)^2 + 0 = 24N-m$.

The system is conservative, so the answers to (a), (b) and (c) are the same.

====================================<>====================================

==================================<>==================================

Problem 4.83 Solve Problem 4.82 if the coefficient of kinetic friction between the mass and surface is $\mu_k = 0.1$.

Solution: (a) $\frac{1}{2}mv_1^2 + \frac{1}{2}ks_1^2 = \frac{1}{2}(12)(2)^2 + 0 = 24 N-m$ (b) The system is not conservative. Let x be the distance the mass moves to the right. Applying work and energy,

$$-\frac{1}{2}k(s_2^2 - s_1^2) - \mu_k mgx = \frac{1}{2}mv_2^2 - \frac{1}{2}mv_1^2: \ -\frac{1}{2}(40)(x^2 - 0) - (0.1)(12)(9.81)x = 0 - \frac{1}{2}(12)(2)^2$$

Solving for , x, $x = s_2 = 0.840 m$. Therefore $\frac{1}{2}mv_2^2 + \frac{1}{2}ks_2^2 = 0 + \frac{1}{2}(40)(0.840)^2 = 14.11 N-m$ (c) Applying work and energy to the reverse motion,

$$-\frac{1}{2}k(s_3^2 - s_2^2) - \mu_k mgx = \frac{1}{2}mv_3^2 - \frac{1}{2}mv_2^2:$$

$$-\frac{1}{2}(40)\left[0 - (0.840)^2\right] - (0.1)(12)(9.81)(0.840) = \frac{1}{2}mv_3^2 - 0.$$

Since $s_3 = 0$, solving this equation gives $\frac{1}{2}mv_3^2 + \frac{1}{2}ks_3^2 = 4.22 N-m.$

==================================<>==================================

Problem 4.84 At the instant shown, the *50-lb* weight is moving downward at *4 ft/s*. Let *d* be the downward displacement of the *50-lb* weight relative to its present position. Use the conservation of energy to determine the magnitude of its velocity as a function of *d*.

Solution: Conservation of energy: $T_1 + V_1 = T_2 + V_2$:

$$\frac{1}{2}(50/32.2)(4)^2 + \frac{1}{2}(10/32.2)(4)^2 + 0 = \frac{1}{2}(50/32.2)v^2 + \frac{1}{2}(10/32.2)v^2 - 50d + 10d$$

Solving for v, $v = \sqrt{16 + 42.9d}$.

==================================<>==================================

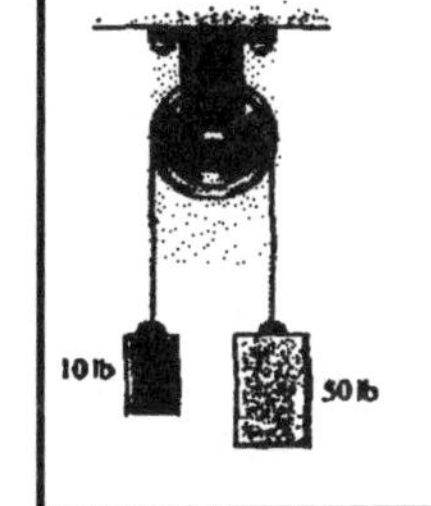

Problem 4.85 The ball is released from rest in position 1. (a) Use conservation of energy to determine the magnitude of its velocity at point 2. (b) Draw graphs of the potential energy, the kinetic energy, and the total energy for values of α from zero to 180°.

Solution The mass of the ball is *m*. Choose the datum point as the lowest point in the path, and the initial point to be position 1. The potential energy at point 1 is $V_1 = mgL$. The potential energy at point 2 is $V = mgL(1 - \sin\alpha)$. The energy condition is $\left(\frac{1}{2}mv^2\right) = V_1 - V = mgL - mgL(1 - \sin\alpha) = mgL\sin\alpha$.

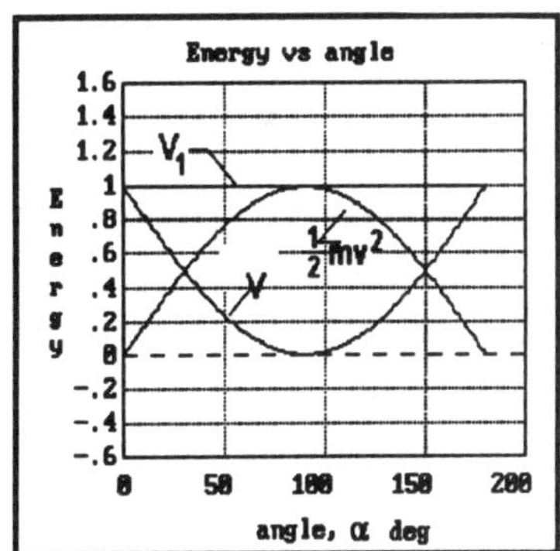

(a) The velocity at position 2 is $\boxed{v = \sqrt{\frac{2(V_1 - V_2)}{m}} = \sqrt{2gL\sin\alpha}}$ (b) The ratios of the energies $\frac{V_1}{mgL}$, $\frac{\left(\frac{1}{2}mv^2\right)}{mgL}$, and $\frac{V}{mgL}$ (equivalent to choosing $mgL = 1$) are graphed as a function of the angle α over the range $0 \le \alpha \le 180^o$.

==================================<>==================================

================================<>================================

Problem 4.86 If the ball is released from rest in position 1, use conservation of energy to determine the initial angle α necessary for it to swing to position 2.

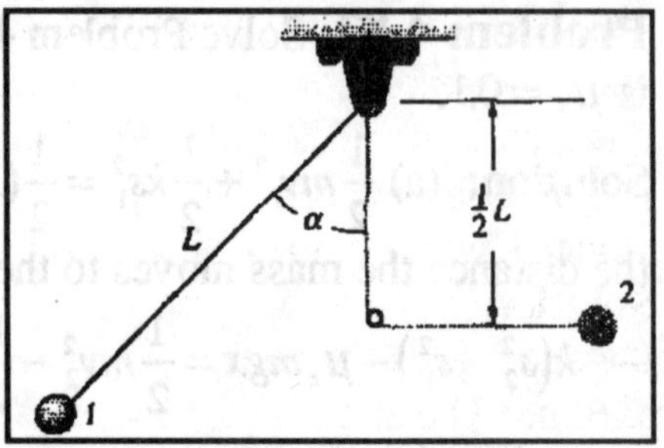

Solution: Choose the lowest point as the datum point. The potential energy in position 1 is $V_1 = mgh_1 = mgL(1-\cos\alpha)$. The potential energy in position 2 is $V_2 = mgh_2 = \frac{mgL}{2}$. From conservation of energy $V_1 = V_2$, and $\cos\alpha = \frac{1}{2}$, $\boxed{\alpha = 60^o}$

================================<>================================

Problem 4.87 The bar is smooth. Use conservation of energy to determine the minimum velocity the 10 *kg* slider must have at A to (a) to reach C, (b) to reach D.

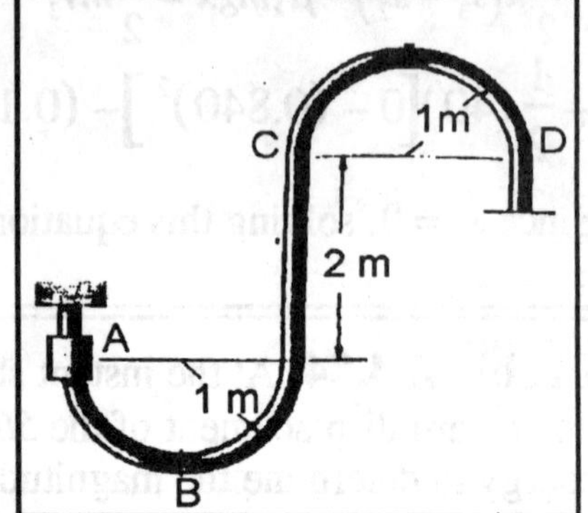

Solution: Choose A as the datum point. (a) Suppose that the slider just reaches C; the velocity is zero and the conservation of energy condition is $\frac{1}{2}mv_A^2 = 2mg$, from which $\boxed{v_A = \sqrt{2g(2)} = 2\sqrt{g} = 6.26 \ m/s}$. (b) To reach D, the slider must reach the top of the bar. Suppose that it reaches the top of the bar with zero velocity. The conservation of energy condition is $\frac{1}{2}mv_A^2 = mg(2+1)$. From which, $\boxed{v_A = \sqrt{2g(3)} = \sqrt{6g} = 7.67 \ m/s}$

================================<>================================

Problem 4.88 In Problem 4.87, what normal force does the bar exert on the slider at B in cases (a) and (b)?

Solution: For case (a), $v_A = 6.26 \ m/s$, and $\frac{1}{2}mv_A^2 = \frac{1}{2}mv_B^2 - mgh$., from which $v_B = \sqrt{v_A^2 + 2g} = 7.67 \ m/s$. From Newton's second law, $\frac{mv_B^2}{R} = F_N - mg$, from which $\boxed{F_N = \frac{mv_B^2}{R} + mg = 686.7 \ N}$, where the radius $R = 1 \ m$. For case (b) $v_A = 7.67 \ m/s$ and from the conservation of energy: $v_B = \sqrt{v_A^2 + 2g} = 8.86 \ m/s$. The normal force is $\boxed{F_N = \frac{mv_B^2}{R} + mg = 882.9 \ N}$

================================<>================================

==================================<>==================================

Problem 4.89 The 10-kg collar starts from rest at position 1 and slides along the bar. The y-axis points upward. The spring constant is $k = 100\ N/m$, and the unstretched length of the spring is 2 m. Use the conservation of energy to determine the collar's velocity when it reaches position 2.

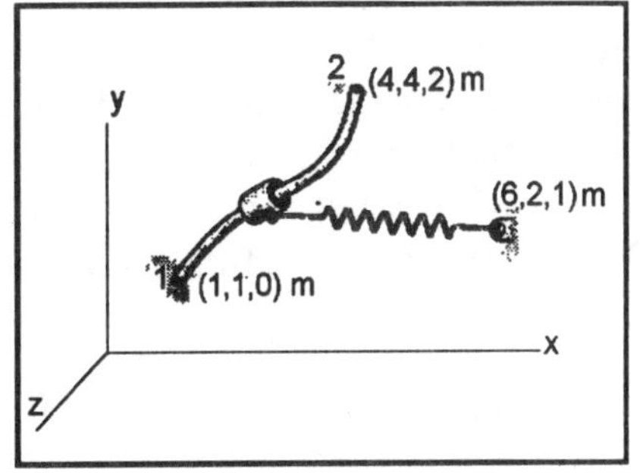

Solution The total energy at any position is the energy carried by the slider plus the energy in the spring. Choose position 1 as the datum. The stretch of the spring is $S_1 = \sqrt{(6-1)^2 + (2-1)^2 + (1-0)^2} - 2 = 3.2\ m$. The energy at position 1 is the potential energy in the spring is $V_1 = \frac{1}{2}kS_1^2 = 510.77\ N\text{-}m$. The stretch of the spring at position 2 is $S_2 = \sqrt{(6-4)^2 + (2-4)^2 + (1-2)^2} - 2 = 1\ m$. From the conservation of energy $\frac{1}{2}kS_1^2 = \frac{1}{2}mv_2^2 + (4-1)mg + \frac{1}{2}kS_2^2$. From which $\boxed{v = \sqrt{\frac{k}{m}(S_1^2 - S_2^2) - 6g} = 5.77\ m/s}$

==================================<>==================================

Problem 4.90 A rock climber of weight W has a rope attached a distance h below him for protection. Suppose that he falls, and assume that the rope behaves like a linear spring with unstretched length h and spring constant $k = \frac{C}{h}$, where C is a constant. Use conservation of energy to determine the maximum force exerted on him by the rope. (Notice that the maximum force is independent of h, which is a reassuring result for climbers--- the maximum force resulting from a long fall is the same as that resulting from a very short one.)

Solution Choose the climber's center of mass before the fall as the datum. The energy of the climber before the fall is zero. As the climber falls, his energy remains the same: $0 = \frac{1}{2}mv^2 - Wy$, where y is positive downward. As the rope tightens, the potential energy stored in the rope becomes $V_{rope} = \frac{1}{2}k(y-2h)^2$. At maximum extension the force on the climber is $F = -\frac{\partial V}{\partial y} = -k(y-2h)$. When the velocity of the falling climber is zero, $0 = -Wy + \frac{1}{2}k(y-2h)^2$, from which: $y^2 + 2by + c = 0$, where $b = -\left(2h + \frac{W}{k}\right)$, and $c = +4h^2$. The solution is $y = \left(2h + \frac{W}{k}\right) \pm \left(\frac{W}{k}\right)\sqrt{\frac{4kh}{W} + 1}$. Substitute: $F = -W\left(1 \pm \sqrt{1 + \frac{4C}{W}}\right)$. The positive sign applies, and the force is $\boxed{F = -W\left(1 + \sqrt{1 + \frac{4C}{W}}\right)}$ (directed upward).

==================================<>==================================

Problem 4.91 The spring constant $k = 700\ N/m$, $m_A = 14kg$, and $m_B = 18kg$. The collar A slides on the smooth horizontal bar. The system is released from rest with the spring unstretched. Use conservation of energy to determine the velocity of the collar A when it has moved *0.2-m* to the right.

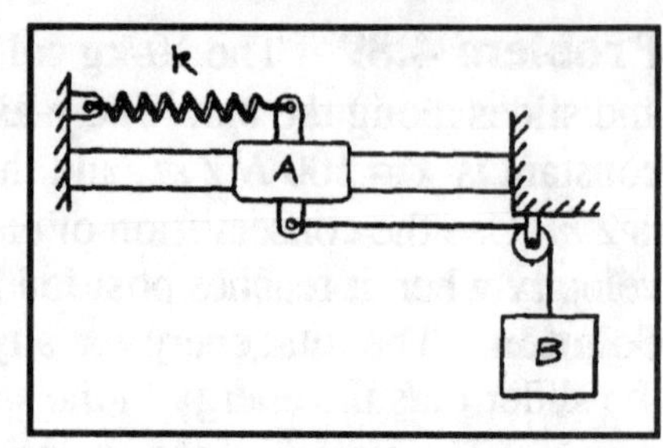

Solution: Let the datum for B be at its initial position

$$T_1 + V_1 = T_2 + V_2:$$

$$0+0=\frac{1}{2}(14)v^2+\frac{1}{2}(18)v^2+\frac{1}{2}(700)(0.2)^2-(18)(9.81)(0.2)$$

Solving, $v = 1.15 m/s$

Problem 4.92 The spring constant $k = 700\ N/m$, $m_A = 14kg$, and $m_B = 18kg$. The collar A slides on the smooth horizontal bar. The system is released from rest with the spring unstretched. Use conservation of energy to determine the velocity of the collar A when it has moved *0.2-m* to the right.

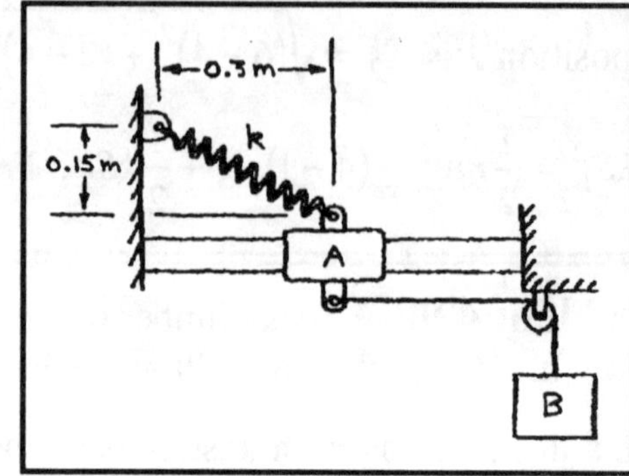

Solution: Let the datum for B be at its initial position. The stretch of the spring when A has moved $0.2m$ to the right is

$$S_2=\sqrt{(0.15)^2+(0.3+0.2)^2}-\sqrt{(0.15)^2+(0.3)^2}=0.187m$$

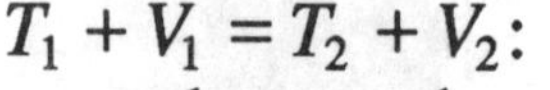

$$T_1 + V_1 = T_2 + V_2:$$

$$0+0=\frac{1}{2}(14)v^2+\frac{1}{2}(18)v^2+\frac{1}{2}(700)(0.187)^2-(18)(9.81)(0.2).$$ Solving, $v = 1.20 m/s$

Problem 4.93 The 5 *lb.* collar starts from rest at A and slides along the semicircular bar. The spring constant is $k = 100\ lb/ft$, and the unstretched length of the string is 1-ft. Use conservation of energy to determine the velocity of the collar at B.

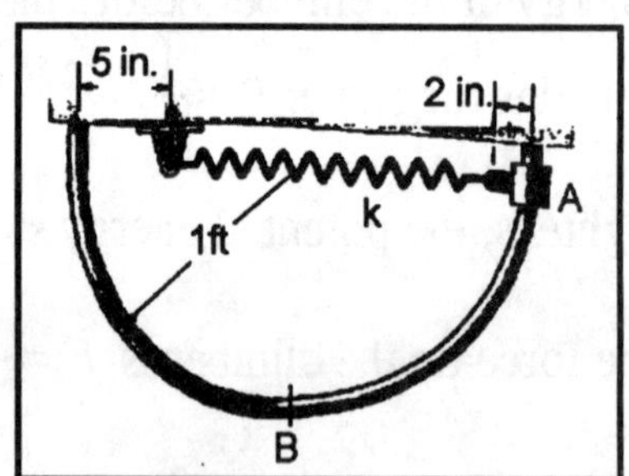

Solution Choose the point B as the datum. The stretch of the spring at A is $S_A = 24-5-2-12 = 5\ in. = \frac{5}{12}\ ft$. The stretch of the spring at B is $S_B = \sqrt{(12-5)^2+12^2}-12 = 1.89\ in. = 0.158\ ft$. The conservation of energy is $W(1)+\frac{1}{2}kS_A^2=\frac{1}{2}mv_B^2+\frac{1}{2}kS_B^2$, from which

$$v_B=\sqrt{2g(1)+\frac{gk}{W}\left(S_A^2-S_B^2\right)}=12.7\ ft/s$$

Problem 4.94 The mass $m = 1\text{-}kg$, the spring constant $k=200\ N/m$, and the unstretched length of the spring is $0.1\text{-}m$. When the system is released from rest in the position shown, the spring contracts, pulling the mass to the right. Use conservation of energy to determine the magnitude of the velocity of the mass when the string and spring are parallel.

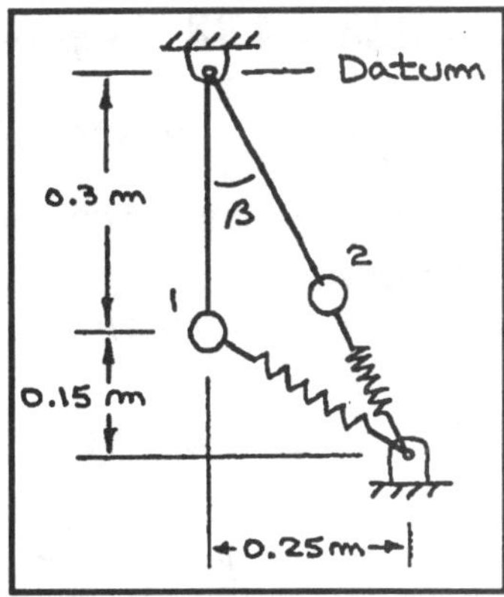

Solution: The stretch of the spring in position 1 is $S_1 = \sqrt{(0.15)^2 + (0.25)^2} - 0.1 = 0.192m$. The stretch in position 2 is $S_2 = \sqrt{(0.3+0.15)^2 + (0.25)^2} - 0.3 - 0.1 = 0.115m$. The angle $\beta = \arctan(0.25/0.45) = 29.1^\circ$. Applying conservation of energy,

$$\frac{1}{2}mv_1^2 + \frac{1}{2}kS_1^2 - mg(0.3) = \frac{1}{2}mv_2^2 + \frac{1}{2}kS_2^2 - mg(0.3\cos\beta):$$

$$0 + \frac{1}{2}(200)(0.192)^2 - (1)(9.81)(0.3) = \frac{1}{2}(1)v_2^2 + \frac{1}{2}(200)(0.115)^2 - (1)(9.81)(0.3\cos 29).$$ Solving,

$v_2 = 1.99 m/s$

Problem 4.95 In Problem 4.94, what is the tension in the string when the string and spring are parallel?

Solution: The free body diagram of the mass is: Newton's second law in the direction normal to the path is $T - kS_2 - mg\cos\beta = ma_n$: $T - (200)(0.115) - (1)(9.81)\cos 29.1^\circ = (1)(v_2^2/0.3)$. We obtain, $T = 44.7N$.

Problem 4.96 The force exerted on an object by a nonlinear spring is $\vec{F} = -\left[k(r - r_o) + q(r - r_o)^3\right]\vec{e}_r$, where k and q are constants and r_o is the unstretched length. Determine the potential energy of the spring in terms of its stretch $S = r - r_o$.

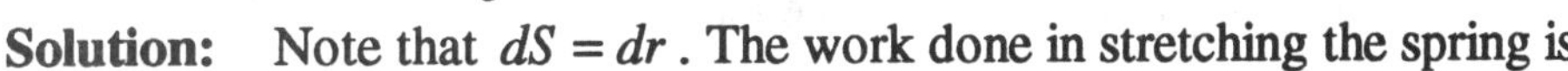

Solution: Note that $dS = dr$. The work done in stretching the spring is

$$V = -\int \vec{F}\cdot d\vec{r} + C = -\int \vec{F}\cdot(dr\vec{e}_r + rd\theta\,\vec{e}_\theta) + C = \int\left[k(r - r_o) + q(r - r_o)^2\right]dr + C,$$

$V = \int\left[kS + qS^3\right]dS + C$. Integrate: $\boxed{V = \frac{k}{2}S^2 + \frac{q}{4}S^4}$, where $C = 0$, since $\vec{F} = 0$ at $S = 0$.

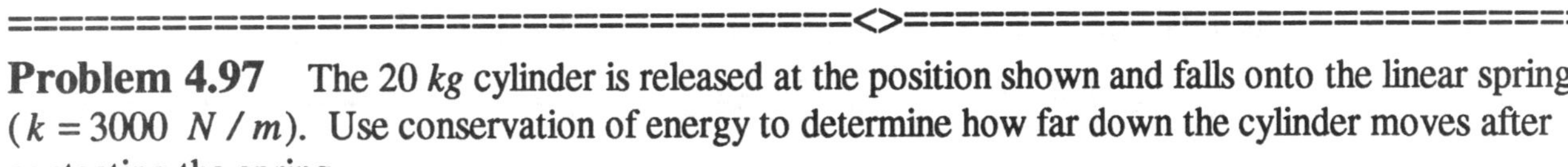

Problem 4.97 The 20 kg cylinder is released at the position shown and falls onto the linear spring ($k = 3000\ N/m$). Use conservation of energy to determine how far down the cylinder moves after contacting the spring.

Solution: Choose the base of the cylinder as a datum. The potential energy of the piston at rest is $V_1 = mg(3.5) = 686.7\ N\text{-}m$. . The conservation of energy condition after the spring has compressed to the point that the piston velocity is zero is $mgh + \frac{1}{2}k(h - 1.5)^2 = mg(3.5)$, where h is the height above the datum. From which $h^2 + 2bh + c = 0$, where $b = -\left(\frac{3}{2} - \frac{mg}{k}\right)$ and $c = 2.25 - \frac{7mg}{k}$. The solution is $h = -b \pm \sqrt{b^2 - c} = 1.95\ m,\ = 0.919\ m$. The value $h = 1.95\ m$ has no physical meaning, since it is above the spring. The downward compression of the spring is $\boxed{S = 1.5 - 0.919 = 0.581\ m}$

Problem 4.98 Suppose that the spring in Problem 4.97 is a nonlinear spring with potential energy $V=\frac{1}{2}kS^2+\frac{1}{4}qS^4$, where $k=3000\ N/m$ and $q=4000\ N/m^3$. What is the velocity of the cylinder when the spring has compressed to 0.5 *m*?

Solution:

Note that $S=1.5-h$ where *h* is the height above the datum, from which $h=1.5-S$. Use the solution to Problem 4.97.

The conservation of energy condition when the spring is being compressed is

$\frac{1}{2}mv^2+V_{spring}+mg(1.5-S)=mg(3.5)$, from which $v=\sqrt{7.0g-2g(1.5-S)-2\frac{V_{spring}}{m}}$.

The potential energy in the spring is $V_{spring}=\frac{1}{2}(3000)(0.5^2)+\frac{1}{4}(4000)(0.5^4)=+437.5\ N\text{-}m$.

Substitute numerical values to obtain $\boxed{v=2.30\ m/s}$

Problem 4.99 The string exerts a force of constant magnitude on the object. Determine the potential energy associated with this force in terms of polar coordinates.

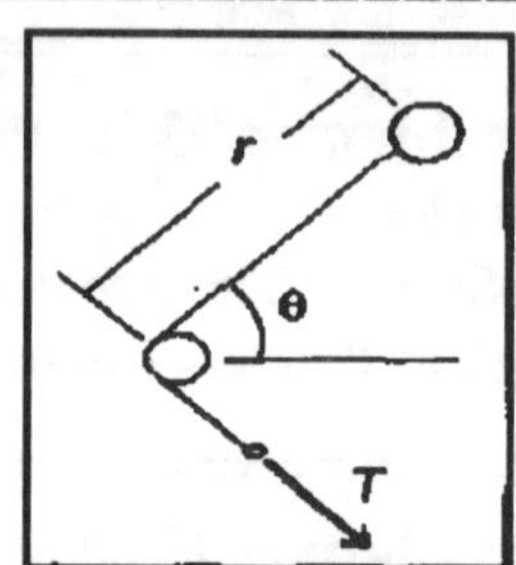

Solution:

The force $\vec{T}=T\vec{e}_r$. The arbitrary path is $d\vec{r}=(dr\,\vec{e}_r+rd\theta\,\vec{e}_\theta)$ The potential energy is $\boxed{V=\int\vec{T}\cdot d\vec{r}=Tr}$, since $\vec{e}_r\cdot\vec{e}_\theta=0$ everywhere.

Problem 4.100 The system is at rest in the position shown, with the 12 *lb.* Collar A is connected to the spring ($k=20\ lb/ft$). When a constant 30 *lb.* force is applied to the cable, what is the velocity of the collar when it has risen 1 ft?

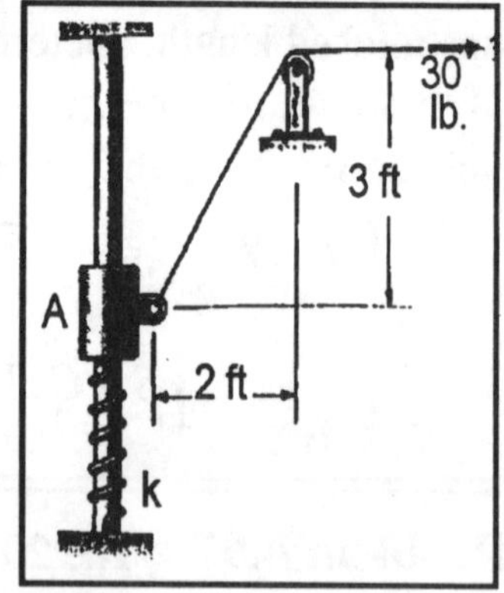

Solution:

Choose the rest position as the datum. At rest, the compression of the spring is $S_1=\frac{-W}{k}=-0.6\,ft$. When the collar rises 1 *ft* the stretch is $S_2=S_1+1=0.4\,ft$ When the collar rises 1 *ft* the constant force on the cable has acted through a distance $s=\sqrt{3^2+2^2}-\sqrt{(3-1)^2+2^2}=0.7771\ ft$. The work done on the system is $U_S=\frac{1}{2}k(S_1^2-S_2^2)-mg(1)+Fs$. From the conservation of energy $U_S=\frac{1}{2}mv^2$ from which $\boxed{v=\sqrt{\frac{k}{m}(S_1^2-S_2^2)-2g+\frac{2Fs}{m}}=8.45\ ft/s}$.

Problem 4.101 The cable extending from A to B engages the arresting hook of the airplane at C. The arresting mechanism maintains the tension in the cable at a constant value T. The airplane's mass is m. (a) Derive the potential energy V associated with the force exerted on the airplane as a function of its displacement s. (b) The airplane is moving at a velocity v_0 when it engages the arresting hook. Use conservation of energy to determine the distance s_0 required to bring it to rest.

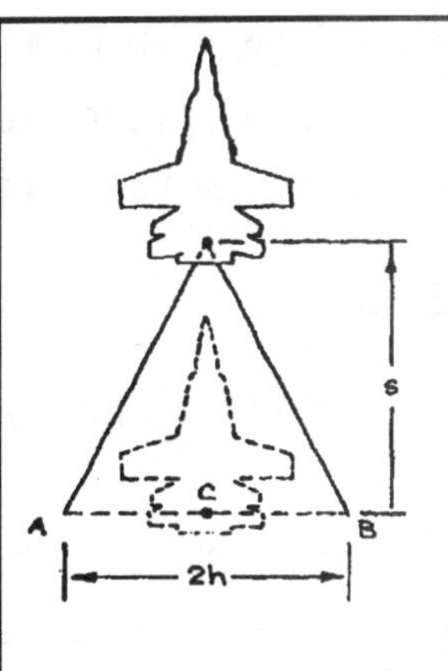

Solution:

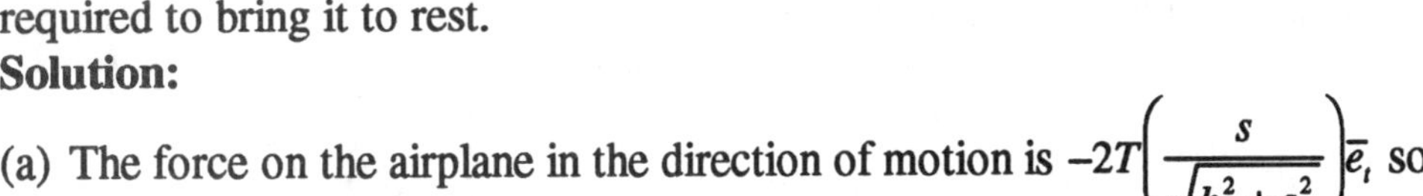

(a) The force on the airplane in the direction of motion is $-2T\left(\frac{s}{\sqrt{h^2+s^2}}\right)\bar{e}_t$ so

$$dV = -\sum \bar{F}\bullet d\bar{r} = -\left[-2T\left(\frac{s}{\sqrt{h^2+s^2}}\right)\bar{e}_t\right]\bullet ds\bar{e}_t = \frac{2Tsds}{\sqrt{h^2+s^2}}.$$

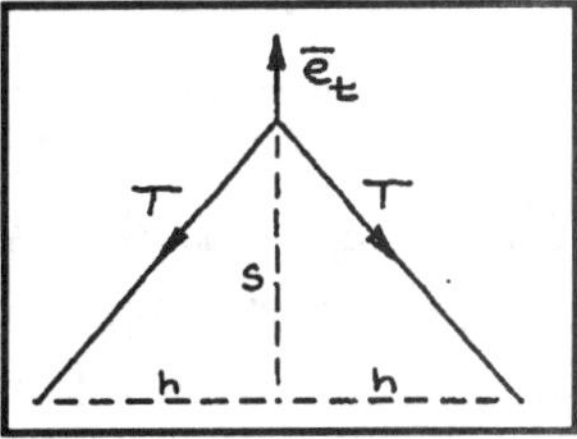

Integrating (introduce a new variable $z^2 = h^2 + s^2$), we obtain $V = 2T\sqrt{h^2+s^2}$

(b) $T_1 + V_1 = T_2 + V_2$; $\frac{1}{2}mv_0^{\,2} + 2Th = 0 + 2T\sqrt{h^2+s_0^2}$ Solving for s_0,

$s_0 = \sqrt{\left[h + mv_0^{\,2}/(4T)\right]^2 - h^2}$.

Problem 4.102 In Example 4.11, assume that the pressure of the gas is related to its volume by $pV^\gamma = constant$, where γ is a constant. (a) Determine the potential energy associated with the force exerted on the projectile in terms of s. (b) If the projectile starts from rest at $s = s_0$ and friction is negligible, what is its velocity as a function of s?

Solution:

(a) The volume $V = sA$, so $p(sA)^\gamma = p_0(s_0A)^\gamma$; $p = p_0(s_0/s)^\gamma$.

(b) The force on the projectile is $\sum \bar{F} = pA\bar{e}_t = p_0A(s_0/s)^\gamma \bar{e}_t$ so

$$dV = -\sum \bar{F}\bullet d\bar{r} = -p_oA(s_0/s)^\gamma \bar{e}_t \bullet ds\bar{e}_t = -p_0A(s_0/s)^\gamma ds.$$

(c) Integrating, we obtain $V = p_0As_0^\gamma s^{1-\gamma}/(\gamma-1)$.

(d) (b) $T_1 + V_1 = T_2 + V_2$: $0 + p_0As_o/(\gamma-1) = \frac{1}{2}mv^2 + p_0As_0^\gamma s^{1-\gamma}/(\gamma-1)$

(e) Solving for v, $v = \sqrt{2p_0As_0^\gamma\left(s_0^{1-\gamma} - s^{1-\gamma}\right)/\left[m(\gamma-1)\right]}$.

Problem 4.103 A satellite at a distance r_o from the center of the earth has a velocity of magnitude v_o. Use conservation of energy to determine the magnitude of its velocity v when it is a distance r from the center of the earth?

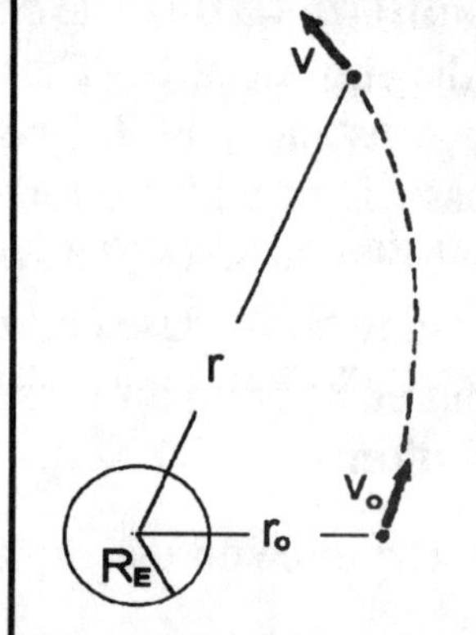

Solution:

From Eq (4.27) the potential energy at a distance r is $V = -\dfrac{mgR_E^2}{r}$.

From conservation of energy at r: $\dfrac{1}{2}mv_o^2 - \dfrac{mgR_E^2}{r_o} = \dfrac{1}{2}mv^2 - \dfrac{mgR_E^2}{r}$.

Solve: $\boxed{v = \sqrt{v_o^2 + 2gR_E^2\left(\dfrac{1}{r} - \dfrac{1}{r_o}\right)}}$

Problem 4.104 Astronomers detect an asteroid 100,000 *km* from the earth moving at 2 *km/s* relative to the center of the earth. If it should strike the earth, use conservation of energy to determine the magnitude of its velocity as it enters the atmosphere. (You can neglect the thickness of the atmosphere in comparison to the earth's 6370 *km* radius.)

Solution:

Use the solution to Problem 4.103. The potential energy at a distance r is $V = -\dfrac{mgR_E^2}{r}$.

The conservation of energy condition at r_o: $\dfrac{1}{2}mv_o^2 - \dfrac{mgR_E^2}{r_o} = \dfrac{1}{2}mv^2 - \dfrac{mgR_E^2}{r}$.

Solve: $v = \sqrt{v_o^2 + 2gR_E^2\left(\dfrac{1}{r} - \dfrac{1}{r_o}\right)}$. Substitute: $r_o = 1\times10^8\ m$, $v_o = 2\times10^3\ m/s$, and $r = 6.37\times10^6\ m$.

The velocity at the radius of the earth is $\boxed{v = 11\ km/s}$

Problem 4.105 A satellite is in an elliptic orbit around the earth. Its velocity at the perigee A is 28,280 *ft/s*. Use conservation of energy to determine its velocity at B. The radius of the earth is 3960 *mi*.

Solution:

Use the solution to Problem 4.104.

The potential energy at a distance r is $V = -\dfrac{mgR_E^2}{r}$.

The conservation of energy condition at r_o:

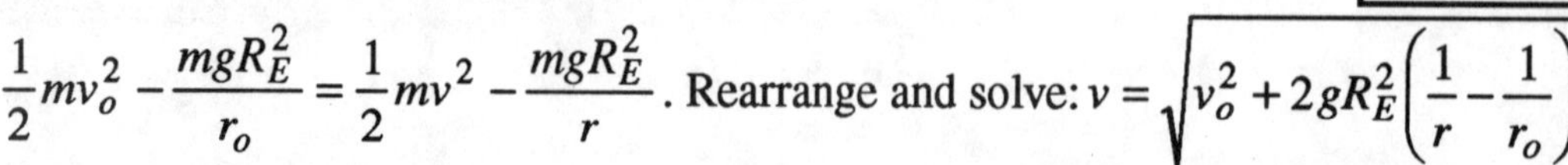

$\dfrac{1}{2}mv_o^2 - \dfrac{mgR_E^2}{r_o} = \dfrac{1}{2}mv^2 - \dfrac{mgR_E^2}{r}$. Rearrange and solve: $v = \sqrt{v_o^2 + 2gR_E^2\left(\dfrac{1}{r} - \dfrac{1}{r_o}\right)}$

Substitute numerical values: The magnitude of the velocity at B is $\boxed{v = 16{,}341\ ft/s}$

================================<>================================

Problem 4.106 For the satellite orbit in Problem 4.105, use conservation of energy to determine the velocity at the apogee C. Using your result, confirm numerically that the velocities at perigee and apogee satisfy the relation $r_A v_A = r_C v_C$.

Solution: Use the solution to Problem 4.105. The potential energy at r is $V = -\frac{mgR_E^2}{r}$.

The conservation of energy condition at r_C is: $\frac{1}{2}mv_o^2 - \frac{mgR_E^2}{r_o} = \frac{1}{2}mv_C^2 - \frac{mgR_E^2}{r_C}$.

Rearrange and solve to get $v_C = \sqrt{v_o^2 + 2gR_E^2\left(\frac{1}{r_C} - \frac{1}{r_o}\right)}$. Substitute:

$r_o = 5000 \text{ mi} = 5000(5280) = 2.64 \times 10^7 \text{ ft}$. $r_C = 15000 \text{ mi} = 7.92 \times 10^7 \text{ ft}$. $v_o = 28{,}280 \text{ ft/s}$, $g = 32.17 \text{ ft/s}^2$, $R_E = 3960(5280) = 2.09 \times 10^7 \text{ ft}$., $g = 32.2 \text{ ft/s}^2$ The velocity at C is $v_C = 9423 \text{ ft/s}$..Using three significant figures $r_C v_C = 7.46 \times 10^{11} \text{ ft}^2/s$, $r_A v_A = 7.46 \times 10^{11} \text{ ft}^2/s$

================================<>================================

Problem 4.107 The *Voyager* and *Galileo* spacecraft have observed volcanic plumes, believed to consist of condensed sulfur or sulfur dioxide gas, above the surface of the Jovian satellite *Io.* The plume observed a volcano named *Prometheus* was estimated to extend *50-km* above the surface. The acceleration due to gravity at the surface is $1.80 \ m/s^2$. Using conservation of energy and neglecting the variation of gravity with height, determine the velocity at which a solid particle would have to be ejected to reach *50-km* above the surface.

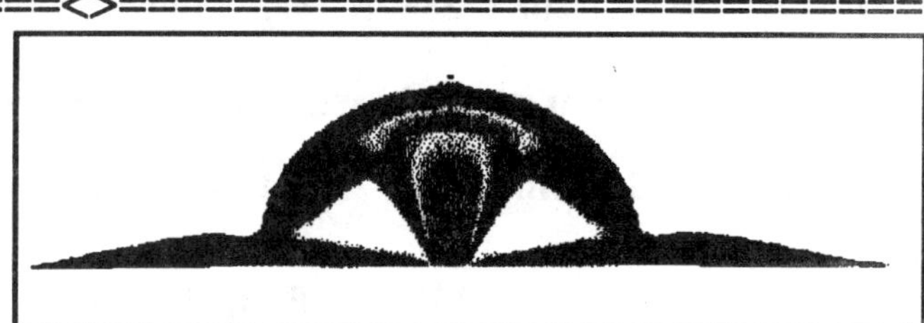

Solution:

Conservation of energy yields: $T_1 + V_1 = T_2 + V_2$: Using the forms for a constant gravity field, we get $\frac{1}{2}mv_1^2 + 0 = 0 + mgy_2$ Evaluating, we get $\frac{1}{2}v_1^2 = (1.8)(50{,}000)$, or $v_1 = 424 m/s$

================================<>================================

Problem 4.108 Solve Problem 4.107 using conservation of energy and accounting for the variation of gravity with height. The radius of *Io* is *1815-km.*

Solution:

Conservation of energy yields: $T_1 + V_1 = T_2 + V_2$: Only the form of potential energy changes from that used in Problem 4.107. Here we get $\frac{1}{2}mv_1^2 - \frac{mgR_I^2}{R_I} = 0 - \frac{mgR_I^2}{r_I}$.

Evaluating, $\frac{1}{2}v_1^2 - \frac{(1.8)(1{,}815{,}000)^2}{1{,}815{,}000} = -\frac{(1.8)(1{,}815{,}000)^2}{1{,}815{,}000 + 50{,}000}$. or $v_1 = 419 m/s$

================================<>================================

================================<>================================

Problem 4.109 The component of the total external force tangent to the path of a 10 kg object moving along the x axis is $\sum F_x = 3x^2\vec{i}\ N$, where x is in meters. At $x = 2\ m$, the object's velocity is $v_x = 4\ m/s$. (a) Use the principle of work and energy to determine its velocity at $x = 6\ m$. (b) Determine the potential energy associated with the force $\sum F_x$ and use the conservation of energy to determine its velocity at $x = 6\ m$.

Solution:

(a) The work done by the force $\vec{F} = 3x^2\vec{i}$ in going from $x = 2$ to $x = 6$ is

$U = \int_2^6 Fdx = \int_2^6 (3x^2)dx = [x^3]_2^6 = 208\ N\text{-}m$. From the principle of work and energy:

$U = \left(\frac{1}{2}mv^2\right)_{x=6} - \left(\frac{1}{2}mv^2\right)_{x=2}$. At $x = 2\ m$, $\left(\frac{1}{2}mv^2\right)_{x=2} = \frac{1}{2}m(4^2) = 80\ N\text{-}m$, from which

$\left(\frac{1}{2}mv^2\right)_{x=6} = 80 + 208 = 288\ N\text{-}m$, from which $\boxed{v = \sqrt{\frac{2(288)}{m}} = 7.59\ m/s}$.

(b) Choose the datum at $x = 0$, and assume that the velocity at $x = 0$ is zero. The potential energy associated with the force is $V = -\int_0^x Fdx = -x^3\ N\text{-}m$. From the conservation of energy:

$\left(\frac{1}{2}mv^2\right)_{x=2} + [V]_{x=2} = 72\ N\text{-}m$. From the conservation of energy, this is also the total energy at

$x = 6\ m$, $\left(\frac{1}{2}mv^2\right)_{x=6} + [V]_{x=6} = 72\ N\text{-}m$, from which $\frac{1}{2}mv^2 = 72 - [V]_{x=6} = 288\ N\text{-}m$. The

velocity is $\boxed{v = \sqrt{\frac{2(288)}{m}} = 7.59\ m/s}$

================================<>================================

Problem 4.110 The potential energy associated with a force acting on an object is $V = 2x^2 - y\ N\text{-}m$, where x and y are in meters. (a) Determine . (b) If the object moves from position 1 to position 2 along the paths A and B, determine the work done by F along those paths.

Solution:

The force is $\boxed{\vec{F} = -\nabla V = -(4x\vec{i} - 1\vec{j})\ N}$. The work done along a path 1,2 is $U_{1,2} = \int_{1,2} \vec{F}\cdot d\vec{r}$

Integrating along path B $\boxed{U_B = \int_0^1 -(4x\vec{i} - \vec{j})|_{y=0}\cdot\vec{i}\,dx - \int_0^1 (4x\vec{i} - \vec{j})|_{x=1}\cdot\vec{j}\,dy = -2 + 1 = -1\ N\text{-}m}$.

Along path A: $\boxed{U_A = \int_0^1 -(4x\vec{i} - \vec{j})|_{x=0}\cdot\vec{j}\,dy + \int_0^1 -(4x\vec{i} - \vec{j})|_{y=1}\cdot\vec{i}\,dx = 1 - 2 = -1\ N\text{-}m}$

================================<>================================

=================================<>=================================

Problem 4.111 An object is subjected to the force $\vec{F} = y\vec{i} - x\vec{j}$ N where x and y are in meters. (a) Show that $\vec{F}$ is not conservative. (b) If the object moves from point 1 to point 2 along the paths A and B shown in Problem 4.110, determine the work done by $\vec{F}$ along each path.

Solution: (a) A necessary and sufficient condition that $\vec{F}$ be conservative is $\nabla \times \vec{F} = 0$.

$$\nabla \times \vec{F} = \begin{bmatrix} \vec{i} & \vec{j} & \vec{k} \\ \frac{\partial}{\partial x} & \frac{\partial}{\partial y} & \frac{\partial}{\partial z} \\ y & -x & 0 \end{bmatrix} = \vec{i}\,0 - \vec{j}\,0 + (-1-1)\vec{k} = -2\vec{k} \neq 0.$$ Therefore $\vec{F}$ is non conservative.

(b) The integral along path B is $\boxed{U_B = \int_0^1 (y\vec{i} - x\vec{j})|_{y=0} \cdot \vec{i}\,dx + \int_0^1 (y\vec{i} - x\vec{j})|_{x=1} \cdot \vec{j}\,dy = 0 - 1 = -1 \ N\text{-}m}$.

Along path A: $\boxed{U_A = \int_0^1 \left(y\vec{i} - x\vec{j}\right)|_{x=0} \cdot \vec{j}\,dy + \int_0^1 \left(y\vec{i} - x\vec{j}\right)|_{y=1} \cdot \vec{i}\,dx = 0 + 1 = +1 \ N\text{-}m}$

=================================<>=================================

Problem 4.112 In terms of polar coordinates, the potential energy associated with the force $\vec{F}$ exerted on an object by a nonlinear spring is $V = \frac{1}{2}k(r - r_o)^2 + \frac{1}{4}q(r - r_o)^4$, where k and q are constants and r_o is the unstretched length of the spring. Determine $\vec{F}$ in terms of polar coordinates.

Solution: The force is given by

$$\boxed{\vec{F} = -\nabla V = -\left(\frac{\partial}{\partial r}\vec{e}_r + \frac{1}{r}\frac{\partial}{\partial \theta}\vec{e}_\theta\right)\left(\frac{1}{2}k(r - r_o)^2 + \frac{1}{4}q(r - r_o)^4\right) = -\left(k(r - r_o) + q(r - r_o)^3\right)\vec{e}_r}$$

=================================<>=================================

Problem 4.113 In terms of polar coordinates, the force exerted on an object by a non-linear spring is $\vec{F} = -\left(k(r - r_o) + q(r - r_o)^3\right)\vec{e}_r$, where k and q are constants and r_o is the unstretched length. Use Eq. (4.36) to show that $\vec{F}$ is conservative.

Solution:

A necessary and sufficient condition that $\vec{F}$ be conservative is $\nabla \times \vec{F} = 0$.

$$\nabla \times \vec{F} = \frac{1}{r}\begin{bmatrix} \vec{e}_r & r\vec{e}_\theta & \vec{e}_z \\ \frac{\partial}{\partial r} & \frac{\partial}{\partial \theta} & \frac{\partial}{\partial z} \\ -\left[k(r - r_o) + q(r - r_o)^3\right] & 0 & 0 \end{bmatrix} = \frac{1}{r}\left[0\vec{e}_r - 0r\vec{e}_\theta + 0\vec{e}_z\right] = 0.$$ $\vec{F}$ is conservative.

=================================<>=================================

==================================<>==================================

Problem 4.114 The potential energy associated with a force $\vec{F}$ acting on an object is $V = -r\sin\theta + r^2\cos^2\theta\ ft\text{-}lb$, where r is in feet. (a) Determine $\vec{F}$. (b) If the object moves from point 1 to point 2 along the circular path, how much work is done by $\vec{F}$?

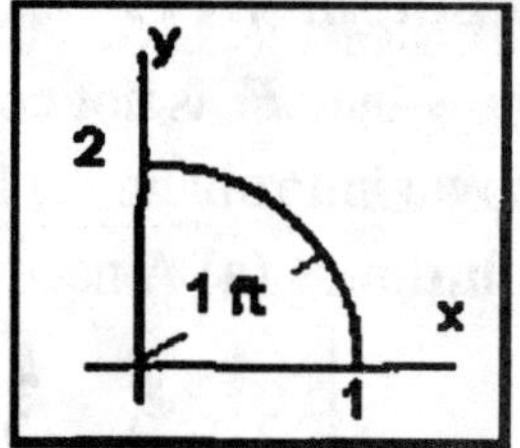

Solution: The force is $\vec{F} = -\nabla V = -\left(\frac{\partial}{\partial r}\vec{e}_r + \frac{1}{r}\frac{\partial}{\partial\theta}\vec{e}_\theta\right)\left(-r\sin\theta + r^2\cos^2\theta\right)$.

$\boxed{\vec{F} = \left(\sin\theta - 2r\cos^2\theta\right)\vec{e}_r + \left(\cos\theta + 2r\sin\theta\cos\theta\right)\vec{e}_\theta}$. The work done is $U_{1,2} = \int_{1,2}\vec{F}\cdot d\vec{r}$, where $d\vec{r} = \vec{e}_r dr + r\vec{e}_\theta d\theta$. Since the path is everywhere normal to $\vec{e}_r$, the radial term does not contribute to the work. The integral is $U_{1,2} = \int_0^{\frac{\pi}{2}}(\cos\theta + 2r\cos\theta\sin\theta)rd\theta = \left[r\sin\theta - r^2\cos^2\theta\right]_0^{\frac{\pi}{2}} = 1 + 1 = 2\ ft\text{-}lb$

Check: Since the force is derivable from a potential, the system is conservative. In a conservative system the work done is $U_{1,2} = -(V_2 - V_1)$, where V_1, V_2 are the potentials at the beginning and end of the path. At $r = 1$, $\theta = 0$, $V_1 = 1\ ft\text{-}lb$. At $r = 1\ ft$. $\theta = \frac{\pi}{2}$, $V_1 = -1$, from which $U_{1,2} = -(V_2 - V_1) = 2\ ft\text{-}lb$.

check.

==================================<>==================================

Problem 4.115 In terms of polar coordinates, the force exerted on an object of mass m by the gravity of a hypothetical two-dimensional planet is $\vec{F} = -\left(\frac{mg_T R_T}{r}\right)\vec{e}_r$, where g_T is the acceleration due to gravity at the surface, R_T is the radius of the planet, and r is the distance from the center of the planet. (a) Determine the potential energy associated with this gravitational force. (b) If the object is given a velocity v_o at a distance r_o, what is the velocity v as a function of r?

Solution:

(a) The potential is $V = -\int \vec{F}\cdot d\vec{r} + C = \int\left(\frac{mg_T R_T}{r}\right)\vec{e}_r\cdot(\vec{e}_r dr) + C = mg_T R_T\ln(r) + C$, where C is the constant of integration. Choose $r = R_T$ as the datum, from which $C = -mg_T R_T\ln(R_T)$, and

$\boxed{V = mg_T R_T\ln\left(\frac{r}{R_T}\right)}$ (*Note:* Alternatively, the choice of $r = 1$ *length-unit* as the datum, from which $C = mg_M R_T\ln(1)$, yields $V = mg_T R_T\ln\left(\frac{r}{1}\right) = mg_T R_T\ln(r)$.)

(b) From conservation of energy, $\frac{1}{2}mv^2 + mg_T R_T\ln\left(\frac{r}{R_T}\right) = \frac{1}{2}mv_o^2 + mg_T R_T\ln\left(\frac{r_o}{R_T}\right)$.

Solve for the velocity $\boxed{v = \sqrt{v_o^2 + 2g_T\ln\left(\frac{r_o}{r}\right)}}$

==================================<>==================================

==================================<>==================================

Problem 4.116 By substituting Eqs. (4.312 into Eq.(4.35), confirm that $\nabla \times \vec{F} = 0$ if $\vec{F}$ is conservative.

Solution:

Eq (4.35) is $\nabla \times \vec{F} = \begin{bmatrix} \vec{i} & \vec{j} & \vec{k} \\ \frac{\partial}{\partial x} & \frac{\partial}{\partial y} & \frac{\partial}{\partial z} \\ -\frac{\partial V}{\partial x} & -\frac{\partial V}{\partial y} & -\frac{\partial V}{\partial z} \end{bmatrix} = \vec{i}\left(-\frac{\partial^2 V}{\partial y \partial z} + \frac{\partial^2 V}{\partial y \partial z}\right) - \vec{j}\left(\frac{\partial^2 V}{\partial x \partial z} - \frac{\partial^2 V}{\partial x \partial z}\right) + \vec{k}\left(\frac{\partial^2 V}{\partial x \partial y} - \frac{\partial^2 V}{\partial x \partial y}\right) = 0$

Thus, **F** is conservative.

==================================<>==================================

Problem 4.117 Determine which of the following are conservative. (a) $\vec{F} = \left(3x^2 - 2xy\right)\vec{i} - x^2\vec{j}$ (b) $\vec{F} = (x - xy^2)\vec{i} + x^2y\vec{j}$. (c) $\vec{F} = \left(2xy^2 + y^3\right)\vec{i} + \left(2x^2y - 3xy^2\right)\vec{j}$.

Solution: Use Eq (4.35)

(a) $\nabla \times \vec{F} = \begin{bmatrix} \vec{i} & \vec{j} & \vec{k} \\ \frac{\partial}{\partial x} & \frac{\partial}{\partial y} & \frac{\partial}{\partial z} \\ 3x^2 - 2xy & -x^2 & 0 \end{bmatrix} = \vec{i}(0) - \vec{j}(0) + \vec{k}(-2x + 2x) = 0$ Force is conservative.

(b) $\nabla \times \vec{F} = \begin{bmatrix} \vec{i} & \vec{j} & \vec{k} \\ \frac{\partial}{\partial x} & \frac{\partial}{\partial y} & \frac{\partial}{\partial z} \\ x - xy^2 & x^2y & 0 \end{bmatrix} = \vec{i}(0) - \vec{j}(0) + \vec{k}(2xy + 2xy) = \vec{k}(4xy) \neq 0$ Force is non-conservative.

(c) $\nabla \times \vec{F} = \begin{bmatrix} \vec{i} & \vec{j} & \vec{k} \\ \frac{\partial}{\partial x} & \frac{\partial}{\partial y} & \frac{\partial}{\partial z} \\ 2xy^2 + y^3 & 2x^2y - 3xy^2 & 0 \end{bmatrix} = \vec{i}(0) - \vec{j}(0) + \vec{k}\left(4xy - 3y^2 - 4xy - 3y^2\right) = \vec{k}\left(-6y^2\right) \neq 0$.

Force is non-conservative.

==================================<>==================================

================================<>================================

Problem 4.118 Determine which of the following forces are conservative.
(a) $\vec{F} = 3r^2 \sin^2 \theta \vec{e}_r + 2r^2 \sin\theta \cos\theta \vec{e}_\theta$, (b) $\vec{F} = (2r\sin\theta - \cos\theta)\vec{e}_r + (r\cos\theta - \sin\theta)\vec{e}_\theta$,
(c) $\vec{F} = \left(\sin\theta + r\cos^2\theta\right)\vec{e}_r + (\cos\theta - r\sin\theta\cos\theta)\vec{e}_\theta$.

Solution:

(a) Use Eq (4.35). $\nabla \times \vec{F} = \frac{1}{r}\begin{bmatrix} \vec{e}_r & r\vec{e}_\theta & \vec{e}_z \\ \frac{\partial}{\partial r} & \frac{\partial}{\partial \theta} & \frac{\partial}{\partial z} \\ 3r^2\sin^2\theta & 2r^3\sin\theta\cos\theta & 0 \end{bmatrix}$.

$= \frac{1}{r}\left(\vec{e}_r(0) - r\vec{e}_\theta(0) + \vec{e}_z\left(6r^2\sin\theta\cos\theta - 6r^2\sin\theta\cos\theta\right)\right) = 0$. The force is conservative.

(b) $\nabla \times \vec{F} = \frac{1}{r}\begin{bmatrix} \vec{e}_r & r\vec{e}_\theta & \vec{e}_z \\ \frac{\partial}{\partial r} & \frac{\partial}{\partial \theta} & \frac{\partial}{\partial z} \\ 2r\sin\theta - \cos\theta & r^2\cos\theta - r\sin\theta & 0 \end{bmatrix}$

$= \frac{1}{r}\left(\vec{e}_r(0) - r\vec{e}_\theta(0) + \vec{e}_z(2r\cos\theta - \sin\theta - 2r\cos\theta - \sin\theta)\right) = \vec{e}_z(-2\sin\theta) \neq 0$.

The force is non-conservative.

(c) $\nabla \times \vec{F} = \frac{1}{r}\begin{bmatrix} \vec{e}_r & r\vec{e}_\theta & \vec{e}_z \\ \frac{\partial}{\partial r} & \frac{\partial}{\partial \theta} & \frac{\partial}{\partial z} \\ \sin\theta + r\cos^2\theta & r\cos\theta - r^2\sin\theta\cos\theta & 0 \end{bmatrix}$.

$= \frac{1}{r}\left(\vec{e}_r(0) - r\vec{e}_\theta(0) + \vec{e}_z(\cos\theta - 2r\sin\theta\cos\theta - \cos\theta + 2r\cos\theta\sin\theta)\right) = 0$. The force is conservative.

================================<>================================

Problem 4.119 The component of the total external force tangent to a 4 *kg* object's path is $\sum F_x = 200 + 2s^2 - 0.2s^3\ N$, where *s* is the position measured along the path in meters. At $s = 0$, the object's velocity is $v = 10\ m/s$. What distance along its path has the object traveled when the velocity reaches 30 *m/s*?

Solution: Choose the datum at $s = 0$. From Eq (4.29) the potential associated with the force is

$V = -\int F ds = -200s - \left(\frac{2}{3}\right)s^3 + \left(\frac{0.2}{4}\right)s^4\ N\text{-}m$. At the point in the path where the velocity is 30 *m/s* the conservation of energy condition is $\frac{1}{2}mv_o^2 = \frac{1}{2}mv_p^2 + V\ N\text{-}m$. Rearrange:

$U(s) = 0 = v_p^2 - v_o^2 - \left(\frac{2}{m}\right)\left(200s + \left(\frac{2}{3}\right)s^3 - \left(\frac{0.2}{4}\right)s^4\right)$. The function $U(s)$ was graphed against *s* to find the zero crossing. The graph is shown. The zero crossing occurs at $\boxed{s = 7.40\ m}$. [*Check*: This is also the value obtained by iteration using **TK Solver Plus.**]

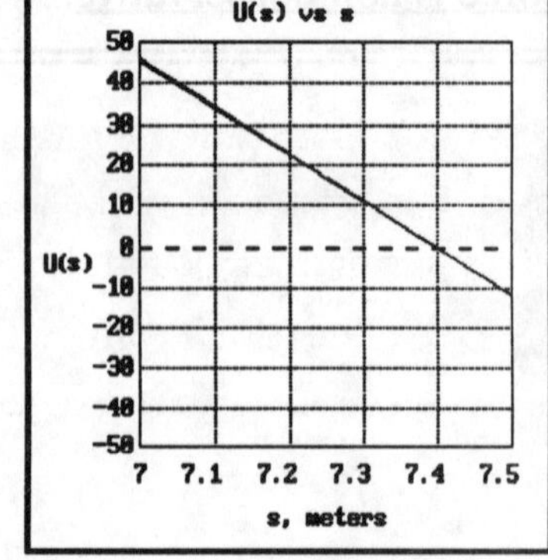

================================<>================================

==============================<>==============================

Problem 4.120 The 6 *kg* collar is released from rest in the position shown. If the spring constant is $k = 4\ kN/m$ and the unstretched length of the spring is 150 *mm*, how far does the mass fall from its original position before rebounding?

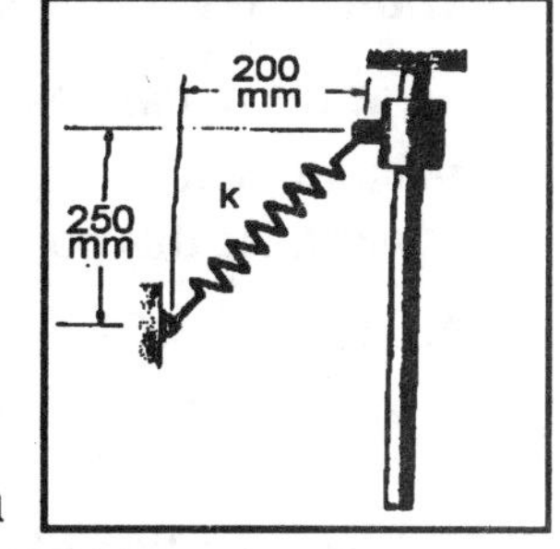

Solution: Choose the initial position as the datum. The deflection of the spring is $S_1 = \sqrt{0.2^2 + 0.25^2} - 0.15\ m$. The potential energy at the initial position is $V_1 = \frac{1}{2}kS_1^2$. Denote the lowest position reached by the collar before rebounding by y, where y is positive downward. At this lowest position the deflection of the spring is $S_2 = \sqrt{0.2^2 + (y - 0.25)^2} - 0.15$. The energy condition at the lowest position is $\frac{1}{2}kS_1^2 = \frac{1}{2}kS_2^2 - mgy$. Substitute and rearrange: $U(y) = 0 = S_2^2 - S_1^2 - \left(\frac{2mg}{k}\right)y$. The function $U(y)$ was graphed against y to find the zero crossing, and this value was then refined by iteration to obtain $\boxed{y = 0.552\ m}$, which is measured from the initial position.

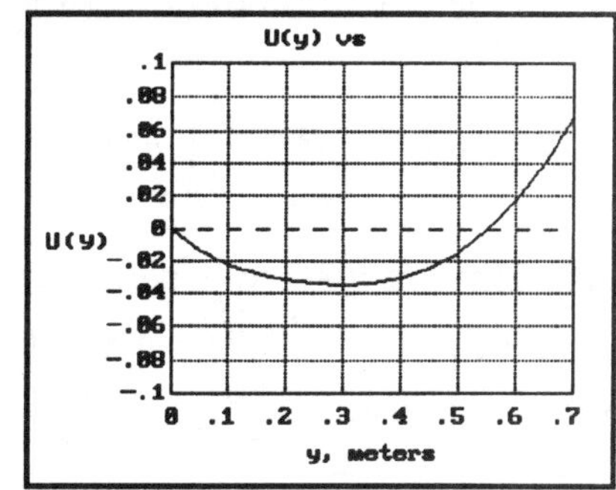

==============================<>==============================

Problem 4.121 How far below its initial position does the collar in Problem 4.120 reach its maximum velocity, and what is the maximum velocity?

Solution: Use the results of the solution to Problem 4.120. The deflection of the spring at the initial position is $S_1 = \sqrt{0.2^2 + 0.25^2} - 0.15\ m$. The potential energy at the initial position is $V_1 = \frac{1}{2}kS_1^2$. Denote the position reached by the collar as it falls by *y*. At this position the deflection of the spring is $S_2 = \sqrt{0.2^2 + (y - 0.25)^2} - 0.15$. The energy at this position is $\frac{1}{2}kS_1^2 = \frac{1}{2}mv^2 + \frac{1}{2}kS_2^2 - mgy$. Collect and rearrange:

$$v = \sqrt{\frac{k}{m}\left(S_1^2 - S_2^2\right) + 2gy}$$

. This is graphed over the range $0 \le y \le 0.55$ to find the maximum velocity.

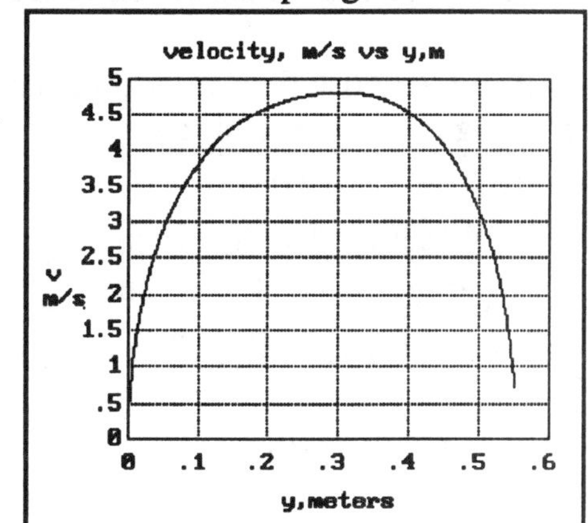

The value of the maximum velocity and the associated value of *y* is returned by an automatic search of the velocity list (array) using the built-in functions for this purpose in **TK Solver Plus.** The result: $\boxed{v_{max} = 4.805\ m/s}$ at $\boxed{y = 0.303\ m}$ (downward).

==============================<>==============================

==================================<>==================================

Problem 4.122 How far below it initial position does the power being transferred to the collar in Problem 4.120 reach its maximum, and what is the maximum power?

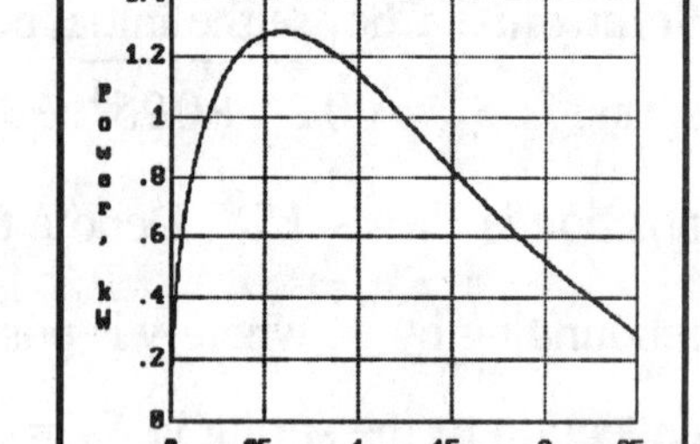

Solution: Use the solution to Problem 4.120. The deflection of the spring at the initial position is (1) $S_1 = \sqrt{0.2^2 + 0.25^2} - 0.15\ m$. Denote the position reached by the collar as it falls by y, where y is positive downward. At this position the deflection of the spring is (2) $S_2 = \sqrt{0.2^2 + (y - 0.25)^2} - 0.15$. From conservation of energy (3) $mv^2/2 = kS_1^2/2 - kS_2^2/2 + mgy$ .From the equation following Eq (4.17) the definition of power transferred is $P = \frac{d}{dt}\left(\frac{1}{2}mv^2\right)$. The derivative is (4) $P = \frac{d}{dt}\left(\frac{1}{2}mv^2\right) = mgv - kS_2\frac{dS_2}{dt}$, where $\frac{dS_2}{dt} = \frac{-(0.25 - y)v}{\sqrt{0.2^2 + (0.25 - y)^2}}$. The four equations are used to generate an array of values of the power over the range $0 \le y \le 0.25\ (m)$.The graph of power transferred as a function of y is shown. From an automatic search of the lists (tables), the maximum power is $\boxed{P_{max} = 1.2776\ kW}$ at a value of $\boxed{y = 0.0587\ m}$ and a velocity of $\boxed{v = 3.1212\ m/s}$

==================================<>==================================

Problem 4.123 The system is released from rest in the position shown. The weights are $W_A = 200\ lb$ and $W_B = 300\ lb$. Neglect friction. Determine the maximum velocity attained by A as it rises.

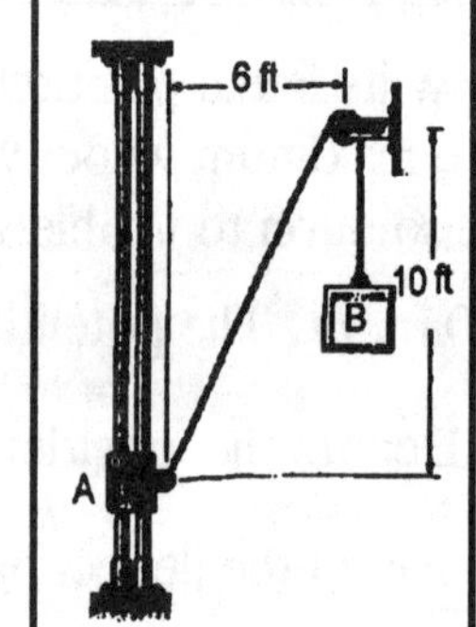

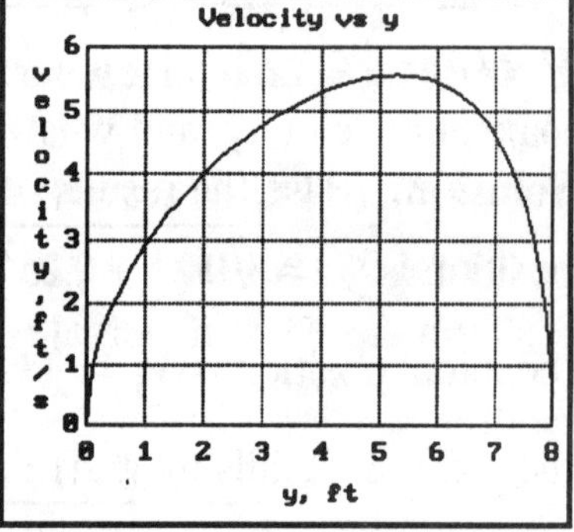

Solution: Choose the datum at the initial point. The weight B moves a distance $h = \sqrt{10^2 + 6^2} - \sqrt{(10 - y)^2 + 6^2}$ downward. The potential energy of weight B is $V_B = -(-W_B)(-h) = -W_B h$. The weight A moves a distance y upward. The potential energy of weight A is $V_A = -(-W_A y) = W_A y$. The energy at the initial point is zero. As the collar rises, the condition of conservation of energy is $0 = \frac{1}{2}\left(\frac{W_A}{g}\right)v_A^2 + \frac{1}{2}\left(\frac{W_B}{g}\right)v_B^2 + V_A + V_B$. The two velocities are related by $v_B = v_A \cos\theta = (10 - y)v_A / \sqrt{(10 - y)^2 + 6^2}$. Substitute into the energy equation:

$$0 = \left(\left(\frac{(10 - y)^2}{(10 - y)^2 + 6^2}\right)\left(\frac{W_B}{g}\right) + \left(\frac{W_A}{g}\right)\right)v_A^2 + 2(W_A y - W_B h).$$

Solve for the velocity: $v_A = \sqrt{\frac{2g(W_B h - W_A y)}{W_B \cos^2\theta + W_A}}$, where $\cos^2\theta = \frac{(10 - y)^2}{(10 - y)^2 + 6^2}$ and $h = \sqrt{10^2 + 6^2} - \sqrt{(10 - y)^2 + 6^2}$. A graph of the velocity is shown.

An automatic search of the lists (tables) used to produce the graph returns a maximum value of the velocity $\boxed{v_{max} = 5.581\ ft/s}$ at the value $\boxed{y = 5.28\ ft}$

==================================<>==================================

Problem 4.124 In Problem 4.123, what maximum height is reached by A relative to its initial position?

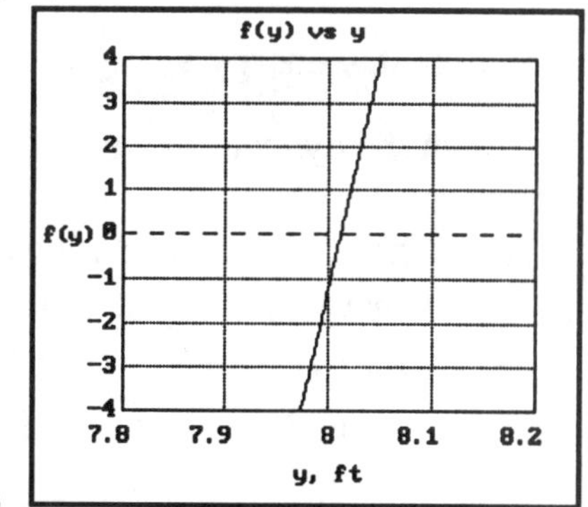

Solution:

At the maximum height the velocity is zero. Use the solution to Problem 4.123: $v_A = \sqrt{\dfrac{2g(W_B y_o - W_A y)}{W_B \cos^2\theta + W_A}}$, from which, if $v = 0$,

(1) $(W_A y - W_B y_o) = 0$, where from the solution to Problem 4.122,

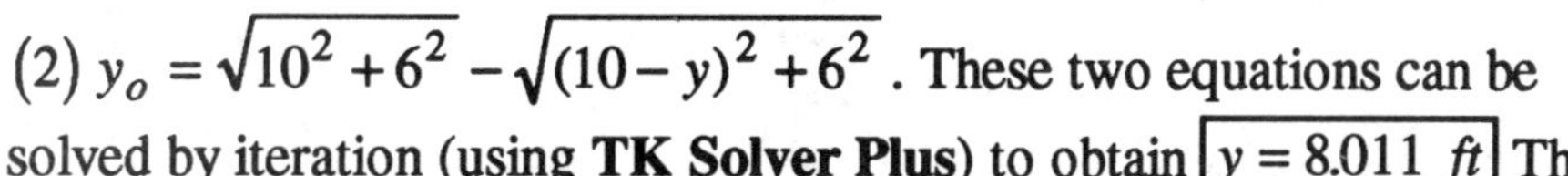

(2) $y_o = \sqrt{10^2 + 6^2} - \sqrt{(10-y)^2 + 6^2}$. These two equations can be solved by iteration (using **TK Solver Plus**) to obtain $\boxed{y = 8.011\ ft}$ This forced iteration was confirmed by a graph of $f(y) = (W_A y - W_B y_o)$ against y, as shown. An automatic search of the table of values confirms that the zero crossing occurs at this value of y.

Problem 4.125 The spring constant k = *2000 N/m*, $m_A = 14 kg$, and $m_B = 18 kg$. The collar A slides on the smooth horizontal bar. The system is released from rest with the spring unstretched. (a) Use conservation of energy to determine the veolcity of the collar A when it has moved a distance x to the right. Draw a graph of the velocity for $0 \le x \le 0.15 m$. Use your graph to estimate the maximum velocity.

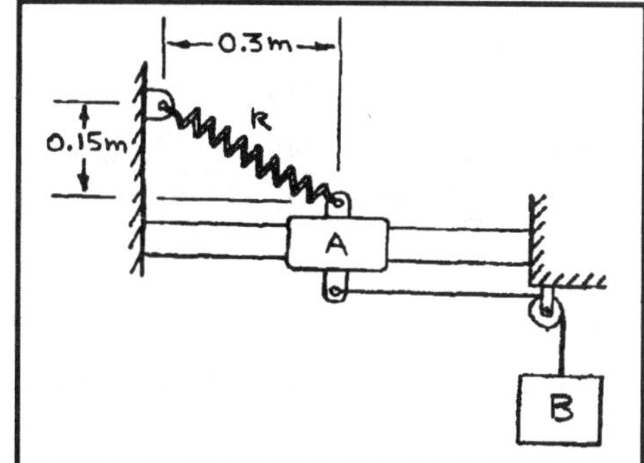

Solution:

The stretch of the spring when A has moved a distance x to the right is $S_2 = \sqrt{(0.15)^2 + (0.3+x)^2} - \sqrt{(0.15)^2 + (0.3)^2}$. Placing the datum for β at its initial position, conservation of energy is

$$T_1 + V_1 = T_2 + V_2:\ 0 + 0 = \frac{1}{2}m_A V^2 + \frac{1}{2}m_B V^2 + \frac{1}{2}kS_2^2 - m_B g x.$$

Solving for V, $V = \sqrt{(2m_B g x. - kS_2^2)/(m_A + m_B)}$.

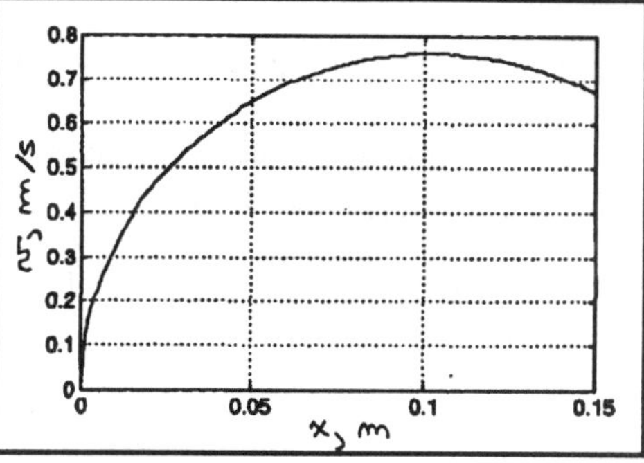

The graph of as a function of x is: By examining the computed results, we estimate the maximum velocity to be $0.760 m/s$ at $x = 0.103 m$.

Problem 4.126 In Problem 4.125, estimate the maximum distance the collar slides to the right.

Solution:

Drawing a graph of v^2 as a function of x, we obtain:

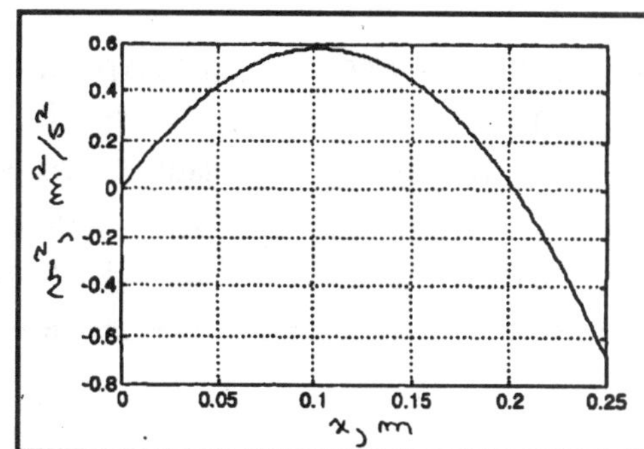

Examining the computed results, we estimate that $v^2 = 0$ (and the maximum distance has been reached) at $x = 0.203 m$.

Problem 4.127 The 16 kg cylinder is released at the position shown and falls onto a nonlinear spring with potential energy $V = \frac{1}{2}kS^2 + \frac{1}{4}qS^4$, where $k = 2400\ N/m$ and $q = 3000\ N/m^3$. Determine how far down the cylinder moves after contacting the spring.

Solution:

Choose the datum at the spring contact position. The potential energy of the piston at the initial position is $V_1 = (-mg)(-h) = mgh$, where $h = 2\ m$. After contact with the spring, the energy condition is $mgh = -mgS + \frac{1}{2}mv^2 + \frac{1}{2}kS^2 + \frac{1}{4}qS^4$, where S is the distance traveled after contact. At the maximum distance the velocity is zero. Rearrange: $f(S) = 0 = -mg(S+h) + \frac{1}{2}kS^2 + \frac{1}{4}qS^4$. The function $f(S)$ was graphed against S to find the zero crossing, and the value was then refined by iteration. The result: $\boxed{S = 0.5305... = 0.531\ m}$

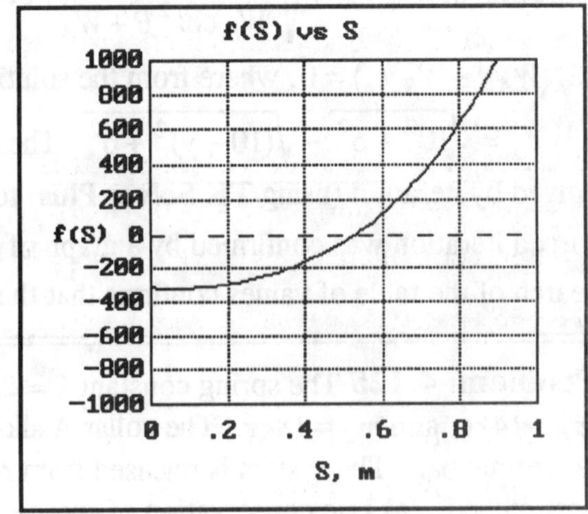

Problem 4.128 In Problem 4.127, what is the maximum velocity attained by the cylinder?

Solution:

Use the solution to Problem 4.127: The conservation of energy condition is $mgh = -mgS + \frac{1}{2}mv^2 + \frac{1}{2}kS^2 + \frac{1}{4}qS^4$, where $h = 2\ m$ and S is the distance traveled after contact with the cylinder. Rearrange:

$$v^2 = 2g(S+h) - \left(\frac{k}{m}S^2 + \frac{q}{2m}S^4\right).$$

The maximum occurs at $\frac{dv^2}{dS} = 0 = 2g - \left(\frac{2k}{m}\right)S - \left(\frac{2q}{m}\right)S^3$

$$= g - \left(\frac{k}{m}\right)S - \left(\frac{q}{m}\right)S^3 = 0.$$

The derivative was graphed as a function of S to determine the zero crossing. The velocity was also graphed against S to demonstrate that the crossing corresponds to a maximum. The refinement of the graphical estimate of the zero crossing yields $\boxed{S = 0.065\ m}$, and the maximum velocity is $\boxed{v = 6.32\ m/s}$.

=================================<>=================================

Problem 4.129 The system shown in Fig. (a) is released from rest. The mass $m = 1\ kg$, the spring constant $k = 200\ N/m$, and the unstretched length of the spring is $0.1\text{-}m$. (a) Use conservation of energy to determine the magnitude of the velocity of the mass as a function of the angle θ between the string and the vertical (Fig. b) and draw a graph of the velocity for $0 \le \theta \le 40°$. (b) Use your graph to estimate the maximum velocity.

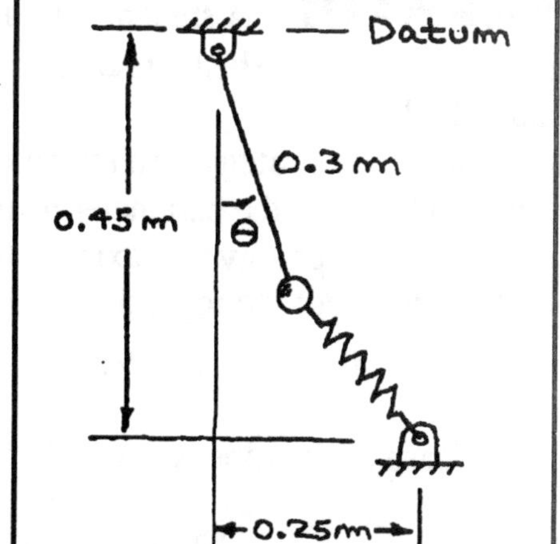

Solution:

In terms of θ , the stretch of the spring is

$$S = \sqrt{(0.25 - 0.3\sin\theta)^2 + (0.45 - 0.3\cos\theta)^2} - 0.1m.$$

Conservation of energy is $T_1 + V_2 = T_2 + V_2$: or

$$0 + \frac{1}{2}kS_1^2 - mg(0.3) = \frac{1}{2}mv^2 + \frac{1}{2}kS_2^2 - mg(0.3)\cos\theta$$

Solving for v,

$$v = \sqrt{(k/m)(S_1^2 - S_2^2) - 2g(0.3)(1-\cos\theta)}.$$

The graph of v as a function of θ is:

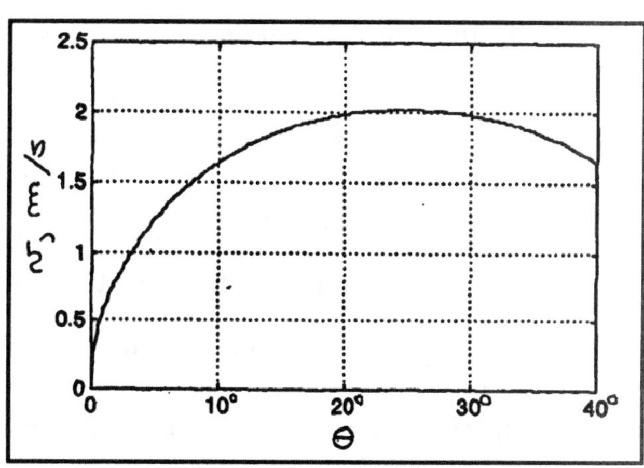

By examining the computed results, we estimate the maximum velocity to be $2.02 m/s$ at $\theta = 25°$.

=================================<>=================================

Problem 4.130 In Problem 4.129, estimate the maximum value of θ reached by the mass.

Solution:

Drawing a graph of v^2 as a function of θ , we obtain:

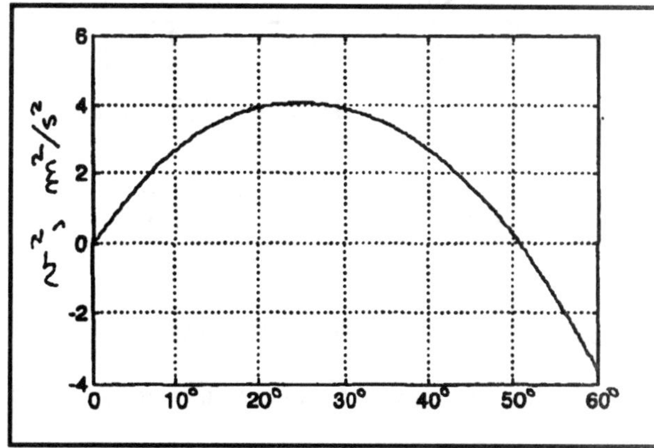

Examining the computed results, we estimate that $v^2 = 0$ (and the maximum distance has been reached) at $\theta = 51°$.

=================================<>=================================

Problem 4.131 A student runs at 15 *ft/s*, grabs a rope, and swings out over a lake. Determine the angle θ at which he should release the rope to maximize the horizontal distance *b*. What is the resulting value of *b*?

Solution The strategy is to express the initial conditions at the point of release in terms of the parameter *b* and the velocity, and then to use the conservation of energy to maximize to determine the velocity. At the point of release of the rope, the components of the velocity are $v_x = v_2 \cos(90 - \theta) = v_2 \sin\theta$, $v_y = v_2 \sin(90^o - \theta) = v_2 \sin\theta$. The path after release of the rope is $y = -\frac{g}{2}t^2 + (v_2 \sin\theta)t + h$, and $x = (v_2 \cos\theta)t + X_o$. The value $X_o = L\sin\theta$ The time of impact is given $y = 0$, from which $t_{imp}^2 + 2bt_{imp} + c = 0$, where $b = -\frac{v_2 \sin\theta}{g}$, $c = -\frac{2h}{g}$, from which, since the time is positive, $t_{imp} = \frac{v_2 \sin\theta}{g}\left(1 + \sqrt{1 + \frac{2gh}{v_2^2 \sin^2\theta}}\right)$ is the time of impact after release. The range $b = (v_2 \cos\theta)(t_{imp}) + L\sin\theta$. Choose the datum at ground level. The energy of the runner at the instant of grabbing the rope is $\frac{1}{2}mv_1^2$. After the rope is seized, the conservation of energy condition is $\frac{1}{2}mv_1^2 = \frac{1}{2}mv_2^2 + mgh$. Rearrange and solve for the velocity: $v_2 = \sqrt{v_1^2 - 2gh}$.

$$b = v_2 \cos\theta\left(\frac{v_2 \sin\theta}{g}\left(1 + \sqrt{1 + \frac{2gh}{v_2^2 \sin^2\theta}}\right)\right) + L\sin\theta.$$

The parameter $h = L(1 - \sin\theta)$, where L is the length of the rope. Rearrange:

$$b = \frac{v_2^2 \cos\theta \sin\theta}{g}\left(1 + \sqrt{1 + \frac{2gL(1 - \sin\theta)}{v_2^2 \sin^2\theta}}\right) + L\sin\theta,$$ where $v_2 = \sqrt{v_1^2 - 2gL(1 - \sin\theta)}$. The only unknown on the right side of the equation is the angle θ. The function $b(\theta)$ was graphed against θ to estimate the maximum. An inspection of tabulated values yielded $\boxed{b = 15.78\ ft}$, at $\boxed{\theta = 24.7^o}$

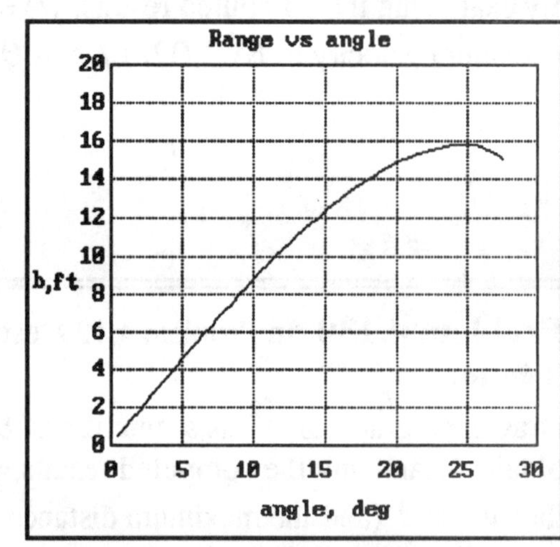

==================================<>==================================

Problem 4.132 The driver of a 3000 *lb.* car moving at 40 *mph* applies an increasing force on the brake pedal. The magnitude of the resulting force exerted on the car by the road is $f = 250 + 6s\ lb$, where *s* is its horizontal position relative to its position when the brakes were applied. Assuming that the car's tires do not slip, determine the distance required for the car to stop (a) by using Newton's second law; (b) by using the principle of work and energy.

Solution (a) Newton's second law: $\left(\frac{W}{g}\right)\frac{dv}{dt} = -f$, where *f* is the force on the car in opposition to the motion. Use the chain rule: $\left(\frac{W}{g}\right)v\frac{dv}{ds} = -f = -(250 + 6s)$. Integrate and rearrange: $v^2 = -\left(\frac{2g}{W}\right)(250s + 3s^2) + C$. At $s = 0$, $v(0) = 40\left(\frac{5280}{3600}\right) = 58.67\ ft/s$, from which $C = (58.67^2) = v_1^2$. The velocity is $v^2 = -\left(\frac{2g}{W}\right)(250s + 3s^2) + v_1^2\ (ft/s)^2$. At $v = 0$, $s^2 + 2bs + c = 0$, where $b = \frac{125}{3} = 41.67$, $c = -\frac{Wv_1^2}{6g} = -53493.6$. The solution: $s = -b \pm \sqrt{b^2 - c} = 193.3\, ft,\ = -276.7\, ft$, from which $\boxed{s = 193.3\ ft}$. (a) Principle of work and energy: The energy of the car when the brakes are first applied is $\frac{1}{2}\left(\frac{W}{g}\right)v_1^2 = 1.0605 \times 10^5\ ft\text{-}lb$. The work done is $U = \int_0^s f ds = -\int_0^s (250 + 6s)ds = -(250s + 3s^2)$. From the principle of work and energy, after the brakes are applied, $U = \frac{1}{2}\left(\frac{W}{g}\right)v_2^2 - \frac{1}{2}\left(\frac{W}{g}\right)v_1^2$.Rearrange: $\frac{1}{2}\left(\frac{W}{g}\right)v_2^2 = \frac{1}{2}\left(\frac{W}{g}\right)v_1^2 - (250s + 3s^2)$. When the car comes to a stop, $v_2 = 0$, from which $0 = \frac{1}{2}\left(\frac{W}{g}\right)v_1^2 - (250s + 3s^2)$. Reduce: $s^2 + 2bs + c = 0$, where $b = \frac{125}{3} = 41.67$, $c = -\frac{Wv_1^2}{6g} = -53493.6$. The solution $s = -b \pm \sqrt{b^2 - c} = 193.3\, ft,\ -276.6\, ft$, from which $\boxed{s = 193.3\, ft}$.

==================================<>==================================

Problem 4.133 Suppose that the car in Problem 4.132 is on wet pavement and the coefficients of friction between the tires and the road are $\mu_s = 0.4$, $\mu_k = 0.35$. Determine the distance for the car to stop.

Solution The initial velocity of the vehicle is $v_1 = 40\ mi/h = 58.67\ ft/s$ (a) Assume that the force $f = 250 + 6s\ lb$ applies until the tire slips. Slip occurs when $f = 250 + 6s = \mu_s W$, from which $s_{slip} = 158.33\, ft$. The work done by the friction force is

$$U_f = \int_0^{s_{slip}} -f ds + \int_{s_{slip}}^{s_{stop}} -\mu_k W = -(250 s_{slip} + 3 s_{slip}{}^2) - \mu_k W(s_{stop} - s_{slip}) = 51458.3 - 1050(s_{stop})$$

.From the principle of work and energy: $U_f = 0 - \left(\frac{1}{2}mv_1^2\right) = -160481\ lb\text{-}ft$, from which $\boxed{s_{stop} = 201.85\ ft}$.

==================================<>==================================

Problem 4.134 An astronaut in a small rocket vehicle (combined mass 450 *kg*) is hovering 100 *m* above the surface of the moon when he discovers he is nearly out of fuel and can only exert the thrust necessary to cause the vehicle to hover for 5 more seconds. He quickly considers two strategies for getting to the surface: (a) Fall 20 *m*, turn on the thrust for 5 *s*, then fall the rest of the way. (b) Fall 40 *m* turn on the thrust for 5 *s* then fall the rest of the way. Which strategy gives him the best chance of surviving? How much work is done by the engine's thrust in each case?

Solution Assume $g = 1.62\ m/s^2$ and that (as in Problem 4.61) the fuel mass is negligible. Since the thruster causes the vehicle to hover, the thrust is $T = mg$. The potential energy at $h_1 = 100\ m$ is $V_1 = mgh$. (a) Consider the first strategy: The energy condition at the end of a 20 *m* fall is $mgh = \frac{1}{2}mv_2^2 + mgh_2$, where $h_2 = h_1 - 20 = 80\ m$, from which $\frac{1}{2}mv_2^2 = mg(h_1 - h_2)$, from which $v_2 = \sqrt{2g(h_1 - h_2)} = 8.05\ m/s$. The work done by the thrust is $U_{thrust} = -\int_{h_2}^{h_3} Fdh = -mg(h_3 - h_2)$, where $F = mg$, acting upward, h_3 is the altitude at the end of the thrusting phase. The energy condition at the end of the thrusting phase is $mgh = \frac{1}{2}mv_3^2 + mgh_3 + U_{thrust}$, from which $mgh = \frac{1}{2}mv_3^2 + mgh_2$. It follows that the velocities $v_3 = v_2 = 8.05\ m/s$, that is, the thruster does not reduce the velocity during the time of turn-on. The height at the end of the thruster phase is $h_3 = h_2 - v_3 t = 80 - (8.04)(5) = 39.75\ m$. The energy condition at the beginning of the free fall after the thruster phase is $\frac{1}{2}mv_3^2 + mgh_3 = 44526.3\ N\text{-}m$, which, by conservation of energy is also the energy at impact: is $\frac{1}{2}mv_4^2 = \frac{1}{2}mv_3^2 + mgh_3 = 43558.3\ N\text{-}m$, from which $\boxed{v_4 = \sqrt{\frac{2(43558.3)}{m}} = 13.9\ m/s}$ at impact. (b) Consider strategy (b): Use the solution above, with $h_2 = h_1 - 40 = 60\ m$ The velocity at the end of the free fall is $v_2 = \sqrt{2g(h_1 - h_2)} = 11.38\ m/s$. The velocity at the end of the thruster phase is $v_3 = v_2$. The height at the end of the thruster phase is $h_3 = h_2 - v_2 t = 3.08\ m$. The energy condition at impact is : $\frac{1}{2}mv_4^2 = \frac{1}{2}mv_3^2 + mgh_3 = 31405\ N\text{-}m$. The impact velocity is

$$\boxed{v_4 = \sqrt{\frac{2(31405)}{m}} = 11.81\ m/s}$$

He should choose strategy (b) since the impact velocity is reduced by $\Delta v = 13.91 - 11.81 = 2.1\ m/s$. The work done by the engine in strategy (a) is $U_{thrust} = \int_{h_3}^{h_3} Fdh = mg(h_3 - h_2) = -29.3\ kN\text{-}m$. The work done by the engine in strategy (b) is $\boxed{U_{thrust} = \int_{h_2}^{h_3} Fdh = mg(h_3 - h_2) = -41.5\ kN\text{-}m}$

==================================<>==================================

Problem 4.135 The coefficients of friction between the 20 *kg* crate and the inclined surface are $\mu_s = 0.24$ and $\mu_k = 0.22$. If the crate starts from rest and the horizontal force $F = 200\ N$, what is the magnitude of the velocity when it has moved 2 *m*?

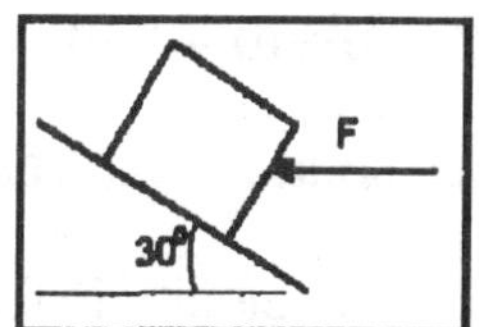

Solution

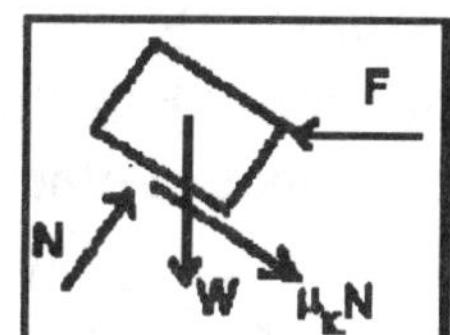

Denote $\theta = 30^o$. Since $F\cos\theta > mg\sin\theta$, the crate moves up the incline. Choose a coordinate system with the *x* axis parallel to the horizontal, *y* positive upward. The normal force is $N = mg\cos\theta + F\sin\theta$. The force parallel to *x* axis is $F_x = -F\cos\theta + mg\sin\theta + \mu_k N$. The work done in moving up the incline is $U = \int_0^{-2} F_x dx = -2F_x$. From the principle of work and energy: $U = \frac{1}{2}mv^2$. Solve:

$$v = \sqrt{\frac{4(F(\cos\theta - \mu_k\sin\theta) - mg(\sin\theta + \mu_k\cos\theta))}{m}} = 1.773\ m/s$$

==================================<>==================================

Problem 4.136 In Problem 4.135, what is the magnitude of the crate's velocity when it has moved 2 *m* if the horizontal force $F = 40\ N$?

Solution

Since $F\cos\theta < mg\sin\theta$ the box slides down the incline. The normal force is $N = mg\cos\theta + F\sin\theta$. The force parallel to the *x* axis is $F_x = -F\cos\theta + mg\sin\theta - \mu_k N$. The work done sliding down the incline is $U = \int_0^2 F_x dx = 2F_x$ From the principle of work and energy: $U = \frac{1}{2}mv^2$.Solve:

$$v = \sqrt{\frac{4(-F\cos\theta + \mu_k\sin\theta) + mg(\sin\theta - \mu_k\cos\theta)}{m}} = 2.082\ m/s$$

==================================<>==================================

Problem 4.137 The Union Pacific *Big Boy* locomotive weighs 1.19 million *lb.*, and the traction force (tangential force) of its drive wheels is 135, 000 *lb*. If you neglect other tangential forces, what distance is required for it to accelerate from zero to 60 *mi/hr*?

Solution

The potential associated with the force is $V = -\int_o^s F ds = -Fs$. The energy at rest is zero. The energy at $v = 60\ mi/hr = 88\ ft/s$ is $0 = \frac{1}{2}\left(\frac{W}{g}\right)v^2 + V$, from which $s = \frac{1}{2}\left(\frac{W}{gF}\right)v^2 = 1061\ ft$

==================================<>==================================

================================<>================================

Problem 4.138 In Problem 4.137, suppose that the total tangential force on the locomotive as it accelerates from zero to 60 *mi/hr* is $F = \left(\frac{F_o}{m}\right)\left(1-\frac{v}{88}\right)$, where $F_o = 135{,}000\ lb$, and m is its mass, and v is its velocity in feet per second. (a) How much work is done in accelerating it to 60 *mi/hr*? (b) Determine its velocity as a function of time.

Solution [Note: F is not a force, but an acceleration, with the dimensions of acceleration.] (a) The work done by the force is equal to the energy acquired by the locomotive in attaining the final speed, in the absence of other tangential forces. Thus the work done by the traction force is

$$U = \frac{1}{2}mv^2 = \frac{1}{2}\left(\frac{W}{g}\right)(88^2) = 1.43\times 10^8\ ft\text{-}lb$$

(b) From Newton's second law $m\frac{dv}{dt} = mF$, from which $\frac{dv}{dt} = \left(\frac{F_o}{m}\right)\left(1-\frac{v}{88}\right)$. Separate variables:

$\frac{dv}{\left(1-\frac{v}{88}\right)} = \left(\frac{F_o}{m}\right)dt$. Integrate: $\ln\left(1-\frac{v}{88}\right) = -\left(\frac{F_o}{88m}\right)t + C_1$. Invert: $v(t) = 88\left(1-Ce^{-\frac{F_o}{88m}t}\right)$. At $t=0$, $v(0)=0$, from which $C=1$. The result: $\boxed{v(t) = 88\left(1-e^{-\frac{gF_o}{88W}t}\right)}$ *Check:* To demonstrate that this is a correct expression, it is used to calculate the work done: Note that $U = \int_0^S mFds = \int_0^T mF\left(\frac{ds}{dt}\right)dt = \int_0^T mFvdt$. For brevity write $K = \frac{gF_o}{88W}$. Substitute the velocity into the force: $mF = F_o\left(1-\frac{v}{88}\right) = F_oe^{-Kt}$. The integral $U = \int_0^T mFvdt = \int_0^T 88F_oe^{-Kt}\left(1-e^{-Kt}\right)dt$

$U = 88F_o\int_0^T\left(e^{-Kt}-e^{-2Kt}\right)dt = -\frac{88F_o}{K}\left[e^{-Kt}-\frac{1}{2}e^{-2Kt}\right]_0^T = -\frac{88F_o}{K}\left[e^{-KT}-\frac{e^{-2KT}}{2}-\frac{1}{2}\right]$. The expression for the velocity is asymptotic in time to the limiting value of 60 *mi/hr*: in strict terms the velocity never reaches 60 *mi/hr*; in practical terms the velocity approaches within a few tenths of percent of 60 *mi/hr* within the first few minutes. Take the limit of the above integral:

$\lim_{T\to\infty}\int_0^T mFvdt = \lim_{T\to\infty} -\frac{88F_o}{K}\left[e^{-KT}-\frac{e^{-2KT}}{2}-\frac{1}{2}\right] = \frac{88F_o}{2K} = \frac{1}{2}\frac{W}{g}(88^2) \equiv \textit{kinetic energy}$, which checks, and confirms the expression for the velocity. *check.*

================================<>================================

=====================================<>=====================================

Problem 4.139 If a car traveling 65 *mi/hr* hits the crash barrier described in Problem 4.9, determine the maximum deceleration of the passengers are subjected to if (a) the car weighs 2500 *lb.*, (b) 5000 *lb.*

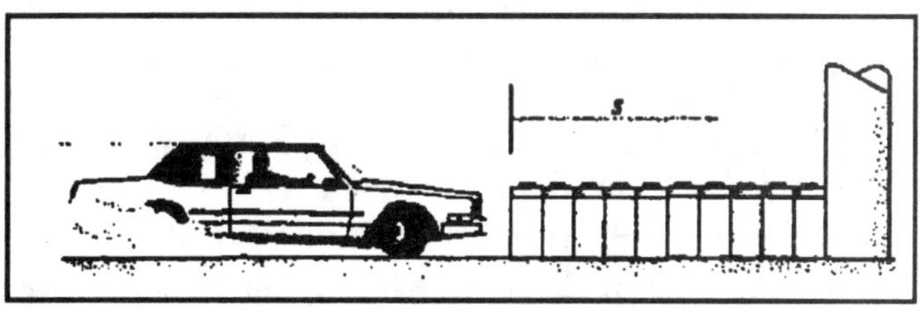

Solution Denote $A = 1000$, $B = 10{,}000$, so that the force exerted on the car is $F = -(A + Bs)\ lb$.

From Newton's second law: $m\frac{dv}{dt} = F = -(A + Bs)$. By inspection, the maximum magnitude of the acceleration occurs at the maximum value of *s*. This maximum is determined as follows: The conservation of energy condition after contact with the barrier is $\frac{1}{2}\left(\frac{W}{g}\right)v_0^2 = \frac{1}{2}\left(\frac{W}{g}\right)v_1^2 + V$, where *V* is associated with the force: $V = -\int_0^s F ds = \left(As + \frac{B}{2}s^2\right)$. Substitute and rearrange: $\frac{1}{2}\left(\frac{W}{g}\right)v_1^2 = \frac{1}{2}\left(\frac{W}{g}\right)v_0^2 - \left(As + \frac{B}{2}s^2\right)$.

When the car comes to a full stop $v_1 = 0$, and $s^2 + 2bs + c = 0$, where $b = \frac{A}{B}$, $c = -\frac{Wv_0^2}{gB}$, and the solution $s = -b \pm \sqrt{b^2 - c} = -\frac{A}{B} \mp \sqrt{\left(\frac{A}{B}\right)^2 + \frac{Wv_0^2}{gB}}$.

Substitute into the expression for the acceleration:

$$\left(\frac{dv}{dt}\right)_{\max} = -\frac{g}{W}\left(A - A \mp \sqrt{A^2 + \frac{BWv_o^2}{g}}\right) = -\sqrt{\frac{g^2A^2}{W^2} + \frac{Bgv_o^2}{W}}$$

. (a) Noting that: $v_o = 65\ mi/hr = 95.3\ ft/s$. For a 2500 *lb.* car, the max deceleration is

$$\boxed{\left|\left(\frac{dv}{dt}\right)_{\max}\right| = 1081.5\ ft/s^2 = 33.6g}$$

(b) For a 5000 *lb.* car

$$\boxed{\left|\left(\frac{dv}{dt}\right)_{\max}\right| = 764.7\ ft/s^2 = 23.8g}$$

=====================================<>=====================================

Problem 4.140 In a preliminary design for a mail sorting machine, parcels moving at 2 *ft/s* slide down a smooth ramp and are brought to rest by a linear spring. What should the spring constant be if you don't want the 10 *lb.* parcel to be subjected to a maximum deceleration greater than 10 g's?

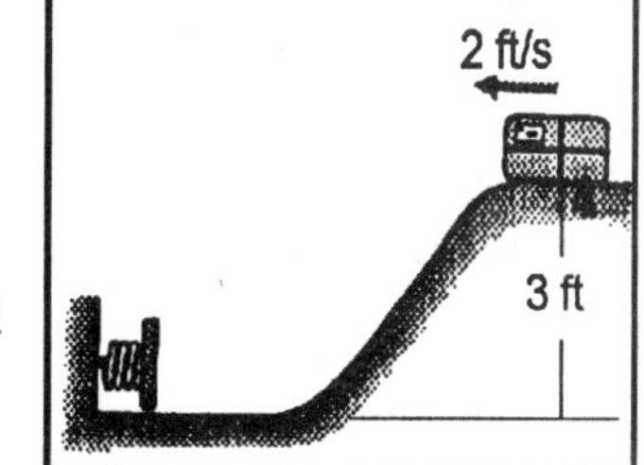

Solution From Newton's second law, the acceleration after contact with the spring is given by: $\frac{W}{g}\left(\frac{dv}{dt}\right) = -F = -kS$, where *k* is the spring constant and *S* is the stretch of the spring. Rearrange: $\left(\frac{dv}{dt}\right) = -\frac{gk}{W}S$. This expression has two unknowns, *k* and *S*. *S* is determined as follows: Choose the bottom of the ramp as the datum. The energy at the top of the ramp is $\frac{1}{2}\left(\frac{W}{g}\right)v_o^2 + V$, where *V* is the potential energy of the package due to gravity: $V = Wh$ where $h = 3\ ft$.

Solution continued on next page

The conservation of energy condition after contact with the spring is $\frac{1}{2}\left(\frac{W}{g}\right)v_0^2 + Wh = \frac{1}{2}\left(\frac{W}{g}\right)v_1^2 + \frac{1}{2}kS^2$. When the spring is fully compressed the velocity is zero, and $S = \sqrt{\frac{W}{gk}v_0^2 + 2\left(\frac{W}{k}\right)h}$. Substitute into the expression for the acceleration: $\left(\frac{dv}{dt}\right) = -\sqrt{k}\sqrt{\frac{gv_o^2}{W} + \frac{2g^2h}{W}}$

(where the negative sign appears because $\frac{dv}{dt} = -10g$), from which $k = \frac{\left(\frac{dv}{dt}\right)^2}{\left(\frac{gv_o^2}{W} + \frac{2g^2h}{W}\right)}$. Substitute

numerical values: $v_o = 2\ ft/s$, $W = 10\ lb$, $h = 3\ ft$, $\left(\frac{dv}{dt}\right) = -10g\ ft/s^2$, from which

$\boxed{k = 163.3\ lb/ft}$

==================================<>==================================

Problem 4.141 When the 1 *kg* collar is in position 1, the tension in the spring is 50 *N*, and the unstretched length of the spring is 260 *mm*. If the collar is pulled to position 2 and released from rest. what is its velocity when it returns to 1?

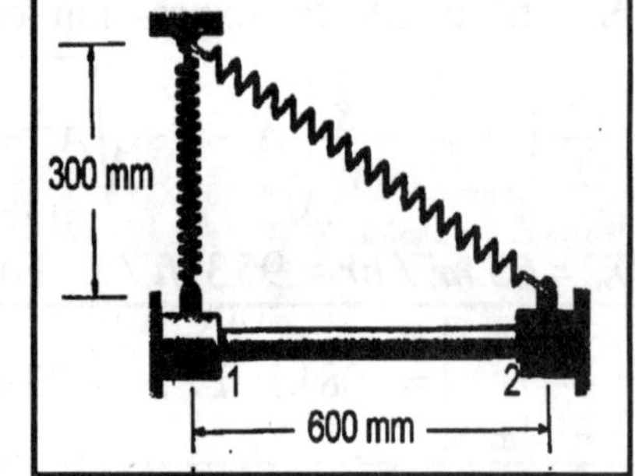

Solution The stretched length of the spring in position 1 is $S_1 = 0.3 - 0.26 = 0.04$ m. The stretched length of the spring in position 2 is $S_2 = \sqrt{0.3^2 + 0.6^2} - 0.26 = 0.411$ m. The spring constant is $k = \frac{50}{S_1} = 1250$ N/m. The potential energy of the spring in position 2 is $\frac{1}{2}kS_2^2$. The potential energy of the spring in position 1 is $\frac{1}{2}kS_1^2$. The energy in the collar at position 1 is $\frac{1}{2}kS_2^2 = \frac{1}{2}mv_1^2 + \frac{1}{2}kS_1^2$, from which

$v_1 = \sqrt{\frac{k}{m}(S_2^2 - S_1^2)} = 14.46$ m/s

==================================<>==================================

==================================<>==================================

Problem 4.142 In Problem 4.141 suppose that the tensions in the spring in positions 1 and 2 are 100 *N* and 400 *N* respectively. (a) What is the spring constant *k*? (b) If the collar is given a velocity of 15 *m/s* at 1, what is its velocity when it reaches 2?

Solution (a) Assume that the dimensions defining locations 1 and 2 remain the same, and that the unstretched length of the spring changes from that given in Problem 4.141. The stretched length of the spring in position 1 is $S_1 = 0.3 - S_o$, and in position 2 is $S_2 = \sqrt{0.3^2 + 0.6^2} - S_o$. The two conditions: $\sqrt{0.6^2 + 0.3^2} - S_o = \frac{400}{k}$, $0.3 - S_o = \frac{100}{k}$. Subtract the second from the first, from which $k = 809$ N/m. Substitute and solve: $S_o = 0.176$ m, and $S_1 = 0.124$ m, $S_2 = 0.494$ m. (b) The energy at the onset of motion at position 1 is $\frac{1}{2}mv_1^2 + \frac{1}{2}kS_1^2$. At position 2: $\frac{1}{2}mv_1^2 + \frac{1}{2}kS_1^2 = \frac{1}{2}mv_2^2 + \frac{1}{2}kS_2^2$, from which

$$\boxed{v_2 = \sqrt{v_1^2 + \frac{k}{m}\left(S_1^2 - S_2^2\right)} = 6.29 \ m/s}$$

==================================<>==================================

Problem 4.143 The 30 *lb.* weight is released from rest with the two springs ($k_A = 30$ *lb/ft*, $k_B = 15$ lb/ft) unstretched. (a) How far does the weight fall before rebounding? (b) What maximum velocity does it attain?

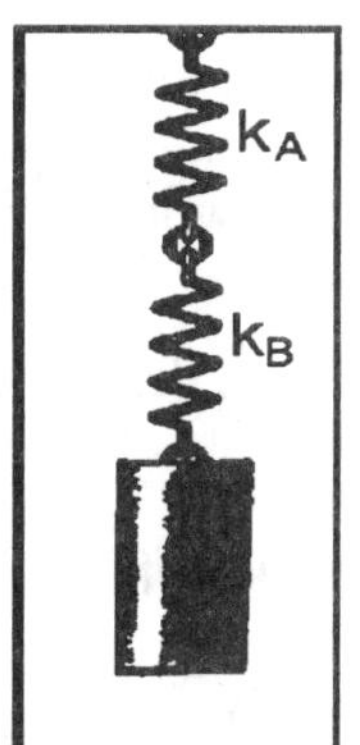

Solution Choose the datum as the initial position. (a) The work done as the weight falls is: for the springs $U_{spring} = \int_0^{-S_A} k_A s ds + \int_0^{-S_B} k_B s ds = -\frac{1}{2}k_A S_A^2 - \frac{1}{2}k_B S_B^2$. For the weight $U_{weight} = \int_0^{-(S_A+S_B)} -W ds = W(S_A + S_B)$. From the principle of work and energy: $U_{springs} + U_{weight} = (mv^2/2)$. At the juncture of the two springs the sum of the forces is $k_A S_A - k_B S_B = 0$, from which $S_B = \frac{k_A}{k_B} S_A$, from which

$-\left(\frac{1}{2}\right)k_A S_A^2\left(1 + \frac{k_A}{k_B}\right) + WS_A\left(1 + \frac{k_A}{k_B}\right) = \left(\frac{1}{2}mv^2\right)$ At the maximum extension the velocity is zero, from which $S_A = \frac{2W}{k_A} = 2$ ft, $S_B = \left(\frac{k_A}{k_B}\right)S_A = 4$ ft. The total fall of the weight is $S_A + S_B = 6$ ft (b) The maximum velocity occurs at $\frac{d}{dS_A}\left(\frac{1}{2}mv^2\right) = \frac{d}{dS_A}(U_{spring} + U_{weight}) = -k_A S_A(1 + \frac{k_A}{k_B}) + W(1 + \frac{k_A}{k_B}) = 0$, from which

$[S_A]_{v\max} = \frac{W}{k_A} = 1$ ft. The maximum velocity is $|v_{\max}| = \left[\sqrt{\frac{2(U_{spring} + U_{weight})}{m}}\right]_{S_A=1} = 9.82$ ft/s *Check:*

Replace the two springs with an equivalent spring of stretch $S = S_A + S_B$, with spring constant k_{eq}, from which $S = \frac{F}{k_A} + \frac{F}{k_B} = \frac{F}{k_{eq}}$. from which $k_{eq} = \frac{F}{S} = \frac{F}{S_A + S_B} = \frac{F}{\frac{F}{k_A} + \frac{F}{k_A}} = \frac{k_A + k_B}{k_A k_B} = 10$ lb/ft.. From conservation of energy $0 = mv^2/2 + k_{eq}S^2/2 - WS$. Set $v = 0$ and solve: $S = 2W/k_{eq} = 6$ ft is the maximum stretch. *check.* The velocity is a maximum when $\frac{d}{dS}\left(\frac{1}{2}mv^2\right) = W - k_{eq}S = 0$, from which $[S]_{v=v_{max}} = 3$ ft, and the maximum velocity is $v = 9.82$ ft/s. *check.*

==================================<>==================================

==================================<>==================================

Problem 4.144 The piston and the load it supports are accelerated upward by the gas in the cylinder. The total weight of and load is 1000 *lb.* The cylinder wall exerts a 50 *lb.* friction force on the piston as it rises. The net force exerted on the piston as it rises is $(p - p_{atm})A$, where *p* is the pressure of the gas, $p_{atm} = 2117 \text{ lb/ft}^2$ is the atmospheric pressure, and $A = 1 \text{ ft}^2$ is the cross sectional area of the piston. Assume that the product of *p* and the volume of the cylinder is constant. When $s = 1$ ft the piston is stationary and $p = 5000 \text{ lb/ft}^2$. What is the velocity at $s = 2$ ft?

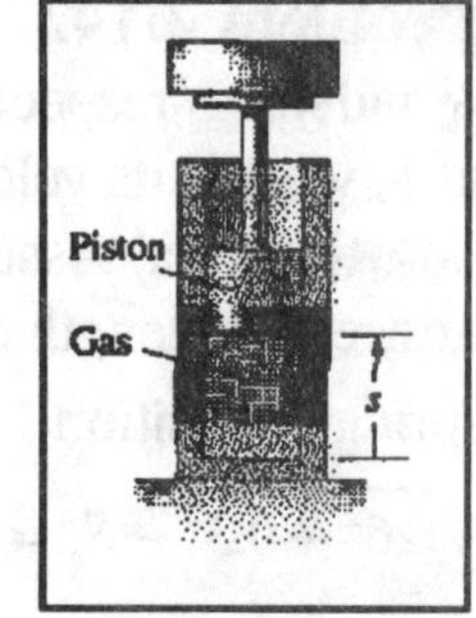

Solution

At the rest position, $p_o A s = p_o V = K$, where $V = 1 \text{ ft}^3$, from which $K = p_o$. Denote the datum: $s_o = 1$ ft. The potential energy of the piston due to the gas pressure after motion begins is

$$V_{gas} = -\int_{s_o}^{s} F ds = -\int_{s_o}^{s} (p - p_{atm}) A ds = p_{atm} A (s - s_o) - \int_{s_o}^{s} pA ds.$$

From which $V_{gas} = p_{atm} A (s - s_o) - K \int_{s_o}^{s} \frac{ds}{s} = p_{atm} A (s - s_o) - K \ln\left(\frac{s}{s_o}\right)$.

The potential energy due to gravity is $V_{gravity} = -\int_{s_o}^{s} (-W) ds = W(s - s_o)$.

The work done by the friction is $U_{friction} = \int_{s_o}^{s} (-f) ds = -f(s - s_o)$., where $f = 50$ lb.

From the principle of work and energy: $U_{friction} = \frac{1}{2}\left(\frac{W}{g}\right) v^2 + V_{gas} + V_{gravity}$

Rearrange: $\frac{1}{2}\left(\frac{W}{g}\right) v^2 = U_{friction} - V_{gas} - V_{gravity}$. At $s = 2$ ft,

$\frac{1}{2}\left(\frac{W}{g}\right) v^2 = -(-1348.7) - (1000) - 50 = 298.7 \text{ ft-lb}$, from which $v = \sqrt{\frac{2(298.7)g}{W}} = 4.38 \text{ ft/s}$

==================================<>==================================

Problem 4.145 When a 22 *Mg* rocket's engine burns out at an altitude of 2 *km*, its velocity is 3 *km/s* and it is traveling at an angle of 60° relative to the horizontal. Neglect the variation in the gravitational force with altitude. (a) If you neglect aerodynamic forces, what is the magnitude of the rocket's velocity when it reaches an altitude of 6 *km*? (b) If the rocket's actual velocity when it reaches an altitude of 6 *km* is 2.8 *km/s*, how much work is done by aerodynamic forces as the rocket moves from 2 *km* to 6 *km* altitude?

Solution Choose the datum to be 2 *km* altitude. (a) The energy is $\frac{1}{2} m v_o^2$ at the datum. The energy condition of the rocket when it reaches 6 *km* is $\frac{1}{2} m v_o^2 = \frac{1}{2} m v^2 + mgh$, where $h = (6 - 2) \times 10^3 = 4 \times 10^3$ m. Rearrange the energy expression: $v^2 = v_o^2 - 2gh$, from which the velocity at 6 *km* is $v = \sqrt{v_o^2 - 2gh} = 2.987 \text{ km/s}$ (b) Define U_{aero} to be the work done by the aerodynamic forces. The energy condition at 6 *km* is $\frac{1}{2} m v_o^2 = \frac{1}{2} m v^2 + mgh - U_{aero}$. Rearrange:

$$U_{aero} = +\frac{1}{2} m v^2 + mgh - \frac{1}{2} m v_o^2 = -1.19 \times 10^{10} \text{ N-m}$$

==================================<>==================================

==================================<>==================================

Problem 4.146 The 12 *kg* collar A is at rest in the position shown at $t = 0$ and is subjected to the tangential force $F = 24 - 12t^2$ N for 1.5 *s*. Neglecting friction, what maximum height *h* does it reach?

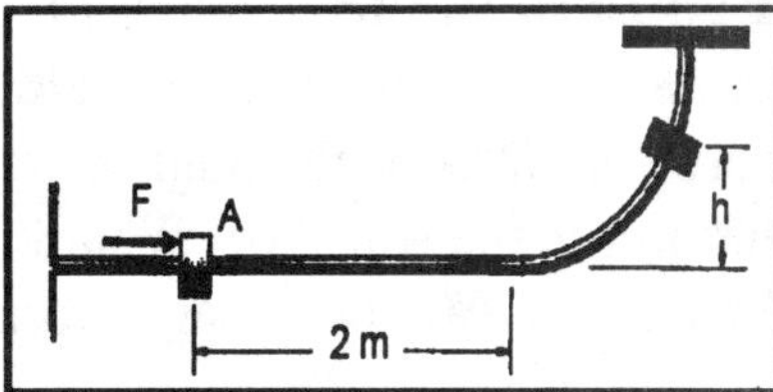

Solution Choose the datum at the initial point. The strategy is to determine the velocity at the end of the 1.5 *s* and then to use work and energy methods to find the height *h*. From Newton's second law: $m\frac{dv}{dt} = F = 24 - 12t^2$. Integrating: $v = \frac{1}{m}\int_0^{1.5}(24 - 12t^2)dt = \left(\frac{1}{m}\right)\left[24t - 4t^3\right]_0^{1.5} = 1.875 \text{ m/s}$. [*Note*: The displacement during this time must not exceed 2 *m*. Integrate the velocity: $s = \left(\frac{1}{m}\right)\int_0^{1.5}(24t - 4t^3)dt = \left(\frac{1}{m}\right)\left[12t^2 - t^4\right]_0^{1.5} = 1.82 \text{ m} < 2 \text{ m}$, so the collar is still at the datum level at the end of 1.5 *s*.] The energy condition as the collar moves up the bar is $\frac{1}{2}mv_0^2 = \frac{1}{2}mv^2 + mgh.$

At the maximum height *h*, the velocity is zero, from which $h = \frac{v_o^2}{2g} = 0.179 \text{ m}$

==================================<>==================================

Problem 4.147 Suppose that in designing a loop for a roller coaster's track, you establish as a safety criterion that at the top of the loop, the normal force exerted on a passenger by the roller coaster should equal 10% of his weight. (That is, the passenger's "effective weight" pressing him down into his seat is 10 % of his weight.) The roller coaster is moving at 62 *ft/s* when it enters the loop. What is the necessary instantaneous radius of curvature at the top of the loop?

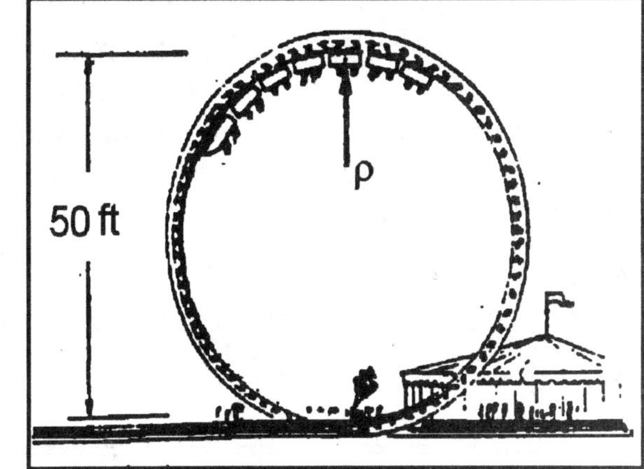

Solution The energy at the top of the loop is $\frac{1}{2}mv_o^2 = \frac{1}{2}mv_{top}^2 + mgh$, where $v_o = 62 \text{ ft/s}$, $h = 50 \text{ ft}$, and $g = 32.2 \text{ ft/s}^2$, from which $v_{top} = \sqrt{v_o^2 - 2gh} = 25 \text{ ft/s}$. From Newton's second law: $m\left(\frac{v_{top}^2}{\rho}\right) = (1.1)mg$, from which $\rho = \frac{v_{top}^2}{1.1g} = 17.7 \text{ ft}$

==================================<>==================================

==================================<>==================================

Problem 4.148 A 180 *lb.* student runs at 15 ft/s, grabs a rope, and swings out over a lake. He releases the rope when his velocity is zero. (a) What is the angle θ when he releases the rope? (b) What is the tension in the rope? (c) What is the maximum tension in the rope?

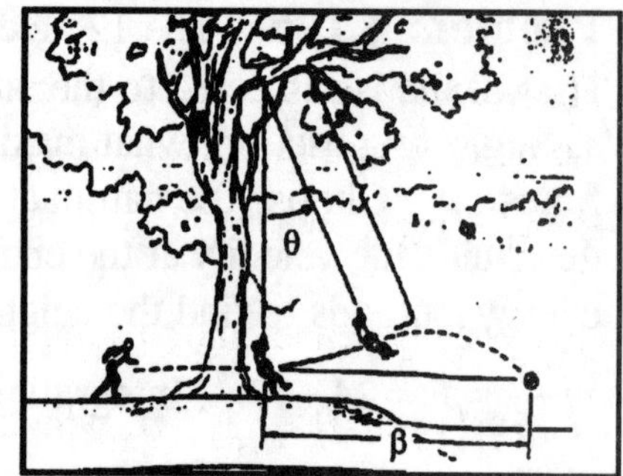

Solution (a) The energy condition after the seizure of the rope is $\frac{1}{2}mv_o^2 = \frac{1}{2}mv^2 + mgL(1-\cos\theta)$, where $v_o = 15$ ft/s, $L = 30$ ft. When the velocity is zero, $v_o^2 = 2gL(1-\cos\theta)$, from which $\cos\theta = 1 - \frac{v_o^2}{2gL} = 0.883$, $\theta = 27.9°$ (b) From the energy equation $v^2 = v_o^2 - 2gL(1-\cos\theta)$. From Newton's second law, $(W/g)(v^2/L) = T - W\cos\theta$, from which $T = \left(\frac{W}{g}\right)\left(\frac{v_o^2}{L}\right) + W\cos\theta = 159.0$ lb. (c) The maximum tension occurs at the angle for which $\frac{dT}{d\theta} = 0 = -2W\sin\theta - W\sin\theta$, from which $\theta = 0$, from which $T_{max} = W\left(\frac{v_o^2}{gL}+1\right) = 222$ lb

==================================<>==================================

Problem 4.149 If the student in Problem 4.148 releases the rope when $\theta = 25°$, what maximum height does he reach relative to his position when he grabs the rope?

Solution Use the solution to Problem 4.126. [The height when he releases the rope is $h_1 = L(1-\cos 25°) = 2.81$ ft.] Before he releases the rope, the total energy is $\frac{1}{2}\left(\frac{W}{g}\right)v_0^2 - WL = \frac{1}{2}\left(\frac{W}{g}\right)v^2 - WL\cos\theta$. Substitute $v_o = 15$ ft/s, $\theta = 25°$ and solve: $v = 6.63$ ft/s. The horizontal component of velocity is $v\cos\theta = 6.01$ ft/s. From conservation of energy: $W(2.81) + \frac{1}{2}m(6.63^2) = Wh + \frac{1}{2}m(6.01^2)$ from which $h = 2.93$ ft

==================================<>==================================

Problem 4.150 A boy takes a running start and jumps on his sled at 1. He leaves the ground at 2 and lands in deep snow at a distance of $b = 25$ ft. How fast was he going at 1?

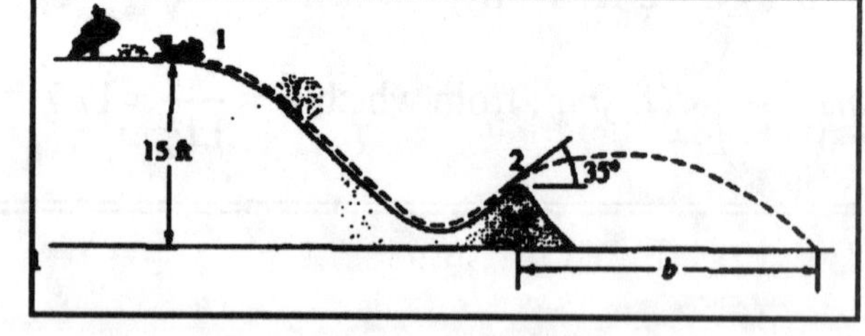

Solution The components of velocity at the point of leaving the ground are $v_y = v_2\sin\theta$ and $v_x = v_2\cos\theta$, where $\theta = 35°$. The path is $y = -\frac{g}{2}t^2 + (v_2\sin\theta)t + h$, where $h = 5$ ft, and $x = (v_2\cos\theta)t$. At impact $y = 0$, from which $t_{impact}^2 + 2bt_{impact} + c = 0$, where $b = -\frac{v_2\sin\theta}{g}$ [not to be confused with the b in the drawing], $c = -\frac{2h}{g}$. From which, since the time is positive, the time of impact is (1) $t_{impact} = \frac{v_2\sin\theta}{g}\left(1+\sqrt{1+\frac{2gh}{v_2^2\sin^2\theta}}\right)$. The range is (2) $x(v_1) = b = (v_2\cos\theta)t_{impact}$.

Solution continued on next page

The velocity v_2 is found in terms of the initial velocity from the energy conditions: Choose the datum at the point where he leaves the ground. The energy after motion begins but before descent is under way is $\frac{1}{2}mv_1^2 + mgh_1$, where h_1 is the height above the point where he leaves the ground, $h_1 = 15 - 5 = 10$ ft. The energy as he leaves the ground is $\frac{1}{2}mv_o^2 + mgh_o = \frac{1}{2}mv_2^2$, from which (3) $v_2 = \sqrt{v_1^2 + 2gh_1}$. The function $x(v_1) = (v_2\cos\theta)t_{impact} - b$, where $b = 25$ ft, was graphed as a function of initial velocity, using the equations (2) and (3) above, to find the zero crossing. The value was refined by iteration to yield $v_1 = 4.72$ ft/s The other values were $v_2 = 25.8$ ft/s, and the time in the air before impact was $t_{impact} = 1.182$ s. *Check:* An analytical solution is found as follows: Combine (1) and (2)
$b = \frac{v_2^2 \sin\theta\cos\theta}{g}\left(1 + \sqrt{1 + \frac{2gh}{v_2^2\sin^2\theta}}\right)$. Invert this algebraically to obtain
$v_2 = b\sqrt{\frac{g}{2\cos\theta(b\sin\theta + h\sin\theta)}} = 25.80$ ft. Use $v_1^2 = v_2^2 - 2g(h_1 - h_2)$, from which $v_1 = 4.724$ ft/s.
check.

==================================<>==================================

Problem 4.151 In Problem 4.150, if the boy starts at 1 going 15 *ft/s*, what distance *b* does he travel through the air?

Solution

Use the solution to Problem 4.150. The distance $b = (v_2\cos\theta)t_{impact}$, where
$t_{impact} = \frac{v_2\sin\theta}{g}\left(1 + \sqrt{1 + \frac{2gh}{v_2^2\sin^2\theta}}\right)$, and $v_2 = \sqrt{v_1^2 + 2gh_1}$. Numerical values are: $h = 5$ ft, $\theta = 35°$, $h_1 = 10$ ft, $v_1 = 15$ ft/s. Substituting, $b = 31.176... = 31.2$ ft

==================================<>==================================

Problem 4.152 The 1 *kg* collar is attached to the linear spring ($k = 500$ N/m) by a string. The collar starts from rest in the position shown, and the initial tension in the spring is 100 *N*. What distance does the collar slide up the smooth bar?

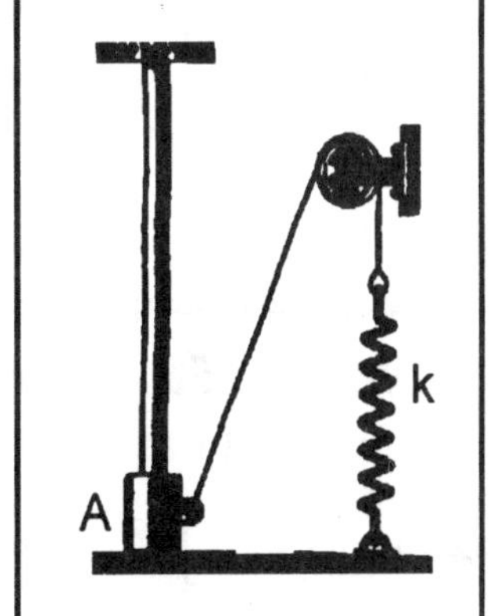

Solution

The deflection of the spring is $S = \frac{100}{k} = 0.2$ m. The potential energy of the spring is $V_{spring} = \frac{1}{2}kS^2$. The energy condition after the collar starts sliding is $V_{spring} = \frac{1}{2}mv^2 + mgh$. At the maximum height, the velocity is zero, from which $h = \frac{V_{spring}}{mg} = \frac{k}{2mg}S^2 = 1.02$ m

==================================<>==================================

Problem 4.153 The masses $m_A = 40kg$ and $m_B = 60kg$. The collar A slides on the smooth horizontal bar. The system is released from rest. Use conservation of energy to determine the velocity of the collar A when it has moved *0.5-m* to the right.

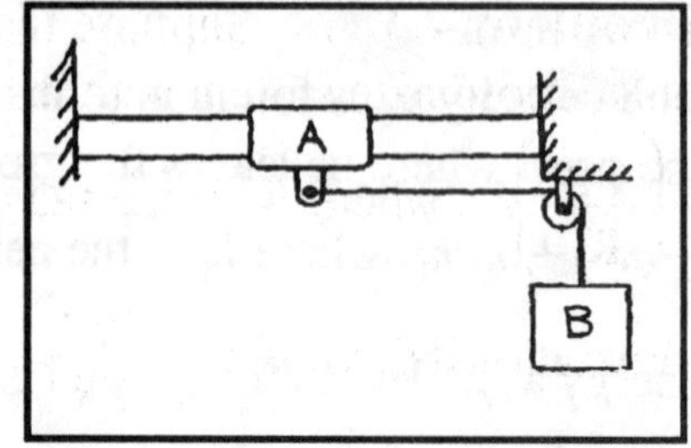

Solution: Placing the datum for B at its initial position, conservation of energy gives $T_1 + V_1 = T_2 + V_2$: Evaluating, we get

$$0+0=\frac{1}{2}(40)v^2+\frac{1}{2}(60)v^2-(60)(9.81)(0.5) \text{ or } v = 2.43 m/s.$$

Problem 4.154 The spring constant is *k = 850 N/m*, $m_A = 40kg$, and $m_B = 60kg$. The collar A slides on the smooth horizontal bar. The system is released from rest in the position shown. Use conservation of energy to determine the velocity of the collar A when it has moved *0.5-m* to the right.

Solution:

Let v_A and v_B be the velocities of A and B when A has moved $0.5m$. The component of $A's$ velocity parallel to the cable equals $B's$ velocity: $v_A \cos 45° = v_B$. $B's$ downward displacement during $A's$ motion is

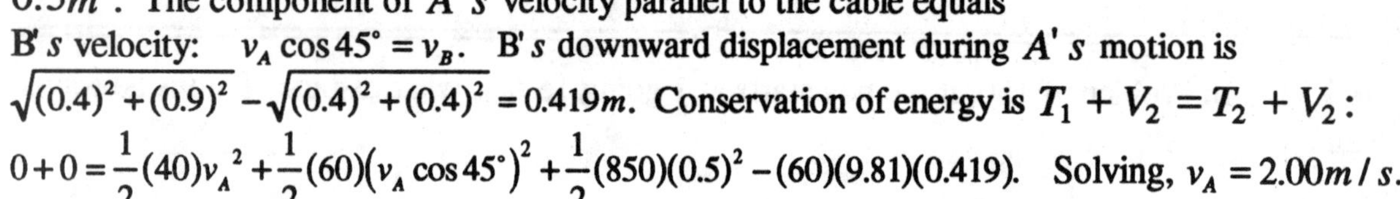

$$\sqrt{(0.4)^2+(0.9)^2}-\sqrt{(0.4)^2+(0.4)^2}=0.419m.$$ Conservation of energy is $T_1 + V_2 = T_2 + V_2$:

$$0+0=\frac{1}{2}(40)v_A{}^2+\frac{1}{2}(60)(v_A\cos 45°)^2+\frac{1}{2}(850)(0.5)^2-(60)(9.81)(0.419).$$ Solving, $v_A = 2.00 m/s$.

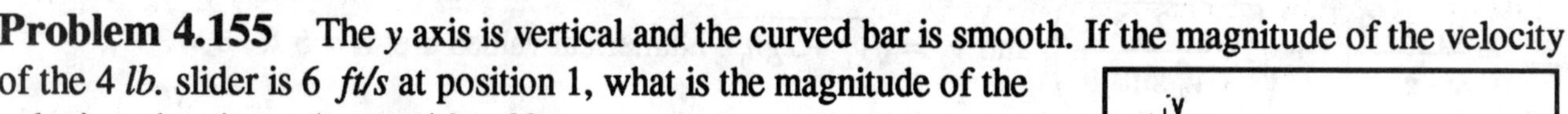

Problem 4.155 The y axis is vertical and the curved bar is smooth. If the magnitude of the velocity of the 4 *lb.* slider is 6 *ft/s* at position 1, what is the magnitude of the velocity when it reaches position 2?

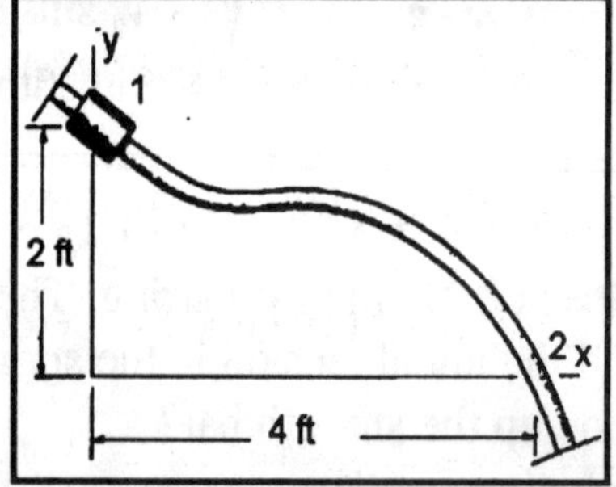

Solution Choose the datum at position 2. At position 2, the energy condition is $\frac{1}{2}\left(\frac{W}{g}\right)v_1^2 + Wh = \frac{1}{2}\left(\frac{W}{g}\right)v_2^2$, where $h = 2$, from which

$$v_2 = \sqrt{v_1^2+2gh} = \sqrt{6^2+2g(2)} = 12.83 \text{ ft/s}$$

Problem 4.156 In Problem 4.155, determine the magnitude of the slider's velocity if it is subjected to the additional force $\vec{F} = 3x\vec{i} - 2\vec{j}$ (lb) during its motion.

Solution $U = \int \vec{F}\cdot d\vec{r} = \int_2^0 (-2)dy + \int_0^4 3xdx = [-2y]_2^0 + \left[\frac{3}{2}x^2\right]_0^4 = 4 + 24 = 28 \text{ ft-lb}$. From the solution to Problem 4.155, the energy condition at position 2 is $\frac{1}{2}\left(\frac{W}{g}\right)v_1^2 + Wh + U = \frac{1}{2}\left(\frac{W}{g}\right)v_2^2 + V$, from which

$$v_2 = \sqrt{v_1^2+2gh+\frac{2g(28)}{W}} = \sqrt{6^2+2g(2)+\frac{2g(28)}{4}} = 24.8 \text{ ft/s}$$

Problem 4.157 Suppose that an object of mass m is beneath the surface of the earth. In terms of a polar coordinate system with its origin at the center of the earth, the gravitational force on the object is $-\left(\frac{\text{mgr}}{\text{R}_\text{E}}\right)\vec{e}_r$, where R_E is the radius of the earth. Show that the potential energy associated with the gravitational force is $V = \frac{mgr^2}{2R_E}$.

Solution

By definition, the potential associated with a force $\vec{F}$ is $\text{V} = -\int \vec{F}\cdot\text{d}\vec{r}$. If $\text{d}\vec{r} = \vec{e}_r\text{dr} + \text{r}\vec{e}_\theta\text{d}\theta$, then

$$\text{V} = -\int\left(-\frac{\text{mgr}}{\text{R}_\text{E}}\right)\vec{e}_r\cdot\vec{e}_r\text{dr} - \int\left(-\frac{\text{mgr}}{\text{R}_\text{E}}\right)\vec{e}_r\cdot\vec{e}_\theta\text{rd}\theta = -\int\left(-\frac{\text{mgr}}{\text{R}_\text{E}}\right)dr = \left(\frac{mgr^2}{2R_E}\right)$$

Problem 4.158 It has been pointed out that if tunnels could be drilled straight through the earth between points on the surface, trains could travel between these points using gravitational force for acceleration and deceleration. (The effects of friction and aerodynamic drag could be minimized by evacuating the tunnels and using magnetically levitated trains.) Suppose that such a train travels from the North Pole to the equator. Determine the magnitude of the train's velocity (a) when it arrives at the equator; (b) when it is halfway from the North Pole to the equator. The radius of the earth is $R_E = 3960$ mi.

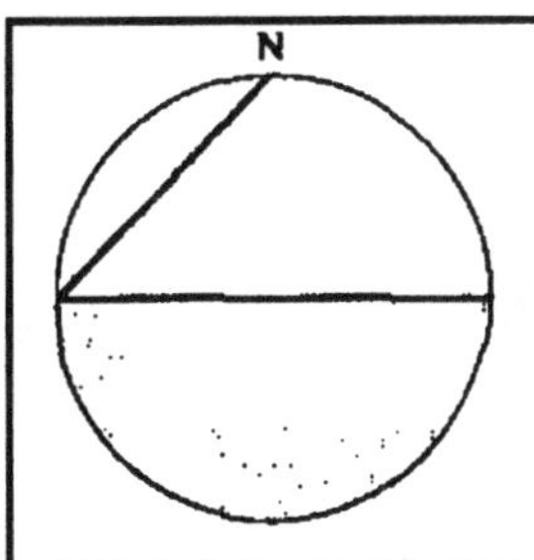

Solution The potential associated with gravity is $V_{gravity} = \frac{mgr^2}{2R_E}$. With an initial velocity at the North Pole of zero, from conservation of energy, at any point in the path $\left(\frac{mgr^2}{2R_E}\right)_{NP} = \frac{1}{2}mv^2 + \left(\frac{mgr^2}{2R_E}\right)$. (a) At the equator, the conservation of energy condition reduces to $\left(\frac{mgR_E}{2}\right) = \frac{1}{2}mv_{EQ}^2 + \left(\frac{mgR_E}{2}\right)$, from which $v_{EQ} = 0$ (b) At the midway point, $r = R_E\sin 45^o = \frac{R_E}{\sqrt{2}}$, and from conservation of energy $\left(\frac{mgR_E}{2}\right) = \frac{1}{2}mv_M^2 + \left(\frac{mgR_E}{4}\right)$, from which $v_M = \sqrt{\frac{gR_E}{2}} = 18{,}339$ ft/s $= 12{,}504$ mi/hr *Note:* Adopting the value $g = 32.2$ ft/s, $v_M = 18348$ ft/s $= 12{,}510$ mi/hr.

Problem 4.159 In Problem 4.137, what is the maximum power transferred to the locomotive during its acceleration?

Solution From Problem 4.137, the drive wheel traction force $F = 135{,}000$ lb is a constant, and the final velocity is $v = 60$ mi/hr $= 88$ ft/s. The power transferred is $P = Fv$, and since the force is a constant, *by inspection* the maximum power transfer occurs at the maximum velocity, from which $P = Fv = (135000)(88) = 11.88\times10^6$ ft-lb/sec $= 21{,}600$ hp

---

==============================<>==============================

Problem 4.160 Just before it lifts off, the 10.5 *Mg* airplane is traveling at 60 *m/s*. The total horizontal force exerted by its engines is 189 *kN* and the plane is accelerating at 15 m/s^2. (a) How much power is being transferred to the plane by its engines? (b) What is the total power being transferred to the plane?

Solution:

(a) The power being transferred by its engines is $P = Fv = (189 \times 10^3)(60) = 1.134 \times 10^7$ Joule / s = 11.34 MW. (b) Part of the thrust of the engines is accelerating the airplane: From Newton's second law, $m\frac{dv}{dt} = T = (10.5 \times 10^3)(15) = 157.5$ kN. The difference $(189 - 157.5) = 31.5$ kN is being exerted to overcome friction and aerodynamic losses.

(b) The total power being transferred to the plane is $P_t = (157.5 \times 10^3)(60) = 9.45$ MW

==============================<>==============================

Problem 4.161 The "Paris Gun", used by Germany in World War I, has a range of 120 *km*, a 37.5 *m* barrel, a muzzle velocity of 1550 *m/s*, and it fired a 120 *kg* shell. (a) If you assume the shell's acceleration to be constant, what maximum power was transferred to it as it traveled along the barrel? (b) What average power was transferred to the shell?

Solution:

From Newton's second law, $m\frac{dv}{dt} = F$, from which, for a constant acceleration, $v = \left(\frac{F}{m}\right)t + C$. At $t = 0$, $v = 0$, from which $C = 0$. The position is $s = \frac{F}{2m}t^2 + C$. At $t = 0$, $s = 0$, from which $C = 0$. At $s = 37.5$ m, $v = 1550$ m / s, from which $F = 3.844 \times 10^6$ N and $t = 4.84 \times 10^{-2}$ s is the time spent in the barrel.

The power is $P = Fv$, and since F is a constant and v varies monotonically with time, the maximum power transfer occurs just before the muzzle exit: $P = F(1550) = 5.96 \times 10^9$ joule / s = 5.96 GW. (b) From Eq.(4.18) the average power transfer is $P_{ave} = \frac{\frac{1}{2}mv_2^2 - \frac{1}{2}mv_1^2}{t} = 2.98 \times 10^9$ W = 2.98 GW

==============================<>==============================

================================<>================================

Problem 5.1 The aircraft carrier *Nimitz* weighs 91,000 tons (A ton is 2000 lb). Suppose that its engines and hydrodynamic drag exert a constant 1,000,000 *lb* decelerating force on it. (a) Use the principle of impulse and momentum to determine how long it requires the ship to come to rest from its top speed of 30 *knots*. (A knot is approximately 6076 *ft/s*.) (b) Use the principle of work and energy to determine the distance the ship travels during the time it takes to come to rest.

Solution: . (a) The mass is $m = 9.1\times10^4(2\times10^3)/32.2 = 5.65\times10^6$ slug. The velocities

$v_1 = \dfrac{30(6076)}{3600} = 50.6\ ft/s$, $v_2 = 0$, From Eq (5.2) the time is given by

$(t_1 - t_2) = \dfrac{mv_2 - mv_1}{-\sum F_{ave}} = 286.2\ s = 4.77\ min$ (b) The work done by the force is

$U = \int -\sum F_{ave} ds = -\sum F_{ave} s$ since the force is constant. From the principle of work and energy,

$U = \dfrac{1}{2}mv_2^2 - \dfrac{1}{2}mv_1^2$, from which $\boxed{s = \dfrac{0 - \frac{1}{2}mv_1^2}{-\sum F_{ave}} = 7245\ ft = 1.37\ mi}$

================================<>================================

Problem 5.2 The 2000 *lb.* drag racer accelerates from rest to 300 *mi/hr* in 6 *s*. (a) What impulse is applied to the car during the 6 *s*? (b) If you assume as a first approximation that the tangential force exerted on the car is constant, what is the magnitude of the force?

Solution: (a) The velocity is $v_1 = 300\left(\dfrac{5280}{3600}\right) = 440\ ft/s$. From Eq (5.1) the impulse is

$\boxed{\int_{t_1}^{t_2} \sum F dt = mv_2 - mv_1 = \left(\dfrac{W}{g}\right)v_2 = 2.73\times10^4\ lb\text{-}s}$. From Eq (5.4) the force is

$\boxed{\sum F = \dfrac{1}{t_2 - t_1}\left(\dfrac{W}{g}\right)v_2 = \dfrac{2.73\times10^6}{6} = 4.555\times10^3\ lb}$

================================<>================================

Problem 5.3 The 21,900 *kg Gloster Saro Protector,* designed for rapid response to airport emergencies, accelerates from rest to 80 *km/hr* in 35 *s*.(a) What impulse is applied to the vehicle during the 35 *s*? (b) If you assume as a first approximation that the tangential force exerted on the vehicle is constant, what is the magnitude of the force? (c) What average power is transferred to the vehicle?

Solution: (a) By definition, the impulse is

$\boxed{\int_{t_1}^{t_2} F dt = mv_2 - mv_1 = 21900\left(\dfrac{80\times10^3}{3600}\right) = 4.867\times10^5\ N-s}$

(b) The average force is $\boxed{F = \dfrac{mv_2}{t_2 - t_1} = 1.39\times10^4\ N}$.

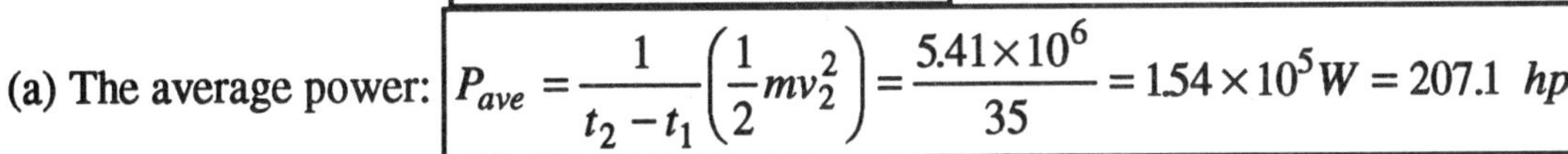

(a) The average power: $\boxed{P_{ave} = \dfrac{1}{t_2 - t_1}\left(\dfrac{1}{2}mv_2^2\right) = \dfrac{5.41\times10^6}{35} = 1.54\times10^5 W = 207.1\ hp}$

================================<>================================

==================================<>==================================

Problem 5.4 The combined weight of the motorcycle and the rider is 300 *lb*. The coefficient of kinetic friction between the motorcycle's tires and the road is $\mu_k = 0.8$. Suppose that the rider starts from rest and spins the rear (drive) wheel. The normal force between the rear wheel and the road is 250 *lb*. (a) What impulse does the friction force on the rear wheel exert in 5 *s*? (b) If you neglect other horizontal forces, what velocity is attained in 5 *s*?

Solution: (a) The force exerted by the tire on the road is $F = \mu_k N = (0.8)(250) = 200\ lb$. The impulse is $\boxed{\int_{t_1}^{t_2} F dt = 200(5) = 1000\ lb\text{-}s}$. (b) The velocity is obtained from the impulse by $mv_2 - mv_1 = \int_{t_1}^{t_2} F dt$, from which, since the cyclist starts from rest,

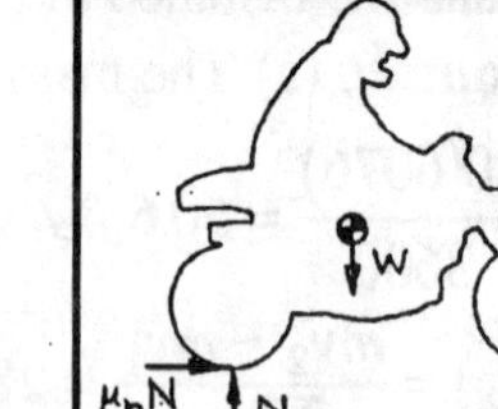

$$\boxed{v_2 = \frac{g(1000)}{W} = 107.3\ ft/s = 73.2\ mi/hr}$$

==================================<>==================================

Problem 5.5 The combined mass of the motorcycle and rider is *160-kg*. The motorcycle starts from rest at $t = 0$. The total horizontal force exerted on it from $t = 0$ to $t = 10\ s$ is $1200e^{-0.1t}$ *(newtons)*.
(a) Determine the impulse exerted from $t = 0$ to $t = 10\ s$.
(b) What is the motorcycle's velocity in *km/hr* at $t = 10\ s$?

Solution: (a) The impulse is $\int_{t_1}^{t_2} F dt = \int_0^{10} 1200e^{-0.1t} dt = 1200\left[\frac{e^{-0.1t}}{-0.1}\right]_0^{10} = 7590 N - s.$

(b) $7590 = mv_2 - mv_1 = (160)(v_2 - 0)$ or $v_2 = 47.4 m/s = 171 km/hr.$

==================================<>==================================

Problem 5.6 In Problem 5.5, what average total horizontal force acts on the motorcycle from $t = 0$ to $t = 10\ s$?

Solution: The average force is given by $\sum F_{av}\Delta t = mv_2 - mv_1$. Evaluating, we get $\sum F_{av}(10) = (160)(47.4 - 0)$, or $\sum F_{av} = 759N.$

==================================<>==================================

Problem 5.7 An astronaut drifts toward a space station at 8 *m/s*. He carries a maneuvering unit (a small hydrogen peroxide rocket) that has an impulse rating of 720 *N-s*. The total mass of the astronaut, his suit, and the maneuvering unit is 120 *kg*. If he uses all of the impulse to slow himself down, what will be his velocity relative to the station?

Solution: By definition, the impulse is $\int_{t_1}^{t_2} F dt = mv_1 - mv_2 = 720\ N\text{-}s$, (since the impulse is being used to slow the astronaut down) from which $\boxed{v_2 = v_1 - \frac{720}{m} = 8 - \frac{720}{120} = 2\ m/s}$ toward the station.

==================================<>==================================

Problem 5.8 The total force on a *20-kg* object is $10t^2\,\mathbf{i}+60t\,\mathbf{j}$ (*newtons*). At $t = 0$, the object's velocity is $8\,\mathbf{i}-4\,\mathbf{j}$ (m/s).
(a) What impulse is applied to the object from $t = 0$ to $t = 4\ s$?
(b) What is the object's velocity at $t = 4\ s$?

Soution:

(a) $\int_{t_1}^{t_2}\sum \bar{F}\,dt=\int_0^4\left(10t^2\bar{i}+60t\bar{j}\right)dt=\left[\frac{10t^3}{3}\bar{i}+30t^2\bar{j}\right]_0^4=213\bar{i}+480\bar{j}(N-s).$

(b) $213\bar{i}+480\bar{j}=m\bar{v}_2-m\bar{v}_1=20v_2-20\left(8\bar{i}-4\bar{j}\right)$, Hence $\bar{v}_2=18.7\bar{i}+20\bar{j}(m/s).$

Problem 5.9 In Problem 5.8, what is the average total force on the object from $t = 0$ to $t = 4\ s$?

Solution: The average force is given by $\sum \bar{F}_{va}\Delta t=m\bar{v}_2-m\bar{v}_1$: Evaluating, we get $\sum F_{av}(4)=20\left(18.7\bar{i}+20\bar{j}\right)-20\left(8\bar{i}-4\bar{j}\right)$, or $\sum F_{av}=53.3\bar{i}+120\bar{j}(N).$

Problem 5.10 The *1-lb* collar A is initially at rest in the position shown on the smooth horizontal bar. At $t = 0$, a force $\mathbf{F}=(t^2/20)\,\mathbf{i}+(t/10)\,\mathbf{j}-(t^3/30)\mathbf{k}$ (*lb*) is applied to the collar, causing it to slide along the bar. What is the velocity of the collar at $t = 2\ s$?

Solution: The impulse applied to the collar is

$\int_{t_1}^{t_2}\sum \bar{F}\,dt=mv_{x2}-mv_{x1}$: Evaluating, we get $\int_0^2\frac{1}{20}t^2dt=(1/32.2)v_{x2}$, or $\left[\frac{1}{60}t^3\right]_0^2=(1/32.2)v_{x2}.$

Hence, $v_{x2}=4.29\,ft/s.$

Problem 5.11 (a) In Problem 5.10, use the principle of impulse and momentum to determine the collar's velocity as a function of time. (b) Use the result of (a) to determine the time at which the collar reaches the right hand end of the bar.

Solution: (a) The impulse acting on the collar is $\int_{t_1}^{t_2}\sum \bar{F}dt=mv_{x2}-mv_{x1}$: or $\int_o^t\frac{1}{20}t^2dt=(1/32.2)v_x-0,$

and we get $v_x=\frac{32.2}{60}t^3\,ft/s.$

(b) Start with $v_x=\frac{dx}{dt}=\frac{32.2}{60}t^3$, Integrate $\int_0^4 dx=\int_0^t\frac{32.2}{60}t^3dt$, and evaluate $4=\left(\frac{32.2}{60}\right)\left(\frac{t^4}{4}\right)$. Hence $t=2.34s.$

Problem 5.12 During the first 5 s of the 32,200 $lb.$ airplane's takeoff roll, the pilot increases the engine's thrust at a constant rate from 5000 $lb.$ to its full thrust of 25,000 $lb.$ (a) What impulse does the thrust exert on the airplane during the 5 s? (b) If you neglect other forces, what total time is required for the airplane to reach its takeoff speed of 150 ft/s?

Solution:

(a) The thrust varies linearly with time for the first 5 s, and then is constant thereafter. For the first 5 s,

$F = 5000 + \frac{20000}{5}t = 5000 + 4000t\ lb$. The impulse is

$$\boxed{\int_{t_1}^{t_2} F dt = \left[5000t + 2000t^2\right]_0^5 = 75{,}000\ lb\text{-}s}$$

(b) The airplane starts from rest. The total impulse is

$\int_{t_1}^{t_2} F dt = \left(\frac{W}{g}\right)v_2 + \left(\left(\frac{W}{g}\right)v_3 - \left(\frac{W}{g}\right)v_2\right) = \left(\frac{W}{g}\right)v_3 = \left(\frac{32200}{g}\right)(150) = 150{,}140\ lb\text{-}s$. The total impulse is also $\int_{0_1}^{5} F_1 dt + \int_5^{t_3} F_2 dt = 75000 + [25000t_3 - 125000] = 25000t_3 - 50000\ lb\text{-}s$. Solve: $\boxed{t_3 = 8.0\ s}$ is the total time required to reach the takeoff speed.

Problem 5.13 The 100 $lb.$ box starts from rest and is subjected to the force shown. If you neglect friction, what is the box's velocity at $t = 8\ s$?

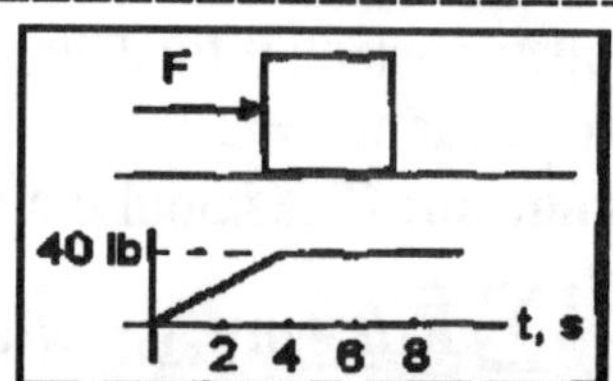

Solution:

The impulse is the area under the force-time curve:

$= \left(\frac{1}{2}\right)(40)(4) + (8-4)40 = 240\ lb\text{-}s$. The impulse is also given by

$240 = \left(\frac{W}{g}\right)v = 3.108v$. Solve: $\boxed{v = 77.2\ ft/s}$

Problem 5.14 Solve Problem 5.13 if the coefficients of friction between the box and the floor are $\mu_s = \mu_k = 0.2$.

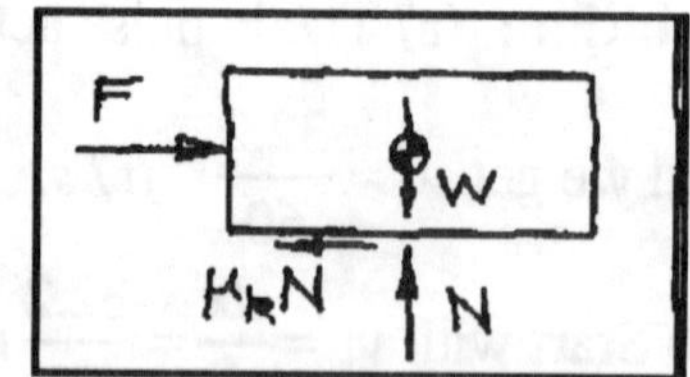

Solution:

The equation for the external force is $F = 10t\ lb\ (0 \le t \le 4)\ s$ and $F = 40\ lb\ (t > 4\ s)$. The box does not start moving until the external force exceeds the static friction: $t > \frac{\mu_s N}{10} = 2\ s$.

The impulse is $\int_{t_1}^{t_2} F dt = \int_2^4 (10t - \mu_k N)dt + \int_4^8 (40 - \mu_k N)dt = \left[5t^2 - 20t\right]_2^4 + [40t - 20t]_4^8 = 100\ lb\text{-}s$.

The velocity is $\boxed{v = \frac{100g}{W} = 32.17\ ft/s}$

===================================<>===================================

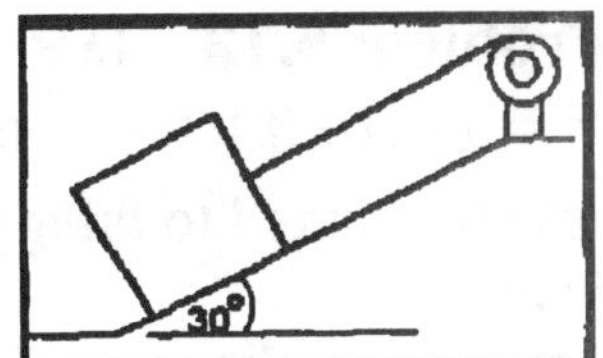

Problem 5.15 The crate has a mass of 120 *kg* and the coefficients of friction between it and the sloping dock are $\mu_s = 0.6$, $\mu_k = 0.5$. The crate starts from rest, and the winch exerts a tension $T = 1220\ N$. (a) What impulse is applied to the crate during the first second of motion? (b) What is the crate's velocity after 1 *s*?

Solution:

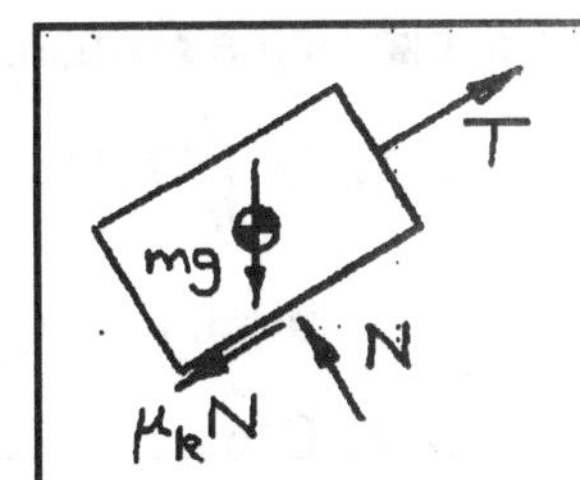

The motion starts only if $T - mg\sin 30^o > \mu_s mg\cos 30^o$, from which $631.4 > 611.7$. The motion indeed starts.

(a) The impulse in the first second is

$$\boxed{\int_{t_1}^{t_2} Fdt = \int_0^1 (T - mg\sin 30^o - mg\mu_k \cos 30^o)dt = 121.7t = 121.7\ N\text{-}s}$$

(b) The velocity is $\boxed{v = \frac{121.7}{120} = 1.01\ m/s}$

===================================<>===================================

Problem 5.16 Solve Problem 5.15 if the crate starts from rest at $t = 0$ and the winch exerts a tension $T = 1220 + 200t\ N$.

Solution:

From the solution to Problem 5.15, motion will start.

(a) The impulse at the end of 1 second is

$$\boxed{\int_{t_1}^{t_2} Fdt = \int_0^1 \left(1220 + 200t - mg\sin 30^o - \mu_k mg\cos 30^o\right)dt = \left[1220 + 100t^2 - 1098.3\right]_0^1 = 221.7\ N\text{-}s}$$

(b) The velocity is $\boxed{v = \frac{221.7}{m} = \frac{221.7}{120} = 1.85\ m/s}$

===================================<>===================================

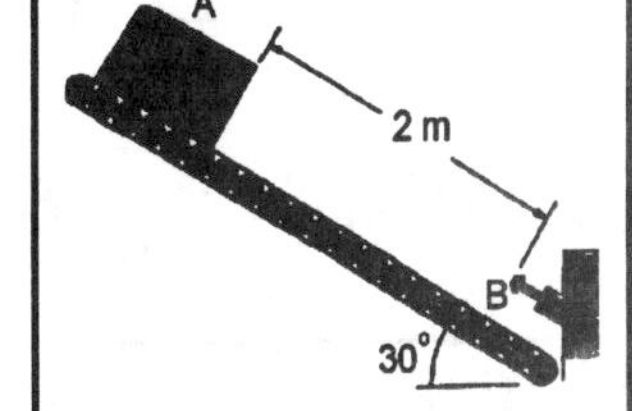

Problem 5.17 In an assembly line process, the 20 *kg* package A starts from rest and slides down the smooth ramp. Suppose that you want to design the hydraulic device B to exert a constant force of magnitude F on the package and bring it to a stop in 0.2 *s*. What is the required force *F*?

Solution:

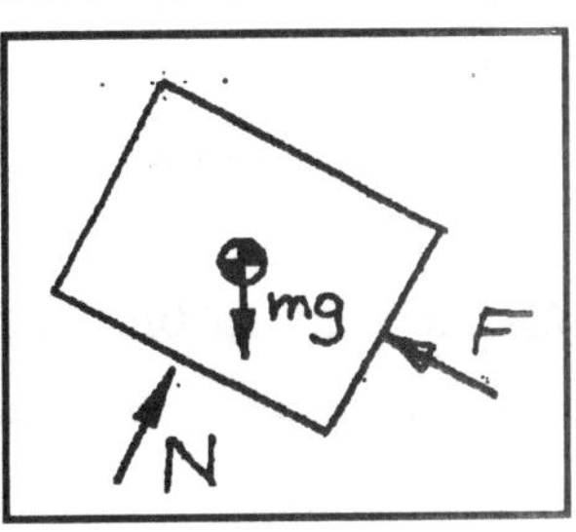

Use conservation of energy to obtain the velocity of the crate at point of contact with device B. $mgh = \frac{1}{2}mv_B^2$, where $h = 2\sin 30^o = 1\ m$, from which $v_B = \sqrt{2g} = 4.43\ m/s$. The impulse to be exerted by B is $\int_{t_1}^{t_2} Fdt = mv_B = 88.6\ N\text{-}s$. The constant force to be applied by device B is

$F - mg\sin 30^o = \frac{88.6}{0.2} = 443\ N$, from which $\boxed{F = 541.\ N}$

===================================<>===================================

===================================<>===================================

Problem 5.18 In Problem 5.17, if the hydraulic device B exerts a force of magnitude of $F = 540(1+0.4t)\ N$ on the package, where t is in seconds measured from the time of first contact, what time is required to bring the package to rest?

Solution:

Assume that the force F is exclusive of the component of the weight of the package exerted at B. The device exerts the resultant force on the package: $F_R = F - mg\sin 30^o$.

The impulse is $\int_{t_1}^{t_2} Fdt = \int_0^t \left(540(1+0.4t) - mg\sin 30^o\right)dt = \left[540\left(t+0.2t^2\right) - mg\sin 30^o t\right] = mv_B$.

From the solution to Problem 5.17, $mv_B = 88.6\ N\text{-}s$. Rearrange and reduce to obtain $t^2 + 2bt + c = 0$, where $b = 2.046$, $c = -0.8203$. The solution: $t = -b \pm \sqrt{b^2 - c} = 0.192\ s$, $= -4.28\ s$. Since negative time has no meaning here, $\boxed{t = 0.192\ s}$ is the time required to bring the package to a stop.

===================================<>===================================

Problem 5.19 In a cathode ray tube an electron (mass $= 9.11\times 10^{-31}\ kg$) is projected at O with velocity $v = 2.2\times 10^7\ m/s$. While it is between the charged plates, the electric field generated by the plates subjects if to a force $\vec{F} = -eE\vec{j}$. The charge of the electron is $e = 1.6\times 10^{-19}\ C\ (coulombs)$, and the electric field strength is $E = 15\sin(\omega t)\ kN/C$, where the circular frequency $\omega = 2\times 10^9\ s^{-1}$. (a) What impulse does the electric field exert on the electron while is it between the plates? (b) What is the velocity of the electron as it leaves the region between the plates?

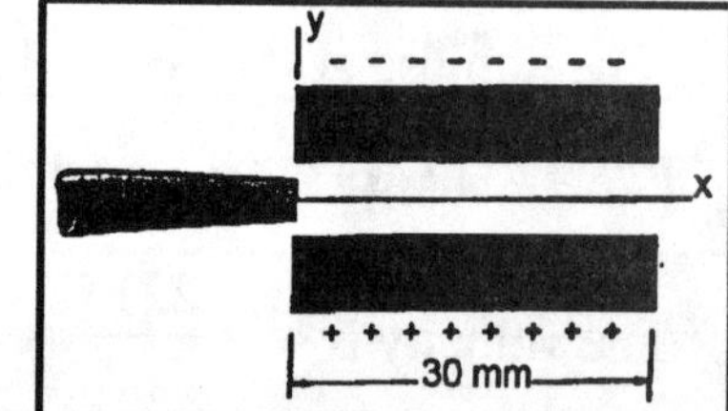

Solution:

The x component of the velocity is unchanged. The time spent between the plates is $t = \dfrac{0.03}{2.2\times 10^7} = 1.36\times 10^{-9}\ s$.

(a) The impulse is $\int_{t_1}^{t_2} Fdt = \int_0^t (-eE)dt = \int_0^t -e\left(15\times 10^3\right)(\sin\omega t)dt = \left[\dfrac{\left(15\times 10^3\right)e}{\omega}\cos\omega t\right]_0^{1.36\times 10^{-9}}$.

$$\boxed{\int_{t_1}^{t_2} Fdt = -2.3\times 10^{-24}\ N-s}$$

The y component of the velocity is $v_y = \dfrac{-2.3\times 10^{-24}}{9.11\times 10^{-31}} = -2.52\times 10^6\ m/s$.

(b) The velocity on emerging from the plates is $\boxed{\vec{v} = 22\times 10^6\vec{i} - 2.5\times 10^6\vec{j}\ m/s}$.

===================================<>===================================

==================================<>==================================

Problem 5.20 The two weights are released from rest. What is the magnitude of their velocity after one-half second?

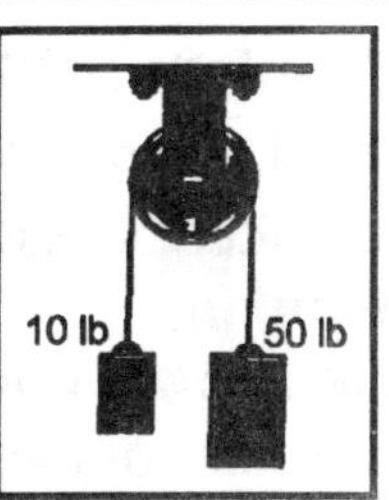

Strategy: Apply the principle of impulse and momentum to each weight individually.

Solution:

Since the pulley is one-to-one, both weights have the same velocity magnitude. The 10 *lb.* weight moves upward. The weights produce a constant force, so that the

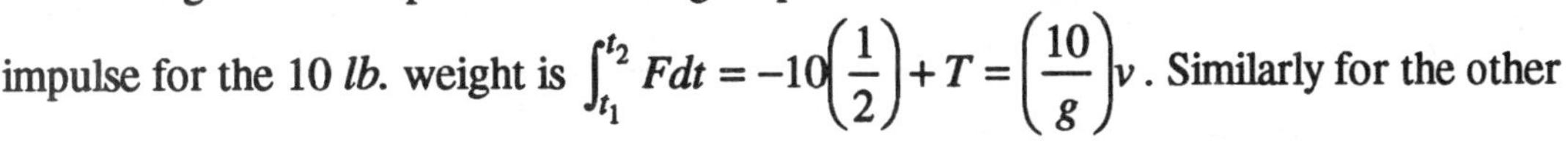

impulse for the 10 *lb.* weight is $\int_{t_1}^{t_2} F dt = -10\left(\frac{1}{2}\right) + T = \left(\frac{10}{g}\right)v$. Similarly for the other

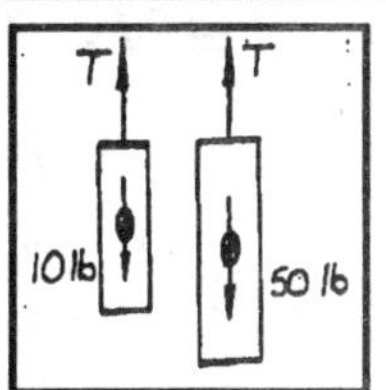

weight: $\int_{t_1}^{t_2} F dt = -50\left(\frac{1}{2}\right) + T = -\left(\frac{50}{g}\right)v$, from which: $\boxed{v = 10.72 \ ft/s}$,

$T = 16.67 \ lb$.

==================================<>==================================

Problem 5.21 The two crates are released from rest. Their masses are $m_A = 40 \ kg$ and $m_B = 30 \ kg$, and the coefficient of kinetic friction between crate A and the inclined surface is $\mu_k = 0.15$. What is the magnitude of their velocity after 1 *s*?

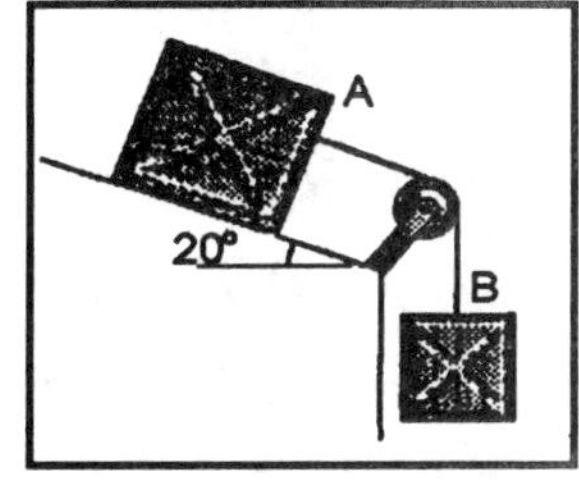

Solution: The force acting to move crate A is

$F_A = T + m_A g(\sin 20^o - \mu_k \cos 20^o) = T + 78.9 \ N$, where T is the tension in the cable. The impulse, since the force is constant, is $(T + 78.9)t = m_A v$. For crate B, $F_B = -T + m_B g = -T + 294.3$. The impulse, since the force is constant, is $(-T + 294.3)t = m_B v$. For $t = 1 \ s$, add and solve: $78.9 + 294.3 = (40 + 30)v$, from which $\boxed{v = 5.33 \ m/s}$

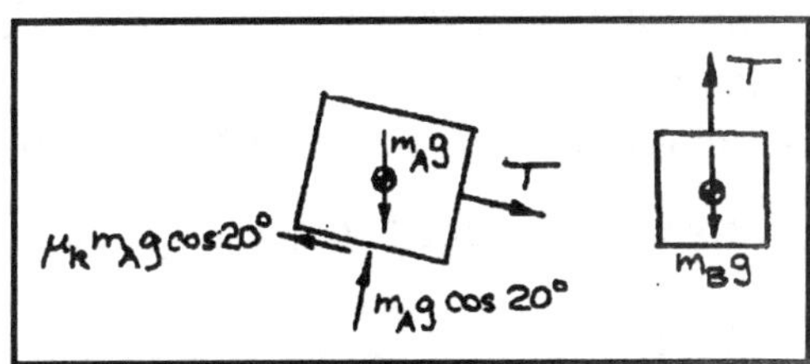

==================================<>==================================

Problem 5.22 The two crates are released from rest. Their masses are $m_A = 20kg$ and $m_B = 80kg$, and the surfaces are smooth. The angle $\theta = 20°$. What is the magnitude of the velocity after *1 s.*?

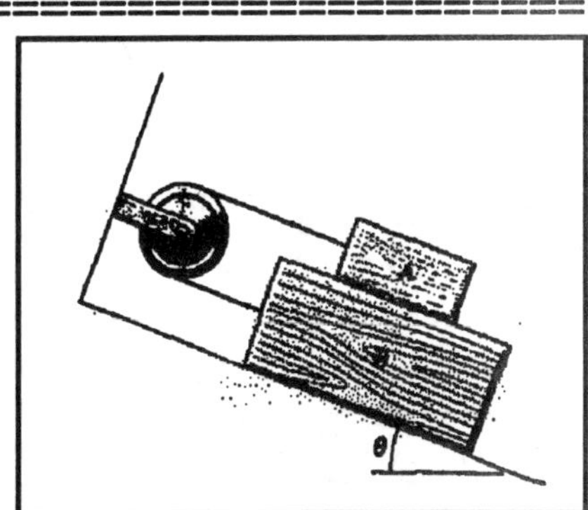

Strategy: Apply the principle of impulse and momentum to each crate individually.

Solution: The free body diagrams are as shown:

Crate B: $\int_{t_1}^{t_2} \sum F_x dt = mv_{x2} - mv_{x1}$:

$$\int_0^1 [(80)(9.81)\sin 20° - T]dt = (80)(v - 0).$$

Crate A: $\int_{t_1}^{t_2} \sum F_x dt = mv_{x2} - mv_{x1}$:

$$\int_0^1 [(20)(9.81)\sin 20° - T]dt = (20)[(-v) - 0]$$

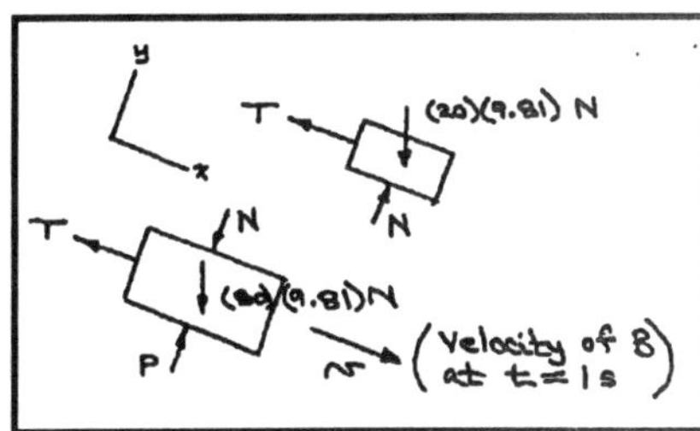

Subtracting the second equation from the first one,

$$\int_0^1 (80 - 20)(9.81)\sin 20° dt = (80 + 20)v.$$

Solving, we get $v = 2.01 m/s$.

==================================<>==================================

Problem 5.23

Problem 5.23 In Problem 5.22, suppose that the coefficient of kinetic friction between the surfaces is $\mu_k = 0.1$. What is the magnitude of the velocity after *1 s*?

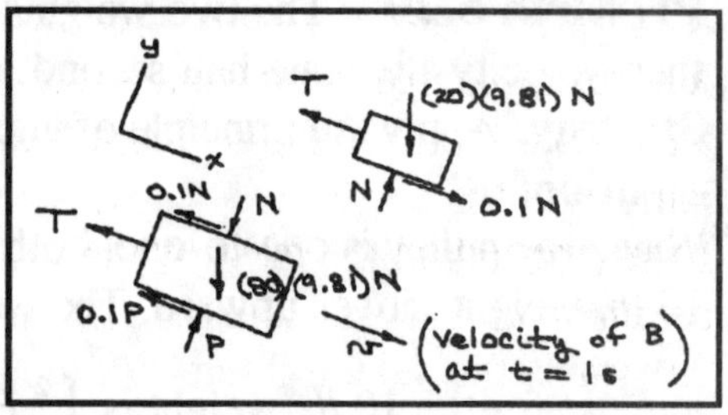

Solution:

The free body diagrams are as shown:

The sums of the forces in the y direction equal zero:

$$\sum F_y = N - (20)(9.81)\cos 20^\circ = 0 \quad N = 184N,$$

$$\sum F_y = P - N - (80)(9.81)\cos 20^\circ = 0 \quad P = 922N$$

Crate B: $\int_{t_1}^{t_2} \sum F_x dt = mv_{x2} - mv_{x1}$:

$$\int_0^1 \left[(80)(9.81)\sin 20^\circ - 0.1P - 0.1N - T\right]dt = (80)(v - 0).$$

Crate A: $\int_{t_1}^{t_2} \sum F_x dt = mv_{x2} - mv_{x1}$: $\int_0^1 \left[(20)(9.81)\sin 20^\circ + 0.1N - T\right]dt = (20)\left[(-v) - 0\right].$

Subtracting Equation (2) from Equation (1), $\int_0^1 \left[(80-20)(9.81)\sin 20^\circ - 0.1P - 0.2N\right]dt = (80+20)v.$

Solving, $v = 0.723 m/s$.

Problem 5.24

Problem 5.24 In Example 5.2, if the range safety officer destroys the booster 1 *s* after it starts rotating, what is its velocity at the time it is destroyed?

Solution:

Following Example 5.2, the vehicle is moving upward at 10 *m/s*, when it begins rotating at $t = 0$. The angular velocity of the booster is $\frac{\pi}{2}$ *rad / s*. The angle between the between the rocket axis and the vertical is $\left(\frac{\pi}{2}\right)t$. The thrust is $T = 1\times 10^6$ *N*, and the mass of the rocket is $m = 90{,}000$ *kg*. The force on the rocket is $\vec{F} = \left(-T\sin\left(\frac{\pi t}{2}\right)\vec{i} + \left(T\cos\left(\frac{\pi t}{2}\right) - mg\right)\vec{j}\right)$. The impulse is

$$\int_0^1 \vec{F}dt = \left[\frac{2T}{\pi}\cos\left(\frac{\pi t}{2}\right)\vec{i} + \left(\frac{2T}{\pi}\sin\left(\frac{\pi t}{2}\right) - mgt\right)\vec{j}\right]_0^1 = -\frac{2T}{\pi}\vec{i} + \left(\frac{2T}{\pi} - mg\right)\vec{j}.$$ The velocity is

$$m\vec{v}_2 - m\vec{v}_1 = -\frac{2T}{\pi}\vec{i} + \left(\frac{2T}{\pi} - mg\right)\vec{j},$$ from which $\vec{v}_2 = 10\vec{j} - \frac{2T}{m\pi}\vec{i} + \left(\frac{2T}{m\pi} - g\right)\vec{j}$, from which

$$\boxed{\vec{v}_2 = -7.07\vec{i} + 7.26\vec{j}}$$

================================<>================================

Problem 5.25 An object of mass m slides with constant velocity v_o on a horizontal table (seen from above in the figure). The object is attached by a string to the fixed point O and is in the position shown, with the string parallel to the x axis, at $t = 0$. (a) Determine the x and y components of the force exerted on the mass by the string as functions of time. (b) Use your results from Part (a) and the principle of impulse and momentum to determine the velocity vector of the mass when it has traveled one-fourth of a revolution about point O.

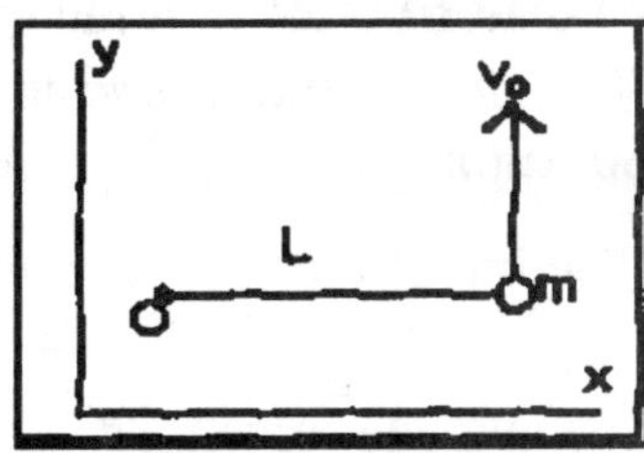

Solution:

(a) From Newton's second law for polar coordinates, $-m\left(\frac{v_o^2}{L}\right) = F_r$. The x and y components of the force are: $\boxed{\vec{F} = \left(-m\left(\frac{v_o^2}{L}\right)\cos\frac{v_o t}{L}\right)\vec{i} - \left(m\left(\frac{v_o^2}{L}\right)\sin\frac{v_o t}{L}\right)\vec{j}}$, . A quarter revolution is $t = \frac{\pi L}{2v_o}$. The impulse is $\int_0^{\frac{\pi L}{2v_o}} \vec{F}dt = -mv_o\vec{i} - mv_o\vec{j}$. The velocity at $t = 0$ is $\vec{v}_1 = \vec{j}v_o$. The velocity at one-quarter revolution is $\boxed{\vec{v}_2 = \vec{v}_1 - v_o\vec{i} - v_o\vec{j} = -v_o\vec{i}}$

================================<>================================

Problem 5.26 At $t = 0$ a 50 *lb.* projectile is given an initial velocity of 40 *ft/s* at 60° above the horizontal. (a) What impulse is applied to the projectile from $t = 0$ to $t = 2\ s$? (b) What is the projectile's velocity at $t = 2\ s$?

Solution:

(a) The force on the projectile is the weight acting downward, $\vec{F} = -W\vec{j}$. The impulse is

$$\boxed{\int_0^2 \vec{F}dt = \left(-W\vec{j}\right)2 = -100\vec{j}\ lb\text{-}s}.$$

(b) The velocity is obtained from $\left(\frac{W}{g}\right)\vec{v}_2 - \left(\frac{W}{g}\right)\vec{v}_1 = -2W\vec{j}$.

The initial velocity is $\vec{v}_1 = \left(40\cos 60^o\right)\vec{i} + \left(40\sin 60^o\right)\vec{j} = 20\vec{i} + 34.64\vec{j}\ \ (ft/s)$, from which

$$\boxed{\vec{v}_2 = \vec{v}_1 - \left(\frac{g}{W}\right)2W\vec{j} = 20\vec{i} + 34.64\vec{j} - 64.4\vec{j} = 20\vec{i} - 29.76\vec{j}\ \ (ft/s)}$$

================================<>================================

=====================================<>=====================================

Problem 5.27 A rail gun, which uses an electromagnetic field to accelerate an object, accelerates a 30 *g* projectile to 5 *km/s* in 0.0005 *s*. What average force is exerted on the projectile?

Solution:

The impulse is $\int_{t_1}^{t_2} Fdt = mv_2 - mv_1 = 0.03(5000) = 150\ N\text{-}s$.

The average force is $\boxed{F_{ave} = \dfrac{150}{0.0005} = 3\times 10^5\ N = 300\ kN}$

=====================================<>=====================================

Problem 5.28 The powerboat is going 50 *mi/hr* when its motor is turned off. In 5 *s* its velocity decreases to 30 *mi/hr*. The boat and its passengers weigh 1800 *lb*. Determine the magnitude of the average force exerted on the boat by hydrodynamic and aerodynamic drag during the 5 *s*.

Solution:

The impulse is $\int_{t_1}^{t_2} Fdt = \left(\dfrac{W}{g}\right)v_2 - \left(\dfrac{W}{g}\right)v_1 = \left(\dfrac{W}{g}\right)(30-50)\left(\dfrac{5280}{3600}\right) = -1640\ lb\text{-}s$.

The average force is $\boxed{F_{ave} = \dfrac{-1640}{5} = -328\ lb}$

=====================================<>=====================================

Problem 5.29 A motorcycle starts from rest at $t = 0$ and travels along a circular track with *400-m* radius. The tangential component of the total force on the motorcycle from $t = 0$ to $t = 30\ s$ is $\sum F_t = 100$ (*newtons*). The combined mass of the motorcycle and rider is *150-kg*. What is the magnitude of the velocity at $t = 30\ s$?

Solution:

The impulse is $\int_{t_1}^{t_2} \sum F_t dt = mv_2 - mv_1$: Evaluating, we get $\int_0^{30} 100dt = (150)(v-0)$. Solving, we find that $v = 20m/s$.

=====================================<>=====================================

Problem 5.30 In Problem 5.29, what is the average of the *normal* component of the total force on the motorcycle from $t = 0$ to $t = 30\ s$?

Strategy:

Use Eq. (5.3) to determine the magnitude of the velocity as a function of time. With the resulting expression, calculate the average value of the total normal force $\sum F_n = mv^2/\rho$.

Solution: The impulse is $\int_{t_1}^{t_2} \sum F_t dt = mv_2 - mv_1$: Evaluating, we get $\int_0^{t} 100dt = (150)(v-0)$, or $v = \dfrac{2}{3}t$. The normal component of the force is $\sum F_n = m\dfrac{v^2}{\rho} = \dfrac{150}{400}\left(\dfrac{2}{3}t\right)^2 = 0.167t^2 N$, and the average normal force is $\sum F_{nav} = \dfrac{1}{t_2 - t_1}\int_{t_1}^{t_2} \sum F_n dt$ or,

$$\sum F_{nav} = \frac{1}{30}\int_0^{30} 0.167t^2 dt = \frac{0.167}{30}\frac{(30)^3}{3} = 50N.$$

=====================================<>=====================================

==================================<>==================================

Problem 5.31 A motorcycle starts from rest at $t = 0$ and travels along a circular track with *400-m* radius. The tangential component of the total force on the motorcycle from $t = 0$ to $t = 30\ s$ is $\sum F_t = (200 - 6t)$ *(newtons)*. The combined mass of the motorcycle and rider is *150-kg*.
(a) What is the magnitude of the velocity at $t = 30\ s$?
(b) What is the average tangential component of the total force from $t = 0$ to $t = 30\ s$?

Solution:

(a) The impulse is $\int_{t_1}^{t_2} \sum F_t dt = mv_2 - mv_1$, or, evaluating, $\int_0^{30} (200 - 6t)dt = (150)(v - 0)$, from which we have $\left[200t - 3t^2\right]_0^{30} = 150v$, Hence, $v = \dfrac{200(30) - (3)(30)^2}{150} = 22 m/s$

(b) The average tangential force is $\Delta t \sum F_{tav} = (150)(22)$, or $\sum F_{tav} = 110N$

==================================<>==================================

Problem 5.32 In Problem 5.31, what is the average of the *normal* component of the total force on the motorcycle from $t = 0$ to $t = 30\ s$?

Solution:

The impulse is $\int_{t_1}^{t_2} \sum F_t dt = mv_2 - mv_1$: or, evaluating, $\int_0^t (200 - 6t)dt = (150)(v - 0)$. Hence, $v = 1.33t - 0.02t^2$. The normal force: $\sum F_n = m\dfrac{v^2}{\rho} = \dfrac{150}{400}(1.33t - 0.02t^2)^2 = 0.667t^2 - 0.02t^3 + 0.00015t^4$ and the average normal force is $\sum F_{nav} = \dfrac{1}{t_2 - t_1}\int_{t_1}^{t_2} \sum F_n dt = \dfrac{1}{30}\int_0^{30} (0.667t^2 - 0.02t^3 + 0.00015t^4)dt$

Hence, $\sum F_{nav} = \dfrac{1}{30}\left[0.667\dfrac{(30)^3}{3} - 0.02\dfrac{(30)^4}{4} + 0.00015\dfrac{(30)^5}{5}\right] = 89.3N$

==================================<>==================================

Problem 5.33 The 77 *kg* skier is traveling at 10 *m/s* at 1, and he goes from 1 to 2 in 0.7 *s*. (a) If you neglect friction and aerodynamic drag, what is the time average of the tangential component of the force exerted on him as he moves from 1 to 2? (b) If his actual velocity is measured at 2 and determined to be 13.1 *m/s*, what is the average of the tangential component of force exerted on him as he moves from 1 to 2?

Solution

(a) The impulse is $\int_{t_1}^{t_2} Fdt = F_{ave}(t_2 - t_1) = mv_2 - mv_1$. If friction and aerodynamic drag are neglected, the velocity at 2 is determined from energy considerations: $\frac{1}{2}mv_2^2 = \frac{1}{2}mv_1^2 + mgh$, from which

$v_2 = \sqrt{v_1^2 + 2gh} = \sqrt{10^2 + 2g(4)} = 13.36\ m/s$. The time average of the tangential force is

$$\boxed{F_{ave} = \frac{mv_2 - mv_1}{(t_2 - t_1)} = \frac{258.7}{0.7} = 369.6\ N}$$

(b) If $v_2 = 13.1\ m/s$, the average tangential force is

$$\boxed{F_{ave} = \frac{238.7}{0.7} = 341\ N}$$

==================================<>==================================

====================================<>====================================

Problem 5.34 In a test of an energy absorbing bumper, a 2800 *lb.* car is driven into a barrier at 5 *mi/hr.* The duration of the impact is 0.4 *s*, and the car bounces back from the barrier at 1 *mi/hr.*
(a) What is the magnitude of the of the average horizontal force exerted on the car during the impact?
(b) (b) What is the average deceleration of the car during the impact?

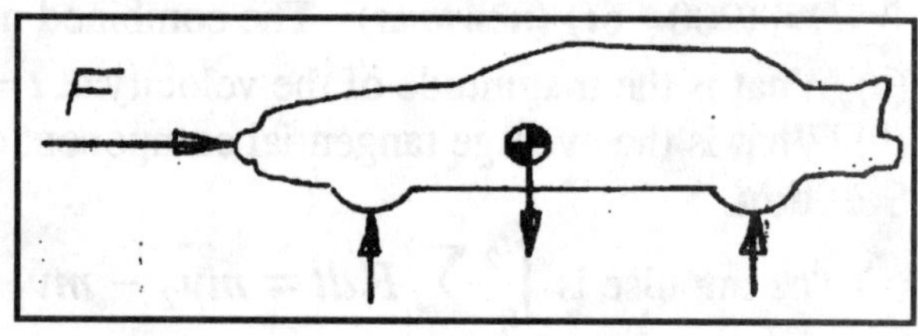

Solution The impulse is $\left(\frac{W}{g}\right)v_2 - \left(\frac{W}{g}\right)v_1 = \int_{t_1}^{t_2} Fdt = F_{ave}(t_2 - t_1)$. The velocity at $t = t_1$ is $v_1 = 5\ mi/hr = 7.33\ ft/s$. The velocity at $t = t_2$ is $v_2 = -1\ mi/hr = -1.47\ ft/s$.

(a)The average horizontal force is $\boxed{F_{ave} = \left(\frac{W}{g}\right)\frac{v_2 - v_1}{(t_2 - t_1)} = -1915\ lb}$

(b) The average deceleration is $\boxed{a_{ave} = \frac{v_2 - v_1}{(t_2 - t_1)} = \frac{-8.8}{0.4} = -22\ ft/s}$

====================================<>====================================

Problem 5.35 A bioengineer, using an instrumented dummy to test a protective mask for a hockey goalie, launches the 170 *g* puck so that it strikes the mask moving horizontally at 40 *m/s*. From photographs of the impact, she estimates the duration to be 0.02 *s* and observes that the puck rebounds at 5 *m/s*. (a) What linear impulse does the puck exert? (b) What is the average value of the impulsive force exerted on the mask by the puck?

Solution (a) The linear impulse is $\int_{t_1}^{t_2} Fdt = F_{ave}(t_2 - t_1) = mv_2 - mv_1$.

The velocities are $v_2 = -5\ m/s$, and $v_1 = 40\ m/s$, from which

$\int_{t_1}^{t_2} Fdt = F_{ave}(t_2 - t_1) = (0.17)(-5 - 40) = -7.65\ N\text{-}s$, where the negative sign means that the force is directed parallel to the negative *x* axis.(b) The average value of the force is $\boxed{F_{ave} = \frac{-7.65}{0.02} = -382.5\ N}$

====================================<>====================================

Problem 5.36 A fragile object dropped onto a hard surface breaks because it is subjected to a large impulsive force. If you drop a 2 *oz* watch from 4 *ft* above the floor, the duration of the impact is 0.001 *s*, and the watch bounces 2 *in.* above the floor, what is the average value of the impulsive force?

Solution The impulse is $\int_{t_1}^{t_2} Fdt = F_{ave}(t_2 - t_1) = \left(\frac{W}{g}\right)(v_2 - v_1)$. The weight of the watch is

$W = \frac{2}{16} = 0.125\ lb$, and its mass is $\left(\frac{W}{g}\right) = 3.88 \times 10^{-3}\ slug$. The velocities are obtained from energy considerations (the conservation of energy in free fall) : $v_1 = \sqrt{2gh} = \sqrt{2(32.2)(4)} = 16.\ ft/s$.

$v_2 = -\sqrt{2gh} = -\sqrt{2(32.2)(2/12)} = -3.28\ ft/s$. The average value of the impulsive force is

$$\boxed{F_{ave} = \frac{(3.88 \times 10^{-3})(-3.28 - 16)}{1 \times 10^{-3}} = -75\ lb}$$

====================================<>====================================

==================================<>==================================

Problem 5.37 A 50 *lb.* projectile is subjected to an impulsive force with a duration of 0.01 *s* that accelerates it from rest to a velocity of 40 *ft/s* at 60° above the horizontal. What is the average value of the impulsive force?

Strategy. Use Eq (5.2) to determine the average total force. To determine the average value of the impulsive force, you must subtract its weight.

Solution The impulse is

$\int_{t_1}^{t_2} \vec{F}dt = \vec{F}_{ave}(t_2 - t_1) = \left(\frac{W}{g}\right)\vec{v}_2 = \left(\frac{W}{g}\right)(20\vec{i} + 34.64\vec{j}) = 31.08\vec{i} + 53.84\vec{j}\ lb\text{-}s$.. The average total force is $\vec{F}_{ave} = \frac{1}{0.01}(31.08\vec{i} + 53.84\vec{j}) = 3108\vec{i} + 5384\vec{j}\ lb$. The average impulsive force is

$$\boxed{\vec{F}_{imp} = \vec{F}_{ave} - (-\vec{j}W) = 3108\vec{i} + 5434\vec{j}\ lb}$$

==================================<>==================================

Problem 5.38 An entomologist measures the motion of a 3 *g* locust during its jump and determines that it accelerates from rest to 3.4 *m/s* in 25 *ms*(milliseconds). The angle of takeoff is 55° above the horizontal. What are the horizontal and vertical components of the average impulsive force exerted by the insect's hind legs during the jump?

Solution

The impulse is $\int_{t_1}^{t_2} \vec{F}dt = \vec{F}_{ave}(t_2 - t_1) = m(\vec{v}_2) = m(3.4\cos 55^o\,\vec{i} + 3.4\sin 55^o\,\vec{j})$, from which

$\vec{F}_{ave}(2.5\times 10^{-2}) = (5.85\times 10^{-3})\vec{i} + (8.36\times 10^{-3})\vec{j}\ N\text{-}s$. The average total force is

$\vec{F}_{ave} = \frac{1}{2.5\times 10^{-2}}\left((5.85\times 10^{-3})\vec{i} + (8.36\times 10^{-3})\vec{j}\right) = 0.234\vec{i} + 0.334\vec{j}\ N$. The impulsive force is

$$\boxed{\vec{F}_{imp} = \vec{F}_{ave} - (-m g\vec{j}) = 0.234\vec{i} + 0.364\vec{j}\ N}$$

==================================<>==================================

Problem 5.39 A 5 *oz* baseball is 3 *ft* above the ground when it is struck by a bat. The horizontal distance to the point where the ball strikes the ground is 180 *ft*. Photographs indicate that the ball was moving approximately horizontally at 100 *ft/s* before it was struck, the duration of the impact was 0.015 *s*, and the ball was traveling at 30° above the horizontal after it was struck.. What was the magnitude of the average impulsive force exerted on the ball?

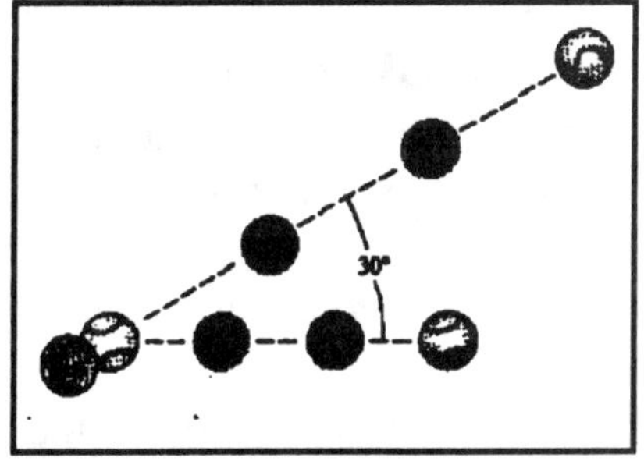

Solution

The impulse is $\int_{t_1}^{t_2} \vec{F}dt = \vec{F}_{ave}(t_2 - t_1) = \left(\frac{W}{g}\right)(\vec{v}_2 - \vec{v}_1)$. The velocity v_2 is determined from the

trajectory. The path is $y = -\frac{gt^2}{2} + (v_2 \sin 30^o)t + y_o$, $x = (v_2\cos 30^o)t$ where v_2 is the magnitude of the velocity at the point of leaving the bat, and $y_o = 3\,ft$. At $x = 180\,ft$, $t = 180/(v_2\cos 30^o)$, and $y = 0$.

Substitute and reduce to obtain $v_2 = 180\sqrt{\frac{g}{2\cos^2 30^o(180\tan 30^o + y_o)}} = 80.62\,ft/s$.

Solution continued on next page

From which: $\vec{F}_{ave} = \left(\frac{1}{0.015}\right)\left(\frac{W}{g}\right)\left(\left(v_2 \cos 30^o\right)\vec{i} + \left(v_2 \sin 30^o\right)\vec{j} - \left(-100\vec{i}\right)\right) = 110\vec{i} + 26.1\vec{j}$. Subtract the weight of the baseball: $\vec{F}_{imp} = \vec{F}_{ave} - \left(-W\vec{j}\right) = 110\vec{i} + 26.42\vec{j}$, from which $\boxed{\left|\vec{F}_{imp}\right| = 113.1\ lb}$

===================================<>===================================

Problem 5.40 The 1 *kg* ball is given a horizontal velocity of 1.2 *m/s* at A. Photographic measurements indicate that $b = 1.2\ m$, $h = 1.3\ m$, and the duration of the bounce at B is 0.1 *s*. What are the components of the average impulsive force exerted on the ball by the floor at B?

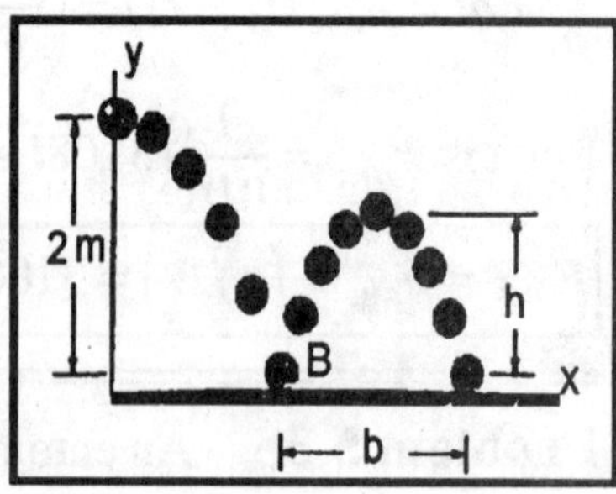

Solution

The impulse is $\int_{t_1}^{t_2} \vec{F}dt = \vec{F}_{ave}(t_2 - t_1) = m(\vec{v}_2 - \vec{v}_1)$. The velocities are determined from the path (from Newton's second law for free fall) : $\vec{v}_1 = 1.2\vec{i} - \sqrt{2gy}\vec{j} = 1.2\vec{i} - 6.26\vec{j}\ (m/s)$. The vertical velocity after the bounce at B is $\sqrt{2gh} = 5.05\ m/s$. The time of flight after the bounce at B is twice the time required to fall a distance h, from which $t = 2\sqrt{\frac{2h}{g}} = 1.03\ s$. The horizontal velocity after the bounce at B is $\frac{b}{t} = 1.17\ \ m/s$, $\vec{v}_2 = 1.2\vec{i} + \sqrt{2gh}\vec{j} = 1.17\vec{i} + 5.05\vec{j}\ \ (m/s)$. From which

$\int_{t_1}^{t_2} \vec{F}dt = \vec{F}_{ave}(t_2 - t_1) = -0.0345\vec{i} + 11.31\vec{j}\ N\text{-}s$. $\vec{F}_{ave} = \frac{1}{0.1}\left(-0.0345\vec{i} + 11.31\vec{j}\right) = \left(-0.345\vec{i} + 113.1\vec{j}\right)\ N$.

Subtract the weight of the ball from the average impulsive force:

$\boxed{\vec{F}_{imp} = \vec{F}_{ave} - \left(-mg\vec{j}\right) = -0.345\vec{i} + 123\vec{j}\ \ N}$

===================================<>===================================

Problem 5.41 At time $t = 0$, the two masses are released from rest on the smooth surface with the spring stretched. Show that at any later time, the velocities of the masses are related by $m_A v_A + m_B v_B = 0$.

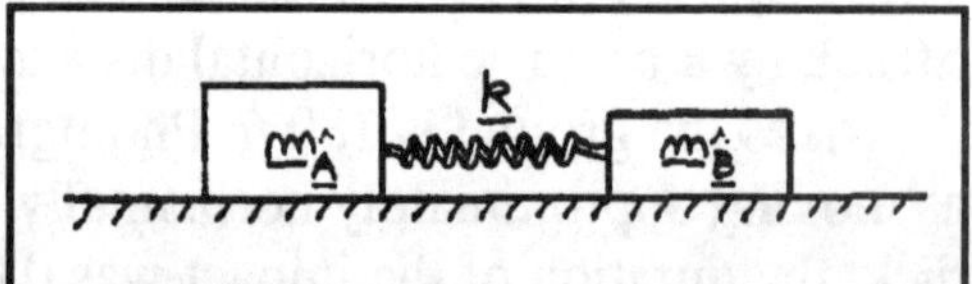

Strategy: Write the principle of impulse and momentum for each mass.

Solution:

The force on mass A is equal and opposite to that on mass B. Hence, we can write $F_A + F_B = 0$ and $\int_0^t F_A dt + \int_0^t F_B dt = \int_0^t 0 dt = 0$. From the principle of impulse and momentum, we can write $\int_0^t F_A dt = m_A \int_0^t dv = m_A v_A$. Similarly, we can write $\int_0^t F_B dt = m_B \int_0^t dv = m_B v_B$. Substituting into the equation above, we get $m_A v_A + m_B v_B = 0$, as was required.

===================================<>===================================

==================================<>==================================

Problem 5.42 In Problem 5.41, $m_A = 40\ kg$, $m_B = 30\ kg$, and $k = 400\ N/m$. The two masses are released from rest on the smooth surface with the spring stretched *1-m*. What are the magnitudes of the velocities of the masses when the spring is unstretched?

Solution:

From the solution of Problem 5.41, (1) $m_A v_A + m_B v_B = 0$: or, evaluating, $40v_A + 30v_B = 0$

Energy is conserved, Thus, (2) $\frac{1}{2}kS^2 = \frac{1}{2}m_A v_A{}^2 + \frac{1}{2}m_B v_B{}^2$. Evaluating, we get

$$\frac{1}{2}(400)(1)^2 = \frac{1}{2}(40)v_A^2 + \frac{1}{2}(30)v_B^2$$

Solving Equations (1) and (2), $|\bar{v}_A| = 2.07 m/s, |\bar{v}_B| = 2.76 m/s$.

==================================<>==================================

Problem 5.43 A girl weighing 100 *lb.* stands at rest on a barge weighing 500 *lb.* She starts running at 10 *ft/s* relative to the barge and runs off the end. Neglect the horizontal force exerted on the barge by the water.

(a) Just before she hits the water, what is the horizontal component of her velocity relative to the water?

(b) What is the velocity of the barge relative to the water while she runs?

Solution:

Denote the velocity of the barge and the girl relative to the water by v_{BW}, v_{GW}.

The linear momentum before and after she starts running is the same, (Eq (5.4)), from which

$$\left(\frac{W_G}{g}\right)v_{GW} + \left(\frac{W_B}{g}\right)v_{BW} = 0.$$

The velocity of the girl relative to the water is $v_{GW} = v_{GB} + v_{BW} = 10 + v_{BW}$, from which

$W_G(10 + v_{BW}) + W_B v_{BW} = 0$. Solve:

(b) $\boxed{v_{BW} = -\frac{W_G 10}{(W_G + W_B)} = -1.67\ ft/s}$ where the negative sign means that the motion is to the left. (a)

The velocity relative to the water before she hits the water is $\boxed{v_{GW} = 10 + v_{BW} = 10 - 1.67 = 8.33\ ft/s}$

==================================<>==================================

Problem 5.44 A 60 *kg* astronaut aboard the space shuttle kicks off toward the center of mass of the 105 *Mg* shuttle at 1 *m/s* relative to the shuttle. He travels 6 *m* relative to the shuttle before coming to rest at the opposite wall. (a) What is the magnitude of the change in the velocity of the shuttle while he is in motion? (b) What is the magnitude of the displacement of the center of mass of the shuttle while he is "in flight"?

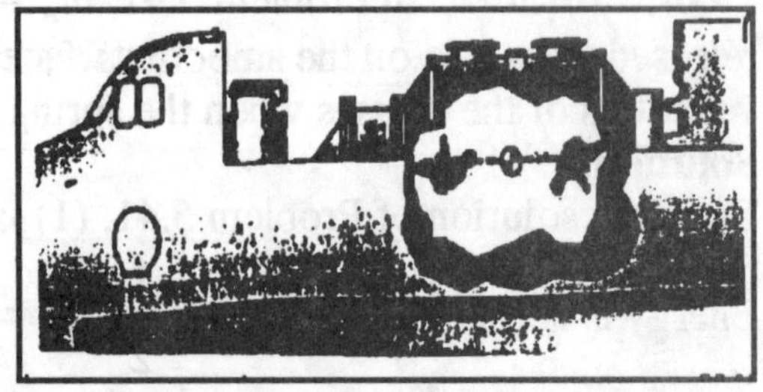

Solution The linear momentum of the system before and after the "kick off" is the same, (Eq (5.5)): $m_A v_A + m_S v_S = (m_A + m_S)v$, where v_A is the velocity of the astronaut relative to the reference system, and v_S is the velocity of the shuttle relative to the reference system. Noting that $v_A = v_{AS} + v_S$, where $v_{AS} = 1\ m/s$ is the velocity of the astronaut relative to the shuttle, the conservation of linear momentum condition becomes: $m_A(v_{AS} + v_S) + m_S v_S = (m_A + m_S)v$. Rearrange and reduce to obtain

$$\boxed{(v - v_S) = \frac{m_A v_{AS}}{m_A + m_S} = 5.71 \times 10^{-4}\ m/s}$$

is the magnitude of the change in the shuttle velocity. (b) The center of mass of the system is the same before and after "kickoff". $m_A r_A + m_S r_S = m_A r'_A + m_S r'_S$, where r_A, r_S and r'_A, r'_S are the center of mass of the astronaut and the shuttle, respectively, relative to the reference system before and after, respectively. Without loss of generality, place the center of mass of the shuttle at the opposite wall, with the origin at the position of that wall, from which $m_A r_A + m_S r_S = m_A(-6)$ and $m_A r'_A + m_S r'_S = (m_A + m_S)r'_S$. From which $m_A(-6) = (m_A + m_S)r'_S$, from which

$$\boxed{|r'_S| = \left|\frac{-6m_A}{m_A + m_S}\right| = 0.00343\ m}$$

is the magnitude of the shift in the center of mass during the "flight".

Problem 5.45 An 80 *lb.* boy sitting stationary in a 20 *lb.* wagon wants to simulate rocket propulsion by throwing a brick out of the wagon. Neglect horizontal forces on the wagon's wheels. If he has three bricks weighing 10 *lb.* each and throws them with a horizontal velocity of 10 *ft/s* relative to the wagon, determine the velocity if (a) he throws the bricks one at a time, (b) if he throws them all at once.

Solution Assume that the bricks are additional weight in the wagon.

(a) *First Brick thrown:* Denote the relative velocity of the thrown brick by $v_{B/W} = 10\ ft/s$. The linear momentum before and after the first brick is thrown is conserved, $\left(\frac{W_B}{g}\right)v_B + \left(\frac{2W_B + W_{Boy} + W_W}{g}\right)v_{W1} = 0$, where $v_B = v_{B/W} + v_{W1} = 10\ ft/s + v_{W1}$ is the velocity of the brick relative to the wagon, from which the velocity of the wagon is

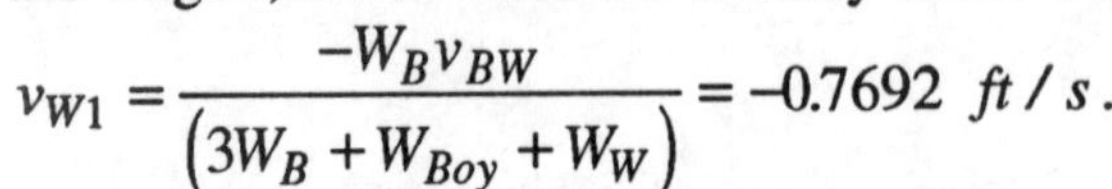

$$v_{W1} = \frac{-W_B v_{BW}}{(3W_B + W_{Boy} + W_W)} = -0.7692\ ft/s.$$

(b) *Second brick thrown:* The conservation of linear momentum condition is

$\left(\frac{W_B}{g}\right)v'_B + \left(\frac{W_B + W_{Boy} + W_W}{g}\right)v_{W2} = \left(\frac{2W_{Boy} + W_W}{g}\right)v_{W1}$, where

$v'_B = v_{B/W} + v_{W2} = 10\ ft/s + v_{W2}$, from which $v_{W2} = v_{W1} - \frac{W_B v_{B/W}}{(2W_B + W_{Boy} + W_W)} = -1.6026\ ft/s$.

Solution continued on next page

Third brick thrown: Carrying out the procedure outlined above, the velocity of the wagon after the third brick is thrown is $v_{W3} = v_{W2} - \dfrac{W_B v_{B/W}}{(W_B + W_{Boy} + W_W)} = -2.512 \ ft/s$

(b) *All three bricks thrown at once:* The conservation of momentum condition is $3\left(\dfrac{W_B}{g}\right)v_B + \left(\dfrac{W_{Boy} + W_W}{g}\right)v_W = 0$, where $v_B = V_{B/W} + V_W = 10 ft/s + V_W$. From which, the velocity of the wagon is $v_W = -\dfrac{3W_B v_{BW}}{(3W_B + W_{Boy} + W_W)} = -2.3077 \ ft/s$ from which $|v_{W3}| > |v_W|$

==================================<>==================================

Problem 5.46 Two railroad cars ($m_A = 1.7 m_B$) collide and become coupled. Car A is full and car B is half full of carbolic acid. When the cars impact, the acid in car B sloshes back and forth violently. (a) Immediately after impact, what is the velocity of the common center of mass of the two cars? (b) A few seconds later, when the sloshing has subsided, what is the velocity of the two cars?

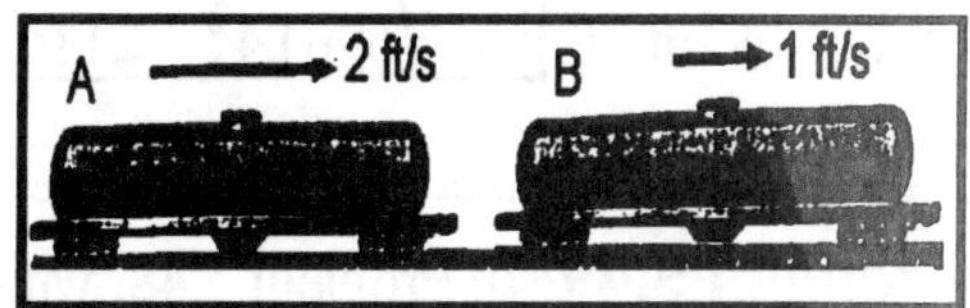

Solution:

(a) The linear momentum of the system is constant: $m_A v_A + m_B v_B = (m_A + m_B)v$, from which

$$v = \frac{m_A v_A + m_B v_B}{(m_A + m_B)} = \frac{m_B(2(1.7) + 1)}{m_B(1 + 1.7)} = \frac{4.4}{2.7} = 1.63 \ ft/s$$

(b) The velocity is the same as in Part (a).

==================================<>==================================

Problem 5.47 In Problem 5.46, if the track slopes one-half degree upward to the right and the cars are initially $10 \, ft$ apart, what is the velocity of their common center of mass immediately after the impact?

Solution:

The impulse due to the weight of the cars moving up the incline is $\int_{t_1}^{t_2} F dt = F_{ave} t_{imp} = -(mg \sin\theta) t_{imp}$.

For car A, $-(m_A g \sin\theta) t_{imp} = m_A (v_{A2} - v_{A1})$, from which $v_{A2} = v_{A1} - (g \sin\theta) t_{imp}$. Similarly for car B, $v_{B2} = v_{B1} - (g \sin\theta) t_{imp}$. The time of impact is determined from the solution to Newton's second law for each car. Choose an origin on the track at $t = 0$ at the instantaneous center of mass of car A. The distance traveled by car A is $s = -\dfrac{g \sin\theta}{2} t^2 + v_{A1} t$. The distance traveled by car B is

$s = -\dfrac{g \sin\theta}{2} t^2 + v_{B1} t + 10$. Impact occurs when the distances are the same, from which

$t_{imp} = \dfrac{10}{v_{A1} - v_{B1}} = 10 \, s$. Substitute into the velocity expressions obtained from the impulse:

$v_{A2} = v_{A1} - (g \sin\theta) t_{imp} = 2 - 10(0.2807) = -0.807 \ ft/s$. $v_{B2} = v_{B1} - (g \sin\theta) t_{imp} = -1.807 \ ft/s$.

Both cars are moving down the incline at time of impact. The linear momentum condition is $m_A v_{A2} + m_B v_{B2} = (m_A + m_B)v$, from which

$$v = \frac{m_A v_{A2} + m_B v_{B2}}{(m_A + m_B)} = \frac{m_B(-0.807(1.7) - 1.807)}{m_B(1 + 1.7)} = -1.18 \ ft/s$$

==================================<>==================================

================================<>================================

Problem 5.48 A 400 *kg* satellite S traveling at 7 *km/s* is hit by a 1 *kg* meteor M traveling at 12 *km/s*. The meteor is embedded in the satellite by the impact. Determine the magnitude of the velocity of their common center of mass after the impact and the angle β between the path of the center of mass and the original path of the satellite.

Solution: The linear momentum condition is $m_S\vec{v}_S + m_M\vec{v}_M = (m_S + m_M)\vec{v}$. The velocity of the satellite is $\vec{v}_S = \left(7\times 10^3\right)\vec{i}$. The velocity of the meteor is $\vec{v}_M = \left(12\times 10^3\right)\left(\vec{i}\cos\left(135^o\right) + \vec{j}\sin(135)\right) = -8485.3\vec{i} + 8485.3\vec{j}$. The velocity of the system center of mass is $\vec{v} = \dfrac{m_S\vec{v}_S + m_M\vec{v}_M}{(m_S + m_M)} = 6961.4\vec{i} + 21.16\vec{j}\ (m/s)$. The magnitude of the velocity is $\boxed{v = \sqrt{6961.4^2 + 21.16^2} = 6961.4\ m/s}$. The angle is $\boxed{\beta = \tan^{-1}\left(\dfrac{21.16}{6961.4}\right) = 0.174^o}$

================================<>================================

Problem 5.49 In Problem 5.48, what would the magnitude of the velocity of the *1-kg* meteor *M* need to be to cause the angle between the original path of the satellite and the path of the center of mass of the combined satellite and meteor after the impact to be $\beta = 0.5^\circ$? What is the magnitude of the velocity of the center of mass after the impact?

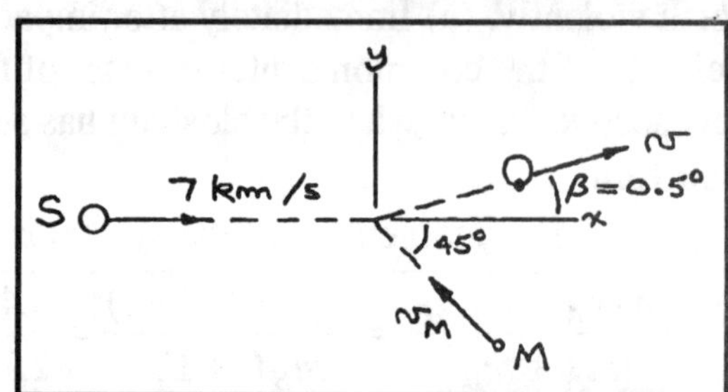

Solution: Conservation of linear momentum yields $(400)(7\vec{i}) + (1)(-v_m\sin 45^\circ\vec{i} + v_m\cos 45^\circ\vec{j}) = (400+1)(v\cos 0.5^\circ i + v\sin 0.5^\circ\vec{j})$.

Equating **i** and **j** components, we get $(400)(7) - v_m\cos 45^\circ = 401v\cos 0.5^\circ$; $v_m\sin 45^\circ = 401v\sin 0.5^\circ$ and solving, we obtain $v_m = 34.26 km/s : v = 6.92 km/s$.

================================<>================================

Problem 5.50 A catapult designed to throw a line to ships in distress throws a *2-kg* projectile. The mass of the catapult is *36-kg* and it rests on a smooth surface. It the velocity of the projectile relative to the earth as it leaves the tube is *50 m/s* at $\theta_0 = 30^\circ$ relative to the horizontal, what is the resulting velocity of the catapult toward the left?

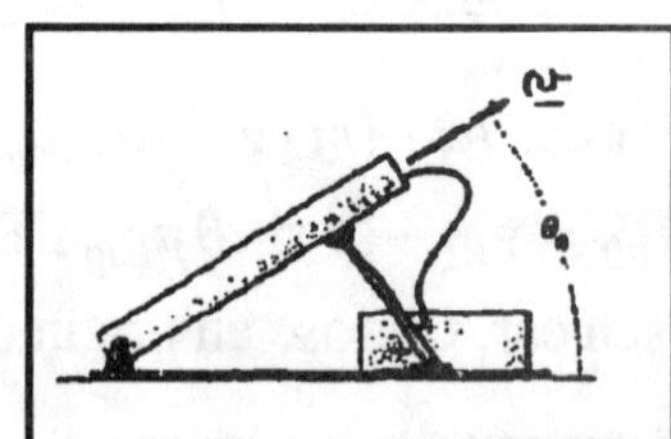

Solution:

Conservation of linear momentum in the horizontal direction yields $0 = m_p v\cos\theta_0 + m_c v_c$:

$0 = (2)(50)\cos 30^\circ + (36)v_c$. Evaluating, we get $0 = (2)(v_c + 50\cos 30^\circ) + (36)v_c$, or $v_c = -2.41 m/s$

================================<>================================

================================<>================================

Problem 5.51 In Problem 5.50, if the velocity of the projectile *relative to the catapult* as it leaves the tube is *50 m/s* at $\theta_0 = 30°$ relative to the horizontal, what is the resulting velocity of the catapult toward the left?

Solution: Conservation of linear momentum yields $0 = m_p(v_c + v\cos\theta_0) + m_c v_c$: Evaluating, we get $0 = (2)(v_c + 50\cos 30°) + (36)v_c$, or $v_c = -2.28 m/s$

================================<>================================

Problem 5.52 A bullet (mass *m*) hits a stationary block of wood (mass m_B) and becomes imbedded in it. The coefficient of kinetic friction between the block and the floor is μ_k. As a result of the impact, the block slides a distance D before stopping. What was the velocity of the bullet?

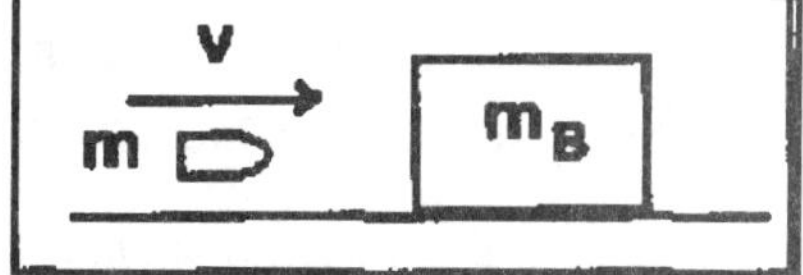

Strategy: First solve the impact problem to determine the velocity of the block and imbedded bullet after the impact in terms of *v*, and then relate the initial velocity of the block and the embedded bullet to the distance D that the block slides.

Solution:

Consider the instant after impact, before the block has moved and friction forces act. The linear momentum condition is $mv = (m + m_B)v_B$, from which $v = \frac{(m + m_B)}{m} v_B$. During the time after impact, the block B begins to move, and friction forces oppose the motion. From Newton's second law of motion and the chain rule: $(m + m_B)v_B \frac{dv_B}{ds} = -\mu_k g(m + m_B)$, from which $v_B^2 = -2\mu_k gs + C$. At $s = D$, $v_B = 0$, from which $v_B^2 = 2\mu_k g(D - s)$. At $s = 0$, the block velocity is $v_B = \sqrt{2\mu_k gD}$. Substitute into the result of the linear momentum condition: $\boxed{v = \left(\frac{m + m_B}{m}\right)\sqrt{2\mu_k gD}}$

================================<>================================

Problem 5.53 A 1 *oz* bullet moving horizontally hits a suspended 100 *lb.* block of wood and becomes embedded in it. If you measure the angle through which the wires supporting the block swing as a result of the impact and determine it to be 7°, what was the bullet's velocity?

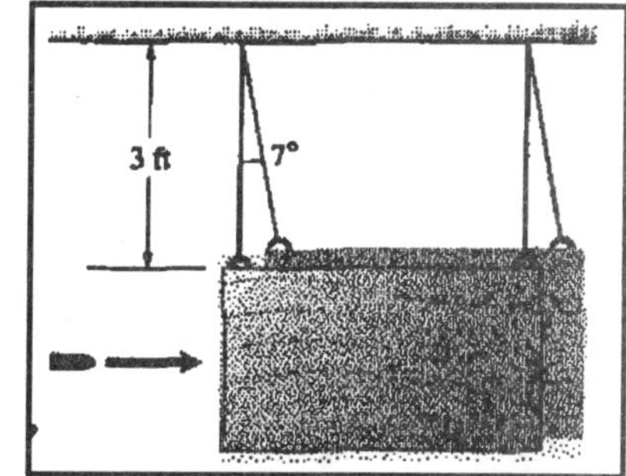

Solution At the instant after impact, before the block has moved, the linear momentum condition is $mv = (m + m_B)v_B$, from which $v = \frac{(m + m_B)}{m} v_B = 1601 v_B$. After impact, and after losses have occurred, energy condition for the block is $\frac{1}{2}(m + m_B)v_B^2 = (W_B + W)(3)(1 - \cos 7^o)$, from which $v_B = \sqrt{2g(3)(1 - \cos 7^o)} = 1.2\ ft/s$, from which $\boxed{v = 1920.4\ ft/s}$

================================<>================================

Problem 5.54 The overhead conveyor drops the 12 *kg* package A into the 1.6 *kg* carton B. The package is tacky" and sticks to the bottom of the carton. If the coefficient of friction between the carton and the horizontal conveyor is $\mu_k = 0.2$, what distance does the carton slide after impact?

Solution

Assume that the height between the overhead conveyor release point and the bottom of the carton is negligible. Consider the instant of time after contact of the package A and B, before carton B has moved, and before friction forces have begun to act. Since the carton can only move parallel to the *x* axis, the linear momentum condition is

$m_A v_A \cos(-26^o) + m_B v_B = (m_A + m_B)v$, from which $v = \dfrac{m_A v_A (0.8988) + m_B v_B}{(m_A + m_B)} = 0.8166\ m/s$. This is the velocity relative to the ground; the velocity relative to the conveyor belt is $v_o = v - v_B = 0.6166\ m/s$. After the impact, the carton B moves and friction forces act. Since the horizontal conveyor moves at constant velocity and the friction force acts on the motion relative to the horizontal conveyor's surface, Newton's second law applies to motion on the horizontal conveyor. Use Newton's second law and the chain rule:. $(m_A + m_B)v\dfrac{dv}{ds} = -\mu_k g(m_A + m_B)$, from which

$v^2 = -2g\mu_k s + C$. At $s = 0$, $v = v_o = 0.6166\ m/s$, from which $v^2 = v_o^2 - 2g\mu_k s$. When the velocity is zero, the distance traveled is $\boxed{s = \dfrac{v_o^2}{2g\mu_k} = 0.09688 \ldots m = 97\ \ mm}$ relative to the horizontal conveyor surface.

Problem 5.55 Suppose you investigate an accident in which a *1300-kg.* car with velocity $\vec{v}_C = 36\bar{j}\ km/hr$ collided with a *5400-kg.* bus with velocity $\vec{v}_B = 20\bar{i}\ \ km/hr$. The vehicles became entangled and remained together after the collision.

(a) What was the velocity of the common center of mass of the two vehicles after collision?

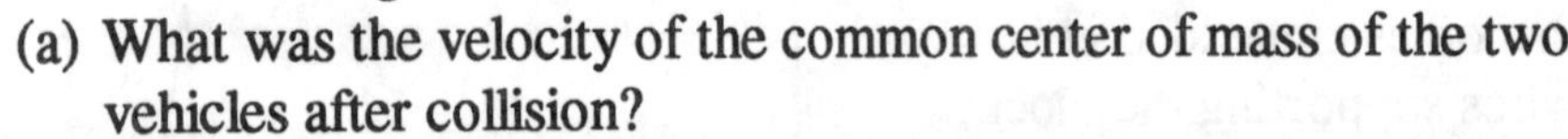

(b) If you estimate the coefficient of friction between the sliding vehicles and the road after the collision to be $\mu_k = 0.4$, what is the approximate final position of the common center of mass relative to its position when the impact occurs?

Solution:

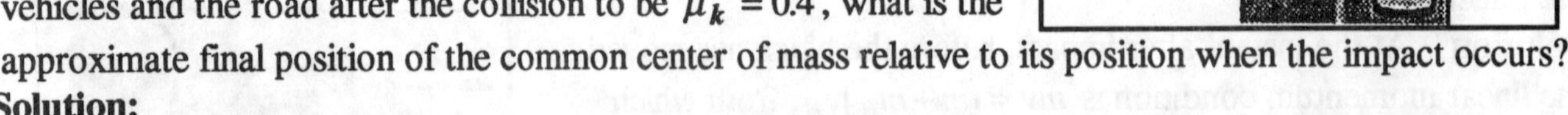

(a) Consider the instant after impact, before the vehicles have started sliding. The linear momentum condition is $m_c\bar{v}_c + m_B\bar{v}_B = (m_c + m_B)\bar{v}$: Evaluating, we get

$(1300)(36j) + (5400)(20\bar{i}) = (1300 + 5400)\bar{v}$, or

$\bar{v} = 16.12\bar{i} + 6.99\bar{j}(km/hr) = 4.48\bar{i} + 1.94\bar{j}(m/s)$

(b) After the vehicles have started sliding, the path is determined by the solution to Newton's second law. Assume that the highway conditions are such that the sliding takes place in a straight line. The friction force during the sliding after the impact is

$f = (1300 + 5400)(9.81)(0.4) = 26{,}300\,N$. Newton's second law in the direction of the

Solution continued on next page

motion of the center of mass after the impact can be written $ma = m\frac{dv}{dt} = m\frac{dv}{ds}v = -f$. Integrating, $m\int_v^0 v dv = -f\int_0^s ds$: $m\frac{v^2}{2} = fs$ The distance traveled is $s = \frac{mv^2}{2f} = \frac{(1300+5400)(4.88)^2}{2(26,300)} = 3.03m$ The x and y distances are $x = 3.03\cos 23.4^\circ = 2.78m$ $y = 3.03\sin 23.4^\circ = 1.21m$

================================<>================================

Problem 5.56 The velocity of the 100 *kg* astronaut A relative to the space station in $40\vec{i} + 30\vec{j}$ (mm/s). The velocity of the 200 *kg* structural member B relative to the station is $-20\vec{i} + 30\vec{j}$ (mm/s). When they approach each other, the astronaut grasps and clings to the structural member. (a) Determine the velocity of their common center of mass when they arrive at the station. (b) Determine the approximate position at which they contact the station.

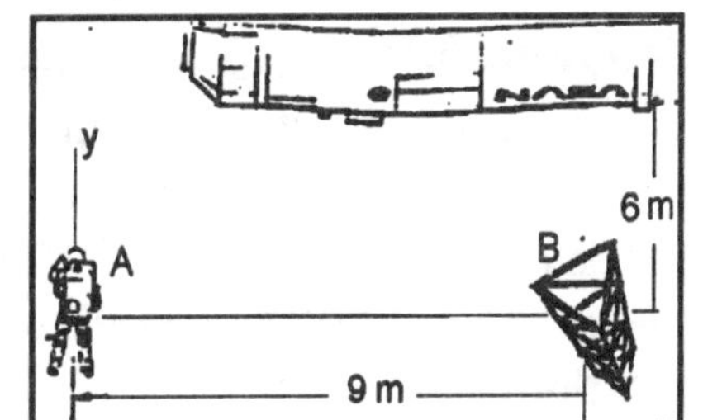

Solution The linear momentum condition is $m_A\vec{v}_A + m_B\vec{v}_B = (m_A + m_B)\vec{v}_{AB}$, from which $\boxed{\vec{v}_{AB} = \frac{m_A\vec{v}_A + m_B\vec{v}_B}{(m_A + m_B)} = 30\vec{j}\ (mm/s)}$ (b) The rendezvous point for the astronaut and the structure is found from: $(0.04)t_{rend} = -(0.02)t_{rend} + 9$, from which the rendezvous time is $t_{rend} = \frac{9}{0.06} = 150\ s$. The coordinates are $x_{rend} = 0.04t_{rend} = 6\ m$, $y_{rend} = 0.03t_{rend} = 4.5\ m$. Since the combined center of mass moves vertically, the coordinates of the point of contact with the space station are $(6,6)$ m, or $\boxed{\vec{r} = 6\vec{i} + 6\vec{j}\ (m)}$

================================<>================================

Problem 5.57 Objects A and B with the same mass m undergo a direct central impact. The velocity of A before the impact is v_A and B is stationary. (a) Determine the velocities of A and B after the impact if it is (a) perfectly plastic ($e = 0$), (b) perfectly elastic ($e = 1$).

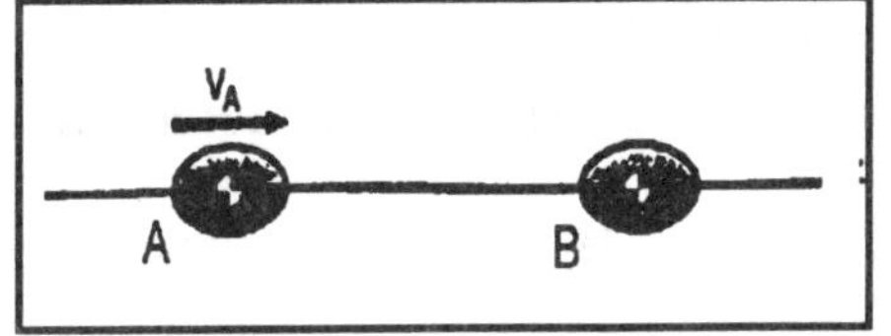

Solution The linear momentum condition is (Eq. (5.8)) $v_A m_A + v_B m_B = m_A v'_A + m_B v'_B$. (a) Since B is stationary, $v_B = 0$. Since $e = 0$, from Eq.(5.14), $v'_A = v'_B$, from which $v_A m_A = (m_A + m_B)v'_A$, and $v'_A = \frac{(m_A)}{(m_A + m_B)}v_A$, from which, for $m_A = m_B = m$, $v'_A = \left(\frac{1}{2}\right)v_A$, $\boxed{v'_B = v'_A}$ (b) For $e = 1$, from Eq (5.14) $v_A - v_B = v'_B - v'_A$. Since B is stationary before impact, the two simultaneous equations are $v_A = v'_B - v'_A$ and $m_A v_A = m_A v'_A + m_B v'_B$. These are solved by the usual (general) procedure of multiplying each equation by a coefficient of the same term appearing in the other equation, and adding the result. The process is repeated for each unknown. The result: $v'_B = v_A\left(\frac{2m_A}{m_A + m_B}\right)$, from which $v'_A = v_A\left(\frac{m_A - m_B}{m_A + m_B}\right)$. For $m_A = m_B = m$, $\boxed{v'_A = 0}$ and $\boxed{v'_B = v_A}$

================================<>================================

================================<>================================

Problem 5.58 In Problem 5.57, if the velocity of B after impact is $0.6v_A$, determine the coefficient of restitution e and the velocity of A after impact.

Solution

From Eq (5.19) the coefficient of restitution is $e = \frac{v'_B - v'_A}{v_A - v_B} = \frac{v'_B - v'_A}{v_A}$ since B is stationary before impact. Substitute the value of e and rearrange, $v'_A = (0.6 - e)v_A$. The conservation of linear momentum condition is $m_A v_A = m_A v'_A + m_B v'_B$, from which $v_A = v'_A + v'_B$, since $m_A = m_B = m$ and B is stationary. Substitute $v'_B = 0.6v_A$, from which $\boxed{v'_A = 0.4v_A}$, and $e = 0.6 - 0.4 = 0.2$

================================<>================================

Problem 5.59 Objects A and B with masses m_A and m_B undergo a direct central impact. (a) If $e = 1$, show that the total kinetic energy after the impact is equal to the total kinetic energy after the impact. (b) If $e = 0$, how much kinetic energy is lost as a result of the collision?

Solution

From Eq (5.19), and from the conservation of linear momentum condition, the two simultaneous equations are $-v'_A + v'_B = ev_A - ev_B$, and $m_A v'_A + m_B v'_B = m_A v_A + m_B v_B$. Write these in canonical form: $a_{11}v'_A + a_{12}v'_B = b_1$, $a_{21}v'_A + a_{22}v'_B = b_2$, for which the solution is $v'_A = \frac{a_{22}b_1}{\det} - \frac{a_{12}b_2}{\det}$,

$v'_B = \frac{-a_{21}b_1}{\det} + \frac{a_{11}b_2}{\det}$, where $\det = a_{11}a_{22} - a_{21}a_{12} = -(m_A + m_B)$ is the determinant of the coefficients.

Substitute and reduce: $v'_A = \left(\frac{1}{m_A + m_B}\right)\left(m_B(1+e)v_B - (em_B - m_A)v_A\right)$,

$v'_B = \left(\frac{1}{m_A + m_B}\right)\left(m_A(1+e)v_A + (m_B - em_A)v_B\right)$.

[*Note:* These results will be referenced in subsequent problem solutions.]

(a) For $e = 1$, these results reduce to $v'_A = -\left(\frac{1}{m_A + m_B}\right)\left((m_B - m_A)v_A - 2m_B v_B\right)$,

$v'_B = \left(\frac{1}{m_A + m_B}\right)\left((m_B - m_A)v_B + 2m_A v_A\right)$. Square both results and substitute into the expression for the kinetic energy of the system after the collision: $\frac{1}{2}m_A(v'_A)^2 + \frac{1}{2}m_B(v'_B)^2$.

After algebraic reduction, $\frac{1}{2}m_A(v'_A)^2 + \frac{1}{2}m_B(v'_B)^2 = \frac{1}{2}m_A v_A^2 + \frac{1}{2}m_B v_B^2$, which demonstrates that the kinetic energy after the collision is equal to the kinetic energy before the collision; thus the <u>total</u> kinetic energy is conserved by a perfectly elastic collision. $e = 1$.

(b) For $e = 0$, $v'_B = \left(\frac{1}{m_A + m_B}\right)(m_A v_A + m_B v_B)$ and $v'_A = v'_B$ [*Check:* This follows from Eq(5.19)].

The kinetic energy after the collision is $\frac{1}{2}(m_A + m_B)(v'_B)^2 = \left(\frac{1}{2}\right)\left(\frac{1}{m_A + m_B}\right)(m_A v_A + m_B v_B)^2$.

Solution continued on next page

The difference in the kinetic energy before and after the collision is
$\frac{1}{2}m_Av_A^2+\frac{1}{2}m_Bv_B^2-\left(\frac{1}{2}\right)\left(\frac{1}{m_A+m_B}\right)(m_Av_A+m_Bv_B)^2$. Carry out the operations and reduce algebraically to obtain for the difference in the kinetic energy: $\boxed{\left(\frac{1}{2}\right)\left(\frac{m_Am_B}{m_A+m_B}\right)(v_A-v_B)^2}$ which is the kinetic energy loss for a completely plastic collision.

====================================<>====================================

Problem 5.60 The *20-lb* weight A and *30-lb* weight B slide on the smooth horizontal bar. Determine their velocities after they collide if the coefficient of restitution is *e = 0.8.*

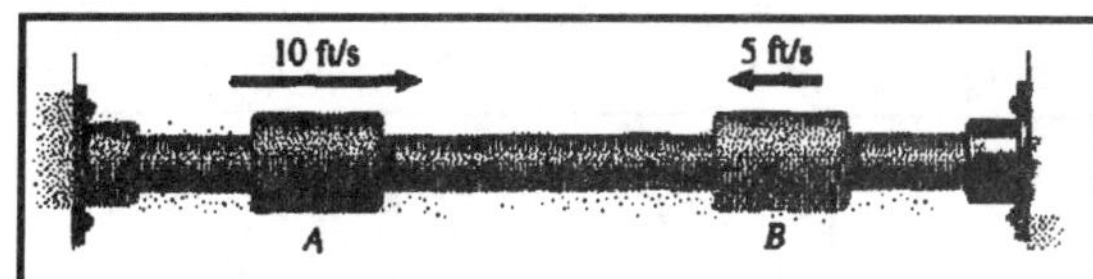

Solution:

Conservation of linear momentum yields: $m_Av_A+m_Bv_B=m_Av_A'+m_Bv_B'$:
Evaluating, we get $(20/32.2)(10)+(30/32.2)(-5)=(20/32.2)v_A'+(30/32.2)v_B'$.

The coefficient of restitution is given by $e=\frac{v_B'-v_A'}{v_A-v_B}$: Evaluating, we get $0.8=\frac{v_B'-v_A'}{10-(-5)}$.

Solving Equations (1) and (2), we get the values $v_A'=-6.2\,ft/s$ $v_B'=5.8\,ft/s$.

====================================<>====================================

Problem 5.61 Two cars with energy absorbing bumpers collide with speeds $v_A=v_B=5\ mi/hr$. Their weights are $W_A=2800\ lb$ and $W_B=4400\ lb$. If the coefficient of restitution of the collision is $e=0.2$, what are the velocities of the cars after the collision?

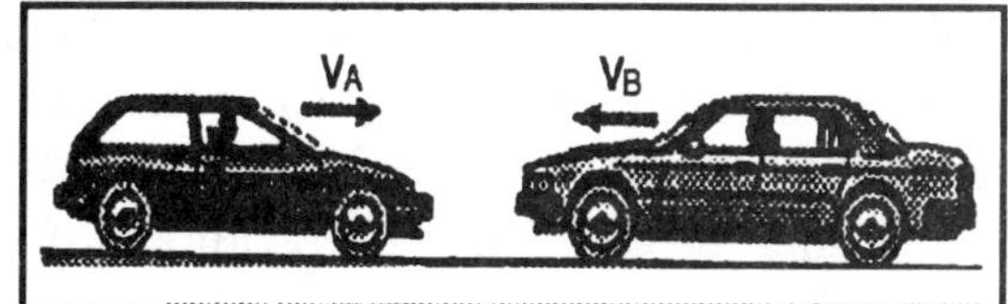

Solution:

Use the results of Problem 5.59. $v_B'=\left(\frac{1}{W_A+W_B}\right)(W_A(1+e)v_A+(W_B-eW_A)v_B)$,

$v_A'=\left(\frac{1}{W_A+W_B}\right)(W_B(1+e)v_B-(eW_B-W_A)v_A)$. Choose the positive direction to the right.

The velocities are $v_A=5\ mi/hr=7.33\ ft/s$, and $v_B=-5\ mi/hr=-7.33\ ft/s$.
Substitute numerical values in the expressions to obtain: $\boxed{v_B'=-0.489\ ft/s}$, $\boxed{v_A'=-3.42\ ft/s}$
(Car B is slowed to 0.489 *ft/s* and car A bounces backwards at 3.42 *ft/s*.)

====================================<>====================================

==================================<>==================================

Problem 5.62 In Problem 5.61, if the duration of the collision is 0.1 *s*, what are the magnitudes of the average accelerations to which the occupants of the two cars are subjected?

Solution

For car A, the impulse is

$\int_{t_1}^{t_2} Fdt = F_{ave}(t_2 - t_1) = \frac{W_A}{g}(v_A - v'_A) = \left(\frac{2800}{32.17}\right)(7.33-(-3.42)) = 936.14\ lb\text{-}s$, from which

$F_{ave} = 9361.4\ lb$. From Newton's second law, $\frac{W_A}{g}\frac{dv}{dt} = -F$. Take the time average of both sides,

$$\frac{1}{t_2-t_1}\left(\frac{W_A}{g}\right)\int_{t_1}^{t_2}\left(\frac{dv}{dt}\right)dt = \frac{1}{t_2-t_1}\left(\frac{W_A}{g}\right)(v_A - v'_A) = \frac{1}{t_2-t_1}\int_{t_1}^{t_2} Fdt = F_{ave} \text{ from which}$$

$$\boxed{\left(\frac{dv}{dt}\right)_{ave} = \left(\frac{g}{W_A}\right)F_{ave} = 107.6\ ft/s^2 = 3.34\ g}$$

For car B the impulse is $\int_{t_1}^{t_2} Fdt = F_{ave}(t_2 - t_1) = \left(\frac{W_B}{g}\right)(v_B - v'_B) = -936.14\ lb\text{-}s$, from which

$F_{ave} = -9361.4\ lb$. From Newton's second law, the magnitude of the average acceleration is

$$\boxed{\left(\frac{dv}{dt}\right)_{ave} = \left(\frac{g}{W_B}\right)F_{ave} = 68.44\ ft/s^2 = 2.13\ g}$$

==================================<>==================================

Problem 5.63 The 10 *kg* mass A is moving at 5 *m/s* when is one meter from the stationary 10 *kg* mass B. The coefficient of kinetic friction between the floor and the two masses is $\mu_k = 0.6$, and the coefficient of restitution of the impact is $e = 0.5$. Determine how far B moves from its initial position as a result of the impact.

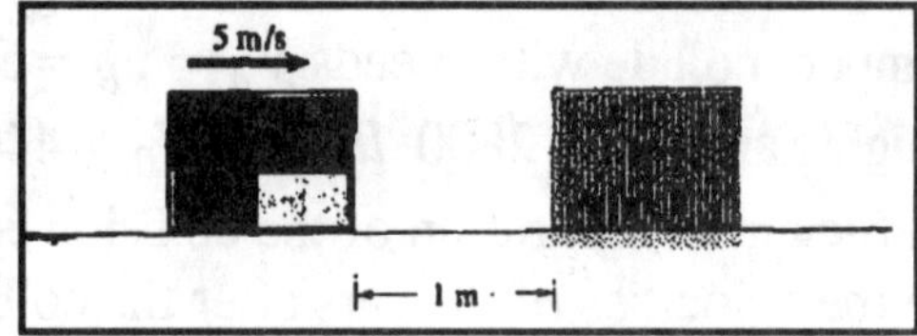

Solution

The strategy is to determine the velocity of A when impact occurs; the velocity of B at the instant after impact but before B has moved, and then determine the distance B will travel.

The work done by friction during the 1 meter travel to impact is

$\int_0^S Fds = \int_0^1 -\mu_k mgds = -(0.6)(9.81)(10) = -58.86\ N\text{-}m$, where the negative sign indicates that the force is in opposition to the motion. The velocity is determined from the change in kinetic energy,

$\frac{1}{2}m_A v_1^2 = \frac{1}{2}m_A v_A^2 + 58.86$, from which the velocity of A at impact is $v_A = \sqrt{v_1^2 - \frac{2(58.86)}{m_A}} = 3.637\ m/s$.

Consider the instant after impact, before B has begun to move. Use the result of the solution to Problem 5.45 for the case when B is stationary and the masses are equal:

$v'_B = \left(\frac{1}{m_A+m_B}\right)(m_A(1+e)v_A + (m_B - em_A)v_B) = \left(\frac{1}{2}\right)(1+e)v_A$, from which $v'_B = 2.73\ m/s$. The work done by friction as B moves is $\int_0^S Fds = \int_0^S -\mu_k m_B gds = -\mu_k m_B gs$. The distance is found from the change in the kinetic energy of B: $\frac{1}{2}m_B v_B^2 = \mu_k m_B gs$. From which $\boxed{s = v_B^2/2\mu_k g = 0.632\ \text{m}}$

==================================<>==================================

=================================<>=================================

Problem 5.64 The kinetic coefficients of friction between the *5-kg* crates A and B and the inclined surface are *0.1* and *0.4*, respectively. The coefficient of restitution between the crates is *e = 0.8*. If the crates are released from rest in the positions shown, what are the magnitudes of their velocities immediately after they collide?

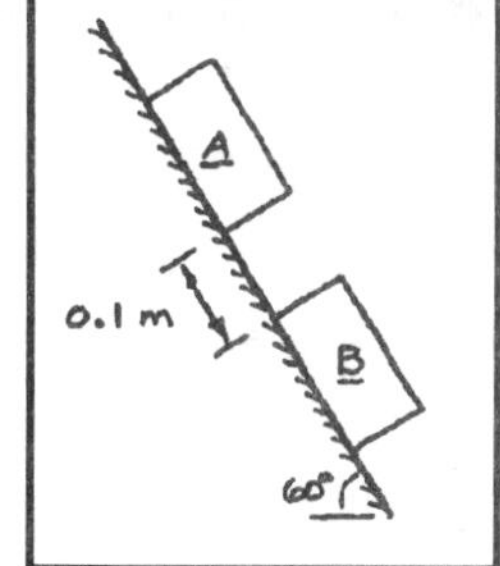

Solution: The free body diagrams of A and B are shown. From the diagram of A, we have $N_A = m_A g\cos 60^\circ = (5)(9.81)\cos 60^\circ = 24.5N$

$\sum F_x = m_A g\sin 60^\circ - 0.1N_A = m_A a_A$. Solving, $(5)(9.81)\sin 60^\circ - 0.1(24.5) = (5)a_A$, $a_A = 8.01m/s^2$. The velocity of A is $v_A = a_A t$ and its position is $x_A = \frac{1}{2}a_A t^2$. From the free body diagram of B, we have $N_B = N_A = 24.5N$ and $\sum F_x = m_A g\sin 60^\circ - 0.4N_B = m_B a_B$. Solving, we

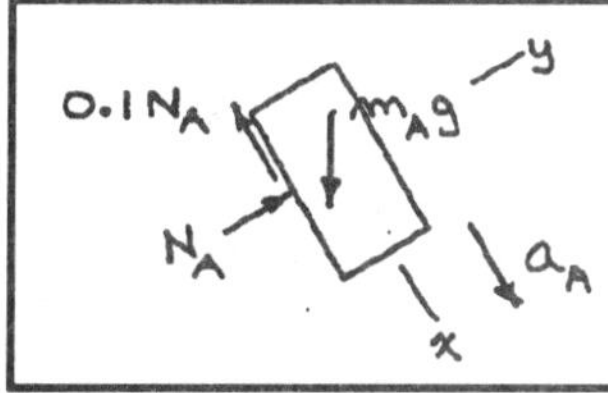

have $(5)(9.81)\sin 60^\circ - 0.4(24.5) = (5)a_B$, or $a_B = 6.53m/s^2$. The velocity of B is $v_B = a_B t$ and its position is $x_B = a_B t^2/2$. To find the time of impact, set $x_A = x_B + 0.1$: $\frac{1}{2}a_A t^2 = \frac{1}{2}a_B t^2 + 0.1$ Solving for t

$t = \sqrt{\frac{2(0.1)}{a_A - a_B}} = \sqrt{\frac{2(0.1)}{8.01 - 6.53}} = 0.369s$. The velocities at impact are

$v_A = a_A t = (8.01)(0.369) = 2.95m/s$ $v_B = a_B t = (6.53)(0.369) = 2.41m/s$ Conservation of linear momentum yields $m_A v_A + m_B v_B = m_A v_A' + m_B v_B'$: $(5)(2.95) + (5)(2.41) = (5)v_A' + (5)v_B'$ (1)

The coefficient of restitution is $e = \frac{v_B' - v_A'}{v_A - v_B}$: Evaluating, we have $0.8 = \frac{v_B' - v_A'}{2.95 - 2.41}$ (2).

Solving equations (1) and (2) $v_A' = 2.46m/s$ $v_B' = 2.90m/s$.

=================================<>=================================

Problem 5.65 Solve Problem 5.64 if crate A has a velocity of *0.2 m/s* down the surface and crate B is at rest when the crates are in the position shown.

Solution:

From the solution of Problem 5.64, A's acceleration is $a_A = 8.01m/s^2$ so A's velocity is

$\int_{0.2}^{v} dv = \int_0^t a_A dt$ $v_A = 0.2 + a_A t$ and its position is $x_A = 0.2t + \frac{1}{2}a_A t^2$ From the solution of Problem 5.64, the acceleration, velocity, and position of B are $a_B = 6.53m/s^2$ $v_B = a_B t$ $x_B = \frac{1}{2}a_B t^2$, At impact, $x_A = x_B + 0.1$; $0.2t + \frac{1}{2}a_A t^2 = \frac{1}{2}a_B t^2 + 0.1$. Solving for t, we obtain $t = 0.257s$ so, $v_A = 0.2 + (8.01)(0.257) = 2.26m/s$; $v_B = (6.53)(0.257) = 1.68m/s$

$m_A v_A + m_B v_B = m_A v_A' + m_B v_B'$: $(5)(2.26) + (5)(1.68) = (5)v_A' + (5)v_B'$ (1); $e = \frac{v_B' - v_A'}{v_A - v_B}$:

$0.8 = \frac{v_B' - v_A'}{2.26 - 1.68}$ (2). Solving Equations (1) and (2), $v_A' = 1.74m/s$ $v_B' = 2.20m/s$.

=================================<>=================================

================================<>================================

Problem 5.66 Suppose you investigate an accident in which a 1300 *kg* car A struck a parked 1200 *kg* car B. All four of Bs wheels were locked, and skid marks indicate it slid 2 *m* after the impact. If you estimate the coefficient of friction between B's tires and the road to $\mu_k = 0.8$ and the coefficient of restitution of the impact to be $e = 0.4$, what was A's velocity just before the impact? (Assume that only one impact occurred.)

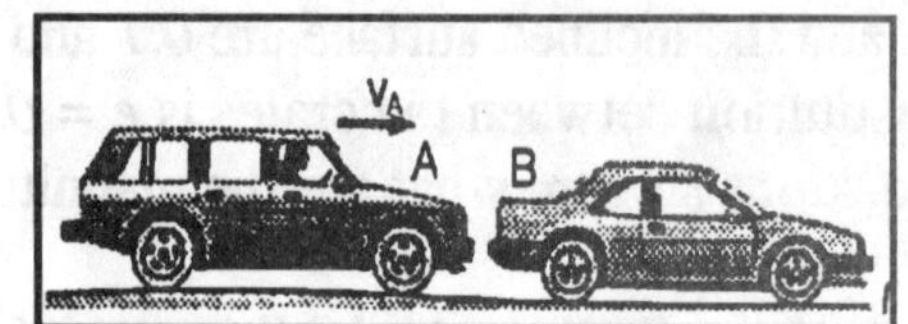

Solution The work done in producing the skid marks is $\int_0^s Fds = \int_0^2 -\mu_k m_B g ds = -2\mu_k m_B g$. This is equal to the kinetic energy of B the instant after impact $\frac{1}{2}m_B(v_B')^2 = 2\mu_k m_B g$, from which $v_B' = \sqrt{4\mu_k g} = 5.6\ m/s$. For B stationary, the conservation of linear momentum condition is $m_A v_A = m_A v_A' + m_B v_B'$, and the coefficient of restitution is $ev_A = v_B' - v_A'$. Solve:
$v_A' = \left(\frac{m_A - em_B}{m_A + m_B}\right)v_A$, $v_B' = \frac{m_A(1+e)}{m_A + m_B}v_A$, from which $\boxed{v_A = \frac{(m_A + m_B)}{m_A(1+e)}v_B' = 7.7\ m/s}$. [*Check.* From the solution to Problem 5.45, for B stationary,

$$v_B' = \left(\frac{1}{m_A + m_B}\right)(m_A(1+e)v_A + (m_B - em_A)v_B) = \frac{m_A}{(m_A + m_B)}(1+e)v_A,$$ from which

$$v_A = \frac{(m_A + m_B)}{m_A(1+e)}v_B' = 7.7\ m/s\ \textit{check.}]$$

================================<>================================

Problem 5.67 Suppose you drop a basketball 5 *ft* above the floor and it bounces to a height of 4 *ft*. If you then throw the basketball downward, releasing it 3 *ft* above the floor moving at 30 *ft/s*, how high does it bounce?

Solution From the definition of the coefficient of restitution (Eq (5.14), for impact on a stationary floor, such that $v_B = v_B' = 0$, $e = -\frac{v_A'}{v_A}$. For a fall of 5 *ft*, the velocity on impact is
$v_A = -\sqrt{2gh} = -\sqrt{2(32.17)5} = -17.94\ ft/s$ (downward). For a rebound of 4 ft,
$v_A' = \sqrt{2gh} = \sqrt{2(32.17)(4)} = 16.04\ ft/s$ (upward), from which $e = 0.894$. For a release at 3 ft with a velocity of 30 *ft/s*, the velocity on impact is $\frac{1}{2}mv_A^2 = \frac{1}{2}m(30)^2 + mg(3)$, from which
$v_A = -\sqrt{30^2 + 2(32.17)(3)} = -33.06\ ft/s$ (downward). The rebound velocity the instant after impact, is
$v_A' = -ev_A = 29.57\ ft/s$ (upward). The height of rebound is obtained from $mgh = \frac{1}{2}m(v_A')^2$, from which $\boxed{h = \frac{(v_A')^2}{2g} = 13.59\ ft}$

================================<>================================

================================<>================================

Problem 5.68 The *1-lb* soccer ball is *3 ft* above the ground when it is kicked upward at *40 ft/s*. If the coefficient of restitution between the ball and the ground is $e = 0.6$, how high above the ground does the ball travel on its first bounce?

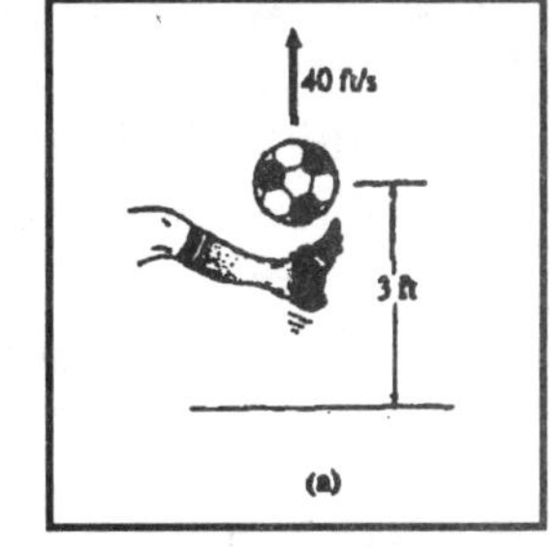

Solution: Applying conservation of energy to the ball's motion before it hits the ground, $\frac{1}{2}mv_1^2 + mgy_1 = \frac{1}{2}mv_2^2 + mgy_2$ Evaluating, we get $\frac{1}{2}(40)^2 + (32.2)(3) = \frac{1}{2}v_2^2 + mg(0)$. Solving, $v_2 = -42.3\,ft/s$.

The coefficient of restitution is given gy $e = 0.6 = \frac{-v_2'}{(-42.3)}$. So the ball's upward velocity after bouncing is $v_2' = 25.4\,ft/s$ Applying conservation of energy to the ball's motion to its highest point, $\frac{1}{2}m(v_2')^2 + mg(0) = \frac{1}{2}m(0)^2 + mgy$, Hence, $y = \frac{(v_2')^2}{2g} = 10.0\,ft$.

================================<>================================

Problem 5.69 If the soccer ball in Problem 5.68 was stationary just before it was kicked and the impact lasted *0.02 s*, what was the average magnitude of the force exerted by the player's foot?

Solution: The impulse is $\sum F_{av}\Delta t = mv_2 - mv_1$: Evaluating, we get $\sum F_{av}(0.02) = (1/32.2)(40)$, resulting in $\sum F_{av} = 62.1\,lb$. This includes the upward force exerted by the player's foot and the downward weight of the ball, so the average upward force exerted by the player is $62.1 + 1 = 63.1\,lb$.

================================<>================================

Problem 5.70 By making measurements directly from the photograph of the bouncing golf ball, estimate the coefficient of restitution.

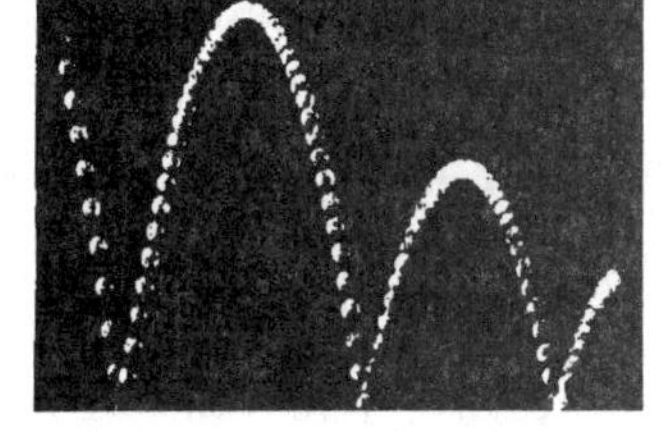

Solution For impact on a stationary surface, the coefficient of restitution is defined to be $e = -v_A'/v_A$. (Since the impact and rebound velocities have opposite signs, e is positive.) (See Eq (5.14)). From the conservation of energy, $\frac{1}{2}m_A(v_A')^2 = m_A gh$, the velocity is proportional to the square root of the rebound height, so that if $h_1,\ h_2,\ \ldots h_N, \ldots$ are successive rebound heights, then an estimate of e is $e = \sqrt{h_{i+1}/h_i}$. Measurements are $h_1 = 5.1\ cm,\ h_2 = 3.1\ cm$, from which

$\boxed{e = \sqrt{3.1/5.1} = 0.78}$

================================<>================================

Problem 5.71 If you throw the golf ball in Problem 5.70 horizontally at 2 *ft/s* and release it 4 ft above the surface, what is the distance between the first two bounces?

Solution The normal velocity at impact is $v_{An} = -\sqrt{2g(4)} = -16.04\ ft/s$ (downward). The rebound normal velocity is (from Eq (5.19)) $v_{An}' = -ev_{An} = -(0.78)(-16.04) = 12.51\ ft/s$ (upward). From the conservation of energy for free fall the first rebound height is $h = (v_{An}')^2/2g = 2.43$ ft. From the solution of Newton's second law for free fall, the time spent between rebounds is twice the time to fall from the maximum height: $t = 2\sqrt{2h/g} = 0.778$ s from which the distance between bounces is:

$\boxed{d = v_o t = 2t = 1.56\,ft}$

================================<>================================

==================================<>==================================

Problem 5.72 In a forging operation, the 100 *lb.* is lifted into position 1 and released from rest. It falls and strikes a workpiece in position 2. If the weight is moving at 15 ft/s immediately before the impact and the coefficient of restitution is $e = 0.3$, what is its velocity immediately after impact.

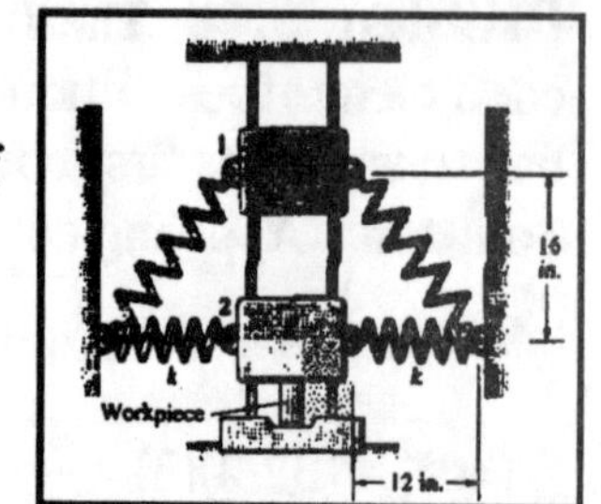

Solution

The strategy is to treat the system as an in-line impact on a rigid, immovable surface. From Eq (5.17) with B's velocity equal to zero: $v'_A = -ev_A$, from which $\boxed{v'_A = -0.6(-15) = 4.5\ ft/s}$

==================================<>==================================

Problem 5.73 In Problem 5.72, suppose that the spring constant is $k = 120\ lb/ft$, the springs are unstretched in position 2, and the coefficient of restitution is $e = 0.2$. Determine the velocity of the weight immediately after impact.

Solution The stretch of each spring is $S = \sqrt{(16/12)^2 + (12/12)^2} - (12/12) = 0.667\ ft$.

From the conservation of energy, $\frac{1}{2}\left(\frac{W}{g}\right)v^2 = 2\left(\frac{1}{2}kS^2\right) + W\left(\frac{16}{12}\right)$, from which the impact velocity is

$v = -\sqrt{\frac{2gk}{W}S^2 + 2g(1.333)} = -10.96\ ft/s$. The velocity immediately after impact is

$\boxed{v' = -ev = 2.19\ ft/s}$

==================================<>==================================

Problem 5.74 A bioengineer studying helmet design strikes a 2.4 *kg* helmet containing a 2 *kg* simulated human head against a rigid surface at 6 *m/s*. The head, being suspended within the helmet, is not immediately affected by the impact of the helmet with the surface and continues to move to the right at 6 *m/s*, so that it then undergoes an impact with the helmet. If the coefficient of restitution of the helmet's impact with the surface is 0.8 and the coefficient of restitution of the following impact of the head and helmet is 0.2, what are the velocities of the helmet and head after their initial interaction?

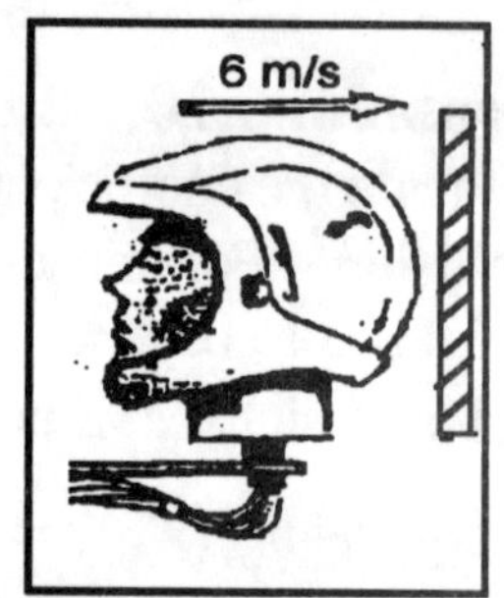

Solution The strategy is to determine the velocity of the helmet after the initial impact (during the rebound), and then to determine the velocity of the head and helmet after the impact of head and helmet. Choose the positive direction to the right. *The initial impact of helmet and surface.* From Eq (5.14), $v'_A = -ev_A$ for rebound from a rigid surface, from which $v'_A = -0.8(6) = -4.8\ m/s$ (to the left). *The impact of head and helmet.* Denote v_A, v'_A to be the before and after velocities of the helmet, and v_B, v'_B to be the before and after velocities of the head. Use the results of the solution of Problem 5.45.

$$v'_A = \left(\frac{1}{m_A + m_B}\right)(m_B(1+e)v_B - (em_B - m_A)v_A),\quad v'_B = \left(\frac{1}{m_A + m_B}\right)(m_A(1+e)v_A + (m_B - em_A)v_B).$$

The initial velocities are $v_A = -4.8\ m/s$, $v_B = 6\ m/s$. The masses are $m_A = 2.4\ kg$, $m_B = 2\ kg$, and the coefficient of restitution is $e = 0.2$. Substitute numerical values to obtain $\boxed{v'_A = 1.09\ m/s}$ for the helmet, and $\boxed{v'_B = -1.07\ m/s}$ for the simulated head.

==================================<>==================================

====================================<>====================================

Problem 5.75 (a) In Problem 5.74, if the duration of the impact of the head with the helmet is 0.008 *s*, what average force is the head subjected to? (b) Suppose that the simulated head alone strikes the surface at 6 *m/s*, the coefficient of restitution is 0.3, and the duration of the impact is 0.002 *s*. What average force is the head subjected to?

Solution (a) The impulse is $\int_{t_1}^{t_2} F dt = F_{ave}(t_2 - t_1) = m_B(v_B - v'_B) = 2(6-(-1.07)) = 14.14\ N\text{-}s$. from which $\boxed{F_{ave} = \frac{14.2}{0.008} = 1{,}767.3\ N}$ (b) From Eq (5.19) the rebound velocity from a rigid surface is $v'_B = -ev_B$, from which $v'_B = -0.3(6) = -1.8\ m/s$ (to the left). The impulse is

$\int_{t_1}^{t_2} F dt = F_{ave}(t_2 - t_1) = m_B(v_B - v'_B) = 2(6-(-1.8)) = 15.6\ N\text{-}s$, from which the average force:

$$\boxed{F_{ave} = \frac{15.6}{0.002} = 7800\ N}$$

====================================<>====================================

Problem 5.76 Two small balls, each of mass *m*, hang from strings of length *L*. The left ball is released from rest at in the position shown. As a result of the first collision, the right ball swings through an angle β. Determine the coefficient of resolution.

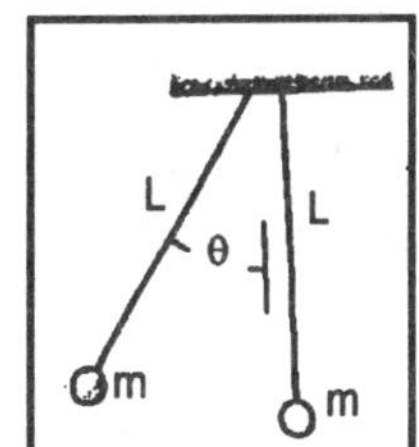

Solution: Assume that when impact occurs, both strings are vertical. The velocity of the ball on the left at impact is determined from the conservation of energy, $\frac{1}{2}mv_A^2 = mgL(1-\cos\theta)$, from which $v_A = \sqrt{2gL(1-\cos\theta)}$. Similarly, the velocity of the ball on the right the instant after impact is determined from the conservation of energy, $v'_B = \sqrt{2gL(1-\cos\beta)}$. For equal masses, and B stationary, the conservation of linear momentum condition is $v_A = v'_A + v'_B$, and the coefficient of restitution is $ev_A = v'_B - v'_A$. Solve: $v'_B = \frac{(1+e)}{2}v_A$, from which

$\boxed{e = \frac{2v'_B - v_A}{v_A} = 2\sqrt{\frac{(1-\cos\beta)}{(1-\cos\theta)}} - 1}$. [*Check*: From the solution to Problem 5.45, for equal masses, and B stationary before impact, $v'_B = \left(\frac{1}{m_A + m_B}\right)(m_A(1+e)v_A + (m_B - em_A)v_B) = \left(\frac{1}{2}\right)(1+e)v_A$., from which

$e = \frac{2v'_B - v_A}{v_A} = 2\sqrt{\frac{(1-\cos\beta)}{(1-\cos\theta)}} - 1$. *check.*]

====================================<>====================================

Problem 5.77 If the duration of the collision in Problem 5.76 is Δt, what is the magnitude of the average force that the balls exert on each other?

Solution: Applying conservation of energy to the motion of the right ball after the impact, $\frac{1}{2}mv^2 = mg(L - L\cos\beta)$ we find that the ball's velocity after the impact is $v = \sqrt{2gL(1-\cos\beta)}$

Therefore, $F_{av}\Delta t = mv = m\sqrt{2gL(1-\cos\beta)}$, so that $F_{av} = (m/\Delta t)\sqrt{2gL(1-\cos\beta)}$.

====================================<>====================================

====================================<>====================================

Problem 5.78 In Example 5.6, suppose that the CSM (A) approaches the Soyuz capsule (B) with velocity $\mathbf{v}_A = 0.05\,\mathbf{i} - 0.002\,\mathbf{j} + 0.007\,\mathbf{k}\ (m/s)$. The docking is unsuccessful, and a spring in the docking collar of the CSM causes the coefficient of restitution of the impact to be $e = 1$. If you treat the collision as an oblique central impact in which the force is parallel to the x-axis, what are the velocities of the centers of mass of the two vehicles afterward?

Solution: The y and z components of the vehicle's velocities are unchanged by the impact. In the x-direction, $m_A v_{ax} = m_A v'_{Ax} + m_b v'_{Bx}$ Substituting in numbers, we get $(18)(0.05) = (18)v'_{Ax} + (6.6)v'_{Bx}$ and $e = \frac{v'_{Bx} - v'_{Ax}}{v_{ax}}$, or $1 = \frac{v'_{Bx} - v'_{Ax}}{0.05}$. Solving, we obtain $v'_{Ax} = 0.0232 m/s$; $v'_{Bx} = 0.0732 m/s$, so $\bar{v}'_A = 0.0232\bar{i} - 0.002\bar{j} + 0.007\bar{k}\,(m/s)$ or $\bar{v}'_B = 0.0732\bar{i}\,(m/s)$

====================================<>====================================

Problem 5.79 A 1 *slug* object A and a 2 *slug* object B undergo an oblique central impact. The coefficient of restitution is $e = 0.8$. Before the impact, $\vec{v}_B = -10\vec{i}\ ft/s$, and after the impact, $\vec{v}'_A = -15\vec{i} + 4\vec{j} + 2\vec{k}$. Determine the velocity of A before the impact and the velocity of B after the impact.

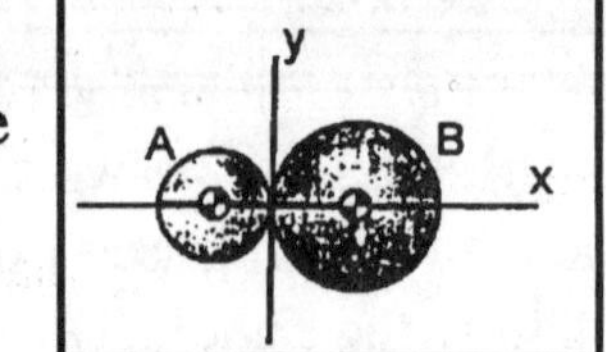

Solution For an oblique central impact along the x axis, the velocities in the y and z directions are unchanged by the impact (see Eq (5.15). The conservation of linear momentum condition for the x axis is $m_A v_{Ax} + m_B v_{Bx} = m_A v'_{Ax} + m_B v'_{Bx}$, with the added condition $e = \frac{v'_{Bx} - v'_{Ax}}{v_A - v_B}$, or, $e v_{Ax} - e v_{Bx} = v'_B - v'_A$.

Solve (see solution to Problem 5.45 for the general procedure used to solve these equations)

$$v'_{Ax} = \left(\frac{1}{m_A + m_B}\right)\left(m_B(1+e)v_{Bx} + (m_A - e m_B)v_{Ax}\right),$$

$$v'_{Bx} = \left(\frac{1}{m_A + m_B}\right)\left(m_A(1+e)v_{Ax} + (m_B - e m_A)v_{Bx}\right).$$ Substitute $m_A = 1\ slug$, $m_B = 2\ slug$, $v'_{Ax} = -15 ft/s$, $v_{Bx} = -10\ ft/s$, and $e = 0.8$. The result: $\boxed{v_{Ax} = 15\ ft/s}$. $\boxed{v'_{Bx} = 5\ ft/s}$.

Collecting terms, and noting that the y and z components of the velocity are unchanged in *an oblique central impact*: $\boxed{\vec{v}_A = 15\vec{i} + 4\vec{j} + 2\vec{k}\ \ (ft/s)}$, is the velocity of A before impact, and $\boxed{\vec{v}'_B = 5\vec{i}\ \ (ft/s)}$ is the velocity of B after impact.

====================================<>====================================

==================================<>==================================

Problem 5.80 The cue gives the cue ball A a velocity parallel to the y axis. It hits the 8 ball B and knocks in straight into the corner pocket. If the magnitude of the velocity of the cue ball just before impact is 2 *m/s* and the coefficient of restitution is $e = 1$, what are the velocity vectors of the two balls just after the impact? (The balls are of equal mass.)

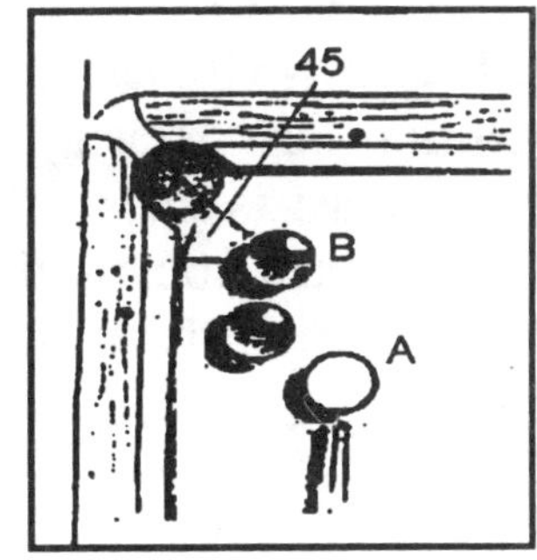

Solution Denote the line from the 8-ball to the corner pocket by BP. This is an oblique central impact about BP. Resolve the cue ball velocity into components parallel and normal to BP. For a 45° angle, the unit vector parallel to BP is $\vec{e}_{BP} = \frac{1}{\sqrt{2}}(-\vec{i} + \vec{j})$, and the unit vector normal to BP is $\vec{e}_{BPn} = \frac{1}{\sqrt{2}}(\vec{i} + \vec{j})$. Resolve the cue ball velocity before impact into components: $\vec{v}_A = v_{AP}\vec{e}_{BP} + v_{APn}\vec{e}_{BPn}$. The magnitudes v_{AP} and v_{APn} are determined from $\sqrt{|v_{AP}\vec{e}_{BP}|^2 + |v_{APn}\vec{e}_{BPn}|^2} = |\vec{v}_A| = 2\ m/s$ and the condition of equality imposed by the 45° angle, from which $v_{AP} = v_{APn} = \sqrt{2}\ m/s$. The cue ball velocity after impact is $\vec{v}'_A = v'_{AP}\vec{e}_{BP} + v_{APn}\vec{e}_{BPn}$, (since the component of $\vec{v}_A$ that is at right angles to BP will be unchanged by the impact). The velocity of the 8-ball after impact is $\vec{v}'_{BP} = v'_{BP}\vec{e}_{BP}$. The unknowns are the magnitudes v'_{BP} and v'_{AP}. These are determined from the conservation of linear momentum along BP and the coefficient of restitution.

$m_A v_{AP} = m_A v'_{AP} + m_B v'_{BP}$, and $1 = \frac{v'_{BP} - v'_{AP}}{v_{AP}}$. For $m_A = m_B$, these have the solution $v'_{AP} = 0$, $v'_{BP} = v_{AP}$, from which $\boxed{\vec{v}'_A = v_{APn}\vec{e}_{BPn} = (\vec{i} + \vec{j})\ (m/s)}$ and $\boxed{\vec{v}'_B = v_{AP}\vec{e}_{BP} = (-\vec{i} + \vec{j})}$

==================================<>==================================

Problem 5.81 In Problem 5.80, what are the velocity vectors of the two balls just after impact if the coefficient of restitution is $e = 0.9$?

Solution Use the results of the solution to Problem 5.80, where the problem is solved as an oblique central impact about the line from the 8-ball to the corner pocket. Denote the line from the 8-ball to the corner pocket by BP. The unit vector parallel to BP is $\vec{e}_{BP} = \frac{1}{\sqrt{2}}(-\vec{i} + \vec{j})$, and the unit vector normal to BP is $\vec{e}_{BPn} = \frac{1}{\sqrt{2}}(\vec{i} + \vec{j})$. Resolve the cue ball velocity before impact into components:

$\vec{v}_A = v_{AP}\vec{e}_{BP} + v_{APn}\vec{e}_{BPn}$, where, from Problem 5.80, $v_{AP} = v_{APn} = \sqrt{2}\ m/s$. The velocity of the 8-ball after impact is $\vec{v}'_{BP} = v'_{BP}\vec{e}_{BP}$. The unknowns are the magnitudes v'_{BP} and v'_{AP}. These are determined from the conservation of linear momentum along BP and the coefficient of restitution.

$m_A v_{AP} = m_A v'_{AP} + m_B v'_{BP}$, and $e = \frac{v'_{BP} - v'_{AP}}{v_{AP}}$. For $m_A = m_B$, these have the solution

$v'_{AP} = \left(\frac{1}{2}\right)(1 - e)v_{AP} = 0.05 v_{AP}$, and $v'_{BP} = \left(\frac{1}{2}\right)(1 + e)v_{AP} = 0.95 v_{AP}$. The result:

$\boxed{\vec{v}'_A = v_{AP}\vec{e}_{BP} + v_{APn}\vec{e}_{BPn} = (-0.05\vec{i} + 0.05\vec{j} + \vec{i} + \vec{j}) = 0.95\vec{i} + 1.05\vec{j}\ m/s}$ $\boxed{\vec{v}'_B = 0.95(-\vec{i} + \vec{j})\ (m/s)}$

==================================<>==================================

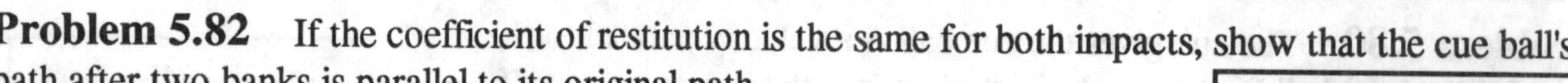

Problem 5.82 If the coefficient of restitution is the same for both impacts, show that the cue ball's path after two banks is parallel to its original path.

Solution The strategy is to treat the two banks as two successive oblique central impacts. Denote the path from the cue ball to the first bank impact as CP1, the path from the first impact to the second as CP2, and the final path after the second bank as CP3. The cue ball velocity along CP1 is

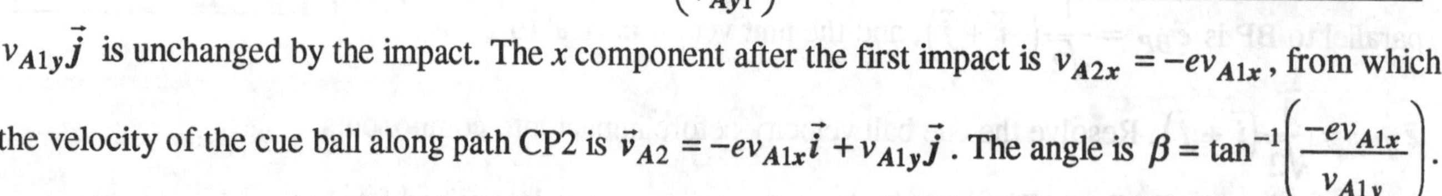

$\vec{v}_{A1} = v_{A1x}\vec{i} + v_{A1y}\vec{j}$, and the angle is $\alpha = \tan^{-1}\left(\dfrac{v_{Ax1}}{v_{Ay1}}\right)$. The component $v_{A1y}\vec{j}$ is unchanged by the impact. The x component after the first impact is $v_{A2x} = -ev_{A1x}$, from which the velocity of the cue ball along path CP2 is $\vec{v}_{A2} = -ev_{A1x}\vec{i} + v_{A1y}\vec{j}$. The angle is $\beta = \tan^{-1}\left(\dfrac{-ev_{A1x}}{v_{A1y}}\right)$.

The x component of the velocity along path CP2 is unchanged after the second impact, and the y component after the second impact is $v_{A3y} = -ev_{A1y}$. The velocity along the path CP3 is

$\vec{v}_{A3} = -ev_{A1x}\vec{i} - ev_{A1y}\vec{j}$, and the angle is $\gamma = \tan^{-1}\left(\dfrac{-ev_{A1x}}{-ev_{A1y}}\right) = \alpha$. The sides of the table at the two banks are at right angles; the angles $\boxed{\alpha = \gamma}$ show that the paths CP1 and CP3 are anti-parallel.

Problem 5.83 The velocity of the 170 *g* hockey puck is $\vec{v}_P = 10\vec{i} - 4\vec{j}$ *(m/s)*. If you neglect the change in the velocity $\vec{v}_S = v_S\vec{j}$ of the stick resulting from the impact and the coefficient of restitution is $e = 0.6$, what should v_S be to send the puck toward the goal?

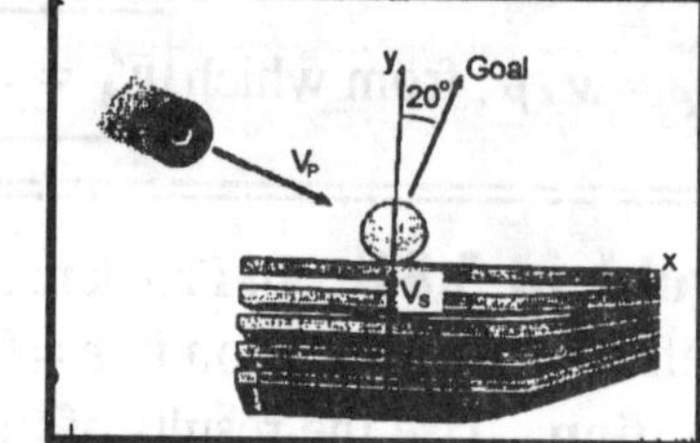

Solution The strategy is to treat the collision as an *oblique central impact* with a moving object of infinite mass. The horizontal component of the puck velocity is unchanged by the impact. The vertical component of the velocity after impact must satisfy the condition $\tan^{-1}(v'_{Px}/v'_{Py}) = \tan^{-1}(10/v'_{Py}) = 20°$, from which the velocity of the puck after impact must be $v'_{Py} = 27.47\ m/s$. Assume for the moment that the hockey stick has a finite mass, and consider only the y component of the puck velocity. The conservation of linear momentum and the definition of the coefficient of restitution are $m_P v_{Py} + m_S v_S = m_P v'_{Py} + m_S v'_S$, and

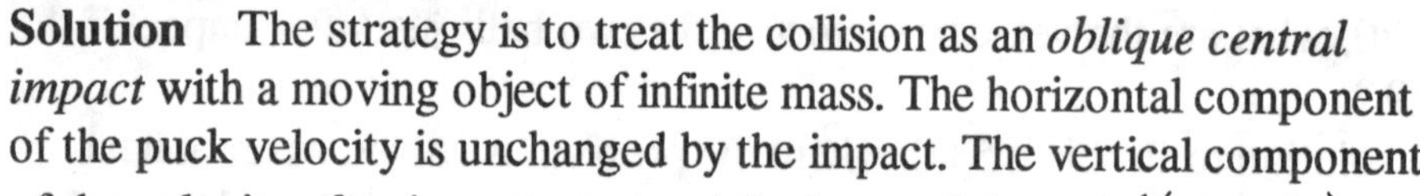

$e = \dfrac{v'_S - v'_{Py}}{v_{Py} - v_S}$. These two simultaneous equations have the solution (see the solution to Problem 5.45) $v'_{Py} = (1/(m_P + m_S))(m_S(1+e)v_S + (m_P - em_S)v_{Py})$. Divide numerator and denominator on the right by m_S and take the limit as $m_S \to \infty$, $v'_{Py} = \lim_{m_S\to\infty}\left(1/\left(\dfrac{m_P}{m_S}+1\right)\right)\left((1+e)v_S + \left(\dfrac{m_P}{m_S} - e\right)v_{Py}\right) = (1+e)v_S - ev_{Py}$.

Substitute the values: $v'_{Py} = 27.47\ m/s$, $e = 0.6$, and $v_{Py} = -4\ m/s$ and solve:

$$\boxed{v_S = \frac{v'_{Py} + ev_{Py}}{(1+e)} = 15.67\ m/s}$$

=================================<>=================================

Problem 5.84 In Problem 5.83, if the stick responds to the impact like an object with the same mass as the puck and the coefficient of restitution is $e = 0.6$, what should v_S be to send the puck towards the goal?

Solution Use the solution to Problem 5.83, where m_S has a finite mass, $v'_{Py} = \left(\frac{1}{m_P + m_S}\right)\left(m_S(1+e)v_S + (m_P - em_S)v_{Py}\right)$. Substitute $m_P = m_S$, $e = 0.6$, $v'_{Py} = 27.47\ m/s$, and $v_{Py} = -4\ m/s$, and solve: $\boxed{v_S = \frac{2v'_{Py} - (1-e)v_{Py}}{(1+e)} = 35.34\ m/s}$

=================================<>=================================

Problem 5.85 The total external force on a 2 *kg* object is $\sum\vec{F} = 2t\vec{i} + 4\vec{j}\ (N)$, where *t* is the time in seconds. At time $t_1 = 0$ its position and velocity are $\vec{r} = 0$, $\vec{v} = 0$. (a) Use Newton's second law to determine the object's position $\vec{r}$ and velocity $\vec{v}$ as functions of time. (b) By integrating $\vec{r} \times \sum\vec{F}$ with respect to time, determine the angular impulse from $t_1 = 0$ to $t_2 = 6\ s$. (c) Use your results from Part (a) to determine the change in the object's angular momentum from $t_1 = 0$ to $t_2 = 6\ s$.

Solution

(a) From Newton's second law, $m\frac{d\vec{v}}{dt} = \sum\vec{F} = 2t\vec{i} + 4\vec{j}$, where $\frac{d\vec{r}}{dt} = \vec{v}$. Integrating,

$\boxed{\vec{v} = \left(\frac{1}{2}\right)\left(t^2\vec{i} + 4t\vec{j}\right)\ (m/s)}$ since $\vec{v} = 0$ at $t = t_1 = 0$. $\boxed{\vec{r} = \left(\frac{1}{2}\right)\left(\frac{t^3}{3}\vec{i} + 2t^2\vec{j}\right)\ (m)}$

(b) From Eq (5.21) the angular impulse is

$$\int_{t_1}^{t_2}\left(\vec{r}\times\sum\vec{F}\right)dt = \left(\frac{1}{m}\right)\int_{t_1}^{t_2}\begin{bmatrix}\vec{i} & \vec{j} & \vec{k}\\ \frac{t^3}{3} & 2t^2 & 0\\ 2t & 4 & 0\end{bmatrix}dt = \vec{k}\left(\frac{1}{m}\right)\int_{t_1}^{t_2}\left(\frac{4}{3} - 4\right)\left(t^3\right)dt = \vec{k}\left(\frac{1}{2}\right)\left[-\frac{2}{3}t^4\right]_0^6$$

$\boxed{= -432\vec{k}\ N\text{-}m\text{-}s}$ (c) The change in the angular momentum is defined to be

$$\vec{H}_2 - \vec{H}_1 = (\vec{r}\times m\vec{v})_2 - (\vec{r}\times m\vec{v})_1, \text{ from which } (\vec{r}\times m\vec{v})_2 = \begin{bmatrix}\vec{i} & \vec{j} & \vec{k}\\ \frac{t^3}{6} & t^2 & 0\\ t^2 & 4t & 0\end{bmatrix}_{t=6} = -432\vec{k}\ kg\text{-}m^2/s,$$

$(\vec{r}\times\vec{v})_1 = (0\times 0) = 0$, from which $\boxed{\vec{H}_2 - \vec{H}_1 = -432\vec{k}\ kg\text{-}m^2/s}$

=================================<>=================================

Problem 5.86 A satellite is in an elliptic orbit around the earth. Its velocity at perigee is *28,280 ft/s*. What is its velocity at apogee.

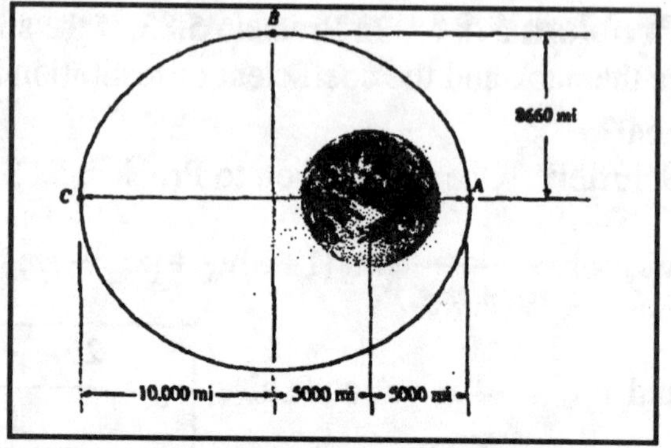

Solution:

Since the velocity has only transverse components at A and C, conservation of angular momentum requires that $r_A v_A = r_c v_c$. Therefore,

$$v_c = \left(\frac{r_A}{r_c}\right) v_A = \left(\frac{5000}{15{,}000}\right)(28{,}280) = 9430\, ft/s$$

Problem 5.87 In Problem 5.86, what are the magnitudes of the radial velocity v_r and the transverse velocity v_θ when the satellite is at point B of its elliptic orbit?

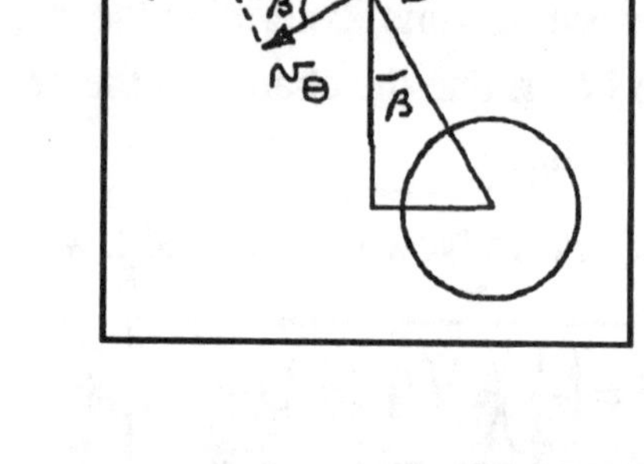

Solution:

We can use conservation of angular momentum to determine the transverse component at B: Thus, $r_B v_\theta = r_A v_A$, and

$$v_\theta = \left(\frac{r_A}{r_B}\right) v_A = \frac{5000}{\sqrt{(5000)^2 + (8660)^2}}(28{,}280) = 14{,}100\, ft/s.$$

From the figure we see that $\dfrac{v_r}{v_\theta} = \tan\beta = \dfrac{5000mi}{8660mi}$, which leads to

$$v_r = \left(\frac{5000}{8660}\right)(14{,}100) = 8160\, ft/s$$

Problem 5.88 The bar rotates *in the horizontal plane* about a smooth pin at the origin. The *2-kg* sleeve C slides on the smooth bar, and the mass of the bar is negligible in comparison to that of the sleeve. The spring constant $k =$ *40 N/m*, and the unstretched length of the spring is *0.8-m*. At $t = 0$, the angular velocity of the bar is $\omega_0 = 6 rad/s$, $r = 0.2m$, and the radial velocity of the sleeve is $v_r = 0$.. What is the angular velocity of the bar when the spring is unstretched?

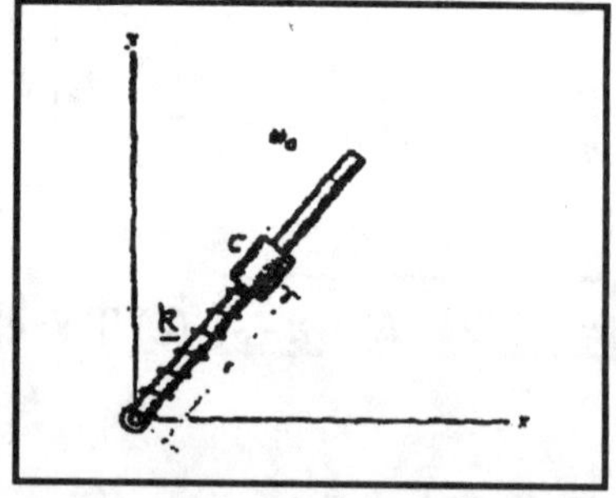

Solution: At $t = 0$, the angular momentum of the sleeve about the origin is $mrv_\theta = mr^2 w = m(0.2)^2(6)$. When the spring is unstretched, it is $mrv_\theta = mr^2\omega = m(0.8)^2\omega$. Therefore $m(0.2)^2(6) = m(0.8)^2\omega$, and $\omega = 0.375 rad/s$.

Problem 5.89 In Problem 5.88, what is the radial velocity of the sleeve when the spring is unstretched?

Solution: From the solution of Problem 5.88, the angular velocity of the bar when the spring is unstretched is $0.375 rad/s$. The kinetic energy of the sleeve is $\frac{1}{2}mv^2 = \frac{1}{2}m(v_r^2 + v_\theta^2) = \frac{1}{2}m(v_r^2 + r^2w^2)$.

Applying conservation of energy, $\frac{1}{2}(2)\left[0 + (0.2)^2(6)^2\right] + \frac{1}{2}(40)(0.2 - 0.8)^2 = \frac{1}{2}(2)\left[v_r^2 + (0.8)^2(0.375)^2\right] + 0$

$\frac{1}{2}m(v_r^2 + r^2w^2) + \frac{1}{2}kS^2 = const$: Solving, $v_r = 2.92 m/s$.

===================================<>===================================

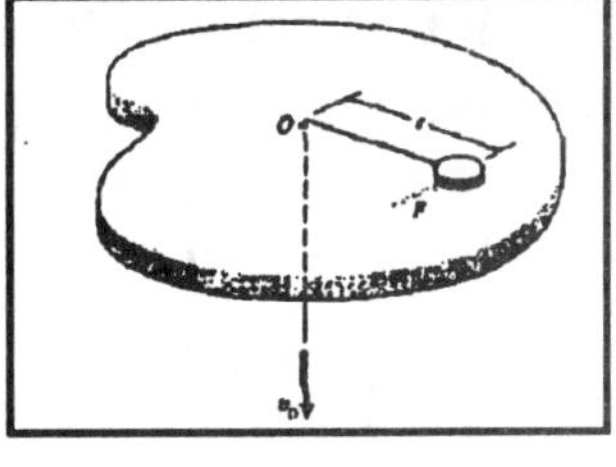

Problem 5.90 In Example 5.7, determine the disk's velocity as a function of time if the force is $F = Ct$, where C is a constant.

Example 5.7: A disk of mass m slides on a smooth horizontal table under the action of a constant transverse force F. The string is drawn through a hole in the table a O at a constant velocity v_o. At $t = 0$, $r = r_o$ and the transverse velocity of the disk is zero. What is the disk's velocity as a function of time?

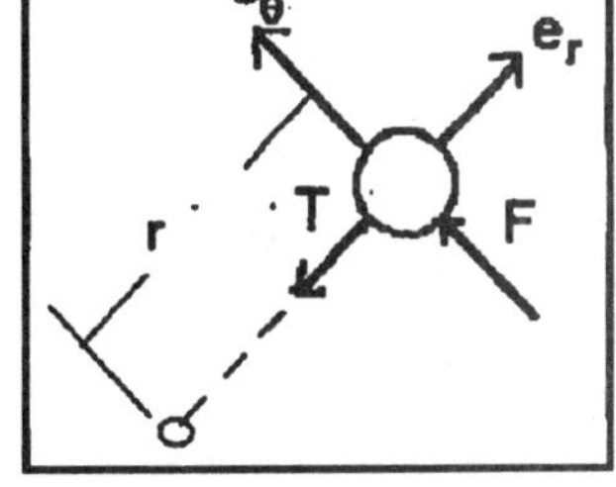

Solution The solution strategy is that followed in Example 5.7. The radial position as a function of time is $r = r_o - v_o t$. Using polar coordinates, the moment about O is $\vec{r} \times \sum \vec{F} = r\vec{e}_r \times (-T\vec{e}_r + F\vec{e}_\theta) = rF(\vec{e}_r \times \vec{e}_\theta) = (r_o - v_o t)Ct\,\vec{e}_z$. The angular momentum is $\vec{H} = (\vec{r} \times m\vec{v}) = r\vec{e}_r \times m(v_r\vec{e}_r + v_\theta\vec{e}_\theta)$

$\vec{H} = mv_\theta r(\vec{e}_r \times \vec{e}_\theta) = mv_\theta(r_o - v_o t)\vec{e}_z$. Substituting into the definition of the angular impulse $\int_{t_1}^{t_2} (\vec{r} \times \sum \vec{F})dt = \vec{H}_2 - \vec{H}_1$, from which $\int_0^t (r_o - v_o t)Ct\,\vec{e}_z dt = mv_\theta(r_o - v_o t)\vec{e}_z$, from which $v_\theta = \dfrac{Ct^2((r_o/2) - (v_o t/3))}{m(r_o - v_o t)}$, and the disk's velocity as a function of time is

$$\boxed{\vec{v} = -v_o\vec{e}_r + \frac{Ct^2\left(\frac{r_o}{2} - \frac{v_o t}{3}\right)}{m(r_o - v_o t)}\vec{e}_\theta}$$

===================================<>===================================

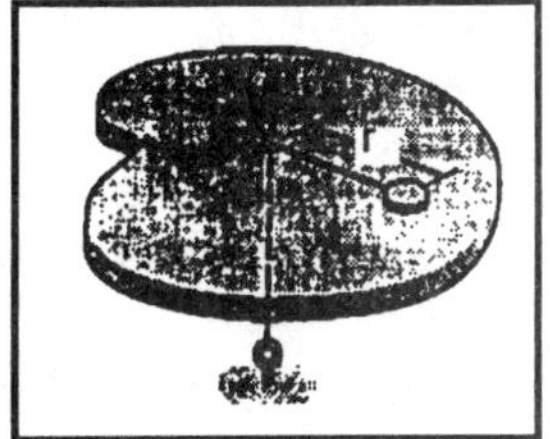

Problem 5.91 A 2 *kg* disk slides on a smooth horizontal table and is connected to an elastic cord whose tension is $T = 6r\ N$, where *r* is the radial position of the disk in meters. If the disk is at $r = 1\ m$ and is given an initial velocity of 4 *m/s* in the transverse direction, what are the magnitudes of the radial and transverse components of its velocity when $r = 2\ m$?

Solution The strategy is to (a) use the principle of conservation of angular momentum to find the transverse velocity and (b) use the conservation of energy to find the radial velocity. The angular momentum the instant after $t = 0$ is $(\vec{r} \times m\vec{v})_o = H_o\vec{e}_z = (mrv_\theta)_o\vec{e}_z$, from which $H_o = 8\ kg\text{-}m^2/s$. In the absence of external transverse forces, the angular momentum impulse vanishes: $\int_{t_1}^{t_2} (\vec{r} \times \sum \vec{F})dt = 0 = \vec{H}_2 - \vec{H}_1$, so that so that $\vec{H}_1 = \vec{H}_2$, that is, the angular momentum is constant. At $r = 2$, $\boxed{v_\theta = \frac{H_o}{mr} = \frac{8}{4} = 2\ m/s}$. From conservation of energy:

$\frac{1}{2}mv_{r_o}^2 + \frac{1}{2}mv_{\theta_o}^2 + \frac{1}{2}kS_o^2 = \frac{1}{2}mv_r^2 + \frac{1}{2}mv_\theta^2 + \frac{1}{2}kS^2$. Solve: $v_r = \sqrt{v_{r_o}^2 + v_{\theta_o}^2 - v_\theta^2 + \left(\frac{k}{m}\right)(S_o^2 - S^2)}$.

Substitute numerical values: Noting $m = 2\ kg$, $v_{r_o} = 0$, $v_{\theta_o} = 4\ m/s$, $k = 6\ N/m$, $r = 2\ m$, $v_\theta = 2\ m/s$, $S_o = 1\ m$, $S = 2\ m$ from which $\boxed{v_r = \sqrt{3}\ m/s}$. The velocity is $\boxed{\vec{v} = 1.732\vec{e}_r + 2\vec{e}_\theta}$

===================================<>===================================

==================================<>==================================

Problem 5.92 In Problem 5.91 determine the maximum value of r reached by the disk.

Solution The maximum value is the stretch of the cord when $v_r = 0$. From the solution to Problem 5.77, $v_r^2 = v_{r_o}^2 + v_{\theta_o}^2 - v_\theta^2 + \left(\frac{k}{m}\right)\left(S_o^2 - S^2\right) = 0$, where $v_\theta = \frac{H_o}{mr}\ m/s$, $v_{r_o}^2 = 0$, $v_{\theta_o} = \frac{H_o}{m(1)}\ m/s$, $S_o = r_o = 1\ m$, $S = r\ m$, and $H_o = 8\ kg\text{-}m^2/s$. Substitute and reduce:

$v_r^2 = 0 = \left(\frac{H_o^2}{m^2}\right)\left(1 - \frac{1}{r^2}\right) + \left(\frac{k}{m}\right)\left(1 - r^2\right)$. Denote $x = r^2$ and reduce to a quadratic canonical form

$x^2 + 2bx + c = 0$, where $b = -\left(\frac{1}{2}\right)\left(\frac{H_o^2}{km} + 1\right) = -3.167$, $c = \frac{H_o^2}{km} = 5.333$. Solve

$r_{1,2}^2 = -b \pm \sqrt{b^2 - c} = 5.333,\ = 1$, from which the greatest positive root is $\boxed{r_{max} = 2.3094... = 2.31\ m}$

[*Check:* This value is confirmed by a graph of the value of $f(r) = \left(\frac{H_o^2}{m^2}\right)\left(1 - \frac{1}{r^2}\right) + \left(\frac{k}{m}\right)\left(1 - r^2\right)$ to find the zero crossing. *check.*]

==================================<>==================================

Problem 5.93 A disk of mass m slides on a smooth horizontal table and is attached to a string that passes through a hole in the table. (a) If the mass moves in a circular path of radius r_o, with transverse velocity v_o, what is the tension T? (b) Starting with the initial condition described in Part (a), the tension is increased in such a way that the string is pulled through the hole at a constant rate until $r = \frac{1}{2} r_o$. Determine T as a function of r while this is taking place. (c) How much work is done on the mass in pulling the string through the hole as described in Part (b) ?

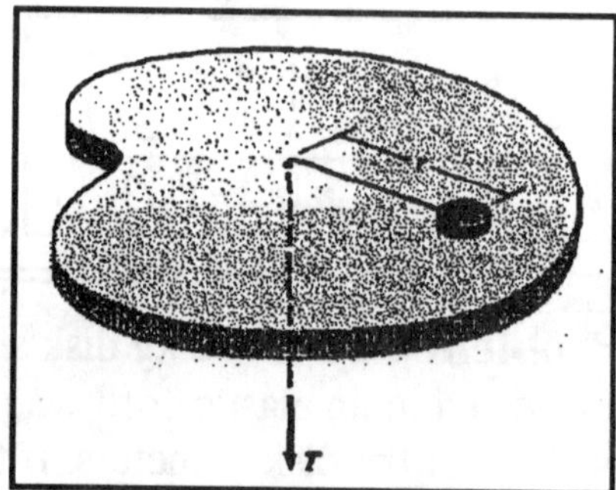

Solution (a) From Newton's second law, $m\left(\frac{d^2r}{dt^2} - r\left(\frac{d\theta}{dt}\right)^2\right) = -T$. With the mass moving in a circle of constant radius r_o, $\frac{d^2r}{dt^2} = 0$, from which $-mr\left(\frac{d\theta}{dt}\right)^2 = -T$. Substitute $r\left(\frac{d\theta}{dt}\right)^2 = \frac{v_o^2}{r}$, to obtain $\boxed{T = mv_o^2 / r_o}$ (b) Assume that "pulled through the hole at constant rate" means that the term $\left|v_r \frac{dv_r}{dr}\right| << \left|\frac{v_\theta^2}{r}\right|$, so that Newton's second law reduces to $-m\frac{v_\theta^2}{r} = -T$. In the absence of external transverse forces $\int_{t_1}^{t_2} \left(\vec{r} \times \sum \vec{F}\right) dt = 0 = \vec{H}_2 - \vec{H}_1$, so that $\vec{H}_1 = \vec{H}_2$; from which the angular momentum is conserved. At the initial condition, the angular momentum is $H_o\vec{e}_z = \left(\vec{r} \times m\vec{v}\right)_o = mr_o v_o \vec{e}_z$, from which $v_\theta = \frac{H_o}{mr}$, where $H_o = mr_o v_o = const$. Newton's second law becomes $\frac{H_o^2}{mr^3} = T$, from which

$$\boxed{T = \frac{mr_o^2 v_o^2}{r^3}}$$

Solution continued on next page

(c) The work done on the mass is $\int_{\frac{r_o}{2}}^{r_o} T dr = \frac{H_o^2}{m}\int_{\frac{r_o}{2}}^{r_o}\frac{dr}{r^3} = -\frac{H_o^2}{2m}\left[\frac{1}{r^2}\right]_{\frac{r_o}{2}}^{r_o} = -\frac{H_o^2}{2m}\left[\frac{1}{r_o^2}-\frac{4}{r_o^2}\right] = \frac{3H_o^2}{2mr_o^2}$

$\boxed{=\frac{3}{2}mv_o^2}$ [*Check*: From the conservation of energy $\left[\frac{1}{2}mv_\theta^2\right]_{r=\frac{r_o}{2}} = \left[\frac{1}{2}mv_\theta^2\right]_{r=r_o} + (\textit{Work done on mass})$.

Rearrange and substitute: $(\textit{Work done on mass}) = \frac{4H_o^2}{2mr_o^2} - \frac{H_o^2}{2mr_o^2} = \frac{3H_o^2}{2mr_o^2} = \frac{3}{2}mv_o^2$, *check.*]

==================================<>==================================

Problem 5.94 In Problem 5.93, how much work is done on the mass in pulling the string through the hole as described in (b)?

Solution: The initial transverse velocity is v_0 so the initial kinetic energy is $T_1 = \frac{1}{2}mv_0^2$ From conservation of angular momentum, the final transverse velocity is $v = \frac{r_0}{(r_0/2)}v_0 = 2v_0$. So the final kinetic energy is $T_2 = \frac{1}{2}m(2v_0)^2 = 2mv_0^2$. The work done is

$V_{12} = T_2 - T_1 = \frac{3}{2}mv_0^2$

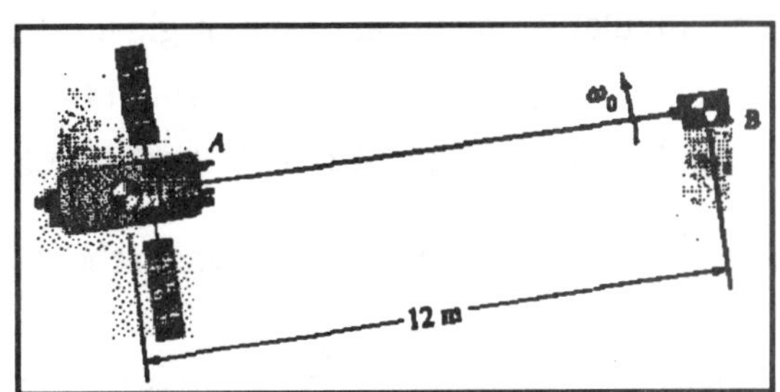

==================================<>==================================

Problem 5.95 Two gravity research satellites ($m_A = 250\ kg$, $m_B = 50\ kg$) are tethered by a cable. The satellite and cable rotate with angular velocity $\omega_o = 0.25$ revolutions per minute. Ground controllers order satellite A to slowly unreel 6 m of additional cable. What is the angular velocity afterward?

Solution The satellite may be rotating in (a) a vertical plane, or (b) in the horizontal plane, or (c) in some intermediate plane. The strategy is to determine the angular velocity for the three possibilities. *Case (a)* Assume that the system rotates in the *x-y* plane, with *y* positive upward. Choose the origin of the coordinates at the center of mass of the system. The distance along the cable from the center of mass to A is $\frac{12m_B}{m_A+m_B} = 2\ m$, from which the distance to B is 10 m.

Assume that both satellites lie on the x axis at $t = 0$. The radius position of satellite A is $\vec{r}_A = -2(\vec{i}\cos\omega_o t + \vec{j}\sin\omega_o t)$, and the radius position of satellite B is $\vec{r}_B = 10(\vec{i}\cos\omega_o t + \vec{j}\sin\omega_o t)$

The acceleration due to gravity is $\vec{W} = -\frac{mgR_E^2}{r^2}\vec{j} = -mg'\vec{j}$, from which $\vec{W}_A = -m_A g'\vec{j}$ and $\vec{W}_B = -m_B g'\vec{j}$ (or, alternatively, . The angular momentum impulse is $\int_{t_1}^{t_2}\left(\vec{r}\times\sum\vec{F}\right)dt = \int_{t_1}^{t_2}\left(\vec{r}_A\times\vec{W}_A + \vec{r}_B\times W_B\right)dt$. Carry out the indicated operations:

Solution is continued on next page

$$\vec{r}_A \times \vec{W}_A = \begin{bmatrix} \vec{i} & \vec{j} & \vec{k} \\ -2(\cos\omega_o t) & -2(\sin\omega_o t) & 0 \\ 0 & -m_A g' & 0 \end{bmatrix} = 2m_A g' \cos\omega_o t\vec{k}.$$

$$\vec{r}_B \times \vec{W}_B = \begin{bmatrix} \vec{i} & \vec{j} & \vec{k} \\ 10(\cos\omega_o t) & 10(\sin\omega_o t) & \\ 0 & -m_B g' & 0 \end{bmatrix} = -10m_B g' \cos\omega_o t\vec{k}.$$

Substitute into the angular momentum impulse: $\int_{t_1}^{t_2}\left(\vec{r} \times \sum \vec{F}\right)dt = 0 = \vec{H}_2 - \vec{H}_1$, from which $\vec{H}_1 = \vec{H}_2$; that is, the angular momentum is conserved. From Newton's second law, the center of mass remains unchanged as the cable is slowly reeled out.

A repeat of the argument above for *any* additional length of cable leads to the same result, namely, the angular momentum is constant, from which the angular momentum is conserved as the cable is reeled out. The angular momentum of the original system is

$\vec{r}_A \times m_A\vec{v} + \vec{r}_B \times m_B\vec{v} = 4m_A\omega_o\vec{k} + 100m_B\omega_o\vec{k} = 157.1\ kg\text{-}m^2/s$, in magnitude, where $\omega_o = 0.026\ rad/s$. After 6 meters is reeled out, the distance along the cable from the center of mass to A is $\dfrac{m_B(6+12)}{m_A + m_B} = 3\ m$, from which the distance to B is 15 m. The new angular velocity when the 6 m is reeled out is $\boxed{\omega = \dfrac{H}{3^2 m_A + 15^2 m_B} = 0.0116\ rad/s = 0.1111\ rpm}$

Case (b): Assume that the system rotates in the *x-z* plane, with y positive upward. As above, choose the origin of the coordinates at the center of mass of the system. Assume that both satellites lie on the *x* axis at $t = 0$. The radius position of satellite A is $\vec{r}_A = -2(\vec{i}\cos\omega_o t + \vec{k}\sin\omega_o t)$, and the radius position of satellite B is $\vec{r}_B = 10(\vec{i}\cos\omega_o t + \vec{k}\sin\omega_o t)$. The force due to gravity is $\vec{W}_A = -m_A g'\vec{j}$ and $\vec{W}_B = -m_B g'\vec{j}$. The angular momentum impulse is $\int_{t_1}^{t_2}\left(\vec{r} \times \sum\vec{F}\right)dt = \int_{t_1}^{t_2}\left(\vec{r}_A \times \vec{W}_A + \vec{r}_B \times \vec{W}_B\right)dt$. Carry out the indicated operations: $\vec{r}_A \times \vec{W}_A = 2m_A g'\cos\omega_o t\vec{i}$, $\vec{r}_B \times \vec{W}_B = -10m_B g'\cos\omega_o t\vec{i}$. Substitute into the angular momentum impulse: $\int_{t_1}^{t_2}\left(\vec{r} \times \sum\vec{F}\right)dt = 0 = \vec{H}_2 - \vec{H}_1$, from which $\vec{H}_1 = \vec{H}_2$; that is, the angular momentum is conserved. By a repeat of the argument given in Case (a), the new angular velocity is $\omega = 0.0116\ rad/s = 0.111\ rpm$.

Case (c): Since the angular momentum is conserved, a repeat of the above for *any orientation of the system relative to the gravity vector* leads to the same result.

================================<>================================

==================================<>==================================

Problem 5.96 An astronaut moves in the x-y plane at the end of a 10 m tether attached to a large space station at O. The total mass of the astronaut and his equipment is 120 kg. (a) What is his angular momentum about O before the tether becomes taut? (b) What is the magnitude of his velocity perpendicular to the tether immediately after the tether becomes taut?

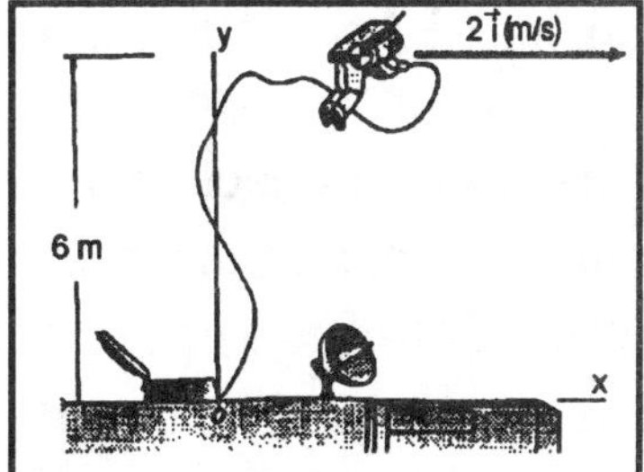

Solution

(a) The angular momentum by definition is

$$(\vec{r}\times m\vec{v}) = \begin{bmatrix} \vec{i} & \vec{j} & \vec{k} \\ 0 & 6 & 0 \\ (120)2 & 0 & 0 \end{bmatrix} = -6(2)(120)\vec{k} = -1440\vec{k}\ kg\text{-}m^2/s.$$

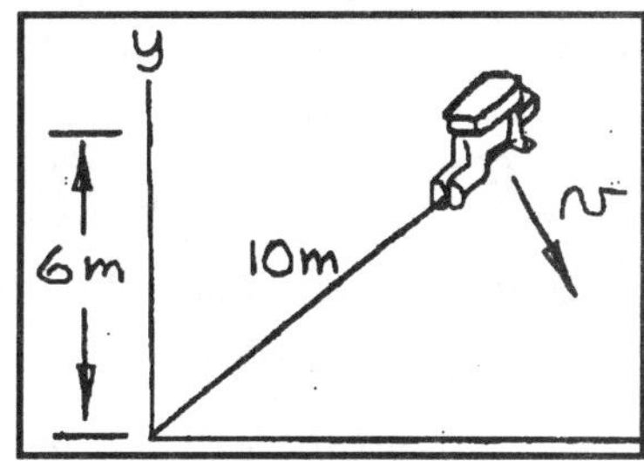

(b)From the conservation of momentum,

(b) $\vec{H} = -1440\vec{k} = -(10)(120)v\vec{k}$, from which

$$\boxed{v = \frac{1440}{(10)(120)} = 1.2\ m/s}$$

==================================<>==================================

Problem 5.97 In Problem 5.96, if the coefficient of restitution of the "impact" that occurs when the astronaut reaches the end of the tether is $e = 0.8$, what are the x and y components of his velocity after the tether becomes taut?

Solution

The strategy is to treat the "impact" as an oblique central impact about the line from O to the astronaut when the tether first tightens, for impact with a "rigid" surface. Since the astronaut is moving parallel to the x axis, the x coordinate of the astronaut when the tether first tightens is $x = \sqrt{10^2 - 6^2} = 8\ m$, from which $\vec{r} = 8\vec{i} + 6\vec{j}\ (m)$ defines the line OA. Project the velocity of the astronaut along the line OA. The unit vector parallel to the line is $\vec{e}_p = \frac{\vec{r}}{|\vec{r}|} = 0.8\vec{i} + 0.6\vec{j}$. The vector normal to the line OA is $\vec{e}_n = 0.8\vec{i} - 0.6\vec{j}$. From $\vec{v}_A = v_{Ap}\vec{e}_p + v_{An}\vec{e}_n$ two simultaneous equations are $2 = 0.8v_{Ap} + 0.6v_{An}$, $0 = 0.6v_{Ap} - 0.8v_{An}$. The solution: $v_{Ap} = 1.6$, $v_{An} = 1.2$, from which the parallel component of the velocity is $\vec{v}_{Ap} = v_{Ap}\vec{e}_p = 1.28\vec{i} + 0.96\vec{j}\ (m/s)$ and the normal component is $\vec{v}_{An} = v_{An}\vec{e}_n = 0.72\vec{i} - 0.96\vec{j}\ (m/s)$. The normal component is unchanged by the "impact". The change in the parallel component is determined from the definition of the coefficient of restitution. Since the velocity of the space station is initially zero and is unchanged by the impact, the definition of the coefficient of restitution (Eq 5.14) reduces to $\vec{v}'_A = -e\vec{v}_{Ap} = -1.024\vec{i} - 0.768\vec{j}$. The velocity of the astronaut after the tether tightens is $\boxed{\vec{v}_{At} = \vec{v}_{An} + \vec{v}'_{Ap} = -0.304\vec{i} - 1.728\vec{j}\ (m/s)}$

==================================<>==================================

================================<>================================

Problem 5.98 A ball suspended from a string that goes through a hole in the ceiling at O moves with velocity v_A in a horizontal circular path of radius r_A. The string is then drawn through the hole until the ball moves with velocity v_B in a horizontal circular path of radius r_B. Use the principle of angular impulse and momentum to show that $r_A v_A = r_B v_B$.
Strategy: Let $\vec{e}$ be a unit vector that is perpendicular to the ceiling. Although this is not a central force problem-- the ball's weight does not point toward O-- you can show the $\vec{e}\cdot(\vec{r}\times\vec{F}) = 0$ so that $\vec{e}\cdot\vec{H}_o$ is conserved.

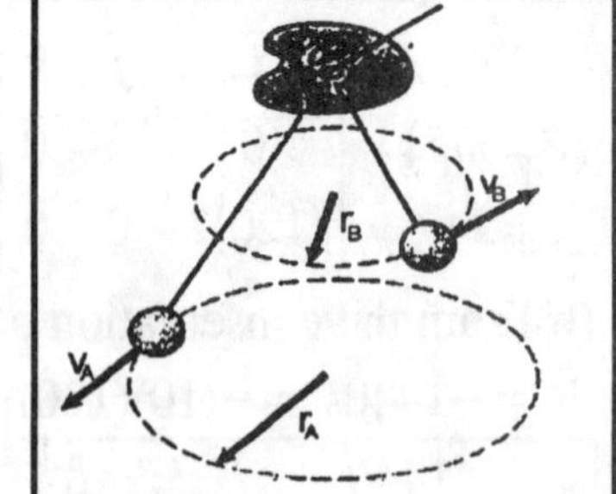

Solution Assume that the motion is in the *x-y* plane, and that the ball lies on the positive *x* axis at $t = 0$. The radius vector $\vec{r}_A = r_A(\vec{i}\cos\omega_A t + \vec{j}\sin\omega_A t)$, where ω_A is the angular velocity of the ball in the path. The velocity is $\vec{v}_A = -\vec{i}r_A\omega_A\sin\omega_A t + \vec{j}r_A\omega_A\cos\omega_A t$ The angular momentum per unit mass about the axis normal to the ceiling is

$$\left(\frac{\vec{r}\times m\vec{v}}{m}\right) = \begin{bmatrix} \vec{i} & \vec{j} & \vec{k} \\ r_A\cos\omega_A t & r_A sin\omega_A t & 0 \\ -r_A\omega_A\sin\omega_A t & r_A\omega_A\cos\omega_A t & 0 \end{bmatrix} = \vec{k}(r_A^2\omega_A)$$

Define the unit vector parallel to this angular momentum vector, $\vec{e} = \vec{k}$. From the principle of angular impulse and momentum, the external forces do not act to change this angular momentum. This is shown as follows:

The external force is the weight, $\vec{W} = -mg\vec{k}$. The momentum impulse is $\int_{t_1}^{t_2}(\vec{r}\times\sum\vec{F})dt = \int_{t_1}^{t_2}(\vec{r}_A\times\vec{W})dt$. Carry out the operation

$$\vec{r}_A\times\vec{W} = \begin{bmatrix} \vec{i} & \vec{j} & \vec{k} \\ r_A\cos\omega_A t & r_A\sin\omega_A t & 0 \\ 0 & 0 & -mg \end{bmatrix} = r_A mg\cos\omega_A t\vec{j}$$

, from which .

$\int_{t_1}^{t_2}\vec{j}r_A mg\cos\omega t dt = -\vec{j}(r_A\omega_A mg)(\sin\omega_A t_2 - \sin\omega_A t_1) = \vec{H}_2 - H_1$. Since this has no component parallel to the unit vector $\vec{e} = \vec{k}$, the angular momentum along the axis normal to the ceiling is unaffected by the weight, that is, the projection of the angular momentum impulse due to the external forces on the unit vector normal to the ceiling is zero $\vec{e}\cdot\vec{H}_2 = \vec{e}\cdot\vec{H}_1 = 0$, hence the *angular momentum normal to the ceiling is conserved.* This result holds true for any length of string, hence $(\vec{r}\times\vec{v})_A = \vec{k}r_A^2\omega_A = (\vec{r}\times\vec{v})_B = \vec{k}r_B^2\omega_B$, from which $r_A^2\omega_A = r_B^2\omega_B$. Since $v_A = r_A\omega_A$, $v_B = r_B\omega_B$, the result can be expressed $\boxed{v_A r_A = v_B r_B}$

================================<>================================

==================================<>==================================

Problem 5.99 The Cheverton fire-fighting and rescue boat can pump 3.8 *kg/s* of water from each of its two pumps at a velocity of 44 *m/s*. If both pumps point in the same direction, what total force do they exert on the boat.

Solution: From Eq (5.25) The force exerted by a mass flow from an object $\vec{F}_f = -\frac{dm_f}{dt}\vec{v}_f$ is exerted on the object, where $\left(\frac{dm_f}{dt}\right)$ is the mass flow rate in *kg/s*, and $\vec{v}_f$ is the exit velocity in *m/s*. The mass flow rate of the two pumps is $2(3.8) = 7.6\ kg/s$. The velocity is 44 *m/s*. The magnitude of the force exerted on the boat is $\boxed{2(3.8) = 7.6\ \ kg/s}$

==================================<>==================================

Problem 5.100 The mass flow rate of water through the nozzle is 1.6 *slug/s*. Determine the magnitude of the horizontal force exerted on the truck by the flow of the water.

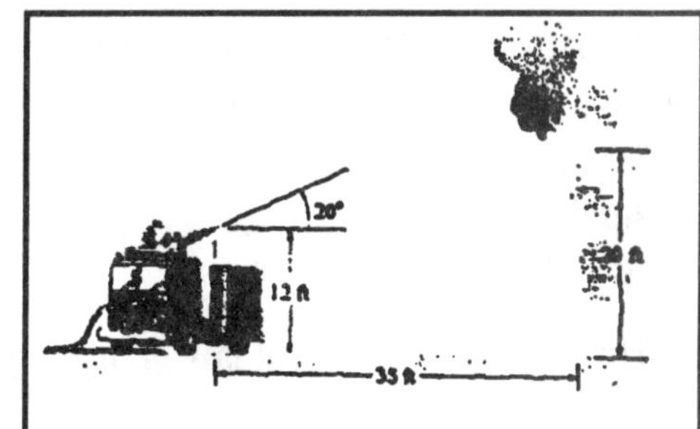

Solution: We must determine the velocity with which the water exits the nozzle. Relative to the end of the nozzle, the x-coordinate of a particle of water is $x = v_0 \cos 20° t$ and the y coordinate is $y = v_0 \sin 20° t - \frac{1}{2}(32.2)t^2$. Setting $x = 35 ft$ and $y = 8 ft$ and eliminating t we obtain $8 = v_0 \sin 20°\left(\frac{35}{v_0 \cos 20°}\right) - \frac{1}{2}(32.2)\left(\frac{35}{v_0 \cos 20°}\right)^2$ From this equation, $v_0 = 68.7 ft/s$. The horizontal force exerted by the flow of water is

$$\frac{dm_f}{dt} v_f \cos 20° = (1.6)(68.7)\cos 20° = 103 lb$$

==================================<>==================================

Problem 5.101 A front end loader moves at 2 *mi/hr* and scoops up 66,000 *lb.* of iron ore in 3 *s*. What horizontal force must its tires exert?

Solution:

The mass flow rate is

$$\left(\frac{dm_f}{dt}\right) = \frac{66,000}{3(32.17)} = 683.87\ slug/s.$$ The velocity is

$v_f = 2\ mi/hr = 2.933\ \ ft/s$. The force the tires must exert of the ground is $\boxed{F = \left(\frac{dm_f}{dt}\right)v_f = 2006\ \ lb}$

==================================<>==================================

==================================<>==================================

Problem 5.102 The snow blower moves at 1 *m/s* and scoops up 750 *kg/s* of snow. Determine the force exerted by the entering flow of snow.

Solution

The mass flow rate is $\left(\frac{dm_f}{dt}\right) = 750\ kg/s$. The velocity is $v_f = 1\ m/s$. The force exerted by the entering flow of snow is $\boxed{F = \left(\frac{dm_f}{dt}\right)v_f = 750(1) = 750\ N}$

==================================<>==================================

Problem 5.103 If you design the snow blower in Problem 5.102 so that it blows snow out at 45° above the horizontal from a port 2 *m* above the ground and the snow lands 20 *m* away, what horizontal force is exerted on the blower by the departing flow of snow?

Solution

The strategy is to use the solution of Newton's second law to determine the exit velocity.

From Newton's second law (ignoring drag) $m_f \frac{dv_y}{dt} = -m_f g$, from which $v_y = -gt + v_f \sin 45^o\ (m/s)$,

$v_x = v_f \cos 45^o\ (m/s)$, and $y = -\frac{g}{2}t^2 + \left(v_f \sin 45^o\right)t + 2\ m$, $x = \left(v_f \cos 45^o\right)t$.

At $y = 0$, the time of impact is $t_{imp}^2 + 2bt_{imp} + c = 0$, where $b = -\frac{v_f \sin 45^o}{g}$, $c = -\frac{4}{g}$.

The solution: $t_{imp} = -b \pm \sqrt{b^2 - c} = \frac{v_f}{\sqrt{2}g}\left(1 \pm \sqrt{1 + \frac{8g}{v_f^2}}\right)$. Substitute:

$$x = 20 = \left(\frac{v_f}{\sqrt{2}}\right)\left(\frac{v_f}{\sqrt{2}g}\right)\left(1 \pm \sqrt{1 + \frac{8g}{v_f^2}}\right).$$

This equation is solved by iteration using **TK Solver Plus** to yield $v_f = 13.36\ m/s$.

The horizontal force exerted on the blower is $\boxed{F = \left(\frac{dm_f}{dt}\right)v_f = 750(13.36)\cos 45^o = 7082.7\ kN}$

==================================<>==================================

=================================<>=================================

Problem 5.104 A nozzle ejects a stream of water horizontally at 40 *m/s* with a mass flow rate of 30 *kg/s*, and the stream is deflected in the horizontal plane by a plate. Determine the force exerted on the plate by the stream in cases (a), (b), and (c).

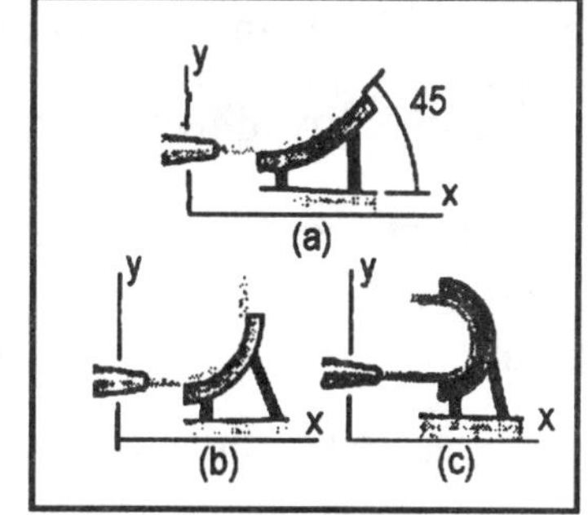

Solution: Apply the strategy used in Example 5.7. The exit velocity is $\vec{v}_{fe} = v_o(\vec{i}\cos\theta + \vec{j}\sin\theta)$ where θ is the total angle of deflection of the stream, and v_o is the magnitude of the stream velocity. The inlet velocity is $\vec{v}_{fi} = v_o\vec{i}$. The force on the plate exerted by the exit stream is in a direction opposite the stream flow, whereas the force exerted on the plate by the inlet stream is in the direction of stream flow. The sum of the forces *exerted on the plate* (see Eq (5.25)) is $\sum\vec{F} = (dm_f/dt)(\vec{v}_{fi} - \vec{v}_{fe})$, from which $\sum\vec{F} = (dm_f/dt)v_o(\vec{i}(1-\cos\theta) - \vec{j}\sin\theta)$. The mass flow is $(dm_f/dt) = 30$ kg/s and the stream velocity is $v_o = 40\ m/s$. For Case(a), $\theta = 45^o$,

$\boxed{\sum\vec{F} = (30)(40)(0.2929\vec{i} - 0.7071\vec{j}) = 351.5\vec{i} - 848.5\vec{j}\ (N)}$. The force is downward to the right. For Case (b) $\theta = 90^o$, $\boxed{\sum\vec{F} = (30)(40)(\vec{i} - \vec{j}) = 1200\vec{i} - 1200\vec{j}}$ For Case (c) $\theta = 180^o$,

$\boxed{\sum\vec{F} = (30)(40)(2\vec{i}) = 2400\vec{i}\ (N)}$

=================================<>=================================

Problem 5.105 A stream of water with velocity $80\vec{i}$ *m/s* and a mass flow of 6 *kg/s* strikes a turbine blade moving with constant velocity $20\vec{i}\ m/s$. (a) What force is exerted on the blade by the water? (b) What is the magnitude of velocity of the water as it leaves the blade?

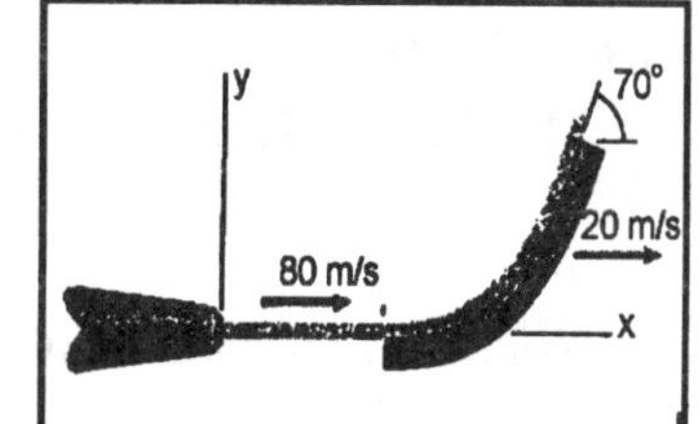

Solution Denote the fixed reference frame as the *nozzle frame,* and the moving blade frame as the *blade rest frame.* Assume that the discharge angle (70°) is referenced to the blade rest frame, so that the magnitude of the stream velocity and the effective angle of discharge in the blade rest frame is not modified by the velocity of the blade. Denote the velocity of the blade by v_B. The inlet velocity of the water relative to the blade is $\vec{v}_{fi} = (v_o - v_B)\vec{i}$ The magnitude $(v_o - v_B)$ is the magnitude of the stream velocity as it flows along the contour of the blade. At exit, the magnitude of the discharge velocity in the blade rest frame is the inlet velocity *in the blade rest frame* $|\vec{v}_{fe}| = (v_o - v_B) = 60\ m/s$, and the vector

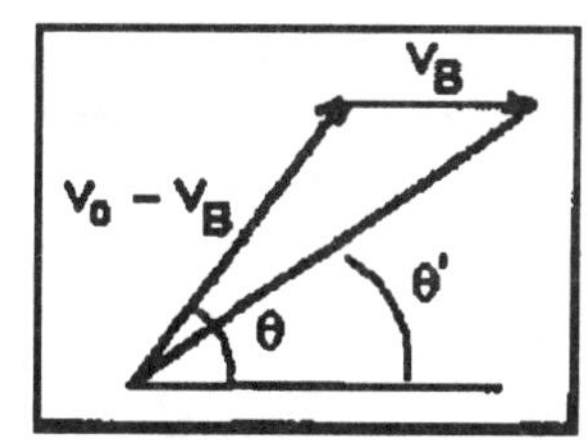

velocity is $\vec{v}_{fe} = v_{fe}(\vec{i}\cos 70^o + \vec{j}\sin 70^o)$ in the blade rest frame. From Eq (5.25) the sum of the forces on the blade is $\sum\vec{F} = (dm_f/dt)((v_o - v_B - v_{fe}\cos 70^o)\vec{i} - \vec{j}(v_{fe}\sin 70^o))$. (a) Substitute numerical values: $(dm_f/dt) = 6$ kg/s, $v_o = 80\ m/s$, $v_B = 20\ m/s$, from which $\boxed{\sum\vec{F} = 236.9\vec{i} - 338.3\vec{j}}$ (b) Assume that the magnitude of the velocity of the water as it leaves the blade is required *in the nozzle frame.* The velocity of the water leaving the blade in the nozzle frame (see vector diagram) is

$\vec{v}_{ref} = 20\vec{i} + 60\cos 70^o\vec{i} + 60\sin 70^o\vec{j} = 40.52\vec{i} + 56.38\vec{j}\ (m/s)$. The magnitude of the velocity is $\boxed{|\vec{v}| = 69.4\ m/s}$.

=================================<>=================================

=================================<>=================================

Problem 5.106 The nozzle A of the lawn sprinkler is located at $(7, -0.5, 0.5)$ *in.*. Water exits each nozzle at 25 *ft/s* with a weight flow rate of 0.5 *lb./s*. The direction cosines of the flow direction from A are $(\frac{1}{\sqrt{3}}, -\frac{1}{\sqrt{3}}, \frac{1}{\sqrt{3}})$. What is the total moment about the *z* axis exerted on the sprinkler by all four nozzles?

Solution:

The unit vector defining the exit velocity is $\vec{e} = \frac{1}{\sqrt{3}}\vec{i} - \frac{1}{\sqrt{3}}\vec{j} + \frac{1}{\sqrt{3}}\vec{k}$.

The position of the exit is $\vec{r} = 7\vec{i} - 0.5\vec{j} + 0.5\vec{k}$. The mass flow rate is

$$\left(\frac{dm_f}{dt}\right) = \frac{0.5}{32.17} = 0.01554 \ slug/s.$$

The moment exerted by the exiting stream is

$$\sum \vec{M} = -\left(\frac{dm_f}{dt}\right) v_f (\vec{r} \times \vec{e}) = -0.3886 \begin{bmatrix} \vec{i} & \vec{j} & \vec{k} \\ 7 & -0.5 & 0.5 \\ \frac{1}{\sqrt{3}} & -\frac{1}{\sqrt{3}} & \frac{1}{\sqrt{3}} \end{bmatrix} = -0.3886(-3.752\vec{j} - 3.752\vec{k}) \ (in.\text{-}lb)$$

$\vec{M} = 1.458\vec{j} + 1.458\vec{k} \ (in.\text{-}lb)$. From symmetry, the total moment about the z axis is

$$\boxed{M_{z\text{-}axis} = 4(1.458) = 5.83 \ in\text{-}lb = 0.486 \ ft\text{-}lb}$$

=================================<>=================================

Problem 5.107 A 45 *kg/s* flow of gravel exits the chute at 2 *m/s* and falls onto a conveyor belt moving at 0.3 *m/s*. Determine the components of the force exerted on the conveyor by the flow of gravel if $\theta = 0$.

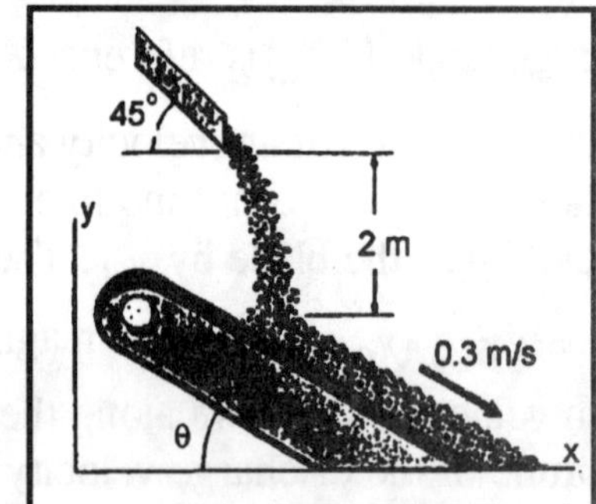

Solution:

The horizontal component of the velocity of the gravel flow is $v_x = 2\cos 45^o = \sqrt{2} \ m/s$. From Newton's second law, (using the chain rule) the vertical component of the velocity is $v_y = -\sqrt{(v\sin 45^o)^2 + 2gh} = -\sqrt{2 + 2(9.81)(2)} = -6.422 \ m/s$. The mass flow rate is $\left(\frac{dm_f}{dt}\right) = 45 \ kg/s$. The force exerted on the belt is

$$\boxed{\sum \vec{F} = \left(\frac{dm_f}{dt}\right)((v_x - 0.3)\vec{i} + v_y\vec{j}) = 50.1\vec{i} - 289\vec{j} \ (N)}$$

=================================<>=================================

==================================<>==================================

Problem 5.108 Solve Problem 5.107 if $\theta = 30^o$.

Solution Use the solution to Problem 5.107 as appropriate. From Problem 5107: $v_x = 2\cos 45^o = \sqrt{2}\ m/s$, $v_y = -\sqrt{(v \sin 45^o)^2 + 2gh} = -\sqrt{2 + 2(9.81)(2)} = -6.422\ m/s$. The velocity of the conveyor belt is $\vec{v}_B = v_{Bx}\vec{i} + v_{By}\vec{j} = 0.3(\vec{i}\cos 30^o - \vec{j}\sin 30^o) = 0.2598\vec{i} - 0.15\vec{j}\ (m/s)$. The magnitude of the velocity of the gravel: $v_{mag} = \sqrt{(v_x - v_{Bx})^2 + (v_y - v_{by})^2} = 6.377\ (m/s)$. The angle of impact: $\beta = \tan^{-1}\left(\frac{v_y - v_{bx}}{v_x - v_{bx}}\right) = -79.6^o$. The force on the belt is

$$\boxed{\sum \vec{F} = \left(\frac{dm_f}{dt}\right)(v_{mag})(\vec{i}\cos\beta + \vec{j}\sin\beta) = 51.9\vec{i} - 282.2\vec{j}\ (N)}$$

==================================<>==================================

Problem 5.109 A toy car is propelled by water that squirts from an internal tank at 3 *m/s* relative to the car. If the mass of the car of the empty car is 1 *kg*, it holds 2 *kg* of water, and you neglect other tangential forces, what is its top speed?

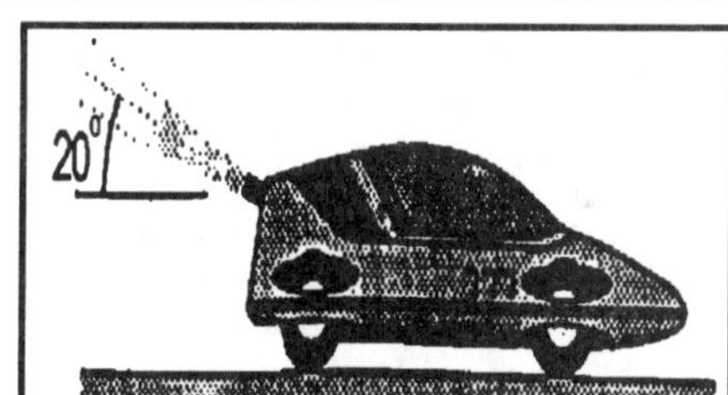

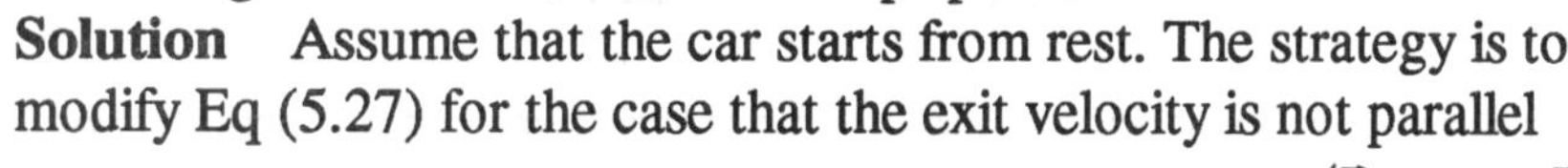

Solution Assume that the car starts from rest. The strategy is to modify Eq (5.27) for the case that the exit velocity is not parallel with the motion. The exit velocity of the jet is $\vec{v}_f = v_f(\vec{i}\cos\theta + \vec{j}\sin\theta) = v_{fx}\vec{i} + v_{fy}\vec{j}$. From Newton's second law,(Eq (5.26)), $\frac{dm_f}{dt}v_{fx} = m\frac{dv}{dt}$. The rate of change of the mass is $\frac{dm}{dt} = -\frac{dm_f}{dt}$, from which $dv = -v_{fx}\frac{dm}{m}$. Integrating, $v - v_o = v_{fx}\ln\left(\frac{m_o}{m}\right)$. Substitute: $v_{fx} = v_f\cos\theta$, $v_o = 0$ $v_f = 3\ m/s$, and $\theta = 20^o$, $m_o = 3\ kg$, $m = 1\ kg$, to obtain $\boxed{v = 3.1\ m/s}$

==================================<>==================================

Problem 5.110 A rocket consists of a 2 *Mg* payload and a 40 Mg booster. Eighty percent of the booster's mass is fuel, and its exhaust velocity is 1 *km/s* . If the rocket starts from rest and you neglect external forces, what velocity will it reach?

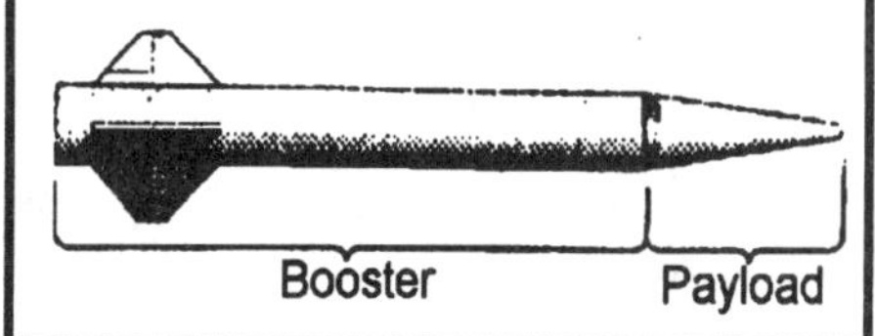

Solution The "empty" mass is $m = 2000 + 0.2(40000) = 10{,}000\ kg$. The "full" mass is $m_o = 2000 + 40000 = 42000\ kg$. The top velocity is (see solution to Problem 5.95 and Eq(5.25)).

$$\boxed{v = v_f \ln\left(\frac{m_o}{m}\right) = 1000\ln\left(\frac{m_o}{m}\right) = 1435.1\ m/s}$$

==================================<>==================================

=====================================<>=====================================

Problem 5.111 A rocket consists of a 2 *Mg* payload and a booster. The booster has two stages whose total mass is 40 *Mg*. Eighty percent of the mass of each stage is fuel. When the fuel of stage 1 is expended, it is discarded and the motor in stage 2 is ignited. The exhaust velocity of both stages is 1 *km/s*. Assume that the rocket starts from rest and neglect external forces. Determine the velocity reached by the rocket if the two stages are of equal mass, and compare your answer to Problem 5.110.

Solution:

The mass at the beginning of the first burn is $m_{o1} = 2000 + 40000 = 42000\ kg$. The "empty" mass at the end of the first burn is

$m_1 = 2000 + \dfrac{40000}{2} + \left(\dfrac{0.2}{2}\right)(40000) = 26000\ kg$. The first booster is discarded, so that the "full" mass at the beginning of the second burn is

$m_{o2} = 2000 + \left(\dfrac{40000}{2}\right) = 22000\ kg$ The "empty" mass at the end of the second burn is

$m_2 = 2000 + (0.2)20000 = 6000\ \ kg$. The top velocity is

$$\boxed{v = v_f \ln\left(\frac{m_{o1}}{m_1}\right) + v_f \ln\left(\frac{m_{o2}}{m_2}\right) = 1000\left(\ln\left(\frac{42000}{26000}\right) + \ln\left(\frac{22000}{6000}\right)\right) = 1778.9\ \ m/s}$$ as compared to

$v = 1435\ m/s$ in Problem 5.110.

=====================================<>=====================================

Problem 5.112 A rocket of initial mass m_0 takes off straight up. Its exhaust velocity v_f and the mass flow rate of its engine $\dot{m}_f = dm_f/dt$ are constant. During the initial part of the flight when aerodynamic drag is negligible, show that the rocket's upward velocity as a function of time is

$$v = v_f \ln\left(\frac{m_0}{m_0 - \dot{m}_f t}\right) - gt.$$

Solution:

Start by adding a gravity term to the second equation in Example 5.10. (Newton's second law). We get $\sum F_x = \dfrac{dm_f}{dt} v_f - mg = m\dfrac{dv_x}{dt}$. Substitute $\dfrac{dm}{dt} = -\dfrac{dm_f}{dt}$ as in the example and divide through by m to get $-v_f\left(\dfrac{1}{m}\right)\dfrac{dm}{dt} - g = \dfrac{dv_x}{dt}$. Integrate with respect to time gives the relation $-v_f \int_0^t \left(\dfrac{1}{m}\right)\dfrac{dm}{dt}dt - \int_0^t g dt = \int_0^t \dfrac{dv_x}{dt}dt$. Simplifying the integrals and setting appropriate limits when we change the variable of integration, we get $-v_f \int_{m_0}^{m} \dfrac{dm}{m} - g\int_0^t dt = \int_0^v dv_x$. Integrating and evaluating at the limits of integration, we get $v = v_f \ln(m_0/m) - gt$. Recalling that $m = m_0 - \dot{m}_f t$ and substituting this in, we get the desired result.

=====================================<>=====================================

====================================<>====================================

Problem 5.113 The mass of the rocket sled in Example 5.9 is *30 slugs.* The only significant force acting on the sled in the direction of its motion is the force exerted by the flow of water entering it. Determine the time and distance required for the sled to decelerate from *1000 ft/s* to *100 ft/s.*

Solution:

From Example 5.9, the decelerating force of the sled is $\rho A v^2 = (1.94 slug/ft^3)(0.1 ft^2)v^2 = 0.194v^2\ lb$ where v is the sled's velocity in ft/s. Therefore $m\frac{dv}{dt} = 30\frac{dv}{dt} = -0.194v^2$.

Integrating, $30\int_{1000}^{100}\frac{dv}{v^2} = -0.194\int_0^t dt$ or $30\left[-\frac{1}{v}\right]_{1000}^{100} = -0.194t$ Solving, $t = 1.39s$.

Applying the chain rule, $30\frac{dv}{dt} = 30\frac{dv}{ds}v = -0.194v^2$. Integrating, $30\int_{1000}^{100}\frac{dv}{v} = -0.194\int_0^s ds$, or

$30\ln\left(\frac{100}{1000}\right) = -0.194s$. Solving, we get $s = 356 ft$

====================================<>====================================

Problem 5.114 Suppose that you grasp the end of a chain that weighs 3 *lb./ft* and lift it straight up off the floor at a constant speed of 2 *ft/s.* (a) Determine the upward force that you must exert as a function of *s*. (b) How much work do you do in lifting the top of the chain to $s = 4\ ft$?

Strategy: Treat the part of the chain you have lifted as an object that is gaining mass.

Solution:

The force is the sum of the "mass flow" reaction and the weight of suspended part of the chain, $F = mg + v\left(\frac{dm}{dt}\right)$. (a) Substitute: $\left(\frac{dm}{dt}\right) = \left(\frac{3}{32.17}\right)v\ slug/s$. The velocity is 2 *ft/s.* The suspended portion of the chain weighs 3*s lb.* The force required to lift the chain is $3s + \left(\frac{3}{32.17}v\right)v = 3s + \frac{3(2^2)}{32.17}$

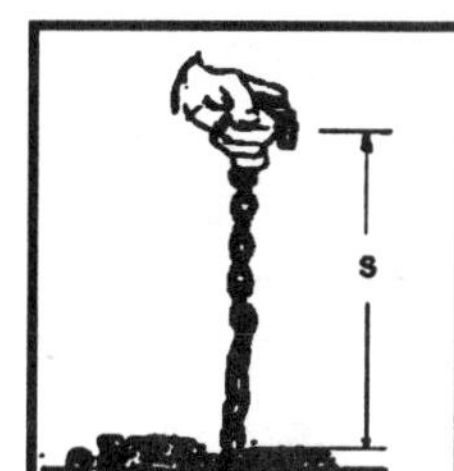

(b) The work done is $\int_0^4 F ds = \left[\left(\frac{12}{g}\right)s + \left(\frac{3}{2}\right)s^2\right]_0^4 = 25.49\ ft\text{-}lb$

====================================<>====================================

Problem 5.115 Solve Problem 5.114, assuming that you lift the end of the chain straight up off the floor with a constant acceleration of $2\ ft/s^2$.

Solution Assume that the velocity is zero at $s = 0$. The mass of the chain currently suspended is $m = \left(\frac{3}{g}\right)s$. Use the solution to Problem 5.114. From Newton's second law, $m\left(\frac{dv}{dt}\right) = F - v\left(\frac{dm}{dt}\right) - mg$. The velocity is expressed in terms of s as follows: The acceleration is constant: $dv/dt = 2\ ft/s^2$. Use the chain rule $v(dv/ds) = 2$. Integrate: $v^2 = 4s$, where it is assumed that the velocity is zero at $s = 0$, from which $F = m\left(\frac{dv}{dt}\right) + 2\sqrt{s}\left(\frac{dm}{dt}\right) + 3s$ Substitute: $m = \left(\frac{3}{g}\right)s$, $\frac{dm}{dt} = \left(\frac{3}{g}\right)v = \left(\frac{6}{g}\right)\sqrt{s}$, from which

$F = \left(\frac{18}{g} + 3\right)s = 3.56s$ (b) The work done is $\int_0^4 F ds = \left[\left(\frac{18}{g} + 3\right)\frac{s^2}{2}\right]_0^4 = 28.47\ ft\text{-}lb$

====================================<>====================================

==========================<>==========================

Problem 5.116 It has been suggested that a heavy chain could be used to gradually stop an airplane that rolls past the end of the runway. A hook attached to the end of the chain engages the planes nose wheel, and the plane drags an increasing length of the chain as it rolls. Let m be the airplane's mass and v_o its initial velocity, and let ρ_L be the mass per unit length of the chain. If you neglect friction and aerodynamic drag, what is the airplane's velocity as a function of s?

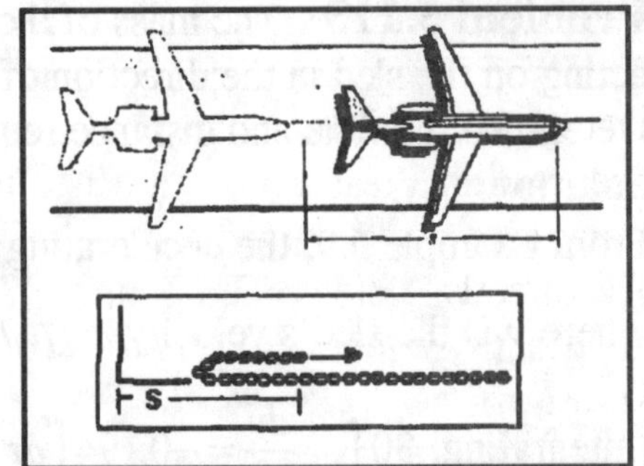

Solution Assume that the chain is laid out lengthwise along the runway, such that the aircraft hook seizes the nearest end as the aircraft proceeds down the runway. As the distance s increases, the length of chain being dragged is $\frac{s}{2}$ (see figure). The mass of the chain being dragged is $\frac{\rho_L s}{2}$ The mass "flow" of the chain is $\frac{d\left(\frac{\rho_L s}{2}\right)}{dt} = \frac{\rho_L v}{2}$. From Newton's second law,

$\left(\frac{\rho_L s}{2} + m\right)\frac{dv}{dt} = -\frac{\rho_L v^2}{2}$. Use the chain rule and integrate: $\left(\frac{\rho_L s}{2} + m\right)\frac{dv}{ds} = -\frac{\rho_L v}{2}$,

$\frac{dv}{v} = -\frac{\rho_L}{2\left(\frac{\rho_L s}{2} + m\right)} ds$, $\ln(v) = -\ln\left(m + \frac{\rho_L s}{2}\right) + C$. For $v = v_o$ at $s = 0$, $C = \ln(m v_o)$, and

$$\boxed{v = \frac{m v_o}{m + \frac{\rho_L s}{2}}}.$$

==========================<>==========================

Problem 5.117 In Problem 5.116 the friction force exerted on the chain by the ground would actually dominate other forces as the distance s increases. If the coefficient of kinetic friction between chain and the ground is μ_k, and you neglect all other forces except the friction force, what is the airplane's velocity as a function of s?

Solution Assume that the chain layout is configured as shown in Problem 5.116. From Problem 5.116, the weight of the chain being dragged at distance s is $\frac{\rho_L g s}{2}$. From Newton's second law

$\left(m + \frac{\rho_L s}{2}\right)\frac{dv}{dt} = -\frac{\mu_k \rho_L g s}{2}$, where only the friction force is considered. Use the chain rule and integrate:

$\left(m + \frac{\rho_L s}{2}\right) v \frac{dv}{ds} = -\frac{\mu_k \rho_L g s}{2}$, $v dv = -\mu_k g \frac{\rho_L s}{2\left(m + \frac{\rho_L s}{2}\right)} ds$, from which

$\frac{v^2}{2} = \frac{-2\mu_k g}{\rho_L}\left(m + \frac{\rho_L s}{2} - m\ln\left(m + \frac{\rho_L s}{m}\right)\right) + C$. For $v = v_o$ at $s = 0$, $C = \frac{v_o^2}{2} + \frac{2\mu_k g m}{\rho_L}(1 - \ln(m))$, from

which, after reduction $\boxed{v^2 = v_o^2 - 2g\mu_k\left(s - \frac{2m}{\rho_L}\ln\left(1 + \frac{\rho_L s}{2m}\right)\right)}$.

==========================<>==========================

==================================<>==================================

Problems 5.118-5.122 are related to Example 5.12.

==================================<>==================================

Problem 5.118 The turbojet engine in Fig 5.24 is being operated on a test stand. The mass flow rate of air entering the compressor is 13.5 *kg/s,* and the mass flow rate of fuel is 0.13 *kg/s.* The effective velocity of air entering the compressor is zero, and the exhaust velocity is 500 *m/s*. What is the thrust of the engine?

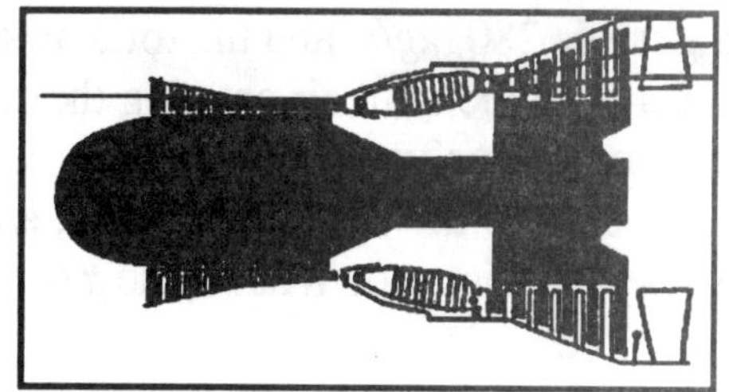

Solution:

The sum of the mass flows is

$\left(\frac{dm}{dt}\right)=\left(\frac{dm_c}{dt}\right)+\left(\frac{dm_f}{dt}\right)=13.5+0.13=13.63\ kg/s$. The inlet velocity is zero, and the exit velocity is 500 *m/s*. The thrust is $\boxed{T=13.63(500)=6{,}815\ N}$

==================================<>==================================

Problem 5.119 Suppose that the engine described in Problem 5.118 is in an airplane flying at 400 *km/hr*. The effective velocity of the air entering the inlet is equal to the airplane's velocity. What is the thrust of the engine?

Solution:

Use the "rest frame" of the engine to determine the thrust. Use the solution to Problem 5.118. The inlet velocity is $v_i=400\left(\frac{10^3}{3600}\right)=111.11\ m/s$. The thrust is $T=\left(\frac{dm_c}{dt}+\frac{dm_f}{dt}\right)500-\left(\frac{dm_c}{dt}\right)111.11$,

$\boxed{T=13.63(500)-(111.11)13.5=5315\ N}$

==================================<>==================================

Problem 5.120 A turbojet engine's thrust reverser causes the exhaust to exit the engine at 20° from the engine centerline. The mass flow rate of air entering the compressor is 3 *slug/s*, and it enters at 200 *ft/s*. The mass flow rate of fuel is 0.1 *slug/s* , and the exhaust velocity is 1200 *ft/s*. What braking velocity does the engine exert on the airplane.

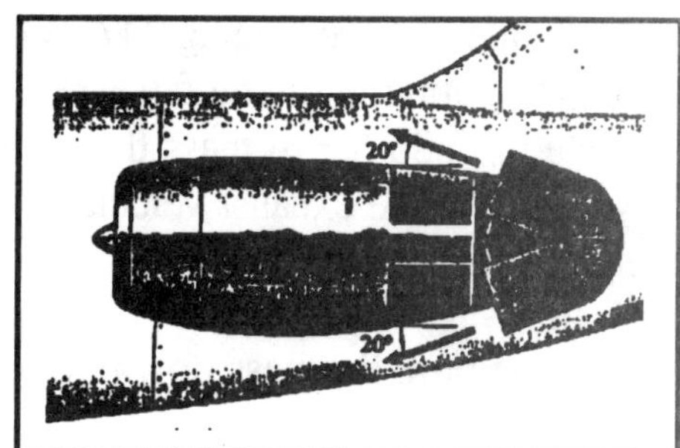

Solution:

The force exerted by the intake flow is

$F_{inlet}=(3)(200)=600\ lb$. The force exerted by the exhaust is

$F_{exhaust}=(3+0.1)(1200)\cos 20^o=3495.7\ lb$. The total braking force is

$\boxed{F_{braking}=F_{inlet}+F_{exhaust}=4095.7\ lb}$.

==================================<>==================================

================================<>================================

Problem 5.121 The 13.6 *Mg* airplane is moving at 400 *km/hr*. The total mass flow rate of air entering the compressors of its turbojet engines is 280 *kg/s*, and the total mass flow rate of fuel is 2.6 *kg/s* . The effective velocity of air entering the compressors is equal to the airplane's velocity, and the exhaust velocity is 480 *m/s*. The ratio of the lift force *L* to the drag force *D* is 6, and the *z* component of the airplane's acceleration is zero. What is the *x* component of its acceleration?

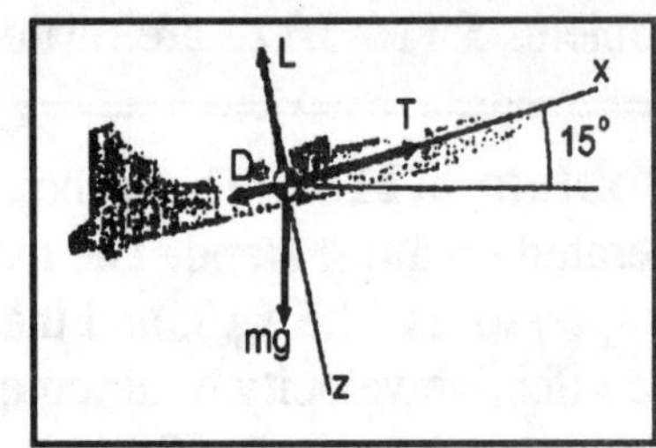

Solution:

The forward velocity is $v = 400\left(\dfrac{10^3}{3600}\right) = 111.11\ m/s$. The thrust is $\vec{T} = ((280+2.6)(480) - (280)111.11)\vec{i} = 104{,}537\vec{i}\ (N)$. The lift force is $\vec{L} = L\vec{k}$. The drag force is $\vec{D} = -D\vec{i}$. The weight is $\vec{W} = g13.6\times 10^3\left(-\vec{i}\sin 15^o - \vec{k}\cos 15^o\right) = -34{,}531\vec{i} - 128{,}870\vec{k}\ (N)$. The sum of the forces $\sum \vec{F} = \vec{T} + \vec{L} + \vec{D} + \vec{W} = (104{,}537 - D - 34{,}531)\vec{i} + (L - 128{,}870)\vec{k}$, from which $L = 128{,}870\ N$. The drag is $D = \dfrac{L}{6} = 21{,}478\ N$. From Newton's second law,

$m\dfrac{dv_x}{dt} = (104{,}537 - 21{,}478 - 34{,}531) = 48{,}528\ N$, from which the acceleration along the *x* axis is

$$\boxed{\frac{dv_x}{dt} = \frac{48{,}528}{13{,}600} = 3.57\ m/s^2}$$

================================<>================================

Problem 5.122 The fan engine in Fig 5.25 is similar to the Pratt and Whitney JT9D-3A engine used on early models of the Boeing 747. When the airplane begins its takeoff run, the velocity of the air entering the compressor and fan is negligible. A mass flow rate of 38.5 *slug/s* enters the fan and is accelerated to 885 *ft/s*. A mass flow rate of 7.7 *slug/s* enters the compressor. The mass flow rate of fuel is 0.23 *slug/s*, and the exhaust velocity is 1190 *ft/s*. (a) What is the bypass ratio? (b) What is the thrust of the engine? If the airplane weighs 500,000 *lb*., what is its initial acceleration? (It has four engines.)

Solution:

(a) By definition, the mass flow rate of air entering the fan to the mass flow rate of air entering the compressor is the bypass ratio, $R_{bypass} = \dfrac{38.5}{7.7} = 5$.

(b) The thrust of the engine is $\boxed{T = (38.5)(885) + (7.7 + 0.23)(1190) = 43{,}509\ lb}$

(c) From Newton's second law, the acceleration from all four engines thrusting is

$$\boxed{\left(\frac{dv}{dt}\right) = \frac{(4)43{,}509(32.17)}{500{,}000} = 11.2\ ft/s^2}$$

================================<>================================

================================<>================================

Problem 5.123 The total external force on a 10 *kg* object is constant and equal to $90\vec{i} - 60\vec{j} + 20\vec{k}$ (N). At $t = 2$ s, the object's velocity is $-8\vec{i} + 6\vec{j}$ (m/s). (a) What impulse is applied to the object from $t = 2$ s to $t = 4$ s? (b) What is the object's velocity at $t = 4$ s?

Solution:

(a)The impulse is $\int_{t_1}^{t_2} \vec{F}dt = 90(4-2)\vec{i} - 60(4-2)\vec{j} + 20(4-2)\vec{k}$ $\boxed{= 180\vec{i} - 120\vec{j} + 40\vec{k} \ (N\text{-}s)}$.

(b) The velocity is $m\vec{v}_2 - m\vec{v}_1 = \int_{t_1}^{t_2} \vec{F}dt$, from which

$$\vec{v}_2 = \vec{v}_1 + \left(\frac{1}{m}\right)\int_{t_1}^{t_2} \vec{F}dt = \left(-8 + \frac{180}{10}\right)\vec{i} + \left(6 - \frac{120}{10}\right)\vec{j} + \left(\frac{40}{10}\right)\vec{k} \boxed{= 10\vec{i} - 6\vec{j} + 4\vec{k} \ (m/s)}$$

================================<>================================

Problem 5.124 The total external force on an object is $\vec{F} = 10t\,\vec{i} + 60\vec{j}$ (lb). At $t = 0$, its velocity is $\vec{v} = 20\vec{j}$ (ft/s). At $t = 12$ s the x component of its velocity is 48 ft/s. (a) What impulse is applied to the object from $t = 0$ to $t = 6$ s? (b) What is its velocity at $t = 6$ s?

Solution:

(a) The impulse is $\int_{t_1}^{t_2} \vec{F}dt = \left[5t^2\right]_0^6 \vec{i} + \left[60t\right]_0^6 \vec{j} = 180\vec{i} + 360\vec{j}$ $(lb\text{-}s)$

(b) The mass of the object is found from the x component of the velocity at 12 s.

$m = \dfrac{\int_{t_1}^{t_2} Fdt}{48 - 0} = \dfrac{5(12^2)}{48} = 15$ $slug$. The velocity at 6 s is $m\vec{v}_2 - m\vec{v}_1 = \int_{t_1}^{t_2} \vec{F}dt$, from which

$$\boxed{\vec{v}_2 = 20\vec{j} + \frac{180}{15}\vec{i} + \frac{360}{15}\vec{j} = 12\vec{i} + 44\vec{j} \ (ft/s)}$$

================================<>================================

Problem 5.125 An aircraft arresting system is used to stop airplanes whose braking systems fail. The system stops a 47.5 *Mg* airplane moving at 80 *m/s* in 9.15 *s*. (a) What impulse is applied to the airplane during the 9.15 *s*? (b) What is the average deceleration to which the passengers are subjected?

Solution: (a) The impulse is $\boxed{\int_{t_1}^{t_2} Fdt = mv_2 - mv_1 = 47500(80) = 3.8 \times 10^6 \ N\text{-}s}$. (b) The average force is $F_{ave} = \dfrac{\int_{t_1}^{t_2} Fdt}{t_2 - t_1} = \dfrac{3.8 \times 10^6}{9.15} = 4.153 \times 10^5$ N. From Newton's second law

$$\boxed{\left(\frac{dv}{dt}\right)_{ave} = \frac{F_{ave}}{47500} = 8.743 \ m/s^2}$$

================================<>================================

Problem 5.126 The 1895 Austrian 150 *mm* howitzer had a 1.94 *m* long barrel, a muzzle velocity of 300 *m/s* and fired a 38 *kg* shell. If the shell took 0.013 *s* to travel the length of the barrel, what average force was exerted on the shell?

Solution: The average force is $\boxed{F_{ave} = \dfrac{\int_{t_1}^{t_2} Fdt}{t_2 - t_1} = \dfrac{(38)(300)}{0.013} = 876{,}923 \ N}$

================================<>================================

Problem 5.127 An athlete throws a shot put weighing 16 *lb*. When he releases it, the shot put is 7 *ft* above the ground and its components of velocity are $v_x = 31 ft/s$, $v_y = 26 ft/s$. (a) If he accelerates the shot put from rest in 0.8 *s* and you assume as a first approximation that the force $\vec{F}$ he exerts on it is constant, use the principle of impulse and momentum to determine the *x* and *y* components of $\vec{F}$. (b) What is the horizontal distance from the point where he releases the shot put to the point where it strikes the ground?

Solution:

(a) Let F_x, F_y be the components of the force exerted by the athlete. The impulse is

$$\int_{t_1}^{t_2} \vec{F} dt = \vec{i}(F_x)(t_2 - t_1) + \vec{j}(F_y - W)(t_2 - t_1) = \vec{i}\left(\frac{W}{g}\right)(31) + \vec{j}\left(\frac{W}{g}\right)(26)$$ from which

$$\boxed{F_x = \left(\frac{W}{g}\right)\left(\frac{31}{0.8}\right) = 19.3\ lb}. \quad \boxed{F_y = \left(\frac{W}{g}\right)\left(\frac{26}{0.8}\right) + W = 32.2\ lb}.$$

(b) From the conservation of energy for the vertical component of the motion, the maximum height reached is $h = \left(\frac{1}{2g}\right)v_y^2 + 7 = 17.5\ ft$. From the solution of Newton's second law for free fall, the time of flight is $t_f = \sqrt{\frac{2(h-7)}{g}} + \sqrt{\frac{2h}{g}} = 1.85\ s$ and the horizontal distance is $\boxed{D = 31 t_f = 57.4\ ft}$

Problem 5.128 The 6000 *lb.* pickup truck A moving at 40 *ft/s* collides with the 4000 *lb.* car moving at 30 *ft/s*. (a) What is the magnitude of the velocity of their common center of mass after the impact? (b) If you treat the collision as a perfectly plastic impact, how much kinetic energy is lost?

Solution:

(a) From the conservation of linear momentum,

$$\left(\frac{W_A}{g}\right)v_A + \left(\frac{W_B}{g}\right)v_B \cos\theta = \left(\frac{W_A + W_B}{g}\right)v_x,$$

$$\left(\frac{W_B}{g}\right)v_B \sin\theta = \left(\frac{W_A + W_B}{g}\right)v_y.$$ Substitute numerical values,

with $\theta = 30^o$ to obtain $v_x = 34.39\ ft/s$, $v_y = 6\ ft/s$, from which $v = \sqrt{v_x^2 + v_y^2} = 34.91\ ft/s$

(b) The kinetic energy before the collision minus the kinetic energy after the collision is the loss in kinetic energy: $\boxed{\frac{1}{2}\left(\frac{W_A}{g}\right)v_A^2 + \frac{1}{2}\left(\frac{W_B}{g}\right)v_B^2 - \frac{1}{2}\left(\frac{W_A + W_B}{g}\right)v^2 = 15724\ ft\text{-}lb}$

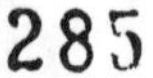

==================================<>==================================

Problem 5.129 Two hockey players ($m_A = 80$ kg, $m_B = 90$ kg) converging on the puck at $x = 0$, $y = 0$ become entangled and fall. Before the collision $\vec{v}_A = 9\vec{i} + 4\vec{j}$ (m/s) and $\vec{v}_B = -3\vec{i} + 6\vec{j}$ (m/s). If the coefficient of kinetic friction between the players and the ice, $\mu_k = 0.1$, what is the approximate position of the players when they stop sliding?

Solution:

The strategy is to determine the velocity of their combined center of mass immediately after the collision using the conservation of linear momentum, and determine the distance using the conservation of energy. The conservation of linear momentum is $9m_A - 3m_B = (m_A + m_B)v_x$, and $4m_A + 6m_B = (m_A + m_B)v_y$, from which $v_x = 2.65$ m/s, and $v_y = 5.06$ m/s, from which $v = \sqrt{v_x^2 + v_y^2} = 5.71$ m/s, and the angle of the path is $\theta = \tan^{-1}\left(\frac{v_y}{v_x}\right) = 62.4^o$. The work done by friction is

$\int_0^s -\mu_k g(m_A + m_B)ds = -\mu_k g(m_A + m_B)s$, where s is the distance the players slide after collision. From the conservation of work and energy $\frac{1}{2}(m_A + m_B)v^2 - \mu_k g(m_A + m_B)s = 0$, from which

$s = \frac{v^2}{2\mu_k g} = 16.6$ m.. The position of the players after they stop sliding is

$$\boxed{\vec{r} = s(\vec{i}\cos\theta + \vec{j}\sin\theta) = 7.7\vec{i} + 14.7\vec{j}\ (m)}$$

==================================<>==================================

Problem 5.130 The cannon weighed 400 *lb.*, fired a cannonball weighing 10 *lb.*, and had a muzzle velocity of 200 *ft/s*. For the 10° elevation shown, determine (a) the velocity of the cannon after it was fired; (b) the distance the cannonball traveled.

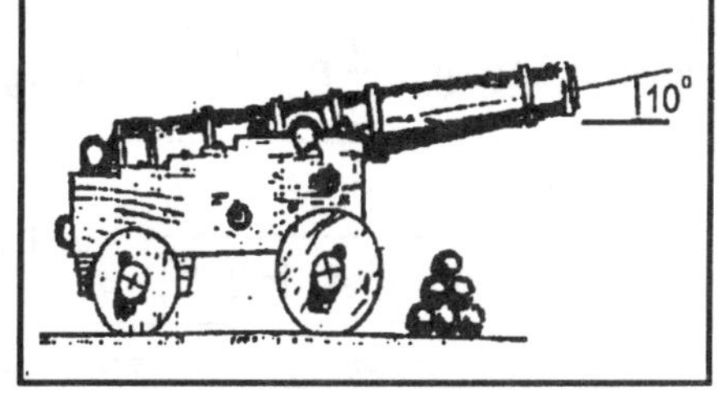

Solution:

Assume that the height of the cannon mouth above the ground is negligible. (a) Relative to a reference frame moving with the cannon, the cannonball's velocity is 200 ft/s at . The conservation of linear momentum condition is

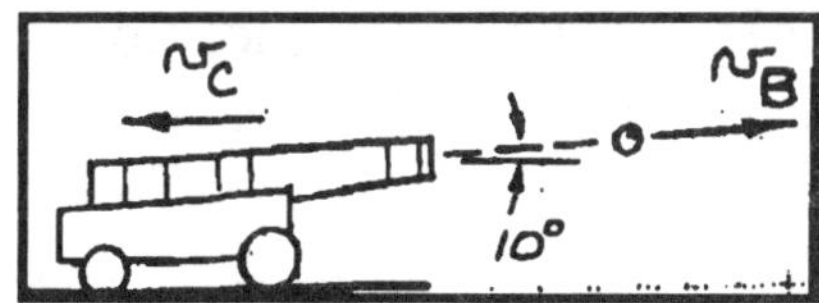

$-m_C v_C + m_B\left(v_B\cos 10^o - v_C\right) = 0$, from which

$v_C = \frac{m_B v_B \cos 10^o}{m_B + m_C} = 4.804$ ft/s. The velocity of the cannon ball relative to the ground is $v_B = \left(200\cos 10^o - v_C\right)\vec{i} + 200\sin 10^o\vec{j} = 192.16\vec{i} + 34.73\vec{j}$ ft/s, from which $v_{ox} = 192.16$ ft/s, $v_{oy} = 34.73$ ft/s. The maximum height is (from the conservation of energy for free fall) $h_{\max} = v_{oy}^2/(2g) = 18.75$ ft, where the height of the cannon mouth above the ground is negligible. The time of flight (from the solution of Newton's second law for free fall) is twice the time required to fall from the maximum height, $t_{flight} = 2\sqrt{\frac{2h_{\max}}{g}} = 2.16$ s. From which the range is

$\boxed{x_{impact} = v_{ox}t_{flight} = 414.5\ ft}$ since v_{ox} is constant during the flight.

==================================<>==================================

=================================<>=================================

Problem 5.131 A 1 *kg* ball moving horizontally at 12 *m/s* strikes a 10 *kg* block. The coefficient of restitution of the impact is $e = 0.6$, and the coefficient of kinetic friction between the block and the inclined surface is $\mu_k = 0.4$. What distance does the block slide before stopping?

Solution:

This an oblique central impact about a line parallel to the inclined surface. The strategy is to determine the velocity of the block immediately after impact using the conservation of linear momentum and the definition of the coefficient of restitution, and then determine the distance the block slides using the principle of work and energy. The component of the velocity of the ball normal to a line parallel to the incline is unchanged by the impact. For the component parallel to the incline the conservation of linear momentum is $m_A v_A + m_B v_B = m_A v'_A + m_B v'_B$.

The definition of the coefficient of restitution is $e = \dfrac{v'_B - v'_A}{v_A - v_B}$ These two equations have the solution (see the solution to Problem 5.59)

$$v'_B = \left(\frac{1}{m_A + m_B}\right)\left(m_A(1+e)v_{AP} + (m_B - e m_A)v_B\right) = \left(\frac{1}{m_A + m_B}\right)\left(m_A(1+e)v_{AP}\right)$$

, where v'_B is the velocity of the block immediately after impact *parallel to the incline*, $v_B = 0$ is the velocity of the block before impact, and $v_{AP} = 12\cos 25^o$ m/s is the component parallel to the incline of the ball's velocity before impact. Substitute numerical values to obtain $v'_B = 1.58$ m/s. The work done by friction is

$\int_0^s F ds = -\mu_k g m_B (\cos 25^o) s$, where s is the distance traveled along the incline. From the principle of work and energy, $\frac{1}{2} m_B v_B'^2 - \mu_k g m_B s \cos 25^o - m_B g s \sin 25^o = 0$, from which

$$\boxed{s = \frac{(v'_B)^2}{2g(\mu_k \cos 25^o + \sin 25^o)} = 0.162\ m}$$

=================================<>=================================

Problem 5.132 A Peace Corps volunteer designs the simple device shown for drilling water wells in remote areas. A 70 *kg* "hammer", such as a section of log or a steel drum partially filled with concrete, is hoisted to $h = 1$ m and allowed to drop onto a protective cap on the section of pipe being pushed into the ground. The mass of the cap and section of pipe is 20 *kg*. Assume that the coefficient of restitution is nearly zero. (a) What is the velocity of the cap and pipe immediately after impact? (b) If the pipe moves 30 *mm* downward when the hammer is dropped, what resistive force was exerted on the pipe by the ground? (Assume that the resistive force is constant during the motion of the pipe.)

Solution: The conservation of momentum principle for the hammer, pipe and cap is $m_H v_H + m_P v_P = (m_H + m_P)v$, where v_H, v_P are the velocity of the hammer and pipe, respectively, before impact, and v is the velocity of their combined center of mass immediately after impact. The

Solution continued on next page

velocity of the hammer before impact (from the conservation of energy for a free fall from height h) is $v_H = \sqrt{2gh} = \sqrt{2(9.81)(1)} = 4.43\ m/s$. Since $v_P = 0$, $\boxed{v = \frac{m_H}{m_H + m_P} v_H = \frac{70}{90}(4.43) = 3.45\ m/s}$. (b)

The work done by the resistive force exerted on the pipe by the cap is $\int_0^s -F ds = -Fs\ N\text{-}m$, where $s = 0.03\ m$. The work and energy principle for the hammer, pipe and cap is

$-Fs + (m_H + m_P)gs = \frac{1}{2}(m_H + m_P)v^2$, from which

$$\boxed{F = \frac{(m_H + m_P)(v^2 + 2gs)}{2s} = \frac{(90)(3.45^2 + 2(9.81)(0.03))}{2(0.03)} = 18{,}686\ N}$$

=================================<>=================================

Problem 5.133 A tugboat ($mass = 40\ Mg$) and a barge ($mass = 160\ Mg$) are stationary with a slack hawser connecting them. The tugboat accelerates to 2 *knots* ($1\ knot = 1852\ m/hr$) before the hawser becomes taut. Determine the velocities of the tugboat and the barge just after the hawser becomes taut (a) if the impact is perfectly plastic ($e = 0$); (b) if the impact is perfectly elastic ($e = 1$). Neglect the forces exerted by the water and the tugboat's engines.

Solution:

The conservation of linear momentum is $m_T v_T + m_B v_B = m_T v'_T + m_B v'_B$, where the prime indicates the velocity when the hawser is taut. The definition of the coefficient of restitution is $e = \frac{v'_B - v'_T}{v_T - v_B}$. The solution of these two equations is (see solution to Problem 5.45 for details)

$$v'_T = \left(\frac{1}{m_T + m_B}\right)(m_B(1+e)v_B + (m_T - e m_B)v_T), \text{ and } v'_B = \left(\frac{1}{m_T + m_B}\right)(m_T(1+e)v_T + (m_B - e m_T)v_B)$$

For $v_B = 0$ these reduce to $v'_T = \left(\frac{1}{m_T + m_B}\right)((m_T - e m_B)v_T)$, and $v'_B = \left(\frac{1}{m_T + m_B}\right)(m_T(1+e)v_T)$.

Case (a) $e = 0$, substitute numerical values $m_T = 4 \times 10^4\ kg$, $m_B = 1.6 \times 10^5\ kg$, $v_T = 2(1852)(1/3600) = 1.03\ m/s$. The result: $\boxed{v'_T = v'_B = 0.823\ m/s = 0.4\ knots}$.

Case (b) $e = 1$. Substitute numerical values to obtain $\boxed{v'_B = 0.8\ knot}$, $\boxed{v'_T = -1.2\ knot}$

=================================<>=================================

=================================<>=================================

Problem 5.134 In Problem 5.133, determine the magnitude of the impulsive force exerted on the tugboat in the two cases if the duration of the "impact" is 4 *s*. Neglect the forces exerted by the water and the tugboat's engines during this period.

Solution: Since the tugboat is stationary at $t = t_1$, the linear impulse is $\int_{t_1}^{t_2} Fdt = F_{ave}(t_2 - t_1) = m_B v'_B$, from which $F_{ave} = m_B v'_B / 4 = (4\times10^4)v'_B$. For Case (a) $v'_B = 0.2058\ m/s$, from which $\boxed{F_{ave} = 8231.1\ N}$. Case (b). $v'_B = 0.4116\ m/s$, $\boxed{F_{ave} = 16{,}462\ N}$

=================================<>=================================

Problem 5.135 The balls are of equal mass *m*. Balls B and C are connected by an unstretched linear spring and are stationary. Ball A moves toward ball B with velocity v_A. The impact with B is perfectly elastic ($e = 1$). Neglect external forces. (a) What is the velocity of the common center of mass of balls B and C immediately after impact? (b) What is the velocity of the common center of mass of balls B and C at time *t* after impact?

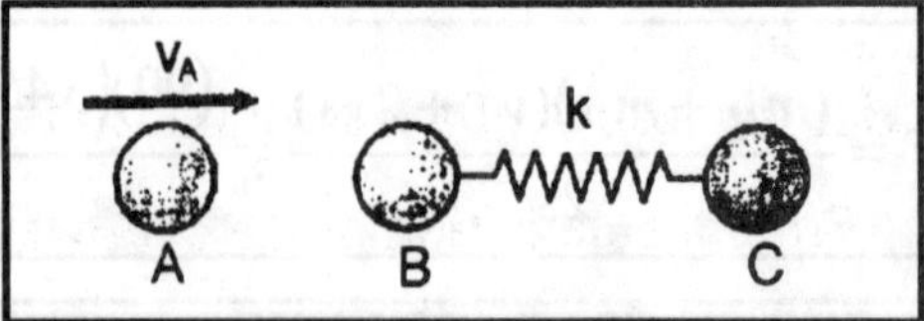

Solution: For a perfectly elastic impact, with B and C stationary, the conservation of linear momentum is $m_A v_A = m_A v'_A + m_B v'_B$, and the coefficient of restitution definition becomes $v_A = v'_B - v'_A$. (a) The solution to these two equations is (see Problem 5.45 for details) $v'_A = 0$, and $v'_B = v_A$. The velocity of the combined mass center of B and C is $v_{BC} = \dfrac{m_B v'_B + m_C v'_C}{m_B + m_C}$. At the instant of impact $v'_C = 0$, from which $\boxed{v_{BC} = v'_B / 2 = v_A / 2}$. (b) Since the external forces are zero, by Newton's second law the velocity remains unchanged at any time *t*: $\boxed{v_{BC} = v_A / 2}$

=================================<>=================================

Problem 5.136 In Problem 5.135, what is the maximum compressive force in the spring as a result of the impact?

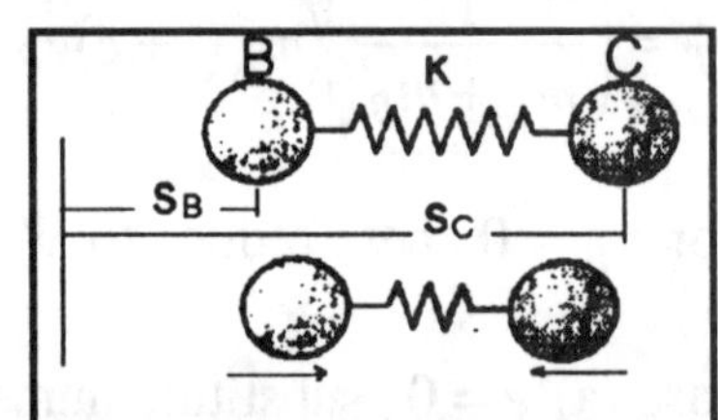

Solution:

The change in external kinetic energy is

$\frac{1}{2}m_A v_A^2 - \frac{1}{2}(m_B + m_C)\left(\frac{v_A}{2}\right)^2 = \frac{1}{2}\left(\frac{mv_A^2}{2}\right)$. Since the impact is elastic, without friction, the change in kinetic energy due to the impact must be stored internally in the system B-C. The mechanism for internal storage is the relative motion of the balls B and C about the common center of mass and the elongation (or compression) of the internal spring. Since the center of mass has a constant velocity, the motion of the balls must be symmetric (equal and opposite) about the center of mass. When the balls are farthest apart (or closest together), their velocity is zero, from which all of the internal energy is stored in the elongation (or compression) of the spring. Therefore the maximum energy in the spring is equal to the change in the external kinetic energy due to the impact, $\frac{1}{2}kS^2 = \frac{1}{2}\left(\frac{mv_A^2}{2}\right)$.

The maximum elongation (or compression) of the spring is $S = \pm\sqrt{\frac{m}{2k}}v_A$. The magnitude of the maximum force is $\boxed{F_{comp} = kS = \sqrt{\frac{mk}{2}}v_A}$.

=================================<>=================================

==================================<>==================================

Problem 5.137 Suppose that you interpret Problem 5.135 as an impact between the ball A and "object" D consisting of the connected balls B and C. (a) What is the coefficient of restitution of the impact between A and D? (b) If you consider the total energy after the impact to be the sum of the kinetic energies $\frac{1}{2}mv_A'^2 + \left(\frac{1}{2}\right)(2m)(v_D')^2$, where v_D' is the velocity of the center of mass after impact, how much energy is "lost" as a result of the impact? (c) How much energy is actually lost as a result of the impact? (This problem is an interesting model for one of the mechanisms for energy loss in impacts between objects. The energy "loss" calculated in Part (b) is transformed into "internal energy"--- the vibrational motions of B and C relative to their common center of mass.)

Solution:

Use the results of the solutions to Problem 5.135 and 5.136 as appropriate.

(a) From the definition of the coefficient of restitution (Eq (5.14)), $e = \frac{v_D' - v_A'}{v_A - v_D}$. From the solution to Problem 5.135, $v_A' = 0$, $v_D = 0$, and $v_D' = \frac{v_A}{2}$, from which $\boxed{e = \frac{\left(\frac{v_A}{2}\right)}{v_A} = \left(\frac{1}{2}\right)}$

(b) From the solution to Problem 5.136, the "loss" in kinetic energy is

$\frac{1}{2}m_A v_A^2 - \frac{1}{2}(m_B + m_C)\left(\frac{v_A}{2}\right)^2 = \frac{1}{2}\left(\frac{mv_A^2}{2}\right)$, or, *one half of the external kinetic energy is "lost".*

(c) None of the energy is lost. As discussed in the solution of Problem 5.135, the energy is stored internally in the object D.

==================================<>==================================

==================================<>==================================

Problem 5.138 A small object starts from rest at A and slides down the smooth ramp. The coefficient of restitution of its impact with the floor is $e = 0.8$. At what height above the floor does it hit the wall?

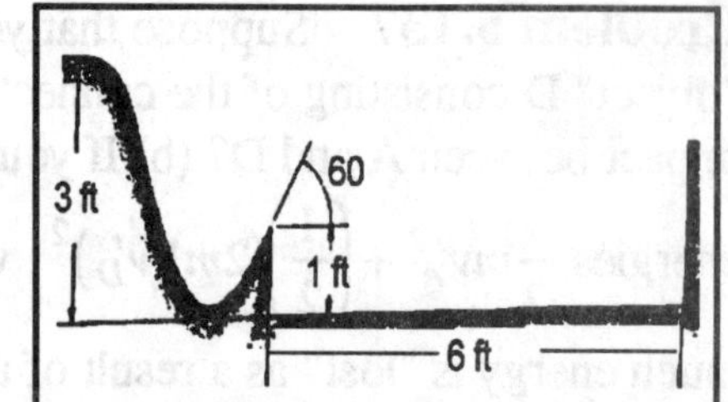

Solution: The impact with the floor is an oblique central impact in which the horizontal component of the velocity is unchanged. The strategy is (a) determine the time between leaving the ramp and impact with the wall, (b) determine the time between the first bounce and impact with the wall (c) from the velocity and height after the first bounce, determine the height at the time of impact with the wall. The steps in this process are:

(1) *The velocities on leaving the ramp.* The velocity at the bottom edge of the ramp is (from the conservation of energy) $v = \sqrt{2gh} = \sqrt{2(32.17)(3-1)} = 11.34\ ft/s$. The horizontal component of the velocity is $v_x = v\cos 60^o = 5.67\ ft/s$. The vertical component of the velocity at the bottom edge of the ramp is $v_y = v\sin 60^o = 9.824\ ft/s$.

(2) *The maximum height after leaving the ramp:* From the conservation of energy, the maximum height reached is $h_{\max} = \dfrac{v_y^2}{2g} + 1 = 2.5\ ft$.

(3) *The velocity and maximum height after the first bounce*: The velocity of the first impact (from the conservation of energy for a free fall) is $v_{yimpact} = -\sqrt{2gh_{\max}} = -12.68\ ft/s$. The vertical velocity after the first bounce (see equation following Eq (5.17)) is $v'_y = -ev_{yimpact} = 0.8(12.68) = 10.15\ ft/s$. The maximum height after the first bounce is $h'_y = \dfrac{(v'_y)^2}{2g} = 1.6\ ft$.

(4) *The time of impact with the wall, the time at the first bounce, and the time between the first bounce and impact with the wall.* The time required to reach the wall is $t_w = \dfrac{6}{v_x} = 1.058\ s$. The time at the first bounce (from a solution to Newton's second law for a free fall) is

$$t_b = \sqrt{\frac{2(h_{\max}-1)}{g}} + \sqrt{\frac{(2)h_{\max}}{g}} = 0.7\ s.$$ The time between the first bounce and wall impact is

$t_{w-b} = t_w - t_b = 0.3582\ s$.

(5) *Is the ball on an upward or downward part of its path when it strikes the wall?* The time required to reach maximum height after the first bounce is $t'_{bh} = \sqrt{\dfrac{2h'_y}{g}} = 0.3154\ s$. Since $t_{w-b} > t'_{bh}$, the ball is on a downward part of its trajectory when it impacts the wall.

(6) *The height at impact with the wall:* The height of impact is (from a solution to Newton's second law for free fall) $\boxed{h_w = h'_y - \frac{g}{2}(t_{w-b} - t'_{bh})^2 = 1.57\ ft}$

==================================<>==================================

===================================<>===================================

Problem 5.139 The cue gives the ball A a velocity of magnitude 3 *m/s*. The angle $\beta = 0$ and the coefficient of restitution of the impact of the cue ball and the 8-ball is $e = 1$. If the magnitude of the 8-ball's velocity after the impact is 0.9 *m/s*, what was the coefficient of restitution of the cue ball's impact with the cushion? (The balls are of equal mass.)

Solution:

The strategy is to treat the two impacts as oblique central impacts. Denote the paths of the cue ball as $P1$ before the bank impact, $P2$ after the bank impact, and $P3$ after the impact with the 8-ball. The velocity of the cue ball is $\vec{v}_{AP1} = 3(\vec{i}\cos 30^o + \vec{j}\sin 30^o) = 2.6\vec{i} + 1.5\vec{j}\ (m/s)$. The x component is unchanged by the bank impact. The y component after impact is $v_{AP2y} = -ev_{AP1y} = -1.5e$, from which the velocity of the cue ball after the bank impact is $\vec{v}_{AP2} = 2.6\vec{i} - 1.5e\vec{j}$. At impact with the 8-ball, the x component is unchanged. The y component after impact is obtained from the conservation of linear momentum and the coefficient of restitution. The two equations are $m_A v_{AP2y} = m_A v_{AP3y} + m_B v_{BP3y}$ and

$1 = \dfrac{v_{BP3y} - v_{AP3y}}{v_{AP2y}}$. For $m_A = m_B$, these equations have the solution $v_{AP3y} = 0$ and $v_{BP3y} = v_{AP2y}$,

from which the velocities of the cue ball and the 8-ball after the second impact are $\vec{v}_{AP3} = 2.6\vec{i}\ (m/s)$, and $\vec{v}_{BP3} = -1.5e\vec{j}\ (m/s)$. The magnitude of the 8-ball velocity is $v_{BP3} = 0.9\ m/s$, from which

$$\boxed{e = \frac{0.9}{1.5} = 0.6}$$

===================================<>===================================

Problem 5.140 What is the solution to Problem 5.139 if the angle $\beta = 10^o$?

Solution:

Use the results of the solution to Problem 5139. The strategy is to treat the second collision as an oblique central impact about the line P when $\beta = 10^o$. The unit vector parallel to the line P is $\vec{e}_P = (\vec{i}\sin 10^o - \vec{j}\cos 10^o) = 0.1736\vec{i} - 0.9848\vec{j}$. The vector normal to the line P is $\vec{e}_{Pn} = 0.9848\vec{i} + 0.1736\vec{j}$. The projection of the velocity $\vec{v}_{AP2} = v_{AP3}\vec{e}_P + v_{AP3n}\vec{e}_{Pn}$. From the solution to Problem 5.63, $\vec{v}_{AP2} = 2.6\vec{i} - 1.5e\vec{j}$, from which the two simultaneous equations for the new components: $2.6 = 0.1736v_{AP3} + 0.9848v_{AP3n}$, and $-1.5e = -0.9848v_{AP3} + 0.1736v_{AP3n}$. (See Problem 5.45 for procedure to solve two-by-two systems.) Solve: $v_{AP3} = 0.4512 + 1.477e$, $v_{AP3n} = 2.561 - 0.2605e$. The component of the velocity normal to the line P is unchanged by impact. The change in the component parallel to P is found from the conservation of linear momentum and the coefficient of restitution: $m_A v_{AP3} = m_A v'_{AP3} + m_B v'_{BP3}$, and $1 = \dfrac{v'_{BP3} - v'_{AP3}}{v_{AP3}}$. For $m_A = m_B$, these equations have the solution $v'_{AP3} = 0$ and $v'_{BP3} = v_{AP3}$. From the value of $v'_{BP3} = 0.9\ m/s$,

$0.9 = 0.4512 + 1.477e$, from which $\boxed{e = \dfrac{0.9 - 0.4512}{1.477} = 0.304}$

===================================<>===================================

===================================<>===================================

Problem 5.141 What is the solution to Problem 5.139 if $\beta = 15^o$ and the coefficient of restitution of the impact between the two balls is $e = 0.9$?

Solution:

Use the solution to Problem 5.139. The strategy is to treat the second collision as an oblique central impact about P when $\beta = 15^o$. The unit vector parallel to the line P is $\vec{e}_P = \left(\vec{i}\sin 15^o - \vec{j}\cos 15^o\right) = 0.2588\vec{i} - 0.9659\vec{j}$. The vector normal to the line P is $\vec{e}_{Pn} = 0.9659\vec{i} + 0.2588\vec{j}$. The projection of the velocity $\vec{v}_{AP2} = v_{AP3}\vec{e}_P + v_{AP3n}\vec{e}_{Pn}$. From the solution to Problem 5.63, $\vec{v}_{AP2} = 2.6\vec{i} - 1.5e\vec{j}$, from which the two simultaneous equations for the new components: $2.6 = 0.2588 v_{AP3} + 0.9659 v_{AP3n}$, and $-1.5e = -0.9659 v_{AP3} + 0.2588 v_{AP3n}$. (See Problem 5.45 for general procedure to solve two-by-two systems.) Solve: $v_{AP3} = 0.6724 + 1.449e$, $v_{AP3n} = 2.51 - 0.3882e$. The component of the velocity normal to the line P is unchanged by impact.. The velocity of the 8-ball after impact is found from the conservation of linear momentum and the coefficient of restitution: $m_A v_{AP3} = m_A v'_{AP3} + m_B v'_{BP3}$, and $e_B = \dfrac{v'_{BP3} - v'_{AP3}}{v_{AP3}}$, where $e_B = 0.9$. For $m_A = m_B$, these equations have the solution $v'_{BP3} = \left(\dfrac{1}{2}\right)(1 + e_B)v_{AP3} = 0.95 v_{AP3}$. From the value of $v'_{BP3} = 0.9\ m/s$, $0.9 = 0.95(0.6724 + 1.449e)$, from which $\boxed{e = 0.189... = 0.19}$

===================================<>===================================

Problem 5.142 A ball is given a horizontal velocity of 3 *m/s* at 2 *m* above the smooth floor. Determine the distance D between its first and second bounces if the coefficient of restitution is $e = 0.6$

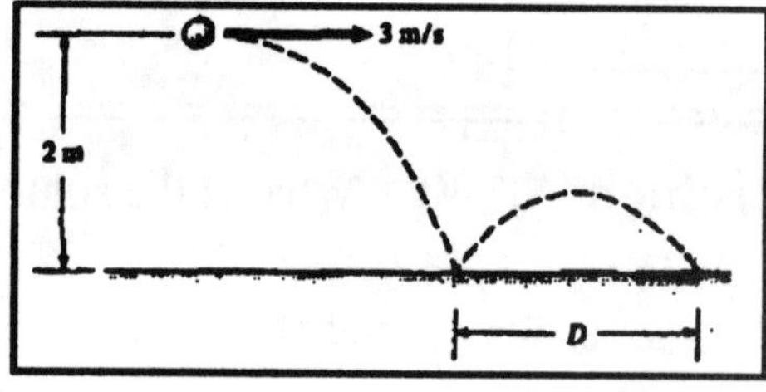

Solution:

The strategy is to treat the impact as an oblique central impact with a rigid surface. The horizontal component of the velocity is unchanged by the impact. The vertical velocity after impact is $v'_{Ay} = -e v_{Ay}$. The vertical velocity before impact is $v_{Ay} = -\sqrt{2gh} = -\sqrt{2(9.81)2} = -6.264\ m/s$, from which $v'_{Ay} = -0.6(-6.264) = 3.759\ m/s$. From the conservation of energy for a free fall, the height of the second bounce is $h' = \dfrac{(v'_{Ay})^2}{2g} = 0.720\ m$. From the solution of Newton's second law for free fall, the time between the impacts is twice the time required to fall from height h, $t = 2\sqrt{\dfrac{2h'}{g}} = 0.7663\ s$, and the distance D is

$$\boxed{D = v_o t = (3)(0.7663) = 2.3\ m}$$

===================================<>===================================

=================================<>=================================

Problem 5.143 A basketball dropped from a height of 4 *ft* rebounds to a height of 3 *ft*. In the "lay-up" shot shown the magnitude of the ball's velocity is 5 *ft/s* and the angles between the velocity vector and the positive coordinate axes are $\theta_x = 42^o$, $\theta_y = 68^o$, and $\theta_z = 124^o$ just before it hits the backboard. What are the magnitudes of its velocity and the angles between its velocity vector and the positive coordinate axes just after it hits the backboard?

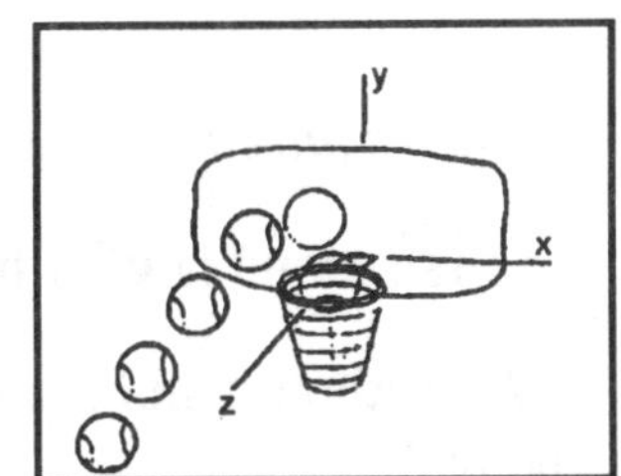

Solution:

The strategy is to treat this as an oblique central impact about a line parallel to the z axis. The coefficient of restitution is $e = \dfrac{v'_A}{v_A} = \sqrt{\dfrac{h'}{h}} = 0.866$. The component of the velocity vector parallel to the z-axis is $\vec{v}_P = v_A(\cos 124^o)\vec{k} = (2)(-0.5592)\vec{k} = -2.796\vec{k}\ \ (ft/s)$. The components normal to the z-axis are $\vec{v}_n = v_A(\cos 42^o)\vec{i} + v_A(\cos 68^o)\vec{j} = 3.716\vec{i} + 1.873\vec{j}\ \ (ft/s)$. The components normal to the z axis are unaffected by the impact. From Eq (5.17), with a rigid backboard, the velocity parallel to the z- axis after rebound is $v'_P = -ev_P = 2.421\ ft/s$. The velocity after impact is $\vec{v}' = 3.716\vec{i} + 1.873\vec{j} + 2.421\vec{k}\ \ (ft/s)$. [*Note:* From this expression, the ball is moving *to the right*, upward, and *outward* after impact with the backboard.] The magnitude of the velocity after impact is $|\vec{v}'| = \sqrt{3.716^2 + 1.873^2 + 2.421^2} = 4.814\ ft/s$

The angles are (using the principal values) $\theta'_x = \cos^{-1}\left(\dfrac{3.716}{4.814}\right) = 39.5^o$ $\theta'_y = \cos^{-1}\left(\dfrac{1.873}{4.814}\right) = 67.1^o$

$\theta'_z = \cos^{-1}\left(\dfrac{2.421}{4.814}\right) = 59.8^o$

=================================<>=================================

Problem 5.144 In Problem 5.143, the basketball's diameter is 9.5 *in.*, the coordinates of the center of basket rim are $x = 0$, $y = 0$, and $z = 12\ in.$, and the backboard is in the *x-y* plane. Determine the *x* and *y* coordinates of the point where ball must hit the backboard so that the center of the ball passes through the center of the basket rim.

Solution:

The strategy is to determine the impact location from the times required for the ball to fall into the basket. From the solution to Problem 5.143 the velocity vector after impact with the board is $\vec{v}' = 3.716\vec{i} + 1.873\vec{j} + 2.421\vec{k}\ \ (ft/s)$, from which the ball is moving upward, to the right, and outward after impact.

The time to travel in the z-direction: The distance the ball travels in the *z* direction is known: location of the basket center less the radius of the ball: $z_{travel} = 12 - 4.25 = 7.75\ in = 0.6458\ ft$. The time required is $t = z_{travel} / v_z = 0.6458 / 2.421 = 0.2668$ s.

The time required to travel in the x- direction is $t = \dfrac{x_{imp}}{v_x} = \dfrac{x_{imp}}{3.716}\ s$, from which

$x_{imp} = v_x t_z = (3.716)(0.2668) = 0.9913\ ft = 11.9\ in.$ Since the ball is traveling to the right toward the origin, the impact point on the board must be negative: $x = -11.9$ in .

Solution continued on next page

Continuation of the solution to Problem 5.144

The time required to travel in the y-direction: The basket is located at $y = 0$. From a solution of Newton's second law for free fall: $0 = -\frac{g}{2}(t_z)^2 + v_y t_z + y_{imp}$, from which $\boxed{y_{imp} = 0.6447\ ft = 7.74\ in.}$.

Check: Although not required by the problem, a useful method of checking results is an Euler integration of the path:

The algorithms: $x[i] = x[i-1] + u[i-1]dt$, $u[i] = u[i-1]$, with $x[0] = -11.9\ in.$, $x[0] = 44.59\ in/s$.

$y[i] = y[i-1] + v[i-1]dt$, $v[i] = v[i-1] - (386.04)dt$, with $y[0] = 7.737\ in.$, $v[0] = 22.477\ in/s$.

$z[i] = z[i-1] + w[i-1]dt$, $w[i] = w[i-1]$, with $z[0] = 12 - 4.75 = 7.75\ in.$, $w[0] = 29.057\ in/s$.

The initial velocities (converted to *in/s*) are obtained from the solution to Problem 5.125, and the initial positions (converted in inches) are obtained from the solution above. Note that the *x*, *y*, *z* paths arrive at the hoop center at the same instant. The *y*- path has an initial *upward* velocity, so that the ball rises a small distance before it starts the downward trajectory. The *z* path starts at $z = 0$ and moves outward a distance of 7.75 *in.*, whereas the *x*, *y* paths start at the coordinates of the impact point on the board and arrive at the origin at the same instant. *check.*

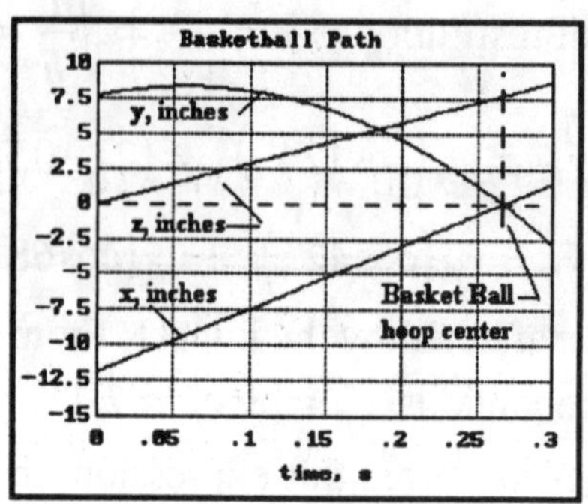

================================<>================================

Problem 5.145 A satellite at $r_o = 10{,}000\ mi$ from the center of the earth is given an initial velocity $v_o = 20{,}000\ ft/s$ in the direction shown. Determine the magnitude of its transverse component of velocity when $r = 20{,}000\ mi$. The radius of the earth is 3960 *mi.*

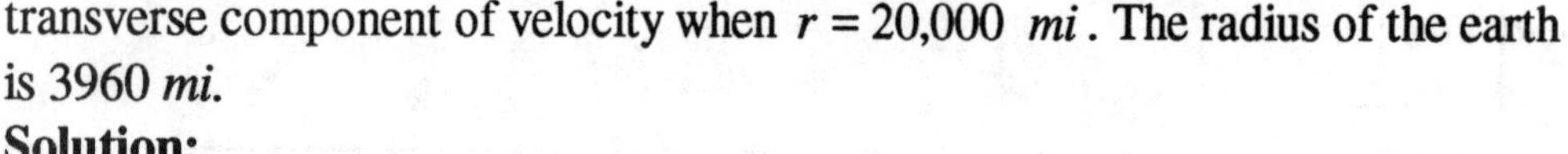

Solution:

By definition,

$$H_o = |\vec{r} \times m\vec{v}| = mr_o v_o \sin 45^o = m(10{,}000)(5280)(20{,}000)\sin 45^o$$

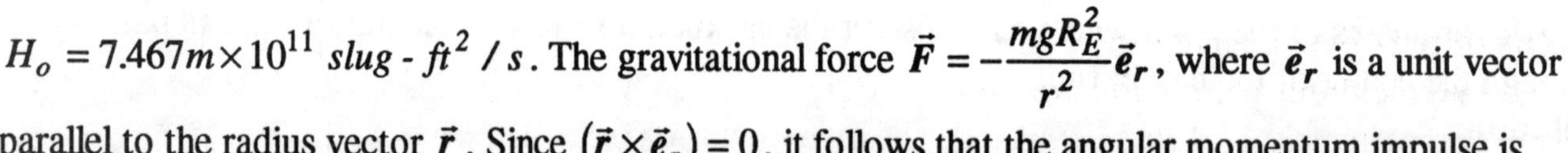

$H_o = 7.467m \times 10^{11}\ slug\text{-}ft^2/s$. The gravitational force $\vec{F} = -\frac{mgR_E^2}{r^2}\vec{e}_r$, where $\vec{e}_r$ is a unit vector parallel to the radius vector $\vec{r}$. Since $(\vec{r} \times \vec{e}_r) = 0$, it follows that the angular momentum impulse is $\int_{t_1}^{t_2} (\vec{r} \times \sum \vec{F})dt = 0 = \vec{H}_1 - \vec{H}_o$, from which $\vec{H}_o = \vec{H}_1$, and the angular momentum is conserved. Thus the angular momentum at the distance 20,000 *mi* is $H_o = mr_1 v_{\theta 1} \sin 90^o$. From which

$$\boxed{v_{\theta 1} = \frac{7.467m \times 10^{11}}{1.056m \times 10^8} = 7071\ ft/s}$$ is the magnitude of the transverse velocity.

================================<>================================

=================================<>=================================

Problem 5.146 In Problem 5.145, determine the magnitudes of the radial and transverse components of the satellite's velocity when $r = 15{,}000\ mi$.

Solution: From the solution to Problem 5.145, the constant angular momentum of the satellite is $H_o = 7.467m \times 10^{11}\ slug\text{-}ft^2/s$. The transverse velocity at $r = 15{,}000\ mi$ is

$$\boxed{v_\theta = \frac{H}{m(15{,}000)(5280)} = 9428.1\ ft/s}$$ From the conservation of energy:

$$\left(\frac{1}{2}mv_o^2\right)_{r=10{,}000} - \frac{mgR_E^2}{r_o} = \left(\frac{1}{2}mv_1^2\right)_{r=15{,}000} - \frac{mgR_E^2}{r_1}, \text{ from which}$$

$$v_1^2 = v_o^2 + 2gR_E^2\left(\frac{1}{r} - \frac{1}{r_o}\right) = (20{,}000)^2 + 2(32.17)(3960)^2\left(\frac{1}{15{,}000} - \frac{1}{10000}\right)(5280)$$

$= 2.224 \times 10^8\ (ft/s)^2$. The radial velocity is $\boxed{v_r = \sqrt{v_1^2 - v_\theta^2} = 11555.8\ ft/s}$

=================================<>=================================

Problem 5.147 The snow is 2 ft deep and weighs 20 *lb./ft*^{2}, the snowplow is 8 *ft* wide, and the truck travels at 5 *mi/hr*. What force does the snow exert on the truck?

Solution:

The velocity of the truck is $v = 5\left(\frac{5280}{3600}\right) = 7.33\ ft/s$ The mass flow is the product of the depth of the snow, the width of the plow, the mass density of the snow, and the velocity of the truck, from which. $\left(\frac{dm_f}{dt}\right) = (2)\left(\frac{20}{g}\right)(8)(14.67) = 72.95\ slug/s$.

The force is $\boxed{F = \left(\frac{dm_f}{dt}\right)v = 534.9\ lb}$

=================================<>=================================

Problem 5.148 An empty 55 *lb.* drum, 3 *ft* in diameter, stands on a set of scales. Water begins pouring into the drum at 1200 *lb./min* from 8 *ft* above the bottom of the drum. The weight density of the water is approximately 62.4 *lb./ft*^{3}. What do the scales read 40 *s* after the water starts pouring?

Solution: The mass flow rate is

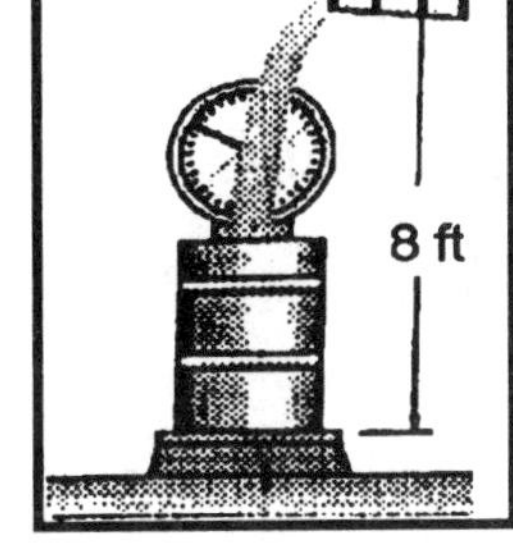

$\left(\frac{dm_f}{dt}\right) = \left(\frac{1200}{g}\right)\left(\frac{1}{60}\right) = 0.6217\ slug/s$ After 40 seconds, the volume of water in the drum is $volume = \frac{1200}{62.4}\left(\frac{40}{60}\right) = 12.82\ ft^3$. The height of water in the drum

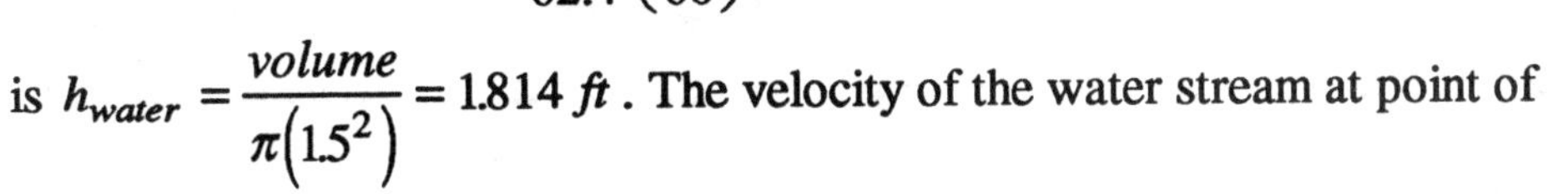

is $h_{water} = \frac{volume}{\pi(1.5^2)} = 1.814\ ft$. The velocity of the water stream at point of impact is (from the conservation of energy for free fall) $v = \sqrt{2g(8 - h_{water})} = 19.95\ ft/s$. The scale reading is $\boxed{W = (volume)(62.4) + 55 + \left(\frac{dm_f}{dt}\right)v = 12.82(62.4) + 55 + 0.6216(19.95) = 867.4\ lb}$

=================================<>=================================

==================================<>==================================

Problem 5.149 The ski boat's propulsive system draws water in at A and expels it at B at 80 *ft/s* relative to the boat. Assume that the water drawn in enters with no horizontal velocity relative to the surrounding water. The maximum mass flow of water through the engine is 2.5 *slug/s*. Hydrodynamic drag exerts a force on the boat of magnitude 1.5*v* *lb.*, where *v* is the boat's velocity in *ft/s*. If you neglect aerodynamic drag, what is the ski boat's maximum velocity?

Solution: Use the boat's "rest" frame of reference. The force exerted by the inlet mass flow on the boat is $F_{inlet} = -\left(\frac{dm_f}{dt}\right)v = -2.5v$. The force exerted by the exiting mass flow on the boat is $F_{exit} = \left(\frac{dm_f}{dt}\right)(80) = 2.5(80)$ The hydrodynamic drag is $F_{drag} = -1.5v$. At top speed the sum of the forces vanishes:

$\sum F = F_{inlet} + F_{exit} + F_{drag} = 0 = -2.5v + 2.5(80) - 1.5v = -4.0v + 2.5(80) = 0$, from which

$$\boxed{v = \frac{2.5(80)}{4.0} = 50\ ft/s}$$

==================================<>==================================

Problem 5.150 The ski boat of Problem 5.149 weighs 2800 *lb*. The mass flow rate of water through its engine is 2.5 *slug/s*, and it starts from rest at $t = 0$. Determine the boat's velocity (a) at $t = 20\ s$, (b) at $t = 60\ s$.

Solution: Use the solution to Problem 5.149: the sum of the forces on the boat is $\sum F = -4v + 2.5(80)\ lb$. From Newton's second law $\frac{W}{g}\left(\frac{dv}{dt}\right) = -4.0v + 200$. Separate variables and integrate: $\frac{dv}{50 - v} = \frac{4g}{W}dt$, from which $\ln(50 - v) = -\frac{4g}{W}t + C$. At $t = 0$, $v = 0$, from which $C = \ln(50)$, from which $\ln\left(1 - \frac{v}{50}\right) = -\frac{4g}{W}t$. Invert: $v(t) = 50\left(1 - e^{-\frac{4g}{W}t}\right)$. Substitute numerical values: $W = 2800\ lb$, $g = 32.17\ ft/s^2$. (a) At $t = 20\ s$, $\boxed{v = 30.1\ ft/s}$, (b) At $t = 60\ s$, $\boxed{v = 46.8\ ft/s}$

==================================<>==================================

Problem 5.151 A crate of mass *m* slides across the smooth floor pulling chain from a stationary pile. The mass per unit length of chain is ρ_L. If the velocity of the crate is v_o at $s = 0$, what is its velocity as a function of *s*?

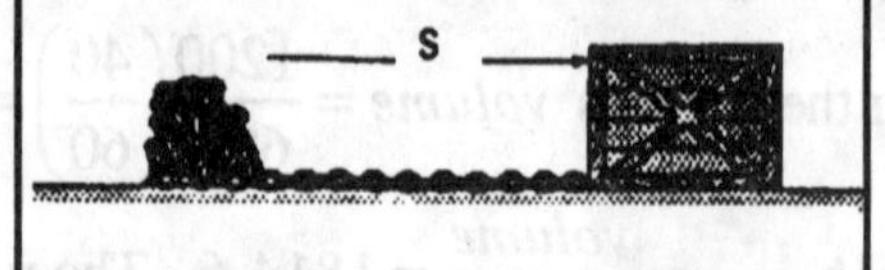

Solution: The "mass flow" of the chain is $\left(\frac{dm_f}{dt}\right) = \rho_L v$. The force exerted by the "mass flow" is $F = \rho_L v^2$. From Newton's second law $(\rho_L s + m)\left(\frac{dv}{dt}\right) = -\rho_L v^2$. Use the chain rule: $(\rho_L s + m)v\frac{dv}{ds} = -\rho_L v^2$. Separate variables and integrate: $\ln(v) = -\ln\left(s + \frac{m}{\rho_L}\right) + C$, from which $C = \ln\left(\frac{mv_o}{\rho_L}\right)$. Reduce and solve: $\boxed{v = \frac{mv_o}{(\rho_L s + m)}}$

==================================<>==================================

=================================<>=================================

Problem 6.1 The disk rotates relative to the coordinate system about the fixed shaft that is coincident with the z-axis. At the instant shown, the disk has a counterclockwise angular velocity of *3 rad/s* and a counterclockwise angular acceleration of $4\ rad/s^2$ What are the x and y components of the velocity and acceleration of point A relative to the coordinate system? .

Solution:

From Figure (a), the velocity is

$\bar{v}_A = r\omega\bar{j} = (2)(3)\bar{j} = 6\bar{j}(ft/s)$

From Figure (b) the acceleration is

$\bar{a}_A = -r\omega^2\bar{i} + r\alpha\bar{j} = -(2)(3)^2\bar{i} + (2)(4)\bar{j}$

$= -18\bar{i} + 8\bar{j}(ft/s^2).$

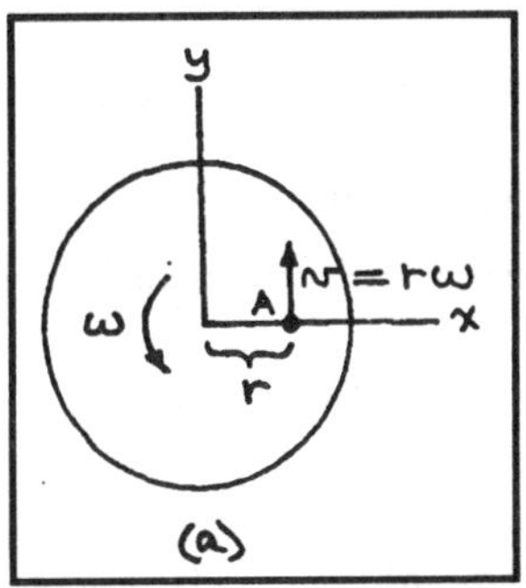

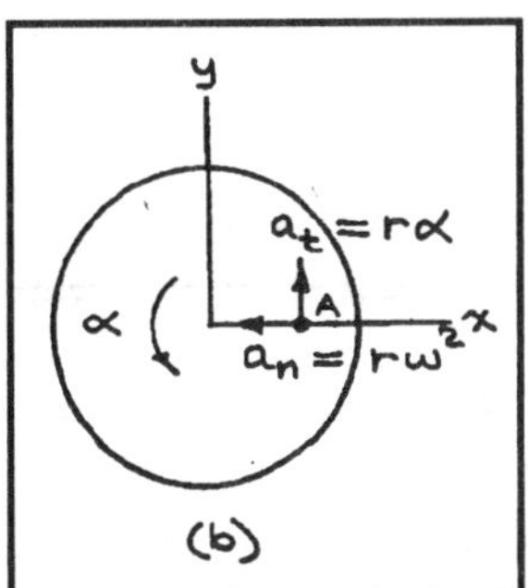

=================================<>=================================

Problem 6.2 If the angular acceleration of the disk in Problem 6.1 is constant, what are the x and y components of the velocity and acceleration of Point A when the disk has rotated 90° relative to the positon shown?

Solution:

The angular acceleration is given by

$\alpha = \dfrac{d\omega}{dt} = 4rad/s^2 \int_3^\omega d\omega = \int_0^t 4dt,$ and

$\omega = \dfrac{d\theta}{dt} = 3 + 4t\ (rad/s),$ and $\int_0^\theta d\theta = \int_0^t (3+4t)dt,$

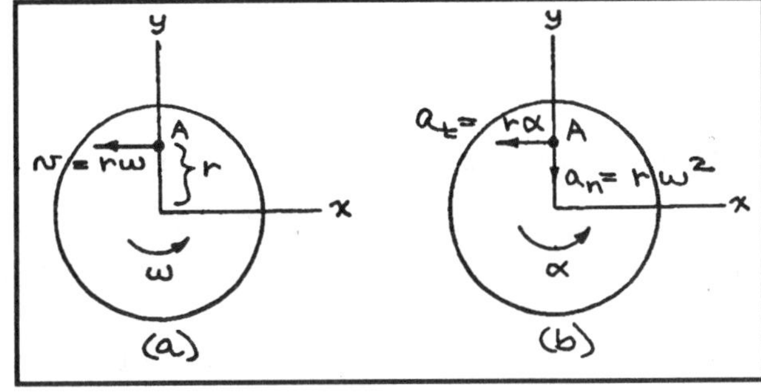

$\theta = 3t + 2t^2$. Setting $\theta = \pi/2rad$ and solving for t, $t = 0.411s$ Then solving for ω, $\theta = 4.64rad/s$ From Figure (a), the velocity is $\bar{v}_A = -r\omega\bar{i} = -(2)(4.64)\bar{i} = -9.29\bar{i}(ft/s)$. From Figure (b) the acceleration is $\bar{a}_A = -r\alpha\bar{i} - r\omega^2\bar{j} = -(2)(4)\bar{i} - (2)(4.64)^2\bar{j} = -8\bar{i} - 43.1\bar{j}(ft/s^2).$

=================================<>=================================

Problem 6.3 The weight A hangs from a rope that is wrapped around the disk. If the weight is moving downward at *4 m/s* , what is the angular velocity of the disk?

Solution:

The magnitude of the velocity of a point on the edge of the disk is

$4m/s = r\omega = (0.1m)\omega$,so $\omega = \dfrac{4}{0.1} = 40rad/s$ clockwise

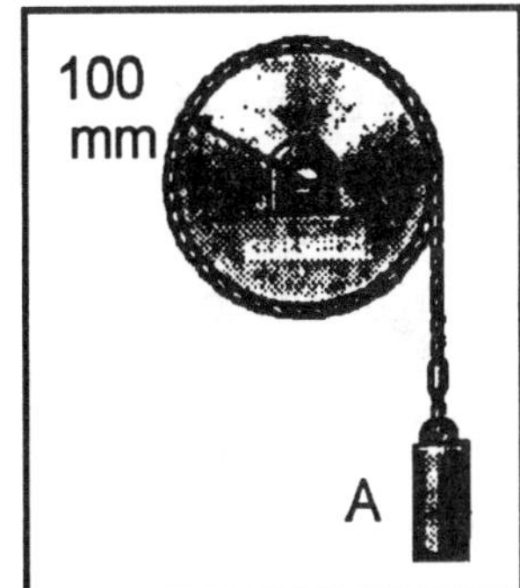

=================================<>=================================

================================<>================================

Problem 6.4 Suppose that the weight A in Problem 6.3 starts from rest at $t = 0$ and falls with a constant acceleration of $2m/s^2$.

(a) What is the angular acceeleration of the disk?

(b) How many revolutions has the disk turned at $t = 1\ s$?

Solution:

(a) The clockwise angular acceleration is $\alpha = \dfrac{2m/s^2}{0.1m} = 20 rad/s^2$ (b)

$\alpha = \dfrac{d\omega}{dt} = 20 rad/s^2 \Rightarrow \int_0^{\omega} d\omega = \int_0^t 20 dt$. Hence, $\omega = \dfrac{d\theta}{dt} = 20t$ and $\int_0^{\theta} d\theta = \int_0^t 20t dt, \theta = 10t^2 rad,$

(b) At $t = 1s$ $\theta = 10 rad = 1.59$ revolutions.

================================<>================================

Problem 6.5 Determine $\dfrac{\omega_B}{\omega_A}$ and $\dfrac{\omega_C}{\omega_A}$.

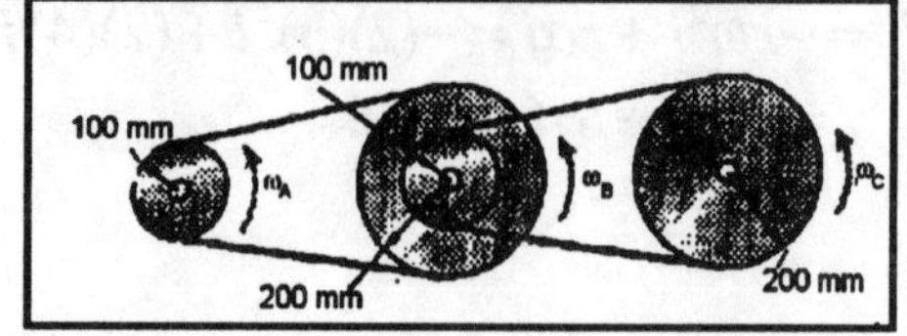

Solution

Denote the pulleys from left to right by A, B, and C. The linear velocity of the rim of the outer pulley at B equals the linear velocity of the rim of A, and the linear velocity of the rim of C is equal to the linear velocity of the rim of the inner pulley at B., from which $v_A = v_{Bo}$, $v_C = v_{Bi}$. Substitute $v = \omega r$, $\omega_A(100) = \omega_B(200)$, from which $\boxed{\dfrac{\omega_B}{\omega_A} = \dfrac{1}{2}}$ $\omega_C(200) = \omega_B(100)$, from which $\dfrac{\omega_C}{\omega_B} = \dfrac{1}{2}$, and $\boxed{\dfrac{\omega_C}{\omega_A} = \dfrac{1}{4}}$

================================<>================================

Problem 6.6 The bicycle's 120 *mm* sprocket wheel turns at 3 rad/s. What is the angular velocity of the 45 *mm* gear?

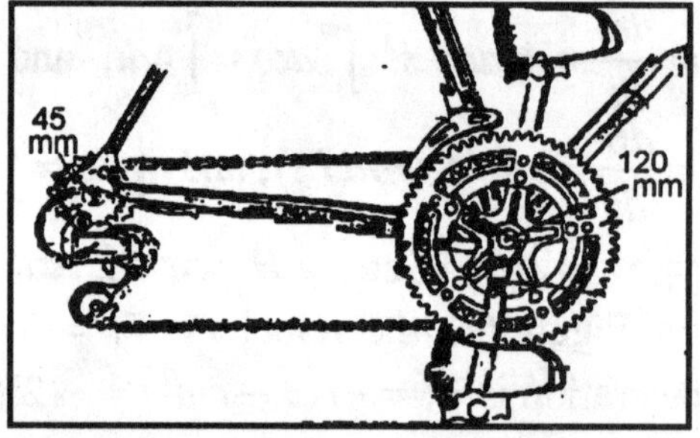

Solution

The linear velocity of the rim of each wheel is the same, $v_{120} = v_{45}$. Substitute $v = \omega r$, $\omega_{120}(120) = \omega_{45}(45)$, from which $\boxed{\omega_{45} = \omega_{120}\left(\dfrac{120}{45}\right) = 3\left(\dfrac{120}{45}\right) = 8\ rad/s}$

================================<>================================

Problem 6.7 The rear wheel of the bicycle in Problem 6.6 has a 330 *mm* radius, and is rigidly attached to the 45 *mm* gear. If the rider turns the pedals which are rigidly attached to the 120 *mm* sprocket wheel, at 1 revolution per second, what is the bicycle's velocity?

Solution

The angular velocity of the 120 *mm* sprocket wheel is $\omega = 1\ rev/s = 2\pi\ rad/s$. Use the solution to Problem 6.6. The angular velocity of the 45 *mm* gear is $\omega_{45} = 2\pi\left(\dfrac{120}{45}\right) = 16.76\ rad/s$. This is also the angular velocity of the rear wheel, from which the velocity of the bicycle is $v = \omega_{45}(330) = 5.53\ m/s$

================================<>================================

=================================<>================================

Problem 6.8 Relative to the coordinate system, the disk rotates about the origin with a constant counterclockwise angular velocity of 10 *rad/s*. What are the velocity and acceleration of points A and B at the instant shown?

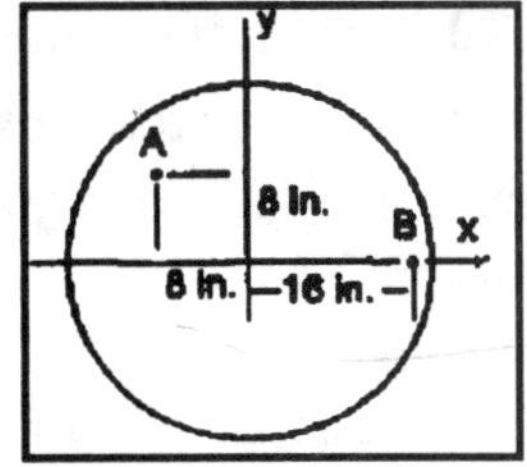

Solution:

The radial distances to A and B are $r_A = \sqrt{8^2 + 8^2} = 11.3in = 0.943ft$

$r_B = 16in = 1.33ft$. The velocity and acceleration of A are shown:

In terms of x and y components, $\bar{v}_A = -\omega r_A \cos 45°\bar{i} - \omega r_A \cos 45°\bar{j} = -(10)(0.943)\cos 45°(\bar{i} + \bar{j})$

$= -6.67(\bar{i} + \bar{j}) ft/s$. $\bar{a}_A = \omega^2 r_A \cos 45°\bar{i} - \omega^2 r_A \cos 45°\bar{j} = (10)^2(0.943)\cos 45°(\bar{i} - \bar{j})$

$= 66.7(\bar{i} - \bar{j}) ft/s$. The velocity and acceleration of B are shown: In terms of x and y components,

$\bar{v}_B = \omega r_B \bar{j} = (10)(1.33)j = 13.3\bar{j}(ft/s)$; $a_B = \omega^2 r_B \bar{i} = -(10)^2(1.33)\bar{i} = -133\bar{i}(ft/s^2)$.

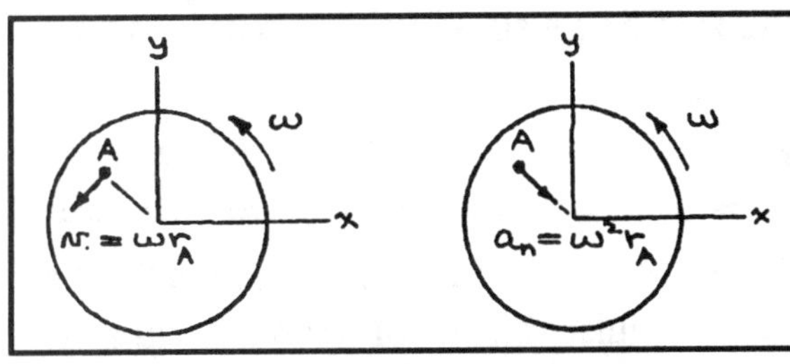

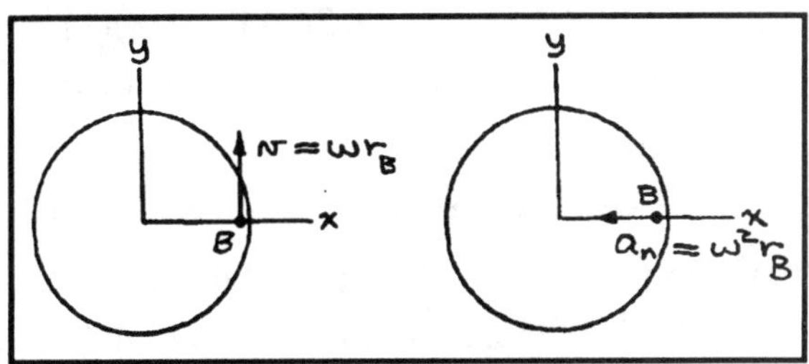

=================================<>================================

Problem 6.9 Suppose that the disk in Problem 6.8 starts from rest in the position shown at $t = 0$ and has a counterclockwise angular acceleration of $6rad/s^2$ relative to the coordinate system. Determine the x and y components of the velocity and acceleration of point B at $t = 1\ s$.

Solution:

The angular acceleration is $\alpha = \dfrac{d\omega}{dt} = 6rad/s^2 = \int_0^{\omega} d\omega = \int_0^t 6dt$. Hence, $\omega = \dfrac{d\theta}{dt} = 6t\ rad/s$. Integrating again, we get $\int_0^{\theta} d\theta = \int_0^t 6tdt, \theta = 3t^2 rad,$

At $t = 1s$ $\omega = 6rad/s$ and $\theta = 3rad = 171.89°$.

The velocity and acceleration of B are shown:

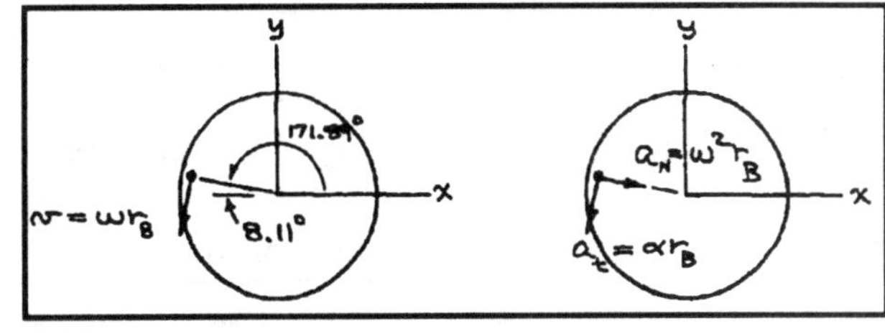

The velocity is $\bar{v}_B = -\omega r_B \sin 8.11°\bar{i} - \omega r_B \cos 8.11°\bar{j}$

$\bar{v}_B = -(6)(16/12)(\sin 8.11°\bar{i} + \cos 8.11°\bar{j})$, or

$\bar{v}_B = -1.13\bar{i} - 7.92\bar{j}(ft/s)$.

The acceleration is $\bar{a}_B = (\omega^2 r_B \cos 8.11° - \alpha r_B \sin 8.11°)\bar{i} - (\omega^2 r_B \sin 8.11° + \alpha r_B \cos 8.11°)\bar{j}$, or

$\bar{a}_B = 46.4\bar{i} - 14.7\bar{j}(ft/s^2)$.

=================================<>================================

==================================<>==================================

Problem 6.10 Suppose the disk in Problem 6.8 starts from rest in the position shown at $t = 0$ and has a constant counterclockwise angular acceleration of $4 rad/s^2$ relative to the coordinate system. Determine the x and y components of the velocity of point A at $t = 2\ s$.

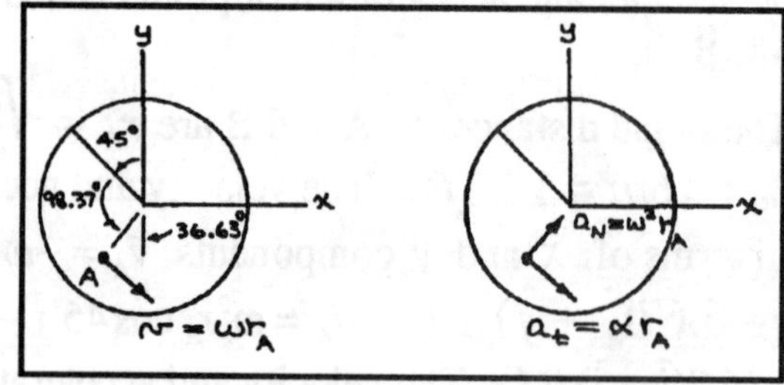

Solution:

The angular acceleration is given by

$\alpha = \frac{d\omega}{dt} = 4 rad/s^2$: $\int_0^{\omega} d\omega = \int_0^{t} 4dt$. Hence,

$\omega = \frac{d\theta}{dt} = 4t\ rad/s$ and $\int_0^{\theta} d\theta = \int_0^{t} 4t dt, \theta = 2t^2 rad,$

At $t = 2s$, $\omega = 8 rad/s$ and $\theta = 8 rad = 458.37°$, or one revolution plus $98.37°$: The velocity is $\bar{v}_A = \omega r_A \cos 36.63° \bar{i} - \omega r_A \sin 36.63° \bar{j} = (8)(11.3/12)(\cos 36.63° \bar{i} - \sin 36.63° \bar{j}) = 6.05\bar{i} - 4.50\bar{j} (ft/s)$.

The acceleration is $\bar{a}_A = (\omega^2 r_A \sin 36.63° + \alpha r_A \cos 36.63°)\bar{i} +$
$(\omega^2 r_A \cos 36.63° - \alpha r_A \sin 36.63°)\bar{j} = 39.0\bar{i} + 46.2\bar{j} (ft/s)$

==================================<>==================================

Problem 6.11 The bracket rotates relative to the coordinate system shown about a fixed shaft that is coincident with the z axis. . If it has a counterclockwise angular velocity of 20 *rad/s* and a clockwise angular acceleration of 200 *rad/s*^{2},. what are the magnitudes of the accelerations of points A and B?

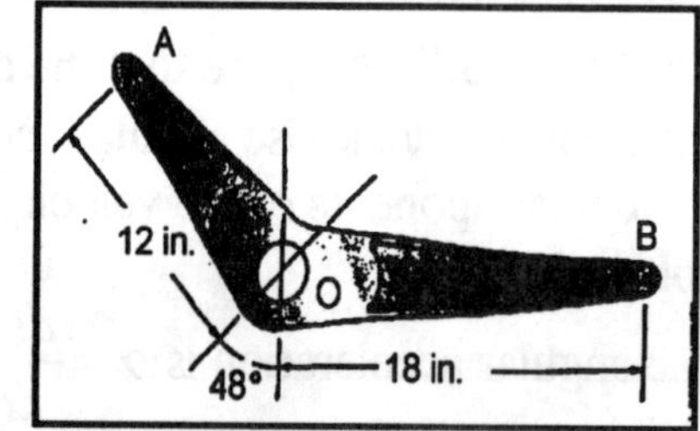

Solution:

The radial distances of A and B from the origin O are $r_A = 12\ in. = 1\ ft$ and $r_B = 18\ in. = 1.5\ ft$.*Point A*: The tangential acceleration is $a_{At} = r\,\alpha = -(1)(200) = -200\ ft/s^2$. The normal acceleration is $a_{An} = \omega^2 r = (20^2)(1) = 400\ ft/s^2$, from which the magnitude is $\boxed{a_A = \sqrt{200^2 + 400^2} = 447.2\ ft/s^2}$ *Point B*: The tangential acceleration is $a_{Bt} = r\,\alpha = (-200)(1.5) = -300\ ft/s^2$. The normal acceleration is $a_{Bn} = \omega^2 r = (20^2)(1.5) = 600\ ft/s^2$, from which the magnitude is $\boxed{a_B = \sqrt{300^2 + 600^2} = 670.8\ ft/s^2}$

==================================<>==================================

Problem 6.12 Consider the bracket in Problem 6.11. If $|\vec{v}_A| = 10\ ft/s$ and $|\vec{a}_A| = 200\ ft/s^2$, what are $|\vec{v}_B|$ and $|\vec{a}_B|$?

Solution:

From Eqs (6.2) and (6.3) $v = \omega\, r$, $a_t = \alpha\, r$, and $a_n = \omega^2 r$, and $a = \sqrt{a_t^2 + a_n^2}$, from which the ratios of the velocities and accelerations relative to the origin at two points on a rigid body are equal to the ratios of the radial distances from the origin to the two points. $\boxed{|\vec{v}_B| = 1.5|\vec{v}_A| = 15\ ft/s}$ and $\boxed{|\vec{a}_B| = 1.5|\vec{a}_A| = 1.5(200) = 300\ ft/s^2}$

==================================<>==================================

====================================<>====================================

Problem 6.13 Consider the bracket in Problem 6.11. If $|\vec{v}_A| = 36\ in/s$ and $|\vec{a}_B| = 600\ in./s^2$, what are $|\vec{v}_B|$ and $|\vec{a}_A|$?

Solution:

From the solution to Problem 6.12, the ratios of the velocities and accelerations relative to the origin at two points on a rigid body are equal to the ratios of the radial distances from the origin to the two points.

$\boxed{|\vec{v}_B| = 1.5(36) = 54\ \ in/s}$, and $\boxed{|\vec{a}_A| = \frac{|\vec{a}_B|}{1.5} = \frac{600}{1.5} = 400\ in/s^2}$

====================================<>====================================

Problem 6.14 The turbine rotates relative to the coordinate system at 30 *rad/s* about a fixed axis coincident with the x axis. What is angular velocity vector?

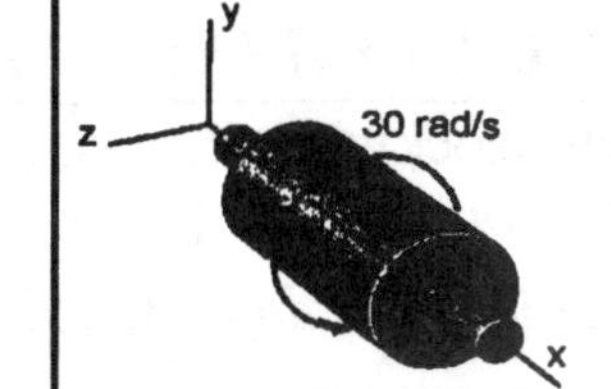

Solution:

The angular velocity vector is parallel to the *x* axis, with magnitude 30 *rad/s*. By the right hand rule, the positive direction coincides with the positive direction of the *x* axis. $\boxed{\vec{\omega} = 30\vec{i}}$

====================================<>====================================

Problem 6.15 The rectangular plate swings in the x-y plane from arms of equal length. Determine the angular velocity of the (a) the rectangular plate, (b) the bar AB.

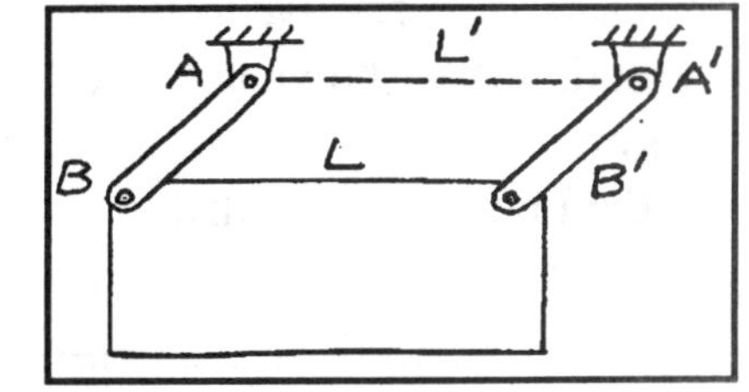

Solution:

Denote the upper corners of the plate by B and B', and denote the distance between these points (the length of the plate) by L. Denote the suspension points by A and A', the distance separating them by L'. By inspection, since the arms are of equal length, and since $L = L'$, the figure $AA'B'B$ is a parallelogram. By definition, the opposite sides of a parallelogram remain parallel, and since the fixed side AA' does not rotate, then BB' cannot rotate, so that the plate does not rotate and $\boxed{\vec{\omega}_{BB'} = 0}$. Similarly, by inspection the angular velocity of the bar AB is $\boxed{\vec{\omega}_{AB} = 10\vec{k}}$, where by the right hand rule the direction is along the positive *z* axis (out of the paper).

====================================<>====================================

Problem 6.16 What are the angular velocity vectors of each bar of the linkage.

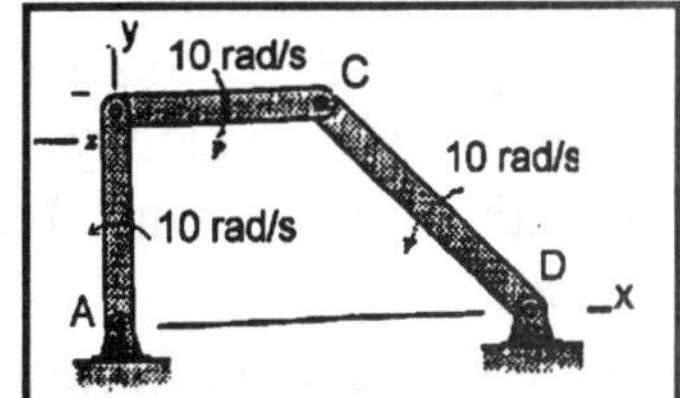

Solution:

The strategy is to use the definition of angular velocity, including the application of the right hand rule. For bar AB the magnitude is 10 *rad/s*, and the right hand rule indicates a direction in the positive *z* direction. (out of the paper). $\boxed{\vec{\omega}_{AB} = 10\vec{k}\ \ rad/s}$. For bar BC, $\boxed{\vec{\omega}_{BC} = -10\vec{k}\ \ rad/s}$.

For bar CD, $\boxed{\vec{\omega}_{CD} = 10\vec{k}\ \ rad/s}$

====================================<>====================================

==================================<>==================================

Problem 6.17 If you model the earth as a rigid body, what is the magnitude of the angular velocity vector ω_E ?.Does ω_E point north or south?

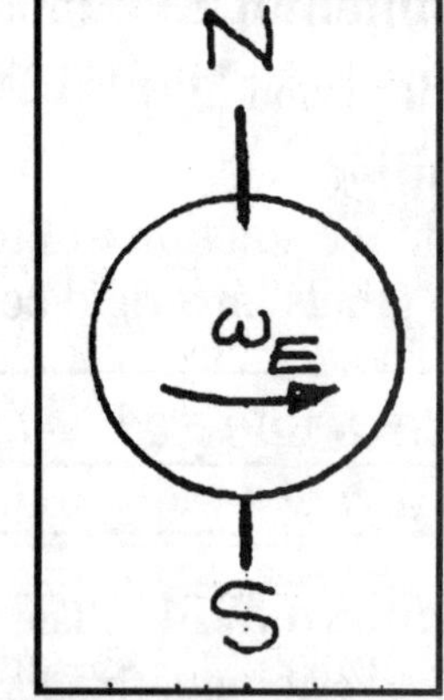

Solution:

With respect to an inertial reference (the "fixed" stars) the earth rotates 2π radians per day relative to the sun plus one revolution per year relative to the fixed direction. Take the length of the year to be 365.242 days. The revolutions per day is $rpd = 1 + \frac{1}{year} = 1.00274\ rev/day = 6.300\ rad/day$

The magnitude of the angular velocity is

$\boxed{\omega_E = 6.300\ rad/day = 7.292\times 10^{-5}\ rad/s}$. The earth turns to the east, or counterclockwise when viewing it from the north pole. From the right hand rule, the *angular velocity vector points north.* [*Note*: If the earth's orbital motion aroud the sun is ignored, then *revolutions per day* = 1, and $\omega_E = 7.27\times 10^{-5}\ rad/s$.]

==================================<>==================================

Problem 6.18 The rigid body rotates with a counterclockwise angular velocity ω about a fixed axis through B that is coincident with the z axis. Determine the x and y components of the velocity of A relative to B by representing it as shown in Fig. 6.11(b).

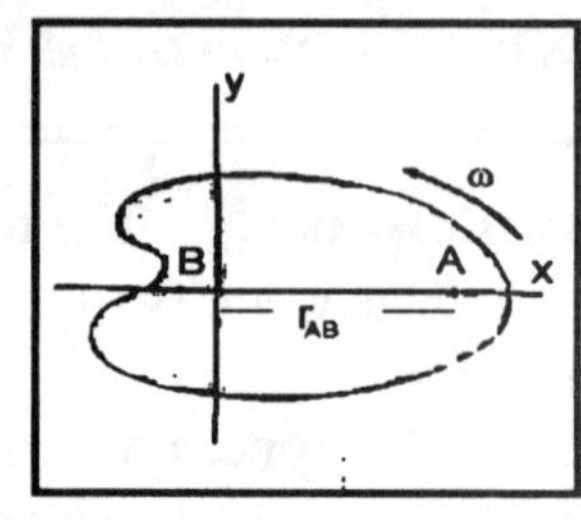

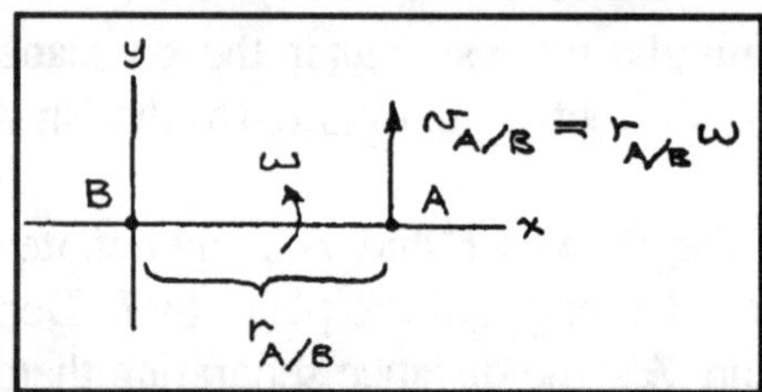

Solution:

From the Figure, we see that

$\bar{v}_{A/B} = r_{A/B}\omega\ \bar{j}$.

==================================<>==================================

Problem 6.19 Consider the rotating rigid body in Problem 6.18. (a) What is the angular velocity vector? (b) Use Eq. (6.5) to determine the velocity of A relative to B.

Solution:

(a) The angular velocity is just $\bar{\omega} = \omega\bar{k}$

(b) Eq. 6.5 gives us $\bar{v}_{A/B} = \bar{\omega}\times\bar{r}_{A/B} = (\omega\bar{k})\times(r_{A/B}\bar{i}) = \omega r_{A/B}\bar{j}$

==================================<>==================================

Problem 6.20 The bar is rotating with a counterclockwise angular velocity of 20 rad/s about the fixed point O. Determine the x and y components of the velocity of A relative to B by representing it as shown in Fig. 6.11(b).

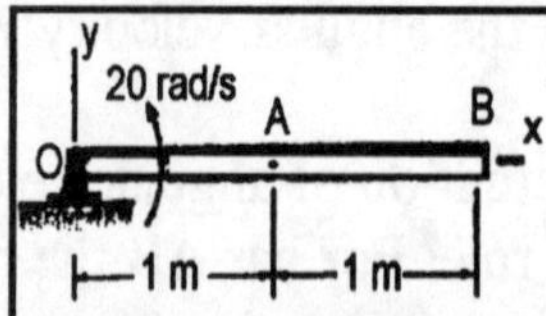

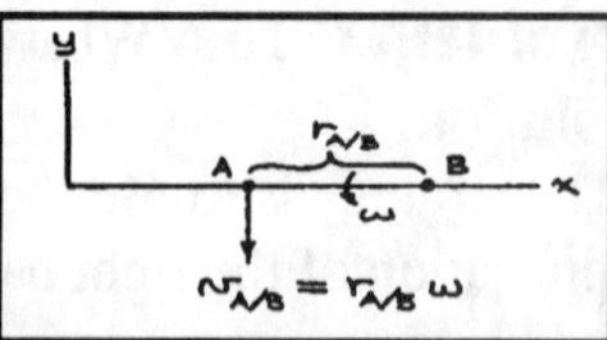

Solution:

From the figure, the relative velocity is given by $\bar{v}_{A/B} = -r_{A/B}\omega\bar{j}$, or $\bar{v}_{A/B} = -(1)(20)\bar{j} = -20\bar{j}(m/s)$.

==================================<>==================================

================================<>================================

Problem 6.21 Consider the bar in Problem 6.20. (a) Use Eq. (6.5) to determine the velocity of A relative to B. (b) Use the result of (a) and Eq. (6.6) to determine the velocity of A relative to the fixed coordinates.

Solution:

(a) $\bar{v}_{A/B} = \bar{\omega} \times \bar{r}_{A/B} = (20\bar{k}) \times (-\bar{i}) = -20\bar{j}(m/s)$.

(b) The velocity of B relative to the fixed coordinate system is $\bar{v}_B = \bar{v}_0 + \bar{\omega} \times \bar{r}_{B/0}$, or $\bar{v}_B = 0 + (20\bar{k}) \times (2\bar{i}) = 40\bar{j}(m/s)$. Therefore, the velocity of A is $\bar{v}_A = \bar{v}_B + \bar{v}_{A/B}$, or $\bar{v}_A = 40\bar{j} - 20\bar{j} = 20\bar{j}(m/s)$

================================<>================================

Problem 6.22 Determine the x and y components of the velocity of point A.

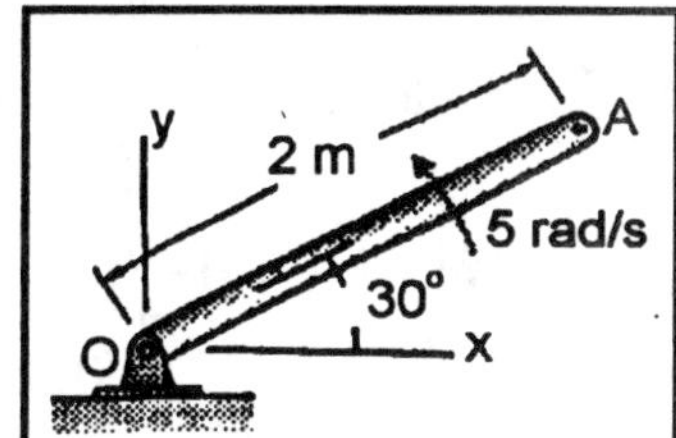

Solution:

The velocity of point A is given by: $\bar{v}_A = \bar{v}_0 + \bar{\omega} \times \bar{r}_{A/o}$.

Hence, $\bar{v}_A = 0 + \begin{vmatrix} \bar{i} & \bar{j} & \bar{k} \\ 0 & 0 & 5 \\ 2\cos 30^\circ & 2\sin 30^\circ & 0 \end{vmatrix} = -10\sin 30^\circ \bar{i} + 10\cos 30^\circ \bar{j}$,

or $\bar{v}_A = -5\bar{i} + 8.66\bar{j}(m/s)$

================================<>================================

Problem 6.23 If the angular velocity of the bar in Problem 6.22 is constant, what are the x and y components of the velocity of Point A *0.1 s* after the instant shown?

Solution: The angular velocity is given by $\omega = \frac{d\theta}{dt} = 5\ rad/s$, $\int_0^\theta d\theta = \int_0^t 5dt$, and $\theta = 5t\ rad$.

At $t = 0.1s$, $\theta = 0.5\ rad = 28.6^\circ$. $\bar{v}_A = \bar{v}_0 + \bar{\omega} \times \bar{r}_{A/o} = 0 + \begin{vmatrix} \bar{i} & \bar{j} & \bar{k} \\ 0 & 0 & 5 \\ 2\cos 58.6^\circ & 2\sin 58.6^\circ & 0 \end{vmatrix}$.

Hence, $\bar{v}_A = -10\sin 58.6^\circ \bar{i} + 10\cos 58.6^\circ \bar{j} = -8.54\bar{i} + 5.20\bar{j}(m/s)$.

================================<>================================

Problem 6.24 The disk is rotating about the z axis at *50 rad/s* in the clockwise direction. Determine the x and y components of the velocities of points A, B, and C.

Solution:

The velocity of A is given by $\bar{v}_A = \bar{v}_0 + \bar{\omega} \times \bar{r}_{A/o}$, or

$\bar{v}_A = 0 + (-50\bar{k}) \times (0.1\bar{j}) = 5\bar{i}\,(m/s)$.

For B, we have

$$\bar{v}_B = \bar{v}_0 + \bar{\omega} \times \bar{r}_{B/o} = 0 + \begin{vmatrix} \bar{i} & \bar{j} & \bar{k} \\ 0 & 0 & -50 \\ 0.1\cos 45^\circ & -0.1\sin 45^\circ & 0 \end{vmatrix} = -3.54\bar{i} - 3.54\bar{j}\,(m/s),$$

For C, we have

$$\bar{v}_c = \bar{v}_0 + \bar{\omega} \times \bar{r}_{c/o} = 0 + \begin{vmatrix} \bar{i} & \bar{j} & \bar{k} \\ 0 & 0 & -50 \\ -0.1\cos 45^\circ & -0.1\sin 45^\circ & 0 \end{vmatrix} = -3.54\bar{i} + 3.54\bar{j}\,(m/s).$$

Problem 6.25 If the angular velocity of the disk in Problem 6.24 is constant, what are the x and y components of the velocity of A *0.02 s* after the instant shown?

Solution: The clockwise angular velocity is $\omega = \dfrac{d\theta}{dt} = 50\,rad/s$. Thus, we get $\int_0^\theta d\theta = \int_0^t 50\,dt$, and $\theta = 50t\ rad$. At $t = 0.02s$ the clockwise rotation is $\theta = 5(0.02) = 1\ rad = 57.3^\circ$. The position of A in this problem is 57.3° counterclockwise from the position shown in the figure in Problem 6.24. The velocity of A is $\bar{v}_A = \bar{v}_0 + \bar{\omega} \times \bar{r}_{A/0}$, or, expanding, we get

$$\bar{v}_A = 0 + \begin{vmatrix} \bar{i} & \bar{j} & \bar{k} \\ 0 & 0 & -50 \\ 0.1\sin 57.3^\circ & 0.1\cos 57.3^\circ & 0 \end{vmatrix}.$$

Hence, $\bar{v}_A = 2.70\bar{i} - 4.21\bar{j}\,(m/s)$.

================================<>================================

Problem 6.26 The car is moving to the right at 100 *km/hr*, and its tires are 600 *mm* in diameter. (a) What is the angular velocity of its tires? (b) Which point on the tire has the largest velocity relative to a reference frame fixed to the road, and what is the magnitude of that velocity?

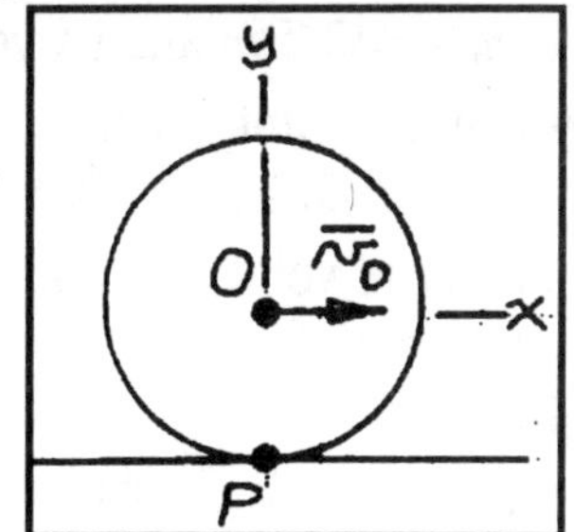

Solution: Choose a coordinate system with the origin O at the hub of the tire and the *x* axis parallel to the direction of travel and *y* upward. (a)The velocity of the point on the rim of the tire in contact with the road is zero. The velocity of the tire hub is $\vec{v}_O = 100\ km/hr = 27.8\ m/s$. The point of contact with the road is stationary, from which

$$\vec{v}_P = 0 = \vec{v}_O + \vec{\omega} \times \vec{r}_{O/P} = 27.8\vec{i} + \begin{bmatrix} \vec{i} & \vec{j} & \vec{k} \\ 0 & 0 & \omega \\ 0 & -R & 0 \end{bmatrix} = 27.8\vec{i} + 0.3\omega\vec{i} = 0,$$ from

which $\boxed{\omega = -\frac{(27.8)}{0.3} = -92.6\ rad/s}$.(b) The radius vector to the upper rim is $\vec{r}_{OU} = +\vec{j}(0.3)\ (m)$. The velocity is $\vec{v}_{\max} = \vec{v}_O + \vec{\omega} \times \vec{r}_{OU} = 27.78\vec{i}\,\omega(0.3)(\vec{k} \times \vec{j}) = 27.8\vec{i} + 0.3\omega \begin{bmatrix} \vec{i} & \vec{j} & \vec{k} \\ 0 & 0 & 1 \\ 0 & 1 & 0 \end{bmatrix} \boxed{= \vec{i}(55.6)\ \ (m/s)}$

================================<>================================

Problem 6.27 The disk rolls on the plane surface. Point A is moving to the right at 6 *ft/s* . (a) What is the angular velocity of the disk? (b) Use Eq (6.6) to determine the x and y components of the velocities of points B, C, and D.

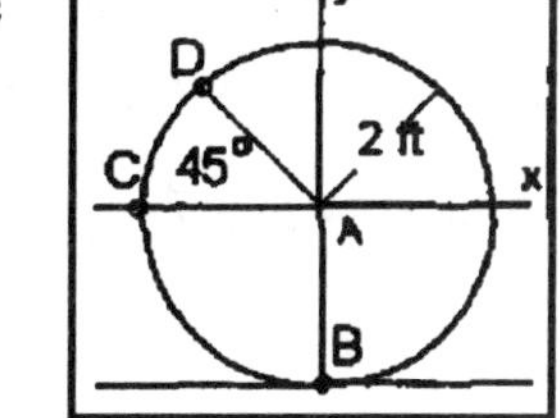

Solution (a) From Eq (6.2), $\omega = \frac{v}{R} = \frac{6}{2} = 3\ rad/s$. The disk rotates in a clockwise direction about the *z* axis, from which $\boxed{\vec{\omega} = -3\vec{k}\ \ (rad/s)}$

(b) The radius vector from A to B is $\vec{r}_{B/A} = -2\vec{j}\ \ (m)$The velocity of the center of the disk A is $\vec{v}_A = 6\vec{i}\ (ft/s)$, from which The velocity of point B is

$$\vec{v}_B = \vec{v}_A + \vec{\omega} \times \vec{r}_{B/A} = 6\vec{i} + 3(2)(\vec{k} \times \vec{j}) = 6\vec{i} + 6\begin{bmatrix} \vec{i} & \vec{j} & \vec{k} \\ 0 & 0 & 1 \\ 0 & 1 & 0 \end{bmatrix}$$ $\boxed{\vec{v}_B = 6\vec{i} - 6\vec{i} = 0}$ The radius vector from A to C is $\vec{r}_{C/A} = -2\vec{i}\ \ (m)$. The velocity of point C is $\vec{v}_C = \vec{v}_A + \vec{\omega} \times \vec{r}_{C/A} = 6\vec{i} + 3(2)(\vec{k} \times \vec{i}) = \begin{bmatrix} \vec{i} & \vec{j} & \vec{k} \\ 0 & 0 & 1 \\ 1 & 0 & 0 \end{bmatrix}$,

$\boxed{\vec{v}_C = 6\vec{i} + 6\vec{j}\ \ (ft/s)}$ The radius vector from A to D is $\vec{r}_{D/A} = 2(-\vec{i}\cos 45^o + \vec{j}\sin 45^o) = \sqrt{2}(-\vec{i} + \vec{j})$. The velocity of point D is

$$\vec{v}_D = \vec{v}_A + \vec{\omega} \times \vec{r}_{D/A} = 6\vec{i} - 3(\sqrt{2})(\vec{k} \times (-\vec{i} + \vec{j})) = 6\vec{i} - 3\sqrt{2}\begin{bmatrix} \vec{i} & \vec{j} & \vec{k} \\ 0 & 0 & 1 \\ -1 & 1 & 0 \end{bmatrix} = 6\vec{i} + 3\sqrt{2}\vec{i} + 3\sqrt{2}\vec{j},$$

$\boxed{\vec{v}_D = 10.24\vec{i} + 4.24\vec{j}\ \ (ft/s)}$

================================<>================================

Problem 6.28 The helicopter is in planar motion in the *x-y* plane. At the instant shown, the position of its center of mass *G* is $x = 2\ m$, $y = 2.5\ m$ and its velocity is $\mathbf{v}_G = 12\,\vec{i} + 4\,\vec{j}\ (m/s)$. The position of point *T* where the tail rotor is mounted is $x = -3.5\ m$, $y = 4.5\ m$. The helicopter's angular veolcity is 0.2 (rad/s) clockwise. What is the velocity of point *T*?

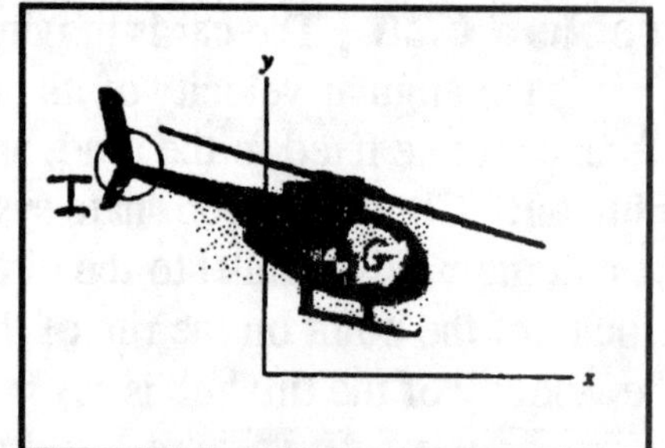

Solution:

The position of T relative to G is $\bar{r}_{T/G} = (-3.5-2)\bar{i} + (4.5-2.5)\bar{j} = -5.5\bar{i} + 2\bar{j}\ (m)$.

The velocity of T is $\bar{v}_T = \bar{v}_G + \bar{\omega} \times \bar{r}_{T/G} = 12\,\bar{i} + 4\,\bar{j} + \begin{vmatrix} \bar{i} & \bar{j} & \bar{k} \\ 0 & 0 & -0.2 \\ -5.5 & 2 & 0 \end{vmatrix} = 12.4\bar{i} + 5.1\bar{j}(m/s)$

Problem 6.29 The bar is in two dimensional motion in the *x-y* plane. The velocity of point A is $8\vec{i}$ *ft/s*. The x component of the velocity of point B is 6 *ft/s*. (a) What is the angular velocity vector of the bar? (b) What is the velocity of point B?

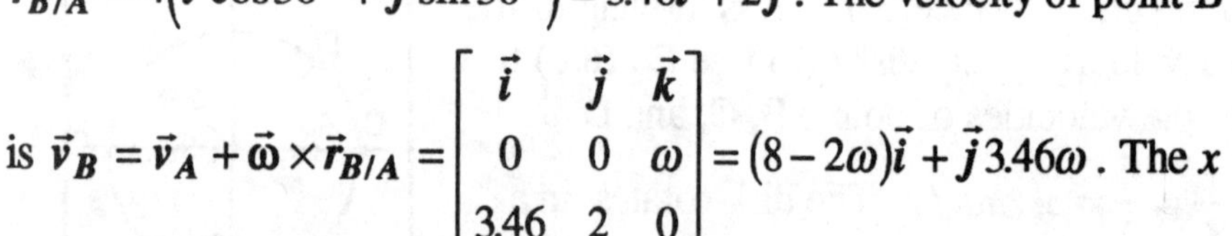

Solution (a) The vector from A to B is $\vec{r}_{B/A} = 4\left(\vec{i}\cos 30^o + \vec{j}\sin 30^o\right) = 3.46\vec{i} + 2\vec{j}$. The velocity of point B

is $\vec{v}_B = \vec{v}_A + \vec{\omega} \times \vec{r}_{B/A} = \begin{bmatrix} \vec{i} & \vec{j} & \vec{k} \\ 0 & 0 & \omega \\ 3.46 & 2 & 0 \end{bmatrix} = (8 - 2\omega)\vec{i} + \vec{j}3.46\omega$. The *x*

component is known, from which $8 - 2\omega = 6$, $\omega = 1\ rad/s$, and $\boxed{\vec{\omega} = 1\vec{k}\ \ rad/s}$. (b) From Part (a), the velocity of point B is $\boxed{\vec{v}_B = 6\vec{i} + 3.46\vec{j}\ \ ft/s}$

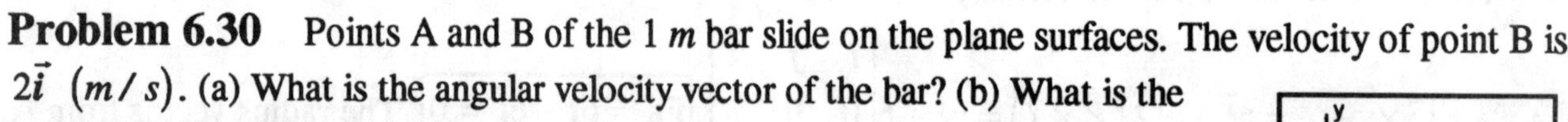

Problem 6.30 Points A and B of the 1 *m* bar slide on the plane surfaces. The velocity of point B is $2\vec{i}\ \ (m/s)$. (a) What is the angular velocity vector of the bar? (b) What is the velocity of Point A?

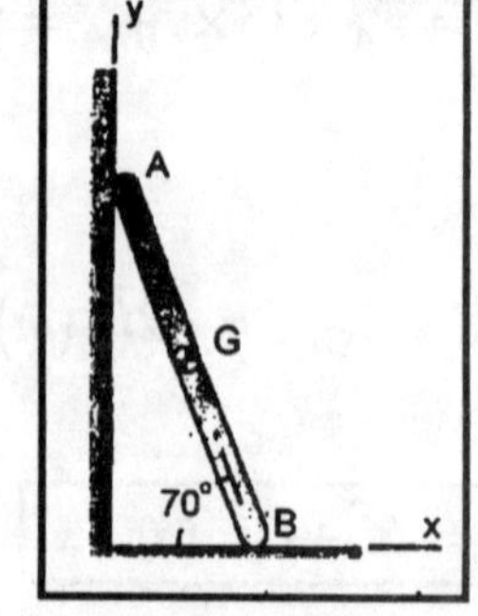

Solution (a) The surfaces constrain the motion of the point at A to motion parallel to the *y* axis. . The vector $\vec{r}_{A/B} = 1\left(-\vec{i}\cos 70^o + \vec{j}\sin 70^o\right) = -0.342\vec{i} + 0.940\vec{j}$. The velocity of point A is

$$\vec{v}_A = \vec{v}_B + \vec{\omega} \times \vec{r}_{A/B} = 2\vec{i} + \begin{bmatrix} \vec{i} & \vec{j} & \vec{k} \\ 0 & 0 & \omega \\ -3.42 & 0.940 & 0 \end{bmatrix} = (-0.940\omega + 2)\vec{i} - 3.42\omega\vec{j}.$$

From the constraint on the motion, $0 = (2 - 0.940\omega)\vec{i}$, from which $\omega = \dfrac{2}{0.940} = 2.13\ rad/s$, and $\boxed{\vec{\omega} = 2.13\vec{k}\ \ rad/s}$. (b) The velocity of point A, from Part(a), $\boxed{\vec{v}_A = -0.342\omega\vec{j} = -0.728\vec{j}\ \ (m/s)}$

==================================<>==================================

Problem 6.31 In Problem 6.30, what is the velocity of the midpoint G of the bar?

Solution:

From Eqs (6.6), we have

$$\vec{v}_G = \vec{v}_B + \vec{w} \times \vec{r}_{G/B} = 2\vec{i} + \frac{1}{2}\begin{bmatrix} \vec{i} & \vec{j} & \vec{k} \\ 0 & 0 & 2.13 \\ 0.342 & 9.40 & 0 \end{bmatrix}, \text{ from which } \vec{v}_G = 1\vec{i} - 0.364\vec{j} \quad (\mathrm{m/s})$$

==================================<>==================================

Problem 6.32 If $\theta = 45°$ and bar OQ is rotating in the counterclockwise direction at *0.2 rad/s*, what is the velocity of the sleeve P?

Solution:

The velocity of Q is $\bar{v}_Q = \bar{v}_0 + \bar{\omega}_{0Q} \times \bar{r}_{Q/0}$. Expanding, we get

$$\bar{v}_Q = 0 + \begin{vmatrix} \bar{i} & \bar{j} & \bar{k} \\ 0 & 0 & 0.2 \\ 2\cos 45° & 2\sin 45° & 0 \end{vmatrix}, \text{ or}$$

$\bar{v}_Q = -0.4\cos 45°\bar{i} + 0.4\cos 45°\bar{j}(ft/s)$.

The velocity of P is $v_P\bar{i} = \bar{v}_Q + \bar{\omega}_{QP} \times \bar{r}_{P/Q}$,

$$\text{or } v_P\bar{i} = -0.4\cos 45°\bar{i} + 0.4\cos 45°\bar{j} + \begin{vmatrix} \bar{i} & \bar{j} & \bar{k} \\ 0 & 0 & \omega_{QP} \\ 2\cos 45° & -2\sin 45° & 0 \end{vmatrix}.$$

Equating $\bar{i}$ and $\bar{j}$ components, $v_P = -0.4\cos 45° + 2\omega_{QP}\cos 45°$, and $0 = 0.4\cos 45° + 2\omega_{QP}\cos 45°$. Solving, we obtain $\omega_{QP} = -0.2 rad/s$ and $v_P = -0.566 ft/s$

==================================<>==================================

Problem 6.33 Consider the system shown in Problem 6.32. If $\theta = 40°$, and the sleeve P is moving to the right at *1.0 ft/s,* what are the angular velocities of the bars *OQ* and *PQ* ?

Solution:

$$\text{The velocity of } Q \text{ is } \bar{v}_Q = \bar{v}_0 + \bar{\omega}_{0Q} \times \bar{r}_{Q/0} = 0 + \begin{vmatrix} \bar{i} & \bar{j} & \bar{k} \\ 0 & 0 & -\omega_{0Q} \\ 2\cos 40° & 2\sin 40° & 0 \end{vmatrix}, \text{ or}$$

$\bar{v}_Q = 2\omega_{0Q}\sin 40°\bar{i} - 2\omega_{0Q}\cos 40°\bar{j}$. The velocity of P is $\bar{v}_P = (1)\bar{i} = \bar{v}_Q + \bar{\omega}_{QP} \times \bar{r}_{P/Q}$. Expanding, we

$$\text{get } \bar{v}_P = 2\omega_{0Q}\sin 40°\bar{i} - 2\omega_{0Q}\cos 40°\bar{j} + \begin{vmatrix} \bar{i} & \bar{j} & \bar{k} \\ 0 & 0 & +\omega_{QP} \\ 2\cos 40° & -2\sin 40° & 0 \end{vmatrix}. \text{ Equating } \bar{i} \text{ and } \bar{j}$$

components, we get $1 = 2\omega_{0Q}\sin 40° + 2\omega_{QP}\sin 40°$ and $0 = -2\omega_{0Q}\cos 40° + 2\omega_{QP}\cos 40°$. Solving, we obtain $\omega_{0Q} = \omega_{QP} = 0.389$ *rad/s*, where $\bar{\omega}_{0Q}$ is clockwise and $\bar{\omega}_{QP}$ is counterclockwise.

==================================<>==================================

Problem 6.34 The bar AB rotates in the counterclockwise direction at 6 *rad/s*. Determine the angular velocity of bar BCD and the velocity of point D.

Solution The constraint imposed at C limits the motion at C to translation parallel to the *x* axis. The point A is fixed.

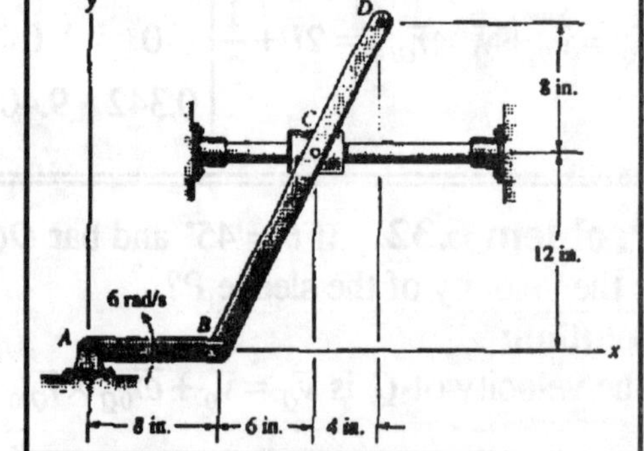

The velocity at B: . The vector from A to B: $\vec{r}_{B/A} = 8\vec{i}$. The

velocity at B: $\vec{v}_B = \vec{v}_A + \vec{\omega} \times \vec{r}_{B/A} = \begin{bmatrix} \vec{i} & \vec{j} & \vec{k} \\ 0 & 0 & 6 \\ 8 & 0 & 0 \end{bmatrix} = 48\vec{j} \ \ (in/s)$.

The velocity at C: The vector BC is $\vec{r}_{C/B} = 6\vec{i} + 12\vec{j} \ \ (in.)$

The velocity at C is

$$\vec{v}_C = \vec{v}_B + \vec{\omega}_{BCD} \times \vec{r}_{C/B} = 48\vec{j} + \begin{bmatrix} \vec{i} & \vec{j} & \vec{k} \\ 0 & 0 & \omega_{BCD} \\ 6 & 12 & 0 \end{bmatrix}, \text{ from which}$$

$\vec{v}_C = -12\omega_{BCD}\vec{i} + (48 + 6\omega_{BCD})\vec{j}$. Since C is constrained to motion parallel to the *x* axis, $(48 + 6\omega_{BCD})\vec{j} = 0$, from which $\boxed{\omega_{BCD} = -8 \ \ rad/s}$.(clockwise).

The velocity at point D: The vector from C to D is $\vec{r}_{D/C} = 10\vec{i} + 20\vec{j} \ \ (in.)$. The velocity of point D is

$$\vec{v}_D = \vec{v}_B + \vec{\omega}_{BCD} \times \vec{r}_{D/B} = 48\vec{j} + \begin{bmatrix} \vec{i} & \vec{j} & \vec{k} \\ 0 & 0 & -8 \\ 10 & 20 & 0 \end{bmatrix} = 48\vec{j} + 160\vec{i} - 80\vec{j}, \text{ from which}$$

$\boxed{\vec{v}_D = 160\vec{i} - 32\vec{j} \ \ (in/s)}$

Problem 6.35 If the crankshaft AB rotates at 6000 *rpm* (revolutions per minute) in the counterclockwise direction, what is the velocity of the piston at the instant shown?

Solution The angle between the crank and the vertical is 40°. The piston is constrained to move in parallel to the y-axis. The angular velocity of the crank is: $\omega_{AB} = \dfrac{6000(2\pi)}{60} = 200\pi \ \ rad/s$. The radius vector of the crankshaft is

$\vec{r}_{B/A} = 36\left(-\vec{i}\sin 40^o + \vec{j}\cos 40^o\right) = -23.1\vec{i} + 27.6\vec{j} \ \ (mm)$. The velocity of the end of the crank is

$$\vec{v}_B = \vec{v}_A + \vec{\omega}_{AB} \times \vec{r}_{B/A} = 0 + \begin{bmatrix} \vec{i} & \vec{j} & \vec{k} \\ 0 & 0 & 200\pi \\ -23.1 & 27.6 & 0 \end{bmatrix}, \text{ from which}$$

$\vec{v}_B = -17327.5\vec{i} - 14539.5\vec{j} \ \ (mm/s)$. The angle of the connecting rod with the horizontal is

$$\theta = 90^o - \sin^{-1}\left(\frac{36\sin 40^o}{125}\right) = 79.33^o.$$

Solution continued on next page

The vector distance from B to C is $\vec{r}_{C/B} = 125(\vec{i}\cos\theta + \vec{j}\sin\theta) = 23.1\vec{i} + 122.8\vec{j}$ (mm). The velocity of point C is $\vec{v}_C = \vec{v}_B + \vec{\omega}_{BC} \times \vec{r}_{BC} = \vec{v}_B + \begin{bmatrix} \vec{i} & \vec{j} & \vec{k} \\ 0 & 0 & \omega_{BC} \\ 23.1 & 122.8 & 0 \end{bmatrix} = \vec{v}_B - 122.8\omega_{BC}\vec{i} + 23.1\omega_{BC}\vec{j}$, from which. $\vec{v}_C = (-17327.5 - \omega_{BC}122.8)\vec{i} + (-14539.3 + \omega_{BC}23.1)\vec{j}$. From the constraint on the piston's motion, $0 = (-17327.5 - \omega_{BC}122.8)\vec{i}$, from which $\omega_{BC} = -\dfrac{17327.5}{122.8} = -141.1$ (rad/s). Substitute:

$$\boxed{\vec{v}_C = (-14539.5 - 141.1(23.1))\vec{j} = -17798\vec{j} \quad (mm/s)}$$

=================================<>=================================

Problem 6.36 Bar AB rotates at 10 *rad/s* in the counterclockwise direction. Determine the angular velocity of bar CD.

Strategy: Since you know the angular velocity of the bar AB, you can determine the velocity of B. Then apply Eq (6.6) to points B and C to obtain an equation for $\vec{v}_C$ in terms of the angular velocity of the bar BC, then apply it to points C and D to obtain an equation for $\vec{v}_C$ in terms of the angular velocity of the bar CD. By equating the two expressions, you will obtain a vector equation in two unknowns: the angular velocities of bars BC and CD.

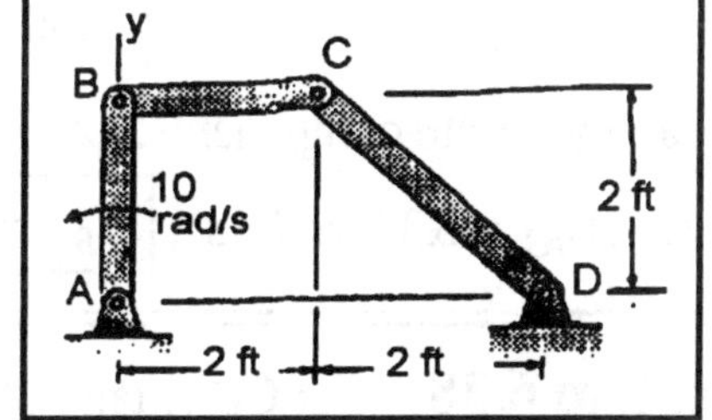

Solution:

The velocity of point B is $\vec{v}_B = \begin{bmatrix} \vec{i} & \vec{j} & \vec{k} \\ 0 & 0 & \omega_{AB} \\ 0 & 2 & 0 \end{bmatrix} = -20\vec{i}$ (ft/s)

The vector BC is $\vec{r}_{C/B} = 2\vec{i}$ (ft). The vector DC is $\vec{r}_{C/D} = -2\vec{i} + 2\vec{j}$ (ft). The velocity of point C is $\vec{v}_C = \vec{v}_B + \vec{\omega}_{BC} \times \vec{r}_{C/B}$. $\vec{v}_C = \vec{v}_B + \vec{\omega} \times \vec{r}_{C/B} = \begin{bmatrix} \vec{i} & \vec{j} & \vec{k} \\ 0 & 0 & \omega_{BC} \\ 0 & 2 & 0 \end{bmatrix} = -20\vec{i} + 2\omega_{BC}\vec{j}$ (ft/s) Similarly, since D is a fixed point: $\vec{v}_C = \vec{\omega}_{CD} \times \vec{r}_{C/D} = \begin{bmatrix} \vec{i} & \vec{j} & \vec{k} \\ 0 & 0 & \omega_{CD} \\ -2 & 2 & 0 \end{bmatrix} = -2\omega_{CD}(\vec{i} + \vec{j})$. Equating the two expressions for the velocity of point C and separating components: $(-20 + 2\omega_{CD})\vec{i} = 0$, and $(2\omega_{BC} + 2\omega_{CD})\vec{j} = 0$. Solve: $\omega_{CD} = 10$ rad/s, $\boxed{\vec{\omega}_{CD} = 10\vec{k} \ (rad/s)}$ and $\omega_{BC} = -\omega_{CD}$ rad/s.

Check: From inspection, the motion of B to the left should cause C to move to the left and down, producing a clockwise rotation of bar BC. Similarly, the motion of C to the left and down should cause bar CD to rotate counterclockwise. Thus the solution satisfies minimal reasonableness tests. *check.*

=================================<>=================================

Problem 6.37 Bar AB rotates at 12 *rad/s* in the clockwise direction. Determine the angular velocities of bars BC and CD.

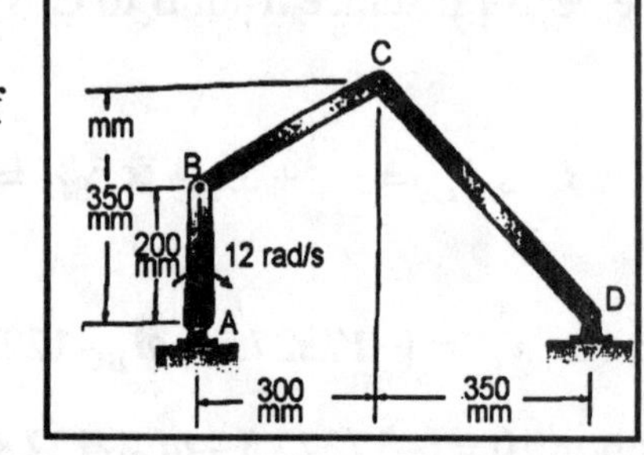

Solution The strategy is analogous to that used in Problem 6.28. The radius vector AB is $\vec{r}_{B/A} = 200\vec{j}\ \ (mm)$. The angular velocity of AB is $\vec{\omega} = -12\vec{k}\ \ (rad/s)$. The velocity of point B is

$$\vec{v}_B = \vec{v}_A + \vec{\omega}\times\vec{r}_{B/A} = 0 + \begin{bmatrix} \vec{i} & \vec{j} & \vec{k} \\ 0 & 0 & -12 \\ 0 & 200 & 0 \end{bmatrix} = 2400\vec{i}\ \ (mm/s).$$

The radius vector BC is $\vec{r}_{C/B} = 300\vec{i} + (350-200)\vec{j} = 300\vec{i} + 150\vec{j}\ \ (mm)$. The velocity of point C is

$$\vec{v}_C = \vec{v}_B + \vec{\omega}_{BC}\times\vec{r}_{C/B} = \vec{v}_B + \begin{bmatrix} \vec{i} & \vec{j} & \vec{k} \\ 0 & 0 & \omega_{BC} \\ 300 & 150 & 0 \end{bmatrix} = (2400 - 150\omega_{BC})\vec{i} + \omega_{BC}300\vec{j}\ \ (mm/s).$$

The radius vector DC is $\vec{r}_{C/D} = -350\vec{i} + 350\vec{j}\ \ (mm)$. The velocity of point C is

$$\vec{v}_C = \vec{v}_D + \vec{\omega}_{CD}\times\vec{r}_{C/D} = 0 + \begin{bmatrix} \vec{i} & \vec{j} & \vec{k} \\ 0 & 0 & \omega_{CD} \\ -350 & 350 & 0 \end{bmatrix} = -350\omega_{CD}\left(\vec{i} + \vec{j}\right).$$

Equate the two expressions for $\vec{v}_C$, and separate components: $(2400 - 150\omega_{BC} + 350\omega_{CD})\vec{i} = 0$, and $(300\omega_{BC} + 350\omega_{CD})\vec{j} = 0$. Solve: $\omega_{BC} = 5.33\ \ rad/s$, $\boxed{\vec{\omega}_{BC} = 5.33\vec{k}\ \ (rad/s)}$. $\omega_{CD} = -4.57\ \ rad/s$, $\boxed{\vec{\omega} = -4.57\vec{k}\ \ (rad/s)}$

Problem 6.38 Bar CD rotates at 2 *rad/s* in the clockwise direction. Determine the angular velocities of bars AB and BC.

Solution The strategy is analogous to that used in Problem 6.36, except that the computation is started with bar CD. Denote the length of CD by $L_{CD} = \dfrac{10\sin 45^o}{\sin 30^o} = \sqrt{2}(10) = 14.14\ in.$. The radius vector DC is

$\vec{r}_{C/D} = L_{CD}\left(-\vec{i}\cos 30^o + \vec{j}\sin 30^o\right) = \left(-12.25\vec{i} + 7.07\vec{j}\right)\ (in.)$. The angular velocity of CD is $\vec{\omega}_{CD} = -2\vec{k}\ \ (rad/s)$. The velocity of point C is

$$\vec{v}_C = \vec{\omega}_{CD}\times\vec{r}_{C/D} = \begin{bmatrix} \vec{i} & \vec{j} & \vec{k} \\ 0 & 0 & -2 \\ -12.25 & 7.07 & 0 \end{bmatrix} = \left(14.14\vec{i} + 24.49\vec{j}\right)\ (in/s).$$

The radius vector BC is $\vec{r}_{C/B} = 12\vec{i}\ \ (in.)$. The velocity of point C is

$$\vec{v}_C = \vec{v}_B + \vec{\omega}_{BC}\times\vec{r}_{C/B} = \vec{v}_B + \begin{bmatrix} \vec{i} & \vec{j} & \vec{k} \\ 0 & 0 & \omega_{BC} \\ 12 & 0 & 0 \end{bmatrix} = \vec{v}_B + 12\omega_{BC}\vec{j}\ \ (in/s).$$

Solution continued on next page

The radius vector AB is $\vec{r}_{B/A} = \frac{10}{\sqrt{2}}(\vec{i}+\vec{j})$ (*in.*). The velocity of point B is

$$\vec{v}_B = \vec{\omega}\times\vec{r}_{B/A} = \begin{bmatrix} \vec{i} & \vec{j} & \vec{k} \\ 0 & 0 & \omega_{AB} \\ \frac{10}{\sqrt{2}} & \frac{10}{\sqrt{2}} & 0 \end{bmatrix} = \left(\frac{10}{\sqrt{2}}\right)\omega_{AB}(-\vec{i}+\vec{j})\ (in/s).$$ Substitute:

$\vec{v}_C = \frac{10}{\sqrt{2}}\omega_{AB}(-\vec{i}+\vec{j}) + 12\omega_{BC}\vec{j}$. Equate the two expressions for $\vec{v}_C$ and separate components:

$0 = \left(14.14 + \frac{10}{\sqrt{2}}\omega_{AB}\right)\vec{i}$, $0 = \left(24.49 - \frac{10}{\sqrt{2}}\omega_{AB} - 12\omega_{BC}\right)\vec{j}$. Solve: $\omega_{AB} = -2\ rad/s$,

$\boxed{\vec{\omega}_{AB} = -2\vec{k}\ (rad/s)}$. $\omega_{BC} = 3.22\ rad/s$, $\boxed{\vec{\omega}_{BC} = 3.22\vec{k}\ (rad/s)}$

==================================<>==================================

Problem 6.39 In Problem 6.38, what is the magnitude of the velocity of the midpoint G of the bar BC?

Solution Use the solution to Problem 6.38. The radius vector from C to G is $\vec{r}_{G/C} = -6\vec{i}$. The velocity of point G is $\vec{v}_G = \vec{v}_C + \vec{\omega}_{BC}\times\vec{r}_{G/C}$. From the solution to Problem 6.38,

$\vec{v}_C = (14.14\vec{i} + 24.49\vec{j})$ (*in./s*), and $\vec{\omega}_{BC} = 3.22\vec{k}\ (rad/s)$, from which

$$\vec{v}_G = 14.14\vec{i} + 24.49\vec{j} + \begin{bmatrix} \vec{i} & \vec{j} & \vec{k} \\ 0 & 0 & 3.22 \\ -6 & 0 & 0 \end{bmatrix} = 14.14\vec{i} + 5.18\vec{j}\ (in/s).$$ The magnitude is

$$\boxed{|\vec{v}_G| = \sqrt{14.14^2 + 5.18^2} = 15.0\ in/s}$$

==================================<>==================================

Problem 6.40 Bar AB rotates at 10 *rad/s* in the counterclockwise direction. Determine the velocity of point E.

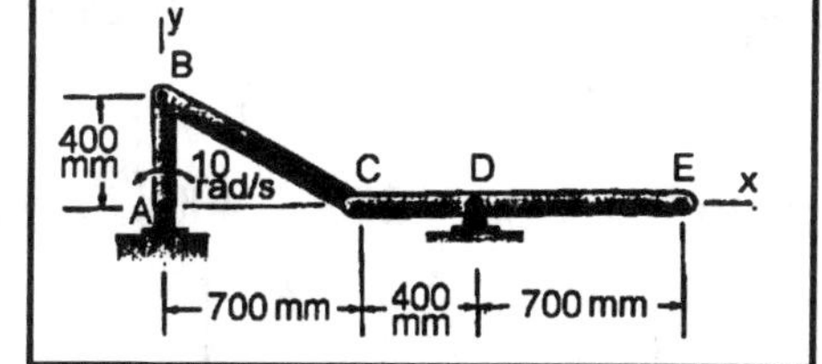

Solution The strategy is analogous to that used in Problem 6.28. The radius vector AB is $\vec{r}_{B/A} = 400\vec{j}\ (mm)$. The angular velocity of bar AB is $\vec{\omega}_{AB} = 10\vec{k}\ (rad/s)$. The velocity of point B is

$$\vec{v}_B = \vec{\omega}_{AB}\times\vec{r}_{AB} = \begin{bmatrix} \vec{i} & \vec{j} & \vec{k} \\ 0 & 0 & 10 \\ 0 & 400 & 0 \end{bmatrix} = -4000\vec{i}\ (mm/s).$$ The radius vector BC is

$\vec{r}_{C/B} = 700\vec{i} - 400\vec{j}\ (mm)$. The velocity of point C is

$$\vec{v}_C = \vec{v}_B + \vec{\omega}_{BC}\times\vec{r}_{C/B} = -4000\vec{i} + \begin{bmatrix} \vec{i} & \vec{j} & \vec{k} \\ 0 & 0 & \omega_{BC} \\ 700 & -400 & 0 \end{bmatrix} = (-4000 + 400\omega_{BC})\vec{i} + 700\omega_{BC}\vec{j}.$$

Solution continued on next page

The radius vector CD is $\vec{r}_{C/D} = -400\vec{i}\ \ (mm)$. The point D is fixed (cannot translate). The velocity at point C is $\vec{v}_C = \vec{\omega}_{CD} \times \vec{r}_{C/D} = \omega_{CD}\left(\vec{k} \times (-400\vec{i})\right) \begin{bmatrix} \vec{i} & \vec{j} & \vec{k} \\ 0 & 0 & \omega_{CD} \\ -400 & 0 & 0 \end{bmatrix} = -400\omega_{CD}\vec{j}$. Equate the two expressions for the velocity at point C, and separate components: $0 = (-4000 + 400\omega_{BC})\vec{i}$, $0 = (700\omega_{BC} + 400\omega_{CD})\vec{j}$. Solve: $\omega_{BC} = 10\ rad/s$, $\omega_{CD} = -17.5\ \ rad/s$.

The radius vector DE is $\vec{r}_{D/E} = 700\vec{i}\ \ (mm)$. The velocity of point E is

$$\vec{v}_E = \vec{\omega}_{CD} \times \vec{r}_{D/E} = \begin{bmatrix} \vec{i} & \vec{j} & \vec{k} \\ 0 & 0 & -17.5 \\ 700 & 0 & 0 \end{bmatrix} \boxed{\vec{v}_E = -12250\vec{j}\ \ (mm/s)}$$

==================================<>==================================

Problem 6.41 Bar AB rotates at 4 *rad/s* in the counterclockwise direction. Determine the velocity of point C.

Solution The strategy is analogous to that used in Problem 6.36. The angular velocity of bar AB is $\vec{\omega} = 4\vec{k}\ \ (rad/s)$. The radius vector AB is $\vec{r}_{B/A} = 300\vec{i} + 600\vec{j}\ \ (mm)$. The velocity of point B is

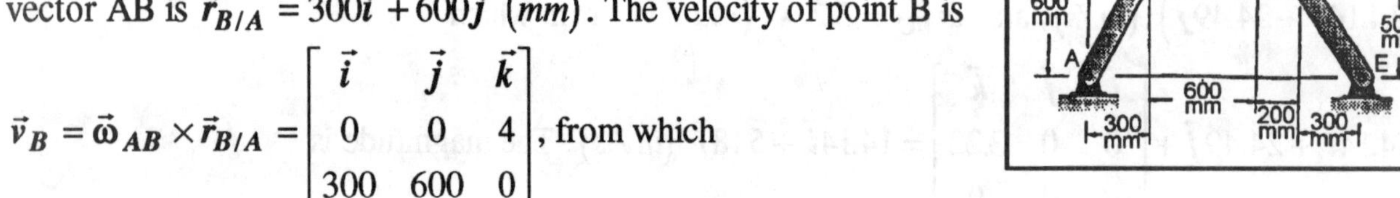

$$\vec{v}_B = \vec{\omega}_{AB} \times \vec{r}_{B/A} = \begin{bmatrix} \vec{i} & \vec{j} & \vec{k} \\ 0 & 0 & 4 \\ 300 & 600 & 0 \end{bmatrix}, \text{ from which}$$

$\vec{v}_B = -2400\vec{i} + 1200\vec{j}\ \ (mm/s)$. The vector radius from B to C is $\vec{r}_{C/B} = 600\vec{i} + (900 - 600)\vec{j} = 600\vec{i} + 300\vec{j}\ \ (mm)$. The velocity of point C is

$$\vec{v}_C = \vec{v}_B + \begin{bmatrix} \vec{i} & \vec{j} & \vec{k} \\ 0 & 0 & \omega_{BC} \\ 600 & 300 & 0 \end{bmatrix} = (-2400 - 300\omega_{BC})\vec{i} + (1200 + 600\omega_{BC})\vec{j}\ \ (mm/s).$$

The radius vector from C to D is $\vec{r}_{D/C} = 200\vec{i} - 400\vec{j}\ \ (mm)$. The velocity of point D is

$$\vec{v}_D = \vec{v}_C + \begin{bmatrix} \vec{i} & \vec{j} & \vec{k} \\ 0 & 0 & \omega_{BC} \\ 200 & -400 & 0 \end{bmatrix} = \vec{v}_C + 400\omega_{BC}\vec{i} + 200\omega_{BC}\vec{j}\ \ (mm/s).$$

The radius vector from E to D is $\vec{r}_{D/E} = -300\vec{i} + 500\vec{j}\ \ (mm)$. The velocity of point D is

$$\vec{v}_D = \vec{\omega}_{DE} \times \vec{r}_{D/E} = \begin{bmatrix} \vec{i} & \vec{j} & \vec{k} \\ 0 & 0 & \omega_{DE} \\ -300 & 500 & 0 \end{bmatrix} = -500\omega_{DE}\vec{i} - 300\omega_{DE}\vec{j}\ \ (mm/s).$$

Equate the expressions for the velocity of point D; solve for $\vec{v}_C$, to obtain one of two expressions for the velocity of point C.

Solution continued on next page

Equate the two expressions for $\vec{v}_C$, and separate components: $0=(-500\omega_{DE}-100\omega_{BC}+2400)\vec{i}$, $0=(1200+300\omega_{DE}+800\omega_{BC})\vec{j}$. Solve $\omega_{DE}=5.51\ rad/s$, $\omega_{BC}=-3.57\ rad/s$. Substitute into the expression for the velocity of point C to obtain $\boxed{\vec{v}_C=-1330\vec{i}-941\vec{j}\ \ (mm/s)}$

==================================<>==================================

Problem 6.42 In the system shown in Problem 6.41, if the magnitude of the velocity of point C is $|\vec{v}_C|=2\ m/s$, what are the magnitudes of the angular velocities of bars AB and DE?

Solution

Use the solution to Problem 6.41.

The velocity of point B is $\vec{v}_B=\vec{\omega}_{AB}\times\vec{r}_{B/A}=\begin{bmatrix}\vec{i} & \vec{j} & \vec{k}\\ 0 & 0 & \omega_{AB}\\ 300 & 600 & 0\end{bmatrix}=-600\omega_{AB}\vec{i}+300\omega_{AB}\vec{j}$.

The velocity of point C is

$$\vec{v}_C=\vec{v}_B+\vec{\omega}_{BC}\times\vec{r}_{C/B}=\begin{bmatrix}\vec{i} & \vec{j} & \vec{k}\\ 0 & 0 & \omega_{BC}\\ 600 & 300 & 0\end{bmatrix}=(-600\omega_{AB}-300\omega_{BC})\vec{i}+(300\omega_{AB}+600\omega_{BC})\vec{j}.$$

Also, $\vec{v}_C=\vec{v}_D+\vec{\omega}\times\vec{r}_{C/D}=\begin{bmatrix}\vec{i} & \vec{j} & \vec{k}\\ 0 & 0 & \omega_{BC}\\ -200 & +400 & 0\end{bmatrix}=\vec{v}_D-400\omega_{BC}\vec{i}-200\omega_{BC}\vec{j}\ \ (mm/s)$, and,

$$\vec{v}_D=\vec{v}_E+\vec{\omega}_{DE}\times\vec{r}_{D/E}=\begin{bmatrix}\vec{i} & \vec{j} & \vec{k}\\ 0 & 0 & \omega_{DE}\\ -300 & 500 & 0\end{bmatrix}=-500\omega_{DE}\vec{i}-300\omega_{DE}\vec{j}\ \ (mm/s).$$

Substitute into the expression for the velocity of point C:

$\vec{v}_C=(-500\omega_{DE}-400\omega_{BC})\vec{i}+(-300\omega_{DE}-200\omega_{BC})\vec{j}$.

The three equations in three unknowns are obtained by taking the magnitude of $\vec{v}_C$, and by equating the two expressions for $\vec{v}_C$ and then separating the components. The results:

$4\times10^6=(600\omega_{AB}+300\omega_{BC})^2+(300\omega_{AB}+600\omega_{BC})^2$, $0=-500\omega_{DE}+600\omega_{AB}-100\omega_{BC}$, $0=+300\omega_{DE}+800\omega_{BC}+300\omega_{AB}$. These equations are conveniently solved by direct iteration, using **TK Solver Plus** (or any equivalent commercial software package). The results are:

$\boxed{\omega_{DE}=6.77\ \ rad/s}$, $\boxed{\omega_{AB}=4.91\ \ rad/s}$, and $\omega_{BC}=-4.38\ \ rad/s$.

[*Check:* To overcome any risks associated with a "brute force" iteration of a non-linear system, an alternate method of solution is as follows: (a) assume a value of ω_{DE}, (b) solve the second and third equations for ω_{AB} and ω_{BC}, and (c) graph the equation

$f(\omega_{AB},\omega_{BC})=-4\times10^6+(600\omega_{AB}+300\omega_{BC})^2+(300\omega_{AB}+200\omega_{BC})^2$ to find the zero crossing. The results are identical with those obtained by direct iteration. *check.*]

==================================<>==================================

==================================<>==================================

Problem 6.43 The horizontal member *ADE* supporting the scoop is stationary. If the link *BD* is rotating in the clockwise direction at *1 rad/s*, what is the angular velocity of the scoop?

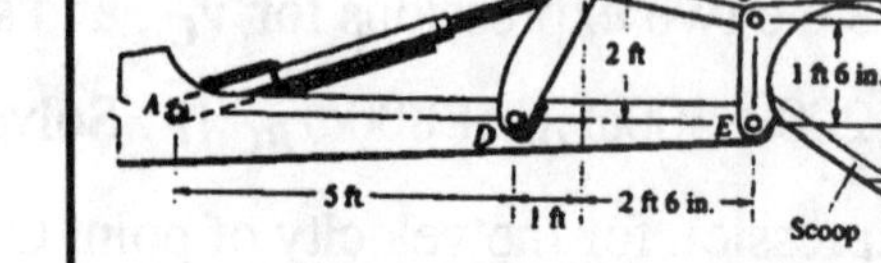

Solution:

The velocity of B is $\vec{v}_B = \vec{v}_D + \vec{\omega}_{BD} \times \vec{r}_{B/D}$. Expanding, we get

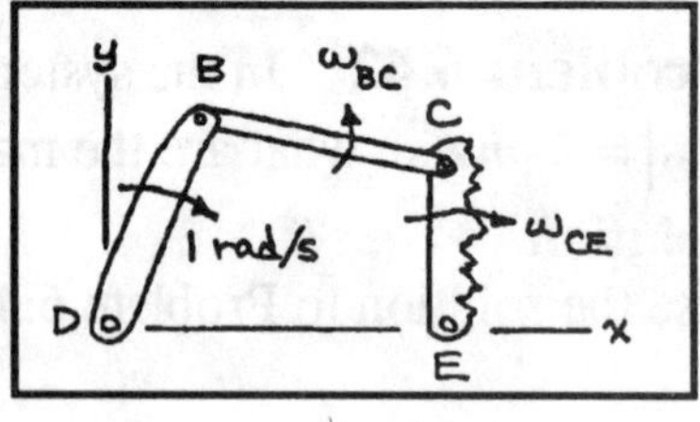

$$\vec{v}_B = 0 + \begin{vmatrix} \bar{i} & \bar{j} & \bar{k} \\ 0 & 0 & -1 \\ 1 & 2 & 0 \end{vmatrix} = 2\bar{i} - 1\bar{j} \ (ft/s)$$ The velocity of C is

$$\vec{v}_C = \vec{v}_B + \vec{\omega}_{BC} \times \vec{r}_{C/B} = 2\bar{i} - 1\bar{j} + \begin{vmatrix} \bar{i} & \bar{j} & \bar{k} \\ 0 & 0 & +\omega_{BC} \\ 2.5 & -0.5 & 0 \end{vmatrix} \quad (1).$$

We can also express the velocity of C as

$$\vec{v}_C = \vec{v}_E + \vec{\omega}_{CE} \times \vec{r}_{C/E} \text{ or } \vec{v}_C = 0 + \begin{vmatrix} \bar{i} & \bar{j} & \bar{k} \\ 0 & 0 & +\omega_{CE} \\ 0 & 1.5 & 0 \end{vmatrix} \quad (2).$$ Equating $\bar{i}$ and $\bar{j}$ components in

Equations (1) and (2) and solving, we obtain $\omega_{BC} = 0.4 rad/s$ and $\omega_{CE} = -1.20 rad/s$.

==================================<>==================================

Problem 6.44 The diameter of the disk is 1 *m*, and the length of the bar AB is 1 *m*. The disk is rolling, and point B slides on the plane surface. Determine the angular velocity of the bar AB and the velocity of point B.

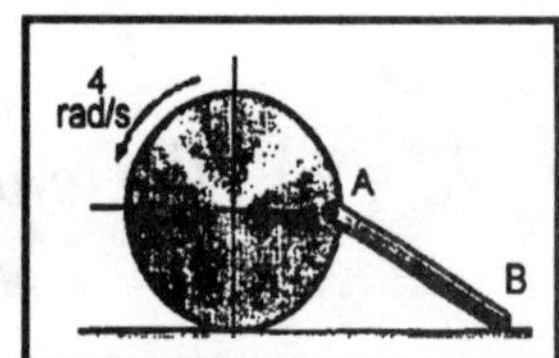

Solution:

Choose a coordinate system with the origin at O, the center of the disk, with *x* axis parallel to the horizontal surface. The point P of contact with the surface is stationary, from which

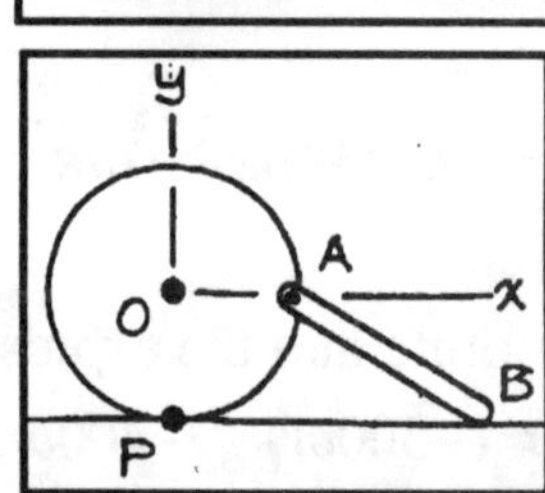

$$\vec{v}_P = 0 = \vec{v}_O + \vec{\mathbf{w}}_O \times -\vec{R} = \vec{v}_O + \begin{bmatrix} \vec{i} & \vec{j} & \vec{k} \\ 0 & 0 & \omega_O \\ 0 & -0.5 & 0 \end{bmatrix} = \vec{v}_O + 2\vec{i},$$

from which $\vec{v}_O = -2\vec{i}$ (m/s). The velocity at A is $\vec{v}_A = \vec{v}_O + \vec{\mathbf{w}}_O \times \vec{r}_{A/O}$.

$$\vec{v}_A = -2\vec{i} + \begin{bmatrix} \vec{i} & \vec{j} & \vec{k} \\ 0 & 0 & \omega_O \\ 0.5 & 0 & 0 \end{bmatrix} = -2\vec{i} + 2\vec{j} \ (\text{m/s}).$$ The vector from B to A is $\vec{r}_{A/B} = -\vec{i}\cos\theta + \vec{j}\sin\theta$ (m),

where $\theta = \sin^{-1} 0.5 = 30^o$. The motion at point B is parallel to the *x* axis. The velocity at A is

$$\vec{v}_A = \text{v}_B\vec{i} + \vec{\mathbf{w}} \times \vec{r}_{A/B} = \begin{bmatrix} \vec{i} & \vec{j} & \vec{k} \\ 0 & 0 & \omega_{AB} \\ -0.866 & 0.5 & 0 \end{bmatrix} = (\text{v}_B - 0.5\omega_{AB})\vec{i} - 0.866\omega_{AB}\vec{j} \ (\text{m/s}).$$

Equate and solve: $(-2 - 0.866\omega_{AB})\vec{j} = 0$, $(\text{v}_B - 0.5\omega_{AB} + 2)\vec{i} = 0$, from which $\vec{\mathbf{w}}_{AB} = -2.31\vec{k}$ rad/s, $\vec{v}_B = -3.15\vec{i}$ (m/s)

==================================<>==================================

==================================<>==================================

Problem 6.45 A motor rotates the circular disk mounted at A, moving the saw back and forth. (The saw is supported by a horizontal slot so that point C moves horizontally.) The radius at AB is 4 *in.* and the link BC is 14 *in.* long. In the position shown, $\theta = 45^o$ and the link BC is horizontal. If the angular velocity of the disk is one revolution per second counterclockwise, what is the velocity of the saw?

Solution The radius vector from A to B is $\vec{r}_{B/A} = 4\left(\vec{i}\cos 45^o + \vec{j}\sin 45^o\right) = 2\sqrt{2}\left(\vec{i}+\vec{j}\right)$ $(in.)$. The angular velocity of B is $\vec{v}_B = \vec{v}_A + \vec{\omega}_{AB} \times \vec{r}_{B/A}$,

$$\vec{v}_B = 0 + 2\pi\left(2\sqrt{2}\right)\begin{bmatrix}\vec{i} & \vec{j} & \vec{k}\\ 0 & 0 & 1\\ 1 & 1 & 0\end{bmatrix} = 4\pi\sqrt{2}\left(-\vec{i}+\vec{j}\right)\ (in/s).$$

The radius vector from B to C is $\vec{r}_{C/B} = \left(4\cos 45^o - 14\right)\vec{i}$. The velocity of point C is

$$\vec{v}_C = \vec{v}_B + \vec{\omega}_{BC} \times \vec{r}_{BC} = \vec{v}_B\begin{bmatrix}\vec{i} & \vec{j} & \vec{k}\\ 0 & 0 & \omega_{BC}\\ 2\sqrt{2}-14 & 0 & 0\end{bmatrix} = -\sqrt{2}4\pi\vec{i} + \left(\left(\sqrt{2}2-14\right)\omega_{BC}+1\right)\vec{j}.$$ The saw is

constrained to move parallel to the x axis, hence $\left(\sqrt{2}2-14\right)\omega_{BC}+1=0$, and the saw velocity is

$\boxed{\vec{v}_S = -4\sqrt{2}\pi\vec{i} = -17.8\vec{i}\ \ (in./s)}$.

==================================<>==================================

Problem 6.46 In Problem 6.45, if the angular velocity of the disk is one revolution per second counter clockwise and $\theta = 270^o$, what is the velocity of the saw?

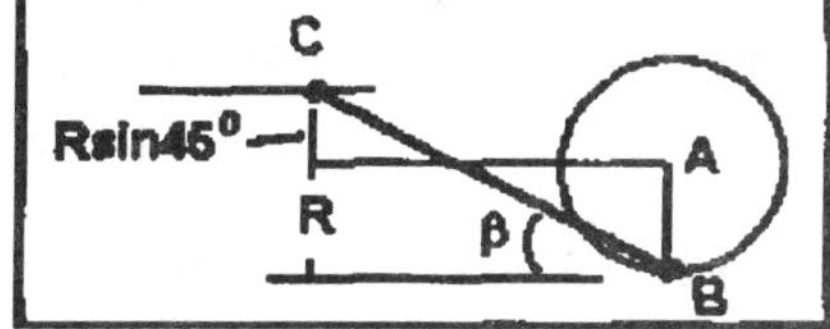

Solution The radius vector from A to B is $\vec{r}_{B/A} = -4\vec{j}$ $(in.)$. The velocity of B is

$$\vec{v}_B = \vec{\omega} \times \vec{r}_{B/A} = 2\pi(-4)\begin{bmatrix}\vec{i} & \vec{j} & \vec{k}\\ 0 & 0 & 1\\ 0 & 1 & 0\end{bmatrix} = 8\pi\vec{i}\ (in/s).$$ The

coordinates of point C are $\left(-14\cos\beta, +4\sin 45^o\right) = \left(-12.22, 2\sqrt{2}\right)$ *in.*, where

$\beta = \sin^{-1}\left(\dfrac{4(1+\sin 45^o)}{14}\right) = 29.19^o$. The coordinates of point B are $(0, -4)$ *in.*. The vector from C to B is $\vec{r}_{C/B} = (-12.22-0)\vec{i} + \left(2\sqrt{2}-(-4)\right)\vec{j} = -12.22\vec{i} + 6.828\vec{j}$ $(in.)$. The velocity at point C is

$$\vec{v}_C = \vec{v}_B + \vec{\omega}_{BC} \times \vec{r}_{C/B} = \vec{v}_B + \begin{bmatrix}\vec{i} & \vec{j} & \vec{k}\\ 0 & 0 & \omega_{BC}\\ 12.22 & 6.828 & 0\end{bmatrix} = (8\pi - 6.828\omega_{BC})\vec{i} + 12.22\omega_{BC}\vec{j}.$$ Since the saw

is constrained to move parallel to the x axis, $12.22\omega_{BC}\vec{j} = 0$, from which $\omega_{BC} = 0$, and the velocity of the saw is $\boxed{\vec{v}_S = 8\pi\vec{i} = 25.1\vec{i}\ \ (in./s)}$ [*Note:* Since the vertical velocity at B reverses direction at $\theta = 270^o$, the angular velocity $\omega_{BC} = 0$ can be determined on physical grounds by inspection , simplifying the solution.]

==================================<>==================================

Problem 6.47 The disks roll on a plane surface. The angular velocity of the left disk is 2 *rad/s* in the clockwise direction. What is the angular velocity of the right disk?

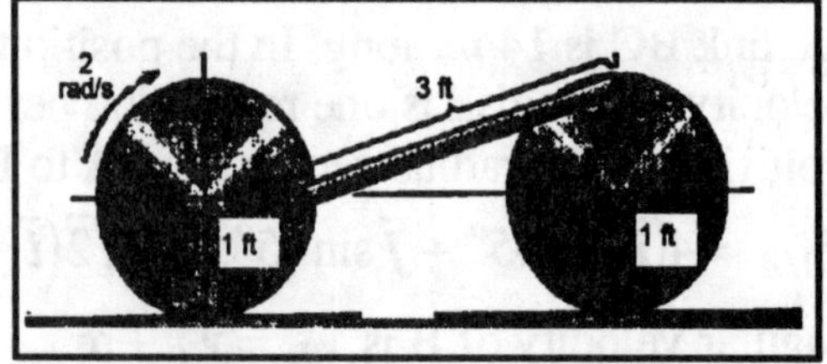

Solution The velocity at the point of contact P of the left disk is zero. The vector from this point of contact to the center of the left disk is $\vec{r}_{O/P} = 1\vec{j}\ (ft)$. The velocity of the center of the left disk is

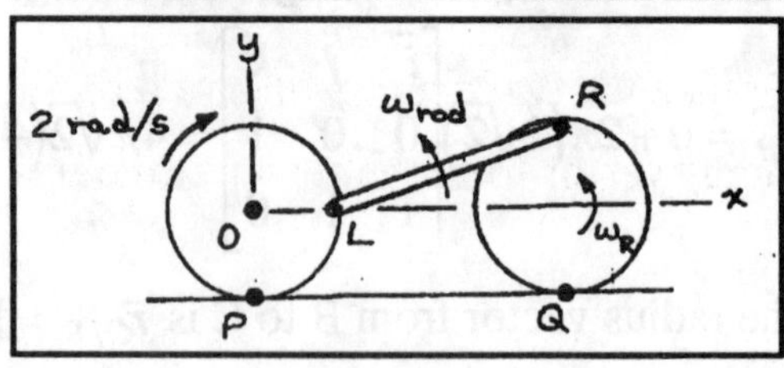

$\vec{v}_O = \vec{\omega} \times \vec{r}_{O/P} = \begin{bmatrix} \vec{i} & \vec{j} & \vec{k} \\ 0 & 0 & -2 \\ 0 & 1 & 0 \end{bmatrix} = 2\vec{i}\ (ft/s)$. The vector from the center of the left disk to the point of attachment of the rod is $\vec{r}_{L/O} = 1\vec{i}\ (ft)$. The velocity of the point of attachment of the rod to the left disk is $\vec{v}_R = \vec{v}_O + \vec{\omega} \times \vec{r}_{L/O} = 2\vec{i} + \begin{bmatrix} \vec{i} & \vec{j} & \vec{k} \\ 0 & 0 & -2 \\ 1 & 0 & 0 \end{bmatrix} = 2\vec{i} - 2\vec{j}\ (ft/s),$. The vector from the point of attachment of the left disk to the point of attachment of the right disk is $\vec{r}_{R/L} = 3(\vec{i}\cos\theta + \vec{j}\sin\theta)\ (ft)$, where $\theta = \sin^{-1}\left(\frac{1}{3}\right) = 19.47^o$. The velocity of the point on attachment on the right disk is

$\vec{v}_R = \vec{v}_L + \vec{\omega}_{rod} \times \vec{r}_{R/L} = \vec{v}_L + \begin{bmatrix} \vec{i} & \vec{j} & \vec{k} \\ 0 & 0 & \omega_{rod} \\ 2.83 & 1 & 0 \end{bmatrix} = (2 - \omega_{rod})\vec{i} + (-2 + 2.83\omega_{rod})\vec{j}\ (ft/s)$. The velocity of point R is also expressed in terms of the contact point Q,

$\vec{v}_R = \vec{\omega}_{RO} \times \vec{r}_{R/O} = \omega_{RO}(2)\begin{bmatrix} \vec{i} & \vec{j} & \vec{k} \\ 0 & 0 & 1 \\ 0 & 1 & 0 \end{bmatrix} = -2\omega_{RO}\vec{i}\ (ft/s)$. Equate the two expressions for the velocity $\vec{v}_R$ and separate components: $(2 - \omega_{rod} + 2\omega_{RO})\vec{i} = 0$, $(-2 + 2.83\omega_{rod})\vec{j} = 0$, from which $\boxed{\vec{\omega}_{RO} = -0.65\vec{k}\ (rad/s)}$ and $\omega_{rod} = 0.707\ rad/s$.

Problem 6.48 The disk rolls on the curved surface. The bar rotates at 10 *rad/s* in the counterclockwise direction. Determine the velocity of point A.

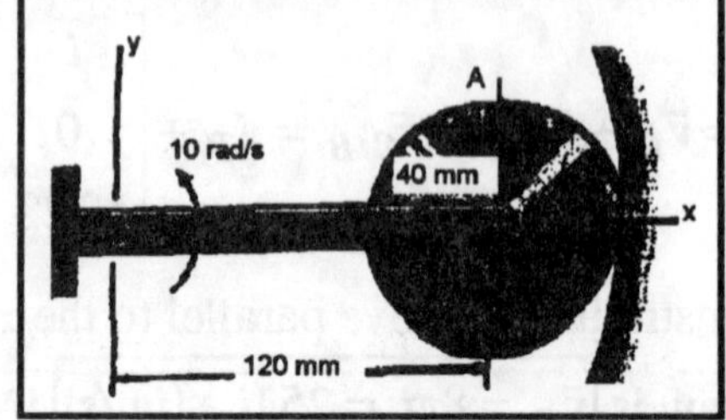

Solution The radius vector from the left point of attachment of the bar to the center of the disk is $\vec{r}_{bar} = 120\vec{i}\ (mm)$. The velocity of the center of the disk is

$\vec{v}_O = \vec{\omega}_{bar} \times \vec{r}_{bar} = 10(120)\begin{bmatrix} \vec{i} & \vec{j} & \vec{k} \\ 0 & 0 & 1 \\ 1 & 0 & 0 \end{bmatrix} = 1200\vec{j}\ (mm/s)$.

Solution continued on next page

The radius vector from the point of contact with the disk and the curved surface to the center of the disk is $\vec{r}_{O/P} = -40\vec{i}\ (mm)$. The velocity of the point of contact of the disk with the curved surface is zero, from which

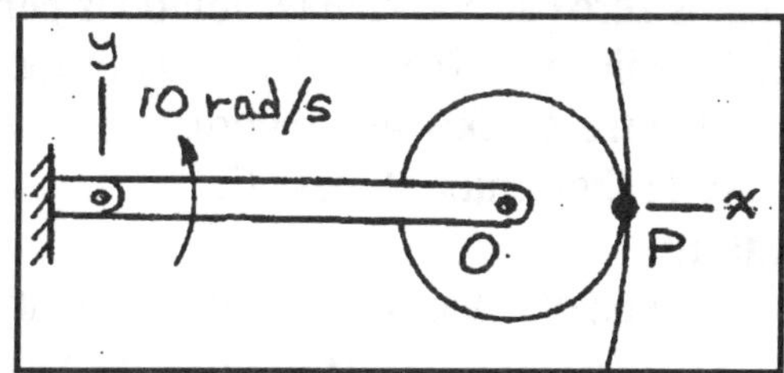

$$\vec{v}_O = \vec{\omega}_O \times \vec{r}_{O/P} = \begin{bmatrix} \vec{i} & \vec{j} & \vec{k} \\ 0 & 0 & \omega_O \\ -40 & 0 & 0 \end{bmatrix} = -40\omega_O \vec{j}.$$

Equate the two expressions for the velocity of the center of the disk and solve: $\omega_O = -30\ rad/s$. The radius vector from the center of the disk to point A is $\vec{r}_{A/O} = 40\vec{j}\ (mm)$. The velocity of point A is

$$\vec{v}_A = \vec{v}_O + \vec{\omega}_O \times \vec{r}_{A/O} = 1200\vec{j} - (30)(40)\begin{bmatrix} \vec{i} & \vec{j} & \vec{k} \\ 0 & 0 & 1 \\ 0 & 1 & 0 \end{bmatrix} \boxed{= 1200\vec{i} + 1200\vec{j}\ (mm/s)}$$

==================================<>==================================

Problem 6.49 If $\omega_{AB} = 2\ rad/s$ and $\omega_{BC} = 4\ rad/s$, what is the velocity of point C, where the excavator's bucket is attached?

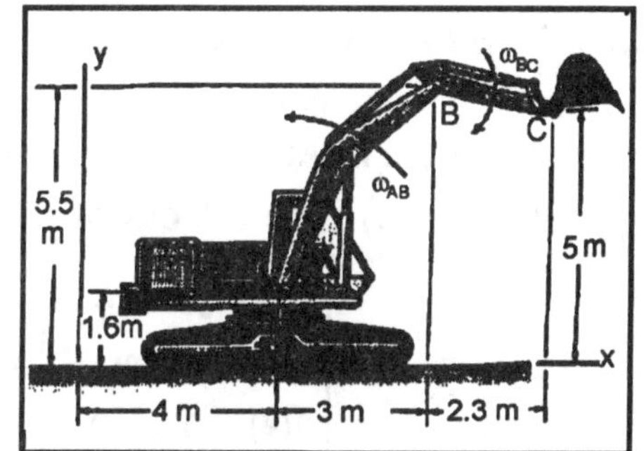

Solution The radius vector AB is $\vec{r}_{B/A} = 3\vec{i} + (5.5 - 1.6)\vec{j} = 3\vec{i} + 3.9\vec{j}\ (m)$. The velocity of point B is

$$\vec{v}_B = \vec{\omega}_{AB} \times \vec{r}_{B/A} = \begin{bmatrix} \vec{i} & \vec{j} & \vec{k} \\ 0 & 0 & 2 \\ 3 & 3.9 & 0 \end{bmatrix} = -7.8\vec{i} + 6\vec{j}\ (m/s).$$

The radius vector BC is $\vec{r}_{C/B} = 2.3\vec{i} + (5 - 5.5)\vec{j} = 2.3\vec{i} - 0.5\vec{j}\ (m)$. The velocity at point C is

$$\vec{v}_C = \vec{v}_B + \vec{\omega}_{BC} \times \vec{r}_{C/B} = -7.8\vec{i} + 6\vec{j} + \begin{bmatrix} \vec{i} & \vec{j} & \vec{k} \\ 0 & 0 & -4 \\ 2.3 & -0.5 & 0 \end{bmatrix} \boxed{= -9.8\vec{i} - 3.2\vec{j}\ (m/s)}$$

==================================<>==================================

Problem 6.50 In Problem 6.49, if $\omega_{AB} = 2\ rad/s$, what clockwise angular velocity ω_{BC} will cause the vertical component of the velocity of point C to be zero? What is the resulting velocity of point C?

Solution Use the solution to Problem 6.49. The velocity of point B is $\vec{v}_B = -7.8\vec{i} + 6\vec{j}\ (m/s)$. The velocity of point C is

$$\vec{v}_C = \vec{v}_B + \vec{\omega}_{BC} \times \vec{r}_{C/B} = -7.8\vec{i} + 6\vec{j} + \begin{bmatrix} \vec{i} & \vec{j} & \vec{k} \\ 0 & 0 & -\omega_{BC} \\ 2.3 & -0.5 & 0 \end{bmatrix},$$

$\vec{v}_C = (-7.8 - 0.5\omega_{BC})\vec{i} + (6 - 2.3\omega_{BC})\vec{j}\ (m/s)$. For the vertical component to be zero,

$\boxed{\omega_{BC} = \frac{6}{2.3} = 2.61\ rad/s\ \ clockwise}$. The velocity of point C is $\boxed{\vec{v}_C = -9.1\vec{i}\ (m/s)}$

==================================<>==================================

================================<>================================

Problem 6.51 The motorcycle's rear wheel is rolling on the ground (the velocity of its point of contact with the ground is zero) at *500 rpm* (revolutions per minute). The wheel's radius is *280 mm.* The body of the motorcycle is rotating in the clockwise direction at *6 rad/s.* Determine the velocity of the center of mass *G.*

Solution:

The rear wheel's clockwise angular velocity is $\omega = (500)(2\pi)/60 = 52.4 rad/s$ so the velocity of the center C of the wheel is $\bar{v}_C = (52.4)(0.28)\bar{i} = 14.7\bar{i}(m/s)$. The velocity of the center of mass is

$$\bar{v}_G = \bar{v}_c + \bar{\omega}\times\bar{r}_{G/c} = 14.7\bar{i} + \begin{vmatrix} \bar{i} & \bar{j} & \bar{k} \\ 0 & 0 & -6 \\ 0.65 & 0.5 & 0 \end{vmatrix} = 17.7\bar{i} - 3.9\bar{j}(m/s).$$

================================<>================================

Problem 6.52 An athlete exercises his arm by raising the mass *m*. The shoulder joint is stationary. The distance AB is 300 *mm*, and the distance BC is 400 *mm*. At the instant shown, $\omega_{AB} = 1\ rad/s$ and $\omega_{BC} = 2\ rad/s$. How fast is the mass *m* rising?

Solution The magnitude of the velocity of the point C parallel to the cable at C is also the magnitude of the velocity of the mass *m*. The radius vector AB is $\vec{r}_{B/A} = 300\vec{i}\ (mm)$. The velocity of point B is

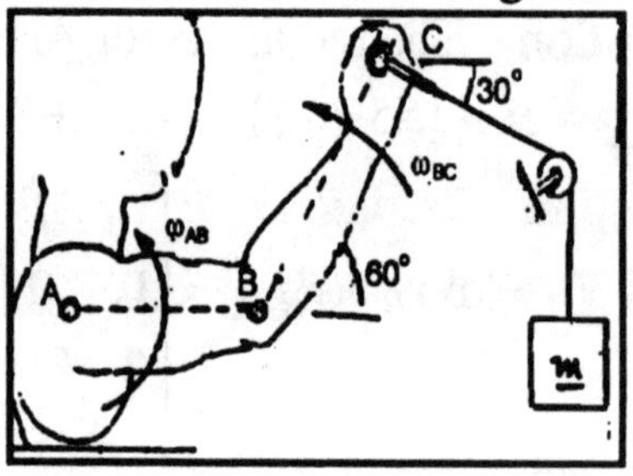

$$\vec{v}_B = \vec{\omega}_{AB}\times\vec{r}_{B/A} = \begin{bmatrix} \vec{i} & \vec{j} & \vec{k} \\ 0 & 0 & \omega_{AB} \\ 300 & 0 & 0 \end{bmatrix} = 300\vec{j}\ (mm/s).$$

The radius vector BC is

$\vec{r}_{C/B} = 400(\vec{i}\cos 60^o + \vec{j}\sin 60^o) = 200\vec{i} + 346.4\vec{j}\ (mm)$. The velocity of point C is

$$\vec{v}_C = \vec{v}_B + \vec{\omega}_{BC}\times\vec{r}_{C/B} = 300\vec{j} + \begin{bmatrix} \vec{i} & \vec{j} & \vec{k} \\ 0 & 0 & 2 \\ 200 & 346.4 & 0 \end{bmatrix} = -692.8\vec{i} + 700\vec{j}\ (mm/s).$$

The unit vector parallel to the cable at C is $\vec{e}_C = -\vec{i}\cos 30^o + \vec{j}\sin 30^o = -0.866\vec{i} + 0.5\vec{j}$. The component of the velocity parallel to the cable at C is $\boxed{\vec{v}_P = \vec{v}_C \cdot \vec{e}_C = 950\ mm/s}$, which is the velocity of the mass *m*.

================================<>================================

====================================<>====================================

Problem 6.53 In Problem 6.52, suppose that the distance AB is 12 *in.*, the distance BC is 16 *in.*, $\omega_{AB} = 0.6\ rad/s$, and the mass *m* is rising at 24 *in/s*. What is the angular velocity ω_{BC} ?

Solution: The radius vector AB is $\vec{r}_{B/A} = 12\vec{i}\ (in.)$.

The velocity at point B is $\vec{v}_B = \vec{\omega}_{AB} \times \vec{r}_{B/A} = \begin{bmatrix} \vec{i} & \vec{j} & \vec{k} \\ 0 & 0 & 0.6 \\ 0 & 12 & 0 \end{bmatrix} = 7.2\vec{j}\ (in/s)$.

The radius vector BC is $\vec{r}_{C/B} = 16(\vec{i}\sin 60^o + \vec{j}\cos 60^o) = 8\vec{i} + 13.9\vec{j}\ (in.)$.

The velocity at C is $\vec{v}_C = \vec{v}_B + \vec{\omega}_{BC} \times \vec{r}_{C/B} = 7.2\vec{j} + \begin{bmatrix} \vec{i} & \vec{j} & \vec{k} \\ 0 & 0 & \omega_{BC} \\ 8 & 13.9 & 0 \end{bmatrix} = -13.9\omega_{BC}\vec{i} + (7.2 + 8\omega_{BC})\vec{j}$.

The unit vector parallel to the cable at C is $\vec{e}_C = -\vec{i}\cos 30^o + \vec{j}\sin 60^o = -0.5\vec{i} + 0.866\vec{j}$. The component of the velocity at C parallel to the cable is
$|\vec{v}_{CP}| = \vec{v}_C \cdot \vec{e}_C = +6.93\omega_{BC} + 6.93\omega_{BC} + 6.24\ (in/s)$. This is also the velocity of the rising mass, from which $13.86\omega_{BC} + 6.24 = 24$, $\boxed{\omega_{BC} = 1.28\ rad/s}$

====================================<>====================================

Problem 6.54 Points B and C are in the *x-y* plane. The angular velocity vectors of the arms AB and BC are $\vec{\omega}_{AB} = -0.2\vec{k}\ (rad/s)$, and $\vec{\omega}_{BC} = 0.4\vec{k}\ (rad/s)$. Determine the velocity of point C.

Solution The radius vector AB is
$\vec{r}_{B/A} = 30(\vec{i}\cos 40^o + \vec{j}\sin 40^o) = 23\vec{i} + 19.3\vec{j}\ (in.)$. The velocity of point B is $\vec{v}_B = \vec{\omega} \times \vec{r}_{B/A} = \begin{bmatrix} \vec{i} & \vec{j} & \vec{k} \\ 0 & 0 & -0.2 \\ 23 & 19.3 & 0 \end{bmatrix} = 3.86\vec{i} - 4.6\vec{j}\ (in/s)$. The radius vector BC is $\vec{r}_{BC} = 35(\vec{i}\cos 30^o - \vec{j}\sin 30^o) = 30.3\vec{i} - 17.5\vec{j}\ (in.)$.

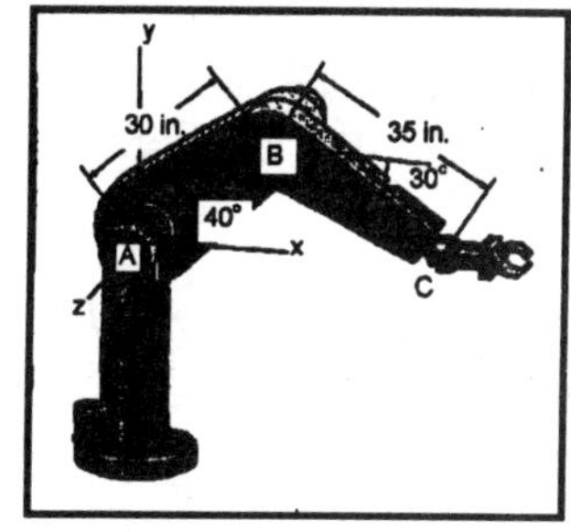

The velocity of point C is $\vec{v}_C = \vec{v}_B + \vec{\omega}_{BC} \times \vec{r}_{BC} = 3.86\vec{i} - 4.6\vec{j} + \omega_{BC}(\vec{k} \times (30.3\vec{i} - 17.5\vec{j}))$

$\boxed{\vec{v}_C = 10.86\vec{i} + 7.53\vec{j}\ (in/s)}$

====================================<>====================================

Problem 6.55 In Problem 6.54, if the velocity at point C is $\vec{v}_C = 10\vec{j}\ (in/s)$, what are the angular velocities of the arms AB and BC?

Solution Use the solution to Problem 6.54. $\vec{r}_{B/A} = 23\vec{i} + 19.3\vec{j}\ (in.)$, $\vec{r}_{BC} = 30.3\vec{i} - 17.5\vec{j}\ (in.)$ The velocity of points B and C: $\vec{v}_B = \vec{\omega} \times \vec{r}_{B/A} = \begin{bmatrix} \vec{i} & \vec{j} & \vec{k} \\ 0 & 0 & \omega_{AB} \\ 23 & 19.3 & 0 \end{bmatrix} = \omega_{AB}(-19.3\vec{i} + 23\vec{j})\ (in/s)$.

$.\vec{v}_C = \vec{v}_B + \vec{\omega}_{BC} \times \vec{r}_{C/B} = \vec{v}_B + \begin{bmatrix} \vec{i} & \vec{j} & \vec{k} \\ 0 & 0 & \omega_{BC} \\ 30.3 & -17.5 & 0 \end{bmatrix} = 10\vec{j}\ (in/s)$

Solution continued on next page

$\vec{v}_C = \omega_{AB}\left(-19.3\vec{i} + 23\vec{j}\right) + \omega_{BC}\left(17.5\vec{i} + 30.3\vec{j}\right) = 10\vec{j}$. Separate components:

$(-19.3\omega_{AB} + 17.5\omega_{BC})\vec{i} = 0$, $(23\omega_{AB} + 30.3\omega_{BC} - 10)\vec{j} = 0$. Solve: $\omega_{AB} = 0.177\ rad/s$,

$\omega_{BC} = 0.195\ rad/s$, from which $\boxed{\vec{\omega}_{AB} = 0.177\vec{k}\ (rad/s)}$, $\boxed{\vec{\omega}_{BC} = 0.195\vec{k}\ (rad/s)}$

===================================<>===================================

Problem 6.56 The link AB of the robot's arm is rotating at *2 rad/s* in the counterclockwise direction, the link BC is rotating at *3 rad/s* in the clockwise direction, and the link CD is rotating at *4 rad/s* in the counterclockwise direction. What is the velocity of point D?

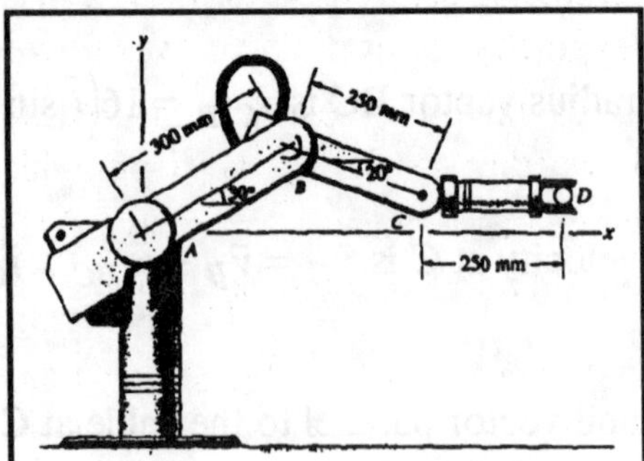

Solution:

The velocity of B is $\bar{v}_B = \bar{v}_A + \bar{\omega}_{AB} \times \bar{r}_{B/A}$, or

$$\bar{v}_B = 0 + \begin{vmatrix} \bar{i} & \bar{j} & \bar{k} \\ 0 & 0 & 2 \\ 0.3\cos 30^\circ & 0.3\sin 30^\circ & 0 \end{vmatrix} = -0.3\bar{i} + 0.520\bar{j}(m/s).$$

The velocity of C is $\bar{v}_C = \bar{v}_B + \bar{\omega}_{BC} \times \bar{r}_{C/B} = -0.3\bar{i} + 0.520\bar{j}(m/s)$, or

$$\bar{v}_C = -0.3\bar{i} + 0.520\bar{j} + \begin{vmatrix} \bar{i} & \bar{j} & \bar{k} \\ 0 & 0 & -3 \\ 0.25\cos 20^\circ & -0.25\sin 20^\circ & 0 \end{vmatrix} = -0.557\bar{i} - 0.185\bar{j}(m/s).$$

The velocity of D is $\bar{v}_D = \bar{v}_C + \bar{\omega}_{CD} \times \bar{r}_{D/C} = -0.557\bar{i} - 0.185\bar{j} + \begin{vmatrix} \bar{i} & \bar{j} & \bar{k} \\ 0 & 0 & 4 \\ 0.25 & 0 & 0 \end{vmatrix}$, or

$\bar{v}_D = -0.557\bar{i} + 0.815\bar{j}(m/s)$

===================================<>===================================

Problem 6.57 Consider the robot shown in Problem 6.56. Link AB is rotating at *2 rad/s* in the counterclockwise direction and link BC is rotating at *3 rad/s* in the clockwise direction. If you want the velocity of point D to be parallel to the x axis, what is the necessary angular velocity of link CD? What is the resulting velocity of point D?

Soluton:

From the solution of Problem 6.56, the velocity of point C is $\bar{v}_C = -0.557\bar{i} - 0.185\bar{j}(m/s)$. Let ω_{CD} be the counterclockwise angular velocity of link CD. The velocity of D is

$$v_D\bar{i} = \bar{v}_C + \bar{\omega}_{CD} \times \bar{r}_{D/C} = -0.557\bar{i} - 0.185\bar{j} + \begin{vmatrix} \bar{i} & \bar{j} & \bar{k} \\ 0 & 0 & \omega_{CD} \\ 0.25 & 0 & 0 \end{vmatrix}.$$ Equating $\bar{i}$ and $\bar{j}$

components, $v_D = -0.557\ m/s$ and $0 = -0.185 + 0.25\omega_{CD}$ we obtain $\omega_{CD} = 0.741 rad/s$ and $v_D = -0.557 m/s$

===================================<>===================================

==================================<>==================================

Problem 6.58 Determine the velocity v_W and the angular velocity of the small pulley.

Solution:

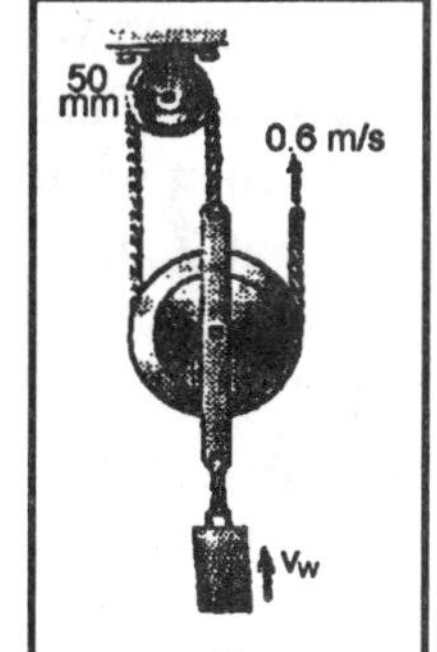

Since the radius of the bottom pulley is not given, we cannot use Eq (6.6) (or the equivalent). The strategy is to use the fact (derived from elementary principles) that the velocity of the center of a pulley is the mean of the velocities of the extreme edges, where the edges lie on a line normal to the motion, *taking into account the directions of the velocities at the extreme edges.* The center rope from the bottom pulley to the upper pulley moves upward at a velocity of v_W. Since the small pulley is fixed, the velocity of the center is zero, and the rope to the left moves downward at a velocity v_W, from which the left edge of the bottom pulley is moving at a velocity v_W downward. The right edge of the bottom pulley moves upward at a velocity of 0.6 *m/s*. The velocity of the center of the bottom pulley is the mean of the velocities at the extreme edges, from which $v_W = \frac{6 - v_W}{2}$. Solve: $\boxed{v_W = \frac{0.6}{3} = 0.2 \; m/s}$. The angular velocity of the small pulley is $\boxed{\omega = \frac{v_W}{r} = \frac{0.2}{0.05} = 4 \;\; rad/s}$

==================================<>==================================

Problem 6.59 Determine the velocity of the block and the angular velocity of the small pulley.

Solution:

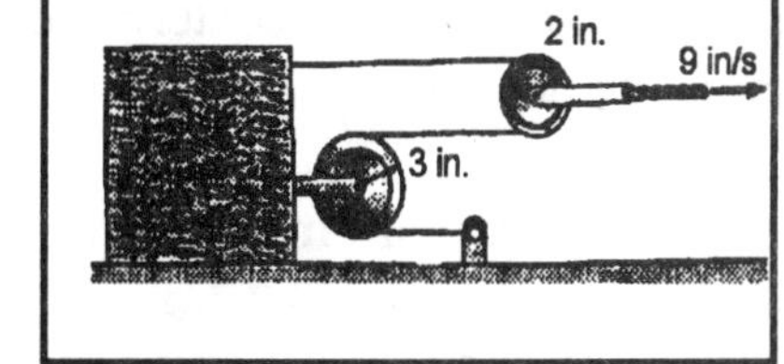

Denote the velocity of the block by v_B. The strategy is to determine the velocities of the extreme edges of a pulley by the determining the velocity of the element of rope in contact with the pulley. The upper rope is fixed to the block, so that it moves to the right at the velocity of the block, from which the upper edge of the small pulley moves to the right at the velocity of the block. The fixed end of the rope at the bottom is stationary, so that the bottom edge of the large pulley is stationary. The center of the large pulley moves at the velocity of the block, from which the upper edge of the bottom pulley moves at twice the velocity of the block (since the velocity of the center is equal to the mean of the velocities of the extreme edges, one of which is stationary) from which the bottom edge of the small pulley moves at twice the velocity of the block. The center of the small pulley moves to the right at 9 *in/s*. The velocity of the center of the small pulley is the mean of the velocities at the extreme edges, from which $9 = \frac{2v_B + v_B}{2} = \frac{3}{2}v_B$, from which $\boxed{v_B = \frac{2}{3}9 = 6 \;\; in/s}$. The angular velocity of small pulley is given by

$$9\vec{i} = 2v_B\vec{i} + \vec{\omega} \times 2\vec{j} = \begin{bmatrix} \vec{i} & \vec{j} & \vec{k} \\ 0 & 0 & \omega \\ 0 & 2 & 0 \end{bmatrix} = 2v_B\vec{i} - 2\omega\vec{i}$$, from which $\boxed{\omega = \frac{12-9}{2} = 1.5 \;\; rad/s}$

==================================<>==================================

================================<>================================

Problem 6.60 The device shown is used in the semiconductor industry to polish silicon wafers. The wafers are placed on the faces of the carriers. The outer and inner rings are then rotated, causing the wafers to move and rotate against an abrasive surface. If the outer ring rotates in the clockwise direction at *7 rpm* (revolutions per minute) and the inner ring rotates in the counterclockwise direction at *12 rpm*, what is the angular velocity of the carriers?

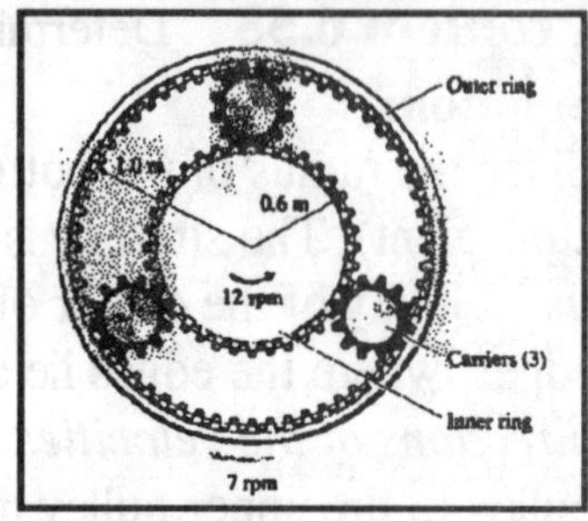

Solution:

The velocity of point *A* is

$\bar{v}_A = \bar{v}_0 + \bar{\omega}_i \times \bar{r}_{A/0} = 0 + (\omega_i \bar{k}) \times (0.6\bar{j}) = -0.6\omega_i \bar{i}$.

The velocity of point *B* is

$\bar{v}_B = \bar{v}_0 + \bar{\omega}_0 \times \bar{r}_{B/0} = 0 + (\omega_0 \bar{k}) \times (1.0\bar{j}) = \omega_0 \bar{i}$. Let $\bar{\omega}_c = \omega_c \bar{k}$ be the angular velocity of the carrier. Then $\bar{v}_B = \bar{v}_A + \bar{\omega}_c \times \bar{r}_{B/A}$: Hence,

$\omega_0 \bar{i} = -0.6\omega_i \bar{i} + (\omega_c \bar{k}) \times (0.4\bar{j})$.

Solving, $\omega_c = -2.5\omega_0 - 1.5\omega_i = -2.5(7rpm) - 1.5(12rpm) = -35.5rpm$

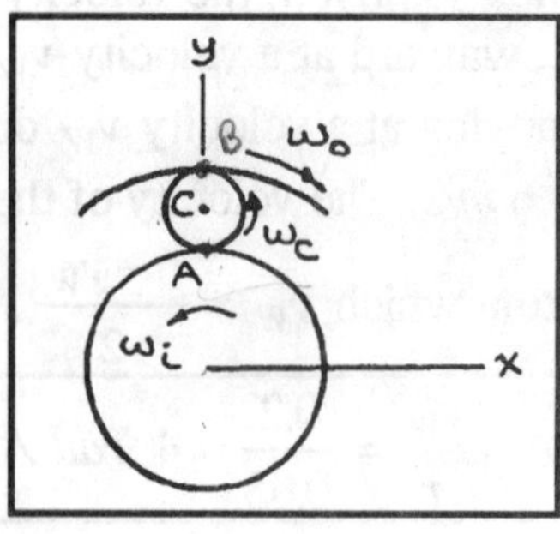

================================<>================================

Problem 6.61 In Problem 6.60, suppose that the outer ring rotates in the clockwise direction at *5 rpm* and you want the centerpiece of the carriers to remain stationary during the polishing process. What is the necessary angular velocity of the inner ring?

Solution:

From the solution of Problem 6.60,

$\bar{v}_A = -0.6\omega_i \bar{i}$ and $\omega_c = -2.5\omega_0 - 1.5\omega_i$ (1)

The velocity of the centerpoint *C* is

$$\bar{v}_C = 0 = \bar{v}_A + \bar{\omega}_C \times \bar{r}_{C/A} = -0.6\omega_i \bar{i} + \begin{vmatrix} \bar{i} & \bar{j} & \bar{k} \\ 0 & 0 & \omega_C \\ 0 & 0.2 & 0 \end{vmatrix} = -(0.6\omega_i \bar{i} + 0.2\omega_c)\bar{i} \quad (2).$$

Eliminating ω_c from Equations (1) and (2) gives $\omega_i = 1.67\omega_o = 1.67(5rpm) = 8.33rpm$.

================================<>================================

==================================<>==================================

Problem 6.62 The ring gear is fixed and the hub and planet gears are bonded together. The connecting rod rotates in the counterclockwise direction at 60 *rpm* (revolutions per minute). Determine the angular velocity of the sun gear and the magnitude of the velocity of point A.

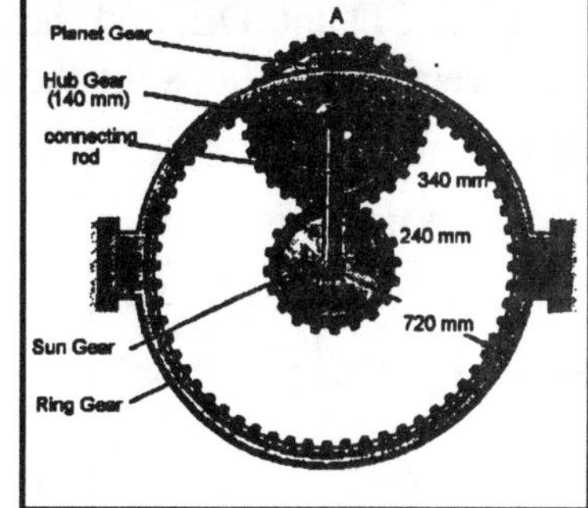

Solution Denote the centers of the sun, hub and planet gears by the subscripts *Sun, Hub*, and *Planet*, respectively. Denote the contact points between the sun gear and the planet gear by the subscript *SP* and the point of contact between the hub gear and the ring gear by the subscript *HR*. The angular velocity of the connecting rod is $\omega_{CR} = 6.28 \ rad/s$. The vector distance from the center of the sun gear to the center of the hub gear is $\vec{r}_{Hub/Sun} = (720-140)\vec{j} = 580\vec{j} \ (mm)$ The velocity of the center of the hub gear is

$$\vec{v}_{Hub} = \vec{\omega}_{CR} \times \vec{r}_{Hub/Sun} = \begin{bmatrix} \vec{i} & \vec{j} & \vec{k} \\ 0 & 0 & 2\pi \\ 0 & 580 & 0 \end{bmatrix} = -3644\vec{i} \ (mm/s)$$

The angular velocity of the hub gear is found from

$$\vec{v}_{HR} = 0 = \vec{v}_{Hub} + \vec{\omega}_{Hub} \times 140\vec{j} = \begin{bmatrix} \vec{i} & \vec{j} & \vec{k} \\ 0 & 0 & \omega_{Hub} \\ 0 & 140 & 0 \end{bmatrix} = -3644\vec{i} - 140\omega_{Hub}\vec{i},$$

from which $\omega_{Hub} = -\dfrac{3644}{140} = -26.03 \ rad/s$. This is also the angular velocity of the planet gear. The linear velocity of point A is

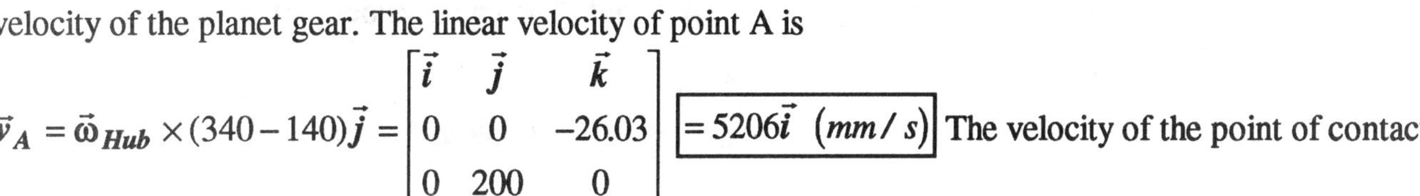

$$\vec{v}_A = \vec{\omega}_{Hub} \times (340-140)\vec{j} = \begin{bmatrix} \vec{i} & \vec{j} & \vec{k} \\ 0 & 0 & -26.03 \\ 0 & 200 & 0 \end{bmatrix} \boxed{= 5206\vec{i} \ (mm/s)}$$

The velocity of the point of contact with the sun gear is

$$\vec{v}_{PS} = \vec{\omega}_{Hub} \times \left(-480\vec{j}\right) = \begin{bmatrix} \vec{i} & \vec{j} & \vec{k} \\ 0 & 0 & -26.03 \\ 0 & -480 & 0 \end{bmatrix} = -12494.6\vec{i} \ (mm/s).$$

The angular velocity of the sun gear is found from

$$\vec{v}_{PS} = -12494.6\vec{i} = \vec{\omega}_{Sun} \times \left(240\vec{j}\right) = \begin{bmatrix} \vec{i} & \vec{j} & \vec{k} \\ 0 & 0 & \omega_{Sun} \\ 0 & 240 & 0 \end{bmatrix} = -240\omega_{sun}\vec{i},$$

from which

$$\boxed{\omega_{Sun} = \frac{12494.6}{240} = 52.06 \ rad/s}$$

==================================<>==================================

Problem 6.63 The large gear is fixed. Bar AB has a counterclockwise angular velocity of 2 *rad/s*. What are the angular velocities of bars CD and DE?

Solution The strategy is to express vector velocity of point D in terms of the unknown angular velocities of CD and DE, and then to solve the resulting vector equations for the unknowns. The vector distance AB is $\vec{r}_{B/A} = 14\vec{j}$ (*in.*) The linear velocity of point B is

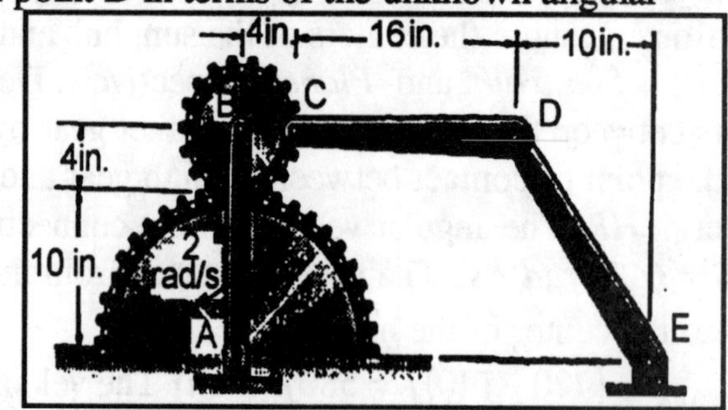

$$\vec{v}_B = \vec{\omega} \times \vec{r}_{B/A} = \begin{bmatrix} \vec{i} & \vec{j} & \vec{k} \\ 0 & 0 & 2 \\ 0 & 14 & 0 \end{bmatrix} = -28\vec{i} \ (in/s).$$ The lower edge of gear B is stationary. The radius vector from the lower edge to B is $\vec{r}_B = 4\vec{j}$ (*in.*), The angular velocity of B is $\vec{v}_B = \vec{\omega}_B \times \vec{r}_B = \begin{bmatrix} \vec{i} & \vec{j} & \vec{k} \\ 0 & 0 & \omega_B \\ 0 & 4 & 0 \end{bmatrix} = -4\omega_B\vec{i} \ (in/s)$, from which

$\omega_B = -\frac{v_B}{4} = 7 \ rad/s$. The vector distance from B to C is $\vec{r}_{C/B} = 4\vec{i}$ (*in.*). The velocity of point C is

$$\vec{v}_C = \vec{v}_B + \vec{\omega}_B \times \vec{r}_{C/B} = -28\vec{i} + \begin{bmatrix} \vec{i} & \vec{j} & \vec{k} \\ 0 & 0 & 7 \\ 4 & 0 & 0 \end{bmatrix} = -28\vec{i} + 28\vec{j} \ (in/s).$$ The vector distance from C to D is

$\vec{r}_{D/C} = 16\vec{i}$ (*in.*), and from E to D is $\vec{r}_{D/E} = -10\vec{i} + 14\vec{j}$ (*in.*). The linear velocity of point D is

$$\vec{v}_D = \vec{v}_C + \vec{\omega}_{CD} \times \vec{r}_{D/C} = -28\vec{i} + 28\vec{j} + \begin{bmatrix} \vec{i} & \vec{j} & \vec{k} \\ 0 & 0 & \omega_{CD} \\ 16 & 0 & 0 \end{bmatrix} = -28\vec{i} + (16\omega_{CD} + 28)\vec{j} \ (in/s).$$ The velocity of point D is also given by

$$\vec{v}_D = \vec{\omega}_{DE} \times \vec{r}_{D/E} = \begin{bmatrix} \vec{i} & \vec{j} & \vec{k} \\ 0 & 0 & \omega_{DE} \\ -10 & 14 & 0 \end{bmatrix} = -14\omega_{DE}\vec{i} - 10\omega_{DE}\vec{j} \ (in/s).$$ Equate components:

$(-28 + 14\omega_{DE})\vec{i} = 0$,

$(16\omega_{CD} + 28 + 10\omega_{DE})\vec{j} = 0$. Solve: $\boxed{\omega_{DE} = 2 \ rad/s}$, $\boxed{\omega_{CD} = -3 \ rad/s}$. The negative sign means a clockwise rotation.

================================<>================================

Problem 6.64 If the bar has a clockwise angular velocity of 10 *rad/s and* $v_A = 20\ m/s$. What are the coordinates of its instantaneous center and the value of v_B?

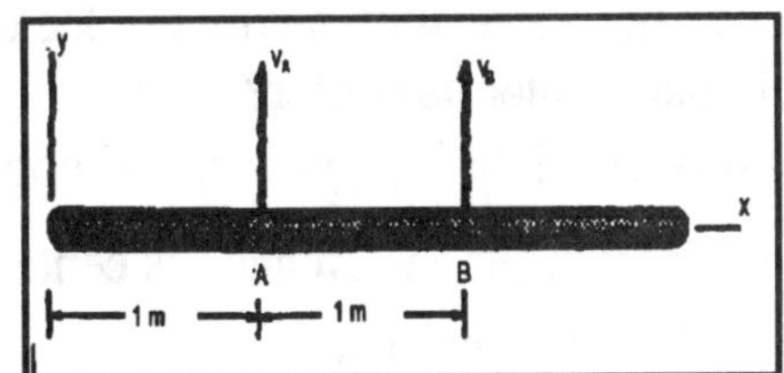

Solution Assume that the coordinates of the instantaneous center are (x_C, y_C), $\vec{\omega} = -\omega\vec{k} = -10\vec{k}$ The distance to point A is $\vec{r}_{A/C} = (1 - x_C)\vec{i} + y_C\vec{j}$. The velocity at A is

$$\vec{v}_A = 20\vec{j} = \vec{\omega} \times \vec{r}_{A/C} = \begin{bmatrix} \vec{i} & \vec{j} & \vec{k} \\ 0 & 0 & -\omega \\ 1 - x_C & y_C & 0 \end{bmatrix} = y_C\omega\vec{i} - \omega(1 - x_C)\vec{j}$$

, from which $y_C\omega\vec{i} = 0$, and $(20 + \omega(1 - x_C))\vec{j} = 0$. Substitute $\omega = 10\ rad/s$ to obtain $y_C = 0$ and $x_C = 3\ m$. The coordinates of the instantaneous center are $\boxed{(3,\ 0)\ (m)}$. The vector distance from C to B is $\vec{r}_{B/C} = (2 - 3)\vec{i} = -\vec{i}\ (m)$. The velocity of point B is

$$\vec{v}_B = \vec{\omega} \times \vec{r}_{B/C} = \begin{bmatrix} \vec{i} & \vec{j} & \vec{k} \\ 0 & 0 & -10 \\ -1 & 0 & 0 \end{bmatrix} = -10(-\vec{j}) \boxed{= 10\vec{j}\ (m/s)}$$

================================<>================================

Problem 6.65 In Problem 6.64, if $v_A = 24\ m/s$ and $v_B = 36\ m/s$, what are the coordinates of its instantaneous center of the bar and its angular velocity?

Solution Let (x_C, y_C) be the coordinates of the instantaneous center. The vectors from the instantaneous center and the points A and B are $\vec{r}_{A/C} = (1 - x_C)\vec{i} + y_C\vec{j}\ \ (m)$ and $\vec{r}_{B/C} = (2 - x_C)\vec{i} + y_C\vec{j}$. The velocity of A is given by

$$\vec{v}_A = 24\vec{j} = \vec{\omega}_{AB} \times \vec{r}_{A/C} = \begin{bmatrix} \vec{i} & \vec{j} & \vec{k} \\ 0 & 0 & \omega_{AC} \\ 1 - x_C & y_C & 0 \end{bmatrix} = -\omega_{AB}y_C\vec{i} + \omega_{AB}(1 - x_C)\vec{j}\ \ (m/s)$$ The velocity of B is

$$\vec{v}_B = 36\vec{j} = \vec{\omega}_{AB} \times \vec{r}_{B/C} = \begin{bmatrix} \vec{i} & \vec{j} & \vec{k} \\ 0 & 0 & \omega_{AB} \\ 2 - x_C & y_C & 0 \end{bmatrix} = -y_C\omega_{AB}\vec{i} + \omega_{AB}(2 - x_C)\vec{j}\ \ (m/s).$$ Separate

components : $24 - \omega_{AB}(1 - x_C) = 0$,

$36 - \omega_{AB}(2 - x_C) = 0$,

$\omega_{AB}y_C = 0$. Solve:

$\boxed{x_C = -1}$, $\boxed{y_C = 0}$, and $\boxed{\omega_{AB} = 12\ \ rad/s}$ *counterclockwise.*

================================<>================================

====================================<>====================================

Problem 6.66 The velocity of point O of the bat is $\vec{v}_B = -6\vec{i} - 1.4\vec{j}\ (ft/s)$ and the bat rotates about the *z* axis with a counterclockwise angular velocity of 4 *rad/s*. What are the *x* and *y* coordinates of the instantaneous center?

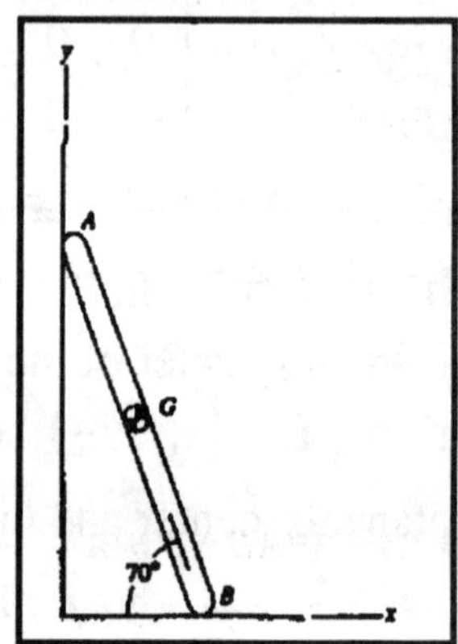

Solution Let (x_C, y_C) be the coordinates of the instantaneous center. The vector from the instantaneous center to point O is $\vec{r}_{O/C} = -x_C\vec{i} - y_C\vec{j}\ (ft)$. The velocity of point O is

$$\vec{v}_O = -6\vec{i} - 1.4\vec{j} = \vec{\omega} \times \vec{r}_{O/C} = \begin{bmatrix} \vec{i} & \vec{j} & \vec{k} \\ 0 & 0 & \omega \\ -x_C & -y_C & 0 \end{bmatrix} = y_C\omega\vec{i} - x_C\omega\vec{j}\ (ft/s).$$ Equate

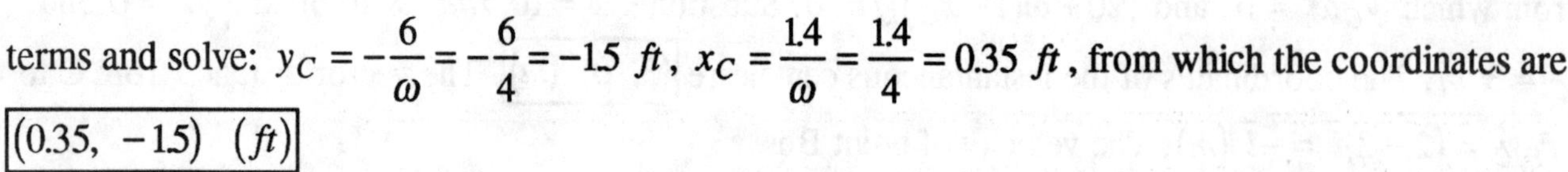

terms and solve: $y_C = -\frac{6}{\omega} = -\frac{6}{4} = -1.5\ ft$, $x_C = \frac{1.4}{\omega} = \frac{1.4}{4} = 0.35\ ft$, from which the coordinates are $\boxed{(0.35,\ -1.5)\ (ft)}$

====================================<>====================================

Problem 6.67 Points A and B of the 1 *m* bar slide on the plane surfaces. The velocity of B is $\vec{v}_B = 2\vec{i}\ (m/s)$. (a) What are the coordinates of the instantaneous center? (b) Use the instantaneous center to determine the velocity at A.

Solution (a) A is constrained to move parallel to the *y* axis, and B is constrained to move parallel to the *x* axis. Draw perpendiculars to the velocity vectors at A and B. From geometry, the perpendiculars intersect at $\boxed{(\cos 70^o, \sin 70^o) = (0.3420, 0.9397)\ (m)}$. (b) The vector from the instantaneous center to point B is

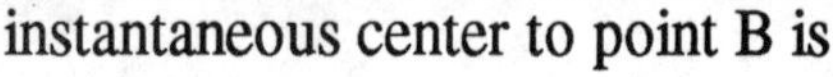

$\vec{r}_{B/C} = \vec{r}_B - \vec{r}_C = 0.3420\vec{i} - 0.3420\vec{i} - 0.9397\vec{j} = -0.9397\vec{j}$ The angular velocity of bar AB is obtained from

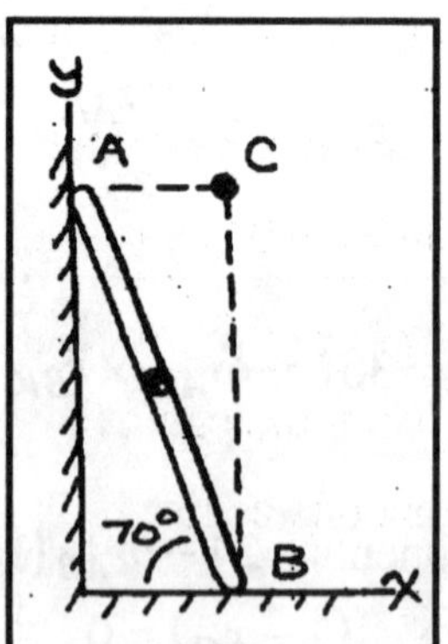

$$\vec{v}_B = 2\vec{i} = \vec{\omega}_{AB} \times \vec{r}_{B/C} = \begin{bmatrix} \vec{i} & \vec{j} & \vec{k} \\ 0 & 0 & \omega_{AB} \\ 0 & -0.9397 & 0 \end{bmatrix} = \omega_{AB}(0.9397)\vec{i},$$ from

which $\omega_{AB} = \frac{2}{0.9397} = 2.13\ rad/s$. The vector from the instantaneous center to point A is $\vec{r}_{A/C} = \vec{r}_A - \vec{r}_C = -0.3420\vec{i}\ (m)$. The velocity at A is

$$\vec{v}_A = \vec{\omega}_{AB} \times \vec{r}_{A/C} = \begin{bmatrix} \vec{i} & \vec{j} & \vec{k} \\ 0 & 0 & 2.13 \\ -0.3420 & 0 & 0 \end{bmatrix} \boxed{= -0.7279\vec{j}\ (m/s)}$$

====================================<>====================================

=================================<>=================================

Problem 6.68 In Problem 6.67, use the instantaneous center to determine the velocity of the bar's midpoint G.

Solution:

The vector to point G is $\vec{r}_{G/O} = (1/2)(0.3420\vec{i} + 0.9397\vec{j}) = 0.1710\vec{i} + 0.4698\vec{j}\ (m)$. From the solution to Problem 6.67, the vector to the instantaneous center is $\vec{r}_C = 0.3420\vec{i} + 0.9397\vec{j}\ (m)$, and $\omega_{AB} = 2.13\ rad/s$. The vector from the instantaneous center to the point G is $\vec{r}_{G/C} = \vec{r}_G - \vec{r}_C = -0.1710\vec{i} - 0.4698\vec{j}\ (m)$. The velocity of point G is

$$\vec{v}_G = \vec{\omega}_{AB} \times \vec{r}_{G/C} = \begin{bmatrix} \vec{i} & \vec{j} & \vec{k} \\ 0 & 0 & 2.13 \\ -0.1710 & -0.4698 & 0 \end{bmatrix} = \boxed{\vec{i} - 0.364\vec{j}\ (m/s)}$$

=================================<>=================================

Problem 6.69 The bar is in two dimensional motion in the *x-y* plane. The velocity of point A is $\vec{v}_A = 8\vec{i}\ (ft/s)$, and B is moving in the direction parallel to the bar. Determine the velocity of B by (a) using Eq (6.6); (b) by using the instantaneous center.

Solution:

(a) The unit vector parallel to the bar is $\vec{e}_{AB} = (\vec{i}\cos 30^o + \vec{j}\sin 30^o) = 0.866\vec{i} + 0.5\vec{j}$. The vector from A to B is $\vec{r}_{B/A} = 4\vec{e}_{AB} = 3.46\vec{i} + 2\vec{j}\ (ft)$. The velocity of point B is

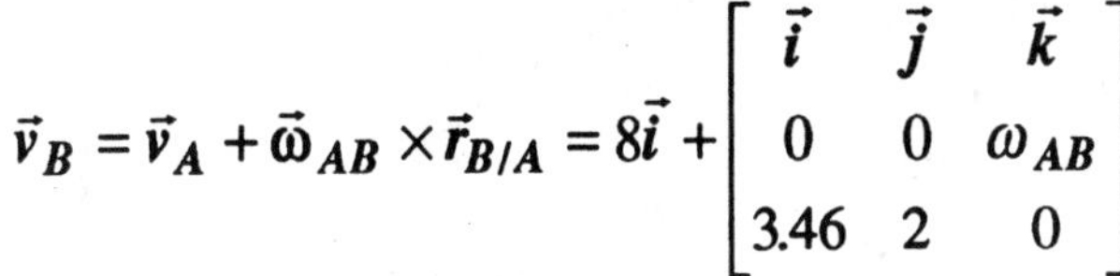

$$\vec{v}_B = \vec{v}_A + \vec{\omega}_{AB} \times \vec{r}_{B/A} = 8\vec{i} + \begin{bmatrix} \vec{i} & \vec{j} & \vec{k} \\ 0 & 0 & \omega_{AB} \\ 3.46 & 2 & 0 \end{bmatrix}$$

$\vec{v}_B = (8 - 2\omega_{AB})\vec{i} + 3.46\omega_{AB}\vec{j}$. But $\vec{v}_B$ is also moving parallel to the bar, $\vec{v}_B = v_B\vec{e}_{AB} = v_B(0.866\vec{i} + 0.5\vec{j})$. Equate, and separate components: $(8 - 2\omega_{AB} - 0.866v_B)\vec{i} = 0$, $(0.346\omega_{AB} - 0.5v_B)\vec{j} = 0$. Solve: $\omega_{AB} = 1\ rad/s$, $v_B = 6.93\ ft/s$, from which $\boxed{\vec{v}_B = v_B\vec{e}_{AB} = 6\vec{i} + 3.46\vec{j}\ (ft/s)}$ (b) Let (x_C, y_C) be the coordinates of the instantaneous center. The vector from the center to A is $\vec{r}_{A/C} = \vec{r}_A - \vec{r}_C = -\vec{r}_C = -x_C\vec{i} - y_C\vec{j}\ (ft)$. The vector from the instantaneous center to B is $\vec{r}_{B/C} = \vec{r}_B - \vec{r}_C = (3.46 - x_C)\vec{i} + (2 - y_C)\vec{j}$. The velocity of point A is

$$\vec{v}_A = 8\vec{i} = \vec{\omega}_{AB} \times \vec{r}_{A/C} = \begin{bmatrix} \vec{i} & \vec{j} & \vec{k} \\ 0 & 0 & \omega_{AB} \\ -x_C & -y_C & 0 \end{bmatrix} = \omega_{AB}y_C\vec{i} - \omega_{AB}x_C\vec{j}\ (ft/s).$$

From which $\boxed{x_C = 0}$, and $\omega_{AB}y_C = 8$. The velocity of point B is

$$\vec{v}_B = v_B\vec{e}_{AB} = \vec{\omega}_{AB} \times \vec{r}_{B/C} = \begin{bmatrix} \vec{i} & \vec{j} & \vec{k} \\ 0 & 0 & \omega_{AB} \\ 3.46 - x_C & 2 - y_C & 0 \end{bmatrix} = -\omega_{AB}(2 - y_C)\vec{i} + \omega_{AB}(3.46 - x_C)\vec{j}.$$

Solution continued on next page

Equate terms and substitute $\omega_{AB} y_C = 8$, and $x_C = 0$, to obtain: $(0.866 v_B + 2\omega_{AB} - 8)\vec{i} = 0$, and $(0.5 v_C - 3.46\omega_{AB})\vec{j} = 0$. These equations are algebraically identical with those obtained in Part (a) above (as can be shown by multiplying all terms by −1). Thus $\omega_{AB} = 1\ rad/s$, $v_B = 6.93\ (ft/s)$, and the velocity of B is that obtained in Part (a) $\boxed{\vec{v}_B = v_B \vec{e}_{AB} = 6\vec{i} + 3.46\vec{j}\ (ft/s)}$.

===================================<>===================================

Problem 6.70 Points A and B of the 4 *ft* bar slide on the plane surfaces. Point B is sliding down the slanted surface at 2 *ft/s*. (a) What are the coordinates of the instantaneous center? (b) Use the instantaneous center to determine the velocity of A.

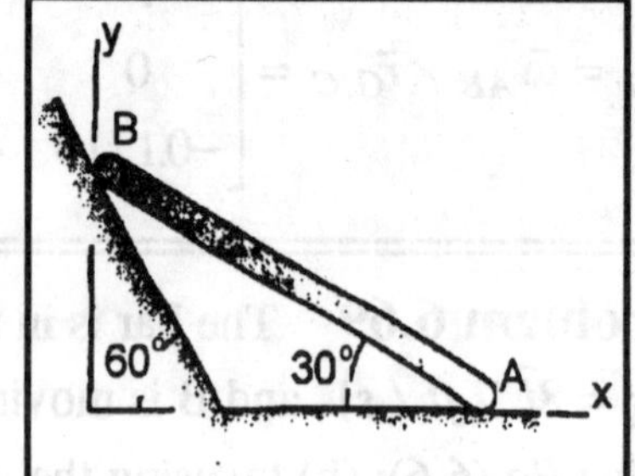

Solution (a) The strategy is to determine the coordinates of the instantaneous center by finding the intersection of perpendiculars to the motion. The unit vector parallel to the slanting surface in the direction of motion of B is $\vec{e}_S = \vec{i}\cos 60^o - \vec{j}\sin 60^o = 0.5\vec{i} - 0.866\vec{j}$. The vector perpendicular to this motion is $\vec{e}_{SP} = 0.866\vec{i} + 0.5\vec{j}$. A point on the line perpendicular to the velocity of B, from point B, is

$\vec{L}_{PB} = L_B(0.866\vec{i} + 0.5\vec{j}) + 4\sin 30^o \vec{j} = 0.866 L_B \vec{i} + (0.5 L_B + 2)\vec{j}$

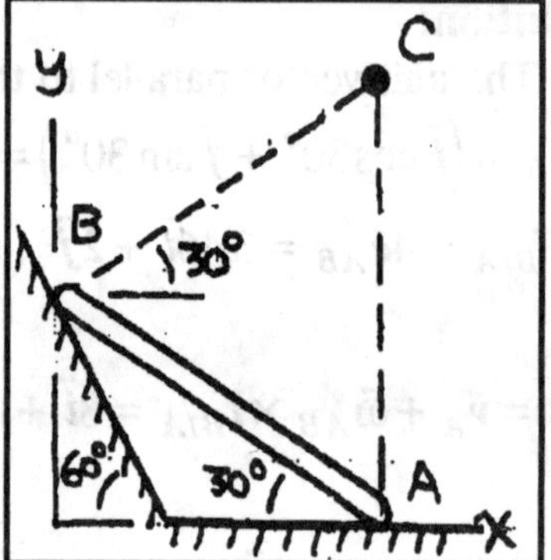

where L_B is the magnitude of the distance to the point along the line, and the height of B has been added to the *y* coordinate. The horizontal distance to the intersection of this line with a perpendicular to the motion of A is: $L_B(0.866) = 4\cos 30^o = 3.46\ in.$, from which $L_B = 4\ ft$. The vertical height of the intercept of this line and the perpendicular to the motion of A is $0.5 L_B + 2 = 4\ ft$. The coordinates of the instantaneous center are $\boxed{(3.46, 4)\ ft}$ (b) The angular velocity of the bar is determined from the known velocity of point B. The vector distance from the instantaneous center is

$\vec{r}_{B/C} = \vec{r}_B - \vec{r}_C = -3.46\vec{i} - (4 - 4\sin 30^o)\vec{j} = -3.46\vec{i} - 2\vec{j}\ (ft)$ The velocity of B is

$$\vec{v}_B = 2\vec{e}_S = \begin{bmatrix} \vec{i} & \vec{j} & \vec{k} \\ 0 & 0 & \omega_{AB} \\ -3.46 & -2 & 0 \end{bmatrix} = \omega_{AB}(2\vec{i} - 3.46\vec{j}).$$ Equate components: $(1 - 2\omega_{AB})\vec{i} = 0$, from which $\omega_{AB} = 0.5\ rad/s$. The vector from the instantaneous center to point A is

$\vec{r}_{A/C} = \vec{r}_A - \vec{r}_C = (3.46\vec{i} - 3.46\vec{i} - 4\vec{j}) = -4\vec{j}\ (ft)$. The velocity of point A is

$$\vec{v}_A = \vec{\omega} \times \vec{r}_{A/C} = \begin{bmatrix} \vec{i} & \vec{j} & \vec{k} \\ 0 & 0 & \omega_{AB} \\ 0 & -4 & 0 \end{bmatrix} = 4\omega_{AB}\vec{i} = \boxed{2\vec{i}\ (ft/s)}$$

===================================<>===================================

==================================<>==================================

Problem 6.71 Use instantaneous centers to determine the horizontal velocity of B.

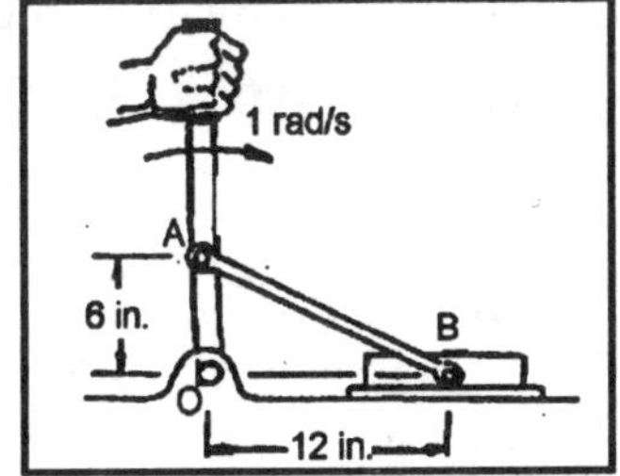

Solution The instantaneous center of OA lies at O, by definition, since O is the point of zero velocity, and the velocity at point A is parallel to the x-axis: $\vec{v}_A = \vec{\omega}_{OA} \times \vec{r}_{A/O} = \begin{bmatrix} \vec{i} & \vec{j} & \vec{k} \\ 0 & 0 & -\omega_{OA} \\ 0 & 6 & 0 \end{bmatrix} = 6\vec{i}\ \ (in/s)$. A line perpendicular to this motion is parallel to the y axis. The point B is constrained to move on the x axis, and a line perpendicular to this motion is also parallel to the y axis. These two lines will not intersect at any finite distance from the origin, hence *at the instant shown the instantaneous center of bar AB is at infinity and the angular velocity of bar AB is zero.* At the instant shown, the bar AB translates only, from which the horizontal velocity of B is the horizontal velocity at A:

$$\boxed{\vec{v}_B = \vec{v}_A = 6\vec{i}\ \ (in/s)}$$

==================================<>==================================

Problem 6.72 When the mechanism in Problem 6.71 is in this position, use instantaneous centers to determine the horizontal velocity of B.

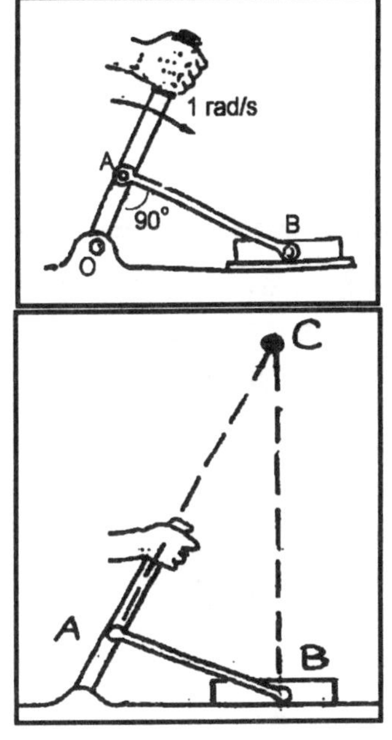

Solution The strategy is to determine the intersection of lines perpendicular to the motions at A and B. The velocity of A is parallel to the bar AB. A line perpendicular to the motion at A will be parallel to the bar OA. From the dimensions given in Problem 6.71, the length of bar AB is $r_{AB} = \sqrt{6^2 + 12^2} = 13.42\ \ in.$. Consider the triangle OAB. The interior angle at B is $\beta = \tan^{-1}\left(\frac{6}{r_{AB}}\right) = 24.1^o$, and the interior angle at O is $\theta = 90^o - \beta = 65.9^o$. The unit vector parallel to the handle OA is $\vec{e}_{OA} = \vec{i}\cos\theta + \vec{j}\sin\theta$, and a point on the line is $\vec{L}_{OA} = L_{OA}\vec{e}_{OA}$, where L_{OA} is the magnitude of the distance of the point from the origin. A line perpendicular to the motion at B is parallel to the y axis. At the intersection of the two lines $L_{OA}\cos\theta = \frac{r_{AB}}{\cos\beta}$, from which $L_{OA} = 36\ \ in.$. The coordinates of the instantaneous center are $(14.7, 32.9)$ $(in.)$. *Check*: From geometry, the triangle OAB and the triangle formed by the intersecting lines and the base are similar, and thus the interior angles are known for the larger triangle. From the law of sines $\frac{L_{OA}}{\sin 90^o} = \frac{r_{OB}}{\sin\beta} = \frac{r_{AB}}{\sin\beta\cos\beta} = 36\ \ in.$, and the coordinates follow immediately from $\vec{L}_{OA} = L_{OA}\vec{e}_{OA}$. *check.* The vector distance from O to A is $\vec{r}_{A/O} = 6(\vec{i}\cos\theta + \vec{j}\sin\theta) = 2.450\vec{i} + 5.478\vec{j}\ \ (in.)$. The angular velocity of the bar AB is determined from the known linear velocity at A.

$$\vec{v}_A = \vec{\omega}_{OA} \times \vec{r}_{A/O} = \begin{bmatrix} \vec{i} & \vec{j} & \vec{k} \\ 0 & 0 & -1 \\ 2.450 & 5.477 & 0 \end{bmatrix} = 5.48\vec{i} - 2.45\vec{j}\ \ (in/s).$$

Solution continued on next page

The vector from the instantaneous center to point A is
$\vec{r}_{A/C} = \vec{r}_{OA} - \vec{r}_C = 6\vec{e}_{OA} - (14.7\vec{i} + 32.86\vec{j}) = -12.25\vec{i} - 27.39\vec{j}$ (*in.*). The velocity at point A is

$$\vec{v}_A = \vec{\omega}_{AB} \times \vec{r}_{A/C} = \begin{bmatrix} \vec{i} & \vec{j} & \vec{k} \\ 0 & 0 & \omega_{AB} \\ -12.25 & -27.39 & 0 \end{bmatrix} = \omega_{AB}\left(27.39\vec{i} - 12.25\vec{j}\right)\ (ft/s)$$. Equate the two

expressions for the velocity at point A and separate components, $5.48\vec{i} = 27.39\omega_{OA}\vec{i}$, $-2.45\vec{j} = -12.25\omega_{OA}\vec{j}$ (one of these conditions is superfluous) and solve to obtain $\omega_{AB} = 0.2\ rad/s$, counterclockwise. [*Check:* The distance OA is 6 *in.* The magnitude of the velocity at A is $\omega_{OA}(6) = (1)(6) = 6\ in/s$. The distance to the instantaneous center from O is $\sqrt{14.7^2 + 32.9^2} = 36\ in.$, and from C to A is $(36-6) = 30\ in.$ from which $30\omega_{AB} = 6\ in/s$, from which $\omega_{AB} = 0.2\ rad/s$. *check.*]. The vector from the instantaneous center to point B is
$\vec{r}_{B/C} = \vec{r}_B - \vec{r}_C = 14.7\vec{i} - \left(14.7\vec{i} + 32.86\vec{j} = -32.86\vec{j}\right)$ (*in.*) The velocity at point B is

$$\vec{v}_B = \vec{\omega}_{AB} \times \vec{r}_{B/C} = \begin{bmatrix} \vec{i} & \vec{j} & \vec{k} \\ 0 & 0 & 0.2 \\ 0 & -32.86 & 0 \end{bmatrix} \boxed{= 6.57\vec{i}\ (in/s)}$$

================================<>================================

Problem 6.73 The angle $\theta = 45°$ and the bar OQ is rotating in the counterclockwise direction at *0.2 rad/s.* Use instantaneous centers to determine the velocity of the sleeve P.

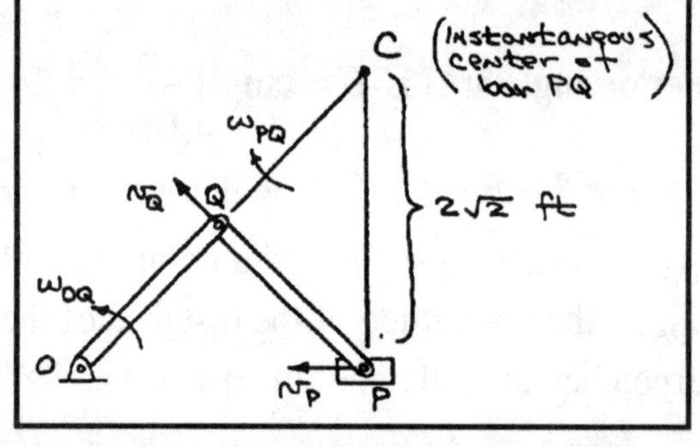

Solution:

The velocity of Q is $v_Q = 2\omega_{0Q} = 2(0.2) = 0.4\,ft/s$.

Therefore $|\overline{\omega}_{PQ}| = \dfrac{v_Q}{2\,ft} = \dfrac{0.4}{2} = 0.2\,rad/s$ (clockwise) and

$|\vec{v}_P| = 2\sqrt{2}\omega_{PQ} = 0.566\,ft/s$ ($\vec{v}_P$ is to the left).

================================<>================================

Problem 6.74 Consider the system shown in Problem 6.73. The angle $\theta = 40°$ and the sleeve P is moving to the right at *1 ft/s.* Use instantaneous centers to determine the angular velocities of the bars *OQ* and *PQ.*

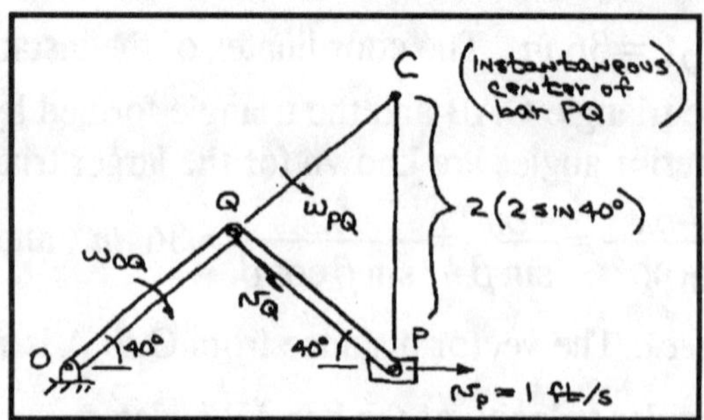

Solution:

From the figure, $\omega_{PQ} = \dfrac{1\,ft/s}{2(2\sin 40°)} = 0.389\,rad/s$. The velocity of Q is $v_Q = 2\omega_{PQ} = 0.778\,ft/s$ so

$$\omega_{0Q} = \frac{v_Q}{2\,ft} = 0.389\,rad/s \text{ (clockwise)}$$

================================<>================================

==================================<>==================================

Problem 6.75 Bar AB rotates at 6 *rad/s* in the clockwise direction. Use the instantaneous centers to determine the angular velocity of bar BC.

Solution:

Choose a coordinate system with origin at A and y axis vertical. Let C' denote the instantaneous center. The instantaneous center for bar AB is the point A, by definition, since A is the point of zero velocity. The vector AB is $\vec{r}_{B/A} = 4\vec{i} + 4\vec{j}$ $(in.)$. The velocity at B is

$$\vec{v}_B = \vec{\omega}_{AB} \times \vec{r}_{B/A} = \begin{bmatrix} \vec{i} & \vec{j} & \vec{k} \\ 0 & 0 & 6 \\ 4 & 4 & 0 \end{bmatrix} = 24\vec{i} - 24\vec{j} \ \ (in/s).$$

The unit vector parallel to AB is also the unit vector perpendicular to the velocity at B, $\vec{e}_{AB} = \frac{1}{\sqrt{2}}(\vec{i} + \vec{j})$. The vector location of a point on a line perpendicular to the velocity at B is $\vec{L}_{AB} = L_{AB}\vec{e}_{AB}$, where L_{AB} is the magnitude of the distance from point A to the point on the line. The vector location of a point on a perpendicular to the velocity at C is $\vec{L}_C = (14\vec{i} + y\vec{j})$ where y is the y-coordinate of the point referenced

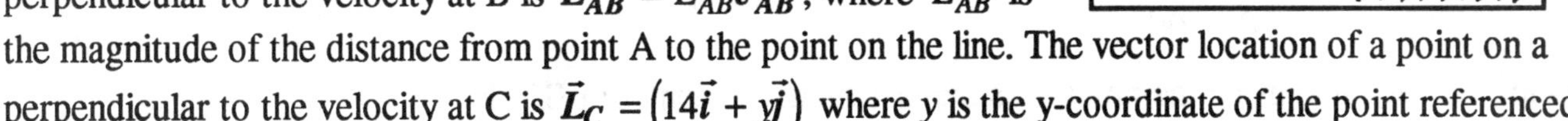

to an origin at A. When the two lines intersect, $\frac{L_{AB}}{\sqrt{2}}\vec{i} = 14\vec{i}$, and $y = \frac{L_{AB}}{\sqrt{2}} = 14$ from which $L_{AB} = 19.8$ *in.*, and the coordinates of the instantaneous center are $(14, 14)$ $(in.)$.

[*Check*: The line AC' is the hypotenuse of a right triangle with a base of 14 *in.* and interior angles of 45°, from which the coordinates of C' are (14, 14) *in. check.*]. The angular velocity of bar BC is determined from the known velocity at B. The vector from the instantaneous center to point B is $\vec{r}_{B/C} = \vec{r}_B - \vec{r}_C = 4\vec{i} + 4\vec{j} - 14\vec{i} - 14\vec{j} = -10\vec{i} - 10\vec{j}$. The velocity of point B is

$$\vec{v}_B = \vec{\omega}_{BC} \times \vec{r}_{B/C} = \begin{bmatrix} \vec{i} & \vec{j} & \vec{k} \\ 0 & 0 & \omega_{BC} \\ -10 & -10 & 0 \end{bmatrix} = \omega_{BC}(10\vec{i} - 10\vec{j}) \ \ (in/s).$$

Equate the two expressions for the velocity:: $24 = 10\omega_{BC}$, from which $\boxed{\omega_{BC} = 2.4 \ rad/s}$,

==================================<>==================================

====================================<>====================================

Problem 6.76 Bar AB rotates at 10 *rad/s* in the counterclockwise direction. Use the instantaneous centers to determine the velocity of point E.

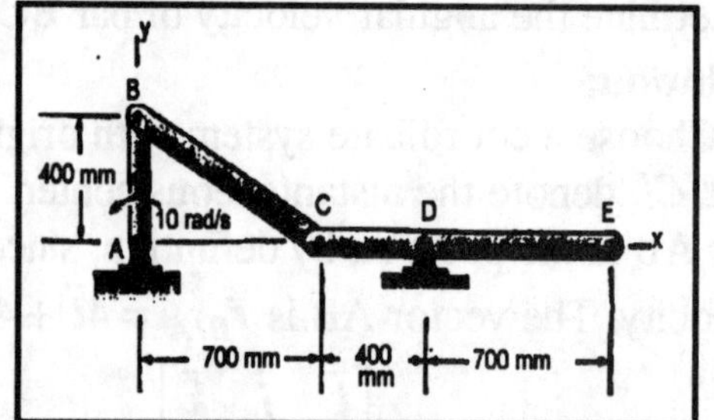

Solution:

Choose a coordinate system with origin at A, with the *x* axis parallel to bar CDE. The instantaneous center of bar AB is point A, and the instantaneous center of bar CDE is point D, by definition, since these are the points with zero velocity. Since AB rotates about A, the velocity of point B will be parallel to the *x* axis, and a line perpendicular to the velocity of B will be parallel to the *y* axis. Since CDE rotates about D, the velocity of point C is parallel to the *y* axis. A line perpendicular to the velocity at C will be parallel to the *x* axis. From inspection, the intersection of these perpendicular lines will be point A. *Thus point A is the instantaneous center for both bar AB and bar BC.* The velocity of point B, using the known angular velocity of bar AB, is

$$\vec{v}_B = \vec{\omega}_{AB} \times \vec{r}_{B/A} = \begin{bmatrix} \vec{i} & \vec{j} & \vec{k} \\ 0 & 0 & 10 \\ 0 & 400 & 0 \end{bmatrix} = -4000\vec{i} \ (mm/s)$$

. The velocity at B, using the unknown angular velocity of BC, and using the point A as the instantaneous center of BC, is

$$\vec{v}_B = \vec{\omega}_{BC} \times \vec{r}_{B/A} = \begin{bmatrix} \vec{i} & \vec{j} & \vec{k} \\ 0 & 0 & \omega_{BC} \\ 0 & 400 & 0 \end{bmatrix} = -400\omega_{BC}\vec{i} \ (mm/s)$$

. Equate the two expressions for the velocity at B, $-4000 = -400\omega_{BC}$, from which, $\omega_{BC} = \omega_{AB} = 10 \ rad/s$. The vector distance from A to C is $\vec{r}_{C/A} = 700\vec{i} \ (mm)$. The velocity of point C is

$$\vec{v}_C = \vec{\omega}_{BC} \times \vec{r}_{C/A} = \begin{bmatrix} \vec{i} & \vec{j} & \vec{k} \\ 0 & 0 & 10 \\ 700 & 0 & 0 \end{bmatrix} = 7000\vec{j} \ (mm/s)$$

. The vector distance from the instantaneous center at D to C is $\vec{r}_{C/D} = -400\vec{i} \ (mm)$. The velocity at point C is

$$\vec{v}_C = \vec{\omega}_{CDE} \times \vec{r}_{C/D} = \begin{bmatrix} \vec{i} & \vec{j} & \vec{k} \\ 0 & 0 & \omega_{CDE} \\ -400 & 0 & 0 \end{bmatrix} = -400\omega_{CDE}\vec{j}$$

Equate the expressions for the velocity at C, $7000 = 400\omega_{CDE}$ from which: $\omega_{CDE} = -17.5 \ rad/s$ clockwise. The vector from D to E is $\vec{r}_{E/D} = 700\vec{i} \ (mm)$. The velocity of point E is

$$\vec{v}_E = \vec{\omega}_{CDE} \times \vec{r}_{E/D} = \begin{bmatrix} \vec{i} & \vec{j} & \vec{k} \\ 0 & 0 & -17.5 \\ 700 & 0 & 0 \end{bmatrix} = \boxed{-12250\vec{j} \ (mm/s)}$$

====================================<>====================================

==============================<>==============================

Problem 6.77 The disks roll on the plane surface. The left disk rotates at 2 *rad/s* in the clockwise direction. Use the instantaneous centers to determine the angular velocities of the bar and the right disk.

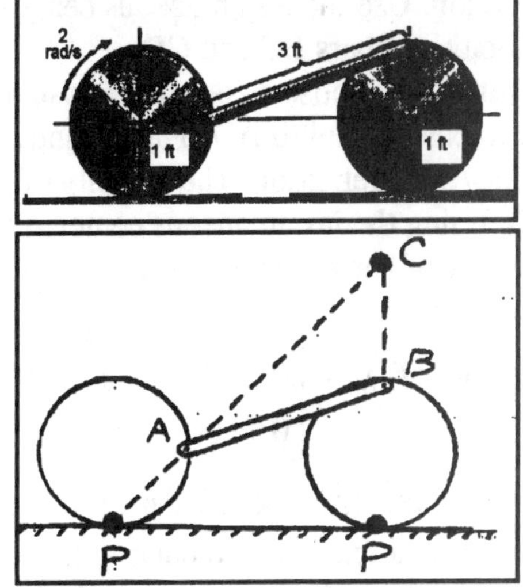

Solution Choose a coordinate system with the origin at the point of contact of the left disk with the surface, and the x axis parallel to the plane surface. Denote the point of attachment of the bar to the left disk by A, and the point of attachment to the right disk by B. The instantaneous center of the left disk is the point of contact with the surface. The vector distance from the point of contact to the point A is $\vec{r}_{A/P} = \vec{i} + \vec{j}$ (ft). The velocity of point A is

$$\vec{v}_A = \vec{\omega}_{LD} \times \vec{r}_{A/P} = \begin{bmatrix} \vec{i} & \vec{j} & \vec{k} \\ 0 & 0 & 2 \\ 1 & 1 & 0 \end{bmatrix} = 2\vec{i} - 2\vec{j} \ (ft/s).$$

The point on a line perpendicular to the velocity at A is $\vec{L}_A = L_A(\vec{i} + \vec{j})$, where L_A is the distance of the point from the origin. The point B is at the top of the right disk, and the velocity is constrained to be parallel to the x axis. A point on a line perpendicular to the velocity at B is $\vec{L}_B = (1 + 3\cos\theta)\vec{i} + y\vec{j}$ (ft), where $\theta = \sin^{-1}\left(\frac{1}{3}\right) = 19.5^o$. At the intersection of these two lines $L_A = 1 + 3\cos\theta = 3.83$ ft, and the coordinates of the instantaneous center of the bar are $(3.83, 3.83)$ (ft). The angular velocity of the bar is determined from the known velocity of point A. The vector from the instantaneous center to point A is $\vec{r}_{A/C} = \vec{r}_A - \vec{r}_C = \vec{i} + \vec{j} - 3.83\vec{i} - 3.83\vec{j} = -2.83\vec{i} - 2.83\vec{j}$ (ft). The velocity of point A is

$$\vec{v}_A = \vec{\omega}_{AB} \times \vec{r}_{A/C} = \begin{bmatrix} \vec{i} & \vec{j} & \vec{k} \\ 0 & 0 & \omega_{AB} \\ -2.83 & -2.83 & 0 \end{bmatrix} = \omega_{AB}(2.83\vec{i} - 2.83\vec{j}) \ (ft/s).$$

Equate the two expressions and solve: $\boxed{\omega_{AB} = \frac{2}{2.83} = 0.7071 \ (rad/s)}$ counterclockwise.

The vector from the instantaneous center to point B is $\vec{r}_{B/C} = \vec{r}_B - \vec{r}_C = (1 + 3\cos\theta)\vec{i} + 2\vec{j} - 3.83\vec{i} - 3.83\vec{j} = -1.83\vec{j}$. The velocity of point B is

$$\vec{v}_B = \vec{\omega}_{AB} \times \vec{r}_{B/A} = \begin{bmatrix} \vec{i} & \vec{j} & \vec{k} \\ 0 & 0 & 0.7071 \\ 0 & -1.83 & 0 \end{bmatrix} = 1.294\vec{i} \ (ft/s).$$

Using the fixed center at point of contact:

$$\vec{v}_B = \vec{\omega}_{RD} \times \vec{r}_{B/P} = \begin{bmatrix} \vec{i} & \vec{j} & \vec{k} \\ 0 & 0 & \omega_{RD} \\ 0 & 2 & 0 \end{bmatrix} = -2\omega_{RD}\vec{i} \ (ft/s).$$

Equate the two expressions for $\vec{v}_B$ and solve: $\boxed{\omega_{RD} = -0.647 \ rad/s}$, clockwise.

==============================<>==============================

Problem 6.78 The bar AB rotates at 12 *rad/s* in the clockwise direction. Use the instantaneous centers to determine the angular velocities of bars BC and CD.

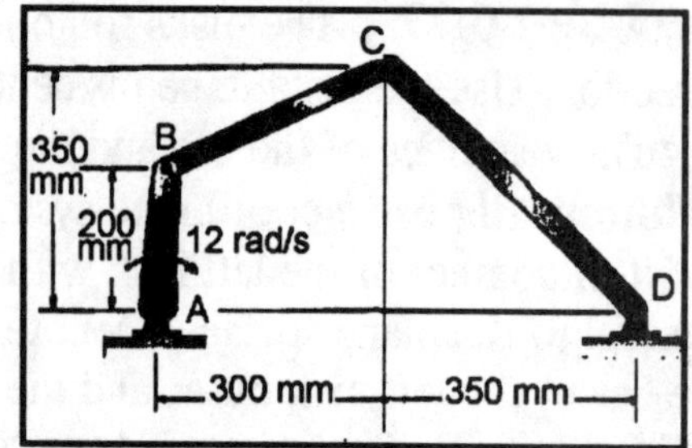

Solution Choose a coordinate system with the origin at A and the *x* axis parallel to AD. The instantaneous center of bar AB is point A, by definition. The velocity of point B is normal to the bar AB. Using the instantaneous center A and the known angular velocity of bar AB the velocity of B is

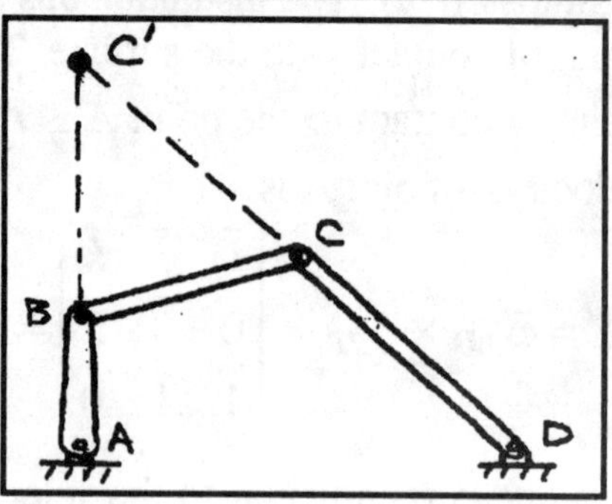

$$\vec{v}_B = \vec{\omega} \times \vec{r}_{B/A} = \begin{bmatrix} \vec{i} & \vec{j} & \vec{k} \\ 0 & 0 & -12 \\ 0 & 200 & 0 \end{bmatrix} = 2400\vec{i} \ \ (mm/s).$$

The unit vector perpendicular to the velocity of B is $\vec{e}_{AB} = \vec{j}$, and a point on a line perpendicular to the velocity at B is $\vec{L}_{AB} = L_{AB}\vec{j} \ \ (mm)$. The instantaneous center of bar CD is point D, by definition. The velocity of point C is constrained to be normal to bar CD. The interior angle at D is 45°, by inspection. The unit vector parallel to DC (and perpendicular to the velocity at C) is $\vec{e}_{DC} = -\vec{i}\cos 45^o + \vec{j}\sin 45^o = \left(\frac{1}{\sqrt{2}}\right)(-\vec{i} + \vec{j})$. The point on a line parallel to DC is $\vec{L}_{DC} = \left(650 - \frac{L_{DC}}{\sqrt{2}}\right)\vec{i} + \frac{L_{DC}}{\sqrt{2}}\vec{j} \ \ (mm)$. At the intersection of these lines $\vec{L}_{AB} = \vec{L}_{DC}$, from which $\left(650 - \frac{L_{DC}}{\sqrt{2}}\right) = 0$ and $L_{AB} = \frac{L_{DC}}{\sqrt{2}}$, from which $L_{DC} = 919.2 \ mm$, and $L_{AB} = 650 \ mm$.

The coordinates of the instantaneous center of bar BC are $(0, \ 650) \ \ (mm)$. Denote this center by C'. The vector from C' to point B is $\vec{r}_{B/C'} = \vec{r}_B - \vec{r}_{C'} = 200\vec{j} - 650\vec{j} = -450\vec{j}$.

The vector from C' to point C is $\vec{r}_{C/C'} = 300\vec{i} + 350j - 650j = 300\vec{i} - 300\vec{j} \ \ (mm)$.

The velocity of point B is $\vec{v}_B = \vec{\omega}_{BC} \times \vec{r}_{B/C'} = \begin{bmatrix} \vec{i} & \vec{j} & \vec{k} \\ 0 & 0 & \omega_{BC} \\ 0 & -450 & 0 \end{bmatrix} = 450\omega_{BC}\vec{i} \ \ (mm/s)$. Equate and solve: $2400 = 450\omega_{BC}$, from which $\boxed{\omega_{BC} = \frac{2400}{450} = 5.33 \ (rad/s)}$. The angular velocity of bar CD is determined from the known velocity at point C. The velocity at C is

$$\vec{v}_C = \vec{\omega}_{BC} \times \vec{r}_{C/C'} = \begin{bmatrix} \vec{i} & \vec{j} & \vec{k} \\ 0 & 0 & 5.33 \\ 300 & -300 & 0 \end{bmatrix} = 1600\vec{i} + 1600\vec{j} \ \ (mm/s).$$

The vector from D to point C is $\vec{r}_{C/D} = -350\vec{i} + 350\vec{j} \ \ (mm)$. The velocity at C is

$$\vec{v}_C = \vec{\omega}_{CD} \times \vec{r}_{C/D} = \begin{bmatrix} \vec{i} & \vec{j} & \vec{k} \\ 0 & 0 & \omega_{CD} \\ -350 & 350 & 0 \end{bmatrix} = -350\omega_{CD}\vec{i} - 350\omega_{CD}\vec{j} \ \ (mm/s).$$

Equate and solve:

$\boxed{\omega_{CD} = -4.57 \ rad/s}$ clockwise.

==================================<>==================================

Problem 6.79 The horizontal member *ADE* supporting the scoop is stationary. The link BD is rotating in the clockwise direction at *1 rad/s*. Use instantaneous centers to determine the angular velocity of the scoop.

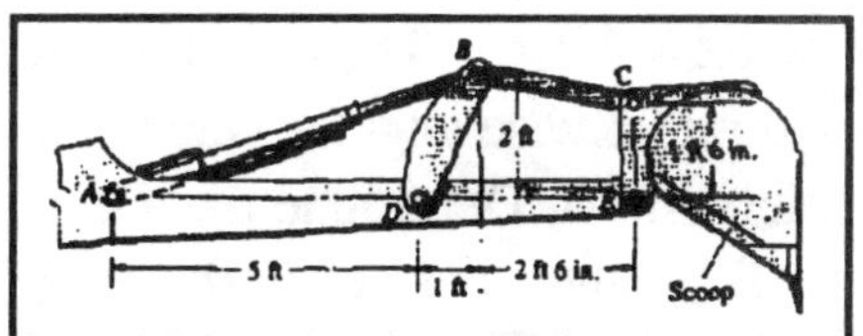

Solution:

The distance from D to B is $r_{BD} = \sqrt{1^2 + 2^2} = 2.24\,ft$. The distance from B to H is $r_{BH} = \dfrac{3.5}{\cos 63.4^\circ} - r_{BD} = 5.59\,ft$, and the distance from C to H is $r_{CH} = 3.5\tan 63.4^\circ - r_{CE} = 5.5\,ft$. The velocity of B is $v_B = r_{BD}\omega_{BD} = (2.24)(1) = 2.24\,ft/s$. Therefore $\omega_{BC} = \dfrac{v_B}{r_{BH}} = \dfrac{2.24}{5.59} = 0.4\,rad/s$. The velocity of C is $v_c = r_{CH}\omega_{BC} = (5.5)(0.4) = 2.2\,ft/s$, so the angular velocity of the scoop is $\omega_{CE} = \dfrac{v_C}{r_{CE}} = \dfrac{2.2}{1.5} = 1.47\,rad/s$

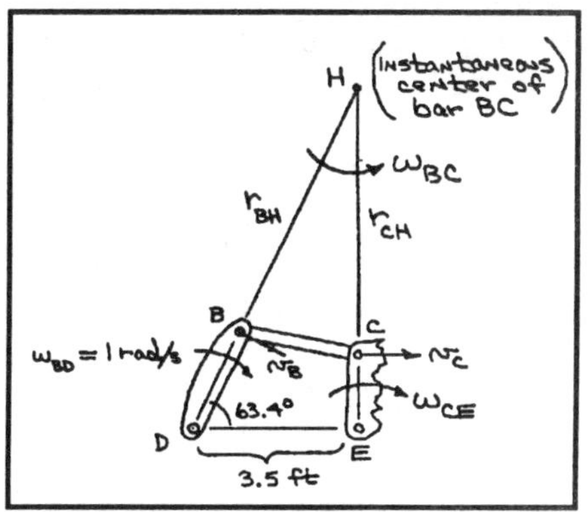

==================================<>==================================

Problem 6.80 Show that if a rigid body in planar motion has two instantaneous centers, it is stationary at that instant.

Solution Let A be any point on the body with coordinates (x, y). Let (x_1, y_1) and (x_2, y_2) be the coordinates of the instantaneous centers. The vector distances from the centers to A are $\vec{r}_{A/1} = (x - x_1)\vec{i} + (y - y_1)\vec{j}$, and $\vec{r}_{A/2} = (x - x_2)\vec{i} + (y - y_2)\vec{j}$. The velocity of point A is $\vec{v}_A = \vec{\omega} \times \vec{r}_{A/1} = \vec{\omega} \times \vec{r}_{A/2}$, from which

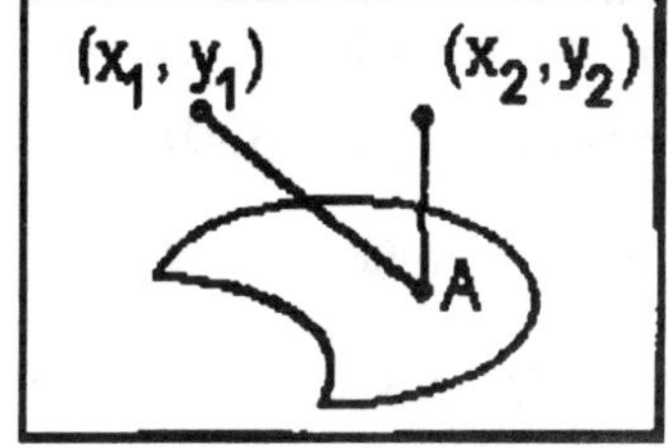

$$\vec{v}_A = \vec{\omega} \times \vec{r}_{A/1} = \begin{bmatrix} \vec{i} & \vec{j} & \vec{k} \\ 0 & 0 & \omega \\ x - x_1 & y - y_1 & 0 \end{bmatrix} = -\omega(y - y_1)\vec{i} + \omega(x - x_1)\vec{j},$$

$$\vec{v}_A = \vec{\omega} \times \vec{r}_{A/2} = \begin{bmatrix} \vec{i} & \vec{j} & \vec{k} \\ 0 & 0 & \omega \\ x - x_2 & y - y_2 & 0 \end{bmatrix} = -\omega(y - y_2)\vec{i} + \omega(x - x_2)\vec{j}.$$

Equate the expressions and separate components: $\omega(y - y_1) = \omega(y - y_2)$, and $\omega(x - x_1) = \omega(x - x_2)$. These two equations reduce to: $\omega y_1 = \omega y_2$, and $\omega x_1 = \omega x_2$. If $\omega \neq 0$, then these imply that $x_1 = x_2$, and $y_1 = y_2$, that is, only one instantaneous center exists. But this violates the initial assumptions, hence the only solution is $\omega = 0$, that is, *the angular velocity must be zero if there is more than one instantaneous center.* (*Note*: This does not rule out a translation of the body; rather it means that the body cannot be rotating.)

==================================<>==================================

Problem 6.81 The rigid body rotates about the *z* axis with counterclockwise angular velocity ω and counterclockwise angular acceleration α. Determine the acceleration of point A relative to point B (a) by using Eq (6.9); (b) by using Eq (6.10).

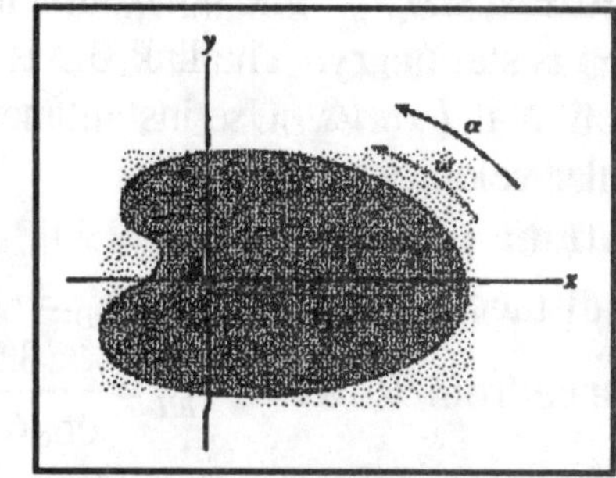

Solution (a) The acceleration of A relative to B is given by Eq (6.9): $\vec{a}_A = \vec{a}_B + \vec{\alpha}\times\vec{r}_{A/B} + \vec{\omega}\times(\vec{\omega}\times\vec{r}_{A/B})$. Expanding the cross products:

$$\vec{\alpha}\times\vec{r}_{A/B} = \begin{bmatrix} \vec{i} & \vec{j} & \vec{k} \\ 0 & 0 & \alpha \\ r_{A/B} & 0 & 0 \end{bmatrix} = \alpha r_{A/B}\vec{j}.$$

$$\vec{\omega}\times(\vec{\omega}\times\vec{r}_{A/B}) = \vec{\omega}\times\begin{bmatrix} \vec{i} & \vec{j} & \vec{k} \\ 0 & 0 & \omega \\ r_{A/B} & 0 & 0 \end{bmatrix} = \begin{bmatrix} \vec{i} & \vec{j} & \vec{k} \\ 0 & 0 & \omega \\ 0 & \omega r_{A/B} & 0 \end{bmatrix} = -\omega^2 r_{A/B}\vec{i} = -\omega^2\vec{r}_{A/B}.$$

Collecting terms $\vec{a}_A = \vec{a}_B + \alpha r_{AB}\vec{j} - \omega^2 r_{A/B}\vec{i}$. Noting that $\vec{a}_B = 0$, $\boxed{\vec{a}_A = \alpha r_{A/B}\vec{j} - \omega^2 r_{A/B}\vec{i}}$ (b) The acceleration of A relative to B is given by Eq (6.10): $\vec{a}_A = \vec{a}_B + \vec{\alpha}\times\vec{r}_{A/B} - \omega^2\vec{r}_{A/B}$. Using the expansions of the cross product above, noting that $\vec{a}_B = 0$, $\boxed{\vec{a}_A = \alpha r_{A/B}\vec{j} - \omega^2 r_{A/B}\vec{i}}$

Problem 6.82 The bar rotates with a counterclockwise angular velocity of 5 *rad/s* and a counterclockwise angular acceleration of 30 *rad/s²*. Determine the acceleration of A (a) by using Eq (6.9); (b) by using Eq (6.10).

Solution (a) Eq (6.9): $\vec{a}_A = \vec{a}_B + \vec{\alpha}\times\vec{r}_{A/B} + \vec{\omega}\times(\vec{\omega}\times\vec{r}_{A/B})$.

Substitute values: $\vec{a}_B = 0$. $\vec{\alpha} = 30\vec{k}\ (rad/s^2)$,

$\vec{r}_{A/B} = 2(\vec{i}\cos 30^o + \vec{j}\sin 30^o) = 1.732\vec{i} + \vec{j}\ (m)$.

$\vec{\omega} = 5\vec{k}\ (rad/s)$. Expand the cross products:

$$\vec{\alpha}\times\vec{r}_{A/B} = \begin{bmatrix} \vec{i} & \vec{j} & \vec{k} \\ 0 & 0 & 30 \\ 1.732 & 1 & 0 \end{bmatrix} = -30\vec{i} + 52\vec{j}\ (m/s^2).$$

$$\vec{\omega}\times\vec{r}_{A/B} = \begin{bmatrix} \vec{i} & \vec{j} & \vec{k} \\ 0 & 0 & 5 \\ 1.732 & 1 & 0 \end{bmatrix} = -5\vec{i} + 8.66\vec{j}\ (m/s).$$

$$\vec{\omega}\times(\vec{\omega}\times\vec{r}_{A/B}) = \begin{bmatrix} \vec{i} & \vec{j} & \vec{k} \\ 0 & 0 & 5 \\ -5 & 8.66 & 0 \end{bmatrix} = -43.3\vec{i} - 25\vec{j}\ (m/s^2).$$

Collect terms:

$\boxed{\vec{a}_A = -73.3\vec{i} + 27\vec{j}\ (m/s^2)}$.(b) Eq(6.10): $\vec{a}_A = \vec{a}_B + \vec{\alpha}\times\vec{r}_{A/B} - \omega^2\vec{r}_{A/B}$. Substitute values, and expand the cross product as in Part (b) to obtain $\boxed{\vec{a}_A = -30\vec{i} + 52\vec{j} - (5^2)(1.732\vec{i} + \vec{j}) = -73.3\vec{i} + 27\vec{j}\ (m/s^2)}$

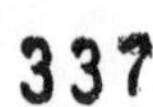

Problem 6.83 The bar rotates with a constant angular velocity of 20 *rad/s* in the counterclockwise direction. (a) Determine the acceleration of point B. (b) Using your result from Part (a) and Eq(6.10), determine the acceleration of point A.

Solution:

(a) Use Eq (6.10). $\vec{a}_B = \vec{a}_O + \vec{\mathbf{a}} \times \vec{r}_{O/B} - \omega^2 \vec{r}_{O/B}$.

Substitute values: $\vec{a}_O = 0$, $\vec{\mathbf{a}} = 0$, $\vec{r}_{B/O} = 2\vec{i}$ (m),

$\omega = 20$ (rad / s), to obtain

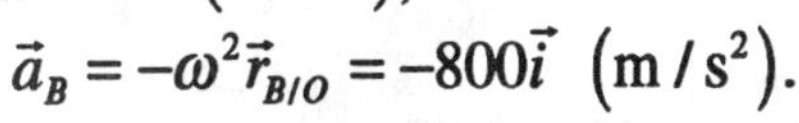

$\vec{a}_B = -\omega^2 \vec{r}_{B/O} = -800\vec{i}$ $\left(\mathrm{m/s^2}\right)$.

(b) Eq (6.10): $\vec{a}_A = \vec{a}_B + \vec{\mathbf{a}} \times \vec{r}_{A/B} - \omega^2 \vec{r}_{A/B}$.

Substitute values: $\vec{a}_B = -800\vec{i}$ $\left(\mathrm{m/s^2}\right)$, $\vec{\mathbf{a}} = 0$, $\vec{r}_{A/B} = -\vec{i}$ (m), $\omega = 20$ (rad / s),

from which $\vec{a}_B = -800\vec{i} + (400)\vec{i} = -400\vec{i}$ $\left(\mathrm{m/s^2}\right)$

Problem 6.84 The helicopter is in planar motion in the *x-y* plane. At the instant shown, the position of its center of mass *G* is *x = 2 m, y = 2.5 m,* its velocity is $\mathbf{v}_G = 12\,\mathbf{i} + 4\,\mathbf{j}\ (m/s)$, and its acceleration is $\mathbf{a}_G = 2\,\mathbf{i} + 3\,\mathbf{j}\ (m/s^2)$. The position of point *T* where the tail rotor is mounted is *x = -3.5 m, y = 4.5 m.* The helicopter's angular velocity is *0.2 rad/s* clockwise, and its angular acceleration is 0.1 rad/s^2 counterclockwise. What is the acceleration of point *T*?

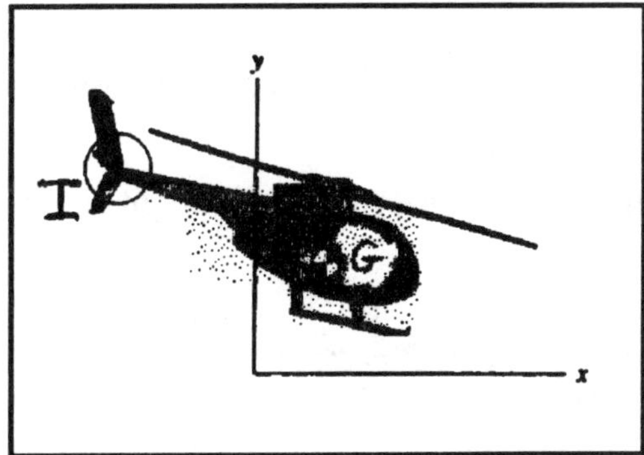

Solution:

The acceleration of *T* is $\bar{a}_T = \bar{a}_G + \bar{\alpha} \times \bar{r}_{T/G} - \omega^2 \bar{r}_{T/G}$;

$$\bar{a}_T = 2\bar{i} + 3\bar{j} + \begin{vmatrix} \bar{i} & \bar{j} & \bar{k} \\ 0 & 0 & 0.1 \\ -5.5 & 2 & 0 \end{vmatrix} - (0.2)^2(-5.5\bar{i} + 2\bar{j}) = 2.02\bar{i} + 2.37\bar{j}(m/s^2).$$

=====================================<>=====================================

Problem 6.85 The disk rolls on the plane surface. The velocity of point A is 6 *m/s* to the right, and its acceleration is 20 m/s^2 to the right. (a) What is the angular acceleration vector of the disk? (b) Determine the accelerations of point B, C and D.

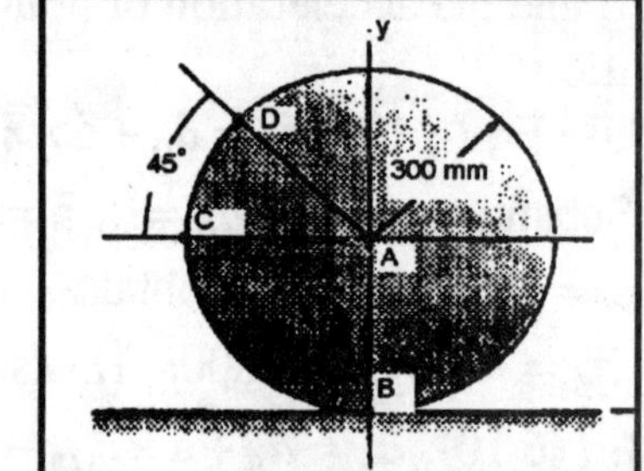

Solution (a) The point of contact P between the disk and the surface is stationary. The distance A to P is $\vec{r}_{P/A} = -\vec{R} = -R\vec{j}$ (m), from which

$$\vec{v}_P = 0 = \vec{v}_A + \vec{\omega} \times \vec{r}_{P/A} = 6\vec{i} + \begin{bmatrix} \vec{i} & \vec{j} & \vec{k} \\ 0 & 0 & \omega \\ 0 & -0.3 & 0 \end{bmatrix} = (6 + 0.3\omega)\vec{i} = 0, \text{ from}$$

which $\omega = -\dfrac{6}{0.3} = -20$ rad/s, $\vec{\omega} = -20\vec{k}$. The instantaneous center of the rolling disk is point B. From $\vec{v}_P = 0 = \vec{v}_A + \vec{\omega} \times \vec{r}_{P/A}$, the velocity of point A is $\vec{v}_A = \vec{\omega} \times \vec{R}$, where $\vec{R} = \vec{j}R$. Differentiating, $\dfrac{d\vec{v}_A}{dt} = \vec{a}_A = \dfrac{d\vec{\omega}}{dt} \times \vec{R} = \vec{\alpha} \times \vec{R}$. By definition $\vec{\alpha} = \dfrac{d\vec{\omega}}{dt} = \dfrac{d(\omega\vec{k})}{dt} = \dfrac{d\omega}{dt}\vec{k} + \omega\dfrac{d\vec{k}}{dt} = \alpha\vec{k}$, since by assumption the disk rolls in a straight line, and $\dfrac{d\vec{k}}{dt} = 0$. From which

$$\vec{a}_A = \frac{d\vec{v}_A}{dt} = \vec{\alpha} \times \vec{R} = \alpha\vec{k} \times \vec{R} = \begin{bmatrix} \vec{i} & \vec{j} & \vec{k} \\ 0 & 0 & \alpha \\ 0 & R & 0 \end{bmatrix} = -\alpha R\vec{i} \ \left(m/s^2\right).$$ Substitute values, $-\alpha R\vec{i} = 20\vec{i}$, from

which $\alpha = -\dfrac{20}{0.3} = -66.7$ $\left(m/s^2\right)$, $\boxed{\vec{\alpha} = -66.7\vec{k} \ \left(m/s^2\right)}$. (b) Use Eq (6.10) to determine the accelerations. The acceleration of point B is

$$\vec{a}_B = \vec{a}_A + \vec{\alpha} \times \vec{r}_{B/A} - \omega^2\vec{r}_{B/A} = 20\vec{i} + \begin{bmatrix} \vec{i} & \vec{j} & \vec{k} \\ 0 & 0 & -66.7 \\ 0 & -3 & 0 \end{bmatrix} - (20)^2(-0.3\vec{j}),$$

$\boxed{\vec{a}_B = 20\vec{i} - 0.3(66.7)\vec{i} + \left(20^2\right)(0.3)\vec{j} = 120\vec{j} \ \left(m/s^2\right)}$ The acceleration of point C is

$$\vec{a}_C = \vec{a}_A + \vec{\alpha} \times \vec{r}_{C/A} - \omega^2\vec{r}_{C/A} = 20\vec{i} + \begin{bmatrix} \vec{i} & \vec{j} & \vec{k} \\ 0 & 0 & -66.7 \\ -0.3 & 0 & 0 \end{bmatrix} - (20)^2(-0.3\vec{i}),$$

$\boxed{\vec{a}_C = 20\vec{i} + 20\vec{j} + \left(20^2\right)(0.3)\vec{i} = 140\vec{i} + 20\vec{j} \ \left(m/s^2\right)}$ The acceleration of point D is $\vec{a}_D = \vec{a}_A + \vec{\alpha} \times \vec{r}_{D/A} - \omega^2\vec{r}_{D/A}$, from which

$$\vec{a}_D = 20\vec{i} + \begin{bmatrix} \vec{i} & \vec{j} & \vec{k} \\ 0 & 0 & -66.7 \\ -0.3\cos 45^o & 0.3\sin 45^o & 0 \end{bmatrix} - \left(20^2\right)(0.3)\left(-\vec{i}\cos 45^o + \vec{j}\sin 45^o\right),$$

$\boxed{\vec{a}_D = 119\vec{i} - 70.7\vec{j} \ \left(m/s^2\right)}$

=====================================<>=====================================

==================================<>==================================

Problem 6.86 The disk rolls on the circular surface with a constant clockwise angular velocity of 1 *rad/s*. What are the accelerations of points A and B?

Solution:

The strategy is to find the accelerations of points A and B relative to the center of the rolling disk, and then determine the accelerations of the center of the rolling disk about the fixed center of the circular surface. The angular acceleration of the disk about its center is zero, since the angular velocity is constant. The acceleration of point A relative to the disk center is $\vec{a}_A = \vec{a}_O - \omega^2 \vec{r}_{A/O} = \vec{a}_O - 1(4\vec{j}) = \vec{a}_O - 4\vec{j}\ (in/s^2)$. The acceleration of point B is

$\vec{a}_B = \vec{a}_O - \omega^2 \vec{r}_{B/O} = \vec{a}_O - \omega^2(-4\vec{j}) = \vec{a}_O + 4\vec{j}\ (in/s)$

The contact point B is stationary.

The velocity of B relative to the center of the disk is

$$\vec{v}_B = 0 = \vec{v}_O + \vec{\omega} \times \vec{r}_{P/O} = \vec{v}_O + \begin{bmatrix} \vec{i} & \vec{j} & \vec{k} \\ 0 & 0 & -1 \\ 0 & -4 & 0 \end{bmatrix} = \vec{v}_O - 4\vec{i}\ (in/s), \text{ from which } \vec{v}_O = 4\vec{i}\ (in/s).$$

A *fixed center* for the center of the rolling disk is the center of the circular surface. Since the acceleration of a fixed center is zero, it can be used to find the accelerations of the center of the disk.

The angular velocity of the center of the disk about the fixed center is found from

$$\vec{v}_O = 4\vec{i} = \vec{\omega}_{OC} \times \vec{r}_{O/C} = \begin{bmatrix} \vec{i} & \vec{j} & \vec{k} \\ 0 & 0 & \omega_{OC} \\ 0 & 12+4 & 0 \end{bmatrix} = -16\omega_{OC}\vec{i}, \text{ from which } \omega_{OC} = -\frac{4}{16} = -0.25\ rad/s,$$

from which $\vec{\omega}_{OC} = -0.25\vec{k}\ (rad/s)$.

The angular acceleration of the center of the disk about the fixed center is zero, since the angular velocity of the rolling disk is constant.

The acceleration of the center of the rolling disk is $\vec{a}_O = -\omega^2(16\vec{j}) = -\vec{j}\ rad/s$, from which the acceleration of point A is $\boxed{\vec{a}_A = -j - 4j = -5\vec{j}\ (in/s^2)}$ and the acceleration of point B is

$\boxed{\vec{a}_B = -\vec{j} + 4\vec{j} = 3\vec{j}\ (in/s^2)}$

==================================<>==================================

=======================================<>=======================================

Problem 6.87 The endpoints of the bar slide on the plane surfaces. Show that the acceleration of the midpoint G is related to the bar's angular velocity and angular acceleration by

$$\vec{a}_G = \frac{1}{2}L\left(\left(\alpha\cos\theta - \omega^2\sin\theta\right)\vec{i} - \left(\alpha\sin\theta - \omega^2\cos\theta\right)\vec{j}\right).$$

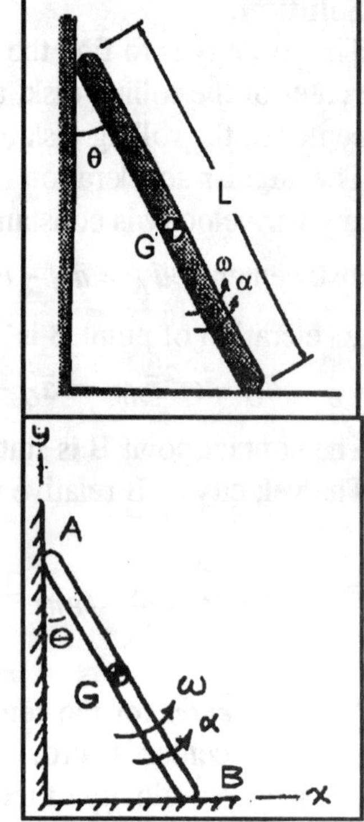

Solution:

Denote the upper point of contact between rod and the wall by A and the lower point of contact between the rod and the floor by B. The strategy is to use Eq (6.10) to find the (a) acceleration of G relative to the points A and B (b) equate the expressions for the acceleration of G to find the accelerations of A and B, and (c) substitute to find the acceleration of G. The constraint that the motion of A is parallel to the *y* axis and the motion of B is parallel to the *x* axis is essential to the solution.

The acceleration of G relative to A is $\vec{a}_G = \vec{a}_A + \vec{\alpha}\times\vec{r}_{G/A} - \omega^2\vec{r}_{G/A}$. The acceleration of G relative to B is $\vec{a}_G = \vec{a}_B + \vec{\alpha}\times\vec{r}_{G/B} - \omega^2\vec{r}_{G/B}$. Substitute:

$\vec{a}_A = a_A\vec{j}$, $\vec{a}_B = a_B\vec{i}$,

$$\vec{r}_{G/A} = \vec{r}_G - \vec{r}_A = \frac{L}{2}\left(\vec{i}\sin\theta + \vec{j}\cos\theta\right) - L\left(\vec{j}\cos\theta\right) = \frac{L}{2}\left(\vec{i}\sin\theta - \vec{j}\cos\theta\right),$$

$$\vec{r}_{G/B} = \vec{r}_G - \vec{r}_B = \frac{L}{2}\left(\vec{i}\sin\theta + \vec{j}\cos\theta\right) - L\left(\vec{i}\sin\theta\right) = \frac{L}{2}\left(-\vec{i}\sin\theta + \vec{j}\cos\theta\right).$$

From which $\vec{a}_G = a_A\vec{j} + \dfrac{\alpha L}{2}\begin{bmatrix}\vec{i} & \vec{j} & \vec{k}\\ 0 & 0 & 1\\ \sin\theta & -\cos\theta & 0\end{bmatrix} - \dfrac{\omega^2 L}{2}\left(\vec{i}\sin\theta - \vec{j}\cos\theta\right)$

$\vec{a}_G = \left(\dfrac{L}{2}\left(\alpha\cos\theta - \omega^2\sin\theta\right)\right)\vec{i} + \left(a_A + \dfrac{L}{2}\left(\alpha\sin\theta + \omega^2\cos\theta\right)\right)\vec{j}$. The acceleration of G in terms of the acceleration of B, $\vec{a}_G = a_B\vec{i} + \dfrac{\alpha L}{2}\begin{bmatrix}\vec{i} & \vec{j} & \vec{k}\\ 0 & 0 & 1\\ -\sin\theta & \cos\theta & 0\end{bmatrix} - \dfrac{\omega^2 L}{2}\left(-\vec{i}\sin\theta + \vec{j}\cos\theta\right)$.

$\vec{a}_G = \left(a_B - \dfrac{L}{2}\left(\alpha\cos\theta - \omega^2\sin\theta\right)\right)\vec{i} - \left(\dfrac{L}{2}\left(\alpha\sin\theta + \omega^2\cos\theta\right)\right)\vec{j}$. Equate the expressions for $\vec{a}_G$,

$$\left(\frac{L}{2}\left(\alpha\cos\theta - \omega^2\sin\theta\right)\right) = \left(a_B - \frac{L}{2}\left(\alpha\cos\theta - \omega^2\sin\theta\right)\right).$$

$\left(a_A + \dfrac{L}{2}\left(\alpha\sin\theta + \omega^2\cos\theta\right)\right) = \left(\dfrac{L}{2}\left(\alpha\sin\theta + \omega^2\cos\theta\right)\right)$. Solve: $a_A = -L\left(\alpha\sin\theta + \omega^2\cos\theta\right)$,

$a_B = L\left(\alpha\cos\theta - \omega^2\sin\theta\right)$. Substitute the expression for the acceleration of the point A into the expression for $\vec{a}_G$, $\boxed{\vec{a}_G = \dfrac{L}{2}\left(\left(\alpha\cos\theta - \omega^2\sin\theta\right)\vec{i} - \left(\alpha\sin\theta + \omega^2\cos\theta\right)\vec{j}\right)}$ which demonstrates the result.

=======================================<>=======================================

================================<>================================

Problem 6.88 The angular velocity and angular acceleration of bar AB are $\omega_{AB} = 2 \ rad/s$ and $\alpha_{AB} = 10 \ rad/s^2$. The dimensions of the rectangular plate are 12 in. by 24 in. What are the angular velocity and angular acceleration of the rectangular plate?

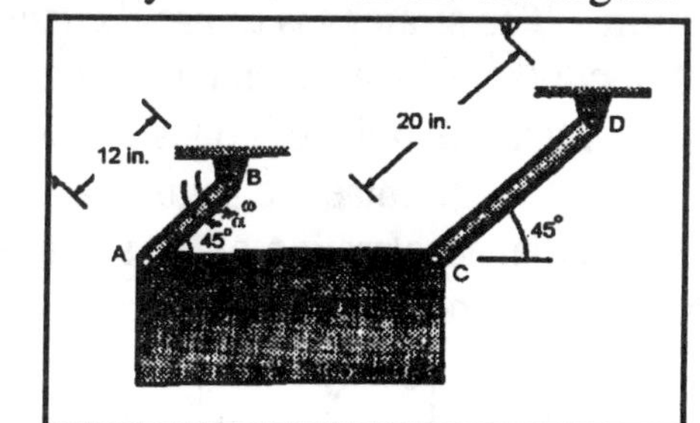

Solution:

The instantaneous center for bar AB is point B, by definition. The instantaneous center for bar CD is point D, by definition. The velocities at points A and C are normal to the bars AB and CD, respectively. However, by inspection these bars are parallel at the instant shown, so that lines perpendicular to the velocities at A and C will never intersect--- the instantaneous center of the plate AC is at infinity, hence *the plate only translates at the instant shown,* and $\boxed{\omega_{AC} = 0}$. If the plate is not rotating, the velocity at every point on the plate must be the same, and in particular, the vector velocity at A and C must be identical. The vector A/B is

$\vec{r}_{A/B} = -\vec{i}\cos 45^o - \vec{j}\sin 45^o = \left(\frac{-1}{\sqrt{2}}\right)(\vec{i} + \vec{j}) \ (ft)$. The velocity at point A is

$\vec{v}_A = \vec{\omega}_{AB} \times \vec{r}_{A/B} = \frac{-\omega_{AB}}{\sqrt{2}}\begin{bmatrix} \vec{i} & \vec{j} & \vec{k} \\ 0 & 0 & 1 \\ 1 & 1 & 0 \end{bmatrix} = \sqrt{2}(\vec{i} - \vec{j}) \ (ft/s)$. The vector C/D is

$\vec{r}_{C/D} = \left(\frac{20}{12}\right)\left(-\vec{i}\cos 45^o - \vec{j}\sin 45^o\right) = -1.179(\vec{i} + \vec{j}) \ (ft)$. The velocity at point C is

$\vec{v}_C = -1.179\omega_{CD}\begin{bmatrix} \vec{i} & \vec{j} & \vec{k} \\ 0 & 0 & 1 \\ 1 & 1 & 0 \end{bmatrix} = 1.179\omega_{CD}(\vec{i} - \vec{j}) \ (ft/s)$. Equate the velocities $\vec{v}_C = \vec{v}_A$, separate components and solve: $\omega_{CD} = 1.2 \ rad/s$. Use Eq (6.10) to determine the accelerations. The acceleration of point A is

$\vec{a}_A = \vec{\alpha}_{AB} \times \vec{r}_{A/B} - \omega_{AB}^2\vec{r}_{A/B} = -\frac{10}{\sqrt{2}}\begin{bmatrix} \vec{i} & \vec{j} & \vec{k} \\ 0 & 0 & 1 \\ 1 & 1 & 0 \end{bmatrix} + \left(\frac{2^2}{\sqrt{2}}\right)(\vec{i} + \vec{j}) = 9.9\vec{i} - 4.24\vec{j} \ (ft/s^2)$. The acceleration of point C relative to point A is

$\vec{a}_C = \vec{a}_A + \vec{\alpha}_{AC} \times \vec{r}_{C/A} = \vec{a}_A + \begin{bmatrix} \vec{i} & \vec{j} & \vec{k} \\ 0 & 0 & \alpha_{AC} \\ 2 & 0 & 0 \end{bmatrix} = 9.9\vec{i} + (2\alpha_{AC} - 4.24)\vec{j} \ (ft/s^2)$. The acceleration of point C relative to point D is $\vec{a}_C = \vec{a}_D + \vec{\alpha}_{CD} \times \vec{r}_{C/D} - \omega_{CD}^2\vec{r}_{C/D}$. Noting $\vec{a}_D = 0$,

$\vec{a}_C = -1.179\alpha_{CD}\begin{bmatrix} \vec{i} & \vec{j} & \vec{k} \\ 0 & 0 & 1 \\ 1 & j & 0 \end{bmatrix} + 1.179\omega_{CD}^2(\vec{i} + \vec{j}) = (1.179\alpha_{CD} + 1.697)\vec{i} + (-1.179\alpha_{CD} + 1.697)\vec{j} \ (ft/s^2)$. Equate the two expressions for the acceleration at point C and separate components:

$(-9.9 + 1.179\alpha_{CD} + 1.697)\vec{i} = 0$, $(2\alpha_{AC} - 4.24 + 1.179\alpha_{CD} - 1.697)\vec{j} = 0$. Solve:

$\boxed{\alpha_{AC} = -1.13 \ (rad/s^2)}$ (clockwise), $\alpha_{CD} = 6.96 \ (rad/s^2)$ (counterclockwise).

================================<>================================

Problem 6.89 The ring gear is stationary, and the sun gear has an angular acceleration of 10 *rad/s*^{2} in the counterclockwise direction. Determine the angular acceleration of the planet gears.

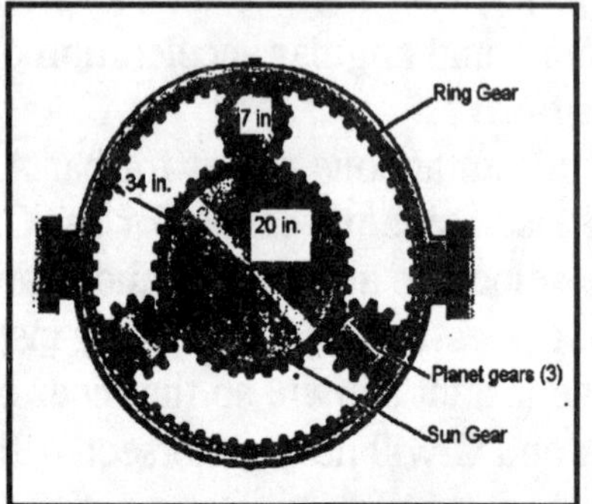

Solution The strategy is to use the tangential acceleration at the point of contact of the sun and planet gears, together with the constraint that the point of contact of the planet gear and ring gear is stationary, to determine the angular acceleration of the planet gear. The tangential acceleration of the sun gear at the point of contact with the top planet gear is

$$\vec{a}_{ST} = \vec{\alpha} \times \vec{R}_S = \begin{bmatrix} \vec{i} & \vec{j} & \vec{k} \\ 0 & 0 & 10 \\ 0 & 20 & 0 \end{bmatrix} = -200\vec{i} \ \left(in/s^2\right).$$

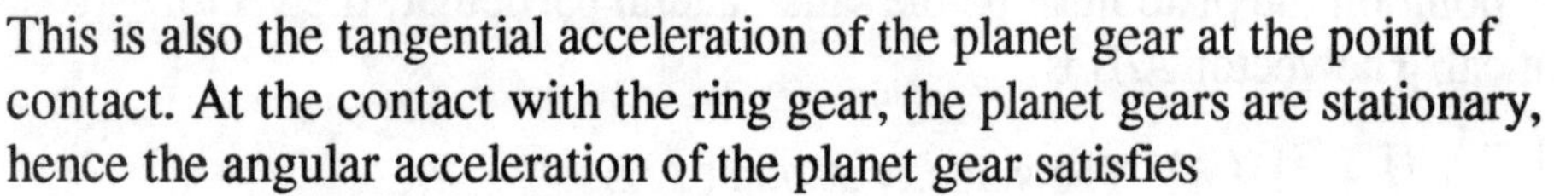

This is also the tangential acceleration of the planet gear at the point of contact. At the contact with the ring gear, the planet gears are stationary, hence the angular acceleration of the planet gear satisfies

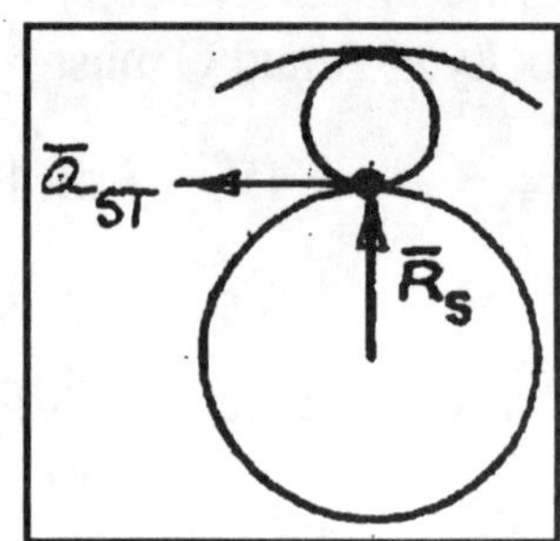

$$\vec{\alpha}_P \times \left(-2\vec{R}_P\right) = \begin{bmatrix} \vec{i} & \vec{j} & \vec{k} \\ 0 & 0 & \alpha_P \\ 0 & -14 & 0 \end{bmatrix} = -200\vec{i} \text{ from which}$$

$$\boxed{\alpha_P = -\frac{200}{14} = -14.29 \ \left(rad/s^2\right)} \text{ (clockwise).}$$

Problem 6.90 The sun gear in Problem 6.89 has a counterclockwise angular velocity of 4 *rad/s* and a clockwise angular acceleration of 12 *rad/s*^{2}. What is the magnitude of the acceleration of the centerpoints of the planet gears?

Solution:

The strategy is to use the tangential velocity and acceleration at the point of contact of the sun and planet gears, together with the constraint that the point of contact of the planet gear and ring gear is stationary, to determine the angular accelerations of the centers of the planet gears. The magnitude of the tangential velocity and tangential acceleration at the point of contact of the sun and a planet gear are, respectively, $v_{SP} = \omega R_S = (4)(20) = 80 \ in/s$, and $a_t = \alpha R_S = 12(20) = 240 \ in/s^2$. . The point of contact of the planet gear and ring gear is stationary. The magnitude of the velocity of the center of the planet gear is the mean of the sum of the velocities at its extreme edges, $v_P = \frac{v_{SP} + 0}{2} = 40 \ (in/s)$. The center of a planet gear moves on a radius of $20 + 7 = 27$ *in.* so the normal acceleration of the center is $a_n = \left(\frac{v_P^2}{27}\right) = 59.62 \ in/s^2$. The magnitude of the acceleration of the center of a planet gear is

$$\boxed{a_P = \sqrt{a_t^2 + a_n^2} = \sqrt{120^2 + 59.26^2} = 133.8 \ in/s^2}$$

===================================<>===================================

Problem 6.91 The 1 *m* diameter disk rolls and point B of the 1 *m* long bar slides on the plane surface. Determine the angular acceleration of the bar and the acceleration of point B

Solution:

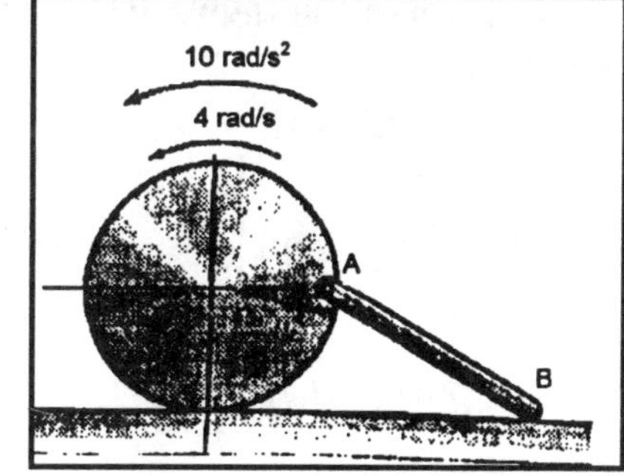

Choose a coordinate system with the origin at O, the center of the disk, with *x* axis parallel to the horizontal surface. The point P of contact with the surface is stationary, from which

$$\vec{v}_P = 0 = \vec{v}_O + \vec{\omega}_O \times -\vec{R} = \vec{v}_O + \begin{bmatrix} \vec{i} & \vec{j} & \vec{k} \\ 0 & 0 & \omega_O \\ 0 & -0.5 & 0 \end{bmatrix} = \vec{v}_O + 2\vec{i}\text{, from}$$

which $\vec{v}_O = -2\vec{i}\ (m/s)$.

The velocity at A is $\vec{v}_A = \vec{v}_O + \vec{\omega}_O \times \vec{r}_{A/O} = -2\vec{i} + \begin{bmatrix} \vec{i} & \vec{j} & \vec{k} \\ 0 & 0 & \omega_O \\ 0.5 & 0 & 0 \end{bmatrix} = -2\vec{i} + 2\vec{j}\ (m/s)$

The motion at point B is constrained to be parallel to the *x* axis. The line perpendicular to the velocity of B is parallel to the *y* axis. The line perpendicular to the velocity at A forms an angle at 45° with the *x* axis. From geometry, the line from A to the fixed center is the hypotenuse of a right triangle with base $\cos 30^o = 0.866$ and interior angles 45°. The coordinates of the fixed center are $(0.5 + 0.866, 0.866) = (1.366, 0.866)$ *in.* The vector from the instantaneous center to the point A is $\vec{r}_{A/C} = \vec{r}_A - \vec{r}_C = -0.866\vec{i} - 0.866\vec{j}\ (m)$. The angular velocity of the bar AB is obtained from

$$\vec{v}_A = \vec{\omega}_{AB} \times \vec{r}_{A/C} = \begin{bmatrix} \vec{i} & \vec{j} & \vec{k} \\ 0 & 0 & \omega_{AB} \\ -0.866 & -0.866 & 0 \end{bmatrix} = 0.866\omega_{AB}\vec{i} - 0.866\omega_{AB}\vec{j}\ (m/s)\text{, from which}$$

$\omega_{AB} = -\dfrac{2}{0.866} = -2.31\ (rad/s)$. The acceleration of the center of the rolling disk is $\vec{a}_C = -\alpha R\vec{i} = -10(0.5)\vec{i} = -5\vec{i}\ (m/s^2)$. The acceleration of point A is

$$\vec{a}_A = \vec{a}_O + \vec{\alpha}_O \times \vec{r}_{A/O} - \omega_O^2\vec{r}_{A/O} = -5\vec{i} + \begin{bmatrix} \vec{i} & \vec{j} & \vec{k} \\ 0 & 0 & \alpha \\ 0.5 & 0 & 0 \end{bmatrix} - 16(0.5)\vec{i}.\ \vec{a}_A = -13\vec{i} + 5\vec{j}\ (m/s^2)\text{. The vector}$$

B/A is $\vec{r}_{B/A} = \vec{r}_B - \vec{r}_A = (0.5 + \cos\theta)\vec{i} - 0.5\vec{j} - 0.5\vec{i} = 0.866\vec{i} - 0.5\vec{j}\ (m)$. The acceleration of point is

$$\vec{a}_B = \vec{a}_A + \vec{\alpha}_{AB} \times \vec{r}_{A/B} - \omega_{AB}^2\vec{r}_{A/B} = -13\vec{i} + 5\vec{j} + \begin{bmatrix} \vec{i} & \vec{j} & \vec{k} \\ 0 & 0 & \alpha_{AB} \\ -\cos\theta & \sin\theta & 0 \end{bmatrix} - \omega_{AB}^2(-\vec{i}\cos\theta + \vec{j}\sin\theta)\text{. The}$$

constraint on B insures that the acceleration of B will be parallel to the *x* axis. Separate components:

$a_B = -13 + 0.5\alpha_{AB} - \omega_{AB}^2(0.866)$,

$0 = 0.5 + 0.866\alpha_{AB} + 0.5\omega_{AB}^2$. Solve: $\boxed{\alpha_{AB} = -8.85\ (rad/s^2)}$, where the negative sign means a clockwise rotation. $\boxed{\vec{a}_B = -22.04\vec{i}\ (m/s^2)}$

===================================<>===================================

===================================<>===================================

Problem 6.92 The angle $\theta = 45^\circ$ and the bar OQ has a constant counterclockwise angular velocity of $2\ rad/s$. What is the acceleration of the sleeve P?

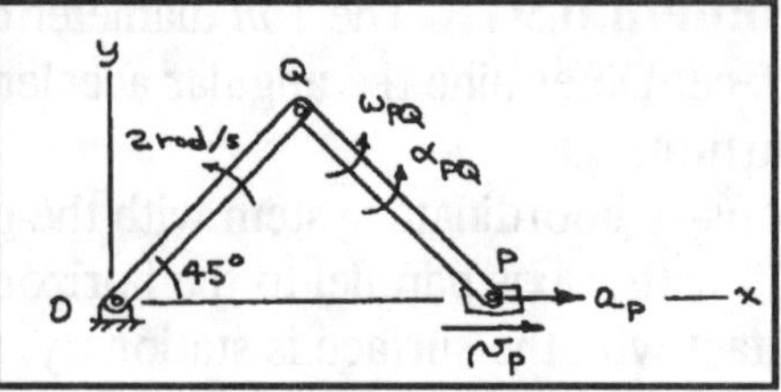

Solution:

The velocity of Q is

$$\bar{v}_Q = \bar{v}_0 + \bar{\omega}_{0Q} \times \bar{r}_{Q/0} = 0 + \begin{vmatrix} \bar{i} & \bar{j} & \bar{k} \\ 0 & 0 & 2 \\ 2\cos 45^\circ & 2\sin 45^\circ & 0 \end{vmatrix}$$

$= -2.83\bar{i} + 2.83\bar{j}\ (ft/s)$

The velocity of P is $\bar{v}_p = \bar{v}_Q + \bar{\omega}_{PQ} \times \bar{r}_{P/Q} = -2.83\bar{i} + 2.83\bar{j} + \begin{vmatrix} \bar{i} & \bar{j} & \bar{k} \\ 0 & 0 & \omega_{PQ} \\ 2\cos 45^\circ & -2\sin 45^\circ & 0 \end{vmatrix}$

Equating $\bar{j}$ components$0 = 2.83 + 2w_{PQ}\cos 45^\circ$, we see that $\omega_{PQ} = -2\ rad/s$.

The acceleration of $Q$is $\bar{a}_Q = \bar{a}_0 + \bar{\alpha}_{0Q} \times \bar{r}_{A/0} - \omega_{0Q}^2 \bar{r}_{Q/0} = 0 + 0 - (2)^2(2\cos 45^\circ \mathbf{i} + 2\sin 45^\circ \mathbf{j})$ or

$\bar{a}_Q = -5.66\ \mathbf{i} - 5.66\ \mathbf{j}\ (ft/s)$

The acceleration $P$of is $a_P\mathbf{i} = \bar{a}_Q + \bar{\alpha}_{PQ} \times \bar{r}_{P/Q} - \omega_{PQ}^2 \bar{r}_{P/Q}$, or

$$a_P\bar{i} = -5.66\,\bar{i} - 5.66\,\bar{j} + \begin{vmatrix} \bar{i} & \bar{j} & \bar{k} \\ 0 & 0 & \alpha_{PQ} \\ 2\cos 45^\circ & -2\sin 45^\circ & 0 \end{vmatrix} - (2)^2(2\cos 45^\circ \bar{i} - 2\sin 45^\circ \bar{j}).$$

Equating the $\bar{i}$ and $\bar{j}$ components, we get $a_P = -5.66 + 2\alpha_{PQ}\cos 45^\circ - 8\sin 45^\circ$, and $0 = -5.66 + 2\alpha_{PQ}\cos 45^\circ + 8\sin 45^\circ$. Solving, we obtain $\alpha_{PQ} = 0$, $a_P = -11.3\ (ft/s^2)$.

===================================<>===================================

Problem 6.93 Consider the system shown in Problem 6.92. If $\theta = 40^\circ$ and the sleeve P is moving to the right with a constant velocity of $5\ ft/s$, what are the angular accelerations of the bars OQ and PQ ?

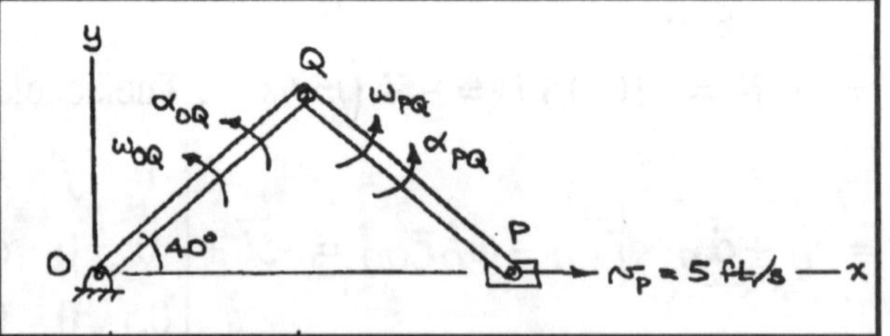

Solution:

The velocity of Q is

$$\bar{v}_Q = \bar{v}_0 + \bar{\omega}_{0Q} \times \bar{r}_{Q/0} = 0 + \begin{vmatrix} \bar{i} & \bar{j} & \bar{k} \\ 0 & 0 & \omega_{0Q} \\ 2\cos 40^\circ & 2\sin 40^\circ & 0 \end{vmatrix} = -2\omega_{0Q}\sin 40^\circ \bar{i} + 2\omega_{0Q}\cos 40^\circ \bar{j}\ .$$

The velocity of P is

$$\bar{v}_p = 5\bar{i} = \bar{v}_Q + \bar{\omega}_{PQ} \times \bar{r}_{P/Q} = -2\omega_{0Q}\sin 40^\circ \bar{i} + 2\omega_{0Q}\cos 40^\circ \bar{j} + \begin{vmatrix} \bar{i} & \bar{j} & \bar{k} \\ 0 & 0 & \omega_{PQ} \\ 2\cos 40^\circ & -2\sin 40^\circ & 0 \end{vmatrix}$$

Equating $\bar{i}$ and $\bar{j}$ components, we get $5 = -2\omega_{0Q}\sin 40^\circ + 2\omega_{PQ}\sin 40^\circ$,and $0 = 2\omega_{0Q}\cos 40^\circ + 2\omega_{PQ}\cos 40^\circ$, to obtain $\omega_{PQ} = -\omega_{0Q} = 1.94 rad/s$.

Solution continued on next page

The acceleration of Q is

$$\bar{a}_Q = \bar{a}_0 + \bar{\alpha}_{0Q} \times \bar{r}_{Q/0} - \omega_{0Q}^2 \bar{r}_{Q/0} = 0 + \begin{vmatrix} \bar{i} & \bar{j} & \bar{k} \\ 0 & 0 & \alpha_{0Q} \\ 2\cos 40^\circ & 2\sin 40^\circ & 0 \end{vmatrix} - (-1.94)^2 (2\cos 40^\circ \bar{i} + 2\sin 40^\circ \bar{j})$$

$\bar{a}_Q = \left[-2\alpha_{0Q}\sin 40^\circ - (1.94)^2\ 2\cos 40^\circ\right]\bar{i} + \left[2\alpha_{0Q}\cos 40^\circ - (1.94)^2\ 2\sin 40^\circ\right]\bar{j}$.

The acceleration P of is $\bar{a}_P = 0 = \bar{a}_Q + \bar{\alpha}_{PQ} \times \bar{r}_{P/Q} - \omega_{PQ}^2 \bar{r}_{P/Q}$, or

$$\bar{a}_P = \left[-2\alpha_{0Q}\sin 40^\circ - (1.94)^2\ 2\cos 40^\circ\right]\bar{i} + \left[2\alpha_{0Q}\cos 40^\circ - (1.94)^2\ 2\sin 40^\circ\right]\bar{j}$$

$$+ \begin{vmatrix} \bar{i} & \bar{j} & \bar{k} \\ 0 & 0 & \alpha_{PQ} \\ 2\cos 40^\circ & -2\sin 40^\circ & 0 \end{vmatrix} - (1.94)^2 (2\cos 40^\circ \bar{i} - 2\sin 40^\circ \bar{j}).$$

Equating the $\bar{i}$ and $\bar{j}$ components to zero

$0 = -2\alpha_{0Q}\sin 40^\circ - (1.94)^2 2\cos 40^\circ + 2\alpha_{PQ}\sin 40^\circ - (1.94)^2 2\cos 40^\circ$, and

$0 = 2\alpha_{0Q}\cos 40^\circ - (1.94)^2 2\sin 40^\circ + 2\alpha_{PQ}\cos 40^\circ + (1.94)^2 2\sin 40^\circ$. Solving, we obtain

$\alpha_{PQ} = -\alpha_{0Q} = 4.51\, rad/s^2$.

==================================<>==================================

Problem 6.94 The angle $\theta = 60^\circ$ and the bar OQ has a constant counterclockwise angular velocity of *2 rad/s*. What is the angular acceleration of the bar PQ?

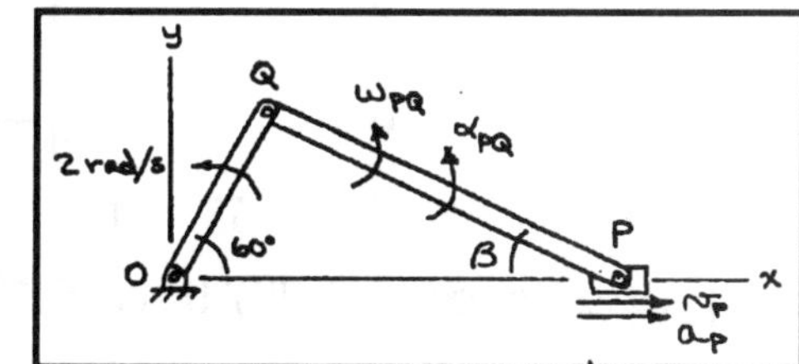

Solution:

By applying the law of sines, the angle $\beta = 25.7^\circ$

The velocity of Q is $\bar{v}_Q = \bar{v}_0 + \bar{\omega}_{0Q} \times \bar{r}_{Q/0}$

$$\bar{v}_Q = 0 + \begin{vmatrix} \bar{i} & \bar{j} & \bar{k} \\ 0 & 0 & 2 \\ 0.2\cos 60^\circ & 0.2\sin 60^\circ & 0 \end{vmatrix} = -0.4\sin 60^\circ \bar{i} + 0.4\cos 60^\circ \bar{j}.$$

The velocity of P is

$$v_P\bar{i} = \bar{v}_Q + \bar{\omega}_{PQ} \times \bar{r}_{P/Q} = -0.4\sin 60^\circ \bar{i} + 0.4\cos 60^\circ \bar{j} + \begin{vmatrix} \bar{i} & \bar{j} & \bar{k} \\ 0 & 0 & \omega_{PQ} \\ 0.4\cos\beta & -0.4\sin\beta & 0 \end{vmatrix}.$$

Equating $\bar{j}$ components, we get $0 = 0.4\cos 60^\circ + 0.4\omega_{PQ}\cos\beta$, and obtain $\omega_{PQ} = -0.555\, rad/s$.

The acceleration of Q is $\bar{a}_Q = \bar{a}_0 + \alpha_{0Q} \times \bar{r}_{Q/0} - \omega_{0Q}{}^2 \bar{r}_{Q/0}$, or

$\bar{a}_Q = 0 + 0 - (2)^2 (0.2\cos 60^\circ \bar{i} + 0.2\sin 60^\circ \bar{j}) = -0.8\cos 60^\circ \bar{i} - 0.8\sin 60^\circ \bar{j}$.

The acceleration of P is

$$a_P\bar{i} = \bar{a}_Q + \bar{\alpha}_{PQ} \times \bar{r}_{P/Q} - \omega_{PQ}^2 \bar{r}_{P/Q} = -0.8\cos 60^\circ \bar{i} - 0.8\sin 60^\circ \bar{j}$$

$$+ \begin{vmatrix} \bar{i} & \bar{j} & \bar{k} \\ 0 & 0 & \alpha_{PQ} \\ 0.4\cos\beta & -0.4\sin\beta & 0 \end{vmatrix} - (-0.555)^2 (0.4\cos\beta\,\bar{i} - 0.4\sin\beta\,\bar{j}).$$

Equating $\bar{j}$ components $0 = -0.8\sin 60^\circ + 0.4\alpha_{PQ}\cos\beta + (0.555)^2 0.4\sin\beta$. Solving, we obtain $\alpha_{PQ} = 1.77\, rad/s^2$.

==================================<>==================================

Problem 6.95 Consider the system shown in Problem 6.94. If $\theta = 55^\circ$ and the sleeve P is moving to the right with a constant velocity of $2\ m/s$, what are the angular accelerations of the bars OQ and PQ?

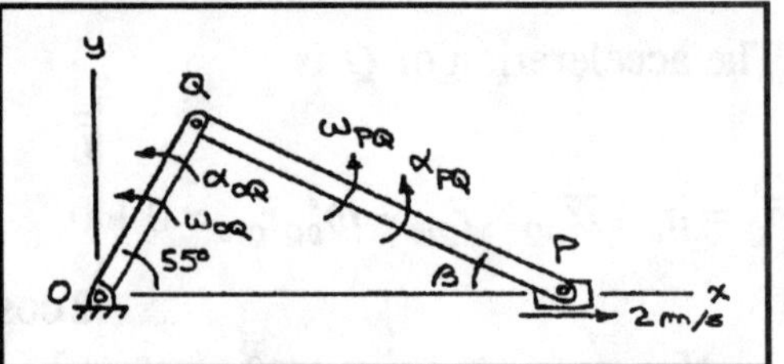

Solution:

By applying the law of sines, the angle $\beta = 24.2^\circ$. The velocity of Q is $\bar{v}_Q = \bar{v}_0 + \bar{\omega}_{0Q} \times \bar{r}_{Q/0}$, or

$$\bar{v}_Q = 0 + \begin{vmatrix} \bar{i} & \bar{j} & \bar{k} \\ 0 & 0 & \omega_{0Q} \\ 0.2\cos 55^\circ & 0.2\sin 55^\circ & 0 \end{vmatrix} = -0.2\omega_{0Q}\sin 55^\circ \bar{i} + 0.2\omega_{0Q}\cos 55^\circ \bar{j}.$$

The velocity of P is

$$\bar{v}_p = 2\bar{i} = \bar{v}_Q + \bar{\omega}_{PQ} \times \bar{r}_{P/Q} = -0.2\omega_{0Q}\sin 55^\circ \bar{i} + 0.2\omega_{0Q}\cos 55^\circ \bar{j} + \begin{vmatrix} \bar{i} & \bar{j} & \bar{k} \\ 0 & 0 & \omega_{PQ} \\ 0.4\cos\beta & -0.4\sin\beta & 0 \end{vmatrix}.$$

Equating $\bar{i}$ and $\bar{j}$ components, $2 = -0.2\omega_{0Q}\sin 55^\circ + 0.4\omega_{PQ}\sin\beta$, and $0 = 0.2\omega_{0Q}\cos 55^\circ + 0.4\omega_{PQ}\cos\beta$, Solving, we obtain $\omega_{0Q} = -9.29\,rad/s$ and $\omega_{PQ} = 2.92\,rad/s$.

The acceleration of Q is $\bar{a}_Q = \bar{a}_0 + \bar{\alpha}_{0Q} \times \bar{r}_{Q/0} - \omega_{0Q}^2 \bar{r}_{Q/0}$

$$\bar{a}_Q = 0 + \begin{vmatrix} \bar{i} & \bar{j} & \bar{k} \\ 0 & 0 & \alpha_{0Q} \\ 0.2\cos 55^\circ & 0.2\sin 55^\circ & 0 \end{vmatrix} -(9.29)^2(0.2\cos 55^\circ \bar{i} + 0.2\sin 55^\circ \bar{j}), \text{ or}$$

$$\bar{a}_Q = \left[-0.2\alpha_{0Q}\sin 55^\circ - (9.29)^2 0.2\cos 55^\circ\right]\bar{i} + \left[0.2\alpha_{0Q}\cos 55^\circ - (9.29)^2\, 0.2\sin 55^\circ\right]\bar{j}.$$

The acceleration P of is $\bar{a}_P = 0 = \bar{a}_Q + \bar{\alpha}_{PQ} \times \bar{r}_{P/Q} - \omega_{PQ}^2 \bar{r}_{P/Q}$, or

$$\bar{a}_P = \bar{a}_Q + \begin{vmatrix} \bar{i} & \bar{j} & \bar{k} \\ 0 & 0 & \alpha_{PQ} \\ 0.4\cos\beta & -0.4\sin\beta & 0 \end{vmatrix} -(2.92)^2(0.4\cos\beta\,\bar{i} - 0.4\sin\beta\,\bar{j}).$$ Equating the $\bar{i}$ and $\bar{j}$ components to zero, $0 = -0.2\alpha_{0Q}\sin 55^\circ - (9.29)^2 0.2\cos 55^\circ + 0.4\alpha_{PQ}\sin\beta - (2.92)^2 0.4\cos\beta$, and $0 = 0.2\alpha_{0Q}\cos 55^\circ - (9.29)^2 0.2\sin 55^\circ + 0.4\alpha_{PQ}\cos\beta + (2.92)^2 0.4\sin\beta$. Solving, we obtain $\alpha_{0Q} = -33.8\,rad/s^2$ $\alpha_{PQ} = 45.5\ rad/s^2$

=================================<>=================================

Problem 6.96 The angular velocity and acceleration of bar AB are $\omega_{AB} = 2\ rad/s$, $\alpha_{AB} = 6\ rad/s^2$. What are the angular velocity and angular acceleration of bar BD?

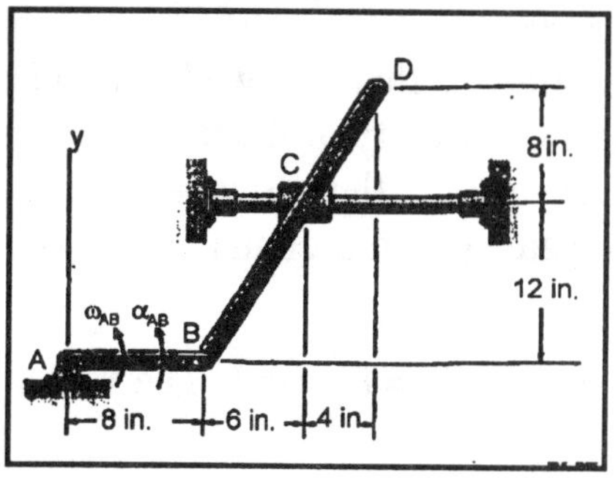

Solution Choose a coordinate system with the origin at A and the x axis parallel to AB. The strategy is to determine the angular velocity of bar BC from the constraint on the motion of C. The accelerations are determined from the angular velocity, the known accelerations of B, and the constraint on the motion of C.

The vector $\vec{r}_{B/A} = \vec{r}_B - \vec{r}_A = 8\vec{i}\ (in.)$. The velocity of point B is

$$\vec{v}_B = \vec{\omega}_{AB} \times \vec{r}_{B/A} = \begin{bmatrix} \vec{i} & \vec{j} & \vec{k} \\ 0 & 0 & 2 \\ 8 & 0 & 0 \end{bmatrix} = 16\vec{j}\ (in/s).$$

The acceleration of point B is

$$\vec{a}_B = \vec{\alpha}_{AB} \times \vec{r}_{B/A} - \omega^2 \vec{r}_{A/B} = \begin{bmatrix} \vec{i} & \vec{j} & \vec{k} \\ 0 & 0 & 6 \\ 8 & 0 & 0 \end{bmatrix} - 4(8\vec{i}) = -32\vec{i} + 48\vec{j}\ (in/s^2).$$

The line perpendicular to the velocity of B is parallel to the x axis, and the line perpendicular to the velocity at C is parallel to the y axis. The intercept is at $(14, 0)\ in.$, which are the coordinates of the instantaneous center of bar BC. Denote the instantaneous center by C''. The vector

$\vec{r}_{B/C'} = \vec{r}_B - \vec{r}_{C'} = (8\vec{i}) - (14\vec{i}) = -6\vec{i}\ (in.)$ The velocity of point B is

$$\vec{v}_B = \vec{\omega}_{BC} \times \vec{r}_{B/C'} = \begin{bmatrix} \vec{i} & \vec{j} & \vec{k} \\ 0 & 0 & \omega_{BC} \\ -6 & 0 & 0 \end{bmatrix} = -6\omega_{BC}\vec{j} = 16\vec{j}\ (in/s).$$

From which

$$\boxed{\omega_{BC} = -\frac{16}{6} = -2.67\ rad/s}$$

(clockwise)

The acceleration of point C is

$$\vec{a}_C = \vec{a}_B + \vec{\alpha}_{BC} \times \vec{r}_{C/C''} - \omega_{BC}^2 \vec{r}_{C/C''} = -32\vec{i} + 48\vec{j} + \begin{bmatrix} \vec{i} & \vec{j} & \vec{k} \\ 0 & 0 & \alpha_{BC} \\ 6 & 12 & 0 \end{bmatrix} - (13.33^2)(6\vec{i} + 12\vec{j}).$$

$\vec{a}_C = -32\vec{i} + 48\vec{j} + \alpha_{BC}(-12\vec{i} + 6\vec{j}) - 42.67\vec{i} - 85.33\vec{j}\ (in/s^2)$. The acceleration of point C is constrained to be parallel to the x axis. Separate components: $a_C = -32 - 12\alpha_{BC} - 42.67$, $0 = 48 + 6\alpha_{BC} - 85.33$. Solve: $\vec{a}_C = -113.2\vec{i}\ (in/s^2)$, $\boxed{\alpha_{BC} = \alpha_{BD} = 6.22\ rad/s^2}$ (counterclockwise).

=================================<>=================================

==================================<>==================================

Problem 6.97 In Problem 6.96, if the angular velocity and acceleration of bar AB are $\omega_{AB} = 2\ rad/s$, $\alpha_{AB} = -10\ rad/s^2$, what are the velocity and acceleration of point D?

Solution:

The strategy is that followed in Problem 6.96, namely to determine the angular velocity of bar BC from the known velocity of B and the constraint on the motion of C. The accelerations are determined from the angular velocity, the known acceleration of B, and the constraint on the motion of C. Choose the coordinate system used in Problem 9.96, with the origin at A and the *x*-axis parallel to AB. The vector $\vec{r}_{B/A} = \vec{r}_B - \vec{r}_A = 8\vec{i}\ (in.)$. The velocity of point B is $\vec{v}_B = \vec{\omega}_{AB} \times \vec{r}_{B/A} = \begin{bmatrix} \vec{i} & \vec{j} & \vec{k} \\ 0 & 0 & 2 \\ 8 & 0 & 0 \end{bmatrix} = 16\vec{j}\ (in/s)$. The acceleration of point B is $\vec{a}_B = \vec{\alpha}_{AB} \times \vec{r}_{B/A} - \omega_{AB}^2 \vec{r}_{A/B} = \begin{bmatrix} \vec{i} & \vec{j} & \vec{k} \\ 0 & 0 & -10 \\ 8 & 0 & 0 \end{bmatrix} - 4(8\vec{i}) = -32\vec{i} - 80\vec{j}\ (in/s^2)$.

The vector $\vec{r}_{B/C} = \vec{r}_C - r_B = (14\vec{i} + 12\vec{j}) - (8\vec{i}) = 6\vec{i} + 12\vec{j}\ (in.)$. From the solution to Problem 9.80, the coordinates of the instantaneous center of bar BC are (14, 0). Denote the instantaneous center by C". The vector $\vec{r}_{B/IC} = \vec{r}_B - \vec{r}_{B/C'} = (8\vec{i}) - (14\vec{i}) = -6\vec{i}\ (in.)$ The velocity of point B is

$$\vec{v}_B = \vec{\omega}_{BC} \times \vec{r}_{B/C'} = \begin{bmatrix} \vec{i} & \vec{j} & \vec{k} \\ 0 & 0 & \omega_{BC} \\ -6 & 0 & 0 \end{bmatrix} = -6\omega_{BC}\vec{j} = 16\vec{j}\ (in/s).$$

From which $\omega_{BC} = -\frac{16}{6} = -2.67\ rad/s$. (clockwise) The velocity of point C is

$$\vec{v}_C = \vec{v}_B + \vec{\omega}_{BC} \times \vec{r}_{C/B} = 16\vec{j} + \begin{bmatrix} \vec{i} & \vec{j} & \vec{k} \\ 0 & 0 & \omega_{BC} \\ 6 & 12 & 0 \end{bmatrix} = -12\omega_{BC}\vec{i} + (16 + 6\omega_{BC})\vec{j} = 32\vec{i}\ (in/s).$$

The acceleration of point C is

$$\vec{a}_C = \vec{a}_B + \vec{\alpha}_{BC} \times \vec{r}_{B/C} - \omega_{BC}^2 \vec{r}_{B/C} = -32\vec{i} - 80\vec{j} + \begin{bmatrix} \vec{i} & \vec{j} & \vec{k} \\ 0 & 0 & \alpha_{BC} \\ 6 & 12 & 0 \end{bmatrix} - (2.67^2)(6\vec{i} + 12\vec{j}).$$

$\vec{a}_C = -32\vec{i} - 80\vec{j} + \alpha_{BC}(-12\vec{i} + 6\vec{j}) - 42.67\vec{i} - 83.33\vec{j}\ (in/s^2)$. The acceleration of point C is constrained to be parallel to the *x* axis. Separate components: $a_C = -32 - 12\alpha_{BC} - 42.67$, $0 = -80 + 6\alpha_{BC} - 83.33$. Solve: $\vec{a}_C = -405.3\vec{i}\ (in/s^2)$, $\alpha_{BC} = \alpha_{BD} = 27.6\ rad/s^2$ (counterclockwise).

Solution continued on next page

The velocity of point D is $\vec{v}_D = \vec{v}_C + \vec{\omega}_{BC} \times \vec{r}_{D/C} = 32\vec{i} + \begin{bmatrix} \vec{i} & \vec{j} & \vec{k} \\ 0 & 0 & 2.67 \\ 4 & 8 & 0 \end{bmatrix} = 53.33\vec{i} - 10.67\vec{j}$ (in/s)

The acceleration of point D is

$$\vec{a}_D = \vec{a}_C + \vec{\alpha}_{BC} \times \vec{r}_{D/C} - \omega_{BC}^2 \vec{r}_{D/C} = -405.3\vec{i} + \begin{bmatrix} \vec{i} & \vec{j} & \vec{k} \\ 0 & 0 & 27.56 \\ 4 & 8 & 0 \end{bmatrix} - (2.67^2)(4\vec{i} + 8\vec{j})$$

$\boxed{\vec{a}_D = -654.2\vec{i} + 53.3\vec{j} \ (in/s^2)}$

==================================<>==================================

Problem 6.98 If $\omega_{AB} = 6$ rad/s and $\alpha_{AB} = 20$ rad/s^2, what are the velocity and acceleration of point C?

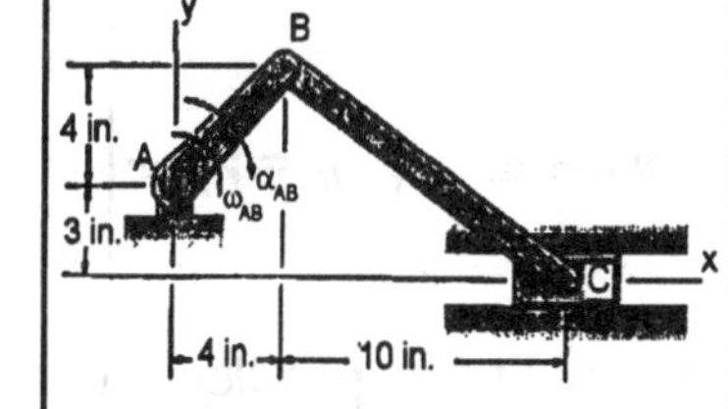

Solution The vector B/A is $\vec{r}_{B/A} = \vec{r}_B - \vec{r}_A = 4\vec{i} + 4\vec{j}$ $(in.)$. The velocity of point B is

$\vec{v}_B = \vec{\omega}_{BA} \times \vec{r}_{B/A} = \begin{bmatrix} \vec{i} & \vec{j} & \vec{k} \\ 0 & 0 & 6 \\ 4 & 4 & 0 \end{bmatrix} = -24(-\vec{i} + \vec{j})$ (in/s). The vector

$\vec{r}_{C/B} = \vec{r}_C - \vec{r}_B = (14\vec{i}) - (4\vec{i} + 7\vec{j}) = 10\vec{i} - 7\vec{j}$ $(in.)$ The velocity of point C in terms of the velocity of B

is $\vec{v}_C = \vec{v}_B + \vec{\omega} \times \vec{r}_{C/B} = 24(\vec{i} - \vec{j}) + \begin{bmatrix} \vec{i} & \vec{j} & \vec{k} \\ 0 & 0 & \omega_{BC} \\ 10 & -7 & 0 \end{bmatrix} = 24(\vec{i} - \vec{j}) + \omega_{BC}(7\vec{i} + 10\vec{j})$. The velocity at point C is constrained to be parallel to the x axis. Separate components: $v_C = 24 + 7\omega_{BC}$, $0 = -24 + 10\omega_{BC}$, from which $\boxed{\vec{v}_C = 40.8\vec{i} \ (in/s)}$, $\omega_{BC} = 2.4$ rad/s. The acceleration of point B is

$$\vec{a}_B = \vec{\alpha}_{AB} \times \vec{r}_{B/A} - \omega_{AB}^2 \vec{r}_{AB} = \begin{bmatrix} \vec{i} & \vec{j} & \vec{k} \\ 0 & 0 & -20 \\ 4 & 4 & 0 \end{bmatrix} - 36(4\vec{i} + 4\vec{j}). = 80\vec{i} - 80\vec{j} - 144\vec{i} - 144\vec{j}.$$

$\vec{a}_B = -64\vec{i} - 224\vec{j}$ (in/s^2) The acceleration of point C in terms of the acceleration of point C:

$$\vec{a}_C = \vec{a}_B + \vec{\alpha}_{BC} \times \vec{r}_{C/B} - \omega_{BC}^2 \vec{r}_{C/B} = -64\vec{i} - 224\vec{j} + \begin{bmatrix} \vec{i} & \vec{j} & \vec{k} \\ 0 & 0 & \alpha_{BC} \\ 10 & -7 & 0 \end{bmatrix} - (\omega_{BC}^2)(10\vec{i} - 7\vec{j}).$$ The

acceleration of C is constrained to be parallel to the x axis. Separate components: $a_C = -64 + 7\alpha_{BC} - 10\omega_{BC}^2$, $0 = -224 + 10\alpha_{BC} + 7\omega_{BC}^2$. Substitute $\omega_{BC} = 2.4$ rad/s and solve:

$\boxed{\vec{a}_C = 6.98\vec{i} \ (in/s^2)}$, $\alpha_{BC} = 18.4$ rad/s^2 (counterclockwise).

==================================<>==================================

Problem 6.99 A motor rotates the circular disk mounted at A, moving the saw back and forth. (The saw is supported by horizontal slot so that point C moves horizontally.) The radius AB is 4 *in.* and the link BC is 14 *in.* long. In the position shown, $\theta = 45^o$ and the link BC is horizontal. If the disk has a constant angular velocity of one revolution per second counterclockwise, what is the acceleration of the saw?

Solution:

The angular velocity of the disk is $\omega_{AB} = 2\pi\ rad/s$. The vector from A to B is $\vec{r}_{B/A} = 4(\vec{i}\cos\theta + \vec{j}\sin\theta)$ (*in.*). The velocity of point B is

$$\vec{v}_B = \vec{\omega}_{AB} \times \vec{r}_{B/A} = \begin{bmatrix} \vec{i} & \vec{j} & \vec{k} \\ 0 & 0 & 2\pi \\ 2\sqrt{2} & 2\sqrt{2} & 0 \end{bmatrix} = 2\pi(-2.83\vec{i} + 2.83\vec{j}).\ \vec{v}_B = -17.8(\vec{i} - \vec{j})\ (in/s).$$

The vector from B to C is $\vec{r}_{C/B} = -14\vec{i}$ (*in.*) The velocity of point C in terms of the velocity of B is

$$\vec{v}_C = \vec{v}_B + \vec{\omega} \times \vec{r}_{C/B} = \vec{v}_B + \begin{bmatrix} \vec{i} & \vec{j} & \vec{k} \\ 0 & 0 & \omega_{BC} \\ -14 & 0 & 0 \end{bmatrix} = -17.8\vec{i} + 17.8\vec{j} - 14\omega_{BC}\vec{j}\ (in/s).$$ The velocity of C is constrained to be parallel to the x axis. Separate components and solve: $\vec{v}_C = -17.8\vec{i}$.

$\omega_{BC} = 1.27\ (rad/s)$. The acceleration of B is $\vec{a}_B = -\omega_{AB}^2(2.83\vec{i} + 2.83\vec{j}) = -111.7(\vec{i} + \vec{j})\ (in/s^2)$.

The acceleration of point C in terms of the acceleration at B:

$$\vec{a}_C = \vec{a}_B + \vec{\alpha}_{BC} \times \vec{r}_{C/B} - \omega_{BC}^2\vec{r}_{C/B} = -111.7(\vec{i} + \vec{j}) + \begin{bmatrix} \vec{i} & \vec{j} & \vec{k} \\ 0 & 0 & \alpha_{BC} \\ -14 & 0 & 0 \end{bmatrix} - (1.27^2)(-14\vec{i}).$$ The acceleration of C is constrained to lie parallel to the x axis. Separate components: $a_C = -111.7 + 14\omega_{BC}^2$, $0 = -111.7 - 14\alpha_{BC}$. Solve: $\boxed{\vec{a}_C = -89\vec{i}\ (in/s^2)}$

================================<>================================

Problem 6.100 In Problem 6.99, if the disk has a constant angular velocity of one revolution per second counterclockwise and $\theta = 180^o$, what is the acceleration of point C?

Solution The angular velocity of the disk is $\omega_{AB} = 2\pi\ rad/s$. The vector location of B is $\vec{r}_B = -4\vec{i}\ (in.)$. The bar BC is level when the angle is $\theta = 45^o$, from which the vector location of point C when $\theta = 180^o$ is $\vec{r}_C = -(14\cos\alpha + 4)\vec{i} + \left(4\sin 45^o\right)\vec{j} = -17.7\vec{i} + 2.83\vec{j}\ (in.)$, where the angle $\alpha = \sin^{-1}\left(\dfrac{4\sin 45^o}{14}\right) = 11.66^o$. The velocity of point B is

$$\vec{v}_B = \vec{\omega}_{AB} \times \vec{r}_{B/C} = \begin{bmatrix} \vec{i} & \vec{j} & \vec{k} \\ 0 & 0 & 2\pi \\ -4 & 0 & 0 \end{bmatrix} = -8\pi\vec{j} = -25.1\vec{j}\ (in/s).$$

The vector $\vec{r}_{C/B} = \vec{r}_C - \vec{r}_B = -13.7\vec{i} + 2.83\vec{j}\ (in.)$, The velocity of point C in terms of the velocity of B is

$$\vec{v}_C = \vec{v}_B + \vec{\omega} \times \vec{r}_{C/B} = \vec{v}_{B+}\begin{bmatrix} \vec{i} & \vec{j} & \vec{k} \\ 0 & 0 & \omega_{BC} \\ -13.7 & 2.83 & 0 \end{bmatrix} = -25.1\vec{j} - 2.83\omega_{BC}\vec{i} - 13.7\omega_{BC}\vec{j}\ (in/s).$$

The velocity of C is constrained to be parallel to the *x* axis.
Separate components and solve: $v_C = 2.83\omega_{BC}$, $0 = -25.1 - 13.7\omega_{BC}$, from which $\vec{v}_C = -6.83\omega_{BC}\vec{i}\ (in/s)$. $\omega_{BC} = -1.833\ rad/s$.

The acceleration of B is $\vec{a}_B = -\omega_{AB}^2(-4\vec{i}) = 157.9\vec{i}\ \left(in/s^2\right)$.

The acceleration of point C in terms of the acceleration at B:

$$\vec{a}_C = \vec{a}_B + \vec{\alpha}_{BC} \times \vec{r}_{C/B} - \omega_{BC}^2\vec{r}_{C/B} = 157.9\vec{i} + \begin{bmatrix} \vec{i} & \vec{j} & \vec{k} \\ 0 & 0 & \alpha_{BC} \\ -13.7 & 2.83 & 0 \end{bmatrix} - \omega_{BC}^2\left(-13.7\vec{i} + 2.83\vec{j}\right)$$

$$\vec{a}_C = 157.9\vec{i} + \alpha_{BC}\left(-2.83\vec{i} - 13.7\vec{j}\right) - \omega_{BC}^2\left(-13.7\vec{i} + 2.83\vec{j}\right).$$

The acceleration of C is constrained to lie parallel to the *x* axis. Separate components:
$a_C = 157.9 - 2.83\alpha_{BC} + 13.7\omega_{BC}^2$, $0 = -13.7\alpha_{BC} - 2.83\omega_{BC}^2$. Solve: $\boxed{\vec{a}_C = 205.9\vec{i}\ \left(in/s^2\right)}$

================================<>================================

===========================<>===========================

Problem 6.101 If $\vec{\omega}_{AB} = 2\vec{k}\ rad/s$, $\vec{\alpha}_{AB} = 2\vec{k}\ rad/s^2$, $\vec{\omega}_{BC} = -1\vec{k}\ rad/s$, and $\vec{\alpha}_{BC} = -4\vec{k}\ rad/s^2$, what is the acceleration of point C, where the scoop of the excavator is attached.?

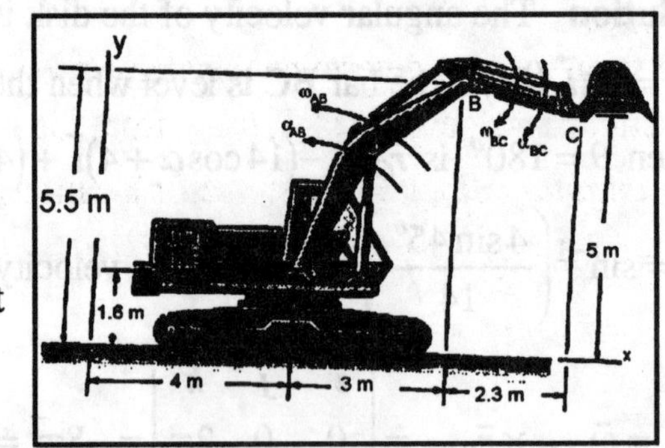

Solution The vector locations of points A, B, C are $\vec{r}_A = 4\vec{i} + 1.6\vec{j}\ (m)$, $\vec{r}_B = 7\vec{i} + 5.5\vec{j}\ (m)$, $\vec{r}_C = 9.3\vec{i} + 5\vec{j}\ (m)$. The vectors $\vec{r}_{B/A} = \vec{r}_B - \vec{r}_A = 3\vec{i} + 3.9\vec{j}\ (m)$, $\vec{r}_{C/B} = \vec{r}_C - \vec{r}_B = 2.3\vec{i} - 0.5\vec{j}\ (m)$. The acceleration of point B is $\vec{a}_B = \vec{\alpha}_{AB} \times \vec{r}_{B/A} - \omega_{AB}^2 \vec{r}_{B/A}$.

$$\vec{a}_B = \begin{bmatrix} \vec{i} & \vec{j} & \vec{k} \\ 0 & 0 & 2 \\ 3 & 3.9 & 0 \end{bmatrix} - (2^2)(3.0\vec{i} + 3.9\vec{j}),$$

$\vec{a}_B = +2(-3.9\vec{i} + 3\vec{j}) - (4)(3.\vec{i} + 3.9\vec{j}) = -19.8\vec{i} - 9.6\vec{j}\ (m/s^2)$. The acceleration of point C in terms of the acceleration at point B is

$$\vec{a}_C = \vec{a}_B + \vec{\alpha}_{BC} \times \vec{r}_{C/B} - \omega_{BC}^2(\vec{r}_{C/B}) = -19.8\vec{i} - 9.6\vec{j} + \begin{bmatrix} \vec{i} & \vec{j} & \vec{k} \\ 0 & 0 & -4 \\ 2.3 & -0.5 & 0 \end{bmatrix} - 1^2(2.3\vec{i} - 0.5\vec{j}),$$

$$\vec{a}_C = -19.8\vec{i} - 9.6\vec{j} - 2\vec{i} - 9.2\vec{j} - 2.3\vec{i} + 0.5\vec{j} = \boxed{-24.1\vec{i} - 18.3\vec{j}\ (m/s^2)}$$

===========================<>===========================

================================<>================================

Problem 6.102 If the velocity of point C of the excavator in Problem 6.101 is $\vec{v}_C = 4\vec{i}\ (m/s)$ and is constant at the instant shown, what are ω_{AB}, α_{AB}, ω_{BC}, α_{BC}?

Solution:

The strategy is to determine the angular velocities ω_{AB}, ω_{BC} from the known velocity at point C, and the angular velocities α_{AB}, α_{BC} from the data that the linear acceleration at point C is constant.

The angular velocities: The vector locations of points A, B, C are $\vec{r}_A = 4\vec{i} + 1.6\vec{j}\ (m)$, $\vec{r}_B = 7\vec{i} + 5.5\vec{j}\ (m)$, $\vec{r}_C = 9.3\vec{i} + 5\vec{j}\ (m)$. The vectors $\vec{r}_{B/A} = \vec{r}_B - \vec{r}_A = 3\vec{i} + 3.9\vec{j}\ (m)$, $\vec{r}_{C/B} = \vec{r}_C - \vec{r}_B = 2.3\vec{i} - 0.5\vec{j}\ (m)$.

The velocity of point B is $\vec{v}_B = \vec{\omega}_{AB} \times \vec{r}_{B/A} = \begin{bmatrix} \vec{i} & \vec{j} & \vec{k} \\ 0 & 0 & \omega_{AB} \\ 3 & 3.9 & 0 \end{bmatrix} = -3.9\omega_{AB}\vec{i} + 3\omega_{AB}\vec{j}$.

The velocity of C in terms of the velocity of B

$$\vec{v}_C = \vec{v}_B + \vec{\omega}_{BC} \times \vec{r}_{C/B} = -3.9\omega_{AB}\vec{i} + 3\omega_{AB}\vec{j} + \begin{bmatrix} \vec{i} & \vec{j} & \vec{k} \\ 0 & 0 & \omega_{BC} \\ 2.3 & -0.5 & 0 \end{bmatrix},$$

$\vec{v}_C = -3.9\omega_{AB}\vec{i} + 3\omega_{AB}\vec{j} + 0.5\omega_{BC}\vec{i} + 2.3\omega_{BC}\vec{j}\ (m/s)$. Substitute $\vec{v}_C = 4\vec{i}\ (m/s)$, and separate components: $4 = -3.9\omega_{AB} + 0.5\omega_{BC}$, $0 = 3\omega_{AB} + 2.3\omega_{BC}$. Solve: $\boxed{\omega_{AB} = -0.8787\ rad/s}$, $\boxed{\omega_{BC} = 1.146\ rad/s}$.

The angular accelerations: The acceleration of point B is

$$\vec{a}_B = \vec{\alpha}_{AB} \times \vec{r}_{B/A} - \omega_{AB}^2\vec{r}_{B/A} = \begin{bmatrix} \vec{i} & \vec{j} & \vec{k} \\ 0 & 0 & \alpha_{AB} \\ 3 & 3.9 & 0 \end{bmatrix} - \left(\omega_{AB}^2\right)\left(3\vec{i} + 3.9\vec{j}\right),$$

$\vec{a}_B = -3.9\alpha_{AB}\vec{i} + 3\alpha_{AB}\vec{j} - 3\omega_{AB}^2\vec{i} - 3.9\omega_{AB}^2\vec{j}\ \left(m/s^2\right)$. The acceleration of C in terms of the acceleration of B is

$$\vec{a}_C = \vec{a}_B + \vec{\alpha}_{BC} \times \vec{r}_{C/B} - \omega_{BC}^2\vec{r}_{C/B} = \vec{a}_B + \begin{bmatrix} \vec{i} & \vec{j} & \vec{k} \\ 0 & 0 & \alpha_{BC} \\ 2.3 & -0.5 & 0 \end{bmatrix} - \omega_{BC}^2\left(2.3\vec{i} - 0.5\vec{j}\right)$$

$\vec{a}_C = \left(-3.9\alpha_{AB} - 3\omega_{AB}^2\right)\vec{i} + \left(3\alpha_{AB} - 3.9\omega_{AB}^2\right)\vec{j} + (0.5\alpha_{BC} - 2.3\omega_{BC}^2)\vec{i} + \left(2.3\alpha_{BC} + 0.5\omega_{BC}^2\right)\vec{j}$. Substitute $\vec{a}_C = 0$ from the conditions of the problem, and separate components:

$0 = -3.9\alpha_{AB} + 0.5\alpha_{BC} - 3\omega_{AB}^2 - 2.3\omega_{BC}^2$, $0 = 3\alpha_{AB} + 2.3\alpha_{BC} - 3.9\omega_{AB}^2 + 0.5\omega_{BC}^2$. Solve: $\boxed{\alpha_{BC} = 2.406\ rad/s^2}$, $\boxed{\alpha_{AB} = -1.06\ rad/s^2}$.

================================<>================================

==================================<>==================================

Problem 6.103 Bar AB rotates in the counterclockwise direction with a constant angular velocity of 10 *rad/s*. What are the angular accelerations of BC and CD?

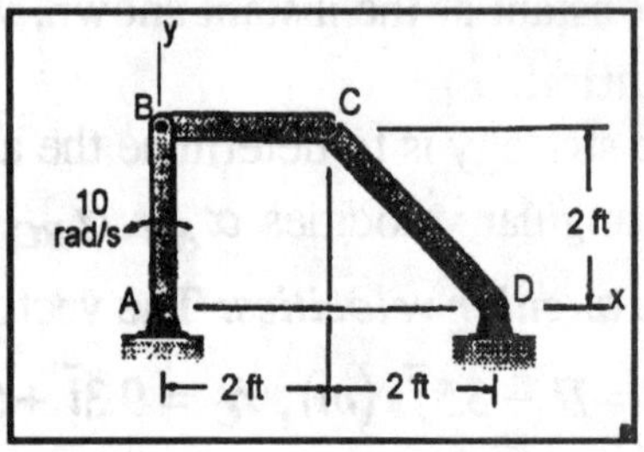

Solution:

The vector locations of A,B,C and D are: $\vec{r}_A = 0$, $\vec{r}_B = 2\vec{j}$ (ft), $\vec{r}_C = 2\vec{i} + 2\vec{j}$ (ft), $\vec{r}_D = 4\vec{i}$ (ft). The vectors $\vec{r}_{B/A} = \vec{r}_B - \vec{r}_A = 2\vec{j}$ (ft). $\vec{r}_{C/B} = \vec{r}_C - \vec{r}_B = 2\vec{i}$ (ft), $\vec{r}_{C/D} = \vec{r}_D - \vec{r}_C = -2\vec{i} + 2\vec{j}$ (ft).

(a) *Get the angular velocities* ω_{BC}, ω_{CD}. The velocity of point B is

$$\vec{v}_B = \vec{\omega}_{AB} \times \vec{r}_{B/A} = \begin{bmatrix} \vec{i} & \vec{j} & \vec{k} \\ 0 & 0 & 10 \\ 0 & 2 & 0 \end{bmatrix} = -20\vec{i} \ (ft/s).$$ The velocity of C in terms of the velocity of B is

$$\vec{v}_C = \vec{v}_B + \vec{\omega}_{BC} \times \vec{r}_{C/B} = -20\vec{i} + \begin{bmatrix} \vec{i} & \vec{j} & \vec{k} \\ 0 & 0 & \omega_{BC} \\ 2 & 0 & 0 \end{bmatrix} = -20\vec{i} + 2\omega_{BC}\vec{j} \ (ft/s).$$ The velocity of C in terms of the velocity of point D $$\vec{v}_C = \vec{\omega}_{CD} \times \vec{r}_{C/D} = \begin{bmatrix} \vec{i} & \vec{j} & \vec{k} \\ 0 & 0 & \omega_{CD} \\ -2 & 2 & 0 \end{bmatrix} = 2\omega_{CD}\left(-\vec{i} - \vec{j}\right) \ (ft/s).$$ Equate the expressions for $\vec{v}_C$ and separate components: $-20 = 2\omega_{CD}$, $2\omega_{BC} = -2\omega_{CD}$. Solve: $\omega_{BC} = -10 \ rad/s$, $\omega_{CD} = 10 \ rad/s$.

(b) *Get the angular accelerations.* The acceleration of point B is

$\vec{a}_B = \vec{\alpha}_{AB} \times \vec{r}_{B/A} - \omega_{AB}^2 \vec{r}_{B/A} = -\omega_{AB}^2(2\vec{j}) = -200\vec{j} \ \left(ft/s^2\right)$.

The acceleration of point C in terms of the acceleration of point B:

$$\vec{a}_C = \vec{a}_B + \vec{\alpha}_{BC} \times \vec{r}_{C/B} - \omega_{BC}^2 \vec{r}_{C/B} = -200\vec{j} + \begin{bmatrix} \vec{i} & \vec{j} & \vec{k} \\ 0 & 0 & \alpha_{BC} \\ 2 & 0 & 0 \end{bmatrix} - \omega_{BC}^2(2\vec{i}).$$

$\vec{a}_C = -200\vec{j} + 2\alpha_{BC}\vec{j} - 200\vec{i} \ \left(ft/s^2\right)$.

The acceleration of point C in terms of the acceleration of point D:

$$\vec{a}_C = \vec{\alpha}_{CD} \times \vec{r}_{C/D} - \omega_{CD}^2 \vec{r}_{C/D} = \begin{bmatrix} \vec{i} & \vec{j} & \vec{k} \\ 0 & 0 & \alpha_{CD} \\ -2 & 2 & 0 \end{bmatrix} - \omega_{CD}^2(-2\vec{i} + 2\vec{j}).$$

$\vec{a}_C = -2\alpha_{CD}\vec{i} - 2\alpha_{CD}\vec{j} + 200\vec{i} - 200\vec{j} \ \left(ft/s^2\right)$. Equate the expressions and separate components: $-200 = -2\alpha_{CD} + 200$, $-200 + 2\alpha_{BC} = -200 - 2\alpha_{CD}$. Solve: $\boxed{\alpha_{BC} = -200 \ rad/s^2}$, $\boxed{\alpha_{CD} = 200 \ rad/s}$, where the negative sign means a clockwise angular acceleration.

==================================<>==================================

==================================<>==================================

Problem 6.104 At the instant shown, bar AB has no angular velocity but has a counterclockwise angular acceleration of 10 rad/s^2. Determine the acceleration of point E.

Solution:

The vector locations of A,B,C and D are: $\vec{r}_A = 0$, $\vec{r}_B = 400\vec{j}$ (mm), $\vec{r}_C = 700\vec{i}$ (mm), $\vec{r}_D = 1100\vec{i}$ (mm). $\vec{r}_E = 1800\vec{i}$ (mm) The vectors $\vec{r}_{B/A} = \vec{r}_B - \vec{r}_A = 400\vec{j}$ (mm).

$\vec{r}_{C/B} = \vec{r}_C - \vec{r}_B = 700\vec{i} - 400\vec{j}$ (mm),

$\vec{r}_{C/D} = \vec{r}_D - \vec{r}_C = -400\vec{i}$ (mm). $\vec{r}_{E/D} = 700\vec{i}$ (mm)

(a) *Get the angular velocities* ω_{BC}, ω_{CD}. The velocity of point B is zero. The velocity of C in terms of the velocity of B is $\vec{v}_C = \vec{v}_B + \vec{\omega}_{BC} \times \vec{r}_{C/B} = \begin{bmatrix} \vec{i} & \vec{j} & \vec{k} \\ 0 & 0 & \omega_{BC} \\ 700 & -400 & 0 \end{bmatrix} = +400\omega_{BC}\vec{i} + 700\omega_{BC}\vec{j}$ (mm/s).

The velocity of C in terms of the velocity of point D

$\vec{v}_C = \vec{\omega}_{CD} \times \vec{r}_{C/D} = \begin{bmatrix} \vec{i} & \vec{j} & \vec{k} \\ 0 & 0 & \omega_{CD} \\ -400 & 0 & 0 \end{bmatrix} = -400\omega_{CD}\vec{j}$ (mm/s). Equate the expressions for $\vec{v}_C$ and separate components: $400\omega_{BC} = 0$, $700\omega_{BC} = -400\omega_{CD}$. Solve: $\omega_{BC} = 0$ rad/s, $\omega_{CD} = 0$ rad/s.

(b) *Get the angular accelerations.* The acceleration of point B is

$\vec{a}_B = \vec{\alpha}_{AB} \times \vec{r}_{B/A} - \omega_{AB}^2 \vec{r}_{B/A} = \begin{bmatrix} \vec{i} & \vec{j} & \vec{k} \\ 0 & 0 & 10 \\ 0 & 400 & 0 \end{bmatrix} = -4000\vec{i}$ (mm/s^2). The acceleration of point C in terms of the acceleration of point B: $\vec{a}_C = \vec{a}_B + \vec{\alpha}_{BC} \times \vec{r}_{C/B} - \omega_{BC}^2 \vec{r}_{C/B} = -4000\vec{i} + \begin{bmatrix} \vec{i} & \vec{j} & \vec{k} \\ 0 & 0 & \alpha_{BC} \\ 700 & -400 & 0 \end{bmatrix}$.

$\vec{a}_C = -4000\vec{i} + 400\alpha_{BC}\vec{i} + 700\alpha_{BC}\vec{j}$ (mm/s^2).

The acceleration of point C in terms of the acceleration of point D:

$\vec{a}_C = \vec{\alpha}_{CD} \times \vec{r}_{C/D} - \omega_{CD}^2 \vec{r}_{C/D} = \begin{bmatrix} \vec{i} & \vec{j} & \vec{k} \\ 0 & 0 & \alpha_{CD} \\ -400 & 0 & 0 \end{bmatrix} = -400\alpha_{CD}\vec{j}$ (mm/s^2). . Equate the expressions and separate components: $-4000 + 400\alpha_{CD} = 0$, $700\alpha_{BC} = -400\alpha_{CD}$.

Solve: $\alpha_{BC} = 10$ rad/s^2, $\alpha_{CD} = -17.5$ rad/s^2, The acceleration of point E in terms of the acceleration of point D is $a_E = \vec{\alpha}_{CD} \times \vec{r}_{E/D} = \begin{bmatrix} \vec{i} & \vec{j} & \vec{k} \\ 0 & 0 & -17.5 \\ 700 & 0 & 0 \end{bmatrix}$ $\boxed{= -12250\vec{j} \ (mm/s^2)}$ (clockwise)

==================================<>==================================

================================<>================================

Problem 6.105 If $\omega_{AB} = 12\ rad/s$ and $\alpha_{AB} = 100\ rad/s^2$, what are the angular accelerations of bars BC and CD?

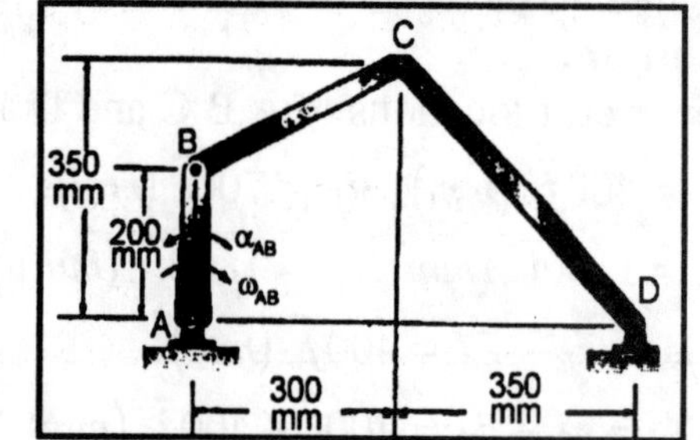

Solution The vector locations of A,B,C and D are: $\vec{r}_A = 0$, $\vec{r}_B = 200\vec{j}\ (mm)$, $\vec{r}_C = 300\vec{i} + 350\vec{j}\ (mm)$, $\vec{r}_D = 650\vec{i}\ (mm)$. The vectors $\vec{r}_{B/A} = \vec{r}_B - \vec{r}_A = 200\vec{j}\ (mm)$.

$\vec{r}_{C/B} = \vec{r}_C - \vec{r}_B = 300\vec{i} + 150\vec{j}\ (mm)$,

$\vec{r}_{C/D} = \vec{r}_D - \vec{r}_C = -350\vec{i} + 350\vec{j}\ (mm)$. $\vec{r}_{E/D} = 700\vec{i}\ (mm)$

(a) *Get the angular velocities* ω_{BC}, ω_{CD}. The velocity of point B is $\vec{v}_B = \vec{\omega}_{AB} \times \vec{r}_{B/A} = \begin{bmatrix} \vec{i} & \vec{j} & \vec{k} \\ 0 & 0 & -12 \\ 0 & 200 & 0 \end{bmatrix} = 2400\vec{i}\ (mm/s)$. The velocity of C in terms of the velocity of B is $\vec{v}_C = \vec{v}_B + \vec{\omega}_{BC} \times \vec{r}_{C/B} = \vec{v}_B + \begin{bmatrix} \vec{i} & \vec{j} & \vec{k} \\ 0 & 0 & \omega_{BC} \\ 300 & 150 & 0 \end{bmatrix} = 2400\vec{i} - 150\omega_{BC}\vec{i} + 300\omega_{BC}\vec{j}\ (mm/s)$. The velocity of C in terms of the velocity of point D

$$\vec{v}_C = \vec{\omega}_{CD} \times \vec{r}_{C/D} = \begin{bmatrix} \vec{i} & \vec{j} & \vec{k} \\ 0 & 0 & \omega_{CD} \\ -350 & 350 & 0 \end{bmatrix} = -350\omega_{CD}\vec{i} - 350\omega_{CD}\vec{j}\ (mm/s)$$

. Equate the expressions for $\vec{v}_C$ and separate components: $2400 - 150\omega_{BC} = -350\omega_{CD}$, $300\omega_{BC} = -350\omega_{CD}$. Solve: $\omega_{BC} = 5.33\ rad/s$, $\omega_{CD} = -4.57\ rad/s$.

(b) *Get the angular accelerations.* The acceleration of point B is

$$\vec{a}_B = \vec{\alpha}_{AB} \times \vec{r}_{B/A} - \omega_{AB}^2 \vec{r}_{B/A} = \begin{bmatrix} \vec{i} & \vec{j} & \vec{k} \\ 0 & 0 & 100 \\ 0 & 200 & 0 \end{bmatrix} - \omega_{AB}^2(200\vec{j}) = -20{,}000\vec{i} - 28{,}800\vec{j}\ \left(mm/s^2\right)$$

. The acceleration of point C in terms of the acceleration of point B:

$$\vec{a}_C = \vec{a}_B + \vec{\alpha}_{BC} \times \vec{r}_{C/B} - \omega_{BC}^2 \vec{r}_{C/B} = \vec{a}_B + \begin{bmatrix} \vec{i} & \vec{j} & \vec{k} \\ 0 & 0 & \alpha_{BC} \\ 300 & 150 & 0 \end{bmatrix} - \omega_{BC}^2(300\vec{i} + 150\vec{j}).$$

$\vec{a}_C = \left(-20{,}000 - 150\alpha_{BC} - 300\omega_{BC}^2\right)\vec{i} + \left(-28{,}800 + 300\alpha_{BC} - 150\omega_{BC}^2\right)\vec{j}\ \left(mm/s^2\right)$. The acceleration of point C in terms of the acceleration of point D:

$$\vec{a}_C = \vec{\alpha}_{CD} \times \vec{r}_{C/D} - \omega_{CD}^2 \vec{r}_{C/D} = \begin{bmatrix} \vec{i} & \vec{j} & \vec{k} \\ 0 & 0 & \alpha_{CD} \\ -350 & 350 & 0 \end{bmatrix} - \omega_{CD}^2(-350\vec{i} + 350\vec{j})$$

Solution continued on next page

$\vec{a}_C = -350\alpha_{CD}\vec{i} - 350\alpha_{CD}\vec{j} + 350\omega_{CD}^2\vec{i} - 350\omega_{CD}^2\vec{j}$ (mm/s^2). Equate the expressions and separate components: $-20{,}000 - 150\alpha_{BC} - 300\omega_{BC}^2 = -350\alpha_{CD} + 350\omega_{CD}^2$, $-28{,}800 + 300\alpha_{BC} - 150\omega_{BC}^2 = -350\alpha_{CD} - 350\omega_{CD}^2$. Solve: $\boxed{\alpha_{BC} = -22.43 \ rad/s^2}$, $\boxed{\alpha_{CD} = 92.8 \ rad/s^2}$, where the negative sign means a clockwise acceleration.

================================<>================================

Problem 6.106 If $\omega_{AB} = 4 \ rad/s$ counterclockwise and $\alpha_{AB} = 12 \ rad/s^2$ counterclockwise, what is the acceleration of point C.

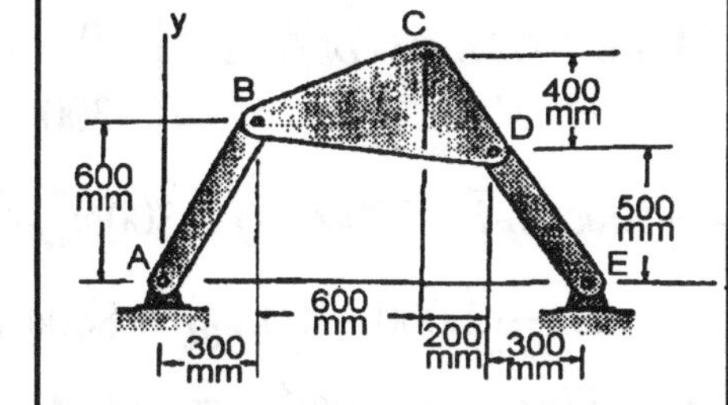

Solution The vector locations of A,B,C and D are: $\vec{r}_A = 0$, $\vec{r}_B = 300\vec{i} + 600\vec{j}$ (mm), $\vec{r}_C = 900\vec{i} + 900\vec{j}$ (mm), $\vec{r}_D = 1100\vec{i} + 500\vec{j}$ (mm), $\vec{r}_E = 1400\vec{i}$ (mm). The vectors $\vec{r}_{B/A} = \vec{r}_B - \vec{r}_A = 300\vec{i} + 600\vec{j}$ (mm). $\vec{r}_{C/B} = \vec{r}_C - \vec{r}_B = 600\vec{i} + 300\vec{j}$ (mm), $\vec{r}_{D/B} = \vec{r}_D - \vec{r}_B = 800\vec{i} - 100\vec{j}$ (mm). $\vec{r}_{D/E} = \vec{r}_D - \vec{r}_E = -300\vec{i} + 500\vec{j}$ (mm)

(a) *Get the angular velocity* ω_{BC}. The velocity of point B is

$$\vec{v}_B = \vec{\omega}_{AB} \times \vec{r}_{B/A} = \begin{bmatrix} \vec{i} & \vec{j} & \vec{k} \\ 0 & 0 & 4 \\ 300 & 600 & 0 \end{bmatrix} = -2400\vec{i} + 1200\vec{j} \ (mm/s).$$

The velocity of D in terms of the velocity of B is

$$\vec{v}_D = \vec{v}_B + \vec{\omega}_{BC} \times \vec{r}_{D/B} = \vec{v}_B + \begin{bmatrix} \vec{i} & \vec{j} & \vec{k} \\ 0 & 0 & \omega_{BC} \\ 800 & -100 & 0 \end{bmatrix}.$$

$\vec{v}_D = -2400\vec{i} + 1200\vec{j} + 100\omega_{BC}\vec{i} + 800\omega_{BC}\vec{j}$ (mm/s). The velocity of D in terms of the velocity of point E

$$\vec{v}_D = \vec{\omega}_{DE} \times \vec{r}_{D/E} = \begin{bmatrix} \vec{i} & \vec{j} & \vec{k} \\ 0 & 0 & \omega_{DE} \\ -300 & 500 & 0 \end{bmatrix} = -500\omega_{DE}\vec{i} - 300\omega_{DE}\vec{j} \ (mm/s).$$

Equate the expressions for $\vec{v}_D$ and separate components: $-2400 + 100\omega_{BC} = -500\omega_{DE}$, $1200 + 800\omega_{BC} = -300\omega_{DE}$. Solve: $\omega_{BC} = -3.57 \ rad/s$, $\omega_{DE} = 5.51 \ rad/s$.

(b) *Get the angular accelerations.* The acceleration of point B is

$$\vec{a}_B = \vec{\alpha}_{AB} \times \vec{r}_{B/A} - \omega_{AB}^2\vec{r}_{B/A} = \begin{bmatrix} \vec{i} & \vec{j} & \vec{k} \\ 0 & 0 & 12 \\ 300 & 600 & 0 \end{bmatrix} - (4^2)(300\vec{i} + 600\vec{j}).$$

$\vec{a}_B = -1200\vec{i} - 600\vec{j}$ (mm/s^2).

Solution continued on next page

The acceleration of point D in terms of the acceleration of point B:

$$\vec{a}_D = \vec{a}_B + \vec{\alpha}_{BC} \times \vec{r}_{D/B} - \omega_{BC}^2 \vec{r}_{D/B} = \vec{a}_B + \begin{bmatrix} \vec{i} & \vec{j} & \vec{k} \\ 0 & 0 & \alpha_{BC} \\ 800 & -100 & 0 \end{bmatrix} - \omega_{BC}^2 (800\vec{i} - 100\vec{j}).$$

$\vec{a}_C = -12{,}000\vec{i} - 6{,}000\vec{j} + 100\alpha_{BC}\vec{i} + 800\alpha_{BC}\vec{j} - 800\omega_{BC}^2\vec{i} + 100\omega_{BC}^2\vec{j}\ (mm/s^2)$. The acceleration of point D in terms of the acceleration of point E:

$$\vec{a}_D = \vec{\alpha}_{DE} \times \vec{r}_{D/E} - \omega_{DE}^2 \vec{r}_{D/E} = \begin{bmatrix} \vec{i} & \vec{j} & \vec{k} \\ 0 & 0 & \alpha_{DE} \\ -300 & 500 & 0 \end{bmatrix} - \omega_{CD}^2 (-300\vec{i} + 500\vec{j}).$$

$\vec{a}_D = -500\alpha_{DE}\vec{i} - 300\alpha_{DE}\vec{j} + 300\omega_{DE}^2\vec{i} + 500\omega_{DE}^2\vec{j}\ (mm/s^2)$. Equate the expressions and separate components: $-12{,}000 + 100\alpha_{BC} - 800\omega_{BC}^2 = -500\alpha_{DE} + 300\omega_{DE}^2$,
$-6{,}000 + 800\alpha_{BC} + 100\omega_{BC}^2 = -300\alpha_{DE} + 500\omega_{DE}^2$. Solve: $\alpha_{BC} = -39.53\ rad/s^2$,
$\alpha_{DE} = 70.51\ rad/s^2$, where the negative sign means a clockwise acceleration.

Get the acceleration of point C. The acceleration of point C in terms of the acceleration of point B is

$$\vec{a}_C = \vec{a}_B + \vec{\alpha}_{BC} \times \vec{r}_{C/B} - \omega_{BC}^2 \vec{r}_{C/B} = \vec{a}_B + \begin{bmatrix} \vec{i} & \vec{j} & \vec{k} \\ 0 & 0 & \alpha_{BC} \\ 600 & 300 & 0 \end{bmatrix} - \omega_{BC}^2 (600\vec{i} + 300\vec{j}),$$

$\vec{a}_C = -12000\vec{i} - 6000\vec{j} - 300\alpha_{BC}\vec{i} + 600\alpha_{BC}\vec{j} - \omega_{BC}^2(600\vec{i} + 300\vec{j})$. Substitute values found above:

$$\boxed{\vec{a}_C = -7777\vec{i} - 33{,}537\vec{j}\ (mm/s^2)}$$

==================================<>==================================

Problem 6.107 In Problem 6.106, if $\omega_{AB} = 6\ rad/s$, and $\alpha_{DE} = 0$, what is the acceleration of point C?

Solution The vector locations of A,B,C and D are: $\vec{r}_A = 0$, $\vec{r}_B = 300\vec{i} + 600\vec{j}\ (mm)$, $\vec{r}_C = 900\vec{i} + 900\vec{j}\ (mm)$, $\vec{r}_D = 1100\vec{i} + 500\vec{j}\ (mm)$, $\vec{r}_E = 1400\vec{i}\ (mm)$. The vectors $\vec{r}_{B/A} = \vec{r}_B - \vec{r}_A = 300\vec{i} + 600\vec{j}\ (mm)$. $\vec{r}_{C/B} = \vec{r}_C - \vec{r}_B = 600\vec{i} + 300\vec{j}\ (mm)$, $\vec{r}_{D/B} = \vec{r}_D - \vec{r}_B = 800\vec{i} - 100\vec{j}\ (mm)$. $\vec{r}_{D/E} = \vec{r}_D - \vec{r}_E = -300\vec{i} + 500\vec{j}\ (mm)$

(a) *Get the angular velocity* ω_{BC}. The velocity of point B is

$$\vec{v}_B = \vec{\omega}_{AB} \times \vec{r}_{B/A} = \begin{bmatrix} \vec{i} & \vec{j} & \vec{k} \\ 0 & 0 & 6 \\ 300 & 600 & 0 \end{bmatrix} = -3600\vec{i} + 1800\vec{j}\ (mm/s).$$

Solution continued on next page

The velocity of D in terms of the velocity of B is $\vec{v}_D = \vec{v}_B + \vec{\omega}_{BC} \times \vec{r}_{D/B} = \vec{v}_B + \begin{bmatrix} \vec{i} & \vec{j} & \vec{k} \\ 0 & 0 & \omega_{BC} \\ 800 & -100 & 0 \end{bmatrix}$.

$\vec{v}_D = -3600\vec{i} + 1800\vec{j} + 100\omega_{BC}\vec{i} + 800\omega_{BC}\vec{j}$ (mm/s). The velocity of D in terms of the velocity of point E, $\vec{v}_D = \vec{\omega}_{DE} \times \vec{r}_{D/E} = \begin{bmatrix} \vec{i} & \vec{j} & \vec{k} \\ 0 & 0 & \omega_{DE} \\ -300 & 500 & 0 \end{bmatrix} = -500\omega_{DE}\vec{i} - 300\omega_{DE}\vec{j}$ (mm/s). Equate the expressions for $\vec{v}_D$ and separate components: $-3600 + 100\omega_{BC} = -500\omega_{DE}$, $1800 + 800\omega_{BC} = -300\omega_{DE}$. Solve: $\omega_{BC} = -5.35$ rad/s, $\omega_{DE} = 8.27$ rad/s.

(b) *Get the angular accelerations.* The acceleration of point B is

$$\vec{a}_B = \vec{\alpha}_{AB} \times \vec{r}_{B/A} - \omega_{AB}^2 \vec{r}_{B/A} = \begin{bmatrix} \vec{i} & \vec{j} & \vec{k} \\ 0 & 0 & \alpha_{AB} \\ 300 & 600 & 0 \end{bmatrix} - \omega_{AB}^2\left(300\vec{i} + 600\vec{j}\right).$$

$\vec{a}_B = -600\alpha_{AB}\vec{i} + 300\alpha_{AB}\vec{j} - 300\omega_{AB}^2\vec{i} - 600\omega_{AB}^2\vec{j}$ $\left(mm/s^2\right)$. The acceleration of point D in terms of the acceleration of point B:

$$\vec{a}_D = \vec{a}_B + \vec{\alpha}_{BC} \times \vec{r}_{D/B} - \omega_{BC}^2 \vec{r}_{D/B} = \vec{a}_B + \begin{bmatrix} \vec{i} & \vec{j} & \vec{k} \\ 0 & 0 & \alpha_{BC} \\ 800 & -100 & 0 \end{bmatrix} - \omega_{BC}^2(800\vec{i} - 100\vec{j}),.$$

$\vec{a}_D = \vec{a}_B + 100\alpha_{BC}\vec{i} + 800\alpha_{BC}\vec{j} - 800\omega_{BC}^2\vec{i} + 100\omega_{BC}^2\vec{j}$ $\left(mm/s^2\right)$. The acceleration of point D in terms of the acceleration of point E: $\vec{a}_D = \vec{\alpha}_{DE} \times \vec{r}_{D/E} - \omega_{DE}^2 \vec{r}_{D/E} = -\omega_{CD}^2(-300\vec{i} + 500\vec{j})$.

$\vec{a}_D = 300\omega_{DE}^2\vec{i} + 500\omega_{DE}^2\vec{j}$ $\left(mm/s^2\right)$. Equate the expressions and separate components:

$-600\alpha_{AB} - 300\omega_{AB}^2 + 100\alpha_{BC} - 800\omega_{BC}^2 = 300\omega_{DE}^2$,

$300\alpha_{AB} - 600\omega_{AB}^2 + 800\alpha_{BC} + 100\omega_{BC}^2 = 500\omega_{DE}^2$. Solve: $\alpha_{BC} = 13.7$ rad/s^2,

$\alpha_{AB} = -88.1$ rad/s^2, where the negative sign means a clockwise acceleration. Substitute to obtain:

$\vec{a}_B = 42{,}057\vec{i} - 48{,}029\vec{j}$ $\left(mm/s^2\right)$.

Get the acceleration of point C. The acceleration of point C in terms of the acceleration of point B is

$$\vec{a}_C = \vec{a}_B + \vec{\alpha}_{BC} \times \vec{r}_{C/B} - \omega_{BC}^2 \vec{r}_{C/B} = \vec{a}_B + \begin{bmatrix} \vec{i} & \vec{j} & \vec{k} \\ 0 & 0 & \alpha_{BC} \\ 600 & 300 & 0 \end{bmatrix} - \omega_{BC}^2\left(600\vec{i} + \vec{j}\right),$$

$\vec{a}_C = 42{,}057\vec{i} - 48{,}029\vec{j} - 300\alpha_{BC}\vec{i} + 600\alpha_{BC}\vec{j} - \omega_{BC}^2\left(600\vec{i} + 300\vec{j}\right)$. Substitute values found above:

$\boxed{\vec{a}_C = 20{,}763\vec{i} - 48{,}395\vec{j} \ \left(mm/s^2\right)}$

==============================<>==============================

================================<>================================

Problem 6.108 If arm AB has a constant clockwise angular velocity of 0.8 *rad/s*, arm BC has a constant angular velocity of 0.2 *rad/s*, and arm CD remains vertical, what is the acceleration of part D?

Solution The constraint that the arm CD remain vertical means that the angular velocity of arm CD is zero. This implies that arm CD translates only, and in a translating, non-rotating element the velocity and acceleration at any point is the same, and the velocity and acceleration of arm CD is the velocity and acceleration of point C. The vectors:

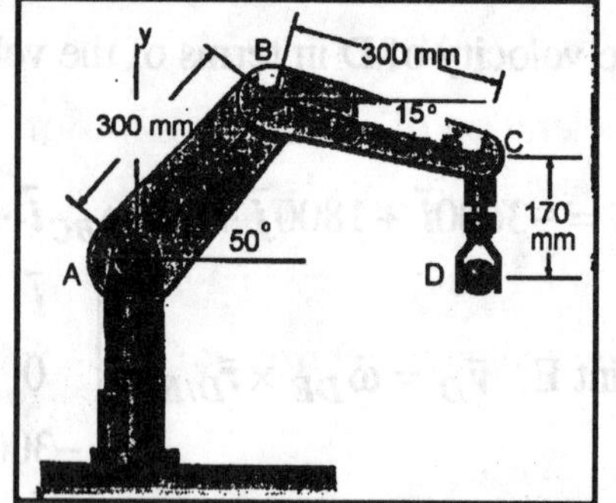

$\vec{r}_{B/A} = 300(\vec{i}\cos 50^o + \vec{j}\sin 50^o) = 192.8\vec{i} + 229.8\vec{j}$ (mm).

$\vec{r}_{C/B} = 300(\vec{i}\cos 15^o - \vec{j}\sin 15^o) = 289.78\vec{i} - 77.6\vec{j}$ (mm). The acceleration of point B is $\vec{a}_B = \vec{\alpha}_{AB} \times \vec{r}_{A/B} - \omega_{AB}^2\vec{r}_{A/B} = -\omega_{AB}^2(192.8\vec{i} + 229.8\vec{j})$ (mm/s^2), since $\alpha_{AB} = 0$. $\vec{a}_B = -123.4\vec{i} - 147.1\vec{j}$ (mm/s) The acceleration of C in terms of the acceleration of B is $\vec{a}_C = \vec{a}_B + \vec{\alpha}_{BC} \times \vec{r}_{C/B} - \omega_{BC}^2\vec{r}_{C/B} = -123.4\vec{i} - 147.1\vec{j} - \omega_{BC}^2(289.8\vec{i} - 77.6\vec{j})$, since $\alpha_{BC} = 0$.

$\vec{a}_C = -135\vec{i} - 144\vec{j}$ (mm/s^2). Since CD is translating: $\boxed{\vec{a}_D = \vec{a}_C = -135\vec{i} - 144\vec{j} \quad (mm/s^2)}$

================================<>================================

Problem 6.109 In Problem 6.108, if arm AB has a constant clockwise angular velocity of 0.8 *rad/s* and you want part CD to have zero velocity and acceleration, what are the necessary angular velocities and accelerations of arms BC and CD?

Solution

Except for numerical values, the solution follows the same strategy as the solution strategies for Problems 6.103 and 6.105. The vectors: $\vec{r}_{B/A} = 300(\vec{i}\cos 50^o + \vec{j}\sin 50^o) = 192.8\vec{i} + 229.8\vec{j}$ (mm).

$\vec{r}_{C/B} = 300(\vec{i}\cos 15^o - \vec{j}\sin 15^o) = 289.78\vec{i} - 77.6\vec{j}$ (mm), $\vec{r}_{C/D} = 170\vec{j}$ (mm). The velocity of point B is $\vec{v}_B = \vec{\omega}_{AB} \times \vec{r}_{B/A} = \begin{bmatrix} \vec{i} & \vec{j} & \vec{k} \\ 0 & 0 & -0.8 \\ 192.8 & 229.8 & 0 \end{bmatrix} = 183.8\vec{i} - 154.3\vec{j}$ (mm/s). The velocity of C in terms of the velocity of B is $\vec{v}_C = \vec{v}_B + \vec{\omega}_{BC} \times \vec{r}_{C/B} = \vec{v}_B + \begin{bmatrix} \vec{i} & \vec{j} & \vec{k} \\ 0 & 0 & \omega_{BC} \\ 289.8 & -77.6 & 0 \end{bmatrix}$.

$\vec{v}_C = 183.9\vec{i} - 154.3\vec{j} + \omega_{BC}(77.6\vec{i} + 289.8\vec{j})$ (mm/s). The velocity of C in terms of the velocity of point D: $\vec{v}_C = \vec{\omega}_{CD} \times \vec{r}_{C/D} = \begin{bmatrix} \vec{i} & \vec{j} & \vec{k} \\ 0 & 0 & \omega_{CD} \\ 0 & 170 & 0 \end{bmatrix} = -170\omega_{CD}\vec{i}$ (mm/s). Equate the expressions for $\vec{v}_C$ and separate components: $183.9 + 77.6\omega_{BC} = -170\omega_{CD}$, $-154.3 + 289.8\omega_{BC} = 0$. Solve: $\boxed{\omega_{BC} = 0.532 \ rad/s}$, $\boxed{\omega_{CD} = -1.325 \ rad/s}$.

Solution continued on next page

Get the angular accelerations. The acceleration of point B is $\vec{a}_B = \vec{\alpha}_{AB} \times \vec{r}_{B/A} - \omega_{AB}^2 \vec{r}_{B/A} = -\omega_{AB}^2(192.8\vec{i} + 229.8\vec{j}) = -123.4\vec{i} - 147.1\vec{j}\ (mm/s^2)$. The acceleration of point C in terms of the acceleration of point B:

$\vec{a}_C = \vec{a}_B + \vec{\alpha}_{BC} \times \vec{r}_{C/B} - \omega_{BC}^2 \vec{r}_{C/B} = \vec{a}_B - \omega_{BC}^2 \vec{r}_{C/B}$.

$\vec{a}_C = -123.4\vec{i} - 147.1\vec{j} + 77.6\alpha_{BC}\vec{i} + 289.8\alpha_{BC}\vec{j} - 289.8\omega_{BC}^2\vec{i} + 77.6\omega_{BC}^2\vec{j}\ (mm/s^2)$.

The acceleration of point C in terms of the acceleration of point D:

$$\vec{a}_C = \vec{\alpha}_{CD} \times \vec{r}_{C/D} - \omega_{CD}^2 \vec{r}_{C/D} = \begin{bmatrix} \vec{i} & \vec{j} & \vec{k} \\ 0 & 0 & \alpha_{CD} \\ 0 & 170 & 0 \end{bmatrix} - \omega_{CD}^2(170\vec{j}).$$

$\vec{a}_C = -170\alpha_{CD}\vec{i} - 170\omega_{CD}^2\vec{j}\ (mm/s^2)$. Equate the expressions and separate components:

$-123.4 + 77.6\alpha_{BC} - 289.8\omega_{BC}^2 = -170\alpha_{CD}$, $-147.1 + 289.8\alpha_{BC} + 77.6\omega_{BC}^2 = 170\omega_{CD}^2$. Solve: $\boxed{\alpha_{BC} = -0.598\ rad/s^2}$, $\boxed{\alpha_{CD} = 1.482\ rad/s^2}$, where the negative sign means a clockwise angular acceleration.

================================<>================================

Problem 6.110 In Problem 6.108, if you want arm CD to remain vertical and you want part D to have velocity $\vec{v}_D = 1.0\vec{i}\ (m/s)$ and zero acceleration, what are the necessary angular velocities and angular accelerations of arms AB and BC?

Solution The constraint that CD remain vertical with zero acceleration means that every point on arm CD is translating, without rotation, at a velocity of 1 *m/s*. This means that the velocity of point C is $\vec{v}_C = 1.0\vec{i}\ (m/s)$, and the acceleration of point C is zero. The vectors:

$\vec{r}_{B/A} = 300(\vec{i}\cos 50^o + \vec{j}\sin 50^o) = 192.8\vec{i} + 229.8\vec{j}\ (mm)$.

$\vec{r}_{C/B} = 300(\vec{i}\cos 15^o - \vec{j}\sin 15^o) = 289.78\vec{i} - 77.6\vec{j}\ (mm)$.

The angular velocities of AB and BC: The velocity of point B is

$$\vec{v}_B = \vec{\omega}_{AB} \times \vec{r}_{B/A} = \begin{bmatrix} \vec{i} & \vec{j} & \vec{k} \\ 0 & 0 & \omega_{AB} \\ 192.8 & 229.8 & 0 \end{bmatrix} = \omega_{AB}(-229.8\vec{i} + 192.8\vec{j})\ (mm/s).$$

. The velocity of C in terms of the velocity of B is $\vec{v}_C = \vec{v}_B + \vec{\omega}_{BC} \times \vec{r}_{C/B} = \vec{v}_B + \begin{bmatrix} \vec{i} & \vec{j} & \vec{k} \\ 0 & 0 & \omega_{BC} \\ 289.8 & -77.6 & 0 \end{bmatrix}$.

$\vec{v}_C = -229.8\omega_{AB}\vec{i} + 192.8\omega_{AB}\vec{j} + \omega_{BC}(77.6\vec{i} + 289.8\vec{j})\ (mm/s)$. The velocity of C is known, $\vec{v}_C = 1000\vec{i}\ (mm/s)$. Equate the expressions for $\vec{v}_C$ and separate components:

$1000 = -229.8\omega_{AB} + 77.6\omega_{BC}$, $0 = 192.8\omega_{AB} + 289.8\omega_{BC}$. Solve: $\boxed{\omega_{AB} = -3.55\ rad/s}$, $\boxed{\omega_{BC} = 2.36\ rad/s}$, where the negative sign means a clockwise angular velocity.

Solution continued on next page

The accelerations of AB and BC: The acceleration of point B is

$$\vec{a}_B = \vec{\alpha}_{AB} \times \vec{r}_{A/B} - \omega_{AB}^2 \vec{r}_{A/B} = \begin{bmatrix} \vec{i} & \vec{j} & \vec{k} \\ 0 & 0 & \alpha_{AB} \\ 192.8 & 229.8 & 0 \end{bmatrix} - \omega_{AB}^2 (192.8\vec{i} + 229.8\vec{j}) \ (mm/s^2), .$$

$\vec{a}_B = \alpha_{AB}(-229.8\vec{i} + 192.8\vec{j}) - \omega_{AB}^2(192.8\vec{i} + 229.8\vec{j}) \ (mm/s^2)$ The acceleration of C in terms of the acceleration of B is

$$\vec{a}_C = \vec{a}_B + \vec{\alpha}_{BC} \times \vec{r}_{C/B} - \omega_{BC}^2 \vec{r}_{C/B} = \vec{a}_B + \begin{bmatrix} \vec{i} & \vec{j} & \vec{k} \\ 0 & 0 & \alpha_{BC} \\ 289.8 & -77.6 & 0 \end{bmatrix} - \omega_{BC}^2 (289.8\vec{i} - 77.6\vec{j}),$$

$\vec{a}_C = \vec{a}_B + \alpha_{BC}(77.6\vec{i} + 289.8\vec{j}) - \omega_{BC}^2(289.8\vec{i} - 77.6\vec{j}) \ (mm/s^2)$. The acceleration of point C is known to be zero. Substitute this value for $\vec{a}_C$, and separate components:

$-229.8\alpha_{AB} - 192.8\omega_{AB}^2 + 77.6\alpha_{BC} - 289.8\omega_{BC}^2 = 0$,

$192.8\alpha_{AB} - 229.8\omega_{AB}^2 + 289.8\alpha_{BC} + 77.6\omega_{BC}^2 = 0$. Solve: $\boxed{\alpha_{AB} = -12.05 \ rad/s^2}$, $\boxed{\alpha_{BC} = 16.53 \ rad/s^2}$, where the negative sign means a clockwise angular acceleration.

===================================<>===================================

Problem 6.111 The link AB of the robot's arm is rotating with a constant counterclockwise angular velocity of *2 rad/s,* and link BC is rotating with a constant clockwise angular velocity of *3 rad/s.* The link CD is rotating at *4 rad/s* in the counterclockwise direction and has a counterclockwise angular acceleration of 6 rad/s^2. What is the acceleration of point D?

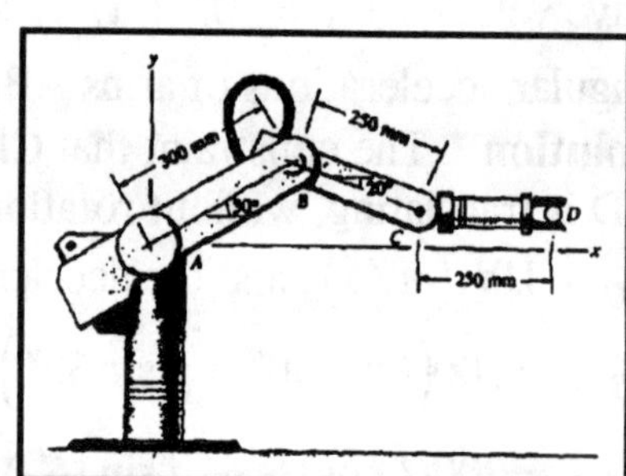

Solution:

The acceleration of B is $\bar{a}_B = \bar{a}_A + \bar{\alpha}_{AB} \times \bar{r}_{B/A} - \omega_{AB}^2 \bar{r}_{B/A}$. Evaluating, we get $\bar{a}_B = 0 + 0 - (2)^2(0.3\cos 30°\mathbf{i} + 0.3\sin 30°\mathbf{j}) = -1.039\ \mathbf{i} - 0.600\ \mathbf{j}\ (m/s^2)$.

The acceleration of C is $\bar{a}_C = \bar{a}_B + \bar{\alpha}_{BC} \times \bar{r}_{C/B} - \omega_{BC}^2 \bar{r}_{C/B}$. Evaluating, we get $\bar{a}_C = -1.039\mathbf{i} - 0.600\mathbf{j} - (3)^2(0.25\cos 20°\mathbf{i} - 0.25\sin 20°\mathbf{j}) = -3.154\ \mathbf{i} + 0.170\ \mathbf{j}\ (m/s^2)$.

The acceleration of D is $\bar{a}_D = \bar{a}_C + \bar{\alpha}_{CD} \times \bar{r}_{D/C} - \omega_{CD}^2 \bar{r}_{D/C}$. Evaluating, we get

$$\bar{a}_D = -3.154\ \mathbf{i} + 0.170\ \mathbf{j} + \begin{vmatrix} \vec{i} & \vec{j} & \vec{k} \\ 0 & 0 & 6 \\ 0.25 & 0 & 0 \end{vmatrix} - (4)^2(0.25\mathbf{i}) = 7.154\ \mathbf{i} + 1.67\ \mathbf{j}\ (m/s^2)$$

===================================<>===================================

==================================<>==================================

Problem 6.112 Consider the robot shown in Problem 6.111. Link AB is rotating with a constant counterclockwise angular velocity of *2 rad/s* and link BC is rotating with a constant clockwise angular velocity of *3 rad/s*. Link CD is rotating at *4 rad/s*. If you want the acceleration of point D to be parallel to the x axis, what is the necessary angular acceleration of link CD? What is the resulting acceleration of point D?

Solution: From the solution of Problem 6.111, the acceleration of C is $\bar{a}_c = -3.154\bar{i} + 0.170\bar{j}(m/s^2)$. The acceleration of D is

$$a_D\bar{i} = \bar{a}_c + \bar{\alpha}_{CD}\times\bar{r}_{D/C} - \omega_{CD}^2\bar{r}_{D/C} = -3.154\bar{i} + 0.170\bar{j} + \begin{vmatrix} \bar{i} & \bar{j} & \bar{k} \\ 0 & 0 & \alpha_{CD} \\ 0.25 & 0 & 0 \end{vmatrix} - (4)^2(0.25\bar{i}).$$

Equating $\bar{i}$ and $\bar{j}$ components, $a_D = -3.154 - 4$, and $0 = 0.170 + 0.25\alpha_{CD}$, we obtain $\alpha_{CD} = -0.678 rad/s^2$ $a_D = -7.15 m/s^2$.

==================================<>==================================

Problem 6.113 The horizontal member *ADE* supporting the scoop is stationary. If the link BD has a clockwise angular velocity of *1 rad/s* and a counterclockwise angular acceleration of 2 rad/s^2, what is the angular acceleration of the scoop?

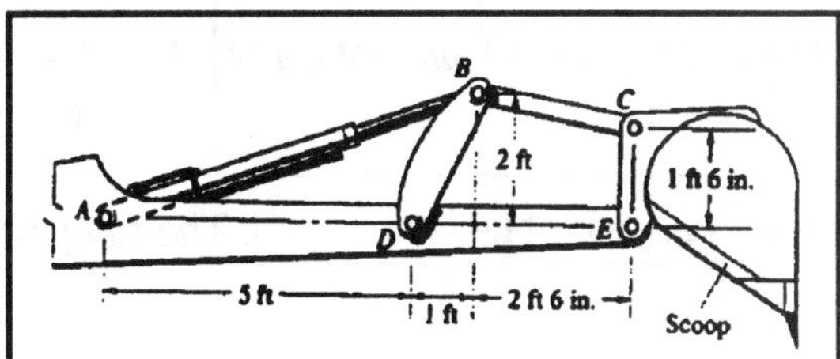

Solution: The velocity of B is

$$\bar{v}_B = \bar{v}_D + \bar{\omega}_{BD}\times\bar{r}_{B/D} = 0 + \begin{vmatrix} \bar{i} & \bar{j} & \bar{k} \\ 0 & 0 & -1 \\ 1 & 2 & 0 \end{vmatrix} = 2\bar{i} - \bar{j}(ft/s).$$

The velocity of C is $\bar{v}_c = \bar{v}_B + \bar{\omega}_{BC}\times\bar{r}_{C/B} = 2\bar{i} - \bar{j} + 0 + \begin{vmatrix} \bar{i} & \bar{j} & \bar{k} \\ 0 & 0 & \omega_{BC} \\ 2.5 & -.5 & 0 \end{vmatrix}$ (1).

We can also express $\bar{v}_c$ as $\bar{v}_c = \bar{v}_E + \bar{\omega}_{CE}\times\bar{r}_{C/E} = 0 + (\omega_{CE}\bar{k})\times(1.5\bar{j}) = -1.5\omega_{CE}\bar{i}$. (2)

Equating $\bar{i}$ and $\bar{j}$ components in Equations (1) and (2) we get $2 + 0.5\omega_{BC} = -1.5\omega_{CE}$, and $-1 + 2.5\omega_{BC} = 0$. Solving, we obtain $\omega_{BC} = 0.400 rad/s$ and $\omega_{CE} = -1.467 rad/s$.

The acceleration of B is $\bar{a}_B = \bar{a}_D + \bar{\alpha}_{BD}\times\bar{r}_{B/D} - \omega_{BD}^2\bar{r}_{B/D}$, or

$$\bar{a}_B = 0 + \begin{vmatrix} \bar{i} & \bar{j} & \bar{k} \\ 0 & 0 & 2 \\ 1 & 2 & 0 \end{vmatrix} - (1)^2(\bar{i} + 2\bar{j}) = -5\bar{i}(ft/s^2).$$

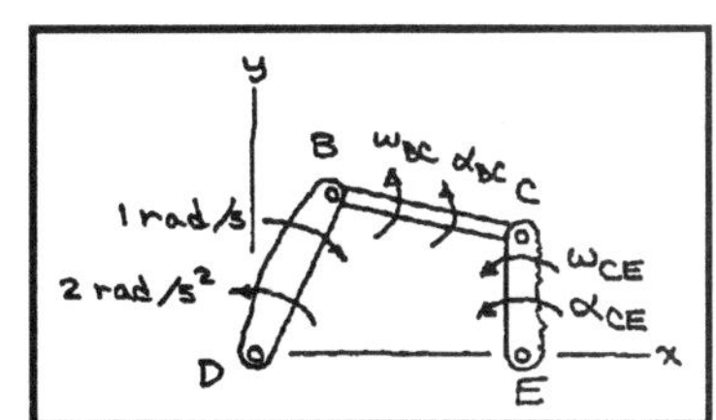

The acceleration of C is $\bar{a}_C = \bar{a}_B + \bar{\alpha}_{BC}\times\bar{r}_{C/B} - \omega_{BC}^2\bar{r}_{C/B}$

$$\bar{a}_C = -5\bar{i} + \begin{vmatrix} \mathrm{i} & \mathrm{j} & \mathrm{k} \\ 0 & 0 & \alpha_{BC} \\ 2.5 & -0.5 & 0 \end{vmatrix} - (0.4)^2(2.5\bar{i} - 0.5\bar{j}).$$ (3) We can also express $\bar{a}_C$ as

$$\bar{a}_c = \bar{a}_E + \bar{\alpha}_{CE}\times\bar{r}_{C/E} - \omega_{CE}^2\bar{r}_{C/E} = 0 + (\alpha_{CE}\bar{k})\times(1.5\bar{j}) - (-1.467)^2(1.5\bar{j}) = -1.5\alpha_{CE}\bar{i} - 3.23\bar{j}.$$ (4)

Equating $\bar{i}$ and $\bar{j}$ components in Equations (3) and (4), we get $-5 + 0.5\alpha_{BC} - (0.4)^2(2.5) = -1.5\alpha_{CE}$, and $2.5\alpha_{BC} + (0.4)^2(0.5) = -3.23$. Solving, we obtain $\alpha_{BC} = -1.32 rad/s^2$ $\alpha_{CE} = 4.04 rad/s^2$.

==================================<>==================================

=====================================<>=====================================

Problem 6.114 The ring gear is fixed, and the hub and planet gears are bonded together. The connecting rod has a counterclockwise angular acceleration of 10 *rad/s*^{2}. Determine the angular acceleration of the planet and sun gears.

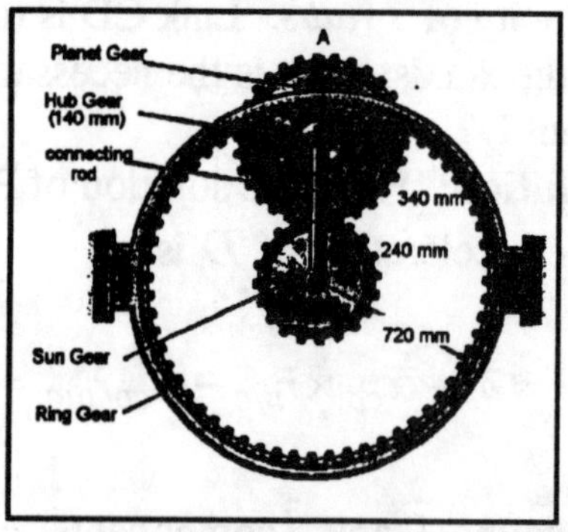

Solution At the point of contact between two gears, the *tangential components* of the accelerations are equal; the radial components of the acceleration are not equal. The tangential component of the acceleration at the end of the connecting rod is

$$\vec{a}_{rod} = \vec{\alpha}_{rod} \times \vec{r}_{rod} = \begin{bmatrix} \vec{i} & \vec{j} & \vec{k} \\ 0 & 0 & 10 \\ 0 & -140 & 0 \end{bmatrix} = -5800\vec{i} \ \left(mm/s^2\right).$$

The tangential acceleration at the point of contact between the ring gear and the hub gear is zero, from which

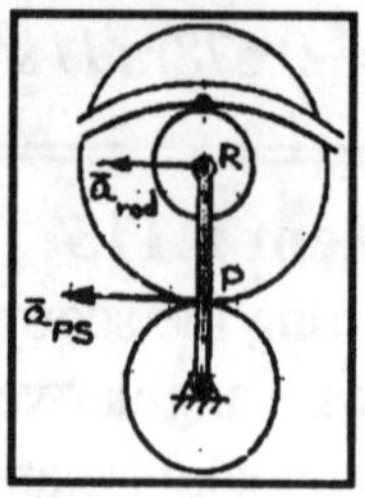

$$0 = \vec{a}_{rod} + \vec{\alpha}_{hub} \times \vec{r}_{hub} = \alpha_{rod} + \begin{bmatrix} \vec{i} & \vec{j} & \vec{k} \\ 0 & 0 & \alpha_{hub} \\ 0 & 140 & 0 \end{bmatrix} = -5800\vec{i} - 140\alpha_{hub}\vec{i},$$

from which $\boxed{\alpha_{hub} = -41.43 \ rad/s^2}$. The tangential acceleration at the interface of the planet and sun gears is

$$\vec{a}_{PS} = \vec{\alpha}_{hub} \times \vec{r}_{P/R} = \begin{bmatrix} \vec{i} & \vec{j} & \vec{k} \\ 0 & 0 & -41.43 \\ 0 & -340 & 0 \end{bmatrix} = -19{,}885.7\vec{i} \ \left(mm/s^2\right).$$

This tangential acceleration is also

$$\vec{a}_{PS} = \vec{\alpha}_{sun} \times \vec{r}_{sun} = \begin{bmatrix} \vec{i} & \vec{j} & \vec{k} \\ 0 & 0 & \alpha_{sun} \\ 0 & 240 & 0 \end{bmatrix} = -240\alpha_{sun}\vec{i} \ \left(mm/s^2\right).$$

Solve:

$\boxed{\alpha_{sun} = 82.86 \ rad/s^2}$

=====================================<>=====================================

===================================<>===================================

Problem 6.115 The connecting rod in Problem 6.114 has a counterclockwise angular velocity of 4 *rad/s* and a clockwise angular acceleration of 12 *rad/s*^{2}. Determine the magnitude of the acceleration at point A.

Solution The velocity of the center of the hub gear is

$\vec{v}_{hub} = \vec{\omega}_{rod} \times \vec{r}_{rod} = \begin{bmatrix} \vec{i} & \vec{j} & \vec{k} \\ 0 & 0 & 4 \\ 0 & 580 & 0 \end{bmatrix} = -2320\vec{i}\ (mm/s)$. The angular velocity of the hub gear is obtained from the fact that the contact between the hub gear and the ring gear is stationary:

$0 = \vec{v}_{rod} + \vec{\omega}_{hub} \times \vec{r}_{hub} = \vec{v}_{rod} + \begin{bmatrix} \vec{i} & \vec{j} & \vec{k} \\ 0 & 0 & \omega_{hub} \\ 0 & 140 & 0 \end{bmatrix} = -2320\vec{i} - 140\omega_{hub}\vec{i}$, from which

$\omega_{hub} = -16.57\ rad/s$. This is also the angular velocity of the planet gear. The *tangential acceleration* of the center of the hub gear is $\vec{a}_{rod} = \vec{\alpha}_{rod} \times \vec{r}_{rod} = \begin{bmatrix} \vec{i} & \vec{j} & \vec{k} \\ 0 & 0 & -12 \\ 0 & 580 & 0 \end{bmatrix} = 6960\vec{i}\ (mm/s^2)$.. The angular acceleration of the hub gear is obtained from $0 = \vec{a}_{rod} + \begin{bmatrix} \vec{i} & \vec{j} & \vec{k} \\ 0 & 0 & \alpha_{hyb} \\ 0 & 140 & 0 \end{bmatrix} = 6960\vec{i} - 140\alpha_{hub}\vec{i}$, from which $\alpha_{hub} = 49.71\ rad/s^2$. This is also the angular acceleration of the planet gear. The acceleration at point A is $\vec{a}_A = \vec{a}_{hub} - \omega_{rod}^2(580\vec{j}) - \alpha_P(340\vec{i}) - \omega_{hub}^2(340\vec{j}) = -9942.9\vec{i} - 102649\vec{j}\ (mm/s^2)$. The magnitude of the acceleration at point A is $\boxed{|\vec{a}_A| = \sqrt{9943^2 + 102649^2} = 103{,}129\ mm/s^2}$

===================================<>===================================

Problem 6.116 The large gear is fixed. The angular velocity and angular acceleration of the bar AB are $\omega_{AB} = 2\ rad/s$, $\alpha_{AB} = 4\ rad/s^2$. Determine the angular acceleration of the bars CD and DE.

Solution The strategy is to express vector velocity of point D in terms of the unknown angular velocities and accelerations of CD and DE, and then to solve the resulting vector equations for the unknowns.

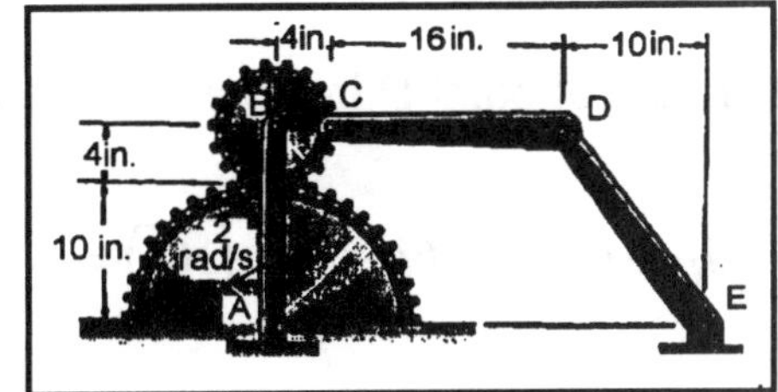

The angular velocities ω_{CD} *and* ω_{DE}. (See solution to Problem 6.51). The linear velocity of point B is

$\vec{v}_B = \vec{\omega}_{AB} \times \vec{r}_{AB} = \begin{bmatrix} \vec{i} & \vec{j} & \vec{k} \\ 0 & 0 & 2 \\ 0 & 14 & 0 \end{bmatrix} = -28\vec{i}\ (in/s)$. The lower edge of gear B is stationary. The velocity of B is also $\vec{v}_B = \vec{\omega}_B \times \vec{r}_B = \begin{bmatrix} \vec{i} & \vec{j} & \vec{k} \\ 0 & 0 & \omega_B \\ 0 & 4 & 0 \end{bmatrix} = -4\omega_B\vec{i}\ (in/s)$.

Solution continued on next page

Equate the velocities $\vec{v}_B$ to obtain the angular velocity of B: $\omega_B = -\frac{v_B}{4} = 7\ rad/s$. The velocity of point C is $\vec{v}_C = \vec{v}_B + \vec{\omega}_B \times \vec{r}_{BC} = -28\vec{i} + \begin{bmatrix} \vec{i} & \vec{j} & \vec{k} \\ 0 & 0 & 7 \\ 4 & 0 & 0 \end{bmatrix} = -28\vec{i} + 28\vec{j}\ (in/s)$. The velocity of point D is

$$\vec{v}_D = \vec{v}_C + \vec{\omega}_{CD} \times \vec{r}_{CD} = -28\vec{i} + 28\vec{j} + \begin{bmatrix} \vec{i} & \vec{j} & \vec{k} \\ 0 & 0 & \omega_{CD} \\ 16 & 0 & 0 \end{bmatrix} = -28\vec{i} + (16\omega_{CD} + 28)\vec{j}\ (in/s)$$. The velocity of point D is also given by $\vec{v}_D = \vec{\omega}_{DE} \times \vec{r}_{ED} = \begin{bmatrix} \vec{i} & \vec{j} & \vec{k} \\ 0 & 0 & \omega_{DE} \\ -10 & 14 & 0 \end{bmatrix} = -14\omega_{DE}\vec{i} - 10\omega_{DE}\vec{j}\ (in/s)$. Equate and separate components : $(-28 + 14\omega_{DE})\vec{i} = 0$, $(16\omega_{CD} + 28 + 10\omega_{DE})\vec{j} = 0$. Solve: $\omega_{DE} = 2\ rad/s$, $\omega_{CD} = -3\ rad/s$. The negative sign means a clockwise rotation.

The angular accelerations. The tangential acceleration of point B is

$$\vec{a}_B = \vec{\alpha}_{AB} \times \vec{r}_{B/A} = \begin{bmatrix} \vec{i} & \vec{j} & \vec{k} \\ 0 & 0 & 4 \\ 0 & 14 & 0 \end{bmatrix} = -56\vec{i}\ \left(in/s^2\right)$$. The tangential acceleration at the point of contact between the gears A and B is zero, from which $\vec{a}_B = \vec{\alpha}_{BC} \times 4\vec{j} = \begin{bmatrix} \vec{i} & \vec{j} & \vec{k} \\ 0 & 0 & \alpha_{BC} \\ 0 & 4 & 0 \end{bmatrix} = -4\alpha_{BC}\vec{i}\ \left(in/s^2\right)$, from which $\alpha_{BC} = 14\ rad/s^2$. The acceleration of point C in terms of the acceleration of point B is

$$\vec{a}_C = \vec{a}_B + \vec{\alpha}_{BC} \times 4\vec{i} - \omega_{BC}^2\left(4\vec{i}\right) = -56\vec{i} + \begin{bmatrix} \vec{i} & \vec{j} & \vec{k} \\ 0 & 0 & 14 \\ 4 & 0 & 0 \end{bmatrix} - 49\left(4\vec{i}\right) = -252\vec{i} + 56\vec{j}\ \left(in/s^2\right)$$. The acceleration of point D in terms of the acceleration of point C is

$$\vec{a}_D = \vec{a}_C + \vec{\alpha}_{CD} \times 16\vec{i} - \omega_{CD}^2\left(16\vec{i}\right) = \vec{a}_C + \begin{bmatrix} \vec{i} & \vec{j} & \vec{k} \\ 0 & 0 & \alpha_{CD} \\ 16 & 0 & 0 \end{bmatrix} - \omega_{CD}^2\left(16\vec{i}\right),$$

$\vec{a}_D = -396\vec{i} + (16\alpha_{CD} + 56)\vec{j}\ \left(in/s^2\right)$. The acceleration of point D in terms of the acceleration of point E is $\vec{a}_D = \begin{bmatrix} \vec{i} & \vec{j} & \vec{k} \\ 0 & 0 & \alpha_{DE} \\ -10 & 14 & 0 \end{bmatrix} - \omega_{DE}^2\left(-10\vec{i} + 14\vec{j}\right) = (40 - 14\alpha_{DE})\vec{i} - (10\alpha_{DE} + 56)\vec{j}\ \left(in/s^2\right)$

Solution continued on next page

Equate the expressions for $\vec{a}_D$ and separate components: $-396 = 40 - 14\alpha_{DE}$, $16\alpha_{CD} + 56 = -10\alpha_{DE} - 56$. Solve: $\boxed{\alpha_{DE} = 31.14\ rad/s^2}$, $\boxed{\alpha_{CD} = -26.46\ rad/s^2}$, where the negative sign means a clockwise angular acceleration.

================================<>================================

Problem 6.117 The bar rotates with a constant counterclockwise angular velocity of 10 *rad/s* and the sleeve slides at 4 *ft/s* relative to the bar. Use Eq (6.11) and the body-fixed coordinate system shown to determine the velocity of A.

Solution Eq (6.11) is $\vec{v}_A = \vec{v}_B + \vec{v}_{Arel} + \vec{\omega} \times \vec{r}_{A/B}$.

Substitute: $$\vec{v}_A = 0 + 4\vec{i} + \begin{bmatrix} \vec{i} & \vec{j} & \vec{k} \\ 0 & 0 & 10 \\ 2 & 0 & 0 \end{bmatrix} \boxed{= 4\vec{i} + 20\vec{j}\ (ft/s)}$$

================================<>================================

Problem 6.118 The sleeve A in Problem 6.117 slides relative to the bar at a constant velocity of 4 *ft/s*. Use Eq (6.15) to determine the acceleration of A.

Solution Eq (6.15) is $\vec{a}_A = \vec{a}_B + \vec{a}_{Arel} + 2\omega \times \vec{v}_{Arel} + \alpha \times \vec{r}_{A/B} - \omega^2 \vec{r}_{A/B}$. Substitute:

$$\vec{a}_A = 0 + 0 + 2\begin{bmatrix} \vec{i} & \vec{j} & \vec{k} \\ 0 & 0 & 10 \\ 4 & 0 & 0 \end{bmatrix} + 0 - 100(2\vec{i}) \boxed{= -200\vec{i} + 80\vec{j}\ \left(ft/s^2\right)}$$

================================<>================================

Problem 6.119 The sleeve C slides at 1 *m/s* relative to bar AB. Use the body-fixed coordinate system shown to determine the velocity of C.

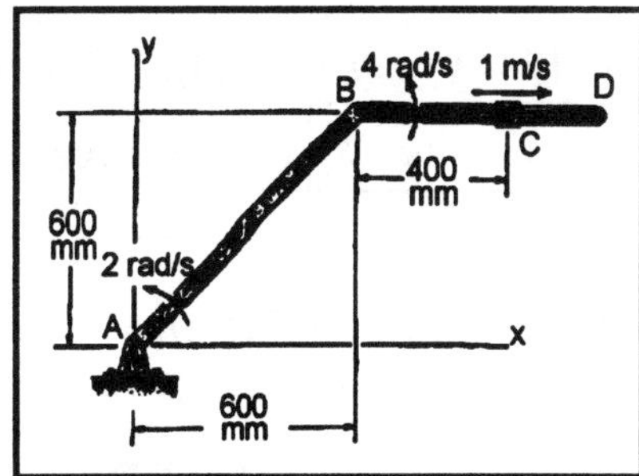

Solution The velocity of point B is

$$\vec{v}_B = \vec{\omega}_{AB} \times \vec{r}_{B/A} = \begin{bmatrix} \vec{i} & \vec{j} & \vec{k} \\ 0 & 0 & 2 \\ 600 & 600 & 0 \end{bmatrix} = -1200(\vec{i} - \vec{j})\ (mm/s).$$

Use Eq (6.11). The velocity of sleeve C is $\vec{v}_C = \vec{v}_B + \vec{v}_{Arel} + \vec{\omega}_{BD} \times \vec{r}_{C/B}$.

$$\vec{v}_C = -1200\vec{i} + 1200\vec{j} + 1000\vec{i} + \begin{bmatrix} \vec{i} & \vec{j} & \vec{k} \\ 0 & 0 & 4 \\ 400 & 0 & 0 \end{bmatrix}.$$

$$\boxed{\vec{v}_C = -200\vec{i} + 2800\vec{j}\ (mm/s)}$$

================================<>================================

==================================<>==================================

Problem 6.120 In Problem 6.119, the angular accelerations of the two bars are zero and the sleeve slides at a constant velocity of 1 *m/s* relative to bar BD. What is the acceleration of A?

Solution From Problem 6.119, $\omega_{AB} = 2\ rad/s$, $\omega_{BC} = 4\ rad/s$. The acceleration of point B is $\vec{a}_B = -\omega_{AB}^2 \vec{r}_{B/A} = -4(600\vec{i} + 600\vec{j}) = -2400\vec{i} - 2400\vec{j}\ (mm/s^2)$. Use Eq (6.15). The acceleration of C is $\vec{a}_C = \vec{a}_B + \vec{a}_{Crel} + 2\vec{\omega}_{BD} \times \vec{v}_{Crel} + \vec{\alpha}_{BD} \times \vec{r}_{C/B} - \omega_{BC}^2 \vec{r}_{C/B}$.

$$\vec{a}_C = -2400\vec{i} - 2400\vec{j} + 2\begin{bmatrix} \vec{i} & \vec{j} & \vec{k} \\ 0 & 0 & \omega_{BD} \\ 1000 & 0 & 0 \end{bmatrix} - \omega_{BD}^2(400\vec{i}),\ \boxed{\vec{a}_C = -8800\vec{i} + 5600\vec{j}\ (mm/s^2)}$$

==================================<>==================================

Problem 6.121 The bar AC has an angular velocity of 2 *rad/s* in the counterclockwise direction that is decreasing at 4 *rad/s*^{2}. The pin at C slides in the slot in bar BD. (a) Determine the angular velocity of bar BD and the velocity of the pin relative to the slot. (b) Determine the angular acceleration of bar BD and the acceleration of the pin relative to the slot.

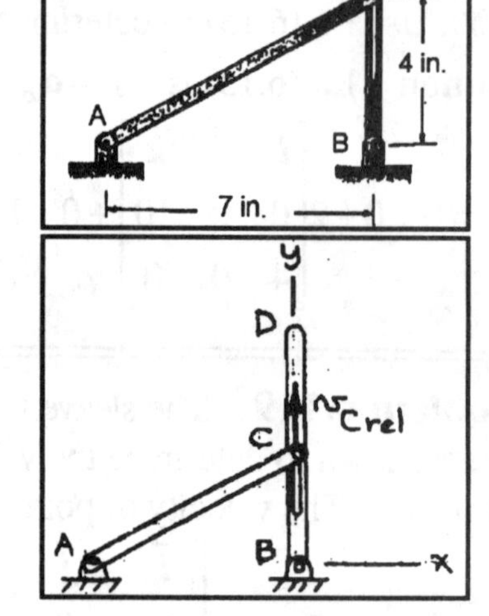

Solution The angular velocity of bar AC is $\omega_{AC} = 4\ rad/s$; the acceleration is $\alpha_{AC} = -4\ rad/s^2$ (a)The velocity of point C is

$$\vec{v}_C = \vec{\omega}_{AC} \times \vec{r}_{C/A} = \begin{bmatrix} \vec{i} & \vec{j} & \vec{k} \\ 0 & 0 & 2 \\ 7 & 4 & 0 \end{bmatrix} = -8\vec{i} + 14\vec{j}\ (in/s).$$

The velocity of point C is also determined from bar BD:

$$\vec{v}_C = \vec{v}_{Crel} + \vec{\omega}_{BD} \times (4\vec{j}) = v_{Crel}\vec{j} - \begin{bmatrix} \vec{i} & \vec{j} & \vec{k} \\ 0 & 0 & \omega_{BD} \\ 4 & 0 & 0 \end{bmatrix}.$$ Substitute:

$-8\vec{i} + 14\vec{j} = v_{Crel}\vec{j} - 4\omega_{BD}\vec{i}$, from which $\boxed{\omega_{BD} = 2\ rad/s}$, $\boxed{\vec{v}_{Crel} = 14\vec{j}\ (in/s)}$ (b) The acceleration of point C using bar AC is

$$\vec{a}_C = \vec{\alpha}_{AC} \times \vec{r}_{C/A} - \omega_{AC}^2 \vec{r}_{C/A} = \begin{bmatrix} \vec{i} & \vec{j} & \vec{k} \\ 0 & 0 & -4 \\ 7 & 4 & 0 \end{bmatrix} - 4(7\vec{i} + 4\vec{j}) = -12\vec{i} - 44\vec{j}\ (in/s^2).$$ The acceleration of C is also given in terms of bar BD:

$$\vec{a}_C = \vec{a}_{Crel} + 2\vec{\omega}_{BD} \times \vec{v}_{Crel} + \vec{\alpha}_{BD} \times \vec{r}_{C/B} - \omega_{BD}^2 \vec{r}_{C/B} = a_{Crel}\vec{j} + \begin{bmatrix} \vec{i} & \vec{j} & \vec{k} \\ 0 & 0 & 4 \\ 0 & 14 & 0 \end{bmatrix} + \begin{bmatrix} \vec{i} & \vec{j} & \vec{k} \\ 0 & 0 & \alpha_{BD} \\ 0 & 4 & 0 \end{bmatrix} - 4(4\vec{j})$$

$= a_{Crel}\vec{j} - 56\vec{i} - 4\alpha_{BD}\vec{i} - 16\vec{j}$. Equate like terms: $-12 = -56 - 4\alpha_{BD}$, $-44 = a_{Crel} - 16$. Solve: $\boxed{\alpha_{BD} = -11\ rad/s^2}$ (clockwise), $\boxed{\vec{a}_{Crel} = -28\vec{j}\ in/s^2}$

==================================<>==================================

================================<>================================

Problem 6.122 In the system shown in Problem 6.121, the velocity of the pin C relative to the slot is 21 *in/s* upward and is decreasing at 42 *in/s*^{2}. What are the angular velocity and acceleration of bar AC?

Solution *The angular velocities:* The velocity of point C is

$$\vec{v}_C = \vec{\omega}_{AC} \times \vec{r}_{C/A} = \begin{bmatrix} \vec{i} & \vec{j} & \vec{k} \\ 0 & 0 & \omega_{AC} \\ 7 & 4 & 0 \end{bmatrix} = -4\omega_{AC}\vec{i} + 7\omega_{AC}\vec{j} \ (in/s).$$

The velocity of point C is also determined from bar BD:

$$\vec{v}_C = \vec{v}_{Crel} + \vec{\omega}_{BD} \times (4\vec{j}) = \vec{v}_{Crel} + \begin{bmatrix} \vec{i} & \vec{j} & \vec{k} \\ 0 & 0 & \omega_{BD} \\ 0 & 4 & 0 \end{bmatrix} = 21\vec{j} - 4\omega_{BD}\vec{i} \ (in/s).$$

Equate like terms: $-4\omega_{AC} = -4\omega_{BD}$, $7\omega_{AC} = 21$ from which $\boxed{\omega_{AC} = 3\ rad/s}$, $\omega_{BD} = 3\ rad/s$.

The angular accelerations: The acceleration of C:

$$\vec{a}_C = \vec{\alpha}_{AC} \times \vec{r}_{C/A} - \omega_{AC}^2(7\vec{i} + 4\vec{j}) = \begin{bmatrix} \vec{i} & \vec{j} & \vec{k} \\ 0 & 0 & \alpha_{AC} \\ 7 & 4 & 0 \end{bmatrix} - (3^2)(7\vec{i} + 4\vec{j}),$$

$$\vec{a}_C = \alpha_{AC}(-4\vec{i} + 7\vec{j}) - 64\vec{i} - 36\vec{j} \ (in/s^2)$$

For bar BD, $\vec{a}_C = -42\vec{j} + 2\vec{\omega}_{BD} \times (21\vec{j}) + \vec{\alpha}_{BD} \times (4\vec{j}) - \omega_{BD}^2(4\vec{j})$.

$$\vec{a}_C = -42\vec{j} + 2\begin{bmatrix} \vec{i} & \vec{j} & \vec{k} \\ 0 & 0 & \omega_{BD} \\ 0 & 21 & 0 \end{bmatrix} + \begin{bmatrix} \vec{i} & \vec{j} & \vec{k} \\ 0 & 0 & \alpha_{BD} \\ 0 & 4 & 0 \end{bmatrix} - \omega_{BD}^2(4\vec{j}) = -42\vec{j} - 126\vec{i} - 4\alpha_{BD}\vec{i} - 36\vec{j}$$

Equate like terms: $-4\alpha_{AC} - 64 = -126 - 4\alpha_{BD}$, $7\alpha_{AC} - 36 = -42 - 36$. Solve: $\boxed{\alpha_{AC} = -6\ rad/s^2}$

================================<>================================

Problem 6.123 In the system shown in Problem 6.121, what should the angular velocity and acceleration of bar AC be if you want the angular velocity and acceleration of bar BD to be 4 *rad/s* counterclockwise and 24 *rad/s*^{2} counterclockwise, respectively?

Solution *Get the angular velocities:* The velocity of point C is

$$\vec{v}_C = \vec{\omega}_{AC} \times \vec{r}_{C/A} = \begin{bmatrix} \vec{i} & \vec{j} & \vec{k} \\ 0 & 0 & \omega_{AC} \\ 7 & 4 & 0 \end{bmatrix} = -4\omega_{AC}\vec{i} + 7\omega_{AC}\vec{j} \ (in/s).$$

The velocity of point C is also determined from bar BD:

$$\vec{v}_C = \vec{v}_{Crel} + \vec{\omega}_{BD} \times (4\vec{j}) = v_{Crel}\vec{j} + \begin{bmatrix} \vec{i} & \vec{j} & \vec{k} \\ 0 & 0 & 4 \\ 0 & 4 & 0 \end{bmatrix} = v_{Crel}\vec{j} - 16\vec{i} \ (in/s).$$

Equate like terms:

$-4\omega_{AC} = -16\vec{i}$, $7\omega_{AC} = v_{Crel}$ from which $\boxed{\omega_{AC} = 4\ rad/s}$, $v_{Crel} = 28\ in/s$.

Solution continued on next page

Get the angular accelerations: The acceleration of C:

$$\vec{a}_C = \vec{\alpha}_{AC} \times \vec{r}_{C/A} - \omega_{AC}^2(7\vec{i} + 4\vec{j}) = \begin{bmatrix} \vec{i} & \vec{j} & \vec{k} \\ 0 & 0 & \alpha_{AC} \\ 7 & 4 & 0 \end{bmatrix} - 112\vec{i} - 64\vec{j},$$

$\vec{a}_C = \alpha_{AC}(-4\vec{i} + 7\vec{j}) - 112\vec{i} - 64\vec{j}$ (in/s^2). For bar BD,

$$\vec{a}_C = a_{Crel}\vec{j} + 2\begin{bmatrix} \vec{i} & \vec{j} & \vec{k} \\ 0 & 0 & \omega_{BD} \\ 0 & 28 & 0 \end{bmatrix} + \begin{bmatrix} \vec{i} & \vec{j} & \vec{k} \\ 0 & 0 & \alpha_{BD} \\ 0 & 4 & 0 \end{bmatrix} - \omega_{BD}^2(4\vec{j}) = a_{Crel}\vec{j} - 224\vec{i} - 96\vec{i} - 64\vec{j}.$$ Equate like

terms: $-4\alpha_{AC} - 112 = -224 - 96$, $7\alpha_{AC} - 64 = a_{Crel} - 64$. Solve: $\boxed{\alpha_{AC} = 52 \ rad/s^2}$

==================================<>==================================

Problem 6.124 Bar AB has an angular velocity of 4 *rad/s* in the clockwise direction. What is the velocity of pin B relative to the slot?

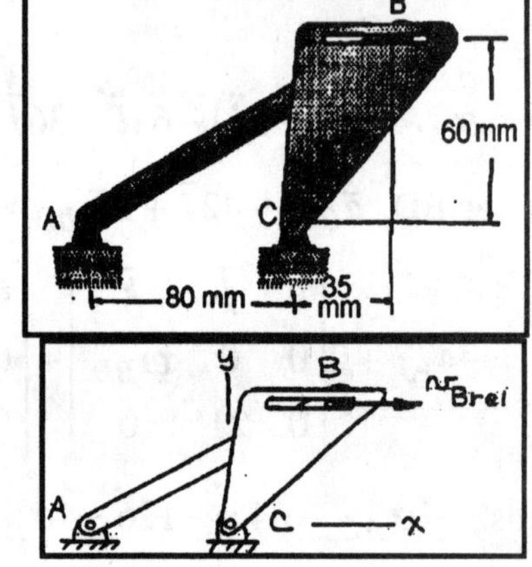

Solution The velocity of point B is

$$\vec{v}_B = \vec{\omega}_{AB} \times \vec{r}_{B/A} = \begin{bmatrix} \vec{i} & \vec{j} & \vec{k} \\ 0 & 0 & \omega_{AB} \\ 115 & 60 & 0 \end{bmatrix} = 240\vec{i} - 460\vec{j} \ (mm/s).$$

The velocity of point B is also determined from bar CB

$\vec{v}_B = \vec{v}_{Brel} + \vec{\omega}_{CB} \times (35\vec{i} + 60\vec{j})$,

$$\vec{v}_B = \vec{v}_{Brel} + \begin{bmatrix} \vec{i} & \vec{j} & \vec{k} \\ 0 & 0 & \omega_{CB} \\ 35 & 60 & 0 \end{bmatrix}$$

$\vec{v}_B = v_{Brel}\vec{i} - 60\omega_{CB}\vec{i} + 35\omega_{CB}\vec{j}$ (mm/s). Equate like terms: $240 = v_{Brel} - 60\omega_{CB}$, $-460 = 35\omega_{CB}$ from which $\omega_{BC} = -13.14 \ rad/s$, $\boxed{v_{Brel} = -548.6 \ mm/s}$

==================================<>==================================

Problem 6.125 In the system shown in Problem 6.124, the bar AB has an angular velocity of 4 *rad/s* in the clockwise direction and an angular acceleration of 10 *rad/s*^{2} in the counterclockwise direction. What is the acceleration of pin B relative to the slot?

Solution:

Use the solution to Problem 6.124, from which $\omega_{BC} = -13.14 \ rad/s$, $v_{Brel} = -548.6 \ mm/s$.

The angular acceleration and the relative acceleration. The acceleration of point B is

$$\vec{a}_B = \vec{\alpha}_{AB} \times \vec{r}_{B/A} - \omega_{AB}^2\vec{r}_{B/A} = \begin{bmatrix} \vec{i} & \vec{j} & \vec{k} \\ 0 & 0 & 10 \\ 115 & 60 & 0 \end{bmatrix} - (16)(115\vec{i} + 60\vec{j}) \ (mm/s^2),$$

$$\vec{a}_B = -600\vec{i} + 1150\vec{j} - 1840\vec{i} - 960\vec{j} = -2240\vec{i} + 190\vec{j} \ (mm/s^2).$$

Solution continued on next page

The acceleration of pin B in terms of bar BC is $\vec{a}_B = a_{Brel}\vec{i} + 2\vec{\omega}_{BC} \times v_{Brel} + \vec{\alpha}_{BC} \times \vec{r}_{B/C} - \omega_{BC}^2 \vec{r}_{B/C}$,

$$\vec{a}_B = a_{Brel}\vec{i} + 2\begin{bmatrix} \vec{i} & \vec{j} & \vec{k} \\ 0 & 0 & -13.1 \\ -548.6 & 0 & 0 \end{bmatrix} + \begin{bmatrix} \vec{i} & \vec{j} & \vec{k} \\ 0 & 0 & \alpha_{BC} \\ 35 & 60 & 0 \end{bmatrix} - (13.14^2)(35\vec{i} + 60\vec{j}).$$

$\vec{a}_B = a_{Brel}\vec{i} + 14{,}419.5\vec{j} + 35\alpha_{BC}\vec{i} - 60\alpha_{BC}\vec{j} - 6045.7\vec{j}$. Equate expressions for $\vec{a}_B$ and separate components: $-2440 = a_{Brel} - 60\alpha_{BC} - 6045.7$, $190 = 14{,}419.6 + 35\alpha_{BC} - 10364.1$. Solve:

$\boxed{\vec{a}_{Brel} = -3021\vec{i}\ (mm/s^2)}$, $\alpha_{BC} = -110.4\ rad/s^2$.

==================================<>==================================

Problem 6.126 Arm AB is rotating at 4 *rad/s* in the clockwise direction. Determine the angular velocity of arm AB and the velocity of point B relative to the slot in arm BC.

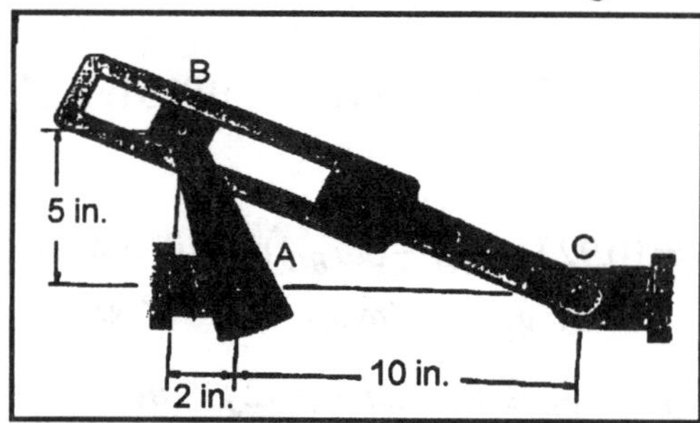

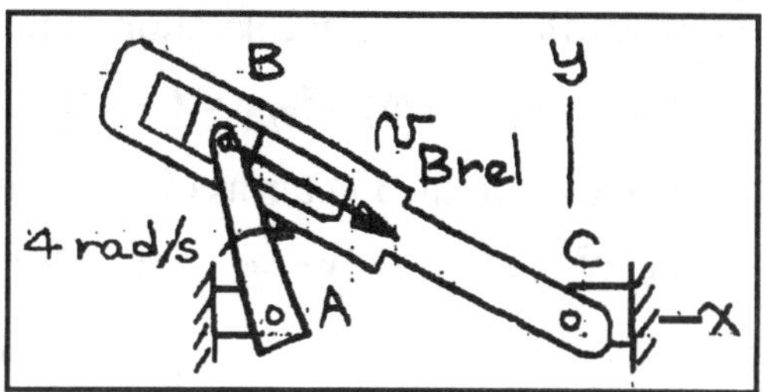

Solution:

The velocity of point B is

$$\vec{v}_B = \vec{\omega}_{AB} \times \vec{r}_{B/A} = \begin{bmatrix} \vec{i} & \vec{j} & \vec{k} \\ 0 & 0 & -4 \\ -2 & 5 & 0 \end{bmatrix} = (20\vec{i} + 8\vec{j}).$$

The unit vector parallel to the slot is

$$\vec{e} = \frac{(12\vec{i} - 5\vec{j})}{\sqrt{12^2 + 5^2}} = 0.9231\vec{i} - 0.3846\vec{j} \text{ (towards C).}$$

The velocity of point B in terms of the rotation of the arm BC is $\vec{v}_B = v_{Brel}\vec{e} + \vec{\omega}_{BC} \times \vec{r}_{B/C}$,

$$\vec{v}_B = 0.9231 v_{Brel}\vec{i} - 0.3846 v_{Brel}\vec{j} + \begin{bmatrix} \vec{i} & \vec{j} & \vec{k} \\ 0 & 0 & \omega_{BC} \\ -12 & 5 & 0 \end{bmatrix}$$

$\vec{v}_B = (0.9231 v_{Brel} - 5\omega_{BC})\vec{i} - (0.3846 v_{Brel} + 12\omega_{BC})\vec{j}$. Equate like terms in the two expressions: $20 = 0.9231 v_{Brel} - 5\omega_{BC}$, $8 = -0.3846 v_{Brel} - 12\omega_{BC}$. Solve: $\boxed{v_{Brel} = 15.4\ (in/s)}$ (towards C),

$\boxed{\omega_{BC} = -1.16\ rad/s}$

==================================<>==================================

==================================<>==================================

Problem 6.127 Arm AB in Problem 6.126 is rotating with a constant velocity of 4 *rad/s* in the clockwise direction. Determine the angular acceleration of arm BC and the acceleration of point B relative to the slot in arm BC.

Solution:

Use the solution to Problem 6.126.

The angular velocities: The velocity of point B is $\vec{v}_B = \vec{\omega}_{AB} \times \vec{r}_{B/A} = \begin{bmatrix} \vec{i} & \vec{j} & \vec{k} \\ 0 & 0 & -4 \\ -2 & 5 & 0 \end{bmatrix} = 20\vec{i} + 8\vec{j} \ (m/s)$.

The unit vector parallel to the slot is $\vec{e} = \dfrac{(12\vec{i} - 5\vec{j})}{\sqrt{12^2 + 5^2}} = 0.9231\vec{i} - 0.3846\vec{j}$ (towards C).

The velocity of point B in terms of the rotation of the arm BC is

$$\vec{v}_B = v_{Brel}\vec{e} + \vec{\omega}_{BC} \times \vec{r}_{B/C} = 0.9231 v_{Brel}\vec{i} - 0.3846 v_{Brel}\vec{j} + \begin{bmatrix} \vec{i} & \vec{j} & \vec{k} \\ 0 & 0 & \omega_{BC} \\ -12 & 5 & 0 \end{bmatrix}.$$

$\vec{v}_B = (0.9231 v_{Brel} - 5\omega_{BC})\vec{i} - (0.3846 v_{Brel} + 12\omega_{BC})\vec{j}$. Equate like terms in the two expressions: $20 = 0.9231 v_{Brel} - 5\omega_{BC}$, $8 = -0.3846 v_{Brel} - 12\omega_{BC}$. Solve: $v_{Brel} = 15.4 \ (in/s)$ (towards C)

$\vec{v}_{Brel} = v_{Brel}\vec{e} = 14.2\vec{i} - 5.92\vec{j} \ (in/s)$, $\omega_{BC} = -1.16 \ rad/s$.

The accelerations: The acceleration of point B is

$\vec{a}_B = \vec{\alpha}_{AB} \times \vec{r}_{B/A} - \vec{\omega}_{AB}^2 \vec{r}_{B/A} = 0 - 16(-2\vec{i} + 5\vec{j}) = 32\vec{i} - 80\vec{j} \ (in/s^2)$, since $\alpha_{AB} = 0$. The acceleration of point B in terms of the acceleration of arm BC is

$\vec{a}_B = a_{Brel}\vec{e} + 2\vec{\omega}_{BC} \times (v_{Brel}\vec{e}) + \vec{\alpha}_{BC} \times \vec{r}_{B/C} - \omega_{BC}^2 \vec{r}_{B/C}$,

$$\vec{a}_B = a_{Brel}(0.9231\vec{i} - 0.3846\vec{j}) + 2\begin{bmatrix} \vec{i} & \vec{j} & \vec{k} \\ 0 & 0 & -1.16 \\ 14.2 & -5.92 & 0 \end{bmatrix} + \begin{bmatrix} \vec{i} & \vec{j} & \vec{k} \\ 0 & 0 & \alpha_{BC} \\ -12 & 5 & 0 \end{bmatrix} - \omega_{BC}^2(-12\vec{i} + 5\vec{j}),$$

$$\vec{a}_B = \left(0.9231 a_{Brel} - 2.32(5.92) - 5\alpha_{BC} + (1.16^2)(12)\right)\vec{i}$$
$$+ \left(-0.3846 a_{Brel} - 2.32(14.2) - 12\alpha_{BC} + (1.16^2)(5)\right)\vec{j}$$

Equate like terms in the two expressions: $32 = 0.9231 a_{Brel} - (2)(1.16)(5.92) - 5\alpha_{BC} + 12(1.16^2)$,

$-80 = -0.3846 a_{Brel} - (2)(1.16)(14.2) - 12\alpha_{BC} - (1.16^2)(5)$. Solve: $\boxed{\alpha_{BC} = 1.9887... = 2 \ rad/s^2}$,

$\boxed{a_{Brel} = 42.8 \ (in/s^2)}$

==================================<>==================================

================================<>================================

Problem 6.128 The angular velocity $\omega_{AC} = 5^o$ per second. Determine the angular velocity of the hydraulic actuator BC and the rate at which it is extending.

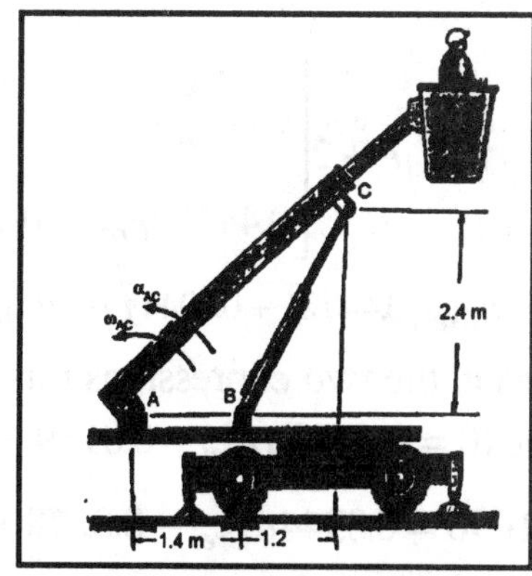

Solution :

The point C effectively slides in a slot in the arm BC. The angular velocity of $\omega_{AC} = 5\left(\frac{\pi}{180}\right) = 0.0873 \ rad/s$. The velocity of point C with respect to arm AC is

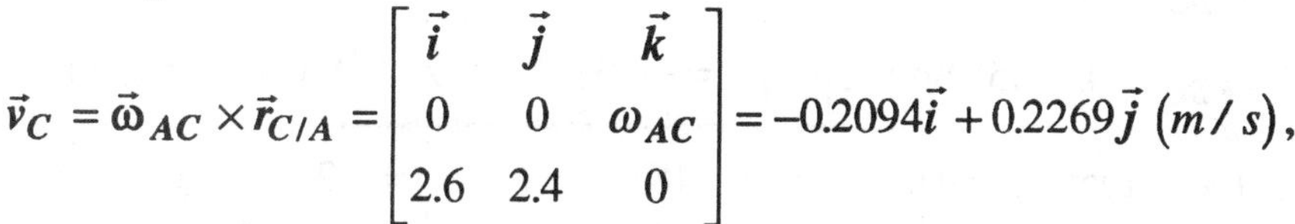

$$\vec{v}_C = \vec{\omega}_{AC} \times \vec{r}_{C/A} = \begin{bmatrix} \vec{i} & \vec{j} & \vec{k} \\ 0 & 0 & \omega_{AC} \\ 2.6 & 2.4 & 0 \end{bmatrix} = -0.2094\vec{i} + 0.2269\vec{j} \ (m/s),$$

The unit vector parallel to the actuator BC is

$$\vec{e} = \frac{1.2\vec{i} + 2.4\vec{j}}{\sqrt{1.2^2 + 2.4^2}} = 0.4472\vec{i} + 0.8944\vec{j}.$$

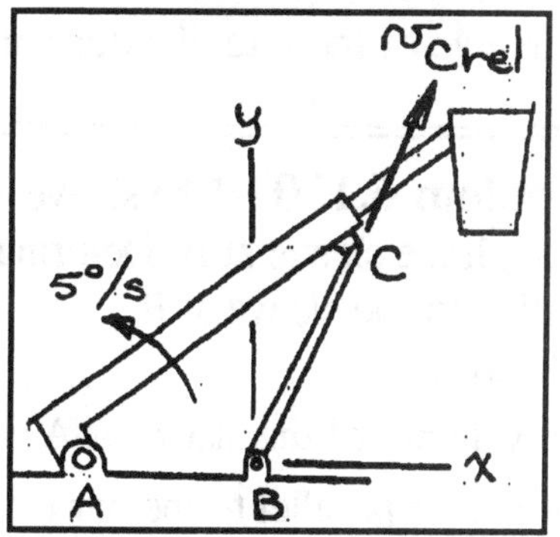

The velocity of point C in terms of the velocity of the actuator is $\vec{v}_C = v_{Crel}\vec{e} + \vec{\omega}_{BC} \times \vec{r}_{C/B}$.

$$\vec{v}_C = v_{Crel}\left(0.4472\vec{i} + 0.8944\vec{j}\right) + \begin{bmatrix} \vec{i} & \vec{j} & \vec{k} \\ 0 & 0 & \omega_{BC} \\ 1.2 & 2.4 & 0 \end{bmatrix}$$

$\vec{v}_C = v_{Crel}\left(0.4472\vec{i} + 0.8944\vec{j}\right) + \omega_{BC}\left(-2.4\vec{i} + 1.2\vec{j}\right)$. Equate like terms in the two expressions: $-0.2094 = 0.4472v_{Crel} - 2.4\omega_{BC}$, $0.2269 = 0.8944v_{Crel} + 1.2\omega_{BC}$. Solve: $\boxed{\omega_{BC} = 0.1076 \ rad/s = 6.167 \ deg/s}$, $\boxed{v_{Crel} = 0.1093 \ (m/s)}$, which is also the velocity of extension of the actuator.

================================<>================================

Problem 6.129 In Problem 6.128, if the angular velocity $\omega_{AC} = 5^o$ per second and the angular acceleration $\alpha_{AC} = -2^o$ per (second^2), determine the angular acceleration of the hydraulic actuator BC and the rate of change of its rate of extension.

Solution:

Use the solution to Problem 6.128 for the velocities: $\omega_{BC} = 0.1076 \ rad/s$, $v_{Crel} = 0.1093 \ (m/s)$. The angular acceleration $\alpha_{AC} = -2\left(\frac{\pi}{180}\right) = -0.03491 \ rad/s^2$. The acceleration of point C is

$$\vec{a}_C = \vec{\alpha}_{AC} \times \vec{r}_{C/A} - \omega_{AC}^2\vec{r}_{C/A} = \begin{bmatrix} \vec{i} & \vec{j} & \vec{k} \\ 0 & 0 & \alpha_{AC} \\ 2.6 & 2.4 & 0 \end{bmatrix} - \omega_{AC}^2\left(2.6\vec{i} + 2.4\vec{j}\right),$$

$\vec{a}_C = \alpha_{AC}\left(-2.4\vec{i} + 2.6\vec{j}\right) - \omega_{AC}^2\left(2.6\vec{i} + 2.4\vec{j}\right) = 0.064\vec{i} - 0.109\vec{j} \ \left(m/s^2\right)$. The acceleration of point C in terms of the hydraulic actuator is $\vec{a}_C = a_{Crel}\vec{e} + 2\vec{\omega}_{BC} \times \vec{v}_{Crel} + \vec{\alpha}_{BC} \times \vec{r}_{C/B} - \omega_{BC}^2\vec{r}_{C/B}$,

Solution continued on next page

$$\vec{a}_C = a_{Crel}\vec{e} + 2\begin{bmatrix} \vec{i} & \vec{j} & \vec{k} \\ 0 & 0 & \omega_{BC} \\ 0.4472v_{Crel} & 0.8944v_{Crel} & 0 \end{bmatrix} + \begin{bmatrix} \vec{i} & \vec{j} & \vec{k} \\ 0 & 0 & \alpha_{BC} \\ 1.2 & 2.4 & 0 \end{bmatrix} - \omega_{BC}^2\left(1.2\vec{i} + 2.4\vec{j}\right)$$

$\vec{a}_C = a_{Crel}\left(0.4472\vec{i} + 0.8944\vec{j}\right) + 2\omega_{BC}\left(-0.0977\vec{i} + 0.0489\vec{j}\right) + \alpha_{BC}\left(-2.4\vec{i} + 1.2\vec{j}\right) - \omega_{BC}^2\left(1.2\vec{i} + 2.4\vec{j}\right)$. Equate like terms in the two expressions for $\vec{a}_C$,

$0.0640 = 0.4472a_{Crel} - 0.0139 - 2.4\alpha_{BC} - 0.0210$,

$-0.1090 = 0.8944a_{Crel} - 0.0278 + 1.2\alpha_{BC} + 0.0105$. Solve: $\boxed{a_{Crel} = -0.0378\ \left(m/s^2\right)}$, which is the rate of change of the rate of extension of the actuator, and $\boxed{\alpha_{BC} = -0.0483\ \left(rad/s^2\right) = -2.77\ deg/s^2}$

==================================<>==================================

Problem 6.130 The sleeve at A slides upward at a constant velocity of 10 *m/s*. The bar AC slides through the sleeve at B. Determine the angular velocity of bar AC and the velocity at which it slides relative to the sleeve at B.

Solution:

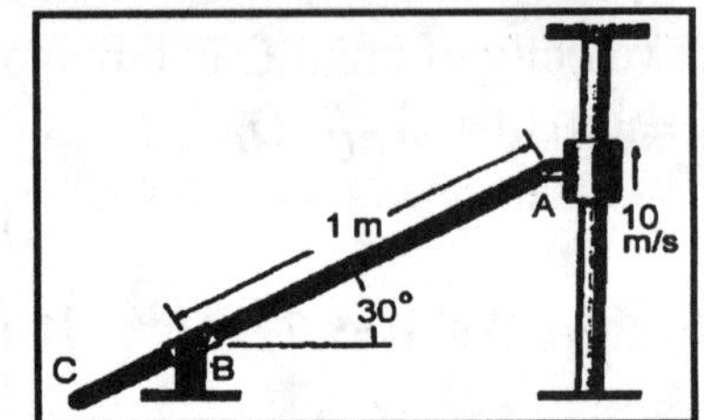

The velocity of the sleeve at A is given to be $\vec{v}_A = 10\vec{j}\ (m/s)$. The unit vector parallel to the bar (toward A) is

$\vec{e} = 1\left(\cos 30^o\,\vec{i} + \sin 30^o\,\vec{j}\right) = 0.866\vec{i} + 0.5\vec{j}$.

Choose a coordinate system with origin at B that rotates with the bar. The velocty at A is

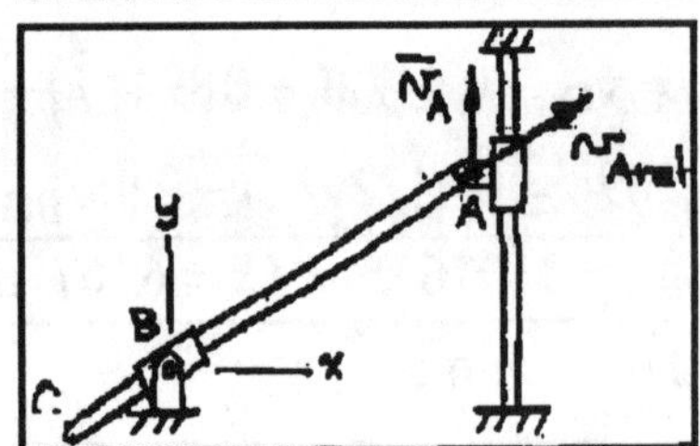

$$\vec{v}_A = \vec{v}_B + v_{Arel}\vec{e} + \vec{\omega}_{AB} \times \vec{r}_{A/B} = 0 + v_{Arel}\vec{e} + \begin{bmatrix} \vec{i} & \vec{j} & \vec{k} \\ 0 & 0 & \omega_{AB} \\ 0.866 & 0.5 & 0 \end{bmatrix}$$

$\vec{v}_A = \left(0.866\vec{i} + 0.5\vec{j}\right)v_{Arel} + \omega_{AB}\left(-0.5\vec{i} + 0.866\vec{j}\right)\ (m/s)$. The given

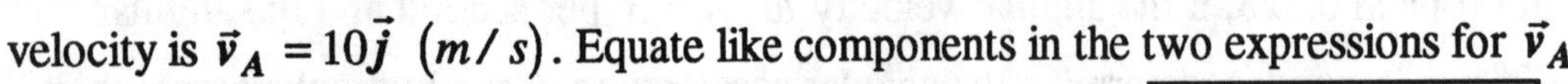

velocity is $\vec{v}_A = 10\vec{j}\ (m/s)$. Equate like components in the two expressions for $\vec{v}_A$:

$0 = 0.866v_{Arel} - 0.5\omega_{AB}$, $10 = 0.5v_{Arel} + 0.866\omega_{AB}$. Solve: $\boxed{\omega_{AB} = 8.66\ rad/s}$ (counterclockwise), $\boxed{v_{Arel} = 5\ m/s}$ from B toward A.

==================================<>==================================

================================<>================================

Problem 6.131 In Problem 6.130, the sleeve at A slides upward at a constant velocity of 10 *m/s*. Determine the angular acceleration of the bar AC and the rate of change of the velocity at which it slides relative to the sleeve at B.

Solution:

Use the solution of Problem 6.130: $\vec{e} = 0.866\vec{i} + 0.5\vec{j}$, $\omega_{AB} = 8.66\ rad/s$, $v_{Arel} = 5\ m/s$. The acceleration of the sleeve at A is given to be zero. The acceleration in terms of the motion of the arm is $\vec{a}_A = 0 = a_{Arel}\vec{e} + 2\vec{\omega}_{AB} \times v_{Arel}\vec{e} + \vec{\alpha}_{AB} \times \vec{r}_{A/B} - \omega_{AB}^2 \vec{r}_{A/B}$.

$$\vec{a}_A = 0 = a_{Arel}\vec{e} + 2v_{Arel}\begin{bmatrix} \vec{i} & \vec{j} & \vec{k} \\ 0 & 0 & \omega_{AB} \\ 0.866 & 0.5 & 0 \end{bmatrix} + \begin{bmatrix} \vec{i} & \vec{j} & \vec{k} \\ 0 & 0 & \alpha_{AB} \\ 0.866 & 0.5 & 0 \end{bmatrix} - \omega_{AB}^2(0.866\vec{i} + 0.5\vec{j})$$

$0 = (0.866\vec{i} + 0.5\vec{j})a_{Arel} - 43.3\vec{i} + 75\vec{j} + \alpha_{AB}(-0.5\vec{i} + 0.866\vec{j}) - 64.95\vec{i} - 37.5\vec{j}$. Separate components:

$0 = 0.866a_{Brel} - 43.3 - 0.5\alpha_{AB} - 64.95$, $0 = 0.5a_{Arel} + 75 + 0.866\alpha_{AB} - 37.5$. Solve:

$\boxed{a_{Arel} = 75\ (m/s^2)}$ (toward A). $\boxed{\alpha_{AB} = -86.6\ rad/s^2}$, (clockwise).

================================<>================================

Problem 6.132 The block A slides up the inclined surface at 2 *ft/s*. Determine the angular velocity of bar BC and the velocity of point C.

Solution The velocity at A is given to be $\vec{v}_A = 2(-\vec{i}\cos 20^o + \vec{j}\sin 20^o) = -1.879\vec{i} + 0.6840\vec{j}\ (ft/s)$. From geometry, the coordinates of point C are $\left(7,\ 2.5\left(\frac{7}{4.5}\right)\right) = (7,\ 3.89)\ (ft)$. The unit vector parallel to the bar (toward A) is $\vec{e} = (7^2 + 3.89^2)^{-\frac{1}{2}}(-7\vec{i} - 3.89\vec{j}) = -0.8742\vec{i} - 0.4856\vec{j}$.

The velocity at A in terms of the motion of the bar is

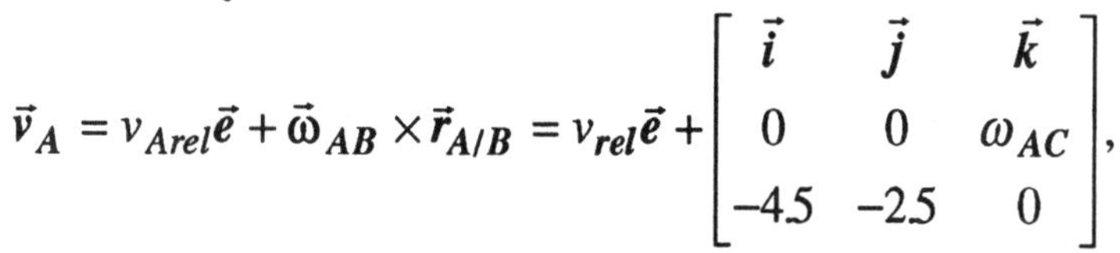

$$\vec{v}_A = v_{Arel}\vec{e} + \vec{\omega}_{AB} \times \vec{r}_{A/B} = v_{rel}\vec{e} + \begin{bmatrix} \vec{i} & \vec{j} & \vec{k} \\ 0 & 0 & \omega_{AC} \\ -4.5 & -2.5 & 0 \end{bmatrix},$$

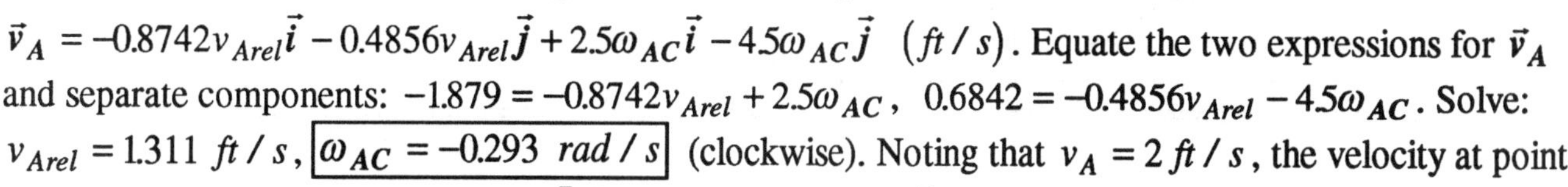

$\vec{v}_A = -0.8742v_{Arel}\vec{i} - 0.4856v_{Arel}\vec{j} + 2.5\omega_{AC}\vec{i} - 4.5\omega_{AC}\vec{j}\ (ft/s)$. Equate the two expressions for $\vec{v}_A$ and separate components: $-1.879 = -0.8742v_{Arel} + 2.5\omega_{AC}$, $0.6842 = -0.4856v_{Arel} - 4.5\omega_{AC}$. Solve: $v_{Arel} = 1.311\ ft/s$, $\boxed{\omega_{AC} = -0.293\ rad/s}$ (clockwise). Noting that $v_A = 2\ ft/s$, the velocity at point C is

$$\vec{v}_C = v_A(-0.8742\vec{i} - 0.4856\vec{j}) + \begin{bmatrix} \vec{i} & \vec{j} & \vec{k} \\ 0 & 0 & -0.293 \\ 2.5 & 3.89-2.5 & 0 \end{bmatrix}, \boxed{\vec{v}_C = -0.738\vec{i} - 1.37\vec{j}\ (ft/s)}.$$

================================<>================================

===================================<>===================================

Problem 6.133 In Problem 6.132 the block A slides up the inclined surface at a constant velocity of 2 *ft/s*. Determine the angular acceleration of bar AC and the acceleration of point C.

Solution *The velocities:* The velocity at A is given to be $\vec{v}_A = 2\left(-\vec{i}\cos 20^o + \vec{j}\sin 20^o\right) = -1.879\vec{i} + 0.6840\vec{j}$ (ft/s). From geometry, the coordinates of point C are $\left(7,\ 2.5\left(\frac{7}{4.5}\right)\right) = (7,\ 3.89)$ (ft). The unit vector parallel to the bar (toward A) is

$\vec{e} = \frac{-7\vec{i} - 3.89\vec{j}}{\sqrt{7^2 + 3.89^2}} = -0.8742\vec{i} - 0.4856\vec{j}$. The velocity at A in terms of the motion of the bar is

$$\vec{v}_A = v_{Arel}\vec{e} + \vec{\omega}_{AC} \times \vec{r}_{A/B} = v_{Arel}\vec{e} + \begin{bmatrix} \vec{i} & \vec{j} & \vec{k} \\ 0 & 0 & \omega_{AC} \\ -4.5 & -2.5 & 0 \end{bmatrix},$$

$\vec{v}_A = -0.8742 v_{Brel}\vec{i} - 0.4856 v_{Brel}\vec{j} + 2.5\omega_{AC}\vec{i} - 4.5\omega_{AC}\vec{j}$ (ft/s). Equate the two expressions and separate components: $-1.879 = -0.8742 v_{Brel} + 2.5\omega_{AC}$, $0.6842 = -0.4856 v_{Brel} - 4.5\omega_{AC}$. Solve: $v_{Brel} = 1.311\ ft/s$, $\omega_{AC} = -0.293\ rad/s$ (clockwise).

The accelerations: The acceleration of block A is given to be zero. In terms of the bar AC, the acceleration of A is $\vec{a}_A = 0 = a_{Arel}\vec{e} + 2\vec{\omega}_{AC} \times v_{Arel}\vec{e} + \vec{\alpha}_{AC} \times \vec{r}_{A/B} - \omega_{AC}^2 \vec{r}_{A/B}$.

$$0 = a_{Arel}\vec{e} + 2\omega_{AC} v_{Arel} \begin{bmatrix} \vec{i} & \vec{j} & \vec{k} \\ 0 & 0 & 1 \\ -0.8742 & -0.4856 & 0 \end{bmatrix} + \begin{bmatrix} \vec{i} & \vec{j} & \vec{k} \\ 0 & 0 & \alpha_{AC} \\ -4.5 & -2.5 & 0 \end{bmatrix} - \omega_{AC}^2\left(-4.5\vec{i} - 2.5\vec{j}\right).$$

$0 = a_{Arel}\vec{e} + 2\omega_{AC} v_{Arel}(-e_y\vec{i} + e_x\vec{j}) + \alpha_{AC}(2.5\vec{i} - 4.5\vec{j}) - \omega_{AC}^2(-4.5\vec{i} - 2.5\vec{j})$. Separate components to obtain: $0 = -0.8742 a_{Arel} - 0.3736 + 2.5\alpha_{AC} + 0.3875$, $0 = -0.4856 a_{Arel} + 0.6742 - 4.5\alpha_{AC} + 0.2153$. Solve: $a_{Arel} = 0.4433$ (ft/s^2) (toward A). $\boxed{\alpha_{AC} = 0.1494\ rad/s^2}$ (counterclockwise). The acceleration of point C is $\vec{a}_C = a_{Arel}\vec{e} + 2\vec{\omega}_{AC} \times \vec{v}_{Arel} + \vec{\alpha}_{AC} \times \vec{r}_{C/B} - \omega_{AC}^2 \vec{r}_{C/B}$

$$\vec{a}_C = a_{Arel}\vec{e} + 2\omega_{AC} v_{Arel} \begin{bmatrix} \vec{i} & \vec{j} & \vec{k} \\ 0 & 0 & 1 \\ e_x & e_y & 0 \end{bmatrix} + \begin{bmatrix} \vec{i} & \vec{j} & \vec{k} \\ 0 & 0 & \alpha_{AC} \\ 2.5 & 3.89 - 2.5 & 0 \end{bmatrix} - \omega_{AC}^2(2.5\vec{i} + (3.89 - 2.5)\vec{j}).$$ Substitute numerical values: $\boxed{\vec{a}_C = -1.184\vec{i} + 0.711\vec{j}\ \left(ft/s^2\right)}$

===================================<>===================================

================================<>================================

Problem 6.134 The angular velocity of the scoop is 1.0 *rad/s* clockwise. Determine the rate at which the hydraulic actuator AB is extending.

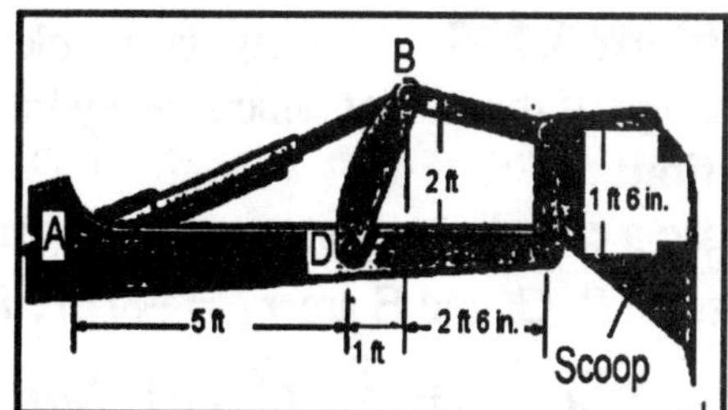

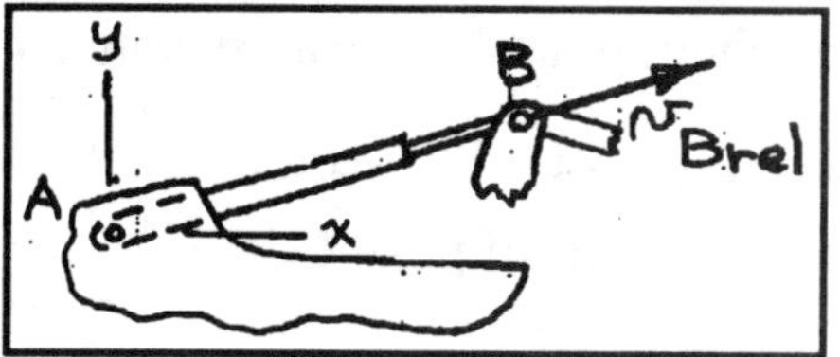

Solution:

The point B slides in the arm AB. The velocity of point C is

$$\vec{v}_C = \vec{\omega}_{scoop} \times (1.5\vec{j}) = \begin{bmatrix} \vec{i} & \vec{j} & \vec{k} \\ 0 & 0 & 1 \\ 0 & 1.5 & 0 \end{bmatrix} = 1.5\vec{i} \ \ (ft/s).$$

Point B is constrained to move normally to the arm DB: The unit vector parallel to DB is $\vec{e}_{DB} = \dfrac{1\vec{i} + 2\vec{j}}{\sqrt{1^2 + 2^2}} = 0.4472\vec{i} + 0.8944\vec{j}$. The unit vector normal to $\vec{e}_{DB}$ is $\vec{e}_{NDB} = 0.8944\vec{i} - 0.4472\vec{j}$, from which the velocity of C in terms of BC is

$$\vec{v}_C = v_B\vec{e}_{NBD} + \vec{\omega}_{BC} \times \vec{r}_{C/B} = v_B(0.8944\vec{i} - 0.4472\vec{j}) + \begin{bmatrix} \vec{i} & \vec{j} & \vec{k} \\ 0 & 0 & \omega_{BC} \\ 2.5 & -0.5 & 0 \end{bmatrix}.$$

$\vec{v}_C = v_B(0.8944\vec{i} - 0.4472\vec{j}) + \omega_{BC}(0.5\vec{i} + 2.5\vec{j})$. Equate terms in $\vec{v}_C$, $1.5 = 0.8944v_B + 0.5\omega_{BC}$, $0 = 0.162v_B + 6\omega_{BC}$. Solve: $\omega_{BC} = 0.2727 \ \ rad/s$, $v_B = 1.525 \ \ ft/s$, from which $\vec{v}_B = v_B\vec{e}_{NDB} = 1.364\vec{i} - 0.6818\vec{j} \ \ (ft/s)$.

The unit vector parallel to the arm AB is $\vec{e}_{AB} = \dfrac{6\vec{i} + 2\vec{j}}{\sqrt{6^2 + 2^2}} = 0.9487\vec{i} + 0.3162\vec{j}$. Choose a coordinate system with with origin at A rotating with arm AB. The velocity of point B is

$$\vec{v}_B = v_{Brel}\vec{e}_{AB} + \vec{\omega}_{AB} \times \vec{r}_{B/A} = v_{Brel}(0.9487\vec{i} + 0.3162\vec{j}) + \begin{bmatrix} \vec{i} & \vec{j} & \vec{k} \\ 0 & 0 & \omega_{AB} \\ 6 & 2 & 0 \end{bmatrix}.$$

$\vec{v}_B = v_{Brel}(0.9487\vec{i} + 0.3162\vec{j}) + \omega_{AB}(-2\vec{i} + 6\vec{j})$. Equate the expressions and separate components: $1.364 = 0.9487v_{Brel} - 2\omega_{AB}$, $-0.6818 = 0.3162v_{Brel} + 6\omega_{AB}$. Solve: $\omega_{AB} = -0.1704 \ \ rad/s$, $\boxed{v_{Brel} = 1.078 \ \ ft/s}$ which is the rate of extension of the actuator.

================================<>================================

====================================<>====================================

Problem 6.135 The angular acceleration of the scoop in Problem 6.134 is zero. Determine the rate of change of the rate at which the hydraulic actuator is extending.

Solution:

Choose a coordinate system with the origin at D and the x axis parallel to ADE. The vector locations of points A, B, C, and E are $\vec{r}_A = -8\vec{i}\ ft$, $\vec{r}_B = 1\vec{i} + 2\vec{j}\ ft$, $\vec{r}_C = 3.5\vec{i} + 1.5\vec{j}\ ft$, $\vec{r}_E = 3.5\vec{i}\ \ ft$. The vector AB is $\vec{r}_{B/A} = \vec{r}_B - \vec{r}_A = 6\vec{i} + 2\vec{j}\ \ (ft)$, $\vec{r}_{B/D} = \vec{r}_B - \vec{r}_D = 1\vec{i} + 2\vec{j}\ \ (ft)$. Assume that the scoop rotates at 1 *rad/s* about point E. The acceleration of point C is $\vec{a}_C = \vec{\alpha}_{Scoop} \times 1.5\vec{j} - \omega_{scoop}^2(1.5\vec{j}) = -1.5\vec{j}\ \ (ft/s^2)$, since $\alpha_{scoop} = 0$. The vector from C to B is $\vec{r}_{B/C} = \vec{r}_B - \vec{r}_C = -2.5\vec{i} + 0.5\vec{j}\ (ft)$. The acceleration of point B in terms of point C is

$$\vec{a}_B = \vec{a}_C + \vec{\alpha}_{BC} \times \vec{r}_{B/C} - \omega_{BC}^2\vec{r}_{B/C} = -1.5\vec{j} + \begin{bmatrix} \vec{i} & \vec{j} & \vec{k} \\ 0 & 0 & \alpha_{BC} \\ -2.5 & +0.5 & 0 \end{bmatrix} - \omega_{BC}^2(-2.5\vec{i} + 0.5\vec{j}),$$ from which

$\vec{a}_B = -(0.5\alpha_{BC} - 2.5\omega_{BC}^2)\vec{i} - (1.5 + 2.5\alpha_{BC} + 0.5\omega_{BC}^2)\vec{j}$. The acceleration of B in terms of D is

$$\vec{a}_B = \vec{a}_D + \vec{\alpha}_{BD} \times \vec{r}_{B/D} - \omega_{BD}^2\vec{r}_{B/D} = \vec{a}_D + \begin{bmatrix} \vec{i} & \vec{j} & \vec{k} \\ 0 & 0 & \alpha_{BD} \\ 1 & 2 & 0 \end{bmatrix} - \omega_{BD}^2(\vec{i} + 2\vec{j}).$$ The acceleration of point D

is zero, from which $\vec{a}_B = -(2\alpha_{BD} + \omega_{BD}^2)\vec{i} + (\alpha_{BD} - 2\omega_{BD}^2)\vec{j}$. Equate like terms in the two expressions for $\vec{a}_B$, $-(0.5\alpha_{BC} - 2.5\omega_{BC}^2) = -(2\alpha_{BD} + \omega_{BD}^2)$, $-(1.5 + 2.5\alpha_{BC} + 0.5\omega_{BC}^2) = (\alpha_{BD} - 2\omega_{BD}^2)$. From the solution to Problem 6.134, $\omega_{BC} = 0.2727\ \ rad/s$, and $v_B = 1.525\ ft/s$. The velocity of point B is normal to the link BD, from which $\omega_{BD} = \dfrac{v_B}{\sqrt{1^2 + 2^2}} = 0.6818\ rad/s$. Substitute and solve for the angular accelerations: $\alpha_{BC} = -0.1026\ \ rad/s^2$, $\alpha_{BD} = -0.3511\ \ rad/s^2$. From which the acceleration of point B is $\vec{a}_B = -(2\alpha_{BD} + \omega_{BD}^2)\vec{i} + (\alpha_{BD} - 2\omega_{BD}^2)\vec{j} = 0.2372\vec{i} - 1.281\vec{j}\ \ (ft/s^2)$.

The acceleration of point B in terms of the arm AB is

$\vec{a}_B = a_{Brel}\vec{e}_{B/A} + 2\vec{\omega}_{AB} \times v_{Brel}\vec{e}_{B/A} + \vec{\alpha}_{AB} \times \vec{r}_{B/A} - \omega_{AB}^2\vec{r}_{B/A}$. From the solution to Problem 6.134: $\vec{e}_{B/A} = 0.9487\vec{i} + 0.3162\vec{j}$, $v_{Brel} = 1.078\ \ ft/s$, $\omega_{AB} = -0.1705\ \ rad/s$. From which $\vec{a}_B = a_{Brel}(0.9487\vec{i} + 0.3162\vec{j}) + 0.1162\vec{i} - 0.3487\vec{j} + \alpha_{AB}(-2\vec{i} + 6\vec{j}) - 0.1743\vec{i} - 0.0581\vec{j}$. Equate the accelerations of point B and separate components: $0.2371 = 0.9487a_{Brel} - 2\alpha_{AB} - 0.0581$, $-1.281 = 0.3162a_{Brel} + 6\alpha_{AB} - 0.4068$. Solve: $\boxed{a_{Brel} = 0.0038\ \ ft/s^2}$, which is the rate of change of the rate at which the actuator is extending.

====================================<>====================================

Problem 6.136 Suppose that the curved bar in Example 6.9 rotates with a counterclockwise angular velocity of 2 *rad/s*. (a) What is the angular velocity of bar AB? (b) What is the velocity of block B relative to the slot?

Solution:

The angle defining the position of B in the circular slot is

$$\beta = \sin^{-1}\left(\frac{350}{500}\right) = 44.4^{o}.$$

The vectors are

$\vec{r}_{B/A} = (500 + 500\cos\beta)\vec{i} + 350\vec{j} = 857\vec{i} + 350\vec{j}\ (mm)$.

$\vec{r}_{B/C} = (-500 + 500\cos\beta)\vec{i} + 350\vec{j}\ (mm)$. The unit vector tangent to the slot at B is given by

$\vec{e}_B = -\sin\beta\vec{i} + \cos\beta\vec{j} = -0.7\vec{i} + 0.714\vec{j}$.

The velocity of B in terms of AB is

$$\vec{v}_B = \vec{\omega}_{AB} \times \vec{r}_{B/A} = \begin{bmatrix} \vec{i} & \vec{j} & \vec{k} \\ 0 & 0 & \omega_{AB} \\ 857 & 350 & 0 \end{bmatrix} = \omega_{AB}\left(-350\vec{i} + 857\vec{j}\right)\ (mm/s).$$

The velocity of B in terms of BC is

$$\vec{v}_B = v_{Brel}\vec{e}_B + \vec{\omega}_{BC} \times \vec{r}_{B/C} = v_{Brel}\left(-0.7\vec{i} + 0.714\vec{j}\right) + \begin{bmatrix} \vec{i} & \vec{j} & \vec{k} \\ 0 & 0 & \omega_{BC} \\ -142.9 & 350 & 0 \end{bmatrix},$$

$$\vec{v}_B = v_{Brel}\left(-0.7\vec{i} + 0.714\vec{j}\right) + \left(-700\vec{i} - 285.8\vec{j}\right)\ (mm/s)$$

Equate the expressions for the velocity of B and separate components: $-350\omega_{AB} = -0.7v_{Brel} - 700$, $-857\omega_{AB} = 0.714v_{Brel} - 285.8$. Solve: (a) $\boxed{\omega_{AB} = -2\ rad/s}$ (clockwise). (b) $\boxed{v_{Brel} = -2000\ mm/s}$ (toward C).

==================================<>==================================

Problem 6.137 Suppose that the curved bar in Example 6.9 has a clockwise angular velocity of 10 4 *rad/s* and a counterclockwise angular acceleration of 10 *rad/s*^{2}. What is the angular acceleration of bar AB?

Solution Use the solution to Problem 6.118 with new data.

Get the velocities: The angle defining the position of B in the circular slot is $\beta = \sin^{-1}\left(\frac{350}{500}\right) = 44.4^o$.

The vectors $\vec{r}_{B/A} = (500 + 500\cos\beta)\vec{i} + 350\vec{j} = 857\vec{i} + 350\vec{j}\ (mm)$.

$\vec{r}_{B/C} = (-500 + 500\cos\beta)\vec{i} + 350\vec{j}\ (mm)$. The unit vector tangent to the slot at B

$\vec{e}_B = -\sin\beta\vec{i} + \cos\beta\vec{j} = -0.7\vec{i} + 0.714\vec{j}$. The component normal to the slot at B is

$\vec{e}_{NB} = \cos\beta\vec{i} + \sin\beta\vec{j} = 0.7141\vec{i} + 0.7\vec{j}$. The velocity of B in terms of AB

$$\vec{v}_B = \vec{\omega}_{AB} \times \vec{r}_{B/A} = \begin{bmatrix} \vec{i} & \vec{j} & \vec{k} \\ 0 & 0 & \omega_{AB} \\ 857 & 350 & 0 \end{bmatrix} = \omega_{AB}\left(-350\vec{i} + 857\vec{j}\right)\ (mm/s).$$ The velocity of B in terms of BC is $$\vec{v}_B = v_{Brel}\vec{e}_B + \vec{\omega}_{BC} \times \vec{r}_{B/C} = v_{Brel}\left(-0.7\vec{i} + 0.714\vec{j}\right) + \begin{bmatrix} \vec{i} & \vec{j} & \vec{k} \\ 0 & 0 & \omega_{BC} \\ -142.9 & 350 & 0 \end{bmatrix},$$

$\vec{v}_B = v_{Brel}\left(-0.7\vec{i} + 0.714\vec{j}\right) + \left(1400\vec{i} + 571.6\vec{j}\right)\ (mm/s)$. Equate the expressions for the velocity of B and separate components: $-350\omega_{AB} = -0.7v_{Brel} + 1400$, $857\omega_{AB} = 0.714v_{Brel} + 571.6$. Solve: $\omega_{AB} = 4\ rad/s$ (counterclockwise). $v_{Brel} = 4000\ mm/s$ (away from C).

Get the accelerations: The acceleration of point B in terms of the AB is

$$\vec{a}_B = \vec{\alpha}_{AB} \times \vec{r}_{B/A} - \omega_{AB}^2\vec{r}_{B/A} = \begin{bmatrix} \vec{i} & \vec{j} & \vec{k} \\ 0 & 0 & \alpha_{AB} \\ 857 & 350 & 0 \end{bmatrix} - \omega_{AB}^2\left(857\vec{i} + 350\vec{j}\right),$$

$\vec{a}_B = \alpha_{AB}\left(-350\vec{i} + 857\vec{j}\right) - 138713\vec{i} - 5665\vec{j}\ \left(mm/s^2\right)$. The acceleration in terms of the arm BC is

$\vec{a}_B = \vec{a}_{Brel} + 2\vec{\omega}_{BC} \times v_{Brel}\vec{e}_B + \vec{\alpha}_{BC} \times \vec{r}_{B/C} - \omega_{BC}^2\vec{r}_{B/C}$. Expanding term by term:

$$\vec{a}_{Brel} = a_{Brel}\vec{e}_B - \left(\frac{v_{Brel}^2}{500}\right)\vec{e}_{NB} = a_{Brel}\left(-0.7\vec{i} + 0.7141\vec{j}\right) - 22{,}852.6\vec{i} - 22{,}400\vec{j}.$$

Other terms: $2\vec{\omega}_{BC} \times v_{Brel}\vec{e}_B = 22852\vec{i} + 22{,}400\vec{j}$, $\vec{\alpha}_{BC} \times \vec{r}_{B/C} = -3500\vec{i} - 1429.3\vec{j}$, $-\omega_{BC}^2\vec{r}_{B/C} = 2286.8\vec{i} - 5600\vec{j}$. Collect terms:

$$\vec{a}_B = a_{Brel}\left(-0.7\vec{i} + 0.7141\vec{j}\right) - 22852.6\vec{i} - 22400\vec{j} + 22852.6\vec{i} + 22400\vec{j} - 3500\vec{i} - 1429.3\vec{j} + 2286.9\vec{i} - 5600\vec{j}$$

$\vec{a}_B = a_{Brel}\left(-0.7\vec{i} + 0.7141\vec{j}\right) - 1213\vec{i} - 7029.3\vec{j}$. Equate the two expressions for the acceleration of B to obtain the two equations: $-350\alpha_{AB} - 13{,}871 = -0.7a_{Brel} - 1213.1$,

$857\alpha_{AB} - 5665 = 0.7141a_{Brel} - 7029.3$. Solve: $a_{Brel} = 29180\ \left(mm/s^2\right)$, $\boxed{\alpha_{AB} = 22.65\ rad/s^2}$ (counterclockwise).

==================================<>==================================

================================<>================================

Problem 6.138 The disk rolls on the plane surface with a counterclockwise angular velocity of 10 *rad/s*. Bar AB slides on the surface of the disk at A. Determine the angular velocity of the bar AB.

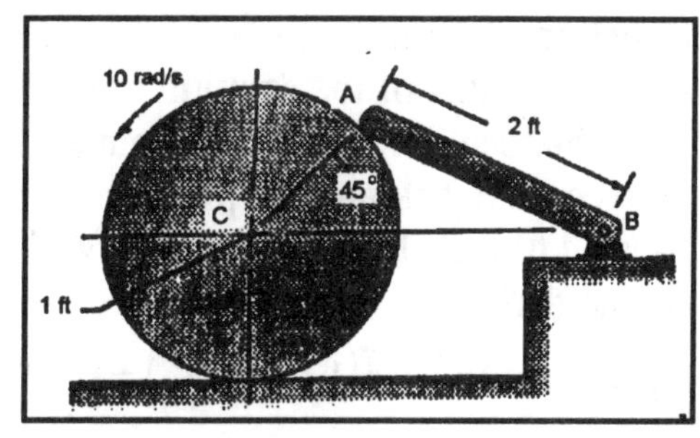

Solution:
Choose a coordinate system with the origin at the point of contact between the disk and the plane surface, with the x axis parallel to the plane surface. Let A be the point of the bar in contact with the disk. The vector location of point A on the disk is $\vec{r}_A = \vec{i}\cos 45^o + \vec{j}(1+\sin 45^o) = 0.707\vec{i} + 1.707\vec{j}$ (ft). The unit vector parallel to the radius of the disk is $\vec{e}_A = \cos 45^o \vec{i} + \sin 45^o \vec{j} = 0.707\vec{i} + 0.707\vec{j}$. The unit vector tangent to the surface of the disk at A is $\vec{e}_{NA} = \vec{i}\sin 45^o - \vec{j}\cos 45^o = 0.707\vec{i} - 0.707\vec{j}$. The angle formed by the bar AB with the horizontal is $\beta = \sin^{-1}\left(\frac{\sin 45^o}{2}\right) = 20.7^o$. The velocity of point A in terms of the motion of bar AB is

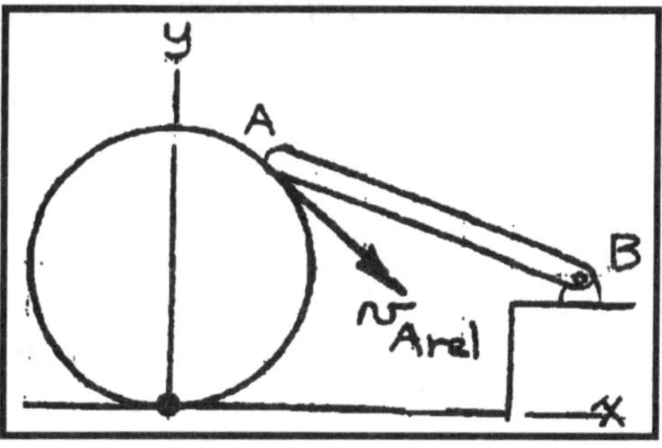

$$\vec{v}_A = \vec{\omega}_{AB} \times \vec{r}_{A/B} = \begin{bmatrix} \vec{i} & \vec{j} & \vec{k} \\ 0 & 0 & \omega_{AB} \\ -2\cos\beta & 2\sin\beta & 0 \end{bmatrix}.\ \vec{v}_A = \omega_{AB}(-0.707\vec{i} - 1.871\vec{j})\ (ft/s).$$

The velocity of point A in terms of the point of the disk in contact with the plane surface is

$$\vec{v}_A = v_{Arel}\vec{e}_{NB} + \vec{\omega}_{disk} \times \vec{r}_A = v_{Arel}(0.707\vec{i} - 0.707\vec{j}) + \begin{bmatrix} \vec{i} & \vec{j} & \vec{k} \\ 0 & 0 & \omega_{disk} \\ 0.707 & 1.707 & 0 \end{bmatrix},$$

$\vec{v}_A = v_{Arel}(0.707\vec{i} - 0.707\vec{j}) + (-17.07\vec{i} + 7.07\vec{j})$. Equate the expressions and separate components: $-0.707\omega_{AB} = 0.707 v_{Arel} - 17.07$, $-1.871\omega_{AB} = -0.707 v_{Arel} + 7.07$. Solve: $v_{Arel} = 20.3\ ft/s$, $\boxed{\omega_{AB} = 3.88\ rad/s}$ (counterclockwise).

================================<>================================

==================================<>==================================

Problem 6.139 In Problem 6.138, the disk rolls on the plane surface with a constant counterclockwise angular velocity of 10 *rad/s*. Determine the angular acceleration of the bar AB.

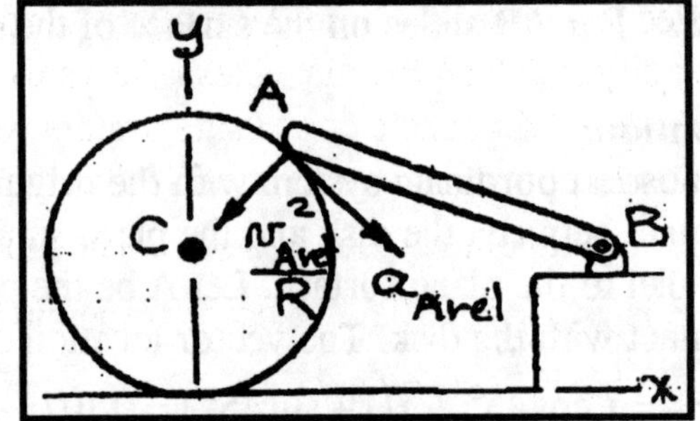

Solution:

Use the results of the solution to Problem 6.138. Choose a coordinate system with the origin at the point of contact between the disk and the plane surface, with the *x* axis parallel to the plane surface. The vector location of point A on the disk is

$\vec{r}_A = \vec{i}\cos 45^o + \vec{j}(1+\sin 45^o) = 0.707\vec{i} + 1.707\vec{j}\ (ft)$. The The unit vector tangent to the surface of the disk at A is $\vec{e}_{NA} = \vec{i}\sin 45^o - \vec{j}\cos 45^o = 0.707\vec{i} - 0.707\vec{j}$.

The angle formed by the bar AB with the horizontal is $\beta = \sin^{-1}(\sin 45^o/2) = 20.7^o$.

Get the velocities: The velocity of point A in terms of the motion of bar AB is

$$\vec{v}_A = \vec{\omega}_{AB} \times \vec{r}_{A/B} = \begin{bmatrix} \vec{i} & \vec{j} & \vec{k} \\ 0 & 0 & \omega_{AB} \\ -2\cos\beta & 2\sin\beta & 0 \end{bmatrix} = \omega_{AB}(-0.707\vec{i} - 1.871\vec{j})\ (ft/s).$$

The acceleration of the center of the disk is zero. The velocity of point A in terms of the center of the disk is

$$\vec{v}_A = v_{Arel}\vec{e}_{NB} + \vec{\omega}_{disk} \times \vec{r}_A = v_{Arel}(0.707\vec{i} - 0.707\vec{j}) + \begin{bmatrix} \vec{i} & \vec{j} & \vec{k} \\ 0 & 0 & \omega_{disk} \\ 0.707 & 1.707 & 0 \end{bmatrix},$$

$\vec{v}_A = v_{Arel}(0.707\vec{i} - 0.707\vec{j}) + (-17.07\vec{i} + 7.07\vec{j})$. Equate the expressions and separate components: $-0.707\omega_{AB} = 0.707 v_{Arel} - 17.07$, $-1.871\omega_{AB} = -0.707 v_{Arel} + 7.07$. Solve: $v_{Arel} = 20.3\ ft/s$, $\omega_{AB} = 3.88\ rad/s$ (counterclockwise).

Get the accelerations: The acceleration of point A in terms of the arm AB is

$$\vec{a}_A = \vec{\alpha}_{AB} \times \vec{r}_{A/B} - \omega_{AB}^2 \vec{r}_{A/B} = \begin{bmatrix} \vec{i} & \vec{j} & \vec{k} \\ 0 & 0 & \alpha_{AB} \\ -1.87 & 0.707 & 0 \end{bmatrix} + 28.15\vec{i} - 10.64\vec{j}\ (ft/s^2),$$

$\vec{a}_A = \alpha_{AB}(-0.707\vec{i} - 1.87\vec{j}) + 28.15\vec{i} - 10.64\vec{j}\ (ft/s^2)$. The acceleration of point A in terms of the disk is $\vec{a}_A = \vec{a}_{Arel} + 2\vec{\omega}_{disk} \times v_{Arel}\vec{e}_{NA} + \vec{\alpha}_{disk} \times \vec{r}_{A/C} - \omega_{disk}^2 \vec{r}_{A/C}$. Expanding term by term: The acceleration $\vec{a}_{Arel}$ is composed of a tangential component and a radial component:

$$\vec{a}_{Arel} = a_{Arel}\vec{e}_{NA} - \left(\frac{v_{Arel}^2}{1}\right)\vec{e}_A = a_{Arel}(0.707\vec{i} - 0.707\vec{j}) - 290.3\vec{i} - 290.3\vec{j}.$$

$2\vec{\omega}_{disk} \times v_{Arel}\vec{e}_{NB} = 286.6\vec{i} + 286.6\vec{j}$, $\vec{\alpha}_{disk} \times \vec{r}_A = 0$, since the acceleration of the disk is zero. $-\omega_{disk}^2 \vec{r}_{A/C} = -70.7\vec{i} - 70.7\vec{j}$. Collect terms and separate components to obtain:

$-0.707\alpha_{AB} + 28.15 = 0.707 a_{Arel} - 290.3 + 286.6 - 70.7$,

$-1.87\alpha_{AB} - 10.64 = -0.707 a_{Arel} - 290.3 + 286.6 - 70.7$. Solve: $a_{Arel} = 80.6\ ft/s^2$,

$\boxed{\alpha_{AB} = 64.6\ rad/s^2}$ (counterclockwise).

==================================<>==================================

==================================<>==================================

Problem 6.140 The bar BC rotates with a counterclockwise angular velocity of 2 *rad/s*. A pin at B slides in a circular slot in the rectangular plate. Determine the angular velocity of the plate and the velocity to which the pin slides relative to the circular slot.

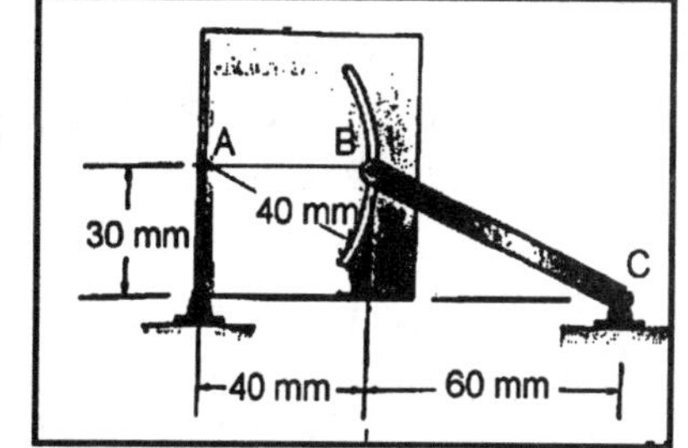

Solution Choose a coordinate system with the origin O at the lower left pin and the x axis parallel to the plane surface. The unit vector parallel to AB is $\vec{e}_{AB} = \vec{i}$. The unit vector tangent to the slot at B is $\vec{e}_{NAB} = \vec{j}$. The velocity of the pin in terms of the motion of BC is $\vec{v}_B = \vec{\omega}_{BC} \times \vec{r}_{B/C}$.

$$\vec{v}_B = \begin{bmatrix} \vec{i} & \vec{j} & \vec{k} \\ 0 & 0 & \omega_{BC} \\ -60 & 30 & 0 \end{bmatrix} = 2(-30\vec{i} - 60\vec{j}) = -60\vec{i} - 120\vec{j} \ (mm/s).$$

The velocity of the pin in terms of the plate is

$$\vec{v}_B = v_{Brel}\vec{j} + \vec{\omega}_{AB} \times \vec{r}_{B/A} = \begin{bmatrix} \vec{i} & \vec{j} & \vec{k} \\ 0 & 0 & \omega_{AB} \\ 40 & 30 & 0 \end{bmatrix} = v_{Brel}\vec{j} + \omega_{AB}(-30\vec{i} + 40\vec{j}) \ (mm/s).$$ Equate the expressions and separate components to obtain $-60 = -30\omega_{AB}$, $-120 = v_{Brel} + 40\omega_{AB}$. Solve: $\vec{v}_{Brel} = -200\vec{j} \ mm/s$, $\boxed{\omega_{AB} = 2 \ rad/s}$ (counterclockwise)

==================================<>==================================

Problem 6.141 The bar BC in Problem 6.140 rotates with a constant counterclockwise angular velocity of 2 *rad/s*. Determine the angular acceleration of the plate.

Solution Choose the same coordinate system as in Problem 6.140. *Get the velocities:* The unit vector parallel to AB is $\vec{e}_{AB} = \vec{i}$. The unit vector tangent to the slot at B is $\vec{e}_{NAB} = \vec{j}$. The velocity of the pin in terms of the motion of BC is $\vec{v}_B = \vec{\omega}_{BC} \times \vec{r}_{B/C} = \begin{bmatrix} \vec{i} & \vec{j} & \vec{k} \\ 0 & 0 & \omega_{BC} \\ -60 & 30 & 0 \end{bmatrix} = (-60\vec{i} - 120\vec{j}) \ (mm/s)$. The velocity of the pin in terms of the plate is

$$\vec{v}_B = v_{Brel}\vec{j} + \begin{bmatrix} \vec{i} & \vec{j} & \vec{k} \\ 0 & 0 & \omega_{AB} \\ 40 & 30 & 0 \end{bmatrix} = v_{Brel}\vec{e}_{NAB} + \vec{\omega}_{AB} \times \vec{r}_{B/O} = v_{Brel}\vec{j} + \omega_{AB}(-30\vec{i} + 40\vec{j}) \ (mm/s).$$ Equate the expressions and separate components to obtain $-60 = -30\omega_{AB}$, $-120 = v_{Brel} + 40\omega_{AB}$. Solve: $\vec{v}_{Brel} = -200\vec{j} \ mm/s$, $\omega_{AB} = 2 \ rad/s$ (counterclockwise). *Get the accelerations:* The acceleration of the pin in terms of the arm BC is

$\vec{a}_B = \vec{\alpha}_{BC} \times \vec{r}_{B/C} - \omega_{BC}^2 \vec{r}_{B/C} = 0 - 4(-60\vec{i} + 30\vec{j}) = 240\vec{i} - 120\vec{j} \ (mm/s^2)$. The acceleration of the pin in terms of the plate AB is $\vec{a}_B = \vec{a}_{Brel} + 2\vec{\omega}_{AB} \times v_{Brel}\vec{e}_{NAB} + \vec{\alpha}_{AB} \times \vec{r}_{B/O} - \omega_{AB}^2 \vec{r}_{B/O}$.

Solution continued on next page

Expand term by term: $\vec{a}_{Brel} = a_{Brel}\vec{e}_{NAB} - \left(\frac{v_{Brel}^2}{40}\right)\vec{e}_{AB} = a_{Brel}\vec{j} - 1000\vec{i}\ \left(mm/s^2\right)$,

$2\vec{\omega}_{AB} \times v_{Brel}\vec{e}_{NAB} = 800\vec{i}\ \left(mm/s^2\right)$. $\vec{\alpha}_{AB} \times \vec{r}_{B/O} = \begin{bmatrix} \vec{i} & \vec{j} & \vec{k} \\ 0 & 0 & \alpha_{AB} \\ 40 & 30 & 0 \end{bmatrix} = \alpha_{BA}\left(-30\vec{i} + 40\vec{j}\right)\ \left(mm/s^2\right)$,

$-\omega_{AB}^2\left(40\vec{i} + 30\vec{j}\right) = -160\vec{i} - 120\vec{j}\ \left(mm/s^2\right)$. Collect terms and separate components to obtain:

$240 = -1000 + 800 - 30\alpha_{BA} - 160$, $-120 = a_{Brel} + 40\alpha_{BA} - 120$. Solve: $a_{Brel} = 800\ mm/s^2$ (upward), $\boxed{\alpha_{AB} = -20\ rad/s^2}$, (clockwise).

==================================<>==================================

Problem 6.142 By taking the derivative of Eq (6.11) and using Eq (6.12), derive Eq (6.13)

Solution Eq (6.11) is $\vec{v}_A = \vec{v}_B + \vec{v}_{Arel} + \vec{\omega} \times \vec{r}_{A/B}$. Eq (6.12) is $\vec{v}_{Arel} = \left(\frac{dx}{dt}\right)\vec{i} + \left(\frac{dy}{dt}\right)\vec{j} + \vec{k}\left(\frac{dz}{dt}\right)\vec{k}$.

Assume that the coordinate system is body fixed and that B is a point on the rigid body, (A is not necessarily a point on the rigid body), such that $\vec{r}_A = \vec{r}_B + \vec{r}_{A/B}$, where $\vec{r}_{A/B} = x\vec{i} + y\vec{j} + z\vec{k}$, and x, y, z are the coordinates of A in body fixed coordinates. Take the derivative of both sides of Eq(6.11):

$\frac{d\vec{v}_A}{dt} = \frac{d\vec{v}_B}{dt} + \frac{d\vec{v}_{Arel}}{dt} + \frac{d\vec{\omega}}{dt} \times \vec{r}_{A/B} + \vec{\omega} \times \frac{d\vec{r}_{A/B}}{dt}$. By definition, $\frac{d\vec{v}_A}{dt} = \vec{a}_A$, $\frac{d\vec{v}_B}{dt} = \vec{a}_B$, and $\frac{d\vec{\omega}}{dt} = \vec{\alpha}$.

The derivative: $\frac{d\vec{v}_{Arel}}{dt} = \frac{d^2x}{dt^2}\vec{i} + \frac{d^2y}{dt^2}\vec{j} + \frac{d^2z}{dt^2}\vec{k} + \frac{dx}{dt}\frac{d\vec{i}}{dt} + \frac{dy}{dt}\frac{d\vec{j}}{dt} + \frac{dz}{dt}\frac{d\vec{k}}{dt}$. Using the fact that the derivative of a unit vector represents a rotation of the unit vector, $\frac{d\vec{i}}{dt} = \vec{\omega} \times \vec{i}$, $\frac{d\vec{j}}{dt} = \vec{\omega} \times \vec{j}$,

$\frac{d\vec{k}}{dt} = \vec{\omega} \times \vec{k}$. Substitute into the derivative: $\frac{d\vec{v}_{Arel}}{dt} = \vec{a}_{Arel} + \vec{\omega} \times \vec{v}_{Arel}$.

Noting $\vec{\omega} \times \frac{d\vec{r}_{A/B}}{dt} = \vec{\omega} \times \left(\frac{dx}{dt}\vec{i} + \frac{dy}{dt}\vec{j} + \frac{dz}{dt}\vec{k}\right) + \vec{\omega} \times \left(x\frac{d\vec{i}}{dt} + y\frac{d\vec{j}}{dt} + z\frac{d\vec{k}}{dt}\right) = \vec{\omega} \times \vec{v}_{Arel} + \vec{\omega} \times (\vec{\omega} \times \vec{r}_{A/B})$.

Collect and combine terms: the derivative of Eq (6.11) is

$\boxed{\vec{a}_A = \vec{a}_B + \vec{a}_{Rel} + 2\vec{\omega} \times \vec{v}_{Arel} + \vec{\alpha} \times \vec{r}_{A/B} + \vec{\omega} \times (\vec{\omega} \times \vec{r}_{A/B})}$, which is Eq (6.13).

==================================<>==================================

==================================<>==================================

Problem 6.143 A merry-go-round rotates at a constant velocity of 0.5 *rad/s*. The person A walks at a constant speed of 1 *m/s* along a radial line. Determine A's velocity and acceleration *relative to the earth* when she is 2 *m* from the center of the merry-go-round, using two methods: (a) Express the velocity and acceleration in terms of polar coordinates. (b) Use Eqs (6.21) and (6.22) to express the velocity and acceleration in terms of a body fixed coordinate system with its *x* axis aligned with the line along which A walks and its *z* axis perpendicular to the merry-go-round.

Solution:

(a) The velocity in polar coordinates is $\boxed{\vec{v}_A = 1\vec{e}_r + 0.5(2)\vec{e}_\theta = \vec{e}_r + \vec{e}_\theta \ (m/s)}$. The acceleration is

$$\vec{a}_A = \left(\left(\frac{d^2 v_A}{dt^2}\right) - r\left(\frac{d\theta}{dt}\right)^2\right)\vec{e}_r + \left(r\frac{d^2\theta}{dt^2} + 2\left(\frac{dr}{dt}\right)\left(\frac{d\theta}{dt}\right)\right)\vec{e}_\theta.$$ Substitute noting

$\frac{d\theta}{dt} = \omega = 0.5 \ rad/s$, $\vec{a}_A = -r\omega^2\vec{e}_r + 2\omega v_A\vec{e}_\theta = -0.5\vec{e}_r + \vec{e}_\theta \ (m/s^2)$.

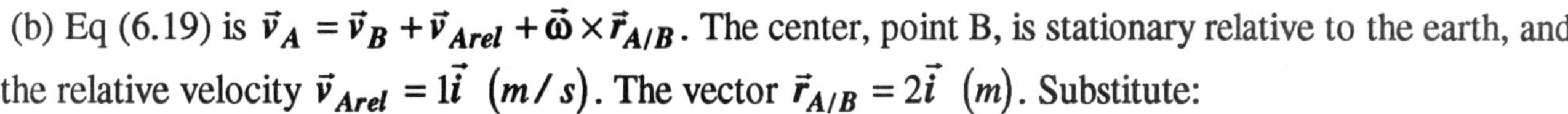

(b) Eq (6.19) is $\vec{v}_A = \vec{v}_B + \vec{v}_{Arel} + \vec{\omega}\times\vec{r}_{A/B}$. The center, point B, is stationary relative to the earth, and the relative velocity $\vec{v}_{Arel} = 1\vec{i} \ (m/s)$. The vector $\vec{r}_{A/B} = 2\vec{i} \ (m)$. Substitute:

$\boxed{\vec{v}_A = \vec{i} + \omega(\vec{k}\times 2\vec{i}) = \vec{i} + \vec{j} \ (m/s)}$. Eq (6.20) is

$\vec{a}_A = \vec{a}_B + \vec{a}_{Arel} + 2\vec{\omega}\times\vec{v}_{Arel} + \vec{\alpha}\times\vec{r}_{A/B} + \vec{\omega}\times(\vec{\omega}\times\vec{r}_{A/B})$. The point B is stationary, $\vec{a}_{Arel} = 0$, and

$\vec{\alpha} = 0$. Substitute: $\vec{a}_A = 0 + 0 + 2\omega(\vec{k}\times\vec{i}) + 0 + \omega^2(\vec{k}\times(\vec{k}\times 2\vec{i})) = 2\omega\vec{j} - 2\omega^2\vec{i} \boxed{= -0.5\vec{i} + \vec{j} \ (m/s^2)}$

==================================<>==================================

Problem 6.144 A disk shaped space station of Radius R rotates with a constant angular velocity ω about the axis perpendicular to the page. Two persons are stationary relative to the station at A and B, and O is the center of the station. Using Eqs (6.21) and (6.22) and the body fixed coordinate system shown, (a) determine A's velocity and acceleration relative to a non rotating reference frame whose origin is at O; (b) determine A's velocity and acceleration relative to a non rotating reference frame whose origin moves with point B.

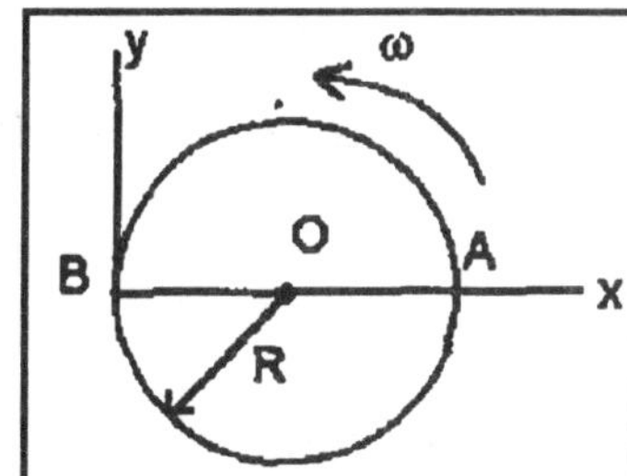

Solution:

(a) The center O is stationary; the velocity of A relative to O is $\vec{v}_{Arel} = 0$, and $\vec{r}_{A/O} = R\vec{i}$. The velocity of A is $\boxed{\vec{v}_A = \vec{v}_O + \vec{v}_{Arel} + \vec{\omega}\times\vec{r}_{A/O} = 0 + 0 + \omega(\vec{k}\times R\vec{i}) = \omega R\vec{j}}$. The acceleration is

$\vec{a}_A = \vec{a}_O + \vec{a}_{Arel} + 2\vec{\omega}\times\vec{v}_{Arel} + \vec{\alpha}\times\vec{r}_{A/O} + \vec{\omega}\times(\vec{\omega}\times\vec{r}_{A/O})$ The accelerations $\vec{a}_O = 0$, $\vec{\alpha} = 0$ and the relative acceleration , $\vec{a}_{Arel} = 0$, from which $\boxed{\vec{a}_B = 0 + 0 + 0 + 0 + \omega^2(\vec{k}\times(\vec{k}\times R\vec{i})) = -\omega^2 R\vec{i}}$

(b) In the non rotating system whose origin moves with B, the velocity is $\vec{v}_A = \vec{v}_B + \vec{v}_{Arel} + \vec{\omega}\times\vec{r}_{A/B}$, and $\vec{v}_B = 0$, $\vec{v}_{Arel} = 0$, $\vec{r}_{A/B} = 2R\vec{i}$, from which $\boxed{\vec{v}_A = 0 + 0 + \omega(\vec{k}\times 2R\vec{i}) = 2R\omega\vec{j}}$. The acceleration is

$\vec{a}_A = \vec{a}_B + \vec{a}_{Arel} + 2\vec{\omega}\times\vec{v}_{Arel} + \vec{\alpha}\times\vec{r}_{A/B} + \vec{\omega}\times(\vec{\omega}\times\vec{r}_{A/B})$. The accelerations $\vec{a}_{Arel} = 0$, $\vec{\alpha} = 0$, from which $\boxed{\vec{a}_A = 0 + 0 + 0 + 0 + \omega^2(\vec{k}\times(\vec{k}\times 2R\vec{i})) = -2R\omega^2\vec{i}}$

==================================<>==================================

====================================<>====================================

Problem 6.145 The metal plate is attached to a fixed ball and socket support at O. The pin A slides in a slot in the plate. At the instant shown, $x_A = 1\ m$, $\frac{dx_A}{dt} = 2\ m/s$, and $\frac{d^2 x_A}{dt^2} = 0$, and the plate's angular velocity and angular acceleration are $\vec{\omega} = 2\vec{k}\ (rad/s)$ and $\vec{\alpha} = 0$. What are the *x,y,z* components of the velocity and acceleration of A relative to a non rotating reference frame that is stationary with respect to O?

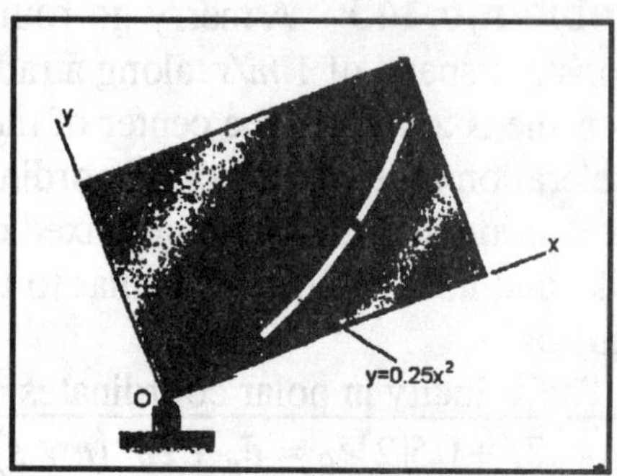

Solution:

The velocity is $\vec{v}_A = \vec{v}_O + \vec{v}_{Arel} + \vec{\omega} \times \vec{r}_{A/O}$. The relative velocity is $\vec{v}_{Arel} = \left(\frac{dx}{dt}\right)\vec{i} + \left(\frac{dy}{dt}\right)\vec{j} + \left(\frac{dz}{dt}\right)\vec{k}$,

where $\frac{dx}{dt} = 2\ m/s$, $\frac{dy}{dt} = \frac{d}{dt} 0.25x^2 = 0.5x\frac{dx}{dt} = 1\ m/s$, $\frac{dz}{dt} = 0$, and

$\vec{r}_{A/O} = x\vec{i} + y\vec{j} + z\vec{j} = \vec{i} + 0.25\vec{j} + 0$, from which

$$\boxed{\vec{v}_A = 2\vec{i} + \vec{j} + \omega\left(\vec{k} \times \left(\vec{i} + 0.25\vec{j}\right)\right) = 2\vec{i} + \vec{j} + 2\left(-0.25\vec{i} + \vec{j}\right) = 1.5\vec{i} + 3\vec{j}\ (m/s)}.$$

The acceleration is $\vec{a}_A = \vec{a}_O + \vec{a}_{Arel} + 2\vec{\omega} \times \vec{v}_{Arel} + \vec{\alpha} \times \vec{r}_{A/O} + \vec{\omega} \times (\vec{\omega} \times \vec{r}_{A/O})$.

Noting $\vec{a}_{Arel} = \left(\frac{d^2x}{dt^2}\right)\vec{i} + \left(\frac{d^2y}{dt^2}\right)\vec{j} + \left(\frac{d^2z}{dt^2}\right)\vec{k}$, where $\left(\frac{d^2x}{dt^2}\right) = 0$, $\frac{d^2y}{dt^2} = \frac{d^2}{dt^2} 0.25x^2 = 0.5\left(\frac{dx}{dt}\right)^2 = 2$, $\left(\frac{d^2z}{dt^2}\right) = 0$.

Substitute: $\vec{a}_A = 2\vec{j} + 2\omega\left(\vec{k} \times \left(2\vec{i} + \vec{j}\right)\right) + \omega^2\left(\vec{k} \times \left(\vec{k} \times \left(\vec{i} + 0.25\vec{j}\right)\right)\right) = \begin{bmatrix} \vec{i} & \vec{j} & \vec{k} \\ 0 & 0 & 4 \\ 2 & 1 & 0 \end{bmatrix} + 4\vec{k} \times \begin{bmatrix} \vec{i} & \vec{j} & \vec{k} \\ 0 & 0 & 1 \\ 1 & 0.25 & 0 \end{bmatrix}$

$$\vec{a}_A = 2\vec{j} - 4\vec{i} + 4\omega\vec{j} + 4\begin{bmatrix} \vec{i} & \vec{j} & \vec{k} \\ 0 & 0 & 1 \\ -0.25 & 1 & 0 \end{bmatrix} = \boxed{-8\vec{i} + 9\vec{j}\ (m/s^2)}$$

====================================<>====================================

==================================<>==================================

Problem 6.146 Suppose that in the instant shown in Problem 6.145, $x_A = 1\,m$, $\frac{dx_A}{dt} = -3\ m/s$, $\frac{d^2x_A}{dt^2} = 4\ m/s^2$, and the plate's angular velocity and angular acceleration are $\vec{\omega} = -4\vec{j} + 2\vec{k}\ (rad/s)$, and $\vec{\alpha} = 3\vec{i} - 6\vec{j}\ (rad/s^2)$. What are the *x,y,z* components of the velocity and acceleration of A relative to a non rotating reference frame that is stationary with respect to O?

Solution:

The velocity is $\vec{v}_A = \vec{v}_O + \vec{v}_{Arel} + \vec{\omega}\times\vec{r}_{A/O}$. The relative velocity is $\vec{v}_{Arel} = \left(\frac{dx}{dt}\right)\vec{i} + \left(\frac{dy}{dt}\right)\vec{j} + \left(\frac{dz}{dt}\right)\vec{k}$, where $\frac{dx}{dt} = -3\ m/s$, $\frac{dy}{dt} = \frac{d}{dt}0.25x^2 = 0.5x\frac{dx}{dt} = -1.5\ m/s$, $\frac{dz}{dt} = 0$, and $\vec{r}_{A/O} = x\vec{i} + y\vec{j} + z\vec{j} = \vec{i} + 0.25\vec{j} + 0$, from which $\vec{v}_A = -3\vec{i} - 1.5\vec{j} + \vec{\omega}\times(\vec{i} + 0.25\vec{j})$.

$$\vec{v}_A = -3\vec{i} - 1.5\vec{j} + \begin{bmatrix} \vec{i} & \vec{j} & \vec{k} \\ 0 & -4 & 2 \\ 1 & 0.25 & 0 \end{bmatrix} = -3\vec{i} - 1.5\vec{j} - 0.5\vec{i} + 2\vec{j} + 4\vec{k}\ \boxed{= -3.5\vec{i} + 0.5\vec{j} + 4\vec{k}\ (m/s)}$$

The acceleration is $\vec{a}_A = \vec{a}_O + \vec{a}_{Arel} + 2\vec{\omega}\times\vec{v}_{Arel} + \vec{\alpha}\times\vec{r}_{A/O} + \vec{\omega}\times(\vec{\omega}\times\vec{r}_{A/O})$. Noting $\vec{a}_{Arel} = \left(\frac{d^2x}{dt^2}\right)\vec{i} + \left(\frac{d^2y}{dt^2}\right)\vec{j} + \left(\frac{d^2z}{dt^2}\right)\vec{k}$, where $\left(\frac{d^2x}{dt^2}\right) = 4\ m/s^2$, $\frac{d^2y}{dt^2} = \frac{d^2}{dt^2}0.25x^2 = 0.5\left(\frac{dx}{dt}\right)^2 + 0.5x\left(\frac{d^2x}{dt^2}\right) = 6.5\ (m/s^2)$, $\left(\frac{d^2z}{dt^2}\right) = 0$, $\vec{\alpha} = 3i - 6\vec{j}\ (rad/s^2)$, $\vec{v}_{Arel} = -3\vec{i} - 1.5\vec{j}$, and from above: $\vec{\omega}\times\vec{r}_{A/O} = -05\vec{i} + 2\vec{j} + 4\vec{k}$. Substitute:

$$\vec{a}_A = 4\vec{i} + 6.5\vec{j} + 2\begin{bmatrix} \vec{i} & \vec{j} & \vec{k} \\ 0 & -4 & 2 \\ -3 & -1.5 & 0 \end{bmatrix} + \begin{bmatrix} \vec{i} & \vec{j} & \vec{k} \\ 3 & -6 & 0 \\ 1 & 0.25 & 0 \end{bmatrix} + \begin{bmatrix} \vec{i} & \vec{j} & \vec{k} \\ 0 & -4 & 2 \\ -0.5 & 2 & 4 \end{bmatrix}.$$

$$\vec{a}_A = 4\vec{i} + 6.5\vec{j} + 2(3\vec{i} + 6\vec{j} - 12\vec{k}) + (6.75\vec{k}) + (-20\vec{i} - \vec{j} - 2\vec{k})\ \boxed{\vec{a}_A = -10\vec{i} - 6.5\vec{j} - 19.25\vec{k}\ (m/s^2)}$$

==================================<>==================================

Problem 6.147 The coordinate system shown is fixed relative to the ship B. At the instant shown, the ship is sailing north at 10 *ft/s* relative to the earth and its angular velocity is 0.02 *rad/s* clockwise. The airplane is flying east at 400 *ft/s* relative to the earth, and its position relative to the ship is $\vec{r}_{A/B} = 2000\vec{i} + 2000\vec{j} + 1000\vec{k}\ (ft)$. If the ship uses its radar to measure the plane's velocity relative to its body fixed coordinates, what is the result?

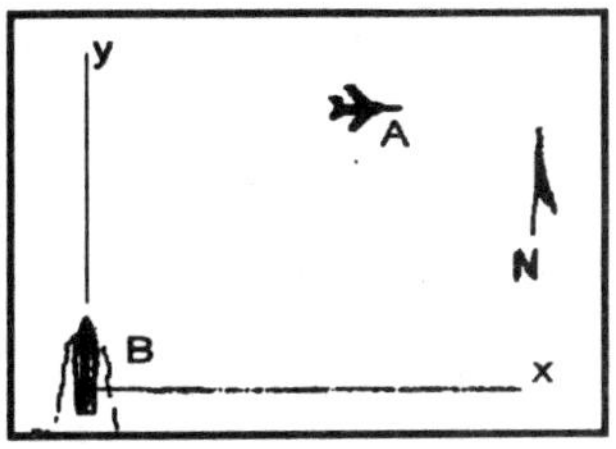

Solution:

The relative velocity is

$$\vec{v}_{Arel} = \vec{v}_A - \vec{v}_B - \vec{\omega}\times\vec{r}_{A/B}, = 400\vec{i} - 10\vec{j} + 0.02(\vec{k}\times(2000\vec{i} + 2000\vec{j} + 100\vec{k}))\ (ft/s)$$

$\boxed{\vec{v}_{Arel} = 360\vec{i} + 30\vec{j}\ (ft/s)}$.

==================================<>==================================

====================================<>====================================

Problem 6.148 The space shuttle is attempting to recover a satellite for repair. At the current time. the satellite's position relative to a coordinate system fixed to the shuttle is $50\vec{i}\ (m)$. The rate gyros on the shuttle indicate that its current angular velocity is $0.05\vec{j}+0.03\vec{k}\ (rad/s)$. The shuttle pilot measures the velocity of the satellite relative to the body fixed coordinate system and determines it to be $-2\vec{i}-1.5\vec{j}+2.5\vec{k}\ (m/s)$. What are the *x,y,z* components of the satellite's velocity relative to a non rotating coordinate system with its origin at the shuttle?

Solution:

The velocity of the satellite is $\vec{v}_A=\vec{v}_B+\vec{v}_{Arel}+\vec{\omega}\times\vec{r}_{A/B}=0-2\vec{i}-1.5\vec{j}+2.5\vec{k}+\begin{bmatrix}\vec{i} & \vec{j} & \vec{k}\\ 0 & 0.05 & 0.03\\ 50 & 0 & 0\end{bmatrix}$

$$\boxed{=-2\vec{i}+1.5\vec{j}+2.5\vec{k}-1.5\vec{j}-2.5\vec{k}=-2\vec{i}\ (m/s)}$$

====================================<>====================================

Problem 6.149 The train on the circular track is traveling at a constant speed of 50 *ft/s* in the direction shown. The train on the straight track is traveling at 20 *ft/s* in the direction shown and is increasing its speed at 2 *ft/s*^{2}. Determine the velocity of passenger A that passenger B observes relative to the coordinate system shown, which is fixed to the car in which B is riding.

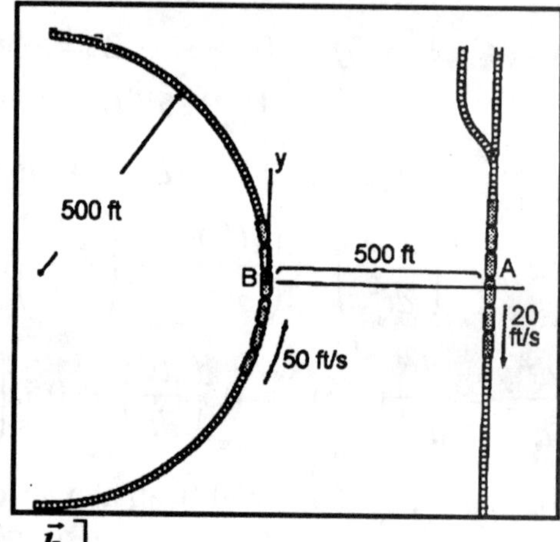

Solution:

The angular velocity of B is $\omega=\dfrac{50}{500}=0.1\ rad/s$.

The velocity of A is $\vec{v}_A=\vec{v}_B+\vec{v}_{Arel}+\vec{\omega}\times\vec{r}_{A/B}$.

At the instant shown, $\vec{v}_A=-20\vec{j}\ (ft/s)$, $\vec{v}_B=+50\vec{j}\ (ft/s)$,

and $\vec{r}_{A/B}=500\vec{i}\ (ft)$, from which $\vec{v}_{Arel}=-20\vec{j}-50\vec{j}-\begin{bmatrix}\vec{i} & \vec{j} & \vec{k}\\ 0 & 0 & 0.1\\ 500 & 0 & 0\end{bmatrix}=-20\vec{j}-50\vec{j}-50\vec{j}$,

$$\boxed{\vec{v}_{Arel}=-120\vec{j}\ (ft/s)}$$

====================================<>====================================

Problem 6.150 In Problem 6.149, determine the acceleration of passenger A that passenger B observes relative to the coordinate system fixed to the car in which B is riding.

Solution:

Use the solution to Problem 6.149: $\vec{v}_{Arel}=-120\vec{j}\ (ft/s)$, $\omega=\dfrac{50}{500}=0.1\ rad/s$, $\vec{r}_{A/B}=500\vec{i}\ (ft)$.

The acceleration of A is $\vec{a}_A=-2\vec{j}\ (ft/s)$,. The acceleration of B is $\vec{a}_B=-500(\omega^2)\vec{i}=-5\vec{i}\ ft/s^2$,

and $\vec{\alpha}=0$, from which $\vec{a}_A=\vec{a}_B+\vec{a}_{Arel}+2\vec{\omega}\times\vec{v}_{Arel}+\vec{\omega}\times(\vec{\omega}\times\vec{r}_{A/B})$. Rearrange:

$\vec{a}_{Arel}=\vec{a}_A-\vec{a}_B-2\vec{\omega}\times\vec{v}_{Arel}-\vec{\omega}\times(\vec{\omega}\times\vec{r}_{A/B})$.

Solution continued on next page

$$\vec{a}_{Rel} = -2\vec{j} + 5\vec{i} - 2\begin{bmatrix} \vec{i} & \vec{j} & \vec{k} \\ 0 & 0 & \omega \\ 0 & -120 & 0 \end{bmatrix} - \omega^2\left(\vec{k} \times \begin{bmatrix} \vec{i} & \vec{j} & \vec{k} \\ 0 & 0 & 1 \\ 500 & 0 & 0 \end{bmatrix}\right).$$

$$\vec{a}_{Rel} = -2\vec{j} + 5\vec{i} - 2(120\omega)\vec{i} - \omega^2\begin{bmatrix} \vec{i} & \vec{j} & \vec{k} \\ 0 & 0 & 1 \\ 0 & 500 & 0 \end{bmatrix}, \boxed{\vec{a}_{Arel} = -14\vec{i} - 2\vec{j}\ \left(ft/s^2\right)}.$$

=================================<>=================================

Problem 6.151 The satellite A is in a circular polar orbit (a circular orbit that intersects to poles). The radius of the orbit is R, and the magnitude of the satellite's velocity relative to a non rotating reference frame with its origin at the center of the earth is v_A. At the instant shown, the satellite is above the equator. A observer B on the equator directly below the satellite measures its motion using the earth-fixed coordinate system shown. What are the velocity and acceleration of the satellite relative B's earth fixed coordinate system? The radius of the earth is R_E and its angular velocity is ω_E.

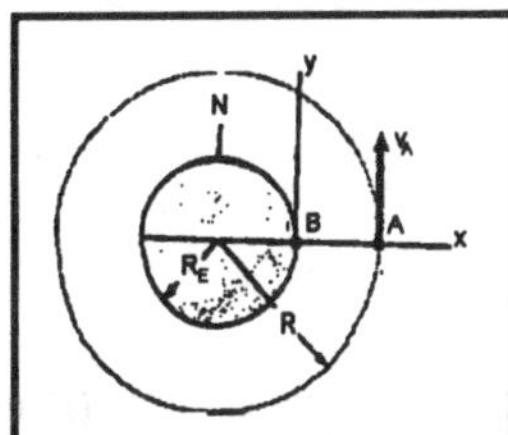

Solution: From the sketch, in the coordinate system shown, the location of the satellite in this system is $\vec{r}_A = (R - R_E)\vec{i}$, from which $\vec{r}_{A/B} = \vec{r}_A - 0 = (R - R_E)\vec{i}$. The angular velocity of the observer is $\vec{\omega}_E = -\omega_E\vec{k}$. The velocity of the observer is $\vec{v}_B = -\omega_E R_E\vec{k}$. The velocity of the satellite is $\vec{v}_A = v_A\vec{j}$. The relative velocity is $\vec{v}_{Arel} = \vec{v}_A - \vec{v}_B - \vec{\omega}_E \times \vec{r}_{A/B}$, .

$$\vec{v}_{Arel} = v_A\vec{j} + \omega_E R_E\vec{k} - (\omega_E)\begin{bmatrix} \vec{i} & \vec{j} & \vec{k} \\ 0 & 1 & 0 \\ R - R_E & 0 & 0 \end{bmatrix} \boxed{= v_A\vec{j} + R\omega_E\vec{k}}.$$ From Eqs (6.26) and (6.27)

$\vec{a}_{Arel} = \vec{a}_A - \vec{a}_B - 2\vec{\omega}_E \times \vec{v}_{Arel} - \vec{\alpha} \times \vec{r}_{A/B} - \vec{\omega}_E \times (\vec{\omega}_E \times \vec{r}_{A/B})$. The accelerations:

$\vec{a}_A = -\omega_A^2 R\vec{i} = -\left(\frac{v_A^2}{R}\right)\vec{i}$, $\vec{a}_B = -\omega_E^2 R_E\vec{i}$, $\vec{\alpha} = 0$, from which

$$\vec{a}_{Arel} = -\left(\frac{v_A^2}{R}\right) + \vec{i}\,\omega_E^2 R_E - 2\begin{bmatrix} \vec{i} & \vec{j} & \vec{k} \\ 0 & \omega_E & 0 \\ 0 & v_A & R\omega_E \end{bmatrix} - \vec{\omega}_E \times \begin{bmatrix} \vec{i} & \vec{j} & \vec{k} \\ 0 & \omega_E & 0 \\ R - R_E & 0 & 0 \end{bmatrix}.$$

$$\vec{a}_{Arel} = -\omega_A^2 R\vec{i} + \omega_E^2 R_E\vec{i} - 2\omega_E^2 R\vec{i} - \begin{bmatrix} \vec{i} & \vec{j} & \vec{k} \\ 0 & \omega_E & 0 \\ 0 & 0 & -\omega_E(R - R_E) \end{bmatrix}$$

$$= -\left(\frac{v_A^2}{R}\right)\vec{i} + \omega_E^2 R_E\vec{i} - 2\omega_E^2 R\vec{i} + \omega_E^2(R - R_E)\vec{i}\ \boxed{\vec{a}_{Arel} = -\left(\left(\frac{v_A^2}{R}\right) + \omega_E^2 R\right)\vec{i}}$$

=================================<>=================================

====================================<>====================================

Problem 6.152 A car A at north latitude L drives north on a north-south highway with constant speed v. The earth's radius is R_E and its angular ω_E. Determine the *x,y,z* components of the car's velocity and acceleration relative to the coordinate system shown, (a) if the coordinate system is earth-fixed, and (b) if the coordinate system does not rotate.

Solution:

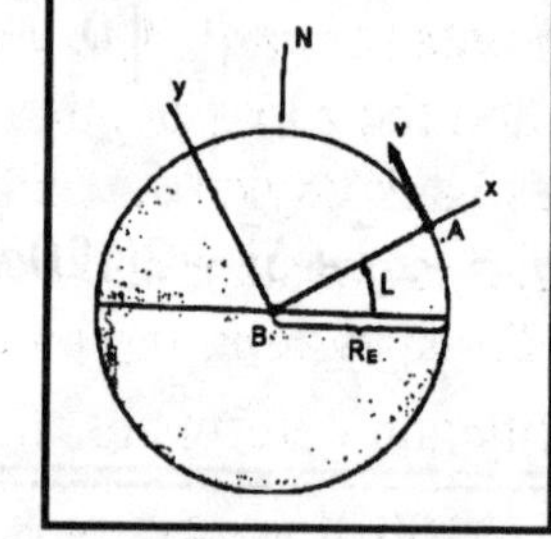

(a) By inspection, the velocity of point A in the fixed coordinate system is $\vec{v}_A = v_{Arel}\vec{j}$. *Check:* This is confirmed as follows: The velocity of the car in the non rotating earth fixed system is $\vec{v}_A = \vec{v}_B + \vec{v}_{Arel} + \vec{\omega} \times \vec{r}_{A/B}$. The velocity $\vec{v}_B = 0$, since the origin is not translating. The rotation of the earth fixed coordinate system shown is $\vec{\omega} = \omega_E \cos L\vec{j}$.

The velocity of the car in the non rotating earth centered system is $\vec{v}_A = v_A\vec{j} - R_E\omega_E \cos L\vec{k}$, from which $\vec{v}_{Arel} = \vec{v}_A - \vec{\omega} \times \vec{r}_{A/B} = v_A\vec{j} - R_E\omega_E \cos L\vec{k} - \begin{bmatrix} \vec{i} & \vec{j} & \vec{k} \\ 0 & \omega_E \cos L & 0 \\ R_E & 0 & 0 \end{bmatrix} = v_A\vec{j}$. *check.* The acceleration of the point A is $\vec{a}_A = \vec{a}_B + \vec{a}_{Arel} + \vec{\alpha} \times \vec{r}_{A/B} - \vec{\omega} \times (\vec{\omega} \times R\vec{i})$, The acceleration in the non rotating earth centered system is $\vec{a}_A = -(v_A^2 / R_E + R_E\omega_E^2 \cos^2 L)\vec{i}$. The angular velocity of rotation of the earth fixed system is $\vec{\omega} = \omega_E \cos L\vec{j}$, and $\vec{\alpha} = 0$, from which

$\vec{a}_{Arel} = -\left(\frac{v_A^2}{R_E} + R_E\omega_E^2 \cos^2 L\right)\vec{i} + (R_E\omega_E^2 \cos^2 L)\vec{i}$, $\boxed{\vec{a}_{Arel} = -\frac{v_A^2}{R_E}\vec{i}}$ (b) The angular velocity of the earth fixed relative to the rotating system is $\vec{\omega} = \omega_E \sin L\vec{i} + \omega_E \cos L\vec{j}$ *rad / s*. The velocity of the car in the non rotating system is $\vec{v}_A - \vec{v}_B = \vec{v}_{Rel} + \vec{\omega} \times \vec{r}_{A/B} = v_A\vec{j} + \begin{bmatrix} \vec{i} & \vec{j} & \vec{k} \\ \omega_E \sin L & \omega_E \cos L & 0 \\ R_E & 0 & 0 \end{bmatrix}$

$\vec{v}_A - \vec{v}_B = v_A\vec{j} - R_E\omega_E \cos L\vec{k}$. The acceleration in the non rotating system is $\vec{a}_A - \vec{a}_B = \vec{a}_{Arel} + 2\vec{\omega} \times \vec{v}_{Arel} + \vec{\omega} \times (\vec{\omega} \times r_{A/B})$. Expanding term by term: From above, the apparent acceleration in the rotating frame is: $\vec{a}_{Arel} = -\left(\frac{v_A^2}{R_E}\right)\vec{i}$, $2\vec{\omega} \times \vec{v}_{Rel} = 2\begin{bmatrix} \vec{i} & \vec{j} & \vec{k} \\ \omega_E \sin L & \omega_E \cos L & 0 \\ 0 & v_A & 0 \end{bmatrix}$,

$2\vec{\omega} \times \vec{v}_{Arel} = (2v_A\omega_E \sin L)\vec{k}$.

$$\vec{\omega} \times (\vec{\omega} \times \vec{r}_{A/O}) = \vec{\omega} \times \begin{bmatrix} \vec{i} & \vec{j} & \vec{k} \\ \omega_E \sin L & \omega_E \cos L & 0 \\ R_E & 0 & 0 \end{bmatrix} = \begin{bmatrix} \vec{i} & \vec{j} & \vec{k} \\ \omega_E \sin L & \omega_E \cos L & 0 \\ 0 & 0 & -R_E\omega_E \cos L \end{bmatrix}$$

Expanding and collecting terms:

$$\boxed{\vec{a}_A = \left(-R_E\omega_E^2 \cos^2 L - \frac{v_A^2}{R_E}\right)\vec{i} + (R_E\omega_E^2 \sin L \cos L)\vec{j} + (2v_A\omega_E \sin L)\vec{k}}$$

====================================<>====================================

===================================<>===================================

Problem 6.153 The airplane B conducts flight tests of a missile. At the instant shown, the airplane is traveling at 200 *m/s* relative to the earth in a circular path of 2000 *m in the horizontal plane.* The coordinate system is fixed relative to the airplane. The *x* axis is tangent to the plane's path and points forward. The *y* axis points out the airplane's right side, and the *z* axis points out the bottom of the airplane. The plane's bank angle (the inclination of the *z* axis from the vertical) is constant and equal to 20°. *Relative to the airplane's coordinate system,* the pilot measures the missile's position and velocity and determines them to be $\vec{r}_{A/B} = 1000\vec{i}\ (m)$ and

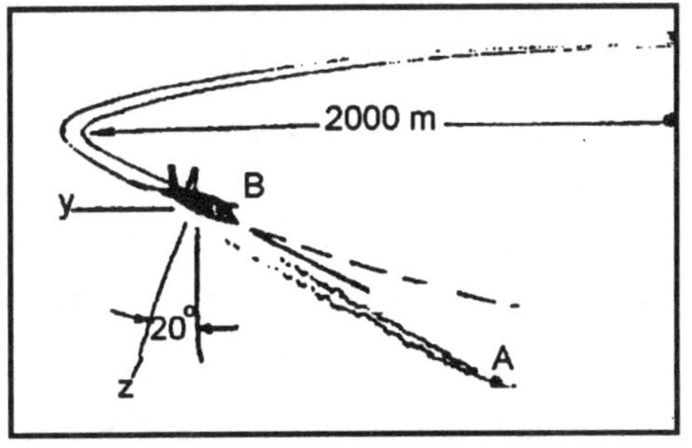

$\vec{v}_{A/B} = 100.0\vec{i} + 94.0\vec{j} + 34.2\vec{k}\ (m/s)$. (a) What are the *x,y,z* components of the airplane's angular velocity vector? b) What are the *x,y,z* components of the missile's velocity relative to the earth?

Solution:

(a) The bank angle is a rotation about the *x* axis; assume that the rotation is counterclockwise, so that the *z* axis is rotated toward the positive *y* axis. The magnitude of the angular velocity is $\omega = \dfrac{200}{2000} = 0.1\ rad/s$. In terms of airplane fixed coordinates,

$\vec{\omega} = 0.1\left(\vec{i}\sin 20^o - \vec{j}\sin 20^o\right)(rad/s)$. $\boxed{\vec{\omega} = 0.03242\vec{j} - 0.0940\vec{k}\ \ rad/s}$

(b) The velocity of the airplane in earth fixed coordinates is

$$\vec{v}_A = \vec{v}_B + \vec{v}_{Arel} + \vec{\omega}\times\vec{r}_{A/B} = 200\vec{i} + 100\vec{i} + 94.0\vec{j} + 34.2\vec{k} + \begin{bmatrix} \vec{i} & \vec{j} & \vec{k} \\ 0 & 0.0342 & -0.940 \\ 1000 & 0 & 0 \end{bmatrix}$$

$\boxed{\vec{v}_A = 300\vec{i} + 94.0\vec{j} + 34.2\vec{k} - 94.0\vec{j} - 34.2\vec{k} = 300\vec{i}\ (m/s)}$

===================================<>===================================

Problem 6.154 To conduct experiments related to long term space flight, engineers construct a laboratory on earth that rotates about the vertical axis at B with a constant angular velocity ω of one revolution every 6 seconds. They establish a laboratory fixed coordinate system with origin at B and the *z* axis upward. An engineer holds an object stationary relative to the laboratory at point A, 3 meters from the axis of rotation, holds an object at rest relative to him, and releases it. At the instant he drops the object, determine its acceleration relative to the laboratory fixed coordinate system, (a) assuming that the laboratory fixed system is inertial; (b) by not assuming that the laboratory system is inertial, but assuming that an earth fixed coordinate system with origin at B is inertial.

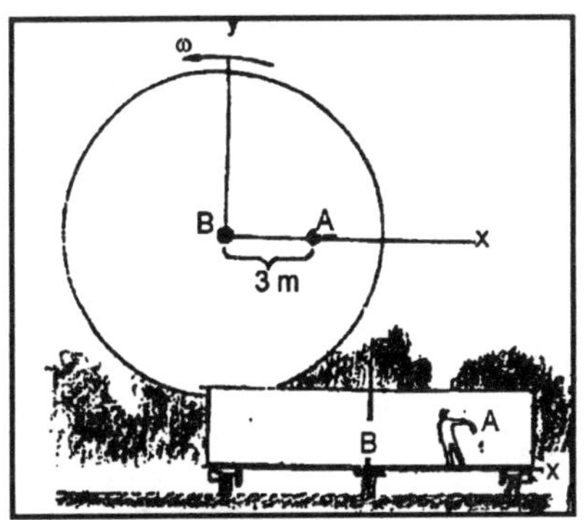

Solution: (a) If the laboratory system is inertial, Newton's second law is $\vec{F} = m\vec{a}$. The only force is the force of gravity; so that as the object free falls the acceleration is $\boxed{-g\vec{k} = -9.81\vec{k}\ \left(m/s^2\right)}$.

If the earth fixed system is inertial, the acceleration observed is the centripetal acceleration and the acceleration of gravity: $\vec{\omega}\times(\vec{\omega}\times\vec{r}_B) - \vec{g}$, where the angular velocity is the angular velocity of the

Solution continues on next page

coordinate system relative to the inertial frame.

$$\vec{\omega}\times(\vec{\omega}\times\vec{r}_{A/B})-\vec{g}=-\left(\frac{2\pi}{6}\right)^2\left(\vec{k}\times\begin{bmatrix}\vec{i} & \vec{j} & \vec{k}\\ 0 & 0 & 1\\ 3 & 0 & 0\end{bmatrix}\right)-9.81\vec{k}=\left(\frac{\pi^2}{3}\right)\vec{i}-9.81\vec{k}\;\boxed{=3.29\vec{i}-9.81\vec{k}\;\left(m/s^2\right)}$$

================================<>================================

Problem 6.155 A disk *lying in the horizontal plane* rotates about a fixed shaft at the origin with constant angular velocity ω. The slider A of mass m moves in a smooth slot in the disk. The spring is unstretched when $x=0$. (a) By using Eq. (6.20) to express Newton's second law in terms of a body-fixed coordinate system, show that the slider's motion is governed by the equation $\frac{d^2x}{dt^2}+\left(\frac{k}{m}-\omega^2\right)x=0$. (b) The Slider is given an initial velocity $\frac{dx}{dt}=v_O$ at $x=0$. Determine its velocity as a function of *x*.

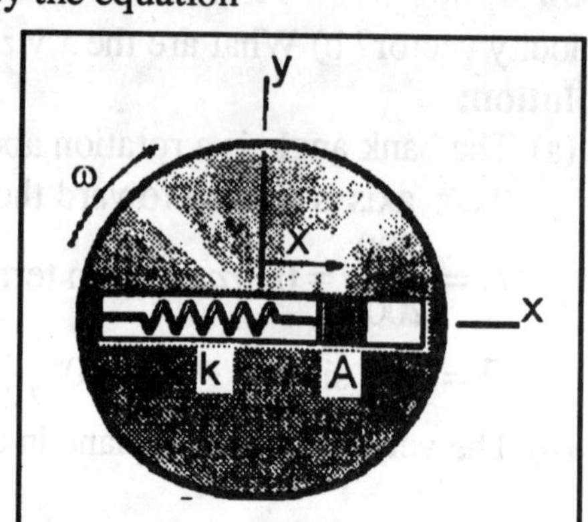

Solution:

Using Eq (6.28):

$$\sum\vec{F}-m\left[\vec{a}_B+2\vec{\omega}\times\vec{v}_{Arel}+\vec{\alpha}\times\vec{r}_{A/B}+\vec{\omega}\times(\vec{\omega}\times\vec{r}_{A/B})\right]=m\vec{a}_{Arel}.$$

The acceleration of the origin is $\vec{a}_B=0$. The relative velocity is zero, and the angular acceleration of the fixed coordinate system is

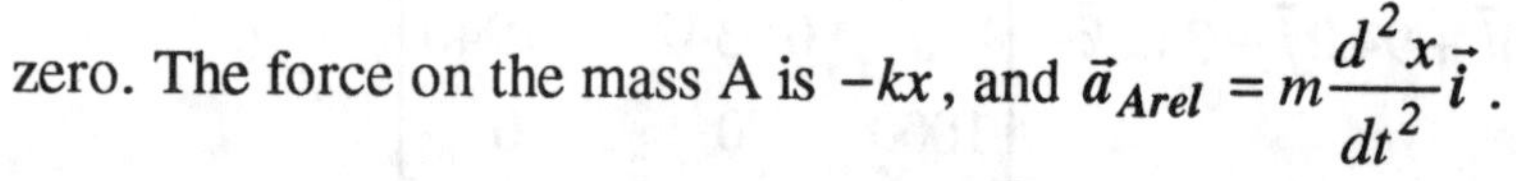

zero. The force on the mass A is $-kx$, and $\vec{a}_{Arel}=m\frac{d^2x}{dt^2}\vec{i}$.

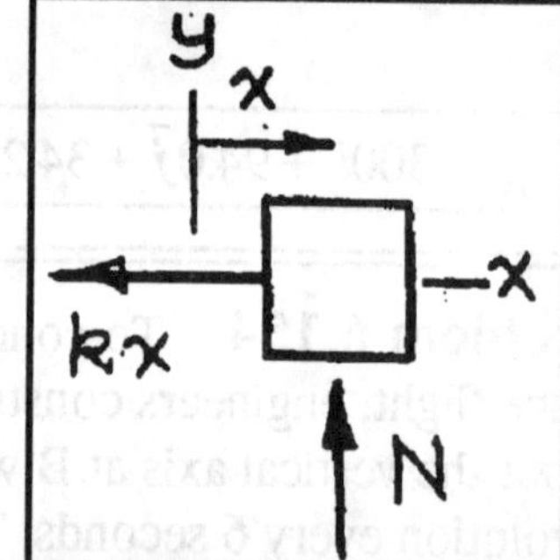

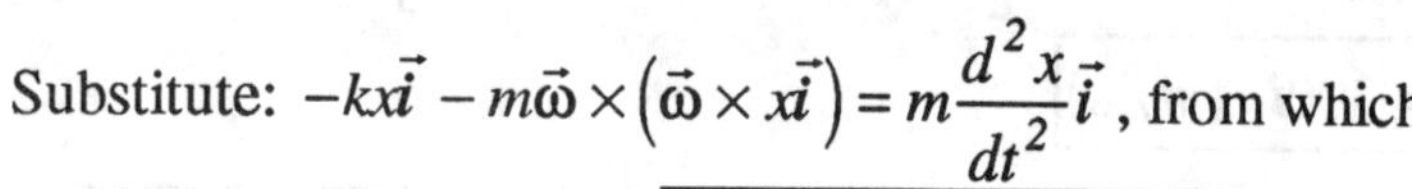

Substitute: $-kx\vec{i}-m\vec{\omega}\times\left(\vec{\omega}\times x\vec{i}\right)=m\frac{d^2x}{dt^2}\vec{i}$, from which

$m\frac{d^2x}{dt^2}+kx-m\omega^2x=0$, $\boxed{\frac{d^2x}{dt^2}+\left(\frac{k}{m}-\omega^2\right)x=0}$ (b) Use the chain rule to write $\frac{d^2x}{dt^2}=\frac{d}{dt}\left(\frac{dx}{dt}\right)=\frac{dv}{dt}=\frac{dv}{dx}\frac{dx}{dt}=v\frac{dv}{dx}$, from which

$v\frac{dv}{dx}=-\left(\frac{k}{m}-\omega^2\right)x$. Integrate: $\frac{v^2}{2}=-\left(\frac{k}{m}-\omega^2\right)\frac{x^2}{2}+C$. At $x=0$, $v=v_0$, from which $C=\frac{v_0^2}{2}$, and

$v^2(x)=-\left(\frac{k}{m}-\omega^2\right)x^2+v_0^2$, $v=\pm\sqrt{v_0^2-\left(\frac{k}{m}-\omega^2\right)x^2}$. Choose the positive sign since it satisfies the condition at $x=0$, $\boxed{v=\sqrt{v_0^2-\left(\frac{k}{m}-\omega^2\right)x^2}}$

================================<>================================

=================================<>=================================

Problem 6.156 Engineers conduct flight tests of a rocket at 30° north latitude. They measure the rocket's motion using an earth fixed coordinate system with x axis upward and the y axis northward. At a particular instant, the mass of the rocket is 4000 kg, its velocity is relative to their coordinate system is $2000\vec{i}+2000\vec{j}\ (m/s)$, and the sum of the forces exerted on the rocket by its thrust, weight and aerodynamic forces is $400\vec{i}+400\vec{j}\ (N)$. Determine the rocket's acceleration relative to their coordinate system (a) assume that their earth fixed coordinate system is inertial; (b) do not assume that their earth fixed coordinate system is inertial.

N
y
x
30°

Solution:

Use Eq (6.22): $\sum\vec{F}-m[\vec{a}_B+2\vec{\omega}\times\vec{v}_{Arel}+\vec{\alpha}\times\vec{r}_{A/B}+\vec{\omega}\times(\vec{\omega}\times\vec{r}_{A/B})]=m\vec{a}_{Arel}$.

(a) If the earth fixed coordinate system is assumed to be inertial, this reduces to $\sum\vec{F}=m\vec{a}_{Arel}$, from which $\vec{a}_{Arel}=\frac{1}{m}\sum\vec{F}=\frac{1}{4000}(400\vec{i}+400\vec{j})\boxed{=0.1\vec{i}+0.1\vec{j}\ (m/s^2)}$ (b) If the earth fixed system is not assumed to be inertial, $\vec{a}_B=-R_E\omega_E^2\cos^2\lambda\vec{i}+R_E\omega_E^2\cos\lambda\sin\lambda\vec{j}$, the angular velocity of the rotating coordinate system is $\vec{\omega}=\omega_E\sin\lambda\vec{i}+\omega_E\cos\lambda\vec{j}\ (rad/s)$. The relative velocity in the earth fixed system is $\vec{v}_{Arel}=2000\vec{i}+2000\vec{j}\ (m/s)$, and $\vec{r}_{A/B}=R_E\vec{i}\ (m)$.

$$2\vec{\omega}\times\vec{v}_{Arel}=2\omega_E\begin{bmatrix}\vec{i} & \vec{j} & \vec{k}\\ \sin\lambda & \cos\lambda & 0\\ 2000 & 2000 & 0\end{bmatrix}=4000\omega_E(\sin\lambda-\cos\lambda)\vec{k}$$

$$\vec{\omega}\times(\vec{\omega}\times\vec{r}_{A/B})=\vec{\omega}\times\begin{bmatrix}\vec{i} & \vec{j} & \vec{k}\\ \omega_E\sin\lambda & \omega_E\cos\lambda & 0\\ R_E & 0 & 0\end{bmatrix}=\vec{\omega}\times(-R_E\omega_E\cos\lambda)\vec{k}$$

$$\vec{\omega}\times(-R_E\omega_E\cos\lambda)\vec{k}=R_E\omega_E^2\cos\lambda\begin{bmatrix}\vec{i} & \vec{j} & \vec{k}\\ \sin\lambda & \cos\lambda & 0\\ 0 & 0 & -1\end{bmatrix}=(-R_E\omega_E^2\cos^2\lambda)\vec{i}+(R_E\omega_E^2\cos\lambda\sin\lambda)\vec{j}$$. Collect

terms, $\vec{a}_{Arel}=+(R_E\omega_E^2\cos^2\lambda)\vec{i}-(R_E\omega_E^2\cos\lambda\sin\lambda)\vec{j}-4000\omega_E(\sin\lambda-\cos\lambda)\vec{k}+0.1\vec{i}+0.1\vec{j}$.

Substitute values: $R_E=6336\times10^3\ m$, $\omega_E=0.73\times10^{-4}\ rad/s$, $\lambda=30^o$,

$\boxed{\vec{a}_{Arel}=0.125\vec{i}+0.0854\vec{j}+0.1069\vec{k}\ (m/s^2)}$ *Note:* The last two terms in the parenthetic expression for $\vec{a}_A$ in $\sum\vec{F}-m[\vec{a}_B+2\vec{\omega}\times\vec{v}_{Arel}+\vec{\alpha}\times\vec{r}_{A/B}+\vec{\omega}\times(\vec{\omega}\times\vec{r}_{A/B})]=m\vec{a}_{Arel}$ can be neglected without significant change in the answers.

=================================<>=================================

Problem 6.157 Consider a point A on the surface of the earth at north latitude L. The radius of the earth is R_E and its angular velocity is ω_E. A plumb bob suspended just above the ground at point A will hang at a small angle β relative to the vertical because of the earth's rotation. Show that β is related to the latitude by

$$\tan\beta = \frac{\omega_E^2 R_E \sin L \cos L}{g - \omega_E^2 R_E \cos^2 L}.$$

Strategy: Use the earth-fixed coordinate system shown, express Newton's second law in the form given by Eq. (6.27).

Solution:

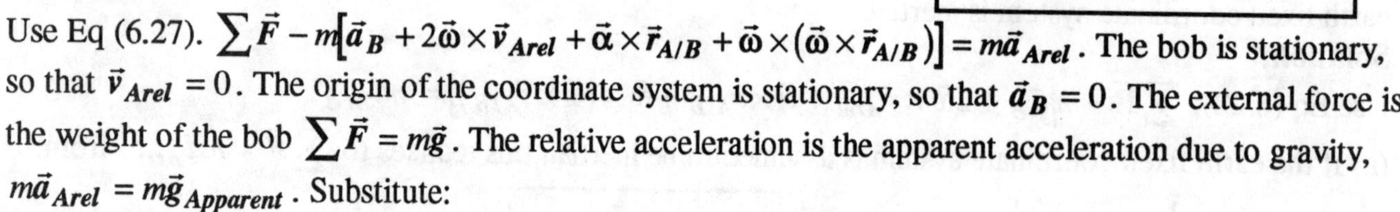

Use Eq (6.27). $\sum \vec{F} - m\left[\vec{a}_B + 2\vec{\omega}\times\vec{v}_{Arel} + \vec{\alpha}\times\vec{r}_{A/B} + \vec{\omega}\times\left(\vec{\omega}\times\vec{r}_{A/B}\right)\right] = m\vec{a}_{Arel}$. The bob is stationary, so that $\vec{v}_{Arel} = 0$. The origin of the coordinate system is stationary, so that $\vec{a}_B = 0$. The external force is the weight of the bob $\sum \vec{F} = m\vec{g}$. The relative acceleration is the apparent acceleration due to gravity, $m\vec{a}_{Arel} = m\vec{g}_{Apparent}$. Substitute:

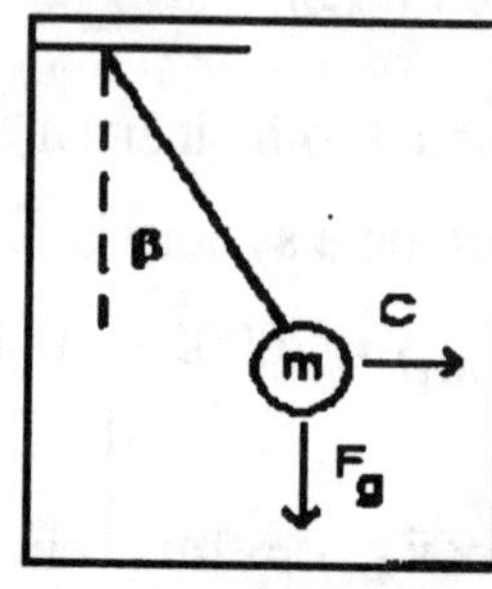

$$\vec{g}_{Apparent} = \vec{g} - \vec{\omega}\times\left(\vec{\omega}\times\vec{r}_{A/B}\right) = g\vec{i} - \vec{\omega}\times\begin{bmatrix} \vec{i} & \vec{j} & \vec{k} \\ \omega_E \sin L & \omega_E \cos L & 0 \\ R_E & 0 & 0 \end{bmatrix}$$

$$\vec{g}_{Apparent} = g\vec{i} - \vec{\omega}\times\left(-R_E\omega_E \cos L\right)\vec{k} = g\vec{i} + R_E\omega_E^2 \cos L \begin{bmatrix} \vec{i} & \vec{j} & \vec{k} \\ \sin L & \cos L & 0 \\ 0 & 0 & -1 \end{bmatrix}$$

$$\vec{g}_{Apparent} = g\vec{i} - \left(R_E\omega_E^2 \cos^2 L\right)\vec{i} + \left(R_E\omega_E^2 \cos L \sin L\right)\vec{j}.$$

The vertical component of the apparent acceleration due to gravity is $g_{vertical} = g - R_E\omega_E^2 \cos^2 L$. The horizontal component of the apparent acceleration due to gravity is $g_{horizontal} = R_E\omega_E^2 \cos L \sin L$. From equation of angular motion, the moments about the bob suspension are $M_{vertical} = (\lambda \sin\beta) m g_{vertical}$ and $M_{horizontal} = (\lambda\cos\beta) m g_{horizontal}$, where λ is the length of the bob, and m is the mass of the bob. In equilibrium, $M_{vertical} = M_{horizontal}$, from which $g_{vertical}\sin\beta = g_{horizontal}\cos\beta$. Substitute and rearrange:

$$\boxed{\tan\beta = \frac{g_{horizontal}}{g_{vertical}} = \frac{R_E\omega_E^2 \cos L \sin L}{g - R_E\omega_E^2 \cos^2 L}}$$

==================================<>==================================

Problem 6.158 Suppose that a space station is in orbit around the earth and two astronauts on the station toss a ball back and forth. They observe that the ball *appears* to travel between them in a straight line at constant velocity. (a) Write Newton's second law for the ball as it travels between them in terms of a non rotating coordinate system that is stationary relative to the station. What is the term $\sum F$? Use the equation to explain the behavior of the ball. (b) Write Newton's second law for the ball as it travels between them in terms of a non rotating coordinate system that is stationary relative to the center of the earth. What is the term $\sum \vec{F}$? Explain the difference between this equation and the one you obtained in part (a).

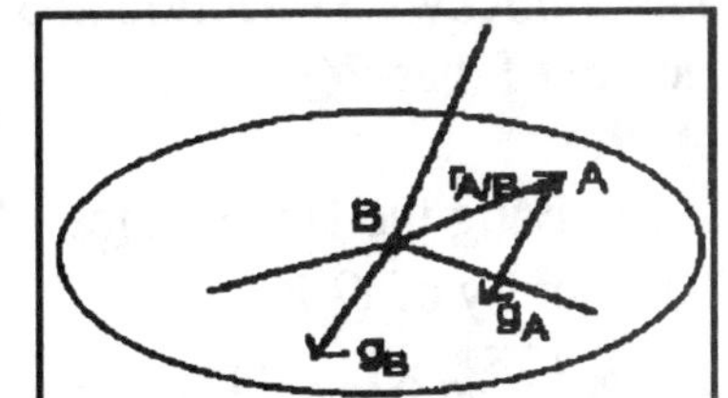

Solution :

An earth centered, non rotating coordinate system can be treated as inertial for analyzing the motions of objects near the earth (See Section 6.2.) Let O be the reference point of this reference frame, and let B be the origin of the non rotating reference frame fixed to the space station, and let A denote the ball. The orbiting station and its contents and the station-fixed non rotating frame are in free fall about the earth (they accelerate relative to the earth due to the earth's gravitational attraction), so that the forces on the ball in the fixed reference frame *exclude the earth's gravitational attraction.* Let $\vec{g}_B$ be the station's acceleration, and let $\vec{g}_A$ be the ball's acceleration relative to the earth due to the earth's gravitational attraction. Let $\sum F$ be the sum of all forces on the ball, *not including the earth's gravitational attraction.* Newton's second law for the ball of mass *m* is $\sum \vec{F} + m\vec{g}_A = m\vec{a}_A = m(\vec{a}_B + \vec{a}_{A/B}) = m\vec{g}_B + m\vec{a}_{A/B}$. Since the ball is within a space station whose dimensions are small compared to the distance from the earth, $\vec{g}_A$ is equal to $\vec{g}_B$ within a close approximation , from which $\sum \vec{F} = m\vec{a}_{A/B}$. The sum of the forces on the ball *not including the force exerted by the earth's gravitational attraction* equals the mass times the ball's acceleration relative to a reference frame fixed with respect to the station. As the astronauts toss the ball back and forth, the only other force on it is aerodynamic drag. Neglecting aerodynamic drag, $\vec{a}_{A/B} = 0$, from which the ball will travel in a straight line with constant velocity. (b) Relative to the earth centered non rotating reference frame, Newton's second law for the ball is $\sum \vec{F} = m\vec{a}_A$ where $\sum \vec{F}$ is the sum of all forces on the ball, including aerodynamic drag *and the force due to the earth's gravitational attraction.* Neglect drag, from which $\vec{a}_A = \vec{g}_A$; the ball's acceleration is its acceleration due to the earth's gravitational attraction, because in this case we are determining the ball's motion relative to the earth. *Note*: An obvious unstated assumption is that the time of flight of the ball as it is tossed between the astronauts is much less than the period of an orbit. Thus the very small acceleration differences $\vec{g}_A - \vec{g}_B$ will have a negligible effect on the path of the ball over the short time interval.

==================================<>==================================

==================================<>==================================

Problem 6.159 If $\theta = 60^\circ$ and bar OQ is rotating in the counterclockwise direction at *5 rad/s*, what is the angular velocity of bar PQ?

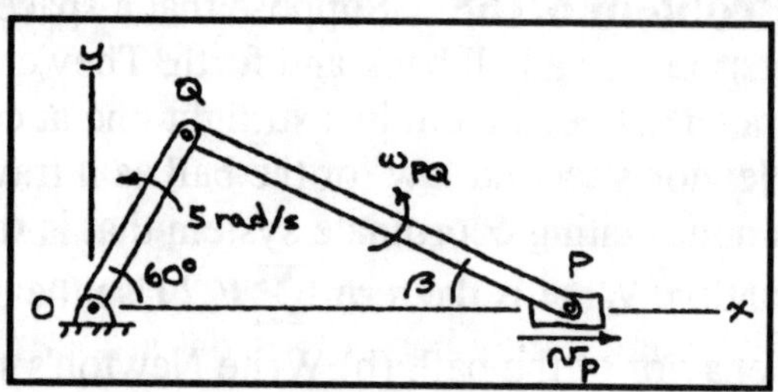

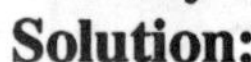

Solution:

By applying the law of sines, $\beta = 25.7^\circ$.

The velocity of Q is

$\bar{v}_Q = \bar{v}_0 + \bar{\omega}_{0Q} \times \bar{r}_{Q/0}$ or

$$\bar{v}_Q = 0 + \begin{vmatrix} \bar{i} & \bar{j} & \bar{k} \\ 0 & 0 & 5 \\ 0.2\cos 60^\circ & 0.2\sin 60^\circ & 0 \end{vmatrix} = -\sin 60^\circ \bar{i} + \cos 60^\circ \bar{j}.$$

The velocity of P is

$$v_p \bar{i} = \bar{v}_Q + \bar{\omega}_{PQ} \times \bar{r}_{P/Q} = -\sin 60^\circ \bar{i} + \cos 60^\circ \bar{j} + \begin{vmatrix} \bar{i} & \bar{j} & \bar{k} \\ 0 & 0 & \omega_{PQ} \\ 0.4\cos\beta & -0.4\sin\beta & 0 \end{vmatrix}.$$

Equating $\bar{i}$ and $\bar{j}$ components $v_P = -\sin 60^\circ + 0.4\omega_{PQ}\sin\beta$, and $0 = \cos 60^\circ + 0.4\omega_{PQ}\cos\beta$.

Solving, we obtain $v_P = -1.11\ m/s$ and $\omega_{PQ} = -1.39 rad/s$

==================================<>==================================

Problem 6.160 Consider the system shown in Problem 6.159. If $\theta = 55^\circ$ and the sleeve P is moving to the left at *2 m/s*, what is the angular velocity of the bar PQ ?

Solution:

By applying the law of sines, $\beta = 24.2^\circ$The velocity of Q is

$$\bar{v}_Q = \bar{v}_0 + \bar{\omega}_{0Q} \times \bar{r}_{Q/0} = 0 + \begin{vmatrix} \bar{i} & \bar{j} & \bar{k} \\ 0 & 0 & \omega_{0Q} \\ 0.2\cos 55^\circ & 0.2\sin 55^\circ & 0 \end{vmatrix} = -0.2\omega_{0Q}\sin 55^\circ \bar{i} + 0.2\omega_{0Q}\cos 55^\circ \bar{j} (1)$$

We can also express $\bar{v}_Q$ as $\bar{v}_Q = \bar{v}_p + \bar{\omega}_{PQ} \times \bar{r}_{Q/P} = -2\bar{i} + \begin{vmatrix} \bar{i} & \bar{j} & \bar{k} \\ 0 & 0 & \omega_{PQ} \\ -0.4\cos\beta & 0.4\sin\beta & 0 \end{vmatrix}.$

Equating $\bar{i}$ and $\bar{j}$ components in Equations (1) and (2), we get $-0.2\omega_{0Q}\sin 55^\circ = -2 - 0.4\omega_{PQ}\sin\beta$, and $0.2\omega_{0Q}\cos 55^\circ = -0.4\omega_{PQ}\cos\beta$. Solving, we obtain $\omega_{0Q} = 9.29 rad/s$ $\omega_{PQ} = -2.92 rad/s$.

==================================<>==================================

=====================================<>=====================================

Problem 6.161 Determine the vertical velocity of the hook and the angular velocity of the small pulley.

Solution:

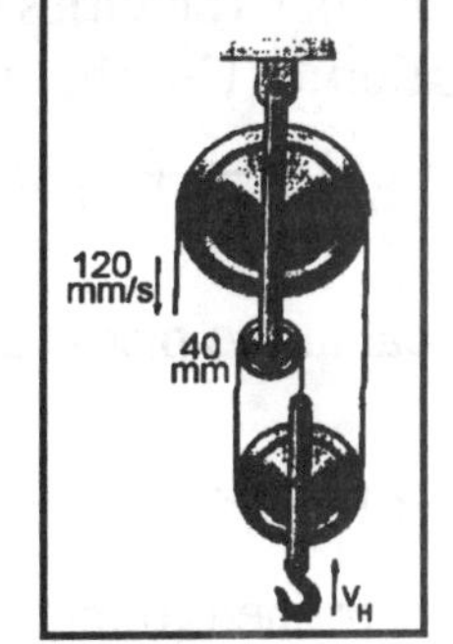

The upper pulley is fixed so that it cannot move, from which the upward velocity of the rope on the right is equal to the downward velocity on the left, and the upward velocity of the rope on the right of the lower pulley is 120 *mm/s.* The small pulley is fixed so that it does not move. The upward velocity on the right of the small pulley is $v_H\ mm/s$, from which the downward velocity on the left is $v_H\ mm/s$. The upward velocity of the center of the bottom pulley is the mean of the difference of the velocities on the right and left, from which $v_H = \frac{120 - v_H}{2}$. Solve, $\boxed{v_H = 40\ mm/s}$. The angular velocity of the small pulley is $\boxed{\omega = \frac{v_H}{R} = \frac{40}{40} = 1\ rad/s}$

=====================================<>=====================================

Problem 6.162 If the crankshaft AB is turning in the counterclockwise direction at 2000 rpm (revolutions per minute), what is the velocity of the piston?

Solution:

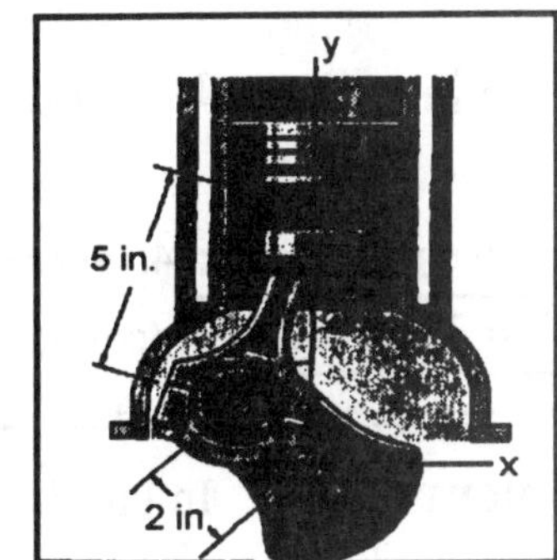

The angle of the crank with the vertical is 45°. The angular velocity of the crankshaft is $\omega = 2000\left(\frac{2\pi}{60}\right) = 209.44\ rad/s$. The vector location of point B (the main rod bearing)

$\vec{r}_B = 2\left(-\vec{i}\sin 45^o + \vec{j}\cos 45^o\right) = 1.414\left(-\vec{i} + \vec{j}\right)\ in.$ The velocity of point B (the main rod bearing) is

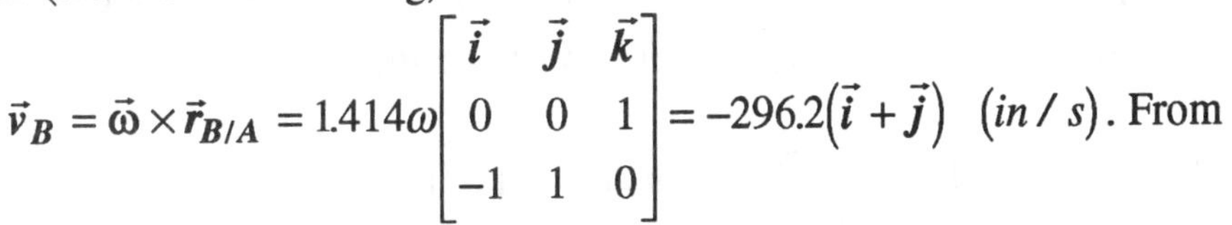

$$\vec{v}_B = \vec{\omega} \times \vec{r}_{B/A} = 1.414\omega\begin{bmatrix} \vec{i} & \vec{j} & \vec{k} \\ 0 & 0 & 1 \\ -1 & 1 & 0 \end{bmatrix} = -296.2\left(\vec{i} + \vec{j}\right)\ (in/s).$$

From the law if sines the interior angle between the connecting rod and the vertical at the piston is obtained from $\frac{2}{\sin\theta} = \frac{5}{\sin 45^o}$, from which $\theta = \sin^{-1}\left(\frac{2\sin 45^o}{5}\right) = 16.43^o$. The location of the piston is

$\vec{r}_C = \left(2\sin 45^o + 5\cos\theta\right)\vec{j} = 6.21\vec{j}\ (in.)$. The vector $\vec{r}_{B/C} = \vec{r}_B - \vec{r}_C = -1.414\vec{i} - 4.796\vec{j}\ (in.)$ The piston is constrained to move along the *y* axis. In terms of the connecting rod the velocity of the point B is

$$\vec{v}_B = \vec{v}_C + \vec{\omega}_{BC} \times \vec{r}_{B/C} = v_C\vec{j} + \begin{bmatrix} \vec{i} & \vec{j} & \vec{k} \\ 0 & 0 & \omega_{BC} \\ -14.14 & -4.796 & 0 \end{bmatrix} = v_C\vec{j} + 4.796\omega_{BC}\vec{i} - 1.414\omega_{BC}\vec{j}\ (in/s).$$

Equate expressions for $\vec{v}_B$ and separate components: $-296.2 = 4.796\omega_{BC}$, $-296.2 = v_C - 1.414\omega_{BC}$.

Solve: $\boxed{\vec{v}_C = -383.5\vec{j}\ (in/s) = -32\vec{j}\ (ft/s)}$, $\omega_{BC} = -61.8\ rad/s$.

=====================================<>=====================================

==============================<>==============================

Problem 6.163 In Problem 6.162, if the piston is moving with velocity $\vec{v}_C = 20\vec{j}\ (ft/s)$, what are the angular velocities of the crankshaft AB and the connecting rod BC?

Solution: Use the solution to Problem 6.162. The vector location of point B (the main rod bearing) $\vec{r}_B = 1.414(-\vec{i}+\vec{j})\ in.$. From the law if sines the interior angle between the connecting rod and the vertical at the piston is $\theta = \sin^{-1}\left(\frac{2\sin 45^o}{5}\right) = 16.43^o$. The location of the piston is

$\vec{r}_C = \left(2\sin 45^o + 5\cos\theta\right)\vec{j} = 6.21\vec{j}\ (in.)$. The vector $\vec{r}_{B/C} = \vec{r}_B - \vec{r}_C = -1.414\vec{i} - 4.796\vec{j}\ (in.)$ The piston is constrained to move along the y axis. In terms of the connecting rod the velocity of the point B is

$$\vec{v}_B = \vec{v}_C + \vec{\omega}_{BC} \times \vec{r}_{B/C} = 240\vec{j} + \begin{bmatrix} \vec{i} & \vec{j} & \vec{k} \\ 0 & 0 & \omega_{BC} \\ -1.414 & -4.796 & 0 \end{bmatrix}$$

$\vec{v}_B = 240\vec{j} + 4.796\omega_{BC}\vec{i} - 1.414\omega_{BC}\vec{j}\ (in/s)$.

In terms of the crank angular velocity, the velocity of point B is

$\vec{v}_B = \vec{\omega}_{AB} \times \vec{r}_{B/A} = 1.414\omega_{AB}\begin{bmatrix} \vec{i} & \vec{j} & \vec{k} \\ 0 & 0 & 1 \\ -1 & 1 & 0 \end{bmatrix} = -1.414\omega_{AB}\left(\vec{i}+\vec{j}\right)\ (in/s)$. Equate expressions and separate components: $4.796\omega_{BC} = -1.414\omega_{AB}$, $240 - 1.414\omega_{BC} = -1.414\omega_{AB}$. Solve:

$\boxed{\omega_{BC} = 38.65\ rad/s}$ (counterclockwise). $\boxed{\omega_{AB} = -131.1\ rad/s = -1251.5\ rpm}$ (clockwise).

==============================<>==============================

Problem 6.164 In Problem 6.162, if the piston is moving with velocity $\vec{v}_C = 20\vec{j}\ (ft/s)$ and its acceleration is zero, what are the angular accelerations of crankshaft AB and connecting rod BC?

Solution:

Use the solution to Problem 6.163 .. $\vec{r}_{B/A} = 1.414(-\vec{i}+\vec{j})\ in.$, $\omega_{AB} = -131.1\ rad/s$, $\vec{r}_{B/C} = \vec{r}_B - \vec{r}_C = -1.414\vec{i} - 4.796\vec{j}\ (in.)$, $\omega_{BC} = 38.65\ rad/s$. For point B,

$$\vec{a}_B = \vec{a}_A + \vec{\alpha}_{AB} \times \vec{r}_{B/A} - \omega_{AB}^2\vec{r}_{B/A} = 1.414\alpha_{AB}\begin{bmatrix} \vec{i} & \vec{j} & \vec{k} \\ 0 & 0 & 1 \\ -1 & 1 & 0 \end{bmatrix} - 1.414\omega_{AB}^2\left(-\vec{i}+\vec{j}\right),$$

$\vec{a}_B = -1.414\alpha_{AB}\left(\vec{i}+\vec{j}\right) + 24291(\vec{i}-\vec{j})\ \left(in/s^2\right)$. In terms of the angular velocity of the connecting rod, $\vec{a}_B = \vec{a}_C + \vec{\alpha}_{BC} \times \vec{r}_{B/C} - \omega_{BC}^2\vec{r}_{B/C}$,

$$\vec{a}_B = \alpha_{BC}\begin{bmatrix} \vec{i} & \vec{j} & \vec{k} \\ 0 & 0 & 1 \\ -1.414 & -4.796 & 0 \end{bmatrix} - \omega_{BC}^2\left(-1.414\vec{i} - 4.796\vec{j}\right)\ \left(in/s^2\right),$$

$\vec{a}_B = 4.796\alpha_{BC}\vec{i} - 1.414\alpha_{BC}\vec{j} + 2112.3\vec{i} + 7163.0\vec{j}\ \left(in/s^2\right)$. Equate expressions and separate components: $-1.414\alpha_{AB} + 24291 = 4.796\alpha_{BC} + 2112.3$, $-1.414\alpha_{AB} - 24291 = -1.414\alpha_{BC} + 7163$.

Solve: $\boxed{\alpha_{AB} = -13{,}605\ rad/s^2}$ (clockwise). $\boxed{\alpha_{BC} = 8636.5\ rad/s^2}$ (counterclockwise).

==============================<>==============================

==================================<>==================================

Problem 6.165 The bar AB rotates at 6 *rad/s* in the counterclockwise direction. Use instantaneous centers to determine the angular velocity of bar BCD and the velocity of point D.

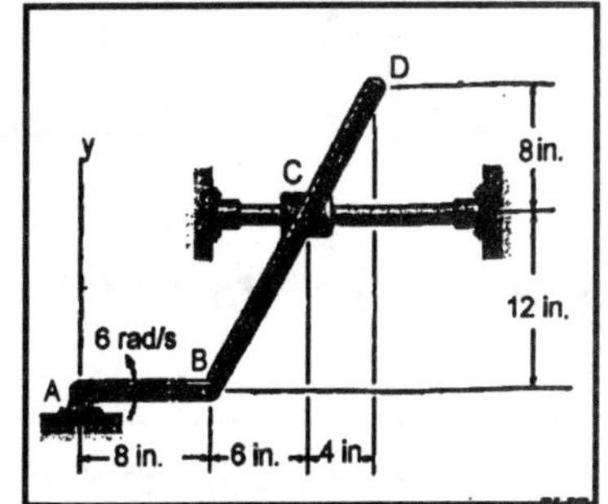

Solution: The strategy is to determine the angular velocity of bar BC from the instantaneous center; using the constraint on the motion of C. The vector $\vec{r}_{B/A} = \vec{r}_B - \vec{r}_A = (8\vec{i} + 4\vec{j}) - 4\vec{j} = 8\vec{i}$ (*in.*). The velocity of point B is $\vec{v}_B = \vec{\omega}_{AB} \times \vec{r}_{B/A} = \omega_{AB}(\vec{k} \times 8\vec{i}) = 48\vec{j}$ (*in/s*). The velocity of point B is normal to the *x* axis, and the velocity of C is parallel to the *x* axis. The instantaneous center of bar BC has the coordinates (14, 0). The vector $\vec{r}_{B/IC} = \vec{r}_B - \vec{r}_{IC} = (8\vec{i} + 4\vec{j}) - (14\vec{i} - 4\vec{j}) = -6\vec{i}$ (*in.*).

The velocity of point B is $\vec{v}_B = \vec{\omega}_{BC} \times \vec{r}_{B/IC} = \begin{bmatrix} \vec{i} & \vec{j} & \vec{k} \\ 0 & 0 & \omega_{BC} \\ -6 & 0 & 0 \end{bmatrix} = -6\omega_{BC}\vec{j} = 48\vec{j}$, from which

$\boxed{\omega_{BC} = -\frac{48}{6} = -8 \ rad/s}$. The velocity of point C is

$\vec{v}_C = \vec{\omega}_{BC} \times \vec{r}_{C/IC} = \begin{bmatrix} \vec{i} & \vec{j} & \vec{k} \\ 0 & 0 & \omega_{BC} \\ 0 & 12 & 0 \end{bmatrix} = 96\vec{i}$ (*in/s*). The velocity of point D is normal to the unit vector parallel to BCD $\vec{e}_{CD} = \frac{6\vec{i} + 12\vec{j}}{\sqrt{6^2 + 12^2}} = 0.4472\vec{i} + 0.8944\vec{j}$. The intersection of the projection of this unit vector with the projection of the unit vector normal to velocity of C is occurs at point C, from which the coordinates of the instantaneous center for the part of the bar CD are (14,12). The instantaneous center is translating at velocity $\vec{v}_C$, from which the velocity of point D is

$\boxed{\vec{v}_D = \vec{v}_C + \vec{\omega}_{BC} \times \vec{r}_{D/ICD} = 96\vec{i} - 8(\vec{k} \times (4\vec{i} + 8\vec{j})) = 160\vec{i} - 32\vec{j} \ (in/s)}$

==================================<>==================================

Problem 6.166 In Problem 6.165, bar AB rotates with a constant angular acceleration of 6 *rad/s* in the counterclockwise direction. Determine the angular acceleration of point D.

Solution:

Use the solution to Problem 6.165. The accelerations are determined from the angular velocity, the known accelerations of B, and the constraint on the motion of C. The vector $\vec{r}_{B/A} = \vec{r}_B - \vec{r}_A = 8\vec{i}$ (*in.*).

The acceleration of point B is $\vec{a}_B = \vec{\alpha}_{AB} \times \vec{r}_{B/A} - \omega^2\vec{r}_{A/B} = 0 - 36(8\vec{i}) = -288\vec{i}$ (in/s^2). From the solution to Problem 6.165, $\omega_{BC} = -8 \ rad/s$. (clockwise).

$\vec{r}_{C/B} = \vec{r}_C - \vec{r}_B = (14\vec{i} + 12\vec{j}) - (8\vec{i}) = 6\vec{i} + 12\vec{j}$ (*in.*). The vector $\vec{r}_{B/C} = -\vec{r}_{C/B}$ The acceleration of point B in terms of the angular acceleration of point C is

$$\vec{a}_B = \vec{a}_C + \vec{\alpha}_{BC} \times \vec{r}_{BC} - \omega_{BC}^2\vec{r}_{B/C} = a_C\vec{i} + \begin{bmatrix} \vec{i} & \vec{j} & \vec{k} \\ 0 & 0 & \alpha_{BC} \\ -6 & -12 & 0 \end{bmatrix} - 64(-6\vec{i} - 12\vec{j}),$$

Solution continued on next page

$\vec{a}_B = a_C\vec{i} + 12\alpha_{BC}\vec{i} - 6\alpha_{BC}\vec{j} + 384\vec{i} + 768\vec{j}$. Equate the expressions and separate components: $-288 = a_C + 12\alpha_{BC} + 384$, $0 = -6\alpha_{BC} + 768$. Solve $\alpha_{BC} = 128 \ rad/s^2$ (counterclockwise), $a_C = -2208 \ in/s^2$. The acceleration of point D is

$$\vec{a}_D = \vec{a}_C + \vec{\alpha}_{BC} \times \vec{r}_{D/C} - \omega_{BC}^2\vec{r}_{D/C} = -2208\vec{i} + \begin{bmatrix} \vec{i} & \vec{j} & \vec{k} \\ 0 & 0 & \alpha_{BC} \\ 4 & 8 & 0 \end{bmatrix} - \omega_{BC}^2(4i + 8\vec{j}).$$

$$\boxed{\vec{a}_D = -2208\vec{i} + (128)(-8\vec{i} + 4\vec{j}) - (64)(4\vec{i} + 8\vec{j}) = -3488\vec{i} \ (in/s^2)}$$

==================================<>==================================

Problem 6.167 Point C is moving to the right at 20 *in/s*. What is the velocity of the midpoint G of bar BC?

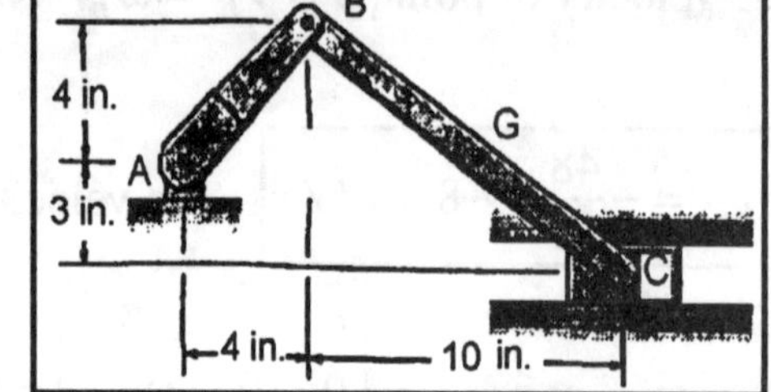

Solution:

Choose a coordinate system with origin at A and *y* vertical. The vector location of a point on a line perpendicular to the velocity at B is $\vec{L}_{AB} = L_{AB}\left(\frac{1}{\sqrt{2}}\right)(\vec{i} + \vec{j})$, where L_{AB} is the magnitude of the distance from point A to the point on the line.

The vector location of a point on a perpendicular to the velocity at C is $\vec{L}_C = (14\vec{i} + y\vec{j})$ where *y* is the y-coordinate of the point referenced to an origin at A. When the two lines intersect, $\frac{L_{AB}}{\sqrt{2}} = 14$, from which $L_{AB} = 19.8 \ in.$, and the coordinates of the instantaneous center C' are $(14, 14)$ $(in.)$. [*Check:* The line AC' is the hypotenuse of a right triangle with base 14 *in.* and interior angles of 45°. *check.*] The angular velocity of bar BC is determined from the known velocity at C.

$$\vec{v}_C = 20\vec{i} = \vec{\omega}_{BC} \times \vec{r}_{C/C'} = \begin{bmatrix} \vec{i} & \vec{j} & \vec{k} \\ 0 & 0 & \omega_{BC} \\ 0 & -17 & 0 \end{bmatrix} = 17\omega_{BC}\vec{i}$$, from which $\omega_{BC} = 1.176 \ rad/s$. The interior angle at C formed by BC is $\theta = \tan^{-1}\left(\frac{7}{10}\right) = 35^o$. The length of the bar BC is

$L_{BC} = \sqrt{7^2 + 10^2} = 12.2 \ in.$. The vector location of point G is

$$\vec{r}_G = \left(14 - \frac{L_{BC}}{2}\cos\theta\right)\vec{i} + \left(-3 + \frac{L_{BC}}{2}\sin\theta\right)\vec{j} = 9\vec{i} + 0.5\vec{j} \ (in.)$$. The vector

$\vec{r}_{G/C} = \vec{r}_G - \vec{r}_C = -5\vec{i} + 3.5\vec{j} \ (in.)$. The velocity at G is

$$\vec{v}_G = \vec{v}_C + \vec{\omega}_{BC} \times \vec{r}_{G/C} = 20\vec{i} + \omega_{BC}\left(\vec{k} \times (-5\vec{i} + 3.5\vec{j})\right) = 20\vec{i} - 3.5\omega_{BC}\vec{i} - 5\omega_{BC}\vec{j} \ (in/s),$$

$$\boxed{\vec{v}_G = 15.88\vec{i} - 5.88\vec{j} \ (in/s)}$$

==================================<>==================================

==================================<>==================================

Problem 6.168 In Problem 6.167, point C is moving to the right with a constant velocity of 20 *in/s*. What is the acceleration of the midpoint G of bar BC?

Solution:

The strategy is the determine the angular acceleration of the bar BC from the known conditions, from which the acceleration of point G is determined. Use the solution to Problem 6.47: The angular velocity of bar BC is $\omega_{BC} = 1.176 \ rad/s$. The velocity of the point B is determined from the known velocity of point C and the known angular velocity of BC:

$$\vec{v}_B = \vec{v}_C + \vec{\omega}_{BC} \times \vec{r}_{B/C} = 20\vec{i} + \begin{bmatrix} \vec{i} & \vec{j} & \vec{k} \\ 0 & 0 & \omega_{BC} \\ -10 & 7 & 0 \end{bmatrix} = 20\vec{i} - 7\omega_{BC}\vec{i} - 10\omega_{BC}\vec{j}.$$

$\vec{v}_B = 11.77\vec{i} - 11.77\vec{j} \ (in/s)$.

The angular velocity of bar AB is determined from the known velocity of B.

$$\vec{v}_B = \vec{\omega}_{AB} \times \vec{r}_{B/A} = \begin{bmatrix} \vec{i} & \vec{j} & \vec{k} \\ 0 & 0 & \omega_{AB} \\ 4 & 4 & 0 \end{bmatrix} = -4\omega_{AB}(i - \vec{j}) \ (in/s)$$ Equate expressions, separate components,

and solve: $\omega_{AB} = -\frac{11.77}{4} = -2.941 \ rad/s$. The acceleration of B relative to A is

$$\vec{a}_B = \vec{\alpha}_{AB} \times \vec{r}_{B/A} - \omega_{AB}^2 \vec{r}_{B/A} = \begin{bmatrix} \vec{i} & \vec{j} & \vec{k} \\ 0 & 0 & \alpha_{AB} \\ 4 & 4 & 0 \end{bmatrix} - \omega_{AB}^2(4\vec{i} + 4\vec{j}), \ \vec{a}_B = 4\alpha_{AB}(-\vec{i} + \vec{j}) - 4\omega_{AB}^2(\vec{i} + \vec{j}).$$

The acceleration of B relative to C is

$$\vec{a}_B = \vec{\alpha}_{BC} \times \vec{r}_{B/C} - \omega_{BC}^2 \vec{r}_{B/C} = \begin{bmatrix} \vec{i} & \vec{j} & \vec{k} \\ 0 & 0 & \alpha_{BC} \\ -7 & -10 & 0 \end{bmatrix} - \omega_{BC}^2(-10\vec{i} + 7\vec{j}).$$

Equate the two expressions and separate the components: $-4\alpha_{AB} - 4\omega_{AB}^2 = -7\alpha_{BC} + 10\omega_{BC}^2$., $4\alpha_{AB} - 4\omega_{AB}^2 = 7\alpha_{AB} - 10\omega_{BC}^2$, where ω_{AB}, and ω_{BC} are known. Solve: $\alpha_{AB} = -4.56 \ rad/s^2$, and $\alpha_{BC} = 4.315 \ rad/s^2$. The vector location of the point G is $\vec{r}_G = 9\vec{i} + 0.5\vec{j} \ (in.)$. The vector $\vec{r}_{G/C} = \vec{r}_G - \vec{r}_C = -5\vec{i} + 3.5\vec{j} \ (in.)$. The acceleration of point C is zero, from which the acceleration of point G relative to C is

$$\vec{a}_G = \vec{\alpha}_{BC} \times \vec{r}_{G/C} - \omega_{BC}^2 \vec{r}_{G/C} = \begin{bmatrix} \vec{i} & \vec{j} & \vec{k} \\ 0 & 0 & \alpha_{BC} \\ -5 & 3.5 & 0 \end{bmatrix} - \omega_{BC}^2(-5\vec{i} + 3.5j),$$

$$\boxed{\vec{a}_G = \alpha_{BC}(-3.5\vec{i} - 5\vec{j}) - \omega_{BC}^2(-5i + 3.5\vec{i}) = -8.18\vec{i} - 26.42\vec{j} \ (in/s^2)}.$$

==================================<>==================================

================================<>================================

Problem 6.169 In Problem 6.167, if the velocity of point C is $\vec{v}_C = 1.0\vec{i}$ (in/s), what are the angular velocity vectors of arms AB and BC.?

Solution:

Use the solution to Problem 6.167: The velocity of the point B is determined from the known velocity of point C and the known velocity of C:

$\vec{v}_B = \vec{v}_C + \vec{\omega}_{BC} \times \vec{r}_{B/C} = 1.0\vec{i} + \begin{bmatrix} \vec{i} & \vec{j} & \vec{k} \\ 0 & 0 & \omega_{BC} \\ -10 & 7 & 0 \end{bmatrix} = 1.0\vec{i} - 7\omega_{BC}\vec{i} - 10\omega_{BC}\vec{j}$. The angular velocity of bar AB is determined from the velocity of B.

$\vec{v}_B = \vec{\omega}_{AB} \times \vec{r}_{B/A} = \begin{bmatrix} \vec{i} & \vec{j} & \vec{k} \\ 0 & 0 & \omega_{AB} \\ 4 & 4 & 0 \end{bmatrix} = -4\omega_{AB}(i - \vec{j})$ (in/s) Equate expressions, separate components,

$1.0 - 7\omega_{BC} = -4\omega_{AB}$, $-10\omega_{BC} = 4\omega_{AB}$. Solve: $\omega_{AB} = -0.147$ rad/s, $\omega_{BC} = 0.0588$ rad/s, from which $\boxed{\vec{\omega}_{AB} = -0.147\vec{k}\ (rad/s)}$. $\boxed{\vec{\omega}_{BC} = 0.0588\vec{k}\ (rad/s)}$

================================<>================================

Problem 6.170 Points B and C are in the *x-y* plane. The angular velocity vectors of arms AB and BC are $\vec{\omega}_{AB} = -0.5\vec{k}$ (rad/s), $\vec{\omega}_{BC} = 2.0\vec{k}$ (rad/s). Determine the velocity of point C.

Solution: The vector

$\vec{r}_{B/A} = 760(\vec{i}\cos 15^o - \vec{j}\sin 15^o) = 734.1\vec{i} - 196.7\vec{j}$ (mm). The vector

$\vec{r}_{C/B} = 900(\vec{i}\cos 50^o + \vec{j}\sin 50^o) = 578.5\vec{i} + 689.4\vec{j}$ (mm). The velocity of point B is

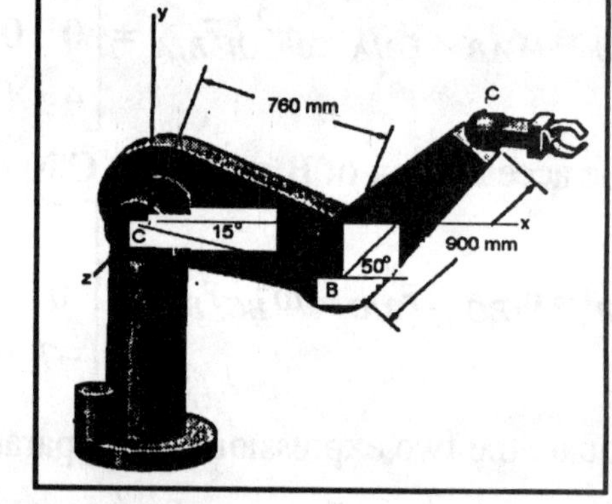

$$\vec{v}_B = \vec{\omega}_{AB} \times \vec{r}_{B/A} = -0.5\begin{bmatrix} \vec{i} & \vec{j} & \vec{k} \\ 0 & 0 & 1 \\ 734.1 & -196.7 & 0 \end{bmatrix}$$

$\vec{v}_B = -98.35\vec{i} - 367.1\vec{j}$ (mm/s). The velocity of point C $\vec{v}_C = \vec{v}_B + \vec{\omega}_{BC} \times \vec{r}_{C/B}$

$= -98.35\vec{i} - 367.1\vec{j} + (2)(\vec{k} \times (578.5\vec{i} + 689.4\vec{j}))$,

$\boxed{\vec{v}_C = -98.35\vec{i} - 367.1\vec{j} - 1378.9\vec{i} + 1157.0\vec{j} = -1477.2\vec{i} + 790\vec{j}\ (mm/s)}$

================================<>================================

=====================================<>=====================================

Problem 6.171 In Problem 6.170, if the velocity vector of point C is $\vec{v}_C = 1.0\vec{i}\ (m/s)$, what are the angular velocity vectors of arms AB and BC?

Solution:

Use the solution to Problem 6.150. The vector

$\vec{r}_{B/A} = 760(\vec{i}\cos 15^o - \vec{j}\sin 15^o) = 734.1\vec{i} - 196.7\vec{j}\ (mm)$. The vector

$\vec{r}_{C/B} = 900(\vec{i}\cos 50^o + \vec{j}\sin 50^o) = 578.5\vec{i} + 689.4\vec{j}\ (mm)$. The velocity of point B is

$$\vec{v}_B = \vec{\omega}_{AB} \times \vec{r}_{B/A} = \begin{bmatrix} \vec{i} & \vec{j} & \vec{k} \\ 0 & 0 & \omega_{AB} \\ 734.1 & -196.7 & 0 \end{bmatrix} = 196.7\omega_{AB}\vec{i} + 734.1\omega_{AB}\vec{j}\ (mm/s)$$. The velocity of

point C is $$\vec{v}_C = \vec{v}_B + \vec{\omega}_{BC} \times \vec{r}_{C/B} = 196.7\omega_{AB}\vec{i} + 734.1\omega_{AB}\vec{j} + \begin{bmatrix} \vec{i} & \vec{j} & \vec{k} \\ 0 & 0 & \omega_{BC} \\ 578.5 & 687.4 & 0 \end{bmatrix},$$

$1000\vec{i} = 196.7\omega_{AB}\vec{i} + 734.1\omega_{AB}\vec{j} - 687.4\omega_{BC}\vec{i} + 578.5\omega_{BC}\vec{j}\ (mm/s)$. Separate components:

$1000 = 196.7\omega_{AB} - 687.4\omega_{BC}$, $0 = 734.1\omega_{AB} + 578.5\omega_{BC}$. Solve: $\boxed{\vec{\omega}_{AB} = 0.933\vec{k}\ (rad/s)}$,

$\boxed{\vec{\omega}_{BC} = -1.184\vec{k}\ (rad/s)}$

=====================================<>=====================================

Problem 6.172 In Problem 6.170, if the angular velocity vectors of arms AB and BC are $\vec{\omega}_{AB} = -0.5\vec{k}\ (rad/s)$, $\vec{\omega}_{BC} = 2.0\vec{k}\ (rad/s)$ and their angular accelerations are $\vec{\alpha}_{AB} = 1.0\vec{k}\ (rad/s^2)$, $\vec{\alpha}_{BC} = 1.0\vec{k}\ (rad/s^2)$, what is the acceleration of point C?

Solution:

Use the solution to Problem 6.170. The vector

$\vec{r}_{B/A} = 760(\vec{i}\cos 15^o - \vec{j}\sin 15^o) = 734.1\vec{i} - 196.7\vec{j}\ (mm)$. The vector

$\vec{r}_{C/B} = 900(\vec{i}\cos 50^o + \vec{j}\sin 50^o) = 578.5\vec{i} + 689.4\vec{j}\ (mm)$. The acceleration of point B is

$$\vec{a}_B = \vec{\alpha}_{AB} \times \vec{r}_{B/A} - \omega_{AB}^2\vec{r}_{B/A} = -0.5\begin{bmatrix} \vec{i} & \vec{j} & \vec{k} \\ 0 & 0 & 1 \\ 734.1 & -196.7 & 0 \end{bmatrix} - (0.5^2)(734.1\vec{i} - 196.7\vec{j})\ (mm/s^2).$$

$\vec{a}_B = 196.7\vec{i} + 734.1\vec{j} - 183.5\vec{i} + 49.7\vec{j} = 13.2\vec{i} + 783.28\vec{j}\ (mm/s^2)$. The acceleration of point C is

$$\vec{a}_C = \vec{a}_B + \vec{\alpha}_{BC} \times \vec{r}_{C/B} - \omega_{BC}^2\vec{r}_{C/B} = \vec{a}_B + \begin{bmatrix} \vec{i} & \vec{j} & \vec{k} \\ 0 & 0 & 1 \\ 578.5 & 689.4 & 0 \end{bmatrix} - (2^2)(578.5\vec{i} + 689.4\vec{j})$$

$\vec{a}_C = 13.2\vec{i} + 783.3\vec{j} - 689.4\vec{i} + 578.5\vec{j} - 2314\vec{i} - 2757.8\vec{j}\ (mm/s^2)$.

$\boxed{\vec{a}_C = -2990\vec{i} - 1396\vec{j}\ (mm/s^2)}$

=====================================<>=====================================

==================================<>==================================

Problem 6.173 In Problem 6.170 if the velocity of point C is $\vec{v}_C = 1.0\vec{i}\ (m/s)$ and $\vec{a}_C = 0$, what are the angular velocity and angular acceleration of arm BC?

Solution:

Use the solution to Problem 6.171. The vector $\vec{r}_{B/A} = 760(\vec{i}\cos 15^o - \vec{j}\sin 15^o) = 734.1\vec{i} - 196.7\vec{j}\ (mm)$. The vector $\vec{r}_{C/B} = 900(\vec{i}\cos 50^o + \vec{j}\sin 50^o) = 578.5\vec{i} + 689.4\vec{j}\ (mm)$. The velocity of point B is

$$\vec{v}_B = \vec{\omega}_{AB} \times \vec{r}_{B/A} = \begin{bmatrix} \vec{i} & \vec{j} & \vec{k} \\ 0 & 0 & \omega_{AB} \\ 734.1 & -196.7 & 0 \end{bmatrix} = 196.7\omega_{AB}\vec{i} + 734.1\omega_{AB}\vec{j}\ (mm/s)$$

. The velocity of point C is

$$\vec{v}_C = \vec{v}_B + \vec{\omega}_{BC} \times \vec{r}_{C/B} = 196.7\omega_{AB}\vec{i} + 734.1\omega_{AB}\vec{j} + \begin{bmatrix} \vec{i} & \vec{j} & \vec{k} \\ 0 & 0 & \omega_{BC} \\ 578.5 & 687.4 & 0 \end{bmatrix}$$

$1000\vec{i} = 196.7\omega_{AB}\vec{i} + 734.1\omega_{AB}\vec{j} - 689.4\omega_{BC}\vec{i} + 578.5\omega_{BC}\vec{j}\ (mm/s)$. Separate components: $1000 = 196.7\omega_{AB} - 689.4\omega_{BC}$, $0 = 734.1\omega_{AB} - 578.5\omega_{BC}$. Solve: $\vec{\omega}_{AB} = 0.933\vec{k}\ (rad/s)$, $\boxed{\vec{\omega}_{BC} = -1.184\vec{k}\ (rad/s)}$.

The acceleration of point B is

$$\vec{a}_B = \vec{\alpha}_{AB} \times \vec{r}_{B/A} - \omega_{AB}^2\vec{r}_{B/A} = \begin{bmatrix} \vec{i} & \vec{j} & \vec{k} \\ 0 & 0 & \alpha_{AB} \\ 734.1 & -196.7 & 0 \end{bmatrix} - (\omega_{AB}^2)(734.1\vec{i} - 196.7\vec{j})\ (mm/s^2).$$

$\vec{a}_B = 196.7\alpha_{AB}\vec{i} + 734.1\alpha_{AB}\vec{j} - 639.0\vec{i} + 171.3\vec{j}\ (mm/s^2)$ The acceleration of point C is

$$\vec{a}_C = \vec{a}_B + \vec{\alpha}_{BC} \times \vec{r}_{C/B} - \omega_{BC}^2\vec{r}_{C/B} = \vec{a}_B + \begin{bmatrix} \vec{i} & \vec{j} & \vec{k} \\ 0 & 0 & \alpha_{BC} \\ 578.5 & 689.4 & 0 \end{bmatrix} - (\omega_{BC}^2)(578.5\vec{i} + 689.4)$$

$$\vec{a}_C = 0 = \alpha_{AB}(196.7\vec{i} + 734.1\vec{j}) + \alpha_{BC}(-689.4\vec{i} + 578.5\vec{j})$$
$$-811.0\vec{i} - 966.5\vec{j} - 639.0\vec{i} + 171.3\vec{j}\ (mm/s^2).$$

Separate components: $196.7\alpha_{AB} - 689.4\alpha_{BC} - 811.3 - 639.0 = 0$,

$734.1\alpha_{AB} + 578.5\alpha_{BC} - 966.5 + 171.3 = 0$. Solve: $\alpha_{AB} = 2.24\ rad/s^2$, $\boxed{\alpha_{BC} = -1.465\ (rad/s^2)}$

==================================<>==================================

==================================<>==================================

Problem 6.174 The crank AB has a constant clockwise angular velocity of 200 *rpm* (revolutions per minute). What are the velocity and acceleration of the piston P?

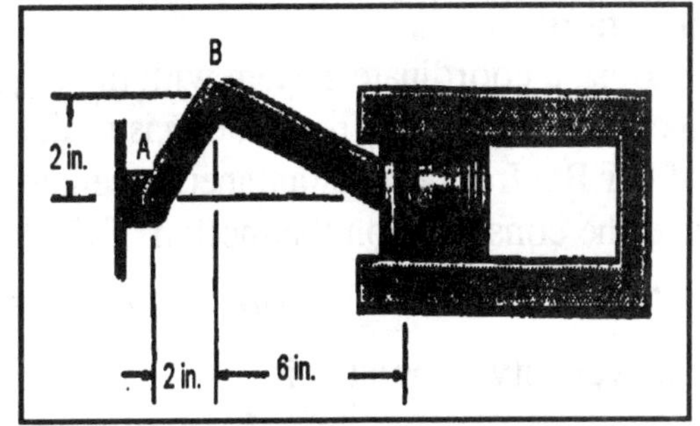

Solution Choose a coordinate system with the origin at A and the *x* axis parallel to AP. The strategy is to determine the angular velocity of bar BP from the instantaneous center; using the angular velocity and the constraint on the motion of P, the velocity acceleration are determined. The angular velocity of crank AB is $\omega_{AB} = 200(2\pi)/60 = 20.94$ (rad/s). The velocity of point B is

$$\vec{v}_B = \vec{\omega}_{AB} \times \vec{r}_{B/A} = \begin{bmatrix} \vec{i} & \vec{j} & \vec{k} \\ 0 & 0 & 20.94 \\ 2 & 2 & 0 \end{bmatrix} = -41.89\left(-\vec{i} + \vec{j}\right) \ (in/s).$$

The instantaneous center for bar BP is at the intersection of the lines parallel to bar AB and the vertical line from P. A point on the line parallel to AB is $\vec{L}_{AB} = \frac{L_{AB}}{\sqrt{2}}\left(\vec{i} + \vec{j}\right)$ $(in.)$. At the intersection, $\frac{L_{AB}}{\sqrt{2}} = 8$, from which of the coordinates of the instantaneous center are $(8, 8)$ $(in.)$. [*Check*: The line from A to the fixed center is the hypotenuse of a right triangle of base 8 *in.* and interior angles of 45°. *check.*] The angular velocity of bar in terms of the instantaneous center is determined from

$$\vec{v}_B = \vec{\omega}_{BP} \times \vec{r}_{B/C} = \begin{bmatrix} \vec{i} & \vec{j} & \vec{k} \\ 0 & 0 & \omega_{BP} \\ -6 & -6 & 0 \end{bmatrix} = 6\omega_{BP}\left(\vec{i} - \vec{j}\right) = 41.89\left(\vec{i} - \vec{j}\right) \ (in/s), \text{ from which}$$

$\omega_{BP} = \frac{41.89}{6} = 6.98$ rad/s. The velocity of point P is $\vec{v}_P = \vec{\omega}_{BP} \times \vec{r}_{P/C} = \begin{bmatrix} \vec{i} & \vec{j} & \vec{k} \\ 0 & 0 & 6.98 \\ 0 & -8 & 0 \end{bmatrix}$

$\boxed{\vec{v}_P = 55.85\vec{i} \ (in/s)}$. The acceleration of point B is

$\vec{a}_B = \vec{\alpha}_{AB} \times \vec{r}_{B/A} - \omega_{AB}^2 \vec{r}_{B/A} = -\omega_{AB}^2 \vec{r}_{B/A} = -877.3\left(\vec{i} + \vec{j}\right) \left(in/s^2\right)$. The acceleration of point P is constrained to be parallel to the *x* axis. The only way this can be true is for the angular acceleration of the bar BP to be non zero.

$$\vec{a}_P = \vec{a}_B + \vec{\alpha}_{BP} \times \vec{r}_{P/B} - \omega_{BP}^2 \vec{r}_{P/B} = \vec{a}_B + \begin{bmatrix} \vec{i} & \vec{j} & \vec{k} \\ 0 & 0 & \alpha_{BP} \\ 6 & -2 & 0 \end{bmatrix} - \omega_{BP}^2\left(6\vec{i} - 2\vec{j}\right).$$

$a_P\vec{i} = -877.3\left(\vec{i} + \vec{j}\right) + \alpha_{BP}\left(2\vec{i} + 6\vec{j}\right) - 292.4\vec{i} + 97.5\vec{j}$. Separate components:

$a_p = -877.3 + 2\alpha_{BP} - 292.4$, $0 = -877.3 + 6\alpha_{BP} + 97.5$. Solve: $\boxed{\vec{a}_P = -909.8\vec{i} \ \left(in/s^2\right)}$,

$\alpha_{BP} = 130$ rad/s^2 (counterclockwise).

==================================<>==================================

==================================<>==================================

Problem 6.175 Bar AB has a counterclockwise angular velocity of 10 *rad/s* and a clockwise angular acceleration of 20 *rad/s* . Determine the angular acceleration of bar BC and the acceleration of point C.

Solution:

Choose a coordinate system with the origin at the left end of the horizontal rod and the *x* axis parallel to the horizontal rod. The strategy is to determine the angular velocity of bar BC from the instantaneous center; using the angular velocity and the constraint on the motion of C, the accelerations are determined. The vector $\vec{r}_{B/A} = \vec{r}_B - \vec{r}_A = (8\vec{i} + 4\vec{j}) - 4\vec{j} = 8\vec{i}$ (*in.*).

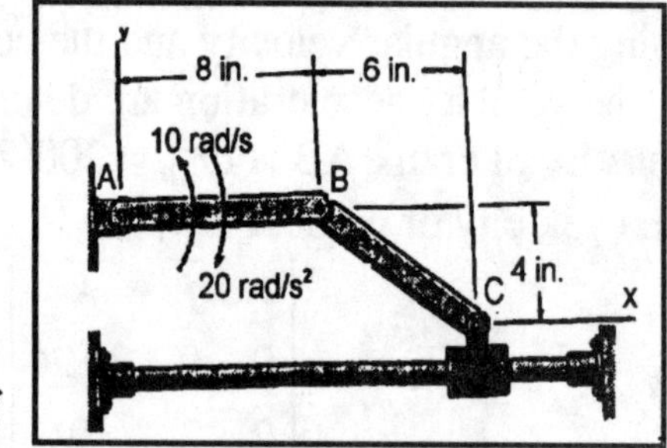

The velocity of point B is

$$\vec{v}_B = \vec{\omega}_{AB} \times \vec{r}_{B/A} = \begin{bmatrix} \vec{i} & \vec{j} & \vec{k} \\ 0 & 0 & 10 \\ 8 & 0 & 0 \end{bmatrix} = 80\vec{j} \ (in/s).$$

The acceleration of point B is

$$\vec{a}_B = \vec{\alpha}_{AB} \times \vec{r}_{B/A} - \omega^2 \vec{r}_{A/B} = \begin{bmatrix} \vec{i} & \vec{j} & \vec{k} \\ 0 & 0 & -20 \\ 8 & 0 & 0 \end{bmatrix} - 100(8\vec{i}) = -800\vec{i} - 160\vec{j} \ (in/s^2).$$

The velocity of point B is normal to the *x* axis, and the velocity of C is parallel to the *x* axis. The line perpendicular to the velocity at B is parallel to the *x*-axis, and the line perpendicular to the velocity at C is parallel to the *y* axis. The intercept is at $(14,4)$, which is the instantaneous center of bar BC. Denote the instantaneous center by C''. The vector $\vec{r}_{B/C''} = \vec{r}_B - \vec{r}_{C''} = (8\vec{i} + 4\vec{j}) - (14\vec{i} - 4\vec{j}) = -6\vec{i}$ (*in.*). The velocity of point B is

$$\vec{v}_B = \vec{\omega}_{BC} \times \vec{r}_{B/IC} = \begin{bmatrix} \vec{i} & \vec{j} & \vec{k} \\ 0 & 0 & \omega_{BC} \\ -6 & 0 & 0 \end{bmatrix} = -6\omega_{BC}\vec{j} = 80\vec{j},$$

from which $\omega_{BC} = -\frac{80}{6} = -13.33 \ rad/s$.

The vector $\vec{r}_{C/B} = \vec{r}_C - \vec{r}_B = (14\vec{i}) - (8\vec{i} + 4\vec{j}) = 6\vec{i} - 4\vec{j}$ (*in.*). The acceleration of point C is

$$\vec{a}_C = \vec{a}_B + \vec{\alpha}_{BC} \times \vec{r}_{C/B} - \omega_{BC}^2 \vec{r}_{C/B} = -800\vec{i} - 160\vec{j} + \begin{bmatrix} \vec{i} & \vec{j} & \vec{k} \\ 0 & 0 & \alpha_{BC} \\ 6 & -4 & 0 \end{bmatrix} - 1066.7\vec{i} + 711.1\vec{j} \ (in/s^2).$$

The acceleration of point C is constrained to be parallel to the *x* axis. Separate components: $a_C = -800 + 4\alpha_{BC} - 1066.7$, $0 = -160 + 6\alpha_{BC} + 711.1$. Solve: $\boxed{\vec{a}_C = -2234\vec{i} \ (in/s^2)}$,

$\boxed{\alpha_{BC} = -91.9 \ rad/s^2}$ (clockwise).

==================================<>==================================

==================================<>==================================

Problem 6.176 The angular velocity of arm AC is 1 *rad/s* counterclockwise. What is the angular velocity of the scoop?

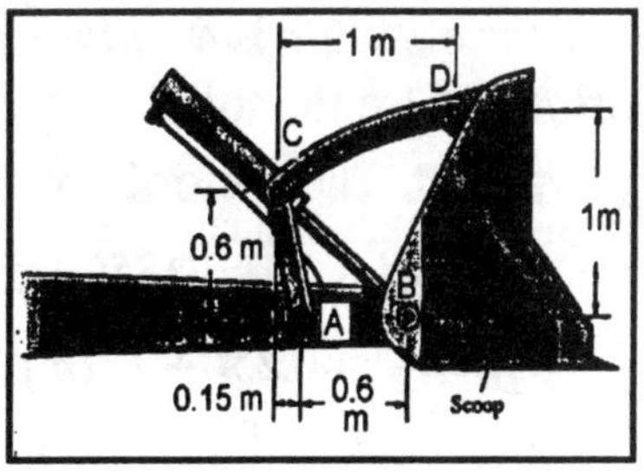

Solution:

Choose a coordinate system with the origin at A and the y axis vertical. The vector locations of B, C and D are $\vec{r}_B = 0.6\vec{i}\ (m)$,. $\vec{r}_C = -0.15\vec{i} + 0.6\vec{j}\ (m)$, $\vec{r}_D = (1-0.15)\vec{i} + 1\vec{j} = 0.85\vec{i} + \vec{j}\ (m)$, from which $\vec{r}_{D/C} = \vec{r}_D - \vec{r}_C = 1\vec{i} + 0.4\vec{j}\ (m)$, and $\vec{r}_{D/B} = \vec{r}_D - \vec{r}_B = 0.25\vec{i} + \vec{j}\ (m)$. The velocity of point C is

$$\vec{v}_C = \vec{\omega}_{AC} \times \vec{r}_{C/A} = \begin{bmatrix} \vec{i} & \vec{j} & \vec{k} \\ 0 & 0 & \omega_{AC} \\ -0.15 & 0.6 & 0 \end{bmatrix} = -0.6\vec{i} - 0.15\vec{j}\ (m/s)$$

The velocity of D in terms of the velocity of C is

$$\vec{v}_D = \vec{v}_C + \vec{\omega}_{CD} \times \vec{r}_{D/C} = -0.6\vec{i} - 0.15\vec{j} + \begin{bmatrix} \vec{i} & \vec{j} & \vec{k} \\ 0 & 0 & \omega_{CD} \\ 1 & 0.4 & 0 \end{bmatrix} = -0.6\vec{i} - 0.15\vec{j} + \omega_{CD}(-0.4\vec{i} + \vec{j})$$

The velocity of point D in terms of the angular velocity of the scoop is

$$\vec{v}_D = \vec{\omega}_{DB} \times \vec{r}_{D/B} = \begin{bmatrix} \vec{i} & \vec{j} & \vec{k} \\ 0 & 0 & \omega_{DB} \\ 0.25 & 1 & 0 \end{bmatrix} = \omega_{DB}(-\vec{i} + 0.25\vec{j})$$

Equate expressions and separate components: $-0.6 - 0.4\omega_{CD} = -\omega_{DB}$, $-0.15 + \omega_{CD} = 0.25\omega_{DB}$. Solve: $\omega_{CD} = 0.333\ rad/s$, $\boxed{\omega_{DB} = 0.733\ rad/s}$ (counterclockwise).

==================================<>==================================

================================<>================================

Problem 6.177 The angular velocity of arm AC in Problem 6.176 is 2 *rad/s* clockwise and its angular acceleration is 4 *rad/s* clockwise. What is the angular acceleration of the scoop?

Solution: Use the solution to Problem 6.154. Choose a coordinate system with the origin at A and the y axis vertical. The vector locations of B, C and D are $\vec{r}_B = 0.6\vec{i}$ (m),. $\vec{r}_C = -0.15\vec{i} + 0.6\vec{j}$ (m), $\vec{r}_D = (1 - 0.15)\vec{i} + 1\vec{j} = 0.85\vec{i} + \vec{j}$ (m), from which $\vec{r}_{D/C} = \vec{r}_D - \vec{r}_C = 1\vec{i} + 0.4\vec{j}$ (m), and $\vec{r}_{D/B} = \vec{r}_D - \vec{r}_B = 0.25\vec{i} + \vec{j}$ (m). The velocity of point C is

$$\vec{v}_C = \vec{\omega}_{AC} \times \vec{r}_{C/A} = \begin{bmatrix} \vec{i} & \vec{j} & \vec{k} \\ 0 & 0 & \omega_{AC} \\ -0.15 & 0.6 & 0 \end{bmatrix} = -1.2\vec{i} - 0.3\vec{j} \ (m/s).$$

The velocity of D in terms of the velocity of C is

$$\vec{v}_D = \vec{v}_C + \vec{\omega}_{CD} \times \vec{r}_{D/C} = -0.6\vec{i} - 0.15\vec{j} + \begin{bmatrix} \vec{i} & \vec{j} & \vec{k} \\ 0 & 0 & \omega_{CD} \\ 1 & 0.4 & 0 \end{bmatrix} = -1.2\vec{i} - 0.3\vec{j} + \omega_{CD}\left(-0.4\vec{i} + \vec{j}\right).$$ The velocity of point D in terms of the angular velocity of the scoop is

$$\vec{v}_D = \vec{\omega}_{DB} \times \vec{r}_{D/B} = \begin{bmatrix} \vec{i} & \vec{j} & \vec{k} \\ 0 & 0 & \omega_{DB} \\ 0.25 & 1 & 0 \end{bmatrix} = \omega_{DB}\left(-\vec{i} + 0.25\vec{j}\right).$$

Equate expressions and separate components: $-1.2 - 0.4\omega_{CD} = -\omega_{DB}$, $-0.3 + \omega_{CD} = 0.25\omega_{DB}$. Solve: $\omega_{CD} = 0.667$ rad/s, $\omega_{DB} = 1.47$ rad/s. The angular acceleration of the point C is

$$\vec{a}_C = \vec{\alpha}_{AC} \times \vec{r}_{C/A} - \omega_{AC}^2\vec{r}_{C/A} = \begin{bmatrix} \vec{i} & \vec{j} & \vec{k} \\ 0 & 0 & \alpha_{AC} \\ -0.15 & 0.6 & 0 \end{bmatrix} - \omega_{AC}^2\left(-0.15\vec{i} + 0.6\vec{j}\right),$$

$\vec{a}_C = 2.4\vec{i} + 0.6\vec{j} + 0.6\vec{i} - 2.4\vec{j} = 3\vec{i} - 1.8\vec{j}$ $\left(m/s^2\right)$. The acceleration of point D in terms of the acceleration of point C is

$$\vec{a}_D = \vec{a}_C + \vec{\alpha}_{CD} \times \vec{r}_{D/C} - \omega_{CD}^2\vec{r}_{D/C} = 3\vec{i} - 1.8\vec{j} + \begin{bmatrix} \vec{i} & \vec{j} & \vec{k} \\ 0 & 0 & \alpha_{CD} \\ 1 & 0.4 & 0 \end{bmatrix} - \omega_{CD}^2\left(\vec{i} + 0.4\vec{j}\right).$$

$\vec{a}_C = \alpha_{CD}\left(-0.4\vec{i} + \vec{j}\right) + 2.56\vec{i} - 1.98\vec{j}$ $\left(m/s^2\right)$. The acceleration of point D in terms of the angular acceleration of point B is

$$\vec{a}_D = \vec{\alpha}_{BD} \times \vec{r}_{D/B} - \omega_{BD}^2\vec{r}_{D/B} = \begin{bmatrix} \vec{i} & \vec{j} & \vec{k} \\ 0 & 0 & \alpha_{CD} \\ 0.25 & 1 & 0 \end{bmatrix} - \omega_{BD}^2\left(0.25\vec{i} + \vec{j}\right).$$

$\vec{a}_D = \alpha_{BD}\left(-\vec{i} + 0.25\vec{j}\right) - 0.538\vec{i} - 2.15\vec{j}$. Equate expressions for $\vec{a}_D$ and separate components: $-0.4\alpha_{CD} + 2.56 = -\alpha_{BD} - 0.538$, $\alpha_{CD} - 1.98 = 0.25\alpha_{BD} - 2.15$. Solve: $\alpha_{CD} = -1.052$ rad/s^2, $\boxed{\alpha_{BD} = -3.51 \ rad/s^2}$, where the negative sign means a clockwise acceleration.

================================<>================================

Problem 6.178 If you want to program the robot so that, at the instant shown the velocity of point D is $\vec{v}_D = 0.2\vec{i} + 0.8\vec{j}$ (m/s) and the angular velocity of arm CD is 0.3 *rad/s* counterclockwise, what are the necessary angular velocities of arms AB and BC?

Solution:

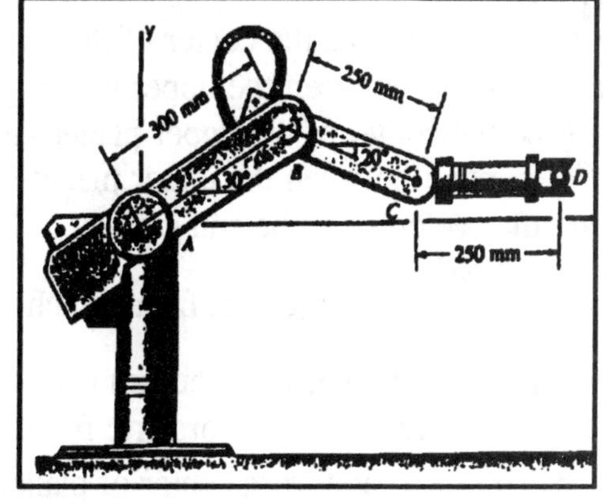

The position vectors are:

$\vec{r}_{B/A} = 300(\vec{i}\cos 30^o + \vec{j}\sin 30^o) = 259.8\vec{i} + 150\vec{j}$ (mm),

$\vec{r}_{C/B} = 250(\vec{i}\cos 20^o - \vec{j}\sin 20^o) = 234.9\vec{i} - 85.5\vec{j}$ (mm).

$\vec{r}_{C/D} = -250\vec{i}$ (mm).

The velocity of the point B is

$$\vec{v}_B = \vec{\omega}_{AB} \times \vec{r}_{B/A} = \begin{bmatrix} \vec{i} & \vec{j} & \vec{k} \\ 0 & 0 & \omega_{AB} \\ 259.8 & 150 & 0 \end{bmatrix}$$

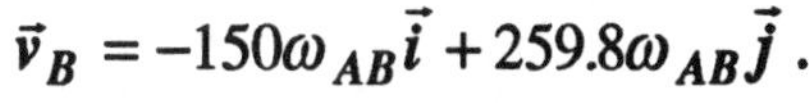

$\vec{v}_B = -150\omega_{AB}\vec{i} + 259.8\omega_{AB}\vec{j}$.

The velocity of point C in terms of the velocity of B is

$$\vec{v}_C = \vec{v}_B + \vec{\omega}_{BC} \times \vec{r}_{C/B} = \vec{v}_B + \begin{bmatrix} \vec{i} & \vec{j} & \vec{k} \\ 0 & 0 & \omega_{BC} \\ 234.9 & -85.5 & 0 \end{bmatrix} = \omega_{BC}(85.5\vec{i} + 234.9\vec{j})$$

$\vec{v}_C = -150\omega_{AB}\vec{i} + 259.8\omega_{AB}\vec{j} + 85.5\omega_{BC}\vec{i} + 234.9\omega_{BC}\vec{j}$ (mm/s).

The velocity of point C in terms of the velocity of point D is

$$\vec{v}_C = \vec{v}_D + \vec{\omega}_{CD} \times \vec{r}_{C/D} = 200\vec{i} + 800\vec{j} + 0.3\begin{bmatrix} \vec{i} & \vec{j} & \vec{k} \\ 0 & 0 & 1 \\ -250 & 0 & 0 \end{bmatrix} = 200\vec{i} + 725\vec{j} \ (mm/s)$$. Equate the expressions for $\vec{v}_C$ and separate components: $-150\omega_{AB} + 85.5\omega_{BC} = 200$, and $259.8\omega_{AB} + 234.9\omega_{BC} = 725$.

Solve: $\boxed{\omega_{AB} = 0.261 \ rad/s}$, $\boxed{\omega_{BC} = 2.797... = 2.80 \ rad/s}$

Problem 6.179 The ring gear is stationary, and the sun gear rotates at 120 *rpm* (revolutions per minute) in the counterclockwise direction. Determine the angular velocity of the planet gears and the magnitude of the velocity of their centerpoints.

Solution:

Denote the point O be the center of the sun gear, point S to be the point of contact between the upper planet gear and the sun gear, point P be the center of the upper planet gear, and point C be the point of contact between the upper planet gear and the ring gear. The angular velocity of the sun gear is

$\omega_S = \dfrac{120(2\pi)}{60} = 4\pi\ rad/s$, from which $\vec{\omega}_S = 4\pi\vec{k}\ (rad/s)$. At the point of contact between the sun gear and the upper planet gear the velocities are equal. The vectors are: from center of sun gear to S is $\vec{r}_{P/S} = 20\vec{j}\ (in.)$, and from center of planet gear to S is $\vec{r}_{S/P} = -7\vec{j}\ (in.)$. The velocities are:

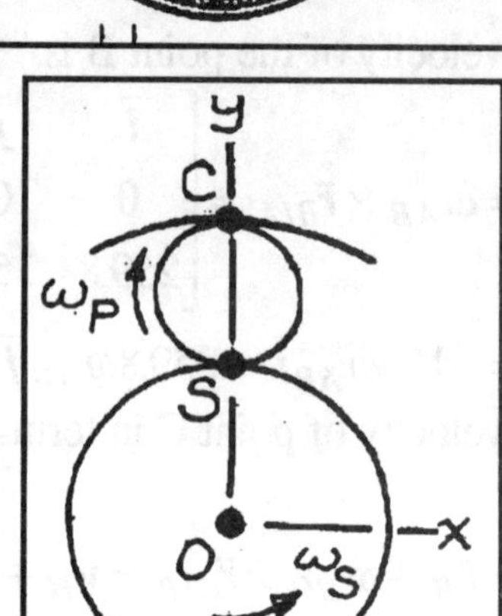

$$\vec{v}_{S/O} = \vec{v}_O + \vec{\omega}_S \times (20\vec{j}) = 0 + \omega_S(20)(\vec{k}\times\vec{j})$$

$\vec{v}_{S/O} = 20\omega_S \begin{bmatrix} \vec{i} & \vec{j} & \vec{k} \\ 0 & 0 & 1 \\ 0 & 1 & 0 \end{bmatrix} = -20(\omega_S)\vec{i} = -251.3\vec{i}\ (in/s)$. From equality of the velocities, $\vec{v}_{S/P} = \vec{v}_{S/O} = -251.3\vec{i}\ (in/s)$. The point of contact C between the upper planet gear and the ring gear is stationary, from which

$\vec{v}_{S/P} = -251.3\vec{i} = \vec{v}_C + \vec{\omega}_P \times \vec{r}_{C/S} = 0 + \begin{bmatrix} \vec{i} & \vec{j} & \vec{k} \\ 0 & 0 & \omega_P \\ -14 & 0 & 0 \end{bmatrix} = 14\omega_P\vec{i} = -251.3\vec{i}$ from which

$\boxed{\omega_P = 17.95\ rad/s}$. The velocity of the centerpoint of the topmost planet gear is

$$\vec{v}_P = \vec{v}_{S/P} + \vec{\omega}_P \times \vec{r}_{P/S} = -251.3\vec{i} + (-17.95)(-7)(\vec{k}\times\vec{j}) = -251.3\vec{i} + 125.65\begin{bmatrix} \vec{i} & \vec{j} & \vec{k} \\ 0 & 0 & 1 \\ 0 & 1 & 0 \end{bmatrix}$$

$\vec{v}_P = -125.7\vec{i}\ (in/s)$ The magnitude is $\boxed{v_{PO} = 125.7\ in./s}$ By symmetry, the magnitudes of the velocities of the centerpoints of the other planetary gears is the same.

===================================<>==================================

Problem 6.180 Arm AB is rotating at 10 *rad/s* in the clockwise direction. Determine the angular velocity of the arm BC and the velocity at which is slides relative to the sleeve at C.

Solution:

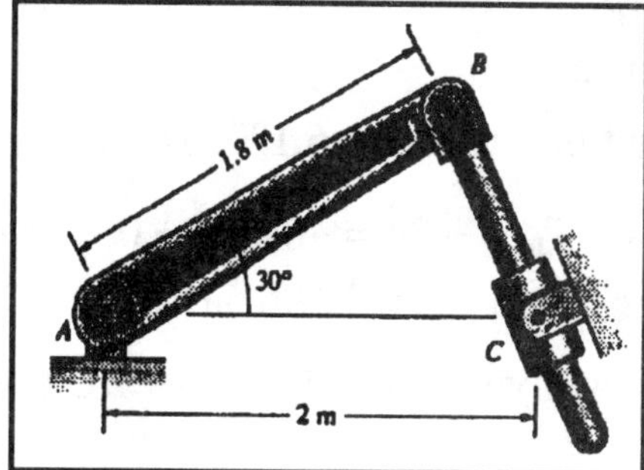

The position vectors are

$\vec{r}_{B/A} = 1.8\left(\vec{i}\cos 30^o + \vec{j}\sin 30^o\right) = 1.56\vec{i} + 0.9\vec{j} \ (m)$.

$\vec{r}_{B/C} = \vec{r}_{B/A} - 2\vec{i} = -0.441\vec{i} + 0.9\vec{j} \ (m)$.

The velocity of point B is

$$\vec{v}_B = \vec{\omega}_{AB} \times \vec{r}_{B/A} = \begin{bmatrix} \vec{i} & \vec{j} & \vec{k} \\ 0 & 0 & -10 \\ 1.56 & 0.9 & 0 \end{bmatrix} = 9\vec{i} - 15.6\vec{j} \ (m/s)$$ The

unit vector from B to C is $\vec{e}_{BC} = \dfrac{-\vec{r}_{B/C}}{|\vec{r}_{B/C}|} = 0.4401\vec{i} - 0.8979\vec{j}$. The relative velocity is parallel to this vector: $\vec{v}_{Crel} = v_{Crel}\vec{e}_{BC} = v_{Crel}\left(0.4401\vec{i} - 0.8979\vec{j}\right) \ (m/s)$

The velocity of B in terms of the velocity of C is

$$\vec{v}_B = \vec{v}_{Rel} + \vec{\omega}_{BC} \times \vec{r}_{B/C} = \vec{v}_{Rel} + \begin{bmatrix} \vec{i} & \vec{j} & \vec{k} \\ 0 & 0 & \omega_{BC} \\ -0.441 & 0.9 & 0 \end{bmatrix},$$

$\vec{v}_B = 0.4401 v_{Crel}\vec{i} - 0.8979 v_{Crel}\vec{j} - 0.9\omega_{BC}\vec{i} - 0.441\omega_{BC}\vec{j} \ (m/s)$.

Equate the expressions for $\vec{v}_B$ and separate components: $9 = 0.4401 v_{Crel} - 0.9\omega_{BC}$, and $-15.6 = -0.8979 v_{Crel} - 0.441\omega_{BC}$.

Solve: $\boxed{v_{Crel} = 17.96 \ m/s}$ (toward C). $\boxed{\omega_{BC} = -1.22 \ rad/s}$ clockwise)

===================================<>==================================

==================================<>==================================

Problem 6.181 In Problem 6.180, arm AB is rotating with an angular velocity of 10 *rad/s* and an angular acceleration of 20 *rad/s²*, both in the clockwise direction. Determine the angular acceleration of arm BC.

Solution:

Use the solution to 6.180. The vector $\vec{r}_{B/A} = 1.8(\vec{i}\cos 30^o + \vec{j}\sin 30^o) = 1.56\vec{i} + 0.9\vec{j}$ (m).

$\vec{r}_{B/C} = \vec{r}_{B/A} - 2\vec{i} = -0.441\vec{i} + 0.9\vec{j}$ (m). The angular velocity: $\omega_{BC} = -1.22$ rad/s, and the relative velocity is $v_{Crel} = 17.96$ m/s.

The unit vector parallel to bar BC is $\vec{e} = 0.4401\vec{i} - 0.8979\vec{j}$

The acceleration of point B is $\vec{a}_B = \vec{\alpha}_{AB} \times \vec{r}_{B/A} - \omega_{AB}^2 \vec{r}_{B/A} = \begin{bmatrix} \vec{i} & \vec{j} & \vec{k} \\ 0 & 0 & \omega_{AB} \\ 1.56 & 0.9 & 0 \end{bmatrix} - \omega_{AB}^2(1.56\vec{i} + 0.9\vec{j})$,

$\vec{a}_B = 18\vec{i} - 31.2\vec{j} - 155.9\vec{i} - 90\vec{j} = -137.9\vec{i} - 121.2\vec{j}$ (m/s^2).

The acceleration of point B in terms of the acceleration of bar BC is

$\vec{a}_B = \vec{a}_{Crel} + 2\vec{\omega}_{BC} \times \vec{v}_{Crel} + \vec{\alpha}_{BC} \times \vec{r}_{B/C} - \omega_{BC}^2 \vec{r}_{B/C}$.

Expanding term by term: $\vec{a}_{Crel} = a_{Crel}(0.4401\vec{i} - 0.8979\vec{j})$ (m/s^2),

$$2\vec{\omega}_{BC} \times \vec{v}_{Crel} = 2v_{Crel}\omega_{BC}\begin{bmatrix} \vec{i} & \vec{j} & \vec{k} \\ 0 & 0 & 1 \\ 0.440 & -0.8979 & 0 \end{bmatrix} = -39.26\vec{i} - 19.25\vec{j} \ (m/s^2),$$

$$\vec{\alpha}_{BC} \times \vec{r}_{B/C} = \begin{bmatrix} \vec{i} & \vec{j} & \vec{k} \\ 0 & 0 & \alpha_{BC} \\ -0.4411 & 0.9 & 0 \end{bmatrix} = \alpha_{BC}(-0.9\vec{i} - 0.4411\vec{j}) \ (m/s^2)$$

$-\omega_{BC}^2(-0.4411\vec{i} + 0.9\vec{j}) = 0.6539\vec{i} - 1.334\vec{j}$ (m/s^2).

Collecting terms, $\vec{a}_B = a_{Crel}(0.4401\vec{i} - 0.8979\vec{j}) - \alpha_{BC}(0.9\vec{i} + 0.4411\vec{j}) - 38.6\vec{i} - 20.6\vec{j}$ (m/s^2).

Equate the two expressions for $\vec{a}_B$ and separate components: $-137.9 = 0.4401a_{Crel} - 0.9\alpha_{BC} - 38.6$, and $-121.2 = -0.8979a_{Crel} - 0.4411\alpha_{BC} - 20.6$.

Solve: $a_{Crel} = 46.6$ m/s^2 (toward C) $\boxed{\alpha_{BC} = 133.1 \ rad/s^2}$

==================================<>==================================

====================================<>====================================

Problem 6.182 The arm AB is rotating with a constant counterclockwise angular velocity of 10 rad/s. Determine the vertical velocity and acceleration of the rack R of the rack and pinion gear.

Solution:

The vectors: $\vec{r}_{B/A} = 6\vec{i} + 12\vec{j}$ (in.). $\vec{r}_{C/B} = 16\vec{i} - 2\vec{j}$ (in.). The velocity of point B is

$$\vec{v}_B = \vec{\omega}_{AB} \times \vec{r}_{B/A} = \begin{bmatrix} \vec{i} & \vec{j} & \vec{k} \\ 0 & 0 & 10 \\ 6 & 12 & 0 \end{bmatrix} = -120\vec{i} + 60\vec{j} \ (in/s).$$

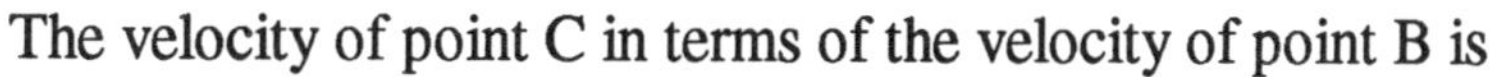

The velocity of point C in terms of the velocity of point B is

$$\vec{v}_C = \vec{v}_B + \vec{\mathbf{w}}_{BC} \times \vec{r}_{C/B} = \vec{v}_B + \begin{bmatrix} \vec{i} & \vec{j} & \vec{k} \\ 0 & 0 & \omega_{BC} \\ 16 & -2 & 0 \end{bmatrix} = -120\vec{i} + 60\vec{j} + \omega_{BC}(2\vec{i} + 16\vec{j}) \ \text{(in/s)}$$

The velocity of point C in terms of the velocity of the gear arm CD is

$$\vec{v}_C = \vec{\omega}_{CD} \times \vec{r}_{C/D} = \begin{bmatrix} \vec{i} & \vec{j} & \vec{k} \\ 0 & 0 & \omega_{CD} \\ -6 & 10 & 0 \end{bmatrix} = -10\omega_{CD}\vec{i} - 6\omega_{CD}\vec{j} \ (in/s).$$

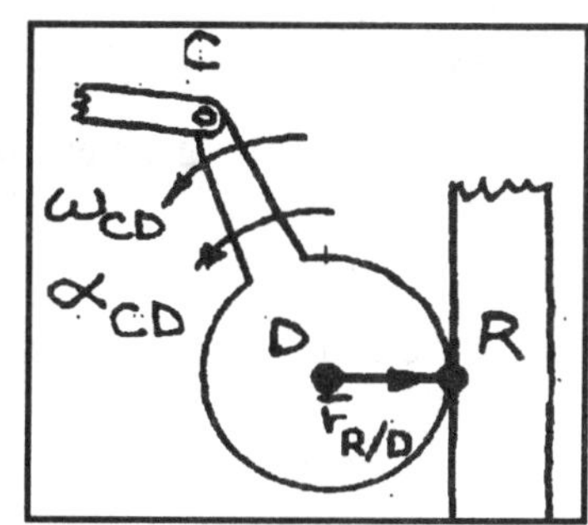

Equate the two expressions for $\vec{v}_C$ and separate components: $-120 + 2\omega_{BC} = -10\omega_{CD}$, $60 + 16\omega_{BC} = -6\omega_{CD}$. Solve: $\omega_{BC} = -8.92 \ rad/s$, $\omega_{CD} = 13.78 \ rad/s$, where the negative sign means a clockwise rotation. The velocity of the rack is $\vec{v}_R = \vec{\omega}_{CD} \times \vec{r}_{R/D} = \begin{bmatrix} \vec{i} & \vec{j} & \vec{k} \\ 0 & 0 & \omega_{CD} \\ 6 & 0 & 0 \end{bmatrix} = 6\omega_{CD}\vec{j}$,

$$\boxed{\vec{v}_R = 82.7\vec{j} \ (in/s) = 6.89\vec{j} \ (ft/s)}$$

The angular acceleration of point B is $\vec{a}_B = -\omega_{AB}^2 \vec{r}_{B/A} = -100(6\vec{i} + 12\vec{j}) = -600\vec{i} - 1200\vec{j} \ (in/s^2)$.

The acceleration of point C is $\vec{a}_C = \vec{a}_B + \vec{\alpha}_{BC} \times \vec{r}_{C/B} - \omega_{BC}^2 \vec{r}_{C/B}$,

$$\vec{a}_C = \vec{a}_B + \begin{bmatrix} \vec{i} & \vec{j} & \vec{k} \\ 0 & 0 & \alpha_{BC} \\ 16 & -2 & 0 \end{bmatrix} - \omega_{BC}^2(16\vec{i} - 2\vec{j}) = \vec{a}_B + 2\alpha_{BC}\vec{i} + 16\alpha_{BC}\vec{j} - \omega_{BC}^2(16\vec{i} - 2\vec{j}).$$ Noting

$\vec{a}_B - \omega_{BC}^2(16\vec{i} - 2\vec{j}) = -600\vec{i} - 1200\vec{j} - 1272.7\vec{i} + 159.1\vec{j} = -1872.7\vec{i} - 1040.9\vec{j}$, from which

$\vec{a}_C = +\alpha_{BC}(2\vec{i} + 16\vec{j}) - 1873\vec{i} - 1041\vec{j} \ (in/s^2)$ The acceleration of point C in terms of the gear arm is

$$\vec{a}_C = \vec{\alpha}_{CD} \times \vec{r}_{C/D} - \omega_{CD}^2 \vec{r}_{C/D} = \begin{bmatrix} \vec{i} & \vec{j} & \vec{k} \\ 0 & 0 & \alpha_{CD} \\ -6 & 10 & 0 \end{bmatrix} - \omega_{CD}^2(-6\vec{i} + 10\vec{j}) \ (in/s^2),$$

$$\vec{a}_C = -10\alpha_{CD}\vec{i} - 6\alpha_{CD}\vec{j} + 1140\vec{i} - 1900\vec{j} \ (in/s^2).$$

Solution continued on next page

Equate expressions for $\vec{a}_C$ and separate components: $2\alpha_{BC} - 1873 = -10\alpha_{CD} + 1140$, $16\alpha_{BC} - 1041 = -6\alpha_{CD} - 1900$. Solve: $\alpha_{CD} = 337.3\ rad/s^2$, and $\alpha_{BC} = -180.2\ rad/s^2$. The acceleration of the rack R is the tangential component of the acceleration of the gear at the point of contact with the rack: $\vec{a}_R = \vec{\alpha}_{CD} \times \vec{r}_{R/D} = \begin{bmatrix} \vec{i} & \vec{j} & \vec{k} \\ 0 & 0 & \alpha_{CD} \\ 6 & 0 & 0 \end{bmatrix} = 6\alpha_{CD}\vec{j}\ (in/s^2)$.

$$\boxed{\vec{a}_R = 2024\vec{j}\ (in/s^2) = 168.7\vec{j}\ (ft/s^2)}$$

==================================<>==================================

Problem 6.183 In Problem 6.182, if the rack R of the rack and pinion gear is moving upward with a constant velocity of 10 *ft/s*, what are the angular velocity and acceleration of bar BC?

Solution:

The constant velocity of the rack R implies that the angular acceleration of the gear is zero, and the angular velocity of the gear is $\omega_{CD} = \frac{120}{6} = 20\ rad/s$. The velocity of point C in terms of the gear angular velocity is $\vec{v}_C = \vec{\omega}_{CD} \times \vec{r}_{C/D} = \begin{bmatrix} \vec{i} & \vec{j} & \vec{k} \\ 0 & 0 & 20 \\ -6 & 10 & 0 \end{bmatrix} = -200\vec{i} - 120\vec{j}\ (in/s)$. The velocity of point B in terms of the velocity of point C is $\vec{v}_B = \vec{v}_C + \vec{\omega}_{BC} \times \vec{r}_{B/C} = \vec{v}_C + \begin{bmatrix} \vec{i} & \vec{j} & \vec{k} \\ 0 & 0 & \omega_{BC} \\ -16 & 2 & 0 \end{bmatrix}$,

$\vec{v}_B = -200\vec{i} - 120\vec{j} - 2\omega_{BC}\vec{i} - 16\omega_{BC}\vec{j}\ (in/s)$. The velocity of point B in terms of the angular velocity of the arm AB is $\vec{v}_B = \vec{\omega}_{AB} \times \vec{r}_{B/A} = \begin{bmatrix} \vec{i} & \vec{j} & \vec{k} \\ 0 & 0 & \omega_{AB} \\ 6 & 12 & 0 \end{bmatrix} = -12\omega_{AB}\vec{i} + 6\omega_{AB}\vec{j}\ (in/s)$. Equate the expressions for $\vec{v}_B$ and separate components $-200 - 2\omega_{BC} = -12\omega_{AB}$, $-120 - 16\omega_{BC} = 6\omega_{AB}$. Solve: $\omega_{AB} = 14.5\ rad/s$, $\boxed{\omega_{BC} = -12.94\ rad/s}$, where the negative sign means a clockwise rotation. The angular acceleration of the point C in terms of the angular velocity of the gear is $\vec{a}_C = \vec{\alpha}_{CD} \times \vec{r}_{C/D} - \omega_{CD}^2\vec{r}_{C/D} = 0 - \omega_{CD}^2(-6\vec{i} + 10\vec{j}) = 2400\vec{i} - 4000\vec{j}\ (in/s^2)$. The acceleration of point B in terms of the acceleration of C is

$$\vec{a}_B = \vec{a}_C + \vec{\alpha}_{BC} \times \vec{r}_{B/C} - \omega_{BC}^2\vec{r}_{B/C} = \vec{a}_C + \begin{bmatrix} \vec{i} & \vec{j} & \vec{k} \\ 0 & 0 & \alpha_{BC} \\ -16 & 2 & 0 \end{bmatrix} - \omega_{BC}^2(-16\vec{i} + 2\vec{j}),$$

$\vec{a}_B = \alpha_{BC}(-2\vec{i} - 16\vec{j}) + 2400\vec{i} - 4000\vec{j} + 2680\vec{i} - 335\vec{j}$.

$\vec{a}_B = -2\alpha_{BC}\vec{i} - 16\alpha_{BC}\vec{j} + 5080\vec{i} - 4335\vec{j}\ (in/s^2)$.

Solution continued on next page

The acceleration of point B in terms of the angular acceleration of arm AB is

$\vec{a}_B = \vec{\alpha}_{AB} \times \vec{r}_{B/A} - \omega_{AB}^2 \vec{r}_{B/A} = \alpha_{AB}\left(\vec{k} \times \left(6\vec{i} + 12\vec{j}\right)\right) - \omega_{AB}^2\left(6\vec{i} + 12\vec{j}\right) \ \left(in/s^2\right)$

$\vec{a}_B = \alpha_{AB}\left(-12\vec{i} + 6\vec{j}\right) - 1263.2\vec{i} - 2526.4\vec{j} \ \left(in/s^2\right)$. Equate the expressions for $\vec{a}_B$ and separate components: $-2\alpha_{BC} + 5080 = -12\alpha_{AB} - 1263.2$, $-16\alpha_{BC} - 4335 = 6\alpha_{AB} - 2526.4$. Solve:

$\alpha_{AB} = -515.2 \ rad/s^2$, $\boxed{\alpha_{BC} = 80.17 \ rad/s^2}$

===================================<>===================================

Problem 6.184 The bar AB has a constant clockwise angular velocity of 2 *rad/s*. The 1 *kg* collar C slides on the smooth horizontal bar. At the instant shown, what is the tension in the cable BC?

Solution:

Choose a coordinate system with the origin at A and the x axis parallel to AC. The vectors: $\vec{r}_B = 1\vec{i} + 2\vec{j} \ (m)$, $\vec{r}_C = 3\vec{i} \ (m)$.

$\vec{r}_{B/A} = \vec{r}_B$, $\vec{r}_{C/B} = \vec{r}_C - \vec{r}_B = -2\vec{i} - 2\vec{j} \ (m)$. The acceleration of point B is $\vec{a}_B = -\omega_{AB}^2 \vec{r}_{B/A} = -4\left(\vec{i} + 2\vec{j}\right) = -4\vec{i} - 8\vec{j} \ \left(m/s^2\right)$.

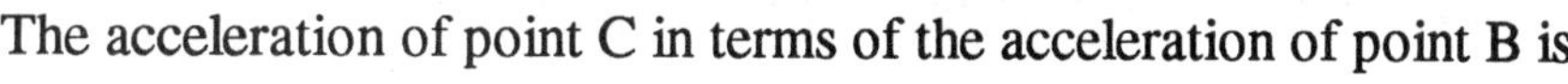

The acceleration of point C in terms of the acceleration of point B is

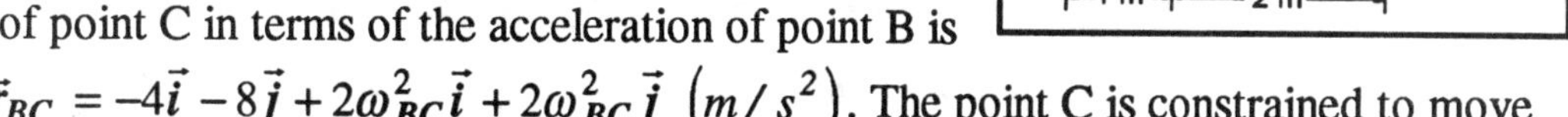

$\vec{a}_C = \vec{a}_B - \omega_{BC}^2 \vec{r}_{BC} = -4\vec{i} - 8\vec{j} + 2\omega_{BC}^2 \vec{i} + 2\omega_{BC}^2 \vec{j} \ \left(m/s^2\right)$. The point C is constrained to move parallel to the x axis, from which $\vec{a}_C = a_C \vec{i} \ \left(m/s^2\right)$. Equate like terms: $a_C = -4 + 2\omega_{BC}^2$, $0 = -8 + 2\omega_{BC}^2$. Solve: $\omega_{BC} = 2 \ rad/s$, $a_C = 0$.

The tension in the cable is $T\cos 45^o = m a_C = 0$, from which $\boxed{T = 0}$

[*Check:* Denote the unit vector from B to C by $\vec{e}_{BC}$. The force at point B is $\vec{F}_B = T\,\vec{e}_{AB}$, where T is the tension in the cable. The moment about A is $\vec{M}_A = \vec{r}_{B/A} \times \vec{F}_B = \vec{r}_{B/A} \times T\vec{e}_{AB}$. From the equation of angular motion $\vec{M}_A = I_A \vec{\alpha}_{AB}$ where I_A is the mass moment of inertia about A, and $\vec{\alpha}_{AB}$ is the angular acceleration of the bar AB. But $\vec{\alpha}_{AB} = 0$ from which $\vec{M}_A = 0$ and $T = 0$. *check.*]

===================================<>===================================

Problem 6.185 An athlete exercises his arm by raising the 8 *kg* mass *m*. The shoulder joint A is stationary. The distance AB is 300 *mm*, the distance BC is 400 *mm*, and the distance from C to the pulley is 340 *mm*. The angular velocities $\omega_{AB} = 1.5 \ rad/s$ and $\omega_{BC} = 2 \ rad/s$ are constant. What is the tension in the cable?

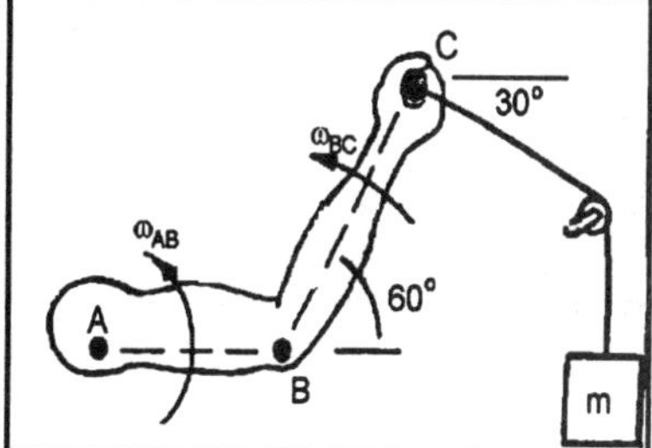

Solution Choose a coordinate system with the origin at A and the x axis parallel to AB. Let P denote the position of the pulley rim. The vectors: $\vec{r}_{B/A} = 300\vec{i} \ (mm)$,

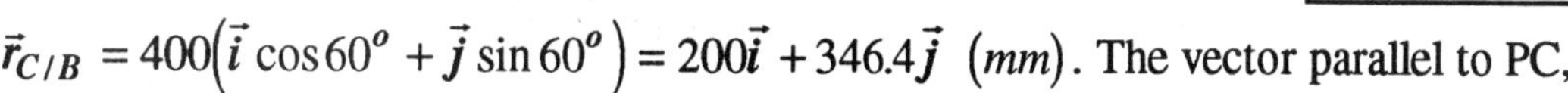

$\vec{r}_{C/B} = 400\left(\vec{i}\cos 60^o + \vec{j}\sin 60^o\right) = 200\vec{i} + 346.4\vec{j} \ (mm)$. The vector parallel to PC,

$\vec{r}_{P/C} = 340\left(\vec{i}\cos 30^o - \vec{j}\sin 30^o\right) = 294.4\vec{i} - 170\vec{j} \ (mm)$. The velocity at point B is

$\vec{v}_B = \vec{\omega}_{AB} \times \vec{r}_{B/A} = \left(300\omega_{AB}\right)\vec{j} = 450\vec{j} \ (mm/s)$.

Solution continued on next page

The velocity at point C is

$$\vec{v}_C = \vec{v}_B + \vec{\omega}_{BC} \times \vec{r}_{C/B} = \vec{v}_B + \begin{bmatrix} \vec{i} & \vec{j} & \vec{k} \\ 0 & 0 & \omega_{BC} \\ 200 & 346.4 & 0 \end{bmatrix} = -728.8\vec{i} + 850\vec{j}\ (mm/s).$$

The velocity at C in terms of the velocity at P is

$$\vec{v}_C = \vec{v}_P + \vec{v}_{Crel} + \vec{\omega}_{PC} \times \vec{r}_{C/P} = 0 + \vec{v}_{Crel} + \begin{bmatrix} \vec{i} & \vec{j} & \vec{k} \\ 0 & 0 & \omega_{PC} \\ -294.4 & 170 & 0 \end{bmatrix}.$$

$$= \left(-v_{Crel}\cos 30^o\right)\vec{i} + \left(v_{Crel}\sin 30^o\right)\vec{j} - 170\omega_{PC}\vec{i} - 294.4\omega_{PC}\vec{j}.$$

Equate like terms: $-728.8 = -v_{Crel}\cos 30^o - 170\omega_{PC}$, $850 = v_{Crel}\sin 30^o - 294.4\omega_{PC}$.

Solve: $v_{Crel} = 412.3\ mm/s$ (toward C), $\omega_{PC} = -2.19\ rad/s$.

The relative velocity is $\vec{v}_{Crel} = -357.1\vec{i} + 206.2\vec{j}\ (mm/s)$ (toward C).

The acceleration of point B is $\vec{a}_B = -\omega_{AB}^2\left(\vec{r}_{B/A}\right) = -675\vec{i}\ \left(mm/s^2\right)$ since $\vec{\alpha}_{AB} = 0$.

The acceleration of point C is $\vec{a}_C = \vec{a}_B - \omega_{BC}^2\vec{r}_{C/B} = -1475\vec{i} - 1385.6\vec{j}\ \left(mm/s^2\right)$, since $\vec{\alpha}_{BC} = 0$.

The acceleration of C in terms of the pulley P is

$\vec{a}_C = \vec{a}_P + \vec{a}_{Crel} + \vec{\alpha}_{PC} \times \vec{r}_{C/P} + 2\vec{\omega}_{PC} \times \vec{v}_{Crel} - \omega_{PC}^2\vec{r}_{C/P}$.

$$\vec{a}_C = \vec{a}_P + a_{Crel}\left(-0.866\vec{i} + 0.5\vec{j}\right) + \begin{bmatrix} \vec{i} & \vec{j} & \vec{k} \\ 0 & 0 & \alpha_{PC} \\ -294.4 & 170 & 0 \end{bmatrix} + 2\begin{bmatrix} \vec{i} & \vec{j} & \vec{k} \\ 0 & 0 & \omega_{PC} \\ -357. & 206.2 & 0 \end{bmatrix} - \omega_{PC}^2\vec{r}_{C/P}$$

$$\vec{a}_C = \vec{a}_P + a_{Crel}\left(-0.866\vec{i} + 0.5\vec{j}\right) - \alpha_{PC}\left(170\vec{i} + 294.4\vec{j}\right) - \left(206.2\vec{i} + 357.1\vec{j}\right)\omega_{PC} - \omega_{PC}^2\left(-294.4\vec{i} + 170\vec{j}\right).$$

Note $\vec{a}_P = 0$.

Equate like terms and solve: : $-1475 = -a_{Crel}(0.866) - 170\alpha_{PC} - 206.2\omega_{PC} + 294.4\omega_{PC}^2$, and $-1385.6 = a_{Crel}(0.5) - 294.4\alpha_{PC} - 357.1\omega_{PC} - 170\omega_{PC}^2$.

Solve: $a_{Crel} = 2210\ mm/s^2$ (toward C), $\alpha_{PC} = 8.35\ rad/s^2$. **The acceleration of the weight is equal to the relative acceleration (toward C) plus the acceleration of gravity, from which the tension is**

$\boxed{T = (a_{Crel} + g)m = 96.16\ N}$

==================================<>==================================

==================================<>==================================

Problem 6.186 The secondary reference frame shown rotates with a constant angular velocity $\overline{\omega} = 2\overline{k}\ rad/s$ relative to a primary reference frame. The point A moves outward along the x axis at a constant rate of *5 m/s*.

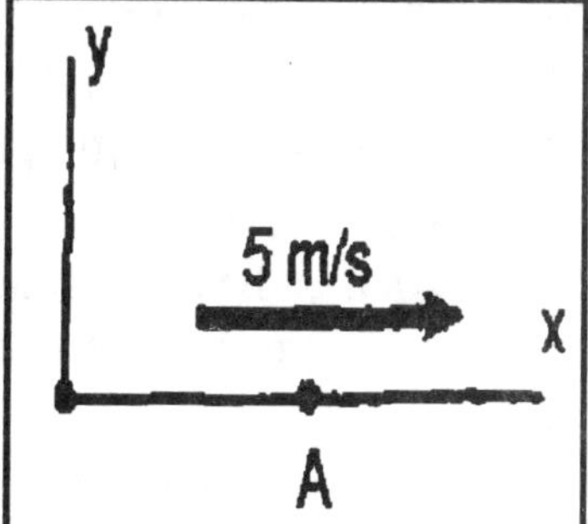

(a) What are the velocity and acceleration of A relative to the secondary reference frame?

(b) If the origin B is stationary relative to the primary reference frame, what are the velocity and acceleration of A relative to the primary reference frame when A is at the position $x = 1\ m$?

Solution:

(a) $\overline{v}_{Arel} = 5\overline{i}\,(m/s)$, $\overline{a}_{Arel} = 0$;

$$\overline{v}_A = \overline{v}_B + \overline{v}_{Arel} + \overline{\omega}\times\overline{r}_{A/B} = 0 + 5\overline{i} + \begin{vmatrix} \overline{i} & \overline{j} & \overline{k} \\ 0 & 0 & 2 \\ 1 & 0 & 0 \end{vmatrix}, \text{ or } \quad \overline{v}_A = 5\overline{i} + 2\overline{j}(m/s)$$

(b) $\overline{a}_A = \overline{a}_B + \overline{a}_{Arel} + 2\overline{\omega}\times\overline{v}_{Arel} + \overline{\alpha}\times\overline{r}_{A/B} - \omega^2\overline{r}_{A/B}$

$$\overline{a}_A = 0 + 0 + 2\begin{vmatrix} \overline{i} & \overline{j} & \overline{k} \\ 0 & 0 & 2 \\ 5 & 0 & 0 \end{vmatrix} + 0 - (2)^2(\overline{i}) = +20\overline{j}\ (m/s^2) + 0 - (2)^2(\overline{i}) = -4\overline{i} + 20\overline{j}(m/s^2)$$

==================================<>==================================

Problem 6.187 The coordinate system shown is fixed relative to the ship B. The ship uses its radar to measure the position of a stationary buoy A and determines it to be $400\,\mathbf{i} + 200\,\mathbf{j}\ (m)$. The ship also measures the velocity of the buoy relative to its body-fixed coordinate system and determines it to be $2\,\mathbf{i} - 8\,\mathbf{j}\ (m/s)$. What are the ship's velocity and angular velocity relative to the earth? (Assume that the ship's velocity is in the direction of the y axis).

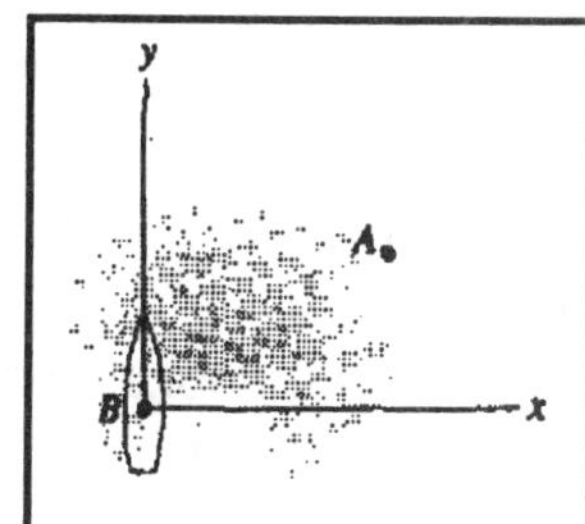

Solution:

$$\overline{v}_A = 0 = \overline{v}_B + \overline{v}_{Arel} + \overline{\omega}\times\overline{r}_{A/B} = v_b\overline{j} + 2\overline{i} - 8\overline{j} + \begin{vmatrix} \overline{i} & \overline{j} & \overline{k} \\ 0 & 0 & \omega \\ 400 & 200 & 0 \end{vmatrix}.$$

Equating $\overline{i}$ and $\overline{j}$ components to zero, $0 = 2 - 200\omega$ $0 = v_B - 8 + 400\omega$ we obtain $\omega = 0.01\ rad/s$ and $v_B = 4m/s$.

==================================<>==================================

=================================<>=================================

Problem 7.1 A refrigerator of mass *m* rests on casters at A and B. Suppose that if you push on it with a horizontal force $\vec{F}$ as shown and that the casters remain on the smooth floor. (a) What is the acceleration of the refrigerator? (b) What normal forces are exerted on the casters at A and B?

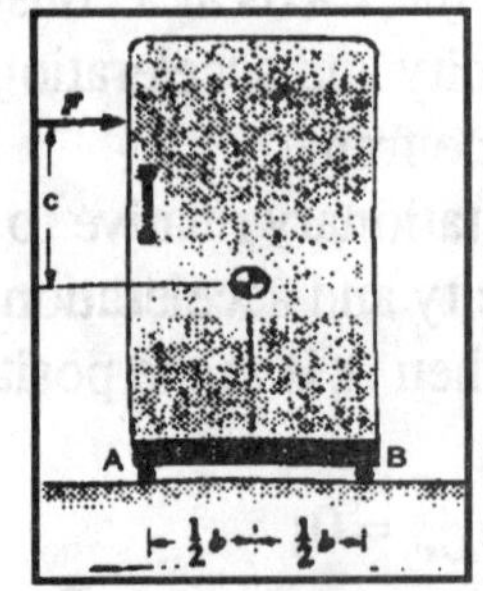

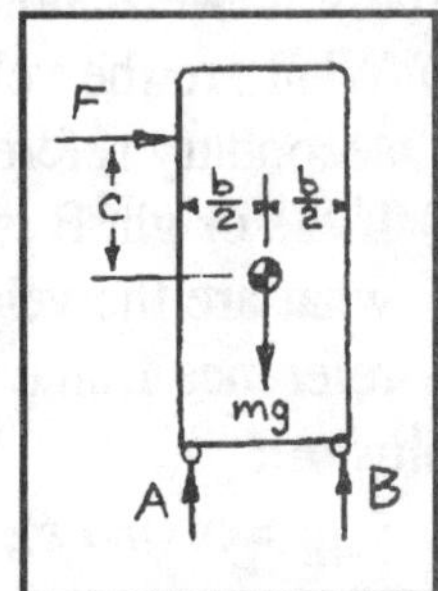

Solution:

(a) From Newton's second law, $\sum F = ma$, from which $\boxed{a = F/m}$.

(b) The sum of the forces in the y direction: $\sum F_y = -mg + A + B = 0$. The sum of moments about the center of mass: $\sum M_{CM} = -cF + \frac{b}{2}B - \frac{b}{2}A = 0$. Solve: $\boxed{A = -\left(\frac{c}{b}\right)F + \frac{mg}{2}}$. $\boxed{B = \left(\frac{c}{b}\right)F + \frac{mg}{2}}$.

=================================<>=================================

Problem 7.2 In Problem 7.1, what is the largest force *F* that you can apply if you want the refrigerator to remain on the floor at A and B?

Solution:

When $F_A = 0$, the caster at A will be on the verge of leaving the floor. From the solution to Problem 7.1, $A = -\left(\frac{c}{b}\right)F + \frac{mg}{2} = 0$, from which $\boxed{F = \left(\frac{b}{2c}\right)mg}$ is the largest force.

=================================<>=================================

Problem 7.3 The 14,000 *lb.* airplane's arresting hook exerts the force F and causes the plane to decelerate at six *g*'s. The horizontal forces exerted by the landing gear are negligible. Determine *F* and the normal forces exerted on the landing gear.

Solution:

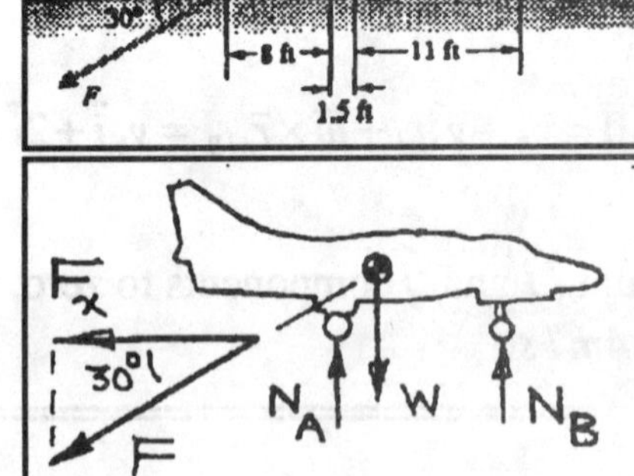

The horizontal component of the force is obtained from Newton's second law: $F_x = \frac{14000}{g} 6g = 84{,}000\ lb$.

The force exerted on the hook is $\boxed{F = \frac{F_x}{\cos 30^o} = 96{,}995\ lb}$.

The sum of the vertical forces is

$\sum F_y = -F\sin 30^o + N_A + N_B - 14000 = 0$.

The line of action of the hook passes through the center of mass, from which the sum of the moments about the center of mass is

$\sum M_{CM} = -1.5 N_A + 11 N_B = 0$. Solve: $\boxed{N_A = 54{,}997\ lb}$, $\boxed{N_B = 7500\ lb}$

=================================<>=================================

Problem 7.4 A student catching a ride to his summer job unwisely supports himself on the back of an accelerating truck by exerting a horizontal force F on the truck's cab at A. Determine the horizontal force he must exert in terms of his weight W, the truck's acceleration a, and the dimensions shown.

Solution:

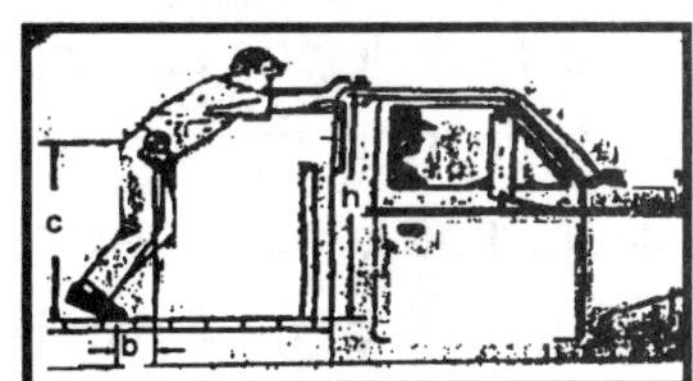

From a free body diagram of the student, four forces are involved: His weight, the normal force *exerted on* his feet, the friction force *exerted on* his feet, the horizontal force F_H *exerted on* his hands by the truck.

From Newton's second law: $\sum F_x = f + F_H = \left(\frac{W}{g}\right)a$, from which

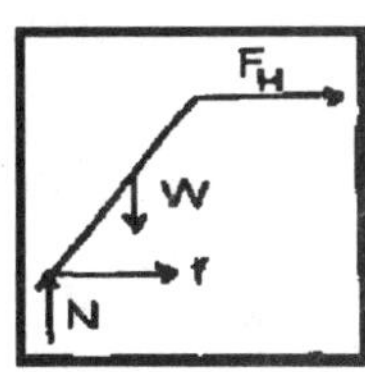

$$f = \left(\frac{W}{g}\right)a - F_H.$$

The sum of the moments about the center of mass:

$$\sum M_{CM} = -bW + cf - (h-c)F_H = 0.$$

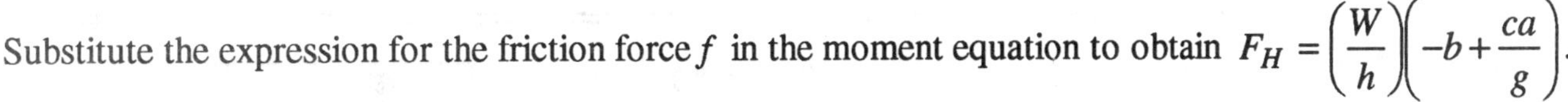

Substitute the expression for the friction force f in the moment equation to obtain $F_H = \left(\frac{W}{h}\right)\left(-b + \frac{ca}{g}\right)$.

The force exerted by the student: $|F| = |F_H| = \left(\frac{W}{h}\right)\left(b - \frac{ca}{g}\right)$

Problem 7.5 The crane moves to the right with constant acceleration, and the 800 kg load moves without swinging. (a) What is the acceleration of the crane and load? (b) What are the tensions in the cables attached at A and B?

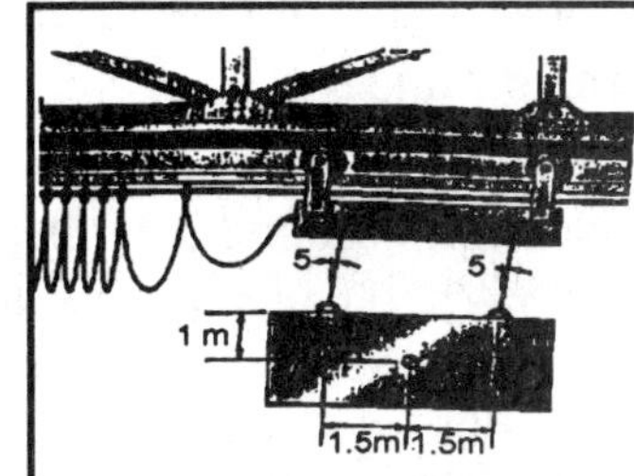

Solution:

From Newton's second law: $F_x = 800a\ N$.

The sum of the forces on the load:

$$\sum F_x = F_A \sin 5^o + F_B \sin 5^o - 800a = 0.$$

$$\sum F_y = F_A \cos 5^o + F_B \cos 5^o - 800g = 0.$$

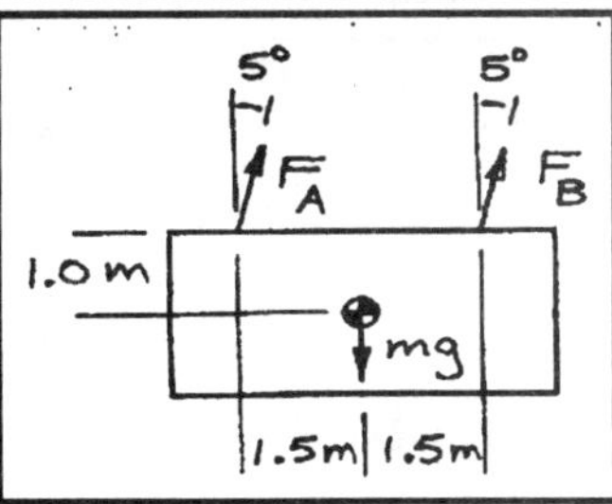

The sum of the moments about the center of mass:

$$\sum M_{CM} = -1.5F_A \cos 5^o + 1.5F_B \cos 5^o - F_A \sin 5^o - F_B \sin 5^o = 0.$$

Solve these three simultaneous equations:

$a = 0.858\ m/s^2$, $F_A = 3709\ N$, $F_B = 4169\ N$

=====================================<>=====================================

Problem 7.6 If the acceleration of the crane in Problem 7.5 suddenly decreases to zero, what are the tensions in the cables attached to at A and B immediately afterward?

Solution:

The linear acceleration in the *x* direction is $a_x = L\alpha\cos 5^o$, where L is the (unknown) length of the cables supporting the load, and α angular acceleration of the support cable. Similarly, the acceleration in the *y* direction is $a_y = L\alpha\sin 5^o$. From Newton's second law, the corresponding forces are ma_x and ma_y. The sum of the forces in the *x* and *y* directions are:

$\sum F_x = F_A \sin 5^o + F_B \sin 5^o = ma_x$.

$\sum F_y = F_A \cos 5^o + F_B \cos 5^o = m(g - a_y)$. The sum of the moments about the center of mass of the load is $\sum M_{CM} = -F_A(1.5\cos 5^o + \sin 5^o) + F_B(1.5\cos 5^o - \sin 5^o) = 0$. Solve these three simultaneous equations in three unknowns: $L\alpha = 0.8547 \ m/s^2$, $\boxed{F_A = 3681.1 \ N}$, and $\boxed{F_B = 4137.1 \ N}$

=====================================<>=====================================

Problem 7.7 The combined mass of the person and bicycle is *m*. The location of their combined center of mass is shown. (a) If they have acceleration *a*, what are the normal forces exerted on the wheels by the ground? (Neglect the horizontal force exerted on the ground by the front wheel.) (b) Based on the results of part (a), what is the largest acceleration that can be achieved without causing the front wheel to leave the ground?

Solution:

From Newton's second law, the horizontal force exerted on the system is $F = ma$. (a) The sum of the forces in the *y* direction: $\sum F_y = N_A + N_B - mg = 0$. The sum of the moments about the center of mass: $\sum M_{CM} = -hF - bN_A + cN_B = 0$. Solve:

$\boxed{N_B = m\left(\frac{bg + ha}{b + c}\right)}$ $\boxed{N_A = m\left(\frac{cg - ha}{b + c}\right)}$ (b) When $N_A = 0$, $\boxed{a = \left(\frac{c}{h}\right)g}$

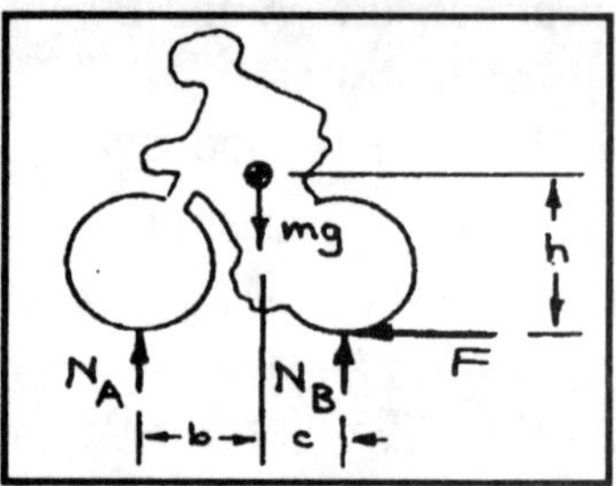

=====================================<>=====================================

Problem 7.8 In Problem 7.7, $b = 615 \ mm$, $c = 445 \ mm$, $h = 985 \ mm$, and $m = 77 \ kg$. If the bicycle is traveling at 6 *m/s* and the person engages the brakes, achieving the largest deceleration for which the rear wheel will not leave the ground, how long does it take the bicycle to stop, and what distance does it travel?

Solution:

Use the solution to Problem 7.7, $N_B = m\left(\frac{bg + ha}{b + c}\right)$, from which, for $N_B = 0$, the acceleration is $a = -bg/h = -0.615(9.81)/0.985 = -6.125 \ \text{m}/\text{s}^2$. The time to stop is $\boxed{t = v/a = 6/6.125 = 0.98 \ \text{s}}$, and the distance traveled is $\boxed{s = \frac{1}{2}at^2 = \frac{6.125(0.98^2)}{2} = 2.94 \ m}$

=====================================<>=====================================

=================================<>=================================

Problem 7.9 The combined mass of the motorcycle and rider is 160 kg. The rear wheel exerts a 400 N horizontal force on the road, and you can neglect the horizontal force exerted by the front wheel. Modeling the motorcycle and its wheels as a rigid body, determine (a) the motorcycle's acceleration; (b) the normal forces exerted on the road by the rear and front wheels.

Solution:

The friction force on the rear wheel is $F = 400\ N$. From Newton's second law, the acceleration is $\boxed{a = \frac{F}{m} = \frac{400}{160} = 2.5\ m/s^2}$. The sum of the forces in the y direction is $\sum F_y = N_A + N_B - mg = 0$. The moment about the center of mass is $\sum M_{CM} = -0.66N_A + (1.5 - 0.66)N_B + 0.66(400) = 0$. Solve these two equations in two unknowns: $\boxed{N_A = 1055\ N}$, $\boxed{N_B = 514.6\ N}$

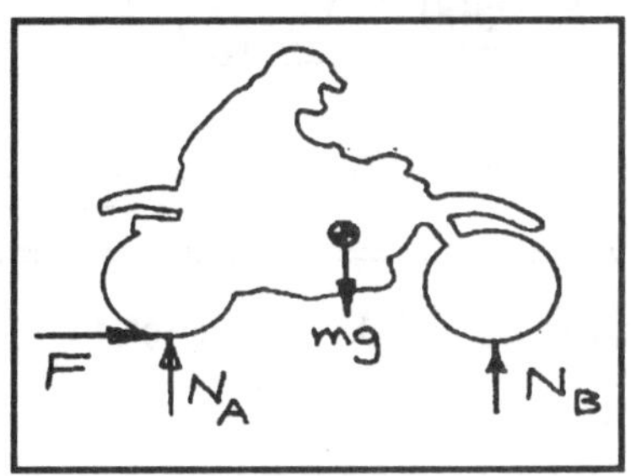

=================================<>=================================

Problem 7.10 In Problem 7.9, the coefficient of kinetic friction between the motorcycle's rear wheel and the road is $\mu_k = 0.8$. If the rider spins the rear wheel, what is the motorcycle's acceleration and what are the normal forces exerted on the road by the rear and front wheels?

Solution:

The friction force on the rear wheel is $F = \mu_k N_A$. The sum of the forces in the y direction is $\sum F_y = N_A + N_B - mg = 0$. The sum of the moments about the center of mass is $\sum M_{CM} = -0.66N_A + (1.5 - 0.66)N_B + 0.66\mu_k N_A = 0$. Solve the two equations in two unknowns: $\boxed{N_A = 1356.4\ N}$, $\boxed{N_B = 213.2\ N}$. The horizontal force exerted by the rear wheel is $F = \mu_k N_A = 1085.2\ N$. From Newton's second law the acceleration is $\boxed{a = \frac{F}{m} = \frac{1085.2}{160} = 6.78\ m/s^2}$

=================================<>=================================

Problem 7.11 During extravehicular activity, an astronaut fires a thruster of his maneuvering unit, exerting a force $T = 14.2\ N$ for one second. It requires 60 seconds from the time the thruster is fired for him rotate through one revolution. If you model the astronaut and maneuvering unit as a rigid body, what is the moment of inertia about their center of mass?

Solution:

The angular velocity is $\omega = \alpha t = \left(\frac{2\pi}{60}\right)\ rad/s$. Since the thruster fires for one second, $\alpha = \frac{2\pi}{60}\ rad/s^2$. The sum of the moments: $\sum M_{CM} = Tr = 14.2(0.3) = 4.26\ N\text{-}m$. From the equation of angular motion $I\alpha = \sum M_{CM} = 4.26\ N\text{-}m$, from which

$$\boxed{I = \frac{4.26}{\left(\frac{2\pi}{60}\right)} = 40.7\ kg\text{-}m^2}$$

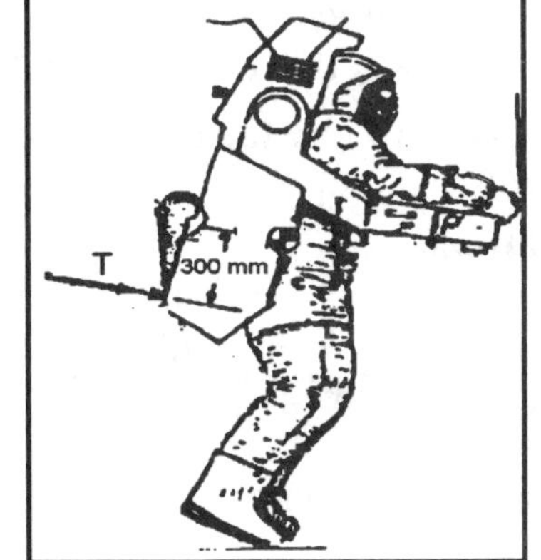

=================================<>=================================

Problem 7.12 The moment of inertia of the helicopter's rotor is 400 *slug-ft²*. If the rotor starts from rest at $t = 0$, the engine exerts a constant torque of 500 *ft-lb.* on the rotor, and the aerodynamic drag is neglected, what is the rotor's angular velocity ω at $t = 6$ seconds?

Solution:

From the equation of angular motion,

$$\alpha = \frac{T}{I} = \frac{500}{400} = 1.25 \ rad/s^2.$$

The angular velocity $\boxed{\omega = \alpha t = (1.25)(6) = 7.5 \ rad/s}$

Problem 7.13 In Problem 7.12, if aerodynamic drag exerts a torque on the helicopter's rotor of magnitude $20\omega^2$ *ft - lb*, what is the rotor's angular velocity at $t = 6$ seconds?

Solution:

The moment on the rotor is $\sum M = 500 - 20\omega^2 \ lb\text{-}ft$.

From the equation of angular motion $\alpha = \frac{500 - 20\omega^2}{400} = 1.25 - 0.05\omega^2 \ rad/s^2$. Denote $a^2 = 1.25$, $b^2 = 0.05$. By definition, $\frac{d\omega}{dt} = \alpha = (a^2 - b^2\omega^2)$.

Separate variables and integrate: $\int \frac{d\omega}{a^2 - b^2\omega^2} = \int dt + C$.

Note that $\frac{1}{a^2 - b^2\omega^2} = \frac{1}{2a}\left(\frac{1}{a + b\omega} + \frac{1}{a - b\omega}\right)$, from which $\left(\frac{1}{2ab}\right)(\ln(a + b\omega) - \ln|(a - b\omega)| = t + C$,

$\ln\left|\frac{a + b\omega}{a - b\omega}\right| = 2abt + 2abC$. Assume $\omega = 0$ at $t = 0$, from which $C = 0$. Invert and solve:

$\frac{a + b\omega}{a - b\omega} = e^{2abt}$, $\omega = \left(\frac{a}{b}\right)\left(\frac{e^{2abt} - 1}{e^{2abt} + 1}\right)$. At $t = 6 \ s$, $\boxed{\omega = 4.53 \ rad/s}$ at 6 seconds.

Problem 7.14 The moment of inertia of the robotic manipulator arm about the y axis is 8 $slug\text{-}ft^2$. The moment of inertia of the 30 *lb.* casting held by the arm about the y' axis is 0.6 $slug - ft^2$. What couple about the y axis is necessary to give the manipulator arm an angular acceleration of 2 *rad/s*?

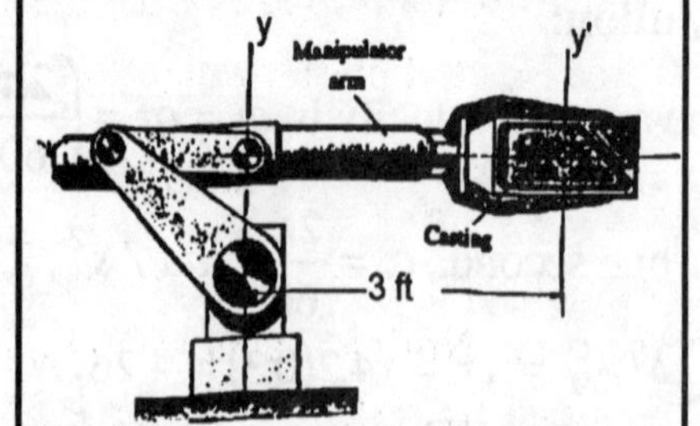

Solution:

The moment of inertia about the y axis is

$$I_{yy} = 8 + 0.6 + (3^2)\left(\frac{30}{g}\right) = 17 \ slug\text{-}ft^2.$$

From the equation of angular motion, $\boxed{T = I_{yy}\alpha = 17(2) = 34 \ ft\text{-}lb}$

Problem 7.15 The gears A and B turn freely on their pin supports. Their moments of inertia are $I_A = 0.002\ kg\text{-}m^2$ and $I_B = 0.006\ kg\text{-}m^2$. They are initially stationary, and at $t = 0$ a constant couple $M = 2\ N\text{-}m$ is applied to gear B. How many revolutions has gear A turned at $t = 4\ s$?

Solution:

Use the usual conventions of directions for angles and angular accelerations. The radius of gear A is 60 *mm*, and the radius of gear B is 90 *mm*. The tangential accelerations at the point of contact are equal: $-r_A\alpha_A = r_B\alpha_B$. From the equation of angular motion

$M - Fr_B = I_B\alpha_B$, and $-Fr_A = I_A\alpha_A$, from which $M + \left(\frac{r_B}{r_A}\right)I_A\alpha_A = I_B\alpha_B$

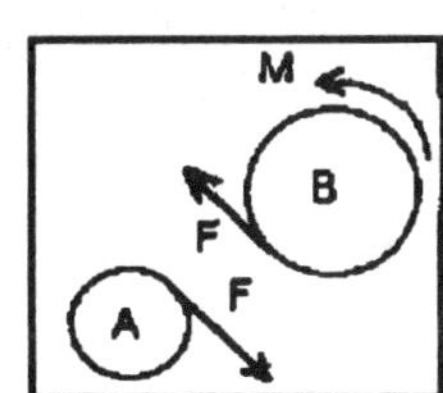

$$2 = -\left(\frac{r_A}{r_B}\right)I_B\alpha_A - \left(\frac{r_B}{r_A}\right)I_A\alpha_A,\ \alpha_A = \frac{-2}{\left(\frac{r_A}{r_B}\right)I_B + \left(\frac{r_B}{r_A}\right)I_A} = -285.7\ rad/s^2$$

The angle of revolution in 4 seconds is $n = \left(\frac{\alpha_A}{2}\right)(4^2) = -2285.7\ rad$, from which

$\boxed{N = \frac{n}{2\pi} = -363.8\ revs}$.

Problem 7.16 The moment of inertia of the pulley is 0.4 *slug-ft*^{2}. Determine the pulley's angular acceleration and the tension in the cable in the two cases.

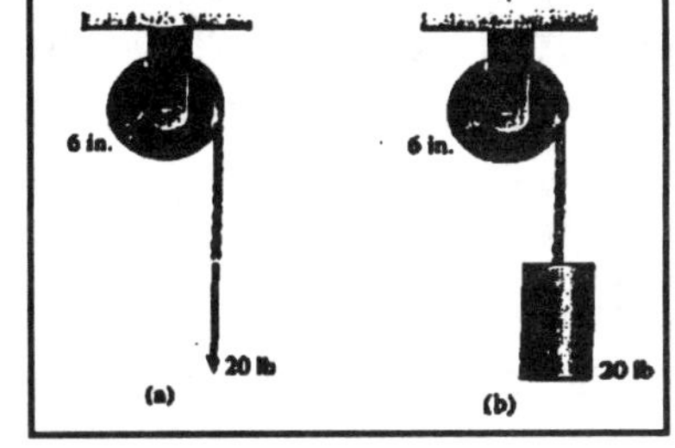

Solution:

Choose a coordinate system with *y* positive upward.

(a) The clockwise moment applied to the pulley by the rope is $M = 0.5(20) = 10\ ft\text{-}lb$. From the equation of angular motion, the clockwise angular acceleration of the pulley is $\boxed{\alpha = \frac{M}{I} = \frac{10}{0.4} = 25\ rad/s^2}$. The tension in the rope is the force on the rope: $\boxed{T = 20\ lb}$.

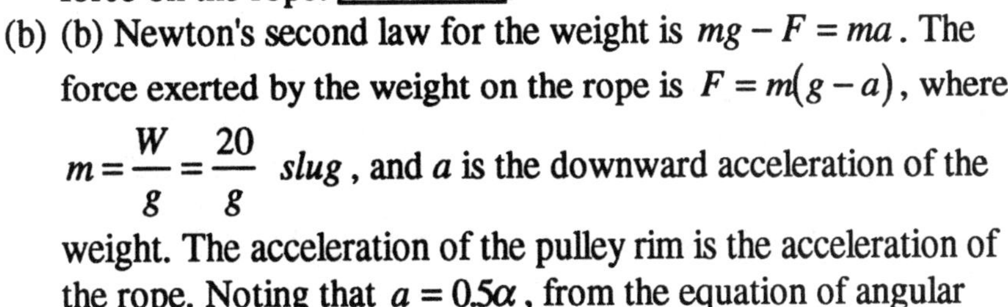

(b) (b) Newton's second law for the weight is $mg - F = ma$. The force exerted by the weight on the rope is $F = m(g - a)$, where $m = \frac{W}{g} = \frac{20}{g}\ slug$, and *a* is the downward acceleration of the weight. The acceleration of the pulley rim is the acceleration of the rope. Noting that $a = 0.5\alpha$, from the equation of angular motion, $I\alpha = 0.5F = 0.5m(g - 0.5\alpha)$, from which

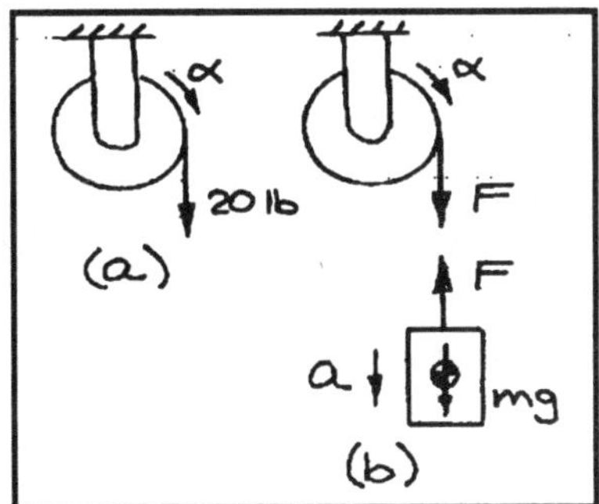

$(I + 0.25m)\alpha = 0.5mg = 10$, and $\alpha = \frac{10}{I + 0.25m} = 18\ rad/s^2$. The force exerted by the weight on the rope is $\boxed{F = W - 0.5m\alpha = 14.4\ lb}$

====================================<>====================================

Problem 7.17 Each box weighs 50 *lb.*, the moment of inertia of the pulley is 0.6 *slug-ft*^{2} and friction can be neglected. If the boxes start from rest at $t = 0$, determine the magnitude of their velocity. and the distance they have moved from their initial position at $t = 1\ s$.

Solution:

Choose a coordinate system with the origin at O and the *x* axis parallel to the plane surface, positive to the right. The usual convention for the direction of pulley angles applies.

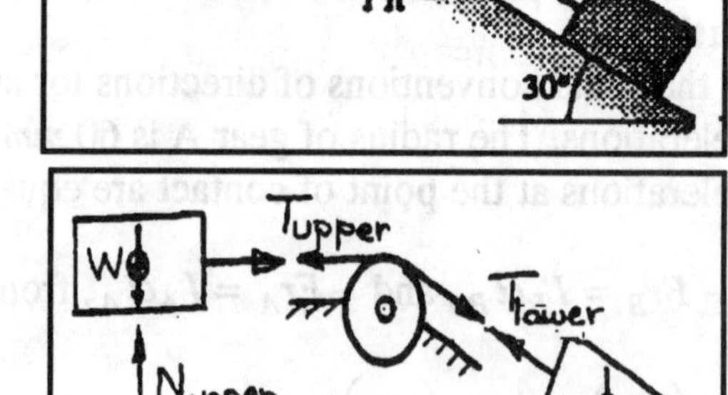

From Newton's second law: $-T_{lower} + W\sin 30^o = \left(\frac{W}{g}\right)a$,

$T_{upper} = \left(\frac{W}{g}\right)a$, and $RT_{upper} - RT_{lower} = I\alpha$.

From kinematics (the pulley has a one foot radius; the symbol *R* is retained for dimensional checks): $a = -(R)\alpha$,

from which $2\left(\frac{W}{g}\right)a - W\sin 30 = -\left(\frac{I}{R^2}\right)a$

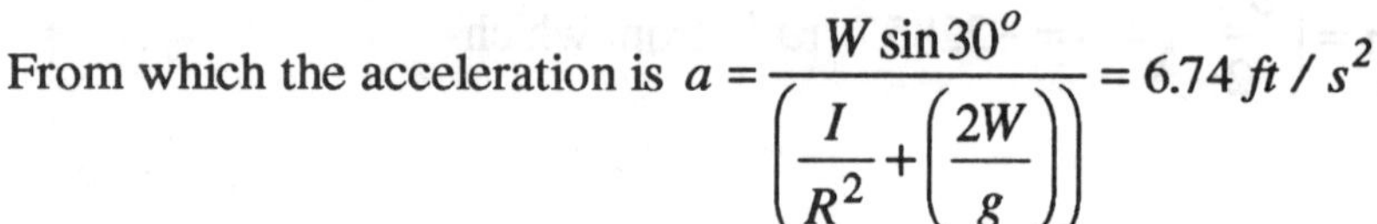

From which the acceleration is $a = \frac{W\sin 30^o}{\left(\frac{I}{R^2} + \left(\frac{2W}{g}\right)\right)} = 6.74\ ft/s^2$.

The velocity after 1 second is $\boxed{v = at = 6.74\ ft/s}$. The distance traveled is $\boxed{s = \frac{a}{2}t^2 = 3.37\ ft}$

====================================<>====================================

Problem 7.18 The slender bar weighs 10 *lb.* and the disk weighs 20 *lb.* The coefficient of kinetic friction between the disk and the horizontal surface is $\mu_k = 0.1$. If the disk has an initial counterclockwise angular velocity of 10 *rad/s*, how long does it take to stop spinning?

Solution:

From the free body diagram of the disk:

$\sum F_x = -\mu N + F_{Bx} = 0$, $\sum F_y = -F_{By} - W_D + N = 0$.

For the bar, the sum of the moments about the left hand support:

$\sum M_A = -(1.5\cos 30^o)W_B + (3\cos 30^o)F_{By} - (3\sin 30^o)F_{Bx} = 0$.

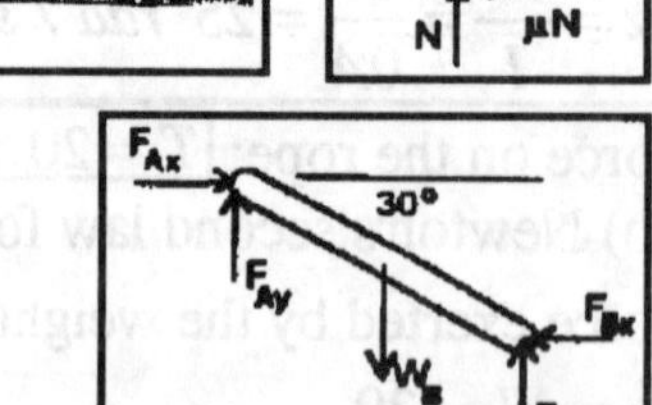

Solve these three equations in three unknowns: $N = 26.53\ lb$, $F_{Bx} = 2.65\ lb$, $F_{By} = 6.53\ lb$, from which $\mu_k N = 2.65\ lb$.

The moment of inertia of a disk about the axis of rotation is $I_D = \frac{1}{2}mR^2 = \frac{W}{2g} = 0.311\ slug\text{-}ft^2$.

From the equation of angular motion, $-\mu_k N(1) = I_D\alpha$, from which $\alpha = -8.54\ rad/s^2$.

The time required for the disk to stop is obtained from $0 = at + \omega$, from which $\boxed{t = \frac{\omega}{\alpha} = 1.17\ s}$

====================================<>====================================

=================================<>=================================

Problem 7.19 In Problem 7.18, how long does it take the disk to stop spinning if it has a clockwise angular velocity of 10 *rad/s* ?

Solution:

Use the solution to Problem 7.18, with the friction force between the surface and the disk reversed in direction in the free body diagrams.

The sum of forces for the disk: $\sum F_x = \mu N + F_{Bx} = 0$, $\sum F_y = -F_{By} - W_D + N = 0$.

The sum of moments for the bar about the left support is

$\sum M_A = -(1.5\cos 30^o)W_B + (3\cos 30^o)F_{By} - (3\sin 30^o)F_{Bx} = 0$.

Solve: $N = 23.6\ lb$, $F_{Bx} = -2.36\ lb$, and $F_{By} = 3.63\ lb$, from which $\mu_k N = 2.36\ lb$.

From the equation of angular motion: $\alpha = \frac{\mu_k N}{I_D} = \frac{2.36}{0.31} = 7.60\ rad/s^2$, from which $\boxed{t = \frac{\omega}{\alpha} = 1.315\ s}$

=================================<>=================================

Problem 7.20 The objects consist of identical 3 *ft*, 10 *lb.* bars welded together. If they are released from rest in the positions shown, what are their angular accelerations and what are the components of the reactions at A at that instant? (The *y* axes are vertical.)

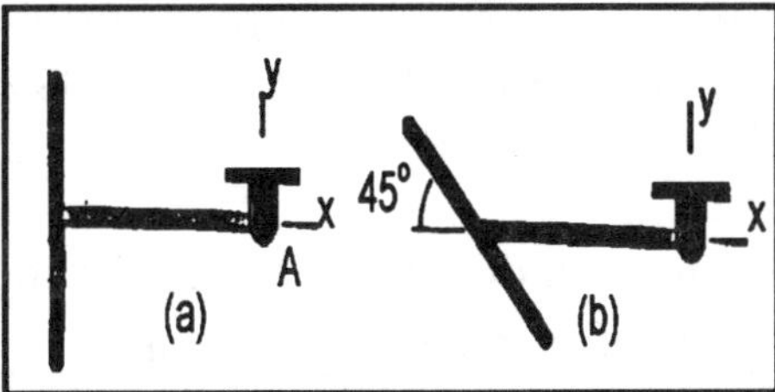

Solution:

The angular velocity $\omega = 0$ at the instant of interest since the system is released from rest. The moment of inertia of a slender rod about the end is $I_{end} = mL^2/3$. The moment of inertia about the center of the slender rod is

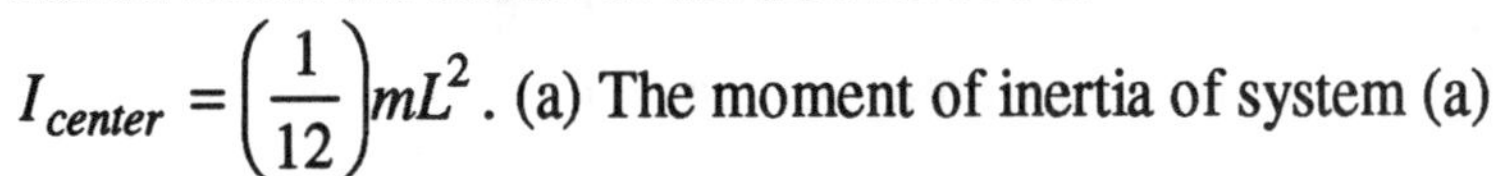

$I_{center} = \left(\frac{1}{12}\right)mL^2$. (a) The moment of inertia of system (a) about A is

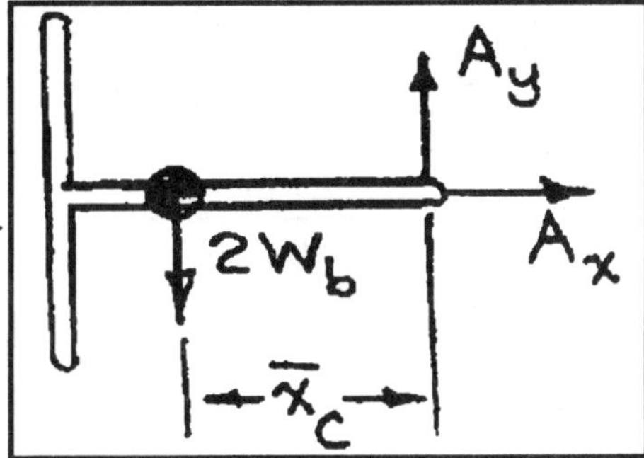

$I_A = mL^2\left(1 + \frac{1}{3} + \frac{1}{12}\right) = \left(\frac{10}{32.17}\right)(3^2)\left(\frac{12+4+1}{12}\right) = 3.96\ slug$ - . The center of mass of the system is

$\bar{x}_C = \frac{m\left(\frac{L}{2}\right) + mL}{2m} = 2.25\ ft$. The moment about A is

$\vec{M} = \vec{r}_{C/A} \times -2m\vec{g} = \begin{bmatrix} \vec{i} & \vec{j} & \vec{k} \\ \bar{x}_C & \bar{y}_C & 0 \\ 0 & -2mg & 0 \end{bmatrix} = -2mg\bar{x}_C\vec{k} = 45\vec{k}\ ft\text{-}lb$. From the equation of angular motion,

$\vec{M} = I_A\vec{\alpha}$, from which $\boxed{\alpha = \frac{M}{I_A} = \frac{45}{3.96} = 11.35\ rad/s^2}$.. The reactions at A are obtained from

Newton's second law: $A_y - 2W_b = \frac{2W_b}{g}a_y$, $A_x = \frac{2W_b}{g}a_x$, where a_x, a_y are the accelerations at the center of mass of the system. From the kinematic relationships: $\vec{a}_{CM} = \vec{\alpha} \times \vec{r}_{C/A} - \omega^2\vec{r}_{C/A}$ from which, since

Solution continued on next page

$\omega = 0$, $\vec{a}_{CM} = \vec{\alpha} \times \vec{r}_{C/A} = \begin{bmatrix} \vec{i} & \vec{j} & \vec{k} \\ 0 & 0 & \alpha \\ -\bar{x}_C & 0 & 0 \end{bmatrix} = -\alpha \bar{x}_C \vec{j} = -25.55\vec{j}\ ft/s^2$. The reaction at A: $A_x = 0$,

$$\boxed{A_y = 2W_b + \left(\frac{2W_b}{g}\right) a_y = 20 + \frac{20}{32.17}(-25.5) = 4.12\ lb}.$$

(b) The moment of inertia of system (b) about A is

$I_A = mL^2\left(1 + \frac{1}{3} + \frac{1}{12}\right) = \left(\frac{10}{32.17}\right)(3^2)\left(\frac{12+4+1}{12}\right) = 3.96\ slug\text{-}ft^2$. The center of mass of the system is $\bar{x}_C = \frac{m(L/2) + mL}{2m} = 2.25$ ft. These are identical with the results of part (a), from which

$\boxed{\alpha = 11.35\ rad/s^2}$, $\boxed{A_y = 2W_b + (2W_b/g)a_y = 4.12\ \text{lb}}$, and $\boxed{A_x = 0}$

==============================<>==============================

Problem 7.21 The object consists of identical 1 *m*, 5 *kg* bars welded together. If it is released from rest in the position shown, what is its angular acceleration and what are the components of the reaction at A at that instant?

Solution:

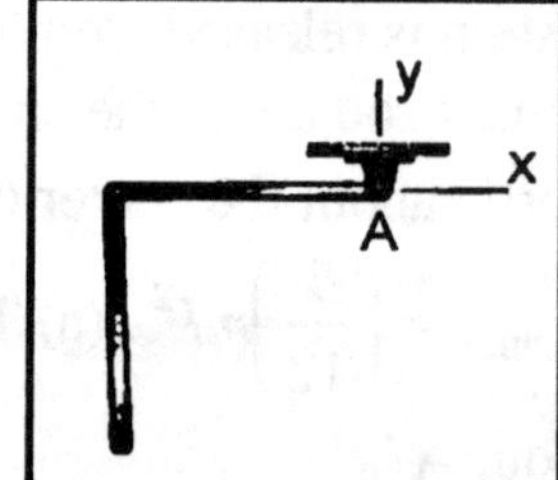

The moment of inertia of a slender rod about one end is $I_{end} = mL^2/3$. The moment of inertia of the system is

$I_A = mL^2\left(\frac{1}{3} + \frac{1}{3} + 1\right) = 5(1^2)\left(\frac{5}{3}\right) = 8.33\ kg\text{-}m^2$. The center of mass of the system is $\bar{x}_C = \frac{-m(L/2) - mL}{2m} = -\frac{3L}{4} = -0.75$ m.

$\bar{y}_C = \frac{-m(L/2) - m0}{2m} = -\frac{L}{4} = -0.25$. The moment about the point A is

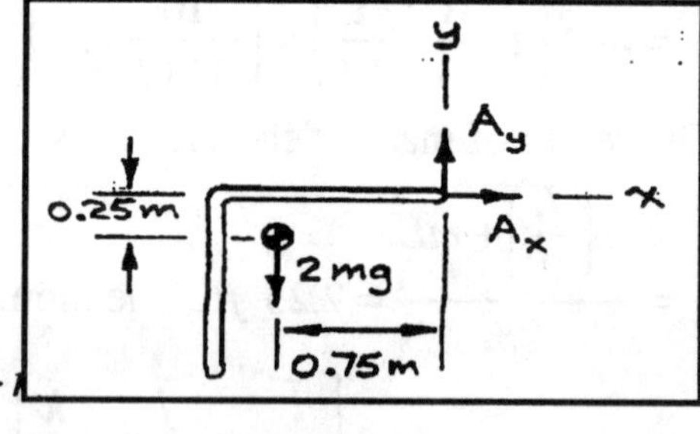

$$\vec{M} = \vec{r}_{C/A} \times 2m\vec{g} = \begin{bmatrix} \vec{i} & \vec{j} & \vec{k} \\ \bar{x}_C & \bar{y}_C & 0 \\ 0 & -2mg & 0 \end{bmatrix} = -2mg\bar{x}_C\vec{k} = 73.58\vec{k}\ N\text{-}m$$

From the equation of angular motion: $\vec{M} = I_A \vec{\alpha}$, from which $\alpha = \frac{M}{I_A} = \frac{73.58}{8.33} = 8.829\ rad/s^2$. The reactions at A are obtained from Newton's second law: $A_x = 2ma_x$, $A_y - 2mg = 2ma_y$, where a_x, a_y are the accelerations of the center of mass. From the kinematics, for $\omega = 0$ since the system is released

from rest: $\vec{a}_{C/A} = \vec{\alpha} \times \vec{r}_{C/A} - \omega^2 \vec{r}_{C/A} = \begin{bmatrix} \vec{i} & \vec{j} & \vec{k} \\ 0 & 0 & \alpha \\ \bar{x}_C & \bar{y}_C & 0 \end{bmatrix} = -\alpha \bar{y}_C \vec{i} + \alpha \bar{x}_C \vec{j} = 4.413\vec{i} - 6.622\vec{j}\ (m/s^2)$..

The reactions: $\boxed{A_x = 2ma_x = 22.07\ N}$ $\boxed{A_y = 2mg + 2ma_y = 31.89\ N}$

==============================<>==============================

=================================<>=================================

Problem 7.22 For what value of x is the horizontal bar's angular acceleration a maximum, and what is the maximum angular acceleration?

Solution:

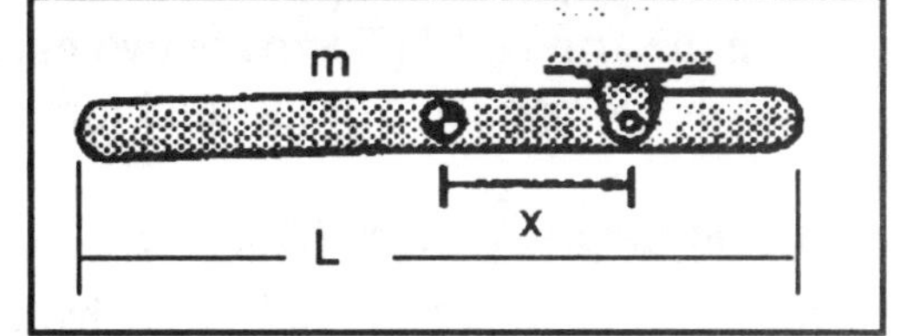

The moment of inertia is $I = mx^2 + \frac{1}{12}mL^2$. The moment about the pin is $M = mgx$. From the equation of angular motion, $I\alpha = mgx$, from which $\alpha = gx\left(x^2 + \frac{L^2}{12}\right)^{-1}$. Take the derivative: $\frac{d\alpha}{dx} = 0 = g\left(x^2 + \frac{L^2}{12}\right)^{-1}\left(1 - 2x^2\left(x^2 + \frac{L^2}{12}\right)^{-1}\right)$. Solve: $\boxed{x = \frac{L}{\sqrt{12}}}$, and $\boxed{\alpha_{\max} = \frac{\sqrt{3}g}{L}}$

=================================<>=================================

Problem 7.23 Model the arm ABC as a single rigid body. Its mass is 300 kg and the moment of inertia about the center of mass is $I = 360\ kg\text{-}m^2$. If point A is stationary and the angular acceleration of the arm is 0.6 rad/s^2 counterclockwise, what force does the hydraulic cylinder exert on the arm at B? (The arm has two hydraulic cylinders, one on each side of the vehicle. You are to determine the total force exerted by the two cylinders.)

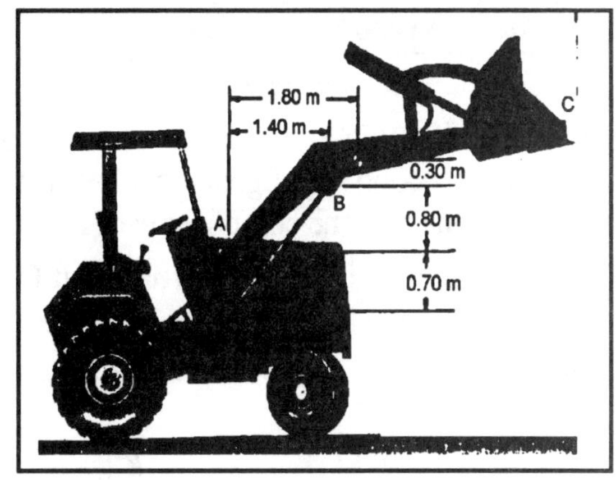

Solution:

The angle of the hydraulic cylinder with the horizontal is $\theta = \tan^{-1}\left(\frac{1.5}{1.4}\right) = 47^o$. The moment of inertia about A is $I_A = 360 + (1.8^2 + 1.1^2)(300) = 1695\ kg\text{-}m^2$. The moment is the sum of the moments exerted by the cylinder and the weight of the arm:

$\vec{M} = \vec{r}_{AB} \times \vec{F} + \vec{r}_{CM} \times m\vec{g}$.

$$\vec{M} = \begin{bmatrix} \vec{i} & \vec{j} & \vec{k} \\ 1.4 & 0.8 & 0 \\ F\cos\theta & F\sin\theta & 0 \end{bmatrix} + \begin{bmatrix} \vec{i} & \vec{j} & \vec{k} \\ 1.8 & 1.1 & 0 \\ 0 & -mg & 0 \end{bmatrix}$$

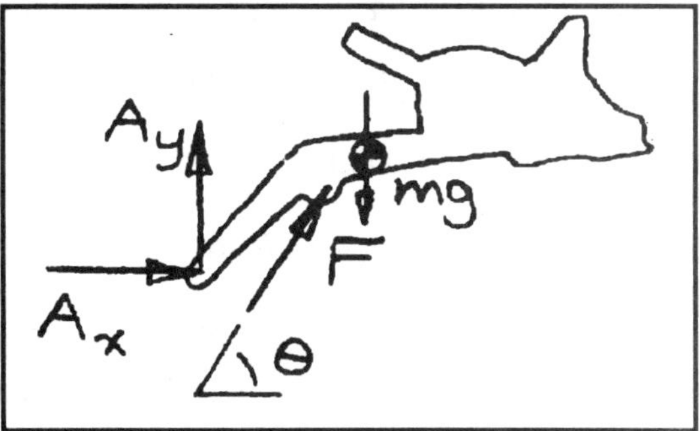

$\vec{M} = (0.4776F - 5297.4)\vec{k}\ (N\text{-}m)$. From the equation of angular motion: $\vec{M} = I_A\vec{\alpha}$, from which $(0.4776F - 5297.4)\vec{k} = 1695(0.6)\vec{k}$, and

$\boxed{F = 13220\ N}$

=================================<>=================================

================================<>================================

Problem 7.24 In Problem 7.23, if the angular acceleration of arm ABC is 0.6 *rad/s*^{2} counterclockwise, and its angular velocity is 1.4 *rad/s* clockwise, what are the components of the force exerted on the arm at A? (There are two pin supports, one on each side of the vehicle. You are to determine the components of the total force exerted by the two supports.)

Solution:

Use the solution to Problem 7.23. The force exerted by the hydraulic cylinder is $F = 13220\ N$. The components of the force exerted by the hydraulic cylinder are $F_x = F\cos\theta = 13220\cos 47^o = 9021\ N$, $F_y = F\sin 47^o = 9665\ N$. From Newton's second law: $A_x + F_x = ma_x$, $A_y - mg + F_y = ma_y$, where a_x, a_y are the components of the acceleration at the center of mass. From kinematics, this acceleration is given by $\vec{a} = \vec{\alpha}\times\vec{r}_{CM/A} - \omega^2\vec{r}_{CM/A} = \begin{bmatrix} \vec{i} & \vec{j} & \vec{k} \\ 0 & 0 & 0.6 \\ 1.8 & 1.1 & 0 \end{bmatrix} - (1.4^2)(1.8\vec{i} + 1.1\vec{j})$. Carry out the indicated operations to obtain: $\vec{a} = -4.188\vec{i} - 1.076\vec{j}\ (m/s^2)$. Substitute into the expressions for Newton's second law relating the forces to the acceleration, and note that $mg = 2943\ N$, to obtain $\boxed{A_x = -10277\ N}$ and $\boxed{A_y = -7044\ N}$

================================<>================================

Problem 7.25 To lower the drawbridge, the gears that raised it are disengaged and a fraction of a second later a second set of gears that lower it are engaged. At the instant the gears that raised it are disengaged, what are the components of force exerted by the bridge on its support at O? The drawbridge weighs 360 *kip*, its moment of inertia about O is $I_O = 1.0\times 10^7\ slug\text{-}ft^2$, and the coordinates of its center of mass at the instant the gears are disengaged are $\bar{x} = 8\ ft$, $\bar{y} = 16\ ft$.

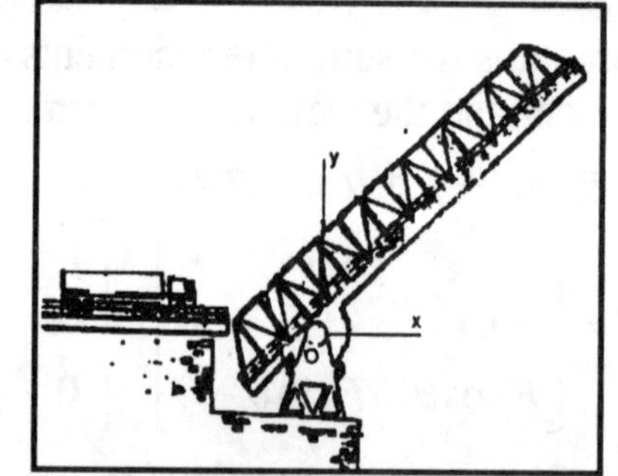

Solution:

The moment about O is

$$\vec{M}_O = \vec{r}_{CM/O}\times\vec{W} = \begin{bmatrix} \vec{i} & \vec{j} & \vec{k} \\ 8 & 16 & 0 \\ 0 & -360000 & 0 \end{bmatrix} = -2{,}880{,}000\vec{k}\ lb\text{-}ft$$

From the equation of angular motion, $\vec{M}_O = I_O\vec{\alpha}$, from which

$$\alpha = \frac{M_O}{I_O} = \frac{-2{,}880{,}000}{1\times 10^7} = -0.288\ rad/s^2.$$

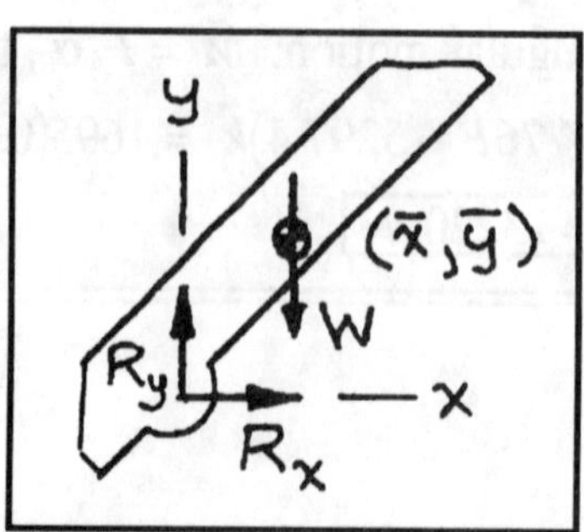

From Newton's second law the reactions are: $R_x = ma_x$, $R_y - W = ma_y$, where a_x, a_y are the accelerations of the center of mass. From kinematics:

$$\vec{a} = \vec{\alpha}\times\vec{r}_{CM/O} = \begin{bmatrix} \vec{i} & \vec{j} & \vec{k} \\ 0 & 0 & -0.288 \\ 8 & 16 & 0 \end{bmatrix} = 4.61\vec{i} - 2.30\vec{j}\ (ft/s^2).$$

Substitute and solve: $\boxed{R_x = 51{,}566\ lb}$, $\boxed{R_y = 334{,}217\ lb}$

================================<>================================

================================<>================================

Problem 7.26 The arm BC has a mass of 12 *kg* and the moment of inertia about its center of mass is 3 $kg\text{-}m^2$. If B is stationary and arm BC has a constant counterclockwise angular velocity of 2 *rad/s* at the instant shown, determine the couple and the components of the force exerted on arm BC at B.

Solution:

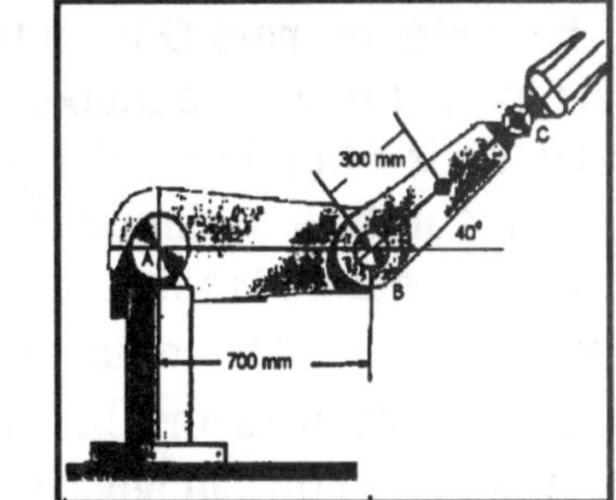

Since the angular acceleration of arm BC is zero, the sum of the moments about the fixed point B must be zero. Let $\vec{M}_B$ be the couple exerted by the support at B. Then

$$\vec{M}_B + \vec{r}_{CM/B} \times m\vec{g} = \vec{M}_B + \begin{bmatrix} \vec{i} & \vec{j} & \vec{k} \\ 0.3\cos 40^o & 0.3\sin 40^o & 0 \\ 0 & -117.2 & 0 \end{bmatrix} = 0.$$

$\boxed{\vec{M}_B = 27.05\vec{k} \quad (N\text{-}m)}$ is the couple exerted at B. From Newton's second law: $B_x = ma_x$, $B_y - mg = ma_y$ where a_x, a_y are the accelerations of the center of mass. From kinematics:

$\vec{a} = \vec{\alpha} \times \vec{r}_{CM/O} - \omega^2 \vec{r}_{CM/O} = -(2^2)(\vec{i}\,0.3\cos 40^o + \vec{j}\,0.3\sin 40^o) = -0.919\vec{i} - 0.771\vec{j} \ (m/s^2)$, where the angular acceleration is zero from the problem statement. Substitute into Newton's second law to obtain the reactions at B: $\boxed{B_x = -11.03 \ N}$, $\boxed{B_y = 108.5 \ N}$.

================================<>================================

Problem 7.27 In Problem 7.26, what are the couple and the components of the force exerted on arm BC if arm AB has a constant clockwise angular velocity of 2 *rad/s* and arm BC has counterclockwise angular velocity of 2 *rad/s* and a clockwise angular acceleration of 4 rad/s^2 at the instant shown?

Solution:

Because the point B is accelerating, the equations of angular motion must be written about the center of mass of arm BC. The vector distances from A to B and B to G, respectively, are $\vec{r}_{B/A} = \vec{r}_B - \vec{r}_A = 0.7\vec{i}$, $\vec{r}_{G/B} = 0.3\cos(40^o)\vec{i} + 0.3\sin(40^o)\vec{j} = 0.2298\vec{i} + 0.1928\vec{j} \ (m)$.

The acceleration of point B is $\vec{a}_B = \vec{\alpha} \times \vec{r}_{B/A} - \omega_{AB}^2 \vec{r}_{B/A} = -\omega_{AB}^2 (0.7\vec{i}) \ (m/s^2)$.

The acceleration of the center of mass is $\vec{a}_G = \vec{a}_B + \vec{\alpha}_{BC} \times \vec{r}_{G/B} - \omega_{BC}^2 \vec{r}_{G/B}$

$$\vec{a}_G = -2.8\vec{i} + \begin{bmatrix} \vec{i} & \vec{j} & \vec{k} \\ 0 & 0 & -4 \\ & 0 & 0 \end{bmatrix} - 0.9193\vec{i} - 0.7713\vec{j} = -2.948\vec{i} - 1.691\vec{j} \ (m/s^2).$$

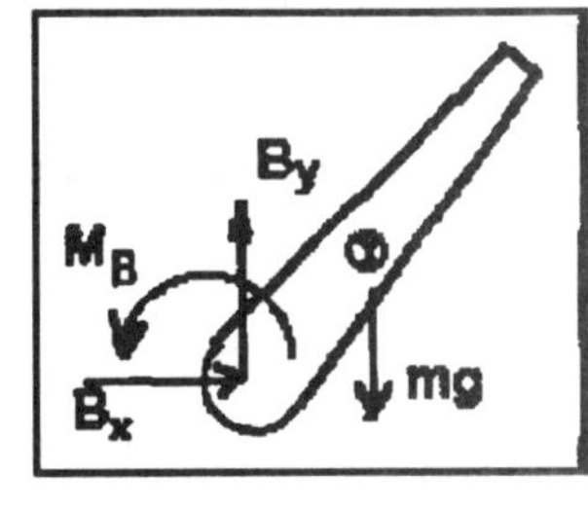

From Newton's second law, $\boxed{B_x = ma_{Gx} = (12)(-2.948) = -35.37 \ N}$,

$B_y - mg = ma_{Gy}$, $\boxed{B_y = (12)(-1.691) + (12)(9.81) = 97.43 \ N}$

From the equation of angular motion, $\vec{M}_G = I\,\vec{\alpha}_{BC}$. The moment about the center of mass is

$$\vec{M}_G = \vec{M}_B + \vec{r}_{B/G} \times \vec{B} = \begin{bmatrix} \vec{i} & \vec{j} & \vec{k} \\ -0.2298 & -0.1928 & 0 \\ -35.37 & 97.43 & 0 \end{bmatrix} = M_B\vec{k} - 29.21\vec{k} \ (N\text{-}m).$$ Note $I = 3 \ kg\text{-}m^2$ and

$\vec{\alpha}_{BC} = -4\vec{k} \ (rad/s^2)$, from which $\boxed{M_B = 29.21 + 3(-4) = 17.21 \ N\text{-}m}$.

================================<>================================

=====================================<>=====================================

Problem 7.28 The space shuttle's attitude control engines exert two forces $F_f = 8kN$ and $F_r = 2kN$. The force vectors ant the center of mass G lie in the x-y plane of the inertial reference frame. The mass of the shuttle is *54,000 kg* and its moment of inertia about the axis through the center of mass that is parallel to the z axis is $4.5\times10^6 kg-m^2$. Determine the acceleration of the center of mass and the angular acceleration. (You can ignore the force on the shuttle due to its weight).

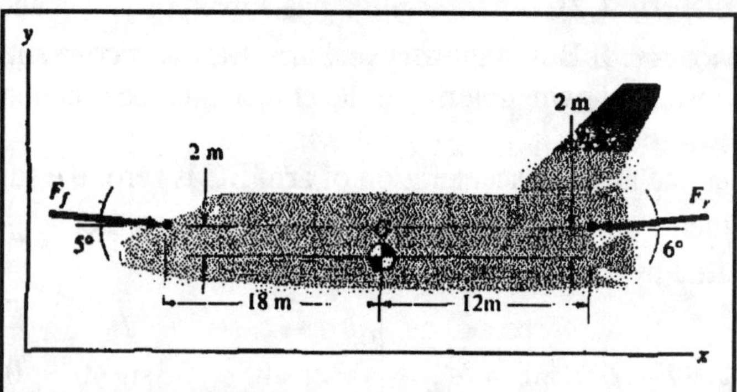

Solution:

Newton's second law is $\sum \bar{F} = (F_f \cos 5^\circ - F_r \cos 6^\circ)\bar{i} - (F_f \sin 5^\circ + F_r \sin 6^\circ)\bar{j} = m\bar{a}$. Setting $F_f = 8000N, F_r = 2000N$ and $m = 54{,}000kg$ and solving for $\bar{a}$,

we obtain $\bar{a} = 0.1108\bar{i} - 0.0168\bar{j}(m/s^2)$

The equation of angular motion is

$\sum M = (18)(F_f \sin 5^\circ) - (2)(F_f \cos 5^\circ) - (12)(F_r \sin 6^\circ) + (2)(F_r \cos 6^\circ) = I\alpha$

where $I = 4.5\times10^6 kg-m^2$.

Solving for α the counterclockwise angular acceleration is $\alpha = -0.000427 rad/s^2$.

=====================================<>=====================================

Problem 7.29 In Problem 7.28, suppose that $F_f = 4kN$ and you want the shuttle's angular acceleration to be zero. Determine the necessary force F_r and the resulting acceleration of the shuttle's center of mass.

Solution:

The total moment about the center of mass must equal zero: $\sum M = (18)(F_f \sin 5^\circ) - (2)(F_f \cos 5^\circ)$

$-(12)(F_r \sin 6^\circ) + (2)(F_r \cos 6^\circ) = 0$

Setting $F_f = 4000N$ and solving $F_r = 2306N$.

From Newton's second law

$\sum \bar{F} = (F_f \cos 5^\circ - F_r \cos 6^\circ)\bar{i} - (F_f \sin 5^\circ + F_r \sin 6^\circ)\bar{j} = 54{,}000\bar{a}$,

we obtain $\bar{a} = 0.0313\bar{i} - 0.0109\bar{j}(m/s^2)$.

=====================================<>=====================================

================================<>================================

Problem 7.30 Points B and C are in the *x-y* plane. At the instant shown, the angular velocity and angular acceleration vectors of arm AB are $\overline{\omega}_{AB} = 0.6\overline{k}\ rad/s$ and $\overline{\alpha}_{AB} = -0.3\overline{k}\ rad/s^2$. The angular velocity and angular acceleration vectors of arm BC are $\overline{\omega}_{BC} = 0.4\overline{k}\ rad/s$ and $\overline{\alpha}_{BC} = 2.0\overline{k}\ rad/s^2$. The center of mass of the *18-kg* arm BC is at the midpoint of the line from B to C, and its moment of inertia about the axis through the center of mass that is parallel to the *z* axis is 1.5 $kg-m^2$. Determine the force and couple exerted on arm BC at B.

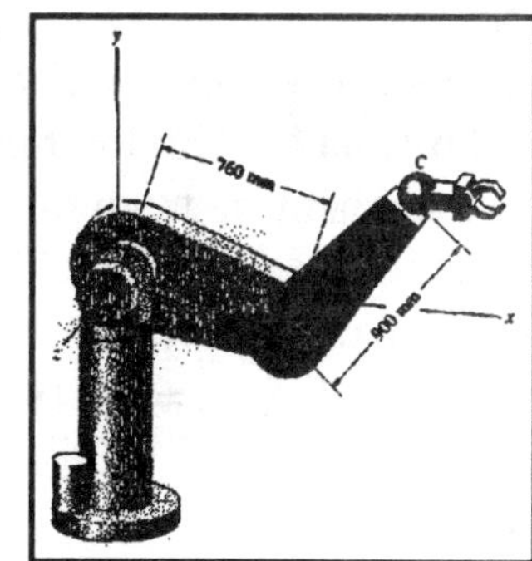

Solution:

The acceleration of point B is $\overline{a}_B = \overline{a}_A + \overline{\alpha}_{AB} \times \overline{r}_{A/B} - \omega_{AB}^2 \overline{r}_{A/B}$ or

$$\overline{a}_B = \begin{vmatrix} \overline{i} & \overline{j} & \overline{k} \\ 0 & 0 & -0.3 \\ 0.76\cos 15^\circ & -0.76\sin 15^\circ & 0 \end{vmatrix} - (0.6)^2(0.76\cos 15^\circ \overline{i} - 0.76\sin 15^\circ \overline{j}) = -0.323\overline{i} - 0.149\overline{j}(m/s^2)$$

The acceleration of the center of mass G of arm BC is $\overline{a}_G = \overline{a}_B + \overline{\alpha}_{BC} \times \overline{r}_{G/B} - \omega_{BC}^2 \overline{r}_{G/B}$

$$\overline{a}_G = -0.323\overline{i} - 0.149\overline{j} + \begin{vmatrix} \overline{i} & \overline{j} & \overline{k} \\ 0 & 0 & 2 \\ 0.45\cos 50^\circ & 0.45\sin 50^\circ & 0 \end{vmatrix}$$

$-(0.4)^2(0.45\cos 50^\circ \overline{i} + 0.45\sin 50^\circ \overline{j})$, or $\overline{a}_G = -1.059\overline{i} + 0.374\overline{j}(m/s^2)$. The free body diagram of arm BC is:

Newton's second law is $\sum \overline{F} = B_x\overline{i} + (B_y - mg)\overline{j} = m\overline{a}_G$:

$B_x\overline{i} + \left[B_y - (18)(9.81)\right]\overline{j} = 18(-1.059\overline{i} + 0.374\overline{j})$.

Solving, we obtain $B_x = -19.1N$ $B_y = 183.3N$.

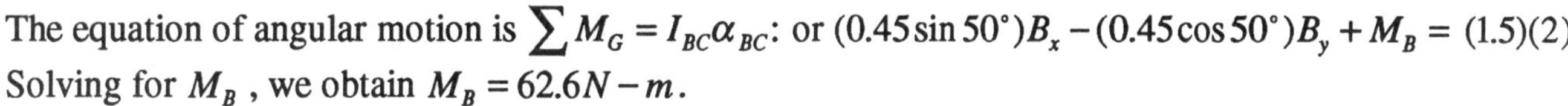

The equation of angular motion is $\sum M_G = I_{BC}\alpha_{BC}$: or $(0.45\sin 50^\circ)B_x - (0.45\cos 50^\circ)B_y + M_B =$ (1.5)(2)

Solving for M_B, we obtain $M_B = 62.6N-m$.

================================<>================================

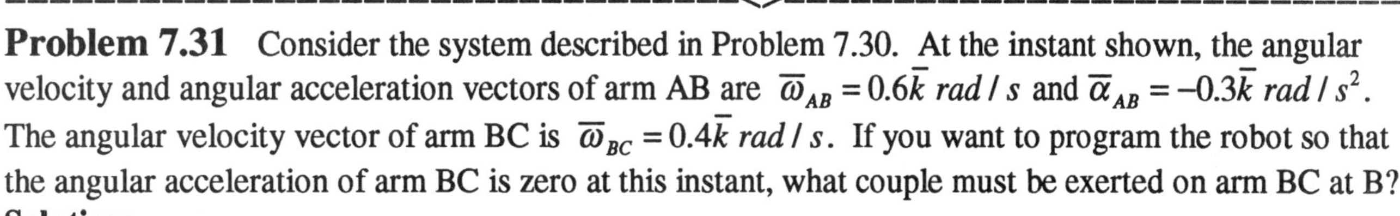

Problem 7.31 Consider the system described in Problem 7.30. At the instant shown, the angular velocity and angular acceleration vectors of arm AB are $\overline{\omega}_{AB} = 0.6\overline{k}\ rad/s$ and $\overline{\alpha}_{AB} = -0.3\overline{k}\ rad/s^2$. The angular velocity vector of arm BC is $\overline{\omega}_{BC} = 0.4\overline{k}\ rad/s$. If you want to program the robot so that the angular acceleration of arm BC is zero at this instant, what couple must be exerted on arm BC at B?

Solution:

From the solution of Problem 7.30, the acceleration of point B is $\overline{a}_B = -0.323\overline{i} - 0.149\overline{j}(m/s^2)$. If $\alpha_{BC} = 0$, the acceleration of the center of mass G of arm BC is

$\overline{a}_G = \overline{a}_B - \omega_{BC}^2 \overline{r}_{G/B} = -0.323\overline{i} - 0.149\overline{j} - (0.4)^2\ (0.45\cos 50^\circ \overline{i} + 0.45\sin 50^\circ \overline{j}) = -0.370\overline{i} - 0.205\overline{j}(m/s^2)$.

From the free body diagram of arm BC in the solution of Problem 7.30, Newton's second law is

$\sum \overline{F} = B_x\overline{i} + (B_y - mg)\overline{j} = m\overline{a}_G$: $B_x\overline{i} + \left[B_y - (18)(9.81)\right]\overline{j} = 18(-0.370\overline{i} - 0.205\overline{j})$.

Solving, we obtain $B_x = -6.65N$, $B_y = 172.90N$. The equation of angular motion is

$\sum M_G = I_{BC}\alpha_{BC} = 0$: $(0.45\sin 50^\circ)B_x - (0.45\cos 50^\circ)B_y + M_B = 0$. Solving for M_B, we obtain $M_B = 52.3N-m$.

================================<>================================

==================================<>==================================

Problem 7.32 The *9000-kg* airplane has just landed. At the instant shown, its angular velocity is zero. Its landing gear are rolling and contact the runway at $x = 10\ m$. The friction force on the wheels is negligible. The coordinates of the airplane's center of mass are $x = 10.50\ m$, $y = 3.00\ m$. The total aerodynamic force is $-26.8\bar{i} + 30.4\bar{j}\ (kN)$, and it effectively acts at the center of pressure located at $x = 10.75\ (m)$, $y = 3.2\ (m)$. The thrust $T = 4.40\ kN$ exerts no moment about the center of mass. The moment of inertia of the airplane about its center of mass is $75{,}000\ kg\text{-}m^2$. Determine the airplane's angular acceleration.

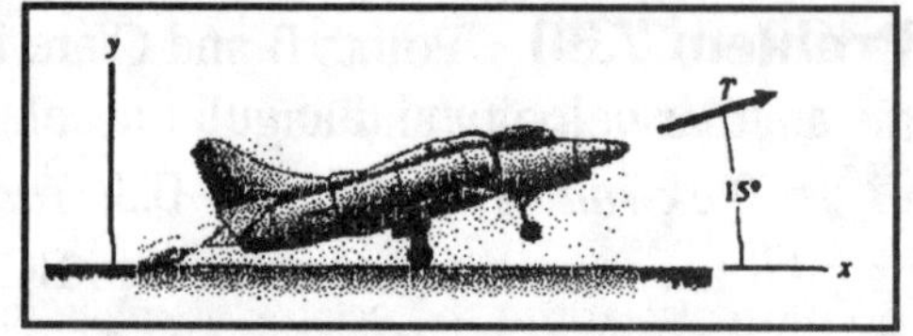

Strategy: Draw a free-body diagram of the airplane including the normal force exerted on the landing gear. To relate the acceleration of the center of mass to the angular acceleration, use the fact that the acceleration of the airplane (treated as a rigid body), is horizontal at the point where the wheels contact the runway.

Solution:

The free body diagram is as shown.

C is the center of pressure and $\bar{A}$ is the aerodynamic force.

Newton's second law is $\sum \bar{F} = \bar{T} + \bar{A} + \bar{N} - mg\bar{j} = m\bar{a}$:

$T\cos 15^\circ \bar{i} + T\sin 15^\circ \bar{j} + A_x\bar{i} + A_y\bar{j} + N\bar{j} - mg\bar{j} = m(a_x\bar{i} + a_y\bar{j})$.

Equating $\bar{i}$ and $\bar{j}$ components, $4400\cos 15^\circ - 26{,}800 = 9000a_x$,

$4400\sin 15^\circ + 30{,}400 + N - (9000)(9.81) = 9000a_y \quad (1)$.

The moment of $\bar{A}$ about the center of mass is $\bar{r}_{C/G} \times \bar{A} = \begin{vmatrix} \bar{i} & \bar{j} & \bar{k} \\ 0.25 & 0.20 & 0 \\ -26{,}800 & 30{,}400 & 0 \end{vmatrix} = 12{,}960\bar{k}\ (N-m)$.

The equation of angular motion is $\sum M = 12{,}960 - (0.5)N = 75{,}000\alpha. \quad (2)$

To relate a_x, a_y and α we express the acceleration of the point of contact P as $a_p\bar{i} = \bar{a}_G + \bar{\alpha} \times \bar{r}_{p/G} =$

$a_x\bar{i} + a_y\bar{j} + \begin{vmatrix} \bar{i} & \bar{j} & \bar{k} \\ 0 & 0 & \alpha \\ -0.5 & -3 & 0 \end{vmatrix}$.

Equating the $\bar{j}$ components, $0 = a_y - 0.5\alpha \quad (3)$

Solving Equations (1) - (3), the angular acceleration is $\alpha = -0.200\ rad/s^2$

==================================<>==================================

Problem 7.33 In Problem 7.32, determine the normal force exerted on the airplane by the runway.

Solution:

From Equations (1) - (3) of the solution of Problem 7.32, we have $N = 55.9\ kN$.

==================================<>==================================

==================================<>==================================

Problem 7.34 A thin ring and a circular disk, each of mass *m* and radius *R*, are released from rest on an inclined surface and allowed to roll a distance D. Determine the ratio of the times required.

Solution Choose a coordinate system with the *x* axis parallel to the inclined surface and the origin at the center of the disk at the instant of release.

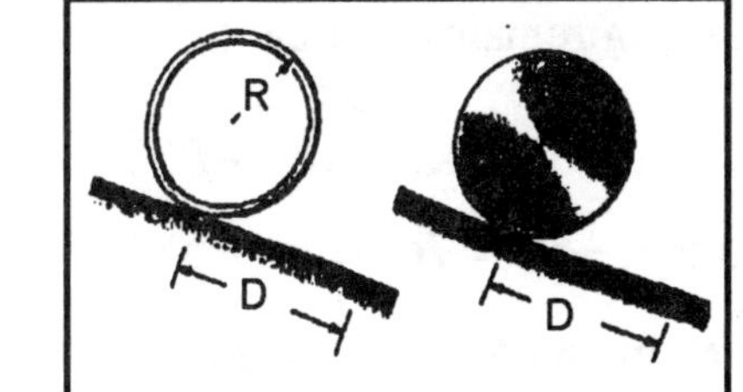

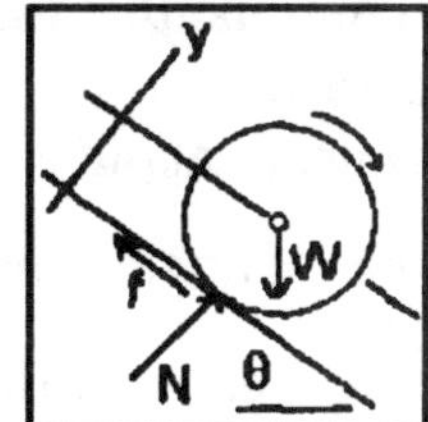

The moment about the center of mass is

$\vec{M} = \vec{R} \times \vec{f} = \begin{bmatrix} \vec{i} & \vec{j} & \vec{k} \\ 0 & -R & 0 \\ -f & 0 & 0 \end{bmatrix} = -Rf\vec{k}$. From the equation of angular motion, $\vec{M} = I\vec{\alpha}$, from which $-Rf = I\alpha$, $f = \frac{-I\alpha}{R}$. From Newton's second law and the free body diagram: $\sum F_x = -f + mg\sin\theta = ma_x$, where a_x is the acceleration of the center of mass. Substitute the expression for the force *f*: $\frac{I\alpha}{R} + mg\sin\theta = ma_x$.

The relation between α and a_x is found from kinematics: the acceleration of the point of contact P with the inclined surface is zero, from which $0 = \vec{a}_{CM} + \overline{\alpha} \times \vec{r}_{P/C} - \omega^2\vec{r}_{PC} = \vec{a}_{CM} + \overline{\alpha} \times (-R)\vec{j} - \omega^2(-R\vec{j})$, and

$\vec{a}_{CM} = \overline{\alpha} \times R\vec{j} + \omega^2 R\vec{j} = \begin{bmatrix} \vec{i} & \vec{j} & \vec{k} \\ 0 & 0 & \alpha \\ 0 & R & 0 \end{bmatrix} - \omega^2 R\vec{j} = -\alpha R\vec{i} - \omega^2 R\vec{j}$. From the constraint on the motion,

$\vec{a}_{CM} = a_x\vec{i}$, from which $a_x = -R\alpha$, or $\alpha = -\frac{a_x}{R}$. Substitute and solve: $a_x = \frac{mg\sin\theta}{\left(\frac{I}{R^2} + m\right)}$. The time required to travel a distance *D* after being released from rest is $t = \sqrt{\frac{2D}{a_x}} = \sqrt{\frac{2D(I + R^2 m)}{R^2 mg\sin\theta}}$. The moment of inertia for a thin ring of radius *R* and mass *m* about the polar axis is $I_{ring} = mR^2$. The time to travel a distance D is $t_{ring} = 2\sqrt{\frac{D}{g\sin\theta}}$. The moment of inertia of a disk of radius *R* and mass *m* about the polar axis is $I_{disk} = \frac{1}{2}mR^2$. The time to travel a distance D is $t_{disk} = \sqrt{\frac{3D}{g\sin\theta}}$. The ratio of the times is $\frac{t_{ring}}{t_{disk}} = \frac{2}{\sqrt{3}} = \sqrt{\frac{4}{3}}$

==================================<>==================================

=================================<>=================================

Problem 7.35 The stepped disk weighs 40 *lb.* and its moment of inertia is $I = 0.2$ slug-ft^2. If it is released from rest, how long does it take the center of the disk to fall three feet? (Assume that the string remains vertical.)

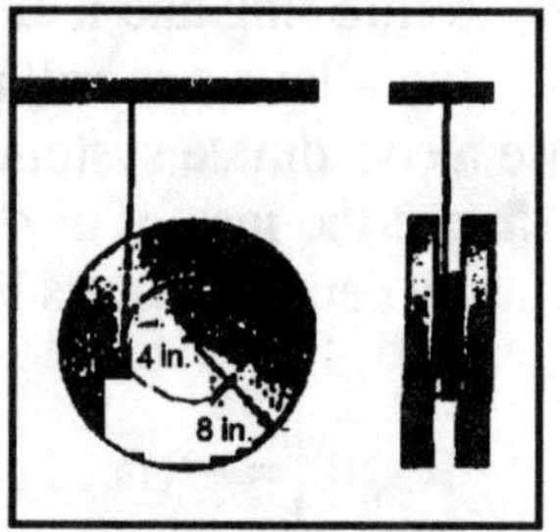

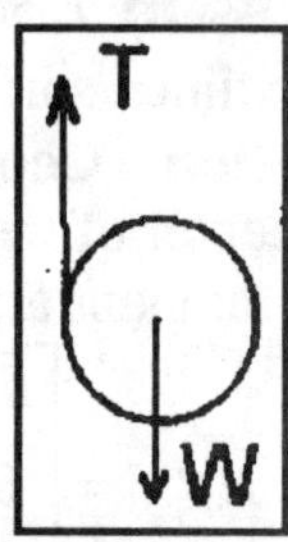

Solution:

The moment about the center of mass is $M = -RT$. From the equation of angular motion: $-RT = I\alpha$, from which $T = -\frac{I\alpha}{R}$.

From the free body diagram and Newton's second law: $\sum F_y = T - W = ma_y$, where a_y is the acceleration of the center of mass. From kinematics: $a_y = R\alpha$. Substitute and solve: $a_y = \frac{W}{\left(\frac{I}{R^2} + m\right)}$. The time required to fall a distance D is $t = \sqrt{\frac{2D}{a_y}} = \sqrt{\frac{2D(I + R^2 m)}{R^2 W}}$. For $D = 3$ ft, $R = \frac{4}{12} = 0.3333$ ft, $W = 40$ lb, $m = \frac{W}{g} = 1.24$ slug, $I = 0.2$ slug-ft^2, $t = 0.676$ s

=================================<>=================================

Problem 7.36 The moment of inertia of the pulley is *I*. The system is released from rest with the spring unstretched. Determine the velocity of the mass as a function of the distance *x* it has fallen.

Strategy: By drawing free-body diagrams of the mass and pulley, determine the acceleration of the mass as a function of the distance it has fallen. The use of the chain rule:

$a = dv/dt = (dv/dx)(dx/dt) = (dv/dt)v$.

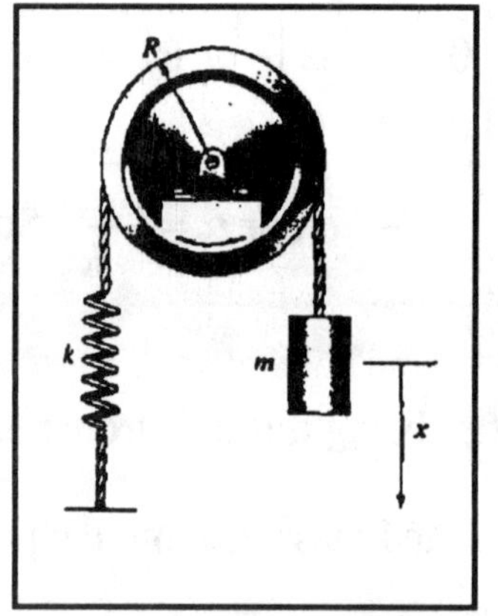

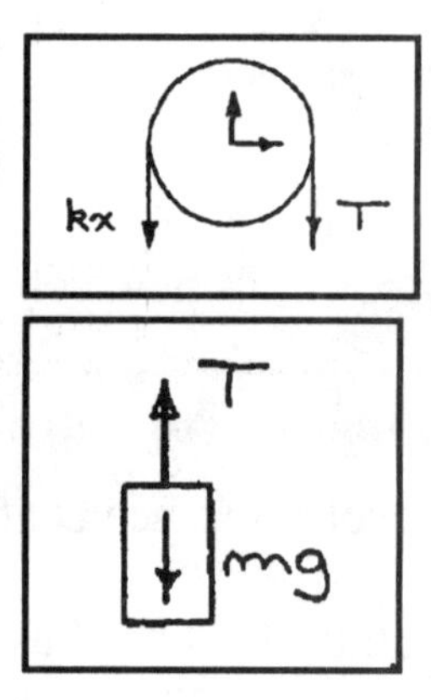

Solution:

The free body diagrams of the pulley and mass are as shown at the right.

Newton's second law for the mass is $\sum F = mg - T = ma$, and the angular equation of motion for the pulley is $\sum M = RT - Rkx = I\alpha = I\left(\frac{a}{R}\right)$. Eliminating T, we obtain $a = \frac{mg - kx}{m + I/R^2}$. Applying the chain rule, $a = \frac{dv}{dx}v = \frac{mg - kx}{m + I/R^2}$, and integrating, $\int_0^v v\,dv = \int_0^x \left(\frac{mg - kx}{m + I/R^2}\right)dx$; $\frac{1}{2}v^2 = \frac{mgx - \frac{1}{2}kx^2}{m + I/R^2}$, we obtain $v = \sqrt{(2mgx - kx^2)/(m + I/R^2)}$.

=================================<>=================================

==================================<>==================================

Problem 7.37 In Problem 7.36, let $R = 100\ mm$, $I = 0.1\ kg-m^2$, $m=5\ kg$, and $k = 135\ N/m$.

Solution:

(a) From the solution of Problem 7.36, the velocity of the mass is zero when $2mgx - kx^2 = 0$ so, $x = 2mg/k = 2(5)(9.81)/135 = 0.727m$.

(b) From the solution of Problem 7.36, the mass's acceleration is $a = \dfrac{mg-kx}{m+I/R^2}$. When $x = 2mg/k$, the acceleration is $a = \dfrac{mg-2mg}{m+I/R^2} = \dfrac{-mg}{m+I/R^2} = \dfrac{-(5)(9.81)}{5+0.1/(0.1)^2} = -3.27m/s^2$

==================================<>==================================

Problem 7.38 The homogeneous disk weighs *100 lb* and its radius is R = *1 ft*. It rolls on the plane surface. The spring constant is k = *100 lb/ft*. If the disk is rolled to the left until the spring is compressed *1 ft* and released from rest, what is its angular acceleration at the instant it is released?

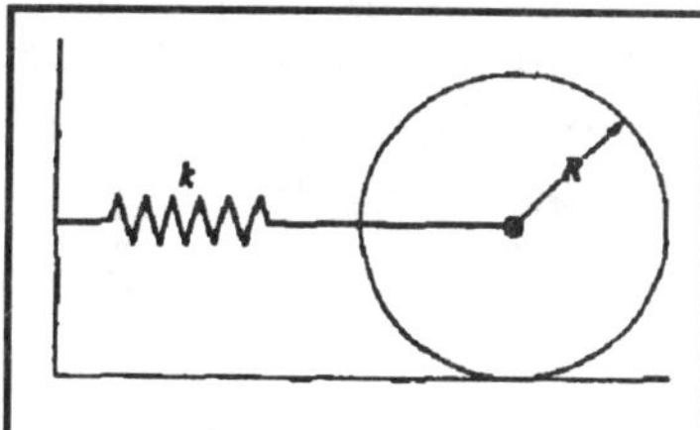

Solution:

The disk's free body diagram at the instant of release is shown.

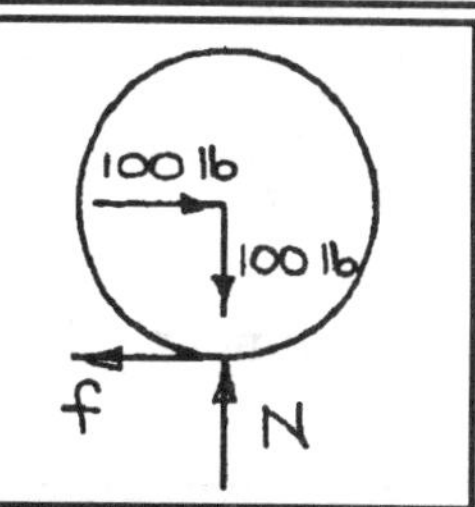

Let a be the acceleration of the center of mass to the right, and let α be the disk's clockwise angular acceleration.

Newton's second law is $\sum F = 100 - f = ma = mR\alpha$ and the equation of angular motion is $\sum M = Rf = I\alpha$. Eliminating f, we obtain $\alpha = \dfrac{100R}{I+mR^2}$ $= \dfrac{200}{3mR}$. Setting $R = 1ft$ and $m = 100/32.2$ slugs, $\alpha = 21.5rad/s^2$.

==================================<>==================================

Problem 7.39 In Problem 7.38, determine the disk's angular velocity when its center has moved *1 ft* to the right of the position in which it was released.

Solution:

Let x be the position of the center of mass relative to its position at release. The free-body diagram is as shown: Let a be the acceleration of the center of mass to the right and α the disk's clockwise angular acceleration. Newton's second law is $\sum F = 100(1-x) - f = ma$, and the equation of angular motion is $\sum M = Rf = I\alpha = I\left(\dfrac{a}{R}\right)$ Eliminating f we obtain $a = \dfrac{100(1-x)}{m+I/R^2}$.

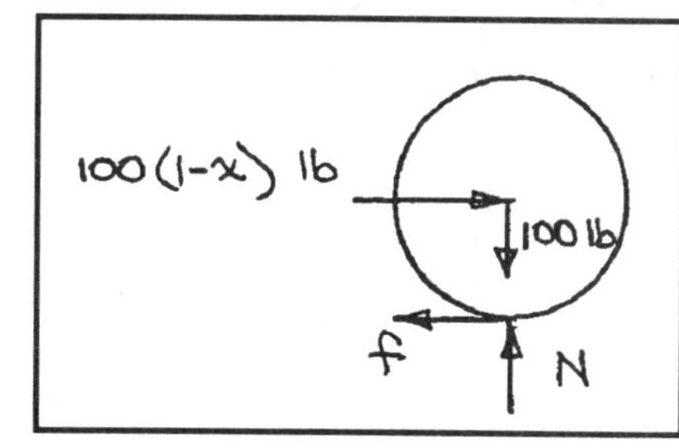

Using the chain rule $a = (dv/dx)v$ and integrating, $\int_0^v v dv = \int_0^1 \dfrac{100(1-x)dx}{m+I/R^2}$ we obtain the velocity of the center: $v = \sqrt{\dfrac{100}{m+I/R^2}} = \sqrt{\dfrac{200}{3m}} = 4.63ft/s$ Therefore $\omega = v/R = 4.63rad/s$.

==================================<>==================================

=================================<>=================================

Problem 7.40 At $t = 0$, a sphere of mass m and radius R ($I = \frac{2}{5}mR^2$) on a flat surface has angular velocity ω_o and the velocity of its center is zero. The coefficient of kinetic friction between the sphere and the surface is μ_k. What is the maximum velocity the center of the sphere will attain, and how long does it take to reach it?

Solution:
At $t = 0$, the sphere is slipping. From Newton's second law: $\sum F_x = \mu_k W = \frac{W}{g} a_x$, from which $a_x = \mu_k g$. Integrating, $v = \mu_k g t$ is velocity of the sphere when the initial velocity is zero. The moment about the center of mass is $M = \mu_k gR$, and $\mu_k gR = I\alpha$, from which $\alpha = \frac{\mu_k WR}{I} = \frac{\mu_k WR}{\frac{2}{5}mR^2} = \frac{5g\mu_k}{2R}$. The angular velocity is

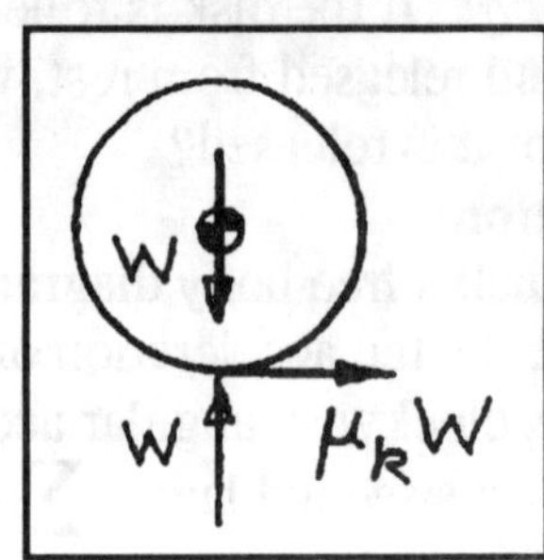

$\omega = \int \alpha \, dt + \omega_o = \frac{5g\mu_k}{2R} t + \omega_o$. When $v = R\omega$, slipping stops. Equate: $-\mu_k g t = \frac{5\mu_k g}{2} t + R\omega_o$. Solve: $t = -\frac{2R\omega_o}{7\mu_k g} = \frac{2R|\omega_o|}{7\mu_k g}$. [*Note*: ω_o is a negative number. (See figure.)]. The velocity: $v = -g\mu_k\left(\frac{2R\omega_o}{7g\mu_k}\right) = \frac{2}{7}R|\omega_o|$

=================================<>=================================

Problem 7.41 A soccer player kicks the ball to a teammate 20 *ft* away. The ball leaves his foot moving parallel to the ground at 20 *ft/s* with no initial angular velocity. The coefficient of kinetic friction between the ball and the grass is $\mu_k = 0.4$. How long does it take the ball to reach his teammate? (The ball is 28 *in.* in circumference and weighs 14 *oz.* Estimate its moment of inertia by using the equation for a thin spherical shell: $I = \frac{2}{3}mR^2$.)

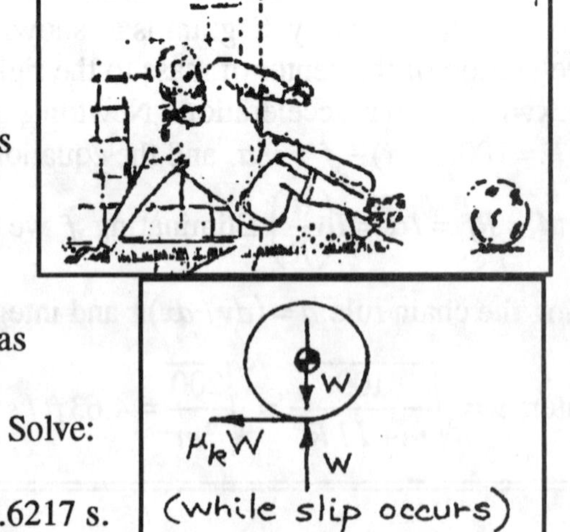

Solution:
The ball is slipping at $t = 0$. From Newton's second law, $\sum F_x = -\mu_k W = ma_x$, from which $a_x = -\mu_k g$. The velocity is $v = -\mu_k g t + v_o$. From Newton's second law $M = I\alpha$, from which $-\mu_k WR = I\alpha$, $\alpha = -\frac{\mu_k WR}{I} = -\frac{3\mu_k g}{2R}$. The angular velocity is $\omega = \int \alpha \, dt = -\frac{3\mu_k g}{2R} t$. When $v = -\omega R$, the ball has stopped slipping and has begun rolling: $-\mu_k g t + v_o = \frac{3\mu_{kg}}{2} t$. Solve: $t = \frac{2v_o}{5\mu_k g}$. For $v_o = 20 \text{ ft/s}$, $\mu_k = 0.4$, $g = 32.17 \text{ ft/s}^2$, $t = 0.6217 \text{ s}$.

Solution continued on next page

The velocity at the time the ball starts rolling is $v = -\mu_k g\left(\dfrac{2v_o}{5\mu_k g}\right) + v_o = v_o\left(1 - \dfrac{2}{5}\right) = \dfrac{3}{5}v_o$. The distance traveled while slipping is $s = \dfrac{a}{2}t^2 + v_o t = -\dfrac{\mu_k g}{2}(0.6217^2) + v_o(0.6217) = 9.947$ ft. The distance remaining is $d = 20 - s = 10.05$ ft. The total time of travel is $t_{total} = \dfrac{d}{v} + t = \dfrac{10.05}{12} + 0.6217 = 1.46$ s

================================<>================================

Problem 7.42 The 100 *kg* cylindrical disk is at rest when the force *F* is applied to a cord wrapped around it. The static and kinetic coefficients of friction between the disk and the surface equal 0.2. Determine the angular acceleration of the disk if (a) $F = 500$ N; (b) $F = 1000$ N.
Strategy: First solve the problem by assuming that the disk does not slip, but rolls on the surface. Determine the friction force and find out if it exceeds the product of the friction coefficient and the normal force. If it does, you must rework the problem assuming that the disk slips.

Solution:

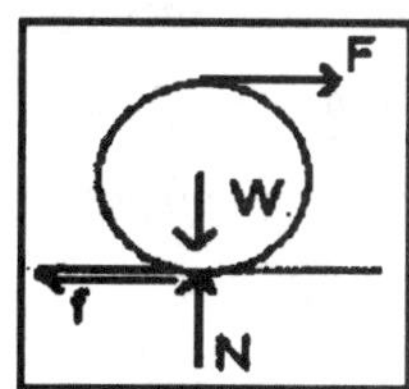

Choose a coordinate system with the origin at the center of the disk in the at rest position, with the *x* axis parallel to the plane surface. The moment about the center of mass is $M = -RF - Rf$, from which $-RF - Rf = I\alpha$.

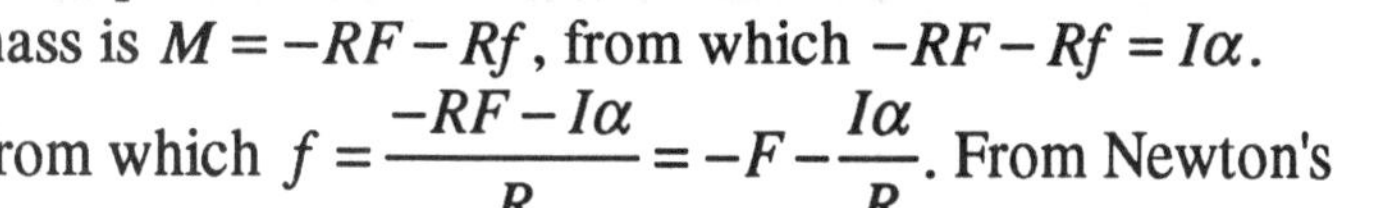

From which $f = \dfrac{-RF - I\alpha}{R} = -F - \dfrac{I\alpha}{R}$. From Newton's second law: $F - f = ma_x$, where a_x is the acceleration of the center of mass. Assume that the disk rolls. At the point of contact $\vec{a}_P = 0$; from which $0 = \vec{a}_G + \vec{\alpha} \times \vec{r}_{P/G} - \omega^2 \vec{r}_{P/G}$.

$$\vec{a}_G = a_x\vec{i} = \vec{\alpha} \times R\vec{j} - \omega^2 R\vec{j} = \begin{bmatrix} \vec{i} & \vec{j} & \vec{k} \\ 0 & 0 & \alpha \\ 0 & R & 0 \end{bmatrix} - \omega^2 R\vec{j} = -R\alpha\vec{i} - \omega^2 R\vec{j}$$, from which $a_y = 0$ and $a_x = -R\alpha$.

Substitute for *f* and solve: $a_x = \dfrac{2F}{\left(m + \dfrac{I}{R^2}\right)}$. (a) For a disk, the moment of inertia about the polar axis is $I = \dfrac{1}{2}mR^2$, from which $a_x = \dfrac{4F}{3m} = \dfrac{2000}{300} = 6.67 \text{ m/s}^2$. (a) For $F = 500$ N the friction force is $f = F - ma_x = -\dfrac{F}{3} = -\dfrac{500}{3} = -167$ N. Note: $-\mu_k W = -0.2mg = -196.2$ N, *the disk does not slip.* The angular velocity is $\alpha = -\dfrac{a_x}{R} = -\dfrac{6.67}{0.3} = -22.22 \text{ rad/s}^2$. (b) For $F = 1000$ N the acceleration is $a_x = \dfrac{4F}{3m} = \dfrac{4000}{300} = 13.33 \text{ m/s}^2$. The friction force is $f = F - ma_x = 1000 - 1333.3 = -333.3$ N. *The drum slips.* The moment equation for slip is $-RF + R\mu_k gm = I\alpha$, from which

$$\alpha = \frac{-RF + R\mu_k gm}{I} = -\frac{2F}{mR} + \frac{2\mu_k g}{R} = -53.6 \text{ rad/s}^2$$

================================<>================================

Problem 7.43 The ring gear is fixed. The mass and moment of inertial of the sun gear are $m_S = 22 slugs$, $I_S = 4400 slug-ft^2$. The mass and moment of inertia of each planet gear are $m_P = 2.7 slugs$, $I_P = 65 slug-ft^2$. If a couple M = *600 ft-lb* is applied to the sun gear, what is its angular acceleration?

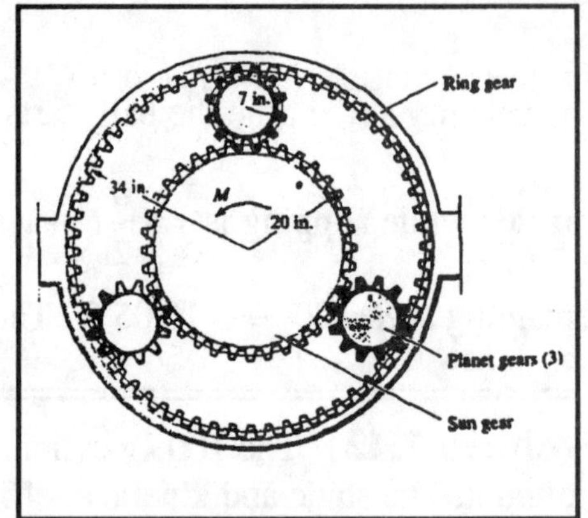

Solution:

The free body diagrams of the sun gear and one of the planet gears are as shown. Let α_s be the counterclockwise angular acceleration of the sun gear and α_P the clockwise angular acceleration of the planet gear. The angular equations of motion are $M - 3r_s T = I_s \alpha_s$ (1), and $r_p T - r_p Q = I_P \alpha_P$ (2).

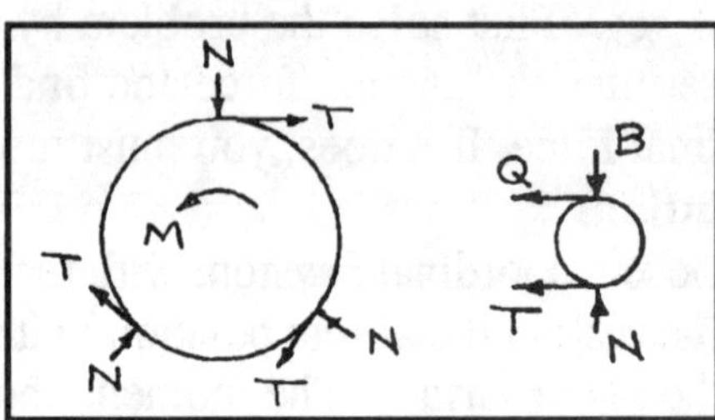

The angular accelerations are related by $r_s \alpha_s = 2 r_P \alpha_P$ (3).

Let a be the tangential acceleration of the center of the planet gear. Newton's second law is $\sum F = T + Q = m_P a = m_P r_P \alpha_P$ (4)

Setting $r_s = (20/12) ft$,

$r_P = (7/12) ft, M = 600 ft-lb, m_P = 2.7 slugs \; I_P = 65 slug-ft^2$

$I_s = 4400 slug-ft^2$ and solving Equations (1) - (4), we obtain $\alpha_s = 0.125 rad/s^2$.

Problem 7.44 In Problem 7.43, what is the magnitude of the tangential force exerted on the sun gear by each planet gear at their point of contact when the *600 ft-lb* couple is applied to the sun gear?

Solution:

From Equations (1) - (4) of the solution of Problem 7.43, the tangential force is $T = 10.1 lb$.

Problem 7.45 The 18 *kg* ladder is released from rest in the position shown. Model it as a slender bar and neglect friction. At the instant of release, determine (a) the angular acceleration; (b) the normal force exerted on the ladder by the floor.

Solution:

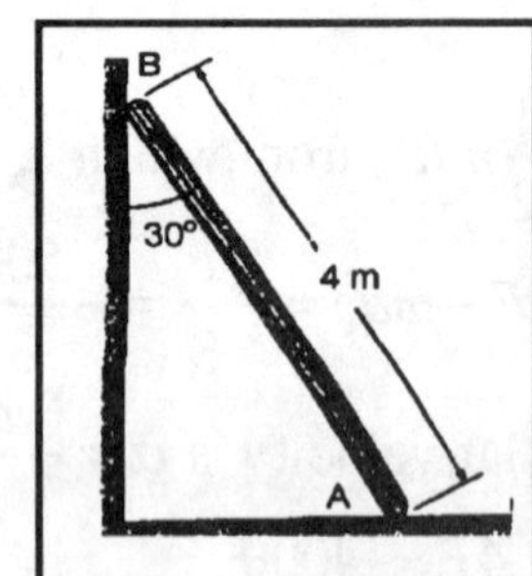

The vector location of the center of mass is $\vec{r}_G = (L/2)\sin 30^\circ \vec{i} + (L/2)\cos 30^\circ \vec{j} = 1\vec{i} + 1.732\vec{j}$ (m). Denote the normal forces at the top and bottom of the ladder by P and N. The vector locations of A and B are $\vec{r}_A = L\sin 30^\circ \vec{i} = 2\vec{i}$ (m), $\vec{r}_B = L\cos 30^\circ \vec{j} = 3.46\vec{j}$ (m). The vectors $\vec{r}_{A/G} = \vec{r}_A - \vec{r}_G = 1\vec{i} - 1.732\vec{j}$ (m), $\vec{r}_{B/G} = \vec{r}_B - \vec{r}_G = -1\vec{i} + 1.732\vec{j}$ (m). The moment about the center of mass is $\vec{M} = \vec{r}_{B/G} \times \vec{P} + \vec{r}_{A/G} \times \vec{N}$,

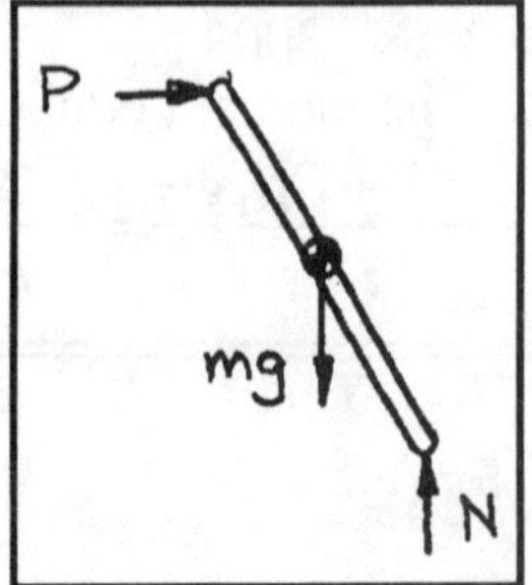

$$\vec{M} = \begin{bmatrix} \vec{i} & \vec{j} & \vec{k} \\ -1 & 1.732 & 0 \\ P & 0 & 0 \end{bmatrix} + \begin{bmatrix} \vec{i} & \vec{j} & \vec{k} \\ 1 & -1.732 & 0 \\ 0 & N & 0 \end{bmatrix} = (-1.732P + N)\vec{k} \text{ (N-m)}.$$ From the equation of angular motion: (1) $-1.732P + N = I\alpha$. From Newton's second law: (2) $P = ma_x$, (3) $N - mg = ma_y$, where a_x, a_y are the accelerations of the center of mass.

Solution continued on next page

From kinematics: $\vec{a}_G = \vec{a}_A + \vec{\alpha} \times \vec{r}_{G/A} - \omega^2 \vec{r}_{G/A}$.
The angular velocity is zero since the system was released from rest,

$$\vec{a}_G = a_A\vec{i} + \begin{bmatrix} \vec{i} & \vec{j} & \vec{k} \\ 0 & 0 & \alpha \\ -1 & 1.732 & 0 \end{bmatrix} = a_A\vec{i} - 1.732\alpha\vec{i} - \alpha\vec{j} = (a_A - 1.732\alpha)\vec{i} - \alpha\vec{j} \ (\mathrm{m/s^2}),$$ from which $a_y = -\alpha$.

Similarly, $\vec{a}_G = \vec{a}_B + \vec{\alpha} \times \vec{r}_{G/B}$, $\vec{a}_G = \vec{a}_B + \begin{bmatrix} \vec{i} & \vec{j} & \vec{k} \\ 0 & 0 & \alpha \\ 1 & -1.732 & 0 \end{bmatrix} = a_B\vec{j} + 1.732\alpha\vec{i} + \alpha\vec{j}$, from which $a_x = 1.732\alpha$.

Substitute into (1), (2) and (3) to obtain three equations in three unknowns:
$-1.732P + N = I\alpha$, $P = m(1.732)\alpha$, $N - mg = -m\alpha$. Solve: (a) $\alpha = 1.84 \ \mathrm{rad/s^2}$, $P = 57.3 \ \mathrm{N}$,
(b) $N = 143.47 \ \mathrm{N}$

===================================<>===================================

Problem 7.46 Suppose that the ladder in Problem 7.45 has a counterclockwise angular velocity of 1.0 *rad/s* in the position shown. Determine (a) the angular acceleration; (b) the normal force exerted on the ladder by the floor.

Solution:

Use the solution to Problem 7.45, from which the three equations are: (1) $-1.732P + N = I\alpha$,
(2) $P = ma_x$, (3) $N - mg = ma_y$. From kinematics: $\vec{a}_G = \vec{a}_A + \vec{\alpha} \times \vec{r}_{G/A} - \omega^2 \vec{r}_{G/A}$,

$$\vec{a}_G = a_A\vec{i} + \begin{bmatrix} \vec{i} & \vec{j} & \vec{k} \\ 0 & 0 & \alpha \\ -1 & 1.732 & 0 \end{bmatrix} - \omega^2(-\vec{i} + 1.732\vec{j}).\ \vec{a}_G = (a_A - 1.732\alpha + 1)\vec{i} - (\alpha + 1.732)\vec{j} \ (\mathrm{m/s^2}),$$ from which $a_y = -\alpha - 1.732$.

Similarly, $\vec{a}_G = \vec{a}_B + \vec{\alpha} \times \vec{r}_{G/B} - \omega^2 r_{G/B}$,

$$\vec{a}_G = \vec{a}_B + \begin{bmatrix} \vec{i} & \vec{j} & \vec{k} \\ 0 & 0 & \alpha \\ 1 & -1.732 & 0 \end{bmatrix} - \omega^2(\vec{i} - 1.732\vec{j}) = a_B\vec{j} + (1.732\alpha - 1)\vec{i} + (\alpha + 1.732)\vec{j},$$ from which

$a_x = 1.732\alpha - 1$. Substitute into (1), (2) and (3) to obtain three equations in three unknowns:
$-1.732P + N = I\alpha$,
$P = m(1.732\alpha - 1)$,
$N - mg = -m(\alpha + 1.732)$. Solve (a) $\alpha = 1.84 \ \mathrm{rad/s^2}$. $P = 39.3 \ \mathrm{N}$. (b) $N = 112.3 \ \mathrm{N}$

===================================<>===================================

================================<>================================

Problem 7.47 Suppose that the ladder in Problem 7.45 has a counterclockwise angular velocity of 1.0 *rad/s* in the position shown and that the coefficient of kinetic friction at the floor and at the wall is $\mu_k = 0.2$. Determine (a) the angular acceleration; (b) the normal force exerted on the ladder by the floor.

Solution:

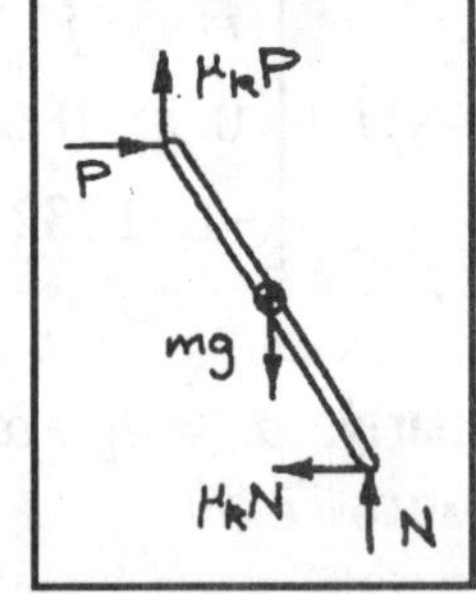

From the solution to Problem 7.45, $\vec{r}_{A/G} = \vec{r}_A - \vec{r}_G = \vec{i} - 1.732\vec{j}$ (m).
$\vec{r}_{B/G} = \vec{r}_B - \vec{r}_G = -\vec{i} + 1.732\vec{j}$ (m) The moment about the center of mass is
$\vec{M} = \vec{r}_{A/G} \times (-\mu_k N\vec{i} + N\vec{j}) + \vec{r}_{B/G} \times (P\vec{i} + \mu_k P\vec{j})$,

$$\vec{M} = \begin{bmatrix} \vec{i} & \vec{j} & \vec{k} \\ 1 & -1.732 & 0 \\ -\mu_k N & N & 0 \end{bmatrix} + \begin{bmatrix} \vec{i} & \vec{j} & \vec{k} \\ -1 & 1.732 & 0 \\ P & \mu_k P & 0 \end{bmatrix} = [N(1 - 1.732\mu_k) - P(\mu_k + 1.732)]\vec{k}$$

The three equations resulting from the application of the equation of angular motion and Newton's second law to the free body diagram are $N(1 - 1.732\mu_k) - P(\mu_k + 1.732) = I\alpha$. $P - \mu_k N = ma_x$. $N - mg + \mu_k P = ma_y$. Use the results of the application of kinematics from Problem 7.34: $a_x = 1.732\alpha - 1$, $a_y = -\alpha - 1.732$. Substitute to obtain three equations in three unknowns.

$N(1 - 1.732\mu_k) - P(\mu_k + 1.732) = I\alpha$,
$P - \mu_k N = m(1.732\alpha - 1)$,
$N - mg + \mu_k P = -m(\alpha + 1.732)$. Solve: (a) $\alpha = 0.808 \text{ rad/s}^2$, $P = 32.1$ N, (b) $N = 124.5$ N

================================<>================================

Problem 7.48 The slender bar weighs 30 *lb.* and the cylindrical disk weighs 20 *lb.* The system is released from rest with the bar horizontal. Determine the bar's angular acceleration at the instant of release if disk and bar are welded together at A.

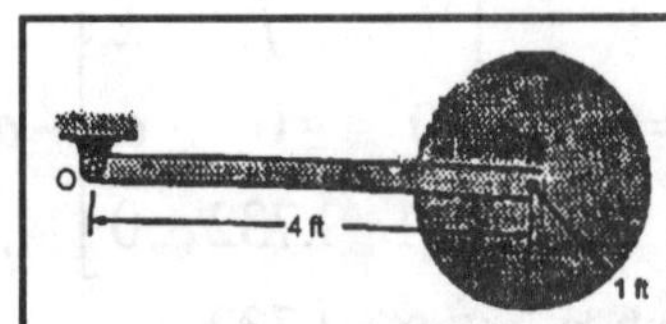

Solution:

The center of mass of the system is $\bar{x}_C = \frac{30(2) + 20(4)}{30 + 20} = 2.8\,ft$. From the equation of angular motion, $-(20 + 30)\bar{x}_C = I\alpha$, since the moment is negative, from which $\alpha = -\frac{140}{I}$. If the bar and disk are welded together, the disk rotates with the bar. The moment of inertia about O is $I = \frac{1}{3}\left(\frac{30}{g}\right)(4^2) + \left(\frac{20}{g}\right)4^2 + \left(\frac{1}{2}\right)\left(\frac{20}{g}\right)(1^2) = 15.23 \text{ slug-ft}^2$, from which $\alpha = -\frac{140}{15.23} = -9.19 \text{ rad/s}^2$

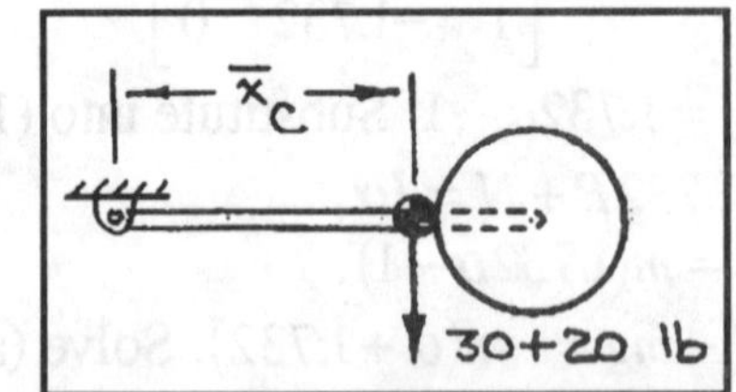

================================<>================================

=================================<>=================================

Problem 7.49 In Problem 7.48, determine the bar's angular acceleration if the bar and disk are pinned together at A.

Solution:

With the bar and disk pinned, there is no couple to cause the disk to rotate, hence the disk can be treated as a point mass. From the solution to Problem 7.48, $\alpha = -\frac{140}{I}$. The moment of inertia is

$I = (1/3)\left(\frac{30}{g}\right)4^2 + \left(\frac{20}{g}\right)4^2 = 14.92$ slug-ft^2, from which $\alpha = -\frac{140}{14.92} = -9.38$ rad/s^2

=================================<>=================================

Problem 7.50 The 0.1 *kg* slender bar and 0.2 *kg* cylindrical disk are released from rest with the bar horizontal. The disk rolls on the curved surface. What is the bar's angular acceleration at that instant?

Solution:

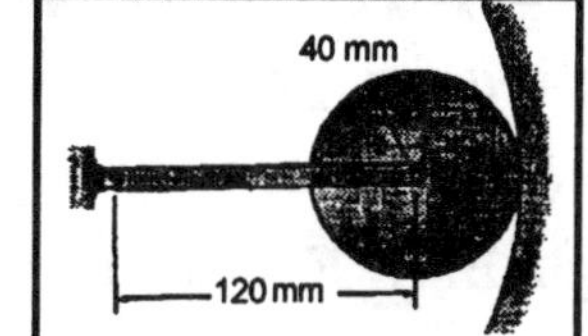

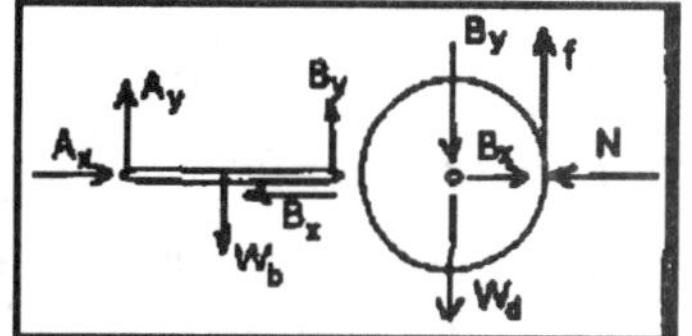

The moment about the center of mass of the disk is $M = fR$, from the equation of angular motion, $Rf = I_d\alpha_d$. From Newton's second law: $f - B_y - W_d = m_d a_{dy}$. Since the disk rolls, the kinematic condition is $a_{dy} = -R\alpha_d$. Combine the expressions and rearrange: $f = I\alpha_d / R$, $I\alpha_d / R - B_y - W_d = m_d a_{dy}$, from which $B_y + W_d = (Rm_d + I_d / R)\alpha_d$. The moment about the center of mass of the bar is $M_b = -\left(\frac{L}{2}\right)A_y + \left(\frac{L}{2}\right)B_y$, from which $-\left(\frac{L}{2}\right)A_y + \left(\frac{L}{2}\right)B_y = I_b\alpha_b$. From Newton's second law `$A_y - W_b + B_y = m_b a_{by}$, where a_{by} is the acceleration of the center of mass of the bar. The kinematic condition for the bar is $\vec{a}_{CM} = \vec{\alpha}_b \times \left(\left(\frac{L}{2}\right)\vec{i}\right) = \left(\frac{L}{2}\right)\alpha_b\vec{j}$, from which $a_{by} = \left(\frac{L}{2}\right)\alpha_b$. Similarly,

$\vec{a}_D = \vec{a}_{CM} + \vec{\alpha}_b \times ((L/2)\vec{i})$, from which $a_{dy} = L\alpha_b$.

From which: $\alpha_d = -L\alpha_b / R$. Substitute to obtain three equations in three unknowns:

$B_y + W_d = \left(Rm_d + \frac{I_d}{R}\right)\left(-\frac{L}{R}\right)\alpha_b$, $-\left(\frac{L}{2}\right)A_y + \left(\frac{L}{2}\right)B_y = I_b\alpha_b$, $A_y - W_b + B_y = m_b\left(\frac{L}{2}\right)\alpha_b$. Substitute known numerical values: $L = 0.12$ m, $R = 0.04$ m, $m_b = 0.1$ kg, $W_b = m_b g = 0.981$ N, $m_d = 0.2$ kg, $W_d = m_d g = 1.962$ N, $I_b = (1/12)m_b(L^2) = 1.2\times10^{-4}$ kg-m^2, $I_d = (1/2)m_d R^2 = 1.6\times10^{-4}$ kg-m^2. Solve: $\alpha_b = -61.3$ rad/s^2, $A_y = 0.368$ N, $B_y = 0.245$ N.

Check: An alternate, analytical solution: from above: $f - B_y - W_d = m_d a_y = -m_d R\alpha_d$, where, from kinematics, $a_y = -R\alpha_d$. From the equation of angular motion, $fR = I_d\alpha_d$. The moment about the left end is $M_A = -(L/2)W_b - (B_y + W_d)L + f(L+R)$. Rearrange $M_A = -(L/2)W_b + (f - W_d - B_y)L + fR$. Substitute into the equation of angular motion: (1) $M_A = -(L/2)W_b + (m_d RL - I_d)(L/R)\alpha_b = I_A\alpha_b$. where, from kinematics (see above) $\alpha_d = -L\alpha_b / R$. The moment of inertia about A for the bar and disk is $I_A = (1/3)m_b L^2 + m_d L^2$. Substitute I_A into (1) and reduce algebraically:

$\alpha_b = -\left(\frac{3W_b}{2m_b L + 3m_d R}\right) = -61.3$ rad/s^2. *check.*

=================================<>=================================

Problem 7.51 The suspended objects A and B weigh *20 lb* and *40 lb*, respectively. The left pulley weighs *16 lb* and its moment of inertia is *0.24 slug-ft²*. The right pulley weighs *6 lb* and its moment of inertia is *0.04 slug-ft²*. If the system is released from rest, what is the acceleration of B?

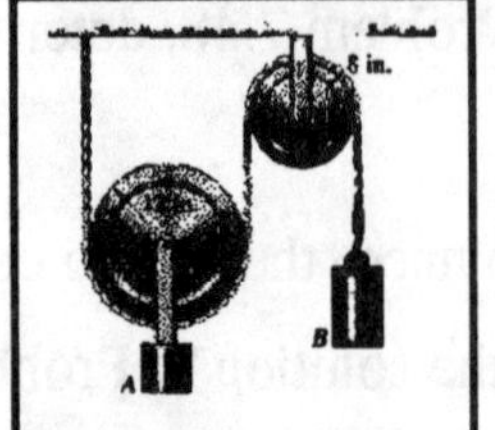

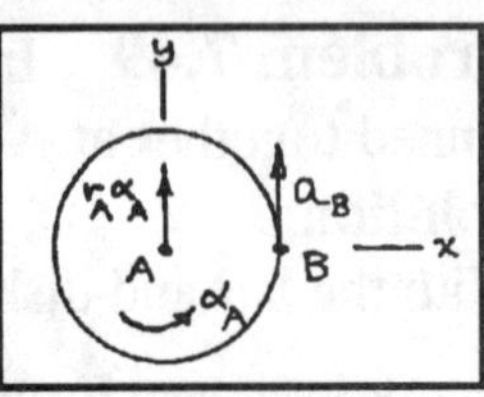

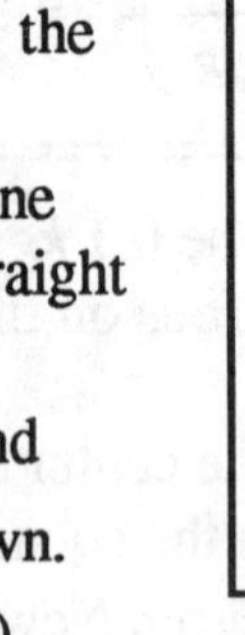

Solution:

Let α_A be the counterclockwise angular acceleration of the left pulley, α_B the clockwise angular acceleration of the right pulley, and a_B the downward acceleration of object B. The left pulley moves as if it was rolling on a plane surface coincident with the left part of the rope, so its center moves in a straight vertical line with upward acceleration $r_A\alpha_A$: From the equation $\bar{a}_B = \bar{a}_A + \alpha_A \times \bar{r}_{B/A} - \omega_A^2 \bar{r}_{B/A}$:, we get $a_B\bar{j} = r_A\alpha_A\bar{j} + (\alpha_A\bar{k} \times r_A\bar{i}) - \omega_A^2 r_A\bar{i}$, and obtain $a_B = 2r_A\alpha_A = 2(12/12)\alpha_A$ (1) The free body diagrams are as shown.

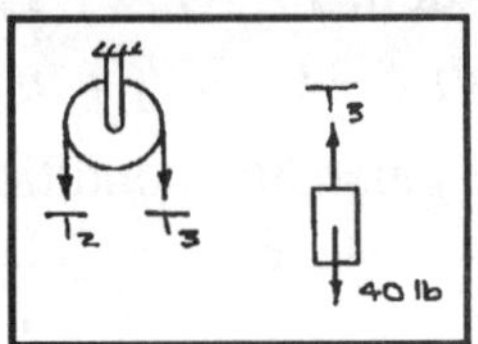

The equation of motion are: Weight B: $\sum F = 40 - T_3 = (40/32.2)a_B$ (2)

Right pulley: $\sum M = (8/12)(T_3 - T_2) = (0.04)\alpha_B = (0.04)a_B/(8/12)$ (3)

Left pulley and weight A: $\sum F = T_1 + T_2 - 36 = (36/32.2)(12/12)\alpha_A$ (4)

$\sum M = (12/12)(T_2 - T_1) = (0.24)\alpha_A$ (5)

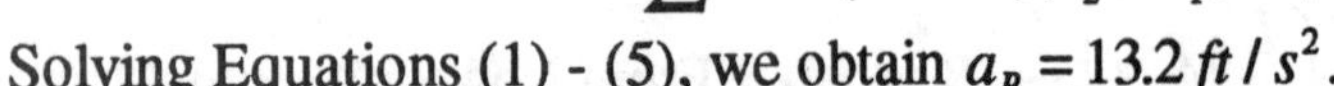

Solving Equations (1) - (5), we obtain $a_B = 13.2\,ft/s^2$.

Problem 7.52 In Problem 7.51, what is the tension in the cable at a point between the two pulleys?

Solution:

Solving Equations (1) - (5) in the solution of Problem 7.51, the tension is $T_2 = 22.5\ lb$.

Problem 7.53 The 4 *lb.* slender bar and 10 *lb.* block are released from rest in the position shown. If friction is negligible, what is the block's acceleration at that instant?

Solution:

Choose a coordinate system with the origin at B and the x axis parallel to the plane surface. Denote $W_{bar} = m_{bar}g = 4$ lb, $W_B = m_B g = 10$ lb.

From Newton's second law for the block: $B_x = m_B a_B$, $N - B_y - W = 0$.

Similarly, for the center of mass of the bar: $-B_x = m_{bar}a_{bx}$, and $B_y - W_b = m_b a_{by}$, where a_{bx}, a_{by} are the accelerations of the center of mass of the bar. The vector location of the center of mass of the bar is

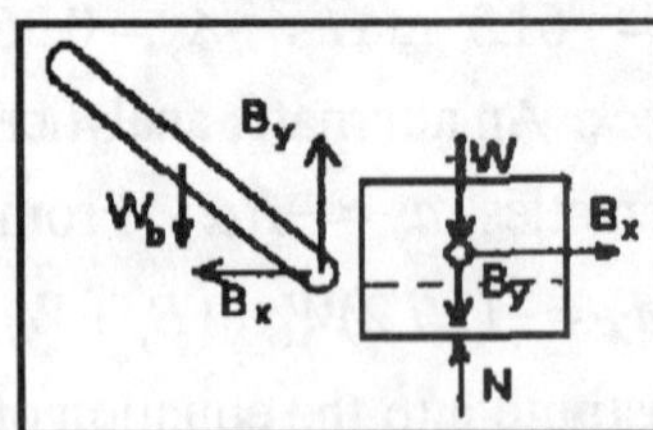

$$\vec{r}_{G/B} = \left(\frac{L}{2}\right)\left(-\vec{i}\cos 45^\circ + \vec{j}\sin 45^\circ\right) = \frac{1}{\sqrt{2}}\left(-\vec{i} + \vec{j}\right).$$

$$\vec{M}_{Gb} = -\vec{r}_{G/B} \times \vec{B} = \frac{1}{\sqrt{2}}\begin{bmatrix} \vec{i} & \vec{j} & \vec{k} \\ 1 & -1 & 0 \\ -B_x & B_y & 0 \end{bmatrix} = \frac{1}{\sqrt{2}}\left(-B_x + B_y\right)\vec{k} \text{ where } L = 2.$$

Solution continued on next page

From the equation of angular motion: $(-B_x + B_y)/\sqrt{2} = I_{bar}\alpha$, where $I_{bar} = m_{bar}L^2/12 = 0.04145$ slug-ft^2. From kinematics: $\vec{a}_{Gbar} = \vec{a}_B + \vec{\alpha}_{bar} \times \vec{r}_{G/B}$. From the constraint on the motion of the block, $\vec{a}_B = a_B\vec{i}$ ft/s^2, from which

$$\vec{a}_{Gbar} = a_B\vec{i} + \begin{bmatrix} \vec{i} & \vec{j} & \vec{k} \\ 0 & 0 & \alpha \\ -\frac{1}{\sqrt{2}} & \frac{1}{\sqrt{2}} & 0 \end{bmatrix} = a_B\vec{i} - \frac{1}{\sqrt{2}}\alpha\vec{i} - \frac{1}{\sqrt{2}}\alpha\vec{j} \text{ ft/s}^2, \text{ from which } a_{bx} = a_B - \frac{1}{\sqrt{2}}\alpha,$$

$a_{by} = -\alpha/\sqrt{2}$. Substitute to obtain five equations in five unknowns: $B_x = ma_B$, $N - B_y - W = 0$, $-B_x = m_{bar}\left(a_B - \frac{\alpha}{\sqrt{2}}\right)$, $B_y - W_b = -\frac{m_b}{\sqrt{2}}\alpha$, $-B_x + B_y = \sqrt{2}I_b\alpha$, where $m_B = \frac{10}{32.17} = 0.3108$ slug, $W_B = 10$ lb, $m_{bar} = \frac{4}{32.17} = 0.1243$ slug, $W_{bar} = 4$ lb, and $I_{bar} = \frac{m_{bar}L^2}{12} = 0.04145$ slug-ft^2. Solve: $\alpha = 19.1$ rad/s^2, $B_x = 1.2$ lb, $B_y = 2.32$ lb, $N = 12.32$ lb, $a_B = 3.86$ ft/s^2.

==================================<>==================================

Problem 7.54 In Problem 7.53 suppose that the velocity of the block is zero and the bar has an angular velocity of 4 *rad/s* at the instant shown. What is the block's acceleration?

Solution:

The strategy is that used in solving Problem 7.53. The difference is in the kinematics: the acceleration of the center of mass of the bar is (using the solution to Problem 7.53) is $\vec{a}_{Gbar} = \vec{a}_B + \vec{\alpha}_{bar} \times \vec{r}_{G/B} - \omega_{bar}^2\vec{r}_{G/B}$,

$$\vec{a}_{Gbar} = a_B\vec{i} + \begin{bmatrix} \vec{i} & \vec{j} & \vec{k} \\ 0 & 0 & \alpha \\ -\frac{1}{\sqrt{2}} & \frac{1}{\sqrt{2}} & 0 \end{bmatrix} - \frac{\omega_{bar}^2}{\sqrt{2}}(-\vec{i} + \vec{j}) = a_B\vec{i} - \frac{1}{\sqrt{2}}\alpha\vec{i} - \frac{1}{\sqrt{2}}\alpha\vec{j} + \frac{16}{\sqrt{2}}\vec{i} - \frac{16}{\sqrt{2}}\vec{j} \text{ ft/s}^2$$

$a_{bx} = a_B - \frac{\alpha}{\sqrt{2}} - \omega^2\left(-\frac{1}{\sqrt{2}}\right) = a_B - \frac{\alpha}{\sqrt{2}} + \frac{16}{\sqrt{2}}$ rad/s^2, and $a_{by} = -\frac{\alpha}{\sqrt{2}} - \omega^2\left(\frac{1}{\sqrt{2}}\right) = -\frac{\alpha}{\sqrt{2}} - \frac{16}{\sqrt{2}}$ rad/s^2.

Using the new kinematic relations, the five equations in five unknowns are: $B_x = m_Ba_B$, $N - B_y - W_B = 0$, $-B_x = m_{bar}\left(a_B - \frac{\alpha}{\sqrt{2}} + \frac{16}{\sqrt{2}}\right)$, $B_y - W_b = m_{bar}\left(-\frac{\alpha}{\sqrt{2}} - \frac{16}{\sqrt{2}}\right)$, $-B_x + B_y = I_b\alpha$, where $m_B = \frac{10}{32.17} = 0.3108$ slug, $W_B = 10$ lb, $m_{bar} = \frac{4}{32.17} = 0.1243$ slug, $W_{bar} = 4$ lb, and $I_{bar} = \frac{m_{bar}L^2}{12} = 0.04145$ slug-ft^2 Solve: $\alpha = 17.2$ rad/s^2, $B_x = 0.075$ lb, $B_y = 1.08$ lb, $N = 11.1$ lb, $a_B = 0.240\,ft/s^2$

==================================<>==================================

Problem 7.55 The 0.4 *kg* slender bar and 1 *kg* disk are released from rest in the position shown. If the disk rolls, what is the bar's angular acceleration?

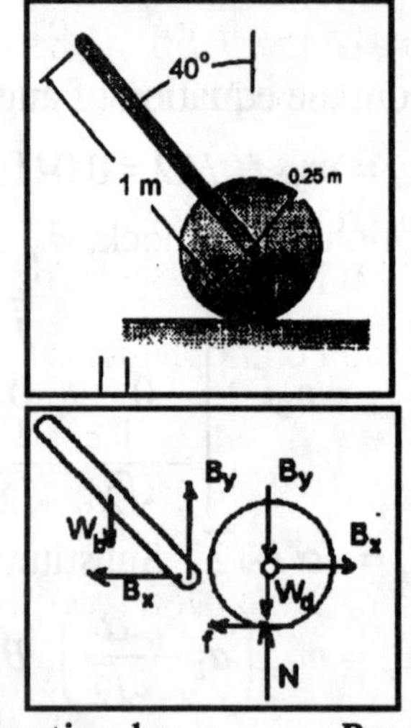

Solution:

Choose a coordinate system with the origin at B and the *x* axis parallel to the plane surface. From Newton's second law applied to the disk, $N - W_d - B_y = 0$, $B_x - f = m_d a_{dx}$. The moment about the mass center of the disk is $M_d = -Rf$, from which $-Rf = I_d\alpha$. From Newton's second law applied to the bar: $-B_x = m_b a_{bx}$, $B_y - W_b = m_b a_{by}$, where a_{bx}, a_{by} are the accelerations of the center of mass of the bar. The vector location of the center of mass of the bar is $\vec{r}_{G/B} = \vec{r}_G - 0 = -0.5\sin 40^o\vec{i} + 0.5\cos 40^o\vec{j} = -0.3214\vec{i} + 0.3830\vec{j}\ (m)$. From the equation of angular motion for the bar: $-0.3830B_x + 0.3214B_y = I_b\alpha_b$,. From the kinematics, the acceleration of the disk is related to the angular acceleration by $a_{dx} = -R\alpha_d$.. The acceleration of the mass center of the bar is $\vec{a}_b = \vec{a}_d + \vec{\alpha}_b \times \vec{r}_{CM/B}$ From the constraint on the motion of the disk, $\vec{a}_d = a_{dx}\vec{i}$. $\vec{a}_b = a_{dx}\vec{i} + \begin{bmatrix} \vec{i} & \vec{j} & \vec{k} \\ 0 & 0 & \alpha_b \\ -0.3214 & 0.3830 & 0 \end{bmatrix} = a_{dx}\vec{i} - 0.3830\alpha\vec{i} - 0.3214\alpha\vec{j}$, from which $a_{bx} = a_{dx} - 0.3830\alpha$, $a_{by} = -0.3214\alpha$. Substitute to obtain six equations in six unknowns:

$N - W_d - B_y = 0$,

$-0.3830B_x + 0.3214B_y = I_b\alpha_b$,

$-B_x = m_b(a_{dx} - 0.3830\alpha_b)$,

$B_y - W_b = m_b(-0.3214\alpha_b)$,

$Rf = \frac{I_d}{R}a_{dx}$,

$B_x - f = m_d a_{dx}$. Substitute known numerical values: $m_b = 0.4\ kg$, $m_d = 1\ kg$,, $L = 1\ m$, $R = 0.25\ m$, $W_b = 3.924\ N$, $W_d = 9.81\ N$, $I_b = \left(\frac{1}{12}\right)mL^2 = 0.0333\ kg\text{-}m^2$, $I_d = \left(\frac{1}{2}\right)mR^2 = 0.03125\ kg\text{-}m^2$. The six equations were solved by iteration using **TK Solver Plus:** $N = 12.39\ N$, $B_y = 2.58\ N$, $B_x = 1.26\ N$, $\boxed{\alpha_b = 10.425\ rad/s^2}$, $a_{dx} = 0.841\ m/s^2$, $f = 0.4203\ N$.

Problem 7.56 The slender bar weighs 20 *lb.* and the crate weighs 80 *lb.* The surface the crate rests on is smooth. If the system is stationary at the instant shown, what couple M will cause the crate to accelerate to the left at 4 ft/s^2 at that instant?

Solution:

The rope is massless and flexible; it can support tension only. The tension components are $T_x = T\cos 45^o = \frac{T}{\sqrt{2}}$, $T_y = T\cos 45^o = \frac{T}{\sqrt{2}}$. From Newton's second law applied to the

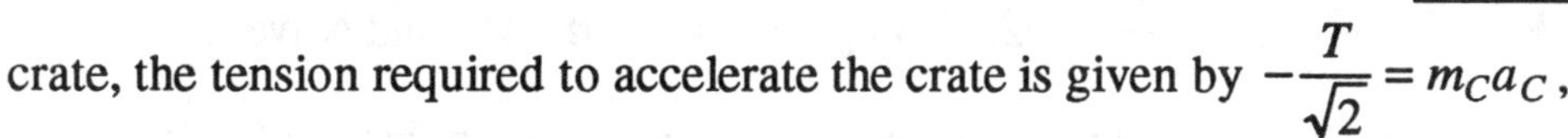

crate, the tension required to accelerate the crate is given by $-\frac{T}{\sqrt{2}} = m_C a_C$,

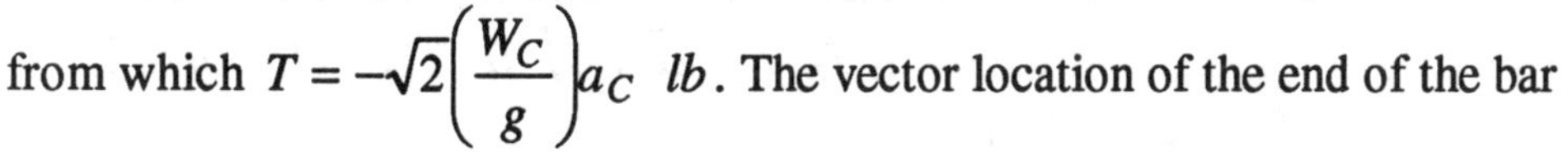

from which $T = -\sqrt{2}\left(\frac{W_C}{g}\right)a_C$ *lb*. The vector location of the end of the bar is $\vec{r}_{B/A} = 2\vec{i} + 4\vec{j}$ *ft*. The moment about the pinned end of the bar is .

$$\vec{M}_A = \vec{M}_{couple} + \vec{r}_{B/A} \times \left(\frac{T}{\sqrt{2}}\vec{i} - \frac{T}{\sqrt{2}}\vec{j}\right) + \vec{r}_{CM} \times \vec{W}_{bar},$$

$.\vec{M} = \vec{M}_{couple} + \begin{bmatrix} \vec{i} & \vec{j} & \vec{k} \\ 2 & 4 & 0 \\ \frac{T}{\sqrt{2}} & -\frac{T}{\sqrt{2}} & 0 \end{bmatrix} + \begin{bmatrix} \vec{i} & \vec{j} & \vec{k} \\ 1 & 2 & 0 \\ 0 & -W_{bar} & 0 \end{bmatrix}$, $\vec{M}_A = (M_{couple} - \sqrt{2}T - 2\sqrt{2}T - W_{bar})\vec{k}$, from

which, in terms of the acceleration of the crate: $M_{couple} + 6\left(\frac{W_C}{g}\right)a_C - W_{bar} = I_{bar}\alpha_{AB}$. From kinematics, the acceleration of the end of the bar in terms of the pinned end is

$$\vec{a}_B = \vec{\alpha}_{AB} \times \vec{r}_{B/A} - \omega_{AB}^2 \vec{r}_{B/A} = \begin{bmatrix} \vec{i} & \vec{j} & \vec{k} \\ 0 & 0 & \alpha \\ 2 & 4 & 0 \end{bmatrix} - \omega_{AB}^2 \vec{r}_{AB} = -4\alpha_{AB}\vec{i} + 2\alpha_{AB}\vec{j} - \omega_{AB}^2(2\vec{i} + 4\vec{j})\ ft/s^2.$$

The acceleration in terms of the motion of the crate is $\vec{a}_B = a_C\vec{i} + \vec{\alpha}_{BC} \times \vec{r}_{B/C} - \omega_{BC}^2 \vec{r}_{B/C}$,

$\vec{a}_B = a_C\vec{i} + \begin{bmatrix} \vec{i} & \vec{j} & \vec{k} \\ 0 & 0 & \alpha_{BC} \\ -4 & 4 & 0 \end{bmatrix} - \omega_{BC}^2(-4\vec{i} + 4\vec{j}) = a_C\vec{i} - 4\alpha_{BC}\vec{i} - 4\alpha_{BC}\vec{j} - \omega_{BC}^2(-4\vec{i} + 4\vec{j})\ ft/s^2$, from which

$-4\alpha_{AB} = a_C - 4\alpha_{BC}$, $2\alpha_{AB} = -4\alpha_{BC}$, where $\omega_{AB}^2 = \omega_{BC}^2 = 0$ from the given conditions.

Solve: $\alpha_{BC} = \frac{a_C}{12}$, $\alpha_{AB} = -\frac{a_C}{6}$. Substitute the kinematic relations into the moment equation and reduce: $M_{couple} = -6\left(\frac{W_C}{g}\right)a_C + W_{bar} - I_{bar}\left(\frac{a_C}{6}\right) = W_{bar} - \left(6\left(\frac{W_C}{g}\right) + \frac{I_{bar}}{6}\right)a_C$. Substitute:

$W_{bar} = 20\ lb$, $W_C = 80\ lb$, $I_{bar} = \left(\frac{1}{3}\right)\left(\frac{W_{bar}}{g}\right)(4^2 + 2^2) = 4.145\ slug\text{-}ft^2$, $a_C = -4\ ft/s^2$, from which $\boxed{M_{couple} = 82.45\ ft\text{-}lb}$

==================================<>==================================

Problem 7.57 Suppose that the slender bar in Problem 7.56 is rotating in the counterclockwise direction at 2 *rad/s* at the instant shown, and that the coefficient of friction between the crate and the horizontal surface is $\mu_k = 0.2$. What couple M will cause the crate to accelerate to the left at 4 *ft/s*^{2} at that instant?

Solution Use the solution to Problem 7.56. From Newton's second law applied to the crate free body diagram, the tension required to accelerate the crate to the left at 4 *ft/s*^{2} is given by the two equations in two unknowns: $N + \frac{T}{\sqrt{2}} - W_C = 0$, $\mu_k N - \frac{T}{\sqrt{2}} = m_C a_C$. Solve: $T = \sqrt{2}\left(\frac{\mu_k W_C - m_C a_C}{1+\mu_k}\right)$. Substitute numerical values $\mu_k = 0.2$, $W_C = 80\ lb$, $m_C = \frac{W_C}{g} = 2.49\ slug$, $a_C = -4\ ft/s^2$, and solve:

$T = 30.58\ lb$, $N = 58.33\ lb$, .(where the latter value is unnecessary to the solution, but useful for numerical checks.). The vector location of the end of the bar is $\vec{r}_{B/A} = 2\vec{i} + 4\vec{j}\ ft$. The velocity of the end of the bar is $\vec{v}_B = \vec{\omega}_{AB} \times \vec{r}_{B/A}$, $\vec{v}_B = \begin{bmatrix} \vec{i} & \vec{j} & \vec{k} \\ 0 & 0 & \omega_{AB} \\ 2 & 4 & 0 \end{bmatrix} = -4\omega_{AB}\vec{i} + 2\omega_{AB}\vec{j}\ ft/s$. The velocity of point B in terms of C is $\vec{v}_B = v_C\vec{i} + \vec{\omega}_{BC} \times \vec{r}_{B/C}$, where $\vec{r}_{B/C} = -4\vec{i} + 4\vec{j}\ ft$.

$\vec{v}_B = v_C\vec{i} + \begin{bmatrix} \vec{i} & \vec{j} & \vec{k} \\ 0 & 0 & \omega_{BC} \\ -4 & 4 & 0 \end{bmatrix} = v_C\vec{i} - 4\omega_{BC}\vec{i} - 4\omega_{BC}\vec{j}\ ft/s$. From the *y* component:

$\omega_{BC} = -\frac{1}{2}\omega_{AB} = -1\ rad/s$. The moment about the pinned end of the bar is

$\vec{M}_A = \vec{M}_{couple} + \vec{r}_{B/A} \times (\frac{T}{\sqrt{2}}\vec{i} - \frac{T}{\sqrt{2}}\vec{j}) + \vec{r}_{CM} \times \vec{W}_{bar}$,

$.\vec{M} = \vec{M}_{couple} + \begin{bmatrix} \vec{i} & \vec{j} & \vec{k} \\ 2 & 4 & 0 \\ \frac{T}{\sqrt{2}} & -\frac{T}{\sqrt{2}} & 0 \end{bmatrix} + \begin{bmatrix} \vec{i} & \vec{j} & \vec{k} \\ 1 & 2 & 0 \\ 0 & -W_{bar} & 0 \end{bmatrix}$, $\vec{M}_A = (M_{couple} - 3\sqrt{2}T - W_{bar})\vec{k}$. From which

$M_{couple} - 3\sqrt{2}T - W_{bar} = I_{bar}\alpha_{AB}$. From kinematics, the acceleration of the end of the bar in terms of the pinned end is

$\vec{a}_B = \vec{\alpha}_{AB} \times \vec{r}_{B/A} - \omega_{AB}^2\vec{r}_{B/A} = \begin{bmatrix} \vec{i} & \vec{j} & \vec{k} \\ 0 & 0 & \alpha_{AB} \\ 2 & 4 & 0 \end{bmatrix} - \omega_{AB}^2\vec{r}_{B/A} = -4\alpha_{AB}\vec{i} + 2\alpha_{AB}\vec{j} - \omega_{AB}^2(2\vec{i} + 4\vec{j})\ ft/s^2$

The acceleration in terms of the motion of the crate is $\vec{a}_B = a_C\vec{i} + \vec{\alpha}_{BC} \times \vec{r}_{B/C} - \omega_{BC}^2\vec{r}_{B/C}$.

Solution continued on next page

$$\vec{a}_B = a_C\vec{i} + \begin{bmatrix} \vec{i} & \vec{j} & \vec{k} \\ 0 & 0 & \alpha_{BC} \\ -4 & 4 & 0 \end{bmatrix} - \omega_{BC}^2(-4\vec{i} + 4\vec{j}) = a_C\vec{i} - 4\alpha_{BC}\vec{i} - 4\alpha_{BC}\vec{j} - \omega_{BC}^2(-4\vec{i} + 4\vec{j})\ ft/s^2$$, from which

$-4\alpha_{AB} + 4\alpha_{BC} = a_C + 3\omega_{AB}^2$, $2\alpha_{AB} + 4\alpha_{BC} = 3\omega_{AB}^2$, where $\omega_{BC}^2 = \left(\frac{1}{4}\right)\omega_{AB}^2$, from above.

Solve: $\alpha_{AB} = -\frac{a_C}{6} = 0.6667\ rad/s^2$, $\alpha_{BC} = \frac{a_C}{12} + \frac{3}{4}\omega_{AB}^2 = 2.667\ rad/s^2$.(where the latter value is unnecessary to the solution, but is useful for numerical checks.) Substitute the kinematic relations into the moment equation and reduce: $M_{couple} = 3\sqrt{2}T + W_{bar} + I_{bar}\left(-\frac{a_C}{6}\right)$. Substitute: $W_{bar} = 20\ lb$,

$I_{bar} = \left(\frac{1}{3}\right)\left(\frac{W_{bar}}{g}\right)(4^2 + 2^2) = 4.145\ slug\text{-}ft^2$, $a_C = -4\ ft/s^2$, $T = 30.58\ lb$, from which

$\boxed{M_{couple} = 152.5\ ft\text{-}lb}$

==================================<>==================================

Problem 7.58 Bar AB rotates with a constant angular velocity of 6 *rad/s* in the counterclockwise direction. The slender bar BCD weighs 10 *lb.* and the collar that the bar BCD is attached to at C weighs 2 *lb.* The *y* axis points upward. Neglecting friction, determine the components of the forces exerted on bar BCD by the pins at B and C at the instant shown.

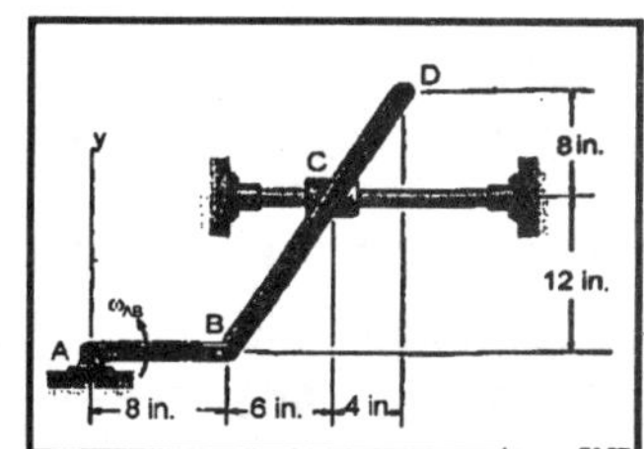

Solution:

The velocity at point B is

$$\vec{v}_B = \vec{\omega}_{AB} \times \vec{r}_{B/A} = \begin{bmatrix} \vec{i} & \vec{j} & \vec{k} \\ 0 & 0 & 6 \\ 8 & 0 & 0 \end{bmatrix} = 48\vec{j}\ in/s = 4\vec{j}\ (ft/s).$$

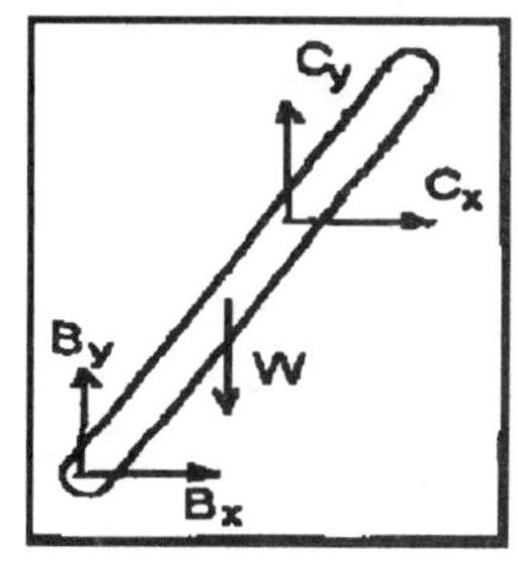

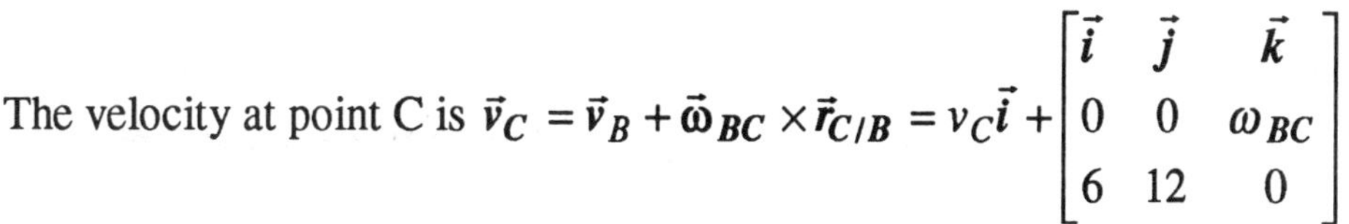

The velocity at point C is $\vec{v}_C = \vec{v}_B + \vec{\omega}_{BC} \times \vec{r}_{C/B} = v_C\vec{i} + \begin{bmatrix} \vec{i} & \vec{j} & \vec{k} \\ 0 & 0 & \omega_{BC} \\ 6 & 12 & 0 \end{bmatrix}$.

$\vec{v}_C = 48\vec{j} - 12\omega_{BC}\vec{i} + 6\omega_{BC}\vec{j}\ in/s$. From the constraint on the collar at C, the *y* component of velocity is zero, from which $48 + 6\omega_{BC} = 0$, $\omega_{BC} = -8\ rad/s$. The acceleration at point B is $\vec{a}_B = \vec{\alpha}_{AB} \times \vec{r}_{B/A} - \omega_{AB}^2\vec{r}_{B/A} = -36(8\vec{i}) = -288\vec{i}\ (in/s^2) = -24\vec{i}\ (ft/s^2)$, since $\vec{\alpha}_{AB} = 0$. The acceleration at point C is $\vec{a}_C = \vec{a}_B + \vec{\alpha}_{BC} \times \vec{r}_{C/B} - \omega_{BC}^2\vec{r}_{C/B}$,

$$\vec{a}_C = -288\vec{i} + \begin{bmatrix} \vec{i} & \vec{j} & \vec{k} \\ 0 & 0 & \alpha_{BC} \\ 6 & 12 & 0 \end{bmatrix} - \omega_{BC}^2(6\vec{i} + 12\vec{j}),$$

$\vec{a}_C = (-288 - 6(8^2) - 12\alpha_{BC})\vec{i} + (6\alpha_{BC} - 12(8^2))\vec{j}\ (in/s^2)$. From the constraint on the collar, $\vec{a}_C = a_C\vec{i}$. Separate components to obtain the two equations in two unknowns: $a_C = -672 - 12\alpha_{BC}$, $0 = -768 + 6\alpha_{BC}$. Solve: $\alpha_{BC} = 128\ rad/s^2$, $a_C = -2208\ in/s^2 = -128\ ft/s^2$.

Solution continued on next page

Continuation of solution to Problem 7.58

The acceleration of the center of mass of the bar is $\vec{a}_G = \vec{a}_B + \vec{\alpha}_G \times \vec{r}_{G/B} - \omega_{BC}^2 \vec{r}_{G/B}$.

$$\vec{a}_G = -288\vec{i} + \begin{bmatrix} \vec{i} & \vec{j} & \vec{k} \\ 0 & 0 & \alpha_{BC} \\ 5 & 10 & 0 \end{bmatrix} - \omega_{BC}^2(5\vec{i} + 10\vec{j}),\ \vec{a}_G = (-288 - 10\alpha_{BC} - 5\omega_{BC}^2)\vec{i} + (5\alpha_{BC} - 10\omega_{BC}^2)\vec{j}.$$

$\vec{a}_G = (-288 - (10)(128) - 5(8^2))\vec{i} + (5(128) - 10(8^2))\vec{j} = -1888\vec{i}\ (in/s^2) = -157.33\vec{i}\ (ft/s^2)$. From Newton's second law and the equation of angular motion applied to the free body diagram of bar BCD and the collar C: for the bar BCD, $B_x + C_x = m_{BCD}a_G$, $B_y + C_y = W$,

$\left(\frac{1}{12}\right)C_y - \left(\frac{5}{12}\right)B_y + \left(\frac{10}{12}\right)B_x - \left(\frac{2}{12}\right)C_x = I_G\alpha_{BC}$, and for the collar C, $-C_x = m_C a_C$, where the units are to be consistent. Solve: $\boxed{C_x = -m_C a_C = \left(\frac{2}{32.17}\right)(184) = 11.44\ lb}$,

$$\boxed{B_x = m_{BCD}a_G - C_x = \left(\frac{10}{32.17}\right)(-157.33) - C_x = -60.3\ lb}$$

$$B_y = -(2)I_G\alpha_{BC} + \left(\frac{10}{6}\right)B_x - \left(\frac{1}{3}\right)C_x + \left(\frac{W_{BCD}}{6}\right) = -\left(\frac{1}{6}\right)\left(\frac{10}{32.17}\right)\frac{(10^2 + 20^2)}{144} + \left(\frac{5}{3}\right)B_x - \left(\frac{1}{3}\right)C_x + \frac{10}{6},$$

$\boxed{B_y = -125.7\ lb}$ $\boxed{C_y = W_{BCD} - B_y = 10 - B_y = 135.7\ lb}$

==================================<>==================================

Problem 7.59 Bar AB weighs 10 *lb.* and bar BC weighs 6 *lb.* If the system is released from rest in the position shown, what are the angular acceleration of bar AB and the normal force exerted by the floor at C at that instant? Neglect friction.

Solution:

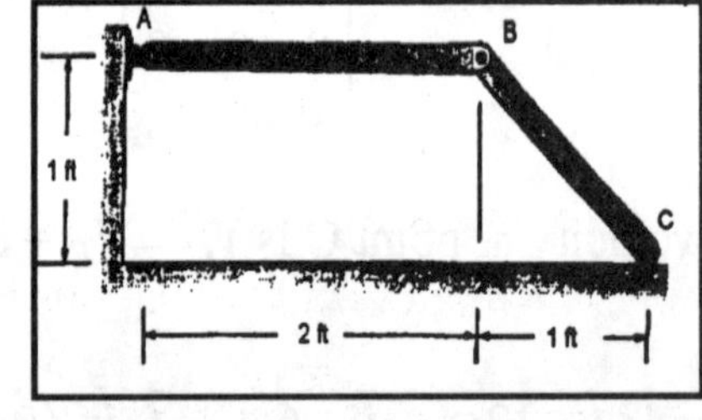

The system starts from rest, from which $\omega_{AB} = 0$, and $\omega_{BC} = 0$. From kinematics, the acceleration of point B is

$\vec{a}_B = \vec{\alpha}_{AB} \times \vec{r}_{B/A} - \omega_{AB}^2 \vec{r}_{B/A}$.,

$$\vec{a}_B = \begin{bmatrix} \vec{i} & \vec{j} & \vec{k} \\ 0 & 0 & \alpha_{AB} \\ 2 & 0 & 0 \end{bmatrix} = 2\alpha_{AB}\vec{j}\ (ft/s^2).$$

In terms of the acceleration of point C, $\vec{a}_B = \vec{a}_C + \vec{\alpha}_{BC} \times \vec{r}_{B/C} - \omega_{BC}^2 \vec{r}_{B/C}$. From the constraint on the motion of point C, $\vec{a}_C = a_C\vec{i}$, from which: $\vec{a}_B = a_C\vec{i} + \begin{bmatrix} \vec{i} & \vec{j} & \vec{k} \\ 0 & 0 & \alpha_{BC} \\ -1 & +1 & 0 \end{bmatrix} = a_C\vec{i} - \alpha_{BC}\vec{i} - \alpha_{BC}\vec{j}$.

The acceleration of point B in terms of the acceleration of the center of mass of bar AB is

$$\vec{a}_B = \vec{a}_{GAB} + \vec{\alpha}_{AB} \times \vec{r}_{B/G} - \omega_{AB}^2 \vec{r}_{B/G} = \vec{a}_{GAB} + \begin{bmatrix} \vec{i} & \vec{j} & \vec{k} \\ 0 & 0 & \alpha_{AB} \\ 1 & 0 & 0 \end{bmatrix} = \vec{a}_{GAB} + \alpha_{AB}\vec{j}\ (ft/s^2).$$

Solution continued on next page

In terms of the acceleration of the center of mass of BC, $\vec{a}_B = \vec{a}_{GBC} + \vec{\alpha}_{BC} \times \vec{r}_{B/GC} - \omega_{BC}^2 \vec{r}_{B/GC}$.

$$\vec{a}_B = \vec{a}_{GBC} + \begin{bmatrix} \vec{i} & \vec{j} & \vec{k} \\ 0 & 0 & \alpha_{BC} \\ -0.5 & 0.5 & 0 \end{bmatrix} = \vec{a}_{GBC} - 0.5\alpha_{BC}\vec{i} - 0.5\alpha_{BC}\vec{j} \ \left(ft/s^2\right).$$

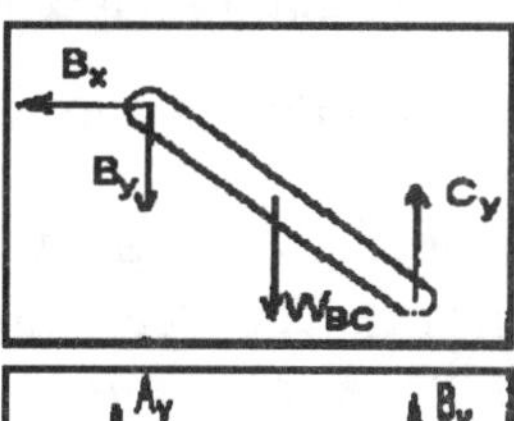

Collecting the kinematic results: $\vec{a}_B = 2\alpha_{AB}\vec{j} \ \left(ft/s^2\right)$,

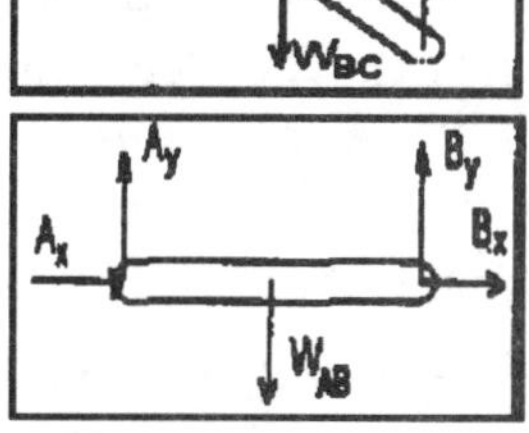

$\vec{a}_B = a_C\vec{i} - \alpha_{BC}\vec{i} - \alpha_{BC}\vec{j} \ \left(ft/s^2\right)$, $\vec{a}_B = \vec{a}_{GAB} + \alpha_{AB}\vec{j} \ \left(ft/s^2\right)$,

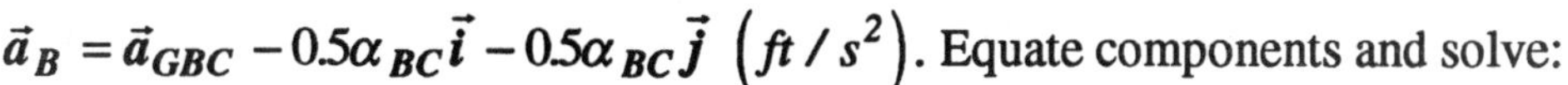

$\vec{a}_B = \vec{a}_{GBC} - 0.5\alpha_{BC}\vec{i} - 0.5\alpha_{BC}\vec{j} \ \left(ft/s^2\right)$. Equate components and solve: $a_{GBCx} = 0$, $a_{GBy} = \alpha_{AB}$, $a_{GCx} = -\alpha_{AB}$, $a_{GCy} = \alpha_{AB}$, and $\alpha_{BC} = -2\alpha_{AB}$. Apply Newton's second law and the equation of angular motion to the free body diagram of bars AB and BC. $A_y + B_y - W_{AB} = m_{AB}a_{GBy}$, $A_x + B_x = 0$, $-A_y + B_y = I_{GB}\alpha_{AB}$, $-B_y + C_y - W_{BC} = m_{BC}a_{GCy}$, $-B_x = m_{BC}a_{GCx}$, $0.5B_x + 0.5C_y + 0.5B_y = I_{GC}\alpha_{BC}$, where $a_{GBy}, a_{GBx}, a_{GCy}, a_{GCx}$ are the accelerations of the center of mass of the two bars, and $I_{GB} = \frac{1}{12}\left(\frac{W_{AB}}{g}\right)2^2 = 0.1036 \ slug\text{-}ft^2$, $I_{GC} = \frac{1}{12}\left(\frac{W_{BC}}{g}\right)(1^2 + 1^2) = 0.0311 \ slug\text{-}ft^2$. Substitute the kinematic results to obtain the six equations in six unknowns: (1) $A_y + B_y - W_{AB} = \left(\frac{W_{AB}}{g}\right)\alpha_{AB}$, (2) $A_x + B_x = 0$, (3) $-A_y + B_y = I_{GB}\alpha_{AB}$, (4) $-B_y + C_y - W_{BC} = \left(\frac{W_{BC}}{g}\right)\alpha_{AB}$, (5) $-B_x = \left(\frac{W_{BC}}{g}\right)(-\alpha_{AB})$, (6) $0.5B_x + 0.5C_y + 0.5B_y = I_{GC}(-2\alpha_{AB})$. Solve by iteration (using **TK Solver Plus**) to obtain: $A_x = 3.27 \ lb$, $A_y = 3.18 \ lb$, $B_x = -3.27 \ lb$, $B_y = 1.36 \ lb$, $\boxed{C_y = 4.09 \ lb}$, $\boxed{\alpha_{AB} = -17.55 \ rad/s^2}$ clockwise. *Check*: These equations have a straight forward analytical solution: From (1) and (3) $B_y = \left(\frac{W_{AB}}{g} + I_{GB}\right)\frac{\alpha_{AB}}{2} + \frac{W_{AB}}{2}$. Equations (4), (5) and (6) collapse to two equations in two unknowns, from which $B_y = -\left(2I_{CG} + \frac{W_{BC}}{g}\right)\alpha_{AB} - \frac{W_{BC}}{2}$. $C_y = -2I_{GC}\alpha_{AB} + \frac{W_{BC}}{2}$. Equate the two expressions for B_y and solve: $\alpha_{AB} = \dfrac{-0.5(W_{BC} + W_{AB})}{(2I_{GC} + 0.5I_{BG}) + \left(\frac{W_{AB}}{2g} + \frac{W_{BC}}{2}\right)} = -17.55 \ rad/s^2$, *check*. Substitute α_{AB} into the expression for C_y: $C_y = 4.091 \ lb$. *check.*

==============================<>==============================

================================<>================================

Problem 7.60 Let the total moment of inertia of the car's two rear wheels and axle be I_R, and let the total moment of inertia of the two front wheels be I_F. The radius of the tires is R, and the total mass of the car including the wheels is m. If the car's engine exerts a torque (couple) T on the rear wheels and the wheels do not slip, show that the car's acceleration is $a = \dfrac{RT}{R^2 m + I_R + I_F}$.

Strategy: Isolate the wheels and draw three free body diagrams.

Solution: The free body diagrams are as shown: We shall write three equations of motion for each wheel and two equations of motion for the body of the car: We shall sum moments about the axles on each wheel.

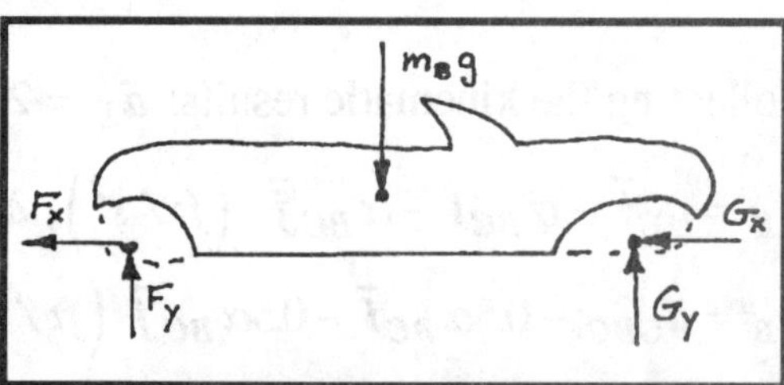

Rear Wheel: $\sum F_x = F_x + f_R = m_R a$,

$\sum F_y = N_R - m_R g - F_y = 0$,

$\sum M_{Raxle} = R f_R - T = I_R \alpha = I_R\left(-\dfrac{a}{R}\right)$

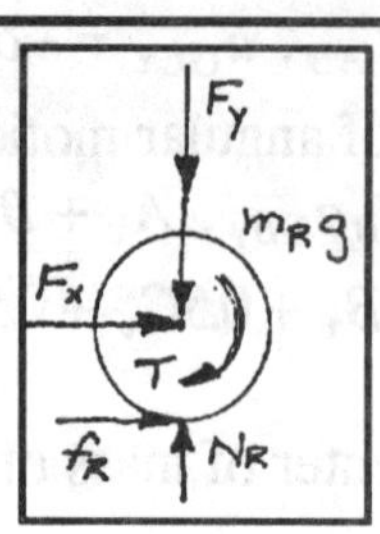

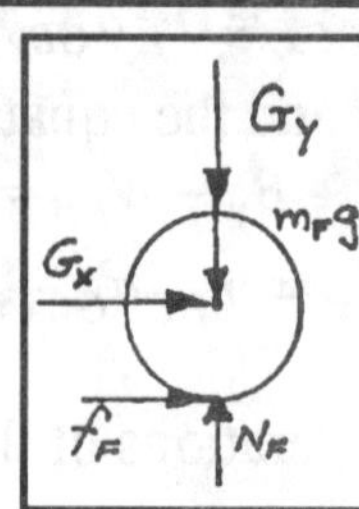

Front Wheel: $\sum F_x = G_x + f_F = m_F a$,

$\sum F_y = N_F - m_F g - G_y = 0$, $\sum M_{Faxle} = R f_F = I_F \alpha = I_F\left(-\dfrac{a}{R}\right)$

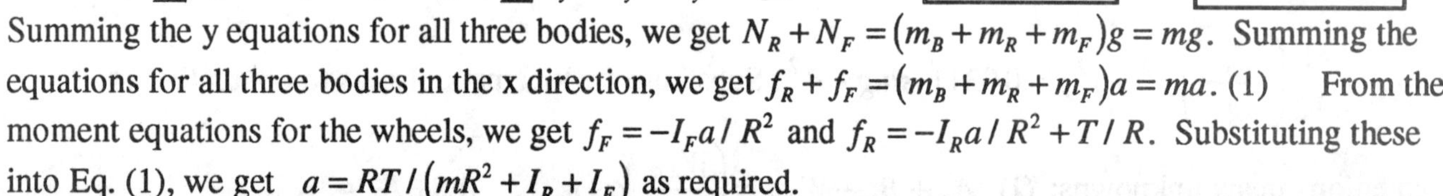

Car Body: $\sum F_x = -F_x - G_x = m_B a$, $\sum F_y = F_y + G_y - m_B g = 0$.

Summing the y equations for all three bodies, we get $N_R + N_F = (m_B + m_R + m_F)g = mg$. Summing the equations for all three bodies in the x direction, we get $f_R + f_F = (m_B + m_R + m_F)a = ma$. (1) From the moment equations for the wheels, we get $f_F = -I_F a / R^2$ and $f_R = -I_R a / R^2 + T/R$. Substituting these into Eq. (1), we get $a = RT/(mR^2 + I_R + I_F)$ as required.

================================<>================================

Problem 7.61 The combined mass of the motorcycle and rider is 160 *kg*. Each 9 *kg* wheel has a 330 *mm* radius and moment of inertia $I = 0.8$ $kg\text{-}m^2$. The engine drives the rear wheel. If the rear wheel exerts a 400 *N* horizontal force on the road and you do not neglect the horizontal force exerted on the road by the front wheel, determine (a) the motorcycle's acceleration; (b) the normal forces exerted on the road by the rear and front wheels. (The location of the center of mass of the motorcycle not including its wheels is shown.) *Strategy:* Isolate the wheels and draw free body diagrams. The motorcycle's engine drives the rear wheel by exerting a couple on it.

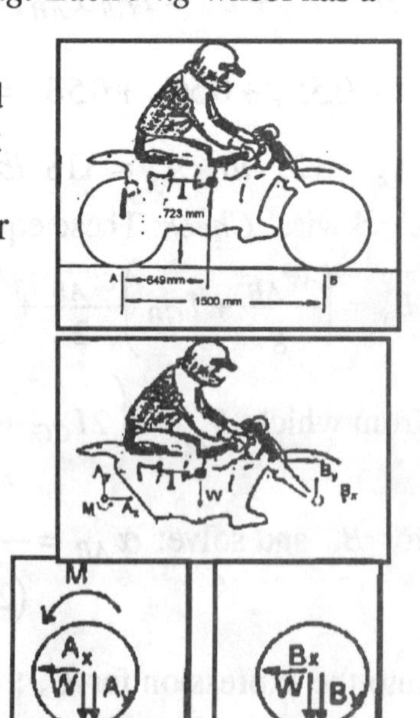

Solution:

Assume that the axles of the motorcycle are directly above the points of contact of the wheels with the road. The mass of the motorcycle and rider exclusive of the wheels is $m_R = 142$ kg. Denote $T_A = -400$ N. Apply Newton's second law and the equation of angular motion to the free body diagram of the wheels. For the rear wheel, $T_A - A_x = m_w a_x$, $M + RT_A = I\alpha_W$. $-A_y - m_w g + N_A = 0$. From kinematics, $a_x = -R\alpha_W$, from which $M + RT_A = -\dfrac{I}{R}a_x$.

Solution continued on next page

For the front wheel, $-RT_B = I\alpha_W$, $-B_x - T_B = m_w a_x$. $-B_y - m_w g + N_B = 0$. From kinematics, $-RT_B = -\frac{I}{R}a_x$. For the frame, choose a coordinate system with the origin at the center of mass and the x axis parallel to the road surface. Since the frame does not rotate, the acceleration of the wheel axle is also the acceleration of the center of mass of the frame. The location of the rear axle is $\vec{r}_A = \left(-0.649\vec{i} - (0.723 - 0.330)\vec{j}\right) = -0.649\vec{i} - 0.393\vec{j}$ (m). The location of the front axle is $\vec{r}_B = (1.50 - 0.649)\vec{i} - 0.393\vec{j} = 0.851\vec{i} - 0.393\vec{j}$ (m). The moment about the center of mass is

$\vec{M}_{cm} = -\vec{M} + \vec{r}_A \times \left(A_x\vec{i} + A_y\vec{j}\right) + \vec{r}_B \times (B_x\vec{i} + B_y\vec{j})$,

$$\vec{M}_{CM} = -\vec{M} + \begin{bmatrix} \vec{i} & \vec{j} & \vec{k} \\ -0.649 & -0.393 & 0 \\ A_x & A_y & 0 \end{bmatrix} + \begin{bmatrix} \vec{i} & \vec{j} & \vec{k} \\ 0.851 & -0.393 & 0 \\ B_x & B_y & 0 \end{bmatrix}$$

$\vec{M}_{CM} = -\vec{M} + \left(-0.649A_y + 0.393A_x + 0.851B_y + 0.393B_x\right)\vec{k} = 0$. Apply Newton's second law to the frame. $A_x + B_x = m_r a_x$. $A_y + B_y - m_r g = 0$, where $m_r = 142$ kg The result is nine equations in nine unknowns: $T_A - A_x = m_w a_x$, $-A_y - m_w g + N_A = 0$, $M + RT_A = -\left(\frac{I}{R}\right)a_x$, $-B_x - T_B = m_w a_x$, $RT_B = \frac{I}{R}a_x$, $-B_y - m_w g + N_B = 0$, $-M + \left(-0.649A_y + 0.393A_x + 0.851B_y + 0.393B_x\right) = 0$, $A_x + B_x = m_r a_x$, $A_y + B_y - m_r g = 0$. Solve by iteration (using **TK Solver Plus**): $A_x = 378.5$ N, $B_x = -39.1$ N, $A_y = 971.1$ N, $B_y = 421.9$ N, (a) $\boxed{a_x = 2.39\ m/s^2}$, $M = -137.8$ N-m, $T_B = 17.6$ N, (b) $\boxed{N_A = 1059.4\ N}$, $\boxed{N_B = 510.2\ N}$

================================<>================================

Problem 7.62 In Problem 7.61, if the front wheel lifts slightly off the road when the rider accelerates, determine (a) the motorcycle's acceleration; (b) the torque exerted by the engine on the rear wheel.

Solution : Use the data and solution in Problem 7.61, *with the assumption that T_A is now unknown.* . With the front wheel off the ground, the *center of mass of the frame and rider and front wheel* is shifted forward from the value used in Problem 7.61 for the frame and rider alone, by an amount

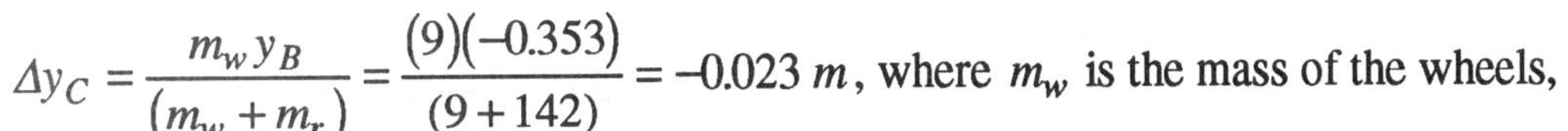

$\Delta x_C = \frac{m_w x_B}{(m_w + m_r)} = \frac{(9)(0.851)}{(142 + 9)} = 0.051\ m$,

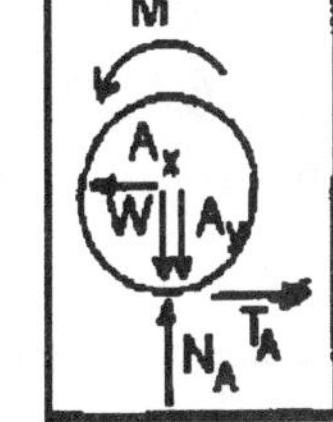

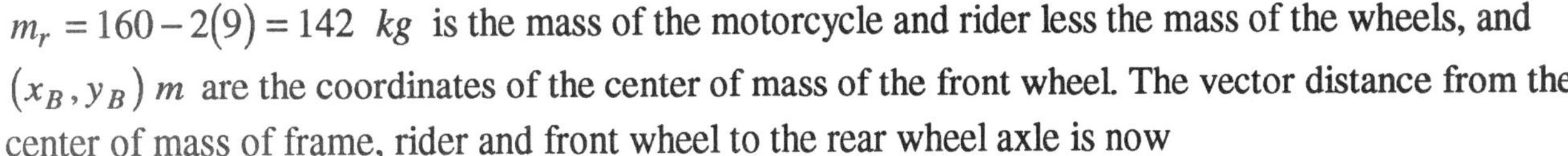

$\Delta y_C = \frac{m_w y_B}{(m_w + m_r)} = \frac{(9)(-0.353)}{(9 + 142)} = -0.023\ m$, where m_w is the mass of the wheels, $m_r = 160 - 2(9) = 142$ kg is the mass of the motorcycle and rider less the mass of the wheels, and (x_B, y_B) m are the coordinates of the center of mass of the front wheel. The vector distance from the center of mass of frame, rider and front wheel to the rear wheel axle is now

Solution continued on next page

$\vec{r}_A = (-0.649 - \Delta x_C)\vec{i} + (-0.393 - \Delta y_C)\vec{j} = -0.7\vec{i} - 0.37\vec{j}$ (m). Apply Newton's second law and the equation of angular motion to the free body diagrams of the wheel and frame: $M + RT_A = I\alpha$, (1) $-A_x + T_A = m_w a_x$, (2) $-A_y - m_w g + N_A = 0$. From kinematics, $a_x = -R\alpha$, from which (3) $M + RT_A = -\frac{I}{R}a_x$. Since the frame does not rotate, the acceleration of the wheel axle is also the acceleration of the center of mass of the frame, rider and front wheel. The moment about this center of mass is

$$\vec{M} = 0 = -\vec{M} + \vec{r}_A \times \left(A_x\vec{i} + A_y\vec{j}\right) = -\vec{M} + \begin{bmatrix} \vec{i} & \vec{j} & \vec{k} \\ -0.7 & -0.37 & 0 \\ A_x & A_y & 0 \end{bmatrix} = -\vec{M} - (0.649A_y - 0.393A_x)\vec{k} = 0$$

, from which (4) $-M - 0.649A_y + 0.393A_x = 0$. (5) $A_x = (m_w + m_r)a_x$, (6) $A_y - (m_w + m_r)g = 0$. By inspection of (2) and (6), A_y and N_A are known quantities: $A_y = (m_r + m_w)g = 1481.3$ N, $N_A = A_y + m_w g = 1569.6$ N. The result is four equations in four unknowns: (1), (3), (4) and (5). Solve by iteration: $A_x = 1409.4$ N, $T_A = 1493.4$ N, $\boxed{a_x = 9.33\ m/s^2}$, $\boxed{M = -515.4\ N\text{-}m}$

==================================<>==================================

Problem 7.63 The moment of inertial of the vertical handle about O is 0.12 $slug-ft^2$. The object B weighs *15-lb* and rests on a smooth surface. The weight of the bar AB is negligible (which means you can treat it as a two-force member). If the person exerts a *0.2-lb* horizontal force on the handle *15-in* above O, what is the resulting angular acceleration of the handle?

Solution:

Let α be the clockwise angular acceleration of the handle. The acceleration of B is: $\bar{a}_B = \bar{a}_A + \bar{\alpha}_{AB} \times \bar{r}_{B/A}$:

$$a_B\bar{i} = (6/12)\alpha\bar{i} + \begin{vmatrix} \bar{i} & \bar{j} & \bar{k} \\ 0 & 0 & \alpha_{AB} \\ 1 & -0.5 & 0 \end{vmatrix}$$

we see that $\alpha_{AB} = 0$ and

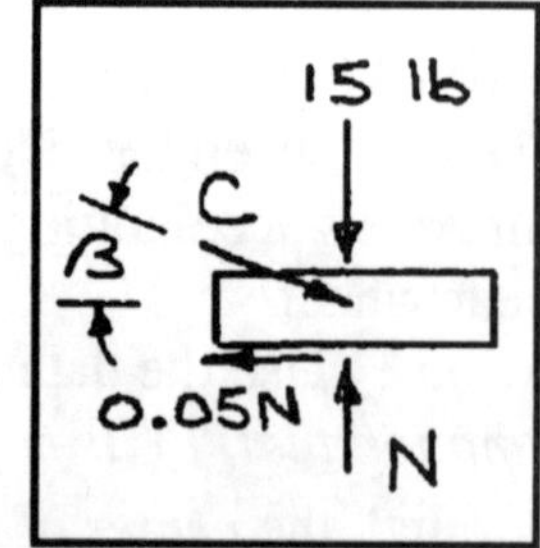

$a_B = (6/12)\alpha$ (1). The free body diagrams of the handle and object B are as shown. Note that $\beta = \arctan(6/12) = 26.6°$. Newton's second law for the object B is $C\cos\beta = (15/32.2)a_B$, (2) The equation of angular motion for the handle is $(15/12)F - (6/12)C\cos\beta = (0.12)\alpha$ (3). Solving Equations (1)-(3) with $F = 0.2lb$, we obtain $\alpha = 1.06 rad/s^2$

==================================<>==================================

==================================<>==================================

Problem 7.64 In Problem 7.63, suppose that the kinetic coefficient of friction between the object B and the horizontal surface is *0.05*. Immediately after B starts slippine, what horizontal force does the person need to exert on the handle *15-in.* above *O* in order that the angular acceleration of the hancle be *1 rad/s* ?

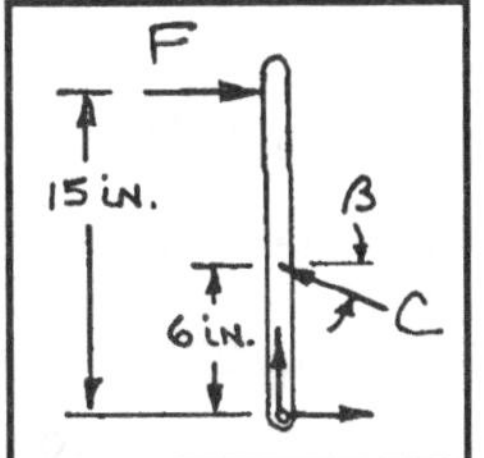

Solution:

Let α be the clockwise angular acceleration of the handle. From the solution of Problem 7.63, the acceleration of B is $a_B = (6/12)\alpha$.(1)

The free body diagrams of the handle and object B are as shown.

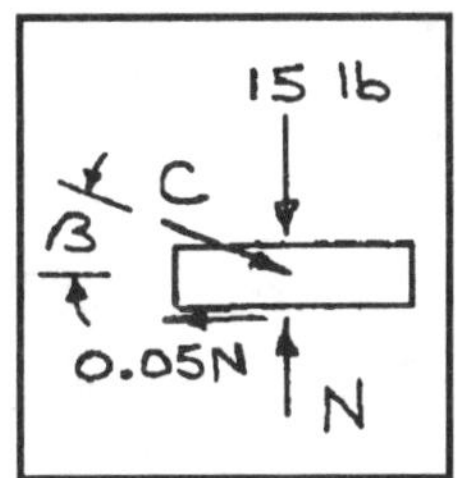

Note that $\beta = 26.6°$.

From the free body diagram of object B, $N - 15 - C\sin\beta = 0$ (2)

$C\cos\beta - 0.05N = (15/32.2)a_B$ (3)

The equation of angular motion for the handle is

$(15/12)F - (6/12)C\cos\beta = (0.12)\alpha$ (4).

Solving Equations (1)-(4) with $\alpha = 1\ rad/s^2$ we obtain $F = 0.499 lb$.

==================================<>==================================

Problem 7.65 Bars *OQ* and *PQ* each weigh *6-lb*. The weight of the collar *P* and friction between the collar and the horizontal bar are negligible. If the system is released from rest with $\theta = 45°$, what are the angular accelerations of the two bars?

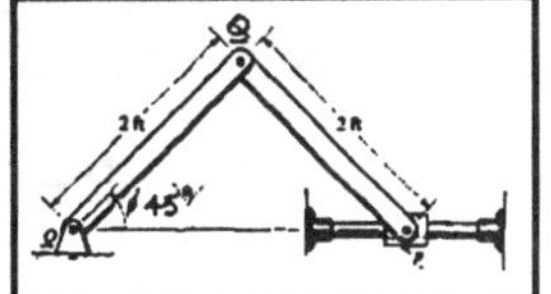

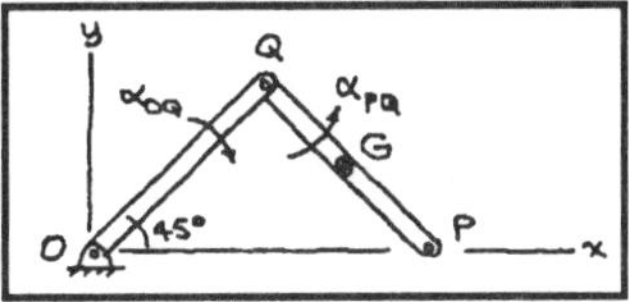

Solutions:

Let α_{OQ} and α_{PQ} be the clockwise angular acceleration of bar OQ and the counterclockwise angular acceleration of bar PQ. The acceleration of Q is

$$\bar{a}_Q = \bar{a}_0 + \bar{\alpha}_{0Q} \times \bar{r}_{Q/0} = \begin{vmatrix} \bar{i} & \bar{j} & \bar{k} \\ 0 & 0 & -\alpha_{OQ} \\ 2\cos 45° & 2\sin 45° & 0 \end{vmatrix} = 2\alpha_{OQ}\sin 45°\bar{i} - 2\alpha_{OQ}\cos 45°\bar{j}.$$

The acceleration of P is $\bar{a}_P = \bar{a}_Q + \bar{\alpha}_{PQ} \times \bar{r}_{P/Q}$

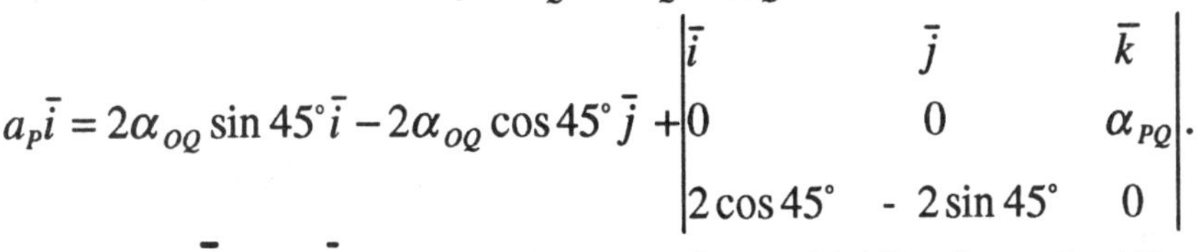

$$a_P\bar{i} = 2\alpha_{OQ}\sin 45°\bar{i} - 2\alpha_{OQ}\cos 45°\bar{j} + \begin{vmatrix} \bar{i} & \bar{j} & \bar{k} \\ 0 & 0 & \alpha_{PQ} \\ 2\cos 45° & -2\sin 45° & 0 \end{vmatrix}.$$

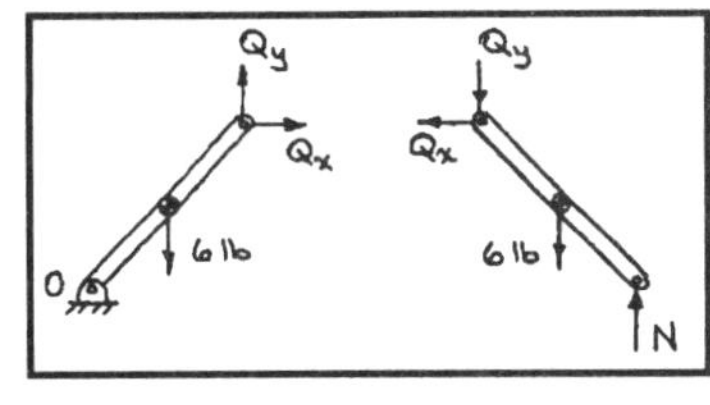

Equating $\bar{i}$ and $\bar{j}$ components, $a_P = 2\alpha_{OQ}\sin 45° + 2\alpha_{PQ}\sin 45°$ (1)

$0 = -2\alpha_{OQ}\cos 45° + 2\alpha_{PQ}\cos 45°$ (2). The acceleration of the center of mass of bar PQ is

$$\bar{a}_G = \bar{a}_Q + \bar{\alpha}_{PQ} \times \bar{r}_{G/Q} = 2\alpha_{OQ}\sin 45°\bar{i} - 2\alpha_{OQ}\cos 45°\bar{j} + \begin{vmatrix} \bar{i} & \bar{j} & \bar{k} \\ 0 & 0 & \alpha_{PQ} \\ \cos 45° & -\sin 45° & 0 \end{vmatrix}.$$ Hence,

$a_{Gx} = 2\alpha_{OQ}\sin 45° + \alpha_{PQ}\sin 45°$ (3); $a_{Gy} = -2\alpha_{OQ}\cos 45° + \alpha_{PQ}\cos 45°$ (4).

From the diagrams:

Solution continued on next page

The equation of angular motion of bar OQ is $\sum M_O = I_O \alpha_{OQ}$:
$Q_x(2\cos 45^\circ) - Q_y(2\cos 45^\circ) + 6\cos 45^\circ = \frac{1}{3}(6/32.2)(2)^2 \alpha_{OQ}$(5).
The equations of motion of bar PQ are
$\sum F_x = -Q_x = (6/32.2)a_{Gx}$ (6)
$\sum F_y = N - Q_y - 6 = (6/32.2)a_{Gy}$(7)
$\sum M = (N + Q_y + Q_x)(\cos 45^\circ) = \frac{1}{12}(6/32.2)(2)^2 \alpha_{PQ}$ (8). Solving Equations (1)-(8), we obtain
$\alpha_{OQ} = \alpha_{PQ} = 6.83 rad/s^2$

================================<>================================

Problem 7.66 In Problem 7.65, what are the angular accelerations of the two bars if the collar P weighs *2 lb*?
Solution:
In the solution of Problem 7.65, the free body diagram of bar PQ has a horizontal component *P* to the left where *P* is the force exerted on the bar by the collar. Equations (6) and (8) become
$\sum F_x = -Q_x - P = (6/32.2)a_{Gx}$ $\sum M = (N - P + Q_y + Q_x)(\cos 45^\circ) = \frac{1}{12}(6/32.2)(2)^2 \alpha_{PQ}$ and the equation of motion for the collar is $P = (2/32.2)a_p$ solving equations (1-9), we obtain
$\alpha_{OQ} = \alpha_{PQ} = 4.88 rad/s^2$.

================================<>================================

Problem 7.67 The *4-kg* slender bar is pinned to *2-kg* sliders at A and B. If friction is negligible and the system is released from rest in the position shown, what is the angular acceleration of the bar at that instant?

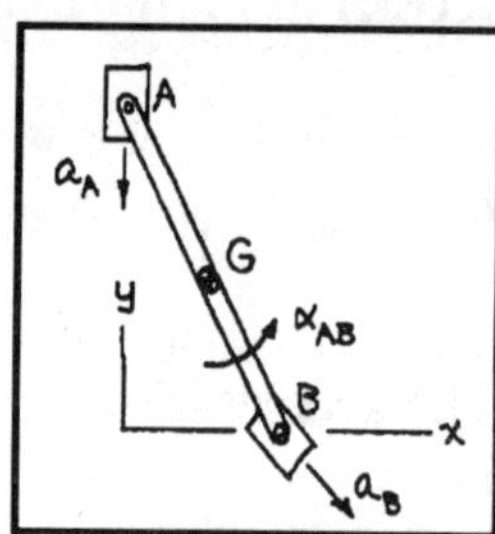

Solution:
Express the acceleration of B in terms of the acceleration of A,
$\bar{a}_B = \bar{a}_A + \bar{\alpha}_{AB} \times \bar{r}_{B/A}$:

$$a_B \cos 45^\circ \bar{i} - a_B \sin 45^\circ \bar{j} = -a_A \bar{j} + \begin{vmatrix} \bar{i} & \bar{j} & \bar{k} \\ 0 & 0 & \alpha_{AB} \\ 0.5 & -1.2 & 0 \end{vmatrix}, \text{ or}$$

$a_B \cos 45^\circ = 1.2\alpha_{AB}$, (1); and $-a_B \sin 45^\circ = -a_A + 0.5\alpha_{AB}$, (2). We express the acceleration of G in terms of the acceleration of A, $\bar{a}_G = \bar{a}_A + \bar{\alpha}_{AB} \times \bar{r}_{G/A}$:

$$\bar{a}_G = a_{Gx}\bar{i} + a_{Gy}\bar{j} = -a_A \bar{j} + \begin{vmatrix} \bar{i} & \bar{j} & \bar{k} \\ 0 & 0 & \alpha_{AB} \\ 0.25 & -0.6 & 0 \end{vmatrix}, \text{ or } a_{Gx} = 0.6\alpha_{AB}, \text{ (3); and}$$

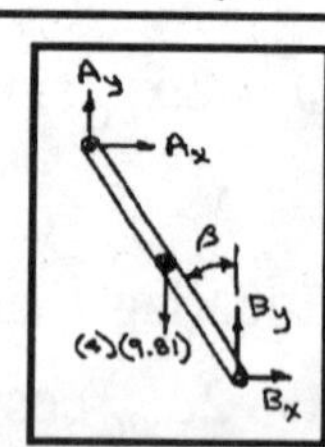

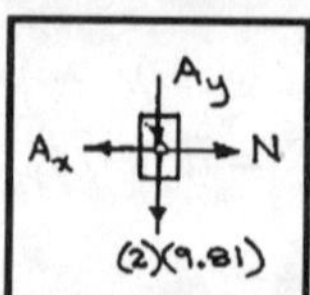

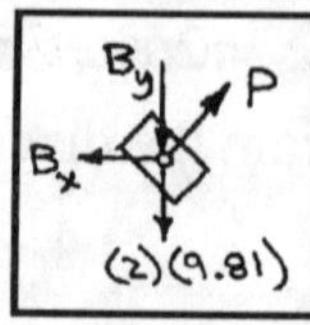

$a_{Gy} = -a_A + 0.25\alpha_{AB}$, (4); The free body diagrams are as shown. The equations of motion are Slider A: $N - A_x = 0$ (5), and $(2)(9.81) + A_y = 2a_A$, (6);
Slider B: $P - [B_x + B_y + (2)(9.81)]\cos 45^\circ = 0$, (7); and
$[(2)(9.81) - B_x + B_y]\cos 45^\circ = 2a_B$, (8);

Solution continued on next page

Bar: $A_x + B_x = 4a_{Gx}$ (9), and $A_y + B_y - (4)(9.81) = 4a_{Gy}$ (10)

$(L/2)\left[(B_x - A_x)\cos\beta + (B_y - A_y)\sin\beta\right] = \frac{1}{12}(4)L^2\alpha_{AB}$ (11), where $L = \sqrt{(0.5)^2 + (1.2)^2}\,m$ and

$\beta \arctan(0.5/1.2) = 22.6°$. Solving Equations (1) - (11), we obtain $\alpha_{AB} = 5.18 rad/s^2$.

==================================<>==================================

Problem 7.68 The mass of the slender bar is m and the mass of the homogenous cylinder is $4m$. The system is released from rest in the position shown. If the disk rolls and the friction between the bar and the horizontal surface is negligible, show that the disk's angular acceleration is $\alpha = \dfrac{6g}{95R}$ counterclockwise.

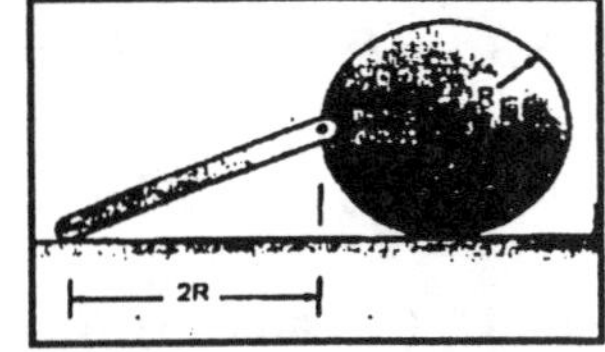

Solution:

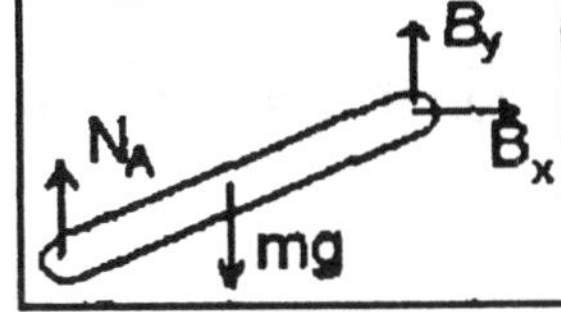

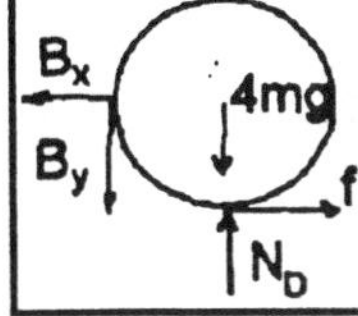

For the bar: The length of the bar is $L = \sqrt{5}R$. Apply Newton's second law to the free body diagram of the bar: $B_x = ma_{Gx}$, $B_y + N_A - mg = ma_{Gy}$, where a_{Gx}, a_{Gy} are the accelerations of the center of mass of the bar. The moment about the bar center of mass is

$$RB_y - RN_A - \frac{R}{2}B_x = I_B\alpha_{AB}.$$

For the disk: Apply Newton's second law and the equation of angular motion to the free body diagram of the disk. $f - B_x = 4ma_{Dx}$, $N_D - 4mg - B_y = 0$, $RB_y + Rf = I_D\alpha_D$

From kinematics: Since the system is released from rest, $\omega_{AB} = \omega_D = 0$. The acceleration of the center of the disk is $\vec{a}_D = -R\alpha_D\vec{i}$ .. The acceleration of point B in terms of the acceleration of the center of the disk is

$$\vec{a}_B = \vec{a}_D + \vec{\alpha}_D \times \vec{r}_{B/D} = \vec{a}_D + \begin{bmatrix} \vec{i} & \vec{j} & \vec{k} \\ 0 & 0 & \alpha_D \\ -R & 0 & 0 \end{bmatrix} = -R\alpha_D\vec{i} - R\alpha_D\vec{j}$$

.. The acceleration of the center of mass of the bar in terms of the acceleration of B is

$$\vec{a}_G = \vec{a}_B + \vec{\alpha}_{AB} \times \vec{r}_{G/B} - \omega_{AB}^2\vec{r}_{G/B} = \vec{a}_B + \begin{bmatrix} \vec{i} & \vec{j} & \vec{k} \\ 0 & 0 & \alpha_{AB} \\ -R & -\frac{R}{2} & 0 \end{bmatrix} = \vec{a}_B + \frac{R\alpha_{AB}}{2}\vec{i} - R\alpha_{AB}\vec{j},$$

$$\vec{a}_G = -R\left(\alpha_D - \frac{\alpha_{AB}}{2}\right)\vec{i} - R(\alpha_D + \alpha_{AB})\vec{j}.$$

The acceleration of the center of mass of the bar in terms of the acceleration of A is

$$\vec{a}_G = \vec{a}_A + \vec{\mathbf{a}}_{AB} \times \vec{r}_{G/A} = \vec{a}_A + \begin{bmatrix} \vec{i} & \vec{j} & \vec{k} \\ 0 & 0 & \alpha_{AB} \\ \mathrm{R} & \frac{\mathrm{R}}{2} & 0 \end{bmatrix} = \vec{a}_A - \frac{\mathrm{R}\alpha_{AB}}{2}\vec{i} + \mathrm{R}\alpha_{AB}\vec{j}.$$

Solution continued on next page

From the constraint on the motion, $\vec{a}_A = a_A\vec{i}$. Equate the expressions for $\vec{a}_G$, separate components and solve: $\alpha_{AB} = -\frac{\alpha_D}{2}$. Substitute to obtain $a_{Gx} = -\frac{5R}{4}\alpha_D$, $a_{Gy} = -\frac{R}{2}\alpha_D$.

Collect the results: (1) $B_x = -\frac{5Rm}{4}\alpha_D$, (2) $B_y + N_A - mg = -\frac{Rm}{2}\alpha_D$,

(3) $RB_y - RN_A - \frac{R}{2}B_x = -\frac{I_B}{2}\alpha_D$, (4) $f - B_x = -4Rm\alpha_D$, (5) $N_D - 4mg - B_y = 0$,

(6) $RB_y + Rf = I_D\alpha_D$.

From (1), (2), and (3) $B_y = \frac{mg}{2} - \left(\frac{9mR}{16} + \frac{I_B}{4R}\right)\alpha_D$. From (1), (4) and (6), $B_y = \left(\frac{I_D}{R} + \frac{21Rm}{4}\right)\alpha_D$.

Equate the expressions for B_y and reduce to obtain $\alpha_D = \left(\frac{mg}{2}\right)\frac{1}{\left(\frac{93Rm}{16} + \frac{I_D}{R} + \frac{I_B}{4R}\right)}$. For a homogenous cylinder of mass 4*m*, $I_D = 2R^2m$. For a slender bar of mass *m* about the center of mass, $I_B = \frac{1}{12}mL^2 = \frac{5}{12}mR^2$. Substitute and reduce: $\boxed{a_D = \frac{6g}{95R}}$.

==================================<>==================================

==================================<>==================================

Problem 7.69 If the disk in Problem 7.68 rolls and the coefficient of kinetic friction between the bar and the horizontal surface is μ_k, what is the disk's angular acceleration at the instant the disk is released?

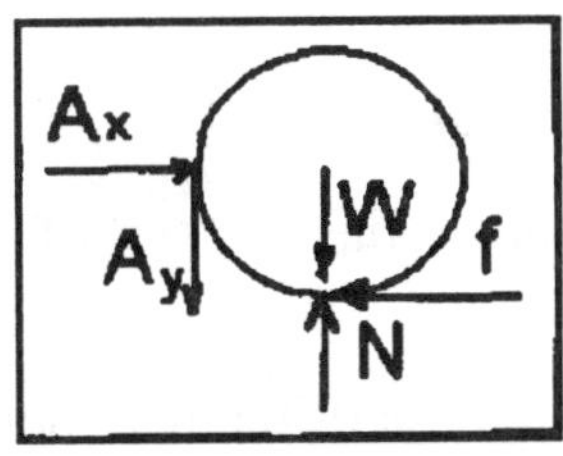

Solution:

Choose a coordinate system with the origin at the point of contact of the disk and the surface, with x axis positive parallel to the plane surface and the y axis positive upward. Denote the end of the bar in contact with the planar surface by B, the point attached to the disk by A, and the center of mass of the bar by G, and the center of the mass of the disk by C. The coordinates of the point A on the sphere are $(-R, R)$, and the coordinates of point B are $(-3R, 0)$.

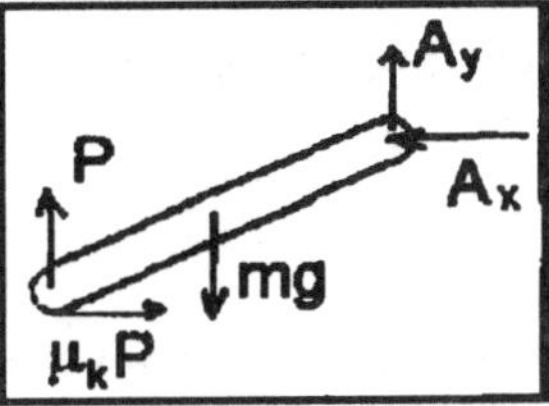

The coordinates of the center of mass G of the bar are $\left(-2R, \frac{1}{2}R\right)$.

The vector distance from A to B is $\vec{r}_{B/A} = \vec{r}_B - \vec{r}_A = -3R\vec{i} - \left(-R\vec{i} + R\vec{j}\right) = -2R\vec{i} - R\vec{j}$.

The vector distance from A to G is $\vec{r}_{G/A} = \vec{r}_G - \vec{r}_A = -2R\vec{i} + \frac{1}{2}R\vec{j} - \left(-R\vec{i} + R\vec{j}\right) = -R\vec{i} - \frac{1}{2}R\vec{j}$.

Kinematics:

The acceleration of the point of contact of the rolling disk with the surface is zero, from which the acceleration of the center of mass of the disk is $0 = \vec{a}_C + \vec{\mathbf{a}} \times \vec{r}_{O/C} = \vec{a}_C + \begin{bmatrix} \vec{i} & \vec{j} & \vec{k} \\ 0 & 0 & \alpha \\ 0 & -R & 0 \end{bmatrix} = \vec{a}_C + R\alpha\vec{i}$, from which (1) $a_{Cx} = -R\alpha$, (2) $a_{Cy} = 0$.

The acceleration of point A on the disk is $\vec{a}_A = \vec{\mathbf{a}} \times \vec{R} = \begin{bmatrix} \vec{i} & \vec{j} & \vec{k} \\ 0 & 0 & \alpha \\ -R & R & 0 \end{bmatrix} = -R\alpha\left(\vec{i} + \vec{j}\right)$.

The acceleration of point B is constrained to be parallel to the x axis, from which

$$\vec{a}_B = a_B\vec{i} = \vec{a}_A + \vec{\mathbf{a}}_{AB} \times \vec{r}_{B/A} = -\alpha R\vec{i} - \alpha R\vec{j} + \begin{bmatrix} \vec{i} & \vec{j} & \vec{k} \\ 0 & 0 & \alpha_{AB} \\ -2R & -R & 0 \end{bmatrix}. \; a_B\vec{i} = (-\alpha R + \alpha_{AB}R)\vec{i} - (\alpha R + 2\alpha_{AB}R)\vec{j},$$

from which (2) $\alpha_{AB} = -\frac{\alpha}{2}$. The acceleration of point G in terms of the acceleration of A is

$$\vec{a}_G = \vec{a}_A + \vec{\mathbf{a}}_{AB} \times \vec{r}_{G/A} = \vec{a}_B + \begin{bmatrix} \vec{i} & \vec{j} & \vec{k} \\ 0 & 0 & \alpha_{AB} \\ -R & -\frac{R}{2} & 0 \end{bmatrix}. \; \vec{a}_G = -R\alpha\left(\vec{i} + \vec{j}\right) + R\alpha_{AB}\left(\frac{1}{2}\vec{i} - \vec{j}\right).$$ Substitute (1),

$\vec{a}_G = -R\alpha\left(\frac{5}{4}\vec{i} + \frac{1}{2}\vec{j}\right)$ from which (3) $a_{Gx} = -\frac{5R\alpha}{4}$, (4) $a_{Gy} = -\frac{R\alpha}{2}$.

Solution continued on next page

Continuation of solution to Problem 7.69

Dynamics:

Denote the moment of inertia of the disk about its center of mass by $I_D = \frac{4mR^2}{2} = 2mR^2$, and the moment of inertia of the bar about its center of mass by $I_B = \frac{m((2R)^2 + R^2)}{12} = \frac{5mR^2}{12}$. Apply Newton's second law and the equation of angular motion to the free body diagram, and apply the identities (1), (2), (3) and (4):

$(1')\ A_y R - fR = I_D\alpha = 2mR^2\alpha,$

$(2')\ f - A_x = (4m)a_{Cx} = 4mR\alpha,$

$(3')\ \mu_k P - A_x = ma_{Gx} = \left(-\frac{5mR}{4}\right)\alpha,$

$(4')\ P + A_y - mg = ma_{Gy} = -\frac{mR\alpha}{2},$

$(5')\ A_x\left(\frac{R}{2}\right) + A_y(R) - PR + \mu_k P\left(\frac{R}{2}\right) = I_B\alpha_{AB} = I_B\left(-\frac{\alpha}{2}\right) = -\frac{5mR^2\alpha}{24}$. These are five equations in the five unknowns A_y, A_x, f, P, α. The steps are: from (2') $f = A_x + 4mR\alpha$. Substitute into (1') to obtain

$(6')\ A_y = A_x + 6mR\alpha$. Substitute into (3'), (4'), (5'), to obtain

$(7')\ \mu_k P - A_x = -\frac{5mR\alpha}{4}$

$(8')\ P + A_x - mg = -\frac{13mR\alpha}{2}$

$(9')\ \frac{3A_x}{2} - P\left(1 - \frac{\mu_k}{2}\right) = -\frac{149mR\alpha}{24}$. Substitute (7') into (8'), (9') to obtain

$P(1+\mu_k) = mg - \frac{31mR\alpha}{4},$

$P(-1+2\mu_k) = -\frac{194}{24}mR\alpha.$

Take the ratio of these two equations to eliminate P $\left(\frac{\mu_k + 1}{2\mu_k - 1}\right) = \frac{-g + \left(\frac{31}{4}\right)R\alpha}{\left(\frac{194}{24}\right)R\alpha}$. Solve:

$$\alpha = \frac{\left(\frac{g}{R}\right)}{\frac{31}{4} - \left(\frac{194}{24}\right)\left(\frac{\mu_k + 1}{2\mu_k - 1}\right)}.$$

Reduce algebraically: $\alpha = \frac{6\left(\frac{g}{R}\right)(1 - 2\mu_k)}{95 - \left(\frac{89}{2}\right)\mu_k}$.

==================================<>==================================

==================================<>==================================

Problem 7.70 The *2-kg* bar rotates *in the horizontal plane* about the smooth pin. The *6-kg* collar A slides on the smooth bar. At the instant shown, *r = 1.2m*, $\omega = 0.4 rad/s$, and the collar is sliding outward at *0.5 m/s* relative to the bar. If you neglect the moment of inertia of the collar (that is, treat the collar as a particle), what is the bar's angular acceleration?

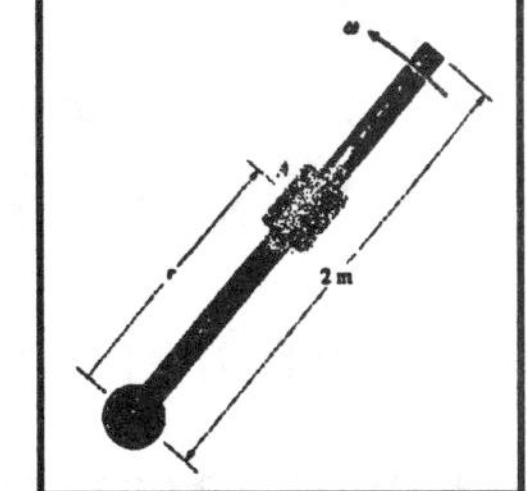

Strategy: Draw individual free-body diagrams for the bar and collar and write Newton's second law for the collar in terms of polar coordinates.

Solution:

Diagrams of the bar and collar showing the force they exert on each other in the horizontal plane are: the bar's equation of angular motion is $\sum M_0 = I_0\alpha$: $-Nr = \frac{1}{3}(2)(2)^2\alpha$ (1)

In polar coordinates, Newton's second law for the collar is

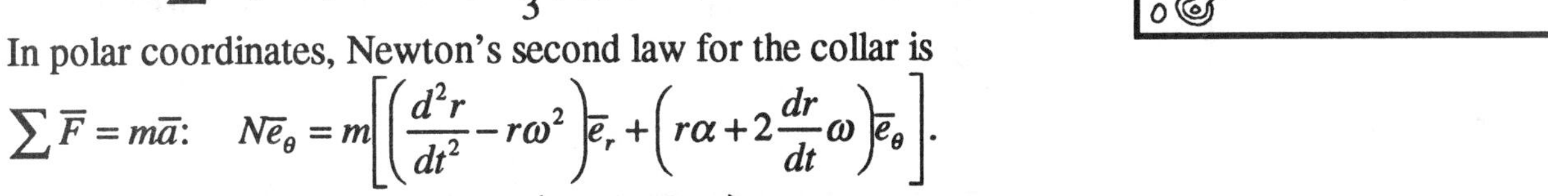

$$\sum \bar{F} = m\bar{a}: \quad N\bar{e}_\theta = m\left[\left(\frac{d^2r}{dt^2} - r\omega^2\right)\bar{e}_r + \left(r\alpha + 2\frac{dr}{dt}\omega\right)\bar{e}_\theta\right].$$

Equating $\bar{e}_\theta$ components, $N = m\left(r\alpha + 2\frac{dr}{dt}\omega\right) = (6)[r\alpha + 2(0.5)(0.4)]$ (2).

Solving Equations (1) and (2) with $r = 1.2m$ gives $\alpha = -0.255 rad/s^2$

==================================<>==================================

Problem 7.71 In Problem 7.70, the moment of inertia of the collar about its center of mass is 0.2 *kg-m*^{2}. Determine the angular acceleration of the bar and compare your answer with the answer to Problem 7.70.

Solution:

Let C be the couple the collar and bar exert on each other: The bar's equation of angular motion is

$$\sum M_0 = I_0\alpha: \quad -Nr - C = \frac{1}{3}(2)(2)^2\alpha \quad (1).$$

The collar's equation of angular motion is $\sum M = I\alpha$: $\quad C = 0.2\alpha$ (2).

From the solution of Problem 7.70, the $\bar{e}_\theta$ component of Newton's second law for the collar is

$N = (6)[r\alpha + 2(0.5)(0.4)]$ (3)

Solving Equations (1) - (3) with $r = 1.2m$ gives $\alpha = -0.250 rad/s^2$.

==================================<>==================================

Problems 7.72 -7.78 are related to Example 7.9.

==================================<>==================================

Problem 7.72 The 3 *Mg* rocket is accelerating upward at 2 *g's*. If you model it as a homogenous bar, what is the magnitude of the axial force at the midpoint?

Solution At the midpoint, the mass above the midpoint is $\frac{m}{2} = \frac{3}{2}$ Mg $= 1500$ kg. Apply Newton's second law to the free body diagram: $P - \left(\frac{m}{2}\right)g = \left(\frac{m}{2}\right)a$, where $a = 2g$. Rearrange:

$P = \left(\frac{m}{2}\right)(g + 2g) = 1500(3)(9.81) = 44{,}145$ N, $P = 44.1$ kN

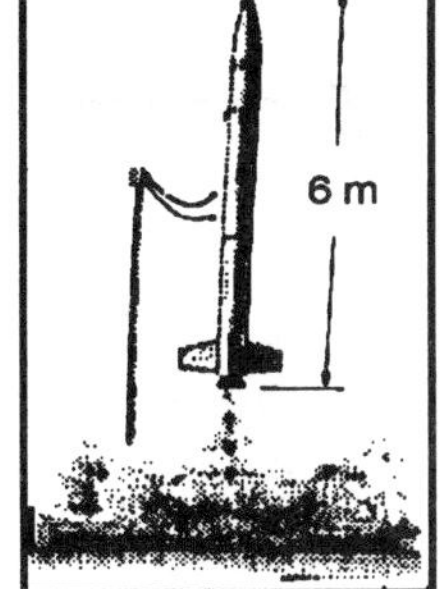

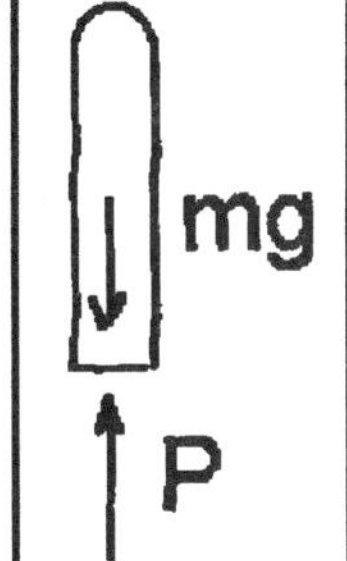

==================================<>==================================

==================================<>==================================

Problem 7.73 The 20 *kg* slender bar is attached to a vertical shaft at A and rotates *in the horizontal plane* with a constant angular velocity of 10 *rad/s.* What is the axial force at the bar's midpoint?

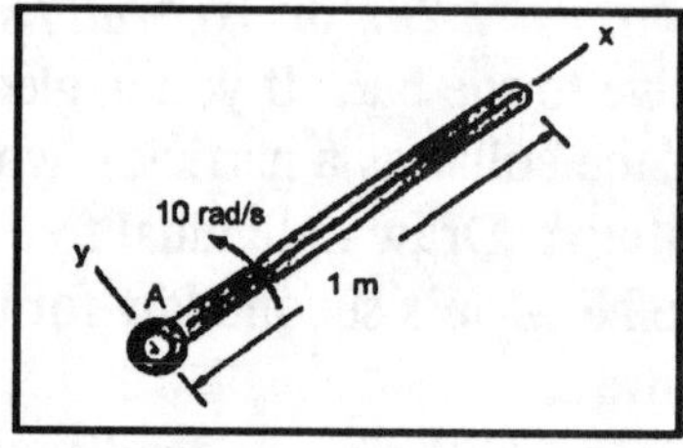

Solution The mass of the outer half of the bar is $\frac{m}{2} = 10$ kg. Apply Newton's law to the free body diagram $-P = \left(\frac{m}{2}\right)a_{CM}$, where a_{CM} is the acceleration of the center of mass of the outer half of the bar.

From kinematics

$$\vec{a}_{CM} = \overline{\omega}\times\left(\overline{\omega}\times\vec{r}_{CM/A}\right) = \overline{\omega}\times\begin{bmatrix} \vec{i} & \vec{j} & \vec{k} \\ 0 & 0 & \omega \\ \frac{3}{4} & 0 & 0 \end{bmatrix} = \begin{bmatrix} \vec{i} & \vec{j} & \vec{k} \\ 0 & 0 & \omega \\ 0 & \frac{3\omega}{4} & 0 \end{bmatrix},\ \vec{a}_{CM} = -\frac{3\omega^2}{4}\vec{i} = -75\vec{i}\ \ \text{m/s}^2\text{, from which}$$

$P = (10)(75) = 750$ N

==================================<>==================================

Problem 7.74 For the rotating bar in Problem 7.73, draw the graph of the axial force as a function of *x*.

Solution:

The vector location of the center of mass of the portion of the bar outboard of x is $\vec{r}_{CM/A} = \left(\frac{1-\text{x}}{2}+\text{x}\right)\vec{i}\ (\text{m}) = \left(\frac{1+\text{x}}{2}\right)\vec{i}\ (\text{m})$. Apply Newton's law to the outboard part of the bar: $-P = m_o a_{CM}$. The mass of the outboard portion is $m_o = m\left(\frac{1-x}{1}\right) = m(1-x)$ kg, from which $-P = m(1-x)a_{CM}$. From kinematics, the acceleration of the center of mass of the outboard part of the bar is

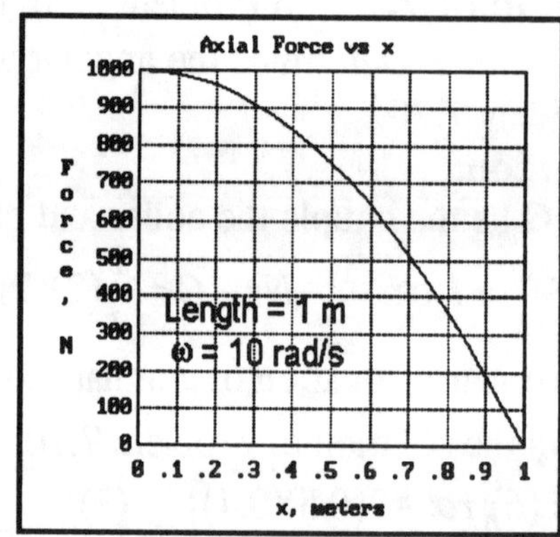

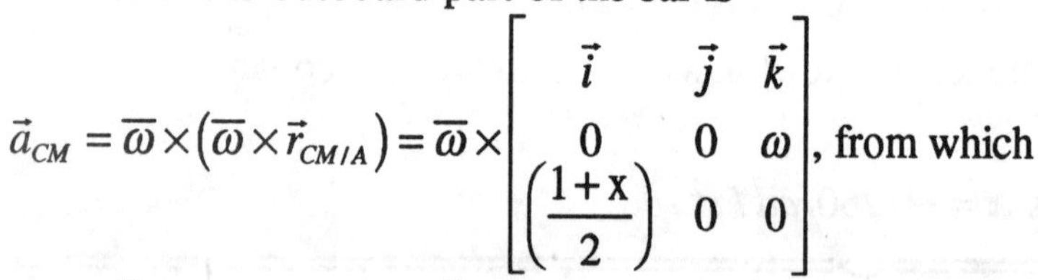

$$\vec{a}_{CM} = \overline{\omega}\times\left(\overline{\omega}\times\vec{r}_{CM/A}\right) = \overline{\omega}\times\begin{bmatrix} \vec{i} & \vec{j} & \vec{k} \\ 0 & 0 & \omega \\ \left(\frac{1+\text{x}}{2}\right) & 0 & 0 \end{bmatrix}\text{, from which}$$

$$\vec{a}_{CM} = \begin{bmatrix} \vec{i} & \vec{j} & \vec{k} \\ 0 & 0 & \omega \\ 0 & \left(\frac{1+\text{x}}{2}\right)\omega & 0 \end{bmatrix} = -\left(\frac{1+\text{x}}{2}\right)\omega^2\vec{i}\ \ P = \frac{m\omega^2}{2}(1-x)(1+x) = 1000\left(1-x^2\right)\text{ N. The graph is shown.}$$

==================================<>==================================

=====================================<>=====================================

Problem 7.75 The 100 *lb.* slender bar AB has a built-in support at A. The *y* axis points upward. Determine the magnitudes of the shear force and bending moment at the bar's midpoint if (a) the support is stationary; (b) the support is accelerating upward at 10 *ft/s*^{2}.

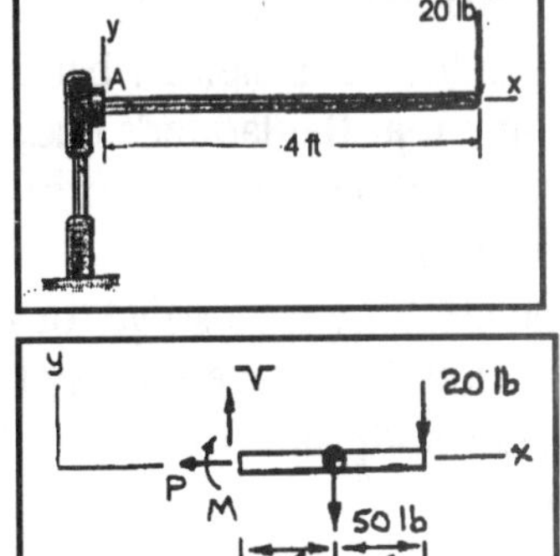

Solution (a) The shear at the midpoint is, by definition, the total load to the right. Thus $V = (50+20) = 70$ lb. The moment is $M = -(50)(1) - 20(2) = -90$ ft-lb. *Check*: The density per unit length is 25 lb/ft. The shear distribution is $V(x) = 120 - 25x$, from which $V(2) = 70$ lb. *check*. The moment is

$$M(x) = \int V(x) + C = 120x - \frac{25}{2}x^2 + C.$$

The constant of integration is found from $M(4) = 0$, from which $C = -280$, from which $M(2) = -90$ lb-ft. *check*. .

(b). Assume that flexure of the bar as point A accelerates is negligible. The mass density is $\frac{25}{g} = 0.778$ slug/ft. The shear is $V(x) = 120 + (0.778(10)(4)) - (25 + 0.778a_y)x = 151.1 - 32.77x$, from which $V(2) = 85.54$ lb.The moment distribution is $M(x) = \int V(x)dx + C = 151.1x - \frac{32.77}{2}x^2 + C$. The constant of integration is determined from $M(4) = 0$, from which $C = -342.2$, and $M(2) = -105.5$ lb-ft. *Check*: The load to the right of the midpoint is $V = 20 + 50 + \frac{50}{g}a_y = 70 + 15.54 = 85.54$ lb. The moment is $M = -20(2) - 25(2) - \frac{50}{g}(1)a_y = -90 - 15.54 = -105.54$ lb-ft. *check*.

=====================================<>=====================================

Problem 7.76 For the bar in Problem 7.75, draw the shear force and bending moment diagrams for the two cases.

Solution Use the solution to Problem 7.75: (a) The shear and moment distributions are

$V(x) = 120 - 25x$ lb,

$$M(x) = 120x - \frac{25}{2}x^2 - 280 \text{ lb-ft}.$$

(b) The shear and moment distributions are

$V(x) = 151.1 - 32.77x$ lb,

$$M(x) = 151.1x - \frac{32.77}{2}x^2 - 342.2 \text{ lb-ft}.$$

The graphs are shown.

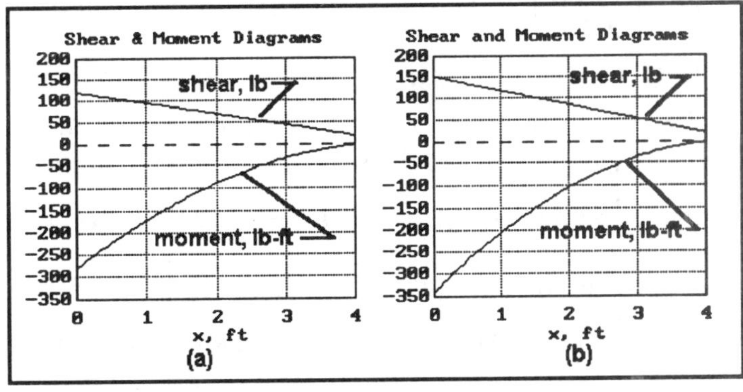

=====================================<>=====================================

==================================<>==================================

Problem 7.77 The 18 *kg* ladder is held in equilibrium in the position shown by the force *F*. Model the ladder as a slender bar and neglect friction. (a) What is the axial force, shear force, and bending moment at the ladder's midpoint? (b) If the force *F* is suddenly removed, what are the axial force, shear force, and bending moment at the ladder's midpoint at that instant?

Solution:

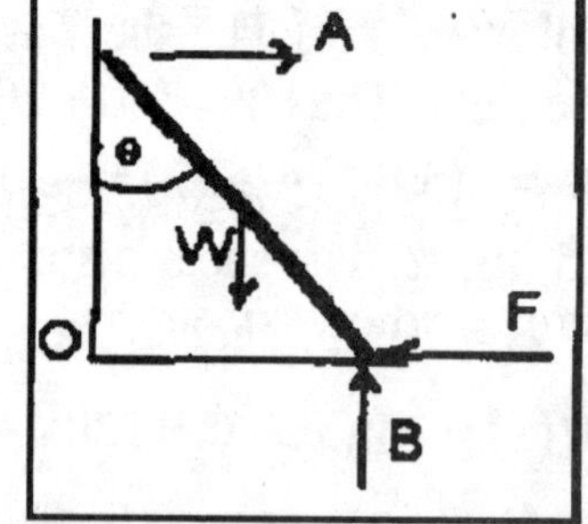

The strategy is to solve for the reactions at the surfaces, and from this solution determine the axial force, shear force, and bending moment at the ladder midpoint, for the static case. The process is repeated for the dynamic case.

(a) *Static case reactions*: Choose a coordinate system with the origin at O and the *x* parallel to the floor. From geometry and the free body diagram of

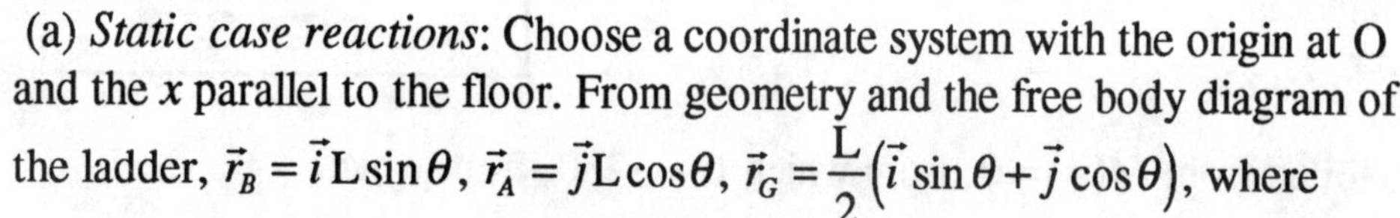

the ladder, $\vec{r}_B = \vec{i}L\sin\theta$, $\vec{r}_A = \vec{j}L\cos\theta$, $\vec{r}_G = \frac{L}{2}(\vec{i}\sin\theta + \vec{j}\cos\theta)$, where $\theta = 30^o$. Apply the static equilibrium conditions to the free body diagram: $-F + A = 0$, $B - W = 0$. The moment about the center of mass of the ladder is $\sum \vec{M}_G = \vec{r}_{B/G} \times (-F\vec{i} + B\vec{j}) + \vec{r}_{A/G} \times A\vec{i}$,

$$\sum \vec{M}_G = \frac{L}{2}\begin{bmatrix} \vec{i} & \vec{j} & \vec{k} \\ \sin\theta & -\cos\theta & 0 \\ -F & B & 0 \end{bmatrix} + \frac{L}{2}\begin{bmatrix} \vec{i} & \vec{j} & \vec{k} \\ -\sin\theta & \cos\theta & 0 \\ A & 0 & 0 \end{bmatrix} = \frac{L}{2}(B\sin\theta - (A+F)\cos\theta)\vec{k} = 0.$$

Substitute numerical values and solve: $B = 176.58$ N, $A = 50.97$ N, $F = 50.97$ N.

Static case axial force, shear force, and bending moment at midpoint: Consider the lower half of the ladder, and note that *from the definition of the bending moment,* $M_{bend} = -M$. Use the definitions and coordinate system for the static case reactions given above. Apply the equilibrium conditions to the free body

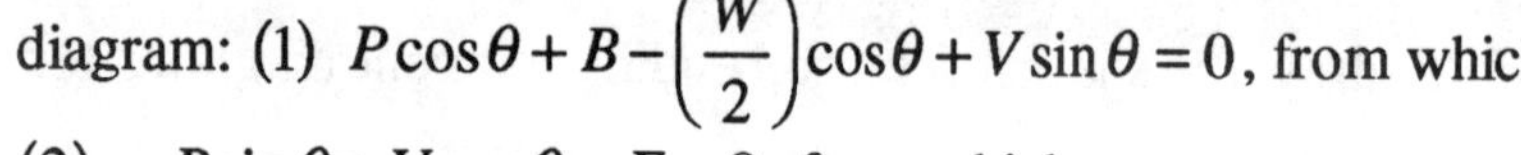

diagram: (1) $P\cos\theta + B - \left(\frac{W}{2}\right)\cos\theta + V\sin\theta = 0$, from which .

(2) $-P\sin\theta + V\cos\theta - F = 0$, from which , ,

(3) $M - \left(\frac{L}{4}\right)(V + F\cos\theta) + \left(\frac{L}{4}\right)B\sin\theta = 0$, from which $P = -101.9$ N, $V = 0$ $M = -44.15$ N-m , $M_{bend} = -M = 44.15$ N-m.

(b) *Dynamic case; the reactions:* The force *F* is zero. From the free body diagram , the application of Newton's second law and the equation of angular motion for the dynamic case yields the three equations:

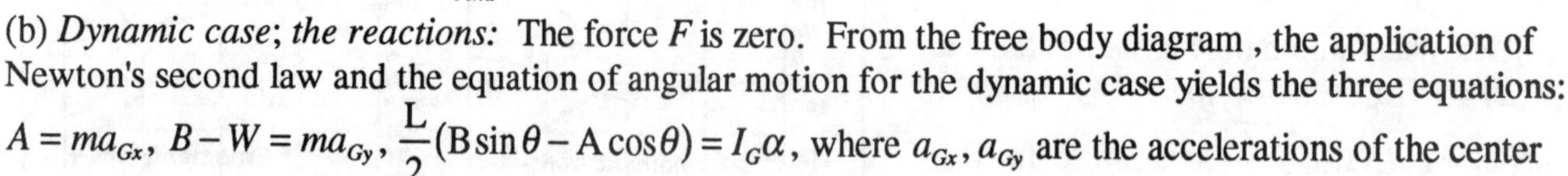

$A = ma_{Gx}$, $B - W = ma_{Gy}$, $\frac{L}{2}(B\sin\theta - A\cos\theta) = I_G\alpha$, where a_{Gx}, a_{Gy} are the accelerations of the center of mass. From the constraint on the motion, the acceleration of points A and B are $\vec{a}_A = a_A\vec{j}$ (m/s^2), $\vec{a}_B = a_B\vec{i}$ (m/s^2), where $\theta = 30^o$. From kinematics, the acceleration of G in terms of the acceleration at A is

$$\vec{a}_G = \vec{a}_A + \mathbf{\vec{a}} \times \vec{r}_{G/A} = \vec{a}_A + \frac{L}{2}\begin{bmatrix} \vec{i} & \vec{j} & \vec{k} \\ 0 & 0 & \alpha \\ +\sin\theta & -\cos\theta & 0 \end{bmatrix} = a_A\vec{j} + \frac{L}{2}(\alpha\cos\theta\vec{i} + \alpha\sin\theta\vec{j})\ (m/s^2),$$

from which $a_{Gx} = \frac{L}{2}\alpha\cos\theta = \sqrt{3}\,\alpha$ m/s^2.

Solution continued on next page

The acceleration of the point G in terms of the acceleration at B is

$$\vec{a}_G = \vec{a}_B + \vec{\mathbf{a}} \times \vec{r}_{G/B} = a_B\vec{i} + \frac{L}{2}\begin{bmatrix} \vec{i} & \vec{j} & \vec{k} \\ 0 & 0 & \alpha \\ -\sin\theta & +\cos\theta & 0 \end{bmatrix} = a_B\vec{i} - \frac{L}{2}(\alpha\cos\theta\vec{i} + \alpha\sin\theta\vec{j}), \text{ from which}$$

$a_{Gy} = -\frac{L}{2}\alpha\sin\theta = -\alpha$. Substitute into the expressions for Newton's laws to obtain the three equations in three unknowns:

$A = m\sqrt{3}\alpha$, $B - W = -m\alpha$, $B\sin\theta - A\cos\theta = \frac{I_G\alpha}{2}$, where $I_G = (1/12)mL^2 = 24 \text{ kg-m}^2$. Solve:

$B = 143.5 \text{ N}$, $A = 57.35 \text{ N}$, $\alpha = 1.84 \text{ rad/s}^2$.

Check: From Example 7.4, $\alpha = \frac{3g}{2L}\sin\theta = 1.84 \text{ rad/s}^2$, *check.*

Axial force, shear force, and bending moment at midpoint: Consider the lower half of the ladder. The vector location of the center of mass is $\vec{r}'_G = \left(\frac{3L}{4}\right)\sin\theta\,\vec{i} + \left(\frac{L}{4}\right)\cos\theta\,\vec{j}$ (m), from which

$\vec{r}'_{G/A} = \frac{3L}{4}(\vec{i}\sin\theta - \vec{j}\cos\theta)$ (m), $\vec{r}'_{G/B} = \left(\frac{L}{4}\right)(-\vec{i}\sin\theta + \vec{j}\cos\theta)$ (m). From kinematics: the acceleration of the midpoint of the lower half in terms of the acceleration at B is

$$\vec{a}'_G = \vec{a}_B + \vec{\mathbf{a}} \times \vec{r}'_{G/B} = \vec{a}_B + \frac{L}{4}\begin{bmatrix} \vec{i} & \vec{j} & \vec{k} \\ 0 & 0 & \alpha \\ -\sin\theta & \cos\theta & 0 \end{bmatrix} = a_B\vec{i} - \frac{L}{4}\alpha(\vec{i}\cos\theta + \vec{j}\sin\theta)\ (\text{m/s}^2). \text{ where } \vec{a}_G \text{ is the}$$

acceleration of the center of mass of the lower half of the ladder, from which

$a'_{Gy} = -\left(\frac{L}{2}\right)\alpha\sin\theta = -\alpha \text{ m/s}^2$. The acceleration of

$$\vec{a}'_G = \vec{a}_A + \vec{\mathbf{a}} \times \vec{r}'_{G/A} = \frac{L}{4}\begin{bmatrix} \vec{i} & \vec{j} & \vec{k} \\ 0 & 0 & \alpha \\ 3\sin\theta & -3\cos\theta & 0 \end{bmatrix} = a_A\vec{j} + \frac{L}{4}(3\alpha\cos\theta\vec{i} + 3\alpha\sin\theta\vec{j})$$

$a'_{Gx} = 3\alpha\cos\theta = \frac{3\sqrt{3}}{2}\alpha \text{ m/s}^2$. Apply Newton's second law and the equation of angular motion to the free body diagram of the lower half of the ladder (see diagram in part (a), with $F = 0$) and use the kinematic relations to obtain: (1′) $P\cos\theta + B - \frac{W}{2} + V\sin\theta = \frac{m}{2}a'_{Gy} = -\frac{m}{4}\alpha$,

(2′) $V\cos\theta - P\sin\theta = \frac{m}{2}a'_{Gx} = \frac{3\sqrt{3}}{4}m\alpha$ The moment about the center of mass is

(3′) $+M - (1)V + (1)B\sin\theta = I'_G\alpha$, where $I'_G = \left(\frac{1}{12}\right)\left(\frac{m}{2}\right)(2^2) = 3 \text{ kg-m}^2$. Solve: $P = -76.46$ N,

$V = 5.518$ N $M = -60.70$ N-m. From the definition of the bending moment,
$M_{Bend} = -M = 60.70$ N-m.

==================================<>=================================

==================================<>==================================

Problem 7.78 For the ladder in Problem 7.77, draw the shear force and bending moment diagrams for the two cases.

Solution Choose a coordinate system as in the solution to Problem 7.77. Use the solution to Problem 7.77 for the reactions at A and B for both cases. Cut the bar at a distance x from the point A, and consider the lower half of the bar, (as shown in the free body diagram), noting that by the definition of the bending moment, $M_{bend} = -M$. The mass density of the ladder per unit length is $\frac{m}{L} = \frac{18}{4} = 4.5 \text{ kg/m}$. The mass of the lower part of the ladder is $m_x = \left(\frac{m}{L}\right)(L-x)$ kg.

(a) *Static Case*: In the solution to Problem 7.77 the reactions B, F were $B = 176.58$ N, $F = 50.97$ N. Apply the static equilibrium conditions to the free body diagram to obtain:

(1) $B - \frac{W(L-x)}{L} + V\sin\theta + P\cos\theta = 0$, (2) $V\cos\theta - P\sin\theta - F = 0$. The moment about the center of mass of the lower part of the ladder is

(3) $M - \left(\frac{L-x}{2}\right)(V + F\cos\theta) + B\left(\frac{L-x}{2}\right)\sin\theta = 0$. These three equations have the solutions:

$$P = \left[W\left(\frac{L-x}{L}\right) - B\right]\cos\theta - F\sin\theta,$$

$$V = \left[W\left(\frac{L-x}{L}\right) - B\right]\sin\theta + F\cos\theta, \text{ and}$$

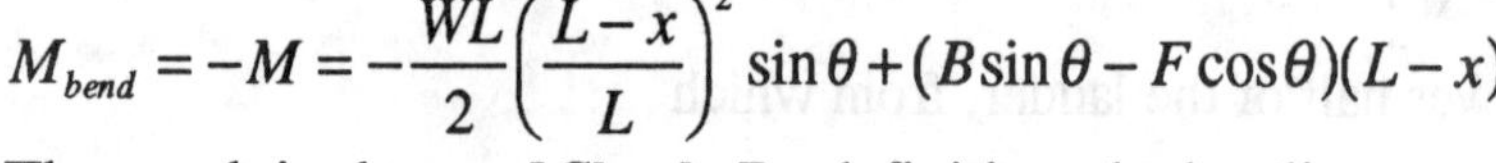

$$M_{bend} = -M = -\frac{WL}{2}\left(\frac{L-x}{L}\right)^2 \sin\theta + (B\sin\theta - F\cos\theta)(L-x)$$

The graph is shown. [*Check:* By definition, the bending moment is the integral of the shear,

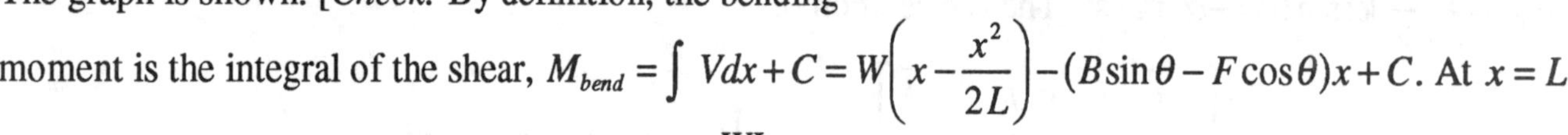

$M_{bend} = \int V dx + C = W\left(x - \frac{x^2}{2L}\right) - (B\sin\theta - F\cos\theta)x + C$. At $x = L$ the bending moment is zero, from which $C = -\frac{WL}{2} + (B\sin\theta - F\cos\theta)L$. Substitute:

$M_{bend} = -\frac{WL}{2}\left(\frac{L-x}{L}\right)^2 \sin\theta + (B\sin\theta - F\cos\theta)(L-x)$. *check.* At $x = 0$, $M_{bend} = 0$, from which $-\frac{WL}{2}\sin\theta + (B\sin\theta - F\cos\theta)L = 0$ *check.*]

(b) *Dynamic case*: Use the solution in Problem 7.77 for the reactions: $B = 143.5$ N, $\alpha = 1.84 \text{ rad/s}^2$. Use the free body diagram and results of (a). The vector distance to the center of mass of the lower part

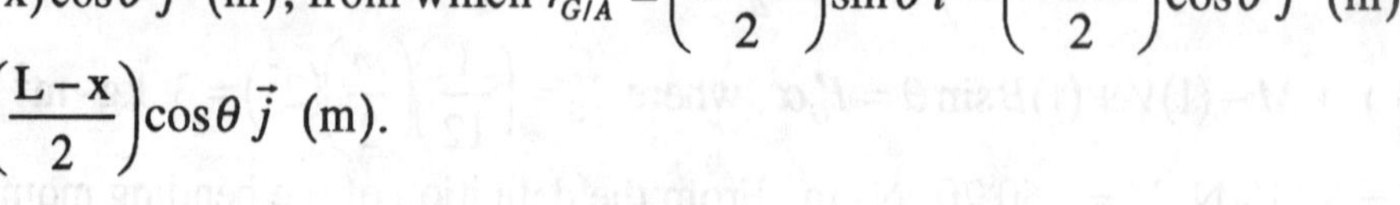

from A is $\vec{r}_G = x\sin\theta\,\vec{i} + (L-x)\cos\theta\,\vec{j}$ (m), from which $\vec{r}_{G/A} = \left(\frac{L+x}{2}\right)\sin\theta\,\vec{i} - \left(\frac{L+x}{2}\right)\cos\theta\,\vec{j}$ (m)

and $\vec{r}_{G/B} = -\left(\frac{L-x}{2}\right)\sin\theta\,\vec{i} + \left(\frac{L-x}{2}\right)\cos\theta\,\vec{j}$ (m).

Solution continued on next page

From kinematics: $\vec{a}_G = \vec{a}_A + \vec{\mathbf{a}} \times \vec{r}_{G/A}$,

$$\vec{a}_G = \vec{a}_A + \left(\frac{L+x}{2}\right)\begin{bmatrix} \vec{i} & \vec{j} & \vec{k} \\ 0 & 0 & \alpha \\ \sin\theta & -\cos\theta & 0 \end{bmatrix},$$

$= a_A\vec{j} + \left(\frac{L+x}{2}\right)(\alpha\cos\theta\vec{i} + \alpha\sin\theta\vec{j})$ (m/s^2) from which $a_{Gx} = \left(\frac{L+x}{2}\right)\alpha\cos\theta$. $\vec{a}_G = \vec{a}_B + \vec{\mathbf{a}} \times \vec{r}_{G/B}$

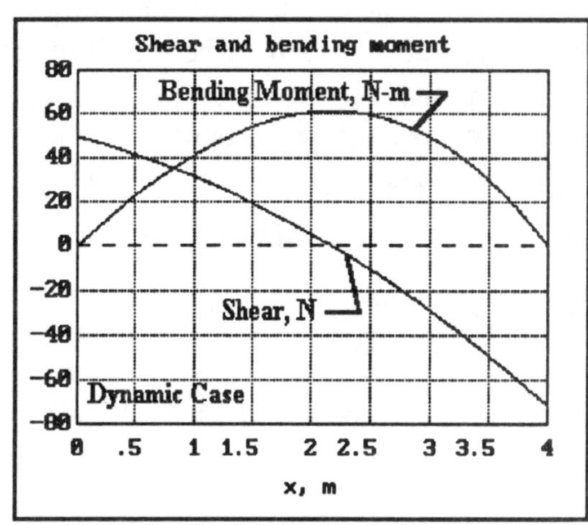

$$\vec{a}_G = a_B\vec{i} + \left(\frac{(L-x)}{2}\right)\begin{bmatrix} \vec{i} & \vec{j} & \vec{k} \\ 0 & 0 & \alpha \\ -\sin\theta & \cos\theta & 0 \end{bmatrix},$$

$= a_B\vec{i} - \frac{(L-x)}{2}(\vec{i}\cos\theta + \vec{j}\sin\theta)$ (m/s^2), from which $a_{Gy} = -\frac{(L-x)}{2}\alpha\sin\theta$. Apply Newton's second law and the equation of angular motion to the free body diagram, and substitute kinematic results to obtain

$$B - \frac{W(L-x)}{L} + V\sin\theta + P\cos\theta = m\left(\frac{L-x}{L}\right)a_{Gy} = -m\left(\frac{(L-x)^2}{2L}\right)\alpha\sin\theta.$$

$V\cos\theta - P\sin\theta = m\left(\frac{L-x}{L}\right)a_{Gx} = m\left(\frac{L-x}{L}\right)\left(\frac{L+x}{2}\right)\alpha\cos\theta$. $M + \left(\frac{L-x}{2}\right)(-V + B\sin\theta) = I_G\alpha$ N-m,

where $I_G = \left(\frac{1}{12}\right)m\left(\frac{L-x}{L}\right)(L-x)^2 = \frac{m(L-x)^3}{12L}$.. The graphs are shown.

==================================<>==================================

Problem 7.79 Continue the calculations presented in Example 7.10, using $\Delta t = 0.1$ s, and determine the ladder's angular position and angular velocity at $t = 0.6$ s and $t = 0.7$ s.

Solution The time was expressed as an array (list) such that $t[i] = t[1] + (i-1)\Delta t$ $(1 \le i \le 8)$, $t[1] = 0$, $\Delta t = 0.1$ s. The first values in the arrays for θ, ω and α are $\theta[1] = 5^\circ = 0.0873$ rad, $\omega[1] = 0$, $\alpha[1] = \frac{3g}{2L}\sin(\theta[1]) = 0.3206$ rad/s^2. The algorithm for integration is

For $i = 2$ to $i = 8$,
$\theta[i] = \theta[i-1] + \omega[i-1]dt$,
$\alpha[i-1] = \frac{3g}{2L}\sin(\theta[i-1])$,
$\omega[i] = \omega[i-1] + \alpha[i-1]dt$,
Next i. The values are tabulated. The first values agree with Example 7.10, as a check.

t, s	θ, rad	ω, rad/s	α, rad/s²
0	.0873	0	.3206
.1	.0873	.0321	.3206
.2	.0905	.0641	.3324
.3	.0969	.0974	.3559
.4	.1066	.1329	.3915
.5	.1199	.1721	.4401
.6	.1371	.2161	.5029
.7	.1587	.2664	.5815

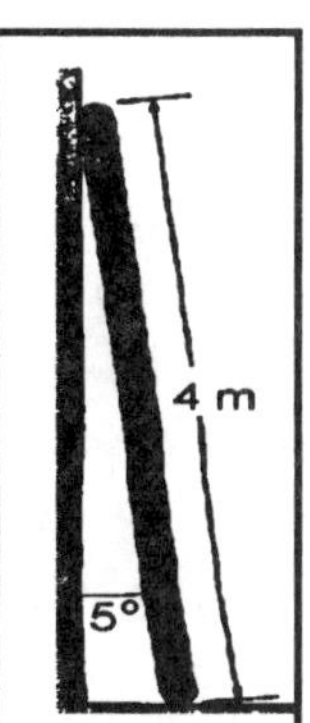

==================================<>==================================

====================================<>====================================

Problem 7.80 The moment of inertia of the helicopter's rotor is 400 slug - ft^2. It starts from rest at $t = 0$, the engine exerts a constant torque of 500 *ft-lb.*, and aerodynamic drag exerts a torque of magnitude $20\omega^2$ $ft - lb$, where ω is the rotor's angular velocity in radians per second. Using $\Delta t = 0.2$ s, determine the rotor's angular position and angular velocity for the first five time steps. Compare your results for the angular velocity with the closed form solution.

Solution The angular acceleration is obtained from the equation of angular motion for the rotor, $I\alpha = T - 20\omega^2$. For convenience, denote $b = 20$. By definition, $\alpha = \frac{d\omega}{dt}$, from which, separating variables, $\frac{d\omega}{T - b\omega^2} = \frac{dt}{I}$. *Make the reasonable assumption that* $T > b\omega^2$ *over the time interval of interest.* Integrate: $\frac{1}{\sqrt{bT}}\tanh^{-1}\left(\sqrt{\frac{b}{T}}\omega\right) = \frac{t}{I} + C$, where C is a constant of integration. Rearrange: $\tanh^{-1}\left(\sqrt{\frac{b}{T}}\omega\right) = \frac{\sqrt{bT}}{I}t + C_1$, from which $\omega = \sqrt{\frac{T}{b}}\tanh\left(\frac{\sqrt{bT}}{I}t\right) + C_2$. When $t = 0$, $\omega = 0$, from which $C_2 = 0$. $\omega = \sqrt{\frac{T}{b}}\tanh\left(\frac{\sqrt{bT}}{I}t\right)$ which is the closed form solution for the angular velocity. *Although not required by the problem*, the closed form for the angular position is a straightforward integration: By definition, $\omega = \frac{d\theta}{dt} = \sqrt{\frac{T}{b}}\tanh\left(\frac{\sqrt{bT}}{I}t\right)$. Integrate: $\theta = \int \omega dt + C = \int \sqrt{\frac{T}{b}}\tanh\left(\frac{\sqrt{bT}}{I}t\right)dt + C$,

$\theta = \frac{I}{b}\ln\left(\cosh\frac{\sqrt{bT}}{I}t\right) + C$ where C is the constant of integration. When $t = 0$, $\cosh(0) = 1$, $\ln(1) = 0$, thus $\theta = 0$ from which $C = 0$. $\theta = \frac{I}{b}\ln\left(\cosh\left(\frac{\sqrt{bT}}{I}t\right)\right)$ is the closed form solution for the angular position. Substitute numerical values: $\omega = 5\tanh(0.25t)$, $\theta = 20\ln(\cosh(0.25t))$. *Check:* $\frac{d\theta}{dt} = \frac{d}{dt}20\ln(\cosh(0.25t)) = \frac{20(0.25)\sinh(0.25t)}{\cosh(0.25t)} = 5\tanh(0.25t)$. *check.* For the numerical integration the time is in an array $t[i] = t[1] + (i-1)\Delta t$ $(1 \le i \le 6)$, where $\Delta t = 0.2$ s, $t[1] = 0$. The initial values are $\theta[1] = 0$, $\omega[1] = 0$, . The numerical integration uses the algorithm:

For $i = 2$ to $i = 6$,
$\theta[i] = \theta[i-1] + \omega[i-1]dt$,
$\alpha[i-1] = 1.25 - 0.05\omega^2[i-1]$,
$\omega[i] = \omega[i-1] + \alpha[i-1]dt$,
Next i. The results are tabulated. The first column is time, in seconds. The second column is the closed form solution for ω, *rad/s.* The third column is the value of ω, *rad/s* obtained from numerical integration. The fourth column is the closed form value of θ, *rads*. The fifth column is the value of θ, *rads*, obtained from the numerical integration. Note that the latter values show poor agreement for the first five steps. (The agreement improves as the number of steps increases.)

t,s	ω_{CF}	ω_{Euler}	θ_{CF}	θ_{Euler}
0	0	0	0	0
.2	.2498	.25	.025	0
.4	.4983	.4994	.0998	.05
.6	.7444	.7469	.2242	.1499
.8	.9869	.9913	.3974	.2993
1	1.225	1.231	.6186	.4975
1.2	1.457	1.466	.8868	.7438

====================================<>====================================

Problem 7.81 In Problem 7.80, draw a graph of the rotor's angular velocity as a function of time from $t = 0$ to $t = 10$ s, comparing the closed form solution, the numerical solution using $\Delta t = 1.0$ s, and the numerical solution using $\Delta t = 0.2$ s.

Solution The graphs are shown. The difference between the closed form and the numerical solution is difficult to see on the scale of the graph for $\Delta t = 0.2$ s, but the closed form is slightly higher than the numerical solution over the entire range. The non-agreement is easy to see for $\Delta t = 1$ s.

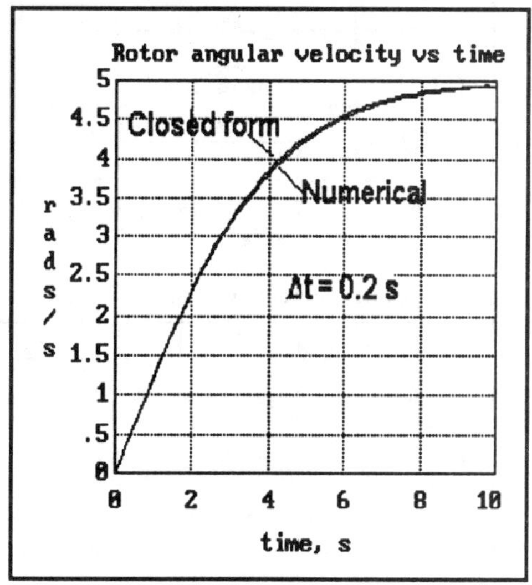

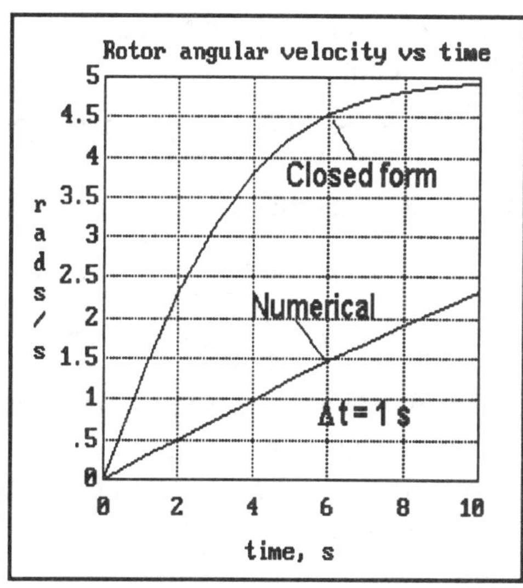

Problem 7.82 The slender 10 *kg* bar is released from rest in the horizontal position shown. Using $\Delta t = 0.1$ s, determine the bar's angular position and angular velocity for the first five time steps.

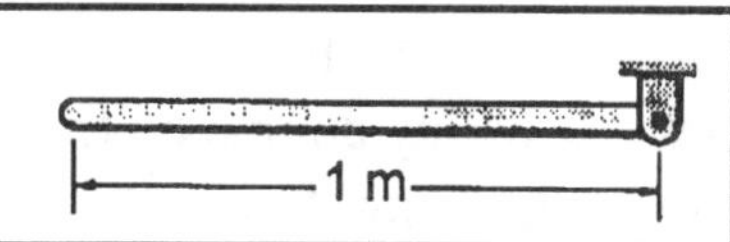

Solution From the equation of angular motion, the moment about the pinned support at A is $W_G\left(\frac{L}{2}\right)\cos\theta = I_A\alpha$, where $I_A = \frac{m}{3}L^2$, $W_G = mg$, from which, by definition, $\alpha = \frac{d\omega}{dt} = \frac{3g}{2L}\cos\theta$.

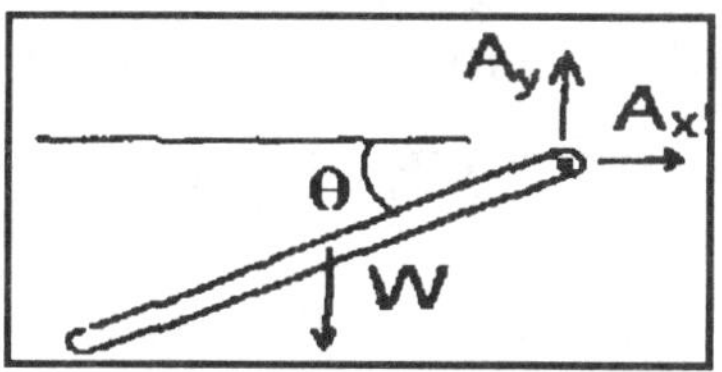

Substitute numerical values: $\frac{d\omega}{dt} = 14.715\cos\theta$.

The algorithm for numerical integration is $\theta[1] = 0$, $\omega[1] = 0$,

For $i = 2$ to $i = 6$,

$\theta[i] = \theta[i-1] + \omega[i-1]dt$,

$\alpha[i-1] = 14.715\cos(\theta[i-1])$,

$\omega[i] = \omega[i-1] + \alpha[i-1]dt$,

Next i. The results are tabulated.

t,s	ω,rad/s	θ, rad
0	0	0
.1	1.472	0
.2	2.943	.1472
.3	4.399	.4415
.4	5.729	.8813
.5	6.665	1.454

====================================<>====================================

Problem 7.83 In Problem 7.82, determine the bar's angular position and angular velocity as functions of time from $t = 0$ to $t = 0.8$ s using $\Delta t = 0.1$ s, $\Delta t = 0.01$ s, and $\Delta t = 0.001$ s. Draw the graphs of the angular velocity as a function of the angular position and for these three cases and compare them with the graph of the closed form solution for the angular velocity as a function of the angular position.

Solution From the solution to Problem 7.82, the angular acceleration is $\alpha = \frac{3g}{2L}\cos\theta$. Use the chain rule and the definition of the angular velocity to obtain $\alpha = \frac{d\omega}{dt} = \frac{d\omega}{d\theta}\frac{d\theta}{dt} = \omega\frac{d\omega}{d\theta} = \frac{3g}{2L}\cos\theta$. Separate variables and integrate. $\omega^2 = \frac{3g}{L}\sin\theta + C$. For $\theta = 0$, $\omega = 0$, from which $C = 0$, and the closed form solution is $\omega = \sqrt{\frac{3g}{L}\sin\theta}$, where the positive sign has been chosen from physical reasoning (the bar is swinging counterclockwise). The algorithm for numerical integration is that given in the solution to Problem 7.82.

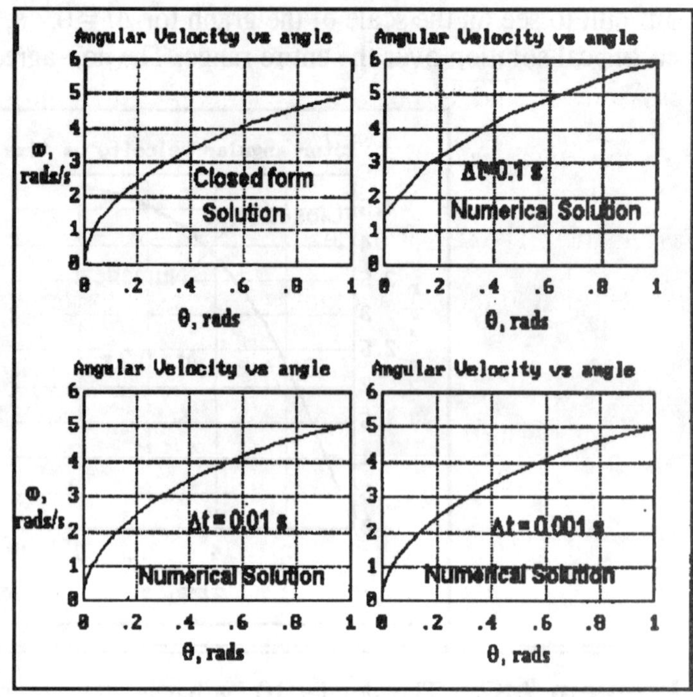

====================================<>====================================

Problem 7.84 In Problem 7.82, suppose that the bar's pin support contains a damping device that exerts a resisting couple on the bar of magnitude $c\omega$ (N-m), where ω is the angular velocity in radians per second. Using $\Delta t = 0.001$ s, draw graphs of the bar's angular velocity as a function of time from $t = 0$ to $t = 0.8$ s for the cases $c = 0$, $c = 2$, $c = 4$, and $c = 8$.

Solution From the application of the equation of angular motion to the bar (see the solution to Problem 7.82) $W_G\left(\frac{L}{2}\right) - c\omega = I_A\alpha$, where $W_G = mg$, $I_A = \frac{m}{3}L^2$ kg-m^2, from which $\frac{d\omega}{dt} = \frac{3g}{2L}\cos\theta - \frac{3c}{mL^2}\omega$. Substitute numerical values: $\frac{d\omega}{dt} = 14.715 - 0.3c\omega$. The algorithm for numerical integration is: $\theta[1] = 0$, $\omega[1] = 0$,

For $i = 2$ to $i = 6$,

$\theta[i] = \theta[i-1] + \omega[i-1]dt$,

$\alpha[i-1] = 14.715\cos(\theta[i-1]) - 0.3c\omega[i-1]$,

$\omega[i] = \omega[i-1] + \alpha[i-1]dt$,

Next i. The graphs are shown.

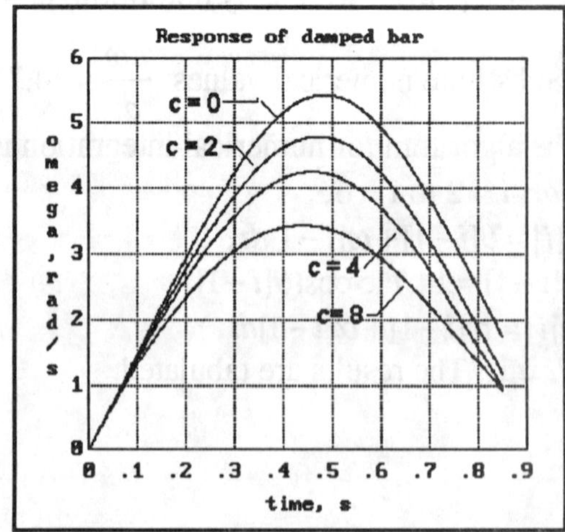

====================================<>====================================

==================================<>==================================

Problem 7.85 The falling ladder in Example 7.10 will lose contact with the wall before it hits the floor. Using $\Delta t = 0.001$ s, estimate the time and the value of the angle between the wall and the ladder when this occurs.

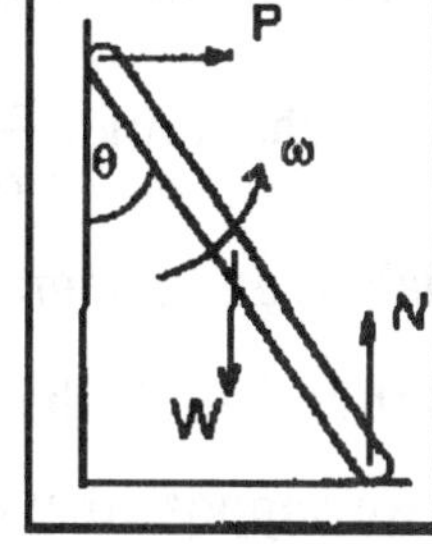

Solution Choose a coordinate system with the origin at the corner of the wall and floor, with the x axis parallel to the floor. Denote the contact point with the wall by P, the contact with the floor by N, and the center of mass by G. From Newton's second law and the equation of angular motion, $P = ma_{Gx}$, $N - mg = ma_y$, and $N\left(\frac{L\sin\theta}{2}\right) - P\left(\frac{L\cos\theta}{2}\right) = I_G\alpha$, where $I_G = \left(\frac{1}{12}\right)mL^2$. From kinematics: The vector distance $\vec{r}_{G/P} = \vec{r}_G - \vec{r}_P = \frac{L}{2}\left(\vec{i}\sin\theta - \vec{j}\cos\theta\right)$. The acceleration of the center of mass in terms of the acceleration at point P is $\vec{a}_G = \vec{a}_P + \vec{\mathbf{a}} \times \vec{r}_{G/P} - \omega^2\vec{r}_{G/A}$,

$$\vec{a}_G = \vec{a}_P + \begin{bmatrix} \vec{i} & \vec{j} & \vec{k} \\ 0 & 0 & \alpha \\ \frac{L\sin\theta}{2} & -\frac{L\cos\theta}{2} & 0 \end{bmatrix} - \omega^2\left(\frac{L\sin\theta}{2}\vec{i} - \frac{L\cos\theta}{2}\vec{j}\right).$$

The constraint on the motion at P is such that $\vec{a}_P = a_P\vec{j}$, from which $a_{Gx} = \frac{L}{2}\left(\alpha\cos\theta - \omega^2\sin\theta\right)$.

The acceleration of the center of mass in terms of the acceleration of point N is

$$\vec{a}_G = \vec{a}_N + \vec{\mathbf{a}} \times \vec{r}_{G/N} - \omega^2\vec{r}_{G/N},\ \vec{a}_G = \vec{a}_N + \begin{bmatrix} \vec{i} & \vec{j} & \vec{k} \\ 0 & 0 & \alpha \\ -\frac{L\sin\theta}{2} & +\frac{L\cos\theta}{2} & 0 \end{bmatrix} - \omega^2\left(-\frac{L\sin\theta}{2}\vec{i} + \frac{L\cos\theta}{2}\vec{j}\right).$$

The constraint on the motion at N is such that $\vec{a}_N = a_N\vec{i}$, from which $a_{Gy} = \frac{L}{2}\left(\alpha\sin\theta + \omega^2\cos\theta\right)$

Collect results and substitute the kinematic relations to obtain:

(1) $P = \frac{mL}{2}\left(\alpha\cos\theta - \omega^2\sin\theta\right)$, (2) $N = mg - \frac{mL}{2}\left(\alpha\sin\theta + \omega^2\cos\theta\right)$,

(3) $N\left(\frac{L\sin\theta}{2}\right) - P\left(\frac{L\cos\theta}{2}\right) = I_G\alpha$. Substitute (1) and (2) into (3) and reduce to obtain

$\alpha = \frac{2mgL\sin\theta}{4I_G + mL^2} = \frac{3g}{2L}\sin\theta$. .The ladder leaves the wall when $P = 0$, when $\alpha\cos\theta = \omega^2\sin\theta$, so that (2) and (3) are not required in the numerical solution.

Substitute numerical values: $m = 18$ kg, $L = 4$ m, to obtain $P = 36\left(\alpha\cos\theta - \omega^2\sin\theta\right)$ N, $\alpha = 3.68\sin\theta$ rad/s^2. A copy of the algorithm used in **TK Solver Plus** is shown. The algorithm is a called procedure, returning the time and angle at which the force exerted by the wall vanishes. These values are: $t = 1.554$ s and $\theta = 0.8455$ rads $= 48.44°$. Although not required by the problem, the graph of the force P against the time is shown.

Algorithm for Problem 7.69

```
Euler(ai,wi,qi,Pi,a,w,q,P ,t,dt;theta,time)

place(a,1):=ai        ;initial values
place(w,1):=wi
place(q,1):=qi
place(P,1):=Pi
for j=2 to length('t)
q[j]:=q[j-1]+w[j-1]*dt          ; theta[j]
w[j]:=w[j-1]+a[j-1]*dt          ; omega[j]
a[j]:=3.67875*sin(q[j])         ; alpha[j]
P[j]:=36*(a[j]*cos(q[j])-(w[j]^2)*sin(q[j]))
if AND (P[j]<>0, P[j]>0) then goto OK
time:=t[j]                      ; stopping time
theta:=q[j]                     ; stopping angle
exit
OK:next j                       ; not ready to stop
```

Solution continued on next page

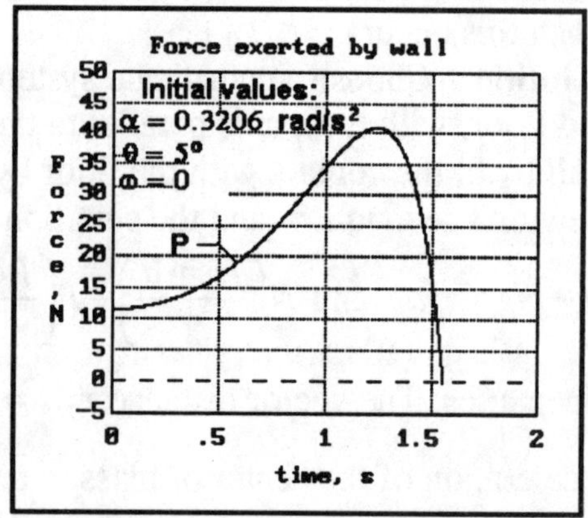

[*Check*: An analytic solution: From $P=0$, $\alpha\cos\theta-\omega^2\sin\theta=0$. Substitute $\alpha=\frac{3g}{2L}\sin\theta$. Noting $\alpha=\frac{d\omega}{dt}$, use the chain rule: $\frac{d\omega}{dt}=\frac{d\omega}{d\theta}\frac{d\theta}{dt}=\omega\frac{d\omega}{d\theta}=\frac{3g}{2L}\sin\theta$. Separate variables and integrate: $\frac{\omega^2}{2}=-\frac{3g}{2L}\cos\theta+C$. Assume that $\omega=0$ at $\theta\rightarrow 0$ (the ladder won't start to fall if $\theta=0$ exactly, but we suppose that θ is very small), from which $C=\frac{3g}{2L}$, from which $\omega=\sqrt{\frac{3g}{L}(1-\cos\theta)}$, where the positive sign is taken because the ladder rotates counterclockwise. Substitute: $\frac{3g}{2L}\sin\theta\cos\theta-\frac{3g}{L}(1-\cos\theta)\sin\theta=0$. Reduce algebraically to obtain $\cos\theta=\frac{2}{3}$ from which $\theta=48.2^{o}$

================================<>================================

Problem 7.86 A torsional spring at A exerts a counterclockwise couple $k\theta$ on the bar, where $k = 20$ *N-m* and θ is in radians. The *2-kg* bar is *1-m* long. At $t = 0$, the bar is released from rest in the horizontal position ($\theta = 0$). Using $\Delta t = 0.01s$, determine the bar's angular position and angular velocity for the first five time steps.

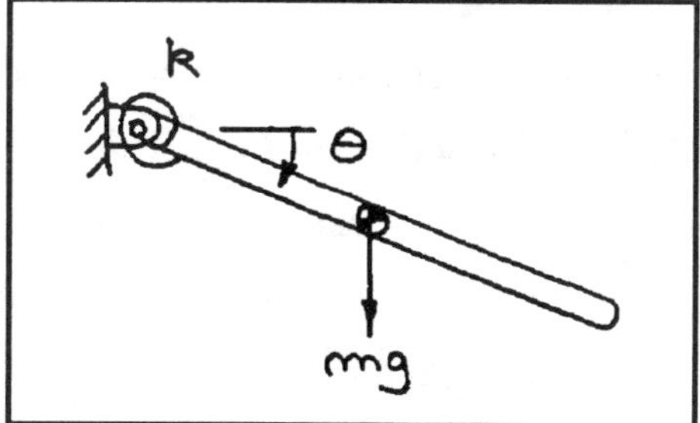

Solution:

The bar's equation of angular motion is $\sum M_0 = I_0\alpha$:: $\quad mg(l/2)\theta - k\cos\theta = \frac{1}{3}ml^2\alpha$ or

$(2)(9.81)(1/2)\cos\theta - 20\theta = \frac{1}{3}(2)(1)^2\alpha$. We see that $\alpha = \frac{dw}{dt} = 14.715\cos\theta - 30\theta$.

Initial Conditions: At $t_0 = 0$, $\theta(t_0) = 0$ and $\omega(t_0) = 0$.

First Time Step: At $t = t_0 + \Delta t = 0.01s$, the angle is $\theta(t_0 + \Delta t) = \theta(t_0) + \omega(t_0)\Delta t = 0$, and the angular velocity is $\omega(t_0 + \Delta t) = \omega(t_0) + \alpha(t_0)\Delta t = [14.715\cos(0) - 30(0)](0.01) = 0.1472 rad/s$.

Second Time Step: At $t = t_0 + 2\Delta t = 0.02s$ the angle is

$\theta(t_0 + 2\Delta t) = \theta(t_0 + \Delta t) + \omega(t_0 + \Delta t)\Delta t = (0.1472)(0.01) = 0.0015 rad$, and the angular velocity is

$\omega(t_0 + 2\Delta t) = \omega(t_0 + \Delta t) + \alpha(t_0 + \Delta t)\Delta t = 0.1472 + [14.715\cos(0) - 30(0)](0.01) = 0.2943 rad/s$.

Continuing, we obtain

t, s	θ, rad	ω, rad/s
0.00	0.0000	0.0000
0.01	0.0000	0.1472
0.02	0.0015	0.2943
0.03	0.0044	0.4410
0.04	0.0088	0.5868
0.05	0.0147	0.7313

Problem 7.87 Using a numerical solution with $\Delta t = 0.001s$, estimate the maximum angle θ reached by the bar in Problem 7.86 when it is released from rest in the horizontal position. At what time after release does the maximum angle occur?

Solution:

Carrying out the numerical solution in the manner described in the solution of Problem 7.86, the resulting graph of θ as a function of time is shown below: By examining the computer results near the maximum, we estimate that the maximum angle $\theta = 0.867 rad$ or $(49.7°)$ occurs at $t = 0.524s$.

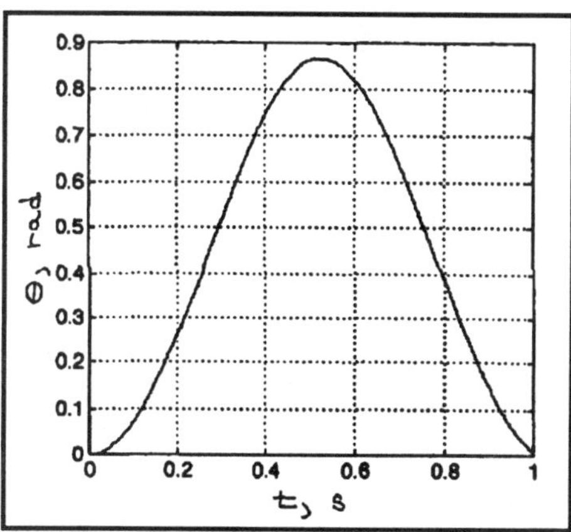

=================================<>=================================

Problem 7.88 The homogenous, slender bar has mass m and length l. Use integration to determine its moment of inertia about the perpendicular axis L_o.

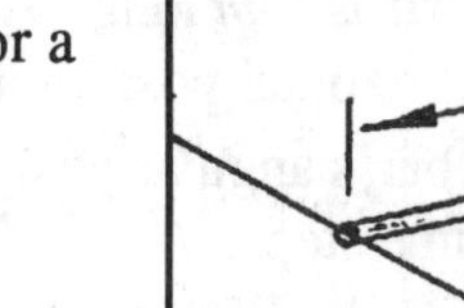

Solution: By definition $I = \int_0^l r^2 dm$. The differential mass for a slender homogenous bar is $dm = \rho A dr$, where ρ in the mass density per unit volume and A is the cross section area. Thus

$$I = \rho A \int_0^l r^2 dr = \rho A \left[\frac{r^3}{3}\right]_0^l = \frac{\rho A l^3}{3} = \frac{ml^2}{3}$$

=================================<>=================================

Problem 7.89 Two homogenous, slender bars, each of mass m and length l, are welded together to form the T-shaped object. Use integration to determine the moment of inertia of the object about the axis through point O that is perpendicular to the bars.

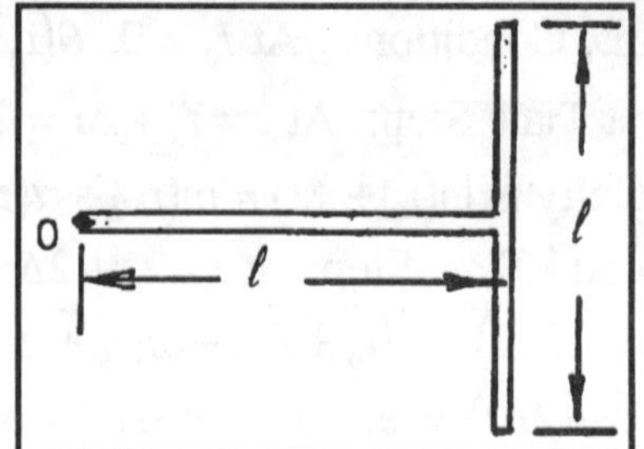

Solution: Divide the object into two pieces, each corresponding to a slender bar of mass m; the first parallel to the y-axis, the second to the x-axis. By definition $I = \int_0^l r^2 dm + \int_m r^2 dm$. For the first bar, the differential mass is $dm = \rho A dr$. Assume that the second bar is very slender, so that the mass is concentrated at a distance l from O. Thus $dm = \rho A dx$, where x lies between the limits $-\frac{l}{2} \le x \le \frac{l}{2}$. The distance to a differential dx is $r = \sqrt{l^2 + x^2}$. Thus the definition becomes $I = \rho A \int_0^l r^2 dr + \rho A \int_{-\frac{l}{2}}^{\frac{l}{2}} (l^2 + x^2) dx$

$$I = \rho A \left[\frac{r^3}{3}\right]_0^l + \rho A \left[l^2 x + \frac{x^3}{3}\right]_{-\frac{l}{2}}^{\frac{l}{2}} = ml^2\left(\frac{1}{3} + 1 + \frac{1}{12}\right) = \frac{17}{12} ml^2$$

=================================<>=================================

Problem 7.90 The homogenous, slender bar has mass m and length l. Use integration to determine the moment of inertia about the axis L.

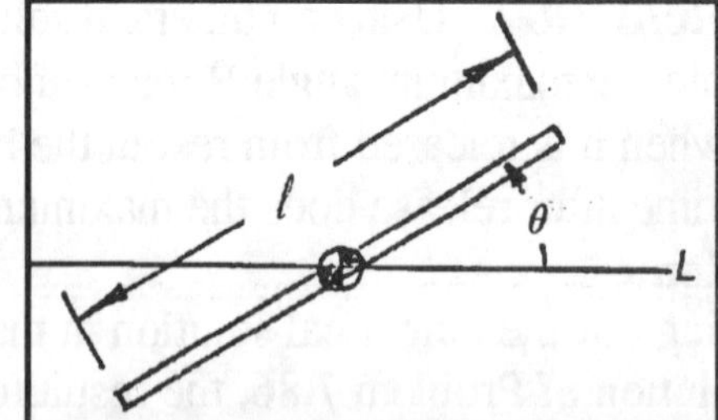

Solution: By definition, $I = \int_m r^2 dm$. Let the differential mass be $dm = \rho A dx$, where x is measured along the bar from $x = -\frac{l}{2}$ to $x = \frac{l}{2}$. The distance from the line L to the bar is r, where $r = x \sin\theta$. Substitute to obtain

$$I = \rho A \sin^2\theta \int_{-\frac{l}{2}}^{\frac{l}{2}} x^2 dx = \rho A \sin^2\theta \left[\frac{x^3}{3}\right]_{-\frac{l}{2}}^{\frac{l}{2}} = \frac{ml^2}{12} \sin^2\theta$$

=================================<>=================================

==================================<>==================================

Problem 7.91 A homogenous, slender bar is bent into a circular ring of mass m and radius R. Determine moment of inertia of the ring (a) about the axis through its center of mass that is perpendicular to the ring; (b) about the axis L.

Solution:

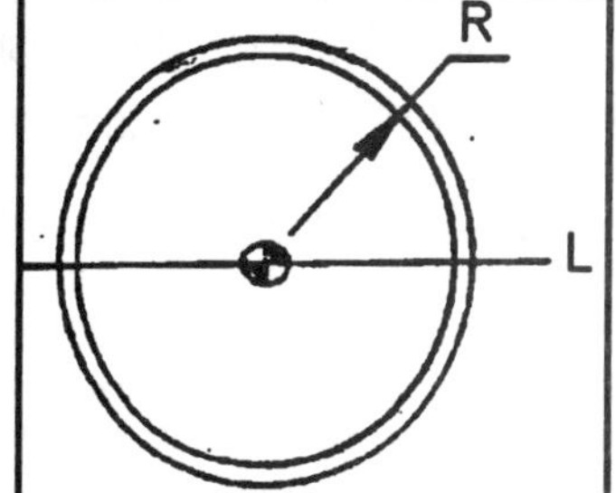

Choose a coordinate system with the origin at the center of the ring and the x-axis parallel to L. The angle θ is measured counterclockwise from x. (a) The definition is $I = \int_m R^2 dm$. The differential mass of the ring is $dm = \rho AR d\theta$, where A is the cross sectional area of the bar. The distance to the ring from the center of mass is a constant, thus $I = \rho AR^3 \int_0^{2\pi} d\theta = 2\pi\rho AR^3 = mR^2$.

(b) The distance to the ring from the line L is y, thus the definition becomes $I = \int_{m_{upper\ half}} y^2 dm + \int_{m_{lower\ half}} y^2 dm = 2\int_{\frac{m}{2}} y^2 dm$, where each half of the *uniform ring* contributes. The differential element of mass is $dm = \rho AR d\theta$. Substitute into the definition $I = 2\rho AR \int_0^{\pi} y^2 d\theta$. Noting that $y^2 = R^2 \sin^2\theta$, from which $I = 2\rho AR^3 \int_0^{\pi} \sin^2\theta d\theta = 2\rho AR^3 \left[\frac{\theta}{2} - \frac{\sin 2\theta}{4}\right]_0^{\pi} = \frac{mR^2}{2}$

==================================<>==================================

Problem 7.92 The homogenous thin plate is of uniform thickness and mass m. Determine its moments of inertia about the x, y, and z axes.

Solution:

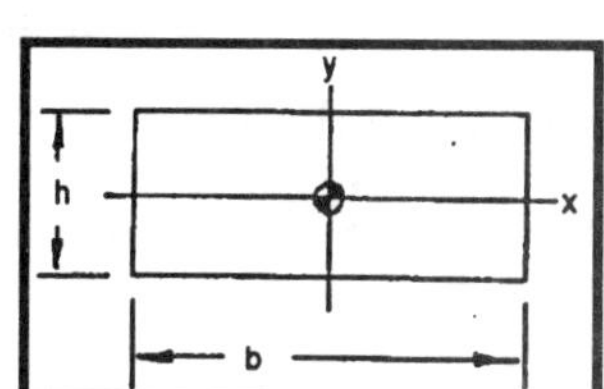

Strategy: The moments of inertia of a thin plate of arbitrary shape are given by Eqs (7.37)-(7.39) in terms of the area moments of inertia of the cross sectional area of the plate. The area moments of inertia can be obtained from Appendix B.

The area moments of inertia are $I_x = \left(\frac{1}{12}\right)bh^3$, $I_y = \left(\frac{1}{12}\right)hb^3$, $I_z = I_x + I_y$. The area mass density of the plate is $\frac{m}{A} = \frac{m}{bh}$, hence the moments of inertia are $I_{(x-axis)} = \frac{m}{bh}I_x = \frac{m}{12}h^2$, $I_{(y-axis)} = \frac{m}{bh}I_y = \frac{m}{12}b^2$

$I_{(z-axis)} = \frac{m}{12}\left(h^2 + b^2\right)$

==================================<>==================================

==============================<>==============================

Problem 7.93 The brass washer is of uniform thickness and mass m. (a) Determine its moments of inertia about the x and z axes. (b) Let $R_i = 0$ and compare your results with the values given in Appendix C for a thin circular plate. (c) Let $R_i \to R_o$ and compare your results with the solutions of Problem 7.73.

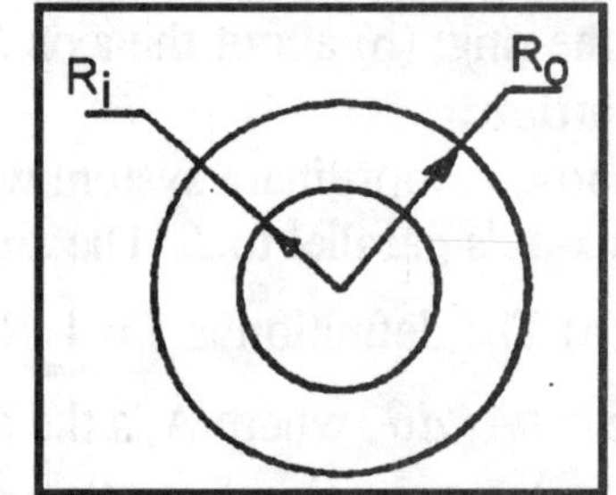

Solution: (a)The area moments of inertia for a circular area are $I_x = I_y = \frac{\pi R^4}{4}$. For the plate with a circular cutout, $I_x = \frac{\pi}{4}(R_o^4 - R_i^4)$

The area mass density is $\frac{m}{A}$, thus for the plate with a circular cut,

$\frac{m}{A} = \frac{m}{\pi(R_o^2 - R_i^2)}$, from which the moments of inertia $I_{(x-axis)} = \frac{m(R_o^4 - R_i^4)}{4(R_o^2 - R_i^2)} = \frac{m}{4}(R_o^2 + R_i^2)$

$I_{(z-axis)} = 2I_{(x-axis)} = \frac{m}{2}(R_o^2 + R_i^2)$. (b) Let $R_i = 0$, to obtain $I_{x-axis} = \frac{m}{4}R_o^2$, $I_{(z-axis)} = \frac{m}{2}R_o^2$, which agrees with table entries. (c) Take the limit as $R_i \to R_o$, $I_{(x-axis)} = \frac{m}{2}R_o^2$, $I_{(z-axis)} = mR_o^2$, which agrees with the solution to Problem 7.73.

==============================<>==============================

Problem 7.94 The homogenous thin plate is of uniform thickness and weighs 20 *lb*. Determine its moment of inertia about the *y* axis.

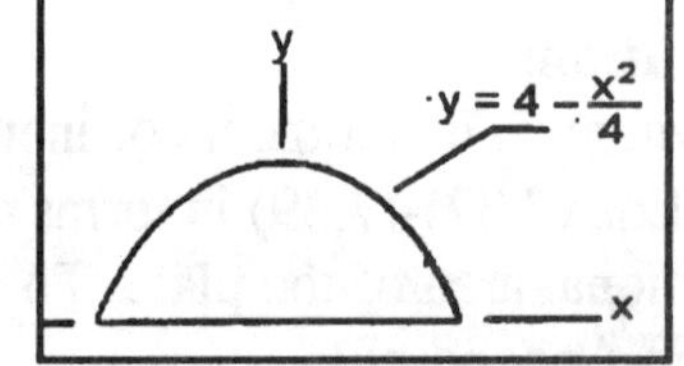

Solution: The definition of the moment of inertia is $I = \int_m r^2 dm$. The distance from the y-axis is *x*, where *x* varies over the range $-4 \le x \le 4$. Let $\tau = \frac{m}{A} = \frac{W}{gA}$ be the area mass density. The mass of an element *ydx* is

$dm = \frac{W}{gA} ydx$. Substitute into the definition: $I_{y-axis} = \frac{W}{gA}\int_{-4}^{4} x^2\left(4 - \frac{x^2}{4}\right)dx$

$= \frac{W}{gA}\left[\frac{4x^3}{3} - \frac{x^5}{20}\right]_{-4}^{+4} = \frac{W}{gA}[68.2667]$. The area is $A = \int_{-4}^{4}\left(4 - \frac{x^2}{4}\right)dx = \left[4x - \frac{x^3}{12}\right]_{-4}^{4} = 21.333 \text{ ft}^2$ The moment of inertia about the y-axis is $I_{(y-axis)} = \frac{W}{g}(3.2) = \frac{20}{32.17}(3.2) = 1.989 \text{ slug-ft}^2$

==============================<>==============================

Problem 7.95 Determine the moment of inertia of the plate in Problem 7.94 about the x-axis.

Solution: The differential mass is $dm = \frac{W}{gA} dydx$. The distance of a mass element from the x-axis is *y*, thus $I = \frac{W}{gA}\int_{-4}^{+4} dx \int_0^{4-\frac{x^2}{4}} y^2 dy = \frac{W}{3gA}\int_{-4}^{+4}\left(4 - \frac{x^2}{4}\right)^3 dx = \frac{W}{3gA}\left[64x - 4x^3 + \frac{3}{20}x^5 - \frac{x^7}{448}\right]_{-4}^{4} = \frac{W}{3gA}[234.057]$.

From the solution to Problem 7.76, $A = 21.333 \text{ ft}^2$. Thus the moment of inertia about the x-axis is

$I_{x-axis} = \frac{W}{3g}\frac{(234.057)}{(21.333)} = \frac{W}{g}(3.657) = 2.27 \text{ slug-ft}^2$.

==============================<>==============================

Problem 7.96 The mass of the object is 10 kg. Its moment of inertia about L_1 is $10\ kg\text{-}m^2$ What is its moment of inertia about L_2? (The three axes are in the same plane.)

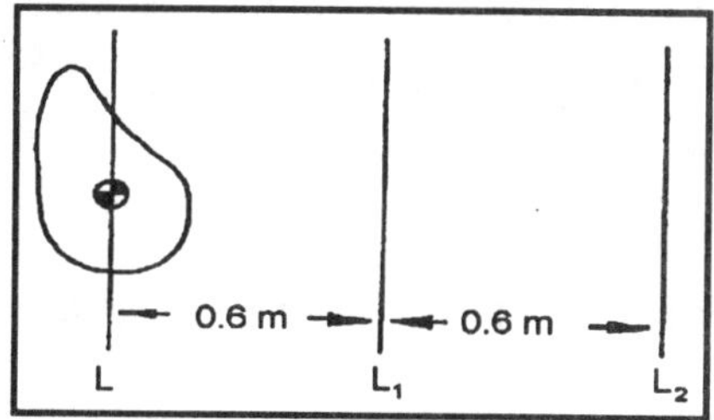

Solution:

The strategy is to use the data to find the moment of inertia about L, from which the moment of inertia about L_2 can be determined. $I_L = -(0.6)^2(10) + 10 = 6.4\ kg\text{-}m^2$, from which $I_{L2} = (1.2)^2(10) + 6.4 = 20.8\ kg\text{-}m^2$

Problem 7.97 An engineer gathering data for the design of a maneuvering unit determines that an astronaut's center of mass is at $x = 1.01\ m$, $y = 0.16\ m$ and that his moment of inertia about the z axis is $105.6\ kg\text{-}m^2$. His mass is 81.6 kg. What is his moment of inertia about the z' axis through his center of mass?

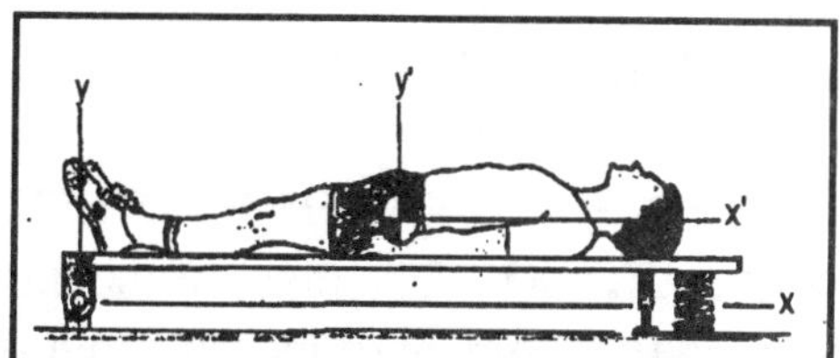

Solution:

The distance from the z' axis to the z axis is $d = \sqrt{x^2 + y^2} = 1.02257\ m$. The moment of inertia about the z' axis is

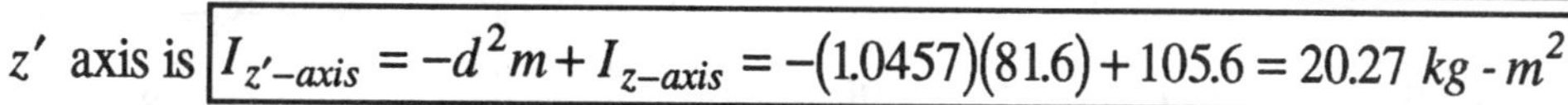

$I_{z'-axis} = -d^2 m + I_{z-axis} = -(1.0457)(81.6) + 105.6 = 20.27\ kg\text{-}m^2$

Problem 7.98 Two homogenous, slender bars, each of mass m and length l, are welded together to form the T-shaped object. Use the parallel axis theorem to determine the moment of inertia of the object about the axis through point O that is perpendicular to the bars.

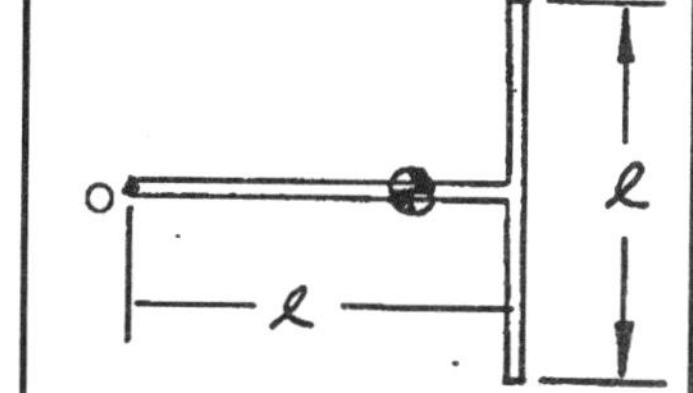

Solution:

Divide the object into two pieces, each corresponding to a bar of mass m. By definition $I = \int_0^l r^2 dm$. For the first bar, the differential mass is $dm = \rho A dr$, from which the moment of inertia about one end is

$$I_1 = \rho A \int_0^l r^2 dr = \rho A \left[\frac{r^3}{3}\right]_0^l = \frac{ml^2}{3}.$$ For the second bar

$$I_2 = \rho A \int_{-\frac{l}{2}}^{\frac{l}{2}} r^2 dr = \rho A \left[\frac{r^3}{3}\right]_{-\frac{l}{2}}^{\frac{l}{2}} = \frac{ml^2}{12}$$ is the moment of inertia about the center of the bar. From the parallel axis theorem, the moment of inertia about O is $I_o = \frac{ml^2}{3} + l^2 m + \frac{ml^2}{12} = \frac{17}{12} ml^2$

==============================<>==============================

Problem 7.99 Use the parallel axis theorem to determine the moment of inertia of the T-shaped object in Problem 7.98 about the axis through the center of mass of the object that is perpendicular to the two bars.

Solution:

The location of the center of mass of the object is $\bar{x} = \dfrac{m\left(\frac{l}{2}\right) + lm}{2m} = \frac{3}{4}l$. Use the results of Problem 7.80 for the moment of inertia of a bar about its center. For the first bar, $I_1 = \left(\frac{l}{4}\right)^2 m + \frac{ml^2}{12} = \frac{7}{48}ml^2$. For the second bar, $I_2 = \left(\frac{l}{4}\right)^2 m + \frac{ml^2}{12} = \frac{7}{48}ml^2$. The composite: $\boxed{I_c = I_1 + I_2 = \frac{7}{24}ml^2}$ *Check:* Use the results of Problem 7.98: $I_c = -\left(\frac{3l}{4}\right)^2 (2m) + \frac{17}{12}ml^2 = \left(\frac{-9}{8} + \frac{17}{12}\right)ml^2 = \frac{7}{24}ml^2$. *check.*

==============================<>==============================

Problem 7.100 The mass of the homogenous, slender bar is 20 *kg*. Determine the moment of inertia about the z axis.

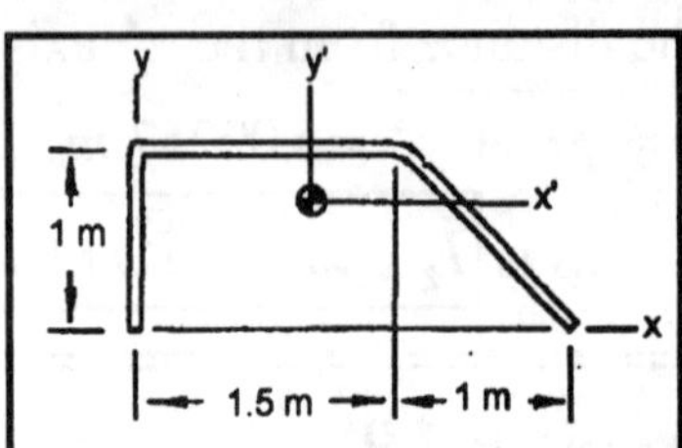

Solution:

Divide the object into three segments. Part (1) is the 1 m bar on the left, Part (2) is the 1.5 m horizontal segment, and Part (3) is the segment on the far right. The mass density is $\rho = \frac{m}{L} = \frac{20}{(1 + 1.5 + \sqrt{2})} = 5.1\ kg/m$

The moments of inertia about the centers of mass and the distances to the centers of mass from the z-axis are:

Part (1) $I_1 = \rho\left(\frac{l_1^{\ 3}}{12}\right) = m_1\frac{l_1^2}{12} = 0.425\ \ kg\text{-}m^2$, $m_1 = \rho 1 = 5.1\ kg$, $d_1 = 0.5\ m$,

Part (2), $I_2 = \rho\frac{l_2^{\ 3}}{12} = m_2\frac{l_2^2}{12} = 1.437\ kg\text{-}m^2$, $m_2 = 7.665\ kg$, $d_2 = \sqrt{0.75^2 + 1^2} = 1.25\ m$

Part (3) $I_3 = \rho\frac{l_3^3}{12} = m_3\frac{(\sqrt{2})^2}{12} = 1.204\ kg\text{-}m^2$, $m_3 = 7.225\ kg$, $d_3 = \sqrt{2^2 + 0.5^2} = 2.062\ m$. The composite: $\boxed{I = d_1^2 m_1 + I_1 + d_2^2 m_2 + I_2 + d_3^2 m_3 + I_3 = 47.02\ kg\text{-}m^2}$

==============================<>==============================

Problem 7.101 Determine the moment of inertia of the bar in Problem 7.100 about the z' axis through its center of mass.

Solution: The center of mass: $\bar{x} = \frac{x_1 m_1 + x_2 m_2 + x_3 m_3}{20} = \frac{0 + 0.75(7.665) + 2(7.225)}{20} = 1.01\ m$.

$\bar{y} = \frac{0.5m_1 + 1m_2 + 0.5m_3}{20} = \frac{0.5(5.1) + 1(7.665) + 0.5(7.225)}{20} = 0.6914\ m$. The distance from the z-axis to the center of mass is $d = \sqrt{\bar{x}^2 + \bar{y}^2} = 1.221\ m$. The moment of inertia about the center of mass:

$\boxed{I_c = -d^2(20) + I_o = 17.19\ kg\text{-}m^2}$

==============================<>==============================

===================================<>===================================

Problem 7.102 The homogeneous slender bar weighs *5 lb*. Determine its moment of inertia about the z axis.

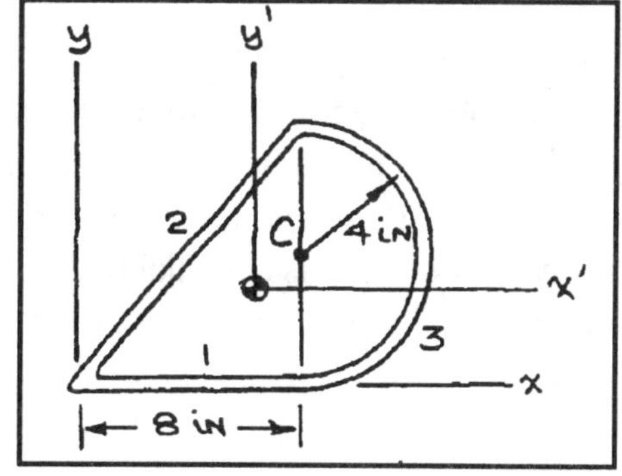

Solution:

The Bar's mass is $m = 5/32.2 slugs$. Its length is $L = L_1 + L_2 + L_3 = 8 + \sqrt{8^2 + 8^2} + \pi(4) = 31.9 in$. The masses of the parts are therefore,

$M_1 = \frac{L_1}{L} m = \left(\frac{8}{31.9}\right)\left(\frac{5}{32.2}\right) = 0.0390 slugs$,

$M_2 = \frac{L_2}{L} m = \left(\frac{\sqrt{2(64)}}{31.9}\right)\left(\frac{5}{32.2}\right) = 0.0551 slugs$,

$M_3 = \frac{L_3}{L} m = \left(\frac{4\pi}{31.9}\right)\left(\frac{5}{32.2}\right) = 0.0612 slugs$. The center of mass of part 3 is located to the right of its center C a distance $2R/\pi = 2(4)/\pi = 2.55 in$ The moment of inertia of part 3 about C is $\int_{m_3} r^2 dm = m_3 r^2 = (0.0612)(4)^2 = 0.979 slug - in^2$. The moment of inertia of part 3 about the center of mass of part 3 is therefore $I_3 = 0.979 - m_3(2.55)^2 = 0.582 slug - in^2$. The moment of inertia of the bar about the z axis is $I_{(z axis)} = \frac{1}{3} m_1 L_1^2 + \frac{1}{3} m_2 L_2^2 + I_3 + m_3\left[(8 + 2.55)^2 + (4)^2\right] = 11.6 slug - in^2 = 0.0803 slug - ft$.

===================================<>===================================

Problem 7.103 Determine the moment of inertia of the bar in Problem 7.102 about the z' axis through the center of mass.

Solution: In the solution of Problem 7.102, it is shown that the moment of inertia of the bar about the z axis is $I_{(z axis)} = 11.6 slug - in^2$. The x and y coordinates of the center of mass coincide with the centroid of the bar: $\bar{x} = \frac{\bar{x}_1 L_1 + \bar{x}_2 L_2 + \bar{x}_3 L_3}{L_1 + L_2 + L_3} = \frac{(4)(8) + (4)\sqrt{8^2 + 8^2} + \left[8 + \frac{2(4)}{\pi}\right]\pi(4)}{8 + \sqrt{8^2 + 8^2} + \pi(4)} = 6.58 in$,

$\bar{y} = \frac{\bar{y}_1 L_1 + \bar{y}_2 L_2 + \bar{y}_3 L_3}{L_1 + L_2 + L_3} = \frac{0 + (4)\sqrt{8^2 + 8^2} + \pi(4)(4)}{8 + \sqrt{8^2 + 8^2} + \pi(4)} = 3.00 in$. The moment of inertia about the z' axis is

$I_{(z' axis)} = I_{(z axis)} - (\bar{x}^2 + \bar{y}^2)\left(\frac{5}{32.2}\right) = 3.44 slug - in^2$.

===================================<>===================================

Problem 7.104 The rocket is used for atmospheric research. Its weight and its moment of inertia about the z-axis through its center of mass (including its fuel) are 10 *kip* and 10200 $slug\text{-}ft^2$, respectively. The rocket's fuel weighs 6000 *lb.*, its center of mass is located at $x = -3\,ft$, $y = 0$, and $z = 0$, and the moment of inertia of the fuel about the axis through the fuel's center of mass parallel to z is 2200 $slug\text{-}ft^2$. When the fuel is exhausted, what is the rocket's moment of inertia about the axis through its new center of mass parallel to z?

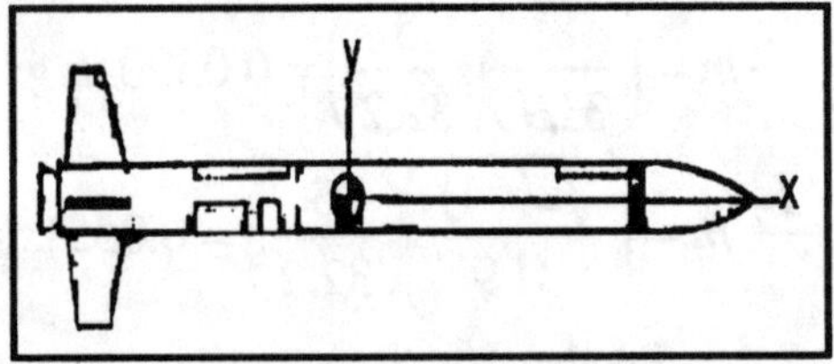

Solution: Denote the moment of inertia of the empty rocket as I_E about a center of mass x_E, and the moment of inertia of the fuel as I_F about a mass center x_F. Using the parallel axis theorem, the moment of inertia of the filled rocket is $I_R = I_E + x_E^2 m_E + I_F + x_F^2 m_F$, about a mass center at the origin ($x_R = 0$). Solve: $I_E = I_R - x_E{}^2 m_E - I_F - x_F{}^2 m_F$. The objective is to determine values for the terms on the right from the data given. Since the filled rocket has a mass center at the origin, the mass center of the empty rocket is found from $0 = m_E x_E + m_F x_F$, from which $x_E = -\left(\frac{m_F}{m_E}\right)x_F$. Using a value of $g = 32\,ft/s^2$, $m_F = \frac{W_F}{g} = \frac{6000}{32} = 186.34\ slug$, $m_E = \frac{(W_R - W_F)}{g} = \frac{10000-6000}{32} = 124.23\ slug$.

From which $x_E = -\left(\frac{186.335}{124.224}\right)(-3) = 4.5\,ft$ is the new location of the center of mass. Substitute:

$$\boxed{I_E = I_R - x_E^2 m_E - I_F - x_F^2 m_F = 10200 - 2515.5 - 2200 - 1677.01 = 3807.5\ slug\text{-}ft^2}$$

Problem 7.105 The mass of the homogenous thin plate is 36 *kg*. Determine its moment of inertia about the x- axis.

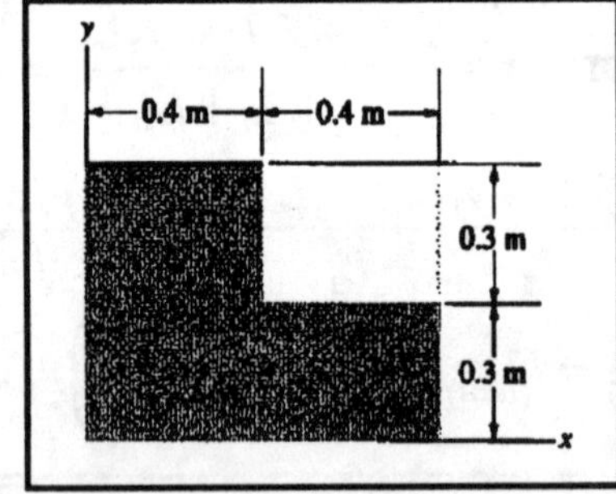

Solution: Divide the plate into two areas: the rectangle 0.4 *m* by 0.6 *m* on the left, and the rectangle 0.4 *m* by 0.3 *m* on the right. The mass density is $\rho = \frac{m}{A}$. The area is $A = (0.4)(0.6) + (0.4)(0.3) = 0.36\ m^2$, from which $\rho = \frac{36}{0.36} = 100\ kg/m^2$. The moment of inertia about the x-axis is

$$\boxed{I_{x-axis} = \rho\left(\frac{1}{3}\right)(0.4)(0.6^3) + \rho\left(\frac{1}{3}\right)(0.4)(0.3)^3 = 3.24\ kg\text{-}m^2}$$

Problem 7.106 Determine the moment of inertia of the plate in Problem 7.105 about the z-axis.

Solution: The basic relation to use is $I_{z-axis} = I_{x-axis} + I_{y-axis}$. The value of I_{x-axis} is given in the solution of Problem 7.105. The moment of inertia about the y-axis using the same divisions as in Problem 7.85 and the parallel axis theorem is

$$I_{y-axis} = \rho\left(\frac{1}{3}\right)(0.6)(0.4)^3 + \rho\left(\frac{1}{12}\right)(0.3)(0.4)^3 + (0.6)^2\rho(0.3)(0.4) = 5.76\ kg\text{-}m^2$$, from which

$$\boxed{I_{z-axis} = I_{x-axis} + I_{y-axis} = 3.24 + 5.76 = 9\ kg\text{-}m^2.}$$

==================================<>==================================

Problem Problem 7.107 The homogenous thin plate weighs 10 *lb*. Determine its moment of inertia about the x-axis.

Solution:

Divide the area into two parts: the lower rectangle 5 *in* by 10 *in* and the upper triangle 5 *in* base and 5 *in* altitude.

The mass density is $\rho = \dfrac{W}{gA}$.

The area is $A = 5(10) + \left(\dfrac{1}{2}\right)5(5) = 62.5\ \text{in}^2$. Using $g = 32\ \text{ft}/\text{s}^2$, the mass density is

$\rho = \dfrac{W}{gA} = 0.005\ \text{slug}/\text{in}^2$. Using the parallel axis theorem, the moment of inertia about the x-axis is

$$I_{x-axis} = \rho\left(\frac{1}{3}\right)(10)(5)^3 + \rho\left(\frac{1}{36}\right)(5)(5^3) + \rho(5+5/3)^2\left(\frac{1}{2}\right)(5)(5) = 4.948\ \text{slug-in}^2 = 0.03436\ \text{slug-ft}^2$$

==================================<>==================================

Problem 7.108 Determine the moment of inertia of the plate in Problem 7.107 about the y-axis.

Solution:

Use the results of the solution in Problem 7.87 for the area and the mass density.

$$I_{y-axis} = \rho\left(\frac{1}{3}\right)5(10^3) + \rho\left(\frac{1}{36}\right)5(5^3) + \rho(5+10/3)^2\left(\frac{1}{2}\right)5(5) = 12.76\ \text{slug-in}^2 = 0.0886\ \text{slug-ft}^2$$

==================================<>==================================

Problem 7.109 The thermal radiator (used to eliminate excess heat from a satellite) can modeled as a homogenous, thin, rectangular plate. Its mass is 5 slugs. Determine its moment of inertia about the x- y- and z- axes.

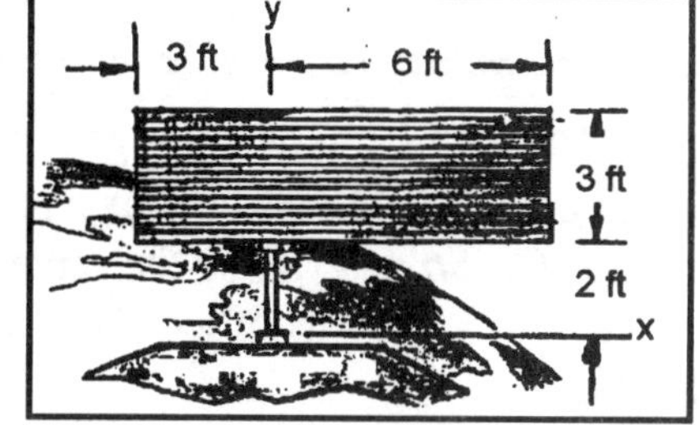

Solution:

The area is $A = 9(3) = 27\ \text{ft}^2$.

The mass density is $\rho = \dfrac{m}{A} = \dfrac{5}{27} = 0.1852\ \text{slugs}/\text{ft}^2$.

The moment of inertia about the centroid of the rectangle is $I_{xc} = \rho\left(\dfrac{1}{12}\right)9(3^3) = 3.75\ \text{slug-ft}^2$,

$I_{yc} = \rho\left(\dfrac{1}{12}\right)3(9^3) = 33.75\ \text{slug-ft}^2$. Use the parallel axis theorem:

$I_{x-axis} = \rho A(2+1.5)^2 + I_{xc} = 65\ \text{slug-ft}^2$, $I_{y-axis} = \rho A(4.5-3)^2 + I_{yc} = 45\ \text{slug-ft}^2$.

$I_{z-axis} = I_{x-axis} + I_{y-axis} = 110\ \text{slug-ft}^2$

==================================<>==================================

==================================<>==================================

Problem 7.110 The mass of the homogenous, thin plate is 2 *kg*. Determine its moment of inertia about the x-axis L_o through point O that is perpendicular to the plate.

Solution:

The strategy is to determine I_x and I_y about point O, from which to determine I_z. The area is $A_1 = \left(\frac{1}{2}\right)80(130) = 5200\ mm^2$,

$A_2 = \pi\left(10^2\right) = 314.16\ mm^2$, $A = A_1 - A_2 = 4885.84\ mm^2$. The mass density is

$\rho = \frac{m}{A} = \frac{2}{A} = 0.0004093\ kg/mm^2$.

The moments of inertia about the centroid of the circle are $I_{xc} = I_{yc} = \rho\left(\frac{1}{4}\right)\pi\left(10^4\right) = 2.849\ kg\text{-}mm^2$.

Use the parallel axis theorem to determine the moments of inertia about the x- and y- axes.

$$I_{y\text{-}axis} = \rho\left(\frac{1}{36}\right)80\left(130^3\right) + \rho A_1\left(\frac{2}{3}130\right)^2 - \rho A_2\left(100^2\right) - I_{yc} = 16697.5\ kg\text{-}mm^2$$

$$I_{x\text{-}axis} = \rho\left(\frac{1}{12}\right)130\left(80^3\right) - \rho A_2\left(30^2\right) - I_{xc} = 2151.5\ kg\text{-}mm^2.$$

$$I_{z\text{-}axis} = I_{x\text{-}axis} + I_{y\text{-}axis} = 18849\ kg\text{-}mm^2 \quad \boxed{I_{z\text{-}axis} = 0.018849\ kg\text{-}m^2}$$

==================================<>==================================

Problem 7.111 The homogenous cone is of mass *m*. Determine its moment of inertia about the z-axis and compare your result with the value in Appendix C. *Strategy:* Use the same approach we used in Example 7.15 to determine the moments of inertia of a homogenous cylinder.

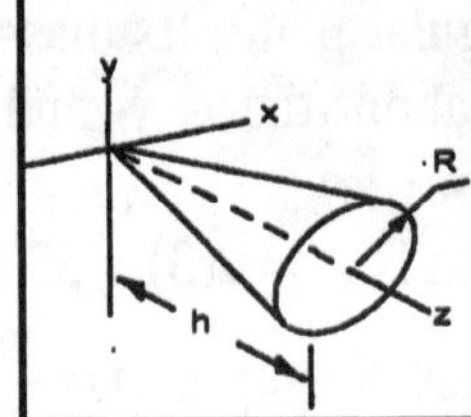

Solution:

The differential mass $dm = \left(\frac{m}{V}\right)\pi r^2 dz = \frac{3m}{R^2 h} r^2\ dz$. The moment of inertia of this disk about the z-axis is $\frac{1}{2}mr^2$. The radius varies with z, $r = \left(\frac{R}{h}\right)z$, from which

$$\boxed{I_{z\text{-}axis} = \frac{3mR^2}{2h^5}\int_0^h z^4 dz = \frac{3mR^2}{2h^5}\left[\frac{z^5}{5}\right]_0^h = \frac{3mR^2}{10}}$$

==================================<>==================================

==================================<>==================================

Problem 7.112 Determine the moments of inertia of the homogenous cone in Problem 7.111 about the x and y axes and compare your results with the values in Appendix C.

Solution: The mass density is $\rho = \frac{m}{V} = \frac{3m}{\pi R^2 h}$ The differential element of mass is $dm = \rho \pi r^2 dz$.. The moment of inertia of this elemental disk about an axis through its center of mass, parallel to the x- and y-axes, is $dI_x = \left(\frac{1}{4}\right) r^2 dm$. Use the parallel axis theorem, $I_x = \int_m \left(\frac{1}{4}\right) r^2 dm + \int_m z^2 dm$. Noting that $r = \frac{R}{h} z$, then $r^2 dm = \rho\left(\frac{\pi R^4}{h^4}\right) z^4 dz$, and $z^2 dm = \rho\left(\frac{\pi R^2}{h^2}\right) z^4 dz$. Substitute: $I_x = \rho\left(\frac{\pi R^4}{4h^4}\right) \int_0^h z^4 dz + \rho\left(\frac{\pi R^2}{h^2}\right) \int_0^h z^4 dz$,

$$I_x = \left(\frac{3mR^2}{4h^5} + \frac{3m}{h^3}\right)\left[\frac{z^5}{5}\right]_0^h = m\left(\frac{3}{20}R^2 + \frac{3}{5}h^2\right) \quad I_y = I_x$$

==================================<>==================================

Problem 7.113 The homogeneous object has the shape of a truncated cone and consists of bronze with a mass density of $\rho = 8200\ kg/m^3$. Determine the moment of inertia about the z axis.

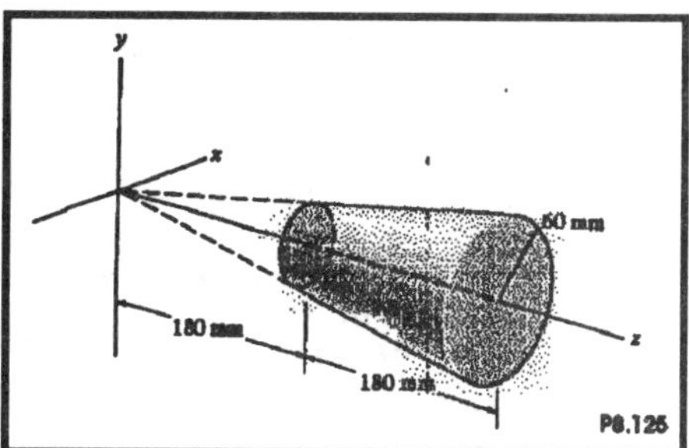

Solution:

Consider an element of the cone consisting of a disk of thickness dz: We can express the radius as a linear function of z $r = az + b$ Using the conditions that $r = 0$ at $z = 0$and $r = 0.06m$ at $z = 0.36m$ to evaluate a and b we find that$r = 0.167z$. From appendix C, the moment of inertia of the element about the z axis is

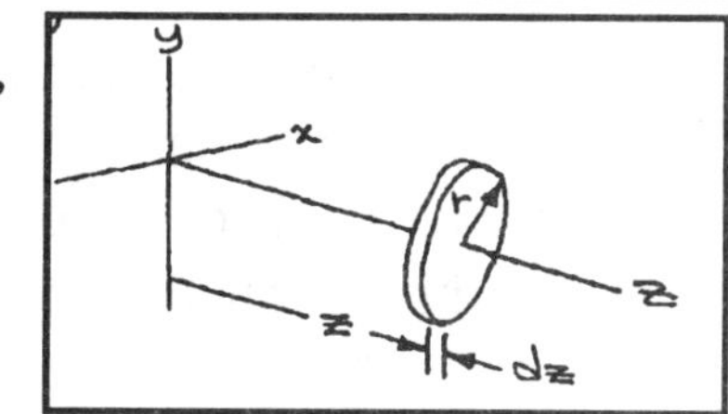

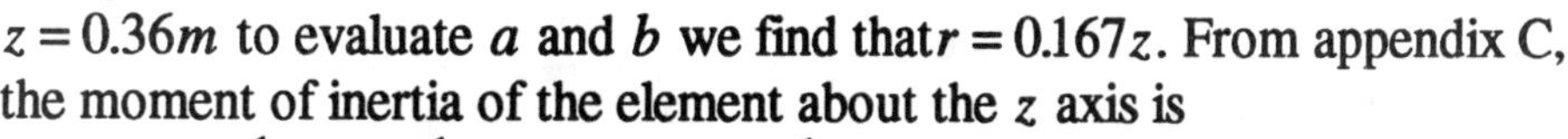

$(I_z)_{element} = \frac{1}{2} mr^2 = \frac{1}{2}\left[\rho(\pi r^2) dz\right] r^2 = \frac{1}{2} \rho\pi(0.167z)^4 dz$. We integrate this result to obtain the mass moment of inertia about the z axis for the cone:

$$I_{(zaxis)} = \int_{0.18}^{0.36} \frac{1}{2} \rho\pi(0.167)^4 \left[\frac{z^5}{5}\right]_{0.18}^{0.36} = \frac{1}{2}(8200)\pi(0.167)^4 \left[\frac{z^5}{5}\right]_{0.18}^{0.36} = 0.0116 kg - m^2.$$

==================================<>==================================

Problem 7.114 Determine the moment of inertia of the object in Problem 7.113 about the x axis.

Solution:

Consider the disk element described in the solution to Problem 7.113. The radius of the laminate is $r = 0.167z$. Using Appendix C and the parallel axis theorem, the moment of inertia of the element about the x axis is $(I_x)_{element} = \frac{1}{4} mr^2 + mz^2 = \frac{1}{4}\left[\rho(\pi r^2) dz\right] r^2 + \left[\rho(\pi r^2) dz\right] z^2$

$= \frac{1}{4} \rho\pi(0.167z)^4 dz + \rho\pi(0.167z)^2 z^2 dz$. Integrating the result,

$$I_{(xaxis)} = \frac{1}{4} \rho\pi(0.167)^4 \int_{0.18}^{0.36} z^4\, dz + \rho\pi(0.167)^2 \int_{0.18}^{0.36} z^4\, dz = 0.844\ kg - m^2.$$

==================================<>==================================

============================<>============================

Problem 7.115 The homogenous parallelepiped is of mass *m*. Determine its moment of inertia about the x, y and z axes and compare your results with the values in Appendix C.

Solution:

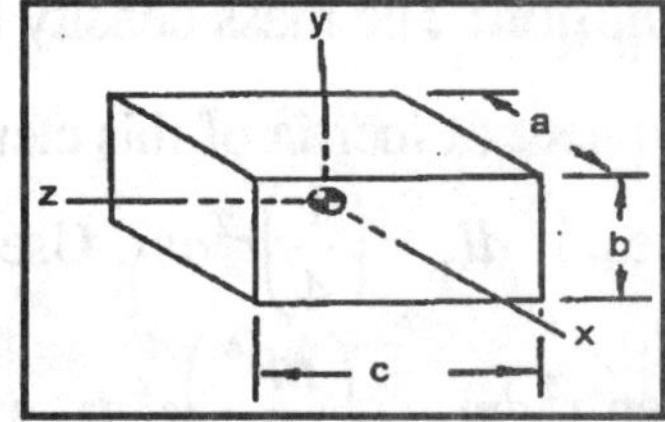

Consider a rectangular slice normal to the x- axis of dimensions b by c and mass dm. The area density of this slice is $\rho = \frac{dm}{bc}$. The moment of inertia about the y axis of the centroid of a thin plate is the product of the area density and the area moment of inertia of the plate:

$dI_y = \rho\left(\frac{1}{12}\right)bc^3$., from which $dI_y = \left(\frac{1}{12}\right)c^2 dm$. By symmetry, the moment of inertia about the z axis is $dI_z = \left(\frac{1}{12}\right)b^2 dm$. Since the labeling of the x- y- and z- axes is arbitrary, $dI_x = dI_z + dI_y$, where the x-axis is normal to the area of the plate. Thus $dI_x = \left(\frac{1}{12}\right)(b^2 + c^2)dm$, from which

$\boxed{I_x = \left(\frac{1}{12}\right)(b^2 + c^2)\int_m dm = \frac{m}{12}(b^2 + c^2)}$. By symmetry, the argument can be repeated for each coordinate, to obtain $\boxed{I_y = \frac{m}{12}(a^2 + c^2)}$ $\boxed{I_z = \frac{m}{12}(b^2 + a^2)}$

============================<>============================

Problem 7.116 The L-shaped machine part is composed of two homogeneous bars. Bar 1 is tungsten alloy with mass density 14000 kg/m^3 and bar 2 is steel with mass density 7800 kg/m^3. Determine its moment of inertia about the x axis.

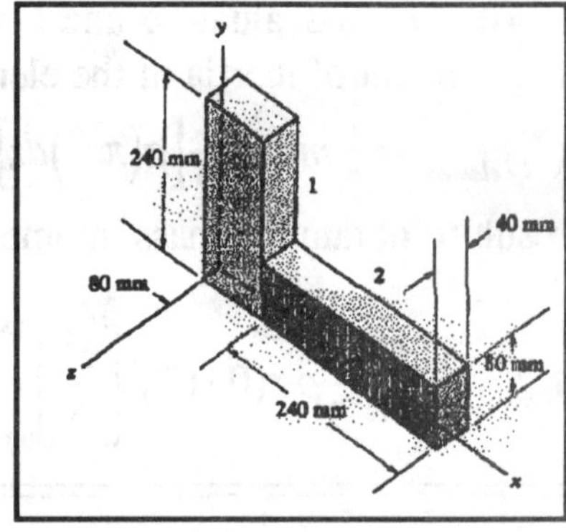

Solution:

The masses of the bars are $m_1 = (14{,}000)(0.24)(0.08)(0.04) = 10.75 kg$ $m_2 = (7800)(0.24)(0.08)(0.04) = 5.99 kg$ Using Appendix C and the parallel axis theorem the moments of inertia of the parts about the x axis are $I_{(xaxis)_1} = \frac{1}{12}m_1\left[(0.04)^2 + (0.24)^2\right] + m_1(0.12)^2 = 0.2079 kg - m^2$,

$I_{(xaxis)_2} = \frac{1}{12}m_2\left[(0.04)^2 + (0.08)^2\right] + m_2(0.04)^2 = 0.0136 kg - m^2$. Therefore

$I_{(xaxis)} = I_{(xaxis)_1} + I_{(xaxis)_2} = 0.221 kg - m^2$

============================<>============================

Problem 7.117 Determine the moment of inertia of the L-shaped machine part in Problem 7.116 about the z axis.

Solution:

From the solution to Problem 7.116, the masses of the parts are $m_1 = 10.75 kg$, $m_2 = 5.99 kg$. Using Appendix C and the parallel axis theorem the moments of inertia of the parts about the z axis are

$I_{(zaxis)1} = (1/12)m_1\left[(0.08)^2 + (0.24)^2\right] + m_1\left[(0.04)^2 + (0.12)^2\right] = 0.229 kg - m^2$,

$I_{(xaxis)2} = \frac{1}{12}m_2\left[(0.24)^2 + (0.08)^2\right] + m_2\left[(0.08 + 0.12)^2 + (0.04)^2\right] = 0.281 kg - m^2$.

Therefore, $I_{(zaxis)} = I_{(zaxis)1} + I_{(zaxis)2} = 0.511 kg - m^2$.

============================<>============================

==================================<>==================================

Problem 7.118 The homogenous ring consists of steel of density $\rho = 15\ slug/ft^3$. Determine the its moment of inertia about the axis L through its center of mass.

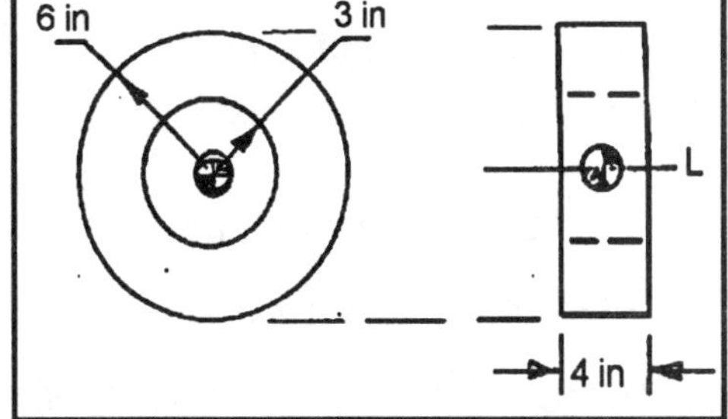

Solution: Consider a solid cylinder of radius *R* and height *h*.. Its mass density is $\rho = \dfrac{m}{V} = \dfrac{m}{\pi R^2 h}$. Consider a thin slice normal to the axis L. The moment of inertia of this slice about the axis L is the product of the area density and the area moment of inertia:

$dI_L = (\rho\, h)\, r^2 dA = (\rho\, h)(2\pi) r^3 dr$, from which

$I_L = (2\rho\, h\pi)\int_0^R r^3 dr = \left(\dfrac{1}{2}\right) mR^2$. From which, for the

outer ring $I_{Lo} = \left(\dfrac{1}{2}\right) m_o R_o{}^2$. For the inner ring $I_{Li} = \left(\dfrac{1}{2}\right) m_i R_i^2$, from which

$I_L = I_{Lo} - I_{Li} = \left(\dfrac{1}{2}\right)\left(m_o R_o^2 - m_i R_i^2\right)$. The mass of the outer cylinder is

$m_o = \rho\pi R_o^2 h = 15\pi\left(0.5^2\right)(1/3) = 3.927\ slugs$. The mass of the inner cylinder is

$m_i = \rho\pi R_i^2 h = 15\pi(0.25^2)(1/3) = 0.982\ slugs$, from which $\boxed{I_L = 0.46\ slug\text{-}ft^2}$

==================================<>==================================

Problem 7.119 The homogenous half cylinder is of mass *m*. Determine its moment of inertia about the axis L though its center of mass.

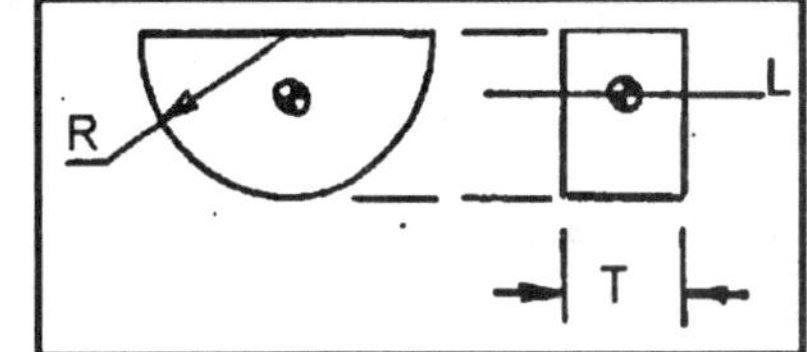

Solution: : The centroid of the half cylinder is located a distance of $\left(\dfrac{4R}{3\pi}\right)$ from the edge diameter. The strategy is to use the parallel axis theorem to treat the moment of inertia of a complete cylinder as the sum of the moments of inertia for the two half cylinders. From Problem 7.96, the moment of inertia about the geometric axis for a cylinder is $I_{cL} = mR^2$, where *m* is one half the mass of the cylinder. By the parallel axis theorem, $I_{cL} = 2\left(\left(\dfrac{4R}{3\pi}\right)^2 m + I_{hL}\right)$. Solve

$$I_{hL} = \left(\frac{I_{cL}}{2} - \left(\frac{4R}{3\pi}\right)^2 m\right) = \left(\frac{mR^2}{2} - \left(\frac{16}{9\pi^2}\right) mR^2\right) = mR^2\left(\frac{1}{2} - \frac{16}{9\pi^2}\right)$$

$$\boxed{= mR^2\left(\frac{1}{2} - \frac{16}{9\pi^2}\right) = 0.31987 mR^2 = 0.32 mR^2}$$

==================================<>==================================

==================================<>==================================

Problem 7.120 The homogeneous machine part is made of aluminum alloy with mass density $\rho = 2800\ kg/m^3$. Determine its moment of inertia about the z axis.

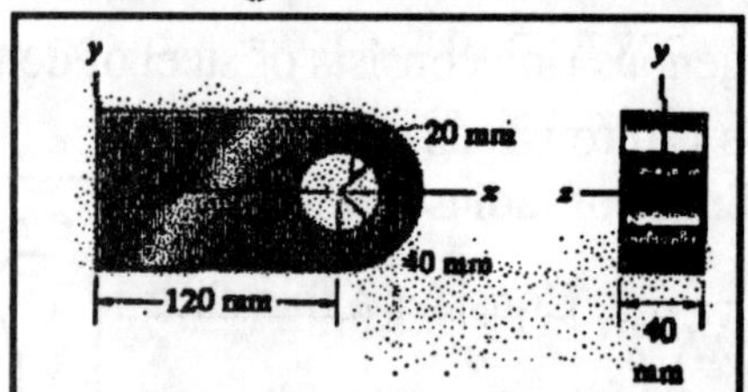

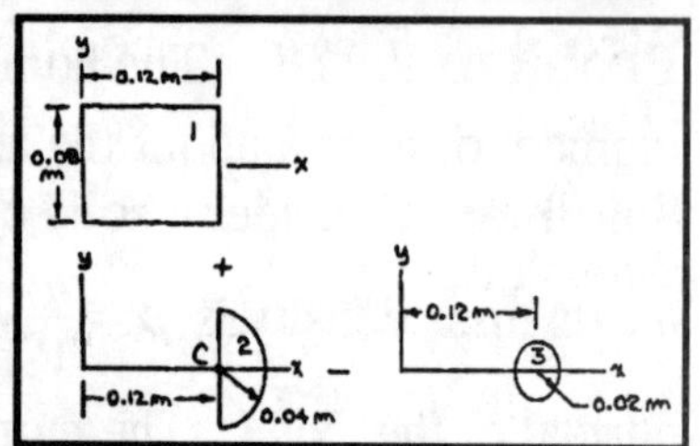

Solution:

We divide the machine part into the 3 parts shown: (The dimension into the page is $0.04m$) The masses of the parts are $m_1 = (2800)(0.12)(0.08)(0.04) = 1.075\ kg$, $m_2 = (2800)\frac{1}{2}\pi(0.04)^2(0.04) = 0.281kg$, $m_3 = (2800)\pi(0.02)^2(0.04) = 0.141kg$. Using Appendix C and the parallel axis theorem the moment of inertia of part 1 about the z axis is $I_{(zaxis)_1} = \frac{1}{12}m_1\left[(0.08)^2 + (0.12)^2\right] + m_1(0.06)^2 = 0.00573kg - m^2$. The moment of inertia of part 2 about the axis through the center C that is parallel to the z axis is $\frac{1}{2}m_2R^2 = \frac{1}{2}m_2(0.04)^2$ The distance along the x axis from C to the center of mass of part 2 is $4(0.04)/(3\pi) = 0.0170m$. Therefore, the moment of inertia of part 2 about the z axis through its center of mass that is parallel to the axis is $\frac{1}{2}m_2(0.04)^2 - m_2(0.0170)^2 = 0.000144kg - m^2$.

Using this result, the moment of inertia of part 2 about the z axis is $I_{(zaxis)2} = 0.000144 + m_2(0.12 + 0.017)^2 = 0.00544kg - m^2$. The moment of inertia of the material that would occupy the hole 3 about the z axis is $I_{(zaxis)3} = \frac{1}{2}m_3(0.02)^2 + m_3(0.12)^2 = 0.00205kg - m^2$.

Therefore, $I_{(zaxis)} = I_{(zaxis)1} + I_{(zaxis)2} - I_{(zaxis)3} = 0.00912kg - m^2$.

==================================<>==================================

Problem 7.121 Determine the moment of inertia of the machine part in Problem 7.120 about the x axis.

Solution:

We divide the machine part into the 3 parts shown in the solution to Problem 7.120. Using Appendix C and the parallel axis theorem, the moments of inertia of the parts about the x axis are:

$$I_{(xaxis)_1} = \frac{1}{12}m_1\left[(0.08)^2 + (0.04)^2\right] = 0.0007168kg - m^2$$

$$I_{(xaxis)_2} = m_2\left[\frac{1}{12}(0.04)^2 + \frac{1}{4}(0.04)^2\right] = 0.0001501kg - m^2$$

$$I_{(xaxis)_3} = m_3\left[\frac{1}{12}(0.04)^2 + \frac{1}{4}(0.02)^2\right] = 0.0000328kg - m^2.$$ Therefore,

$$I_{(xaxis)} = I_{(xaxis)_1} + I_{(xaxis)_2} - I_{(xaxis)_3} = 0.000834kg - m^2.$$

==================================<>==================================

================================<>================================

Problem 7.122 The object shown consists of steel of density $\rho = 7800\ kg/m^3$ of width $w = 40\ mm$. Determine the moment of inertia about the axis L_O.

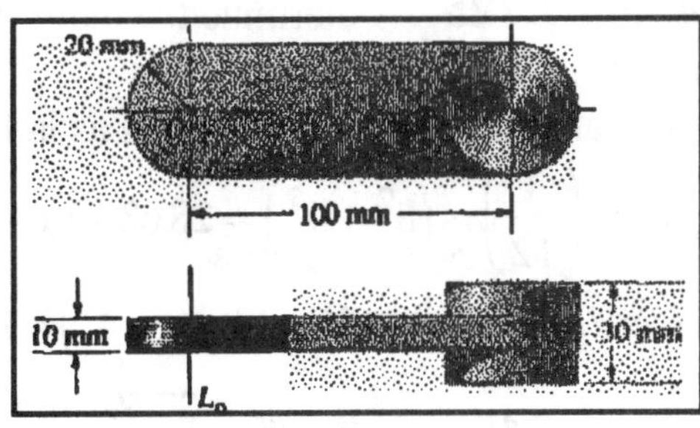

Solution:

Divide the object into four parts: Part (1) The semi-cylinder of radius $R = 0.02\ m$, height $h_1 = 0.01\ m$.
Part (2): The rectangular solid $L = 0.1\ m$ by $h_2 = 0.01\ m$ by $w = 0.04\ m$. Part (3): The semi-cylinder of radius $R = 0.02\ m$, $h_1 = 0.01\ m$ Part (4) The cylinder of radius $R = 0.02\ m$, height $h = 0.03\ m$.

Part (1) $m_1 = \dfrac{\rho\pi R^2 h_1}{2} = 0.049\ kg$, $I_1 = \dfrac{m_1 R^2}{4} = 4.9\times10^{-6}\ kg\text{-}m^2$, Part (2): $m_2 = \rho w L h_2 = 0.312\ kg$, $I_2 = (1/12)m_2(L^2 + w^2) + m_2(L/2)^2 = 0.00108\ \text{kg-m}^2$.

Part (3) $m_3 = m_1 = 0.049\ kg$, $I_3 = -\left(\dfrac{4R}{3\pi}\right)^2 m_2 + I_1 + m_3\left(L - \dfrac{4R}{3\pi}\right)^2 = 0.00041179\ kg\ m^2$. Part (4) $m_4 = \rho\pi R^2 h = 0.294\ kg$, $I_4 = \left(\dfrac{1}{2}\right)m_4\left(R^2\right) + m_4 L^2 = 0.003\ kg\ m^2$. The composite:

$$\boxed{I_{Lo} = I_1 + I_2 - I_3 + I_4 = 0.003674\ kg\ m^2}$$

================================<>================================

Problem 7.123 Determine the moment of inertia of the object in Problem 7.122 about the axis through the center of mass of the object parallel to L_O.

Solution:

The center of mass is located relative to L_O is given by

$$\bar{x} = \frac{m_1\left(-\dfrac{4R}{3\pi}\right) + m_2(0.05) - m_3\left(0.1 - \dfrac{4R}{3\pi}\right) + m_4(0.1)}{m_1 + m_2 - m_3 + m_4} = 0.066\ m,$$

$$\boxed{I_c = -\bar{x}^2 m + I_{Lo} = -0.00265 + 0.00367 = 0.00102\ kg\ m^2}$$

================================<>================================

Problem 7.124 The thick plate consists of steel with a density of $\rho = 15\ slug/ft^3$. Determine its moment of inertia about the z- axis.

Solution:

Divide the object into three parts: Part (1) the rectangle 8 *in* by 16 *in*, Parts (2) & (3) the cylindrical cut outs. Part (1): $m_1 = \rho 8(16)(4) = 4.444\ slugs$. $I_1 = (1/12)m_1\left(16^2 + 8^2\right) = 118.52\ \text{slug in}^2$. Part (2):

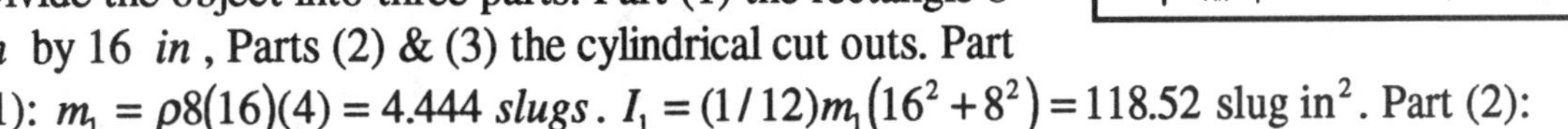

$m_2 = \rho\pi\left(2^2\right)(4)\left(\dfrac{1}{12^3}\right) = 0.4363\ slug$, $I_2 = \dfrac{m_2\left(2^2\right)}{2} + m_2\left(4^2\right) = 7.854\ slug\ in^2$. Part (3):

$m_3 = m_2 = 0.4363\ slugs$, $I_3 = I_2 = 7.854\ slug\text{-}in^2$. The composite: $I_{z-axis} = I_1 - 2I_2 = 102.81\ slug\text{-}in^2$

$$\boxed{I_{z-axis} = 0.714\ slug\text{-}ft^2}$$

================================<>================================

Problem 7.125 Determine the moment of inertia of the object in Problem 7.124 about the x- axis.

Solution:

Use the same divisions of the object as in Problem 7.124. Part (1):

$$I_{1x\text{-}axis} = \left(\frac{1}{12}\right) m_1 \left(8^2 + 4^2\right) = 28.63 \ slug\text{-}in^2,$$

Part (2): $I_{2x\text{-}axis} = (1/12) m_2 \left(3(2^2) + 4^2\right) = 1.018$ slug-in^2. The composite:

$$I_{x\text{-}axis} = I_{1x\text{-}axis} - 2I_{2x\text{-}axis} = 27.59 \ slug \ in^2 \ \boxed{= 0.1916 \ slug \ ft^2}$$

Problem 7.126 The airplane is at the beginning of its takeoff run. Its weight is 1000 *lb.* and the initial thrust *T* exerted by its engine is 300 *lb.* Assume that the thrust is horizontal, and neglect the tangential forces exerted on its wheels. (a) If the acceleration of the airplane remains constant, how long will it take to reach its takeoff speed of 80 *mi/hr*? (b) Determine the normal force exerted on the forward landing gear at the beginning of the takeoff run.

Solution:

The acceleration under constant thrust is

$a = \frac{T}{m} = \frac{300(32.17)}{1000} = 9.65 \ ft/s^2$. The time required to reach $80 \ mph = 117.33 \ ft/s$ is $\boxed{t = \frac{v}{a} = \frac{117.33}{9.65} = 12.16 \ s}$ The sum

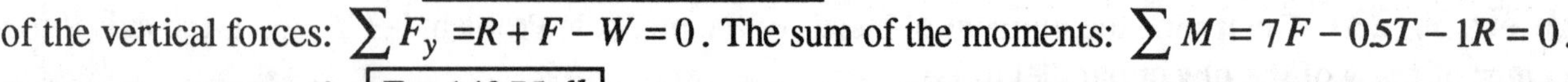

of the vertical forces: $\sum F_y = R + F - W = 0$. The sum of the moments: $\sum M = 7F - 0.5T - 1R = 0$.

Solve: $R = 856.25 \ lb$, $\boxed{F = 143.75 \ lb}$

Problem 7.127 The pulleys can turn freely on their pin supports. Their moments of inertia are $I_A = 0.002 \ kg\text{-}m^2$, $I_B = 0.036 \ kg\text{-}m^2$, and $I_C = 0.032 \ kg\text{-}m^2$. They are initially stationary, and at $t = 0$ a constant $M = 2 \ N\text{-}m$ is applied at pulley A. What is the angular velocity of pulley C and how many revolutions has it turned at $t = 2 \ s$?

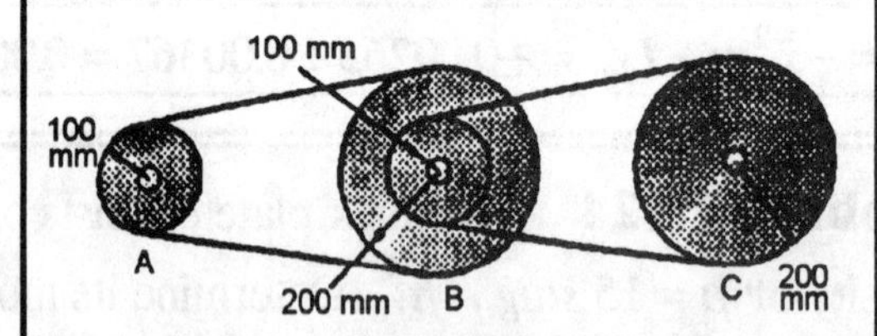

Solution:

Denote the upper and lower belts by the subscripts *U* and *L*. Denote the difference in the *tangential* component of the tension in the belts by $\Delta T_A = T_{LA} - T_{UA}$, $\Delta T_B = T_{LB} - T_{UB}$. From the equation of angular motion: $M + R_A \Delta T_A = I_A \alpha_A$, $-R_{B1} \Delta T_A + R_{B2} \Delta T_B = I_B \alpha_B$, $-R_C \Delta T_B = I_C \alpha_C$. From kinematics, $R_A \alpha_A = R_{B1} \alpha_B$, $R_{B2} \alpha_B = R_C \alpha_C$, from which

$$\alpha_A = \frac{R_{B1} R_C}{R_A R_{B2}} \alpha_C = \frac{(0.2)(0.2)}{(0.1)(0.1)} = 4\alpha_C,$$

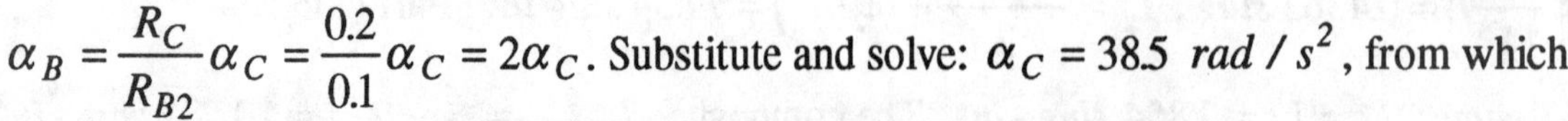

$\alpha_B = \frac{R_C}{R_{B2}} \alpha_C = \frac{0.2}{0.1} \alpha_C = 2\alpha_C$. Substitute and solve: $\alpha_C = 38.5 \ rad/s^2$, from which

$\boxed{\omega_C = \alpha_C t = 76.9 \ rad/s}$ $\boxed{N = \theta\left(\frac{1}{2\pi}\right) = \frac{\alpha_C}{4\pi}\left(2^2\right) = 12.2 \ revolutions}$

Problem 7.128 A 2 *kg* box is subjected to a 40 *N* horizontal force. Neglect friction. (a) If the box remains on the floor, what is its acceleration? (b) Determine the range of values of *c* for which the box will remain on the floor when the force is applied.

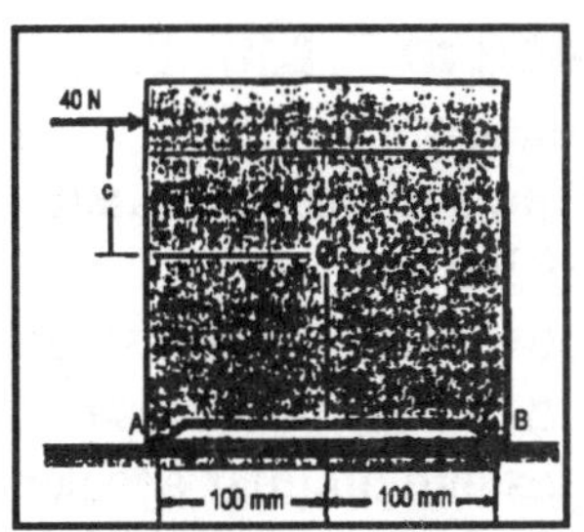

Solution (a) From Newton's second law, $40 = (2)a$, from which $\boxed{a = \frac{40}{2} = 20\ m/s^2}$. (b) The sum of forces: $\sum F_y = A + B - mg = 0$. The sum of the moments about the center of mass:

$\sum M = 0.1B - 0.1A - 40c = 0$. Substitute the value of B from the first equation into the second equation and solve for *c*: $c = \frac{(0.1)mg - (0.2)A}{40}$

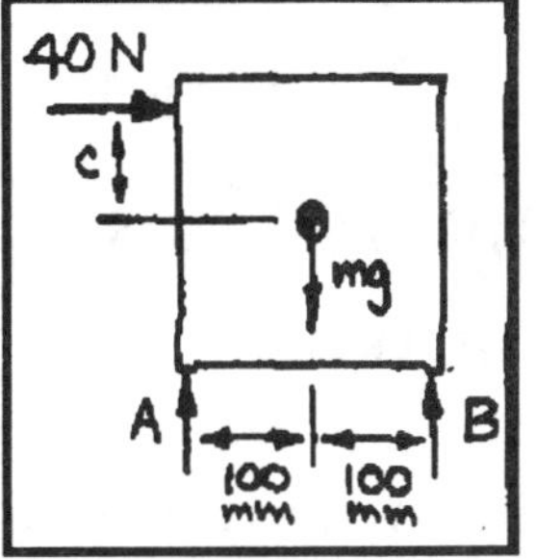

The box leg at A will leave the floor as $A \le 0$, from which $\boxed{c \le \frac{(0.1)(2)(9.81)}{40} \le 0.0495\ m}$ for values of $A \ge 0$.

=================================<>=================================

Problem 7.129 The slender, 2 *slug* bar AB is 3 *ft* long. It is pinned to the cart at A and leans against it at B. (a) If the acceleration of the cart is $a = 20\ ft/s^2$, what normal force is exerted on the bar by the cart at B? (b) What is the largest acceleration *a* for which the bar will remain in contact with the surface at B?

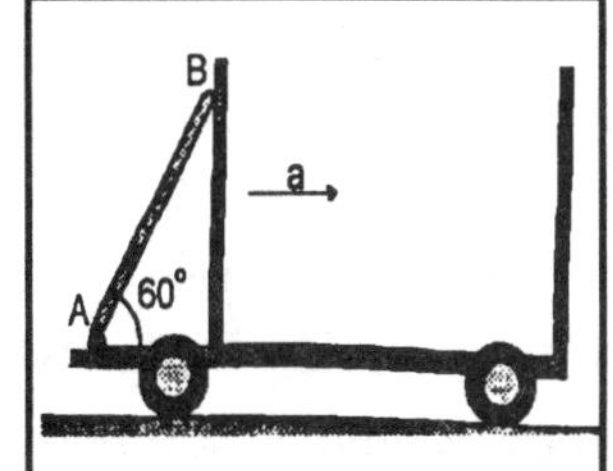

Solution:

Newton's second law applied to the center of mass of the bar yields

$-B + A_x = ma_{Gx}$, $A_y - W = ma_{Gy}$,

$-A_y\left(\frac{L\cos\theta}{2}\right) + (B + A_x)\left(\frac{L\sin\theta}{2}\right) = I_G\alpha$, where a_{Gx}, a_{Gy} are the accelerations of the center of mass. From kinematics, ,

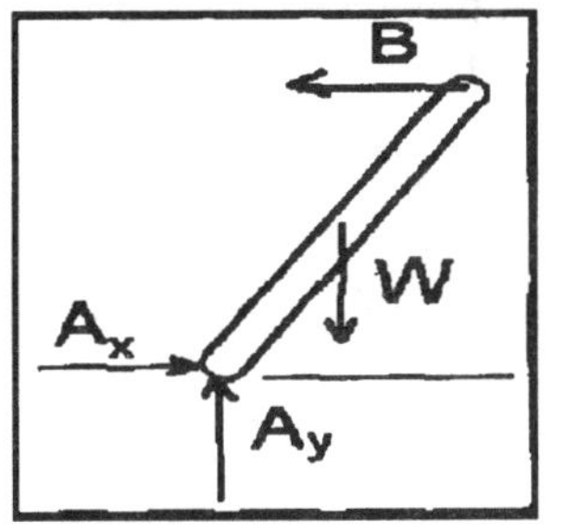

$\vec{a}_G = \vec{a}_A + \vec{\alpha} \times \vec{r}_{G/A} - = \omega_{AB}^2 \vec{r}_{G/A} = 20\vec{i}\ ft/s^2$ where $\alpha = 0, \omega_{AB} = 0$ *so long as the bar is resting on the cart at B and is pinned at A*. Substitute the kinematic relations to obtain three equations in three unknowns:

$-B + A_x = ma$, $A_y - W = 0$, $-A_y\left(\frac{L\cos\theta}{2}\right) + (B + A_x)\left(\frac{L\sin\theta}{2}\right) = 0$.

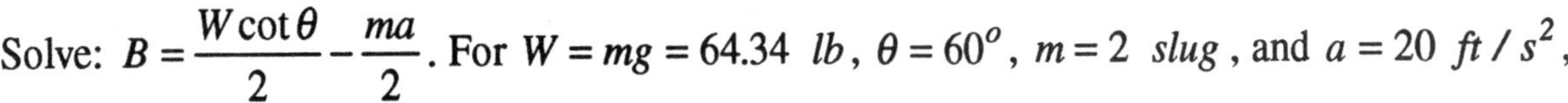

Solve: $B = \frac{W\cot\theta}{2} - \frac{ma}{2}$. For $W = mg = 64.34\ lb$, $\theta = 60^o$, $m = 2\ slug$, and $a = 20\ ft/s^2$, $B = -1.43\ lb$, from which *the bar has moved away from the cart at point B.* (b) The acceleration that produces a zero normal force is $\boxed{a = g\cot\theta = 18.57\ ft/s^2}$.

=================================<>=================================

Problem 7.130

Problem 7.130 To determine a 4.5 *kg* tire's moment of inertia, an engineer lets it roll down an inclined surface. If it takes 3.5 seconds to start from rest and roll three meters down the surface, what is the tire's moment of inertia about its center of mass?

Solution:

Choose a coordinate system with the origin at the center of the disk in its at rest position and the *x* axis parallel to the plane surface. The usual sign conventions for angles and angular acceleration apply. From Newton's second law: $mg\sin\theta - f = ma$, $-Rf = I\alpha$. From kinematics, $a = -R\alpha$. From dynamics, $s = \frac{a}{2}t^2$, from which $a = \frac{2s}{t^2}$. Substitute to obtain two equations in two unknowns:

$mg\sin\theta - f = \frac{2ms}{t^2}$, $Rf = \left(\frac{2ms}{Rt^2}\right)I$. Solve: $f = 9.22$ N (to the left, as shown in the free body diagram), $I = 2.05\ \text{kg-m}^2$

Problem 7.131 Pulley A weighs 4 *lb.*, $I_A = 0.060\ \text{slug-ft}^2$, and $I_B = 0.014\ \text{slug-ft}^2$. If the system is released from rest, what distance will the 16 *lb.* weight fall in one-half second?

Solution:

The strategy is to apply Newton's second law and the equation of angular motion to the free body diagrams. Denote the rightmost weight by $W_R = 16$ lb, the mass by $m_R = 0.4974$ slug, and the leftmost weight by $W_L = 4 + 8 = 12$ lb, and the mass by $m_L = 0.3730$ slug. $R_B = 8$ in. is the radius of pulley B, $I_B = 0.014\ \text{slug-ft}^2$, and $R_A = 12$ in. is the radius of pulley A, and $I_A = 0.060\ \text{slug-ft}^2$. Choose a coordinate system with the *y* axis positive upward.

The 16 lb. weight: (1) $T_1 - W_R = m_R a_{Ry}$.

Pulley B: The center of the pulley is constrained against motion, and the acceleration of the rope is equal (except for direction) on each side of the pulley. (2) $-R_B T_1 + R_B T_2 = I_B \alpha_B$. From kinematics, (3) $a_{Ry} = R_B \alpha_B$.

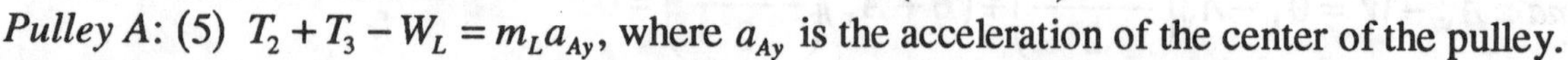

Combine (1), (2) and (3) and reduce: (4) $T_2 = W_L + \left(\frac{I_B}{R_B^2} + m_R\right)a_{Ry}$

Pulley A: (5) $T_2 + T_3 - W_L = m_L a_{Ay}$, where a_{Ay} is the acceleration of the center of the pulley.

(6) $-R_A T_3 + R_A T_2 = I_A \alpha_A$. From the kinematics of pulley A, the acceleration of the left side of the pulley is zero, so that the acceleration of the right side relative to the left side is

$$\vec{a}_{right} = -a_{Ry}\vec{j} = \vec{a}_{left} + \vec{\mathbf{a}}_A \times (2R_A\vec{i}) = \begin{bmatrix} \vec{i} & \vec{j} & \vec{k} \\ 0 & 0 & \alpha_A \\ 2R_A & 0 & 0 \end{bmatrix} = 0 + 2R_A\alpha_A\vec{j}, \text{ from which}$$

(7) $a_{Ry} = -2R_A\alpha_A$, where the change in direction of the acceleration of the 16 *lb.* weight across pulley B is taken into account. Similarly, the acceleration of the right side relative to the acceleration of the center

Solution continued on next page

of the pulley is $\vec{a}_{Aright} = -a_{Ry}\vec{j} = \vec{a}_A + \mathbf{a}_A \times (R_A\vec{i}) = \vec{a}_A + R_A\alpha_A\vec{j}$, from which (8) $a_{Ay} = -\frac{a_{Ry}}{2}$. Combine (5), (6), (7) and (8) and reduce to obtain (9) $T_2 = \frac{W_A}{2} - \left(\frac{I_A}{4R_A^2} + \frac{m_A}{4}\right)a_y$.

The total system: Equate (4) and (9) (the two expressions for T_2) and solve:

$$a_{Ry} = \frac{\left(\frac{W_L}{2} - W_R\right)}{\left(+\frac{I_B}{R_B^2} + m_R + \frac{I_A}{4R_A^2} + \frac{m_L}{4}\right)}.$$

Substitute numerical values: $a_{Ry} = -15.7 \text{ ft/s}^2$. The distance that the 16 *lb.* weight will fall in one-half second is $s = \frac{a_{Ry}}{2}t^2 = \frac{-15.7}{8} = -1.96 \text{ ft}$

==================================<>==================================

Problem 7.132 Model the excavator's arm ABC as a single rigid body. Its mass is 1200 *kg,* and the moment of inertia *about its center of mass* is $I = 3600 \text{ kg-m}^2$. If point A is stationary, the angular velocity of the arm is zero, and the angular acceleration of the arm is 1.0 *rad/s²* counterclockwise, what force does the vertical hydraulic cylinder exert on the arm at B?

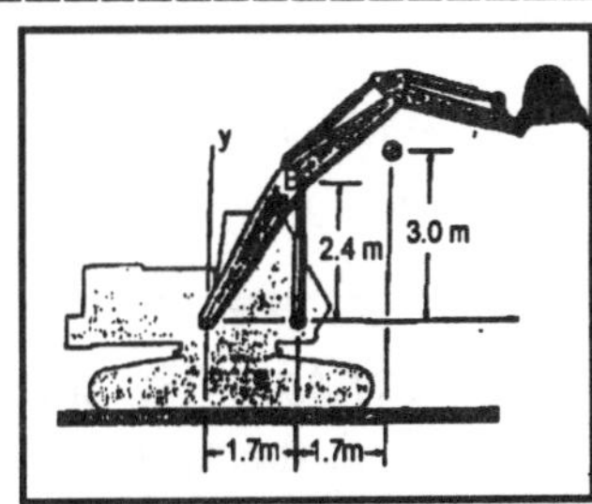

Solution:

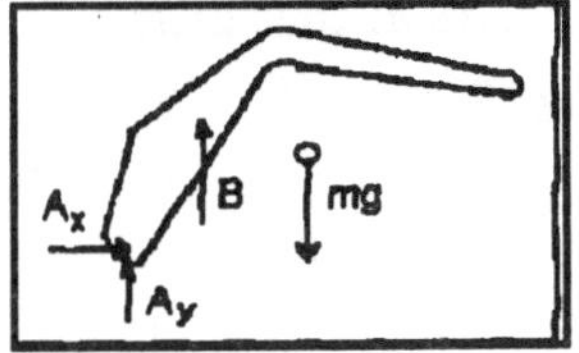

The distance from A to the center of mass is $d = \sqrt{(3.4)^2 + (3)^2} = 4.53 \text{ m}$.
The moment of inertia about A is $I_A = I + d^2m = 28{,}270 \text{ kg-m}^2$. From the equation of angular motion: $1.7B - 3.4mg = I_A\alpha$.
Substitute $\alpha = 1.0 \text{ rad/s}^2$, to obtain $B = 40{,}170 \text{ N}$.

==================================<>==================================

Problem 7.133 In Problem 7.132, if the angular acceleration of arm ABC is 1.0 *rad/s²* counterclockwise and its angular velocity is 2.0 *rad/s,* what are the components of the force exerted on the arm at A?

Solution:

The acceleration of the center of mass is

$$\vec{a}_G = \mathbf{a} \times \vec{r}_{G/A} - \omega^2\vec{r}_{G/B} = \begin{bmatrix} \vec{i} & \vec{j} & \vec{k} \\ 0 & 0 & \alpha \\ 3.4 & 3 & 0 \end{bmatrix} - \omega^2(3.4\vec{i} + 3\vec{j}) = -16.6\vec{i} - 8.6\vec{j} \text{ m/s}^2.$$

From Newton's second law: $A_x = ma_{Gx} = -19{,}900 \text{ N}$, $A_y + B - mg = ma_{Gy}$. From the solution to Problem 7.132, $B = 40{,}170 \text{ N}$, from which $A_y = -38{,}720 \text{ N}$

==================================<>==================================

==================================<>==================================

Problem 7.134 To decrease the elevation of the stationary 200 *kg* ladder, the gears that raised it are disengaged and a fraction of a second later a second set of gears that lower it are engaged. At the instant the gears that raised it are disengaged, what is the ladder's angular acceleration and what are the components of force exerted on the ladder by its support at O? The moment of inertia of the ladder about O is $I_O = 14{,}000 \text{ kg-m}^2$, and the coordinates of its center of mass at the instant the gears are disengaged are $\bar{x} = 3 \text{ m}, \bar{y} = 4 \text{ m}$.

Solution:

The moment about O, $-mg\bar{x} = I_O\alpha$, from which

$\alpha = -\dfrac{(200)(9.81)(3)}{14{,}000} = -0.420 \text{ rad/s}^2$. The acceleration of the center of mass

is $\vec{a}_G = \mathbf{\vec{a}} \times \vec{r}_{G/O} - \omega^2\vec{r}_{G/O} = \begin{bmatrix} \vec{i} & \vec{j} & \vec{k} \\ 0 & 0 & \alpha \\ 3 & 4 & 0 \end{bmatrix} = -4\alpha\vec{i} + 3\alpha\vec{j}$

$\vec{a}_G = 1.68\vec{i} - 1.26\vec{j} \ (\text{m/s}^2)$. From Newton's second law: $F_x = ma_{Gx} = 336.3 \text{ N}$, $F_y - mg = ma_{Gy}$, from which $F_y = 1709.7 \text{ N}$

==================================<>==================================

Problem 7.135 The slender bars each weigh 4 *lb.* and are 10 *in.* long. The homogenous plate weighs 10 *lb.* If the system is released from rest in the position shown, what is the angular acceleration of the bars at that instant?

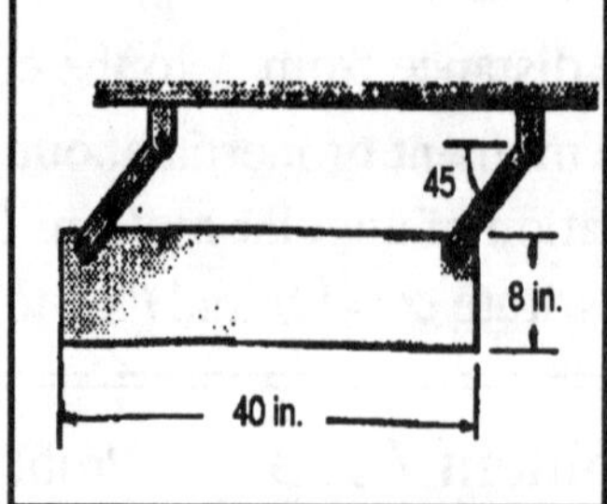

Solution:

From geometry, the system is a parallelogram, so that the plate translates without rotating, so that the acceleration of every point on the plate is the same.

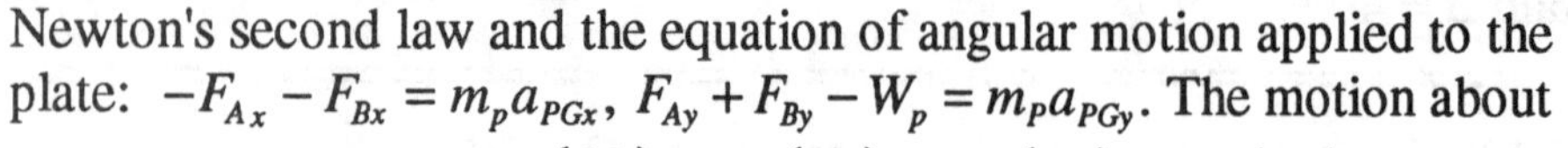

Newton's second law and the equation of angular motion applied to the plate: $-F_{Ax} - F_{Bx} = m_p a_{PGx}$, $F_{Ay} + F_{By} - W_p = m_p a_{PGy}$. The motion about the center of mass: $-F_{Ay}\left(\dfrac{20}{12}\right) + F_{Ax}\left(\dfrac{4}{12}\right) + F_{Bx}\left(\dfrac{4}{12}\right) + F_{By}\left(\dfrac{20}{12}\right) = I_P\alpha = 0$.

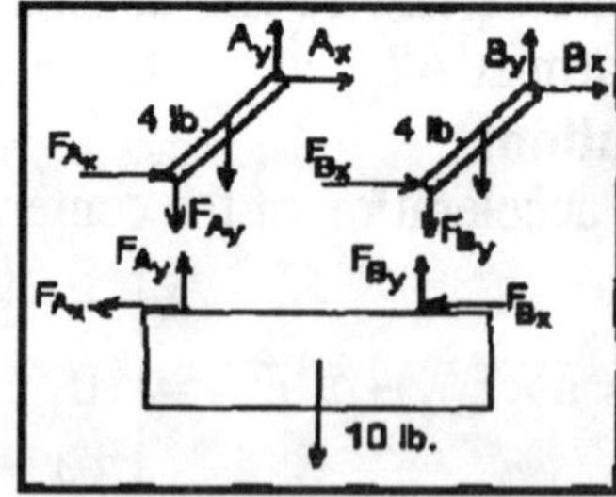

Newton's second law for the bars: $-F_{Ay} + A_y - W_B = m_B a_{BGy}$,

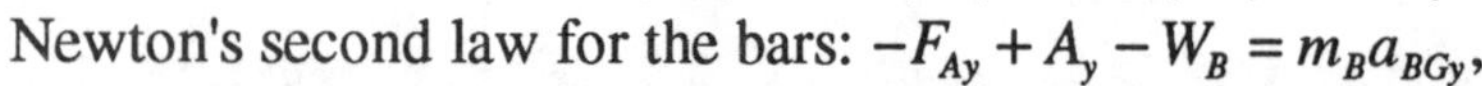

$-F_{Ax} + A_x = m_B a_{BGx}$. $-F_{By} + B_y - W_B = m_B a_{BGy}$. $-F_{Bx} + B_x = m_B a_{BGx}$. The angular acceleration about the center of mass:

$F_{Ax}\left(\dfrac{5}{12}\right)\cos\theta + F_{Ay}\left(\dfrac{5}{12}\right)\sin\theta - A_x\left(\dfrac{5}{12}\right)\cos\theta + A_y\left(\dfrac{5}{12}\right)\sin\theta = I_B\alpha$,

$F_{Bx}\left(\dfrac{5}{12}\right)\cos\theta + F_{By}\left(\dfrac{5}{12}\right)\sin\theta - B_x\left(\dfrac{5}{12}\right)\cos\theta + B_y\left(\dfrac{5}{12}\right)\sin\theta = I_B\alpha$. From kinematics: the acceleration of the center of mass of the bars in terms of the acceleration at point A is

$$\vec{a}_{BG} = \mathbf{\vec{a}} \times \vec{r}_{G/A} - \omega^2\vec{r}_{G/A} = \begin{bmatrix} \vec{i} & \vec{j} & \vec{k} \\ 0 & 0 & \alpha \\ -\dfrac{5}{12}\cos\theta & -\dfrac{5}{12}\sin\theta & 0 \end{bmatrix} = \frac{5}{12}\sin\theta\alpha\,\vec{i} - \frac{5}{12}\cos\alpha\,\vec{j} \ (\text{ft/s}^2).$$ From which

$a_{BGx} = \left(\dfrac{5}{12}\right)\sin\theta\,\alpha$, $a_{BGy} = -\left(\dfrac{5}{12}\right)\cos\theta\,\alpha$, since $\omega = 0$ upon release.

Solution continued on next page

The acceleration of the plate:

$$\vec{a}_P = \vec{\mathbf{a}} \times \vec{r}_{P/A} - \omega^2 \vec{r}_{P/A} = \begin{bmatrix} \vec{i} & \vec{j} & \vec{k} \\ 0 & 0 & \alpha \\ -\frac{10}{12}\cos\theta & -\frac{10}{12}\sin\theta & 0 \end{bmatrix} = \frac{10}{12}\sin\theta\alpha\,\vec{i} - \frac{10}{12}\cos\theta\alpha\,\vec{j}\ \left(\text{ft}/\text{s}^2\right).$$ From which

$a_{Px} = \left(\frac{10}{12}\right)\sin\theta\,\alpha$, $a_{Py} = -\left(\frac{10}{12}\right)\cos\theta\,\alpha$.

Substitute to obtain the nine equations in nine unknowns:

(1) $-F_{Ax} - F_{Bx} = \left(\frac{10}{12}\right)m_p \sin\theta\,\alpha$, (2) $F_{Ay} + F_{By} - W_p = -\left(\frac{10}{12}\right)m_P \cos\theta\,\alpha$,

(3) $-20F_{Ay} + 4F_{Ax} + 20F_{By} + 4F_{Bx} = 0$, (4) $-F_{Ay} + A_y - W_B = -\left(\frac{5}{12}\right)m_B \cos\theta\alpha$,

(5) $F_{Ax} + A_x = \left(\frac{5}{12}\right)m_B \sin\theta\alpha$, (6) $F_{Ax}\sin\theta + F_{Ay}\cos\theta - A_x\sin\theta + A_y\cos\theta = \left(\frac{12}{5}\right)I_B\alpha$,

(7) $F_{Bx} + B_x = \left(\frac{5}{12}\right)m_B \sin\theta\,\alpha$, (8) $-F_{By} + B_y - W_B = -\left(\frac{5}{12}\right)m_B \cos\theta\,\alpha$,

(9) $F_{Bx}\cos\theta + F_{By}\sin\theta - B_x\cos\theta + B_y\sin\theta = \left(\frac{12}{5}\right)I_B\alpha$. The number of equations and number of unknowns can be reduced by combining equations, but here the choice is to solve the system by iteration using **TK Solver Plus.** The results: $F_{Ax} = -2.21$ lb, $F_{Ay} = 1.68$ lb, $F_{Bx} = -3.32$ lb, $A_x = 3.32$ lb, $A_y = 4.58$ lb, $B_x = 4.42$ lb, $B_y = 5.68$ lb. $\alpha = 30.17\ \text{rad}/\text{s}^2$.

=====================================<>=====================================

Problem 7.136 A slender bar of mass m is released from rest in the position shown. The static and kinetic friction coefficients of friction at the floor and the wall have the same value μ. If the bar slips, what is its angular acceleration at the instant of release?

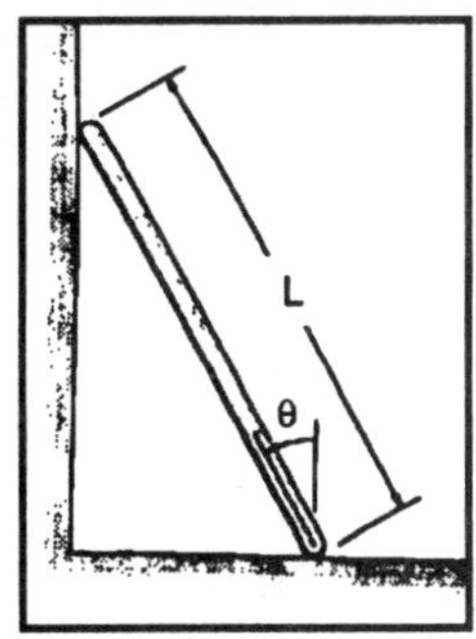

Solution Choose a coordinate system with the origin at the intersection of wall and floor, with the x axis parallel to the floor. Denote the points of contact at wall and floor by P and N respectively, and the center of mass of the bar by G. The vector locations are $\vec{r}_N = \vec{i}\,\mathrm{L}\sin\theta$, $\vec{r}_P = \vec{j}\,\mathrm{L}\cos\theta$, $\vec{r}_G = \frac{\mathrm{L}}{2}\left(\vec{i}\sin\theta + \vec{j}\cos\theta\right)$. From Newton's second law: $P - \mu N = ma_{Gx}$, $N + \mu P - mg = ma_{Gy}$, where a_{Gx}, a_{Gy} are the accelerations of the center of mass. The moment about the center of mass is $\vec{M}_G = \vec{r}_{P/G} \times \left(\mathrm{P}\vec{i} + \mu\mathrm{P}\vec{j}\right) + \vec{r}_{N/G} \times \left(\mathrm{N}\vec{j} - \mu\mathrm{N}\vec{i}\right)$:

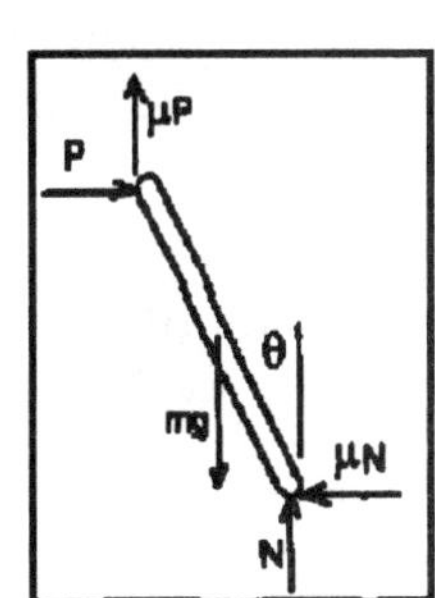

$$\vec{M}_G = \frac{\mathrm{PL}}{2}\begin{bmatrix} \vec{i} & \vec{j} & \vec{k} \\ -\sin\theta & \cos\theta & 0 \\ 1 & \mu & 0 \end{bmatrix} + \frac{\mathrm{NL}}{2}\begin{bmatrix} \vec{i} & \vec{j} & \vec{k} \\ \sin\theta & -\cos\theta & 0 \\ -\mu & 1 & 0 \end{bmatrix}.$$

$\vec{M}_G = -\left(\frac{\mathrm{PL}}{2}\right)(\cos\theta + \mu\sin\theta)\vec{k} + \left(\frac{\mathrm{NL}}{2}\right)(\sin\theta - \mu\cos\theta)\vec{k}$ From the equation of angular motion, $-\left(\frac{\mathrm{PL}}{2}\right)(\cos\theta + \mu\sin\theta) + \left(\frac{\mathrm{NL}}{2}\right)(\sin\theta - \mu\cos\theta) = \mathrm{I}_\mathrm{B}\alpha$

Solution continued on next page

From kinematics: Assume that at the instant of slip the angular velocity $\omega = 0$. The acceleration of the center of mass in terms of the acceleration at point N is

$$\vec{a}_G = \vec{a}_N + \vec{\mathbf{a}} \times \vec{r}_{G/N} - \omega^2 \vec{r}_{G/.N} = \mathrm{a}_N \vec{i} + \begin{bmatrix} \vec{i} & \vec{j} & \vec{k} \\ 0 & 0 & \alpha \\ -\frac{\mathrm{L}\sin\theta}{2} & \frac{\mathrm{L}\cos\theta}{2} & 0 \end{bmatrix} \vec{a}_G = \left(\mathrm{a}_N - \frac{\alpha\mathrm{L}\cos\theta}{2}\right)\vec{i} + \left(-\frac{\alpha\mathrm{L}\sin\theta}{2}\right)\vec{j},$$

from which $a_{Gy} = -\frac{L\sin\theta}{2}\alpha$. The acceleration of the center of mass in terms of the acceleration at point P is $\vec{a}_G = \vec{a}_P + \vec{\mathbf{a}} \times \vec{r}_{G/P}$.

$$\vec{a}_G = \vec{a}_P + \vec{\mathbf{a}} \times \vec{r}_{G/P} - \omega^2 \vec{r}_{G/P} = \mathrm{a}_P \vec{j} + \begin{bmatrix} \vec{i} & \vec{j} & \vec{k} \\ 0 & 0 & \alpha \\ \frac{\mathrm{L}\sin\theta}{2} & -\frac{\mathrm{L}\cos\theta}{2} & 0 \end{bmatrix}, \vec{a}_G = \left(\frac{\alpha\mathrm{L}\cos\theta}{2}\right)\vec{i} + \left(\mathrm{a}_P + \frac{\alpha\mathrm{L}\sin\theta}{2}\right)\vec{j}, \text{ from}$$

which $a_{Gx} = \frac{L\cos\theta}{2}\alpha$. Substitute to obtain the three equations in three unknowns,

(1) $P - \mu N = \frac{mL\cos\theta}{2}\alpha$, (2) $\mu P + N = -\frac{mL\sin\theta}{2}\alpha + mg$.

(3) $-\frac{PL}{2}(\cos\theta + \mu\sin\theta) + \frac{NL}{2}(\sin\theta - \mu\cos\theta) = I_B\alpha$. Solve the first two equations for *P* and *N*:

$P = \frac{mL}{2(1+\mu^2)}(\cos\theta - \mu\sin\theta)\alpha + \frac{\mu mg}{(1+\mu^2)}$. $N = -\frac{mL}{2(1+\mu^2)}(\sin\theta + \mu\cos\theta)\alpha + \frac{mg}{(1+\mu^2)}$. Substitute the first two equations into the third, and reduce to obtain

$\alpha\left[I_B + \frac{mL^2}{4}\left(\frac{1-\mu^2}{1+\mu^2}\right)\right] = \frac{mgL}{2}\left(\frac{1-\mu^2}{1+\mu^2}\right)\sin\theta - mgL\left(\frac{\mu}{1+\mu^2}\right)\cos\theta$. Substitute $I_B = \left(\frac{1}{12}\right)mL^2$, reduce, and

solve: $\alpha = \frac{\left(3(1-\mu^2) - 6\mu\cos\theta\right)g}{(2-\mu^2)L}$

==================================<>==================================

Problem 7.137 Each of the go-cart's front wheels weighs 5 *lb.* and has a moment of inertia of 0.01 *slug-ft*^{2}. The two rear wheels and rear axle form a single rigid body weighing 40 *lb.* and having a moment of inertia of 0.1 *slug-ft*^{2}. The total weight of the go-cart and rider is 240 *lb.* (The location of the center of mass of the go-cart and driver *not including* the front wheels or the rear wheels and axle is shown.) If the engine exerts a torque of 12 *ft-lb.* on the rear axle, what is the go-cart's acceleration?

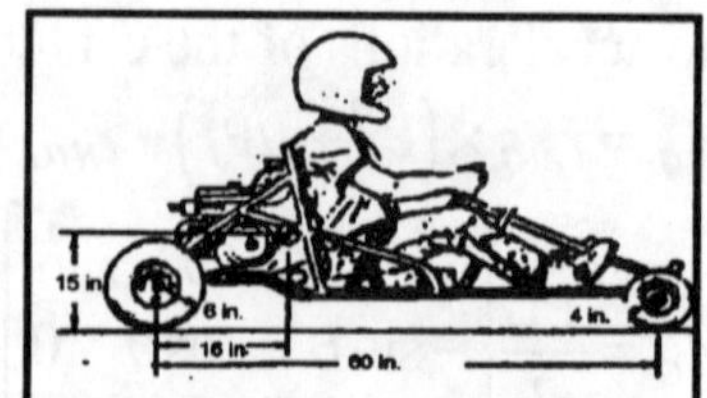

Solution:

Apply Newton's second law and the equation of angular motion to the wheel assemblies:

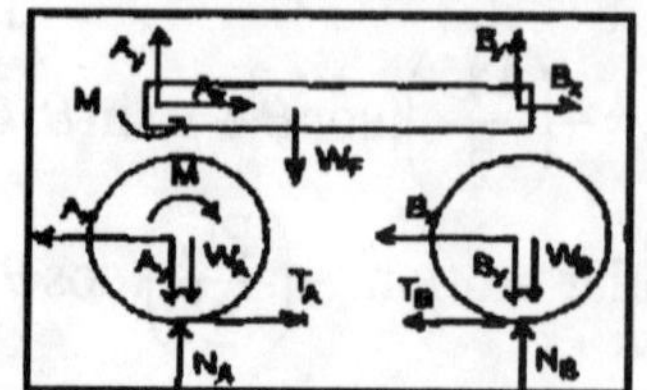

The rear wheel assembly: (1) $T_A - A_x = m_A a_x$, (2) $-A_y - m_A g + N_A = 0$, $-M + R_A T_A = I_A \alpha_A$. From kinematics, $a_x = -R_A \alpha_A$, from which

(3) $-M + R_A T_A = -\frac{I_A}{R_A} a_x$.

Solution continued on next page

For the front wheel assembly:

(4) $-B_x - T_B = m_B a_x$,

(5) $-B_y + m_B g + N_B = 0$. $-R_B T_B = I_B \alpha_B$. From kinematics, $a_x = -R_B \alpha_B$, from which

(6) $-R_B T_B = -\dfrac{I_B}{R_B} a_x$.

For the frame: Choose an origin at the center of mass with the x axis parallel to the road. Since the frame does not rotate, the acceleration of the wheel connections is the acceleration of the center of mass. The location of the rear wheel assembly is $\vec{r}_{A/G} = -\left(\dfrac{16}{12}\right)\vec{i} - \left(\dfrac{15-6}{12}\right)\vec{j} = -1.33\vec{i} - 0.75\vec{j}$ (ft). The location of the front wheel is $\vec{r}_{B/G} = (\dfrac{60-4}{12})\vec{i} - \left(\dfrac{15-4}{12}\right)\vec{j} = 4.67\vec{i} - 9.17\vec{j}$ (ft). The moment about the center of mass is $\vec{M}_G = \vec{M} + \vec{r}_{A/G} \times \vec{A} + \vec{r}_{B/G} \times \vec{B} = \vec{M} + \begin{bmatrix} \vec{i} & \vec{j} & \vec{k} \\ -1.33 & -0.75 & 0 \\ \mathrm{A_x} & \mathrm{A_y} & 0 \end{bmatrix} + \begin{bmatrix} \vec{i} & \vec{j} & \vec{k} \\ 4.67 & -0.917 & 0 \\ \mathrm{B_x} & \mathrm{B_y} & 0 \end{bmatrix} = 0$, from which

(7) $M_G = M - 1.33A_y + 0.75A_x + 4.67B_y + 0.917B_x = 0$. Apply Newton's second law to the frame:

(8) $A_x + B_x = m_f a_x$,

(9) $A_y + B_y - m_f g = 0$. The result is nine equations in nine unknowns. Substitute numerical values: $m_A = \left(\dfrac{40}{32.17}\right) = 1.24$ slug, $m_B = \dfrac{10}{32.17} = 0.311$ slug., $m_f = \dfrac{240-40-10}{32.17} = 5.91$ slug. $M = 12$ ft-lb, $R_A = 0.5$ ft, $R_B = 0.33$ ft, $I_A = 0.1$ slug-ft^2, $I_B = 0.01$ slug-ft^2. Solve by iteration: $a_x = 3.0$ ft/s^2, $T_A = 22.8$ lb, $T_B = 0.27$ lb, $A_x = 19.0$ lb, $A_y = 152$ lb, $B_x = -1.2$ lb, $B_y = 38$ lb, $N_A = 192$ lb, $N_B = 48$ lb. [*Check:* An alternate approach: Denote I_R to be the moment of inertia of the rear wheel and axle assembly, *about the point of contact with the road.* $I_R = 0.1 + \left(\dfrac{40}{32.17}\right)\left(\dfrac{1}{2}\right) = 0.0411$ slug-ft^2. Denote I_F to be the moment of inertia of the front wheels *about the point of contact with the road.* $I_F = 0.01 + \left(\dfrac{5}{32.17}\right)\left(\dfrac{1}{3}\right) = 0.0273$ slug-ft^2. Apply the equation of angular motion to the rear assembly,

(1) $I_R \alpha = M + A_x\left(\dfrac{1}{2}\right)$, where $M = -12$ ft-lb is the torque applied by the engine, and A_x is the (unknown) drive thrust at the center of mass of the rear wheels(see free body diagram.). From kinematics, (2) $a = \left(\dfrac{1}{3}\right)\alpha$, where a is the acceleration of the center of mass of the rear wheels.

Combining: for the cart and rider (without the wheels) (3) $A_x - 2B_x = \left(\dfrac{190}{32.17}\right)a$ where B_x is the force on the front wheel centers (see free body diagram.). For one front wheel, (4) $I_F \alpha = -B_x\left(\dfrac{1}{3}\right)$. Combine these four equations: $\left(\dfrac{190}{32.17} + \dfrac{I_R}{(0.5)^2} + \dfrac{2I_F}{(0.3333)^2}\right)a = -\left(\dfrac{M}{0.5}\right)$. from which $a = 2.98$ ft/s^2. *check*].

==============================<>==============================

Problem 7.138 Bar AB rotates with a constant angular velocity of 10 *rad/s* in the counterclockwise direction. The masses of the slender bars BC and CDE are 2 *kg* and 3.6 *kg* respectively. The *y* axis points upward. Determine the components of the forces exerted on bar BC by the pins at B and C at the instant shown.

Solution The velocity of point B is

$$\vec{v}_B = \vec{\mathbf{w}}_{AB} \times \vec{r}_B = \begin{bmatrix} \vec{i} & \vec{j} & \vec{k} \\ 0 & 0 & 10 \\ 0 & 0.4 & 0 \end{bmatrix} = -0.4(10)\vec{i} = -4\vec{i} \ \ (\mathrm{m/s}).$$

The velocity of point C is

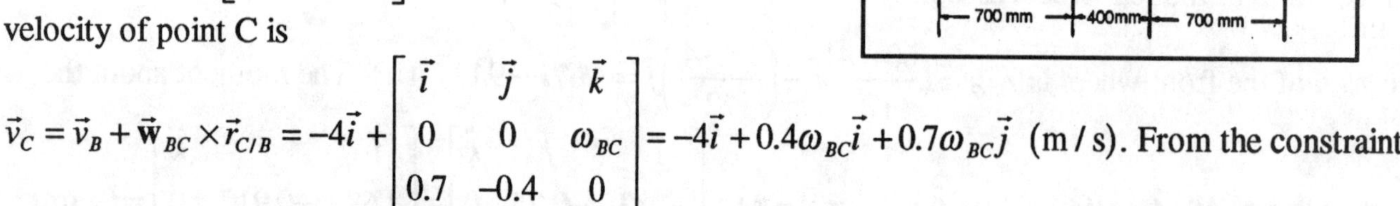

$$\vec{v}_C = \vec{v}_B + \vec{\mathbf{w}}_{BC} \times \vec{r}_{C/B} = -4\vec{i} + \begin{bmatrix} \vec{i} & \vec{j} & \vec{k} \\ 0 & 0 & \omega_{BC} \\ 0.7 & -0.4 & 0 \end{bmatrix} = -4\vec{i} + 0.4\omega_{BC}\vec{i} + 0.7\omega_{BC}\vec{j} \ \ (\mathrm{m/s}).$$

From the constraint on the motion at point C, $\vec{v}_C = \mathrm{v}_C\vec{j}$. Equate components: $0 = -4 + 0.4\omega_{BC}$, $v_C = 0.7\omega_{BC}$, from which $\omega_{BC} = 10 \ \mathrm{rad/s}$, $v_C = 7 \ \ \mathrm{m/s}$. The velocity at C in terms of the angular velocity ω_{CDE},

$$\vec{v}_C = \vec{v}_D + \vec{\mathbf{w}}_{CDE} \times \vec{r}_{C/D} = 0 + \begin{bmatrix} \vec{i} & \vec{j} & \vec{k} \\ 0 & 0 & \omega_{CDE} \\ -0.4 & 0 & 0 \end{bmatrix} = -0.4\omega_{CDE}\vec{j},$$

from which $\omega_{CDE} = -\dfrac{7}{0.4} = -17.5 \ \ \mathrm{rad/s}$.

The acceleration of point B is $\vec{a}_B = -\omega_{AB}^2\vec{r}_B = -(10^2)(0.4)\vec{j} = -40\vec{j} \ \ (\mathrm{m/s^2})$.

The acceleration at point C is $\vec{a}_C = \vec{a}_B + \vec{\mathbf{a}}_{BC} \times \vec{r}_{C/B} - \omega_{BC}^2\vec{r}_{C/B}$.

$$\vec{a}_C = -40\vec{j} + \begin{bmatrix} \vec{i} & \vec{j} & \vec{k} \\ 0 & 0 & \alpha_{BC} \\ 0.7 & -0.4 & 0 \end{bmatrix} - \omega_{BC}^2\left(0.7\vec{i} - 0.4\vec{j}\right) \ (\mathrm{m/s^2}).$$

$$\vec{a}_C = +\left(0.4\alpha_{BC} - 0.7\omega_{BC}^2\right)\vec{i} + \left(-40 + 0.7\alpha_{BC} + 0.4\omega_{BC}^2\right)\vec{j} \ (\mathrm{m/s^2}).$$

The acceleration in terms of the acceleration at D is

$$\vec{a}_C = \begin{bmatrix} \vec{i} & \vec{j} & \vec{k} \\ 0 & 0 & \alpha_{CDE} \\ -0.4 & 0 & 0 \end{bmatrix} - \omega_{CDE}^2\left(-0.4\vec{i}\right) = -0.4\alpha_{CDE}\vec{j} + 0.4\omega_{CDE}^2\vec{i}.$$

Equate components and solve:

$\alpha_{BC} = 481.25 \ \ \mathrm{rad/s^2}$, $\alpha_{CDE} = -842.19 \ \ \mathrm{rad/s^2}$. The acceleration of the center of mass of BC is

$$\vec{a}_G = -40\vec{j} + \begin{bmatrix} \vec{i} & \vec{j} & \vec{k} \\ 0 & 0 & \alpha_{BC} \\ 0.35 & -0.2 & 0 \end{bmatrix} - \omega_{BC}^2\left(0.35\vec{i} - 0.2\vec{j}\right),$$

from which $\vec{a}_G = 61.25\vec{i} + 148.44\vec{j} \ \ (\mathrm{m/s^2})$

The equations of motion:

$B_x + C_x = m_{BC}a_{Gx}$,

$B_y + C_y - m_{BC}g = m_{BC}a_{Gy}$, where the accelerations a_{Gx}, a_{Gy} are known. The moment equation,

$0.35C_y + 0.2C_x - 0.2B_x - 0.35B_y = I_{BC}\alpha_{BC}$, where α_{BC}, is known, and

$I_{BC} = \left(\dfrac{1}{12}\right)m_{BC}L_{BC}^2 = 0.1083 \ \ \mathrm{kg\text{-}m^2}$, $\ 0.4C_y - 0.15m_{CE}g = I_D\alpha_{CE}$, where

Solution continued on next page

$I_D = \left(\frac{1}{12}\right) m_{CE} L_{CE}^2 + (0.15)^2 m_{CE} = 0.444 \text{ kg-m}^2$, is the moment of inertia about the pivot point D, and 0.15 m is the distance between the point D and the center of mass of bar CDE. Solve these four equations in four unknowns by iteration:
$B_x = -1958.8$ N, $B_y = 1238.1$ N, $C_x = 2081.3$ N, $C_y = -921.6$ N.

==================================<>==================================

Problem 7.139 At the instant shown, the arms of the robotic manipulator have the constant counterclockwise angular velocities $\omega_{AB} = -0.5$ rad/s, $\omega_{BC} = 2$ rad/s, and $\omega_{CD} = 4$ rad/s. The mass of arm CD is 10 kg, and the center of mass is at its midpoint. At this instant, what force and couple are exerted on arm CD at C?

Solution:

The relative vector locations of B, C, and D are

$\vec{r}_{B/A} = 0.3(\vec{i}\cos 30^\circ + \vec{j}\sin 30^\circ) = 0.2598\vec{i} + 0.150\vec{j}$ (m),

$\vec{r}_{C/B} = 0.25(\vec{i}\cos 20^\circ - \vec{j}\sin 20^\circ) = 0.2349\vec{i} - 0.08551\vec{j}$ (m),

$\vec{r}_{D/C} = 0.25\vec{i}$ (m).

The acceleration of point B is

$\vec{a}_B = -\omega_{AB}^2 \vec{r}_{B/A} = -(0.5^2)(0.3\cos 30^\circ \vec{i} + 0.3\sin 30^\circ \vec{j})$,

$\vec{a}_B = -0.065\vec{i} - 0.0375\vec{j}$ (m/s^2). The acceleration at point C is

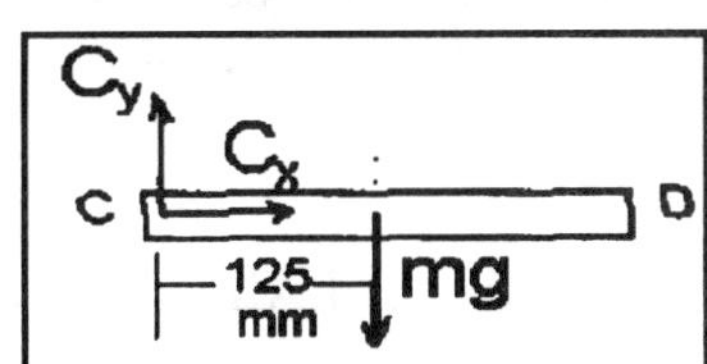

$\vec{a}_C = \vec{a}_B - \omega_{BC}^2 \vec{r}_{C/B} = \vec{a}_B - \omega_{BC}^2(0.2349\vec{i} - 0.08551\vec{j})$.

$\vec{a}_C = -1.005\vec{i} + 0.3045\vec{j}$ (m/s^2).

The acceleration of the center of mass of CD is $\vec{a}_G = \vec{a}_C - \omega_{CD}^2(0.125\vec{i})$ (m/s^2), from which

$\vec{a}_G = -3.005\vec{i} + 0.3045\vec{j}$ (m/s^2).

For the arm CD the three equations of motion in three unknowns are $C_y - m_{CD} g = m_{CD} a_{Gy}$, $C_x = m_{CD} a_{Gx}$, $M - 0.125 C_y = 0$, which have the direct solution:

$C_y = 101.15$ N, $C_x = -30.05$ N. $M = 12.64$ N-m, where the negative sign means a direction opposite to that shown in the free body diagram.

==================================<>==================================

==================================<>==================================

Problem 7.140 Each bar is 1 *m* in length and has a mass of *4 kg*. The inclined surface is smooth. If the system is released from rest in the position shown, what are the angular accelerations of the bars at that instant?

Solution For convenience, denote $\theta = 45^o$, $\beta = 30^o$, and $L = 1$ m. The acceleration of point A is

$$\vec{a}_A = \mathbf{a}_{OA} \times \vec{r}_{A/O} = \begin{bmatrix} \vec{i} & \vec{j} & \vec{k} \\ 0 & 0 & \alpha_{OA} \\ \mathrm{L}\cos\theta & \mathrm{L}\sin\theta & 0 \end{bmatrix}.$$

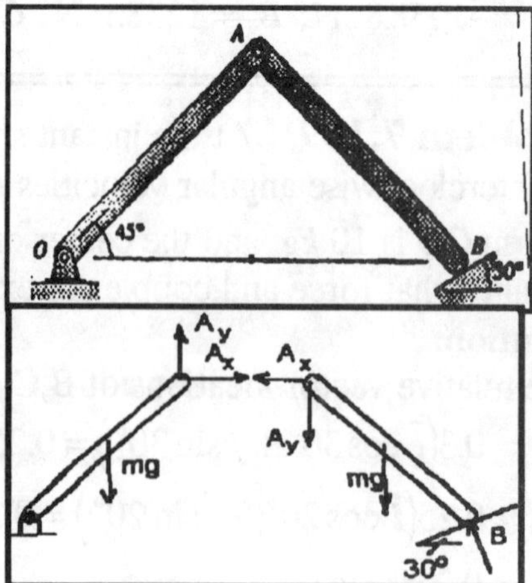

$\vec{a}_A = \alpha_{OA}\left(-\vec{i}\,\mathrm{L}\sin\theta + \vec{j}\mathrm{L}\cos\theta\right)$ $(\mathrm{m/s^2})$. The acceleration of A is also given by $\vec{a}_A = \vec{a}_B + \mathbf{a}_{AB} \times \vec{r}_{A/B}$.

$$\vec{a}_A = \vec{a}_B + \begin{bmatrix} \vec{i} & \vec{j} & \vec{k} \\ 0 & 0 & \alpha_{AB} \\ -\mathrm{L}\cos\theta & \mathrm{L}\sin\theta & 0 \end{bmatrix}.$$

$\vec{a}_A = \vec{a}_B - \vec{i}\,\alpha_{AB}\mathrm{L}\sin\theta - \vec{j}\,\alpha_{AB}\mathrm{L}\cos\theta$ $(\mathrm{m/s^2})$. From the constraint on the motion at B, Equate the expressions for the acceleration of A to obtain the two equations:

(1) $-\alpha_{OA}\mathrm{L}\sin\theta = a_B\cos\beta - \alpha_{AB}L\sin\theta$, and (2) $\alpha_{OA}L\cos\theta = a_B\sin\beta - \alpha_{AB}L\cos\theta$. The acceleration of

the center of mass of AB is $\vec{a}_{GAB} = \vec{a}_A + \mathbf{a}_{AB} \times \vec{r}_{GAB/A} = \vec{a}_A + \begin{bmatrix} \vec{i} & \vec{j} & \vec{k} \\ 0 & 0 & \alpha_{AB} \\ \frac{\mathrm{L}\cos\theta}{2} & -\frac{\mathrm{L}\sin\theta}{2} & 0 \end{bmatrix}$,

$\vec{a}_{GAB} = \vec{a}_A + \dfrac{\mathrm{L}\alpha_{AB}}{2}\sin\theta\vec{i} + \dfrac{\alpha_{AB}\mathrm{L}}{2}\cos\theta\vec{j}$ $(\mathrm{m/s^2})$, from which

(3) $a_{GABx} = -\alpha_{OA}L\sin\theta + \dfrac{L\alpha_{AB}}{2}\sin\theta$ $(\mathrm{m/s^2})$, (4) $a_{GABy} = \alpha_{OA}\mathrm{L}\cos\theta + \dfrac{\mathrm{L}\alpha_{AB}}{2}\cos\theta$.

The equations of motion for the bars: for the pin supported left bar:

(5) $A_yL\cos\theta - A_xL\sin\theta - mg\left(\dfrac{L}{2}\right)\cos\theta = I_{OA}\alpha_{OA}$, where $I_{OA} = \left(\dfrac{mL^2}{3}\right) = \dfrac{4}{3}$ kg-m^2.

The equations of motion for the right bar: (6) $-A_x - B\sin\beta = ma_{GABx}$, (7) $-A_y - mg + B\cos\beta = ma_{GABy}$,

(8) $A_y\left(\dfrac{L}{2}\right)\cos\theta + A_x\left(\dfrac{L}{2}\right)\sin\theta + B\left(\dfrac{L}{2}\right)\sin\theta\,\cos\beta - B\left(\dfrac{L}{2}\right)\cos\theta\,\sin\beta = I_{GAB}\alpha_{AB}$, where

$I_{GAB} = \left(\dfrac{1}{12}\right)mL^2 = \left(\dfrac{1}{3}\right)$ kg-m^2.These eight equations in eight unknowns are solved by iteration:

$A_x = -19.27$ N, $A_y = 1.15$ N, $\alpha_{OA} = 0.425$ rad/s^2, $\alpha_{AB} = -1.59$ rad/s^2, $B = 45.43$ N,

$a_{GABx} = -0.8610$ m/s^2, $a_{GABy} = -0.2601$ m/s^2

==================================<>==================================

====================================<>====================================

Problem 7.141 At the instant the system in Problem 7.140 is released, what is the magnitude of the force exerted on the bar OA by the support at A?

Solution The acceleration of the center of mass of the bar OA is

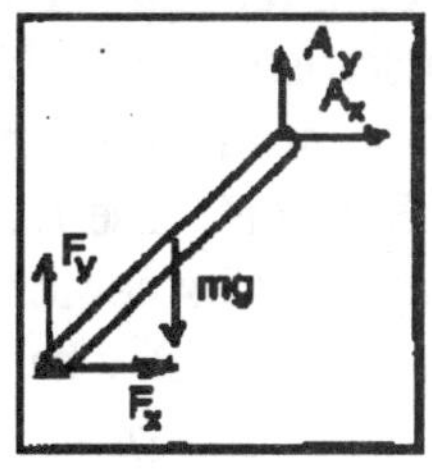

$$\vec{a}_{GOA} = \mathbf{a}_{OA} \times \vec{r}_{G/A} = \vec{a}_A + \begin{bmatrix} \vec{i} & \vec{j} & \vec{k} \\ 0 & 0 & \alpha_{OA} \\ \dfrac{L\cos\theta}{2} & \dfrac{L\sin\theta}{2} & 0 \end{bmatrix},$$

$$\vec{a}_{GOA} = -\frac{L\sin\theta}{2}\alpha_{OA}\vec{i} + \frac{L\cos\theta}{2}\alpha_{OA}\vec{j} \ (\mathrm{m/s^2}).$$

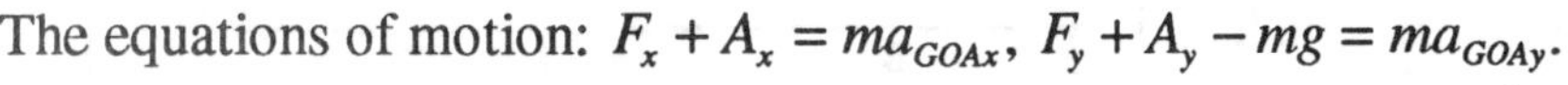

The equations of motion: $F_x + A_x = ma_{GOAx}$, $F_y + A_y - mg = ma_{GOAy}$.

Use the solution to Problem 7.140: $\theta = 45^o$, $\alpha_{OA} = 0.425 \ \mathrm{rad/s^2}$, $A_x = -19.27 \ \mathrm{N}$, $m = 4 \ \mathrm{kg}$, from which $F_x = 18.67 \ \mathrm{N}$, $F_y = 38.69 \ \mathrm{N}$, from which $|\vec{F}| = \sqrt{F_x^2 + F_y^2} = 42.96 \ \mathrm{N}$

====================================<>====================================

Problem 7.142 The fixed ring gear lies in the *horizontal plane.* The hub and planet gears are bonded together. The mass and moment of inertia of the combined hub and planet gears are $m_{HP} = 130 \ \mathrm{kg}$ and $I_{HP} = 130 \ \mathrm{kg\text{-}m^2}$. The moment of inertia of the sun gear is $I_s = 60 \ \mathrm{kg\text{-}m^2}$. The mass of the connecting rod is 5 *kg,* and it can be modeled as a slender bar. If a 1 *kN-m* counterclockwise moment is applied to the sun gear, what is the resulting angular acceleration of the bonded hub and planet gears?

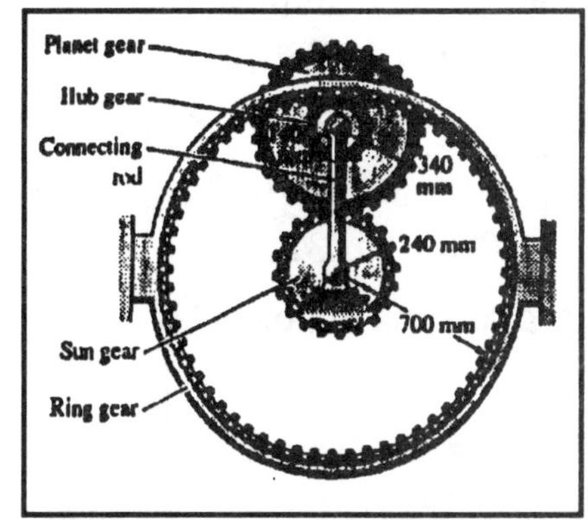

Solution:

The moment equation *for the sun gear* is (1) $M - 0.24F = I_S\alpha_S$.

For the hub and planet gears:

(2) $(0.48)\alpha_{HP} = -0.24\alpha_S$, (3) $F - Q - R = m_{HP}(0.14)(-\alpha_{HP})$,

(4) $(0.14)Q + 0.34F = I_{HP}(-\alpha_{HP})$.

For the connecting rod: (5) $(0.58)R = I_{CR}\alpha_{CR}$, where

$$I_{CR} = \left(\frac{1}{3}\right)m_{CR}(0.58^2) = 0.561 \ \mathrm{kg\text{-}m^2}.$$

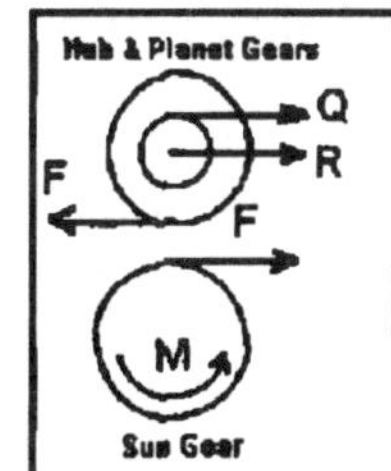

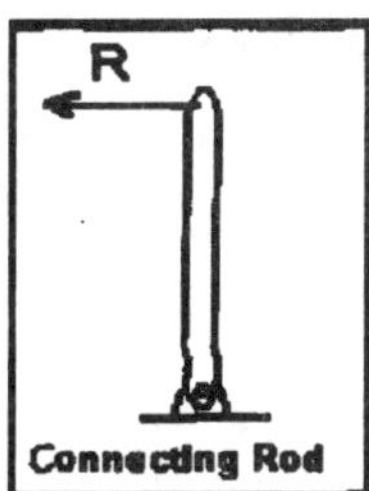

(6) $(0.58)\alpha_{CR} = -(0.14)\alpha_{HP}$. These six equations in six unknowns are solved by iteration: $F = 1482.7 \ \mathrm{N}$, $\alpha_S = 10.74 \ \mathrm{rad/s^2}$, $\alpha_{HP} = -5.37 \ \mathrm{rad/s^2}$, $Q = 1383.7 \ \mathrm{N}$, $R = 1.25 \ \mathrm{N}$, $\alpha_{CR} = 1.296 \ \mathrm{rad/s^2}$.

====================================<>====================================

==================================<>==================================

Problem 7.143 The system is stationary at the instant shown. The net force exerted on the piston by the exploding fuel air mixture and friction is 5 *kN* to the left. A clockwise couple $M = 200$ N-m acts on the crank AB. The moment of inertia of the crank about A is 0.0003 *kg-m*^{2}. The mass of the connecting rod BC is 0.36 *kg*, and its center of mass is 40 *mm* from B on the line from B to C. The connecting rod's moment of inertia about its center of mass is 0.0004 *kg-m*^{2}. The mass of the piston is 4.6 *kg*. What is the piston's acceleration at this instant? (Neglect the gravitational force on the crank and connecting rod.)

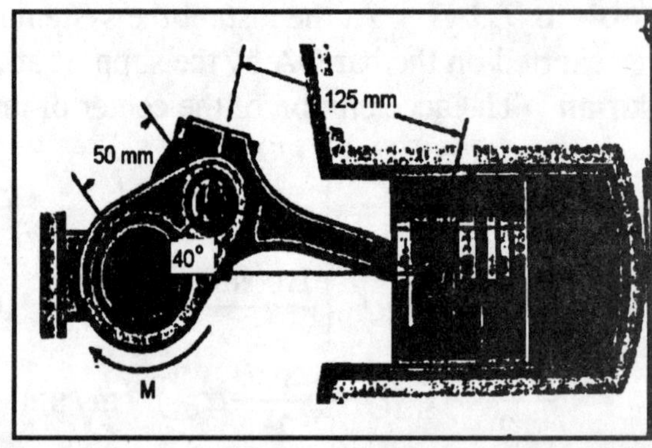

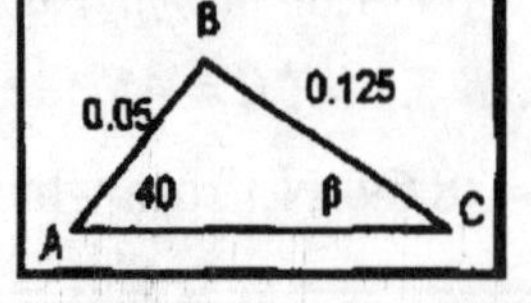

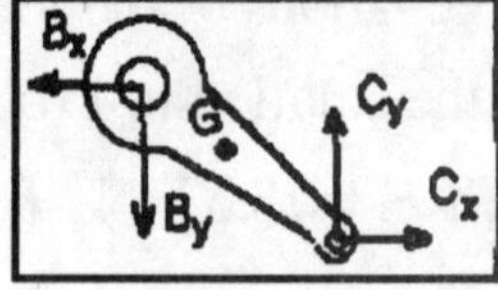

Solution:

From the law of sines: $\dfrac{\sin\beta}{0.05} = \dfrac{\sin 40^\circ}{0.125}$, from which $\beta = 14.9^\circ$. The vectors $\vec{r}_{B/A} = 0.05(\vec{i}\cos 40^\circ + \vec{j}\sin 40^\circ)\vec{r}_{B/A} = 0.0383\vec{i} + 0.0321\vec{j}$ (m).

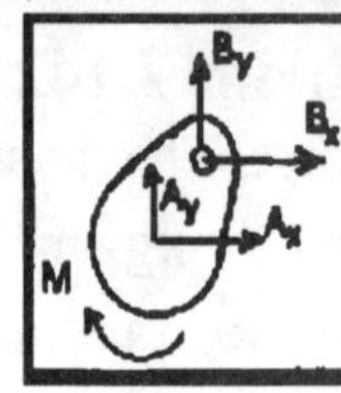

$\vec{r}_{B/C} = 0.125(-\vec{i}\cos\beta + \vec{j}\sin\beta)$ (m). $\vec{r}_{B/C} = -0.121\vec{i} + 0.321\vec{j}$ (m).

The acceleration of point B is

$\vec{a}_B = \mathbf{\vec{a}}_{AB} \times \vec{r}_{B/A} - \omega_{AB}^2 \vec{r}_{B/A}$,

$\vec{a}_B = \begin{bmatrix} \vec{i} & \vec{j} & \vec{k} \\ 0 & 0 & \alpha_{AB} \\ 0.0383 & 0.0321 & 0 \end{bmatrix} - \omega_{AB}^2(0.0383\vec{i} + 0.0321\vec{j})\ (\mathrm{m/s^2})$. The acceleration of point B in terms of the acceleration of point C is

$\vec{a}_B = \vec{a}_C + \mathbf{\vec{a}}_{BC} \times \vec{r}_{B/C} = a_C\vec{i} + \begin{bmatrix} \vec{i} & \vec{j} & \vec{k} \\ 0 & 0 & \alpha_{BC} \\ -0.121 & 0.0321 & 0 \end{bmatrix} - \omega_{BC}^2(-0.121\vec{i} + 0.0321\vec{j})\ (\mathrm{m/s^2})$. Equate the two expressions for the acceleration of point B, note $\omega_{AB} = \omega_{BC} = 0$, and separate components:

(1) $-0.05\alpha_{AB}\sin 40^\circ = a_C - 0.125\alpha_{BC}\sin\beta$, (2) $0.05\alpha_{AB}\cos 40^\circ = -0.125\alpha_{BC}\cos\beta$.

The acceleration of the center of mass of the connecting rod is $\vec{a}_{GCR} = \vec{a}_C + \mathbf{\vec{a}}_{BC} \times \vec{r}_{GCR/C} - \omega_{BC}^2 \vec{r}_{GCR/C}$,

$\vec{a}_{GCR} = a_C\vec{i} + \begin{bmatrix} \vec{i} & \vec{j} & \vec{k} \\ 0 & 0 & \alpha_{BC} \\ -0.085\cos\beta & 0.085\sin\beta & 0 \end{bmatrix} - \omega_{BC}^2(-0.085\cos\beta\vec{i} + 0.085\sin\beta\vec{j})\ (\mathrm{m/s^2})$, from which

(3) $a_{GCRx} = a_C - 0.085\alpha_{BC}\sin\beta\ (\mathrm{m/s^2})$,

(4) $a_{GCRy} = -0.085\alpha_{BC}\cos\beta\ (\mathrm{m/s^2})$.

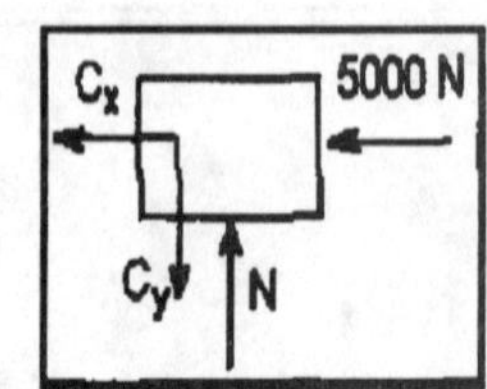

The equations of motion *for the crank:*

(5) $B_y(0.05\cos 40^\circ) - B_x(0.05\sin 40^\circ) - M = I_A\alpha_{AB}$

For the connecting rod:

(6) $-B_x + C_x = m_{CR}a_{GCRx}$

(7) $-B_y + C_y = m_{CR}a_{GCRy}$ (8) $C_y(0.085\cos\beta) + C_x(0.085\sin\beta) + B_x(0.04\sin\beta) + B_y(0.04\cos\beta) = I_{GCR}\alpha_{BC}$

Solution continued on next page

For the piston:

(9) $-C_x - 5000 = m_p a_C$. These nine equations in nine unknowns are solved by iteration:

$\alpha_{AB} = 1255.7 \text{ rad/s}^2$, $\alpha_{BC} = -398.2 \text{ rad/s}^2$, $a_{GCRx} = -44.45 \text{ m/s}^2$, $a_{GCRy} = 32.71 \text{ m/s}^2$,

$B_y = 1254.6 \text{ N}$, $B_x = -4739.5 \text{ N}$, $C_x = -4755.5 \text{ N}$, $C_y = 1266.3 \text{ N}$, $a_C = -53.15 \text{ m/s}^2$.

==================================<>==================================

Problem 7.144 If the crank AB in Problem 7.143 has a counterclockwise angular velocity of 2000 *rpm* (revolutions per minute) at the instant shown, what is the piston's acceleration?

Solution:

The angular velocity of AB is $\omega_{AB} = 2000\left(\frac{2\pi}{60}\right) = 209.44 \text{ rad/s}$.The angular velocity of the connecting rod BC is obtained from the expressions for the velocity at point B and the known value of ω_{AB}:

$$\vec{v}_B = \vec{w}_{AB} \times \vec{r}_{B/A} = \begin{bmatrix} \vec{i} & \vec{j} & \vec{k} \\ 0 & 0 & \omega_{AB} \\ 0.05\cos 40^o & 0.05\sin 40^o & 0 \end{bmatrix}. \ \vec{v}_B = -0.05\sin 40^o \omega_{AB}\vec{i} + 0.05\cos 40^o \omega_{AB}\vec{j} \ (\text{m/s}).$$

$$\vec{v}_B = \text{v}_C\vec{i} + \begin{bmatrix} \vec{i} & \vec{j} & \vec{k} \\ 0 & 0 & \omega_{BC} \\ -0.125\cos\beta & 0.125\sin\beta & 0 \end{bmatrix}, \ \vec{v}_B = \text{v}_C\vec{i} - 0.125\sin\beta\omega_{BC}\vec{i} - 0.125\cos\beta\omega_{BC}\vec{j} \ (\text{m/s}).$$

From the $\vec{j}$ component, $0.05\cos 40^o \omega_{AB} = -0.125\cos\beta\omega_{BC}$, from which $\omega_{BC} = -66.4 \text{ rad/s}$.

The nine equations in nine unknowns obtained in the solution to Problem 7.119 are:

(1) $-0.05\alpha_{AB}\sin 40^o - 0.05\omega_{AB}^2\cos 40^o = \text{a}_C - 0.125\alpha_{BC}\sin\beta + 0.125\omega_{BC}^2\cos\beta$,

(2) $0.05\alpha_{AB}\cos 40^o - 0.05\omega_{AB}^2\sin 40^o = -0.125\alpha_{BC}\cos\beta - 0.125\omega_{BC}^2\sin\beta$,

(3) $a_{GCRx} = a_C - 0.085\alpha_{BC}\sin\beta + 0.085\omega_{BC}^2\cos\beta \ (\text{m/s}^2)$,

(4) $a_{GCRy} = -0.085\alpha_{BC}\cos\beta - 0.085\omega_{BC}^2\sin\beta \ (\text{m/s}^2)$,

(5) $B_y(0.05\cos 40^o) - B_x(0.05\sin 40^o) - M = I_A\alpha_{AB}$,

(6) $-B_x + C_x = m_{CR}a_{GCRx}$.

(7) $-B_y + C_y = m_{CR}a_{GCRy}$,

(8) $C_y(0.085\cos\beta) + C_x(0.085\sin\beta) + B_x(0.04\sin\beta) + B_y(0.04\cos\beta) = I_{GCR}\alpha_{BC}$.

(9) $-C_x - 5000 = m_p a_C$. These nine equations in nine unknowns are solved by iteration:

$\alpha_{AB} = -39{,}386.4 \text{ rad/s}^2$, $\alpha_{BC} = 22{,}985.9 \text{ rad/s}^2$, $a_{GCRx} = -348.34 \text{ m/s}^2$, $a_{GCRy} = -1984.5 \text{ m/s}^2$,

$B_y = 1626.7 \text{ N}$, $B_x = -3916.7 \text{ N}$, $C_x = -4042.1 \text{ N}$, $C_y = 912.25 \text{ N}$, $\boxed{a_C = -208.25 \ (m/s^2)}$

==================================<>==================================

CHAPTER 8 - Solutions

====================================<>====================================

Problem 8.1 A main landing gear wheel of a Boeing 747 weighs 240 *lb* and has a moment of inertia of 17 $slug\text{-}ft^2$, and has a radius of 2 *ft*. If the airplane is moving at 245 *ft/s* and the wheel rolls, what is the wheel's kinetic energy?

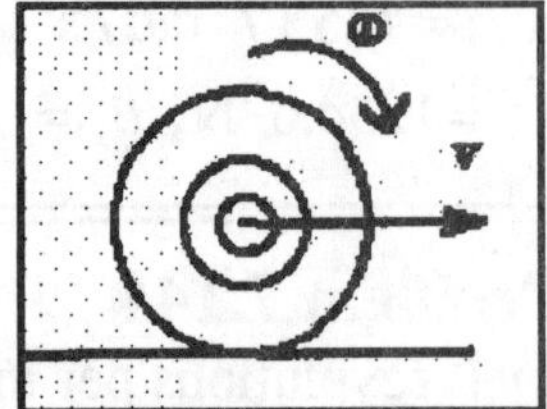

Solution:

The angular velocity of the rolling wheel is $\omega = \frac{v}{r} = \frac{245}{2} = 122.5\ rad/s$. The kinetic energy of rotation is $T_{rot} = \left(\frac{1}{2}\right)I\omega^2 = \frac{(17)(122.5)^2}{2} = 127{,}553.1\ \text{slug-ft}^2/s^2$. The kinetic energy of translation is $T = \left(\frac{1}{2}\right)\left(\frac{W}{g}\right)v^2 = \frac{240(245^2)}{2(32.17)} = 223{,}904.3\ \text{slug-ft}^2/s^2$. The sum is

$$\boxed{T = T_{rot} + T_{trans} = 351{,}457.4\ \text{slug-ft}^2/s^2}$$

====================================<>====================================

Problem 8.2 The *8-kg* slender bar is released from rest in the horizontal position. When it has fallen to the position shown, its angular velocity is 3.226 *rad/s.* (a) Use the given value of the angular velocity to determine the bar's kinetic energy. (b) How much work is done by the bar's weight as it falls to the position shown?

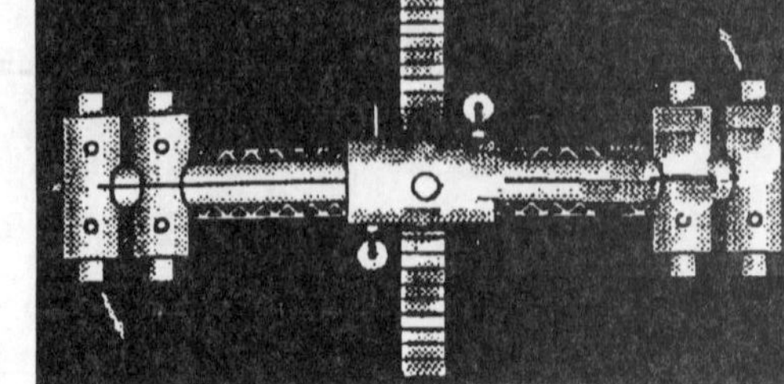

Solution:

From Appendix C, the bar's moment of inertia about the point 0 where it is pinned is $I_0 = (1/3)ml^2 = (1/3)(8)(2)^2 = 10.7 kg - m^2$

(a) From Equation (8.13), the bar's kinetic energy is

$$T = \frac{1}{2}I_0\omega^2 = \frac{1}{2}(10.7)(3.226)^2 = 55.5N - m$$

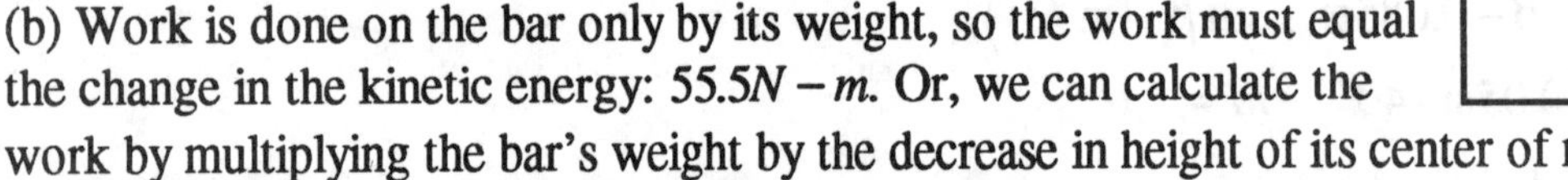

(b) Work is done on the bar only by its weight, so the work must equal the change in the kinetic energy: $55.5N - m$. Or, we can calculate the work by multiplying the bar's weight by the decrease in height of its center of mass: $V_{12} = mg(l/2)\sin 45° = (8)(9.81)(1)\sin 45° = 55.5N - m$.

====================================<>====================================

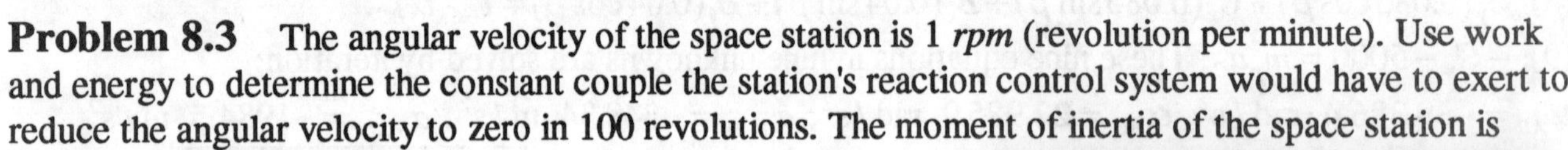

Problem 8.3 The angular velocity of the space station is 1 *rpm* (revolution per minute). Use work and energy to determine the constant couple the station's reaction control system would have to exert to reduce the angular velocity to zero in 100 revolutions. The moment of inertia of the space station is $I = 1.5 \times 10^{10}\ kg\text{-}m^2$.

Solution:

From the principle of work and energy: $U = T_1 - T_2$. Since $T_2 = 0$, $U = T_1$. The work done is related to the constant couple by $U = \int_0^\theta M\,d\theta = M\theta$, where $\theta = 100(2\pi) = 628.32\ rads$. The initial kinetic energy of the space station is $T_1 = \left(\frac{1}{2}\right)I\omega^2 = \frac{1.5\times10^{10}}{2}\left(\frac{2\pi}{60}\right)^2 = 82246703 = 8.225\times10^7\ kg\text{-}m^2/s^2$, from which

$$\boxed{M = T_1/\theta = 1.31\times10^5\ \text{N-m}}$$

====================================<>====================================

=================================<>=================================

Problem 8.4 Determine the average power transferred from the space station in Problem 8.3 as the control system reduces the angular velocity to zero.

Solution:

From the solution of Problem 8.3, the station's initial angular velocity is $\omega_1 = 0.105 rad/s$ and the constant decelerating couple is 131 $kN-m$. From the equation of angular motion,

$-131{,}000 = I\alpha = I\dfrac{d\omega}{dt}$. Integrating $\int_{t_1}^{t_2} -131{,}000 dt = I\int_{\omega_1}^{\omega_2} d\omega$, we obtain the interval of time:

$t_2 - t_1 = \dfrac{I(\omega_2-\omega_1)}{-131{,}000} = \dfrac{(1.5\times 10^{10})(-0.105)}{-131{,}000} = 12{,}000 s$. From the solution of Problem 8.3, the station's initial kinetic energy is $T_1 = 8.22\times 10^7 N-m$. The average power transferred from the station is

$$P_{av} = \frac{8.22\times 10^7}{12{,}000} = 6860 N-m/s$$

=================================<>=================================

Problem 8.5 The moment of inertia of the helicopter's rotor is 400 *slug-ft*^{2}. If the rotor starts from rest, the engine exerts a constant torque of 500 *ft-lb* on it, and the aerodynamic drag is neglected, use the principle of work and energy to determine how many revolutions the rotor must turn to reach an angular velocity of 2 revolutions per second.

Solution:

From the principle of work and energy: $U = T_2 - T_1$. Since the rotor starts from rest, $T_1 = 0$. The work done is $U = \int_0^\theta M\, d\theta = M\theta = 500\theta\ ft\text{-}lb$. The kinetic energy of the rotor is

$T_2 = \left(\dfrac{1}{2}\right)I\omega^2 = \dfrac{400}{2}(2(2\pi))^2 = 31{,}582.7\ slug\text{-}ft^2/s^2$, from which $500\theta = 31{,}582.7$, from which

$\boxed{\theta = 63.17\ rads = 10.05\ revolutions}$

=================================<>=================================

Problem 8.6 What average power is transferred to the rotor in Problem 8.5 in accelerating it from rest to 2 revolutions per second?

Solution:

From Newton's second law, the constant torque produces a constant acceleration. The angular velocity under constant acceleration is $\omega = \int \alpha dt + C = \alpha t$, since the rotor starts from rest. The angular travel is

$\theta = \int \alpha\, t\, dt + C = \dfrac{\alpha}{2}t^2\ rads$, from which $t^2 = \dfrac{2\theta}{\alpha} = \dfrac{2\theta}{\left(\dfrac{\omega}{t}\right)}$, from which $t = \dfrac{2\theta}{\omega}$. The average power, by

definition, is $P_{ave} = \dfrac{T_2}{t} = \dfrac{I\omega^2}{2\left(\dfrac{2\theta}{\omega}\right)} = \dfrac{I\omega^3}{4\theta} = \dfrac{(400)(4\pi)^3}{4(63.17)}$, $\boxed{P_{ave} = 3141.59\ ft\text{-}lb/s = 5.71\ hp}$ [*Check:*

$P_{ave} = (torque,\ ft-lb)\times(\omega_{ave},\ rad/s)$,

$P_{ave} = (500\,ft\text{-}lb)\left(\dfrac{4\pi - 0}{2}\ rad/s\right) = 1000\pi\ ft\text{-}lb/s = 5.71\ hp.\ check.$]

=================================<>=================================

Problem 8.7 During extravehicular activity, an astronaut activates two thrusters of her maneuvering unit, exerting equal and opposite forces $T = 2N$. The moment of inertia of the astronaut and her equipment about their center of mass is 45 *kg-m*^{2}. Using the principle of work and energy, determine her rate of rotation in revolutions per second when she has rotated one-fourth of a revolution from her initial orientation.

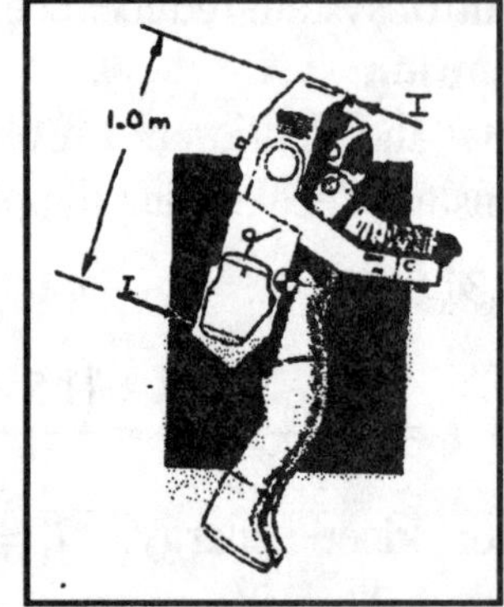

Solution:

The couple exerted is $M = (2N)(1m) = 2N-m$, so the work done in one-fourth revolution is $U_{12} = M\theta = (2)\frac{1}{4}(2\pi) = \pi\ N-m$. From the principle of work and energy $U_{12} = (1/2)I\omega_2^2$, or $\pi = (1/2)(45)\omega_2^2$, so her angular velocity is $\omega_2 = 0.374 rad/s = 0.0595 rev/s$.

Problem 8.8 What average power is transmitted to the astronaut in Problem 8.7 as she rotates one-fourth of a revolution from her initial orientation?

Solution:

The couple exerted is $M = 2N-m$. From the solution of Problem 8.7, the work done is $U_{12} = \pi\ N-m$ and her final angular velocity is $\omega_2 = 0.374 rad/s$. From the equation of angular motion $M = I\alpha = I\frac{d\omega}{dt}$. Integrating, $\int_{t_1}^{t_2} M dt = I\int_{\omega_1}^{\omega_2} d\omega$ we obtain the interval of time:

$t_2 - t_1 = \frac{I(\omega_2 - \omega_1)}{M} = \frac{(45)(0.374)}{2} = 8.41\ s$. From Equation (8.22), the average power is

$P_{av} = \frac{U_{12}}{t_2 - t_1} = \frac{\pi}{8.41} = 0.374W$.

Problem 8.9 A slender bar of mass *m* is released from rest in the horizontal position shown. Determine its angular velocity when it is vertical (a) by using the principle of work and energy, (b) by using conservation of energy.

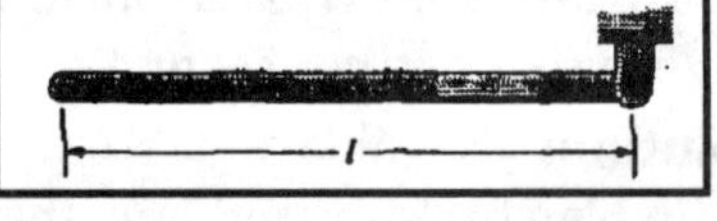

Solution : (a) Choose a coordinate system with *y* positive downward. From the principle of work and energy: $U = T_2 - T_1$. Since the bar is released from rest, $T_1 = 0$. The work done is

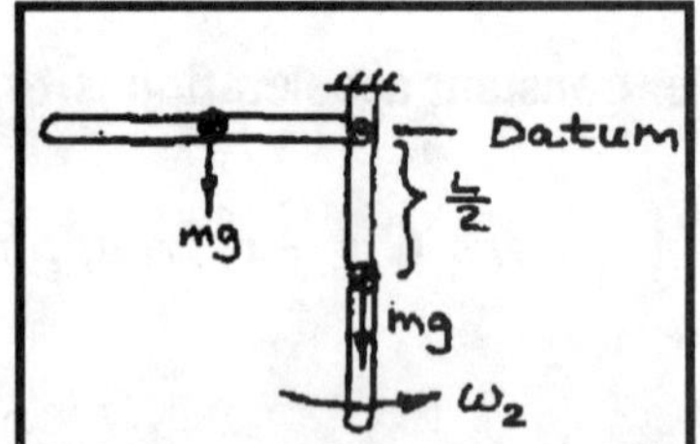

$U = \int_{y_1}^{y_2} F\,dy = \int_0^{\frac{L}{2}} mg\,dy = mg\left(\frac{L}{2}\right)$. The kinetic energy is

$T_2 = (1/2)I\omega_2^2$, where $I = (1/3)mL^2$, from which

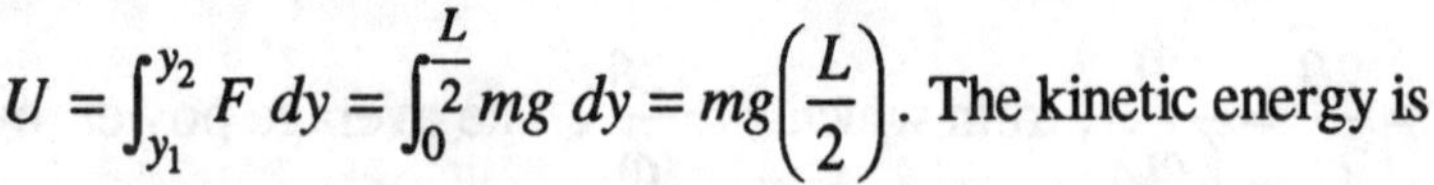

$$\omega_2 = \sqrt{\frac{2U}{I}} = \sqrt{\frac{mgL}{\left(\frac{1}{3}\right)mL^2}} = \sqrt{\frac{3g}{L}}$$

(b) Choose the datum at the level of the pin. The potential is $V = -\int_{y_1}^{y_2} F\,dy$, from which $V_1 = 0$, and $V_2 = -mg\int_0^{\frac{L}{2}} dy = -mgL/2$. From the conservation of energy, $V_1 + T_1 = V_2 + T_2$, from which $0 + 0 = V_2 + \left(\frac{1}{2}\right)I\omega_2^2$, from which

$$\omega_2 = \sqrt{\frac{-2V_2}{I}} = \sqrt{\frac{3g}{L}}$$

Problem 8.10 The moment of inertia of the pulley is 0.4 *slug-ft*^{2}. The pulley starts from rest. For both cases, use the principle of work and energy to determine the pulley's angular velocity when it has turned one revolution.

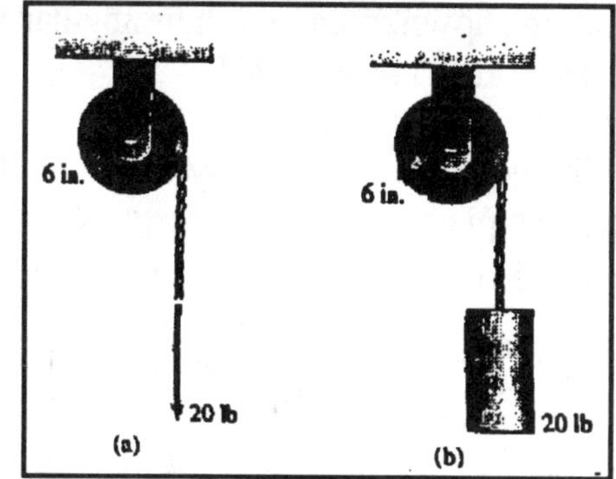

Solution From the principle of work and energy: $U = T_2 - T_1$, where $T_1 = 0$, since the pulley starts from rest. (a) The work done is $U = \int_0^{2\pi} Fr\,d\theta = 2\pi Fr$. The kinetic energy of the pulley is $T_2 = \left(\frac{1}{2}\right) I\omega_2^2$, where $I = 0.4$ $slug$ - ft^2, from which $\boxed{\omega_2 = \sqrt{\frac{4\pi Fr}{I}} = \sqrt{\frac{4\pi(20)(0.5)}{0.4}} = 17.72 \ rad/s}$.(b) The work done is $U = \int_0^{2\pi} Wr\,d\theta = 2\pi Wr$, and the kinetic energy is $T_2 = \left(\frac{1}{2}\right) I\omega_2^2 + \left(\frac{1}{2}\right)\frac{W}{g} v^2$. Noting that $v = r\omega$, $2\pi Wr = \left(\left(\frac{1}{2}\right) I + \left(\frac{1}{2}\right)\frac{W}{g} r^2\right)\omega_2^2$ from which $\boxed{\omega_2 = \sqrt{\frac{4\pi Wr}{I + \left(\frac{W}{g}\right) r^2}} = 15.0 \ rad/s}$. [*Check*: Compare with the solution to Problem 7.16. Apply chain rule: $\frac{d\omega_2}{dt} = \omega_2 \frac{d\omega_2}{d\theta} = \alpha$, from which, for $\omega_2(0) = 0$, (a) $\omega_2 = \sqrt{2\alpha\theta} = \sqrt{2(25)(2\pi)} = 17.72 \ rad/s$.(b) $\omega_2 = \sqrt{2(18)(2\pi)} = 15.0 \ rad/s$. *check.*]

Problem 8.11 The object consists of identical 1 *m*, 5 *kg* bars welded together. If it is released from rest in the position shown, what is the angular velocity when the bar attached to A is vertical?

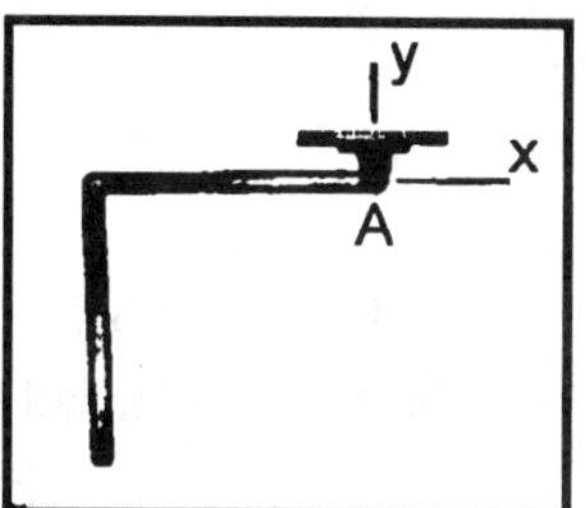

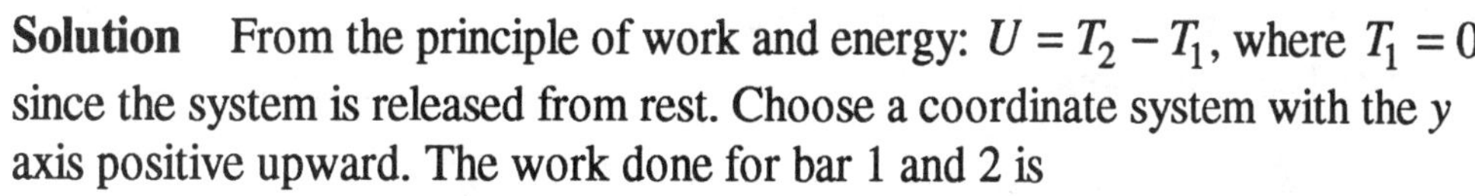

Solution From the principle of work and energy: $U = T_2 - T_1$, where $T_1 = 0$ since the system is released from rest. Choose a coordinate system with the y axis positive upward. The work done for bar 1 and 2 is

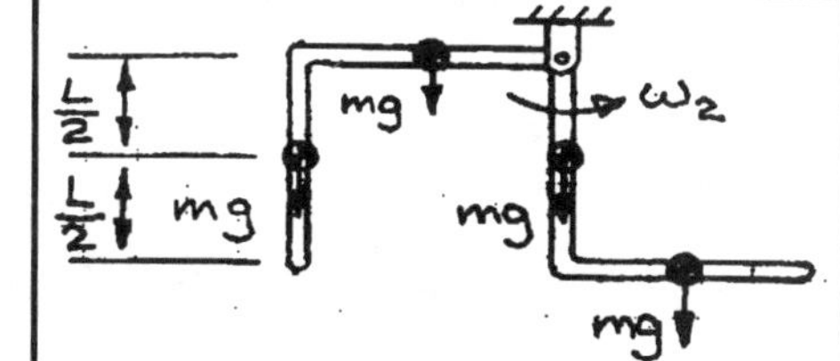

$$U = \left(\int_{y_1}^{y_2} F\,dy\right)_1 + \left(\int_{y_1}^{y_2} F\,dy\right)_2 = -\int_0^{\frac{L}{2}} mg\,dy - \int_0^{\frac{L}{2}} mg\,dy.$$

$U = 2mg\left(\frac{L}{2}\right) = mgL$. The kinetic energy is $T = \left(\frac{1}{2}\right) I_A \omega_2^2$, where I_A is the moment of inertia about point A.

$I_A = I_1 + I_2 + md^2$, where $I_1 = \frac{mL^2}{3}$, $I_2 = \frac{mL^2}{12}$, $d = \sqrt{L^2 + \left(\frac{L}{2}\right)^2}$, from which

$I_A = mL^2\left(\frac{1}{3} + \frac{1}{12} + \frac{5}{4}\right) = \frac{5mL^2}{3}$. Thus $mgL = \left(\frac{1}{2}\right)\left(\frac{5mL^2}{3}\right)\omega_2^2$, from which $\boxed{\omega_2 = \sqrt{\frac{6g}{5L}} = 3.431 \ rad/s}$

===================================<>===================================

Problem 8.12 The objects consist of identical 3 *ft*, 10 *lb* bars welded together. If they are released from rest in the position shown, what are their angular velocities when the bars attached at A are vertical?

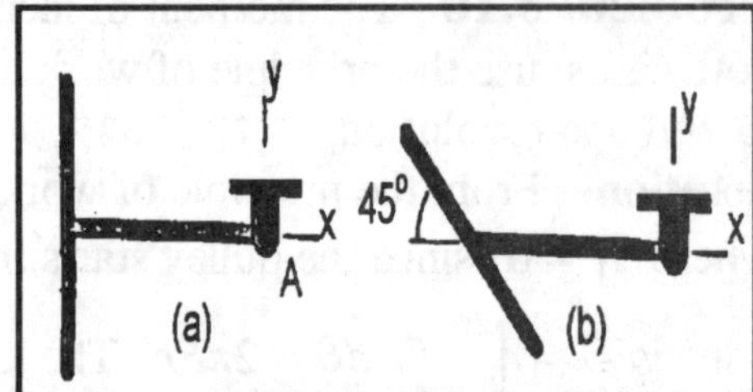

Solution:

Choose the coordinate system with the *y* axis upward. From the principle of work and energy: $U = T_2 - T_1$, where $T_1 = 0$, since the objects are released from rest. (a) The work done is

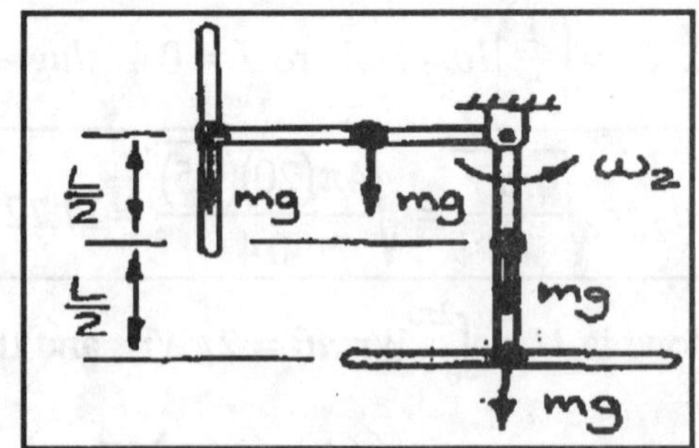

$$U = \left(\int_{y_1}^{y_2} F\,dy\right)_1 + \left(\int_{y_1}^{y_2} F\,dy\right)_2 = -\int_0^{\frac{L}{2}} W dy - \int_0^{L} W dy = \frac{3}{2}WL.$$

The kinetic energy is $T_2 = \left(\frac{1}{2}\right) I_A \omega_2^2$, where

$$I_A = \left(\frac{WL^2}{3g}\right) + \left(\frac{WL^2}{12g}\right) + \left(\frac{W}{g}\right)L^2 = \frac{17WL^2}{12g}, \text{ from which}$$

$\frac{3WL}{2} = \left(\frac{1}{2}\right)\frac{17WL^2}{12g}\omega_2^2$, from which $\boxed{\omega_2 = \sqrt{\frac{36g}{17L}} = 4.765\ rad/s}$. (b) The expressions for the work and the kinetic energy are algebraically identical for the two objects, so that the angular velocities are the same: $\boxed{\omega_2 = 4.765\ rad/s}$

===================================<>===================================

Problem 8.13 The 8 *kg* slender bar is released from rest in the horizontal position. When it has fallen to the position shown, what are the *x* and *y* components of force exerted on the bar by the pin support at A?

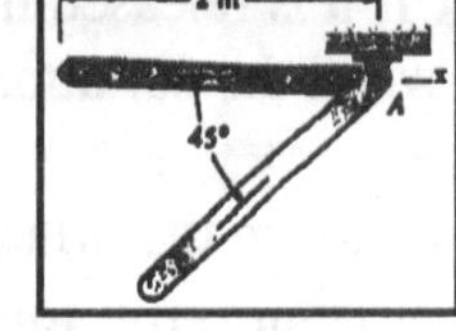

Solution:

Denote $\theta = 45^o$, $L = 2\ m$. Choose a coordinate system with *y* positive upward. From the principle of work and energy: $U = T_2 - T_1$, where $T_1 = 0$ since the bar is released from rest. The work done is

$U = \int_{y_1}^{y_2} F\,dy = -\int_0^{\frac{L\sin\theta}{2}} mg\,dy = \frac{mgL\sin\theta}{2}$. The kinetic energy is $T_2 = \left(\frac{1}{2}\right) I_A \omega_2^2$, where

$I_A = \left(\frac{1}{3}\right) mL^2$, from which $\frac{mgL\sin\theta}{2} = \left(\frac{1}{2}\right)\frac{mL^2}{3}\omega_2^2$. Solve: $\omega_2 = \sqrt{\frac{3g\sin\theta}{L}} = 3.23\ rad/s$. *Check:* For $\theta = 90^o$, this should agree with the solution to Problem 8.7. *check.* Denote the angular acceleration of the bar by α, from which the transverse acceleration is $a_t = \alpha r = \left(\frac{L}{2}\right)\alpha$, and the normal acceleration is $a_n = \omega_2^2 r = \left(\frac{L}{2}\right)\omega_2^2$.

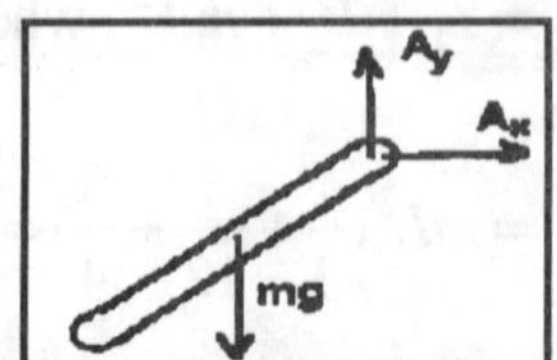

From Newton's second law and the equation of angular motion for the bar:

(1) $A_x \cos\theta + A_y \sin\theta - mg\sin\theta = ma_n = m\omega_2^2(L/2)$,

(2) $A_x \sin\theta - A_y \cos\theta + mg\cos\theta = ma_t = m\alpha(L/2)$,

Solution continued on next page

(3) $mgL\cos\theta = I_A\alpha = \dfrac{mL^2}{3}\alpha$. These three equations in three unknowns are solved by iteration:

$\boxed{A_x = 88.29\ N}$, $\boxed{A_y = 107.91\ N}$, $\alpha = 5.20\ rad/s^2$. *Check:* An analytical solutions is readily obtained:

From the third equation: $\alpha = \dfrac{3g\cos\theta}{2L} = 5.20\ rad/s^2$. *check.* Substitute the expressions for α and for

$\omega_2^2 = \dfrac{3g\sin\theta}{L}$ into the first two equations to obtain: $A_x\sin\theta + A_y\cos\theta = \dfrac{5}{2}mg\sin\theta$,

$A_x\cos\theta - A_y\sin\theta = -\dfrac{1}{4}mg\cos\theta$. Solve: $A_y = \left(\dfrac{11}{4}\right)mg\sin\theta\cos\theta = 107.91\ N$,

$A_x = mg\left(\dfrac{5}{2}\sin^2\theta - \dfrac{1}{4}\cos^2\theta\right) = 88.29\ N$. *check.*

=================================<>=================================

Problem 8.14 The slender bar is released from rest in the position shown. (a) Use conservation of energy to determine the angular velocity when the bar is vertical. (b) For what value of x is the angular velocity determined in part (a) a maximum?

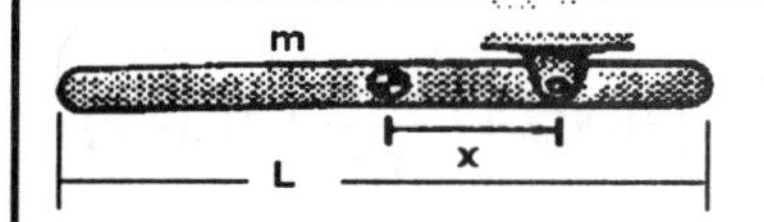

Solution:

Choose the datum at the level of the pin. From the conservation of energy, $T + V = const.$, where T is the kinetic energy and V is the total potential energy (see Eq 8.19). By definition, $V_1 = 0$ at the datum, and $T_1 = 0$ at the datum because the bar is released from rest there, from which $T_2 + V_2 = 0$ at any other position. The change in total potential energy in terms of the work done is $U = V_1 - V_2 = -V_2$ (see equation preceding Eq (8.19)). The work done is $U = \int_{s_1}^{s_2} F\,ds = -\int_0^{-x} mg ds = mgx = -V_2$.

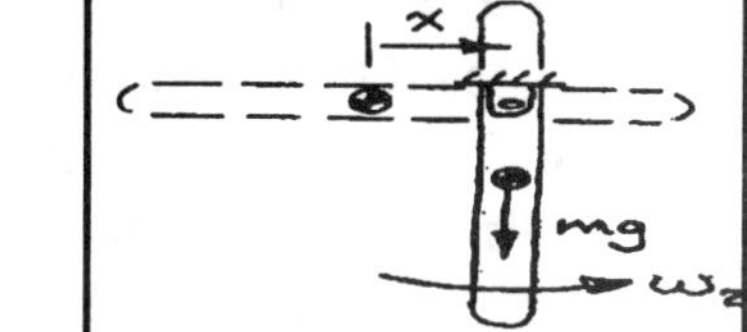

The kinetic energy is $T_2 = \dfrac{1}{2}I\omega_2^2$, where $I = \dfrac{mL^2}{12} + mx^2$, from which $-mgx + \left(\dfrac{1}{2}\right)\left(\dfrac{mL^2}{12} + mx^2\right)\omega_2^2 = 0$. (a) Solve

$$\omega_2^2 = \frac{2gx}{\left(\frac{1}{12}L^2 + x^2\right)},\quad \omega_2 = \sqrt{\frac{2gx}{\left(\frac{1}{12}L^2 + x^2\right)}}$$

(b) Take the derivative to find the maximum:

$$\frac{d\omega_2^2}{dx} = 0 = \frac{2g}{\left(\frac{1}{12}L^2 + x^2\right)} - \frac{2gx}{\left(\frac{1}{12}L^2 + x^2\right)^2}(2x) = 0,$$ from which $\boxed{x = \dfrac{L}{\sqrt{12}}}$

=================================<>=================================

====================================<>====================================

Problem 8.15 The gears can turn freely on their pin supports. Their moments of inertia are $I_A = 0.002\ kg\text{-}m^2$ and $I_B = 0.006\ kg\text{-}m^2$. They are at rest when a constant couple $M = 2\ N\text{-}m$ is applied to gear B. Neglecting friction, use principle of work and energy to determine the angular velocities of the gears when gear A has turned 100 revolutions.

Solution:

From the principle of work and energy: $U = T_2 - T_1$, where $T_1 = 0$ since the system starts from rest. The work done is $U = \int_0^{\theta_B} M d\theta = M\theta_B$.

Gear B rotates in a positive direction; gear A rotates in a negative direction, $\theta_A = -2\pi(100) = -200\pi\ rad$. The angle traveled by the gear B is $\theta_B = -\dfrac{r_A}{r_B}\theta_A = -\left(\dfrac{0.06}{0.09}\right)(-200\pi) = 418.9\ rad$, from which

$U = M\theta_B = 2(418.9) = 837.76\ N\text{-}m$.

The kinetic energy is $T_2 = \left(\dfrac{1}{2}\right)I_A\omega_A^2 + \left(\dfrac{1}{2}\right)I_B\omega_B^2$, where $\omega_B = -\left(\dfrac{r_A}{r_B}\right)\omega_A$, from which

$$U = \left(\frac{1}{2}\right)\left(0.002 + \left(\frac{0.06}{0.09}\right)^2(0.006)\right)\omega_A^2 = 837.76\ N\text{-}m, \text{ from which}$$

$\boxed{\omega_A = \pm\sqrt{359{,}039} = -599.2\ rad/s}$, and $\boxed{\omega_B = -\left(\dfrac{0.06}{0.09}\right)\omega_A = 399.5\ rad/s}$

====================================<>====================================

Problem 8.16 The moments of inertia of gears A and B are $I_A = 0.02 kg\text{-}m^2$ and $I_B = 0.09 kg\text{-}m^2$. Gear A is connected to a torsional spring with constant *k=12 N-m/rad.* If gear B is given an initial counterclockwise angular velocity of *10 rad/s* with the torsional spring unstretched, through what maximum counterclockwise angle does gear B rotate?

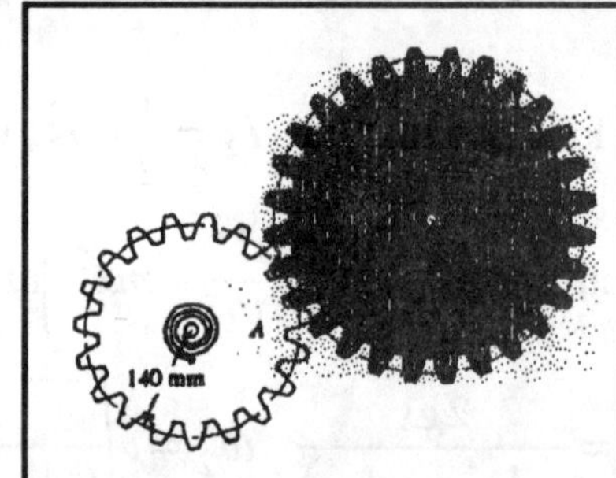

Solution:

The counterclockwise velocity of B and clockwise velocity of A are related by $0.2\omega_B = 0.14\omega_A$.

Applying conservation of energy,

$$\frac{1}{2}I_A\omega_{A1}^2 + \frac{1}{2}I_B\omega_{B1}^2 = \frac{1}{2}k\theta_{A2}^2 : \frac{1}{2}(0.02)\left[\left(\frac{0.2}{0.14}\right)(10)\right]^2 + \frac{1}{2}(0.09)(10)^2$$

$= \dfrac{1}{2}(12)\theta_{A2}^2$ Solving, we obtain $\theta_{A2} = 1.04 rad$ so $\theta_{B2} = \dfrac{0.14}{0.2}\theta_{A2} = 0.731 rad = 41.9°$.

====================================<>====================================

==================================<>==================================

Problem 8.17 The pulleys can turn freely on their pin supports. Their moments of inertia are $I_A = 0.002\ kg\text{-}m^2$, $I_B = 0.036\ kg\text{-}m^2$, and $I_C = 0.032\ kg\text{-}m^2$. They are stationary when a constant couple $M = 2\ N\text{-}m$ is applied to pulley A. What is the angular velocity of pulley A when it has turned 10 revolutions?

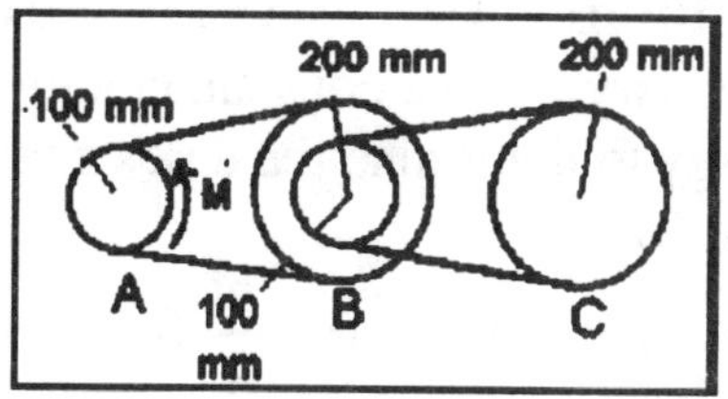

Solution:

All pulleys rotate in a positive direction: $\omega_C = \left(\frac{0.1}{0.2}\right)\omega_B$, $\omega_B = \left(\frac{0.1}{0.2}\right)\omega_A$, from which $\omega_C = \left(\frac{0.1}{0.2}\right)^2 \omega_A$. From the principle of work and energy: $U = T_2 - T_1$, where $T_1 = 0$ since the pulleys start from a stationary position. The work done is $U = \int_0^{\theta_A} M d\theta = 2(2\pi)(10) = 40\pi\ N\text{-}m$. The kinetic energy is $T_2 = \left(\frac{1}{2}\right)I_A\omega_A^2 + \left(\frac{1}{2}\right)I_B\omega_B^2 + \left(\frac{1}{2}\right)I_C\omega_C^2 = \left(\frac{1}{2}\right)\left(I_A + \left(\frac{0.1}{0.2}\right)^2 I_B + \left(\frac{0.1}{0.2}\right)^4 I_C\right)\omega_A^2$, from which $T_2 = 0.0065\omega_A^2$, and $U = 40\pi = 0.0065\omega_A^2$. Solve: $\boxed{\omega_A = \sqrt{\frac{40\pi}{0.0065}} = 139.0\ rad/s}$

==================================<>==================================

Problem 8.18 Model the arm ABC as a rigid body. Its mass is 300 *kg*, and the moment of inertia about its center of mass is $I = 360\ kg\text{-}m^2$. Starting from rest with its center of mass 2 *m* above the ground (position 1), the hydraulic cylinders push arm ABC upward. When it is in the position shown (position 2) its counterclockwise angular velocity is 1.4 *rad/s*. How much work do the hydraulic cylinders do on the arm in moving it from position 1 to position 2?

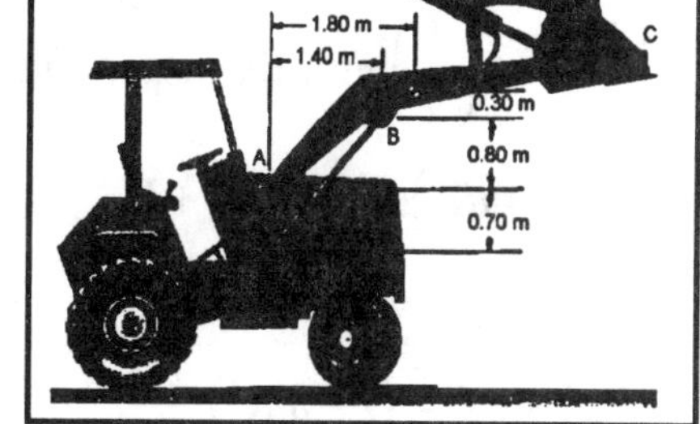

Solution:

From the principle of work and energy: $U = T_2 - T_1$, where $T_1 = 0$ since the system starts from rest. The work done is $U = U_{cylinders} - mg(h_2 - h_1)$, and the kinetic energy is $T_2 = \left(\frac{1}{2}\right)I_A\omega^2$, from which $U_{cylinders} - mg(h_2 - h_1) = \left(\frac{1}{2}\right)I_A\omega^2$

In position 1, $h_1 = 2\ m$ above the ground. In position 2, $h_2 = 2.25 + 0.8 + 0.3 = 3.35\ m$. The distance from A to the center of mass is $d = \sqrt{1.8^2 + 1.1^2} = 2.11\ m$, from which $I_A = I + md^2 = 1695\ kg\text{-}m^2$. Substitute: $U_{cylinders} - 3973.05 = 1661.1$, from which $\boxed{U_{cylinders} = 5634.2\ N\text{-}m}$

==================================<>==================================

================================<>================================

Problem 8.19 The mass of the homogenous cylindrical disk is *m*, and its radius is *R*. The disk is stationary when a constant clockwise couple *M* is applied to it. Use work and energy to determine the disk's angular velocity when it has rolled a distance *b*.

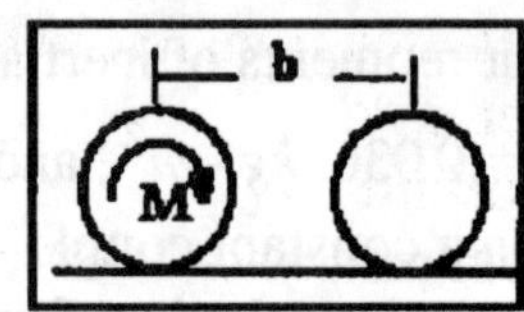

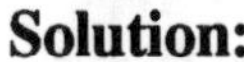

Solution:

The angle rolled is $\theta = \frac{b}{R}$, from which $U = M\theta = \frac{Mb}{R}$. The kinetic energy is

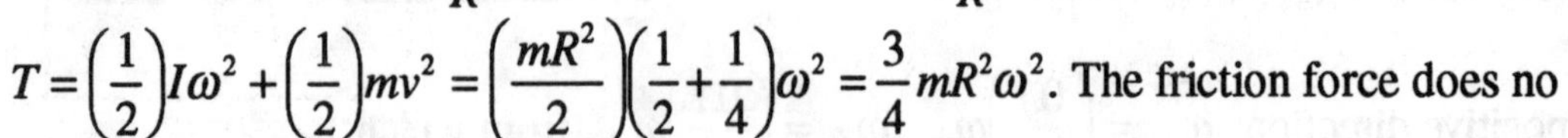

$T = \left(\frac{1}{2}\right)I\omega^2 + \left(\frac{1}{2}\right)mv^2 = \left(\frac{mR^2}{2}\right)\left(\frac{1}{2}+\frac{1}{4}\right)\omega^2 = \frac{3}{4}mR^2\omega^2$. The friction force does no work since the velocity of the point of contact is zero. From the principle of work and energy: $U = T_2$, from which $\omega = -\sqrt{\frac{4Mb}{3mR^3}}$ (clockwise).

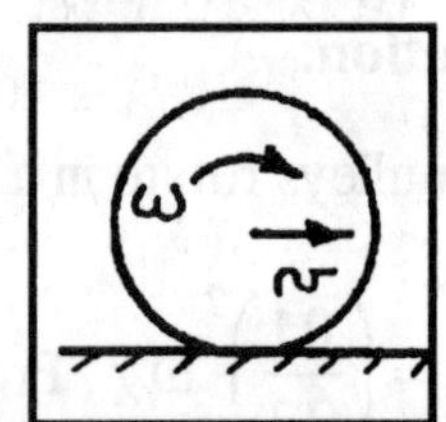

================================<>================================

Problem 8.20 A disk of mass *m* and moment of inertia *I* starts from rest on an inclined surface and is subjected to a constant clockwise couple *M*. Assuming that it rolls, what is the angular velocity of the disk when it has moved a distance *b*?

Solution:

Choose a coordinate system with the *y* axis positive upward. From the principle of work and energy: $U = T_2 - T_1$, where $T_1 = 0$ since the disk starts from rest. The angular distance traveled down the incline is $\theta = \frac{b}{R}$, from which the work done by the negative couple is $U_{couple} = -\int_0^{-\theta} Md\theta = M\theta = \frac{Mb}{R}$. The work done by the weight is $U_{weight} = \int_0^{-h} -mgdh = mgh = mgb\sin\beta$. The kinetic energy is $T_2 = \left(\frac{1}{2}\right)I\omega^2 + \left(\frac{1}{2}\right)mv^2$. By inspection $v = R\omega$, from which $T_2 = \left(\frac{1}{2}\right)(I + mR^2)\omega^2$. $U_{couple} + U_{weight} = T_2$. Substitute:

$$M\frac{b}{R} + mgb\sin\beta = (I + mR^2)\frac{\omega^2}{2}, \text{ from which } \omega = -\sqrt{\frac{2b\left(\frac{M}{R} + mg\sin\beta\right)}{(I + mR^2)}}$$

(clockwise). [*Check*: Note that $v = |\omega|R$, and compare with Example 8.1, for case $M = 0$ (see solution for v in middle of page 377 of text). *check.*]

================================<>================================

====================================<>====================================

Problem 8.21 The stepped disk weighs 40 *lb*, and its mass moment of inertia is $I = 0.2\ \text{slug-ft}^2$. If it is released from rest, what is its angular velocity when the center of the disk has fallen 3 *ft*?

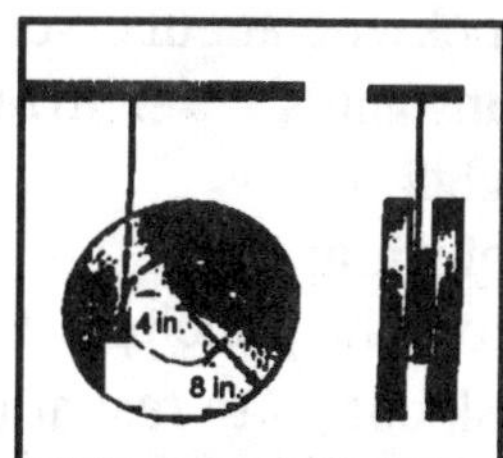

Solution:

From the principle of work and energy: $U = T_2 - T_1$, where $T_1 = 0$ since the disk is released from rest. The work done by the weight is $U_{weight} = \int^{-3} -W dh = 3W = 120\ \text{ft-lb}$. The kinetic energy is $T_2 = \left(\frac{1}{2}\right)I\omega^2 + \left(\frac{1}{2}\right)\frac{W}{g}v^2$. By inspection, $v = \frac{\omega}{3}$, from which,

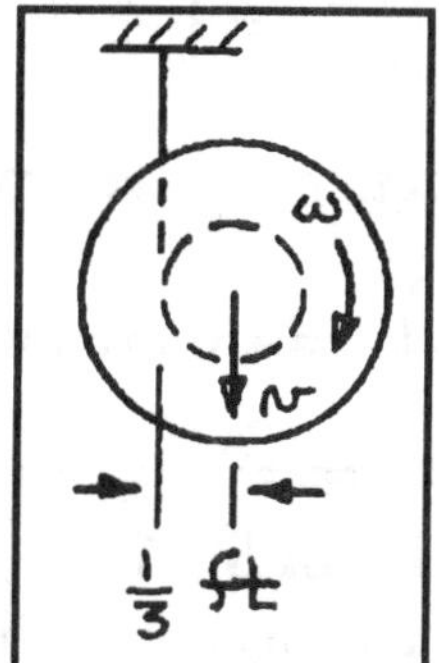

$T_2 = \left(\frac{1}{2}\right)\left(I + \frac{W}{3^2 g}\right)\omega^2 = 0.1691\omega^2$. $U_{weight} = T_2$, from which $120 = 0.1691\omega^2$, from which $|\omega| = 26.64\ \text{rad/s}$.

[*Check*: Note that for constant angular acceleration, starting from rest, $\omega = \alpha\, t$. *check.*]

====================================<>====================================

Problem 8.22 The 100 *kg* homogenous cylindrical disk is at rest when the force $F = 500\ \text{N}$ is applied to a cord wrapped around it, causing the disk to roll. Use the principle of work and energy to determine the disk's angular velocity when it has turned one revolution.

Solution:

From the principle of work and energy: $U = T_2 - T_1$, where $T_1 = 0$ since the disk is at rest initially. The distance traveled in one revolution by the center of the disk is $s = 2\pi R = 0.6\pi\ \text{m}$. As the cord unwinds, the force F acts through a distance of $2s$.

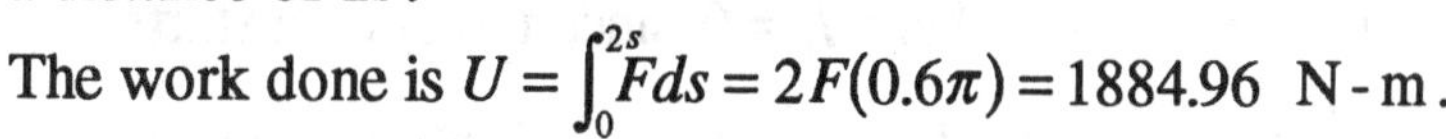
The work done is $U = \int_0^{2s} F ds = 2F(0.6\pi) = 1884.96\ \text{N-m}$.

The kinetic energy is $T_2 = \left(\frac{1}{2}\right)I\omega^2 + \left(\frac{1}{2}\right)mv^2$, where $I = \frac{1}{2}mR^2$, and $v = R\omega$, from which

$T_2 = \frac{3}{4}mR^2\omega^2 = 6.75\omega^2$. $U = T_2$, $1884.96 = 6.75\omega^2$, from which $\omega = -16.71\ \text{rad/s}$ (clockwise).

====================================<>====================================

==============================<>==============================

Problem 8.23 The 1 *slug* homogenous cylindrical disk is given a clockwise angular velocity of 2 *rad/s* with the spring unstretched. The spring constant is $k = 3\ \text{lb/ft}$. If the disk rolls, how far will its center move to the right?

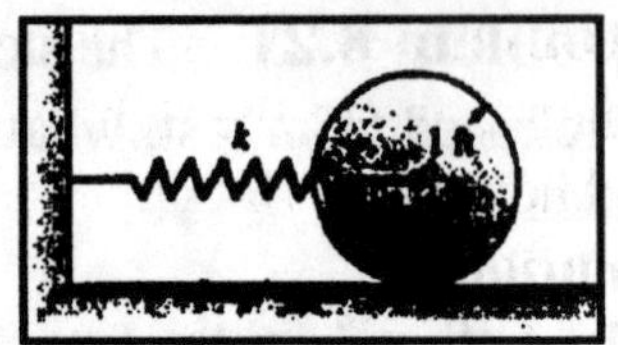

Solution:

From the principle of work and energy: $U = T_2 - T_1$, where $T_2 = 0$ since the disk comes to rest at point 2. The work done by the spring is $U = \int_0^S F ds = \int_0^S -ksds = -\frac{1}{2}kS^2$. The kinetic energy is $T_1 = \frac{1}{2}I\omega^2 + \frac{1}{2}mv^2$. By inspection $v = \omega R$, from which $T_1 = \left(\frac{1}{2}\right)\left(\frac{m}{2}R^2 + mR^2\right)\omega^2 = \frac{3}{4}mR^2\omega^2 = 0.75(2^2) = 3\ \text{ft-lb}$, $U = -T_1$, $-\left(\frac{3}{2}\right)S^2 = -3$, from which $S = \sqrt{2} = 1.414\ \text{ft}$

==============================<>==============================

Problem 8.24 The 22 *kg* platen P rests on four roller bearings. The roller bearings can be modeled as 1 *kg* homogenous cylinders with 30 *mm* radii. The platen is stationary and the spring (900 *N/m*) is unstretched when a constant horizontal force $F = 100\ \text{N}$ is applied as shown. What is the platen's velocity when it has moved 200 *mm* to the right.

Solution:

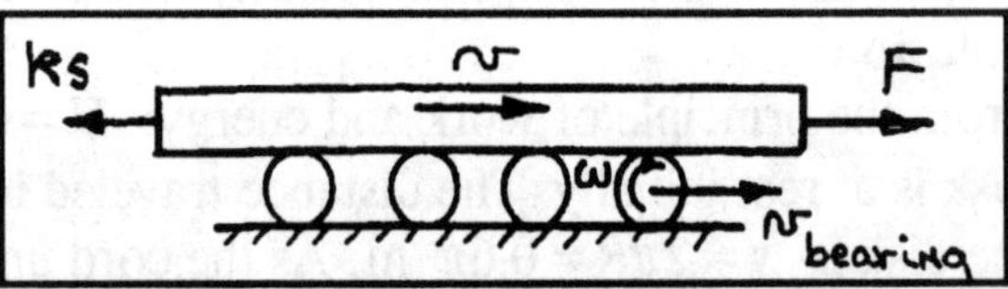

Denote the distance moved by the platen (and the spring stretch) by *S*. The roller bearings rotate with an angular velocity ω, and the center of mass of each translates with a velocity $v_{bearing} = R\omega$. The platen moves with a velocity $v = 2v_{bearing}$. From the principle of work and energy: $U = T_2 - T_1$, where $T_1 = 0$ since the system starts from rest. The work done by the spring is $U_{spring} = \int_0^S F ds = -\int_0^S ks\, ds = -\frac{1}{2}kS^2 = -18\ \text{N-m}$. The work done by the force is $U_{force} = \int_0^S F ds = FS = 20\ \text{N-m}$. The kinetic energy of the platen is $T_{platen} = \frac{1}{2}m_{platen}v^2 = 11v^2$.

The kinetic energy of each ball bearing is $T_{bearing} = \left(\frac{1}{2}\right)I\omega^2 + \left(\frac{1}{2}\right)m_{bearing}v_{bearing}^2 = \left(\frac{3}{4}m_{bearing}\right)v_{bearing}^2 = \frac{3}{16}m_{bearing}v^2 = 0.1875v^2$, where $I = \frac{m_{bearing}R^2}{2}$, $m_{bearing} = 1\ \text{kg}$, and $\omega = \frac{v_{bearing}}{R}$, and $v_{bearing} = \frac{v}{2}$ has been used. The principle of work and energy: $U = U_{force} + U_{spring}$, $T_2 = T_{platen} + 4T_{bearing}$, from which $U_{force} + U_{spring} = T_{platen} + 4T_{bearing}$. Substitute: $20 - 18 = 11v^2 + 0.75v^2$, from which $v = 0.4126\ \text{m/s}$

==============================<>==============================

====================================<>====================================

Problem 8.25 Consider the system described in Problem 8.24. (a) What maximum distance does the platen move to the right when the force is applied? (b) What maximum velocity does the platen achieve, and how far to the right has the platen moved to the right when it occurs?

Solution:

(a) At the maximum distance the velocity is zero. From the solution to Problem 8.24,

$FS_{max} - \left(\frac{1}{2}\right)kS^2_{max} = 0$, from which $S_{max} = \frac{2F}{k} = 0.222$ m (where the other solution $S_{max} = 0$ has been discarded).

(b) From the solution to Problem 8.24, $FS - \frac{1}{2}kS^2 = \frac{1}{2}m_{platen}v^2 + (4)\left(\frac{3}{16}\right)m_{bearing}v^2$, from which

$$v^2 = \frac{FS - \frac{1}{2}kS^2}{\left(\frac{1}{2}m_{platen} + \frac{3}{4}m_{bearing}\right)}. \quad \frac{d(v^2)}{dS} = F - kS = 0, \text{ from which } S_{v-max} = \frac{F}{k} = \frac{1}{9} = 0.1111 \text{ m, and}$$

$v_{max} = 0.6876$ m/s.

====================================<>====================================

Problem 8.26 The rules of a soapbox derby specify the required combined weight of the car and driver and the radius of the wheels. A young contestant designing her car considers two possibilities: (a) use heavy wheels; (b) use light wheels, making up the weight by adding ballast. Analyze this problem using the principle of work and energy, and explain the advice you would give her.

Solution:

[*Discussion:* Suppose that the total work done is given (that is, the downhill course is fixed). Neglecting friction, this total work is divided into a part that increases *translational* kinetic energy and a part that increases *rotational* kinetic energy, respectively, of the car and driver. Since the *final speed is is dependent only on the final translational kinetic energy*, a design policy that increases the final speed will reduce the fraction of the work that goes into rotational kinetic energy. The part of the total work done to provide translational kinetic energy is the same no matter how the mass is arranged, but the part of the total work done to provide rotational kinetic energy depends on how much mass is in the wheels. Thus, *the mass of the wheels should be reduced. End Discussion.]*

From the principle of work and energy: $U = T_2 - T_1$, where $T_1 = 0$ since the car starts from rest. Treat the wheels as homogenous disks, so that $I_{wheel} = \frac{m_{wheel}}{2}R^2$ for each wheel.

Let the required combined weight be $W_{required} = m_{required}g$.

The work done as the car rolls down the hill is $U = W_{required}h$, where h is the change in height.

This is a fixed amount of work, independent of the design.

The kinetic energy of the car is $T_2 = \frac{1}{2}(m_{required})v^2 + 4\left(\frac{1}{2}\right)I_{wheel}\omega^2 = \left(\frac{1}{2}\right)m_{required}v^2 + m_{wheel}v^2$, , where

$\omega = \frac{v}{R}$ has been used. From $U = T_2$, solve for the velocity: $v^2 = \frac{m_{required}gh}{\left(\frac{m_{required}}{2} + m_{wheel}\right)}$.

Solution continued on next page

Denote the ratio $\chi = \frac{m_{wheel}}{m_{required}}$. Rearrange, from which

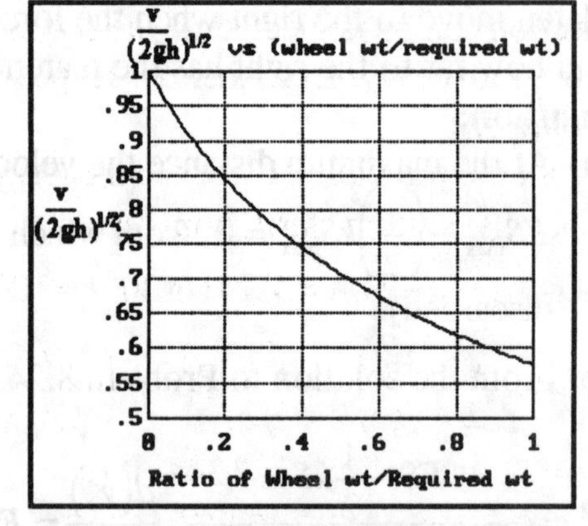

$v = \sqrt{\frac{2gh}{(1+2\chi)}}$ is a decreasing monotone function of χ with a greatest value of $v = \sqrt{2gh}$ for $\chi \to 0$, and a least value $v = \sqrt{\frac{2gh}{3}}$ for $\chi \to 1$.(Since $m_{required}$ is the required mass of driver, car and wheels, the practical range of χ is somewhere within the open range $(0 < \chi < 1)$, with the values $\chi = 0$ and $\chi = 1$ unattainable.) A graph of $\frac{v}{\sqrt{2gh}}$ against χ over the range $0 < \chi < 1$ is shown. Conclusion: For the maximum velocity, χ should be as small as possible: *Use light wheels.*

==================================<>==================================

Problem 8.27 The total moment of inertia of the car's two rear wheels and axle is I_R, and the total moment of inertia of the two front wheels is I_F. The radius of the tires is R, and the total mass of the car, including the wheels, is m. The car is moving at velocity v_0 when the driver applies the brakes. If the car's brakes exert a constant retarding couple M on each wheel and the tires do not slip, determine the car's velocity as a function of the distance s from the point where the brakes were applied.

Solution:

When the car rolls a distance s, the wheels roll through an angle s/R so the work done by the brakes is $V_{12} = -4M(s/R)$. Let m_F be the total mass of the two front wheels, m_R the mass of the rear wheels and axle, and m_c the remainder of the car's mass. When the car is moving at velocity v its total kinetic energy is $T = \frac{1}{2}m_c v^2 + \frac{1}{2}m_R v^2 + \frac{1}{2}I_R(v/R)^2 + \frac{1}{2}m_F v^2 + \frac{1}{2}I_F(v/R)^2 = \frac{1}{2}\left[m + (I_R + I_F)/R^2\right]v^2$.

From the principle of work and energy, $V_{12} = T_2 - T_1$:

$$-4M(s/R) = \frac{1}{2}\left[m + (I_R + I_F)/R^2\right]v^2 - \frac{1}{2}\left[m + (I_R + I_F)/R^2\right]v_0^2$$

Solving for v, we get $v = \sqrt{v_0^2 - 8Ms/\left[Rm + (I_R + I_F)/R\right]}$.

==================================<>==================================

Problem 8.28 In Problem 8.27, $I_R = 0.24 kg-m^2$, $I_F = 0.20 kg-m^2$, $R = 0.30$ *m*, and $m = 1480$ *kg*. If the car's brakes exert a constant retarding couple $M = 650$ *N-m* on each wheel and the tires do not slip, determine the distance s required for the car to come to a stop if it is moving at *100 km/hr*.

Solution:

From the solution of Problem 8.27, the car's velocity when it has moved a distance s is $v = \sqrt{v_0^2 - 8Ms/\left[Rm + (I_R + I_F)/R\right]}$. Setting $v = 0$ and solving for s we obtain $s = \left[Rm + (I_R + I_F)/R\right]v_0^2/(8M)$.

The car's initial velocity is $v_0 = 100{,}000/3600 = 27.8 m/s$

So, $s = [(0.3)(1480) + (0.24 + 0.2)/0.3](27.8)^2/[8(650)] = 66.1 m$.

==================================<>==================================

==================================<>==================================

Problem 8.29 Each box weighs 50 *lb*, the mass moment of inertia of the pulley is 0.6 *slug-ft*^{2}, and friction can be neglected. If the boxes start from rest, determine the magnitude of their velocity when they have moved 4 *ft* from their initial position.

Solution:

From the principle of work and energy: $U = T_2 - T_1$, where $T_1 = 0$ since the system starts from rest. The work done is $U = \int_0^{-4\sin 30} -W dh = W(4\sin 30^o) = 100 \text{ ft-lb}$. The kinetic energy is

$$T_2 = 2\left(\frac{1}{2}\right)\left(\frac{W}{g}\right)v^2 + \left(\frac{1}{2}\right)I\omega^2 = \left(\frac{1}{2}\right)\left(\frac{W}{g} + I\right)v^2 = 1.854v^2,$$ where $\omega = \frac{v}{R} = \frac{v}{1}$ has been used. Substitute into $U = T_2$ and solve: $v = 7.344 \text{ ft/s}$

[*Discussion*: Compare with the solution to Problem 7.17. In Problem 7.17 the time was given and Newton's laws were an "efficient" way to solve for v and x. Here the displacement x is given, and energy methods are the most efficient in solving for v. (This observation forms the basis for two common, useful strategies.) *End*.]

==================================<>==================================

Problem 8.30 The slender bar weighs 30 *lb*, and the cylindrical disk weighs 20 *lb*. The system is released from rest with the bar horizontal. Determine the magnitude of the bar's angular velocity when it is vertical if the bar and disk are welded together at A.

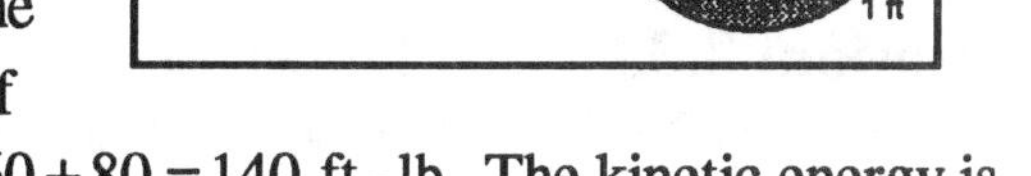

Solution:

From the principle of work and energy: $U = T_2 - T_2$, where $T_1 = 0$ since the system starts from rest. The work done is $U = W_{bar}h_1 + W_{disk}h_2$, where $h_1 = 2$ ft is the change in height of the center of mass of the bar, and $h_2 = 4$ ft is the change in height of the center of mass of the disk, from which $U = (30)2 + (20)4 = 60 + 80 = 140 \text{ ft-lb}$. The kinetic energy is

$$T_2 = \left(\frac{1}{2}\right)I\omega^2,$$ where $I = \left(\frac{1}{3}\right)\left(\frac{W_{bar}}{g}\right)L^2 + \left(\frac{W_{disk}R^2}{2g}\right) + \left(\frac{W_{disk}}{g}\right)L^2 = 15.23 \text{ slug-ft}^2$. Substitute into $U = T_2$ and solve: $|\omega| = 4.288 \text{ rad/s}$

==================================<>==================================

Problem 8.31 In Problem 8.30, determine the magnitude of the bar's angular velocity when it has reached the vertical position if the bar and disk are connected by a smooth pin at A.

Solution:

The disk does not rotate since there are no moments acting on it. From the solution to Problem 8.30, the work done is $U = 140 \text{ ft-lb}$. The kinetic energy is $T_2 = \left(\frac{1}{2}\right)\left(\frac{W_{bar}L^2}{3}\right)\omega^2 + \left(\frac{1}{2}\right)\left(\frac{W}{g}\right)v^2$. Note $v = \omega L$, from which $U = T_2 = 7.46\omega^2$ and $\omega = 4.33 \text{ rad/s}$

==================================<>==================================

Problem 8.32 The 100 *lb* crate is pulled up the inclined surface by the winch. The coefficient of kinetic friction between the crate and the surface is $\mu_k = 0.4$. The mass moment of inertia of the drum on which the cable is wound, including the cable wound on the drum, is $I_A = 3\text{ slug-ft}^2$. The motor exerts a constant couple $M = 40\text{ ft-lb}$ on the drum. If the crate starts from rest, use the principle of work and energy to determine its velocity when it has moved 2 *ft*.

Solution:

From the principle of work and energy: $U = T_2 - T_1$, where $T_1 = 0$, since the system starts from rest. The work done by the weight of the crate is $U_{weight} = \int_0^{2\sin\theta} -W ds = -W(2\sin\theta) = -68.40\text{ ft-lb}$. The work done by friction is $U_{friction} = \int_0^2 -\mu_k N ds = -\mu_k N(2)$. The normal force is $N = W\cos\theta$, from which $U_f = -\mu_k W\cos\theta(2) = -75.175\text{ ft-lb}$. The drum rotates $\theta_{drum} = \frac{2}{0.5} = 4\text{ rad}$, from which the work done by the drum is $U_{drum} = 40(4) = 160\text{ ft-lb}$, from which $U = U_{weight} + U_{friction} + U_{drum} = 16.42\text{ ft-lb}$. The kinetic energy is

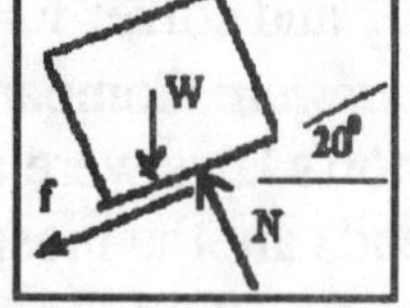

$$T_2 = \left(\frac{1}{2}\right)\left(\frac{W}{g}\right)v^2 + \left(\frac{1}{2}\right)I_A\omega^2 = \left(\frac{1}{2}\right)\left(\frac{W}{g} + \frac{I_A}{(0.5^2)}\right)v^2 = 7.55v^2,$$ where $\omega = \frac{v}{0.5}$ has been used. Substitute into $U = T_2$; solve: $v = 1.474\text{ ft/s}$ [*Check:* Compare with Example 7.2. Apply the chain rule, $\frac{dv}{dt} = v\frac{dv}{dx} = a$, from which, for $v_o = 0$, $v = \sqrt{v_o^2 + 2\alpha\, x} = \sqrt{2(0.544)2} = 1.475\text{ ft/s}$. *check.*]

Problem 8.33 The 2 *ft* slender bars each weigh 4 *lb*, and the rectangular plate weighs 20 *lb*. If the system is released from rest in the position shown, what is the velocity of the plate when the bars are vertical?

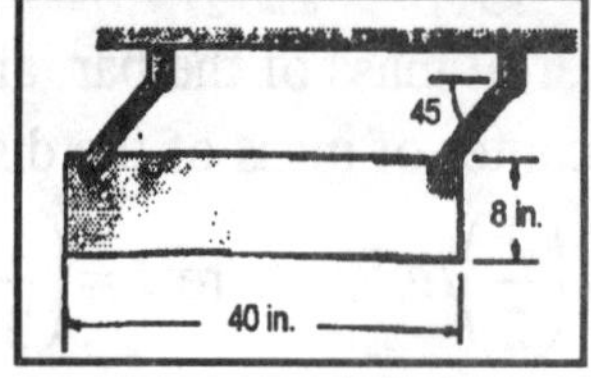

Solution :

The work done by the weights: $U = 2W_{bar}h + W_{plate}2h$, where

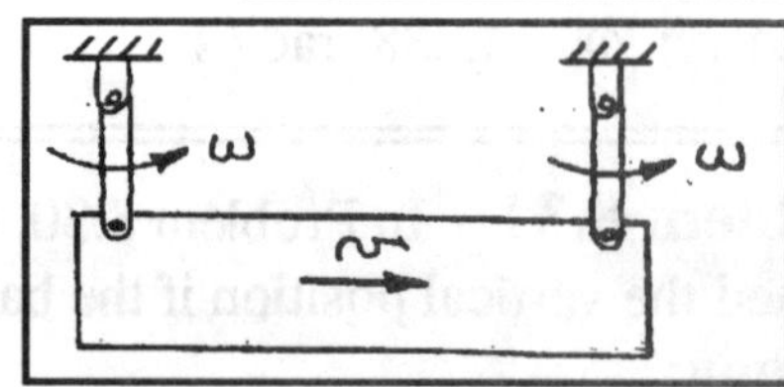

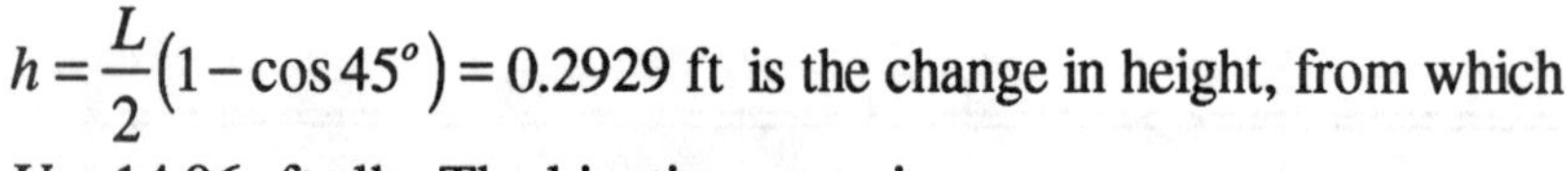

$h = \frac{L}{2}(1 - \cos 45^o) = 0.2929\text{ ft}$ is the change in height, from which

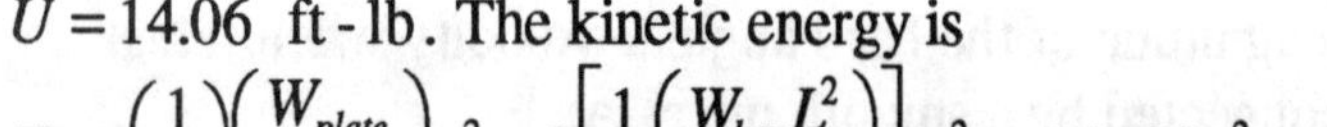

$U = 14.06\text{ ft-lb}$. The kinetic energy is

$$T_2 = \left(\frac{1}{2}\right)\left(\frac{W_{plate}}{g}\right)v^2 + 2\left[\frac{1}{2}\left(\frac{W_{bar}L^2}{3g}\right)\right]\omega^2 = 0.3523v^2,$$ where $\omega = \frac{v}{L}$ has been used. Substitute into $U = T_2$ and solve: $v = 6.32\text{ ft/s}$

==================================<>==================================

Problem 8.34 The slender bar has mass *m* and length *l*. A torsional spring with constant *k* is attached to the bar at the pin support. The spring is unstretched when the bar is vertical. If the bar is released from rest in the position shown, what is its angular velocity as a function of the angle θ between the bar's axis and the vertical?

Solution:

Placing the datum at the level of the pin support, the potential energy is $V = mg\frac{\ell}{2}\cos\theta + \frac{1}{2}k\theta^2$ The bar's kinetic energy is $T = \frac{1}{2}\left(\frac{1}{3}m\ell^2\right)\omega^2$. Conservation of energy requires that $mg\frac{\ell}{2}\cos\theta_0 + \frac{1}{2}k\theta_0^2 = mg\frac{\ell}{2}\cos\theta + \frac{1}{2}k\theta^2 + \frac{1}{6}m\ell^2\omega^2$. Solving for ω we obtain

$\omega = \pm\sqrt{(3/\ell)\left[g(\cos\theta_0 - \cos\theta) + (k/m\ell)(\theta_0^2 - \theta^2)\right]}$

==================================<>==================================

Problem 8.35 The unstretched length of the spring is 1.5 *m*, and its constant is $k = 50\text{ N/m}$. When the 15 *kg* slender bar is horizontal, its angular velocity is 0.1 *rad/s*. What is its angular velocity when it is in the position shown?

Solution:

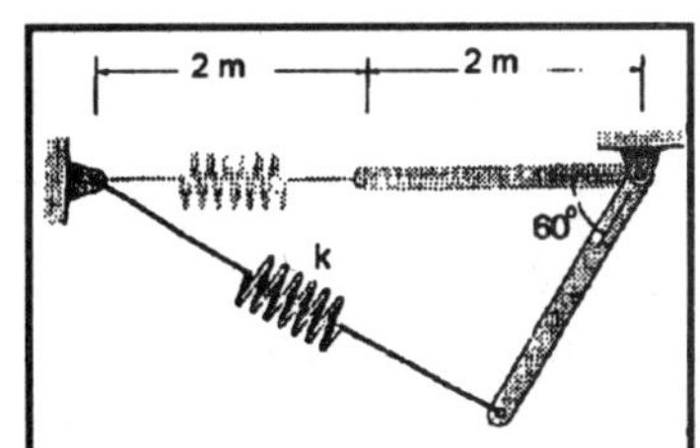

Choose the datum at the initial position. From the conservation of energy $V_1 + T_1 = V_2 + T_2$. The initial potential energy of the spring is $V_1 = -\int_0^{S_1} -ksds = \left(\frac{1}{2}\right)kS_1^2 = \left(\frac{1}{2}\right)(50)(2-1.5)^2 = 6.25\text{ N-m}$. The initial kinetic energy of the bar is $T_1 = \left(\frac{1}{2}\right)\left(\frac{m(2^2)}{3}\right)\omega_1^2 = 10\omega_1^2 = 0.1\text{ N-m}$.

The potential energy of the weight of the bar is $V_{2weight} = -\int_0^{-\sin 60^o} mgdh = -mg\sin 60^o = -127.44\text{ N-m}$.

The potential energy in the spring is $V_{2spring} = -\int_0^S -ksds = \frac{1}{2}kS^2$, where $S = \sqrt{2^2 + 4^2 - 2(2)(4)\cos 60^o} - 1.5 = 1.964\text{ m}$. [*Check*: The triangle is a right triangle, so the Pythagorean theorem also applies. *check*].

From which $V_{2spring} = 96.44\text{ N-m}$, $V_2 = \left(V_{2bar} + V_{2spring}\right) = -31\text{ N-m}$.. The kinetic energy in the final position is $T_2 = \left(\frac{1}{2}\right)\left(\frac{m(2^2)}{3}\right)\omega_2^2 = 10\omega_2^2$. Substitute: $V_1 + T_1 = V_2 + T_2$ and solve: $\omega_2 = 1.932\text{ rad/s}$

==================================<>==================================

Problem 8.36 Pulley A weighs 4 *lb*, $I_A = 0.060$ slug-ft^2, and $I_B = 0.014$ slug-ft^2. If the system is released from rest, what is the velocity of the 16 *lb* weight when it has fallen 2 *ft*?

Solution :

From the principle of work and energy: $U = T_2 - T_1$, where $T_1 = 0$, since the system is at rest initially. The work done by the weight is $U_{weight} = \int_0^{-h} -Wdh = Wh = 16(2) = 32$ ft-lb. The work done by the weight of the pulley A is $U_{pulley} = \int_0^{h_p} -W_p dh = -12h_p$.

By inspection, is the height change of pulley A is one half the drop of the weight, from which $U_{pulley} = -12(1) = -12$ ft-lb, from which $U = U_{weight} + U_{pulley} = 20$ ft-lb.

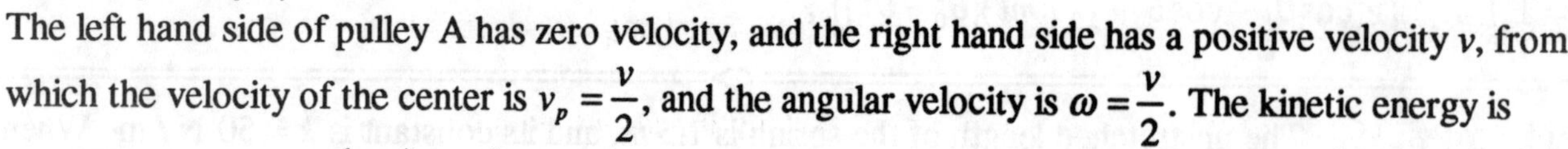

The left hand side of pulley A has zero velocity, and the right hand side has a positive velocity v, from which the velocity of the center is $v_p = \frac{v}{2}$, and the angular velocity is $\omega = \frac{v}{2}$. The kinetic energy is $T_2 = \left(\frac{1}{2}\right)\frac{W}{g}v^2 + \left(\frac{1}{2}\right)\left(\frac{W_p}{g}\right)\left(\frac{v}{2}\right)^2 + \left(\frac{1}{2}\right)I_A\omega_A^2 + \left(\frac{1}{2}\right)I_B\omega_B^2 = 0.3183v^2$, where $\omega_A = \frac{v}{2}$, $\omega_B = \frac{v}{\left(\frac{8}{12}\right)}$ has been used. Substitute into $U = T_2$ and solve: $v = 7.927$ ft/s.

Problem 8.37 The 18 *kg* ladder is released from rest with $\theta = 10^\circ$. The wall and floor are smooth. Modeling the ladder is a slender bar, use conservation of energy to determine its angular velocity when $\theta = 40^\circ$.

Solution:

Choose the datum at floor level. The potential energy at the initial position is $V_1 = mg\left(\frac{L}{2}\right)\cos 10^\circ$. At the final position, $V_2 = mg\left(\frac{L}{2}\right)\cos 40^\circ$.

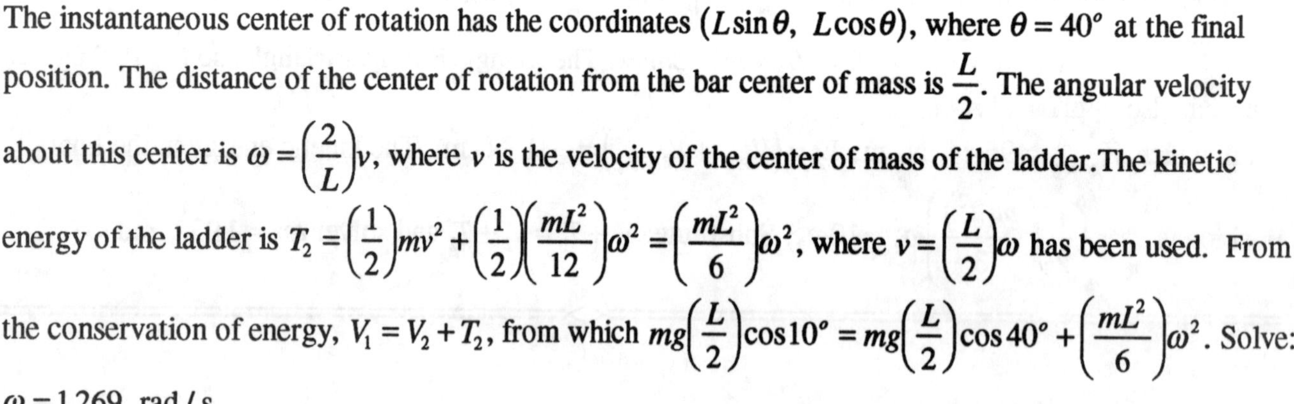

The instantaneous center of rotation has the coordinates $(L\sin\theta,\ L\cos\theta)$, where $\theta = 40^\circ$ at the final position. The distance of the center of rotation from the bar center of mass is $\frac{L}{2}$. The angular velocity about this center is $\omega = \left(\frac{2}{L}\right)v$, where v is the velocity of the center of mass of the ladder. The kinetic energy of the ladder is $T_2 = \left(\frac{1}{2}\right)mv^2 + \left(\frac{1}{2}\right)\left(\frac{mL^2}{12}\right)\omega^2 = \left(\frac{mL^2}{6}\right)\omega^2$, where $v = \left(\frac{L}{2}\right)\omega$ has been used. From the conservation of energy, $V_1 = V_2 + T_2$, from which $mg\left(\frac{L}{2}\right)\cos 10^\circ = mg\left(\frac{L}{2}\right)\cos 40^\circ + \left(\frac{mL^2}{6}\right)\omega^2$. Solve: $\omega = 1.269$ rad/s

===================================<>===================================

Problem 8.38 The spring attached to the slender bar of mass m is unstretched when $\theta = 0^{\circ}$. If the bar falls from rest in the vertical position, what is its angular velocity as a function of θ? Friction is negligible.

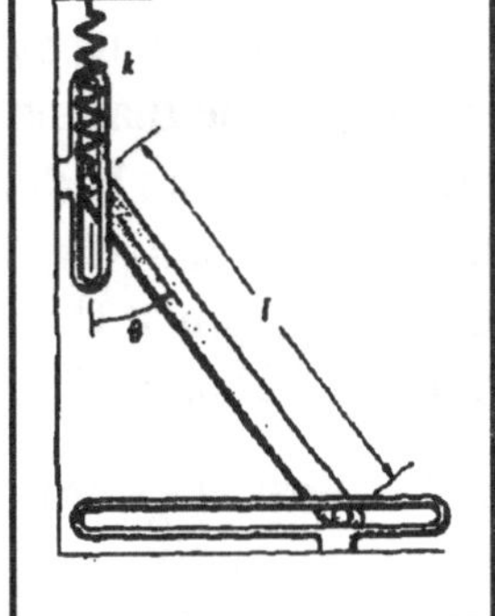

Solution:

In terms of θ, the stretch of the spring is $S = \ell - \ell\cos\theta = \ell(1-\cos\theta)$ Placing the datum at the bottom of the bar, the potential energy is

$V = mg\frac{\ell}{2}\cos\theta + \frac{1}{2}k\ell^2(1-\cos\theta)^2$. The bar's kinetic energy is

$T = \frac{1}{2}mv^2 + \frac{1}{2}I\omega^2$, where v is the magnitude of the velocity of the center of mass. By using the location of the bar's instantaneous center, we see that $v = (\ell/2)\omega$ so the kinetic energy is $T = \frac{1}{2}m(\ell/2)^2\omega^2 + \frac{1}{2}\left(\frac{1}{12}m\ell^2\right)\omega^2 = \frac{1}{6}m\ell^2\omega^2$. Applying conservation of energy,

$\frac{1}{2}mg\ell = \frac{1}{2}mg\ell\cos\theta + \frac{1}{2}k\ell^2(1-\cos\theta)^2 + \frac{1}{6}m\ell^2\omega^2$ solving for ω, we obtain

$$\omega = \pm\sqrt{3\left[(g/\ell)(1-\cos\theta) - (k/m)(1-\cos\theta)^2\right]}$$

===================================<>===================================

Problem 8.39 Consider the system shown inProblem 8.38. The mass and length of the bar are $m = 4\ kg$ and $l = 1.2\ m$. The spring constant is $k = 180\ N/m$. If the bar is released from rest in the position $\theta = 10^{\circ}$, what is its angular velocity when it has fallen to $\theta = 20^{\circ}$?

Solution:

From the solution of Problem 8.38, the potential energy of the system is

$V = \frac{1}{2}mg\ell\cos\theta + \frac{1}{2}k\ell^2(1-\cos\theta)^2$ and its kinetic energy is $T = \frac{1}{6}m\ell^2\omega^2$.

We equate the total energy at $\theta = 10^{\circ}$ to the total energy at $\theta = 20^{\circ}$:

$$\frac{1}{2}(4)(9.81)(1.2)\cos 10^{\circ} + \frac{1}{2}(180)(1.2)^2(1-\cos 10^{\circ})^2 = \frac{1}{2}(4)(9.81)(1.2)\cos 20^{\circ} + \frac{1}{2}(180)(1.2)^2(1-\cos 20^{\circ})^2 + \frac{1}{6}(4)(1.2)^2\omega^2.$$

Solving for ω, we obtain $\omega = 0.804 rad/s$.

===================================<>===================================

==================================<>==================================

Problem 8.40 The 4 *kg* slender bar is pinned to a 2 *kg* slider at A and to a 4 *kg* homogenous cylindrical disk at B. Neglect the friction force on the slider and assume that the disk rolls. If the system is released from rest with $\theta = 60^o$, what is the bar's angular velocity when $\theta = 0$?

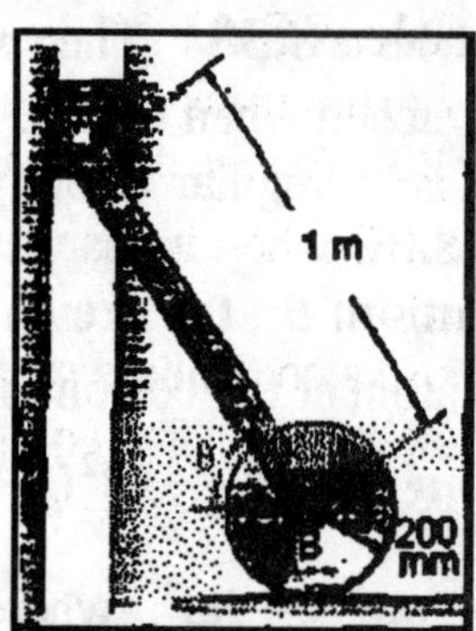

Solution:

Choose the datum at $\theta = 0$. The instantaneous center of the bar has the coordinates $(L\cos\theta,\ L\sin\theta)$ (see figure), and the distance from the center of mass of the bar is $\frac{L}{2}$, from which the angular velocity about the bar's instantaneous center is $v = \left(\frac{L}{2}\right)\omega$, where v is the velocity of the center of mass. The velocity of the slider is $v_A = \omega L\cos\theta$, and the velocity of the disk is $v_B = \omega L\sin\theta$ The potential energy of the system is

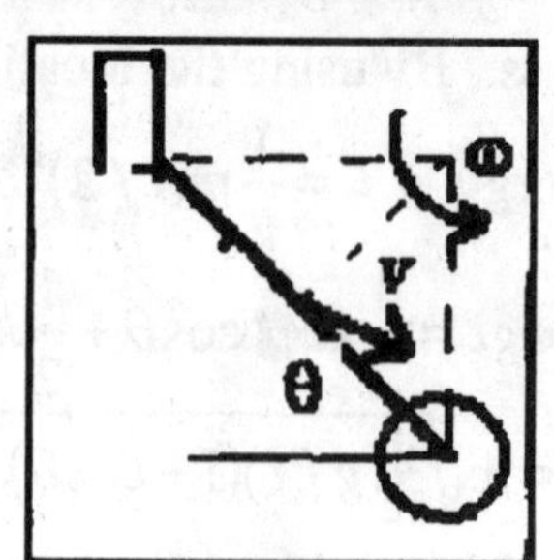

$V_1 = m_A gL\sin\theta_1 + mg\left(\frac{L}{2}\right)\sin\theta_1$. At the datum, $V_2 = 0$. The kinetic energy is

$$T_2 = \left(\frac{1}{2}\right)m_A v_A^2 + \left(\frac{1}{2}\right)mv^2 + \left(\frac{1}{2}\right)\left(\frac{mL^2}{12}\right)\omega^2 + \left(\frac{1}{2}\right)m_B v_B^2 + \left(\frac{1}{2}\right)\left(\frac{m_B R^2}{2}\right)\left(\frac{v_B}{R}\right)^2,$$

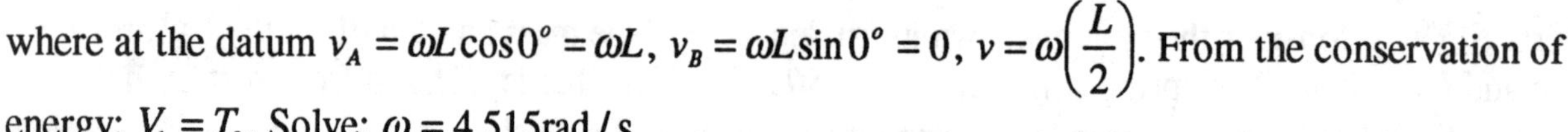

where at the datum $v_A = \omega L\cos 0^o = \omega L$, $v_B = \omega L\sin 0^o = 0$, $v = \omega\left(\frac{L}{2}\right)$. From the conservation of energy: $V_1 = T_2$. Solve: $\omega = 4.515 \text{rad/s}$

==================================<>==================================

Problem 8.41 If the system in Problem 8.40 is released from rest with $\theta = 80^o$, what is the bar's angular velocity when $\theta = 20^o$?

Solution;

Choose the datum at $\theta = 0^o$. Denote $\theta_1 = 80^o$ and $\theta_2 = 20^o$.From the solution to Problem 8.38,

$V_1 = m_A gL\sin\theta_1 + mg\left(\frac{L}{2}\right)\sin\theta_1$ and $V_2 = m_A gL\sin\theta_2 + mg\left(\frac{L}{2}\right)\sin\theta_2$.

At the datum $v_A = \omega L\cos\theta_2$, $v_B = \omega L\sin\theta_2$, and $v = \omega\left(\frac{L}{2}\right)$. The kinetic energy is

$$T_2 = \left(\frac{1}{2}\right)m_A v_A^2 + \left(\frac{1}{2}\right)mv^2 + \left(\frac{1}{2}\right)\left(\frac{mL^2}{12}\right)\omega^2 + \left(\frac{1}{2}\right)m_B v_B^2 + \left(\frac{1}{2}\right)\left(\frac{m_B R^2}{2}\right)\left(\frac{v_B}{R}\right)^2.$$

Substitute into $V_1 = V_2 + T_2$ and solve: $\omega = 3.643\ \text{rad/s}$

==================================<>==================================

================================<>================================

Problem 8.42 The system is in equilibrium in the position shown. The mass of the slender bar ABC is 6 *kg*, the mass of the bar BD is 3 *kg*, and the mass of the slider at C is 1 *kg*. The spring constant is $k = 200\ N/m$. If a constant 100 *N* downward force is applied at A, what is the angular velocity of the bar ABC when it has rotated 20° from its initial position?

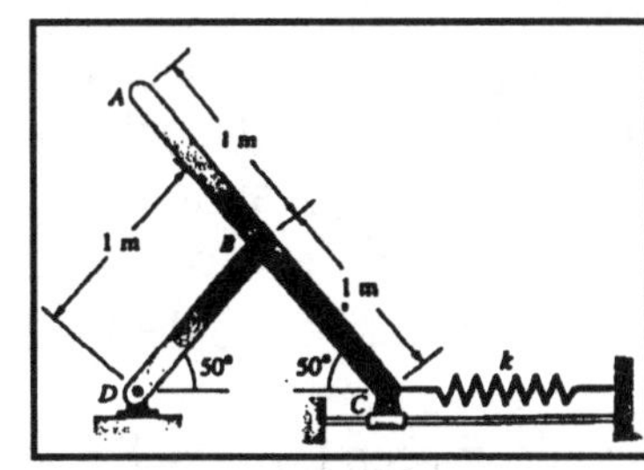

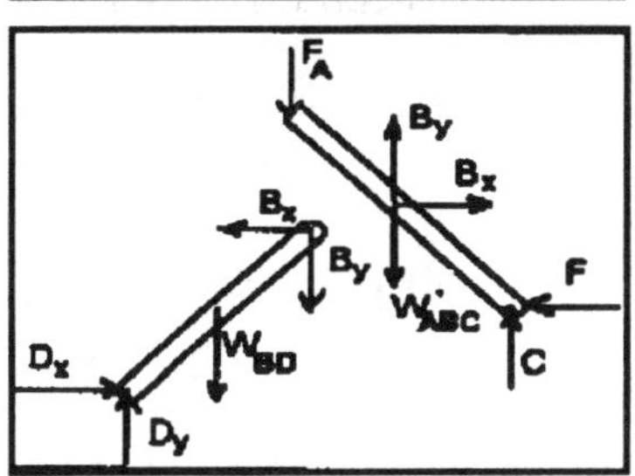

Solution:

Choose a coordinate system with the origin at D and the *x* axis parallel to DC. The equilibrium conditions for the bars: for bar BD, $\sum F_x = -B_x + D_x = 0$, $\sum F_y = -B_y - m_{BD}g + D_y = 0$.

$\sum M_D = B_x \sin 50^o - \left(B_y + \frac{m_{BD}g}{2}\right)\cos 50^o = 0$. For the bar ABC, $\sum F_x = B_x - F = 0$, $\sum F_y = -F_A + C - m_{ABC}g + B_y = 0$.

$\sum M_C = (2F_A - B_y + m_{ABC}g)\cos 50^o - B_x \sin 50^o = 0$. At the initial position $F_A = 0$. The solution: $B_x = 30.87\ N$, $D_x = 30.87\ N$, $B_y = 22.07\ N$, $D_y = 51.5\ N$, $F = 30.87\ N$, $C = 36.79\ N$. [*Note:* Only the value $F = 30.87\ N$ is required for the purposes of this problem.] The initial stretch of the spring is

$S_1 = \frac{F}{k} = \frac{30.87}{200} = 0.154\ m$. The distance D to C is $2\cos\theta$, so that the final stretch of the spring is

$S_2 = S_1 + \left(2\cos 30^o - 2\cos 50^o\right) = 0.601\ m$. From the principle of work and energy: $U = T_2 - T_1$, where $T_1 = 0$ since the system starts from rest. The work done is $U = U_{force} + U_{ABC} + U_{BD} + U_{spring}$. The height of the point A is $2\sin\theta$, so that the change in height is $h = 2\left(\sin 50^o - \sin 50^o\right)$, and the work done by the applied force is $U_{force} = \int_0^h F_A dh = 100(2\sin 50^o - 2\sin 30^o) = 53.2\ N\text{-}m$. The height of the center of mass of bar BD is $\frac{\sin\theta}{2}$, so that the work done by the weight of bar BD is

$U_{BD} = \int_0^h -m_{BD}g dh = -\frac{m_{BD}g}{2}\left(\sin 30^o - \sin 50^o\right) = 3.91\ N\text{-}m$.

The height of the center of mass of bar ABC is $\sin\theta$, so that the work done by the weight of bar ABC is $U_{ABC} = \int_0^h -m_{ABC}g dh = -m_{ABC}g\left(\sin 30^o - \sin 50^o\right) = 15.66\ N\text{-}m$. The work done by the spring is

$U_{spring} = \int_{S_1}^{S_2} -ksds = -\frac{k}{2}(S_2^2 - S_1^2) = -33.72\ N\text{-}m$. Collecting terms, the total work: $U = 39.07\ N\text{-}m$.

The bars form an isosceles triangle, so that the changes in angle are equal; by differentiating the changes, it follows that the angular velocities are equal. The distance D to C is $x_{DC} = 2\cos\theta$, from which $v_C = -2\sin\theta\,\omega$, since D is a stationary. The kinetic energy is

$T_2 = \left(\frac{1}{2}\right)I_{BD}\omega^2 + \left(\frac{1}{2}\right)I_{ABC}\omega^2 + \left(\frac{1}{2}\right)m_{ABC}v_{ABC}^2 + \left(\frac{1}{2}\right)m_C v_C^2 = 5\omega^2$, where $I_{BD} = \frac{m_{BD}}{3}\left(1^2\right)$,

$I_{ABC} = \frac{m_{ABC}}{12}\left(2^2\right)$, $v_{ABC} = (1)\omega$, $v_C = -2\sin\theta\,\omega$. Substitute into $U = T_2$ and solve:

$\boxed{\omega = 2.795\ rad/s}$

================================<>================================

=================================<>=================================

Problem 8.43 Bar AB weighs 10 *lb* and bar BC weighs 6 *lb*. If the system is released from rest in the position shown, what are the angular velocities of the bars at the instant before joint B hits the smooth floor?

Solution:

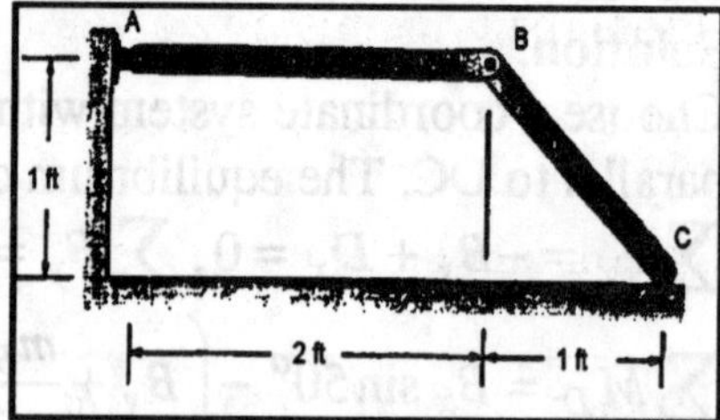

The strategy is to use kinematics to determine (from the velocities at B and C) the ratio of the unknown angular velocities of AB and BC. Choose a coordinate system with the origin at the juncture of the wall and the floor and the *x* axis parallel to the floor. The coordinates of A and B the instant before the bar hits the floor are (0, 1) and $\left(\sqrt{2^2-1^2}=\sqrt{3},\ 0\right)$. The vector from A to B is

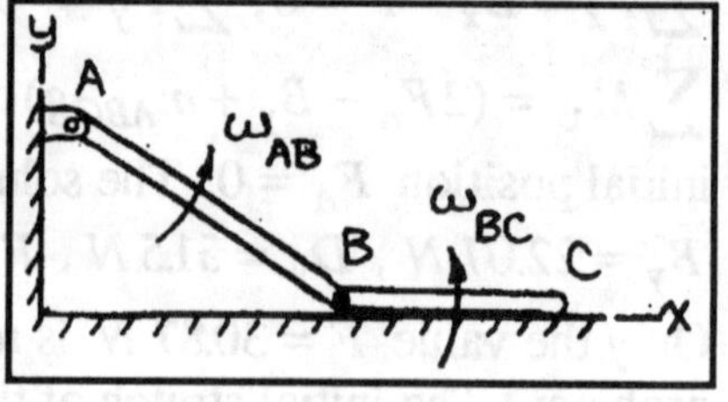

$\vec{r}_{AB}=\sqrt{3}\vec{i}-\vec{j}\ (ft)$. The velocity of point B is

$$\vec{v}_B=\vec{\omega}_{AB}\times\vec{r}_{AB}=\begin{bmatrix}\vec{i} & \vec{j} & \vec{k}\\ 0 & 0 & \omega_{AB}\\ \sqrt{3} & -1 & 0\end{bmatrix}=(\vec{i}+\sqrt{3}\vec{j})\omega_{AB}\ (ft/s).$$

At the instant before point B hits the floor the vector $\vec{r}_{BC}=\sqrt{2}\,\vec{i}$.

The velocity of point C at that instant is $\vec{v}_C=\vec{v}_B+\vec{\omega}_{BC}\times\vec{r}_{BC}=(\vec{i}+\sqrt{3}\vec{j})\omega_{AB}+\begin{bmatrix}\vec{i} & \vec{j} & \vec{k}\\ 0 & 0 & \omega_{BC}\\ \sqrt{2} & 0 & 0\end{bmatrix}$

$\vec{v}_C=\left(\vec{i}\,\omega_{AB}+\left(\sqrt{3}\omega_{AB}+\sqrt{2}\omega_{BC}\right)\vec{j}\right)\ (ft/s)$. The *y* component of the velocity of point C is zero, from which . $\sqrt{3}\omega_{AB}+\sqrt{2}\omega_{BC}=0$, and $\omega_{BC}=-(\sqrt{3})\omega_{AB}/\sqrt{2}$.

The velocity of the center of mass of bar BC the instant before the point B hits the floor is

$$\vec{v}_G=\vec{v}_B+\vec{\omega}_{BC}\times\vec{r}_{BG}=(\vec{i}+\sqrt{3}\vec{j})\omega_{AB}+\begin{bmatrix}\vec{i} & \vec{j} & \vec{k}\\ 0 & 0 & \omega_{BC}\\ \frac{\sqrt{2}}{2} & 0 & 0\end{bmatrix},$$

$\vec{v}_G=\left(\vec{i}\,\omega_{AB}+\left(\sqrt{3}\omega_{AB}+\frac{\omega_{BC}}{\sqrt{2}}\right)\vec{j}\right)=\left(\vec{i}+\frac{\sqrt{3}}{2}\vec{j}\right)\omega_{AB}\ (ft/s)$. From the principle of work and energy: $U=T_2-T_1$, where $T_1=0$ since the system is released from rest. The work done by the weight of the bars is $U_{AB}=\int_0^{-h_{AB}}-W_{AB}\,dh=W_{AB}\,h_{AB}$, $U_{BC}=\int_0^{-h_{BC}}-W_{BC}\,dh=W_{BC}\,h_{BC}$, where $h_{AB}=0.5\,ft$, $h_{BC}=0.5\,ft$, from which $U=U_{AB}+U_{BC}=5+3=8\ ft\text{-}lb$. The kinetic energy is

$T=\left(\frac{1}{2}\right)\left(\frac{W_{AB}r_{AB}^2}{3g}\right)\omega_{AB}^2+\left(\frac{1}{2}\right)\left(\frac{W_{BC}r_{BC}^2}{12g}\right)\omega_{BC}^2+\left(\frac{1}{2}\right)\left(\frac{W_{BC}}{g}\right)v_G^2=0.3937\omega_{AB}^2$, where $\omega_{BC}=-\sqrt{\frac{3}{2}}\omega_{AB}$,

$|v_G|=\left|\sqrt{1+\frac{3}{4}}\omega_{AB}\right|=\left|\sqrt{1.75}\omega_{AB}\right|$ have been used. Substitute into $U=T_2$ and solve:

$\boxed{\omega_{AB}=-4.5075...=-4.51\ rad/s}$ (clockwise), $\boxed{\omega_{BC}=5.52\ rad/s}$ (counterclockwise).

=================================<>=================================

================================<>================================

Problem 8.44 If bar AB in Problem 8.43 is rotating at 1 *rad/s* in the clockwise direction at the instant shown, what are the angular velocities of the bars at the instant before the joint B hits the smooth floor?

Solution:

From the principle of work and energy: $U = T_2 - T_1$. From the solution to Problem 8.43, $U = 8\ ft\text{-}lb$, and $T_2 = 0.3937\omega_{AB}^2$, at the instant before the point B hits the smooth floor.

The initial kinetic energy is $T_1 = \left(\frac{1}{2}\right)\left(\frac{W_{AB}r_{AB}^2}{3g}\right)\omega_{1AB}^2 + \left(\frac{1}{2}\right)\left(\frac{W_{BC}r_{BC}^2}{12g}\right)\omega_{1BC}^2 + \left(\frac{1}{2}\right)\left(\frac{W_{BC}}{g}\right)v_{1G}^2$, where $\omega_{1AB} = -1\ rad/s$, and ω_{1BC}, v_{1G} are found from the kinematics.

The velocity of point B initially is $\vec{v}_{1B} = \vec{\omega} \times \vec{r}_{1AB} = \begin{bmatrix} \vec{i} & \vec{j} & \vec{k} \\ 0 & 0 & \omega_{1AB} \\ 2 & 0 & 0 \end{bmatrix} = \vec{j}2\omega_{1AB} = -2\vec{j}\ (ft/s)$.

The velocity of point C is

$\vec{v}_{1C} = \vec{v}_{1B} + \vec{\omega}_{1BC} \times \vec{r}_{1BC} = -2\vec{j} + \begin{bmatrix} \vec{i} & \vec{j} & \vec{k} \\ 0 & 0 & \omega_{1BC} \\ 1 & -1 & 0 \end{bmatrix} = \vec{i}\omega_{1BC} - (2 - \omega_{1BC})\vec{j}\ (ft/s)$. Since the y component of the velocity of C is zero, $2 - \omega_{1BC} = 0$, from which $\omega_{1BC} = 2\ rad/s$.

The velocity of the center of mass of the bar BC is

$\vec{v}_{1G} = \vec{v}_{1B} + \vec{\omega}_{1BC} \times \vec{r}_{1BG} = -2\vec{j} + \begin{bmatrix} \vec{i} & \vec{j} & \vec{k} \\ 0 & 0 & 2 \\ \frac{1}{2} & -\frac{1}{2} & 0 \end{bmatrix} = -\vec{i} - \vec{j}\ (ft/s)$.

The initial kinetic energy is $T_1 = 0.4559\ ft\text{-}lb$, where $\omega_{1AB} = -1\ rad/s$, $\omega_{1BC} = 2\ rad/s$, and $|v_{1G}| = \sqrt{2}\ rad/s$ have been used.

Substitute into $U = T_2 - T_1$, and solve: $\boxed{\omega_{AB} = -4.634... = -4.63\ rad/s}$ (clockwise) and $\boxed{\omega_{BC} = 5.6757... = 5.68\ rad/s}$ (counterclockwise).

================================<>================================

==================================<>==================================

Problem 8.45 Each bar is of mass m and length l. The spring is unstretched when $\theta = 0°$. If the system is released freom rest with the bars vertical, determine the angular velocity of the lower bar as a function of θ.

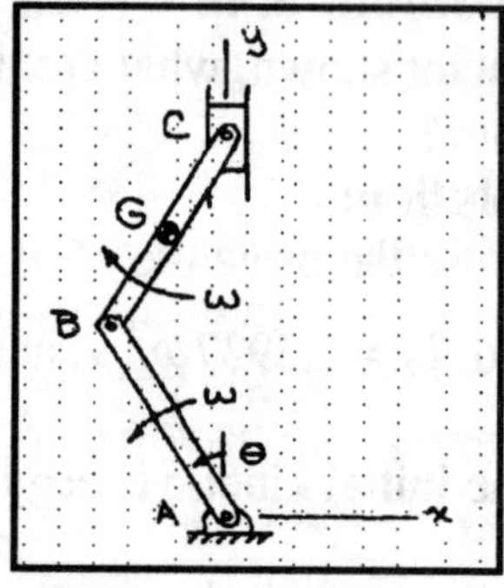

Solution:

In terms of θ , the stretch of the spring is

$S = 2\ell - 2\ell\cos\theta = 2\ell(1-\cos\theta)$.

Placing the datum at the pin support, the potential energy associated with the weights of the the two bars is

$$mg\left(\frac{1}{2}\ell\cos\theta\right) + mg\left(\frac{3}{2}\ell\cos\theta\right) = 2mg\ell\cos\theta,$$

The potential energy of the system is $V = 2mg\ell\cos\theta + 2k\ell^2(1-\cos\theta)^2$ .To determine the kinetic energy of the system, we need to evaluate the velocity of the center of mass of the upper bar. The velocity of B is $\bar{v}_B = \bar{v}_A + \begin{vmatrix} \bar{i} & \bar{j} & \bar{k} \\ 0 & 0 & \omega \\ -\ell\sin\theta & \ell\cos\theta & 0 \end{vmatrix} = -\omega\ell\cos\theta\bar{i} - \omega\ell\sin\theta\bar{j}$, so the velocity of the center of mass G is

$$\bar{v}_G = \bar{v}_B + \bar{\omega}_{BC}\times\bar{r}_{G/B} = -\omega\ell\cos\theta\bar{i} - \omega l\sin\theta\bar{j} + \begin{vmatrix} \bar{i} & \bar{j} & \bar{k} \\ 0 & 0 & -\omega \\ \frac{\ell}{2}\sin\theta & \frac{\ell}{2}\cos\theta & 0 \end{vmatrix} = -\omega\frac{\ell}{2}\cos\theta\bar{i} - \frac{3}{2}\omega\ell\sin\theta\bar{j}.$$

The kinetic energy of the system is $T = \frac{1}{2}I_A\omega^2 + \frac{1}{2}m|\bar{v}_G|^2 + \frac{1}{2}I_G\omega^2$, or

$$T = \frac{1}{2}\left(\frac{1}{3}m\ell^2\right)\omega^2 + \frac{1}{2}m\left[\left(-\frac{1}{2}\omega\ell\cos\theta\right)^2 + \left(-\frac{3}{2}\omega\ell\sin\theta\right)^2\right] + \frac{1}{2}\left(\frac{1}{12}m\ell^2\right)\omega^2 = \left(\frac{1}{3}+\sin^2\theta\right)m\ell^2\omega^2.$$

Conservation of energy requires that $2mg\ell = 2mg\ell\cos\theta + 2k\ell^2 2(1-\cos\theta)^2 + \left(\frac{1}{3}+\sin^2\theta\right)m\ell^2\omega^2$.

Solving for ω , we obtain $\omega = \pm\sqrt{2[(g/\ell)(1-\cos\theta)-(k/m)(1-\cos\theta)^2]/\left(\frac{1}{3}+\sin^2\theta\right)}$

==================================<>==================================

==================================<>==================================

Problem 8.46 The system starts from rest with the crank AB vertical. A constant couple M exerted on the crank causes it to rotate in the clockwise direction, comporssing the gas in the cylinder. Let s be the displacement (in meters) of the piston to the right relative to its initial position. The net force toward the left exerted on the piston by atmospheric pressure and the gas in the cylinder is $350/(1-10s)$ N. The moment of inertia of the crank about A is 0.0003 $kg\text{-}m^2$. The mass of the connecting rod BC is $0.36kg$, and its center of mass is at its midpoint. The connecting rod's moment of inertia about its center of mass is $0.0004kg\text{-}m^2$. The mass of the piston is 4.6 kg. If the clockwise angular velocity of the crank AB is 200 rad/s when it has rotated 90° from its initial position, what is M? (Neglect the work done by the weights of the crank and connecting rod).

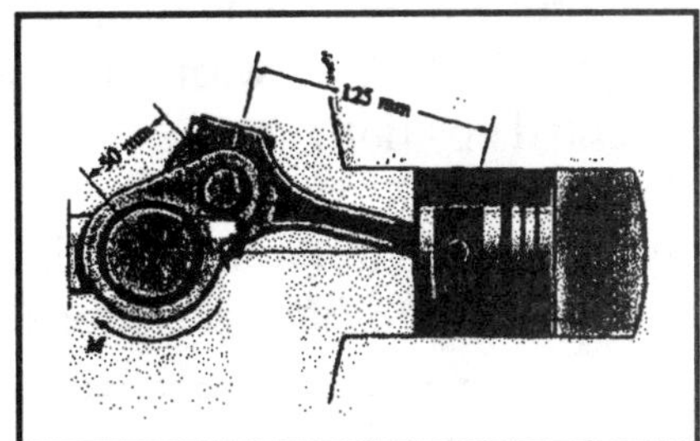

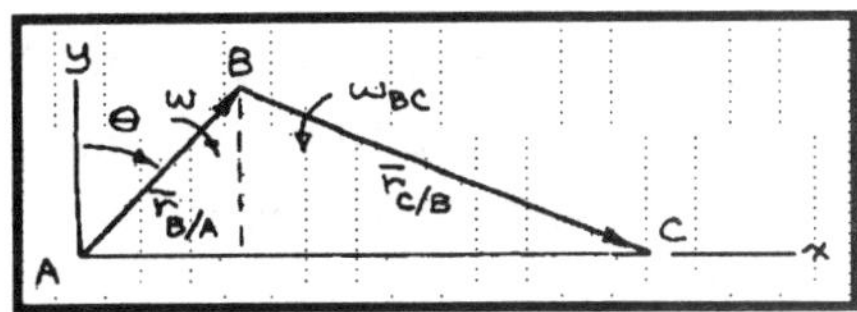

Solution: As the crank rotates through an angle θ the work done by the couple is $M\theta$. As the piston moves to the right a distance s, the work done on the piston by the gas is $-\int_0^s \frac{350ds}{1-10s}$. Letting $1-10s=10z$, this is $\int_{0..1}^{0.1-s} 35\frac{dz}{z} = 35\ell n(1-10s)$. The total work is $U_{12} = M\theta + 35\ell n(1-10s)$. From the given dimensions of the crank and connecting rod, the vector components are $r_{B/Ax} = 0.05\sin\theta$, (1) $r_{B/Ay} = 0.05\cos\theta$ (2); $r_{C/Bx} = \sqrt{(0.125)^2-(0.05\cos\theta)^2}$ (3); $r_{C/By} = -0.05\cos\theta$ (4). The distance s that the piston moves to the right is $s = r_{B/Ax} + r_{C/Bx} - \sqrt{(0.125)^2-(0.05)^2}$ (5). The velocity of B is

$$\bar{v}_B = \bar{v}_A + \begin{vmatrix} \bar{i} & \bar{j} & \bar{k} \\ 0 & 0 & -\omega \\ r_{B/Ax} & r_{B/Ay} & 0 \end{vmatrix} = \omega r_{B/Ay}\bar{i} - \omega r_{B/Ax}\bar{j}.$$

The velocity of C is $\bar{v}_C = \bar{v}_B + \bar{\omega}_{BC} \times \bar{r}_{C/B}$:

$$v_C\bar{i} = \omega r_{B/Ay}\bar{i} - \omega r_{B/Ax}\bar{j} + \begin{vmatrix} \bar{i} & \bar{j} & \bar{k} \\ 0 & 0 & \omega_{BC} \\ r_{c/Bx} & r_{c/By} & 0 \end{vmatrix}.$$

Equating $\bar{i}$ and $\bar{j}$ components, we can solve for v_c and ω_{BC} in terms of ω: $v_C = \left[r_{B/Ay} - r_{C/By}\left(r_{B/Ax}/r_{C/Bx}\right)\right]\omega$ (6); $\omega_{BC} = \left(r_{B/Ax}/r_{C/Bx}\right)\omega$ (7). The velocity of the center of mass G (the midpoint) of the connecting rod BC is

$$\bar{v}_G = \bar{v}_B + \bar{\omega}_{BC} \times \frac{1}{2}\bar{r}_{C/B} = \omega r_{B/Ay}\bar{i} - \omega r_{B/Ax}\bar{j} + \begin{vmatrix} \bar{i} & \bar{j} & \bar{k} \\ 0 & 0 & \omega_{BC} \\ \frac{1}{2}r_{C/Bx} & \frac{1}{2}r_{C/By} & 0 \end{vmatrix}, \text{ or}$$

$v_C = \left(\omega r_{B/Ay} - \frac{1}{2}\omega_{BC}r_{C/By}\right)\bar{i} - \left(\omega r_{B/Ax} - \frac{1}{2}\omega_{BC}r_{C/Bx}\right)\bar{j}$ (8). Let I_A be the moment of inertia of the crank about A, I_{BC} the moment of inertia of the connecting rod about its center of mass, m_{BC} the mass of the connecting rod, and m_p the mass of the piston. The principle of work and energy is: $U_{12} = T_2 - T_1$: $M\theta + 35\ell n(1-10s) = \frac{1}{2}I_A\omega^2 + \frac{1}{2}I_{BC}\omega_{BC}^2 + \frac{1}{2}m_{BC}\left|\bar{v}_G\right|^2 + \frac{1}{2}m_p v_C^2$ (9). Solving this equation for M and using Equation s(1) - (8) with $\omega = 200rad/s$ and $\theta = \pi/2rad$, we obtain $M = 28.2N-m$.

==================================<>==================================

==============================<>==============================

Problem 8.47 In Problem 8.46, if the system starts from rest with the crank AB vertical and the couple *M =40 N-m*, what is the clockwise angular velocity of the crank AB when it has rotated 45° from its initial position?

Solution:

In the solution to Problem 8.46, we substitute Equations (1) - (8) into Equation (9), set $M = 40N-m$ and $\theta = \pi/4\ rad$ and solve for ω obtaining $\omega = 49.6 rad/s$.

==============================<>==============================

Problem 8.48 The *8 kg* slender bar is released from rest in the horizontal position. When it has fallen to the position shown, its angular velocity is 3.226 *rad/s*. Determine the bar's angular momentum (a) about its center of mass; (b) about point A.

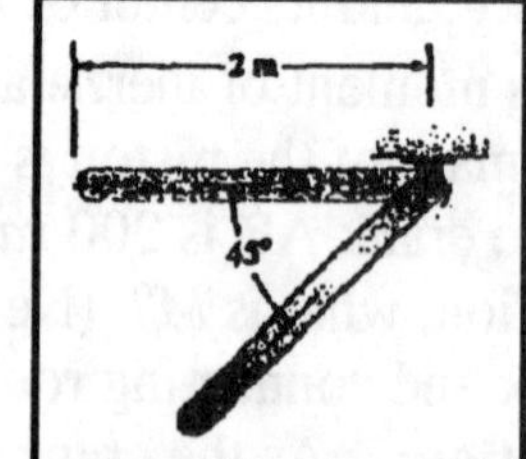

Solution:

The bar's moment of inertia about its center of mass is

$$I = \frac{1}{12}m\ell^2 = \frac{1}{12}(8)(2)^2 = 2.67 kg-m^2.$$

(a) The angular momentum about the center of mass is $H = I\omega = (2.67)(3.226) = 8.60 kg-m^2/s$.

(b) The angular momentum about A (the pinned end) is $H_A = (\bar{r}\times m\bar{v})\bullet\bar{k} + I\omega$ where $\bar{v}$ is the velocity of the center of mass and $\bar{r}$ is the position of the center of mass relative to A. Instead of using the cross product, we can calculate the "moment" of the linear momentum about A:

$$I_A = \frac{1}{3}m\ell^2 = 10.67\ kg-m^2,\quad H = I_A\omega = 34.41\ kg-m^2/s.$$

==============================<>==============================

Problem 8.49 The moments of inertia of the pulley is 0.4 *slug-ft²*. The pulley starts from rest at $t = 0$. For both cases, use momentum principles to determine the pulley's angular velocity at $t = 1\,s$.

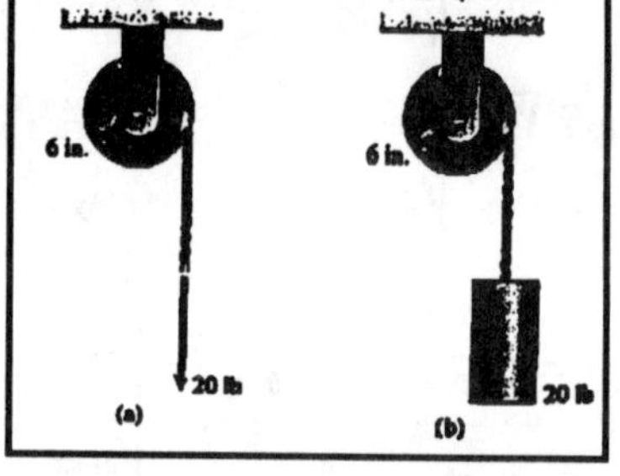

Solution:

Choose *y* coordinate positive upward. The force and weight act downward. (a) The angular momentum-impulse for the pulley is

$\int_{t_1}^{t_2}\sum M\,dt = -\int_0^1 FRdt = -20(0.5)(1) = I\omega_2 - I\omega_1$. Since the pulley starts from rest, $\omega_1 = 0$, from which $-10 = (0.4)\omega_2$, and

$\boxed{\omega_2 = -\frac{10}{0.4} = -25\ \mathrm{rad/s}}$ (clockwise). (b) Denote the tension in the cable by *T*. The angular momentum-impulse for the pulley is

$\int_{t_1}^{t_2}\sum M\,dt = -\int_0^1 TR\,dt = I\omega_2 - I\omega_1$, where $\omega_1 = 0$. The linear-impulse for the cylinder is

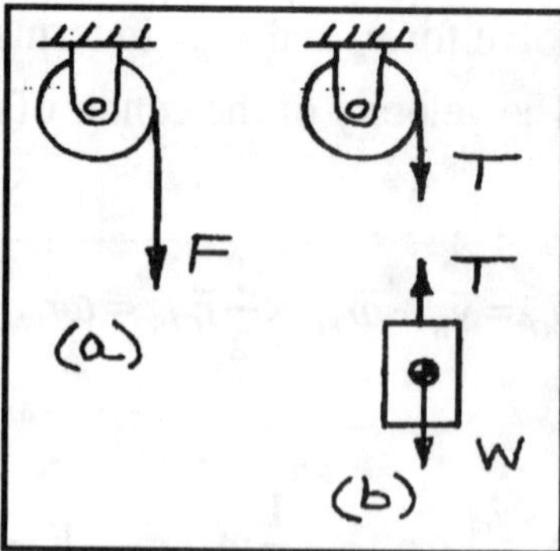

$\int_{t_1}^{t_2}\sum F\,dt = -\int_0^1 (W-T)\,dt = \left(\frac{W}{g}\right)v_2 - \left(\frac{W}{g}\right)v_1$, where $v_1 = 0$. By inspection, $v_2 = 0.5\omega_2$, from which the two equations: $-T(0.5) = I\omega_2$, $T = W + \frac{W}{g}(0.5\omega_2)$. Solve:

$\omega_2 = \frac{-2gW}{4gI + W}$, $\boxed{\omega_2 = -18.0\ rad/s}$ (clockwise).

==============================<>==============================

Problem 8.50

Problem 8.50 An astronaut fires a thruster of his maneuvering unit, exerting a force $T = 2(1+t)\ N$, where t is in seconds. The combined mass of the astronaut and his equipment is 122 *kg*, and the moment of inertia about their center of mass is 45 *kg-m*^{2}. Modeling the astronaut and his equipment as a rigid body, use the principle of angular impulse and angular momentum to determine how long it takes for his angular velocity to reach 0.1 *rad/s*.

Solution :

From the principle of impulse and angular momentum,

$\int_{t_1}^{t_2} \sum M\, dt = I\omega_2 - I\omega_1$, where $\omega_1 = 0$, since the astronaut is initially stationary. The normal distance from the thrust line to the center of mass is $R = 0.3\ m$, from which

$\int_0^{t_2} 2(1+t)(R)\, dt = I\omega_2$. $0.6\left(t_2 + \dfrac{t_2^2}{2}\right) = 45(0.1)$. Rearrange: $t_2^2 + 2bt_2 + c = 0$, where $b = 1$, $c = -15$.

Solve: $t_2 = -b \pm \sqrt{b^2 - c} = 3, \ = -5$. Since the negative solution has no meaning here, $\boxed{t_2 = 3\ s}$

Problem 8.51

Problem 8.51 The maneuvering unit in Problem 8.50 exerts an impulsive force T of 0.2 *s* duration, giving the astronaut a counterclockwise angular velocity of one revolution per minute. (a) What is the average value of the impulsive force? (b) What is the magnitude of the change in the velocity of his center of mass?

Solution:

(a) From the principle of moment impulse and angular momentum, $\int_{t_1}^{t_2} \sum M\, dt = I\omega_2 - I\omega_1$, where $\omega_1 = 0$ since the astronaut is initially stationary. The angular velocity is $\omega_2 = \dfrac{2\pi}{60} = 0.1047\ rad/s$, from which $\int_0^{0.2} TR\, dt = I\omega_2$, from which $T(0.3)(0.2) = 45(\omega_1)$,

$$\boxed{T = \frac{45\omega_2}{0.06} = 78.54\ N}$$

(b) From the principle of impulse and linear momentum $\int_{t_1}^{t_2} \sum F\, dt = m(v_2 - v_1)$ where $v_1 = 0$ since the astronaut is initially stationary. $\int_0^{0.2} T\, dt = mv_2$, from which $\boxed{v_2 = \dfrac{T(0.2)}{m} = 0.1288... = 0.129\ m/s}$

====================================<>====================================

Problem 8.52 A flywheel attached to an electric motor is initially at rest. At $t = 0$, the motor exerts a couple $M = 200e^{-0.1t}$ N - m on the flywheel. The moment of inertia of the flywheel is 10 kg - m^2. (a) What is the flywheel's angular velocity at $t = 10$ s? (b) What maximum angular velocity will the flywheel attain?

Solution:

(a) From the principle of moment impulse and angular momentum, $\int_{t_1}^{t_2} \sum M\, dt = I\omega_2 - I\omega_1$, where $\omega_1 = 0$, since the motor starts from rest. $\int_0^{10} 200e^{-0.1t}\, dt = \frac{200}{0.1}\left[-e^{-0.1t}\right]_0^{10} = 2000\left[1 - e^{-1}\right]$, $= 1264.2$ N - m - s. From which $\boxed{\omega_2 = \frac{1264.24}{10} = 126.4 \ rad/s}$.

(b) An inspection of the angular impulse function shows that the angular velocity of the flywheel is an increasing monotone function of the time, so that the greatest value occurs as $t \to \infty$.

$$\boxed{\omega_{2\max} = \lim_{t\to\infty} \frac{200}{(10)(0.1)}\left[1 - e^{-0.1t}\right] \to 200 \ rad/s}$$

[*Note:* If the maximum (minimum) of a function occurs at the end of an interval, it is strictly speaking *the greatest (least) value* of the function in the interval. However, since the engineer's interest in extreme values is most often independent of where the extremes occur, the authors do not distinguish between maximum (minimum) and greatest (least) values in the text. *End of Note.*)

====================================<>====================================

Problem 8.53 The main landing gear of a Boeing 747 has a moment of inertia of 17 *slug-ft*^{2} and a 2 *ft* radius. The airplane is moving at 245 *ft/s* when it touches down. Suppose that you measure the skid marks where the plane touches down and find that they are 30 *ft* long. Assuming that the airplane's velocity and the normal force are constant while the wheel skids, use the principle of angular impulse and angular momentum to estimate the friction force exerted on the wheel while it skids.

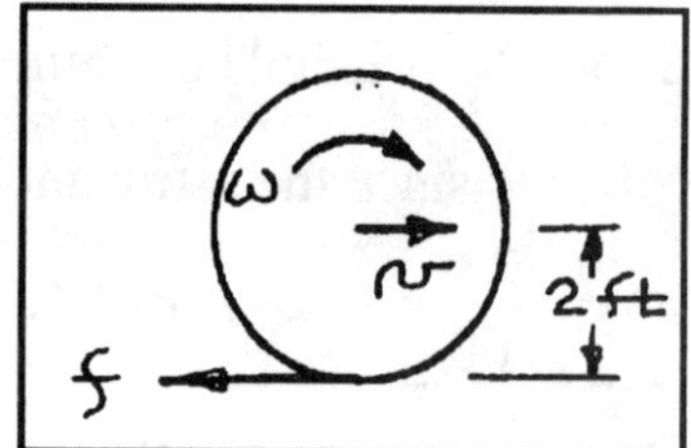

Solution:

From the principle of moment impulse and angular momentum, $\int_{t_1}^{t_2} \sum M\, dt = I\omega_2 - I\omega_1$, where $\omega_1 = 0$, since the wheel is initially stationary. Denote the friction force by *f*. $\int_{t_1}^{t_2} 2f\, dt = I\omega_2$, from which $f = \frac{I\omega_2}{R(t_2 - t_1)}$. The angular velocity is $\omega_2 = \frac{v}{R} = \frac{245}{2} = 122.5$ rad/s, since the velocity is constant over the interval; $t_2 - t_1 = \frac{s}{v} = \frac{30}{245} = 0.1224$ s, from which $\boxed{f = 8503.5 \ lb}$

====================================<>====================================

==================================<>==================================

Problem 8.54 The force a club exerts on a 1.62 *oz* golf ball is shown. The ball is 1.68 *in.* in diameter and can be modeled as a homogenous sphere. The club is in contact for 0.0006 *s*, and the magnitude of the velocity of the ball's center of mass after it is struck is 160 *ft/s*. What is the ball's angular velocity after it is struck?

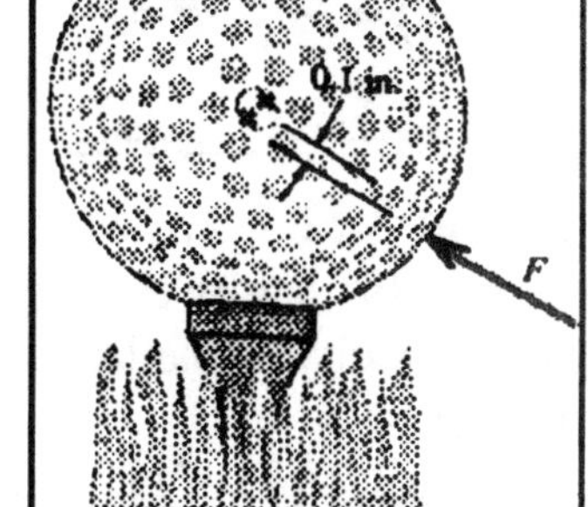

Solution:

The strategy is to use the principle of linear impulse-momentum to determine the magnitude of the force, from which the angular velocity can be found from the principle of moment impulse and angular momentum. From the principle of impulse and linear momentum $\int_{t_1}^{t_2} \sum F\, dt = m(v_2 - v_1)$ where $v_1 = 0$ since the system is initially stationary, from which $F(t_2 - t_1) = mv_2$, From the principle of moment impulse and angular momentum, $\int_{t_1}^{t_2} \sum M\, dt = I(\omega_2 - \omega_1)$, where $\omega_1 = 0$ since the ball is initially stationary, from which $F(d)(t_1 - t_2) = I\omega_2$, from which $\omega_2 = \frac{Fd(t_2 - t_1)}{I} = \frac{mdv_2}{I}$. Substitute $m = \frac{1.62}{16(32.17)} = 3.15 \times 10^{-3}$ *slug*, $R = \frac{1.68}{2(12)} = 0.07$ *ft*, $I = \frac{2mR^2}{5} = 6.17 \times 10^{-6}$ *slug - ft*^{2}, $d = \frac{0.1}{12} = 0.14$ *ft*, $v_2 = 160$ *ft / s*, from which

$\boxed{\omega_2 = 680.3 \text{ rad / s}}$

==================================<>==================================

Problem 8.55 The suspended 8 *kg* slender bar is subjected to a horizontal impulsive force at B. The average value of the force is 1000 *N*, and its duration is 0.03 *s*. If the force causes the bar to swing to the horizontal position before coming to a stop, what is the distance *h*?

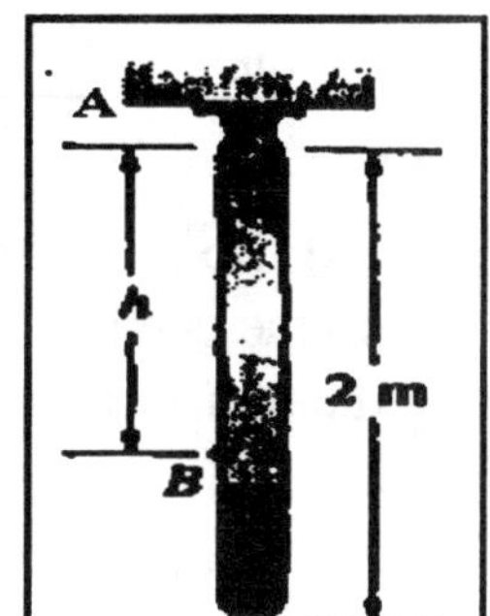

Solution:

The strategy is to use the principle of work and energy to determine the bar's angular velocity the instant after impact, and to use the principle of moment impulse and angular momentum to determine the distance *h*.

From the principle of work and energy, $U = T_2 - T_1$, where $T_2 = 0$, since the bar comes to rest at the horizontal position.

Choose a coordinate system with the origin at A and the y axis positive upward.

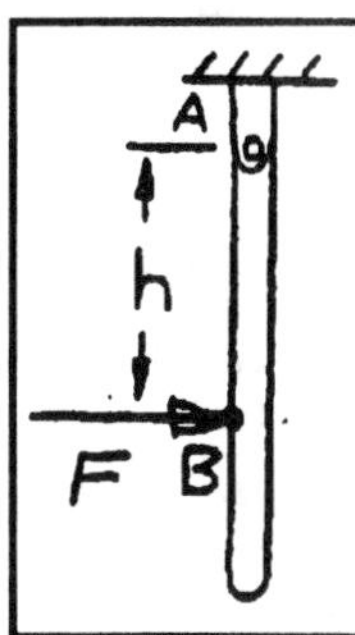

The work done by the weight of the bar is $U = \int_0^{\frac{L}{2}} -mg\, dh = -mg\left(\frac{L}{2}\right)$.

Denote the angular velocity the instant after impact by ω_2.

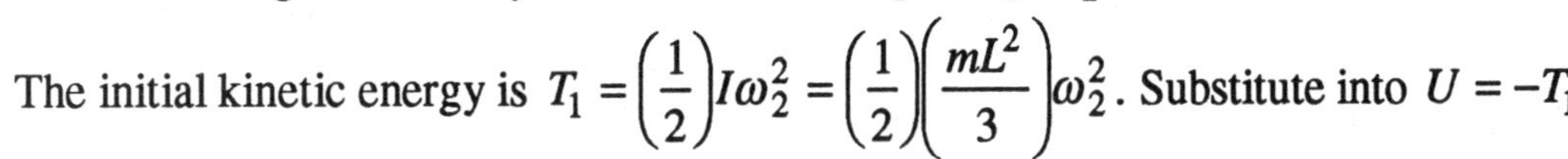

The initial kinetic energy is $T_1 = \left(\frac{1}{2}\right)I\omega_2^2 = \left(\frac{1}{2}\right)\left(\frac{mL^2}{3}\right)\omega_2^2$. Substitute into $U = -T_1$ and solve: $\omega_2 = \sqrt{\frac{3g}{L}} = \sqrt{\frac{3(9.81)}{2}} = 3.84$ *rad / s*.

From the principle of angular impulse and angular momentum,

Solution continued on next page

$\int_{t_1}^{t_2} \sum M\, dt = I(\omega_2 - \omega_1)$, where $\omega_1 = 0$ since the bar is initially at rest, from which

$$Fh(t_2 - t_1) = I\omega_2,\ h = \frac{I\omega_2}{F(t_2 - t_1)} = \frac{\left(\frac{mL^2}{3}\right)\omega_2}{F(t_2 - t_1)} = \frac{8(2^2)(3.84)}{(3)(1000)(0.03)} \boxed{= 1.36\ m}$$

================================<>================================

Problem 8.56 For what value of the distance h in Problem 8.55 will no average horizontal force be exerted on the bar by the support A when the horizontal impulsive force is applied at B? What is the angular velocity of the bar just after the impulsive force is applied?

Solution:

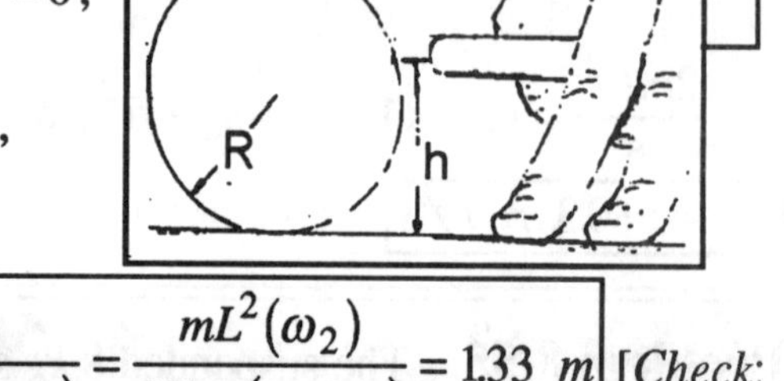

From the principle of impulse and linear momentum $\int_{t_1}^{t_2} A_x dt + \int_{t_1}^{t_2} F_{ave} dt = m(v_2 - v_1)$ where $v_1 = 0$ since the system is initially at rest. From the principle of moment impulse and angular momentum, $\int_{t_1}^{t_2} F_{ave} h\, dt = I(\omega_2 - \omega_1)$, where $\omega_1 = 0$. For $A_x = 0$, and $v_2 = \left(\frac{L}{2}\right)\omega_2$, the two equations: $F_{ave}(t_2 - t_1) = m\omega_2\left(\frac{L}{2}\right)$, $hF_{ave}(t_2 - t_1) = I\omega_2$. Solve: ,

$$\boxed{\omega_2 = \frac{F_{ave}(t_2 - t_1)}{m} = \frac{(1000)(0.03)}{8} = 3.75\ rad/s}\ \boxed{h = \frac{I\omega_2}{F_{ave}(t_2 - t_1)} = \frac{mL^2(\omega_2)}{3F_{ave}(t_2 - t_1)} = 1.33\ m}$$ [*Check*:

================================<>================================

Problem 8.57 The force exerted on the cue ball by the cue is horizontal. Determine the value of h for which the cue ball rolls without slipping. (Assume that the average friction force exerted on the ball during the impact is zero.)

Solution:

From the principle of moment impulse and angular momentum, $\int_{t_1}^{t_2} (h - R)F dt = I(\omega_2 - \omega_1)$, where $\omega_1 = 0$ since the ball is initially stationary. From the principle of impulse and linear momentum $\int_{t_1}^{t_2} F\, dt = m(v_2 - v_1)$ where $v_1 = 0$ since the ball is initially stationary. *Since the ball rolls*, $v_2 = R\omega_2$, from which the two equations: $(h - R)F(t_2 - t_2) = I\omega_2$, $F(t_2 - t_1) = mR\omega_2$. The ball is a homogenous sphere, from which $I = \frac{2}{5}mR^2$. Substitute: $(h - R)mR\omega_2 = \frac{2mR^2}{5}\omega_2$. Solve:

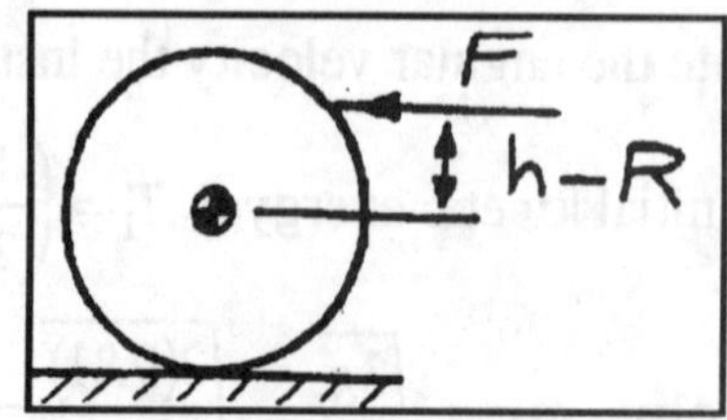

$$\boxed{h = \left(\frac{7}{5}\right)R}$$

================================<>================================

====================================<>==================================

Problem 8.58 Two gravity research satellites ($m_A = 250kg$, $I_A = 350kg - m^2$, $m_B = 50kg$, $I_B = 16kg - m^2$) are tethered by a cable. The satellites and cable rotate with angular velocity $\omega_0 = 0.25rpm$. Ground controllers order satellite A to slowly unreel *6 m* of additional cable. What is the anguar velocity afterward?

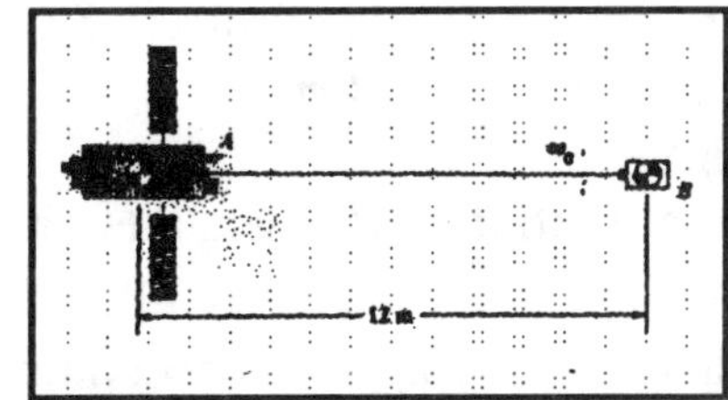

Solution:

The initial distance from A to the common center of mass is $\bar{x}_0 = \dfrac{(0)(250)+(12)(50)}{250+50} = 2m$. The final distance from A to the common center of mass is $\bar{x} = \dfrac{(0)(250)+(18)(50)}{250+50} = 3m$. The total angular momentum about the center of mass is conserved.

$\bar{x}_0 m_A(\bar{x}_0\omega_0) + I_A\omega_0 + (12-\bar{x}_0)m_B[(12-\bar{x}_0)\omega_0] + I_B\omega_0 = \bar{x}m_A(\bar{x}\omega) + I_A\omega + (18-\bar{x})m_B[(18-\bar{x})\omega] + I_B\omega$:

or $(2)(250)(2)\omega_0 + 350\omega_0 + (10)(50)(10)\omega_0 + 16\omega_0 = (3)(250)(3)\omega + 350\omega + (15)(50)(15)\omega + 16\omega$.

We obtain $\omega = 0.459\omega_0 = 0.115rpm.$

====================================<>==================================

Problem 8.59 Solve Problem 8.58 by treating the satellites as particles (that is, neglect their moments of inertia I_A and I_B), and compare your answer to that of Problem 8.58.

Solution:

Setting I_A and I_B equal to zero in the solution of Problem 8.58, we obtain $\omega = 0.444\omega_0 = 0.111rpm.$

====================================<>==================================

Problem 8.60 The *2-kg* bar is *1-m* in length. It rotates in the horizontal plane about the smooth pin. The *6-kg* collar slides on the smooth bar. The unstretched length of the spring is *0.2-m*, and its spring constant is *k = 10 N/m*. At the instant shown, the angular velocity of the bar is $\omega_0 = 2rad/s$, the distance from the pin to the collar is *r = 0.6 m*, and the radial velocity of the collar is zero. Use conservations of angular momentum to determine the bar's angular velocity when the distance from the pin to the collar is *r = 0.8 m*. Neglect the moment of inertia of the collar about its center of mass; that is, treat the collar as a particle.

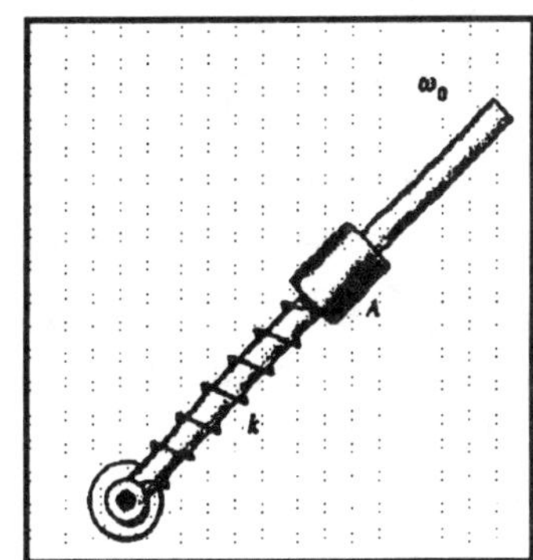

Solution:

Let $r_0 = 0.6m$ be the radial distance to the collar at the initial instant. The total angular momentum of the bar and collar about the pin is conserved:

$\left(\frac{\ell}{2}\right)m_{bar}\left(\frac{\ell}{2}\right)\omega_0 + I_{bar}\omega_0 + r_0 m_{collar}(r_0\omega_0) = \left(\frac{\ell}{2}\right)m_{bar}\left(\frac{\ell}{2}\right)\omega + I_{bar}\omega + r m_{collar}(r\omega)$:

$(0.5)(2)(0.5)\omega_0 + \frac{1}{12}(2)(1)^2\omega_0 + (0.6)(6)(0.6)\omega_0 = (0.5)(2)(0.5)\omega + \frac{1}{12}(2)(1)^2\omega + (0.8)(6)(0.8)\omega.$

Solving, we obtain $\omega = 0.627\omega_0 = 1.25rad/s$

====================================<>==================================

==================================<>==================================

Problem 8.61 In Problem 8.60, what is the collar's radial velocity when the distance from the pin to the collar is $r = 0.8\ m$?

Solution:

In the solution of Problem 8.60, it was shown that when $r = 0.8m$ the angular velocity is $\omega = 1.25 rad/s$. Let I_0 be the bar's moment of inertia about the pinned end. To determine the collar's radial velocity, we use conservation of energy.

$$\frac{1}{2}k(r_0 - 0.2)^2 + \frac{1}{2}I_0\omega_0^2 + \frac{1}{2}m_{collar}(r_o\omega_o)^2 = \frac{1}{2}k(r-0.2)^2 + \frac{1}{2}I_0\omega^2 + \frac{1}{2}m_{collar}\left[(r\omega)^2 + v_r^2\right]$$

$$\frac{1}{2}(10)(0.6-0.2)^2 + \frac{1}{2}\left[\frac{1}{3}(2)(1)^2\right](2)^2 + \frac{1}{2}(6)(0.6)^2(2)^2 = \frac{1}{2}(10)(0.8-0.2)^2 + \frac{1}{2}\left[\frac{1}{3}(2)(1)^2\right](1.25)^2 + \frac{1}{2}(6)\left[(0.8)^2(1.25)^2 + v_r^2\right]$$

Solving, we obtain $v_r = 0.608 m/s$.

==================================<>==================================

Problem 8.62 In Problem 8.60, suppose that the moment of inertia of the collar about its center of mass is 0.2 $kg\text{-}m^2$, Determine the bar's angular velocity when the distance from the pin to the collar is $r = 0.8\ m$ and compare your answer with that of Problem 8.60.

Solution:

We must include the collar's angular momentum about its center of mass in the equation of conservation of angular momentum.

$$\left(\frac{\ell}{2}\right)m_{bar}\left(\frac{\ell}{2}\right)\omega_0 + I_{bar}\omega_0 + r_0 m_{collar}(r_0\omega_0) + I_{collar}\omega_0 = \left(\frac{\ell}{2}\right)m_{bar}\left(\frac{\ell}{2}\right)\omega + I_{bar}\omega + r m_{collar}(r\omega) + I_{collar}\omega:$$

Inserting numbers, we get

$$(0.5)(2)(0.5)\omega_0 + \frac{1}{12}(2)(1)^2\omega_0 + (0.6)(6)(0.6)\omega + 0.2\omega = (0.5)(2)(0.5)\omega + \frac{2}{12}\omega + (0.8)(6)(0.8)\omega + (0.2)\omega$$

Solving, we obtain $\omega = 0.643\omega_0 = 1.29 rad/s$

==================================<>==================================

Problem 8.63 The circular bar is welded to the vertical shafts, which can rotate freely in bearings at A and B. Let I be the moment of inertia of the circular bar and shafts about the vertical axis. The circular bar has an initial angular velocity ω_0 and the mass m is released in the position shown with no velocity relative to the bar. Determine the angular velocity of the circular bar as a function of the angle β between the vertical and the position of the mass. Neglect the moment of inertia of the mass about its center of mass: that is, treat the mass as a particle.

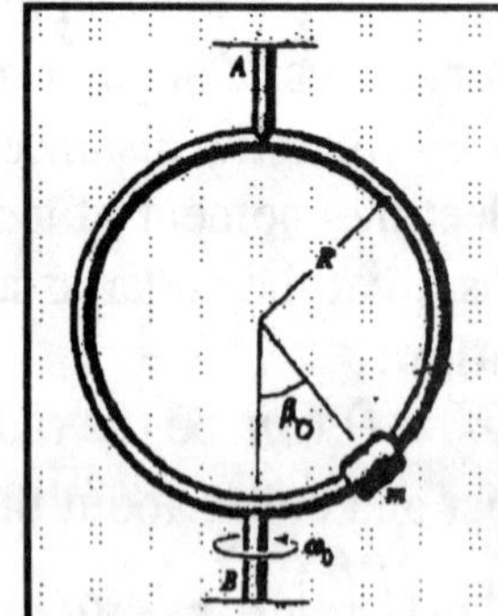

Solution:

Angular momentum about the vertical axis is conserved:

$$I\omega_o + (R\sin\beta_0)m(R\sin\beta_0)\omega_0 = I\omega + (R\sin\beta)m(R\sin\beta)\omega.$$

Solving for ω, $\omega = \left(\dfrac{I + mR^2\sin^2\beta_0}{I + mR^2\sin^2\beta}\right)\omega_0.$

==================================<>==================================

====================================<>====================================

Problem 8.64 The 2 *kg* slender bar starts from rest in the vertical position and falls, striking the smooth surface at P. The coefficient of restitution of the impact is $e = 0.5$. When the bar rebounds, through what angle relative to the horizontal will it rotate? *Strategy*: Use the coefficient of restitution to relate the bar's velocity at P just after the impact to its value just before the impact.

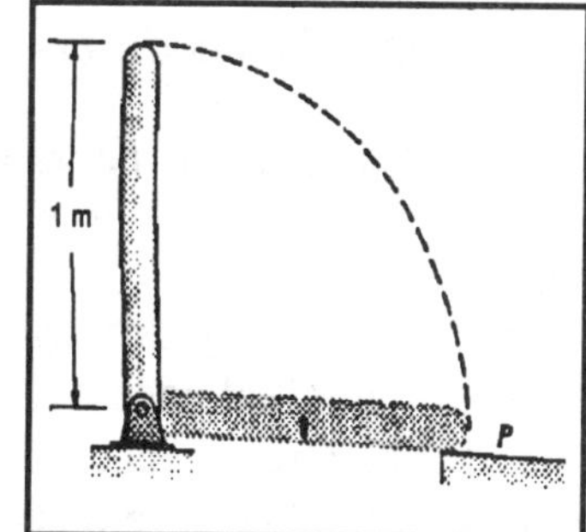

Solution :

Choose a coordinate system with the origin at the center of mass of the bar at the initial position, and the *y* axis positive upward. The strategy is to use the principle of work and energy to determine the angular velocity of the bar the instant before impact; use the coefficient of restitution to determine the bar's velocity the instant after impact, and use the principle of work and energy to determine the maximum angle reached after rebound. From the principle of work and energy, $U = T_2 - T_1$, where $T_1 = 0$, since the bar starts from rest. The work done by gravity is $U = \int_{h_1}^{h_2} -mg dh = -mg\left(-\frac{L}{2} - 0\right) = mg\left(\frac{L}{2}\right)$ $N\text{-}m$

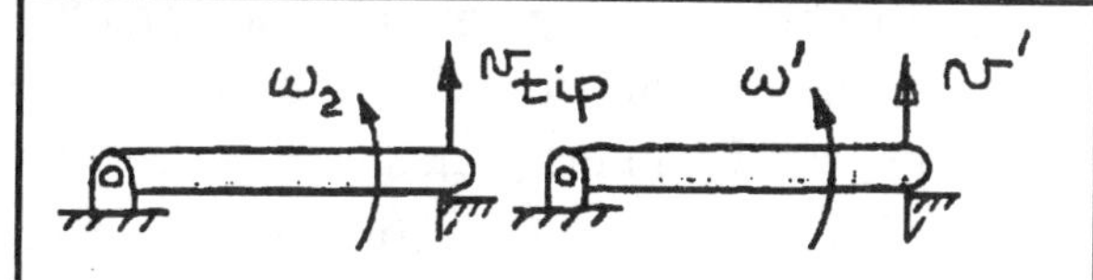

The kinetic energy is $T_2 = \frac{1}{2} I\omega_2^2 = \frac{1}{2}\left(\frac{mL^2}{3}\right)\omega_2^2$. Substitute into $U = T_2$, to obtain

$mg\left(\frac{L}{2}\right) = \left(\frac{1}{2}\right)\left(\frac{mL^2}{3}\right)\omega_2^2$, from which $\omega_2 = -\sqrt{\frac{3g}{L}} = -5.42$ rad/s. The linear velocity at the end of the bar is $v_{tip} = L\omega = -5.42$ m/s. From the definition of the coefficient of restitution (see Eq (8.46)), since the surface P is stationary the upward velocity at the end of the bar after rebound is $v' = -(0.5)(v_{tip}) = 2.71$ m/s, and the angular velocity after rebound is $\omega' = \frac{v'}{L} = 2.71$ rad/s. From the principle of work and energy $U = T_2 - T_1$, where $T_2 = 0$ since the bar comes to rest at the maximum angle. The work done is $U = \int_{h_1}^{h_2} -mg\, dh = -mg(h_2 - 0) = -mgh_2$. The kinetic energy is

$T_1 = \left(\frac{1}{2}\right) I\omega'^2 = \left(\frac{1}{2}\right)\left(\frac{mL^2}{3}\right)\omega'^2$. Substitute into $U = -T_1$ and solve: $h_2 = \frac{L}{6g}\omega'^2 = 0.125$ m, which is the maximum height of the center of mass of the bar after rebound. The associated angle is

$$\boxed{\theta_{\max} = \sin^{-1}\left(\frac{2h_2}{L}\right) = 14.48^\circ}$$

====================================<>====================================

==================================<>==================================

Problem 8.65 The slender bar of mass *m* falls from rest in the position shown and hits the smooth projection at A. The coefficient of restitution is *e*. Show that the velocity of the center of mass of the bar is zero immediately after the impact if $b^2 = \frac{eL^2}{12}$.

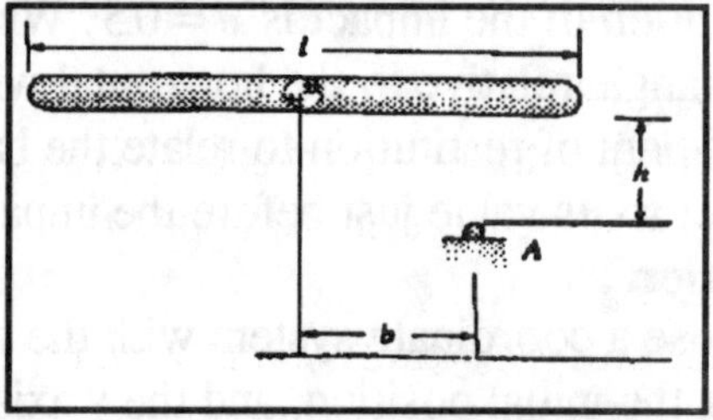

Solution:

Choose a coordinate system with the origin at the center of mass of the bar at the initial position, and the *y* axis positive downward. The strategy is to use the principle of work and energy to find the linear velocity of the center of mass of the bar the instant before impact: use the coefficient of restitution to determine the rebound velocity at the point of impact, and the conservation of angular momentum about A to determine the velocity of the center of mass after impact. From the principle of work and energy $U = T_2 - T_1$, where $T_1 = 0$ since the bar falls from rest. The work done is

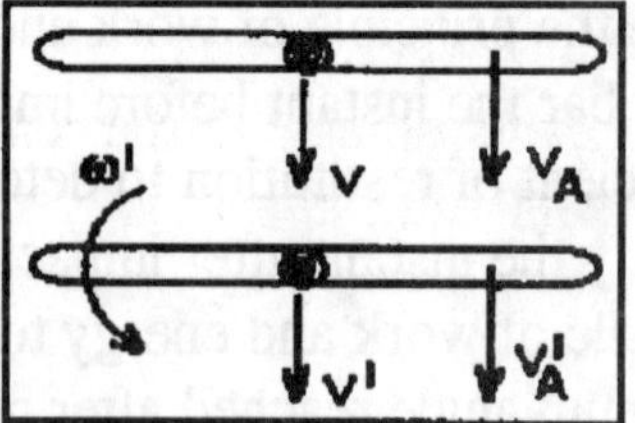

$U = \int_{h_1}^{h_2} mg\, dh = mg(h-0) = mgh$. The kinetic energy is $T_2 = \left(\frac{1}{2}\right)mv^2$,

from which $v = \sqrt{2gh}$. The angular momentum of the bar about A is conserved:

$bmv = bmv' + \left(\frac{mL^2}{12}\right)\omega'$. The velocity at A after rebound is $v'_A = -ev = -e\sqrt{2gh}$. From kinematics

$v' = v'_A + b\omega'$. If $v' = 0$, $\omega' = \frac{e\sqrt{2gh}}{b}$. Substitute into $bmv = \left(\frac{mL^2}{12}\right)\omega'$ and solve: $\boxed{b^2 = \frac{eL^2}{12}}$

==================================<>==================================

Problem 8.66 In Problem 8.65, if $m = 2$ kg, $L = 1$ m, $b = 350$ mm, $h = 200$ mm, and the coefficient of restitution of the impact is $e = 0.4$, determine the bar's angular velocity after the impact.

Solution:

From the solution to Problem 8.65, the velocity of the center of mass before impact is $v = \sqrt{2gh}$, and the *y* component of velocity at point A after impact is $v'_A = -ev = -e\sqrt{2gh}$. From kinematics, $v' = v'_A + b\omega'$. The angular momentum before and after the impact is conserved: $bmv = bmv' + I\omega'$. Substitute to obtain $bmv = -bmev + (b^2 m + I)\omega'$, from which

$\boxed{\omega' = \frac{(1+e)bmv}{(b^2 m + I)} = \frac{(1+e)b\sqrt{2gh}}{(b^2 + L^2/12)} = 4.72 \text{ rad/s}}$. [*Note:* See the solution to Problem 8.60.]

==================================<>==================================

Problem 8.67 If the duration of the impact described in Problem 8.66 is 0.02 *s*, what average force is exerted on the bar by the projection at A?

Solution:

From the principle of impulse and linear momentum for the bar, : $\int_{t_1}^{t_2} \sum F\, dt = -F(t_2 - t_1) = mv' - mv$,

from which $F = -\frac{m(v'-v)}{(t_2 - t_1)}$. From the solution to Problem 8.66, $v = \sqrt{2gh} = 1.98$ m/s,

$v' = v'_A + b\omega' = -e\sqrt{2gh} + b(4.72) = 0.8581$ m/s, from which $\boxed{F = 112.3 \text{ N}}$

==================================<>==================================

Problem 8.68 Wind causes the 600-ton ship to drift sideways at 1 *ft/s* and strike the stationary quay at P. The ship's moment of inertia about its center of mass is $3\times10^{8}\ slug\text{-}ft^{2}$, and the coefficient of restitution of the impact is 0.2. What is the ship's angular velocity after the impact?

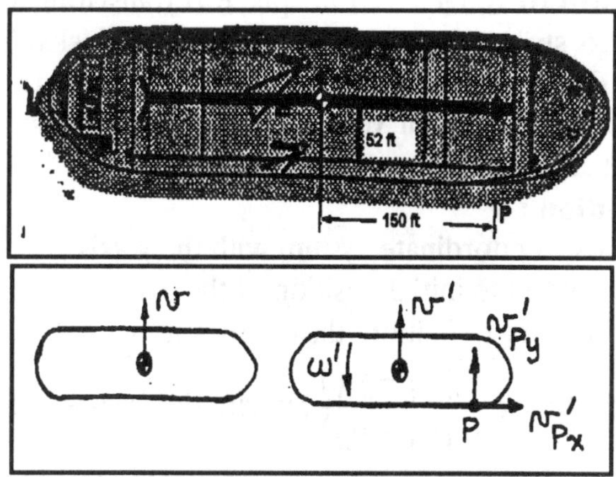

Solution:

Choose a coordinate system with the *x* axis parallel to the keel the instant before impact, and the *y* axis positive upward. Denote $b = 150\,ft$, $c = 52\,ft$. The vector from point P to the center of mass the instants before and after impact is

$\vec{r} = -150\vec{i} + 52\vec{j} = -b\vec{i} + c\vec{j}\ (ft)$. The angular momentum about P is conserved:

$$(\vec{r}\times m\vec{v}) = (\vec{r}\times m\vec{v}') + I\vec{\omega}' = \begin{bmatrix} \vec{i} & \vec{j} & \vec{k} \\ -b & c & 0 \\ 0 & mv & 0 \end{bmatrix} = \begin{bmatrix} \vec{i} & \vec{j} & \vec{k} \\ -b & c & 0 \\ 0 & mv' & 0 \end{bmatrix} + I\omega'\vec{k},\ (-bmv)\vec{k} = (-bmv' + I\omega')\vec{k}.$$

Since the quay is stationary before and after impact, from Eq (8.46) the magnitude of the *y* component of the ship's velocity at P after the impact is $v'_{Py} = -ev$. From kinematics:

$$\vec{v}' = \vec{v}'_P + \vec{\omega}\times\vec{r} = \vec{v}'_P + \begin{bmatrix} \vec{i} & \vec{j} & \vec{k} \\ 0 & 0 & \omega' \\ -b & +c & 0 \end{bmatrix} = v'_{Px}\vec{i} + v'_{Py}\vec{j} - c\omega'\vec{i} - b\omega\vec{j}.$$ The *y* component is

$v' = v'_{Py} - b\omega'$.

Substitute into the angular momentum to obtain: $-b(1+e)mv = (b^2 m + I)\omega'$, (where $v = -1\,ft/s$ is the velocity the instant before impact). Solve: $\omega' = -\dfrac{b(1+e)mv}{I + b^2 m}$.

$$\boxed{\omega' = -\frac{(1.2\times10^{6})150(1+0.2)(-1)}{(32.17)\left(3\times10^{8} + 150^{2}\left(\dfrac{1.2\times10^{6}}{32.17}\right)\right)} = 0.00589\ rad/s},\ v' = -0.6840\ ft/s.$$

Problem 8.69 In Problem 8.68, if the duration of the ship's impact with the quay is 10 *s*, what is the average value of the force exerted on the ship by the impact?

Solution:

From the principle of impulse and linear momentum, $\int_{t_1}^{t_1}\sum F dt = F_{ave}(t_2 - t_1) = m(v' - v)$, from which

$F_{ave} = \dfrac{m(v' - v)}{(t_2 - t_1)}$. From the solution to Problem 8.68, $v = -1\ ft/s$, $v' = -0.6840\ ft/s$, from which

$$\boxed{F_{ave} = \left(\frac{1.2\times10^{6}}{32.17}\right)\frac{(-0.6840+1)}{10} = 1178.7\ lb},$$ which is *the average external force exerted on the ship.*

==================================<>==================================

Problem 8.70 A 1 *lb* sphere A translating at 20 *ft/s* strikes the end of a stationary 10 *lb* slender bar B. The bar is pinned to a fixed support at O. What is the angular velocity of the bar after the impact if the sphere adheres to the bar?

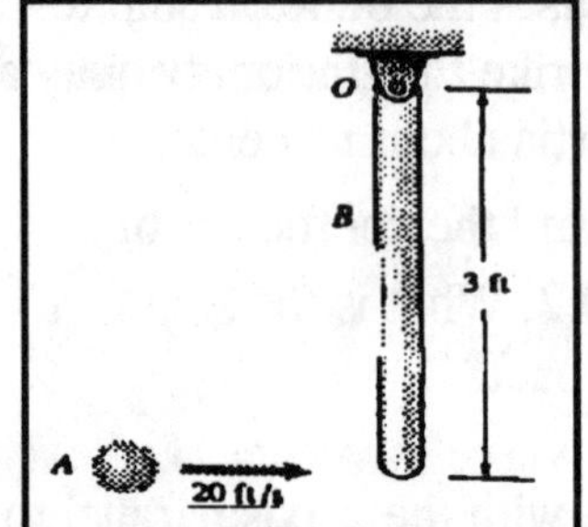

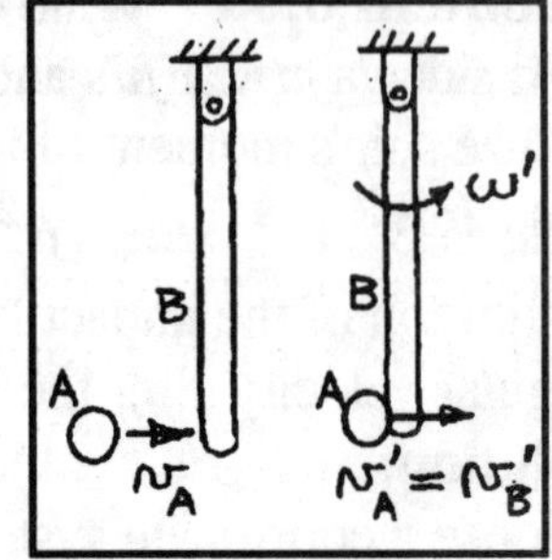

Solution :

Choose a coordinate system with the *y* axis parallel to the initial position of the bar, positive upward. By definition, the coefficient of restitution $e = \frac{(v'_B - v'_A)}{(v_A - v_B)}$, where the primes indicate values after the impact; unprimed values are before impact, and the velocities are normal to the bar. (See Eq (8.46)). The sphere adheres to the bar, from which $v'_A = v'_B$, from which $e = 0$. The angular momentum about O is conserved: $m_A L v_A = m_A L v'_A + I_O \omega'$. From kinematics for a pin supported bar, $v'_B = L\omega' = v'_A$. Substitute into the angular momentum to obtain: $m_A L v_A = \left(m_A L^2 + I_O\right)\omega'$.

Solve: $\omega' = \frac{m_A L v_A}{\left(m_A L^2 + \frac{m_B L^2}{3}\right)} = \frac{v_A}{\left(1 + \frac{m_B}{3 m_A}\right) L}$, or $\boxed{\omega' = \frac{20}{\left(1 + \frac{10}{3}\right)(3)} = 1.538 \ rad/s}$

==================================<>==================================

=====================================<>=====================================

Problem 8.71 In Problem 8.70, determine the velocity of the smooth sphere and the angular velocity of the bar after the impact if the coefficient of restitution is $e = 0.8$.

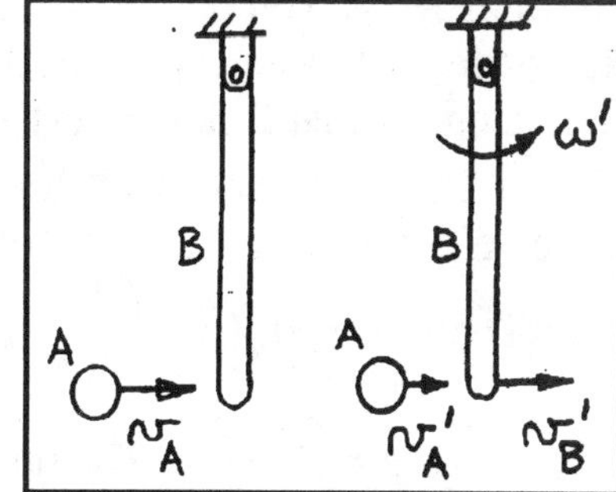

Solution:

Choose a coordinate system with the y axis parallel to the initial position of the bar, positive upward. From Eq (8.50), the coefficient of restitution $e = \frac{(v_B' - v_A')}{(v_A - v_B)}$, where the primes indicate values after the impact; unprimed values are before impact, and the velocities are normal to the bar. Since the bar is initially stationary, $v_B = 0$, from which $(v_B' - v_A') = e v_A$. The angular momentum about O is conserved, $m_A L v_A = m_A L v_A' + I_O \omega'$. From kinematics for a pinned bar, $v_B' = L\omega'$. From the definition of the coefficient of restitution, $v_B' = v_A' + e v_A$, from which $v_A' = L\omega' - e v_A$. Substitute into the angular momentum expression to obtain:

$m_A L(1+e)v_A = (m_A L^2 + I_O)\omega'$. Solve: $\omega' = \frac{m_A L(1+e)v_A}{(m_A L^2 + I_O)}$. [*Check:* See the solutions to Problems 8.66 and 8.68, and Example 8.7. *check.*] $\omega' = \frac{(1+e)v_A}{\left(1 + \frac{m_B}{3m_A}\right)L} = \frac{(1+0.8)(20)}{\left(1+\frac{10}{3(1)}\right)(3)} = 2.77 \text{ rad/s}$.

$v_A' = L\omega' - e v_A = (3)(2.77) - (0.8)(20) = -7.69 \text{ ft/s}$

=====================================<>=====================================

Problem 8.72 The 1 *kg* sphere is traveling at 10 *m/s* when it strikes the and of the 4 *kg* stationary slender bar B. If the sphere adheres to the bar, what is the bar's angular velocity after the impact?

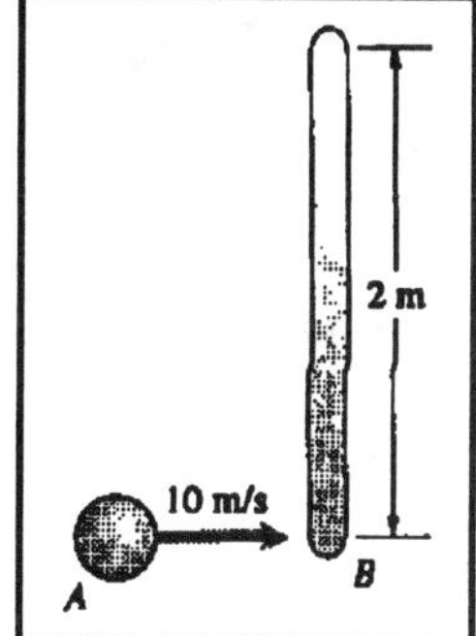

Solution:

The linear momentum is conserved: $m_A v_A = m_A v_A' + m_B v_{CM}'$, where v_{CM}' is the velocity of the center of mass of the bar after impact, and v_A, v_A' are the velocities of the sphere before and after impact. The angular momentum about the point of impact is conserved: $0 = -\left(\frac{L}{2}\right)m_B v_{CM}' + I_{CM}\omega'$. From Eq (8.50) (see solution to Problem 8.70) if the sphere adheres to the bar, $v_A' = v_{tip}'$, and $e = 0$.

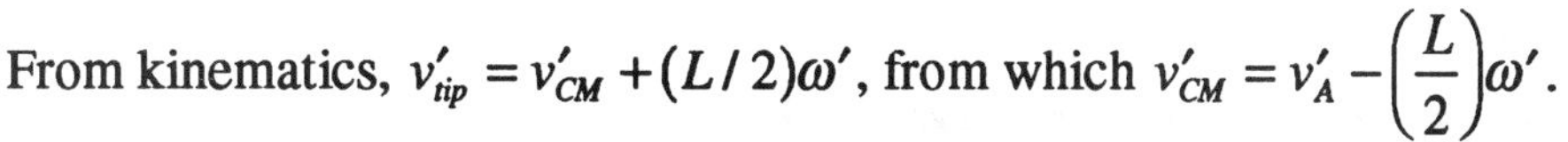

From kinematics, $v_{tip}' = v_{CM}' + (L/2)\omega'$, from which $v_{CM}' = v_A' - \left(\frac{L}{2}\right)\omega'$.

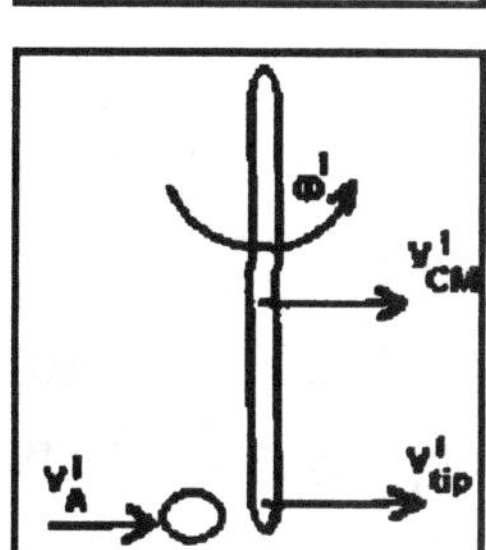

Substitute into the expressions for the conservation of momentum to obtain the two equations in two unknowns: $m_A v_A = (m_A + m_B)v_A' - (L/2)m_B\omega'$, and

$0 = -\left(\frac{L}{2}\right)m_B v_A' + \left(\frac{m_B L^2}{4} + I_{CM}\right)\omega'$. Note $I_{CM} = \frac{m_B L^2}{12}$. Solve:

$\omega' = \frac{6 m_A v_A}{(4m_A + m_B)L} = 3.75 \text{ rad/s}$

=====================================<>=====================================

==============================<>==============================

Problem 8.73 In Problem 8.72, what is the bar's angular velocity after impact if the coefficient of restitution is $e = 0.5$?

Solution:

The strategy is that followed in the solution to Problem 8.72. The linear momentum is conserved: $m_A v_A = m_A v_A' + m_B v_{CM}'$, where v_{CM}' is the velocity of the center of mass of the bar after impact, and v_A, v_A' are the velocities of the sphere before and after impact. The angular momentum about the point of impact is conserved: $0 = -\left(\frac{L}{2}\right) m_B v_{CM}' + I_{CM}\omega'$. By definition, (see Eq (8.46)) the coefficient of restitution is $e = \frac{v_{tip}' - v_A'}{v_A - v_{tip}}$. The tip is stationary before impact, from which $v_{tip}' - v_A' = e v_A$. From kinematics, $v_{tip}' = v_{CM}' + \left(\frac{L}{2}\right)\omega'$, from which $v_{CM}' = v_A' + e v_A - \left(\frac{L}{2}\right)\omega'$. Substitute into the expressions for the conservation of momentum to obtain the two equations in two unknowns: $(m_A - e m_B) v_A = (m_A + m_B) v_A' - \frac{m_B L}{2}\omega'$, and $\frac{m_B L e v_A}{2} = -\frac{L}{2} m_B v_A' + \left(\frac{m_B L^2}{4} + I_{CM}\right)\omega'$, where $I_{CM} = \frac{m_B L^2}{12}$.

Solve by iteration (using **TK Solver Plus**) to obtain: $\omega' = 5.625$ rad/s and $v_A' = 2.5$ m/s.

==============================<>==============================

Problem 8.74 In Problem 8.72, determine the kinetic energy of the sphere and the bar before and after the impact if (a) $e = 0.5$, (b) $e = 1.0$.

Solution:

Assume that it is the total kinetic energy of both sphere and bar before impact, and the total kinetic energy of both after impact, that is required, and not the kinetic energy of each separately.

Before impact: $T_{sphere} = (1/2) m_A v_A^2 = 50$ N-m, $T_{bar} = 0$, from which $T_{total} = T_{sphere} + T_{bar} = 50$ N-m

After impact: For the sphere: $T_{sphere}' = \left(\frac{1}{2}\right) m_A v_A'^2$. For the bar: $T_{bar}' = \left(\frac{1}{2}\right) m_B v_{CM}'^2 + \left(\frac{1}{2}\right) I_{CM}(\omega')^2$, from which $T_{total}' = \left(\frac{1}{2}\right) m_A v_A'^2 + \left(\frac{1}{2}\right) m_B v_{CM}'^2 + \left(\frac{1}{2}\right) I_{CM}(\omega')^2$. From the solution to Problem 8.64, the two equations in two unknowns: $(m_A - e m_B) v_A = (m_A + m_B) v_A' - \frac{m_B L}{2}\omega'$, and $\frac{m_B L e v_A}{2} = -\frac{L}{2} m_B v_A' + \left(\frac{m_B L^2}{4} + I_{CM}\right)\omega'$, where $I_{CM} = \frac{m_B L^2}{12} = 1.333$ kg-m^2, and $v_{CM}' = v_A' + e v_A - \left(\frac{L}{2}\right)\omega'$. (a) From the solution to Problem 8.64, for $e = 0.5$ $\omega' = 5.625$ rad/s, $v_A' = 2.5$ m/s, from which $v_{CM}' = 1.875$ m/s. Substitute into the kinetic energy expression to obtain: $T_{total}' = 31.25$ N-m (b) For $e = 1.0$, the solution is $\omega' = 7.5$ rad/s, $v_A' = 0$, and $v_{CM}' = 2.5$ m/s, from which $T_{total}' = 50$ N-m

==============================<>==============================

====================================<>================================

Problem 8.75 The 5 *oz* ball is translating with velocity $v_A = 80$ ft/s perpendicular to the bat just before impact. The player is swinging the 31 *oz* bat with angular velocity $\omega = 6\pi$ rad/s before the impact. Point C is the bat's instantaneous center both before and after the impact. The distances $b = 14$ in., and $\bar{y} = 26$ in. The bat's moment of inertia about its center of mass is $I_B = 0.033$ slug-ft^2. The coefficient of restitution is $e = 0.6$, and the duration of the impact is 0.008 s. Determine the magnitude of the velocity of the ball after the impact and the average force A_x exerted on the bat by the player during the impact if (a) $d = 0$, (b) $d = 3$ in., (c) $d = 8$ in.

Solution:

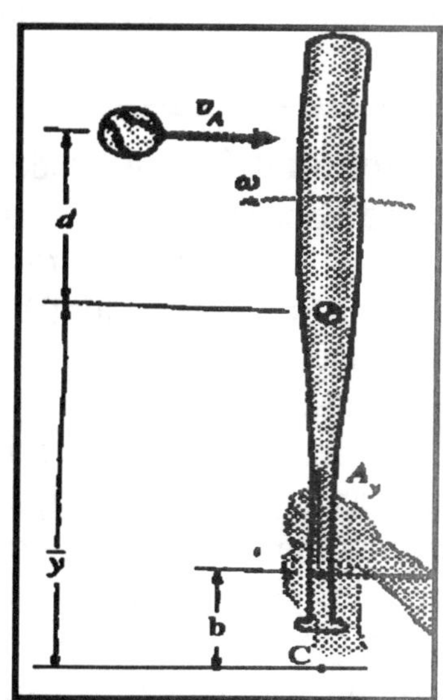

By definition, the coefficient of restitution is (1) $e = \dfrac{v'_P - v'_A}{v_A - v_P}$.

The angular momentum about A is conserved:
(2) $m_A v_A(d + \bar{y} - b) + m_B v_B(\bar{y} - b) - I_B\omega =$
$m_A v'_A(d + \bar{y} - b) + m_B v'_B(\bar{y} - b) - I_B\omega'$. From kinematics, the velocities about the instantaneous center: (3) $v_P = -\omega(\bar{y} + d)$, (4) $v_B = -\omega\bar{y}$,
(5) $v'_P = -\omega'(\bar{y} + d)$, (6) $v'_B = -\omega'\bar{y}$. Since ω, $\bar{y}$, and d are known, v_B and v_P are determined from (3) and (4), and these six equations in six unknowns reduce to four equations in four unknowns. v'_P, v'_B, v'_A, and ω'. Further reductions may be made by substituting (5) and (6) into (1) and (2); however here the remaining four unknowns were solved by iteration for values of $d = 0$, $d = 3$ in. $= 0.25$ ft, $d = 8$ in. $= 0.667$ ft. The reaction at A is determined from the principle of angular impulse-momentum applied about the point of impact:

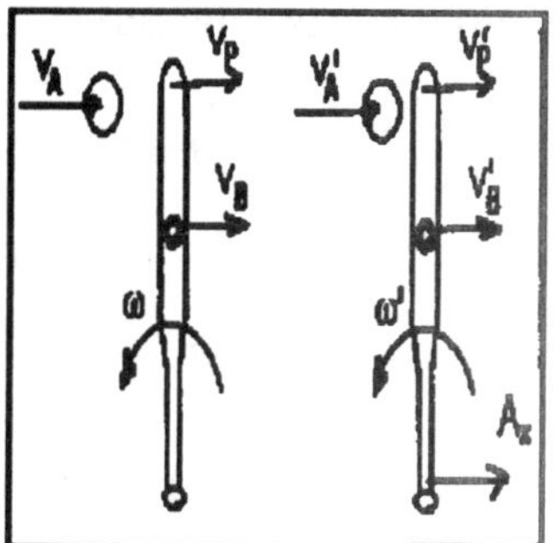

(7) $\int_{t_1}^{t_2} A_x(d + \bar{y} - b)dt = (dm_B v'_B + I_B\omega') - (dm_B v_B + I_B\omega)$, where
$t_2 - t_1 = 0.08$ s. Using (4) and (5), the reaction is

$A_x = \dfrac{(I_B - dm_B\bar{y})}{(d + \bar{y} - b)(t_2 - t_1)}(\omega' - \omega)$, where the unknown, ω', is determined from the solution of the first six equations. The values are tabulated:

d, in.	v'_A, ft/s	A_x, lb	v'_P, ft/s	v'_B, ft/s	ω', rad/s
0	–91.29	–41.98	–18.79	–18.79	8.672
3	–90.32	–0.477	–14.98	–13.43	6.20
8	–86.7	+66.84	–6.64	–5.08	2.34

Only the values in the first two columns are required for the problem; the other values are included for checking purposes. *Note:* The reaction reverses between $d = 3$ in. and $d = 8$ in., which means that the point of zero reaction occurs in this interval.

====================================<>================================

================================<>================================

Problem 8.76 In Problem 8.75, show that the force A_x is zero if $d = \dfrac{I_B}{m_B \bar{y}}$ where m_B is the mass of the bat.

Solution :

From the solution to Problem 8.75, the reaction is $A_x = \dfrac{(I_B - dm_B\bar{y})}{(d+\bar{y}-b)(t_2-t_1)}(\omega'-\omega)$. Since $(\omega'-\omega) \neq 0$, the condition for zero reaction is $I_B - dm_B\bar{y} = 0$, from which $d = \dfrac{I_B}{\bar{y}m_B}$.

================================<>================================

Problem 8.77 A slender bar of mass m is released from rest in the horizontal position at a height h above a peg (Fig (a)). A small hook at the end of the bar engages the peg, and the bar swings from the peg (Fig (b)). What minimum height h is necessary for the bar to swing 270° from its position when it engages the peg?

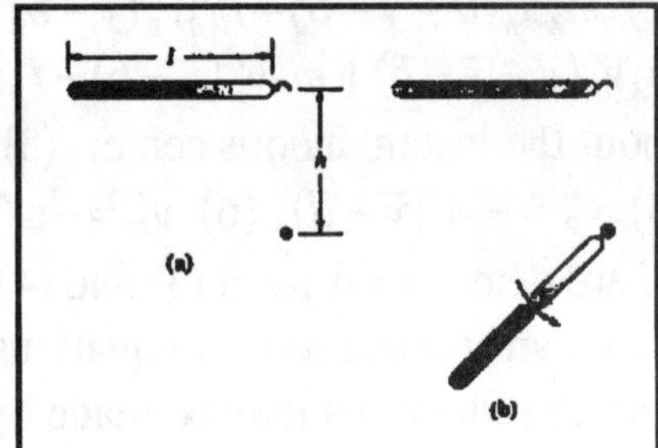

Solution:

The strategy is to use the principle of work and energy to find the velocity of the bar when it engages the peg; the conservation of angular momentum to determine the angular velocity of the bar, and the principle of work and energy to determine the angle through which the bar rotates. Choose a coordinate system with the x axis parallel to the bar in the initial position, and the y axis positive upward. The principle of work and energy: $U = T_2 - T_1$, where $T_1 = 0$ since the bar is released from rest. The work done is $U = \int_{h_1}^{h_2} -mgdh = -mg(h_2 - h_1) = mgh$. The kinetic energy is $T_2 = \left(\frac{1}{2}\right)mv^2$. Substitute and solve: $v = -\sqrt{2gh}$, where the negative sign is chosen to conform with the choice of coordinate system. The conservation of angular momentum about the peg is $-\left(\frac{L}{2}\right)mv = -\left(\frac{L}{2}\right)mv' + I_B\omega'$, where $I_B = \frac{mL^2}{12}$. From kinematics, $v' = -\left(\frac{L}{2}\right)\omega'$. Substitute to obtain $-\left(\frac{L}{2}\right)mv = \left(\frac{mL^2}{4} + I_B\right)\omega'$, from which $\omega' = -\frac{3}{2L}v = \frac{3\sqrt{2gh}}{2L}$. From the principle of work and energy: $U = T_2 - T_1$, where $T_2 = 0$ since the bar comes to rest at $\theta = 270^\circ$. The work done is $U = \int_{h_1'}^{h_2'} -mgdh = -mg(h_2' - h_1') = -mg\left(\frac{L}{2}\right)$. The kinetic energy is $T_1 = \left(\frac{1}{2}\right)mv'^2 + \left(\frac{1}{2}\right)I_B\omega'^2$. Use the kinematic condition and the expression for ω' to obtain $T_1 = \frac{1}{2}\left(\frac{mL^2}{4} + I_B\right)\omega'^2 = \frac{1}{2}\left(\frac{mL^2}{4} + I_B\right)\left(\frac{9gh}{2L^2}\right)$. Substitute into $U = -T_1$ and solve: $h = \frac{2L}{3}$.

================================<>================================

==================================<>==================================

Problem 8.78 Is energy conserved in Problem 8.77? If not, how much energy is lost?

Solution:

Choose the datum at the point of contact with the peg. The potential energy in the initial position is $V = -U = mgh = \frac{2}{3}mgL$, where $h = \frac{2}{3}L$ is the solution to Problem 8.77. The bar is at rest at the $\theta = 270^o$, from which the potential energy (relative to the same datum) is $V' = -U' = mg\left(\frac{L}{2}\right)$. Since $V \neq V'$, *energy is not conserved.*. The loss in energy is $\Delta V = \frac{2mgL}{3} - \frac{mgL}{2} = \frac{mgL}{6}$

==================================<>==================================

Problem 8.79 A wheel that can be modeled as a 1 *slug* homogenous cylindrical disk rolls at 10 *ft/s* on a horizontal surface toward a 6 *in.* step. If the wheel remains in contact with the step and does not slip while rolling up onto it, what is the wheel's velocity once it is on the step?

Solution:

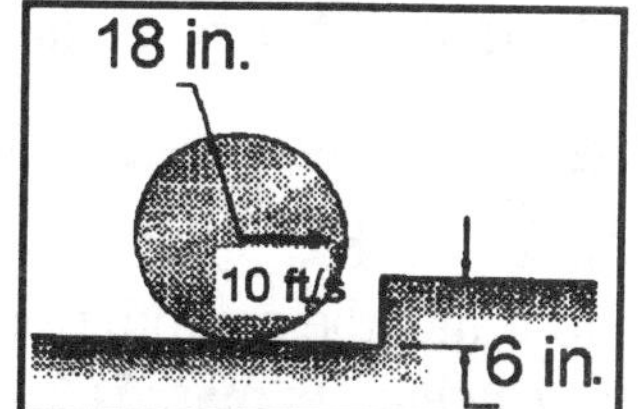

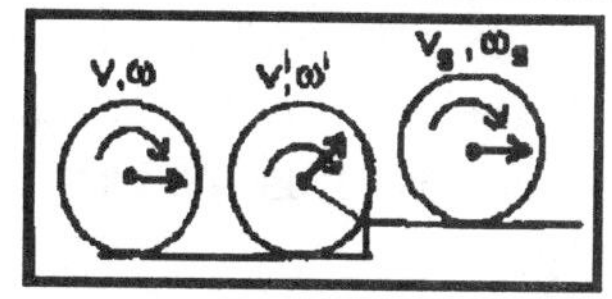

Choose a coordinate system with the *x* axis parallel to the plane surface, with *y* positive upward. Denote the velocity and angular velocity before impact with the step by v, ω, the instant after impact with the step by v', ω', and after the disk has climbed the step by v_S, ω_S. Denote the vector from the point of impact to the center of mass at the instant of impact by $\vec{r}$, the height of the step by *h*, and the radius of the disk by *R*. The strategy is to use the conservation of angular momentum to determine v', ω' the instant after impact, and the principle of work and energy to determine v_S, ω_S.

The conservation of angular momentum about the point of contact: $\mathrm{I}_B\vec{\omega} + (\vec{r} \times \mathrm{m}\vec{v}) = \mathrm{I}_B\vec{\omega}' + (\vec{r} \times \mathrm{m}\vec{v}')$. Before contact, the perpendicular distance from the point of contact to the velocity vector of the disk is $(R-h)$. After contact, as the disk climbs the step, the center of mass moves in an arc of radius R about the fixed point of contact with the step, from which the velocity vector of the center of mass is perpendicular to the vector $\vec{r}$. From the definition of the cross product, for motion in the *x-y* plane: $(\vec{r} \times \mathrm{m}\vec{v}) \cdot \vec{k} = (\mathrm{R} - \mathrm{h})\mathrm{mv}$, $(\vec{r} \times \mathrm{m}\vec{v}') \cdot \vec{k} = \mathrm{Rmv}'$, and $\mathrm{I}\vec{\mathrm{w}}' \cdot \vec{k} = \mathrm{I}\omega'$. From kinematics, $\omega = \frac{v}{R}$, $\omega' = \frac{v'}{R}$. For a homogenous disk $I_B = \frac{mR^2}{2}$. Substitute to obtain $v' = \left(\frac{1}{3}\right)\left(1 + 2\frac{(R-h)}{R}\right)v = 7.778\ \text{ft/s}$. *The principle of work and energy:* $U = T_2 - T_1$. The work done climbing the step is $U = \int_0^h -mgdh = -mgh$. The kinetic energy at the instant after contact is $T_1 = \left(\frac{1}{2}\right)I_D\omega'^2 + \left(\frac{1}{2}\right)mv'^2$. Use the kinematic relation to obtain $T_1 = \left(\frac{I_D}{R^2} + m\right)\frac{v'^2}{2}$. Similarly, $T_2 = \left(\frac{I_D}{R^2} + m\right)\frac{v_S^2}{2}$, where $\omega_S = \frac{v_S}{R}$ has been used. Substitute into $U = T_2 - T_1$ and solve: $v_S = \sqrt{v'^2 - \frac{2mR^2gh}{I_D + R^2m}} = \sqrt{v'^2 - \frac{4gh}{3}} = 6.25\ \text{ft/s}$

==================================<>==================================

==================================<>==================================

Problem 8.80 In Problem 8.79, what is the minimum velocity the wheel must have rolling toward the step in order to climb up onto to it?

Solution:

From the solution to Problem 8.79, the velocity the instant after impact is $v' = \left(\frac{1}{3}\right)\left(1+2\frac{(R-h)}{R}\right)v$, and the velocity after climbing the step is $v_S = \sqrt{v'^2 - \frac{4gh}{3}}$. Suppose that $v_S = 0$, from which the minimum velocity after impact is $v'_{\min} = \sqrt{\frac{4gh}{3}}$. Substitute to obtain the minimum rolling velocity:

$v_{\min} = \dfrac{2\sqrt{3gh}}{1+\dfrac{2(R-h)}{R}} = 5.95$ ft/s [*Side comment*: If $h = \frac{3}{2}R$, $v_{\min} \to \infty$, and the disk cannot climb the step. For the range of step heights $R \le h < \frac{3}{2}R$ it *does climb the step.*]

==================================<>==================================

Problem 8.81 The slender bar is shown just before it hits the smooth floor. The length of the bar is *1 m* and its mass is *2 kg*. The bar's angular velocity is zero and it is moving downward at *4 m/s*. If the coefficient of restitution of the impact is *e = 0.2,* what is the bar's angular velocity after the impact?

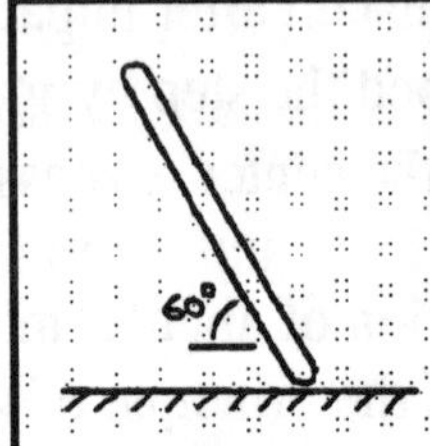

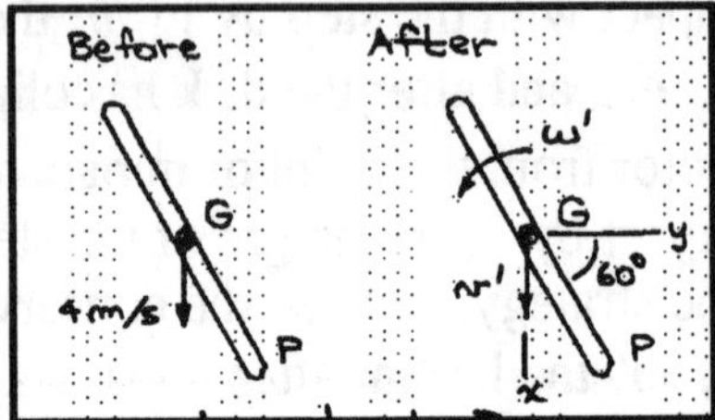

Solution:

No horizontal force is exerted by the floor, so the center of mass is still moving vertically after the impact. Angular momentum about *P* is conserved:

$(0.5)\cos 60°(2)(4) = (0.5)\cos 60°(2)v' + \left[\frac{1}{12}(2)(1)^2\right]\omega'$ (1).

The velocity of the point of impact *P* after the impact is

$$\bar{v}'_P = \bar{v}'_G + \bar{\omega}' \times \bar{r}_{P/G} = v'\bar{i} + \begin{vmatrix} \bar{i} & \bar{j} & \bar{k} \\ 0 & 0 & \omega' \\ 0.5\sin 60° & 0.5\cos 60° & 0 \end{vmatrix} = (v' - 0.5\omega'\cos 60°)\bar{i} + 0.5\omega'\sin 60°\bar{j}$$

The normal components at *P* are related by the coefficient of restitution:

$e = 0.2 = \dfrac{-v'_{Px}}{v_{Px}} = \dfrac{-(v' - 0.5\omega'\cos 60°)}{4}$ (2) Solving Equations (1) and (2), we obtain $v' = 1.26 m/s$ $\omega' = 8.23 rad/s$.

==================================<>==================================

==================================<>==================================

Problem 8.82 If the duration of the impact in Problem 8.81 is *0.03 s*, what average force is exerted on the bar by the floor?

Solution:

From the solution of Problem 8.81, the downward velocity of the center of mass after the impact is $v' = 1.26 m/s$. From Equation (8.26), $(t_2 - t_1)\sum \overline{F}_{av} = mv'\overline{i} - mv\overline{i}$: $(0.03)\sum \overline{F}_{av} = (2)(1.26)\overline{i} - (2)(4)\overline{i}$. The force is $\sum \overline{F}_{av} = -183\overline{i}N$, or $183N$ upward.

==================================<>==================================

Problem 8.83 In Problem 8.81, suppose that the center of mass of the bar is moving downward at *4 m/s* and that the bar has a clockwise angular velocity ω just before it hits the floor. What value of ω would cause the bar to have no angular velocity after the impact?

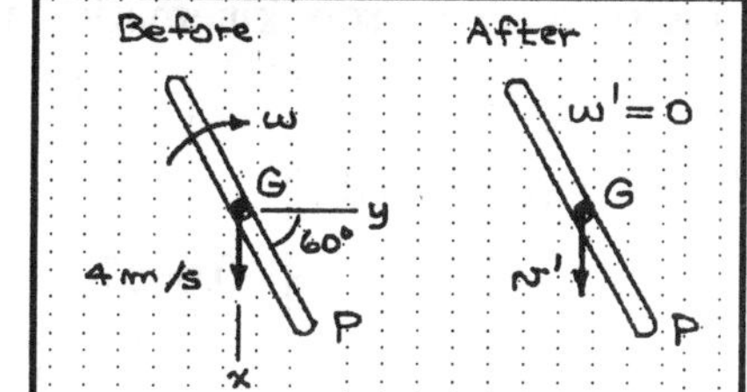

Solution:

Angular momentum about P is conserved:

$$(0.5)\cos 60^\circ(2)(4) + \left[(1/12)(2)(1)^2\right](-\omega) = (0.5)\cos 60^\circ(2)v' \quad (1)$$

The velocity of the point of impact P before the impact is

$$\overline{v}_P = \overline{v}_G + \overline{\omega}\times\overline{r}_{P/G} = 4\overline{i} + \begin{vmatrix} \overline{i} & \overline{j} & \overline{k} \\ 0 & 0 & -\omega \\ 0.5\sin 60^\circ & 0.5\cos 60^\circ & 0 \end{vmatrix} = (4 + 0.5\omega\cos 60^\circ)\overline{i} - (0.5\omega\sin 60^\circ)\overline{j}.$$

The normal velocity components at P are related by the coefficient of restitution:

$$e = 0.2 = \frac{-v'_{Px}}{v_{Px}} = \frac{-v'_{Px}}{4 + 0.5\omega\cos 60^\circ} \quad (2) \ .$$

Solving Eqs. (1) and (2), we obtain $v' = -1.65 m/s$, $\omega = 16.9 rad/s$

==================================<>==================================

================================<>================================

Problem 8.84 During her parallel bars routine, the velocity of the 90 *lb* gymnast's center of mass is $4\vec{i} - 10\vec{j}$ (ft / s) and her angular velocity is zero just as she grasps the bar at A. In the position shown, her moment of inertia about her center of mass is 1.8 slug - ft^2. If she stiffens her shoulders and legs so that she can be modeled as a rigid body, what is the velocity of her center of mass and her angular velocity just after she grasps the bar?

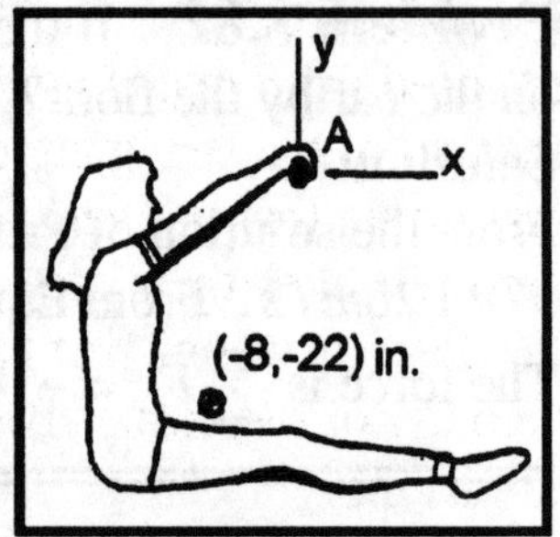

Solution:

Denote $\vec{r} = -8\vec{i} - 22\vec{j} = -0.667\vec{i} - 1.833\vec{j}$ (ft).

The conservation of angular momentum about A is $\vec{r}\times m\vec{v} = \vec{r}\times m\vec{v}' + I\vec{w}'$, since the initial angular velocity is zero. From kinematics, $\vec{v}_G' = \vec{v}_A' + \vec{w}'\times\vec{r}$. Reduce to get

$$\vec{v}_G' = \vec{v}_A' + \vec{w}'\times\vec{r} = 0 + \begin{bmatrix} \vec{i} & \vec{j} & \vec{k} \\ 0 & 0 & \omega' \\ -0.6667 & -1.833 & 0 \end{bmatrix} = (1.833\omega')\vec{i} - (0.667\omega')\vec{j}, \text{ from which } v_{Gx}' = 1.833\omega',$$

$$v_{Gy}' = -0.6667\omega'. \ (\vec{r}\times m\vec{v}) = \begin{bmatrix} \vec{i} & \vec{j} & \vec{k} \\ -0.6667 & -1.833 & 0 \\ 4 & -10 & 0 \end{bmatrix} = 14m\vec{k},$$

$$\vec{r}\times m\vec{v}_G' = \begin{bmatrix} \vec{i} & \vec{j} & \vec{k} \\ -0.6667 & -1.833 & 0 \\ 1.833m\omega' & -0.6667m\omega' & 0 \end{bmatrix} = \left(0.6667^2 + 1.833^2\right)m\omega'\vec{k} = (1.951)^2 m\omega'\vec{k}, \ I\vec{w}'\cdot\vec{k} = 1.8\omega'.$$

Substitute: $14m = \left((1.951)^2 m + 1.8\right)\omega'$. From which $\omega' = \dfrac{14}{(1.951)^2 + \dfrac{32.17(1.8)}{90}} = 3.146$ rad / s,

$v' = 1.951\omega' = 6.14$ ft / s. The velocity vector is $\vec{v}' = v'\vec{e}_P$, where $\vec{e}_P$ is a unit vector perpendicular to $\vec{r}$. The unit vector parallel to $\vec{r}$ is $\vec{e}_r = \dfrac{\vec{r}}{|\vec{r}|} = -0.3417\vec{i} - 0.9398\vec{j}$. From the requirement that $\vec{e}_P\cdot\vec{e}_r = 0$, the two unit vectors perpendicular to $\vec{e}_r$ are constructed by inspection: $\vec{e}_P = \pm 0.9398\vec{i} \mp 0.3417\vec{j}$. The center of mass rotates counterclockwise about point A, (see Figure), from which the unit vector of interest is $\vec{e}_P = +0.9398\vec{i} - 0.3417\vec{j}$, from which $\vec{v}' = 5.77\vec{i} - 2.1\vec{j}$ (ft / s).

================================<>================================

====================================<>===================================

Problem 8.85 The 20 *kg* homogenous rectangular plate is released from rest (Fig a) and falls 200 *mm* before coming to the end of the string attached to the corner A (Fig b). Assuming that the vertical component of the velocity of A is zero just after the plate reaches the end of the string, determine the angular velocity of the plate and the magnitude of the velocity of the corner B at that instant.

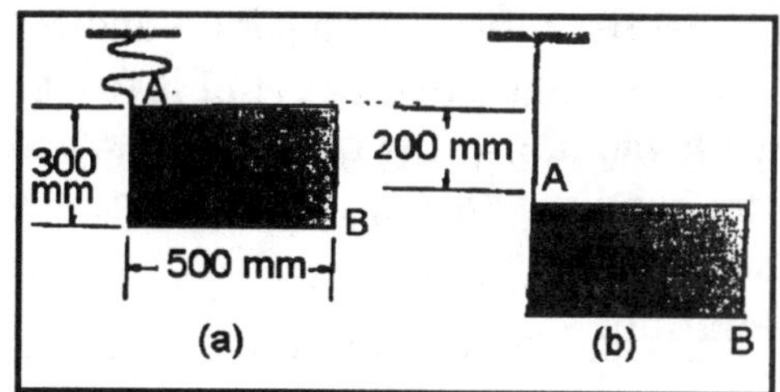

Solution:

Choose a coordinate system with the x axis parallel to top edge of the plate in Fig (a), with the y axis positive upward.

Denote the vector distance from the corner A to the center of mass of the plate by $\vec{r} = a\vec{i} - b\vec{j}$, where $a = 0.25$ m, $b = 0.15$ m. From the principle of work and energy, $U = T_2 - T_1$, where $T_1 = 0$ since the plate is released from rest. The work done is $U = \int_0^h -mgdh = -mgh$, where $h = -0.2$ m. The kinetic energy is $T_2 = (1/2)mv^2$, from which $v = -\sqrt{-2gh} = -1.981$ m, where the negative sign on the square root is chosen to conform to the choice of coordinates. From conservation of linear momentum in the horizontal direction, $mv_{Gx} = mv'_{Gx}$, from which $v_{Gx} = v'_{Gx} = 0$, that is, *the horizontal component of the velocity of the center of mass is zero the instant after the string tightens.* From kinematics, the velocity of the center of mass is $\vec{v}'_G = \vec{v}'_A + \vec{w}' \times \vec{r}$.

$\vec{v}'_G = \vec{v}'_A + \vec{w}' \times \vec{r} = \vec{i}\,v'_{Ax} + \vec{j}\,v'_{Ay} + \begin{bmatrix} \vec{i} & \vec{j} & \vec{k} \\ 0 & 0 & \omega' \\ a & -b & 0 \end{bmatrix} = \vec{i}(v'_{Ax} + b\omega') + \vec{j}(v'_{Ay} + a\omega')$. Since $v'_{Gy} = 0$, $v'_{Gx} = 0$, then

$0 = v'_{Ax} + b\omega'$, and $v'_{Gy} = a\omega'$. The angular momentum about A is conserved:

$(\vec{r} \times m\vec{v}_G) = (\vec{r} \times m\vec{v}'_G) + I_G\vec{w}'$. Substitute and reduce:

$\vec{r} \times m\vec{v}_G = \begin{bmatrix} \vec{i} & \vec{j} & \vec{k} \\ a & -b & 0 \\ 0 & mv & 0 \end{bmatrix} = (amv)\vec{k}$, $\vec{r} \times m\vec{v}'_G = \begin{bmatrix} \vec{i} & \vec{j} & \vec{k} \\ a & -b & 0 \\ 0 & ma\omega' & 0 \end{bmatrix} = a^2 m\omega' \vec{k}$, $I_G\vec{w}' = \left(\dfrac{(2a)^2 + (2b)^2}{12}\right) m\omega' \vec{k}$.

Collect terms and substitute into the conservation of angular momentum expression to obtain

$amv = \left[a^2 m + \left((2a)^2 + (2b)^2\right)(1/12)m\right]\omega'$, from which

$\omega' = 3av/(4a^2 + b^2) = -3a\sqrt{-2gh}/(4a^2 + b^2) = -5.452$ rad/s, (clockwise),

$\vec{v}'_G = +(a\omega')\vec{j} = -1.363\vec{j}$ (m/s), and $v'_{Ax} = -b\omega' = 0.818$ m/s. The velocity of the corner B is obtained

from $\vec{v}'_B = \vec{v}'_G + \vec{w}' \times (a\vec{i} - b\vec{j}) = 0 + \begin{bmatrix} \vec{i} & \vec{j} & \vec{k} \\ 0 & 0 & \omega' \\ a & -b & 0 \end{bmatrix} = (b\omega')\vec{i} + (v'_{Gy} + a\omega')\vec{j}$, $\vec{v}_B = -0.818\vec{i} - 2.73\vec{j}$ (m/s),

from which $|\vec{v}_B| = 2.85$ m/s

====================================<>===================================

Problem 8.86 Two bars A and B are each 2 *m* in length and each has a mass of 4 *kg*. In Fig (a), bar A has no angular velocity and is moving to the right at 1 *m/s*, and bar B is stationary. If the bars bond together at impact, (Fig b), what is their angular velocity ω' after the impact?

Solution:

Linear momentum is conserved, $m_A\bar{v}_A + m_B\bar{v}_B = m_A\bar{v}_A{}' + m_B\bar{v}_B{}'$, or

$(4)(1)+0 = 4v_A{}' + 4v_B{}'$ (1).

The moment of inertia of each bar about its center of mass is

$I = \frac{1}{12}m\ell^2 = \frac{1}{12}(4)(2)^2 = 1.33 kg-m^2$. Angular momentum about any point is conserved.

Choosing the center of mass of A, $0 = -\ell m_B v_B{}' + I\omega' + I\omega'$, or $0 = -(2)(4)v_B{}' + 2(1.33)\omega'$ (2).

The velocities of the centers of mass are related by $\bar{v}_B{}' = \bar{v}_A{}' + \bar{\omega}' \times \bar{r}_{B/A}$: or

$$v_B{}'\bar{i} = v_A{}'\bar{i} + \begin{vmatrix} \bar{i} & \bar{j} & \bar{k} \\ 0 & 0 & \omega' \\ 0 & \ell & 0 \end{vmatrix}.$$

Hence $v_B{}' = v_A{}' - (2)\omega'$ (3).

Solving Equations (1) - (3), we obtain $\omega' = 0.375 rad/s$, $v_A{}' = 0.875 m/s$, and $v_B{}' = 0.125 m/s$.

Problem 8.87 In Problem 8.86, what is the velocity of the center of mass just after impact?

Solution:

From the solution of Problem 8.86, $v_A{}' = 0.875 m/s$ to the right and $v_B' = 0.125 m/s$ so $v_{CM} = m(0.875) + m(0.125) = 0.5 m/s$ to the right. Alternatively, use conservation of linear momentum, i.e., $4kg(1m/s) = 8kg(v_{CM})$, which results in $v_{CM} = 0.5 m/s$ to the right as before.

Problem 8.88 In Problem 8.86, if the bars do not bond together on impact and the coefficient of restitution is $e = 0.8$, what are the angular velocities of the bars after impact?

Solution:

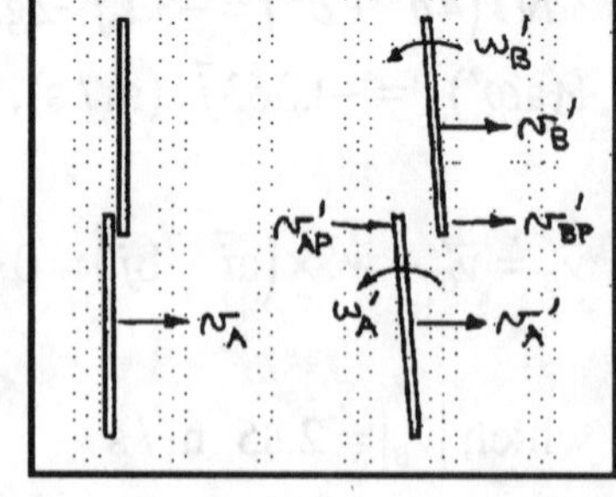

Linear momentum is conserved,

$m_A v_A = m_A v_A{}' + m_B v_B{}'$ or $(4)(1) = 4v_A{}' + 4v_B{}'$ (1).

The angular momentum of each bar about the point of contact is conserved.

$\left(\frac{\ell}{2}\right)m_A v_A = \left(\frac{\ell}{2}\right)m_A v_A{}' + I\omega_A{}'$, or $(1)(4)(1) = (1)(4)v_A{}' + 1.33\omega_A{}'$ (2).

$0 = -\left(\frac{\ell}{2}\right)m_B v_B{}' + I\omega_B{}'$ $0 = -(1)(4)v_B{}' + 1.33\omega_B{}'$ (3).

The velocities at the point of impact are related by the coefficient of restitution: $e = \frac{v_{Bp}{}' - v_{AP}{}'}{v_A}$ (4)

The kinematic relationships between the velocities of the centers of mass and the velocities at the point

Solution continued on next page

of contact are $\bar{v}_{AP}' = \bar{v}_A' + \bar{\omega}_A' \times \bar{r}_{P/A}$:

$$v_{AP}'\bar{i} = v_A'\bar{i} + \begin{vmatrix} \bar{i} & \bar{j} & \bar{k} \\ 0 & 0 & \omega_A' \\ 0 & \ell/2 & 0 \end{vmatrix}, \text{ or } v_{AP}' = v_A' - (1)\omega_A' \ (5).$$

and $\bar{v}_{BP}' = \bar{v}_A' + \bar{\omega}_A' \times \bar{r}_{P/B}$:

$$v_{BP}'\bar{i} = +v_B'\bar{i} + \begin{vmatrix} \bar{i} & \bar{j} & \bar{k} \\ 0 & 0 & \omega_B' \\ 0 & -\ell/2 & 0 \end{vmatrix} \text{ or } v_{BP}' = v_B' + (1)\omega_B' \ (6).$$

Solving Equations (1) - (6), we obtain $\omega_A' = \omega_B' = 0.675 rad/s$.

================================<>================================

==============================<>==============================

Problem 8.89 The horizontal velocity of the airplane is 50 *m/s*, its vertical velocity (rate of descent) is 2 *m/s*, and its angular velocity is zero. The mass of the airplane is 12 *Mg* and the moment of inertia about its center of mass is $1\times10^5 \ kg\text{-}m^2$. When the rear wheels touch the runway, they remain in contact with it. Neglecting the horizontal force exerted on the wheels by the runway, determine the airplane's angular velocity just after it touches down.

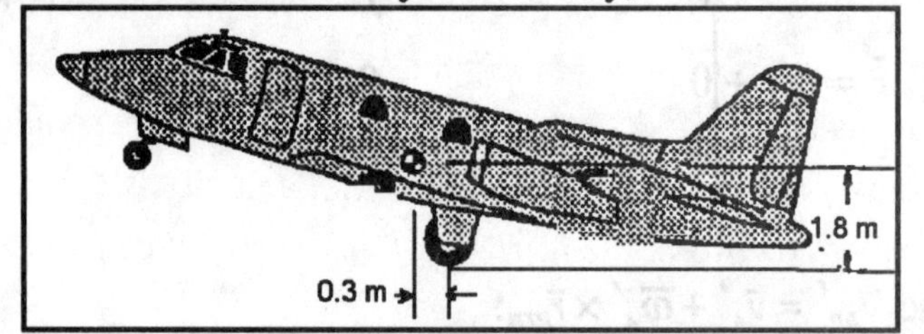

Solution:

Choose a coordinate system with the *x* axis parallel to the ground and *y* positive upward. Denote the vector distance from the rear wheel touchdown point P to the center of mass by $\vec{r} = -0.3\vec{i} + 1.8\vec{j}\ (m)$. Denote the vector velocity of the center of mass before and after touchdown by $\vec{v}_G, \vec{v}'_G$. The angular momentum about P is conserved: $\vec{r}\times m\vec{v}_G = \vec{r}\times m\vec{v}'_G + I\vec{\omega}'$. The initial velocity is $\vec{v}_G = -50\vec{i} - 2\vec{j}\ (m/s)$. From the conservation of linear momentum, the *horizontal velocity of the center of mass is unchanged by touchdown*; that is, $mv'_{Gx} = mv_{Gx}$, from which $v'_{Gx} = v_{Gx} = -50\ m/s$.

From kinematics, the velocity of the center of mass after contact is $\vec{v}'_G = \vec{v}'_P + \vec{\omega}'\times\vec{r}$. Since the wheels remain in contact with the ground after touchdown, $v'_{Py} = 0$. Expanding terms:

$$\vec{v}'_G = \vec{v}'_P + \begin{bmatrix} \vec{i} & \vec{j} & \vec{k} \\ 0 & 0 & \omega' \\ -0.3 & 1.8 & 0 \end{bmatrix},$$

$$v'_G = (v'_{Px} - 1.8\omega')\vec{i} - (0.3\omega')\vec{j}\ (m/s).$$

From the conservation of linear momentum, $v'_{Gx} = v_{Gx} = -50\ m/s$, from which $v'_{Px} = -50 + 1.8\omega'$, and

$$v'_{Gy} = -0.3\omega'.\ (\vec{r}\times m\vec{v}_G) = \begin{bmatrix} \vec{i} & \vec{j} & \vec{k} \\ -0.3 & 1.8 & 0 \\ -50m & -2m & 0 \end{bmatrix} = (0.6m + 50(1.8)m)\vec{k} = 90.6m\vec{k},$$

$$(\vec{r}\times m\vec{v}'_G) = \begin{bmatrix} \vec{i} & \vec{j} & \vec{k} \\ -0.3 & 1.8 & 0 \\ -50m & -0.3m\omega' & 0 \end{bmatrix} = \left((0.3)^2 m\omega' + 50(1.8)m\right)\vec{k}.$$

Substitute into the expression for the conservation of angular momentum and reduce:

$$\boxed{\omega' = \frac{0.6}{0.3^2 + \frac{I}{m}} = \frac{0.6}{0.3^2 + 8.333} = 0.0712\ rad/s},$$

from which $\vec{v}'_G = -50\vec{i} - 0.0214\vec{j}\ (m/s)$, $\vec{v}'_P = -49.87\vec{i}\ (m/s)$.

==============================<>==============================

====================================<>====================================

Problem 8.90 Determine the angular velocity of the airplane in Problem 8.89 just after it touches down if its wheels don't stay in contact with the runway and the coefficient of restitution is $e = 0.4$.

Solution:

Use the results of the solution to Problem 8.89 as appropriate. Choose a coordinate system with the *x* axis parallel to the ground and *y* positive upward. Denote the vector distance from the rear wheel touchdown point P to the center of mass by $\vec{r} = -0.3\vec{i} + 1.8\vec{j}\ (m)$. Denote the vector velocity of the center of mass before and after touchdown by $\vec{v}_G, \vec{v}'_G$. The angular momentum about P is conserved: $\vec{r} \times m\vec{v}_G = \vec{r} \times m\vec{v}'_G + I\vec{\omega}'$. The initial velocity is $\vec{v}_G = -50\vec{i} - 2\vec{j}\ (m/s)$. From the conservation of linear momentum, the *horizontal velocity of the center of mass is unchanged by touchdown*; that is, $mv'_{Gx} = mv_{Gx}$, from which $v'_{Gx} = v_{Gx} = -50\ m/s$. By definition, $e = \dfrac{v'_{Py} - v'_{Ry}}{v_{Ry} - v_{Py}}$, where the primed values are after contact and the unprimed values are before contact. The values $v_{Ry} = v'_{Ry} = 0$ for the runway surface, from which $v'_{Py} = -ev_{Py}$. The value $v_{Py} = -2\ m/s$ is the descent velocity, from which $v'_{Py} = 2e = 0.8\ m/s$. From kinematics, the velocity of the center of mass after contact is $\vec{v}'_G = \vec{v}'_P + \vec{\omega}' \times \vec{r}$. Expanding terms:

$$\vec{v}'_G = \vec{v}'_P + \begin{bmatrix} \vec{i} & \vec{j} & \vec{k} \\ 0 & 0 & \omega' \\ -0.3 & 1.8 & 0 \end{bmatrix} = (v'_{Px} - 1.8\omega')\vec{i} + (v'_{Py} - 0.3\omega')\vec{j}\ (m/s).$$

From the conservation of linear momentum, $v'_{Gx} = v_{Gx} = -50\ m/s$, from which $v'_{Px} = -50 + 1.8\omega'$, and $v'_{Gy} = v'_{Py} - 0.3\omega' = 0.8 - 0.3\omega'$.

$$(\vec{r} \times m\vec{v}_G) = \begin{bmatrix} \vec{i} & \vec{j} & \vec{k} \\ -0.3 & 1.8 & 0 \\ -50m & -2m & 0 \end{bmatrix} = (0.6m + 50(1.8)m)\vec{k} = 90.6m\vec{k},$$

$$(\vec{r} \times m\vec{v}'_G) = \begin{bmatrix} \vec{i} & \vec{j} & \vec{k} \\ -0.3 & 1.8 & 0 \\ -50m & (0.8 - 0.3\omega')m & 0 \end{bmatrix} = (-(0.3)(0.8 - 0.3\omega')m + 50(1.8)m)\vec{k}.$$ Substitute into the expression for the conservation of angular momentum and reduce:

$$\boxed{\omega' = \frac{0.84}{\left(0.3^2 + \dfrac{I}{m}\right)} = 0.0997\ rad/s},$$

and $\vec{v}'_G = -50\vec{i} + 0.770\vec{j}\ (m/s)$, $\vec{v}'_P = -49.82\vec{i} + 0.8\vec{j}\ (m/s)$

====================================<>====================================

===================================<>===================================

Problem 8.91 While attempting to drive on an icy street for the first time, a student skids his 1260 *kg* car (a) into the university president's unoccupied 2700 *kg* Rolls Royce Corniche (b). The point of impact is P. Assume that the impacting surfaces are smooth and parallel to the y axis, and the coefficient of restitution of the impact is $e = 0.5$. The moments of inertia of the cars about their centers of mass are $I_A = 2400 kg-m^2$ and $I_B = 7600 kg-m^2$.

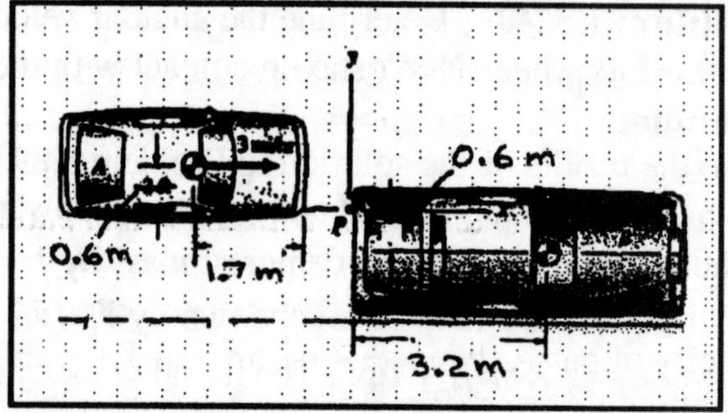

Determine the angular velocities of the cars and the velocities of their centers of mass after the collision.

Solution:

Car A's initial velocity is $v_A = \dfrac{5000}{3600} = 1.39 m/s$ The force of the impact is parallel to the x-axis so the cars are moving in the x direction after the collision. Linear momentum is conserved:

$m_A v_A = m_A v_A' + m_B v_B'$ (1) The angular momentum of each car about the point of impact is conserved.

$-0.6 m_A v_A = -0.6 m_A v_A' + I_A \omega_A'$ (2) $0 = 0.6 m_B v_B' + I_B \omega_B'$ (3). The velocity of car A at the point of impact after the collision is $\bar{v}_{AP}' = \bar{v}_A' + \bar{\omega}_A' \times \bar{r}_{P/A} = v_A' \bar{i} + \begin{vmatrix} \bar{i} & \bar{j} & \bar{k} \\ 0 & 0 & \omega_A' \\ 1.7 & -0.6 & 0 \end{vmatrix}$

$\bar{v}_{AP}' = (v_A' + 0.6\omega_A')\bar{i} + 1.7\omega_A' \bar{j}$. The corresponding equation for car B is

$\bar{v}_{BP}' = \bar{v}_B' + \bar{\omega}_B' \times \bar{r}_{P/B} = v_B' \bar{i} + \begin{vmatrix} \bar{i} & \bar{j} & \bar{k} \\ 0 & 0 & \omega_B' \\ -3.2 & 0.6 & 0 \end{vmatrix}$, or $\bar{v}_{BP}' = (v_B' - 0.6\omega_B')\bar{i} - 3.2\omega_B' \bar{j}$

The x-components of the velocities at P are related by the coefficient of restitution:

$e = \dfrac{(v_B' - 0.6\omega_B') - (v_A' + 0.6\omega_A')}{v_A}$ (4). Solving Equations (1) - (4), we obtain

$v_A' = 0.174 m/s, v_B' = 0.567 m/s$ $\omega_A' = -0.383 rad/s, \omega_B' = -0.121 rad/s$.

===================================<>===================================

Problem 8.92 The student in Problem 8.91 claimed he was moving at 5 *km/hr* prior to the collision, but the investigating police estimate that the center of mass of the Rolls-Royce was moving at 1.7 *m/s* after the collision. What was the student's actual speed?

Solution:

Setting $v_B' = 1.7 m/s$ in Equations (1) - (4) of the solution of Problem 8.91 and treating v_A as an unknown, we obtain $v_A = 4.17 m/s = 15.0 km/hr$.

===================================<>===================================

====================================<>====================================

Problem 8.93 Each slender bar is 48 *in.* long and weighs 20 *lb.* Bar A is released in the horizontal position as shown. The bars are smooth and the coefficient of restitution of their impact is $e = 0.8$. Determine the angle through which B swings afterward.

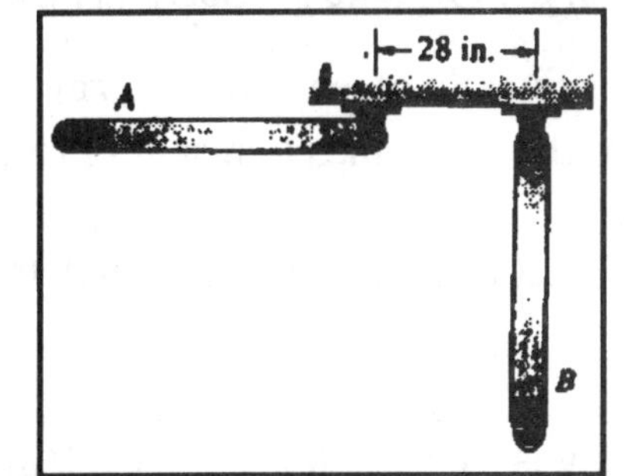

Solution:

Choose a coordinate system with the x axis parallel to bar A in the initial position, and the y axis positive upward. The strategy is (a) from the principle of work and energy, determine the angular velocity of bar A the instant before impact with B; (b) from the definition of the coefficient of restitution, determine the value of e in terms of the angular velocities of the two bars at the instant after impact; (c) from the principle of angular impulse-momentum, determine the relation between the angular velocities of the two bars after impact; (d) from the principle of work and energy, determine the angle through which bar B swings.

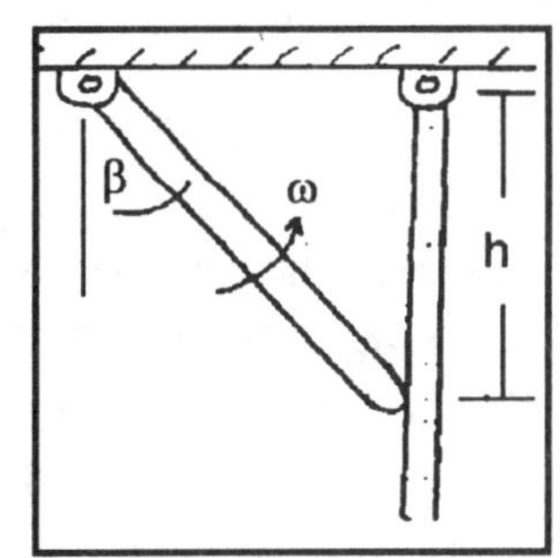

The angular velocity of bar A before impact: The angle of swing of bar A is $\beta = \sin^{-1}\left(\frac{28}{48}\right) = 35.69^o$. Denote the point of impact by P. The point of impact is $-h = -L\cos\beta = -3.25\ ft$. The change in height of the center of mass of bar A is $-\frac{L}{2}\cos\beta = -\frac{h}{2}$. From the principle of work and energy, $U = T_2 - T_1$, where $T_1 = 0$, since the bar is released from rest. The work done by the weight of bar A is $U = \int_0^{\frac{h}{2}} -Wdh = \frac{Wh}{2}$. The kinetic energy the bar is $T_2 = \left(\frac{1}{2}\right) I_A \omega_A^2$, from which

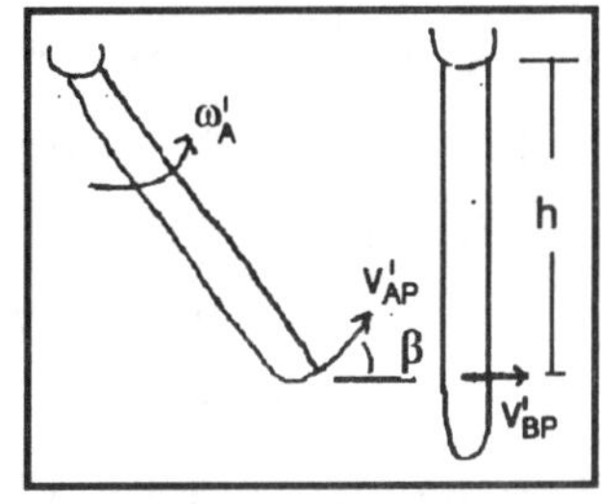

$$\omega_A = \sqrt{\frac{Wh}{I_A}} = \sqrt{\frac{3g\cos\beta}{L}} = 4.43\ rad/s, \text{ where } I_A = \frac{mL^2}{3}.$$

The angular velocities at impact: By definition, the coefficient of restitution is $e = \frac{v'_{BPx} - v'_{APx}}{v_{APx} - v_{BPx}}$. Bar B is at rest initially, from which $v_{BPx} = 0$, $v'_{BPx} = v'_{BP}$, and $v'_{APx} = v'_{AP}\cos\beta$, $v_{APx} = v_{AP}\cos\beta$, from which $e = \frac{v'_{BP} - v'_{AP}\cos\beta}{v_{AP}\cos\beta}$. The angular velocities are related by $v'_{BP} = h\omega'_B$, $v'_{AP} = L\omega'_A$, $v_{AP} = L\omega_A$, from which $e = \frac{h\omega'_B - (L\cos\beta)\omega'_A}{(L\cos\beta)\omega_A} = \frac{\omega'_B - \omega'_A}{\omega_A}$, from which (1) $\omega'_B - \omega'_A = e\omega_A$

The force reactions at P: From the principle of angular impulse-momentum,

$\int_{t_1}^{t_2} Fhdt = Fh(t_2 - t_1) = I_B(\omega'_B - 0)$, and

$\int_{t_1}^{t_2} -F(L\cos\beta)dt = -F(L\cos\beta)(t_2 - t_1) = I_A(\omega'_A - \omega_A)$. Divide the second equation by the first: : $\frac{-L\cos\beta}{h} = -1 = \frac{\omega'_A - \omega_A}{\omega'_B}$, from which

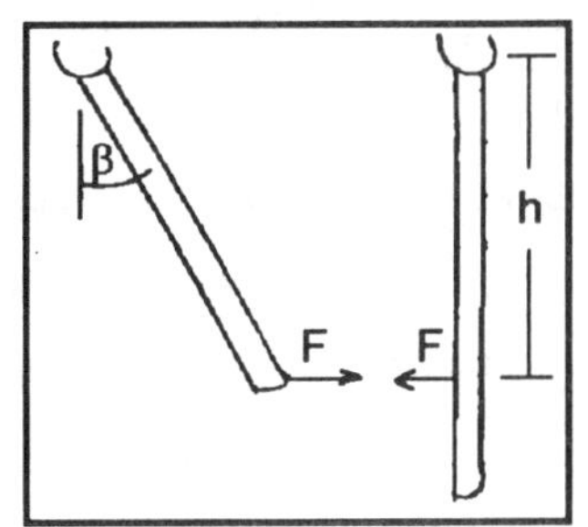

Solution Continued on next page

(2) $\omega'_B + \omega'_A = \omega_A$. Solve (1) and (2): $\omega'_A = \dfrac{(1-e)}{2}\omega_A$, $\omega'_B = \dfrac{(1+e)}{2}\omega_A$.

The principle of work and energy: From the principle of work and energy, $U = T_2 - T_1$, where $T_2 = 0$, since the bar comes to rest after rotating through an angle γ. The work done by the weight of bar B as its center of mass rotates through the angle γ is $U = \int_{\frac{L}{2}}^{\frac{L\cos\gamma}{2}} -W_B dh = -W\left(\frac{L}{2}\right)(1-\cos\gamma)$. The kinetic energy is $T_1 = \left(\frac{1}{2}\right)I_B\omega_B'^2 = \left(\frac{1}{8}\right)I_B(1+e)^2\omega_A^2 = \dfrac{I_B(1+e)^2(3g\cos\beta)}{8L} = \dfrac{WL(1+e)^2\cos\beta}{8}$. Substitute into $U = -T_1$ to obtain $\cos\gamma = 1 - \dfrac{(1+e)^2\cos\beta}{4} = 0.3421$, from which $\boxed{\gamma = 70^o}$

==================================<>==================================

Problem 8.94 The *Apollo* CSM (A) approaches the *Soyuz* Space Station (B). The mass of the *Apollo* is $m_A = 18\ Mg$, and the moment of inertia about the axis through the center of mass parallel to the z axis is $I_A = 114\ Mg\text{-}m^2$. The mass of the *Soyuz* is $m_B = 6.6\ Mg$, and the moment of inertia about the axis through its center of mass parallel to the z axis is $I_B = 70\ Mg\text{-}m^2$. The *Soyuz* is stationary relative to the reference frame shown and the CSM approaches with velocity $\vec{v}_A = 0.2\vec{i} + 0.05\vec{j}\ (m/s)$ and no angular velocity. What is their angular velocity after docking?

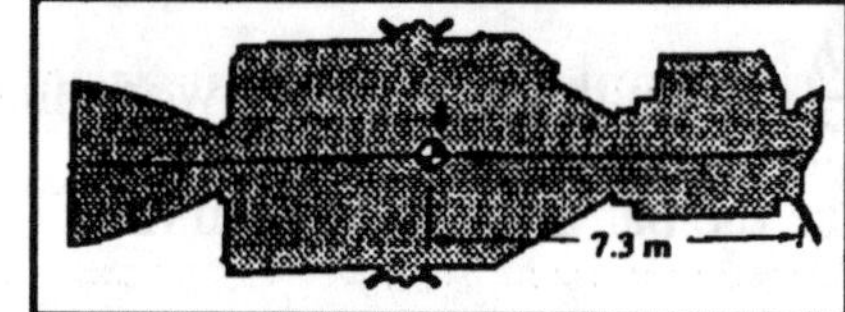

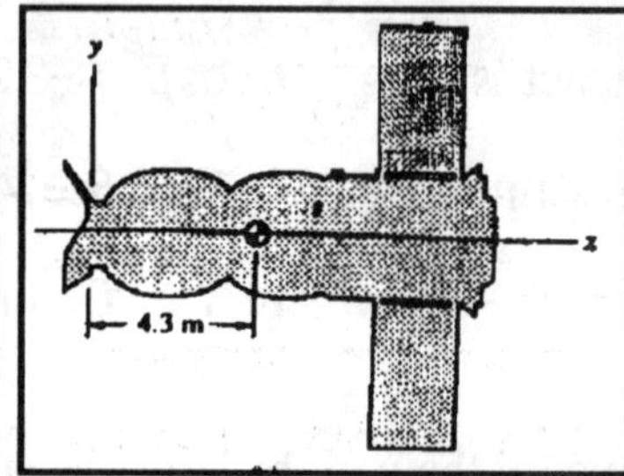

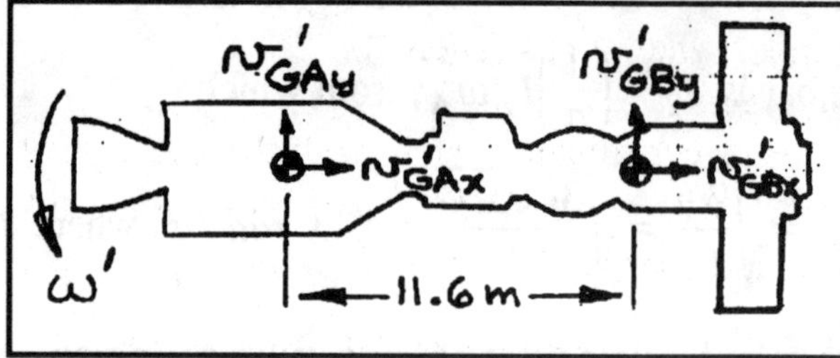

Solution:

The docking port is at the origin on the *Soyuz,* and the configuration the instant after contact is that the centers of mass of both spacecraft are aligned with the x axis. Denote the docking point of contact by P. (P is a point on each spacecraft, and by assumption, lies on the x-axis.) The linear momentum is conserved: $m_A\vec{v}_{GA} = m_A\vec{v}'_{GA} + m_B\vec{v}'_{GB}$, from which $m_A(0.2) = m_A v'_{GAx} + m_B v'_{GBx}$, and (1) $m_A(0.05) = m_A v'_{GAy} + m_B v'_{GBy}$. Denote the vectors from P to the centers of mass by $\vec{r}_{P/GA} = -7.3\vec{i}\ (m)$, and $\vec{r}_{P/GB} = +4.3\vec{i}\ (m)$. The angular momentum about the origin is conserved: $\vec{r}_{P/GA} \times m_A\vec{v}_{GA} = \vec{r}_{P/GA} \times m_A\vec{v}'_{GA} + I_A\vec{\omega}'_A + \vec{r}_{P/GB} \times m_B\vec{v}'_{GB}$. Denote the vector distance from the center of mass of the *Apollo* to the center of mass of the *Soyuz* by $\vec{r}_{B/A} = 11.6\vec{i}\ (m)$. From kinematics, the instant after contact: $\vec{v}'_{GB} = \vec{v}'_{GA} + \vec{\omega}' \times \vec{r}_{B/A}$. Reduce:

$$\vec{v}'_{GB} = \vec{v}'_{GA} + \begin{bmatrix} \vec{i} & \vec{j} & \vec{k} \\ 0 & 0 & \omega' \\ 11.6 & 0 & 0 \end{bmatrix} = v'_{GAx}\vec{i} + \left(v'_{GAy} + 11.6\omega'\right)\vec{j}$$, from which $v'_{GBx} = v'_{GAx}$,

Solution continued on next page

(2) $v'_{GBy} = v'_{GAy} + 11.6\omega'$. $\vec{r}_{P/GA} \times m_A\vec{v}_{GA} = \begin{bmatrix} \vec{i} & \vec{j} & \vec{k} \\ -7.3 & 0 & 0 \\ 0.2m_A & 0.05m_A & 0 \end{bmatrix} = (-0.365m_A)\vec{k}$,

$$\vec{r}_{P/GA} \times m_A\vec{v}'_{GA} = \begin{bmatrix} \vec{i} & \vec{j} & \vec{k} \\ -7.3 & 0 & 0 \\ m_A v'_{GAx} & m_A v'_{GAy} & 0 \end{bmatrix} = \left(-7.3 m_A v'_{GAy}\right)\vec{k}.$$

$$\vec{r}_{P/GB} \times m_B\vec{v}'_{GB} = \begin{bmatrix} \vec{i} & \vec{j} & \vec{k} \\ 4.3 & 0 & 0 \\ m_B v'_{GAx} & m_A\left(v'_{GAy} + 11.6\omega'\right) & 0 \end{bmatrix} = 4.3m_B\left(v'_{GAy} + 11.6\omega'\right)\vec{k}.$$

Collect terms and substitute into conservation of angular momentum expression, and reduce: $-0.365m_A = (-7.3m_A + 4.3m_B)v'_{GAy} + (49.9m_B + I_A + I_B)\omega'$. From (1) and (2), $v'_{GAy} = \dfrac{0.05m_A - 11.6m_B\omega'}{m_A + m_B}$. These two equations in two unknowns can be further reduced by substitution and algebraic reduction, but here they have been solved by iteration: $\boxed{\omega' = -0.003359 \ \ rad/s}$, $v'_{GAy} = 0.04704 \ \ m/s$, from which $v'_{GBy} = v'_{GAy} + 11.6\omega' = 0.00807 \ \ m/s$

====================================<>====================================

Problem 8.95 The moment of inertia of the disk is 0.2 *kg-m*^{2}. The system is released from rest. Use the principle of work and energy to determine the velocity of the 10 *kg* cylinder when it has fallen 1 *m*.

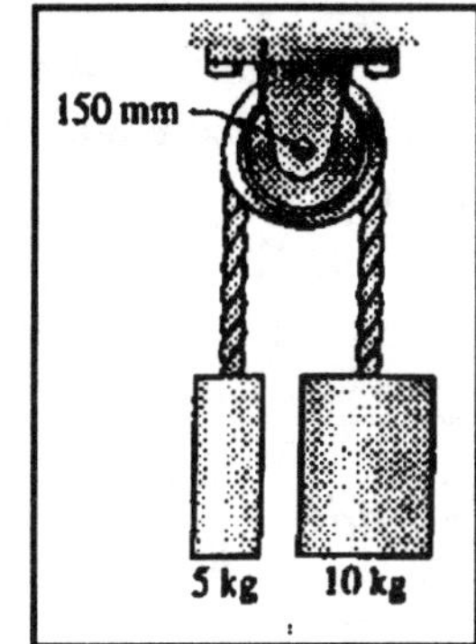

Solution :

Choose a coordinate system with the *y* axis positive upward. Denote $m_L = 5 \ \ kg$, $m_R = 10 \ \ kg$, $h_R = -1 \ m$, $R = 0.15 \ \ m$. From the principle of work and energy, $U = T_2 - T_1$ where $T_1 = 0$ since the system is released from rest. The work done by the left hand weight is $U_L = \int_0^{h_L} -m_L g dh = -m_L g h_L$.

The work done by the right hand weight is $U_L = \int_0^{h_R} -m_R g dh = -m_R g h_R$.

Since the pulley is one-to-one, $h_L = -h_R$, from which $U = U_L + U_R = (m_L - m_R)gh_R$. The kinetic energy is $T_2 = \left(\frac{1}{2}\right)I_P\omega^2 + \left(\frac{1}{2}\right)m_L v_L^2 + \left(\frac{1}{2}\right)m_R v_R^2$. Since the pulley is one-to-one, $v_L = -v_R$. From kinematics $\omega = \dfrac{v_R}{R}$, from which $T_2 = \left(\frac{1}{2}\right)\left(\frac{I_B}{R^2} + m_L + m_R\right)v_R^2$.

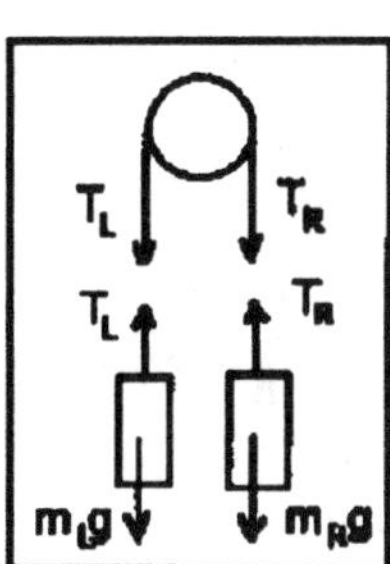

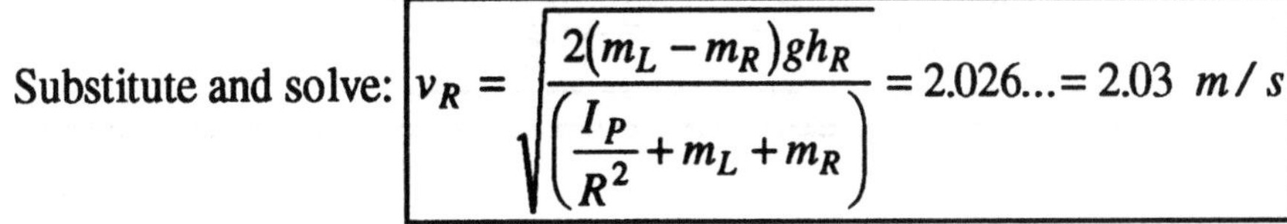

Substitute and solve: $\boxed{v_R = \sqrt{\dfrac{2(m_L - m_R)gh_R}{\left(\dfrac{I_P}{R^2} + m_L + m_R\right)}} = 2.026... = 2.03 \ \ m/s}$

====================================<>====================================

====================================<>====================================

Problem 8.96 Use momentum principles to determine the velocity of the 10 *kg* cylinder in Problem 8.95 one second after the system is released from rest.

Solution:

Use the coordinate system and notations of Problem 8.95. From the principle of linear impulse-momentum for the left hand weight: $\int_{t_1}^{t_2}(T_L - m_L g)dt = m_L(v_{L2} - v_{L1}) = m_L v_{L2}$, since $v_{L1} = 0$, from which (1) $T_L(t_2 - t_1) = m_L g(t_2 - t_1) + m_L v_2$. For the right hand weight, $\int_{t_1}^{t_2}(T_R - m_R g)dt = m_R v_{R2}$, from which (2) $T_R(t_2 - t_1) = m_R g(t_2 - t_1) + m_R v_{R2}$. From the principle of angular impulse-momentum for the pulley: $\int_{t_1}^{t_2}(T_L - T_R)Rdt = I_P\omega_2$, from which (3) $(T_L - T_R)(t_2 - t_1) = \frac{I_P}{R}\omega_2$.

Substitute (1) and (2) into (3): $(m_L - m_R)g(t_2 - t_1) + m_L v_{L2} - m_R v_{R2} = \frac{I_P}{R}\omega_2$. Since the pulley is one to one, $v_{L2} = -v_{R2}$. From kinematics: $\omega_2 = \frac{v_{R2}}{R}$, from which

$$v_{R2} = -\frac{(m_R - m_L)g(t_2 - t_1)}{\frac{I_P}{R^2} + m_L + m_R} = -2.05\ m/s$$

====================================<>====================================

Problem 8.97 Arm BC has a mass of 12 *kg* and the moment of inertia about its center of mass is 3 *kg-m*^{2}. Point B is stationary. Arm BC is initially aligned with the (horizontal) *x* axis, and a constant couple *M* is applied at B causes it to rotate upward. When it is in the position shown, its counterclockwise angular velocity is 2 *rad/s*. Determine *M*.

Solution:

Assume that the arm BC is initially stationary. Denote $R = 0.3\ m$. From the principle of work and energy, $U = T_2 - T_1$, where $T_1 = 0$. The work done is

$$U = \int_0^{\theta} Md\theta + \int_0^{R\sin\theta} -mgdh = M\theta - mgR\sin\theta$$. The angle is

$$\theta = 40\left(\frac{\pi}{180}\right) = 0.6981\ rad.$$

The kinetic energy is $T_2 = \left(\frac{1}{2}\right)mv^2 + \left(\frac{1}{2}\right)I_{BC}\omega^2$. From kinematics, $v = R\omega$, from which

$$T_2 = \left(\frac{1}{2}\right)(mR^2 + I_{BC})\omega^2.$$

Substitute into $U = T_2$ and solve: $M = \frac{(mR^2 + I_{BC})\omega^2 + 2mgR\sin\theta}{2\theta} = 44.2\ N\text{-}m$

====================================<>====================================

================================<>================================

Problem 8.98 The cart is stationary when a constant force F is applied to it. What will be its velocity when it has rolled a distance b? The mass of the body of the cart is m_C, and each of the four wheels has mass m, radius R, and moment of inertia I.

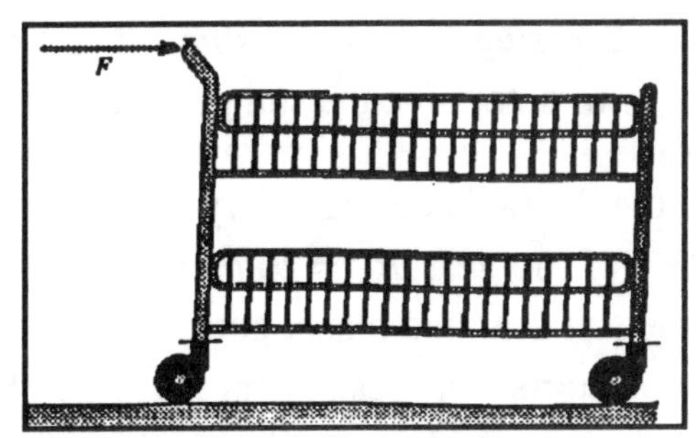

Solution:

From the principle of work and energy, $U = T_2 - T_1$, where $T_1 = 0$. The work done is $U = \int_0^b F dx = Fb$. The kinetic energy is $T_2 = \left(\frac{1}{2}\right) m_C v^2 + 4\left(\frac{1}{2}\right) m v^2 + 4\left(\frac{1}{2}\right) I\omega^2$. From kinematics, $v = R\omega$, from which

$$T_2 = \left(\frac{m_C}{2} + 2m + 2\frac{I}{R^2}\right) v^2.$$

Substitute into $U = T_2$ and solve: $v = \sqrt{\dfrac{2Fb}{m_C + 4m + 4\left(\dfrac{I}{R^2}\right)}}$

================================<>================================

Problem 8.99 Each pulley has moment of inertia $I = 0.003\ kg\text{-}m^2$, and the mass of the belt is 0.2 kg. If constant couple $M = 4\ N\text{-}m$ is applied to the bottom pulley, what will its angular velocity be when it has turned 10 revolutions?

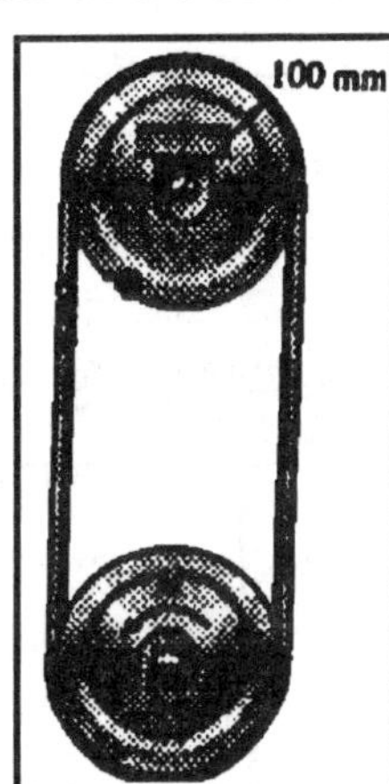

Solution:

Assume that the system is initially stationary.

From the principle of work $U = T_2 - T_1$, where $T_1 = 0$.

The work done is $U = \int_0^\theta M d\theta = M\theta$, where the angle is $\theta = 10(2\pi) = 62.83\ rad$.

The kinetic energy is $T_2 = \left(\frac{1}{2}\right) m_{belt} v^2 + 2\left(\frac{1}{2}\right) I_{pulley}\omega^2$.

From kinematics, $v = R\omega$, where $R = 0.1\ m$, from which

$T_2 = \left(\frac{1}{2}\right)\left(R^2 m_{belt} + 2I_{pulley}\right)\omega^2$. Substitute into $U = T_2$ and solve:

$$\omega = \sqrt{\frac{2M\theta}{\left(R^2 m_{belt} + 2I_{pulley}\right)}} = 250.7\ rad/s$$

================================<>================================

Problem 8.100 The ring gear is fixed. The mass and moment of inertia of the sun gear are $m_S = 22\ slugs$, $I_S = 4400\ slug\text{-}ft^2$. The mass and moment of inertia of each planet gear are $m_P = 2.7\ slugs$, $I_P = 65\ slug\text{-}ft^2$. A couple $M = 600\ ft\text{-}lb$ is applied to the sun gear. Use work and energy to determine the sun gear's angular velocity after it has turned 100 revolutions:

Solution:

Denote the radius of planetary gear, $R_P = 7\ in. = 0.5833\ ft$, the radius of sun gear $R_S = 20\ in. = 1.667\ ft$, and angular velocities of the sun gear and planet gear by ω_S, ω_P. Assume that the system starts from rest. From the principle of work and energy $U = T_2 - T_1$, where $T_1 = 0$. The work done is $U = \int_0^\theta M d\theta = M\theta$, where the angle is $\theta = (100)(2\pi) = 200\pi = 628.3\ rad$. The kinetic energy is $T_2 = \left(\frac{1}{2}\right) I_S \omega_S^2 + 3\left(\left(\frac{1}{2}\right) m_P v_P^2 + \left(\frac{1}{2}\right) I_P \omega_P^2\right)$. The velocity of the outer radius of the sun gear is $v_S = R_S \omega_S$. The velocity of the center of mass of the planet gears is the average velocity of the velocity of the sun gear contact and the ring gear contact, $v_P = \frac{v_S + v_R}{2} = \frac{v_S}{2}$, since $v_R = 0$. The angular

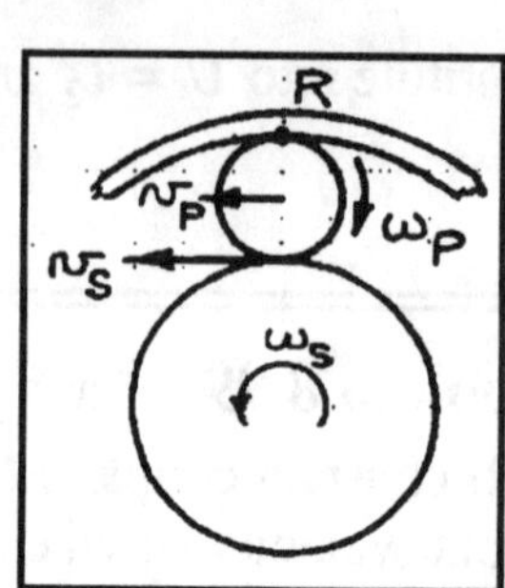

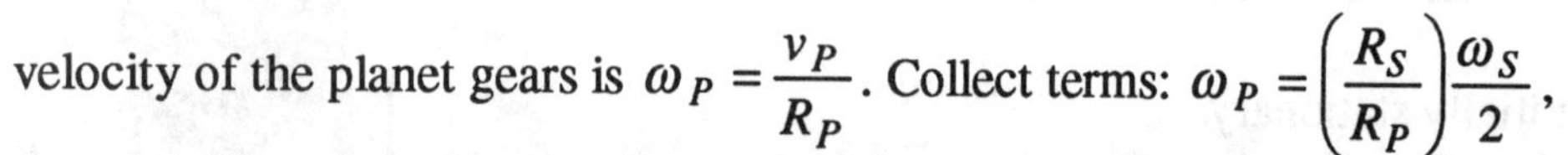

velocity of the planet gears is $\omega_P = \frac{v_P}{R_P}$. Collect terms: $\omega_P = \left(\frac{R_S}{R_P}\right)\frac{\omega_S}{2}$, $v_P = \frac{R_S}{2}\omega_S$. Substitute into $U = T_2$ and solve:

$$\omega_S = \sqrt{\frac{2M\theta}{\left(I_S + \frac{3}{4}\left(m_P R_S^2 + I_P\left(\frac{R_S}{R_P}\right)^2\right)\right)}} = 12.53\ rad/s$$

==================================<>==================================

Problem 8.101 The moments of inertia of gears A and B are $I_A = 0.014\ slug\text{-}ft^2$, and $I_B = 0.100\ slug\text{-}ft^2$. Gear A is connected to a torsional spring with constant $k = 0.2\ ft\text{-}lb/rad$. If the spring is unstretched, and the surface supporting the 5 *lb* weight is removed, what is the weight's velocity when it has fallen 3 *in.*?

Solution:

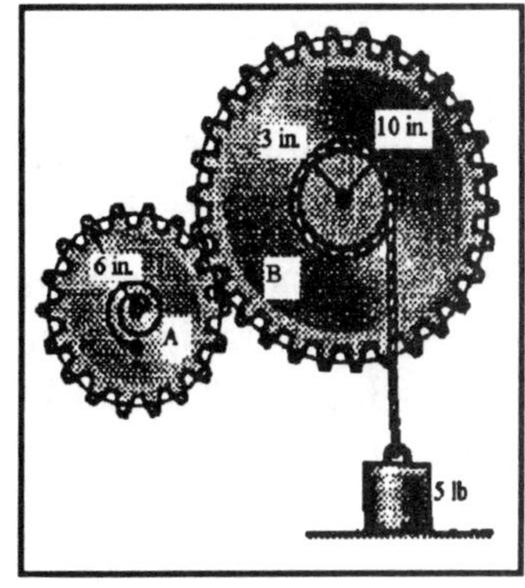

Denote $W = 5\ lb$, $s = 3\ in. = 0.25\ ft$ is the distance the weight falls, $r_B = 10\ in. = 0.833\ ft$, $r_A = 6\ in. = 0.5\ ft$, $r'_B = 3\ in. = 0.25\ ft$, are the radii of the gears and pulley. Choose a coordinate system with *y* positive upward. From the conservation of energy $T + V = const.$ Choose the datum at the initial position, such that $V_1 = 0$, $T_1 = 0$, from which $T_2 + V_2 = 0$ at any position 2. The gear B rotates in a negative direction and the gear A rotates in a positive direction. By inspection, $\theta_B = -\frac{s}{r'_B} = -\frac{0.25}{0.25} = -1\ rad$,

$\theta_A = -\left(\frac{r_B}{r_A}\right)\theta_B = 1.667\ rad$, $v = r'_B\omega_B = 0.25\omega_B$, $\omega_A = -\left(\frac{r_B}{r_A}\right)\omega_B = 1.667\omega_B$. The moment exerted by the spring is negative, from which the potential energy in the spring is $V_{spring} = -\int_0^{\theta_A} M d\theta = \int_0^{\theta_A} k\theta d\theta = \frac{1}{2}k\theta_A^2 = 0.2778\ ft\text{-}lb$. The force due to the weight is negative, from which the potential energy of the weight is $V_{weight} = -\int_0^{-s}(-W)dy = -Ws = -1.25\ ft\text{-}lb$. The kinetic energy of the system is $T_2 = \left(\frac{1}{2}\right)I_A\omega_A^2 + \left(\frac{1}{2}\right)I_B\omega_B^2 + \left(\frac{1}{2}\right)\left(\frac{W}{g}\right)v^2$. Substitute: $T_2 = 1.1888v^2$, from which $V_{spring} + V_{weight} + 1.1888v^2 = 0$. Solve $\boxed{v = 0.9043\ ft/s}$ downward.

==================================<>==================================

Problem 8.102 Consider the system in Problem 8.101. (a) What maximum distance does the weight fall when the supporting system is removed? (b) What maximum velocity does the weight achieve?

Solution :

Use the solution to Problem 8.101: $V_{spring} + V_{weight} + 1.1888v^2 = 0$.

$V_{spring} = -\int_0^{\theta_A} M d\theta = \int_0^{\theta_A} k\theta d\theta = \frac{1}{2}k\theta_A^2 = 4.444s^2$, $V_{weight} = -\int_0^{-s}(-W)dy = -Ws$, from which $4.4444s^2 - 5s + 1.1888v^2 = 0$. (a) The maximum travel occurs when $v = 0$, from which $\boxed{s_{max} = \frac{5}{4.4444} = 1.125\ ft}$ (where the other solution $s_{max} = 0$ is meaningless here). (b) The maximum velocity occurs at $\frac{dv^2}{ds} = 0 = 2\frac{4.4444}{1.1888}s - \frac{5}{1.1888} = 0$, from which $s_{v-max} = 0.5625\ ft$. This is indeed a maximum, since $\frac{d^2(v^2)}{ds^2} = 2\left(\frac{4.444}{1.1888}\right) > 0$, and $\boxed{|v_{max}| = 1.0876\ ft/s}$ (downward).

==================================<>==================================

Problem 8.103 Each of the go-cart's front wheels weighs 5 *lb* and has a mass moment of inertia of 0.01 *slug-ft*^{2}. The two rear wheels and rear axle form a single rigid body weighing 40 *lb* and having a mass moment of inertia of 0.1 *slug-ft*^{2}. The total weight of the rider and go-cart, including its wheels, is 240 *lb*. The go-cart starts from rest, its engine exerts a constant torque of 15 *ft-lb* on the rear axle, and its wheels do not slip. If you neglect friction and aerodynamic drag, how fast is it moving when it has traveled 50 *ft*?

Solution:

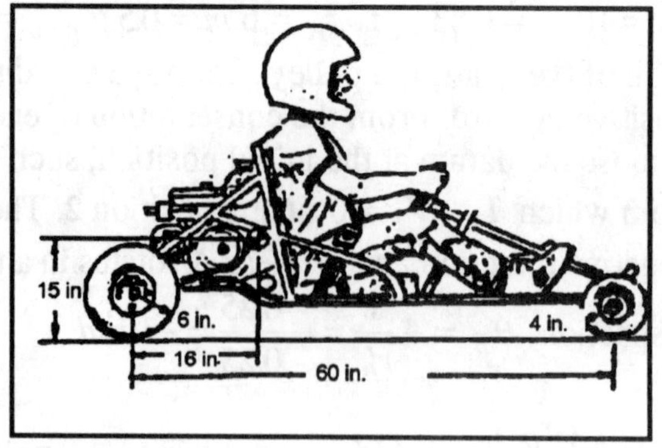

From the principle of work and energy: $U = T_2 - T_1$, where $T_1 = 0$, since the go-cart starts from rest. Denote the rear and front wheels by the subscripts A and B, respectively. The radius of the rear wheels $r_A = 6$ in. $= 0.5$ ft. The radius of the front wheels is $r_B = 4$ in. $= 0.3333$ ft. The rear wheels rotate through an angle $\theta_A = \dfrac{50}{0.5} = 100$ rad, from which the work done is $U = \int_0^{\theta_A} M d\theta = M\theta_A = 15(100)$ ft-lb. The kinetic energy is $T_2 = \left(\dfrac{1}{2}\right)\dfrac{W}{g}v^2 + \left(\dfrac{1}{2}\right)I_A\omega_A^2 + 2\left(\dfrac{1}{2}\right)I_B\omega_B^2$ (for two front wheels). The angular velocities are related to the go-cart velocity by $\omega_A = \dfrac{v}{r_A} = 2v$, $\omega_B = \dfrac{v}{r_B} = 3v$, from which $T_2 = 4.020v^2$ ft-lb. Substitute into $U = T_2$ and solve: $v = 19.32$ ft/s.

Problem 8.104 Determine the maximum power and the average power transferred to the go-cart in Problem 8.103 by its engine.

Solution The maximum power is $P_{\max} = M\omega_{A\max}$, where $\omega_{A\max} = \dfrac{v_{\max}}{R}$. From which $P_{\max} = \dfrac{Mv_{\max}}{R}$. Under constant torque, the acceleration of the go-cart is constant, from which the maximum velocity is the greatest value of the velocity, which will occur at the end of the travel. From the solution to Problem 8.27, $v_{\max} = 19.32$ ft/s, from which $P_{\max} = \dfrac{Mv_{\max}}{r_A} = \dfrac{15(19.32)}{0.5} = 579.5$ ft-lb/s. The average power is $P_{ave} = \dfrac{U}{t}$. From the solution to Problem 8.27, $U = 1500$ ft-lb. Under constant acceleration, $v = at$, and $s = \dfrac{1}{2}at^2$, from which $s = \dfrac{1}{2}vt$, and $t = \dfrac{2s}{v} = \dfrac{100}{19.32} = 5.177$ s, from which

$$P_{ave} = \frac{U}{t} = \frac{1500}{5.177} = 289.74 \text{ ft-lb/s}$$

=================================<>=================================

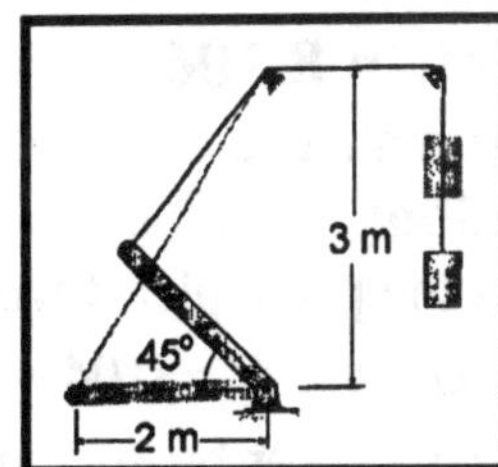

Problem 8.105 The system starts from rest with the 4 *kg* slender bar horizontal. The mass of the suspended cylinder is 10 *kg*. What is the bar's angular velocity when it is in the position shown?

Solution:

From the principle of work and energy: $U = T_2 - T_1$, where $T_1 = 0$ since the system starts from rest. The change in height of the cylindrical weight is found as follows: By inspection, the distance between the end of the bar and the pulley when the bar is in the horizontal position is

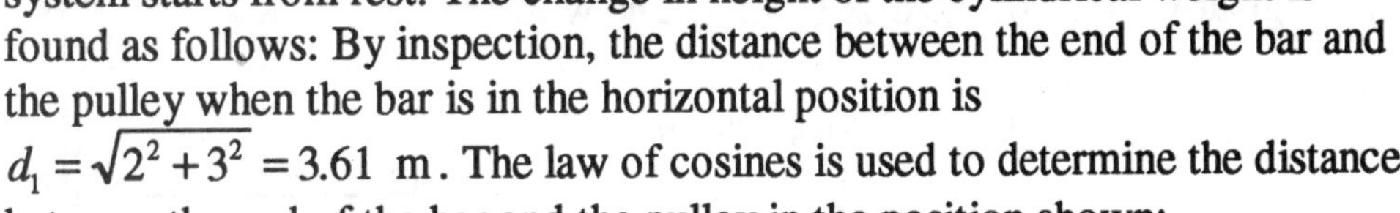

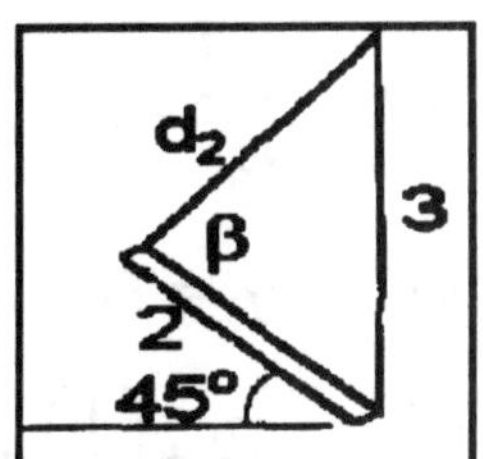

$d_1 = \sqrt{2^2 + 3^2} = 3.61 \text{ m}$. The law of cosines is used to determine the distance between the end of the bar and the pulley in the position shown:

$d_2 = \sqrt{2^2 + 3^2 - 2(2)(3)\cos 45^\circ} = 2.125 \text{ m}$, from which $h = d1 - d2 = 1.481 \text{ m}$.

The work done by the cylindrical weight is

$U_{cylinder} = \int_0^{-h} -m_c g ds = m_c g h = 145.3 \text{ N-m}$. The work done by the weight of the bar is $U_{bar} = \int_0^{\cos 45^o} -m_b g dh = -m_b g \cos 45^o = -27.75 \text{ N-m}$, from which $U = U_{cylinder} + U_{bar} = 117.52 \text{ N-m}$. From the sketch, (which shows the final position) the component of velocity normal to the bar is $v \sin\beta$, from which $2\omega = v\sin\beta$. From the law of sines:

$\sin\beta = \left(\frac{3}{d_2}\right)\sin 45^o = 0.9984$. The kinetic energy is $T_2 = \left(\frac{1}{2}\right)m_c v^2 + \left(\frac{1}{2}\right)I\omega_b^2$, from which

$T_2 = 22.732\omega^2$, where $v = \left(\frac{2}{\sin\beta}\right)\omega$ and $I = \frac{m(2^2)}{3}$ has been used. Substitute into $U = T_2$ and solve:

$\omega = 2.274 \text{ rad/s}$

=================================<>=================================

Problem 8.106 The 0.1 *kg* slender bar and 0.2 *kg* cylindrical disk are released from rest with the bar horizontal. The disk rolls on the curved surface. What is the bar's angular velocity when it is vertical?

Solution:

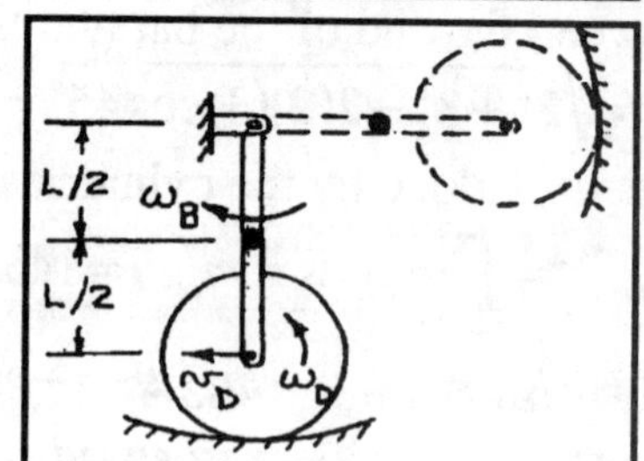

From the principle of work and energy, $U = T_2 - T_1$, where $T_1 = 0$. Denote $L = 0.12\ m$, $R = 0.04\ m$, the angular velocity of the bar by ω_B, the velocity of the disk center by v_D, and the angular velocity of the disk by ω_D. The work done is $U = \int_0^{-\frac{L}{2}} -m_B g dh + \int_0^{-L} -m_D g dh = \left(\frac{L}{2}\right) m_B g + L m_D g$. From kinematics, $v_D = L\omega_B$, and $\omega_D = \frac{v_D}{R}$. The kinetic energy is

$T_2 = \left(\frac{1}{2}\right) I_B \omega_B^2 + \left(\frac{1}{2}\right) m_D v_D^2 + \left(\frac{1}{2}\right) I_D \omega_D^2$. Substitute the kinematic relations to obtain $T_2 = \left(\frac{1}{2}\right)\left(I_B + m_D L^2 + I_D \left(\frac{L}{R}\right)^2\right)\omega_B^2$, where

$I_B = \frac{m_B L^2}{3}$, $I_D = \frac{m_D R^2}{2}$, from which $T_2 = \left(\frac{1}{2}\right)\left(\frac{m_B}{3} + \frac{3m_D}{2}\right) L^2 \omega_B^2$. Substitute into $U = T_2$ and solve: $\boxed{\omega_B = \sqrt{\frac{6g(m_B + 2m_D)}{(2m_B + 9m_D)L}} = 11.07\ rad/s}$.

Problem 8.107 A slender bar of mass *m* is released from rest in the vertical position and is allowed to fall. Neglecting friction and assuming that it remains in contact with the floor and wall, determine its angular velocity as a function of θ.

Solution:

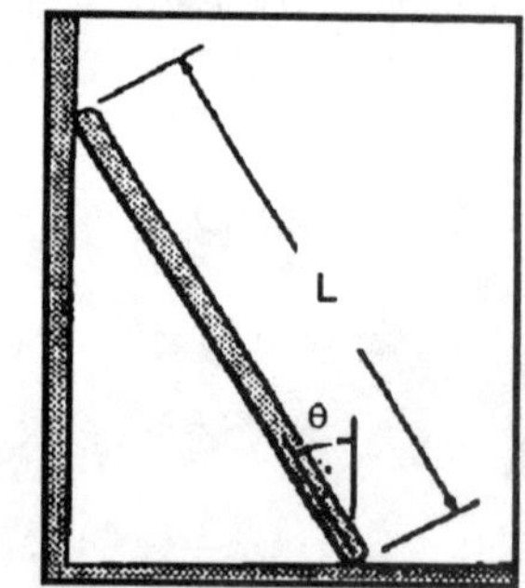

The strategy is (a) to use the kinematic relations to determine the relation between the velocity of the center of mass and the angular velocity about the instantaneous center, and (b) the principle of work and energy to obtain the angular velocity of the bar. *The kinematics:* Denote the angular velocity of the bar about the instantaneous center by ω. The coordinates of the instantaneous center of rotation of the bar are $(L\sin\theta, L\cos\theta)$. The coordinates of the center of mass of the bar are $\left(\frac{L}{2}\sin\theta, \frac{L}{2}\cos\theta\right)$.

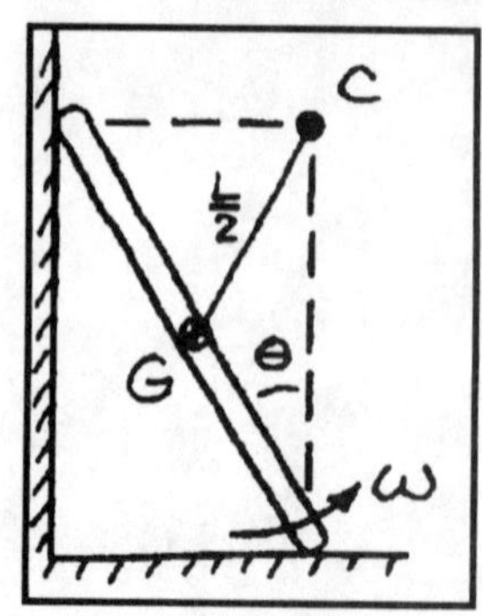

The vector distance from the instantaneous center to the center of mass is $\vec{r}_{G/C} = -\frac{L}{2}(\vec{i}\sin\theta + \vec{j}\cos\theta)$. The velocity of the center of mass is

$$\vec{v}_G = \vec{\omega} \times \vec{r}_{G/C} = \begin{bmatrix} \vec{i} & \vec{j} & \vec{k} \\ 0 & 0 & \omega \\ -\frac{L}{2}\sin\theta & -\frac{L}{2}\cos\theta & 0 \end{bmatrix} = \frac{\omega L}{2}(\vec{i}\cos\theta - \vec{j}\sin\theta),$$

Solution continued on next page

from which $|\vec{v}_G| = \frac{\omega L}{2}$.

The principle of work and energy: $U = T_2 - T_1$ where $T_1 = 0$. The work done by the weight of the bar is $U = \int_{\frac{L}{2}}^{\frac{L}{2}\cos\theta} -mgdh = -mg\left(\frac{L}{2}\cos\theta - \frac{L}{2}\right) = mg\left(\frac{L}{2}\right)(1-\cos\theta)$. The kinetic energy is

$T_2 = \left(\frac{1}{2}\right)mv_G^2 + \left(\frac{1}{2}\right)I_B\omega^2$. Substitute $v_G = \frac{\omega L}{2}$ and $I_B = \frac{mL^2}{12}$ to obtain $T_2 = \frac{mL^2\omega^2}{6}$. Substitute into $U = T_2$ and solve: $\boxed{\omega = \sqrt{\frac{3g(1-\cos\theta)}{L}}}$.[*Check*: Compare with the equation (see bottom of page 343 of text) used in the Discussion of Example 7.10, with $\cos 5^o$ replaced by $\cos 0^o$. *check.*]

==================================<>==================================

Problem 8.108 The 4 *kg* slender bar is pinned to 2 *kg* sliders at A and B. If friction is negligible and the system starts from rest in the position shown, what is the bar's angular velocity when the slider at A has fallen 0.5 *m*?

Solution:

Choose a coordinate system with the origin at the initial position of A and the y axis positive upward. The strategy is (a) to determine the distance that B has fallen and the center of mass of the bar has fallen when A falls 0.5 *m*, (b) use the coordinates of A, B, and the center of mass of the bar and the constraints on the motion of A and B to determine the kinematic relations, and (c) use the principle of work and energy to determine the angular velocity of the bar.

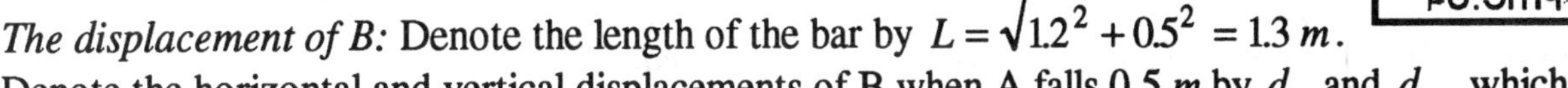

The displacement of B: Denote the length of the bar by $L = \sqrt{1.2^2 + 0.5^2} = 1.3\ m$.

Denote the horizontal and vertical displacements of B when A falls 0.5 *m* by d_x and d_y, which are in the ratio $\frac{d_y}{d_x} = \tan 45^o = 1$, from which $d_x = d_y = d$. The vertical distance between A and B is reduced by the distance 0.5 *m* and increased by the distance d_y, and the horizontal distance between A and B is increased by the distance d_x, from which $L^2 = (1.2 - 0.5 + d_y)^2 + (0.5 + d_x)^2$. Substitute $d_x = d_y = d$ and $L = 1.3\ m$ and reduce to obtain $d^2 + 2bd + c = 0$, where $b = 0.6$, and $c = -0.475$. Solve: $d_{1,2} = -b \pm \sqrt{b^2 - c} = 0.3138\ m,\ = -1.514\ m$, from which only the positive root is meaningful.

The final position coordinates: The coordinates of the initial position of the center of mass of the bar are $(x_{G1}, y_{G1}) = \left(\frac{L}{2}\sin\theta_1,\ -\frac{L}{2}\cos\theta_1\right) = (0.25, -0.6)\ (m)$, where $\theta_1 = \sin^{-1}\left(\frac{0.5}{L}\right) = 22.61^o$ is the angle of the bar relative to the vertical. The coordinates of the final position of the center of mass of the bar are $(x_{G2}, y_{G2}) = \left(\frac{L}{2}\sin\theta_2,\ -0.5 - \frac{L}{2}\cos\theta_2\right) = (0.4069,\ -1.007)$, where $\theta_2 = \sin^{-1}\left(\frac{0.5+d}{L}\right) = 38.75^o$.

The vertical distance that the center of mass falls is $h = y_{G2} - y_{G1} = -0.4069\ m$.

Solution continued on next page

The coordinates of the final positions of A and B are, respectively $(x_{A2}, y_{A2}) = (0, -0.5)$, and $(x_{B2}, y_{B2}) = (0.5 + d, -(1.2 + d)) = (0.8138, -1.514)$. The vector distance from A to B is $\vec{r}_{B/A} = (x_{B2} - x_{A2})\vec{i} + (y_{B2} - y_{A2})\vec{j} = 0.8138\vec{i} - 1.014\vec{j}\ (m)$. *Check:* $|\vec{r}_{B/A}| = L = 1.3\ m$. *check.* The vector distance from A to the center of mass is

$\vec{r}_{G/A} = (x_{G2} - x_{A2})\vec{i} + (y_{G2} - y_{A2})\vec{j} = 0.4069\vec{i} - 0.5069\vec{j}\ (m)$. *Check:* $|\vec{r}_{G/A}| = \frac{L}{2} = 0.65\ m$. *check.*

The kinematic relations: From kinematics, $\vec{v}_B = \vec{v}_A + \vec{\omega} \times \vec{r}_{B/A}$. The slider A is constrained to move vertically, and the slider B moves at a 45° angle, from which $\vec{v}_A = -v_A\vec{j}\ (m/s)$, and

$$\vec{v}_B = (v_B \cos 45^o)\vec{i} - (v_B \sin 45^o)\vec{j}.\ \vec{v}_B = \vec{v}_A + \begin{bmatrix} \vec{i} & \vec{j} & \vec{k} \\ 0 & 0 & \omega \\ 0.8138 & -1.014 & 0 \end{bmatrix} = \vec{i}(1.014\omega) + \vec{j}(-v_A + 0.8138\omega),$$

from which (1) $v_B = \left(\frac{1.014}{\cos 45^o}\right)\omega = 1.434\omega$, and (2) $v_A = v_B \sin 45^o + 0.8138\omega = 1.828\omega$. The velocity of the center of mass of the bar is

$$\vec{v}_G = \vec{v}_A + \vec{\omega} \times \vec{r}_{G/A} = -v_A\vec{j} + \begin{bmatrix} \vec{i} & \vec{j} & \vec{k} \\ 0 & 0 & \omega \\ 0.4069 & -0.5069 & 0 \end{bmatrix} = (0.5069\omega)\vec{i} + (-v_A + 0.4069\omega)\vec{j},$$

$\vec{v}_G = 0.5069\omega\,\vec{i} - 1.421\omega\,\vec{j}\ (m/s)$, from which (3) $v_G = 1.508\omega$

The principle of work and energy: From the principle of work and energy, $U = T_2 - T_1$, where $T_1 = 0$.

The work done is $U = \int_0^{-d_1} -m_B g dh + \int_0^{-0.5} -m_A g dh + \int_0^{-h} -m_{bar} g dh$,

$U = m_B g(0.3138) + m_A g(0.5) + m_{bar} g(0.4069) = 31.93\ N\text{-}m$.

The kinetic energy is $T_2 = \left(\frac{1}{2}\right)m_B v_B^2 + \left(\frac{1}{2}\right)m_A v_A^2 + \left(\frac{1}{2}\right)m_{bar} v_G^2 + \left(\frac{1}{2}\right)I_{bar}\omega^2$, substitute $I_{bar} = \frac{m_{bar}L^2}{12}$,

and (1), (2)and (3) to obtain $T_2 = 10.23\omega^2$. Substitute into $U = T_2$ and solve: $\boxed{\omega = \sqrt{\frac{31.93}{10.23}} = 1.77\ rad/s}$

==<>==

==================================<>==================================

Problem 8.109 A homogenous hemisphere of mass m is released from rest in the position shown. If it rolls on the horizontal surface, what is its angular velocity when its flat surface is horizontal?

Solution:

Choose a coordinate system with the x axis parallel to the plane surface and the y axis positive upward. The moment of inertia of the hemisphere about the z axis passing through center of mass is $I_{zz} = \frac{83}{320} mR^2 = 0.259 mR^2$. The change in height of the center of mass is $h = R - \left(1 - \frac{3}{8}\right)R = \frac{3R}{8}$. The work done is $U = \int_0^{-\frac{3R}{8}} -mg\,dh = mg\left(\frac{3}{8}\right)R$. The kinetic energy is $T_2 = \left(\frac{1}{2}\right)mv^2 + \left(\frac{1}{2}\right)I_{zz}\omega^2$.

From kinematics, $v = \left(1 - \frac{3}{8}\right)R\omega = \frac{5}{8}R\omega$. Substitute to obtain

$T_2 = \left(\frac{1}{2}\right)\left(m\left(\frac{5R}{8}\right)^2 + \frac{83}{320}mR^2\right)\omega^2 = \left(\frac{1}{2}\right)\left(\frac{25}{64} + \frac{83}{320}\right)mR^2\omega^2 = \frac{13}{40}mR^2\omega^2$. Substitute into $U = T_2$, $\frac{3}{8}mgR = \left(\frac{13}{40}\right)mR^2\omega^2$, and solve: $\boxed{\omega = \sqrt{\frac{15g}{13R}}}$

==================================<>==================================

Problem 8.110 What normal force is exerted on the hemisphere in Problem 8.95 by the horizontal surface at the instant its flat surface is horizontal?

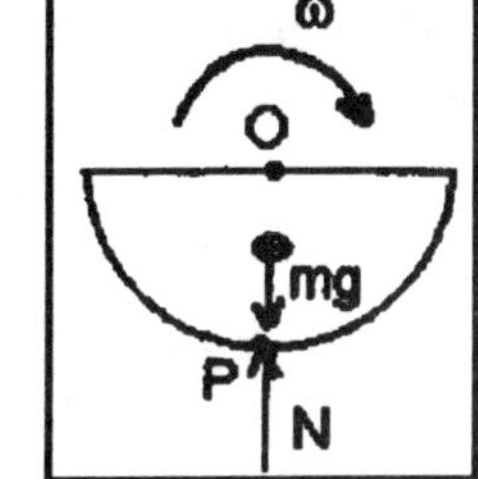

Solution: .

Assume that the hemisphere rocks back and forth in an oscillatory motion. As the hemisphere rocks, it rolls on the surface, so that the point O remains *a constant distance* R *from the floor*, from which the *vertical* velocity and acceleration of the point O is zero. The *angular velocity* $\frac{d\theta}{dt} = \omega$ varies smoothly from zero at the extreme left to a maximum when the flat surface is horizontal, back to zero at the extreme right, and then reverses. The derivative of the angular velocity (the angular acceleration) passes through zero as the angular velocity passes through its maximum, from which *the angular acceleration is zero when the hemisphere is horizontal.* The vector distance from the center of mass to point O is $\vec{r}_{G/O} = -\left(\frac{3}{8}\right)R\vec{j}$, from which $\vec{a}_G = \vec{a}_O + \vec{\alpha} \times \vec{r}_{G/O} - \omega^2 \vec{r}_{G/O}$, $\vec{a}_G = a_{Ox}\vec{i} + \omega^2\left(\frac{3}{8}\right)R\vec{j}$, and

$a_{Gy} = \frac{3}{8}R\omega^2$.

From the solution to Problem 8.95, $\omega^2 = \frac{15g}{13R}$, from which $a_{Gy} = \frac{45g}{104}$. From Newton's second law, $ma_{Gy} = N - mg$, from which $\boxed{N = mg\left(1 + \frac{45}{104}\right) = mg\left(\frac{149}{104}\right) = 1.433mg}$

==================================<>==================================

==============================<>==============================

Problem 8.111 The slender bar rotates freely *in the horizontal plane* about a vertical shaft at O. The bar weighs 20 *lb* and its length is 6 *ft*. The slider A weighs 2 *lb*. If the bar's angular velocity is $\omega = 10\ rad/s$ and the radial component of the velocity of A is zero when $r = 1\ ft$, what is the angular velocity of the bar when $r = 4\ ft$? (The moment of inertia of the slider A about its center of mass is negligible; that is, treat A as a particle.)

Solution:

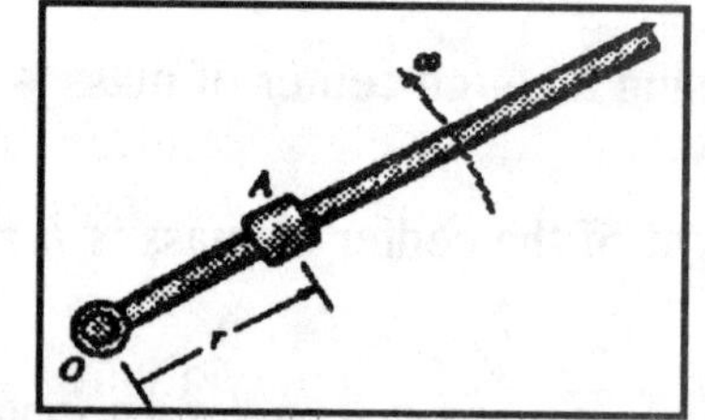

From the definition of angular momentum, only *the radial position* of the slider need be taken into account in applying the principle of the conservation of angular momentum; that is, *the radial velocity* of the slider at $r = 4\ ft$ does not change the angular momentum of the bar. From the conservation of angular momentum:

$I_{bar}\omega_1 + r_{1A}^2\left(\frac{W_A}{g}\right)\omega_1 = I_{bar}\omega_2 + r_{2A}^2\left(\frac{W_A}{g}\right)\omega_2$. Substitute numerical values:

$$I_{bar} + r_{1A}^2\left(\frac{W_A}{g}\right) = \frac{W_{bar}}{3g}L^2 + r_{1A}^2\left(\frac{W_A}{g}\right) = 7.52\ slug\text{-}ft^2.$$

$I_{bar} + r_{21A}^2\left(\frac{W_A}{g}\right) = \frac{W_{bar}}{3g}L^2 + r_{2A}^2\left(\frac{W_A}{g}\right) = 8.46\ slug\text{-}ft^2$, from which

$$\boxed{\omega_2 = \left(\frac{7.52}{8.46}\right)\omega_1 = 8.897... = 8.90\ rad/s}$$

==============================<>==============================

Problem 8.112 A satellite is deployed with angular velocity $\omega = 1\ rad/s$. (Fig, a). Two internally stored antennas that span the diameter of the satellite are then extended. and the satellite's angular velocity decreases to ω'.(Fig. b)- By modeling the satellite as a 500 *kg* sphere of 1.2-*m* radius and each antenna as a 10-kg slender bar, determine ω'.

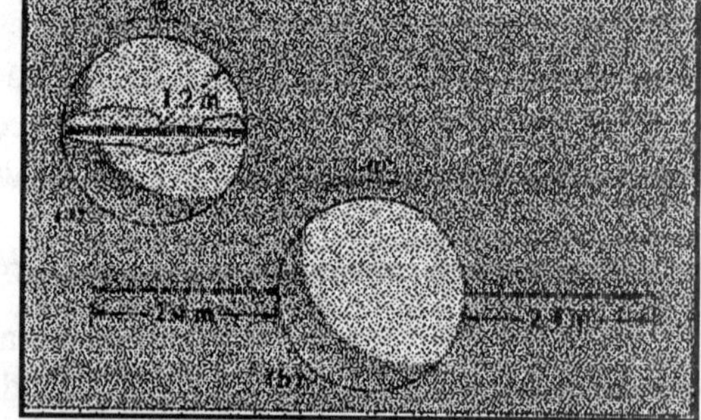

Solution:

Assume (I) in configuration (a) the antennas are folded inward, each lying on a line passing so near the center of the satellite that the distance from the line to the center can be neglected; (II) when extended, the antennas are entirely external to the satellite. Denote $R = 1.2\ m$, $L = 2R = 2.4\ m$. The moment of inertia of the antennas about the center of mass of the satellite in configuration (a) is $I_{ant-folded} = 2\left(\frac{mL^2}{12}\right) = 9.6\ kg\text{-}m^2$. The moment of inertia of the antennas about the center of mass of the satellite in configuration (b) is

$$I_{ant-ext} = 2\left(\frac{mL^2}{12}\right) + 2\left(R + \frac{L}{2}\right)^2 m = 124.8\ kg\text{-}m^2.$$

The moment of inertia of the satellite is $I_{sphere} = \frac{2}{5}m_{sphere}R^2 = 288\ kg\text{-}m^2$. The angular momentum is conserved, $I_{sphere}\omega + I_{ant-folded}\omega = I_{sphere}\omega' + I_{ant-ext}\omega'$, where $I_{ant-folded}$, $I_{ant-ext}$ *are for both antennas,* from which $297.6\omega = 412.8\omega'$. Solve $\boxed{\omega' = 0.721\ rad/s}$

==============================<>==============================

====================================<>====================================

Problem 8.113 An engineer decides to control the angular velocity of a satellite by deploying small masses attached to cables. If the angular velocity of the satellite in configuration (a) is 4 *rpm*, determine the distance *d* in configuration (b) that will cause the angular velocity to be 1 *rpm*. The moment of inertia of the satellite is $I = 500\ kg\text{-}m^2$ and each mass is 2 *kg*. (Assume that the cables and masses rotate with the same angular velocity as the satellite. Neglect the masses of the cables and the mass moments of inertia of the masses about their centers of mass.)

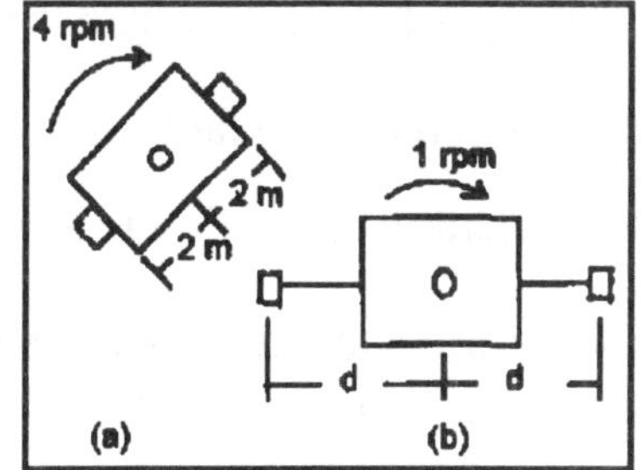

Solution:

From the conservation of angular momentum, $\left(I + 2m\left(2^2\right)\right)\omega_1 = \left(I + 2md^2\right)\omega_2$.

Solve: $\boxed{d = \sqrt{\dfrac{(I + 8m)\left(\dfrac{\omega_1}{\omega_2}\right) - I}{2m}} = 19.8\ m}$

====================================<>====================================

Problem 8.114 A homogenous cylindrical disk of mass *m* rolls on the horizontal surface with angular velocity ω. If it does not slip or leave the slanted surface when it comes in contact with it, what is the angular velocity ω' of the disk immediately afterward?

Solution:

The velocity of the center of mass of the disk is parallel to the surface before and after contact. The angular momentum about the point of contact is conserved. $mvR\cos\beta + I\omega = mv'R + I\omega'$. From kinematics, $v = R\omega$, and $v' = R\omega'$. Substitute into the angular moment condition to obtain:

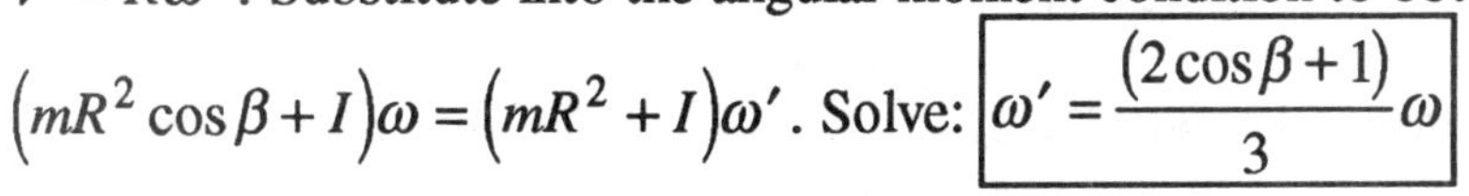

$\left(mR^2\cos\beta + I\right)\omega = \left(mR^2 + I\right)\omega'$. Solve: $\boxed{\omega' = \dfrac{(2\cos\beta + 1)}{3}\omega}$

====================================<>====================================

Problem 8.115 The 10 *lb* slender bar falls from rest in the vertical position and hits the smooth projection at B. The coefficient of restitution of the impact is $e = 0.6$, the duration of the impact is 0.1 *s*, and $b = 1\,ft$. Determine the average force exerted on the bar at B as a result of the impact.

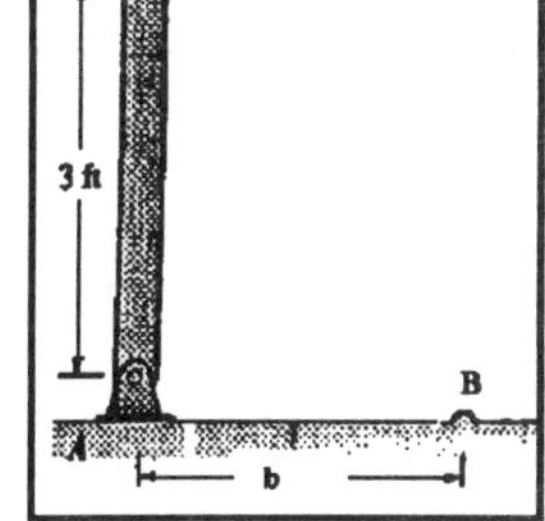

Solution:

Choose a coordinate system with the origin at A and the *x* axis parallel to the plane surface, and *y* positive upward. The strategy is to (a) use the principle of work and energy to determine the velocity before impact, (b) the coefficient of restitution to determine the velocity after impact, (c) and the principle of angular impulse-momentum to determine the average force of impact.

From the principle of work and energy, $U = T_2 - T_1$, where $T_1 = 0$. The center of mass of the bar falls a distance $h = \dfrac{L}{2}$. The work done by the weight of the bar is $U = \int_{\frac{L}{2}}^{0}(-mg)dh = mg\left(\dfrac{L}{2}\right)$. The kinetic

Solution continued on next page

energy is $T_2 = \left(\frac{1}{2}\right) I\omega^2$, where $I = \frac{mL^2}{3}$. Substitute into $U = T_2$ and solve: $\omega = -\sqrt{\frac{3g}{L}}$, where the negative sign on the square root is chosen to be consistent with the choice of coordinates. By definition, the coefficient of restitution is $e = \frac{v'_B - v'_A}{v_A - v_B}$, where v_A, v'_A are *the velocities of the bar at a distance b from A* before and after impact. Since the projection B is stationary before and after the impact, $v_B = v'_B = 0$, from which $v'_A = -ev_A$. From kinematics, $v_A = b\omega$, and $v'_A = b\omega'$, from which $\omega' = -e\omega$. The principle of angular impulse-momentum about the point A is $\int_0^{t_2 - t_1} bF_B dt = (I\omega' - I\omega) = \frac{mL^2}{3}(\omega' - \omega)$, where F_B is the force exerted on the bar by the projection at B, from which $bF_B(t_2 - t_1) = -\frac{mL^2}{3}(1+e)\omega$. Solve: $F_B = \frac{mL^2(1+e)}{3b(t_2 - t_1)}\sqrt{\frac{3g}{L}} = 84.63\ lb$.

================================<>================================

Problem 8.116 In Problem 8.115 determine the value of b for which the average force exerted on the bar at A as a result of the impact is zero.

Solution:

From the principle of linear impulse-momentum, $\int_{t_1}^{t_2} \sum F dt = m(v'_G - v_G)$, where v_G, v'_G are the velocities of the center of mass of the bar before and after impact, and $\sum F = F_A + F_B$ are the forces exerted on the bar at A and B. From kinematics, $v'_G = \left(\frac{L}{2}\right)\omega'$, $v_G = \left(\frac{L}{2}\right)\omega$. From the solution to Problem 8.99, $\omega' = -e\omega$, from which $F_A + F_B = -\frac{mL(1+e)}{2(t_2 - t_1)}\omega$. If the reaction at A is zero, then

$F_B = -\frac{mL(1+e)}{2(t_2 - t_1)}\omega$. From the solution to Problem 8.99, $F_B = -\frac{mL^2(1+e)}{3b(t_2 - t_1)}\omega$. Substitute and solve:

$b = \frac{2}{3}L = 2\ ft$

================================<>================================

==================================<>==================================

Problem 8.117 The 1 *kg* sphere is moving at 2 *m/s* when it strikes the end of the 2 *kg* slender bar B. If the velocity of the sphere after the impact is 0.8 *m/s* to the right, what is the coefficient of restitution?

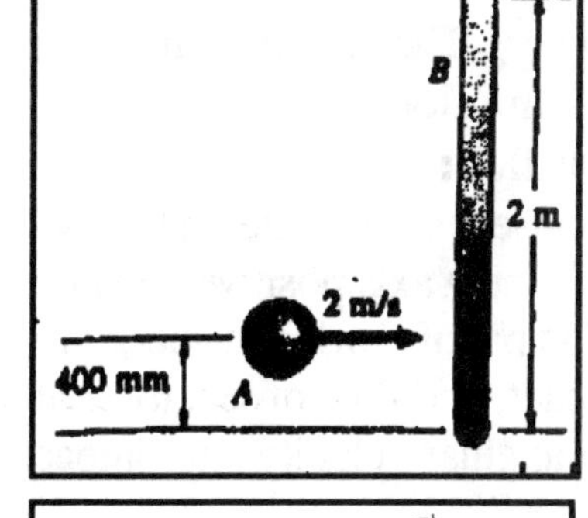

Solution:

Denote the distance of the point of impact P from the end of the bar by $d = 0.4\ m$. The linear momentum is conserved:

(1) $m_A v_A = m_A v_A' + m_B v_{CM}'$, where v_{CM}' is the velocity of the center of mass of the bar after impact, and v_A, v_A' are the velocities of the sphere before and after impact. By definition, the coefficient of restitution is $e = \dfrac{v_P' - v_A'}{v_A - v_P}$, where v_P, v_P' are the velocities of the bar at the point of impact. The point P is stationary before impact, from which (2) $v_P' - v_A' = e v_A$.

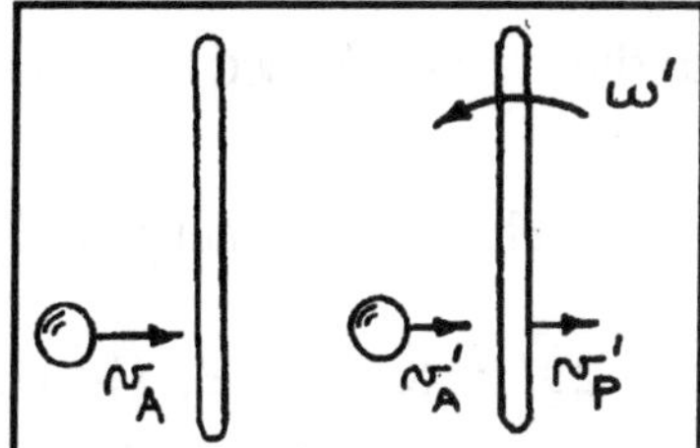

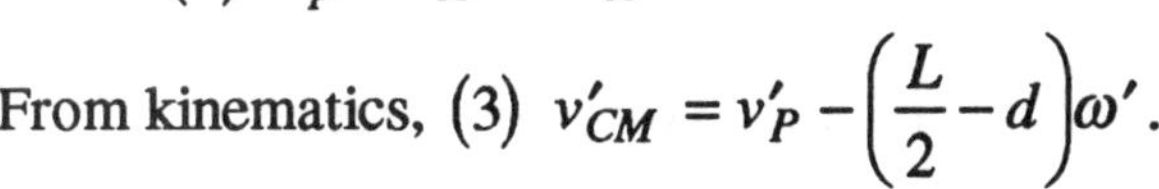

From kinematics, (3) $v_{CM}' = v_P' - \left(\dfrac{L}{2} - d\right)\omega'$.

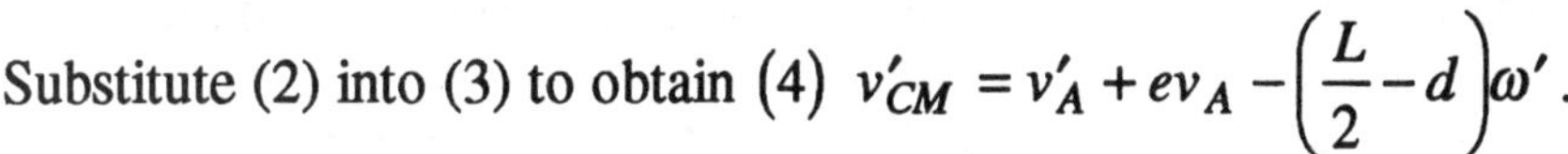

Substitute (2) into (3) to obtain (4) $v_{CM}' = v_A' + e v_A - \left(\dfrac{L}{2} - d\right)\omega'$.

Substitute (4) into (1) to obtain (5) $(m_A - e m_B) v_A = (m_A + m_B) v_A' - \left(\dfrac{L}{2} - d\right) m_B \omega'$. The angular momentum of the bar about the point of impact is conserved: (6) $0 = -\left(\dfrac{L}{2} - d\right) m_B v_{CM}' + I_{CM}\omega'$.

Substitute (3) into (6) to obtain, (7) $\left(\dfrac{L}{2} - d\right) m_B (v_A' + e v_A) = \left(I_{CM} + \left(\dfrac{L}{2} - d\right)^2 m_B\right)\omega'$, where

$I_{CM} = \dfrac{m_B L^2}{12}$. Solve the two equations (5) and (7) for the two unknowns to obtain: $\omega' = 1.08\ rad/s$ and $\boxed{e = 0.224}$.

==================================<>==================================

==================================<>==================================

Problem 8.118 The slender bar is released from rest in the position shown in Fig(a) and falls a distance $h = 1\,ft$. When the bar hits the floor, its tip is supported by a depression and remains on the floor (Fig b.). The length of the bar is 1 *ft* and its weight is 4 *oz.* What is the angular velocity ω just after it hits the floor?

Solution:

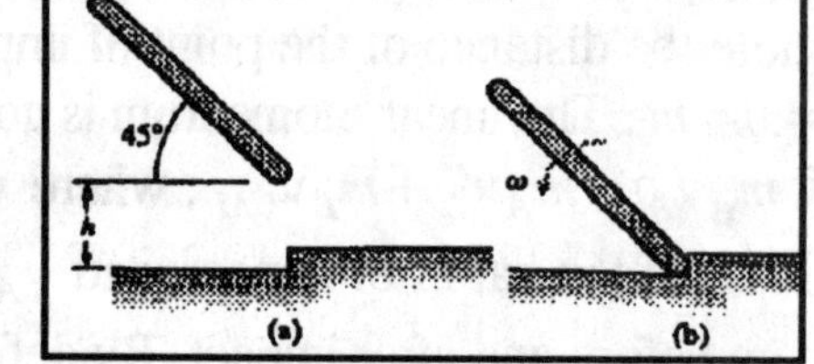

Choose a coordinate system with the *x* axis parallel to the surface, with the *y* axis positive upward. The strategy is to use the principle of work and energy to determine the velocity just before impact, and the conservation of angular momentum to determine the angular velocity after impact. From the principle of work and energy, $U = T_2 - T_1$, where $T_1 = 0$ since the bar is released from rest. The work done is $U = \int_{h_1}^{h_2} -mg dh = -mg(h_2 - h_1) = -mgh$, where $h = 1\,ft$. The kinetic energy is $T_2 = \left(\frac{1}{2}\right) mv^2$. Substitute into $U = T_2 - T_1$ and solve: $v = \sqrt{-2gh} = 8.02\,ft/s$. The conservation of angular momentum about the point of impact is $(\vec{r} \times m\vec{v}) = (\vec{r} \times m\vec{v}') + I_B \vec{\omega}'$. At the instant before the impact, the perpendicular distance from the point of impact to the center of mass velocity vector is $\left(\frac{L}{2}\right)\cos 45^o$. After impact, the center of mass moves in an arc of radius $\left(\frac{L}{2}\right)$ about the point of impact so that the perpendicular distance from the point of impact to the velocity vector is $\left(\frac{L}{2}\right)$. From the definition of the cross product, for motion in the *x*, *y* plane, $(\vec{r} \times m\vec{v}) \cdot \vec{k} = \left(\frac{L}{2}\cos 45^o\right) mv$, $(\vec{r} \times m\vec{v}') \cdot \vec{k} = \left(\frac{L}{2}\right) mv'$, and $I_B \vec{\omega}' \cdot \vec{k} = I_B \omega'$. From kinematics, $v' = \frac{L}{2}\omega'$. Substitute to obtain: $\boxed{\omega' = \frac{mv(\cos 45^o)}{\left(\frac{mL}{2} + \frac{2I_B}{L}\right)} = \frac{3v}{2\sqrt{2}L} = \frac{3\sqrt{-gh}}{2L} = 8.51 \ rad/s}$

==================================<>==================================

===============================<>===============================

Problem 8.119 The slender bar is released from rest with $\theta = 45^o$ and falls a distance $h = 1\ m$ onto the smooth floor. The length of the bar is 1 *m* and its mass is 2 *kg*. If the coefficient of restitution of the impact is $e = 0.4$, what is the bar's angular velocity just after it hits the floor?

Solution Choose a coordinate system with the *x* axis parallel to the plane surface and *y* positive upward. The strategy is to (a) use the principle of work and energy to obtain the velocity the instant before impact, (b) use the definition of the coefficient of restitution to find the velocity just after impact, (c) get the angular velocity-velocity relations from kinematics and (d) use the principle of the conservation of angular momentum about the point of impact to determine the angular velocity.

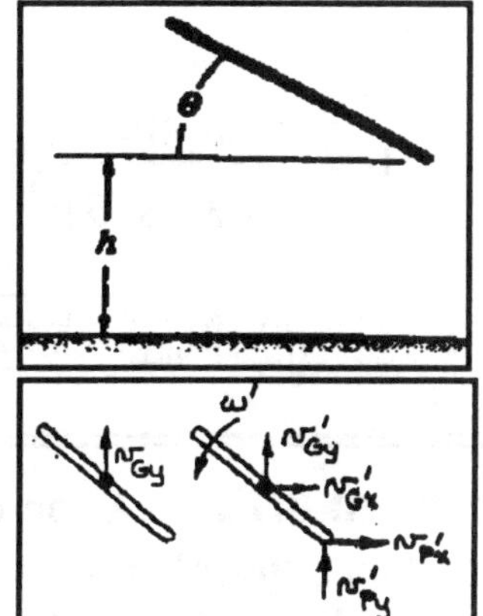

From the principle of work and energy, $U = T_2 - T_1$, where $T_1 = 0$. The center of mass of the bar also falls a distance *h* before impact. The work done is $U = \int_0^{-h} -mgdh = mgh$. The kinetic energy is $T_2 = \left(\frac{1}{2}\right)mv_{Gy}^2$.

Substitute into $U = T_2$ and solve: $v_{Gy} = -\sqrt{2gh}$, where the negative sign on the square root is chosen to be consistent with the choice of coordinates.

Denote the point of impact on the bar by P. From the definition of the coefficient of restitution, $e = \frac{v'_{Py} - v'_{By}}{v_{By} - v_{Py}}$. Since the floor is stationary before and after impact, $v_B = v'_B = 0$, and

(1) $v'_{Py} = -ev_{Py} = -ev_{Gy}$. From kinematics: $\vec{v}'_P = \vec{v}'_G + \vec{\omega}' \times \vec{r}_{P/G}$, where $\vec{r}_{P/G} = \left(\frac{L}{2}\right)(\vec{i}\cos\theta - \vec{j}\sin\theta)$,

from which (2) $\vec{v}'_{Py} = \vec{v}'_{Gy} + \left(\frac{L}{2}\right)\begin{bmatrix} \vec{i} & \vec{j} & \vec{k} \\ 0 & 0 & \omega' \\ \cos\theta & -\sin\theta & 0 \end{bmatrix} = v'_{Gy}\vec{j} + \left(\frac{L}{2}\right)(\vec{i}\,\omega'\sin\theta + \vec{j}\,\omega'\cos\theta)$. Substitute

(1) into (2) to obtain (3) $v'_{Gy} = -ev_{Gy} - \left(\frac{L}{2}\right)\omega'\cos\theta$. The angular momentum is conserved about the

point of impact: (4) $-\left(\frac{L}{2}\right)\cos\theta\, mv_{Gy} = -\left(\frac{L}{2}\right)\cos\theta\, mv'_{Gy} + I_G\omega'$. Substitute (3) into (4) to obtain

$-\frac{:L}{2}m\cos\theta\,(1+e)v_G = \left(I_G + \left(\frac{L}{2}\right)^2 m\cos^2\theta\right)\omega'$. Solve: $\omega' = -\frac{mL(1+e)\cos\theta}{2\left(I_G + \left(\frac{L\cos\theta}{2}\right)^2 m\right)}v_{Gy}$,

$$\boxed{\omega' = -\frac{6(1+e)\cos\theta}{(1+3\cos^2\theta)}(-\sqrt{2gh}) = 10.52\ rad/s}$$

where $I_G = \frac{mL^2}{12}$ has been used.

===============================<>===============================

================================<>================================

Problem 8.120 In Problem 8.119, determine the angle θ for which the angular velocity of the bar just after it hits the floor is a maximum. What is the maximum angular velocity?

Solution From the solution to Problem 8.119, $\omega' = \dfrac{6(1+e)\cos\theta}{\left(1+3\cos^2\theta\right)}\sqrt{2gh}$. Take the derivative:

$$\frac{d\omega'}{d\theta} = 0 = \frac{-6(1+e)\sin\theta}{\left(1+3\cos^2\theta\right)}\sqrt{2gh} + \frac{6(1+e)(6)\cos^2\theta\sin\theta}{\left(1+3\cos^2\theta\right)^2}\sqrt{2gh} = 0,$$ from which $3\cos^2\theta_{max} - 1 = 0$,

$\cos\theta_{max} = \sqrt{\dfrac{1}{3}}$, $\boxed{\theta_{max} = 54.74^o}$, and $\boxed{\omega'_{max} = 10.74\ rad/s}$.

================================<>================================

Problem 8.121 A nonrotating slender bar A moving with velocity v_o strikes a stationary slender bar B. Each bar has mass m and length L. If the bars adhere when they collide, what is their angular velocity after the impact?

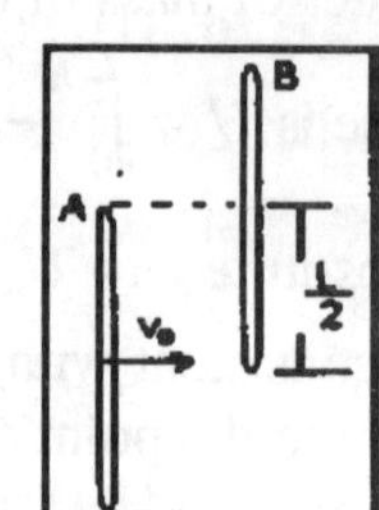

Solution:

From the conservation of linear momentum, $mv_o = mv'_A + mv'_B$.

From the conservation of angular momentum about the mass center of A

$$0 = I_B\vec{\omega}'_A + I_B\vec{\omega}'_B + (\vec{r}\times m\vec{v}'_B) = (I_B\omega'_A + I_B\omega'_B)\vec{k} + \begin{bmatrix} \vec{i} & \vec{j} & \vec{k} \\ 0 & \frac{L}{2} & 0 \\ v'_B & 0 & 0 \end{bmatrix},$$

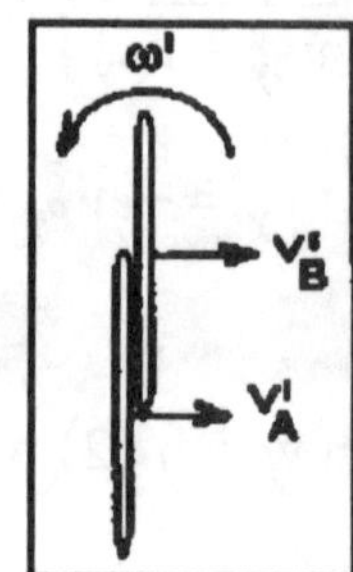

$$0 = (I_B\omega'_A + I_B\omega'_B)\vec{k} - m\left(\frac{L}{2}\right)v'_B\vec{k}.$$

Since the bars adhere, $\omega'_A = \omega'_B$, from which $2I_B\omega' = m\left(\dfrac{L}{2}\right)v'_B$.

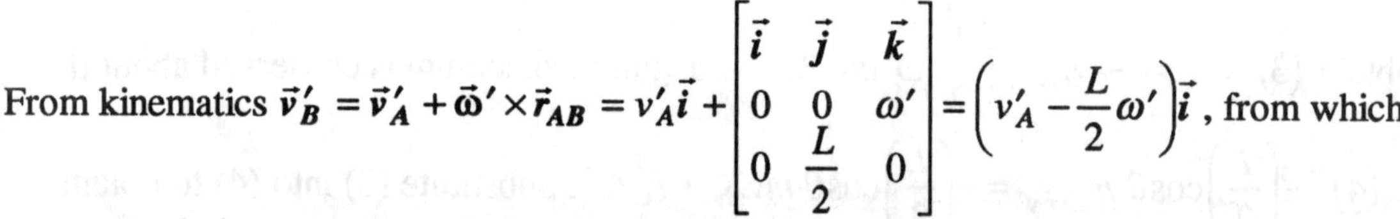

From kinematics $\vec{v}'_B = \vec{v}'_A + \vec{\omega}'\times\vec{r}_{AB} = v'_A\vec{i} + \begin{bmatrix} \vec{i} & \vec{j} & \vec{k} \\ 0 & 0 & \omega' \\ 0 & \frac{L}{2} & 0 \end{bmatrix} = \left(v'_A - \dfrac{L}{2}\omega'\right)\vec{i}$, from which

$v'_B = v'_A - \left(\dfrac{L}{2}\right)\omega'$. Substitute into the expression for conservation of linear momentum to obtain

$v'_B = \dfrac{v_o - \left(\frac{L}{2}\right)\omega'}{2}$. Substitute into the expression for the conservation of angular momentum to obtain:

$\omega' = \dfrac{\left(\frac{mL}{2}\right)v_o}{4I_B + \frac{mL^2}{4}}$. $I_B = \dfrac{mL^2}{12}$, from which $\boxed{\omega' = \dfrac{6v_o}{7L}}$

================================<>================================

==================================<>==================================

Problem 8.122 An astronaut approaches a nonrotating satellite at $1.0\vec{i}$ (m/s) relative to the satellite. Her mass is 136 kg, and the mass moment of inertia about the axis through her center of mass parallel to the z axis is 45 $kg\text{-}m^2$. The mass of the satellite is 450 kg and the mass moment of inertia about the z axis is 675 $kg\text{-}m^2$. At the instant she attaches to the satellite and begins moving with it, the position of her center of mass is $(-1.8, -0.9, 0)$ (m). The axis of rotation of the satellite after she attaches is parallel to the z axis. What is their angular velocity?

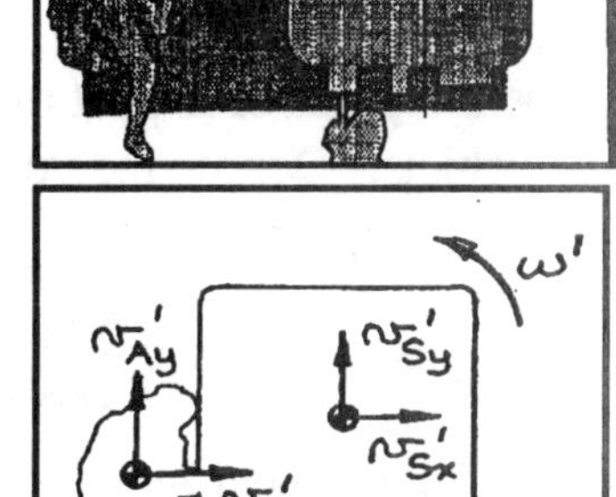

Solution:

Choose a coordinate system with the origin at the center of mass of the satellite, and the y axis positive upward. The linear momentum is conserved: (1) $m_A v_{Ax} = m_A v'_{Ax} + m_S v'_{Sx}$, (2) $0 = m_A v'_{Ay} + m_S v'_{Sy}$. The angular momentum about the center of mass of the satellite is conserved: $\vec{r}_{A/S} \times m_A \vec{v}_A = \vec{r}_{A/S} \times m_A \vec{v}'_A + (I_A + I_S)\vec{\omega}'$, where $\vec{r}_{A/S} = -1.8\vec{i} - 0.9\vec{j}$ (m), and $\vec{v}_A = 1.0\vec{i}$ (m/s).

$$\begin{bmatrix} \vec{i} & \vec{j} & \vec{k} \\ -1.8 & -0.9 & 0 \\ m_A v_{Ax} & 0 & 0 \end{bmatrix} = \begin{bmatrix} \vec{i} & \vec{j} & \vec{k} \\ -1.8 & -0.9 & 0 \\ m_A v'_{Ax} & m_A v'_{Ay} & 0 \end{bmatrix} + (I_A + I_S)\vec{\omega}', \text{ from which}$$

(3) $0.9 m_A v_{Ax} = -1.8 m_A v'_{Ay} + 0.9 m_A v'_{Ax} + (I_A + I_S)\omega'$. From kinematics:

$$\vec{v}'_A = \vec{v}'_S + \vec{\omega}' \times \vec{r}_{A/S} = \vec{v}'_S + \begin{bmatrix} \vec{i} & \vec{j} & \vec{k} \\ 0 & 0 & \omega' \\ -1.8 & -0.9 & 0 \end{bmatrix} = (v'_{Sx} + 0.9\omega')\vec{i} + (v'_{Sy} - 1.8\omega')\vec{j}, \text{ from which}$$

(4) $v'_{Ax} = v'_{Sx} + 0.9\omega'$, (5) $v'_{Ay} = v'_{Sy} - 1.8\omega'$. With $v_{Ax} = 1.0$ m/s, these are five equations in five unknowns. The number of equations can be reduced further, but here they are solved by iteration (using **TK Solver Plus**) to obtain: $v'_{Ax} = 0.289$ m/s, $v'_{Sx} = 0.215$ m/s, $v'_{Ay} = -0.114$ m/s, $v'_{Sy} = 0.0344$ m/s, $\boxed{\omega' = 0.0822 \ rad/s}$

==================================<>==================================

Problem 8.123 In Problem 8.122, suppose that the design parameters of the satellite's control system require that its angular velocity not exceed 0.02 *rad/s*. If the astronaut is moving parallel to the x axis and the position of her center of mass when she attaches is $(-1.8, -0.9, 0)$ (m), what is the maximum relative velocity at which she should approach the satellite?

Solution:

From the solution to Problem 8.122, the five equations are:

(1) $m_A v_{Ax} = m_A v'_{Ax} + m_S v'_S$,

(2) $0 = m_A v'_{Ay} + m_S v'_{Sy}$

(3) $0.9 m_A v_{Ax} = -1.8 m_A v'_{Ay} + 0.9 m_A v'_{Ax} + (I_A + I_S)\omega'$

(4) $v'_{Ax} = v'_{Sx} + 0.9\omega'$,

(5) $v'_{Ay} = v'_{Sy} - 1.8\omega'$. With $\omega' = 0.02$ rad/s, these five equations have the solutions:

$\boxed{v_{Ax} = 0.243 \ m/s}$, $v'_{Ax} = 0.070$ m/s, $v'_{Sx} = 0.052$ m/s, $v'_{Ay} = -0.028$ m/s, $v'_{Sy} = 0.008$ m/s.

==================================<>==================================

==================================<>==================================

Problem 8.124 A 2800 *lb* car skidding on ice strikes a concrete abutment at 3 *mi/hr*. The car's moment of inertia about its center of mass is 1800 *slug-ft*^{2}. Assume that the impacting surfaces are smooth and parallel to the *y* axis and that the coefficient of restitution of the impact is $e = 0.8$. What is the car's angular velocity and the velocity of its center of mass after the impact?

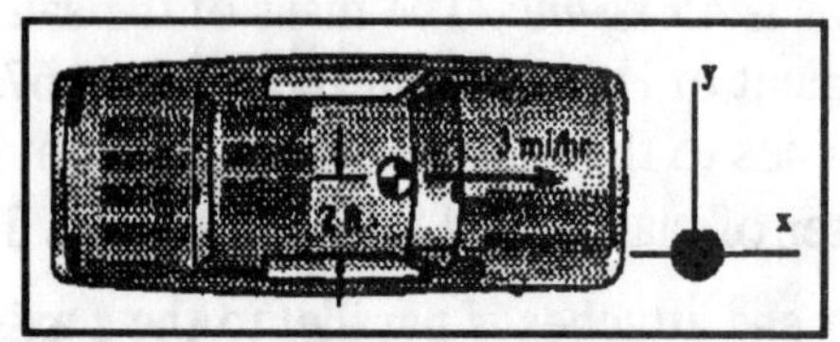

Solution:

Let P be the point of impact on with the abutment. (P is located on the vehicle.) Denote the vector from P to the center of mass of the vehicle by $\vec{r} = a\vec{i} + 2\vec{j}$, where *a* is unknown. The velocity of the vehicle is $v_{Gx} = 3\left(\frac{5280}{3600}\right) = 4.4\ ft/s$. From the conservation of linear momentum in the *y* direction, $mv_{Gy} = mv'_{Gy}$, from which $v'_{Gy} = v_{Gy} = 0$. Similarly,

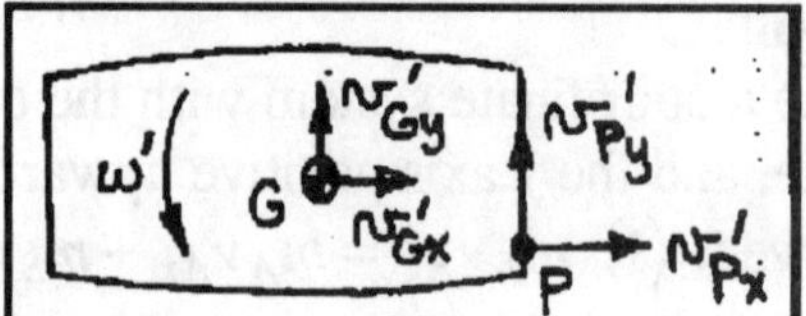

$v_{Ay} = v'_{Ay} = 0$. By definition, $e = \frac{v'_{Px} - v'_{Ax}}{v_{Ax} - v_{Px}}$. *Assume that the abutment does not yield under the impact.* $v_{Ax} = v'_{Ax} = 0$, from which $v'_{Px} = -ev_{Px} = -ev_{Gx} = -(0.8)(4.4) = -3.52\ ft/s$. The conservation of angular momentum about P is $\vec{r} \times m\vec{v}_G = \vec{r} \times m\vec{v}'_G + I\vec{\omega}'$. From kinematics, $\vec{v}'_G = \vec{v}_P + \vec{\omega}' \times \vec{r}$.

Reduce: $\vec{v}'_G = v'_{Px}\vec{i} + \vec{\omega}' \times \vec{r} = \begin{bmatrix} \vec{i} & \vec{j} & \vec{k} \\ 0 & 0 & \omega' \\ a & 2 & 0 \end{bmatrix} = (v'_{Px} - 2\omega')\vec{i} + a\omega\vec{j} = (-ev_{Gx} - 2\omega')\vec{i} + a\omega\vec{j}$, from which

$v'_{Gx} = -(ev_{Gx} + 2\omega')$, $\vec{r} \times m\vec{v}_G = \begin{bmatrix} \vec{i} & \vec{j} & \vec{k} \\ a & 2 & 0 \\ 4.4m & 0 & 0 \end{bmatrix} = -8.8m\vec{k}$.

$\vec{r} \times m\vec{v}'_G = \begin{bmatrix} \vec{i} & \vec{j} & \vec{k} \\ a & 2 & 0 \\ -(ev_{Gx} + 2\omega')m & 0 & 0 \end{bmatrix} = 2(ev_{Gx} + 2\omega')m\vec{k}$, $I\vec{\omega}' = I\omega'\vec{k}$.

Substitute into the expression for the conservation of angular momentum to obtain

$\omega' = \frac{-15.84}{\left(2^2 + \frac{I}{m}\right)} = \frac{-15.84}{(4 + 20.68)} = -0.642\ rad/s$. The velocity of the center of mass is

$\vec{v}'_G = v'_{Gx}\vec{i} = -(ev_{Gx} + 2\omega')\vec{i} = -2.236\vec{i}\ (m/s)$

==================================<>==================================

==================================<>==================================

Problem 8.125 A 170 *lb* receiver jumps vertically to receive a pass and is stationary at the instant he catches the ball. At the same instant he is hit at P by a 180 *lb* linebacker moving horizontally at 15 *ft/s*. The wide receiver's mass moment of inertia about his center of mass is 7 *slug-ft*^{2}.If you model the players as rigid bodies and assume that the coefficient of restitution is $e = 0$, what is the wide receiver's angular velocity immediately after the impact?

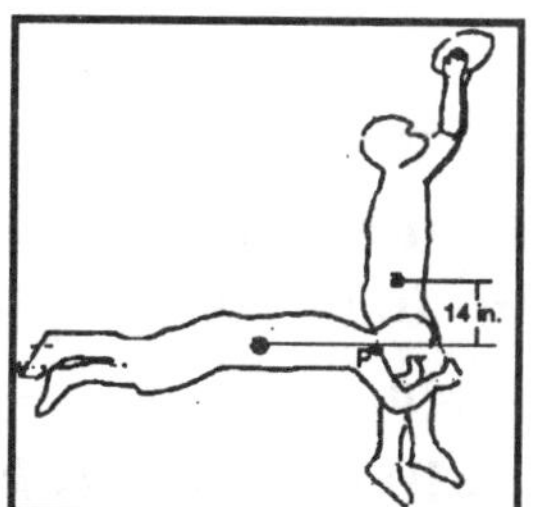

Solution:

Denote the receiver by the subscript B, and the tackler by the subscript A. Denote $d = \left(\frac{14}{12}\right) = 1.17\ ft$. The conservation of linear momentum: $(1)\ m_A v_A = m_A v_A' + m_B v_B'$. From the definition, $e = \frac{v_{BP}' - v_A'}{v_A - v_{BP}}$. Since $v_{BP} = 0$, $(2)\ e v_A = v_{BP}' - v_A'$. The angular momentum is conserved about point P: $(3)\ 0 = -m_B d v_B' + I_B \omega_B'$. From kinematics: $(4)\ v_{BP}' = v_B' + d\omega_B'$.

The solution: For $e = 0$, from (1), $v_A' = v_A - \left(\frac{m_B}{m_A}\right) v_B'$.

From (2) and (4) $v_A' = v_B' + d\omega_B'$. From (3) $v_B' = \left(\frac{I_B}{d m_B}\right)\omega'$. Combine these last two equations and solve: $\boxed{\omega_B' = \frac{v_A}{\frac{I_B}{d m_B}\left(1 + \frac{m_B}{m_A}\right) + d} = 4.445... = 4.45\ rad/s}$. *Check*: The four equations in four unknowns are solved by iteration: for $e = 0$, $v_A' = 10.23\ ft/s$, $v_B' = 5.047\ ft/s$, $v_{BP}' = 10.23\ ft/s$, $\omega_B' = 4.445\ rad/s$. *check.*

==================================<>==================================

Problem 9.1 Relative to a primary reference frame, a right body's angular velocity is $\bar{\omega} = 200\,\bar{i} + 900\,\bar{j} - 600\bar{k}\ (rad/s)$. The position of its center of mass relat ive to the origin O of the reverence frame is $\bar{r}_G = 6\,\bar{i} + 6\,\bar{j} + 2\bar{k}\ (m)$, and the velocity of its center of mass is $\bar{v}_G = 100\,\bar{i} + 80\,\bar{j} - 60\bar{k}\ (m/s)$. What is the velocity of point A of the rigid body whose position relative to O is $\bar{r}_A = 5.8\,\bar{i} + 6.4\,\bar{j} + 1.6\bar{k}\ (m)$.

Solution:

The position of point A relative to the center of mass G is

$\bar{r}_{A/G} = \bar{r}_A - \bar{r}_G = 5.8\bar{i} + 6.4\bar{j} + 1.6\bar{k} - (6\bar{i} + 6\bar{j} + 2\bar{k}) = -0.2\bar{i} + 0.4\bar{j} - 0.4\bar{k}(m)$.

From Equation (9.1) the velocity of point A relative to the primary reference frame is

$$\bar{v}_A = \bar{v}_G + \bar{\omega}\times\bar{r}_{A/G} = 100\bar{i} + 80\bar{j} - 60\bar{k} + \begin{vmatrix} \bar{i} & \bar{j} & \bar{k} \\ 200 & 900 & -600 \\ -0.2 & 0.4 & -0.4 \end{vmatrix}$$

$\bar{v}_A = -20\bar{i} + 280\bar{j} + 200\bar{k}(m/s)$

Problem 9.2 The angular acceleration of the rigid body in Problem 9.1 is $\bar{\alpha} = 8000\,\bar{i} - 8000\,\bar{j} - 4000\bar{k}\ (rad/s^2)$ and the acceleration of its center of mass is zero. What is the acceleration of point A?

Solution:

From Equation (9.2), the acceleration of A relative in the primary reference frame is $\bar{a}_A = \bar{a}_G + \bar{\alpha}\times\bar{r}_{A/G} + \bar{\omega}\times(\bar{\omega}\times\bar{r}_{A/G})$ The position vector is $\bar{r}_{A/G} = \bar{r}_A - \bar{r}_G = -0.2\bar{i} + 0.4\bar{j} - 0.4\bar{k}(m)$ and

$$\bar{\omega}\times\bar{r}_{A/G} = \begin{vmatrix} \bar{i} & \bar{j} & \bar{k} \\ 200 & 900 & -600 \\ -0.2 & 0.4 & -0.4 \end{vmatrix} = -120\bar{i} + 200\bar{j} + 260\bar{k}(m/s).$$ Therefore

$$\bar{a}_A = 0 + \begin{vmatrix} \bar{i} & \bar{j} & \bar{k} \\ 8000 & -8000 & -4000 \\ -0.2 & 0.4 & -0.4 \end{vmatrix} + \begin{vmatrix} \bar{i} & \bar{j} & \bar{k} \\ 200 & 900 & -600 \\ -120 & 200 & 260 \end{vmatrix}$$

$\bar{a}_A = 358{,}800\bar{i} + 24{,}000\bar{j} + 149{,}600\bar{k}(m/s^2)$.

Problem 9.3 The airplane's rate gyros indicate that its angular velocity is $\vec{w} = 4.0\vec{i} + 6.4\vec{j} + 0.2\vec{k}$ (rad / s). What is the velocity relative to the center of mass of point A with coordinates (8,2,2) *ft*?

Solution:

The velocity of point A relative to G is

$$\vec{v}_{A/G} = \vec{w}\times\vec{r}_{A/G} = \begin{bmatrix} \vec{i} & \vec{j} & \vec{k} \\ 4.0 & 6.4 & 0.2 \\ 8 & 2 & 2 \end{bmatrix},\ \vec{v}_{A/G} = 12.4\vec{i} - 6.4\vec{j} - 43.2\vec{k}\ \text{(ft/s)}$$

==================================<>==================================

Problem 9.4 The rate gyros of the airplane in Problem 9.3 indicate that its angular acceleration is $\vec{\mathbf{a}} = -4\vec{i} + 12\vec{j} + 2\vec{k}$ (rad/s^2). What is the acceleration of point A relative to the center of mass?

Solution:

The acceleration of point A relative to G is $\vec{a}_{A/G} = \vec{\mathbf{a}} \times \vec{r}_{A/G} + \vec{\mathbf{w}} \times (\vec{\mathbf{w}} \times \vec{r}_{A/G})$. Assume that

$$\vec{a}_{A/G} = \begin{bmatrix} \vec{i} & \vec{j} & \vec{k} \\ -4 & 12 & 2 \\ 8 & 2 & 2 \end{bmatrix} + \vec{\mathbf{w}} \times \begin{bmatrix} \vec{i} & \vec{j} & \vec{k} \\ 4.0 & 6.4 & 0.2 \\ 8 & 2 & 2 \end{bmatrix} = (20\vec{i} + 24\vec{j} - 104\vec{k}) + \begin{bmatrix} \vec{i} & \vec{j} & \vec{k} \\ 4.0 & 6.4 & 0.2 \\ 12.4 & -6.4 & -43.2 \end{bmatrix}.$$

$$\vec{a}_{A/G} = -255.2\vec{i} + 199.28\vec{j} - 208.96\vec{k} \quad (\text{ft/s}^2)$$

==================================<>==================================

Problem 9.5 Relative to the reference frame shown, the rectangular parallelepiped is rotating about a fixed axis through points A and B. Its direction of rotation is clockwise when the axis of rotation is viewed from point A toward B. (a) What is its angular velocity vector $\vec{\mathbf{w}}$? (b) what are the velocities of points C and D?

Solution:

The angular velocity is positive when viewed along the line AB, since it is rotating in the direction of a right handed screw. (a) The coordinates of A are $(0, 0, 10)$ in. The coordinates of B are $(10, 5, 0)$ in. The axis AB is parallel to the vector

$\vec{r}_{B/A} = (10-0)\vec{i} + (5-0)\vec{j} + (0-10)\vec{k}$,

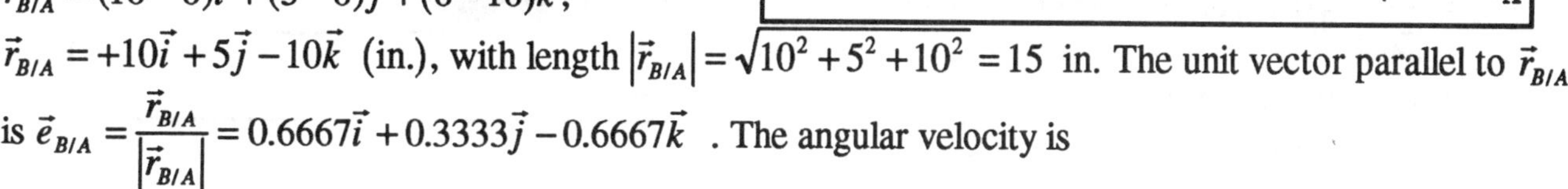

$\vec{r}_{B/A} = +10\vec{i} + 5\vec{j} - 10\vec{k}$ (in.), with length $|\vec{r}_{B/A}| = \sqrt{10^2 + 5^2 + 10^2} = 15$ in. The unit vector parallel to $\vec{r}_{B/A}$ is $\vec{e}_{B/A} = \dfrac{\vec{r}_{B/A}}{|\vec{r}_{B/A}|} = 0.6667\vec{i} + 0.3333\vec{j} - 0.6667\vec{k}$. The angular velocity is

$\vec{\mathbf{w}}_{B/A} = |\vec{\mathbf{w}}|\vec{e}_{B/A} = 20\vec{i} + 10\vec{j} - 20\vec{k}$ (rad/s).

(b) Choose point A as the reference, since it lies on the axis of rotation and therefore has zero velocity. The coordinates of point C are $(0, 5, 10)$ in.. The line AC is parallel to the vector $\vec{r}_{C/A} = (0-0)\vec{i} + (5-0)\vec{j} + (10-10)\vec{k} = 5\vec{j}$ (in.). The velocity of point C relative to A is

$$\vec{v}_{C/A} = v\vec{\mathbf{w}} \times \vec{r}_{C/A} = \begin{bmatrix} \vec{i} & \vec{j} & \vec{k} \\ 20 & 10 & -20 \\ 0 & 5 & 0 \end{bmatrix},$$

$\vec{v}_{C/A} = 100\vec{i} + 100\vec{k}$ (in/s). The coordinates of point D are $(10, 5, 10)$ in. The line AD is parallel to the vector $\vec{r}_{D/A} = (10-0)\vec{i} + (5-0)\vec{j} + (10-10)\vec{k} = 10\vec{i} + 5\vec{j}$ (in.).

The velocity of point D relative to A is $\vec{v}_{D/A} = \vec{v}_A + \vec{\mathbf{w}} \times \vec{r}_{D/A} = 0 + \begin{bmatrix} \vec{i} & \vec{j} & \vec{k} \\ 20 & 10 & -20 \\ 10 & 5 & 0 \end{bmatrix}$,

$\vec{v}_{D/A} = 100\vec{i} - 200\vec{j}$ (in/s)

==================================<>==================================

Problem 9.6 If the angular velocity of the rectangular parallelepiped in Problem 9.5 is constant, what is the acceleration of point C?

Solution:

As in the solution to Problem 9.5, choose point A as the reference, since it lies on the axis of rotation and therefore has zero acceleration.

The acceleration of point C relative to A is $\vec{a}_{C/A} = \vec{\mathbf{a}} \times \vec{r}_{C/A} + \vec{\mathbf{w}} \times (\vec{\mathbf{w}} \times \vec{r}_{C/A})$.

For constant angular velocity $\vec{\mathbf{w}}$, the angular acceleration is zero, $\vec{\mathbf{a}} = \dfrac{d\vec{\mathbf{w}}}{dt} = 0$. From the solution to Problem 9.5, $\vec{r}_{C/A} = 5\vec{j}$ (in.), from which $\vec{a}_{C/A} = \vec{\mathbf{w}} \times \begin{bmatrix} \vec{i} & \vec{j} & \vec{k} \\ 20 & 10 & -20 \\ 0 & 5 & 0 \end{bmatrix} = \begin{bmatrix} \vec{i} & \vec{j} & \vec{k} \\ 20 & 10 & -20 \\ 100 & 0 & 100 \end{bmatrix}$,

$\vec{a}_{C/A} = 1000\vec{i} - 4000\vec{j} - 1000\vec{k}$ (in/s^2)

Problem 9.7 Relative to the reference frame shown, the turbine is rotating about a fixed axis coincident with the line OA. (a) What is its angular velocity vector? (b) What is the velocity of the point of the turbine with the coordinates $(3, 2, 2)$ m?

Solution:

(a) The line OA is parallel to the vector

$\vec{r}_{A/O} = 7\vec{i} + 4\vec{j} + 4\vec{k}$ (m).

The unit vector parallel to OA is

$\vec{e}_{A/O} = \dfrac{\vec{r}_{A/O}}{|\vec{r}_{A/O}|} = 0.7778\vec{i} + 0.4444\vec{j} + 0.4444\vec{k}$.

The angular velocity vector is

$\vec{\mathbf{w}} = -|\vec{\mathbf{w}}|\vec{e}_{A/O} = -700\vec{i} - 400\vec{j} - 400\vec{k}$ rad/s, where the angular velocity is negative because the rotation is in the opposite direction of the rotation of a right handed screw, when viewed along the line OA.

(b) Use the point O as a reference, since the velocity of point O is zero. The line from O to the point with coordinates (3, 2, 2) is parallel to the vector $\vec{r}_{P/O} = 3\vec{i} + 2\vec{j} + 2\vec{k}$ (m).

The velocity of point P is $\vec{v}_{P/O} = \vec{\mathbf{w}} \times \vec{r}_{P/O} = \begin{bmatrix} \vec{i} & \vec{j} & \vec{k} \\ -700 & -400 & -400 \\ 3 & 2 & 2 \end{bmatrix}$, $\vec{v}_{P/O} = 200\vec{j} - 200\vec{k}$ (m/s).

===============================<>===============================

Problem 9.8 The 900 *rad/s* angular velocity of the turbine is decreasing at 100 *rad/s*^{2}. (a) What is the turbine's angular acceleration vector? (b) What is the acceleration of the point on the turbine with coordinates $(3, 2, 2)$ m?

Solution:

(a) The angular velocity is negative, so the angular acceleration is positive: $\vec{\mathbf{a}} = 100\vec{e}_{A/O}$ rad / s^2. From the solution to Problem 9.7, the unit vector parallel to the line OA is $\vec{e}_{A/O} = 0.7778\vec{i} + 0.4444\vec{j} + 0.4444\vec{k}$. The angular acceleration vector is $\vec{\mathbf{a}} = 100\vec{e}_{A/O} = 77.78\vec{i} + 44.44\vec{j} + 44.44\vec{k}$ rad / s^2. (b) Use the point O as a reference, since it has zero linear acceleration. From the solution to Problem 9.7, the line OP is parallel to the vector $\vec{r}_{P/O} = 3\vec{i} + 2\vec{j} + 2\vec{k}$ (m). The angular acceleration of point P is

$$\vec{a}_{P/O} = \vec{\mathbf{a}} \times \vec{r}_{P/O} + \vec{\mathbf{w}} \times (\vec{\mathbf{w}} \times \vec{r}_{P/O}).\ \vec{a}_{P/O} = \begin{bmatrix} \vec{i} & \vec{j} & \vec{k} \\ 77.78 & 44.44 & 44.44 \\ 3 & 2 & 2 \end{bmatrix} + \vec{\mathbf{w}} \times \begin{bmatrix} \vec{i} & \vec{j} & \vec{k} \\ -700 & -400 & -400 \\ 3 & 2 & 2 \end{bmatrix},$$

$$\vec{a}_{P/O} = 0\vec{i} - 22.22\vec{j} + 22.22\vec{k} + \begin{bmatrix} \vec{i} & \vec{j} & \vec{k} \\ -700 & -400 & -400 \\ 0 & 200 & -200 \end{bmatrix},\ \vec{a}_{P/O} = 160{,}000\vec{i} - 139{,}996\vec{j} - 139{,}978\vec{k}\ \left(\mathrm{m/s^2}\right)$$

===============================<>===============================

Problem 9.9 The disk of radiius R is supported by the vertical shaft. Realtive to an earth-fixed reference frame, the shaft rotates with constant angular velocity ω_0. The disk rotates with constant angular velocity ω_d relative to the shaft. Determine: (a) the disk's angular velocity $\overline{\omega}$ relative to the earth-fixed reference frame; (b) the velocity of point A of the disk relative to the earth-fixed reference frame.

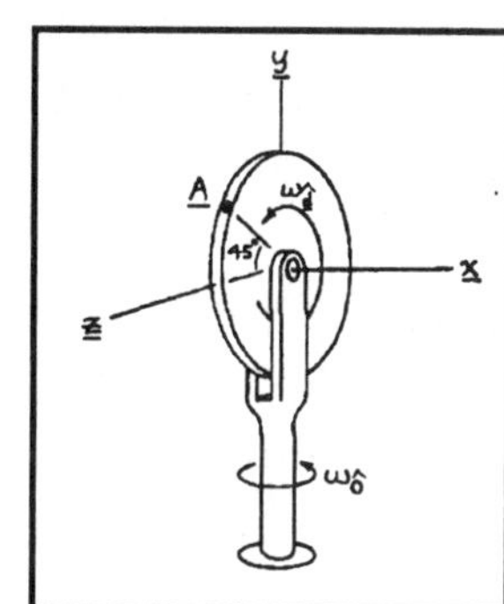

Strategy: (a) Assume that the y axis remains vertical and the x axis remains perpendicular to the disk. Determine $\overline{\omega}$ by expressing it as the sum of the angular velocity $\overline{\Omega}$ of the coordinate system and the angular velocity $\overline{\omega}_{rel}$ of the disk relative to the secondary coordinate system. (b) Use Eq (9.1) to express $\overline{v}_A$ in terms of the stationary center of the disk.

Solution:

(a) The angular velocity of the coordinate system is $\overline{\Omega} = \omega_0 \overline{j}$ The angular velocity of the disk relative to the coordinate system is $\overline{\omega}_{rel} = \omega_d \overline{i}$. So the angular velocity to the disk relative to an earth-fixed reference frame is $\overline{\omega} = \overline{\Omega} + \overline{\omega}_{rel} = \omega_d \overline{i} + \omega_o \overline{j}$.

(b) The position of point A relative to the origin 0 is $\overline{r}_{A/0} = R\sin 45^\circ \overline{j} + R\cos 45^\circ \overline{k}$.
The velocity of point A is

$$\overline{v}_A = \overline{v}_0 + \overline{\omega} \times \overline{r}_{A/0} = 0 + \begin{vmatrix} \overline{i} & \overline{j} & \overline{k} \\ \omega_d & \omega_0 & 0 \\ 0 & R\sin 45^\circ & R\cos 45^\circ \end{vmatrix}, \text{ or}$$

$$\overline{v}_A = \omega_0 R\cos 45^\circ \overline{i} - \omega_d R\cos 45^\circ \overline{j} + \omega_d R\sin 45^\circ \overline{k}.$$

===============================<>===============================

================================<>================================

Problem 9.10 For the disk in Problem 9.9, determine: (a) the disk's angular acceleration $\bar{\alpha}$ relative to the earth-fixed reference frame; (b) the acceleration of point A of the disk.

Solution:

(a) From Equation (9.4) and the solution of Problem 9.9, the disk's angular acceleration relative to an earth-fixed reference frame is

$$\bar{\alpha} = \frac{d\omega_x}{dt}\bar{i} + \frac{d\omega_y}{dt}\bar{j} + \frac{d\omega_z}{dt}\bar{k} + \bar{\Omega}\times\bar{\omega} = \begin{vmatrix} \bar{i} & \bar{j} & \bar{k} \\ 0 & \omega_0 & 0 \\ \omega_d & \omega_0 & 0 \end{vmatrix} = -\omega_0\omega_d\bar{k}.$$

(b) The position of point A relative to the origin 0 is $\bar{r}_{A/0} = R\sin 45^\circ\bar{j} + R\cos 45^\circ\bar{k}$. In the solution of Problem 9.9 it was shown that $\bar{\omega}\times\bar{r}_{A/0} = \omega_0 R\cos 45^\circ\bar{i} - \omega_d R\cos 45^\circ\bar{j} + \omega_d R\sin 45^\circ\bar{k}$

The acceleration of point A is $\bar{a}_A = \bar{a}_0 + \bar{\alpha}\times\bar{r}_{A/0} + \bar{\omega}\times(\bar{\omega}\times\bar{r}_{A/0})$ Hence,

$$\bar{a}_A = 0 + \begin{vmatrix} \bar{i} & \bar{j} & \bar{k} \\ 0 & 0 & -\omega_0\omega_d \\ 0 & R\sin 45^\circ & R\cos 45^\circ \end{vmatrix} + \begin{vmatrix} \bar{i} & \bar{j} & \bar{k} \\ \omega_d & \omega_0 & 0 \\ \omega_0 R\cos 45^\circ & -\omega_d R\cos 45^\circ & \omega_d R\cos 45^\circ \end{vmatrix}, \text{ or}$$

$$\bar{a}_A = 2\omega_0\omega_d R\sin 45^0\bar{i} - \omega_d^2 R\sin 45^\circ\bar{j} - (\omega_d^2 + \omega_0^2)R\cos 45^\circ\bar{k}$$

================================<>================================

Problem 9.11 The base of the dish antenna is rotating at 1 *rad/s*. The angle $\theta = 30^o$ and is increasing at 20 deg/s. (a) What are the components of the antenna's angular velocity vector $\vec{w}$ in terms of the body-fixed coordinates shown? (b) What is the velocity of the point of the antenna with coordinates $(2, 2, -2)$ m?

Solution:

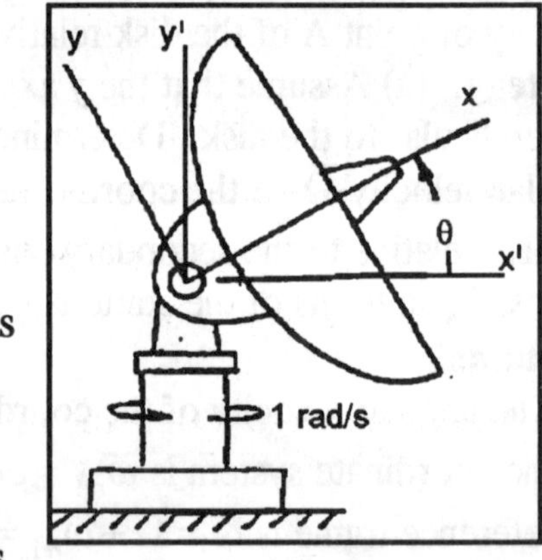

Let x', y', z' denote the fixed coordinates shown, where z' is directed positively out of the paper. The angular velocity in the primed coordinates is $\vec{w} = \omega'\vec{k}' = 1\vec{k}'$ (rad/s). The rate of change of the angle in the primed coordinates is $\frac{d\theta}{dt}\vec{k}' = 20\vec{k}'$ (deg/s) $= 0.3491\vec{k}'$ rad/s where the angular velocity is positive because the rotation is in the direction of a right handed screw. The angular velocity in the body fixed coordinates is the projection of the angular velocity in the primed coordinates onto the rotated coordinates (x, y), plus the angular velocity about the z' axis, which coincides with the z axis: $\vec{w} = \omega'\sin\theta\,\vec{i} + \omega'\cos\theta\,\vec{j} + 0.3491\vec{k}$, $\vec{w} = 0.5\vec{i} + 0.866\vec{j} + 0.3491\vec{k}$ (rad/s). (b) Choose the origin of the coordinates as the reference point, since it is on the axis of rotation and the velocity is zero. The line OP is parallel to the vector $\vec{r}_{P/O} = 2\vec{i} + 2\vec{j} - 2\vec{k}$ (m).

The velocity of the point P relative to O is $\vec{v}_{P/O} = \vec{w}\times(\vec{r}_{P/O}) = \begin{bmatrix} \vec{i} & \vec{j} & \vec{k} \\ 0.5 & 0.866 & 0.3491 \\ 2 & 2 & -2 \end{bmatrix}$,

$$\vec{v}_{P/O} = -2.430\vec{i} + 1.698\vec{j} - 0.7320\vec{k} \text{ (m/s)}$$

================================<>================================

====================================<>====================================

Problem 9.12 The circular disk rotates with angular velocity ω_d relative to the horizontal bar, the horizontal bar rotates with angular velocity ω_b about the z axis, that the vertical shaft rotates with angular velocity veolcity ω_0. In terms of components in the coordinate system shown, determine: (a) the angular velocity $\bar{\omega}_{disk}$ of the disk; (b) the veolcity of point P, which is the uppermost point of the disk at the present instant.

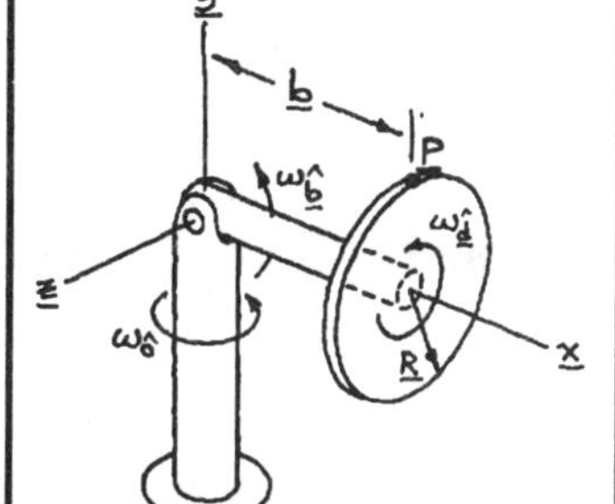

Solution:

Assume that the coordinate system is body fixed with respect to the horizontal bar.

(a) The angular velocity of the coordinate system is $\bar{\Omega} = \omega_o\bar{j} + \omega_b\bar{k}$. The angular velocity of the disk relative to the coordinate system is $\bar{\omega}_{rel} = \bar{\omega}_d\bar{i}$, so the angular velocity of the disk is $\bar{\omega}_{disk} = \bar{\Omega} + \bar{\omega}_{rel} = \omega_d\bar{i} + \omega_0\bar{j} + \omega_b\bar{k}$.

(b) The position of the center of the disk A relative to the origin 0 is $\bar{r}_{A/0} = b\bar{i}$. Since the bar is stationary relative to the coordinate system, its angular velocity is $\bar{\omega}_{bar} = \bar{\Omega}$.

The velocity of A is $\bar{v}_A = \bar{v}_0 + \bar{\omega}_{bar} \times \bar{r}_{A/0}$; or

$$\bar{v}_A = 0 + \begin{vmatrix} \bar{i} & \bar{j} & \bar{k} \\ 0 & \omega_0 & \omega_b \\ b & 0 & 0 \end{vmatrix} = \omega_b b\bar{j} - \omega_0 b\bar{k}.$$

The velocity of point P is

$$\bar{v}_p = \bar{v}_A + \bar{\omega}_{disk} \times \bar{r}_{P/A} = \omega_b b\bar{j} - \omega_0 b\bar{k} + \begin{vmatrix} \bar{i} & \bar{j} & \bar{k} \\ \omega_d & \omega_0 & \omega_b \\ 0 & R & 0 \end{vmatrix}, \text{ or}$$

$$\bar{v}_P = -\omega_b R\bar{i} + \omega_b b\bar{j} + (\omega_d R - \omega_0 b)\bar{k}.$$

====================================<>====================================

Problem 9.13 The angular velocities given in Problem 9.12 are constant. Determine: (a) the angular acceleration $\bar{\alpha}_{disk}$ of the disk; (b) the acceleration of point P.

Solution:

(a) From Equation (9.4) and the solution of Problem 9.12, the angular acceleration of the disk is

$$\bar{\alpha}_{disk} = \bar{\Omega} \times \bar{\omega}_{disk} = \begin{vmatrix} \bar{i} & \bar{j} & \bar{k} \\ 0 & \omega_0 & \omega_b \\ \omega_d & \omega_0 & \omega_b \end{vmatrix} = \omega_b\omega_d\bar{j} - \omega_0\omega_d\bar{k}.$$

(b) The position of the center of the disk A relative to the origin 0 is $\bar{r}_{A/0} = b\bar{i}$.

The angular acceleration of the coordinate system is zero, so the acceleration of point A is

$$\bar{a}_A = \bar{a}_0 + \bar{\omega}_{bar} \times (\bar{\omega}_{bar} \times \bar{r}_{A/0}) = 0 + \begin{vmatrix} \bar{i} & \bar{j} & \bar{k} \\ 0 & \omega_0 & \omega_b \\ 0 & \omega_b b & -\omega_0 b \end{vmatrix} = -(\omega_0^2 + \omega_b^2)b\bar{i}.$$

The acceleration of point P is $\bar{a}_p = \bar{a}_A + \bar{\alpha}_{disk} \times \bar{r}_{P/A} + \bar{\omega}_{disk} \times (\bar{\omega}_{disk} \times \bar{r}_{P/A})$.

Solution continued on next page

From the solution of Problem 9.12, $\overline{\omega}_{disk} \times \overline{r}_{P/A} = \begin{vmatrix} \overline{i} & \overline{j} & \overline{k} \\ \omega_d & \omega_0 & \omega_b \\ 0 & R & 0 \end{vmatrix} = -\omega_b R\overline{i} + \omega_d R\overline{k}$.

Therefore $\overline{a}_P = -(\omega_0^2 + \omega_b^2)b\overline{i} + \begin{vmatrix} \overline{i} & \overline{j} & \overline{k} \\ 0 & \omega_b\omega_d & -\omega_0\omega_d \\ 0 & R & 0 \end{vmatrix} + \begin{vmatrix} \overline{i} & \overline{j} & \overline{k} \\ \omega_d & \omega_0 & \omega_b \\ -\omega_b R & 0 & \omega_d R \end{vmatrix}$

$\overline{a}_P = \left[-(\omega_0^2 + \omega_b^2)b + 2\omega_0\omega_d R\right]\overline{i} - (\omega_b^2 + \omega_d^2)R\overline{j} + \omega_0\omega_b R\overline{k}$.

================================<>================================

Problem 9.14 The bent bar is rigidly attached to the vertical shaft, which rotates with constant angular veolcity ω_0. The circular disk is pinned to the bent bar and rotates with constant angular velocity ω_d relative to the bar. (a) Determine the disk's angular velocity $\overline{\omega}_{disk}$.
(b) Determine the velocity of point P, whith is the uppermost point on the circular disk at the present instant.

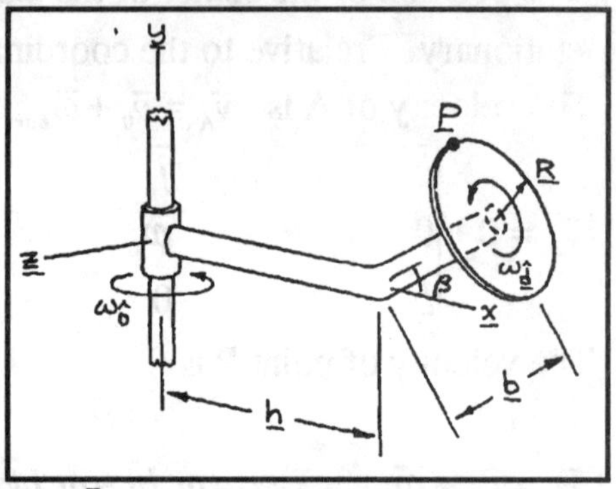

Solution:

(a) Assume that the coordinate system is body-fixed with respect to the bent bar. The angular velocity of the coordinate system is $\overline{\Omega} = \omega_o\overline{j}$. The angular velocity of the disk relative to the coordinate system is $\overline{\omega}_{rel} = \omega_d \cos\beta\overline{i} + \omega_d \sin\beta\overline{j}$, so the angular velocity of the disk is $\overline{\omega}_{disk} = \overline{\Omega} + \overline{\omega}_{rel} = \omega_d\cos\beta\overline{i} + (\omega_0 + \omega_d\sin\beta)\overline{j}$.

(b) The angular velocity of the bent bar is equal to the angular velocity of the coordinate system: $\overline{\omega}_{bar} = \overline{\Omega}$. The position of the center of the disk A relative to the origin 0 is $\overline{r}_{A/0} = (h + b\cos\beta)\overline{i} + b\sin\beta\overline{j}$. The velocity of point A is $\overline{v}_A = \overline{v}_0 + \overline{\omega}_{bar} \times \overline{r}_{A/0}$

$$\overline{v}_A = 0 + \begin{vmatrix} \overline{i} & \overline{j} & \overline{k} \\ 0 & \omega_0 & 0 \\ h + b\cos\beta & b\sin\beta & 0 \end{vmatrix} = -\omega_0(h + b\cos\beta)\overline{k}.$$

The position of point P relative to the center A is $\overline{r}_{P/A} = -R\sin\beta\overline{i} + R\cos\beta\overline{j}$. The velocity of point P is

$$\overline{v}_P = \overline{v}_A + \overline{\omega}_{disk} \times \overline{r}_{P/A} = -\omega_0(h + b\cos\beta)\overline{k} + \begin{vmatrix} \overline{i} & \overline{j} & \overline{k} \\ \omega_d\cos\beta & \omega_0 + \omega_d\sin\beta & 0 \\ -R\sin\beta & R\cos\beta & 0 \end{vmatrix}, \text{ or}$$

$$\overline{v}_P = \left[\omega_d R - \omega_0(h + b\cos\beta - R\sin\beta)\right]\overline{k}.$$

================================<>================================

================================<>================================

Problem 9.15 In Problem 9.14, determine (a) the disk's angular acceleration $\overline{\alpha}_{disk}$; (b) the acceleration of point P.

Solution:

(a) From Equation (9.4) and the solution of Problem 9.14, the disk's angular acceleration

$$\overline{\alpha}_{disk} = \overline{\Omega} \times \overline{\omega}_{disk} = \begin{vmatrix} \overline{i} & \overline{j} & \overline{k} \\ 0 & \omega_0 & 0 \\ \omega_d \cos\beta & \omega_0 + \omega_d \sin\beta & 0 \end{vmatrix} = -\omega_0 \omega_d \cos\beta \overline{k}.$$

(b) The position of the center of the disk A relative to the origin 0 is $\overline{r}_{A/0} = (h + b\cos\beta)\overline{i} + b\sin\beta\overline{j}$. The angular velocity of the bent bar equals the angular velocity of the coordinate system, $\overline{\omega}_{bar} = \overline{\Omega} = \omega_0 \overline{j}$, and the angular acceleration of the coordinate system is zero, so the acceleration of point A is $\overline{a}_A = \overline{a}_0 + \overline{\omega}_{bar} \times (\overline{\omega}_{bar} \times \overline{r}_{A/0})$, or

$$\overline{a}_A = 0 + \begin{vmatrix} \overline{i} & \overline{j} & \overline{k} \\ 0 & \omega_0 & 0 \\ 0 & 0 & -\omega_0 (h + b\cos\beta) \end{vmatrix} = -\omega_0^2 (h + b\cos\beta)\overline{i}.$$

The acceleration of point P is $\overline{a}_P = \overline{a}_A + \overline{\alpha}_{disk} \times \overline{r}_{P/A} + \overline{\omega}_{disk} \times (\overline{\omega}_{disk} \times \overline{r}_{P/A})$.

The position of P relative to A is $\overline{r}_{P/A} = -R\sin\beta\overline{i} + R\cos\beta\overline{j}$. From the solution of Problem 9.14, $\overline{\omega}_{disk} = \omega_d \cos\beta\overline{i} + (\omega_0 + \omega_d \sin\beta)\overline{j}$, so

$$\overline{\omega}_{disk} \times \overline{r}_{P/A} = \begin{vmatrix} \overline{i} & \overline{j} & \overline{k} \\ \omega_d \cos\beta & \omega_0 + \omega_d \sin\beta & 0 \\ -R\sin\beta & R\cos\beta & 0 \end{vmatrix} = (\omega_d R + \omega_0 R \sin\beta)\overline{k}.$$

Therefore the acceleration of point P is

$$\overline{a}_p = -\omega_0^2 (h + b\cos\beta)\overline{i} + \begin{vmatrix} \overline{i} & \overline{j} & \overline{k} \\ 0 & 0 & -\omega_0 \omega_d \cos\beta \\ -R\sin\beta & R\cos\beta & 0 \end{vmatrix} + \begin{vmatrix} \overline{i} & \overline{j} & \overline{k} \\ \omega_d \cos\beta & \omega_0 + \omega_d \sin\beta & 0 \\ 0 & 0 & \omega_d R + \omega_0 R \sin\beta \end{vmatrix}$$

or

$$\overline{a}_P = \left[-\omega_0^2 (h + b\cos\beta) + (\omega_0^2 + \omega_d^2) R\sin\beta + 2\omega_0 \omega_d R\right]\overline{i} - \omega_d^2 R\cos\beta\overline{j}$$

================================<>================================

=================================<>=================================

Problem 9.16 The gyroscope's circular frame rotates about the vertical axis at 2 rad / s in the counterclockwise direction when viewed from above. The 2.4 *in.* diameter wheel rotates relative to the frame at 10 *rad/s*. Determine the velocities of points A and B relative to the origin.

Solution:

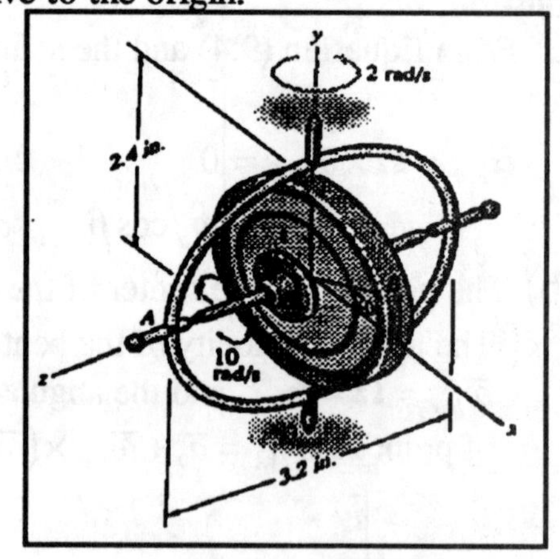

Denote the magnitude of the angular velocity about the vertical axis by $\Omega = 2 \text{ rad/s}$, the magnitude of the angular velocity of the wheel relative to the frame by $\omega = 10 \text{ rad/s}$, and the radius of the wheel by $R = 1.2$ in.. The angular velocity vectors are $\vec{\mathbf{\Omega}} = \Omega \vec{j}$, and $\vec{\mathbf{w}}_{rel} = \omega \vec{k}$. The angular velocity vector of the wheel is $\vec{\mathbf{w}} = \vec{\mathbf{\Omega}} + \vec{\mathbf{w}}_{rel} = 2\vec{j} + 10\vec{k}$ (rad / s). The line OA is parallel to the vector $\vec{r}_{A/O} = \frac{3.2}{2}\vec{k} = 1.6\vec{k}$ (in.). The velocity of point A is

$$\vec{v}_{A/O} = \vec{\mathbf{w}} \times \vec{r}_{A/O} = \begin{bmatrix} \vec{i} & \vec{j} & \vec{k} \\ 0 & 2 & 10 \\ 0 & 0 & 1.6 \end{bmatrix}, \quad \vec{v}_{A/O} = 3.2\vec{i} \text{ (in/s)}.$$

The line OB is

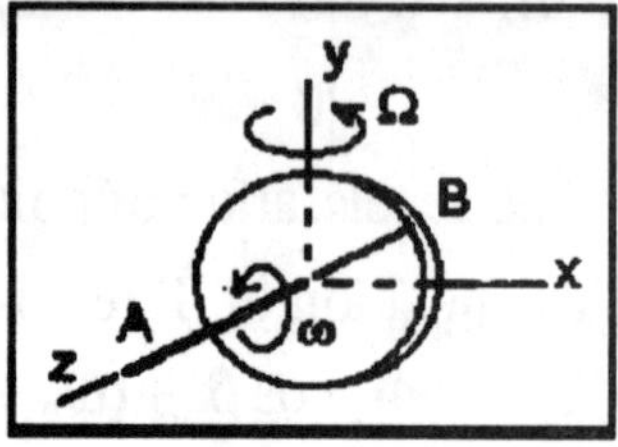

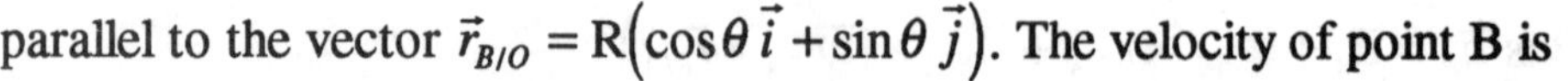

parallel to the vector $\vec{r}_{B/O} = \mathrm{R}(\cos\theta\,\vec{i} + \sin\theta\,\vec{j})$. The velocity of point B is

$$\vec{v}_{B/O} = \vec{\mathbf{w}} \times \vec{r}_{B/O} = \begin{bmatrix} \vec{i} & \vec{j} & \vec{k} \\ 0 & 2 & 10 \\ \mathrm{R}\cos\theta & \mathrm{R}\sin\theta & 0 \end{bmatrix},$$

$$\vec{v}_{B/O} = -4.10\vec{i} + 11.28\vec{j} - 2.26\vec{k} \text{ (in/s)}$$

=================================<>=================================

Problem 9.17 If the angular velocities of the frame and wheel of the gyroscope in Problem 9.16 are constant, what are the accelerations of points A and B relative to the origin?

Solution:

The angular acceleration vector for the wheel is $\vec{\mathbf{a}} = \frac{d\vec{\mathbf{w}}}{dt} + \vec{\mathbf{\Omega}} \times \vec{\mathbf{w}}_{rel} = \frac{d}{dt}(\Omega\vec{j} + \omega_{rel}\vec{k}) + \vec{\mathbf{\Omega}} \times \vec{\mathbf{w}}_{rel} = \vec{\mathbf{\Omega}} \times \vec{\mathbf{w}}_{rel}$.

The acceleration of point A is $\vec{a}_{A/O} = \vec{\mathbf{a}} \times \vec{r}_{A/O} + \vec{\mathbf{w}} \times (\vec{\mathbf{w}} \times \vec{r}_{A/O})$. From the solution to Problem 9.16, $\vec{r}_{A/O} = 1.6\vec{k}$ (in.). Expanding term by term:

$$\vec{\mathbf{a}} \times \vec{r}_{A/O} = (\vec{\mathbf{\Omega}} \times \vec{\mathbf{w}}_{rel}) \times \vec{r}_{A/O} = \begin{bmatrix} \vec{i} & \vec{j} & \vec{k} \\ 0 & 2 & 0 \\ 0 & 0 & 10 \end{bmatrix} \times \vec{r}_{A/O} = \begin{bmatrix} \vec{i} & \vec{j} & \vec{k} \\ 20 & 0 & 0 \\ 0 & 0 & 1.6 \end{bmatrix} = -32\vec{j}.$$

$$\vec{\mathbf{w}} \times (\vec{\mathbf{w}} \times \vec{r}_{A/O}) = \vec{\mathbf{w}} \times \begin{bmatrix} \vec{i} & \vec{j} & \vec{k} \\ 0 & 2 & 10 \\ 0 & 0 & 1.6 \end{bmatrix} = \begin{bmatrix} \vec{i} & \vec{j} & \vec{k} \\ 0 & 2 & 10 \\ 3.2 & 0 & 0 \end{bmatrix} = 32\vec{j} - 6.4\vec{k}.$$

Collect terms: $\vec{a}_{A/O} = -6.4\vec{k} \text{ (in/s}^2\text{)}$

The acceleration of point B is $\vec{a}_{B/O} = \vec{\mathbf{a}} \times \vec{r}_{B/O} + \vec{\mathbf{w}} \times (\vec{\mathbf{w}} \times \vec{r}_{B/O})$. From the solution to Problem 9.12, $\vec{r}_{B/O} = \mathrm{R}(\cos\theta\,\vec{i} + \sin\theta\,\vec{j})$. Expanding term by term:

Solution continued on next page

$$\vec{\mathbf{a}}\times\vec{r}_{B/O}=\vec{\mathbf{\Omega}}\times\vec{\mathbf{w}}_{rel}\times\vec{r}_{B/O}=\begin{bmatrix}\vec{i} & \vec{j} & \vec{k}\\ 20 & 0 & 0\\ R\cos\theta & R\sin\theta & 0\end{bmatrix}=20R\sin\theta\vec{k}=8.208\vec{k},$$

$$\vec{\mathbf{w}}\times(\vec{\mathbf{w}}\times\vec{r}_{B/O})=\vec{\mathbf{w}}\times\begin{bmatrix}\vec{i} & \vec{j} & \vec{k}\\ 0 & 2 & 10\\ R\cos\theta & R\sin\theta & 0\end{bmatrix}=\begin{bmatrix}\vec{i} & \vec{j} & \vec{k}\\ 0 & 2 & 10\\ -10R\sin\theta & 10R\cos\theta & -2R\cos\theta\end{bmatrix}$$

$$\vec{\mathbf{w}}\times(\vec{\mathbf{w}}\times\vec{r}_{B/O})=-117.27\vec{i}-41.04\vec{j}+8.208\vec{k}.$$

Collect terms, $\vec{a}_{B/O}=-117.3\vec{i}-41.04\vec{j}+16.42\vec{k}\ (\text{in}/\text{s}^2)$

==================================<>==================================

Problem 9.18 Relative to an earth-fixed reference frame, the manipulator rotates about the vertical axis with angular velocity $\omega_y=0.1\ (rad/s)$. The y axis of the secondary coordinate system remains vertical, and the x axis rotates with the manipulator so that points A, B, and C remain in the *x-y* plane. The angular velocities of the arms AB and BC, *relative to the secondary coordinate system*, are $-0.2\bar{k}(rad/s)$ and $0.4\bar{k}(rad/s)$, respectively. (a) What is the angular velocity $\bar{\omega}_{BC}$ of arm BC relative to the earth-fixed reference, and (b) What is the velocity of point C relative to the earth-fixed reference?

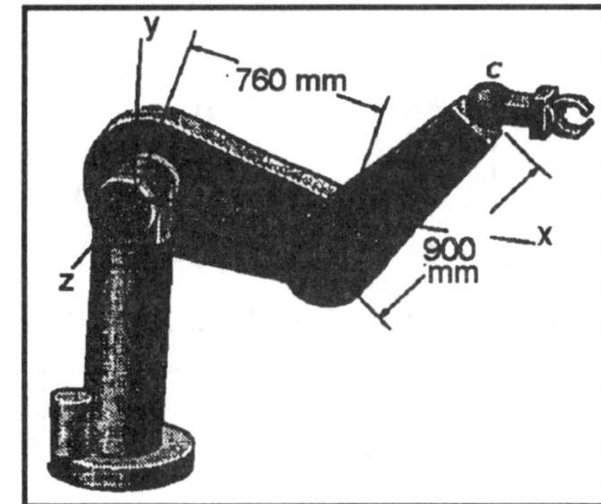

Solution:

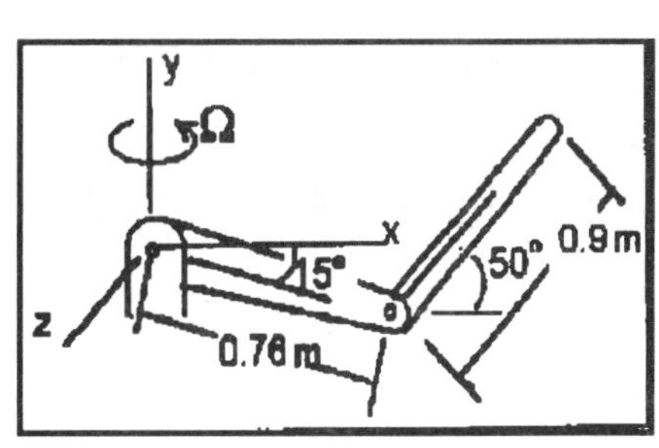

(a) The angular velocity of the coordinate system is $\bar{\Omega}=\omega_y\bar{j}=0.1\bar{j}(rad/s)$. The angular velocity of arm BC relative to the coordinate system is $(\bar{\omega}_{BC})_{rel}=0.4\bar{k}(rad/s)$, so the angular velocity of arm BC relative to the earth-fixed reference frame is $\bar{\omega}_{BC}=\bar{\Omega}+(\bar{\omega}_{BC})_{rel}=0.1\bar{j}+0.4\bar{k}(rad/s)$. The angular velocity of arm AB relative to the coordinate system is $(\bar{\omega}_{AB})_{rel}=-0.2\bar{k}(rad/s)$, so $\bar{\omega}_{AB}=\bar{\Omega}+(\bar{\omega}_{AB})_{rel}=0.1\bar{j}-0.2\bar{k}(rad/s)$. The velocity of point B is

$$\bar{v}_B=\bar{v}_A+\bar{\omega}_{AB}\times\bar{r}_{B/A}=0+\begin{vmatrix}\bar{i} & \bar{j} & \bar{k}\\ 0 & 0.1 & -0.2\\ 0.76\cos15^\circ & -0.76\sin15^\circ & 0\end{vmatrix}$$

$$\bar{v}_B=-(0.2)(0.76)\sin15^\circ\bar{i}-(0.2)(0.76)\cos15^\circ\bar{j}-(0.1)(0.76)\cos15^\circ\bar{k}$$

$$\bar{v}_B=-0.0393\bar{i}-0.1468\bar{j}-0.0734\bar{k}(m/s).$$

(b) The velocity of point C is

$$\bar{v}_c=\bar{v}_B+\bar{\omega}_{BC}\times\bar{r}_{C/B}=-0.0393\bar{i}-0.1468\bar{j}-0.0734\bar{k}+\begin{vmatrix}\bar{i} & \bar{j} & \bar{k}\\ 0 & 0.1 & 0.4\\ 0.9\cos50^\circ & 0.9\sin50^\circ & 0\end{vmatrix},\text{ or}$$

$$\bar{v}_C=-0.315\bar{i}+0.085\bar{j}-0.131\bar{k}(m/s).$$

==================================<>==================================

=================================<>=================================

Problem 9.19 The angular velocity of the manipulator in Problem 9.18 about the vertical axis is constant. The angular acceleration of the arms AB and BC relative to the secondary coordinate system are zero. What is the acceleration of point C relative to the earth-fixed reference frame?

Solution:

From Equation (9.4) and the solution of Problem 9.18, the angular acceleration of arm AB is

$$\overline{\alpha}_{AB}=\overline{\Omega}\times\overline{\omega}_{AB}=\begin{vmatrix}\overline{i} & \overline{j} & \overline{k}\\ 0 & 0.1 & 0\\ 0 & 0.1 & -0.2\end{vmatrix}=-0.02\overline{i}\left(rad/s^2\right).$$

The angular acceleration of arm BC is $\overline{\alpha}_{BC}=\overline{\Omega}\times\overline{\omega}_{BC}=\begin{vmatrix}\overline{i} & \overline{j} & \overline{k}\\ 0 & 0.1 & 0\\ 0 & 0.1 & 0.4\end{vmatrix}=0.04\overline{i}\left(rad/s^2\right).$

The acceleration of point B is $\overline{a}_B=\overline{a}_A+\overline{\alpha}_{AB}\times\overline{r}_{B/A}+\overline{\omega}_{AB}\times\left(\overline{\omega}_{AB}\times\overline{r}_{B/A}\right)$.

From the solution of Problem 9.18, $\overline{\omega}_{AB}\times\overline{r}_{B/A}=-0.0393\overline{i}-0.1468\overline{j}-0.0734\overline{k}(m/s)$, so

$$\overline{a}_B=0+\begin{vmatrix}\overline{i} & \overline{j} & \overline{k}\\ -0.02 & 0 & 0\\ 0.76\cos 15^\circ & -0.76\sin 15^\circ & 0\end{vmatrix}+\begin{vmatrix}\overline{i} & \overline{j} & \overline{k}\\ 0 & 0.1 & -0.2\\ -0.0393 & -0.1468 & -0.0734\end{vmatrix}, \text{ or}$$

$\overline{a}_B=-0.036711\overline{i}+0.00787\overline{j}+0.00787\overline{k}(m/s)$.

The acceleration of point C is $\overline{a}_C=\overline{a}_B+\overline{\alpha}_{BC}\times\overline{r}_{C/B}+\overline{\omega}_{BC}\times\left(\overline{\omega}_{BC}\times\overline{r}_{C/B}\right)$.

The term $\overline{\omega}_{BC}\times\overline{r}_{C/B}$ is $\overline{\omega}_{BC}\times\overline{r}_{C/B}=\begin{vmatrix}\overline{i} & \overline{j} & \overline{k}\\ 0 & 0.1 & 0.4\\ 0.9\cos 50^\circ & 0.9\sin 50^\circ & 0\end{vmatrix}=-0.2758\overline{i}+0.2314\overline{j}-0.0579\overline{k}(m/s)$,

so the acceleration of point C is

$$\overline{a}_c=-0.036711\overline{i}+0.00787\overline{j}+0.00787\overline{k}+\begin{vmatrix}\overline{i} & \overline{j} & \overline{k}\\ 0.04 & 0 & 0\\ 0.9\cos 50^\circ & 0.9\sin 50^\circ & 0\end{vmatrix}$$

$$+\begin{vmatrix}\overline{i} & \overline{j} & \overline{k}\\ 0 & 0.1 & 0.4\\ -0.2758 & 0.2314 & -0.0579\end{vmatrix}, \text{ or}$$

$\overline{a}_C=-0.135\overline{i}-0.102\overline{j}+0.063\overline{k}\left(m/s^2\right)$.

=================================<>=================================

Problem 9.20 The cone's curved surface rolls on the horizontal surface. The x axis of the secondary coordinate system remains coincident with the cone's axis and the z axis remains horizontal. The z axis has a constant angular velocity ω_0 in the horizontal plane. In terms of components in the secondary coordinate system, determine (a) the angular velocity $\overline{\Omega}$ of the secondary coordinate system; (b) the angular velocity $\overline{\omega}_{rel}$ of the cone relative to the secondary coordinate system; (c) the cone's angular velocity $\overline{\omega}$ relative to the primary reference frame.

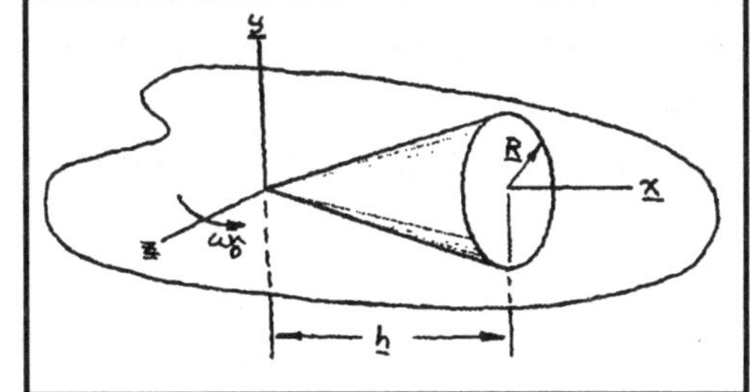

Solution:

Let β be the angle between the y axis and the vertical, which is also the angle between the x axis and the horizontal plane.

(a) The angular velocity of the coordinate system is $\overline{\Omega} = \omega_0 \sin\beta \bar{i} + \omega_0 \cos\beta \bar{j}$. Noting that $\sin\beta = R/\sqrt{h^2+R^2}$ and $\cos\beta = h/\sqrt{h^2+R^2}$, we have $\overline{\Omega} = \dfrac{\omega_0}{\sqrt{h^2+R^2}}\left(R\bar{i} + h\bar{j}\right)$.

(b) Relative to the secondary coordinate system, the cone rotates about the x axis, so we can write its angular velocity relative to the coordinate system as $\overline{\omega}_{rel} = \omega_{rel}\bar{i}$. We can determine ω_{rel} by using the fact that the velocity of the cone is zero where it contacts the surface. The velocity of the point with coordinates $(h,-R,0)$ is $\overline{\omega}\times\left(h\bar{i} - R\bar{j}\right) = \left(\overline{\Omega} + \overline{\omega}_{rel}\right)\times\left(h\bar{i} - R\bar{j}\right) = 0$:

Hence, $\begin{vmatrix} \bar{i} & \bar{j} & \bar{k} \\ \dfrac{\omega_0 R}{\sqrt{h^2+R^2}} + \omega_{rel} & \dfrac{\omega_0 h}{\sqrt{h^2+R^2}} & 0 \\ h & -R & 0 \end{vmatrix} = 0$. Solving this equation for ω_{rel}, we obtain

$$\overline{\omega}_{rel} = \omega_{rel}\bar{i} = \frac{-\omega_0}{\sqrt{h^2+R^2}}\left(R + \frac{h^2}{R}\right)\bar{i}.$$

(c) The cone's velocity is $\overline{\omega} = \overline{\Omega} + \overline{\omega}_{rel} = \dfrac{\omega_0}{\sqrt{h^2+R^2}}\left(-\dfrac{h^2}{R}\bar{i} + h\bar{j}\right)$.

Problem 9.21 In Problem 9.20, determine the velocity relative to the primary reference frame of the point on the base of the cone with coordinates $x = h$, $y = 0$, $z = R$.

Solution:

From the solution of Problem 9.20, the cone's angular velocity is $\overline{\omega} = -\dfrac{\omega_0 h^2/R}{\sqrt{h^2+R^2}}\bar{i} + \dfrac{\omega_0 h}{\sqrt{h^2+R^2}}\bar{j}$. The position of the point relative to the origin 0 is $\bar{r}_{P/O} = h\bar{i} + R\bar{k}$, so its velocity is

$$\bar{v}_P = \bar{v}_0 + \overline{\omega}\times\bar{r}_{P/O} = 0 + \begin{vmatrix} \bar{i} & \bar{j} & \bar{k} \\ -\dfrac{\omega_0 h^2/R}{\sqrt{h^2+R^2}} & \dfrac{\omega_0 h}{\sqrt{h^2+R^2}} & 0 \\ h & 0 & R \end{vmatrix} = \frac{\omega_0 h}{\sqrt{h^2+R^2}}\left(R\bar{i} + h\bar{j} - h\bar{k}\right).$$

Problem 9.22 In Problem 9.20, determine: (a) the cone's angular acceleration $\overline{\alpha}$ relative to the primary reference frame, (b) the acceleration relative to the primary reference frame of the point on the base of the cone with coordinates $x = h,\ y = 0,\ z = R$.

Solution:

(a) From Equation (9.4) and the solution of Problem 9.20, the cone's angular acceleration is $\overline{\alpha} = \overline{\Omega}\times\overline{\omega}$

$$0+\begin{vmatrix} \overline{i} & \overline{j} & \overline{k} \\ \frac{\omega_0 R}{\sqrt{h^2+R^2}} & \frac{\omega_0 h}{\sqrt{h^2+R^2}} & 0 \\ -\frac{\omega_0 h^2/R}{\sqrt{h^2+R^2}} & \frac{\omega_0 h}{\sqrt{h^2+R^2}} & 0 \end{vmatrix} = \frac{\omega_0^2 h}{R}\overline{k}.$$

(b) The position of the point relative to the origin 0 is $\overline{r}_{P/O} = h\overline{i} + R\overline{k}$. From the solution of Problems 9.20 and 9.21, $\overline{\omega} = \frac{\omega_0}{\sqrt{h^2+R^2}}\left(-\frac{h^2}{R}\overline{i} + h\overline{j}\right)$ and $\overline{\omega}\times\overline{r}_{P/O} = \frac{\omega_0 h}{\sqrt{h^2+R^2}}\left(R\overline{i} + h\overline{j} - h\overline{k}\right)$.

(c) The acceleration of the point is $\overline{a}_P = \overline{a}_0 + \overline{\alpha}\times\overline{r}_{P/O} + \overline{\omega}\times\left(\overline{\omega}\times\overline{r}_{P/O}\right)$, or

$$\overline{a}_P = 0+\begin{vmatrix} \overline{i} & \overline{j} & \overline{k} \\ 0 & 0 & \frac{\omega_0^2 h}{R} \\ h & 0 & R \end{vmatrix} + \begin{vmatrix} \overline{i} & \overline{j} & \overline{k} \\ -\frac{\omega_0 h^2/R}{\sqrt{h^2+R^2}} & \frac{\omega_0 h}{\sqrt{h^2+R^2}} & 0 \\ \frac{\omega_0 hR}{\sqrt{h^2+R^2}} & \frac{\omega_0 h^2}{\sqrt{h^2+R^2}} & -\frac{\omega_0 h^2}{\sqrt{h^2+R^2}} \end{vmatrix}, \text{ or}$$

$$\overline{a}_P = -\frac{\omega_0^2 h^2}{h^2+R^2}\left[h\overline{i} - R\overline{j} + \left(R + \frac{h^2}{R}\right)\overline{k}\right].$$

==================================<>==================================

Problem 9.23 A tilted cylinder of length ℓ and radius R undergoes a steady motion in which one end rolls on the plane surface while the center of the cylinder remains stationary relative to the surface. The z axis of the secondary coordinate system remains coincident with the cylinder's axis and the y axis remains horizontal. The angular velocity of the y axis in the horizontal plane is ω_0. In terms of components in the secondary coordinate system, determine: (a) the angular velocity $\overline{\Omega}$ of the secondary coordinate system, (b) the angular velocity $\overline{\omega}_{rel}$ of the cylinder relative to the secondary coordinate system, and (c) the cone's angular velocity $\overline{\omega}$ relative to the primary coordinate system.

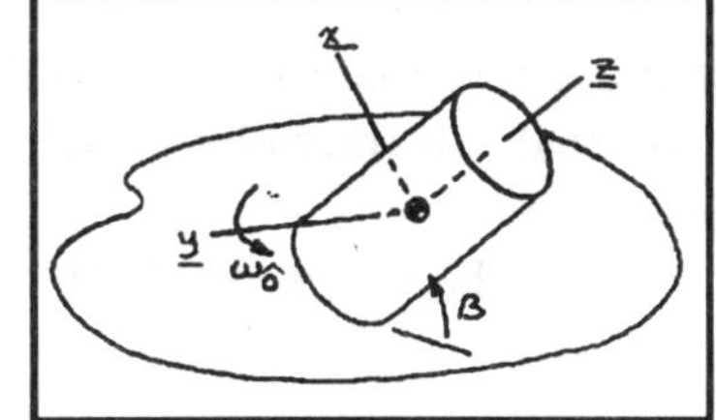

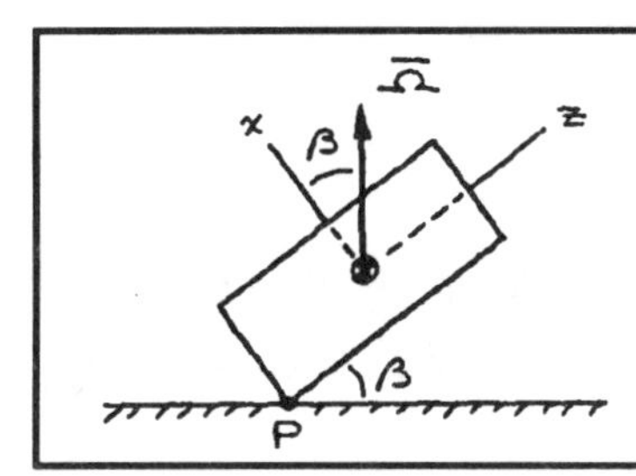

Solution:

(a) The angular velocity of the coordinate system is
$\overline{\Omega} = \omega_0 \cos\beta \bar{i} + \omega_0 \sin\beta \bar{k}$.

(b) Relative to the coordinate system, the cylinder rotates about the z axis, so we can express its angular velocity relative to the coordinate system as $\overline{\omega}_{rel} = \omega_{rel}\bar{k}$.

To determine ω_{rel}, we will use the fact that the velocities of the origin 0 and the point P in contact with the floor are zero. The position of P relative to 0 is $\bar{r}_{P/0} = -R\bar{i} - \frac{\ell}{2}\bar{k}$.

Therefore

$$\bar{v}_p = 0 = \bar{v}_0 + \overline{\omega}\times\bar{r}_{P/0} = 0 + \left(\overline{\Omega} + \overline{\omega}_{rel}\right)\times\bar{r}_{P/0} = 0 + \begin{vmatrix} \bar{i} & \bar{j} & \bar{k} \\ \omega_0\cos\beta & 0 & \omega_0\sin\beta + \omega_{rel} \\ -R & 0 & -\frac{\ell}{2} \end{vmatrix}.$$

Solving this equation for ω_{rel}, we obtain $\overline{\omega}_{rel} = \omega_0\left(\frac{\ell}{2R}\cos\beta - \sin\beta\right)\bar{k}$.

(c) The cylinder's angular velocity is $\overline{\omega} = \overline{\Omega} + \overline{\omega}_{rel} = \omega_0\cos\beta\left(\bar{i} + \frac{\ell}{2R}\bar{k}\right)$.

==================================<>==================================

Problem 9.24 In Problem 9.23, determine the velocity relative to the primary reference frame of the point on the upper base of the cylinder with coordinates $x = R$, $y = 0$, $z = \ell/2$.

Solution:

From the solution of Problem 9.23, the cylinder's angular velocity is $\overline{\omega} = \omega_0\cos\beta\left(\bar{i} + \frac{\ell}{2R}\bar{k}\right)$. The center of the cylinder (the origin 0 the secondary coordinate system) is stationary, so the velocity of the point Q with coordinates $(R, 0, \ell/2)$ is

$$\bar{v}_Q = \bar{v}_0 + \overline{\omega}\times\bar{r}_{Q/0} = 0 + \begin{vmatrix} \bar{i} & \bar{j} & \bar{k} \\ \omega_0\cos\beta & 0 & \omega_0\frac{\ell}{2R}\cos\beta \\ R & 0 & \ell/2 \end{vmatrix} = 0$$

==================================<>==================================

==================================<>==================================

Problem 9.25 In Problem 9.23, determine (a) the cylinder's angular acceleration $\overline{\alpha}$ relative to the primary reference frame, (b) the acceleration relative to the primary reference frame of the point on the upper base of the cylinder with coordinates $x = R,\ y = 0,\ z = \ell/2$.

Solution:

(a) From the Equation (9.4) and the solution of Problem 9.23, the cylinder's angular acceleration is

$$\overline{\alpha} = \overline{\Omega}\times\overline{\omega} = \begin{vmatrix} \overline{i} & \overline{j} & \overline{k} \\ \omega_0\cos\beta & 0 & \omega_0\sin\beta \\ \omega_0\cos\beta & 0 & \omega_0\frac{\ell}{2R}\cos\beta \end{vmatrix},\ \text{or } \overline{\alpha} = \omega_0^2\cos\beta\left(\sin\beta - \frac{\ell}{2R}\cos\beta\right)\overline{j}.$$

(b) The center 0 of the cylinder is stationary so the acceleration of the point Q with coordinates $(R, 0, \ell/2)$ is $\overline{a}_Q = \overline{\alpha}\times\overline{r}_{Q/0} + \overline{\omega}\times(\overline{\omega}\times\overline{r}_{Q/0})$. From the solution of Problem 9.24, $\overline{\omega}\times\overline{r}_{Q/0} = 0$ so

$$\overline{a}_Q = \overline{\alpha}\times\overline{r}_{Q/0} = \begin{vmatrix} \overline{i} & \overline{j} & \overline{k} \\ 0 & \omega_0^2\cos\beta\left(\sin\beta - \frac{\ell}{2R}\cos\beta\right) & 0 \\ R & 0 & \ell/2 \end{vmatrix},\ \text{or}$$

$$a_Q = \omega_0^2\cos\beta\left(\sin\beta - \frac{\ell}{2R}\cos\beta\right)\left(\frac{\ell}{2}\overline{i} - R\overline{k}\right).$$

==================================<>==================================

Problem 9.26 The bar AB is connected by ball and socket joints to the edge of the horizontal circular disk A and to a collar that slides on the vertical bar at B. The disk rotates with constant angular velocity $\omega_d = 4\ rad/s$ and the angular velocity of bar AB about its axis is zero. Determine the velocity of the collar at B and the angular velocity of bar AB.

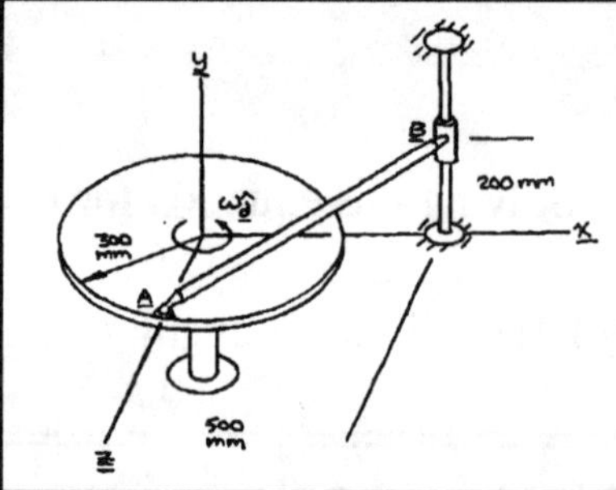

Solution:

The angular velocity of the disk is $\overline{\omega}_{disk} = \omega_d\overline{j} = 4\overline{j}(rad/s)$, so the velocity of A is $\overline{v}_A = \overline{v}_0 + \overline{\omega}_{disk}\times\overline{r}_{A/0} = 0 + \begin{vmatrix} \overline{i} & \overline{j} & \overline{k} \\ 0 & 4 & 0 \\ 0 & 0 & 0.3 \end{vmatrix} = 1.2\overline{i}\,(m/s).$

Let the bar's angular velocity be $\overline{\omega}_{bar} = \omega_x\overline{i} + \omega_y\overline{j} + \omega_z\overline{k}$. We can express the velocity of B as $\overline{v}_B = v_B\overline{j}$, so $\overline{v}_B = \overline{v}_A + \overline{\omega}_{bar}\times\overline{r}_{B/A}$; $\quad v_B\overline{j} = 1.2\overline{i} + \begin{vmatrix} \overline{i} & \overline{j} & \overline{k} \\ \omega_x & \omega_y & \omega_z \\ 0.5 & 0.2 & -0.3 \end{vmatrix}$. Equating $\overline{i}$, $\overline{j}$ and $\overline{k}$ components, we obtain $0 = 1.2 - 0.3\omega_y - 0.2\omega_z$ (1), $v_B = 0.5\omega_z + 0.3\omega_x$ (2), and $0 = 0.2\omega_x - 0.5\omega_y$ (3).

Since the angular velocity of the bar about its axis is zero, i.e., $\overline{r}_{B/A}\bullet\overline{\omega}_{bar} = 0$: or, expanding $0.5\omega_x + 0.2\omega_y - 0.3\omega_z = 0$ (4). Solving Equations (1) - (4), we obtain $\overline{v}_B = 3\overline{j}\,(m/s)$

$\overline{\omega}_{bar} = 2.37\overline{i} + 0.95\overline{j} + 4.58k\,(rad/s)$.

==================================<>==================================

==================================<>==================================

Problem 9.27 In Problem 9.26, the angular acceleration of bar AB about its axis is zero. Determine the acceleration of the collar at B and the angular acceleration of bar AB.

Solution:

The angular acceleration of the disk is zero, so the acceleration of A is $\bar{a}_A = \bar{\omega}_{disk} \times \left(\bar{\omega}_{disk} \times \bar{r}_{A/0}\right)$.

From the solution of Problem 9.26, $\bar{a}_A = \begin{vmatrix} \bar{i} & \bar{j} & \bar{k} \\ 0 & 4 & 0 \\ 1.2 & 0 & 0 \end{vmatrix} = -4.8\bar{k}\left(m/s^2\right)$.

Let the bar's angular acceleration be $\bar{\alpha}_{bar} = \alpha_x \bar{i} + \alpha_y \bar{j} + \alpha_z \bar{k}$.

The acceleration of B is $\bar{a}_B = \bar{a}_A + \bar{\alpha}_{bar} \times \bar{r}_{B/A} + \bar{\omega}_{bar} \times \left(\bar{\omega}_{bar} \times \bar{r}_{B/A}\right)$.

In the solution of Problem 9.26 it was determined that $\bar{\omega}_{bar} = 2.37\bar{i} + 0.95\bar{j} + 4.58\bar{k}(rad/s)$ so

$\omega_{bar} \times \bar{r}_{B/A} = \begin{vmatrix} \bar{i} & \bar{j} & \bar{k} \\ 2.37 & 0.95 & 4.58 \\ 0.5 & 0.2 & -0.3 \end{vmatrix} = -1.2\bar{i} + 3\bar{j}(m/s)$. The acceleration of B is

$a_B\bar{j} = -4.8\bar{k} + \begin{vmatrix} \bar{i} & \bar{j} & \bar{k} \\ \alpha_x & \alpha_y & \alpha_z \\ 0.5 & 0.2 & -0.3 \end{vmatrix} + \begin{vmatrix} \bar{i} & \bar{j} & \bar{k} \\ 2.37 & 0.95 & 4.58 \\ -1.2 & 3 & 0 \end{vmatrix}$. Equating $\bar{i}$, $\bar{j}$ and $\bar{k}$ components yields $0 = -0.3\alpha_y - 0.2\alpha_z - 13.74$ (1), $a_B = 0.5\alpha_z + 0.3\alpha_x - 5.49$ (2), and $0 = -4.8 + 0.2\alpha_x - 0.5\alpha_y + 8.24$ (3)

The angular acceleration of the bar about its axis is zero, so $\bar{r}_{B/A} \bullet \bar{\alpha}_{bar} = 0$: Hence, $0.5\alpha_x + 0.2\alpha_y - 0.3\alpha_z = 0$ (4)

Solving Equations (1) - (4), we obtain $\bar{a}_B = -45\bar{j}\left(m/s^2\right)$ and $\bar{\alpha}_{bar} = -33.00\bar{i} - 6.32\bar{j} - 59.21\bar{k}\left(rad/s^2\right)$.

==================================<>==================================

Problem 9.28 The inertial matrix of a rigid body in terms of a body-fixed coordinate system with its origin at the center of mass is $[I] = \begin{bmatrix} 8.8 & 4.0 & 0 \\ 4.0 & 6.6 & 0 \\ 0 & 0 & 15.4 \end{bmatrix}$ *(kg-m²)*.

If the rigid body's angular velocity is $\bar{\omega} = -6\bar{i} + 2\bar{j} - 2\bar{k}$ *(rad/s)*, what is its angular momentum about its center of mass?

Solution:

The angular momentum is $\begin{bmatrix} H_x \\ H_y \\ H_z \end{bmatrix} = \begin{bmatrix} 8.8 & 4.0 & 0 \\ 4.0 & 6.6 & 0 \\ 0 & 0 & 15.4 \end{bmatrix} \begin{bmatrix} -6 \\ 2 \\ -2 \end{bmatrix} = \begin{bmatrix} -44.8 \\ -10.8 \\ -30.8 \end{bmatrix} (kg - m^2/s)$

==================================<>==================================

Problem 9.29 What is the moment of inertia of the rigid body in Problem 9.28 about the axis that passes through the origin and the point (1, 2, -2) *m*?

Solution:

A unit vector parallel to the axis is $\bar{e} = \dfrac{\bar{i} + 2\bar{j} - 2\bar{k}}{\sqrt{(1)^2 + (2)^2 + (-2)^2}} = 0.333\bar{i} + 0.667\bar{j} - 0.667\bar{k}$. From Eq (9.19), the moment of inertia is $I_0 = (8.8)e_x^2 + (6.6)e_y^2 + (15.4)e_z^2 - 2(-4)e_x e_y = 12.5(kg - m^2)$.

==================================<>==================================

================================<>================================

Problem 9.30 A rigid body rotates about a fixed point O. Its inertia matrix in terms of a body fixed coordinate system with its origin at O is $[I]=\begin{bmatrix}1 & -1 & 0\\ -1 & 5 & 1\\ 0 & 1 & 7\end{bmatrix}$ $slug\text{-}ft^2$. If the rigid body's angular velocity is $\vec{\omega}=6\vec{i}+6\vec{j}-4\vec{k}\ (rad/s)$, what is its angular momentum about O?

Solution: .

The angular momentum is $\begin{bmatrix}H_{Ox}\\ H_{Oy}\\ H_{Oz}\end{bmatrix}=\begin{bmatrix}1 & -1 & 0\\ -1 & 5 & 1\\ 0 & 1 & 7\end{bmatrix}\begin{bmatrix}6\\ 6\\ -4\end{bmatrix}=\begin{bmatrix}6-6+0\\ -6+30-4\\ 0+6-28\end{bmatrix}=\begin{bmatrix}0\\ 20\\ -22\end{bmatrix}$ $slug\text{-}ft^2/s$.

In terms of the unit vectors $\vec{i},\vec{j},\vec{k}$, $\boxed{\vec{H}=20\vec{j}-22\vec{k}\ slug\text{-}ft^2/s}$

================================<>================================

Problem 9.31 What is the moment of inertia of the rigid body in Problem 9.30 about the axis that passes through the origin and the point $(-1,5,2)$ ft?

Solution: The unit vector parallel to the line through (0,0,0) and $(-1,5,2)$ ft is

$\vec{e}=\dfrac{-1\vec{i}+5\vec{j}+2\vec{k}}{\sqrt{1^2+5^2+4^2}}=-0.1826\vec{i}+0.9129\vec{j}+0.36521\vec{k}$. From Problem 9.20, the inertia matrix of the rigid body is $[I]=\begin{bmatrix}1 & -1 & 0\\ -1 & 5 & 1\\ 0 & 1 & 7\end{bmatrix}=\begin{bmatrix}I_{xx} & -I_{xy} & -I_{xz}\\ -I_{xy} & I_{yy} & -I_{yz}\\ -I_{xz} & -I_{yz} & I_{zz}\end{bmatrix}$, From Eq (9.17), the new moment of inertia about the line through (0,0,0) and $(-1,5,2)$ ft is $I_O=1e_x^2+5e_y^2+7e_z^2+2(-1)(e_xe_y)+2(0)(e_xe_z)+2(1)(e_ye_z)$

$\boxed{I_O=6.133\ slug\text{-}ft^2}$

================================<>================================

Problem 9.32 The mass of the homogenous slender bar is 6 kg. Determine its moments and products of inertia in terms of the coordinate system shown.

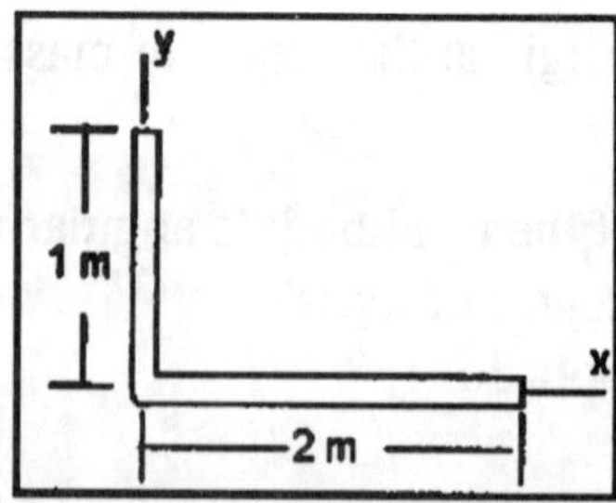

Solution: One strategy (not the simplest, see *Check* note following the solution) is to determine the moment of inertia matrix for each element of the bar, and then to use the parallel axis theorem to transfer each to the coordinate system shown.

(a) *The vertical element Oy of the bar.* The mass density per unit volume is $\rho=\dfrac{6}{3A}\ kg/m^3$, where A is the (unknown) cross section of the bar, from which $\rho A=2\ kg/m$. The element of mass is $dm=\rho AdL$, where dL is an element of length. The mass of the vertical element is $m_v=\rho AL_v=2\ kg$, where $L_v=1\ m$. From Appendix C the moment of inertia about an x' axis passing through the center of mass is $I_{x'x'}^{(1)}=\dfrac{m_vL_v^2}{12}=0.1667\ kg\text{-}m^2$.

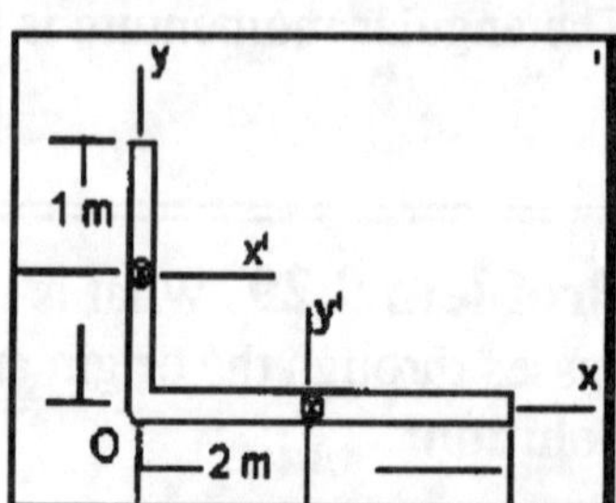

Solution continued on next page

Continuation of solution to Problem 9.32

Since the bar is slender, $I^{(1)}_{y'y'} = 0$. $I^{(1)}_{z'z'} = \dfrac{m_v L_v^2}{12} = 0.1667\ kg\text{-}m^2$. Since the bar is slender, the products of inertia vanish: $I^{(1)}_{x'y'} = \int_m x'y'dm = 0$, $I^{(1)}_{x'z'} = \int_m x'z'dm = 0$, $I^{(1)}_{y'z'} = \int_m y'z'dm = 0$, from which the inertia matrix for the element *Oy* about the x' axis is $\left[I^{(1)}\right] = \begin{bmatrix} 0.1667 & 0 & 0 \\ 0 & 0 & 0 \\ 0 & 0 & 0.1667 \end{bmatrix} kg\text{-}m^2$.

The horizontal element Ox of the bar: The mass of the horizontal element is $m_h = \rho A L_h = 4\ kg$, where $L_h = 2\ m$. From Appendix C the moments and products of inertia about the y' axis passing through the center of mass of the horizontal element are: $I^{(2)}_{x'x'} = 0$, $I^{(2)}_{y'y'} = \dfrac{m_h L_h^2}{12} = 1.333\ kg\text{-}m^2$, $I^{(2)}_{z'z'} = \dfrac{m_h L_h^2}{12} = 1.333\ kg\text{-}m^2$. Since the bar is slender, the cross products of inertia about the y' axis through the center of mass of the horizontal element of the bar vanish: $I^{(2)}_{x'y'} = 0$, $I^{(2)}_{x'z'} = 0$, $I^{(2)}_{y'z'} = 0$.

The inertia matrix is $\left[I^{(2)}\right] = \begin{bmatrix} 0 & 0 & 0 \\ 0 & 1.333 & 0 \\ 0 & 0 & 1.333 \end{bmatrix} kg\text{-}m^2$.

Use the parallel axis theorem to transfer the moment of inertia matrix to the origin O:

For the vertical element the coordinates of the center of mass O are $(d_x, d_y, d_z) = (0, 0.5, 0)\ m$. Use the parallel axis theorem (see Eq (9.16)).

$I^{(1)}_{xx} = I^{(1)}_{x'x'} + (d_y^2 + d_z^2)m_v = \dfrac{m_v L_v^2}{12} + m_v(0.5^2) = 0.6667\ kg\text{-}m^2$. $I^{(1)}_{yy} = I^{(1)}_{y'y'} + (d_x^2 + d_z^2)m_v = 0$.

$I^{(1)}_{zz} = I^{(1)}_{z'z'} + (d_x^2 + d_y^2)mv = 0.6667\ kg\text{-}m^2$. The products of inertia are $I^{(1)}_{xy} = I^{(1)}_{x'y'} + d_x d_y \rho A(1) = 0$, $I^{(1)}_{xz} = I^{(1)}_{x'z'} + d_x d_z \rho A(1) = 0$, $I^{(1)}_{yz} = I^{(1}_{y'z'} + d_y d_z \rho A(1) = 0$. The inertia matrix for the is transferred vertical element: $\left[I^{(1)}\right] = \begin{bmatrix} \dfrac{\rho A}{3} & 0 & 0 \\ 0 & 0 & 0 \\ 0 & 0 & \dfrac{\rho A}{3} \end{bmatrix}$.

For the horizontal element, the coordinates of the center of mass relative to O are $(d_x, d_y, d_z) = (1, 0, 0)\ m$. From the parallel axis theorem, $I^{(2)}_{xx} = I^{(2)}_{x'x'} + (d_y^2 + d_z^2)m_h = 0$.

$I^{(2)}_{yy} = I^{(2)}_{y'y'} + (d_x^2 + d_z^2)m_h = 5.333\ kg\text{-}m^2$. $I^{(2)}_{zz} = I^{(2)}_{z'z'} + (d_x^2 + d_y^2)m_h = 5.333\ kg\text{-}m^2$. By inspection, the products of inertia vanish. The inertia matrix is $\left[I^{(2)}\right] = \begin{bmatrix} 0 & 0 & 0 \\ 0 & 5.333 & 0 \\ 0 & 0 & 5.333 \end{bmatrix}$.

Solution continued on next page

Sum the two inertia matrices: $[I]_O = \left[I^{(1)}\right] + \left[I^{(2)}\right] = \begin{bmatrix} 0.6667 & 0 & 0 \\ 0 & 5.333 & 0 \\ 0 & 0 & 6 \end{bmatrix} kg\text{-}m^2$.

[*Check*: The moment of inertia in the coordinate system shown can be derived by inspection by taking the moment of inertia of each element about the origin: From Appendix C the moments of inertia about the origin of the slender bars are $I_{xx} = \dfrac{m_v L_v^3}{3}$, $I_{yy} = \dfrac{m_h L_h^2}{3}$, and $I_{zz} = I_{xx} + I_{yy}$, where the subscripts v and h denote the vertical and horizontal bars respectively. Noting that the masses are $m_v = \dfrac{mL_v}{L_v + L_h}$, $m_h = \dfrac{mL_h}{L_v + L_h}$, the moment of inertia matrix becomes:

$$[I] = \begin{bmatrix} \dfrac{6(1)^3}{3(3)} & 0 & 0 \\ 0 & 0 & \dfrac{6(2)^3}{3(3)} \\ 0 & 0 & 6.0 \end{bmatrix} = \begin{bmatrix} 0.6667 & 0 & 0 \\ 0 & 5.333 & 0 \\ 0 & 0 & 6 \end{bmatrix} kg\text{-}m^2. \textit{ check.}]$$

==================================<>==================================

Problem 9.33 Consider the bar in Problem 9.32. (a) Determine its moments and products of inertia in terms of a parallel coordinate system $x'\ y'\ z'$ with its origin at the bar's center of mass. (b) If the bar is rotating with angular velocity $\vec{\omega} = 4\vec{i}\ (rad/s)$, what is the angular momentum about its center of mass?

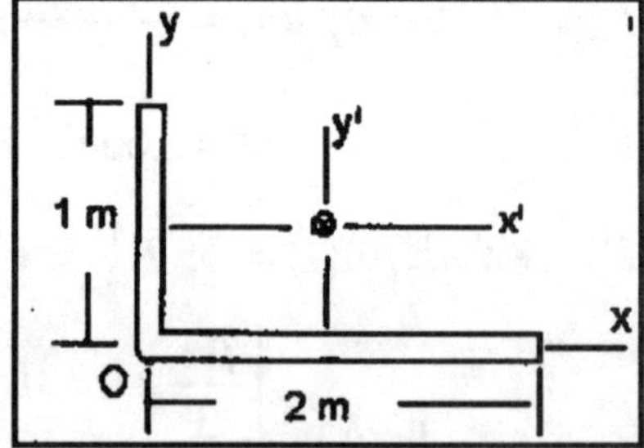

Solution: (a) From the solution to Problem 9.32, the mass of the vertical member is $m_v = 2\ kg$, with a center of mass located at $(0, 0.5, 0)$. The mass of the horizontal member is $m_h = 4\ kg$, with a center of mass located at $(1, 0, 0)$. The coordinates of the center of mass of the bar in the x, y, z system are

$x' = \dfrac{m_v(0) + m_h(1)}{m_v + m_h} = 0.6667\ m$, $y' = \dfrac{m_v(0.5) + m_h(0)}{m_v + m_h} = 0.1667\ m$, $z' = \dfrac{m_v(0) + m_h(0)}{m_v + m_h} = 0$, from which $(d_x, d_y, d_z) = (0.6667, 0.3333, 0)$. From the solution to Problem 9.32, the inertia matrix for the coordinates originating at O is $[I]_O = \begin{bmatrix} 0.6667 & 0 & 0 \\ 0 & 5.333 & 0 \\ 0 & 0 & 6 \end{bmatrix} kg\text{-}m^2$. Algebraically rearrange Eq (9.18) to obtain the moments and products of inertia about the parallel axis passing through the center of mass of the bar when the moments and products of inertia in the system originating at O are known:

Solution continued on next page

$$\boxed{I_{x'x'}^{(G)} = I_{xx}^{(0)} - \left(d_y^2 + d_x^2\right)m = 0.5 \ kg\text{-}m^2}.$$

$$\boxed{I_{y'y'}^{(G)} = I_{yy}^{(O)} - \left(d_x^2 + d_z^2\right)m = 2.667 \ kg\text{-}m^2}.$$

$$I_{z'z'}^{(G)} = I_{zz}^{(O)} - \left(d_x^2 + d_y^2\right)m = 3.167 \ kg\text{-}m^2$$

$$\boxed{I_{x'y'}^{(G)} = I_{xy}^{(0)} - d_x d_y m = -0.6667 \ kg\text{-}m^2},$$

$\boxed{I_{x'z'}^{(G)} = I_{xz}^{(O)} - d_x d_z m = 0}$, $\boxed{I_{y'z'}^{(G)} = I_{yz}^{(O)} - d_y d_z m = 0}$, from which the inertia matrix in the system originating at the center of mass of the bar is

$$[I]_G = \begin{bmatrix} I_{x'x'} & -I_{x'y'} & -I_{x'z'} \\ -I_{x'y'} & I_{y'y'} & -I_{y'z'} \\ -I_{x'z'} & -I_{y'z'} & I_{z'z'} \end{bmatrix} = \begin{bmatrix} 0.5 & 0.6667 & 0 \\ 0.6667 & 2.667 & 0 \\ 0 & 0 & 3.167 \end{bmatrix} kg\text{-}m^2.$$

(b) The angular momentum is $[H] = \begin{bmatrix} 0.5 & 0.6667 & 0 \\ 0.6667 & 2.667 & 0 \\ 0 & 0 & 3.167 \end{bmatrix}\begin{bmatrix} 4 \\ 0 \\ 0 \end{bmatrix} = \begin{bmatrix} 2.0 \\ 2.667 \\ 0 \end{bmatrix}$. In terms of the unit vectors $\vec{i}, \vec{j}, \vec{k}$, $\boxed{\vec{H} = 2.0\vec{i} + 2.667\vec{j} \ kg\text{-}m^2/s}$

==================================<>==================================

Problem 9.34 The 4 *kg* thin rectangular plate lies in the *x-y* plane. Determine its moments and products of inertia in terms of the coordinates system shown.

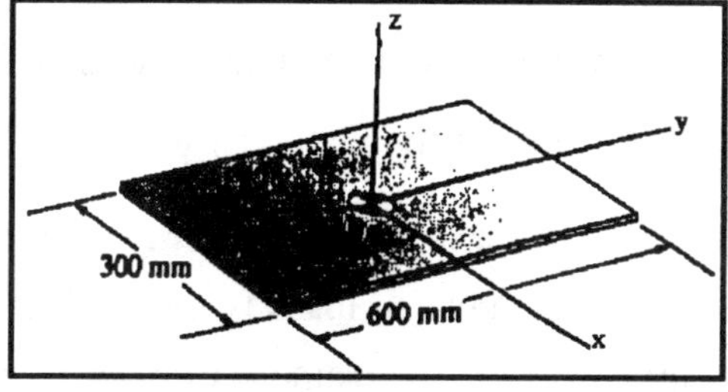

Solution : Denote $h = 0.6\ m$, $b = 0.3\ m$. The mass density per unit volume is $\rho = \frac{m}{AT} = \frac{4}{hbT} = \frac{22.22}{T}\ kg/m^3$, where T is the (unknown) thickness of the plate. The moments of inertia are:

$$I_{xx} = \int_m \left(y^2 + z^2\right)dm = \rho Tb\int_{-\frac{h}{2}}^{+\frac{h}{2}} y^2 dy = \frac{\rho Tb}{3}\left[y^3\right]_{-\frac{h}{2}}^{+\frac{h}{2}} = \frac{\rho Tbh^3}{12} = \frac{mh^2}{12}.$$

$$I_{yy} = \int_m \left(x^2 + z^2\right)dm = \rho Th\int_{-\frac{b}{2}}^{+\frac{b}{2}} x^2 dx = \frac{mb^2}{12}.$$

$$I_{zz} = \int_m (x^2 + y^2)dm = \rho Th\int_{-\frac{b}{2}}^{+\frac{b}{2}} x^2 dx + \rho Tb\int_{-\frac{h}{2}}^{+\frac{b}{2}} y^2 dy = \frac{m}{12}(h^2 + b^2).$$

$$I_{xy} = \int_m xy\ dm = \rho T\int_{-\frac{h}{2}}^{+\frac{h}{2}} y dy\int_{-\frac{b}{2}}^{+\frac{b}{2}} x dx = \rho T\left[\frac{y^2}{2}\right]_{-\frac{h}{2}}^{\frac{h}{2}}\left[\frac{x^2}{2}\right]_{-\frac{b}{2}}^{+\frac{b}{2}} = 0.\ I_{xz} = \int_m xz\ dm = 0,\ I_{yz} = \int_m yz\ dm = 0.$$

[*Check*: Compare these results with Appendix C. *check.*]

Solution continued on next page

. From which the inertia matrix is $[I]=\begin{bmatrix}\frac{mh^2}{12} & 0 & 0\\ 0 & \frac{mb^2}{12} & 0\\ 0 & 0 & \frac{m}{12}(h^2+b^2)\end{bmatrix}=\begin{bmatrix}0.12 & 0 & 0\\ 0 & 0.03 & 0\\ 0 & 0 & 0.15\end{bmatrix} kg\text{-}m^2$

===================================<>===================================

Problem 9.35 If the plate in Problem 9.34 is rotating with angular velocity $\vec{\omega}=6\vec{i}+4\vec{j}-2\vec{k}$ (rad/s), what is its angular momentum about its center of mass?

Solution: The angular momentum is $[H]=\begin{bmatrix}0.12 & 0 & 0\\ 0 & 0.03 & 0\\ 0 & 0 & 0.15\end{bmatrix}\begin{bmatrix}6\\4\\-2\end{bmatrix}=\begin{bmatrix}0.72\\0.12\\-0.3\end{bmatrix} kg\text{-}m^2/s$, from which

$\boxed{\vec{H}=0.72\vec{i}+0.12\vec{j}-0.3\vec{k}\ \ kg\text{-}m^2/s}$

===================================<>===================================

Problem 9.36 The 30 *lb* thin triangular plate lies in the *x-y* plane. Determine its moments and products of inertia in terms of the coordinate system shown.

Solution : Denote $h=4\ ft$, $b=6\ ft$. The mass density per unit volume is $\rho=\left(\frac{30}{32.17}\right)\frac{2}{bhT}=\frac{0.07771}{T}\ slug/ft^3$, where *T* is the (unknown) thickness of the thin plate. The equation of the line defining the upper boundary of the plate are $y=\frac{h}{b}x$. For an element of area dA, the element of mass is $dm=\rho T dx dy$. For a horizontal elemental strip *dy*, the element of mass is $dm=\rho T(b-x)dy$. For a vertical elemental strip *dx*, the element of mass is $dm=\rho T y dx$. The mass of the plate is $m=\frac{\rho T b h}{2}$. The moments and products of inertia (see Eq (9.11)) are: $I_{xx}=\int_m(y^2+z^2)dm=\rho T\int_0^h y^2(b-x)dy$,

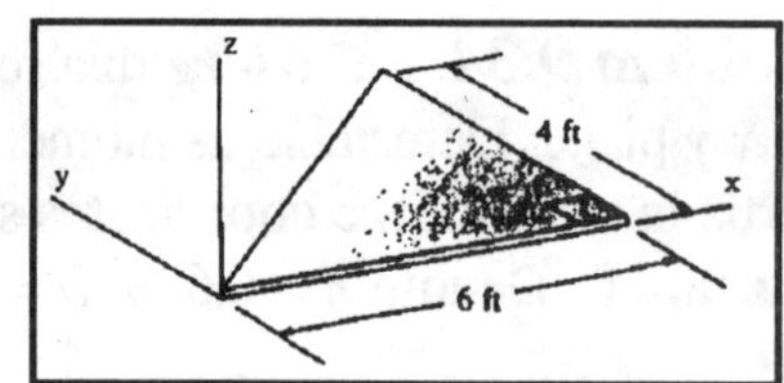

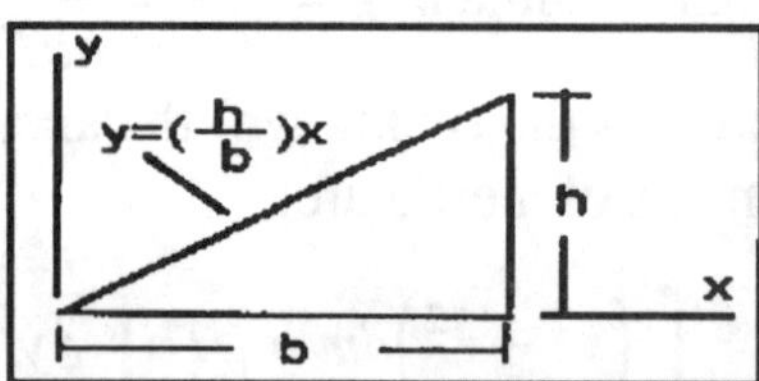

$I_{xx}=\rho T b\int_0^h y^2 dy-\rho T\left(\frac{b}{h}\right)\int_0^h y^3 dy=\frac{mh^2}{6}$, $\boxed{I_{xx}=2.487\ slug\text{-}ft^2}$.

$I_{yy}=\int_m(x^2+z^2)dm=\rho T\int_0^b x^2 y dx=\frac{\rho T h}{b}\int_0^b x^3 dx=\frac{\rho T h b^3}{4}=\frac{mb^2}{2}$, $\boxed{I_{yy}=16.79\ slug\text{-}ft^2}$.

$I_{zz}=\int_m(x^2+y^2)dm=I_{xx}+I_{yy}=\frac{m}{6}(h^2+3b^2)$, $\boxed{I_{zz}=19.28\ slug\text{-}ft^2}$.

$I_{xy}=\int_m xy dm=\rho T\int_0^b x dx\int_0^{\frac{h}{b}x} y dy=\frac{\rho T}{2}\int_0^b\left(\frac{h^2}{b^2}\right)x^3 dx=\frac{mhb}{4}=5.595\ slug\text{-}ft^2$ $\boxed{I_{xz}=\int_m xz\,dm=0}$,

$\boxed{I_{yz}=\int_m yz\,dm=0}$.

===================================<>===================================

=================================<>=================================

Problem 9.37 Consider the triangular plate in Problem 9.26. (a) Determine the moments and products of inertia in terms of a parallel coordinate system $x'\ y'\ z'$ with its origin at the plate's center of mass. (b) If the plate is rotating with angular velocity $\vec{\omega} = 20\vec{i} - 12\vec{j} + 16\vec{k}\ (rad/s)$, what is the angular momentum about its center of mass?

Solution : The coordinates of the center of mass of the triangle are $x' = \left(\frac{2}{3}\right)6 = 4\ ft$, $y' = \left(\frac{1}{3}\right)4 = 1.333\,ft$, $z' = 0$, from which $(d_x, d_y, d_z) = (4, 1.333, 0)$. The mass of the plate is $m = \frac{W}{g} = \frac{30}{32.17} = 0.9325\ slug$. From the solution to Problem 9.36, the inertia matrix of the triangular plate in the x, y, z coordinates is $[I] = \begin{bmatrix} I_{xx} & -I_{xy} & -I_{xz} \\ -I_{yx} & I_{yy} & -I_{yz} \\ -I_{zx} & -I_{zy} & I_{zz} \end{bmatrix} = \begin{bmatrix} 2.487 & -5.595 & 0 \\ -5.59 & 16.79 & 0 \\ 0 & 0 & 19.27 \end{bmatrix}\ slug\text{-}ft^2$.

Rearrange Eq (9.18) algebraically to yield the moments and products of inertia about the center of mass, when the moments and products in the system originating at O are known:

$$\boxed{\begin{aligned} I_{x'x'} &= I_{xx} - (d_y^2 + d_z^2)m = 0.8290\ slug\text{-}ft^2 \\ I_{y'y'} &= I_{yy} - (d_x^2 + d_z^2)m = 1.865\ slug\text{-}ft^2 \\ I_{z'z'} &= I_{zz} - (d_x^2 + d_y^2)m = 2.694\ slug\text{-}ft^2 \\ I_{x'y'} &= I_{xy} - d_x d_y m = 0.6218\ slug\text{-}ft^2 \\ I_{x'z'} &= I_{xz} - d_x d_z m = 0 \\ I_{y'z'} &= I_{yz} - d_y d_z m = 0 \end{aligned}}.$$

(b) The inertia matrix is $[I] = \begin{bmatrix} I_{xx} & -I_{xy} & -I_{xz} \\ -I_{yx} & I_{yy} & -I_{yz} \\ -I_{zx} & -I_{zy} & I_{zz} \end{bmatrix} = \begin{bmatrix} 0.8290 & -0.6218 & 0 \\ -0.6218 & 1.865 & 0 \\ 0 & 0 & 2.694 \end{bmatrix}\ slug\text{-}ft^2$.

The angular momentum is $[H] = \begin{bmatrix} 0.8290 & -0.6218 & 0 \\ -0.6218 & 1.865 & 0 \\ 0 & 0 & 2.694 \end{bmatrix}\begin{bmatrix} 20 \\ -12 \\ 16 \end{bmatrix} = \begin{bmatrix} 24.04 \\ -34.82 \\ 43.11 \end{bmatrix}\ slug\text{-}ft^2/s$.

In terms of unit vectors: $\boxed{\vec{H} = 24.04\vec{i} - 34.81\vec{j} + 43.11\vec{k}\ \ slug\text{-}ft^2/s}$.

=================================<>=================================

==================================<>==================================

Problem 9.38 Determine the inertial matrix of the 2.4-*kg* steel plate in terms of the coordinate system shown.

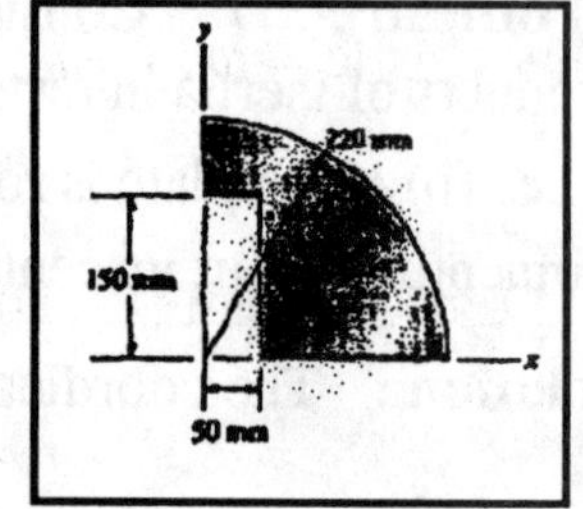

Solution:

Equation (9.15) gives the plate's moments and products of inertia in terms of the moments and product of inertia of its area. Treating the area as a quarter-circle using Appendix B, the moments and products of inertia of the area are

$$I_x = \frac{1}{16}\pi(0.22)^4 - \frac{1}{3}(0.05)(0.15)^3 = 0.000404 m^4$$

$$I_y = \frac{1}{16}\pi(0.22)^4 - \frac{1}{3}(0.15)(0.05)^3 = 0.000454 m^4$$

$$I_{xy}^A = \frac{1}{8}(0.22)^4 - \frac{1}{4}(0.05)^2(0.15)^2 = 0.000279 m^4.$$

The area is $A = \frac{1}{4}\pi(0.22)^2 - (0.05)(0.15) = 0.0305 m^2$.

The moments of inertia of the plate are

$$I_{xx} = \frac{m}{A} I_x = 0.0318 kg - m^2,$$

$$I_{yy} = \frac{m}{A} I_y = 0.0357 kg - m^2 ,$$

$$I_{zz} = I_{xx} + I_{yy} = 0.0674 kg - m^2 ,$$

$$I_{xy} = \frac{m}{A} I_{xy}^A = 0.0219 kg - m^2 \text{ , and } I_{yz} = I_{zx} = 0$$

==================================<>==================================

Problem 9.39 Consider the steel plate in Problem 9.38. (a) Determine its moments and products of inertia in terms of a parallel coordinate system x'y'z' with its origin at the plate's center of mass.

Solution:

(a) The x and y coordinates of the center of mass coincide with the centroid of the area:

$$A_1 = \frac{1}{4}\pi(0.22)^2 = 0.0380 m^2, \qquad A_2 = (0.05)(0.15) = 0.0075 m^2$$

$$\bar{x} = \frac{\frac{4(0.22)}{3\pi} A_1 - (0.025) A_2}{A_1 - A_2} = 0.1102 m, \quad \bar{y} = \frac{\frac{4(0.22)}{3\pi} A_1 - (0.075) A_2}{A_1 - A_2} = 0.0979 m.$$

Using the results of the solution of Problem 9.38 and the parallel axis theorems,

$$I_{x'x'} = I_{xx} - m\bar{y}^2 = 0.00876 kg - m^2;$$

$$I_{y'y'} = I_{yy} - m\bar{x}^2 = 0.00655 kg - m^2;$$

$$I_{z'z'} = I_{x'x'} + I_{y'y'} = 0.01531 kg - m^2$$

$$I_{x'y'} = I_{xy} - m\bar{x}\bar{y} = -0.00396 kg - m^2 \text{ , and } I_{y'z'} = I_{z'x'} = 0.$$

The angular momentum is $\begin{bmatrix} H_{x'} \\ H_{y'} \\ H_{z'} \end{bmatrix} = \begin{bmatrix} I_{x'x'} & -I_{x'y'} & 0 \\ -I_{x'y'} & I_{y'y'} & 0 \\ 0 & 0 & I_{z'z'} \end{bmatrix} \begin{bmatrix} 20 \\ 10 \\ -10 \end{bmatrix} = \begin{bmatrix} 0.215 \\ 0.145 \\ -0.153 \end{bmatrix} (kg - m^2 / s)$

==================================<>==================================

Problem 9.40 The slender bar of mass m is parallel to the x axis. If the angular velocity about the point O is $\vec{\omega} = \omega_y \vec{j} + \omega_z \vec{k}$, what is the bar's angular momentum (a) about its center of mass? (b) about O?

Solution:

(a) From Appendix C and by inspection, the moments and products of inertia about the center of mass of the bar are: $I_{xx} = 0,$, $I_{yy} = I_{zz} = \dfrac{mL^2}{12}$, $I_{xy} = I_{xz} = I_{yz} = 0$.

The angular momentum about its center of mass is

$$[H]_G = \begin{bmatrix} I_{xx} & -I_{xy} & -I_{xz} \\ -I_{yx} & I_{yy} & -I_{yz} \\ -I_{zx} & -I_{zy} & I_{zz} \end{bmatrix} \begin{bmatrix} 0 \\ \omega_y \\ \omega_z \end{bmatrix} = \begin{bmatrix} 0 & 0 & 0 \\ 0 & \dfrac{mL^2}{12} & 0 \\ 0 & 0 & \dfrac{mL^2}{12} \end{bmatrix} \begin{bmatrix} 0 \\ \omega_y \\ \omega_z \end{bmatrix}$$

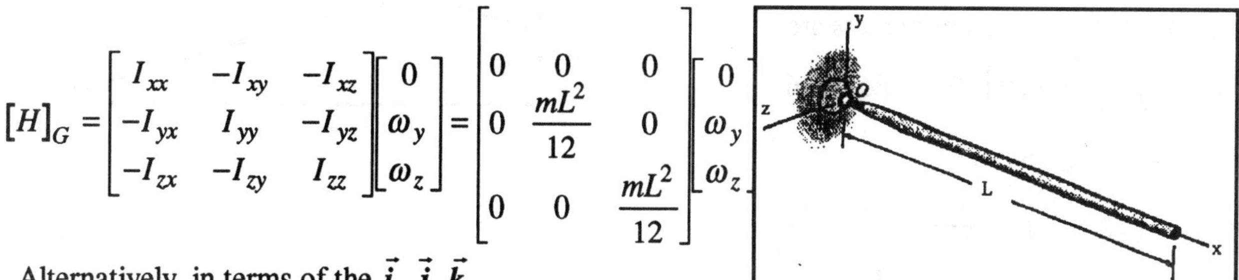

. Alternatively, in terms of the $\vec{i}, \vec{j}, \vec{k}$,

$$\boxed{\vec{H}_G = \frac{mL^2}{12}\left(\omega_y \vec{j} + \omega_z \vec{k}\right)}.$$

(b) From Appendix C and by inspection, the moments and products of inertia about O are $I_{xx} = 0$, $I_{yy} = I_{zz} = \dfrac{mL^2}{3}$, $I_{xy} = I_{xz} = I_{yz} = 0$

The angular momentum about O is

$$[H]_O = \begin{bmatrix} I_{xx} & -I_{xy} & -I_{xz} \\ -I_{yx} & I_{yy} & -I_{yz} \\ -I_{zx} & -I_{zy} & I_{zz} \end{bmatrix} \begin{bmatrix} 0 \\ \omega_y \\ \omega_z \end{bmatrix} = \begin{bmatrix} 0 & 0 & 0 \\ 0 & \dfrac{mL^2}{3} & 0 \\ 0 & 0 & \dfrac{mL^2}{3} \end{bmatrix} \begin{bmatrix} 0 \\ \omega_y \\ \omega_z \end{bmatrix} = \begin{bmatrix} 0 \\ \dfrac{mL^2\omega_y}{3} \\ \dfrac{mL^2\omega_z}{3} \end{bmatrix}$$

, or, alternatively, in terms of the unit vectors $\vec{i}, \vec{j}, \vec{k}$, $\boxed{\vec{H}_O = \dfrac{mL^2}{3}\left(\omega_y \vec{j} + \omega_z \vec{k}\right)}$

Problem 9.41 The slender bar of mass *m* is parallel to the *x* axis. If the coordinate system is body fixed and its angular velocity about the fixed point O is $\vec{\omega} = \omega_y \vec{j}$, what is the angular momentum about O?

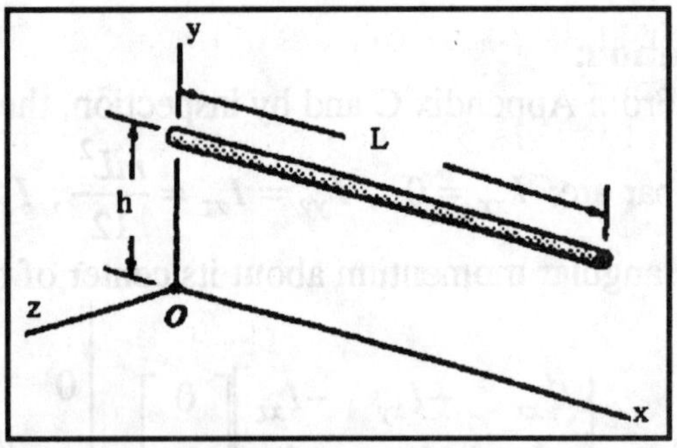

Solution:

From Appendix C and by inspection, the moments and products of inertia about the center of mass of the bar are:

$I_{x'x'} = 0,\ I_{y'y'} = I_{z'z'} = \frac{mL^2}{12},\ I_{x'y'} = I_{x'z'} = I_{y'z'} = 0.$

The coordinates of the center of mass are $(d_x, d_y, d_z) = \left(\frac{L}{2}, h, 0\right)$. From Eq (9.16),

$I_{xx} = I_{x'x'} + (d_y^2 + d_z^2)m = mh^2$

$I_{yy} = I_{y'y'} + (d_x^2 + d_z^2)m = \frac{mL^2}{12} + \frac{mL^2}{4} = \frac{mL^2}{3}$

$I_{zz} = I_{z'z'} + (d_x^2 + d_y^2)m = m\left(h^2 + \frac{L^2}{3}\right),\ I_{xy} = I_{x'y'} + d_x d_y m = 0 + \frac{mLh}{2},\ I_{xz} = I_{x'z'} + d_x d_z m = 0$

$I_{yz} = I_{y'z'} + d_y d_z m = 0$

The angular momentum about O is

$$[H]_O = \begin{bmatrix} I_{xx} & -I_{xy} & -I_{xz} \\ -I_{yx} & I_{yy} & -I_{yz} \\ -I_{zx} & -I_{zy} & I_{zz} \end{bmatrix} \begin{bmatrix} 0 \\ \omega_y \\ 0 \end{bmatrix} = \begin{bmatrix} mh^2 & -\frac{mLh}{2} & 0 \\ -\frac{mLh}{2} & \frac{mL^2}{3} & 0 \\ 0 & 0 & m\left(h^2 + \frac{L^2}{3}\right) \end{bmatrix} \begin{bmatrix} 0 \\ \omega_y \\ 0 \end{bmatrix} = \begin{bmatrix} -\frac{mLh\omega_y}{2} \\ \frac{mL^2\omega_y}{3} \\ 0 \end{bmatrix}$$. Alternatively,

$$\boxed{\vec{H}_O = -\frac{mLh}{2}\omega_y \vec{i} + \frac{mL^2}{3}\omega_y \vec{j}}$$

==================================<>==================================

Problem 9.42 In Example 9.5 the moments and products of inertia of the object consisting of the booms AB and BC were determined in terms of the coordinate system shown in Fig 9.21. Determine the moments and products of inertia of the object in terms of a parallel coordinate system x', y', z' with its origin at the center of mass of the object.

Solution :

From Example 9.5, the inertia matrix for the two booms in the x, y, z system is

$$[I]_O = \begin{bmatrix} I_{xx} & -I_{xy} & -I_{xz} \\ -I_{yx} & I_{yy} & -I_{yz} \\ -I_{zx} & -I_{zy} & I_{zz} \end{bmatrix}$$

$$= \begin{bmatrix} 19{,}200 & 86{,}400 & 0 \\ 86{,}400 & 1{,}036{,}800 & 0 \\ 0 & 0 & 1{,}056{,}000 \end{bmatrix} kg\text{-}m^2.$$

The mass of

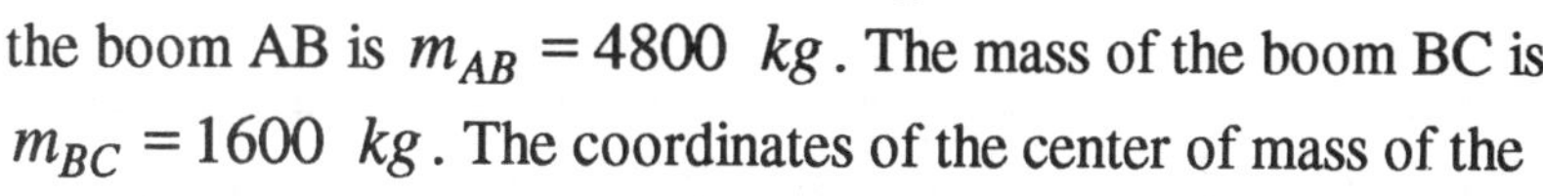

the boom AB is $m_{AB} = 4800 \ kg$. The mass of the boom BC is $m_{BC} = 1600 \ kg$. The coordinates of the center of mass of the two booms are

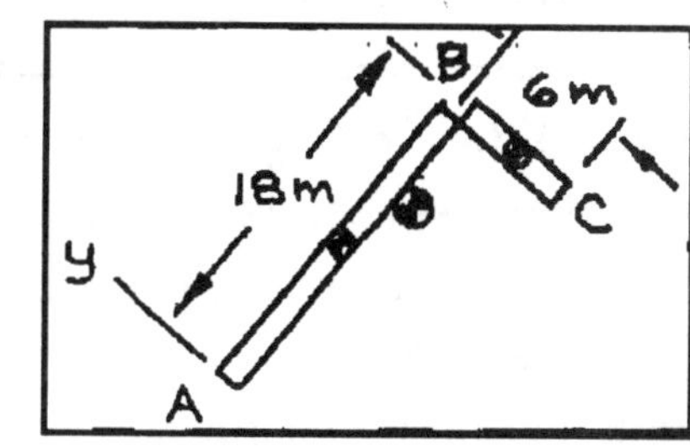

$$x' = \frac{m_{AB}\frac{L}{2} + m_{BC}L}{m_{AB} + m_{BC}} = 11.25 \ m.$$

$$y' = \frac{m_{AB}(0) + m_{BC}(-3)}{m_{AB} + m_{BC}} = -0.75 \ m.$$

$z' = 0$, from which $(d_x, d_y, d_z) = (11.25, -0.75, 0) \ m$. Algebraically rearrange Eq (9.18) to obtain the moments and products of inertia about the parallel axis passing through the center of mass of the two booms when the moments and products of inertia in the x, y, z system are known:

$$\boxed{I_{x'x'}^{(G)} = I_{xx}^{(0)} - (d_y^2 + d_z^2)m = 15600 \ kg\text{-}m^2}.$$

$$\boxed{I_{y'y'}^{(G)} = I_{yy}^{(O)} - (d_x^2 + d_z^2)m = 226800 \ kg\text{-}m^2}.$$

$$\boxed{I_{z'z'}^{(G)} = I_{zz}^{(O)} - (d_x^2 + d_y^2)m = 242400 \ kg\text{-}m^2}$$

$$\boxed{I_{x'y'}^{(G)} = I_{xy}^{(0)} - d_x d_y m = -32400 \ kg\text{-}m^2}, \boxed{I_{x'z'}^{(G)} = 0}, \boxed{I_{y'z'}^{(G)} = 0}.$$

The inertia matrix for the x', y', z' system is

$$[I]_G = \begin{bmatrix} 15{,}600 & 32{,}400 & 0 \\ 32{,}400 & 226{,}800 & 0 \\ 0 & 0 & 242{,}400 \end{bmatrix} kg\text{-}m^2.$$

==================================<>==================================

===================================<>===================================

Problem 9.43 Suppose that the crane described in Example 9.5 undergoes a rigid body rotation about the vertical axis at 0.1 *rad/s* in the counterclockwise direction when viewed from above. (a) What is its angular velocity vector? (b) What is the angular momentum of the object consisting of the boom AB and BC *about its center of mass?*

Solution:

The unit vector parallel to vertical axis in the x', y', z' system is

$\vec{e} = \vec{i}\sin 50^o + \vec{j}\cos 50^o = 0.7660\vec{i} + 0.6428\vec{j}$. The angular velocity vector is

$\boxed{\vec{\omega} = (0.1)\vec{e} = 0.07660\vec{i} + 0.06428\vec{j}}$ From the inertia matrix given in the solution of Problem 9.42, the angular moment about the center of mass is

$$[H]_G = \begin{bmatrix} 15{,}600 & 32{,}400 & 0 \\ 32{,}400 & 226{,}800 & 0 \\ 0 & 0 & 242{,}400 \end{bmatrix}\begin{bmatrix} 0.07660 \\ 0.06428 \\ 0 \end{bmatrix} = \begin{bmatrix} 3277.7 \\ 17060.4 \\ 0 \end{bmatrix} kg\text{-}m^2/s \text{, or,}$$

$\boxed{\vec{H}_G = 3277.7\vec{i} + 17060.4\vec{j}}$

===================================<>===================================

Problem 9.44 A 3 *kg* slender bar is rigidly attached to a 2 *kg* thin circular disk. In terms of the body fixed coordinate system shown, the angular velocity of the composite object is $\vec{\omega} = 100\vec{i} - 4\vec{j} + 6\vec{k}$ (rad/s). What is the object's angular momentum about its center of mass?

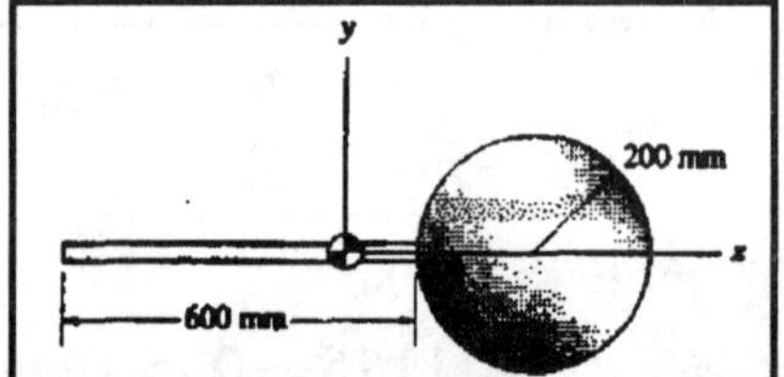

Solution: Choose a coordinate system *x, y, z* originating at the left end of the bar. From Appendix C and by inspection, the inertia matrix for the bar about its left end is

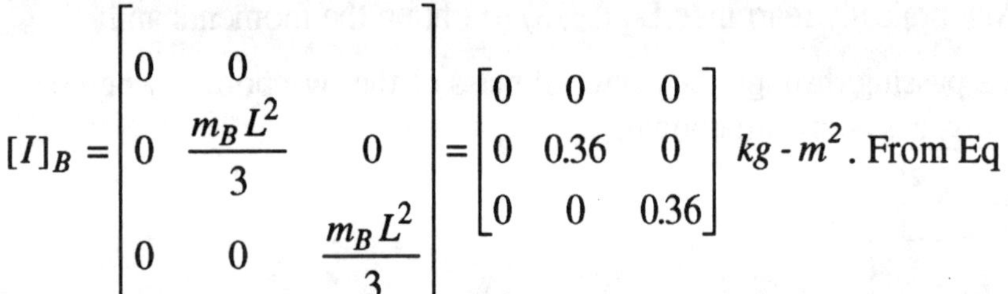

$$[I]_B = \begin{bmatrix} 0 & 0 & 0 \\ 0 & \frac{m_B L^2}{3} & 0 \\ 0 & 0 & \frac{m_B L^2}{3} \end{bmatrix} = \begin{bmatrix} 0 & 0 & 0 \\ 0 & 0.36 & 0 \\ 0 & 0 & 0.36 \end{bmatrix} kg\text{-}m^2.$$

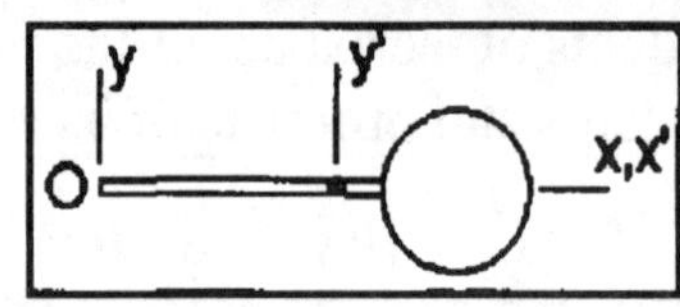

From Eq (9.16) the inertia matrix of the disk about the left end of the bar is

$$[I]_D = \begin{bmatrix} \frac{m_D R^2}{4} & 0 & 0 \\ 0 & \frac{m_D R^2}{4} & 0 \\ 0 & 0 & \frac{m_D R^2}{2} \end{bmatrix} + \begin{bmatrix} 0 & 0 & 0 \\ 0 & m_D(L+R)^2 & 0 \\ 0 & 0 & m_D(L+R)^2 \end{bmatrix} = \begin{bmatrix} 0.02 & 0 & 0 \\ 0 & 1.3 & 0 \\ 0 & 0 & 1.32 \end{bmatrix} kg\text{-}m^2.$$

The inertia matrix of the composite is $[I]_{left\_end} = [I]_B + [I]_D = \begin{bmatrix} 0.02 & 0 & 0 \\ 0 & 1.66 & 0 \\ 0 & 0 & 1.68 \end{bmatrix}$. The coordinates of the center of mass of the composite in the *x, y, z* system are $x' = \frac{m_B(L/2) + m_D(R+L)}{m_B + m_D} = 0.5$ m. $y' = 0$, $z' = 0$, from which $(d_x, d_y, d_z) = (0.5, 0, 0)$ m.

Solution continued on next page

Rearrange Eq (9.16) to yield the inertia matrix in the x', y', z' system when the inertia matrix in the x, y, z system is known:

$$[I]_G = \begin{bmatrix} 0.02 & 0 & 0 \\ 0 & 1.66 & 0 \\ 0 & 0 & 1.68 \end{bmatrix} - \begin{bmatrix} 0 & 0 & 0 \\ 0 & d_x^2(m_B+m_D) & 0 \\ 0 & 0 & d_x^2(m_B+m_D) \end{bmatrix} = \begin{bmatrix} 0.02 & 0 & 0 \\ 0 & 0.41 & 0 \\ 0 & 0 & 0.43 \end{bmatrix} kg\text{-}m^2.$$

The angular momentum is

$$[H]_G = \begin{bmatrix} 0.02 & 0 & 0 \\ 0 & 0.41 & 0 \\ 0 & 0 & 0.43 \end{bmatrix}\begin{bmatrix} 100 \\ -4 \\ 6 \end{bmatrix} = \begin{bmatrix} 2 \\ -1.64 \\ 2.58 \end{bmatrix}, \boxed{\vec{H} = 2\vec{i} - 1.64\vec{j} + 2.58\vec{k}\ kg\text{-}m^2/s}$$

==================================<>==================================

Problem 9.45 The mass of the homogenous slender bar is m. If the bar rotates with angular velocity $\vec{\omega} = \omega_O(24\vec{i} + 12\vec{j} - 6\vec{k})$, what is its angular momentum about its center of mass?

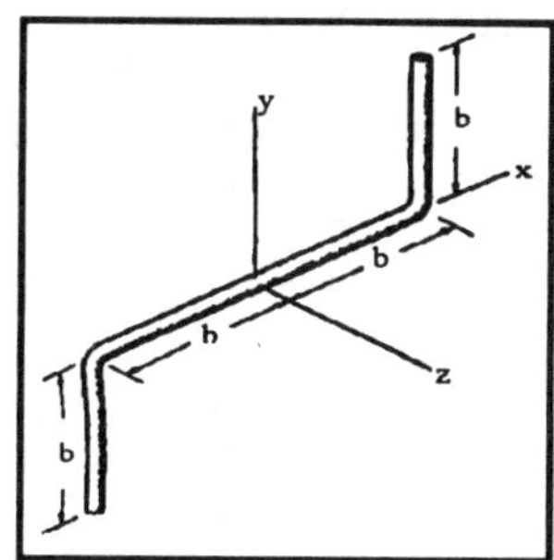

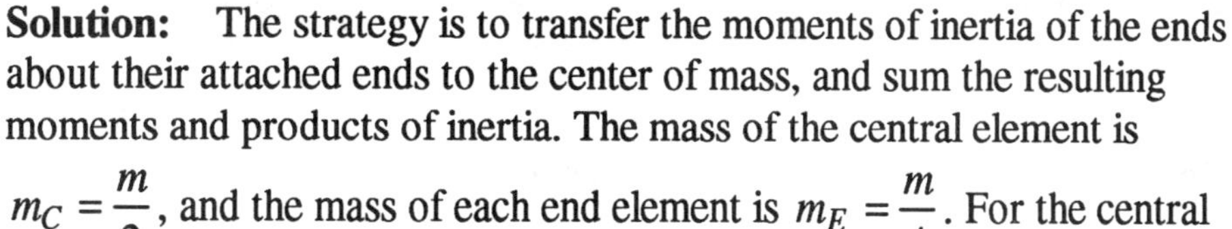

Solution: The strategy is to transfer the moments of inertia of the ends about their attached ends to the center of mass, and sum the resulting moments and products of inertia. The mass of the central element is $m_C = \frac{m}{2}$, and the mass of each end element is $m_E = \frac{m}{4}$. For the central element about its center of mass: $I_{xx} = 0$, $I_{yy} = I_{zz} = \frac{m_C(2b)^2}{12} = \frac{mb^2}{6}$, $I_{xy} = I_{xz} = I_{yz} = 0$. For each end element about its center of mass: $I_{xx} = \frac{m_E b^2}{12} = \frac{mb^2}{48}$, $I_{yy} = 0$, $I_{zz} = \frac{m_E b^2}{12} = \frac{mb^2}{48}$, $I_{xy} = I_{xz} = I_{yz} = 0$. The coordinates of the center of mass (at the origin) relative to the center of mass of the left end is $(d_x, d_y, d_z) = \left(-b, -\frac{b}{2}, 0\right)$, and from the right end $(d_x, d_y, d_z) = \left(b, \frac{b}{2}, 0\right)$. From Eq (9.18), the moments of inertia of each the end pieces about the center of mass (at the origin) are

$$I_{x'x'}^{(G)} = I_{xx}^{(L)} + \left(d_y^2 + d_z^2\right)m_E = \frac{mb^2}{48} + \frac{m_E b^2}{4} = \frac{mb^2}{12}. \quad I_{y'y'}^{(G)} = I_{yy}^{(L)} + \left(d_x^2 + d_z^2\right)m_E = \frac{mb^2}{4}.$$

$$I_{z'z'}^{(G)} = I_{zz}^{(L)} + \left(d_x^2 + d_y^2\right)m_E = \frac{mb^2}{48} + m_E\left(b^2 + \frac{b^2}{4}\right) = \frac{mb^2}{3}, \quad I_{x'y'}^{(G)} = I_{xy}^{(L)} + d_x d_y m_E = mb^2/8, \quad I_{x'z'}^{(G)} = 0,$$

$I_{y'z'}^{(G)} = 0$. The sum of the matrices:

$$[I]_G = [I]_{GC} + 2[I]_{end} = mb^2\begin{bmatrix} 0.1667 & -0.25 & 0 \\ -0.25 & 0.6667 & 0 \\ 0 & 0 & 0.8333 \end{bmatrix}.$$

The angular momentum about the center of mass is

$$[H]_G = \omega_o mb^2\begin{bmatrix} 0.1667 & -0.25 & 0 \\ -0.25 & 0.6667 & 0 \\ 0 & 0 & 0.8333 \end{bmatrix}\begin{bmatrix} 24 \\ 12 \\ -6 \end{bmatrix} = \omega_o mb^2\begin{bmatrix} 1 \\ 2 \\ -5 \end{bmatrix}. \boxed{\vec{H} = \omega_o mb^2\left(\vec{i} + 2\vec{j} - 5\vec{k}\right)}$$

==================================<>==================================

==================================<>==================================

Problem 9.46 The 8 kg homogenous slender bar has ball and socket supports at A and B. (a) What is the bar's moment of inertia about the axis AB? (b) If the bar rotates about the axis AB at 4 *rad/s,* what is the magnitude of its angular moment about its axis of rotation?

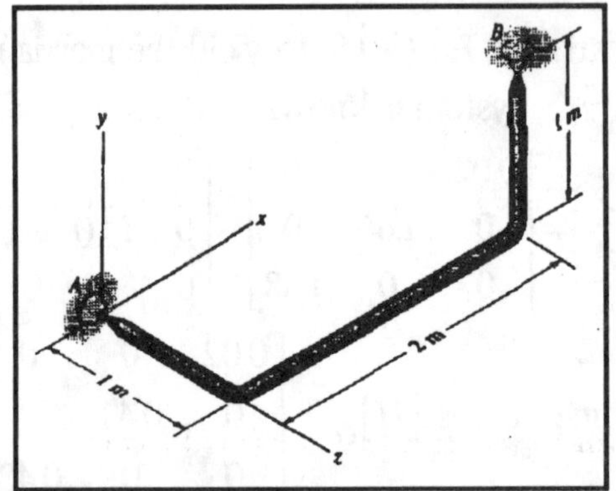

Solution:

Divide the bar into three elements: the central element, and the two end element. The strategy is to find the moments and products of inertia in the *x, y, z* system shown, and then to use Eq (9.17) to find the moment of inertia about the axis AB. Denote the total mass of the bar by $m = 8\ kg$, the mass of each end element by $m_E = \frac{m}{4} = 2\ kg$, and the mass of the central element by $m_C = \frac{m}{2} = 4\ kg$.

The left end element: The moments and products of inertia about point A are:

$$I_{xx}^{(LA)} = \frac{m_E(1^2)}{3} = \frac{m}{12} = 0.6667\ kg\text{-}m^2,\ I_{yy}^{(LA)} = \frac{m_E(1^2)}{3} = \frac{m}{12} = 0.6667\ kg\text{-}m^2,\ I_{zz}^{(LA)} = 0,$$

$$I_{xy}^{(LA)} = I_{xz}^{(LA)} = I_{yz}^{(LA)} = 0.$$

The right end element: The moments and products of inertia about its center of mass are

$$I_{xx}^{(RG)} = \frac{m_E(1^2)}{12} = \frac{m}{48} = 0.1667\ kg\text{-}m^2,\ I_{yy}^{(RG)} = 0,\ I_{zz}^{(RG)} = \frac{m_E(1^2)}{12} = \frac{m}{48} = 0.1667\ kg\text{-}m^2,$$

$I_{xy}^{(RG)} = I_{xz}^{(RG)} = I_{yz}^{(RG)} = 0$. The coordinates of the center of mass of the right end element are $(d_x, d_y, d_z) = (2, 0.5, 1)$. From Eq (9.18), the moments and products of inertia in the *x, y, z* system are

$$I_{xx}^{(RA)} = I_{xx}^{(RG)} + (d_y^2 + d_z^2)\frac{m}{4} = 2.667\ kg\text{-}m^2,\ I_{yy}^{(RA)} = I_{yy}^{(RG)} + (d_x^2 + d_z^2)\frac{m}{4} = 10.0\ kg\text{-}m^2$$

$$I_{zz}^{(RA)} = I_{zz}^{(RG)} + (d_x^2 + d_y^2)\frac{m}{4} = 8.667\ kg - m^2,\ I_{xy}^{(RA)} = I_{xy}^{(RG)} + d_x d_y \frac{m}{4} = 2.0\ kg\text{-}m^2$$

$$I_{xz}^{(RA)} = I_{xz}^{(RG)} + d_x d_z \frac{m}{4} = 4.0\ kg\text{-}m^2,\ I_{yz}^{(RA)} = I_{yz}^{(RG)} + d_y d_z \frac{m}{4} = 1.0\ kg\text{-}m^2$$

Sum the two inertia matrices: $[I]_{RLA} = [I]_{RA} + [I]_{LA} = \begin{bmatrix} 3.333 & -2 & -4 \\ -2 & 10.67 & -1 \\ -4 & -1 & 8.667 \end{bmatrix} kg\text{-}m^2$, where the negative signs are a consequence of the definition of the inertia matrix.

The central element: The moments and products of inertia of the central element about its center of mass are: $I_{xx}^{(CG)} = 0$, $I_{yy}^{(CG)} = \frac{m_C(2^2)}{12} = \frac{m}{6} = 1.333\ kg\text{-}m^2$, $I_{zz}^{(CG)} = \frac{m_C(2^2)}{12} = \frac{m}{6} = 1.333\ kg\text{-}m^2$.

$I_{xy}^{(CG)} = I_{xz}^{(CG)} = I_{yz}^{(CG)} = 0$. The coordinates of the center of mass of the central element are $(d_x, d_y, d_z) = (1, 0, 1)$.

Solution continued on next page

From Eq (9.18) the moments and products of inertia in the *x, y, z* system are:

$$I_{xx}^{(CA)} = I_{xx}^{(CG)} + \left(d_y^2 + d_z^2\right)\frac{m}{2} = 4 \ kg\text{-}m^2, \ I_{yy}^{(CA)} = I_{yy}^{(CG)} + \left(d_x^2 + d_z^2\right)\frac{m}{2} = 9.333 \ kg\text{-}m^2$$

$$I_{zz}^{(CA)} = I_{zz}^{(CG)} + (d_x^2 + d_y^2)\frac{m}{2} = 5.333 \ kg\text{-}m^2, \ I_{xy}^{(CA)} = I_{xy}^{(CG)} + d_x d_y \frac{m}{2} = 0,$$

$$I_{xz}^{(CA)} = I_{xz}^{(CG)} + d_x d_z \frac{m}{2} = 4 \ kg\text{-}m^2, \ I_{yz}^{(CA)} = I_{yz}^{(CG)} + d_y d_z \frac{m}{2} = 0$$

Sum the inertia matrices: $[I]_A = [I]_{RLA} + [I]_{CA} = \begin{bmatrix} 7.333 & -2 & -8 \\ -2 & 20.00 & -1 \\ -8 & -1 & 14..00 \end{bmatrix} kg\text{-}m^2$.

(a) The moment of inertia about the axis AB: The distance AB is parallel to the vector $\vec{r}_{AB} = 2\vec{i} + 1\vec{j} + 1\vec{k} \ (m)$. The unit vector parallel to $\vec{r}_{AB}$ is $\vec{e}_{AB} = \frac{\vec{r}_{AB}}{|\vec{r}_{AB}|} = 0.8165\vec{i} + 0.4082\vec{j} + 4082\vec{k}$.

From Eq (9.17), the moment of inertia about the AB axis is

$I_{AB} = I_{xx}^{(A)} e_x^2 + I_{yy}^{(A)} e_y^2 + I_{zz}^{(A)} e_z^2 - 2I_{xy}^{(A)} e_x e_y - 2I_{xz}^{(A)} e_x e_z - 2I_{yz}^{(A)} e_y e_z$, $\boxed{I_{AB} = 3.556 \ kg\text{-}m^2}$. (b) The angular momentum about the AB axis is $\boxed{H_{AB} = I_{AB}\omega = 3.556(4) = 14.22 \ kg\text{-}m^2/s}$

==================================<>==================================

Problem 9.47 The 8 *kg* homogenous slender bar in Problem 9.46 is released from rest in the position shown. (The *x-z* plane is horizontal.) At that instant, what is the bar's angular acceleration about the axis AB?

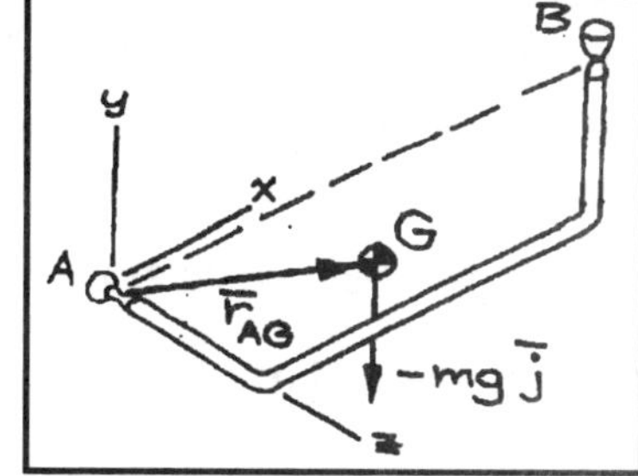

Solution: The center of mass of the bar has the coordinates:

$$x_G = \frac{1}{m}\left(\left(\frac{m}{4}\right)0 + \left(\frac{m}{2}\right)(1) + \left(\frac{m}{4}\right)(2)\right) = 1 \ m.$$

$$y_G = \frac{1}{m}\left(\left(\frac{m}{4}\right)(0) + \left(\frac{m}{2}\right)(0) + \left(\frac{m}{4}\right)(0.5)\right) = 0.125 \ m.$$

$$z_G = \frac{1}{m}\left(\left(\frac{m}{4}\right)(0.5) + \left(\frac{m}{2}\right)(1) + \left(\frac{m}{4}\right)(1)\right) = 0.875 \ m.$$

The line from A to the center of mass is parallel to the vector $\vec{r}_{AG} = \vec{i} + 0.125\vec{j} + 0.875\vec{k} \ (m)$. From the solution to Problem 9.46 the unit vector parallel to the line AB is $\vec{e}_{AB} = 0.8165\vec{i} + 0.4082\vec{j} + 0.4082\vec{k}$. The magnitude of the moment about line AB due to the weight is

$$\vec{e}_{AB} \cdot [\vec{r}_{AB} \times (-mg\vec{j})] = \begin{vmatrix} 0.8165 & 0.4082 & 0.4082 \\ 1.000 & 0.125 & 0.875 \\ 0 & -78.48 & 0 \end{vmatrix} = 24.03 \ N\text{-}m.$$

From the solution to Problem 9.46, $I_{AB} = 3.556 \ kg\text{-}m^2$. From the equation of angular motion about axis AB, $M_{AB} = I_{AB}\alpha$, from which

$$\boxed{a = \frac{24.03}{3.556} = 6.75 \ rad/s^2}.$$

==================================<>==================================

==================================<>==================================

Problem 9.48 In terms of a coordinate system x', y',z' with its origin at the center of mass, the inertia of a rigid is $[I'] = \begin{bmatrix} 20 & 10 & -10 \\ 10 & 60 & 0 \\ -10 & 0 & 80 \end{bmatrix}$ $kg\text{-}m^2$. Determine the principal moments of inertia and unit vectors parallel to the corresponding principal axes.

Solution:

Principal Moments of Inertia: The moments and products: $I_{x'x'} = 20\ kg\text{-}m^2$, $I_{y'y'} = 60\ kg\text{-}m^2$, $I_{z'z'} = 80\ kg\text{-}m^2$, $I_{x'y'} = -10\ kg\text{-}m^2$, $I_{x'z'} = 10\ kg\text{-}m^2$, $I_{y'z'} = 0$. From Eq (9.21), the principal values are the roots of the cubic, $AI^3 + BI^2 + CI + D = 0$, where $A = +1$, $B = -(I_{x'x'} + I_{y'y'} + I_{z'z'}) = -160$, $C = (I_{x'x'}I_{y'y'} + I_{y'y'}I_{z'z'} + I_{z'z'}I_{x'x'} - I_{x'y'}^2 - I_{x'z'}^2 - I_{y'z'}^2) = 7400$, $D = -(I_{x'x'}I_{y'y'}I_{z'z'} - I_{x'x'}I_{y'z'}^2 - I_{y'y'}I_{x'z'}^2 - I_{z'z'}I_{x'y'}^2 - 2I_{x'y'}I_{y'z'}I_{x'z'}) = -82000$. The function $f(I) = AI^3 + BI^2 + CI + D$ is graphed to find the zero crossings, and these values are refined by iteration. The graph is shown. The principal moments of inertia are: $\boxed{I_1 = 16.153\ kg\text{-}m^2}$, $\boxed{I_2 = 62.10\ kg\text{-}m^2}$, $\boxed{I_3 = 81.75\ kg\text{-}m^2}$. *Check:* A numerical check: by theorem (see any text on linear algebra), the sum of the principal values of an inertia matrix is equal to the sum of the diagonal elements of the matrix: $I_1 + I_2 + I_3 = I_{x'x'} + I_{y'y'} + I_{z'z'} = 160$. *check.*

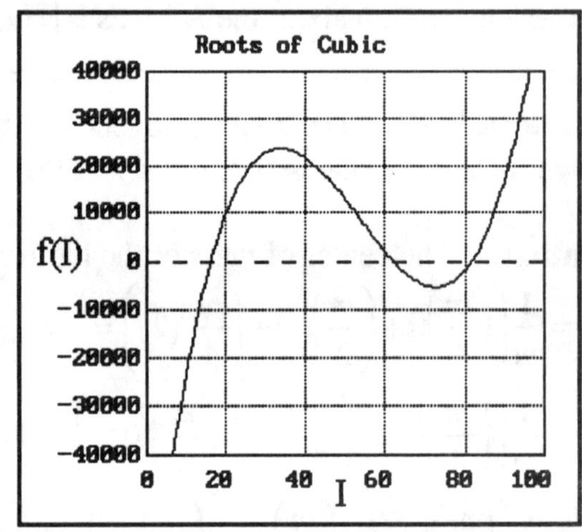

[*Discussion:* The iterative method built into **TK Solver Plus** was used here. However, an algorithm for iteration (the Newton-Raphson algorithm) suitable for a programmable hand calculator is as follows: If $I_j^{(i)}$ is the ith trial guess for the jth root, (estimated from a graph of $f(I)$ against *I*), the $(i+1)^{th}$ guess is improved by

$$I_j^{(i+1)} = I_j^{(i)} - \frac{f\left(I_j^{(i)}\right)}{f'(I_j^{(i)})}$$

where the prime indicates the first derivative. The method works well if the initial guess is close to the correct value, and if distinct principal values are well separated. (We don't have to worry about imaginary principal values: the inertia matrix is symmetrical, and by theorem the principal values of a symmetric matrix are real. (A heuristic "proof" is based on the observation that the moment of inertia is a real, positive number or zero so the principal values are real, positive or zero.) The method loses robustness if two principal values are distinct but very close together; but this can be overcome by expanding the scales of the graph to obtain better first guesses. *End of discussion.*]

Principal Axes: From Eq (9.22), the characteristic vectors parallel to the principal axes are given by the solution to $V_{x'}^{(j)} = (I_{y'y'} - I_j)(I_{z'z'} - I_j) - I_{y'z'}^2$

$$V_{y'}^{(j)} = I_{x'y'}(I_{z'z'} - I_j) + I_{x'z'}I_{y'z'}$$

$$V_{z'}^{(j)} = I_{x'z'}(I_{y'y'} - I_j) + I_{x'y'}I_{y'z'}.$$

Solution continued on next page

For $I_1 = 16.153$, $\vec{V}^{(1)} = 2799.5\vec{i} - 638.47\vec{j} + 438.47\vec{k}$, and the unit vector parallel to $\vec{V}^{(1)}$ is $\vec{e}_1 = \dfrac{\vec{V}^{(1)}}{\left|\vec{V}^{(1)}\right|}$,

$\boxed{\vec{e}_1 = 0.9638\vec{i} - 0.2198\vec{j} + 0.1510\vec{k}}$ For $I_2 = 62.097$, $\vec{V}^{(2)} = -37.55\vec{i} - 179.0\vec{j} - 20.97\vec{k}$,

$\boxed{\vec{e}_2 = -0.2039\vec{i} - 0.9723\vec{j} - 0.1139\vec{k}}$. For $I_3 = 81.75$, $\vec{V}^{(3)} = 38.06\vec{i} + 17.50\vec{j} - 217.5\vec{k}$,

$\boxed{\vec{e}_3 = 0.1718\vec{i} + 0.0790\vec{j} - 0.9820\vec{k}}$. *Check:* Numerical check: The unit vectors parallel to the principal axes must be orthogonal: $\vec{e}_1 \cdot \vec{e}_2 = -2 \times 10^{-15}$, $\vec{e}_1 \cdot \vec{e}_3 = -4 \times 10^{-15}$, where the computation is done with full precision, and not with the four significant figures. *check.*

[*Discussion:* Define the characteristic matrix by $\begin{bmatrix} (I_{xx} - I) & -I_{xy} & -I_{xz} \\ -I_{yx} & (I_{yy} - I) & -I_{yz} \\ -I_{zx} & -I_{zy} & (I_{zz} - I) \end{bmatrix}$ By definition (see any text on linear algebra) the characteristic vectors satisfy the matrix equation

$\begin{bmatrix} (I_{xx} - I) & -I_{xy} & -I_{xz} \\ -I_{yx} & (I_{yy} - I) & -I_{yz} \\ -I_{zx} & -I_{zy} & (I_{zz} - I) \end{bmatrix} \begin{bmatrix} V_x \\ V_y \\ V_z \end{bmatrix} = 0$. This is a set of 3-by-3 simultaneous homogenous equations, and by theorem a unique solution exists if and only if the determinant of the characteristic matrix is zero. The determinant is the characteristic polynomial $(I_{xx} - I)A_{11} - I_{xy}A_{12} - I_{xz}A_{13} = 0$, (which reduces to cubic in Eq (9.21)) where A_{ij} is the determinant (called the cofactor) formed by striking the $i-th$ row and $j-th$ column from the characteristic matrix, that is, $A_{11} = (-1)^{1+1}\begin{bmatrix} (I_{yy} - I) & -I_{yz} \\ -I_{yz} & (I_{zz} - I) \end{bmatrix}$,

$A_{12} = (-1)^{1+2}\begin{bmatrix} -I_{xy} & -I_{yz} \\ -I_{xz} & (I_{zz} - I) \end{bmatrix}$, $A_{13} = (-1)^{1+3}\begin{bmatrix} -I_{xy} & (I_{yy} - I) \\ -I_{xz} & -I_{yz} \end{bmatrix}$.

Expand the matrix equation for the characteristic vectors term by term to obtain:

$(I_{xx} - I)V_x - I_{xy}V_y - I_{xy}V_z = 0$,

$-I_{xy}V_x + (I_{yy} - I)V_y - I_{yz} = 0$,

$-I_{xz}V_x - I_{yz}V_y + (I_{zz} - I)V_z = 0$. By comparison with the expansion of the determinant, and noting that the cofactors are symmetric ($A_{ij} = A_{ji}$) these equations are satisfied if the components of the characteristic vector are equal to the cofactors of the characteristic matrix, that is, if $V_x = A_{11}$, $V_y = A_{12}$, $V_z = A_{13}$. Everything looks neat to this point, *except that a cofactor may vanish, or two (or three) characteristic values may be equal; in general any cofactor may be zero and any two (or three) principal values may be equal without affecting the logical validity of any of these equations.* An example of a zero cofactor occurs when all of the products of inertia for one axis in the x', y', z' system are zero. Two equal principal values corresponds to symmetry in two dimensions, such as a straight slender rod about its center of mass.

Solution continued on next page

While the equations are logically valid, the solutions may be nonsense. A zero result for a characteristic vector is meaningless, since in three dimensional space there must be three orthogonal principal axes defined by three non-zero unit vectors. Similarly, two equal principal values will yield two equal characteristic vectors, which is meaningless, since two principal axes cannot be parallel. However, there are two constraints that must be taken into account: (a) the components of the characteristic vectors must be orthogonal to one another, and (b) the characteristic vectors must be orthogonal to one another. We can discard a meaningless component, or a duplicate vector, and replace them with ones determined from the orthogonality constraint.
To summarize: In general, Eq (9.20) may fail to yield meaningful components for one or more characteristic vectors. These meaningless components are to be discarded, and the orthogonality constraints imposed to obtain a solution. If at least one meaningful characteristic vector can be found from Eq (9.20), then a complete specification of the principal axes is possible. If no meaningful component can be obtained from Eq (9.20), then the orientation of the principal axes is entirely arbitrary, subject only to the orthogonality constraints (e.g., a homogenous sphere). (See discussion on page 439 of the text.) *End of discussion.*]

==================================<>==================================

Problem 9.49 For the steel plate and coordinate system shown in Problem 9.38, determine the principal moments of inertia and unit vectors parallel to the corresponding principal axes. Draw a sketch of the plate showing the principal axes.

Solution:

The moments and products of inertia were calculated in the solution of Problem 9.38. Denoting the coordinate system by $x'y'z'$, the value are $I_{x'x'} = 0.0318 kg-m^2$, $I_{y'y'} = 0.0357 kg-m^2$, $I_{z'z'} = 0.0674 kg-m^2$, $I_{x'y'} = 0.0219 kg-m^2$, and $I_{y'z'} = I_{z'x'} = 0$. Substituting these values into Equation (9.21) the principal moments of inertia are roots of the equation $I^3 - 0.134880 I^2 + 0.005201 I - 0.000044 = 0$. The graph of the left side of this equation as a function of I shows the 3 roots: By examining the calculated results, we find that the roots are $I_1 = 0.0117 kg-m^2$ $I_2 = 0.0557 kg-m^2$, $I_3 = 0.0674 kg-m^2$. From Equation (9.22), for each principal moment of inertia the vector with components $V_{x'} = (0.0357 - I)(0.0674 - I)$ $V_{y'} = 0.0219(0.0674 - I)$ $V_{z'} = 0$ is parallel to the corresponding principal axis. Determining the vector for $I = I_1$ and dividing it by its magnitude yields the unit vector $\bar{e}_1 = 0.738\bar{i} + 0.675\bar{j}$. The result for $I = I_2$ is $\bar{e}_2 = -0.675\bar{i} + 0.738\bar{j}$. This procedure does not yield a vector for $I = I_3$, but we know that the third principal axis is perpendicular to the first two, so $\bar{e}_3 = \bar{k}$. The principal axes are shown:

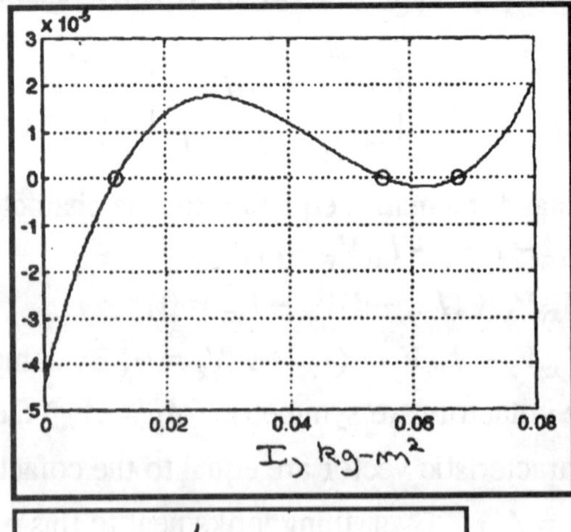

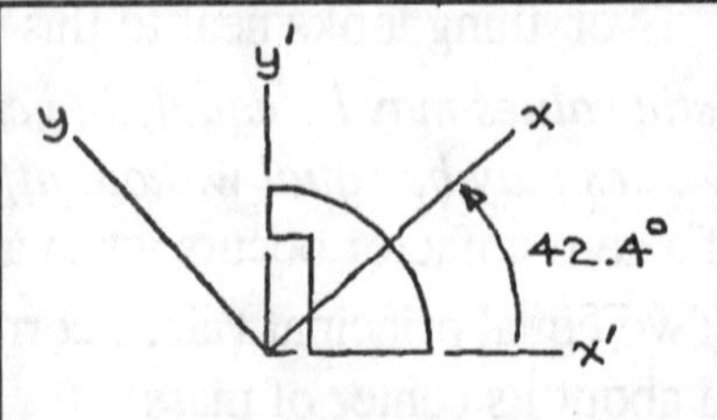

Note: Here we used calculated values from the solution to Problem 9.38 without rounding. If you use the rounded values, your numbers may differ very slightly from the numbers given in this solution.

==================================<>==================================

=================================<>=================================

Problem 9.50 The 1 *kg*, 1 *m* long slender bar lies in the *x-y* plane. Its moment of inertia matrix is

$$[I'] = \begin{bmatrix} \frac{\sin^2\beta}{12} & -\frac{\sin\beta\cos\beta}{12} & 0 \\ -\frac{\sin\beta\cos\beta}{12} & \frac{\cos^2\beta}{12} & 0 \\ 0 & 0 & \frac{1}{12} \end{bmatrix}.$$

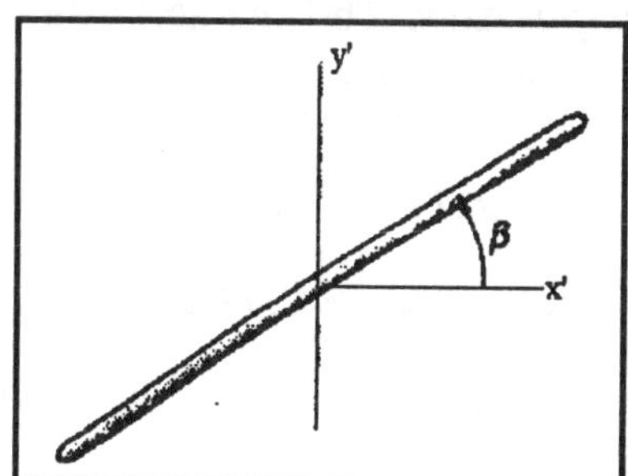

Use Eq (9.19) and (9.20) to determine the principal moments of inertia and unit vectors parallel to the corresponding principal axes.

Solution:

[*Preliminary Discussion:* The moment of inertia about an axis coinciding with the slender rod is zero; it follows that one principal value will be zero, and the associated principal axis will coincide with the slender bar. Since the moments of inertia about the axes normal to the slender bar will be equal, there will be two equal principal values, and Eq (9.20) will fail to yield unique solutions for the associated characteristic vectors. However the problem can be solved by inspection: the unit vector parallel to the axis of the slender rod will be $\vec{e}_1 = \vec{i}\cos\beta + \vec{j}\sin\beta$. A unit vector orthogonal to $\vec{e}_1$ is $\vec{e}_2 = -\vec{i}\sin\beta + \vec{j}\cos\beta$. A third unit vector orthogonal to these two is $\vec{e}_3 = \vec{k}$. The solution based on Eq (9.20) must agree with these preliminary results. (See discussion under solution to Problem 9.38.) *End of Discussion.*]

Principal Moments of Inertia: The moments and products: $I_{x'x'} = \frac{\sin^2\beta}{12}$, $I_{y'y'} = \frac{\cos^2\beta}{12}$, $I_{z'z'} = \frac{1}{12}$, $I_{x'y'} = +\frac{\sin\beta\cos\beta}{12}$, $I_{x'z'} = 0$, $I_{y'z'} = 0$.

From Eq (9.19), the principal values are the roots of the cubic, $AI^3 + BI^2 + CI + D = 0$.

The coefficients are: $A = +1$, $B = -(I_{x'x'} + I_{y'y'} + I_{z'z'}) = -\left(\frac{\sin^2\beta}{12} + \frac{\cos^2\beta}{12} + \frac{1}{12}\right) = -\frac{1}{6}$,

$$C = \left(I_{x'x'}I_{y'y'} + I_{y'y'}I_{z'z'} + I_{z'z'}I_{x'x'} - I_{x'y'}^2 - I_{x'z'}^2 - I_{y'z'}^2\right),$$

$$C = \frac{\sin^2\beta\cos^2\beta}{144} + \frac{\cos^2\beta}{144} + \frac{\sin^2\beta}{144} - \frac{\sin^2\beta\cos^2\beta}{144} = \frac{1}{144}.$$

$$D = -\left(I_{x'x'}I_{y'y'}I_{z'z'} - I_{x'x'}I_{y'z'}^2 - I_{y'y'}I_{x'z'}^2 - I_{z'z'}I_{x'y'}^2 - 2I_{x'y'}I_{y'z'}I_{x'z'}\right),$$

$$D = -\left(\frac{\sin^2\beta\cos^2\beta}{12^3} - \frac{\sin^2\beta\cos^2\beta}{12^3}\right) = 0.$$

The cubic equation reduces to $I^3 - \left(\frac{1}{6}\right)I^2 + \left(\frac{1}{144}\right)I = \left(I^2 - \left(\frac{1}{6}\right)I + \left(\frac{1}{144}\right)\right)I = 0$. By inspection, the least root is $\boxed{I_1 = 0}$. The other two roots are the solution of the quadratic $I^2 + 2bI + c = 0$ where $b = -\frac{1}{12}$, $c = \frac{1}{144}$, from which $I_{1,2} = -b \pm \sqrt{b^2 - c} = \frac{1}{12}$, from which $\boxed{I_2 = \frac{1}{12}}$, $\boxed{I_3 = \frac{1}{12}}$

Solution continued on next page

Principal axes: The characteristic vectors parallel to the principal axes are obtained from Eq (9.22),

$$V_{x'}^{(j)} = (I_{y'y'} - I_j)(I_{z'z'} - I_j) - I_{y'z'}^2$$

$$V_{y}^{(j)} = I_{x'y'}(I_{z'z'} - I_j) + I_{x'z'}I_{y'z'}$$

$$V_{z'}^{(j)} = I_{x'z'}(I_{y'y'} - I_j) + I_{x'y'}I_{y'z'}.$$

For the first root, $I_1 = 0$, $\vec{V}^{(1)} = \frac{\cos^2\beta}{144}\vec{i} + \frac{\cos\beta\sin\beta}{144}\vec{j} = \frac{\cos\beta}{144}(\cos\beta\vec{i} + \sin\beta\vec{j})$.

The magnitude: $\left|\vec{V}^{(1)}\right| = \frac{|\cos\beta|}{144}\sqrt{\cos^2\beta + \sin^2\beta} = \frac{|\cos\beta|}{144}$, and the unit vector is $\vec{e}_1 = \text{sgn}(\cos\beta)(\cos\beta\vec{i} + \sin\beta\vec{j})$, where $\text{sgn}(\cos\beta)$ is equal to the sign of $\cos\beta$.

Without loss of generality, β can be restricted to lie in the first or fourth quadrants, from which $\boxed{\vec{e}_1 = (\cos\beta\vec{i} + \sin\beta\vec{j})}$ For $I_2 = I_3 = \left(\frac{1}{12}\right)$, $\vec{V}^{(2)} = 0$, and $\vec{V}^{(3)} = 0$, from which *the equation* Eq (9.22) *fails for the repeated principal values, and the characteristic vectors are to determined from the condition of orthogonality with* $\vec{e}_1$. From the preliminary discussion, $\boxed{\vec{e}_2 = -\vec{i}\sin\beta + \vec{j}\cos\beta}$ and $\boxed{\vec{e}_3 = 1\vec{k}}$.

====================================<>====================================

Problem 9.51 The mass of a homogenous thin plate is 3 *slugs*. For a coordinate system with its origin at O, determine the principal moments of inertia and unit vectors parallel to the corresponding principal axes.

Solution:

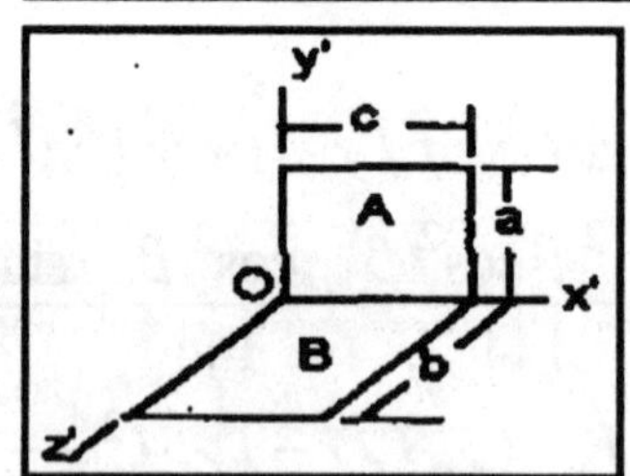

Divide the plate into A and B sheets, as shown. Denote $m = 3$ *slugs*, $a = 2\,ft$, $b = 4\,ft$, and $c = 3\,ft$. The mass of plate A is $m_A = \frac{6m}{18} = \frac{m}{3} = 1\ slug$. The mass of plate B is $m_B = \frac{12m}{18} = 2\ slugs$.

The coordinates of the center of mass of A are

$$\left(d_x^{(A)}, d_y^{(A)}, d_z^{(A)}\right) = \left(\frac{c}{2}, \frac{a}{2}, 0\right) = (1.5, 1, 0)\ ft.$$

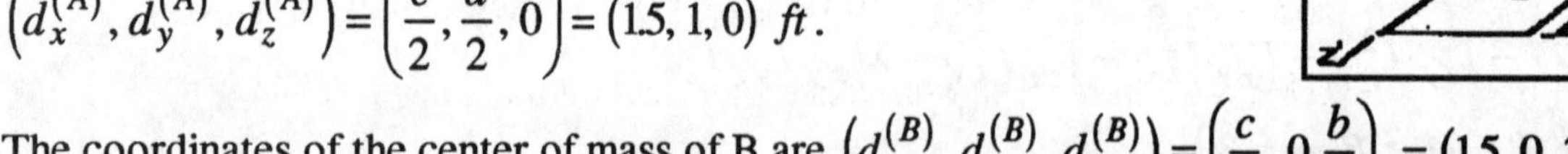

The coordinates of the center of mass of B are $\left(d_x^{(B)}, d_y^{(B)}, d_z^{(B)}\right) = \left(\frac{c}{2}, 0, \frac{b}{2}\right) = (1.5, 0, 2)\,ft$.

Principal Values: From Appendix C, the moments and products of inertia for plate A are

$$I_{x'x'}^{(A)} = \frac{m_A a^2}{12} + \left(d_y^2 + d_z^2\right)m_A = 1.333\ slug\text{-}ft^2,\ I_{y'y'}^{(A)} = \frac{m_A c^2}{12} + \left(d_x^2 + d_z^2\right)m_A = 3\ slug\text{-}ft^2,$$

$$I_{z'z'}^{(A)} = \frac{m_A\left(c^2 + a^2\right)}{12} + \left(d_x^2 + d_y^2\right)m_A = 4.333\ slug\text{-}ft^2,\ I_{x'y'}^{(A)} = d_x d_y m_A = 1.5\ slug\text{-}ft^2,\ I_{x'z'}^{(A)} = 0,$$

$$I_{y'z'}^{(A)} = 0.$$

Solution continued on next page

The moments and products of inertia for plate B are $I_{x'x'}^{(B)} = \frac{m_B b^2}{12} + \left(d_y^2 + d_z^2\right)m_B = 10.67 \ slug\text{-}ft^2$,

$I_{y'y'}^{(B)} = \frac{m_B\left(c^2 + b^2\right)}{12} + \left(d_x^2 + d_z^2\right)m_B = 16.67 \ slug\text{-}ft^2$,

$I_{z'z'}^{(B)} = \frac{m_B c^2}{12} + \left(d_x^2 + d_y^2\right)m_B = 6 \ slug\text{-}ft^2$, $I_{x'y'}^{(B)} = 0$,

$I_{x'z'}^{(B)} = d_x d_z m_B = 6 \ slug\text{-}ft^2$, $I_{y'z'}^{(B)} = 0$. The inertia matrix is the sum of the two matrices:

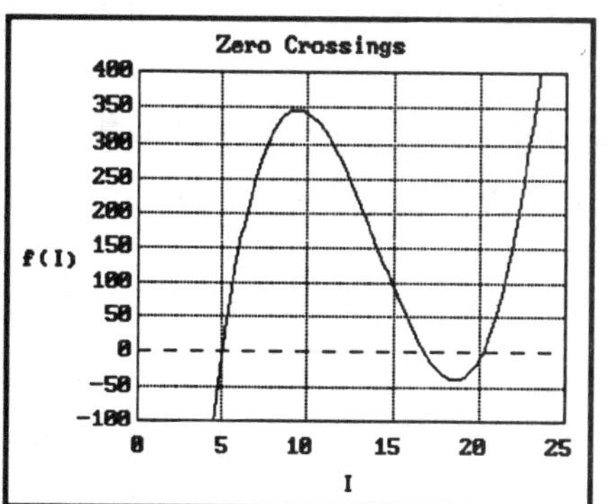

$$[I'] = \begin{bmatrix} I_{x'x'} & -I_{x'y'} & -I_{x'z'} \\ -I_{y'x'} & I_{y'y'} & -I_{y'z'} \\ -I_{z'x'} & -I_{z'y'} & I_{z'z'} \end{bmatrix} = \begin{bmatrix} 12 & -1.5 & -6 \\ -1.5 & 19.67 & 0 \\ -6 & 0 & 10.33 \end{bmatrix}.$$ From Eq (9.19), the principal values are the roots of the cubic, $AI^3 + BI^2 + CI + D = 0$, where $A = +1$, $B = -\left(I_{x'x'} + I_{y'y'} + I_{z'z'}\right) = -42$, $C = \left(I_{x'x'}I_{y'y'} + I_{y'y'}I_{z'z'} + I_{z'z'}I_{x'x'} - I_{x'y'}^2 - I_{x'z'}^2 - I_{y'z'}^2\right) = 524.97$, $D = -\left(I_{x'x'}I_{y'y'}I_{z'z'} - I_{x'x'}I_{y'z'}^2 - I_{y'y'}I_{x'z'}^2 - I_{z'z'}I_{x'y'}^2 - 2I_{x'y'}I_{y'z'}I_{x'z'}\right) = --1707.4$. The function $f(I) = AI^3 + BI^2 + CI + D$ is graphed to determine the zero crossings, and the values refined by iteration. The graph is shown. The principal values are $\boxed{I_1 = 5.042 \ slug\text{-}ft^2}$, $\boxed{I_2 = 16.79 \ slug\text{-}ft^2}$, $\boxed{I_3 = 20.17 \ slug\text{-}ft^2}$. *Check:* A numerical check: the sum of the principal values must be equal to the sum of the diagonal elements of the inertia matrix: $I_1 + I_2 + I_3 = I_{x'x'} + I_{y'y'} + I_{z'z'} = 42$. *check.*

Principal axes: The characteristic vectors parallel to the principal axes are obtained from Eq (9.20),

$V_{x'}^{(j)} = (I_{y'y'} - I_j)(I_{z'z'} - I_j) - I_{y'z'}^2$

$V_{y'}^{(j)} = I_{x'y'}(I_{z'z'} - I_j) + I_{x'z'}I_{y'z'}$

$V_{z'}^{(j)} = I_{x'z'}(I_{y'y'} - I_j) + I_{x'y'}I_{y'z'}$. For $I_1 = 5.042$, $\vec{V}^{(1)} = 77.38\vec{i} + 7.937\vec{j} + 87.75\vec{k}$, and $\boxed{\vec{e}_1 = 0.6599\vec{i} + 0.06768\vec{j} + 0.7483\vec{k}}$. For $I_2 = 16.79$, $\vec{V}^{(2)} = -18.57\vec{i} - 9.687\vec{j} - 17.25\vec{k}$, and $\boxed{\vec{e}_2 = -0.6843\vec{i} - 0.3570\vec{j} + 0.6358\vec{k}}$. For $I_3 = 20.17$, $\vec{V}^{(3)} = 4.911\vec{i} - 14.75\vec{j} - 2.997\vec{k}$, and $\vec{e}_3 = +0.3102\vec{i} - 0.9316\vec{j} - 0.1893\vec{k}$. *Check:* The unit vectors are orthogonal: $\vec{e}_1 \cdot \vec{e}_2 = -4 \times 10^{-14}$, $\vec{e}_2 \cdot \vec{e}_3 = -2 \times 10^{-13}$ (at full 15 digit precision). *check.*

==================================<>==============================

===================================<>===================================

Problem 9.52 The moment of inertia matrix of a rigid body in terms of a body fixed coordinate system with its origin at the center of mass is $[I]=\begin{bmatrix}8.8 & 4.0 & 0\\ 4.0 & 6.6 & 0\\ 0 & 0 & 15.4\end{bmatrix}$ *kg-m*^{2}.

If the rigid body's angular velocity is $\overline{\omega}=-6\overline{i}+2\overline{j}-2\overline{k}$ *(rad/s)*, and its angular acceleration is zero, what are the components of the total moment about its center of mass?

Solution:

the components of the total moment are given by Equation (9.30) with $\overline{\Omega}=\overline{\omega}$

$$\begin{bmatrix}\sum M_x\\ \sum M_y\\ \sum M_z\end{bmatrix}=\begin{bmatrix}0 & 2 & 2\\ -2 & 0 & 6\\ -2 & -6 & 0\end{bmatrix}\begin{bmatrix}8.8 & 4.0 & 0\\ 4.0 & 6.6 & 0\\ 0 & 0 & 15.4\end{bmatrix}\begin{bmatrix}-6\\ 2\\ -2\end{bmatrix} \text{ or}$$

$$\begin{bmatrix}\sum M_x\\ \sum M_y\\ \sum M_z\end{bmatrix}=\begin{bmatrix}0 & 2 & 2\\ -2 & 0 & 6\\ -2 & -6 & 0\end{bmatrix}\begin{bmatrix}-44.8\\ -10.8\\ -30.8\end{bmatrix}=\begin{bmatrix}-83.2\\ -95.2\\ 154.4\end{bmatrix}(N-m)$$

===================================<>===================================

Problem 9.53 If the total moment about the center of mass of the rigid body in Problem 9.52 is zero, what are the components of its angular acceleration?

Solution:

Equation (9.30) can be solved for the components of the angular acceleration.

$$\begin{bmatrix}0\\ 0\\ 0\end{bmatrix}=\begin{bmatrix}8.8 & 4.0 & 0\\ 4.0 & 6.6 & 0\\ 0 & 0 & 15.4\end{bmatrix}\begin{bmatrix}d\omega_x/dt\\ d\omega_y/dt\\ d\omega_z/dt\end{bmatrix}+\begin{bmatrix}0 & 2 & 2\\ -2 & 0 & 6\\ -2 & -6 & 0\end{bmatrix}\begin{bmatrix}8.8 & 4.0 & 0\\ 4.0 & 6.6 & 0\\ 0 & 0 & 15.4\end{bmatrix}\begin{bmatrix}-6\\ 2\\ -2\end{bmatrix}.$$

From the $\overline{i}$, $\overline{j}$ and $\overline{k}$ components of this equation, we obtain

$$8.8\frac{d\omega_x}{dt}+4\frac{d\omega_y}{dt}-83.2=0$$

$$4\frac{d\omega_x}{dt}+6.6\frac{d\omega_y}{dt}-95.2=0$$

$$15.4\frac{d\omega_z}{dt}+154.4=0.$$

Solving, we obtain $d\omega_x/dt=4rad/s^2$ $d\omega_y/dt=12rad/s^2$, and $d\omega_z/dt=-10.0rad/s^2$

===================================<>===================================

==================================<>==================================

Problem 9.54 A rigid body rotates about a fixed point O. Its inertia matrix in terms of a body fixed coordinate system with its origin at O is $[I]=\begin{bmatrix}1 & -1 & 0\\ -1 & 5 & 1\\ 0 & 1 & 7\end{bmatrix}$ *slug - ft*^{2}. If the rigid body's angular velocity is $\vec{\omega}=6\vec{i}+6\vec{j}-4\vec{k}\ (rad/s)$ and its angular acceleration is zero, what are the components of the moment about O?

Solution:

Use Eq (9.26)., which reduces to
$$\begin{bmatrix}\sum M_{ox}\\ \sum M_{oy}\\ \sum M_{oz}\end{bmatrix}=\begin{bmatrix}0 & -\Omega_z & \Omega_y\\ \Omega_z & 0 & -\Omega_x\\ -\Omega_y & \Omega_x & 0\end{bmatrix}\begin{bmatrix}I_{xx} & -I_{xy} & -I_{xz}\\ -I_{yx} & I_{yy} & -I_{yz}\\ -I_{zx} & -I_{zy} & I_{zz}\end{bmatrix}\begin{bmatrix}\omega_x\\ \omega_y\\ \omega_z\end{bmatrix}.$$

The coordinate system is rotating with angular velocity $\vec{\omega}$, from which $\vec{\Omega}=\vec{\omega}$, from which
$$\begin{bmatrix}\sum M_{ox}\\ \sum M_{oy}\\ \sum M_{oz}\end{bmatrix}=\begin{bmatrix}0 & 4 & 6\\ -4 & 0 & -6\\ -6 & 6 & 0\end{bmatrix}\begin{bmatrix}1 & -1 & 0\\ -1 & 5 & 1\\ 0 & 1 & 7\end{bmatrix}\begin{bmatrix}6\\ 6\\ -4\end{bmatrix}=\begin{bmatrix}-4 & 26 & 46\\ -4 & -2 & -42\\ -12 & 36 & 6\end{bmatrix}\begin{bmatrix}6\\ 6\\ -4\end{bmatrix}=\begin{bmatrix}-52\\ 132\\ 120\end{bmatrix},$$

$\boxed{\vec{M}=-52\vec{i}+132\vec{j}+120\vec{k}\ lb\text{-}ft}$

==================================<>==================================

Problem 9.55 If the total moment about O due to forces and couple acting on the rigid body in Problem 9.54 is zero, what are the components of the angular acceleration?

Solution: From the solution to Problems 9.53 and 9.54, the accelerations are given by the matrix equation: $\begin{bmatrix}0\\ 0\\ 0\end{bmatrix}=\begin{bmatrix}1 & -1 & 0\\ -1 & 5 & 1\\ 0 & 1 & 7\end{bmatrix}\begin{bmatrix}\alpha_x\\ \alpha_y\\ \alpha_z\end{bmatrix}+\begin{bmatrix}-52\\ 132\\ 120\end{bmatrix}$. Carry out the matrix multiplication to obtain the three simultaneous equations:

$\alpha_x-\alpha_y+0=52$,

$-\alpha_x+5\alpha_y+\alpha_z=-132$,

$0+\alpha_y+7\alpha_z=-120$. Solve: $\boxed{\vec{\alpha}=35.70\vec{i}-16.30\vec{j}-14.81\vec{k}}$

==================================<>==================================

======================================<>======================================

Problem 9.56 At $t = 0$, the stationary rectangular plate of mass m is subjected to the force F perpendicular to the plate. No other external forces or couples act on the plate. What is the magnitude of the acceleration of point A at $t = 0$?

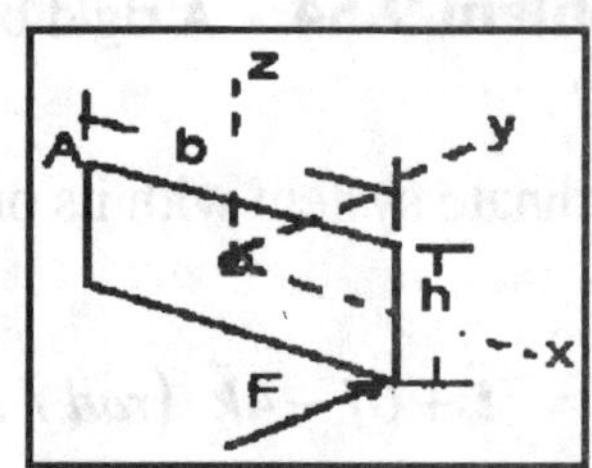

Solution:

The strategy is to write Euler's Equations (Eqs (9.30)) in terms of the principal axes of the plate. For principal axes the cross products vanish and the inertia matrix becomes a diagonal matrix, which markedly simplifies the equations. Choose a coordinate system fixed to the plate, with the origin at the center of mass of the plate, with the plate in the x-z plane. From Appendix C, the inertia matrix for the plate about its center of mass in the x, y, z is

$$[I]=\begin{bmatrix} \dfrac{mh^2}{12} & 0 & 0 \\ 0 & \dfrac{m}{12}\left(h^2+b^2\right) & 0 \\ 0 & 0 & \dfrac{mb^2}{12} \end{bmatrix}$$

. The vector distance between the application of the force and the center of mass is $\vec{r}_{F/G} = (b/2)\vec{i} - (h/2)\vec{k}$. The force vector is $\vec{F} = F\vec{j}$. The moment about the center of mass is

$$\sum \vec{M} = \vec{r}_{F/G} \times \vec{F} = \begin{bmatrix} \vec{i} & \vec{j} & \vec{k} \\ \dfrac{b}{2} & 0 & -\dfrac{h}{2} \\ 0 & F & 0 \end{bmatrix} = \left(\frac{h}{2}F\right)\vec{i} + \left(\frac{b}{2}F\right)\vec{j}.$$

Eq (9.29) is

$$\begin{bmatrix} \sum M_{ox} \\ \sum M_{oy} \\ \sum M_{oz} \end{bmatrix} = \begin{bmatrix} I_{xx} & -I_{xy} & -I_{xz} \\ -I_{yx} & I_{yy} & -I_{yz} \\ -I_{zx} & -I_{zy} & I_{zz} \end{bmatrix} \begin{bmatrix} \alpha_x \\ \alpha_y \\ \alpha_z \end{bmatrix} + \begin{bmatrix} 0 & -\Omega_z & \Omega_y \\ \Omega_z & 0 & -\Omega_x \\ -\Omega_y & \Omega_x & 0 \end{bmatrix} \begin{bmatrix} I_{xx} & -I_{xy} & -I_{xz} \\ -I_{yx} & I_{yy} & -I_{yz} \\ -I_{zx} & -I_{zy} & I_{zz} \end{bmatrix} \begin{bmatrix} \omega_x \\ \omega_y \\ \omega_z \end{bmatrix}$$

. Since the coordinate system is fixed to the plate, and the plate is initially at rest, $\vec{\Omega} = \vec{\omega} = 0$, and Eq (9.25) reduces to

$$\begin{bmatrix} \dfrac{h}{2}F \\ 0 \\ +\dfrac{b}{2}F \end{bmatrix} = \begin{bmatrix} \dfrac{mh^2}{12} & 0 & 0 \\ 0 & \dfrac{m}{12}\left(h^2+b^2\right) & 0 \\ 0 & 0 & \dfrac{mb^2}{12} \end{bmatrix} \begin{bmatrix} \alpha_x \\ \alpha_y \\ \alpha_z \end{bmatrix}$$

, from which $\vec{\alpha}_G = \left(\dfrac{6F}{mh}\right)\vec{i} + \left(\dfrac{6F}{mb}\right)\vec{k}\ \left(rad/s^2\right)$. From Newton's second law, the linear acceleration of the center of mass is $\vec{a}_G = (F/m)\vec{j}$. The vector distance from the center of mass to the point A is $\vec{r}_{A/G} = -(b/2)\vec{i} + (h/2)\vec{k}$.

The acceleration of point A is $\vec{a}_A = \vec{a}_G + \vec{\alpha}_G \times \vec{r}_{A/G} + \vec{\omega} \times \left(\vec{\omega} \times \vec{r}_{A/G}\right)$. Since $\vec{\omega} = 0$,

$$\vec{a}_A = \left(\frac{F}{m}\right)\vec{j} + \begin{bmatrix} \vec{i} & \vec{j} & \vec{k} \\ \dfrac{6F}{mh} & 0 & +\dfrac{6F}{mb} \\ -\dfrac{b}{2} & 0 & \dfrac{h}{2} \end{bmatrix}, \quad \boxed{\vec{a}_A = -\left(\frac{5F}{m}\right)\vec{j}}$$

======================================<>======================================

==================================<>==================================

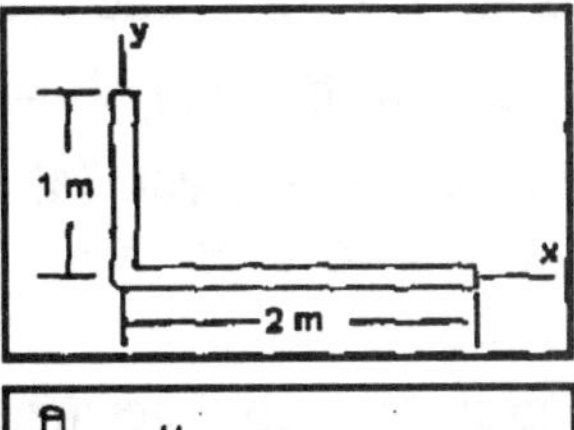

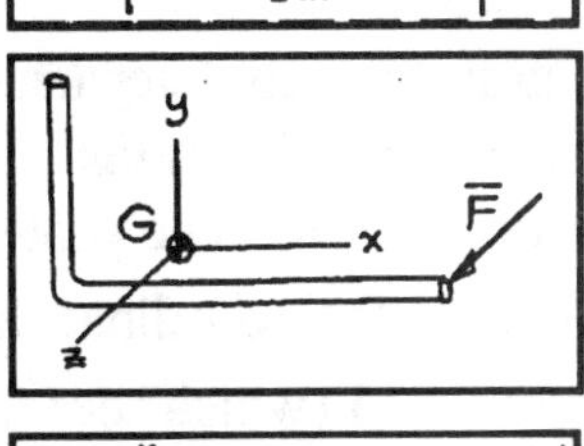

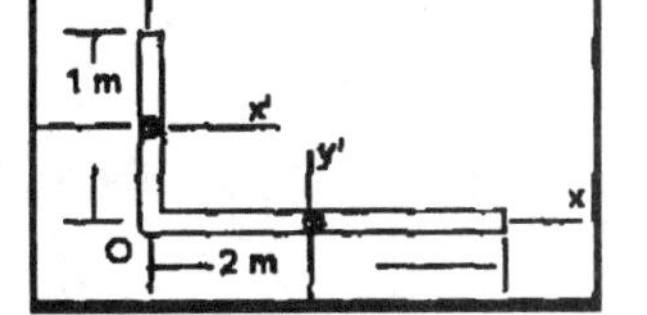

Problem 9.57 The mass of the homogenous slender bar is 6 *kg*. At $t=0$, the stationary bar is subjected to the force $\vec{F}=12\vec{k}\ (N)$ at the point $x=2$, $y=0$. No other external forces or couples act on the bar. (a) What is the bar's angular acceleration at $t=0$? (b) What is the acceleration of the point $x=2\ m$, $y=0$, at $t=0$.

Solution:

The strategy is to determine moments and products of inertia in a parallel system passing through the center of mass of the bar; and then to use Euler's Equations to determine the bar's angular acceleration, from which, with Newton's second law, the acceleration of a point on the bar can be determined.

(a) Choose a coordinate system with the origin at O and the *y* axis positive upward and the *x* axis parallel to the horizontal element of the bar. From the definition of the center of mass, the coordinates of the center of mass of the bar in the x, y, z system originating at O are $x=0.6667\ m$, $y=0.1667\ m$, $z=0$, and the inertia matrix in the parallel system originating at the center of mass of the bar is $[I]_G=\begin{bmatrix} I_{x'x'} & -I_{x'y'} & -I_{x'z'} \\ -I_{x'y'} & I_{y'y'} & -I_{y'z'} \\ -I_{x'z'} & -I_{y'z'} & I_{z'z'} \end{bmatrix}=\begin{bmatrix} 0.5 & 0.6667 & 0 \\ 0.6667 & 2.667 & 0 \\ 0 & 0 & 3.167 \end{bmatrix}$ $kg\text{-}m^2$. Since the coordinate system is fixed on the bar and the system is initially at rest, $\vec{\Omega}=\vec{\omega}=0$, and Eq (9.26) reduces to

$$\begin{bmatrix} \sum M_{ox} \\ \sum M_{oy} \\ \sum M_{oz} \end{bmatrix}=\begin{bmatrix} 0.5 & 0.6667 & 0 \\ 0.6667 & 2.667 & 0 \\ 0 & 0 & 3.167 \end{bmatrix}\begin{bmatrix} \alpha_x \\ \alpha_y \\ \alpha_z \end{bmatrix}.$$

The vector distance from the center of mass to the point of application of the force, $\vec{r}_{F/G}=(2-0.6667)\vec{i}+(0-0.1667)\vec{j}=1.333\vec{i}-0.1667\vec{j}\ (m)$.

The moment about the center of mass is $\vec{M}_G=\vec{r}_{F/G}\times\vec{F}=\begin{bmatrix} \vec{i} & \vec{j} & \vec{k} \\ 1.333 & -0.1667 & 0 \\ 0 & 0 & 12 \end{bmatrix}=-2\vec{i}-16\vec{j}\ N\text{-}m$.

Carry out the matrix multiplication to obtain the simultaneous equations: $0.5\alpha_x+0.6667\alpha_y=-2$, $0.6667\alpha_x+2.667\alpha_y=-16$, and $\boxed{\alpha_z=0}$.

Solve: $\boxed{\vec{\alpha}_G=6.0\vec{i}-7.5\vec{j}\ \left(rad/s^2\right)}$.

(b) From Newton's second law, the acceleration of the center of mass is $\vec{a}_G=\frac{12}{6}\vec{k}=2\vec{k}\ \left(m/s^2\right)$. The acceleration of the point (2, 0, 0) is

$$\boxed{\vec{a}_P=\vec{a}_G+\vec{\alpha}_G\times\vec{r}_{F/G}=2\vec{k}+\begin{bmatrix} \vec{i} & \vec{j} & \vec{k} \\ 6.0 & -7.5 & 0 \\ 1.333 & -0.1667 & 0 \end{bmatrix}=11\vec{k}\ \left(m/s^2\right)}$$

==================================<>==================================

==================================<>==================================

Problem 9.58 The mass of the homogenous slender bar is 1.2 *slug*. At $t=0$, the stationary bar is subjected to the force $\vec{F}=2\vec{i}+4\vec{k}$ (lb) at the point $x=1\ ft$, $y=1\ ft$. No other external forces or couples act on the bar. (a) What is the bar's angular acceleration at $t=0$? (b) What is the acceleration of the point $x=-1\ ft$, $y=-1\ ft$ at $t=0$?

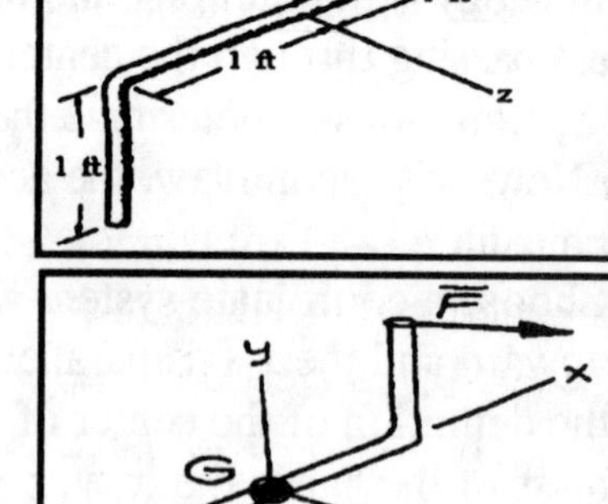

Solution:

The strategy is to determine moments and products of inertia, and then to use Euler's Equations to determine the bar's angular acceleration. Denote $b=1\ ft$. The center of mass of the composite slender bar is (0, 0, 0) in the coordinate system shown. From the solution to Problem 9.35 (which see) the moment of inertia of the bar about the center of mass is

$$[I]_G = mb^2\begin{bmatrix} 0.1667 & -0.25 & 0 \\ -0.25 & 0.6667 & 0 \\ 0 & 0 & 0.8333 \end{bmatrix} = \begin{bmatrix} 0.2 & -0.3 & 0 \\ -0.3 & 0.8 & 0 \\ 0 & 0 & 1 \end{bmatrix} \ slug\text{-}ft^2.$$

(a) The vector from the center of mass to the point of application of the force is $\vec{r}_{F/G}=\vec{i}+\vec{j}$ (ft). The moment about the center of mass is

$$\vec{M}=\vec{r}_{F/G}\times\vec{F}=\begin{bmatrix} \vec{i} & \vec{j} & \vec{k} \\ 1 & 1 & 0 \\ 2 & 0 & 4 \end{bmatrix}=4\vec{i}-4\vec{j}-2\vec{k}\ (lb\text{-}ft).$$ Since the

coordinate system is fixed to the slender bar, and the bar is initially stationary, $\vec{\Omega}=\vec{\omega}=0$, and Eq (9.29)

reduces to $\begin{bmatrix} 4 \\ -4 \\ -2 \end{bmatrix} = \begin{bmatrix} 0.2 & -0.3 & 0 \\ -0.3 & 0.8 & 0 \\ 0 & 0 & 1 \end{bmatrix}\begin{bmatrix} \alpha_x \\ \alpha_y \\ \alpha_z \end{bmatrix}$. The resulting three simultaneous equations:

$0.2\alpha_x-0.3\alpha_y=4, -0.3\alpha_x+0.8\alpha_y=-4$, and $\alpha_z=-2$. Solve:

$\boxed{\vec{\alpha}_G=28.57\vec{i}+5.714\vec{j}-2\vec{k}\ \left(rad/s^2\right)}$ (b) From Newton's second law, $m\vec{a}_G=\vec{F}$. where

$\vec{a}_G=\left(\frac{1}{m}\right)\vec{F}=\left(\frac{1}{1.2}\right)\left(2\vec{i}+4\vec{k}\right)=1.667\vec{i}+3.333\vec{k}\ \left(m/s^2\right)$. Noting $\vec{r}_{P/G}=-\vec{i}-\vec{j}\ (ft)$, the acceleration of the point P is

$$\vec{a}_P=\vec{a}_G+\vec{\alpha}\times\vec{r}_{P/G}=1.667\vec{i}+3.333\vec{k}+\begin{bmatrix} \vec{i} & \vec{j} & \vec{k} \\ 28.57 & 5.714 & -2 \\ -1 & -1 & 0 \end{bmatrix}=\boxed{-0.3333\vec{i}+2\vec{j}-19.52\vec{k}\ \left(ft/s^2\right)}$$

==================================<>==================================

==================================<>==================================

Problem 9.59 In terms of the coordinate system shown, the inertia matrix of the 2.4-*kg* steel plate is

$$[I]=\begin{bmatrix} 0.0318 & -0.0219 & 0 \\ -0.0219 & 0.0357 & 0 \\ 0 & 0 & 0.0674 \end{bmatrix} \ (kg\text{-}m^2).$$

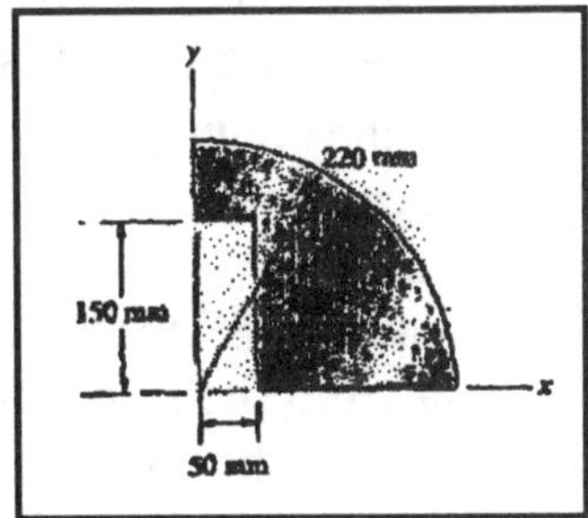

The angular velocity of the plate is $\overline{\omega}=6.4\overline{i}+8.2\overline{j}+14.0\overline{k}$ *(rad/s)* and its angular acceleration is $\overline{\alpha}=60\overline{i}+40\overline{j}-120\overline{k}$ *(rad/s²)*. What are the components of the total moment exerted on the plate about its center of mass?

Solution:

In the solution of Problem 9.39, the location of the center of mass, $\overline{x}=0.1102\ (m)$, $\overline{y}=0.0979\ (m)$ and the moments of inertia in terms of a parallel coordinate system with its origin at the center of mass were determined: $I_{x'x'}=0.00876\ (kg-m^2)$, $I_{y'y'}=I_{yy}=0.00655\ (kg-m^2)$, $I_{z'z'}=0.01531\ (kg-m^2)$ $I_{x'y'}=-0.00396\ (kg-m^2)$ $I_{y'z'}=I_{z'x'}=0$.

The components of the total moment are given by Equation (9.30) with $\overline{\Omega}=\overline{\omega}$:

$$\begin{bmatrix} \sum M_x \\ \sum M_y \\ \sum M_z \end{bmatrix}=\begin{bmatrix} 0.00876 & 0.00396 & 0 \\ 0.00396 & 0.00655 & 0 \\ 0 & 0 & 0.01531 \end{bmatrix}\begin{bmatrix} 60 \\ 40 \\ -120 \end{bmatrix}$$

$$+\begin{bmatrix} 0 & -14.0 & 8.2 \\ 14.0 & 0 & -6.4 \\ -8.2 & 6.4 & 0 \end{bmatrix}\begin{bmatrix} 0.00876 & 0.00396 & 0 \\ 0.00396 & 0.00655 & 0 \\ 0 & 0 & 0.01531 \end{bmatrix}\begin{bmatrix} 6.4 \\ 8.2 \\ 14.0 \end{bmatrix}=\begin{bmatrix} 1.335 \\ 0.367 \\ -2.057 \end{bmatrix}(N-m)$$

==================================<>==================================

Problem 9.60 At t=0, the plate in Problem 9.59 is stationary and is subjected to a force $\overline{F}=-10\overline{k}\ (N)$ at the point (220, 0, 0) *mm*. No other forces or couples act on the plate. At that instant, determine (a) the acceleration of the plate's center of mass; (b) the plate's angular acceleration.

Solution:

(a) From Newton's second law, $\sum\overline{F}=m\overline{a}$: $-10\overline{k}=2.4\,\overline{a}$, and the acceleration of the center of mass is $\overline{a}=-4.17\overline{k}(m/s^2)$.

(b) From the solution of Problem 9.39, the center of mass is at $\overline{x}=0.1102\ (m)$, $\overline{y}=0.0979\ (m)$. Therefore, the moment of the force about the center of mass is

$$\sum\overline{M}=[(0.22-0.1102)\overline{i}-0.0979\overline{j}]\times(-10\overline{k})=0.979\overline{i}+1.098\overline{j}\ (N-m).$$

Equation (9.30) is $\begin{bmatrix} 0.979 \\ 1.098 \\ 0 \end{bmatrix}=\begin{bmatrix} 0.00876 & 0.00396 & 0 \\ 0.00396 & 0.00655 & 0 \\ 0 & 0 & 0.01531 \end{bmatrix}\begin{bmatrix} d\omega_x/dt \\ d\omega_y/dt \\ d\omega_z/dt \end{bmatrix}$.

Solving these equations, we obtain $\overline{\alpha}=d\overline{\omega}/dt=49.5\overline{i}+137.7\overline{j}\ (rad/s^2)$

==================================<>==================================

================================<>================================

Problem 9.61 A 3 *kg* slender bar is rigidly attached to a 2 *kg* thin circular disk. In terms of the body-fixed coordinate system shown, the angular velocity of the composite object is $\vec{w} = 100\vec{i} - 4\vec{j} + 6\vec{k}$ (rad / s) and its angular acceleration is zero. What are the components of the moment exerted on the object about its center of mass?

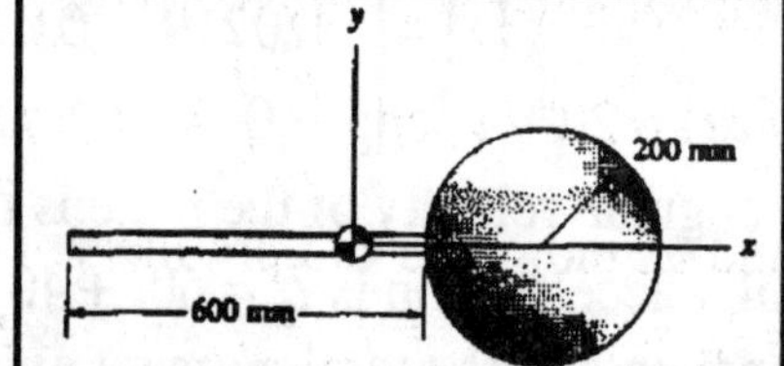

Solution:

Choose an *x, y, z* coordinate system with the origin at O and the *x* axis parallel to the slender rod, as shown. From the solution to Problem 9.34, the coordinates of the center of mass in the *x, y, z* system are $(0.5, 0, 0)$, and the inertia matrix about a parallel coordinate system with origin at the center of mass is:

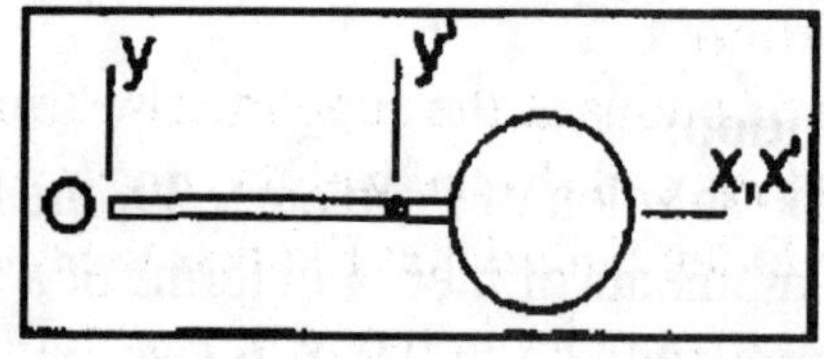

$[I]_G = \begin{bmatrix} 0.02 & 0 & 0 \\ 0 & 0.41 & 0 \\ 0 & 0 & 0.43 \end{bmatrix}$ kg - m^2. Since the coordinate system is body fixed, $\vec{\Omega} = \vec{w}$, and Eq (9.29) reduces to $\begin{bmatrix} \sum M_{ox} \\ \sum M_{oy} \\ \sum M_{oz} \end{bmatrix} = \begin{bmatrix} 0 & -6 & -4 \\ 6 & 0 & -100 \\ 4 & 100 & 0 \end{bmatrix} \begin{bmatrix} 0.02 & 0 & 0 \\ 0 & 0.41 & 0 \\ 0 & 0 & 0.43 \end{bmatrix} \begin{bmatrix} 100 \\ -4 \\ 6 \end{bmatrix}$,

$$\begin{bmatrix} \sum M_{ox} \\ \sum M_{oy} \\ \sum M_{oz} \end{bmatrix} = \begin{bmatrix} 0 & -2.46 & -1.72 \\ 0.12 & 0 & -43 \\ 0.08 & 41 & 0 \end{bmatrix} \begin{bmatrix} 100 \\ -4 \\ 6 \end{bmatrix} = \begin{bmatrix} -0.48 \\ -246 \\ -156 \end{bmatrix} \text{ N-m}, \ \vec{M}_o = -0.48\vec{i} - 246\vec{j} - 156\vec{k} \text{ N-m}$$

================================<>================================

Problem 9.62 At $t = 0$, the composite object in Problem 9.61 is stationary and is subjected to the moment $\sum \vec{M} = -10\vec{i} + 10\vec{j}$ (N - m) about the center of mass. No other forces or couples act on the object. What are the components of angular accelerations at $t = 0$?

Solution:

From the solution to Problem 9.44, the inertia matrix in terms of the parallel coordinate system with origin at the center of mass is $[I]_G = \begin{bmatrix} 0.02 & 0 & 0 \\ 0 & 0.41 & 0 \\ 0 & 0 & 0.43 \end{bmatrix}$ kg - m^2.

Since the coordinates are body-fixed and the object is stationary at $t = 0$, $\vec{\Omega} = \vec{w} = 0$, and Eq (9.26) reduces to: $\begin{bmatrix} -10 \\ 10 \\ 0 \end{bmatrix} = \begin{bmatrix} 0.02 & 0 & 0 \\ 0 & 0.41 & 0 \\ 0 & 0 & 0.43 \end{bmatrix} \begin{bmatrix} \alpha_x \\ \alpha_y \\ \alpha_z \end{bmatrix} = \begin{bmatrix} 0.02\alpha_x \\ 0.41\alpha_y \\ 0.43\alpha_z \end{bmatrix}$. Solve: $\vec{a} = -500\vec{i} + 24.39\vec{j}$ (rad / s^2)

================================<>================================

==================================<>==================================

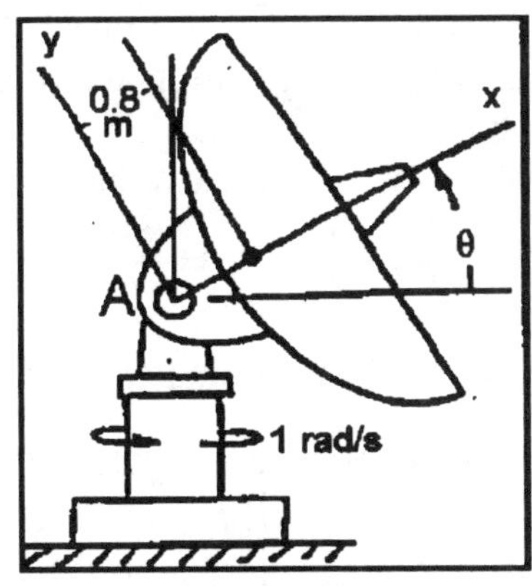

Problem 9.63 The base of the dish antenna is rotating with a constant angular velocity of 1 *rad/s*. The angle $\theta = 30^o$, $(d\theta / dt) = 20^o / s$, and $\frac{d^2\theta}{dt^2} = -40^o / s^2$. The mass of the antenna is 280 *kg*, and its moments and products of inertia in *kg-m*^{2} are $I_{xx} = 140$, $I_{yy} = I_{zz} = 220$, $I_{xy} = I_{xz} = I_{yz} = 0$. Determine the couple exerted on the antenna by its support at this instant.

Solution:

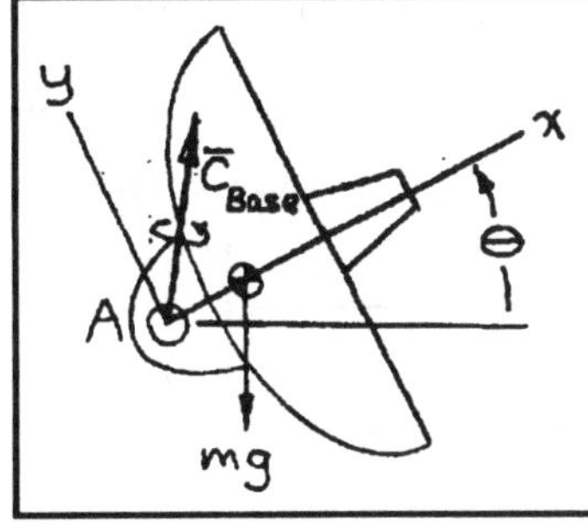

The reactions at the support arise from (a) the Euler moments about the point A, and (b) the weight unbalance due to the offset center of mass. *The Euler Equations:* Express the reactions in the *x, y, z* system. The angular velocity in the *x, y, z* system $\vec{\mathbf{w}} = \vec{i}\sin\theta + \vec{j}\cos\theta + (d\theta / dt)\vec{k} = 0.5\vec{i} + 0.866\vec{j} + 0.3491\vec{k}$ (rad / s). The angular acceleration is $\vec{\mathbf{a}} = \frac{d\vec{\mathbf{w}}}{dt} = (\vec{i}\cos\theta - \vec{j}\sin\theta)\frac{d\theta}{dt} + \frac{d^2\theta}{dt^2}\vec{k}$,

$\vec{\mathbf{a}} = 0.3023\vec{i} - 0.1745\vec{j} - 0.6981\vec{k}$ $(\text{rad} / \text{s}^2)$. Since the coordinates are body fixed $\vec{\Omega} = \vec{\mathbf{w}}$, and Eq (9.25) is

$$\begin{bmatrix} \sum M_{ox} \\ \sum M_{oy} \\ \sum M_{oz} \end{bmatrix} = \begin{bmatrix} 140 & 0 & 0 \\ 0 & 220 & 0 \\ 0 & 0 & 220 \end{bmatrix}\begin{bmatrix} 0.302 \\ -0.1745 \\ -0.6981 \end{bmatrix} + \begin{bmatrix} 0 & -0.3491 & 0.866 \\ 0.3491 & 0 & -0.5 \\ -0.866 & 0.5 & 0 \end{bmatrix}\begin{bmatrix} 140 & 0 & 0 \\ 0 & 220 & 0 \\ 0 & 0 & 220 \end{bmatrix}\begin{bmatrix} 0.5 \\ 0.866 \\ 0.3491 \end{bmatrix}.$$

$$\begin{bmatrix} \sum M_{ox} \\ \sum M_{oy} \\ \sum M_{oz} \end{bmatrix} = \begin{bmatrix} 42.32 \\ -38.39 \\ -153.6 \end{bmatrix} + \begin{bmatrix} 0 & -76.79 & 190.52 \\ 48.87 & 0 & -110 \\ -121.2 & 110 & 0 \end{bmatrix}\begin{bmatrix} 0.5 \\ 0.866 \\ 0.3491 \end{bmatrix} = \begin{bmatrix} 42.32 \\ -38.39 \\ -153.6 \end{bmatrix} + \begin{bmatrix} 0 \\ -13.96 \\ 34.64 \end{bmatrix} = \begin{bmatrix} 42.32 \\ -52.36 \\ -118.9 \end{bmatrix},$$

$\vec{M}_o = 42.32\vec{i} - 52.36\vec{j} - 118.9\vec{k}$ N-m

The unbalance exerted by the offset center of mass: The weight of the antenna acting through the center of mass in the *x, y, z* system is $\vec{W} = \text{mg}(-\vec{i}\sin\theta - \vec{j}\cos\theta) = -1373.4\vec{i} - 2378.8\vec{j}$ (N).

The vector distance to the center of mass is in the *x, y, z* system is $\vec{r}_{G/O} = 0.8\vec{i}$ (m).The moment exerted by the weight is $\vec{M}_W = \vec{r}_{G/O} \times \vec{W} = \begin{bmatrix} \vec{i} & \vec{j} & \vec{k} \\ 0.8 & 0 & 0 \\ -1373.4 & -2378.8 & 0 \end{bmatrix} = -1903.0\vec{k}$.

The couple exerted by the base:

$$\begin{bmatrix} C_x \\ C_y \\ C_z \end{bmatrix} = \begin{bmatrix} \sum M_x \\ \sum M_y \\ \sum M_z \end{bmatrix} - \begin{bmatrix} 0 \\ 0 \\ -1903.0 \end{bmatrix} = \begin{bmatrix} 42.32 \\ -52.36 \\ -118.9 \end{bmatrix} + \begin{bmatrix} 0 \\ 0 \\ +1903.0 \end{bmatrix} = \begin{bmatrix} 42.32 \\ -52.36 \\ 1784.1 \end{bmatrix}(N-m),$$

$\vec{C}_{Base} = 42.35\vec{i} - 52.36\vec{j} + 1784.1\vec{k}$ (N-m)

==================================<>==================================

Problem 9.64 A thin triangular plate of mass m is supported by a ball and socket at O. If it is held in the horizontal position and released from rest, what are the components of its angular acceleration at that instant?

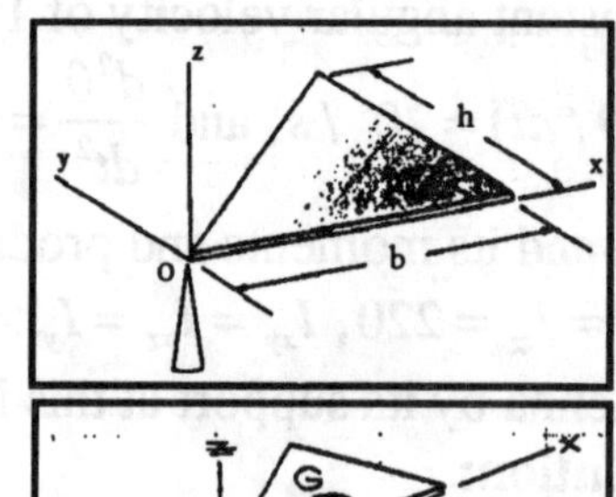

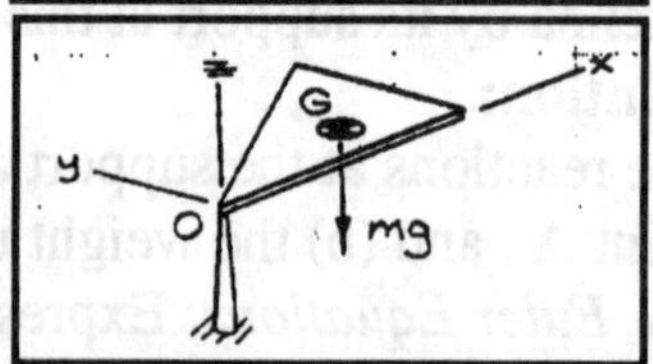

Solution From the solution to Problem 9.26, the moments of inertia are: $I_{xx}=\frac{mh^2}{6}$, $I_{yy}=\frac{mb^2}{2}$, $I_{zz}=\frac{m}{6}\left(h^2+3b^2\right)$, $I_{xy}=\frac{mhb}{4}$, $I_{xz}=0$, $I_{yz}=0$. The coordinates of the center of mass are $\left(\frac{2b}{3},\frac{h}{3},0\right)$. The moment about the origin exerted by the weight of the plate is

$$\vec{M}=\vec{r}_{G/O}\times\vec{W}=\begin{bmatrix}\vec{i} & \vec{j} & \vec{k}\\ \frac{2b}{3} & \frac{h}{3} & 0\\ 0 & 0 & -mg\end{bmatrix},\ \vec{M}=-\frac{mgh}{3}\vec{i}+\frac{2mgb}{3}\vec{j}.$$

The coordinates are body-fixed and the plate is initially stationary, from which $\boldsymbol{\Omega}=\vec{\mathbf{w}}=0$, and Eq (9.25) reduces to

$$\begin{bmatrix}-\frac{mgh}{3}\\ \frac{2mgb}{3}\\ 0\end{bmatrix}=\begin{bmatrix}\frac{mh^2}{6} & -\frac{mbh}{4} & 0\\ -\frac{mbh}{4} & \frac{mb^2}{2} & 0\\ 0 & 0 & \frac{m}{6}\left(h^2+3b^2\right)\end{bmatrix}\begin{bmatrix}\alpha_x\\ \alpha_y\\ \alpha_z\end{bmatrix}.$$

Carry out the matrix multiplication to obtain: $\frac{mh^2}{6}\alpha_x-\frac{mbh}{4}\alpha_y=-\frac{mgh}{3}$, $-\frac{mbh}{4}\alpha_x+\frac{mb^2}{2}\alpha_y=\frac{2mgb}{3}$, and $\alpha_z=0$. Solve: $\alpha_x=0$, $\alpha_y=\frac{4g}{3b}$, from which $\vec{\mathbf{a}}=+\left(\frac{4g}{3b}\right)\vec{j}\ \left(\text{rad/s}^2\right)$

Problem 9.65 Determine the force exerted on the triangular plate in Problem 9.64 by the ball and socket support at the instant of release.

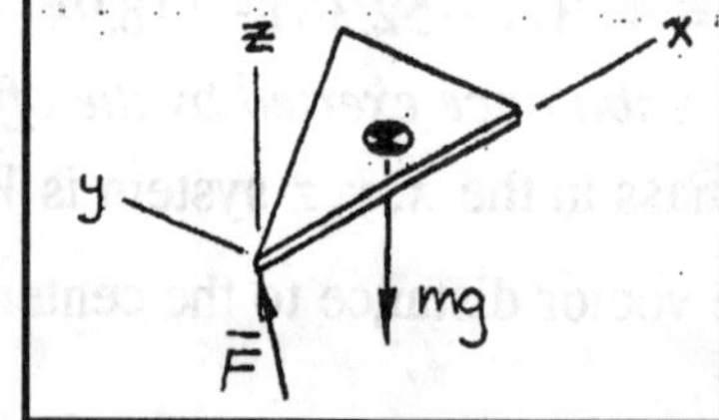

Solution From the solution to Problem 9.64, the vector from O to the center of mass is $\vec{r}_{G/O}=\left(\frac{2b}{3}\right)\vec{i}+\left(\frac{h}{3}\right)\vec{j}$, and the angular acceleration about the point O is $\vec{\mathbf{a}}=\frac{4g}{3b}\vec{j}\ \left(\text{rad/s}^2\right)$. The acceleration of the center of mass is $\vec{a}_G=\vec{a}_O+\vec{\mathbf{a}}\times\vec{r}_{G/O}+\vec{\mathbf{w}}\times\left(\vec{\mathbf{w}}\times\vec{r}_{G/O}\right)=\vec{\mathbf{a}}\times\vec{r}_{G/O}$.

$$\vec{a}_G=\begin{bmatrix}\vec{i} & \vec{j} & \vec{k}\\ 0 & \frac{4g}{3b} & 0\\ \frac{2b}{3} & \frac{h}{3} & 0\end{bmatrix}=-\left(\frac{8g}{9}\right)\vec{k}.$$

From Newton's second law: $m\vec{a}_G=\vec{F}-mg\vec{k}$, from which

$$\vec{F}=\left(-m\left(\frac{8g}{9}\right)+mg\right)\vec{k}=\left(\frac{mg}{9}\right)\vec{k}.$$

====================================<>====================================

Problem 9.66 In Problem 9.64, the mass of the plate is 5 *kg*, $b = 900$ mm, and $h = 600$ mm. If the plate is released in the horizontal position with angular velocity $\vec{w} = 4\vec{i}$ (rad / s), what are the components of its angular acceleration at that instant?

Solution:

From the solution to Problem 9.36, the inertia matrix is

$$\begin{bmatrix} \frac{mh^2}{6} & -\frac{mbh}{4} & 0 \\ -\frac{mbh}{4} & \frac{mb^2}{2} & 0 \\ 0 & 0 & \frac{m}{6}(h^2+3b^2) \end{bmatrix} = \begin{bmatrix} 0.3 & -0.675 & 0 \\ -0.675 & 2.025 & 0 \\ 0 & 0 & 2.325 \end{bmatrix}$$

kg - m^2. From the solution to Problem 9.54, the moment about point O is $\vec{M} = -\frac{\text{mgh}}{3}\vec{i} + \frac{2\text{mgb}}{3}\vec{j} = -9.81\vec{i} + 29.43\vec{j}$ (N - m).

The coordinate system is body-fixed, from which $\vec{\Omega} = \vec{w}$, and Eq (9.26) reduces to

$$\begin{bmatrix} -9.81 \\ 29.43 \\ 0 \end{bmatrix} = \begin{bmatrix} 0.3 & -0.675 & 0 \\ -0.675 & 2.025 & 0 \\ 0 & 0 & 2.325 \end{bmatrix} \begin{bmatrix} \alpha_x \\ \alpha_y \\ \alpha_z \end{bmatrix} + \begin{bmatrix} 0 & 0 & 0 \\ 0 & 0 & -4 \\ 0 & 4 & 0 \end{bmatrix} \begin{bmatrix} 0.3 & -0.675 & 0 \\ -0.675 & 2.025 & 0 \\ 0 & 0 & 2.325 \end{bmatrix} \begin{bmatrix} 4 \\ 0 \\ 0 \end{bmatrix}.$$

$$\begin{bmatrix} -9.81 \\ 29.43 \\ 0 \end{bmatrix} = \begin{bmatrix} 0.3 & -0.675 & 0 \\ -0.675 & 2.025 & 0 \\ 0 & 0 & 2.325 \end{bmatrix} \begin{bmatrix} \alpha_x \\ \alpha_y \\ \alpha_z \end{bmatrix} + \begin{bmatrix} 0 & 0 & 0 \\ 0 & 0 & -9.3 \\ -2.7 & 8.1 & 0 \end{bmatrix} \begin{bmatrix} 4 \\ 0 \\ 0 \end{bmatrix},$$

$$\begin{bmatrix} -9.81 \\ 29.43 \\ 0 \end{bmatrix} = \begin{bmatrix} 0.3 & -0.675 & 0 \\ -0.675 & 2.025 & 0 \\ 0 & 0 & 2.325 \end{bmatrix} \begin{bmatrix} \alpha_x \\ \alpha_y \\ \alpha_z \end{bmatrix} + \begin{bmatrix} 0 \\ 0 \\ -10.8 \end{bmatrix}.$$

Carry out the matrix multiplication to obtain

$0.3\alpha_x - 0.675\alpha_y = -9.81$, $-0.675\alpha_x + 2.025\alpha_y = 29.43$, and $0 + 0 + 2.325\alpha_z = 10.8$.

Solve: $\vec{\mathbf{a}} = 14.53\vec{j} + 4.645\vec{k}$ (rad/s^2)

====================================<>====================================

Problem 9.67 A subassembly of a space station can be modeled as two rigidly connected slender bars, each of mass 5 *Mg*. The subassembly is not rotating at $t = 0$ when a reaction control motor exerts a force $\vec{F} = 400\vec{k}$ (N) at B. What is the acceleration of point A at that instant?

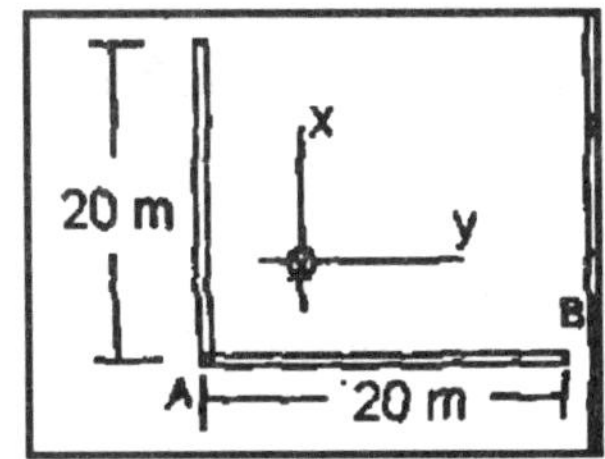

Solution:

Choose a x', y', z' coordinate system with the origin at A and the x' axis parallel to the parallel to the horizontal bar, and a parallel *x*, *y*, *z* system with origin at the center of mass.

The Euler Equations: The center of mass in the x', y', z' system has the coordinates

$x_G = \frac{10(5000) + 0(5000)}{10000} = 5$ m, $y_G = \frac{10(5000) + 0(5000)}{10000} = 5$ m, $z_G = 0$, from which $(d_x, d_y, d_z) = (5, 5, 0)$ m.

From Appendix C, the moments and products of inertia of each bar about A are $I_{xx}^A = I_{yy}^A = \frac{mL^2}{3}$, $I_{zz}^A = I_{xx}^A + I_{yy}^A = \frac{2mL^2}{3}$, $I_{xy}^A = I_{xz}^A = I_{yz}^A = 0$, where $m = 5000$ kg, and $L = 20$ m.

Solution continued on next page

The moment of inertia matrix is $\left[I^A\right]=\begin{bmatrix}0.6667 & 0 & 0\\ 0 & 0.6667 & 0\\ 0 & 0 & 1.333\end{bmatrix}\text{Mg-m}^2$.

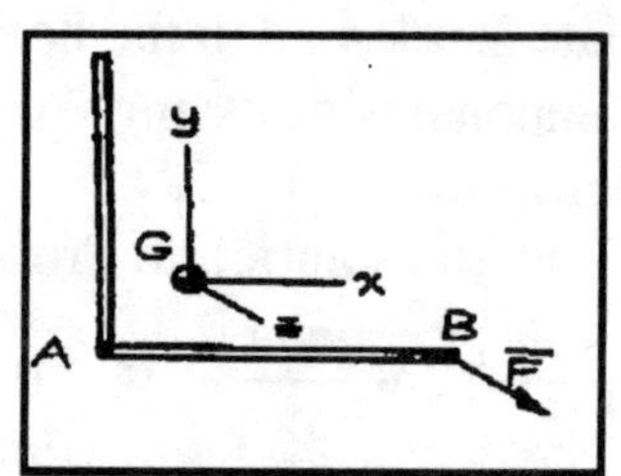

From the parallel axis theorem, Eq (9.18), the moments and products of inertia about the center of mass are:

$I_{xx}=I_{xx}^A-\left(d_z^2+d_y^2\right)(2m)=0.4167\ \text{Mg-m}^2$,

$I_{yy}=I_{yy}^A-\left(d_x^2+d_z^2\right)(2m)=0.4167\ \text{Mg-m}^2$, $I_{zz}=I_{zz}^A-\left(d_x^2+d_y^2\right)(2m)=0.8333\ \text{Mg-m}^2$.

$I_{xy}=I_{xy}^A-d_xd_y(2m)=-0.2500\ \text{Mg-m}^2$, from which the inertia matrix is

$$[I]=\begin{bmatrix}0.4167 & 0.2500 & 0\\ 0.2500 & 0.4167 & 0\\ 0 & 0 & 0.8333\end{bmatrix}\text{Mg-m}^2$$

[*Check*: From Appendix C, the moments and products of inertia about the center of mass of the two bars taken separately are: *Vertical Bar:* $I_{x'x'}^{(G)}=\frac{mL^2}{12}$, $I_{y'y'}^{(G)}=0$, $I_{z'z'}^{(G)}=\frac{mL^2}{12}$, $I_{x'y'}^{(G)}=I_{x'z'}^{(G)}=I_{y'z'}^{(G)}=0$. The distances from the center of mass of the vertical bar to the center of mass of the object are $\left(d_x,d_y,d_z\right)=(5,-5,0)$. The parallel axis theorem (Eq (9.18)): $I_{xx}^{(v)}=\frac{mL^2}{12}+\left(d_y^2\right)m=2.917\times10^5$. $I_{yy}^{(v)}=\left(d_x^2\right)m=1.25\times10^5$, $I_{zz}^{(v)}=\frac{mL^2}{12}+\left(d_x^2+d_y^2\right)m=4.167\times10^5$, $I_{xy}^{(v)}=\left(d_xd_y\right)m=-1.25\times10^5$, $I_{xz}^{(v)}=I_{yz}^{(v)}=0$. *Horizontal Bar:* $I_{x'x'}^{(G)}=0$, $I_{y'y'}^{(G)}=\frac{mL^2}{12}$, $I_{z'z'}^{(G)}=\frac{mL^2}{12}$, $I_{x'y'}^{(G)}=I_{x'z'}^{(G)}=I_{y'z'}^{(G)}=0$. The distances to the center of mass of the object: $\left(d_x,d_y,d_z\right)=(-5,5,0)$. The parallel axis theorem (Eq (9.18)): . $I_{xx}^{(h)}=\left(d_y^2\right)m=1.25\times10^5$, $I_{yy}^{(h)}=\frac{mL^2}{12}+\left(d_x^2\right)m=2.917\times10^5$, $I_{zz}^{(h)}=\frac{mL^2}{12}+\left(d_x^2+d_y^2\right)m=4.167\times10^5$, $I_{xy}^{(h)}=\left(d_xd_y\right)m=-1.25\times10^5$, $I_{xz}^{(h)}=I_{yz}^{(h)}=0$. Sum the values: $I_{xx}=I_{xx}^{(v)}+I_{xx}^{(h)}=4.167\times10^5\ \text{kg-m}^2$. $I_{yy}=I_{yy}^{(v)}+I_{yy}^{(h)}=4.167\times10^5\ \text{kg-m}^2$. $I_{zz}=I_{zz}^{(v)}+I_{zz}^{(h)}=8.333\times10^5\ \text{kg-m}^2$. $I_{xy}=I_{xy}^{(v)}+I_{xy}^{(h)}=-2.5\times10^5\ \text{kg-m}^2$. $I_{xz}=I_{yz}=0$. The inertia matrix in the *x*, *y*, *z* system is $[I]=\begin{bmatrix}4.167\times10^5 & 2.5\times10^5 & 0\\ 2.5\times10^5 & 4.167\times10^5 & 0\\ 0 & 0 & 8.333\times10^5\end{bmatrix}\text{kg-m}^2$. *check.*]

The vector distance from the center of mass to the point B is $\vec{r}_{B/G}=(20-5)\vec{i}+(0-5)\vec{j}=15\vec{i}-5\vec{j}$ (m).The moment about the center of mass is

$$\vec{M}_G=\vec{r}_{B/G}\times\vec{F}=\begin{bmatrix}\vec{i} & \vec{j} & \vec{k}\\ 15 & -5 & 0\\ 0 & 0 & 400\end{bmatrix}=-2000\vec{i}-6000\vec{j}\ \text{(N-m)}.$$

The coordinates are body-fixed, and the object is initially stationary, from which $\vec{\Omega}=\vec{w}=0$, and Eq (9.30) reduces to

$$\begin{bmatrix}-2000\\ -6000\\ 0\end{bmatrix}=[I]=\begin{bmatrix}4.167\times10^5 & 2.5\times10^5 & 0\\ 2.5\times10^5 & 4.167\times10^5 & 0\\ 0 & 0 & 8.333\times10^5\end{bmatrix}\begin{bmatrix}\alpha_x\\ \alpha_y\\ \alpha_z\end{bmatrix}.$$

Solution continued on next page

Carry out the matrix multiplication to obtain: $4.167\times10^5\alpha_x + 2.5\times10^5\alpha_y = -2000$, $2.5\times10^5\alpha_x + 4.167\times10^5\alpha_y = -6000$, and $\alpha_z = 0$. Solve: $\vec{\mathbf{a}} = 0.006\vec{i} - 0.018\vec{j}$ (rad/s^2).

Newton's second law: The acceleration of the center of mass of the object from Newton's second law is $\vec{a}_G = \left(\frac{1}{2\text{m}}\right)\vec{F} = 0.04\vec{k}$ (m/s^2).

The acceleration of point A: The vector distance from the center of mass to the point A is $\vec{r}_{A/G} = -5\vec{i} - 5\vec{j}$ (m). The acceleration of point A is $\vec{a}_A = \vec{a}_G + \vec{\mathbf{a}}\times\vec{r}_{A/G} + \vec{\mathbf{w}}\times(\vec{\mathbf{w}}\times\vec{r}_{A/G})$. Since the object is initially stationary, $\vec{\mathbf{w}} = 0$.

$$\vec{a}_A = \vec{a}_G + \vec{\mathbf{a}}\times\vec{r}_{A/G} = 0.04\vec{k} + \begin{bmatrix} \vec{i} & \vec{j} & \vec{k} \\ 0.006 & -0.018 & 0 \\ -5 & -5 & 0 \end{bmatrix} = -0.08\vec{k}\ (\text{m/s}^2),\ \vec{a}_A = -0.08\vec{k}\ (\text{m/s}^2)$$

==================================<>==================================

Problem 9.68 If the subassembly described in Problem 9.67 rotates about the x axis at a constant rate of one revolution every 10 minutes, what is the magnitude of the couple its reaction control system must exert on it?

Solution :

From Problem 9.67, the inertia matrix is $[I] = \begin{bmatrix} 4.167\times10^5 & 2.5\times10^5 & 0 \\ 2.5\times10^5 & 4.167\times10^5 & 0 \\ 0 & 0 & 8.333\times10^5 \end{bmatrix}$ kg-m^2.

For rotation at a constant rate, the angular acceleration is zero, $\vec{\mathbf{a}} = 0$. The body-fixed coordinate system rotates with angular velocity Eq (9.29) reduces to:

$$\begin{bmatrix} \sum M_{ox} \\ \sum M_{oy} \\ \sum M_{oz} \end{bmatrix} = \begin{bmatrix} 0 & 0 & 0 \\ 0 & 0 & -0.01047 \\ 0 & 0.01047 & 0 \end{bmatrix} \begin{bmatrix} 4.167\times10^5 & 2.5\times10^5 & 0 \\ 2.5\times10^5 & 4.167\times10^5 & 0 \\ 0 & 0 & 8.333\times10^5 \end{bmatrix} \begin{bmatrix} 0.01047 \\ 0 \\ 0 \end{bmatrix},$$

$$\begin{bmatrix} \sum M_{ox} \\ \sum M_{oy} \\ \sum M_{oz} \end{bmatrix} = \begin{bmatrix} 0 & 0 & 0 \\ 0 & 0 & -8726.7 \\ 2618.0 & 4363.3 & 0 \end{bmatrix} \begin{bmatrix} 0.01047 \\ 0 \\ 0 \end{bmatrix} = \begin{bmatrix} 0 \\ 0 \\ 27.42 \end{bmatrix} \text{N-m},\ |\vec{M}_o| = 27.42\ (\text{N-m})$$

==================================<>==================================

Problem 9.69 The thin circular disk of radius R and mass m is attached rigidly to the vertical shaft. The disk is slanted at an angle β relative to the horizontal plane. The shaft rotates with constant angular velocity ω_O. What is the magnitude of the couple exerted on the disk by the shaft?

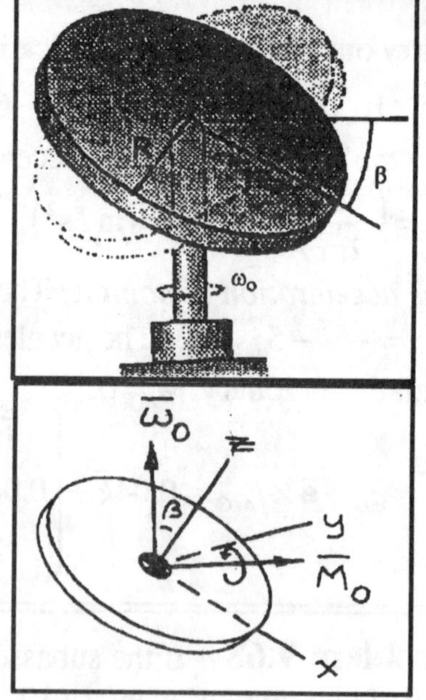

Solution:

Choose a coordinate system fixed to the disk, with the origin at the center of the disk and the x axis along the upper and lower extremes of the disk, with positive x toward the lower extreme. Both x and y axes lie in the plane of the disk. In this system, the angular velocity is $\vec{w}_O = \omega_o(-\vec{i}\sin\beta + \vec{k}\cos\beta +)$ (rad / s). The moments and products of inertia are $I_{xx} = I_{yy} = \frac{mR^2}{4}$, $I_{zz} = \frac{mR^2}{2}$, $I_{xy} = I_{xz} = I_{yz} = 0$. At constant rate of rotation, $\vec{a} = 0$. The body-fixed coordinate system rotates with angular velocity $\vec{\Omega} = \vec{w}$. The Eq (9.25) reduces to

$$\begin{bmatrix} \sum M_{ox} \\ \sum M_{oy} \\ \sum M_{oz} \end{bmatrix} = \begin{bmatrix} 0 & -\omega_O\cos\beta & 0 \\ \omega_O\cos\beta & 0 & \omega_O\sin\beta \\ 0 & -\omega_O\sin\beta & 0 \end{bmatrix} \begin{bmatrix} \frac{mR^2}{4} & 0 & 0 \\ 0 & \frac{mR^2}{4} & 0 \\ 0 & 0 & \frac{mR^2}{2} \end{bmatrix} \begin{bmatrix} -\omega_O\sin\beta \\ 0 \\ \omega_O\cos\beta \end{bmatrix}$$

$$, \begin{bmatrix} \sum M_{ox} \\ \sum M_{oy} \\ \sum M_{oz} \end{bmatrix} = \begin{bmatrix} 0 & -\frac{m\omega_O R^2}{4}\cos\beta & 0 \\ \frac{m\omega_O R^2}{4}\cos\beta & 0 & \frac{m\omega_O R^2}{2}\sin\beta \\ 0 & -\frac{m\omega_O R^2}{4}\sin\beta & 0 \end{bmatrix} \begin{bmatrix} -\omega_O\sin\beta \\ 0 \\ \omega_O\cos\beta \end{bmatrix} = \begin{bmatrix} 0 \\ \frac{m\omega_O^2 R^2}{4}\sin\beta\cos\beta \\ 0 \end{bmatrix}, \text{ from}$$

which $\left|\vec{M}_O\right| = \frac{m\omega_O^2 R^2\cos\beta\sin\beta}{4}$

=================================<>=================================

Problem 9.70 A slender bar of mass m and length L is welded to a horizontal shaft that rotates with constant angular velocity ω_O. Determine the magnitude of the force $\vec{F}$ and couple $\vec{C}$ exerted on the bar by the shaft. (Write the equations of motion in terms of the body fixed coordinate system shown.)

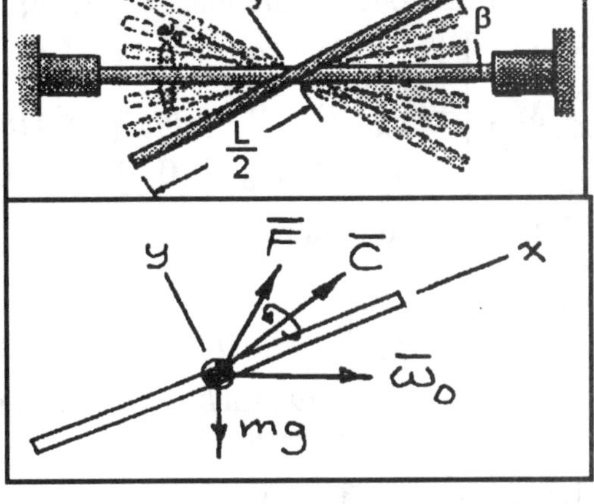

Solution:

The force: The force is the weight of the bar, with constant direction. In the coordinate system shown, the force has a rotation relative to the coordinate system, but has constant magnitude: $\boxed{|\vec{F}| = mg}$.

The couple: In terms of the coordinate system shown, the angular velocity is $\vec{\omega}_O = \omega_O(\vec{i}\cos\beta - \vec{j}\sin\beta)$. The moments and products of inertia for the slender bar are: $I_{xx} = 0$, $I_{yy} = \dfrac{mL^2}{12}$, $I_{zz} = \dfrac{mL^2}{12}$, $I_{xy} = I_{xz} = I_{yz} = 0$. The body-fixed coordinate system rotates with angular velocity $\vec{\Omega} = \vec{\omega}_o$. At a constant rate of rotation, $\vec{\alpha} = 0$. Eq (9.25) reduces to

$$\begin{bmatrix} \sum M_{ox} \\ \sum M_{ox} \\ \sum M_{oy} \end{bmatrix} = \begin{bmatrix} 0 & 0 & -\omega_O\sin\beta \\ 0 & 0 & -\omega_O\cos\beta \\ \omega_O\sin\beta & \omega_O\cos\beta & 0 \end{bmatrix} \begin{bmatrix} 0 & 0 & 0 \\ 0 & I_{yy} & 0 \\ 0 & 0 & I_{zz} \end{bmatrix} \begin{bmatrix} \omega_O\cos\beta \\ -\omega_O\sin\beta \\ 0 \end{bmatrix}.$$

Carry out the matrix multiplication to obtain

$$\begin{bmatrix} \sum M_{ox} \\ \sum M_{oy} \\ \sum M_{oz} \end{bmatrix} = \begin{bmatrix} 0 \\ 0 \\ -\dfrac{m\omega_O^2 L^2}{12}\sin\beta\cos\beta \end{bmatrix},$$

from which $\boxed{|\vec{C}| = \dfrac{m\omega_O^2 L^2}{12}\sin\beta\cos\beta}$

=================================<>=================================

Problem 9.71 A slender bar of mass m and length L is welded to a horizontal shaft that rotates with constant angular velocity ω_O. Determine the magnitude of the couple $\vec{C}$ exerted on the bar by the shaft. (Write the equations of motion in terms of the body fixed coordinate system shown.)

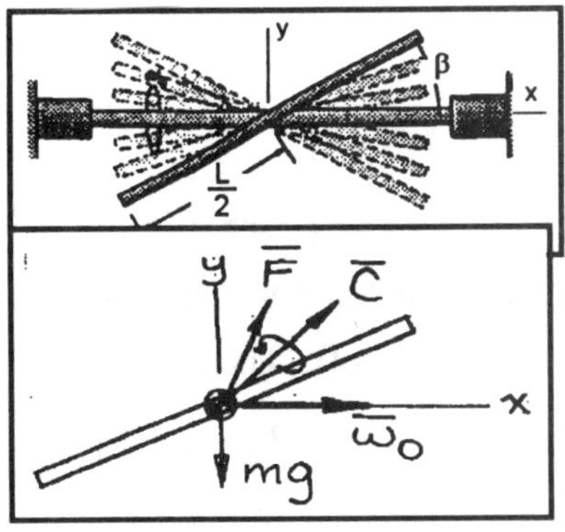

Solution:

In terms of the coordinate system shown, the body-fixed coordinate system rotates with angular velocity $\vec{\Omega} = \vec{\omega}_o = \omega_o\vec{i}$ (rad/s). The moments and products of inertia for the slender bar at an angle β (see solution to Problem 9.50) are: $I_{xx} = \dfrac{mL^2\sin^2\beta}{12}$, $I_{yy} = \dfrac{mL^2}{12}\cos^2\beta$, $I_{zz} = \dfrac{mL^2}{12}$,

$I_{xy} = \dfrac{mL^2}{12}\sin\beta\cos\beta$. $I_{xz} = I_{yz} = 0$.

The body-fixed coordinate system rotates with angular velocity $\vec{\Omega} = \vec{\omega}_o$.

Solution continued on next page

At a constant rate of rotation, $\vec{\alpha} = 0$. Eq (9.26) reduces to

$$\begin{bmatrix} \sum M_{ox} \\ \sum M_{ox} \\ \sum M_{oy} \end{bmatrix} = \frac{mL^2}{12} \begin{bmatrix} 0 & 0 & 0 \\ 0 & 0 & -\omega_o \\ 0 & \omega_o & 0 \end{bmatrix} \begin{bmatrix} \sin^2\beta & -\sin\beta\cos\beta & 0 \\ -\sin\beta\cos\beta & \cos^2\beta & 0 \\ 0 & 0 & 1 \end{bmatrix} \begin{bmatrix} \omega_o \\ 0 \\ 0 \end{bmatrix}$$

. Carry out the matrix multiplication to obtain

$$\begin{bmatrix} \sum M_{ox} \\ \sum M_{oy} \\ \sum M_{oz} \end{bmatrix} = \frac{mL^2}{12} \begin{bmatrix} 0 & 0 & 0 \\ 0 & 0 & -\omega_o \\ -\omega_o \sin\beta\cos\beta & \omega_o \cos^2\beta & 0 \end{bmatrix} \begin{bmatrix} \omega_o \\ 0 \\ 0 \end{bmatrix} = \frac{mL^2}{12} \begin{bmatrix} 0 \\ 0 \\ -\omega_o^2 \sin\beta\cos\beta \end{bmatrix}$$

, from which

$$\boxed{|\vec{C}| = \frac{m\omega_o^2 L^2}{12} \sin\beta\cos\beta}$$

==<>==

Problem 9.72 The slender bar of length L and mass m is pinned to the vertical shaft at O. The vertical shaft rotates with a constant angular velocity ω_O. Show that the value of ω_O necessary for the bar to remain at a constant angle β relative to the vertical is $\omega_O = \sqrt{\dfrac{3g}{2L\cos\beta}}$.

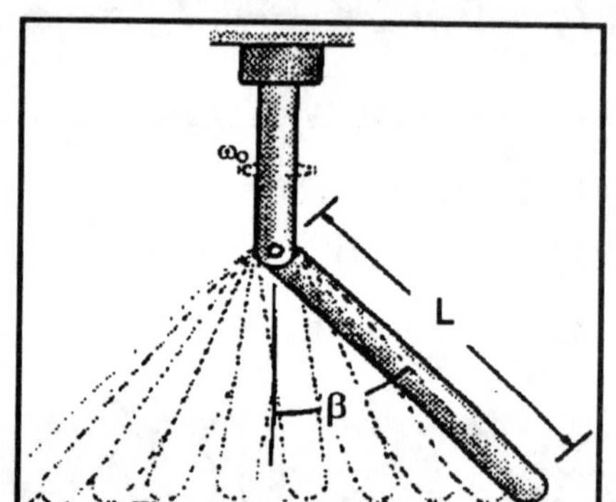

Solution :

This is motion about a fixed point (see page 490 of text) so Eq (9.25) is applicable. Choose a body-fixed x, y, z coordinate system with the origin at O, the positive x axis parallel to the slender bar, and z axis out of the page. The angular velocity of the vertical shaft is $\vec{\Omega} = \omega_o(-\vec{i}\cos\beta + \vec{j}\sin\beta)$. The vector from O to the center of mass of the bar is $\vec{r}_{G/O} = (L/2)\vec{i}$. The weight is $\vec{W} = mg(\vec{i}\cos\beta - \vec{j}\sin\beta)$. The moment about the point O is

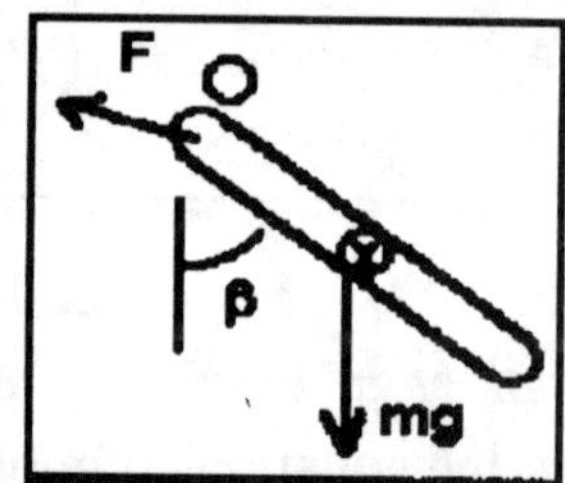

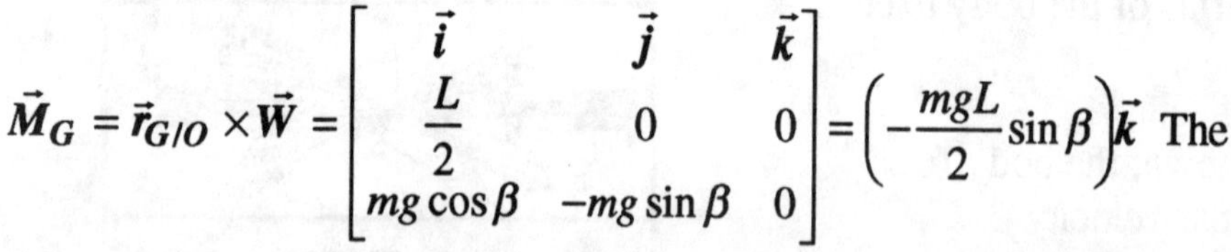

$$\vec{M}_G = \vec{r}_{G/O} \times \vec{W} = \begin{vmatrix} \vec{i} & \vec{j} & \vec{k} \\ \frac{L}{2} & 0 & 0 \\ mg\cos\beta & -mg\sin\beta & 0 \end{vmatrix} = \left(-\frac{mgL}{2}\sin\beta\right)\vec{k}$$

The moments and products of inertia about O in the x, y, z, system are $I_{xx} = 0$, $I_{yy} = I_{zz} = mL^2/3$, $I_{xy} = I_{xz} = I_{yz} = 0$. The body-fixed coordinate system rotates with angular velocity $\vec{\Omega} = \vec{\omega} = \omega_O(-\vec{i}\cos\beta + \vec{j}\sin\beta)$. Eq (9.25) reduces to

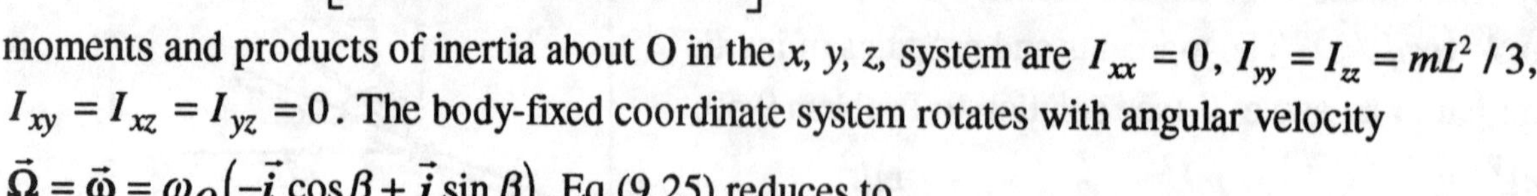

$$\begin{bmatrix} M_{Ox} \\ M_{Oy} \\ M_{Oz} \end{bmatrix} = \begin{bmatrix} 0 & 0 & \omega_o \sin\beta \\ 0 & 0 & \omega_o \cos\beta \\ -\omega_o \sin\beta & -\omega_o \cos\beta & 0 \end{bmatrix} \begin{bmatrix} 0 & 0 & 0 \\ 0 & \frac{mL^2}{3} & 0 \\ 0 & 0 & \frac{mL^2}{3} \end{bmatrix} \begin{bmatrix} -\omega_o \cos\beta \\ \omega_o \sin\beta \\ 0 \end{bmatrix}.$$

Solution continued on next page

Carry out the matrix multiplication,

$$\begin{bmatrix} M_{Ox} \\ M_{Oy} \\ M_{Oz} \end{bmatrix} = \begin{bmatrix} 0 & 0 & \dfrac{\omega_o m L^2 \sin\beta}{3} \\ 0 & 0 & \dfrac{\omega_o m L^2 \cos\beta}{3} \\ 0 & -\dfrac{\omega_o m L^2 \cos\beta}{3} & 0 \end{bmatrix} \begin{bmatrix} -\omega_o \cos\beta \\ \omega_o \sin\beta \\ 0 \end{bmatrix} = \begin{bmatrix} 0 \\ 0 \\ -\dfrac{\omega_o^2 m L^2 \cos\beta \sin\beta}{3} \end{bmatrix}$$

. Equate the z components: $M_{Oz} = -\dfrac{\omega_O^2 m L^2 \cos\beta \sin\beta}{3}$. The pin-supported joint at O cannot support a couple, $\vec{C} = 0$, from which $\vec{M}_O = \vec{M}_G$. $-\dfrac{mgL}{2}\sin\beta = -\dfrac{\omega_O^2 m L^2 \cos\beta \sin\beta}{3}$ Assume that $\beta \neq 0$, from which $\sin\beta \neq 0$, and the equation can be solved for $\boxed{\omega_O = \sqrt{\dfrac{3g}{2L\cos\beta}}}$

===================================<>===================================

Problem 9.73 The vertical shaft rotates with constant angular velocity ω_O. The 35° angle between the edge of the 10 *lb* thin rectangular plate pinned to the shaft and the shaft remains constant. Determine ω_O.

Solution:

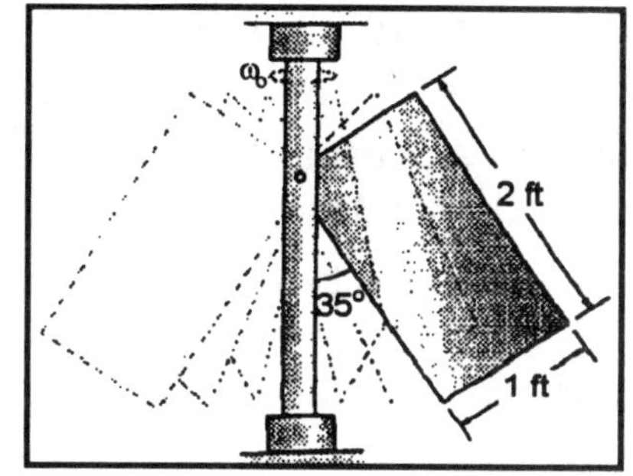

This is motion about a fixed point, and Eq (9.25) is applicable. Choose an *x, y, z* coordinate system with the origin at the pinned joint O and the *x* axis parallel to the lower edge of the plate, and the *y* axis parallel to the upper narrow edge of the plate. Denote $\beta = 35^o$. The plate rotates with angular velocity $\vec{\omega} = \omega_o(-\vec{i}\cos\beta + \vec{j}\sin\beta) = \omega_o(-0.8192\vec{i} + 0.5736\vec{j})(rad/s)$ T

he vector from the pin joint to the center of mass of the plate is $\vec{r}_{G/O} = \vec{i} + (0.5)\vec{j}$ (ft). The weight of the plate is $\vec{W} = 10(\vec{i}\cos\beta - \vec{j}\sin\beta) = 8.192\vec{i} - 5.736\vec{j}$ (lb). The moment about the center of mass is

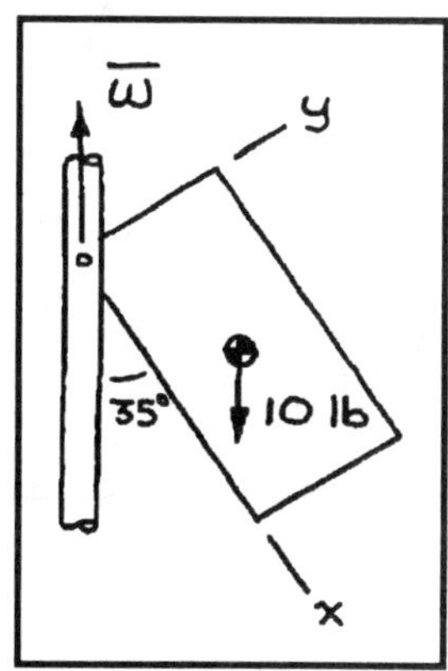

$$\vec{M}_G = \vec{r}_{G/O} \times \vec{W} = \begin{bmatrix} \vec{i} & \vec{j} & \vec{k} \\ 1 & 0.5 & 0 \\ -0.5639 & -0.8054 & 0 \end{bmatrix} = -0.9832\vec{k}\ (ft\text{-}lb).$$

From Appendix C, the moments and products of inertia of a thin plate about O are $I_{xx} = \dfrac{mh^2}{3} = 0.1036\ slug\text{-}ft^2$, $I_{yy} = \dfrac{mb^2}{3} = 0.4145\ slug\text{-}ft^2$, $I_{zz} = \dfrac{m}{3}(h^2 + b^2) = 0.5181\ slug\text{-}ft^2$, $I_{xy} = \dfrac{mbh}{4} = 0.1554\ slug\text{-}ft^2$, $I_{xz} = I_{yz} = 0$. At a constant rate of rotation, the angle $\beta = 35^o = const$, $\vec{\alpha} = 0$.

Solution continued on next page

The body-fixed coordinate system rotates with angular velocity
$\vec{\Omega} = \vec{\omega} = \omega_O\left(-\vec{i}\cos\beta + \vec{j}\sin\beta\right) = -0.8191\omega_O\vec{i} + 0.5736\omega_O\vec{j}\ \ (rad/s)$, and Eq (9.26) reduces to:

$$\begin{bmatrix} M_{ox} \\ M_{oy} \\ M_{oz} \end{bmatrix} = \omega_O^2 \begin{bmatrix} 0 & 0 & 0.5736 \\ 0 & 0 & 0.8191 \\ -0.5736 & -0.8191 & 0 \end{bmatrix} \begin{bmatrix} 0.1036 & -0.1554 & 0 \\ -0.1554 & 0.4145 & 0 \\ 0 & 0 & 0.5181 \end{bmatrix} \begin{bmatrix} -0.8191 \\ 0.5736 \\ 0 \end{bmatrix}$$

. Carry out the matrix operations to obtain: $\begin{bmatrix} M_{ox} \\ M_{oy} \\ M_{oz} \end{bmatrix} = \omega_o^2 \begin{bmatrix} 0 \\ 0 \\ -0.1992 \end{bmatrix}$. The pin support cannot support a couple, from which

$M_{Oz} = M_{Gz}$, $-9.832 = -\omega_o^2(0.1992)$, from which $\boxed{\omega_o = 7.025\ rad/s}$

=================================<>=================================

Problem 9.74 A thin circular disk of mass m mounted on a horizontal shaft rotates relative to the shaft with constant angular velocity ω_d. The horizontal shaft is rigidly attached to the vertical shaft rotating with constant angular velocity ω_O. Determine the magnitude of the couple exerted on the disk by the horizontal shaft.

Solution:

This is rotation about a fixed point, so Eq (9.26) applies. Choose a body-fixed *x, y, z* coordinate system with the origin at the center of mass of the disk and the *y, z* axes in the plane of the disk, and a parallel x', y', z' system with the z' axis parallel to the vertical shaft. With the rates of rotation constant, the accelerations are zero. The body-fixed coordinate system rotates with angular velocity $\vec{\Omega} = \omega_O\vec{k}\ (rad/s)$ and $\vec{\omega} = \omega_d\vec{i} + \omega_o\vec{k}\ (rad/s)$, and

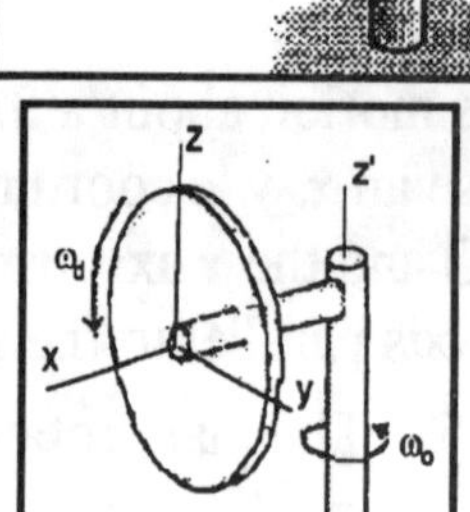

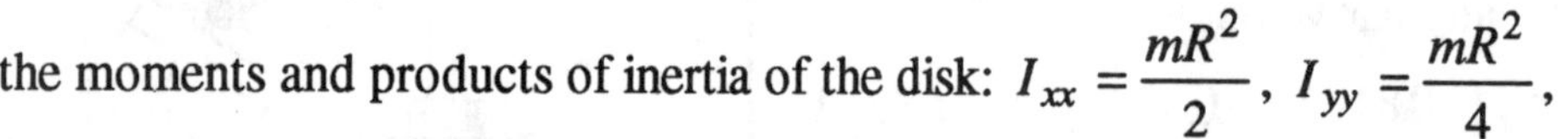

the moments and products of inertia of the disk: $I_{xx} = \dfrac{mR^2}{2}$, $I_{yy} = \dfrac{mR^2}{4}$,

$I_{zz} = \dfrac{mR^2}{4}$, $I_{xy} = I_{xz} = I_{yz} = 0$, Eq (9.25) becomes:

$$\begin{bmatrix} M_x \\ M_y \\ M_z \end{bmatrix} = mR^2\omega_O \begin{bmatrix} 0 & -1 & 0 \\ 1 & 0 & 0 \\ 0 & 0 & 0 \end{bmatrix} \begin{bmatrix} 0.5 & 0 & 0 \\ 0 & 0.25 & 0 \\ 0 & 0 & 0.25 \end{bmatrix} \begin{bmatrix} \omega_d \\ 0 \\ \omega_o \end{bmatrix} = mR^2\omega_O\omega_d \begin{bmatrix} 0 \\ 0.5 \\ 0 \end{bmatrix}$$

, from which

$\boxed{|\vec{M}| = 0.5mR^2\omega_O\omega_d}$

=================================<>=================================

==================================<>==================================

Problem 9.75 The thin triangular plate has ball and socked supports at A and B. The y axis is vertical. If the plate rotates with constant velocity ω_o, what are the horizontal components of the reactions on the plate at A and B?

Solution:

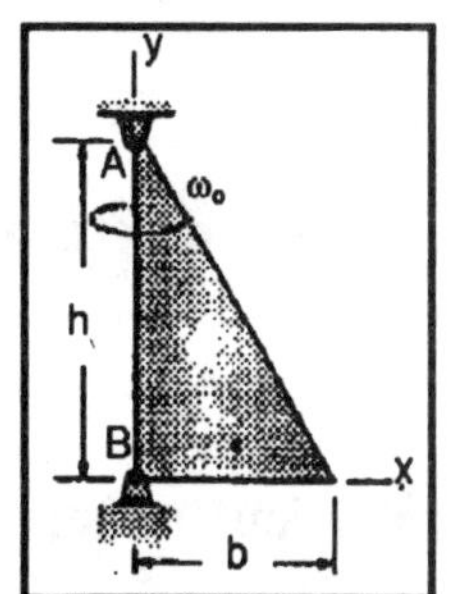

This is rotation about a fixed axis, and Eq (9.25) applies. The horizontal reactions at A and B are the sum of (a) the moments as determined by the Euler Equations, and (b) the horizontal reactions due to the weight of the plate.

The Euler Equations: Denote the mass density by ρ mass units per unit volume. An element of mass of the thin plate is $dm = \rho T dA$, where T is the (unknown) thickness. The area moments and products for a triangle about the origin O are determined as in the solution to Problem 9.26: The equation of the line defining the upper boundary of the plate are $y = h\left(1-\frac{x}{b}\right)$. For an element of area dA, the element of mass is $dm = \rho T dx dy$. For a horizontal elemental strip dy, the element of mass is $dm = \rho T x dy = \rho T(b - \frac{b}{h}y)dy$, For a vertical elemental strip dx, the element of mass is $dm = \rho T y dx = \rho T h\left(1-\frac{x}{b}\right)dx$.The mass of the plate is $m = \frac{\rho T b h}{2}$. The moments and products of inertia (see Eq (9.9)) are:

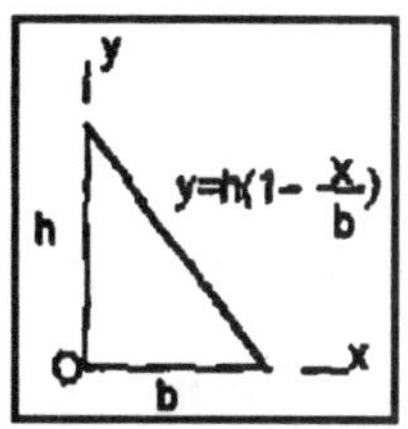

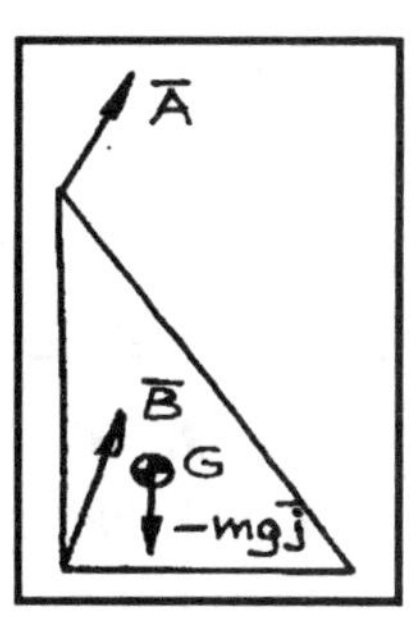

$$I_{xx} = \int_m \left(y^2 + z^2\right)dm = \rho T b \int_0^h y^2\left(1-\frac{y}{h}\right)dy = \frac{mh^2}{6},$$

$$I_{yy} = \int_m (x^2 + z^2)dm = \rho T h \int_0^b \left(1-\frac{x}{b}\right)x^2 dx = \frac{mb^2}{6},$$

$$I_{zz} = \int_m \left(x^2 + y^2\right)dm = I_{xx} + I_{yy} = \frac{m}{6}\left(h^2 + b^2\right),$$

$$I_{xy} = \int_m xy dm = \rho T \int_0^b x dx \int_0^{h\left(1-\frac{x}{b}\right)} y dy = \frac{\rho T h^2}{2}\int_0^b \left(1-\frac{x}{b}\right)^2 x dx = \frac{mhb}{12}$$

$\boxed{I_{xz} = \int_m xz\, dm = 0}$, $\boxed{I_{yz} = \int_m yz\, dm = 0}$.

The acceleration is zero, and the body-fixed coordinate system rotates with angular velocity $\vec{\Omega} = \vec{\omega} = -\omega_O \vec{j}$. Eq (9.30) becomes

$$\begin{bmatrix} M_x \\ M_y \\ M_z \end{bmatrix} = \omega_o^2 \begin{bmatrix} 0 & 0 & +1 \\ 0 & 0 & 0 \\ -1 & 0 & 0 \end{bmatrix}\begin{bmatrix} I_{xx} & -I_{xy} & 0 \\ -I_{yx} & I_{yy} & 0 \\ 0 & 0 & I_{zz} \end{bmatrix}\begin{bmatrix} 0 \\ 1 \\ 0 \end{bmatrix} = \omega_o^2 \begin{bmatrix} 0 \\ 0 \\ I_{xy} \end{bmatrix},$$ from which $M_z = \frac{mhb}{12}\omega_o^2$.

The weight of the plate is $\vec{W} = -mg\vec{j}$

Solution continued on next page

the center of mass from B is $\vec{r}_{G/B} = \frac{b}{3}\vec{i} + \frac{h}{3}\vec{j}$, from which the acceleration is

$$\vec{a}_G = \vec{a}_B + \vec{\omega}\times(\vec{\omega}\times\vec{r}_{G/B}) = \vec{\omega}\times\begin{bmatrix}\vec{i} & \vec{j} & \vec{k}\\ 0 & -\omega_o & 0\\ \frac{b}{3} & \frac{h}{3} & 0\end{bmatrix} = \begin{bmatrix}\vec{i} & \vec{j} & \vec{k}\\ 0 & -\omega_o & 0\\ 0 & 0 & -\frac{b\omega_o}{3}\end{bmatrix} = -\frac{b\omega_o^2}{3}\vec{i}$$

. [*Check:* The perpendicular distance from the axis of rotation to the center of mass is $d = \frac{b}{3}$, from which the acceleration is $\omega_o^2 d$ toward the axis of rotation. *check.*] Let $\vec{A}$, $\vec{B}$ be the reaction forces at A and B. From Newton's second law, $m\vec{a}_G = \vec{A} + \vec{B} + \vec{W}$, from which $\vec{A} + \vec{B} = m\vec{a}_G - \vec{W} = -m\left(\frac{b}{3}\right)\omega_o^2\vec{i} + mg\vec{j}$, from which the three scalar equations are obtained, (1) $A_x + B_x = -\frac{mb}{3}\omega_o^2$, $A_y + B_y = mg$, $A_z + B_z = 0$, of which only the first is necessary here. The external forces in the z direction and couples about the x axis are zero, from which $\boxed{A_z = B_z = 0}$. The moment about point B is (2) $M_z = -hA_x - \frac{b}{3}mg = -\frac{mhb}{12}\omega_o^2$. Solve (1) and (2) $\boxed{A_x = -\frac{m\omega_o^2 b}{12} - \frac{mgb}{3h}}$. $\boxed{B_x = -\frac{m\omega_o^2 b}{4} + \frac{mgb}{3h}}$

==================================<>==================================

Problem 9.76 The 10 *lb* thin circular disk is rigidly attached to the 12 *lb* slender horizontal shaft. The disk and shaft rotate about the axis of the shaft with constant angular velocity $\omega_d = 20\ rad/s$. The entire assembly rotates about the vertical axis with constant angular velocity $\omega_o = 4\ rad/s$. Determine the components of the force and couple exerted on the horizontal shaft by the disk.

Solution:

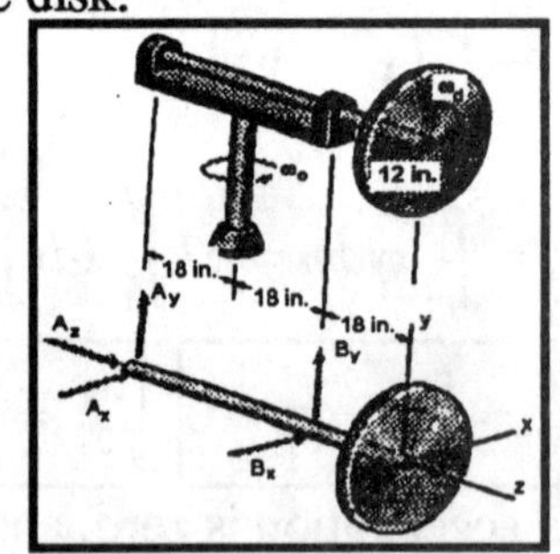

The shaft is $L = 1.5 + 1.5 + 1.5 = 4.5\ ft$ long. The mass of the disk is $m_D = \frac{10}{32.17} = 0.3108\ slug$.

The reaction of the shaft to the disk: The moments and products of inertia of the disk are: $I_{xx} = I_{yy} = \frac{m_D R^2}{4} = 0.07771\ slug\text{-}ft^2$, $I_{zz} = \frac{m_D R^2}{2} = 0.1554\ slug\text{-}ft^2$, $I_{xy} = I_{xz} = I_{yz} = 0$. The rotation rate is constant, $\vec{\alpha} = \frac{d\vec{\omega}}{dt} = 0$. The body-fixed coordinate system rotates with angular velocity $\vec{\Omega} = \omega_o\vec{j}\ (rad/s)$, and $\vec{\omega} = \omega_o\vec{j} + \omega_d\vec{k}\ (rad/s)$. Eq (9.25) reduces to:

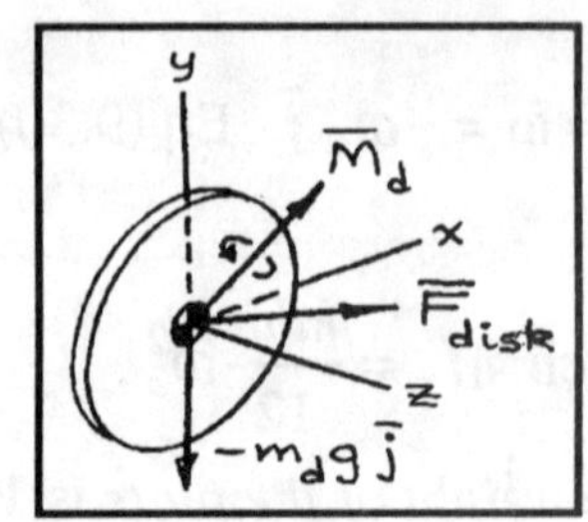

$$\begin{bmatrix}M_{dx}\\ M_{dy}\\ M_{dz}\end{bmatrix} = \omega_d\begin{bmatrix}0 & 0 & 1\\ 0 & 0 & 0\\ -1 & 0 & 0\end{bmatrix}\begin{bmatrix}I_{xx} & 0 & 0\\ 0 & I_{yy} & 0\\ 0 & 0 & I_{zz}\end{bmatrix}\begin{bmatrix}0\\ \omega_o\\ \omega_d\end{bmatrix} = \omega_o\omega_d\begin{bmatrix}I_{zz}\\ 0\\ 0\end{bmatrix}.$$

Solution continued on next page

The total moment exerted by the disk is $\vec{M}_d = \omega_o \omega_d \dfrac{m_d R^2}{2} \vec{i} = 12.43\vec{i}$ $(ft\text{-}lb)$. The reaction on the shaft by the disk is $\boxed{\vec{M}_s = -\vec{M}_d = -12.43\vec{i}\ ft\text{-}lb}$

The reaction of the shaft to the acceleration of the disk: The attachment point of the column to the shaft has the coordinates $(0, 0, -3)$ ft, from which the vector distance from the attachment point to the disk is $\vec{r}_{D/P} = 3\vec{k}$ ft. The acceleration of the disk is $\vec{a}_D = \vec{a}_P + \vec{\alpha} \times \vec{r}_{D/P} + \vec{\Omega} \times (\vec{\Omega} \times \vec{r}_{P/D}) = \vec{\Omega} \times (\vec{\Omega} \times \vec{r}_{P/D})$

$$\vec{a}_D = \vec{\Omega} \times \begin{bmatrix} \vec{i} & \vec{j} & \vec{k} \\ 0 & \omega_o & 0 \\ 0 & 0 & 3 \end{bmatrix} = \begin{bmatrix} \vec{i} & \vec{j} & \vec{k} \\ 0 & \omega_o & \\ 3\omega_o & 0 & 0 \end{bmatrix} = -3\omega_o^2 \vec{k} = -48\vec{k}\ (ft/s^2).$$ From Newton's second law,

$m_d \vec{a}_D = \vec{F}_{disk} + \vec{W} = \vec{F}_{disk} - W_d \vec{j}$, from which the external force on the disk is:

$\vec{F}_{disk} = m_d g\vec{j} - 3m_d \omega_o^2 \vec{k} = 10\vec{j} - 14.92\vec{k}$ (lb). The external force on the shaft is

$\boxed{\vec{F}_{shaft} = -\vec{F}_{disk} = -10\vec{j} + 14.92\vec{k}\ (lb)}$

==================================<>==================================

Problem 9.77 In Problem 9.76, determine the reactions exerted on the horizontal shaft by the two bearings.

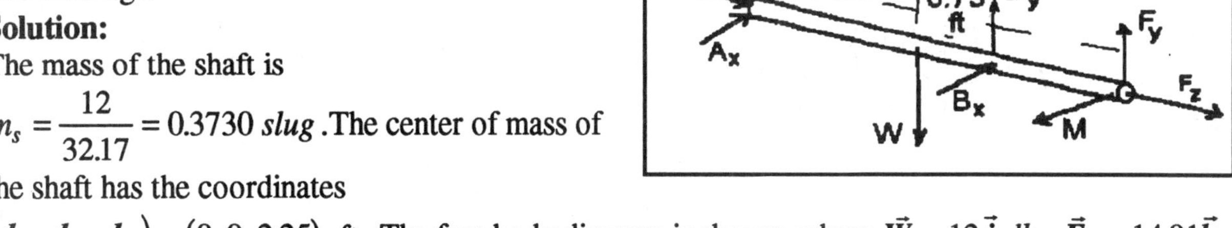

Solution:

The mass of the shaft is $m_s = \dfrac{12}{32.17} = 0.3730$ *slug*. The center of mass of the shaft has the coordinates $(d_x, d_y, d_z) = (0, 0, 2.25)$ ft. The free body diagram is shown, where $\vec{W} = 12\vec{j}$ lb, $\vec{F}_z = 14.91\vec{k}$ lb, $\vec{F}_y = -10\vec{j}$ lb, and $\vec{M} = -12.42\vec{i}$ $ft\text{-}lb$. $\vec{F}_z$, $\vec{F}_y$ and $\vec{M}$ are given by the solution to Problem 9.76, and A_x, A_y, A_z, and B_x, B_y are to be determined. Denote the point of attachment to the shaft by O. The acceleration of the center of mass of the shaft is

$\vec{a}_G = \vec{a}_O + \vec{\alpha} \times \vec{r}_{G/O} - \omega_o^2 \vec{r}_{G/O} = -\omega_o^2 (0.75)\vec{k} = -12\vec{k}$ ft/s^2, since the supporting shaft is rotating at a constant rate. From Newton's second law applied to the free body diagram: $m_s a_G = A_z + F_z$, $A_x + B_x = 0$, $A_y + B_y - W + F_y = 0$. The sum of the moments about the bar's center of mass is $0.75B_y - 2.25A_y + 2.25F_y + M = 0$, and $0.75B_x - 2.25A_x = 0$. These five equations in five unknowns are solved to obtain $\boxed{A_x = 0}$, $\boxed{A_y = 2.14\ lb}$, $\boxed{A_z = -19.38\ lb}$, $\boxed{B_x = 0}$, $\boxed{B_y = 19.86\ lb}$

==================================<>==================================

Problem 9.78 The thin rectangular plate is attached to the rectangular frame by pins. The frame rotates with constant angular velocity ω_o. Show that $\dfrac{d^2\beta}{dt^2} = -\omega_o^2 \sin\beta\cos\beta$.

Solution Assume that the only external moment applied to the object is the moment required to maintain a constant rotation ω_o about the axis of rotation. Denote this moment by $\vec{M}_o$. In the *x, y, z* system $\vec{M}_o = M_o(-\vec{i}\sin\beta + \vec{k}\cos\beta)$, from which $M_x = -M_o\sin\beta$, $M_y = 0$, $M_z = M_o\cos\beta$. From Appendix C, in the *x, y, z* system the moments and products of inertia of the plate are $I_{xx} = \dfrac{mh^2}{12}$, $I_{yy} = \dfrac{mb^2}{12}$, $I_{zz} = \dfrac{m}{12}(h^2 + b^2)$, $I_{xy} = I_{xz} = I_{yz} = 0$. The plate is attached to the frame by pins, so the assumption is that the plate is free to rotate about the *y*-axis. The body-fixed coordinate system rotates with angular velocity $\vec{\Omega} = \vec{\omega} = -\vec{i}\omega_o\sin\beta + \vec{j}\left(\dfrac{d\beta}{dt}\right) + \vec{k}\omega_o\cos\beta$ (rad/s), where $\dfrac{d\beta}{dt}$ is the angular velocity about the *y*-axis. For $\omega_o = const.$ *for all time*, the derivative $\dfrac{d\omega_o}{dt} = 0$, and the acceleration is $\vec{\alpha} = -\vec{i}\omega_o\cos\beta\left(\dfrac{d\beta}{dt}\right) + \vec{j}\left(\dfrac{d^2\beta}{dt^2}\right) - \vec{k}\omega_o\sin\beta\left(\dfrac{d\beta}{dt}\right)$. Eq (9.26) becomes

$$\begin{bmatrix} -M_o\sin\beta \\ 0 \\ M_o\cos\beta \end{bmatrix} = \begin{bmatrix} I_{xx} & 0 & 0 \\ 0 & I_{yy} & 0 \\ 0 & 0 & I_{zz} \end{bmatrix}\begin{bmatrix} \alpha_x \\ \alpha_y \\ \alpha_z \end{bmatrix} + \begin{bmatrix} 0 & -\omega_o\cos\beta & \dfrac{d\beta}{dt} \\ \omega_o\cos\beta & 0 & \omega_o\sin\beta \\ -\dfrac{d\beta}{dt} & -\omega_o\sin\beta & 0 \end{bmatrix}\begin{bmatrix} I_{xx} & 0 & 0 \\ 0 & I_{yy} & 0 \\ 0 & 0 & I_{zz} \end{bmatrix}\begin{bmatrix} -\omega_o\sin\beta \\ \dfrac{d\beta}{dt} \\ \omega_o\cos\beta \end{bmatrix}$$

$$\begin{bmatrix} -M_o\sin\beta \\ 0 \\ M_o\cos\beta \end{bmatrix} = \begin{bmatrix} I_{xx}\alpha_x \\ I_{yy}\alpha_y \\ I_{zz}\alpha_z \end{bmatrix} + \begin{bmatrix} 0 & -\omega_o I_{yy}\cos\beta & I_{zz}\dfrac{d\beta}{dt} \\ \omega_o I_{xx}\cos\beta & 0 & \omega_o I_{zz}\sin\beta \\ -I_{xx}\dfrac{d\beta}{dt} & -\omega_o I_{yy}\sin\beta & 0 \end{bmatrix}\begin{bmatrix} -\omega_o\sin\beta \\ \dfrac{d\beta}{dt} \\ \omega_o\cos\beta \end{bmatrix},$$

$$\begin{bmatrix} -M_o\sin\beta \\ 0 \\ M_o\cos\beta \end{bmatrix} = \begin{bmatrix} -\omega_o I_{xx}\cos\beta\left(\dfrac{d\beta}{dt}\right) \\ I_{yy}\left(\dfrac{d\beta}{dt}\right) \\ -\omega_o I_{zz}\sin\beta\left(\dfrac{d\beta}{dt}\right) \end{bmatrix} + \begin{bmatrix} \omega_o I_{xx}\cos\beta\left(\dfrac{d\beta}{dt}\right) \\ \omega_o^2 I_{yy}\cos\beta\sin\beta \\ \omega_o(I_{xx} - I_{yy})\sin\beta\left(\dfrac{d\beta}{dt}\right) \end{bmatrix} = \begin{bmatrix} 0 \\ I_{yy}\alpha_y + \omega_o^2 I_{yy}\cos\beta\sin\beta \\ -2\omega_o I_{yy}\sin\beta\left(\dfrac{d\beta}{dt}\right) \end{bmatrix}$$ where

$I_{zz} = I_{xx} + I_{yy}$ has been used. The *y*-component is $I_{yy}\alpha_y + \omega_o^2 I_{yy}\cos\beta\sin\beta = 0$, from which

$$\boxed{\frac{d^2\beta}{dt^2} = -\omega_o^2\cos\beta\sin\beta}.$$

==================================<>==================================

Problem 9.79 The axis of the right circular cone of mass *m*, height *h*, and radius *R* spins about the vertical axis with constant angular velocity ω_o. Its center of mass remains stationary and its base rolls on the floor. Show that the angular velocity ω_o necessary for this motion is $\omega_o = \sqrt{\frac{10g}{3R}}$. *Strategy:* Let the *z* axis remain aligned with the axis of the cone and the *x* axis remain vertical.

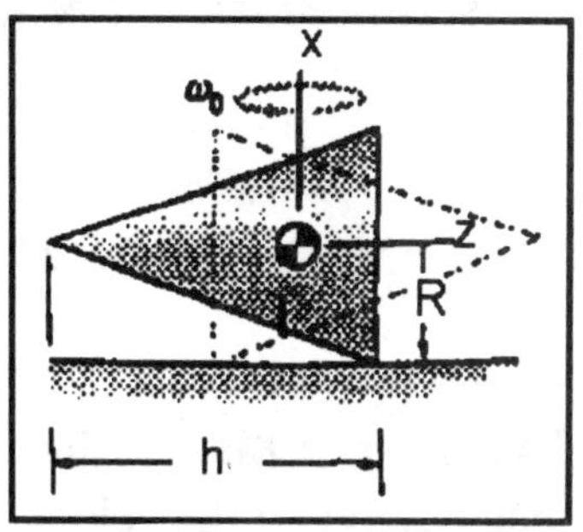

Solution:

This a problem of general motion, and Eq (9.30) applies. The vector distance from the center of mass to the base of the cone is $\vec{r}_{B/G} = \frac{h}{4}\vec{k}$ (see Appendix C). The angular velocity of rotation of the body fixed coordinate system is $\vec{\Omega} = \omega_o \vec{i}$. The velocity of the center of the base is

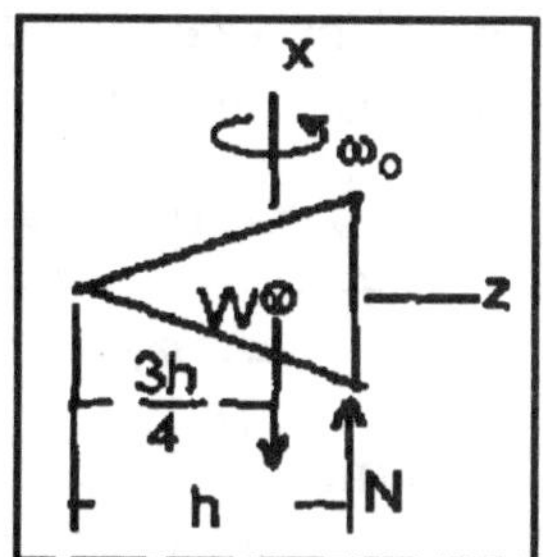

$$\vec{v}_B = \vec{\Omega} \times \vec{r}_{B/G} = \begin{bmatrix} \vec{i} & \vec{j} & \vec{k} \\ \omega_o & 0 & 0 \\ 0 & 0 & \frac{h}{4} \end{bmatrix} = -\frac{\omega_o h}{4}\vec{j}.$$ Let the spin rate about the *z* axis be $\dot{\phi}$, so that the angular velocity, from which $\vec{\omega} = \vec{\Omega} + \dot{\phi}\vec{k}$. The point of contact with the surface is stationary, and the velocity of the center of the base of the cone is $\vec{v} = -\frac{\omega_o h}{4}\vec{j}$, from which

$$0 = \vec{v} + \vec{\omega} \times \left(-R\vec{i}\right) = \begin{bmatrix} \vec{i} & \vec{j} & \vec{k} \\ \omega_o & 0 & \dot{\phi} \\ -R & 0 & 0 \end{bmatrix} = \left(-\frac{\omega_o h}{4} + R\dot{\phi}\right)\vec{j} = 0,$$ from which $\dot{\phi} = -\frac{\omega_o h}{4R}$. The center of mass of the cone is at a zero distance from the axis of rotation, from which the acceleration of the center of mass is zero. The angular velocity about the *z*-axis, $\vec{\omega} = \omega_o \vec{i} - \frac{\omega_o h}{4R}\vec{k}$ (rad/s). The weight of the cone is $\vec{W} = -mg\vec{i}$. The reaction of the floor on the cone is $\vec{N} = -\vec{W}$. The moment about the center of mass exerted by the weight is

$$\vec{M}_G = \vec{r}_{B/G} \times \vec{N} = \begin{bmatrix} \vec{i} & \vec{j} & \vec{k} \\ 0 & 0 & \frac{h}{4} \\ mg & 0 & 0 \end{bmatrix} = +\left(\frac{mgh}{4}\right)\vec{j}.$$ The moments and products of inertia of a cone about its center of mass in the *x, y, z* system are, from Appendix C,

$$I_{xx} = I_{yy} = m\left(\frac{3}{80}h^2 + \frac{3}{20}R^2\right),\ I_{zz} = \frac{3mR^2}{10},\ I_{xy} = I_{xz} = I_{yz} = 0.$$ Since the rotation rate is constant, and the *z* axis remains horizontal, the angular acceleration is zero, $\frac{d\vec{\omega}_o}{dt} = 0$. The body-fixed coordinate system rotates with angular velocity $\vec{\Omega} = \omega_o \vec{i}$, and $\vec{\omega} = \omega_o \vec{i} - \frac{h\omega_o}{4R}\vec{k}$.

Solution continued on next page

Eq (9.26) becomes: $\begin{bmatrix} M_{Gx} \\ M_{Gy} \\ M_{Gz} \end{bmatrix} = \omega_o^2 \begin{bmatrix} 0 & 0 & 0 \\ 0 & 0 & -1 \\ 0 & 1 & 0 \end{bmatrix} \begin{bmatrix} I_{xx} & 0 & 0 \\ 0 & I_{yy} & 0 \\ 0 & 0 & I_{zz} \end{bmatrix} \begin{bmatrix} 1 \\ 0 \\ -\dfrac{h}{4R} \end{bmatrix} = \omega_o^2 \begin{bmatrix} 0 \\ \dfrac{hI_{zz}}{4R} \\ 0 \end{bmatrix}$. For equilibrium,

$M_{oy} = M_{Gy}$, from which $\dfrac{mgh}{4} = \dfrac{3mhR}{40}\omega_o^2$. Solve $\boxed{\omega_o = \sqrt{\dfrac{10g}{3R}}}$

==================================<>==================================

Problem 9.80 A thin circular disk of radius R and mass m rolls along a circular path of radius r. The magnitude v of the velocity of the center of the disk and the angle θ between the disk's axis and the vertical are constants. Show that v satisfies the equation

$$v^2 = \frac{\left(\dfrac{2}{3}\right) g \cot\theta (r - R\cos\theta)^2}{r - \dfrac{5}{6} R\cos\theta}.$$

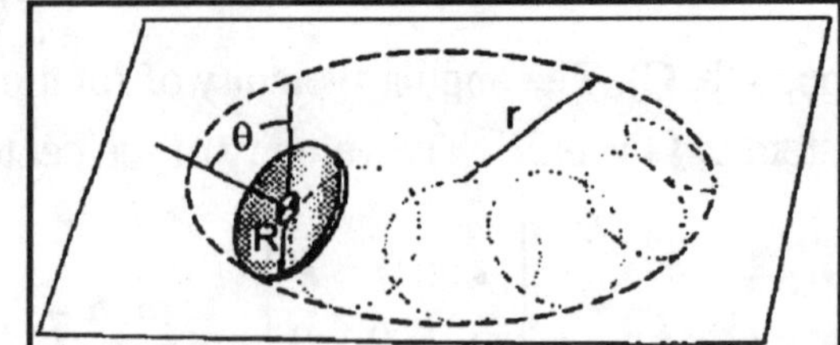

Solution:

This is a problem of general motion, and Eq(9.30) applies. The strategy is to use the Euler Equations find the velocity that will produce an external moment that equals the sum of the external moments exerted by the floor on the wheel. For the purpose of assigning directions to the angular velocities, assume that the disk rolls around the path in a counterclockwise direction. Choose a coordinate system with the origin at the center of the wheel, with the z-axis aligned with the spin of the wheel, and the x axis into the paper.

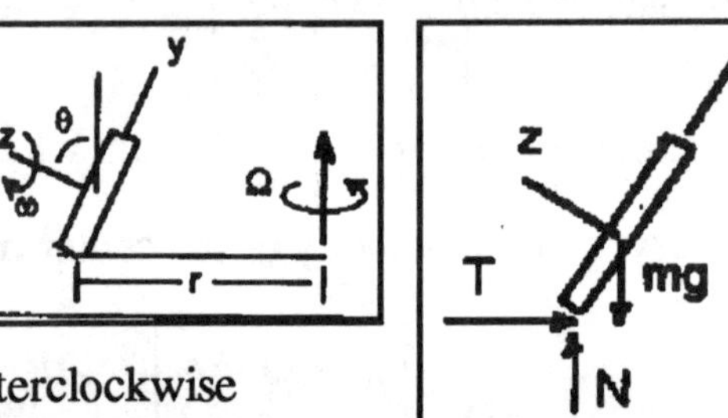

The distance from the center of the path to the center of mass is $r_{G/O} = (r - R\cos\theta)$.The vector distance is $\vec{r}_{G/O} = r_{G/O}(-\vec{j}\cos\theta + \vec{k}\sin\theta)$.Denote the spin rate of the wheel by ω_o, from which the angular velocity of the wheel is $\vec{\omega} = \vec{\Omega} + \omega_o\vec{k}$. The angular velocity of the center of mass of the wheel around the path is $\vec{\Omega} = \Omega_o(\vec{i}\cos\theta + \vec{j}\sin\theta)$ (rad/s). The velocity of the center of mass of the wheel is

$$\vec{v} = \vec{\Omega} \times \vec{r}_{G/O} = \Omega_o\, r_{G/O} \begin{bmatrix} \vec{i} & \vec{j} & \vec{k} \\ 0 & \sin\theta & \cos\theta \\ 0 & -\cos\theta & \sin\theta \end{bmatrix} = \Omega_o r_{G/O}\vec{i},$$

from which $\Omega_o = \dfrac{v}{r_{G/O}}$. Since the point P of contact of the wheel with the floor is stationary, the velocity of the center of the wheel is also given by

$$\vec{v}_P = 0 = \vec{v}_G + \vec{\omega} \times \vec{r}_{P/G} = \Omega_o r_{G/O}\vec{i} + \begin{bmatrix} \vec{i} & \vec{j} & \vec{k} \\ 0 & \Omega_o\sin\theta & \omega_o + \Omega_o\cos\theta \\ 0 & -R & 0 \end{bmatrix},$$

where $\vec{r}_{P/G} = -R\vec{j}$, from which

$\Omega_o r_{G/O} + R(\omega_o + \Omega_o\cos\theta) = 0$, from which $\omega_o = -\Omega_o\left(\dfrac{r}{R}\right)$. In terms of the coordinate systems shown, the body-fixed coordinate system rotates with angular velocity $\vec{\Omega}$, and

$\vec{\omega} = \vec{\Omega} + \omega_o\vec{k} = \Omega_o\left(\vec{j}\sin\theta - \left(\dfrac{r}{R} - \cos\theta\right)\vec{k}\right)$. The acceleration of the center of mass of the

Solution continued on next page

wheel is $\vec{a}_G = \vec{\Omega}\times(\vec{\Omega}\times\vec{r}_{G/O}) = \vec{\Omega}\times\Omega_o r_{G/O}\begin{bmatrix}\vec{i} & \vec{j} & \vec{k}\\ 0 & \sin\theta & \cos\theta\\ 0 & -\cos\theta & \sin\theta\end{bmatrix} = \Omega_o^2 r_{G/O}\begin{bmatrix}\vec{i} & \vec{j} & \vec{k}\\ 0 & \sin\theta & \cos\theta\\ 1 & 0 & 0\end{bmatrix}$,

$\vec{a}_G = \Omega_o^2 r_{G/O}(\vec{j}\cos\theta - \vec{k}\sin\theta)$. The normal force is $\vec{N} = -\vec{W} = mg(\vec{j}\sin\theta + \vec{k}\cos\theta)$. From Newton's second law , $m\vec{a}_G = \vec{T} + \vec{N} + \vec{W} = \vec{T}$, from which $\vec{T} = m\Omega_o^2 r_{G/O}(+\vec{j}\cos\theta - \vec{k}\sin\theta)$ (see figure). The moment about the center of mass is

$\vec{M}_G = \vec{r}_{P/G}\times(\vec{N}+\vec{T}) = Rmg\begin{bmatrix}\vec{i} & \vec{j} & \vec{k}\\ 0 & -1 & 0\\ 0 & \sin\theta & \cos\theta\end{bmatrix} + mR\Omega_o^2 r_{G/O}\begin{bmatrix}\vec{i} & \vec{j} & \vec{k}\\ 0 & -1 & 0\\ 0 & \cos\theta & -\sin\theta\end{bmatrix}$, from which

$\vec{M}_G = mR(\Omega_o^2 r_{G/O}\sin\theta - g\cos\theta)\vec{i} = M_x\vec{i}$. The moments and products of inertia of the disk in the coordinate system shown are $I_{xx} = \frac{mR^2}{4}$, $I_{yy} = \frac{mR^2}{4}$, $I_{zz} = \frac{mR^2}{2}$, $I_{xy} = I_{xz} = I_{yz} = 0$.

The Eq (9.30) becomes: $\begin{bmatrix}M_{ox}\\ M_{oy}\\ M_{oz}\end{bmatrix} = \Omega_o^2\begin{bmatrix}0 & -\cos\theta & \sin\theta\\ \cos\theta & 0 & 0\\ -\sin\theta & 0 & 0\end{bmatrix}\begin{bmatrix}I_{xx} & 0 & 0\\ 0 & I_{yy} & 0\\ 0 & 0 & I_{zz}\end{bmatrix}\begin{bmatrix}0\\ \omega_y\\ \omega_z\end{bmatrix}$. From which, for the x component: $M_{Ox} = -mR\Omega_o^2\sin\theta\left(\frac{r}{2} - \frac{R\cos\theta}{4}\right)$. This moment is the sum of the couple and the external moment exerted by reaction of the floor on the wheel, $\vec{M}_o = \vec{C} + \vec{M}_G$. A rolling point cannot support a couple, $\vec{C} = 0$, from which $\vec{M}_o = \vec{M}_G$. Substitute and reduce:

$-mgR\cos\theta + mR\Omega_o^2 r_{G/O}\sin\theta = mR\Omega_o^2\sin\theta\left(\frac{R\cos\theta}{4} - \frac{r}{2}\right)$, from which $\Omega_o^2 = \dfrac{-g\cot\theta}{\left(\frac{5R}{4}\right)\cos\theta - \left(\frac{3r}{2}\right)}$

.Substitute $\Omega_o^2 = \frac{v^2}{r_{G/O}^2} = \frac{v^2}{(r - R\cos\theta)^2}$ and reduce: $\boxed{v^2 = \left(\frac{2}{3}\right)g\cot\theta\,\dfrac{(r-R\cos\theta)^2}{r - \left(\frac{5}{6}\right)R\cos\theta}}$

[*Note:* See Section 244, page 196 of the Dover reprint (1955) of the Sixth Edition (1905) of Routh's Dynamics of a System of Rigid Bodies,. (This problem appears as Exercise 4, page 197.) See also page 363 of J. L. Merriam's Dynamics, 2nd Edition, 1975.]

==================================<>==================================

=================================<>=================================

Problem 9.81 The vertical shaft rotates with a constant angular velocity ω_o, causing the grinding mill to roll on the horizontal surface. Assume that point P of the mill is stationary at the instant shown, and that the force N exerted on the mill by the surface is perpendicular to the surface and acts at P. The mass of the mill is m and its moments and products of inertia in terms of the coordinate system shown are $I_{xx} = I_{yy}$, I_{zz}, and $I_{xy} = I_{xz} = I_{yz} = 0$. Determine N.

Solution:

The origin of the coordinate system O is a fixed point. Assume for purposes of assigning directions to the angular velocities that the mill moves in a counterclockwise direction viewed from above the point O. The positive x axis points into the paper. The body-fixed coordinate system rotates with the angular velocity of the vertical shaft: *Projected onto the body-fixed coordinates* $\vec{\Omega} = \omega_o(\vec{j}\sin\beta - \vec{k}\cos\beta)$. [*Check:* When the shaft is horizontal, the angular velocity vector is parallel to the y axis, with no component parallel to the z axis, as expected. *check.*] The point O is fixed, from which the velocity of the mill center of mass relative to O is

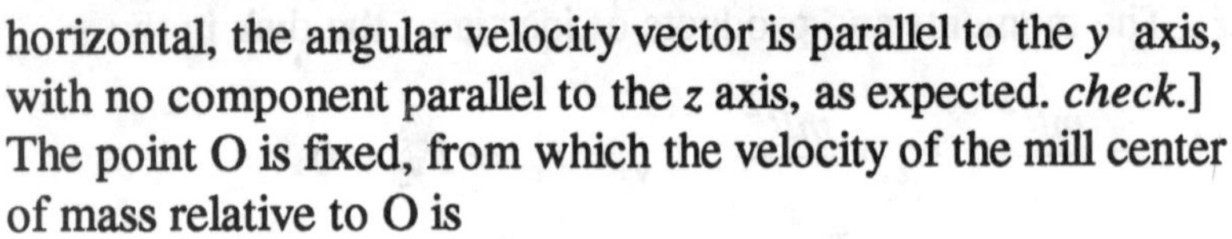

$$\vec{v}_G = \vec{\omega}_o \times \vec{r}_{G/O} = \begin{bmatrix} \vec{i} & \vec{j} & \vec{k} \\ 0 & 0 & \omega_o \\ 0 & b\sin\beta & -b\cos\beta \end{bmatrix} = -\omega_o b\sin\beta\,\vec{i}.$$

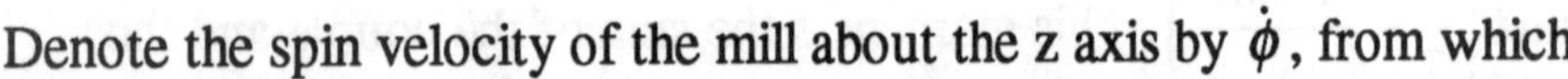

Denote the spin velocity of the mill about the z axis by $\dot{\phi}$, from which $\vec{\omega}_m = \vec{\Omega} + \dot{\phi}\vec{k} = \omega_o(\vec{j}\sin\beta + (\dot{\phi} - \omega_o\cos\beta)\vec{k})$. The point P is stationary, from which the velocity of the mill center of mass relative to O is

$$\vec{v}_G = 0 + \vec{\omega}_m \times \vec{r}_{G/P} = \begin{bmatrix} \vec{i} & \vec{j} & \vec{k} \\ 0 & \omega_o\sin\beta & \dot{\phi} - \omega_o\cos\beta \\ 0 & -r & 0 \end{bmatrix} = r(\dot{\phi} - \omega_o\cos\beta).$$ Equate and solve:

$\dot{\phi} = \omega_o(\cos\beta - \frac{b}{r}\sin\beta)$. [*Check:* When the shaft is horizontal (see Problem 9.77), the spin of the mill is $-\frac{b}{r}\omega_o$, as to be expected. *check.*]

The angular velocity of the mill is $\vec{\omega}_m = \vec{\Omega} + \dot{\phi}\vec{k} = \omega_o\left(\vec{j}\sin\beta - \vec{k}\frac{b}{r}\sin\beta\right)$

The moment about the point O: The weight of the mill is $\vec{W} = -mg\vec{j}$. $\vec{M}_O = \vec{r}_{O/P} \times \vec{N} + \vec{r}_{O/G} \times \vec{W}$,

$$\vec{M}_O = N\begin{bmatrix} \vec{i} & \vec{j} & \vec{k} \\ 0 & -r & b \\ 0 & \sin\beta & -\cos\beta \end{bmatrix} 0 - mg\begin{bmatrix} \vec{i} & \vec{j} & \vec{k} \\ 0 & 0 & b \\ 0 & \sin\beta & -\cos\beta \end{bmatrix}$$

$\vec{M}_O = (N(r\cos\beta - b\sin\beta) + mgb\sin\beta)\vec{i} = M_{ox}\vec{i}$.

Solution continued on next page

[*Check*: The perpendicular distance from the vertical shaft to the center of mass is $b\sin\beta$, from which the moment about O exerted by the weight about the x axis is is counterclockwise looking toward the origin, from which $M_x^{(W)} = mgb\sin\beta$. *check*.

From geometry, the perpendicular distance from the vertical shaft to the point P is $d_N = b\sin\beta - r\cos\beta$. The moment exerted by N about the x axis is clockwise looking toward the origin, from which $M_x^{(N)} = -d_N N = N(r\cos\beta - b\sin\beta)$. The sum of the two moments is the total moment about the x axis. *check*.]

The Euler Equations: Since the point O is a fixed point, Eq (9.26) applies. The angular velocity $\omega_o = const$ and the angle β is constant, the angular accelerations are zero. Eq (9.26) becomes:

$$\begin{bmatrix} M_{ox} \\ M_{oy} \\ M_{oz} \end{bmatrix} = \begin{bmatrix} 0 & -\Omega_z & \Omega_y \\ \Omega_z & 0 & 0 \\ -\Omega_y & 0 & 0 \end{bmatrix} \begin{bmatrix} I_{xx} & 0 & 0 \\ 0 & I_{yy} & 0 \\ 0 & 0 & I_{zz} \end{bmatrix} \begin{bmatrix} 0 \\ \omega_y \\ \omega_z \end{bmatrix} = \begin{bmatrix} -\Omega_z I_{yy}\omega_y + \Omega_y I_{zz}\omega_z \\ 0 \\ 0 \end{bmatrix}.$$

Substitute $\Omega_z = -\omega_o\cos\beta$, $\Omega_y = \omega_o\sin\beta$, $\omega_y = \omega_o\sin\beta$, $\omega_z = -\omega_o\dfrac{b}{r}\sin\beta$.,

$M_{ox} = \omega_o^2\left(I_{yy}\cos\beta\sin\beta - I_{zz}\left(\dfrac{b}{r}\right)\sin^2\beta\right)$. Substitute for M_{ox} and solve:

$$\boxed{N = \frac{\omega_o^2\left(I_{yy}\cos\beta\sin\beta - I_{zz}\left(\frac{b}{r}\right)\sin^2\beta\right) - mgb\sin\beta}{r\cos\beta - b\sin\beta}}.$$

==================================<>==================================

Problem 9.82 The view of an airplane's landing gear looking from behind the airplane is shown in Fig. (a). The radius of the wheel is 300 *mm* and its moment of inertia is 2 *kg-m*^{2}. The airplane takes off at 30 *m/s*. After takeoff, the landing gear retracts by rotating toward the right side of the airplane as shown in Fig.(b). Determine the magnitude of the couple exerted by the wheel on its support.

Solution:

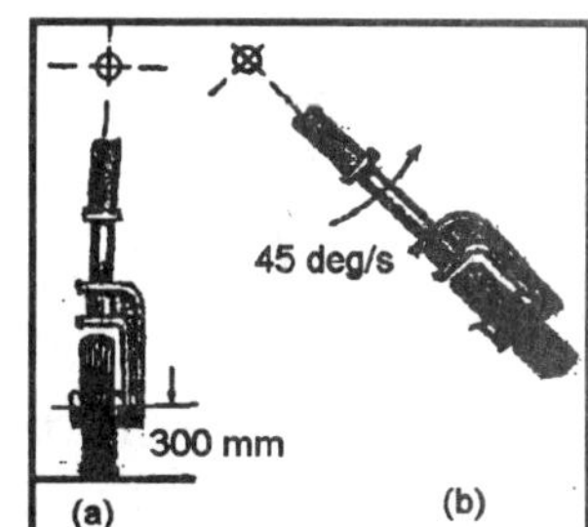

Choose a coordinate system with the origin at the center of mass of the wheel and the z axis aligned with the carriage, as shown.

Assume that the angular velocities are constant, so that the angular accelerations are zero. The moments and products of inertia of the wheel are $I_{xx} = mR^2/2 = 2\text{ kg-m}^2$, from which $m = 44.44\ kg$.

$I_{yy} = I_{zz} = mR^2/4 = 1\text{ kg-m}^2$. The angular velocities are

$\vec{\Omega} = -(45(\pi/180))\vec{j} = -0.7853\vec{j}$ rad/s.

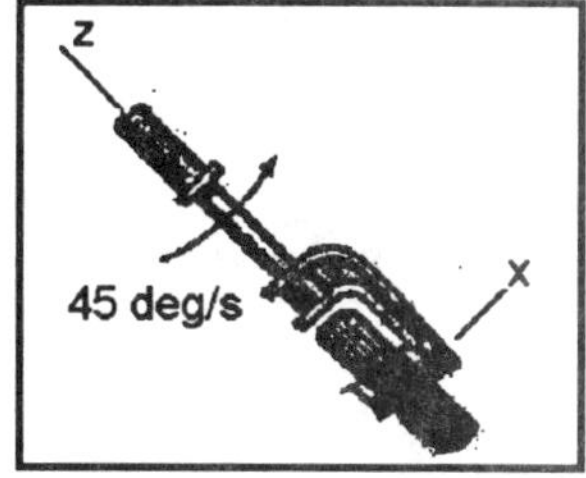

$\vec{\omega} = -\left(\dfrac{v}{R}\right)\vec{i} + \vec{\Omega} = -\left(\dfrac{30}{0.3}\right)\vec{i} - 0.7853\vec{j} = -100\vec{i} - 0.7853\vec{j}$. Eq (9.30)

becomes $$\begin{bmatrix} M_x \\ M_y \\ M_z \end{bmatrix} = \begin{bmatrix} 0 & 0 & \Omega_y \\ 0 & 0 & 0 \\ -\Omega_y & 0 & 0 \end{bmatrix} \begin{bmatrix} I_{xx} & 0 & 0 \\ 0 & I_{yy} & 0 \\ 0 & 0 & I_{zz} \end{bmatrix} \begin{bmatrix} \omega_x \\ \omega_y \\ 0 \end{bmatrix} = \begin{bmatrix} 0 \\ 0 \\ -\Omega_y\omega_x I_{xx} \end{bmatrix},$$

from which $\vec{M}_o = -\Omega_y\omega_x I_{xx}\vec{k}$. Substitute: $\boxed{|\vec{M}| = (0.7854)(100)2 = 157.1\ N\text{-}m}$

==================================<>==================================

===================================<>===================================

Problem 9.83 If the rider turns to his left, will the couple exerted on the motorcycle by its wheels tend to cause the motorcycle to turn lean toward the rider's left side or his right side?

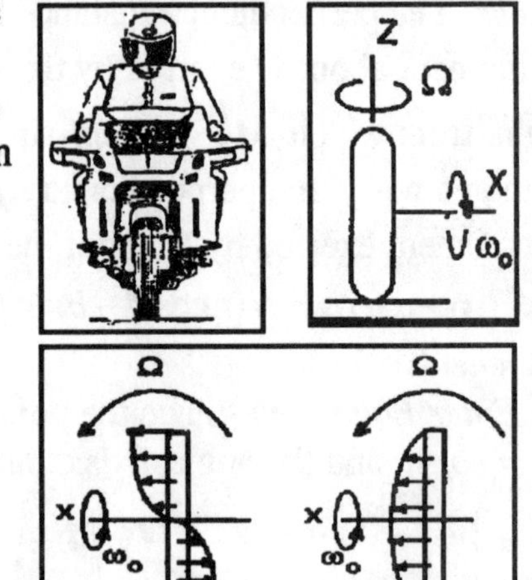

Solution Choose a coordinate system as shown in the front view, with y positive into the paper. The Eqs (9.26) in condensed notation are $\sum \vec{M} = \frac{d\vec{H}}{dt} + \vec{\Omega} \times \vec{H}$. For $\frac{d\vec{H}}{dt} = 0$, $\sum \vec{M} = \vec{\Omega} \times \vec{H}$. If the rider turns to his left, the angular velocity is $\vec{\Omega} = +\Omega\vec{k}$ rad/s. The angular momentum is $\vec{H} = H_x\vec{i} + H_z\vec{k}$, where $H_x > 0$. The cross product $\vec{\Omega} \times \vec{H} = \begin{bmatrix} \vec{i} & \vec{j} & \vec{k} \\ 0 & 0 & +\Omega \\ H_x & 0 & H_z \end{bmatrix} = +\Omega H_x \vec{j}$. For a left turn the moment about y is positive, causing the cycle <u>to lean to the left</u>. [*Check*: This result is supported by a heuristic argument based on the acceleration of a mass particle on the rim of the wheel. Figs (a) and (b) show a *top view* of the front wheel, (as seen by the rider). The motion of the cycle is forward, along the negative y axis. As the wheel turns left with a constant angular velocity Ω, *the x component of velocity* of a mass particle on the rim of the wheel first has a negative value, and then as the wheel rolls, becomes less negative, passes through zero, and then reaches a maximum positive value at the forward point on the wheel, where it begins to decrease. The envelope of the x component of velocity is graphed in Fig (a). For constant rotation, the velocity is a sine function of time (as measured from the top of the wheel). *The x component of particle acceleration* associated with this change in velocity is a cosine function of time, and is graphed in Fig (b). The acceleration reaches a maximum at the top of the wheel *toward the left.* Thus the acceleration has the same direction as the velocity induced by Ω, shifted by 90° around the wheel. From Newton's second law, the moment about the y axis causes the cycle to *lean to the left. check.*]

===================================<>===================================

Problem 9.84 By substituting the components of $\vec{H}_o$ from Eqs (9.7) into Eq (9.24), derive Eqs (9.25).

Solution: The Eqs (9.7) are (page 472 of text)

$H_{ox} = I_{xx}\omega_x - I_{xy}\omega_y - I_{xz}\omega_z$

$H_{oy} = -I_{yx}\omega_x + I_{yy}\omega_y - I_{yz}\omega_z$

$H_{oz} = -I_{zx}\omega_x - I_{zy}\omega_y + I_{zz}\omega_z$. The products of inertia are symmetric, $I_{xy} = I_{yx}$, $I_{xz} = I_{zx}$, $I_{yz} = I_{zy}$ (See Eqs (9.9).) The matrix notation simplifies bookkeeping:

(1) $\begin{bmatrix} H_{ox} \\ H_{oy} \\ H_{oz} \end{bmatrix} = \begin{bmatrix} I_{xx} & -I_{xy} & -I_{xz} \\ -I_{yx} & I_{yy} & -I_{yz} \\ -I_{zx} & -I_{zy} & I_{zz} \end{bmatrix} \begin{bmatrix} \omega_x \\ \omega_y \\ \omega_z \end{bmatrix}$. (See Eq (9.10).)

Eq (9.24) is $\sum \vec{M}_o = \left(\frac{dH_{ox}}{dt}\right)\vec{i} + \left(\frac{dH_{oy}}{dt}\right)\vec{j} + \left(\frac{dH_{oz}}{dt}\right)\vec{k} + \vec{\Omega} \times \vec{H}_o$, which can be written in condensed notation as $\sum \vec{M}_o = \frac{d\vec{H}_o}{dt} + \vec{\Omega} \times \vec{H}_o$. Term by term, substituting $\vec{H}_o$,

Solution continued on next page

$$\frac{d\vec{H}_o}{dt}=\frac{d}{dt}\begin{bmatrix} I_{xx} & -I_{xy} & -I_{xz} \\ -I_{yx} & I_{yy} & -I_{yz} \\ -I_{zx} & -I_{zy} & I_{zz} \end{bmatrix}\begin{bmatrix} \omega_x \\ \omega_y \\ \omega_z \end{bmatrix}=\frac{d}{dt}([I]\vec{w})=\frac{d[I]}{dt}\vec{w}+[I]\frac{d\vec{w}}{dt}.$$

The moments and products of inertia are constants (do not vary with time), from which $\frac{d[I]}{dt}=0$, and

$$(2)\ \frac{d\vec{H}_o}{dt}=[I]\frac{d\vec{w}}{dt}.\ \vec{\Omega}\times\vec{H}_o=\begin{bmatrix} \vec{i} & \vec{j} & \vec{k} \\ \Omega_x & \Omega_y & \Omega_z \\ H_{ox} & H_{oy} & H_{oz} \end{bmatrix}=\begin{bmatrix} \Omega_y H_{oz}-\Omega_z H_{oy} \\ -(\Omega_x H_{oz}-\Omega_z H_{ox}) \\ \Omega_x H_{oy}-\Omega_y H_{ox} \end{bmatrix}.$$

The vector result can be written as the product of an skew-symmetric matrix and a vector. The skew-symmetric matrix is constructed by inspection: $\vec{\Omega}\times\vec{H}_o=\begin{bmatrix} 0 & -\Omega_z & \Omega_y \\ \Omega_z & 0 & -\Omega_x \\ -\Omega_y & \Omega_x & 0 \end{bmatrix}\begin{bmatrix} H_{ox} \\ H_{oy} \\ H_{oz} \end{bmatrix}$. Substitute (1) above: $\vec{\Omega}\times\vec{H}_o=\begin{bmatrix} 0 & -\Omega_z & \Omega_y \\ \Omega_z & 0 & -\Omega_x \\ -\Omega_y & \Omega_x & 0 \end{bmatrix}\begin{bmatrix} I_{xx} & -I_{xy} & -I_{xz} \\ -I_{yx} & I_{yy} & -I_{yz} \\ -I_{zx} & -I_{zy} & I_{zz} \end{bmatrix}\begin{bmatrix} \omega_x \\ \omega_y \\ \omega_x \end{bmatrix}$. Collect terms:

$$\sum\vec{M}_o=\begin{bmatrix} I_{xx} & -I_{xy} & -I_{xz} \\ -I_{yx} & I_{yy} & -I_{yz} \\ -I_{zx} & -I_{zy} & I_{zz} \end{bmatrix}\begin{bmatrix} \frac{d\omega_x}{dt} \\ \frac{d\omega_y}{dt} \\ \frac{d\omega_z}{dt} \end{bmatrix}+\begin{bmatrix} 0 & -\Omega_z & \Omega_y \\ \Omega_z & 0 & -\Omega_x \\ -\Omega_y & \Omega_x & 0 \end{bmatrix}\begin{bmatrix} I_{xx} & -I_{xy} & -I_{xz} \\ -I_{yx} & I_{yy} & -I_{yz} \\ -I_{zx} & -I_{zy} & I_{zz} \end{bmatrix}\begin{bmatrix} \omega_x \\ \omega_y \\ \omega_z \end{bmatrix}$$

(See Eq (9.25).) In condensed notation: $\sum\vec{M}_o=[I]\frac{d\vec{w}}{dt}+[\Omega][I]\vec{w}$. Carry out the multiplication:

$$[I]\frac{d\vec{w}}{dt}=\begin{bmatrix} I_{xx}\frac{d\omega_x}{dt}-I_{xy}\frac{d\omega_y}{dt}-I_{xz}\frac{d\omega_z}{dt} \\ -I_{xy}\frac{d\omega_x}{dt}+I_{yy}\frac{d\omega_y}{dt}-I_{yz}\frac{d\omega_z}{dt} \\ -I_{xz}\frac{d\omega_x}{dt}-I_{yz}\frac{d\omega_y}{dt}I_{zz}\frac{d\omega_z}{dt} \end{bmatrix}$$

, where symmetry of the inertia matrix , $I_{xy}=I_{yx}$, $I_{xz}=I_{zx}$, $I_{yz}=I_{zy}$ has been used.

$$[I]\vec{w}=\begin{bmatrix} I_{xx} & -I_{xy} & -I_{xz} \\ -I_{yx} & I_{yy} & -I_{yz} \\ -I_{zx} & -I_{zy} & I_{zz} \end{bmatrix}\begin{bmatrix} \omega_x \\ \omega_y \\ \omega_z \end{bmatrix}=\begin{bmatrix} I_{xx}\omega_x-I_{xy}\omega_y-I_{xz}\omega_z \\ -I_{xy}\omega_x+I_{yy}\omega_y-I_{yz}\omega_z \\ -I_{xz}\omega_x-I_{yz}\omega_y+I_{zz}\omega_z \end{bmatrix},$$

$$[\Omega][I]\vec{w}=\begin{bmatrix} 0 & -\Omega_z & \Omega_y \\ \Omega_z & 0 & -\Omega_x \\ -\Omega_y & \Omega_x & 0 \end{bmatrix}\begin{bmatrix} I_{xx}\omega_x-I_{xy}\omega_y-I_{xz}\omega_z \\ -I_{xy}\omega_x+I_{yy}\omega_y-I_{yz}\omega_z \\ -I_{xz}\omega_x-I_{yz}\omega_y+I_{zz}\omega_z \end{bmatrix}$$

Solution continued on next page

$$,[\Omega][I]\vec{w}=\begin{bmatrix} -\Omega_z(-I_{xy}\omega_x+I_{yy}\omega_y-I_{yz}\omega_z)+\Omega_y(-I_{xz}\omega_x-I_{yz}\omega_y+I_{zz}\omega_z) \\ \Omega_z(I_{xx}\omega_x-I_{xy}\omega_y-I_{xz}\omega_z)-\Omega_x(-I_{xz}\omega_x-I_{yz}\omega_y+I_{zz}\omega_z) \\ -\Omega_y(I_{xx}\omega_x-I_{xy}\omega_y-I_{xz}\omega_z)+\Omega_x(-I_{xy}\omega_x+I_{yy}\omega_y-I_{yz}\omega) \end{bmatrix}.$$

Collect terms.

$$\sum M_{ox}=I_{xx}\frac{d\omega_x}{dt}-I_{xy}\frac{d\omega_y}{dt}-I_{xz}\frac{d\omega_z}{dt}-\Omega_z(-I_{xy}\omega_x+I_{yy}\omega_y-I_{yz}\omega_z)+\Omega_y(-I_{xz}\omega_x-I_{yz}\omega_y+I_{zz}\omega_z)$$

$$\sum M_{oy}=-I_{xy}\frac{d\omega_x}{dt}+I_{yy}\frac{d\omega_y}{dt}-I_{yz}\frac{d\omega_z}{dt}+\Omega_z(I_{xx}\omega_x-I_{xy}\omega_y-I_{xz}\omega_z)-\Omega_x(-I_{xz}\omega_x-I_{yz}\omega_y+I_{zz}\omega_z)$$

$$\sum M_{oz}=-I_{xz}\frac{d\omega_x}{dt}-I_{yz}\frac{d\omega_y}{dt}+I_{zz}\frac{d\omega_z}{dt}-\Omega_y(I_{xx}\omega_x-I_{xy}\omega_y-I_{xz}\omega_z)+\Omega_x(-I_{xy}\omega_x+I_{yy}\omega_y-I_{yz}\omega)$$

which are the Eqs (9.25). This completes the demonstration.

==================================<>==================================

Problem 9.85 A ship has a turbine engine. The spin axis of the axisymmetric turbine is horizontal and aligned with the ship's longitudinal axis. The turbine rotates at 10,000 *rpm* (revolutions per minute). Its moment of inertia about its spin axis is 1000 *kg-m*^{2}. If the ship turns at a constant rate of 20 degrees per minute, what is the magnitude of the moment exerted on the ship by the turbine? (*Strategy:* Treat the turbine's motion as steady precession with nutation angle $\theta = 90^o$.)

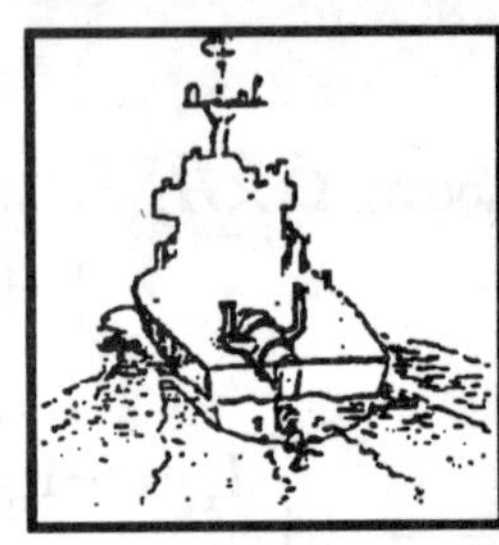

Solution:

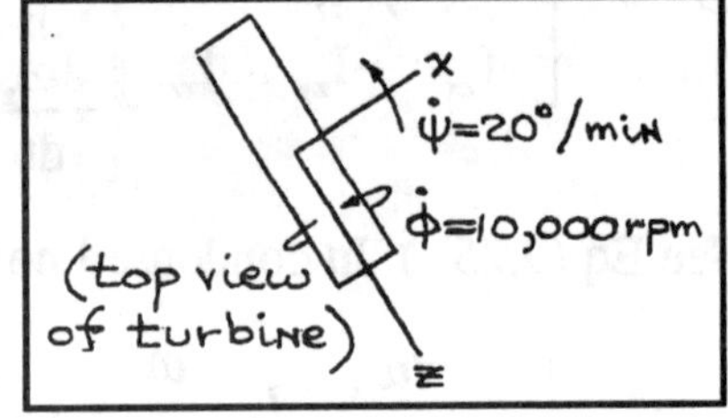

Choose a coordinate system with the z axis parallel to the axis of the turbine, and y positive upward. From Eq (9.39),

$\sum M_x=(I_{zz}-I_{xx})\dot{\psi}^2\sin\theta\cos\theta+I_{zz}\,\dot{\phi}\,\dot{\psi}\sin\theta$, where

$I_{xx}=\frac{1}{2}I_{zz}=500\ \text{kg-m}^2$

$\dot{\psi}=20\left(\frac{\pi}{180}\right)\left(\frac{1}{60}\right)=0.005818\ \text{(rad/s)},$

$\dot{\phi}=10000(2\pi/60)=1047.2\ \text{rad/s},\ \theta=90^o.M_x=6092.3\ \text{N-m}$

==================================<>==================================

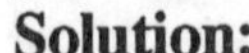

Problem 9.86 The center of the car's wheel A travels in a circular path about O at 15 *mi/hr*. The wheel's radius is 1 *ft* and its moment of inertia about its axis of rotation is 0.8 *slug-ft*^{2}. What is the magnitude of the total external moment about the wheel's center of mass? (*Strategy:* Treat the wheel's motion as steady precession with nutation angle $\theta = 90^o$.

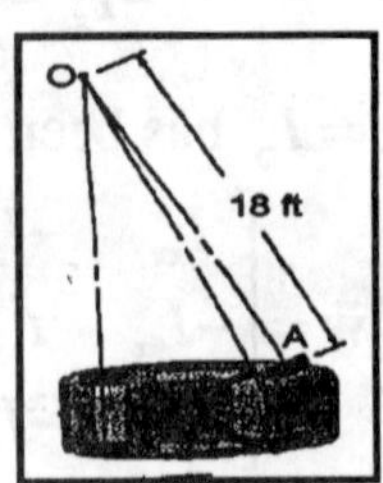

Solution:

From Eq (9.38) $\sum M_x=(I_{zz}-I_{xx})\dot{\psi}^2\sin\theta\cos\theta+I_{zz}\,\dot{\phi}\,\dot{\psi}\sin\theta$, where the spin is

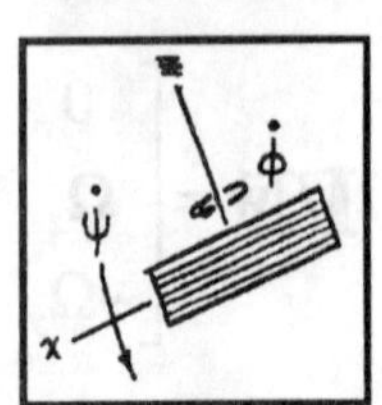

$\dot{\phi}=\frac{v}{R}=\frac{15}{18}\left(\frac{5280}{3600}\right)=\frac{22}{18}=1.222\ \text{rad/s}$, the precession rate is

$\dot{\psi}=\frac{v}{R_w}=\frac{22}{1}=22\ \text{rad/s}$, and the nutation angle is $\theta=90^o$.

Using $I_{xx}=\frac{1}{2}I_{zz}=0.4\ \text{slug-ft}^2$, from which $M_x=21.51\ \text{ft-lb}$

==================================<>==================================

=====================================<>=====================================

Problem 9.87 Solve Problem 9.74 by treating the motion as steady precession.

Solution:

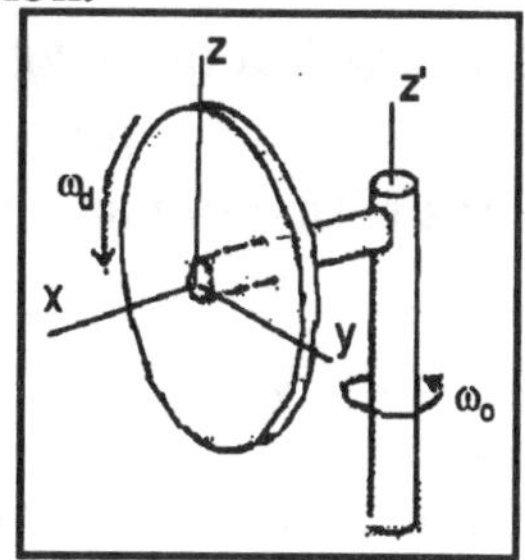

In Problem 9.74 a thin circular disk of mass m is mounted on a horizontal shaft and rotates relative to the shaft with constant angular velocity ω_d. The horizontal shaft is rigidly attached to the vertical shaft rotating with constant angular velocity ω_o. The magnitude of the couple exerted on the disk by the horizontal shaft is to be determined. The nutation angle is $\theta = 90^o$. The precession rate is $\dot{\psi} = \omega_o$, and the spin rate is $\dot{\phi} = \omega_d$. The moments and products of inertia of the disk: $I_{xx} = I_{zz} = \dfrac{mR^2}{2}$, $I_{yy} = \dfrac{mR^2}{4}$, $I_{xy} = I_{xz} = I_{yz} = 0$.

Eq (9.38) is $M_y = (I_{xx} - I_{yy})\dot{\psi}^2 \sin\theta\cos\theta + I_{xx}\dot{\psi}\dot{\phi}\sin\theta$, from which $M_y = \dfrac{mR^2}{2}\omega_o\omega_d$.

=====================================<>=====================================

Problem 9.88 Solve Problem 9.79 by treating the motion as steady precession.

Solution:

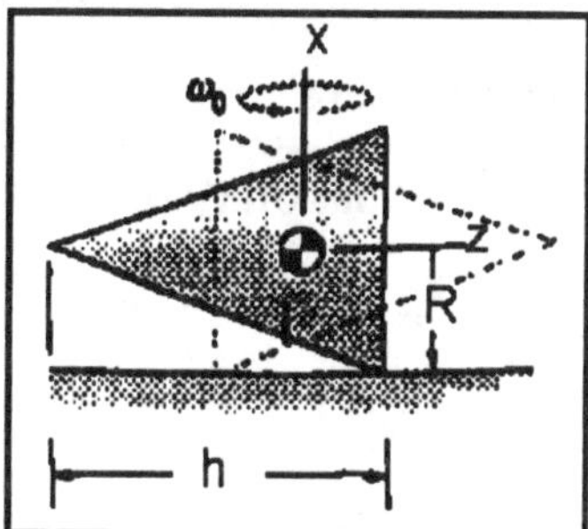

In Problem 9.79, the axis of the right circular cone of mass m, height h, and radius R spins about the vertical axis with constant angular velocity ω_o. Its center of mass remains stationary and its base rolls on the floor. The angular velocity ω_o necessary for this motion is to be determined. The strategy is to determine the angular velocity required to balance the moment due to the weight of the cone. The nutation angle is $\theta = 90^o$. The precession rate is $\dot{\psi} = \omega_o$. The distance from the center of mass to the base of the cone is $d = \dfrac{h}{4}$. The velocity of the center of the base is $v = -\dfrac{\omega_o h}{4}$, from which the spin axis is the z axis and the spin rate is $\dot{\phi} = \dfrac{v}{R} = -\dfrac{\omega_o h}{4R}$.

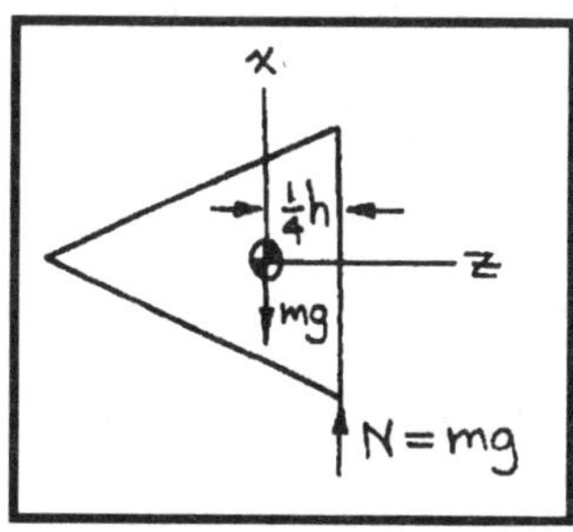

The moments and products of inertia of a cone in the x, y, z system are, from Appendix C, $I_{xx} = I_{yy} = m\left(\dfrac{3}{80}h^2 + \dfrac{3}{20}R^2\right)$, $I_{zz} = \dfrac{3mR^2}{10}$, $I_{xy} = I_{xz} = I_{yz} = 0$. Eq (9.39) is $M_y = (I_{zz} - I_{yy})\dot{\psi}^2 \sin\theta\cos\theta + I_{zz}\dot{\psi}\dot{\phi}\sin\theta$, from which $M_y = -\dfrac{3mhR}{40}\omega_o^2$. The weight of the cone is $W = -\text{mg}$ along the negative x axis. The moment is $M_y^{(mg)} = \dfrac{mgh}{4}$. The sum of the moments is zero $\sum M_y = M_y + M_y^{(mg)} = 0$ from which $\omega_o = \sqrt{\dfrac{10g}{3R}}$:

=====================================<>=====================================

==================================<>==================================

Problem 9.89 The bent bar is rigidly attached to the vertical shaft, which rotates with constant angula velocity ω_0. The thin circular disk of mass m and radius R is pinned to the bent bar and rotates with constant angular velocity ω_d relative to the bar. Determine the magnitudes of the force and couple exerted on the disk by the bar.

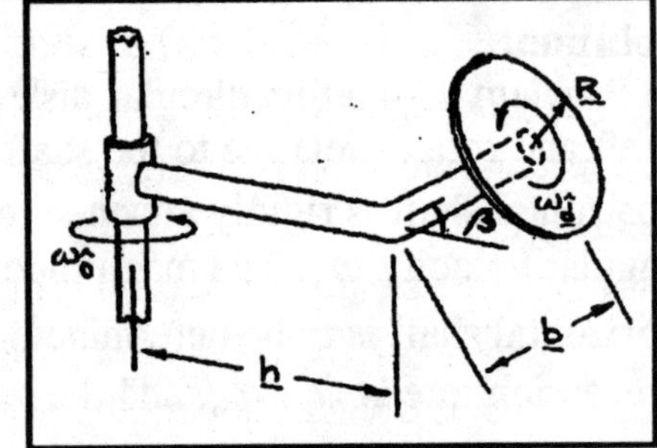

Solution:

(a) The center of mass of the disk moves in a horizontal circular path of radius $h+b\cos\beta$ with angular velocity ω_0. The acceleration normal to the circular path is $a_N=\omega_0^2(h+b\cos\beta)$, so the bar exerts a horizontal force of magnitude $ma_N=m\omega_0^2(h+b\cos\beta)$. The bar also exerts on upward force equal to the weight of the disk, so the magnitude of the total force is $\sqrt{(ma_N)^2+(mg)^2}=m\sqrt{\omega_0^4(h+b\cos\beta)^2+g^2}$.

(b) By orienting a coordinate system as shown, with the z axis normal to the disk and the x axis horizontal, the disk is in steady precession with precession rate $\dot\psi=\omega_0$, spin rate $\dot\phi=\omega_d$, and nutation angle $\theta=\dfrac{\pi}{2}-\beta$. The plate's moments of inertia are $I_{xx}=I_{yy}=\dfrac{1}{4}mR^2$, $I_{zz}=\dfrac{1}{2}mR^2$.

From Equation (9.41), the magnitude of the moment is

$$(I_{zz}-I_{xx})\dot\psi^2\sin\theta\cos\theta+I_{zz}\dot\phi\dot\psi\sin\theta=\frac{1}{4}mR^2\omega_0^2\sin\left(\frac{\pi}{2}-\beta\right)\cos\left(\frac{\pi}{2}-\beta\right)+\frac{1}{2}mR^2\omega_d\omega_0\sin\left(\frac{\pi}{2}-\beta\right)$$

$$=R^2\omega_0 m\left(\frac{1}{4}\omega_0\cos\beta\sin\beta+\frac{1}{2}\omega_d\cos\beta\right)$$

==================================<>==================================

Problem 9.90 In Problem 9.89, determine the value of the angular velocity ω_d which causes no couple to be exerted on the disk by the bar.

Solution:

From the result for the magnitude of the moment in the solution of Problem 9.88 the moment equals zero if $\dfrac{1}{4}\omega_0\sin\beta+\dfrac{1}{2}\omega_d=0$, so $\omega_d=-\dfrac{1}{2}\omega_0\sin\beta$

==================================<>==================================

Problem 9.91 A thin circular disk undergoes moment-free steady precession. The z axis is perpendicular to the disk. Show that the disk's precession rate is $\dot\psi=-\dfrac{2\dot\phi}{\cos\theta}$. (Notice that when the nutation angle is small, the precession rate is approximately two times the spin rate.)

Solution:

Moment free steady precession is described by Eq(9.44), $(I_{zz}-I_{xx})\dot\psi\cos\theta+I_{zz}\dot\phi=0$, where $\dot\psi$ is the precession rate, $\dot\phi$ is the spin rate, and θ is the nutation angle. For a thin circular disk, the moments and products of inertia are $I_{xx}=I_{yy}=\dfrac{mR^2}{4}$, $I_{zz}=\dfrac{mR^2}{2}$,

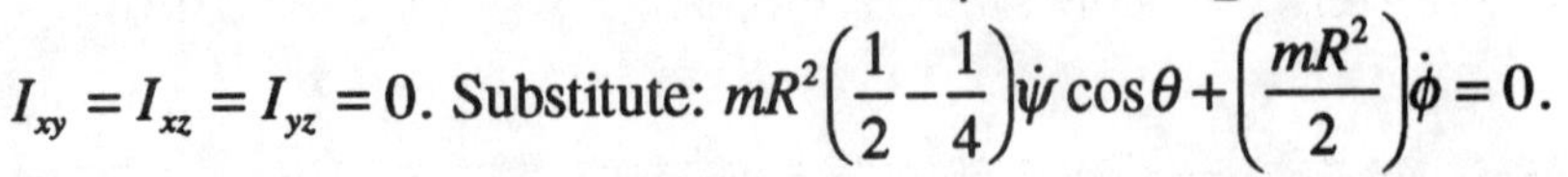

$I_{xy}=I_{xz}=I_{yz}=0$. Substitute: $mR^2\left(\dfrac{1}{2}-\dfrac{1}{4}\right)\dot\psi\cos\theta+\left(\dfrac{mR^2}{2}\right)\dot\phi=0.$

Reduce, to obtain $\dot\psi=-\dfrac{2\dot\phi}{\cos\theta}$. When the nutation angle is small, $\theta\to0$, $\cos\theta\to1$, and $\dot\psi\cong-2\dot\phi$.

==================================<>==================================

Problem 9.92 The rocket is in moment-free steady precession with nutation angle $\theta = 40^o$ and spin rate $\dot{\phi} = 4$ revolutions per second. Its moments of inertia are $I_{xx} = 10{,}000\ \text{kg-m}^2$, and $I_{zz} = 2{,}000\ \text{kg-m}^2$. What is the rocket's precession rate $\dot{\psi}$ in revolutions per second?

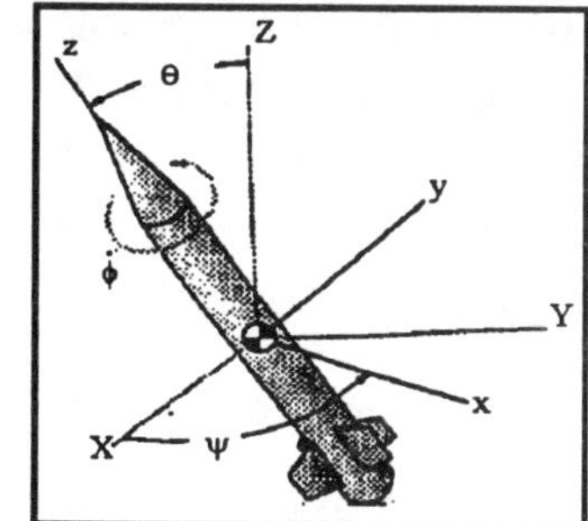

Solution:

Moment-free steady precession is described by Eq (9.44), $(I_{zz} - I_{xx})\dot{\psi}\cos\theta + I_{zz}\dot{\phi} = 0$, where $\dot{\psi}$ is the precession rate, $\dot{\phi}$ is the spin rate, and θ is the nutation angle. Solve for the precession rate: $\dot{\psi} = \dfrac{I_{zz}\dot{\phi}}{(I_{xx} - I_{zz})\cos\theta} = 1.305\ \text{rev/s}$

Problem 9.93 Sketch the body and space cones for the motion of the rocket in Problem 9.92.

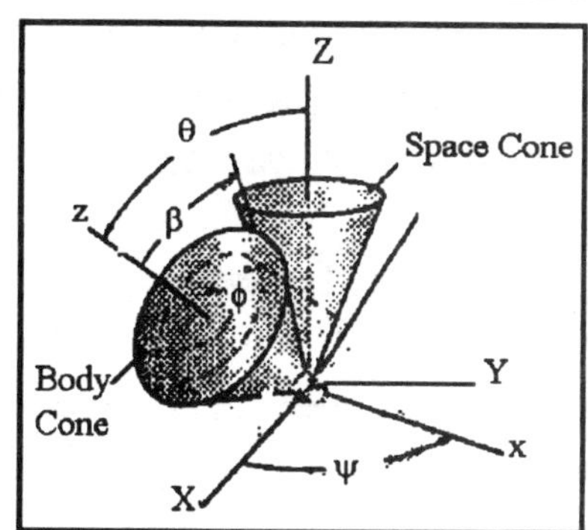

Solution:

The angle β defined by $\beta = \tan^{-1}\left[\left(\dfrac{I_{zz}}{I_{xx}}\right)\tan\theta\right] = 9.53^\circ < \theta$

satisfies the condition $\beta < \theta$ (see discussion on pages 509 –510). The body cone with an axis along the z axis, rolls on a space cone with axis on the Z axis. The result is shown.

Problem 9.94 The top is in steady precession with nutation angle $\theta = 15^\circ$ and precession rate $\dot{\psi} = 1$ revolution per second. The mass of the top is 8×10^{-4} slug, its center of mass is 1 *in.* from the point, and its moments of inertia are $I_{xx} = 6\times10^{-6}\ \text{slug-ft}^2$ and $I_{zz} = 2\times10^{-6}\ \text{slug-ft}^2$. What is the spin rate $\dot{\phi}$ of the top in revolutions per second?

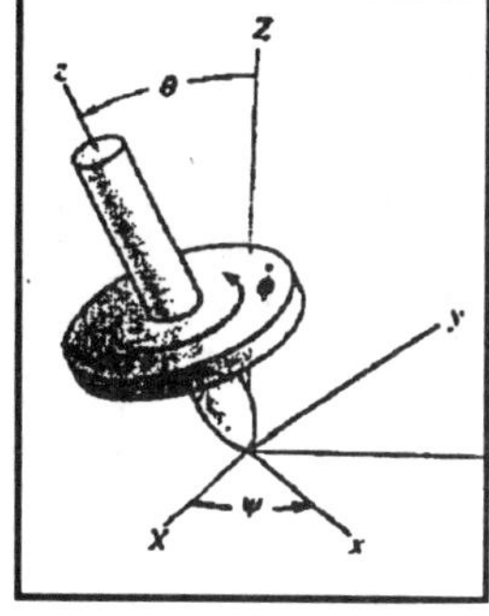

Solution:

The steady precession is not moment-free, since the weight of the top exerts a moment $M_x = \left(\dfrac{mg}{12}\right)\sin\theta$. The motion of a spinning top is described by Eq (9.44),

$mgh = (I_{zz} - I_{xx})\,\dot{\psi}^2\cos\theta + I_{zz}\dot{\psi}\dot{\phi}$, where $\dot{\psi}$ is the rate of precession, $\dot{\phi}$ is the spin rate, and θ is the nutation angle and $h = \left(\dfrac{1}{12}\right)$ ft is the distance from the point to the center of mass. Solve:

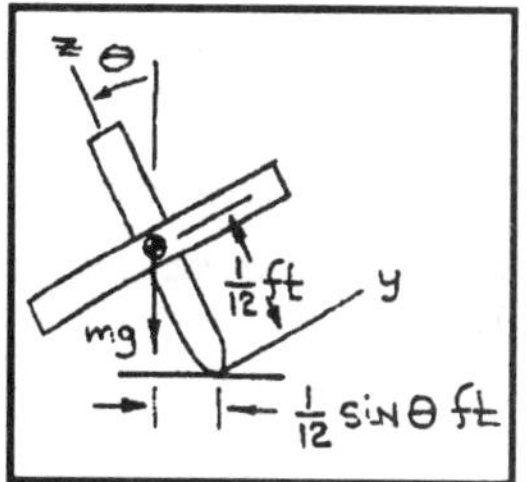

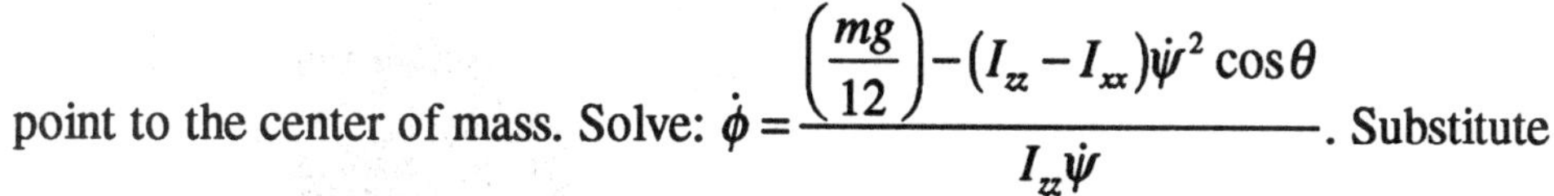

$\dot{\phi} = \dfrac{\left(\dfrac{mg}{12}\right) - (I_{zz} - I_{xx})\dot{\psi}^2\cos\theta}{I_{zz}\dot{\psi}}$. Substitute numerical values (using $\dot{\psi} = 2\pi\ \text{rad/s}$ for dimensional consistency) to obtain $\dot{\phi} = 182.8\ \text{rad/s}$, from which $\dot{\phi} = 29.1\ \text{rev/s}$

Problem 9.95 The top described in Problem 9.94 has a spin rate $\dot{\phi} = 15$ revolutions per second. Draw a graph of the precession rate (in revolutions per second) as a function of the nutation angle θ for values of θ from zero to 45°.

Solution:

The behavior of the top is described in Eq (9.44), $mgh = (I_{xx} - I_{yy})\dot{\psi}^2 \cos\theta + I_{xx}\dot{\psi}\dot{\phi}$, where $\dot{\psi}$ is the rate of precession, $\dot{\phi}$ is the spin rate, and θ is the nutation angle and $h = (1/12)$ ft is the distance from the point to the center of mass. Rearrange: $(I_{zz} - I_{xx})\dot{\psi}^2 \cos\theta + I_{zz}\dot{\psi}\dot{\phi} - mgh = 0$.

The velocity of the center of the base is $v = -\dfrac{\omega_o h}{4}$, from which the spin axis is the z axis and the spin *rate is* $\dot{\phi} = \dfrac{v}{R} = -\dfrac{\omega_o h}{4R}$.

The solution, $\dot{\psi}_{1,2} = -b \pm \sqrt{b^2 - c}$.

The two solutions, which are real over the interval, are graphed as a function of θ over the range $0 \le \theta \le 45^\circ$. The graph is shown. (Before plotting, convert $\dot{\phi}$ to *rad/s,* and then divide $\dot{\psi}_{1,2}$ by 2π to obtain *rev/s.*)

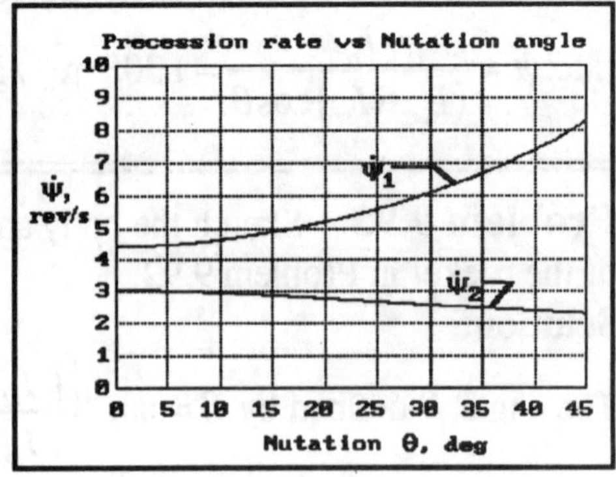

Problem 9.96 The rotor of a tumbling gyroscope can be modeled as being in moment free steady precession. Its moments of inertia are $I_{xx} = I_{yy} = 0.04\ \text{kg-m}^2$, $I_{zz} = 0.18\ \text{kg-m}^2$. Its spin rate is $\dot{\phi} = 1500$ *rpm* and its nutation angle is $\theta = 20^\circ$. (a) What is its precession rate in rpm? (b) Sketch the body and space cones.

Solution:

(a) The motion in moment-free, steady precession is described by Eq (9.43), $(I_{zz} - I_{xx})\dot{\psi}\cos\theta + I_{zz}\dot{\phi} = 0$, where, $\dot{\psi}$ is the precession rate, $\dot{\phi} = 1500$ rpm is the spin rate, and $\theta = 20^\circ$ is the nutation angle.

Solve: $\dot{\psi} = -\dfrac{I_{zz}\dot{\phi}}{(I_{zz} - I_{xx})\cos\theta} = -2052.3$ rpm.

Z
z
θ
β
Space Cone
Y
Body Cone
X
ψ

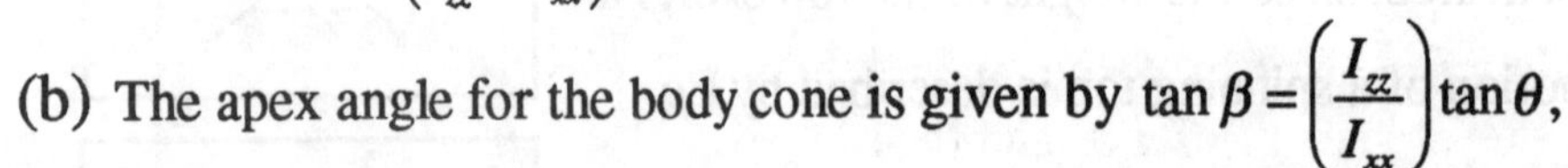

(b) The apex angle for the body cone is given by $\tan\beta = \left(\dfrac{I_{zz}}{I_{xx}}\right)\tan\theta$, from which $\beta = 58.6^\circ$. Since $\beta > \theta$, the space cone lies inside the body cone as in Figure (e) on page 509 of the text.

Problem 9.97 A satellite can be modeled as an 800 *kg* cylinder 4 *m* in length and 2 *m* in diameter. If the nutation angle is $\theta = 20^o$ and the spin rate is one revolution per second, what is the satellite's precession rate $\dot{\psi}$ in revolutions per second?

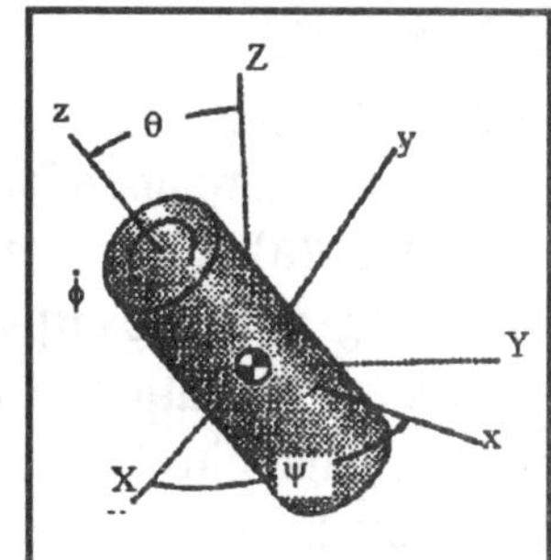

Solution:

From Appendix C, the moments and products of inertia of a homogenous cylinder are $I_{xx} = I_{yy} = m\left(\frac{L^2}{12} + \frac{R^2}{4}\right) = 1267 \text{ kg-m}^2$, $I_{zz} = mR^2/2 = 400 \text{ kg-m}^2$. $I_{xy} = I_{xz} = I_{yz} = 0$. The angular motion of an axisymmetric moment free object in steady precession is described by Eq (9.45), $(I_{zz} - I_{xx})\dot{\psi}\cos\theta + I_{zz}\dot{\phi} = 0$, where $\dot{\psi}$ is the precession rate, $\theta = 20^o$ is the nutation angle, and $\dot{\phi} = 1$ rps is the spin rate. Solve:

$$\dot{\psi} = -\frac{I_{zz}\dot{\phi}}{(I_{zz} - I_{xx})\cos\theta} = 0.49 \text{ rps}$$

Problem 9.98 Solve Problem 9.81 by treating the motion as a steady precession.

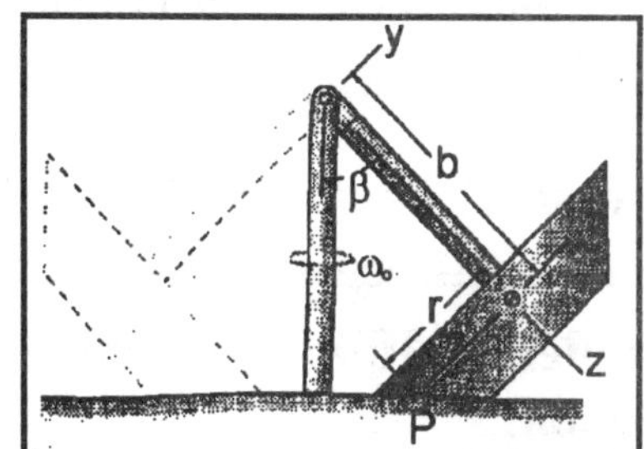

Solution :

The vertical shaft rotates with a constant angular velocity ω_o, causing the grinding mill to roll on the horizontal surface. The point P of the mill is stationary at the instant shown, and the force *N* exerted on the mill by the surface is perpendicular to the surface and acts at P. The force *N* is to be determined. The description of the motion is given by Eq (9.41), $\sum M_x = (I_{zz} - I_{xx})\dot{\psi}^2 \sin\theta\cos\theta + I_{zz}\dot{\phi}\dot{\psi}\sin\theta$.

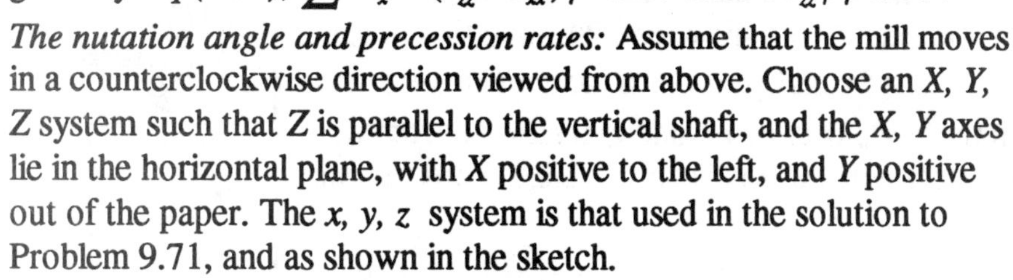

The nutation angle and precession rates: Assume that the mill moves in a counterclockwise direction viewed from above. Choose an *X, Y, Z* system such that *Z* is parallel to the vertical shaft, and the *X, Y* axes lie in the horizontal plane, with *X* positive to the left, and *Y* positive out of the paper. The *x, y, z* system is that used in the solution to Problem 9.71, and as shown in the sketch.

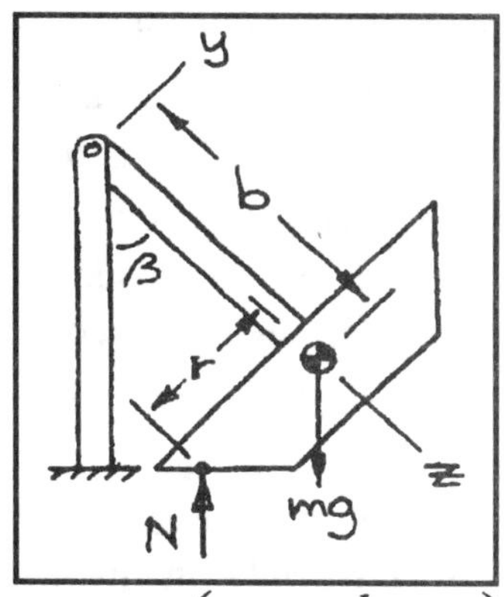

Apply the definitions of the nutation angle and precession rates: the nutation angle of the mill is the angle between the Z axis and the *z*-axis, $\theta = 180^o - \beta$. The precession rate is the angular velocity about the Z axis, $\dot{\psi} = \omega_o$. The spin rate is given in the solution to Problem 9.81: $\dot{\phi} = \omega_o\left(\cos\beta - \frac{b}{r}\sin\beta\right)$.

The moment about the point O. From the solution to Problem 9.81, the moment about the fixed point O is $\vec{M}_O = \left(\text{N}(r\cos\beta - b\sin\beta) + mgb\sin\beta\right)\vec{i} = \text{M}_{ox}\vec{i}$. Substitute the moment into Eq (9.41) to obtain:

$$N = \frac{(I_{zz} - I_{xx})\dot{\psi}^2 \sin\theta\cos\theta + I_{zz}\dot{\phi}\dot{\psi}\sin\theta - mgb\sin\beta}{r\cos\beta - b\sin\beta},$$ Substitute: $\sin\theta = \sin\beta$, $\cos\theta = -\cos\beta$,

$$\dot{\phi} = \omega_o\left(\cos\beta - \frac{b}{r}\sin\beta\right),\ \dot{\psi} = \omega_o,\ I_{xx} = I_{yy},\ N = \frac{\omega_o^2\left[I_{zz}\left(\frac{b}{r}\right)\sin^2\beta - I_{yy}\sin\beta\cos\beta\right] + mgb\sin\beta}{b\sin\beta - r\cos\beta}.$$

[*Check:* Compare with the solution to Problem 9.81. *check.*]

===============================<>===============================

Problem 9.99 Solve Problem 9.82 by treating the motion as steady precession.

Solution:

The view of an airplane's landing gear looking from behind the airplane is shown in Fig. (a). The radius of the wheel is 300 *mm* and its moment of inertia is 2 *kg-m*^{2}. The airplane takes off at 30 *m/s*. After takeoff, the landing gear retracts by rotating toward the right side of the airplane as shown in Fig.(b). The magnitude of the couple exerted by the wheel on its support is to be determined.

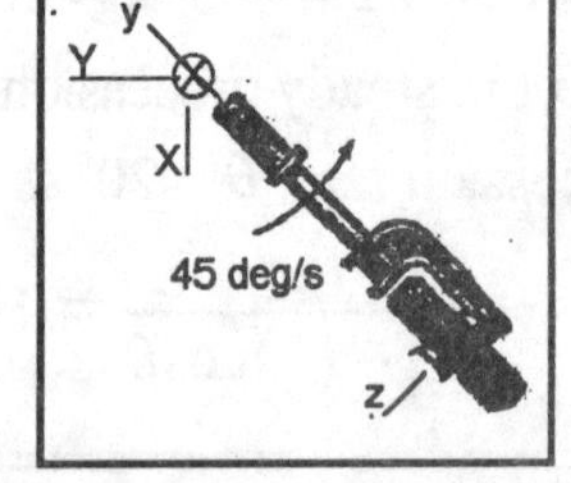

Choose *X, Y, Z* with the *Z* axis parallel to the runway, *X* perpendicular to the runway, and *Y* parallel to the runway. Choose the *x, y, z* coordinate system with the origin at the center of mass of the wheel and the *z* axis aligned with the direction of the axis of rotation of the wheel and the *y* axis positive upward. The Eq (9.41) is $\sum M_x = (I_{zz} - I_{xx})\dot{\psi}^2 \sin\theta\cos\theta + I_{zz}\dot{\phi}\dot{\psi}\sin\theta$.

The nutation angle and rates of precession: The nutation angle is the angle between Z and *z*, $\theta = 90^o$. The precession angle is the angle between the *X* and *x*, which is increasing in value, from which $\dot{\psi} = 45^o / s = 0.7853$ rad / s. The spin vector is aligned with the *z* axis, from which $\dot{\phi} = \left(\frac{V}{R}\right) = \left(\frac{30}{0.3}\right) = 100$ rad / s. The moments and products of inertia of the wheel are $I_{zz} = mR^2 / 2 = 2$ kg-m^2. The moment is $M_x = I_{zz}\dot{\psi}\,\dot{\phi}\sin 90^o = 2(0.7854)(100) = 157.1$ N-m

===============================<>===============================

Problem 9.100 Solve Problem 9.83 by treating the motion as steady precession.

Solution:

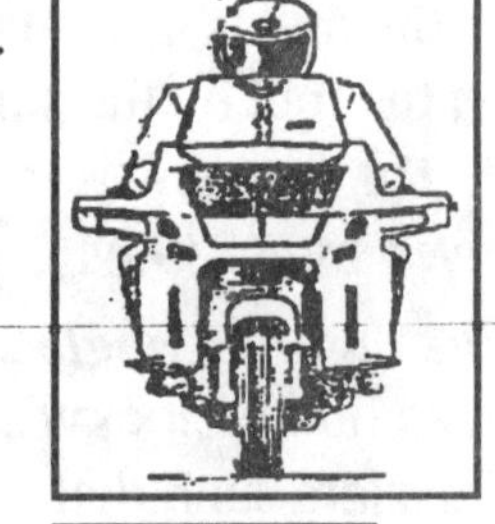

Choose an *X, Y, Z* coordinate system with the origin at O and the Z axis vertical, as shown, an *x, y, z* system with *z* coincident with the direction of wheel spin, and positive *y* vertical. The *X, x* axes are coincident, into the paper. The nutation angle is the angle between Z and *z*, $\theta = 90^o$. The spin angle $\dot{\phi} = \omega_o$ is the angular velocity of the wheel. The precession angle is the angle between X and *x*, measured from X. Thus if the rider turns to his left, the precession angle is positive, and the precession rate is $\dot{\psi} = \Omega$. For steady precession, Eq (9.41) is $\sum M_x = (I_{zz} - I_{xx})\dot{\psi}^2 \sin\theta\cos\theta + I_{zz}\dot{\phi}\dot{\psi}\sin\theta$, from which $\sum M_x = I_{zz}\Omega\omega_o$, and the moment has the same sign as Ω. If the rider turns to the left, $\Omega > 0$, and the moment about the x axis is positive, causing the rider to lean to the left. Coversely, if the rider turns to the right, the moment is negative, causing the rider to lean to the right. [*Note*: See Problem 9.83.]

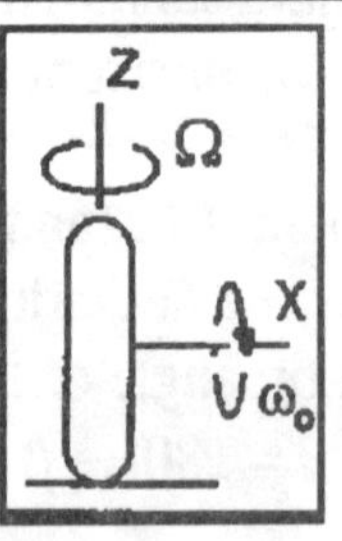

===============================<>===============================

==================================<>==================================

Problem 9.101 Suppose that you are testing a car and use accelerometers and gyroscopes to measure its Eulerian angles and their derivatives relative to a reference coordinate system. At a particular instant, $\psi = 15^o$, $\theta = 4^o$, $\phi = 15^o$, the rates of change of the Eulerian angles are zero, and their second derivatives are $\ddot{\psi} = 0$, $\ddot{\theta} = 1 \text{ rad/s}^2$, and $\ddot{\phi} = -0.5 \text{ rad/s}^2$. The car's principal moments of inertia in *kg-m²* $I_{xx} = 2200$, $I_{yy} = 480$, and $I_{zz} = 2600$. What are the components of the total moment about the car's center of mass?

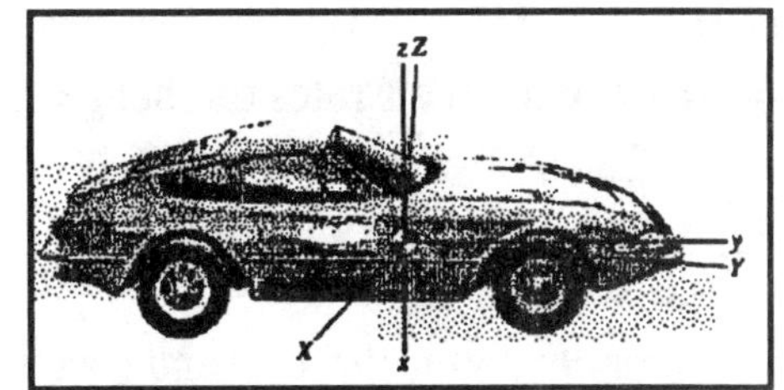

Solution:

The description of the motion of an arbitrarily shaped object is given by the Eqs (9.48).
The Eqs (9.48) are :

$$M_x = I_{xx}\left(\ddot{\psi}\sin\theta\sin\phi + \ddot{\theta}\cos\phi + \dot{\psi}\dot{\theta}\cos\theta\sin\phi + \dot{\psi}\dot{\phi}\sin\theta\cos\phi - \dot{\theta}\dot{\phi}\sin\phi\right)$$

$$-(I_{yy} - I_{zz})(\dot{\psi}\sin\theta\cos\phi - \dot{\theta}\sin\phi)(\dot{\psi}\cos\theta + \dot{\phi})$$

$$M_y = I_{yy}\left(\ddot{\psi}\sin\theta\cos\phi - \ddot{\theta}\sin\phi + \dot{\psi}\dot{\theta}\cos\theta\cos\phi - \dot{\psi}\dot{\phi}\sin\theta\sin\phi - \dot{\theta}\dot{\phi}\cos\phi\right)$$

$$-(I_{zz} - I_{xx})(\dot{\psi}\sin\theta\sin\phi + \dot{\theta}\cos\phi)(\dot{\psi}\cos\theta + \dot{\phi})$$

$$M_z = I_{zz}\left(\ddot{\psi}\cos\theta + \ddot{\phi} - \dot{\psi}\dot{\theta}\sin\theta\right)$$

$$-(I_{xx} - I_{yy})(\dot{\psi}\sin\theta\sin\phi + \dot{\theta}\cos\phi)(\dot{\psi}\sin\theta\cos\phi - \dot{\theta}\sin\phi)$$

Substitute $\ddot{\psi} = 0$, $\dot{\psi} = \dot{\theta} = \dot{\phi} = 0$, to obtain $M_x = I_{xx}\ddot{\theta}\sin\phi$, $M_y = -I_{yy}\ddot{\theta}\sin\phi$, $M_z = I_{zz}\ddot{\phi}$

Substitute values: $M_x = 2125 \text{ N-m}$, $M_y = -124.2$, $M_z = -1300 \text{ N-m}$

==================================<>==================================

Problem 9.102 If the Eulerian angles and their second derivatives of the car described in Problem 9.101 have the given values but their rates of change are $\dot{\psi} = 0.2 \text{ rad/s}$, $\dot{\theta} = -2 \text{ rad/s}$, and $\dot{\phi} = 0$, what are the components of the total moments about the car's center of mass?

Solution:

Use Eqns (9.48),

$$M_x = I_{xx}\left(\ddot{\psi}\sin\theta\sin\phi + \ddot{\theta}\cos\phi + \dot{\psi}\dot{\theta}\cos\theta\sin\phi + \dot{\psi}\dot{\phi}\sin\theta\cos\phi - \dot{\theta}\dot{\phi}\sin\phi\right)$$

$$-(I_{yy} - I_{zz})(\dot{\psi}\sin\theta\cos\phi - \dot{\theta}\sin\phi)(\dot{\psi}\cos\theta + \dot{\phi})$$

$$M_y = I_{yy}\left(\ddot{\psi}\sin\theta\cos\phi - \ddot{\theta}\sin\phi + \dot{\psi}\dot{\theta}\cos\theta\cos\phi - \dot{\psi}\dot{\phi}\sin\theta\sin\phi - \dot{\theta}\dot{\phi}\cos\phi\right)$$

$$-(I_{zz} - I_{xx})(\dot{\psi}\sin\theta\sin\phi + \dot{\theta}\cos\phi)(\dot{\psi}\cos\theta + \dot{\phi})$$

$$M_z = I_{zz}\left(\ddot{\psi}\cos\theta + \ddot{\phi} - \dot{\psi}\dot{\theta}\sin\theta\right)$$

$$-(I_{xx} - I_{yy})(\dot{\psi}\sin\theta\sin\phi + \dot{\theta}\cos\phi)(\dot{\psi}\sin\theta\cos\phi - \dot{\theta}\sin\phi)$$

Substitute values: $M_x = 2122.5 \text{ N-m}$, $M_y = -155.4 \text{ N-m}$, $M_z = 534.0 \text{ N-m}$

==================================<>==================================

==================================<>==================================

Problem 9.103 Suppose that the Eulerian angles of the car described in Problem 9.101 are $\psi = 40^o$, $\theta = 20^o$, $\phi = 5^o$, their rates of change are zero, and the components of the total moment about the car's center of mass are $\sum M_x = -400$ N-m, $\sum M_y = 200$ N-m, $\sum M_z = 0$. What are the *x*, *y*, and *z* components of the car's acceleration?

Solution:

The moments about the *x*, *y* and *z* axes are given by Eq (9.26), From Eqs(9.34) and (9.35), the angular velocities are zero. From Problem 9.101, the off-diagonal elements of $[I]$ are zero.

Eq (9.26) becomes: $\begin{bmatrix} -400 \\ 200 \\ 0 \end{bmatrix} = \begin{bmatrix} 2200 & 0 & 0 \\ 0 & 480 & 0 \\ 0 & 0 & 2600 \end{bmatrix} \begin{bmatrix} \alpha_x \\ \alpha_y \\ \alpha_z \end{bmatrix}$.

Solve: $\alpha_x = -0.1818 \text{ rad/s}^2$, $\alpha_y = 0.417 \text{ rad/s}^2$, $\alpha_z = 0$.

Check: An alternative solution uses Eq (9.48), which simplifies when $\dot{\psi} = \dot{\phi} = \dot{\theta} = 0$ to $M_x = I_{xx}(\ddot{\psi}\sin\theta\sin\phi + \ddot{\theta}\cos\phi)$, $M_y = I_{yy}(\ddot{\psi}\sin\theta\cos\phi - \ddot{\theta}\sin\phi)$, $M_z = I_{zz}(\ddot{\psi}\cos\theta + \ddot{\phi})$. These three simultaneous equations have the solutions,

$$\ddot{\theta} = \frac{M_x}{I_{xx}}\cos\phi - \frac{M_y}{I_{yy}}\sin\phi = -0.2174 \text{ rad/s}^2,$$

$$\ddot{\psi} = \left(\frac{M_{xx}}{I_{xx}}\right)\frac{\sin\phi}{\sin\theta} + \left(\frac{M_y}{I_{yy}}\right)\frac{\cos\phi}{\sin\theta} = 1.167 \text{ rad/s}^2,$$

$$\ddot{\phi} = (M_z / I_{zz}) - \ddot{\psi}\cos\theta = -1.097 \text{ rad/s}^2.$$

From Eq(9.47), when $\dot{\psi} = \dot{\phi} = \dot{\theta} = 0$: $\frac{d\omega_x}{dt} = \ddot{\psi}\sin\theta\sin\phi + \ddot{\theta}\cos\phi = -0.1818 \text{ rad/s}^2$ *check.*

$\frac{d\omega_y}{dt} = \ddot{\psi}\sin\theta\cos\phi - \ddot{\theta}\sin\phi = 0.417 \text{ rad/s}$ *check.* $\frac{d\omega_z}{dt} = \ddot{\psi}\cos\phi + \ddot{\phi} = 0$ *check.*

==================================<>==================================

Problem 9.104 The circular disk remains perpendicular to the horizontal shaft and rotates relative to it with angular velocity ω_d. The horizontal shaft is rigidly attached to a vertical shaft rotating with angular velocity ω_O. (a) What is the disk's angular velocity vector $\vec{\omega}$? (b) What is the velocity of point A on the disk?

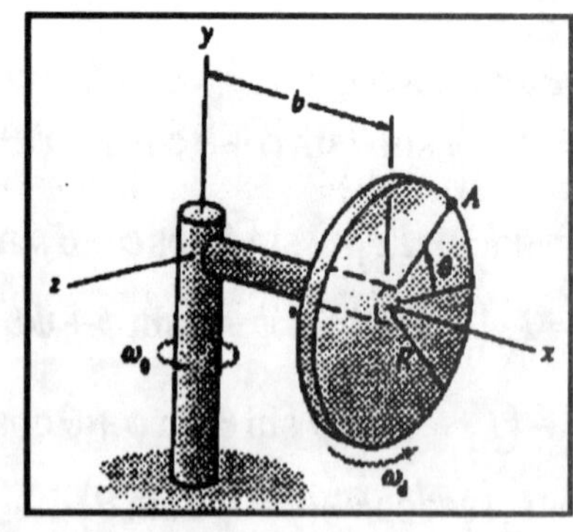

Solution: (a) The angular velocity vector of the of the coordinate system is $\vec{\omega}_O = \omega_O \vec{j}$. The angular velocity vector of the disk relative to the coordinate system is $\vec{\omega}_{d/O} = \omega_d \vec{i}$. Therefore the disk's angular velocity is $\vec{\omega} = \vec{\omega}_O + \vec{\omega}_{d/O} = \omega_d \vec{i} + \omega_O \vec{j}$ (rad/s)

The point A has the coordinates $(b, R\sin\theta, -R\cos\theta)$, with the vector location relative to O of $\vec{r}_{A/O} = b\vec{i} + R\sin\theta\,\vec{j} - R\cos\theta\,\vec{k}$.

(a) The velocity of point A is $\vec{v}_{A/O} = \vec{\omega} \times \vec{r}_{A/O} = 0 + \begin{bmatrix} \vec{i} & \vec{j} & \vec{k} \\ \omega_d & \omega_O & 0 \\ b & R\sin\theta & -R\cos\theta \end{bmatrix}$,

$$\boxed{\vec{v}_{A/O} = (-R\omega_O\cos\theta)\vec{i} + (R\omega_d\cos\theta)\vec{j} + (R\omega_d\sin\theta - b\omega_O)\vec{k}}$$

==================================<>==================================

=================================<>=================================

Problem 9.105 If the angular velocities ω_d and ω_O in Problem 9.104 are constant, what is the acceleration of point A of the disk?

Solution:

(See Figure in solution to Problem 9.104.) The angular acceleration of the disk is given by

$$\frac{d\vec{\omega}}{dt}=\frac{d}{dt}\left(\omega_d\vec{i}+\omega_O\vec{j}\right)+\vec{\omega}_O\times\vec{\omega}_d=0+\begin{bmatrix}\vec{i}&\vec{j}&\vec{k}\\0&\omega_O&0\\\omega_d&0&0\end{bmatrix}=-\omega_O\omega_d\vec{k}$$

. The velocity of point A relative to O is $\vec{a}_{A/O}=\vec{\alpha}\times\vec{r}_{A/O}+\vec{\omega}\times\left(\vec{\omega}\times\vec{r}_{A/O}\right)=\left(-\omega_O\omega_d\right)\left(\vec{k}\times\vec{r}_{A/O}\right)+\vec{\omega}\times\left(\vec{\omega}\times\vec{r}_{A/O}\right)$. Term by term:

$$-\omega_O\omega_d\left(\vec{k}\times\vec{r}_{A/O}\right)=-\omega_O\omega_d\begin{bmatrix}\vec{i}&\vec{j}&\vec{k}\\0&0&1\\b&R\sin\theta&-R\cos\theta\end{bmatrix},\quad -\omega_O\omega_d\left(\vec{k}\times\vec{r}_{A/O}\right)=\omega_O\omega_dR\sin\theta\,\vec{i}-\omega_O\omega_db\vec{j}.$$

$$\vec{\omega}\times\left(\vec{\omega}\times\vec{r}_{A/O}\right)=\vec{\omega}\times\begin{bmatrix}\vec{i}&\vec{j}&\vec{k}\\\omega_d&\omega_O&0\\b&R\sin\theta&-R\cos\theta\end{bmatrix}=\begin{bmatrix}\vec{i}&\vec{j}&\vec{k}\\\omega_d&\omega_O&0\\-R\omega_O\cos\theta&R\omega_d\cos\theta&R\omega_d\sin\theta-b\omega_O\end{bmatrix}.$$

$\vec{\omega}\times\left(\vec{\omega}\times\vec{r}_{A/O}\right)=\left(R\omega_d\sin\theta-b\omega_O\right)\left(\omega_o\vec{i}-\omega_d\vec{j}\right)+\left(R\cos\theta\right)\left(\omega_d^2+\omega_O^2\right)\vec{k}$. Collecting terms:

$$\vec{a}_{A/O}=\left(2R\omega_O\omega_d\sin\theta-b\omega_O^2\right)\vec{i}-\left(R\omega_d^2\sin\theta\right)\vec{j}+\left(R\omega_d^2\cos\theta+R\omega_O^2\cos\theta\right)\vec{k}$$

=================================<>=================================

Problem 9.106 The cone is connected by a ball and socket joint at its vertex to a 100 *mm* post. The radius of its base is 100 *mm*, and the base rolls on the floor. The velocity of the center of the base is $\vec{v}_C=2\vec{k}\ (m/s)$. (a) What is the cone's angular velocity vector $\vec{\omega}$? (b) What is the velocity of point A?

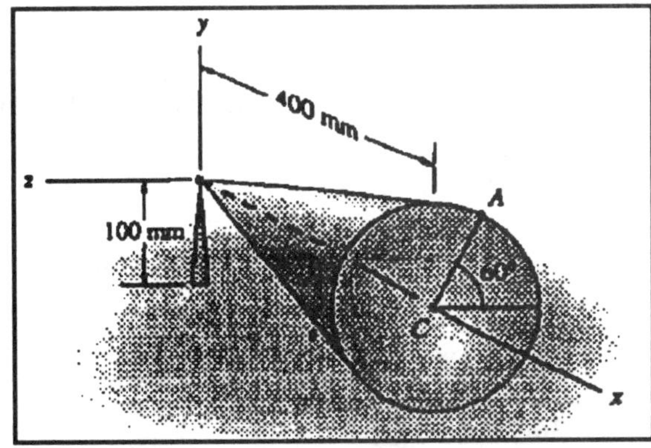

Solution Denote $\theta=60^o$, $h=0.1\ m$, $L=0.4\ m$, $R=0.1\ m$. (a) The strategy is to express the velocity of the center of the base of the cone, point C, in terms of the known (zero) velocities of O and P to formulate simultaneous equations for the angular velocity vector components.

The line OC is parallel to the vector $\vec{r}_{C/O}=\vec{i}L\ (m)$.

The line PC is parallel to the vector $\vec{r}_{C/P}=R\vec{j}\ (m)$.

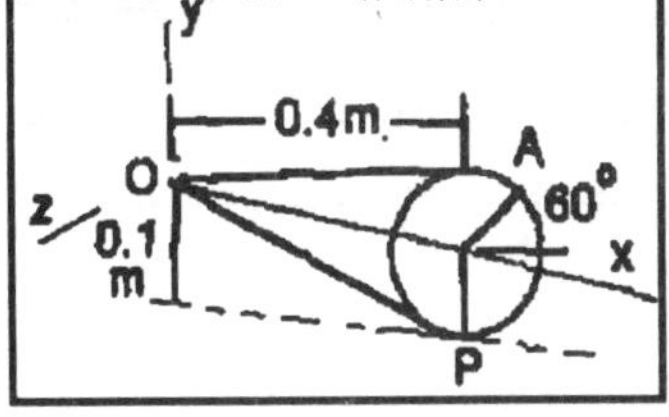

The velocity of the center of the cone is given by the two expressions:

$\vec{v}_C=\vec{v}_O+\vec{\omega}\times\vec{r}_{C/O}$,

and $\vec{v}_C=\vec{v}_P+\vec{\omega}\times\vec{r}_{C/P}$, where $\vec{v}_C=2\vec{k}\ (m/s)$, and $\vec{v}_O=\vec{v}_P=0$.

Solution continued on next page

Expanding: $\vec{v}_C = \vec{\omega} \times \vec{r}_{C/O} = \begin{bmatrix} \vec{i} & \vec{j} & \vec{k} \\ \omega_x & \omega_y & \omega_z \\ L & 0 & 0 \end{bmatrix} = \omega_z L\vec{j} - \omega_y L\vec{k}$.

$\vec{v}_C = \vec{\omega} \times \vec{r}_{C/P} = \begin{bmatrix} \vec{i} & \vec{j} & \vec{k} \\ \omega_x & \omega_y & \omega_Z \\ 0 & R & 0 \end{bmatrix} = -R\omega_Z\vec{i} + R\omega_x\vec{k}$. Solve by inspection: $\omega_x = \frac{v_C}{R} = 20\ rad/s$,

$\omega_y = -\frac{v_C}{L} = -5\ m/s$, $\omega_z = 0$. $\boxed{\vec{\omega} = 20\vec{i} - 5\vec{j}\ rad/s}$

(b) The line OA is parallel to the vector $\vec{r}_{A/O} = \vec{i}L + \vec{j}R\sin\theta - \vec{k}R\cos\theta$. The velocity of point A is given by: $\vec{v}_A = \vec{v}_O + \vec{\omega} \times \vec{r}_{A/O}$, where $\vec{v}_O = 0$. $\vec{v}_A = \begin{bmatrix} \vec{i} & \vec{j} & \vec{k} \\ 20 & -5 & 0 \\ L & R\sin\theta & -R\cos\theta \end{bmatrix}$,

$\boxed{\vec{v}_A = 0.25\vec{i} + 1\vec{j} + 3.732\vec{k}\ \ (m/s)}$

==================================<>==================================

Problem 9.107 The bar AB is connected by ball and socket joints to collars sliding on the fixed bars CD and EF. The bar EF is parallel to the y axis. At the present instant the collar at A has velocity $\bar{v}_A = 20\bar{k}(ft/s)$ and the angular velocity of bar AB about its axis is zero. Determine the velocity of the collar at B and the angular velocity of bar AB.

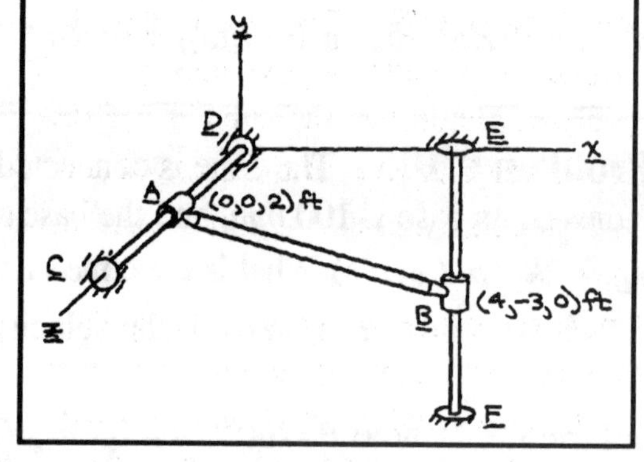

Solution:

We can express the velocity of B and the bar's angular velocity as $\bar{v}_B = v_B\bar{j}$, $\bar{\omega} = \omega_x\bar{i} + \omega_y\bar{j} + \omega_z\bar{k}$.

The velocity of B is $\bar{v}_B = \bar{v}_A + \bar{\omega} \times \bar{r}_{B/A}$: Hence, $v_B\bar{j} = 20\bar{k} + \begin{vmatrix} \bar{i} & \bar{j} & \bar{k} \\ \omega_x & \omega_y & \omega_z \\ 4 & -3 & -2 \end{vmatrix}$.

Equating $\bar{i}, \bar{j}$ and $\bar{k}$ components, we obtain $0 = -2\omega_y + 3\omega_z$ (1), $v_B = 4\omega_z + 2\omega_x$ (2), and $0 = 20 - 3\omega_x - 4\omega_y$(3).

The angular velocity of the bar about its axis is zero, so $\bar{r}_{B/A} \bullet \bar{\omega} = 4\omega_x - 3\omega_y - 2\omega_z = 0$(4)

Solving Equations (1) - (4) yields $\bar{v}_B = 13.3\bar{j}(ft/s)$, $\bar{\omega} = 2.99\bar{i} + 2.76\bar{j} + 1.84\bar{k}(rad/s)$.

==================================<>==================================

================================<>================================

Problem 9.108 The collar in Problem 9.107 has acceleration $\bar{a}_A = -40\bar{k}\ (ft/s^2)$ and the angular acceleration of bar AB about its axis is zero. Determine the acceleration of the collar B and the angular acceleration of bar AB.

Solution:

We start with the solution to Problem 9.107 where we knew that $\bar{r}_{B/A} = 4\,\bar{i} - 3\,\bar{j} - 2\bar{k}\ (ft)$ and $\bar{\omega} = 2.99\bar{i} + 2.76\bar{j} + 1.84\bar{k}(rad/s)$. The acceleration of point B, given the acceleration of point A, is $\bar{a}_B = \bar{a}_A + \bar{\alpha}\times\bar{r}_{B/A} + \bar{\omega}\times(\bar{\omega}\times\bar{r}_{B/A})$. We know that $\bar{a}_A = -40\bar{k}\ (ft/s)$ and $\bar{a}_B = a_{By}\bar{j}$. From the information found in problem 9.107, we can calculate $(\bar{\omega}\times r_{B/A}) = 0\bar{i} + 13.33\bar{j} - 20\bar{k}$ *(ft/s)*, and $\bar{\omega}\times(\bar{\omega}\times\bar{r}_{B/A}) = -79.7\bar{i} + 59.8\bar{j} + 39.8\bar{k}$ *(ft/s²)*. Substituting the known information into the equation for the acceleration of B, we get $a_{By}\bar{j} = -40\bar{k} + \begin{vmatrix} \bar{i} & \bar{j} & \bar{k} \\ \alpha_x & \alpha_y & \alpha_z \\ 4 & -3 & -2 \end{vmatrix} + (-79.7\bar{i} + 59.8\bar{j} + 39.8\bar{k})$. Separating this into component equations, we get $0 = -2\alpha_y + 3\alpha_z - 79.7$, $a_{By} = 4\alpha_z + 2\alpha_x + 59.8$, and $0 = -40 - 3\alpha_x - 4\alpha_y + 39.8$. Finally, we have the requirement that the angular acceleration about the axis AB is zero. This requires that $\bar{\alpha}\bullet\bar{r}_{B/A} = 0 = 4\alpha_x - 3\alpha_y - 2\alpha_z$. Solving the four scalar equations simultaneously, we get $a_{By} = 165.9\,ft/s$, and $\bar{\alpha} = 7.31\bar{i} - 5.52\bar{j} + 22.89\bar{k}$ *(rad/s²)*.

================================<>================================

Problem 9.109 The mechanism shown is a type of universal joint called a yoke and spider. The axis *L* lies in the *x-y* plane. Determine the angular velocity ω_L and the angular velocity ω_S of the cross-shaped "spider" in terms of the angular velocity ω_R at the instant shown.

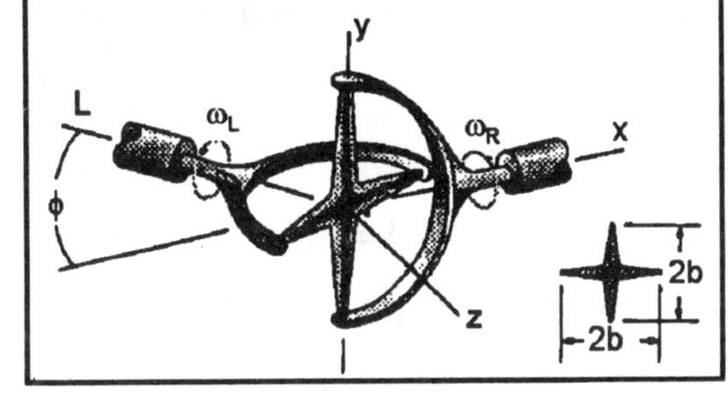

Solution:

Denote the center of mass of the spider by point O, and denote the line coinciding with the vertical arms of the spider (the *y* axis) by $P'P$, and the line coinciding with the horizontal arms by $Q'Q$. The line $P'P$ is parallel to the vector $\vec{r}_{P/O} = b\vec{j}$. The angular velocity of the right hand yoke is positive along the *x* axis, $\vec{\omega}_R = \omega_R\vec{i}$, from which the angular velocity $\vec{\omega}_L$ is positive toward the right, so that $\vec{\omega}_L = \omega_L(\vec{i}\cos\phi + \vec{k}\sin\phi)$. The velocity of the point P on the extremities of the line P'P is $\vec{v}_P = \vec{v}_O + \vec{\omega}_R\times\vec{r}_{P/O}$, where $\vec{v}_O = 0$, from which

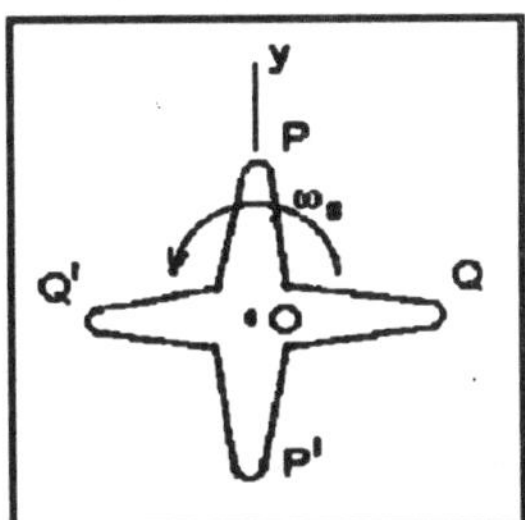

$\vec{v}_P = \begin{bmatrix} \vec{i} & \vec{j} & \vec{k} \\ \omega_R & 0 & 0 \\ 0 & b & 0 \end{bmatrix} = b\omega_R\vec{k}$. The velocity v_P is also given by

$\vec{v}_P = \vec{v}_O + \vec{\omega}_S\times\vec{r}_{P/O} = \begin{bmatrix} \vec{i} & \vec{j} & \vec{k} \\ \omega_{Sx} & \omega_{Sy} & \omega_{Sz} \\ 0 & b & 0 \end{bmatrix} = -\vec{i}b\omega_{Sz} + \vec{k}b\omega_{Sx}$, from which $\omega_{Sz} = 0$, $\omega_{Sx} = \omega_R$.

Solution continued on next page

$\vec{v}_P = \vec{v}_O + \vec{\omega}_S \times \vec{r}_{P/O} = \begin{bmatrix} \vec{i} & \vec{j} & \vec{k} \\ \omega_{Sx} & \omega_{Sy} & \omega_{Sz} \\ 0 & b & 0 \end{bmatrix} = -\vec{i}b\omega_{Sz} + \vec{k}b\omega_{Sx}$, from which $\omega_{Sz} = 0$, $\omega_{Sx} = \omega_R$. The line $Q'Q$ is parallel to the vector $\vec{r}_{Q/O} = \vec{i}b\sin\phi - \vec{k}b\cos\phi$.

The velocity of the point Q is $\vec{v}_Q = \vec{v}_O + \vec{\omega}_S \times \vec{r}_{Q/O} = 0 + \begin{bmatrix} \vec{i} & \vec{j} & \vec{k} \\ \omega_{Sx} & \omega_{Sy} & 0 \\ b\sin\phi & 0 & -b\cos\phi \end{bmatrix}$,

$\vec{v}_Q = -\vec{i}(\omega_{Sy} b\cos\phi) + \vec{j}(\omega_{Sx} b\cos\phi) - \vec{k}(\omega_{Sy} b\sin\phi)$. The velocity $\vec{v}_Q$ is also given by

$\vec{v}_Q = \vec{v}_O + \vec{\omega}_L \times \vec{r}_{Q/O}$., from which $\vec{v}_Q = \begin{bmatrix} \vec{i} & \vec{j} & \vec{k} \\ \omega_L\cos\phi & 0 & \omega_L\sin\phi \\ b\sin\phi & 0 & -b\cos\phi \end{bmatrix} = \vec{j}b\omega_L(\cos^2\phi + \sin^2\phi) = \vec{j}b\omega_L$,

from which $\omega_{Sy} = 0$, from which $\omega_S = \omega_R$, $\boxed{\vec{\omega}_S = \omega_R\vec{i}}$, $\boxed{\omega_L = \omega_{Sx}\cos\phi = \omega_R\cos\phi}$, *Check:* The line PQ is parallel to the vector $\vec{r}_{Q/P} = \vec{r}_{Q/O} - \vec{r}_{P/O} = \vec{i}b\sin\phi - b\vec{j} - \vec{k}b\cos\phi$. The velocity of the point Q

is $\vec{v}_Q = \vec{v}_P + \vec{\omega}_S \times \vec{r}_{Q/P} = b\omega_R\vec{k} + \begin{bmatrix} \vec{i} & \vec{j} & \vec{k} \\ \omega_{Sx} & 0 & \omega_{Sz} \\ b\sin\phi & -b & -b\cos\phi \end{bmatrix}$.

From which $\vec{v}_Q = \vec{i}b\omega_{Sz} + \vec{j}(b\omega_{Sx}\cos\phi + b\omega_{Sz}\sin\phi) + (b\omega_R - b\omega_{Sx})\vec{k}$. But also, $\vec{v}_Q = \vec{j}b\omega_L$, from which $(b\omega_R - b\omega_{Sx}) = 0$, $\omega_{Sz} = 0$, from which $\omega_{Sx} = \omega_R$, $\omega_L = \omega_{Sx}\cos\phi + \omega_{Sz}\sin\phi = \omega_{Sx}\cos\phi$. Since the line L and the x axis lie in the x-z plane, $\omega_{Sy} = 0$, from which $\vec{\omega}_S = \omega_R\vec{i}$, and $\omega_L = \omega_R\cos\phi$. *check.*

==================================<>==================================

Problem 9.110 The inertia matrix of a rigid body in terms of a body fixed coordinate system with its origin at the center of mass is $[I] = \begin{bmatrix} 4 & 1 & -1 \\ 1 & 2 & 0 \\ -1 & 0 & 6 \end{bmatrix}$ $kg\text{-}m^2$. The rigid body's angular velocity is $\vec{\omega} = 10\vec{i} - 5\vec{j} + 10\vec{k}$ (rad/s). What is its angular momentum about its center of mass?

Solution :

The angular momentum is $\begin{bmatrix} H_{Ox} \\ H_{Oy} \\ H_{Oz} \end{bmatrix} = \begin{bmatrix} 4 & 1 & -1 \\ 1 & 2 & 0 \\ -1 & 0 & 6 \end{bmatrix} \begin{bmatrix} 10 \\ -5 \\ 10 \end{bmatrix} = \begin{bmatrix} 40-5-10 \\ 10-10+0 \\ -10+0+60 \end{bmatrix} = \begin{bmatrix} 25 \\ 0 \\ 50 \end{bmatrix}$ $kg\text{-}m^2/s$ In terms of the unit vectors $\vec{i}, \vec{j}, \vec{k}$, $\boxed{\vec{H} = 25\vec{i} + 50\vec{k} \ kg\text{-}m^2/s}$

==================================<>==================================

==================================<>==================================

Problem 9.111 What is the moment of inertia of the rigid body in Problem 9.110 about the axis that passes through the origin and the point $(4,-4,7)$ *m*? *Strategy*: Determine the components of a unit vector parallel to the axis and use Eq (9.17).

Solution:

The unit vector parallel to the line passing through (0,0,0) and $(4,-4,7)$ is

$\vec{e} = \dfrac{4\vec{i} - 4\vec{j} + 7\vec{k}}{\sqrt{4^2 + 4^2 + 7^2}} = 0.4444\vec{i} - 0.4444\vec{j} + 0.7778\vec{k}$. The inertia matrix in Problem 9.110 is

$[I] = \begin{bmatrix} 4 & 1 & -1 \\ 1 & 2 & 0 \\ -1 & 0 & 6 \end{bmatrix} = \begin{bmatrix} I_{xx} & -I_{xy} & -I_{xz} \\ -I_{xy} & I_{yy} & -I_{yz} \\ -I_{xz} & -I_{yz} & I_{zz} \end{bmatrix}$, where advantage is taken of the symmetric property of the inertia matrix. From Eq (9.19), the new moment of inertia about the line through (0,0,0) and $(4,-4,7)$ is

$I_O = 4e_x^2 + 2e_y^2 + 6e_z^2 + 2(1)(e_x e_y) + 2(-1)(e_x e_z) + 2(0)(e_y e_z)$, $\boxed{I_O = 3.728 \ kg\text{-}m^2}$

==================================<>==================================

Problem 9.112 Determine the inertia matrix of the 0.6 *slug* thin plate in terms of the coordinate system shown.

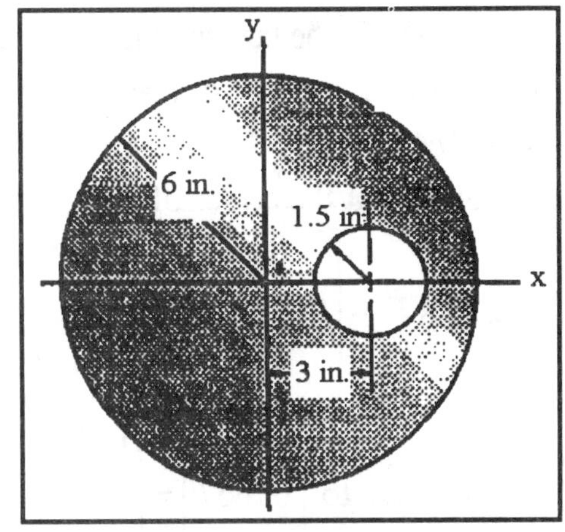

Solution:

The strategy is to determine the moments and products for a solid thin plate about the origin, and then subtract the moments and products of the cut-out. The mass density is

$\rho = \dfrac{0.6}{\left(\pi(0.5^2) - \pi\left(\frac{3}{24}\right)^2\right)T} = \dfrac{0.8149}{T}$ $slug/ft^3$, from which

$\rho T = 0.8149 \ slug/ft^2$, where T is the (unknown) thickness of the plate. From Appendix C and by inspection,

the moments and products of inertia for a thin plate of radius R are: $I_{xx} = I_{yy} = \dfrac{mR^2}{4}$, $I_{zz} = \dfrac{mR^2}{2}$,

$I_{xy} = I_{xz} = I_{yz} = 0$. For a 6 *in.* radius *solid thin plate*, $m = \rho T\pi(0.5^2) = 0.64 \ slug$.

$I_{xx} = I_{yy} = 0.04 \ slug\text{-}ft^2$, $I_{zz} = 0.08 \ slug\text{-}ft^2$, $I_{xy} = I_{xz} = I_{yz} = 0$.

The coordinates of the 1.5 *in.* radius cut-out are $(d_x, d_y, d_x) = (3,0,0)$. The mass removed by the cut-out is $m_C = \rho T\pi R_C^2 = 0.04 \ slug$. The moments and products of inertia of the cut-out are

$I_{xx}^C = \dfrac{m_2 R_C^2}{4} = 1.563 \times 10^{-4} \ slug\text{-}ft^2$, $I_{yy}^C = \dfrac{m_C R_C^2}{4} + \left(\dfrac{3}{12}\right)^2 m_C = 2.656 \times 10^{-3} \ slug\text{-}ft^2$,

$I_{zz}^C = \dfrac{m_C R_C^2}{2} + d_x^2 m_C = 2.813 \times 10^{-3} \ slug - ft^2$. $I_{xy}^C = 0$, $I_{xz}^C = 0$, $I_{yz}^C = 0$.

Solution continued on next page

The inertia matrix of the plate with the cut-out is

$$[I]_O = \begin{bmatrix} 0.04 & 0 & 0 \\ 0 & 0.04 & 0 \\ 0 & 0 & 0.08 \end{bmatrix} - \begin{bmatrix} 1.563\times 10^{-4} & 0 & 0 \\ 0 & 2.656\times 10^{-3} & 0 \\ 0 & 0 & 2.813\times 10^{-3} \end{bmatrix},$$

$$\boxed{[I]_O = \begin{bmatrix} 0.03984 & 0 & 0 \\ 0 & 0.03734 & 0 \\ 0 & 0 & 0.0772 \end{bmatrix} \textit{slug - ft}^2}$$

================================<>================================

Problem 9.113 At $t = 0$, the plate in Problem 9.112 has angular velocity $\vec{\omega} = 10\vec{i} + 10\vec{j}$ (rad/s) and is subjected to the force $\boldsymbol{F} = -10\vec{k}$ (lb) acting at the point $(0, 6, 0)$ *in.* No other forces or couples act on the plate. What are the components of its angular acceleration at that instant?

Solution:

The coordinates of the center of mass are $(-0.01667, 0, 0)$ *ft*.

The vector from the center of mass to the point of application of the force is $\vec{r}_{F/G} = 0.01667\vec{i} + 0.5\vec{j}$ (ft).

The moment about the center of mass of the plate is

$$\vec{M}_G = \vec{r}_{F/G} \times \vec{F} = \begin{bmatrix} \vec{i} & \vec{j} & \vec{k} \\ 0.01667 & 0.5 & 0 \\ 0 & 0 & -10 \end{bmatrix} = -5\vec{i} + 0.1667\vec{j} \ (lb\text{-}ft).$$

Eq (9.30) reduces to
$$\begin{bmatrix} -5 \\ 0.1667 \\ 0 \end{bmatrix} = \begin{bmatrix} 0.03984 & 0 & -0 \\ 0 & 0.03718 & 0 \\ 0 & 0 & 0.07702 \end{bmatrix} \begin{bmatrix} \alpha_x \\ \alpha_y \\ \alpha_z \end{bmatrix} + \begin{bmatrix} 0 \\ 0 \\ -0.267 \end{bmatrix}.$$

Carry out the matrix multiplication to obtain the three equations:

$0.03984\alpha_x = -5$, $0.03718\alpha_y = 0.1667$, $0.07702\alpha_z - 0.267 = 0$.

Solve: $\boxed{\vec{\alpha} = -125.5\vec{i} + 4.484\vec{j} + 3.467\vec{k} \ rad/s^2}$

================================<>================================

==================================<>==================================

Problem 9.114 The inertia matrix of a rigid body in terms of a body-fixed coordinate system with its origin at the center of mass is $[I]=\begin{bmatrix}4 & 1 & -1\\ 1 & 2 & 0\\ -1 & 0 & 6\end{bmatrix}\ kg\text{-}m^2$. If the body's angular velocity is $\vec{\omega}=10\vec{i}-5\vec{j}+10\vec{k}\ (rad/s)$ and its angular acceleration is zero, what are the components of the total moment about its center of mass?

Solution: Use general motion, Eq (9.29),

$$\begin{bmatrix}\sum M_{ox}\\ \sum M_{oy}\\ \sum M_{oz}\end{bmatrix}=\begin{bmatrix}I_{xx} & -I_{xy} & -I_{xz}\\ -I_{yx} & I_{yy} & -I_{yz}\\ -I_{zx} & -I_{zy} & I_{zz}\end{bmatrix}\begin{bmatrix}\alpha_x\\ \alpha_y\\ \alpha_z\end{bmatrix}+\begin{bmatrix}0 & -\Omega_z & \Omega_y\\ \Omega_z & 0 & -\Omega_x\\ -\Omega_y & \Omega_x & 0\end{bmatrix}\begin{bmatrix}I_{xx} & -I_{xy} & -I_{xz}\\ -I_{yx} & I_{yy} & -I_{yz}\\ -I_{zx} & -I_{zy} & I_{zz}\end{bmatrix}\begin{bmatrix}\omega_x\\ \omega_y\\ \omega_z\end{bmatrix}$$ with

$\vec{\alpha}=\dfrac{d\vec{\omega}}{dt}=0$. The coordinate system is rotating with angular velocity $\vec{\omega}$, from which $\vec{\Omega}=\vec{\omega}$. Eq (9.25)

reduces to $$\begin{bmatrix}\sum M_{ox}\\ \sum M_{oy}\\ \sum M_{oz}\end{bmatrix}=\begin{bmatrix}0 & -10 & -5\\ 10 & 0 & -10\\ 5 & 10 & 0\end{bmatrix}\begin{bmatrix}4 & 1 & -1\\ 1 & 2 & 0\\ -1 & 0 & 6\end{bmatrix}\begin{bmatrix}10\\ -5\\ 10\end{bmatrix}=\begin{bmatrix}-5 & -20 & -30\\ 50 & 10 & -70\\ 30 & 25 & -5\end{bmatrix}\begin{bmatrix}10\\ -5\\ 10\end{bmatrix}= N\text{-}m,$$

$\boxed{\vec{M}=-250\vec{i}-250\vec{j}+125\vec{k}\ N\text{-}m}$

==================================<>==================================

Problem 9.115 If the total moment about the center of mass of the rigid body in Problem 9.114 is zero, what are the components of its angular acceleration?

Solution: Use general motion, Eq (9.25), with the moment components equated to zero,

$$\begin{bmatrix}0\\ 0\\ 0\end{bmatrix}=\begin{bmatrix}I_{xx} & -I_{xy} & -I_{xz}\\ -I_{yx} & I_{yy} & -I_{yz}\\ -I_{zx} & -I_{zy} & I_{zz}\end{bmatrix}\begin{bmatrix}\alpha_x\\ \alpha_y\\ \alpha_z\end{bmatrix}+\begin{bmatrix}0 & -\Omega_z & \Omega_y\\ \Omega_z & 0 & -\Omega_x\\ -\Omega_y & \Omega_x & 0\end{bmatrix}\begin{bmatrix}I_{xx} & -I_{xy} & -I_{xz}\\ -I_{yx} & I_{yy} & -I_{yz}\\ -I_{zx} & -I_{zy} & I_{zz}\end{bmatrix}\begin{bmatrix}\omega_x\\ \omega_y\\ \omega_z\end{bmatrix}$$ .Use the numerical results from the solution to Problem 9.42: :

$$\begin{bmatrix}0\\ 0\\ 0\end{bmatrix}=\begin{bmatrix}4 & 1 & -1\\ 1 & 2 & 0\\ -1 & 0 & 6\end{bmatrix}\begin{bmatrix}\alpha_x\\ \alpha_y\\ \alpha_z\end{bmatrix}+\begin{bmatrix}0 & -10 & -5\\ 10 & 0 & -10\\ 5 & 10 & 0\end{bmatrix}\begin{bmatrix}4 & 1 & -1\\ 1 & 2 & 0\\ -1 & 0 & 6\end{bmatrix}\begin{bmatrix}10\\ -5\\ 10\end{bmatrix},\ \begin{bmatrix}0\\ 0\\ 0\end{bmatrix}=\begin{bmatrix}4 & 1 & -1\\ 1 & 2 & 0\\ -1 & 0 & 6\end{bmatrix}\begin{bmatrix}\alpha_x\\ \alpha_y\\ \alpha_z\end{bmatrix}+\begin{bmatrix}-250\\ -250\\ 125\end{bmatrix}.$$

Carry out the matrix multiplication to obtain the three simultaneous equations in the unknowns:

$4\alpha_x+\alpha_y-\alpha_z=250$,

$\alpha_x+2\alpha_y+0=250$,

$-\alpha_x+0+6\alpha_z=-125$. Solve: $\boxed{\vec{\alpha}=31.25\vec{i}+109.4\vec{j}-15.63\vec{k}\ (rad/s^2)}$

==================================<>==================================

==<>==

Problem 9.116 The slender bar of length L and mass m is pinned to the L-shaped bar at O. The L-shaped bar rotates about the vertical axis with a constant angular velocity ω_o. Determine the values of ω_o necessary for the bar to remain at a constant angle β relative to the vertical.

Solution:

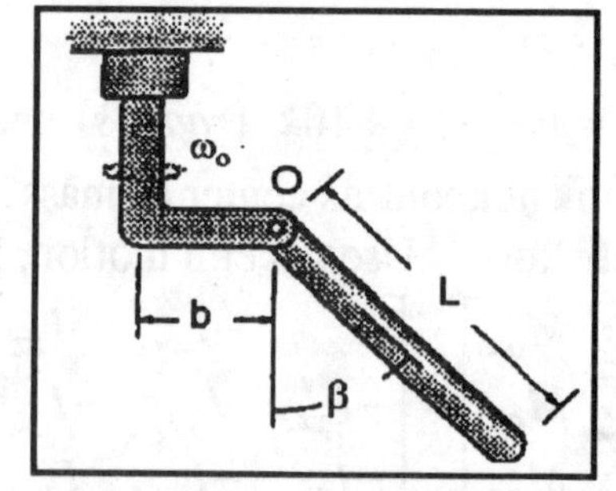

Since the point O is not fixed, this is general motion, in which Eq (9.30) applies. Choose a coordinate system with the origin at O and the *x* axis parallel to the slender bar.

The moment exerted by the bar. Since axis of rotation is fixed, the acceleration must be taken into account in determining the moment. The vector distance from the axis of rotation in the coordinates system shown is

$\vec{r}_O = b(\vec{i}\cos(90^o-\beta)+\vec{j}\sin(90^o-\beta)) = b(\vec{i}\sin\beta+\vec{j}\cos\beta)$. The vector distance to the center of mass of the slender bar is

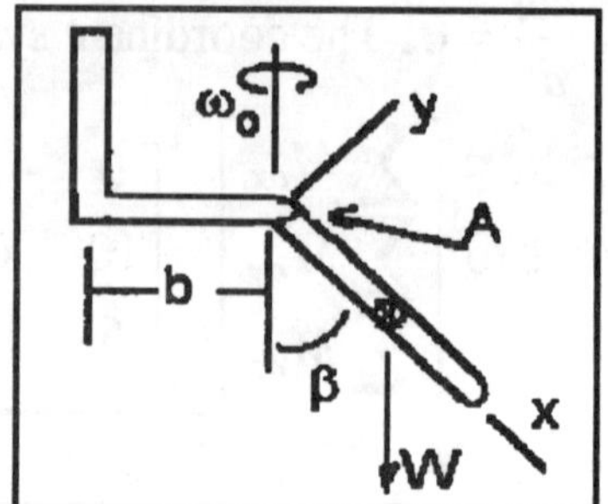

$\vec{r}_{G/O} = \vec{i}\left(\dfrac{L}{2}\right)$. The angular velocity is a constant and the coordinate system is rotating with an angular velocity $\vec{\Omega} = \omega_o(-\vec{i}\cos\beta+\vec{j}\sin\beta)$.

The acceleration of the center of mass relative to O is

$$\vec{a}_G = \vec{\Omega}\times(\vec{\Omega}\times(\vec{r}_O+\vec{r}_{G/O})) = \vec{\Omega}\times\begin{bmatrix}\vec{i} & \vec{j} & \vec{k}\\ -\omega_o\cos\beta & \omega_o\sin\beta & 0\\ b\sin\beta+\dfrac{L}{2} & b\cos\beta & 0\end{bmatrix},$$

$$\vec{a}_G = \begin{bmatrix}\vec{i} & \vec{j} & \vec{k}\\ -\omega_o\cos\beta & \omega_o\sin\beta & 0\\ 0 & 0 & -\omega_o b-\dfrac{\omega_o L\sin\beta}{2}\end{bmatrix} = a_{Gx}\vec{i}+a_{Gy}\vec{j}$$

$$\vec{a}_G = -\omega_o^2\left(+b\sin\beta+\frac{L}{2}\sin^2\beta\right)\vec{i} - \omega_o^2\left(b\cos\beta+\frac{L}{2}\sin\beta\cos\beta\right)\vec{j}.$$

From Newton's second law, $m\vec{a}_G = \vec{A}+\vec{W}$, from which $\vec{A} = m\vec{a}_G - \vec{W}$.

The weight is $\vec{W} = mg(\vec{i}\cos\beta - \vec{j}\sin\beta)$.

The moment about the center of mass is

$$\vec{M}_G = \vec{r}_{O/G}\times\vec{A} = \vec{r}_{O/G}\times(m\vec{a}_G-\vec{W}) = \begin{bmatrix}\vec{i} & \vec{j} & \vec{k}\\ -\dfrac{L}{2} & 0 & 0\\ ma_{Gx}-mg\cos\beta & ma_{Gy}+mg\sin\beta & 0\end{bmatrix}.$$

$$\vec{M}_G = +\left(\frac{m\omega_o^2 bL}{2}\cos\beta - \frac{mgL}{2}\sin\beta + \frac{m\omega_o^2 L^2}{4}\sin\beta\cos\beta\right)\vec{k} = M_z\vec{k}$$

Solution continued on next page

The Euler Equations:

The moments of inertia of the bar about the center of mass are

$I_{xx} = 0$, $I_{yy} = I_{zz} = \frac{mL^2}{12}$, $I_{xy} = I_{xz} = I_{yz} = 0$.

Eq (9.30) becomes:

$$\begin{bmatrix} M_x \\ M_y \\ M_z \end{bmatrix} = \omega_o^2 \begin{bmatrix} 0 & 0 & +\sin\beta \\ 0 & 0 & \cos\beta \\ -\sin\beta & -\cos\beta & 0 \end{bmatrix} \begin{bmatrix} 0 & 0 & 0 \\ 0 & \frac{mL^2}{12} & 0 \\ 0 & 0 & \frac{mL^2}{12} \end{bmatrix} \begin{bmatrix} -\cos\beta \\ \sin\beta \\ 0 \end{bmatrix}.$$

Carry out the matrix multiplication: $\begin{bmatrix} M_x \\ M_y \\ M_z \end{bmatrix} = \begin{bmatrix} 0 \\ 0 \\ -\frac{\omega_o^2 mL^2}{12}\cos\beta\sin\beta \end{bmatrix}$.

Substitute: $\frac{m\omega_o^2 bL}{2}\cos\beta - \frac{mgL}{2}\sin\beta + \frac{m\omega_o^2 L^2}{4}\sin\beta\cos\beta = -\omega_o^2\frac{mL^2}{12}\sin\beta\cos\beta$.

Solve: $\boxed{\omega_o = \sqrt{\frac{g\sin\beta}{\left(\frac{2}{3}\right)L\sin\beta\cos\beta + b\cos\beta}}}$

==================================<>==================================

================================<>================================

Problem 9.117 A slender bar of length L and mass m is rigidly attached to the center of a thin disk of radius R and mass m. The composite object undergoes a motion in which the bar rotates in the horizontal plane with constant angular velocity ω_o about the center of mass of the composite object and the disk rolls on the floor. Show that $\omega_o = 2\sqrt{\frac{g}{R}}$.

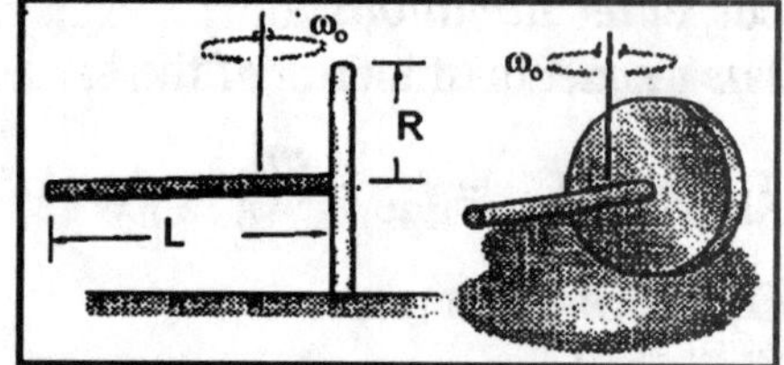

Solution:

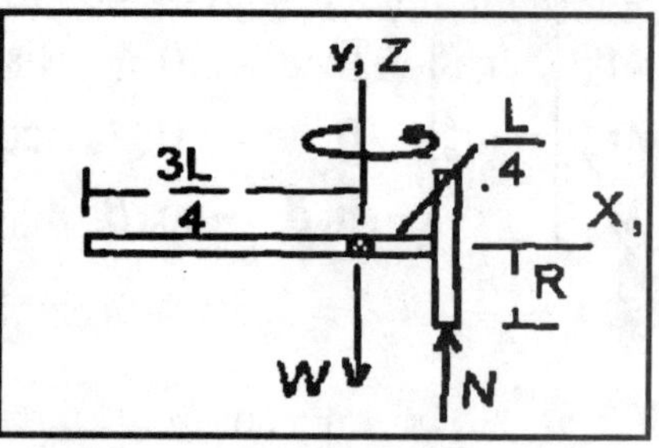

Measuring from the left end of the slender bar, the distance to the center of mass is $d_G = \dfrac{\left(\frac{L}{2}\right)m + mL}{2m} = \dfrac{3L}{4}$. Choose an X, Y, Z coordinate system with the origin at the center of mass, the Z axis parallel to the vertical axis of rotation and the X axis parallel to the slender bar. Choose an x, y, z coordinate system with the origin at the center of mass, the z axis parallel to the slender bar, and the y axis parallel to the Z axis. By definition, the nutation angle is the angle between Z and z, $\theta = 90^o$. The precession rate is the rotation about the Z axis, $\dot{\psi} = \omega_o \; rad/s$. The velocity of the center of mass of the disk is $v_G = (L/4)\dot{\psi}$, from which the spin rate is $\dot{\phi} = \dfrac{v}{R} = \left(\dfrac{L}{4R}\right)\dot{\psi}$.

From Eq (9.41), the moment about the x-axis is

$$M_x = (I_{zz} - I_{xx})\dot{\psi}^2 \sin\theta\cos\theta + I_{zz}\dot{\psi}\dot{\phi}\sin\theta = I_{zz}\dot{\phi}\dot{\psi} = \frac{\omega_o^2 L}{4R} I_{zz}.$$

The moment of inertia is $I_{zz} = \dfrac{mR^2}{2}$, from which $M_x = \left(\dfrac{mRL}{8}\right)\omega_o^2$.

The normal force acting at the point of contact is $N = 2mg$.

The moment exerted about the center of mass $M_G = N\dfrac{L}{4} = mg\left(\dfrac{L}{2}\right)$.

Equate the moments: $mg\left(\dfrac{L}{2}\right) = m\left(\dfrac{RL}{8}\right)\omega_o^2$, from which $\boxed{\omega_o = 2\sqrt{\dfrac{g}{R}}}$.

================================<>================================

==================================<>==================================

Problem 9.118 The thin plate of mass m spins about a vertical axis with the plane of the plate perpendicular to the floor. The corner of the plate at O rests in an indentation so that it remains at the same point on the floor. The plate rotates with constant angular ω_o and the angle β is constant. (a) Show that the angular velocity ω_o is related to the angle β by $\dfrac{h\omega_o^2}{g} = \dfrac{2\cos\beta - \sin\beta}{\sin^2\beta - 2\sin\beta\cos\beta - \cos^2\beta}$.

Solution:

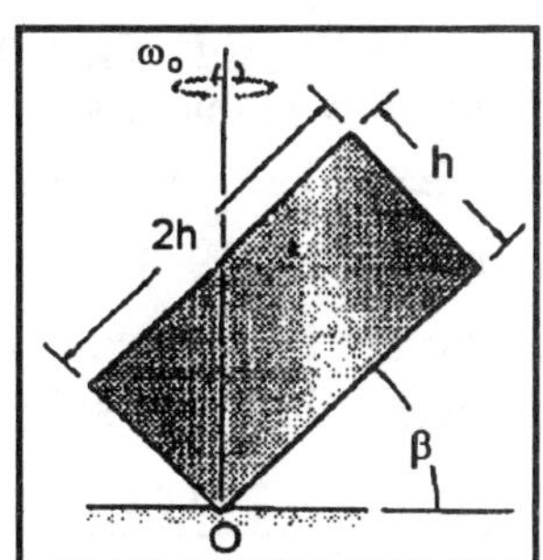

Choose a body-fixed coordinate system with its origin at the fixed point O and the axes aligned with the plate's edges. Using the moments of intertia for a rectangular area, $I_{xx} = \dfrac{mh^2}{3}$, $I_{yy} = \dfrac{4mh^2}{3}$, $I_{zz} = I_{xx} + I_{yy} = \dfrac{5mh^2}{3}$, $I_{xy} = \dfrac{mh^2}{2}$, $I_{xz} = I_{yz} = 0$. The plate's angular velocity is $\vec{\omega} = \omega_o \sin\beta\vec{i} + \omega_o \cos\beta\vec{j}$, and the moment about O due to the plate's weight is $\sum M_x = 0$, $\sum M_y = 0$,. $\sum M_z = \dfrac{h}{2} mg\sin\beta - hmg\cos\beta$.

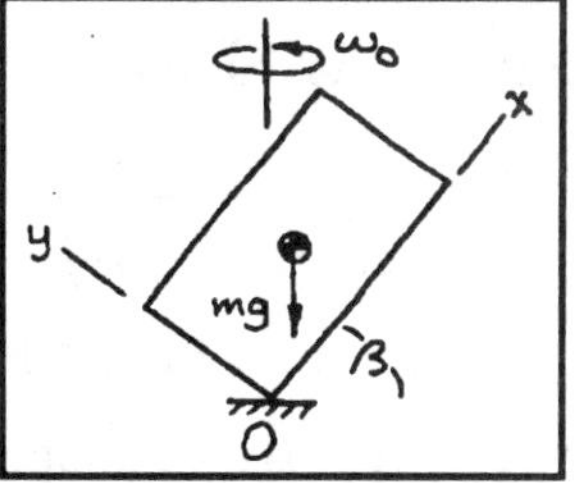

Choose a coordinate system with the origin at the center of mass of the plate and the x axis parallel to the right lower edge of the plate and the y axis parallel to the left lower edge of the plate, as shown. The body fixed coordinate system rotates with angular velocity $\vec{\omega} = \omega_o(\vec{j}\sin\beta + \vec{k}\cos\beta)$.

From Eq (9.26),

$$\begin{bmatrix} 0 \\ 0 \\ M_z \end{bmatrix} = \begin{bmatrix} 0 & 0 & \omega_o\cos\beta \\ 0 & 0 & -\omega_o\sin\beta \\ -\omega_o\cos\beta & \omega_o\sin\beta & 0 \end{bmatrix} \begin{bmatrix} \dfrac{mh^2}{3} & -\dfrac{mh^2}{2} & 0 \\ -\dfrac{mh^2}{2} & \dfrac{4mh^2}{3} & 0 \\ 0 & 0 & \dfrac{5m^2}{3} \end{bmatrix} \begin{bmatrix} \omega_o\cos\beta \\ \omega_o\cos\beta \\ 0 \end{bmatrix}.$$

Expand, $M_z = hmg\left(\dfrac{\sin\beta}{2} - \cos\beta\right) = mh^2\omega_o^2\left(-\dfrac{\sin\beta\cos\beta}{3} + \dfrac{\cos^2\beta}{2} - \dfrac{\sin^2\beta}{2} + \dfrac{4\sin\beta\cos\beta}{3}\right)$

Solve: $\boxed{\dfrac{\omega_o^2 h}{g} = \dfrac{2\cos\beta - \sin\beta}{\sin^2\beta - 2\sin\beta\cos\beta - \cos^2\beta}}$

(b) The perpendicular distance from the axis of rotation to the center of mass of the plate is

$d = \left|\vec{r}_{O/G} \times \dfrac{\vec{\Omega}}{|\vec{\Omega}|}\right| = \left\| \begin{bmatrix} \vec{i} & \vec{j} & \vec{k} \\ 0 & -\dfrac{h}{2} & -h \\ 0 & \cos\beta & \sin\beta \end{bmatrix} \right\| = h\left(\cos\beta - \dfrac{\sin\beta}{2}\right)$. If this distance is zero, $\beta = \tan^{-1}(2) = 63.43^o$, the accelerations of the center of mass and the external moments are zero (see equations above, where for convenience the term $\cos\beta - \dfrac{\sin\beta}{2}$ has been kept as a factor) and the plate is balanced.

Solution continued on next page

The *angular velocity of rotation is zero* (the plate is stationary) *if* $\beta = \tan^{-1}(2) = 63.435^o$, since the numerator of the right hand term in the boxed expression vanishes (the balance at this point would be very unstable, since an infinitesimally small change in β would induce a destabilizing moment.). [*Note:* From the discussion of the *existence* of conditions for dynamic stability given on page 165, Art. 214c, (pages 164-165) of the Dover reprint (1955) of the Sixth Edition (1905) of Routh's Dynamics of a System of Rigid Bodies, one would presume (without certainty) that it is possible for the plate to be dynamically stable under rotation.]

==================================<>==================================

Problem 9.119 In Problem 9.118, determine the range of values of the angle β for which the plate will remain in the steady motion described.

Solution From the solution to Problem 9.118,

$$\frac{\omega_o^2 h}{g} = \frac{2\cos\beta - \sin\beta}{\sin^2\beta - 2\sin\beta\cos\beta - \cos^2\beta}$$. The angular velocity is a real number, from which $\omega_o^2 \geq 0$, from which

$$f(\beta) = \frac{2\cos\beta - \sin\beta}{\sin^2\beta - 2\cos\beta\sin\beta - \cos^2\beta} \geq 0$$. A graph of

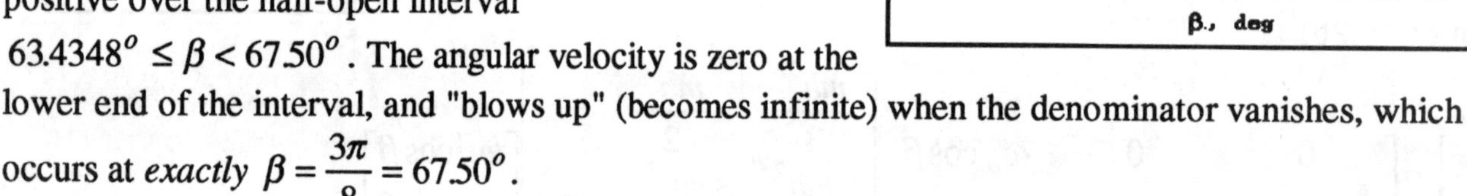

$f(\beta)$ for values of $0 \leq \beta \leq 90^o$ is shown. The function is positive over the half-open interval $63.4348^o \leq \beta < 67.50^o$. The angular velocity is zero at the lower end of the interval, and "blows up" (becomes infinite) when the denominator vanishes, which occurs at *exactly* $\beta = \frac{3\pi}{8} = 67.50^o$.

==================================<>==================================

==================================<>==================================

Problem 9.120 Arm BC has a mass of 12 *kg*, and its moments and products of inertia in terms of the coordinate system shown are $I_{xx} = 0.03\ kg\text{-}m^2$, $I_{yy} = I_{zz} = 4\ kg\text{-}m^2$, $I_{xy} = I_{xz} = I_{yz} = 0$. At the instant shown, arm AB is rotating in the horizontal plane with a constant angular velocity of 1 *rad/s* in the counterclockwise direction viewed from above. Relative to arm AB, BC is rotating about the *z* axis with a constant angular velocity of 2 *rad/s*. Determine the force and couple on arm BC at B.

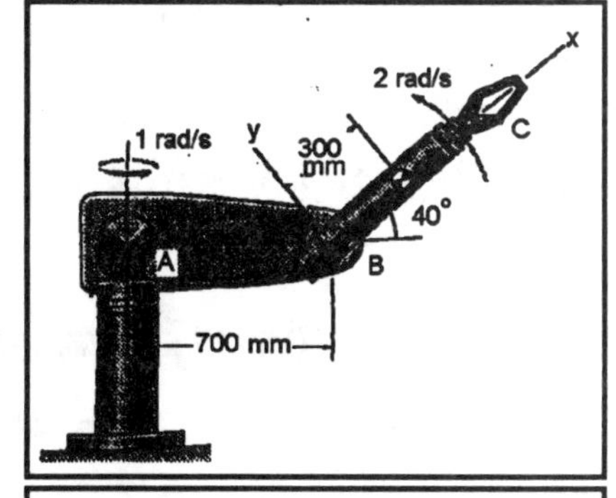

Solution [*Preliminary discussion:* This a general motion problem, and Eq (9.29) applies. The arm BC and the body-fixed coordinates attached to BC rotate relative to the arm AB, and AB is rotating, from which *the angular acceleration of BC is not a constant.* The magnitude of the velocity of rotation of AB is $\Omega_{AB} = 1\ rad/s$ is constant, and the magnitude of the rotation of BC about the z axis $\frac{d\beta}{dt} = 2\ rad/s$ is constant. However, the angular velocity of the body fixed coordinates is $\vec{\omega}_{BC} = \left(\frac{d\beta}{dt}\right)\vec{k} + \vec{\Omega}_{AB}$, from which $\vec{\alpha} = \frac{d\vec{\omega}_{BC}}{dt} + \vec{\omega}_{BC} \times \vec{\omega}_{BC} = \frac{d\vec{\Omega}}{dt}$. This angular acceleration must be taken into account in determining the force and the couple exerted on arm AB by BC. *end discussion.*]

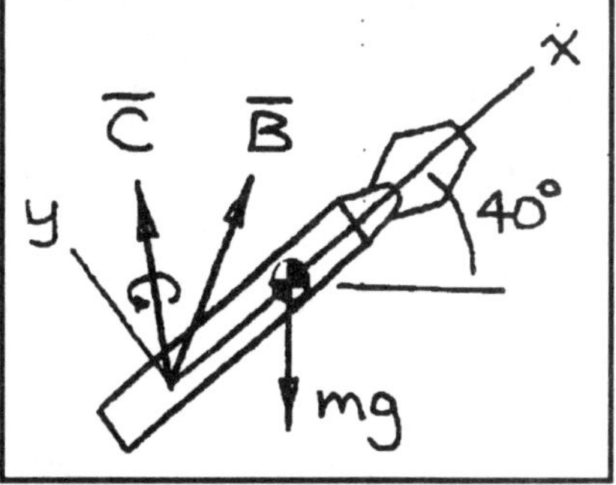

Choose an *x, y,* z axis as shown, with origin at the joint B. Denote $\beta = 40^o$. The angular velocity of arm AB is $\vec{\Omega}_{AB} = 1(\vec{i}\sin\beta + \vec{j}\cos\beta) = 0.6428\vec{i} + 0.7660\vec{j}\ rad/s$. The angular velocity of the arm BC is

$\vec{\omega}_{BC} = 2\vec{k} + \vec{\Omega}_{AB} = 0.6428\vec{i} + 0.7660\vec{j} + 2\vec{k}\ \ rad/s$.

The acceleration of the point B. The vector distance from A to the point B is $\vec{r}_{B/A} = 0.7(\vec{i}\cos\beta - \vec{j}\sin\beta) = 0.5362\vec{i} - 0.4500\vec{j}\ \ (m)$. The acceleration of the point B is

$$\vec{a}_B = \vec{a}_A + \vec{\Omega}_{AB} \times (\vec{\Omega}_{AB} \times \vec{r}_{B/A}) = \vec{\Omega}_{AB} \times \begin{bmatrix} \vec{i} & \vec{j} & \vec{k} \\ 0.6438 & 0.7660 & 0 \\ 0.5362 & -0.4500 & 0 \end{bmatrix},$$

$$\vec{a}_B = \begin{bmatrix} \vec{i} & \vec{j} & \vec{k} \\ 0.6438 & 0.7660 & 0 \\ 0 & 0 & -0.7 \end{bmatrix} = -0.5362\vec{i} + 0.4500\vec{j}\ \ (m/s^2)$$

The acceleration of the center of mass of arm BC. The angular acceleration of the arm BC is

$$\vec{\alpha}_{BC} = \frac{d\vec{\Omega}}{dt} = \cos\beta\frac{d\beta}{dt}\vec{i} - \sin\beta\frac{d\beta}{dt}\vec{j} = 1.532\vec{i} - 1.286\vec{j}\ \ (rad/s^2)$$

The vector distance from B to the center of mass of arm BC is $\vec{r}_{G/B} = 0.3\vec{i}\ (m)$. The acceleration of the center of mass of arm BC is $\vec{a}_G = \vec{a}_B + \vec{\alpha} \times \vec{r}_{G/B} + \vec{\omega}_{BC} \times (\vec{\omega}_{BC} \times \vec{r}_{G/B})$,

Solution continued on next page

Hence, $\vec{a}_G = \vec{a}_B + \begin{bmatrix} \vec{i} & \vec{j} & \vec{k} \\ 1.532 & -1.286 & 0 \\ 0.3 & 0 & 0 \end{bmatrix} + \vec{\omega}_{BC} \times \begin{bmatrix} \vec{i} & \vec{j} & \vec{k} \\ 0.6427 & 0.7660 & 2 \\ 0.5362 & -0.4500 & 0 \end{bmatrix}$

$$\vec{a}_G = -0.5362\vec{i} + 0.4500\vec{j} + 0.3856\vec{j} + \begin{bmatrix} \vec{i} & \vec{j} & \vec{k} \\ 0.6428 & 0.7660 & 2 \\ 0 & 0.6 & -0.2298 \end{bmatrix}.$$

$$\vec{a}_G = -1.912\vec{i} + 0.5977\vec{j} + 0.7713\vec{k} \ \left(m/s^2\right).$$

The force at the joint B. From Newton's second law: $m\vec{a}_G = \vec{B} + \vec{W}$, from which $\vec{B} = m\vec{a}_G - \vec{W}$. The weight is $\vec{W} = -mg\left(\vec{i}\sin\beta + \vec{j}\cos\beta\right) = -75.67\vec{i} - 90.18\vec{j} \ (N)$, from which

$\boxed{\vec{B} = m\vec{a}_G - \vec{W} = 52.72\vec{i} + 97.35\vec{j} + 9.256\vec{k} \ (N)}$ is the reaction at joint B.

The moment about the center of mass:

$$\vec{M}_G = -\vec{r}_{G/B} \times \vec{B} = \begin{bmatrix} \vec{i} & \vec{j} & \vec{k} \\ -0.3 & 0 & 0 \\ 57.72 & 97.35 & 9.256 \end{bmatrix} = 2.776\vec{j} - 29.20\vec{k} \ (N\text{-}m).$$

The Euler Equations: The body fixed coordinates are rotating with angular velocity $\vec{\omega}_{BC}$. The moment of inertia about the center of mass of the arm BC is $I_{x'x'} = 0.03 \ kg\text{-}m^2$,

$I_{y'y'} = I_{z'z'} = 4 - \left(0.3^2\right)12 = 2.92 \ kg\text{-}m^2$, $I_{x'y'} = I_{x'z'} = I_{y'z'} = 0$. Eq (9.29) becomes

$$\begin{bmatrix} M_{Ox} \\ M_{Oy} \\ M_{Oz} \end{bmatrix} = \begin{bmatrix} 0.03 & 0 & 0 \\ 0 & 2.92 & 0 \\ 0 & 0 & 2.92 \end{bmatrix} \begin{bmatrix} 1.532 \\ -1.286 \\ 0 \end{bmatrix} + \begin{bmatrix} 0 & 2 & \cos\beta \\ 2 & 0 & -\sin\beta \\ -\cos\beta & \sin\beta & 0 \end{bmatrix} \begin{bmatrix} 0.03 & 0 & 0 \\ 0 & 2.92 & 0 \\ 0 & 0 & 2.92 \end{bmatrix} \begin{bmatrix} \sin\beta \\ \cos\beta \\ 2 \end{bmatrix}$$

$\begin{bmatrix} M_{Ox} \\ M_{Oy} \\ M_{Oz} \end{bmatrix} = \begin{bmatrix} 0.04596 \\ -3.75 \\ 0 \end{bmatrix} + \begin{bmatrix} 0 & -5.84 & 2.237 \\ 0.06 & 0 & -1.877 \\ -0.0230 & 1.877 & 0 \end{bmatrix} \begin{bmatrix} 0.6428 \\ 0.7660 \\ 2 \end{bmatrix} = \begin{bmatrix} 0.0460 \\ -7.469 \\ 1.423 \end{bmatrix} (N - m)$. The couple acting on the arm BC is $\vec{C} = \vec{M}_O - \vec{M}_G$, or $\begin{bmatrix} C_x \\ C_y \\ C_z \end{bmatrix} = \begin{bmatrix} 0.0460 \\ -7.469 \\ 1.423 \end{bmatrix} - \begin{bmatrix} 0 \\ 2.777 \\ -29.21 \end{bmatrix} = \begin{bmatrix} 0.0460 \\ -10.25 \\ 30.63 \end{bmatrix}$,

$\boxed{\vec{C} = 0.0460\vec{i} - 10.25\vec{j} + 30.63\vec{k} \ (N\text{-}m)}$

================================<>================================

Problem 9.121

Suppose that you throw a football in a wobble spiral with a nutation angle of 25°. The football's moments of inertia are $I_{xx} = I_{yy} = 0.003 \ slug\text{-}ft^2$, $I_{zz} = 0.001 \ slug\text{-}ft^2$. If the spin rate is $\dot{\phi} = 4$ revolutions per second, what is the magnitude of the precession rate (the rate at which it wobbles)?

Solution :

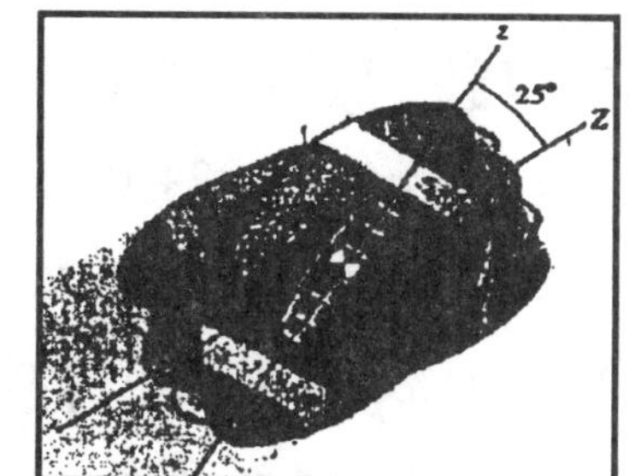

This is modeled as moment-free, steady precession of an axisymmetric object. From Eq (9.43) $(I_{zz} - I_{xx})\dot{\psi}\cos\theta + I_{zz}\dot{\phi} = 0$, from which

$$\dot{\psi} = -\frac{I_{zz}\dot{\phi}}{(I_{zz} - I_{xx})\cos\theta}.$$ Substitute

$$\dot{\psi} = -\frac{(0.001)(4)}{(0.001 - 0.003)\cos 25^o} = 2.21 \ rev/s.$$ $\boxed{|\dot{\psi}| = 2.21 \ rev/s}$

Problem 9.122

Sketch the body and space cones for the motion of the football in Problem 9.121.

Solution:

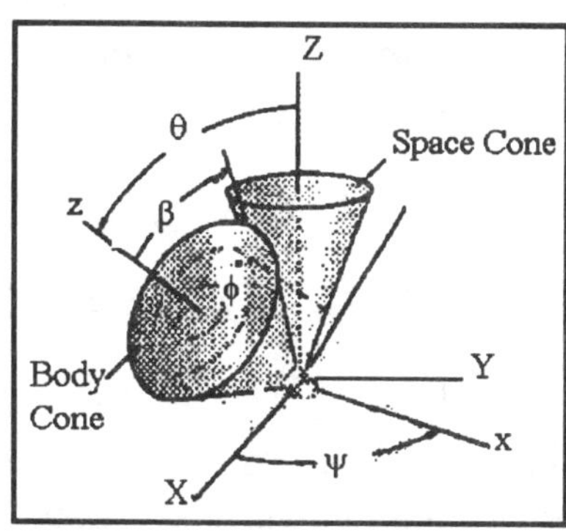

The angle β is given by $\tan\beta = \left(\frac{I_{zz}}{I_{xx}}\right)\tan\theta$, from which $\beta = 8.8^o$.

$\beta < \theta$, and the body cone revolves outside the space cone. The sketch is shown.

Problem 9.123

The mass of the homogenous thin plate is 1 kg. For a coordinate system with its origin at O, determine the principal moments of inertia and the directions of the unit vectors parallel to the corresponding principal axes.

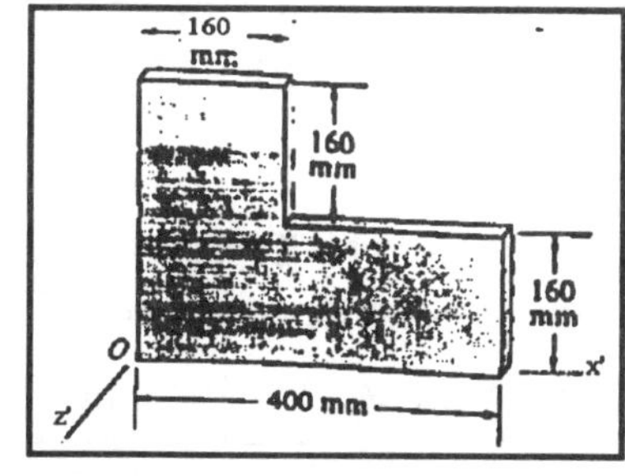

Solution The moment of inertia is determined by the strategy of determining the moments and products of a larger plate, and then subtracting the moments and products of inertia of a cutout, as shown in the sketch. Denote $h = 0.32 \ m$, $b = 0.4 \ m$, $c = 0.16 \ m$, $d = b - c = 0.24 \ m$ The area of the plate is $A = hb - cd = 0.0896 \ m^2$. Denote the mass density by $\rho \ kg/m^3$. The mass is $\rho AT = 1 \ kg$, from which $\rho T = \frac{1}{A} = 11.16 \ kg/m^2$, where T is the (unknown) thickness.

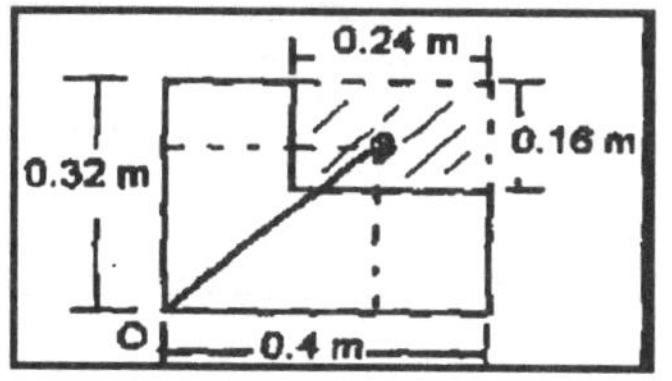

The moments and products of inertia of the large plate: The moments and products of inertia about O are (See Appendix C) $I_{xx}^{(p)} = \frac{\rho Tbh^3}{3}$,

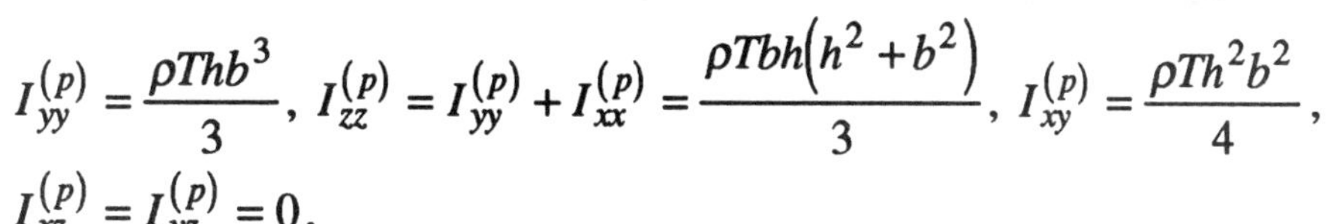

$$I_{yy}^{(p)} = \frac{\rho Thb^3}{3}, \quad I_{zz}^{(p)} = I_{yy}^{(p)} + I_{xx}^{(p)} = \frac{\rho Tbh(h^2 + b^2)}{3}, \quad I_{xy}^{(p)} = \frac{\rho Th^2b^2}{4},$$

$I_{xz}^{(p)} = I_{yz}^{(p)} = 0$.

Solution continued on next page

The moments and products of inertia for the cutout: The moments and products of inertia about the center of mass of the cutout are: $I_{xx}^{(c)} = \dfrac{\rho T d c^3}{12}$, $I_{yy}^{(c)} = \dfrac{\rho T c d^3}{12}$, $I_{zz}^{(c)} = \dfrac{\rho T c d\left(c^2 + d^2\right)}{12}$, $I_{xy}^{(c)} = I_{xz}^{(c)} = I_{yz}^{(c)} = 0$. The distance from O to the center of mass of the cutout is $(d_x, d_y, 0) = (0.28, 0.24, 0)$ m. The moments and products of inertia about the point O are $I_{xx}^{(0)} = I_{xx}^{(c)} + \rho T c d\left(d_y^2\right)$, $I_{yy}^{(0)} = I_{yy}^{(c)} + \rho T c d\left(d_x^2\right)$, $I_{zz}^{(0)} = I_{zz}^{(c)} + \rho T c d\left(d_x^2 + d_y^2\right)$, $I_{xy}^{(0)} = I_{xy}^{(c)} + \rho T c d\left(d_x d_y\right)$, $I_{xz}^{(0)} = I_{yz}^{(0)} = 0$.

The moments and products of inertia of the object: The moments and products of inertia about O are $I_{xx} = I_{xx}^{(p)} - I_{xx}^{(0)} = 0.02316\ kg\text{-}m^2$, $I_{yy} = I_{yy}^{(p)} - I_{yy}^{(0)} = 0.04053\ kg\text{-}m^2$, $I_{zz} = I_{zz}^{(p)} - I_{zz}^{(0)} = 0.06370\ kg\text{-}m^2$, $I_{xy} = I_{xy}^{(p)} - I_{xy}^{(0)} = 0.01691\ kg\text{-}m^2$, $I_{xz} = I_{yz} = 0$.

The principal moments of inertia. The principal values are given by the roots of the cubic, $AI^3 + BI^2 + CI + D = 0$, where $A = 1$, $B = -(I_{xx} + I_{yy} + I_{zz}) = -0.1274$,

$C = I_{xx}I_{yy} + I_{yy}I_{zz} + I_{xx}I_{zz} - I_{xy}^2 - I_{xz}^2 - I_{yz}^2 = 4.71 \times 10^{-3}$,

$D = -\left(I_{xx}I_{yy}I_{zz} - I_{xx}I_{yz}^2 - I_{yy}I_{xz}^2 - I_{zz}I_{xy}^2 - 2I_{xy}I_{yz}I_{xz}\right) = -4.158 \times 10^{-5}$.

The function $f(I) = AI^3 + BI^2 + CI + D$ is graphed to get an estimate of the roots, and these estimates are refined by iteration. The graph is shown. The refined values of the roots are $I_1 = 0.01283\ kg\text{-}m^2$, $I_2 = 0.05086\ kg\text{-}m^2$, $I_3 = 0.06370\ kg\text{-}m^2$.

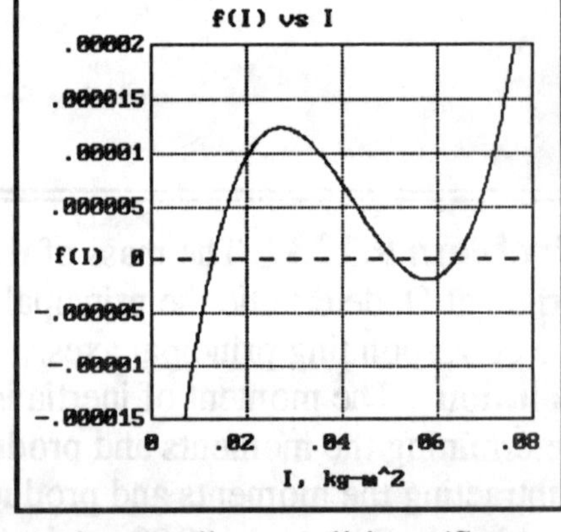

The principal axes. The principal axes are obtained from a solution of the equations

$V_x = (I_{yy} - I)(I_{zz} - I) - I_{yz}^2$

$V_y = I_{xy}(I_{zz} - I) + I_{xz}I_{yz}$

$V_z = I_{xz}(I_{yy} - I) + I_{xz}I_{yz}$. Since $I_{xz} = I_{yz} = 0$, $V_z = 0$, and the solution fails for this axis, and the vector is to be determined from the orthogonality condition. (See discussion under solution to Problem 9.38).

Solving for V_x, V_y, the unit vectors are: for $I = I_1$, $V_x^{(1)} = 0.00141$, $V_y^{(1)} = 8.6 \times 10^{-4}$, from which the unit vectors are $\boxed{\vec{e}_1 = 0.8535\vec{i} + 0.5212\vec{j}}$. For $I = I_2$, $V_x^{(2)} = -1.325 \times 10^{-4}$, $V_y^{(2)} = 2.170 \times 10^{-4}$, from which $\boxed{\vec{e}_2 = -0.5212\vec{i} + 0.8535\vec{j}}$ The third unit vector is determined from orthogonality conditions: $\boxed{\vec{e}_3 = \vec{k}}$.

==================================<>==================================

=================================<>=================================

Problem 9.124 The airplane's principal moments of inertia in *slug-ft*^{2} are $I_{xx} = 8000$, $I_{yy} = 48{,}000$, and $I_{zz} = 50{,}000$. (a) The airplane begins in the reference position shown and maneuvers into the orientation $\psi = \theta = \phi = 45^o$. Draw a sketch showing its orientation relative to the *X, Y, Z* system. (b) If the airplane is in the orientation described in Part (a), the rates of change of the Eulerian angles are $\dot{\psi} = 0$, $\dot{\theta} = 0.2\ rad/s$, and $\dot{\phi} = 0.2\ rad/s$, and their second time derivatives are zero, what are the components of the total moment about the airplane's center of mass?

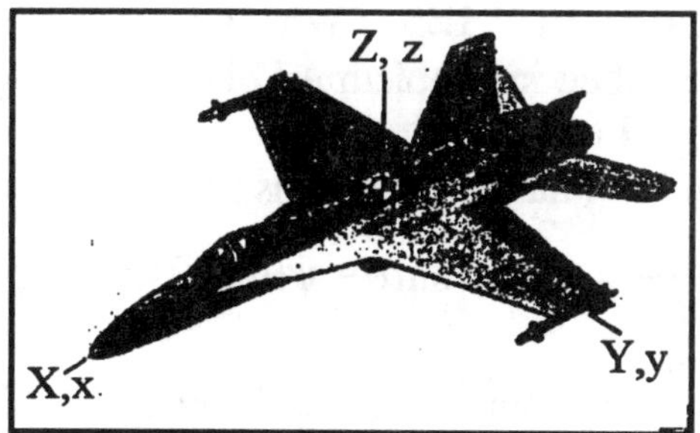

Solution (a)

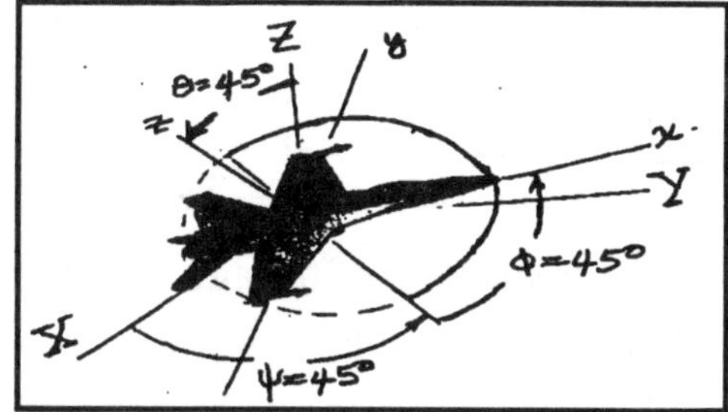

(b) The Eqs (9.46) apply.

$$M_x = I_{xx}(\ddot{\psi}\sin\theta\sin\phi + \ddot{\theta}\cos\phi + \dot{\psi}\dot{\theta}\cos\theta\sin\phi + \dot{\psi}\dot{\phi}\sin\theta\cos\phi - \dot{\theta}\dot{\phi}\sin\phi)$$
$$-(I_{yy} - I_{zz})(\dot{\psi}\sin\theta\cos\phi - \dot{\theta}\sin\phi)(\dot{\psi}\cos\theta + \dot{\phi})$$

$$M_y = I_{yy}(\ddot{\psi}\sin\theta\cos\phi - \ddot{\theta}\sin\phi + \dot{\psi}\dot{\theta}\cos\theta\cos\phi - \dot{\psi}\dot{\phi}\sin\theta\sin\phi - \dot{\theta}\dot{\phi}\cos\phi)$$
$$-(I_{zz} - I_{xx})(\dot{\psi}\sin\theta\sin\phi + \dot{\theta}\cos\phi)(\dot{\psi}\cos\theta + \dot{\phi})$$

$$M_z = I_{zz}(\ddot{\psi}\cos\theta + \ddot{\phi} - \dot{\psi}\dot{\theta}\sin\theta)$$
$$-(I_{xx} - I_{yy})(\dot{\psi}\sin\theta\sin\phi + \dot{\theta}\cos\phi)(\dot{\psi}\sin\theta\cos\phi - \dot{\theta}\sin\phi)$$

Substitute $I_{xx} = 8000$, $I_{yy} = 48{,}000$, and $I_{zz} = 50{,}000$, in *slug-ft*^{2}, and $\dot{\psi} = 0$, $\dot{\theta} = 0.2\ rad/s$, and $\dot{\phi} = 0.2\ rad/s$, and $\ddot{\psi} = \ddot{\theta} = \ddot{\phi} = 0$. The moments: $\boxed{M_x = -282.84\ ft\text{-}lb}$, $\boxed{M_y = -2545.6\ ft\text{-}lb}$, $\boxed{M_z = -800\ ft\text{-}lb}$

=================================<>=================================

==================================<>==================================

Problem 9.125 What are the *x, y,* and *z* components of the angular acceleration of the airplane described in Problem 9.124?

Solution:

The angular accelerations are related to the moments by Eq (9.47),

$$\frac{d\omega_x}{dt} = \ddot{\psi}\sin\theta\sin\phi + \dot{\psi}\dot{\theta}\cos\theta\sin\phi + \dot{\psi}\dot{\phi}\sin\theta\cos\phi + \ddot{\theta}\cos\phi - \dot{\theta}\dot{\phi}\sin\phi$$

$$\frac{d\omega_y}{dt} = \ddot{\psi}\sin\theta\cos\phi + \dot{\psi}\dot{\theta}\cos\theta\cos\phi - \dot{\psi}\dot{\phi}\sin\theta\sin\phi - \ddot{\theta}\sin\phi - \dot{\theta}\dot{\phi}\cos\phi$$

$\frac{d\omega_z}{dt} = \ddot{\psi}\cos\theta - \dot{\psi}\dot{\theta}\sin\theta + \ddot{\phi}$. Substitute $I_{xx} = 8000$, $I_{yy} = 48{,}000$, and $I_{zz} = 50{,}000$, in *slug-ft*², and $\dot{\psi} = 0$, $\dot{\theta} = 0.2\ rad/s$, and $\dot{\phi} = 0.2\ rad/s$, and $\ddot{\psi} = \ddot{\theta} = \ddot{\phi} = 0$, to obtain $\boxed{\alpha_x = -0.0283\ rad/s^2}$, $\boxed{\alpha_y = -0.0283\ rad/s}$, $\boxed{\alpha_z = 0}$

==================================<>==================================

Problem 9.126 If the orientation of the airplane in Problem 9.124 is $\psi = 45^o$, $\theta = 60^o$, $\phi = 45^o$, the rates of change of the Eulerian angles are $\dot{\psi} = 0$, $\dot{\theta} = 0.2\ rad/s$, and $\dot{\phi} = 0.1\ rad/s$, and the components of the total moment about the center of mass are $\sum M_x = 400\ ft\text{-}lb$, $\sum M_y = 1200\ ft\text{-}lb$, and $\sum M_z = 0$, what are the *x, y,* and *z* components of the airplane's angular acceleration?

Solution:

The strategy is to solve the Eqs (9.48) by iteration for the values $\ddot{\theta}$, $\ddot{\phi}$, and $\ddot{\psi}$, and then to use Eqs (9.47) to determine the angular accelerations.

$$M_x = I_{xx}\left(\ddot{\psi}\sin\theta\sin\phi + \ddot{\theta}\cos\phi + \dot{\psi}\dot{\theta}\cos\theta\sin\phi + \dot{\psi}\dot{\phi}\sin\theta\cos\phi - \dot{\theta}\dot{\phi}\sin\phi\right)$$
$$-\left(I_{yy} - I_{zz}\right)\left(\dot{\psi}\sin\theta\cos\phi - \dot{\theta}\sin\phi\right)\left(\dot{\psi}\cos\theta + \dot{\phi}\right)$$

$$M_y = I_{yy}\left(\ddot{\psi}\sin\theta\cos\phi - \ddot{\theta}\sin\phi + \dot{\psi}\dot{\theta}\cos\theta\cos\phi - \dot{\psi}\dot{\phi}\sin\theta\sin\phi - \dot{\theta}\dot{\phi}\cos\phi\right)$$
$$-\left(I_{zz} - I_{xx}\right)\left(\dot{\psi}\sin\theta\sin\phi + \dot{\theta}\cos\phi\right)\left(\dot{\psi}\cos\theta + \dot{\phi}\right)$$

$$M_z = I_{zz}\left(\ddot{\psi}\cos\theta + \ddot{\phi} - \dot{\psi}\dot{\theta}\sin\theta\right)$$
$$-\left(I_{xx} - I_{yy}\right)\left(\dot{\psi}\sin\theta\sin\phi + \dot{\theta}\cos\phi\right)\left(\dot{\psi}\sin\theta\cos\phi - \dot{\theta}\sin\phi\right)$$

Substitute numerical values and solve iteratively to obtain, $\ddot{\phi} = -0.03266\ rad/s^2$, $\ddot{\theta} = 0.01143\ rad/s^2$, $\ddot{\psi} = 0.09732\ rad/s^2$. These are to be used (with the other data) in Eqs (9.47),

$$\frac{d\omega_x}{dt} = \ddot{\psi}\sin\theta\sin\phi + \dot{\psi}\dot{\theta}\cos\theta\sin\phi + \dot{\psi}\dot{\phi}\sin\theta\cos\phi + \ddot{\theta}\cos\phi - \dot{\theta}\dot{\phi}\sin\phi$$

$$\frac{d\omega_y}{dt} = \ddot{\psi}\sin\theta\cos\phi + \dot{\psi}\dot{\theta}\cos\theta\cos\phi - \dot{\psi}\dot{\phi}\sin\theta\sin\phi - \ddot{\theta}\sin\phi - \dot{\theta}\dot{\phi}\cos\phi$$

$$\frac{d\omega_z}{dt} = \ddot{\psi}\cos\theta - \dot{\psi}\dot{\theta}\sin\theta + \ddot{\phi}.$$

Substitute, to obtain: $\boxed{\alpha_x = 0.05354\ rad/s^2}$, $\boxed{\alpha_y = 0.03737\ rad/s^2}$, $\boxed{\alpha_z = 0.016\ rad/s^2}$

==================================<>==================================

CHAPTER 10 - Solutions

==================================<>==================================

Problem 10.1 Confirm that $x = A\sin\omega t + B\cos\omega t$, where A and B are arbitrary constants, satisfies Eq (10.4).

Solution:

Eq (10.4) is $\frac{d^2x}{dt^2} + \omega^2 x = 0$. To confirm that $x = A\sin\omega t + B\cos\omega t$ satisfies this equation, do a substitution. The first derivative is $\frac{dx}{dt} = \omega A\cos\omega t - \omega B\sin\omega t$. The second derivative is $\frac{d^2x}{dt^2} = -\omega^2 A\sin\omega t - \omega^2 B\cos\omega t = -\omega^2(A\sin\omega t + B\cos\omega t) = -\omega^2 x$. Substitute: $-\omega^2 x + \omega^2 x = 0$, which confirms that the equation is satisfied.

==================================<>==================================

Problem 10.2 Confirm that $x = E\sin(\omega t - \phi)$, where E and ϕ are arbitrary constants, satisfies Eq (10.4).

Solution:

Eq (10.4) is $\frac{d^2x}{dt^2} + \omega^2 x = 0$. Take the derivatives and substitute: $\frac{dx}{dt} = \omega E\cos(\omega t - \phi)$, $\frac{d^2x}{dt^2} = -\omega^2 E\sin(\omega t - \phi) = -\omega^2 x$, from which Eq (10.4) is satisfied. [*Check*: Use the sine-of the sum of -angles-dentity $E\sin(\omega t - \phi) = E\cos\phi\sin\omega t - E\sin\phi\cos\omega t$. Let $A = E\cos\phi$, $B = -E\sin\phi$, from which $E\sin(\omega t - \phi) = A\sin\omega t + B\cos\omega t$, and use the solution to Problem 10.1. *check.*]

==================================<>==================================

Problem 10.3 (a) Show that $x = G\cos(\omega t - \psi)$, where G and ψ are arbitrary constants, satisfies Eq (10.4). (b) Determine the constants A and B in the form of the solution given by Eq (10.5) in terms of the constants G and ψ.

Solution:

(a) Eq (10.4) is $\frac{d^2x}{dt^2} + \omega^2 x = 0$. To confirm that $x = G\cos(\omega t - \psi)$ satisfies this equation, do a substitution. The first derivative is $\frac{dx}{dt} = -\omega G\sin(\omega t - \psi)$. The second derivative is $\frac{d^2x}{dt^2} = -\omega^2 G\cos(\omega t - \psi) = -\omega^2 x$. Substitute: $-\omega^2 x + \omega^2 x = 0$, which confirms that the equation is satisfied. (b) Eq (10.5) is $x = A\sin\omega t + B\cos\omega t$. Use the cosine of the sum of angles identity $G\cos(\omega t - \psi) = G\cos\psi\sin\omega t + G\sin\psi\cos\omega t$. Compare to Eq (10.5): $\boxed{A = G\cos\psi}$, $\boxed{B = G\sin\psi}$

==================================<>==================================

==================================<>==================================

Problem 10.4 The position of a vibrating system is $x=(1/\sqrt{2})\sin\omega t-(1/\sqrt{2})\cos\omega t$ ft. (a) Determine the amplitude E of the vibration and the angle ϕ in degrees. (b) Draw a sketch of x for values of ωt from zero to 4π radians, showing the angle ϕ.

Solution: (a) Use the sine of the sum of angles identity:

$E\sin(\omega t-\phi)=E\cos\phi\sin\omega t-E\sin\phi\cos\omega t$. Denote $A=E\cos\phi=\frac{1}{\sqrt{2}}$, $B=E\sin\phi=\left(\frac{1}{\sqrt{2}}\right)$.

The amplitude is defined by Eq (10.8): $|x|_{amplitude}=\sqrt{A^2+B^2}=E$. [*Note*: This is the definition of the "peak amplitude", so-called to distinguish it from other useful definitions of the amplitude found, for example, in alternating current theory.] From the sine-of-the-sum-of-angles identity,

$E=\sqrt{(E\sin\phi)^2+(E\cos\phi)^2}$, from which $\boxed{E=\sqrt{(1/\sqrt{2})^2+(1/\sqrt{2})^2}=1 \text{ ft}}$ The angle ϕ is given by

$\tan\phi=\frac{E\sin\phi}{E\cos\phi}=\frac{(1/\sqrt{2})}{(1/\sqrt{2})}=1$, from which

$\boxed{\phi=\tan^{-1}(1)=45^o}$

(b) The graph is shown. The start of the function $x=\sin\omega t$ is *delayed* by the angle $\phi=45^o$, as shown on the graph.

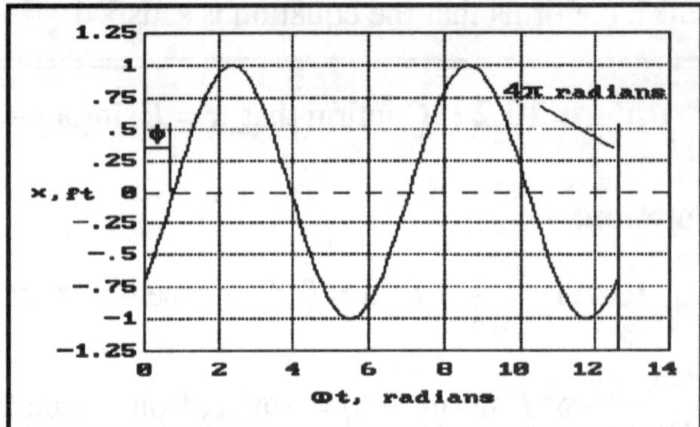

==================================<>==================================

Problem 10.5 The position of a vibrating system is $x=-\sqrt{2}\sin\omega t+\sqrt{2}\cos\omega t$ m. (a) Determine the amplitude E of the vibration and the angle ϕ in degrees. (b) Draw a sketch of x for values of ωt from zero to 4π radians, showing the angle ϕ.

Solution : From the sine of the sum of angles identity $E\sin(\omega t-\phi)=E\cos\phi\sin\omega t-E\sin\phi\cos\omega t$, from which, by comparison, $E\cos\phi=-\sqrt{2}$, $E\sin\phi=-\sqrt{2}$. The amplitude is

$\boxed{E=\sqrt{(\sqrt{2})^2+(\sqrt{2})^2}=2\ m}$, and $\tan\phi=\frac{E\sin\phi}{E\cos\phi}=\frac{-\sqrt{2}}{-\sqrt{2}}=1$, from which $\boxed{\phi=\tan^{-1}(1)+180^o=225^o}$.

[*Discussion.* The quadrant is determined by the signs of the sine and cosine. This is illustrated in the table.

sign of $\cos\phi$	sign of $\sin\phi$	ϕ is in Quadrant	$\phi=\tan^{-1}\left(\frac{\sin\phi}{\cos\phi}\right)+\theta$
+	+	1st	$\theta=0$
+	−	4th	$\theta=360^o$
−	+	2nd	$\theta=180^o$
−	−	3rd	$\theta=180^o$

Solution continued on next page

If angles in the 4th quadrant are to be written as negative angles, this reduces to the algorithm: *if* $\cos\phi < 0$, *then* $\phi = \tan^{-1}\left(\frac{\sin\phi}{\cos\phi}\right) + 180^o$ *else* $\phi = \tan^{-1}\left(\frac{\sin\phi}{\cos\phi}\right)$. (This algorithm will be referenced in later problem solutions, if required.) *End discussion.*]

(b) The position is $x = 2\sin\left(\omega\, t - \frac{5\pi}{4}\right)$. The graph is shown. The start of the function $x = 2\sin\omega\, t$ is *delayed* by the angle $\phi = \frac{5\pi}{4}\ rad = 225^o$.

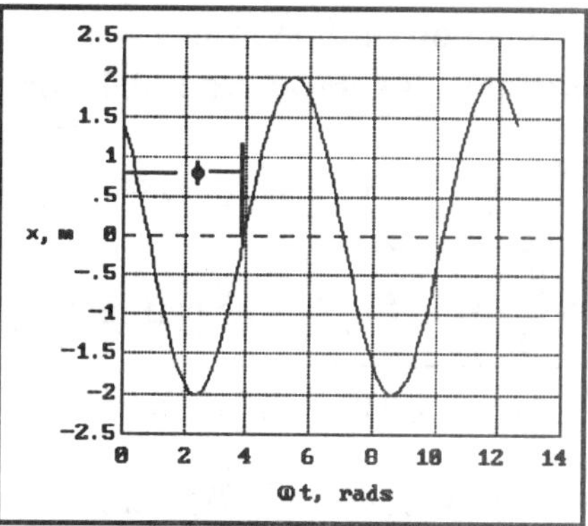

================================<>================================

Problem 10.6 The mass $m = 10\ kg$ and $k = 90\ N/m$. The coordinate x measures the displacement of the mass relative to its equilibrium position. At $t = 0$ the mass is released from rest in the position $x = 0.1\ m$. (a) Determine the period and natural frequency of the resulting vibrations. (b) Determine x as a function of time.

Solution: The equation of motion is given by Eq (10.1), $\frac{d^2x}{dt^2} + \frac{k}{m} = 0$. The canonical form (see Eq (10.4)) is $\frac{d^2x}{dt^2} + \omega^2 x = 0$, with the solution given by Eq (10.6), $x(t) = E\sin(\omega t - \phi)$, where $\omega = \sqrt{\frac{k}{m}}$. (a) For motion described by Eq (10.6) the period is defined as the time required for the angle $\omega t - \phi$ to increase by 2π radians from which $\omega(t+\tau) - \phi = \omega t + 2\pi - \phi$, where τ is the period. Solve: $\tau = 2\pi/\omega$. The natural frequency by definition is $f = 1/\tau$. Substitute numerical values: $\omega = \sqrt{\frac{90}{10}} = 3\ rad/s$. The period is $\boxed{\tau = \frac{2\pi}{3} = 2.094\ s}$. The natural frequency is $\boxed{f = 1/\tau = 0.4775\ \text{Hz}}$. (b) The position is $x(t) = E\sin(\omega t - \phi)$, and the velocity is $\frac{dx}{dt} = \omega E\cos(\omega t - \phi)$. The system is released from rest at $x(0) = 0.1\ m$, from which $x(0) = 0.1 = -E\sin\phi$, and $\frac{dx(0)}{dt} = \omega E\cos\phi = 0$. Solve: $\phi = -\frac{\pi}{2}$, and $E = 0.1\ m$, from which $\boxed{x(t) = E\sin(\omega t - \phi) = 0.1\sin(3t + \frac{\pi}{2}) = 0.1\cos(3t)\ m}$ [*Check*: The solution given in Eq (10.5) is a more convenient form for applying initial conditions: The position is $x(t) = A\sin\omega t + B\cos\omega t$, and the velocity is $\frac{dx(t)}{dt} = \omega A\cos\omega t - \omega B\sin\omega t$, from which $x(0) = 0.1 = B$, $\frac{dx(0)}{dt} = 0 = \omega A$, from which $A = 0$, $B = 0.1\ m$, and $x(t) = B\cos\omega t = 0.1\cos(3t)$. *check.*]

================================<>================================

==============================<>==============================

Problem 10.7 The suspended object weighs 30 *lb* and $k = 20 \ lb / ft$. At $t = 0$, the displacement of the object *relative to its equilibrium position* is $\bar{x} = 0.25 \ ft$ and it is moving downward at $1.5 \ ft / s$. (a) Determine the period and natural frequency of the resulting vibrations. (b) Determine $\bar{x}$ as a function of time. (c) Draw a graph of $\bar{x}$ as a function of time from $t = 0$ to $t = 3 \ s$.

Solution: (a) From Eq (10.4), the canonical form of the equation of motion is $\frac{d^2\bar{x}}{dt^2} + \omega^2\bar{x} = 0$, where $\omega = \sqrt{\frac{gk}{W}} = \sqrt{\frac{32.17(20)}{30}} = 4.631 \ rad / s$. The period is $\boxed{\tau = \frac{2\pi}{\omega} = 1.357 \ s}$. The natural frequency is $\boxed{f = \frac{1}{\tau} = 9.7371 \ Hz}$. (b) The position is $\bar{x}(t) = A\sin\omega t + B\cos\omega t$. The velocity is

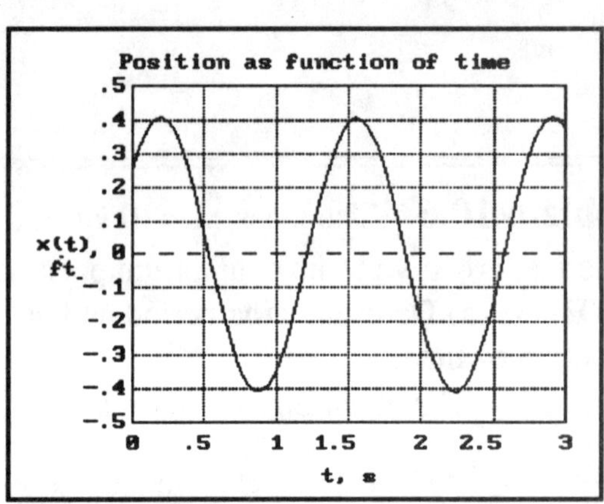

$\frac{d\bar{x}(t)}{dt} = \omega A\cos\omega t - \omega B\sin\omega t$. At $t = 0$,

$\bar{x}(0) = 0.25 = B \ ft$, $\frac{d\bar{x}(0)}{dt} = 1.5 = \omega A$, where, from the figure, x is positive *downward*, from which

$A = \frac{1.5}{\omega} = 0.3239 \ ft$, $B = 0.25 \ ft$, from which

$x(t) = 0.3239\sin 4.631t + 0.25\cos 4.631t$.

(c) The graph is shown. The amplitude is

$|x|_{max} = \sqrt{A^2 + B^2} = 0.4092 \ ft$

==============================<>==============================

Problem 10.8 Determine the natural frequency of vibration of the mass relative to its equilibrium position.

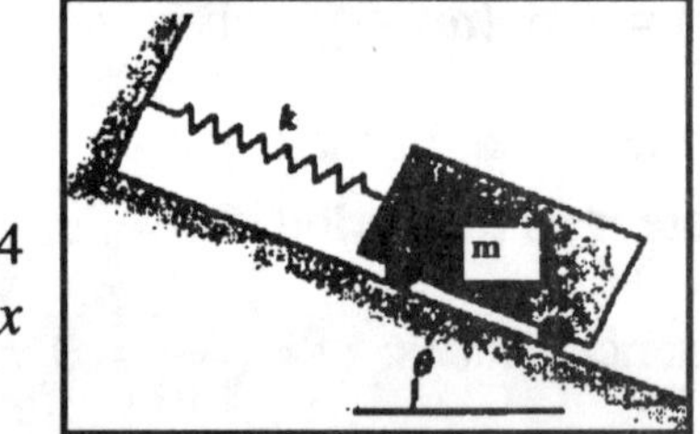

Solution: The strategy is demonstrate that the second order differential equation of motion for motion *about the equilibrium point* reduces to the canonical form. (See discussion at the bottom of page 484 of the text.) Choose a coordinate system with origin at the wall and the x axis parallel to the inclined surface, positive to the right. The forces parallel to x acting on the mass are $\sum F_x = -kx + mg\sin\theta$. From Newton's second law $m(d^2x / dt^2) = -kx + mg\sin\theta$.

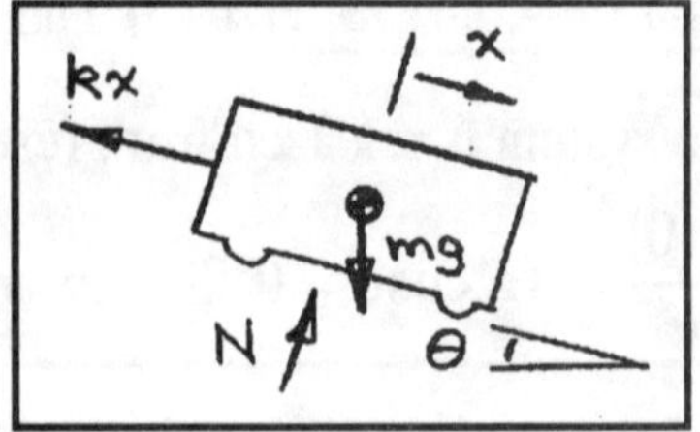

[*Discussion*: (See the discussion on page 508 of the text.) The *general solution* is the sum of two solutions, $x = x_h + x_p$, where x_h, x_p satisfy the *homogenous equation* $m\frac{d^2x_h}{dt^2} + kx_h = 0$ and the *non-homogenous equation* $m\frac{d^2x_p}{dt^2} + kx_p = mg\sin\theta$, respectively. x_h is called the *homogenous solution*, and x_p is called the *particular solution*.

Solution continued on next page

[*Note:* Unlike many older differential equations texts, which use the term *particular integral*, most modern texts use the nomenclature used here, for example, see page 4 of Bronson's Modern Introductory Differential Equations, Schaum's Outline Series, 1973.]
If the non-homogenous term is a constant, that is, if $mg \sin\theta$ is *not a function of x or t*, (as is the case here) then two significant consequences are: (1) the particular solution may be found by setting the derivatives of x to zero in the non-homogenous equation, and solving the result for x_p, and (2) the particular solution x_p is the *equilibrium value* of x, which may be used as a reference value about which the motion takes place. *End discussion.*].
The particular solution is found by setting the acceleration to zero, $-kx_p + mg\sin\theta = 0$, from which $x_p = \frac{mg\sin\theta}{k}$. Define $\bar{x} = (x - x_p)$, and note that $\frac{d^2\bar{x}}{dt^2} = \frac{d^2 x}{dt^2}$, since $x_p = const$. Substitute and rearrange algebraically: $m\frac{d^2\bar{x}}{dt^2} + k\bar{x} = -kx_p + mg\sin\theta = 0$, since by definition the right hand side vanishes. In canonical form (see Eq (10.4)): $\frac{d^2\bar{x}}{dt^2} + \omega^2\bar{x} = 0$, where $\omega = \sqrt{\frac{k}{m}}$. The natural frequency is $f = \frac{\omega}{2\pi}$, from which $\boxed{f = \frac{1}{2\pi}\sqrt{\frac{k}{m}}}$.

==================================<>==================================

Problem 10.9 The thin rectangular plate is attached to the rectangular frame by pins. The frame rotates with conatant angular velocity ω_0. The angle β between the z axis of the body-fixed coordinate system and the vertical is governed by the equation $\frac{d^2\beta}{dt^2} = -\omega_0^2 \sin\beta\cos\beta$. Determine the natural frequency of small vibrations of the plate relative to its horizontal position.

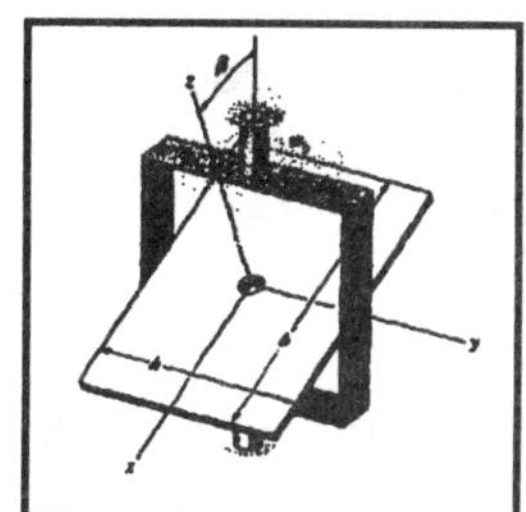

Solution:

For small values of $\beta, \sin\beta \cong \beta$ and $\cos\beta \cong 1$, and the equation of motion is $\frac{d^2\beta}{dt^2} + \omega_0^2\beta = 0$. This is of the form of Equation (10.4) with $\omega = \omega_0$. From Equation (10.10), the natural frequency is $f = \omega_0 / 2\pi$.

==================================<>==================================

Problem 10.10 The rectangular frame in Problem 10.9 rotates with a constant angular velocity $\omega_0 = 6$ (rad/s). At $t = 0$, the angle $\beta = 0.01$ (rad) and $d\beta/dt = 0$. Determine β as a function of time.

Solution:

The equation of motion for small vibrations is $\frac{d^2\beta}{dt^2} + \omega_0^2\beta = 0$. Comparing this equation to Equation (10.4), we can write the solution (10.5) as $\beta = A\sin\omega_0 t + B\cos\omega_0 t$ (1) Taking the time derivative, $\frac{d\beta}{dt} = A\omega_0\cos\omega_0 t - B\omega_0\sin\omega_0 t$ (2) At $t = 0$, $\beta = 0.01 rad$, and $d\beta/dt = 0$. Substituting these conditions into Equations (1) and (2), we obtain $0.01 = B$ and $0 = A\omega_0$.
Therefore the solution is $\beta = 0.01\cos\omega_0 t = 0.01\cos(6t)$ rad.

==================================<>==================================

==============================<>==============================

Problem 10.11 A 200-lb "bungee jumper" jumps from a bridge above a river. The bungee cord has an unstretched length of 60 *ft*, and it stretches an additional 40 *ft* before he rebounds. If you model the cord as a linear spring, what is the period of his vertical oscillations?

Solution:

The total distance that the jumper falls is $h = 60 + 40 = 100\ ft$. The change in potential energy from the action of gravity is Wh. The gain in potential energy in the bungee is $\frac{1}{2}kx^2$. Since the kinetic energy is zero at the end of the drop (the velocity is zero), the total change in the energy is zero, $\frac{1}{2}kx^2 - Wh = 0$, from which $k = \frac{2Wh}{x^2} = \frac{2(200)(100)}{\left(40^2\right)} = 25\ lb/ft$. The period is $\tau = 2\pi\sqrt{\frac{m}{k}} = 2\pi\sqrt{\frac{200}{32.17(25)}} = 3.133\ s$

==============================<>==============================

Problem 10.12 The total mass of the pistion and the load it supports is 90 *kg*. In terms of the distance *s* in meters, the net upward force exerted on the piston by the gas in the cylinder and atmospheric pressure is $(780/s) - 3000$ (*newtons*). Determine the frequency of small vibrations of the piston and load relative to their equilibruim position.

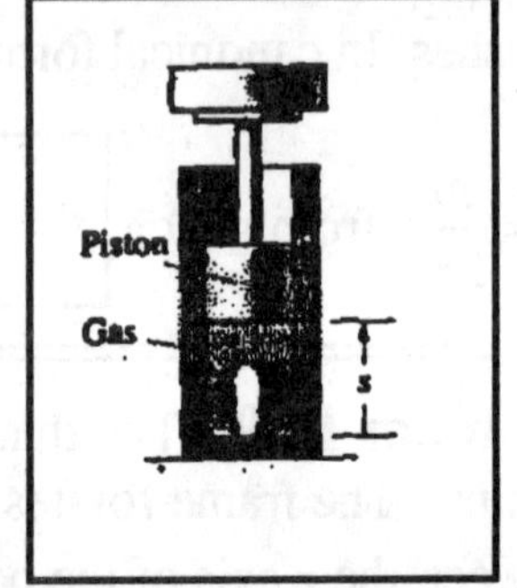

Solution:

Newton's second law for the piston and load is $\sum F_y = ma_y$, or $780s^{-1} - 3000 - (90)(9.81) = 90\frac{d^2s}{dt^2}$. When the system is in equilibrium, $\sum F_y = 0$. Solving for the equilibrium position $s = s_e$, we obtain $s_e = 0.201m$. Let $\tilde{s} = s - s_e$ be the displacement relative to the equilibrium position. Note that $780s_e^{-1} - 3000 - (90)(9.81) = 0$. Expanding $780s^{-1} = \frac{780}{(s_e + \tilde{s})}$ as a Taylor series in terms of $\tilde{s}$, we obtain $\frac{780}{(s_e + \tilde{s})} = \frac{780}{(s_e + \tilde{s})}\Big|_{\tilde{s}=0} + \left[\frac{d}{d\tilde{s}}\left(\frac{780}{s_e + \tilde{s}}\right)\right]_{\tilde{s}=0}\tilde{s} + \ldots = \frac{780}{s_e} - \frac{780}{s_e^2}\tilde{s} + \ldots$. Using this result, we can write the equation of motion as $90\frac{d^2\tilde{s}}{dt^2} = -\frac{780}{s_e^2}\tilde{s}$ or $\frac{d^2\tilde{s}}{dt^2} + 215\tilde{s} = 0$. Comparing this to Equation (10.4), we see that $\omega^2 = 215$ so the natural frequency is $f = \frac{\omega}{2\pi} = 2.33Hz$.

==============================<>==============================

==================================<>==================================

Problem 10.13 The piston and load in Problem 10.12 are displaced downward 2 *mm* from their equilibrium positon and released from rest. Determine their position relative to the equilibrium position as a function of time.

Solution:

From the solution of Problem 10.12, the equation describing the motion relative to the equilibrium position is $\frac{d^2\tilde{s}}{dt^2}+215\tilde{s}=0$ We write the solution in the form $\tilde{s}=A\sin\sqrt{215}t+B\cos\sqrt{215}t$ The time derivative is $\frac{d\tilde{s}}{dt}=(A\sqrt{215})\cos[\sqrt{215}t]-(B\sqrt{215})\sin[\sqrt{215}t]$. At $t=0$, $\tilde{s}=-0.002m$ and $d\tilde{s}/dt=0$.
Using these conditions to determine A and B, we obtain $A=0, B=-0.002m$ so
$\tilde{s}=-0.002\cos[14.7t]\ m=-2\cos[14.7t]\ mm.$

==================================<>==================================

Problem 10.14 The pendulum consists of a homogenous 1 *kg* disk attached to a 0.2 *kg* slender bar. What is the natural frequency of small vibrations of the pendulum?

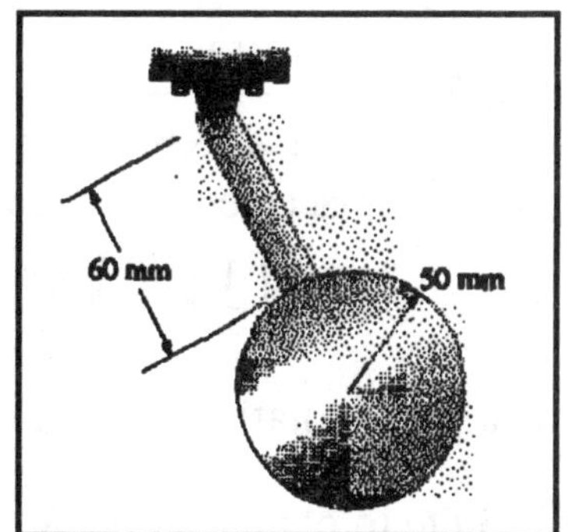

Solution:

The pivot is a fixed point. The center of mass of the combination is located a distance L from the pivot, parallel to the slender bar, where

$L=\frac{0.2(0.03)+1(0.05+0.06)}{0.2+1}=0.09667\ m$. Let θ be the angle from the vertical. From the equation of angular motion: $I\alpha=M$, where

$M=-mgL\sin\theta$, where $m=1.2\ kg$, from which $I\frac{d^2\theta}{dt^2}=-mgL\sin\theta$,

$\frac{d^2\theta}{dt^2}+\frac{mgL}{I}\sin\theta=0$. For small oscillations $\sin\theta\to\theta$, so that the

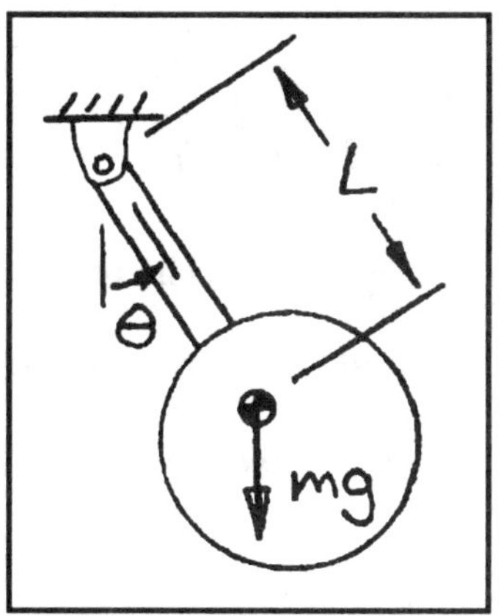

canonical form (see Eq (10.4)) is $\frac{d^2\theta}{dt^2}+\omega^2\theta=0$, where $\omega=\sqrt{\frac{mgL}{I}}$.

The moment of inertia of the slender bar about the pivot point is

$I_b=\frac{0.2(0.06)^2}{12}+0.2\left(\frac{0.06}{2}\right)^2=0.00024\ kg\text{-}m^2$. The moment of inertia of the disk about the pivot is

$I_d=\frac{1(0.05)^2}{2}+1(0.05+0.06)^2=0.01335\ kg\text{-}m^2$, from which $I=I_b+I_d=0.01359\ kg\text{-}m^2$. The natural frequency for small amplitude oscillations is

$$\boxed{f=\frac{\omega}{2\pi}=\frac{1}{2\pi}\sqrt{\frac{mgL}{I}}=\frac{1}{2\pi}\sqrt{\frac{1.2(9.81)(0.09667)}{0.01359}}=1.456\ Hz}$$

==================================<>==================================

================================<>================================

Problem 10.15 The homogenous disk weighs 100 *lb* and its radius is $R = 1$ *ft*. It rolls on the plane surface. The spring constant is $k = 100$ *lb / ft*. (a) Determine the natural frequency of vibrations of the disk relative to its equilibrium position. (b) At $t = 0$ the spring is unstretched and the disk has a clockwise angular velocity of 2 *rad/s*. What is the amplitude of the resulting vibrations of the disk and what is the angular velocity of the disk when $t = 3\ s$?

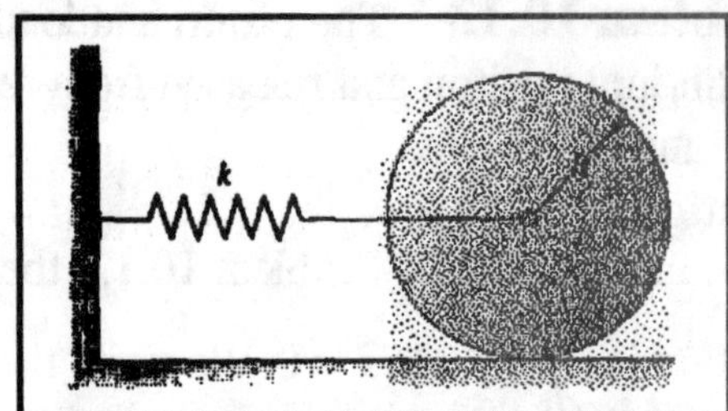

Solution:

The kinetic energy of the disk is the sum of its energy of translation and its energy of rotation, $T = \left(\frac{1}{2}\right)\left(\frac{W}{g}\right)\left(\frac{dx}{dt}\right)^2 + \left(\frac{1}{2}\right)I\left(\frac{d\theta}{dt}\right)^2$. The moment of inertia of the disk about its center is $I = \frac{mR^2}{2}$, from which $T = \left(\frac{1}{2}\right)\left(\frac{W}{g}\right)\left(\frac{dx}{dt}\right)^2 + \left(\frac{1}{4}\right)\left(\frac{W}{g}\right)R^2\left(\frac{d\theta}{dt}\right)^2$. The potential energy of the spring is $V = \frac{1}{2}kx^2$, where x is the displacement from the unstretched spring position. The system is conservative, from which $T + V = const. = \left(\frac{1}{2}\right)\left(\frac{W}{g}\right)\left(\frac{dx}{dt}\right)^2 + \left(\frac{1}{4}\right)\left(\frac{W}{g}\right)R^2\left(\frac{d\theta}{dt}\right)^2 + \frac{1}{2}kx^2$.

The angle of rotation of the disk is related to the displacement of the center of the disk by $x = -R\theta$. Substitute to obtain $\frac{3}{4}\left(\frac{W}{g}\right)R^2\left(\frac{d\theta}{dt}\right)^2 + \frac{1}{2}kR^2\theta^2 = const.$ Take the time derivative $R^2\left(\frac{d\theta}{dt}\right)\left(\frac{3}{2}\left(\frac{W}{g}\right)\left(\frac{d^2\theta}{dt^2}\right) + k\theta\right) = 0$. (a) The time derivative has two possible solutions $\frac{d\theta}{dt} = 0$ or $\frac{d^2\theta}{dt^2} + \frac{2gk}{3W}\theta = 0$. The first is of no interest (the system is not at rest, and *the points* at which the system has zero velocity can be ignored since θ and its first derivative are continuous, by theorem) from which the canonical form (see Eq (10.4)) of the equation of motion is $\frac{d^2\theta}{dt^2} + \omega^2\theta = 0$, where $\omega = \sqrt{\frac{2gk}{3W}}$. The natural frequency is $\boxed{f = \frac{\omega}{2\pi} = \frac{1}{2\pi}\sqrt{\frac{2gk}{3W}} = \frac{1}{2\pi}\sqrt{\frac{2(100)(32.17)}{300}} = 0.7371\ Hz}$ (b) The solution to the equation of motion is $\theta = A\sin\omega t + B\cos\omega t$, where $\omega = \sqrt{\frac{2gk}{3W}} = 4.631\ rad/s$. At $t = 0$, $\frac{d\theta}{dt} = -2\ rad/s$, and $x = 0$, from which $\theta = 0$. Substitute, $[\theta]_{t=0} = 0 = B$, $\left[\frac{d\theta}{dt}\right]_{t=0} = -2 = \omega A$, from which $B = 0$ and $A = \frac{-2}{\omega} = \frac{-2}{2\pi f} = -0.4319\ rad$. The solution is $\theta = A\sin\omega t = -0.4319\sin(4.631\,t)$, from which the amplitude is $\boxed{|\theta|_{max} = 0.4319\ rad}$. At $t = 3\ s$ the angular velocity is

$$\boxed{\left[\frac{d\theta}{dt}\right]_{t=3} = [A\omega\cos\omega t]_{t=3} = -2\cos(4.631(3)) = -0.4832\ rad/s}$$

================================<>================================

==================================<>==================================

Problem 10.16 The radius of the disk is $R = 100\ mm$, and its moment of inertia is $I = 0.1\ \ kg\text{-}m^2$. The mass $m = 5\ kg$, and the spring constant is $k = 135\ \ N/m$. The cable does not slip relative to the disk. The coordinate x measures the displacement of the mass relative to the position in which the spring is unstretched. (a) What is the natural frequency of vertical vibrations of the mass relative to its equilibrium position? (b) Determine x as a function of time if the system is released from rest with $x = 0$.

Solution:

The system is conservative. The kinetic energy of the disk and the weight is $T = \frac{1}{2}I\left(\frac{d\theta}{dt}\right)^2 + \frac{1}{2}m\left(\frac{dx}{dt}\right)^2$. The potential energy is the sum of the energy stored in the spring and the change in the height of the mass,

$V = \frac{1}{2}kx^2 - mgx$. $T + V = const = \frac{1}{2}I\left(\frac{d\theta}{dt}\right)^2 + \frac{1}{2}m\left(\frac{dx}{dt}\right)^2 + \frac{1}{2}kx^2 - mgx$.

From kinematics, $-R\theta = x$. Substitute:

$$\frac{1}{2}\left(\left(\frac{I}{R^2}\right) + m\right)\left(\frac{dx}{dt}\right)^2 + \frac{1}{2}kx^2 - mgx = const.$$

Take the time derivative: $\left(\frac{dx}{dt}\right)\left[\left(\frac{I}{R^2} + m\right)\left(\frac{d^2x}{dt^2}\right) + kx - mg\right] = 0$. This has two possible solutions,

$\left(\frac{dx}{dt}\right) = 0$ or $\left(\frac{I}{R^2} + m\right)\left(\frac{d^2x}{dt^2}\right) + kx - mg = 0$. The first can be ignored (see solution to Problem 10.15)

from which the canonical form (see Eq (10.4)) is $\frac{d^2x}{dt^2} + \omega^2 x = F$, where $\omega = R\sqrt{\frac{k}{I + R^2 m}}$, and

$F = \frac{mgR^2}{\left(I + R^2 m\right)}$. (a) The period of the motion is $\boxed{\tau = \frac{2\pi}{\omega} = \left(\frac{2\pi}{R}\right)\sqrt{\frac{I + R^2 m}{k}} = 2.094\ \ s}$. The natural frequency is $\boxed{f = 1/\tau = 0.4775\ \ \text{Hz}}$.(b) (See discussion on page 508 of text, and discussion under the solution to Problem 10.10.) The general solution is the sum of the *homogenous solution*

$x_h = A\sin\omega t + B\sin\omega t$, and the *particular solution* x_p, which is the solution to $\frac{d^2x_p}{dt^2} + \omega^2 x_p = F$.

Since the non-homogenous term *F* is a constant, *independent of position or time*, x_p can be found by setting the acceleration to zero and solving, from which $x_p = \frac{F}{\omega^2} = \frac{mg}{k}$. The general solution is

$x(t) = x_h + x_p = A\sin\omega t + B\cos\omega t + \frac{mg}{k}$. For this class of problem, where F is a constant independent of position or time, the particular solution is also the equilibrium position of the weight, about which the oscillations take place.

Apply the initial conditions to the solution $x(t) = A\sin\omega t + B\cos\omega t + \frac{mg}{k}$. At $t = 0$, $x = 0$, and $\frac{dx}{dt} = 0$

from which $B = -\frac{mg}{k}$, $A = 0$, from which $\boxed{x(t) = \frac{mg}{k}(1 - \cos\omega t) = 0.3633(1 - \cos 3t)\ \ (m)}$

==================================<>==================================

==================================<>==================================

Problem 10.17 If the spring constant is $k = 30\ lb/ft$ and the moment of inertia of the pulley is negligible, what is the frequency of vertical vibrations of the weights relative to their equilibrium positions?

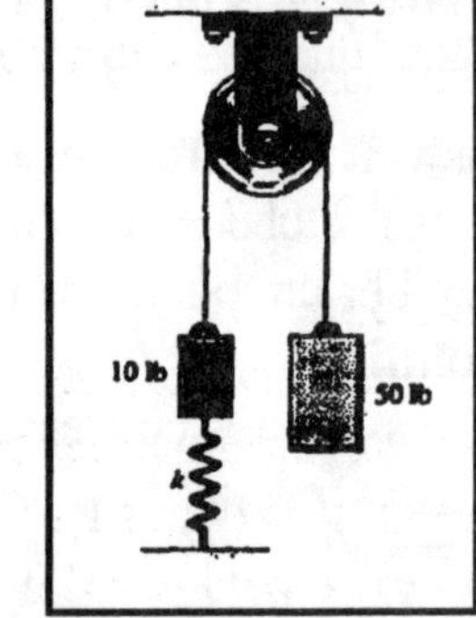

Soluton:

Let x be the downward displacement of the $50-lb$ weight relative to its position when the spring is unstretched. The kinetic energy of the system is

$T=\frac{1}{2}\left(\frac{50}{32.2}\right)\left(\frac{dx}{dt}\right)^2+\frac{1}{2}\left(\frac{10}{32.2}\right)\left(\frac{dx}{dt}\right)^2$ and its potential energy is

$V=\frac{1}{2}(30)x^2-50x+10x$. Since

$\frac{d}{dt}(T+V)=\frac{1}{2}\left(\frac{60}{32.2}\right)2\frac{dx}{dt}\frac{d^2x}{dt^2}+\frac{1}{2}(30)2x\frac{dx}{dt}-50\frac{dx}{dt}+10\frac{dx}{dt}=0$, we obtain the equation of motion

$\frac{60}{32.2}\frac{d^2x}{dt^2}+30x=40$. (1) When the system is in equilibrium, $d^2x/dt^2=0$, so the equilibrium position x_e is $x_e=\frac{40}{30}ft$

Let $\tilde{x}=x-x_e$ be the position relative to equilibrium. Substituting $x=\tilde{x}+x_e$ into Equation (1), the equation of motion becomes $\frac{d^2\tilde{x}}{dt^2}+16.1\tilde{x}=0$ This is of the form of Equation (10.4) with $\omega^2=16.1$, so the natural frequency is $f=\frac{\omega}{2\pi}=0.639Hz$.

==================================<>==================================

Problem 10.18 In Problem 10.17, the spring constant is $k = 30\ lb/ft$, the radius of the pulley is $0.5\ m$, and the moment of inertia of the pulley is $0.25\ slug\text{-}ft^2$. At $t = 0$, the weights are released from rest with the spring unstretched. Determine the position of the $10\text{-}lb$ weight relative to its position at $t = 0$ as a function of time.

Solution:

Let x be the upward displacement of the $10-lb$ weight relative to its position when the spring is unstretched. The angular velocity of the pulley is $(dx/dt)/0.5rad/s$ therefore the kinetic energy of the system is $T=\frac{1}{2}\left(\frac{50}{32.2}\right)\left(\frac{dx}{dt}\right)^2+\frac{1}{2}\left(\frac{10}{32.2}\right)\left(\frac{dx}{dt}\right)^2+\frac{1}{2}(0.25)\frac{1}{(0.5)^2}\left(\frac{dx}{dt}\right)^2$, and its potential energy is

$V=\frac{1}{2}(30)x^2-50x+10x$ By setting $\frac{d}{dt}(T+V)=0$, we obtain the equation of motion

$2.86\frac{d^2x}{dt^2}+30x=40$. (1) When the system is in equilibrium, $d^2x/dt^2=0$, so the equilibrium position x_e is $x_e=\frac{40}{30}=1.33ft$. Let $\tilde{x}=x-x_e$ be the position relative to equilibrium. Substituting $x=\tilde{x}+x_e$ into Eq (1), the equation of motion becomes $\frac{d^2\tilde{x}}{dt^2}+\omega^2\tilde{x}=0$ where $\omega^2=10.48$. At $t=0, x=0$ and $dx/dt=0$, so $\tilde{x}=-x_e=-1.33ft$ and $d\tilde{x}/dt=0$. Writing the solution of Equation (2) in the form $\tilde{x}=A\sin\omega t+B\cos\omega t$ and applying the initial conditions, we obtain $\tilde{x}=-1.33\cos[\sqrt{10.48}t]$. Therefore $x=\tilde{x}+x_e=-1.33\cos[3.24t]+1.33=1.33(1-\cos[3.24t])ft$.

==================================<>==================================

==================================<>==================================

Problem 10.19 A homogenous disk of mass m and radius r rolls on a curved surface of radius R. Show that the natural frequency of small vibrations of the disk relative to its equilibrium position is $f=\frac{1}{\pi}\sqrt{\frac{g}{6(R-r)}}$.

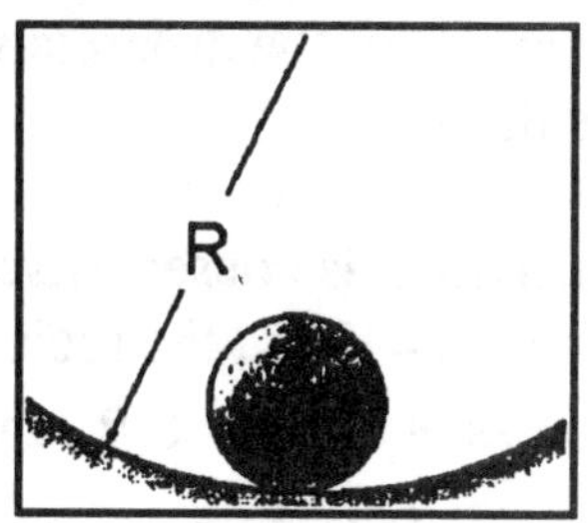

Solution:

Choose a coordinate system with origin at the center of mass of the disk at the equilibrium position (the lowest position on the curved surface) with the y axis positive upward. Denote the angular rotation of the disk by α, and the angle between the y axis and R by θ. The kinetic energy of the disk is the sum of the kinetic energies of translation and rotation: $T=\frac{1}{2}mv^2+\frac{1}{2}I\left(\frac{d\alpha}{dt}\right)^2$, where v is the velocity of the center of mass of the disk. The potential energy is the change in height of the disk: $V=mgh$, where y is the change in height of the center of mass of the disk. Since the system is conservative, $T+V=const.$ From kinematics: the velocity of the center of mass is $v=(R-r)\left(\frac{d\theta}{dt}\right)$. The point of contact with the surface has zero velocity, from which the velocity of the center of mass is $v=-r\left(\frac{d\alpha}{dt}\right)$, from which $\frac{d\alpha}{dt}=-\frac{(R-r)}{r}\left(\frac{d\theta}{dt}\right)$. The height measured from the center of mass at the lowest position is $h=(R-r)(1-\cos\theta)$. For a homogenous disk, the mass moment of inertia about the axis of rotation is $I=\frac{mr^2}{2}$. Substitute the kinematic relations and reduce:

$$\frac{m}{2}\left[\frac{3}{2}(R-r)^2\right]\left(\frac{d\theta}{dt}\right)^2+mg(R-r)(1-\cos\theta)=const.$$

Take the time derivative and divide by the mass: $\left(\frac{d\theta}{dt}\right)\left[\left(\frac{3}{2}(R-r)^2\right)\left(\frac{d^2\theta}{dt^2}\right)+g(R-r)\sin\theta\right]=0$. Ignore the possible solution $\left(\frac{d\theta}{dt}\right)=0$, (see solution to Problem 10.15) from which the equation of motion is $\frac{d^2\theta}{dt^2}+\frac{2g}{3(R-r)}\sin\theta=0$. If the amplitude of the oscillation is small, $\sin\theta\to\theta$, and the canonical form (see Eq (10.4)) of the equation of motion is $\frac{d^2\theta}{dt^2}+\omega^2\theta=0$, where $\omega=\sqrt{\frac{2g}{3(R-r)}}=2\sqrt{\frac{g}{6(R-r)}}$. The natural frequency of the vibration is $\boxed{f=\frac{\omega}{2\pi}=\frac{1}{\pi}\sqrt{\frac{g}{6(R-r)}}}$, which completes the demonstration.

==================================<>==================================

================================<>================================

Problem 10.20 The slender bar has roller supports at its ends and is at rest in circular depression with an 8 *ft* radius. What is the frequency of small vibrations of the disk relative to its equilibrium position?

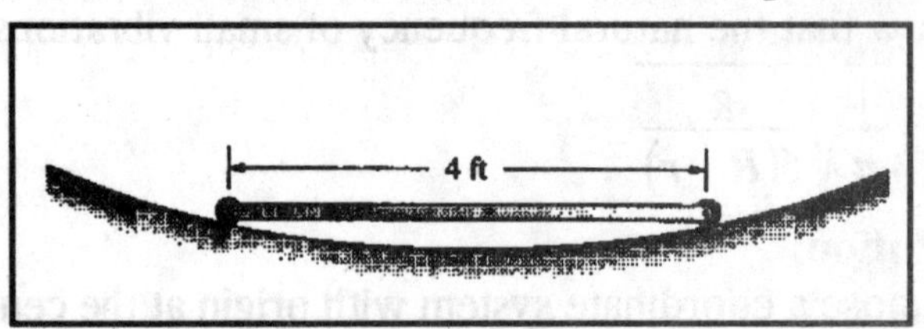

Solution:

The system is conservative. Denote the length of the bar by $L = 4\ ft$, the radius of the circular depression by $R = 8\ ft$, and the fixed angle $\theta = \sin^{-1}\left(\frac{L}{2R}\right) = 14.48^o$. The distance to the center of mass from the fixed center O is $D = R\cos\theta = 7.746\ ft$. Denote the angle of rotation (from the vertical axis) about the fixed center O by α. The kinetic energy of the bar is $T = \frac{1}{2}mv^2 + \frac{1}{2}I\left(\frac{d\alpha}{dt}\right)^2$, where v is the velocity of the center of mass. The potential energy is related to the height of the center of mass of the bar: $V = mgD(1-\cos\alpha)$. From kinematics, $v = D\left(\frac{d\alpha}{dt}\right)$. The moment of inertia of the bar about the center of mass is $I = \frac{mL^2}{12}$. $T + V = const. = \frac{1}{2}mD^2\left(\frac{d\alpha}{dt}\right)^2 + \frac{1}{2}\frac{mL^2}{12}\left(\frac{d\alpha}{dt}\right)^2 + mgD(1-\cos\alpha)$. Take the time derivative and reduce: $\left(\frac{d\alpha}{dt}\right)\left[m\left(D^2 + \frac{L^2}{12}\right)\left(\frac{d^2\alpha}{dt^2}\right) + mgD\sin\alpha\right] = 0$. Ignore the possible solution $\frac{d\alpha}{dt} = 0$, (see the solution to Problem 10.15) from which $\left(D^2 + \frac{L^2}{12}\right)\left(\frac{d^2\alpha}{dt^2}\right) + gD\sin\alpha = 0$. For small amplitude vibrations, $\sin\alpha \to \alpha$, and the canonical form (see Eq (10.4)) is $\frac{d^2\alpha}{dt^2} + \omega^2\alpha = 0$, where

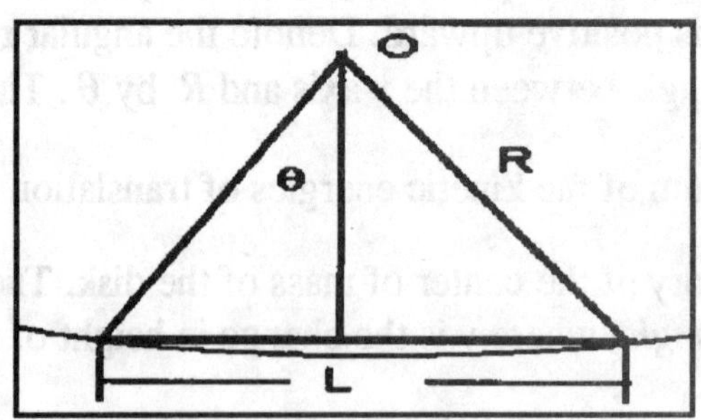

$\omega = \sqrt{\frac{Dg}{D^2 + \frac{L^2}{12}}} = 2.016\ rad/s$, from which the natural frequency is $\boxed{f = \frac{\omega}{2\pi} = 0.3208\ Hz}$

================================<>================================

================================<>================================

Problem 10.21 A slender bar of mass *m* and length *L* is pinned to a fixed support as shown. A torsional spring of spring constant *k* attached to the bar at the support is unstretched when the bar is vertical. Show that the equation governing small vibrations of the bar from its vertical equilibrium position is $\frac{d^2\theta}{dt^2}+\omega^2\theta=0$, where $\omega^2=\frac{\left(k-\frac{1}{2}mgL\right)}{\frac{1}{3}mL^2}$.

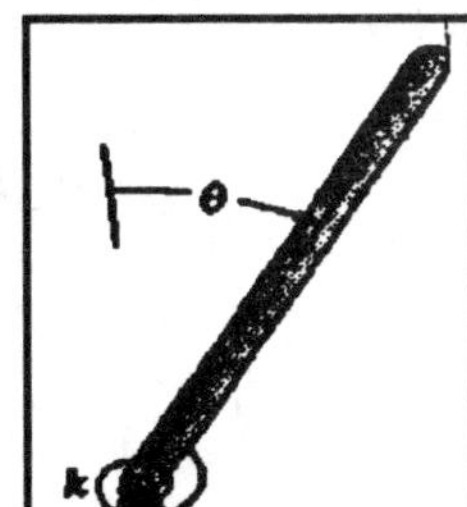

Solution:
The system is conservative. The pivot is a fixed point. The moment of inertia about the fixed point is $I=mL^2/3$. The kinetic energy of the motion of the bar is $T=\frac{1}{2}I\left(\frac{d\theta}{dt}\right)^2$. The potential energy is the sum of the energy in the spring and the gravitational energy associated with the change in height of the center of mass of the bar, $V=\frac{1}{2}k\theta^2-\frac{mgL}{2}(1-\cos\theta)$. For a conservative system, $T+V=const.=\frac{1}{2}I\left(\frac{d\theta}{dt}\right)^2+\frac{1}{2}k\theta^2-\frac{mgL}{2}(1-\cos\theta)$. Take the time derivative and reduce: $\left(\frac{d\theta}{dt}\right)\left[\frac{mL^2}{3}\left(\frac{d^2\theta}{dt^2}\right)+k\theta-\frac{mgL}{2}\sin\theta\right]=0$. Ignore the possible solution $\frac{d\theta}{dt}=0$, from which $\frac{mL^2}{3}\left(\frac{d^2\theta}{dt^2}\right)+k\theta-\frac{mgL}{2}\sin\theta=0$. For small amplitude vibrations $\sin\theta\to\theta$, and the canonical form (see Eq (10.4)) of the equation of motion is $\frac{d^2\theta}{dt^2}+\omega^2\theta=0$, where $\boxed{\omega^2=\frac{k-\frac{mgL}{2}}{\frac{1}{3}mL^2}}$

================================<>================================

Problem 10.22 The conditions of the slender bar in Problem 10.21 are $t=0$, $\theta=0$ and $\frac{d\theta}{dt}=\dot{\theta}_O$.

(a) If $k>\frac{1}{2}mgL$, show that θ is given as a function of time by $\theta=\frac{\dot{\theta}_O}{\omega}\sin\omega t$, where $\omega^2=\frac{\left(k-\frac{mgL}{2}\right)}{\left(\frac{1}{3}mL^2\right)}$.

(b) If $k<\frac{1}{2}mgL^2$, show that θ is given as a function of time by $\theta=\frac{\dot{\theta}_O}{2h}\left(e^{ht}-e^{-ht}\right)$, where $h^2=\frac{\left(\frac{1}{2}mgL-k\right)}{\left(\frac{1}{3}mL^2\right)}$. *Strategy*: To do part (b), seek a solution of the equation of motion of the form $x=Ce^{\lambda t}$, where C and λ are constants.

Solution continued on next page

Continuation of solution to Problem 10.22

Solution: Write the equation of motion in the form $\frac{d^2\theta}{dt^2} + p^2\theta = 0$, where $p = \sqrt{\frac{k - \frac{mgL}{2}}{\frac{1}{3}mL^2}}$.

Define $\omega = \sqrt{\frac{\left(k - \frac{mgL}{2}\right)}{\left(\frac{1}{3}mL^2\right)}}$ if $k > \frac{mgL}{2}$, and $h = \sqrt{\frac{\left(\frac{1}{2}mgL - k\right)}{\left(\frac{1}{3}mL^2\right)}}$, if $k < \frac{mgL}{2}$, from which $p = \omega$ if $k > \frac{mgL}{2}$, and $p = ih$, if $k < \frac{mgL}{2}$, where $i = \sqrt{-1}$. Assume a solution of the form $\theta = A\sin pt + B\cos pt$. The time derivative is $\frac{d\theta}{dt} = pA\cos pt - pB\sin pt$. Apply the initial conditions at $t = 0$, to obtain $B = 0$, and $A = \frac{\dot{\theta}_o}{p}$, from which the solution is $\theta = \frac{\dot{\theta}_o}{p}\sin pt$. (a) Substitute: if $k > \frac{mgL}{2}$, $\boxed{\theta = \frac{\dot{\theta}_o}{\omega}\sin\omega t}$. (b) If $k < \frac{mgL}{2}$, $\theta = \frac{\dot{\theta}_o}{ih}\sin(iht)$. From the definition of the hyperbolic sine, (see any mathematics handbook, for example, see **8.1**, page 26, and **8.74** page 31 of Spiegel's <u>Mathematical Handbook of Formulas and Tables</u>, Schaum's Outline Series.) $\frac{\sinh(ht)}{h} = \frac{\sin(iht)}{ih} = \frac{1}{2h}\left(e^{ht} - e^{-ht}\right)$, from which the solution is $\boxed{\theta = \frac{\dot{\theta}_o}{2h}\left(e^{ht} - e^{-ht}\right)}$. [*Check*: An alternate solution for part (b) based on the suggested strategy is: For $k < \frac{mgL}{2}$ write the equation of motion in the form $\frac{d^2\theta}{dt^2} - h^2\theta = 0$, and assume a general solution of the form $\theta = Ce^{\lambda t} + De^{-\lambda t}$. Substitute into the equation of motion to obtain $\left(\lambda^2 - h^2\right)\theta = 0$, from which $\lambda = \pm h$, and the solution is $\theta = Ce^{ht} + De^{-ht}$, where the positive sign is taken without loss of generality. The time derivative is $\frac{d\theta}{dt} = hCe^{ht} - hDe^{-ht}$. Apply the initial conditions at $t = 0$ to obtain the two equations: $0 = C + D$, and $\dot{\theta}_o = hC - hD$. Solve: $C = \frac{\dot{\theta}_o}{2h}$, $D = -\frac{\dot{\theta}_o}{2h}$, from which the solution is $\theta = \frac{\dot{\theta}_o}{2h}(e^{ht} - e^{-ht})$. *check.*]

==================================<>==================================

==================================<>==================================

Problem 10.23 Engineers use the device shown to measure an astronaut's moment of inertia. The horizontal board is pinned at O and supported by the linear spring with constant $k = 12\ kN/m$. When the astronaut is not present, the frequency of small vibrations of the board about O is measured and determined to be $6.0\ Hz$. When the astronaut is lying on the board as shown, the frequency of small vibrations of the board about O is $2.8\ Hz$. What is the astronaut's moment of inertia about the z axis?

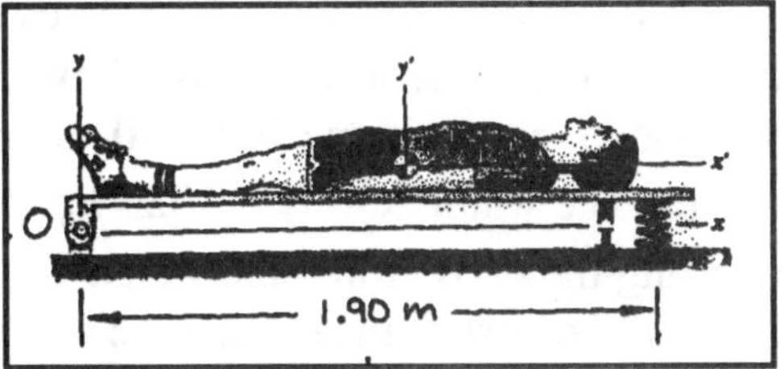

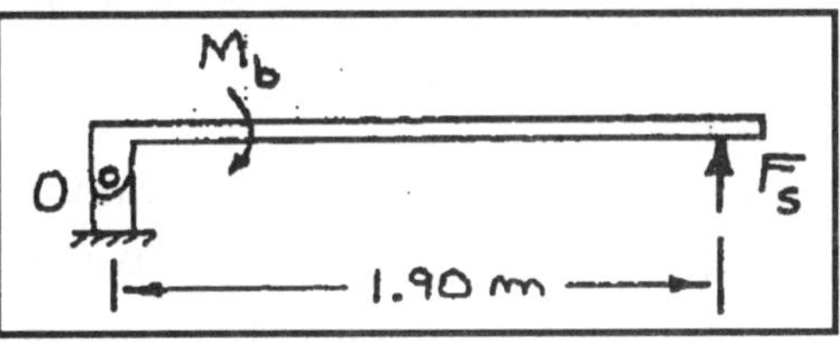

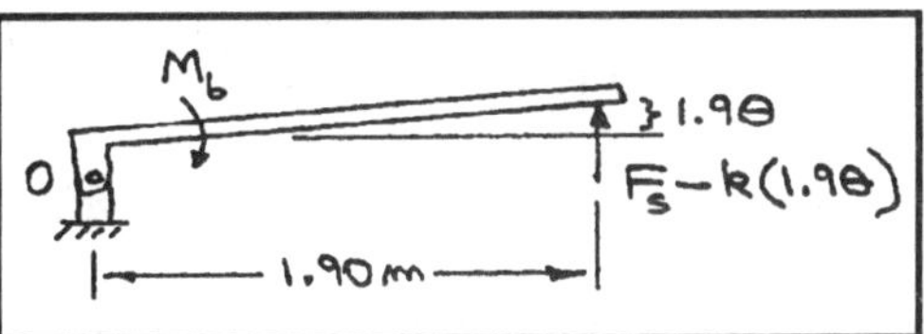

Soluton:

When the astronaut is not present: Let F_s be the spring force and M_b be the moment about 0 due to the board's weight when the system is in equilibrium. The moment about 0 equals zero, $\sum M_{(pt0)} = (1.9)F_s - M_b = 0$ (1). When the system is in motion and displaced by a small counterclockwise angle θ, the spring force decreases to $F_s - k(1.9\ \theta)$: The equation of angular motion about 0 is

$\sum M_{(pt0)} = (1.9)[F_s - k(1.9\theta)] - M_b = I_b \dfrac{d^2\theta}{dt^2}$, where I_b is the moment of inertial of the board about the z axis. Using Equation (1), we can write the equation of angular motion as $\dfrac{d^2\theta}{dt^2} + \omega_1^2\theta = 0$, where

$\omega_1^2 = \dfrac{k(1.9)^2}{I_b} = \dfrac{(12{,}000)(1.9)^2}{I_b}$. We know that $f_1 = \omega_1 / 2\pi = 6Hz$, so $\omega_1 = 12\pi = 37.7 rad/s$ and we can solve for I_b: $I_b = 30.48 kg - m^2$.

When the astronaut is present: Let F_s be the spring force and M_{ba} be the moment about 0 due to the weight of the board and astronaut when the system is in equilibrium. The moment about 0 equals zero, $\sum M_{(pt0)} = (1.9)F_s - M_{ba} = 0$ (2). When the system is in motion and displaced by a small counterclockwise angle θ, the spring force decreases to $F_s - k(1.9\ \theta)$: The equation of angular motion about 0 is $\sum M_{(pt0)} = (1.9)[F_s - k(1.9\ \theta)] - M_{ba} = (I_b + I_a)\dfrac{d^2\theta}{dt^2}$, where I_a is the moment of inertia of the astronaut about the z axis. Using Equation (2), we can write the equation of angular motion as $\dfrac{d^2\theta}{dt^2} + \omega_2^2\theta = 0$, where $\omega_2^2 = \dfrac{k(1.9)^2}{I_b + I_a} = \dfrac{(12{,}000)(1.9)^2}{I_b + I_a}$. In this case $f_2 = \omega_2 / 2\pi = 2.8Hz$, so $\omega_2 = 2.8(2\pi) = 17.59 rad/s$. Since we know I_b, we can determine I_a, obtaining $I_a = 109.48\ kg - m^2$.

==================================<>==================================

Problem 10.24 In Problem 10.23, the astronaut's center of mass is at $x = 1.01\ m$, $y = 0.16m$, *and his mass is 81.6 kg*. What is his moment of inertia about the z' axis through his center of mass?

Solution:

From the solution of Problem 10.23, his moment of inertial about the *zaxis* is $I_z = 109.48 kg - m^2$.
From the parallel-axis theorem, $I_{z'} = I_z - (d_x^2 + d_y^2)m = 109.48 - [(1.01)^2 + (0.16)^2](81.6) = 24.15\ kg - m^2$.

==================================<>==================================

Problem 10.25

Problem 10.25 A floating sonobuoy (sound measuring device) is in equilibrium in the vertical position shown. (Its center of mss is low enough that it is stable in this position.) It is a 10 kg cylinder 1 m in length and 125 mm in diameter. The water density is 1025 kg/m^3, and the buoyancy force supporting the buoy equals the weight of the water that would occupy the volume of the part of the cylinder below the surface. If you push the sonobuoy slightly deeper and release it, what is the natural frequency of the resulting vertical vibrations?

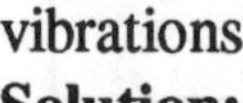

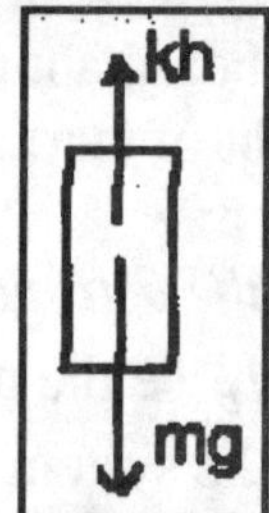

Solution:

Choose a coordinate system with y positive downward. Denote the volume beneath the surface by $V = \pi R^2 d$, where $R = 0.0625\ m$. The density of the water is $\rho = 1025\ kg/m^3$. The weight of the displaced water is $W = \rho V g$, from which the buoyancy force is $F = \rho V g = \pi \rho R^2 g d$. By definition, the spring constant is $k = \frac{\partial F}{\partial d} = \pi \rho R^2 g = 123.4\ N/m$. If h is a positive change in the immersion depth from equilibrium, the force on the sonobuoy is $\sum F_y = -kh + mg$, where the the negative sign is taken because the "spring force" kh opposes the positive motion h. From Newton's second law, $m\frac{d^2h}{dt^2} = mg - kh$. The canonical form (see Eq (10.4)) is $\frac{d^2h}{dt^2} + \omega^2 h = g$, where $\omega = \sqrt{\frac{k}{m}} = 3.513\ rad/s$. The natural frequency is

$$\boxed{f = \frac{\omega}{2\pi} = 0.5591\ Hz}.$$

Problem 10.26

Problem 10.26 A disk rotates about a vertical axis with constant angular velocity Ω. (The plane of the disk is horizontal.) A mass m slides in a smooth slot in the disk and is attached to a spring with constant k. The distance from the center of the disk to the mass when the spring is unstretched is r_o. Show that if $\frac{k}{m} > \Omega^2$, the natural frequency of vibration of the mass is

$$f = \frac{1}{2\pi}\sqrt{\frac{k}{m} - \Omega^2}.$$

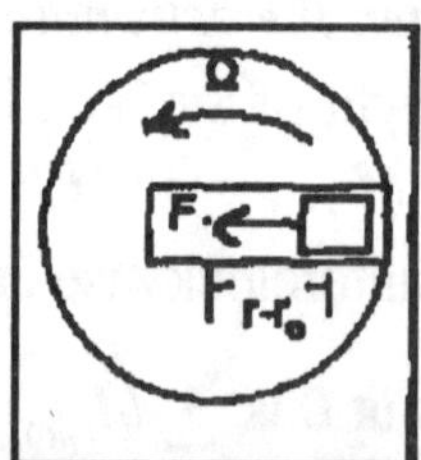

Solution:

The force on the mass is $\sum F = -k(r - r_o)$. From Newton's second law,

$$ma_r = m\left(\frac{d^2r}{dt^2} - r\left(\frac{d\theta}{dt}\right)^2\right) = m\left(\frac{d^2r}{dt^2} - r\Omega^2\right) = -k(r - r_o).$$ In canonical form (see Eq (10.4)),

$\frac{d^2r}{dt^2} + \omega^2 r = \frac{kr_o}{m}$, where $\omega = \sqrt{\frac{k}{m} - \Omega^2}$. If ω is a real number (that is, if $\frac{k}{m} > \Omega^2$), the natural frequency is $\boxed{f = \frac{1}{2\pi}\sqrt{\frac{k}{m} - \Omega^2}}$

=====================================<>=====================================

Problem 10.27 Suppose that at $t = 0$, the mass described in Problem 10.26 is located at $r = r_o$ and the radial velocity is $\frac{dr}{dt} = 0$. Determine the position r of the mass as a function of time.

Solution:

From the solution to Problem 10.26, the equation of motion is $\frac{d^2 r}{dt^2} + \omega^2 r = \frac{k r_o}{m}$, where, $\omega = \sqrt{\frac{k}{m} - \Omega^2}$. The solution is the sum $r(t) = r_h + r_p$, where $r_c(t) = A \sin\omega t + B\cos\omega t$, and $r_p = \frac{k r_o}{m\omega^2} = \frac{k r_o}{k - m\Omega^2}$, where the latter is obtained by setting the acceleration to zero, since the non-homogenous term $\frac{k r_o}{m}$ is not a function of *r* or *t*. The general solution is $r(t) = r_h(t) + r_p(t) = A\sin\omega t + B\cos\omega t + \frac{k r_o}{m\omega^2}$. The velocity is $\frac{dr}{dt} = \omega A\cos\omega t - \omega B \sin\omega t$. Apply the initial conditions to obtain $r_o = B + \frac{k r_o}{m\omega^2}$, and $A = 0$. Solve: $B = r_o\left(1 - \frac{k}{m\omega^2}\right) = r_o\left(\frac{1}{1 - \frac{k}{m\Omega^2}}\right)$, from which

$$\boxed{r = \frac{r_o}{\left(1 - \frac{k}{m\Omega^2}\right)}\cos\omega t + \frac{k r_o}{m\omega^2} = r_o\left(\frac{\cos\omega t}{1 - \frac{k}{m\Omega^2}} + \frac{1}{1 - \frac{m\Omega^2}{k}}\right)}$$

[*Check*: As $\frac{m\Omega^2}{k} \to 1$, the solution diverges asymptotically, as expected. *check*: If the angular velocity is zero, this reduces to $r(t) = r_o\left(1 + \cos\sqrt{\frac{k}{m}}\, t\right)$, as expected, since r_o is the unstretched spring position. *check.* An alternate form is obtained by noting that the equilibrium position for non-zero Ω is $r_e = r_p = \frac{k r_o}{m\omega^2}$, from which $B = r_o - r_e$,, $A = 0$, and the solution is $r(t) = (r_o - r_e)\cos\omega t + r_e$, where $\omega = \sqrt{\frac{k}{m} - \Omega^2}$.*check.*]

=====================================<>=====================================

Problem 10.28 A homogenous 100 *lb* disk with radius $R = 1\ ft$ is attached to two identical cylindrical steel bars of length $L = 1\ ft$. The relation between the moment *M* exerted on the disk and one of the bars and the angle of rotation θ of the disk is $M = \frac{GJ}{L}\theta$, where *J* is the polar moment of inertia of the cross section of the bar and $G = 1.7 \times 10^9\ lb/ft^2$ is the shear modulus of the steel. Determine the required radius of the bars if the natural frequency of vibrations of the disk is to be 10 *Hz*.

Solution:

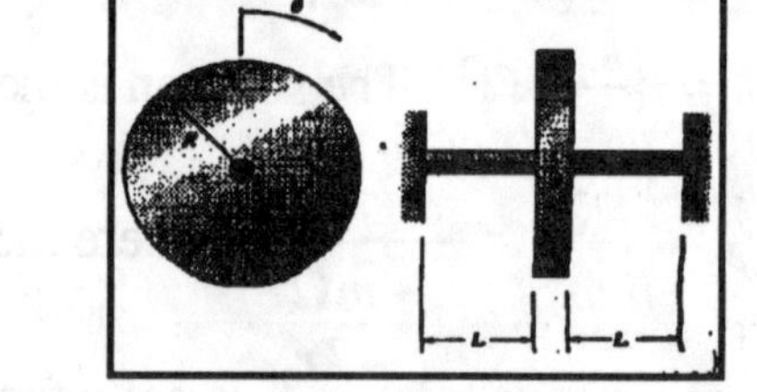

The moment exerted by two bars on the disk is $M = k\theta$, where the spring constant is $k = \frac{\partial M}{\partial \theta} = \frac{2GJ}{L}$. The polar moment of the cross section of a bar is $J = \frac{\pi r^4}{2}$, from which $k = \frac{\pi r^4 G}{L}$.

From the equation of angular motion, $I\frac{d^2\theta}{dt^2} = -k\theta$. The moment of inertia of the disk is $I = \frac{W}{2g}R_d^2$, from which $\frac{d^2\theta}{dt^2} + \omega^2\theta = 0$, where $\omega = \sqrt{\frac{2\pi G g r^4}{WLR_d^2}} = \left(\frac{r^2}{R_d}\right)\sqrt{\frac{2\pi G g}{WL}}$.

Solve: $r = \sqrt{R_d\sqrt{\frac{WL}{2\pi G g}}\omega} = \sqrt{R_d\sqrt{\frac{2\pi WL}{Gg}}f}$. Substitute numerical values: $R_d = 1\,ft$, $L = 1\,ft$, $W = 100\ lb$, $G = 1.7 \times 10^9\ lb/ft^2$, $g = 32.17\ ft/s^2$, from which $\boxed{r = 0.03274\ ft = 0.393\ in.}$

Problem 10.29 The moments of inertia of gears A and B are $I_A = 0.025\ kg\text{-}m^2$ and $I_B = 0.100\ kg\text{-}m^2$. Gear A is attached to a torsional spring with constant $k = 10\ N\text{-}m/rad$. What is the natural frequency of small angular vibrations of the gears?

Solution:

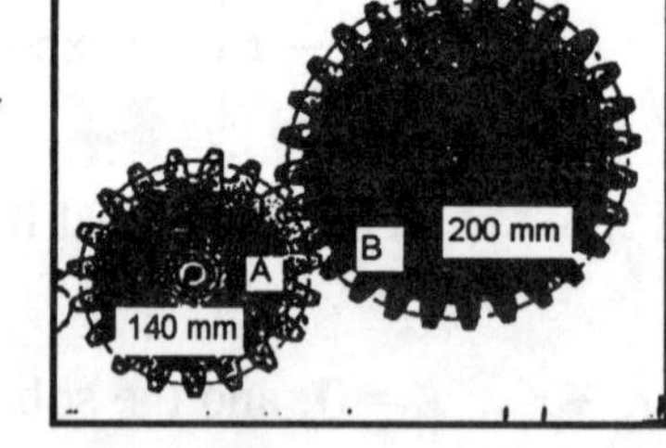

The system is conservative. Denote the rotation velocities of A and B by $\dot{\theta}_A$, and $\dot{\theta}_B$ respectively. The kinetic energy of the gears is $T = \frac{1}{2}I_A\dot{\theta}_A^2 + \frac{1}{2}I_B\dot{\theta}_B^2$. The potential energy of the torsional spring is $V = \frac{1}{2}k\theta_A^2$. $T + V = const = \frac{1}{2}I_A\dot{\theta}_A^2 + \frac{1}{2}I_B\dot{\theta}_B^2 + \frac{1}{2}k\theta_A^2$. From kinematics, $\dot{\theta}_B = -\left(\frac{R_A}{R_B}\right)\dot{\theta}_A$. Substitute, define the *generalized mass moment of inertia* by

$M = (I_A + \left(\frac{R_A}{R_B}\right)^2 I_B) = 0.074\ kg\text{-}m^2$, and take the time derivative: $\left(\frac{d\theta_A}{dt}\right)\left(M\left(\frac{d^2\theta_A}{dt^2}\right) + k\theta_A\right) = 0$.

Ignore the possible solution $\left(\frac{d\theta_A}{dt}\right) = 0$, from which $\frac{d^2\theta_A}{dt^2} + \omega^2\theta_A = 0$, where $\omega = \sqrt{\frac{k}{M}} = 11.62\ rad/s$. The natural frequency is $\boxed{f = \frac{\omega}{2\pi} = 1.850\ Hz}$.

====================================<>====================================

Problem 10.30 At $t = 0$ the torsional spring in Problem 10.29 is unstretched and gear B has a counterclockwise angular velocity of 2 *rad/s*. Determine the counterclockwise angular position of gear B relative to its equilibrium position as a function of time.

Solution:

It is convenient to express the motion in terms of gear A, since the equation of motion of gear A is given in the solution to Problem 10.29: $\frac{d^2\theta_A}{dt^2} + \omega^2\theta_A = 0$, where $M = (I_A + \left(\frac{R_A}{R_B}\right)^2 I_B) = 0.074\ kg\text{-}m^2$, $\omega = \sqrt{\frac{k}{M}} = 11.62\ rad/s$. Assume a solution of the form $\theta_A = A\sin\omega t + B\cos\omega t$. Apply the initial conditions, $\theta_A = 0$, $\dot{\theta}_A = -\left(\frac{R_B}{R_A}\right)\dot{\theta}_B = -2.857\ rad/s$, from which $B = 0$, $A = \frac{\dot{\theta}_A}{\omega} = -0.2458$, and $\theta_A = -0.2458\sin(11.62\ t)\ rad$, and $\boxed{\theta_B = -\left(\frac{R_A}{R_B}\right)\theta_A = 0.1720\sin(11.62\ t)\ rad}$

====================================<>====================================

Problem 10.31 Each slender bar is of mass m and length L. Determine the natural frequency of small vibrations of the system.

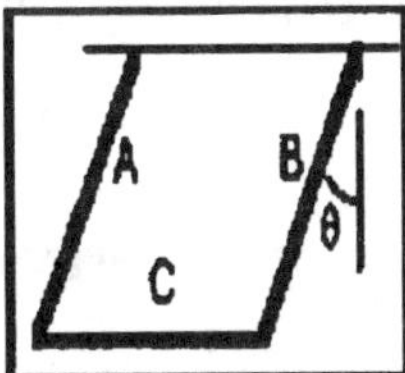

Solution:

The system is conservative. Label the bars as shown. The pivots of bars A and B are fixed points. The mechanism is a parallelogram, so that the bars A and B remain parallel, and bar C remains horizontal during motion. The moments of inertia about the fixed point is the same for the bars A and B. The kinetic energy is $T = \frac{1}{2}I\dot{\theta}^2 + \frac{1}{2}I\dot{\theta}^2 + \frac{1}{2}mv_C^2 = \frac{1}{2}\left(2I\dot{\theta}^2 + mv_C^2\right)$. The potential energy is $V = \frac{mgL}{2}(1-\cos\theta) + \frac{mgL}{2}(1-\cos\theta) + mgL(1-\cos\theta) = 2mgL(1-\cos\theta)$. The system is conservative, $T + V = const. = \frac{1}{2}(2I\dot{\theta}^2 + mv_C^2) + 2mgL(1-\cos\theta)$. From kinematics, $v_C = L\cos\theta(\dot{\theta})$. Substitute: $\frac{1}{2}\left(2I + mL^2\cos^2\theta\right)\dot{\theta}^2 + 2mgL(1-\cos\theta) = const.$ For small angles: $\cos^2\theta \to 1$, $(1-\cos\theta) \to \frac{\theta^2}{2}$, from which $\frac{1}{2}\left(2I + mL^2\right)\dot{\theta}^2 + mgL\theta^2 = const.$ Take the time derivative: $\dot{\theta}^2\left[\left(2I + mL^2\right)\frac{d^2\theta}{dt^2} + 2mgL\theta\right] = 0.$

Ignore the possible solution $\dot{\theta} = 0$, from which $\frac{d^2\theta}{dt^2} + \omega^2\theta = 0$, where $\omega = \sqrt{\frac{2mgL}{2I + mL^2}}$. The moment of inertia of a slender bar about one end (the fixed point) is $I = \frac{mL^2}{3}$, from which $\omega = \sqrt{\frac{6g}{5L}}$, and the natural frequency is $\boxed{f = \frac{\omega}{2\pi} = \frac{1}{2\pi}\sqrt{\frac{6g}{5L}}}$

====================================<>====================================

Problem 10.32 The masses of the slender bar and the homogenous disk are m and m_d, respectively. The spring is unstretched when $\theta = 0$. Assume that the disk rolls on the horizontal surface. (a) show that the motion is governed by the equation
$\left(\frac{1}{3}+\frac{3m_d}{2m}\cos^2\theta\right)\frac{d^2\theta}{dt^2}-\frac{3m_d}{2m}\sin\theta\cos\theta\left(\frac{d\theta}{dt}\right)^2-\frac{g}{2L}\sin\theta+\frac{k}{m}(1-\cos\theta)\sin\theta=0$. (b) if the system is in equilibrium at the angle $\theta=\theta_e$, and $\tilde{\theta}=\theta-\theta_e$, show that the equation governing small vibrations relative to the equilibrium position is

$$\left(\frac{1}{3}+\frac{3m_d}{2m}\cos^2\theta_e\right)\frac{d^2\tilde{\theta}}{dt^2}+\left[\frac{k}{m}(\cos\theta_e-\cos^2\theta_e+\sin^2\theta_e)-\frac{g}{2L}\cos\theta_e\right]\tilde{\theta}=0.$$

Solution:

(See Example 10.2.) The system is conservative. (a) The kinetic energy is $T=\frac{1}{2}I\dot{\theta}^2+\frac{1}{2}mv^2+\frac{1}{2}I_d\dot{\theta}_d^2+\frac{1}{2}m_dv_d^2$, where $I=\frac{mL^2}{12}$ is the moment of inertia of the bar about its center of mass, v is the velocity of the center of mass of the bar, $I_d=\frac{mR^2}{2}$ is the polar moment of inertia of the disk, and v_d is the velocity of the center of mass of the disk. The height of the center of mass of the bar is $h=\frac{L\cos\theta}{2}$, and the stretch of the spring is $S=L(1-\cos\theta)$, from which the potential energy is $V=\frac{mgL}{2}\cos\theta+\frac{1}{2}kL^2(1-\cos\theta)^2$. The system is conservative, $T+V=const\,\dot{\theta}l_d=\frac{v}{R}=\frac{L\cos\theta}{R}\dot{\theta}$. Choose a coordinate system with the origin at the pivot point and the x axis parallel to the lower surface. The instantaneous center of rotation of the bar is located at $(L\sin\theta, L\cos\theta)$.

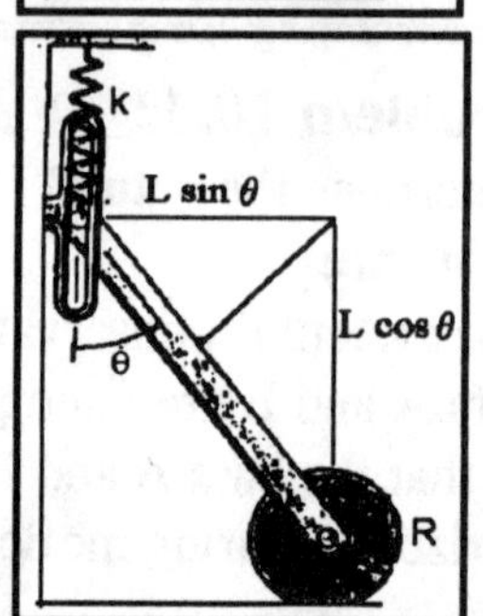

The center of mass of the bar is located at $\left(\frac{L}{2}\sin\theta, \frac{L}{2}\cos\theta\right)$.

The distance from the center of mass to the center of rotation is

$r=\sqrt{\left(L-\frac{L}{2}\right)^2\sin^2\theta+\left(L-\frac{L}{2}\right)^2\cos^2\theta}=\frac{L}{2}$, from which $v=\frac{L}{2}\dot{\theta}$. The velocity of the center of mass of the disk is $v_d=(L\cos\theta)\dot{\theta}$. The angular velocity of the disk is $\dot{\theta}l_d=\frac{v}{R}=\frac{L\cos\theta}{R}\dot{\theta}$. Substitute and reduce:

$\frac{1}{2}\left(\frac{1}{3}+\frac{3m_d}{2m}\cos^2\right)\dot{\theta}^2+\frac{g}{2L}\cos\theta+\frac{k}{2m}(1-\cos\theta)^2=const.$ Take the time derivative:

$\dot{\theta}\left[\left(\frac{1}{3}+\frac{3m_d}{2m}\cos^2\theta\right)\frac{d^2\theta}{dt^2}-\frac{3m_d}{2m}\sin\theta\cos\theta\left(\frac{d\theta}{dt}\right)^2-\frac{g}{2L}\sin\theta+\frac{k}{m}(1-\cos\theta)\sin\theta\right]=0$, from which

$$\left(\frac{1}{3}+\frac{3m_d}{2m}\cos^2\theta\right)\frac{d^2\theta}{dt^2}-\frac{3m_d}{2m}\sin\theta\cos\theta\left(\frac{d\theta}{dt}\right)^2-\frac{g}{2L}\sin\theta+\frac{k}{m}(1-\cos\theta)\sin\theta=0$$

Solution continued on next page

(b) The non-homogenous term is $\dfrac{g}{2L}$, as can be seen by dividing the equation of motion by $\sin\theta \neq 0$,

$$\frac{\left(\frac{1}{3}+\frac{3m_d}{2m}\cos^2\theta\right)}{\sin\theta}\frac{d^2\theta}{dt^2}-\frac{3m_d}{2m}\cos\theta\left(\frac{d\theta}{dt}\right)^2+\frac{k}{m}(1-\cos\theta)=\frac{g}{2L}.$$

Since the non-homogenous term is independent of time and angle, the equilibrium point can be found by setting the acceleration and the velocity terms to zero, $\cos\theta_e = 1-\dfrac{mg}{2Lk}$. [*Check*: This is identical to Eq (10.15) in Example 10.2, as expected. *check*.]. Denote $\tilde\theta = \theta - \theta_e$. For small angles:

$\cos\theta = \cos\tilde\theta\cos\theta_e - \sin\tilde\theta\sin\theta_e \to \cos\theta_e - \tilde\theta\sin\theta_e$.

$\cos^2\theta \to \cos^2\theta_e - 2\tilde\theta\sin\theta_e\cos\theta_e$.

$\sin\theta = \sin\tilde\theta\cos\theta_e + \cos\tilde\theta\sin\theta_e \to \tilde\theta\cos\theta_e + \sin\theta_e$.

$(1-\cos\theta)\sin\theta \to (1-\cos\theta_e)\sin\theta_e + \tilde\theta(\cos\theta_e - \cos\theta_e + \sin^2\theta_e)$.

$\sin\theta\cos\theta \to \tilde\theta(\cos^2\theta_e - \sin^2\theta_e)$. Substitute and reduce:

$$(1)\left(\frac{1}{3}+\frac{3m_d}{2m}\cos^2\theta\right)\frac{d^2\theta}{dt^2}\to\left(\frac{1}{3}+\frac{3m_d}{m}\tilde\theta\sin\theta_e\cos\theta_e\right)\frac{d^2\tilde\theta}{dt^2}\to\left(\frac{1}{3}+\frac{3m_d}{2m}\cos^2\theta_e\right)\frac{d^2\tilde\theta}{dt^2}.$$

$$(2)=-\frac{3m_d}{2m}\cos\theta\sin\theta\left(\frac{d\theta}{dt}\right)^2\to-\frac{3m_d}{2m}\tilde\theta\sin\theta_e\cos\theta_e\left(\frac{d\tilde\theta}{dt}\right)^2\to 0.$$

$$(3)-\frac{g}{2L}\sin\theta\to-\frac{g}{2L}\tilde\theta\cos\theta_e-\frac{g}{2L}\sin\theta_e.$$

$$(4)\frac{k}{m}(1-\cos\theta)\sin\theta\to\frac{k}{m}(1-\cos\theta_e)\sin\theta_e+\frac{k}{m}\tilde\theta(\cos\theta_e-\cos^2\theta_e+\sin^2\theta_e),$$

where the terms $\tilde\theta\dfrac{d^2\tilde\theta}{dt^2}\to 0$, $\tilde\theta\left(\dfrac{d\tilde\theta}{dt}\right)^2\to 0$, and terms in $\tilde\theta^2$ have been dropped.

Collect terms in (1) to (4) and substitute into the equations of motion:

$$\frac{\left(\frac{1}{3}+\frac{3m_d}{2m}\cos^2\theta_e\right)}{\sin\theta_e}\frac{d^2\tilde\theta}{dt^2}+\left[\frac{k}{m}\frac{(\cos\theta_e-\cos^2\theta_e+\sin^2\theta_e)}{\sin\theta_e}-\frac{g}{2L}\frac{\cos\theta_e}{\sin\theta_e}\right]=\left[\frac{g}{2L}-\frac{k}{m}(1-\cos\theta_e)\right].$$

The term on the right $\dfrac{g}{2L}-\dfrac{k}{m}(1-\cos\theta_e)=0$, as shown by substituting the value $\cos\theta_e = 1-\dfrac{mg}{2Lk}$, from which

$$\left(\frac{1}{3}+\frac{3m_d}{2m}\cos^2\theta_e\right)\frac{d^2\tilde\theta}{dt^2}+\left[\frac{k}{m}(\cos\theta_e-\cos^2\theta_e+\sin^2\theta_e)-\frac{g}{2L}\cos\theta_e\right]\tilde\theta=0$$

is the equation of motion for small amplitude oscillations about the equilibrium point.

==================================<>==================================

==================================<>==================================

Problem 10.33 The masses of the bar and disk in Problem 10.32 are $m = 2$ kg and $m_d = 4$ kg, respectively. The dimensions $L = 1$ m and $R = 0.28$ m, and the spring constant is $k = 70$ N/m. (a) Determine the angle θ_e at the system is in equilibrium. (b) The system is at rest in the equilibrium position, and the disk is given a clockwise angular velocity of 0.1 rad/s. Determine θ as a function of time.

Solution:

From the solution to Problem 10.32, the static equilibrium angle is $\theta_e = \cos^{-1}\left(1 - \frac{mg}{2kL}\right) = 30.7^o = 0.5358$ rad. (b) The canonical form (see Eq (10.4)) of the equation of motion is $\frac{d^2\tilde{\theta}}{dt^2} + \omega^2\tilde{\theta} = 0$, where $\omega = \sqrt{\frac{\frac{k}{m}(\cos\theta_e - \cos^2\theta_e + \sin^2\theta_e) - \frac{g}{2L}\cos\theta_e}{\left(\frac{1}{3} + \frac{3m_d}{2m}\cos^2\theta_e\right)}} = 1.891$ rad/s. Assume a solution of the form $\tilde{\theta} = \theta - \theta_e = A\sin\omega t + B\cos\omega t$, from which $\theta = \theta_e + A\sin\omega t + B\cos\omega t$. From the solution to Problem 10.34 the angular velocity of the disk is $\dot{\theta}_d = \frac{v_d}{R} = \frac{(L\cos\theta_e)}{R}\dot{\theta}$. The initial conditions are $t = 0$, $\theta = \theta_e$, $\dot{\theta} = \frac{R\dot{\theta}_d}{L\cos\theta_e} = \frac{(0.1)(0.28)}{0.86} = 0.03256$ rad/s, from which $B = 0$, $A = 0.03256/\omega = 0.01722$, from which the solution is $\theta = 0.5358 + 0.01722\sin(1.891t)$.

==================================<>==================================

================================<>================================

Problem 10.34 The mass of each slender bar is 1 *kg*. If the natural frequency of small vibrations of the system is 0.935 *Hz*, what is the mass of the object A?

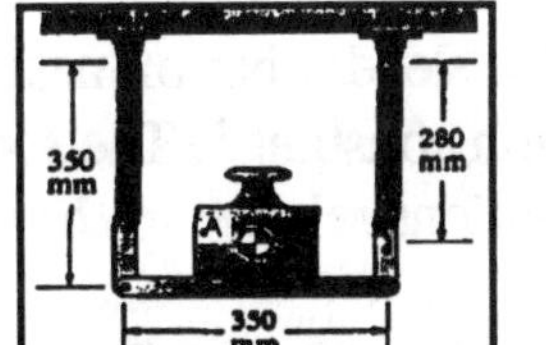

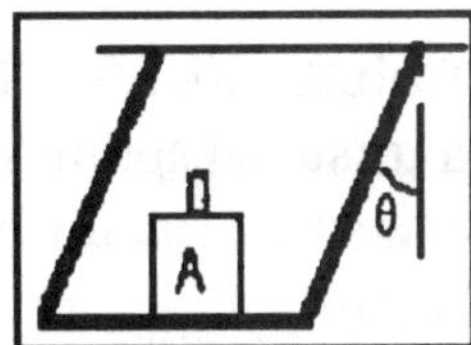

Solution:

The system is conservative. (See the solution to Problem 10.31.) Denote $L = 0.350\ m$, $L_A = 0.280\ m$, $m = 1\ kg$, and M the mass of A. The moments of inertia about the fixed point is the same for the two vertical bars. The kinetic energy is $T = \frac{1}{2}I\dot{\theta}^2 + \frac{1}{2}I\dot{\theta}^2 + \frac{1}{2}mv^2 + \frac{1}{2}Mv_A^2$, where v is the velocity of the center of mass of the lower bar and v_A is the velocity of the center of mass of A, from which $T = \frac{1}{2}\left(2I\dot{\theta}^2 + mv^2 + Mv_A^2\right)$. The potential energy is $V = \frac{mgL}{2}(1-\cos\theta) + \frac{mgL}{2}(1-\cos\theta) + mgL(1-\cos\theta) + MgL_A(1-\cos\theta)$,

$V = (MgL_A + 2mgL)(1-\cos\theta)$. $T + V = const. = \frac{1}{2}(2I\dot{\theta}^2 + mv^2 + Mv_A^2) + (MgL_A + 2mgL)(1-\cos\theta)$.

From kinematics, $v = L\cos\theta(\dot{\theta})$, and $v_A = L_A\cos\theta(\dot{\theta})$. Substitute:

$\frac{1}{2}\left(2I + \left(mL^2 + ML_A^2\right)\cos^2\theta\right)\dot{\theta}^2 + (MgL_A + 2mgL)(1-\cos\theta) = const.$ For small angles: $\cos^2\theta \to 1$, $(1-\cos\theta) \to \frac{\theta^2}{2}$, from which $\frac{1}{2}\left(2I + mL^2 + ML_A^2\right)\dot{\theta}^2 + \left(\frac{MgL_A}{2} + mgL\right)\theta^2 = const.$ Take the time derivative: $\dot{\theta}\left[\left(2I + mL^2 + ML_A^2\right)\frac{d^2\theta}{dt^2} + (MgL_A + 2mgL)\theta\right] = 0$. Ignore the possible solution $\dot{\theta} = 0$, from which $\frac{d^2\theta}{dt^2} + \omega^2\theta = 0$, where $\omega = \sqrt{\frac{2mgL + MgL_A}{2I + mL^2 + ML_A^2}}$. The moment of inertia of a slender bar about one end (the fixed point) is $I = \frac{mL^2}{3}$, from which $\omega = \sqrt{\frac{g(2+\eta)}{\frac{5}{3}L + \eta L_A}}$, where $\eta = \frac{ML_A}{mL}$. The natural frequency is $f = \frac{\omega}{2\pi} = \frac{1}{2\pi}\sqrt{\frac{3g(2+\eta)}{5L + 3\eta L_A}} = 0.935\ Hz$. Solve: $\eta = \frac{5L\omega^2 - 6g}{3\left(g - L_A\omega^2\right)} = 3.502$, from which

$$\boxed{M = \frac{mL}{L_A}(3.502) = 4.378\ kg}$$

================================<>================================

===================================<>===================================

Problem 10.35 The slender bar of mass *m* and length *L* is held in equilibrium in the position shown by a torsional spring with constant *k*. The spring is unstretched when the bar is vertical. Determine the frequency of small vibrations relative to the equilibrium position shown.

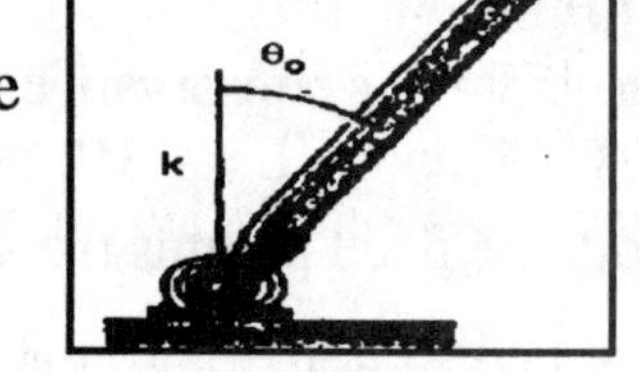

Solution:

(See the solution to Problem 10.21.). The system is conservative. Denote the *static equilibrium angle* by θ_o. The pivot is a fixed point. From statics, $\sum M = -k\theta_o + (mgL/2)\sin\theta_o = 0$. The kinetic energy is $T = (1/2)I\dot{\theta}^2$. The potential energy is $V = (1/2)k\theta^2 - (mgL/2)(1-\cos\theta)\theta$.

Denote the displacement from the static equilibrium point by $\eta = \theta - \theta_o$, from which the potential energy is $V = (1/2)k\eta^2 + k\eta\theta_o + (1/2)k\theta_o^2 - (mgL/2)(1 - \cos\eta\cos\theta_o + \sin\eta\sin\theta_o)\theta$, . $T + V = const = (1/2)I\dot{\eta}^2 + (1/2)k\eta^2 + k\eta\theta_o + (1/2)k\theta_o^2 - (mgL/2)(1 - \cos\eta\cos\theta_o + \sin\eta\sin\theta_o)$. Take the time derivative: $\dot{\eta}\left[I(d^2\eta/dt^2) + k\eta + k\theta_o - (mgL/2)(\sin\eta\cos\theta_o + \cos\eta\sin\theta_o)\right] = 0$. Ignore the solution $\dot{\eta} = 0$. For small angles, $\sin\eta \to \eta$, and $\cos\eta \to 1$. The moment of inertia about the axis is $I = mL^2/3$. The equation of motion (see Eq (10.4)) is $(d^2\eta/dt^2) + \omega^2\eta = (3/(mL^2))(-k\theta_o + (mgL/2)\sin\theta_o)$, where, from the definition of θ_o, the right hand side vanishes, and $\omega = \sqrt{\left(k - \frac{mgL}{2}\cos\theta_o\right)\left(\frac{mL^2}{3}\right)^{-1}}$. The natural frequency is

$$\boxed{f = \frac{\omega}{2\pi} = \frac{1}{2\pi}\sqrt{\frac{k - \frac{mgL\cos\theta_o}{2}}{(mL^2/3)}}}$$

===================================<>===================================

Problem 10.36 (a) What are the natural frequency and period of the spring mass system described in example 10.3? (b) What are the natural frequency and period if the damping element is removed?

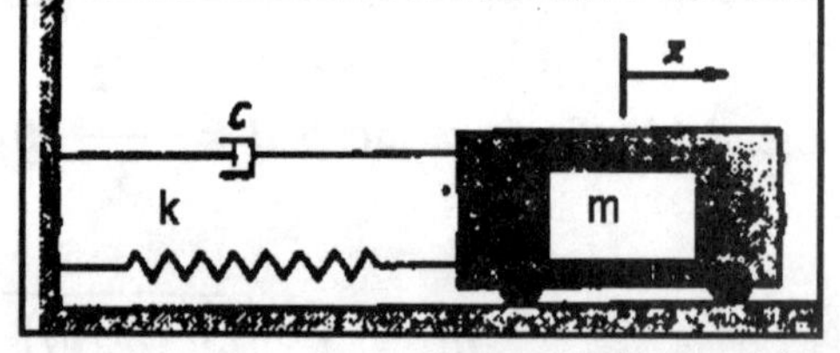

Solution:

The mass is $m = 2\ kg$, the spring constant is $k = 8\ N/m$, and the damping constant is $c = 1\ N\text{-}s/m$. The sum of the *horizontal* forces on the mass are $\sum F_x = -kx - c\frac{dx}{dt}$. From Newton's second law: $m\frac{d^2x}{dt^2} + c\frac{dx}{dt} + kx = 0$. The canonical form of the equation of motion (see Eq (10.16)) is $\frac{d^2x}{dt^2} + 2d\frac{dx}{dt} + \omega^2 = 0$, where $d = \frac{c}{2m} = 0.25\ rad/s$, and $\omega = \sqrt{(k/m)} = 2$ rad/s. Since $\omega^2 > d^2$, the damping is sub-critical. (a) The natural frequency is (see Eq (10.18)) $\boxed{f_d = \frac{\omega_d}{2\pi} = \frac{1}{2\pi}\sqrt{\omega^2 - d^2} = 0.3158\ Hz}$. The period is $\boxed{\tau_d = \frac{1}{f_d} = 3.166\ s}$. (b) The natural frequency and period if the damping element is removed are $\boxed{f = \frac{\omega}{2\pi} = \frac{1}{2\pi}\sqrt{\frac{k}{m}} = 0.3183\ Hz}$, $\boxed{\tau = 1/f = 3.142\ s}$.

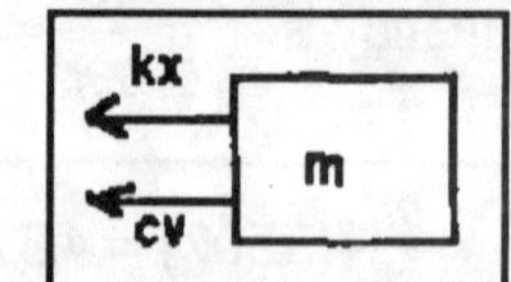

===================================<>===================================

================================<>================================

Problem 10.37 (a) What value of c is necessary for the stepped disk in Example 10.4 to be critically damped? (b) If c equals the value you determined in part (a) and the disk is released from rest with the spring unstretched, determine the position of the center of the disk relative to its equilibrium point as a function of time.

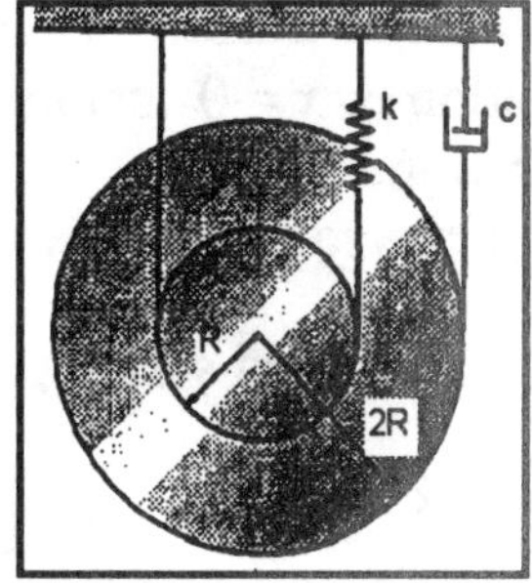

Solution :

The disk weighs 40 lb, $R = 1\ ft$, and $k = 10\ lb/ft$. From Example 10.4, the equation of motion is shown to be $4m\frac{d^2x}{dt^2} + 9c\frac{dx}{dt} + 4kx = 0$. The equilibrium position is determined by setting the acceleration and velocity to zero (since the non-homogenous term mg is not a function of time or position) from which $\tilde{x} = x - mg/(4k)$. Substitute (see Example 10.4) to obtain the canonical form $\frac{d^2\tilde{x}}{dt^2} + 2d\frac{d\tilde{x}}{dt} + \omega^2\tilde{x} = 0$, where $d = \frac{9c}{8m}$ and $\omega^2 = \frac{k}{m}$. (a) For critical damping $d^2 = \omega^2$, from which $\boxed{c = \frac{8m}{9}\sqrt{\frac{k}{m}} = \frac{8}{9}\sqrt{\frac{Wk}{g}} = \frac{8}{9}\sqrt{\frac{40(10)}{32.17}} = 3.134\ lb\text{-}s/ft}$. (b) Since the system is critically damped, from Eq (10.25) the solution has the form $\tilde{x} = Ae^{dt} + Bte^{-dt}$, where $d = 2.836\ rad/s$. The initial conditions are $[x]_{t=0}$, $\left[\frac{dx}{dt}\right]_{t=0} = 0$ from which $[\tilde{x}]_{t=0} = -x_e = -\frac{mg}{4k} = -1\ ft$ and $\left[\frac{d\tilde{x}}{dt}\right]_{t=0} = 0$ from which $A = -x_e = -1$, and $B = Ad = -d$. The solution is $\boxed{\tilde{x} = -e^{-dt}(1 + dt) = -e^{-2.836t}(1 + 2.836t)\ ft}$.

================================<>================================

Problem 10.38 The damping constant of the damped spring-mass oscillator is $c = 20\ N\text{-}s/m$. What are the period and natural frequency of the system? Compare them to the period and natural frequency if the system is undamped.

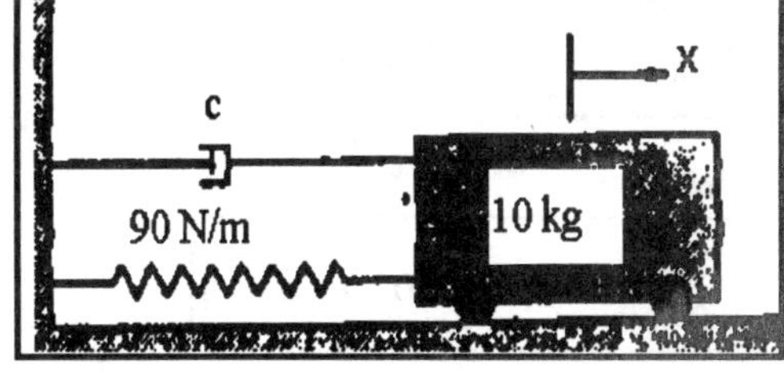

Solution:

From the solution to Problem 10.36 the canonical form of the equation of motion about the unstretched spring position is $\frac{d^2x}{dt^2} + 2d\frac{dx}{dt} + \omega^2 x = 0$, where $d = \frac{c}{2m} = 1\ rad/s$ and $\omega = \sqrt{\frac{k}{m}} = \sqrt{\frac{90}{10}} = 3\ rad/s$. The damping is sub-critical, since $d^2 < \omega^2$. The natural frequency is $\boxed{f_d = \frac{\omega_d}{2\pi} = \frac{1}{2\pi}\sqrt{\frac{k}{m} - d^2} = 0.4502\ Hz}$. The period is $\boxed{\tau_d = \frac{1}{f_d} = 2.221\ s}$. The undamped natural frequency and period are $\boxed{f = \frac{1}{2\pi}\sqrt{\frac{k}{m}} = 0.4775\ Hz}$ and $\boxed{\tau = \frac{1}{f} = 2.094\ s}$.

================================<>================================

=====================================<>=====================================

Problem 10.39 At $t = 0$, the position of the mass in Problem 10.38 relative to its equilibrium position is $x = 0$ and its velocity is 1 *m/s* to the right. Determine *x* as a function of time.

Solution:

From the solution to Problem 10.38, the canonical form (see Eq 10.16)) of the equation of motion about the unstretched spring position is $\frac{d^2x}{dt^2} + 2d\frac{dx}{dt} + \omega^2 x = 0$, where $d = \frac{c}{2m} = 1\ rad/s$ and $\omega^2 = \frac{k}{m} = 9\ rad/s^2$. The system is sub-critically damped, since $d^2 < \omega^2$. Assume a solution of the form $x(t) = e^{-dt}(A\sin\omega t + B\cos\omega t)$, where $\omega_d = \sqrt{\omega^2 - d^2} = 2.828\ rad/s$. The initial conditions are $[x]_{t=0} = 0$, $\left[\frac{dx}{dt}\right]_{t=0} = +1\ m/s$, from which $B = 0$, $A = \frac{1}{\omega_d}$, and the position is given by

$$x(t) = \frac{e^{-dt}\sin\omega_d t}{\omega_d} = 0.3536e^{-t}\sin(2.828t)$$

=====================================<>=====================================

Problem 10.40 In Problem 10.38, what value of damping constant *c* will cause the amplitude of the vibration of the system to decrease to one-half its value in 10 *s*?

Solution:

From Eq. (10.19) assume a solution of the form $x = e^{-dt}(A\sin\omega_d t + B\cos\omega_d t) = e^{-dt}E\sin(\omega_d t + \phi)$, where $\omega_d = \sqrt{\omega^2 - d^2} = 2.828\ rad/s$, $d = \frac{c}{2m} = 1\ rad/s$, $E = \sqrt{A^2 + B^2}$, and $\phi = \tan^{-1}\left(\frac{B}{A}\right)$. By analogy with the undamped solution, (see the solution to Problem 10.4), define *the peak amplitude of the vibration at any time t* to be $|x(t)|_{amplitude} = e^{-dt}E$. The peak amplitude at $t + 10\ s$ is

$|x(t+10)|_{amplitude} = e^{-d(t+10)}E$, and the ratio of the two amplitudes is

$\frac{|x(t+10)|_{amplitude}}{|x(t)|_{amplitude}} = \frac{e^{-d(t+10)}}{e^{-dt}} = e^{-d(10)} = 0.5$, from which $d = -\frac{1}{10}\ln(0.5)$, from which

$$c = -\frac{2m}{10}\ln(0.5) = 1.386\ N\text{-}s/m$$

=====================================<>=====================================

==================================<>==================================

Problem 10.41 At $t=0$, the position of the mass in Problem 10.38 is $x=0$ and it has a velocity of $1\ m/s$ to the right. Determine x as a function of time if c has twice the value necessary for the system to be critically damped.

Solution:

From the solution to Problem 10.38, the canonical form of the equation of motion is $\frac{d^2x}{dt^2}+2d\frac{dx}{dt}+\omega^2 x=0$, where $d=\frac{c}{2m}$, $m=10\ kg$, and $\omega^2=9\ (rad/s)^2$. Critical damping occurs when $d^2=\omega^2$, from which $c=4\sqrt{mk}=120\ N\text{-}s/m$ *for twice critical damping*. From Eq (10.24) the solution is $x=Ce^{-(d-h)t}+De^{-(d+h)t}$, where $d=\frac{c}{2m}=6\ rad/s$ and $h=\sqrt{d^2-\omega^2}=5.196\ rad/s$.

Apply the initial conditions, from which $C+D=0$, and $C-D=\frac{1}{h}$. Solve: $C=0.09622$, $D=-0.09622$, from which $\boxed{x=0.09622\left(e^{-0.8038t}-e^{-11.20t}\right)\ m}$

==================================<>==================================

Problem 10.42 For small vertical displacements of the tire and wheel, the motion of the car's suspension can be modeled by the damped spring-mass oscillator in Fig 10.9 with *m = 36 kg,* and *k = 22 kN/m.* Determine the value of the damping constant *c* that must be provided by the suspension's shock absorber to achieve critical damping.

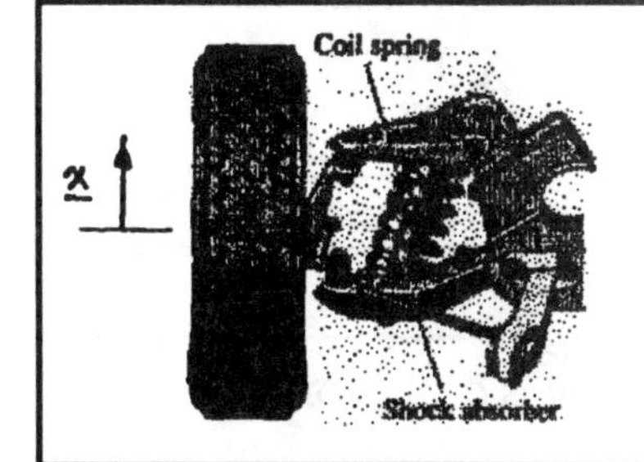

Solution:

Critical damping occurs when $\omega=d$ in Equation (10.16). For the spring -mass oscillator in Figure 10.9, $\omega=\sqrt{k/m}$ and $d=c/2m$ so $\sqrt{\frac{k}{m}}=\frac{c}{2m}$, or $\sqrt{\frac{22{,}000}{36}}=\frac{c}{2(36)}$.

We obtain $c=1780$ *N-s/m.* .

==================================<>==================================

Problem 10.43 The motion of the car's suspension shown in Problem 10.42 can be modeled by the damped spring-mass oscillator in Fig. 10.9 with *m = 36 kg,, k = 22 kN/m* and *c = 2.2 kN-s/m.* Assume that no external forces act on the tire and wheel. At *t = 0,* the spring is unstretched and the tire and wheel are given a velocity $dx/dt=10m/s$. Determine the position *x* as a function of time.

Solution:

Calculating ω and d, we obtain $\omega=\sqrt{\frac{k}{m}}=\sqrt{\frac{22{,}000}{36}}=24.72 rad/s$, and

$d=\frac{c}{2m}=\frac{2200}{2(36)}=30.56 rad/s$. Since $d>\omega$, the motion is supercritically damped and Equation (10.24) is the solution, where $h=\sqrt{d^2-\omega^2}=17.96 rad/s$. Equation (10.24) is $x=Ce^{-(d-h)t}+De^{-(d+h)t}$, or $x=Ce^{-12.6t}+De^{-48.5t}$. The time derivative is $\frac{dx}{dt}=-12.6Ce^{-12.6t}-48.5De^{-48.5t}$.

At $t=0, x=0$ and $dx/dt=10m/s$: Hence, $0=C+D$ and $10=-12.6C-48.5D$.

Solving for C and D we obtain $C=0.278m, D=-0.278m$.

The solution is $x=0.278\left(e^{-12.6t}-e^{-48.5t}\right)m.$

==================================<>==================================

==================================<>==================================

Problem 10.44 The homogenous slender bar is 4 *ft* long and weighs 10 *lb*. Aerodynamic drag and friction at the support exert a resisting moment on the bar of magnitude $0.5\left(\frac{d\theta}{dt}\right)$ $ft\text{-}lb$, where $\frac{d\theta}{dt}$ is the angular velocity of the bar in *rad/s*. (a) What are the period and natural frequency of small vibrations of the bar? (b) How long does it take for the amplitude of vibration to decrease to one-half its initial value?

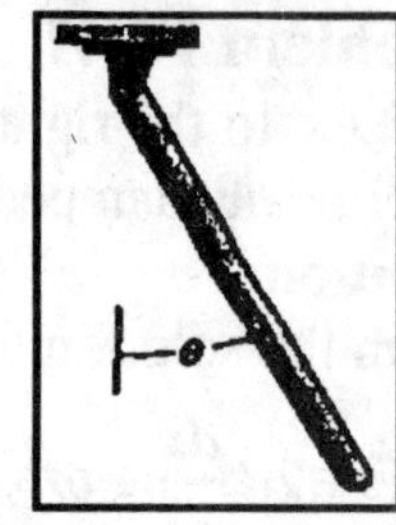

Solution:

The pivot is a fixed point. The moment about the pivot is $\sum M = -0.5\frac{d\theta}{dt} - W\left(\frac{L}{2}\right)\sin\theta$. From the equation of angular motion, $I\frac{d^2\theta}{dt^2} = \sum M$. The moment of inertia about the end of the bar is $I = \left(\frac{W}{g}\right)\frac{L^2}{3}$, from which the equation of motion is $\frac{d^2\theta}{dt^2} + \frac{3g}{2WL^2}\frac{d\theta}{dt} + \frac{3g}{2L}\sin\theta = 0$.

For small amplitude vibrations $\sin\theta \to \theta$, from which the canonical form (see Eq (10.16)) is $\frac{d^2\theta}{dt^2} + 2d\frac{d\theta}{dt} + \omega^2\theta = 0$, where $d = \frac{3g}{4WL^2} = 0.1508\ rad/s$ and $\omega^2 = \frac{3g}{2L} = 12.06\ (rad/s)^2$. (a) The period is $\boxed{\tau_d = \frac{2\pi}{\omega_d} = \frac{2\pi}{\sqrt{\omega^2 - d^2}} = 1.811\ s}$. The natural frequency is $\boxed{f_d = \frac{1}{\tau_d} = 0.5523\ Hz}$. (b) The solution is of the form $\theta = e^{-dt}E\sin(\omega_d t - \phi)$. By analogy with the undamped solution amplitude, (see the solution to Problem 10.4) *define the peak amplitude* by $|\theta|_{amplitude} = Ee^{-dt}$. The time for the peak amplitude to decrease to one-half value is $\frac{|\theta(T)|_{amplitude}}{|\theta(0)|_{amplitude}} = 0.5 = e^{-dT}$, from which

$$\boxed{T = -\frac{1}{d}\ln(0.5) = 4.597\ s}$$

==================================<>==================================

Problem 10.45 If the bar in Problem 10.44 is displaced a small angle θ_o and released from rest at $t = 0$, what is θ as a function of time?

Solution:

From the solution to Problem 10.44, the canonical form (Eq (10.16)) of the equation of motion is $\frac{d^2\theta}{dt^2} + 2d\frac{d\theta}{dt} + \omega^2\theta = 0$, where $d = \frac{3g}{4WL^2} = 0.1508\ rad/s$ and $\omega^2 = \frac{3g}{2L} = 12.06\ (rad/s)^2$. Assume a solution of the form $\theta = e^{-dt}E\sin(\omega_d t - \phi) = e^{-dt}(A\sin\omega_d t + B\cos\omega_d t)$. Apply the initial conditions: $t = 0$, $[\theta]_{t=0} = \theta_o$, $\left[\frac{d\theta}{dt}\right]_{t=0} = 0$, from which $B = \theta_o$, and $0 = -d\theta_o + \omega_d A$, from which $A = \frac{d}{\omega_d}\theta_o$.

The solution, $\theta(t) = \theta_o e^{-dt}\left(\frac{d}{\omega_d}\sin\omega_d t + \cos\omega_d t\right)$ $\boxed{= \theta_o e^{-0.1508t}(0.04346\sin(3.470\,t) + \cos(3.470\,t))}$

==================================<>==================================

================================<>================================

Problem 10.46 The radius of the pulley is $R = 100\ mm$ and its moment of inertia is $I = 0.1\ kg\text{-}m^2$. The mass $m = 5\ kg$, and the spring constant is $k = 135\ N/m$. The cable does not slip relative to the pulley. The coordinate x measures the displacement of the mass relative to the position in which the spring is unstretched. Determine x as a function of time if $c = 60\ N\text{-}s/m$ and the system is released from rest with $x = 0$.

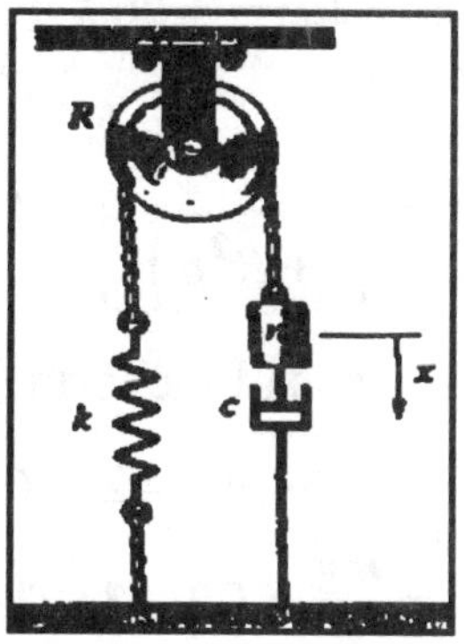

Solution:

Denote the angular rotation of the pulley by θ. The moment on the pulley is $\sum M = R(kx) - RF$, where F is the force acting on the right side of the pulley. From the equation of angular motion for the pulley, $I\frac{d^2\theta}{dt^2} = Rkx - RF$, from which $F = -\frac{I}{R}\frac{d^2\theta}{dt^2} + kx$. The force on the mass is $-F + f + mg$, where the friction force $f = -c\frac{dx}{dt}$ acts in opposition to the velocity of the mass. From Newton's second law for the mass,

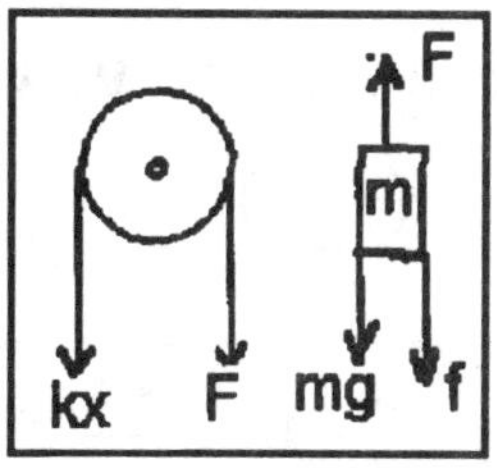

$m\frac{d^2x}{dt^2} = -F - c\frac{dx}{dt} + mg = \frac{I}{R}\frac{d^2\theta}{dt^2} - kx - c\frac{dx}{dt} + mg$. From kinematics, $\theta = -\frac{x}{R}$, from which the equation of motion for the mass is

$\left(\frac{I}{R^2} + m\right)\frac{d^2x}{dt^2} + c\frac{dx}{dt} + kx = mg$. The canonical form (see Eq (10.16)) of the equation of motion is $\frac{d^2x}{dt^2} + 2d\frac{dx}{dt} + \omega^2 x = \frac{R^2 mg}{I + R^2 m}$, where $d = \frac{cR^2}{2(I + R^2 m)} = 2\ rad/s$, $\omega^2 = \frac{kR^2}{(I + R^2 m)} = 9\ (rad/s)^2$.

The damping is sub-critical, since $d^2 < \omega^2$. The solution is the sum of the solution to the homogenous equation of motion, of the form $x_c = e^{-dt}(A\sin\omega_d t + B\cos\omega_d t)$, where $\omega_d = \sqrt{\omega^2 - d^2} = 2.236\ rad/s$, and the solution to the non-homogenous equation, of the form $x_p = \frac{mgR^2}{(I + R^2 m)\omega^2} = \frac{mg}{k} = 0.3633$. (The particular solution x_p is obtained by setting the acceleration and velocity to zero and solving, since the non-homogenous term mg is not a function of time or position.) The solution is $x = x_c + x_p = e^{-dt}(A\sin\omega_d t + B\cos\omega_d t) + \frac{mg}{k}$. Apply the initial conditions: at $t = 0$, $x = 0$, $\frac{dx}{dt} = 0$, from which $0 = B + \frac{mg}{k}$, and $0 = -d[x_c]_{t=0} + \omega_d A = -dB + \omega_d A$, from which $B = -0.3633$, $A = \frac{dB}{\omega_d} = -0.3250$, and

$$x(t) = e^{-dt}\left(-\frac{dmg}{k\omega_d}\sin\omega_d - \left(\frac{mg}{k}\right)\cos\omega_d t\right) + \left(\frac{mg}{k}\right)$$

$$\boxed{x(t) = e^{-2t}(-0.3250\sin(2.236\,t) - 0.3633\cos(2.236\,t)) + 0.3633\ (m)}$$

================================<>================================

================================<>================================

Problem 10.47 For the system described in Problem 10.46, determine x as a function of time if $c = 120\ N\text{-}s/m$ and the system is released from rest with $x = 0$.

Solution:

From the solution to Problem 10.46 the canonical form of the equation of motion is

$\frac{d^2x}{dt^2} + 2d\frac{dx}{dt} + \omega^2 x = \frac{R^2 mg}{I + R^2 m}$, where $d = \frac{cR^2}{2(I + R^2 m)} = 4\ rad/s$, and

$\omega^2 = \frac{kR^2}{(I + R^2 m)} = 9\ (rad/s)^2$. The system is supercritically damped, since $d^2 > \omega^2$. The homogenous solution is of the form (see Eq (10.24)) $x_c = Ce^{-(d-h)t} + De^{-(d+h)t}$, where

$h = \sqrt{d^2 - \omega^2} = 2.646\ rad/s$, $(d - h) = 0.8307$, $(d + h) = 6.646$. The particular solution is

$x_p = \frac{mg}{k} = 0.3633$. The solution is $x(t) = x_c + x_{p_c} = Ce^{-(d-h)t} + De^{-(d+h)t} + \frac{mg}{k}$. Apply the initial conditions, at $t = 0$, $x = 0$, $\frac{dx}{dt} = 0$, from which $0 = C + D + \frac{mg}{k}$, and $0 = -(d - h)C - (d + h)D$. Solve:

$C = -\frac{(d + h)}{2h}\left(\frac{mg}{k}\right)$, and $D = \frac{(d - h)}{2h}\left(\frac{mg}{k}\right)$, from which

$\boxed{x(t) = \frac{mg}{k}\left(1 - \frac{(d + h)}{2h}e^{-(d-h)t} + \frac{(d - h)}{2h}e^{-(d+h)t}\right) = 0.3633(1 - 1.256e^{-1.354t} + 0.2559e^{-6.646t})\ (m)}$. [*Check:* An alternate form of the solution is obtained by combining factors:

$x(t) = 0.3633 - 0.4563e^{-1.354t} + 0.0929e^{-6.646t}$. *check.*]

================================<>================================

Problem 10.48 For the system described in Problem 10.46, choose the value of c so that the system is critically damped and determine x as a function of time if the system is released from rest with $x = 0$.

Solution:

From the solution to Problem 10.46, the canonical form (see Eq (10.16)) of the equation of motion is

$\frac{d^2x}{dt^2} + 2d\frac{dx}{dt} + \omega^2 x = \frac{R^2 mg}{I + R^2 m}$, where $d = \frac{cR^2}{2(I + R^2 m)}$, and $\omega^2 = \frac{kR^2}{(I + R^2 m)} = 9\ (rad/s)^2$. For critical damping, $d^2 = \omega^2$, from which $d = 3\ rad/s$. The homogenous solution is (see Eq (10.25))

$x_c = Ce^{-dt} + Dte^{-dt}$ and the particular solution is $x_p = \frac{mg}{k} = 0.3633\ m$. The solution:

$x(t) = x_c + x_p = Ce^{-dt} + Dte^{-dt} + \frac{mg}{k}$. Apply the initial conditions at $t = 0$, $x = 0$, $\frac{dx}{dt} = 0$, from which

$C + \frac{mg}{k} = 0$, and $-dC + D = 0$. Solve: $C = -\frac{mg}{k} = -0.3633$, $D = dC = -1.09$. The solution is

$x(t) = \frac{mg}{k}(1 - e^{-dt}(1 + dt))$, $\boxed{x(t) = 0.3633(1 - e^{-3t}(1 + 3t))\ m}$ [*Check:* Combining factors,

$x = 0.3633 - 0.3633e^{-3t} - 1.09te^{-3t}\ m$. *check.*]

================================<>================================

==================================<>==================================

Problem 10.49 The spring constant is $k = 30$ *lb/ft* and the damping constant is $c = 3.5$ *lb-s/ft*. The raduis of the pulley is *o.5 ft*, and its moment of inertia is *0.25 slug-ft*^{2}. What is the frequency of small vibrations of the system relative to its equilibrium position?

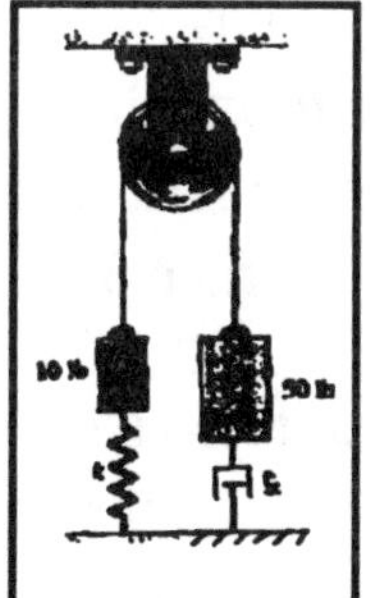

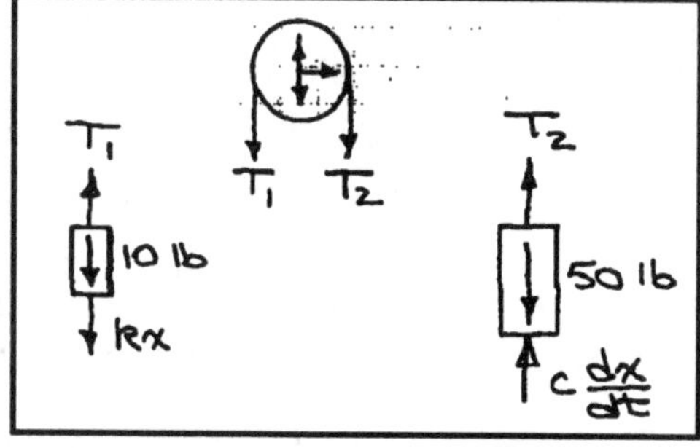

Solution:

Let x be the upward displacement of the $10-lb$ weight relative to its position when the spring is unstretched. The free body diagrams of the pulley and the two weights are as shown. We write Newton's second law for the *10-lb* and *50-lb* weights, $T_1 - 10 - kx = \left(\frac{10}{32.2}\right)\frac{d^2x}{dt^2}$ (1), and $50 - T_2 - c\frac{dx}{dt} = \left(\frac{50}{32.2}\right)\frac{d^2x}{dt^2}$ (2) and write the equation of angular motion for the pulley, $(0.5)T_2 - (0.5)T_1 = (0.25)\left(\frac{1}{0.5}\right)\frac{d^2x}{dt^2}$ (3). Solving Eqns (1) and (2) for T_1 and T_2 and substituting the results into Eq (3), we obtain the equation of motion $\left(\frac{60}{32.2}+1\right)\frac{d^2x}{dt^2}+c\frac{dx}{dt}+kx = 40$ or $\frac{d^2x}{dt^2}+1.22\frac{dx}{dt}+10.48x = 13.97$. When the system is in equilibrium, $d^2x/dt^2 = dx/dt = 0$ so the equilibrium position x_e is $x_e = \frac{13.97}{10.48} = 1.33 ft$. Defining $\tilde{x} = x - x_e$, the equation of motion expressed in terms of $\tilde{x}$ is $\frac{d^2\tilde{x}}{dt^2}+1.22\frac{d\tilde{x}}{dt}+10.48\tilde{x} = 0$ Comparing this equation to Equation (10.16), we see that $d = \frac{1.22}{2} = 0.611 rad/s$, $\omega = \sqrt{10.48} = 3.24 rad/s$. The system is subcritically damped with frequency $f_d = \frac{\omega_d}{2\pi} = \frac{\sqrt{\omega^2 - d^2}}{2\pi} = 0.506 Hz$.

==================================<>==================================

Problem 10.50 The system described in Problem 10.49 is released from rest with the spring unstretched. Determine the position of the *10-lb* weight relative to its position at $t = 0$ as a function of time.

Solution:

In the solution to Problem 10.49 it was shown that the upward displacement $\tilde{x}$ of the $10-lb$ weight relative to its equilibrium position is governed by the equation $\frac{d^2\tilde{x}}{dt^2}+1.22\frac{d\tilde{x}}{dt}+10.48\tilde{x} = 0$, that $d = 0.611 rad/s, \omega = 3.24 rad/s$ and the position when the spring is unstretched is $\tilde{x} = -1.33 ft$. The system is subcritically damped, so the motion is described by Equation (10.19) with $\omega_d = \sqrt{\omega^2 - d^2} = 3.18 rad/s$ The solution is $\tilde{x} = e^{-0.611t}(A\sin 3.18t + B\cos 3.18t)$. The time derivative is $\frac{d\tilde{x}}{dt} = -0.611e^{-0.611t}(A\sin 3.18t + B\cos 3.18t) + e^{-0611t}(3.18A\cos 3.18t - 3.18B\sin 3.18t)$.

At $t = 0, \tilde{x} = -1.33 ft$ and $d\tilde{x}/dt = 0$. Thus, $-1.33 = B$, $0 = -0.611B + 3.18A$. Therefore $B = -1.33, A = -0.256$ and we obtain $\tilde{x} = e^{-0.611t}(-0.256\sin 3.18t - 1.33\cos 3.18t)$. The position of the weight relative to its position at $t = 0$ is $x = 1.33 + \tilde{x} = 1.33 + e^{-0.611}(-0.256\sin 3.18t - 1.33\cos 3.18t)$.

==================================<>==================================

================================<>================================

Problem 10.51 The homogenous disk weighs 100 *lb* and its radius is $R = 1\ ft$. It rolls on the plane surface. The spring constant is $k = 100\ lb/ft$ and the damping constant is $c = 3\ lb\text{-}s/ft$. Determine the natural frequency of small vibrations of the disk relative to its equilibrium position.

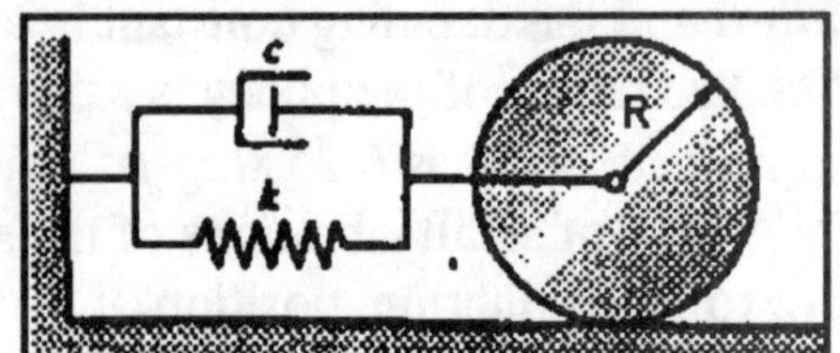

Solution:

Choose a coordinate system with the origin at the center of the disk and the positive *x* axis parallel to the floor. Denote the angle of rotation by θ. The *horizontal* forces acting on the disk are $\sum F = -kx - c\frac{dx}{dt} + f$.

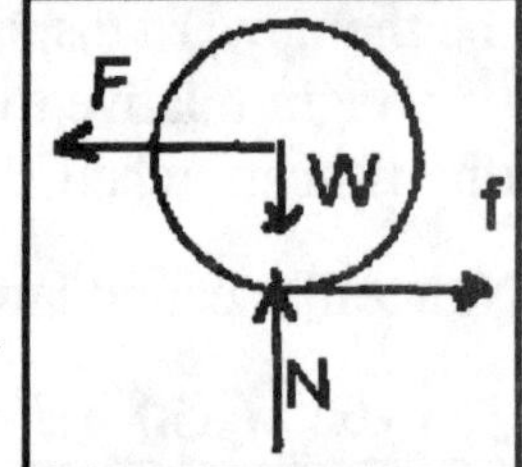

From Newton's second law, $m\frac{d^2x}{dt^2} = \sum F = -kx - c\frac{dx}{dt} + f$. The moment about the center of mass of the disk is $\sum M = Rf$. From the equation of angular motion, $I\frac{d^2\theta}{dt^2} = Rf$, from which $f = \frac{I}{R}\frac{d^2\theta}{dt^2}$, where the polar moment of inertia is $I = \frac{mR^2}{2} = 1.554\ slug\text{-}ft^2$. Substitute: $m\frac{d^2x}{dt^2} = -kx - c\frac{dx}{dt} + \frac{I}{R}\frac{d^2\theta}{dt^2}$. From kinematics, $\theta = -\frac{x}{R}$, from which the equation of motion is $\left(m + \frac{I}{R^2}\right)\frac{d^2x}{dt^2} + c\frac{dx}{dt} + kx = 0$. The canonical form (see Eq (10.16)) is $\frac{d^2x}{dt^2} + 2d\frac{dx}{dt} + \omega^2 x = 0$, where $d = \frac{cR^2}{2(mR^2 + I)} = \frac{c}{3m} = 0.3217\ rad/s$, and $\omega^2 = \frac{kR^2}{(I + R^2 m)} = \frac{2k}{3m} = 21.45\ (rad/s)^2$. The damping is sub-critical, since $d^2 < \omega^2$. The natural frequency is $\boxed{f_d = \frac{1}{2\pi}\sqrt{\omega^2 - d^2} = 0.7353\ Hz}$

================================<>================================

Problem 10.52 In Problem 10.51 the spring is unstretched at $t = 0$ and the disk has a clockwise angular velocity of 2 *rad/s*. What is the angular velocity of the disk when $t = 3\ s$?

Solution:

From the solution to Problem 10.51, the canonical form (see Eq (10.16)) of the equation of motion is $\frac{d^2x}{dt^2} + 2d\frac{dx}{dt} + \omega^2 x = 0$, where $d = \frac{c}{3m} = 0.3217\ rad/s$, and $\omega^2 = \frac{2k}{3m} = 21.45\ (rad/s)^2$. The system is sub-critically damped, so that the solution is of the form $x = e^{-dt}(A\sin\omega_d t + B\cos\omega_d t)$, where $\omega_d = \sqrt{\omega^2 - d^2} = 4.620\ rad/s$. Apply the initial conditions: $x_o = 0$, and from kinematics, $\dot{\theta}_o = -\dot{x}_o/R = -2$ rad/s, from which $\dot{x}_o = 2\ ft/s$, from which $B = 0$, and $A = \dot{x}_o/\omega_d = 0.4329$. The solution is $x(t) = e^{-dt}\left(\frac{\dot{x}_o}{\omega_d}\right)\sin\omega_d t$, and $\dot{x}(t) = -dx + \dot{x}_o e^{-dt}\cos\omega_d t$. At $t = 3\ s$, $x(3) = 0.1586\ ft$, and $\dot{x}(3) = 0.1577\ ft/s$ From kinematics, $\theta(t) = -\frac{\dot{x}(t)}{R}$. At $t = 3\ s$, $\boxed{\dot{\theta}(3) = -0.1577\ rad/s}$ clockwise.

================================<>================================

================================<>================================

Problem 10.53 The moment of inertia of the stepped disk is I. Let θ be the angular displacement of the disk relative to its position when the spring is unstretched. Show that the equation governing θ is identical in form to Eq (10.16), where $d = \frac{R^2 c}{2I}$, and $\omega^2 = \frac{4R^2 k}{I}$.

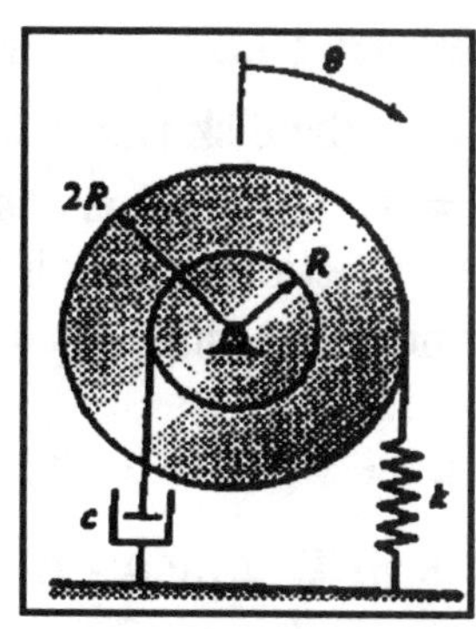

Solution:

The sum of the moments about the center of the stepped disk is $\sum M = -R\left(Rc\frac{d\theta}{dt}\right) - 2R(2Rk\theta)$. From the equation of angular motion $\sum M = I\frac{d^2\theta}{dt^2}$. The equation of motion is $I\frac{d^2\theta}{dt^2} + R^2 c\frac{d\theta}{dt} + 4R^2 k\theta = 0$. The cannonical form is $\boxed{\frac{d^2\theta}{dt^2} + 2d\frac{d\theta}{dt} + \omega^2\theta = 0}$, where $d = \frac{R^2 c}{2I}$, $\omega^2 = \frac{4R^2 k}{I}$. [*Check:* Let x be the stretch of the spring, positive upward. The sum of the moments about the center of the stepped disk is $\sum M = fR - 2Rkx$, where the damping force $f = -c\frac{dx}{dt}$ acts in opposition to the linear velocity. From the equation of angular motion, $I\frac{d^2\theta}{dt^2} = \sum M = -Rc\frac{dx}{dt} - 2Rkx$. From kinematics, $\theta = \frac{x}{2R}$, for the stretch of the spring, and $\frac{d\theta}{dt} = \frac{1}{R}\frac{dx}{dt}$, for the linear velocity at the damper, from which the equation of motion is $I\frac{d^2\theta}{dt^2} + R^2 c\frac{d\theta}{dt} + 4R^2 k\theta = 0$. The canonical form (see Eq (10.16)) of the equation of motion is $\frac{d^2\theta}{dt^2} + 2d\frac{d\theta}{dt} + \omega^2\theta = 0$, where $d = \frac{R^2 c}{2I}$ and $\omega^2 = \frac{4R^2 k}{I}$. *check*]

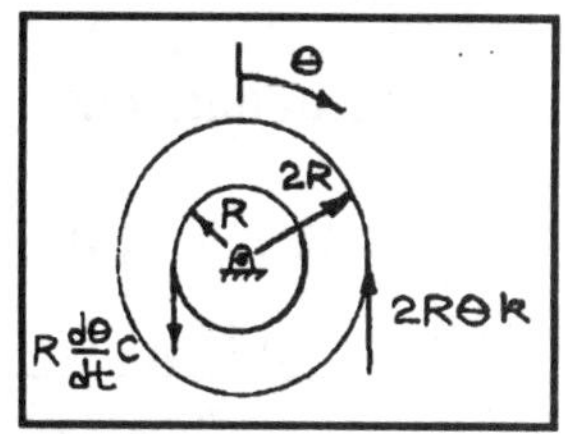

================================<>================================

================================<>================================

Problem 10.54 In Problem 10.53, the radius $R = 250\ mm$, $k = 150\ N/m$, and the moment of inertia of the disk is $I = 2\ kg\text{-}m^2$. (a) What value of c will cause the system to be critically damped? b) At $t = 0$, the spring is unstretched and the clockwise angular velocity of the disk is 10 *rad/s*. Determine θ as a function of time if the system is critically damped. (c) Using the result of part (b), determine the maximum resulting angular displacement of the disk and the time at which is occurs.

Solution:

From the solution to Problem 10.53, the canonical form (see Eq (10.16)) of the equation of motion is $\frac{d^2\theta}{dt^2} + 2d\frac{d\theta}{dt} + \omega^2\theta = 0$, where $d = \frac{R^2 c}{2I}$ and $\omega^2 = \frac{4R^2 k}{I}$.

(a) For critical damping, $d^2 = \omega^2$, from which $\boxed{c = \frac{4}{R}\sqrt{kI} = 277.1\ N\text{-}s/m}$, and $d = 4.330\ rad/s$.

(b) The solution is of the form (see Eq (10.25)) $\theta = Ce^{-dt} + Dte^{-dt}$. Apply the initial conditions: $\theta_o = 0$, $\dot{\theta}_o = -10\ rad/s$, from which $C = 0$, and $D = -10$. The solution is

$\boxed{\theta(t) = \dot{\theta}_o te^{-dt} = -10te^{-4.330t}\ rad/s}$

(c) The maximum (or minimum) value of the angular displacement is obtained from $\frac{d\theta}{dt} = 0 = -10e^{-4.330t}(1 - 4.33t) = 0$, from which the maximum/minimum occurs at

$\boxed{t_{\max} = \frac{1}{4.330} = 0.2309\ s}$. The angle is $[\theta]_{t=t_{\max}} = -0.8496\ rad$ *clockwise.* [*Note:* If the clockwise direction of the angular velocity is taken to be *positive* (see figure in Problem 10.53), then the results are changed in sign, $\theta(t) = \dot{\theta}_o te^{-dt} = 10te^{-4.330t}\ rad/s$, $[\theta]_{t=t_{\max}} = 0.8496\ rad$ at $t = 0.231\ s$.]

================================<>================================

Problem 10.55 The moments of inertia of gears A and B are $I_A = 0.025\ kg\text{-}m^2$ and $I_B = 0.100\ kg\text{-}m^2$. Gear A is connected to a torsional spring with constant $k = 10\ N\text{-}m/rad$. The bearing supporting gear B incorporates a damping element that exerts a resisting moment on gear B of $2\left(\frac{d\theta_B}{dt}\right)\ N\text{-}m$, where $\frac{d\theta_B}{dt}$ is the angular velocity of gear B in *rad/s*. What is the frequency of small angular vibrations of the gears?

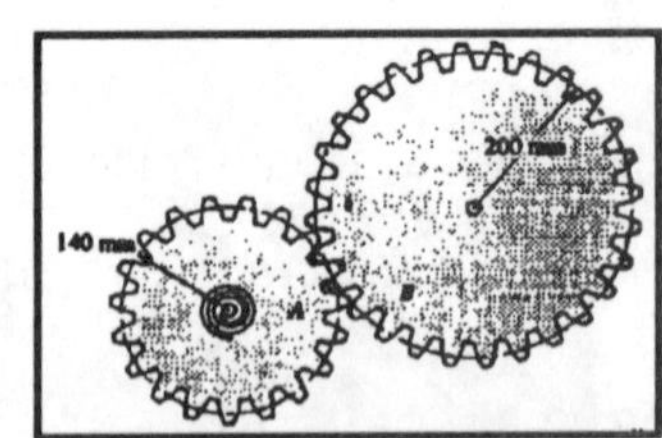

Solution:

(See the solution to Problem 10.28.) The sum of the moments on gear A is $\sum M = -k\theta_A + R_A F$, where the moment exerted by the spring opposes the angular displacement θ_A. From the equation of angular motion, $I_A \frac{d^2\theta_A}{dt^2} = \sum M = -k\theta_A + R_A F$, from which

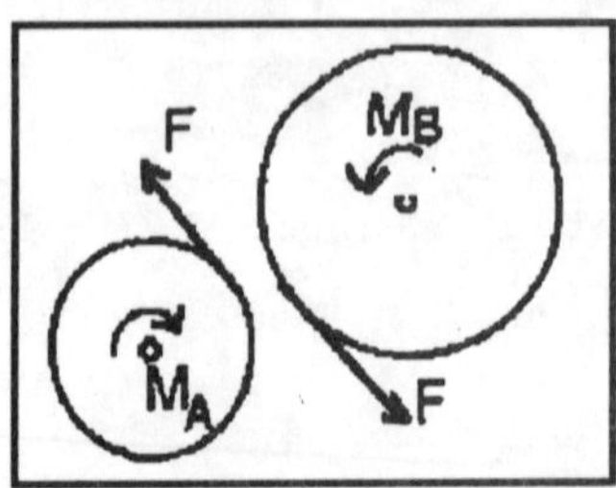

$F = \left(\frac{I_A}{R_A}\right)\frac{d^2\theta_A}{dt^2} + \left(\frac{k}{R_A}\right)\theta_A$. The sum of the moments acting on gear B is $\sum M = -2\frac{d\theta_B}{dt} + R_B F$, where the moment exerted by the damping element opposes the angular velocity of B.

Solution continued on next page

From the equation of angular motion applied to B, $I_B \frac{d^2\theta_B}{dt^2} = \sum M = -2\frac{d\theta_B}{dt} + R_B F$. Substitute the expression for F, $I_B \frac{d^2\theta_B}{dt^2} + 2\frac{d\theta_B}{dt} - \left(\frac{R_B}{R_A}\right)\left(I_A \frac{d^2\theta_A}{dt^2} + k\theta_A\right) = 0$. From kinematics, $\theta_A = -\left(\frac{R_B}{RA}\right)\theta_B$, from which the equation of motion for gear B is $\left(I_B + \left(\frac{R_B}{R_A}\right)^2 I_A\right)\frac{d^2\theta_B}{dt^2} + 2\frac{d\theta_B}{dt} + \left(\frac{R_B}{R_A}\right)^2 k\theta_B = 0$.

Define the *generalized mass moment of inertia* by $M = I_B + \left(\frac{R_B}{R_A}\right)^2 I_A = 0.1510\ kg\text{-}m^2$,

The canonical form (see Eq (10.16)) of the equation of motion is $\frac{d^2\theta_B}{dt^2} + 2d\frac{d\theta_B}{dt} + \omega^2\theta_B = 0$, where $d = \frac{1}{M} = 6.622\ rad/s$, and $\omega^2 = \left(\frac{R_B}{R_A}\right)^2 \frac{k}{M} = 135.1\ (rad/s)^2$. The system is sub critically damped, since $d^2 < \omega^2$, from which $\omega_d = \sqrt{\omega^2 - d^2} = 9.555\ rad/s$, from which the frequency of small vibrations is $\boxed{f_d = \frac{\omega_d}{2\pi} = 1.521\ Hz}$

==============================<>==============================

Problem 10.56 At $t = 0$, the torsional spring in Problem 10.55 is unstretched and gear B has a counterclockwise angular velocity of 2 *rad/s*. Determine the counterclockwise angular position of gear B relative to its equilibrium position as a function of time.

Solution:

From the solution to Problem 10.55, the canonical form (see Eq (10.16)) of the equation of motion for gear B is $\frac{d^2\theta_B}{dt^2} + 2d\frac{d\theta_B}{dt} + \omega^2\theta_B = 0$, where $M = I_B + \left(\frac{R_B}{R_A}\right)^2 I_A = 0.1510\ kg\text{-}m^2$,

$d = \frac{1}{M} = 6.622\ rad/s$, and $\omega^2 = \left(\frac{R_B}{R_A}\right)^2 \frac{k}{M} = 135.1\ (rad/s)^2$. The system is sub critically damped, since $d^2 < \omega^2$, from which $\omega_d = \sqrt{\omega^2 - d^2} = 9.555\ rad/s$. The solution is of the form $\theta_B(t) = e^{-dt}(A\sin\omega_d t + B\cos\omega_d t)$. Apply the initial conditions, $[\theta_B]_{t=0} = 0$, $[\dot\theta_B]_{t=0} = 2\ rad/s$, from which $B = 0$, and $A = \frac{2}{\omega_d} = 0.2093$, from which the solution is

$$\boxed{\theta_B(t) = e^{-6.622t}(0.2093\sin(9.555t))}$$

==============================<>==============================

==================================<>==================================

Problem 10.57 For the case of critically damped motion, confirm that the expression $x = Ce^{-dt} + Dte^{-dt}$ is a solution of Eq (10.16).

Solution :

Eq (10.16) is $\frac{d^2x}{dt^2} + 2d\frac{dx}{dt} + \omega^2 x = 0$. We show that the expression is a solution by substitution. The individual terms are: (1) $x = e^{-dt}(C + Dt)$, (2) $\frac{dx}{dt} = -dx + De^{-dt}$,

(3) $\frac{d^2x}{dt^2} = -d\frac{dx}{dt} - dDe^{-dt} = d^2x - 2dDe^{-dt}$.

Substitute: $\frac{d^2x}{dt^2} + 2d\frac{dx}{dt} + \omega^2 x = \left(d^2x - 2De^{-dt}\right) + 2d\left(-dx + De^{-dt}\right) + \omega^2 x = 0$.

Reduce:

$$\frac{d^2x}{dt^2} + 2d\frac{dx}{dt} + \omega^2 x = (-d^2 + \omega^2)x = 0.$$

This is true if $d^2 = \omega^2$, which is the definition of a critically damped system. [*Note:* Substitution leading to an identity shows that $x = Ce^{-dt} + Dte^{-dt}$ *is a solution.* It *does not* prove that it is *the only solution.*]

==================================<>==================================

Problem 10.58 The mass $m = 2$ *slug* and $k = 200$ *lb / ft*. Let *x* be the position of the mass relative to its position when the spring is unstretched. The force $F(t) = 36\sin 8t$ *lb*. (a) Determine the particular solution. (b) At $t = 0$, $x = 1$ *ft* and the velocity of the mass is zero. Determine *x* as a function of time.

Solution:

The horizontal forces on the mass are $\sum F = -kx + F(t)$.

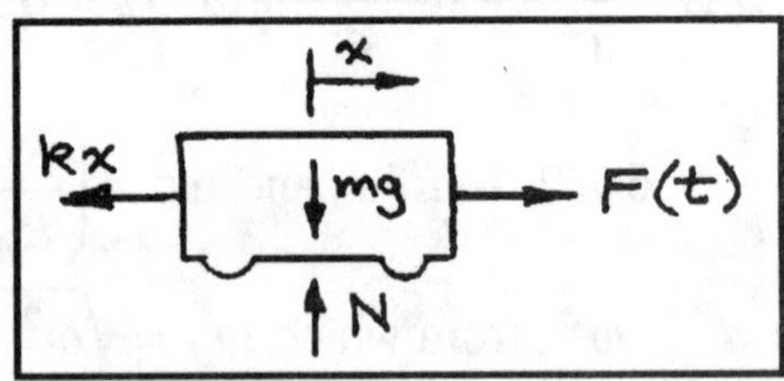

From Newton's second law, $m\frac{d^2x}{dt^2} = \sum F = -kx + F(t)$. For $d = 0$, the canonical form of the equation of motion (see Eq (10.26)) is $\frac{d^2x}{dt^2} + \omega^2 x = a(t)$, where

$\omega^2 = \frac{k}{m} = 100\ (rad / s)^2$, and

$a(t) = \frac{F(t)}{m} = \frac{F_o}{m}\sin\omega_o t = 18\sin 8t$.

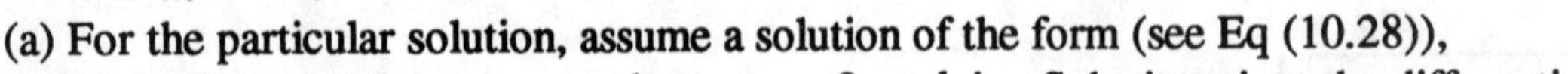

(a) For the particular solution, assume a solution of the form (see Eq (10.28)), $x_p = A_p\sin\omega_o t + B_p\cos\omega_o t$, where $\omega_o = 8$ *rad / s*. Substitute into the differential equation.

$$\frac{d^2x}{dt^2} + \omega^2 x = A_p(\omega^2 - \omega_o^2)\sin\omega_o t + B_p(\omega^2 - \omega_o^2)\cos\omega_o t = \frac{F_o}{m}\sin\omega_o t.$$

Equate like coefficients to obtain $B_p = 0$, $A_p = \frac{F_o}{m\left(\omega^2 - \omega_o^2\right)} = 0.5$, from which the particular solution is

Solution continued on next page

$$x_p(t) = \frac{F_o}{m(\omega^2 - \omega_o^2)} \sin\omega_o t = 0.5\sin 8t \ ft$$

(b) The solution is $x(t) = x_h(t) + x_p(t)$, where the solution to the homogenous equation is

$x_h = A\sin\omega t + B\cos\omega t$.

Apply the initial conditions $x_o = 1$, $\dot{x}_o = 0$ to obtain $B = 1$, $A = -\left(\frac{\omega_o}{\omega}\right)A_p = -0.4$ (since $x_p(0) = 0$, and $\dot{x}_p(0) = \omega_o A_p$). The solution: $x(t) = -0.4\sin 10t + \cos 10t + 0.5\sin 8t \ ft$

==================================<>==================================

Problem 10.59 Consider the spring-mass oscillator in Problem 10.58. The mass is $m = 2$ *slugs* and $k = 200$ *lb/ft*. The mass is initially stationary with the spring unstretched. At $t = 0$, a force $F(t) = 200 - 80t^2$ (*lb*) is applied to the mass. (a) What is the steady-state (particular) solution? (b) Determine the position of the mass as a function of time.

Solution:

Writing Newton's second law for the mass, the equation of motion is

$F(t) - kx = m\frac{d^2x}{dt^2}$: or $200 - 80t^2 - 200x = 2\frac{d^2x}{dt^2}$, which we can write as $\frac{d^2x}{dt^2} + 100x = 100 - 40t^2$. (1)

(a) We seek a particular solution of the form $x_p = A_0 + A_1 t + A_2 t^2$.

Substituting it into Equation (1), we can write the resulting equation as

$(2A_2 + 100A_0 - 100) + 100A_1 t + (100A_2 + 40)t^2 = 0$, which is satisfied if

$2A_2 + 100A_0 - 100 = 0, A_1 = 0$, and $100A_2 + 40 = 0$.

From these equations we obtain $A_0 = 1.008, A_1 = 0, A_2 = -0.4$ so the particular solution is

$x_p = 1.008 - 0.4t^2 \ ft$

(b) $\omega = \sqrt{k/m} = 10 rad/s$, so the homogeneous solution is $x_h = A\sin 10t + B\cos 10t$.

The general solution is $x = x_h + x_p = A\sin 10t + B\cos 10t + 1.008 - 0.4t^2$.

The time derivative is $\frac{dx}{dt} = 10A\cos 10t - 10B\sin 10t - 0.8t$.

At $t = 0, x = 0$ and $dx/dt = 0$: Hence, $0 = B + 1.008$, and $0 = 10A$.

Thus $A = 0$ and $B = -1.008$ and the solution is

$x = -1.008\cos 10t + 1.008 - 0.4t^2 \ ft$

==================================<>==================================

==================================<>==================================

Problem 10.60 The damped spring-mass oscillator is initially stationary with the spring unstretched. At $t = 0$, a constant force $F(t) = 6$ (*newtons*) is applied to the mass. (a) What is the steady-state (particular) solution? (b) Determine the position of the mass as a function of time.

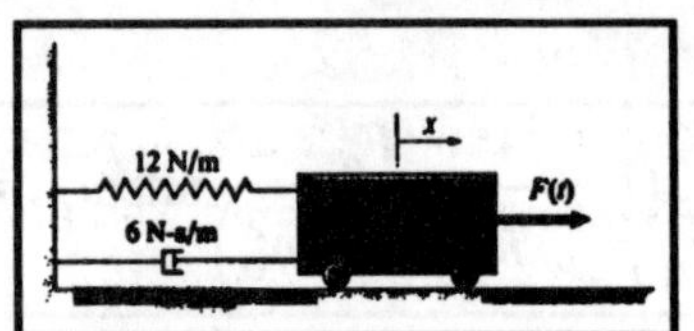

Solution:

Writing Newton's second law for the mass, the equation of motion is $F(t) - c\dfrac{dx}{dt} - kx = m\dfrac{d^2x}{dt^2}$

$F(t) - 6\dfrac{dx}{dt} - 12x = 3\dfrac{d^2x}{dt^2}$, which we can write as $\dfrac{d^2x}{dt^2} + 2\dfrac{dx}{dt} + 4x = \dfrac{F(t)}{3}$. (1)

(a) If $F(t) = 6N$, we seek a particular solution of the form $x_p = A_0$, a constant. Substituting it into Equation (1), we get $4x_p = \dfrac{F(t)}{3} = 2$ and obtain the particular solution: $x_p = 0.5m$.

(b) Comparing equation (1) with Equation (10.26) we see that $d = 1 rad/s$ and $\omega = 2 rad/s$. The system is subcritically damped and the homogeneous solution is given by Equation (10.19) with $\omega_d = \sqrt{\omega^2 - d^2} = 1.73 rad/s$ The general solution is
$x = x_h + x_p = e^{-t}(A\sin 1.73t + B\cos 1.73t) + 0.5m.$
The time derivative is
$\dfrac{dx}{dt} = -e^{-t}(A\sin 1.73t + B\cos 1.73t) + e^{-t}(1.73A\cos 1.73t - 1.73B\sin 1.73t)$. At $t = 0, x = 0$, and $dx/dt = 0$ $0 = B + 0.5$, and $0 = -B + 1.73A$. We see that $B = -0.5$ and $A = -0.289$ and the solution is $x = e^{-t}(-0.289\sin 1.73t - 0.5\cos 1.73t) + 0.5m.$

==================================<>==================================

Problem 10.61 The damped spring-mass oscillator shown in Problem 10.60 is initially stationary with the spring unstretched. At $t = 0$, a force $F(t) = 6\cos[1.6t]$ (*newtons*) is applied. (a) What is the steady-state (particular) solution? (b) Determine the position of the mass as a function of time.

Solution:

Writing Newton's second law for the mass, the equation of motion can be written as
$\dfrac{d^2x}{dt^2} + 2\dfrac{dx}{dt} + 4x = \dfrac{F(t)}{3} = 2\cos 1.6t$,

(a) Comparing this equation with Equation (10.26), we see that $d = 1 rad/s$, $\omega = 2 rad/s$, and the forcing function is $a(t) = 2\cos 1.6t$. This forcing function is of the form of Equation (10.27), with $a_0 = 0, b_0 = 2$ and $\omega_o = 1.6$. Substituting these values into Equation (10.30), the particular solution is $x_p = 0.520\sin 1.6t + 0.234\cos 1.6t$ (m).

(b) The system is subcritically damped so the homogeneous solution is given by Equation (10.19) with $\omega_d = \sqrt{\omega^2 - d^2} = 1.73 rad/s$. The general solution is
$x = x_h + x_p = e^{-t}(A\sin 1.73t + B\cos 1.73t) + 0.520\sin 1.6t + 0.234\cos 1.6t$. The time derivative is
$$\frac{dx}{dt} = -e^{-t}(A\sin 1.73t + B\cos 1.73t) + e^{-t}(1.73A\cos 1.73t - 1.73B\sin 1.73t)$$
$$+(1.6)(0.520)\cos 1.6t - (1.6)(0.234)\sin 1.6t$$
At $t = 0, x = 0$ and $dx/dt = 0$: $0 = B + 0.234$ $0 = -B + 1.73A + (1.6)(0.520)$. Solving, we obtain $A = -0.615$ and $B = -0.234$, so the solution is
$x = e^{-t}(-0.615\sin 1.73t - 0.234\cos 1.73t) + 0.520\sin 1.6t + 0.234\cos 1.6t$ (m).

==================================<>==================================

==============================<>==============================

Problem 10.62 The disk has moment of inertia $I = 3\ kg\text{-}m^2$. It rotates about a fixed shaft and is attached to a torsional spring with constant $k = 20\ N\text{-}m/rad$. At $t = 0$, the angle $\theta = 0°$, the angular velocity is $(d\theta/dt) = 4 rad/s$, and the disk is subjected to a couple $M(t) = 10\sin[2t]\ N\text{-}m$. Determine θ as a function of time.

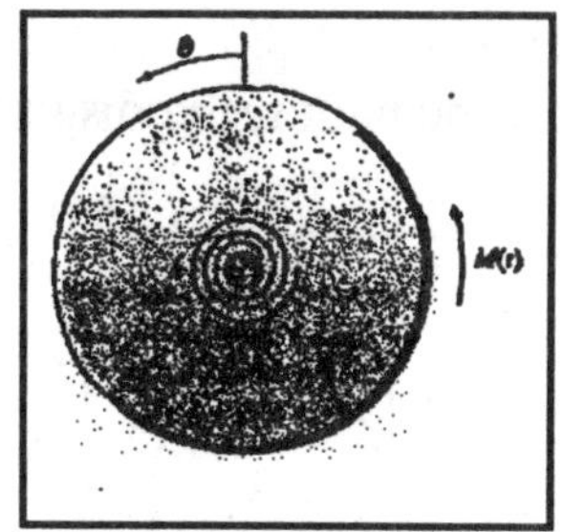

Solution: The equation of angular motion for the disk is $M(t) - k\theta = I\dfrac{d^2\theta}{dt^2}$: or $10\sin 2t - (20)\theta = 3\dfrac{d^2\theta}{dt^2}$, or , rewriting in standard form, we have $\dfrac{d^2\theta}{dt^2} + \dfrac{20}{3}\theta = \dfrac{10}{3}\sin 2t$

(a) Comparing this equation with equation (10.26), we see that $d = 0$, $\omega = \sqrt{20/3} = 2.58 rad/s$ and the forcing function is $a(t) = \dfrac{10}{3}\sin 2t$. This forcing function is of the form of Equation (10.27), with $a_0 = 10/3, b_0 = 0$ and $\omega_0 = 2$.

Substituting these values into Equation (10.30), the particular solution is $\theta_p = 1.25\sin 2t$.

(b) The general solution is $\theta = \theta_h + \theta_p = A\sin 2.58t + B\cos 2.58t + 1.25\sin 2t$

The time derivative is $\dfrac{d\theta}{dt} = 2.58A\cos 2.58t - 2.58B\sin 2.58t + 2.50\cos 2t$.

At $t = 0, \theta = 0$ and $d\theta/dt = 4 rad/s$, $0 = B$, and $4 = 2.58A + 2.50$

Solving, we obtain $A = 0.581$ and $B = 0$

The solution is $\theta = 0.581\sin 2.58t + 1.25\sin 2t\ (rad)$

==============================<>==============================

Problem 10.63 The stepped disk weighs 20 *lb* and its moment of inertia is $I = 0.6\ slug\text{-}ft^2$. It rolls on the horizontal surface. The disk is initially stationary with the spring unstretched, and at $t = 0$ a constant force of $F = 10\ lb$ is applied as shown. Determine the position of the center of the disk as a function of time.

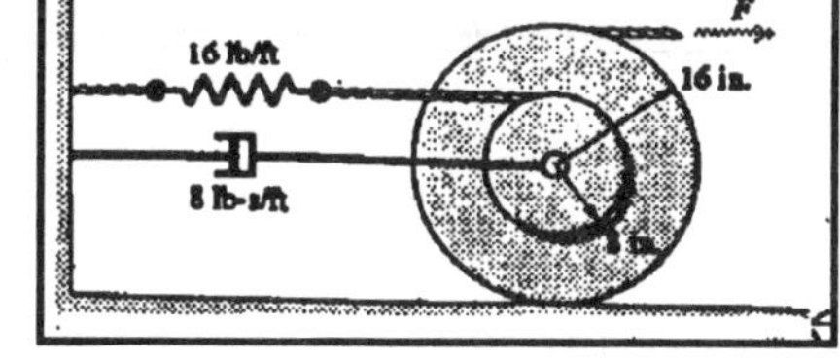

Solution:

The strategy is to apply the free body diagram to obtain equations for both θ and x, and then to eliminate one of these. An essential element in the strategy is the determination of the stretch of the spring. Denote $R = 8\ in. = 0.6667\ ft$, and the stretch of the spring by S. Choose a coordinate system with the positive x axis to the right. The sum of the moments about the center of the disk is $\sum M_C = RkS + 2Rf - 2RF$. From the equation of angular motion, $I\dfrac{d^2\theta}{dt^2} = \sum M_C = RkS + 2Rf - 2RF$. Solve for the reaction at the floor: $f = \dfrac{I}{2R}\dfrac{d^2\theta}{dt^2} - \dfrac{k}{2}S + F$. The sum of the horizontal forces:

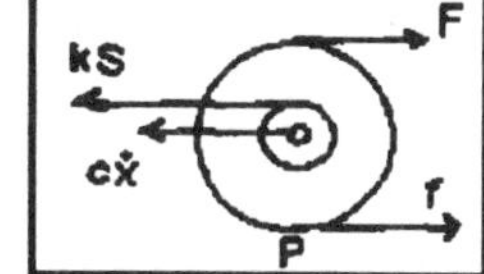

$\sum F_x = -kS - c\dfrac{dx}{dt} + F + f$. From Newton's second law: $m\dfrac{d^2x}{dt^2} = \sum F_x = -kS - c\dfrac{dx}{dt} + F + f$.

Substitute for f and rearrange: $m\dfrac{d^2x}{dt^2} + \dfrac{I}{2R}\dfrac{d^2\theta}{dt^2} + c\dfrac{dx}{dt} + \dfrac{3}{2}kS = 2F$.

Solution continued on the next page

Continuation of the solution to Problem 10.63

From kinematics, the displacement of the center of the disk is $x = -2R\theta$. The stretch of the spring is the amount wrapped around the disk plus the translation of the disk, $S = -R\theta - 2R\theta = -3R\theta = \frac{3}{2}x$.

Substitute: $\left(m + \frac{I}{(2R)^2}\right)\frac{d^2x}{dt^2} + c\frac{dx}{dt} + \left(\frac{3}{2}\right)^2 kx = 2F$. Define the *generalized mass* by

$M = m + \frac{I}{(2R)^2} = 0.9592$ *slug*. The canonical form (see Eq (10.26)) of the equation of motion is

$\frac{d^2x}{dt^2} + 2d\frac{dx}{dt} + \omega^2 x = a$, where $d = \frac{c}{2M} = 4.170$ rad/s, $\omega^2 = \left(\frac{3}{2}\right)^2 \frac{k}{M} = 37.53$ $(rad/s)^2$, and

$a = \frac{2F}{M} = 20.85$ ft/s^2. The particular solution is found by setting the acceleration and velocity to zero

and solving: $x_p = \frac{a}{\omega^2} = \frac{8F}{9k} = 0.5556$ ft. Since $d^2 < \omega^2$, the system is sub-critically damped, so the

homogenous solution is $x_h = e^{-dt}(A\sin\omega_d t + B\cos\omega_d t)$, where $\omega_d = \sqrt{\omega^2 - d^2} = 4.488$ rad/s. The

complete solution is $x = e^{-dt}(A\sin\omega_d t + B\cos\omega_d t) + \frac{8F}{9k}$. Apply the initial conditions: at $t = 0$,

$x_o = 0$, $\dot{x}_o = 0$, from which $B = -\frac{8F}{9k} = -0.5556$, and $A = \frac{dB}{\omega_d} = -0.5162$. Adopting

$g = 32.17$ ft/s^2 the solution is

$$x = \frac{8F}{9k}\left[1 - e^{dt}\left(\frac{d}{\omega_d}\sin\omega_d t + \cos\omega_d t\right)\right] \boxed{= 0.5556 - e^{-4.170t}(0.5162\sin 4.488t + 0.5556\cos 4.488t)\ ft},$$

===================================<>===================================

==================================<>==================================

Problem 10.64 An electric motor is bolted to a metal table. When the motor is on, it causes the tabletop to vibrate horizontally. Assume that the legs of the table behave like linear springs and neglect damping. The total weight of the motor and table top is 150 *lb*. When the motor is not on, the frequency of horizontal vibration is 5 *Hz*. When the motor is running at 600 *rpm* (revolutions per minute), the amplitude of the horizontal vibrations is 0.01 *in.* What is the magnitude of the oscillatory force exerted on the table by the motor at this speed?

Solution:

For $d = 0$, the canonical form (see Eq (10.26)) of the equation of motion is $\frac{d^2x}{dt^2} + \omega^2 x = a(t)$, where

$\omega^2 = (2\pi f)^2 = (10\pi)^2 = 986.96\ (rad/s)^2$, and

$a(t) = \frac{F(t)}{m} = \frac{gF(t)}{W}$. The forcing frequency is

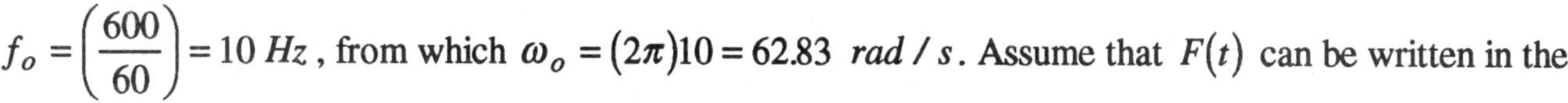

$f_o = \left(\frac{600}{60}\right) = 10\ Hz$, from which $\omega_o = (2\pi)10 = 62.83\ rad/s$. Assume that $F(t)$ can be written in the form $F(t) = F_o \sin\omega_o t$. From Eq (10.31), the amplitude of the oscillation is

$x_o = \frac{a_o}{\omega^2 - \omega_o^2} = \frac{gF_o}{W(\omega^2 - \omega_o^2)}$. Solve for the magnitude: $|F_o| = \frac{W}{g}\left|(\omega^2 - \omega_o^2)\right| x_o$ Substitute numerical values: $W = 150\ lb$, $g = 32.17\ ft/s^2$, $\left|(\omega^2 - \omega_o^2)\right| = 2960.9\ (rad/s)^2$, $x_o = 0.01\ in. = 8.33 \times 10^{-4}\ ft$, from which $\boxed{F_o = 11.5\ lb}$

==================================<>==================================

Problem 10.65 The moments of inertia of gears A and B are $I_A = 0.014\ slug\text{-}ft^2$ and $I_B = 0.100\ slug\text{-}ft^2$. Gear A is connected to a torsional spring with constant $k = 2\ ft\text{-}lb/rad$. The system is in equilibrium at $t = 0$ when it is subjected to an oscillatory force $F(t) = 4\sin 3t\ lb$. What is the downward displacement of the 5 *lb* weight as a function of time?

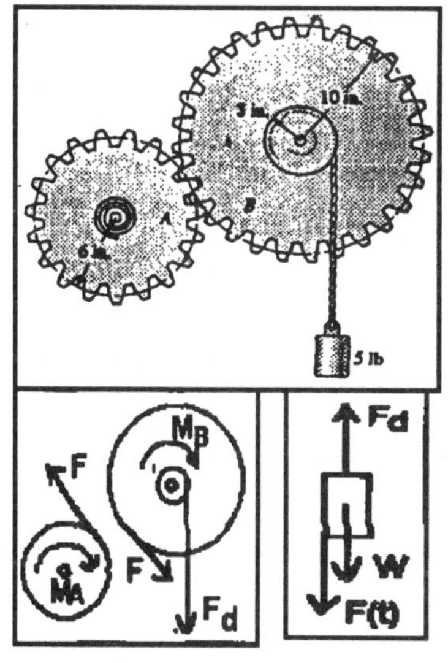

Solution:

(See the solution to Problem 10.55.) Choose a coordinate system with the *x* axis positive upward. The sum of the moments on gear A is $\sum M = -k\theta_A + R_A F$. From Newton's second law,

$I_A \frac{d^2\theta_A}{dt^2} = \sum M = -k\theta_A + R_A F$, from which

$F = \left(\frac{I_A}{R_A}\right)\frac{d^2\theta_A}{dt^2} + \left(\frac{k}{R_A}\right)\theta_A$. The sum of the moments acting on gear B is $\sum M = R_B F - R_W F_d$. From the equation of angular motion applied to gear B,

$I_B \frac{d^2\theta_B}{dt^2} = \sum M_B = R_B F - R_W F_d$.

Solution continued on next page

Substitute for F to obtain the equation of motion for gear B:

$$I_B \frac{d^2\theta_B}{dt^2} - \left(\frac{R_B}{R_A}\right) I_A \frac{d^2\theta_A}{dt^2} - \left(\frac{R_A}{R_B}\right) k\theta_A - R_W F_d = 0.$$

Solve: $F_d = \left(\frac{I_B}{R_W}\right)\frac{d^2\theta_B}{dt^2} - \left(\frac{R_B}{R_A R_W}\right)\frac{d^2\theta_A}{dt^2} - \left(\frac{R_B}{R_A R_W}\right) k\theta_A$. The sum of the forces on the weight are $\sum F = +F_d - W - F(t)$. From Newton's second law applied to the weight, $\left(\frac{W}{g}\right)\frac{d^2x}{dt^2} = F_d - W - F(t)$.

Substitute for F_d, and rearrange to obtain the equation of motion for the weight:

$\frac{W}{g}\frac{d^2x}{dt^2} - \frac{I_B}{R_W}\frac{d^2\theta_B}{dt^2} + \frac{R_B I_A}{R_W R_A}\frac{d^2\theta_A}{dt^2} + \frac{R_B}{R_W R_A} k\theta_A = -W - F(t)$. From kinematics, $\theta_A = -\left(\frac{R_B}{RA}\right)\theta_B$, and $x = -R_W\theta_B$, from which $\frac{d^2\theta_B}{dt^2} = -\frac{1}{R_W}\frac{d^2x}{dt^2}$, $\frac{d^2\theta_A}{dt^2} = -\left(\frac{R_B}{R_A}\right)\frac{d^2\theta_B}{dt^2} = \frac{R_B}{R_W R_A}\frac{d^2x}{dt^2}$, and

$\theta_A = \frac{R_B}{R_W R_A} x$. Define $\eta = \frac{R_B}{R_W R_A} = 6.667\ ft^{-1}$, and the *generalized mass* by

$M = \frac{W}{g} + \frac{I_B}{(R_W)^2} + \left(\frac{R_B}{R_W R_A}\right)^2 I_A = 2.378\ slug$, from which the equation of motion for the weight *about the unstretched spring position* is: $M\frac{d^2x}{dt^2} + \left(k\eta^2\right)x = -W - F(t)$. For $d = 0$, the canonical form (see Eq (10.16)) of the equation of motion is $\frac{d^2x}{dt^2} + \omega^2 x = a(t)$, where $\omega^2 = \frac{\eta^2 k}{M} = 37.39\ (rad/s)^2$. [*Check*: This agrees with the result in the solution to Problem 10.53, as it should, since nothing has changed except for the absence of a damping element. *check*.] The non-homogenous terms are $a(t) = \frac{W}{M} + \frac{F(t)}{M}$.

Since $\frac{W}{M}$ is not a function of *t* or *x*, $x_{pw} = -\frac{W}{\omega^2 M} = -\frac{W}{\eta^2 k} = -0.05625\ ft$, which is the equilibrium point. Make the transformation $\tilde{x}_p = x_p - x_{pw}$. The equation of motion about the equilibrium point is $\frac{d^2\tilde{x}}{dt^2} + \omega^2\tilde{x} = -\frac{4\sin 3t}{M}$ Assume a solution of the form $\tilde{x}_p = A_p \sin 3t + B_p \cos 3t$. Substitute:

$\left(\omega^2 - 3^2\right)\left(A_p \sin 3t + B_p \cos 3t\right) = -\frac{3\sin 3t}{M}$, from which $B_p = 0$, and

$A_p = -\frac{4}{M\left(\omega^2 - 3^2\right)} = -0.05927\ ft$. The particular solution is

$\tilde{x}_p = -\frac{4}{M\left(\omega^2 - 3^2\right)}\sin 3t = -0.05927 \sin 3t\ ft$. The solution to the homogenous equation is

$x_h = A_h \sin\omega t + B_h \cos\omega t$, and the complete solution is

Solution continued on next page

Continuation of solution to Problem 10.65

$\tilde{x}(t) = A_h \sin\omega t + B_h \cos\omega t - \frac{4}{M(\omega^2 - 3^2)}\sin 3t$. Apply the initial conditions: at $t = 0$, $x_o = 0$, $\dot{x}_o = 0$, from which $0 = B$, $0 = \omega A_h - \frac{12}{M(\omega^2 - 3^2)}$, from which $A_h = \frac{12}{\omega M(\omega^2 - 3^2)} = 0.02908$.

The complete solution for vibration about the equilibrium point is :

$\tilde{x}(t) = \frac{4}{M(\omega^2 - \omega_o^2)}\left(\frac{3}{\omega}\sin\omega t - \sin 3t\right) = 0.02908 \sin 6.114t - 0.05927 \sin 3t\ ft$. The downward travel is the negative of this: $\boxed{\tilde{x}_{down} = -0.02908 \sin 6.114t + 0.05927 \sin 3t\ ft}$

=================================<>=================================

Problem 10.66 A 1.5 *kg* cylinder is mounted on a "sting" in a wind tunnel with the cylinder axis transverse to the flow direction. When there is no flow, a 10 *N* vertical force applied to the cylinder causes it to deflect 0.15 *mm*. When air flows in the wind tunnel, vortices subject the cylinder to alternating lateral forces. The velocity of the air is 5 *m/s*, the distance between vortices is 80 *mm*, and the magnitude of the lateral forces is 1 *N*. If you model the lateral forces by the oscillatory function $F(t) = 1.0 \sin\omega_o t\ N$, what is the amplitude of the steady state lateral motion of the sphere?

Solution:

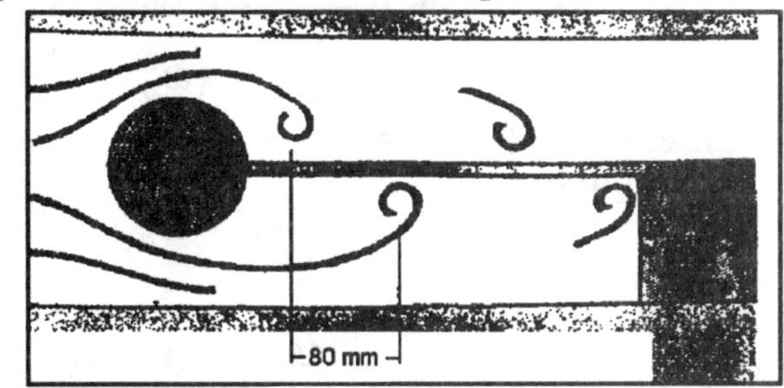

The time interval between the appearance of an upper vortex and a lower vortex is $\delta t = \frac{0.08}{5} = 0.016\ s$, from which the period of a *sinusoidal-like* disturbance is $\tau = 2(\delta t) = 0.032\ s$, from which $f_o = \frac{1}{\tau} = 31.25\ Hz$. [*Check:* Use the physical relationship between frequency, wavelength and velocity of propagation of a *small amplitude sinusoidal wave*, $\lambda f = v$. The wavelength of a traveling sinusoidal disturbance is the distance between two peaks or two troughs, or twice the distance between adjacent peaks and troughs, $\lambda = 2(0.08) = 0.16\ m$, from which the frequency is $f_o = \frac{v}{\lambda} = 31.25\ Hz$.*check*] The circular frequency is $\omega_o = 2\pi f_o = 196.35\ rad/s$. The spring constant of the sting is $k = \frac{F}{\delta} = \frac{10}{0.00015} = 66667\ N/m$. The natural frequency of the sting-cylinder system is $f = \frac{1}{2\pi}\sqrt{\frac{k}{m}} = \frac{1}{2\pi}\sqrt{\frac{66667}{1.5}} = 33.55\ Hz$, from which $\omega = 2\pi f = 210.82\ rad/s$. For $d = 0$, the canonical form (see Eq (10.16)) of the equation of motion is $\frac{d^2x}{dt^2} + \omega^2 x = a(t)$, where $a(t) = \frac{F}{m} = \frac{1}{1.5}\sin\omega_o t = 0.6667 \sin 196.3t$. From Eq (10.31) the amplitude is

$\boxed{E = \frac{a_o}{\omega^2 - \omega_o^2} = \frac{0.6667}{5091.3} = 1.132 \times 10^{-4}\ m}$ [*Note*: This is a small deflection (113 *microns*) but the associated aerodynamic forces may be significant to the tests (e.g. $F_{amp} = 7.5\ N$), since the sting is stiff. Vortices may cause undesirable noise in sensitive static aerodynamic loads test measurements.]

=================================<>=================================

=====================================<>=====================================

Problem 10.67 Show that the amplitude of the particular solution given by Eq (10.31) is a maximum when the frequency of the oscillatory forcing function is $\omega_o = \sqrt{\omega^2 - 2d^2}$.

Solution:

Eq (10.31) is $E_p = \dfrac{\sqrt{a_o^2 + b_o^2}}{\sqrt{(\omega^2 - \omega_o^2)^2 + 4d^2\omega_o^2}}$. Since the numerator is a constant, rearrange:

$\eta = \dfrac{E_p}{\sqrt{a_o^2 + b_o^2}} = \left[(\omega^2 - \omega_o^2)^2 + 4d^2\omega_o^2\right]^{-\frac{1}{2}}$. The maximum (or minimum) is found from $\dfrac{d\eta}{d\omega_o} = 0$.

$\dfrac{d\eta}{d\omega_o} = -\dfrac{1}{2}\dfrac{4\left[-(\omega^2 - \omega_o^2)(\omega_o) + 2d^2\omega_o\right]}{\left[(\omega^2 - \omega_o^2)^2 + 4d^2\omega_o^2\right]^{\frac{3}{2}}} = 0$, from which $-(\omega^2 - \omega_o^2)(\omega_o) + 2d^2\omega_o = 0$. Rearrange, $(\omega_o^2 - \omega^2 + 2d^2)\omega_o = 0$. Ignore the possible solution $\omega_o = 0$, from which $\boxed{\omega_o = \sqrt{\omega^2 - 2d^2}}$. Let $\tilde{\omega}_o = \sqrt{\omega^2 - 2d^2}$ be the maximizing value. To show that η is indeed a maximum, take the second

derivative: $\left[\dfrac{d^2\eta}{d\omega_o^2}\right]_{\omega_o = \tilde{\omega}_o} = \dfrac{3}{4}\dfrac{\left[(4\omega_o)(\omega_o^2 - \omega^2 + 2d^2)\right]^2_{\omega_o = \tilde{\omega}_o}}{\left[(\omega^2 - \omega_o^2)^2 + 2d^2\omega_o^2\right]^{\frac{5}{2}}_{\omega_o = \tilde{\omega}_o}} - \dfrac{1}{2}\dfrac{4\left[3\omega_o^2 - \omega^2 + 2d^2\right]_{\omega_o = \tilde{\omega}_o}}{\left[(\omega^2 - \omega_o^2)^2 + 2d^2\omega_o^2\right]^{\frac{3}{2}}_{\omega_o = \tilde{\omega}_o}}$, from

which $\left[\dfrac{d^2\eta}{dt^2}\right]_{\omega_o = \tilde{\omega}_o} = -\dfrac{4\omega_o^2}{\left[2\omega^2 d^2\right]^{\frac{3}{2}}} < 0$, which demonstrates that it is a maximum.

=====================================<>=====================================

==================================<>==================================

Problem 10. 68 A sonobuoy (sound measuring device) floats in a standing wave tank. It is a cylinder of mass m and cross sectional area A. The water density is ρ, and the buoyancy force supporting the buoy equals the weight of the water that would occupy the volume of the part of the cylinder below the surface. When the water in the tank is stationary, the buoy is in equilibrium in the vertical position shown in Fig (a). Waves are then generated in the tank, causing the depth of the water at the sonobuoy's position *relative to its original depth* to be $d = d_o \sin\omega_o t$. Let y be the sonobuoy's vertical position relative to its original position. Show that the sonobuoy's vertical position is governed by the equation $\frac{d^2y}{dt^2}+\left(\frac{A\rho g}{m}\right)y=\left(\frac{A\rho g}{m}\right)d_o \sin\omega_o t$.

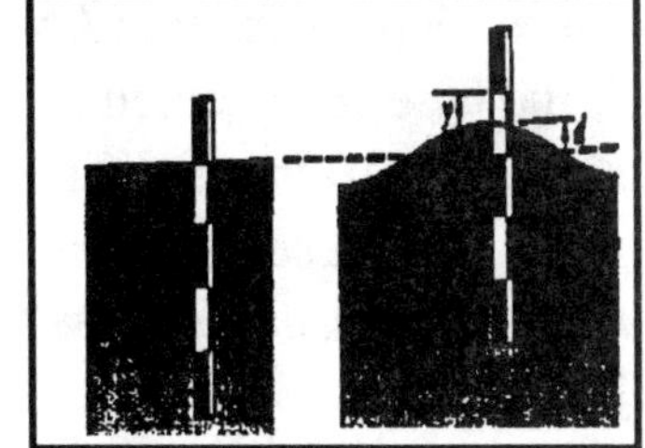

Solution:

The volume of the water displaced at equilibrium is $V = Ah$ where A is the cross-sectional area, and h is the equilibrium immersion depth. The weight of water displaced is $\rho V g = \rho g A h$, so that the buoyancy force is $F_b = \rho A g h$.

The sum of the vertical forces is $\sum F_y = \rho g A h - mg = 0$ at equilibrium, where m is the mass of the buoy. By definition, the spring constant is $\frac{\partial F_b}{\partial h} = k = \rho g A$. For any displacement δ of the immersion depth from the equilibrium depth, the net vertical force on the buoy is $\sum F_y = -\rho g A(h+\delta) + mg = -\rho A g\delta = -k\delta$, since h is the equilibrium immersion depth. As the waves are produced, $\delta = y - d$, where $d = d_o \sin\omega_o t$, from which $\sum F_y = -k(y-d)$. From Newton's second law, $m\frac{d^2y}{dt^2} = -k(y-d)$, from which $m\frac{d^2y}{dt^2} + ky = kd$. Substitute:

$$\boxed{\frac{d^2y}{dt^2}+\left(\frac{\rho g A}{m}\right)y=\left(\frac{\rho g A}{m}\right)d_o \sin\omega_o t}$$

==================================<>==================================

Problem 10.69 Suppose that the mass of the sonobuoy in Problem 10.68 is $m = 10\ kg$. Its diameter is $125\ mm$, and the water density is $\rho = 1025\ kg/m^3$. If the water depth is $d = 0.1\sin 2t\ m$, what is the magnitude of the steady state vertical vibrations of the sonobuoy?

Solution: From the solution to Problem 10.68, $\frac{d^2y}{dt^2}+\left(\frac{\rho g A}{m}\right)y=\left(\frac{\rho g A}{m}\right)d_o \sin\omega_o t$. The canonical form is $\frac{d^2y}{dt^2}+\omega^2 y = a(t)$, where $\omega^2 = \frac{\rho\pi d^2 g}{4m} = \frac{1025\pi(0.125^2)(9.81)}{4(10)} = 12.34\ (rad/s)^2$, and $a(t) = \omega^2(0.1)\sin 2t$. From Eq (10.31), the amplitude of the steady state vibrations is

$$\boxed{E_p = \frac{\omega^2(0.1)}{\left|(\omega^2 - 2^2)\right|} = 0.1480\ m}$$

==================================<>==================================

==<>==

Problems 10.70-10.73 are related to Example 10.7.

==<>==

Problem 10.70 The mass in Fig 10.21 weighs 50 *lb*. The spring constant is $k = 200\ lb/ft$, and $c = 10\ lb\text{-}s/ft$. If the base is subjected to an oscillatory displacement x_i of amplitude 10 *in.*, and circular frequency $\omega_i = 15\ rad/s$, what is the resulting steady-state amplitude of the displacement of the mass relative to the base?

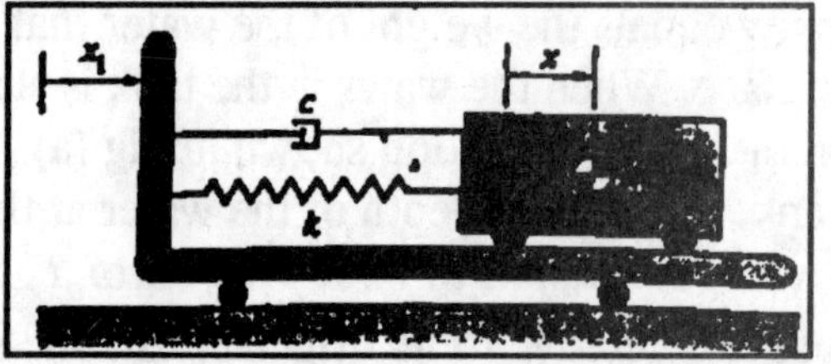

Solution:

The canonical form of the equation of motion (see Eq (10.26) and Example 10.7) is

$\frac{d^2x}{dt^2} + 2d\frac{dx}{dt} + \omega^2 x = a(t)$ where $d = \frac{c}{2m} = \frac{gc}{2W} = 3.217\ rad/s$, $\omega^2 = \frac{kg}{W} = 128.7\ (rad/s)^2$, and

(see Eq (10.38) $a(t) = -\frac{d^2x_i}{dt^2} = x_i\omega_i^2 \sin(\omega_i t - \phi)$. The displacement of the mass relative to its base is

$$E_p = \frac{\omega_i^2 x_i}{\sqrt{(\omega^2 - \omega_i^2)^2 + 4d^2\omega_i^2}} = \frac{(15^2)(10)}{\sqrt{(11.34^2 - 15^2)^2 + 4(3.217^2)(15^2)}} = 16.50\ in.$$

==<>==

Problem 10.71 The mass in Fig. 10.21 is 100 *kg*. The spring constant is $k = 4\ N/m$, and $c = 24\ N\text{-}s/m$. The base is subjected to an oscillatory displacement of circular frequency $\omega_i = 0.2\ rad/s$. The steady-state amplitude of the displacement of the mass relative to the base is measured and determined to be 200 *mm*. What is the magnitude of the displacement of the base?

Solution:

From Example 10.7 and the solution to Problem 10.70 the canonical form of the equation of motion is

$\frac{d^2x}{dt^2} + 2d\frac{dx}{dt} + \omega^2 x = a(t)$ where $d = \frac{c}{2m} = 0.12\ rad/s$, $\omega^2 = \frac{k}{m} = 0.04\ (rad/s)^2$, and (see Eq

(10.38) $a(t) = -\frac{d^2x_i}{dt^2} = x_i\omega_i^2 \sin(\omega_i t - \phi)$. The displacement of the mass relative to its base is

$$E_p = 0.2 = \frac{\omega_i^2 x_i}{\sqrt{(\omega^2 - \omega_i^2)^2 + 4d^2\omega_i^2}} = \frac{(0.2^2)x_i}{\sqrt{(0.2^2 - 0.2^2)^2 + 4(0.12^2)(0.2^2)}} = 0.8333x_i\ m,$$ from which

$$x_i = \frac{0.2}{0.8333} = 0.24\ m$$

==<>==

=================================<>=================================

Problem 10.72 A team of engineering students builds the simple seismograph shown. The coordinate x_i measures the local horizontal ground motion. The coordinate x measures the position of the of the mass relative to the frame of the seismograph. The spring is unstretched when $x = 0$. The mass $m = 1\ kg$, $k = 10\ N/m$, and $c = 2\ N\text{-}s/m$. Show that the seismograph is initially stationary and at $t = 0$ it is subjected to an oscillatory ground motion $x_i = 10\sin 2t\ mm$. What is the amplitude of the steady state response of the mass?

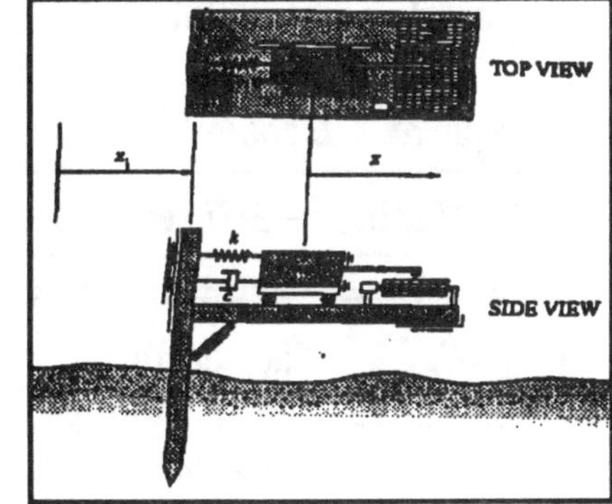

Solution:

From Example 10.7 and the solution to Problem 10.70 the canonical form of the equation of motion is

$$\frac{d^2x}{dt^2} + 2d\frac{dx}{dt} + \omega^2 x = a(t) \text{ where } d = \frac{c}{2m} = 1\ rad/s,$$

$\omega^2 = \frac{k}{m} = 10\ (rad/s)^2$, and (see Eq (10.38)

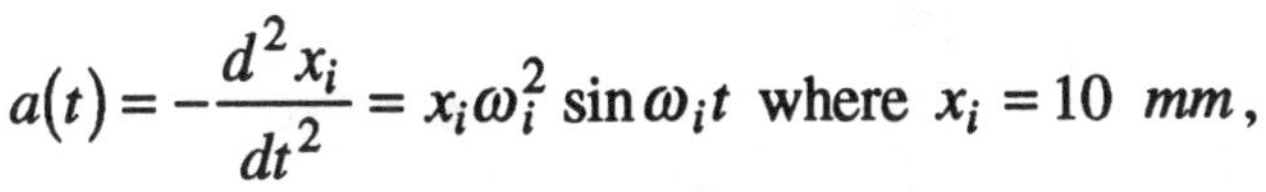

$$a(t) = -\frac{d^2x_i}{dt^2} = x_i\omega_i^2 \sin\omega_i t \text{ where } x_i = 10\ mm,$$

$\omega_i = 2\ rad/s$. The amplitude of the steady state response of the mass relative to its base is

$$\boxed{E_p = \frac{\omega_i^2 x_i}{\sqrt{(\omega^2 - \omega_i^2)^2 + 4d^2\omega_i^2}} = \frac{(2^2)(10)}{\sqrt{(3.162^2 - 2^2)^2 + 4(1^2)(2^2)}} = 5.547\ mm}$$

=================================<>=================================

Problem 10.73 In Problem 10.72, determine the position x of the mass relative to the base as a function of time.

Solution :

From Example 10.7 and the solution to Problem 10.70 the canonical form of the equation of motion is

$$\frac{d^2x}{dt^2} + 2d\frac{dx}{dt} + \omega^2 x = a(t) \text{ where } d = \frac{c}{2m} = 1\ rad/s,\ \omega^2 = \frac{k}{m} = 10\ (rad/s)^2, \text{ and}$$

$a(t) = -\frac{d^2x_i}{dt^2} = x_i\omega_i^2 \sin\omega_i t$ where $x_i = 10\ mm$, $\omega_i = 2\ rad/s$. From a comparison with Eq(10.27) and Eq (10.30), the particular solution is $x_p = A_p \sin\omega_i t + B_p \cos\omega_i t$, where

$$A_p = \frac{(\omega^2 - \omega_i^2)\omega_i^2 x_i}{(\omega^2 - \omega_i^2)^2 + 4d^2\omega_i^2} = 4.615.\ B_p = -\frac{2d\omega_i^3 x_i}{(\omega^2 - \omega_i^2)^2 + 4d^2\omega_i^2} = -3.077.$$

[*Check*: Assume a solution of the form $x_p = A_p \sin\omega_i t + B_p \cos\omega_i t$. Substitute into the equation of motion:

$$\left[(\omega^2 - \omega_i^2)A_p - 2d\omega_i B_p\right]\sin\omega_i t + \left[(\omega^2 - \omega_i^2)B_p + 2d\omega_i Ap\right]\cos\omega_i t = x_i\omega_i^2 \sin\omega_i t.$$

Equate like coefficients: $(\omega^2 - \omega_i^2)A_p - 2d\omega_i B_p = x_i\omega_i^2$, and $2d\omega_i A_p + (\omega^2 - \omega_i^2)B_p = 0$. Solve:

$$A_p = \frac{(\omega^2 - \omega_i^2)\omega_i^2 x_i}{(\omega^2 - \omega_i^2) + 4d^2\omega_i^2},\ B_p = \frac{-2d\omega_i^3 x_i}{(\omega^2 - \omega_i^2)^2 + 4d^2\omega_i^2}, \textit{ check.}]$$

Solution continued on next page

Since $d^2 < \omega^2$, the system is sub-critically damped, and the homogenous solution is $x_h = e^{-dt}(A\sin\omega_d t + B\cos\omega_d t)$, where $\omega_d = \sqrt{\omega^2 - d^2} = 3\ rad/s$. The complete solution is $x = x_h + x_p$. Apply the initial conditions: at $t = 0$, $x_o = 0$, $\dot{x}_o = 0$, from which $0 = B + B_p$, and $0 = -dB + \omega_d A + \omega_i A_p$. Solve: $B = -B_p = 3.077$, $A = \frac{dB}{\omega_d} - \frac{\omega_i A_p}{\omega_d} = -2.051$ The solution,

$$x = e^{-dt}(A\sin\omega t + B\cos\omega t) + A_p\sin\omega_i t + B_p\cos\omega_i t$$

$$\boxed{x = -e^{-t}(2.051\sin 3t - 3.077\cos 3t) + 4.615\sin 2t - 3.077\cos 2t\ mm}$$

=================================<>=================================

Problem 10.74 The coordinate x measures the displacement of the mass relative to the position in which the spring is unstretched. The mass is given; the initial conditions are: $t = 0$: $x = 0.1\ m$, $\frac{dx}{dt} = 0$.
(a) Determine the position of the mass as a function of time. (b) Draw graphs of the position and velocity of the mass as functions of time for the first 5 s of motion.

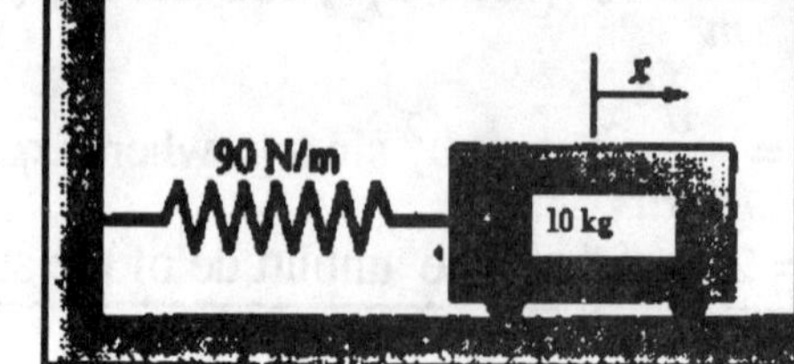

Solution:

The canonical equation of motion is

$$\frac{d^2x}{dt^2} + \omega^2 x = 0,\text{ where}$$

$\omega = \sqrt{\frac{k}{m}} = \sqrt{\frac{90}{10}} = 3\ rad/s$. (a) The position is $x(t) = A\sin\omega t + B\cos\omega t$, and the velocity is $\frac{dx(t)}{dt} = \omega A\cos\omega t - \omega B\sin\omega t$. At $t = 0$, $x(0) = 0.1 = B$, and $\frac{dx(0)}{dt} = 0 = \omega A$, from which $A = 0$, $B = 0.1\ m$, and $\boxed{x(t) = 0.1\cos 3t}$, $\frac{dx(t)}{dt} = -0.3\sin 3t$. (b) The graphs are shown.

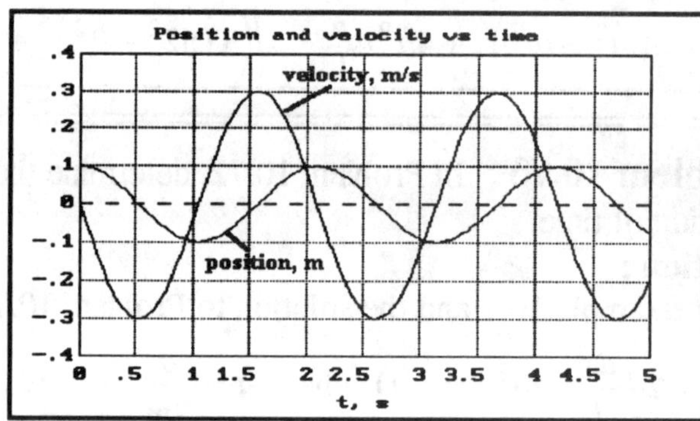

=================================<>=================================

==================================<>====================================

Problem 10.75 When $t = 0$, the mass in Problem 10.74 is in the position in which the spring is unstretched and has a velocity of 0.3 *m/s* to the right. Determine the position of the mass as functions of time and the amplitude of vibration; (a) by expressing the solution in the form given by Eq (10.5); (b) by expressing the solution in the form given by Eq (10.6).

Solution:

From Eq (10.4) the canonical form of the equation of motion is $\frac{d^2x}{dt^2} + \omega^2 x = 0$, where

$\omega = \sqrt{\frac{k}{m}} = \sqrt{\frac{90}{10}} = 3\ rad/s$.

(a) From Eq (10.5) the position is $x(t) = A\sin\omega t + B\cos\omega t$, and the velocity is $\frac{dx(t)}{dt} = \omega A\cos\omega t - \omega B\sin\omega t$. At $t = 0$, $x(0) = 0 = B$, and $\frac{dx(0)}{dt} = 0.3 = \omega A\ \ m/s$, from which $B = 0$ and $A = \frac{0.3}{\omega} = 0.1\ \ m$. The position is $\boxed{x(t) = 0.1\sin 3t\ \ m}$. The amplitude is $\boxed{|x(t)|_{\max} = 0.1\ \ m}$

(b) From Eq (10.6) the position is $x(t) = E\sin(\omega t - \phi)$, and the velocity is $\frac{dx(t)}{dt} = \omega E\cos(\omega t - \phi)$. At $t = 0$, $x(0) = -E\sin\phi$, and the velocity is $\frac{dx(0)}{dt} = 0.3 = \omega E\cos\phi$. Solve: $\phi = 0$, $E = \frac{0.3}{\omega} = 0.1$, from which $\boxed{x(t) = 0.1\sin 3t}$. The amplitude is $\boxed{|x(t)| = E = 0.1\ \ m}$

==================================<>====================================

Problem 10.76 A homogenous disk of mass *m* and radius *R* rotates about a fixed shaft and is attached to a torsional spring with constant *k*. (The torsional spring exerts a restoring moment of magnitude $k\theta$, where θ is the angle of rotation of the disk relative to its position in which the spring is unstretched.) Show that the period of rotational vibrations of the disk is $\tau = \pi R\sqrt{\frac{2m}{k}}$.

Solution :

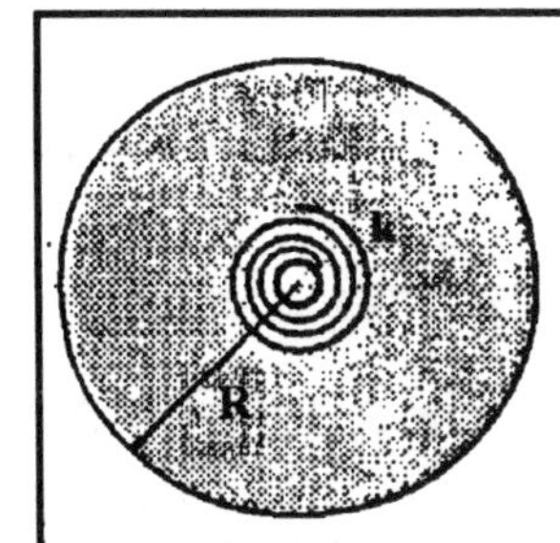

From the equation of angular motion , the equation of motion is $I\alpha = M$, where $M = -k\theta$, from which $I\frac{d^2\theta}{dt^2} + k\theta = 0$, and the canonical form (see Eq (10.4)) is $\frac{d^2\theta}{dt^2} + \omega^2\theta = 0$, where $\omega = \sqrt{\frac{k}{I}}$. For a homogenous disk the moment of inertia about the axis of rotation is $I = \frac{mR^2}{2}$, from which

$\omega = \sqrt{\frac{2k}{mR^2}} = \frac{1}{R}\sqrt{\frac{2k}{m}}$. The period is $\boxed{\tau = \frac{2\pi}{\omega} = 2\pi R\sqrt{\frac{m}{2k}} = \pi R\sqrt{\frac{2m}{k}}}$

==================================<>====================================

==================================<>==================================

Problem 10.77 Assigned to determine the moments of inertia of astronaut candidates, an engineer attaches a horizontal platform to a vertical steel bar. The moment of inertia of the platform about L is 7.5 $kg\text{-}m^2$, and the natural frequency of torsional oscillations of the unloaded platform is 1 Hz. With an astronaut candidate in the position shown, the natural frequency of torsional oscillations is 0.520 Hz. What is the candidate's moment of inertia about L?

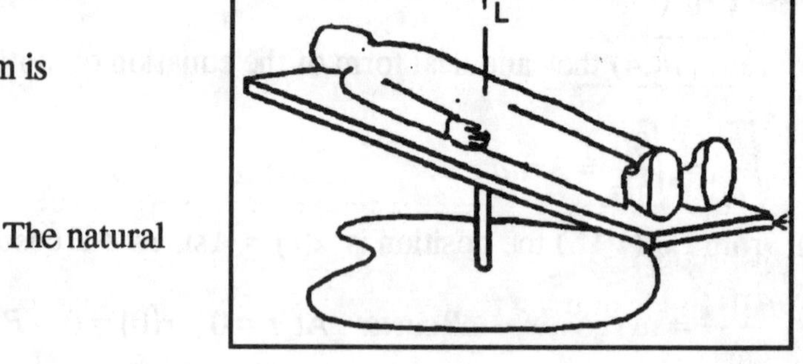

Solution:

The natural frequency of the unloaded platform is

$f = \frac{\omega}{2\pi} = \frac{1}{2\pi}\sqrt{\frac{k}{I}} = 1\, Hz$, from which

$k = (2\pi f)^2 I = (2\pi)^2 7.5 = 296.1\ N\text{-}m/rad$. The natural frequency of the loaded platform is

$f_1 = \frac{\omega_1}{2\pi} = \frac{1}{2\pi}\sqrt{\frac{k}{I_1}} = 0.520\, Hz$, from which

$I_1 = \left(\frac{1}{2\pi f_1}\right)^2 k = 27.74\ kg\text{-}m^2$, from which $\boxed{I_A = I_1 - I = 20.24\ kg\text{-}m^2}$

==================================<>==================================

Problem 10.78 The 22 kg platen P rests on four roller bearings. The roller bearings can be modeled as 1 kg homogenous cylinders with 30 mm radii, and the spring constant is $k = 900\ N/m$. What is the natural frequency of horizontal vibrations of the platen relative to its equilibrium position?

Solution:

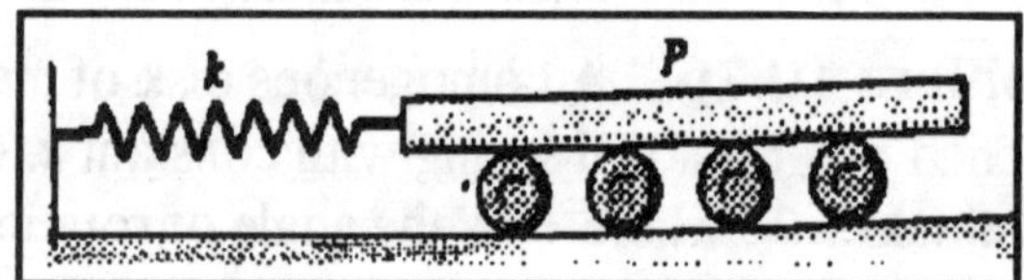

The kinetic energy is the sum of the kinetic energies of translation of the platen P and the roller bearings, and the kinetic energy of rotation of the roller bearings. Denote references to the platen by the subscript P and references to the ball bearings by the subscript B: $T = \frac{1}{2}m_P\left(\frac{dx_P}{dt}\right)^2 + \frac{1}{2}(4m_B)\left(\frac{dx_B}{dt}\right)^2 + \frac{1}{2}(4I_B)\left(\frac{d\theta}{dt}\right)^2$. The potential energy is the energy stored in the spring: $V = \frac{1}{2}kx_P^2$. From kinematics, $-R\theta = x_B$ and $x_B = \frac{x_P}{2}$.

Since the system is conservative $T + V = const.$ Substitute the kinematic relations and reduce:

$\left(\frac{1}{2}\right)\left(m_P + m_B + \frac{I_B}{R^2}\right)\left(\frac{dx_P}{dt}\right)^2 + \left(\frac{1}{2}\right)kx_P^2 = const.$. Take the time derivative:

$\left(\frac{dx}{dt}\right)\left[\left(m_P + m_B + \frac{I_B}{R^2}\right)\left(\frac{d^2x_P}{dt^2}\right) + kx\right] = 0$. This has two possible solutions, $\left(\frac{dx}{dt}\right) = 0$ or

$\left(m_P + m_B + \frac{I_B}{R^2}\right)\left(\frac{d^2x_P}{dt^2}\right) + kx_P = 0$. The first can be ignored (see solution to Problem 10.15) from

Solution continued on next page

which the canonical form of the equation of motion is $\frac{d^2 x_P}{dt^2} + \omega^2 x_P = 0$, where

$\omega = R\sqrt{\frac{k}{R^2(m_P + m_B) + I_B}}$. For a homogenous cylinder, $I_B = \frac{m_B R^2}{2}$, from which

$\omega = \sqrt{\frac{k}{m_P + \frac{3}{2} m_B}} = 6.189\ rad/s$. The natural frequency is $\boxed{f = \frac{\omega}{2\pi} = 0.9849\ Hz}$.

====================================<>====================================

Problem 10.79 At $t = 0$ the platen described in Problem 10.78 is 0.1 *m* to the left of its equilibrium position and is moving to the right at 2 *m/s*. What are the platen's position and velocity at $t = 4\ s$.

Solution:

The position is $x(t) = A\sin\omega t + B\cos\omega t$, and the velocity is $\frac{dx}{dt} = \omega A\cos\omega t - \omega B\sin\omega t$, where, from the solution to Problem 10.78, $\omega = 6.189\ rad/s$. At $t = 0$, $x(0) = -0.1\ m$, and $\left[\frac{dx}{dt}\right]_{t=0} = 2\ m/s$, from which $B = -0.1\ m$, $A = \frac{2}{\omega} = 0.3232\ m$. The position and velocity are

$x(t) = 0.3232\sin(6.189t) - 0.1\cos(6.189t)\ (m)$, $\frac{dx}{dt} = 2\cos(6.189t) + 0.6189\sin(6.189t)\ (m/s)$. At

$\boxed{x(4) = -0.2124\ m}$, $\boxed{\left[\frac{dx}{dt}\right]_{t=4} = 1.630\ m/s}$.

====================================<>====================================

==================================<>==================================

Problem 10.80 The moments of inertia of gears A and B are $I_A = 0.014\ slug\text{-}ft^2$ and $I_B = 0.100\ slug\text{-}ft^2$. Gear A is attached to a torsional spring with constant $k = 2\ ft\text{-}lb/rad$. What is the natural frequency of small angular vibrations of the gears relative to their equilibrium position?

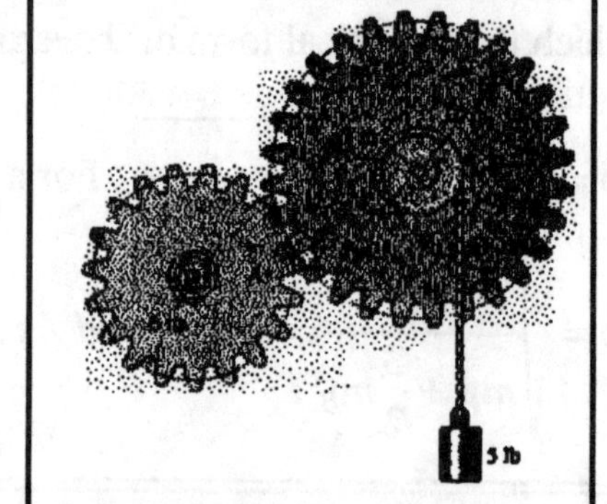

Solution :

The system is conservative. The strategy is to determine the equilibrium position from the equation of motion about the unstretched spring position. Choose a coordinate system with the y axis positive upward. Denote $R_A = 6\ in. = 0.5\ ft$, $R_B = 10\ in. = 0.8333\ ft$, and $R_M = 3\ in. = 0.25\ ft$, and $W = 5\ lb$. The kinetic energy of the system is $T = \frac{1}{2} I_A \dot{\theta}_A^2 + \frac{1}{2} I_B \dot{\theta}_B^2 + \frac{1}{2}\frac{W}{g} v^2$, where $\dot{\theta}_A, \dot{\theta}_B$ are the angular velocities of gears A and B respectively, and v is the velocity of the 5 lb weight. The potential energy is the sum of the energy stored in the spring plus the energy gain due to the increase in the height of the 5 lb weight: $V = \frac{1}{2} k\theta_A^2 + Wy$. From kinematics, $v = R_M \dot{\theta}_B$, $\dot{\theta}_B = -\left(\frac{R_A}{R_B}\right)\dot{\theta}_A$, $y = R_M \theta_B = -R_M\left(\frac{R_A}{R_B}\right)\theta_A$. Substitute,

$$T + V = const. = \left(\frac{1}{2}\right)\left[I_A + \left(\frac{R_A}{R_B}\right)^2 I_B + \left(\frac{W}{g}\right)\left(R_M^2\right)\left(\frac{R_A}{R_B}\right)^2\right]\dot{\theta}_A^2 + \left(\frac{1}{2}\right)k\theta_A^2 - \left(R_M\right)\left(\frac{R_A}{R_B}\right)\theta_A W.$$

Define the *generalized mass moment of inertia* by $M = I_A + \left(\frac{R_A}{R_B}\right)^2 I_B + \frac{W}{g}\left(R_M^2\right)\left(\frac{R_A}{R_B}\right)^2 = 0.05350\ slug\text{-}ft^2$, and take the time derivative: $\dot{\theta}_A\left[M\left(\frac{d^2\theta_A}{dt^2}\right) + k\theta_A - W\left(R_M\right)\left(\frac{R_A}{R_B}\right)\right] = 0.$

Ignore the possible solution $\dot{\theta}_A = 0$, to obtain $\frac{d^2\theta_A}{dt^2} + \omega^2\theta_A = F$, where $F = \frac{WR_M}{M}\left(\frac{R_A}{R_B}\right)$, and $\omega = \sqrt{\frac{k}{M}} = 6.114\ rad/s$, is the equation of motion *about the unstretched spring position.* Note that $\frac{F}{\omega^2} = \frac{W}{k} R_M\left(\frac{R_A}{R_B}\right) = 0.375\ rad$ is the equilibrium position of θ_A, obtained by setting the acceleration to zero (since the non-homogenous term F is a constant). Make the change of variable: $\tilde{\theta} = \theta_A - \frac{F}{\omega^2}$, from which the canonical form (see Eq(10.4))of the equation of motion *about the equilibrium point* is $\frac{d^2\tilde{\theta}}{dt^2} + \omega^2\tilde{\theta} = 0$, and the natural frequency is $\boxed{f = \frac{\omega}{2\pi} = 0.9732\ H_z}$.

==================================<>==================================

====================================<>====================================

Problem 10.81 The 5 *lb* weight in Problem 10.80 is raised 0.5 *in.* from its equilibrium position and released from rest at $t = 0$. Determine the counterclockwise angular position of gear *B* relative to its equilibrium position as a function of time.

Solution:

From the solution to Problem 10.80, the equation of motion for gear A is $\frac{d^2\theta_A}{dt^2} + \omega^2\theta_A = F$, where

$$M = I_A + \left(\frac{R_A}{R_B}\right)^2 I_B + \frac{W}{g}\left(R_M^2\right)\left(\frac{R_A}{R_B}\right)^2 = 0.05350 \ slug\text{-}ft^2,\ \omega = \sqrt{\frac{k}{M}} = 6.114 \ rad/s, \text{ and}$$

$$F = \frac{WR_M\left(\frac{R_A}{R_B}\right)}{M} = 14.02 \ rad/s^2.$$

As in the solution to Problem 10.80, the *equilibrium angular position* θ_A associated with the equilibrium position of the weight is not the angular position associated with the unstretched spring but rather: $[\theta_A]_{eq} = \frac{F}{\omega^2} = 0.375 \ rad$. Make the change of variable: $\tilde{\theta}_A = \theta_A - [\theta_A]_{eq}$, from which the canonical form (see Eq(10.4))of the equation of motion *about the equilibrium point* is $\frac{d^2\tilde{\theta}_A}{dt^2} + \omega^2\tilde{\theta}_A = 0$. Assume a solution of the form $\tilde{\theta}_A = A\sin\omega t + B\cos\omega t$. The displacement from the equilibrium position is , from kinematics,

$\tilde{\theta}_A(t=0) = \left(\frac{R_B}{R_A}\right)\theta_A = -\frac{1}{R_M}\left(\frac{R_B}{R_A}\right)y(t=0) = -0.2778 \ rad$, from which the initial conditions are

$\tilde{\theta}_A(t=0) = -0.2778 \ rad$ and $\left[\frac{d\tilde{\theta}_A}{dt}\right]_{t=0} = 0$, from which $B = -0.2778$, and $A = 0$. The angular position of gear A is $\theta_a = -0.2778\cos(6.114\,t) \ rad$, from which the angular position of gear B is

$$\boxed{\tilde{\theta}_B = -\left(\frac{R_A}{R_B}\right)\tilde{\theta}_A = 0.1667\cos(6.114\,t) \ rad}$$

about the equilibrium position. [*Check*: At $t = 0$, $y = R_M\theta_B(t=0) = +0.04167 \ ft = +0.5 \ in.$ *check*:]

====================================<>====================================

=================================<>=================================

Problem 10.82 The mass of the slender bar is *m*. The spring is unstretched when the bar is vertical. The light collar slides on the smooth vertical bar so that the spring remains horizontal. Determine the natural frequency of small vibrations of the bar.

Solution:

The system is conservative. Denote the angle between the bar and the vertical by θ. The base of the bar is a fixed point. The kinetic energy of the bar is $T = \frac{1}{2} I \left(\frac{d\theta}{dt}\right)^2$. Denote the datum by $\theta = 0$. The potential energy is the result of the change in the height of the center of mass of the bar from the datum and the stretch of the spring, $V = -\frac{mgL}{2}(1-\cos\theta) + \frac{1}{2}k(L\sin\theta)^2$. The system is conservative, $T + V = const$ $= \frac{1}{2} I \left(\frac{d\theta}{dt}\right)^2 + \frac{kL^2}{2}\sin^2\theta - \frac{mgL}{2}(1-\cos\theta)$. Take the time derivative:

$$\left(\frac{d\theta}{dt}\right)\left[I\frac{d^2\theta}{dt^2} + kL^2 \sin\theta\cos\theta - \frac{mgL}{2}\sin\theta \right] = 0.$$ From which

$I\frac{d^2\theta}{dt^2} + kL^2 \sin\theta\cos\theta - \frac{mgL}{2}\sin\theta = 0$. For small angles, $\sin\theta \to \theta$, $\cos\theta \to 1$. The moment of inertia about the fixed point is $I = \frac{mL^2}{3}$, from which (see Eq (10.4)) $\frac{d^2\theta}{dt^2} + \omega^2\theta = 0$, where $\omega = \sqrt{\frac{3k}{m} - \frac{3g}{2L}}$.

The natural frequency is $\boxed{f = \frac{\omega}{2\pi} = \frac{1}{2\pi}\sqrt{\frac{3k}{m} - \frac{3g}{2L}}}$

=================================<>=================================

Problem 10.83 The homogenous hemisphere of radius R and mass *m* rests on a level surface. If you rotate the hemisphere slightly from its equilibrium position and release it, what is the natural frequency of its vibrations?

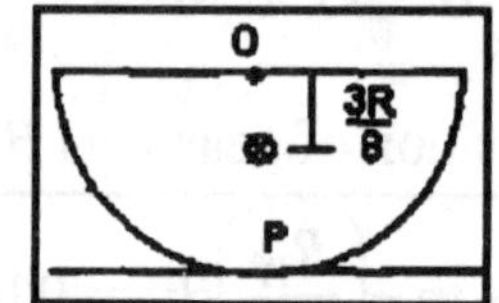

Solution:

The system is conservative. The distance from the center of mass to point O is $h = 3R/8$. Denote the angle of rotation about P by θ. Rotation about P causes the center of mass to rotate relative to the radius center OP, suggesting the analogy with a pendulum suspended from O. The kinetic energy is $T = (1/2)I_p\dot{\theta}^2$. The potential energy is $V = mgh(1-\cos\theta)$, where $h(1-\cos\theta)$ is the increase in height of the center of mass. $T + V = const = I_p\dot{\theta}^2 + mgh(1-\cos\theta)$. Take the time derivative:

$\dot{\theta}\left[I_P \frac{d^2\theta}{dt^2} + mgh\sin\theta \right] = 0$, from which $\frac{d^2\theta}{dt^2} + \frac{mgh}{I_P}\sin\theta = 0$. For small angles $\sin\theta \to \theta$, and the moment of inertia about P is $I_P = I_{CM} + m(R-h)^2 = \frac{83}{320}mR^2 + \left(\frac{5}{8}\right)^2 mR^2 = \frac{13}{20}mR^2$, from which

$\frac{d^2\theta}{dt^2} + \omega^2\theta = 0$, where $\omega = \sqrt{\frac{20(3)g}{13(8)R}} = \sqrt{\frac{15g}{26R}}$. The natural frequency is $\boxed{f = \frac{1}{2\pi}\sqrt{\frac{15g}{26R}}}$

=================================<>=================================

Problem 10.84 The frequency of the spring mass oscillator is measured and determined to be 4 *Hz*. The spring mass oscillator is then immersed in oil, and its frequency is determined to be 3.8 *Hz*. What is the logarithmic decrement of vibrations of the mass when the oscillator is immersed in oil?

Solution :

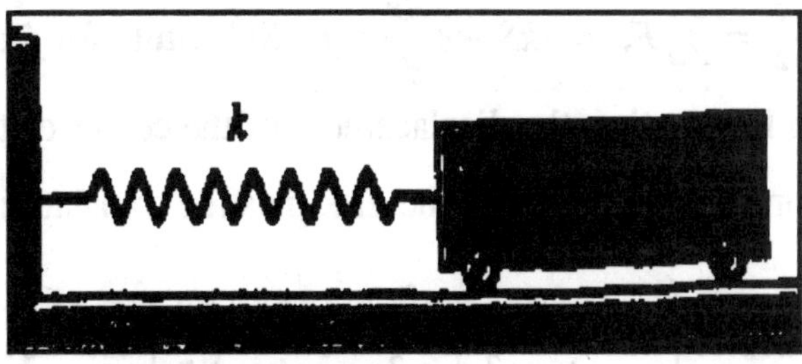

The logarithm decrement is defined by $\delta = d\tau$, where

$d = \sqrt{\omega^2 - \omega_d^2} = 2\pi\sqrt{4^2 - 3.8^2} = 7.848 \ rad/s$,

and $\tau = \dfrac{2\pi}{\omega_d} = 2.632 \ s$. ($\omega$ and ω_d are defined by Eq (10.16) and Eq (10.18)). The logarithmic decrement is $\boxed{\delta = d\tau = 2.0652}$.

Problem 10.85 Consider the oscillator immersed in oil described in Problem 10.84. If the mass is displaced 0.1 *m* to the right of its equilibrium position and released from rest, what is its position relative to the equilibrium position as a function of time?

Solution:

The mass and spring constant are unknown. The canonical form of the equation of motion (see Eq (10.16)) is $\dfrac{d^2x}{dt^2} + 2d\dfrac{dx}{dt} + \omega^2 x = 0$, where, from the solution to Problem 10.84, $d = 7.848 \ rad/s$, and $\omega = 2\pi(4) = 25.13 \ rad/s$. The solution is of the form (see Eq (10.19)) $x = e^{-dt}(A\sin\omega_d t + B\cos\omega_d t)$, where $\omega_d = 2\pi(3.8) = 23.88 \ rad/s$. Apply the initial conditions: at $t = 0$, $x_o = 0.1 \ m$, and $\dot{x}_o = 0$, from which $B = x_o$, and $0 = -dB + \omega_d A$, from which $A = \dfrac{dx_o}{\omega_d}$, from which the solution is

$$x = x_o e^{-dt}\left(\frac{d}{\omega_d}\sin\omega_d + \cos\omega_d\right) \boxed{= 0.1e^{-7.848t}(0.3287\sin 23.88t + \cos 23.88t)}$$

Problem 10.86 The stepped disk weighs 20 *lb*, and its moment of inertia is $I = 0.6 \ slug\text{-}ft^2$. It rolls on the horizontal surface. If $c = 8 \ lb\text{-}s/ft$, what is the frequency of small vibrations?

Solution :

(See the solution to Problem 10.63). The strategy is to apply the free body diagram to obtain equations for both θ and x, and then to eliminate one of these. An essential element in the strategy is the determination of the stretch of the spring. Denote $R = 8 \ in. = 0.6667 \ ft$, and the stretch of the spring by S. Choose a coordinate system with the positive x axis to the right. The sum of the moments about the center of the disk is

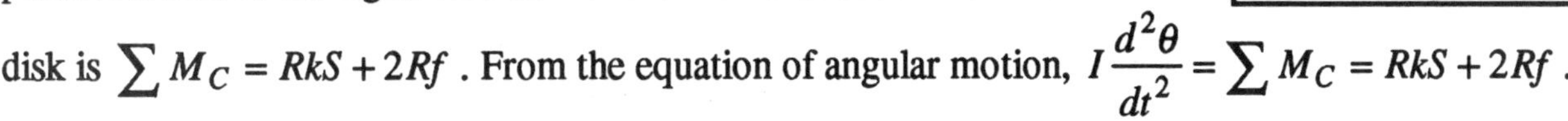

$\sum M_C = RkS + 2Rf$. From the equation of angular motion, $I\dfrac{d^2\theta}{dt^2} = \sum M_C = RkS + 2Rf$.

Solve for the reaction at the floor: $f = \dfrac{I}{2R}\dfrac{d^2\theta}{dt^2} - \dfrac{k}{2}S$.

Solution continued on next page

Continuation of solution to Problem 10.86

The sum of the horizontal forces: $\sum F_x = -kS - c\frac{dx}{dt} + f$. From Newton's second law:

$m\frac{d^2x}{dt^2} = \sum F_x = -kS - c\frac{dx}{dt} + f$. Substitute for f and rearrange: $m\frac{d^2x}{dt^2} + \frac{I}{2R}\frac{d^2\theta}{dt^2} + c\frac{dx}{dt} + \frac{3}{2}kS = 0$

From kinematics, the displacement of the center of the disk is $x = -2R\theta$. The stretch of the spring is the amount wrapped around the disk plus the translation of the disk, $S = -R\theta - 2R\theta = -3R\theta = \frac{3}{2}x$.

Substitute: $\left(m + \frac{I}{(2R)^2}\right)\frac{d^2x}{dt^2} + c\frac{dx}{dt} + \left(\frac{3}{2}\right)^2 kx = 0$. Define the *generalized mass* by

$M = m + \frac{I}{(2R)^2} = 0.9592\ \ slug$. The canonical form (see Eq (10.16)) of the equation of motion is

$\frac{d^2x}{dt^2} + 2d\frac{dx}{dt} + \omega^2 x = 0$, where $d = \frac{c}{2M} = 4.170\ \ rad/s$, $\omega^2 = \left(\frac{3}{2}\right)^2 \frac{k}{M} = 37.53\ \ (rad/s)^2$. The

natural circular frequency is $\omega_d = \sqrt{\omega^2 - d^2} = 4.488\ rad/s$, and the natural frequency is

$$\boxed{f_d = \frac{\omega_d}{2\pi} = 0.7143\ Hz}$$

================================<>================================

Problem 10.87 The stepped disk described in Problem 10.86 is initially in equilibrium, and at $t = 0$ it is given a clockwise angular velocity of 1 *rad/s*. Determine the position of the center of the disk relative to its equilibrium position as a function of time.

Solution:

From the solution to 10.86, the equation of motion is $\frac{d^2x}{dt^2} + 2d\frac{dx}{dt} + \omega^2 x = 0$, where

$d = 4.170\ \ rad/s$, $\omega^2 = 37.53\ \ (rad/s)^2$. The natural circular frequency is

$\omega_d = \sqrt{\omega^2 - d^2} = 4.488\ rad/s$. Since $d^2 < \omega^2$, the system is sub-critically damped. The solution is of the form $x = e^{-dt}(A\sin\omega_d t + B\cos\omega_d t)$. Apply the initial conditions: at $t = 0$, $\theta_o = 0$, and $\dot{\theta}_o = -1\ \ rad/s$. From kinematics, $\dot{x}_o = -2R\dot{\theta}_o = 2R\ ft$. Substitute, to obtain $B = 0$ and

$A = \frac{\dot{x}_o}{\omega_d} = 0.2971$, and the position of the center of the disk is $\boxed{x = 0.2971e^{-4.170t}\sin 4.488t}$

================================<>================================

==================================<>==================================

Problem 10.88 The stepped disk described in Problem 10.86 is initially in equilibrium, and at $t = 0$ it is given a clockwise angular velocity of 1 *rad/s*. Determine the position of the center of the disk relative to its equilibrium position as a function of time if $c = 16\ lb\text{-}s/ft$.

Solution:

From the solution to Problem 10.86, the canonical form (see Eq (10.16)) of the equation of motion is $\frac{d^2x}{dt^2} + 2d\frac{dx}{dt} + \omega^2 x = 0$, where $d = \frac{c}{2M} = 8.340\ rad/s$, $\omega^2 = \left(\frac{3}{2}\right)^2 \frac{k}{M} = 37.53\ (rad/s)^2$. Since $d^2 > \omega^2$, the system is supercritically damped. The solution is of the form (see Eq (10.24)) $x = e^{-dt}(Ce^{ht} + De^{-ht})$, where $h = \sqrt{d^2 - \omega^2} = 5.659\ rad/s$.

Apply the initial conditions: Apply the initial conditions: at $t = 0$, $\theta_o = 0$, and $\dot{\theta}_o = -1\ rad/s$. From kinematics, $\dot{x}_o = -2R\dot{\theta}_o = 2R\ ft$. Substitute, to obtain $0 = C + D$ and $\dot{x}_o = -(d-h)C - (d+h)D$.

Solve: $C = \frac{x_o}{2h} = 0.1178$, $D = -\frac{x_o}{2h} = -0.1178$, from which the position of the center of the disk is

$$x = 0.1179e^{-8.340t}\left(e^{5.659t} - e^{-5.659t}\right) = 0.1179\left(e^{-2.680t} - e^{-14.00t}\right)\ ft$$

==================================<>==================================

Problem 10.89 The 22 *kg* platen P rests on four roller bearings. The roller bearings can be modeled as 1 *kg* homogenous cylinders with 30 *mm* radii, and the spring constant is $k = 900\ N/m$. The platen is subjected to a force $F(t) = 100\sin 3t\ N$. What is the magnitude of the platen's steady state horizontal vibration?

Solution:

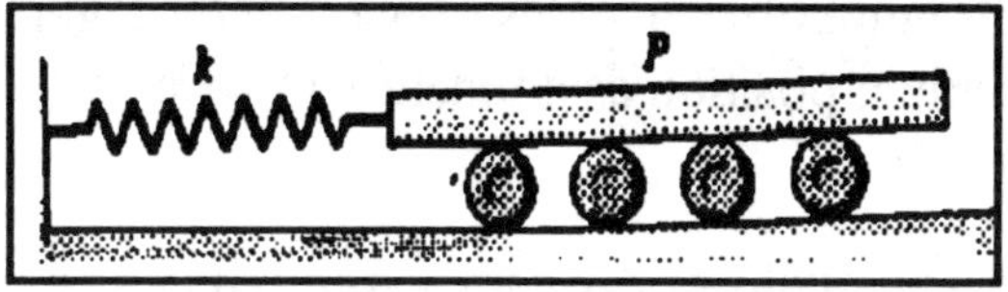

Choose a coordinate system with the origin at the wall and the *x* axis parallel to the plane surface. Denote the roller bearings by the subscript *B* and the platen by the subscript *P*.

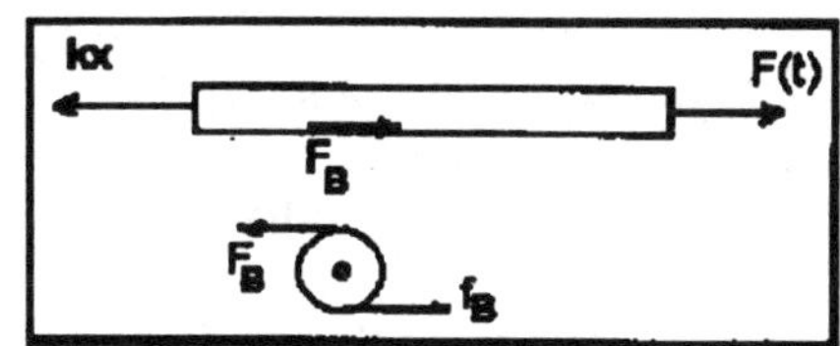

The roller bearings: The sum of the moments about the mass center of a roller bearing is $\sum M_{B-cm} = +RF_B + Rf_B$. From Newton's second law: $I_B \frac{d^2\theta}{dt^2} = RF_B + Rf_B$. Solve for the reaction at the floor: $f_B = \frac{I_B}{R}\frac{d^2\theta}{dt^2} - F_B$. The sum of the *horizontal* forces on each roller bearing: $\sum F_x = -F_B + f_P$. From Newton's second law $m_B \frac{d^2x_B}{dt^2} = -F_B + f_B$, where x_B is the translation of the center of mass of the roller bearing. Substitute f_P, $m_B \frac{d^2x_B}{dt^2} = \frac{I_B}{R}\frac{d^2\theta}{dt^2} - 2F_B$. From kinematics, $\theta_B = -\frac{x_B}{R}$, from which $\left(m_B + \frac{I_B}{R^2}\right)\frac{d^2x_B}{dt^2} = -2F_B$.

The platen: The sum of the forces on the platen are $\sum F_P = -kx + 4F_B + F(t)$.

Solution continued on next page

From Newton's second law, $m_P \frac{d^2 x_P}{dt^2} = -kx_p + 4F_B + F(t)$. Substitute for F_B and rearrange:

$m_p \frac{d^2 x_P}{dt^2} + kx_P + 2\left(m_B + \frac{I_B}{R^2}\right)\frac{d^2 x_B}{dt^2} = F(t)$. From kinematics, $x_B = \frac{x_P}{2}$, from which

$\left(m_P + m_B + \frac{I_B}{R^2}\right)\frac{d^2 x_P}{dt^2} + kx_P = F(t)$. [*Check*: The homogenous equation agrees with that in the solution to Problem 10.17 (which was obtained by an alternate method), as it should. *check.*]

For a homogenous cylinder, $I_B = \frac{m_B R^2}{2}$, from which the *generalized mass* is defined

$M = m_p + \frac{3}{2}m_B = 23.5\ kg$. For $d = 0$, the canonical form of the equation of motion (see Eq (10.26) is

$\frac{d^2 x_P}{dt^2} + \omega^2 x_p = a(t)$, where $\omega^2 = 38.30\ (rad/s)^2$, and $a(t) = \frac{F(t)}{M} = 4.255 \sin 3t\ (m/s^2)$. The amplitude of the steady state motion is given by Eq (10.31) $\boxed{E_p = \frac{4.255}{(\omega^2 - 3^3)} = 0.1452\ m}$

===================================<>===================================

Problem 10.90 At $t = 0$ the platen described in Problem 10.89 is 0.1 m to the right of its equilibrium position and is moving to the right at 2 *m/s*. Determine the platen's position relative to its equilibrium position as a function of time.

Solution :

From the solution to Problem 10.81, the equation of motion is $\frac{d^2 x_P}{dt^2} + \omega^2 x_p = a(t)$, where

$\omega^2 = 38.30\ (rad/s)^2$, and $a(t) = \frac{F(t)}{M} = 4.255 \sin 3t\ (m/s^2)$. The solution is in the form $x = x_h + x_p$, where the homogenous solution is of the form $x_h = A \sin \omega t + B \cos \omega t$ and the particular solution x_p is given by Eq (10.30), with $d = 0$ and $b_o = 0$. The result: $x = A \sin \omega t + B \cos \omega t + \frac{a_o}{(\omega^2 - \omega_o^2)} \sin \omega_o t$,

where $a_o = 4.255\ m$, $\omega = 6.189\ rad/s$, and $\omega_o = 3\ rad/s$. Apply the initial conditions: at $t = 0$, $x_o = 0.1\ m$, and $\dot{x}_o = 2\ m/s$, from which $B = 0.1$, and $A = \frac{2}{\omega} - \left(\frac{\omega_o}{\omega}\right)\frac{a_o}{(\omega^2 - \omega_o^2)} = 0.2528$, from which $\boxed{x = 0.2528 \sin 6.189t + 0.1 \cos 6.189t + 0.1452 \sin 3t\ \ m}$

===================================<>===================================

==================================<>==================================

Problem 10.91 The moments of inertia of gears A and B are $I_A = 0.014\ slug\text{-}ft^2$ and $I_B = 0.100\ slug\text{-}ft^2$. Gear A is connected to a torsional spring with constant $k = 2\ ft\text{-}lb/rad$ .m The bearing supporting gear B incorporates a damping element that exerts a resisting moment on gear B of magnitude $1.5\left(\frac{d\theta_B}{dt}\right)\ ft\text{-}lb$, where $\frac{d\theta_B}{dt}$ is the angular velocity of gear B in *rad/s*. What is the frequency of small angular vibrations of the gears?

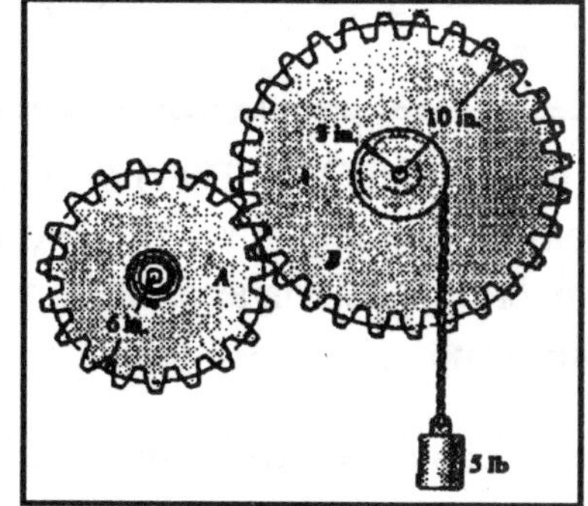

Solution :

(See the solution to Problem 10.29.) Choose a coordinate system with the x axis positive downward. The sum of the moments on gear A is $\sum M = -k\theta_A + R_A F$, where the moment exerted by the spring opposes the angular displacement θ_A. From the equation of angular motion, $I_A \frac{d^2\theta_A}{dt^2} = \sum M = -k\theta_A + R_A F$, from which

$$F = \left(\frac{I_A}{R_A}\right)\frac{d^2\theta_A}{dt^2} + \left(\frac{k}{R_A}\right)\theta_A.$$

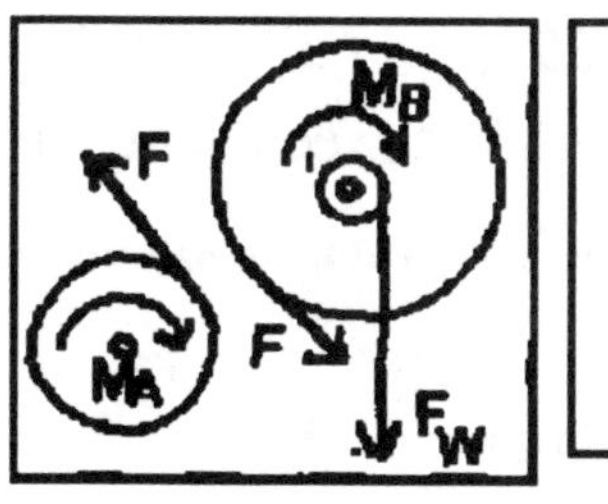

The sum of the moments acting on gear B is $\sum M = -1.5\frac{d\theta_B}{dt} + R_B F - R_W F_W$, where $W = 5\ lb$, and the moment exerted by the damping element opposes the angular velocity of B. From the equation of angular motion applied to B, $I_B \frac{d^2\theta_B}{dt^2} = \sum M = -1.5\frac{d\theta_B}{dt} + R_B F - R_W F_W$.

The sum of the forces on the weight are $\sum F = +F_W - W$. From Newton's second law applied to the weight, $\left(\frac{W}{g}\right)\frac{d^2x}{dt^2} = F_W - W$, from which $F_W = \left(\frac{W}{g}\right)\frac{d^2x}{dt^2} + W$. Substitute for F and F_W to obtain the equation of motion for gear B:

$$I_B \frac{d^2\theta_B}{dt^2} + 1.5\frac{d\theta_B}{dt} - \left(\frac{R_B}{R_A}\right)\left(I_A \frac{d^2\theta_A}{dt^2} + k\theta_A\right) - R_W\left(\left(\frac{W}{g}\right)\frac{d^2x}{dt^2} + W\right) = 0$$

. From kinematics, $\theta_A = -\left(\frac{R_B}{RA}\right)\theta_B$, and $x = -R_W\theta_B$, from which $M\ \frac{d^2\theta_B}{dt^2} + 1.5\frac{d\theta_B}{dt} + \left(\frac{R_B}{R_A}\right)^2 k\theta_B = R_W W$, where

$$M = I_A + \left(\frac{R_B}{R_A}\right)^2 I_A + R_W^2\left(\frac{W}{g}\right) = 0.1486\ slug\text{-}ft^2$$

The canonical form of the equation of motion is

$$\frac{d^2\theta_B}{dt^2} + 2d\frac{d\theta_B}{dt} + \omega^2\theta_B = P\text{, where } d = \frac{1.5}{2M} = 5.047\ rad/s,\ \omega^2 = \frac{\left(\frac{R_B}{R_A}\right)^2 k}{M} = 37.39\ (rad/s)^2,$$

Solution continued on next page

and $P = \dfrac{R_W W}{M} = 8.412\ (rad/s)^2$. The system is sub critically damped, since $d^2 < \omega^2$, from which $\omega_d = \sqrt{\omega^2 - d^2} = 3.452\ rad/s$, and the natural frequency is $\boxed{f_d = \dfrac{\omega_d}{2\pi} = 0.5493\ Hz}$

==========================<>==========================

Problem 10.92 The 5 *lb* weight in Problem 10.91 is raised 0.5 *in.* from its equilibrium position and released from rest at $t = 0$. Determine the counterclockwise angular position of gear B relative to its equilibrium position as a function of time.

Solution:

From the solution to Problem 10.91, $\dfrac{d^2\theta_B}{dt^2} + 2d\dfrac{d\theta_B}{dt} + \omega^2\theta_B = P$, where $d = 5.047\ rad/s$, $\omega^2 = 37.39\ (rad/s)^2$, and $P = 8.412\ (rad/s)^2$. Since the non homogenous term P is independent of time and angle, the equilibrium position is found by setting the acceleration and velocity to zero in the equation of motion and solving: $\theta_{eq} = \dfrac{P}{\omega^2}$. Make the transformation $\tilde{\theta}_B = \theta_B - \theta_{eq}$, from which, by substitution, $\dfrac{d^2\tilde{\theta}_B}{dt^2} + 2d\dfrac{d\tilde{\theta}_B}{dt} + \omega^2\tilde{\theta} = 0$, is the equation of motion about the equilibrium point.

Since $d^2 < \omega^2$, the system is sub critically damped, from which the solution is $\tilde{\theta}_B = e^{-dt}(A\sin\omega_d t + B\cos\omega_d t)$. Apply the initial conditions: $x_o = -0.5\ in. = -0.04167\ ft$, from which $\left[\tilde{\theta}_B\right]_{t=0} = -\dfrac{x_o}{R_w} = 0.1667\ rad$, $\dfrac{d\tilde{\theta}_B}{dt} = 0$, from which $B = 0.1667$, $A = \dfrac{d(0.1667)}{\omega_d} = 0.2437$. The solution is $\boxed{\tilde{\theta}(t) = e^{-5.047t}(0.2437\sin(3.452t) + 0.1667\cos(3.452t))}$

==========================<>==========================

Problem 10.93 The base and mass are initially stationary. The base is subjected to a vertical displacement of $h = \sin\omega_i t$ relative to its original position. What is the magnitude of the resulting steady state vibration of the mass m relative to its base?

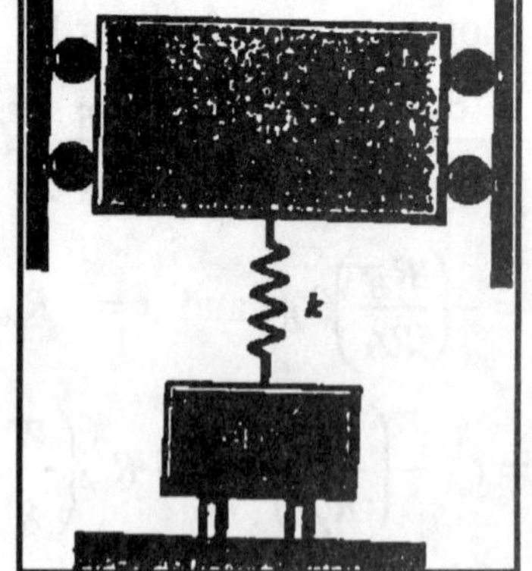

Solution:

From Eq (10.26), for $d = 0$, $\dfrac{d^2x}{dt^2} + \omega^2 x = a(t)$, where $\omega^2 = \dfrac{k}{m}$, and $a(t) = \omega_i^2 h\sin\omega_i t$. From Eq (10.31), the steady state amplitude is

$$\boxed{E_p = \frac{\omega_i^2 h}{(\omega^2 - \omega_i^2)} = \frac{\omega_i^2 h}{\left(\frac{k}{m} - \omega_i^2\right)}}$$

==========================<>==========================

==================================<>==================================

Problem 10.94 The mass of the trailer, not including its wheels and axle, is *m*, and the spring suspension is *k*. To analyze its behavior, and engineer assumes that the height of the road surface relative to its mean height is $h = \sin\left(\frac{2\pi x}{\lambda}\right)$. Assume that the trailer's wheels remain on the road and its horizontal component of velocity is *v*. Neglect the damping due to the suspension's shock absorbers. (a) Determine the magnitude of the trailer's vertical steady state vibration *relative to the road surface.* (b) At what velocity *v* does resonance occur?

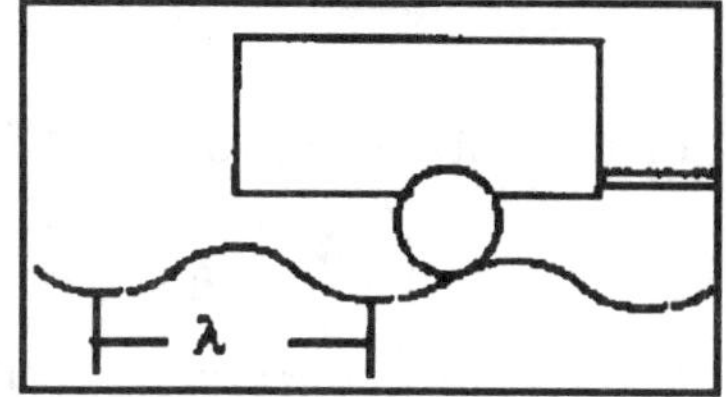

Solution:

Since the wheels and axle act as a base that moves with the disturbance, this is analogous to the transducer problem (Example 10.7). For a *constant velocity* the distance $x = \int_0^t v dt = vt$, from which the movement of the axle-wheel assembly as a function of time is $h_f(t) = h\sin(\omega_o t)$, where $\omega_o = \frac{2\pi v}{\lambda}$ rad/s. [*Check*: The velocity of the disturbing "waves" in the road relative to the trailer is *v*. Use the physical relationsip between frequency, wavelength and velocity of propagation $\lambda f = v$. The wavelength of the road disturbance is λ, from which the forcing function frequency is $f_o = \frac{v}{\lambda}$, and the circular frequency is $\omega_o = 2\pi f_o = \frac{2\pi v}{\lambda}$. *check.*] The forcing function on the spring-mass oscillator (that is, the trailer body and spring) is (see Example 10.7) $F(t) = -m\frac{d^2 h_f(t)}{dt^2} = m\left(\frac{2\pi v}{\lambda}\right)^2 h\sin\left(\frac{2\pi v}{\lambda}t\right)$. For $d = 0$, the canonical form of the equation of motion (see Eq (10.26) and Example 10.7) is

$\frac{d^2 y}{dt^2} + \omega^2 y = a(t)$, where $\omega = \sqrt{\frac{k}{m}} = 8.787$ rad/s, and

$a(t) = \frac{F(t)}{m} = \left(\frac{2\pi v}{\lambda}\right)^2 h\sin\left(\frac{2\pi v}{\lambda}t\right) = a_o\left(\frac{2\pi v}{\lambda}\right)^2 \sin\left(\frac{2\pi v}{\lambda}t\right)$, where $a_o = h$. (a) The magnitude of the steady state amplitude of the motion relative to the wheel-axle assembly is given by Eq (10.31) and the equations in Example 10.7 for $d = 0$ and $b_o = 0$,

$$E_p = \frac{\left(\frac{2\pi v}{\lambda}\right)^2 h}{\left(\frac{k}{m} - \left(\frac{2\pi v}{\lambda}\right)^2\right)}$$

(b) Resonance, by definition, occurs when the denominator vanishes, from which $v = \frac{\lambda}{2\pi}\sqrt{\frac{k}{m}}$

==================================<>==================================

==================================<>==================================

Problem 10.95 The trailer in Problem 10.94, not including its wheels and axle, weighs 1000 *lb.* The spring constant of its suspension is $k = 2400 \ lb/ft$, and the damping coefficient due to its shock absorbers is $200c = 200 \ lb\text{-}s/ft$. The road surface parameters are $h = 2$ *in.* and $\lambda = 8 \ ft$. The trailer's horizontal velocity is $v = 6 \ mi/hr$. Determine the magnitude of the trailer's vertical steady state vibration relative to the road surface: (a) neglecting the damping due to the shock absorbers, (b) not neglecting the damping.

Solution:

(a) From the solution to Problem 10.94, for zero damping, the steady state amplitude relative to the road surface is $E_p = \dfrac{\left(\frac{2\pi v}{\lambda}\right)^2 h}{\left(\frac{k}{m} - \left(\frac{2\pi v}{\lambda}\right)^2\right)}$. Substitute numerical values: $v = 6 \ mi/hr = 8.8 \ ft/s$, $\lambda = 8 \, ft$, $k = 2400 \ lb/ft$, $m = \dfrac{W}{g} = 31.08 \ slug$, to obtain $\boxed{E_p = 0.2704 \ ft = 3.25 \ in.}$

(b) From Example 10.7, and the solution to Problem 10.94, the canonical form of the equation of motion is $\dfrac{d^2y}{dt^2} + 2d\dfrac{dy}{dt} + \omega^2 y = a(t)$, where $d = \dfrac{c}{2m} = \dfrac{200}{2(31.08)} = 3.217 \ rad/s$, $\omega = 8.787 \ rad/s$, and $a(t) = a_o\left(\dfrac{2\pi v}{\lambda}\right)^2 \sin\left(\dfrac{2\pi v}{\lambda}t\right)$, where $a_o = h$. The system is sub-critically damped, from which $\omega_d = \sqrt{\omega^2 - d^2} = 8.177 \ rad/s$. From Example 10.7, the magnitude of the steady state response is

$$\boxed{E_p = \frac{\left(\frac{2\pi v}{\lambda}\right)^2 h}{\sqrt{\left(\omega^2 - \left(\frac{2\pi v}{\lambda}\right)^2\right)^2 + 4d^2\left(\frac{2\pi v}{\lambda}\right)^2}} = 0.149 \ ft = 1.791 \ in.}$$

==================================<>==================================

====================================<>====================================

Problem 10.96 A disk with moment of inertia I rotates about a fixed shaft and is attached to a torsional spring with constant k. The angle θ measures the angular position of the disk relative to its position when the spring is unstretched. The disk is initially stationary with the spring unstretched. At $t = 0$, a time dependent moment $M(t) = M_o\left(1 - e^{-t}\right)$ is applied to the disk, where M_o is a constant. Show that the angular position of the disk is

$$\theta = \frac{M_o}{I}\left[-\frac{1}{\omega\left(1+\omega^2\right)}\sin\omega t - \frac{1}{\omega^2\left(1+\omega^2\right)}\cos\omega t + \frac{1}{\omega^2} - \frac{1}{\left(1-\omega^2\right)}e^{-t}\right].$$

Strategy: To determine the particular solution, seek a solution of the form $\theta_p = A_p + B_p e^{-t}$, where A_p and B_p are constants you must determine.

Solution:

The sum of the moments on the disk are $\sum M = -k\theta + M(t)$. From the equation of angular motion, $I\frac{d^2\theta}{dt^2} = -k\theta + M(t)$. For $d = 0$ the canonical form is (see Eq (10.26)) is $\frac{d^2\theta}{dt^2} + \omega^2\theta = a(t)$, where $\omega = \sqrt{\frac{k}{I}}$, and $a(t) = \frac{M_o}{I}(1 - e^{-t})$. (a) For the particular solution, assume a solution of the form $\theta_p = A_p + B_p e^{-t}$. Substitute into the equation of motion,

$\frac{d^2\theta_p}{dt^2} + \omega^2\theta_p = B_p e^{-t} + \omega^2(A_p + B_p e^{-t}) = \frac{M_o}{I}(1 - e^{-t})$. Rearrange:

$\omega^2 A_p + B_p\left(1+\omega^2\right)e^{-t} = \frac{M_o}{I} - \frac{M_o}{I}e^{-t}$. Equate like coefficients: $\omega^2 A_p = \frac{M_o}{I}$, $B_p\left(1+\omega^2\right) = -\frac{M_o}{I}$,

from which $A_p = \frac{M_o}{\omega^2 I}$, and $B_p = -\frac{M_o}{\left(1+\omega^2\right)I}$. The particular solution is $\theta_p = \frac{M_o}{I}\left[\frac{1}{\omega^2} - \frac{e^{-t}}{\left(1+\omega^2\right)}\right]$.

The system is undamped. The solution to the homogenous equation has the form $\theta_h = A\sin\omega t + B\cos\omega t$. The trial solution is:

$\theta = \theta_h + \theta_p = A\sin\omega t + B\cos\omega t + \frac{M_o}{I}\left(\frac{1}{\omega^2} - \frac{e^{-t}}{\left(1+\omega^2\right)}\right)$. Apply the initial conditions: at $t = 0$, $\theta_o = 0$ and $\dot{\theta}_o = 0$: $\theta_o = 0 = B + \frac{M_o}{I}\left(\frac{1}{\omega^2} - \frac{1}{\left(1+\omega^2\right)}\right)$, $\dot{\theta}_o = 0 = \omega A + \frac{M_o}{I}\left(\frac{1}{\left(1+\omega^2\right)}\right)$, from which

$A = -\frac{M_o}{I}\left(\frac{1}{\omega(1+\omega^2)}\right)$, $B = -\frac{M_o}{I}\left(\frac{1}{\omega^2} - \frac{1}{\left(1+\omega^2\right)}\right) = -\frac{M_o}{I}\left(\frac{1}{\omega^2\left(1+\omega^2\right)}\right)$, and the complete solution is

$$\boxed{\theta = \frac{M_o}{I}\left[-\frac{1}{\omega\left(1+\omega^2\right)}\sin\omega t - \frac{1}{\omega^2\left(1+\omega^2\right)}\cos\omega t + \frac{1}{\omega^2} - \frac{1}{\left(1+\omega^2\right)}e^{-t}\right]}.$$

====================================<>====================================

Problem [illegible] A disk with moment of inertia I rotates about a fixed shaft and is attached to a torsional spring with constant k. The angle θ measures the angular position of the disk relative to its position when the spring is unstretched. The disk is initially stationary with the spring unstretched. At $t = 0$, a time-dependent moment $M(t) = M_0\left(1 - e^{-t}\right)$ is applied to the disk, where M_0 is a constant. Show that the angular position of the disk is

[illegible]

Strategy: To determine the particular solution, seek a solution of the form $\theta_p = A_p + B_p e^{-t}$, where A_p and B_p are constants you must determine.

Solution:

The sum of the moments on the disk are $\sum M = -k\theta + M(t)$. From the equation of angular motion, $I\frac{d^2\theta}{dt^2} = -k\theta + M(t)$. For $d = 0$ the canonical form is Eq [illegible] $\frac{d^2\theta}{dt^2} + \omega^2\theta = [illegible]$, where [illegible] (a) For the particular solution, assume a solution of the form [illegible] Substitute into the equation of motion:

[illegible] $\frac{M_0}{I}\left(1 - e^{-t}\right)$. Rearrange:

[illegible] Equate like coefficients: [illegible] $= \frac{M_0}{I}$, [illegible] $= -\frac{M_0}{I}$, from which [illegible]. The particular solution is $\theta_p =$ [illegible]

[illegible] solution to the homogeneous equation is of the form [illegible]. The total solution is

$\theta = \theta_h + \theta_p = A \sin\omega t + B \cos\omega t +$ [illegible]. Apply the initial conditions: at $t = 0$, $\theta_0 = 0$

and [illegible], from which [illegible]

[illegible], and the complete solution

[illegible]